S0-AZR-901

PEARSON

INDIANA

Chemistry

TEACHER'S EDITION

Antony C. Wilbraham • Dennis D. Staley • Michael S. Matta • Edward L. Waterman

Boston, Massachusetts • Chandler, Arizona • Glenview, Illinois • Upper Saddle River, New Jersey

---

Front matter credits: **T5** Carlos Alvarez/iStockphoto; **T7** dra schwartz/iStockphoto; **T8** sidsnapper/iStockphoto; **T14** Dmitriy Melnikov/iStockphoto.

Additional credits appear on pages R156–R157, which constitute an extension of this copyright page.

ISBN-13: 978-0-13-252584-8
ISBN-10: 0-13-252584-4

2 3 4 5 6 7 8 9 10 V052 15 14 13 12 11

Table of Contents

Overview of Teacher's Edition

Contents in Brief

PEARSON Chemistry

INDIANA

"The integration of problem-solving support, technology, and real-world application in *Pearson Chemistry* is above and beyond any program I've seen in my years teaching."

– Mary Chuboff, Athens Academy

Pearson Chemistry for Indiana – All the Elements for Success

The *Pearson Chemistry for Indiana* program combines proven content with cutting-edge digital support and a variety of hands-on learning opportunities to help ensure success in your chemistry classroom. *Pearson Chemistry* provides tools to help you engage and motivate your students, as well as offers support for all the levels and types of learners in your classroom.

A Program Backed by Research

In developing *Pearson Chemistry for Indiana*, the use of research is a guiding, central construct. Research indicated key elements of a textbook program that ensure students' success: support for reading and mathematics in science, consistent opportunities for inquiry, and an ongoing assessment strand.

1 Exploratory Needs Assessment

(Quantitative and Qualitative)

Key research events include —

- Teacher interviews
- Classroom observations
- Mail surveys
- Reviews of educational research

2 Formative Prototype Development and Field Testing

(Quantitative and Qualitative)

Key research events include —

- Field testing of prototypes
- Classroom observations
- Teacher reviews
- Supervisor reviews
- Educator advisory panel

3 Summative Validation Research

(Experimental and Quasi-Experimental Study Designs & Qualitative Research)

Key research events include —

- Pre-publication learner verification research
- Post-publication efficacy studies
- Classroom observations
- Effect size studies

Build a Framework for Learning with Understanding by Design

Understanding by Design® (UbD), developed by Grant Wiggins and Jay McTighe, offers a framework for creating understanding, not by accident or happenstance, but deliberately by design. UbD is a disciplined way of thinking about the design of curriculum, instruction, and assessment that moves teaching from covering the content to ensuring understanding. In the Teacher's Edition of *Pearson Chemistry for Indiana*, we've incorporated Understanding by Design principles into every lesson.

1 Plan for Understanding

The goal of teaching chemistry is not to teach students a bunch of facts about chemistry, but to develop a deeper understanding. Chemistry is the study of the composition of matter and the changes that matter undergoes. How do you help your students understand chemistry on this scale?

In *Pearson Chemistry for Indiana*, your students are introduced to Big Ideas, Essential Questions, and Key Questions in the Student Edition. The Teacher's Edition builds on these, providing questions to help students uncover a deeper understanding.

"The big-idea questions signal that education is not just about learning 'the answer' but about learning how to learn."

— Grant Wiggins, Ed.D., Harvard University

CHAPTER 9

What's Online

CHEM TUTOR Students access guided, step-by-step tutorials for naming and writing formulas for compounds and for calculating mass ratios.

PROBLEM-SOLVING HELP Students can practice key problem-solving skills in an online problem set.

MATH HELP Identify the students who struggle with math by assigning an online math skills diagnostic test. These students can then improve and practice math skills using the MathXL tutorial system.

VIRTUAL LAB Students go on an animated virtual lab tour in which ionic compounds are explored in a simulated laboratory environment.

... communicate their ideas and results of investigations verbally and in written form using tables, graphs, diagrams, and photographs.

- Use analogies and models (mathematical and physical) to simplify and represent systems that are difficult to understand or directly experience due to their size, time scale, or complexity, and recognize the limitations of analogies and models.

Standard 3: Bonding and Molecular Structure

C.3.4 Draw structural formulas for and name simple molecules.

C.3.5 Predict chemical formulas for ionic compounds given their name and vice versa.

262 Chapter 9

9 Chemical Names and Formulas

INSIDE:
- 9.1 Naming Ions
- 9.2 Naming and Writing Formulas for Ionic Compounds
- 9.3 Naming and Writing Formulas for Molecular Compounds
- 9.4 Naming and Writing Formulas for Acids and Bases
- 9.5 The Laws Governing How ...

BIG IDEAS
- ELECTRONS AND THE STRUCTURE OF ATOMS
- BONDING AND INTERACTIONS

Essential Questions:
1. *How does the periodic table help you determine the names and formulas of ions and compounds?*
2. *What is the difference between an ionic and a molecular compound?*

CHEMYSTERY

Cucumber Blunder

Tara's friend, Ellie, raves about the delicious lime pickles that Tara's grandma makes. As a surprise, Tara decided to make a batch for Ellie. She got the recipe from her grandma. It called for "1 cup of lime for every 3 pounds of cucumbers."

When the pickles were ready, Tara proudly presented the gift to Ellie. She watched Ellie's face as she bit into a pickle. After the first bite, Ellie made an awful face. "What's wrong?" ...

INDIANA'S ACADEMIC STANDARDS FOR SCIENCE

Bonding and Molecular Structure C.3.4, C.3.5. See lessons for details.

Focus on ELL

1 CONTENT AND LANGUAGE Write the the chapter title on the board, and underline the words *Names* and *Formulas*. Ask students to predict the difference between a chemical name and a chemical formula.

BEGINNING: LOW/HIGH Rewrite the lesson titles containing a conjugate as two sentences. Read the new lesson titles aloud.

INTERMEDIATE

LOW Present previously learned examples of names and formulas of simple compounds.

HIGH Predict how naming compounds differs from writing formulas for compounds.

ADVANCED: LOW/HIGH Use information from previous chapters to predict what is likely to vary between naming ions, ionic compounds, and molecular compounds. Begin a list of naming rules for each.

Introduce the Chapter

REAL-WORLD CONNECTION Use this activity to draw attention to information students can obtain by looking at the chemical names of ingredients in a variety of household products.

Activity *Display a selection of household products or just their packaging labels. Examples include soap, detergent, toothpaste, shampoo, ibuprofen, bathroom cleaner, windshield wiper fluid, sunscreen, and food. Make sure that at least one chemical name is displayed on the label and that it indicates the elements in the compound.* Have students work with a partner to examine the ingredients listed on each package and record the chemical names with elements they recognize. **Ask** What can you learn from the chemical names? *(the elements in the compounds)* **Ask** Would knowing the chemical formula of an ingredient help you to know more, less, or about the same amount of information? Explain. *(More; the formula would show the proportion of the elements and possibly the type of bonding.)*

UbD Understanding by Design

Students are building toward naming and writing formulas for compounds using their knowledge of atoms, ions, and bonding.

PERFORMANCE GOALS At the end of Chapter 9, students will be able to answer the essential questions by applying their knowledge of ions, ionic compounds, and molecular compounds. Students will also be able to name and write formulas for compounds.

ESSENTIAL QUESTIONS Read the essential questions aloud. **Ask** What is an ion? *(an atom or group of atoms that has a positive or negative charge)* **Ask** How is an ion different from a molecule? *(An ion has a charge, but a molecule is neutral.)*

BIG IDEA Use the photo of an artist's paints to help students connect to the concepts they will learn in this chapter. Have students study the photograph. Have students imagine that the artist holding the brush needs to buy a new tube of green paint to finish a picture, but his supplier is out of his favorite brand, and none of the other manufacturers use the same common name for that particular hue. **Ask** How could the artist make sure he gets an exact match? *(Answers will vary.)* Explain that all paint pigments have official color index name codes, such as PG7, which stands for Pigment Green 7. The artist can use the color index name code to make sure he buys an exact match.

Next, have students read the caption. **Ask** What is unique about transition elements? *(Many transition elements form more than one ion.)* Point out that early chemists faced the same challenges identifying chemical compounds with these elements as paint manufacturers do naming paint colors. Explain that, in this chapter, students will discover how chemists make sure that each chemical compound is identified by a unique name and formula.

CHEMYSTERY Have students read over the CHEMystery. Connect the CHEMystery to the Big Ideas of Electrons and the Structure of Atoms and Bonding and Interactions by having students describe what information they get by reading the chemical formula or chemical name of a compound. For example, write "water" and "H_2O" on the board, and point out that the formula gives information about the composition of water that the term "water" does not supply. Ask students to predict why the lime pickles Tara made tasted funny. As a hint, write "lime" on the board, and ask students how much information they can get from just the name.

Chemical Names and Formulas 263

Each chapter is represented by one or more Big Ideas that can be explored using an Essential Question. Each lesson is also accompanied by Key Questions that are tied to the Essential Questions. If students can answer these questions, you know they have grasped important concepts in chemistry.

2 Set Assessment Goals

A critical aspect of any UbD lesson plan is thinking backwards from the desired goal of enduring understandings to what you will accept as evidence of those understandings. The following features of the Teacher's Edition will help you identify desired accomplishments and plan appropriate assessment goals.

- Performance Goals, Chapter Opener—highlights activities by which students can demonstrate a broader understanding of chemistry concepts.
- Evidence of Understanding, Lesson Opener—alternative assessments that help you determine if your students can make meaning from what they read in a lesson.
- Performance Tasks, Chapter Study Guide—suggestions for both a Summative and a Transfer Task to that ensure that students grasp chapter content.

UbD

Understanding by Design

Students are building toward naming and writing formulas for compounds using their knowledge of atoms, ions, and bonding.

PERFORMANCE GOALS At the end of Chapter 9, students will be able to answer the essential questions by applying their knowledge of ions, ionic compounds, and molecular compounds. Students will also be able to name and write formulas for compounds.

ESSENTIAL QUESTIONS Read the essential questions aloud. **Ask** What is an ion? *(an atom or group of atoms that has a positive or negative charge)* **Ask** How is an ion different from a molecule? *(An ion has a charge, but a molecule is neutral.)*

3 Teach the Concepts

In addition to being goal-focused, a UbD lesson plan should include interim assessment tasks to assess students' progress through a lesson. The Teacher's Edition includes **Check for Understanding** boxes that can be used at a specific point in a lesson. These include a variety of informal assessment strategies—for example, follow-up probes, visual representations, question boxes, one-minute responses, and hand signals. Also included are suggestions for adjusting your instruction to address students' misunderstandings or misconceptions mid-lesson.

UbD **Check for Understanding**

The Essential Question How can you determine the charges of monoatomic ions?

Assess students' comprehension by asking them to show you how they use the groups in the periodic table to determine the charge on any given monatomic ion. Students should properly use the terms *cations* and *anions* in their explanation.

ADJUST INSTRUCTION If students are having trouble answering, provide them a clean copy of the periodic table. Instruct them to use colored pencils to color code the groups within the table to help them determine the charges. Direct students to revisit the text following the *Cations* and *Anions* subheads.

Integrate Technology in Your Chemistry Classroom

PearsonChem.com is the next generation of digital instruction, offering complete online Student and Teacher Editions, a comprehensive teacher center, digital lessons, interactive animations, simulations, and assessments in one place. Featuring the latest in digital instructional technology, **PearsonChem.com** integrates the key concepts from the text and brings them alive online.

Enhanced Classroom Instruction

- A sophisticated classroom management system allows you to organize your class reports and track student progress easily.
- Online Teacher Center contains assignable activities, homework, quizzes, and tests with automatic grading capabilities.

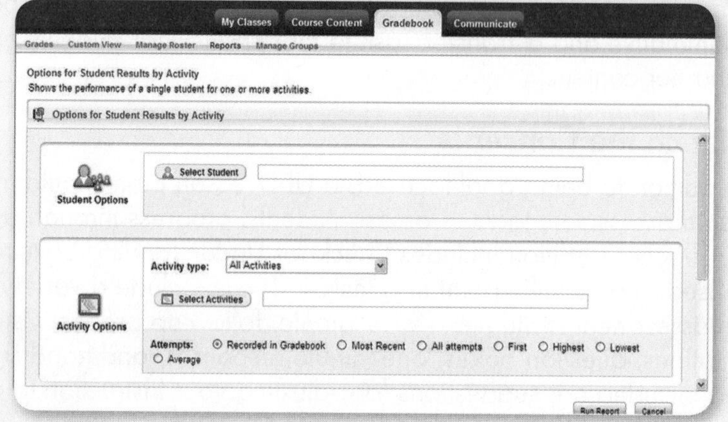

Flexible Tools to Customize Your Instruction

- Hundreds of editable labs in various formats (traditional, small-scale, probeware, lab practicals) allow you to choose the labs that best fit your teaching style.
- PowerPoint presentations provide you with pre-set lesson overview slides that you can use as is or change with the click of a button.
- Editable assessment resources let you pick the level and type of chapter or unit test that fits your classroom needs.
- Student Study worksheets in an editable format give you control over adding or changing anything you might want to in the Reading and Study Workbook.

Integration of Chemistry and Technology

Each lesson opener directs you and your students to the online resources available. Go to the Online Student Edition with Audio.

Use Math Tutorial to review key math skills that apply to the chapter.

Get extra practice with Online Problem Sets.

See art from the textbook brought to life as interactive animations or simulations.

Review key chapter concepts in a real-world context with Concepts in Action.

Use Chem Tutorial to reinforce the problem-solving approach step by step.

Directed Virtual Lab demonstrations allow students to conduct labs with no safety issues and clean up at the click of a button.

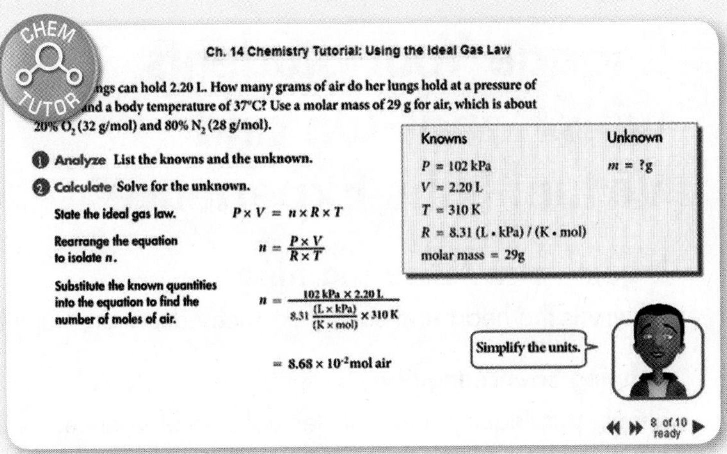

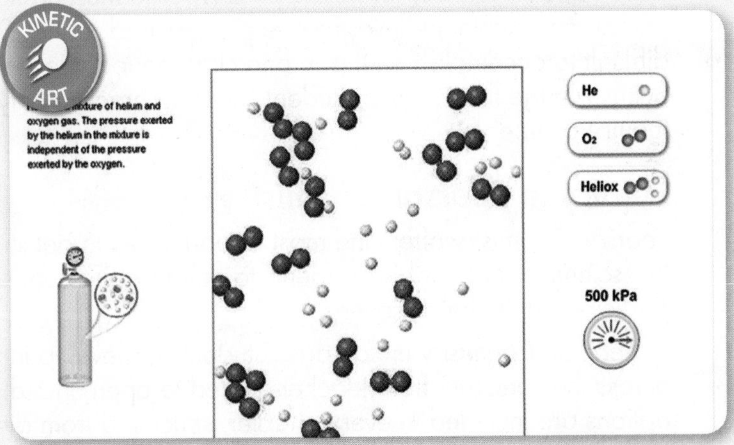

*my*PearsonTraining.com

Go to MyPearsonTraining.com for Professional Development on your own time.

Engage Your Students with Hands-On and Virtual Labs Experience

Inquiry and Active Learning

Inquiry is the heart and soul of science education, but what is it?

Defining Science Inquiry

Simply put, inquiry is the intellectual side of science. It's thinking like a scientist—being inquisitive, asking why, and searching for answers. The National Science Education Content Standards define inquiry as the process in which students begin with a question, design an investigation, gather evidence, formulate an answer to the original question, and communicate the investigative process and results. Since it's often difficult to accomplish all this in one class period, the standards also acknowledge that at times students need to practice only one or two inquiry components.

Inquiry in *Pearson Chemistry*

Pearson Chemistry offers the most opportunities to get students to think like scientists and enables students to enhance their understanding by participating in the discovery.

Pearson Chemistry encourages students to develop inquiry skills across the spectrum from teacher-guided to open-ended. Flexible lab options are included in every chapter, structured from directed to open-ended—providing the flexibility to address all types of learners and accommodate your class time and equipment requirements.

In the Student Edition

Quick Labs

Quick Labs provide students the opportunity to discover science with a variety of investigations through open-ended inquiry. These labs deepen student understanding of Key Concepts and Big Ideas.

Chapter Labs

Chapter-level Small-Scale Labs offer longer, more rigorous investigations that connect back to the Big Idea of the chapter. Each Small-Scale Lab includes extensions for further analysis and/or opportunities to design follow-up experiments.

Ideal Gases and Real Gases

Under what conditions are real gases most likely to differ from ideal gases?

An ideal gas is one that follows the gas laws at all conditions of pressure and temperature. Such a gas would have to conform precisely to the assumptions of kinetic theory. Its particles could have no volume, and there could be no attraction between particles in the gas. As you probably suspect, there is no gas for which these assumptions are true. So an ideal gas does not exist. Nevertheless, at many conditions of temperature and pressure, a real gas behaves very much like an ideal gas.

The particles in a real gas do have volume, and there are attractions between the particles. Because of these attractions, a gas can condense, or even solidify, when it is compressed or cooled. For example, if water vapor is cooled below 100°C at standard atmospheric pressure, it condenses to a liquid. The behavior of other real gases is similar, although lower temperatures and greater pressures may be required. Such conditions are required to produce the liquid nitrogen in Figure 14.14. **Real gases differ most from an ideal gas at low temperatures and high pressures.**

Figure 14.14 Liquid Nitrogen
A lab technician places a cell sample into an insulated tank containing liquid nitrogen. Nitrogen boils at −196°C.

Quick Lab

Carbon Dioxide From Antacid Tablets

Purpose To measure the amount of carbon dioxide gas given off when antacid tablets dissolve in water

Materials
- 6 effervescent antacid tablets
- 3 rubber balloons (spherical)
- plastic medicine dropper
- water
- clock or watch
- metric tape measure
- graph paper
- water
- pressure sensor (optional)

Procedure

1. Break six antacid tablets into small pieces. Keep the pieces from each tablet in a separate pile. Put the pieces from one tablet into the first balloon. Put the pieces from two tablets into a second balloon. Put the pieces from three tablets into a third balloon. **CAUTION** *If you are allergic to latex, do not handle the balloons.*
2. After you use the medicine dropper to squirt about 5 mL of cold water into each balloon, immediately tie off each balloon.
3. Shake the balloons to mix the contents. Allow the contents to warm to room temperature.
4. Measure and record the circumference of each balloon several times during the next 20 minutes.
5. Use the maximum circumference of each balloon to calculate its volume. (*Hint:* For the volume of a sphere, use $V = \frac{4}{3}\pi r^3$ and r = circumference/2π.)

Analyze and Conclude

1. Graph Make a graph of volume versus number of tablets. Use your graph to describe the relationship between the number of tablets used and the volume of the balloon.
2. Calculate Assume that the balloon is filled with carbon dioxide gas at 20°C and standard pressure. Calculate the mass and the number of moles of CO_2 in each balloon at maximum inflation.
3. Analyze Data If a typical antacid tablet contains 2.0 g of sodium hydrogen carbonate, how many moles of CO_2 should one tablet yield? Compare this theoretical value with your results.

The Behavior of Gases 467

Small-Scale Lab

Diffusion

Purpose
To infer diffusion of a gas by observing color changes during chemical reactions

Materials
- clear plastic cup or petri dish
- reaction surface
- dropper bottles containing bromothymol blue, hydrochloric acid, and sodium hydrogen sulfite
- ruler
- cotton swab
- NaOH, NH_4Cl (optional)

Procedure

1. Use the plastic cup or petri dish to draw the large circle shown below on a sheet of paper.

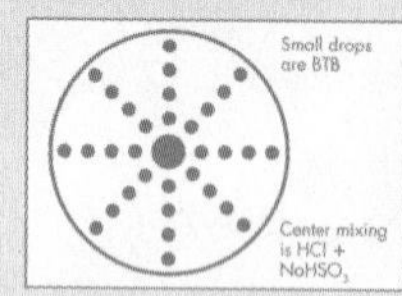

2. Place a reaction surface over the grid and add small drops of bromothymol blue (BTB) in the pattern shown by the small circles. Make sure the drops do not touch one another.
3. Mix one drop each of hydrochloric acid (HCl) and sodium hydrogen sulfite ($NaHSO_3$) in the center of the pattern.
4. Place the cup or petri dish over the grid and observe what happens.
5. If you plan to do Activity 1 in the You're the Chemist section, don't dispose of your materials yet.

Analyze

1. Observe Describe in detail the changes you observed in the drops of BTB over time. Draw pictures to illustrate the changes.
2. Describe Draw a series of pictures showing how one of the BTB drops might look over time if you could view the drop from the side.
3. Explain The BTB changed even though you added nothing to it. If the mixture in the center circle produced a gas, would this explain the change in the drops of BTB? Use kinetic theory to explain your answer.
4. Describe Translate the following word equation into a balanced chemical equation: Sodium hydrogen sulfite reacts with hydrochloric acid to produce sulfur dioxide gas, water, and sodium chloride.

You're the Chemist

The following activities allow you to develop your own procedures and analyze the results.

1. Analyze Data Carefully absorb the center mixture of the original experiment onto a cotton swab and replace it with one drop of NaOH and one drop of NH_4Cl. Describe what happens and explain in terms of kinetic theory. Ammonium chloride reacts with sodium hydroxide to produce ammonia gas, water, and sodium chloride. Write and balance a chemical equation to describe this reaction.
2. Design an Experiment Design an experiment to observe the effect of the size of the BTB drops on the rate at which they change. Explain your results in terms of kinetic theory.

The Behavior of Gases 475

On PearsonChem.com

Online Editable Labs

A wealth of labs are available to you online in an editable format, which allows you to make the labs your own and tailor them to fit your needs. Labs available include traditional labs, small-scale labs, lab practicals, and probeware labs. *Pearson Chemistry*'s online lab program allows you the flexibility to select the labs that you want and make the necessary changes to fit your classroom needs.

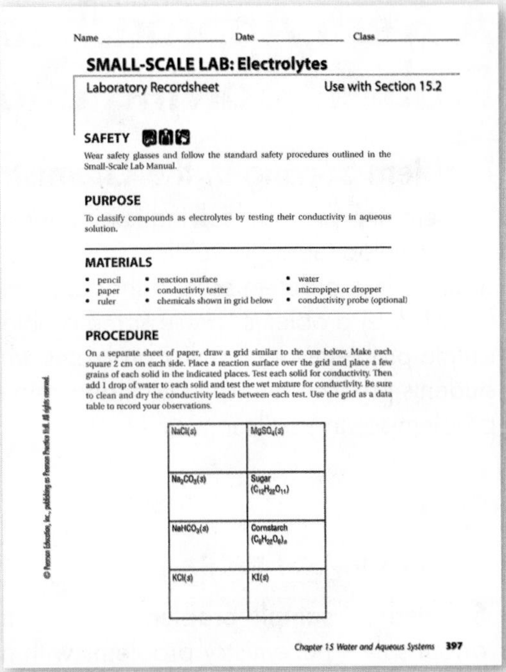

Name ____________ Date ____________ Class ____________

SMALL-SCALE LAB: Electrolytes

Laboratory Recordsheet — Use with Section 15.2

SAFETY

Wear safety glasses and follow the standard safety procedures outlined in the Small-Scale Lab Manual.

PURPOSE

To classify compounds as electrolytes by testing their conductivity in aqueous solution.

MATERIALS

- pencil
- paper
- ruler
- reaction surface
- conductivity tester
- chemicals shown in grid below
- water
- micropipet or dropper
- conductivity probe (optional)

PROCEDURE

On a separate sheet of paper, draw a grid similar to the one below. Make each square 2 cm on each side. Place a reaction surface over the grid and place a few grains of each solid in the indicated places. Test each solid for conductivity. Then add 1 drop of water to each solid and test the wet mixture for conductivity. Be sure to clean and dry the conductivity leads between each test. Use the grid as a data table to record your observations.

$NaCl(s)$	$MgSO_4(s)$
$Na_2CO_3(s)$	Sugar ($C_{12}H_{22}O_{11}$)
$NaHCO_3(s)$	Cornstarch $(C_6H_{10}O_5)_n$
$KCl(s)$	$KI(s)$

Chapter 15 Water and Aqueous Systems 397

On CD-ROM

Virtual ChemLab

The Virtual ChemLab CD-ROM helps chemistry concepts come to life in a hands-on lab environment simulated on the computer screen, rendering safety and equipment needs obsolete. The Virtual ChemLab Workbook provides an additional 30 labs that are correlated to existing labs in the Student Edition and Laboratory Manual.

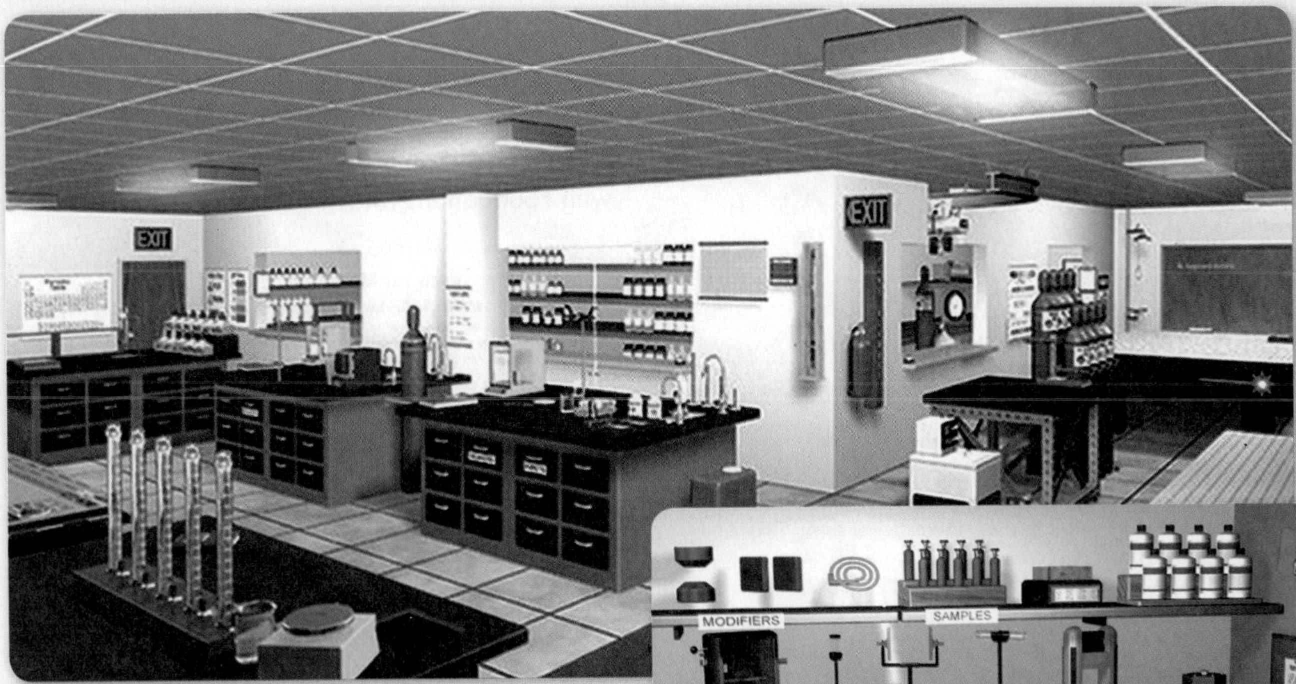

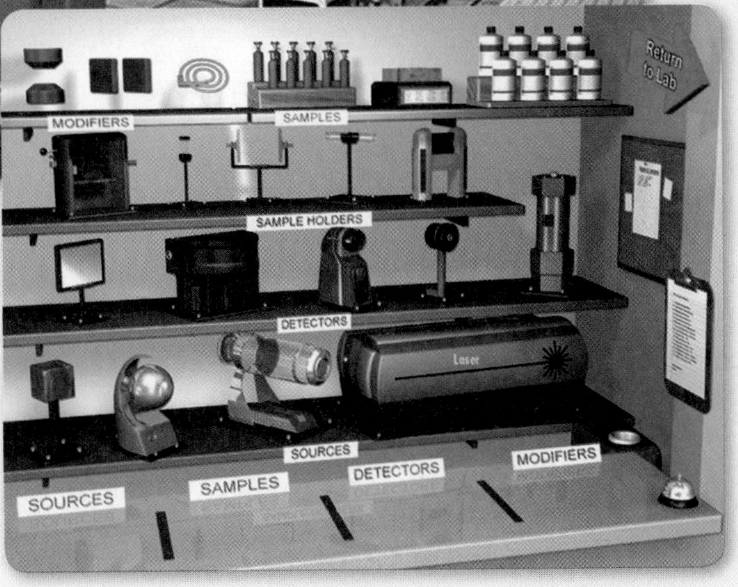

Virtual ChemLab offers a powerful lab simulation environment for performing investigations involving gas properties, calorimetry, titrations, qualitative analysis, and quantum chemistry.

A Fresh Approach to Problem-Solving and Math Support

Problem Solving in the Chemistry Curriculum

In chemistry, problem solving is an important set of skills that students must learn and then develop through practice. In *Pearson Chemistry for Indiana*, students are taught an approach to problem solving that applies to both numeric (quantitative) and conceptual (qualitative) problems. These same problem-solving skills can be used to solve relatively simple problems or complex problems whose solutions require multiple steps. As students practice and gain confidence in their abilities, they will be able to apply their problem-solving skills in new situations and trust that their answers make sense.

"By having the skills of understanding and utilizing problem solving at their disposal, students can approach brand new situations and trust that their results make sense."

– Dennis D. Staley, *Pearson Chemistry* co-author

In the Student Edition

Stepped-out sample problems show your students how to go about solving chemistry problems with hints along the way.

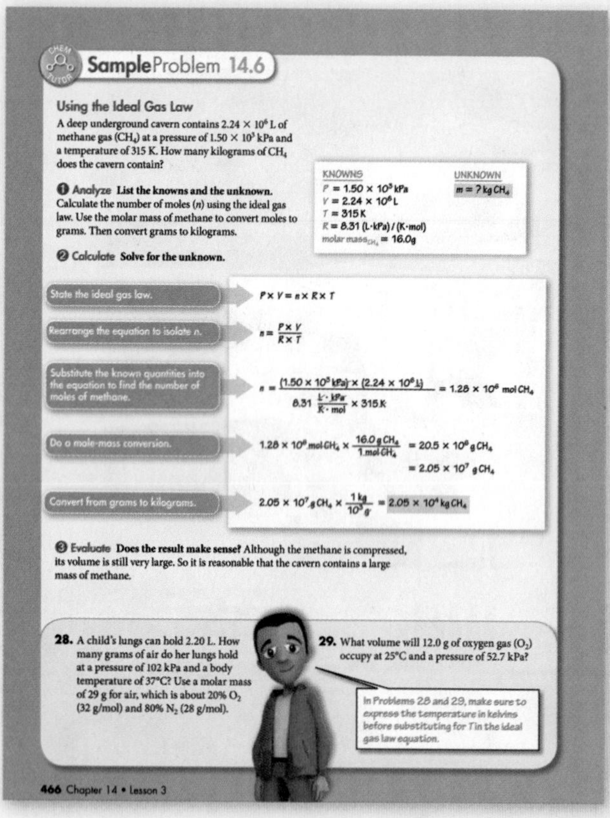

SampleProblem 14.6

Using the Ideal Gas Law

A deep underground cavern contains 2.24×10^6 L of methane gas (CH_4) at a pressure of 1.50×10^3 kPa and a temperature of 315 K. How many kilograms of CH_4 does the cavern contain?

1 Analyze List the knowns and the unknown. Calculate the number of moles (n) using the ideal gas law. Use the molar mass of methane to convert moles to grams. Then convert grams to kilograms.

KNOWNS
$P = 1.50 \times 10^3$ kPa
$V = 2.24 \times 10^6$ L
$T = 315$ K
$R = 8.31$ (L·kPa)/(K·mol)
molar mass$_{CH_4}$ = 16.0g

UNKNOWN
$m = ?$ kg CH_4

2 Calculate Solve for the unknown.

State the ideal gas law. $P \times V = n \times R \times T$

Rearrange the equation to isolate n. $n = \frac{P \times V}{R \times T}$

Substitute the known quantities into the equation to find the number of moles of methane. $n = \frac{(1.50 \times 10^3\ \text{kPa}) \times (2.24 \times 10^6\ \text{L})}{8.31\ \frac{\text{L}\cdot\text{kPa}}{\text{K}\cdot\text{mol}} \times 315\ \text{K}} = 1.28 \times 10^6\ \text{mol}\ CH_4$

Do a mole-mass conversion. $1.28 \times 10^6\ \text{mol}\ CH_4 \times \frac{16.0\ \text{g}\ CH_4}{1\ \text{mol}\ CH_4} = 20.5 \times 10^6\ \text{g}\ CH_4 = 2.05 \times 10^7\ \text{g}\ CH_4$

Convert from grams to kilograms. $2.05 \times 10^7\ \text{g}\ CH_4 \times \frac{1\ \text{kg}}{10^3\ \text{g}} = 2.05 \times 10^4\ \text{kg}\ CH_4$

3 Evaluate Does the result make sense? Although the methane is compressed, its volume is still very large. So it is reasonable that the cavern contains a large mass of methane.

28. A child's lungs can hold 2.20 L. How many grams of air do her lungs hold at a pressure of 102 kPa and a body temperature of 37°C? Use a molar mass of 29 g for air, which is about 20% O_2 (32 g/mol) and 80% N_2 (28 g/mol).

29. What volume will 12.0 g of oxygen gas (O_2) occupy at 25°C and a pressure of 52.7 kPa?

In Problems 28 and 29, make sure to express the temperature in kelvins before substituting for T in the ideal gas law equation.

466 Chapter 14 • Lesson 3

In the Teacher's Edition

Teacher support for problem-solving and math skills is provided in the Teacher's Edition wrap with **Foundations for Math** features.

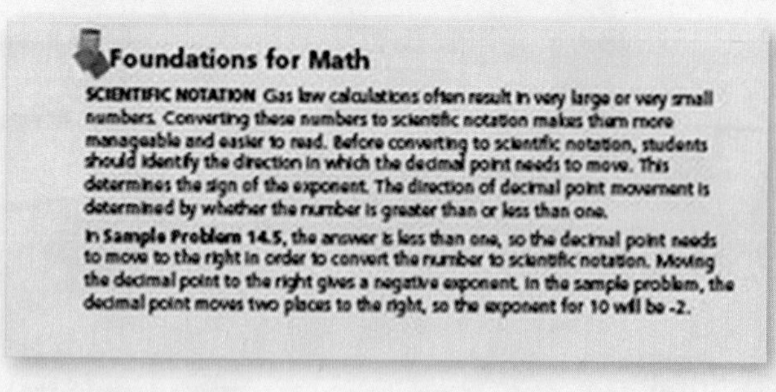

Foundations for Math

SCIENTIFIC NOTATION Gas law calculations often result in very large or very small numbers. Converting these numbers to scientific notation makes them more manageable and easier to read. Before converting to scientific notation, students should identify the direction in which the decimal point needs to move. This determines the sign of the exponent. The direction of decimal point movement is determined by whether the number is greater than or less than one.

In **Sample Problem 14.5**, the answer is less than one, so the decimal point needs to move to the right in order to convert the number to scientific notation. Moving the decimal point to the right gives a negative exponent. In the sample problem, the decimal point moves two places to the right, so the exponent for 10 will be -2.

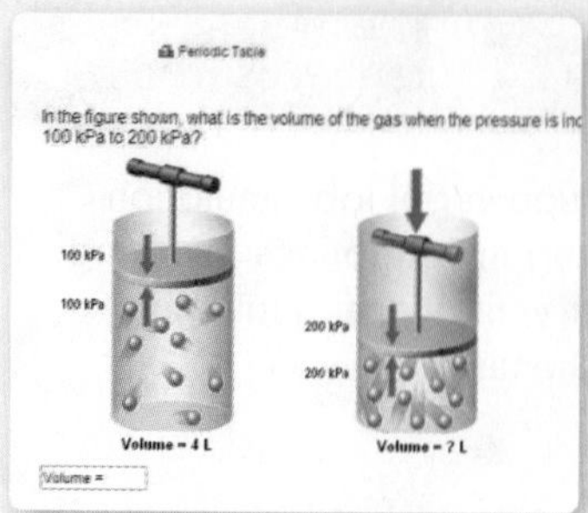

Online

Online Problem Sets give your students extra practice in solving problems similar to those you'll find in your textbook.

Math in the Science Classroom

Why should students concern themselves with mathematics in your chemistry class? Good science requires good data from which to draw conclusions. Technology enhances the ability to measure in a variety of ways. Often scientists must measure large amounts of data, and thus an aim of analysis is to reduce the data to a summary that makes sense and is consistent with established norms of communication—i.e., mathematics. Mathematics is the language used by scientists to model change in the world.

In the Student Edition

Math Tune-Up pages at the end of each chapter help your students quickly reference how to solve a problem—and are also a great way to review before a test!

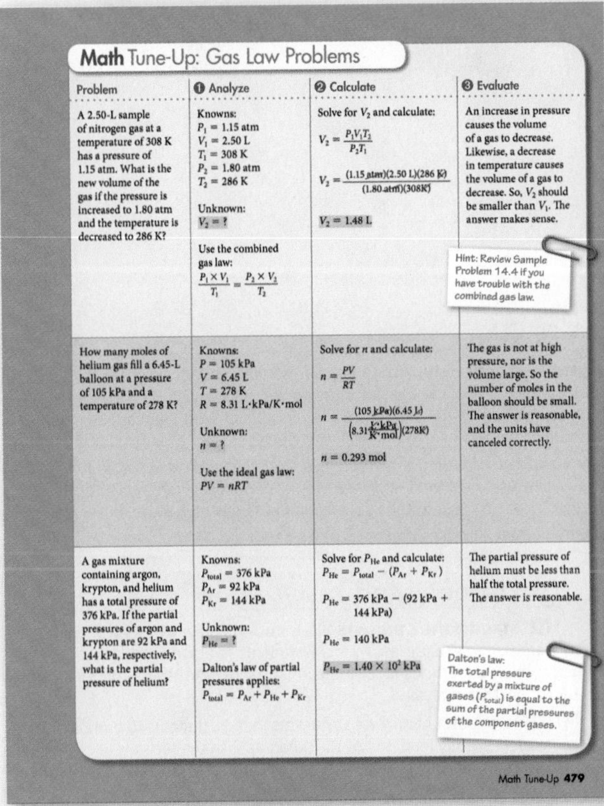

Math Tune-Up: Gas Law Problems

Problem	❶ Analyze	❷ Calculate	❸ Evaluate
A 2.50-L sample of nitrogen gas at a temperature of 308 K has a pressure of 1.15 atm. What is the new volume of the gas if the pressure is increased to 1.80 atm and the temperature is decreased to 286 K?	Knowns: $P_1 = 1.15$ atm, $V_1 = 2.50$ L, $T_1 = 308$ K, $P_2 = 1.80$ atm, $T_2 = 286$ K Unknown: $V_2 = ?$ Use the combined gas law: $\frac{P_1 \times V_1}{T_1} = \frac{P_2 \times V_2}{T_2}$	Solve for V_2 and calculate: $V_2 = \frac{P_1V_1T_2}{P_2T_1}$ $V_2 = \frac{(1.15\text{ atm})(2.50\text{ L})(286\text{ K})}{(1.80\text{ atm})(308\text{ K})}$ $V_2 = 1.48$ L	An increase in pressure causes the volume of a gas to decrease. Likewise, a decrease in temperature causes the volume of a gas to decrease. So, V_2 should be smaller than V_1. The answer makes sense. Hint: Review Sample Problem 14.4 if you have trouble with the combined gas law.
How many moles of helium gas fill a 6.45-L balloon at a pressure of 105 kPa and a temperature of 278 K?	Knowns: $P = 105$ kPa, $V = 6.45$ L, $T = 278$ K, $R = 8.31$ L·kPa/K·mol Unknown: $n = ?$ Use the ideal gas law: $PV = nRT$	Solve for n and calculate: $n = \frac{PV}{RT}$ $n = \frac{(105\text{ kPa})(6.45\text{ L})}{\left(8.31\frac{\text{L·kPa}}{\text{K·mol}}\right)(278\text{ K})}$ $n = 0.293$ mol	The gas is not at high pressure, nor is the volume large. So the number of moles in the balloon should be small. The answer is reasonable, and the units have canceled correctly.
A gas mixture containing argon, krypton, and helium has a total pressure of 376 kPa. If the partial pressures of argon and krypton are 92 kPa and 144 kPa, respectively, what is the partial pressure of helium?	Knowns: $P_{total} = 376$ kPa, $P_{Ar} = 92$ kPa, $P_{Kr} = 144$ kPa Unknown: $P_{He} = ?$ Dalton's law of partial pressures applies: $P_{total} = P_{Ar} + P_{He} + P_{Kr}$	Solve for P_{He} and calculate: $P_{He} = P_{total} - (P_{Ar} + P_{Kr})$ $P_{He} = 376\text{ kPa} - (92\text{ kPa} + 144\text{ kPa})$ $P_{He} = 140$ kPa $P_{He} = 1.40 \times 10^2$ kPa	The partial pressure of helium must be less than half the total pressure. The answer is reasonable. Dalton's law: The total pressure exerted by a mixture of gases (P_{total}) is equal to the sum of the partial pressures of the component gases.

Math Tune-Up 479

Online

Math Tutorial animations review key math skills that are needed to solve the chapter problems your students will see in their textbook.

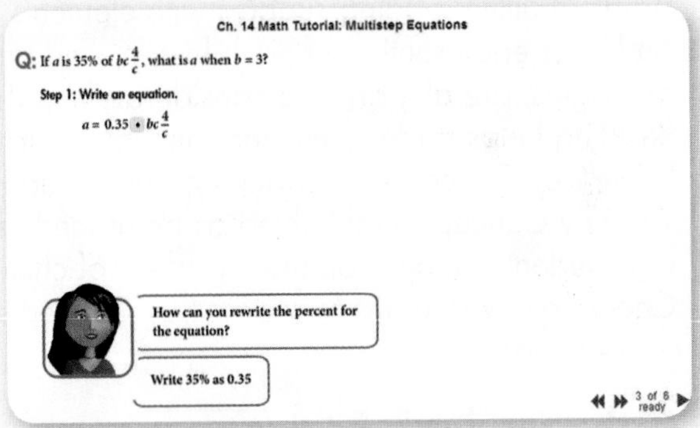

In the Workbook

In the Reading and Study Workbook, students can find extra skill-building strategies to enhance the math support provided in the Student Edition.

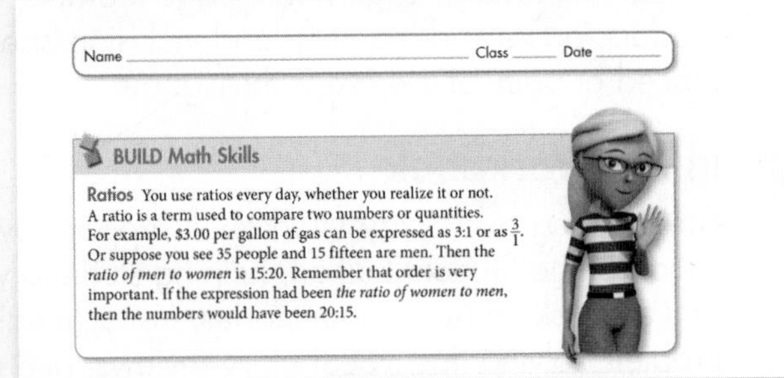

Name ______________ Class ______ Date ______

BUILD Math Skills

Ratios You use ratios every day, whether you realize it or not. A ratio is a term used to compare two numbers or quantities. For example, \$3.00 per gallon of gas can be expressed as 3:1 or as $\frac{3}{1}$. Or suppose you see 35 people and 15 fifteen are men. Then the *ratio of men to women* is 15:20. Remember that order is very important. If the expression had been *the ratio of women to men*, then the numbers would have been 20:15.

In the Teacher's Edition

Math support in the Teacher's Edition is geared toward the math objectives on high-stakes tests. Teaching notes provide you with strategies for alternative instruction and additional practice.

Meeting Individual Needs of Your Classroom

What's different about each of the students in your classroom? Just about everything. A typical classroom is comprised of students whose differences include culture, language, interest, motivation, and knowledge.

Here's how *Pearson Chemistry* meets individual needs:

In the Student Edition

Considerate Text Structure

Ample, built-in reading support with elements before, during, and after each section of each lesson, and frequent use of analogies, are all part of a considerate text. The Guide for Reading helps students master concepts by introducing them to the Key Concepts, Vocabulary, and a Reading Strategy. The Key Concepts highlighted as boldfaced sentences encourage students to focus on the big ideas of chemistry. Lesson Checks allow students to review Key Concepts at the end of every lesson.

In the Teacher's Edition

Chapter Planning Guides label activities by levels of difficulty, meeting the needs of all students. All program resources are identified by levels for Conceptual, for Standard, and for Honors.

The Teacher's Edition includes Differentiated Instructional Support Ideas. In order to facilitate, these practical ideas for visual learning, English language learners, less proficient readers, advanced learners, and inclusion/special needs are placed at point-of-use areas throughout.

In the Workbook

The Reading and Study Workbook supports every lesson of the Student Edition, helping students with reading and math skills. Lesson summaries, instructional graphics, problem-solving tips, vocabulary reviews, and extension activities help contribute to a better understanding of core chemistry concepts.

Focus on ELL

1 CONTENT AND LANGUAGE Ask students to state the key questions in th words. Have pairs make and use flashcards for key vocabulary. They shoul word on one side and a definition, visual, and sentence with the word on Allow students with limited English to use their native languages for supp

2 FRONTLOAD THE LESSON Have students discuss games they've played t inflatable balls. Ask them what differences they've noticed, if any, when b underinflated. Direct them to the information in Chemistry & You.

3 COMPREHENSIBLE INPUT Explain Figure 14.5 and restate information in as need board.

Differentiated Instruction

L1 STRUGGLING STUDENTS Have students draw a balloon and use circle represent gas particles. Then ask students to draw a smaller balloon with number of particles to illustrate what happens to the gas particles in the when it's compressed.

ELL ENGLISH LANGUAGE LEARNERS Assign students to groups of three student silently read one of the paragraphs in the sections titled Amount Volume, and Temperature. Ask students to cover the page and restate to one fact they read.

L3 ADVANCED STUDENTS Ask students to explain why a fully inflated p balloon appears underinflated after moving from a warm place to a cool

Inform Instruction with Comprehensive Assessment

A variety of assessment opportunities in the *Pearson Chemistry for Indiana* program helps you monitor student progress, evaluate content mastery, and ensure student success on high-stakes tests.

Assessment Monitors Learning

Lesson Checks at the end of each lesson call on students to think critically and apply vocabulary. Expanded chapter assessments offer students plenty of opportunity to master the concepts. (Solutions to asterisked assessment problems can be found in Appendix E of the Student Edition.) In addition, Standardized Test Prep pages in your state test format provide students practice on high-stakes exams. Additional assessment opportunities are offered in the Progress Monitoring Assessments book.

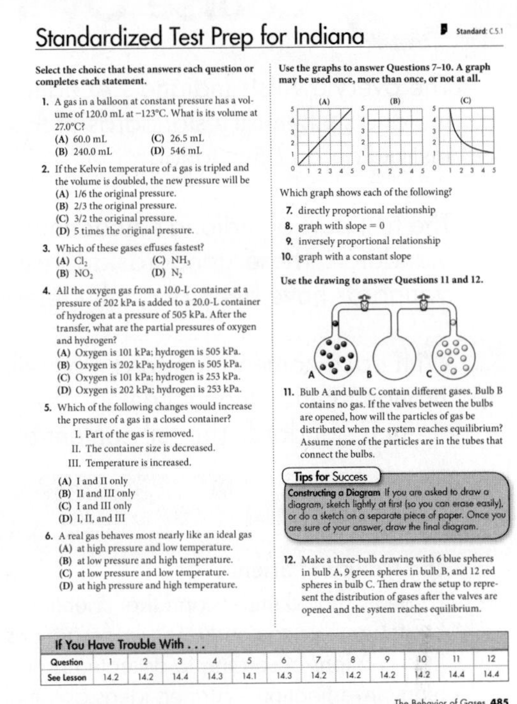

Standardized Test Prep for Indiana

Standard: C.5.1

Select the choice that best answers each question or completes each statement.

1. A gas in a balloon at constant pressure has a volume of 120.0 mL at −123°C. What is its volume at 27.0°C?
   (A) 60.0 mL (C) 26.5 mL
   (B) 240.0 mL (D) 546 mL
2. If the Kelvin temperature of a gas is tripled and the volume is doubled, the new pressure will be
   (A) 1/6 the original pressure.
   (B) 2/3 the original pressure.
   (C) 3/2 the original pressure.
   (D) 5 times the original pressure.
3. Which of these gases effuses fastest?
   (A) Cl_2 (C) NH_3
   (B) NO_2 (D) N_2
4. All the oxygen gas from a 10.0-L container at a pressure of 202 kPa is added to a 20.0-L container of hydrogen at a pressure of 505 kPa. After the transfer, what are the partial pressures of oxygen and hydrogen?
   (A) Oxygen is 101 kPa; hydrogen is 505 kPa.
   (B) Oxygen is 202 kPa; hydrogen is 505 kPa.
   (C) Oxygen is 101 kPa; hydrogen is 253 kPa.
   (D) Oxygen is 202 kPa; hydrogen is 253 kPa.
5. Which of the following changes would increase the pressure of a gas in a closed container?
   I. Part of the gas is removed.
   II. The container size is decreased.
   III. Temperature is increased.
   (A) I and II only
   (B) II and III only
   (C) I and III only
   (D) I, II, and III
6. A real gas behaves most nearly like an ideal gas
   (A) at high pressure and low temperature.
   (B) at low pressure and high temperature.
   (C) at low pressure and low temperature.
   (D) at high pressure and high temperature.

Use the graphs to answer Questions 7–10. A graph may be used once, more than once, or not at all.

(A) (B) (C)

Which graph shows each of the following?

7. directly proportional relationship
8. graph with slope = 0
9. inversely proportional relationship
10. graph with a constant slope

Use the drawing to answer Questions 11 and 12.

A B C

11. Bulb A and bulb C contain different gases. Bulb B contains no gas. If the valves between the bulbs are opened, how will the particles of gas be distributed when the system reaches equilibrium? Assume none of the particles are in the tubes that connect the bulbs.

Tips for Success

Constructing a Diagram If you are asked to draw a diagram, sketch lightly at first (so you can erase easily), or do a sketch on a separate piece of paper. Once you are sure of your answer, draw the final diagram.

12. Make a three-bulb drawing with 6 blue spheres in bulb A, 9 green spheres in bulb B, and 12 red spheres in bulb C. Then draw the setup to represent the distribution of gases after the valves are opened and the system reaches equilibrium.

If You Have Trouble With . . .

Question	1	2	3	4	5	6	7	8	9	10	11	12
See Lesson	14.2	14.2	14.4	14.3	14.1	14.3	14.2	14.2	14.2	14.2	14.4	14.4

The Behavior of Gases 485

Each chapter assessment concludes with standardized test practice.

Editable Online Assessment Resource

Ensure that your students are mastering content with this brand-new assessment program! Assessments include leveled Unit Tests and leveled Chapter Tests to help you prepare your classroom for end-of-year testing.

Online Assessment and Remediation

PearsonChem.com provides diagnostic and Benchmark tests with automatic scoring, class reports, and remediation for students.

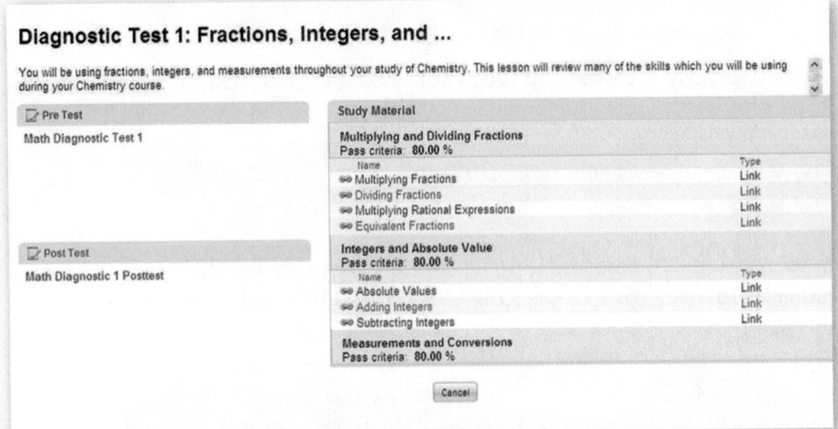

Indiana Exam*View* Assessment Suite CD-ROM

Exam*View*® Computer Test Bank CD-ROM provides you access to thousands of modifiable test questions, hundreds of which contain algorithmically-generated variable data. Create and print custom tests in minutes that correlate to your state standards with this test bank that is also compatible with QuickTake software.

Course Overview

The overview lists Indiana's Academic Standards for Chemistry I. Use this chart to see the coverage of Indiana's Chemistry standards in this book. With the help of this chart, you can find solid, fully developed instruction on any standard.

The first part of Indiana's Academic Standards for Science consists of a set of eleven Nature of Science Standards. These standards address scientific methods and scientific thinking. In this front matter, these standards have been given the reference numbers NoS.1 through NoS.11.

With each content standard, you will find a Summary Phrase that is used to reference that standard throughout the Student Edition. The phrase is a convenient tool to help student's remember the basic content of each standard. The full content of each standard is listed in this chart.

Standard	Where You'll Find It
The Nature of Science Students should understand that scientific knowledge is gained from observation of natural phenomena and experimentation, by designing and conducting investigations guided by theory, and by evaluating and communicating the results of those investigations according to accepted procedures. Thus, scientific knowledge is scientists' best explanations for the data from many investigations. Further, ideas about objects in the microscopic world that we cannot directly sense are often understood in terms of concepts developed to understand objects in the macroscopic world that we can see and touch. In the science classroom student work should align with this process of science and should be guided by the following principles. These should be woven throughout the daily work that students are doing when learning the content presented in the standard indicators.	
NoS.1 Develop explanations based on reproducible data and observations gathered during laboratory investigations.	All Quick Labs Chemistry & You: pp. 208, 355, 744 All Small-Scale Labs
NoS.2 Recognize that their explanations must be based both on their data and other known information from investigations of others.	Small-Scale Lab: p. 545
NoS.3 Clearly communicate their ideas and results of investigations verbally and in written form using tables, graphs, diagrams, and photographs.	Lesson: 1.3 Quick Lab: pp. 180, 328, 354, 571, 896 All Small-Scale Labs
NoS.4 Regularly evaluate the work of their peers and in turn have their work evaluated by their peers.	Quick Lab: p. 109
NoS.5 Apply standard techniques in laboratory investigations to measure physical quantities in appropriate units and convert known quantities to other units as necessary.	Lessons: 12.1, 12.2, 12.3 Quick Lab: pp. 328, 571 Small-Scale Lab: pp. 92, 149, 184, 324, 545, 583
NoS.6 Use analogies and models (mathematical and physical) to simplify and represent systems that are difficult to understand or directly experience due to their size, time scale, or complexity, and recognize the limitations of analogies and models.	Lessons: 9.1, 12.1 Quick Lab: p. 230, 778 Chemistry & You: p. 316 Small-Scale Lab: pp. 184, 787
NoS.7 Focus on the development of explanatory models based on their observations during laboratory investigations.	Lessons: 4.1, 4.2, 4.3 Chemistry & You: p. 316 Small-Scale Lab: p. 184
NoS.8 Explain that the body of scientific knowledge is organized into major theories, which are derived from and supported by the results of many experiments, and allow us to make testable predictions.	Lessons: 1.1, 4.1, 4.2, 4.3

Course Overview *(continued)*

NoS.9 Recognize that new scientific discoveries often lead to a re-evaluation of previously accepted scientific knowledge and of commonly held ideas.	Lessons: 4.1, 4.2, 4.3, 5.1 Chemistry & You: p. 133
NoS.10 Describe how scientific discoveries lead to the development of new technologies, and conversely how technological advances can lead to scientific discoveries through new experimental methods and equipment.	Lesson: 5.3 Chemistry & You: pp. 133, 368, 736
NoS.11 Explain how scientific knowledge can be used to guide decisions on environmental and social issues.	Lesson: 1.2 Chemistry & You: pp. 52, 440, 476

Standard	Summary Phrase	Where You'll Find It
Standard 1: Properties and States of Matter **Core Standard:** Describe the nature of physical and chemical properties and changes of matter. **Core Standard:** Compare and contrast states of matter at the molecular level.		
C.1.1 Based on physical properties, differentiate between pure substances and mixtures	Pure substance or mixture?	Lessons: 2.1, 2.2 Quick Lab: p. 39 Chemistry & You: p. 52 Small-Scale Lab: p. 51
C.1.2 Observe and describe chemical and physical properties of different types of matter and designate them as either extensive or intensive.	Extensive and intensive properties	Lesson: 2.1
C.1.3 Recognize observable indicators of chemical changes.	Recognize chemical changes	Lesson: 2.4 Quick Lab: p. 699 CHEMystery: Ch 2
C.1.4 Describe physical and chemical changes at the molecular level.	Physical and chemical changes	Lessons: 2.1, 2.4, 11.1, 11.2, 11.3 Small-Scale Lab: p. 51
C.1.5 Describe the characteristics of solids, liquids, and gases and state changes at the molecular level.	States of matter	Lessons: 2.1, 13.1, 13.2, 13.2, 13.4 CHEMystery: Ch 13
C.1.6 Explain and apply the law of conservation of mass as it applies to chemical changes.	Conservation of mass: chemical changes	Lessons: 2.4, 12.1, 12.2 Chemistry & You: p. 355
C.1.7 Define density and distinguish between materials based on densities. Perform calculations involving density.	Density	Lesson: 3.2 Small-Scale Lab: p. 92

Standard	Summary Phrase	Where You'll Find It
Standard 2: Atomic Structure and the Periodic Table **Core Standard:** Describe how the properties and arrangement of the subatomic particles contributes to the structure of the atom. **Core Standard:** Describe how the structure of the periodic table reflects the numbers of electrons and protons and the configuration of the electrons in an atom.		
C.2.1 Describe how models of atomic structure changed over time based on available experimental evidence and understand the current model of atomic structure.	Modern atomic theory: historic development	Lessons: 4.1, 4.2, 4.3, 5.1, 5.2, 5.3
C.2.2 Describe how the subatomic particles (protons, neutrons, and electrons) contribute to the structure of an atom and recognize that the particles within the nucleus are held together against the electrical repulsion of the protons.	Subatomic particles and atomic structure	Lessons: 4.2, 4.3, 5.1, 5.2, 5.3, 22.1

C.2.3 Determine the number of protons, neutrons, and electrons in isotopes and in those isotopes that comprise a specific element. Relate these numbers to atomic number and mass number.	Isotopes, atomic number, and atomic mass	Lessons: 4.2, 4.3
C.2.4 Calculate the average atomic mass of an element from isotopic abundance data.	Average atomic mass	Lesson: 4.3
C.2.5 Write the electron configuration of an element and relate this to its position on the periodic table.	Write an element's electron configuration	Lesson: 6.2
C.2.6 Use the periodic table and electron configurations to determine an element's number of valence electrons and chemical and physical properties.	Determine valence electrons and properties	Lessons: 6.1, 6.2, 6.3, 7.1 Quick Lab: p. 180 Small-Scale Lab: pp. 184, 200
C.2.7 Compare and contrast nuclear reactions with chemical reactions. For nuclear reactions, describe how the fusion and fission processes transform elements present before the reaction into elements present after the reaction.	Compare and contrast nuclear reactions with chemical reactions	Lessons: 25.1, 25.2, 25.3, 25.4
C.2.8 Understand that the radioactive decay process is random for any given atom, but that this property leads to a predictable and measurable exponential decay of a sample of radioactive material. On this basis, calculate the half-life of a radioactive sample given its molarity and its decay rate.	Radioactive decay	Lesson: 25.2 CHEMystery: Ch 25 Chemistry & You: p. 892 Small-Scale Lab: p. 887
Standard	**Summary Phrase**	**Where You'll Find It**

Standard 3: Bonding and Molecular Structure
Core Standard: Describe how the distribution of electrons within an atom determines its interactions with other atoms.
Core Standard: Describe the attractive forces between molecules and their effect on chemical and physical properties.

C.3.1 Describe, compare, and contrast the characteristics of the interactions between atoms in ionic and covalent compounds.	Interactions between atoms in ionic and covalent compunds	Lessons: 7.2, 8.1, 8.2, 8.3, 8.4 CHEMystery: Ch 7
C.3.2 Compare and contrast how ionic and covalent compounds form.	Formation of ionic and covalent compounds	Lessons: 7.2, 8.1, 8.2, 8.3, 8.4
C.3.3 Compare and contrast ionic, covalent network, metallic and polar and non-polar molecular crystals with respect to constituent particles, strength of bonds, melting and boiling points, and conductivity; provide examples of each type.	Molecular and atomic attractions	Lessons: 7.1, 7.2, 7.3, 8.1, 8.2, 8.3, 8.4, 13.2, 13.3, 15.1 Chemistry & You: p. 208
C.3.4 Draw structural formulas for and name simple molecules.	Structural formulas and names	Lessons: 8.3, 9.3, 9.4, 9.5, 22.1, 22.2, 22.3, 22.4 Chemistry & You: pp. 803, 821
C.3.5 Write chemical formulas for ionic compounds given their name and vice versa.	Write chemical formulas	Lessons: 9.1, 9.2

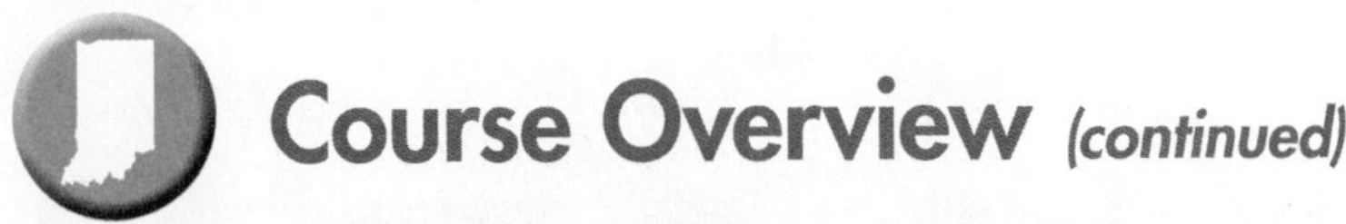

Course Overview *(continued)*

Standard	Summary Phrase	Where You'll Find It
Standard 4: Reactions and Stoichiometry **Core Standard:** Use balanced chemical equations and the mole concept to determine the quantities of reactants and products.		
C.4.1 Predict products of simple reactions such as synthesis, decomposition, single replacement and double replacement.	Predict products	Lessons: 11.1, 11.2, 11.3 CHEMystery: Ch 11
C.4.2 Balance chemical equations using the law of conservation of mass and use them to describe chemical reactions.	Balance chemical equations	Lessons: 11.1, 12.1, 23.1 Quick Lab: pp. 354, 404 Chemistry & You: p. 355 Small-Scale Lab: pp. 374, 399, 475
C.4.3 Use the mole concept to determine the number of moles and number of atoms or molecules in samples of elements and compounds, given mass of the sample.	The mole concept	Lessons: 10.1, 10.2, 10.3, 12.1, 12.2, 12.3 Quick Lab: p. 404 CHEMystery: Ch 10 Small-Scale Lab: pp. 324, 399
C.4.4 Using a balanced chemical equation, calculate the quantities of reactants needed and products made in a chemical reaction that goes to completion.	Calculate reactants and products	Lessons: 12.2, 12.3 Quick Lab: p. 404 Small-Scale Lab: p. 399
C.4.5 Describe, classify and give examples of various kinds of reactions-synthesis (combination), decomposition, single displacement, double displacement and combustion.	Classify types of reactions	Lessons: 11.2, 11.3, 23.1, 23.2, 23.3 Chemistry & You: p. 368
C.4.6 Determine oxidation states and identify the substances apparently gaining and losing electrons in redox reactions.	Redox reactions	Lessons: 20.1, 20.2, 20.3 CHEMystery: Ch 20 Chemistry & You: p. 700, 716
C.4.7 Perform calculations to determine percent composition by mass of a compound or mixture when given the formula.	Percent composition, empirical formulas, and molecular formulas	Lesson: 10.3 CHEMystery: Ch 10 Quick Lab: p. 328 Chemistry & You: p. 532

Standard	Summary Phrase	Where You'll Find It
Standard 5: Behavior of Gases **Core Standard:** Using the kinetic molecular theory, describe and explain the behavior of ideal gases. **Core Standard:** Examine the relationship between moles, volume, pressure, and temperature for ideal gases, using the ideal gas equation of state $pv = nRT$.		
C.5.1 Use kinetic molecular theory to explain changes in gas volumes, pressure, moles, and temperature.	Kinetic molecular theory	Lessons: 14.1, 14.2, 14.3 Quick Lab: p. 467
C.5.2 Using the ideal gas equation of state, pv =nRT, calculate the change in one variable when another variable is changed and the others are held constant.	Ideal gas equation	Lesson: 14.3
C.5.3 Given the equation for a chemical reaction involving one or more gases as reactants and/or products, calculate the volumes of gas assuming the reaction goes to completion and the ideal gas law holds.	Reactant and product calculations: gases	Lessons: 12.2, 12.3 Quick Lab: p. 467

Standard	Summary Phrase	Where You'll Find It
Standard 6: Thermochemistry **Core Standard:** Recognize that chemical reactions result in either the release or absorption of energy. **Core Standard:** Apply the law of conservation of energy.		
C.6.1 Explain that atoms and molecules that make up matter are in constant motion and that this motion increases as thermal energy increases.	Explain that atoms and molecules are in constant motion	Lessons: 13.1, 13.2, 13.3
C.6.2 Distinguish between the concepts of temperature and heat flow in macroscopic and microscopic terms.	Distinguish between temperature and heat flow	Lessons: 13.1, 13.2, 14.2
C.6.3 Solve problems involving heat flow and temperature changes, using known values of specific heat and/or phase change constants (latent heats).	Heat flow and temperature change problems	Lessons: 17.1, 17.2, 17.3, 17.4 Quick Lab: p. 571
C.6.4 Classify chemical reactions and phase changes as exothermic or endothermic.	Exothermic and endothermic reactions	Lessons: 17.3, 17.2 CHEMystery: Ch 17 Chemistry & You: p. 576

Standard	Summary Phrase	Where You'll Find It
Standard 7: Solutions **Core Standard:** Describe the composition and characteristics of solutions. **Core Standard:** Identify the factors that qualitatively affect solubility, reaction rates and dynamic equilibrium.		
C.7.1 Describe the composition and properties of types of solutions.	Composition and properties of solutions	Lessons: 15.1, 15.2, 15.3, 16.1, 16.2, 16.3, 16.4
C.7.2 Explain how temperature, pressure, and polarity of the solvent affect the solubility of a solute.	Factors affecting solubility	Lessons: 16.1, 18.4
C.7.3 Describe the concentration of solutes in solution in terms of molarity. Perform calculations using molarity, mass, and volume.	Molarity	Lesson: 16.4
C.7.4 Prepare a specific volume of a solution of a given molarity when provided with a known solute.	Prepare a solution	Lesson: 16.2 Small-Scale Lab: p. 545
C.7.5 Explain how the rate of a reaction is qualitatively affected by changes in concentration, temperature, surface area, and the use of a catalyst.	Factors affecting rate of reaction	Lessons: 18.1, 18.2
C.7.6 Write equilibrium expressions for reversible reactions.	Reversible reactions	Lesson: 18.3

Standard	Summary Phrase	Where You'll Find It
Standard 8: Acids and Bases **Core Standard:** Use acid-base definitions to identify acids and bases given their formulas and reactions. **Core Standard:** Explain the meaning of the value indicated by the pH scale in terms of the hydrogen ion concentration for any aqueous solution.		

Standard	Summary Phrase	Where You'll Find It
C.8.1 Use Arrhenius and Bronsted-Lowry models to classify substances as acids or bases.	Arrhenius and Bronsted-Lowry	Lesson: 19.1
C.8.2 Describe the characteristic properties of acids and bases.	Properties of acids and bases	Lessons: 19.1, 19.2, 19.3, 19.4, 19.5
C.8.3 Compare and contrast the dissociation and strength of acids and bases in solution.	Dissociation and strength of acids and bases	Lesson: 19.3
C.8.4 Given the hydronium (H3O+) ion concentration in a solution, calculate the pH, and vice versa. Explain the meaning of these values.	Calculate pH and hydronium ion concentration	Lesson: 19.2
C.8.5 For an acid-base titration, calculate the concentration of an unknown solution.	Titration calculations	Lesson: 19.4

Standard	Summary Phrase	Where You'll Find It
Standard 9: Organic Chemistry and Biochemistry **Core Standard:** Describe the unique nature of carbon atoms demonstrated by their ability to bond to one another and other elements, forming countless carbon-based substances and macromolecules.		
C.9.1 Use structural formulas to illustrate carbon atoms' ability to bond covalently to form many different substances.	Use structural formulas: carbon atoms	Lessons: 22.1, 22.2, 22.3, 22.4
C.9.2 Identify the variety of molecular types formed by the covalent bonding of carbon atoms and describe the typical properties of these molecular types.	Identify the variety of carbon molecules	Lessons: 22.1, 22.2, 22.3, 22.4, 23.1

Lesson-by-Lesson Correlation and Pacing Guide

This chart helps you plan your instruction time. The specific Academic Standards covered in each lesson are provided. The Fast Track column indicates those lessons, features, and labs required to provide in-depth coverage of Indiana's Academic Standards for Chemistry I. Lessons, features, and labs providing supplementary information are indicated under Enrichments. Chapter pacing helps you plan how much time to spend teaching and assessing each chapter.

Chapters/Lessons	Academic Standards for Science	Fast Track	Enrichments
Chapter 1 Introduction to Chemistry			5 periods, 2.5 blocks
1.1 The Scope of Chemistry	NoS.8		•
1.2 Chemistry and You	NoS.11		•
Chemistry & You			•
1.3 Thinking Like a Scientist	NoS.1, NoS.3		•
Small-Scale Lab	NoS.1		•
1.4 Problem Solving in Chemistry			•
Chapter 2 Matter and Change			5 periods, 2.5 blocks
2.1 Properties of Matter	C.1.1, C.1.2, C.1.4, C.1.5	•	
2.2 Mixtures	C.1.1	•	
2.3 Elements and Compounds	NoS.1		•
2.4 Chemical Reactions	C.1.3, C.1.4, C.1.6	•	
Small-Scale Lab	C.1.1, C.1.4, NoS.1	•	
Chemistry & You	C.1.1, NoS.11		•

Chapters/Lessons	Academic Standards for Science	Fast Track	Enrichments
Chapter 3 Scientific Measurement			10 periods, 5 blocks
3.1 Using and Expressing Measurements	NoS.1		•
Chemistry & You			•
3.2 Units of Measurement	C.1.7	•	
Chemistry & You			•
3.3 Solving Conversion Problems			•
Small-Scale Lab	C.1.7, NoS.1, NoS.5	•	
Chapter 4 Atomic Structure			5 periods, 2.5 blocks
4.1 Defining the Atom	C.2.1, NoS.7, NoS.8, NoS.9	•	
4.2 Structure of the Nuclear Atom	C.2.1, C.2.2, C.2.3, NoS.1, NoS.4, NoS.7, NoS.8, NoS.9	•	
Chemistry & You			•
4.3 Distinguishing Among Atoms	C.2.1, C.2.2, C.2.3, C.2.4, NoS.7, NoS.8, NoS.9	•	
Small-Scale Lab	NoS.1		•
Chapter 5 Electrons in Atoms			5 periods, 2.5 blocks
5.1 Revising the Atomic Model	C.2.1, C.2.2, NoS.9	•	
Chemistry & You	NoS.9, NoS.10		•
5.2 Electron Arrangement in Atoms	C.2.1, C.2.2	•	
5.3 Atomic Emission Spectra and the Quantum Mechanical Model	C.2.1, C.2.2, NoS.1, NoS.10	•	
Chemistry & You			•
Small-Scale Lab	NoS.1, NoS.5		•

Lesson-by-Lesson Correlation and Pacing Guide *(continued)*

Chapters/Lessons	Academic Standards for Science	Fast Track	Enrichments
Chapter 6 The Periodic Table			5 periods, 2.5 blocks
6.1 Organizing the Elements	C.2.6	•	
Chemistry & You			•
6.2 Classifying the Elements	C.2.5, C.2.6	•	
6.3 Periodic Trends	C.2.6, NoS.1, NoS.3	•	
Chemistry & You			•
Small-Scale Lab	C.2.6, NoS.1, NoS.5, NoS.6, NoS.7	•	
Chapter 7 Ionic and Metallic Bonding			10 periods, 5 blocks
7.1 Ions	C.2.6, C.3.3	•	
Small-Scale Lab	C.2.6, NoS.1		•
7.2 Ionic Bonds and Ionic Compounds	C.3.1, C.3.2, C.3.3, NoS.1	•	
Chemistry & You	C.3.3, NoS.1		•
7.3 Bonding in Metals	C.3.3	•	
Chapter 8 Covalent Bonding			10 periods, 5 blocks
8.1 Molecular Compounds	C.3.1, C.3.2, C.3.3	•	
8.2 The Nature of Covalent Bonding	C.3.1, C.3.2, C.3.3, NoS.1, NoS.6	•	
Chemistry & You			•
8.3 Bonding Theories	C.3.1, C.3.2, C.3.3, C.3.4	•	

Chapters/Lessons	Academic Standards for Science	Fast Track	Enrichments
Chapter 8 Covalent Bonding *(continued)*			10 periods, 5 blocks
8.4 Polar Bonds and Molecules	C.3.1, C.3.2, C.3.3	•	
Small-Scale Lab	NoS.1		•
Chapter 9 Chemical Names and Formulas			12 periods, 6 blocks
9.1 Naming Ions	C.3.5, NoS.6	•	
Chemistry & You			•
9.2 Naming and Writing Formulas for Ionic Compounds	C.3.5, NoS.1	•	
9.3 Naming and Writing Formulas for Molecular Compounds	C.3.4	•	
Chemistry & You			•
9.4 Naming and Writing Formulas for Acids & Bases	C.3.4	•	
9.5 The Laws Governing How Compounds Form	C.3.4	•	
Small-Scale Lab	NoS.1		•
Chapter 10 Chemical Quantities			8 periods, 4 blocks
10.1 The Mole: A Measurement of Matter	C.4.3	•	
Chemistry & You	NoS.6, NoS.7		•
10.2 Mole-Mass and Mole-Volume Relationships	C.4.3	•	
Small-Scale Lab	C.4.3, NoS.1, NoS.5	•	
10.3 Percent Composition and Chemical Formulas	C.4.3, C.4.7, NoS.1, NoS.3, NoS.5	•	
Chemistry & You			•

Lesson-by-Lesson Correlation and Pacing Guide *(continued)*

Chapters/Lessons	Academic Standards for Science	Fast Track	Enrichments
Chapter 11 Chemical Reactions			8 periods, 4 blocks
11.1 Describing Chemical Reactions	C.1.4, C.4.1, C.4.2, NoS.1, NoS.3	•	
Chemistry & You	C.1.6, C.4.2, NoS.1		•
11.2 Types of Chemical Reactions	C.1.4, C.4.1, C.4.5	•	
Chemistry & You	C.4.5, NoS.10		•
11.3 Reactions in Aqueous Solution	C.1.4, C.4.1, C.4.5	•	
Small-Scale Lab	C.4.2, NoS.1	•	
Chapter 12 Stoichoimetry			8 periods, 4 blocks
12.1 The Arithmetic of Equations	C.1.6, C.4.2, C.4.3, NoS.5, NoS.6	•	
12.2 Chemical Calculations	C.1.6, C.4.3, C.4.4, C.5.3, NoS.5	•	
Chemistry & You			•
Small-Scale Lab	C.4.2, C.4.3, C.4.4, NoS.1	•	
12.3 Limiting Reagent and Percent Yield	C.4.2, C.4.3, C.4.4, C.5.3, NoS.1, NoS.5	•	
Chemistry & You			•
Chapter 13 States of Matter			5 periods, 2.5 blocks
13.1 The Nature of Gases	C.1.5, C.6.1, C.6.2	•	
13.2 The Nature of Liquids	C.1.5, C.3.3, C.6.1, C.6.2	•	
13.3 The Nature of Solids	C.1.5, C.3.3, C.6.1	•	
Small-Scale Lab	NoS.1		•
13.4 Changes of States	C.1.5, NoS.1	•	

Chapters/Lessons	Academic Standards for Science	Fast Track	Enrichments
Chemistry & You	NoS.11		
Chapter 14 The Behavior of Gases		5 periods, 2.5 blocks	
14.1 The Properties of Gases	C.5.1	●	
Chemistry & You			●
14.2 The Gas Laws	C.5.1, C.6.2	●	
14.3 Ideal Gases	C.5.1, C.5.2, C.5.3, NoS.1	●	
14.4 Gases: Mixtures and Movements			●
Small-Scale Lab	C.4.2, NoS.1	●	
Chemistry & You	NoS.11		●
15 Water and Aqueous Systems		5 periods, 2.5 blocks	
15.1 Water and Its Properties	C.3.3, C.7.1, NoS.1	●	
15.2 Homogeneous Aqueous Systems	C.7.1	●	
Chemistry & You			●
15.3 Heterogeneous Aqueous Systems	C.7.1	●	
Small-Scale Lab	NoS.1		●
Chapter 16 Solutions		6 periods, 3 blocks	
16.1 Properties of Solutions	C.7.1, C.7.2, NoS.1	●	
16.2 Concentrations of Solutions	C.7.1, C.7.4	●	
Chemistry & You	C.4.7		●

Lesson-by-Lesson Correlation and Pacing Guide *(continued)*

Chapters/Lessons	Academic Standards for Science	Fast Track	Enrichments
16.3 Colligative Properties	C.7.1	•	
16.4 Calculations Involving Colligative Properties	C.7.1, C.7.3	•	
Small-Scale Lab	C.7.4, NoS.1, NoS.2, NoS.5	•	
Chapter 17 Thermochemistry		9 periods, 4.5 blocks	
17.1 The Flow of Energy	C.6.3	•	
17.2 Measuring and Expressing Enthalpy Changes	C.6.3, C.6.4	•	
17.3 Heat in Changes of State	C.6.3, C.6.4, NoS.1, NoS.3, NoS.5	•	
Chemistry & You	C.6.4		•
17.4 Calculating Heats of Reaction	C.6.3	•	
Small-Scale Lab	NoS.1, NoS.5		•
Chapter 18 Reaction Rates and Equilibrium		9 periods, 4.5 blocks	
18.1 Rates of Reaction	C.7.5, NoS.1	•	
Chemistry & You			•
18.2 The Progress of Chemical Reactions	C.7.5	•	
18.3 Reversible Reactions and Equilibrium	C.7.6	•	
18.4 Solubility Equilibrium	C.7.2	•	
18.5 Entropy and Free Energy			•

Chapters/Lessons	Academic Standards for Science	Fast Track	Enrichments
Small-Scale Lab	NoS.1		•
Chapter 19 Acids, Bases, and Salts			7 periods, 3.5 blocks
19.1 Acid-Base Theories	C.8.1, C.8.2	•	
19.2 Hydrogen Ions and Acidity	C.8.2, C.8.4, NoS.1	•	
Chemistry & You			•
19.3 Strengths of Acids and Bases	C.8.2, C.8.3	•	
Small-Scale Lab	NoS.1		•
19.4 Neutralization Reactions	C.8.2, C.8.5	•	
Chemistry & You			•
19.5 Salts in Solution	C.8.2	•	
Chemistry & You			
Chapter 20 Oxidation-Reduction Reactions			3 periods, 1.5 blocks
20.1 The Meaning of Oxidation and Reduction	C.1.3, C.4.6, NoS.1	•	
Chemistry & You	C.4.6		•
20.2 Oxidation Numbers	C.4.6	•	
20.3 Describing Redox Equations	C.4.6	•	
Chemistry & You	C.4.6		•
Small-Scale Lab	NoS.1		•

Lesson-by-Lesson Correlation and Pacing Guide *(continued)*

Chapters/Lessons	Academic Standards for Science	Fast Track	Enrichments
Chapter 21 Electrochemistry			5 periods, 2.5 blocks
21.1 Electrochemical Cells			●
21.2 Half-Cells and Cell Potentials			●
Chemistry & You	NoS.1		●
21.3 Electrolytic Cells	NoS.1		●
Small-Scale Lab	NoS.1		●
Chapter 22 Hydrocarbon Compounds			10 periods, 5 blocks
22.1 Hydrocarbons	C.2.2, C.3.4, C.9.1, C.9.2	●	
Chemistry & You	NoS.10		●
22.2 Unsaturated Hydrocarbons	C.3.4, C.9.1, C.9.2	●	
Chemistry & You			●
22.3 Isomers	C.3.4, C.9.1, C.9.2, NoS.1, NoS.6	●	
22.4 Hydrocarbon Rings			●
22.5 Hydrocarbons From Earth's Crust			●
Chemistry & You			●
Small-Scale Lab	NoS.1, NoS.6		●
Chapter 23 Functional Groups			5 periods, 2.5 blocks
23.1 Introduction to Functional Groups	C.4.2, C.4.5, C.9.2		●
Chemistry & You	C.3.4		●

Chapters/Lessons	Academic Standards for Science	Fast Track	Enrichments
23.2 Alcohols, Ethers, and Amines	C.4.5		•
23.3 Carbonyl Compounds	C.4.5, NoS.1		•
Chemistry & You	C.3.4		•
23.4 Polymers			•
Small-Scale Lab	NoS.1		•
Chapter 24 The Chemistry of Life			10 periods, 5 blocks
24.1 A Basis for Life			•
24.2 Carbohydrates			•
24.3 Amino Acids and Their Polymers			•
Small-Scale Lab	NoS.1		•
24.4 Lipids			•
Chemistry & You			•
24.5 Nucleic Acids	NoS.1		•
24.6 Metabolism			•
Chapter 25 Nuclear Chemistry			5 periods, 2.5 blocks
25.1 Nuclear Radiation	C.2.7	•	
25.2 Nuclear Transformations	C.2.7, C.2.8	•	
25.3 Fission and Fusion	C.2.7	•	
25.4 Radiation in Your Life	C.2.7, NoS.1, NoS.3	•	

Course Planning Options

Most chemistry teachers are faced with the challenge of educating students of varied academic abilities. It takes careful planning to map out a method by which students can best be exposed to the major topics in the high school chemistry curriculum.

Prentice Hall Chemistry is organized in a way that can meet the needs of a variety of students and teachers. The textbook is comprehensive enough to satisfy the most comprehensive high school chemistry curriculum, while remaining readable, interesting, and accessible to less-academically motivated students. In addition, the division of chapters into manageable numbered sections provides today's teachers with the flexibility needed to teach classes comprised of students with diverse learning levels.

Pages T20 and T21 contain section planning options to assist teachers in designing lessons that meet the needs of all students. Symbols are used as follows.

The Conceptual Curriculum

In recent years, more students have been required to take chemistry, either in a mainstream setting or in a conceptual chemistry course. Thus, classroom populations have more students with below-average academic abilities in reading comprehension and math skills. The conceptual curriculum allows for a more concrete approach to ensure these students achieve their maximum potential. Course options are designated with symbols as follows.

● **Concrete concepts**—These sections typically describe fundamental (concrete) chemical concepts without relying on highly theoretical (abstract) or mathematical principles. These topics will help students focus on qualitative aspects of chemistry.

❍ **More abstract concepts or math/problem-solving**—These sections offer a combination of concrete topics plus theoretically abstract topics and/or mathematically challenging problem solving. In these sections, the emphasis on theoretical aspects or the degree of problem solving can be determined by the teacher.

Some sections have no symbol because they are beyond the scope of a typical conceptual curriculum and do not include concepts that are prerequisites for later chapters.

The Standard Curriculum

The standard curriculum is designed for students considered average in ability and achievement levels. These students should have little difficulty reading and understanding most of the material presented, and they will have sufficient math skills for tackling the problems in the textbook. Course options are designated with symbols as follows.

■ **Core content**—These sections are suitable for all students in a standard curriculum. A course that includes most of these sections will prepare students for other science courses and for becoming scientifically literate citizens.

❑ **Extension topics**—These sections may be covered as time allows or according to the interest of the teacher and the needs and/or problem-solving skills of the students.

Some sections have no symbol because a teacher would not have sufficient time to cover the topic well, or because the material is too challenging or abstract, or is not usually required in state or national curricula. Assume if there is no symbol that the section is not a prerequisite for later chapters.

The Honors Curriculum

The standard curriculum is designed to be used for students considered above average in their reading comprehension and general scholastic abilities. These students should be able to master most of the content in the textbook with less time and effort. Course options are designated with symbols as follows.

▲ **Core honors content**—These sections are suitable for all students in an honors curriculum. A course including these sections will prepare students for advanced science courses and for college coursework.

△ **Options to accelerate**—These sections may be covered more rapidly using homework assignments. Students may only need to review this material because they have studied it in a previous course or a previous chapter.

		C	S	H
1	**Introduction to Chemistry**			
1.1	Chemistry	●	■	△
1.2	Chemistry Far and Wide	●	■	△
1.3	Thinking Like a Scientist	●	■	△
1.4	How to Study for Chemistry	●	■	△
2	**Matter and Change**			
2.1	Matter	●	■	△
2.2	Mixtures	●	■	▲
2.3	Elements and Compounds	●	■	▲
2.4	Chemical Reactions	●	■	▲
3	**Scientific Measurement**			
3.1	Measurements and their Uncertainty	○	■	▲
3.2	The International System of Units	●	■	△
3.3	Density	●	■	▲
4	**Atomic Structure**			
4.1	Defining the Atom	●	■	△
4.2	Structure of the Atom	●	■	▲
4.3	Distinguishing Among Atoms	○	■	▲
5	**Electrons in Atoms**			
5.1	Models of the Atom	●	■	▲
5.2	Electron Arrangement in Atoms	○	■	▲
5.3	Physics and the Quantum Mechanical Model	○		▲
6	**The Periodic Table**			
6.1	Organizing the Elements	●	■	▲
6.2	Classifying the Elements	●	■	▲
6.3	Periodic Trends	●	■	▲
7	**Ionic and Metallic Bonding**			
7.1	Ions	●	■	▲
7.2	Ionic Bonds and Ionic Compounds	●	■	▲
7.3	Bonding in Metals	●	■	▲

		C	S	H
8	**Covalent Bonding**			
8.1	Molecular Compounds	●	■	▲
8.2	The Nature of Covalent Bonding	●	■	▲
8.3	Bonding Theories		❑	▲
8.4	Polar Bonds and Molecules	●	■	▲
9	**Chemical Names and Formulas**			
9.1	Naming Ions	●	■	▲
9.2	Naming and Writing Formulas for Ionic Compounds	●	■	▲
9.3	Naming and Writing Formulas for Molecular Compounds	●	■	▲
9.4	Naming and Writing Formulas for Acids and Bases	●	■	▲
9.5	The Laws Governing Formulas and Names	●	■	▲
10	**Chemical Quantities**			
10.1	The Mole: A Measurement of Matter	○	■	▲
10.2	Mole–Mass and Mole–Volume Relationships	○	■	▲
10.3	Percent Composition and Chemical Formulas		■	▲
11	**Chemical Reactions**			
11.1	Describing Chemical Reactions	●	■	▲
11.2	Types of Chemical Reactions	●	■	▲
11.3	Reactions in Aqueous Solutions	○	■	▲
12	**Stoichiometry**			
12.1	The Arithmetic of Equations	●	■	△
12.2	Chemical Calculations		■	▲
12.3	Limiting Reagent and Percent Yield		❑	▲
13	**States of Matter**			
13.1	The Nature of Gases	○	■	▲
13.2	The Nature of Liquids	●	■	▲
13.3	The Nature of Solids	○	■	▲
13.4	Changes of State		❑	▲

Conceptual Curriculum
- ● Concrete concepts
- ○ More abstract concepts or math/problem-solving

The Standard Curriculum
- ■ Core content
- ❑ Extension topics

The Honors Curriculum
- ▲ Core honors content
- △ Options to accelerate

		C	S	H
14	**The Behavior of Gases**			
14.1	Properties of Gases	●	■	△
14.2	The Gas Laws	○	■	▲
14.3	Ideal Gases		■	▲
14.4	Gases: Mixtures and Movements	○	■	▲
15	**Water and Aqueous Systems**			
15.1	Water and Its Properties	●	■	▲
15.2	Homogeneous Aqueous Systems	●	■	▲
15.3	Heterogenous Aqueous Systems		❑	△
16	**Solutions**			
16.1	Properties of Solutions	●	■	▲
16.2	Concentrations of Solutions	○	■	▲
16.3	Colligative Properties of Solutions		■	▲
16.4	Calculations Involving Colligative Properties		❑	▲
17	**Thermochemistry**			
17.1	The Flow of Energy—Heat and Work	○	■	▲
17.2	Measuring and Expressing Enthalpy Changes	○	■	▲
17.3	Heat in Changes of State	○	❑	▲
17.4	Calculating Heats of Reaction			▲
18	**Reaction Rates and Equilibrium**			
18.1	Rates of Reaction	●	■	▲
18.2	Reversible Reactions and Equilibrium	○	❑	▲
18.3	Solubility Product		❑	▲
18.4	Calculating Entropy and Free Energy			▲
18.5	The Progress of Chemical Reactions			▲
19	**Acids and Bases**			
19.1	Acid-Base Theories	●	■	△
19.2	Hydrogen Ions and Acidity	○	■	▲
19.3	Strengths of Acids and Bases		❑	▲
19.4	Salts in Solution		❑	▲

		C	S	H
20	**Oxidation-Reduction Reactions**			
20.1	The Meaning of Oxidation and Reduction	●	■	▲
20.2	Oxidation Numbers		■	▲
20.3	Balancing Redox Equations		❑	▲
21	**Electrochemistry**			
21.1	Electrochemical Cells	●	■	▲
21.2	Half-Cells and Cell Potentials	○	❑	▲
21.3	Electrolytic Cells	○	■	▲
22	**Hydrocarbon Compounds**			
22.1	Hydrocarbons	●	■	▲
22.2	Unsaturated Hydrocarbons	○	■	▲
22.3	Isomerism		❑	▲
22.4	Hydrocarbon Rings	○	❑	▲
22.5	Hydrocarbons from the Earth	●	■	△
23	**Functional Groups and Organic Reactions**			
23.1	Introduction to Functional Groups	○	■	▲
23.2	Alcohols and Ethers		❑	▲
23.3	Carbonyl Compounds		❑	▲
23.4	Polymerization		❑	▲
24	**The Chemistry Of Life**			
24.1	A Strategy for Life	●	■	△
24.2	Carbohydrates		❑	▲
24.3	Amino Acids and Their Polymers		❑	▲
24.4	Lipids		❑	▲
24.5	Nucleic Acids		❑	▲
24.6	Metabolism		❑	▲
25	**Nuclear Chemistry**			
25.1	Nuclear Radiation	●	■	▲
25.2	Nuclear Transformations	○	■	▲
25.3	25.3 Fission and Fusion of Atomic Nuclei	●	■	▲
25.4	25.4 Radiation in Your Life	●	■	△

Conceptual Curriculum
- ● Concrete concepts
- ○ More abstract concepts or math/problem-solving

The Standard Curriculum
- ■ Core content
- ❑ Extension topics

The Honors Curriculum
- ▲ Core honors content
- △ Options to accelerate

Materials Lists

Master Solutions List

The following list gives directions for preparing all the chemical solutions necessary for doing all the Small-Scale Labs in **Pearson Chemistry**. The most efficient way to organize your small-scale lab is to make up the appropriate aqueous solutions, then use and replace them as needed. Store each solution in a plastic reagent bottle. Depending on the labs you do, some solutions will have to be replaced several times during the year; others will last the entire year. Most solutions are 0.1*M* or 0.2*M*. In no case is any solution of greater concentration than 1.0*M*. Except for sodium chloride, the labs do not require you to prepare multiple solutions for the same chemical.

For example, all the experiments that use sodium hydroxide use 0.5*M* NaOH.

Label each reagent bottle with the chemical formula, the concentration, the number of grams of solute per volume of solution, and the date. When the bottle is empty, simply replenish it by following the directions on the label, and change the date. When making dilute acid solutions from concentrated acids, always add acid to water slowly and carefully. The Teacher's Edition of the Small-Scale Chemistry Laboratory Manual has a combined list of solutions for the labs in the Student Edition and the Small-Scale Chemistry Laboratory Manual.

Solution	Preparation	Labs
alizarin yellow R, 0.02%	50 mg AYR in 250 mL	670
aluminum chloride, 0.2*M*	12.1 g $AlCl_3 \cdot 6H_2O$ in 250 mL	200
ammonia, 1.0*M*	67 mL of 15*M* NH_3 in 1.0 L **Caution! Using proper ventilation, add ammonia to water carefully and slowly!**	436, 508, 604
borax, 4%	10 g $Na_2B_4O_7 \cdot 10H_2O$ in 250 mL	828
bromocresol green*, 0.04%	100 mg BCG in 14.3 mL 0.01*M* NaOH; dilute to 250 mL	670
bromophenol blue*, 0.04%	100 mg BPB in 14.9 mL 0.01*M* NaOH; dilute to 250 mL	670
bromothymol blue*, 0.04%	100 mg BTB in 16.0 mL 0.01*M* NaOH; dilute to 250 mL	435, 475, 670, 752
calcium chloride, 0.5*M*	13.9 g $CaCl_2$ in 250 mL	200, 295, 374
copper(II) sulfate, 0.2*M*	12.5 g $CuSO_4 \cdot 5H_2O$ in 250 mL	51, 200, 295, 752, 849
ethanoic acid, 0.8*M* CH_3COOH	undiluted white vinegar	254, 435, 508, 717
hydrochloric acid, 1.0*M*	82 mL of 12*M* HCl in 1.0 L **Caution! Always add acid to water carefully and slowly!**	399, 508, 717, 849
hydrogen peroxide, 3% H_2O_2	undiluted household supply	51
iron(III) chloride, 0.1*M*	6.8 g $FeCl_3 \cdot 6H_2O$ in 25 mL of 1.0*M* NaCl; dilute to 250 mL	200, 295
lead(II) nitrate, 0.2*M*	16.6 g $Pb(NO_3)_2$ in 250 mL	295, 374
magnesium sulfate, 0.2*M*	6.0 g $MgSO_4$ in 250 mL	200, 295
metacresol purple*, 0.04%	100 mg MCP in 26.2 mL 0.01*M* NaOH; dilute to 250 mL	670

Master Solutions List *(continued)*

Solution	Preparation	Labs
methyl orange, 0.05%	125 mg MO in 250 mL	670
nitric acid, 1.0*M*	63 mL of 15.8*M* HNO_3 in 1.0 L **Caution! Always add acid to water carefully and slowly!**	508, 717
nickel(II) sulfate, 0.2*M*	2.94 g $NiSO_4$ in 100 mL	200
pH buffer, Solution A	14.9 g boric acid and 12.6 g citric acid in 1.2 L	670
pH buffer, Solution B	45.6 g $Na_3PO_4 \cdot 12H_2O$ in 1.2 L	670
phenolphthalein, 0.1%	250 mg phenolphthalein in 250 mL of 70% 2-propanol	670
potassium bromide, 0.2*M*	6.0 g KBr in 250 mL	752
potassium iodide, 0.1*M*	4.2 g KI in 250 mL	51
2-propanol (70% isoproponal)	undiluted rubbing alcohol	254, 508
silver nitrate, 0.05*M*	2.1 g $AgNO_3$ in 250 mL	200, 295, 374
sodium carbonate, 1.0*M*	26.5 g Na_2CO_3 in 250 mL	200, 295, 374
sodium chloride, 0.1%	1.0 g NaCl in 1.0 L	254
sodium chloride, 1.0*M*	14.6 g NaCl in 250 mL	200, 374, 752
sodium hydrogen sulfite, 1.0*M*	4.2 g $NaHSO_3$ in 250 mL	475
sodium hydroxide, 0.5*M*	20.0 g NaOH in 1.0 L	200, 295, 374, 399, 508, 849
sodium hypochlorite, 1%	50 mL liquid bleach (NaClO) in 200 mL	51
sodium phosphate, 0.1*M*	9.5 g $Na_3PO_4 \cdot 12H_2O$ in 250 mL	200, 295, 374
sodium sulfate, 0.2*M*	7.1 g Na_2SO_4 in 250 mL	295, 374, 752
starch, 20%	50 mL liquid starch in 200 mL	51, 752
sulfuric acid, 1.0*M*	56 mL of 18*M* H_2SO_4 in 1.0 L **Caution! Always add acid to water carefully and slowly!**	508, 717
thymol blue*, 0.04%	100 mg TB in 21.5 mL 0.01*M* NaOH; dilute to 250 mL	399, 670
zinc chloride, 0.2*M*	6.8 g $ZnCl_2$ in 250 mL	200

*If the indicator is available as the sodium salt, dissolve directly in 250 mL of deionized or distilled water.

Master Solids List

For each solid that can be poured, fill six 30- to 60-mL plastic pill vials. Use a 1/4-inch office punch to make a hole in each lid. Cover the holes with plastic tape. Label and store each set in a plastic ice cube tray.

Name	Labs
ammonium chloride, NH_4Cl	635
calcium carbonate, $CaCO_3$	324
calcium chloride, $CaCl_2$	435, 635
guar gum	828
iron, Fe (staples)	717
magnesium, Mg	717
magnesium sulfate, $MgSO_4$	508
potassium chloride, KCl	508, 635
potassium iodide, KI	508
sodium carbonate, Na_2CO_3	508
sodium chloride, NaCl (table salt)	51, 324, 508, 635
sodium chloride, NaCl (lab grade)	545
sodium hydrogen carbonate, $NaHCO_3$ (baking soda)	399, 508, 635
sodium phosphate, Na_3PO_4	635
sucrose (table sugar)	508, 545
zinc, Zn (galvanized nails)	717

Master Equipment List

Name	Labs
balance and weighing paper	92, 120, 324, 399, 545, 583, 849
beaker, small	324
chromatography paper	254
colored filters	149
conductivity tester	508
diffraction grating	149
electrolysis device	752
Petri dish, plastic	435, 475
pipet	51, 92, 349, 545, 828
spot plate, 96-well	184
volumetric flask, 50-mL plastic	545
well strip, 8-well	92

Master Household Products List

Store liquids in their original containers.

Name	Labs
alcohol thermometer	635
aluminum cans	92
aluminum foil	583
antacid tablets	51
baking powder	399
beverages, assorted	508
candles, votive	583
candy (colored)	120, 254
cereal	51
cereal box	149
construction paper, colored	51
cornstarch	508, 828
cotton swabs	475
cups, plastic	92, 120, 254, 399, 635, 828
dice	92
eggs	849
food coloring	254
glue, white	828
graph paper	435, 887
ice, cubes or crushed	435, 635
marking pens, fine line	51, 184
matches, safety	583
medicine droppers	51, 200, 508
milk, powdered	849
modeling clay	787
pennies (pre- and post-1982)	92, 717, 887
pickle juice	508
powdered drinks	254
reaction surface	51, 200, 295, 374, 475, 508, 670, 717, 752
scissors	149, 184, 254
spoons, plastic	324, 635, 828
starch, liquid laundry	828
straws	184, 399, 828
toothpicks	254, 787

Quick Lab and On Your Own Materials List

Material	Quick Lab	On Your Own
air freshener, solid	437	
aluminum foil	750	
ammonia, household	662	
ammonia, 6*M* 100 mL 15*M* NH_3 in 250 mL **Caution! Using proper ventilation, add ammonia to water carefully and slowly!**	818	
antacid tablets, effervescent, varied brands	467	
baking soda ($NaHCO_3$)	207, 662	335
balance, centigram	328, 404	
ball-and-stick molecular model kit	778	
balloons, round, different colors	404, 467	
batteries, D-cell	207	
9-volt battery	750	
beakers, 100-mL	354	
bell wire	207	
bleach, powder	699	
borax	662	
bowls		671
box	109	
Bunsen burner	142, 328, 600	
calcium carbonate ($CaCl_2$)		671
calcium chloride ($CaCl_2{\cdot}2H_2O$)	328	
calcium chloride, 0.5*M* 13.9 g $CaCl_2$ in 250 mL	142	
can, 6-oz and 16-oz	238	
carbonated beverage	662	
carbonated water		671
cardboard, thick, cut in strips	437	
paper towel roll	856	
clock or watch	467	
cloth, clean white	662	
coat hanger	238	

Material	Quick Lab	On Your Own
colored fabric	699	
copper(II) chloride, 0.1*M* 3.4 g $CuCl_2$ in 250 mL	142	
copper strip		744
copper(II) sulfate ($CuSO_4{\cdot}5H_2O$)	328	
cornstarch	519	
cotton swabs	142	
cups, 8-oz plastic, clear	17, 39, 207, 437, 589, 662	
cups, paper		355
cup, foam	571	
dish or other shallow container	437, 491	
dishwashing liquid	17, 491, 662	
distilled water	207, 519	
dropper bottles, plastic	279	
duct tape	896	
Erlenmeyer flasks, 250-mL	404	
ethanol	818	
filter paper	39, 750	
flashlight	519, 896	
flower petals, different colors	699	
glasses, drinking		681
glucose solution	819	
graduated cylinders, 10-mL, 100-mL, and 250-mL	404, 571	
grape juice	699	
graph paper	180, 238 467, 896	

Continues

Material	Quick Lab	On Your Own
grass stain on piece of white fabric	699	
grocery bag, plastic	238	
heat-resistant surface	600	
hydrochloric acid, 1.0*M* 20.5 mL of 12*M* HCl in 250 mL **Caution! Always add acid to water carefully and slowly!**	404	
hydrogen peroxide, 3% H_2O_2	699	
ice	437, 571	
index cards, 3 inch by 5 inch	72	
iodine solution, 1% I_2 in 2% KI 2.5 g I_2 plus 5.0 g KI in 250 mL	699	
iron(III) chloride, 0.1*M* 6.8 g $FeCl_3 \cdot 6H_2O$ in 250 mL	279	
jar, cylindrical with lid	519, 662	
knife	662	
laundry detergent	662	
lead(II) nitrate, 0.2*M* 16.6 g $Pb(NO_3)_2$ in 250 mL	279	
lemon		744
lemon juice	662	355
lithium chloride, 0.1*M* 1.1 g LiCl in 250 mL	142	
magnesium ribbon	404	
marble	109	
masking tape	207, 519	671, 681
measuring cups	17, 662	208, 681
measuring spoons	354	681
magnesium sulfate		208
medicine droppers	467, 662 699, 818	355

Material	Quick Lab	On Your Own
metric ruler, meter stick	39, 72, 238 662, 856, 896	
milk	662	
milk of magnesia	662	
mouthwash	662	
nails, iron finishing	750	
nickel coin	750	
oxalic acid solution, 1% 2.5 g in 250 mL	699	
paper clips, metal	238, 491	
paper towels		671
pencil	39	
pennies	750	
pens, colored marking, green included	39, 856	
permanent markers		671, 681
pH test strips		681
pie plates		208
plastic foam shapes	109	
plastic wrap	39, 662	
poster board, white	896	
potassium permanganate, 0.05*M* 0.40 g $KMnO_4$ in 250 mL	699	
propanone	818	

Continues

Quick Lab and On Your Own Materials List *(continued)*

Material	Quick Lab	On Your Own
reaction surface	750	
red cabbage leaves	662	
rubber band, No. 25 (about 2 inches long)	238, 491	
rubbing alcohol	39	
rusty water	699	
sandpaper, fine		744
scissors	896	
sea salt		208, 681
shampoo	662	
silver nitrate, 5% 12.5 g $AgNO_3$ in 250 mL	818	
silver nitrate, 0.05*M* 2.1 g $AgNO_3$ in 250 mL	279	
sodium carbonate, 1.0*M* 26.5 g Na_2CO_3 in 250 mL	279	
sodium chloride (NaCl)	17, 207, 662	
sodium chloride, 0.1*M* 0.6 g NaCl in 100 mL	142	
sodium hydrogen carbonate ($NaHCO_3$)	354, 519	
sodium hydroxide, 1*M* 10 g NaOH in 250 mL	818	
sodium hydroxide, 0.5*M* 5 g NaOH in 250 mL	279	
sodium hypochlorite (NaClO), household liquid bleach	699	
sodium phosphate, 0.1*M* 9.5 g $Na_3PO_4 \cdot 12H_2O$ in 250 mL	279	
sodium sulfate ($Na_2SO_4 \cdot 10H_2O$)	328, 750	
sodium thiosulfate solution, 0.2*M* 12.4 g $Na_2S_2O_3 \cdot 5H_2O$ in 250 mL	699	
spatula	328	
spoon	662	355
spot plate	699	

Material	Quick Lab	On Your Own
steel wool (0000 weight)	600	
stirring rod	519	208
straw, drinking	17	
sucrose (table sugar)	17, 207	
tape measure, metric	467	
tape, transparent	39, 662	
tarnished silver item	354	
teaspoon	17, 519, 662	
test tube holder	328	
test tube rack	142, 279, 328, 818	
test tubes, medium	328	
test tubes, small	142, 279, 818	
thermometer	600	
thumbtack	856	
tissue paper	600	
tongs	142, 600	
toothpaste	662	
toothpicks, wooden	856	
vegetable oil	491	
vinegar, white (CH_3COOH)	207, 662	355, 671
water, distilled	207	681
wire for flame tests (10-cm length of nichrome wire)	142	
zinc strip		744

Lab Safety

Chemistry teachers are no longer exempt from successful prosecution for accidents that occur in their classrooms. That threat, coupled with our increased awareness of the insidious effects of chemical exposure, explains the demand for safety education, particularly in science classrooms.

When it comes to laboratory chemistry, the safety of students and teachers is the primary concern of Prentice Hall. To that end, both the traditional Laboratory Manual and the Small-Scale Chemistry Laboratory Manual give detailed safety information for the student and the teacher at the beginning of each lab and in the front of each laboratory manual. In the Student Edition, the labs display appropriate safety icons, which are defined for the students in Appendix D.

Safety in the Chemistry Laboratory

The following information about safety issues in the high school chemistry laboratory is intended to be used by classroom chemistry teachers. This reference section is intended to be accessible to the teacher for planning demonstrations and preparing for lab. The material included here is by no means comprehensive. Prentice Hall takes no responsibility for the completeness of this information or for the implementation of these suggestions.

The safety of the students in a high school laboratory is traditionally assigned to the teacher. But the teacher teaches safety awareness by holding each student accountable for the safety of all other participants in the lab, including the teacher. Have students sign a safety contract at the beginning of the year. There is a sample contract in each laboratory manual. Of course, the teacher will most effectively teach safety awareness by modeling safe practices and by making a discussion of safety conerns the first priority for all lab activities. The following suggestions are intended to help you establish a program of safety awareness for staff and students.

Education

- Put together a safety committee for your school. Include faculty, students, and administrators.
- Schedule regular meetings of the safety committee to discuss issues, accident reports, and the results of lab inspections.
- Implement a safety education program for staff and students.
- Require students to know the lab safety rules, assess this by giving a safety quiz, and have the students sign a safety contract, saying that they know the rules and will comply with them. Send the contract home with a set of rules and require that one parent or guardian sign the contract, having read the rules.
- Give a safety grade with each lab, based on each student's behavior during the lab. Penalize unsafe behavior by removing students from the lab.
- Encourage school staff to become certified in CPR and first aid.
- Require that science staff read any of several publications on safety in the high school science classroom.

Management

- Report all accidents to the safety committee.
- Periodically inspect laboratories and report findings to the safety committee.
- Establish an emergency plan and review it periodically with students and staff.
- Keep phone numbers for the fire department, police department, local ambulance, and poison control center next to every phone in a lab or stockroom.
- Display warning signs and safety posters.
- Prohibit student access to stockrooms.
- Require students to keep aisles clear and counters clean. There should be two exits in a laboratory or stockroom, and they should be kept free of obstacles.
- Keep a current chemical inventory. Update the chemical inventory often. Include a Material Safety Data Sheet (MSDS) for each chemical.
- Label all chemicals with the date and possible hazards. Establish a policy for disposing of and buying chemicals.
- In prelab discussions, stress safety issues relevant to each specific lab.

Lab Techniques

- Do not allow any student or staff member to work alone in the laboratory.
- Do not leave any experiment unattended.
- Do not eat, smoke, or drink in the lab.
- Do not store food in chemical refrigerators. Store chemicals only in chemical refrigerators.
- Never taste a chemical in the laboratory.
- Always add acids to water, never water to acids.
- When pipetting, use a suction bulb. Never use your mouth.
- When heating the contents of a test tube, direct the mouth of the test tube away from yourself and away from others.

Personal Safety

- Always wear goggles in lab. Require students and anyone entering the lab to wear goggles.
- Provide lab coats and aprons for all participants in the lab.
- Take into account student health problems such as allergies or asthma.
- Do not allow untied long hair, loose clothing, bare feet, open-toed or -heeled shoes, or dangling jewelry to be worn in the lab.
- Strongly discourage the wearing of contact lenses in lab. In the event of a chemical splash into the eye, the lenses may hold chemicals in contact with the eye, potentially causing greater injury.

Equipment

- Make available the following items in all labs and demonstration areas: an ABC fire extinguisher, a smoke alarm, a safety shower, an eye-wash faucet, a fume hood.
- Maintain and check all safety equipment regularly.
- Keep a first-aid kit stocked and available.
- Keep flammable chemicals in an approved fireproof cabinet.
- Attach compressed gas cylinders to counters or walls. Move them on dollies on which they are secured.
- Use safety shields when there is a possibility of explosion, implosion, chemical splashing, or fire.
- Use hot plates in lieu of open flames as often as possible.
- Use dispensing devices such as syringes, spigots, or droppers, rather than having students pour chemicals from their containers.
- Be prepared for spills, especially acid spills. Spill-control devices are available from suppliers. To avoid mercury spills, use alcohol thermometers.

Storing Chemicals

- Store chemicals in stockrooms, not in the lab.
- Do not store chemicals alphabetically. Use an approved system of chemical storage, and post legends in the stockroom.
- Store only the chemicals you will use during a school year.

Disposal

- Dispose of broken glass in a dedicated container kept out of the way and clearly labeled.
- Dispose of chemicals in the following ways:

  Neutralize acids and bases and flush down the drain.

  Flush light metal salts and their solutions down the drain with excess water.

  Precipitate and filter heavy metals. Evaporate heavy metal solutions down to salt. Keep heavy metals salts in labeled containers and dispose of them by following local regulations.

  When in doubt, check with local authorities about local regulations.

- Have a disposal plan for each lab, and include directions for disposal in your prelab discussion.

PEARSON

Chemistry

INDIANA

Antony C. Wilbraham • Dennis D. Staley • Michael S. Matta • Edward L. Waterman

Boston, Massachusetts • Chandler, Arizona • Glenview, Illinois • Upper Saddle River, New Jersey

Print Components

Indiana Student Edition

Indiana Teacher's Edition

Reading and Study Workbook

Reading and Study Workbook, Teacher's Edition

Indiana Progress Monitoring Assessments

Indiana Standardized Test Preparation Workboook

Technology Components

Online Student Edition

Online Teacher's Edition

Math Support:

- Math Diagnostic and Remediation
- Online Problem Sets

Tutorials:

- ChemTutor
- MathTutor

Animations:

- Concepts in Action Animations
- Kinetic Art Animations
- Directed Virtual ChemLabs

Videos:

- Untamed Science® Videos

Editable Resources:

- Lab Manual
- Small-Scale Lab Manual
- Probeware Lab Manual
- Lab Practicals
- Reading and Study Workbook
- Assessment Workbook
- PowerPoint® Presentations

CD-ROM and DVD-ROM

Indiana Exam*View*® CD-ROM

Indiana Classroom Resource DVD-ROM

Virtual ChemLab CD-ROM

Front matter credits: **ii** Superstock.
Additional credits appear on pages R156–R157, which constitute an extension of this copyright page.

ISBN-13: 978-0-13-252579-4
ISBN-10: 0-13-252579-8

2 3 4 5 6 7 8 9 10 V011 15 14 13 12 11

About the Authors

Dennis Staley earned a Master of Science in chemistry at Southern Illinois University Edwardsville. As an Instructor in the Department of Chemistry and the Office of Science and Math Education at Southern Illinois University Edwardsville, he taught high school and college-level chemistry and also led teacher workshops. Mr. Staley has been writing high school and college-level chemistry textbooks for over 30 years. He currently enjoys gardening, bicycling, sharing hands-on science with kids, and traveling to visit his grandchildren.

Antony Wilbraham spent the majority of his career at Southern Illinois University Edwardsville, where he currently holds the position of Emeritus Professor of Chemistry. He is a member of several professional societies. For more than 30 years, he has been writing high school and college-level chemistry textbooks and has published extensively in scientific journals. Professor Wilbraham enjoys traveling, woodworking, gardening, and making toys for his granddaughters.

Michael Matta earned a Bachelor of Science in chemistry at the University of Dayton and a Doctor of Philosophy in chemistry at Indiana University. He spent most of his career at Southern Illinois University Edwardsville, where he was most recently an Emeritus Professor. Dr. Matta developed and wrote high school and college-level chemistry textbooks and related ancillaries for over 30 years and published extensively in scientific journals. He was a member of several professional societies. In his spare time, he enjoyed woodworking, watercolor painting, and playing with his six grandchildren.

Michael Matta died shortly after the onset of this program. While he is greatly missed by his many friends throughout the chemistry community, his coauthors remain inspired by his visionary dedication to education, and they are committed to searching for even better ways to engage students in the wonders of chemistry.

Edward Waterman taught chemistry and advanced placement chemistry from 1976 to 2007 at Rocky Mountain High School in Fort Collins, Colorado. He now conducts workshops for teachers on inquiry, differentiation, small-scale chemistry, AP chemistry and virtual chemistry laboratory. He also presents photo-essay lectures about the natural history of molecules, engaging the general public in the appreciation for and understanding of chemistry. Mr. Waterman holds a Bachelor of Science degree in chemistry from Montana State University and a Master of Science degree in chemistry from Colorado State University. In his free time, he enjoys exploring wild places in the Rocky Mountains and on the Colorado Plateau by hiking, kayaking, and cross-country skiing.

Consultants/Reviewers

Grant Wiggins, Ed.D. Dr. Wiggins is a coauthor of *Understanding by Design®*, a philosophy of instructional design. The Understanding by Design methodology is a disciplined way of thinking about curriculum design, assessment, and instruction that moves teaching from covering the content to ensuring understanding.

BIGIDEA Big Ideas are one of the core components of the Understanding by Design methodology in **Pearson Chemistry.** These Big Ideas, such as The Mole and Quantifying Matter, establish a conceptual framework for the program. Look for opportunities throughout each chapter to link back to the Big Ideas.

The Association for Supervision of Curriculum Development (ASCD), publisher of the "Understanding by Design Handbook" co-authored by Grant Wiggins and registered owner of the trademark "Understanding by Design", has not authorized, approved or sponsored this work and is in no way affiliated with Pearson or its products.

Teacher Advisory Board

Linda Dearth-Monroe
Warren Central High School
Indianapolis, Indiana

Jason Gilley
Cypress Creek High School
Orlando, Florida

Kenneth A. Greathouse
Parkway Central High School
Chesterfield, Missouri

Paul Holloman
Rocky Mount High School
Rocky Mount, North Carolina

George "Rod" Larsen
West Orange High School
Winter Garden, Florida

Stephanie C. LeGrone
Mary G. Montgomery High School
Semmes, Alabama

Christopher Schrempp
Los Osos High School
Rancho Cucamonga, California

Content Reviewers

Matthew Asplund, Ph.D.
Department of Chemistry and Biochemistry
Brigham Young University
Provo, Utah

Regina M. Barrier
Western Outreach Coordinator
The Science House
North Carolina State University
Lenoir, North Carolina

J. Phillip Bowen, Ph.D.
Department of Chemistry and Biochemistry
University of North Carolina
Greensboro, North Carolina

Alison J. Frontier, Ph.D.
Department of Chemistry
University of Rochester
Rochester, New York

David J. Merkler, Ph.D.
Department of Chemistry
University of South Florida
Tampa, Florida

Gregory S. Owens, Ph.D.
Department of Chemistry
University of Utah
Salt Lake City, Utah

Eric T. Sevy, Ph.D.
Department of Chemistry and Biochemistry
Brigham Young University
Provo, Utah

William H. Steinecker, Ph.D.
Miami University
Oxford, Ohio

Harry A. Stern, Ph.D.
Department of Chemistry
University of Rochester
Rochester, New York

Mark E. Welker, Ph.D.
Department of Chemistry
Wake Forest University
Winston-Salem, North Carolina

Teacher Reviewers

Jeff Bilyeu
West Linn High School
West Linn, Oregon

Mary Chuboff
Athens Academy
Athens, Georgia

Linda Dearth-Monroe
Warren Central High School
Indianapolis, Indiana

Jason Gilley
Cypress Creek High School
Orlando, Florida

Stella Glogover
Head-Royce School
Oakland, California

Paul Holloman
Rocky Mount High School
Rocky Mount, North Carolina

Laura McGregor
Marist School
Atlanta, Georgia

Nancy Monson
West Linn High School
West Linn, Oregon

Daniel R. Mullaney
Walpole High School
Walpole, Massachusetts

Michael Roadruck, Ph.D.
Department of Chemistry
University of Toledo
Toledo, Ohio

Michelle Tindall
Birmingham Groves High School
Beverly Hills, Michigan

Safety Consultant

Kenneth R. Roy, Ph.D.
Director of Science and Safety
Glastonbury Public Schools
Glastonbury, Connecticut

All the Elements for Success

The new **Pearson Chemistry** program, created specifically for Indiana, combines proven content with cutting-edge digital support and a variety of lab investigations to help ensure your success in chemistry.

In your new program, you'll find:

- **Personalized learning** opportunities to support your unique learning style on your path to mastery of the Indiana Academic Standards.
- **Connections to the real world** relate abstract concepts and processes to your every day life.
- **Cutting-edge technology** that is integrated throughout the program providing options for you to interact with the content in multiple ways.
- **Rich lab explorations** and **study support** provide numerous opportunities to practice and reinforce essential chemistry skills.

In the following pages, you'll see just a few of the elements contained in your **Pearson Chemistry** for Indiana program that will lead to your success!

Big Ideas ensure that the lesson is centered around the most important concepts of each chapter. Essential Questions that support the Big Idea organize and set the focus for the chapter.

22
Hydrocarbon Compounds

INSIDE:
- 22.1 Hydrocarbons
- 22.2 Unsaturated Hydrocarbons
- 22.3 Isomers
- 22.4 Hydrocarbon Rings
- 22.5 Hydrocarbons From Earth's Crust

PearsonChem.com

VIRTUAL LAB

Turkmenistan has some of the world's largest deposits of natural gas. This natural gas deposit in the city of Darvaza has been burning for over three decades!

760

BIGIDEA

CARBON CHEMISTRY

Essential Questions:
1. *How are hydrocarbons named?*
2. *What are the general properties of hydrocarbons?*

CHEMYSTERY

Nose for Hire

Walking home from school one day, Anthony spotted a poster soliciting participants for a smell test. The poster offered participants "$50 for less than an hour of your time." So he decided to go for it.

When Anthony reported for the study, after filling out some paperwork, a researcher asked him to smell two chemical samples. The first one smelled like a freshly cut orange. The second had an odor that reminded him of pine trees.

Anthony was curious. "What am I smelling?" he asked. "Limonene," the researcher answered. "What about the second one?" Anthony asked. The researcher gave the same reply: "Limonene." Anthony was perplexed. How could two substances, both with the name limonene, smell so different?

► Connect to the **BIGIDEA** As you read about hydrocarbons, think about what could make this phenomenon possible.

INDIANA'S ACADEMIC STANDARDS FOR SCIENCE

Atomic Structure and the Periodic Table C.2.2; **Bonding and Molecular Structure** C.3.4; **Organic Chemistry and Biochemistry** C.9.1, C.9.2. See lessons for details.

CHEMystery features present mini-mysteries that take the Big Idea a step further into a practical context. Once you study the chapter, you'll have the opportunity to solve the mystery and understand how it connects to the Big Idea.

Personalized Learning

Each student approaches learning in a variety of ways. **Pearson Chemistry** for Indiana provides all the tools to support your unique style to help you master the Indiana Academic Standards. Sample chemistry problems and math support are just a few examples.

Math Tune-Up pages at the end of each chapter help you quickly reference how to solve a problem—and are also a great way to review before your test!

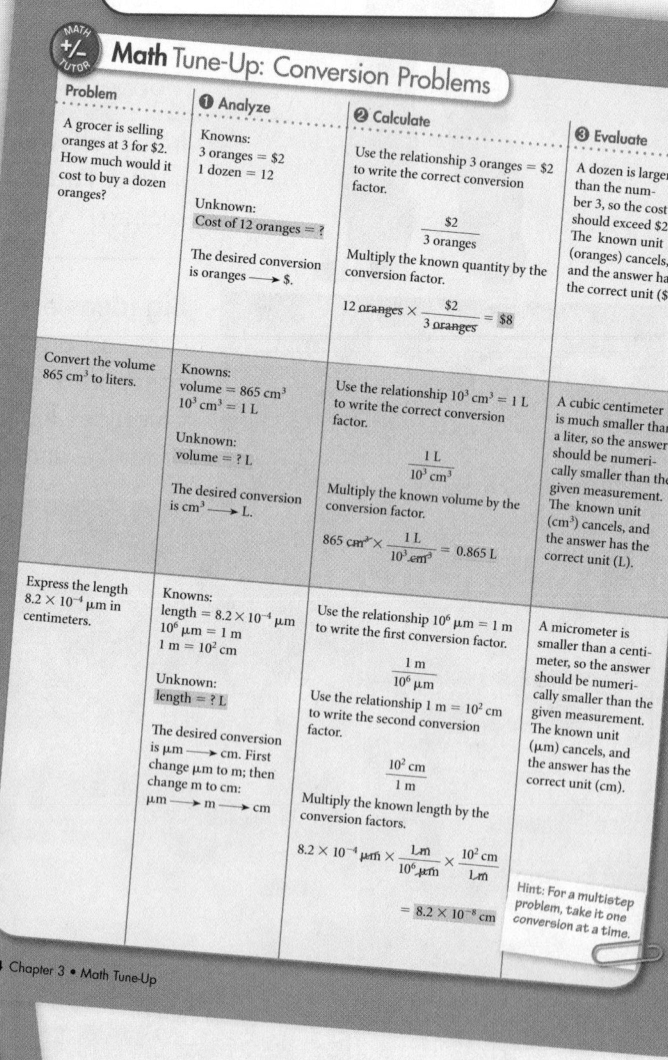

MATH +/- TUTOR **Math** Tune-Up: Conversion Problems

Problem	❶ Analyze	❷ Calculate	❸ Evaluate
A grocer is selling oranges at 3 for \$2. How much would it cost to buy a dozen oranges?	Knowns: 3 oranges = \$2; 1 dozen = 12. Unknown: Cost of 12 oranges = ? The desired conversion is oranges ⟶ \$.	Use the relationship 3 oranges = \$2 to write the correct conversion factor. $\frac{\$2}{3 \text{ oranges}}$ Multiply the known quantity by the conversion factor. $12 \text{ oranges} \times \frac{\$2}{3 \text{ oranges}} = \$8$	A dozen is larger than the number 3, so the cost should exceed \$2. The known unit (oranges) cancels, and the answer has the correct unit (\$).
Convert the volume 865 cm^3 to liters.	Knowns: volume = 865 cm^3; $10^3\ cm^3 = 1\ L$. Unknown: volume = ? L. The desired conversion is cm^3 ⟶ L.	Use the relationship $10^3\ cm^3 = 1\ L$ to write the correct conversion factor. $\frac{1\ L}{10^3\ cm^3}$ Multiply the known volume by the conversion factor. $865\ cm^3 \times \frac{1\ L}{10^3\ cm^3} = 0.865\ L$	A cubic centimeter is much smaller than a liter, so the answer should be numerically smaller than the given measurement. The known unit (cm^3) cancels, and the answer has the correct unit (L).
Express the length $8.2 \times 10^{-4}\ \mu m$ in centimeters.	Knowns: length = $8.2 \times 10^{-4}\ \mu m$; $10^6\ \mu m = 1\ m$; $1\ m = 10^2\ cm$. Unknown: length = ? L. The desired conversion is μm ⟶ cm. First change μm to m; then change m to cm: μm ⟶ m ⟶ cm	Use the relationship $10^6\ \mu m = 1\ m$ to write the first conversion factor. $\frac{1\ m}{10^6\ \mu m}$ Use the relationship $1\ m = 10^2\ cm$ to write the second conversion factor. $\frac{10^2\ cm}{1\ m}$ Multiply the known length by the conversion factors. $8.2 \times 10^{-4}\ \mu m \times \frac{1\ m}{10^6\ \mu m} \times \frac{10^2\ cm}{1\ m} = 8.2 \times 10^{-8}\ cm$	A micrometer is smaller than a centimeter, so the answer should be numerically smaller than the given measurement. The known unit (μm) cancels, and the answer has the correct unit (cm).

Hint: For a multistep problem, take it one conversion at a time.

94 Chapter 3 • Math Tune-Up

Sample Problems guide you step-by-step through complicated chemistry problems. Remember to look for the orange button to get more practice problems.

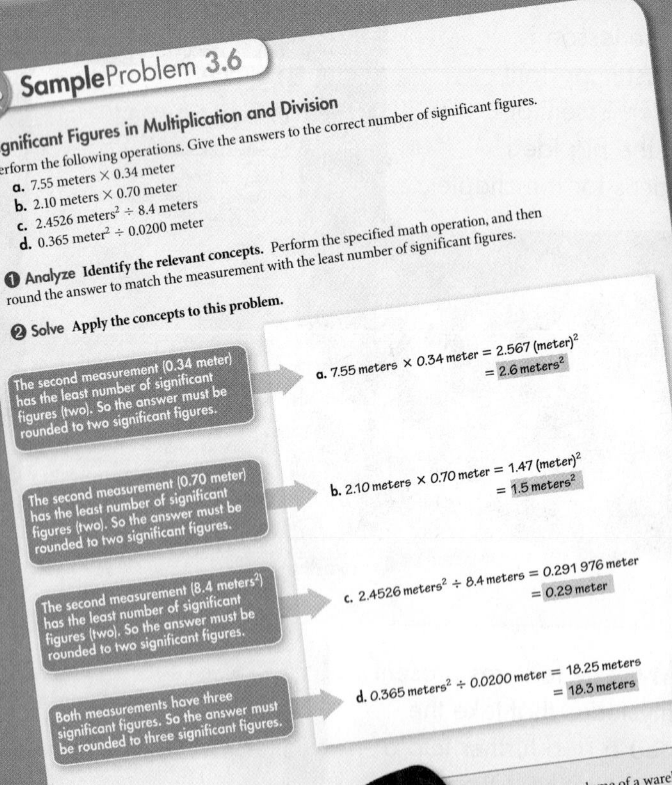

CHEM TUTOR **Sample**Problem 3.6

Significant Figures in Multiplication and Division

Perform the following operations. Give the answers to the correct number of significant figures.

a. 7.55 meters × 0.34 meter
b. 2.10 meters × 0.70 meter
c. 2.4526 $meters^2$ ÷ 8.4 meters
d. 0.365 $meter^2$ ÷ 0.0200 meter

❶ **Analyze Identify the relevant concepts.** Perform the specified math operation, and then round the answer to match the measurement with the least number of significant figures.

❷ **Solve Apply the concepts to this problem.**

The second measurement (0.34 meter) has the least number of significant figures (two). So the answer must be rounded to two significant figures.

a. 7.55 meters × 0.34 meter = 2.567 $(meter)^2$ = 2.6 $meters^2$

The second measurement (0.70 meter) has the least number of significant figures (two). So the answer must be rounded to two significant figures.

b. 2.10 meters × 0.70 meter = 1.47 $(meter)^2$ = 1.5 $meters^2$

The second measurement (8.4 $meters^2$) has the least number of significant figures (two). So the answer must be rounded to two significant figures.

c. 2.4526 $meters^2$ ÷ 8.4 meters = 0.291 976 meter = 0.29 meter

Both measurements have three significant figures. So the answer must be rounded to three significant figures.

d. 0.365 $meters^2$ ÷ 0.0200 meter = 18.25 meters = 18.3 meters

10. Solve each problem. Give your answers to the correct number of significant figures and in scientific notation.
a. 8.3 meters × 2.22 meters
b. 8432 $meters^2$ ÷ 12.5 meters
c. 35.2 seconds × $\frac{1 \text{ minute}}{60 \text{ seconds}}$

11. Calculate the volume of a warehouse that has measured dimensions of 22.4 meters by 11.3 meters by 5.2 meters. (Volume = $l \times w \times h$)

In Problem 11, the measurement with the fewest significant figures is 5.2 meters. What does this tell you?

Scientific Measurement 71

Connecting Chemistry to your World

Chemistry connections are all around you. **Pearson Chemistry** for Indiana provides examples of chemistry in daily life, connecting content to familiar examples and experiences in your world.

Chemistry & You engaging questions begin each lesson to help spark your curiosity and guide your learning.

22.1 Hydrocarbo

C.2.2 Subatomic particles and atomic structure; C.3.4 Structural formulas and names; C.9.1 Use structural formulas: carbon atoms; C.9.2 Identify the variety of carbon molecules.

CHEMISTRY & YOU

Q: *Why are some fossil fuels gases, some liquids, and some solids?* The gasoline used to fuel this motorcycle is a liquid at STP. So are the diesel fuel used in trucks and buses, and the kerosene used in lanterns. Other fuels are gases or solids. For example, the fuel used in a furnace might be natural gas or a solid such as coal. All these fuels contain mixtures of compounds called hydrocarbons. In this lesson, you will learn about the structure and properties of hydrocarbons.

Key Questions

Why does a carbon atom form four covalent bonds?

What are two possible arrangements of carbon atoms in an alkane?

Vocabulary

- hydrocarbon
- alkane
- straight-chain alkane
- homologous series
- condensed structural formula
- substituent
- alkyl group
- branched-chain alkane

Organi and Hy

Why d

Fewer than 2 could synthe compounds these comp thought tha bon compo refuted thi to synthesi organic ch pounds, r

Introduc erties of longer th compour pounds at least t contain are met

Me times on dec and te of dig

Chemistry & You features throughout your book showcase exciting real world applications of chemistry in technology, the environment, and in numerous careers.

CHEMISTRY & YOU: TECHNOLOGY

Bioremediation

Oil and water don't mix. You may have witnessed this fact watching footage of an oil spill in the news. Oil spills can lead to the deaths of seabirds and marine mammals and can contaminate soil and drinking water.

One tool being used to clean up spilled oil is a relatively new technology called bioremediation. The technology uses "oil-eating" microbes—particularly bacteria—to remedy the spill. For these microbes, the hydrocarbons in crude oil are not a contaminant but a food source. During the digestion process, harmful hydrocarbons are converted to less harmful products—mainly carbon dioxide and water. Bioremediation is a safe, simple, and relatively inexpensive method of dealing with spilled oil. However, the process takes time to work. In addition, it is usually only effective on residual oil on shorelines, after part of the spill has been removed by other means.

Take It Further

1. **Describe** Two hydrocarbons found in crude oil spills are methylbenzene and methylcyclopentane. Draw structural formulas for these two compounds.
2. **Research a Problem** Another technology called a dispersion agent is often used to remedy oil spills. Research this technology and compare it to bioremediation.

HIGH-TECH . . . AND ALIVE The degradation of petroleum in the marine environment is carried out by diverse microorganisms, including the *Pseudomonas* species shown here.

784 Chapter 22 • Chemistry & You

Integrating Chemistry and Technology

Pearson Chemistry for Indiana comes alive online with numerous ways to practice, tutors to step through chemistry and math problems, online labs, interactive art, animations and much more to expand your learning beyond the classroom. The buttons throughout the text direct you to exciting and helpful online activities on PearsonChem.com.

Solutions

INSIDE:

- **16.1** Properties of Solutions
- **16.2** Concentrations of Solutions
- **16.3** Colligative Properties of Solutions
- **16.4** Calculations Involving Colligative Properties

PearsonChem.com

516

River water contains many dissolved ions, including so calcium, magnesium, chlor and sulfate ions.

Concepts in Action animations provide an overview of a key chapter concept using real-world contexts and concrete examples and analogies.

Chem Tutorial animations give you extra problem-solving help by demonstrating the steps shown in one or more selected Sample Problems from the chapter.

Kinetic Art animations help you grasp abstract concepts by bringing art from your textbook to life online.

Math Tutorial animations review key math skills that are needed to solve the chapter problems you'll see in your textbook. It shows you a 2–3 minute video that will walk you through the steps to solve a specific type of problem.

Online Problem Sets give you extra practice solving problems similar to those you'll find in your textbook.

Directed Virtual Lab activities guide you through Virtual ChemLab investigations designed to explain and reinforce chapter concepts in a virtual and realistic lab environment.

Dalton's Law

Heliox is a mixture of helium and oxygen gas. The pressure exerted by the helium in the mixture is independent of the pressure exerted by the oxygen.

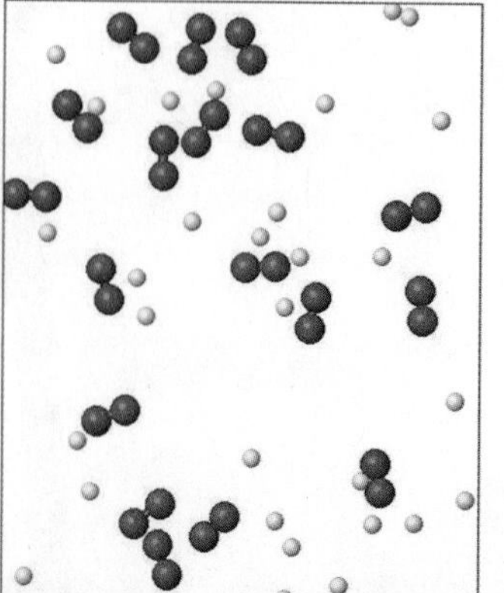

He

O_2

Heliox

500 kPa

Pearson is pleased to partner with **Molecular Workbench** for all Kinetic Art animations.

Lab Investigations and Study Tools

Laboratory Investigations and Study Tools extend your understanding of chemistry concepts by providing you with hands-on practice and extra study support in the areas you need most.

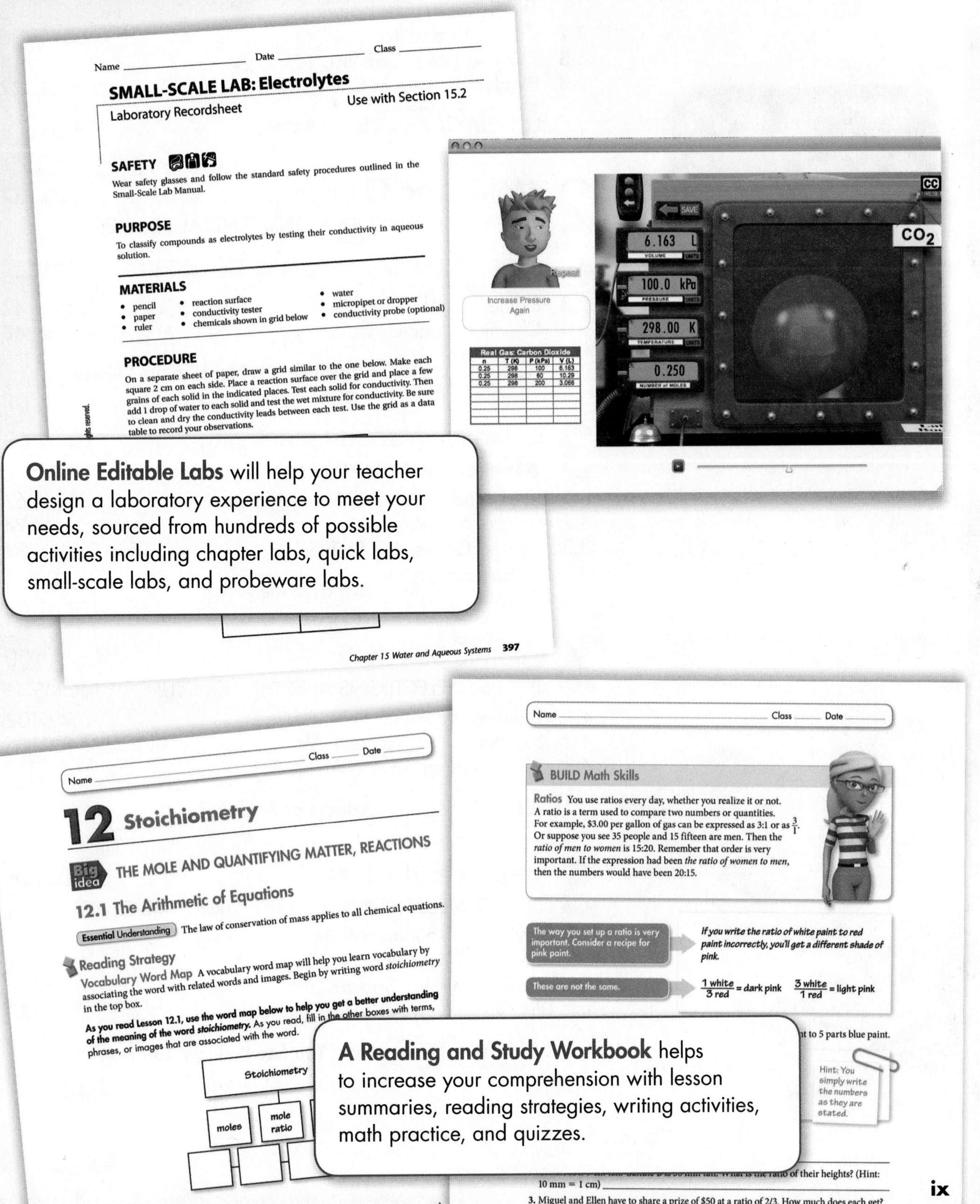

Contents

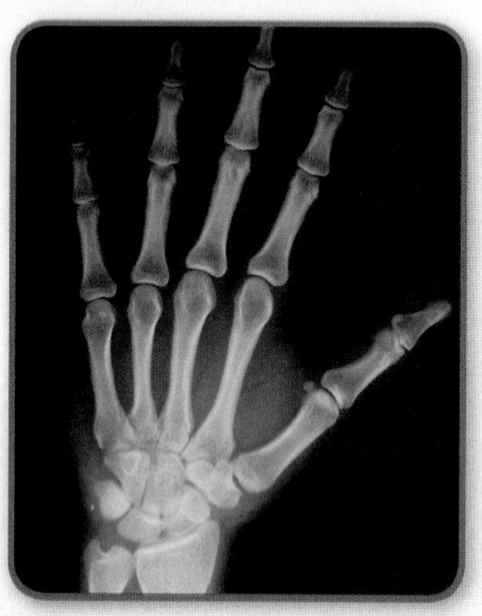

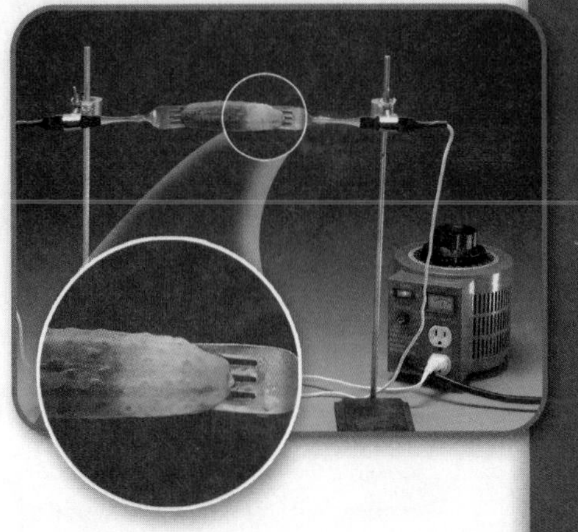

CHEMYSTERY

Enhance your understanding by exploring a chemical mystery that connects to each chapter's Big Idea. For example, you'll learn how less expensive jewelry can appear to be made of pure gold.

PearsonChem.com

What's Online

Your chemistry book comes alive online at **PearsonChem.com.** Anytime you spot one of these icons on a page, you can visit **PearsonChem.com** for an online activity, tutorial, or practice problem that helps reinforce the concepts introduced in the book.

Small-Scale Lab

Use readily available materials and easy procedures to produce reliable lab results.

Quick Lab

Apply chemistry concepts and skills with these quick, effective hands-on opportunities.

Probe or sensor versions available in the Probeware Lab Manual.

InterpretGraphs

Visualize chemical data in graphical form and improve your critical-thinking skills.

InterpretData

Organize and interpret data while building critical-thinking skills.

CHEMISTRY & YOU

Feature Pages

Learn more about how chemistry applies to real-world situations. You'll read about the societal and environmental impact of chemical technologies, and survey some interesting careers that apply chemistry. Some features include fun experiments that you can do on your own or with classmates.

CHEMISTRY & YOU: GREEN CHEMISTRY

Carbon Footprints

To measure a footprint, you might use units such as centimeters or inches. But what about a carbon footprint? A carbon footprint is a measure of how much greenhouse gas is released into the atmosphere by a person, country, or industry. Greenhouse gases, such as carbon dioxide (CO_2) and methane (CH_4), are gases that contribute to global warming.

Any activity that involves the burning of fossil fuels results in carbon dioxide emissions. Car travel, air travel, home heating and cooling, and electricity usage all add to an individual's carbon footprint. Your own carbon footprint is the total mass of CO_2 that you put into the atmosphere over the course of a year. This quantity can be expressed in metric tons (t) of CO_2 per year. A metric ton equals 1000 kg. So, the units of your carbon footprint can be abbreviated as: t CO_2/yr or 10^3 kg CO_2/yr.

CARBON COSTS Your choices affect the size of your carbon footprint. For example, using a clothes dryer consumes electricity, but hanging wet laundry on a clothesline does not. The things you buy also contribute to your carbon footprint. Not only is energy required to make these goods, but the goods themselves (such as TVs) may consume energy.

FOOTPRINT UNITS The carbon footprint of fresh produce can be expressed in g CO_2 per serving. Cars require different units: kg CO_2 per gallon of gasoline. For planes, the units are kg CO_2 per passenger mile.

Take It Further

1. Calculate A car emits 8.6 kg of CO_2 per gallon of unleaded gas. How much CO_2 is produced if the car burns 2.5 gal of fuel?

2. Infer What factors do you think determine the carbon footprint of an apple? Why might the carbon footprints of two apples in the same store differ substantially?

Chemistry & You 83

CHEMISTRY & YOU: CAREERS

Arson Investigator

Arson is a dangerous crime in which a person starts a fire with the intent to cause property damage or to harm another person. The person responsible, an arsonist, often uses an accelerant to initiate the fire. Gasoline and lighter fluid—both ignitable fluids composed of hydrocarbons—are two of the most commonly used accelerants. If arson is suspected, an arson investigator carefully examines the fire scene and looks for evidence of a crime.

Arson investigators often collect debris from the scene to take back to the laboratory to analyze for accelerants. Investigators frequently analyze the samples with an instrument called a gas chromatograph. The output of the analysis is sometimes called a "fingerprint" because, like a human fingerprint, each accelerant shows a characteristic pattern. With the knowledge of the molecules in each accelerant, the investigator can identify the accelerant from the fingerprint. Knowing which accelerant was used in the crime can help police narrow the search for the arsonist.

WHO'S YOUR PARTNER? Trained dogs with keen noses are often used to guide the evidence collection.

ACCELERANTS Gasoline and lighter fluid are complex mixtures of hydrocarbons. Each peak in the chromatograph represents a different chemical compound.

Take It Further

1. Describe One of the hydrocarbons found in gasoline is the branched-chain alkane 3-methylhexane. Write the condensed structural formula for this compound.

Sample Problem

Take advantage of these stepped-out problems to guide your solving process.

1 Introduction to Chemistry Planning Guide

Introducing the BIGIDEA: CHEMISTRY AS THE CENTRAL SCIENCE

Chemistry overlaps with all of the other sciences—physics, biology, astronomy, geology, environmental science, and others.

IASS	Lessons and Objectives	Print Resources		
		For the Student	For the Teacher	
	1.1 The Scope of Chemistry p 2–5 **1.1.1** Explain why the scope of chemistry is so vast. **1.1.2** Identify five traditional areas of study in chemistry. **1.1.3** Identify the central themes of chemistry.	Reading and Study Workbook Lesson 1. 1 Lesson Assessment 1.1 p 5		
	1.2 Chemistry and You p 6–11 **1.2.1** Identify three general reasons to study chemistry. **1.2.2** Identify some outcomes of modern research in chemistry.	Reading and Study Workbook Lesson 1.2 Lesson Assessment 1.2 p 11	Teacher Demo, p 7: Explaining the Natural World Class Activity, p 11: Composition of the Sun	
	1.3 Thinking Like a Scientist p 14–19 **1.3.1** Describe how Lavoisier helped transform chemistry. **1.3.2** Identify three steps in the scientific method. **1.3.3** Explain the role collaboration and communication play in science.	Reading and Study Workbook Lesson 1.3 Lesson Assessment 1.3 p 19 Quick Lab: Bubbles! p 17 Small-Scale Lab: Laboratory Safety p 20–21	Class Activity, p 15: Invisible Ink	
	1.4 Problem Solving in Chemistry p 22–26 **1.4.1** Identify the general approach to solving a problem. **1.4.2** Describe the steps for solving numeric problems. **1.4.3** Describe the steps for solving nonnumeric problems.	Reading and Study Workbook Lesson 1.4 Lesson Assessment 1.4 p 26	Teacher Demo, p 25: Fit an Ice Cube in a Soda Bottle	

Assessing the BIGIDEA: CHEMISTRY AS THE CENTRAL SCIENCE

	Essential Questions 1. Why is it important to study chemistry? 2. How do chemists solve problems?	Study Guide p 27 STP p 31 Reading and Study Workbook Self-Check and Vocabulary Review Chapter 1		

	PearsonChem.com	
	Editable Worksheets	Online Activities
	Teaching Resources, Lesson 1.1 Review, Interpret Graphs	1.1 Lesson Overview What Would Happen in a World Without Chemistry?
	Teaching Resources, Lesson 1.2 Review	1.2 Lesson Overview
	Teaching Resources, Lesson 1.3 Review Lab 1: Observing and Inferring Small-Scale Lab Manual Lab 1: Making Observations of Matter Small-Scale Lab Manual Lab 20: Absorption of Water by Paper Towels: A Consumer Lab	1.3 Lesson Overview Molecular Models Laboratory Safety
	Teaching Resources, Lesson 1.4 Review	1.4 Lesson Overview Estimating Walking Time
	Core Teaching Resources (includes editable worksheets) • Lesson Reviews • Practice Problems • Interpret Graphs • Vocabulary Review • Chapter Quizzes and Tests • Lab Record Sheets	Chapter 1 Problem Set Equations

Materials List

For the Student

Quick Lab p 17
- 3 plastic drinking cups
- liquid dish detergent
- measuring cup and spoons
- water
- table sugar
- table salt

For the Teacher

Teacher Demo p 7
- slides that show a wide variety of events or structures
- projector

Class Activity p 15
- 250 mg of phenolphthalein
- 250 mL of 70% solution 2–propanol
- small artist's paintbrush
- sheets of copy paper or white butcher paper
- household glass cleaner containing ammonia

Teacher Demo p 25
- bowl of ice cubes
- empty narrow neck (soda) bottle

Additional Digital Resources

Online Student Edition
Online Teacher's Edition
ExamView *Assessment Suite CD-ROM*
Classroom Resources DVD

CHAPTER 1

What's Online

CHEMISTRY TUTORIAL Students can work step-by-step with the online tutorial to solve estimation problems by accessing the Chapter 1 Chemistry Tutorial.

ONLINE PROBLEM SET Additional practice problems about general chemistry concepts can be found in the Chapter 1 online Problem Set.

MATH TUTORIAL The online Math Tutorial guides students through how to solve basic math equations.

DIRECTED VIRTUAL LAB Encourage students to review laboratory safety procedures using the animated virtual lab tour for Chapter 1. The animation is followed by analysis questions.

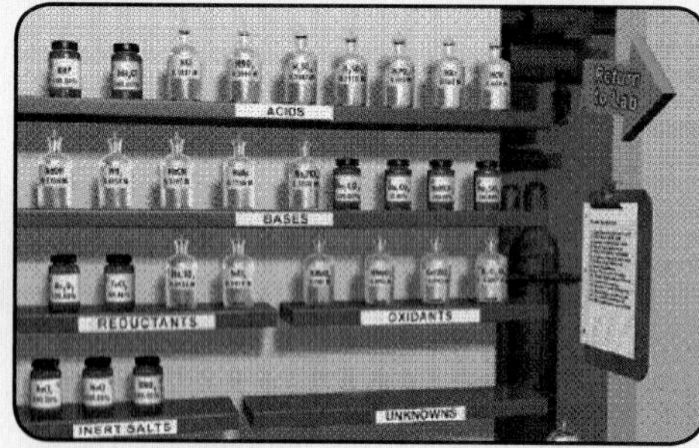

KINETIC ART Have students compare models that chemists use to visualize the structures of molecules. The animation is followed by questions aimed at further understanding of the concept.

CONCEPTS IN ACTION Students can watch an interactive animation showing chemistry applications in the real world. They can also learn about the different fields of chemistry and see examples of applied and pure chemistry.

IN ACADEMIC STANDARDS for SCIENCE

Nature of Science

- Develop explanations based on reproducible data and observations gathered during laboratory investigations.
- Clearly communicate their ideas and results of investigations verbally and in written form using tables, graphs, diagrams, and photographs.
- Explain that the body of scientific knowledge is organized into major theories, which are derived from and supported by the results of many experiments, and allow us to make testable predictions.
- Explain how scientific knowledge can be used to guide decisions on environmental and social issues.

1 Introduction to Chemistry

INSIDE:

PearsonChem.com

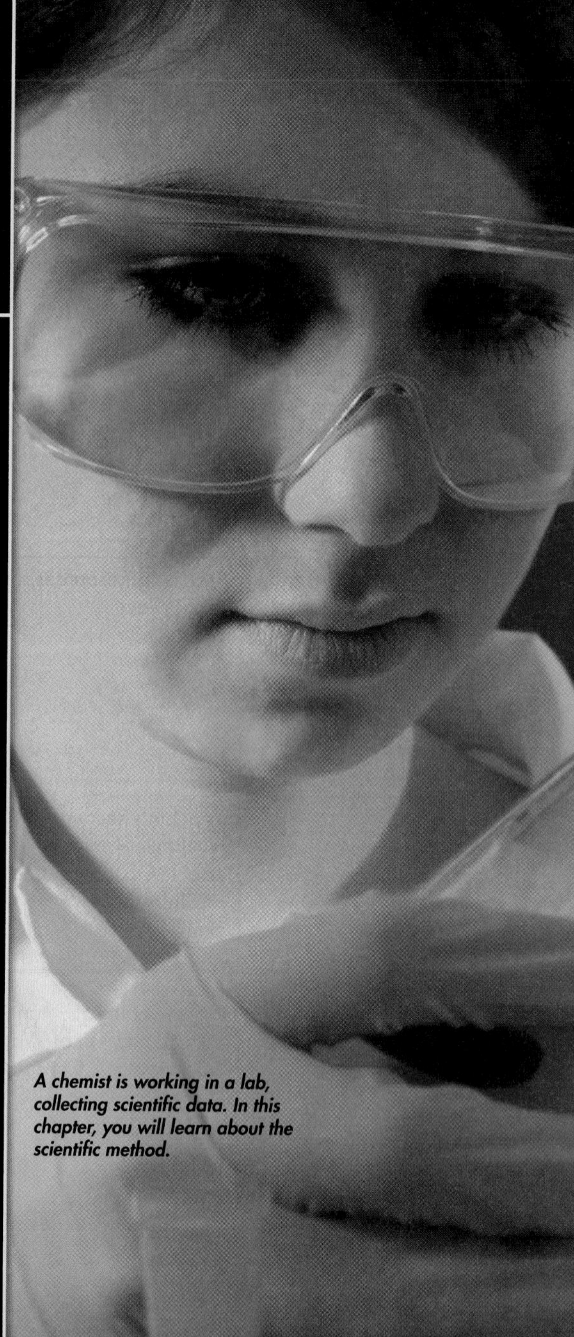

A chemist is working in a lab, collecting scientific data. In this chapter, you will learn about the scientific method.

Focus on ELL

1 CONTENT AND LANGUAGE Have student volunteers read the chapter and lesson titles out loud, and assist them with pronunciation as necessary. Clarify for the students that the words *scope* and *problem* have both common and scientific meanings. Have students work in groups to create Frayer models for each word, including definitions, parts of speech, characteristics, examples, and nonexamples.

BEGINNING

LOW Create a drawing or symbol for each word, and present a native-language definition if you are literate in students' native language.

HIGH Write the word, and point to it while saying it out loud. Have students repeat the word after you.

INTERMEDIATE: LOW/HIGH Brainstorm examples and nonexamples and present to the class.

ADVANCED: LOW/HIGH Have students look up definitions and rewrite in their own words.

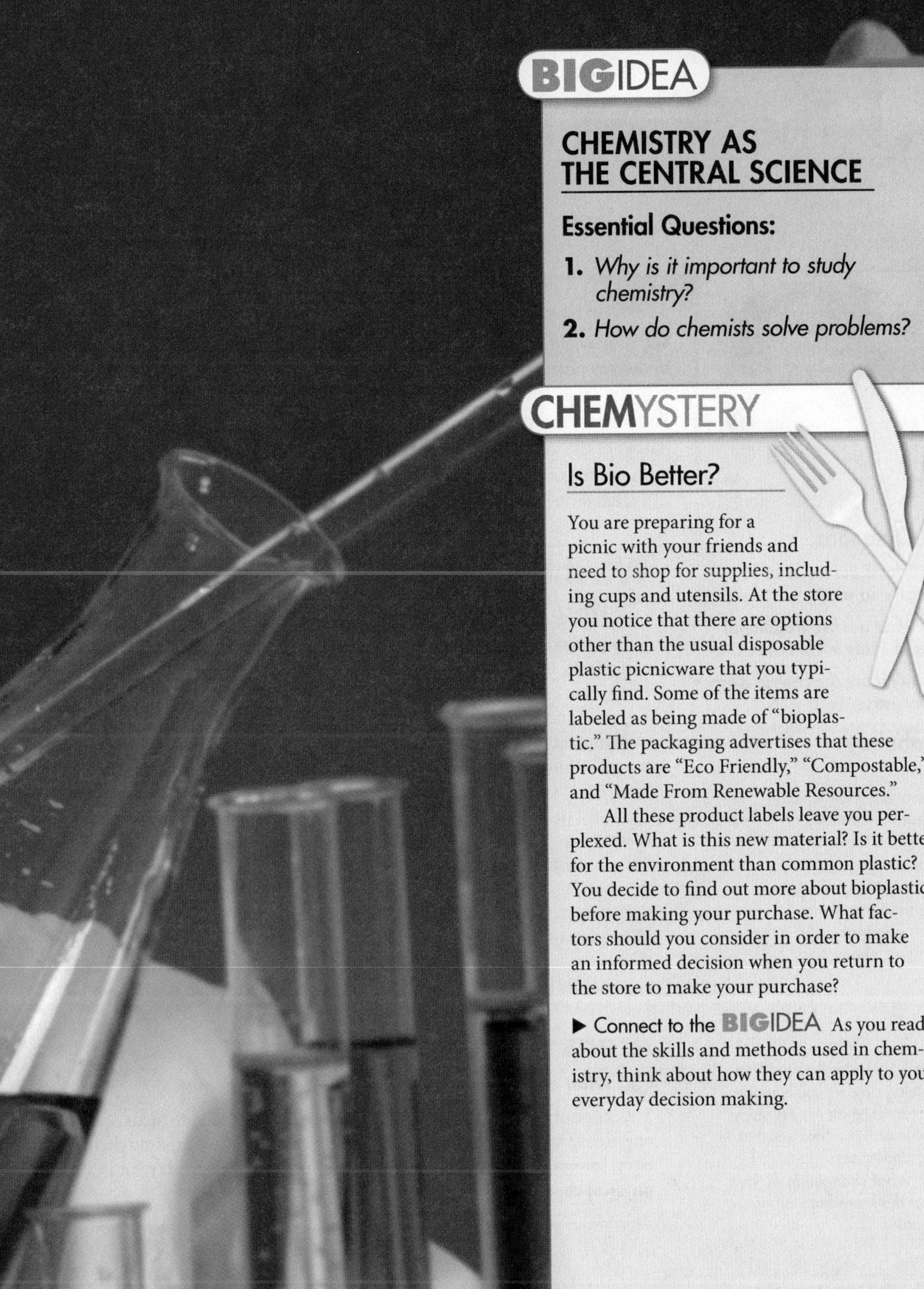

CHAPTER 1

UbD

Understanding by Design

Students are building toward an understanding of matter, its properties, and its changes using problem solving in chemistry.

PERFORMANCE GOALS At the end of Chapter 1, students will be able to answer the essential questions by applying their knowledge of the scope of chemistry.

ESSENTIAL QUESTIONS Read the essential questions aloud. **Ask** Why is it important to study so many different subjects in school. **Ask** Why is it useful to learn problem-solving skills?

BIGIDEA Use the photo of the chemist to help students connect to this chapter's concepts. Activate prior knowledge by asking students if they have ever made something that required reading instructions. **Ask** How could you make or fix something if there were no instructions or guidelines? Point out that chemists launch the problem-solving method by learning as much as they can about an unknown and then using that information to plan experiments and draw conclusions that are useful.

CHEMYSTERY Have students read over the CHEMystery. Connect the CHEMystery to the Big Idea by explaining that all branches of science are related. Have students consider the steps they can take to solve the problem of which plastic picnic supplies to buy. Point out that the problem can be solved in different ways. Have students predict what some of the words in the CHEMystery paragraph mean, such as *bioplastic* and *Eco Friendly*. As a hint, tell students to think about the prefixes *bio-* and *eco-*.

Introduce the Chapter

CONNECTIONS TO EVERYDAY LIFE Students often question whether the subjects they study in school relate to them. Use the activity to help students understand ways chemistry is important in their everyday lives.

Activity Set up stations with different items: (1) a cup of water, powdered drink mix, and a spoon; (2) a flashlight; (3) a candle, a match with a matchbox; and (4) a sliced apple that is turning brown. Assign a student to each station. As the class watches, have the student at Station 1 stir the drink mix into the water. **Ask** What do you think chemistry has to do with the drink mix? *(Sample answers: Chemistry explains how the substances mix. Chemistry is responsible for the color and the flavor.)* Have the student at Station 2 turn the flashlight on and off and then remove the batteries. **Ask** What do you think the flashlight and batteries have to do with chemistry? *(Sample answer: Chemistry explains how the batteries provide energy to the flashlight.)* Continue to follow the same process at the other two stations, asking how chemistry relates to the items. If time allows, have students choose other items in the classroom and discuss possible connections to chemistry.

UntamedScience™

What Good Can Chemistry Do?
Students and chemists explain why they love chemistry and how they hope to use what they learn to make the world a better place.

LESSON 1.1

Key Objectives

1.1.1 **EXPLAIN** why the scope of chemistry is so vast.

1.1.2 **IDENTIFY** five traditional areas of study in chemistry.

1.1.3 **IDENTIFY** the central themes of chemistry.

Additional Resources

Reading and Study Workbook, Lesson 1.1
Available Online or on Digital Media:
- Teaching Resources, Lesson 1.1 Review, Interpreting Graphics

PearsonChem.com

Engage

CHEMISTRY & YOU Direct students' attention to the photograph and text that open the lesson. **Ask** In what ways does society benefit from chemists looking for the positive in a perceived difficult or dangerous situation? *(Sample answers: medical discoveries, safer automobiles, healthier food products)*

Build Background

Ask students to state what immediately comes to mind when they hear the word *chemistry*. Generate a list on the board of students' responses, and categorize the named things as living or nonliving and visible or not visible. Use this activity as an introduction to matter.

1.1 The Scope of Chemistry

CHEMISTRY & YOU

Q: *Why might this creature interest you if you were a chemist?* Fugu, also known as puffer fish, is a sushi delicacy that can also be lethal. Puffer fish contain a powerful toxin that can kill an adult a few hours after ingestion. Sushi chefs who prepare fugu must be specially trained because any contamination of the toxin-free areas of the fish can be deadly. Recently this toxin has been put to good use, as scientists have discovered that a purified form of it can treat severe pain in cancer patients.

Key Questions

Why is the scope of chemistry so vast?

What are five traditional areas of study in chemistry?

What are the central themes of chemistry?

Vocabulary

- matter
- chemistry
- organic chemistry
- inorganic chemistry
- biochemistry
- analytical chemistry
- physical chemistry
- pure chemistry
- applied chemistry

What Is Chemistry?

Why is the scope of chemistry so vast?

Look around you. This book you are reading, the chair you sit in, and the computer you use are all made of matter. Matter is the general term for all the things that can be described as materials, or "stuff." **Matter** is anything that has mass and occupies space. The trees, the water, and the buildings you see in Figure 1.1 are all examples of matter. However, you don't have to be able to see something for it to qualify as matter. The air you breathe is an example of matter that you cannot see with the naked eye.

Have you ever wondered how some creatures can survive deep in the ocean where there is no light? Why some foods taste sweet and some taste bitter? Chemistry answers these questions and the many other questions you may have about the world you live in. **Chemistry** is the study of the composition of matter and the changes that matter undergoes. **Chemistry affects all aspects of life and most natural events because all living and nonliving things are made of matter.** Chemistry is also known as the central science, because it is fundamental to the understanding of the other sciences.

Areas of Study

What are five traditional areas of study in chemistry?

The scope of chemistry is vast, so individual chemists tend to focus on one area of study. **Five traditional areas of study are organic chemistry, inorganic chemistry, biochemistry, analytical chemistry, and physical chemistry.**

Figure 1.1 Matter Around You
Everything around you is made of matter. Chemistry is the study of matter and the changes that matter undergoes.
Infer *What changes in matter do you think are happening in this photo?*

Learn more about the areas of chemistry online.

IN ACADEMIC STANDARDS for SCIENCE

- Explain that the body of scientific knowledge is organized into major theories, which are derived from and supported by the results of many experiments, and allow us to make testable predictions.

Focus on ELL

1 CONTENT AND LANGUAGE Help students actively learn to pronounce and understand the lesson vocabulary by showing them various word-analysis strategies, such as identifying and translating pronounceable word parts within a word and identifying and interpreting prefixes, suffixes, and roots within multisyllabic words.

2 FRONTLOAD THE LESSON Begin the lesson by dividing the class into three groups. Assign each group of students one of the key questions, and have them rephrase their question with simpler words. For the group with Key Question 1, have students predict reasons for why chemistry is so vast. For the other two groups, have students create a main-ideas-and-details chart that the class can complete as the material is presented.

3 COMPREHENSIBLE INPUT Slowly read aloud the captions in Figures 1.1 and 1.2, emphasizing key points by altering your tone. Enhance learning of both of these concepts by setting up an outdoor activity in which students utilize their five senses.

Most chemicals found in organisms contain carbon. Therefore, organic chemistry was originally defined as the study of the carbon-based chemicals found in organisms. Today, with few exceptions, **organic chemistry** is defined as the study of all chemicals containing carbon. The study of chemicals that, in general, do not contain carbon is called **inorganic chemistry.** Many inorganic chemicals are found in non-living things, such as rocks. The study of processes that take place in living organisms is **biochemistry.** These processes include muscle contraction and digestion. The area of study that focuses on the composition of matter is **analytical chemistry.** A task that would fall into this area of chemistry is measuring the level of carbon dioxide in the atmosphere. **Physical chemistry** is the area that deals with the mechanism, rate, and energy transfer that occurs when matter undergoes a change.

The boundaries between the five areas are not firm. A chemist is likely to be working in more than one area of chemistry at any given time. For example, an organic chemist uses analytical chemistry to determine the composition of an organic chemical. Figure 1.1 shows examples of the types of research different chemists do.

Some chemists do research on fundamental aspects of chemistry. This type of research is sometimes called pure chemistry. **Pure chemistry** is the pursuit of chemical knowledge for its own sake. The chemist doesn't expect that there will be any immediate practical use for the knowledge. However, most chemists do research that is designed to answer a specific question. **Applied chemistry** is research that is directed toward a practical goal or application. In practice, pure chemistry and applied chemistry are often linked. Pure research can lead directly to an application, but an application can exist before research is done to explain how it works.

CHEMISTRY & YOU

Q: *Why would you study a puffer fish if you were a biochemist? If you were an organic chemist?*

Analytical Chemistry
An analytical chemist might test the air for the presence of pollutants.

Physical Chemistry
A physical chemist might study factors that affect the rate of photosynthesis in trees.

Inorganic Chemistry
An inorganic chemist might develop metal materials that provide strong structural parts for buildings.

Biochemistry
A biochemist might study how the energy used for the contraction of muscles is produced and stored.

Organic Chemistry
An organic chemist might develop new lightweight plastics for flying disks.

Foundations for Reading

BUILD VOCABULARY Explain that the given definition of the term *matter* is one of the definitions used in everyday conversation. Point out that for most scientific terms with common meanings, however, the scientific definitions are almost always more specific than common usage.

READING STRATEGY Have students write a summary of this lesson. Verify that the summaries answer the lesson's key questions.

Explain

What Is Chemistry?

APPLY CONCEPTS Create a two-column table on the board. Label one column *Matter* and the other column *Nonmatter*. Tell students to think of one example that falls under each column head and to write their examples in the table. Review the table with the class once all students have participated. Explain incorrect placement of examples, and clear up any confusion about students' understanding of matter.

Areas of Study

START A CONVERSATION As a class, discuss if students feel that they have any connection to chemistry in their daily lives. **Ask** What daily activities do you do that may involve a chemical process or contact with chemicals? *(Sample answers: eating, brushing teeth, doing laundry, digesting food, using makeup, wearing deodorant, riding in a car that uses gasoline)* **Ask** Can you think of four items or activities that do not involve chemistry? *(It will be hard to find any that do not; for all responses, discuss reasons why it does involve chemistry.)*

CHEMISTRY & YOU *Sample answer:* A biochemist might study how the toxin is produced. An organic chemist might study the structure of the toxin.

Differentiated Instruction

LPR LESS PROFICIENT READERS Lead the students on a walk around the school or campus. Show students objects that can be used as examples to convey vocabulary or concepts in the lesson. Allow students to collect a couple of items that can be posted in their vocabulary notebooks.

ELL ELL STUDENTS The lesson includes many photographs to illustrate the concepts presented. Create short, simple descriptions about each image that students can write in their notebooks. As you work through the presentation, ask students questions that they can answer using yes/no or one- or two-word responses or by using gestures.

L3 ADVANCED STUDENTS Challenge students to conduct research on the traditional areas of study in chemistry. Encourage students to set up an interview with a chemist at a nearby college or local research laboratory. They could also search for information online. Ask students to share their research with the class.

Answers

FIGURE 1.1 Students are most likely to note that the leaves are changing color and the dog's muscles are contracting.

LESSON 1.1

Explain

Big Ideas in Chemistry

USE VISUALS Direct students to Figure 1.2 and ask for a show of hands of students who have roasted marshmallows over a campfire. Pair these students and ask them to write step-by-step directions for roasting a marshmallow; have them include a description of the characteristics of the marshmallow at each step. Have the students read or act out their directions for the class. Assess the understanding of the students who did not have to write the directions by passing out a marshmallow to each student and having them inspect it. **Ask** Why does the marshmallow change colors during roasting? *(Sample answer: When heated, some substances within the marshmallow darken.)* What about the marshmallow prevents it from initially falling off of the stick? *(Sample answer: It has a spongy, solid consistency.)* What causes the marshmallow to lose its stability on the stick after it is heated? *(Sample answer: When heated, the marshmallow slowly begins to melt.)*

Misconception Alert

Based on how the term *chemical* is used in the media, and how it is defined in most dictionaries, students may assume that the term *chemical* refers exclusively to harmful materials. Explain that the word *chemical* describes all types of matter, including life-sustaining substances, such as water and oxygen. Explain that it is not a fair assumption to think that all chemicals are dangerous. Tell students that you will alert them when potentially harmful chemicals are used in the lab.

Big Ideas in Chemistry

What are the central themes of chemistry?

This book contains many ideas in the science of chemistry. One of the goals of your course in chemistry is to help you understand these ideas so you can use them to explain real situations that you may encounter in your life, such as the one shown in Figure 1.2. Fortunately, most of the topics of interest in chemistry are connected by a relatively few organizing principles, or "big ideas." **Some of chemistry's big ideas are as follows: chemistry as the central science, electrons and the structure of atoms, bonding and interactions, reactions, kinetic theory, the mole and quantifying matter, matter and energy, and carbon chemistry.**

BIGIDEA Chemistry As the Central Science Chemistry overlaps with all of the other sciences. Many physicists, biologists, astronomers, geologists, environmental scientists, and others use chemistry in their work.

BIGIDEA Electrons and the Structure of Atoms Carbon, oxygen, and copper are all examples of elements. Elements are composed of particles called atoms, and every atom contains a nucleus and one or more electrons. The type of products obtained in a chemical reaction is largely determined by the electrons in the reacting chemicals.

BIGIDEA Bonding and Interactions Most elements exist in chemical compounds, which are collections of two or more elements held together by relatively strong attractive forces. These forces, called chemical bonds, greatly influence the properties of compounds. Weak bonds between the particles of an element or compound can also contribute to the properties of the material.

Figure 1.2 Big Ideas
The big ideas in chemistry can help you understand the world around you. For example, all matter is made up of atoms, which are held together in compounds by chemical bonds. The fire is a result of a chemical reaction between the carbon-containing compounds in the wood and the oxygen in the air. The fire gives off energy in the form of heat and light. The gas particles in the air around the fire begin to move faster as the air heats up.
Predict *Marshmallows are made up of mostly sugar, a carbon-containing compound. What do you think happens when the sugar is heated by the fire?*

UbD Check for Understanding

BIGIDEA CHEMISTRY IS THE CENTRAL SCIENCE Write on the board the sentence below. Instruct students to copy the sentence onto a slip of paper and to fill in the blanks using words or pictures for a specific example. For struggling students, provide the hint that there are many correct answers.

One reason why chemistry is known as the central science is that it relates ______________ to ______________.

(Sample answers: natural science/geology; life science/medicine; volcano gases/volcano eruption; composition of stars/astronomy; illness/medicine; materials of science/safety)

ADJUST INSTRUCTION If students are having difficulty with this concept, have students reread the last sentence of the What is Chemistry? section and apply it to Figure 1.1.

BIGIDEA Reactions Chemical reactions involve processes in which reactants produce products. When you strike a match, the compounds in the head of the match combine with oxygen in the air to produce a flame. New compounds, along with light and heat, are formed. The compounds in the match head and oxygen are the reactants, and the new compounds are the products. Chemical reactions are important to the chemistry of living and nonliving things.

BIGIDEA Kinetic Theory The particles in matter are in constant motion. The ways in which these motions vary with changes in temperature and pressure determine whether a substance will be a solid, liquid, or gas.

BIGIDEA The Mole and Quantifying Matter In many aspects of chemistry, it is vital to know the amount of material with which you are dealing. In conducting a chemical reaction, you would want to use just the right amount of the reacting material so none is wasted. This measurement is possible using the mole, the chemist's invaluable unit for specifying the amount of material. Other concepts in chemistry also rely on the mole unit.

BIGIDEA Matter and Energy Every chemical process uses or produces energy, often in the form of heat. The heat changes that occur in chemical reactions are easy to measure. Changes in a quantity called free energy allow you to predict whether a reaction will actually occur under the given conditions.

BIGIDEA Carbon Chemistry There are about 10 million carbon-containing compounds, with new ones being prepared each day. Many of these compounds, including plastics and synthetic fibers, are produced from petroleum. Carbon compounds are the basis of life in all living organisms.

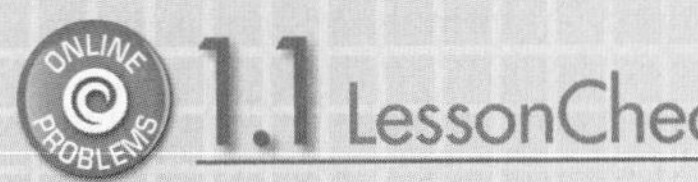

1.1 LessonCheck

1. **Explain** Why does chemistry affect all aspects of life and most natural events?
2. **List** Name the five traditional areas into which chemistry can be divided.
3. **Review** What are the "big ideas" of chemistry?
4. **Describe** What is the relationship between pure and applied chemistry?
5. **Infer** Why might a geologist ask an analytical chemist to help identify the minerals in a rock?
6. **Apply Concepts** Workers digging a tunnel through a city find some ancient pots decorated with geometric designs. Which of the following tasks might they ask a chemist to do? Explain.
   a. Determine the materials used to make the pots.
   b. Explain what the designs on the pots represent.
   c. Recommend how to store the pots to prevent further damage.

BIGIDEA CHEMISTRY AS THE CENTRAL SCIENCE

7. Why would a student who wants to be a doctor need to study chemistry?

Lesson Check Answers

1. Living and nonliving things are made of matter, and chemistry is the study of matter.
2. organic chemistry, analytical chemistry, biochemistry, physical chemistry, and inorganic chemistry
3. chemistry as the central science; electrons and the structure of atoms; bonding and interactions; reactions; kinetic theory; the mole and classifying matter; matter and energy, carbon chemistry
4. Pure research can lead directly to an application; an application can exist before research is done to explain how it works.
5. An analytical chemist focuses on the composition of matter.
6. a and c
7. **BIGIDEA** *Sample answer:* A doctor needs to understand reactions that take place in the body as well as factors that can interfere with these processes.

Extend

Connect to MEDICINE

Point out that medicine is one discipline which often relies on applied chemistry. Tell students that one area of medical interest is the use of animal venom as a source of therapeutic drugs. Challenge students to research vampire bats, death stalker scorpions, cone snails, or poison dart frogs to learn about recent discoveries. Have students present to the class the reasons why their animal produces venom and the properties of the venom that sparked the interest in medical research. Students should also be prepared to discuss the successes and failures of the research and show how this type of research exemplifies the Big Idea of chemistry being the central science.

Evaluate

Informal Assessment

Assess students' understanding of the science of chemistry by asking them to define *chemistry* in their own words. *(Chemistry is the study of the composition of matter and the changes it undergoes.)* **Ask** What do chemists do? *(Chemists study the composition and behavior of matter.)* Then have students complete the 1.1 Lesson Check.

Reteach

Direct students' attention to Figure 1.2. Beginning with the Big Idea that chemistry is a central science, challenge students to identify why this photograph illustrates each Big Idea. For example, explain that this photograph illustrates that chemistry is the central science because food scientists used chemistry to develop the recipe for the marshmallows, and textile scientists used chemistry to develop the dyes in the clothes.

Answers

FIGURE 1.2 Students are likely to say that the sugar reacts. (Some students may know that caramel forms when sugar is heated.)

LESSON 1.2

Key Objectives

1.2.1 IDENTIFY three general reasons to study chemistry.

1.2.2 IDENTIFY some outcomes of modern research in chemistry.

Additional Resources

Reading and Study Workbook, Lesson 1.2

Available Online or on Digital Media:

- Teaching Resources, Lesson 1.2 Review

PearsonChem.com

Engage

CHEMISTRY & YOU As a class, discuss past and current space exploration studies and benefits. **Ask** What do you think scientists hoped to gain by launching a telescope out of Earth's atmosphere into space? *(Sample answer: evidence of life on other planets)*

Activate Prior Knowledge

Encourage students to think of ways that technology might benefit chemical research, and vice versa. Explain that, in this lesson, they will explore the roles that chemistry plays in society and why the study of chemistry is important.

IN ACADEMIC STANDARDS for SCIENCE

- Explain how scientific knowledge can be used to guide decisions on environmental and social issues.

1.2 Chemistry and You

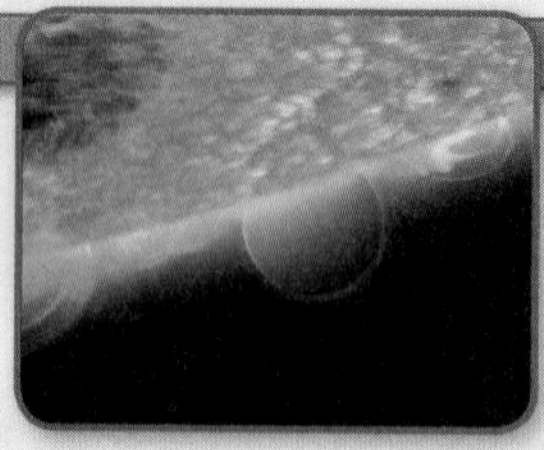

CHEMISTRY & YOU

Q: *How is chemistry used to study worlds other than your own?* The Hubble Space Telescope has provided detailed views of celestial objects. Scientists who know chemistry have also used the telescope to discover water and compounds containing carbon on a planet located 63 light years from Earth. Such compounds are necessary for life on Earth. However, the planet, designated HD189733b, is much too hot to support life. Perhaps the Hubble Space Telescope or its successors will someday be used to find evidence of life on planets with atmospheres more like that of Earth.

Key Questions

What are three general reasons to study chemistry?

What are some outcomes of modern research in chemistry?

Vocabulary

- technology

Figure 1.3 Chemistry and Food
When cut apples are exposed to air, a chemical reaction takes place, which causes the color to change to brown.

Why Study Chemistry?

What are three general reasons to study chemistry?

You may not realize it, but chemistry can answer many questions you have about the world around you. Should you use hot water or cold water to remove a grass stain from a shirt? How could you prepare for a career in nursing, firefighting, or journalism? If your local government wanted to build a solid waste incinerator in your town, what questions would you ask about the project? **Chemistry can be useful in explaining the natural world, preparing people for career opportunities, and producing informed citizens.**

Explaining the Natural World You were born with a curiosity about your world. Chemistry can help you satisfy your natural desire to understand how things work. For example, chemistry can be seen in all aspects of food preparation. Chemistry can explain why cut apples, such as the one shown in Figure 1.3, turn brown upon exposure to air. It can explain why the texture of eggs changes from runny to firm as eggs are boiled. Chemistry can explain why water expands as it freezes, why sugar dissolves faster in hot water than in cold water, and why yeast makes bread dough rise. After you study this textbook, you will know the answers to these questions and many more.

Preparing for a Career Being a chemist can be rewarding. Chemists contribute to society in many ways. In this book, you will find features on careers that require knowledge of chemistry. Some of the choices may surprise you. You do not need to have the word *chemist* in your job title to benefit from understanding chemistry. For example, a reporter may be asked to interview a chemist to gather background for a story. Turf managers have the important task of keeping the grass on golf courses, lawns, and soccer fields, such as the one shown in Figure 1.4a, healthy. This job requires an understanding of soil chemistry. Figure 1.4b shows a firefighter, who must know which chemicals to use to fight different types of fires.

Focus on ELL

1 CONTENT AND LANGUAGE Ask students to state the key questions in their own words. Explain that the wording of each question indicates that both questions will have more than one answer. Then, have students locate the boldfaced answers to each question. Show students that commas are used to separate the series of answers to each question in a single sentence. Demonstrate how to rewrite each answer series as a numbered list that can be used as the framework for an outline.

2 FRONTLOAD THE LESSON Have students take a picture walk through the lesson, and ask students to identify any visual displays that they find difficult to read or interpret. Use Figure 1.6 to explain how to read a graph.

3 COMPREHENSIBLE INPUT Stage a debate on the pros and cons of polylactic acid (PLA) plastics. Divide the class into two groups of mixed proficiency: one for PLA and one against PLA. Prior to the debate, have each group create drawings to depict entries in the table on page 9 that reflect the group's position and to use as visual aids.

Figure 1.4 Careers
Many careers require a knowledge of chemistry. **a.** Turf managers must know how the soil and other conditions affect grass. **b.** Firefighters must choose the correct chemicals to extinguish different types of fires.
Infer ***What are some factors that may affect the health and appearance of the grass on a soccer field?***

Being an Informed Citizen Industry, private foundations, and the federal and state governments all provide funds for scientific research. The availability of funding can influence the direction of research. Those who distribute funds have to balance the importance of a goal against the cost. Areas of research often compete for funds because there is limited money available.

For example, space exploration research could not take place without federal funding. Critics argue that the money spent on space exploration would be better spent on programs such as cancer research. Those who support space exploration point out that NASA research has led to the development of many items used on Earth. These include smoke detectors, scratch-resistant plastic lenses, heart monitors, and flat-screen televisions. What if all the money spent on space exploration was used to find a cure for cancer? Are there enough valid avenues of research to take advantage of the extra funding? Would there be qualified scientists to do the research?

Like the citizen shown in Figure 1.5, you will need to make choices that will influence the direction of scientific research. You may vote directly on some issues through ballot initiatives or indirectly through the officials you elect. You may speak at a public hearing, write a letter to the editor, or sign a petition. When it comes to scientific research, there is no one correct answer. However, knowledge of chemistry and other sciences can help you evaluate the data presented, arrive at an informed opinion, and take appropriate action.

Figure 1.5 Voting
Through voting, citizens have a say in the decisions their government makes. Those decisions include how much money to provide for scientific research.

Differentiated Instruction

LPR LESS PROFICIENT READERS Refer students to the graph in Figure 1.6. Direct students to read the title of the graph, the *x*-axis label and scale, and the *y*-axis label and scale. Ask students to describe the pattern in the bar lengths from 1960 to 2007. Explain that a graph displays information visually rather than in words.

ELL ENGLISH LANGUAGE LEARNERS Prior to students reading the "Pros & Cons" chart on page 9, explain that *pro* is a Latin word meaning "for or on behalf of" and *con* is derived from the Latin word *contra*, meaning "against." Try using visual cues such as thumbs up and thumbs down to represent for and against.

L3 ADVANCED STUDENTS Ask students to determine if they think the growth of plastic waste in Figure 1.6 could be represented by a linear function, quadratic function, or exponential function. Ask them to explain. *(Linear function; the rate of change in the different intervals is approximately the same. The data do not show an exponential pattern but approximate a straight line.)*

Foundations for Reading

BUILD VOCABULARY As a class, brainstorm a list of words to describe technology. Guide students to use the list to paraphrase the definition of *technology*.

READING STRATEGY Have students create and use a web to visually display their answers to the key questions. Explain that a web is a flexible outline for organizing and prioritizing information.

Explain

Why Study Chemistry?

START A CONVERSATION Ask Why did you decide to take this chemistry course? *(Sample answers: I needed a science credit; I want to attend college; I want to be a scientist.)* **Ask** After previewing this lesson, what do you think is a benefit of studying chemistry that you did not expect when you signed up for this course? *(Sample answers: a greater appreciation for research and discovery, an awareness of technological advances, an understanding of how my actions affect the environment)*

Extend

Connect to JOURNALISM

Have students scan newspapers for current news stories related to chemistry. Have students cut out or print one article and prepare a short presentation on how the content of the article relates to chemistry.

Explore

Teacher Demo

PURPOSE To provide a visual display of how chemistry plays a role in a variety of natural processes

MATERIALS slides or computer images showing events or structures such as a volcanic eruption, a forest fire, a limestone cave, and lightning

ADVANCE PREPARATION Create a slide show of images for which you can make a connection to chemistry without having to supply details that students will study later.

PROCEDURE Run through the set of slides once without discussion. Discuss the common thread among the images. Then go through the slides a second time and explain the connection to chemistry.

Answers

FIGURE 1.4 *Sample answer:* amount of rain, amount of sunlight, amount of use of the field for practice and games

LESSON 1.2

Explain

Chemistry, Technology, and Society

USE MODELS Point out that in the modern world, manufactured items often are made from materials that are created by chemists rather than those that are natural. Show students an assortment of classroom electronics, such as a computer, a calculator, and an electronic balance or pH meter. **Ask** What is similar about the construction materials used in all of these electronic devices? *(Sample answers: They are primarily made of plastic.)* **Ask** What properties of plastic make it a useful material for these electronic devices? *(Sample answers: lightweight, moldable, strong, durable, resistant to water and heat, inexpensive)*

USE VISUALS Direct students to the graph in Figure 1.6. **Ask** How much plastic waste do you think was produced prior to 1950? *(Sample answer: about half of what is shown in 1960, so about 60,000 pounds)* **Ask** What led to your conclusion? *(The value for each decade is approximately double that of the previous decade.)* **Ask** What do you think accounted for the increase of plastic waste after 1960? *(Sample answer: increased use of plastic as a material for multiple items)*

CRITICAL THINKING Refer students back to the previous conversation regarding plastics. Engage students in a discussion about the pervasiveness of plastic-encased electronics in today's society and people's tendency to upgrade their electronics at an ever-increasing rate. Encourage students to speculate on how the increase in plastic packaging waste is driven not by convenience but by human nature and the desire to have the newest and most technologically advanced products.

Interpret Graphs

Figure 1.6 Plastics packaging includes soft-drink and water bottles, milk and water jugs, and other plastic containers.

a. Read Graphs How much plastics packaging was in U.S. waste in 2007?

b. Calculate How much more plastics packaging was in U.S. waste in 2000 than in 1990?

c. Predict Do you think that the amount of plastics packaging in U.S. waste will increase or decrease in the next 10 years? Explain.

Chemistry, Technology, and Society

What are some outcomes of modern research in chemistry?

You have probably heard the term *high tech* used when describing the latest gadgets and inventions, such as computers that are the size of your watch, or cars that can drive themselves. However, you may not realize that many of the basic items you use every day, such as the sneakers you wear or the cereal you eat, are products of technology. **Technology** is the means by which a society provides its members with those things needed and desired. Technology allows humans to do some things more quickly or with less effort. It also allows people to do things that would be otherwise impossible, such as traveling to the moon. **Modern research in chemistry can lead to technologies that aim to benefit the environment, conserve and produce energy, improve human life, and expand our knowledge of the universe.**

Materials and the Environment Chemists don't just study matter—they also use what they know about the structure and properties of matter to make new materials with different or improved properties. Take plastics, for example. If you have ever consumed bottled water, eaten a salad with a disposable fork, or put on your helmet before riding your bike, you have used plastics. Chemistry has played a large role in developing plastics for different uses.

Most plastics are made using petrochemicals, which are chemical products derived from petroleum. Although plastics are a part of most of our daily lives, there is concern that their use is taking a toll on the environment and on natural resources. The supply of petrochemicals is limited, and the manufacture of plastics uses large amounts of energy. Unrecycled plastics end up in landfills, where they remain for hundreds of years. Figure 1.6 shows the amount of plastics packaging in U.S. waste. Understandably, there has been a demand for plastics that are better for the environment. Figure 1.7 describes a new technology that offers an alternative to petroleum-based plastics.

Biodegradable Insulation

Air makes a good barrier to the movement of thermal energy as long as the air is not moving. In fiberglass, air is trapped by a network of thin fibers that are woven together. In foam insulation, the air pockets are completely enveloped by the solid that makes up the framework of the foam. SEAgel insulation is made of agar, which is derived from seaweed. A gelatin-like mixture of agar and water is freeze-dried to remove the water. What is left is a honeycomb of dried agar filled with air. SEAgel has a density approximately equal to that of air. It is also biodegradable.

Figure 1.7 Bioplastic
Polylactic acid (PLA) is known as a bioplastic. Made from corn, PLA can be used to manufacture many items that are typically made out of petroleum-based plastics, including cups, utensils, containers and packaging for food, and bags.

Pros & Cons

Advantages of PLA	Disadvantages of PLA
✔ **Made from renewable resources** PLA can be made from corn and other crops, which can be grown year after year.	✘ **Melts at low temperatures** Since PLA melts at temperatures lower than other common plastics, it doesn't have as many applications.
✔ **Less energy** The production of PLA uses less energy than the production of conventional plastics.	✘ **Made from corn** The fields used to grow corn for PLA could be used to grow food for Earth's growing population. There is also concern that the corn used to produce PLA is genetically modified.
✔ **Better for the environment** Fewer greenhouse gases, such as carbon dioxide, are produced in the production of PLA than in the production of other plastics. Also, PLA can be broken down into compounds found in nature.	✘ **Difficult to dispose of** PLA must be taken to special compost facilities in order to ensure that the material breaks down. Products made from PLA cannot be recycled with other plastics.

LESSON 1.2

MAKE A CONNECTION Explain to students that composting is the disposal of food waste and plant remains that relies on aerobic decomposition to break materials down into soil-like particles. Point out that this breakdown occurs naturally when leaves pile up on the forest floor, eventually decay, and return to the soil. Because of the rich nutrients in the soil that is produced, people started using composting as a method that not only reduces the amount of garbage sent to landfills but also produces nutrient-rich soil to add to houseplants, gardens, and landscapes.

CRITICAL THINKING Ask In what ways will society benefit from using PLA rather than petroleum-based plastics in products? *(Sample answers: reduction of plastic in landfills, reduction of energy costs for production)* **Ask** Describe other ways in which chemistry has played a role in benefiting the environment. *(Guide students to think of energy-efficient home designs, such as green roofs, solar panels, and LED lighting.)*

UbD Check for Understanding

The Essential Question Why is it important to study chemistry?

Ask students to write a one-minute summary that answers this essential question.

ADJUST INSTRUCTION If students are having difficulty answering this question, have students reread the lesson. Then generate a discussion on the importance of having not only an informed citizenry but also informed and knowledgeable individuals in industry and government.

Answers

INTERPRET GRAPHS

a. 13,630,000 tons
b. 4,290,000 tons
c. *Sample answer:* It is likely to increase unless there is a dramatic change in how items are packaged.

LESSON 1.2

Explain

START A CONVERSATION Discuss with students the difference between renewable energy and nonrenewable energy. Ask a volunteer to explain why the U.S. is trying to develop alternative ways to obtain energy other than oil. **Ask** What types of alternative energy do you think are most promising? *(Sample answers: solar power, nuclear power, biodiesel, wind power, hydropower, hydrogen fuel cells)* Explain to students that they will be learning more about the chemistry of nuclear power, fossil fuels, alternative fuels, and fuel cells as the year progresses.

USE VISUALS Have students study Figure 1.8. Tell students that one of the applications of this technology was a worldwide research endeavor called the Human Genome Project, which ran from 1990 to 2003. Scientists involved in this project successfully used the technology to identify and sequence the genes that make up the human genome–which totals about 30,000 genes. They also aimed to develop analytical tools that could be transferred to the private sector and address any ethical, legal, and social issues (ELSI). These developments are utilized in today's biotechnology research. The chemistry of DNA will be explored in Chapter 24.

Extend

Connect to MEDICINE

Point out that it is one thing to produce new effective medicines and medical devices and quite a different challenge to test them, get them approved, and produce them in quantity. Have students work in small groups. Tell them to research the steps involved in getting approval for a new medicine or device from the FDA and how much time is usually required. Or assign each group a prescription medication that is currently on the market and ask them to trace it backward through the approval stages.

Energy The needs of any modern society require energy to power homes, factories, and transportation. With population growth and more industrialization around the globe, the demand for energy is on the rise. There are only two ways to meet the demand for energy—conserve it or produce more of it. Chemistry plays an essential role in both of these options.

Gasoline-electric hybrid cars play a substantial role in the conservation of energy. They have greater fuel efficiencies than gasoline-powered vehicles. Hybrids use both a gasoline engine and a set of batteries to run the car. A knowledge of chemistry was necessary to develop these batteries. In an effort to produce more energy, sustainable energy sources are important to consider. Unlike fossil fuels, the sun is a renewable energy source. Chemists help design materials that collect energy from the sun that is then converted to electricity.

Medicine and Biotechnology Chemistry supplies the medicines, materials, and technology that doctors use to treat their patients. Biochemists work with biologists and doctors to understand the structure of matter found in the human body and the chemical changes that occur in cells.

There are more than 10,000 prescription drugs, which have been designed to treat various conditions including infections, high blood pressure, and depression. Other drugs, such as aspirin and antacids, can be sold without a prescription. Many drugs are effective because they interact in a specific way with chemicals in cells. Chemists who develop these drugs must have knowledge of the structure and function of these target chemicals in order to design safe and effective drugs. Chemistry can also develop materials to repair or replace body parts. Diseased arteries can be replaced with plastic tubes. Artificial hips and knees made from metals and plastics can replace worn-out joints and allow people to walk again without pain.

Figure 1.8a shows a model of a small piece of DNA. Segments of DNA, called genes, store the information that controls changes that take place in cells. Biotechnology applies science to the production of biological products or processes. It uses techniques that can alter the DNA in living organisms. It may depend on the transfer of genes from one organism to another. When genes from humans are inserted into bacteria, the bacteria act as factories that produce chemicals of importance to humans, such as insulin. Production takes place in large versions of the bioreactors shown in Figure 1.8b.

Figure 1.8 Biotechnology
The discovery of the structure of DNA led to the development of biotechnology. **a.** This computer graphics model shows a small segment of DNA. **b.** The conditions in a bioreactor are controlled so that the bacteria produce as much of the product as possible.

Aspirin

In the late 1940s, Dr. Lawrence Craven observed that aspirin-containing gum, which was used to relieve pain after the removal of tonsils, caused excessive bleeding in children. He hypothesized that aspirin prevented the blood from clotting. He then began to prescribe aspirin to prevent heart attacks. In 1985, the FDA approved aspirin for patients who had suffered a heart attack. In 1996, the FDA proposed using aspirin during a suspected heart attack. In 1997, an advisory committee recommended daily low doses of aspirin for people at high risk of a heart attack. Currently aspirin is still used and recommended by doctors for this purpose.

The Universe Scientists assume that the methods used to study Earth can be applied to other objects in the universe. To study the universe, chemists gather data from afar and analyze matter that is brought back to Earth.

In the early 1800s, scientists began to study the composition of stars by analyzing the light they transmitted to Earth. In 1868, Pierre Janssen discovered a gas on the sun's surface that was not known on Earth. Joseph Norman Lockyer named the gas helium from the Greek word *helios,* meaning "sun." In 1895, William Ramsay discovered helium on Earth.

The moon and the planets do not emit light, so scientists must use other methods to gather data about these objects. The methods used depend on matter brought back to Earth by astronauts or on probes that can analyze matter in space. Chemists have analyzed more than 850 pounds of moon rocks that were brought back to Earth. The rocks were similar to rocks formed by volcanoes on Earth, suggesting that vast oceans of molten lava once covered the moon's surface. Figure 1.9 is a drawing of the robotic vehicle *Opportunity.* The vehicle was designed to determine the chemical composition of rocks and soil on Mars. Data collected at the vehicle's landing site indicated that the site was once drenched with water.

Figure 1.9 Space Exploration With help from NASA, chemists study matter from other bodies in the solar system. This drawing shows the robotic vehicle *Opportunity* on the surface of Mars.

Q: *How can chemistry be used to find evidence of life on other planets?*

1.2 LessonCheck

8. **List** What are three reasons for studying chemistry?
9. **Review** How has modern research in chemistry impacted society?
10. **Describe** How do chemists study the universe?
11. **Form an Opinion** Do the advantages of substituting the bioplastic PLA for conventional plastics outweigh the disadvantages? Would you use products made out of PLA? Why or why not?
12. **Explain** How can a knowledge of chemistry help you be a more informed citizen?

BIGIDEA
CHEMISTRY AS THE CENTRAL SCIENCE

13. A friend tells you that she doesn't think it is important to learn chemistry. What would be your response?

Lesson Check Answers

8. explaining the natural world, preparing for a career, being an informed citizen.
9. Students should cite some of the examples from the lesson.
10. They gather data from afar and analyze matter brought back to Earth.
11. Students should use data from Figure 1.7 to support their opinions.
12. A possible answer is that knowledge of chemistry helps a citizen evaluate data and arrive at an informed opinion about a public issue that involves technology.
13. **BIGIDEA** Students will likely focus on the three general reasons presented in the lesson. They may also note how central chemistry is to the friend's life.

Explore

Class Activity

Have students use the Internet to locate data on the composition of the sun. Have them explain how scientists can know the composition of the sun without collecting matter from the sun. Ask students to explain why it would be difficult to present the data as a pie chart or bar graph.

CHEMISTRY & YOU Chemists can use a probe to detect evidence of water in rocks or soil; they can analyze light absorbed by a planet's atmosphere to detect water vapor.

Extend

Connect to LITERATURE

Explain to the class how public perception and expectations of science have changed dramatically over the last two hundred years. Advances in technology can lead to increased public concern about the role that science plays in shaping the future. These societal concerns are reflected in the work of 19th- and 20th-century novelists such as H. G. Wells, Edward Bellamy, Jules Verne, George Orwell, and Mary Shelley. Have students select a novel by one of these novelists and critically examine it to determine how the author addressed the role of science, technology, and society in his or her era. Students should be prepared to share their findings with the class.

Evaluate

Informal Assessment

Ask students to name examples of recent technological developments that were made possible by the study of chemistry. List the technologies on the board. Challenge students to identify whether each application benefits the environment, conserves and produces energy, improves the environment, or expands our knowledge of the universe. Then have students complete the 1.2 Lesson Check.

Reteach

Point out that being aware of the way chemistry affects the United States, other countries, and the universe will help students to better understand, benefit from, and contribute to the advances of the modern world in which they live. Ask students to generate a list of what is essential in their lives. Then have them look at their lists and determine three items that provide benefits because of chemistry.

CHEMISTRY & YOU

CHEMISTRY & YOU The results from an experiment do not always match the predicted outcome or hypothesis and in many cases may only lead to more questions. However, there are times that the unexpected results lead to exciting new ideas about future experiments with slightly modified variables.

Have students read the feature and take note of the corresponding photos. Point out that these examples highlight how a failed initial hypothesis resulted in products that improve everyday life. Emphasize the significance of the scientist not giving up after the "accident."

Ask What are some other uses that may have come out of these discoveries? *(Sample answers: sports balls, shoe soles, skateboard wheels, stain-resistant fabrics, lubricants, plumbing tape, no-residue tape, new artificial sweeteners, diet soda)*

Accidental Chemistry

Typically, chemists direct their research toward a practical goal or application. However, sometimes scientists accidentally stumble upon a discovery that they didn't intend to find.

Polytetrafluoroethylene In 1938, Roy J. Plunkett was researching new refrigerants, which are compounds used in refrigerators and air conditioners. In preparation for an experiment, Plunkett stored a compound called tetrafluoroethylene gas (TFE) in cylinders at a low temperature overnight.

When Plunkett was ready to use the TFE the next day, he found that none of the gas came out. He opened a cylinder to find a white, waxy solid that was very slippery and did not react with any other chemicals.

The substance that Plunkett accidentally made, called polytetrafluoroethylene (PTFE) has many uses that you may be familiar with, such as insulation for electrical cables, soil and stain repellents for fabrics, and coatings for nonstick cookware.

Saccharin In 1879, Constantine Fahlberg was looking for new uses for coal tar and forgot to wash his hands after working one day in the lab. When he went home to eat dinner that night he noticed that his bread tasted sweet. He realized that the sweet taste was coming from his unwashed hands.

The substance on his hands was saccharin, which is the oldest artificial sweetener. Saccharin was initially sold in tablet form from the late 1890s to the 1940s. In 1957, it was introduced in granulated form.

Sugar Substitute

21st Century Learning To be successful in the 21st century, students need skills and learning experiences that extend beyond subject matter mastery. This project helps students build the following 21st-Century Skills: *Creativity and Innovation; Communication and Collaboration; Information, Media, and Technology Literacy; Initiative and Self-Direction;* and *Social Skills.*

UNDER PRESSURE Scientists must secure funding for their research. Government and corporate financial backers require regular project updates and proof of progress before approving additional funding. Pose the following challenge to your students. They are to form research groups of three to five students each and choose one of these four experiments as the group's project. They will prepare a 15-20 minute presentation for their financial backers (the class) on the status of their project. The presentation will be made via an Internet meeting program from their "lab" (an unused classroom or computer lab). It is their job to convince the backers to continue funding their project, despite the outcome of their original research. Presentations should include a summary of the purpose of the original research, a graphic presentation of the results, a proposal for exploring the possible uses of the discovery, and a proposed budget.

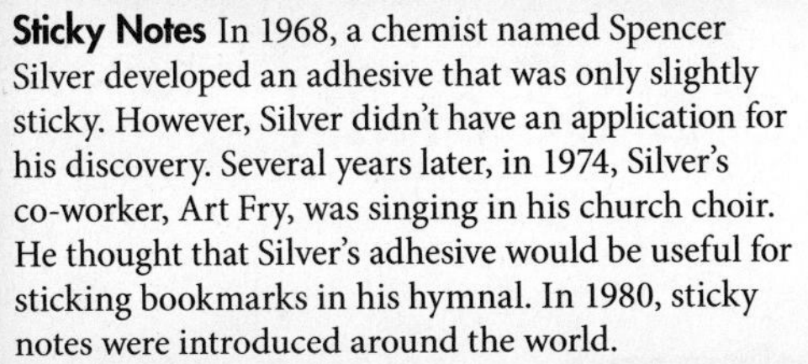

Sticky Notes In 1968, a chemist named Spencer Silver developed an adhesive that was only slightly sticky. However, Silver didn't have an application for his discovery. Several years later, in 1974, Silver's co-worker, Art Fry, was singing in his church choir. He thought that Silver's adhesive would be useful for sticking bookmarks in his hymnal. In 1980, sticky notes were introduced around the world.

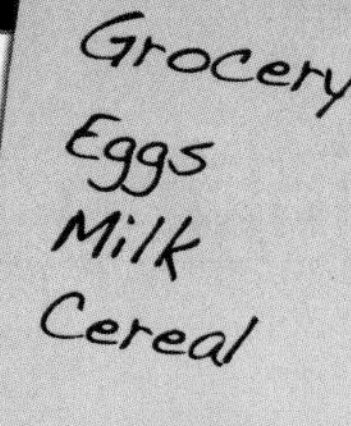

Vulcanized Rubber When rubber was introduced in the early 1830s, it was not very useful. It froze in the winter and melted into a glue-like substance in the summer. Charles Goodyear, a hardware merchant from Philadelphia was on a quest to improve the properties of the substance. After countless failed experiments, Goodyear began using sulfur in his rubber formulations. In the winter of 1839, a chunk of his rubber-sulfur formulation landed on top of a hot stove. Instead of melting, the rubber charred. The mixture had hardened but it was still elastic.

Thomas Hancock, an Englishman, saw a sample of Goodyear's rubber and re-invented it in 1843. He named the processes "vulcanization," after Vulcan, the Roman god of fire.

Take It Further

1. Identify Coal tar contains carbon. What type of chemistry was the focus of Fahlberg's research?

2. Classify Were the inventors in these examples performing pure or applied research when they made their discoveries? Explain.

3. Infer When Plunkett couldn't get the TFE gas out of the cylinders, he weighed them and found that they weighed the same as they did the night before. What was the significance of this finding?

4. Form an Opinion Are the inventions described examples of technology? Why or why not?

Differentiated Instruction

L1 LESS PROFICIENT READERS Display visual examples of each product, and use the images as you discuss each discovery.

ELL ENGLISH LANGUAGE LEARNERS For each of the four products shown in the pictures, have samples available for students to touch, smell, and observe.

L3 ADVANCED STUDENTS Assign students one of the four products featured, and have them generate a flowchart outlining the steps they would have expected the researcher to follow.

Explain

START A CONVERSATION Explore the initial disappointment the researchers must have felt by relating it to an experience in which the school, class, or student had to deal with an unexpected outcome. Ask students to consider what would have happened had the researchers simply stopped at that point, rather than continuing to experiment. **Ask** What step of the scientific method did these researchers pursue that resulted in an amazing discovery? *(They modified the initial hypothesis and designed new experiments.)*

Extend

Encourage students to identify examples of discoveries they have discussed in other science courses that could be described as "accidental". Ask students to create a timeline noting how and when one such discovery was made, and showing past and current uses, applications, or subsequent discoveries.

Connect to LANGUAGE ARTS

Have students write a factual yet humorous narrative about the way in which Sir Isaac Newton discovered gravity. Students may write from the perspective of either an unseen observer or Newton himself.

Answers

TAKE IT FURTHER

1. organic chemistry
2. They were doing applied chemistry because they were focused on solving a specific problem.
3. Because the mass was constant, no matter had entered the cylinder or leaked out of the cylinder.
4. They are examples of technology because they provide society with things that are useful.

LESSON 1.3

Key Objectives

1.3.1 DESCRIBE how Lavoisier transformed chemistry.

1.3.2 IDENTIFY three steps in the scientific method.

1.3.3 EXPLAIN the role collaboration and communication play in science.

Additional Resources

Reading and Study Workbook, Lesson 1.3

Available Online or on Digital Media:

- Teaching Resources, Lesson 1.3 Review
- Laboratory Manual, Lab 1
- Small-Scale Chemistry Laboratory Manual, Lab 1

PearsonChem.com

Engage

CHEMISTRY & YOU Have students study the photograph and read the text that opens the lesson. **Ask** What made Fleming different from other scientists who had seen this mold? *(Fleming recognized the importance of this discovery. He assumed that the mold had released a chemical that prevented the growth of bacteria.)*

Activate Prior Knowledge

Have students draw a T-chart in their notebooks. In the first column, they should write what they already know about the scientific method. In the second column, they should add the new information they learn as they proceed with the lesson.

IN ACADEMIC STANDARDS for SCIENCE

- Develop explanations based on reproducible data and observations gathered during laboratory investigations.
- Clearly communicate their ideas and results of investigations verbally and in written form using tables, graphs, diagrams, and photographs.

1.3 Thinking Like a Scientist

CHEMISTRY & YOU

Q: *How do you think Alexander Fleming tested his hypothesis?* In 1928, Alexander Fleming, a Scottish scientist, noticed that a bacteria he was studying did not grow in the presence of a yellow-green mold. Other scientists had made the same observation, but Fleming was the first to recognize its importance. He assumed that the mold had released a chemical that prevented the growth of the bacteria. That chemical was penicillin, which can kill a wide range of harmful bacteria.

Key Questions

How did Lavoisier help to transform chemistry?

What are the steps in the scientific method?

What role do collaboration and communication play in science?

Vocabulary

- scientific method
- observation
- hypothesis
- experiment
- independent variable
- dependent variable
- model
- theory
- scientific law

An Experimental Approach to Science

How did Lavoisier help to transform chemistry?

The word *chemistry* comes from the word *alchemy*. Long before there were chemists, alchemists were studying matter. Alchemy arose independently in many regions of the world. It was practiced in China and India as early as 400 B.C. In the eighth century, Arabs brought alchemy to Spain, and from there it spread quickly to other parts of Europe.

You may have heard that alchemists were concerned with searching for a way to change other metals, such as lead, into gold. Although alchemists did not succeed with this quest, the work they did spurred the development of chemistry. Alchemists developed the tools and techniques for working with chemicals. For example, alchemists developed processes for separating mixtures and purifying chemicals. They designed equipment that is still used today, including beakers, flasks, tongs, funnels, and the mortar and pestle, which is shown in Figure 1.10. What they did not do was provide a logical set of explanations for the changes in matter that they observed. Chemists would accomplish that task many years later.

Figure 1.10 Mortar and Pestle
Pharmacists still use a bowl-shaped mortar and club-shaped pestle to mix drugs for patients. The mortar and pestle in this photograph are made of porcelain, which is a hard material.
Infer *What may be some other uses of a mortar and pestle?*

Focus on ELL

1 CONTENT AND LANGUAGE Preview the vocabulary terms by pronouncing each term and having students repeat it. Allow students time to add the terms to their vocabulary notebooks, with definitions, parts of speech, and examples or sketches. Encourage struggling students to add native language equivalents.

2 FRONTLOAD THE LESSON Have students preview and predict what the lesson is about by reading the first sentence of each major paragraph and looking at the visual displays. Once the lesson has been read, predictions can be confirmed or denied.

3 COMPREHENSIBLE INPUT Use Figure 1.14 to explain how to read a flowchart, and provide students with a copy of a flowchart that explains each step of the scientific method. Use the Quick Lab to model the scientific method.

By the 1500s in Europe, there was a shift from alchemy to science. Science flourished in Britain in the 1600s, partly because King Charles II was a supporter of the sciences. With his permission, some scientists formed the Royal Society of London for the Promotion of Natural Knowledge. The scientists met to discuss scientific topics and conduct experiments. The society's aim was to encourage scientists to base their conclusions about the natural world on experimental evidence, not on philosophical debates.

In France, Antoine-Laurent Lavoisier did work in the late 1700s that would revolutionize the science of chemistry. **Lavoisier helped to transform chemistry from a science of observation to the science of measurement that it is today.** To make careful measurements, Lavoisier designed a balance that could measure mass to the nearest 0.0005 gram.

One of the many things Lavoisier accomplished was to settle a long-standing debate about how materials burn. The accepted explanation was that materials burn because they contain phlogiston, which is released into the air as a material burns. To support this explanation, scientists had to ignore the evidence that metals can gain mass as they burn. By the time Lavoisier did his experiments, he knew that there were two main gases in air—oxygen and nitrogen. Lavoisier was able to show that oxygen is required for a material to burn. Lavoisier's wife Marie Anne, shown in Figure 1.11, helped with his scientific work. She made drawings of his experiments and translated scientific papers from English.

Figure 1.11 Antoine Lavoisier
This portrait of Antoine Lavoisier and his wife Marie Anne was painted by Jacques Louis David in 1788.

The Scientific Method

What are the steps in the scientific method?

Scientists have a powerful tool that they use to produce valuable results. Like all scientists, the biochemist shown in Figure 1.12 is using the scientific method to solve difficult problems. The **scientific method** is a logical, systematic approach to the solution of a scientific problem. **Steps in the scientific method include making observations, proposing and testing hypotheses, and developing theories.**

Making Observations The scientific method is useful for solving many kinds of problems. Suppose you try to turn on a flashlight and you notice that it does not light. When you use your senses to obtain information, you make an **observation.** An observation can lead to a question: What is wrong with the flashlight?

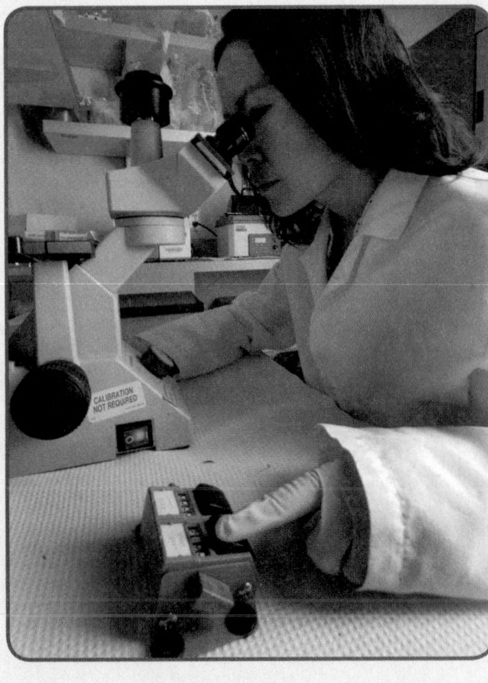

Figure 1.12 Observing With a Microscope
Observation is an essential step in the scientific method.

How Oxygen Got Its Name

The ancient Greeks thought that flammable objects contained the element fire, which George Stahl (1660–1734) named *phlogiston*. During burning, phlogiston transferred to the air. Phlogiston-rich air, now called nitrogen, did not support burning; objects burned brightly in phlogiston-poor air.

Lavoisier measured the mass of metals before and after heating in a closed container. He showed that the mass gained by the metal was lost by the air. Thus, the process of burning involved a gain of matter, not a loss of phlogiston. Lavoisier named the portion of air that supported combustion oxygen.

Foundations for Reading

BUILD VOCABULARY When discussing the two types of variables, explain that one meaning of *manipulate* is "to manage or control" and one meaning of *respond* is "to answer or act in turn."

Explain

An Experimental Approach to Science

MAKE A CONNECTION Students often think that an experiment is a failure if they do not get the expected or right results. As students perform experiments, help them analyze results that do not fit a hypothesis or vary widely from those of other students. Often, you can identify experimental errors that explain deviation. Also point out that scientists can gain important insights from "failed" experiments.

Explore

Class Activity

PURPOSE Students make hypotheses based on a set of observations.

MATERIALS phenolphthalein, 70% solution of 2-propanol, small artist's paintbrush, sheets of copy paper or white butcher paper, household glass cleaner containing ammonia

ADVANCE PREPARATION Prepare a solution of phenolphthalein in alcohol (250 mg of phenolphthalein in 250 mL of 70% 2-propanol). Use the brush to letter messages on sheets of paper. Try messages such as "CHEM-IS-TRY" and "This is a LABOR-atory, not a lab-ORATORY." Allow the paper to dry until the messages are invisible. Then post the messages around the room in well-ventilated areas.

PROCEDURE At intervals, spray each sheet with the glass cleaner. Ask students to hypothesize why the pink messages appear, disappear after a few minutes, and then reappear when the paper is sprayed again.

EXPECTED OUTCOME Students are likely to infer that something in the cleaner caused a reversible change to something on the paper. They may infer that the material in the cleaner is volatile.

Answers

FIGURE 1.10 *Sample answers:* to grind herbs, grains, or other food items; to grind pigments for homemade oil paints or watercolors

Explain

The Scientific Method

USE VISUALS Direct students to Figure 1.14 and discuss each step. Then discuss how the steps in the scientific method allow scientists to support theories. Ask students to explain why they think some pathways in the flowchart are cyclic rather than straight. *(Cyclic pathways occur where steps may occur more than once, as with the processes of developing hypotheses and conducting experiments to test them.)*

Misconception Alert

Students may think the steps in the scientific method must be done in a particular order every time. Guide students to understand that the steps may or may not be repeated, depending on the problem.

Also, explain to students that some disciplines resemble science, but they are not science. They are not science because the ideas related to these disciplines are not tested using the scientific method. These disciplines are called pseudoscience. Astrology and phrenology are two examples of pseudoscience.

Extend

Connect to ASTRONOMY

A theory in astronomy, called the Big Bang theory, states that the universe began billions of years ago in an explosion. The Big Bang theory also states that the universe is still expanding. Although the Big Bang cannot be observed directly, its effects can be observed.

Scientists have gathered a great deal of evidence that supports the Big Bang theory. This evidence includes the red shifts of distant galaxies, which indicates that the universe is still expanding. Scientists have also observed cosmic microwave background radiation. Scientists believe this radiation was produced during the Big Bang. Have students research how the Big Bang theory was developed and then make a flowchart, similar to the one in Figure 1.14.

Figure 1.13 Computer Models
This scientist is using a computer to model complex molecules, which are difficult to study with experiments alone.

Testing Hypotheses If you guess that the batteries in the flashlight are dead, you are making a hypothesis. A **hypothesis** is a proposed explanation for an observation. You can test your hypothesis by putting new batteries in the flashlight. Replacing the batteries is an **experiment,** a procedure that is used to test a hypothesis. If the flashlight lights, you can be fairly certain that your hypothesis was true. What if the flashlight does not work after you replace the batteries? A hypothesis is useful only if it accounts for what is actually observed. When experimental data does not fit a hypothesis, the hypothesis must be changed. A new hypothesis might be that the light bulb is burnt out. An experiment to test this new hypothesis is to replace the bulb.

When you design experiments, you deal with variables, or factors that can change. The variable that you change during an experiment is the **independent variable,** also called the manipulated variable. The variable that is observed during the experiment is the **dependent variable,** also called the responding variable. If you keep other factors that can affect the experiment from changing during the experiment, you can relate any change in the dependent variable to changes in the independent variable. For the results of an experiment to be accepted, the experiment must produce the same result no matter how many times it is repeated, or by whom. This is why scientists are expected to publish a description of their procedures along with their results.

Sometimes the experiment a scientist must perform to test a hypothesis is difficult or impossible. For example, atoms and molecules, which are some of the smallest units of matter, cannot be easily seen. In these situations, scientists often turn to models to gain more understanding of a problem. A **model** is a representation of an object or event. Figure 1.13 shows a scientist working with computer models of complex biological molecules. Chemists may also use models to study chemical reactions and processes.

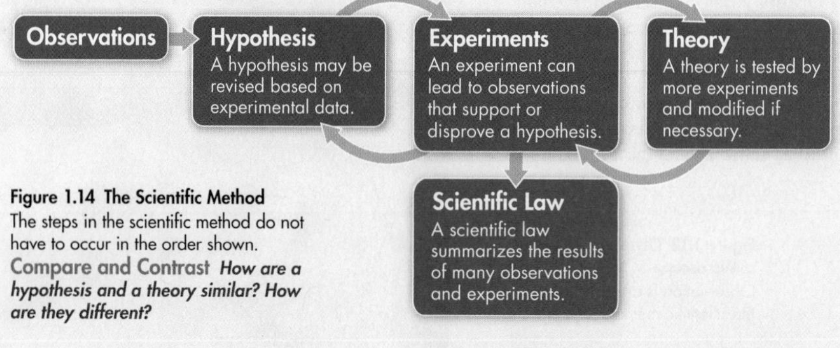

Figure 1.14 The Scientific Method
The steps in the scientific method do not have to occur in the order shown.
Compare and Contrast *How are a hypothesis and a theory similar? How are they different?*

Differentiated Instruction

L1 STRUGGLING STUDENTS Explain to students that if they make a mistake in the lab, the scientific method allows them to repeat their experiments as many times as needed, as long as they record their work.

ELL ENGLISH LANGUAGE LEARNERS Have English learners work with a partner to describe, act out, or illustrate in a drawing the different steps that make up the scientific method. Challenge students to identify how they used each step to solve problems in the past.

L3 ADVANCED STUDENTS Have students state the constants, the independent variable, and the dependent variable in the Quick Lab experiment.

Developing Theories Figure 1.14 shows how the steps of the scientific method fit together. Once a hypothesis meets the test of repeated experimentation, it may be raised to a higher level of ideas. It may become a theory. A **theory** is a well-tested explanation for a broad set of observations. Some of the theories in chemistry are very useful because they help you form mental pictures of objects or processes that cannot be seen. Other theories allow you to predict the behavior of matter.

When scientists say that a theory can never be proved, they are not saying that a theory is unreliable. They are simply leaving open the possibility that a theory may need to be changed at some point in the future to explain new observations or experimental results.

Scientific Laws Figure 1.14 shows how scientific experiments can lead to laws as well as theories. A **scientific law** is a concise statement that summarizes the results of many observations and experiments. In Chapter 14, you will study laws that describe how gases behave. One law describes the relationship between the volume of a gas in a container and its temperature. If all other variables are kept constant, the volume of the gas increases as the temperature increases. The law doesn't try to explain the relationship it describes. That explanation requires a theory.

Q: *What was Alexander Fleming's hypothesis? How could he test his hypothesis?*

Quick Lab

Purpose To test the hypothesis that bubble making can be affected by adding sugar or salt to a bubble-blowing mixture

Materials

- 3 plastic drinking cups
- measuring cup and spoons
- liquid dish detergent
- water
- table sugar
- table salt
- drinking straw

Bubbles!

Procedure

1. Label three drinking cups 1, 2, and 3. Measure and add one teaspoon of liquid dish detergent to each cup.
2. Use the measuring cup to add two-thirds cup of water to each drinking cup. Then swirl the cups to form a clear mixture. **CAUTION** *Wipe up any spills immediately so that no one will slip and fall.*
3. Add a half teaspoon of table sugar to cup 2 and a half teaspoon of table salt to cup 3. Swirl each cup for one minute.
4. Dip the drinking straw into cup 1, remove it, and blow gently into the straw to make the largest bubble you can. Practice making bubbles until you feel you have reasonable control over your bubble production.
5. Repeat Step 4 with the mixtures in cups 2 and 3.

Analyze and Conclude

1. **Observe** Did you observe any differences in your ability to produce bubbles using the mixtures in cup 1 and cup 2?
2. **Observe** Did you observe any differences in your ability to produce bubbles using the mixtures in cup 1 and cup 3?
3. **Draw Conclusions** What can you conclude about the effects of table sugar and table salt on your ability to produce bubbles?
4. **Design an Experiment** Propose another hypothesis related to bubble making. Design an experiment to test your hypothesis.

Focus on ELL

4 LANGUAGE PRODUCTION Have students work in groups of four to complete the lab. Make sure each group has ELLs of varied language proficiencies so that more proficient students can help less proficient ones. Have students work according to their proficiency level.

BEGINNING: LOW/HIGH Model this lab and use student volunteers to help with each step. For example, ask one student to label the cups and another student to pour the detergent.

INTERMEDIATE: LOW/HIGH Provide a picture for each step of the procedure to depict what is supposed to be done. Provide a chart for answering questions 1-4.

ADVANCED: LOW/HIGH Ask students to compare and contrast the use of salt and sugar for making bubbles.

CHEMISTRY & YOU Fleming's hypothesis was that mold released a chemical that prevented the growth of bacteria. He could test his hypothesis with other types of bacteria to see if the mold had the same effect on them.

Explore

Quick Lab

OBJECTIVE By completing this activity, students will be able to test the hypothesis that bubble making can be affected by adding sugar or salt to the bubble-blowing mixture.

PREP TIME 5 minutes

CLASS TIME 15 minutes

SAFETY Remind students to be careful not to draw any liquid into their mouth through the straw.

EXPECTED OUTCOME Students conclude that sugar has no effect on bubble production and that salt prevents bubble production.

ANALYZE AND CONCLUDE

1. no
2. Yes; no bubbles formed from the liquid in cup 3.
3. Sugar has no effect on bubble production, but salt stops it completely.
4. Answers might include examining the effect of temperature or dilution of the bubble-making mixture. For example, diluting the mixture can reverse the salt effect.

FOR ENRICHMENT Facilitate a class discussion about the experiments that students proposed for Question 4. Discuss whether or not the experiments will work. Perform two of the experiments and discuss the results. Emphasize the need to learn from "failed" experiments.

Answers

FIGURE 1.14 A hypothesis and a theory are both explanations. A hypothesis is only a proposed explanation; a theory is a well-tested explanation for a broad set of observations. When a hypothesis meets the test of repeated experimentation, it may be become a theory.

Explain

Collaboration and Communication

CRITICAL THINKING Explain that students can learn about research from news reports, from specialized journals, and from the Internet. The more reliable information people have, the better able they are to effectively address public issues related to science and technology. Stress that students should not avoid reading about science, but they should look for reliable sources and approach the news with a certain amount of healthy skepticism. **Ask** How can the Internet help people learn about advances in science? *(Anyone can access information on the Internet.)* **Ask** What is one disadvantage of getting information from the Internet? *(The information is not always reliable.)* Note that sites with *.gov*, *.edu*, and *.org* domains are usually the most reliable sources, especially those sites that are maintained by federal and state science agencies, colleges and universities, and professional organizations.

Have students conduct an Internet search on a scientific topic of their choice and print the first page of the results. Demonstrate how to locate information to verify a site's credibility, such as professional affiliations or expert contributors or authors. Then, have students check each source on their printout and note whether they believe the site to be a reliable or unreliable source of scientific information.

Extend

Have students use the Internet to research examples of successful collaborative projects that involved a multidisciplinary approach to solving a scientific problem. For instance, the series of NASA missions to the moon in the 1970s was a collaborative project. Have students orally report their findings to the class; tell students to be prepared for questions.

Figure 1.15 Teamwork
For a volleyball team to win, the players must work together.

Collaboration and Communication

What role do collaboration and communication play in science?

No matter how talented the players on a team may be, one player cannot ensure victory for the team. Individuals must collaborate, or work together, for the good of the team. Think about the volleyball players in Figure 1.15. In volleyball, the person who spikes the ball depends on the person who sets the ball. Unless the ball is set properly, the spiker will have limited success. Many sports recognize the importance of collaboration by keeping track of assists. During a volleyball game, the players also communicate with one another so it is clear who is going to do which task. Strategies that are successful in sports can work in other fields, such as science. **When scientists collaborate and communicate with one another, they increase the likelihood of a successful outcome.**

Collaboration Scientists choose to collaborate for different reasons. For example, some research problems are so complex that no one person could have all the knowledge, skills, and resources to solve the problem. It is often necessary to bring together individuals from different disciplines. Each scientist will typically bring different knowledge and, perhaps, a different approach to a problem. Just talking with a scientist from another discipline may provide insights that are helpful.

There may be a practical reason for collaboration. For example, an industry may give a university funding for pure research in an area of interest to the industry. Scientists at the university get the equipment and financing required to do the research. In exchange, the scientists provide ideas and expertise. The industry may profit from its investment by marketing applications based on the research.

Collaboration isn't always a smooth process. Conflicts can arise about use of resources, amount of work, who is to receive credit, and when and what to publish. Like the students in Figure 1.16, you will likely work in pairs or on a team in the laboratory. If so, you may face some challenges. However, you can also experience the benefits of a successful collaboration.

Figure 1.16 Lab Partners
Working in pairs or in a group can be challenging, but it can also be rewarding.
Apply Concepts ***What steps in the scientific method are these students using?***

UbD Check for Understanding

What role do collaboration and communication play in science?

Assess students' understanding of the roles of collaboration and communication in science by asking for thumbs-up or thumbs-down gestures in response to the statement *I understand the importance of collaboration and communication in science and can explain it.*

ADJUST INSTRUCTION If students use a thumbs-down response, ask students to identify what aspects of collaboration or communication are unclear. Provide a targeted review of those aspects with students.

Communication The way scientists communicate with each other and with the public has changed over the centuries. In earlier centuries, scientists exchanged ideas through letters. They also formed societies to discuss the latest work of their members. When societies began to publish journals, scientists could use the journals to keep up with new discoveries.

Today, many scientists, like those in Figure 1.17, work as a team. They can communicate face to face. They also can exchange ideas with other scientists by e-mail, by phone, and at local and international conferences. Scientists still publish their results in scientific journals, which are the most reliable source of information about new discoveries. Most journals are now published online and are readily accessible. Articles are published only after being reviewed by experts in the author's field. Reviewers may find errors in experimental design or challenge the author's conclusions. This review process is good for science because work that is not well founded is usually not published.

The Internet is a major source of information. One advantage of the Internet is that anyone can get access to information. One disadvantage is that anyone can post information on the Internet without first having that information reviewed. To judge the reliability of information you find on the Internet, you have to consider the source. This same advice applies to articles in newspapers and magazines or the news you receive from television. If a media outlet has a reporter who specializes in science, chances are better that a report will be accurate.

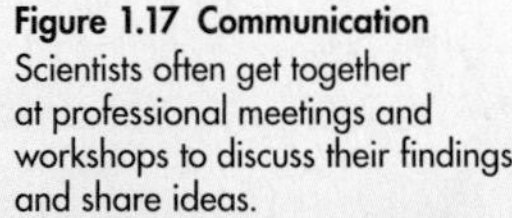

Figure 1.17 Communication
Scientists often get together at professional meetings and workshops to discuss their findings and share ideas.

1.3 LessonCheck

14. **Review** How did Lavoisier revolutionize the science of chemistry?
15. **List** Name three steps in the scientific method.
16. **Explain** Why are collaboration and communication important in science?
17. **Describe** What did alchemists contribute to the development of chemistry?
18. **Explain** How did Lavoisier's wife help him to communicate the results of his experiments?
19. **Describe** What process takes place before an article is published in a scientific journal?
20. **Explain** Why is it important for scientists to publish a description of their procedures along with the results of their experiments?
21. **Infer** Why should a hypothesis be developed before experiments take place?
22. **Compare** What is the difference between a theory and a hypothesis?
23. **Classify** In Chapter 2, you will learn that matter is neither created nor destroyed in any chemical change. Is this statement a theory or a law? Explain your answer.

BIGIDEA
CHEMISTRY AS THE CENTRAL SCIENCE

24. Do the steps in the scientific method always need to be followed in order? Explain.

LESSON 1.3

Evaluate

Informal Assessment

Ask students to identify an everyday type of problem that could potentially be solved by the scientific method. Then have them explain how they would use the scientific method to solve that problem. Then have students complete the 1.3 Lesson Check.

Reteach

Students may mistakenly infer that a theory grows into a law by constant testing and refinement. Or they may confuse a theory with a hypothesis. A hypothesis differs from a theory in that a hypothesis can be formed on the basis of a single set of observations. It cannot become a theory unless it explains a broad set of observations obtained from repeated experimentation. Remind students that theories can, and do, change as new information becomes available through experimentation.

Answers

FIGURE 1.16 making observations

Lesson Check Answers

14. He helped transform chemistry from a science of observation to a science of measurement.
15. *Sample answers:* making observations, testing hypotheses, and developing theories
16. They help increase the likelihood of a successful outcome.
17. They developed the tools and techniques for working with chemicals.
18. She made drawings of his experiments and translated scientific papers.
19. Articles are reviewed by experts in the author's field of research.
20. so that other scientists can repeat the experiments and confirm the results
21. It guides the design of the experiments.
22. A theory is a well-tested explanation of a broad set of observations; a hypothesis is a proposed explanation for an observation.
23. A law; it is not an explanation.
24. **BIGIDEA** *Sample answer:* No; for example, most experiments do not lead directly to the development of or revision of a theory.

SMALL-SCALE LAB

Explore

Small-Scale Lab

OBJECTIVE After completing this activity, students will be able to demonstrate their knowledge of safe laboratory practices.

CLASS TIME 40 minutes

SKILLS FOCUS Observing, communicating results

TEACHING TIPS Students need to read Appendix C to answer the questions in the lab. Use this activity as part of an orientation in which you present your rules for working in the laboratory. Provide a floor plan of the room and have students record the location of safety equipment such as an eyewash fountain or a fire extinguisher. After you discuss the safety rules, have students sign a safety contract. For an example, see the safety contract for the small-scale chemistry labs.

ANSWERS TO QUESTIONS

Stand back, notify your teacher, and warn other students.

Wash your hands thoroughly with soap and water.

Wear safety goggles at all times when working in the lab.

Tie back long hair and loose clothing. Never reach across a lit burner. Keep flammable materials away from the flame.

It isn't always appropriate to dispose of chemicals by flushing them down the drain. Follow your teacher's instructions for disposal.

Tell your teacher and nearby classmates. Dispose of the glass as instructed by your teacher.

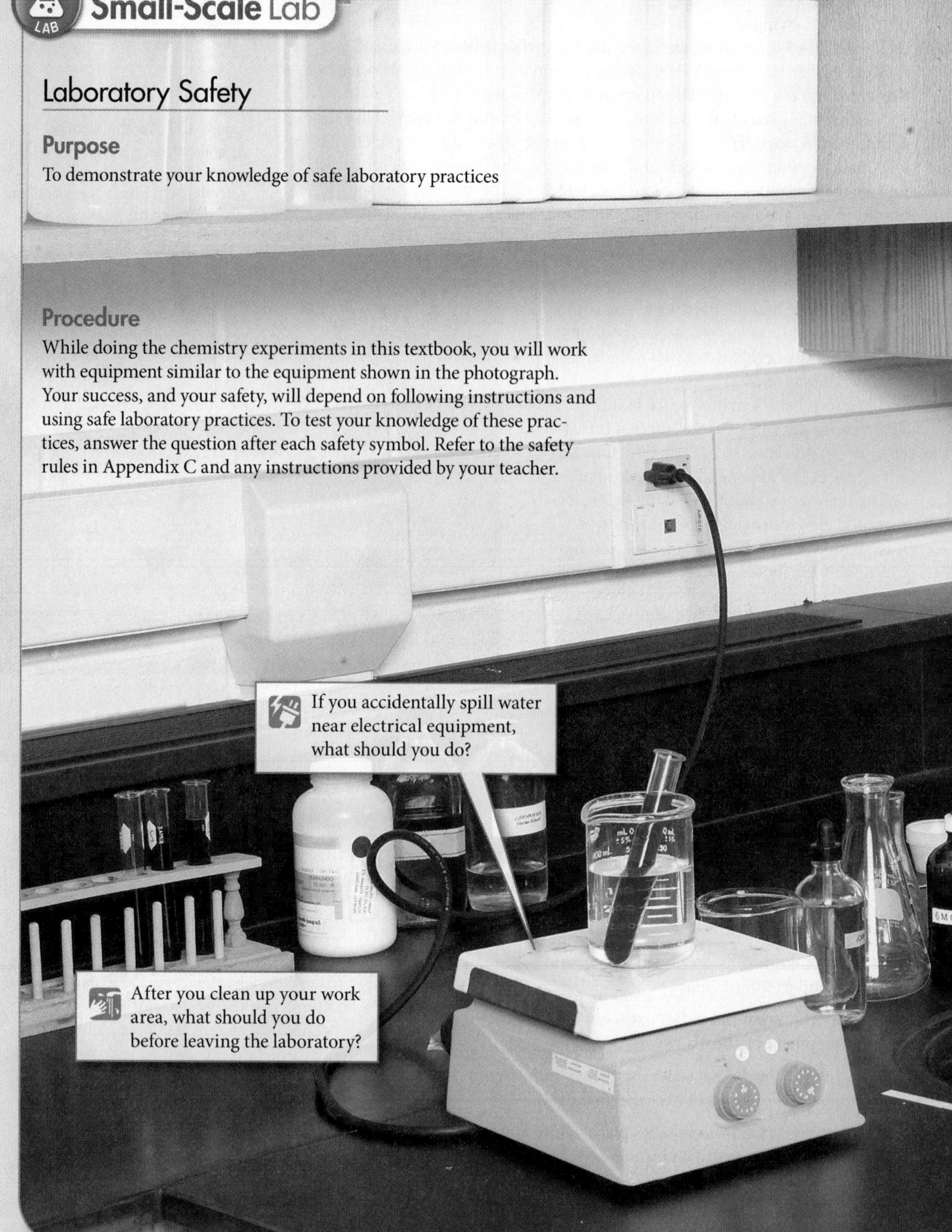

Small-Scale Lab

Laboratory Safety

Purpose

To demonstrate your knowledge of safe laboratory practices

Procedure

While doing the chemistry experiments in this textbook, you will work with equipment similar to the equipment shown in the photograph. Your success, and your safety, will depend on following instructions and using safe laboratory practices. To test your knowledge of these practices, answer the question after each safety symbol. Refer to the safety rules in Appendix C and any instructions provided by your teacher.

If you accidentally spill water near electrical equipment, what should you do?

After you clean up your work area, what should you do before leaving the laboratory?

Focus on ELL

4 LANGUAGE PRODUCTION Have students work in groups of four to complete the lab. Make sure each group has ELLs of varied language proficiencies so that more proficient students can help less proficient ones. Have students work according to their proficiency level.

BEGINNING: LOW/HIGH Provide true or false pictures that represent each question on the safety test. For instance, create a picture of a student pouring a chemical into the sink. Students can circle a "thumbs up" or "thumbs down" to indicate if this is a good or bad idea.

INTERMEDIATE: LOW/HIGH Show students how to use the appendix in the textbook.

ADVANCED: LOW/HIGH Have students help students with lower proficiencies with the written answers and predictions.

When should safety goggles be worn?

What precautions should you take when working near an open flame?

Is it always appropriate to dispose of chemicals by flushing them down the sink?

What should you do if glassware breaks?

SMALL-SCALE LAB

Background

The small-scale chemistry experiments in this book are designed to help you teach students important chemical principles, not just process. For most experiments, the procedure is short and simple. In many cases, students are asked to construct a grid, place a small-scale reaction surface over the grid, do the experiment, and record their results in a similar grid. Sometimes the students are told to mark the grid with black X's. These X's provide black-and-white backgrounds against which students can observe reaction mixtures.

The YOU'RE THE CHEMIST activities ask students to apply what they learned in the initial experiment. Some of these activities could be used for performance-based assessment.

FOR ENRICHMENT Have students look at the safety rules in Appendix C. Have each student choose five of the rules in Appendix C and make a list of what things could happen if each rule is not followed. Organize students into groups of three and have them share their answers with each other.

Laboratory Hazard Symbols

The four-color diamond pictured on page R50 of the text is a familiar sight in every laboratory stockroom and chemical cabinet. The hazard diamond was originally developed by the National Fire Protection Association (NFPA) in response to federal requirements that information about hazardous materials must be visibly posted and easily accessible to all workers. The diamond is divided into four smaller diamonds: blue, red, yellow, and white. The colors symbolize four areas of safety concern. The blue diamond provides information on potential dangers to human health. The red diamond provides information on the flammability of the material. Yellow provides information on the reactivity of the substance. Each of these three diamonds is occupied by a number ranging from 0 to 4, with 0 representing the lowest level of danger and 4 the highest. The white diamond is reserved for special hazards. Instead of numbers, the NFPA system uses two symbols in this diamond—W and OX—to indicate hazards of particular interest to firefighters. The W, often seen with a horizontal line through it, provides information about reactivity with water, and OX indicates that the material is an oxidizer. The official NFPA hazard diamond uses only the W and OX symbols in the white diamond; other versions may have an expanded list that indicates whether a substance is an acid (ACID), a base (ALK), corrosive (COR), or radioactive (universal radioactive warning symbol).

LESSON 1.4

Key Objectives

1.4.1 IDENTIFY the general approach to solving a problem.

1.4.2 DESCRIBE the steps for solving numeric problems.

1.4.3 DESCRIBE the steps for solving nonnumeric problems.

Additional Resources

Reading and Study Workbook, Lesson 1.4

Available Online or on Digital Media:

- Teaching Resources, Lesson 1.4 Review

PearsonChem.com

Engage

CHEMISTRY & YOU Have students study the photograph and read the text that opens the lesson. **Ask** Describe your approach to solving a crossword puzzle. *(Sample answer: I try to answer the first five clues across and the first five clues down. I continue like this until I get enough of the puzzle filled in; then I begin to randomly answer the clues.)* **Ask** What is the most effective way you have learned to solve a problem? *(Approaches vary depending on the given problem; effective approaches are generally organized and well planned.)*

Activate Prior Knowledge

Ask students to share the steps of the problem-solving plan they have learned in mathematics courses. Explain that scientists use a similar plan to approach the process of solving chemistry problems.

IN ACADEMIC STANDARDS for SCIENCE

1.4 Problem Solving in Chemistry

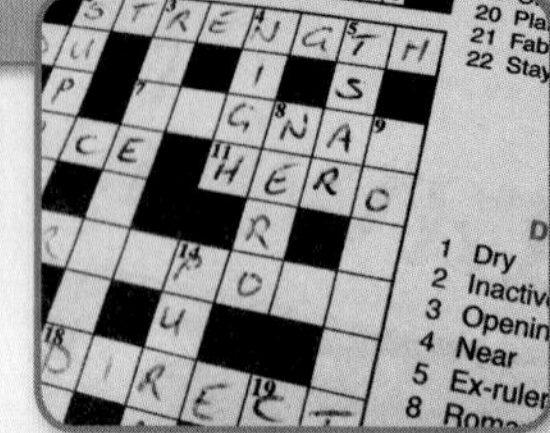

CHEMISTRY & YOU

Q: *How does having a plan make solving problems easier?* Have you ever tried to solve a crossword puzzle? If so, you may have found it helpful to develop a strategy before you begin. For example, you may try to fill in all the "down" clues before attempting the "across" clues. Or, you may first try to complete the fill-in-the-blank clues before moving on to the more difficult clues. In chemistry, it is helpful to develop a strategy to solve both numeric and non-numeric problems.

Key Questions

What is a general approach to solving a problem?

What are the steps for solving numeric problems?

What are the steps for solving nonnumeric problems?

Skills Used in Solving Problems

What is a general approach to solving a problem?

Problem solving is a skill you use all the time. You are in a supermarket. Do you buy a name brand or the store brand of peanut butter? Do you buy the 1-liter bottle or the 2-liter bottle of a carbonated beverage? Do you choose the express line if there are five customers ahead of you or the non-express line with a single shopper who has a lot of items?

When you solve a problem, you may have a data table, a graph, or another type of visual to refer to. The shopper in Figure 1.18 is reading the label on a container while trying to decide whether to buy the item. She may need to avoid certain ingredients because of a food allergy. She may also want to know the number of Calories per serving.

The skills you use to solve a word problem in chemistry are not that different from those you use while shopping, cooking, or planning a party. **Effective problem solving always involves developing a plan and then implementing that plan.**

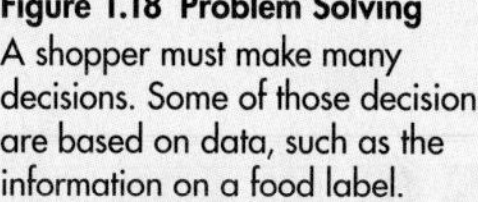

Figure 1.18 Problem Solving
A shopper must make many decisions. Some of those decisions are based on data, such as the information on a food label.

Focus on ELL

1 CONTENT AND LANGUAGE Present the objectives of the lesson by paraphrasing or using visual aids. Write the words *numeric* and *nonnumeric* on the board. **Ask** What word are you familiar with that sounds similar to *numeric*? *(number)* Explain that *numeric* is an adjective meaning "indicating a number." Tell students that the objectives tell them that they will be learning to solve problems with and without numbers.

2 FRONTLOAD THE LESSON Preview Figures 1.19 and 1.21 and Sample Problems 1.1 and 1.2. Show students how each sample problem is broken down into the steps presented in the corresponding figure. Point out that each sample problem contains hints to help students during the solution process.

3 COMPREHENSIBLE INPUT For Sample Problem 1.1, model writing down the knowns and unknowns while reading through the problem. When explaining Sample Problem 1.2, demonstrate how to use a table to organize the known information and to assist in eliminating some possibilities.

Solving Numeric Problems

What are the steps for solving numeric problems?

Most word problems in chemistry require math because measurement is such an important part of chemistry. The techniques used in this book to solve numeric problems are conveniently organized into a three-step, problem-solving approach. This approach has been shown to be very helpful and effective. We recommend that you follow this approach when working on numeric problems in this textbook. **The steps for solving a numeric word problem are analyze, calculate, and evaluate.** Figure 1.19 summarizes the three-step process, and Sample Problem 1.1 on the next page shows how the steps work in solving a numeric problem.

READING SUPPORT

Build Reading Skills: ***Main Ideas*** Under the heading Solving Numeric Problems, there are three main ideas presented as subheads. ***What are two details that support each main idea?***

❶ Analyze To solve a word problem, you must first determine where you are starting from (identify what is known) and where you are going (identify what is unknown). What is known may be a measurement or an equation that shows a relationship between measurements. If you expect the answer (the unknown) to be a number, you need to determine what unit(s) the answer should have before you do any calculations.

After you identify the known and the unknown, you need to make a plan for using what is known to arrive at the unknown. Planning is at the heart of successful problem solving. As part of planning, you might draw a diagram that helps you visualize a relationship between the known and the unknown. You might need to use a table or graph to identify data or to identify a relationship between a known quantity and the unknown. You may need to select an equation that you can use to calculate the unknown.

❷ Calculate If you make an effective plan, doing the calculations is usually the easiest part of the process. For some problems, you will have to convert a measurement from one unit to another. For other problems, you may need to rearrange an equation before you can solve for an unknown. You will be taught these math skills as needed.

❸ Evaluate After you calculate an answer, you should evaluate it. Is the answer reasonable? Does it make sense? If not, reread the word problem. Did you copy the data correctly? Did you choose the right equations?

Check that your answer has the correct unit(s) and the correct number of significant figures. You may need to use scientific notation in your answer. You will study significant figures and scientific notation in Chapter 3.

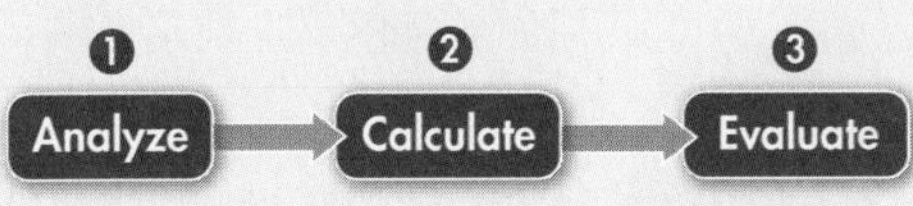

Figure 1.19 Solving Numeric Problems
This flowchart summarizes the steps for solving a numeric problem.
Identify ***In which step do you make a plan for getting from what is known to what is unknown?***

Foundations for Reading

BUILD VOCABULARY Students may benefit from a reminder that certain key words and phrases in each word problem indicate the unknown quantity and its units. These include *how much*, *determine*, *what is*, *find*, and *how long*. Have students generate a list of key words that indicate an operational symbol, such as *add*, *remove*, and *total*.

READING STRATEGY Some students are convinced they cannot do word problems because the problems are too hard. Tell students to examine one sentence at a time to locate information that will help them solve the problem and list what is already known, such as identities and quantities of substances or experimental conditions.

Explain

Skills Used in Solving Problems

MAKE A CONNECTION Explain that the memorization of facts is a relatively small part of learning chemistry. A person who succeeds in chemistry has become a good problem solver. Encourage students to share problem-solving methods and techniques. Discuss the "supermarket" problems. Stress that for these problems, unlike most problems in a chemistry course, answers can vary and still be correct.

Solving Numeric Problems

START A CONVERSATION Point out that in the laboratory, as well as during students' normal routines, students are often presented with more data than is needed to solve a problem. Explain that they will need to sort essential data from extraneous data. In cases with more than the required data, sorting the data into knowns and unknowns helps students to determine the answer to "What data do I need to solve this problem?"

Differentiated Instruction

LPR LESS PROFICIENT READERS Use short, simple words and examples to restate the three steps of solving numeric problems. Have students use a bulleted list to rewrite the steps.

ELL ELL Create a two-column chart relating common key words and mathematical operation symbols. Clarify misinterpretations involving the fraction bar and the division bar. Guide students to understand that those two symbols mean the same thing. State the various words used to read a numerical sentence involving those symbols.

L3 ADVANCED STUDENTS Have students create a list of all the various types of problem-solving plans they have learned throughout previous mathematics and science courses. Ask the students to share and explain each method to the class. Encourage students to use skits, visual aids, and examples to present each method. Students should also note that more than one method can be used to solve a problem.

Answers

READING SUPPORT

1. Determine where you are starting from and where you are going. Then, make a plan.
2. You may need to convert a measurement from one unit to another. You may need to rearrange an equation before solving for an unknown.
3. If the answer does not make sense, reread the problem. Check that the answer has the correct units.

FIGURE 1.19 Step 1, analyze

LESSON 1.4

Explain

Estimating Walking Time

APPLY CONCEPTS As a class, rewrite the problem to exclude extraneous information, such as *It is a nice day*. Then list the knowns and unknowns on the board as well as the steps needed to calculate the answer. You may wish to use chalk or markers of different colors for the knowns and unknowns to make the process clearer. Stress the importance of evaluating whether the result makes sense. Students should make an initial mental estimate to compare with the final result displayed by a calculator.

Sample Practice Problem

A student maps out a 25-block jogging loop through the downtown streets of Indianapolis. She states that she ran the loop one time in about 15 minutes.

- **a.** How many miles did the student run? *(2.5 miles)*
- **b.** How long did it take the student to run one mile? *(6 minutes)*
- **c.** The next day the student runs for 90 minutes. She states that she ran the loop 4 times. How many miles did she run? What was her running rate in miles per minute and miles per hour? *(10 miles; about 0.11 mile per minute; about 6.7 miles per hour)*

Estimating Walking Time

You are visiting Indianapolis for the first time. It is a nice day, so you decide to walk from the Indiana State Capital to the Murat Centre for an afternoon performance. According to the map in Figure 1.20 on the next page, the shortest route from the capital to the theater is eight blocks. How many minutes will the trip take if you can walk one mile in 20 minutes? Assume that ten blocks equals one mile.

1 Analyze **List the knowns and the unknown.**
This problem is an example of what is typically called a conversion problem. In a conversion problem, one unit of measure (in this case, blocks) must be expressed in a different unit (in this case, minutes).

Divide the distance to be traveled (in blocks) by the number of blocks in one mile to get the distance of the trip in miles. Then multiply the number of miles by the time it takes to walk one mile.

KNOWNS
distance to be traveled = 8 blocks
walking speed = 1 mile/20 minutes
1 mile = 10 blocks

UNKNOWN
time of trip = ? minutes

2 Calculate **Solve for the unknown.**

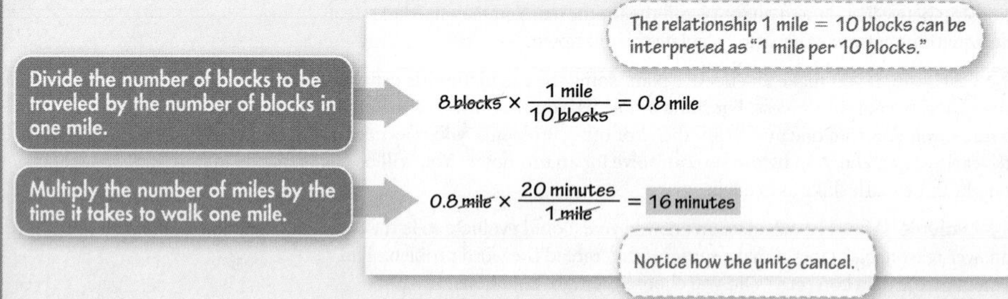

3 Evaluate **Does the result make sense?** The answer seems reasonable, 16 minutes to walk eight blocks. The answer has the correct unit. The relationships used are correct.

25. There is an ice cream shop six blocks north of your hotel. How many minutes will it take to walk there and back? Use the information in the sample problem.

In Problem 25, you must account for both the distance to and from the ice cream shop.

26. Using the information in the sample problem, how many blocks can be walked in 48 minutes?

In Problem 26, first determine how many miles can be walked in 48 minutes. Then, convert miles to blocks.

Foundations for Math

CONVERSION PROBLEMS Tell students that this is the most common type of numerical problem they will see in an introductory chemistry course.

IN SAMPLE PROBLEM 1.1, show students how the units in the numerator and the denominator of Step 2 cancel. It may be helpful to show 8 blocks as a fraction, with 8 blocks over 1. Make sure students write units at every step of the problems. Students need to understand that if their answers do not indicate units, the answers are not considered correct. (NOTE: Conversion problems are the most common type of numeric problem in an introductory chemistry course. They will be discussed in detail in Chapter 3. In that chapter, students will learn how to combine the two calculate steps in Sample Problem 1.1 into a single step.)

Figure 1.20 Reading a Map
Refer to this map of Indianapolis, Indiana, while you do Sample Problem 1.1.

Solving Nonnumeric Problems

What are the steps for solving nonnumeric problems?

Not every word problem in chemistry requires calculations. Some problems ask you to apply the concepts you are studying to a new situation. To solve a nonnumeric problem, you still need to identify what is known and what is unknown. Most importantly, you still need to make a plan for getting from the known to the unknown. If your answer is not a number, you do not need to check the units, make an estimate, or check your calculations.

The three-step problem-solving approach is modified for nonnumeric problems. **The steps for solving a nonnumeric problem are analyze and solve.** Problem-solving methods, such as drawing a diagram, creating a flowchart, or building a model, may be useful. Figure 1.21 summarizes the process, and Sample Problem 1.2 on the next page shows how the steps work in an actual problem.

CHEMISTRY & YOU

Q: *Pick a game you like to play or a type of puzzle you enjoy solving. How does having a plan make playing the game or solving the puzzle easier?*

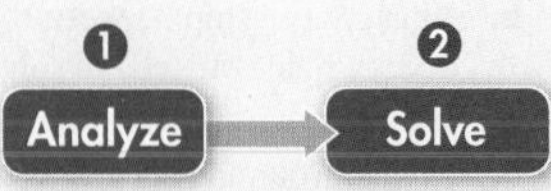

Figure 1.21 Solving Nonnumeric Problems
This flowchart summarizes the steps for solving a nonnumeric problem.
Compare ***With a nonnumeric problem, why is the second step called solve, rather than calculate?***

UbD Check for Understanding

The Essential Question How do chemists solve problems?

Have students write a one-minute statement describing the steps to follow when solving a nonnumeric problem and a numeric problem.

ADJUST INSTRUCTION If students are having difficulty with the processes for solving these two types of problems, refer students back to Figures 1.19 and 1.21.

Explain

Solving Nonnumeric Problems

MAKE A CONNECTION Discuss using classification schemes as tools for organizing information. **Ask** What classification schemes have you encountered in your daily life? *(Sample answer: the ways that books are organized in stores or in libraries)*

Explore

Teacher Demo

PURPOSE Students suggest different approaches for solving a problem.

MATERIALS bowl of ice cubes, empty narrow-neck bottle

PROCEDURE Place a bowl of ice cubes and an empty soda bottle in front of the class. Explain that the problem is to transfer the ice to the inside of the bottle. Have students analyze the problem and suggest different approaches.

EXPECTED OUTCOME Students may suggest crushing the ice so that pieces are small enough to fit through the mouth of the bottle; or they may suggest melting the ice, pouring the water into the bottle, and placing the bottle in a freezer.

CHEMISTRY & YOU Answers will vary, but should include specific examples of how having a plan helped.

Answers

25. 24 minutes
26. 24 blocks

FIGURE 1.21 Solving a nonnumeric problem usually does not require any calculations.

LESSON 1.4

Explain

Sample Practice Problem

How would Manny's schedule change if Physical Education was offered only during 7th period? (*History would have to be moved to 4th period.*)

Evaluate

Informal Assessment

To determine student understanding of the three-step problem-solving approach, ask students to suggest ways to evaluate an answer. *(Reread the problem to be sure the answer supplies the unknown; round off the numbers and do a quick estimate.)* Then have students complete the 1.4 Lesson Check.

Reteach

Have students try working in pairs to solve problems. One student thinks aloud while trying to solve a problem. The other keeps a careful record of the process. Then they reverse roles and work on another problem.

SampleProblem 1.2

Scheduling Classes

Manny needs to schedule his classes for next year. The school day is divided into seven periods, and he must take Algebra II, Art, Chemistry, English, History, and Physical Education. Using the information from the course catalog, find a way for Manny to schedule all his classes.

- ✔ Algebra II is offered during either 1st or 2nd period.
- ✔ Art is offered during 2nd period only.
- ✔ Chemistry is offered during either 3rd or 6th period.
- ✔ English is offered during either 6th or 7th period.
- ✔ History is offered during either 4th or 7th period.
- ✔ Physical Education is offered during 4th period only.
- ✔ Lunch is scheduled for 5th period.

❶ **Analyze** **Identify the relevant concepts.**
Manny must take lunch during 5th period. Art is only available during 2nd period. Physical Education is only available during 4th period.

❷ **Solve** **Apply the concepts to this problem.**

Place lunch, Art, and Physical Education into Manny's schedule.

Manny's Schedule	
1st period	
2nd period	Art
3rd period	
4th period	P.E.
5th period	Lunch
6th period	
7th period	

Since Art is only available during 2nd period, Manny must take Algebra II during 1st period. Since P.E. is only available during 4th period, he must take History during 7th period.

Fit the remainder of the classes into Manny's schedule.

Manny's Schedule	
1st period	Algebra II
2nd period	Art
3rd period	Chemistry
4th period	P.E.
5th period	Lunch
6th period	English
7th period	History

27. How would Manny's schedule change if Art was available during 1st period, instead of 2nd period?

28. Would Manny's schedule change if Algebra II was available during 1st, 2nd, and 3rd periods? Explain.

1.4 LessonCheck

29. **Review** What are the two general steps in successful problem solving?

30. **List** What are the three steps for solving numeric problems?

31. **List** What are the two steps for solving nonnumeric problems?

32. **Compare and Contrast** How are the processes for solving numeric and nonnumeric problems similar? In what way are they different?

33. **Calculate** Read the following conversion problem, and then answer the questions. "There are 3600 seconds in an hour. How many seconds are there in one day?"

a. Identify the known and the unknown.
b. What relationship between the known and unknown do you need to solve the problem?
c. Calculate the answer to the problem.
d. Evaluate your answer and explain why your answer makes sense.

Lesson Check Answers

29. Develop a plan and implement the plan.

30. The three steps are analyze, calculate, and evaluate.

31. analyze and solve

32. In both cases, the solver analyzes the problem, makes a plan, and carries out the plan. Problems with numeric answers require that the answers be evaluated to see if they are reasonable.

33. **a.** known: 3600 s = 1 h; unknown: ? s = 1 day
b. 24 h = 1 day
c. 3600 s/h × 24 h/day = 86,400 s/day
d. 86,400 seconds in one day seems reasonable in relationship to 3600 seconds in one hour. The answer has the correct units, and the relationship used is correct.

1 Study Guide

BIGIDEA
CHEMISTRY AS THE CENTRAL SCIENCE

It is important to study chemistry because chemistry is fundamental to the understanding of the other sciences. Chemistry is relevant to many professions. A knowledge of chemistry can help you become an informed citizen. Chemists use the scientific method to solve problems and develop theories about the natural world.

1.1 The Scope of Chemistry

Chemistry affects all aspects of life and most natural events because all living and nonliving things are made of matter.

Five traditional areas of study are organic chemistry, inorganic chemistry, biochemistry, analytical chemistry, and physical chemistry.

Some of chemistry's big ideas are as follows: chemistry as the central science, electrons and the structure of atoms, bonding and interactions, reactions, kinetic theory, the mole and quantifying matter, matter and energy, and carbon chemistry.

- matter (2)
- chemistry (2)
- organic chemistry (3)
- inorganic chemistry (3)
- biochemistry (3)
- analytical chemistry (3)
- physical chemistry (3)
- pure chemistry (3)
- applied chemistry (3)

1.2 Chemistry and You

Chemistry can be useful in explaining the natural world, preparing people for career opportunities, and producing informed citizens.

Modern research in chemistry can lead to technologies that aim to benefit the environment, conserve and produce energy, improve human life, and expand our knowledge of the universe.

- technology (8)

1.3 Thinking Like a Scientist

Lavoisier helped to transform chemistry from a science of observation to the science of measurement that it is today.

Steps in the scientific method include making observations, proposing and testing hypotheses, and developing theories.

When scientists collaborate and communicate with one another, they increase the likelihood of a successful outcome.

- scientific method (15)
- observation (15)
- hypothesis (16)
- experiment (16)
- independent variable (16)
- dependent variable (16)
- model (16)
- theory (17)
- scientific law (17)

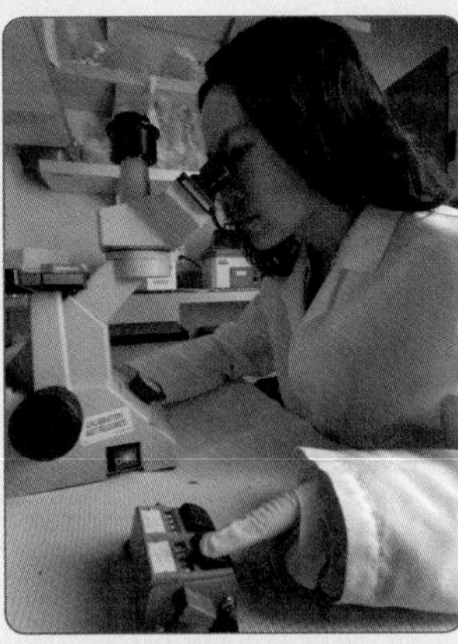

1.4 Problem Solving in Chemistry

Effective problem solving always involves developing a plan and then implementing that plan.

The steps for solving a numeric word problem are analyze, calculate, and evaluate.

The steps for solving a nonnumeric problem are analyze and solve.

Study Guide

Review and Assessment Resources

CHEMISTRY TUTORIAL Encourage students to review solving estimation problems using the animated step-by-step tutorial for Chapter 1. Each Chemistry Tutorial is followed by a similar practice problem, where the student is guided through the problem-solving process.

ONLINE PROBLEM SET Have students practice more general chemistry problems by using the Chapter 1 online Problem Set.

MATH TUTORIAL Struggling students can use a tutorial on solving equations as a quick review. Ask students to answer the guided practice problem that follows the Math Tutorial.

DIRECTED VIRTUAL LAB Have students complete the Virtual Lab *Laboratory Safety* to help reinforce safety procedures for the lab. In the post-lab exercise, students will draw conclusions based on their observations.

KINETIC ART Direct students to the Kinetic Art animation *Molecular Models*, and have them answer the questions that accompany the animation.

Study Tip

STUDY WITH A FRIEND Encourage students to find a study partner. Tell them to review each section and quiz each other on the key concepts. Suggest that they separately create a summary for each section and then compare notes.

Answers

27. Algebra II would have to move to the second period.

28. The schedule would not need to change unless the schedule for some other class changed.

Focus on ELL

5 ASSESS UNDERSTANDING Assign ELL students one of the Performance Tasks as a group for an alternative assessment. Provide guidelines for the assignment and a grading rubric.

BEGINNING

LOW Create the visual representation for the assignment.

HIGH Find specific text information needed for the assignment.

INTERMEDIATE: LOW/HIGH Provide the written explanation for the visuals.

ADVANCED: LOW/HIGH Present the assignment to the class.

UbD Performance Tasks

IN THE FIELDS Ask students to identify one important discovery and some problems researchers faced during the process of discovery. Tell students to create an oral presentation with visual aids to be presented to the class.

DEBATE Have partners use a newspaper or other media to find a scientific topic that is commonly debated. One student should take the pro side and the other, the con. Each pair of students should research facts and scientists' opinions for support, create guidelines for the debate (such as time limits), and debate the topic in front of the class.

Evaluate

Answers

LESSON 1.1

34. Although air is "invisible," it has mass and occupies space.

35. Chemistry is concerned with the changes that matter undergoes.

36. Organic chemistry is the study of chemicals that contain carbon; inorganic chemistry is the study of chemicals that do not contain carbon.

37. applied research because the research is directed toward a practical goal or application

LESSON 1.2

38. A firefighter needs to know which chemicals to use to fight different types of fires; knowledge of chemistry will help a reporter gather information during an interview with a chemist.

39. Possible products include cups and utensils; containers and packaging for food; and bags.

40. *Sample answer:* Chemists develop medicines to treat conditions such as high blood pressure; materials to replace damaged body parts, and technology such as ways to alter genes.

41. to understand the structure of matter found in cells and chemical changes that occur in cells

42. by analyzing the light they transmit to Earth

LESSON 1.3

43. They based their conclusions on experimental evidence.

44. the scientific method

45. to test a hypothesis

46. c

47. The independent variable is changed during an experiment; the dependent variable is observed during an experiment.

48. Repeat the experiment. If you get the same result, you must propose a new hypothesis.

49. A scientific law summarizes the results of many experiments; a theory explains them.

50. to share knowledge across disciplines and resources between industry and academia

LESSON 1.4

51. b, c, and d

52. developing a plan and implementing that plan

53. Step 1, analyze

54. 54 games (one third of 162)

55. 4320 times an hour, 103,680 times a day

56. 12 days

UNDERSTAND CONCEPTS

57. Answers may vary, but possible answers are 1c, 2d, 3e, 4b, and 5a.

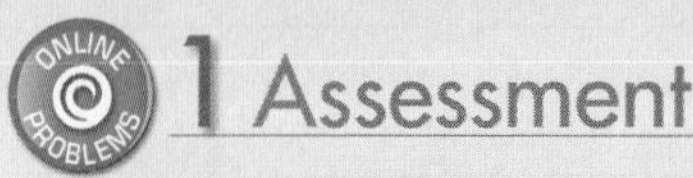

1 Assessment

★ Solutions appear in Appendix E

Lesson by Lesson

1.1 The Scope of Chemistry

34. Explain why air is classified as matter.

★**35.** The Japanese characters for chemistry literally mean "change study." Why are these appropriate characters to represent chemistry?

36. Describe the main difference between inorganic chemistry and organic chemistry.

★**37.** Is a scientist who is studying cancer with the goal of finding an effective treatment doing pure research or applied research?

1.2 Chemistry and You

★**38.** Why would a firefighter or a reporter need to understand chemistry?

39. What are some products that could be made from the bioplastic polylactic acid (PLA)?

40. How do chemists help doctors treat patients?

41. What is the overall goal of biochemists who work in the field of medicine?

★**42.** How can scientists study the composition of distant stars?

1.3 Thinking Like a Scientist

43. What did the scientists who founded the Royal Society of London have in common with Lavoisier?

★**44.** What is the most powerful tool that any scientist can have?

45. What is the purpose of an experiment?

★**46.** Which of the following is not a part of the scientific method?

a. hypothesis
b. experiment
c. guess
d. theory

47. How do an independent variable and a dependent variable differ?

★**48.** You perform an experiment and get unexpected results. According to the scientific method, what should you do next?

★**49.** Explain how the results of many experiments can lead to both a theory and a scientific law.

50. List two general reasons why scientists are likely to collaborate.

1.4 Problem Solving in Chemistry

★**51.** Identify the statements that correctly describe good problem solvers.

a. read a problem only once
b. check their work
c. look up missing facts
d. look for relationships among the data

52. What do effective problem-solving strategies have in common?

53. In which step of the three-step problem-solving approach for numeric problems is a problem-solving strategy developed?

★**54.** On average, a baseball team wins two out of every three games it plays. How many games will this team lose in a 162-game season?

55. If your heart beats at an average rate of 72 times per minute, how many times will your heart beat in an hour? In a day?

★**56.** How many days would it take you to count a million pennies if you could count one penny each second?

Understand Concepts

★**57.** Match each area of chemistry with a numbered statement.

a. physical chemistry
b. organic chemistry
c. analytical chemistry
d. inorganic chemistry
e. biochemistry

(1) measure the level of lead in blood
(2) study non-carbon-based chemicals in rocks
(3) investigate changes that occur as food is digested in the stomach
(4) study carbon-based chemicals in coal
(5) explain the energy transfer that occurs when ice melts

Use this photograph of the runners to answer Questions 58 and 59.

*58. Explain how chemistry has affected the ability of these athletes to compete.

59. What type of chemist might study how an athlete uses energy during a competition? Give a reason for your answer.

60. Explain why chemistry might be useful in a career you are thinking of pursuing.

*61. A doctor examines a patient's sore throat and suggests that the patient has strep throat. She takes a sample to test for the bacteria that cause strep throat. What parts of the scientific method is the doctor applying?

*62. You perform an experiment and find that the results do not agree with an accepted theory. Should you conclude that you made an error in your procedure? Explain.

63. A student is planning a science fair project called "Does Temperature Affect How High a Basketball Can Bounce?"
   a. Based on the project title, identify the independent variable and the dependent variable.
   b. Name at least two factors that would need to be kept constant during the experiment.

64. Describe a situation in which you used at least two steps in the scientific method to solve a problem.

*65. Pure water freezes at 0°C. A student wanted to test the effect of adding table salt to the water. The table shows the data that was collected.

Effect of Salt on Freezing Point of Water

Salt Added	Freezing Point
5 g	−4.8°C
10 g	−9.7°C
15 g	−15.1°C
20 g	−15.0°C

   a. What was the independent variable?
   b. What was the dependent variable?
   c. Why must the volume of water be the same for each test?
   d. Based on the data, the student hypothesized, "As more salt is added to water, the temperature at which water freezes decreases." Is this hypothesis supported by the data? Explain.

*66. In the time a person on a bicycle travels 4 miles, a person in a car travels 30 miles. Assuming a constant speed, how far will the car travel while the bicycle travels 40 miles?

Think Critically

67. **Compare and Contrast** How is the study of chemistry similar to the study of a language? How is it different?

68. **Infer** Comment on the idea that science accepts what works and rejects what does not work.

69. **Apply Concepts** You are asked to design an experiment to answer the question: "Which paper towel is the best?"
   a. What is the independent variable in your experiment?
   b. List three possible dependent variables that could be used to define "best."
   c. Pick one of the dependent variables and rewrite the question as a hypothesis.
   d. List at least five factors that must be kept constant when you test the hypothesis.

*70. **Compare** Important discoveries in science are sometimes the result of an accident. Louis Pasteur said, "Chance favors the prepared mind." Explain how both of these statements can be true.

Answers

58. One possible answer is development of the materials to produce an artificial limb.

59. Students are likely to choose biochemist because biochemistry is the study of processes that take place in organisms. However, a student might choose physical chemist because physical chemistry includes the study of energy transfer as matter undergoes a change.

60. Answers will vary but should demonstrate an understanding that chemistry is the study of matter and the changes it undergoes.

61. The doctor's hypothesis is that the sore throat is the result of bacteria that cause strep throat. She tests the hypothesis by testing a sample for the presence of the bacteria.

62. Your experiment may be correct, but your hypothesis may be wrong. You should reexamine your hypothesis and repeat the experiment.

63. **a.** The independent variable is the temperature of the basketball. The dependent variable is the height of the basketball's bounce.
   b. The method for dropping the basketball before it bounces; the method for measuring the height of the bounce.

64. Answers will vary but should reflect knowledge of the steps in a scientific method including making observations and testing hypotheses.

65. **a.** amount of salt added
   b. freezing point
   c. Changing the volume of the water would also change the concentration of the salt solution and affect the results.
   d. Yes, to a point. The freezing point appears to be leveling off at about 15°C.

66. 300 miles

THINK CRITICALLY

67. One possible answer is that with both areas of study, students slowly build up a vocabulary and a set of concepts that relate the new terms. Both areas rely on a systematic approach—grammar versus the scientific method. Chemistry deals with matter and words do not qualify as matter. Science is independent of a specific culture; language varies from culture to culture.

68. A possible answer is that scientists accept hypotheses that are supported by experimental results and reject hypotheses that are not supported by experimental results.

69. **a.** the brand of paper towel
   b. Possible answers include strongest, most absorbent, strongest when wet, largest area per sheet, and easiest to tear from roll.
   c. *Sample answer:* The best towel will absorb the most liquid.
   d. Possible answers include size of paper towel sample, type of liquid, amount of liquid, method for measuring absorbency, method for measuring strength.

70. A person who is educated in the theories and practice of chemistry would be more likely to recognize the significance of an accidental discovery and have the means and motivation to develop that accidental discovery into an important scientific contribution.

CHEMYSTERY

START A CONVERSATION Have students read over the CHEMystery. As a class, discuss the factors students identified at the beginning of the chapter as important ones to consider before purchasing PLA utensils. **Ask** Did any of the information explained here answer your questions about the utensil material? *(Answers will vary.)* Ask students to tell whether or not they'd purchase this type of utensil, and explain how they made their decisions. Challenge students to find out if any composting facilities for PLA are located in your community.

CHEMYSTERY ANSWERS

80. Possible factors in addition to the environmental issues are durability, appearance, cost, and availability.

81. **BIG**IDEA *Sample answer:* A knowledge of chemistry might affect the priority given to environmental concerns.

*71. **Calculate** Four beakers have a total weight of 2.0 lb. Each beaker weighs 0.5 lb. Describe two different methods you could use to calculate the weight of two beakers. Then try both methods and compare the answers.

*72. **Apply Concepts** Explain what is wrong with the statement, "Theories are proven by experiments."

*73. **Interpret Diagrams** The air you breathe is composed of about 20% oxygen and 80% nitrogen. Use your problem-solving skills to decide which drawing best represents a sample of air. Explain your choice.

a.

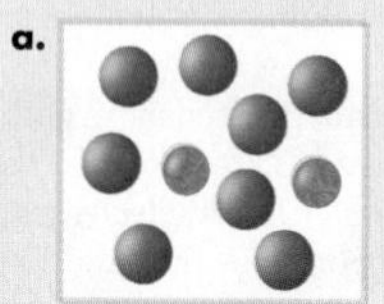

c.

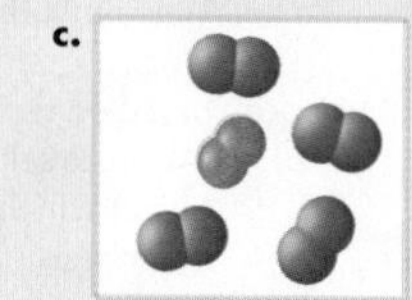

b.

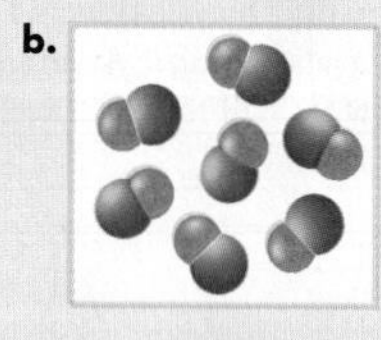

d.

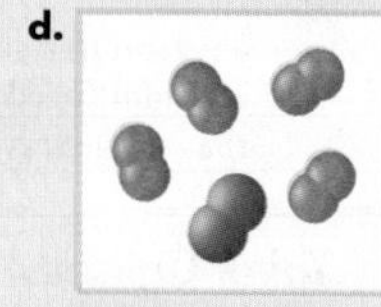

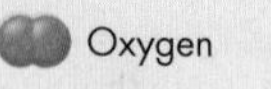

Oxygen   Nitrogen

Enrichment

74. **Calculate** A certain ball, when dropped from any height, bounces one-half the original height. If the ball was dropped from a height of 60 inches and allowed to bounce freely, what is the total distance the ball has traveled when it hits the ground for the third time? Assume the ball bounces straight up and down.

*75. **Calculate** Eggs are shipped from a farm to a market by truck. They are packed 12 eggs to a carton and 20 cartons to a box. Four boxes are placed in each crate. Crates are stacked on a truck 5 crates wide, 6 crates deep, and 5 crates high. How many eggs are on the truck?

76. **Analyze Data** An oil tanker containing 4,000,000 barrels of oil is emptied at the rate of 5000 gallons per minute. What information do you need to figure out how long it would take to empty the tanker?

*77. **Calculate** A crate of envelopes sells for $576.00. A package of envelopes contains 250 envelopes. Six packages are packed inside a carton. Twelve cartons are packed in a box. Eight boxes are packed in a crate.

a. What does a package of envelopes cost?

b. What fact given in the problem was not needed to calculate the answer?

Write About Science

78. **Explain** Pick one activity that you can do faster or with less effort because of technology. Write a paragraph in which you describe the activity, identify the technology, and explain how the technology affects the activity.

79. **Relate Cause and Effect** Write a paragraph explaining how you can learn about the research that is done by scientists. Then explain how this information could help you be an informed citizen.

CHEMYSTERY

Is Bio Better?

When you return home from the store, you go online and search for "bioplastics." You learn that the products you found at the store were most likely made from polylactic acid (PLA). You like that the PLA products come from natural resources, such as corn, and that less energy is used in making PLA than other plastics. However, you are concerned that it would be difficult to find a facility that would be able to compost the cups and utensils when you and your friends are finished with them.

*80. **Relate Cause and Effect** What factors will affect your decision as to whether to purchase the picnic products made from PLA? Explain.

81. **Connect to the BIGIDEA** How would a knowledge of chemistry help you make an informed decision?

Answers

71. Divide the weight of 4 beakers by 2 or multiply the weight of 1 beaker by 2 (2.0 lb/2 or 0.5 lb × 2 = 1 lb).

72. A theory can never be proven. It is a well-tested explanation of a broad set of observations. A theory may need to be changed in the future to explain new observations.

73. c; the 1 to 4 ratio of oxygen to nitrogen corresponds to the composition of air.

ENRICHMENT

74. 150 in.

75. 144,000 eggs

76. the number of gallons in a barrel

77. **a.** $1.00 per package
b. number of envelopes in a package

WRITE ABOUT SCIENCE

78. Answers will vary but should reflect an understanding of the definition of technology.

79. Students are likely to refer to multiple forms of media. They should also acknowledge the role that citizens play in influencing the direction of scientific research.

Standardized Test Prep for Indiana

Select the choice that best answers each question or completes each statement.

1. The branch of chemistry that studies chemicals containing carbon is _____ chemistry.
   (A) physical
   (B) inorganic
   (C) analytical
   (D) organic
2. An analytical chemist is most likely to
   (A) explain why paint is stirred before it is used.
   (B) explain what keeps paint attached to the steel frame of an automobile.
   (C) identify the type of paint chips found at the scene of a hit-and-run accident.
   (D) investigate the effect of leaded paint on the development of a young child.
3. Chemists who work in the biotechnology field are most likely to work with
   (A) X-ray technicians.
   (B) geologists.
   (C) physicians.
   (D) physicists.

Respond to each statement in Questions 4–6.

4. Someone who wears contact lenses does not have to wear safety goggles in the lab.
5. Eating food that is left over from an experiment is an alternative to discarding the food.
6. For a student who has read the procedure, the teacher's pre-lab instructions are unnecessary.

Use the flowchart to answer Question 7.

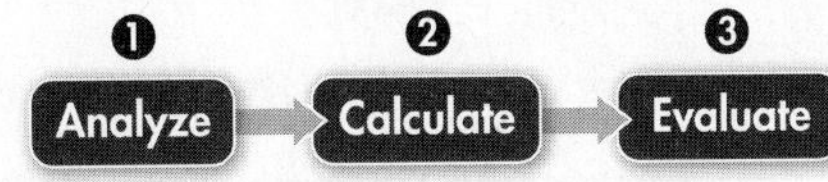

7. What should you do before you calculate an answer to a numeric problem, and what should you do after you calculate the answer?

Use this paragraph to answer Questions 8–10.

(A) One day, your car doesn't start. **(B)** You say, "The battery is dead!" **(C)** Your friend uses a battery tester and finds that the battery has a full charge. **(D)** Your friend sees rust on the battery terminals. **(E)** Your friend says, "Maybe rust is causing a bad connection in the electrical circuit, preventing the car from starting." **(F)** Your friend cleans the terminals, and the car starts.

8. Which statements are observations?
9. Which statements are hypotheses?
10. Which statements describe experiments?

Tips for Success

True-False Questions When the word *because* is placed between two statements, you must first decide if the statements are both true, both false, or if one statement is true and the other false. If both are true, you must decide if the second statement is a correct explanation for the first.

For each question, there are two statements. Decide whether each statement is true or false. Then decide whether Statement II is a correct explanation for Statement I.

	Statement I		Statement II
11.	A hypothesis may be rejected after an experiment.	BECAUSE	Experiments are used to test hypotheses.
12.	Theories help you make mental models of objects that cannot be seen.	BECAUSE	Theories summarize the results of many observations and experiments.
13.	All Internet sites that provide scientific information are equally reliable.	BECAUSE	All information on these sites is reviewed by qualified scientists.

If You Have Trouble With . . .

Question	1	2	3	4	5	6	7	8	9	10	11	12	13
See Lesson	1.1	1.1	1.2	1.3	1.3	1.3	1.4	1.3	1.3	1.3	1.3	1.3	1.3

STP Answers

1. d
2. c
3. c
4. Contact lenses increase the possibility of eye damage because chemicals can become trapped behind a lens. Safety goggles also protect the eyes from flying debris.
5. No food should be eaten in the laboratory because food may become contaminated with a toxic substance.
6. The teacher may modify the instructions to suit available equipment or chemicals. The teacher will stress any safety precautions. Accidents are more likely to happen when instructions or warnings are ignored.
7. Analyze the question and the given information; evaluate the answer against a reasonable estimate and check the units.
8. A and D
9. B and E
10. C and F
11. true, true, correct explanation
12. true, false
13. false, false

2 Matter and Change Planning Guide

Introducing the BIGIDEA: CHEMISTRY AS THE CENTRAL SCIENCE

Chemistry is the central science because it overlaps with all of the other sciences.

IASS	Lessons and Objectives	Print Resources		
		For the Student	For the Teacher	
C.1.1, C.1.2, C.1.4, C.1.5	**2.1** Properties of Matter p 34–37 **2.1.1** Explain why all samples of a substance have the same intensive properties. **2.1.2** Identify the three states of matter. **2.1.3** Classify physical changes.	Reading and Study Workbook Lesson 2. 1 Lesson Assessment 2.1 p 37	Teacher Demo, p 35: Volume and Mass Teacher Demo, p 36: States of Chocolate Class Activity, p 36: Comparing Solids and Liquids	
C.1.1	**2.2** Mixtures p 38–41 **2.2.1** Explain how mixtures are classified. **2.2.2** Explain how mixtures can be separated.	Reading and Study Workbook Lesson 2.2 Lesson Assessment 2.2 p 41 Quick Lab: Separating Mixtures p 39	Teacher Demo, p 40: Metallic Breakfast	
	2.3 Elements and Compounds p 42–47 **2.3.1** Explain the difference between an element and a compound. **2.3.2** Distinguish between a substance and a mixture. **2.3.3** Describe what chemists use to represent elements and compounds. **2.3.4** Explain how a periodic table is useful.	Reading and Study Workbook Lesson 2.3 Lesson Assessment 2.3 p 47	Teacher Demo, p 43: Decomposition of Sugar Class Activity, p 44: Substances	
C.1.1, C.1.3, C.1.4, C.1.6	**2.4** Chemical Reactions p 48–50 **2.4.1** Describe what happens during a chemical change. **2.4.2** Identify four possible clues that a chemical change has taken place. **2.4.3** Describe how the mass of the reactants and the mass of the products of a chemical reaction are related.	Reading and Study Workbook Lesson 2.4 Lesson Assessment 2.4 p 50 Small-Scale Lab: 1 + 2 + 3 = BLACK! p 51	Teacher Demo, p 49: Identifying a Chemical Change	

Assessing the BIGIDEA: CHEMISTRY AS THE CENTRAL SCIENCE

	Essential Questions 1. What properties are used to describe matter? 2. How can matter change its form?	Study Guide p 54 STP p 59 Reading and Study Workbook Self-Check and Vocabulary Review Chapter 2		

	PearsonChem.com	
	Editable Worksheets	Online Activities
	Teaching Resources, Lesson 2.1 Review	2.1 Lesson Overview States of Matter
	Teaching Resources, Lesson 2.2 Review	2.2 Lesson Overview Distillation
	Teaching Resources, Lesson 2.3 Review Lab 26: Distillation Lab 32: Introduction to Chromatography Lab Practical 16-1: Chromatography	2.3 Lesson Overview
	Teaching Resources, Lesson 2.4 Review Lab 2: Physical and Chemical Change Lab 3: Observing a Chemical Reaction Small-Scale Lab Manual Lab 2: A Study of Chemical Changes Lab Practical 2-1: Chemical and Physical Properties	2.4 Lesson Overview Cooking Up Physical and Chemical Changes Chemical Reactions
	Core Teaching Resources (includes editable worksheets) • Lesson Reviews • Practice Problems • Interpret Graphs • Vocabulary Review • Chapter Quizzes and Tests • Lab Record Sheets	Chapter 2 Problem Set

Materials List

For the Student

Quick Lab p 39
- green marking pen
- filter paper strip
- metric ruler
- clear plastic tape
- pencil
- rubbing alcohol
- clear plastic drinking cup
- clear plastic wrap

Small Scale Lab p 51
- metric ruler
- reaction surface
- 0.1*M* KI
- 0.2*M* $CuSO_4$
- 3% H_2O_2
- 1% NaClO
- liquid starch
- cereal
- pipette
- medicine droppers
- spatulas
- iodized and noniodized salt
- antacid tablets
- colored marking pens
- foods

For the Teacher

Teacher Demo p 35
- plastic syringe (50 mL or larger)
- balance that reads to 0.01 g
- Erlenmeyer flask with tightly fitting 1-hole rubber stopper
- funnel
- water

Class Activity p 36
- foam balls or miniature marshmallows

Teacher Demo p 36
- piece of solid chocolate
- small beaker
- large beaker
- boiling water

Teacher Demo p 40
- iron-fortified breakfast cereal
- 400-mL beaker
- distilled water
- magnetic stirrer with stirring bar

Teacher Demo p 43
- powdered sugar
- 100-mL beaker
- 18*M* sulfuric acid
- glass stirring rod

Class Activity p 44
- 20 index cards
- masking tape

Teacher Demo p 49
- Bunsen burner
- match
- tongs
- magnesium ribbon
- 4 test tubes
- solutions of 1*M* $AgNO_3$, 0.1*M* NaCl, 0.1*M* K_2CrO_4, and 3*M* H_2SO_4
- mossy zinc
- marble chip
- spatulas
- cobalt blue glass filters

Additional Digital Resources

Online Student Edition
Online Teacher's Edition
Exam*View Assessment Suite CD-ROM*
Classroom Resources DVD

CHAPTER 2

What's Online

ONLINE PROBLEM SET Additional practice problems about mixtures and chemical change can be found in the Chapter 2 online Problem Set.

DIRECTED VIRTUAL LAB Encourage students to review different types of chemical reactions using the animated virtual lab for Chapter 2. The animation is followed by analysis questions, where the student is guided through the lab concepts.

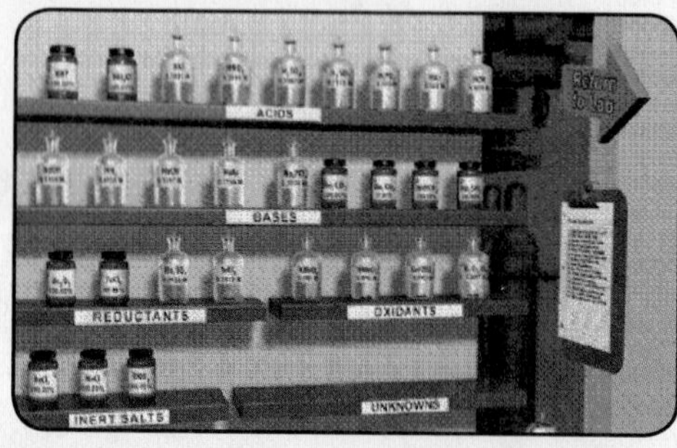

KINETIC ART Have students watch animations about the states of matter and distillation. Each animation is followed by analysis questions to guide further understanding of the concepts.

CONCEPTS IN ACTION Students can watch an interactive animation about physical and chemical changes in the world around them and learn to distinguish between them. The animation is followed by questions to help the student practice identifying physical and chemical changes in common food items.

2 Matter and Change

INSIDE:

PearsonChem.com

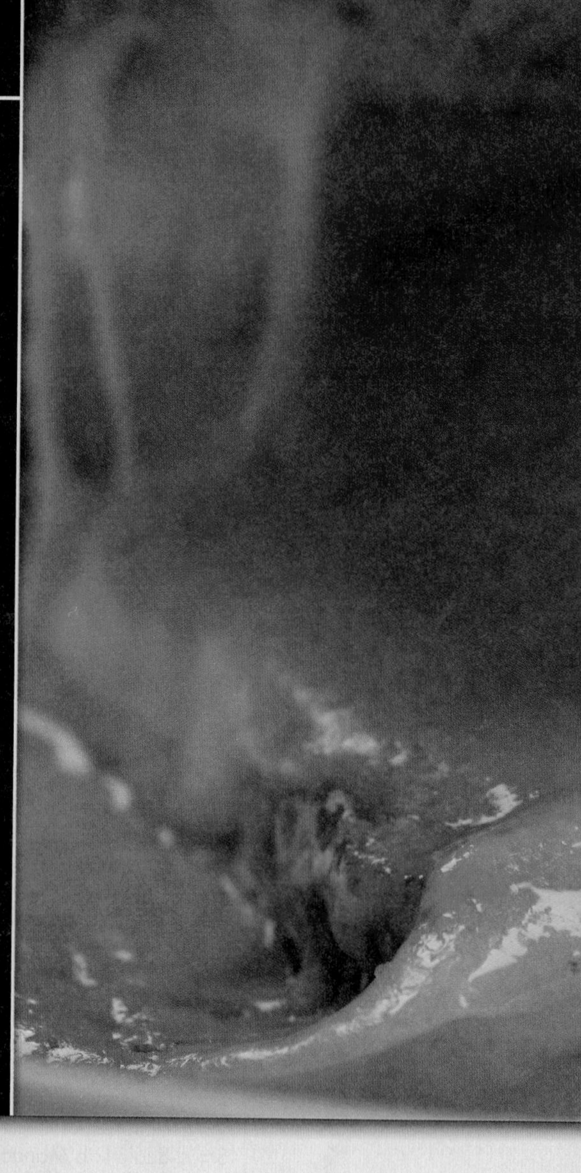

As a grilled cheese sandwich cooks, physical and chemical changes are taking place. Melting cheese is a physical change. Toasting bread is a chemical change.

32

IN ACADEMIC STANDARDS for SCIENCE

Nature of Science

- Develop explanations based on reproducible data and observations gathered during laboratory investigations.
- Explain how scientific knowledge can be used to guide decisions on environmental and social issues.

Standard 1: Properties and States of Matter

C.1.1 Based on physical properties, differentiate between pure substances and mixtures

Also covered C.1.2, C.1.3, C.1.4, C.1.5, C.1.6

Focus on ELL

1 CONTENT AND LANGUAGE Write the word *change* on the board and pronounce it for students. Ask volunteers to define this term in their own words. Then explain that the title refers to the fact that matter is capable of undergoing change, and that these changes in matter can be either physical.

BEGINNING: LOW/HIGH Provide students with a simplified definition of change that they can write in their vocabulary notebooks. Allow students to include a native language definition as well.

INTERMEDIATE: LOW/HIGH Have students write several sentences to demonstrate an understanding of the definition.

ADVANCED: LOW/HIGH Have students identify an example of change in the classroom and write a brief paragraph describing the change. Review the paragraphs to assess students' understanding of the definition of *change*.

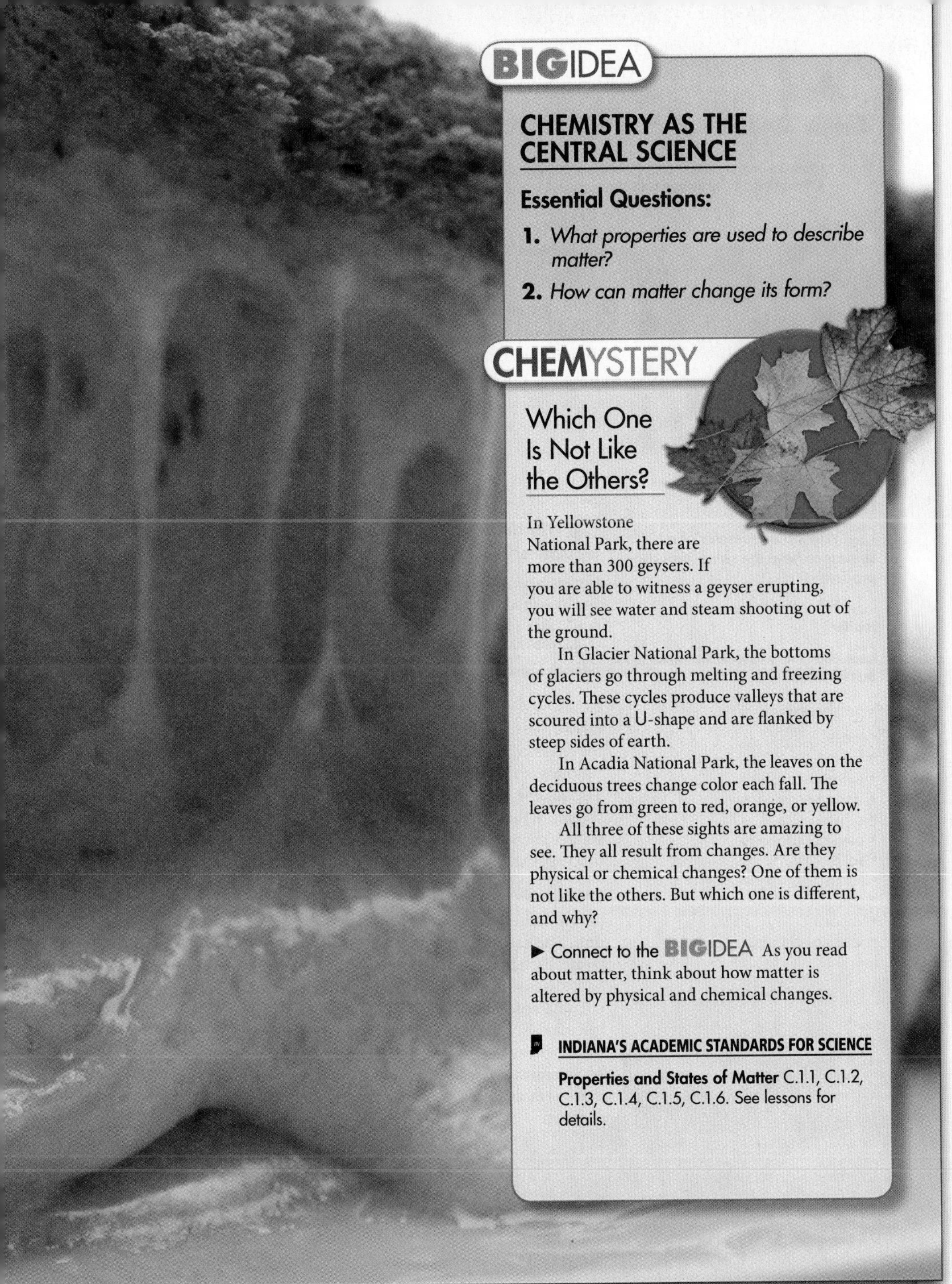

BIGIDEA

CHEMISTRY AS THE CENTRAL SCIENCE

Essential Questions:

1. *What properties are used to describe matter?*
2. *How can matter change its form?*

CHEMYSTERY

Which One Is Not Like the Others?

In Yellowstone National Park, there are more than 300 geysers. If you are able to witness a geyser erupting, you will see water and steam shooting out of the ground.

In Glacier National Park, the bottoms of glaciers go through melting and freezing cycles. These cycles produce valleys that are scoured into a U-shape and are flanked by steep sides of earth.

In Acadia National Park, the leaves on the deciduous trees change color each fall. The leaves go from green to red, orange, or yellow.

All three of these sights are amazing to see. They all result from changes. Are they physical or chemical changes? One of them is not like the others. But which one is different, and why?

▶ Connect to the **BIG**IDEA As you read about matter, think about how matter is altered by physical and chemical changes.

INDIANA'S ACADEMIC STANDARDS FOR SCIENCE

Properties and States of Matter C.1.1, C.1.2, C.1.3, C.1.4, C.1.5, C.1.6. See lessons for details.

Introduce the Chapter

CONNECT TO THE REAL WORLD Students observe changes around them on a daily basis. Use the following activities to help students think about differences in changes.

Activity 1 *You'll need a beaker with ice cubes.* Point out that one change students might observe regularly is the melting of ice. **Ask** Does the composition change as the ice melts? *(No; it is frozen water before and liquid water afterward.)*

Activity 2 *You'll need a small ball of cookie dough and a baked cookie made from the dough.* Have students examine the dough and the cookie. Point out that the cookie was baked from the same batch of dough. **Ask** How does the baked cookie differ from the cookie dough? *(Sample answers: The cookie is dry and crisp and has a crumbly texture; the dough is moist and sticky and has a smooth texture.)* **Ask** Do you think the composition of the cookie dough changes as it bakes? Explain. *(The composition most likely changes; otherwise the cookie would have the same physical properties as the uncooked dough.)*

UbD

Understanding by Design

Students are building toward describing matter and change by using Chemistry as the Central Science.

PERFORMANCE GOALS At the end of Chapter 2, students will be able to answer the essential questions by applying their knowledge of matter and change.

ESSENTIAL QUESTIONS Read the essential questions aloud. Display a piece of paper and have students brainstorm a list of its properties. *(Sample answers: flat, solid, white, easy to bend and tear)* **Ask** What are some ways we can change the paper's form? *(Sample answers: burning, folding, dyeing.)*

BIGIDEA Use the photo of the grilled cheese sandwich to help students connect to this chapter's concepts. Activate prior knowledge by having them describe their experiences with grilled cheese sandwiches. **Ask** What are some ways that properties of the bread and cheese change as the sandwich is grilled? *(The bread browns and becomes crispy on the outside; the cheese melts.)* **Ask** Do you think these changes are permanent? *(The cheese will resolidiy, but the bread cannot change back to its original color or texture.)*

CHEMYSTERY Connect the CHEMystery to the Big Idea by reminding students that chemistry is the study of the composition of matter and its changes. Have students think about how matter changes when a geyser erupts, the bottom of a glacier melts and refreezes, and leaves change color. Ask them to predict which change is different. Also have them predict the difference between a physical change and a chemical change.

UntamedScience™

A New "Roll" for Old Tires

Haley from Untamed Science takes us to a research facility to learn how chemists are using microwaves to recycle old tires.

LESSON 2.1

Key Objectives

2.1.1 EXPLAIN why all samples of a substance have the same intensive properties.

2.1.2 IDENTIFY three states of matter.

2.1.3 CLASSIFY physical changes.

Additional Resources

Reading and Study Workbook, Lesson 2.1

Available Online or on Digital Media:

- Teaching Resources, Lesson 2.1 Review

PearsonChem.com

Engage

CHEMISTRY & YOU Have students read the opening text. **Ask** Do you know of any other substance that shares the property of transparency with glass? *(Plastic)* **Ask** Why do you think glass was chosen for the aquarium window rather than plastic? Glass has physical properties that make it a better choice for such a large window than plastic, such as transparency and strength.

Activate Prior Knowledge

Ask students to list the three types of matter they encounter every day. *(solids, liquids, and gases)* Have students visualize the structure of each type of matter when viewed under a microscope. Have students compare their mental pictures or actual sketches with the drawings in Figure 2.3.

IN ACADEMIC STANDARDS for SCIENCE

C.1.1 Based on physical properties, differentiate between pure substances and mixtures.

C.1.2 Observe and describe chemical and physical properties of different types of matter and designate them as either extensive or intensive.

C.1.4 Describe physical and chemical changes at the molecular level.

C.1.5 Describe the characteristics of solids, liquids, and gases and state changes at the molecular level.

2.1 Properties of Matter

C.1.1 Pure substance or mixture? **C.1.2** Extensive and intensive properties; **C.1.5** Physical and chemical changes; **C.1.5** States of matter

CHEMISTRY & YOU

Q: *Why are windows made of glass?* When you think of a window, you probably think of something that you can look through. Most windows are made of glass and are transparent, meaning you can see through them. If you found a piece of broken glass on the ground, you would probably recognize it as glass. It is hard, yet easy to shatter, and it is heat resistant. In this lesson, you will learn how properties can be used to classify and identify matter.

Key Questions

Why do all samples of a substance have the same intensive properties?

What are three states of matter?

How can physical changes be classified?

Vocabulary

- mass
- volume
- extensive property
- intensive property
- substance
- physical property
- solid
- liquid
- gas
- vapor
- physical change

Describing Matter

Why do all samples of a substance have the same intensive properties?

Understanding matter begins with observation, and what you observe when you look at a particular sample of matter is its properties. Is a solid shiny or dull? Does a liquid flow quickly or slowly? Is a gas odorless, or does it have a smell? Properties used to describe matter can be classified as extensive or intensive properties.

Extensive Properties Recall that matter is anything that has mass and takes up space. The **mass** of an object is a measure of the amount of matter the object contains. The mass of the basketball in Figure 2.1 is greater than the mass of the golf ball. There is also a difference in the volume of the balls. The **volume** of an object is a measure of the space occupied by the object. The volume of the basketball is greater than the volume of the golf ball. Mass and volume are both examples of extensive properties. An **extensive property** is a property that depends on the amount of matter in a sample.

Intensive Properties Basketballs may appear to be all the same. But, there are properties to consider when selecting a basketball besides mass and volume. The outer covering may be made of leather, rubber, or a synthetic composite. Each of these materials has different properties which make the basketballs suitable for different playing situations. For example, leather balls are suitable for indoor play but not outdoor play. Leather balls absorb water and dirt more than rubber balls. Absorbancy is an example of an intensive property. An **intensive property** is a property that depends on the type of matter in a sample, not the amount of matter.

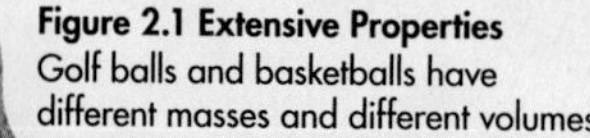

Figure 2.1 Extensive Properties
Golf balls and basketballs have different masses and different volumes.

34

Focus on ELL

1 CONTENT AND LANGUAGE Assign small groups of mixed proficiencies one content vocabulary term each. Groups should define the term, create a visual example and a non-linguistic representation, and identify any common misinterpretations of the assigned term. Have each group present its word to the class.

2 FRONTLOAD THE LESSON Ask students to state the types of materials used for various features of their residence or classroom, such as windows, walls, and doors. Have students group the materials based on their similarities. Explain that the groupings are an example of classification of matter and that students have used the properties of these materials to classify them. Revisit the activity after the lesson to see if students wish to use their newfound knowledge of properties to change their classifications.

3 COMPREHENSIBLE INPUT Illustrate that substances can be classified in more than one way on the basis of their properties by having students create Venn diagrams to classify the substances listed in Table 2.1. Ask volunteers to share their diagrams with the class.

Identifying a Substance Each object in Figure 2.2 has a different chemical makeup, or composition. The soda can is mainly aluminum. The watering can is mainly copper. Matter that has a uniform and definite composition is called a **substance.** Aluminum and copper are examples of substances, which are also referred to as *pure substances.* **Every sample of a given substance has identical intensive properties because every sample has the same composition.**

Aluminum and copper have some properties in common, but there are differences besides their distinctive colors. Aluminum is highly reflective and is often used in silver paints. Pure copper can scratch the surface of pure aluminum because copper is harder than aluminum. Copper is better than aluminum as a conductor of heat or electric current. Copper and aluminum are both malleable, which means they can be hammered into sheets without breaking. Hardness, color, conductivity, and malleability are examples of physical properties. A **physical property** is a quality or condition of a substance that can be observed or measured without changing the substance's composition.

Table 2.1 lists physical properties for some substances. The states of the substances are given at room temperature. (Although scientists use room temperature to refer to a range of temperatures, in this book it will be used to refer to a specific temperature, 20°C.) Physical properties can help chemists identify substances. For example, a colorless substance that was found to boil at 100°C and melt at 0°C would likely be water. A colorless substance that boiled at 78°C and melted at −117°C would definitely not be water. Based on Table 2.1, it would likely be ethanol.

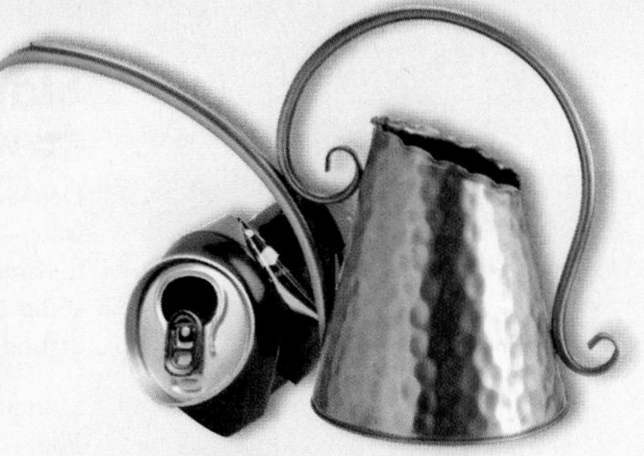

Figure 2.2 Aluminum and Copper
This soda can is made almost entirely of a thin sheet of aluminum. The watering can is made of copper, which has been hammered to give its textured appearance.
Analyze Data ***Which of the properties listed in Table 2.1 could not be used to distinguish copper from aluminum?***

CHEMISTRY & YOU

Q: *Glass is often used to make windows, while copper is often used in electrical wires. What properties of glass make it a desirable material to use for windows?*

Interpret Data

Physical Properties of Some Substances

Substance	State	Color	Melting point (°C)	Boiling point (°C)
Neon	Gas	Colorless	−249	−246
Oxygen	Gas	Colorless	−218	−183
Chlorine	Gas	Greenish-yellow	−101	−34
Ethanol	Liquid	Colorless	−117	78
Mercury	Liquid	Silvery-white	−39	357
Bromine	Liquid	Reddish-brown	−7	59
Water	Liquid	Colorless	0	100
Sulfur	Solid	Yellow	115	445
Aluminum	Solid	Silver	660	2519
Sodium chloride	Solid	White	801	1413
Gold	Solid	Yellow	1064	2856
Copper	Solid	Reddish-yellow	1084	2562

Table 2.1 A substance can be described and identified by its physical properties.

a. Identify Which property can most easily distinguish chlorine from the other gases?

b. Identify A colorless liquid boils at 40°C. Is the substance water? Why or why not?

c. Calculate Which of the liquid substances has the highest melting point? The lowest boiling point?

d. Draw Conclusions Which of the properties would be the most helpful in identifying an unknown substance?

Differentiated Instruction

L3 ADVANCED STUDENTS Ask students to identify two states of matter other than solid, liquid, and gas. Have students prepare a report describing the general characteristics of these states and the conditions under which matter is likely to exist in these states. Students should explain why these states of matter are not common on Earth. *(Plasmas are likely at extremely high temperatures; Bose-Einstein condensates are possible at extremely low temperatures.)*

LPR LESS PROFICIENT READERS Provide steps to follow for reading and extracting information from a reference table. Then have students practice reading Table 2.1 by posing simple questions such as, *What is the boiling point of ethanol?*

L1 SPECIAL NEEDS STUDENTS Help students understand that air is matter by giving each student an air-filled balloon tied with a ribbon. Show or pass around an unfilled balloon. Untie each ribbon and slowly release the air against the student's skin. Explain that a balloon expands because air takes up space.

Foundations for Reading

BUILD VOCABULARY To help students distinguish the word *extensive* from *intensive,* explain that the prefix *in-* means "within." The related word *extend* often refers to an increase in a dimension, such as length, area, or time.

READING STRATEGY Have students strategically read the opening paragraph of the Describing Matter section. Then teach them how to use the self-questioning strategy to frontload their understanding of the content. Guide students to look for the answers to their questions as they read the rest of the section.

Explore

Describing Matter

Teacher Demo

PURPOSE Students will observe that air takes up space and has mass.

MATERIALS plastic syringe (50 mL or larger), balance that reads to 0.01 g, Erlenmeyer flask with tightly fitting 1-hole rubber stopper, funnel, water

PROCEDURE Have students predict whether or not air takes up space and has mass. Determine the mass of the empty syringe. Then determine the mass when the syringe is full of air. Subtract the two masses to find the mass of air in the syringe. Fit the stopper and funnel into the Erlenmeyer flask. The equipment must be airtight. Quickly pour water through the funnel.

EXPECTED OUTCOME The mass of the air will be about 0.06 g. Some water will run into the flask (air can be compressed) but not very much because air takes up the space in the flask.

Explain

CHEMISTRY & YOU Some of the physical properties that make glass a good material for windows are that glass has a high melting point making it heat resistant, it is transparent, and it is hard.

USE VISUALS Direct students to Table 2.1. **Ask** What is the relationship between the state of a substance at room temperature and its boiling point? *(Substances that are gases at room temperature have boiling points below 20°C; liquids and solids have boiling points above 20°C.)* **Ask** Are the properties listed in the table intensive or extensive? *(intensive)*

Answers

FIGURE 2.2 State; both are solids.

INTERPRET DATA

a. color
b. No; water boils at 100°C.
c. water; bromine
d. melting point or boiling point

LESSON 2.1

Explore

States of Matter

Student Activity

PURPOSE Students will construct models of a solid and a liquid.

MATERIALS foam balls or miniature marshmallows

PROCEDURE Use Figure 2.3 as a guide for the three-dimensional models.

EXPECTED OUTCOME The particles in each model should be closely packed, but the arrangement in the solid should be orderly.

Teacher Demo

PURPOSE Students will observe three states of matter when chocolate is heated.

MATERIALS piece of solid chocolate, small beaker, large beaker, boiling water

SAFETY Use a thermal mitt while handling the boiling water.

PROCEDURE Show students pieces of solid chocolate. Place the pieces in a small beaker surrounded by boiling water in a larger beaker. After a few minutes, **Ask** How many physical states can you discern in the mixture? *(three)*

EXPECTED OUTCOME Three states can be discerned as the chocolate melts: solid, liquid, and vapor. Remind students that their ability to smell the chocolate means that some components of the mixture must have vaporized.

States of Matter

What are three states of matter?

Depending on the circumstances, you use three different words to refer to water—water, ice, and steam. Water, which is a common substance, exists in three different physical states. So can most other substances. **Three states of matter are solid, liquid, and gas.** Certain characteristics that can distinguish these three states of matter are summarized in Figure 2.3.

Solids A **solid** is a form of matter that has a definite shape and volume. The shape of a solid doesn't depend on the shape of its container. The particles in a solid are packed tightly together, often in an orderly arrangement, as shown in Figure 2.3a. As a result, solids are almost incompressible; that is, it is difficult to squeeze a solid into a smaller volume. In addition, solids expand only slightly when heated.

Liquids Look at Figure 2.3b. The particles in a liquid are in close contact with one another, but the arrangement of particles in a liquid is not rigid or orderly. Because the particles in a liquid are free to flow from one location to another, a liquid takes the shape of the container in which it is placed. However, the volume of the liquid doesn't change as its shape changes. The volume of a liquid is fixed or constant. Thus, a **liquid** is a form of matter that has an indefinite shape, flows, yet has a fixed volume. Liquids are almost incompressible, but they tend to expand slightly when heated.

Figure 2.3
The arrangement of particles is different in solids, liquids, and gases.
Relate Cause and Effect ***Use the arrangements of their particles to explain the general shape and volume of solids and gases.***

See states of matter animated online. KINETIC ART

Solid In a solid, the particles are packed closely together in a rigid arrangement.

Liquid In a liquid, the particles are close together, but they are free to flow past one another.

Gas In a gas, the particles are relatively far apart and can move freely.

Professional Development Note: Thixotropic Materials

Changes of state are typically associated with changes in temperature. However, the state of matter can also be affected by other variables. Thixotropic materials are solidlike materials that liquefy when subjected to shearing forces. For example, many paints are thixotropic; they thin out when brushed on a surface and thicken when the brush strokes stop, thus keeping the paint from sliding off the wall! A shearing force has an opposite effect on quicksand. Quick movements "thicken" the quicksand and make it much more difficult for a person or animal trapped in it to move. (A shearing force causes two layers to slide in opposite directions along a plane of contact.)

Gases Like a liquid, a gas takes the shape of its container. But unlike a liquid, a gas can expand to fill any volume. A **gas** is a form of matter that takes both the shape and volume of its container. Look back at Figure 2.3c. As shown in the model, the particles in a gas are usually much farther apart than the particles in a liquid. Because of the space between particles, gases are easily compressed into a smaller volume.

The words *vapor* and *gas* are sometimes used interchangeably. But there is a difference. The term *gas* is used for substances, like oxygen, that exist in the gaseous state at room temperature. (*Gaseous* is the adjective form of *gas*.) **Vapor** describes the gaseous state of a substance that is generally a liquid or solid at room temperature, as in *water vapor.*

Physical Changes

How can physical changes be classified?

The melting point of gallium metal is 30°C. Figure 2.4 shows how heat from a person's hand can melt a sample of gallium. The shape of the sample changes during melting as the liquid begins to flow, but the composition of the sample does not change. Melting is a physical change. During a **physical change,** some properties of a material change, but the composition of the material does not change.

Figure 2.4 Physical Change
The silvery substance in the photograph is gallium, which has a melting point of 30°C.
Infer ***What can you infer about the temperature of the hand holding the gallium?***

Words such as *boil, freeze, melt,* and *condense* are used to describe physical changes. So are words such as *break, split, grind, cut,* and *crush.* However, there is a difference between these two sets of words. Each set describes a different type of physical change. **Physical changes can be classified as reversible or irreversible.** Melting is an example of a reversible physical change. If a sample of liquid gallium is cooled below its melting point, the liquid will become a solid. All physical changes that involve a change from one state to another are reversible. Cutting hair, filing nails, and cracking an egg are examples of irreversible physical changes.

2.1 LessonCheck

ONLINE PROBLEMS

1. **Explain** Explain why all samples of a given substance have the same intensive properties.
2. **Identify** Name three states of matter.
3. **Describe** Describe the two categories used to classify physical changes.
4. **Identify** Name two categories used to classify properties of matter.
5. **Interpret Tables** Which property in Table 2.1 can most easily distinguish sodium chloride from the other solids?
6. **Compare and Contrast** In what way are liquids and gases alike? In what way are liquids and solids different?
7. **Explain** Is the freezing of mercury a reversible or irreversible physical change? Explain your answer.
8. **Explain** Explain why samples of platinum and copper can have the same extensive properties but not the same intensive properties.

BIGIDEA
CHEMISTRY AS THE CENTRAL SCIENCE

9. How would understanding the properties of matter be helpful in other fields of study besides chemistry?

Lesson Check Answers

1. Every sample of a given substance has the same chemical composition.
2. solid, liquid, gas
3. Physical changes are either reversible or irreversible. Reversible changes can be "undone," or reversed. Irreversible changes cannot be undone.
4. intensive and extensive properties
5. Color; sodium chloride is the only white solid listed.
6. Liquids and gases both have an indefinite shape. The shape of a solid is definite; the shape of a liquid is indefinite.
7. The freezing of mercury is reversible because solid mercury can be melted.
8. Samples of platinum and copper can have the same mass and volume (extensive properties). They cannot have the same set of intensive properties because they have different chemical compositions.
9. **BIGIDEA** Answers will vary; possible answers include: choosing materials for automobile parts; combining and cooking foods in a restaurant; handling products for cleaning and maintaining swimming pools.

Explain

Physical Changes

USE VISUALS Direct students to Figure 2.4. Have students study the photograph. Refer them to page R10 in the Elements Handbook for data on the properties of gallium. **Ask** Is the melting of gallium a reversible or irreversible change? *(reversible)* **Ask** What is the physical state of gallium at room temperature? *(solid)*

Evaluate

Informal Assessment

Assess students' understanding of states of matter by having them consider a lit candle. **Ask** How are three states of matter involved when a candle burns? Explain. *(Solid wax has a definite shape and volume; the melted liquid wax is shapeless and runs down the side of the candle; the vapor that forms above the wick has no shape.)* Then have students complete the 2.1 Lesson Check.

Reteach

If possible, bring samples of some of the substances listed in Table 2.1 to class. Describe the physical properties of selected samples and have students practice identifying them by referring to Table 2.1.

Answers

FIGURE 2.3 Solids have a definite shape and volume because their particles are in a rigid, orderly arrangement; gases have no definite shape or volume because their particles are relatively far apart and can move freely.

FIGURE 2.4 Because gallium is shown in the liquid state, the temperature of the hand must be greater than 30°C.

LESSON 2.2

Key Objectives

2.2.1 **EXPLAIN** how mixtures are classified.

2.2.2 **EXPLAIN** how mixtures can be separated.

Additional Resources

Reading and Study Workbook, Lesson 2.2

Available Online or on Digital Media:

- Teacher Resources, Lesson 2.2 Review

PearsonChem.com

Engage

CHEMISTRY & YOU Have students study the photograph and read the opening text. Explain that the coffee in the cup is also a mixture. **Ask** Can you use a filter to separate the water from the other components of the coffee? Explain. *(no, because the particles are too small)* Point out that not all mixtures can be separated with the same method. Tell students to keep this in mind as they read the lesson.

Activate Prior Knowledge

Review the concepts from Lesson 2.1, directing students' attention back to Table 2.1. Emphasize that the substances in that table are pure substances and always have the same composition. In this lesson, students will focus on mixtures. Substances can be combined in different ratios to form mixtures, so a mixture doesn't always have the same composition.

Foundations for Reading

BUILD VOCABULARY Have students look up the term *homogenize* and explain how this process could be used to turn a heterogeneous mixture into one that is more homogeneous.

ACADEMIC STANDARDS for SCIENCE

C.1.1 Based on physical properties, differentiate between pure substances and mixtures

2.2 Mixtures

C.1.1 Pure substance or mixture?

CHEMISTRY & YOU

Q: *Why aren't there coffee grounds in a cup of coffee?* Coffee is often brewed by mixing hot water with ground coffee beans. But when people drink coffee, the grounds are usually not in their mug. In this lesson, you will learn how to classify and separate mixtures.

Key Questions

How can mixtures be classified?

How can mixtures be separated?

Vocabulary

- mixture
- heterogeneous mixture
- homogeneous mixture
- solution
- phase
- filtration
- distillation

Classifying Mixtures

How can mixtures be classified?

A salad bar, like the one in Figure 2.5, provides a range of items, such as lettuce, tomatoes, cheese, and green peppers. Customers choose which items to use in their salads and how much of each item to use. So each salad mixture has different types and amounts of components. A **mixture** is a physical blend of two or more components.

Most samples of matter are mixtures. Some mixtures are easier to recognize than others. You can easily recognize chicken noodle soup as a mixture of chicken, noodles, and broth. Recognizing air as a mixture of gases is more difficult. But the fact that air can be drier or more humid shows that the amount of one component of air—water vapor—can vary. Chicken noodle soup and air represent two different types of mixtures. **Based on the distribution of their components, mixtures can be classified as heterogeneous mixtures or as homogeneous mixtures.**

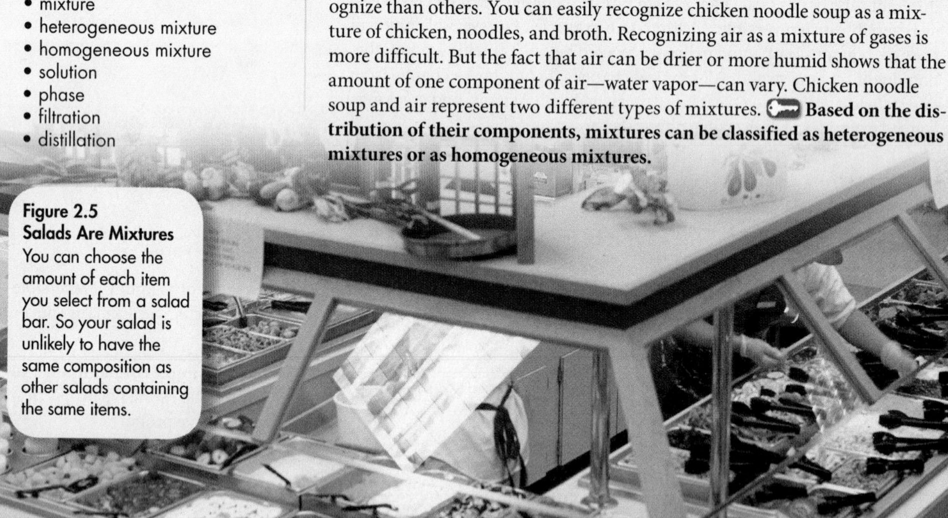

Figure 2.5
Salads Are Mixtures
You can choose the amount of each item you select from a salad bar. So your salad is unlikely to have the same composition as other salads containing the same items.

Focus on ELL

1 CONTENT AND LANGUAGE Write the word *mixture* on the board. Have students brainstorm examples of mixtures. Create a list of their ideas. Have students use a dictionary to look up the various uses of the word *mixture*. Once the scientific definition is determined, revisit the list of examples; point to each example and ask the class if it should stay on the list.

2 FRONTLOAD THE LESSON Generate a discussion on the seasonings used in students' favorite foods. Identify seasonings that are blends rather than single herbs or spices. Explain that the blends are examples of mixtures, because they are made by combining two or more spices and/or herbs. Have students predict how this fact explains why different brands of the same spice blend often taste quite different.

3 COMPREHENSIBLE INPUT Perform as many teacher demos and class activities as possible to expose students to a variety of approaches to separating mixtures. Use crumpled and torn paper to demonstrate the difference between reversible and irreversible physical changes.

Quick Lab

Purpose To separate a mixture using paper chromatography

Materials
- green marking pen
- filter paper strip
- metric ruler
- clear plastic tape
- pencil
- rubbing alcohol
- clear plastic drinking cup
- clear plastic wrap

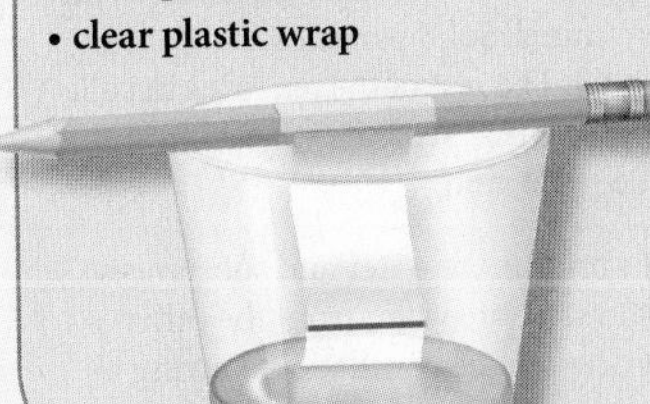

Separating Mixtures

Procedure

1. Use the marking pen to draw a line across a strip of filter paper, as shown in the drawing. The line should be 2 cm from one end of the strip.

2. Tape the unmarked end of the filter paper to the center of a pencil so that the strip hangs down when the pencil is held horizontally, as shown in the diagram below.

3. Working in a well-ventilated room, pour rubbing alcohol into a plastic cup to a depth of 1 cm.

4. Rest the pencil on the rim of the cup so that the ink end of the strip touches the rubbing alcohol but does not extend below its surface. Use plastic wrap to cover the top of the cup.

5. Observe the setup for 15 minutes.

Analyze and Conclude

1. Identify How did the appearance of the filter paper change during the procedure?

2. Analyze Data What evidence is there that green ink is a mixture?

3. Apply Concepts How could you use this procedure to identify an unknown type of green ink?

Heterogeneous Mixtures In chicken noodle soup, the ingredients in the soup are not evenly distributed throughout the mixture. There is likely to be different amounts of chicken and noodles in each spoonful. A mixture in which the composition is not uniform throughout is a **heterogeneous mixture.**

Homogeneous Mixtures The substances in the olive oil in Figure 2.6 are evenly distributed throughout the mixture. So, olive oil doesn't look like a mixture. The same is true for vinegar. Vinegar is a mixture of water and acetic acid, which dissolves in the water. Olive oil and vinegar are homogeneous mixtures. A **homogeneous mixture** is a mixture in which the composition is uniform throughout. Another name for a homogeneous mixture is a **solution.** Many solutions are liquids. But some are gases, like air, and some are solids, like stainless steel, which is a mixture of iron, chromium, and nickel.

The term **phase** is used to describe any part of a sample with uniform composition and properties. By definition, a homogeneous mixture consists of a single phase. A heterogeneous mixture consists of two or more phases. When oil and vinegar are mixed, they form a heterogeneous mixture with two layers, or phases. As shown in Figure 2.6, the oil phase floats on the water, or vinegar, phase.

Figure 2.6 Homogeneous Mixtures
Olive oil and vinegar are homogeneous mixtures. The substances in these mixtures are evenly distributed. When olive oil and vinegar are mixed, they form a heterogeneous mixture with two distinct phases.

Focus on ELL

4 LANGUAGE PRODUCTION Have students work in groups of four to complete the lab. Make sure each group has ELLs of varied language proficiencies so that more proficient students can help less proficient ones. Have students work according to their proficiency level.

BEGINNING: LOW/HIGH Have students draw a picture of their observations.

INTERMEDIATE: LOW/HIGH Allow students to orally explain their observations.

ADVANCED: LOW/HIGH Have students assist students with lower English proficiencies with the calculations and answering the questions.

Explain

Classifying Mixtures

MAKE A CONNECTION Bring orange juice or liquid salad dressing to class and compare the properties of these mixtures to those of pure water or pure NaCl. Explain that mixtures are variable in composition.

Explore

Quick Lab

OBJECTIVE After completing this activity, students will be able to separate the components of a mixture by using paper chromatography.

SKILLS FOCUS Observing, inferring, drawing conclusions

PREP TIME 15 minutes

CLASS TIME 25 minutes

ADVANCE PREPARATION Cut the paper strips in advance to save time. Strips of paper toweling can be used in place of filter paper.

SAFETY Rubbing alcohol is poisonous and flammable. It is also an irritant when inhaled. Keep containers covered and away from heat. If the room is not well ventilated, use a fume hood.

EXPECTED OUTCOME Bands of color will separate on the filter paper.

EXTENSION Students could repeat the lab by using different brands of markers, different colors, or water in place of rubbing alcohol.

ANALYZE AND CONCLUDE

1. Bands of colors appeared as the alcohol moved up the paper.
2. The bands of colors indicate that green ink is a mixture.
3. The color pattern of the unknown ink could be compared with color patterns from known types of green ink.

FOR ENRICHMENT Students can research R_f values of different dyes and how they are calculated (ratio of distance traveled by dye to distance traveled by solvent). They can then quantify this lab by calculating the R_f values of the dyes in the ink.

Explain

Separating Mixtures

USE VISUALS Direct students' attention to Figure 2.8. Review the distillation process, and describe the components of the apparatus. **Ask** What might be an advantage to having a long condenser in a distillation apparatus? *(There would be more surface area on which the vapor can condense.)* **Ask** Is distilled water that contains dissolved gases a pure substance? *(No, distilled water is not pure unless the dissolved gases are removed.)*

CHEMISTRY & YOU Filtration; the brewed coffee passes through the filter paper, but the ground coffee beans do not.

Explore

Teacher Demo

PURPOSE Students will observe the separation of iron filings from iron-fortified breakfast cereal.

MATERIALS iron-fortified breakfast cereal, 400-mL beaker, distilled water, magnetic stirrer with stirring bar

SAFETY Remind students not to eat the cereal.

PROCEDURE Place a stirring bar in a 400-mL beaker. Add about 30 g of cereal to the beaker and add distilled water until the beaker is about half full. Using a magnetic stirrer, mix gently for about 20 minutes. Retrieve the stirring bar and observe the black iron filings attached to it.

EXPECTED OUTCOME Iron filings will cover the stirring bar. They are added to cereal as an iron supplement. Explain that stomach acid changes the iron into a form the body can use.

Figure 2.7 Filtration
A filter is used to separate ground coffee beans from brewed coffee. This process is a type of filtration.

Q: *Brewing coffee is a mixture of ground coffee beans and water. What process is used to separate ground coffee beans from brewed coffee?*

Separating Mixtures

How can mixtures be separated?

If you have a salad containing an ingredient you don't like, you can use a fork to remove the pieces of the unwanted ingredient. Many mixtures are not as easy to separate. To separate a mixture of olive oil and vinegar, for example, you could decant, or pour off, the oil layer. Or you might cool the mixture until the oil turned solid. The first method takes advantage of the fact that oil floats on water. The second method takes advantage of a difference in the temperatures at which the olive oil and vinegar freeze. **Differences in physical properties can be used to separate mixtures.**

Filtration The coffee filter in Figure 2.7 can separate ground coffee beans from brewed coffee. The liquid brewed coffee passes through the paper filter, but the solid coffee grounds cannot pass through the filter. Filter paper used in a laboratory is similar to coffee filters. Filter paper is often placed in a funnel. Then the mixture is poured into the funnel. Solid particles that cannot pass through the filter remain in the funnel. The rest of the particles in solution pass through the filter paper. The process that separates a solid from the liquid in a heterogeneous mixture is called **filtration.**

Distillation Tap water is a homogeneous mixture of water and substances that are dissolved in the water. One way to separate water from the other components in tap water is through a process called distillation. During a **distillation,** a liquid is boiled to produce a vapor that is then condensed into a liquid. Figure 2.8 shows an apparatus that can be used to perform a small-scale distillation.

As water in the distillation flask is heated, water vapor forms, rises in the flask, and passes into a glass tube in the condenser. The tube is surrounded by cold water, which cools the vapor to a temperature at which it turns back into a liquid. The liquid water is collected in a second flask. The solid substances that were dissolved in the water remain in the distillation flask because their boiling points are much higher than the boiling point of water.

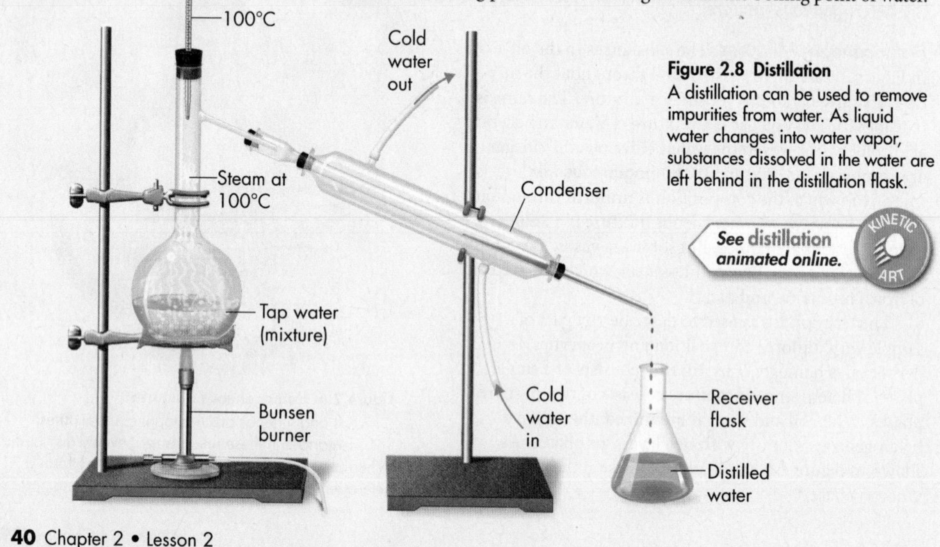

Figure 2.8 Distillation
A distillation can be used to remove impurities from water. As liquid water changes into water vapor, substances dissolved in the water are left behind in the distillation flask.

See distillation animated online. KINETIC ART

Distillation

A working distillation apparatus was described in the writings of Maria of Alexandria, an alchemist who lived and worked nearly two thousand years ago. The city of Alexandria, located on the Nile River in North Africa, was a world center of science and culture at that time. Maria of Alexandria is also credited with inventing other chemical apparatus, such as the water bath. To this day the water bath is often referred to as the *bain marie* in her honor.

SampleProblem 2.1

Separating a Heterogeneous Mixture

How could a mixture of aluminum nails and iron nails be separated?

1 Analyze **Identify the relevant concepts.** In order to identify how to separate aluminum and iron nails, the properties of both aluminum and iron must be known.

2 Solve **Apply concepts to this situation.**

List the properties of each substance in the mixture.

Aluminum	Iron
• metal	• metal
• gray color	• gray color
• doesn't dissolve in water	• doesn't dissolve in water
• not attracted to magnet	• attracted to magnet

Identify a property that can be used to separate different substances from each other.

The ability to be attracted by a magnet is a property that iron and aluminum do not share. You could use a magnet to remove the iron nails from a mixture of iron and aluminum.

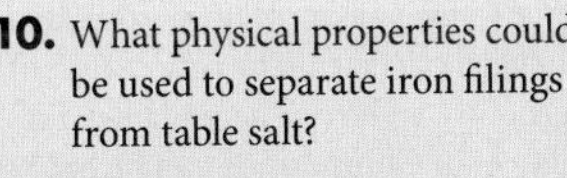

10. What physical properties could be used to separate iron filings from table salt?

11. Air is mainly a mixture of nitrogen and oxygen, with small amounts of other gases such as argon and carbon dioxide. What property could you use to separate the gases in air?

2.2 LessonCheck

12. **Identify** How are mixtures classified?

13. **List** What type of properties can be used to separate mixtures?

14. **Explain** Explain the term *phase* as it relates to homogeneous and heterogeneous mixtures.

15. **Classify** Classify each of the following as a homogeneous or heterogeneous mixture.

- **a.** food coloring
- **b.** ice cubes in liquid water
- **c.** mouthwash
- **d.** mashed, unpeeled potatoes

16. **Compare and Contrast** How are a substance and a solution similar? How are they different?

17. **Apply Concepts** In general, when would you use filtration to separate a mixture? When would you use distillation to separate a mixture?

18. **Explain** Describe a procedure that could be used to separate a mixture of sand and table salt.

BIGIDEA
CHEMISTRY AS THE CENTRAL SCIENCE

19. Give three examples of when you have separated mixtures at home.

Lesson Check Answers

12. as heterogeneous or homogeneous

13. differences in physical properties

14. A phase is any part of a sample with uniform composition. A homogeneous mixture has one phase; a heterogeneous mixture has two or more phases.

15. **a.** homogeneous **b.** heterogeneous **c.** homogeneous **d.** heterogeneous

16. Both have a uniform composition throughout. A substance has a definite composition; a solution has a variable composition.

17. Filtration separates solids from a liquid in a heterogeneous mixture. Distillation can separate a liquid from substances dissolved in the liquid.

18. Add water to dissolve the salt. Filter the mixture to remove the sand. Evaporate the water from the liquid to isolate the solid salt.

19. **BIGIDEA** Answers will vary; examples of mixtures at home include: sorting laundry, draining cooked pasta, putting away bags of groceries.

Sample Practice Problem

Describe how you could separate metal paper clips and plastic paper clips in a bucket of paper clips. *(A magnet could be used to remove the metal paper clips.)*

Evaluate

Informal Assessment

Have students identify five items that fit each of the following categories:

1. substance
2. homogeneous mixture
3. heterogeneous mixture
4. solution

Have students select three items, one each from categories 2, 3, and 4. For each item selected, students should outline a method for separating the components. Then have students complete the 2.2 Lesson Check.

Reteach

Explain that distillation can be used to separate a mixture of gases. Show students the drawing of the fractional distillation of liquid air, on page R20 of the Elements Handbook. **Ask** What physical property is used to separate the gases? *(boiling point)*

Answers

10. Iron is magnetic; table salt is not. Table salt will dissolve in water; iron will not.

11. By lowering the temperature to below the boiling point of each gas, you could condense each substance and separate the gases.

LESSON 2.3

Key Objectives

2.3.1 EXPLAIN the difference between an element and a compound.

2.3.2 DISTINGUISH between a substance and a mixture.

2.3.3 DESCRIBE what chemists use to represent elements and compounds.

2.3.4 EXPLAIN how a periodic table is useful.

Additional Resources

Reading and Study Workbook, Lesson 2.3

Available Online or on Digital Media:

- Teaching Resources, Lesson 2.3 Review

PearsonChem.com

Engage

CHEMISTRY & YOU Have students examine the photo and read the opening text. Ask students who have eaten or smelled burned toast to describe its properties. Explain that the burned taste is due to a chemical reaction between sugar and amino acids, which are chemical compounds in the bread.

Activate Prior Knowledge

Review heterogeneous and homogeneous mixtures. **Ask** What are two methods that can be used to separate mixtures? *(filtration and distillation)* **Ask** Do these methods change the composition of the substances in the mixtures? *(no)* Tell students to keep this in mind as they read the lesson.

IN ACADEMIC STANDARDS for SCIENCE

- Develop explanations based on reproducible data and observations gathered during laboratory investigations.

2.3 Elements and Compounds

CHEMISTRY & YOU

Q: *Why does burned toast taste so bad?* Bread that is toasted to a nice golden brown makes for a tasty addition to breakfast. But most people would agree that bread that is cooked so long that it is burned and black is not tasty.

Key Questions

How are elements and compounds different?

How can substances and mixtures be distinguished?

What do chemists use to represent elements and compounds?

Why is a periodic table useful?

Vocabulary

- element
- compound
- chemical change
- chemical symbol
- periodic table
- period
- group

Distinguishing Elements and Compounds

How are elements and compounds different?

Substances can be classified as elements or compounds. An **element** is the simplest form of matter that has a unique set of properties. Oxygen and hydrogen are two of the more than 100 known elements. A **compound** is a substance that contains two or more elements chemically combined in a fixed proportion. For example, carbon, oxygen, and hydrogen are chemically combined in the compound sucrose. Sometimes sucrose is called table sugar to distinguish it from other sugar compounds. In every sample of sucrose, there are twice as many hydrogen particles as oxygen particles. The proportion of hydrogen particles to oxygen particles in sucrose is fixed. There is a key difference between elements and compounds. **Compounds can be broken down into simpler substances by chemical means, but elements cannot.**

Breaking Down Compounds Physical methods that are used to separate mixtures cannot be used to break a compound into simpler substances. Boil liquid water and you get water vapor, not the oxygen and hydrogen that water contains. Dissolve a sugar cube in water and you still have sucrose, not oxygen, carbon, and hydrogen. This result does not mean that sucrose or water cannot be broken down into simpler substances. But the methods must involve a chemical change.

Figure 2.9 Chemical Changes
When table sugar is heated, it goes through a series of chemical changes. The final products of these changes are solid carbon and water vapor.

Focus on ELL

1 CONTENT AND LANGUAGE Write the words *element*, *compound*, *change* on the board and pronounce them for the students. Explain that these words have both common and scientific meanings. Provide students with the common definitions and have students write them in their vocabulary notebooks. Remind students to leave room to add the scientific definitions as they discover them in the lesson.

2 FRONTLOAD THE LESSON Write a series of recognizable two-letter abbreviations, such as those used for U.S. states, on the board. Ask for a show of hands of students who can readily identify one or more of the abbreviations. Explain that an abbreviation is a form of symbol. Have students consider how the use of symbols in chemistry might make it easier for scientists to communicate with each other.

3 COMPREHENSIBLE INPUT Guide students to use a think-aloud process as they use the flowchart in Figure 2.11 to classify various types of matter. Suggest possible lines of thought, such as, *Is the matter a pure substance or a mixture of substances?* and *Can the matter be physically separated or chemically separated?*

A **chemical change** is a change that produces matter with a different composition than the original matter. Heating is one of the processes used to break down compounds into simpler substances. The layer of sugar in Figure 2.9 is heated in a skillet until it breaks down into solid carbon and water vapor. Can the substances that are produced also be broken down?

There is no chemical process that will break down carbon into simpler substances because carbon is an element. Heat will not cause water to break down, but electricity will. When an electric current passes through water, oxygen gas and hydrogen gas are produced. The following diagram summarizes the overall process.

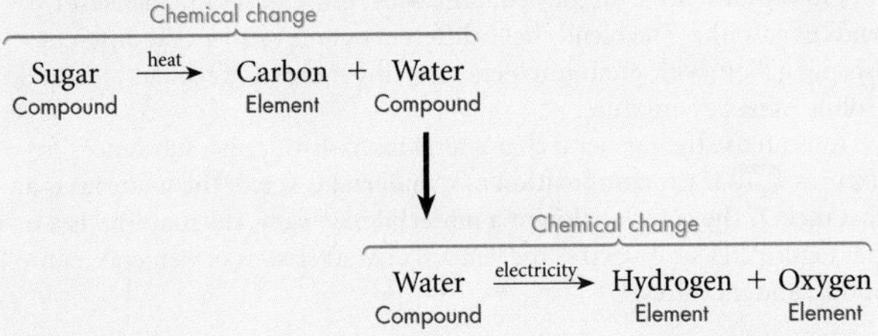

CHEMISTRY & YOU

Q: *What happens to the compounds in bread when it is overcooked that causes the changes to the taste of the bread?*

Properties of Compounds In general, the properties of compounds are quite different from those of their component elements. Sugar is a sweet-tasting, white solid, but carbon is a black, tasteless solid. Hydrogen is a gas that burns in the presence of oxygen—a colorless gas that supports burning. The product of this chemical change is water, a liquid that can stop materials from burning. Figure 2.10 shows samples of table salt (sodium chloride), sodium, and chlorine. When the elements sodium and chlorine combine chemically to form sodium chloride, there is a change in composition and a change in properties. Sodium is a soft, gray metal. Chlorine is a pale yellow-green poisonous gas. Sodium chloride is a white solid.

Figure 2.10 Properties of Compounds and Their Elements
Compounds and the elements from which they form have different properties. The elements sodium and chlorine have different properties from each other and from the compound sodium chloride.
Observe ***Based on the photographs, describe two physical properties of sodium and two of chlorine.***

Sodium is stored under oil to keep it from reacting with oxygen or water vapor in air. Sodium vapor produces the light in some street lamps.

Chlorine is used to make compounds that kill harmful organisms in swimming pools.

Sodium chloride (commonly known as table salt) is a compound used to season or preserve food.

Differentiated Instruction

LPR LESS PROFICIENT READERS Have students list the key questions in their notebooks, leaving space after each question. As they read the lesson, they should list terms and phrases or create drawings that give the information needed for the answers. Check entries for accuracy.

L3 ADVANCED STUDENTS Other forms of energy besides electricity can be used to bring about the decomposition of water into hydrogen and oxygen. Ask students to consider whether heat could be used for this purpose. Tell them that early chemists considered water to be an element because it would not decompose when heated. Have students research the conditions necessary to decompose water using heat. Ask students to find what hazards, if any, might preclude the routine use of this method.

Foundations for Reading

BUILD VOCABULARY Compare the common meanings of *element, elemental,* and *elementary* in relation to the specific use of *element* in chemistry. In the broadest sense, the terms refer to fundamentals, first principles, or basics.

Explain

Distinguishing Elements and Compounds

USE VISUALS Have students study Figure 2.9. **Ask** What causes the chemical changes? *(the addition of heat)* **Ask** How do the photos and caption convince you that sugar is a compound and not an element? *(The photos and caption show that sugar can be broken down into simpler substances: solid carbon and water vapor. Elements cannot be broken down into simpler substances.)*

CHEMISTRY & YOU When bread overcooks, the compounds in the bread undergo a chemical change and break down into other compounds. The compounds produced have a different taste and appearance.

Explore

Teacher Demo

PURPOSE Students will observe the decomposition of sugar by acid.

MATERIALS 100-mL beaker, powdered sugar, 18*M* sulfuric acid, glass stirring rod

SAFETY Perform the demo in a fume hood or in a well-ventilated area and be sure to wear safety goggles.

CAUTION *This is an extremely exothermic reaction.*

PROCEDURE Place about 25 g of powdered sugar in a 100-mL beaker. Carefully add about 10–15 mL of 18*M* sulfuric acid and stir rapidly with a glass rod.

DISPOSAL Place the cooled beaker and its contents in a large beaker and add water. Neutralize with NaOH and pour the liquid down the drain, flushing with excess water. Throw the solid residue in the trash.

EXPECTED OUTCOME The mixture turns dark, and a column of carbon rises out of the beaker.

Answers

FIGURE 2.10 sodium: gray color, shiny, solid at room temperature; chlorine: yellow-green color, gas at room temperature

LESSON 2.3

Explain

Distinguishing Substances and Mixtures

USE VISUALS Have students examine the flowchart in Figure 2.11. **Ask** Can compounds be separated into their component elements by physical processes? *(No, they must be separated by chemical processes.)* Have students suggest two other examples for each of the four classifications and explain why each example matches the definition of its classification.

Explore

Substances

PURPOSE Students will use an inductive approacth to establish the definition of *substance*.

MATERIALS 20 index cards, masking tape

PROCEDURE Make signs with the number and name of each of these items: 1 oxygen, 2 neon, 3 apple, 4 sand, 5 iron, 6 water, 7 air, 8 paint, 9 sodium chloride, 10 sucrose, 11 carbon dioxide, 12 granite, 13 laundry detergent, 14 citric acid, 15 cereal, 16 salad, 17 salad dressing, 18 copper, 19 ocean water, 20 gold. Use masking tape to create two columns on the board or wall. Place items 1 and 2 in Column A and items 3 and 4 in Column B. Ask students to think about the criteria for each column. Then place item 5 in Column A. Students may have to change their criteria. Place item 6 in Column A. Ask students what they think the criteria are now. Continue placing one item at a time in a column until students understand the criteria. Everything in Column A has a uniform and definite composition.

EXPECTED OUTCOME Column A is for substances: 1, 2, 5, 6, 9, 10, 11, 14, 18, 20. Column B is for mixtures: 3, 4, 7, 8, 12, 13, 15, 16, 17, 19. Make sure students understand the difference between the uniform composition of a substance and the composition of a mixture. The composition of the substance is fixed. The composition of the mixture can vary.

Distinguishing Substances and Mixtures

How can substances and mixtures be distinguished?

Deciding whether a sample of matter is a substance or a mixture based solely on appearance can be difficult. After all, homogeneous mixtures and substances will both appear to contain only one kind of matter. Sometimes you can decide by considering whether there is more than one version of the material in question. For example, you can buy whole milk, low-fat milk, no-fat milk, light cream, or heavy cream. From this information, you can conclude that milk and cream are mixtures. You might infer that these mixtures differ in the amount of fat they contain. Most gas stations offer at least two blends of gasoline. The blends have different octane ratings and different costs per gallon, with premium blends costing more than regular blends. So gasoline must be a mixture.

You can use their general characteristics to distinguish substances from mixtures. **If the composition of a material is fixed, the material is a substance. If the composition of a material may vary, the material is a mixture.** Figure 2.11 summarizes the general characteristics of elements, compounds, and mixtures.

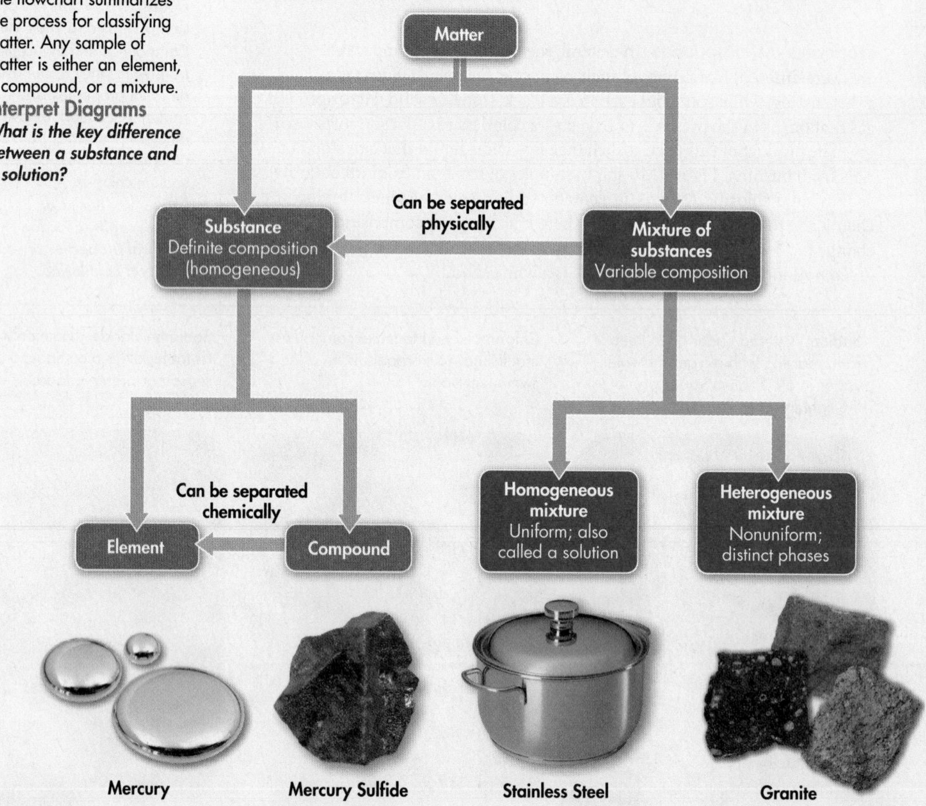

Figure 2.11 Element, Compound, or Mixture? The flowchart summarizes the process for classifying matter. Any sample of matter is either an element, a compound, or a mixture.
Interpret Diagrams *What is the key difference between a substance and a solution?*

UbD Check for Understanding

How can substances and mixtures be distinguished?

To assess students' understanding of the difference between substances and mixtures, ask students to use a thumbs-up gesture if they are confident in their ability to distinguish between the two classifications of matter. Instruct students to use a thumbs-down gesture to indicate continuing confusion about the difference between the two classifications.

ADJUST INSTRUCTION If students indicate discomfort with their level of understanding, review Figure 2.11 with students. Then provide students with additional practice in using the flowchart to identify a group of samples as either substances or mixtures.

SampleProblem 2.2

Classifying Materials

When a certain blue-green solid is heated, a colorless gas and a black solid form. All three materials are substances. Is it possible to classify these substances as elements or compounds?

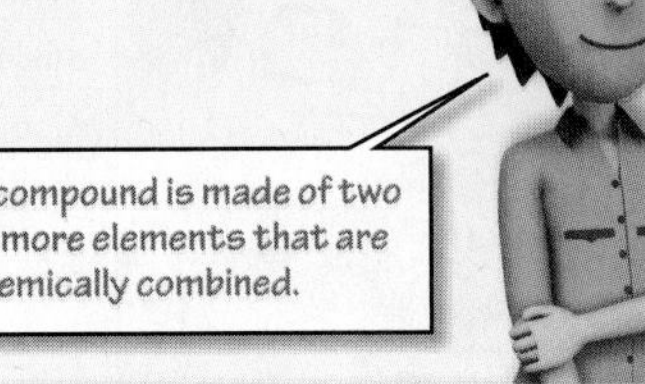

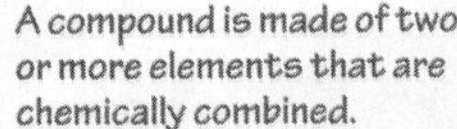

1 Analyze Identify the relevant concepts. A compound can be broken down into simpler substances by a chemical change, but an element cannot. Heating can cause a chemical change.

2 Solve Apply concepts to this situation.

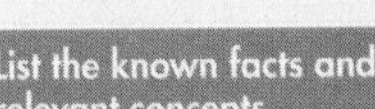

- A blue-green solid is heated.

Determine if the substances are elements or compounds.

- A colorless gas and a black solid appear.

Before heating, there was one substance. After heating, there were two substances. The blue-green solid must be a compound. Based on the information given, it isn't possible to know if the colorless gas and the black solid are elements or compounds.

20. Liquid A and Liquid B are clear liquids. They are placed in open containers and allowed to evaporate. When evaporation is complete, there is a white solid in container B but no solid in container A. From these results, what can you infer about the two liquids?

21. A clear liquid in an open container is allowed to evaporate. After three days, a solid is left in the container. Was the clear liquid an element, a compound, or a mixture? How do you know?

Symbols and Formulas

What do chemists use to represent elements and compounds?

The common names *water* and *table salt* do not provide information about the chemical composition of these substances. Also, words are not ideal for showing what happens to the composition of matter during a chemical change. **Chemists use chemical symbols to represent elements, and chemical formulas to represent compounds.**

Using symbols to represent different kinds of matter is not a new idea. For thousands of years, alchemists used symbols, such as the ones in Figure 2.12, to represent elements. The symbols used today for elements are based on a system developed by a Swedish chemist, Jöns Jacob Berzelius (1779–1848). He based his symbols on the Latin names of elements. Each element is represented by a one- or two-letter **chemical symbol.** The first letter of a chemical symbol is always capitalized. When a second letter is used, it is lowercase.

Figure 2.12 Element Symbols The symbols used to represent elements have changed over time. Alchemists and the English chemist John Dalton (1766–1844) both used drawings to represent chemical elements. Today, elements are represented by one- or two-letter symbols.

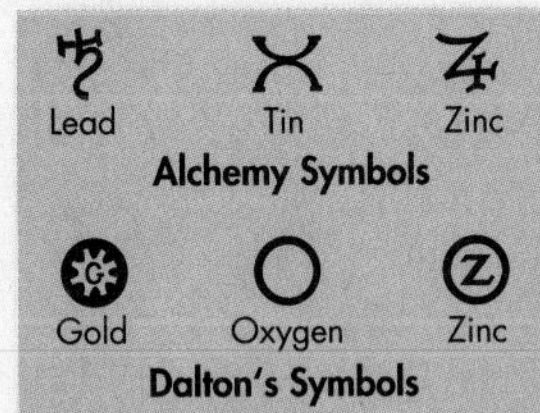

Explain

Sample Practice Problem

Have students classify each of the following types of matter as an element, a compound, or a mixture and provide reasons for their answers.

a. Epsom salt (magnesium sulfate) *(Compound; it is composed of at least two elements.)*

b. hydrogen peroxide (H_2O_2) *(Compound; it is composed of two elements.)*

c. mercury (Hg) *(Element; its symbol shows that it is composed of only one kind of matter.)*

d. sweetened iced tea *(Mixture; it has a composition that can vary.)*

Symbols and Formulas

APPLY CONCEPTS Point out that the use of chemical symbols is an example of the "international language of chemistry." Discuss why people all over the world use the same set of chemical symbols. Show how symbols are used in chemical formulas. Emphasize that a compound is always made up of the same elements in the same proportions. Encourage students to begin a symbol reference sheet and place it in their notebooks where it can be accessed easily—for instance, inside the front cover.

Differentiated Instruction

L1 SPECIAL NEEDS STUDENTS Work with students in creating a large game board out of Figure 2.11. Use large print and symbols for words and colored pathways for the flow options. When asking students to classify matter, have them start with a game piece at the top of the game board and follow along the pathways, pausing at each classification. Use the light from a flashlight or gently clap your hands to signal that students are on the correct path. Tell students that when the light is off or when they no longer hear clapping, they are not on the right path.

ELL ENGLISH LANGUAGE LEARNERS Have students make flashcards to help them learn the names and symbols of at least 30 elements. If an element has a different common name in a student's native country, have the student include that information on the flashcard.

Answers

FIGURE 2.11 The composition of a substance is fixed; the composition of a solution may vary.

20. Liquid A is probably a substance. Liquid B is a mixture.

21. The liquid was not an element because a solid was left when the liquid evaporated. A physical process, such as evaporation, cannot be used to break down a compound. Therefore, the liquid was a mixture.

Explain

The Periodic Table—A Preview

USE VISUALS Direct students to examine Table 2.2 and Figure 2.13. Have students locate the elements given in Table 2.2 on the periodic table. **Ask** Why do some elements have symbols that are similar to the spelling of their names (for example, helium, He) and others have symbols that seem to have nothing to do with names (for example, sodium, Na)? *(Only some element names are derived from their English names, and others are derived from their Latin names, Greek names, the names of scientists who discovered the elements, or other names.)* **Ask** Are the elements in alphabetical order on the periodic table? *(no)* **Ask** What do the numbers above the element symbols represent? *(atomic numbers)* Encourage students to familiarize themselves with the names and the symbols of the elements, especially those in the first four rows of the periodic table. Explain that in Chapter 6 students will explore the periodic table in depth.

Extend

Connect to HISTORY

Have students explore the history of science by asking them to work collaboratively in creating a timeline for the discovery of the chemical elements. Encourage students to use the timeline to detect shifts in naming conventions. Ask students to predict whether the timeline of element discovery has anything to do with the location of the elements on the periodic table.

Table 2.2

Symbols and Latin Names for Some Elements

Name	Symbol	Latin name
Sodium	Na	*natrium*
Potassium	K	*kalium*
Antimony	Sb	*stibium*
Copper	Cu	*cuprum*
Gold	Au	*aurum*
Silver	Ag	*argentum*
Iron	Fe	*ferrum*
Lead	Pb	*plumbum*
Tin	Sn	*stannum*

If the English name and the Latin name of an element are similar, the symbol will appear to have been derived from the English name. Examples include Ca for calcium, N for nitrogen, and S for sulfur. Table 2.2 shows examples of elements for which the symbols do not match the English names. Chemical symbols provide a shorthand way to write the chemical formulas of compounds. The symbols for hydrogen, oxygen, and carbon are H, O, and C. The formula for water is H_2O. The formula for sucrose, or table sugar, is $C_{12}H_{22}O_{11}$. Subscripts in chemical formulas tell you how many of each type of element are in the compound. For example, the subscript 2 in H_2O indicates that there are always two parts of hydrogen for each part of oxygen in water. Because a compound has a fixed composition, the formula for a compound is always the same.

The Periodic Table—A Preview

Why is a periodic table useful?

All the known elements are organized in a special table called the periodic table. A **periodic table** is an arrangement of elements in which the elements are separated into groups based on a set of repeating properties. **The periodic table allows you to easily compare the properties of one element (or a group of elements) to another element (or group of elements).**

Figure 2.13 shows the most commonly used form of the modern periodic table, sometimes called the long form. Each element is identified by its symbol placed in a square. The elements are listed in order from left to right and top to bottom by atomic number, a number that is unique to each element. The atomic number of the element is shown centered above the symbol. You will learn more about atomic numbers in Chapter 4. Hydrogen (H), the lightest element, is in the top left corner. Helium (He), atomic number 2, is at the top right. Lithium (Li), atomic number 3, is at the left end of the second row.

Each horizontal row of the periodic table is called a **period.** There are seven periods in the periodic table. The number of elements per period ranges from 2 (hydrogen and helium) in Period 1 to 32 in Period 6. Within a period, the properties of the elements vary as you move across the period. This pattern of properties then repeats as you move to the next period.

Origins of Element Names

Many elements were named for the people who discovered them or the places where they were discovered. Some elements were given descriptive names taken from classical Latin or Greek. Others were named for figures in mythology. Polonium is named for Poland, the native land of Marie Curie, who also discovered radium. Californium was discovered at the University of California. The name *chlorine* comes from the Greek *chloros*, meaning "greenish-yellow"; it is a greenish-yellow gas. The name *calcium* is derived from the Latin *calx*, meaning "lime." Calcium is a major component of limestone.

Each vertical column of the periodic table is called a **group**, or family. Elements within a group have similar chemical and physical properties. Note that each group is identified by a number and the letter A or B. For example, Group 2A contains the elements beryllium (Be), magnesium (Mg), calcium (Ca), strontium (Sr), barium (Ba), and radium (Ra). You will learn more about specific trends in the periodic table in Chapter 6.

Figure 2.13 The Periodic Table
Elements are arranged in the modern periodic table in order of atomic number.
Interpret Diagrams ***How many elements are in Period 2? In Group 2A?***

	1A	2A	3B	4B	5B	6B	7B	8B	8B	8B	1B	2B	3A	4A	5A	6A	7A	8A
1	1 H																	2 He
2	3 Li	4 Be											5 B	6 C	7 N	8 O	9 F	10 Ne
3	11 Na	12 Mg											13 Al	14 Si	15 P	16 S	17 Cl	18 Ar
4	19 K	20 Ca	21 Sc	22 Ti	23 V	24 Cr	25 Mn	26 Fe	27 Co	28 Ni	29 Cu	30 Zn	31 Ga	32 Ge	33 As	34 Se	35 Br	36 Kr
5	37 Rb	38 Sr	39 Y	40 Zr	41 Nb	42 Mo	43 Tc	44 Ru	45 Rh	46 Pd	47 Ag	48 Cd	49 In	50 Sn	51 Sb	52 Te	53 I	54 Xe
6	55 Cs	56 Ba	71 Lu	72 Hf	73 Ta	74 W	75 Re	76 Os	77 Ir	78 Pt	79 Au	80 Hg	81 Tl	82 Pb	83 Bi	84 Po	85 At	86 Rn
7	87 Fr	88 Ra	103 Lr	104 Rf	105 Db	106 Sg	107 Bh	108 Hs	109 Mt	110 Ds	111 Rg	112 Cn	113 Uut	114 Uuq	115 Uup	116 Uuh		118 Uuo

57 La	58 Ce	59 Pr	60 Nd	61 Pm	62 Sm	63 Eu	64 Gd	65 Tb	66 Dy	67 Ho	68 Er	69 Tm	70 Yb
89 Ac	90 Th	91 Pa	92 U	93 Np	94 Pu	95 Am	96 Cm	97 Bk	98 Cf	99 Es	100 Fm	101 Md	102 No

2.3 LessonCheck

ONLINE PROBLEMS

22. Compare How is a compound different from an element?

23. Compare How can you distinguish a substance from a mixture?

24. Identify What are chemical symbols and chemical formulas used for?

25. Explain What makes the periodic table such a useful tool?

26. Identify Name two methods that can be used to break down compounds into simpler substances.

27. Classify Classify each of these samples of matter as an element, a compound, or a mixture.
a. table sugar **c.** cough syrup
b. tap water **d.** nitrogen

28. Identify Write the chemical symbol for each of the following elements:
a. lead **c.** silver **e.** hydrogen
b. oxygen **d.** sodium **f.** aluminum

29. Identify Name the chemical elements represented by the following symbols:
a. C **c.** K **e.** Fe
b. Ca **d.** Au **f.** Cu

30. Identify What elements make up the pain reliever acetaminophen, chemical formula $C_8H_9O_2N$? Which element is present in the greatest proportion by number of particles?

31. Identify Name two elements that have properties similar to those of the element calcium (Ca).

Lesson Check Answers

22. Compounds can be broken down into simpler substances by chemical means, but elements cannot.

23. A substance has a fixed composition. The composition of a mixture may vary.

24. Chemical symbols are used to represent elements. Chemical formulas are used to represent compounds.

25. It allows you to compare the properties of the elements.

26. heating or an electric current

27. a. compound **c.** mixture
b. mixture **d.** element

28. a. Pb **c.** Ag **e.** H
b. O **d.** Na **f.** Al

29. a. carbon **c.** potassium **e.** iron
b. calcium **d.** gold **f.** copper

30. Carbon, hydrogen, oxygen, and nitrogen; hydrogen is present in the greatest proportion by number of atoms.

31. Any two: beryllium (Be), magnesium (Mg), strontium (Sr), barium (Ba), radium (Ra).

Evaluate

Informal Assessment

Ask students to explain in their own words why a chemical change might indicate that a substance is a compound. Then have students complete the 2.3 Lesson Check.

Reteach

Have students review their notes on elements, compounds, substances, and mixtures. Then, as a class, read through the flowchart in Figure 2.11. Have the class brainstorm examples of each classification and confirm their examples by using the flowchart. For any elements or compounds, have students use Table 2.2 or Figure 2.13 to identify the symbols or names for the elements or the elements that make up the compounds.

Answers

FIGURE 2.13 8; 6

LESSON 2.4

Key Objectives

2.4.1 DESCRIBE what happens during a chemical change.

2.4.2 IDENTIFY four possible clues that a chemical change has taken place.

2.4.3 DESCRIBE how the mass of the reactants and the mass of the products of a chemical reaction are related.

Additional Resources

Reading and Study Workbook, Lesson 2.4

Available Online or on Digital Media:

- Laboratory Manual, Labs 2 and 3
- Small-Scale Laboratory Manual, Lab 2

PearsonChem.com

Engage

CHEMISTRY & YOU Have students read the opening text and study the photograph. **Ask** What evidence do you see in the photograph that a chemical reaction has occurred? *(The burning part of the match has changed color.)*

Activate Prior Knowledge

Before students begin to read this lesson, review the concepts of physical change and chemical change. Have students predict why these concepts might be related to chemical reactions.

IN ACADEMIC STANDARDS for SCIENCE

C.1.3 Recognize observable indicators of chemical changes.

C.1.4 Describe physical and chemical changes at the molecular level.

C.1.6 Explain and apply the law of conservation of mass as it applies to chemical processes.

2.4 Chemical Reactions

C.1.3 Recognize chemical changes; **C.1.4** Physical and chemical changes; **C.1.6** Conservation of mass: chemical changes

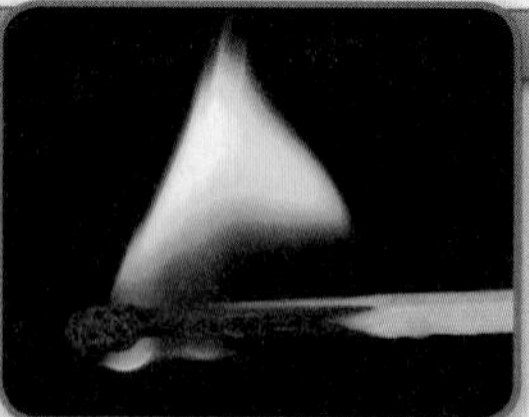

CHEMISTRY & YOU

Q: *What happened to the match?* Matches are often used to light candles on a cake. A match is usually lit at the tip and then burns down the match. So you better be quick, or your fingers will be burned by the lit match. A lit match is different than an unlit match. In this lesson, you will learn to recognize whether the burning match is a chemical change or physical change.

Key Questions

What always happens during a chemical change?

What are four possible clues that a chemical change has taken place?

How are the mass of the reactants and the mass of the products of a chemical reaction related?

Vocabulary

- chemical property
- chemical reaction
- reactant
- product
- precipitate
- law of conservation of mass

Chemical Changes

What always happens during a chemical change?

The compound formed when iron rusts is iron oxide (Fe_2O_3). Words such as *burn, rot, rust, decompose, ferment, explode,* and *corrode* usually signify a chemical change. The ability of a substance to undergo a specific chemical change is called a **chemical property.** Iron is able to combine with oxygen to form rust. So the ability to rust is a chemical property of iron. Chemical properties can be used to identify a substance. But chemical properties can be observed only when a substance undergoes a chemical change.

Figure 2.14 compares a physical change and a chemical change that can occur in a sample of charcoal. When charcoal is broken into smaller pieces, the change is a physical change. The substances present before the change are the same substances present after the change, although the charcoal pieces are not as large. Recall that during a physical change, the composition of matter never changes. **During a chemical change, the composition of matter always changes.** When the charcoal is heated and burned, a chemical change occurs. The substances in charcoal react with oxygen in the air to form other substances.

A chemical change is also called a chemical reaction. One or more substances change into one or more new substances during a **chemical reaction.** A substance present at the start of the reaction is a **reactant.** A substance produced in the reaction is a **product.** In the burning of charcoal, carbon and oxygen are the main reactants, and carbon dioxide is the main product.

Figure 2.14
Physical and Chemical Changes
Charcoal is used as a fuel in charcoal grills. **a.** Breaking the charcoal into smaller pieces is a physical change. **b.** Burning the charcoal is a chemical change.

READING SUPPORT

Build Study Skills: *Preview Visuals* Before you start reading the lesson, preview the visuals in Figure 2.16. Then write two questions you have about the visuals. After you finish reading, answer your questions.

Focus on ELL

1 CONTENT AND LANGUAGE Write the word *reaction* on the board. Explain that this word has both common and scientific meanings. State some nonscientific examples of the word. Then have students predict the scientific meaning of the word.

2 FRONTLOAD THE LESSON Prepare a beaker of boiling water on a hot plate. Prepare a second beaker with ethanoic acid. Add a small quantity of sodium hydrogen carbonate to the ethanoic acid. Have students identify the similarities and differences between the events occurring in each beaker.

3 COMPREHENSIBLE INPUT For Figure 2.15, rewrite each caption as a word equation to reinforce that a chemical reaction is taking place in each photograph. Assist students in answering the question associated with Figure 2.16 by asking them to list the similarities and differences between the two photos.

Figure 2.15 Chemical Change
Clues to chemical change often have practical applications.

Learn more about physical and chemical changes *online.* CONCEPTS IN ACTION

Production of a Gas
Bubbles of carbon dioxide gas form when an antacid tablet is dropped into a glass of water.

Color Change
When a test strip is dipped in a solution, the color change is used to determine the pH of the solution.

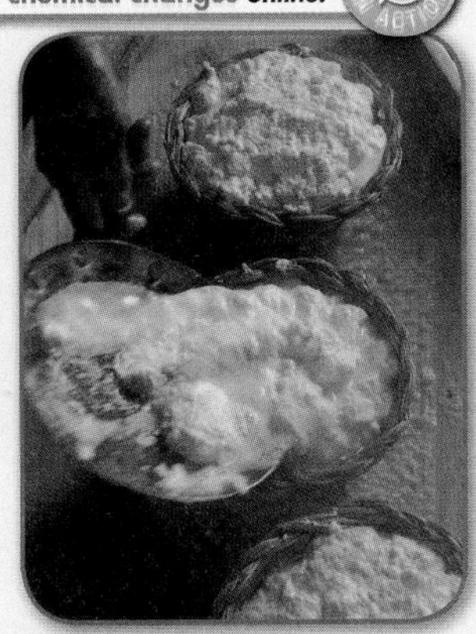

Formation of a Precipitate
One step in the production of cheese is a reaction that causes milk to separate into solid curds and liquid whey.

Recognizing Chemical Changes

What are four possible clues that a chemical change has taken place?

How can you tell whether a chemical change has taken place? There are four clues that can serve as a guide. **Possible clues to chemical change include a transfer of energy, a change in color, the production of a gas, or the formation of a precipitate.**

Every chemical change involves a transfer of energy. For example, energy stored in natural gas is used to cook food. When the methane in natural gas chemically combines with oxygen in the air, energy is given off in the form of heat and light. Some of this energy is transferred to and absorbed by food that is cooking over a lit gas burner. The energy causes chemical changes to take place in the food. The food may change color and brown as it cooks, which is another clue that chemical changes are occurring.

You can observe two other clues to chemical change while cleaning a bathtub. The ring of soap scum that can form in a bathtub is an example of a precipitate. A **precipitate** is a solid that forms and settles out of a liquid mixture. Some bathroom cleaners that you can use to remove soap scum start to bubble when you spray them on the scum. The bubbles are produced because a gas is released during the chemical change that is taking place in the cleaner.

If you observe a clue to chemical change, you cannot be certain that a chemical change has taken place. The clue may be the result of a physical change. For example, energy is always transferred when matter changes from one state to another. Bubbles form when you boil water or open a carbonated drink. The only way to be sure that a chemical change has occurred is to test the composition of a sample before and after the change. Figure 2.15 shows examples of practical situations in which different clues to chemical change are visible.

Q: *Are the changes that happen to a burning match chemical changes or physical changes? How do you know?*

Focus on ELL

4 LANGUAGE PRODUCTION Have students work in groups of four to complete the lab on page 51. Make sure each group has ELLs of varied language proficiencies so that more proficient students can help less proficient ones. Have students work according to their proficiency level.

BEGINNING: LOW/HIGH Provide students with colored markers to use to indicate their observations.

INTERMEDIATE: LOW/HIGH Provide students with a checklist they can use to better guide them through the procedure. Have students verbally respond to the Analyze questions.

Foundations for Reading

BUILD VOCABULARY Ask a student volunteer to explain the term *precipitation*. Point out that just as precipitation falls from the sky, a *precipitate* "falls" out of solution.

READING STRATEGY Preview the lesson by having students examine the photographs and diagrams and read the related captions. Encourage students to predict the answers to the questions presented in the captions.

Explain

Chemical Changes

CHEMISTRY & YOU Chemical changes; clues include a transfer of energy (heat is produced), production of a gas (smoke is produced), and a color change (the match darkens).

Explore

Recognizing Chemical Changes

PURPOSE Students will practice identifying chemical changes.

MATERIALS Bunsen burner, match, tongs, magnesium ribbon, 4 test tubes, solutions of 0.1*M* $AgNO_3$, 0.1*M* NaCl, 0.1*M* K_2CrO_4, and 3*M* H_2SO_4, mossy zinc, marble chip, spatulas, cobalt blue glass filters

SAFETY For Step c, wear gloves to avoid stains on your skin from silver nitrate.

CAUTION Students should not look at burning magnesium without the cobalt blue glass filters.

PROCEDURE As you do each step, have students identify the clues for chemical change. Emphasize that the clues point to a chemical change but do not confirm that a change has taken place.

a. Light a Bunsen burner. *(Heat and light are produced.)*

b. Using tongs, hold a piece of magnesium ribbon in a burner flame until it ignites. Remove from heat and observe. *(The product is a white powder; heat and light are produced.)*

c. Put 5 mL of silver nitrate solution in each of two test tubes. Add a small amount of sodium chloride solution to one tube and potassium chromate solution to the other. *(color change, formation of a precipitate)*

d. Put 5 mL of sulfuric acid in each of two test tubes. Add a piece of mossy zinc to one tube and a marble chip ($CaCO_3$) to the other. *(formation of a gas)*

LESSON 2.4

Explain

Conservation of Mass

USE VISUALS Direct students' attention to Figure 2.16. Explain that the photos are before and after views of a reaction. **Ask** Why are the containers covered? *(The covers prevent evaporation of the solutions and loss of mass.)*

Evaluate

Informal Assessment

Write this word equation on the board:

methane + oxygen → carbon dioxide + water + energy

Have students identify the reactants and products and one possible clue for a chemical change. Then have students complete the 2.4 Lesson Check.

Reteach

Write a word equation on the board for the reaction of magnesium and oxygen to produce magnesium oxide. **Ask** How is mass conserved in this reaction? *(The difference between the mass of the magnesium and the mass of the magnesium oxide is the mass of the oxygen with which the magnesium combines.)*

Conservation of Mass

How are the mass of the reactants and the mass of the products of a chemical reaction related?

When wood burns, substances in the wood combine with oxygen from the air. As the wood burns, a sizable amount of matter is reduced to a small pile of ashes. The reaction seems to involve a reduction in the amount of matter. But appearances can be deceiving. **During any chemical reaction, the mass of the products is always equal to the mass of the reactants.** Two of the products of burning wood—carbon dioxide gas and water vapor—are released into the air. When the mass of these gases is considered, the amount of matter is unchanged. Careful measurements show that the total mass of the reactants (wood and the oxygen consumed) equals the total mass of the products (carbon dioxide, water vapor, and ash).

Mass also holds constant during physical changes. For example, when 10 grams of ice melt, 10 grams of liquid water are produced. Similar observations have been recorded for all chemical and physical changes studied. The scientific law that reflects these observations is the law of conservation of mass. The **law of conservation of mass** states that in any physical change or chemical reaction, mass is conserved. Mass is neither created nor destroyed. The conservation of mass is more easily observed when a change occurs in a closed container, as in Figure 2.16.

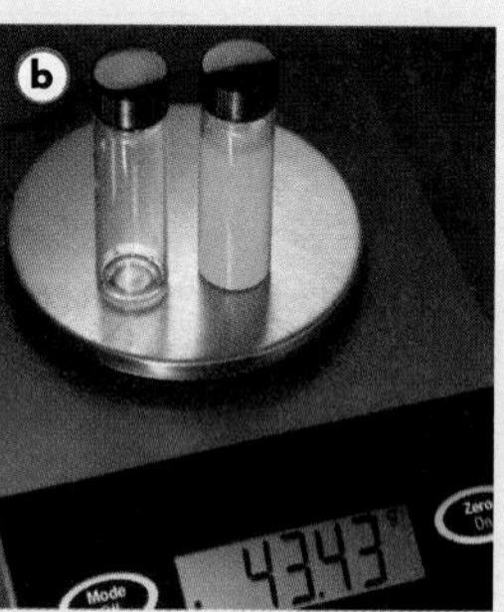

Figure 2.16
Conservation of Mass
When the liquids in **a.** are mixed, they react. The products are shown in **b.** None of the products are gases.
Analyze Data ***How do you know that a reaction took place and that mass was conserved during the reaction?***

2.4 LessonCheck

32. **Explain** How does a chemical change affect the composition of matter?

33. **List** Name four possible clues that a chemical change has taken place.

34. **Compare** In a chemical reaction, how does the mass of the reactants compare with the mass of the products?

35. **Compare** What is the main difference between physical changes and chemical changes?

36. **Classify** Classify the following changes as physical or chemical changes.
a. Water boils.
b. Salt dissolves in water.
c. Milk turns sour.
d. A metal rusts.

37. **Explain** According to the law of conservation of mass, when is mass conserved?

38. **Calculate** Hydrogen and oxygen react chemically to form water. How much water would form if 4.8 grams of hydrogen reacted with 38.4 grams of oxygen?

Lesson Check Answers

32. The composition of matter always changes during a chemical change.

33. a transfer of energy, a change in color, the production of a gas, the formation of a precipitate

34. The mass of the products is always equal to the mass of the reactants.

35. In a physical change, the chemical composition of a substance does not change. In a chemical change, the chemical composition of the reactants changes as one or more products form.

36. **a.** physical **c.** chemical
b. physical **d.** chemical

37. Mass is conserved in every physical change or chemical reaction.

38. 43.2 g

Small-Scale Lab

1 + 2 + 3 = Black!

Purpose

To make macroscopic observations of chemical reactions and use them to solve problems

Materials

- paper
- metric ruler
- reaction surface
- materials shown in grid
- pipette, medicine droppers, and spatulas

Procedure

1. Draw two copies of the grid on separate sheets of paper. Make each square in the grid 2 cm on each side.

2. Place a reaction surface over one of the grids. Use the second grid as a data table to record your observations.

3. Use the column and row labels to determine which materials belong in each square. Depending on the material, add one drop, one piece, or a few grains.

4. Stir each mixture by forcing air from an empty pipette as directed by your teacher.

	NaClO	H_2O_2	$CuSO_4$
KI			
KI + Starch			
KI + Paper			
KI + Cereal			

Analyze

Using your experimental data, record the answers to the following questions below your data table.

1. Describe What color is a mixture of sodium hypochlorite (NaClO) and potassium iodide (KI)?

2. Describe What happens when you mix NaClO, KI, and starch?

3. Compare What do NaClO, H_2O_2, and $CuSO_4$ have in common?

4. Compare What substance is found in both paper and cereal? How do you know?

5. Predict If you used NaClO instead of $CuSO_4$ in reactions other than the reaction with KI and starch, would you expect the results to always be identical? Explain your answer.

You're the Chemist

The following small-scale activities allow you to develop your own procedures and analyze the results.

1. Analyze Data NaClO is a bleaching agent. Such agents are used to whiten clothes and remove stains. Use different color marker pens to draw several lines on a piece of white paper. Add one drop of NaClO to each line. What happens? Try inventing a technique that you can use to make "bleach art."

2. Design an Experiment Design and carry out an experiment to see which foods contain starch.

3. Design an Experiment Read the label on a package of iodized salt. How much KI does iodized salt contain? Design an experiment to demonstrate the presence of KI in iodized salt and its absence in salt that is not iodized.

4. Design an Experiment Antacid tablets often contain starch as a binder to hold the ingredients in the tablet together. Design and carry out an experiment to explore various antacid tablets to see if they contain starch.

Small-Scale Lab

OBJECTIVE In this activity, students will begin to familiarize themselves with small-scale equipment and methodology. They will also observe and analyze some chemical reactions

SKILLS FOCUS Observing, classifying

PREP TIME 1 hour

CLASS TIME 40 minutes

MATERIALS paper, metric ruler, reaction surface, pipette, dropper, spatulas, KI, liquid starch, cereal, NaClO, H_2O_2, $CuSO_4$, foods, iodized and non-iodized salt, antacid tablets, color marker pens

ADVANCE PREPARATION

Solution	Preparation
0.1*M* KI	4.2 g KI in 250 mL water
0.2*M* $CuSO_4$	12.5 g $CuSO_4 \cdot 5H_2O$ in 250 mL water
1% NaClO	50 mL household bleach in 200 mL water
3% H_2O_2	Use undiluted household hydrogen peroxide
20% starch	50 mL liquid starch in 200 mL water

TEACHING TIPS Demonstrate how to use a pipette as a stirrer. This information, as well as a discussion of reaction surfaces, can be found in the *Small-Scale Chemistry Laboratory Manual.* Explain that using a mixture of KI and NaClO to test for starch is an example of qualitative analysis.

EXPECTED OUTCOME

	NaClO	H_2O_2	$CuSO_4$
KI	yellow	yellow	brown ppt
KI + starch	black	black	black
KI + paper	black	black	black
KI + cereal	black	black	black

FOR ENRICHMENT Have students design and carry out an experiment to quantify the amount of KI in iodized salt.

ANALYZE AND CONCLUDE

1. yellow
2. The mixture turns a blue-black color.
3. They all turn a mixture of KI and starch black.
4. Starch; both turn blue-black, which suggests the presence of starch.
5. The results may be the same in reactions that are similar to the one with KI and starch but different in other reactions.

YOU'RE THE CHEMIST

1. The color fades. A picture can be drawn with colored ink. Areas can be treated with NaClO to bleach parts of the picture.
2. Add KI + NaClO to various foods. A black color indicates the presence of starch.
3. Most table salt contains 0.01% KI. Wet only a portion of a small pile of salt with starch. Add $CuSO_4$ or H_2O_2. A black color indicates the presence of KI.
4. If an antacid tablet contains starch, it will turn black when treated with KI + NaClO.

Answers

FIGURE 2.16 Change of color and formation of a precipitate; the balance reading did not change.

CHEMISTRY & YOU

CHEMISTRY & YOU Draw students' attention to the container of recyclables in the photo. Point out that most communities and businesses now actively participate in recycling, particularly of glass, paper, and plastic products. Explain that recycling guidelines vary from municipality to municipality, depending upon how the garbage/recyclables collection process is set up in their community. Some cities might mandate recycling as a part of citywide garbage pickup. Other communities might rely on private haulers to collect garbage and recyclables.

Engage students in a discussion of the types of recycling they commonly do in their homes or workplaces, and create a list of the types of recyclables their hauler will accept. For example, some waste management authorities will accept only glass, metal cans, certain plastics, and newspapers. Others might accept these items as well as pasteboard, cardboard, and magazines. Still other communities accept all of the previously mentioned items as well as used motor oil, lead batteries, and appliances.

Pose the following question to students: *What factors determine the types of materials that a community waste management authority or private hauler collects for recycling?*

You may need to guide students in the following ways:

- Some recycled materials have stronger market demands than other materials.
- Certain materials, such as used motor oil and lead batteries, are or contain environmental hazards.
- Equipment upgrades are often necessary if a new type of recyclable is added to the list or if the collection method changes from drop-off centers to curbside recycling.

CHEMISTRY & YOU: GREEN CHEMISTRY

Recycled Mixtures

You may be used to sorting recyclables into separate bins. But now, a new process known as single-stream recycling allows you to put all recyclables into a single container. The separation is then done at a plant known as a Material Recovery Facility (MRF), which uses machines to separate paper, glass, metal, and plastics based on their properties.

1. FEEDER Unsorted paper, glass, plastic, and metal are loaded onto a conveyer belt. In some locations, workers pull any mixed-in trash off the belt and discard it.

2. SCREENING MACHINES A series of fast, rotating rubber blades, called star screens, separate out cardboard, newspaper, and other paper from the rest of the material. The plastic, glass, and metals fall through the screens onto another conveyer belt.

3. ELECTROMAGNET Heavy-duty electro-magnets grab items made of ferrous metals, such as iron and steel, off the conveyer belt.

4. EDDY CURRENT A rotating magnetic field creates an eddy current that repels nonmagnetic metals, such as aluminum, and ejects them from the rest of the materials.

5. GLASS CRUSHER The remaining materials move through another series of screens designed to break glass and remove it from the stream.

6. OPTICAL SCANNER Optical scanners use infrared technology to recognize different types or colors of plastics and can trigger blasts of air to separate the plastics into bins.

52

21st Century Learning To be successful in the 21st century, students need skills and learning experiences that extend beyond subject matter mastery. This project helps students build the following 21st Century Skills: *Global Awareness; Financial, Economic, Business and Entrepreneurial Literacy; Critical Thinking and Problem Solving; Communication and Collaboration; Information Literacy; Media Literacy; Productivity and Accountability.*

BUILD A COMPOST PILE Pose the following challenge to your students. *A rural community in your state has received a grant to begin a recycling program, and approaches your company to develop a recycling plan.* Form a "company" of four to six students to develop a plan that is suited to a rural community with a widely dispersed population and present your plan as an audio slideshow on digital media. Plans should include a determination as to whether curbside or drop-off recycling is most appropriate for the community, a comparison of the types of sorting systems available, a list of the most accessible markets for recycled materials, and a determination of the financial and environmental impact of enacting the plan.

Explain

START A CONVERSATION Engage students in a conversation of the different methods of recycling. Have them consider the steps needed to recycle different materials. Point out the following:

- Soda cans are an alloy (homogeneous mixture) of metals, but the tops do not have the same percentages of metals as the rest of the can. After the can is melted, more of some metals must be added before using the recycled metal to make new cans.
- One of the misconceptions of recycling is that paper products degrade quickly in a landfill. In fact, all products degrade slowly because it is an anaerobic (low oxygen) environment.
- Yard trimmings should not be placed in garbage that will go to landfills because they take up so much room. Instead, yard trimmings are easily composted.

Extend

Connect to BIOLOGY Remind students that decomposers are nature's recyclers. Explain to students that vermicomposting is a type of composting that uses worms to break down yard and food wastes. Explain that worms placed in a compost pile eat and digest the materials, and then excrete a nutrient-rich soil. Have students determine the physical and chemical changes in the materials. *(Physical change: During digestion, the food passes through the worm's gizzard where stones the worm has previously ingested grind the food into smaller pieces. Chemical change: The food then moves into the intestine where it is chemically broken down.)*

Differentiated Instruction

L1 ENGLISH LANGUAGE LEARNERS Have students translate the words *recycle, compost, plastic,* and *landfill* into their native language and teach the class how to pronounce each word. As a class use creative ways such as act it out or photographs to teach the meaning of these words.

LPR LESS PROFICIENT READERS Divide the class into groups of 3 to 4 students. Have students independently read a paragraph or caption on the page and then discuss it with others in their group. Then have them independently write a sentence summarizing what they have read. Repeat for the remaining paragraphs and captions.

L3 ADVANCED STUDENTS Provide students with a list of plastic recycling codes. Have students choose one type of plastic from the list and research the specific method used to recycle it. Then have them make a small poster explaining the method. Students should include details about physical and chemical changes used in the recycling process as well as an economic summary outlining cost associated with the process.

Answers

TAKE IT FURTHER

1. Magnetism, color, conductivity.
2. Single-stream recycling makes recycling easier and more convenient, so people may be more likely to recycle. Single-stream recycling centers may accept more types of materials for recycling.

STUDY GUIDE

Evaluate

Review and Assessment Materials

CHAPTER PROBLEM SET Have students practice more mixture and reaction problems using the Chapter 2 online Problem Set.

DIRECTED VIRTUAL LAB Have students complete the Virtual Lab *Chemical Reactions* to help reinforce the properties of chemical changes. In the post-lab exercise, students will analyze the data and draw conclusions based on the results of the experiment.

CONCEPTS IN ACTION If students have trouble understanding the difference between physical and chemical changes, ask them to view the Concepts in Action animation, which presents the concepts in a familiar context. Have students complete the questions after the animation.

Study Tip

ORGANIZE NEW INFORMATION Have students learn new vocabulary terms by making vocabulary flashcards. Students should write a vocabulary term on one side of a card and its definition on the other side. The cards can be used by students to quiz each other.

Understanding by Design

Performance Tasks

SECRET MATTER Direct students to create a secret mixture and record the items used in the mixture. Have students exchange mixtures with a partner. Students should determine a method to separate their partner's mixture into its various components. Have students submit their findings and written descriptions of the substance properties and the processes used to separate the mixture, along with a copy of the ingredient list for the mixture they created.

MATTER OF FACT! Have students design and carry out a science experiment about physical and chemical changes. Students should submit a written report that includes the purpose, the materials, the procedure, the expected outcome, safety issues, and an Analyze and Conclude question section. The procedure should demonstrate at least one physical change and at least one chemical change.

2 Study Guide

BIGIDEA
CHEMISTRY AS THE CENTRAL SCIENCE

Physical properties, such as melting point and boiling point, and chemical properties, such as whether a substance will corrode or burn, are used to describe matter. Matter may be made of elements or compounds. Elements and compounds are pure substances but can be physically combined to make heterogenous or homogeneous mixtures. These different forms of matter may undergo physical or chemical changes.

2.1 Properties of Matter

Every sample of a given substance has identical intensive properties because every sample has the same composition.

Three states of matter are solid, liquid, and gas.

Physical changes can be classified as reversible or irreversible.

- mass (34)
- volume (34)
- extensive property (34)
- intensive property (34)
- substance (35)
- physical property (35)
- solid (36)
- liquid (36)
- gas (37)
- vapor (37)
- physical change (37)

2.2 Mixtures

Mixtures can be classified as heterogeneous mixtures or as homogeneous mixtures, based on the distribution of their components.

Differences in physical properties can be used to separate mixtures.

- mixture (38)
- heterogeneous mixture (39)
- homogeneous mixture (39)
- solution (39)
- phase (39)
- filtration (40)
- distillation (40)

2.3 Elements and Compounds

Compounds can be broken down into simpler substances by chemical means, but elements cannot.

If the composition of a material is fixed, the material is a substance. If the composition may vary, the material is a mixture.

Chemists use chemical symbols to represent elements, and chemical formulas to represent compounds.

The periodic table allows you to easily compare the properties of one element (or a group of elements) to another element (or group of elements).

- element (42)
- compound (42)
- chemical change (43)
- chemical symbol (45)
- periodic table (46)
- period (46)
- group (47)

2.4 Chemical Reactions

During a chemical change, the composition of matter always changes.

Four possible clues to chemical change include a transfer of energy, a change in color, the production of a gas, or the formation of a precipitate.

During any chemical reaction, the mass of the products is always equal to the mass of the reactants.

- chemical property (48)
- chemical reaction (48)
- reactant (48)
- product (48)
- precipitate (49)
- law of conservation of mass (50)

FOCUS ON ELL

5 ASSESS UNDERSTANDING Have ELL students keep a content log to track their progress during this chapter. Provide suggestions for the log and follow up often with students to determine the material that needs further explanation.

BEGINNING: LOW/HIGH Provide students with two colors of small sticky notes, one color indicating they understand a topic, the other color indicating that they do not understand the topic. Have students mark each major section with the appropriately colored sticky note. Review areas of concern with students.

INTERMEDIATE: LOW/HIGH Provide students with a two-column chart with headings that read *What I Understand* and *What I Do Not Understand*. Have students complete the chart by using symbols, page numbers, simple sentences, or their native languages.

ADVANCED: LOW/HIGH Group students in small topic circles and lead a discussion of the common areas of struggle. Allow extra time for ELL students to explain or to show the concepts requiring additional help.

2 Assessment

★ Solutions appear in Appendix E

Lesson by Lesson

2.1 Properties of Matter

39. Describe the difference between an extensive property and an intensive property and give an example of each.

40. List three physical properties of copper.

41. Name two physical properties that could be used to distinguish between water and ethanol.

★42. Name one physical property that could not be used to distinguish chlorine from oxygen.

43. What is the physical state of each of these materials at room temperature?

a. gold
b. gasoline
c. oxygen
d. neon
e. olive oil
f. sulfur
g. mercury

★44. Fingernail-polish remover (mostly acetone) is a liquid at room temperature. Would you describe acetone in the gaseous state as a vapor or a gas? Explain your answer.

45. Compare the arrangements of individual particles in solids, liquids, and gases.

46. Use Table 2.1 to identify four substances that undergo a physical change if the temperature is reduced from 50°C to −50°C. What is the physical change that takes place in each case?

★47. Explain why sharpening a pencil is a different type of physical change than freezing water to make ice cubes.

2.2 Mixtures

48. What is the difference between homogeneous mixtures and heterogeneous mixtures?

49. How many phases does a solution have? Explain your answer.

★50. Classify each of the following as a homogeneous or heterogeneous mixture.

a. chocolate-chip ice cream
b. green ink
c. cake batter
d. cooking oil
e. granite rock
f. salt water
g. paint
h. a silver ring

51. What is the goal of a distillation? Describe briefly how this goal is accomplished.

2.3 Elements and Compounds

52. How could you distinguish an element from a compound?

★53. Classify the following materials as an element, compound, or mixture. Give reasons for your answers.

a. table salt (NaCl)
b. salt water
c. sodium (Na)

54. Describe the relationship between the three items in each of the following groups. Identify each item as an element, compound, or mixture.

a. hydrogen, oxygen, and water
b. nitrogen, oxygen, and air
c. sodium, chlorine, and table salt
d. carbon, water, and table sugar

55. Name the elements found in each of the following compounds.

a. ammonia (NH_3)
b. potassium oxide (K_2O)
c. sucrose ($C_{12}H_{22}O_{11}$)
d. calcium sulfide (CaS)

56. Not all element names come from English or Latin words. The symbol for tungsten is W from the German word *wolfram*. The symbol for mercury is Hg from the Greek word *hydragyrum*. Use the symbols W and Hg to explain the system of symbols for elements.

★57. What does the formula H_2O tell you about the composition of water?

54. a. Hydrogen and oxygen are the elements that make up the compound water.
b. Nitrogen and oxygen are elements present in the mixture air.
c. Sodium and chlorine are elements in the compound sodium chloride (table salt).
d. The element carbon and the compound water are the final products of heating table sugar.

55. a. nitrogen, hydrogen **b.** potassium, oxygen **c.** carbon, hydrogen, oxygen **d.** calcium, sulfur

56. In W, the single letter is capitalized. In Hg, the first letter is a capital and the second letter is lowercase.

57. The compound water contains two parts hydrogen to one part oxygen.

Evaluate

ASSESSMENT

Answers

LESSON 2.1

39. An extensive property depends on the amount of matter; an intensive property depends on the type of matter. Mass and volume are extensive properties. Color and hardness are intensive properties.

40. Answers could include reddish-yellow color, conductor of heat and electricity, malleable, melting point of 1084°C, and boiling point of 2562°C.

41. melting point and boiling point

42. State; both are gases.

43. a. solid **b.** liquid **c.** gas **d.** gas **e.** liquid **f.** solid **g.** liquid

44. A vapor; the term *vapor* is used to refer to the gaseous state of a substance that normally exists as a liquid or solid at room temperature.

45. The particles in a solid are packed tightly together in an orderly arrangement. Particles in a liquid are in close contact, but not in a orderly arrangement. The particles in a gas are relatively far apart.

46. Chlorine condenses. Mercury, bromine, and water freeze.

47. Sharpening a pencil is an irreversible change. Making ice cubes is a reversible change.

LESSON 2.2

48. Heterogeneous mixtures have a nonuniform composition with two or more phases. Homogeneous mixtures have a uniform composition.

49. One; Solutions are homogeneous mixtures with uniform composition throughout.

50. a. heterogeneous **b.** homogeneous
c. depends on how well the batter is mixed
d. homogeneous **e.** heterogeneous
f. homogeneous **g.** heterogeneous
h. homogeneous

51. The goal is to separate the components of a solution. The solution is boiled to produce a vapor, which is condensed to a liquid. Dissolved solids are left behind.

LESSON 2.3

52. Compounds can be separated by chemical means into simpler substances. Elements cannot.

53. a. Compound; formula indicates it's made of more than one element, but only one compound so it's not a mixture. **b.** Mixture; made of two compounds, salt and water. **c.** Element; formula indicates it's made of only one element.

ASSESSMENT

Evaluate

Answers

58. Sample answer: The table is set up so that chemical properties of elements recur at regular intervals.

LESSON 2.4

59. When heated, sulfur and iron react and form iron sulfide. The composition of the reactants in a chemical change is different from the composition of the products. In a physical change, the chemical composition of a sample doesn't change.

60. **a.** physical **b.** chemical (color change) **c.** chemical (production of a gas) **d.** physical

61. chemical property

62. 18 g of water

UNDERSTAND CONCEPTS

63. ethanol

64. neon

65. sulfur

66. sulfur

67. Mass is an extensive property, which depends only on the amount of matter in the sample, not on the composition of the sample.

68. Malleability is an intensive property, which depends on the type of matter in a sample.

69. If no temperature is specified, we can assume that the state is being given at room temperature.

70. The particles in solids are packed tightly together. The particles in a gas are spaced relatively far apart.

71. kitchen, mixtures; park, mixtures

72. **a.** heterogeneous mixture
b. compound
c. homogeneous mixture
d. heterogeneous mixture

73. **a.** physical **c.** physical **e.** chemical
b. physical **d.** physical

74. **a.** Both are elements and solids at room temperature; they are a different color and have different melting points and boiling points.
b. Both are clear liquids at room temperature; distilled water is a compound and saltwater is a mixture.
c. Both are white, solid compounds. Sugar is composed of carbon, hydrogen, and oxygen. Salt is composed of sodium and chlorine.

75. **a.** element **c.** mixture **e.** mixture
b. compound **d.** element **f.** compound

58. Look up the word *periodic* in the dictionary. Propose a reason for the naming of the periodic table.

2.4 Chemical Reactions

59. Use the word equation below to explain how a chemical change differs from a physical change.

$$\text{iron} + \text{sulfur} \xrightarrow{\text{heat}} \text{iron sulfide}$$

***60.** Classify each of the following as a physical or chemical change. For any chemical change, list at least one clue to support your answer.
a. A copper wire is bent.
b. Charcoal burns in a grill.
c. Bread dough rises when yeast is added.
d. Sugar dissolves in water.

61. Which type of property cannot be observed without changing the composition of a substance?

***62.** When ammonium nitrate (NH_4NO_3) explodes, the products are nitrogen, oxygen, and water. When 40 grams of ammonium nitrate explode, 14 grams of nitrogen and 8 grams of oxygen form. How many grams of water form?

Understand Concepts

Use the data table to answer Questions 63–66.

Substance	Color	Melting point (°C)	Boiling point (°C)
Bromine	Red-brown	−7	59
Chlorine	Green-yellow	−101	−34
Ethanol	Colorless	−117	78
Mercury	Silvery-white	−39	357
Neon	Colorless	−249	−246
Sulfur	Yellow	115	445
Water	Colorless	0	100

63. Which colorless substance is a liquid at −30°C?

64. Which colorless substance is a gas at 60°C?

65. Which substance is a solid at 7°C?

***66.** As the temperature rises, which solid will melt before mercury boils?

***67.** Explain why mass cannot be used as a property to identify a sample of matter.

68. Is malleability an extensive property or an intensive property? Explain.

69. The state of a substance can change when the substance is heated or cooled. So what does it mean to say that a certain substance is a solid, liquid, or gas?

***70.** Use the arrangement of particles in solids and gases to explain why solids are not as easy to compress as gases.

71. You are standing in a kitchen and then in the middle of a park. When you view your surroundings in each location, do you see mostly elements, compounds, or mixtures?

72. Identify each of the following items as a mixture or compound. Classify the mixtures as homogeneous or heterogeneous.
a. raw egg
b. ice
c. gasoline
d. blood

73. Classify the following properties of the element silicon as chemical or physical properties:
a. blue-gray color
b. brittle
c. doesn't dissolve in water
d. melts at 1410°C
e. reacts vigorously with fluorine

74. How are the items in each of the following pairs similar? How are they different?
a. copper and silver
b. distilled water and salt water
c. table sugar and table salt

75. Identify each of the following as an element, compound or mixture.
a. iron
b. distilled water
c. laundry detergent
d. sulfur
e. chicken broth
f. sodium fluoride

*76. Describe clues you might observe during the following events that could support the conclusion that a chemical change is occurring.
  a. An antacid tablet is dropped into water.
  b. A ring of scum forms around a bathtub.
  c. Iron rusts.
  d. A firecracker explodes.
  e. Bubbles form when hydrogen peroxide is poured onto an open wound.
  f. A hamburger cooks.

77. In photograph A, a coil of zinc metal is in a solution of sulfuric acid. In photograph B, a yellow solution of sodium chromate is being added to a colorless solution of silver nitrate. What clues in the photographs indicate that a chemical change is probably occurring?

A.

B.

78. Classify each of the following as a chemical change or a physical change.
  a. Plastic drink bottles are burned in an incinerator to generate electricity.
  b. Digesting a cereal bar.
  c. Water in a rain puddle evaporates.
  d. Slicing a tomato for a sandwich.
  e. Plastic drink bottles are recycled to make fiberfill for ski jackets.

79. Explain why the production of a gas does not always mean that a chemical reaction has occurred.

*80. The wax seems to disappear as a candle burns. How can the law of conservation of mass apply to this reaction?

Think Critically

*81. **Apply Concepts** Devise a way to separate sand from a mixture of charcoal, sand, sugar, and water.

82. **Sequence** Assume that water, mercury, and gallium are all at 40°C. As the temperature drops, which substance will freeze first? Which will be the last to freeze?

83. **Apply Concepts** A change in odor can also be a clue that a chemical change has occurred. Describe at least one situation in which you might be likely to detect such a change in odor in a kitchen.

84. **Explain** Explain why this statement is false. "Because there is no change in composition during a physical change, the appearance of the substance will not change."

*85. **Interpret Graphs** The mass of the elements iron and oxygen in four samples of a rust-colored substance was measured in grams (g). The amount of iron and oxygen in each sample is shown on the graph.

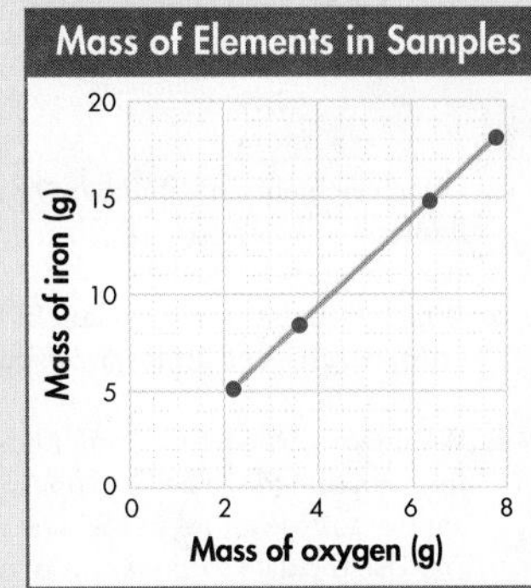

  a. Do you think all four samples are the same compound? Explain.
  b. Another sample of similar material was found to contain 9.9 grams of iron and 3.4 grams of oxygen. Is this sample the same substance as the other four samples? Explain.

86. **Explain** When powdered iron is left exposed to the air, it rusts. Explain why the mass of the rust is greater than the mass of the powdered iron.

87. **Explain** Discuss the statement "A gas requires a container, but a solid is its own container."

Evaluate

Answers

76. **a.** Gas is produced.
b. formation of a precipitate
c. color and texture change
d. energy change, odor change
e. Gas is produced.
f. energy change, color change, odor change

77. In photograph A, bubbles indicate the production of a gas. In photograph B, there is a color change and a precipitate.

78. **a.** chemical **c.** physical **e.** physical
b. chemical **d.** physical

79. A gas may be released during a physical change. For example, bubbles form when water boils.

80. The wax appears to disappear because the products of the reaction—carbon dioxide and water vapor—are colorless gases.

THINK CRITICALLY

81. Add sufficient water to dissolve all of the sugar. Separate the charcoal and sand from the sugar water by filtration. Large pieces of charcoal could be separated on the basis of color. Small pieces of charcoal could be burned.

82. Gallium will freeze first; mercury will freeze last.

83. Smelling something burning is one possible answer.

84. The appearance of a substance will change during a change of state, which is a physical change.

85. **a.** Yes; because the graph is a straight line, the proportion of iron to oxygen is a constant, which is true for a compound.
b. No; a point for the values given wouldn't fall on the line. The mass ratio of iron to oxygen is different.

86. Iron rusts when it reacts with oxygen in the air to form an oxide (Fe_2O_3). The mass of the rust is the sum of the mass of the iron and the mass of the oxygen that combined with the iron.

87. A gas expands to fill any space; a gas has no shape or volume without a container. A solid has a definite shape and volume; a solid doesn't need a container to maintain its shape and volume.

ASSESSMENT

CHEMYSTERY **START A CONVERSATION** After students have read through the CHEMystery, call on volunteers to describe how matter changes in each of the three scenarios. **Ask** How can you determine whether a change is physical or chemical? *(You consider whether the composition of matter is the same before and after the change. If the composition changes, matter has changed chemically.)*

Have students think about water boiling to form water vapor. **Ask** Is the formation of bubbles always a sign that a physical change is occurring? *(no)* **Ask** What is an example of bubbles occurring during a chemical change? *(Sample answer: Bubbles might form during the production of a gas, as when vinegar and baking soda react.)*

Have students think about leaves changing color in the fall. **Ask** Is color change always a sign that a chemical change is occurring? *(no)* **Ask** What is an example of a color change during a physical change? *(Sample answer: You can change the color of your room by painting it.)* Point out that even though paint is a chemical substance, the color change of the wall is not the result of a chemical reaction between the new paint and the old paint. The change occurs because a new mixture of pigments is applied to the wall.

CHEMYSTERY ANSWERS

94. The physical changes are reversible. The water in the geyser can change from steam to liquid water and the water in the glacier can freeze and melt.

95. **BIGIDEA** If rangers have knowledge of physical and chemical changes then they will have a better understanding of the changes they see occurring in the park. This might help them assess whether the changes are normal or possibly harmful.

Enrichment

*** 88. Interpret Graphs** Five elements make up 98% of the mass of the human body. These elements are oxygen (61%), carbon (23%), hydrogen (10.0%), nitrogen (2.6%), and calcium (1.4%). Compare these data with those in the pie graph below, which shows the five most abundant elements by mass in Earth's crust, oceans, and atmosphere.

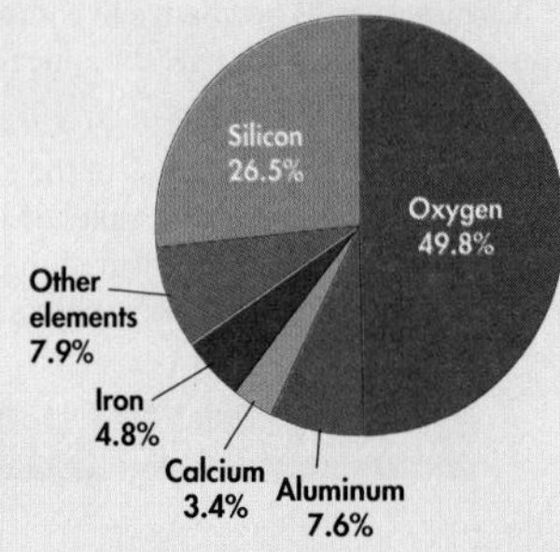

a. Which elements are abundant both in the human body and Earth's crust, oceans, and atmosphere?
b. Which elements are abundant in Earth's crust, oceans, and atmosphere, but not in the human body?
c. Would you expect the compounds found in the human body to be the same as or different from those found in rocks, seawater, and air? Use the data to explain your answer.

89. Evaluate Each day you encounter some chemical changes that are helpful and some that are harmful to humans or the environment. Cite three examples of each type. For each example, list the clues that identified the change as a chemical change.

90. Interpret Tables Use Table 2.1 on page 35 to answer this question.
a. Which substances in the table are in the liquid state at 125°C?
b. Use the physical properties of one of these substances to explain how you figured out the answer to Question 90a.
c. The substances in the table are listed in order of increasing melting point. Propose another way that these data could be arranged.

Write About Science

91. Explain Write a paragraph in support of this statement: "Dry tea is a mixture, not a substance." Include at least two pieces of evidence to support your argument.

92. Explain Lavoisier proposed the law of conservation of mass in 1789. Write a paragraph describing, in general, what Lavoisier must have done before he proposed this law. Use what you have learned about the scientific method.

93. Connect to the BIGIDEA Compare elements and compounds by saying how they are alike. Contrast elements and compounds by describing how they are different.

CHEMYSTERY

Which One Is Not Like the Others?

The eruption of geysers in Yellowstone National Park is caused by a physical change. Underground water is heated to temperatures hot enough to turn the water into steam. This steam causes an increase in pressure underground. In the cases of geysers, the steam is at first unable to escape. But, when the pressure reaches a critical level, an eruption occurs, producing the geyser.

The shaping of valleys by melt-refreeze cycles is also a physical change. The melting and refreezing of water is a physical change, and as the earth underneath the glacier is broken apart and moved, this is also a physical change.

The changing of the color of leaves is a chemical change. As the temperature and hours of sunlight change in the fall, chemical changes occur. Therefore, leaves changing color, a chemical change, is not like the physical changes of geysers erupting or glaciers moving.

94. Identify Are the physical changes in the mystery reversible or irreversible physical changes? Explain your answer.

95. Connect to the BIGIDEA Why would it be important for rangers at national parks to have knowledge of physical and chemical changes?

ENRICHMENT

88. **a.** oxygen and calcium
b. silicon, aluminum, and iron
c. Different; the second most abundant element in Earth's crust, silicon, is not present in the human body, and the second most abundant element in the human body, carbon, is not among the most abundant elements in Earth's crust. If the elements are different, then the compounds must also be different.

89. Many answers are possible. Cooking food (color change, odor change) is an example of a helpful change; the burning of gasoline in a gasoline engine (production of a gas, energy change) is an example of a harmful change.

90. **a.** mercury and sulfur
b. Sulfur melts at 113°C and boils at 445°C. Between 113°C and 445°C, it exists as a liquid. Mercury melts at −39°C and boils at 357°C. Between these temperatures, it exists as a liquid.
c. Possibilities include: by color, by boiling point, or in alphabetical order.

WRITE ABOUT SCIENCE

91. Answers will vary. Possible evidence includes the fact that only some substances in the tea dissolve in water, and tea is sold both with and without caffeine.

92. Before he proposed the law of conservation of mass, Lavoisier would have done a series of experiments in which he measured the masses of reactants and products.

93. Elements and compounds are alike in that they both are substances of fixed composition. They are different in that elements cannot be separated into simpler substances through chemical changes.

Standardized Test Prep for Indiana

Standards: C.1.1, C.1.5

Select the choice that best answers each question or completes each statement.

1. Which of the following is not a chemical change?
   (A) paper being shredded
   (B) steel rusting
   (C) charcoal burning
   (D) a newspaper yellowing in the sun
2. Which phrase best describes an apple?
   (A) heterogeneous mixture
   (B) homogeneous compound
   (C) heterogeneous substance
   (D) homogeneous mixture
3. Which element is paired with the wrong symbol?
   (A) sulfur, S
   (B) potassium, P
   (C) nitrogen, N
   (D) calcium, Ca
4. Which of these properties could not be used to distinguish between table salt and table sugar?
   (A) boiling point
   (B) melting point
   (C) density
   (D) color
5. The state of matter characterized by a definite volume and an indefinite shape is a
   (A) solid.
   (B) liquid.
   (C) mixture.
   (D) gas.

The lettered choices below refer to Questions 6–9. A lettered choice may be used once, more than once, or not at all.

(A) compound
(B) heterogeneous mixture
(C) element
(D) homogeneous mixture

Which description correctly identifies each of the following materials?

6. air
7. carbon monoxide
8. zinc
9. mushroom pizza

Tips for Success

Using Models To answer some test questions, you will be asked to use visual models. At first the models may look very similar. Decide which information will help you answer the question. The number of particles, their colors, or their shapes may or may not be important.

Use the atomic windows to answer Question 10.

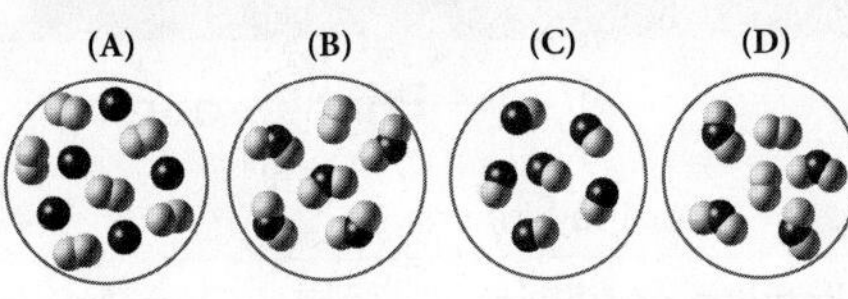

10. The species in window A react. Use the law of conservation of mass to determine which window best represents the reaction products.

Use the data table to answer Questions 11–14.

Mass of magnesium (g)	Mass of oxygen (g)	Mass of magnesium oxide (g)
5.0	3.3	8.3
6.5	(a)	10.8
13.6	9.0	(b)
(c)	12.5	31.5

11. Magnesium metal burns vigorously in oxygen to produce the compound magnesium oxide. Use the law of conservation of mass to identify the masses labeled (a), (b), and (c) in the table.
12. Use the data in the completed table to construct a graph with mass of magnesium on the x-axis and mass of magnesium oxide on the y-axis.
13. How many grams of magnesium oxide form when 8.0 g of magnesium are burned?
14. How many grams of magnesium and oxygen react to form 20.0 g of magnesium oxide?

If You Have Trouble With . . .

Question	1	2	3	4	5	6	7	8	9	10	11	12	13	14
See Lesson	2.4	2.2	2.3	2.1	2.1	2.3	2.3	2.3	2.3	2.4	2.4	2.4	2.4	2.4

STP Answers

1. a
2. a
3. b
4. d
5. b
6. D
7. A
8. C
9. B
10. b
11. **(a)** is 4.3 g; **(b)** is 22.6 g; **(c)** is 19.0 g
12.

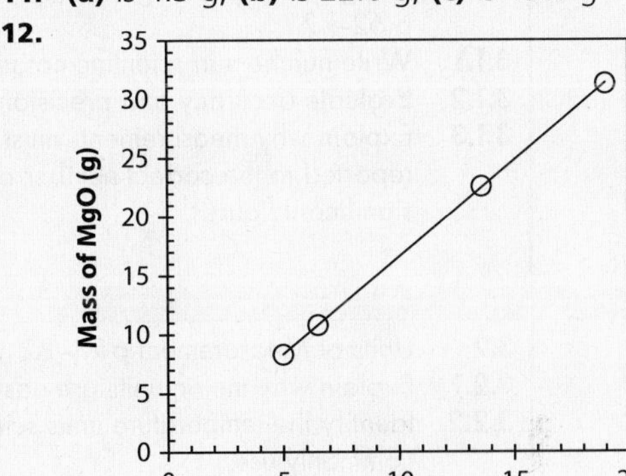

13. 13.3 g magnesium oxide
14. 12.1 g magnesium and 7.9 g oxygen

3 Scientific Measurement Planning Guide

Introducing the BIGIDEA: QUANTIFYING MATTER

In many aspects of chemistry, it is vital to know the amount of material with which you are dealing.

IASS	Lessons and Objectives	Print Resources		
		For the Student	For the Teacher	
	3.1 Using and Expressing Measurements p 62–72 **3.1.1** Write numbers in scientific notation. **3.1.2** Evaluate accuracy and precision. **3.1.3** Explain why measurements must be reported to the correct number of significant figures.	Reading and Study Workbook Lesson 3.1 Lesson Assessment 3.1 p 72 Quick Lab: Accuracy and Precision p 72	Class Activity, p 64: Precision and Accuracy Class Activity, p 67: Olympic Times Class Activity, p 68: Significant Zeros	
C.1.7	**3.2** Units of Measurement p 74–82 **3.2.1** Explain why metric units are easy to use. **3.2.2** Identify the temperature units scientists commonly use. **3.2.3** Calculate the density of a substance.	Reading and Study Workbook Lesson 3.2 Lesson Assessment 3.2 p 82	Teacher Demo, p 76: Volume Measurements Class Activity, p 77: Mass of a Penny Teacher Demo, p 81: Density Calculations	
C.1.7	**3.3** Solving Conversion Problems p 84–91 **3.3.1** Explain what happens when a measurement is multiplied by a conversion factor. **3.3.2** Describe the kinds of problems that can be easily solved using dimensional analysis.	Reading and Study Workbook Lesson 3.3 Lesson Assessment 3.3 p 91 Small-Scale Lab: Now What Do I Do? p 92	Class Activity, p 85: Expanding a Recipe Class Activity, p 89: Sports Stats	

Assessing the BIGIDEA: QUANTIFYING MATTER

	Essential Questions 1. How do scientists express the degree of uncertainty in their measurements? 2. How is dimensional analysis used to solve problems?	Study Guide p 93 Math Tune-Up p 94 STP p 99 Reading and Study Workbook Self-Check and Vocabulary Review Chapter 3		

	PearsonChem.com	
	Editable Worksheets	Online Activities
		3.1 Lesson Overview Significant Figures in Multiplication and Division Precision in Measurement
	Small-Scale Lab Manual Lab 3: Design and Construction of a Small-Scale Balance Small-Scale Lab Manual Lab 4: Design and Construction of a Set of Standardized Masses Small-Scale Lab Manual Lab 5: Massing Activities for a Small-Scale Lab Lab 4: Mass, Volume, and Density Lab Practical 3-1: Basic Measurement Lab Practical 3-2: Density	3.2 Lesson Overview SI Units! Exactly! Converting Between Temperature Scales The Density of Solids and Liquids
		3.3 Lesson Overview Conversion Factors Using Dimensional Analysis Using Density as a Conversion Factor Converting Ratios of Units
	Core Teaching Resources (includes editable worksheets) • Lesson Reviews • Practice Problems • Interpret Graphs • Vocabulary Review • Chapter Quizzes and Tests • Lab Record Sheets	Chapter 3 Problem Set Converting Units Scientific Notation

Materials List

For the Student

Quick Lab p 72
- 3 × 5 in. index cards
- metric ruler

Small-Scale Lab p 92
- meter stick
- balance
- pair of dice
- aluminum can
- calculator
- small-scale pipet
- a pre- and post-1982 penny
- 8-well strip
- plastic cup

For the Teacher

Class Activity p 64
- a small object, such as a lead fishing weight
- triple-beam balance

Class Activity p 67
- almanacs or Internet access

Class Activity p 68
- scientific literature
- index cards

Class Activity p 76
- sets of Erlenmeyer flasks
- buret
- graduated cylinder, beaker
- volumetric flask
- water
- food coloring

Class Activity p 77
- balance with a precision of at least 0.01 g
- sets of 10 pennies each separated according to minting dates between 1970 and the present (Be sure that sets include pre- and post-1982 minting dates.)

Class Activity p 81
- 3 or 4 different-sized cubes of a material such as wood, metal, or marble
- metric ruler
- balance

Class Activity p 85
- recipe
- lists of equivalents and conversions among the following measurements: teaspoon, tablespoon, 1/4 cup, 1/2 cup, and 1 cup

Class Activity p 89
- media guides containing vital statistics, such as heights and weights, of sports players

Additional Digital Resources

Online Student Edition
Online Teacher's Edition
Exam*View Assessment Suite CD-ROM*
Classroom Resources DVD

CHAPTER 3

What's Online

CHEMISTRY TUTORIAL Students can work step-by-step with the online tutorial to learn how to solve significant figure and conversion problems by accessing the Chapter 3 Chemistry Tutorials.

ONLINE PROBLEM SET Additional practice problems about measurements and conversions can be found in the Chapter 3 online Problem Set.

MATH TUTORIALS Online Math Tutorials will help guide students through converting units and numbers to scientific notation.

DIRECTED VIRTUAL LAB Encourage students to review calculating the density of solids and liquids using the animated virtual lab for Chapter 3. The animation is followed by analysis questions.

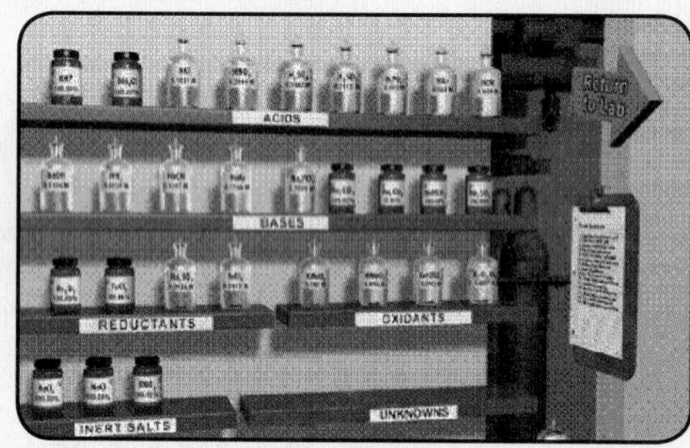

KINETIC ART Have students watch animations about precision in measurement and conversion factors. Each animation is followed by analysis questions.

CONCEPTS IN ACTION Students can watch an interactive animation about the international system of units. The animation is followed by questions.

IN ACADEMIC STANDARDS for SCIENCE

Nature of Science

- Clearly communicate their ideas and results of investigations verbally and in written form using tables, graphs, diagrams, and photographs.
- Apply standard techniques in laboratory investigations to measure physical quantities in appropriate units and convert known quantities to other units as necessary.

Standard 1: Properties and States of Matter

C.1.7 Define density and distinguish between materials based on densities. Perform calculations involving density.

3 Scientific Measurement

INSIDE:

- **3.1** Using and Expressing Measurements
- **3.2** Units of Measurement
- **3.3** Solving Conversion Problems

PearsonChem.com

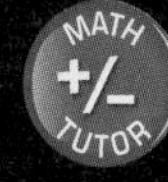

A surveyor in Antarctica uses a device called a theodolite to measure the landscape for a future airstrip.

60

Focus on ELL

1 CONTENT AND LANGUAGE Read the title of the chapter aloud to the class. List reasons why the word *scientific* is in front of the word *measurement*. Discuss what it means if a measurement is *not* scientific. Have students document the discussion with a KWL chart.

BEGINNING: LOW/HIGH Have students identify tools used to measure distance, mass, volume, and time and order the tools based on their accuracy for small measurements.

INTERMEDIATE: LOW/HIGH Have students relate the measures provided with various tools to a familiar object, such as: the mass of a paper clip is about one gram.

ADVANCED: LOW/HIGH Have students practice identifying measurements as a length, a mass, a time, a temperature, or a volume.

UbD

Understanding by Design

Students are building toward understanding how to use scientific measurement as a method of quantifying matter.

PERFORMANCE GOALS At the end of Chapter 3, students will be able to answer the essential questions by applying their knowledge of scientific measurement.

ESSENTIAL QUESTIONS Read the essential questions aloud. **Ask** When you make a measurement, what are some possible sources of uncertainty? *(Sample answer: Your instrument might have increments that are too large for an accurate measurement.)* **Ask** Given the meanings of the terms "dimension" and "analysis," what do you think "dimensional analysis" is? *("Dimension" is a unit of measurement. "Analysis" is trying to explain something by examining its parts. "Dimensional analysis" is trying to explain something by using its units.)*

BIGIDEA Use the photo of the surveyor to help students connect to this chapter's concepts. Activate prior knowledge by discussing units of measurement. Explain that the surveyor in the photo is using the theodolite to make very accurate measurements of angles. **Ask** Why is it important for his measurements to be so accurate? *(Sample answer: The airstrip needs to be located on a flat surface for safety.)*

CHEMYSTERY Connect the CHEMystery to the Big Idea by engaging students in a discussion of the units they might use for different types of measurements. Point out that quantified descriptions of distances need to identify a unit to avoid ambiguity. Have students suggest ways that they might know for sure whether the distances on the sign are in kilometers or miles.

BIGIDEA

QUANTIFYING MATTER

Essential Questions:

1. *How do scientists express the degree of uncertainty in their measurements?*
2. *How is dimensional analysis used to solve problems?*

CHEMYSTERY

Just Give Me a Sign

While traveling in a foreign country, you happen to get lost, as many tourists do. But then you spot these signs along the road. If you know the distance to your destination, you can find your way. However, in the signs shown here, the distances are listed as numbers with no units attached. For example, is Preston 8 kilometers away or 8 miles away? Is there any way to know for sure?

▶ Connect to the **BIGIDEA** As you read the chapter, try to familiarize yourself with common metric units used in science.

INDIANA'S ACADEMIC STANDARDS FOR SCIENCE

Properties and States of Matter C.1.7. See lessons for details.

Introduce the Chapter

IDENTIFYING PRECONCEPTIONS Students may think that measurements are either correct or incorrect. Use the activity to help them realize that measurements can be correct but have degrees of uncertainty.

Activity Divide the class into groups of two or three students, and provide each group with a meterstick and a metric ruler. Have different groups measure and record the length of a wall to the nearest meter, decimeter, centimeter, or millimeter. Afterward, have groups compare their measurements. **Ask** Were the measurements for each type of unit the same? *(Answers will vary.)* **Ask** Which units do you think gave measurements closest to the actual length of the wall? *(millimeters)*

Untamed Science™

Being Precisely Accurate

Suze visits a manufacturing plant to learn how accuracy and precision are vital in producing quality products.

LESSON 3.1

Key Objectives

3.1.1 WRITE numbers in scientific notation.

3.1.2 EVALUATE accuracy and precision.

3.1.3 EXPLAIN why measurements must be reported to the correct number of significant figures.

Additional Resources

- Reading and Study Workbook, Lesson 3.1

Core Teaching Resources, Lesson 3.1 Review

PearsonChem.com

Engage

CHEMISTRY & YOU Have students study the photograph and read the text that opens the section. **Ask** How do you think scientists ensure measurements are accurate and precise? *(Acceptable answers include that scientists make multiple measurements by using the most precise equipment available. They use samples with known values to check the reliability of the equipment.)*

Activate Prior Knowledge

Discuss the various everyday activities that involve measuring, as well as the tools used to make those measurements. Ask a volunteer to be measured for height. Hand out a tape measure, a yardstick, and a metric ruler to three students. Have these students measure the height of the volunteer with their tools and state aloud their measures. **Ask** Which measurement appears to be the most accurate? Why? *(Answers will vary based on the techniques used.)*

IN ACADEMIC STANDARDS for SCIENCE

- Develop explanations based on reproducible data and observations gathered during laboratory investigations.

3.1 Using and Expressing Measurements

CHEMISTRY & YOU

Q: *How do you measure a photo finish?* You probably know that a 100-meter dash is timed in seconds. But if it's a close finish, measuring each runner's time to the nearest second will not tell you who won. That's why sprint times are often measured to the nearest hundredth of a second (0.01 s). Chemistry also requires making accurate and often very small measurements.

Key Questions

How do you write numbers in scientific notation?

How do you evaluate accuracy and precision?

Why must measurements be reported to the correct number of significant figures?

Vocabulary

- measurement
- scientific notation
- accuracy • precision
- accepted value
- experimental value
- error • percent error
- significant figures

Scientific Notation

How do you write numbers in scientific notation?

Everyone makes and uses measurements. A **measurement** is a quantity that has both a number and a unit. Your height (66 inches), your age (15 years), and your body temperature (37°C) are examples of measurements.

Measurements are fundamental to the experimental sciences. For that reason, it is important to be able to make measurements and to decide whether a measurement is correct. In chemistry, you will often encounter very large or very small numbers. A single gram of hydrogen, for example, contains approximately 602,000,000,000,000,000,000,000 hydrogen atoms. The mass of an atom of gold is 0.000 000 000 000 000 000 000 327 gram. Writing and using such large and small numbers is cumbersome. You can work more easily with these numbers by writing them in scientific notation.

In **scientific notation,** a given number is written as the product of two numbers: a coefficient and 10 raised to a power. For example, the number 602,000,000,000,000,000,000,000 can be written in scientific notation as 6.02×10^{23}. The coefficient in this number is 6.02. The power of 10, or exponent, is 23. **In scientific notation, the coefficient is always a number greater than or equal to one and less than ten. The exponent is an integer.** A positive exponent indicates how many times the coefficient must be multiplied by 10. A negative exponent indicates how many times the coefficient must be divided by 10. Figure 3.1 shows a magnified view of a human hair, which has a diameter of about 0.00007 m, or 7×10^{-5} m.

When writing numbers greater than ten in scientific notation, the exponent is positive and equals the number of places that the original decimal point has been moved to the left.

$$6{,}300{,}000. = 6.3 \times 10^{6} \qquad 94{,}700. = 9.47 \times 10^{4}$$

Numbers less than one have a negative exponent when written in scientific notation. The value of the exponent equals the number of places the decimal has been moved to the right.

$$0.000\,008 = 8 \times 10^{-6} \qquad 0.00736 = 7.36 \times 10^{-3}$$

Figure 3.1 Just a Hair
A hair's width expressed in meters is a very small measurement.

$0.00007\ \text{m} = 7 \times 10^{-5}\ \text{m}$

Decimal point moves 5 places to the right. Exponent is −5

Focus on ELL

1 CONTENT AND LANGUAGE Review the abbreviations used in the lesson. Supply the class with a listing for each abbreviation and its meaning. Read aloud each abbreviation and have students repeat.

2 FRONTLOAD THE LESSON Put together a display of common and uncommon measuring tools that students can examine prior to the lesson. Ask students to group the tools on the basis of the type of measurement for which they believe the tools are used: distance, time, volume, and so forth. Have students predict which tools in each group are the most and least accurate. Revisit students' predictions after completing the lesson.

3 COMPREHENSIBLE INPUT When discussing Figure 3.2, assist students in completing the associated analogy by allowing them to test their answers with an unzeroed balance.

Multiplication and Division To multiply numbers written in scientific notation, multiply the coefficients and add the exponents.

$$(3 \times 10^4) \times (2 \times 10^2) = (3 \times 2) \times 10^{4+2} = 6 \times 10^6$$

$$(2.1 \times 10^3) \times (4.0 \times 10^{-7}) = (2.1 \times 4.0) \times 10^{3+(-7)} = 8.4 \times 10^{-4}$$

To divide numbers written in scientific notation, divide the coefficients and subtract the exponent in the denominator from the exponent in the numerator.

$$\frac{3.0 \times 10^5}{6.0 \times 10^2} = \left(\frac{3.0}{6.0}\right) \times 10^{5-2} = 0.5 \times 10^3 = 5.0 \times 10^2$$

Addition and Subtraction If you want to add or subtract numbers expressed in scientific notation and you are not using a calculator, then the exponents must be the same. In other words, the decimal points must be aligned before you add or subtract the numbers. For example, when adding 5.4×10^3 and 8.0×10^2, first rewrite the second number so that the exponent is a 3. Then add the numbers.

$$\begin{aligned}(5.4 \times 10^3) + (8.0 \times 10^2) &= (5.4 \times 10^3) + (0.80 \times 10^3)\\ &= (5.4 + 0.80) \times 10^3\\ &= 6.2 \times 10^3\end{aligned}$$

Sample Problem 3.1

Using Scientific Notation

Solve each problem and express the answer in scientific notation.
a. $(8.0 \times 10^{-2}) \times (7.0 \times 10^{-5})$ **b.** $(7.1 \times 10^{-2}) + (5 \times 10^{-3})$

1 Analyze Identify the relevant concepts. To multiply numbers in scientific notation, multiply the coefficients and add the exponents. To add numbers in scientific notation, the exponents must match. If they do not, then adjust the notation of one of the numbers.

2 Solve Apply the concepts to this problem.

Multiply the coefficients and add the exponents.

$$\begin{aligned}\textbf{a. } (8.0 \times 10^{-2}) \times (7.0 \times 10^{-5}) &= (8.0 \times 7.0) \times 10^{-2+(-5)}\\ &= 56 \times 10^{-7}\\ &= 5.6 \times 10^{-6}\end{aligned}$$

Rewrite one of the numbers so that the exponents match. Then add the coefficients.

$$\begin{aligned}\textbf{b. } (7.1 \times 10^{-2}) + (5 \times 10^{-3}) &= (7.1 \times 10^{-2}) + (0.5 \times 10^{-2})\\ &= (7.1 + 0.5) \times 10^{-2}\\ &= 7.6 \times 10^{-2}\end{aligned}$$

1. Solve each problem and express the answer in scientific notation.
a. $(6.6 \times 10^{-8}) + (5.0 \times 10^{-9})$
b. $(9.4 \times 10^{-2}) - (2.1 \times 10^{-2})$

2. Calculate the following and write your answer in scientific notation:

$$\frac{6.6 \times 10^6}{(8.8 \times 10^{-2}) \times (2.5 \times 10^3)}$$

Foundations for Math

CALCULATIONS INVOLVING SCIENTIFIC NOTATION Tell students that solving problems in chemistry often involves numbers written in scientific notation. Explain that when multiplying or dividing numbers written in scientific notation, the exponents do not have to be the same. However, when adding or subtracting without a calculator, the exponents must be the same. Point out that another way to combine numbers written in scientific notation is to write both numbers in standard form first, perform the operation, and then convert the resulting answer back to scientific notation.

In **Sample Problem 3.1b**, each number can be written in standard form first. Some students may find this easier than changing the exponent for one of the numbers to make them match.

$$7.1 \times 10^{-2} = 0.071$$

$$5 \times 10^{-3} = 0.005$$

Now add: $0.071 + 0.005 = 0.076 = 7.6 \times 10^{-2}$

Foundations for Reading

BUILD VOCABULARY Have students write definitions of the words *accurate* and *precise* in their own words. As they read the text, have students compare the definitions with those of *accuracy* and *precision* given in the text.

READING STRATEGY After reading each section, create a bulleted list summarizing the meaning of the key words or summarizing the rules to follow. Have students compare their lists with others. Break the class into small groups and have each group consolidate their individual efforts into a group summary.

Explain

Scientific Notation

USE VISUALS Have students study Figure 3.1. **Ask** How is the exponent of a number expressed in scientific notation related to the number of places the decimal point is moved to the right in a number smaller than 1? *(They are equal.)*

Sample Practice Problem

Calculate the following and express your answers in scientific notation.

1. $(3.0 \times 10^{-3}) \times (2.5 \times 10^{-4}) =?$ *(7.5×10^{-7})*
2. $(4.2 \times 10^{-3}) + (7.0 \times 10^{-4}) =?$ *(4.9×10^{-3})*

Answers

1. **a.** 7.1×10^{-8} **b.** 7.3×10^{-2}
2. 3.0×10^4

Explain

Accuracy, Precision, and Error

USE VISUALS Have students inspect Figure 3.2. **Ask** If one dart in the third figure were closer to the bull's-eye, what would happen to the accuracy? *(The accuracy would increase.)* **Ask** What would happen to the precision? *(The precision would increase.)* **Ask** What is the operational definition of error implied by this figure? *(The error is the distance between the dart and the bull's-eye.)*

Explore

Class Activity

PURPOSE Students will explore the concepts of precision and accuracy.

MATERIALS a small object (such as a lead fishing weight), triple-beam balance

PROCEDURE Place the object and a triple-beam balance in a designated area. Set a deadline by which each student will have measured the mass of the object. After everyone has had an opportunity, have students compile a summary of all the measurements. Illustrate precision by having the students find the average and compare their measurement to it.

EXPECTED OUTCOME Measured values should be similar, but not necessarily identical, for all students.

USE ANALOGIES Guide students to think of other analogies that may be useful in explaining precision vs. accuracy, such as casting a fishing line, pitching horseshoes, or marching in a precision marching band.

Accuracy, Precision, and Error

How do you evaluate accuracy and precision?

Your success in the chemistry lab and in many of your daily activities depends on your ability to make reliable measurements. Ideally, measurements should be both correct and reproducible.

Accuracy and Precision Correctness and reproducibility relate to the concepts of accuracy and precision, two words that mean the same thing to many people. In chemistry, however, their meanings are quite different. **Accuracy** is a measure of how close a measurement comes to the actual or true value of whatever is measured. **Precision** is a measure of how close a series of measurements are to one another, irrespective of the actual value. **To evaluate the accuracy of a measurement, the measured value must be compared to the correct value. To evaluate the precision of a measurement, you must compare the values of two or more repeated measurements.**

Darts on a dartboard illustrate accuracy and precision in measurement. Let the bull's-eye of the dartboard in Figure 3.2 represent the true, or correct, value of what you are measuring. The closeness of a dart to the bull's-eye corresponds to the degree of accuracy. The closer it comes to the bull's-eye, the more accurately the dart was thrown. The closeness of several darts to one another corresponds to the degree of precision. The closer together the darts are, the greater the precision and the reproducibility.

Figure 3.2 Accuracy vs. Precision
The distribution of darts illustrates the difference between accuracy and precision.
Use Analogies ***Which outcome describes a scenario in which you properly measure an object's mass three times using a balance that has not been zeroed?***

Good Accuracy, Good Precision
Closeness to the bull's-eye indicates a high degree of accuracy. The closeness of the darts to one another indicates high precision.

Poor Accuracy, Good Precision
Precision is high because of the closeness of grouping—thus, the high level of reproducibility. But the results are inaccurate.

Poor Accuracy, Poor Precision
The darts land far from one another and from the bull's-eye. The results are both inaccurate and imprecise.

Determining Error Suppose you use a thermometer to measure the boiling point of pure water at standard pressure. The thermometer reads 99.1°C. You probably know that the true or accepted value of the boiling point of pure water at these conditions is actually 100.0°C.

Striving for Scientific Accuracy

The French chemist Antoine-Laurent Lavoisier worked hard to establish the importance of accurate measurement in scientific inquiry. Lavoisier devised an experiment to test the popular idea that when water was heated, it could turn into earth. For 100 days, Lavoisier boiled water in a glass flask constructed to allow steam to condense without escaping. He weighed the water and the flask separately before and after boiling. He found that the mass of the water had not changed. The flask, however, lost a small mass equal to the sediment he found in the bottom of it. Lavoisier proved that the sediment was not earth, but part of the flask etched away by the boiling water.

There is a difference between the **accepted value,** which is the correct value for the measurement based on reliable references, and the **experimental value,** the value measured in the lab. The difference between the experimental value and the accepted value is called the **error.**

$$\text{Error} = \text{experimental value} - \text{accepted value}$$

Error can be positive or negative, depending on whether the experimental value is greater than or less than the accepted value. For the boiling-point measurement, the error is 99.1°C − 100.0°C, or −0.9°C.

The magnitude of the error shows the amount by which the experimental value differs from the accepted value. Often, it is useful to calculate the relative error, or percent error. The **percent error** of a measurement is the absolute value of the error divided by the accepted value, multiplied by 100%.

$$\text{Percent error} = \frac{|\text{error}|}{\text{accepted value}} \times 100\%$$

READING SUPPORT

Build Reading Skills: *Inference* As you read, try to identify some of the factors that cause experimental error. ***What factors might result in inaccurate measurements? What factors might result in imprecise measurements?***

SampleProblem 3.2

Calculating Percent Error

The boiling point of pure water is measured to be 99.1°C. Calculate the percent error.

1 Analyze **List the knowns and unknown.** The accepted value for the boiling point of pure water is 100°C. Use the equations for error and percent error to solve the problem.

KNOWNS	UNKNOWN
Experimental value = 99.1°C	Percent error = ?
Accepted value = 100.0°C	

2 Calculate **Solve for the unknown.**

Start with the equation for percent error.

$$\text{Percent error} = \frac{|\text{error}|}{\text{accepted value}} \times 100\%$$

Substitute the equation for error, and then plug in the known values.

$$\text{Percent error} = \frac{|\text{experimental value} - \text{accepted value}|}{\text{accepted value}} \times 100\%$$

$$= \frac{|99.1°\text{C} - 100.0°\text{C}|}{100.0°\text{C}} \times 100\%$$

$$= \frac{0.9\cancel{°\text{C}}}{100.0\cancel{°\text{C}}} \times 100\% = 0.9\%$$

3 Evaluate **Does the result make sense?** The experimental value was off by about 1°C, or $\frac{1}{100}$ of the accepted value (100°C). The answer makes sense.

3. A student measures the depth of a swimming pool to be 2.04 meters at its deepest end. The accepted value is 2.00 m. What is the student's percent error?

Think about it: Using the absolute value of the error means that percent error will always be a positive value.

Explain

START A CONVERSATION Review the concept of absolute value. **Ask** What is the meaning of a *positive* error? *(The measured value is greater than the accepted value.)* **Ask** What is the meaning of a *negative* error? *(The measured value is less than the accepted value.)* Explain that the absolute value of the error is a positive value that describes the difference between the measured value and the accepted value, but not which is greater.

Sample Practice Problem

Diamonds and other gemstones are measured in carats. The accepted value for the weight of a carat is 0.2 gram. A jeweler measured the weight of a carat of opals to be 0.192 gram. What is the percent error? *(4%)*

Foundations for Math

ABSOLUTE VALUE Remind students that the absolute value of a number is its distance from zero on a number line and is always written as a positive number. So, |4.2| = 4.2 and |−4.2| = 4.2.

In **Sample Problem 3.2**, the numerator in the fraction of the formula is the absolute value of the error. Regardless of whether the experimental value or the accepted value is larger, the percent error will always be a positive value.

Answers

READING SUPPORT Inaccurate measurements may be due to uncalibrated instruments or incorrect experimental procedure; imprecise measurements may be due to poor instrument operation, variable experimental conditions, or errors in reading output values.

FIGURE 3.2 poor accuracy, good precision

3. 2%

LESSON 3.1

Explain

Significant Figures

START A CONVERSATION Point out that the concept of significant figures applies only to measured quantities. If students ask why an estimated digit is considered significant, tell them a significant figure is one that is known to be reasonably reliable. A careful estimate fits this definition.

USE VISUALS Direct students' attention to Figure 3.3. Point out that when calibration marks on an instrument are spaced very close together (e.g., on certain thermometers and graduated cylinders), it is sometimes more practical to estimate a measurement to the nearest half of the smallest calibrated increment, rather than to the nearest tenth.

CRITICAL THINKING As students inspect Figure 3.4, model the use of the top meter stick by pointing out that one can be certain that the width of the door is between 0 and 1 m, and one can say that the actual width is closer to 1 m. Thus, one can estimate the width as 0.8 m. Similarly, using the middle meter stick, one can say with certainty that the width is between 70 and 80 cm. Because the width is very close to 80 cm, one should estimate the width as 77 cm or 0.77 m. Have students study the bottom meter stick and use similar reasoning to describe the measurement and estimation process. **Ask** If the bottom meter stick were divided into 0.001 m intervals, as are most meter sticks, what would be the estimated width of the door in meters? *(Acceptable answers range from 0.7715 to 0.7724 m.)* In millimeters? *(771.5 to 772.4 mm)*

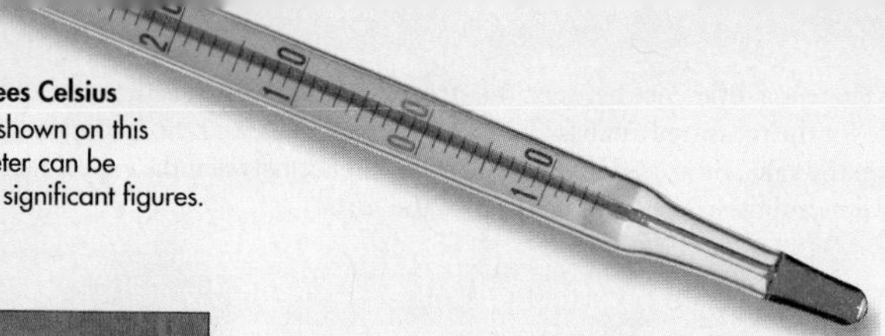

Figure 3.3 Degrees Celsius
The temperature shown on this Celsius thermometer can be reported to three significant figures.

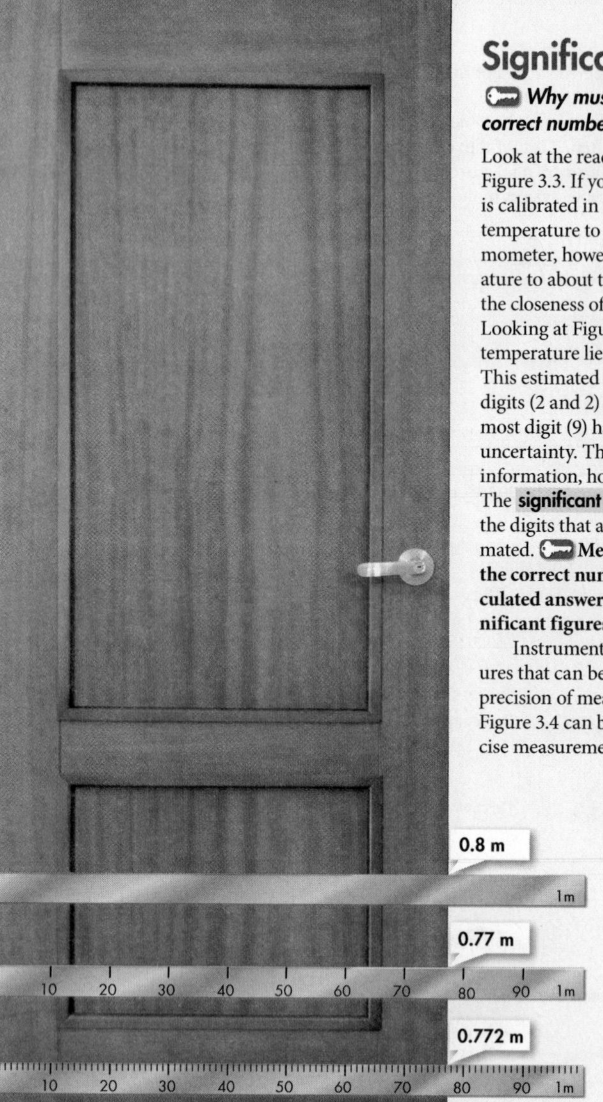

Significant Figures

Why must measurements be reported to the correct number of significant figures?

Look at the reading of the thermometer shown in Figure 3.3. If you use a liquid-filled thermometer that is calibrated in 1°C intervals, you can easily read the temperature to the nearest degree. With the same thermometer, however, you can also estimate the temperature to about the nearest tenth of a degree by noting the closeness of the liquid inside to the calibrations. Looking at Figure 3.3, suppose you estimate that the temperature lies between 22°C and 23°C, at 22.9°C. This estimated number has three digits. The first two digits (2 and 2) are known with certainty. But the rightmost digit (9) has been estimated and involves some uncertainty. These reported digits all convey useful information, however, and are called significant figures. The **significant figures** in a measurement include all of the digits that are known, plus a last digit that is estimated. **Measurements must always be reported to the correct number of significant figures because calculated answers often depend on the number of significant figures in the values used in the calculation.**

Instruments differ in the number of significant figures that can be obtained from their use and thus in the precision of measurements. The three meter sticks in Figure 3.4 can be used to make successively more precise measurements.

More on precision in measurements online. KINETIC ART

Figure 3.4 Increasing Precision
Three differently calibrated meter sticks are used to measure a door's width. A meter stick calibrated in 0.1-m (1 dm) intervals is more precise than one calibrated in a 1-m interval but less precise than one calibrated in 0.01-m (1 cm) intervals. **Measure** ***How many significant figures are reported in each measurement?***

UbD Check for Understanding

The Essential Question How do scientists express their uncertainty in measurement?

Assess students' understanding of the Essential Question by asking students to complete the following statement:

A measured value's uncertainty is expressed by its _____ digit. *(final; last; rightmost)*

ADJUST INSTRUCTION If students are having difficulty answering, refer them to Figure 3.4. Point out that the rightmost digit in each measurement is an estimated value. The fact that the rightmost digit is estimated gives the digit its uncertainty.

Determining Significant Figures in Measurements To determine whether a digit in a measured value is significant, you need to apply the following rules.

1. Every nonzero digit in a reported measurement is assumed to be significant.

24.7 meters
0.743 meter
714 meters

Each of these measurements has three significant figures.

2. Zeros appearing between nonzero digits are significant.

7003 meters
40.79 meters
1.503 meters

Each of these measurements has four significant figures.

3. Leftmost zeros appearing in front of nonzero digits are not significant. They act as placeholders. By writing the measurements in scientific notation, you can eliminate such placeholding zeros.

0.0071 meter $= 7.1 \times 10^{-3}$ meter
0.42 meter $= 4.2 \times 10^{-1}$ meter
0.000099 meter $= 9.9 \times 10^{-5}$ meter

Each of these measurements has only two significant figures.

4. Zeros at the end of a number and to the right of a decimal point are always significant.

43.00 meters
1.010 meters
9.000 meters

Each of these measurements has four significant figures.

5. Zeros at the rightmost end of a measurement that lie to the left of an understood decimal point are not significant if they serve as placeholders to show the magnitude of the number.

300 meters (one significant figure)
7000 meters (one significant figure)
27,210 meters (four significant figures)

The zeros in these measurements are not significant.

If such zeros were known measured values, however, then they would be significant. Writing the value in scientific notation makes it clear that these zeros are significant.

300 meters $= 3.00 \times 10^{2}$ meters (three significant figures)

The zeros in this measurement are significant.

6. There are two situations in which numbers have an unlimited number of significant figures. The first involves counting. A number that is counted is exact.

23 people in your classroom

This measurement is a counted value, so it has an unlimited number of significant figures.

The second situation involves exactly defined quantities such as those found within a system of measurement.

60 min = 1 hr
100 cm = 1 m

Each of these numbers has an unlimited number of significant figures.

Explore

PURPOSE Students will explore how similar measurements from different eras may vary in precision.

MATERIALS Almanacs or Internet access

PROCEDURE Have students look up the winning times for the men's and women's 100-meter dashes at the 1948 and 2008 Olympic Games. **Ask** Why do the more recently recorded race times contain more digits to the right of the decimal? *(The technology used for timekeeping improved to allow for more precise measurements.)*

EXPECTED OUTCOME Students should find that the race times from 1948 were recorded to the nearest tenth of a second. The race times from 2008 were recorded to the nearest hundredth of a second.

Differentiated Instruction

LPR LESS PROFICIENT READERS Have students write in their own words the rules for determining the number of significant digits. Help them if necessary. Direct them to make several measurements, and then use their rules to correctly determine the correct number of significant digits in the measurements.

L1 STRUGGLING STUDENTS Create one set of flash cards for each rule. Write three numerical values on each card—two values that follow the rule and one that does not follow the rule. Have students identify the two values that follow the rule and explain why the third values does not.

ELL ADVANCED LEARNERS Have students determine how a measurement, such as area or volume, can be precise without being accurate.

Answers

FIGURE 3.4 0.8 m, one significant figure; 0.77 m, two significant figures; 0.772 m, three significant figures

LESSON 3.1

Explore

Class Activity

PURPOSE Students will practice applying the rules governing the significance of zeros in measurements.

MATERIALS textbook and scientific literature, index cards

PROCEDURE Have students search their textbooks and other sources for length, mass, volume, or temperature measurements that contain zeros. Have them include some examples written in scientific notation. Ask them to write each measurement on the front of an index card; on the back of each card, have them write (1) all the rules governing the significance of zeros that apply to the measurement, and (2) the number of significant figures in the measurement. Have pairs of students exchange index cards and agree on the appropriateness of the rules and the answers.

EXPECTED OUTCOME Students should be able to apply correctly rules 2–5 listed on page 67.

Explain

DRAW A CONCLUSION Explain that due to rounding, there will often be discrepancies between actual values and calculated values derived from measurements. Ask students to draw a conclusion about whether following the rules for significant figures introduces error in calculated values. *(Although rounding errors are generally small, they should be acknowledged when performing calculations.)*

CHEMISTRY & YOU The time of 9.98 seconds has three significant figures, and the time of 10.05 seconds has four significant figures. Both times have the same precision because they are measured to the nearest 0.01 second.

Sample Practice Problem

How many significant figures does the number 103,400 have? *(4)*

Sample Problem 3.3

Counting Significant Figures in Measurements

How many significant figures are in each measurement?

a. 123 m
b. 40,506 mm
c. 9.8000×10^4 m
d. 22 meter sticks
e. 0.070 80 m
f. 98,000 m

Make sure you understand the rules for counting significant figures (on the previous page) before you begin, okay?

1 Analyze **Identify the relevant concepts.** The location of each zero in the measurement and the location of the decimal point determine which of the rules apply for determining significant figures. These locations are known by inspecting each measurement value.

2 Solve **Apply the concepts to this problem.**

Apply the rules for determining significant figures. All nonzero digits are significant (rule 1). Use rules 2 through 6 to determine if the zeros are significant.

a. three (rule 1)
b. five (rule 2)
c. five (rule 4)
d. unlimited (rule 6)
e. four (rules 2, 3, 4)
f. two (rule 5)

4. Count the significant figures in each measured length.
a. 0.057 30 meter
b. 8765 meters
c. 0.000 73 meter
d. 40.007 meters

5. How many significant figures are in each measurement?
a. 143 grams
b. 0.074 meter
c. 8.750×10^{-2} gram
d. 1.072 meters

CHEMISTRY & YOU

Q: *Suppose that the winner of a 100-meter dash finishes the race in 9.98 seconds. The runner in second place has a time of 10.05 seconds. How many significant figures are in each measurement? Is one measurement more accurate than the other? Explain your answer.*

Significant Figures in Calculations Suppose you use a calculator to find the area of a floor that measures 7.7 meters by 5.4 meters. The calculator would give an answer of 41.58 square meters. However, each of the measurements used in the calculation is expressed to only two significant figures. As a result, the answer must also be reported to two significant figures (42 m^2). In general, a calculated answer cannot be more precise than the least precise measurement from which it was calculated. The calculated value must be rounded to make it consistent with the measurements from which it was calculated.

Rounding To round a number, you must first decide how many significant figures the answer should have. This decision depends on the given measurements and on the mathematical process used to arrive at the answer. Once you know the number of significant figures your answer should have, round to that many digits, counting from the left. If the digit immediately to the right of the last significant digit is less than 5, it is simply dropped and the value of the last significant digit stays the same. If the digit in question is 5 or greater, the value of the digit in the last significant place is increased by 1.

Foundations for Math

ZEROS IN SIGNIFICANT FIGURES Remind students that a number may contain both significant and nonsignificant zeros. In this situation, more than one rule must be applied to determine which zeros are significant and which zeros are not significant. In **Sample Problem 3.3e,** the number 0.07080 contains both significant and nonsignificant zeros. Rewriting this number in scientific notation makes it easier to determine which zeros are significant figures, by removing the nonsignificant zeros from immediate view:

$$0.07080 = 7.080 \times 10^{-2}$$

Now it is obvious that the zeros to the left of the 7 act as placeholders and are not significant, while the remaining two zeros *are* significant because of their location.

Sample Problem 3.4

Rounding Measurements

Round off each measurement to the number of significant figures shown in parentheses. Write the answers in scientific notation.

a. 314.721 meters (four)
b. 0.001 775 meter (two)
c. 8792 meters (two)

1 Analyze **Identify the relevant concepts.** Using the rules for determining significant figures, round the number in each measurement. Then apply the rules for expressing numbers in scientific notation.

2 Solve **Apply the concepts to this problem.**

Starting from the left, count the first four digits that are significant. The arrow points to the digit immediately following the last significant digit.

a. 314.721 meters
↑
2 is less than 5, so you do not round up.
314.7 meters = 3.147×10^2 meters

Starting from the left, count the first two digits that are significant. The arrow points to the digit immediately following the second significant digit.

b. 0.001 775 meters
↑
7 is greater than 5, so round up.
0.0018 meter = 1.8×10^{-3} meter

Starting from the left, count the first two digits that are significant. The arrow points to the digit immediately following the second significant digit.

c. 8792 meters
↑
9 is greater than 5, so round up.
8800 meters = 8.8×10^3 meters

6. Round each measurement to three significant figures. Write your answers in scientific notation.

a. 87.073 meters
b. 4.3621×10^8 meters
c. 0.01552 meter
d. 9009 meters
e. 1.7777×10^{-3} meter
f. 629.55 meters

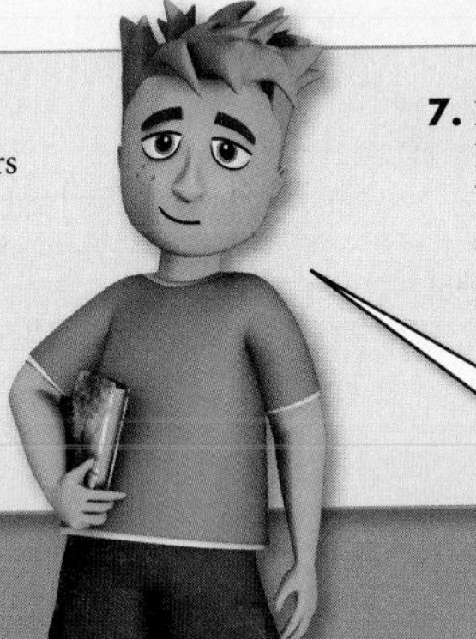

7. Round each measurement in Problem 6 to one significant figure. Write each of your answers in scientific notation.

If you're already familiar with rounding numbers, you can skip to Sample Problems 3.5 and 3.6.

Foundations for Math

COMMON ERROR Stress to students that rounding to a given number of significant figures is not necessarily the same thing as rounding to that many decimal places. For example, if 7.376 is rounded to two significant figures, the result is 7.4. Students often confuse this process with rounding to two decimal places, in which case the result would be 7.38. This number has three significant figures, not two.

In **Sample Problem 3.4b,** none of the zeros are significant. Stress that the answer is not rounded to two decimal places. Rather, the number will be rounded to four decimal places because the second significant figure (7) is in the ten-thousandths place.

Explain

Significant Figures

SUMMARIZE As a class, summarize the rules for significant figures in a bulleted list or a fishbone map. Students should state the rules in their own words and provide an example for each rule.

Misconception Alert

Some students may think that because they use a calculator, the results are shown with the proper number of significant figures. Explain that this is not the case, and even scientific calculators or graphing calculators do not round answers to the correct number of significant figures.

Sample Practice Problem

Round each measurement to two significant figures. Write your answer in scientific notation.

a. 94.592 grams *(9.5×10^1 g)*
b. 2.4332×10^3 grams *(2.4×10^3 g)*
c. 0.007438 grams *(7.4×10^{-3} g)*
d. 54,752 grams *(5.5×10^4 g)*
e. 6.0289×10^{-3} grams *6.0×10^{-3} g)*
f. 405.11 grams *(4.1×10^2 g)*

Answers

4. **a.** 4 **c.** 2
b. 4 **d.** 5

5. **a.** 3 **c.** 4
b. 2 **d.** 4

6. **a.** 8.71×10^1 m
b. 4.36×10^8 m
c. 1.55×10^{-2} m
d. 9.01×10^3 m
e. 1.78×10^{-3} m
f. 6.30×10^2 m

7. **a.** 9×10^1 m
b. 4×10^8 m
c. 2×10^{-2} m
d. 9×10^3 m
e. 2×10^{-3} m
f. 6×10^2 m

Explain

START A CONVERSATION The rules for rounding calculated numbers can be compared with the old adage, "A chain is only as strong as its weakest link." Explain that an answer cannot be more precise than the least precise value used to calculate the answer. **Ask** In addition and subtraction, what is the least precise value? (*The measurement with the fewest digits to the right of the decimal point.*) **Ask** In multiplication and division, what is the least precise value? (*The measurement with the fewest significant figures.*) If students wonder why addition and subtraction rules differ from multiplication and division rules, point out that in addition and subtraction of measurements, the measurements are of the same property, such as length or volume. However, in the multiplication and division of measurements, new quantities or properties are being described, such as speed (length ÷ time), area (length × length), or density (mass ÷ volume).

Misconception Alert

Students may argue that making one measurement of a dimension, such as length, is adequate. **Ask** What possible errors may occur when making only one length measurement? (*Acceptable answers include misreading the ruler or not holding the ruler parallel to the length of the object.*)

Sample Practice Problem

What is the total mass of three rock samples that have measured masses of 20.72 grams, 24.8 grams, and 17.35 grams? (*62.9 grams*)

Addition and Subtraction The answer to an addition or subtraction calculation should be rounded to the same number of decimal places (not digits) as the measurement with the least number of decimal places. Sample Problem 3.5 gives examples of rounding in addition and subtraction.

Multiplication and Division In calculations involving multiplication and division (such as those in Sample Problem 3.6), you need to round the answer to the same number of significant figures as the measurement with the least number of significant figures. The position of the decimal point has nothing to do with the rounding process when multiplying and dividing measurements. The position of the decimal point is important only in rounding the answers of addition or subtraction problems.

Sample Problem 3.5

Significant Figures in Addition and Subtraction

Perform the following addition and subtraction operations. Give each answer to the correct number of significant figures.

a. 12.52 meters + 349.0 meters + 8.24 meters

b. 74.626 meters − 28.34 meters

1 Analyze **Identify the relevant concepts.** Perform the specified math operation, and then round the answer to match the measurement with the least number of decimal places.

2 Solve **Apply the concepts to this problem.**

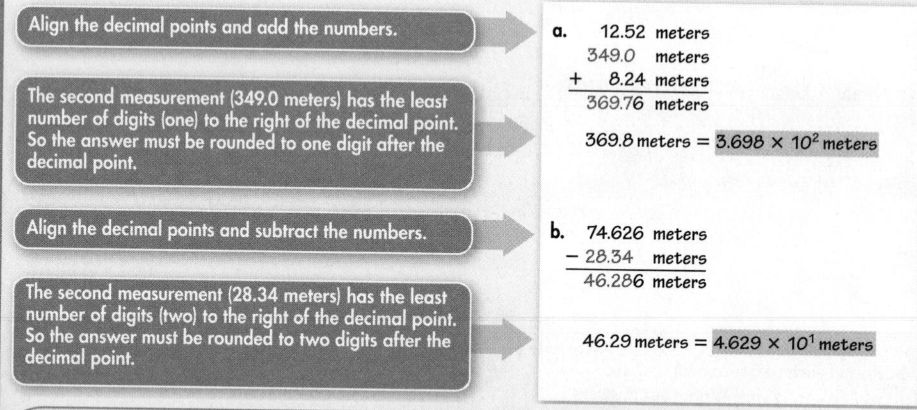

8. Perform each operation. Express your answers to the correct number of significant figures.

a. 61.2 meters + 9.35 meters + 8.6 meters

b. 9.44 meters − 2.11 meters

c. 1.36 meters + 10.17 meters

d. 34.61 meters − 17.3 meters

9. Find the total mass of three diamonds that have masses of 14.2 grams, 8.73 grams, and 0.912 gram.

Foundations for Math

ADDITION AND SIGNIFICANT FIGURES Explain to students that when performing calculations that involve measurements with significant figures, the final result must be rounded to be consistent with the measurements. When adding and subtracting, if the measurements have different numbers of significant figures, the final result should be rounded to the same number of decimal places as the number with the least number of decimal places.

In **Sample Problem 3.5,** the first step should be to identify the measurement with the least number of decimal places. Encourage students to underline or circle this measurement. The result of the calculation should be rounded to the same number of decimal places (not digits) as this measurement.

Sample Problem 3.6

Significant Figures in Multiplication and Division

Perform the following operations. Give the answers to the correct number of significant figures.

a. 7.55 meters × 0.34 meter
b. 2.10 meters × 0.70 meter
c. 2.4526 meters^2 ÷ 8.4 meters
d. 0.365 meter^2 ÷ 0.0200 meter

1 Analyze **Identify the relevant concepts.** Perform the specified math operation, and then round the answer to match the measurement with the least number of significant figures.

2 Solve **Apply the concepts to this problem.**

The second measurement (0.34 meter) has the least number of significant figures (two). So the answer must be rounded to two significant figures.

a. 7.55 meters × 0.34 meter = 2.567 (meter)^2
= 2.6 meters^2

The second measurement (0.70 meter) has the least number of significant figures (two). So the answer must be rounded to two significant figures.

b. 2.10 meters × 0.70 meter = 1.47 (meter)^2
= 1.5 meters^2

The second measurement (8.4 meters) has the least number of significant figures (two). So the answer must be rounded to two significant figures.

c. 2.4526 meters^2 ÷ 8.4 meters = 0.291 976 meter
= 0.29 meter

Both measurements have three significant figures. So the answer must be rounded to three significant figures.

d. 0.365 meters^2 ÷ 0.0200 meter = 18.25 meters
= 18.3 meters

10. Solve each problem. Give your answers to the correct number of significant figures and in scientific notation.
a. 8.3 meters × 2.22 meters
b. 8432 meters^2 ÷ 12.5 meters
c. 35.2 seconds × $\frac{1 \text{ minute}}{60 \text{ seconds}}$

11. Calculate the volume of a warehouse that has measured dimensions of 22.4 meters by 11.3 meters by 5.2 meters. (Volume = $l \times w \times h$)

In Problem 11, the measurement with the fewest significant figures is 5.2 meters. What does this tell you?

Focus on ELL

4 LANGUAGE PRODUCTION Have students work in pairs to complete the lab. Make sure each pair has ELLs of varied language proficiencies, so that more proficient students can help less proficient ones. Have students work according to their proficiency level.

BEGINNING: LOW/HIGH Underline the action verbs in each step of the procedure. Work with a partner to follow each step.

INTERMEDIATE: LOW/HIGH Use a chart to keep track of the different measurements needed for the lab.

ADVANCED: LOW/HIGH With the help of a partner, answer #2 of the Analyze and Conclude section of the lab.

Explain

Sample Practice Problem

A small rectangular container has measured dimensions of 4.25 inches by 8.5 inches by 1.75 inches. What is the volume of the container? *(63 cubic inches)*

Evaluate

Informal Assessment

Write the following sets of measurements on the board.
(1) 70°C, 70°C, 80°C
(2) 77°C, 78°C, 78°C
(3) 80.25°C, 84.50°C, 86.00°C

Ask The temperature of a liquid under similar conditions was measured every ten minutes by three different students. Which student had the most precise measuring instrument? *(Set 3 is the most precise because the measurements have the greatest number of significant figures.)* **Ask** What would have to be known to determine which set is the most accurate? *(the accepted value of the liquid's boiling point)*

Reteach

Use Figure 3.4 to reteach the method of correctly recording the number of significant figures in a measurement. Then have students convert each measurement into scientific notation. *(8×10^{-1} m, 7.7×10^{-1} m, 7.72×10^{-1} m)*

Answers

8. **a.** 79.2 m **b.** 7.33 m **c.** 11.53 m **d.** 17.3 m
9. 23.8 g
10. **a.** 1.8×10^1 m^2
b. 6.75×10^2 m
c. 5.87×10^{-1} min
11. 1.3×10^3 m^3

LESSON 3.1

Explore

Quick Lab

OBJECTIVES After completing this activity, students will be able to measure length with accuracy and precision, apply rules for rounding answers calculated from measurements, and determine experimental error and express it as percent error.

SKILLS FOCUS Measuring, calculating

PREP TIME 5 minutes

CLASS TIME 15 minutes

MATERIALS 3 inch × 5 inch index cards, metric rulers

TEACHING TIPS Emphasize that students should use an interior, marked line, such as 10.0 cm, as the initial point, instead of the end of the ruler, which may be damaged.

EXPECTED OUTCOME Measured values should be similar, but not necessarily identical, for all students.

ANALYZE AND CONCLUDE

1. Four for length; three for width
2. See Expected Outcome.
3. Significant digits for rounded-off answers are area, 3, and perimeter, 4. Some students may not round to the proper number of digits.
4. Errors of ±0.03 cm are acceptable. Such errors yield percent errors of 0.2% for length and 0.4% for width.

FOR ENRICHMENT Have students devise methods of calculating the volume of one card. Point out that measuring the thickness of one card with a ruler would be very inaccurate. **Ask** How might the measurement of the thickness of the card be improved? *(Use a more precise instrument, such as a micrometer, or measure the thickness of a stack of cards and divide by the number of cards.)* Have students determine the thickness of one card and calculate its volume. Using the class average of the calculated volumes, have each student determine the percent error using the average as the accepted value.

Quick Lab

Purpose To measure the dimensions of an object as accurately and precisely as possible and to apply rules for rounding answers calculated from the measurements

Materials

- 3-inch × 5-inch index card
- metric ruler

Accuracy and Precision

Procedure

1. Use a metric ruler to measure in centimeters the length and width of an index card as accurately as you can. The hundredths place in your measurement should be estimated.
2. Calculate the area ($A = l \times w$) and the perimeter [$P = 2 \times (l + w)$] of the index card. Write both your unrounded answers and your correctly rounded answers on the chalkboard.

Analyze and Conclude

1. **Identify** How many significant figures are in your measurements of length and of width?
2. **Compare** How do your measurements compare with those of your classmates?
3. **Explain** How many significant figures are in your calculated value for the area? In your calculated value for the perimeter? Do your rounded answers have as many significant figures as your classmates' measurements?
4. **Evaluate** Assume that the correct (accurate) length and width of the card are 12.70 cm and 7.62 cm, respectively. Calculate the percent error for each of your two measurements.

3.1 LessonCheck

ONLINE PROBLEMS

12. **Review** How can you express a number in scientific notation?
13. **Review** How are accuracy and precision evaluated?
14. **Explain** Why must a given measurement always be reported to the correct number of significant figures?
15. **Calculate** A technician experimentally determined the boiling point of octane to be 124.1°C. The actual boiling point of octane is 125.7°C. Calculate the error and the percent error.
16. **Evaluate** Determine the number of significant figures in each of the following measurements:
    a. 11 soccer players
    b. 0.070 020 meter
    c. 10,800 meters
    d. 0.010 square meter
    e. 5.00 cubic meters
    f. 507 thumbtacks
17. **Calculate** Solve the following and express each answer in scientific notation and to the correct number of significant figures.
    a. $(5.3 \times 10^4) + (1.3 \times 10^4)$
    b. $(7.2 \times 10^{-4}) \div (1.8 \times 10^3)$
    c. $10^4 \times 10^{-3} \times 10^6$
    d. $(9.12 \times 10^{-1}) - (4.7 \times 10^{-2})$
    e. $(5.4 \times 10^4) \times (3.5 \times 10^9)$

BIGIDEA QUANTIFYING MATTER

18. Write a brief paragraph explaining the differences between the accuracy, precision, and error of a measurement.

Lesson Check Answers

12. Write the number as a product of two numbers: a coefficient greater than or equal to one and less than ten, and 10 raised to an integer power.
13. Accuracy compares the measured value to the correct value. Precision compares more than one measurement.
14. The significant figures in a calculated answer depend on the number of significant figures of the measurements and the mathematical operation used in the calculation.
15. error = −1.6°C; percent error = 1.3%
16. a. unlimited d. 2
    b. 5 e. 3
    c. 3 f. unlimited
17. a. 6.6×10^4 d. 8.65×10^{-1}
    b. 4.0×10^{-7} e. 1.9×10^{14}
    c. 10^7
18. **BIGIDEA** Accuracy compares a measured value to an accepted value of the measurement; precision compares a measured value to a set of measurements made under similar conditions; and error is the difference between the measured and accepted values.

Watch What You Measure

Just because you live in a digital age doesn't mean that you no longer have to do things by hand. In fact, manually measuring quantities remains an important everyday skill in a number of professions and activities. For example, chefs measure volumes of ingredients in cups (C) or liters (L). Tailors use a tape measure calibrated in inches (in. or ″) to measure length, while biologists use metric rulers or calipers calibrated in centimeters (cm). A ship's navigator uses a sextant to measure the angle between the sun and the horizon. The angle is expressed in degrees (°) and minutes (′).

The next time you make a measurement in lab, keep in mind that lots of other measurers are rounding and noting significant figures, just like you are.

Take It Further

1. **Measure** What is the measured height of the tomato shown above? How many significant figures does your answer have?
2. **Identify** What are some other activities that involve measurements done by hand? What units and measuring tools are used?

CHEMISTRY & YOU Have students study the photographs closely. Point out that the most important decision these professionals make when they begin their measurement is not whether to use a manual tool or a technologically advanced tool—the most important decision is to determine the correct tool for the task. The choice of tool helps determine the overall accuracy of the measurement. Explain to students that the flexible cloth tape measure used by the tailor is divided into increments of 1/16 of an inch. Pose the following question to students: *Imagine that the only tool the tailor had available was a wooden ruler divided into 1-inch increments. How might the use of the wooden ruler affect the outcome of the custom-tailored suit the tailor is constructing?* You may need to assist students in the following ways:

- Custom-tailored clothing is designed to fit the body measurements of the intended wearer.
- Many of the measurements a tailor makes involve curves or circumferences.
- Typical nonretractable cloth tape measures range in length from 60 to 96 inches.

Extend

Connect to ARCHITECTURE

Explain to students that one field that makes use of manual systems of measurement is architecture. Discuss how manual tools are utilized by an architect during both the design stages of a project (such as making blueprints or producing models) and by the various tradesmen during the implementation of the architect's design and blueprints. Have students think of a specific example of an architectural project and write a brief paragraph discussing how, during each stage of construction, various manual systems of measurement might be utilized to create a finished product.

Differentiated Instruction

L1 STRUGGLING STUDENTS Have students produce a list of several tools used to manually take measurements. The list should include both the name of the tool and what quantity it is used to measure.

ELL ENGLISH LANGUAGE LEARNERS Have students print out pictures of tools commonly used to measure and arrange them on a poster board. Students should write a description of the tools and people who might use such tools for their work or in everyday lives.

L3 ADVANCED STUDENTS Have students choose a trade that commonly utilizes computerized measuring tools and research how the systems of measurement have evolved over time in that trade. Encourage students to be creative in what format they use to display their findings.

Answers

TAKE IT FURTHER

1. *Sample answer:* 32.5 mm (three significant figures)
2. Answers may vary; possibilities include measuring length of fabric (yards and yardsticks), measuring weight of a letter (ounces and a scale), measuring weight of produce in a store (pounds and a scale).

LESSON 3.2

Key Objectives

3.2.1 **EXPLAIN** why metric units are easy to use.

3.2.2 **IDENTIFY** the temperature units scientists commonly use.

3.2.3 **CALCULATE** the density of a substance.

Additional Resources

- Reading and Study Workbook, Lesson 3.2
- Core Teaching Resources, Lesson 3.2 Review
- Laboratory Manual, Lab 4

Small-Scale Lab Manual, Labs 3, 4, 5

PearsonChem.com

Engage

CHEMISTRY & YOU Have students study the weather forecast and read the text that opens the section. **Ask** Does the forecast show measurements? *(No; the forecast shows only numbers. A measurement contains both a number and a unit.)*

Activate Prior Knowledge

As a class, create an itemized list of units of measures learned in mathematics and other science courses. Have students classify these measures as linear, squared, cubic, and rate measurements.

Foundations for Reading

BUILD VOCABULARY Have students predict the meanings of the SI prefixes *deci-*, *centi-*, and *milli-* from their knowledge of the words *decimal*, *century*, and *millennium*.

IN ACADEMIC STANDARDS for SCIENCE

C.1.7 Define density and distinguish between materials based on densities. Perform calculations involving density.

3.2 Units of Measurement

C.1.7 Density

CHEMISTRY & YOU

Q: *What's the forecast for tomorrow—hot or cold?* In the weather forecast shown here, the temperatures are in degrees, but without a temperature scale. Will the high temperature tomorrow be 28°C, which is very warm? Or 28°F, which is very cold? Without the correct units, you can't be sure. When you make a measurement, you must assign the correct units to the number. Without the units, it's impossible to communicate the measurement clearly to others.

Key Questions

What makes metric units easy to use?

What temperature units do scientists commonly use?

What determines the density of a substance?

Vocabulary

- International System of Units (SI)
- meter (m) • liter (L)
- kilogram (kg) • gram (g)
- weight • energy
- joule (J) • calorie (cal)
- temperature • Celsius scale
- Kelvin scale • absolute zero
- density

Using SI Units

What makes metric units easy to use?

All measurements depend on units that serve as reference standards. The standards of measurement used in science are those of the metric system. The metric system is important because of its simplicity and ease of use. **All metric units are based on multiples of 10. As a result, you can convert between units easily.** The metric system was originally established in France in 1795. The **International System of Units** (abbreviated **SI,** after the French name, *Le Système International d'Unités*) is a revised version of the metric system. The SI was adopted by international agreement in 1960. There are seven SI base units, which are listed in Table 3.1. From these base units, all other SI units of measurement can be derived. Derived units are used for measurements such as volume, density, and pressure.

All measured quantities can be reported in SI units. Sometimes, however, non-SI units are preferred for convenience or for practical reasons. In this textbook you will learn about both SI and non-SI units.

Table 3.1

SI Base Units

Quantity	SI base unit	Symbol
Length	meter	m
Mass	kilogram	kg
Temperature	kelvin	K
Time	second	s
Amount of substance	mole	mol
Luminous intensity	candela	cd
Electric current	ampere	A

Focus on ELL

1 CONTENT AND LANGUAGE For 30 seconds, have students call out what they think of when they hear the words *units of measurement*. Record every response in a list and use the list to create a wordle. Pass out a copy of the wordle to each student. Ask students to deduce the type of measurement with which the class is most and least familiar.

2 FRONTLOAD THE LESSON Ask students to share the common units of measurement used in the country they are from. If the common units are not metric, have the students compare them to metric units.

3 COMPREHENSIBLE INPUT Make some of the items (listed in tables 3.3, 3.4, and 3.5) that represent the dimensions commonly stated in metric units available for students to see and touch. Have each item labeled with the unit that it represents.

Table 3.2

Commonly Used Metric Prefixes			
Prefix	Symbol	Meaning	Factor
mega	M	1 million times larger than the unit it precedes	10^6
kilo	k	1000 times larger than the unit it precedes	10^3
deci	d	10 times smaller than the unit it precedes	10^{-1}
centi	c	100 times smaller than the unit it precedes	10^{-2}
milli	m	1000 times smaller than the unit it precedes	10^{-3}
micro	μ	1 million times smaller than the unit it precedes	10^{-6}
nano	n	1 billion times smaller than the unit it precedes	10^{-9}
pico	p	1 trillion times smaller than the unit it precedes	10^{-12}

Units of Length Size is an important property of matter. In SI, the basic unit of length, or linear measure, is the **meter (m).** All measurements of length can be expressed in meters. (The length of a page in this book is about one fourth of a meter.) For very large and very small lengths, however, it may be more convenient to use a unit of length that has a prefix. Table 3.2 lists the prefixes in common use. For example, the prefix *milli-* means 1/1000 (one-thousandth), so a millimeter (mm) is 1/1000 of a meter, or 0.001 m. A hyphen (-) measures about 1 mm.

For large distances, it is usually most appropriate to express measurements in kilometers (km). The prefix *kilo-* means 1000, so 1 km equals 1000 m. A standard marathon distance race of about 42,000 m is more conveniently expressed as 42 km (42 × 1000 m). Table 3.3 summarizes the relationships among metric units of length.

button diameter = 1 cm

dime thickness = 1 mm

Table 3.3

Metric Units of Length			
Unit	Symbol	Relationship	Example
Kilometer	km	1 km = 10^3 m	length of about five city blocks ≈ 1 km
Meter	m	base unit	height of doorknob from the floor ≈ 1 m
Decimeter	dm	10^1 dm = 1 m	diameter of large orange ≈ 1 dm
Centimeter	cm	10^2 cm = 1 m	diameter of shirt button ≈ 1 cm
Millimeter	mm	10^3 mm = 1 m	thickness of dime ≈ 1 mm
Micrometer	μm	10^6 μm = 1 m	diameter of bacterial cell ≈ 1 μm
Nanometer	nm	10^9 nm = 1 m	thickness of RNA molecule ≈ 1 nm

SI Units of Distance and Time

In 1790, with the establishment of the metric system, the French became the first to adopt measurement standards that were close to being precise. The meter was defined as one ten-millionth of the distance from the equator to the North Pole along the meridian that passes through Paris. The second was defined as 1/86,400 of the average day. The definitions of these and other base units have since become more precise. The meter is now defined as the distance traveled by light in a vacuum in 1/299,792,458 of a second. The second is now defined in terms of the number of cycles of radiation given off by a specific isotope of the element cesium.

READING STRATEGY Have students look at Tables 3.1, 3.3, 3.4, and 3.5. Tell students to read the title of each table and the column heads before looking at the information in the table. Have students write in their vocabulary notebook any words from these tables with which they are unfamiliar.

Explain

Using SI Units

MAKE A CONNECTION Explain that with the exception of those working in the sciences, the United States is the only country *not* using the metric system as a routine method of measurement. **Ask** What are some real-world examples of applications in the United States in which the metric system is the commonly accepted unit of measure? *(Acceptable answers include 35-mm film; meter distances in track and field; the dimensions of tools (wrenches) and hardware (bolts); power rating in watts of appliances, and the occasional road sign.)*

USE VISUALS Have students examine Table 3.2. Discuss with students how SI prefixes are always in increments of ten and can be expressed using scientific notation. **Ask** What does the prefix *kilo-* mean? *(The prefix* kilo- *means 1000 times larger than the unit it precedes.)* **Ask** How is this prefix represented in scientific notation? *(10^3)* **Ask** How much larger is the prefix *centi-* compared to the prefix *micro-?* Show students how to make this comparison by writing the following on the board:

$$\frac{\text{centi-}}{\text{micro-}} = \frac{10^{-2}}{10^{-6}} = 10^{-2-(-6)} = 10^{(-2+6)} = 10^4$$

Remind students that when dividing numbers in scientific notation, the exponent of the denominator is subtracted from the exponent of the numerator. Convey the importance of knowing how to interpret an answer. For instance, point out that the difference of 10^4 means that the *centi-* is a measure that is 10,000 times larger than the *micro-*.

Explain

Using SI Units

APPLY CONCEPTS Point out that data can be reported accurately in various related units (mm, cm, m), but it is best to choose the unit closest in scale to the object being measured—a unit that may or may not be an SI base unit. Share with the class the following:

- the unit of mass most often used by chemists is the gram—not the kilogram. Explain that kilograms of chemicals are too large for most laboratory experiments.
- the density of a gas is typically measured in grams per liter; the densities of most solids and liquids are typically measured in grams per cubic centimeter.
- the milliliter is the most common unit of volume used by laboratory chemists. The cubic meter is much too large a volume for most laboratory experiments.

Explore

Teacher Demo

PURPOSE Students will be introduced to laboratory equipment used to measure volume.

MATERIALS sets of Erlenmeyer flasks, buret, graduated cylinder, beaker, and volumetric flask; water; food coloring

SAFETY Have students wear goggles to emphasize safety in the laboratory.

PROCEDURE Before the demonstration, fill the glassware with water and add a few drops of food coloring. Place the sets at various stations around the classroom. Assign groups of students to each station. As students inspect the glassware, explain that beakers, Erlenmeyer flasks, and graduated cylinders are used to measure approximate volumes. Volumetric flasks and burets allow measurements that are more precise. **Ask** What is the appropriate instrument to use to measure a large volume, such as 1 L, accurately? *(Students should choose the volumetric flask.)*

EXPECTED OUTCOME Students should be able to select the appropriate glassware when presented with the volumetric parameters for an experiment.

Figure 3.5 Volumetric Units
The volume of 20 drops of liquid from a medicine dropper is about 1 mL. This is the same volume as that of a sugar cube, which is 1 cm on each edge; 1 cm^3 = 1 mL. A liter bottle has a volume of 1 L, or 1000 mL.
Describe ***What is the volume of a 2-L bottle in cubic centimeters?***

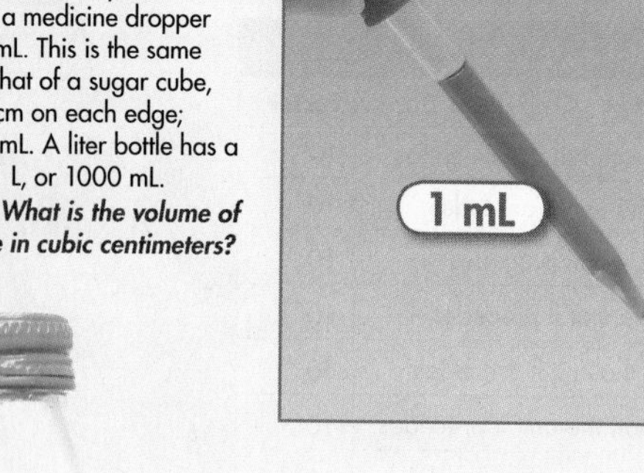

Units of Volume The space occupied by any sample of matter is called its volume. You calculate the volume of any cubic or rectangular solid by multiplying its length by its width by its height. The unit for volume is thus derived from units of length. The SI unit of volume is the amount of space occupied by a cube that is 1 m along each edge. This volume is a cubic meter (m^3). An automatic dishwasher has a volume of about 1 m^3.

A more convenient unit of volume for everyday use is the liter, a non-SI unit. A **liter (L)** is the volume of a cube that is 10 centimeters (10 cm) along each edge (10 cm × 10 cm × 10 cm = 1000 cm^3 = 1 L). A decimeter (dm) is equal to 10 cm, so 1 L is also equal to 1 cubic decimeter (dm^3). A smaller non-SI unit of volume is the milliliter (mL); 1 mL is 1/1000 of a liter. Thus, there are 1000 mL in 1 L. Because 1 L is defined as 1000 cm^3, 1 mL and 1 cm^3 are the same volume. The units milliliter and cubic centimeter are thus used interchangeably. Figure 3.5 gives you some idea of the relative sizes of a liter and a milliliter. Table 3.4 summarizes the relationships among common metric units of volume.

There are many devices for measuring liquid volumes, including graduated cylinders, pipets, burets, volumetric flasks, and syringes. Note that the volume of any solid, liquid, or gas will change with temperature (although the change is much more dramatic for gases). Consequently, accurate volume-measuring devices are calibrated at a given temperature—usually 20 degrees Celsius (20°C), which is about normal room temperature.

Table 3.4

Metric Units of Volume

Unit	Symbol	Relationship	Example
Liter	L	base unit	quart of milk ≈ 1 L
Milliliter	mL	10^3 mL = 1 L	20 drops of water ≈ 1 mL
Cubic centimeter	cm^3	1 cm^3 = 1 mL	cube of sugar ≈ 1 cm^3
Microliter	µL	10^6 µL = 1 L	crystal of table salt ≈ 1 µL

Differentiated Instruction

L1 STRUGGLING STUDENTS Recreate Table 3.2 on the board, adding a fifth column that shows the factor of each prefix as a whole number or fraction. For example, *mega-* represents a factor of 1,000,000 and *deci-* represents a factor of 1/10.

LPR LESS PROFICIENT READERS Have students search magazines and newspapers for examples of SI units. Have them report seasonal day/night temperature ranges for their countries of origin.

L3 ADVANCED STUDENTS Have students research the efforts to make the metric system standard in the United States. Who has promoted the change? What progress has been made? What obstacles remain?

Table 3.5

Metric Units of Mass			
Unit	**Symbol**	**Relationship**	**Example**
Kilogram (base unit)	kg	1 kg $= 10^3$ g	small textbook $\approx$ 1 kg
Gram	g	1 g $= 10^{-3}$ kg	dollar bill $\approx$ 1 g
Milligram	mg	10^3 mg = 1 g	ten grains of salt $\approx$ 1 mg
Microgram	μg	10^6 μg = 1 g	particle of baking powder $\approx$ 1 μg

Units of Mass The mass of an object is measured in comparison to a standard mass of 1 **kilogram (kg),** which is the basic SI unit of mass. A kilogram was originally defined as the mass of 1 L of liquid water at 4°C. A cube of water at 4°C measuring 10 cm on each edge would have a volume of 1 L and a mass of 1000 grams (g), or 1 kg. A **gram (g)** is 1/1000 of a kilogram; the mass of 1 cm^3 of water at 4°C is 1 g. The relationships among units of mass are shown in Table 3.5.

You can use a platform balance to measure the mass of an object. The object is placed on one side of the balance, and standard masses are added to the other side until the balance beam is level. The unknown mass is equal to the sum of the standard masses. Laboratory balances range from very sensitive instruments with a maximum capacity of only a few milligrams to devices for measuring quantities in kilograms. An analytical balance is used to measure objects of less than 100 g and can determine mass to the nearest 0.0001 g (0.1 mg).

Weight is a force that measures the pull on a given mass by gravity. Weight, a measure of force, is different from mass, which is a measure of the quantity of matter. The weight of an object can change with its location. For example, an astronaut on the surface of the moon weighs one sixth of what he weighs on Earth. The reason for this difference is that the force of Earth's gravity is about six times greater than that of the moon. The astronaut in Figure 3.6 is in free fall as he orbits Earth and is therefore weightless. Although it's possible for an object to become weightless, it can never become massless.

Figure 3.6 Weightlessness
An astronaut in orbit is weightless, but not massless. The astronaut's mass remains constant regardless of location or motion.

Units of Energy The capacity to do work or to produce heat is called **energy.** Like any other quantity, energy can be measured. The SI unit of energy is the **joule (J),** named after the English physicist James Prescott Joule (1818–1889). A common non-SI unit of energy is the calorie. One **calorie (cal)** is the quantity of heat that raises the temperature of 1 g of pure water by 1°C. Conversions between joules and calories can be carried out using the following relationships:

$$1\ \text{J} = 0.2390\ \text{cal} \qquad 1\ \text{cal} = 4.184\ \text{J}$$

In this book, you will see energy values expressed in both joules and calories, as well as kilojoules (kJ) and kilocalories (kcal). A kilojoule is 1000 joules; a kilocalorie is 1000 calories.

UbD Check for Understanding

BIGIDEA **QUANTIFYING MATTER** Assess students' understanding of the appropriate choice of units for a given measurement by listing the following measurements on the board: liquid volume, body temperature, height, and body mass. Ask students to identify the appropriate SI unit for the measurement and indicate whether the unit is a base unit or derived unit. *(liquid volume: liter, derived; body temperature: Kelvin, base; height: meter, base; body mass: kilogram, base)*

ADJUST INSTRUCTION If students are still struggling with the concept of derived units, have students reread the first paragraph of the Using SI Units section, and then review the derivation of the liter for volume as an example of how an SI unit can be used to calculate a common non-SI unit.

Explore

Class Activity

PURPOSE Students will practice weighing groups of objects.

MATERIALS balance with a precision of at least 0.01 g, sets of 10 pennies each separated according to minting dates between 1970 and the present. Make sure that sets include pre- and post-1982 minting dates.

SAFETY Have student wear goggles when weighing the pennies.

PROCEDURE Have students weigh three five-penny samples from each set of pennies and record their data. Have students calculate the average mass of a penny and plot the average mass versus the year it was minted.

EXPECTED OUTCOME Students should observe slight variations in mass from year to year due to chance error, and a dramatic decrease in mass between 1982 and 1983. Note that pennies minted before 1982 are mostly copper; pennies minted after 1982 are mostly zinc, which is less dense than copper. Ask students to propose a hypothesis for their findings. *(Acceptable answers include a change in the composition of the penny.)* Instruct students to keep their graph, results, and hypothesis for the discussion of density on page 80.

Answers

FIGURE 3.5 2000 cm^3

Explain

Temperature Scales

USE VISUALS Have students inspect the thermometers in Figure 3.7. **Ask** Which thermometer is more precise? *(the bulb thermometer)* **Ask** Which is the least precise? *(the bimetallic strip thermometer)* **Ask** Why do you think a chef would choose the less precise thermometer for determining when the turkey is done? *(The range of the bimetallic strip thermometer is greater than that of the bulb thermometer and more appropriate for an oven environment.)*

Temperature Scales

What temperature units do scientists commonly use?

When you hold a glass of hot water, the glass feels hot because heat transfers from the glass to your hand. When you hold an ice cube, it feels cold because heat transfers from your hand to the ice cube. **Temperature** is a measure of how hot or cold an object is. An object's temperature determines the direction of heat transfer. When two objects at different temperatures are in contact, heat moves from the object at the higher temperature to the object at the lower temperature. In Chapter 13, you will learn how the temperature of an object is related to the energy and motion of particles.

Almost all substances expand with an increase in temperature and contract as the temperature decreases. (A very important exception is water.) These properties are the basis for the common bulb thermometer. The liquid in the thermometer expands and contracts more than the volume of the glass, producing changes in the column height of liquid. Figure 3.7 shows two different types of thermometers.

Several temperature scales with different units have been devised. **Scientists commonly use two equivalent units of temperature, the degree Celsius and the kelvin.** The Celsius scale of the metric system is named after the Swedish astronomer Anders Celsius (1701–1744). It uses two readily determined temperatures as reference temperature values: the freezing point and the boiling point of water. The **Celsius scale** sets the freezing point of water at 0°C and the boiling point of water at 100°C. The distance between these two fixed points is divided into 100 equal intervals, or degrees Celsius (°C).

Another temperature scale used in the physical sciences is the Kelvin, or absolute, scale. This scale is named for Lord Kelvin (1824–1907), a Scottish physicist and mathematician. On the **Kelvin scale,** the freezing point of water is 273.15 kelvins (K), and the boiling point is 373.15 K. Notice that with the Kelvin scale, the degree sign is not used.

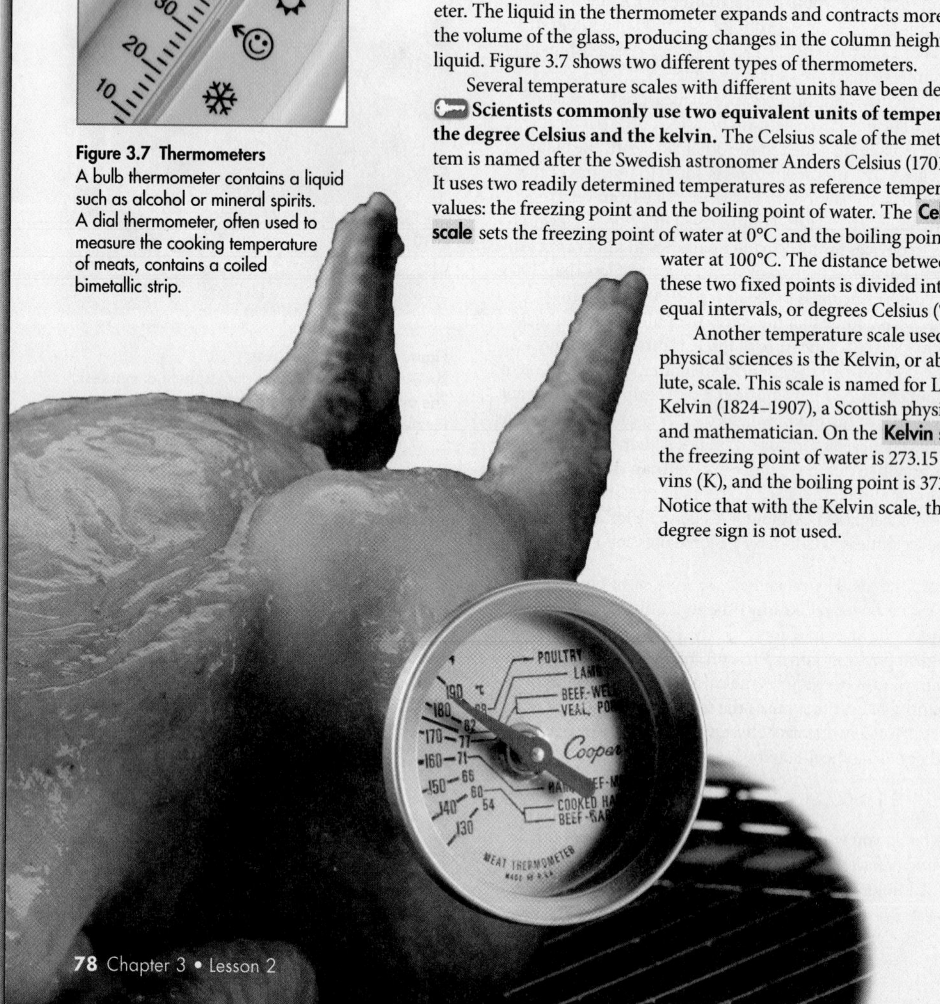

Figure 3.7 Thermometers
A bulb thermometer contains a liquid such as alcohol or mineral spirits. A dial thermometer, often used to measure the cooking temperature of meats, contains a coiled bimetallic strip.

Cryogenics

William Francis Giauque was awarded the 1949 Nobel Prize in chemistry for his research regarding the properties of matter at temperatures close to absolute zero. The study of phenomena at extremely low temperatures is known as cryogenics. Giauque's findings made it possible to calculate and predict chemical equilibria and the behavior of some chemical reactions.

Prior to Giauque's research, the coldest temperature achieved in a laboratory was 0.8 K in 1910 by another Nobel laureate, Heike Kamerlingh Onnes. By 2000, the coldest temperature achieved in a laboratory was 500 pK, or 0.000 000 000 5 K. In comparison, the lowest temperature known outside a laboratory is about 2.6 K, which is the estimated background temperature of deep space.

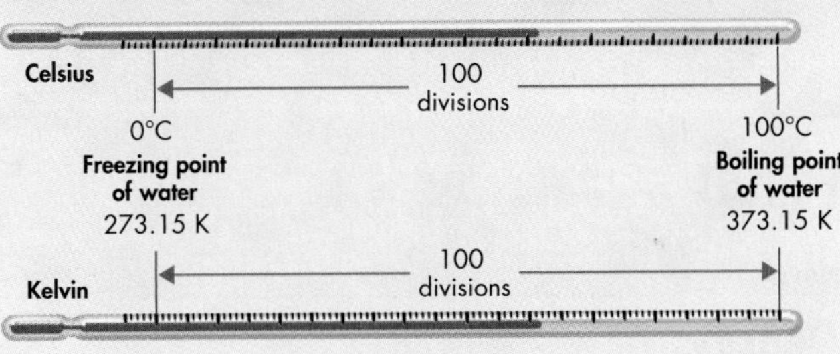

Figure 3.8 Temperature Scales
A 1°C change on the Celsius scale is equal to a 1 K change on the Kelvin scale.
Interpret Diagrams *What is a change of 10°C equivalent to on the Kelvin scale?*

Figure 3.8 compares the Celsius and Kelvin scales. A change of one degree on the Celsius scale is equivalent to one kelvin on the Kelvin scale. The zero point on the Kelvin scale, 0 K, or **absolute zero,** is equal to −273.15°C. For problems in this text, you can round −273.15°C to −273°C. Because one degree on the Celsius scale is equivalent to one kelvin on the Kelvin scale, converting from one temperature to another is easy. You simply add or subtract 273, as shown in the following equations:

$$K = °C + 273$$
$$°C = K - 273$$

CHEMISTRY & YOU

Q: *In a few countries, such as the United States, metric units are not commonly used in everyday measurements. What temperature units are used for a typical weather forecast in the United States? What about for a country that uses the metric system, such as Australia or Japan?*

SampleProblem 3.7

Converting Between Temperature Scales

Normal human body temperature is 37°C. What is this temperature in kelvins?

1 Analyze List the known and the unknown.
Use the known value and the equation K = °C + 273 to calculate the temperature in kelvins.

KNOWN
Temperature in °C = 37°C

UNKNOWN
Temperature in K = ? K

2 Calculate Solve for the unknown.

Substitute the known value for the Celsius temperature into the equation and solve.

$$K = °C + 273 = 37 + 273 = 310\text{ K}$$

3 Evaluate Does the result make sense? You should expect a temperature in this range, since the freezing point of water is 273 K and the boiling point of water is 373 K; normal body temperature is between these two values.

19. The element silver melts at 960.8°C and boils at 2212°C. Express these temperatures in kelvins.

20. Liquid nitrogen boils at 77.2 K. What is this temperature in degrees Celsius?

Foundations for Math

CHECK ANSWERS When converting between the Kelvin and Celsius scales, the conversion itself is straightforward. However, students often become confused as to when the conversion factor should be added and when it should be subtracted. Remind students that the value for a temperature in the Kelvin scale is always greater numerically than its equivalent on the Celsius scale.

In **Sample Problem 3.7**, 310 is numerically greater than 37, so the conversion is correct.

Explain

Temperature Scales

USE VISUALS Direct students to Figure 3.8. Have students compare the freezing point of water on the Celsius and Kelvin temperature scales *(0°C and 273.15 K)* **Ask** What temperature scale has its zero point at absolute zero? *(the Kelvin scale)* **Ask** Why do you think scientists use more than one temperature scale? *(Acceptable answers include that one scale may be more convenient than the other depending on what is being measured; for example, extremely low temperatures are all negative numbers when expressed in degrees Celsius but positive when expressed in kelvins.)*

Extend

Connect to PHYSICS

Ask students to determine a way to represent a temperature given in degrees Fahrenheit to the Kelvin scale. (Convert the temperature from °F to °C using the formula $°C = \frac{5}{9}(°F - 32)$ and then convert the temperature from °C to K.) Explain that absolute zero is defined as precisely 0 K on the Kelvin scale and as −273.15° on the Celsius scale. Ask students to research if there is another unit for temperature that has a relationship to the Fahrenheit scale that is similar to the relationship between the Celsius scale and the Kelvin scale. *(Absolute zero is precisely equivalent to 0 R on the Rankine scale, which is the Fahrenheit equivalent of the Kelvin scale. 0 R = −459.67° on the Fahrenheit scale.)*

CHEMISTRY & YOU In the United States, temperatures are typically reported in degrees Fahrenheit (°F). In countries that have adopted the metric system, temperatures are reported in degrees Celsius (°C).

Sample Practice Problem

At 10,000 feet above sea level, water boils at 90°C. What is this temperature in kelvins? *(363 K)*

Answers

FIGURE 3.8 10 K
19. melting point: 1234 K; boiling point: 2485 K
20. −196°C

Explain

Density

USE VISUALS Direct students to the photograph in Figure 3.9. **Ask** What property of a cranberry allows it to float on the water rather than sink to the bottom? *(Cranberries are less dense than water.)* **Ask** What measurements would you need to make to determine whether an object would float in water? *(Measure the object's volume and mass; then compute its density and compare the result to the density of water.)*

USE VISUALS Direct students' attention to Figure 3.10. Point out that the samples in the figure are cubes. **Ask** What would be the volume of a 10-g *sphere* of each of the substances shown? *(Li: 19 cm^3; H_2O: 10 cm^3; Al: 3.7 cm^3; and Pb: 0.88 cm^3)* Point out that even though the shape of the sample changed, its density did not. As long as the samples have the same mass and the same density, their shape has no bearing on their volume.

MAKE A CONNECTION If students performed the *Mass of a Penny* activity, have them refer to the explanation of the results. **Ask** How can you use density to test the hypothesis that the composition of the penny changed? *(Determine the volume and mass of a pre- and post-1982 penny. Calculate the density of each and compare them. If the densities changed, the composition of the penny must have changed.*

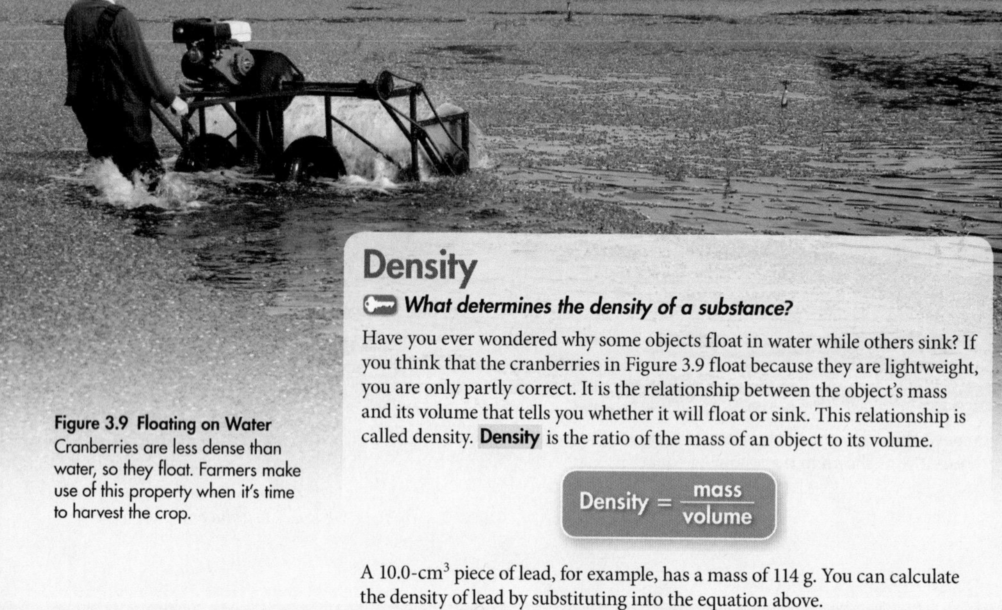

Figure 3.9 Floating on Water
Cranberries are less dense than water, so they float. Farmers make use of this property when it's time to harvest the crop.

Density

What determines the density of a substance?

Have you ever wondered why some objects float in water while others sink? If you think that the cranberries in Figure 3.9 float because they are lightweight, you are only partly correct. It is the relationship between the object's mass and its volume that tells you whether it will float or sink. This relationship is called density. **Density** is the ratio of the mass of an object to its volume.

$$\text{Density} = \frac{\text{mass}}{\text{volume}}$$

A 10.0-cm^3 piece of lead, for example, has a mass of 114 g. You can calculate the density of lead by substituting into the equation above.

$$\frac{114\text{ g}}{10.0\text{ cm}^3} = 11.4\text{ g/cm}^3$$

Note that when mass is measured in grams, and volume in cubic centimeters, density has units of grams per cubic centimeter (g/cm^3). The SI unit of density is kilograms per cubic meter (kg/m^3).

Figure 3.10 compares the density of four substances: lithium, water, aluminum, and lead. Why does each 10-g sample have a different volume? The volumes vary because the substances have different densities. **Density is an intensive property that depends only on the composition of a substance, not on the size of the sample.** With a mixture, density can vary because the composition of a mixture can vary.

Figure 3.10 Comparing Densities
A 10-g sample of pure water has less volume than 10 g of lithium, but more volume than 10 g of lead or 10 g of aluminum. The faces of the cubes are shown actual size.
Predict *Which of the solids shown will sink in water?*

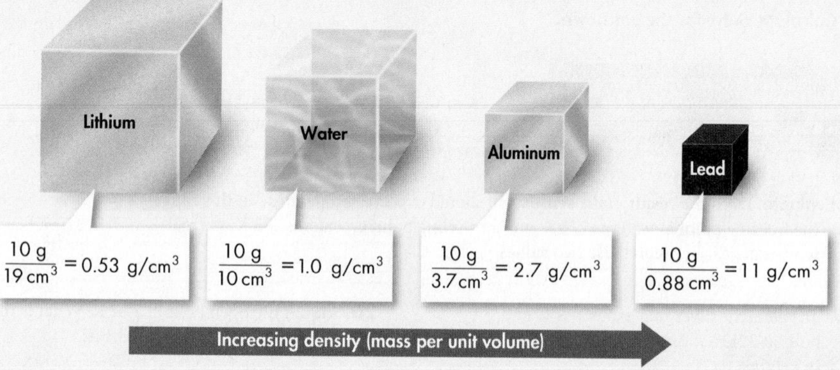

Differentiated Instruction

ELL ENGLISH LANGUAGE LEARNERS Have students find photos of objects containing substances from Table 3.6. They could label each photo with the substance's name in multiple languages as well as the density, whose units are part of a universal language.

LPR LESS PROFICIENT READERS Using context clues, have students orally describe in their own words, or with an analogy, the meaning of the term *intensive property*. **Ask** What other properties may be intensive? *(Acceptable answers include temperature and color.)*

L3 ADVANCED STUDENTS Point out that the specific gravity is the ratio of the density of a material to that of water. Have students design and carry out an experiment for determining the specific gravity of two different liquids. As a hint, you could suggest that they build and use a hydrometer in their experiment. Ask students to prepare a lab report detailing their research.

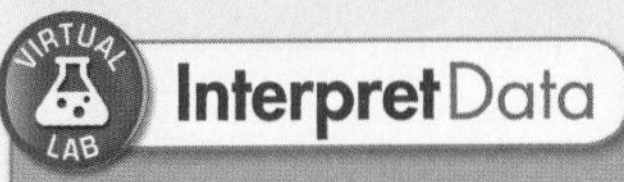

Interpret Data

Densities of Some Common Materials			
Solids and Liquids		Gases	
Material	Density at 20°C (g/cm^3)	Material	Density at 20°C (g/L)
Gold	19.3	Chlorine	2.95
Mercury	13.6	Carbon dioxide	1.83
Lead	11.3	Argon	1.66
Aluminum	2.70	Oxygen	1.33
Table sugar	1.59	Air	1.20
Corn syrup	1.35–1.38	Nitrogen	1.17
Water (4°C)	1.000	Neon	0.84
Corn oil	0.922	Ammonia	0.718
Ice (0°C)	0.917	Methane	0.665
Ethanol	0.789	Helium	0.166
Gasoline	0.66–0.69	Hydrogen	0.084

Table 3.6 Density is the mass per unit volume of a material.

a. Compare Why do you think the densities of the gases are reported in units that are different from those used for the densities of the solids and liquids?

b. Predict Would a balloon filled with carbon dioxide sink or rise in air? Explain.

c. Infer Why are the densities of corn syrup and gasoline expressed as a range of values?

Note the units here: Densities of the solids and liquids are expressed in g/cm^3. Densities of the gases are expressed in g/L.

What do you think will happen if corn oil is poured into a container of water? Using Table 3.6, you can see that the density of corn oil is less than the density of water. For that reason, the oil floats on top of the water. Figure 3.11 shows different liquids forming distinct layers in a container due to differences in density. For example, the corn syrup (colored red), sinks below the water (colored green) because the density of corn syrup is greater than the density of water.

You have probably seen a helium-filled balloon rapidly rise to the ceiling when it is released. Whether a gas-filled balloon will sink or rise when released depends on how the density of the gas compares with the density of air. Helium is less dense than air, so a helium-filled balloon rises. The densities of various gases are listed in Table 3.6.

What happens to the density of a substance as its temperature increases? Experiments show that the volume of most substances increases as the temperature increases. Meanwhile, the mass remains the same despite the temperature and volume changes. Remember that density is the ratio of an object's mass to its volume. So if the volume changes with temperature (while the mass remains constant), then the density must also change with temperature. The density of a substance generally decreases as its temperature increases. As you will learn in Chapter 15, water is an important exception. Over a certain range of temperatures, the volume of water increases as its temperature decreases. Ice, or solid water, floats because it is less dense (0.917 g/cm^3) than liquid water (1.000 g/cm^3).

Figure 3.11 Liquid Layers Because of differences in density, the liquids separate into layers.

Compare *Is the blue-colored liquid more or less dense than water?*

Professional Development Note

Specific Gravity

Specific gravity is the ratio of the density of a substance to the density of a reference substance.

$$\text{Specific gravity} = \frac{\text{density of substance}}{\text{density of reference substance}}$$

Specific gravity is expressed without units. The reference substance for solids and liquids is usually water at 4°C. The density of water at that temperature is 1.0000 g/mL.

Physicians use the measured specific gravity of a patient's urine to help diagnose certain diseases, such as diabetes. Auto mechanics measure the specific gravity of antifreeze in radiators during tune-ups.

Explore

Class Activity

PURPOSE Students will calculate density from a mass-volume graph.

MATERIALS 3 or 4 different-sized cubes of a material such as wood, metal, or marble; metric ruler; balance

PROCEDURE Have student volunteers measure the masses of the cubes, measure the lengths of the cubes, and calculate their volumes. On the board or an overhead projector, record the masses and volumes. Construct a mass-volume graph (mass on the *y*-axis) and plot the data. Remind students how to calculate the slope of the line using rise over run. **Ask** What does the slope of this line indicate? *(Elicit that the slope of the line is the ratio of mass/volume and has the units of g/cm³, which is the unit of density.)*

EXPECTED OUTCOME The mass and volume data should lie along a straight line. **Ask** What is the ratio of mass to volume of a substance? *(density)*

Answers

FIGURE 3.10 aluminum and lead

FIGURE 3.11 more dense

INTERPRET DATA a. Gases at ambient temperature and pressure take up much more volume than liquids or solids. If g/cm^3 were used as the unit for gases, the reported values would be very small numbers. **b.** Sink; carbon dioxide is more dense than air. **c.** Corn syrup and gasoline are mixtures with variable composition, so they have variable density.

LESSON 3.2

Explain

Density

Sample Practice Problem

What is the density of an object whose mass is 142.5 grams and whose volume is 7.5 cm^3? *(19 g/cm³)*

Extend

Remind students of other equations they have manipulated, such as the formula for the area of a triangle, and have students state the properties of equality they used to solve the formula for a specified variable. Then ask students to show how the density equation can be manipulated to determine the mass or volume of an object, if the density of a substance is known. *(mass/density = volume; volume × density = mass)*

Evaluate

Informal Assessment

Have students identify the most common unit of density and explain why density is a derived SI unit rather than a base unit. *(g/cm³; it is a derived unit because it is calculated from units of mass and length. A gram is 10^{-3} kilogram, and a centimeter is 10^{-2} meter. Kilograms and meters are base SI units.)* **Ask** How can density be used to determine whether a metal washer is aluminum or zinc? *(Measure and compare the density of the washer to the accepted density values of aluminum and zinc.)* Then have students complete the 3.2 Lesson Check.

Reteach

Review Tables 3.3, 3.4, 3.5, and 3.6 with students. Review the SI base unit listed in each table, pronunciations of the derived units, and the use of the prefixes to determine the numerical relationships between the base unit and the derived units.

Sample Problem 3.8

Calculating Density

A copper penny has a mass of 3.1 g and a volume of 0.35 cm^3. What is the density of copper?

1 Analyze List the knowns and the unknown. Use the known values and the equation for density to solve the problem.

KNOWNS	UNKNOWN
mass = 3.1 g	density = ? g/cm^3
volume = 0.35 cm^3	

2 Calculate Solve for the unknown.

Start with the equation for density.

$$\text{Density} = \frac{\text{mass}}{\text{volume}}$$

Substitute the known values for mass and volume and then calculate.

$$\text{Density} = \frac{3.1\ g}{0.35\ cm^3} = 8.8571\ g/cm^3 = 8.9\ g/cm^3$$

The calculated answer must be rounded to two significant figures.

3 Evaluate Does the result make sense? A piece of copper with a volume of about 0.3 cm^3 has a mass of about 3 grams. About three times that volume of copper, 1 cm^3, should have a mass three times larger, about 9 grams. This estimate is close to the calculated result.

21. A student finds a shiny piece of metal that she thinks is aluminum. In the lab, she determines that the metal has a volume of 245 cm^3 and a mass of 612 g. Calculate the density. Is the metal aluminum?

22. A bar of silver has a mass of 68.0 g and a volume of 6.48 cm^3. What is the density of silver?

3.2 LessonCheck

ONLINE PROBLEMS

23. Review Why are metric units easy to use?

24. Identify What temperature units do scientists commonly use?

25. Review What determines density?

26. Identify Write the name and symbol of the SI units for mass, length, volume, and temperature.

27. Define Write the symbol and meaning of each prefix below.

a. *milli-* **b.** *nano-* **c.** *deci-* **d.** *centi-*

28. List Arrange the following units in order from largest to smallest: m^3, mL, cL, μL, L, dL.

29. Calculate What is the volume of a paperback book 21 cm tall, 12 cm wide, and 3.5 cm thick?

30. Compare State the difference between mass and weight.

31. Calculate Surgical instruments may be sterilized by heating at 170°C for 1.5 hr. Convert 170°C to kelvins.

32. Calculate A weather balloon is inflated to a volume of 2.2×10^3 L with 374 g of helium. What is the density of helium in grams per liter?

33. Apply Concepts A 68-g bar of gold is cut into three equal pieces. How does the density of each piece compare to the density of the original gold bar?

34. Interpret Data Look up the densities of the elements in Group 1A on page R2. Which Group 1A elements are less dense than pure water at 4°C?

35. Explain How does density vary with temperature?

Lesson Check Answers

23. All metric units are based on multiples of 10, which makes them easy to use.

24. the degree Celsius and the kelvin

25. Density is an intensive property that depends only on the composition of a substance, not on the size of the sample. Density = mass/volume

26. mass, kilogram (kg); length, meter (m); volume, cubic meter (m^3); temperature, kelvin (K)

27. **a.** m; 10^{-3} of the unit
b. n; 10^{-9} of the unit
c. d; 10^{-1} of the unit
d. c; 10^{-2} of the unit

28. m^3, L, dL, cL, mL, μL

29. 8.8×10^2 cm^3

30. Mass is a measure of the amount of matter in an object. Weight is a measure of the force of gravity on an object.

31. 443 K

32. 1.7×10^{-1} g/L

33. All the densities are equal.

34. Li, Na, and K are less dense than water.

35. Density generally decreases when temperature increases.

Carbon Footprints

To measure a footprint, you might use units such as centimeters or inches. But what about a carbon footprint? A carbon footprint is a measure of how much greenhouse gas is released into the atmosphere by a person, country, or industry. Greenhouse gases, such as carbon dioxide (CO_2) and methane (CH_4), are gases that contribute to global warming.

Any activity that involves the burning of fossil fuels results in carbon dioxide emissions. Car travel, air travel, home heating and cooling, and electricity usage all add to an individual's carbon footprint. Your own carbon footprint is the total mass of CO_2 that you put into the atmosphere over the course of a year. This quantity can be expressed in metric tons (t) of CO_2 per year. A metric ton equals 1000 kg. So, the units of your carbon footprint can be abbreviated as: t CO_2/yr or 10^3 kg CO_2/yr.

CARBON COSTS Your choices affect the size of your carbon footprint. For example, using a clothes dryer consumes electricity, but hanging wet laundry on a clothesline does not. The things you buy also contribute to your carbon footprint. Not only is energy required to make these goods, but the goods themselves (such as TVs) may consume energy.

FOOTPRINT UNITS The carbon footprint of fresh produce can be expressed in g CO_2 per serving. Cars require different units: kg CO_2 per gallon of gasoline. For planes, the units are kg CO_2 per passenger mile.

Take It Further

1. **Calculate** A car emits 8.6 kg of CO_2 per gallon of unleaded gas. How much CO_2 is produced if the car burns 2.5 gal of fuel?
2. **Infer** What factors do you think determine the carbon footprint of an apple? Why might the carbon footprints of two apples in the same store differ substantially?

21st Century Learning To be successful in the 21st century, students need skills and learning experiences that extend beyond subject matter mastery. The following project helps students build the following 21st Century Skills: *Global Awareness; Financial, Economic, Business, and Entrepreneurial Literacy; Creativity and Innovation; Communication and Collaboration; Information, Media, and Technology Skills; Initiative and Self-Direction; and Productivity and Accountability.*

Divide the class into groups of three to five students. Pose the following challenge to students: *Estimate your individual personal carbon footprint for a day. As a group, compile your results and prepare a graphic representation to depict the activities that contribute the most to the personal carbon footprint of your group.* Students should include possible strategies they could implement to reduce that amount.

CHEMISTRY & YOU The idea of a *carbon footprint* is a relatively new concept, and it is not always easy to understand just what determines an object's or an individual's personal carbon footprint. Explain to students first how a particular object's carbon footprint is determined. Take a common classroom item such as a chair as an example. Explain how each step in production and transportation of that object adds to its carbon footprint. For example, consider the CO_2 released by the factory while the chair is produced and the CO_2 released by the delivery truck that transported the chair from the factory to the warehouse to the school.

After explaining factors that contribute to an individual object's carbon footprint, talk to students about how those objects then contribute to every individual's own carbon footprint. Pose the following question to students: *Which would contribute more to your personal carbon footprint, driving a car to school or riding a bus?* You may need to assist students in the following ways:

- A bus produces more CO_2 per mile.
- A bus carries significantly more people per trip.
- Sharing a ride with other people reduces a person's individual carbon footprint.

Extend

Connect to ECONOMICS

A major issue in the world today concerning carbon footprints is that with current technologies, it often costs significantly more to create a product with a low carbon footprint. For example, an LED-type bulb has a lower carbon footprint than a similar incandescent bulb because it uses less energy and lasts significantly longer. However, an LED bulb is more expensive to produce than its incandescent counterpart.

Have students work in groups of three to discover examples in the world around them where the issue of price versus carbon footprint might be an issue, and report their findings to the class.

Answers

21. density = 2.50 g/cm^3; no
22. 10.5 g/cm^3

TAKE IT FURTHER

1. 22 kg CO_2
2. Factors might include the amount of fertilizer used to help the apple tree grow and the amount of energy used to harvest, package, and ship the apples. Two apples in the same store might come from different orchards and require different amounts of energy to grow, harvest, package, and ship.

LESSON 3.3

Key Objectives

3.3.1 EXPLAIN what happens when a measurement is multiplied by a conversion factor.

3.3.2 DESCRIBE the kinds of problems that can be easily solved using dimensional analysis.

Additional Resources

- Reading and Study Workbook, Lesson 3.3
- Core Teaching Resources, Lesson 3.3 Review

PearsonChem.com

Engage

CHEMISTRY & YOU Have students examine the photograph that opens the section. **Ask** Has anyone ever noticed a chart or table in a bank or in the newspaper relating the values of foreign currency to the U.S. dollar? *(yes or no)* Explain that these are tables are called currency conversion tables. These are the daily values that allow people to relate one currency to another. **Ask** How could you know for certain which amount of money is worth more—75 euros or 75 British pounds? *(Convert these values to a familiar currency—U.S. dollars.)*

Activate Prior Knowledge

Provide various types of measuring devices to the class. Tell students to select one device and record the length of their shoe. Then ask them to create a chart and record the length of their shoe in inches, feet, yards, millimeters, centimeters, and meters. As a class, discuss how students determined the measurements. Begin a list of conversions on the word wall.

3.3 Solving Conversion Problems

CHEMISTRY & YOU

Q: *How can you convert U.S. dollars to euros?* Perhaps you have traveled to another country or are planning to do so. If so, you know (or will soon discover) that different countries have different currencies. As a tourist, exchanging money is an important part of having a good trip. After all, you often must use cash to pay for meals, transportation, and souvenirs. Because each country's currency compares differently with the U.S. dollar, knowing how to convert currency units correctly is essential. Conversion problems are readily solved by a problem-solving approach called dimensional analysis.

Key Questions

What happens when a measurement is multiplied by a conversion factor?

What kinds of problems can you solve using dimensional analysis?

Vocabulary

- conversion factor
- dimensional analysis

Conversion Factors

What happens when a measurement is multiplied by a conversion factor?

If you think about any number of everyday situations, you will realize that a quantity can usually be expressed in several different ways. For example, consider the monetary amount $1.

1 dollar = 4 quarters = 10 dimes = 20 nickels = 100 pennies

These are all expressions, or measurements, of the same amount of money. The same thing is true of scientific quantities. For example, consider a distance that measures exactly 1 meter.

1 meter = 10 decimeters = 100 centimeters = 1000 millimeters

These are different ways to express the same length.

Whenever two measurements are equivalent, a ratio of the two meaurements will equal 1, or unity. For example, you can divide both sides of the equation 1 m = 100 cm by 1 m or by 100 cm.

$$\frac{1\text{m}}{1\text{m}} = \frac{100\text{ cm}}{1\text{ m}} = 1 \quad \text{or} \quad \frac{1\text{ m}}{100\text{ cm}} = \frac{100\text{ cm}}{100\text{ cm}} = 1$$

conversion factors

The ratios 100 cm/1 m and 1 m/100 cm are examples of conversion factors. A **conversion factor** is a ratio of equivalent measurements. The measurement in the numerator (on the top) is equivalent to the measurement in the denominator (on the bottom). The conversion factors shown above are read "one hundred centimeters per meter" and "one meter per hundred centimeters."

Focus on ELL

1 CONTENT AND LANGUAGE Present academic vocabulary for this lesson, such as *analyze*, *calculate*, and *evaluate*. Pronounce the words and have students repeat them. Provide explanations, examples, and visuals of each word so students are aware of what tasks lie ahead when they encounter these words.

2 FRONTLOAD THE LESSON Preview the conversion factors and Figure 3.12 in the text, and point out that conversion factors are fractions that contain both numbers and units of measurement. Briefly review the basics of working with fractions, including terminology and equivalent fractions.

3 COMPREHENSIBLE INPUT When writing problems on the board, assign a different color to each unique unit of measurement in the problem. This will assist students in understanding that a unit in the numerator can only be cancelled by an equivalent unit in the denominator.

1 meter

100 centimeters

Smaller number → $\frac{1 \text{ m}}{100 \text{ cm}}$ ← Larger unit

Larger number → ← Smaller unit

Figure 3.12 Conversion Factor
The two parts of a conversion factor, the numerator and the denominator, are equal.

See conversion factors animated online.

Figure 3.12 illustrates another way to look at the relationships in a conversion factor. Notice that the smaller number is part of the measurement with the larger unit. That is, a meter is physically larger than a centimeter. The larger number is part of the measurement with the smaller unit.

Conversion factors are useful in solving problems in which a given measurement must be expressed in some other unit of measure. **When a measurement is multiplied by a conversion factor, the numerical value is generally changed, but the actual size of the quantity measured remains the same.** For example, even though the numbers in the measurements 1 g and 10 dg (decigrams) differ, both measurements represent the same mass. In addition, conversion factors within a system of measurement are defined quantities or exact quantities. Therefore, they have an unlimited number of significant figures and do not affect the rounding of a calculated answer.

Here are some additional examples of pairs of conversion factors written from equivalent measurements. The relationship between grams and kilograms is 1000 g = 1 kg. The conversion factors are

$$\frac{1000 \text{ g}}{1 \text{ kg}} \text{ and } \frac{1 \text{ kg}}{1000 \text{ g}}$$

Figure 3.13 shows a scale that can be used to measure mass in grams or kilograms. If you read the scale in terms of grams, you can convert the mass to kilograms by multiplying by the conversion factor 1 kg/1000 g.

The relationship between nanometers and meters is given by the equation 10^9 nm = 1 m. The possible conversion factors are

$$\frac{10^9 \text{ nm}}{1 \text{ m}} \text{ and } \frac{1 \text{ m}}{10^9 \text{ nm}}$$

Common volumetric units used in chemistry include the liter and the microliter. The relationship 1 L = 10^6 μL yields the following conversion factors:

$$\frac{1 \text{ L}}{10^6 \text{ μL}} \text{ and } \frac{10^6 \text{ μL}}{1 \text{ L}}$$

Based on what you have learned about metric prefixes, you should easily be able to write conversion factors that relate equivalent metric quantities.

Figure 3.13 Measuring Mass
This scale is calibrated to measure mass to the nearest 20 g.
Interpret Photos *What is the scale showing in grams? In kilograms?*

Monetary Exchange Rates

The conversion of chemical units is similar to the exchange of currency. Americans who travel outside the United States must exchange U.S. dollars for foreign currency at a given rate of exchange. These exchange rates vary from day to day. The daily exchange rates affect all international monetary transactions. Each time one type of money is exchanged for another, the current exchange rate serves as a conversion factor. International currency traders keep track of exchange rates 24 hours a day through a linked computer network.

Foundations for Reading

BUILD VOCABULARY Have students write definitions of *conversion factor* and *dimensional analysis* in their own words. *(Sample answers: a ratio of equivalent measurements used to convert a quantity from one unit to another; a technique of problem-solving that uses the units that are part of a measurement to solve the problem.)*

Explain

Conversion Factors

USE VISUALS Have students inspect Figure 3.12. Emphasize that a conversion factor relates two equivalent measurements. **Ask** What two parts does every measurement have? *(a number and a unit)* Point out that if this is so, then every conversion factor must contain two numbers and two units so that one number and its unit equal another number and its unit.

Explore

Class Activity

PURPOSE Students will use dimensional analysis to convert common units.

MATERIALS copies of a recipe, lists of equivalents and conversions among the following measurements:

teaspoon, tablespoon, 1/4 cup, 1/2 cup, and 1 cup

(These lists are found in most cookbooks.)

PROCEDURE Distribute the recipe and the conversion list to pairs of students. Explain that they must rewrite the recipe so that it can feed six times the number of serving sizes suggested by the recipe. Point out that it would be tedious to have to measure out a particular ingredient (pick out one) in teaspoons or tablespoons six times, so students must rewrite the recipe in appropriately larger units. After students have rewritten the recipe, have student pairs exchange and compare recipes.

EXPECTED OUTCOME Students should use the conversion lists to write simple conversion factors, such as 3 teaspoons/1 tablespoon, and then rewrite the recipe using larger measurements.

Answers

FIGURE 3.13 The scale is showing 220 grams, or 0.22 kg.

LESSON 3.3

Explain

Dimensional Analysis

START A CONVERSATION Explain that dimensional analysis is an extremely powerful problem-solving tool. Learning this method requires extra effort on the part of students. They must often use multiple conversion factors. The extra effort can be justified because the proper manipulation of the units assures accurate manipulation of the numbers.

Emphasize that students can use dimensional analysis as a tool for solving nearly all of the problems they encounter in chemistry. Their first question about any quantity should be, "What are the units of this quantity?" By comparing the units of various quantities in a problem, students can discover whether they need to perform any unit conversions before proceeding.

Sample Practice Problem

Mary worked 43,200 seconds this week. How many hours did Mary work this week? *(12 hours)*

Dimensional Analysis

What kinds of problems can you solve using dimensional analysis?

Some problems are best solved using algebra. For example, converting a kelvin temperature to Celsius can be done by using the equation °C = K − 273. Many problems in chemistry are conveniently solved using dimensional analysis. **Dimensional analysis** is a way to analyze and solve problems using the units, or dimensions, of the measurements. The best way to explain this technique is to use it to solve an everyday situation, as in Sample Problem 3.9.

As you read Sample Problem 3.10, you might see how the same problem could be solved algebraically but is more easily solved using dimensional analysis. In either case, you should choose the problem-solving method that works best for you. Try to be flexible in your approach to problem solving, as no single method is best for solving every type of problem.

SampleProblem 3.9

Using Dimensional Analysis

How many seconds are in a workday that lasts exactly eight hours?

1 Analyze **List the knowns and the unknown.**
To convert time in hours to time in seconds, you'll need two conversion factors. First you must convert hours to minutes: h ⟶ min. Then you must convert minutes to seconds: min ⟶ s. Identify the proper conversion factors based on the relationships 1 h = 60 min and 1 min = 60 s.

KNOWNS	UNKNOWN
time worked = 8 h	seconds worked = ? s
1 hour = 60 min	
1 minute = 60 s	

2 Calculate **Solve for the unknown.**

The first conversion factor is based on 1 h = 60 min. The unit hours must be in the denominator so that the known unit will cancel.

$$\frac{60\ \text{min}}{1\ \text{h}}$$

The second conversion factor is based on 1 min = 60 s. The unit minutes must be in the denominator so that the desired units (seconds) will be in your answer.

$$\frac{60\ \text{s}}{1\ \text{min}}$$

Before you do the actual arithmetic, it's a good idea to make sure that the units cancel and that the numerator and denominator of each conversion factor are equal to each other.

Multiply the time worked by the conversion factors.

$$8\ \cancel{\text{h}} \times \frac{60\ \cancel{\text{min}}}{1\ \cancel{\text{h}}} \times \frac{60\ \text{s}}{1\ \cancel{\text{min}}} = 28{,}800\ \text{s} = 2.8800 \times 10^4\ \text{s}$$

3 Evaluate **Does the result make sense?** The answer has the desired unit (s). Since the second is a small unit of time, you should expect a large number of seconds in 8 hours. The answer is exact since the given measurement and each of the conversion factors is exact.

36. How many minutes are there in exactly one week?

37. How many seconds are in an exactly 40-hour work week?

Foundations for Math

CONVERSION FACTORS Point out that a conversion factor is a ratio of equivalent *measurements*–that is, the numerator should contain both a number and a unit, and the denominator should contain both a number and a unit. For example, the relationship 60 minutes = 1 hour yields the following conversion factors:

$$\frac{60\ \text{minutes}}{1\ \text{hour}} \text{ and } \frac{1\ \text{hour}}{60\ \text{minutes}}$$

Knowing the unit of the given quantity and the desired unit of the answer dictates which conversion factor to use. To convert from hours to minutes, the denominator of the conversion factor must contain hours in order for the known unit to cancel, so you would use 60 min/1 h. Note that **Sample Problem 3.9** involves two conversions (from hours to minutes, and from minutes to seconds), each requiring its own conversion factor.

In **Sample Problem 3.9**, note that two such expressions are used, $\frac{60\ \text{minutes}}{1\ \text{hour}}$ and $\frac{60\ \text{seconds}}{1\ \text{minute}}$. This reflects the fact that for each equivalent relationship there are two possible conversion factors that can be written.

Sample Problem 3.10

Using Dimensional Analysis

The directions for an experiment ask each student to measure 1.84 g of copper (Cu) wire. The only copper wire available is a spool with a mass of 50.0 g. How many students can do the experiment before the copper runs out?

1 Analyze **List the knowns and the unknown.**
From the known mass of copper, use the appropriate conversion factor to calculate the number of students who can do the experiment. The desired conversion is mass of copper ⟶ number of students.

KNOWNS
mass of copper available = 50.0 g Cu
Each student needs 1.84 grams of copper.

UNKNOWN
number of students = ?

2 Calculate **Solve for the unknown.**

The experiment calls for 1.84 grams of copper per student. Based on this relationship, you can write two conversion factors.

$$\frac{1.84\text{ g Cu}}{1\text{ student}} \text{ and } \frac{1\text{ student}}{1.84\text{ g Cu}}$$

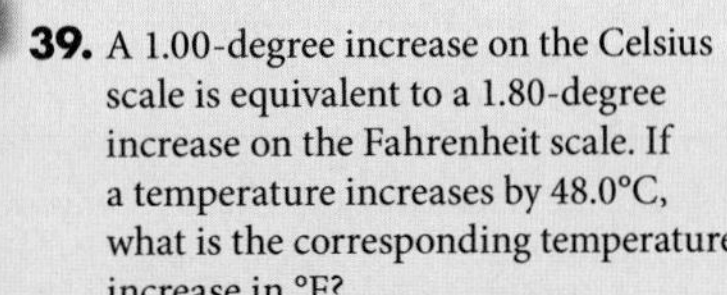

Because the desired unit for the answer is students, use the second conversion factor. Multiply the mass of copper by the conversion factor.

$$50.0\text{ g Cu} \times \frac{1\text{ student}}{1.84\text{ g Cu}} = 27.174\text{ students} = 27\text{ students}$$

3 Evaluate **Does the result make sense?** The unit of the answer (students) is the one desired. You can make an approximate calculation using the following conversion factor.

$$\frac{1\text{ student}}{2\text{ g Cu}}$$

Multiplying the above conversion factor by 50 g Cu gives the approximate answer of 25 students, which is close to the calculated answer.

38. An experiment requires that each student use an 8.5-cm length of magnesium ribbon. How many students can do the experiment if there is a 570-cm length of magnesium ribbon available?

Here's a tip: The equalities needed to write a particular conversion factor may be given in the problem. In other cases, you'll need to know or look up the necessary equalities.

39. A 1.00-degree increase on the Celsius scale is equivalent to a 1.80-degree increase on the Fahrenheit scale. If a temperature increases by 48.0°C, what is the corresponding temperature increase in °F?

40. An atom of gold has a mass of 3.271×10^{-22} g. How many atoms of gold are in 5.00 g of gold?

Foundations for Math

ROUNDING Explain to students that some problems require a more logical rounding technique than using significant figures or decimal places. For example, money is typically rounded to the hundredths place because our money system includes pennies. The number of buses needed to transport a certain number of students on a field trip would be rounded UP to the next whole bus to make sure all students are accommodated. Have students think of other examples where rounding would not follow the conventional rules.

In **Sample Problem 3.10,** logically there is no such thing as 0.174 student, so the result of the calculation is rounded to a whole number. Explain that the number is rounded DOWN (unlike in the bus example above) because there is not enough copper for another student to do the experiment.

ADDRESS MISCONCEPTIONS Students may think that because a conversion factor equals unity, it does not matter whether the conversion factor or its reciprocal is used in a calculation. Remind students that a given pair of equivalent measurements yields two different conversion factors, only one of which can be used to obtain the correct conversion.

Sample Practice Problem

An experiment requires that each student use 8 grams of baking soda. If a full container of baking soda holds 340 grams, how many students can do the experiment? *(42 students)*

Answers

36. 1.0080×10^4 min
37. 1.44000×10^5 s
38. 67 students
39. 86.4°F
40. 1.53×10^{22} atoms

LESSON 3.3

Explain

START A CONVERSATION Explain that measurements are often made using one unit and then converted into a related unit before being used in calculations. For example, students might measure volume in liters or milliliters in the laboratory, but express it as cubic centimeters in a calculation. Explain to the students that conversions are done using conversion factors. Emphasize that these conversion factors are ratios of *equivalent physical quantities,* such as 1 mL/1 cm^3.

MAKE A CONNECTION Reassure students that they are more familiar with conversions than they may realize. Point out examples of everyday conversions, such as converting money from cents to dollars and converting time from minutes to hours. Start out by giving them practice with everyday examples.
Ask A chicken needs to be cooked 20 minutes for each pound it weighs. How long should the chicken be cooked if it weighs 4.5 pounds? *(4.5 lb × 20 min/lb = 90 min; 90 min × 1 h/60 min = 1.5 h. Most students will automatically relate 90 minutes to 1.5 hours. This may help them become comfortable with the process.)* If students are having difficulty with conversion factors, provide a more tactile environment for students to discover these relationships.

Divide the class in half and have each group challenge the other to write the conversion factor given two related units. Remind them that each conversion factor can appear in two forms depending on which value they put in the numerator.

CHEMISTRY & YOU Sample answer (the conversion factor will vary with the exchange rate): \$50 × (€1/\$1.360) = €36.76

Sample Practice Problem

Make the following conversions.

a. 0.045 L to cubic centimeters *(4.5×10^1 cm^3)*
b. 14.3 mg to grams *(1.43×10^{-2} g)*
c. 0.0056 m to micrometers *(5.6×10^3 μm)*
d. 0.035 cm to millimeters *(3.5×10^{-1} mm)*

CHEMISTRY & YOU

Q: *Look up the exchange rate between U.S. dollars and euros on the Internet. Write a conversion factor that allows you to convert from U.S. dollars to euros. How many euros could you buy with \$50?*

Simple Unit Conversions In chemistry, as in everyday life, you often need to express a measurement in a unit different from the one given or measured initially. **Dimensional analysis is a powerful tool for solving conversion problems in which a measurement with one unit is changed to an equivalent measurement with another unit.** Sample Problems 3.11 and 3.12 walk you through how to solve simple conversion problems using dimensional analysis.

Sample Problem 3.11

Converting Between Metric Units

Express 750 dg in grams. (Refer to Table 3.2 if you need to refresh your memory of metric prefixes.)

1 Analyze List the knowns and the unknown.
The desired conversion is decigrams ⟶ grams. Multiply the given mass by the proper conversion factor.

KNOWNS	UNKNOWN
mass = 750 dg	mass = ? g
1 g = 10 dg	

2 Calculate Solve for the unknown.

Use the relationship 1 g = 10 dg to write the correct conversion factor. $\frac{1\,g}{10\,dg}$

Multiply the known mass by the conversion factor. $750\,\cancel{dg} \times \frac{1\,g}{10\,\cancel{dg}} = 75\,g$

Note that the known unit (dg) is in the denominator and the unknown unit (g) is in the numerator.

3 Evaluate Does the result make sense? Because the unit gram represents a larger mass than the unit decigram, it makes sense that the number of grams is less than the given number of decigrams. The answer has the correct unit (g) and the correct number of significant figures.

41. Using tables from this chapter, convert the following:
a. 0.044 km to meters
b. 4.6 mg to grams
c. 0.107 g to centigrams

42. Convert the following:
a. 15 cm^3 to liters
b. 7.38 g to kilograms
c. 6.7 s to milliseconds
d. 94.5 g to micrograms

Foundations for Math

MORE ON METRIC CONVERSIONS Explain to students that it is possible to convert "through" the basic metric unit in order to quickly go from very large units to very small units or vice versa. This requires more than one conversion factor in the dimensional analysis calculation, each of which involves the basic metric unit. Demonstrate this example: How many centigrams are there in 0.8 kilograms?

$$0.8\,\cancel{kg} \times \frac{1000\,\cancel{g}}{1\,\cancel{kg}} \times \frac{100\,cg}{1\,\cancel{g}} = 80{,}000\,cg$$

Point out that the desired unit (cg) is placed in the numerator of the last conversion factor. In **Sample Problem 3.11**, point out that the initial unit is placed in the denominator of the conversion factor and the desired unit is placed in the numerator.

Sample Problem 3.12

Using Density as a Conversion Factor

What is the volume of a pure silver coin that has a mass of 14 g? The density of silver (Ag) is 10.5 g/cm^3.

1 Analyze List the knowns and the unknown. You need to convert the mass of the coin into a corresponding volume. The density gives you the following relationship between volume and mass: 1 cm^3 Ag = 10.5 g Ag. Multiply the given mass by the proper conversion factor to yield an answer in cm^3.

KNOWNS	UNKNOWN
mass = 14 g	volume of coin = ? cm^3
density of silver = 10.5 g/cm^3	

2 Calculate Solve for the unknown.

Use the relationship 1 cm^3 Ag = 10.5 g Ag to write the correct conversion factor.

$$\frac{1\ cm^3\ Ag}{10.5\ g\ Ag}$$

Notice that the known unit (g) is in the denominator and the unknown unit (cm^3) is in the numerator.

Multiply the mass of the coin by the conversion factor.

$$14\ \cancel{g\ Ag} \times \frac{1\ cm^3\ Ag}{10.5\ \cancel{g\ Ag}} = 1.3\ cm^3\ Ag$$

3 Evaluate Does the result make sense? Because a mass of 10.5 g of silver has a volume of 1 cm^3, it makes sense that 14.0 g of silver should have a volume slightly larger than 1 cm^3. The answer has two significant figures because the given mass has two significant figures.

43. Use dimensional analysis and the given densities to make the following conversions:
- **a.** 14.8 g of boron to cm^3 of boron. The density of boron is 2.34 g/cm^3.
- **b.** 4.62 g of mercury to cm^3 of mercury. The density of mercury is 13.5 g/cm^3.

44. Rework the preceding problems by applying the following equation:

$$\text{Density} = \frac{\text{mass}}{\text{volume}}$$

45. What is the mass, in grams, of a sample of cough syrup that has a volume of 50.0 cm^3? The density of cough syrup is 0.950 g/cm^3.

Density can be used to write two conversion factors. To figure out which one you need, consider the units of your given quantity and the units needed in your answer.

Explore

Class Activity

PURPOSE Students will use dimensional analysis to convert between English and metric units.

MATERIALS Internet access or copies of media guides containing vital statistics, such as heights and weights, of players on a sports team (These guides are available from local sports franchises.)

PROCEDURE Distribute the media guides and assign each group of students a set of players. Ask the group to convert heights and weights into heights and masses expressed in meters and kilograms, respectively. Have students document their approach, including dimensional analysis expressions, conversion factors, and calculations.

EXPECTED OUTCOMES Students should use conversion factors, such as 2.54 cm/1 inch and 454 g/1 lb, to convert their measurements.

Sample Practice Problem

The density of mineral oil is 0.835 g/cm^3. What is the mass in grams of a sample of mineral oil that has a volume of 2.50 cm^3? *(2.09 grams)*

Differentiated Instruction

ELL ENGLISH LANGUAGE LEARNERS Provide as much class time as possible for students to work on problem assignments in cooperative learning groups of varying proficiencies. Encourage students with high proficiency to assist students with lower proficiencies by reading the problem out loud.

LPR LESS PROFICIENT READERS Students may benefit from a reminder that certain key words and phrases in each word problem indicate the unknown quantity and its units. Some of these phrases are: *how much*, *what is*, *how long*, *determine*, and *find*.

L3 ADVANCED STUDENTS Direct students' attention to Tables 3.3, 3.4, and 3.5 in Lesson 3.2. Have students express the ratio in scientific notation between the largest and smallest units listed in each table. *(Table 3.3: 1×10^{12}; Table 3.4: 1×10^6; Table 3.5: 1×10^9)*

Answers

41. a. 44 m
b. 4.6×10^{-3} g
c. 10.7 cg

42. a. 1.5×10^{-2} L
b. 7.38×10^{-3} kg
c. 6.7×10^3 ms
d. 9.45×10^7 μg

43. a. 6.32 cm^3 b. 0.342 cm^3

44. See answers for Problem 43.

45. 47.5 g

LESSON 3.3

Explain

Sample Practice Problem

How many deciliters are in 6.5 kiloliters? *(6.5×10^4 dL)*

Multistep Problems Many complex tasks in your life are best handled by breaking them down into smaller, manageable parts. For example, if you were cleaning a car, you might first vacuum the inside, then wash the exterior, then dry the exterior, and finally put on a fresh coat of wax. Similarly, many complex word problems are more easily solved by breaking the solution down into steps.

When converting between units, it is often necessary to use more than one conversion factor. Sample Problems 3.13 and 3.14 illustrate the use of multiple conversion factors.

Sample Problem 3.13

Converting Between Metric Units

The diameter of a sewing needle is 0.073 cm. What is the diameter in micrometers?

1 Analyze List the knowns and the unknown. The desired conversion is centimeters ⟶ micrometers. The problem can be solved in a two-step conversion. First change centimeters to meters; then change meters to micrometers: centimeters ⟶ meters ⟶ micrometers.

KNOWNS	UNKNOWN
diameter = 0.073 cm = 7.3×10^{-2} cm	diameter = ? μm
10^2 cm = 1 m	
1 m = 10^6 μm	

2 Calculate Solve for the unknown.

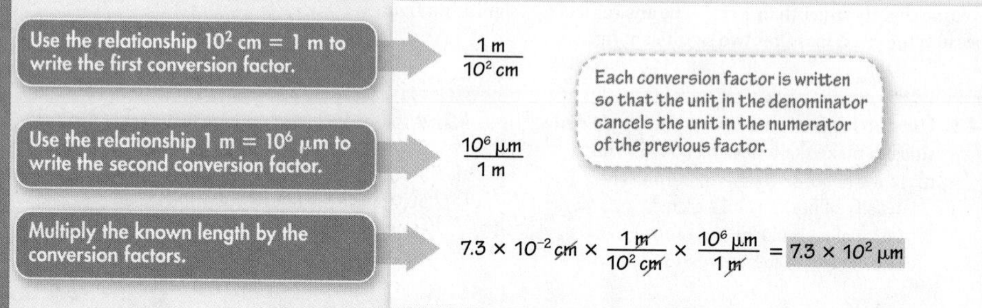

3 Evaluate Does the result make sense? Because a micrometer is a much smaller unit than a centimeter, the answer should be numerically larger than the given measurement. The units have canceled correctly, and the answer has the correct number of significant figures.

46. The radius of a potassium atom is 0.227 nm. Express this radius in the unit centimeters.

47. The diameter of Earth is 1.3×10^4 km. What is the diameter expressed in decimeters?

Focus on ELL

4 LANGUAGE PRODUCTION Have students work in pairs to complete the Small-Scale Lab on page 92. Make sure each pair has ELLs of varied language proficiencies, so that more proficient students can help less proficient ones. Have students work according to their proficiency level.

BEGINNING: LOW/HIGH Provide students a detailed step-by-step procedure to follow using pictures, words, and symbols. Be sure to show a balance with a question mark on top of it to indicate that students are to find the mass. Convey that the same procedure is to be followed three times.

INTERMEDIATE: LOW/HIGH Have accelerated students perform the procedure for finding the mass of a drop of water using verbal questioning of what to do next. Have students mimic this procedure for finding the mass of the pre-1982 penny and the post-1982 penny.

ADVANCED: LOW/HIGH Have students read out loud the questions in the Analyze and You're the Chemist sections. Then assist and/or edit the work of classmates with lower language proficiency.

Sample Problem 3.14

Converting Ratios of Units

The density of manganese, a metal, is 7.21 g/cm^3. What is the density of manganese expressed in units of kg/m^3?

1 Analyze **List the knowns and the unknown.**
The desired conversion is g/cm^3 ⟶ kg/m^3. The mass unit in the numerator must be changed from grams to kilograms: g ⟶ kg. In the denominator, the volume unit must be changed from cubic centimeters to cubic meters: cm^3 ⟶ m^3. Note that the relationship 10^6 cm^3 = 1 m^3 was derived by cubing the relationship 10^2 cm = 1 m. That is, $(10^2 \text{ cm})^3 = (1 \text{ m})^3$, or 10^6 cm^3 = 1 m^3.

KNOWNS
density of manganese = 7.21 g/cm^3
10^3 g = 1 kg
10^6 cm^3 = 1 m^3

UNKNOWN
density of manganese = ? kg/m^3

2 Calculate **Solve for the unknown.**

Multiply the known density by the correct conversion factors.

$$\frac{7.21 \text{ g}}{1 \text{ cm}^3} \times \frac{1 \text{ kg}}{10^3 \text{ g}} \times \frac{10^6 \text{ cm}^3}{1 \text{ m}^3} = 7.21 \times 10^3 \text{ kg/m}^3$$

3 Evaluate **Does the result make sense?**
Because the physical size of the volume unit m^3 is so much larger than cm^3 (10^6 times), the calculated value of the density should be larger than the given value even though the mass unit is also larger (10^3 times). The units cancel, the conversion factors are correct, and the answer has the correct ratio of units.

48. Gold has a density of 19.3 g/cm^3. What is the density in kilograms per cubic meter?

49. There are 7.0×10^6 red blood cells (RBCs) in 1.0 mm^3 of blood. How many red blood cells are in 1.0 L of blood?

3.3 LessonCheck

50. Review What happens to the numerical value of a measurement that is multiplied by a conversion factor? What happens to the actual size of the quantity?

51. Review What types of problems can be solved using dimensional analysis?

52. **Identify** What conversion factor would you use to convert between these pairs of units?
a. minutes to hours
b. grams to milligrams
c. cubic decimeters to milliliters

53. **Calculate** Make the following conversions. Express your answers in scientific notation.
a. 14.8 g = ? μg
b. 3.72 g = ? kg
c. 66.3 L = ? cm^3
d. 7.5×10^4 J = ? kJ
e. 3.9×10^5 mg = ? dg
f. 2.1×10^{-4} dL = ? μL

54. **Calculate** What is the mass, in kilograms, of 14.0 L of gasoline? (Assume that the density of gasoline is 0.680 g/cm^3.)

55. **Apply Concepts** Light travels at a speed of 3.00×10^{10} cm/s. What is the speed of light in kilometers/hour?

Lesson Check Answers

50. The numerical value (and the unit) changes; the actual size does not change.
51. conversion problems
52. **a.** 1 hour / 60 min
b. 10^3 mg / 1 g
c. 10^3 mL / 1 dm^3
53. **a.** 1.48×10^7 μg
b. 3.72×10^{-3} kg
c. 6.63×10^4 cm^3
d. 7.5×10^1 kJ
e. 3.9×10^3 dg
f. 2.1×10^1 μL
54. 9.52 kg
55. 1.08×10^9 km/h

Explain

Sample Practice Problem

In physics, acceleration is usually given in units of m/s^2. The acceleration of a falling object due to gravity is 9.8 m/s^2. What is the acceleration due to gravity in km/min^2? *(35 km/min^2)*

Evaluate

Informal Assessment

To determine students' grasp of conversion factors, ask students to orally explain the relationship between the numerator and the denominator of any measurement conversion factor. *(They are equivalent so that the ratio of numerator to denominator equals 1.)* Then have students complete the 3.3 Lesson Check.

Reteach

Model the conversion of 2 L to 2000 mL. Suggest that students check the answer by explaining that when using a conversion factor, such as 1 L = 1000 mL, the measurement expressed with the smaller unit (mL) should have a larger number associated with it than the measurement expressed with the larger unit (L).

Answers

46. 2.27×10^{-8} cm
47. 1.3×10^8 dm
48. 1.93×10^4 kg/m^3
49. 7.0×10^{12} RBC/L

SMALL-SCALE LAB

Small-Scale Lab

OBJECTIVE After completing this activity, students will be able to solve problems in divergent ways.

SKILLS FOCUS measuring, calculating, evaluating, designing experiments

PREP TIME 1 hour

MATERIALS Calculators, small-scale pipets, meter sticks, water, mass balances, pre- and post-1982 pennies, dice, 8-well strips, aluminum cans, plastic cups

ADVANCE PREP A day before doing the lab, obtain soda cans from the cafeteria. Wash and let air-dry overnight.

CLASS TIME 40 minutes

EXPECTED OUTCOME Students should find that the mass of the pre-1982 penny is 3.11 g and the mass of the post-1982 penny is 2.50 g.

ANALYZE

1. 19 mg
2. 0.019 cm^3; 0.019 mL; 19 μL
3. 1000 mg/cm^3; 1000 mg/mL
4. 2.95 g Cu; 0.16 g Zn
5. 0.060 g Cu; 2.44 g Zn
6. The new penny is mostly zinc, which has a lower density than copper.

YOU'RE THE CHEMIST

1. at 90°, mass of 1 drop: 0.0242 g
   at 45°, mass of 1 drop: 0.0218 g
   at 0°, mass of 1 drop: 0.019 g
   Pipets give different results.
2. The pipet is easiest to control at 90°. Expel the air bubble so that the first drop will be the same size as the rest.
3. Find the mass of the can and divide by the density of aluminum. Sample answer: mass of one can: 14.77 g; density of Al: 2.70 g/cm^3; $V = 5.47$ cm^3
4. (1) Measure the mass before and after you fill the can with water. Use the mass and density of water to find the volume.
   (2) Measure the height and radius and calculate volume. $V = \pi r^2 h$ (Can is not a perfect cylinder.)
   (3) Read label: 12 oz = 355 mL
5. Sample answer:
   $V = 16.5$ m × 3.0 m × 12.8 m = 630 m^3 × 1000 L/m^3 = 630,000 L. Assume 30 people with an average weight of 130 lb (1 kg = 2.2 lb) and a density of about 1.0 kg/L. Volume of 30 people = 30 × 130 lb × 1 kg/2.2 lb × 1 L/1.0 kg = 1800 L. The volume of 30 chairs, 15 tables, and 2 desks is about that of 30 people or 1800 L. The volume of people and furniture is 3600 L. % error = (3600 L/630,000 L)(100%) = 0.57%.

Small-Scale Lab

Now What Do I Do?

Purpose

To solve problems by making accurate measurements and applying mathematics

Materials

- pencil
- paper
- meter stick
- balance
- pair of dice
- aluminum can
- calculator
- small-scale pipet
- water
- a pre-1982 penny
- a post-1982 penny
- 8-well strip
- plastic cup

Procedure

1. Determine the mass, in grams, of one drop of water. To do this, measure the mass of an empty cup. Add 50 drops of water from a small-scale pipet to the cup and measure its mass again. Subtract the mass of the empty cup from the mass of the cup with water in it. To determine the average mass in grams of a single drop, divide the mass of the water by the number of drops (50). Repeat this experiment until your results are consistent.

2. Determine the mass of a pre-1982 penny and a post-1982 penny.

Analyze

Using your experimental data, record the answers to the following questions.

1. Calculate What is the average mass of a single drop of water in milligrams? (1 g = 1000 mg)

2. Calculate The density of water is 1.00 g/cm^3. Calculate the volume of a single drop in cm^3 and mL. (1 mL = 1 cm^3) What is the volume of a drop in microliters (μL)? (1000 μL = 1 mL)

3. Calculate What is the density of water in units of mg/cm^3 and mg/mL? (1 g = 1000 mg)

4. Calculate Pennies made before 1982 consist of 95.0% copper and 5.0% zinc. Calculate the mass of copper and the mass of zinc in the pre-1982 penny.

5. Calculate Pennies made after 1982 are made of zinc with a thin copper coating. They are 97.6% zinc and 2.4% copper. Calculate the mass of copper and the mass of zinc in the newer penny.

6. Explain Why does one penny have less mass than the other?

You're the Chemist

The following small-scale activities allow you to develop your own procedures and analyze the results.

1. Design an Experiment Design an experiment to determine if the size of drops varies with the angle at which they are delivered from the pipet. Try vertical (90°), horizontal (0°), and halfway between (45°). Repeat until your results are consistent.

2. Analyze Data What is the best angle to hold a pipet for ease of use and consistency of measurement? Explain. Why is it important to expel the air bubbles before you begin the experiment?

3. Design an Experiment Make the necessary measurements to determine the volume of aluminum used to make an aluminum soda can. *Hint:* Look up the density of aluminum in your textbook.

4. Design an Experiment Design and carry out some experiments to determine the volume of liquid that an aluminum soda can will hold.

5. Design an Experiment Measure a room and calculate the volume of air it contains. Estimate the percent error associated with not taking into account the furniture in the room.

6. Design an Experiment Make the necessary measurements and do the necessary calculations to determine the volume of a pair of dice. First, ignore the volume of the dots on each face, and then account for the volume of the dots. What is your error and percent error when you ignore the holes?

6. *Sample answer:* If die measures 1.55 cm on a side:
   $V = (1.55\ cm)^3 = 3.72\ cm^3$
   A die has 21 holes that are hemispheres with a radius of 0.20 cm.
   V of hemisphere = $(2/3)\pi r^3 = 0.017\ cm^3$
   V of 21 hemispheres = 0.36 cm^3
   V of die = 3.72 cm^3 – 0.36 cm^3 = 3.36 cm^3
   Error = 0.36 cm^3
   % error = (0.36 cm^3/3.36 cm^3)(100%) = 11%

3 Study Guide

BIGIDEA QUANTIFYING MATTER

Scientists express the degree of uncertainty in their measurements and calculations by using significant figures. In general, a calculated answer cannot be more precise than the least precise measurement from which it was calculated. Dimensional analysis is a problem-solving method that involves analyzing the units of the given measurement and the unknown to plan a solution.

3.1 Using and Expressing Measurements

- In scientific notation, the coefficient is always a number greater than or equal to one and less than ten. The exponent is an integer.
- To evaluate accuracy, the measured value must be compared to the correct value. To evaluate the precision of a measurement, you must compare the values of two or more repeated measurements.
- Measurements must always be reported to the correct number of significant figures because calculated answers often depend on the number of significant figures in the values used in the calculation.

- measurement (62)
- scientific notation (62)
- accuracy (64)
- precision (64)
- accepted value (65)
- experimental value (65)
- error (65)
- percent error (65)
- significant figures (66)

Key Equations

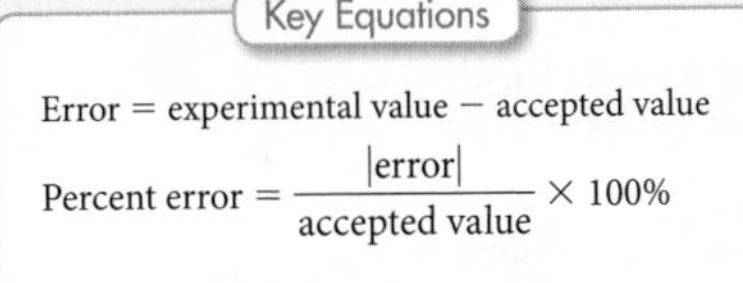

$$\text{Error} = \text{experimental value} - \text{accepted value}$$

$$\text{Percent error} = \frac{|\text{error}|}{\text{accepted value}} \times 100\%$$

3.2 Units of Measurement

- All metric units are based on multiples of 10. As a result, you can convert between units easily.
- Scientists commonly use two equivalent units of temperature, the degree Celsius and the kelvin.
- Density is an intensive property that depends only on the composition of a substance, not on the size of the sample.

- International System of Units (SI) (74)
- meter (m) (75)
- liter (L) (76)
- kilogram (kg) (77)
- gram (g) (77)
- weight (77)
- energy (77)
- joule (J) (77)
- calorie (cal) (77)
- temperature (78)
- Celsius scale (78)
- Kelvin scale (78)
- absolute zero (79)
- density (80)

Key Equations

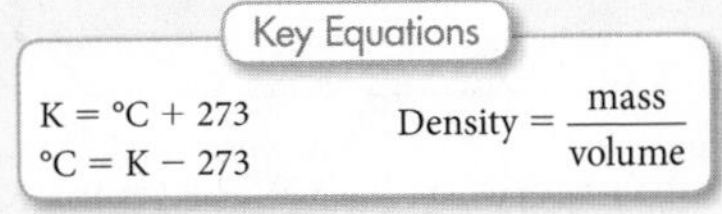

$$K = °C + 273$$

$$°C = K - 273$$

$$\text{Density} = \frac{\text{mass}}{\text{volume}}$$

3.3 Solving Conversion Problems

- When a measurement is multiplied by a conversion factor, the numerical value is generally changed, but the actual size of the quantity measured remains the same.
- Dimensional analysis is a powerful tool for solving conversion problems in which a measurement with one unit is changed to an equivalent measurement with another unit.

- conversion factor (84)
- dimensional analysis (86)

Focus on ELL

5 ASSESS UNDERSTANDING Have students keep a content area log. Consider using a two-column format with the headings *What I understand* and *What I do not understand.* Follow up with students on the items in the *What I do not understand* column.

BEGINNING

LOW Give students a list of examples, vocabulary, and concepts with boxes that they can check for *understand* or *do not understand*.

HIGH Students can copy vocabulary, sample problem numbers, and key questions into either column of the chart.

INTERMEDIATE: LOW/HIGH Ask students to specify the step(s) in the sample problems that they do not understand.

ADVANCED: LOW/HIGH Ask students to review the content area logs of those students with lower proficiencies. If there is an area that they understand but another student does not, encourage them to reteach the concept to that student.

Evaluate

Review and Assessment Materials

CHEMISTRY TUTORIAL Encourage students to review solving significant figure and conversion problems using the animated step-by-step tutorials for Chapter 3. Each Chemistry Tutorial is followed by a similar practice problem, where the student is guided through the problem-solving process.

CHAPTER PROBLEM SET Have students practice more unit conversion and scientific measurement problems using the Chapter 3 online Problem Set.

MATH TUTORIALS Struggling students can use tutorials on converting units and scientific notation as quick reviews. Ask students to answer the guided practice problem that follows each Math Tutorial.

DIRECTED VIRTUAL LAB Have students complete the Virtual Lab *The Density of Solids and Liquids* to help reinforce the concepts of precise measurement of mass and volume. In the post-lab exercise, students will analyze the data and draw conclusions based on the results of the experiment.

KINETIC ART Direct students to the Kinetic Art animations *Precision in Measurement* and *Conversion Factors*, and have them answer the questions that accompany each animation.

CONCEPTS IN ACTION If students have trouble understanding standard units of measurement ask them to view the Concepts in Action animation, which presents the material as applied to a familiar context. Have students complete the questions after the animation.

Study Tip

CHOOSE A QUIET PLACE Tell students that being able to concentrate while studying will increase their comprehension. Therefore, a quiet place where they will not be distracted can help them achieve a better grasp of the subject material and will make their study time more efficient.

MATH TUNE-UP

Math Review

Example and sample practice problems pertaining to scientific measurements and conversions can be found on the following pages:

LESSON 3.1 Using Scientific Notation—page 63; Calculating Percent Error—page 65; Determining Significant Figures in Measurements—page 66; Counting Significant Figures in Measurements—page 68; Rounding Measurements—page 69; Significant Figures in Addition and Subtraction—page 70; Significant Figures in Multiplication and Division—page 70

LESSON 3.2 Converting Between Temperature Scales—page 79; Calculating Density—page 82

LESSON 3.3 Using Dimensional Analysis—page 86 and page 87; Converting Between Metric Units—page 88; Using Density as a Conversion Factor—page 89; Converting Between Metric Units—page 90; Converting Ratios of Units—page 91

Math Tune-Up: Conversion Problems

Problem	❶ Analyze	❷ Calculate	❸ Evaluate
A grocer is selling oranges at 3 for \$2. How much would it cost to buy a dozen oranges?	Knowns: 3 oranges = \$2 1 dozen = 12 Unknown: Cost of 12 oranges = ? The desired conversion is oranges ⟶ \$.	Use the relationship 3 oranges = \$2 to write the correct conversion factor. $\frac{\$2}{3 \text{ oranges}}$ Multiply the known quantity by the conversion factor. $12 \text{ oranges} \times \frac{\$2}{3 \text{ oranges}} = \$8$	A dozen is larger than the number 3, so the cost should exceed \$2. The known unit (oranges) cancels, and the answer has the correct unit (\$).
Convert the volume 865 cm^3 to liters.	Knowns: volume = 865 cm^3 10^3 cm^3 = 1 L Unknown: volume = ? L The desired conversion is cm^3 ⟶ L.	Use the relationship $10^3 \text{ cm}^3 = 1 \text{ L}$ to write the correct conversion factor. $\frac{1 \text{ L}}{10^3 \text{ cm}^3}$ Multiply the known volume by the conversion factor. $865 \text{ cm}^3 \times \frac{1 \text{ L}}{10^3 \text{ cm}^3} = 0.865 \text{ L}$	A cubic centimeter is much smaller than a liter, so the answer should be numerically smaller than the given measurement. The known unit (cm^3) cancels, and the answer has the correct unit (L).
Express the length 8.2×10^{-4} μm in centimeters.	Knowns: length = 8.2×10^{-4} μm 10^6 μm = 1 m 1 m = 10^2 cm Unknown: length = ? L The desired conversion is μm ⟶ cm. First change μm to m; then change m to cm: μm ⟶ m ⟶ cm	Use the relationship $10^6 \text{ μm} = 1 \text{ m}$ to write the first conversion factor. $\frac{1 \text{ m}}{10^6 \text{ μm}}$ Use the relationship $1 \text{ m} = 10^2 \text{ cm}$ to write the second conversion factor. $\frac{10^2 \text{ cm}}{1 \text{ m}}$ Multiply the known length by the conversion factors. $8.2 \times 10^{-4} \text{ μm} \times \frac{1 \text{ m}}{10^6 \text{ μm}} \times \frac{10^2 \text{ cm}}{1 \text{ m}}$ $= 8.2 \times 10^{-8} \text{ cm}$	A micrometer is smaller than a centimeter, so the answer should be numerically smaller than the given measurement. The known unit (μm) cancels, and the answer has the correct unit (cm). Hint: For a multistep problem, take it one conversion at a time.

Performance Tasks

OUTER SPACE Develop an original way to show the advantages and disadvantages of NASA eliminating the need for uncertainty in measurements.

SCALE MODELING Create a three-dimensional model to physically show the relationship between the prefixes given in Table 3.2 to scale. Present the model to the class.

3 Assessment

* Solutions appear in Appendix E

Lesson by Lesson

3.1 Using and Expressing Measurements

56. Three students made multiple weighings of a copper cylinder, each using a different balance. Describe the accuracy and precision of each student's measurements if the correct mass of the cylinder is 47.32 g.

Mass of Cylinder (g)			
	Colin	Lamont	Kivrin
Weighing 1	47.13	47.45	47.95
Weighing 2	47.94	47.39	47.91
Weighing 3	46.83	47.42	47.89
Weighing 4	47.47	47.41	47.93

57. How many significant figures are in each underlined measurement?

a. 60 s = 1 min
b. 47.70 g of copper
c. 1 km = 1000 m
d. 25 computers
e. 9 innings in a baseball game
f. 0.0950 m of gold chain

58. Round off each of these measurements to three significant figures.

a. 98.473 L
b. 0.000 763 21 cg
c. 57.048 m
d. 12.17°C
e. $0.007\,498\,3 \times 10^4$ mm
f. 1764.9 mL

***59.** Round off each of the answers correctly.

a. 8.7 g + 15.43 g + 19 g = 43.13 g
b. 4.32 cm × 1.7 cm = 7.344 cm^2
c. 853.2 L − 627.443 L = 225.757 L
d. 38.742 m^2 ÷ 0.421 m = 92.023 75 m
e. 5.40 m × 3.21 m × 1.871 m = 32.431 914 m^3
f. 5.47 m^3 + 11 m^3 + 87.300 m^3 = 103.770 m^3

***60.** Express each of the rounded-off answers in Problems 58 and 59 in scientific notation.

61. How are the *error* and the *percent* error of a measurement calculated?

3.2 Units of Measurement

62. Write the SI base unit of measurement for each of these quantities.

a. time
b. length
c. temperature
d. mass
e. energy
f. amount of substance

***63.** Order these units from smallest to largest: cm, μm, km, mm, m, nm, dm, pm. Then give each measurement in terms of meters.

64. Measure each of the following dimensions using a unit with the appropriate prefix.

a. the height of this letter I
b. the width of Table 3.3
c. the height of this page

65. State the relationship between degrees Celsius and kelvins.

***66.** The melting point of silver is 962°C. Express this temperature in kelvins.

67. What equation is used to determine the density of an object?

68. Would the density of a person be the same on the surface of Earth and on the surface of the moon? Explain.

***69.** A shiny, gold-colored bar of metal weighing 57.3 g has a volume of 4.7 cm^3. Is the bar of metal pure gold?

70. Three balloons filled with neon, carbon dioxide, and hydrogen are released into the atmosphere. Using the data in Table 3.6 on page 81, describe the movement of each balloon.

3.3 Solving Conversion Problems

71. What is the name given to a ratio of two equivalent measurements?

72. What must be true for a ratio of two measurements to be a conversion factor?

73. How do you know which unit of a conversion factor must be in the denominator?

Evaluate

Answers

LESSON 3.1

56. Colin: inaccurate and imprecise; Lamont: accurate and precise; Kivrin: inaccurate and precise

57. **a.** infinite **b.** 4 **c.** infinite **d.** infinite **e.** infinite **f.** 3

58. **a.** 98.5 L **b.** 0.000 763 cg **c.** 57.0 m **d.** 12.2°C **e.** 0.00750×10^4 mm **f.** 1760 mL

59. **a.** 43 g **b.** 7.3 cm^2 **c.** 225.8 L **d.** 92.0 m **e.** 32.4 m^3 **f.** 104 m^3

60. (58) **a.** 9.85×10^1 L **b.** 7.63×10^{-4} cg **c.** 5.70×10^1 m **d.** 1.22×10^1°C **e.** 7.50×10^1 mm **f.** 1.76×10^3 mL; (59) **a.** 4.3×10^1 g **b.** 7.3×10^0 cm^2 **c.** 2.258×10^2 L **d.** 9.20×10^1 m **e.** 3.24×10^1 m^3 **f.** 1.04×10^2 m^3

61. The error is the difference between the experimental value and the accepted value. The percent error is the error divided by the accepted value multiplied by 100.

LESSON 3.2

62. **a.** second **b.** meter **c.** kelvin **d.** kilogram **e.** joule **f.** mole

63. pm, nm, μm, mm, cm, dm, m, km; 1 km = 10^3 m, 1 dm = 10^{-1} m, 1 cm = 10^{-2} m, 1 mm = 10^{-3} m, 1 μm = 10^{-6} m, 1 nm = 10^{-9} m, 1 pm = 10^{-12} m

64. **a.** 2.4 mm **b.** 14.10 cm **c.** 29.32 cm

65. Degrees Celsius equals kelvins minus 273.

66. 1235 K

67. density = mass/volume

68. Yes; neither mass nor volume of a solid or liquid changes appreciably with location.

69. No; the density of the metal bar is 12 g/cm^3, but the density of gold is 19 g/cm^3.

70. The carbon dioxide–filled balloon would sink. The neon and hydrogen-filled balloons would rise, the hydrogen-filled one at a much faster rate.

LESSON 3.3

71. conversion factor

72. They must equal one another.

73. The unit of the conversion factor in the denominator must be identical to the unit in the given measurement or the previous conversion factor.

Evaluate

Answers

74. **a.** 1.57 s **b.** 4.27×10^4 mL **c.** 2.61×10^{-4} mm **d.** 6.5×10^2 dm **e.** 6.42×10^{-3} kg **f.** 8.25×10^9 ng

75. **a.** 7.3 µL/s **b.** 78.6 mg/mm^2 **c.** 1.54 g/cm^3

76. 10^6 mL

77. **a.** 2.83×10^2 mg **b.** 0.283 g **c.** 2.83×10^{-4} kg **d.** 6.6 g **e.** 6.6×10^2 cg **f.** 6.6×10^{-3} kg **g.** 2.8×10^{-1} mg **h.** 2.8×10^{-2} cg **i.** 2.8×10^{-7} kg

UNDERSTAND CONCEPTS

78. improper calibration or improper use of the measuring device

79. e, d, c, f, a, b

80. **a.** accurate and precise **b.** inaccurate and precise **c.** inaccurate and imprecise

81. °F = 1.8°C + 32

82. germanium

83. 0.69–0.789 g/cm^3

84. Sink; the ball's density is 0.802 g/cm^3, which is greater than the density of gasoline (0.66–0.69 g/cm^3, Table 3.6).

85. 1 g/10^2 cg, 10^2 cg/1 g, 1 g/10^3 mg, 10^3 mg/1 g, 10^2 cg/10^3 mg, 10^3 mg/10^2 cg

86. 81.3 kg

87. 4.20×10^4 cm^3

88. 73 g

89. 0.804 g/cm^3

90. 3.6 min lost

91. 0.92 kg/L

***74.** Make the following conversions:

a. 157 cs to seconds
b. 42.7 L to milliliters
c. 261 nm to millimeters
d. 0.065 km to decimeters
e. 642 cg to kilograms
f. 8.25×10^2 cg to nanograms

***75.** Make the following conversions:

a. 0.44 mL/min to microliters per second
b. 7.86 g/cm^2 to milligrams per square millimeter
c. 1.54 kg/L to grams per cubic centimeter

76. How many milliliters are contained in 1 m^3?

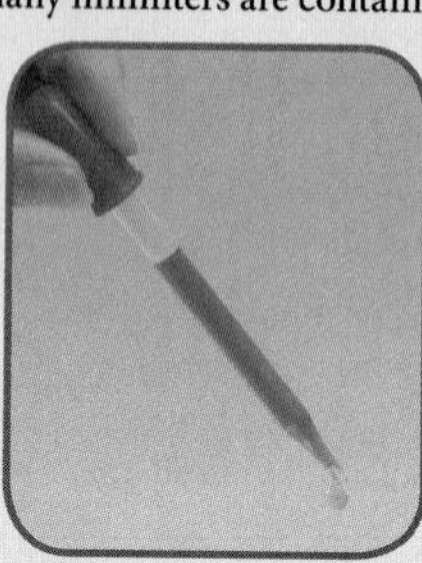

***77.** Complete this table so that all the measurements in each row have the same value.

mg	g	cg	kg
a. ____	b. ____	28.3	c. ____
6.6×10^3	d. ____	e. ____	f. ____
g. ____	2.8×10^{-4}	h. ____	i. ____

Understand Concepts

78. List two possible reasons for reporting precise, but inaccurate, measurements.

79. Rank these numbers from smallest to largest.

a. 5.3×10^4
b. 57×10^3
c. 4.9×10^{-2}
d. 0.0057
e. 5.1×10^{-3}
f. 0.0072×10^2

80. Comment on the accuracy and precision of these basketball free-throw shooters.

a. 99 of 100 shots are made.
b. 99 of 100 shots hit the front of the rim and bounce off.
c. 33 of 100 shots are made; the rest miss.

81. Fahrenheit is a third temperature scale. Plot the data in the table and use the graph to derive an equation for the relationship between the Fahrenheit and Celsius temperature scales.

Example	°C	°F
Melting point of selenium	221	430
Boiling point of water	100	212
Normal body temperature	37	98.6
Freezing point of water	0	32
Boiling point of chlorine	−34.6	−30.2

82. Which would melt first, germanium with a melting point of 1210 K or gold with a melting point of 1064°C?

83. A piece of wood floats in ethanol but sinks in gasoline. Give a range of possible densities for the wood.

84. A plastic ball with a volume of 19.7 cm^3 has a mass of 15.8 g. Would this ball sink or float in a container of gasoline?

85. Write six conversion factors involving these units of measure: 1 g = 10^2 cg = 10^3 mg.

***86.** A 2.00-kg sample of bituminous coal is composed of 1.30 kg of carbon, 0.20 kg of ash, 0.15 kg of water, and 0.35 kg of volatile (gas-forming) material. Using this information, determine how many kilograms of carbon are in 125 kg of this coal.

***87.** The density of dry air measured at 25°C is 1.19×10^{-3} g/cm^3. What is the volume of 50.0 g of air?

88. What is the mass of a cube of aluminum that is 3.0 cm on each edge? The density of aluminum is 2.7 g/cm^3.

***89.** A flask that can hold 158 g of water at 4°C can hold only 127 g of ethanol at the same temperature. What is the density of ethanol?

***90.** A watch loses 0.15 s every minute. How many minutes will the watch lose in 1 day?

***91.** A tank measuring 28.6 cm by 73.0 mm by 0.72 m is filled with olive oil. The oil in the tank has a mass of 1.38×10^4 g. What is the density of olive oil in kilograms per liter?

92. Alkanes are a class of molecules that have the general formula C_nH_{2n+2}, where n is an integer (whole number). The table below gives the boiling points for the first five alkanes with an odd number of carbon atoms. Using the table, construct a graph with the number of carbon atoms on the x-axis.

Boiling point (°C)	Number of carbon atoms
−162.0	1
−42.0	3
36.0	5
98.0	7
151.0	9

a. What are the approximate boiling points for the C_2, C_4, C_6, and C_8 alkanes?
b. Which of these nine alkanes are gases at room temperature (20°C)?
c. How many of these nine alkanes are liquids at 350 K?
d. What is the approximate increase in boiling point per additional carbon atom in these alkanes?

*93. Earth is approximately 1.5×10^8 km from the sun. How many minutes does it take light to travel from the sun to Earth? The speed of light is 3.0×10^8 m/s.

*94. The average density of Earth is 5.52 g/cm³. Express this density in units of kg/dm³.

95. How many kilograms of water (at 4°C) are needed to fill an aquarium that measures 40.0 cm by 20.0 cm by 30.0 cm?

Think Critically

96. **Explain** Is it possible for an object to lose weight but at the same time not lose mass? Explain your answer.

*97. **Calculate** One of the first mixtures of metals, called amalgams, used by dentists for tooth fillings, consisted of 26.0 g of silver, 10.8 g of tin, 2.4 g of copper, and 0.8 g of zinc. How much silver is in a 25.0 g sample of this amalgam?

*98. **Calculate** A cheetah can run 112 km/h over a 100-m distance. What is this speed in meters per second?

99. **Evaluate** You are hired to count the number of ducks on three northern lakes during the summer. In the first lake, you estimate 500,000 ducks, in the second 250,000 ducks, and in the third 100,000 ducks. You write down that you have counted 850,000 ducks. As you drive away, you see 15 ducks fly in from the south and land on the third lake. Do you change the number of ducks that you report? Justify your answer.

100. **Describe** What if ice were more dense than water? It would certainly be easier to pour water from a pitcher of ice cubes and water. Can you think of situations of more consequence?

101. **Graph** Plot these data that show how the mass of sulfur increases with an increase in volume. Determine the density of sulfur from the slope of the line.

Volume of sulfur (cm³)	Mass of sulfur (g)
11.4	23.5
29.2	60.8
55.5	115
81.1	168

102. **Analyze Data** At 20°C, the density of air is 1.20 g/L. Nitrogen's density is 1.17 g/L. Oxygen's density is 1.33 g/L.
a. Will balloons filled with oxygen and balloons filled with nitrogen rise or sink in air?
b. Air is mainly a mixture of nitrogen and oxygen. Which gas is the main component? Explain.

92. a. C_2 = −90°C, C_4 = 0°C, C_6 = 70°C, C_8 = 125°C
b. C_1 through C_4
c. three
d. From C_1 through C_9, the increase is approximately 39°C/additional carbon. Over the range C_3 through C_9, the increase is approximately 32°C/additional carbon.
93. 8.3 min
94. 5.52 kg/dm³
95. 24.0 kg of water

THINK CRITICALLY

96. Yes, the mass of an object is constant. The weight of an object varies with location.
97. 16.3 g
98. 31.1 m/s
99. You do not change your estimate. Counting the extra 15 ducks would suggest greater precision than was used in the estimate.
100. Answers will vary. Lakes would freeze solid from the bottom up; aquatic life would be destroyed; possible climate changes would occur.
101.

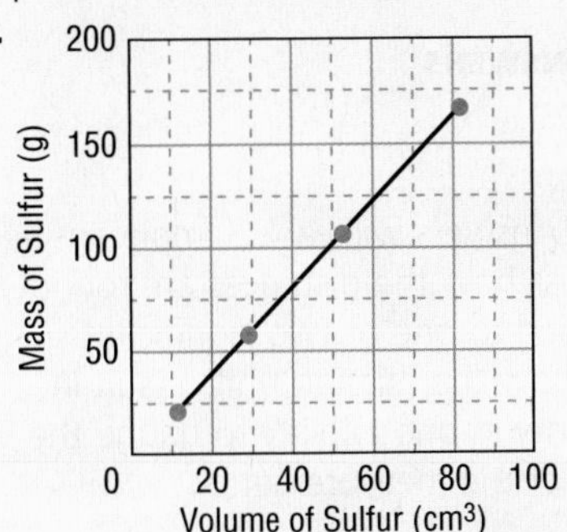

density of sulfur = 2.1 g/cm³
102. a. The oxygen-filled balloon will sink; the nitrogen-filled balloon will rise.
b. Nitrogen; the "weighted" average density of air is closer to the density of nitrogen.

CHEMYSTERY

START A CONVERSATION After students have read through the CHEMystery, discuss the importance of quantifying a measure with units. **Ask** What problems can occur if people have to rely on the customs of the city they are in to interpret the measure of a value? *(Answers will vary.)* **Ask** What are some units used to measure distance? *(Sample answer: centimeter, mile, foot, kilometer)* **Ask** Do you think if you know a measurement is linear, such as distance, that then it is okay not to label the quantity with a unit? Explain. *(no, because the measurement could have been made in any one of several metric or nonmetric scales)* Explain to students that measured values should be of help to other people, not a hindrance. Point out that a value expressed without a unit can have a great effect on something else. Encourage students to make a concerted effort to write the units of a value throughout a calculation as well as when the final answer is computed.

CHEMYSTERY ANSWERS

111. 7.2 km

112. 43 minutes

113. **BIG**IDEA Answers will vary. Sample answers might discuss how the distances are rounded and expressed to one or two significant figures, or how the missing units result in an uncertainty in one's ability to gauge the magnitude of the measurements.

Concept Challenge

*103. **Calculate** The mass of a cube of iron is 355 g. Iron has a density of 7.87 g/cm^3. What is the mass of a cube of lead that has the same dimensions?

*104. **Calculate** Sea water contains 8.0×10^{-1} cg of the element strontium per kilogram of sea water. Assuming that all the strontium could be recovered, how many grams of strontium could be obtained from one cubic meter of sea water? Assume the density of sea water is 1.0 g/mL.

105. **Calculate** The density of dry air at 20°C is 1.20 g/L. What is the mass of air, in kilograms, of a room that measures 25.0 m by 15.0 m by 4.0 m?

106. **Graph** Different volumes of the same liquid were added to a flask on a balance. After each addition of liquid, the mass of the flask with the liquid was measured. Graph the data using mass as the dependent variable. Use the graph to answer these questions.

Volume (mL)	Mass (g)
14	103.0
27	120.4
41	139.1
55	157.9
82	194.1

a. What is the mass of the flask?
b. What is the density of the liquid?

*107. **Predict** A 34.5-g gold nugget is dropped into a graduated cylinder containing water. By how many milliliters does the measured volume increase if the nugget is completely covered by water? The density of water is 1.0 g/mL. The density of gold is 19.3 g/cm^3.

108. **Predict** Equal amounts of mercury, water, and corn oil are added to a beaker. Use Table 3.6 to help you anwer the following questions.
a. Describe the arrangement of the layers of liquids in the beaker.
b. A small sugar cube is added to the beaker. Describe its location.
c. What change will occur to the sugar cube over time?

Write About Science

109. **Describe** For one of the topics below, write a short paragraph that identifies both metric and nonmetric units that are commonly used to communicate information.
a. measurements used in cooking
b. measurements used in sports
c. measurements used in transportation

110. **Connect to the BIGIDEA** Explain how the three-step problem-solving approach defined in Chapter 1 (*Analyze, Calculate, Evaluate*) applies to problems that involve dimensional analysis.

CHEMYSTERY

Just Give Me a Sign

The road signs point to locations in England. Although England has adopted metric units for many everyday quantities, distances shown on road signs are not among them. The road signs above list distances in miles, a nonmetric unit. Speed-limit signs in England are typically expressed in miles per hour—also nonmetric. However, in the same country, gasoline is sold by metric units of volume (liters), fabric is measured in metric units of area (square meters), and the local weather report uses metric units of temperature (°C).

111. **Calculate** The relationship between kilometers and miles (mi) is 1 km = 0.621 mi. How far it is to Chipping in kilometers?

112. **Calculate** Suppose you encounter the road signs above while cycling. If your average speed is 18 km/h, how many minutes will it take you to reach Preston?

113. **Connect to the BIGIDEA** Describe two ways in which the road signs above might be considered examples of "uncertainty in measurement."

Answers

ENRICHMENT

103. Volume of iron = 45.1 cm^3
Mass of lead = 510 g Pb

104. 8.0 g Sr

105. 1.8×10^3 kg

106. a. 85 g **b.** 1.3 g/mL

107. 1.79 mL

108. a. corn oil on top of water on top of mercury
b. The density of sugar is greater than the density of water and less than the density of mercury; it floats between the layers of mercury and water.
c. The sugar cube will dissolve in the water over time.

WRITE ABOUT SCIENCE

109. Answers will vary. Students should be comfortable with recognizing metric and nonmetric units in any given measurement and have an intuitive idea of their magnitude.

110. Answers will vary. Students should understand that solving dimensional analysis problems uses the same general approach as any other chemistry problem, including setting up the problem, performing calculations, and checking the results for sense.

Standardized Test Prep for Indiana

Standard: C.1.7

Select the choice that best answers each question or completes each statement.

1. Which of these series of units is ordered from smallest to largest?
   (A) μg, cg, mg, kg
   (B) mm, dm, m, km
   (C) μs, ns, cs, s
   (D) nL, mL, dL, cL
2. Which answer represents the measurement 0.00428 g rounded to two significant figures?
   (A) 4.28×10^3 g
   (B) 4.3×10^3 g
   (C) 4.3×10^{-3} g
   (D) 4.0×10^{-3} g
3. An over-the-counter medicine has 325 mg of its active ingredient per tablet. How many grams does this mass represent?
   (A) 325,000 g
   (B) 32.5 g
   (C) 3.25 g
   (D) 0.325 g
4. If 10^4 μm = 1 cm, how many $\mu m^3 = 1\ cm^3$?
   (A) 10^4
   (B) 10^6
   (C) 10^8
   (D) 10^{12}
5. If a substance contracts when it freezes, its
   (A) density will remain the same.
   (B) density will increase.
   (C) density will decrease.
   (D) change in density cannot be predicted.

For Questions 6–7, identify the known and the unknown. Include units in your answers.

6. The density of water is 1.0 g/mL. How many deciliters of water will fill a 0.5-L bottle?
7. A graduated cylinder contains 44.2 mL of water. A 48.6-g piece of metal is carefully dropped into the cylinder. When the metal is completely covered with water, the water rises to the 51.3-mL mark. What is the density of the metal?

Tips for Success

Interpret Diagrams Before you answer questions about a diagram, study the diagram carefully. Ask: What is the diagram showing? What does it tell me?

Use the diagrams below to answer Questions 8 and 9.

The atomic windows represent particles of the same gas occupying the same volume at the same temperature. The systems differ only in the number of gas particles per unit volume.

a. 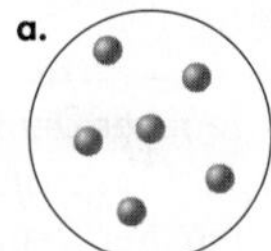

b. 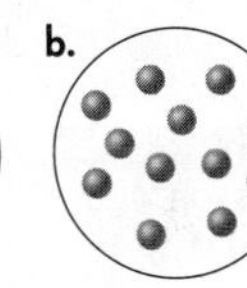

c. 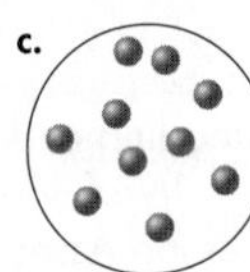

8. List the windows in order of decreasing density.
9. Compare the density of the gas in window (a) to the density of the gas in window (b).

For each question, there are two statements. Decide whether each statement is true or false. Then decide whether Statement II is a correct explanation for Statement I.

	Statement I		Statement II
10.	There are five significant figures in the measurement 0.00450 m.	BECAUSE	All zeros to the right of a decimal point in a measurement are significant.
11.	Precise measurements will always be accurate measurements.	BECAUSE	A value that is measured 10 times in a row must be accurate.
12.	A temperature in kelvins is always numerically larger than the same temperature in degrees Celsius.	BECAUSE	A temperature in kelvins equals a temperature in degrees Celsius plus 273.

If You Have Trouble With . . .

Question	1	2	3	4	5	6	7	8	9	10	11	12
See Lesson	3.2	3.1	3.3	3.3	3.2	3.3	3.2	3.2	3.2	3.1	3.1	3.2

STP Answers

MULTIPLE CHOICE

1. B
2. C
3. D
4. D
5. B

OPEN RESPONSE

6. known: volume in liters; unknown: volume in deciliters
7. knowns: mass of metal, initial and final readings of the volume; unknown: density of metal
8. b, c, a
9. The density in *a* is one-half the density in *b*.
10. false, false
11. false, false
12. true, true, correct explanation

4 Atomic Structure Planning Guide

Introducing the BIGIDEA: ELECTRONS AND THE STRUCTURE OF ATOMS

Elements are composed of particles called atoms, and every atom contains a nucleus and one or more electrons.

IASS	Lessons and Objectives	Print Resources		
		For the Student	For the Teacher	
C.2.1	**4.1** Defining the Atom p 102–104 **4.1.1** Explain how Democritus and John Dalton described atoms. **4.1.2** Identify instruments used to observe individual atoms.	Reading and Study Workbook Lesson 4.1 Lesson Assessment 4.1 p 104		
C.2.1, C.2.2, C.2.3	**4.2** Structure of the Nuclear Atom p 105–109 **4.2.1** Identify three types of subatomic particles. **4.2.2** Describe the structure of atoms according to the Rutherford model.	Reading and Study Workbook Lesson 4.2 Lesson Assessment 4.2 p 109 Quick Lab: Using Inference: The Black Box p 109	Teacher Demo, p 106: Observing Cathode Rays Teacher Demo, p 107: Atomic Model Timeline	
C.2.1, C.2.2, C.2.3, C.2.4	**4.3** Distinguishing Among Atoms p 112–119 **4.3.1** Explain what makes one element different from another **4.3.2** Explain how isotopes of an element differ. **4.3.3** Calculate the atomic mass of an element.	Reading and Study Workbook Lesson 4.3 Lesson Assessment 4.3 p 119 Small-Scale Lab: The Atomic Mass of "Candium" p 120	Class Activity, p 114: Applications of Isotopes	

Assessing the BIGIDEA: ELECTRONS AND THE STRUCTURE OF ATOMS

	Essential Questions 1. What components make up an atom? 2. How are atoms of one element different from atoms of another element?	Study Guide p 121 STP p 125 Reading and Study Workbook Self-Check and Vocabulary Review Chapter 4		

	PearsonChem.com	
	Editable Worksheets	Online Activities
	Teaching Resources, Lesson 4.1 Review	4.1 Lesson Overview
	Teaching Resources, Lesson 4.2 Review Lab 5: Atomic Structure: Rutherford's Experiment	4.2 Lesson Overview Cathode Ray Rutherford's Experiment Rutherford's Experiment
	Teaching Resources, Lesson 4.3 Review Lab Practical 4-1: Modeling Atomic Structure Small-Scale Lab Manual Lab 6: Isotopes and Atomic Mass	4.3 Lesson Overview Calculating Atomic Mass Determining the Composition of an Atom If Atoms Were Pennies, Then Banks Would Be Huge
	Core Teaching Resources (includes editable worksheets) • Lesson Reviews • Practice Problems • Interpret Graphs • Vocabulary Review • Chapter Quizzes and Tests • Lab Record Sheets	Chapter 4 Problem Set Weighted Averages

Materials List

For the Student

Quick Lab p 109
- box containing a regularly shaped object fixed in place
- loose marble

Small-Scale Lab p 120
- mass balance
- coated candies (3 different brands)
- small plastic cups or containers

For the Teacher

Teacher Demo p 106
- magnet
- cathode ray tube

Additional Digital Resources

Online Student Edition
Online Teacher's Edition
Exam*View Assessment Suite CD-ROM*
Classroom Resources DVD
4.2 Virtual Chem Lab 4: Thomson Cathode-Ray Tube Experiment
Virtual Chem Lab 5: Millikan Oil Drop Experiment
Virtual Chem Lab 6: Atomic Structure: Rutherford's Experiment

What's Online

Students access guided step-by-step tutorials for solving various atomic structure problems.

Students can practice key problem-solving skills in an online problem set.

Identify the students who struggle with math by assigning an online math skills diagnostic test. These students can then improve and practice math skills using the MathXL tutorial system.

Students go into a virtual lab tour in which key experiments in the history of atomic theory are studied in a simulated laboratory environment.

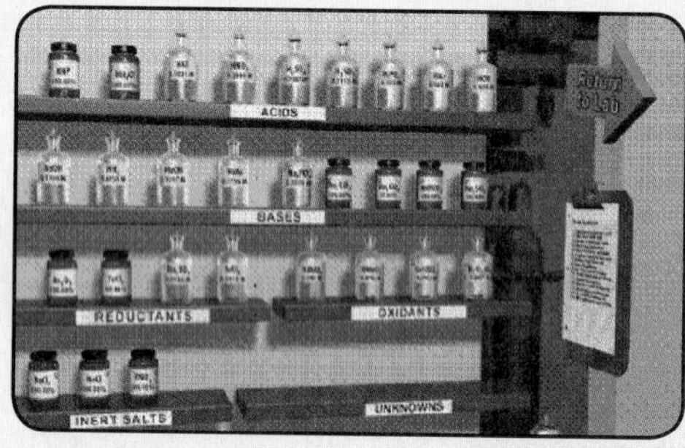

Student watch animations of a selected figures from the chapter followed by questions to check for understanding.

Students watch an overview of a key chapter concept using real-world contexts and concrete examples and analogies. Each activity includes an interactive animation followed by analysis questions.

IN ACADEMIC STANDARDS for SCIENCE

Nature of Science

- Regularly evaluate the work of their peers and in turn have their work evaluated by their peers.
- Explain that the body of scientific knowledge is organized into major theories, which are derived from and supported by the results of many experiments, and allow us to make testable predictions.
- Recognize that new scientific discoveries often lead to a re-evaluation of previously accepted scientific knowledge and of commonly held ideas.

Standard 2: Atomic Structure and the Periodic Table

C.2.1 Describe the historical development of the modern atomic theory.

Also covered C.2.2, C.2.3, C.2.4

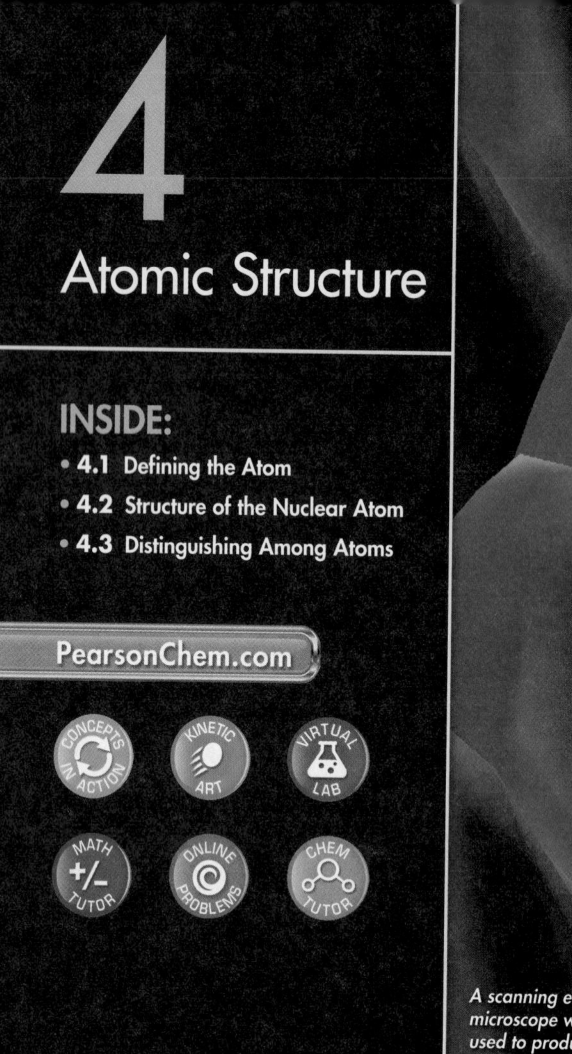

4 Atomic Structure

INSIDE:

- **4.1** Defining the Atom
- **4.2** Structure of the Nuclear Atom
- **4.3** Distinguishing Among Atoms

PearsonChem.com

A scanning electron microscope was used to produce this color-enhanced image of nickel atoms.

Focus on ELL

1 CONTENT AND LANGUAGE Dissect the chapter title into two parts. Begin with the word *structure*. Give examples of real world objects or situations that have a structure. Explain that this chapter is about the structure of *atoms*. As a class brainstorm the meaning of an atom. Have students identify any other words in the lesson titles that they do not know. Add these words to the classroom word wall.

BEGINNING

LOW Write the unfamiliar words on the board and say them out loud.

HIGH Write the plural, noun, and adjective forms of the word *atom*.

INTERMEDIATE

LOW Have students use an English dictionary to find the meaning of the unfamiliar words and write the definitions in their notebooks.

HIGH Have students write an original sentence using each word.

ADVANCED: LOW/HIGH Have students discuss the meaning or synonyms of the unfamiliar words with an elbow partner.

BIGIDEA

ELECTRONS AND THE STRUCTURE OF ATOMS

Essential Questions:

1. *What components make up an atom?*
2. *How are atoms of one element different from atoms of another element?*

CHEMYSTERY

Artifact or Artifake?

Crystal skulls are shaped like a human skull and carved from quartz crystal. Crystal skulls are thought to have originated from pre-Columbian Central American cultures. If so, then crystal skulls would have been carved several hundred or even thousands of years ago. They would probably have been carved using primitive stone, wooden, and bone tools.

Although crystal skulls are displayed in museums throughout the world, none of them were found in an actual archaeological dig. This unusual circumstance has led to some debate about the history of the skulls. People have questioned whether crystal skulls were ever carved by people from ancient civilizations. Are these sculptures true artifacts that were carved in the pre-Columbian era, or are they just fakes?

▶ Connect to the **BIGIDEA** As you read about the structure of atoms, think about how scientists could identify whether a crystal skull is from an ancient civilization or is just a fake.

INDIANA'S ACADEMIC STANDARDS FOR SCIENCE

Atomic Structure and the Periodic Table C.2.1, C.2.2, C.2.3, C.2.4

CHAPTER 4

UbD

Understanding by Design

Students are building toward an understanding of the characteristics of individual atoms and the structural differences between different types of atoms.

PERFORMANCE GOALS At the end of Chapter 4, students will be able to answer the essential questions by applying their knowledge of the structure of the nuclear atom.

ESSENTIAL QUESTIONS Read the essential questions aloud. **Ask** Why do you think it is important to understand the structure of the atom? *(The structure of the atom determines the properties of matter.)* **Ask** Can you predict ways that one atom might differ from another atom?

BIGIDEA Activate students' prior knowledge. Use the photo of the atoms to help students connect to this chapter's concepts. Explain that a scanning-electron microscope (SEM) produced this image of a sample of matter. **Ask** What information does this picture tell you about the sample? *(The atoms have a similar appearance.)* **Ask** What can you infer about the sample from this image? *(All of the nickel atoms look alike.)*

CHEMYSTERY Have students read over the CHEMystery. Connect the CHEMystery to the Big Idea by having students consider differences in markings caused by carving the crystals. Carving the crystals with primitive tools would leave rough and uneven markings. Ask students to predict what a microscopic inspection of the crystals would show if more advanced carving methods had been used. Then have students tell how this microscopic inspection could be made. **Ask** How do you think this technology is used to investigate the crystal skull mystery?

Introduce the Chapter

IDENTIFY MISCONCEPTIONS Students may think of atoms as either solid spheres or composed of smaller particles that are close together. Use the activity to help students develop a clearer understanding of atomic structure.

Activity *You will need several small balls of modeling clay, a piece of foam board, and about fifteen straight pins with ball heads.* Place one clay ball and the foam board on a table. Cluster about 10 of the pins in the center of the foam board to model an atomic nucleus. Insert the other pins around the rest of the foam board to model electrons. **Ask** Which of these is the best model for an atom? *(the pins)* **Ask** What makes the ball of clay a poor model? *(It is solid.)*

Gently push the balls of clay close together into a sphere. Explain that each ball of clay represents a particle in the atom. Have students compare this clay sphere to the pin model. **Ask** Which model is a better representation of an atom? *(pin model)* Why? *(The pin model represents tiny particles with a large amount of space between them. The clay model has no space between the particles.)* Explain that in this chapter, students will begin their study of atomic structure.

What Can Bones Tell Us?
Jonas recreates how chemistry helped detectives investigate a crime.

LESSON 4.1

Key Objectives

4.1.1 EXPLAIN how Democritus and John Dalton described atoms.

4.1.2 IDENTIFY instruments used to observe individual atoms.

Additional Resources

Reading and Study Workbook, Lesson 4.1
Available Online or on Digital Media:
- Teaching Resources, Lesson 4.1 Review

PearsonChem.com

Engage

CHEMISTRY & YOU Ask students to think of objects that require experimental data in order to "picture" them, either because they are small or inaccessible. *(Sample answers: objects in deep space or deep underground)*

Build Background

Explain that early philosophers and scientists developed models of the atom to help explain the nature of matter. Tell students that in the same way that they might use a globe to learn about Earth, they can use an atomic model to learn about atoms. **Ask** Why do people use models? *(to study things too large, too small, or too complex to easily see or understand)* What models do you use or have you used? *(Sample answers: subway maps or weather maps)* Tell students they will learn more about atomic models in Lessons 4.2 and 5.1.

IN ACADEMIC STANDARDS for SCIENCE

- Focus on the development of explanatory models based on their observations during laboratory investigations.
- Explain that the body of scientific knowledge is organized into major theories, which are derived from and supported by the results of many experiments, and allow us to make testable predictions.
- Recognize that new scientific discoveries often lead to a re-evaluation of previously accepted scientific knowledge and of commonly held ideas.

Also covered C.2.1

4.1 Defining the Atom

C.2.1 Modern atomic theory: historic development

CHEMISTRY & YOU

Q: *How do you study something that you cannot see?* It is sometimes fun to try to figure out what is inside a present before opening it. You could look at the shape or weight of the box. Or maybe you would shake the box a little to find out if anything moved around or made noise inside the box. Similar to how you might study a giftwrapped present, scientists often study things that cannot be seen with the naked eye. In this lesson, you will learn how scientists obtained information about the atoms that they couldn't see.

Key Questions

How did the concept of the atom change from the time of Democritus to the time of John Dalton?

What instruments are used to observe individual atoms?

Vocabulary

- atom
- Dalton's atomic theory

READING SUPPORT

Build Vocabulary: *Word Origins* *Atom* comes from the Greek word *atomos,* meaning "indivisible." ***How does the word origin of atom relate to Dalton's atomic theory?***

Early Models of the Atom

How did the concept of the atom change from the time of Democritus to the time of John Dalton?

Using your unaided eyes, you cannot see the tiny fundamental particles that make up matter. Yet, all matter is composed of such particles, which are called atoms. An **atom** is the smallest particle of an element that retains its identity in a chemical reaction.

The concept of the atom intrigued a number of early scholars. Although these philosophers and scientists could not observe individual atoms, they still were able to propose ideas about the structure of atoms.

Democritus's Atomic Philosophy The Greek philosopher Democritus (460 B.C.–370 B.C.) was among the first to suggest the existence of atoms. **Democritus reasoned that atoms were indivisible and indestructible.** Although Democritus's ideas agreed with later scientific theory, they did not explain chemical behavior. They also lacked experimental support because Democritus's approach was not based on the scientific method.

Dalton's Atomic Theory The real nature of atoms and the connection between observable changes and events at the atomic level were not established for more than 2000 years after Democritus's death. The modern process of discovery regarding atoms began with John Dalton (1766–1844), an English chemist and schoolteacher. **By using experimental methods, Dalton transformed Democritus's ideas on atoms into a scientific theory.** Dalton studied the ratios in which elements combine in chemical reactions.

Focus on ELL

1 CONTENT AND LANGUAGE Direct students' attention to the words *indivisible* and *indestructible* on this page. Have students identify the roots of these words. (divisible *and* destructible) Have students look up the definitions of all four words in the dictionary. Then, have students use the differences between the definitions of *indivisible* and *indestructible* and their root words to identify the meaning of the prefix *in-*. (*in-* means "not") Have students brainstorm other words with this prefix that also have meanings opposite those of their roots.

2 FRONTLOAD THE LESSON Have students draw a compare/contrast table with one column labeled *Democritus* and another labeled *Dalton*. Tell students they will use their tables to track the similarities and differences between the two men's beliefs about the atom as they work through the lesson.

3 COMPREHENSIBLE INPUT Read aloud the four parts of Dalton's theory, demonstrating each point with molecular models. Use an iron bar, iron filings, and a magnet to reinforce the concept that different forms of an element all have the same properties because they are made of the same atoms.

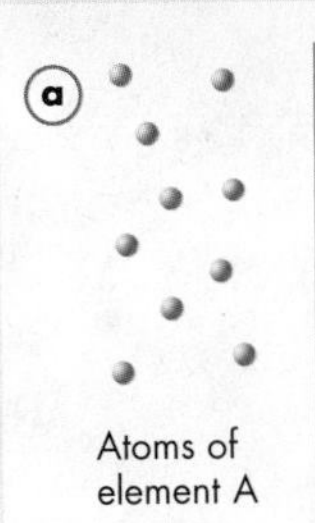

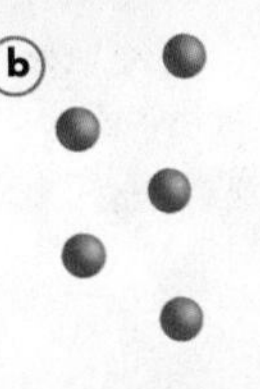

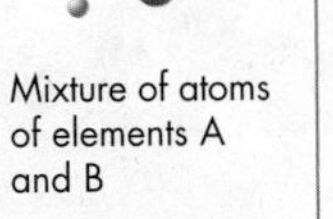

Figure 4.1 Dalton's Atomic Theory
According to Dalton's atomic theory, an element is composed of only one kind of atom, and a compound is composed of particles that are chemical combinations of different kinds of atoms.
Interpret Diagrams ***How does a mixture of atoms of different elements differ from a compound?***

Based on the results of his experiments, Dalton formulated hypotheses and theories to explain his observations. The result of his work is known as **Dalton's atomic theory,** which includes the ideas illustrated in Figure 4.1 and listed below.

1. All elements are composed of tiny indivisible particles called atoms.
2. Atoms of the same element are identical. The atoms of any one element are different from those of any other element.
3. Atoms of different elements can physically mix together or can chemically combine in simple whole-number ratios to form compounds.
4. Chemical reactions occur when atoms are separated from each other, joined, or rearranged in a different combination. Atoms of one element, however, are never changed into atoms of another element as a result of a chemical reaction.

CHEMISTRY & YOU

Q: *How was John Dalton able to study atoms even though he couldn't observe them directly? What evidence did he use to formulate his atomic theory?*

Sizing up the Atom

What instruments are used to observe individual atoms?

The liquid mercury in Figure 4.2 illustrates Dalton's concept of the atom. Whether the size of the drop of mercury is large or small, all drops have the same properties because they are all made of the same kind of atoms.

A coin the size of a penny and composed of pure copper (Cu) is another example. If you were to grind the copper coin into a fine dust, each speck in the small pile of shiny red dust would still have the properties of copper. If by some means you could continue to make the copper dust particles smaller, you would eventually come upon a particle of copper that could no longer be divided and still have the chemical properties of copper. This final particle is an atom.

Atoms are very small. A pure copper coin the size of a penny contains about 2×10^{22} atoms. By comparison, Earth's population is only about 7×10^{9} people. There are about 3×10^{12} times as many atoms in the coin as there are people on Earth. If you could line up 100,000,000 copper atoms side by side, they would produce a line only 1 cm long!

Figure 4.2 Drops of Mercury
This petri dish contains drops of liquid mercury. Every drop, no matter its size, has the same properties. Even if you could make a drop the size of one atom, it would still have the chemical properties of mercury.

LESSON 4.1

Foundations for Reading

BUILD VOCABULARY Discuss the meaning of the term *scientific theory*. *(an explanation of the way the world works, based on observation)*

READING STRATEGY Ask students to think about these questions as they read this chapter.

- What could you do if someone asked you to describe something that is too small to see?
- How would you find out what it looked like?

Explain

Early Models of the Atoms

START A CONVERSATION Explain that John Dalton's work, published in 1808, became the basis for the modern atomic theory. This model represented the atom as a simple sphere with no internal structure. Point out how experimental data have been used to test and refine atomic theory over time.

USE VISUALS Direct students' attention to Figure 4.1. **Ask** How many atoms of A and of B form one particle in Figure 4.1d? *(2 A and 1 B)* Do you think that a mixture, shown in Figure 4.1c, always leads to compounds, shown in Figure 4.1d? Why? *(No; some atoms may not readily combine.)*

CHEMISTRY & YOU Dalton used experimental methods to mix elements and observe how they combined with each other. He studied the ratios by which elements combine with each other to propose what is now known as Dalton's atomic theory.

UbD Check for Understanding

BIGIDEA ELECTRONS AND THE STRUCTURE OF ATOMS Assess students' understanding of Dalton's atomic theory. **Ask** How might Dalton have used his observations of chemical reactions to develop his atomic theory? *(Based on his understanding of chemical reactions, he would most likely know that elements are present in compounds in fixed proportions, and so might hypothesize that the atoms of these elements were also present in fixed proportions. He also might have hypothesized that because the same elements are present in reactants and products and no new elements appeared as a result of the reaction that atoms can be rearranged, separated, or joined but that no atoms are changed into atoms of a new element.)*

ADJUST INSTRUCTION If students are having difficulty understanding Dalton's atomic theory, review the four parts of the theory and Figure 4.1.

Answers

READING SUPPORT Dalton's atomic theory states that atoms are indivisible particles. "Indivisible" is the meaning of *atomos*.

FIGURE 4.1 In a mixture of atoms of different elements, each element in the mixture retains its chemical properties. In a compound, different elements have been combined to form a new substance with chemical properties different from its component elements. Atoms are separated, joined, or rearranged.

LESSON 4.1

Explain

Sizing up the Atom

USE VISUALS Point out that the image in Figure 4.3 is greatly magnified. Explain that a nanometer is 1×10^{-9} meters. **Ask** If the nanocars are about 2 nm across, about how many atoms wide are the nanocars if the average width of the atoms is 1.0×10^{-10} m? *(about 20 atoms wide)*

Evaluate

Informal Assessment

Have students evaluate and criticize the following statements according to Dalton's theory.

- "All atoms are identical." *(False. Dalton said: "All atoms of a given element are identical.")*
- "Chemical reactions occur when atoms of one element change into atoms of another element." *(False. Chemical reactions occur when atoms are separated, joined, or rearranged. The elemental identity of atoms does not change during chemical reactions.)*

Then have students complete the 4.1 Lesson Check.

Reteach

Review Dalton's model of the atom. Discuss how a scanning electron microscope might be used to support or contest Dalton's theory, referring students back to the opening photo as needed. *(No matter how small the sample, you can see that that there are many similar atoms in each sample. You could also see where atoms are and whether they change after a chemical reaction.)*

Figure 4.3 Model of a Nanocar
These nanocars are each made of a single molecule. Each nanocar is only about 2 nanometers across. A light-activated paddle wheel on the car propels the car so it can move. The arrow represensents the direction the nanocar moves.

Learn more about the size of the atom online.

The radii of most atoms fall within the range of 5×10^{-11} m to 2×10^{-10} m. Does seeing individual atoms seem impossible? **Despite their small size, individual atoms are observable with instruments such as scanning electron microscopes.** In scanning electron microscopes, a beam of electrons is focused on the sample. Electron microscopes are capable of much higher magnifications than light microscopes.

With the help of electron microscopes, individual atoms can even be moved around and arranged in patterns. The ability to move individual atoms holds future promise for the creation of atomic-sized electronic devices, such as circuits and computer chips. An example of a device made from individual atoms is the nanocar shown in Figure 4.3. This atomic-scale, or "nanoscale," technology could become essential to future applications in medicine, communications, solar energy, and space exploration.

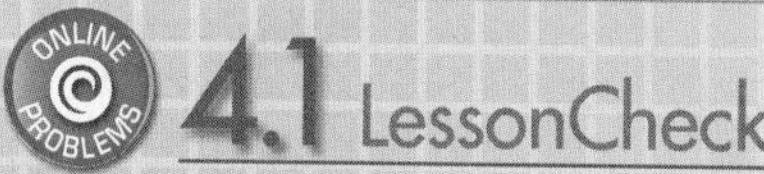

4.1 LessonCheck

1. **Review** How did Democritus characterize atoms?
2. **Explain** How did Dalton advance the atomic philosophy proposed by Democritus?
3. **Identify** What instrument can be used to observe individual atoms?
4. **Explain** In your own words, explain the main ideas of Dalton's atomic theory.
5. **Evaluate** Explain why the ideas on atoms proposed by Dalton constitute a theory, while the ideas proposed by Democritus do not.
6. **Identify** What is the range of the radii of most atoms in nanometers (nm)?
7. **Calculate** A sample of copper with a mass of 63.5 g contains 6.02×10^{23} atoms. Calculate the mass of a single copper atom.

BIGIDEA ELECTRONS AND THE STRUCTURE OF ATOMS

8. According to Dalton's theory, is it possible to convert atoms of one element into atoms of another? Explain.

Lesson Check Answers

1. as indivisible and indestructible
2. by using experimental methods
3. a scanning electron microscope
4. Answers should include the ideas that all matter is composed of atoms; atoms of different elements differ; and chemical change involves a rearrangement of atoms.
5. Democritus's ideas were not based on experimental results and did not explain chemical behavior. Dalton's ideas were empirically based and did explain chemical behavior; his experiments showed that the ratios in which elements combined were whole numbers.
6. 5×10^{-2} nm to 2×10^{-1} nm
7. 1.05×10^{-22} g
8. **BIGIDEA** Atoms of one element are never changed into atoms of another element as a result of a chemical reaction.

4.2 Structure of the Nuclear Atom

C.2.1 Modern atomic theory: historic development; C.2.2 Subatomic particles and atomic structure; C.2.3 Isotopes, atomic number, and atomic mass

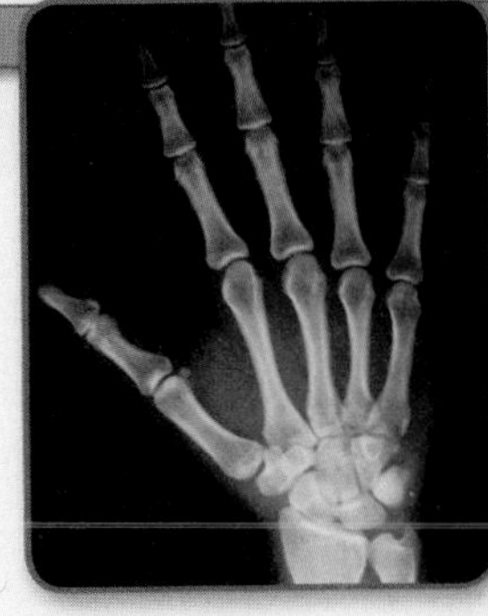

CHEMISTRY & YOU

Q: *You can X-ray a person's hand to see inside it—but how can you see inside an atom?* You may have seen X-rays like the one of the hand shown here. Doctors often use X-rays to see bones and other structures that cannot be seen through the skin. Scientists tried to figure out what was inside an atom without being able to see inside the atom. In this lesson, you will learn the methods scientists used to "see" inside an atom.

Key Questions

What are three kinds of subatomic particles?

How can you describe the structure of the nuclear atom?

Vocabulary

- electron
- cathode ray
- proton
- neutron
- nucleus

Subatomic Particles

What are three kinds of subatomic particles?

Much of Dalton's atomic theory is accepted today. One important change, however, is that atoms are now known to be divisible. They can be broken down into even smaller, more fundamental particles, called subatomic particles. **Three kinds of subatomic particles are electrons, protons, and neutrons.**

Electrons In 1897, the English physicist J. J. Thomson (1856–1940) discovered the electron. **Electrons** are negatively charged subatomic particles. Thomson performed experiments that involved passing electric current through gases at low pressure. He sealed the gases in glass tubes fitted at both ends with metal disks called electrodes. The electrodes were connected to a source of electricity, as shown in Figure 4.4. One electrode, the anode, became positively charged. The other electrode, the cathode, became negatively charged. The result was a glowing beam, or **cathode ray,** that traveled from the cathode to the anode.

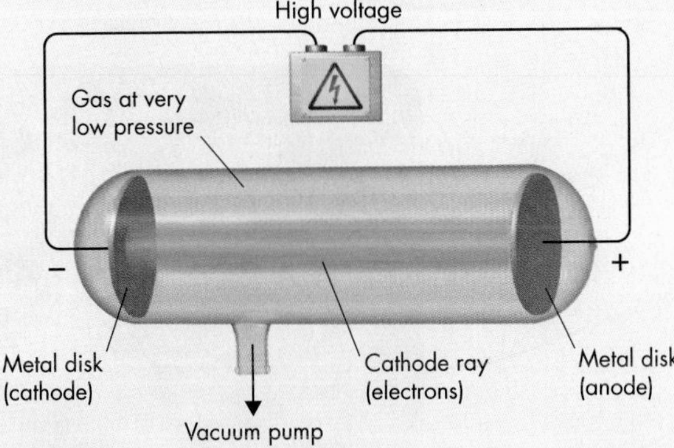

Figure 4.4
Cathode-Ray Tube
In a cathode-ray tube, electrons travel as a ray from the cathode (−) to the anode (+). Televisions used to be made with a specialized type of cathode-ray tube.

Focus on ELL

1 CONTENT AND LANGUAGE Write the term *subatomic* on the board. Explain that the prefix *sub-* means "below" or "a part of." Have students brainstorm the literal meaning of *subatomic. (below atomic)* Now write the words *electron*, *proton*, and *neutron* on the board. Explain that in these three words, the suffix *-on* means "subatomic particle."

2 FRONTLOAD THE LESSON Have students preview Figure 4.4 and identify any similarities and differences between the parts of a cathode ray tube and parts of a battery. Explain that, just as a battery generates a stream of electrons that flow through an electrical circuit, a cathode-ray tube generates a stream of electrons that travel as a beam.

3 COMPREHENSIBLE INPUT Draw Table 4.1 on the board, using a contrasting color to write the mathematical signs in the *Symbols* column. Use the same color to write the values of the relative charges. Point out the relative mass of an electron, and ask students to locate the text where this figure is referenced.

Key Objectives

4.2.1 IDENTIFY three types of subatomic particles.

4.2.2 DESCRIBE the structure of atoms according to the Rutherford atomic model.

Additional Resources

Reading and Study Workbook, Lesson 4.2
Available Online or on Digital Media:
- Teaching Resources, Lesson 4.2 Review
- Laboratory Manual, Lab 5
- Virtual Chemistry Laboratory Manual, Labs 4–6

PearsonChem.com

Engage

CHEMISTRY & YOU Have students study the opening photo and read the accompanying text. **Ask** How could you determine what your hand looks like under the skin without dissecting it? *(X-rays, CT scans, or MRI scans)* Explain that scientists use technology to "see" inside atoms, just as doctors use it to "see" inside the body.

Build Background

Ask Do you have an electronic item in your home that uses a CRT? *(Sample answers: a television or a computer monitor)* Explain that television and computer CRTs have a source of rays at the small end. The rays are projected on the large receiving end to create the display.

IN ACADEMIC STANDARDS for SCIENCE

- Focus on the development of explanatory models based on their observations during laboratory investigations.
- Explain that the body of scientific knowledge is organized into major theories, which are derived from and supported by the results of many experiments, and allow us to make testable predictions.

C.2.3 Determine the number of protons, neutrons, and electrons in isotopes and in those isotopes that comprise a specific element. Relate these numbers to atomic number and mass number.

Also covered C.2.1, C.2.2

LESSON 4.2

Foundations for Reading

BUILD VOCABULARY Have students think of words that start with the same word parts as the three subatomic particles described in this section: neutron, electron, and proton. *(Sample answer: neutral, electric, protein)* Words that students already know may help them remember the meaning of new terms.

Explore

Teacher Demo

PURPOSE To demonstrate a cathode-ray tube and observe properties of cathode rays

MATERIALS cathode-ray tube, magnet

PROCEDURE Demonstrate a cathode-ray tube in class. Use a magnet to deflect the beam of particles. Review the components of a cathode-ray tube, and discuss the connection to television picture tubes and computer monitors.

EXPECTED OUTCOME Students should be able to explain how the CRT works and see how the cathode ray is deflected by a magnetic field.

Explain

Subatomic Particles

USE VISUALS Have students study Table 4.1 and compare the masses and charges of the three elementary particles. Students should recognize from this comparison that the mass of an atom is due mainly to the number of protons and neutrons that it has. Point out that the assigned charges for protons and electrons are relative charges. **Ask** What particles make up most of the mass of an atom? *(protons and neutrons)* **Ask** Why are relative charges and mass useful in talking about subatomic particles? *(The actual values are unwieldy numbers.)* The absolute charge on an electron is 1.602177×10^{-19} coulombs.

Thomson found that a cathode ray is deflected by electrically charged metal plates, as in Figure 4.5a. A positively charged plate attracts the cathode ray, while a negatively charged plate repels it. Thomson knew that opposite charges attract and like charges repel, so he hypothesized that a cathode ray is a stream of tiny negatively charged particles moving at high speed. Thomson called these particles corpuscles; later they were named electrons.

To test his hypothesis, Thomson set up an experiment to measure the ratio of an electron's charge to its mass. He found this ratio to be constant. Also, the charge-to-mass ratio of electrons did not depend on the kind of gas in the cathode-ray tube or the type of metal used for the electrodes. Thomson concluded that electrons are a component of the atoms of all elements.

The U.S. physicist Robert A. Millikan (1868–1953) carried out experiments to find the quantity of an electron's charge. In his oil-drop experiment, Millikan suspended negatively charged oil droplets between two charged plates. He then changed the voltage on the plates to see how this affected the droplets' rate of fall. From his data, he found that the charge on each oil droplet was a multiple of 1.60×10^{-19} coulomb, meaning this must be the charge of an electron. Using this charge value and Thomson's charge-to-mass ratio of an electron, Millikan calculated an electron's mass. Millikan's values for electron charge and mass are similar to those accepted today. An electron has one unit of negative charge, and its mass is 1/1840 the mass of a hydrogen atom.

Figure 4.5
Thomson's Experiment
a. Thomson found that cathode rays are attracted to metal plates that have a positive electrical charge. **b.** A cathode ray can also be deflected by a magnet.
Infer *If a cathode ray is attracted to a positively charged plate, what can you infer about the charge of the particles that make up the cathode ray?*

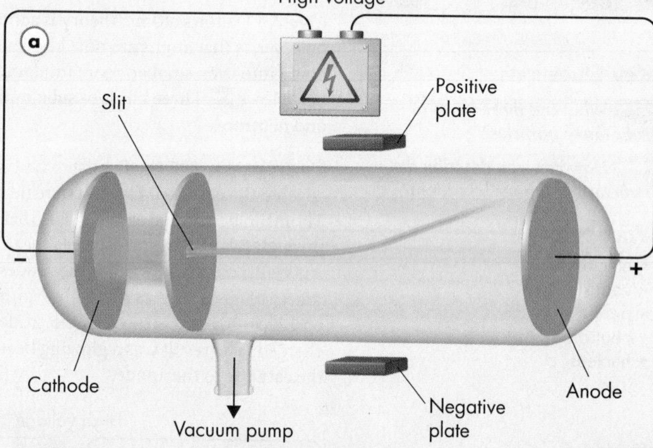

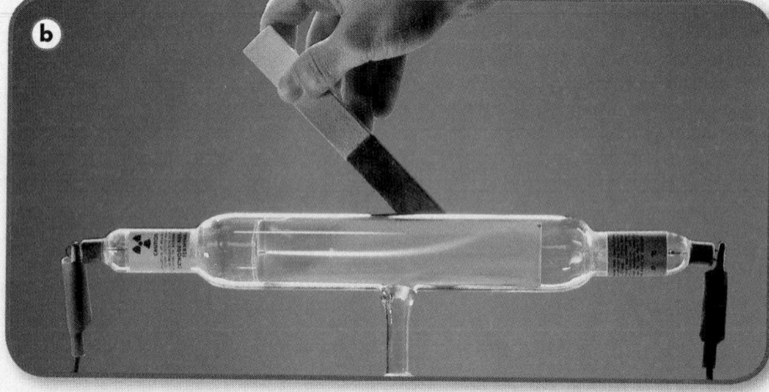

Differentiated Instruction

LPR LESS PROFICIENT READERS Direct students to use Figures 4.4 and 4.5 to describe J. J. Thomson's experiment with cathode-ray tubes in their own words.

L1 SPECIAL NEEDS STUDENTS Prior to explaining subatomic particles, organize students into small groups. Give each group a pair of inflated rubber balloons (be aware of any latex allergies in the class) and a swatch of wool cloth. Have students rub the balloons vigorously with the wool. Have students test the balloons against different objects and against each other to discover how opposite charges attract and like charges repel. Use this activity as a lead in to Thompson's cathode-ray experiment.

L3 ADVANCED STUDENTS Have students use the Internet or library to find the original papers for the discoveries described in this chapter and write a report on what they have learned.

Table 4.1

Properties of Subatomic Particles				
Particle	Symbol	Relative charge	Relative mass (mass of proton = 1)	Actual mass (g)
Electron	e^-	1−	1/1840	9.11×10^{-28}
Proton	p^+	1+	1	1.67×10^{-24}
Neutron	n^0	0	1	1.67×10^{-24}

Protons and Neutrons If cathode rays are electrons given off by atoms, what remains of the atoms that have lost the electrons? For example, after a hydrogen atom (the lightest kind of atom) loses an electron, what is left? You can think through this problem using four simple ideas about matter and electric charges. First, atoms have no net electric charge; they are electrically neutral. (One important piece of evidence for electrical neutrality is that you do not receive an electric shock every time you touch something!) Second, electric charges are carried by particles of matter. Third, electric charges always exist in whole-number multiples of a single basic unit; that is, there are no fractions of charges. Fourth, when a given number of negatively charged particles combines with an equal number of positively charged particles, an electrically neutral particle is formed.

Considering all of this information, it follows that a particle with one unit of positive charge should remain when a typical hydrogen atom loses an electron. Evidence for such a positively charged particle was found in 1886, when Eugen Goldstein (1850–1930) observed a cathode-ray tube and found rays traveling in the direction opposite to that of the cathode rays. He called these rays canal rays and concluded that they were composed of positive particles. Such positively charged subatomic particles are called **protons.** Each proton has a mass about 1840 times that of an electron.

In 1932, the English physicist James Chadwick (1891–1974) confirmed the existence of yet another subatomic particle: the neutron. **Neutrons** are subatomic particles with no charge but with a mass nearly equal to that of a proton. Table 4.1 summarizes the properties of these subatomic particles. Although protons and neutrons are exceedingly small, theoretical physicists believe that they are composed of yet smaller subnuclear particles called *quarks.*

The Atomic Nucleus

How can you describe the structure of the nuclear atom?

When subatomic particles were discovered, scientists wondered how the particles were put together in an atom. This question was difficult to answer, given how tiny atoms are. Most scientists—including J. J. Thomson, discoverer of the electron—thought it likely that electrons were evenly distributed throughout an atom filled uniformly with positively charged material. In Thomson's atomic model, known as the "plum-pudding model," electrons were stuck into a lump of positive charge, similar to raisins stuck in dough. This model of the atom turned out to be short-lived, however, due to the work of a former student of Thomson, Ernest Rutherford (1871–1937), shown in Figure 4.6.

Figure 4.6 Ernest Rutherford Born in New Zealand, Rutherford was awarded the Nobel Prize in Chemistry in 1908. His portrait appears on the New Zealand $100 bill.

Explore

Teacher Demo

PURPOSE To trace the history of atomic models and examine the role of the scientific method in the development of such models

MATERIALS Library or Internet access

PROCEDURE Have students create a timeline that traces the development of the atomic model. Have them note the data that led to an existing model being changed.

EXPECTED OUTCOME Students' timelines should list at least some of the atomic models shown on page 133. Students should be able to explain how certain scientific discoveries (e.g., Rutherford's gold foil experiment) resulted in the revision of the prevailing atomic model at the time.

Extend

Explain to students that by 1887, the British scientist William Crookes knew that metal atoms contained negatively charged particles. He used a cathode-ray tube containing hydrogen gas at low pressure and discovered that hydrogen contains positive charges. Ask students to research the following: What conclusion about the mass of the atom and the mass of its protons was derived from Crookes's findings? *(The mass of an atom is greater than its proton content.)* Which subatomic particle accounted for this additional mass? *(the neutron)*

UbD Check for Understanding

The Essential Question What structures make up an atom?

Give each student an index card. Assess students' knowledge of subatomic particles by reading out loud the description of each of the three kinds of particles. Students should write the names of the particles on the card in the order in which they were stated. Then ask students to indicate where each particle is located within the atom. *(electron, proton, and neutron; protons and neutrons are located inside the nucleus; electrons are located outside)*

ADJUST INSTRUCTION If students are having difficulty remembering which particles are located inside the nucleus, have them review Rutherford's atomic model by drawing and labeling an atom.

Answers

FIGURE 4.5 The particles are negatively charged.

LESSON 4.2

Explain

USE AN ANALOGY Explain to students that the ratio of the size of the nucleus to the size of an atom is about 10^{-5}. Discuss how small the nucleus is compared to the entire atom. Explain that if a housefly sitting on second base in a baseball stadium represented the nucleus of an atom, the rest of the atom would be the size of the stadium.

Evaluate

Informal Assessment

Have each student come to your desk and orally state in one minute or less one of the discoveries by Thomson, Millikan, or Rutherford. Then ask them to describe how the discovery led to the current understanding of atomic structure. Then have students complete the 4.2 Lesson Check.

Reteach

Have students review Table 4.1. Ask them to create a table or diagram to compare the characteristics of electrons, protons, and neutrons.

CHEMISTRY & YOU Scientists observed the behavior of atoms in different experimental conditions. This gave them clues as to what is inside an atom without actually "seeing" inside an atom.

Q: *How did scientists "see" inside an atom to determine the structures that are inside an atom?*

Rutherford's Gold-Foil Experiment In 1911, Rutherford and his co-workers at the University of Manchester, England, wanted to test the existing plum-pudding model of atomic structure. So, they devised the gold-foil experiment. Their test used alpha particles, which are helium atoms that have lost their two electrons and have a double positive charge because of the two remaining protons. In the experiment, illustrated in Figure 4.7, a narrow beam of alpha particles was directed at a very thin sheet of gold foil. According to the prevailing theory, the alpha particles should have passed easily through the gold, with only a slight deflection due to the positive charge thought to be spread out in the gold atoms.

Rutherford's results were that most alpha particles went straight through the gold foil, or were slightly deflected. However, what was surprising is that a small fraction of the alpha particles bounced off the gold foil at very large angles. Some even bounced straight back toward the source. Rutherford later recollected, "This is almost as incredible as if you fired a 15-inch shell at a piece of tissue paper and it came back and hit you."

The Rutherford Atomic Model Based on his experimental results, Rutherford suggested a new theory of the atom. He proposed that the atom is mostly empty space, thus explaining the lack of deflection of most of the alpha particles. He concluded that all the positive charge and almost all the mass are concentrated in a small region that has enough positive charge to account for the great deflection of some of the alpha particles. He called this region the nucleus. The **nucleus** is the tiny central core of an atom and is composed of protons and neutrons.

Figure 4.7 Rutherford's Experiment
Rutherford's gold-foil experiment yielded evidence of the atomic nucleus.

a. Rutherford and his co-workers aimed a beam of alpha particles at a sheet of gold foil surrounded by a fluorescent screen. Most of the particles passed through the foil with no deflection at all. A few particles were greatly deflected.

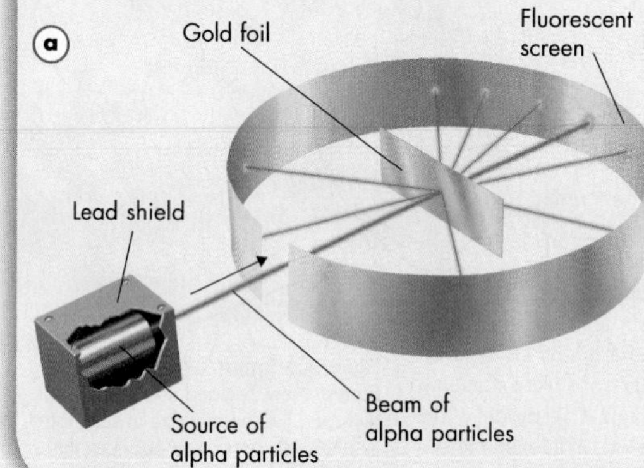

b. Rutherford concluded that most of the alpha particles pass through the gold foil because the atom is mostly empty space. The mass and positive charge are concentrated in a small region of the atom. Rutherford called this region the nucleus. Particles that approach the nucleus closely are greatly deflected.

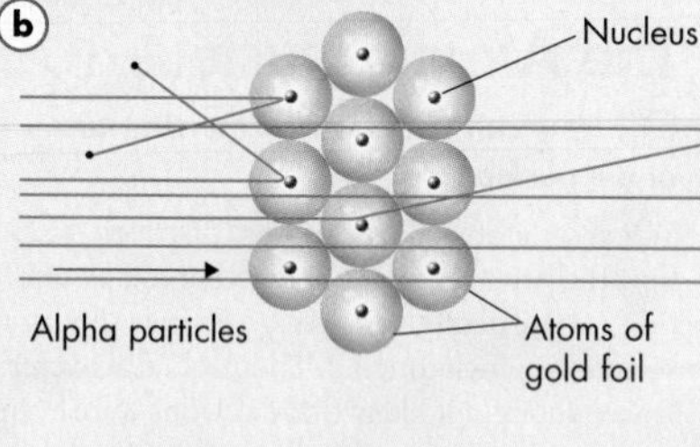

See Rutherford's gold-foil experiment *animated online.* KINETIC ART

Focus on ELL

4 LANGUAGE PRODUCTION Have students work with partners to complete the Quick Lab. Make sure each pairing has ELLs of varied language proficiencies, so that a more proficient student can help a less proficient one. Have students work according to their proficiency level.

BEGINNING: LOW/HIGH Model this lab for these students. Show them your data, sketch, and conclusions. Explain your method for guessing what is in the box.

INTERMEDIATE: LOW/HIGH Ask students to brainstorm what they think they may observe, prior to performing each step. Allow students to orally present their results.

ADVANCED: LOW/HIGH Ask students to present to the class their sketch, data, and interpretation of the object in one of the boxes.

The Rutherford atomic model is known as the nuclear atom. **In the nuclear atom, the protons and neutrons are located in the positively charged nucleus. The electrons are distributed around the nucleus and occupy almost all the volume of the atom.** According to this model, the nucleus is tiny and densely packed compared with the atom as a whole. If an atom were the size of a football stadium, the nucleus would be about the size of a marble.

Although it was an improvement over Thomson's model of the atom, Rutherford's model turned out to be incomplete. In Chapter 5, you will learn how the Rutherford atomic model had to be revised in order to explain the chemical properties of elements.

Quick Lab

Purpose To determine the shape of a fixed object inside a sealed box without opening the box

Materials

- box containing a regularly shaped object fixed in place and a loose marble

Using Inference: The Black Box

Procedure

1. Do not open the box.
2. Manipulate the box so that the marble moves around the fixed object.
3. Gather data (clues) that describe the movement of the marble.
4. Sketch a picture of the object in the box, showing its shape, size, and location within the box.
5. Repeat this activity with a different box containing a different object.

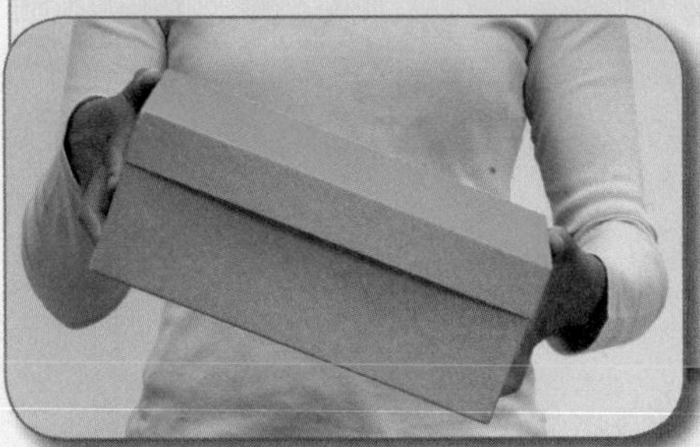

Analyze and Conclude

1. **Compare** Find a classmate who had a box with the same letter as yours, and compare your findings.
2. **Apply Concepts** Think about the experiments that have contributed to a better understanding of the atom. Which experiment does this activity remind you of?

4.2 LessonCheck

9. **Review** What are three types of subatomic particles?
10. **Explain** How does the Rutherford model describe the structure of atoms?
11. **Review** What are the charges and relative masses of the three main subatomic particles?
12. **Explain** Describe Thomson's and Millikan's contributions to atomic theory.
13. **Compare and Contrast** Compare Rutherford's expected outcome of the gold-foil experiment with the actual outcome.
14. **Analyze Data** What experimental evidence led Rutherford to conclude that an atom is mostly empty space?
15. **Compare and Contrast** How did Rutherford's model of the atom differ from Thomson's?

LESSON 4.2

Explore

PURPOSE The students will determine the shape of the hidden object by analyzing the rebound paths of a marble rolled at the object.

SKILLS FOCUS Observing, inferring

PREP TIME 5 minutes

CLASS TIME 10 minutes

MATERIALS box containing a regularly shaped object fixed in place and a loose marble

ADVANCE PREP Cut geometric shapes—such as a triangle, circle, or L, from a sheet of 1-inch plastic foam.

EXPECTED OUTCOME Students' inferences may or may not be different for the same object.

ANALYZE AND CONCLUDE

1. See Expected Outcome.
2. The activity simulates the strategy that Rutherford used to probe the structure of metal atoms. Like the students, Rutherford and his coworkers were also faced with the problem of identifying properties of an object not visible to the eye.

FOR ENRICHMENT Make a more challenging black box for students who have an easy time with the simple boxes. Put two single objects in one box, or a single object with a complex shape.

Lesson Check Answers

9. protons, neutrons, and electrons
10. A positively charged nucleus surrounded by electrons, which occupy most of the volume.
11. proton, positive charge, relative mass = 1; electron, negative charge, relative mass = 1/1840; neutron, no charge, relative mass = 1
12. Thomson passed an electric current through sealed glass tubes filled with gases. The resulting glowing beam consisted of tiny negatively charged particles moving at high speed. Thomson concluded that electrons must be parts of the atoms of all elements. Millikan determined the charge and mass of the electron.
13. Rutherford expected all the alpha particles to pass straight through with little deflection. He found that most alpha particles passed straight through, but some particles were deflected at very large angles—and some even bounced straight back.
14. The great majority of the alpha particles passed straight through the gold foil.
15. Rutherford's atomic model described the atom as having a positively charged, dense nucleus that is tiny compared to the atom as a whole. In Thomson's plum-pudding model, electrons were stuck in a chunk of positive charge.

CHEMISTRY & YOU

Electron Microscopy

CHEMISTRY & YOU Electron microscopes are capable of much greater levels of magnification than light microscopes, but their real power lies in their resolution. Resolution refers to the degree of detail that is revealed by a microscope. The greater a microscope's resolution, the more detail that can be seen in the images it transmits. Light microscopes rely on the natural reflective, refractive, and absorbent behaviors of light to make objects visible under magnification. However, these same behaviors can inhibit the degree of resolution, regardless of the magnitude of magnification. Thus, the resolution of a typical compound light microscope is about 200 nm. In contrast, some SEMs can achieve a resolution of 5 nm at high magnification.

Pose the following question to students: *How does the behavior of light inhibit the ability of a light microscope to deliver images with high resolution?* You may need to assist students in the following ways:

- Waves of visible light occur in wavelengths from 400 nm to 700 nm.
- The light provided to a light microscope by its mirror or electric light source are unfocused and spread out from the light source, while that electron beam in an SEM is highly focused and narrow.
- Light waves reflected by and transmitted through the object are subject to interference from each other and from incident light from both the source and from the microscope's surroundings.
- Some light waves transmitted through the sample may be directed away from the lens of the microscope.

CHEMISTRY & YOU: TECHNOLOGY

Electron Microscopy

Within 30 years of J. J. Thomson's discovery of the electron, scientists were studying how to produce images of objects by using an electron beam. In 1931, German scientists Ernst Ruska and Max Knoll built the first electron microscope. There are two types of electron microscopes, scanning electron microscopes (SEM) and transmission electron microscopes (TEM). The images shown here are from SEMs. In an SEM, a beam of electrons is focused down to a very small diameter and scanned across the sample. Most materials eject electrons when the electron beam hits them. The location of the ejected electrons is detected and used to produce an image.

A typical light microscope is capable of magnifying an object 1000 times. An electron microscope can magnify an object over 100,000 times. Another advantage of electron microscopes is their higher resolution. Resolution is the ability to differentiate two objects that are very close to each other. So, an electron microscope has the ability to produce a clearer image than a light microscope at the same magnification. Electron microscopes do not produce color images. The color images shown here have false color that has been added to the images. Electron microscopes are useful in chemistry, but also in other fields, such as archeology, pharmacology and quality assurance testing.

SEM This microscope is a scanning electron microscope. The image from the microscope is seen by using a computer screen.

BIOLOGY This diatom is a single-celled organism that lives in the water. The image shown above is a dust mite on a piece of fabric.

110

21st Century Learning To be successful in the 21st century, students need skills and learning experiences that extend beyond subject matter mastery. This project helps students build the following 21st Century Learning: *Creativity and Innovation*; *Communication and Collaboration*; *Information, Media*, and *Technology Literacy*; *Initiative and Self-Direction*; and *Social Skills*.

Pose the following challenge to your students. Your school has invited four different specialists in electron microscope technology to participate in a roundtable discussion during a college fair. The specialists are there to answer questions and convey information about the uses of electron microscopes and the types of careers that utilize this type of technology.

Ask for four student volunteers to be the team of four "experts" in electron microscopy in the following fields: biology, forensics, biochemistry, and food science. The remaining students will form four interview panels. Each panel will select one of the four specialists to interview. The interview panels should prepare a set of questions and their expected answers on the specific uses of electron microscopy in the expert's field, as well as pros and cons of its use. The team of "experts" will prepare a set of expected questions and their answers on which to base their preparation for their roles.

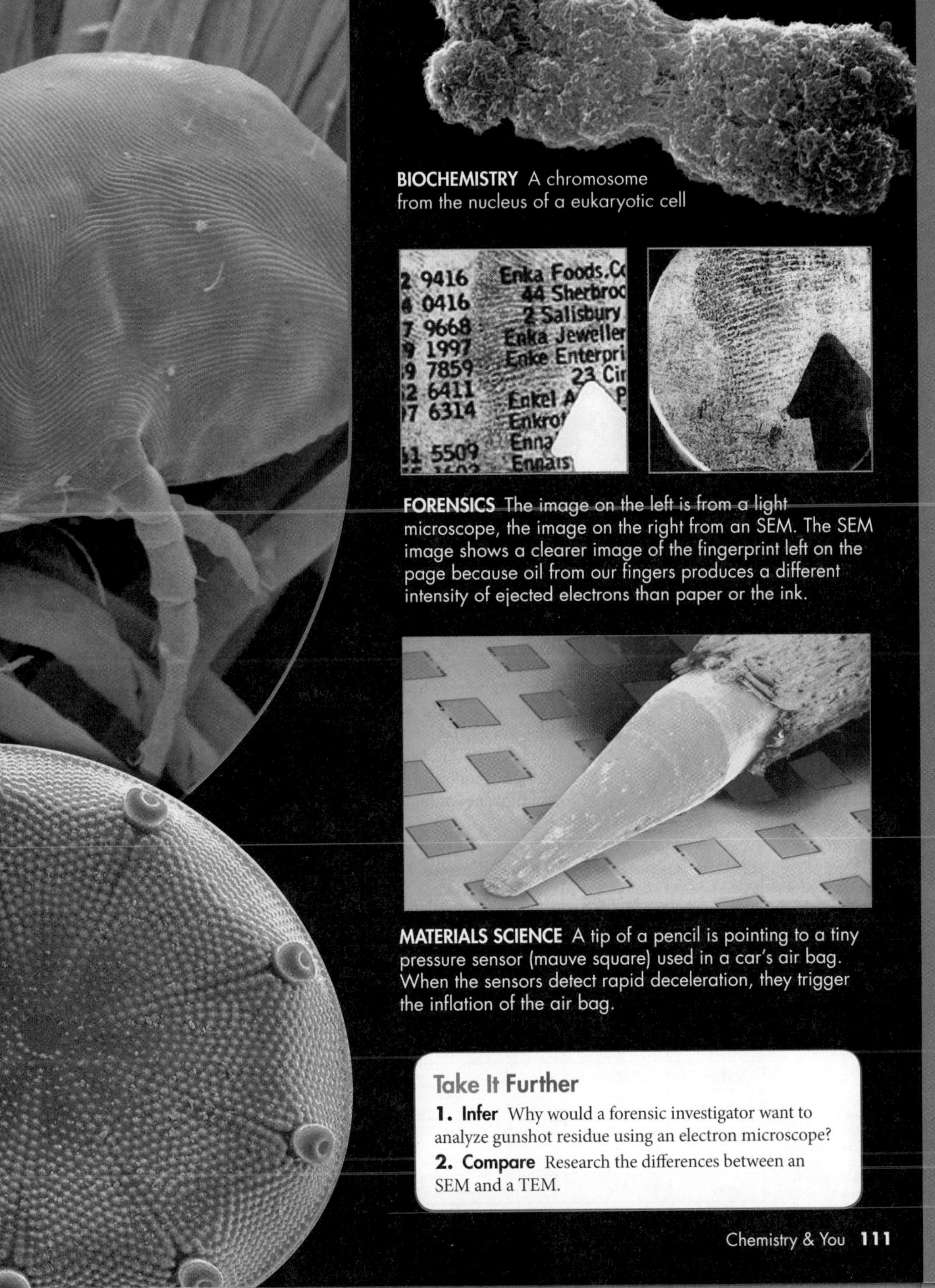

BIOCHEMISTRY A chromosome from the nucleus of a eukaryotic cell

FORENSICS The image on the left is from a light microscope, the image on the right from an SEM. The SEM image shows a clearer image of the fingerprint left on the page because oil from our fingers produces a different intensity of ejected electrons than paper or the ink.

MATERIALS SCIENCE A tip of a pencil is pointing to a tiny pressure sensor (mauve square) used in a car's air bag. When the sensors detect rapid deceleration, they trigger the inflation of the air bag.

Take It Further

1. **Infer** Why would a forensic investigator want to analyze gunshot residue using an electron microscope?
2. **Compare** Research the differences between an SEM and a TEM.

Chemistry & You 111

Explain

USE VISUALS Have students examine the image of the scientists working with the electron microscope. Point out how a person sees an image made by the electron microscope; the scientists in the photo are not looking into eyepieces, but see a projection on a computer screen. Explain that the type of electron microscope a person uses is typically dependent on the type of information he or she needs to know. A microscope is typically selected based on its resolution, magnification, depth of field, field of view, illumination method, degree of automation, and type of image produced.

Extend

Connect to TECHNOLOGY

Emphasize the time period that the electron microscope was invented and the technology associated with the apparatus. Tell students that cathode ray tubes, magnets, electron beams, and vacuum chambers are all integral parts of the electron microscope. In pairs, have students research other technological advances that came about using the same or similar components. *(Televisions, hand held radios, X-ray machines and lasers all were invented around the same time as the electron microscope and as a result of the same technology.)*

Differentiated Instruction

L1 STRUGGLING STUDENTS Have students draw pictures of an ordinary object as they imagine it would appear under an ordinary microscope and then as they imagine it would appear under an electron microscope.

ELL ENGLISH LANGUAGE LEARNERS Have magnifying glasses available in class, as well as a compound microscope. Allow students time to examine and use both tools.

L3 ADVANCED STUDENTS Assign students to write a summary explaining how an electron microscope provides a far greater degree of magnification and a better resolution than an ordinary compound microscope. Summary should include how Thomson's research on cathode rays propelled electron microscope technology.

Answers

TAKE IT FURTHER

1. An electron microscope would allow scientists to see details in the gunshot residue that might not be visible by other microscopes.
2. An SEM is a scanning electron microscope. In an SEM, an electron beam is passed over the surface of an object. An SEM produces an image of the surface of on an object. A TEM is a transmission electron microscope. In a TEM, an electron beam is directed through an object, producing an image of what is inside an object.

LESSON 4.3

Key Objectives

4.3.1 EXPLAIN what makes elements and isotopes different from each other.

4.3.2 EXPLAIN how isotopes of an element differ.

4.3.3 CALCULATE the atomic mass of an element.

Additional Resources

Reading and Study Workbook, Lesson 4.3

Available Online or on Digital Media:

- Teaching Resources, Lesson 4.3 Review
- Small-Scale Chemistry Manual, Lab 6

PearsonChem.com

Engage

CHEMISTRY & YOU Direct students to look at the photograph of the dogs and to read the corresponding text. **Ask** What characteristics can you use to classify different dogs? *(Sample answers: height, weight, color, hair length, tail, etc.)* Point out that each dog in the picture is a different variety, called a *breed*, but they are all dogs. Tell students to keep this distinction in mind as they work through the lesson.

Activate Prior Knowledge

Briefly review what students know about subatomic particles and Dalton's atomic theory. Have a student volunteer go to the board, and draw an atom based on the description given by his or her classmates. Have the volunteer label the different parts of the atom with important characteristics, such as charge.

IN ACADEMIC STANDARDS for SCIENCE

- Focus on the development of explanatory models based on their observations during laboratory investigations.
- Explain that the body of scientific knowledge is organized into major theories, which are derived from and supported by the results of many experiments, and allow us to make testable predictions.

C.2.4 Calculate the average atomic mass of an element from its isotopic abundance data.

Also covered C.2.1, C.2.2, C.2.3

4.3 Distinguishing Among Atoms

C.2.1 Modern atomic theory: historic development; **C.2.2** Subatomic particles and atomic structure; **C.2.3** Isotopes, atomic number, and atomic mass; **C.2.4** Average atomic mass.

CHEMISTRY & YOU

Q: *How can there be different varieties of atoms?* Some things exist in many different varieties. For example, dogs can differ in many ways, such as color, size, ear shape, and length of hair. Just as there are many types of dogs, atoms come in different varieties, too.

Key Questions

What makes one element different from another?

How do isotopes of an element differ?

How do you calculate the atomic mass of an element?

Vocabulary

- atomic number • mass number
- isotope • atomic mass unit (amu)
- atomic mass

Atomic Number and Mass Number

What makes one element different from another?

Atoms are composed of protons, neutrons, and electrons. Protons and neutrons make up the nucleus. Electrons surround the nucleus. How, then, are atoms of hydrogen, for example, different from atoms of oxygen?

Atomic Number Look at Table 4.2. Notice that a hydrogen atom has one proton, but an oxygen atom has eight protons. **Elements are different because they contain different numbers of protons.** An element's **atomic number** is the number of protons in the nucleus of an atom of that element. Since all hydrogen atoms have one proton, the atomic number of hydrogen is 1. All oxygen atoms have eight protons, so the atomic number of oxygen is 8. The atomic number identifies an element. For each element listed in Table 4.2, the number of protons equals the number of electrons. Remember that atoms are electrically neutral. Thus, the number of electrons (negatively charged particles) must equal the number of protons (positively charged particles).

Table 4.2

Atoms of the First Ten Elements

Name	Symbol	Atomic number	Protons	Neutrons*	Mass number	Electrons
Hydrogen	H	1	1	0	1	1
Helium	He	2	2	2	4	2
Lithium	Li	3	3	4	7	3
Beryllium	Be	4	4	5	9	4
Boron	B	5	5	6	11	5
Carbon	C	6	6	6	12	6
Nitrogen	N	7	7	7	14	7
Oxygen	O	8	8	8	16	8
Fluorine	F	9	9	10	19	9
Neon	Ne	10	10	10	20	10

* Number of neutrons in the most abundant isotope. Isotopes are introduced later in Lesson 4.3.

Focus on ELL

1 CONTENT AND LANGUAGE Direct students' attention to the second paragraph under the Mass Number heading on the following page. Read the second sentence aloud. Explain that this sentence describes a mathematical relationship. Have students identify the words that relate to mathematical symbols. Then demonstrate how to rewrite the sentence using these signs to create a word equation. Point out that this technique can be used whenever they are having difficulty understanding similar sentences. Have students practice identifying and rewriting the other descriptions of mathematical relationships in the paragraph.

2 FRONTLOAD THE LESSON Ask students to preview Table 4.2 and the Periodic Table of Elements. Show students how the numbers on the Periodic Table of Elements correspond to the values in Table 4.2.

3 COMPREHENSIBLE INPUT As you write, use a black marker to write the element symbol, a red marker to represent the protons, and a blue marker to represent the number of electrons. Continue to use the corresponding color throughout the calculation. Use green to represent the neutrons.

SampleProblem 4.1

Understanding Atomic Number

The element nitrogen (N) has an atomic number of 7. How many protons and electrons are in a neutral nitrogen atom?

1 Analyze **Identify the relevant concepts.** The atomic number gives the number of protons, which in a neutral atom equals the number of electrons.

2 Solve **Apply the concepts to this problem.**

Identify the atomic number. Then use the atomic number to find the number of protons and electrons. → The atomic number of nitrogen is 7. So, a neutral nitrogen atom has 7 protons and 7 electrons.

16. How many protons and electrons are in each atom?

a. fluorine (atomic number = 9)
b. calcium (atomic number = 20)
c. aluminum (atomic number = 13)
d. potassium (atomic number = 19)

17. Complete the table.

Element	Atomic number	Protons	Electrons
S	16	a. ___	b. ___
V	c. ___	23	d. ___
e. ___	f. ___	g. ___	5

Mass Number Most of the mass of an atom is concentrated in its nucleus and depends on the number of protons and neutrons. The total number of protons and neutrons in an atom is called the **mass number.** For example, a helium atom has two protons and two neutrons, so its mass number is 4. A carbon atom has six protons and six neutrons, so its mass number is 12.

If you know the atomic number and mass number of an atom of any element, you can determine the atom's composition. The number of neutrons in an atom is the difference between the mass number and atomic number.

Number of neutrons = mass number − atomic number

Table 4.2 shows that a fluorine atom has an atomic number of 9 and a mass number of 19. Since the atomic number equals the number of protons, which equals the number of electrons, a fluorine atom has nine protons and nine electrons. The mass number of fluorine is equal to the number of protons plus the number of neutrons. So the fluorine atom has ten neutrons, which is the difference between the mass number and the atomic number ($19 - 9 = 10$).

The composition of any atom can be represented in shorthand notation using the atomic number and mass number, as in Figure 4.8. The chemical symbol for gold, Au, appears with two numbers written to its left. The atomic number is the subscript. The mass number is the superscript. You can also refer to atoms by using the mass number and the name of the element. For example, $^{197}_{79}Au$ may be written as gold-197.

Figure 4.8 Chemical Symbol
Au is the chemical symbol for gold.
Apply Concepts ***How many electrons does a gold atom have?***

Differentiated Instruction

LPR LESS PROFICIENT READERS Have students make a list of familiar elements and describe at least one real world use for each element. Have students combine their lists into a master list on the chalkboard.

L1 STRUGGLING STUDENTS Have students recreate Table 4.2 in their notebooks, omitting the Neutrons column and the Mass number column. Guide students to see the pattern between the atomic number of an element and its corresponding number of protons and electrons.

ELL ENGLISH LANGUAGE LEARNERS Pair students of limited proficiency with more advanced ELL students. Have student pairs create a poster showing how to interpret the superscript and subscript associated with an element. Display the posters in the classroom for reference.

Foundations for Reading

BUILD VOCABULARY Have students use the vocabulary for this section to build a concept map that links and relates the vocabulary terms.

READING STRATEGY Allot time at the beginning of the lesson to remind students how to read a table. Review each table in this lesson, except for the periodic table. Point out that students should read the title and the headings of each row and column and take note of units. Relate the cells of each table to a matrix or to a spreadsheet. Students need to be able to read the text, locate the appropriate table, and extract the necessary information.

Explain

Atomic Number and Mass Number

USE VISUALS Have students look at Table 4.2. Point out that the atomic number is equal to the number of protons for each element. **Ask** Why must the number of electrons equal the number of protons for each element? *(Atoms are electrically neutral.)* **Ask** What is the relationship between the number of protons and the number of neutrons? *(The number of neutrons tends to rise with the number of protons.)*

Sample Practice Problems

A. Identify the element that has 6 protons and 6 neutrons. *(carbon)*

B. What is the atomic number of Lithium (Li)? How many electrons and protons does this element have? *(atomic number = 3, 3 electrons, 3 protons)*

APPLY CONCEPTS Explain that the mass number of an element is defined as the total number of protons and neutrons in the element. Explain that chemists have arbitrarily assigned a value of one atomic mass unit to represent the mass of one twelfth of a carbon-12 atom. **Ask** How do you find the number of neutrons in an atom? *(Subtract the atomic number from the mass number.)*

Answers

FIGURE 4.8: 79 electrons

16. **a.** 9 protons and 9 electrons
b. 20 protons and 20 electrons
c. 13 protons and 13 electrons
d. 19 protons and 19 electrons

17. **a.** 16 **e.** B
b. 16 **f.** 5
c. 23 **g.** 5
d. 23

LESSON 4.3

Explain

CHEMISTRY & YOU Atoms of different elements have different numbers of protons. Isotopes of an element have the same number of protons, but different numbers of neutrons.

Misconception Alert

Students are used to seeing subscripts and superscripts in mathematics class. Be sure to emphasize how subscripts and superscripts are used in chemical symbols. When representing elements with shorthand notation, the superscripts and subscripts are located to the left of a chemical element, and the numbers indicate important information about the element.

Sample Practice Problems

A. How many neutrons are in the each atom?

a. $^{40}_{20}\text{Ca}$ *(20)* **b.** $^{204}_{81}\text{Te}$ *(123)* **c.** $^{127}_{53}\text{I}$ *(74)*

B. Use Table 4.2 to express the composition of each atom below in shorthand form.

a. oxygen-16 *($^{16}_{8}O$)* **c.** nitrogen-14 *($^{14}_{7}N$)*

b. helium-4 *($^{4}_{2}He$)*

Explore

Student Activity

PURPOSE To learn about practical applications of isotopes

MATERIALS Library or Internet access

PROCEDURE Have students use the library or Internet to find an isotope of an element that has a practical or everyday use.

EXPECTED OUTCOME Students' research will most likely focus on applications of radioisotopes such as carbon-14 (used in archaeological dating), americium-241 (used in smoke alarms), iodine-131 (used in the treatment of thyroid disorders), and cobalt-60 (used in the treatment of some cancers). Point out that the instability of these isotopes is what makes them useful. Radioisotopes and radioactivity are discussed in Chapter 25.

Sample Problem 4.2

Determining the Composition of an Atom

How many protons, electrons, and neutrons are in each atom?

a. $^{9}_{4}\text{Be}$ **b.** $^{20}_{10}\text{Ne}$ **c.** $^{23}_{11}\text{Na}$

KNOWNS	UNKNOWNS
• Beryllium (Be)	number of:
atomic number = 4	protons = ?
mass number = 9	electrons = ?
• Neon (Ne)	neutrons = ?
atomic number = 10	
mass number = 20	
• Sodium (Na)	
atomic number = 11	
mass number = 23	

1 Analyze **List the knowns and the unknowns.** Use the definitions of atomic number and mass number to calculate the numbers of protons, electrons, and neutrons.

2 Calculate **Solve for the unknowns.**

Use the atomic number to find the number of protons.

atomic number = number of protons
a. 4 b. 10 c. 11

Use the atomic number to find the number of electrons.

atomic number = number of electrons
a. 4 b. 10 c. 11

Use the mass number and atomic number to find the number of neutrons.

number of neutrons = mass number − atomic number
a. number of neutrons = 9 − 4 = 5
b. number of neutrons = 20 − 10 = 10
c. number of neutrons = 23 − 11 = 12

3 Evaluate **Do the results make sense?** For each atom, the mass number equals the number of protons plus the number of neutrons. The results make sense.

18. How many neutrons are in each atom?

a. $^{80}_{35}\text{Br}$ **b.** $^{32}_{16}\text{S}$ **c.** $^{108}_{47}\text{Ag}$ **d.** $^{207}_{82}\text{Pb}$

19. Use Table 4.2 to express the composition of each atom below in shorthand form.

a. carbon-12 **c.** beryllium-9
b. boron-11 **d.** oxygen-16

CHEMISTRY & YOU

Q: *How are atoms of one element different from the atoms of another element? How are isotopes of the same element different?*

Isotopes

How do isotopes of an element differ?

Figure 4.9 shows that there are three different kinds of neon atoms. How do these atoms differ? All have the same number of protons (10) and electrons (10), but they each have different numbers of neutrons. **Isotopes** are atoms that have the same number of protons but different numbers of neutrons. **Because isotopes of an element have different numbers of neutrons, they also have different mass numbers.** Despite these differences, isotopes are chemically alike because they have identical numbers of protons and electrons, which are the subatomic particles responsible for chemical behavior. Remember the dogs at the beginning of the lesson. Their color or size doesn't change the fact that they are all dogs. Similarly, the number of neutrons in isotopes of an element doesn't change which element it is because the atomic number doesn't change.

UbD Check for Understanding

The Essential Question How are atoms of one element different from atoms of another element?

Have students create an acrostic poem using either *atomic number* or *mass number* that answers this Essential Question. Poems should demonstrate that students understand that the number of protons is a key determinant of difference between elements, but should also make note of differences in numbers of electrons and neutrons, and the fact that most of the atom's mass is concentrated in its nucleus.

ADJUST INSTRUCTION If students are having problems answering the Essential Question, have them review and analyze the differences between the atoms in the elements listed in Table 4.2.

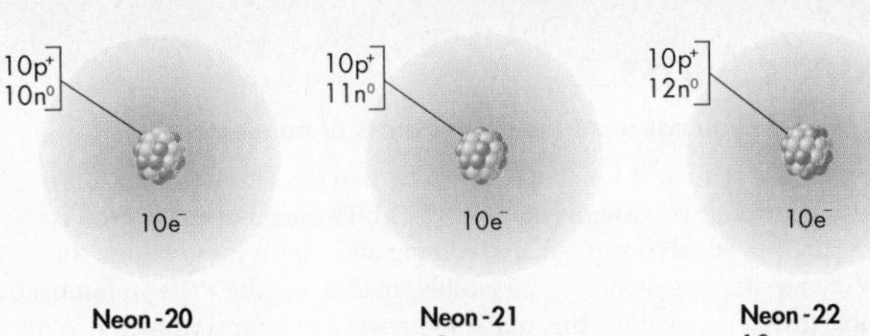

Neon-20
10 protons
10 neutrons
10 electrons

Neon-21
10 protons
11 neutrons
10 electrons

Neon-22
10 protons
12 neutrons
10 electrons

Figure 4.9 Isotopes
Neon-20, neon-21, and neon-22 are three isotopes of neon.
Compare and Contrast *How are these isotopes different? How are they similar?*

There are three known isotopes of hydrogen. Each isotope of hydrogen has one proton in its nucleus. The most common hydrogen isotope has no neutrons. It has a mass number of 1 and is called hydrogen-1 ($^{1}_{1}H$) or hydrogen. The second isotope has one neutron and a mass number of 2. It is called either hydrogen-2 ($^{2}_{1}H$) or deuterium. The third isotope has two neutrons and a mass number of 3. This isotope is called hydrogen-3 ($^{3}_{1}H$) or tritium.

SampleProblem 4.3

Writing Chemical Symbols for Isotopes

Diamonds are a naturally occurring form of elemental carbon. Two stable isotopes of carbon are carbon-12 and carbon-13. Write the symbol for each isotope using superscripts and subscripts to represent the mass number and the atomic number.

1 Analyze Identify the relevant concepts. Isotopes are atoms that have the same number of protons but different numbers of neutrons. The composition of an atom can be expressed by writing the chemical symbol, with the atomic number as a subscript and the mass number as a superscript.

2 Solve Apply the concepts to this problem.

Step	Result
Use Table 4.2 to identify the symbol and the atomic number for carbon.	The symbol for carbon is C. The atomic number of carbon is 6.
Look at the name of the isotope to find the mass number.	For carbon-12, the mass number is 12. For carbon-13, the mass number is 13.
Use the symbol, atomic number, and mass number to write the symbol of the isotope.	For carbon-12, the symbol is $^{12}_{6}C$. For carbon-13, the symbol is $^{13}_{6}C$.

20. Three isotopes of oxygen are oxygen-16, oxygen-17, and oxygen-18. Write the symbol for each, including the atomic number and mass number.

21. Three chromium isotopes are chromium-50, chromium-52, and chromium-53. How many neutrons are in each isotope, given that chromium has an atomic number of 24?

Nuclear Chemistry

Isotopes and nuclides are both similar in the fact that they have negligible effects on the chemical properties of the element. Chemical reactivity depends on the electron structure of an atom, not its nuclear makeup. In nuclear technology, however, the emphasis is on the neutron number because of its drastic effect on the *nuclear properties* of the element. Because neutrons help bind protons together in the nucleus, the ratio of protons to neutrons is a major factor in determining stability. It is not uncommon for a nuclide to be unstable. Isotopes with unstable nuclides are referred to as radioisotopes. In order for these radioisotopes to gain stability, they do not gain or lose electrons, they undergo changes in the nucleus that involve large amounts of energy. This is the stepping stone into nuclear chemistry.

Explain

Isotopes

START A CONVERSATION Explain that isotopes of an atom all have the same number of protons and electrons. **Ask** What do you know about subatomic particles that explains why different isotopes of an atom exhibit almost identical chemical behavior? *(Sample answer: The electrons, not the neutrons, determine the atom's chemical properties.)*

Misconception Alert

Many students associate the term *isotope* with radioactivity. Reinforce the point that an isotope is simply one form of an element. Explain that isotopes can be stable or unstable, and that only unstable isotopes are radioactive. Tell students that they will explore radioactivity and nuclear chemistry in more detail in Chapter 25.

Sample Practice Problems

A. The element cesium has numerous isotopes, including these four: cesium-126, cesium-129, cesium-131, and cesium-132. Given that cesium has an atomic number of 55, how many neutrons are in each of these isotopes? *(71,74, 76, 77)*

B. How does the number of neutrons in carbon-14 differ from the number of neutrons in carbon-12 and carbon-13? *(Carbon-14 has one more neutron than carbon-13, and two more neutrons than carbon-12.)*

Answers

FIGURE 4.9 They have different numbers of neutrons but the same number of protons.

18. **a.** 45 **b.** 16 **c.** 61 **d.** 125

19. **a.** $^{12}_{6}C$ **b.** $^{11}_{5}B$ **c.** $^{9}_{4}Be$ **d.** $^{16}_{8}O$

20. $^{16}_{8}O$, $^{17}_{8}O$, $^{18}_{8}O$

21. Chromium-50 has 26 neutrons; chromium-52 has 28 neutrons; chromium-53 has 29 neutrons.

Explain

Atomic Mass

USE VISUALS Direct students to Table 4.3. Point out that the average atomic masses listed in Table 4.3 are based on the masses of stable isotopes and their percent abundance in Earth's crust. Have students study the average atomic masses in the table. **Ask** Which elements exist predominantly as one natural isotope? *(those with atomic masses closest to a whole number)* **Ask** Which element has substantial amounts of each of its natural isotopes? *(chlorine)*

MAKE A CONNECTION Students may ask why the masses (in amu) of most of the isotopes in Table 4.3 are not whole-number values like the mass numbers that they are used to seeing. Explain that although the mass number of carbon-12 exactly equals its mass in amu, this is generally not the case for other isotopes due to the mass defect. Point out that the mass defect is the difference between the mass of a nucleus and the sum of the masses of its component protons and neutrons.

Mass defect also varies with different elements. As a result the masses of isotopes other than carbon-12 in amu (a unit based on the mass of a carbon-12 atom) will generally not be whole numbers. For example, the mass of hydrogen-2 is 2.0141 amu; the mass of oxygen-16 is 15.995 amu. (*NOTE:* Assessment problem 79 asks students to calculate the mass defect of an atom in grams by comparing the actual mass of the atom to the sum of the masses of the atom's component protons, neutrons, and electrons.)

Atomic Mass

How do you calculate the atomic mass of an element?

A glance back at Table 4.1 on page 107 shows that the actual mass of a proton or a neutron is very small (1.67×10^{-24} g). The mass of an electron is 9.11×10^{-28} g, which is negligible in comparison. Given these values, the mass of even the largest atom is incredibly small. Since the 1920s, it has been possible to determine these tiny masses by using a mass spectrometer. With this instrument, the mass of a fluorine atom was found to be 3.155×10^{-23} g, and the mass of an arsenic atom was found to be 1.244×10^{-22} g. Such data about the actual masses of individual atoms can provide useful information, but in general, these values are inconveniently small and impractical to work with. Instead, it is more useful to compare the relative masses of atoms using a reference isotope as a standard. The reference isotope chosen is carbon-12. This isotope of carbon has been assigned a mass of exactly 12 atomic mass units. An **atomic mass unit (amu)** is defined as one twelfth of the mass of a carbon-12 atom. Using these units, a helium-4 atom has one third the mass of a carbon-12 atom. On the other hand, a nickel-60 atom has five times the mass of a carbon-12 atom.

Interpret Data

Natural Percent Abundance of Stable Isotopes of Some Elements

Name	Symbol	Natural percent abundance	Mass (amu)	Atomic mass
Hydrogen	$^{1}_{1}H$ $^{2}_{1}H$ $^{3}_{1}H$	99.985 0.015 negligible	1.0078 2.0141 3.0160	1.0079
Helium	$^{3}_{2}He$ $^{4}_{2}He$	0.0001 99.9999	3.0160 4.0026	4.0026
Carbon	$^{12}_{6}C$ $^{13}_{6}C$	98.89 1.11	12.000 13.003	12.011
Nitrogen	$^{14}_{7}N$ $^{15}_{7}N$	99.63 0.37	14.003 15.000	14.007
Oxygen	$^{16}_{8}O$ $^{17}_{8}O$ $^{18}_{8}O$	99.759 0.037 0.204	15.995 16.995 17.999	15.999
Sulfur	$^{32}_{16}S$ $^{33}_{16}S$ $^{34}_{16}S$ $^{36}_{16}S$	95.002 0.76 4.22 0.014	31.972 32.971 33.967 35.967	32.06
Chlorine	$^{35}_{17}Cl$ $^{37}_{17}Cl$	75.77 24.23	34.969 36.966	35.453

Table 4.3 The atomic mass of an element is calculated using the percent abundance and mass of its isotopes.
a. Identify Which isotope of oxygen is the most abundant?
b. Describe How could you use the atomic mass of helium to determine which isotope of helium is most abundant?

Hint: The natural percent abundance of hydrogen-3 is "negligible" because the amount of naturally occurring hydrogen-3 is so small that it doesn't affect the atomic mass of hydrogen.

UbD Check for Understanding

How do you calculate the atomic mass of an element?

To assess students' understanding of atomic mass, ask students to write a few sentences, create an example, or write a step-by-step explanation of how to determine the atomic mass of an element. Students' products should indicate an understanding of how the natural abundance of each isotope affects the atomic weight of an element.

ADJUST INSTRUCTION To reinforce this concept, perform a probability experiment in your class. Survey 10 students in your class. Ask them to select one sandwich from tuna, turkey, or jelly, one fruit from apple or pear, and one drink from milk or water. Tally the results and create a weighted probability tree diagram. Point out how each selection of the tree is weighted. Show students how this is similar to understanding weighted average mass.

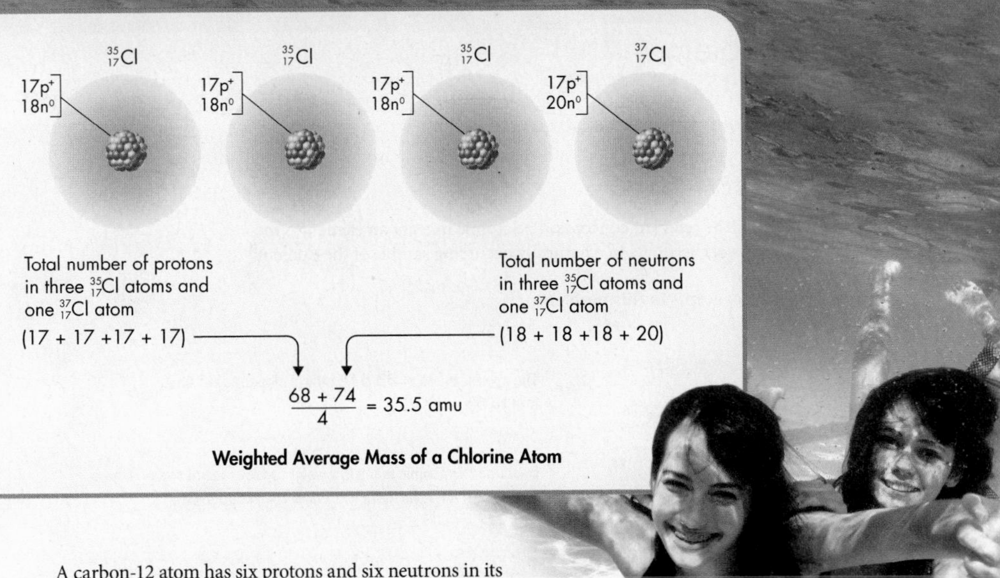

A carbon-12 atom has six protons and six neutrons in its nucleus, and its mass is set as 12 amu. The six protons and six neutrons account for nearly all of this mass. Therefore, the mass of a single proton or a single neutron is about one twelfth of 12 amu, or about 1 amu. Because the mass of any single atom depends mainly on the number of protons and neutrons in the nucleus of the atom, you might predict that the atomic mass of an element should be a whole number. However, that is not usually the case.

In nature, most elements occur as a mixture of two or more isotopes. Each isotope of an element has a fixed mass and a natural percent abundance. Consider the three isotopes of hydrogen discussed earlier in this section. According to Table 4.3, almost all naturally occurring hydrogen (99.985 percent) is hydrogen-1. The other two isotopes are present in trace amounts. Notice that the atomic mass of hydrogen listed in Table 4.3 (1.0079 amu) is very close to the mass of hydrogen-1 (1.0078 amu). The slight difference takes into account the larger masses, but much smaller amounts, of the other two isotopes of hydrogen.

Now consider the two stable isotopes of chlorine listed in Table 4.3: chlorine-35 and chlorine-37. If you calculate the arithmetic mean of these two masses ((34.969 amu + 36.966 amu)/2), you get an average atomic mass of 35.968 amu. However, this value is higher than the actual value of 35.453. To explain this difference, you need to know the natural percent abundance of the isotopes of chlorine. Chlorine-35 accounts for 75 percent of the naturally occurring chlorine atoms; chlorine-37 accounts for only 25 percent. See Figure 4.10. The **atomic mass** of an element is a weighted average mass of the atoms in a naturally occurring sample of the element. A weighted average mass reflects both the mass and the relative abundance of the isotopes as they occur in nature.

Figure 4.10 Isotopes of Chlorine
Chlorine is a reactive element used to disinfect swimming pools. Chlorine occurs as two isotopes: chlorine-35 and chlorine-37. Because there is more chlorine-35 than chlorine-37 in nature, the atomic mass of chlorine, 35.453 amu, is closer to 35 than to 37.
Evaluate ***How does a weighted average differ from an arithmetic mean?***

USE VISUALS Direct students to Figure 4.10. Explain that atomic mass is the weighted average mass of the atoms in a naturally occurring sample of the element. Relate students' knowledge of calculating the mean of a data set or calculating weighted probability to help them understand how this concept is related to chemistry.

Carbon-14 Dating

All living organisms contain carbon-12 and carbon-14 in a fixed ratio. After an organism dies, this ratio changes as the carbon-14 decays. Paleontologists and archaeologists use this fact to establish the age of fossils and ancient artifacts.

Answers

TABLE 4.3

a. oxygen-16

b. Helium has two isotopes, helium-3 and helium-4. The atomic mass of helium is very close to 4 amu, therefore helium-4 must be most abundant.

FIGURE 4.10 In an arithmetic mean, all the numbers in the calculation are weighted equally. A weighted average takes into account the varying weights of the numbers in the data set.

LESSON 4.3

Explain

MAKE A CONNECTION Explain to the class that some teachers evaluate a student's performance based on the weighted average for different work. For example, homework and quizzes might be worth 20%, a term paper might be worth 30%, and an exam might be worth 50%. Have students calculate a grade for a student who receives an 84 on homework and quizzes, a 79 on the term paper, and an 86 on the exam. *(The student would receive a final grade of 83.5 for the grading period.)* Then, explain your own weighting strategy for grading, and have students identify which items have the most influence over the grade that appears on their grade reports.

Sample Practice Problem

Argon has three isotopes with mass numbers 36, 38, and 40, respectively. Which of these isotopes is the most abundant? *(argon-40)*

Sample Problem 4.4

Understanding Relative Abundance of Isotopes

The atomic mass of copper is 63.546 amu. Which of copper's two isotopes is more abundant: copper-63 or copper-65?

1 Analyze **Identify the relevant concepts.** The atomic mass of an element is the weighted average mass of the atoms in a naturally occurring sample of the element.

2 Solve **Apply the concepts to this problem.**

Step	
Compare the atomic mass to the mass of each isotope.	The atomic mass of 63.546 amu is closer to 63 than it is to 65.
Determine the most abundant isotope based on which isotope's mass is closest to the atomic mass.	Because the atomic mass is a weighted average of the isotopes, copper-63 must be more abundant than copper-65.

22. Boron has two isotopes: boron-10 and boron-11. Which is more abundant, given that the atomic mass of boron is 10.81 amu?

23. There are three isotopes of silicon; they have mass numbers of 28, 29, and 30. The atomic mass of silicon is 28.086 amu. Comment on the relative abundance of these three isotopes.

Now that you know that the atomic mass of an element is a weighted average of the masses of its isotopes, you can determine atomic mass based on relative abundance. To do this, you must know three things: the number of stable isotopes of the element, the mass of each isotope, and the natural percent abundance of each isotope.  **To calculate the atomic mass of an element, multiply the mass of each isotope by its natural abundance, expressed as a decimal, and then add the products.** The resulting sum is the weighted average mass of the atoms of the element as they occur in nature. You can calculate the atomic masses listed in Table 4.3 based on the given masses and natural abundances of the isotopes for each element.

For example, carbon has two stable isotopes: carbon-12, which has a natural abundance of 98.89 percent, and carbon-13, which has a natural abundance of 1.11 percent. The mass of carbon-12 is 12.000 amu; the mass of carbon-13 is 13.003 amu. The atomic mass of carbon is calculated as follows:

$$\text{Atomic mass of carbon} = (12.000 \text{ amu} \times 0.9889) + (13.003 \text{ amu} \times 0.0111)$$
$$= (11.867 \text{ amu}) + (0.144 \text{ amu})$$
$$= 12.011 \text{ amu}$$

Foundations for Math

PERCENTAGES Tell students they will need to have a basic understanding of percents in order to calculate atomic mass. Atomic mass is a weighted average that takes into account natural percent abundance.

In **Sample Problem 4.5,** show students that the values given for natural abundance total 100%. In order to use these values to correctly calculate the atomic mass, remind students how to convert a percent to a decimal. Tell students to check the reasonableness of their answer by comparing it to the mass of the isotopes. The atomic mass of the element should be close to the mass of the isotope with the greatest percent of abundance.

Sample Problem 4.5

Calculating Atomic Mass

Element X has two naturally occurring isotopes. The isotope with a mass of 10.012 amu (^{10}X) has a relative abundance of 19.91 percent. The isotope with a mass of 11.009 amu (^{11}X) has a relative abundance of 80.09 percent. Calculate the atomic mass of element X.

To find all the knowns, change the percent abundance to decimals. A percent is a shorthand way of expressing a fraction whose denominator is 100. 19.91% is equivalent to 19.91/100 or 0.1991.

1 Analyze **List the knowns and the unknown.** The mass each isotope contributes to the element's atomic mass can be calculated by multiplying the isotope's mass by its relative abundance. The atomic mass of the element is the sum of these products.

KNOWNS
- isotope ^{10}X:
  mass = 10.012 amu
  relative abundance = 19.91% = 0.1991
- isotope ^{11}X:
  mass = 11.009 amu
  relative abundance = 80.09% = 0.8009

UNKNOWN
atomic mass of X = ?

2 Calculate **Solve for the unknown.**

Use the atomic mass and the decimal form of the percent abundance to the find the mass contributed by each isotope.

for ^{10}X: $10.012 \text{ amu} \times 0.1991 = 1.993 \text{ amu}$
for ^{11}X: $11.009 \text{ amu} \times 0.8009 = 8.817 \text{ amu}$

Add the atomic mass contributions for all the isotopes.

For element X, atomic mass $= 1.993 \text{ amu} + 8.817 \text{ amu} = 10.810 \text{ amu}$

3 Evaluate **Does the result make sense?** The calculated value is closer to the mass of the more abundant isotope, as would be expected.

24. The element copper has naturally occurring isotopes with mass numbers of 63 and 65. The relative abundance and atomic masses are 69.2% for mass = 62.93 amu, and 30.8% for mass = 64.93 amu. Calculate the atomic mass of copper.

25. Calculate the atomic mass of bromine. The two isotopes of bromine have atomic masses and relative abundance of 78.92 amu (50.69%) and 80.92 amu (49.31%).

4.3 LessonCheck

26. **Explain** What distinguishes the atoms of one element from the atoms of another?

27. **Compare and Contrast** How do the isotopes of a given element differ from one another?

28. **Explain** How is atomic mass calculated?

29. **Identify** What equation tells you how to calculate the number of neutrons in an atom?

30. **Compare** How is atomic number different from mass number?

31. **Use Models** What does the number represent in the isotope platinum-194?

32. **Explain** The atomic masses of elements are generally not whole numbers. Explain why.

33. **Identify** Which of argon's three isotopes is most abundant: argon-36, argon-38, or argon-40? (*Hint:* the atomic mass of argon is 39.948 amu.)

34. **Calculate** List the number of protons, neutrons, and electrons in each pair of isotopes.
a. $^{6}_{3}Li$, $^{7}_{3}Li$ **b.** $^{42}_{20}Ca$, $^{44}_{20}Ca$ **c.** $^{78}_{34}Se$, $^{80}_{34}Se$

Lesson Check Answers

26. Atoms of different elements contain different numbers of protons.

27. They have different mass numbers and different numbers of neutrons.

28. For each isotope, multiply its atomic mass by its percent abundance; then add the products.

29. mass number – atomic number = number of neutrons

30. Atomic number corresponds to the number of protons. Mass number represents the total number of protons and neutrons.

31. Mass number, $^{194}_{78}Pt$

32. The atomic mass is the weighted average of the masses of its isotopes.

33. argon-40

34. **a.** lithium-6: 3 p^+, 3 e^-, 3 n^0 ; lithium-7: 3 p^+, 3 e^-, 4 n^0
b. calcium-42: 20 p^+, 20 e^-, 22 n^0; calcium-44: 20 p^+, 20 e^-, 24 n^0
c. selenium-78: 34 p^+, 34 e^-, 44 n^0; selenium-80: 34 p^+, 34 e^-, 46 n^0

Sample Practice Problem

The element antimony (Sb) has naturally occurring isotopes with mass numbers of 121 and 123. The relative abundance and atomic masses are 57.21% for mass = 120.90 amu, and 42.79% for mass = 122.90 amu. Calculate the atomic mass of antimony. *(121.76 amu)*

Evaluate

Informal Assessment

Write the symbols for isotopes of an element not described in the chapter. **Ask** How does changing the value of the subscript change the chemical properties of the atom? *(The subscript designates the number of protons in the atoms of that isotope. Changing the number of protons would change the chemical identity of the isotope to that of another element.)* Then have students complete the 4.3 Lesson Check.

Reteach

Review the concept of weighted averages. Work through the following calculations. If 75% of chlorine atoms are ^{35}Cl species and 25% are ^{37}Cl species, this implies that for a sample of 100 atoms, 75 atoms are ^{35}Cl and 25 atoms are ^{37}Cl species. The combined masses of these atoms would be (75 × 35 amu) + (25 × 37 amu) = 3550 amu for 100 atoms, or 35.5 amu for one atom.

Answers

22. boron-11

23. Silicon-28 must be by far the most abundant. The other two isotopes must be present in very small amounts.

24. 63.5 amu

25. 79.90 amu

SMALL-SCALE LAB

Explore

PURPOSE Students make measurements to calculate the relative abundances of three types of candy in a mixture and use their data to calculate the atomic mass of a candium particle.

MATERIALS mass balance, coated candies (3 different brands), small plastic cups or containers

ADVANCE PREP Prepare in advance a large mixture of the three candies and half fill a clean 3.5-ounce plastic cup for each student. Each sample will contain about 50 total pieces.

SAFETY Discourage students from eating the candies after the experiment. Contamination can easily occur in a lab even if you have taken every precaution to keep the candy free of contamination.

TEACHING TIPS This lab is similar to the longer Small-Scale Lab "Isotopes and Atomic Mass" found in the Small-Scale Chemistry Laboratory Manual.

EXPECTED OUTCOME Sample data are listed below.

Total Mass: 13.16 g; 13.83 g; 15.40 g; 42.39 g (total).

Number: 15; 13; 20; 48 (total).

Average mass (grams): 0.8773 g; 1.064 g; 0.7700 g; 0.8831 g (total).

Relative abundance: 0.3125; 0.2708; 0.4167; 1.000 (total).

Percent abundance: 31.25%; 27.08%; 41.67%; 100% (total).

Relative mass: 0.2742 g; 0.2883 g; 0.3208 g; 0.8833 g (total).

ANALYZE

1–5. See Expected Outcome.

6. Percent abundance is parts per hundred. Relative abundance is parts per one, or the decimal form of percent. The individual percent abundances add up to 100. The individual relative abundances add up to 1.

7. Relative abundance tells you the decimal fraction of particles.

8. The total in row 3 is an average that ignores the relative abundances of particles. The total in row 6 is a weighted average that best represents atomic mass because it considers differences in mass and abundance among the particles.

9. Another student might not have had the same relative abundance of each candy.

YOU'RE THE CHEMIST

1. Any differences are probably due to small variations in the numbers of each kind of candy in the samples, which affects the relative abundances.

2. The larger the samples, the better the results with any of the methods. Mass is likely to provide better results than volume.

Small-Scale Lab

The Atomic Mass of "Candium"

Purpose

To analyze the isotopes of "candium" and to calculate its atomic mass

Materials

- sample of candium
- balance

Procedure

Obtain a sample of "candium" that contains three different brands of round, coated candy. Treat each brand of candy as an isotope of candium. Separate the three isotopes into groups labeled A, B, and C, and measure the mass of each isotope. Count the number of atoms in each sample. Make a table similar to the one below to record your measured and calculated data.

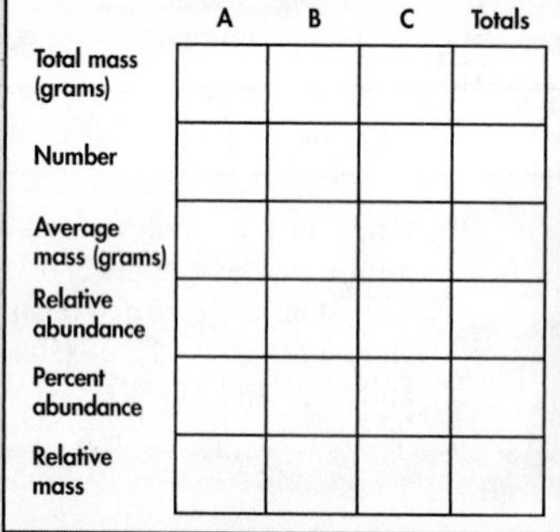

	A	B	C	Totals
Total mass (grams)				
Number				
Average mass (grams)				
Relative abundance				
Percent abundance				
Relative mass				

Analyze

Using the experimental data, record the answers to the following questions below your data table.

1. Calculate Calculate the average mass of each isotope by dividing its total mass by the number of particles of that isotope.

2. Calculate Calculate the relative abundance of each isotope by dividing its number of particles by the total number of particles.

3. Calculate Calculate the percent abundance of each isotope by multiplying the relative abundance from Step 2 by 100.

4. Calculate Calculate the relative mass of each isotope by multiplying its relative abundance from Step 2 by its average mass.

5. Calculate Calculate the weighted average mass of all candium particles by adding the relative masses. This weighted average mass is the atomic mass of candium.

6. Explain What is the difference between percent abundance and relative abundance? What is the result when you total the individual relative abundances? The individual percent abundances?

7. Identify The percent abundance of each kind of candy tells you how many of each kind of candy there are in every 100 particles. What does relative abundance tell you?

8. Analyze Data Compare the total values for rows 3 and 6 in the table. Explain why the totals differ and why the value in row 6 best represents atomic mass.

9. Analyze Data Explain any differences between the atomic mass of your candium sample and that of your neighbor. Explain why the difference would be smaller if larger samples were used.

You're the Chemist

The following small-scale activity allows you to develop your own procedures and analyze the results.

1. Analyze Data Determine the atomic mass of a second sample of candium. How does it compare with the first? Suggest reasons for any differences between the samples.

Focus on ELL

4 LANGUAGE PRODUCTION Have students work in groups of three to complete the lab. Make sure each group has ELLs of varied language proficiencies, so that more proficient students can help less proficient ones. Have students work according to their proficiency level.

BEGINNING: LOW/HIGH Paraphrase the procedural steps and group students with higher language proficiency partners.

INTERMEDIATE: LOW/HIGH Have students brainstorm what they think they may observe, prior to performing each step. Allow students to orally present their results.

ADVANCED: LOW/HIGH Have students write a bulleted list stating the procedure, their findings, and their conclusions.

4 Study Guide

BIGIDEA ELECTRONS AND THE STRUCTURE OF ATOMS

Atoms are the smallest particles of an element that still have the chemical properties of that element. Atoms have positively charged protons and neutral neutrons inside a nucleus, and negatively charged electrons outside the nucleus. Atoms of the same element have the same number of protons, which is equal to an atom's atomic number. But atoms of the same element can have different numbers of neutrons. Atoms of the same element with different numbers of neutrons are isotopes.

4.1 Defining the Atom

Democritus reasoned that atoms were indivisible and indestructible. By using experimental methods, Dalton transformed Democritus's ideas on atoms into a scientific theory.

Scientists can observe individual atoms by using instruments such as scanning electron microscopes.

- atom (102)
- Dalton's atomic theory (103)

4.2 Structure of the Nuclear Atom

Three kinds of subatomic particles are electrons, protons, and neutrons.

In the nuclear atom, the protons and neutrons are located in the nucleus. The electrons are distributed around the nucleus and occupy almost all the volume of the atom.

- electron (105)
- cathode ray (105)
- proton (107)
- neutron (107)
- nucleus (108)

4.3 Distinguishing Among Atoms

Elements are different because they contain different numbers of protons.

Because isotopes of an element have different numbers of neutrons, they also have different mass numbers.

To calculate the atomic mass of an element, multiply the mass of each isotope by its natural abundance, expressed as a decimal, and then add the products.

- atomic number (112)
- mass number (113)
- isotope (114)
- atomic mass unit (amu) (116)
- atomic mass (117)

Key Equation

$$\text{number of neutrons} = \text{mass number} - \text{atomic number}$$

Evaluate

Review and Assessment Resources

CHEMISTRY TUTORIAL Encourage students to review solving atomic composition problems using the animated step-by-step tutorials for Chapter 4. Each Chemistry Tutorial is followed by a similar practice problem, where the student is guided through the problem-solving process.

CHAPTER PROBLEM SET Have students practice more atomic structure problems using the Chapter 4 online Problem Set.

MATH TUTORIAL Struggling students can use a tutorial on weighted averages as a quick review. Ask students to answer the guided practice problem that follows the Math Tutorial.

DIRECTED VIRTUAL LAB Have students complete the Virtual Lab *Rutherford's Experiment* to help reinforce the concept of the atomic nucleus. In the post-lab exercise, students will analyze the data and draw conclusions based on the results of the experiment.

KINETIC ART Direct students to the Kinetic Art animations *Cathode Ray* and *Rutherford's Gold-Foil Experiment*, and have them answer the questions that accompany each animation.

Study Tip

ORGANIZE NEW INFORMATION Tell students: Use a reading strategy that you practiced in this chapter to organize the subject matter in a way that helps you study. For example, if you have trouble remembering all the new vocabulary, try listing the new terms and defining them in your own words.

UbD Performance Tasks

IT'S "ELEMENT"ARY Have students choose three different elements from the Periodic Table of Elements. For each element, ask students to do the following: write the symbol for the element; determine the atomic number, the number of protons, the number of electrons, and the number of neutrons. Then have students look at the atomic mass and identify two possible isotopes of each element. Finally, ask students to compare their elements with those of a partner. Students can present their information on a poster, slide show, or as a pamphlet.

SHOWING IN 3D Create a three dimensional model of one of the elements on Periodic Table of Elements. The numbers of protons, neutrons, and electrons should be clearly represented. Use the model to demonstrate two different isotopes of the element. Present the model to the class and explain how to determine the mass number.

ASSESSMENT

Evaluate

ANSWERS

LESSON 4.1

35. The smallest particle of an element that still has the properties of that element.

36. Democritus's ideas were not helpful in explaining chemical behavior because they lacked experimental support.

37. Dalton would agree with all four statements because they all fit his atomic theory.

38. The atoms are separated, joined, and rearranged.

LESSON 4.2

39. **a.** A beam of electrons (cathode rays) is deflected by an electric field toward the positively charged plate.

b. The cathode rays were always composed of electrons regardless of the metal used in the electrodes or the gas used in the cathode-ray tube.

40. repel

41. The mass of the proton and neutron are equal; protons are positively charged and neutrons are neutral.

42. Atoms are neutral: number of protons = number of electrons. Loss of an electron means that the number of p^+ is greater than the number of e^-, so the remaining particle is positively charged.

43. The electrons were stuck in a lump of positive charge.

44. He did not expect alpha particles to be deflected at a large angle, and he did not expect so many to pass right through.

45. positive

46. protons and neutrons (Rutherford suspected there was something in the nucleus in addition to protons—but didn't know them as neutrons.)

LESSON 4.3

47. It has equal numbers of positively charged protons and negatively charged electrons.

48. the number of protons in the nucleus

49. **a.** 15 **b.** 42 **c.** 13 **d.** 48 **e.** 24 **f.** 82

50. The atomic number is the number of protons. The mass number is the sum of the protons and neutrons.

51. **a.** 19 **b.** 9 **c.** 14 **d.** 29 **e.** 22 **f.** 22 **g.** 25 **h.** 30

52. *Sample answers (any 2):* mass numbers, atomic masses, number of neutrons, relative abundance

53. 6.94

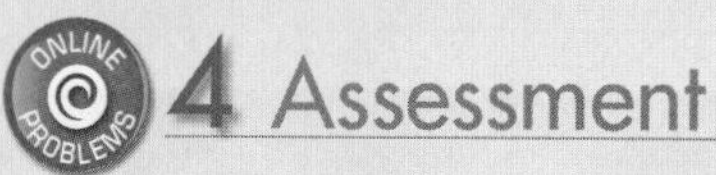

4 Assessment

★ Solutions appear in Appendix E

Lesson by Lesson

4.1 Defining the Atom

35. What is an atom?

36. What were the limitations of Democritus's ideas about atoms?

37. With which of these statements would John Dalton have agreed in the early 1800s? For each, explain why or why not.

- **a.** Atoms are the smallest particles of matter.
- **b.** The mass of an iron atom is different from the mass of a copper atom.
- **c.** Every atom of silver is identical to every other atom of silver.
- **d.** A compound is composed of atoms of two or more different elements.

38. Use Dalton's atomic theory to describe how atoms interact during a chemical reaction.

4.2 Structure of the Nuclear Atom

39. What experimental evidence did Thomson have for each statement?

- **a.** Electrons have a negative charge.
- **b.** Atoms of all elements contain electrons.

★40. Would you expect two electrons to attract or repel each other?

41. How do the charge and mass of a neutron compare to the charge and mass of a proton?

42. Why does it make sense that if an atom loses electrons, it is left with a positive charge?

43. Describe the location of the electrons in Thomson's "plum-pudding" model of the atom.

★44. How did the results of Rutherford's gold-foil experiment differ from his expectations?

45. What is the charge, positive or negative, of the nucleus of every atom?

46. In the Rutherford atomic model, which subatomic particles are located in the nucleus?

4.3 Distinguishing Among Atoms

47. Why is an atom electrically neutral?

48. What does the atomic number of each atom represent?

49. How many protons are in the nuclei of the following atoms?

- **a.** phosphorus (P)
- **b.** molybdenum (Mo)
- **c.** aluminum (Al)
- **d.** cadmium (Cd)
- **e.** chromium (Cr)
- **f.** lead (Pb)

50. What is the difference between the mass number and the atomic number of an atom?

★51. Complete the following table.

Atomic number	Mass number	Number of protons	Number of neutrons
9	a. ____	b. ____	10
c. ____	d. ____	14	15
e. ____	47	f. ____	25
g. ____	55	25	h. ____

52. Name two ways that isotopes of an element differ.

★53. Lithium has two isotopes, lithium-6 (atomic mass = 6.015, relative abundance = 7.5%) and lithium-7 (atomic mass = 7.016, relative abundance = 92.5%). Calculate the atomic mass of lithium.

Understand Concepts

★54. How can there be more than 1000 different atoms when there are only about 100 different elements?

55. What data must you know about the isotopes of an element to calculate the atomic mass of the element?

56. How is an average mass different from a weighted average mass?

57. What is the atomic mass of an element?

58. Characterize the size of an atom.

59. Compare the size and density of an atom with its nucleus.

★60. You are standing on the top of a boron-11 nucleus. Describe the numbers and kinds of subatomic particles you see looking down into the nucleus, and those you see looking out from the nucleus.

UNDERSTAND CONCEPTS

54. because of the existence of isotopes

55. which isotopes exist, their masses, and their natural percent abundance

56. Average mass is the arithmetic mean. Weighted average mass considers both the masses and their relative abundance.

57. The atomic mass is the weighted average of the masses of all the isotopes.

58. very, very tiny—but larger than protons and electrons

59. The nucleus is very small and very dense compared with the atom.

60. 5 protons and 6 neutrons in the nucleus; 5 electrons outside the nucleus

61. What parts of Dalton's atomic theory no longer agree with the current picture of the atom?

62. Millikan measured the quantity of charge carried by an electron. How did he then calculate the mass of an electron?

63. How is the atomic mass of an element calculated from isotope data?

*64. The four isotopes of lead are shown below, each with its percent by mass abundance and the composition of its nucleus. Using these data, calculate the approximate atomic mass of lead.

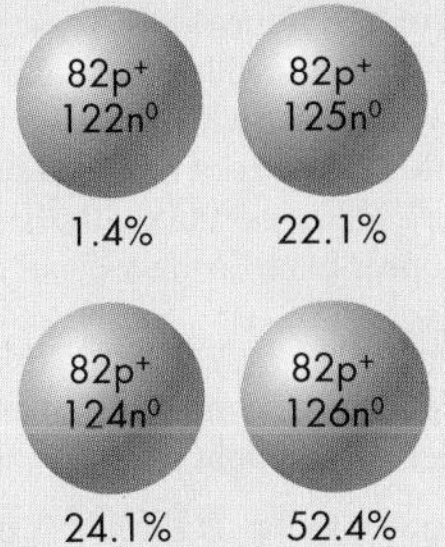

65. Dalton's atomic theory was not correct in every detail. Should this be taken as a criticism of Dalton as a scientist? Explain.

*66. The following table shows some of the data collected by Rutherford and his colleagues during their gold-foil experiment.

Angle of deflection (degrees)	Number of deflections
5	8,289,000
10	502,570
15	120,570
30	7800
45	1435
60	477
75	211
>105	198

a. What percentage of the alpha particle deflections were 5° or less?
b. What percentage of the deflections were 15° or less?
c. What percentage of the deflections were 60° or greater?

*67. Using the data for nitrogen listed in Table 4.3, calculate the weighted average atomic mass of nitrogen. Show your work.

68. What characteristics of cathode rays led Thomson to conclude that the rays consisted of negatively charged particles?

69. If you know the atomic number and mass number of an atom of an element, how can you determine the number of protons, neutrons, and electrons in that atom?

70. What makes isotopes of the same element chemically alike?

71. If isotopes are chemically alike, but physically different, propose which subatomic particles are responsible for determining an element's chemical reactivity.

Think Critically

72. **Interpret Diagrams** The diagram below shows gold atoms being bombarded with fast-moving alpha particles.

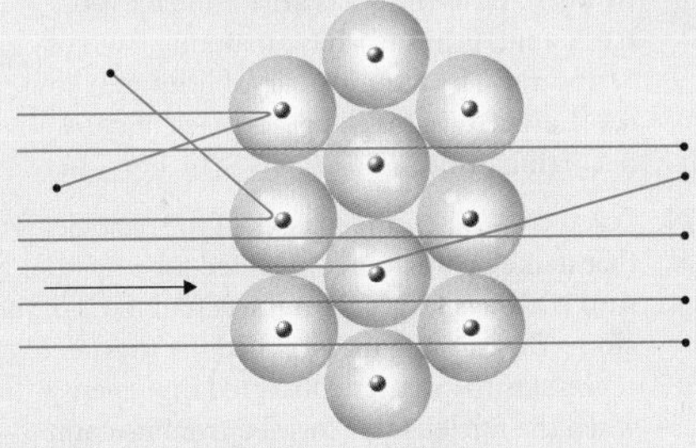

a. The large yellow spheres represent gold atoms. What do the small gray spheres represent?
b. List at least two characteristics of the small gray spheres.
c. Which subatomic particle cannot be found in the area represented by the gray spheres?

73. **Evaluate and Revise** How could you modify Rutherford's experimental procedure to determine the relative sizes of different nuclei?

ASSESSMENT

61. All atoms of the same element are not identical (isotopes). The atom is not the smallest particle of matter.
62. He used the quantity of charge value and the charge-to-mass ratio measured by Thomson.
63. The masses of isotopes in a sample of the element are averaged, based on relative abundance. The result is the element's atomic mass.
64. 207 amu
65. No; in general he proposed a valid theory in line with the experimental evidence available to him.
66. **a.** 92.90% **b.** 99.89% **c.** 0.00993%
67. $^{14}_{7}$N: 14.003 amu; 99.63%
$^{15}_{7}$N: 15.000 amu; 0.37%
average atomic mass = 14.01 amu
68. They were attracted to a positively charged plate.
69. Atomic number is the same as the number of protons and electrons; mass number minus atomic number equals number of neutrons.
70. Because they have identical numbers of protons, they also have identical numbers of electrons; electrons are the subatomic particles that are responsible for chemical behavior.
71. electrons

THINK CRITICALLY

72. **a.** the nucleus of an atom;
b. very small volume; almost all the mass of the atom; high density; positive charge;
c. electron
73. Change the metal used as a target and account for differences in deflection patterns.

Focus on ELL

5 ASSESS UNDERSTANDING Create a formative assessment to determine the students' level of mastery of the concepts. If the level is low, reteach the concepts that the students found to be the most difficult.

BEGINNING
LOW Ask students to label a diagram of an atom.
HIGH Ask students to create their own diagram of an atom.

INTERMEDIATE
LOW Paraphrase or reread the directions to the formative assessment.
HIGH Ask students to compare their answers on the formative assessment to a partner's answers. Have them discuss any discrepancies.

ADVANCED: LOW/HIGH Review the formative assessment with students. Have them give you feedback on the topics they understood the best and the topics that they still need help with.

ASSESSMENT

Explain

CHEMYSTERY

START A CONVERSATION Have students read over the CHEMystery. As a class discuss the findings obtained by the use of the electron microscope. Discuss whether or not students came to this conclusion during the study of this chapter. Have students share what they learned from the activities in the Chemistry and Technology Feature to help other students answer the CHEMystery questions. Have students highlight what they know about electrons and the structure of atoms that can help them to understand what the researchers must have seen in the microscope.

CHEMYSTERY ANSWERS

82. Electron microscopes have a higher resolution than light microscopes so smaller aspects of an object can be differentiated using an electron microscope.

83. **BIG**IDEA The function of the electron microscope is based on the properties of electrons. Without the knowledge of electrons and other subatomic particles, electron microscopes might not have been invented.

***74. Explain** Rutherford's atomic theory proposed a dense nucleus surrounded by very small electrons. This structure implies that atoms are composed mainly of empty space. If all matter is mainly empty space, why is it impossible to walk through walls or pass your hand through your desk?

75. Explain This chapter illustrates the scientific method in action. What happens when new experimental results cannot be explained by the existing theory?

76. Apply Concepts The law of conservation of mass was introduced in Chapter 2. Use Dalton's atomic theory to explain this law.

77. Infer Diamond and graphite are both composed of carbon atoms. The density of diamond is 3.52 g/cm^3. The density of graphite is 2.25 g/cm^3. In 1955, scientists successfully made diamond from graphite. Using the relative densities, consider what happens at the atomic level when this change occurs. Then suggest how this synthesis may have been accomplished.

***78. Calculate** Lithium has two naturally occurring isotopes. Lithium-6 has an atomic mass of 6.015 amu; lithium-7 has an atomic mass of 7.016 amu. The atomic mass of lithium is 6.941 amu. What is the percentage of naturally occurring lithium-7?

***79. Calculate** When the masses of the particles that make up an atom are added together, the sum is always larger than the actual mass of the atom. The missing mass, called the mass defect, represents the matter converted into energy when the nucleus was formed from its component protons and neutrons. Calculate the mass defect of a chlorine-35 atom by using the data in Table 4.1. The actual mass of a chlorine-35 atom is 5.81×10^{-23} g.

Write About Science

80. Communicate Explain how Rutherford's gold-foil experiment yielded new evidence about atomic structure. *Hint:* First, describe the setup of the experiment. Then, explain how Rutherford interpreted his experimental data.

81. Connect to the BIGIDEA Choose two atoms from Table 4.2. Compare and contrast the structure of the two atoms.

CHEMYSTERY

Artifacts or Artifakes?

There are currently no crystal skulls that have been proven to be from ancient civilizations. The Linnean Society of London is a research institute specializing in taxonomy and natural history. They have used electron microscopy to view the surface of crystal skulls, including a crystal skull that is part of the Smithsonian collection. Images from the scanning electron microscope reveal circular patterns on the surface of the skulls. These patterns indicate the skulls were likely carved using a modern carving device with a rotary wheel. Ancient civilizations would not have had such devices. Therefore, all the crystal skulls that are known today appear to be "artifakes," not artifacts.

82. Infer Why would electron microscopes be able to provide more information about an object than a light microscope?

83. Connect to the BIGIDEA How has knowledge of atomic structure aided in the development of the electron microscope?

Cumulative Review

84. How does a scientific law differ from a scientific theory?

85. Classify each as an element, a compound, or a mixture.

a. sulfur
b. salad oil
c. newspaper
d. orange

***86.** Oxygen and hydrogen react explosively to form water. In one reaction, 6 g of hydrogen combines with oxygen to form 54 g of water. How much oxygen was used?

87. An aquarium measures 54.0 cm × 31.10 m × 80.0 cm. How many cubic centimeters of water will this aquarium hold?

***88.** What is the mass of 4.42 cm^3 of platinum? The density of platinum is 22.5 g/cm^3.

If You Have Trouble With . . .					
Question	84	85	86	87	88
See Chapter	1	2	2	3	3

Answers

74. The following are reasonable hypotheses: The space in an individual atom is large relative to the volume of the atom, but very small relative to an object the size of a hand. There are many layers of atoms in a wall or a desk. The space that exists is distributed evenly throughout the solid, similar to the distribution of air pockets in foam insulation.

75. The theory must be modified and then retested.

76. In a chemical change, atoms are not created or destroyed; they are rearranged.

77. Because diamond is more dense than graphite, pressure could be used to "squeeze" the carbon atoms closer together.

78. 92.5%

79. 4×10^{-25} g

WRITE ABOUT SCIENCE

80. In Rutherford's gold-foil experiment, a narrow beam of alpha particles was directed at a very thin sheet of gold foil. Most of the alpha particles passed straight through the gold atoms without being deflected, but a small fraction bounced off the foil at very large angles. To explain his results, Rutherford proposed that an atom is mostly empty space, and that the positive charge and most of the mass of the atom are concentrated in a very small region. Rutherford's experiment yielded evidence of the nucleus, and led to an improved atomic model known as the nuclear atom.

Standardized Test Prep for Indiana

Standards: C.2.2, C.2.3, C.2.4

Select the choice that best answers each question or completes each statement.

1. The smallest particle of an element that retains its identity in a chemical reaction is a
   (A) proton.
   (B) neutron.
   (C) atom.
   (D) compound.

2. Which of these descriptions is *incorrect*?
   (A) proton: positive charge, in nucleus, mass of $\approx$1 amu
   (B) electron: negative charge, mass of $\approx$0 amu, in nucleus
   (C) neutron: mass of $\approx$1 amu, no charge

3. Thallium has two isotopes, thallium-203 and thallium-205. Thallium's atomic number is 81, and its atomic mass is 204.38 amu. Which statement about the thallium isotopes is true?
   (A) There is more thallium-203 in nature.
   (B) Atoms of both isotopes have 81 protons.
   (C) Thallium-205 atoms have fewer neutrons.
   (D) The most common atom of thallium has a mass of 204.38 amu.

4. Which atom is composed of 16 protons, 16 electrons, and 16 neutrons?
   (A) $^{48}_{16}S$ (C) $^{32}_{16}S$
   (B) $^{16}_{32}Ge$ (D) $^{16}_{32}S$

Use the art to answer Question 5.

5. How many nitrogen-14 atoms (^{14}N) would you need to place on the right pan to balance the three calcium-42 atoms (^{42}Ca) on the left pan of the "atomic balance" below? Describe the method you used to determine your answer, including any calculations.

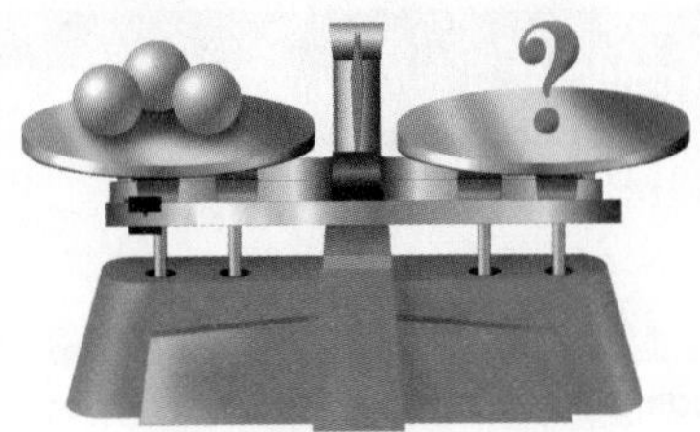

Tips for Success

Connectors Sometimes two phrases in a true/false question are connected by a word such as *because* or *therefore*. These words imply a relationship between one part of the sentence and another. Statements that include such words can be false even if both parts of the statement are true by themselves.

For each question below, there are two statements. Decide whether each statement is true or false. Then decide whether Statement II is a correct explanation for Statement I.

	Statement I		Statement II
6.	Every aluminum-27 atom has 27 protons and 27 electrons.	BECAUSE	The mass number of aluminum-27 is 27.
7.	Isotopes of an element have different atomic masses.	BECAUSE	The nuclei of an element's isotopes contain different numbers of protons.
8.	An electron is repelled by a negatively charged particle.	BECAUSE	An electron has a negative charge.
9.	In an atom, the number of neutrons is generally equal to or greater than the number of protons.	BECAUSE	The mass number is generally equal to or greater than the atomic number.

If You Have Trouble With . . .

Question	1	2	3	4	5	6	7	8	9
See Lesson	4.1	4.2	4.3	4.3	4.3	4.3	4.3	4.2	4.2

STP Answers

1. C
2. B
3. B
4. C
5. 9; three ^{42}Ca atoms have an approximate mass of 3 × 42 = 126 amu; one ^{14}N atom has an approximate mass of 14 amu; 126/14 = 9 ^{14}N atoms with an approximate mass of 126 amu.
6. false, true
7. true, false
8. true, true, correct explanation
9. false, true

Answers

81. Answer will vary. Boron and carbon have different atomic numbers and a different number of protons, with boron having 5 protons and an atomic number of 5 and carbon having 6 protons and an atomic number of 6. Both boron and carbon have 6 neutrons. Boron has 5 electrons, while carbon has 6 electrons.

CUMULATIVE REVIEW

84. Scientific law describes a natural phenomenon but does not explain it. Scientific theory attempts to explain why experiments give certain results.
85. **a.** element **c.** mixture
    b. mixture **d.** mixture
86. 48 g
87. 1.34×10^7 cm^3
88. 99.5 g

5 Electrons in Atoms Planning Guide

Introducing the BIGIDEA: ELECTRONS AND THE STRUCTURE OF ATOMS

Elements are composed of particles called atoms, and every atom contains a nucleus and one or more electrons.

IASS	Lessons and Objectives	Print Resources: For the Student	Print Resources: For the Teacher	
C.2.1, C.2.2	**5.1** Revising the Atomic Model p 128–132 **5.1.1** Describe what Bohr proposed in his model of the atom. **5.1.2** Describe what the quantum mechanical model determines about the electrons in an atom. **5.1.3** Explain how sublevels of principal energy levels differ.	Reading and Study Workbook Lesson 5.1 Lesson Assessment 5.1 p 132	Teacher Demo, p 129: Quantized Energy Teacher Demo, p 130: Energy and Energy Levels Teacher Demo, p 131: The Shapes of Orbitals	
C.2.1, C.2.2	**5.2** Electron Arrangement in Atoms p 134–137 **5.2.1** List the three rules for writing the electron configurations of elements.	Reading and Study Workbook Lesson 5.2 Lesson Assessment 5.2 p 137	Class Activity, p 135: Writing Electron Configurations	
C.2.1, C.2.2	**5.3** Atomic Emission Spectra and the Quantum Mechanical Model p 138–148 **5.3.1** Explain what causes atomic emission spectra. **5.3.2** Describe how Einstein explained the photoelectric effect. **5.3.3** Explain how the frequencies of light are related to changes in electron energies. **5.3.4** Distinguish between quantum mechanics and classical mechanics.	Reading and Study Workbook Lesson 5.3 Lesson Assessment 5.3 p 148 Quick Lab: Flame Tests p 142 Small-Scale Lab: Atomic Emission Spectra p 149	Student Activity, p 139: Black Box Discovery Class Activity, p 141: Properties of Waves Class Activity, p 147: Atomic Scientists	

Assessing the BIGIDEA: ELECTRONS AND THE STRUCTURE OF ATOMS

	Essential Questions 1. How does the quantum mechanical model describe the arrangement of electrons in atoms? 2. What happens when electrons in atoms absorb or release energy?	Study Guide p 150 Math Tune-Up p 151 STP p 157 Reading and Study Workbook Self-Check and Vocabulary Review Chapter 5		

	PearsonChem.com	
	Editable Worksheets	Online Activities
	Teaching Resources, Lesson 5.1 Review	5.1 Lesson Overview Electron Orbitals
	Teaching Resources, Lesson 5.2 Review	5.2 Lesson Overview Writing Electron Configurations
	Teaching Resources, Lesson 5.3 Review Lab 6: Flame Tests for Metals Lab 7: Energies of Electrons Lab 8: Introduction to the Spectrophotometer Small-Scale Lab Manual Lab 7: Design and Construction of a Quantitative Spectroscope Small-Scale Lab Manual Lab 8: Visible Spectra and the Nature of Light and Color Lab Practical 5-1: Flame Tests Lab Practical 5-2: Spectroscopy with a Spectrophotometer	5.3 Lesson Overview A Probability Swim in Lake S Orbital Calculating the Wavelength of Light Flame Tests The Hydrogen Emission Spectrum
	Core Teaching Resources (includes editable worksheets) • Lesson Reviews • Practice Problems • Interpret Graphs • Vocabulary Review • Chapter Quizzes and Tests • Lab Record Sheets	Chapter 5 Problem Set Probability

Materials List

For the Student

Small-Scale Lab p 149
- cereal box
- diffraction grating
- tape
- ruler
- scissors
- black construction paper
- white notebook paper

Quick Lab p 142
- Bunsen burner
- 6 small test tubes
- 6 cotton swabs
- 0.1*M* NaCl, 0.5*M* $CaCl_2$, 0.1*M* LiCl, 0.1*M* $CuCl_2$, 0.1*M* KCl, unknown solution

For the Teacher

Teacher Demo p 129
- a trumpet or trombone
- a player

Teacher Demo p 130
- three balls: red, blue, and yellow
- step stool with three steps

Teacher Demo p 131
- three metal or wooden skewers
- three foam balls (3 or 4 inches in diameter)

Class Activity p 139
- several sealed boxes containing common items such as rubber bands, golf, tennis, or table-tennis balls
- cans of different sizes, and different shapes cut from blocks of wood or foam

Class Activity p 141
- paper and pencil
- ruler

Additional Digital Resources

Online Student Edition
Online Teacher's Edition
Exam*View Assessment Suite CD-ROM*
Classroom Resources DVD
5.3 Virtual Chem Lab 1: Flame Tests for Metals
Virtual Chem Lab 7: Atomic Emissions Spectra
Virtual Chem Lab 8: Photoelectric Effect
Virtual Chem Lab 9: Diffraction Experiments
Virtual Chem Lab 10: Electronic State Energy Levels

CHAPTER 5

What's Online

CHEMISTRY TUTORIAL Students can work step-by-step with the online tutorial to learn how to write electron configurations and calculate wavelengths of different colors of light by accessing the Chapter 5 Chemistry Tutorials.

ONLINE PROBLEM SET Have students practice more electron property problems in the Chapter 5 online Problem Set.

MATH TUTORIAL Encourage students to practice probability problems using the animated step-by-step tutorial for Chapter 5. The tutorial is followed by a similar practice problem, where the student is guided through the problem-solving process.

DIRECTED VIRTUAL LAB Encourage students to review the characteristics of different element flame tests using the animated virtual lab for Chapter 5. The animation is followed by analysis questions.

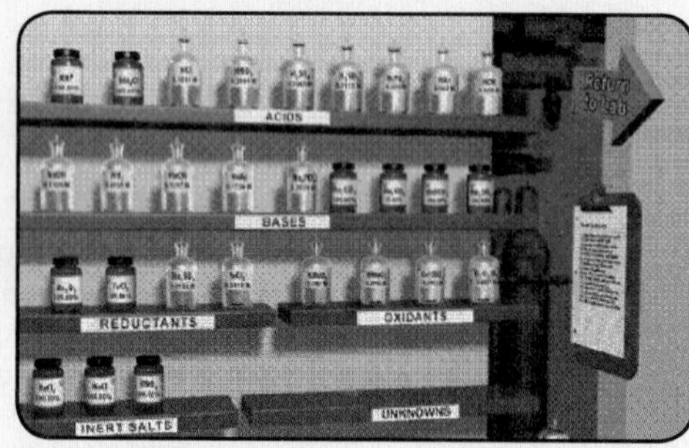

KINETIC ART Have students watch animations about electron orbitals and the hydrogen emission spectrum. Each animation is followed by analysis questions.

CONCEPTS IN ACTION Students can watch an interactive animation about orbitals to understand probability regions around nuclei. The animation is followed by analysis questions.

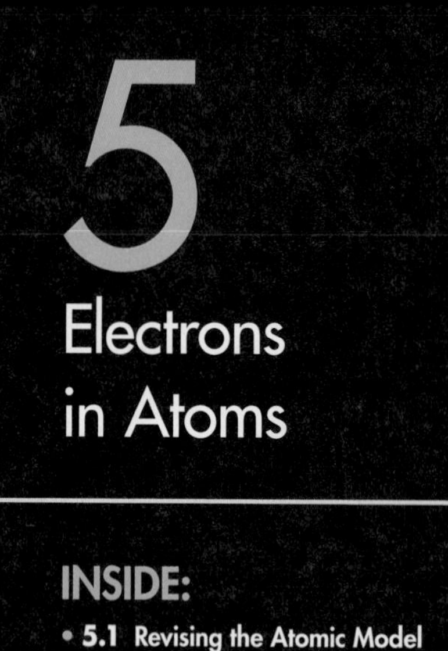

5 Electrons in Atoms

INSIDE:

- **5.1** Revising the Atomic Model
- **5.2** Electron Arrangement in Atoms
- **5.3** Atomic Emission Spectra and the Quantum Mechanical Model

PearsonChem.com

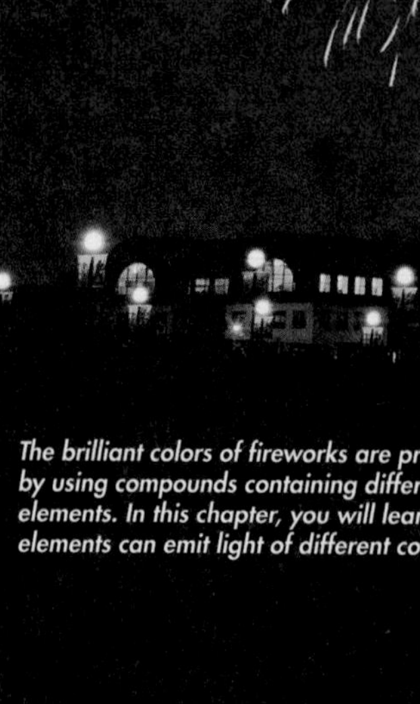

The brilliant colors of fireworks are produced by using compounds containing different elements. In this chapter, you will learn how elements can emit light of different colors.

126

IN ACADEMIC STANDARDS for SCIENCE

Nature of Science

- Recognize that new scientific discoveries often lead to a re-evaluation of previously accepted scientific knowledge and of commonly held ideas.
- Describe how scientific discoveries lead to the development of new technologies, and conversely how technological advances can lead to scientific discoveries through new experimental methods and equipment.

Also covered C.2.1, C.2.2

Focus on ELL

1 CONTENT AND LANGUAGE Begin the class by reviewing the three lesson titles. Focus students' attention on the words *atom, model,* and *electron.* Ask students to recall definitions based on what they learned in chapter 4. Write the words *quantum* and *spectra* on the board. Pair students and have them create a word map for each word, which should include terms, phrases, or drawings that are associated with the word. Have each pair share their word maps with the class.

BEGINNING: LOW/HIGH Write each word on a separate card and create a drawing on the back. Add these words to the word wall.

INTERMEDIATE

LOW Paraphrase the Essential Questions.

HIGH Summarize the CHEMystery story by writing the main idea of the story.

ADVANCED: LOW/HIGH Compare the Essential Questions to the lesson titles. Predict which lessons will contain the answers to the Essential Questions.

CHAPTER 5

BIG IDEA

ELECTRONS AND THE STRUCTURE OF ATOMS

Essential Questions:

1. *How does the quantum mechanical model describe the arrangement of electrons in atoms?*
2. *What happens when electrons in atoms absorb or release energy?*

CHEMYSTERY

Now You See It... Now You Don't

Young Liam loves to go outside at night and gaze at the stars. For his birthday, his parents gave him glow-in-the-dark star stickers so that he could look at the stars from the comfort of his bed.

That night, Liam looked at the constellations he had created on his ceiling while he drifted off to sleep. A few hours later, Liam woke up very disturbed. The stars were no longer glowing even though it was still dark in his room. He turned on his bedroom light and ran down the hall to wake his parents. However, when Liam brought his parents back to his room and turned off the light, the stars were glowing again. Why did the stars cease to glow and then light up again later?

▶ Connect to the BIGIDEA As you read about electrons in atoms, think about how glow-in-the-dark stickers might work.

INDIANA'S ACADEMIC STANDARDS FOR SCIENCE

Atmoic Structure and the Periodic Table C.2.1, C.2.2. See lessons for details.

Introduce the Chapter

CONNECTIONS TO EVERYDAY LANGUAGE Students have probably heard the expression *quantum jump* or *quantum leap*. In everyday language, this expression is often used to describe large and sudden changes. For example, a large and sudden increase in a person's income may be described as a *quantum jump*. Use the activity to introduce students to ideas about quantum effects.

Activity You'll need to identify stairs where students can walk up and down. Have two or three students at a time walk up several steps. Then have students walk back down the steps. **Ask** How was your height above the ground limited by the steps? *(I could only move distances that corresponded to the heights of the steps.)* Explain that the term *quantum jump* in science refers to small, discrete changes in energy. **Ask** How is the scientific meaning of *quantum jump* similar to the everyday use of the term? How is it different? *(Both refer to discrete changes, but the everyday use often refers to large changes, whereas the science use refers to tiny changes.)*

UbD

Understanding by Design

Students are building toward an understanding of atoms by studying the behavior of electrons.

PERFORMANCE GOALS At the end of Chapter 5, students will be able to answer the essential questions by applying their knowledge of the quantum mechanical model of the atom.

ESSENTIAL QUESTIONS Read the essential questions aloud. Explain to students that one of the fundamental ideas of quantum mechanics is describing electron position using probability rather than precise calculation. **Ask** Why do you think we use probability? *(We can't precisely calculate an electron's location or speed.)* **Ask** How do you think electrons in an atom might gain or release energy?

BIGIDEA Use the photo of the fireworks to help students connect to this chapter's concepts. Activate students' prior knowledge about fireworks. Encourage them to describe the colors and patterns. **Ask** Why do you think the photo of fireworks opens this chapter about electrons and atoms? *(The fireworks display is due to changes in the electron configurations of atoms.)*

CHEMYSTERY Have students read over the CHEMystery. Then activate their prior knowledge by asking if they have ever seen glow-in-the-dark stars. Connect the CHEMystery to the Big Idea by reminding students that light is a form of energy. Have them consider what the gradual decrease in light indicates about the stickers' energy. **Ask** What clues in the story indicate why the stars stop glowing and then start glowing again later? *(The stars glow after they have been exposed to light. They gradually glow less and less after the light is turned off.)*

The Chemistry of Fireworks
Erica visits a chemistry professor who specializes in finding the best chemicals for brilliant fireworks colors.

LESSON 5.1

Key Objectives

5.1.1 DESCRIBE what Bohr proposed in his model of the atom.

5.1.2 DESCRIBE what the quantum mechanical model determines about the electrons in an atom.

5.1.3 EXPLAIN how sublevels of principal energy levels differ.

Additional Resources

Reading and Study Workbook, Lesson 5.1
Available Online or on Digital Media:
- Teaching Resources, Lesson 5.1

PearsonChem.com

Engage

CHEMISTRY & YOU Have students read the opening text. Explain that models are often used to represent something that is very large or very small. **Ask** How are models useful? *(You can test a model to see if your ideas or design are correct.)*

Build Background

Compare the use of atomic models to develop the atomic theory to the use of physical or computer-generated models to design, develop, and test cars.

Foundations for Reading

BUILD VOCABULARY Explain that the word *orbital* was coined by scientists to describe the space in which an electron has a high probability of being found. It is derived from the word *orbit*, the path around the nucleus that electrons were thought to follow.

IN ACADEMIC STANDARDS for SCIENCE

- Recognize that new scientific discoveries often lead to a re-evaluation of previously accepted scientific knowledge and of commonly held ideas.

C.2.1 Describe how models of atomic structure changed over time based on available experimental evidence and understand the current model of atomic structure.

Also covered C.2.2.

5.1 Revising the Atomic Model

C.2.1 Modern atomic theory: historic development; **C.2.2** Subatomic particles and atomic structure

CHEMISTRY & YOU

Q: *Why do scientists use mathematical models to describe the position of electrons in atoms?* Wind tunnels and models are often used to simulate the forces from the moving air on a design. Shown here is a life-sized model of a speed skier. It is a physical model. However, not all models are physical. In fact, the current model of the atom is a mathematical model.

Key Questions

What did Bohr propose in his model of the atom?

What does the quantum mechanical model determine about the electrons in an atom?

How do sublevels of principal energy levels differ?

Vocabulary

- energy level
- quantum
- quantum mechanical model
- atomic orbital

Energy Levels in Atoms

What did Bohr propose in his model of the atom?

Thus far, the atomic model presented in this textbook has considered atoms as consisting of protons and neutrons making up a nucleus surrounded by electrons. After discovering the atomic nucleus, Rutherford used existing ideas about the atom and proposed an atomic model in which the electrons move around the nucleus like the planets move around the sun.

Limitations of Rutherford's Atomic Model Rutherford's atomic model explained only a few simple properties of atoms. It could not explain the chemical properties of elements. For example, Rutherford's model could not explain why metals or compounds of metals give off characteristic colors when heated in a flame. It also could not explain why an object such as the iron scroll shown in Figure 5.1 first glows dull red, then yellow, and then white when heated to higher and higher temperatures. Explaining what leads to the chemical properties of elements required a model that better described the behavior of electrons in atoms.

Figure 5.1 Glowing Metal
Rutherford's model failed to explain why objects change color when heated. As the temperature of this iron scroll is increased, it first appears black, then red, then yellow, and then white. The observed behavior could be explained only if the atoms in the iron gave off light in specific amounts of energy. A better atomic model was needed to explain this observation.

Focus on ELL

1 CONTENT AND LANGUAGE Begin the lesson by drawing students' attention to the word *quantum* in the vocabulary list, and have them read the Reading Support feature on page 129. Have students brainstorm the answer to the feature's associated question. Point out that just as a quantity of matter can be measured, so can a quantum of energy.

2 FRONTLOAD THE LESSON Ask students to look at Figure 5.2 and discuss with a partner the differences between the two ladders.

3 COMPREHENSIBLE INPUT Create a compare and contrast chart for students. Have students compare each type of electron model. Encourage students to draw pictures to help them.

The Bohr Model In 1913, Niels Bohr (1885–1962), a young Danish physicist and a student of Rutherford, developed a new atomic model. He changed Rutherford's model to incorporate newer discoveries about how the energy of an atom changes when the atom absorbs or emits light. He considered the simplest atom, hydrogen, which has one electron. **Bohr proposed that an electron is found only in specific circular paths, or orbits, around the nucleus.**

Each possible electron orbit in Bohr's model has a fixed energy. The fixed energies an electron can have are called **energy levels.** The fixed energy levels of electrons are somewhat like the rungs of the ladder in Figure 5.2a. The lowest rung of the ladder corresponds to the lowest energy level. A person can climb up or down the ladder by stepping from rung to rung. Similarly, an electron can move from one energy level to another. A person on the ladder cannot stand between the rungs. Similarly, the electrons in an atom cannot exist between energy levels. To move from one rung to another, a person climbing the ladder must move just the right distance. To move from one energy level to another, an electron must gain or lose just the right amount of energy. A **quantum** of energy is the amount of energy required to move an electron from one energy level to another energy level. The energy of an electron is therefore said to be quantized.

READING SUPPORT

Build Vocabulary: *Latin Word Origins* Quantum comes from the Latin word *quantus*, meaning "how much." ***What other commonly used English word comes from this root?***

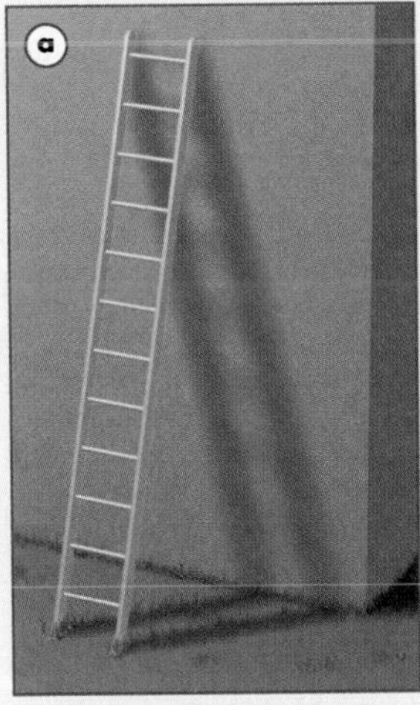

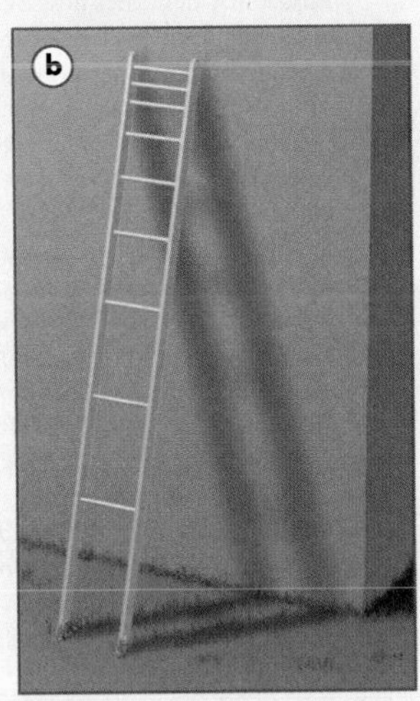

Figure 5.2 Energy Levels
The rungs of a ladder are somewhat like the energy levels in Bohr's model of the atom. **a.** In an ordinary ladder, the rungs are equally spaced. **b.** The energy levels in atoms are unequally spaced, like the rungs in this unusual ladder. The higher energy levels are closer together.
Compare ***For the ladder in b, compare the amount of energy it would take to move from the first rung to the second rung with the amount of energy it would take to move from the second rung to the third rung.***

The amount of energy an electron gains or loses in an atom is not always the same. Like the rungs of the strange ladder in Figure 5.2b, the energy levels in an atom are not equally spaced. The higher energy levels are closer together. It takes less energy to climb from one rung to another near the top of the ladder in Figure 5.2b, where the rungs are closer. Similarly, the higher the energy level occupied by an electron, the less energy it takes the electron to move from that energy level to the next higher energy level.

The Bohr model provided results in agreement with experiments using the hydrogen atom. However, the Bohr model failed to explain the energies absorbed and emitted by atoms with more than one electron.

Development of Wave Mechanics

Schrödinger's equation accounted mathematically for de Broglie's discovery of the wave-like properties of the electron. These developments led to a major break away from the Bohr model of the atom. Schrödinger extended de Broglie's work by considering the movement of a particle in an electromagnetic field, which established an area of physics known as wave mechanics.

Explain

Energy Levels in Atoms

START A CONVERSATION Prepare students for the development of atomic theory by asking questions about their understanding of what they already know about atomic theory. **Ask** What three major subatomic particles comprise an atom? *(electron, proton, and neutron)* **Ask** What electrical charges are associated with each particle? *(electron, 1–; proton, 1+; neutron, 0)* Have students describe the structure of the nuclear atom in terms of the locations of each of the subatomic particles. *(Every atom, except hydrogen, consists of a small dense nucleus composed of protons and neutrons that accounts for most of the mass of the atom. Hydrogen is an exception because it only has one proton in its nucleus.)* **Ask** Do electrons contribute to the mass of an atom? *(Electrons occupy most of its volume but do not contribute significantly to the actual mass of the atom.)*

USE VISUALS Call students' attention to Figure 5.2. **Ask** What are the rungs in the ladders compared to? *(the energy levels in atoms)* **Ask** How is the ladder in picture b different from the ladder in picture a? *(The rungs are not evenly spaced. They get closer together nearer to the top.)*

Explore

Teacher Demo

PURPOSE Students witness an example of quantized energy.

MATERIALS trumpet or trombone and a player

PROCEDURE Have a student blow into a trumpet or trombone. Challenge the student to produce as many different notes as he or she can without depressing any valves.

EXPECTED OUTCOMES It is not possible to play the entire scale without changing the valve positions. In the open position, no matter how much energy is put in, the instrument will only accept certain specific amounts of energy and produce certain notes.

Answers

READING SUPPORT The word quantity has the same root as quantum and means "a definite amount or number."

FIGURE 5.2 Less energy is required to move between the second and third rungs than between the first and second rungs.

Explore

Teacher Demo

PURPOSE Students observe an analogy for energy transitions in atoms.

MATERIALS 1 red, 1 yellow, and 1 blue ball, step stool with 3 steps

PROCEDURE Stand on the first step of the step stool. Jump down to the floor while throwing the red ball to a student catcher. Stand on the second step and repeat the process throwing the yellow ball. Repeat jumping from the third step while throwing the blue ball. **Ask** What analogy are we trying to show you? *(The different colored balls are different amounts of energy that correspond to the energy emitted when electrons drop from excited energy levels to a lower level. Blue must be the largest amount of energy and red the smallest.)*

Explain

The Quantum Mechanical Model

MAKE A CONNECTION Draw a dartboard on the board with concentric circles of decreasing value. Draw 20 or so small solid dots and explain that the dots represent the dart holes made by two dart players. Make sure the dart holes are closer to the bull's eye than the edge. *(The analogy will not work for professional dartboards.)* Have students describe the distribution of the holes. *(There are more holes closer to the bull's-eye than near the edge.)*

Ask How is the distribution of holes analogous to the distribution of electrons in an atom? *(The probability diminishes with distance from the nucleus, approaching zero as distance approaches infinity.)*

Ask If the dartboard is a model for an atom, what are its major failures? *(It is a 2-dimensional model for a 3-dimensional atom; the distribution of holes may not decrease uniformly.)* Redraw the solid points of the dart holes so that the holes do decrease uniformly. **Ask** What are the probabilities for two holes on a circle at a specified distance from the center? *(The probabilities are equal.)* This rule holds true for electrons that are equidistant from the nucleus.

The Quantum Mechanical Model

What does the quantum mechanical model determine about the electrons in an atom?

The Rutherford model and the Bohr model of the atom described the path of a moving electron as you would describe the path of a large moving object. Later theoretical calculations and experimental results were inconsistent with describing electron motion this way. In 1926, the Austrian physicist Erwin Schrödinger (1887–1961) used these calculations and results to devise and solve a mathematical equation describing the behavior of the electron in a hydrogen atom. The modern description of the electrons in atoms, the **quantum mechanical model,** came from the mathematical solutions to the Schrödinger equation.

Like the Bohr model, the quantum mechanical model of the atom restricts the energy of electrons to certain values. Unlike the Bohr model, however, the quantum mechanical model does not specify an exact path the electron takes around the nucleus. **The quantum mechanical model determines the allowed energies an electron can have and how likely it is to find the electron in various locations around the nucleus of an atom.**

Probability describes how likely it is to find an electron in a particular location around the nucleus of an atom. If you placed three red marbles and one green marble into a box and then picked a marble without looking, the probability of picking the green marble would be one in four, or 25 percent. This percentage means that if you put the four marbles in a box and picked one, and repeated this many times, you would pick a green marble in 25 percent of your tries.

The quantum mechanical model description of how electrons move around the nucleus is similar to a description of how the blades of a windmill rotate. The windmill blades in Figure 5.3a have some probability of being anywhere in the blurry region they produce in the picture, but you cannot predict their exact locations at any instant. In the quantum mechanical model of the atom, the probability of finding an electron within a certain volume of space surrounding the nucleus can be represented as a fuzzy cloudlike region, as shown in Figure 5.3b. The cloud is more dense where the probability of finding the electron is high and is less dense where the probability of finding the electron is low. There is no boundary to the cloud because there is a slight chance of finding the electron at a considerable distance from the nucleus. Therefore, attempts to show probabilities as a fuzzy cloud are usually limited to the volume in which the electron is found 90 percent of the time. To visualize an electron probability cloud, imagine that you could mold a sack around the cloud so that the electron was inside the sack 90 percent of the time. The shape of the sack would then give you a picture of the shape of the cloud.

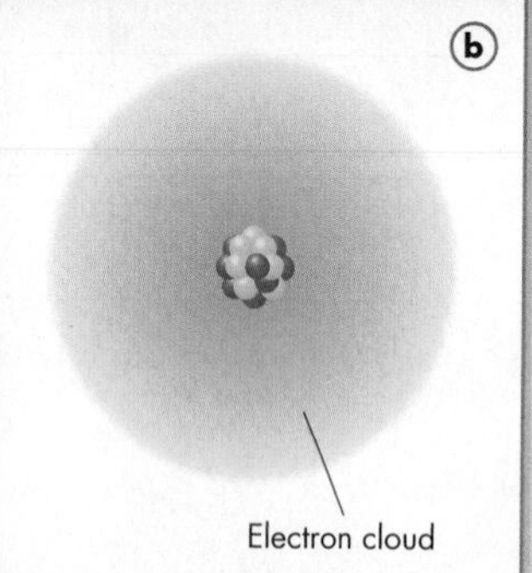

Figure 5.3 Electron Cloud
The electron cloud of an atom can be compared to a photograph of spinning windmill blades. **a.** The windmill blades are somewhere in the blurry region they produce in this picture, but the picture does not tell you their exact positions at any instant. **b.** Similarly, the electron cloud of an atom represents the locations where an electron is likely to be found, but it is not possible to know where an electron is in the cloud at any instant.

Differentiated Instruction

L1 LESS PROFICIENT READERS Have students expand Table 5.1 on page 132 to include a column with the heading *Number of orbitals* (n^2) and another headed *Number of electrons* ($2n^2$). Have students fill in the rows. Make sure students do not confuse an orbital with any predictable physical location. Concentrate primarily on *s* and *p* orbitals since these are key to understanding the behavior of representative elements.

ELL ENGLISH LANGUAGE LEARNERS Have students make a poster to show the evolution of the model of the atom. Have them start with Dalton's model and continue through all the modifications to the current quantum mechanical model. Have them include enough information to make the model clear and understandable.

L3 ADVANCED STUDENTS Have students research de Broglie's work on the wave-particle nature of matter and write a short report that explains how particles and light both produce diffraction patterns.

Atomic Orbitals

How do sublevels of principal energy levels differ?

Solutions to the Schrödinger equation give the energies, or energy levels, an electron can have. For each energy level, the Schrödinger equation also leads to a mathematical expression, called an **atomic orbital,** describing the probability of finding an electron at various locations around the nucleus. An atomic orbital is represented pictorially as a region of space in which there is a high probability of finding an electron.

The energy levels of electrons in the quantum mechanical model are labeled by principal quantum numbers (n). These numbers are assigned the values $n = 1, 2, 3, 4$, and so forth. For each principal energy level greater than 1, there are several orbitals with different shapes and at different energy levels. These energy levels within a principal energy level constitute energy sublevels. **Each energy sublevel corresponds to one or more orbitals of different shapes. The orbitals describe where an electron is likely to be found.**

Different atomic orbitals are denoted by letters. As shown in Figure 5.4a, s orbitals are spherical, and p orbitals are dumbbell-shaped. The probability of finding an electron at a given distance from the nucleus in an s orbital does not depend on direction because of its spherical shape. The three kinds of p orbitals have different orientations in space. Figure 5.4b shows the shapes of d orbitals. Four of the five kinds of d orbitals have cloverleaf shapes. The shapes of f orbitals are more complicated than the shapes of d orbitals.

a Shapes of *s* and *p* orbitals

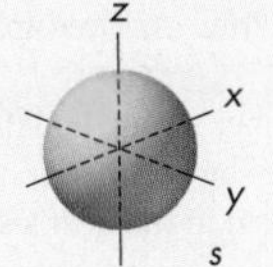

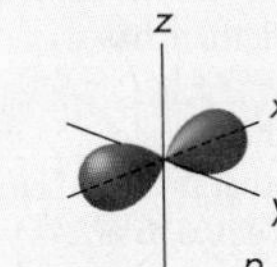

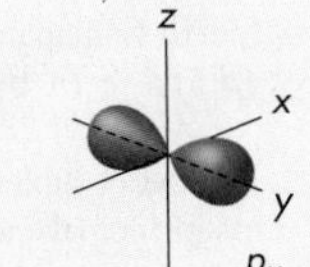

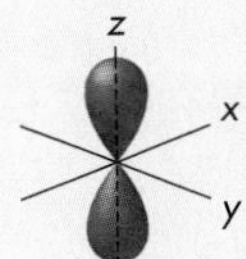

b Shapes of *d* orbitals

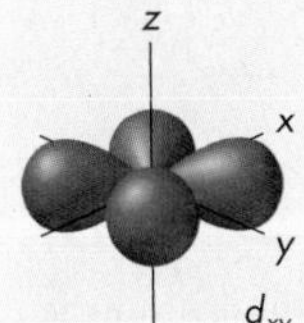

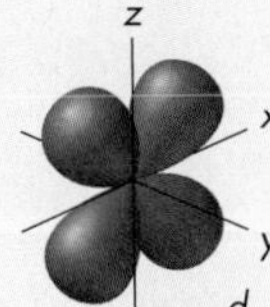

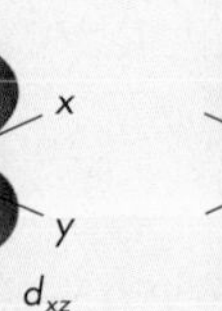

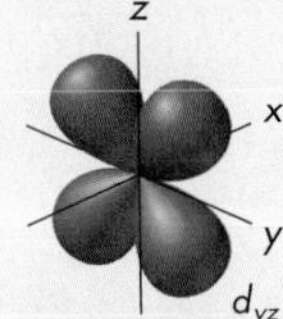

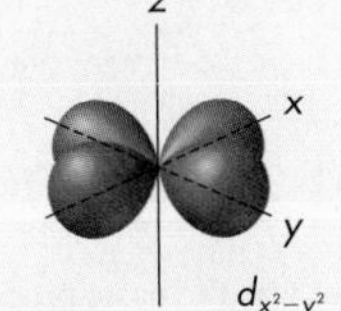

Figure 5.4 Atomic Orbitals
Solutions to the Schrödinger equation give rise to atomic orbitals. **a.** For a given principal energy level greater than 1, there is one *s* orbital and three *p* orbitals. **b.** Four of the five *d* orbitals have the same shape but different spatial orientations.
Interpret Diagrams ***How are the orientations of the d_{xy} and $d_{x^2-y^2}$ orbitals similar? How are they different?***

See atomic orbitals animated online. KINETIC ART

UbD Check for Understanding

The Essential Question How does the quantum mechanical model describe the arrangement of electrons in atoms?

Assess students' knowledge about the quantum mechanical model by asking them to fill in the blanks of the following prompt:

Analogy: The quantum mechanical model is like a __________ because ________.

(Sample answer: windmill, propeller blade, fan blade; it is difficult to predict what region a blade will be in, although you know that the blade is within a certain space.)

ADJUST INSTRUCTION If students are having trouble answering, have them reread the fourth paragraph of The Quantum Mechanical Model section for review. Then ask them to complete the analogy again.

Explore

Teacher Demo

PURPOSE Students observe the shapes of *s* and *p* orbitals and their orientation in a three-dimensional coordinate system.

MATERIALS 3 metal or wooden skewers, 3 foam balls (3 or 4 inches in diameter)

PROCEDURE Push three skewers through a ball at right angles to each other and as close to the ball's center as possible. **Ask** What type of orbital does this represent? *(s)* **Ask** What is its shape? *(Students may say it is circular. Make the point that it is spherical.)* Identify the *x*-, *y*-, and *z*-axes. Push a skewer through two balls to represent a *p* orbital. Tell students that the skewer can represent any one of the three axes. **Ask** Where along the axis is the atomic nucleus? *(at the point between the two balls)*

Explain

Atomic Orbitals

INTERPRET DIAGRAMS Draw students' attention to Figure 5.4. Present the orbital concept as an outgrowth of the quantum mechanical model and describe the shapes of the *s* and *p* orbitals. Stress that the quantum mechanical model predicts the shapes of the various orbitals. The fact that experiments affirm the model's predictions is validation that the model is correct. Remind students that an orbital is only a region of mathematical probability, not a concrete item that can be seen or felt. An orbital does not have the sharp boundaries that models and diagrams seem to indicate.

Answers

FIGURE 5.4 Both lie in the *xy* plane. The lobes of the d_{xy} orbital lie between the *x* and *y* axes. Those of the $d_{x^2-y^2}$ orbital lie along the *x*- and *y*-axes.

Extend

USE VISUALS Refer students to Table 5.1. Explain that every atom has an infinite number of energy levels, but no element, when it is in its ground state, requires more than seven energy levels to describe the probability regions, *or* orbitals, where its electrons can be found. **Ask** How many sublevels are in level 5? *(5)* in level 6? *(6)* in level 7? *(7)* As a class, discuss determine the types of sublevels that are in level 5. *(5s, 1 orbital; 5p, 3 orbitals; 5d, 5 orbitals; 5f, 7 orbitals; and 5g, 9 orbitals)*

Ask What is the mathematical relationship between type (s, p, d, f, g, and so on) and number of orbitals? *(1, 3, 5, 7, 9, and so on)*. **Ask** How many orbitals relate to level 1? *(1)* level 2? *(4)* level 3? *(9)* level 4? *(16)* **Ask** Do you see a pattern? If so, describe the pattern. *(Yes; The number of orbitals is the square of the energy level,* n^2*.)* **Ask** How many orbitals would you expect to relate to level 7? *(49)*

Stress that orbitals are mathematical probability regions where electrons may be located. The equation for determining how many electrons can relate to each energy level is $2n^2$ where *n* represents the energy level.

CHEMISTRY & YOU The physical models of electron motion were based on describing the paths of electrons in the same way as the paths of a large moving object and were inconsistent with theoretical calculations and experimental results.

Evaluate

Informal Assessment

Create two to three true or false questions about each model explained in the lesson. Divide the class into two groups. State one of the true or false questions and allow groups time to deliberate. Tell groups to have a designated person state the answer and the reasoning for the answer. Assign one point for each correct answer and reason. The group with the most points wins. Then have students complete the 5.1 Lesson Check.

Reteach

Review the following information. Energy levels are designated with the letter *n*. Each energy level has as many sublevels as the number of the energy level. When $n = 2$, there are 2 sublevels. Each energy level relates to n^2 orbitals. When $n = 3$, there are 3 sublevels and 9 orbitals ($1s + 3p + 5d = 9$). Each orbital contains up to 2 electrons. The maximum number of electrons at each energy level is $2n^2$. Thus, energy level 3 has up to 18 electrons.

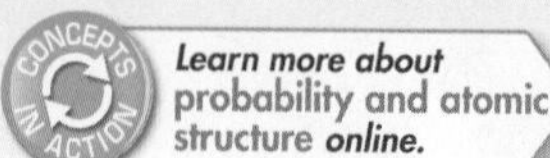

Table 5.1

Summary of Principal Energy Levels and Sublevels

Principal energy level	Number of sublevels	Type of sublevel	Maximum number of electrons
$n = 1$	1	1*s* (1 orbital)	2
$n = 2$	2	2*s* (1 orbital), 2*p* (3 orbitals)	8
$n = 3$	3	3*s* (1 orbital), 3*p* (3 orbitals), 3*d* (5 orbitals)	18
$n = 4$	4	4*s* (1 orbital), 4*p* (3 orbitals), 4*d* (5 orbitals), 4*f* (7 orbitals)	32

Q: *Previous models of the atom were physical models based on the motion of large objects. Why do scientists no longer use physical models to describe the motion of electrons?*

As shown in Table 5.1, the numbers and types of atomic orbitals depend on the principal energy level. The lowest principal energy level ($n = 1$) has only one sublevel, called 1*s*. The second principal energy level ($n = 2$) has two sublevels, 2*s* and 2*p*. The 2*p* sublevel is of higher energy than the 2*s* sublevel and consists of three *p* orbitals of equal energy. Thus the second principal energy level has four orbitals (one 2*s* and three 2*p* orbitals).

The third principal energy level ($n = 3$) has three sublevels. These are called 3*s*, 3*p*, and 3*d*. The 3*d* sublevel consists of five *d* orbitals of equal energy. Thus the third principal energy level has nine orbitals (one 3*s*, three 3*p*, and five 3*d* orbitals).

The fourth principal energy level ($n = 4$) has four sublevels, called 4*s*, 4*p*, 4*d*, and 4*f*. The 4*f* sublevel consists of seven *f* orbitals of equal energy. The fourth principal energy level, then, has sixteen orbitals (one 4*s*, three 4*p*, five 4*d*, and seven 4*f* orbitals).

As shown in the table, the principal quantum number always equals the number of sublevels within that principal energy level. The number of orbitals in a principal energy level is equal to n^2. As you will learn in the next lesson, a maximum of two electrons can occupy an orbital. Therefore, the maximum number of electrons that can occupy a principal energy level is given by the formula $2n^2$.

5.1 LessonCheck

ONLINE PROBLEMS

1. **Review** What was the basic proposal in the Bohr model of the atom?
2. **Describe** What does the quantum mechanical model determine about electrons in atoms?
3. **Review** How do two sublevels of the same principal energy level differ from each other?
4. **Describe** How can electrons in an atom move from one energy level to another?
5. **Explain** The energies of electrons are said to be quantized. Explain what this means.
6. **Apply Concepts** How many orbitals are in the following sublevels?
   - **a.** 3*p* sublevel
   - **b.** 2*s* sublevel
   - **c.** 4*p* sublevel
   - **d.** 3*d* sublevel
   - **e.** 4*f* sublevel

BIGIDEA ELECTRONS AND THE STRUCTURE OF ATOMS

7. How do the Bohr model and the quantum mechanical model differ in the way they describe the arrangement of electrons in atoms?

Lesson Check Answers

1. An electron is found only in specific circular paths or orbits around the nucleus.
2. It determines the allowed energy levels an electron can have and the likelihood of finding an electron in various locations around the nucleus.
3. The sublevels have different shapes.
4. by losing or gaining just the right amount of energy—a quantum
5. In an atom, the electrons can have certain fixed energy levels. To move from one energy level to another requires the emission or absorption of an exact amount of energy, or quantum. Thus the energy of the electron is said to be quantized.
6. **a.** 3 **b.** 1 **c.** 3 **d.** 5 **e.** 7
7. **BIGIDEA** The Bohr model limits electrons to specific circular paths. The quantum mechanical model expresses the probability of finding an electron in a given location within the electron cloud based on its current energy level.

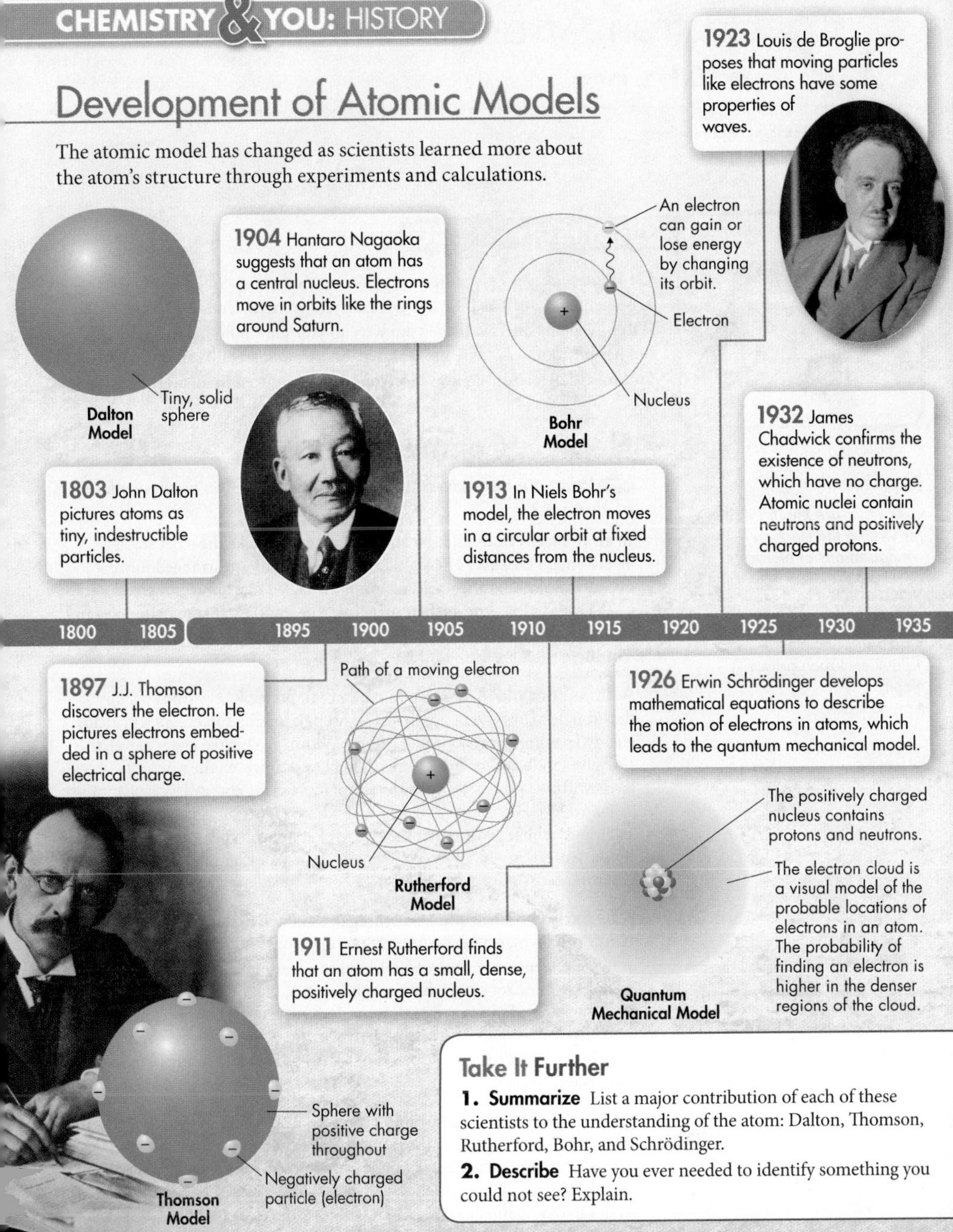

CHEMISTRY & YOU: HISTORY

Development of Atomic Models

The atomic model has changed as scientists learned more about the atom's structure through experiments and calculations.

Take It Further

1. **Summarize** List a major contribution of each of these scientists to the understanding of the atom: Dalton, Thomson, Rutherford, Bohr, and Schrödinger.
2. **Describe** Have you ever needed to identify something you could not see? Explain.

Chemistry & You 133

CHEMISTRY & YOU Have students examine the timeline. Point out that it took over 90 years from the time that Dalton first proposed his atomic theory for the discovery of the electron to occur. As with many other important scientific discoveries, the development of a critical piece of technology led to major revisions of a long-standing theory as one new discovery led quickly to another. For example, the development of the microscope led to the discovery of the cell. The development of x-ray crystallography was necessary before the structure of DNA could be determined. In the case of the atomic model, the critical technological advance was the development of the cathode-ray tube. **Ask** Why was the development of the cathode-ray tube a turning point in the development of the modern atomic model?

Extend

Connect to LANGUAGE ARTS

Explain that some of the greatest works of literature written in this time period, such as *Frankenstein* and *Dr. Jekyl and Mr. Hyde,* were inspired by scientific discoveries of the day. Literature sometimes influences the field of science as well, as in the case of the naming of the subatomic particle known as a *quark*. Have students research the identification of the quark as a subatomic particle and write a brief paragraph about its discovery.

Differentiated Instruction

L1 STRUGGLING STUDENTS Have students create their own model of an atom based on any of the scientists discussed in the text. The model need not be elaborate but should include labeling of the various parts and a brief description of the model.

LPR LESS PROFICIENT READERS Have students recreate the timeline shown using a poster board. Tell students to add information and diagrams that are helpful in their understanding of the development of the atomic model.

L3 ADVANCED STUDENTS Have students research early literary works of science fiction and their authors. Tell students to write their findings on note cards; each note card should state the title of the work, the author, the date of publication, and the basis of the scientific concepts that provided the foundation of the story.

Answers

1. *Sample Answer:* Dalton recognized that atoms of the same element are identical. Thomson discovered the electron. Rutherford proposed that an atom has a nucleus. Bohr proposed that electrons orbited the nucleus in circular paths. Schrödinger devised and solved a mathematical equation that described the behavior of an electron in a hydrogen atom.
2. Answers will vary.

LESSON 5.2

Key Objectives

5.2.1 LIST the three rules for writing the electron configurations of elements.

Additional Resources

Reading and Study Workbook, Lesson 5.2

Available Online or on Digital Media:

- Teaching Resources, Lesson 5.2 Review
- Small-Scale Chemistry Laboratory Manual, Lab 7

PearsonChem.com

Engage

CHEMISTRY & YOU Have students study the photograph and read the text that opens the lesson. **Ask** What role do energy and stability play in the way that electrons are configured in an atom? *(In the most stable atoms, electrons occupy the lowest possible energy orbitals.)*

Build Background

As a class, discuss processes familiar to students that follow a set of rules or guidelines. Discuss whether these rules always have to be followed, or if there are exceptions to the rules. *(Guide discussion toward rules for driving, security line at the airport, school rules, grammar rules, etc.)*

Foundations for Reading

BUILD VOCABULARY Explain that a configuration is an arrangement of parts. An electron configuration is the arrangement of electrons in orbitals around the nucleus of an atom.

IN ACADEMIC STANDARDS for SCIENCE

C.2.1 Describe how models of atomic structure changed over time based on available experimental evidence and understand the current model of atomic structure.

C.2.2 Describe how the subatomic particles (protons, neutrons, and electrons) contribute to the structure of an atom and recognize that the particles within the nucleus are held together against the electrical repulsion of the protons.

5.2 Electron Arrangement in Atoms

C.2.1 Modern atomic theory: historic development; C.2.2 Subatomic particles and atomic structure

CHEMISTRY & YOU

Q: *What makes the electron configuration of an atom stable?* Unstable arrangements, such as the yoga position shown here, tend to become more stable by losing energy. If the yogi were to fall, she would have less energy, but her position would be more stable. Energy and stability play an important role in determining how electrons are configured in an atom.

Key Question

What are the three rules for writing the electron configurations of elements?

Vocabulary

- electron configuration
- aufbau principle
- Pauli exclusion principle
- spin
- Hund's rule

Electron Configurations

What are the three rules for writing the electron configurations of elements?

In an atom, electrons and the nucleus interact to make the most stable arrangement possible. The ways in which electrons are arranged in various orbitals around the nuclei of atoms are called **electron configurations.** **Three rules—the aufbau principle, the Pauli exclusion principle, and Hund's rule—tell you how to find the electron configurations of atoms.** The three rules are as follows.

Aufbau Principle According to the **aufbau principle,** electrons occupy the orbitals of lowest energy first. The orbitals for any sublevel of a principal energy level are always of equal energy. Within a principal energy level, the *s* sublevel is always the lowest-energy sublevel. However, the range of energy levels within a principal energy level can overlap the energy levels of another principal level. Look at the aufbau diagram in Figure 5.5. Each box represents an atomic orbital. Notice that the filling of atomic orbitals does not follow a simple pattern beyond the second energy level. For example, the 4*s* orbital is lower in energy than a 3*d* orbital.

Pauli Exclusion Principle According to the **Pauli exclusion principle,** an atomic orbital may describe at most two electrons. For example, either one or two electrons can occupy an *s* orbital or a *p* orbital. To occupy the same orbital, two electrons must have opposite spins; that is, the electron spins must be paired. **Spin** is a quantum mechanical property of electrons and may be thought of as clockwise or counterclockwise. A vertical arrow indicates an electron and its direction of spin (↑ or ↓). An orbital containing paired electrons is written as [↑↓].

Hund's Rule According to **Hund's rule,** electrons occupy orbitals of the same energy in a way that makes the number of electrons with the same spin direction as large as possible. For example, three electrons would occupy three orbitals of equal energy as follows: [↑][↑][↑]. Electrons then occupy each orbital so that their spins are paired with the first electron in the orbital.

Focus on ELL

1 CONTENT AND LANGUAGE Write the vocabulary terms on the board. Divide the class into groups and ask them to find the paragraph or section of the lesson where the definitions to each term can be found. They should notice that the vocabulary is highlighted in the text.

2 FRONTLOAD THE LESSON Point out that this lesson is about the arrangement of electrons in an atom and the rules that determine where each electron can be found. Have students brainstorm other types of arrangements they have encountered that are governed by rules.

3 COMPREHENSIBLE INPUT Have students create a two-column chart that allows them to copy the three rules for writing the electron configuration of elements in the first column and draw an example or two that they can label in the second column. A third column can be added for a native language explanation of each rule if desired.

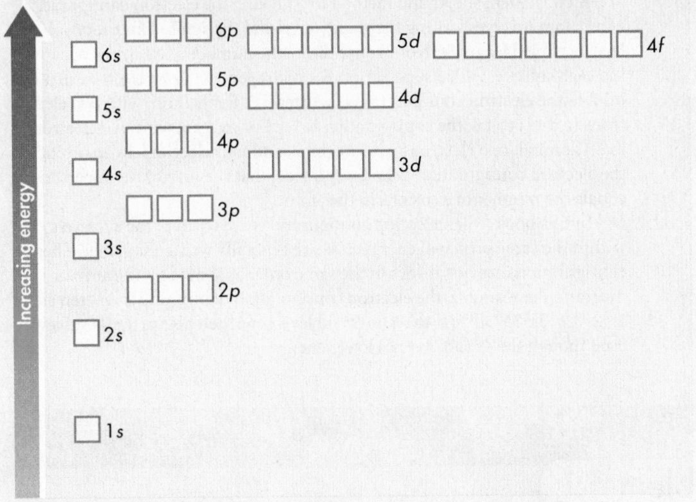

6s, 6p, 5d, 4f
5s, 5p, 4d
4s, 4p, 3d
3s, 3p
2s, 2p
1s
Increasing energy

Figure 5.5 Aufbau Diagram This aufbau diagram shows the relative energy levels of the various atomic orbitals. Orbitals of greater energy are higher on the diagram.
Interpret Tables ***Which is of higher energy, a 4d orbital or a 5s orbital?***

Look at the orbital filling diagrams of the atoms listed in Table 5.2. An oxygen atom contains eight electrons. The orbital of lowest energy, *1s*, has one electron, then a second electron of opposite spin. The next orbital to fill is *2s*. It also has one electron, then a second electron of opposite spin. One electron then occupies each of the three *2p* orbitals of equal energy. The remaining electron now pairs with an electron occupying one of the *2p* orbitals. The other two *2p* orbitals remain only half filled, with one electron each.

Table 5.2

Electron Configurations of Selected Elements

Element	1s	2s	$2p_x$	$2p_y$	$2p_z$	3s	Electron configuration
H	↑						$1s^1$
He	↑↓						$1s^2$
Li	↑↓	↑					$1s^2 2s^1$
C	↑↓	↑↓	↑	↑			$1s^2 2s^2 2p^2$
N	↑↓	↑↓	↑	↑	↑		$1s^2 2s^2 2p^3$
O	↑↓	↑↓	↑↓	↑	↑		$1s^2 2s^2 2p^4$
F	↑↓	↑↓	↑↓	↑↓	↑		$1s^2 2s^2 2p^5$
Ne	↑↓	↑↓	↑↓	↑↓	↑↓		$1s^2 2s^2 2p^6$
Na	↑↓	↑↓	↑↓	↑↓	↑↓	↑	$1s^2 2s^2 2p^6 3s^1$

Differentiated Instruction

LPR LESS PROFICIENT READERS Have students copy Table 5.2 with the boxes and last column left blank. Then fill in the arrows and electron configurations together to help them become familiar with the order in which sublevels fill when writing electron configurations.

L1 STRUGGLING STUDENTS Encourage students to draw the aufbau diagram (Figure 5.5) on a 5 x 7 in. card and keep it handy when writing electron configurations. Also encourage them to make flash cards, each with the name and atomic number of an element on one side and its electron configuration on the other side. Have partners use the flash cards to quiz each other.

L3 ADVANCED STUDENTS Direct students to a college text and ask them to name and explain the four quantum numbers. *(The principle quantum number, n, is the energy level of the orbital; the azimuthal quantum number, l, defines the shape of the orbital; the magnetic quantum number, m, describes the orientation of the orbital in space; and the spin quantum number, m_s, denotes the direction of electron spin, +1/2 or –1/2.)*

READING STRATEGY Encourage students to apply SQ3R as they read this section. Model how to employ it for the first two pages.

Explain

Electron Configurations

INTERPRET DIAGRAMS Direct students' attention to Figure 5.5. **Ask** Do all the orbitals at energy level *n* = 4 have the same energy? *(No, 4p orbitals have higher energy than 4s, and 4d have higher energy than 4p.)* **Ask** Which has higher energy, 4s or 3*d*? *(3d)*

START A CONVERSATION Develop the electron configurations for several of the simpler elements. Introduce each rule governing the process as needed. Begin with hydrogen. Use the aufbau diagram to explain that electrons enter orbitals of lowest energy first. Show how the orbital notation *($1s^1$)* describes the energy level, the orbital, and the number of electrons. Repeat the process for helium. Then continue with lithium, beryllium, and boron. Apply the Pauli exclusion principle to explain why additional orbitals must be used. When you reach carbon, explain and apply Hund's rule. Complete the configurations for the second period elements.

Explore

Class Activity

PURPOSE Students gain practice in determining the correct order for filling orbitals.

PROCEDURE Have students work with a partner to develop the electron configurations for the third period elements. When the exercise is complete, ask students to compare the configurations of second and third period elements in preparation for future discussions of the periodic table.

Answers

FIGURE 5.5 4*d*

LESSON 5.2

Explain

Electron Configurations

MAKE A CONNECTION Students may ask how two negative electrons can occupy the same orbital, or why an orbital never contains more than two electrons. The answer lies in the quantum property known as spin. A spinning electron acts as a tiny magnet, with a north pole at one end and a south pole at the other. If electrons with opposite spins enter one orbital, the attraction between their opposite magnetic poles counteracts some of the electrical repulsive force. If a third electron tries to enter the orbital, its spin will always be the same as one of the existing electrons, and it will be repelled.

CHEMISTRY & YOU Electrons fill the orbitals with the lowest energy first (the aufbau principle). The 2*p* sublevel of oxygen has less energy than the 3*s* sublevel.

Sample Practice Problem

What are the electron configurations for atoms of the following elements? How many unpaired electrons does each atom have?

a. neon *($1s^22s^22p^6$; no unpaired electrons)*

b. phosphorus *($1s^22s^22p^63s^23p^3$; three unpaired electrons)*

Q: *Explain why the correct electron configuration of oxygen is $1s^2 2s^2 2p^4$ and not $1s^2 2s^2 2p^3 3s^1$.*

A convenient shorthand method for showing the electron configuration of an atom involves writing the energy level and the symbol for every sublevel occupied by an electron. You indicate the number of electrons occupying each sublevel with a superscript. For hydrogen, with one electron in a 1*s* orbital, the electron configuration is written $1s^1$. For helium, with two electrons in a 1*s* orbital, the configuration is $1s^2$. For oxygen, with two electrons in a 1*s* orbital, two electrons in a 2*s* orbital, and four electrons in 2*p* orbitals, the electron configuration is $1s^2 2s^2 2p^4$. Note that the sum of the superscripts equals the number of electrons in the atom.

In this book, when electron configurations are written, the sublevels within the same principal energy level are generally written together. These configurations are not always in the same order as shown on the aufbau diagram. For example, the electron configuration of bromine is written as $1s^2 2s^2 2p^6 3s^2 3p^6 3d^{10} 4s^2 4p^5$. The 3*d* sublevel is written before the 4*s* sublevel, even though the 4*s* sublevel has lower energy.

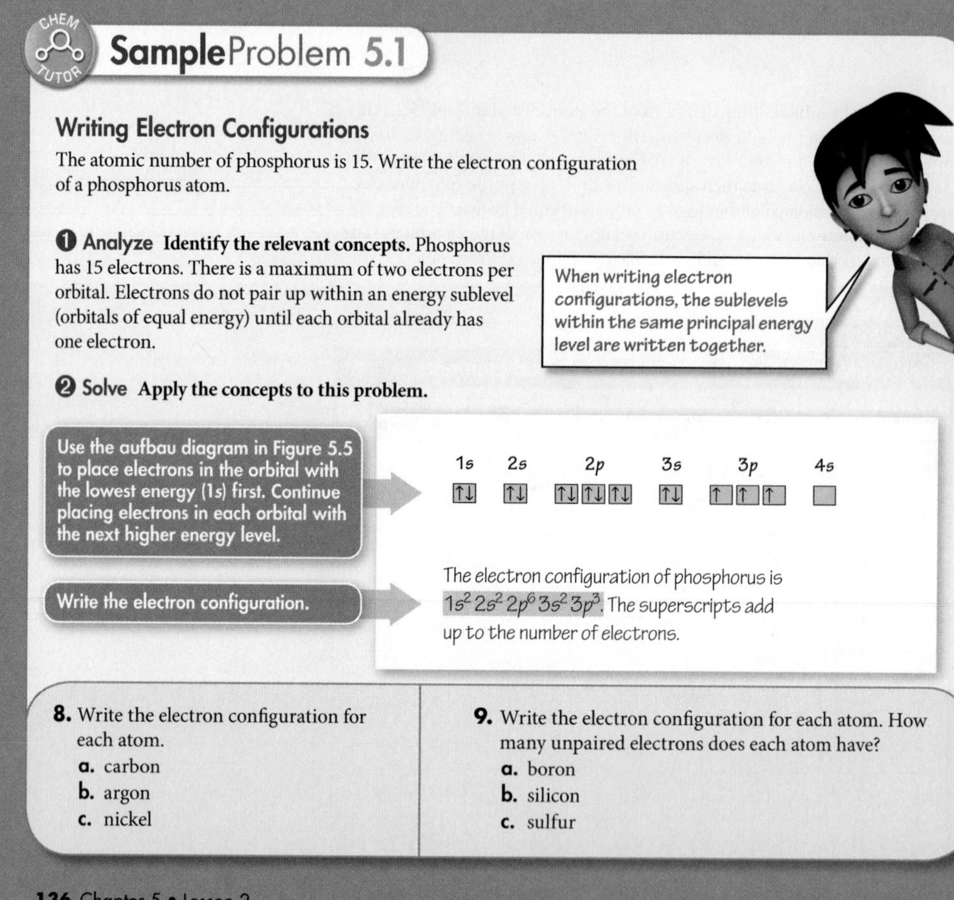

136 Chapter 5 • Lesson 2

UbD Check for Understanding

BIGIDEA ELECTRONS AND THE STRUCTURE OF ATOMS Give each student an index card. Assess students' knowledge of writing electron configurations of elements by reading out loud the description of each of the three rules. Students should write the name of the rule on the card in the order in which they were stated. *(the aufbau principle, the Pauli exclusion principle, and Hund's rule)*

ADJUST INSTRUCTION If students are having difficulty answering the question, have them re-read the previous page then repeat the exercise.

Exceptional Electron Configurations Copper, which is shown in Figure 5.6, has an electron configuration that is an exception to the aufbau principle. You can obtain correct electron configurations for the elements up to vanadium (atomic number 23) by following the aufbau diagram for orbital filling. If you were to continue in that fashion, however, you would assign chromium and copper the following incorrect configurations.

$$\text{Cr } 1s^2 2s^2 2p^6 3s^2 3p^6 3d^4 4s^2$$
$$\text{Cu } 1s^2 2s^2 2p^6 3s^2 3p^6 3d^9 4s^2$$

The correct electron configurations are as follows:

$$\text{Cr } 1s^2 2s^2 2p^6 3s^2 3p^6 3d^5 4s^1$$
$$\text{Cu } 1s^2 2s^2 2p^6 3s^2 3p^6 3d^{10} 4s^1$$

These arrangements give chromium a half-filled *d* sublevel and copper a filled *d* sublevel. Filled energy sublevels are more stable than partially filled sublevels. Some actual electron configurations differ from those assigned using the aufbau principle because although half-filled sublevels are not as stable as filled sublevels, they are more stable than other configurations. This tendency overcomes the small difference between the energies of the 3*d* and 4*s* sublevels in copper and chromium.

At higher principal quantum numbers, energy differences between some sublevels (such as 5*f* and 6*d*, for example) are even smaller than in the chromium and copper examples. As a result, there are other exceptions to the aufbau principle. Although it is worth knowing that exceptions to the aufbau principle occur, it is more important to understand the general rules for determining electron configurations in the many cases in which the aufbau principle applies.

Figure 5.6 Copper
Copper is a shiny metal that can be molded into different shapes. The electron configuration of copper does not follow the aufbau principle.

5.2 LessonCheck

10. List What are the three rules for writing the electron configurations of elements?

11. Sequence Use Figure 5.5 to arrange the following sublevels in order of decreasing energy: 2*p*, 4*s*, 3*s*, 3*d*, and 3*p*.

12. Explain Why do the actual electron configurations for some elements differ from those assigned using the aufbau principle?

13. Infer Why does one electron in a potassium atom go into the fourth energy level instead of squeezing into the third energy level along with the eight already there?

14. Apply Concepts The atomic number of arsenic is 33. What is the electron configuration of an arsenic atom?

Lesson Check Answers

10. aufbau principle, Pauli exclusion principle, Hund's rule

11. 3*d*, 4*s*, 3*p*, 3*s*, 2*p*

12. Half-filled sublevels and filled sublevels are more stable than other configurations.

13. The 3*s* and 3*p* orbitals are already filled, so the last electron must go to the next higher energy orbital, which is 4*s*.

14. $1s^2 2s^2 2p^6 3s^2 3p^6 3d^{10} 4s^2 4p^3$

Explain

Exceptional Electron Configurations

SUMMARIZE Have students write a summary statement about why there are exceptions to the filling rules among the transition elements. *(Exceptions can be explained by the atom's tendency to keep its energy as low as possible. These exceptions help explain the unexpected chemical behavior of transition elements.)*

Extend

Connect to PHYSICS

The Pauli exclusion principle applies to all four quantum numbers of an electron. Therefore, no two electrons can have exactly the same set of quantum numbers. Refer students to the Internet or a college text to find a table summarizing the quantum numbers. Have students demonstrate the application of the Pauli exclusion principle on all quantum numbers by writing out the possible combinations of quantum numbers for a first or second period element.

Evaluate

Informal Assessment

Make a set of small cards, each with the symbol and atomic number of an element. Choose elements from the first five periods of the periodic table. Have students choose a card and write the electron configuration for the element. Then have students complete the 5.2 Lesson Check.

Reteach

Review the three rules for writing electron configurations. Then demonstrate each rule in action by writing the electron configuration for a third period (non-transition) element.

Answers

8. **a.** $1s^2 2s^2 2p^2$
b. $1s^2 2s^2 2p^6 3s^2 3p^6$
c. $1s^2 2s^2 2p^6 3s^2 3p^6 3d^8 4s^2$

9. **a.** $1s^2 2s^2 2p^1$; one unpaired electron
b. $1s^2 2s^2 2p^6 3s^2 3p^2$; two unpaired electrons
c. $1s^2 2s^2 2p^6 3s^2 3p^4$; two unpaired electrons

LESSON 5.3

Key Objectives

5.3.1 EXPLAIN what causes atomic emission spectra.

5.3.2 DISTINGUISH between quantum mechanics and classical mechanics.

5.3.3 EXPLAIN how the frequencies of emitted light are related to changes in electron energies.

5.3.4 DISTINGUISH between quantum mechanics and classical mechanics.

Additional Resources

Reading and Study Workbook, Lesson 5.3
Available Online or on Digital Media:
- Teaching Resources, Lesson 5.3 Review
- Laboratory Manual, Lab 6-8
- Small-Scale Chemistry Laboratory Manual, Lab 8
- Virtual Chemistry Laboratory Manual, Labs 1, 7-10

PearsonChem.com

Engage

CHEMISTRY & YOU Have students study the photograph and read the opening text. **Ask** What causes atoms to emit light when excited by an electric current? *(Energy absorbed by electrons is emitted as light.)*

Activate Prior Knowledge

Have students draw a **KWL Chart** in their notebooks as you draw one on the board. Together fill in the **K** column with what they know about light. In groups, have them fill in the **W** column. Have students fill in the **L** column as they read.

IN ACADEMIC STANDARDS for SCIENCE

- Develop explanations based on reproducible data and observations gathered during laboratory investigations.
- Describe how scientific discoveries lead to the development of new technologies, and conversely how technological advances can lead to scientific discoveries through new experimental methods and equipment.

Also covered C.2.1, C.2.2

5.3 Atomic Emission Spectra and the Quantum Mechanical Model

C.2.1 Modern atomic theory: historic development; **C.2.2** Subatomic particles and atomic structure

FAMOUS BURGERS

CHEMISTRY & YOU

Q: *What gives gas-filled lights their colors?* If you walk in the evening along a busy street lined with shops and theaters, you are likely to see lighted advertising signs. The signs are formed from glass tubes bent in various shapes. An electric current passing through the gas in each glass tube makes the gas glow with its own characteristic color.

Key Questions

What causes atomic emission spectra?

How did Einstein explain the photoelectric effect?

How are the frequencies of light emitted by an atom related to changes of electron energies?

How does quantum mechanics differ from classical mechanics?

Vocabulary

- amplitude • wavelength
- frequency • hertz
- electromagnetic radiation
- spectrum
- atomic emission spectrum
- Planck's constant
- photoelectric effect • photon
- ground state
- Heisenberg uncertainty principle

Light and Atomic Emission Spectra

What causes atomic emission spectra?

The previous sections in this chapter introduced you to some ideas about how electrons in atoms are arranged in orbitals, each with a particular energy level. You also learned how to write electron configurations for atoms. You will now get a closer look into what led to the development of Schrödinger's equation and the quantum mechanical model of the atom.

The Nature of Light Rather curiously, the quantum mechanical model grew out of the study of light. Isaac Newton (1642–1727) tried to explain what was known about the behavior of light by assuming that light consists of particles. By the year 1900, however, there was enough experimental evidence to convince scientists that light consists of waves. Figure 5.7 illustrates some of the properties of waves. As shown, each complete wave cycle starts at zero on the y-axis, increases to its highest value, passes through zero to reach its lowest value, and returns to zero again. The **amplitude** of a wave is the wave's height from zero to the crest, as shown in Figure 5.7. The **wavelength,** represented by λ (the Greek letter lambda), is the distance between the crests. The **frequency,** represented by ν (the Greek letter nu), is the number of wave cycles to pass a given point per unit of time. The units of frequency are usually cycles per second. The SI unit of cycles per second is called the **hertz** (Hz). A hertz can also be expressed as a reciprocal second (s^{-1}).

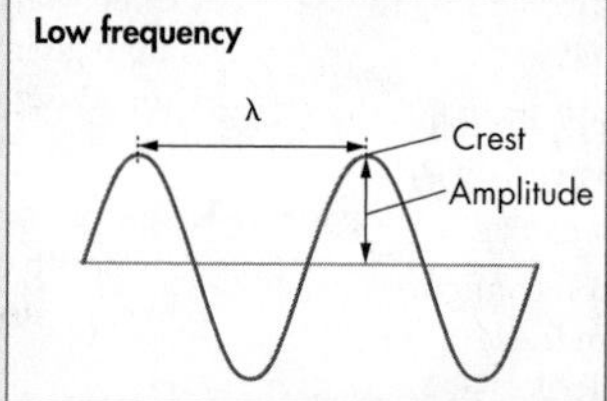

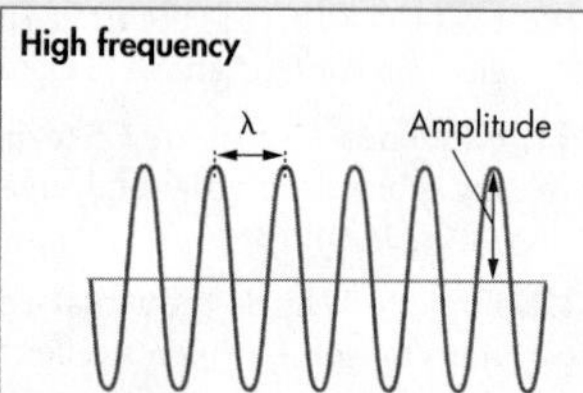

Figure 5.7 Light Waves
The frequency (ν) and wavelength (λ) of light waves are inversely related. As the wavelength decreases, the frequency increases.

Focus on ELL

1 CONTENT AND LANGUAGE Write the content vocabulary that appears in the lesson, such as *amplitude, wavelength, frequency,* and *hertz* on the board. Pronounce each word and ask students to repeat. Clarify the part of speech and discuss any cognates.

2 FRONTLOAD THE LESSON Ask students for examples of where they have seen a neon sign in or around town. Have students discuss how they think the signs produce different colored lights. Direct them to the information in the opening *Chemistry & You* text and ask them to predict how an electric current might affect the atoms in these gases.

3 COMPREHENSIBLE INPUT Provide students with a step-by-step explanation of Sample Problem 5.2. Work out another example similar to Sample Problem 5.2 and have students follow the same procedure. Speak clearly and slowly and use block arrows to step out the process.

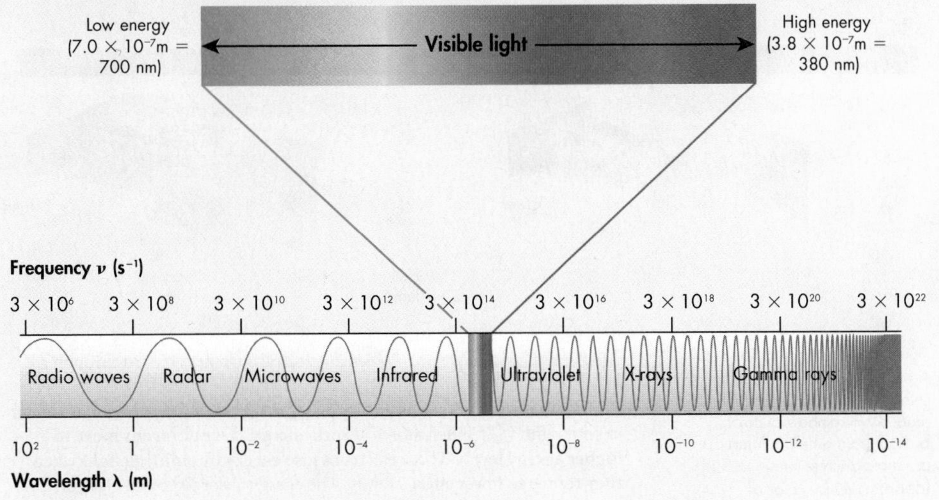

Figure 5.8 Electromagnetic Spectrum
The electromagnetic spectrum consists of radiation over a broad range of wavelengths. The visible light portion is very small. It is in the 10^{-7} m wavelength range and 10^{15} Hz (s^{-1}) frequency range.
Interpret Diagrams ***What types of nonvisible radiation have wavelengths close to those of red light? To those of blue light?***

The product of frequency and wavelength equals a constant (*c*), the speed of light.

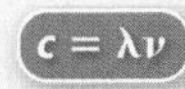

$$c = \lambda\nu$$

The wavelength and frequency of light are inversely proportional to each other. As the wavelength of light increases, the frequency decreases.

According to the wave model, light consists of electromagnetic waves. **Electromagnetic radiation** includes radio waves, microwaves, infrared waves, visible light, ultraviolet waves, X-rays, and gamma rays. All electromagnetic waves travel in a vacuum at a speed of 2.998×10^{8} m/s.

The sun and incandescent light bulbs emit white light, which consists of light with a continuous range of wavelengths and frequencies. As you can see from Figure 5.8, the wavelength and frequency of each color of light are characteristic of that color. When sunlight passes through a prism, the different wavelengths separate into a **spectrum** of colors. A rainbow is an example of this phenomenon. Each tiny droplet of water acts as a prism to produce a spectrum. Each color blends into the next in the order red, orange, yellow, green, blue, and violet. As can be seen in Figure 5.8, red light has the longest wavelength and the lowest frequency in the visible spectrum.

Light as a Particle

In 1666, Sir Isaac Newton (1642–1727) carried out important experiments with light that led to the proposal that light consists of tiny particles called corpuscles. Twelve years later, a Dutch scientist, Christian Huygens, suggested a wave theory to explain the properties of light. For more than 200 years, scientists argued about these seemingly contradictory theories. By the year 1900, most scientists had finally accepted that light could be described as a wave. Paradoxically, they soon had to accept that light could, at the same time, be described as a particle.

Foundations for Reading

BUILD VOCABULARY The word *emission* is from the Latin *emittere* meaning "to send out" The word *spectrum* is Latin for "appearance." Explain that an atomic emission spectrum allows the light sent out from an atom to be seen. **Ask** What is an emissary? *(a person sent out to represent another person)*

Explain

Light and Atomic Emission Spectra

USE VISUALS Direct students' attention to Figure 5.7. Tell students to assume that both waves are traveling at the same speed. **Ask** Which wave would have more complete wave cycles passing a point in a given time interval? *(The wave on the right.)* **Ask** Which wave has the higher frequency? *(The wave on the right.)*

INTERPRET DIAGRAMS Direct students' attention to Figure 5.8. **Ask** What are some types of radiation that are not visible? *(infrared, ultraviolet, X-rays, gamma rays)* **Ask** Are any of these invisible rays more energetic than visible light, and if so, what evidence do you have? *(UV causes suntan or sunburns, X-rays penetrate flesh, and gamma rays can penetrate bones.)* **Ask** What do all the rays in the electromagnetic spectrum have in common? *(All travel at the speed of light.)*

Explore

Student Activity

PURPOSE Students discover information about objects without seeing them.

MATERIALS several sealed boxes containing different common items such as rubber bands, golf, tennis, or table-tennis balls, cans of different sizes, and different shapes cut from blocks of wood or foam

PROCEDURE Pass the boxes around the room and record student observations and guesses about what is inside each "black box." Encourage students to move the boxes and listen to what happens. Point out that they are adding kinetic energy and observing how the hidden object responds. Point out that this method is analogous to the way scientists add energy to atoms and observe how they respond.

EXPECTED OUTCOME If the items are carefully selected, students will be surprised at how much they can determine without opening the boxes. This activity could also be done as a group activity with groups reporting what they find.

Answers

FIGURE 5.8 infrared, ultraviolet

LESSON 5.3

Explain

Atomic Emission Spectra

USE VISUALS Direct students' attention to Figure 5.9. **Ask** What is the difference between the spectrum of light from an ordinary light bulb and the spectrum from a helium lamp? *(The light bulb spectrum shows a continuous display of all the colors of visible light. The helium lamp shows discrete bands of only certain frequencies of light.)* Explain that the frequencies correspond to transitions of electrons in the helium atom.

Now direct students' attention to Figure 5.10. The atoms release the absorbed electrical energy in discrete packets with definite frequencies. Each different frequency is observed as a line in the sodium or mercury vapor emission spectrum. Scientists who were trying to understand the inner architecture of atoms recognized a relationship between the emission spectrum of an atom and its atomic structure.

Ask Why is hydrogen a good model for studying the correlation between emission spectrum and atomic structure? *(Hydrogen, with only one electron, has the simplest electron configuration and thus the simplest emission spectrum.)*

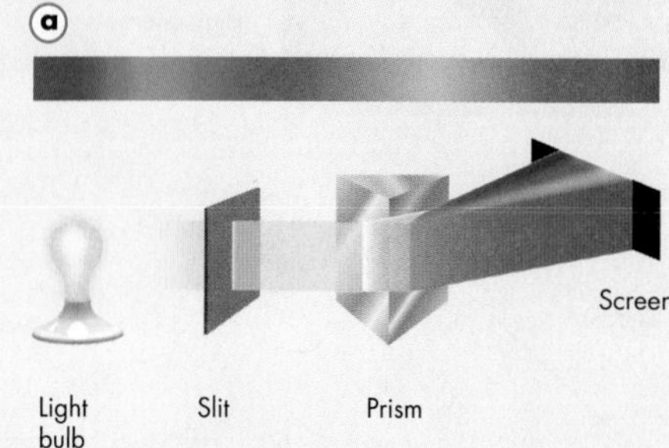

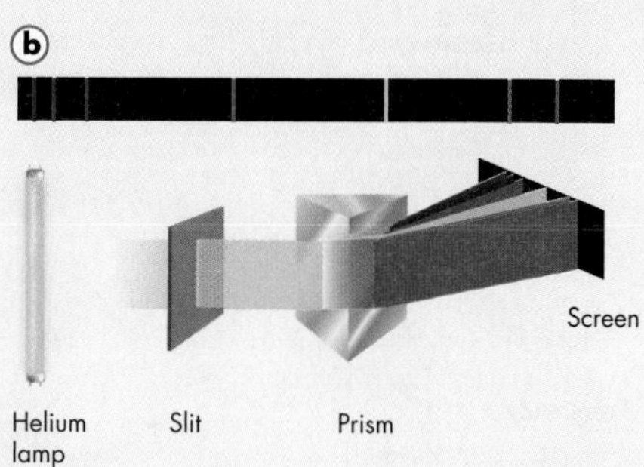

Figure 5.9 Comparing Spectra
A prism separates light into the colors it contains. **a.** White light produces a rainbow of colors. **b.** Light from a helium lamp produces discrete lines.
Identify *Which color of the rainbow has the highest frequency?*

Atomic Emission Spectra When an electric current is passed through a gaseous element, or through the vapor of a liquid or solid element, the electrons of the atoms of the gas or vapor are energized. This energy causes them to emit light. **When atoms absorb energy, their electrons move to higher energy levels. These electrons lose energy by emitting light when they return to lower energy levels.** The energy absorbed by an electron for it to move from its current energy level to a higher energy level is identical to the energy of the light emitted by the electron as it drops back to its original energy level. Figure 5.9a shows the visible spectrum of white light. Notice that all the wavelengths of visible light are blurred together as in a rainbow. However, when the light emitted by the energized electrons of a gaseous element is passed through a prism, as shown in Figure 5.9b, the spectrum consists of a limited number of narrow lines of light. The wavelengths of these spectral lines are characteristic of the element, and they make up the **atomic emission spectrum** of the element.

Each spectral line in an atomic emission spectrum of an element corresponds to exactly one wavelength of light emitted by the electrons of that element. Figure 5.9b shows the visible portion of the atomic emission spectrum of helium.

The atomic emission spectrum of each element is like a person's fingerprint. Just as no two people have the same fingerprints, no two elements have the same atomic emission spectrum. In the same way that fingerprints identify people, atomic emission spectra are useful for identifying elements. Figure 5.10 shows the characteristic colors emitted by sodium and by mercury. Much of the knowledge about the composition of the universe comes from studying the atomic emission spectra of the stars, which are hot glowing bodies of gases.

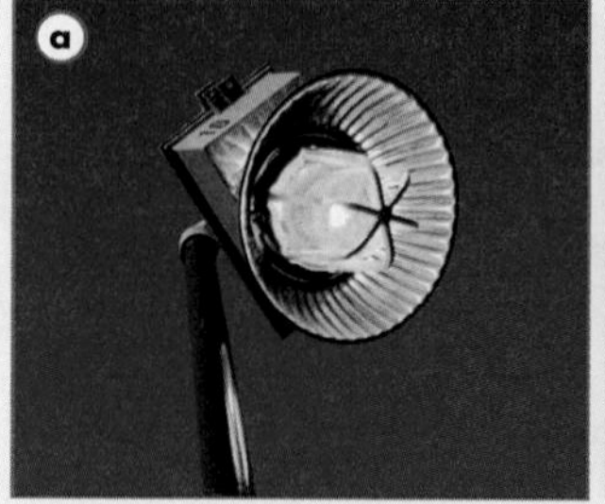

Figure 5.10
Atomic Emission Spectra
No two elements have the same atomic emission spectrum.
a. Sodium vapor lamps produce a yellow glow. **b.** Mercury vapor lamps produce a blue glow.

Differentiated Instruction

LPR LESS PROFICIENT READERS Create a bulleted list of the important information in the text following the Atomic Emission Spectra paragraph.

L1 SPECIAL NEEDS STUDENTS When introducing the formulas and units in this section, be sure to review mathematical operations for multiplication and division using scientific notation. Also be sure to show that $1/10 = 0.10 = 10^{-1}$. This will help students understand the units used for frequency: cycles per second = cycles/s = s^{-1}, or hertz (Hz).

L3 ADVANCED STUDENTS Have students design a system for refracting light and projecting a rainbow on a wall or screen. Depending on their setup, they may need a prism and a bright light source.

Calculating the Wavelength of Light

Calculate the wavelength of the yellow light emitted by a sodium lamp if the frequency of the radiation is 5.09×10^{14} Hz (5.09×10^{14}/s).

1 Analyze List the knowns and the unknown. Use the equation $c = \lambda\nu$ to solve for the unknown wavelength.

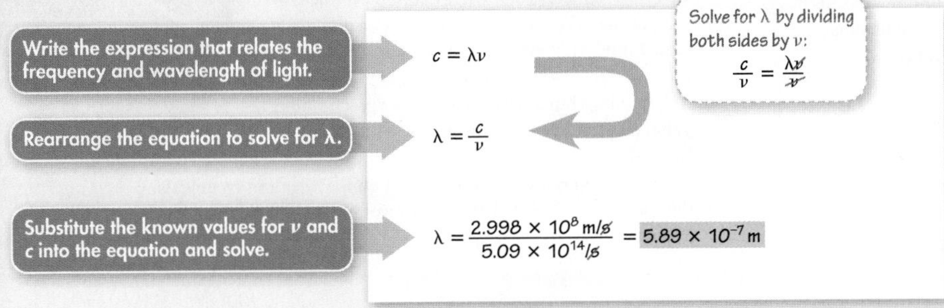

2 Calculate Solve for the unknown.

3 Evaluate Does the result make sense? The magnitude of the frequency is much larger than the numerical value of the speed of light, so the answer should be much less than 1. The answer should have three significant figures.

15. What is the wavelength of radiation with a frequency of 1.50×10^{13} Hz? Does this radiation have a longer or shorter wavelength than red light?

16. What is the frequency of radiation with a wavelength of 5.00×10^{-8} m? In what region of the electromagnetic spectrum is this radiation?

In Problem 15, solve for wavelength.

In Problem 16, solve for frequency.

Foundations for Math

ALGEBRAIC EQUATIONS Tell students that solving problems in chemistry often involves manipulating an equation to come up with a formula for a particular unknown. Students need to use the inverse properties of multiplication, division, addition, and subtraction in order to solve for the unknown variable.

In **Sample Problem 5.2**, the value for ν (nu) is given and c is a constant. Point out that the equation needs to be rearranged in order to isolate λ (lambda). Show students that they can do this by dividing both sides of the equation by ν (nu).

Explore

Class Activity

PURPOSE Students gain familiarity with wavelength, frequency, and amplitude.

PROCEDURE Have students draw a horizontal line approximately 5 inches long on a sheet of paper. Then ask them to use the line as a midpoint to draw two waves, superimposed, that have the same amplitudes but different frequencies. Have them draw another horizontal line and two waves with the same wavelength but different amplitudes.

Sample Practice Problem

What is the frequency of radiation which has a wavelength of 4.00×10^{-5} cm? *(7.49×10^{12} s^{-1})* In what region of the electromagnetic spectrum is this radiation? *(infrared)*

Answers

Figure 5.9 violet

15. 2.00×10^{-5} m; longer wavelength than red light

16. 6.00×10^{15} s^{-1}; ultraviolet

Explore

Quick Lab

OBJECTIVE Students use flame tests to determine the identity of the cation in an unknown solution.

SKILLS FOCUS Observe, interpret, identify

PREP TIME 30 minutes

CLASS TIME 20 minutes

MATERIALS Bunsen burner, 6 small test tubes, test tube rack, tongs, 6 cotton swabs, 0.1*M* NaCl, 0.5*M* $CaCl_2$, 0.1*M* LiCl, 0.1*M* $CuCl_2$, 0.1*M* KCl, unknown solution

ADVANCE PREP

Solution	Preparation
0.1 *M* NaCl	0.6 g NaCl in 100 mL water
0.5 *M* $CaCl_2$	5.5 g $CaCl_2$ in 100 mL water
0.1 *M* LiCl	0.5 g LiCl in 100 mL water
0.1 *M* $CuCl_2$	1.3 g $CuCl_2$ in 100 mL water
0.1 *M* KCl	0.8 g KCl in 100 mL water

SAFETY Students should wear safety goggles and a lab apron and observe all safety procedures.

EXPECTED OUTCOME

Na^+	bright yellow
Ca^{2+}	orange-red
Li^+	bright crimson red
Cu^{2+}	blue-green
K^+	light purple

ANALYZE AND CONCLUDE

1. Answer depends on the unknown.
2. Yes; because the composition of every atom is different, the amount of energy needed for transitions between energy sublevels is different. These energy differences are reflected in the different line spectra.
3. copper (green), lithium (red), or calcium (orange-red)
4. red, lithium; yellow, sodium

FOR ENRICHMENT Mix two unknown solutions. Ask students to identify the two metal ions.

Purpose To determine the identity of the metal in an unknown solution based on its characteristic color in a flame

Materials

- 6 small test tubes
- sodium chloride (NaCl) solution
- calcium chloride ($CaCl_2$) solution
- lithium chloride (LiCl) solution
- copper(II) chloride ($CuCl_2$) solution
- potassium chloride (KCl) solution
- unknown solution
- 6 cotton swabs
- gas burner

Flame Tests

Procedure

1. Make a two-column data table. Label the columns Metal and Flame Color. Enter the metal's name for each solution in the first column.
2. Label each of five test tubes with the name of a solution; label the sixth tube Unknown. Add 1 mL of each solution to the appropriately labeled test tube.
3. Dip one of the cotton ends of a cotton swab into the sodium chloride solution and then hold it briefly in the burner flame. Record the color of the flame. Do not leave the swab in the flame too long or the plastic will melt.
4. Repeat Step 3 for each of the remaining solutions using a new cotton swab each time.
5. Perform a flame test with the unknown solution. Note the color of the flame.

Analyze and Conclude

1. **Identify** What is the metal in the unknown?
2. **Draw Conclusions** Each solution produces a unique color. Would you expect this result based on the modern view of the atom? Explain.
3. **Analyze Data** Some commercially available fireplace logs burn with a red and/or green flame. What elements could be responsible for these colored flames?
4. **Predict** Aerial fireworks contain gunpowder and chemicals that produce colors. What element would you include to produce crimson red? Yellow?

The Quantum Concept and Photons

How did Einstein explain the photoelectric effect?

According to the laws of classical physics, the atomic emission spectrum of an element should be continuous. Thus, classical physics does not explain the emission spectra of atoms, which consist of lines.

The Quantization of Energy Recall the iron scroll in Figure 5.1 that changed color when heated. In 1900, the German physicist Max Planck (1858–1947) was trying to describe why such a body first appears black, then red, then yellow, and then white as its temperature increases. Planck found that he could explain the color changes if he assumed that the energy of a body changes only in small discrete units, or quanta. Planck showed mathematically that the amount of radiant energy (E) of a single quantum absorbed or emitted by a body is proportional to the frequency of radiation (ν).

$$E \propto \nu \text{ or } E = h\nu$$

Focus on ELL

4 LANGUAGE PRODUCTION Have students work in groups of four to complete the lab. Make sure each group has ELLs of varied language proficiencies, so that more proficient students can help less proficient ones. Have students work according to their proficiency level.

BEGINNING

LOW Model this lab and ask students to say the color that each metal produces. Ask students to discuss the color of the unknown metal.

HIGH Provide students with a chart of the solutions and then a column for them to write the color produced.

INTERMEDIATE: LOW/HIGH Have students predict which metal represents their unknown and compare the result with their prediction. Allow students to orally present their results.

ADVANCED: LOW/HIGH Ask students to complete the Analyze and Conclude questions with written responses.

The constant (h), which has a value of 6.626×10^{-34} J·s, (J is the joule, the SI unit of energy) is called **Planck's constant.** The energy of a quantum equals $h\nu$. A small energy change involves the emission or absorption of low-frequency radiation. A large energy change involves the emission or absorption of high-frequency radiation.

The Photoelectric Effect A few years after Planck presented his theory on the quantization of energy, scientists began to use it to explain many experimental observations that could not be explained by classical physics. In 1905, Albert Einstein (1879–1955), then a patent examiner in Bern, Switzerland, used Planck's quantum theory to explain the photoelectric effect, which is illustrated in Figure 5.11. In the **photoelectric effect,** electrons are ejected when light shines on a metal. Not just any frequency of light will cause the photoelectric effect. For example, red light will not cause potassium to eject electrons, no matter how intense the light. Yet a very weak yellow light shining on potassium begins the effect.

The photoelectric effect could not be explained by classical physics. Although classical physics correctly described light as a form of energy, it assumed that under weak light of any wavelength, an electron in a metal should eventually collect enough energy to be ejected. The photoelectric effect presented a serious problem for the classical wave theory of light.

To explain the photoelectric effect, Einstein proposed that light could be described as quanta of energy that behave as if they were particles. These light quanta are called **photons.** The energy of photons is quantized according to the equation $E = h\nu$. Einstein recognized that there is a threshold value of energy below which the photoelectric effect does not occur. According to $E = h\nu$, all the photons in a beam of monochromatic light (light of only one frequency) have the same energy. If the frequency, and therefore the energy, of the photons is too low, then no electrons will be ejected. It does not matter whether a single photon or a steady stream of low-energy photons strikes an electron in the metal. Only if the frequency of light is above the threshold frequency will the photoelectric effect occur.

Einstein's theory that light behaves as a stream of particles explains the photoelectric effect and many other observations. However, light also behaves as waves in other situations. Therefore, we must consider that light possesses both wavelike and particle-like properties.

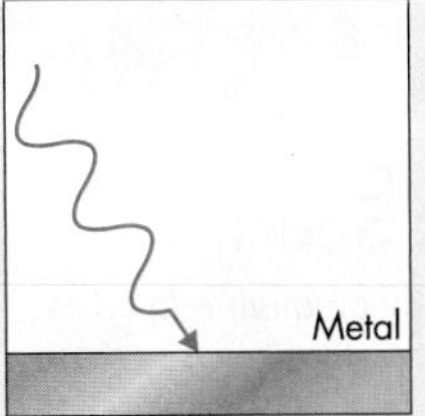

No electrons are ejected because the frequency of the light is below the threshold frequency.

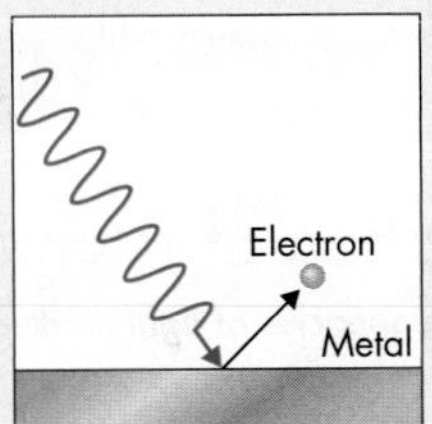

If the light is at or above the threshold frequency, electrons are ejected.

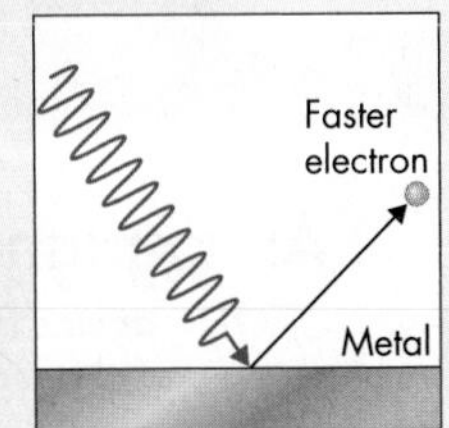

If the frequency is increased, the ejected electrons will travel faster.

Figure 5.11
Photoelectric Effect
Einstein explained the photoelectric effect by proposing that light behaves as particles.
Predict ***What will happen if ultraviolet light shines on the metal?***

UbD Check for Understanding

The Essential Question What happens when electrons in atoms absorb or release energy?

Have students write a brief one-minute summary that explains what they know about the absorption and release of energy by electrons in atoms. Summaries should reference the change in energy levels of the electrons and the photoelectric effect.

ADJUST INSTRUCTION If students are having difficulty writing their summaries, briefly review how atomic emission spectra are generated. Connect this concept to the photoelectric effect.

Explain

The Quantum Concept and Photons

APPLY CONCEPTS Determine by a show of hands how familiar students are with solar-powered landscaping lights or photoelectric cells. Explain that landscaping lights use photoelectric cells to convert solar (light) energy to electrical energy. The specially treated material used to make photoelectric cells has a positive and a negative side. When light strikes this material, electrons are emitted. The electrons move toward the positive side of the material in a current. This current is stored in batteries until the lights come on at dusk.

Extend

Connect to PHYSICS

Max Planck (1858–1947) showed that energy could be absorbed or emitted by a body only in quanta whose energy is given by the equation $E = h \times \nu$, where h is a proportionality constant called Planck's constant. Planck's proposal was revolutionary. Everyday experience had led people to believe that no limitation existed for the smallness of permissible energy changes in a system. For example, appearances would lead you to believe that thermal energy may be continuously supplied to heat liquid water to any temperature between 0°C and 100°C. Actually, the water temperature increases by infinitesimally small steps that occur as individual molecules absorb quanta of energy. Have students research Max Planck and write a summary about his other scientific discoveries.

Answers

FIGURE 5.11 an electron will be ejected from the metal

Extend

Connect to PHYSICS

Explain that the word *laser* is an acronym for **light amplification by the stimulated emission of radiation.** Encourage students to find out how coherent light—light with one wavelength—is produced. *(When a photon of a particular frequency strikes an atom, the photon stimulates the atom to emit a photon of the same frequency. The original photon is unaffected. Now two photons of the same frequency can collide with other atoms and stimulate them to produce identical photons. The process goes on building up an aggregate of photons all with the same frequency.)* Have students investigate what substances are stimulated to create laser light and what wavelengths correspond to each substance. *(Sample answers: Some typical lasers are nitrogen, $\lambda = 337$ nm; helium-cadmium, $\lambda = 441.6$ nm; argon, $\lambda = 476.5$ and 488.0 nm; krypton, $\lambda = 534.5$ nm; ruby, $\lambda = 694.3$ nm; neon, $\lambda = 632,8$ nm; gallium arsenide, $\lambda = 840–930$ nm, depending on temperature; neodynium, $\lambda = 1040$ nm; carbon dioxide, $\lambda = 10,600$ nm)*

Sample Practice Problem

What is the energy of a photon of red light with a frequency of 4.30×10^{14}/s? *(2.85×10^{-19} J)*

SampleProblem 5.3

Calculating the Energy of a Photon

What is the energy of a photon of microwave radiation with a frequency of 3.20×10^{11}/s?

KNOWNS
frequency (ν) = 3.20×10^{11}/s
$h = 6.626 \times 10^{-34}$ J·s

UNKNOWN
energy (E) = ? J

1 Analyze List the knowns and the unknown. Use the equation $E = h\nu$ to calculate the energy of the photon.

2 Calculate Solve for the unknown.

Write the expression that relates the energy of a photon of radiation and the frequency of the radiation.

$$E = h\nu$$

Substitute the known values for ν and h into the equation and solve.

$$E = (6.626 \times 10^{-34}\ \text{J}\cdot\text{s}) \times (3.20 \times 10^{11}/\text{s})$$
$$= 2.12 \times 10^{-22}\ \text{J}$$

3 Evaluate Does the result make sense? Individual photons have very small energies, so the answer seems reasonable.

In Problem 18, use the equation $c = \lambda\nu$ to calculate the frequency of light from the wavelength. Then, calculate the energy.

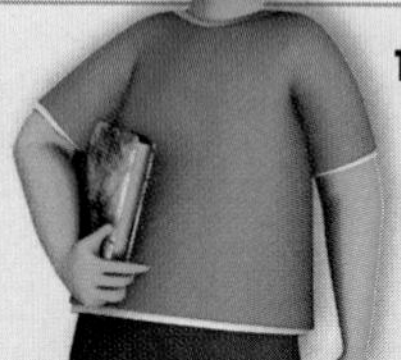

17. Calculate the energy of a quantum of radiant energy with a frequency of 5.00×10^{11}/s.

18. The threshold photoelectric effect in tungsten is produced by light of a wavelength 260 nm. Give the energy of a photon of this light in joules.

An Explanation of Atomic Spectra

How are the frequencies of light emitted by an atom related to changes of electron energies?

Atomic emission spectra were known before Bohr proposed his model of the hydrogen atom. Bohr applied quantum theory to electron energy levels in atoms to explain the atomic emission spectrum of hydrogen. Bohr's model not only explained why the atomic emission spectrum of hydrogen consists of specific frequencies of light, but it also predicted specific values of these frequencies that agreed with the experimental results.

Foundations for Math

SCIENTIFIC NOTATION Review with students that a number written in proper scientific notation form must have a coefficient greater than or equal to 1 and less than 10, a base of 10 and an exponent to indicate the number of decimal places that the decimal needs to be moved to change the number to standard notation. To find the product of numbers written in scientific notation, you multiply the coefficients, keep the base the same, and add the exponents.

In **Sample Problem 5.3**, the Associative Property of Multiplication can be used to find the product of $h \times \nu$. Begin by regrouping the terms:

$$E = h\nu = (6.626 \cdot 10^{-34}\ \text{J} \cdot \text{s}) \cdot (3.20 \cdot 10^{11}/\text{s})$$
$$= (6.626 \times 3.20)(10^{-34} \times 10^{11})\ \text{J}$$

Next, multiply the coefficients, add the exponents, and simplify:

$$= 21.2 \times 10^{-34+11}\ \text{J} = 21.2 \times 10^{-23}\ \text{J}$$

Finally, move the decimal one place to the left and increase the exponent by 1:

$$= 2.12 \times 10^{-22}\ \text{J}$$

In the Bohr model, the lone electron in the hydrogen atom can have only certain specific energies. When the electron has its lowest possible energy, the atom is in its **ground state.** In the ground state, the principal quantum number (n) is 1. Excitation of the electron by absorbing energy raises the atom to an excited state with $n = 2, 3, 4, 5$, or 6, and so forth. A quantum of energy in the form of light is emitted when the electron drops back to a lower energy level. The emission occurs in a single step, called an electronic transition. Bohr already knew that this quantum of energy E is related to the frequency ν of the emitted light by the equation $E = h\nu$. **The light emitted by an electron moving from a higher to a lower energy level has a frequency directly proportional to the energy change of the electron.** Therefore, each transition produces a line of a specific frequency in the spectrum.

Figure 5.12 shows the three groups of lines in the emission spectrum of hydrogen atoms. The lines at the ultraviolet end of the hydrogen spectrum are the Lyman series. These lines are due to the transitions of electrons from higher energy levels to the lowest energy level, $n = 1$. The lines in the visible spectrum are the Balmer series. These lines result from transitions from higher energy levels to $n = 2$. These transitions generally involve a smaller change in electron energy than transitions to $n = 1$. Transitions to $n = 3$ from higher energy levels produce the Paschen series. The energy changes of the electron are generally smaller still. The lines are in the infrared range. Spectral lines for the transitions from higher energy levels to $n = 4$ and $n = 5$ also exist. Note that the spectral lines in each group become more closely spaced at increased values of n because the energy levels become closer together. There is an upper limit to the frequency of emitted light for each set of lines because an electron with enough energy completely escapes the atom.

Bohr's model explained the atomic emission spectrum of hydrogen but not the emission spectra of atoms with more than one electron. Also, it did not help in understanding how atoms bond to form molecules. Eventually the quantum mechanical model displaced the Bohr model of the atom.

Q: *The glass tubes in lighted signs contain helium, neon, argon, krypton, or xenon gas, or a mixture of these gases. Why do the colors of the light depend on the gases that are used?*

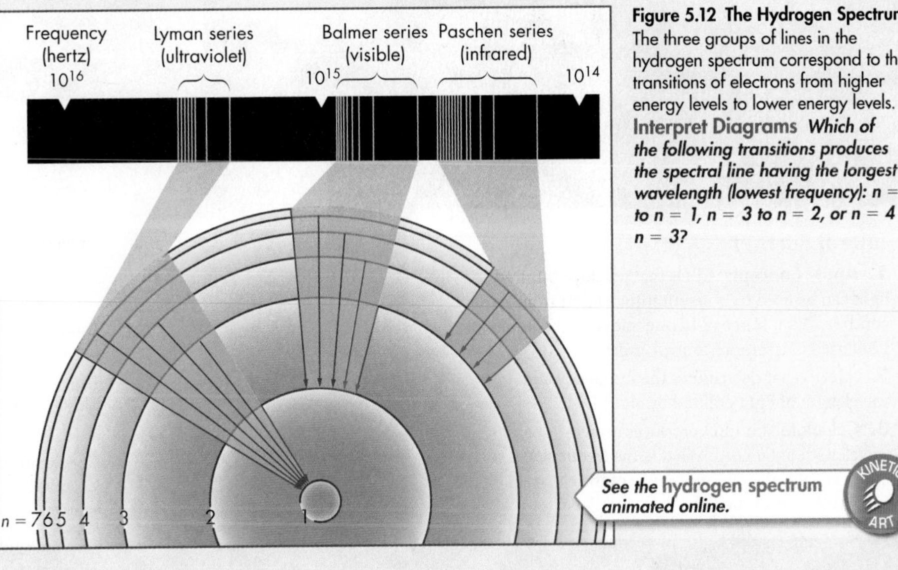

Figure 5.12 The Hydrogen Spectrum The three groups of lines in the hydrogen spectrum correspond to the transitions of electrons from higher energy levels to lower energy levels.
Interpret Diagrams ***Which of the following transitions produces the spectral line having the longest wavelength (lowest frequency): $n = 2$ to $n = 1$, $n = 3$ to $n = 2$, or $n = 4$ to $n = 3$?***

See the hydrogen spectrum animated online. KINETIC ART

UbD Check for Understanding

How are the frequencies of light emitted by an atom related to changes of electron energies?

Assess students' understanding of how atomic spectra occur by having students rephrase the key answer in their own words. *(Sample answer: When an electron moves from a higher to a lower energy level, the frequency of the light emitted will be in direct proportion to the electron's change in energy.)*

ADJUST INSTRUCTION If students are struggling to understand this concept, review the difference between direct and indirect proportionality. Then rewrite Planck's equation on the board to isolate frequency. Ask students to identify what would happen to frequency if E is increased. *(frequency would increase)* **Ask** What happens to frequency if E decreases? *(frequency would decrease)* Point out that the two values are directly proportional because both increase or both decrease. If they were indirectly proportional, then an increase in one would cause a decrease in the other.

Explain

An Explanation of Atomic Spectra

CHEMISTRY & YOU Each gas emits light at a different wavelength. This wavelength determines the characteristic color of light associated with a particular gas.

USE VISUALS Direct students' attention to Figure 5.12. **Ask** Suppose an electron, in its ground state at energy level one, absorbs enough energy to jump to level two. What type of radiation will it emit when it returns to the ground state? *(UV radiation)* **Ask** If you observed a hydrogen gas discharge tube through a diffraction grating, could you see the line corresponding to this emission? *(No)* Why not? *(The human eye cannot detect radiation in the UV range.)* **Ask** Which series of lines could you detect? *(the Balmer series, which has frequencies in the visible region of the spectrum)*

Have students compare the energy of the Paschen and Balmer series. *(The Paschen series has lower energy.)* **Ask** What do you notice about the spacing of the energy levels from $n = 1$ to $n = 7$? *(The levels are not evenly spaced. The lines get closer as the distance from the nucleus increases.)*

Point out that each set of lines is a converging series that approaches the ionization energy of hydrogen.

CRITICAL THINKING Point out to students that astronomers are able to identify the composition of stars that are at the farthest reaches of the universe, even though no technology exists that allows them to take actual samples of the gases that make up these stars. Have students use what they have learned about atomic emission spectra to explain how scientists can identify these gases.

Answers

17. 3.31×10^{-22} J
18. 7.6×10^{-19}

FIGURE 5.12 $n = 4$ to $n = 3$

CHEMISTRY & YOU

CHEMISTRY & YOU Call students' attention to the cut-out diagram of the LED. Explain how an LED works by first defining what a "diode" is: a small semiconductor that allows electrons to flow in only one direction across a material when a direct current is applied. Then explain that the light emitted from a diode is the result of the electrons from the atoms moving from the negative pole of the direct current to the positive pole. The light that is emitted is in the form of particles called "photons," which are small particles that have energy and momentum, but no mass.

Remind students that earlier they learned of another use of semiconductors: photoelectric cells. **Ask** How do photoelectric cells work? *(Sunlight causes the semiconductor material to emit electrons, which form a current by travelling between the negatively and positively charged sides of the semiconductor material.)* **Ask** What type of energy conversion occurs in photoelectric cells? *(light energy is converted to electrical energy)* Point out that in LEDs, the opposite conversion is taking place—electrical energy is being converted to light energy.

Extend

Connect to ECONOMICS

Lead students in a discussion about the efficiency of LED lighting. Discuss the respective amounts of energy needed to illuminate an LED bulb, a fluorescent bulb, and an incandescent bulb all of the same wattage output. Discuss the varying life expectancy of the three types of bulbs.

Have students prove mathematically that LED bulbs use less energy and cost less to operate than incandescent lights. Provide students with the following formulas: *Energy = power × time*, and *Cost of energy use = Energy × Cost of energy per unit*. Tell students to use $0.10 per kilowatt-hours for the cost of energy per unit, and to submit a source supporting the initial cost of the light. Have students research the amount of energy used per hour for a 60-watt incandescent bulb and an LED bulb providing the same illumination, and calculate the cost of illuminating each bulb for 24 hours per day for one year. Compare the costs and determine which type of bulb is the most cost effective.

Ask students to predict how the cost and efficiency of a fluorescent light compares to the cost and efficiency of incandescent and LED lights.

CHEMISTRY & YOU: TECHNOLOGY

Light Emitting Diodes

Although they are small, you may have seen light emitting diodes, or LEDs, several times today. These tiny light bulbs may form the numbers on your digital clock, light up your watch, or illuminate the traffic light you stopped at on your way to school. You may have even watched television on a giant screen made out of LEDs.

Light from a typical incandescent bulb is generated when the filament inside the bulb is heated. Light from an LED is generated in a different way. A diode is made out of two materials with different properties. The electrons in one of the materials are at a higher energy level than the electrons in the other material. When a voltage, supplied by a battery or other power supply, is applied to the diode, electrons flow across the boundary between the two materials. The electrons at the higher energy level flow into the other material, fall to a lower energy level, and emit light.

LEDs last much longer than incandescent light bulbs because there is no filament to burn out. Also, a large amount of the energy that is used to produce light in an incandescent bulb is wasted as heat. However, LEDs produce very little heat, so they use less energy and cost less to operate than incandescent lights.

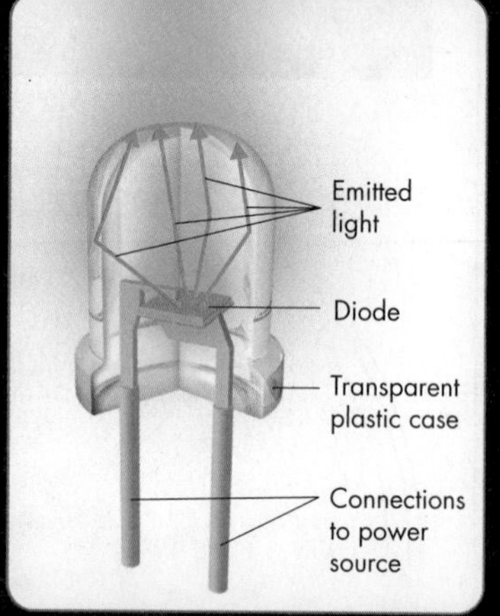

Take It Further

1. Apply Concepts LEDs that produce infrared light can be used to transmit information from remote controls. What range of frequencies of light would an LED used in a remote control emit?

2. Infer What determines the frequency and wavelength of light emitted by an LED?

3. Calculate An LED produces orange light with a wavelength of 605 nm. What is the frequency of this light? What is the energy change of the electrons in the LED as they emit the light?

21st Century Learning To be successful in the 21st century, students need skills and learning experiences that extend beyond subject matter mastery. The following project helps students build the following 21st Century Skills: *Financial, Economic, Business, and Entrepreneurial Literacy*; *Creativity and Innovation*; *Communication and Collaboration*; *Information, Media, and Technology Literacy*; *Initiative and Self-Direction*; and *Productivity and Accountability*.

Pose the following challenge to the students: *Conduct a cost-benefit analysis of upgrading the lighting in your home from incandescent to LED. Submit your report in electronic form, including any spreadsheets and graphs that were generated as a result of your analysis.*

You may need to guide the students in the following ways:

- A cost-benefit analysis quantifies the benefits, or positive aspects, of a project and compares this quantity to the total costs, or negative aspects, of the project.
- Electrical customers are billed in cents per kWh (kilowatt-hour). A rate schedule is available from the local electrical utility.

Quantum Mechanics

How does quantum mechanics differ from classical mechanics?

In 1924, Louis de Broglie (1892–1987), a French graduate student, asked an important question: Given that light behaves as waves and particles, can particles of matter behave as waves? De Broglie referred to the wavelike behavior of particles as matter waves. His reasoning led him to a mathematical expression for the wavelength of a moving particle.

The Wavelike Nature of Matter The proposal that matter moves in a wavelike way would not have been accepted unless experiments confirmed its validity. Only three years later, experiments by Clinton Davisson and Lester Germer at Bell Labs in New Jersey did just that. The two scientists had been studying the bombardment of metals with beams of electrons. They noticed that the electrons reflected from the metal surface produced curious patterns. The patterns were like those obtained when X-rays (which are electromagnetic waves) reflect from metal surfaces. The electrons, which were believed to be particles, were reflected as if they were waves! De Broglie was awarded the Nobel prize for his work on the wave nature of matter. Davisson also received the Nobel prize for his experiments demonstrating the wave nature of electrons.

Today, the wavelike properties of beams of electrons are useful in viewing objects that cannot be viewed with an optical microscope. The electrons in an electron microscope have much smaller wavelengths than visible light. These smaller wavelengths allow a much clearer enlarged image of a very small object, such as the pollen grain in Figure 5.13, than is possible with an ordinary microscope.

De Broglie's equation predicts that all moving objects have wavelike behavior. Why are you unable to observe the effects of this wavelike motion for ordinary objects like baseballs and trains? The answer is that the mass of the object must be very small in order for its wavelength to be large enough to observe. For example, a 50-gram golf ball traveling at 40 m/s (about 90 mi/h) has a wavelength of only 3×10^{-34} m, which is much too small to detect experimentally. On the other hand, an electron has a mass of only 9.11×10^{-28} g. If it were moving at a velocity of 40 m/s, it would have a wavelength of 2×10^{-5} m, which is comparable to infrared radiation and is readily measured.

De Broglie's prediction that matter exhibits both wave and particle properties set the stage for a new way of describing the motions of subatomic particles and atoms. The newer theory is called quantum mechanics; the older theory is called classical mechanics. **Classical mechanics adequately describes the motions of bodies much larger than atoms, while quantum mechanics describes the motions of subatomic particles and atoms as waves.**

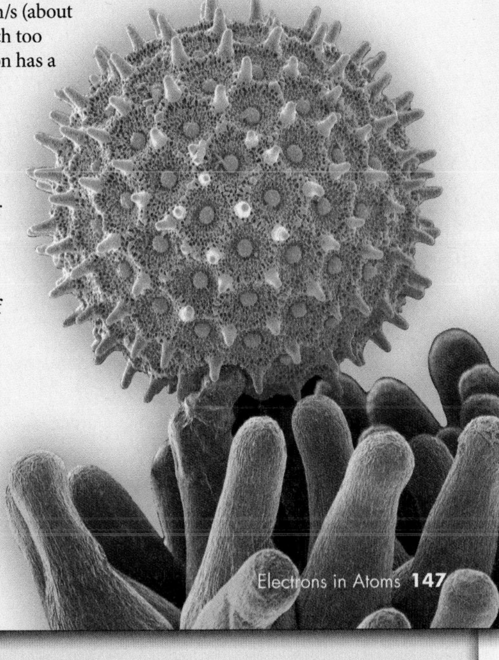

Figure 5.13 Electron Micrograph
An electron microscope can produce sharp images of a very small object, such as this pollen grain, because of the small wavelength of a moving electron compared with that of light.

A Scientific Controversy

Albert Einstein and Niels Bohr had different views of quantum mechanics. Although Einstein accepted quantum mechanical theory as the best explanation available at the time, he was convinced the theory was incomplete. He believed in the existence of "hidden variables" which, if known, would eliminate problems such as wave-particle duality and uncertainty. Bohr, on the other hand, urged physicists to accept quantum mechanics without qualification.

Explain

Quantum Mechanics

INTERPRET DATA The wave-particle behavior of light was difficult for scientists trained in classical physics to accept. However, it became difficult to dispute because it provided an explanation for a previously mysterious phenomenon called the photoelectric effect. In the photoelectric effect, it was shown that metals eject electrons called photoelectrons when light shines on them. But not just any frequency of light will cause the photoelectric effect. Red light ($\nu = 4.3 \times 10^{14}$ s^{-1} to 4.6×10^{14} s^{-1}), for example, will not cause the ejection of photoelectrons from potassium, no matter how intense the light or for how long the light radiates the metal surface. But even a weak yellow light ($\nu = 5.1 \times 10^{14}$ s^{-1} to 5.2×10^{14} s^{-1}) shining on potassium causes the emission of photoelectrons. Thus, photons of red light do not have sufficient energy to cause the ejection of electrons, but photons of yellow light do.

Explore

Class Activity

ATOMIC SCIENTISTS Numerous physicists contributed to modern theories of light and atoms. Dalton, Becquerel, Roentgen, Rutherford, Thomson, Millikan, Planck, Einstein, de Broglie, Heisenberg, Schrödinger, Bohr, Chadwick, Balmer, Huygens, Newton, and Foucault all made contributions in one of the most productive periods of scientific discovery.

PURPOSE Students acquire a deeper understanding and appreciation for the achievements of these scientists.

PROCEDURE Have students work in teams of two or three. Assign scientists to each group and ask students to prepare a brief oral discussion highlighting how the work of the scientists contributed to the understanding of light or the atom. Encourage students to consider the work of other scientists they encounter in their research. Contributions at the periphery can help students understand the interactive and collaborative nature of scientific research.

EXPECTED OUTCOME Class presentations allow students to share information.

Answers

TAKE IT FURTHER

1. 3×10^{14} to 3×10^{12} s^{-1}
2. The size of the energy change of the transition from a higher energy level to a lower energy level determines the frequency and wavelength of light emitted by an LED.
3. $\nu = 4.96 \times 10^{14}$ s^{-1}; $E = 3.29 \times 10^{-19}$ J

LESSON 5.3

Explain

USE AN ANALOGY Tell students that, according to quantum mechanics, scientists cannot observe an object without changing the object in the process. The change produced by making an observation is the basis for the uncertainty principle. Present the following analogy: Suppose you want to find out if a car is in a long tunnel. In quantum mechanics, the only sort of experiment you could do would be to send another car into the tunnel and listen for a crash. Although it is possible to detect the presence of a car with this method, it is obvious that the car will be changed by the crash.

Misconception Alert

Refer students to Figure 5.14 and the supporting text. Often students think that Heisenberg's principle is a technological limitation of the time in which it was discovered. Point out that this principle is in fact based on a physical limit—the level of precision that can be achieved when measuring the momentum of a particle.

Evaluate

Informal Assessment

In small groups, have students draw and label wave diagrams that illustrate the relationship between wavelength and frequency. Ask students to explain the origin of the lines in the Lyman, Balmer, and Paschen series in the atomic emission spectrum of hydrogen. *(Each line arises from light emitted by an atom as electrons move from a higher energy level to a lower one. The Lyman series is in the ultraviolet region of the spectrum, the Balmer is in the visible, and the Paschen is in the infrared.)*

Reteach

Use the illustrations in this lesson as a basis for review. Starting with Figure 5.7 and continuing to Figure 5.14, question students about what is being illustrated and what is significant about it.

The Heisenberg Uncertainty Principle German physicist Werner Heisenberg examined another feature of quantum mechanics. The **Heisenberg uncertainty principle** states that it is impossible to know both the velocity and the position of a particle at the same time. This limitation is critical when dealing with small particles such as electrons, but it does not matter for ordinary-sized objects such as cars or airplanes.

Consider how you determine the location of an object. To locate a set of keys in a dark room you can use a flashlight. You see the keys when the light bounces off them and strikes your eyes. To locate an electron, you might strike it with a photon of light, as shown in Figure 5.14. However, the electron has such a small mass that striking it with a photon affects its motion in a way that cannot be predicted accurately. The very act of measuring the position of the electron changes its velocity, making its velocity uncertain.

The discovery of matter waves led the way for Schrödinger's quantum mechanical description of electrons in atoms. Schrödinger's theory leads to the concept of electron orbitals and includes the uncertainty principle.

Figure 5.14 Heisenberg Uncertainty Principle
According to the Heisenberg uncertainty principle, it is impossible to know exactly both the velocity and the position of a particle at the same time.

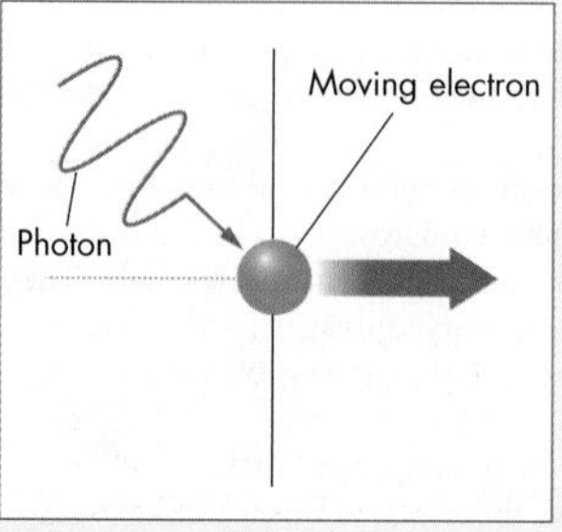

Before collision A photon strikes an electron during an attempt to observe the electron's position.

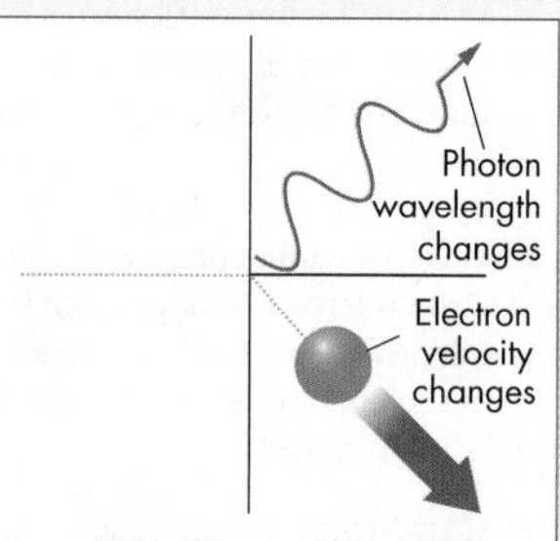

After collision The impact changes the electron's velocity, making it uncertain.

5.3 LessonCheck

19. **Describe** What is the origin of the atomic emission spectrum of an element?
20. **Review** What was Einstein's explanation for the photoelectric effect?
21. **Explain** How is the change in electron energy related to the frequency of light emitted in electronic transitions?
22. **Explain** How does quantum mechanics differ from classical mechanics?
23. **Sequence** Arrange the following in order of decreasing wavelength.
    a. infrared radiation from a heat lamp
    b. dental X-rays
    c. signal from a shortwave radio station
24. **Calculate** A hydrogen lamp emits several lines in the visible region of the spectrum. One of these lines has a wavelength of 6.56×10^{-5} cm. What is the frequency of this radiation?
25. **Calculate** What is the energy of a photon of blue light with a wavelength of 460 nm?

BIGIDEA
ELECTRONS AND THE STRUCTURE OF ATOMS

26. When a strontium compound is heated in a flame, red light is produced. When a barium compound is heated in a flame, yellow-green light is produced. Explain why these colors are emitted.

Lesson Check Answers

19. Electrons in atoms absorb energy as they move to higher energy levels and then lose the energy by emitting it as light as they drop back.
20. He proposed that light could be described as quanta of energy that behave as if they were particles.
21. The light emitted in an electronic transition from a higher to a lower energy level has a frequency that is directly proportional to the energy change of the electron.
22. Quantum mechanics describes the motions of atoms and subatomic particles; classical mechanics describes the motions of larger bodies.
23. *c, a, b*
24. $4.57 \times 10^{14}\ s^{-1}$
25. 4.32×10^{-19} J
26. **BIGIDEA** The colors, when separated by a spectroscope, are the colors of the identifying lines for strontium and barium and arise from electrons moving from higher energy levels to lower energy levels.

Small-Scale Lab

Atomic Emission Spectra

Purpose

To build a spectroscope and use it to measure the wavelengths, frequencies, and energies of atomic emission lines

Materials

- black construction paper
- tape
- cereal box
- scissors
- white notebook paper
- diffraction grating
- ruler

Procedure

Tape together two 2.0 cm × 10 cm strips of black construction paper so that they are parallel and form a narrow slit about 2 mm wide. Remove the top of a cereal box and tape the construction paper slit as shown. Cover the rest of the opening with white notebook paper. Cut a square hole (approximately 2 cm per side) and tape a diffraction grating over the hole as shown. Point the spectroscope toward a fluorescent light. Tape up any light leaks. Your lab partner should mark the exact positions of all the colored emission lines you see on the notebook paper. Measure the distances between the violet line and the other lines you have marked.

Diffraction grating
Black construction paper
Cereal box
Notebook paper
Slit
Spectrum appears here

Analyze and Conclude

1. Observe List the number of distinct lines that you see as well as their colors.

2. Measure Each line you see has a wavelength. The prominent violet line has a wavelength of 436 nm and the prominent green line has a wavelength of 546 nm. How many mm apart are these lines on the paper? By how many nm do their wavelengths differ? How many nanometers of wavelength are represented by each millimeter you measured?

3. Calculate Using the nm/mm value you calculated and the mm distance you measured for each line from the violet reference line, calculate the wavelengths of all the other lines you see.

4. Calculate Use the wavelength value of each line to calculate its frequency given that $\nu = c/\lambda$ where $c = 2.998 \times 10^{17}$ nm/s (2.998×10^{8} m/s).

5. Calculate The energy E of a quantum of light an atom emits is related to its frequency ν by $E = h\nu$. Use the frequency value for each line and $h = 6.626 \times 10^{-34}$ J·s to calculate its corresponding energy.

You're the Chemist

1. Design an Experiment Design and carry out an experiment to measure the longest and shortest wavelengths you can see in daylight. Use your spectroscope to observe light from daylight reflected off a white piece of paper. **CAUTION** *Do not look directly at the sun!* Describe the differences in daylight and fluorescent light.

2. Design an Experiment Design and carry out an experiment to determine the effect of colored filters on the spectrum of fluorescent light or daylight. For each filter, tell which colors are transmitted and which are absorbed.

3. Analyze Data Use your spectroscope to observe various atomic emission discharge tubes provided by your teacher. Note and record the lines you see and measure their wavelengths.

Focus on ELL

4 LANGUAGE PRODUCTION Consider restating the procedure of the lab into numbered steps or directions. Have students work in groups of four to complete the lab. Make sure each group has ELLs of varied language proficiencies, so that more proficient students can help less proficient ones. Have students work according to their proficiency level.

BEGINNING: LOW/HIGH Paraphrase orally the procedural steps and group students with higher language proficiency partners.

INTERMEDIATE: LOW/HIGH Have students brainstorm what they think they may observe, prior to performing each step. Allow students to orally present their results.

ADVANCED: LOW/HIGH Have students write a couple of paragraphs explaining their procedure, findings, and results.

Explore

OBJECTIVE After completing this activity, students will be able to build a simple spectroscope and use it to determine wavelength, frequency, and energy of emission lines.

PREP TIME 10 minutes

CLASS TIME 20 minutes

TEACHING TIPS Have a completed spectroscope for students to examine. Have students bring in their own cereal boxes. Use holographic diffraction gratings of about 700 lines per millimeter. The diffraction grating needs to be positioned so that the lines on the grating are parallel to the slit. Very narrow slits do not let in enough light to see a bright spectrum.

EXPECTED OUTCOME Students calculate the wavelength, frequency, and energy of emissions from a fluorescent light.

ANALYZE

1. Most fluorescent lights display 5 lines: violet, blue, green, yellow and red. All fluorescent lights display violet, green and yellow. Older lights will display only two distinct violet and blue lines. A diffuse yellow line is also evident.
2. A typical box will yield a measurement of 30 mm between the violet and green lines; 546 – 436 = 110 nm; 110 nm/30 mm = 3.67 nm/mm
3. Blue: 436 nm + (16 mm × 3.67 nm/mm) = 495 nm; Green: 436 nm + (30 mm × 3.67 nm/mm) = 546 nm; Yellow: 436 nm + (42 × 3.67 nm/mm) = 590 nm; Red: 436 nm + (50 × 3.67 nm/mm) = 620 nm
4. Violet: $\nu = 3.00 \times 10^{17}$ nm/s436 nm $= 6.88 \times 10^{14}\ s^{-1}$; Blue: $\nu = 3.00 \times 10^{17}$ nm/s495 nm $= 6.06 \times 10^{14}\ s^{-1}$; Green: $\nu = 3.00 \times 10^{17}$ nm/s/546 nm $= 5.49 \times 10^{14}\ s^{-1}$; Yellow: $\nu = 3.00 \times 10^{17}$ nm/s/590 nm $= 5.08 \times 10^{14}\ s^{-1}$; Red: $\nu = 3.00 \times 10^{17}$ nm/s/620 nm $= 4.84 \times 10^{14}\ s^{-1}$
5. Violet: 6.63×10^{-34} J · s × $6.88 \times 10^{14}\ s^{-1} = 4.56 \times 10^{-19}$ J; Blue: 6.63×10^{-34} J · s × $6.06 \times 10^{14}\ s^{-1} = 4.02 \times 10^{-19}$ J; Green: 6.63×10^{-34} J · s × $5.49 \times 10^{14}\ s^{-1} = 3.64 \times 10^{-19}$ J; Yellow: 6.63×10^{-34} J · s × $5.08 \times 10^{14}\ s^{-1} = 3.37 \times 10^{-19}$ J; Red: 6.63×10^{-34} J · s × $4.84 \times 10^{14}\ s^{-1} = 3.21 \times 10^{-19}$ J

YOU'RE THE CHEMIST

1. Sunlight has more red and more violet and lacks the distinct mercury emission lines of fluorescent light. Typical measurements are: Violet line to edge of red = 59 mm; 436 nm + (59 mm × 3.67 nm/mm) = 653 nm
Violet line to edge of violet = – 4 mm; 436 nm – (4 × 3.67) nm/mm = 421 nm
2. Translucent colored file tabs from an office supply store work well as filters. A yellow filter transmits green, yellow, orange and red and absorbs violet and blue. A green filter transmits blue, green and yellow and absorbs violet, orange and red. A blue filter transmits violet, blue and green and absorbs yellow, orange and red. A red filter transmits yellow, orange and red and absorbs violet, blue and green.
3. Answers will vary. Provide students with gas discharge tubes containing hydrogen and noble gases.

Evaluate

Review and Assessment Resources

CHEMISTRY TUTORIAL Encourage students to review writing electron configurations and calculating the wavelength of light using the animated step-by-step tutorials for Chapter 5. Each Chemistry Tutorial is followed by a similar practice problem, where the student is guided through the problem-solving process.

CHAPTER PROBLEM SET Have students practice more electron configuration problems using the Chapter 5 online Problem Set.

DIRECTED VIRTUAL LAB Have students complete the Virtual Lab *Flame Tests* to help reinforce the concept of emission spectra of elements. In the post-lab exercise, students will analyze the data and draw conclusions based on the results of the experiment.

KINETIC ART Direct students to the Kinetic Art animations *Electron Orbitals* and *The Hydrogen Emission Spectrum*, and have them answer the questions that accompany each animation.

Study Tip

SUMMARIZE Summarizing requires students to identify key ideas and state them briefly in their own words. Tell students that they will remember the content of an entire section better even if they summarize only a portion of the section.

5 Study Guide

BIGIDEA ELECTRONS AND THE STRUCTURE OF ATOMS

The quantum mechanical model of the atom comes from the solutions to the Schrödinger equation. Solutions to the Schrödinger equation give the energies an electron can have and the atomic orbitals, which describe the regions of space where an electron may be found. Electrons can absorb energy to move from one energy level to a higher energy level. When an electron moves from a higher energy level back down to a lower energy level, light is emitted.

5.1 Revising the Atomic Model

Bohr proposed that an electron is found only in specific circular paths, or orbits, around the nucleus.

The quantum mechanical model determines the allowed energies an electron can have and how likely it is to find the electron in various locations around the nucleus of an atom.

Each energy sublevel corresponds to one or more orbitals of different shapes. The orbitals describe where an electron is likely to be found.

- energy level (129)
- quantum (129)
- quantum mechanical model (130)
- atomic orbital (131)

5.2 Electron Arrangement in Atoms

Three rules—the aufbau principle, the Pauli exclusion principle, and Hund's rule—tell you how to find the electron configurations of atoms.

- electron configuration (134)
- aufbau principle (134)
- Pauli exclusion principle (134)
- spin (134)
- Hund's rule (134)

5.3 Atomic Emission Spectra and the Quantum Mechanical Model

When atoms absorb energy, their electrons move to higher energy levels. These electrons lose energy by emitting light when they return to lower energy levels.

To explain the photoelectric effect, Einstein proposed that light could be described as quanta of energy that behave as if they were particles.

The light emitted by an electron moving from a higher to a lower energy level has a frequency directly proportional to the energy change of the electron.

Classical mechanics adequately describes the motions of bodies much larger than atoms, while quantum mechanics describes the motions of subatomic particles and atoms as waves.

- amplitude (138)
- wavelength (138)
- frequency (138)
- hertz (138)
- electromagnetic radiation (139)
- spectrum (139)
- atomic emission spectrum (140)
- Planck's constant (143)
- photoelectric effect (143)
- photon (143)
- ground state (145)
- Heisenberg uncertainty principle (148)

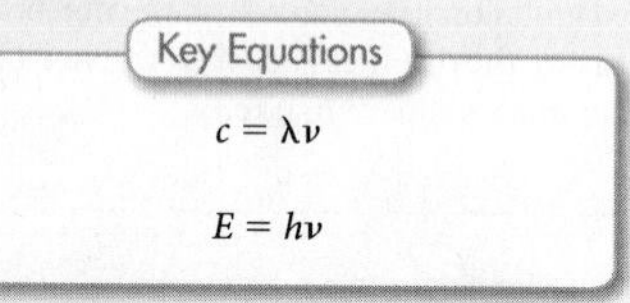

UbD Performance Tasks

MODEL MAKERS: Have students compare and contrast the three different electron models in a five paragraph essay. Students should include the scientist associated with each model, the reason why each model was accepted, and the reasons why a new model was needed.

SEEING THE LIGHT: In a computer slideshow presentation, explain the atomic emission spectrum. Then have students describe how they would make use of the spectrum to create either a colored light sign or a fireworks display.

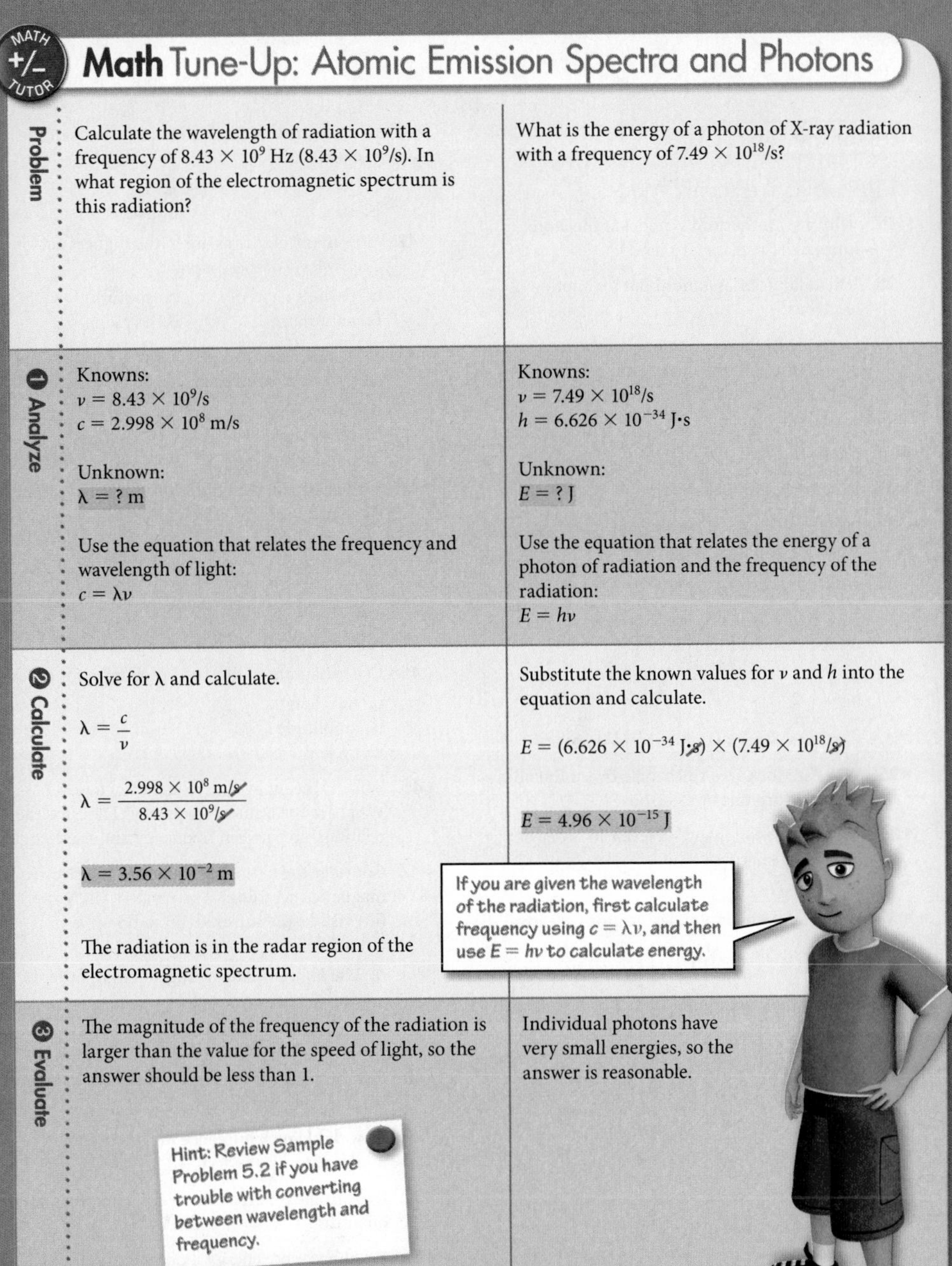

Math Tune-Up: Atomic Emission Spectra and Photons

Problem	Calculate the wavelength of radiation with a frequency of 8.43×10^9 Hz (8.43×10^9/s). In what region of the electromagnetic spectrum is this radiation?	What is the energy of a photon of X-ray radiation with a frequency of 7.49×10^{18}/s?
1 Analyze	Knowns: $\nu = 8.43 \times 10^9$/s $c = 2.998 \times 10^8$ m/s Unknown: $\lambda = ?$ m Use the equation that relates the frequency and wavelength of light: $c = \lambda\nu$	Knowns: $\nu = 7.49 \times 10^{18}$/s $h = 6.626 \times 10^{-34}$ J·s Unknown: $E = ?$ J Use the equation that relates the energy of a photon of radiation and the frequency of the radiation: $E = h\nu$
2 Calculate	Solve for λ and calculate. $\lambda = \frac{c}{\nu}$ $\lambda = \frac{2.998 \times 10^8 \text{ m/s}}{8.43 \times 10^9\text{/s}}$ $\lambda = 3.56 \times 10^{-2}$ m The radiation is in the radar region of the electromagnetic spectrum.	Substitute the known values for ν and h into the equation and calculate. $E = (6.626 \times 10^{-34} \text{ J·s}) \times (7.49 \times 10^{18}\text{/s})$ $E = 4.96 \times 10^{-15}$ J
3 Evaluate	The magnitude of the frequency of the radiation is larger than the value for the speed of light, so the answer should be less than 1.	Individual photons have very small energies, so the answer is reasonable.

Math Tune-Up 151

Math Review

Example and sample practice problems for the various electron problems can be found on the following pages:

LESSON 5.2 Writing Electron Configurations—page 136

LESSON 5.3 Calculating the Wavelength of Light—page 141; Calculating the Energy of a Photon—page 144

Focus on ELL

5 ASSESS UNDERSTANDING Preview the questions in the chapter assessment. Have students underline the words they don't understand. Review these words by referring to the word wall, bilingual glossary, or dictionary.

BEGINNING: LOW/HIGH Allow students to use their glossary during the assessment.

INTERMEDIATE

LOW Paraphrase or reread the assessment questions.

HIGH Encourage students to skip the questions they are unsure of and to go back to those once they have answered the questions they know.

ADVANCED: LOW/HIGH Encourage students to use good test taking strategies, like eliminating obvious distracters.

Evaluate

Answers

LESSON 5.1

27. He used existing ideas about the atom to propose that electrons move around the nucleus like the planets move around the sun.

28. that electrons traveled in circular paths around the nucleus

29. In Rutherford's model, negatively charged electrons surround a dense, positively charged nucleus. In Bohr's model, the electrons are assigned to concentric circular orbits of fixed energy.

30. An electron is found 90% of the time inside this boundary.

31. a region in space around the nucleus in which there is a high probability of finding an electron

32. The 1*s* orbital is spherical. The 2*s* orbital is spherical with a diameter larger than that of the 1*s* orbital. The three 2*p* orbitals are dumbbell shaped and oriented at right angles to each other.

33. 3

34. **a.** 1 **b.** 2 **c.** 3 **d.** 4

LESSON 5.2

35. Electrons occupy the lowest possible energy levels. An atomic orbital can hold at most two electrons. One electron occupies each of a set of orbitals with equal energies before any pairing of electrons occurs.

36. 2*s*, 3*p*, 4*s*, 3*d*

37. **b** and **c**

38. **a.** 2 **c.** 6 **e.** 6 **g.** 2
b. 2 **d.** 14 **f.** 10 **h.** 6

39. The *p* orbitals in the third quantum level have three electrons.

40. **a.** $1s^22s^22p^3$
b. $1s^22s^22p^5$
c. $1s^22s^22p^63s^2$
d. $1s^22s^22p^63s^23p^63d^{10}4s^24p^6$

41. **a.** $1s^22s^22p^63s^1$
b. $1s^22s^22p^63s^23p^64s^1$
c. $1s^22s^22p^63s^23p^63d^{10}4s^24p^64d^{10}5s^25p^5$
d. $1s^22s^22p^6$

42. **a.** 2 **b.** 3 **c.** 1 **d.** 6

43. **a.** 8 **b.** 8 **c.** 8

44. **a.** $1s^22s^22p^63s^23p^63d^{10}4s^24p^4$
b. $1s^22s^22p^63s^23p^63d^24s^2$
c. $1s^22s^22p^63s^23p^63d^34s^2$
d. $1s^22s^22p^63s^23p^64s^2$

5 Assessment

★ Solutions appear in Appendix E

Lesson by Lesson

5.1 Revising the Atomic Model

27. Why was Rutherford's model of the atom known as the planetary model?

★28. What did Bohr assume about the motion of electrons?

29. Describe Rutherford's model of the atom and compare it with the model proposed by his student Niels Bohr.

★30. What is the significance of the boundary of an electron cloud?

31. What is an atomic orbital?

32. Sketch 1*s*, 2*s*, and 2*p* orbitals using the same scale for each.

★33. How many orbitals are in the 2*p* sublevel?

★34. How many sublevels are contained in each of these principal energy levels?
a. $n = 1$ **c.** $n = 3$
b. $n = 2$ **d.** $n = 4$

5.2 Electron Arrangement in Atoms

★35. What are the three rules that govern the filling of atomic orbitals by electrons?

★36. Arrange the following sublevels in order of increasing energy:
3*d*, 2*s*, 4*s*, 3*p*.

★37. Which of these orbital designations are invalid?
a. 4*s* **c.** 3*f*
b. 2*d* **d.** 3*p*

38. What is the maximum number of electrons that can go into each of the following sublevels?
a. 2*s* **e.** 3*p*
b. 4*s* **f.** 3*d*
c. 4*p* **g.** 5*s*
d. 4*f* **h.** 5*p*

★39. What is meant by $3p^3$?

40. Write electron configurations for the elements that are identified by these atomic numbers:
a. 7 **c.** 12
b. 9 **d.** 36

41. Give electron configurations for atoms of these elements:
a. Na **c.** I
b. K **d.** Ne

★42. How many electrons are in the highest occupied energy level of these atoms?
a. barium **c.** sodium
b. aluminum **d.** oxygen

43. How many electrons are in the second energy level of an atom of each element?
a. chlorine
b. phosphorus
c. potassium

★44. Write electron configurations for atoms of these elements:
a. selenium **c.** vanadium
b. titanium **d.** calcium

5.3 Atomic Emission Spectra and the Quantum Mechanical Model

45. Use a diagram to illustrate each term for a wave.
a. wavelength
b. amplitude
c. cycle

46. What is meant by the frequency of a wave? What are the units of frequency? Describe the relationship between frequency and wavelength.

★47. Consider the following regions of the electromagnetic spectrum: (i) ultraviolet, (ii) X-ray, (iii) visible, (iv) infrared, (v) radio wave, (vi) microwave.
a. Use Figure 5.8 to arrange them in order of decreasing wavelength.
b. How does this order differ from that of decreasing frequency?

48. List the colors of the visible spectrum in order of increasing wavelength.

49. How did Planck influence the development of modern atomic theory?

★50. Explain the difference between a photon and a quantum.

★51. What has more energy, a photon of infrared light or a photon of ultraviolet light?

LESSON 5.3

45. diagrams similar to those in Figure 5.7

46. Frequency is the number of wave cycles that pass a given point per unit time. Frequency units are cycles/s or s^{-1} or Hertz. Wavelength and frequency are inversely related.

47. **a.** v, vi, iv, iii, i, ii
b. It is the reverse.

48. Violet, indigo, blue, green, yellow, orange, red

49. He showed mathematically that the amount of radiant energy (*E*) of a single quantum absorbed or emitted by a body is proportional to the frequency of radiation (ν).

50. A quantum is a discrete amount of energy. Photons are light quanta.

51. A photon of ultraviolet light has a higher frequency (smaller wavelength) than a photon of infrared light. Therefore, a photon of ultraviolet light has more energy than a photon of infrared light.

*52. What is the energy of a photon of green light with a frequency of 5.80×10^{14}/s?

*53. Explain the difference between the energy lost or gained by an atom according to the laws of classical physics and according to the quantum model of the atom.

*54. What happens when a hydrogen atom absorbs a quantum of energy?

55. The transition of electrons from higher energy levels to the $n = 2$ energy level results in the emission of light from hydrogen atoms. In what part of the spectrum is the emitted light, and what is the name given to this transition series?

Understand Concepts

56. Give the symbol for the atom that corresponds to each electron configuration.
   a. $1s^22s^22p^63s^23p^6$
   b. $1s^22s^22p^63s^23p^63d^{10}4s^24p^64d^75s^1$
   c. $1s^22s^22p^63s^23p^63d^{10}4s^24p^64d^{10}4f^75s^25p^65d^16s^2$

*57. Write the electron configuration for an arsenic atom. Calculate the total number of electrons in each energy level and state which energy levels are not full.

58. How many paired electrons are there in an atom of each element?
   a. helium
   b. sodium
   c. boron
   d. oxygen

*59. An atom of an element has two electrons in the first energy level and five electrons in the second energy level. Write the electron configuration for this atom and name the element. How many unpaired electrons does an atom of this element have?

*60. Give the symbols and names of the elements that correspond to these configurations of an atom.
   a. $1s^22s^22p^63s^1$
   b. $1s^22s^22p^3$
   c. $1s^22s^22p^63s^23p^2$
   d. $1s^22s^22p^4$
   e. $1s^22s^22p^63s^23p^64s^1$
   f. $1s^22s^22p^63s^23p^63d^24s^2$

61. What is the maximum number of electrons that can be found in any orbital of an atom?

62. Suppose your favorite AM radio station broadcasts at a frequency of 1150 kHz. What is the wavelength, in centimeters, of the radiation from the station?

*63. A mercury lamp, such as the one below, emits radiation with a wavelength of 4.36×10^{-7} m.

   a. What is the wavelength of this radiation in centimeters?
   b. In what region of the electromagnetic spectrum is this radiation?
   c. Calculate the frequency of this radiation.

64. Sodium vapor lamps are used to illuminate streets and highways. The very bright light emitted by these lamps is actually due to two closely spaced emission lines in the visible region of the electromagnetic spectrum. One of these lines has a wavelength of 5.890×10^{-7} m, and the other line has a wavelength of 5.896×10^{-7} m.
   a. What are the wavelengths of these radiations in centimeters?
   b. Calculate the frequencies of these radiations.
   c. In what region of the visible spectrum do these lines appear?

*65. What will happen if the following occur?
   a. Monochromatic light shining on cesium metal is just above the threshold frequency.
   b. The intensity of the light increases, but the frequency remains the same.
   c. Monochromatic light of a shorter wavelength is used.

*66. Calculate the energy of a photon of red light with a wavelength of 6.45×10^{-5} cm. Compare your answer with the answer to Question 52. Is red light of higher or lower energy than green light?

ASSESSMENT

Answers

52. 3.84×10^{-19} J

53. Classical physics views energy changes as continuous. In the quantum concept, energy changes occur in tiny discrete units called quanta.

54. The electron of the hydrogen atom is raised (excited) to a higher energy level.

55. visible spectrum, Balmer series

UNDERSTAND CONCEPTS

56. **a.** Ar **b.** Ru **c.** Gd

57. $1s^22s^22p^63s^23p^63d^{10}4s^24p^3$; level 1, 2; level 2, 8; level 3, 18; level 4, 5; The fourth energy level is not filled.

58. **a.** 2 **b.** 10 **c.** 4 **d.** 6

59. $1s^22s^22p^3$ nitrogen; 3

60. **a.** Na, sodium
b. N, nitrogen
c. Si, silicon
d. O, oxygen
e. K, potassium
f. Ti, titanium

61. 2

62. 2.61×10^4 cm

63. **a.** 4.36×10^{-5} cm **b.** visible **c.** $6.88 \times 10^{14} s^{-1}$

64. **a.** 5.890×10^{-5} cm and 5.896×10^{-5} cm
b. 5.090×10^{14} s^{-1} (Hz) and 5.085×10^{14} s^{-1} (Hz)
c. yellow

65. **a.** Electrons with a low velocity will be emitted.
b. More electrons will be emitted but with a low velocity.
c. Electrons will be emitted with a higher velocity.

66. 3.08×10^{-19} J; Red light is lower energy than green light.

ASSESSMENT

Evaluate

Answers

67. It is not possible to know both the position and the velocity of a particle at the same time.

68. The frequency is inversely proportional to the wavelength, so if the frequency increases by a factor of 1.5, the wavelength will decrease by a factor of 1.5.

69. emits: a; absorbs: b, c and d

70. The outermost electron of sodium absorbs photons of wavelength 589 nm as it jumps to a higher energy level.

71. 1.57×10^{-24} J

72. **a.** 1.7×10^{-21} J **b.** 3.9×10^{-19} J
c. 1.4×10^{-17} J; The energy of the photon of light increases as its wavelength decreases.

THINK CRITICALLY

73. An orbit confines the electron to a fixed circular path around the nucleus; an orbital is a region around the nucleus in which electrons are likely to be found.

74. **a.** fluorine **b.** germanium **c.** vanadium

75. Answers will vary. Some students may note that radio waves have the lowest energy in the electromagnetic spectrum, and thus would not be energetic enough to cook food. Others may reason that if microwaves cook food faster than infrared radiation, then radio waves would cook food even faster.

76. Answers will vary. The model of the atom is based on the abstract idea of probability. Light is considered a particle and a wave at the same time. Atoms and light cannot be compared to familiar objects or observations because humans cannot experience atoms or photons directly and because matter and energy behave differently at the atomic level than at the level humans can observe directly.

77. **a.** Electrons in $2p$ boxes should not be paired—there should be one electron in each.
b. Magnesium has 12 electrons. Two more electrons need to be added to $3s$.

78. **a.** $n = 1$ level
b. $n = 4$ level
c. $n = 4$ level
d. $n = 1$ level

79. **a.** potassium, excited state, valence electron has been promoted from $4s$ to $5p$
b. potassium, ground state, correct electron configuration
c. impossible configuration, $3p$ orbitals can hold a maximum of 6 electrons, not 7

80. The electrons obey Hund's rule.

67. State the Heisenberg uncertainty principle.

68. Describe how the wavelength of a wave changes if the frequency of the wave is multiplied by 1.5.

★69. Indicate whether each of the following electron transitions emits energy or requires the absorption of energy.

a. $3p$ to $3s$ **c.** $2s$ to $2p$
b. $3p$ to $4p$ **d.** $1s$ to $2s$

★70. White light is viewed in a spectroscope after passing through sodium vapor too cool to emit light. The spectrum is continuous except for a dark line at 589 nm. How can you explain this observation? (*Hint:* Recall from Sample Problem 5.2 that the atomic emission spectrum of sodium exhibits a strong yellow line at 589 nm.)

71. You use a microwave oven to heat your dinner. The frequency of the radiation is $2.37 \times 10^9\ s^{-1}$. What is the energy of one photon of this radiation?

★72. Calculate the following energies:

a. One photon of infrared radiation, if $\lambda = 1.2 \times 10^{-4}$ m.
b. One photon of visible radiation, if $\lambda = 5.1 \times 10^{-7}$ m.
c. One photon of ultraviolet radiation, if $\lambda = 1.4 \times 10^{-8}$ m.

What do the answers indicate about the relationship between the energy of light and its wavelength?

Think Critically

★73. Compare Explain the difference between an orbit in the Bohr model and an orbital in the quantum mechanical model of the atom.

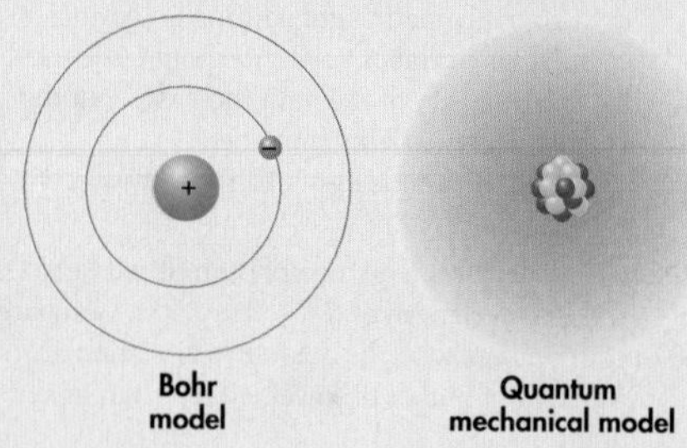

Bohr model | Quantum mechanical model

74. Apply Concepts Identify the elements whose electrically neutral atoms have the following electron configurations.

a. $1s^22s^22p^5$
b. $1s^22s^22p^63s^23p^63d^{10}4s^24p^2$
c. $1s^22s^22p^63s^23p^63d^34s^2$

★75. Predict Traditional cooking methods make use of infrared radiation (heat). Microwave radiation cooks food faster. Could radio waves be used for cooking? Explain.

76. Draw Conclusions Think about the currently accepted models of the atom and of light. In what ways do these models seem strange to you? Why are these models not exact or definite?

77. Evaluate and Revise Orbital diagrams for the ground states of two elements are shown below. Each diagram shows something that is incorrect. Identify the error in each diagram and then draw the correct diagram.

a. Nitrogen

$1s$	$2s$	$2p$
↑↓	↑↓	↑↓ ↑ ☐

b. Magnesium

$1s$	$2s$	$2p$	$3s$
↑↓	↑↓	↑↓ ↑↓ ↑↓	☐

★78. Infer Picture two hydrogen atoms. The electron in the first hydrogen atom is in the $n = 1$ level. The electron in the second atom is in the $n = 4$ level.

a. Which atom has the ground state electron configuration?
b. Which atom can emit electromagnetic radiation?
c. In which atom is the electron in a larger orbital?
d. Which atom has the lower energy?

★79. Infer Which of the following is the ground state of an atom? Which is its excited state? Which is an impossible electron configuration? Identify the element and briefly explain your choices.

a. $1s^22s^22p^63s^23p^65p^1$
b. $1s^22s^22p^63s^23p^64s^1$
c. $1s^22s^22p^63s^23p^7$

80. Relate Cause and Effect Why do electrons occupy equal energy orbitals singly before beginning to pair up?

Enrichment

★81. Graph The energy of a photon is related to its frequency and its wavelength.

Energy of photon (J)	Frequency (s^{-1})	Wavelength (cm)
3.45×10^{-21}	ν_1____	5.77×10^{-3}
2.92×10^{-20}	ν_2____	6.82×10^{-4}
6.29×10^{-20}	ν_3____	3.16×10^{-4}
1.13×10^{-19}	ν_4____	1.76×10^{-4}
1.46×10^{-19}	ν_5____	1.36×10^{-4}
3.11×10^{-19}	ν_6____	6.38×10^{-5}

a. Complete the table above.
b. Plot the energy of the photon (y-axis) versus the frequency (x-axis).
c. Determine the slope of the line.
d. What is the significance of this slope?

82. Calculate The average distance between Earth and Mars is about 2.08×10^8 km. How long would it take to transmit television pictures from Mars to Earth?

★83. Calculate Bohr's atomic theory can be used to calculate the energy required to remove an electron from an orbit of a hydrogen atom or an ion (an atom that has lost or gained electrons) containing only one electron. This number is the ionization energy for that atom or ion. The formula for determining the ionization energy (E) is

$$E = Z^2 \times \frac{k}{n^2}$$

where Z is the atomic number, k is 2.18×10^{-18} J, and n is the energy level. What is the energy required to eject an electron from a hydrogen atom when the electron is in the ground state ($n = 1$)? In the second energy level? How much energy is required to eject a ground state electron from the species Li^{2+} (a lithium atom that has lost two electrons)?

84. Draw Conclusions In a photoelectric experiment, a student shines light on the surface of a metal. The frequency of the light is greater than the threshold frequency of the metal. The student observes that after a long time, the maximum energy of the ejected electrons begins to decrease. Explain this observation.

Write About Science

★85. Explain Write a brief description of how trying to place two bar magnets pointing in the same direction alongside each other is like trying to place two electrons into the same orbital.

86. Connect to the BIGIDEA The late 1800s and early 1900s were significant times for the rapid development of chemistry. Bohr improved on Rutherford's model of the atom, then Schrödinger developed the quantum mechanical model of the atom. Explain why a model of the atom is crucial to understanding chemistry and in explaining the behavior of matter.

CHEMYSTERY

Now You See It... Now You Don't

Liam eventually realized that his star stickers would always stop glowing after a period of time. He discovered that he could "recharge" the stickers by turning on the lights. After he turned off the lights, the stars would glow again. However, after a few hours, the stars would eventually stop glowing.

Glow-in-the-dark objects contain compounds that react with light. When these objects are exposed to light, the electrons in the compounds absorb energy and become excited. As the electrons drop back down to a lower energy level, they emit light. This process, called phosphorescence, occurs more slowly in the compounds contained in glow-in-the-dark objects than in other compounds.

87. Infer Do Liam's glow-in-the-dark stars glow when the lights are on? Explain.

★88. Connect to the BIGIDEA Light emitted from an incandescent light bulb is in the visible region of the electromagnetic spectrum (300 nm to 700 nm). What does this information tell you about the energy of the photons absorbed by the electrons in glow-in-the dark objects?

CHEMISTRY & YOU

SUMMARIZE After students have read through the CHEMystery, call on volunteers to summarize how the behavior of electrons helps explain the glow-in-the-dark stars in Liam's bedroom. **Ask** Why do the stars glow in the dark for only a short period of time? **Answer:** *(The stars glow after they have been exposed to light because the atoms absorbed energy. This resulted in the electrons moving into higher energy levels. When it is dark, the electrons lose energy and return to a lower energy level, resulting in a gradual fading of the light.)* **Ask** What caused the stars to start glowing again? *(The light in the room was turned on which allowed the atoms in the stars to absorb energy. The electrons moved into higher energy levels, resulting in the star's ability to glow in the dark once again.)*

CHEMYSTERY ANSWERS

87. No. As long as the lights are on the atoms in the stars are absorbing energy, and the electrons remain in their higher energy levels. Once the lights are turned off, the atoms are no longer absorbing energy and the electrons can drop to lower energy levels, emitting light in the process.

88. BIGIDEA The light emitted from an incandescent bulb has wavelengths from 300 nm to 700 nm, which corresponds to a frequency range of about 4×10^{14} s^{-1} to 1×10^{15} s^{-1}. This means that the energy absorbed by the photons is in the range of about 3×10^{-19} J to 7×10^{-19} J.

Answers

ENRICHMENT

81. a. Frequency(s^{-1}): 5.20×10^{12}; 4.40×10^{13}; 9.49×10^{13}; 1.70×10^{14}; 2.20×10^{14}; 4.70×10^{14}

b.

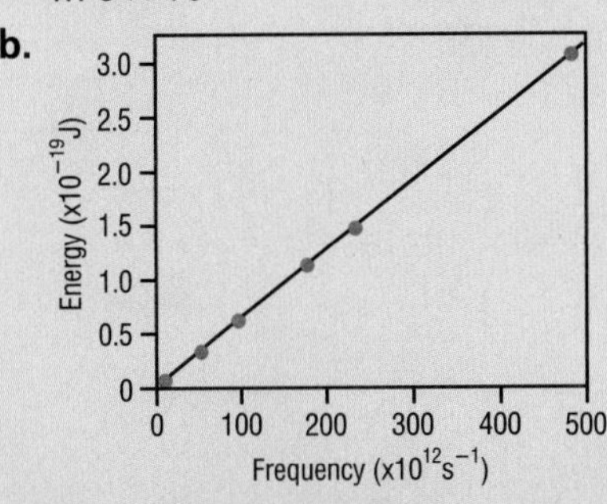

c. 6.63×10^{-34} joule • second.

d. The slope is Planck's constant.

82. 6.94×10^2 s

83. Hydrogen atom ($Z = 1$), $n = 1$: 2.18×10^{-18} J; Hydrogen atom ($Z = 1$), $n = 2$: 5.45×10^{-19} J; Li^{2+} ion ($Z = 3$), $n = 1$: 1.96×10^{-17} J

84. Answers will vary but should reflect an understanding that the maximum energy of an ejected electron is directly proportional to frequency.

WRITE ABOUT SCIENCE

85. Two magnets would push each other apart. In the same way, electrons with the same spin would push apart and be unable to occupy the same orbital.

86. Answers will vary but should reflect an understanding that atomic structure determines the chemical and physical properties of matter, and that these properties determine how matter behaves.

Evaluate

Answers

CUMULATIVE REVIEW

89. **a** and **b** are heterogeneous; **c** is homogeneous

90. Answers will vary but could include: water is lost as steam and burned meat gives off carbon dioxide.

91. A compound has constant composition; the composition of a mixture can vary.

92. a heterogeneous mixture

93. 7.7×10^{-5} μm

94. 18.9 cm^3

95. the piece of lead

96. **a.** 3.9×10^{-5} kg
b. 7.84×10^{2} L
c. 8.30×10^{-2} g
d. 9.7×10^{6} ng

97. **a** and **b** are exact.

98. Mass remains the same; weight decreases because gravity on the moon is less than gravity on Earth.

99. 8.92 g/cm^3

100. 154 g, 1.54×10^{-1} kg

101. Helium gas is less dense than air.

102. Accuracy is a measure of how close the value is to the true value; precision is a measure of how close a series of measurements are to one another.

103. **a.** 55 protons, 55 electrons
b. 47 protons, 47 electrons
c. 48 protons, 48 electrons
d. 34 protons, 34 electrons

104. a

105. Neon-20 has 10 neutrons; neon-21 has 11 neutrons.

106. The value 35.453 amu is a weighted average. Its calculation is based on the percentage natural abundance of two isotopes, chlorine-35 and chlorine-37.

Cumulative Review

*__89.__ Classify each of the following as homogeneous or heterogeneous:
a. a page of this textbook
b. a banana split
c. the water in bottled water

90. Hamburger undergoes a chemical change when cooked on a grill. All chemical changes are subject to the law of conservation of mass. Yet, a cooked hamburger will weigh less than the uncooked meat patty. Explain.

*__91.__ Homogeneous mixtures and compounds are both composed of two or more elements. How do you distinguish between a homogeneous mixture and a compound?

92. The photo shows a magnified view of a piece of granite. Is granite a substance or a mixture?

*__93.__ The diameter of a carbon atom is 77 pm. Express this measurement in μm.

94. A silver bar has a mass of 368 g. What is the volume, in cm^3, of the bar? The density of silver is 19.5 g/cm^3.

*__95.__ Which has more mass, a 28.0-cm^3 piece of lead or a 16.0-cm^3 piece of gold? The density of lead is 11.3 g/cm^3; the density of gold is 19.3 g/cm^3.

96. Express the following measurements in scientific notation.
a. 0.000039 kg
b. 784 L
c. 0.0830 g
d. 9,700,000 ng

97. Which of these quantities or relationships are exact?
a. 10 cm = 1 dm
b. There are 9 baseball players on the field.
c. A diamond has a mass of 12.4 g.
d. The temperature is 21°C.

98. A one-kilogram steel bar is brought to the moon. How are its mass and its weight each affected by this change in location? Explain.

*__99.__ When a piece of copper with a mass of 36.4 g is placed into a graduated cylinder containing 20.00 mL of water, the water level rises to 24.08 mL, completely covering the copper. What is the density of copper?

100. The density of gold is 19.3 g/cm^3. What is the mass, in grams, of a cube of gold that is 2.00 cm on each edge? In kilograms?

*__101.__ A balloon filled with helium will rise upward when released. What does this result show about the relative densities of helium and air?

*__102.__ Explain the difference between the accuracy of a measurement and the precision of a measurement.

103. Give the number of protons and electrons in each of the following:
a. Cs
b. Ag
c. Cd
d. Se

104. Which of these was an essential part of Dalton's atomic model?
a. indivisible atoms
b. electrons
c. atomic nuclei
d. neutrons

*__105.__ How do neon-20 and neon-21 differ from each other?

106. The mass of an atom should be very nearly the sum of the masses of its protons and neutrons. The mass of a proton and the mass of a neutron are each very close to 1 amu. Why is the atomic mass of chlorine, 35.453 amu, so far from a whole number?

If You Have Trouble With . . .

Question	89	90	91	92	93	94	95	96	97	98	99	100	101	102	103	104	105	106
See Chapter	2	2	2	2	3	3	3	3	3	3	3	3	3	3	4	4	4	4

Standardized Test Prep for Indiana

Standard: C.2.2

Select the choice that best answers each question or completes each statement.

Tips for Success

Eliminate Wrong Answers If you don't know which response is correct, start by eliminating those you know are wrong. If you can rule out some choices, you'll have fewer left to consider and you'll increase your chances of choosing the correct answer.

1. Select the correct electron configuration for silicon, atomic number 14.
   (A) $1s^22s^22p^23s^23p^23d^24s^2$
   (B) $1s^22s^22p^43s^23p^4$
   (C) $1s^22s^62p^6$
   (D) $1s^22s^22p^63s^23p^2$

2. Which two orbitals have the same shape?
   (A) 2*s* and 2*p*
   (B) 2*s* and 3*s*
   (C) 3*p* and 3*d*
   (D) More than one is correct.

3. Which of these statements characterize the nucleus of every atom?
   I. It has a positive charge.
   II. It is very dense.
   III. It is composed of protons, electrons, and neutrons.
   (A) I and II only
   (B) II and III only
   (C) I and III only
   (D) I, II, and III

4. As the wavelength of light increases,
   (A) the frequency increases.
   (B) the speed of light increases.
   (C) the energy decreases.
   (D) the intensity increases.

5. In the third energy level of an atom,
   (A) there are two energy sublevels.
   (B) the *f* sublevel has 7 orbitals.
   (C) there are three *s* orbitals.
   (D) a maximum of 18 electrons are allowed.

The lettered choices below refer to Questions 6–10. A lettered choice may be used once, more than once, or not at all.

(A) s^2p^6 (B) s^2p^2 (C) s^2 (D) s^4p^1 (E) s^2p^4

Which configuration is the configuration of the highest occupied energy level for each of these elements?

6. sulfur
7. germanium
8. beryllium
9. krypton
10. strontium

Use the drawings to answer Questions 11–14. Each drawing represents an electromagnetic wave.

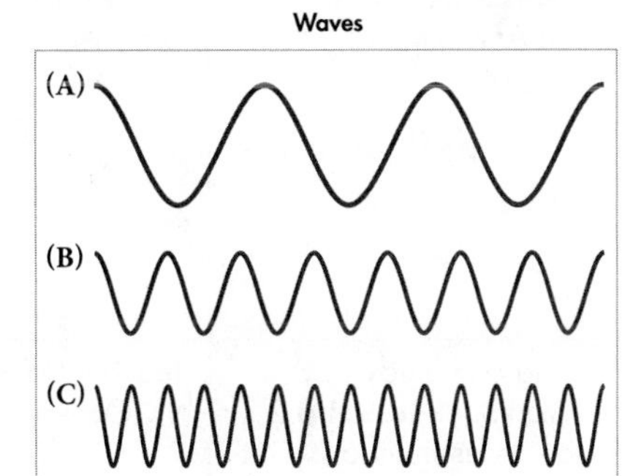

11. Which wave has the longest wavelength?
12. Which wave has the highest energy?
13. Which wave has the lowest frequency?
14. Which wave has the highest amplitude?

Write a short essay to answer Question 15.

15. Explain the rules that determine how electrons are arranged around the nuclei of atoms.

If You Have Trouble With . . .

Question	1	2	3	4	5	6	7	8	9	10	11	12	13	14	15
See Lesson	5.2	5.1	5.1	5.3	5.1	5.2	5.2	5.2	5.2	5.2	5.3	5.3	5.3	5.3	5.2

STP Answers

1. D
2. B
3. A
4. C
5. D
6. E
7. B
8. C
9. A
10. C
11. A
12. C
13. A
14. A
15. According to the aufbau principle, electrons enter orbitals of lowest energy first. According to the Pauli exclusion principle, an orbital may contain at most two electrons. According to Hund's rule, one electron will enter each orbital of equal energy before electrons begin to pair up.

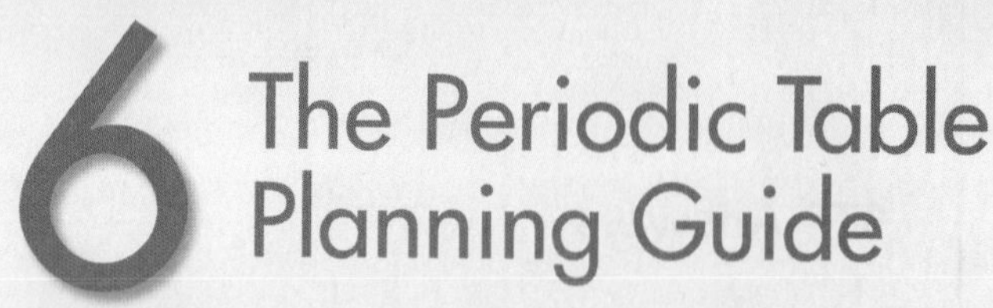

6 The Periodic Table Planning Guide

Introducing the BIGIDEA: ELECTRONS AND THE STRUCTURE OF ATOMS

Elements are composed of particles called atoms, and every atom contains a nucleus and one or more electrons.

IASS	Lessons and Objectives	Print Resources	
		For the Student	For the Teacher
C.2.6	**6.1** Organizing the Elements p 160–166 **6.1.1** Explain how chemists began to organize the known elements. **6.1.2** Describe how Mendeleev organized his periodic table. **6.1.3** Describe how the modern periodic table is organized. **6.1.4** Identify three broad classes of elements.	Reading and Study Workbook Lesson 6.1 Lesson Assessment 6.1 p 166	Teacher Demo, p 161: Organizing Elements Class Activity, p 165: Name the Element
C.2.5, C.2.6	**6.2** Classifying the Elements p 167–173 **6.2.1** List the types of information that can be displayed in a periodic table. **6.2.2** Classify elements based on electron configuration.	Reading and Study Workbook Lesson 6.2 Lesson Assessment 6.2 p 173	Teacher Demo, p 171: Observing Differences
C.2.6	**6.3** Periodic Trends p 174–182 **6.3.1** Describe trends among elements for atomic size. **6.3.2** Explain how ions form. **6.3.3** Describe periodic trends for first ionization energy, ionic size, and electronegativity.	Reading and Study Workbook Lesson 6.3 Lesson Assessment 6.3 p 182 Quick Lab: Periodic Trends in Ionic Radii p 180 Small-Scale Lab: Periodicity in Three Dimensions p 184	Class Activity, p 176: Listing Elements Class Activity, p 177: Effective Nuclear Charge and Electron Shielding Teacher Demo, p 178: Predicting Reactivity Teacher Demo, p 179: Trends in Ionic Size

Assessing the BIGIDEA: ELECTRONS AND THE STRUCTURE OF ATOMS

	Essential Questions 1. What information does the periodic table provide? 2. How can periodic trends be explained?	Study Guide p 185 STP p 191 Reading and Study Workbook Self-Check and Vocabulary Review Chapter 6	

	PearsonChem.com	
	Editable Worksheets	Online Activities
	Teaching Resources, Lesson 6.1 Review	6.1 Lesson Overview Heat Energy and Metals The Periodic Table of Musical Instruments
	Teaching Resources, Lesson 6.2 Review	6.2 Lesson Overview Periodic Table Tour Using Energy Sublevels to Write Electron Configurations
	Teaching Resources, Lesson 6.3 Review Lab 9: Periodic Properties Small-Scale Lab Manual Lab 9: A Periodic Table Logic Problem Lab Practical 6-1: Density and the Periodic Table Lab Practical 6-2: Solubility and the Periodic Table	6.3 Lesson Overview Trends in the Periodic Table
	Core Teaching Resources (includes editable worksheets) • Lesson Reviews • Practice Problems • Interpret Graphs • Vocabulary Review • Chapter Quizzes and Tests • Lab Record Sheets	Chapter 6 Problem Set Trend Lines

Materials List

For the Student

Quick Lab p 180
- graph paper

Small-Scale Lab p 184
- 96-well spot plate
- 20 straws
- scissors
- metric ruler
- permanent fine-line marker

For the Teacher

Teacher Demo p 161
- samples of elements
- photographs of elements

Teacher Demo p 171
- 0.2*M* HCl
- 6 large test tubes
- test tube rack
- small pieces of clean magnesium, tin, and copper
- matches
- wood splint

Teacher Demo p 178
- 20 mL 1*M* HCl
- two 50-mL beakers
- overhead projector
- 20 cm of magnesium ribbon
- 1 g of calcium

Teacher Demo p 179
- washers or other small circular items
- smaller item to represent nucleus
- overhead projector

Additional Digital Resources

Online Student Edition
Online Teacher's Edition
Exam*View Assessment Suite CD-ROM*
Classroom Resources DVD

CHAPTER 6

What's Online

CHEMISTRY TUTORIAL Students can work step-by-step with the online tutorial to write electron configurations by accessing the Chapter 6 Chemistry Tutorial.

ONLINE PROBLEM SET Additional practice problems about the periodic table can be found in the Chapter 6 online Problem Set.

MATH TUTORIAL The online Math Tutorial will help students understand trend lines.

DIRECTED VIRTUAL LAB Encourage students to use the animated virtual lab for Chapter 6 to learn about heat capacities of metals. The animation is followed by guided analysis questions.

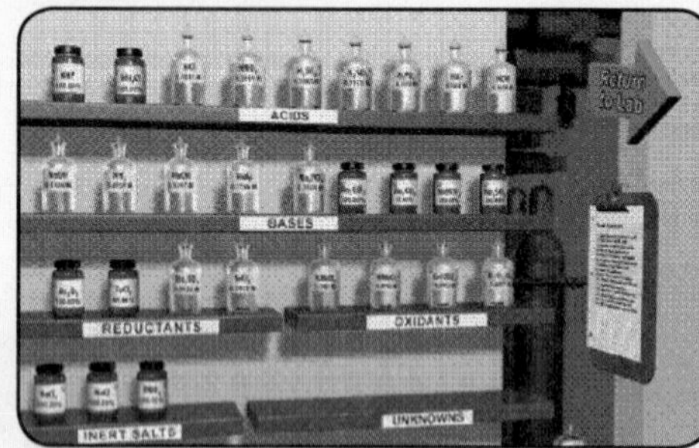

KINETIC ART Have students watch animations about the organization of the periodic table and properties of different groups. Each animation is followed by questions.

CONCEPTS IN ACTION This interactive animation compares the organization of the periodic table to the organization of instruments of an orchestra. The animation is followed by questions about periodic table organization.

IN ACADEMIC STANDARDS for SCIENCE

Nature of Science

- Use analogies and models (mathematical and physical) to simplify and represent systems that are difficult to understand or directly experience due to their size, time scale, or complexity, and recognize the limitations of analogies and models.
- Focus on the development of explanatory models based on their observations during laboratory investigations.

Also covered C.2.5, C.2.6.

6 The Periodic Table

INSIDE:

- **6.1** Organizing the Elements
- **6.2** Classifying the Elements
- **6.3** Periodic Trends

PearsonChem.com

These beads are organized by color and shape. In the periodic table, elements are organized into groups with similar properties.

158

Focus on ELL

1 CONTENT AND LANGUAGE Ask students to create a KWL chart. Have them complete the first column by writing everything they already know about the periodic table. Then have them use the Essential Questions and the CHEMystery to identify what they want to know. At the end of each lesson, have them identify what they learned.

BEGINNING: LOW/HIGH Allow students to work in pairs and use pictures or symbols to represent what they already know and make lists for what they want to know.

INTERMEDIATE: LOW/HIGH Encourage students to copy the information from the essential questions onto their chart.

ADVANCED: LOW/HIGH Tell students to read the CHEMystery feature and determine at least one thing that they want to learn from the reading.

BIGIDEA

ELECTRONS AND THE STRUCTURE OF ATOMS

Essential Questions:

1. *What information does the periodic table provide?*
2. *How can periodic trends be explained?*

CHEMYSTERY

Made in the USA

Only 90 elements in the periodic table are known to occur naturally. You might ask then, "Why are there more than a hundred elements listed in the periodic table?"

Between the years of 1940 and 1958, nine of these "unnatural" elements were discovered by the American scientist Glenn Seaborg and his colleagues at the University of California at Berkeley. Later in 1974, Seaborg and his team discovered yet another element. Several other "unnatural" elements were discovered by Russian and German scientists.

How do these elements compare to other elements in the periodic table?

▶ Connect to the BIGIDEA As you read about the periodic table, keep an eye out for elements that do not occur naturally.

INDIANA'S ACADEMIC STANDARDS FOR SCIENCE

Solutions C.2.5, C.2.6. See lessons for details.

Introduce the Chapter

IDENTIFYING PRECONCEPTIONS Most students have probably heard of the periodic table, but they may have a limited understanding of the information it contains. Use the activity below to assess students' level of knowledge about the periodic table.

Activity Divide the class into small groups. Provide each group with a sheet of paper on which the names of twenty of the most familiar elements of the periodic table are printed. Challenge students to organize the list into a table using any criteria they prefer. Each group should then present and explain the organization of their list to the class. Afterward, describe to students some of the ways they will learn in this chapter that the periodic table is organized and is useful.

CHAPTER 6

UbD

Understanding by Design

Students are building toward understanding the periodic table by applying a knowledge of electrons and the structure of atoms.

PERFORMANCE GOALS At the end of Chapter 6, students will be able to answer the essential questions by applying their knowledge of the periodic table.

ESSENTIAL QUESTIONS Read the essential questions aloud. **Ask** What are some ways that elements are different from each other? *(Their atoms have different numbers of protons, and they have different electron configurations.)*

BIGIDEA Use the photo of sorted beads to help students connect to this chapter's concepts. Activate prior knowledge by discussing with students the benefits of sorting things. Have them imagine how much harder it would be to find one of the beads in the photo if they were not sorted. Then point out that there are so many elements with different properties, sorting them helps you understand how they are alike and different.

CHEMYSTERY Have students read over the CHEMystery. Connect the CHEMystery to the Big Idea by explaining that the "unnatural" elements were discovered because scientists were able to predict their existence by studying the atomic structure of known elements. Encourage students to predict how the "unnatural" elements are different from the other elements. As a hint, suggest that they consider what property an element might have that would make it unlikely to occur in nature.

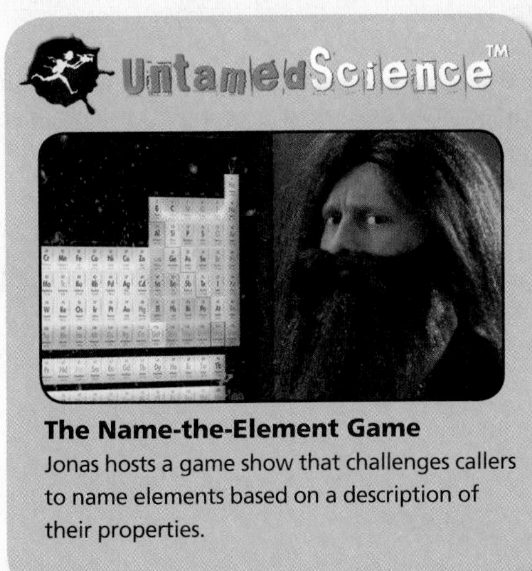

The Name-the-Element Game

Jonas hosts a game show that challenges callers to name elements based on a description of their properties.

Key Objectives

6.1.1 EXPLAIN how chemists began to organize the known elements.

6.1.2 DESCRIBE how Mendeleev organized his periodic table.

6.1.3 DESCRIBE how the modern periodic table is organized.

6.1.4 IDENTIFY three broad classes of elements.

Additional Resources

Reading and Study Workbook, Lesson 6.1

Available Online or on Digital Media:

- Teaching Resources, Lesson 6.1 Review

PearsonChem.com

Engage

CHEMISTRY & YOU Have students look at the opening photograph and read the opening text. Have students list the types of characteristics that might be used to organize a group of cards as discussed in the text. Discuss as a class how these categories can then be subdivided.

Access Prior Knowledge

Have students recall what makes one element different from another. *(They contain different numbers of protons.)* **Ask** What is the term used to identify the number of protons in an element? *(atomic number)* Have students keep this in mind as they read this section.

IN ACADEMIC STANDARDS for SCIENCE

C.2.6 Use the periodic table and electron configurations to determine an element's number of valence electrons and chemical and physical properties.

6.1 Organizing the Elements

C.2.6 Determine valence electrons and properties

CHEMISTRY & YOU

Q: *How can you organize and classify elements?* If you have ever played a card game, then you have probably organized your cards. Maybe you classified them by color or number. Elements can also be classified. In this lesson, you will learn how elements are arranged in the periodic table and what that arrangement reveals about the elements.

Key Questions

How did chemists begin to organize the known elements?

How did Mendeleev organize his periodic table?

How is the modern periodic table organized?

What are three broad classes of elements?

Vocabulary

- periodic law
- metal
- nonmetal
- metalloid

Searching for an Organizing Principle

How did chemists begin to organize the known elements?

A few elements, including copper, silver, and gold, have been known for thousands of years. Yet, there were only 13 elements identified by the year 1700. Chemists suspected that other elements existed. They had even assigned names to some of these elements, but they were unable to isolate the elements from their compounds. As chemists began to use scientific methods to search for elements, the rate of discovery increased. In one decade (1765–1775), chemists identified five new elements, including three colorless gases, hydrogen, nitrogen, and oxygen. Was there a limit to the number of elements? How would chemists know when they had discovered all the elements? To begin to answer these questions, chemists needed to find a logical way to organize the elements.

Early chemists used the properties of elements to sort them into groups. In 1829, a German chemist, J. W. Dobereiner (1780–1849), published a classification system. In his system, the known elements were grouped into triads. A triad is a set of three elements with similar properties. The elements in Figure 6.1 formed one triad. Chlorine, bromine, and iodine may look different, but they have very similar chemical properties. For example, they react easily with metals. Unfortunately, all the known elements could not be grouped into triads.

Figure 6.1 Triad in Dobereiner's System
Chlorine, bromine, and iodine formed one triad. These elements have similar chemical properties.

Dobereiner noted a pattern in his triads. One element in each triad tended to have properties with values that fell midway between those of the other two elements. For example, the average of the atomic masses of chlorine and iodine is [(35.453 1 + 126.90)/2] or 81.18 amu. This value is close to the atomic mass of bromine, which is 79.904 amu.

Focus on ELL

1 CONTENT AND LANGUAGE Have students create a four-column chart. In the first column, have them copy the key questions and the vocabulary terms. Have students title the second column *Answer/Definition*, the third column, *Where can I find it?*, and the fourth column, *Do I know it?* As you proceed through the lesson, encourage students to fill in columns two and three. At the end of the lesson, ask students to self-assess their knowledge by filling in the fourth column; use this column to determine whether there is a need for reteaching.

2 FRONTLOAD THE LESSON Have students consider how information is organized on their MP3 players. For example, ask students how their playlists are organized and what information is available about each song. Explain that the periodic table lists all the elements by atomic number but that the resulting arrangement of the table groups the elements in different ways. Have students keep this in mind as they proceed with the lesson.

но въ ней, мнѣ кажется, уже ясно выражается примѣнимость выставляемаго мною начала ко всей совокупности элементовъ, пай которыхъ извѣстенъ съ достовѣрностію. На этотъ разъ я и желалъ преимущественно найдти общую систему элементовъ. Вотъ этотъ опытъ:

			Ti=50	Zr=90	?=180.
			V=51	Nb=94	Ta=182.
			Cr=52	Mo=96	W=186.
			Mn=55	Rh=104,4	Pt=197,4
			Fe=56	Ru=104,4	Ir=198.
			Ni=Co=59	Pl=106,6	Os=199.
H=1			Cu=63,4	Ag=108	Hg=200.
	Be=9,4	Mg=24	Zn=65,2	Cd=112	
	B=11	Al=27,4	?=68	Ur=116	Au=197?
	C=12	Si=28	?=70	Sn=118	
	N=14	P=31	As=75	Sb=122	Bi=210
	O=16	S=32	Se=79,4	Te=128?	
	F=19	Cl=35,5	Br=80	I=127	
Li=7	Na=23	K=39	Rb=85,4	Cs=133	Tl=204
		Ca=40	Sr=87,6	Ba=137	Pb=207.
		?=45	Ce=92		
		?Er=56	La=94		
		?Yt=60	Di=95		
		?In=75,6	Th=118?		

Figure 6.2
Mendeleev's Periodic Table
In this early version of the periodic table, Mendeleev (shown on the stamp below) arranged elements with similar properties in the same row.
Identify ***Which element is grouped with chlorine (Cl), bromine (Br), and (I) iodine?***

Mendeleev's Periodic Table

How did Mendeleev organize his periodic table?

From 1829 to 1869, different systems for organizing the elements were proposed, but none of them gained wide acceptance. In 1869, a Russian chemist and teacher, Dmitri Mendeleev, published a table of the elements. Later that year, a German chemist, Lothar Meyer, published a nearly identical table. Mendeleev was given more credit than Meyer because he published his table first and because he was better able to explain its usefulness.

Mendeleev developed his table while working on a textbook for his students. He needed a way to show the relationships among more than 60 elements. He wrote the properties of each element on a separate note card. This approach allowed him to move the cards around until he found an organization that worked. The organization he chose was a periodic table. Elements in a periodic table are arranged into groups based on a set of repeating properties. **Mendeleev arranged the elements in his periodic table in order of increasing atomic mass.**

Figure 6.2 is an early version of Mendeleev's periodic table. Look at the column that starts with Ti = 50. Notice the two question marks between the entries for zinc (Zn) and arsenic (As). Mendeleev left these spaces in his table because he knew that bromine belonged with chlorine and iodine. He predicted that elements would be discovered to fill those spaces, and he predicted what their properties would be based on their locations in the table. The elements between zinc and arsenic were gallium and germanium, which were discovered in 1875 and 1886, respectively. There was a close match between the predicted properties and the actual properties of these elements. This match helped convince scientists that Mendeleev's periodic table was a powerful tool.

Three-Dimensional Periodic Attempts

A table of atomic weights, along with letters to symbolize all known elements, was proposed in 1828, by Jöns Jakob Berzelius. In 1862, French geologist A. E. Beguyer de Chancourtois made use of the relationship between the elements and their atomic weights by representing their relationship on a geometric model. The model involved a listing of elements wrapped around a cylinder in specified intervals so that similar elements lined up. Although the model contained flaws, de Chancourtois is one of the first scientists to visually realize a recurring pattern among elemental properties.

Foundations for Reading

BUILD VOCABULARY Point out that *metal, nonmetal,* and *metalloid* all have the same root. Discuss the meanings of the suffix *-oid* (resembling) and prefix *non-* (not).

READING STRATEGY Have students write down the following terms at the top of a sheet of paper: *groups, periods, periodic law, periodic table, repeating properties, metals, nonmetals,* and *metalloids.* Instruct students to generate a concept map relating these terms as they progress through the chapter.

Explore

Searching for an Organizing Principle

Teacher Demo

PURPOSE To provide students with an understanding of how elements are organized.

MATERIALS samples of elements, or photographs of elements

PROCEDURE Display the samples of the elements or the photographs the elements. Ask students to use properties of the elements to organize them and to explain the organization.

Explain

Mendeleev's Periodic Table

USE VISUALS Have students study Mendeleev's version of the periodic table in Figure 6.2. Show students how the elements with similar chemical properties were arranged in rows. Point out the gaps in the table where Mendeleev left room for elements that had not yet been discovered.

APPLY CONCEPTS Emphasize that Mendeleev did not make the first periodic table, but he did make the first table that gained wide acceptance. Explain that these tables were true periodic tables since they arranged the elements according to periodicity of their chemical and physical properties. Bring awareness to the fact that protons are used to define elements, but at the point in time when these tables were created, the proton had yet to be discovered.

Answers

FIGURE 6.2 F

LESSON 6.1

Explain

Today's Periodic Table

USE VISUALS Direct students' attention to Figure 6.3. This is the first of three periodic tables in the chapter. The information in this first table is limited so that students can see how all the elements fit in the table and can focus on atomic number as an organizing principle.

Have students compare Figures 6.2 and 6.3. **Ask** How is the modern periodic table shown here similar to Mendeleev's periodic table? *(In both tables, elements are arranged in rows and columns. Mendeleev's periodic table ordered the elements according to atomic mass. Modern periodic tables order elements according to atomic number.)* **Ask** What does an element's position in the table indicate about its properties? *(The properties of an element are similar to those of other elements in the same group.)*

INTERPRET DIAGRAMS Display Figure 6.3 on an overhead. To emphasize the concept that elements in the same group, or column, of the periodic table have similar properties, isolate one group with two sheets of paper. Describe some physical and chemical properties of the elements in that group. Then, point to the group on the far right, and explain that all of these elements are gases that do not normally form compounds. Have students compare these elements to those in other groups.

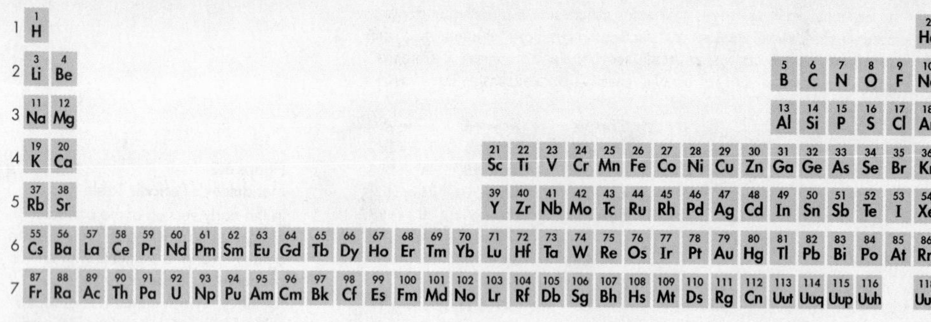

Figure 6.3 Modern Periodic Table
In the modern periodic table, the elements are arranged in order of increasing atomic number.
Interpret Diagrams *How many elements are in the second period?*

Today's Periodic Table

How is the modern periodic table organized?

The atomic mass of iodine (I) is 126.90. The atomic mass of tellurium (Te) is 127.60. In a periodic table based on atomic mass, iodine should come before tellurium since iodine has a smaller atomic mass than tellurium does. However, based on its chemical properties, iodine belongs in a group with bromine and chlorine. Mendeleev broke his rule about placing elements in strict order of atomic mass and placed tellurium before iodine in his periodic table. He assumed that the atomic masses for iodine and tellurium were incorrect, but they were not. A similar problem occurred with other pairs of elements. The problem wasn't with the atomic masses but with using atomic mass to organize the periodic table.

Mendeleev developed his table before scientists knew about the structure of atoms. He didn't know that the atoms of each element contain a unique number of protons. Recall that the number of protons is the atomic number. In 1913, a British physicist, Henry Moseley, determined an atomic number for each known element. Tellurium's atomic number is 52 and iodine's is 53, so it makes sense for iodine to come after tellurium in the periodic table. **In the modern periodic table, elements are arranged in order of increasing atomic number.**

READING SUPPORT

Build Vocabulary: *Word Origins* *Periodic* comes from the Greek roots *peri*, meaning "around" and *hodos*, meaning "path." In a periodic table, properties repeat from left to right across each period. ***If the Greek word metron means "measure," what does perimeter mean?***

The elements in Figure 6.3 are arranged in order of atomic number, starting with hydrogen, which has atomic number 1. There are seven rows, or periods, in the table. Period 1 has 2 elements, Period 2 has 8 elements, Period 4 has 18 elements, and Period 6 has 32 elements. Each period corresponds to a principal energy level. There are more elements in higher numbered periods because there are more orbitals in higher energy levels. (Recall the rules you studied in Chapter 5 for how electrons fill orbitals.)

The properties of the elements within a period change as you move across a period from left to right. However, the pattern of properties within a period repeats as you move from one period to the next. This pattern gives rise to the **periodic law:** When elements are arranged in order of increasing atomic number, there is a periodic repetition of their physical and chemical properties. The arrangement of the elements into periods has an important consequence. Elements that have similar chemical and physical properties end up in the same column in the periodic table.

Differentiated Instruction

LPR LESS PROFICIENT READERS Using a large poster-size display of the periodic table, trace one period across the table. On the display, note some physical and chemical properties of each element in that period. Then have students research the properties of each element in the period below the one you just highlighted and write a comparison statement comparing the properties of both periods.

ELL ENGLISH LANGUAGE LEARNERS To help students understand the term "group" on the periodic table, assign the term the down-arrow symbol. Convey that a *group* of elements means the same as a *column* of elements. To help students understand the term "period" on the periodic table, assign the term the right-arrow symbol. Convey that a *period* of elements means the same as a *row* of elements. Other words such as *vertical* and *horizontal* can also be used in place of *group* and *period*.

L3 STRUGGLING STUDENTS Review Lesson 4.3 to help students understand how the atomic numbers and the atomic mass of elements are important to the modern periodic table.

CHEMISTRY & YOU: EVERYDAY MATTER

"Elemental" Trivia

Did you know that the stench of a skunk's spray is largely due to compounds that contain the element sulfur, or that rubies are red because of small amounts of chromium? Discover more fun facts about other elements as you read this page.

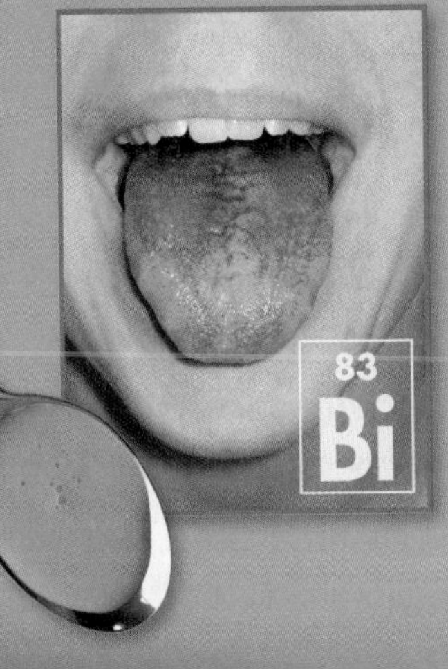

BISMUTH A compound containing bismuth is commonly used to treat indigestion. The bismuth can combine with sulfur in saliva and temporarily turn a person's tongue black!

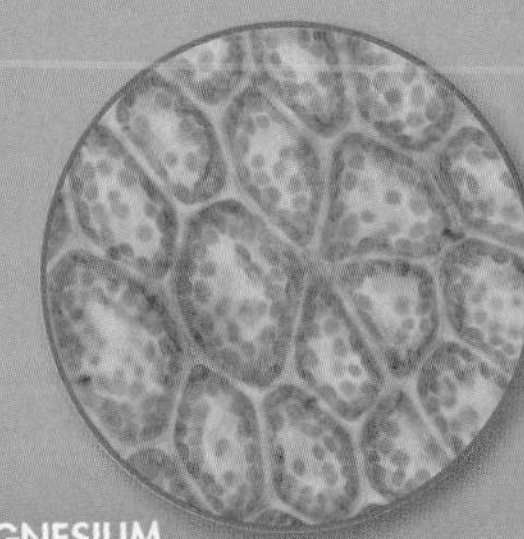

12 **Mg**

MAGNESIUM Magnesium is a component of chlorophyll, the green pigment in plants that allows photosynthesis to occur.

ANTIMONY The element antimony is often used to increase the hardness and strength of pewter figurines.

SEABORGIUM In 1974, this element was created by a team of scientists that included Glenn T. Seaborg. It was the first element to be named after a living person.

1 **H**

HYDROGEN Hydrogen is the most abundant element in the universe and, among other places, is found in stars and auroras.

Take It Further

1. Explain In elemental form, antimony (Sb) and bismuth (Bi) are both brittle, crystalline solids at room temperature. They are also poor conductors of heat and electricity. How does the periodic law support this observation? Use Figure 6.3 to explain your answer.

2. Classify Look ahead to Figure 6.4. Use the figure to classify the five elements above as metals, nonmetals, or metalloids.

21st Century Learning The following project involves the following 21st Century Skills: *Creativity and Innovation, Communication and Collaboration, Information Literacy, Media Literacy, Flexibility and Adaptability, Initiative and Self-Direction, Social and Cross-Cultural Skills, Productivity and Accountability, and Leadership and Responsibility.*

ELEMENT GAME Pose the following challenge to students. *You have been hired to create an educational game. Form a team of 3–5 people to create a new board game to teach facts about the elements.*

- Research and make a list of at least 30 element fun facts. Be sure each fact is something interesting. Keep a copy of source materials for each fact.
- Create a board game that can be used to learn about chemical elements. Be sure the design is engaging, creative, and durable. Design and create all the parts for the game, and build a prototype.
- Test your game by having other groups play it, and then complete a survey about the game. Redesign the game as needed based on feedback.

CHEMISTRY & YOU Lead students in discussing the photos and reading aloud the trivia facts about the elements. Encourage students to describe ways that these elements affect their everyday lives. Point out that some elements have interesting stories behind their names. In some cases, a name is derived from Greek or Latin words early scientists felt were representative of the element. Some elements have interesting facts about their uses or their discoveries. Note that the International Union for Pure and Applied Chemistry (IUPAC) now requires elements to be named after a concept in mythology, a person, a place or country, a mineral, a property, or a scientist.

Have students examine the periodic table. Pose the following challenge to students: *How many elements can you find that are named after places? People? Greek or Latin words? Properties? Minerals?* Provide students with the following examples to get them started:

- Places: californium, after the state
- People: copernicum, after Copernicus
- Mineral: beryllium, after a gemstone
- Property: chlorine, for its color
- Scientist: seaborgium, after one of its discoverers
- Mythological concept: thorium, after the Scandinavian god of war

Extend

Connect to ASTRONOMY

Have students look at the periodic table and identify the four elements named after planets.

- Mercury was named after the planet Mercury, but its symbol, Hg, is derived from the Greek word *hydrargyros,* meaning "liquid silver."
- Uranium is named after the planet Uranus.
- Plutonium is named after the dwarf planet Pluto.
- Neptunium is named after the planet Neptune.

Answers

FIGURE 6.3 8

READING SUPPORT One meaning of *perimeter* is "a line or boundary around an area." It comes from *peri-*, "around," and *meter*, "measure."

TAKE IT FURTHER

1. Antimony and bismuth are in the same column in the periodic table.
2. metals: bismuth, magnesium, and seaborgium; nonmetal: hydrogen; metalloid: antimony

Explain

Metals, Nonmetals, and Metalloids

USE VISUALS Have students examine Figure 6.4. Explain the color key and use the color coding to help identify the sections of the table that correspond to metals, nonmetals, and metalloids. Emphasize the periodic trend across each period from metals to nonmetals. **Ask** Which class of elements do most of the elements belong to? *(metals)* **Ask** What do the three numbers at the top of each column represent? *(different systems for numbering the groups)* Explain that the numbers in blue were used initially, but over time the International Union for Pure and Applied Chemistry changed to the red numbers. The most recent change, to the black system, eliminated the use of A and B subgroups and established a 1 to 18 numbering system. Note that both red and black systems are in use today, but that the system shown in blue is no longer used.

USE AN ANALOGY Use a monthly calendar to show students how the concepts of a monthly calendar are similar in theory to that of the periodic table: The same progression of days occurs each week. Then explain how the calendar and periodic table differ: Although the length of a day is based on a natural event (Earth's rotation), what happens on a Monday or Saturday depends on human decisions, not on an underlying natural principle.

Metals | Metalloids | Nonmetals

	1 IA 1A	2 IIA 2A	3 IIIA 3B	4 IVA 4B	5 VA 5B	6 VIA 6B	7 VIIA 7B	8	9 VIIIA 8B	10	11 IB 1B	12 IIB 2B	13 IIIB 3A	14 IVB 4A	15 VB 5A	16 VIB 6A	17 VIIB 7A	18 VIIIB 8A
1	1 H																	2 He
2	3 Li	4 Be											5 B	6 C	7 N	8 O	9 F	10 Ne
3	11 Na	12 Mg											13 Al	14 Si	15 P	16 S	17 Cl	18 Ar
4	19 K	20 Ca	21 Sc	22 Ti	23 V	24 Cr	25 Mn	26 Fe	27 Co	28 Ni	29 Cu	30 Zn	31 Ga	32 Ge	33 As	34 Se	35 Br	36 Kr
5	37 Rb	38 Sr	39 Y	40 Zr	41 Nb	42 Mo	43 Tc	44 Ru	45 Rh	46 Pd	47 Ag	48 Cd	49 In	50 Sn	51 Sb	52 Te	53 I	54 Xe
6	55 Cs	56 Ba	71 Lu	72 Hf	73 Ta	74 W	75 Re	76 Os	77 Ir	78 Pt	79 Au	80 Hg	81 Tl	82 Pb	83 Bi	84 Po	85 At	86 Rn
7	87 Fr	88 Ra	103 Lr	104 Rf	105 Db	106 Sg	107 Bh	108 Hs	109 Mt	110 Ds	111 Rg	112 Cn	113 Uut	114 Uuq	115 Uup	116 Uuh		118 Uuo

57 La	58 Ce	59 Pr	60 Nd	61 Pm	62 Sm	63 Eu	64 Gd	65 Tb	66 Dy	67 Ho	68 Er	69 Tm	70 Yb
89 Ac	90 Th	91 Pa	92 U	93 Np	94 Pu	95 Am	96 Cm	97 Bk	98 Cf	99 Es	100 Fm	101 Md	102 No

Figure 6.4 Classifying Elements
Periodic tables are sometimes color-coded to classify certain types of elements. This periodic table classifies elements as metals (yellow), nonmetals (blue), and metalloids (green).

Metals, Nonmetals, and Metalloids

What are three broad classes of elements?

Most periodic tables are laid out like the one in Figure 6.4. Notice some elements from Periods 6 and 7 are placed beneath the table. This arrangement makes the periodic table more compact. It also reflects an underlying structure of the periodic table, which you will study in Lesson 6.2. Each column, or group, in this table has three labels. Scientists in the United States primarily use the labels shown in red. Scientists in Europe use the labels shown in blue.

For scientists to communicate clearly, they need to agree on the standards they will use. The International Union of Pure and Applied Chemistry (IUPAC) is an organization that sets standards for chemistry. In 1985, IUPAC proposed a new system for labeling groups in the periodic table. They numbered the groups from left to right 1 through 18 (the black labels in Figure 6.4). The large periodic table in Figure 6.9 includes the IUPAC system and the system used in the United States.

Dividing the elements into groups is not the only way to classify them based on their properties. The elements can be grouped into three broad classes based on their general properties. **Three classes of elements are metals, nonmetals, and metalloids.** Across a period, the properties of elements become less metallic and more nonmetallic.

Figure 6.5 Metals
The metals aluminum, copper, and iron have many important uses. The properties of the metal determine how it is used.

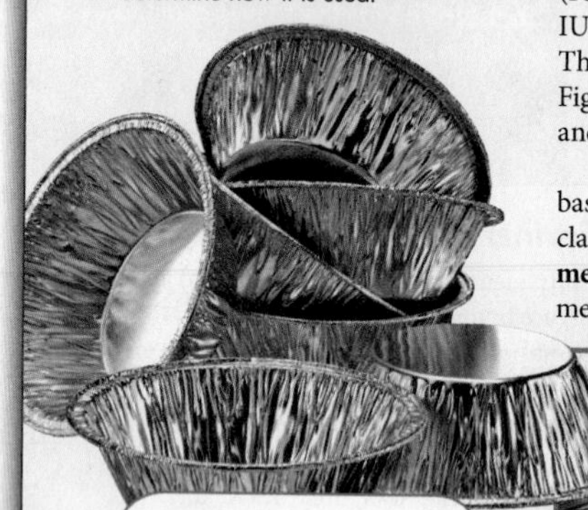

Aluminum (Al)
Aluminum is one of the metals that can be shaped into a thin sheet, or foil.

Copper (Cu)
Copper is ductile and second only to silver as a conductor of electric current. The copper used in electrical cables must be 99.99 percent pure.

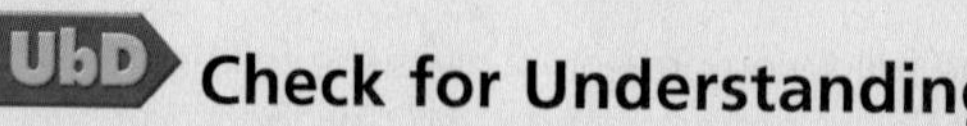

Check for Understanding

How is the modern periodic table organized?

Assess students' understanding of the organization of the periodic table by having them write a one-minute explanation of how the periodic table of elements used today is organized. Explanations should describe the main reason and/or benefit for this organizational structure. *(Today's periodic table is organized by atomic number. This allows for an easier understanding of an element's properties due to the patterns and trends in the groups and periods.)*

ADJUST INSTRUCTION If students are struggling in their responses, direct them to read the text under the heading Today's Periodic Table and to create a list of facts. Then, allow students to use their lists to revise their explanations.

Metals The number of yellow squares in Figure 6.4 shows that most elements are metals—about 80 percent. **Metals** are generally good conductors of heat and electric current. A freshly cleaned or cut surface of a metal will have a high luster, or sheen. The sheen is caused by the metal's ability to reflect light. All metals are solids at room temperature, except for mercury (Hg). Many metals are ductile, meaning that they can be drawn into wires. Most metals are malleable, meaning that they can be hammered into thin sheets without breaking. Figure 6.5 shows how the properties of metals can determine how metals are used.

Nonmetals In Figure 6.4, blue is used to identify the nonmetals. With the exception of hydrogen, these elements are in the upper-right corner of the periodic table. There is a greater variation in physical properties among nonmetals than among metals. Most nonmetals are gases at room temperature, including the main components of air—nitrogen and oxygen. A few are solids, such as sulfur and phosphorus. One nonmetal, bromine, is a dark-red liquid. Some examples of nonmetals are shown in Figure 6.6.

The variation among nonmetals makes it difficult to describe one set of general properties that will apply to all nonmetals. However, nonmetals tend to have properties that are opposite to those of metals. In general, **nonmetals** are poor conductors of heat and electric current. Carbon, in the form of graphite, is an exception to this rule. Solid nonmetals tend to be brittle, meaning that they will shatter if hit with a hammer.

Figure 6.6 Nonmetals
The properties of nonmetals vary.

Carbon (C) and Phosphorus (P)
A diamond, which is composed of carbon, is very hard. Some match heads are coated with phosphorus, a brittle solid.

Iron (Fe)
Cloud Gate in Chicago, Illinois, is covered in stainless steel, which contains iron and chromium (Cr). The steel is shiny, malleable, and strong. It also resists rusting.

Test properties of metals *online.* VIRTUAL LAB

LESSON 6.1

Explore

Class Activity

Have each student choose an element without revealing the choice to other students. Each student should write a short description of the chosen element to read to the class. Ask the other students to identify the element from the description. Encourage students to be as specific as possible in their descriptions. Example: bromine is a reddish-brown liquid.

Extend

Connect to METALLURGY

Aluminum has the best strength-to-weight ratio of any commonly used metal, and it weighs about one-third as much as an equivalent volume of steel. Have students determine four other properties of aluminum that make it a better option to use in a product than other commonly used metals. *(Properties of aluminum: low density, relatively high strength, corrosion resistant even without surface protection, nonstaining, high conductivity, nonmagnetic, etc.)*

Differentiated Instruction

LPR LESS PROFICIENT READERS Make a chart on the board of the three types of elements. Have students complete the chart by adding details about the general characteristics of each type of element. As a class, compare and contrast metals, nonmetals, and metalloids.

L1 SPECIAL NEEDS STUDENTS Provide a large periodic table mat and colored chips or blocks to use as markers. Have students place a marker(s) on the element(s) that is being discussed. Allow students to use these tactile objects during assessments.

L3 ADVANCED STUDENTS Have students research semiconductors; students should define the term and note where the elements that are used for the application are located in the periodic table.

LESSON 6.1

Explain

CHEMISTRY & YOU You could use the periodic table to classify the elements according to physical state (solid, liquid, gas) or whether or not they are found in nature.

CRITICAL THINKING Point out silicon's position in the periodic table and explain that silicon is a metalloid, an element having both metallic and nonmetallic properties. **Ask** In terms of conductance, how can you distinguish metals from nonmetals? *(Metals are good conductors of heat and electricity, and nonmetals are poor conductors of heat and electricity.)* **Ask** Name other elements that are classified as metalloids. *(Chemists do not always agree on which elements to classify as metalloids. Some include polonium (Po) and astatine (At) as metalloids. Some classify polonium as a metal and astatine as a nonmetal.)*

Evaluate

Informal Assessment

Have students explain how Mendeleev contributed to the development of today's periodic table. **Ask** Why did Mendeleev leave gaps in his periodic table? *(He predicted that more elements would one day be discovered, and he was correct.)* **Ask** What do elements in the same group have in common? *(They have similar properties.)* Have students name a metal, a metalloid, and a nonmetal using the name and symbol of the element. *(Answers will vary.)* Then have students complete the 6.1 Lesson Check.

Reteach

Direct students' attention to Figure 6.2 and provide them with a printout of the modern table. Remind students that Mendeleev's rows are the equivalent of the modern groups, so students may wish to rotate their periodic table to correspond to Mendeleev's table. Point out that Mendeleev's practice of classification on the basis of increasing atomic mass had its problems. Have students locate thallium (Tl) and nickel (Ni) in both tables and compare their placement in each table. Point out that when classified on the basis of atomic mass, Tl is grouped with the alkali metals, even though its properties are much less metallic than those of other elements in Group 1A. Note that Ni overlaps with cobalt (Co) in Mendeleev's table because at that time, their atomic masses were thought to be equal. By organizing on the basis of atomic numbers, the modern table avoids these types of problems.

Q: *All of the known elements are listed in the periodic table. What are different ways you could use the periodic table to classify elements?*

Metalloids There is a heavy stair-step line in Figure 6.4 that separates the metals from the nonmetals. Most of the elements that border this line are shaded green. These elements are metalloids. A **metalloid** generally has properties that are similar to those of metals and nonmetals. Under some conditions, a metalloid may behave like a metal. Under other conditions, it may behave like a nonmetal. The behavior often can be controlled by changing the conditions. For example, like most nonmetals, pure silicon is a poor conductor of electric current. However, if a small amount of boron is mixed with silicon, the mixture is a good conductor of electric current. Silicon can be cut into wafers and used to make computer chips. Silicon is also present as the compound silicon dioxide in glass items like the one in Figure 6.7.

Figure 6.7 Metalloid Computer chips and glass are two common items that contain metalloids.

Silicon (Si) Glass contains silicon in the compound silicon dioxide. Skilled artisans prepare molten glass, and then shape it into the desired form as it cools.

6.1 LessonCheck

1. **Explain** How did chemists begin the process of organizing elements?
2. **Identify** What property did Mendeleev use to organize his periodic table?
3. **Explain** How are elements arranged in the modern periodic table?
4. **List** Name the three broad classes of elements.
5. **Classify** Identify each element as a metal, a metalloid, or a nonmetal.
   a. gold
   b. silicon
   c. sulfur
   d. barium
6. **Compare** Which of the following sets of elements have similar physical and chemical properties?
   a. oxygen, nitrogen, carbon, boron
   b. strontium, magnesium, calcium, beryllium
   c. nitrogen, neon, nickel, niobium
7. **Identify** Name two elements that have properties similar to those of the element sodium.

BIGIDEA ELECTRONS AND THE STRUCTURE OF ATOMS

8. Why is atomic number better than atomic mass for organizing the elements in the periodic table?

Lesson Check Answers

1. Chemists used the properties of elements to sort them into groups.
2. atomic mass
3. in order of increasing atomic number
4. metals, metalloids, nonmetals
5. **a.** metal **b.** metalloid **c.** nonmetal **d.** metal
6. b
7. Possible answers (any two): lithium, potassium, rubidium, cesium, or francium
8. **BIGIDEA** It is a good way to organize elements because atomic number is unique for each element, while other properties, such as atomic mass, can vary for atoms of an element.

6.2 Classifying the Elements

C.2.5 Write an element's electron configuration; C.2.6 Determine valence electrons and properties

CHEMISTRY & YOU

Q: *What can you learn about each element from the periodic table?* Many people carry a form of identification, such as a driver's license. An ID contains information specific to a particular person, such as the person's name, address, height, eye color, and weight. The periodic table contains a square for each element that supplies information about that element. In this lesson, you will learn the types of information that are usually listed in a periodic table.

Key Questions

What information can be displayed in a periodic table?

How can elements be classified based on electron configurations?

Vocabulary

- alkali metal
- alkaline earth metal
- halogen
- noble gas
- representative element
- transition metal
- inner transition metal

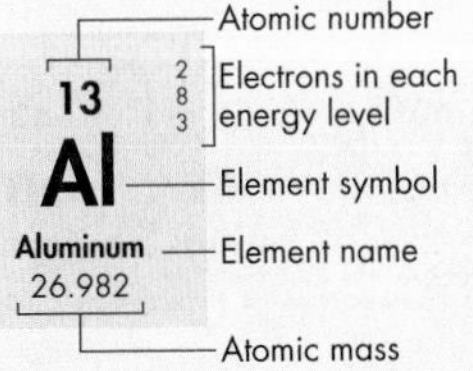

Figure 6.8 Periodic Table Square
This is the element square for aluminum from the periodic table in Figure 6.9.
Interpret Diagrams *What does the data in the square tell you about the structure of an aluminum atom?*

Reading the Periodic Table

What information can be displayed in a periodic table?

The periodic table is a very useful tool in chemistry. **The periodic table usually displays the symbols and names of the elements, along with information about the structure of their atoms.** Figure 6.8 shows one square from the detailed periodic table of the elements in Figure 6.9 on page 168. In the center of the square is the symbol for aluminum (Al). The atomic number for aluminum (13) is above the symbol. The element name and atomic mass are below the symbol. There is also a vertical column with the numbers 2, 8, and 3, which indicate the number of electrons in each occupied energy level of an aluminum atom.

The symbol for aluminum is printed in black because aluminum is a solid at room temperature. In Figure 6.9, the symbols for gases are in red. The symbols for the two elements that are liquids at room temperature, mercury and bromine, are in blue. The symbols for some elements in Figure 6.9 are printed in grey. These elements are not found in nature. In Chapter 25, you will learn how scientists produce these elements.

The background colors in the squares are used to distinguish groups of elements in the periodic table. For example, two shades of orange are used for the metals in Groups 1A and 2A. The elements in Group 1A are called **alkali metals.** The elements in Group 2A are called **alkaline earth metals.** The name alkali comes from the Arabic *al aqali,* meaning "the ashes." Wood ashes are rich in compounds of the alkali metals sodium and potassium. Some groups of nonmetals also have special names. The nonmetals of Group 7A are called **halogens.** The name *halogen* comes from the combination of the Greek word *hals,* meaning "salt," and the Latin word *genesis,* meaning "to be born." There is a general class of compounds called salts, which include the compound called table salt. Chlorine, bromine, and iodine, the most common halogens, can be prepared from their salts.

Key Objectives

6.2.1 DESCRIBE the information in a periodic table.

6.2.2 CLASSIFY elements based on electron configuration.

Additional Resources

Reading and Study Workbook, Lesson 6.2

Available Online or on Digital Media:

- Teaching Resources, Lesson 6.2 Review
- Laboratory Manual, Lab 9

LESSON 6.2

PearsonChem.com

Engage

CHEMISTRY & YOU After students read the opening paragraph, have them brainstorm other examples of small items that contain lots of information (e.g., coins, train schedules, digital watch face, and nutrient labels). As a class, discuss the information that each of these items provides.

Access Prior Knowledge

Have students review what they learned about atomic numbers, atomic mass, and chemical symbols in Chapter 4. Then, review what they learned about the periodic table in the previous lesson, and determine how this information relates to the information from Chapter 4.

Focus on ELL

1 CONTENT AND LANGUAGE Use a deck of cards to clarify the difference between the words *organize* and *classify.* Show students how they can classify the cards into suits or colors, but organizing the cards is different.

2 FRONTLOAD THE LESSON Display pictures of coins from various countries and in different denominations. Be sure that the images show both sides of each coin. Ask students to list the types of information found on each coin and consider how that information can be used to identify each coin. Tells students that the squares of the periodic table also contain several important pieces of information that can be used to identify each element.

3 COMPREHENSIBLE INPUT When discussing Figure 6.13, project the figure on the board. Refer students back to the electron configurations for the elements in Groups 1A–8A, and then fill in these elements in their appropriate blocks on the figure to reinforce the relationship between electron configuration and element groups.

IN ACADEMIC STANDARDS for SCIENCE

C.2.5 Write the electron configuration of an element and relate this to its position on the periodic table.

C.2.6 Use the periodic table and electron configurations to determine an element's number of valence electrons and chemical and physical properties.

Answers

FIGURE 6.8 There are 13 protons in the nucleus and 13 electrons in the three occupied energy levels.

Foundations for Reading

BUILD VOCABULARY Have students use the LINCS strategy. In LINCS exercises, students **L**ist what they know about each term, **I**magine a picture that describes the term, **N**ote a reminding "sound-alike" word, **C**onnect the terms to the sound-alike word by making up a short story, and then perform a brief **S**elf-test.

READING STRATEGY After students preview the lesson, have them predict the relationship between an element's electron configuration and its position in the periodic table. Students can revise their predictions after they read the lesson.

Explain

Reading the Periodic Table

USE VISUALS Have students examine Figure 6.8 on the previous page. Remind students what atomic mass represents. Make sure that students understand what the vertical column of numbers on the right-hand side of each block represents. Refer students back to the discussion of electron configuration in Chapter 5 for review, if needed.

Make sure students understand that each column in the table is a group, and at the top of each column is a number that indicates the group number. **Ask** What elements are in Group 4B? *(Ti, Zr, Hf, Rf)*

USE VISUALS Have students examine the periodic table in Figure 6.9. Discuss as a class the information that can extracted from the table. **Ask** Why are the symbols for bromine and mercury displayed in blue ink? *(They are liquids.)* **Ask** What series of elements come between Groups 2A and 3B in the periodic table? *(the lanthanide and actinide elements)*

SUMMARIZE Have students write summary statements for the information that can or cannot be extracted from the periodic table for the following: element name, atomic number, atomic mass, and physical state at room temperature.

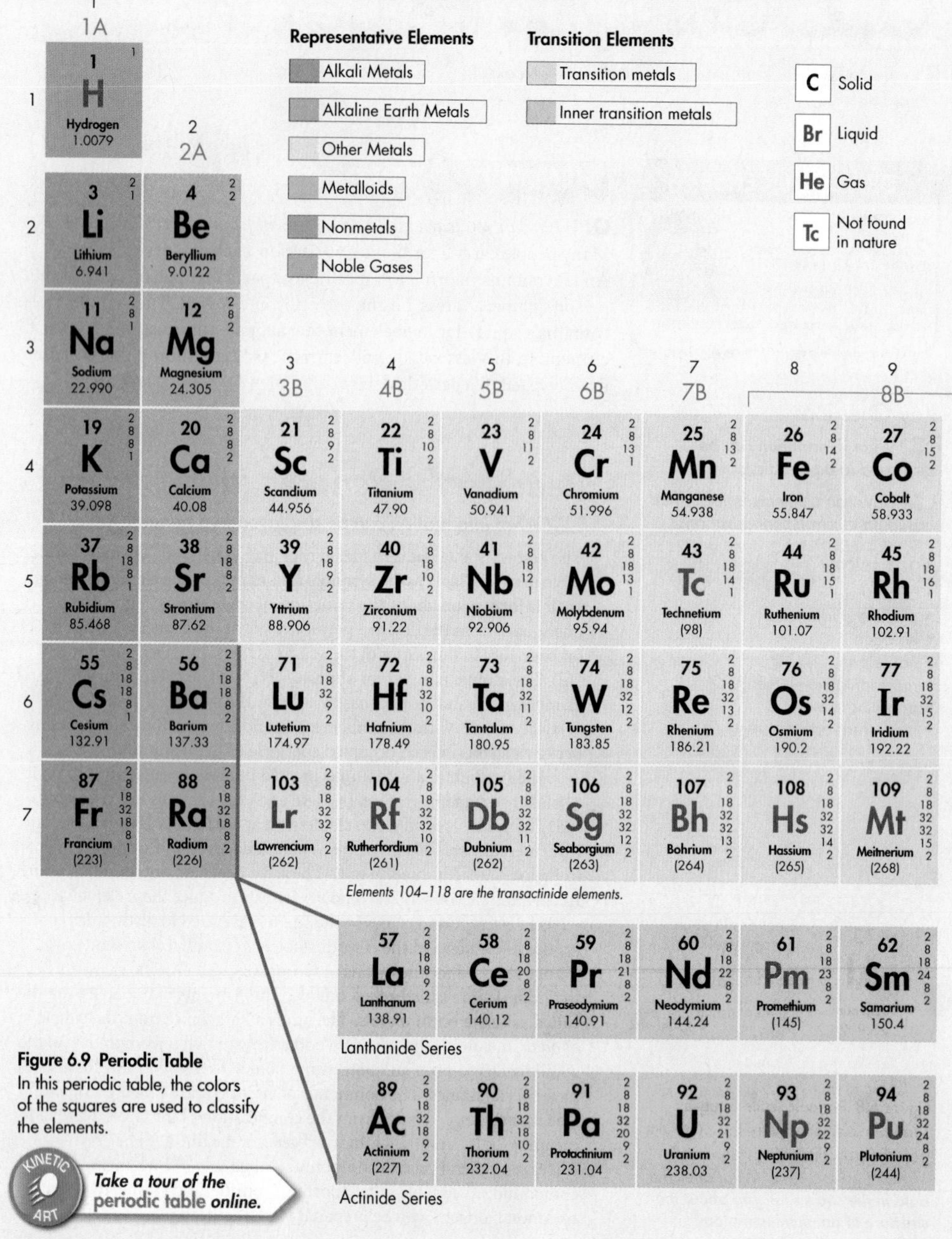

Figure 6.9 Periodic Table
In this periodic table, the colors of the squares are used to classify the elements.

Professional Development Note: Discovery and Naming of New Elements

Recently, a laboratory in Darmstadt, Germany, reported the discovery of the element with atomic number 117. This element will temporarily go by the name ununseptium, which is based on the Latin root for the number 117. Element names follow rules set by the International Union of Pure and Applied Chemistry (IUPAC). An official name will be assigned to element 117 after IUPAC officially recognizes its discovery. The element's official name will be chosen from names proposed by the scientists who discovered the element.

Often several years pass before an official name is assigned to a newly discovered element. For example, the most recently named element, copernicium, formerly unubium, was reported in 1996. It was officially named about 14 years later, in 2010.

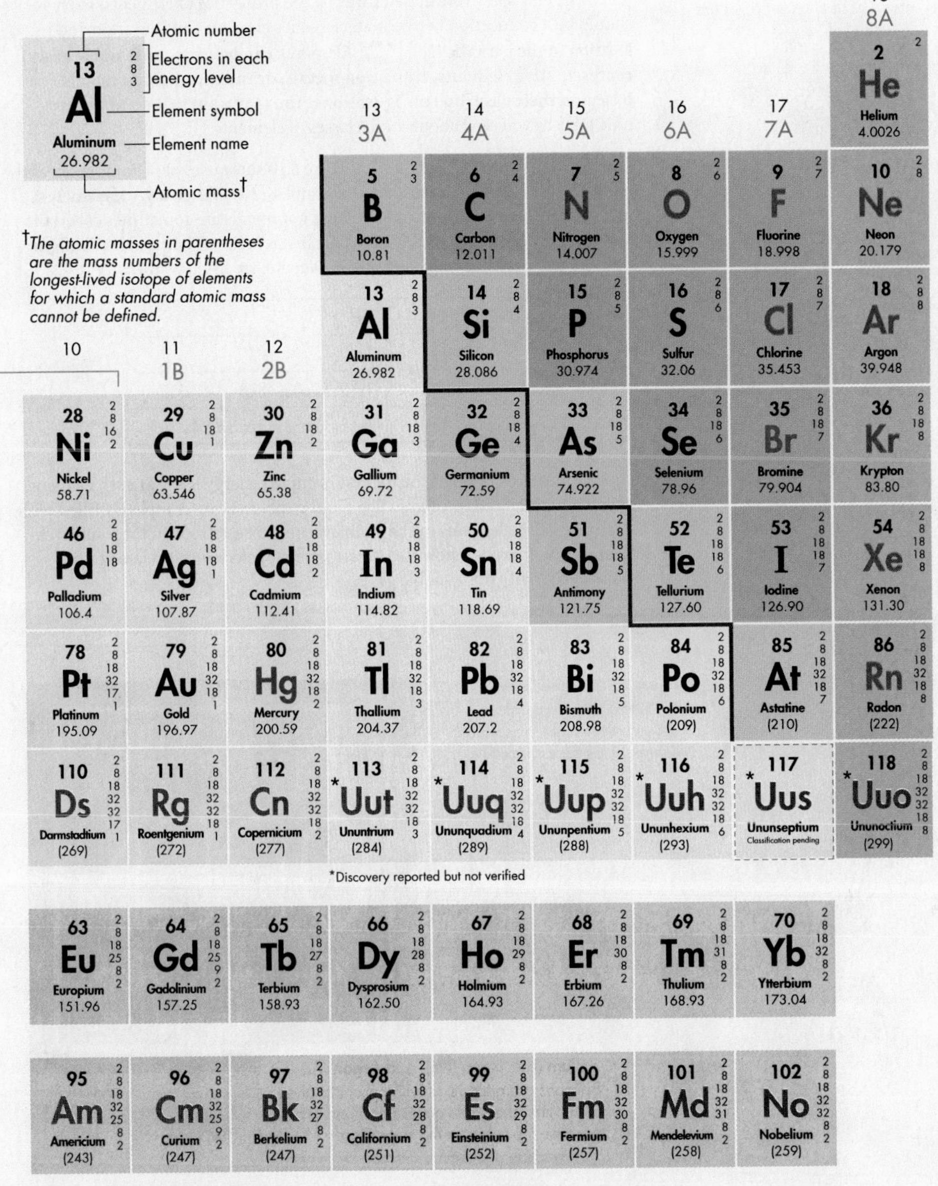

DRAW CONCLUSIONS Have students examine the periodic table and determine the accuracy of this statement: Atomic mass always increases as atomic number increases. *(The trend is generally true, but there are exceptions. For example, the atomic numbers of Co, Ni, and Cu increase by one unit (27, 28, 29), but Ni has the lowest average atomic mass of these three elements.)*

Extend

Connect to PHYSIOLOGY

Have students start a list of elements found in the human body that are essential for metabolism. Explain that many elements are present in trace amounts but are, nevertheless, essential for survival. Have students research elements such as copper, chromium, iodine, and manganese to find out more about their biological roles. For further enrichment, suggest that students look into micronutrients.

Differentiated Instruction

LPR LESS PROFICIENT READERS Explain that reading a table involves reading the title first, followed by reading the key, and then by examining the data presented in the table. Review each aspect of the color key used in the periodic table.

L1 SPECIAL NEEDS STUDENTS Have students use an online annotated periodic table. Students can click on an element to see or hear information and properties about the element.

L3 ADVANCED STUDENTS Mendeleev predicted that certain elements that were missing from his table would have specific properties. Have students research Mendeleev's predictions for ekaaluminum (gallium), ekaboron (scandium), and ekasilicon (germanium) and compare them to the known properties of each element. Have students create a table demonstrating the similarities and differences between Mendeleev's predictions and what we now know about each element's properties.

The Periodic Table 169

Explain

Electron Configurations in Groups

APPLY CONCEPTS Lead a class discussion on electron configurations of noble gases and representative elements. Select some elements and have students write out the electron configurations for those elements. Have students compare the electron configurations for all the elements in a single group. Ask students to identify similarities. *(Noble gases are sometimes classified as representative elements because they are in the p block of elements.)*

CHEMISTRY & YOU From the periodic table, you can learn several things about a particular element—whether the element is a solid, liquid, or gas at room temperature; how many electrons and protons it has; its atomic mass; whether it is a metal, metalloid, or nonmetal; its electron configuration; and its name and symbol.

CHEMISTRY & YOU

Q: *What can you learn about each element from the periodic table?*

Electron Configurations in Groups

How can elements be classified based on electron configurations?

Electrons play a key role in determining the properties of elements, so there should be a connection between an element's electron configuration and its location in the periodic table. **Elements can be sorted into noble gases, representative elements, transition metals, or inner transition metals based on their electron configurations.** You may want to refer to the periodic table as you read about these classes of elements.

The Noble Gases The photos in Figure 6.10 show uses of helium, neon, and argon. Helium, neon, and argon are examples of **noble gases,** the elements in Group 8A of the periodic table. These nonmetals are sometimes called the inert gases because they rarely take part in a reaction. The electron configurations for the first four noble gases in Group 8A are listed below.

Helium (He)	$1s^2$
Neon (Ne)	$1s^2 2s^2 2p^6$
Argon (Ar)	$1s^2 2s^2 2p^6 3s^2 3p^6$
Krypton (Kr)	$1s^2 2s^2 2p^6 3s^2 3p^6 3d^{10} 4s^2 4p^6$

Look at the description of the highest occupied energy level for each element, which is highlighted in yellow. The *s* and *p* sublevels are completely filled with electrons—two electrons in the *s* sublevel and six electrons in the *p* sublevel. Chapter 7 will explain how this arrangement of electrons is related to the relative inactivity of the noble gases.

Figure 6.10 Noble Gases
The noble gases are unique because their highest occupied energy levels are completely filled.
Infer *Helium-filled balloons rise in air. What does this tell you about the density of helium?*

Helium (He), Neon (Ne), and Argon (Ar)
Balloons are often filled with helium to give them "lift." The noble gases neon and argon produce the colors in this neon sign.

UbD Check for Understanding

The Essential Question What information does the periodic table provide?

Assess students' knowledge about the periodic table by having students use a thumbs-up sign to indicate if the information listed in each statement below can be found in the periodic table, and a thumbs-down sign if it cannot be found there.

- The date an element was discovered *(no)*
- An element with properties similar to neon *(yes)*
- The boiling point of magnesium *(no)*
- The atomic mass of nitrogen *(yes)*
- An element that is a good conductor of heat *(yes)*

ADJUST INSTRUCTION If students are having trouble answering, have them re-examine Figures 6.8 and 6.9. Call out an element and have students write down all of the information about the element and what the information is based on.

The Representative Elements Figure 6.11 shows the portion of the periodic table containing Groups 1A through 7A. Elements in Groups 1A through 7A are often referred to as **representative elements** because they display a wide range of physical and chemical properties. Some elements in these groups are metals, some are nonmetals, and some are metalloids. Most of them are solids, but a few are gases at room temperature, and one, bromine, is a liquid.

In atoms of representative elements, the *s* and *p* sublevels of the highest occupied energy level are not filled. Look at the electron configurations for lithium, sodium, and potassium below. In atoms of these Group 1A elements, there is only one electron in the highest occupied energy level. The electron is in an *s* sublevel.

Lithium (Li)	$1s^22s^1$
Sodium (Na)	$1s^22s^22p^63s^1$
Potassium (K)	$1s^22s^22p^63s^23p^64s^1$

In atoms of the Group 4A elements carbon, silicon, and germanium, there are four electrons in the highest occupied energy level.

Carbon (C)	$1s^22s^22p^2$
Silicon (Si)	$1s^22s^22p^63s^23p^2$
Germanium (Ge)	$1s^22s^22p^63s^23p^63d^{10}4s^24p^2$

For any representative element, its group number equals the number of electrons in the highest occupied energy level.

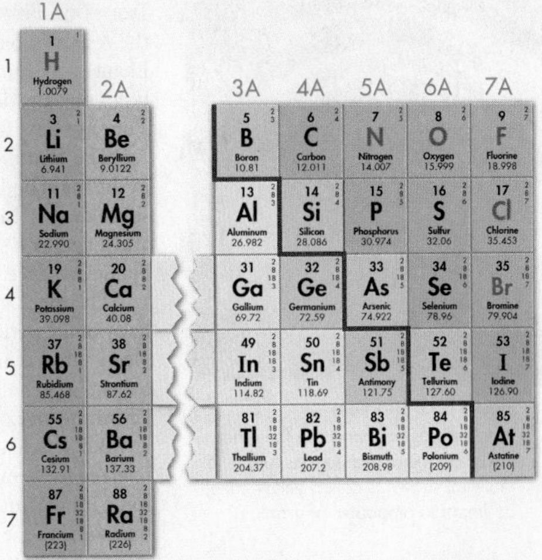

Figure 6.11 Representative Elements
Some of the representative elements exist in nature as elements. Others are found only in compounds.

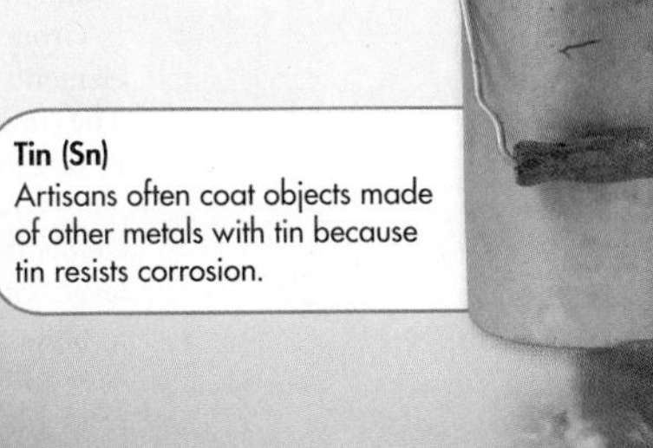

Tin (Sn)
Artisans often coat objects made of other metals with tin because tin resists corrosion.

Lithium (Li)
Batteries in many electronic devices, including this personal vehicle, use lithium to generate electrical energy.

Sulfur (S)
Some volcanos release high amounts of sulfur vapors. The sulfur cools and is deposited as a solid yellow powder.

Explain

USE VISUALS Direct students' attention to Figure 6.11. **Ask** What pattern do you notice in the electron configuration for Group 1A elements? *(The s-level is not completely filled.)* **Ask** Predict the pattern for the electron configuration for Group 2A elements. *(Answers will vary.)*

Explore

Teacher Demo

PURPOSE Students observe differences in the reactivity of magnesium, tin, and copper.

MATERIALS 0.2*M* HCl; 6 large test tubes; test tube rack; small pieces of clean magnesium, tin, or copper; matches; wood splint

SAFETY Wear goggles for this demo.

PROCEDURE Place three large test tubes in a test tube rack. To each test tube, add a small, clean piece of a different metal—magnesium, tin, or copper. Carefully add some of the 0.2*M* HCl to each test tube and invert another test tube over it. Point out the appearance of bubbles. After a while, carefully ignite the hydrogen in the inverted test tube from the magnesium or tin reaction.

EXPECTED OUTCOME The magnesium and tin react with the HCl, producing hydrogen. The hydrogen ignites with an explosive pop. The copper shows no sign of reaction.

Differentiated Instruction

L1 STRUGGLING STUDENTS Divide the class into small groups. Have each group write electron configurations for an *s*-block element, a *p*-block element, a *d*-block element, and an *f*-block element. Avoid assigning any of the *d*-block elements with unpredictable configurations. Ask students to explain the method they used to determine specific electron configurations.

LPR LESS PROFICIENT READERS Prior to the teacher demo, have students place markers on the elements magnesium, tin, copper, and hydrogen on their periodic table. Discussing one element at a time, explain what is needed in regard to gaining or losing an electron in order to achieve the nearest noble gas electron configuration. Have students predict which element is most likely to react and which element is least likely to react with HCl. Discuss the predictions after the demonstration.

L3 ADVANCED STUDENTS Have students write an expression for the pattern they notice in the outer shell electron configuration for the noble gases. *(ns^2np^6, where n = the number of the period, n >1)*

Answers

FIGURE 6.10 Helium is less dense than air.

LESSON 6.2

Explain

Electron Configurations in Groups

Misconception Alert

Students may think that elements whose names are familiar are always more abundant on Earth than less familiar elements. Use the following data to show students why calling inner transition metals "rare-earth elements" is misleading. In Earth's crust, there is 1.7 ppb of osmium, 37 ppb of platinum, 67 ppb of mercury, 1700 ppb of tantalum, 6000 ppb of thorium, 10,000 ppb of lead, 60,000 ppb of cerium, and 34,000 ppb of lanthanum. Both platinum and mercury, which should be familiar elements to students, are present in much smaller quantities than thorium, cerium, and lanthanum—elements with which most students will not be familiar.

USE VISUALS Have students examine Figure 6.13. Emphasize the relationship between the position of an element in the periodic table and the element's atomic structure. **Ask** What do all the elements in Group 1A have in common? *(They all have one electron in the* s *orbital in the highest occupied energy level.)* **Ask** What do all the elements in Group 3A have in common? *(They all have one electron in a* p *orbital in the highest occupied energy level.)* To help students understand the instructions for writing electron configurations for elements in the *d* and *f* blocks, review the aufbau diagram in Lesson 5.2.

Extend

Connect to PHYSICS

Direct students' attention to Figure 6.12. Explain that uranium is a heavy metal element that can be used as a source of concentrated energy. Briefly describe how electricity can be generated from uranium in nuclear reactors. Have students write a persuasion paper on the risks or the rewards of increasing the production of nuclear power as an alternative energy source in the United States.

Figure 6.12 Uranium
Nuclear power plants use the inner transition metal uranium as fuel. The material shown is called yellowcake, an impure compound of uranium.

Transition Elements In the periodic table, the B group elements separate the A groups on the left side of the table from the A groups on the right side. Elements in the B groups are referred to as transition elements. There are two types of transition elements—transition metals and inner transition metals. They are classified based on their electron configurations.

The **transition metals** are the Group B elements that are usually displayed in the main body of a periodic table. Copper, silver, gold, and iron are transition metals. In atoms of a transition metal, the highest occupied *s* sublevel and a nearby *d* sublevel contain electrons. These elements are characterized by the presence of electrons in *d* orbitals.

The **inner transition metals** are the elements that appear below the main body of the periodic table. In atoms of these elements, the highest occupied *s* sublevel and a nearby *f* sublevel generally contain electrons. The inner transition metals are characterized by the presence of electrons in *f* orbitals. Uranium, an example of an inner transition metal, is shown in Figure 6.12.

Before scientists knew much about inner transition metals, people referred to them as rare-earth elements. This name is misleading because some inner transition metals are more abundant than other elements. Notice that some of the inner transition metals are not found in nature. These elements were prepared in laboratories using methods presented in Chapter 25.

Blocks of Elements If you consider both the electron configurations and the positions of the elements in the periodic table, another pattern emerges. In Figure 6.13, the periodic table is divided into sections, or blocks, that correspond to the highest occupied sublevels. The *s* block contains the elements in Groups 1A and 2A and the noble gas helium. The *p* block contains the elements in Groups 3A, 4A, 5A, 6A, 7A, and 8A, with the exception of helium. The transition metals belong to the *d* block, and the inner transition metals belong to the *f* block.

You can use Figure 6.13 to help determine electron configurations of elements. Each period on the periodic table corresponds to a principal energy level. Suppose an element is located in Period 3. You know that the *s* and *p* sublevels in energy levels 1 and 2 are filled with electrons. You then read across Period 3 from left to right to complete the configuration. For transition elements, electrons are added to a *d* sublevel with a principal energy level that is one less than the period number. For the inner transition metals, the principal energy level of the *f* sublevel is two less than the period number. This procedure gives the correct electron configurations for most atoms.

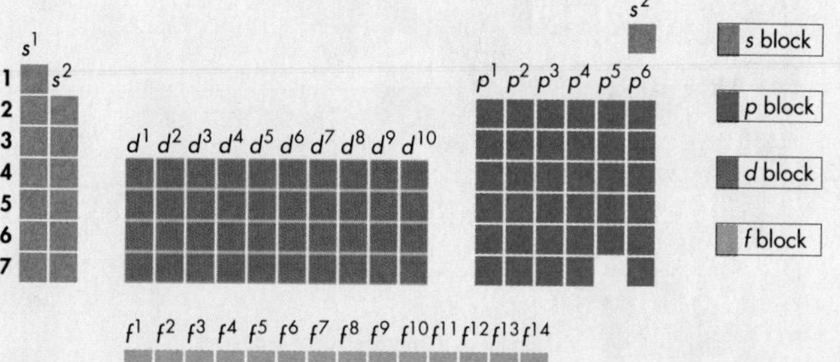

Figure 6.13 Electron Configurations
This diagram classifies elements into blocks according to sublevels that are filled or are filling with electrons.
Interpret Diagrams *In the highest occupied energy level of a halogen atom, how many electrons are in the* p *sublevel?*

172 Chapter 6 • Lesson 2

Using Shorthand Notation

The electron configurations of noble gases can be used to produce a shorthand notation for other electron configurations. The symbol of a noble gas in brackets represents its electron configuration. The configuration of sodium can be written as [Ne] $3s^1$. The electron configuration for zinc can be written as [Ar] $3d^{10}4s^2$.

Using Energy Sublevels to Write Electron Configurations

Use Figure 6.9 and Figure 6.13 to write the electron configuration for nickel (Ni).

1 Analyze **Identify the relevant concepts.** For all elements, the atomic number is equal to the total number of electrons. For a representative element, the highest occupied energy level is the same as the number of the period in which the element is located. You can tell how many electrons are in this energy level from the group in which the element is located.

2 Solve **Apply the concepts to this problem.**

Use Figure 6.9 to identify where the atom is in the periodic table and the number of electrons in the atom. → Nickel is located in the fourth period and has 28 electrons.

Use Figure 6.13 to determine the electron configuration. → In nickel, the *s* and *p* sublevels in the first three energy levels are full, so the configuration begins with $1s^22s^22p^63s^23p^6$. Next is $4s^2$ and $3d^8$. Put it all together: $1s^22s^22p^63s^23p^63d^84s^2$

9. Use Figure 6.9 and Figure 6.13 to write the electron configurations of the following elements:
a. carbon
b. strontium
c. vanadium

10. List the symbols for all the elements whose electron configurations end as follows. *Note*: Each *n* represents an energy level.
a. ns^2np^1
b. ns^2np^5
c. $ns^2np^6nd^2(n+1)s^2$

Remember that the principal energy level number for elements in the *d* block is always one less than the period number.

6.2 LessonCheck

11. Identify What types of information can be included in a periodic table?

12. List Into what four classes can elements be sorted based on their electron configurations?

13. Explain Why do the elements potassium and sodium have similar chemical properties?

14. Classify Identify each element as an alkali metal, an alkaline earth metal, or a halogen:
a. barium
b. chlorine
c. lithium
d. beryllium

15. Classify Based on the following electron configurations, identify each element as a representative element, transition metal, or noble gas.
a. $1s^22s^22p^63s^23p^63d^{10}4s^24p^6$
b. $1s^22s^22p^63s^23p^63d^64s^2$
c. $1s^22s^22p^63s^23p^2$

16. Describe How many electrons are in the highest occupied energy level of an element in Group 5A?

17. Identify Which of these elements are transition metals: Cu, Sr, Cd, Au, Al, Ge, Co?

Lesson Check Answers

11. symbols and names of the elements, atomic number and atomic mass, information about electron configuration
12. representative elements, noble gases, transition metals, and inner transition metals
13. They are in the same group and have the same number of electrons in the highest occupied energy level.
14. **a.** alkaline earth metal
b. halogen
c. alkali metal
d. alkaline earth metal
15. **a.** noble gas
b. transition metal
c. representative element
16. 5
17. Cu, Cd, Au, Co

Explain

Sample Practice Problem

Write the electron configuration for the following elements.
a. fluorine *([He] $2s^22p^5$)*
b. chromium *([Ar] $3d^54s^1$)*

Evaluate

Informal Assessment

Call out pairs of elements in the same group and have students write their electron configuration. If desired, this activity can be made into a game by dividing the class into groups that compete to see who writes the correct configuration first. Then have students complete the 6.2 Lesson Check.

Reteach

To reinforce the relationship between configurations and positions in the periodic table, have students work backward from electron configurations to identify their associated elements. Work with students to identify and locate the elements in the periodic table. Ask students to explain which parts of a configuration proved most useful for determining the identity.

Answers

FIGURE 6.13 five
9. **a.** $1s^22s^22p^2$
b. $1s^22s^22p^63s^23p^63d^{10}4s^24p^65s^2$
c. $1s^22s^22p^63s^23p^63d^34s^2$
10. **a.** B, Al, Ga, In, Tl
b. F, Cl, Br, I, At
c. Ti, Zr, Hf, Rf

Key Objectives

6.3.1 DESCRIBE trends among elements for atomic size.

6.3.2 EXPLAIN how ions form.

6.3.3 DESCRIBE trends for first ionization energy, ionic size, and electronegativity.

Additional Resources

Reading and Study Workbook, Lesson 6.3

Available Online or on Digital Media:

- Teaching Resources, Lesson 6.3 Review
- Small-Scale Chemistry Laboratory Manual, Lab 9

PearsonChem.com

Engage

CHEMISTRY & YOU Have students read the opening paragraph. As a class, discuss the seasonal trends in weather typical for your region of the country. **Ask** How is knowing the trends in weather for a specific region helpful? *(Sample answer: It can help in determining the type of plants that will grow well in your yard.)* Have students consider how knowing trends in elemental properties might be helpful to scientists.

Access Prior Knowledge

Ask student volunteers to summarize what they have learned about the organization of the periodic table. Have students compare and contrast properties and characteristics of periods with those of groups.

IN ACADEMIC STANDARDS for SCIENCE

- Develop explanations based on reproducible data and observations gathered during laboratory investigations.
- Clearly communicate their ideas and results of investigations verbally and in written form using tables, graphs, diagrams, and photographs.

C.2.6 Use the periodic table and electron configurations to determine an element's number of valence electrons and chemical and physical properties.

6.3 Periodic Trends

C.2.6 Determine valence electrons and properties

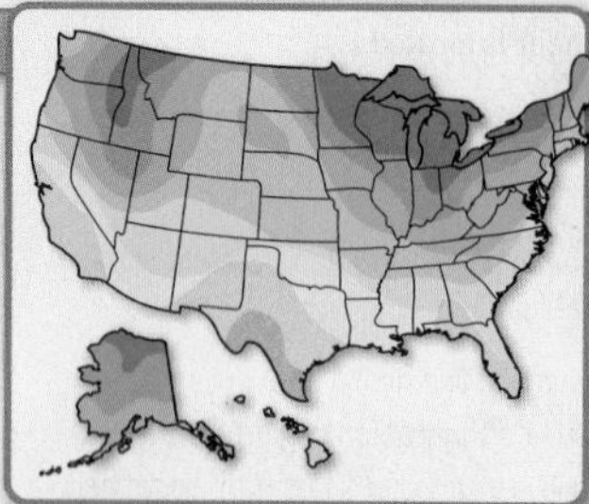

CHEMISTRY & YOU

Q: *How are trends in the weather similar to trends in the properties of elements?* Although the weather changes from day to day. The weather you experience is related to your location on the globe. For example, Florida has an average temperature that is higher than Minnesota's. Similarly, a rain forest receives more rain than a desert. These differences are attributable to trends in the weather. In this lesson, you will learn how a property such as atomic size is related to the location of an element in the periodic table.

Key Questions

What are the trends among the elements for atomic size?

How do ions form?

What are the trends among the elements for first ionization energy, ionic size, and electronegativity?

Vocabulary

- atomic radius
- ion
- cation
- anion
- ionization energy
- electronegativity

Trends in Atomic Size

What are the trends among the elements for atomic size?

One way to think about atomic size is to look at the units that form when atoms of the same element are joined to one another. These units are called molecules. Figure 6.14 shows models of molecules (molecular models) for seven nonmetals. Because the atoms in each molecule are identical, the distance between the nuclei of these atoms can be used to estimate the size of the atoms. This size is expressed as an atomic radius. The **atomic radius** is one half of the distance between the nuclei of two atoms of the same element when the atoms are joined.

The distances between atoms in a molecule are extremely small. So the atomic radius is often measured in picometers (pm). Recall that there are one trillion, or 10^{12}, picometers in a meter. The molecular model of iodine in Figure 6.14 is the largest. The distance between the nuclei in an iodine molecule is 280 pm. Because the atomic radius is one half the distance between the nuclei, a value of 140 pm (280/2) is assigned as the radius of the iodine atom. **In general, atomic size increases from top to bottom within a group and decreases from left to right across a period.**

Figure 6.14 Atomic Radii
This diagram compares the atomic radii of seven nonmetals.

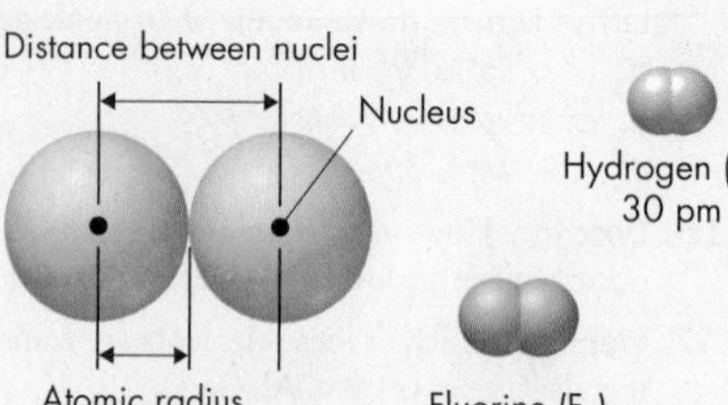

Hydrogen (H_2) 30 pm
Oxygen (O_2) 66 pm
Nitrogen (N_2) 70 pm
Fluorine (F_2) 62 pm
Chlorine (Cl_2) 102 pm
Bromine (Br_2) 120 pm
Iodine (I_2) 140 pm

Focus on ELL

1 CONTENT AND LANGUAGE Have students write the lesson title *Periodic Trends* in their vocabulary notebook. Have students determine the common definitions of each word and write them in their notebook. Have students predict the lesson content based on these definitions and what they have learned previously about the periodic table.

2 FRONTLOAD THE LESSON Provide four sets of photos of fashion trends from the 1950s, the 1970s, the 1990s, and the current decade. Ask students to identify the fashion trends for each decade. Explain that a trend is a pattern over time. Then ask students if they notice any similarities between clothes in the pictures. Explain that in fashion, and in element properties, trends repeat periodically.

3 COMPREHENSIBLE INPUT Play "The Elements" song written by Tom Lehler. Use this as a tool for engaging students' interest in the content. Point out that, even though the element names are never repeated, sections of the music are repeated. Similarly, trends in the periodic table are repeated in each period.

Interpret Graphs

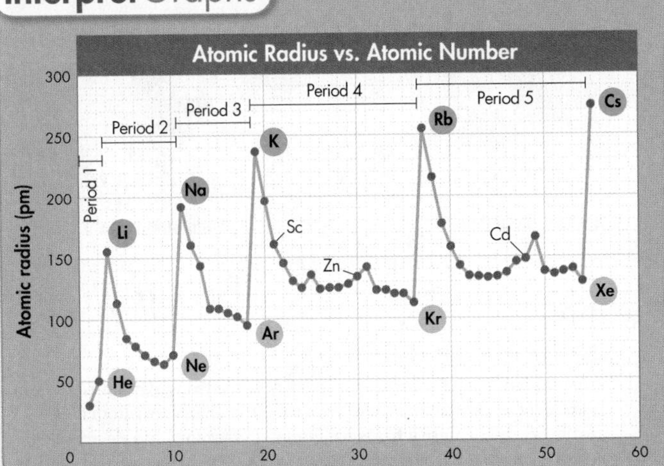

Figure 6.15 This graph plots atomic radius versus atomic number for elements with atomic numbers from 1 to 55.
a. Read Graphs Which alkali metal has an atomic radius of 238 pm?
b. Draw Conclusions Based on the data for alkali metals and noble gases, how does atomic size change within a group?
c. Predict Is an atom of barium, atomic number 56, smaller or larger than an atom of cesium (Cs)?

Group Trends in Atomic Size Look at the data for the alkali metals and noble gases in Figure 6.15. The atomic radius within these groups increases as the atomic number increases. This increase is an example of a trend.

As the atomic number increases within a group, the charge on the nucleus increases and the number of occupied energy levels increases. These variables affect atomic size in opposite ways. The increase in positive charge draws electrons closer to the nucleus. The increase in the number of occupied orbitals shields electrons in the highest occupied energy level from the attraction of protons in the nucleus. The shielding effect is greater than the effect of the increase in nuclear charge, so the atomic size increases.

Period Trends in Atomic Size Look again at Figure 6.15. With increasing atomic number, each element has one more proton and one more electron than the preceding element. Across a period, the electrons are added to the same principal energy level. The shielding effect is constant for all the elements in a period. The increasing nuclear charge pulls the electrons in the highest occupied energy level closer to the nucleus, and the atomic size decreases. Figure 6.16 summarizes the group and period trends in atomic size.

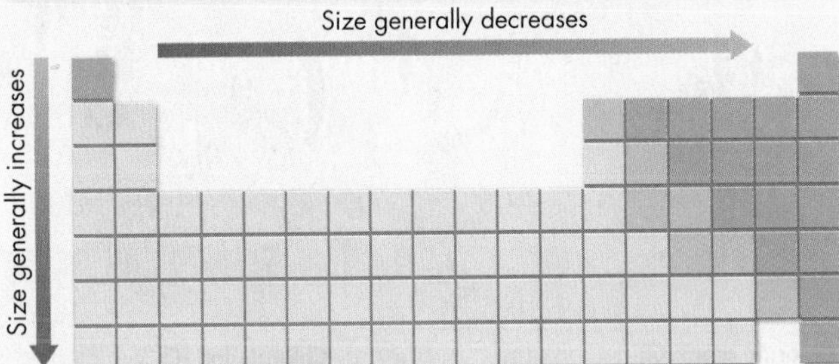

Figure 6.16 Trends in Atomic Size The size of atoms tends to decrease from left to right across a period and increase from top to bottom within a group.
Predict ***If a halogen and an alkali metal are in the same period, which one will have the larger radius?***

See periodic trends animated online.

Elements and the Big Bang

At the time of the Big Bang, the temperature was many billions of degrees. Neutrons, protons, and electrons may have formed within 10^{-4} second after the Big Bang, and the lightest nuclei formed within 3 minutes. Matter was in the form of plasma, a sea of positive nuclei and negative electrons. It took an estimated 500,000 years for electrons and nuclei to cool enough to form atoms. According to the Big Bang theory, Earth, with its wealth of chemical elements, formed from the debris of supernova explosions.

Foundations for Reading

BUILD VOCABULARY Have students make a compare/contrast table for *cation* and *anion*, and use the table to decide which type of ion an element is likely to form.

READING STRATEGY Tell students that they will be learning about trends related to the location of elements in the periodic table. Guide students to read the visuals throughout the lesson closely, as the visuals summarize the trends described in the text.

Explain

Trends in Atomic Size

USE VISUALS Guide students' attention to Figure 6.14. **Ask** What is the main reason why a scientist cannot measure the diameter of a single atom? *(because an atom does not have a sharply defined border)* Discuss how measuring the distance between nuclei solves this problem. *(NOTE: In Chapter 8 there are formal definitions of* molecule *and* diatomic molecule. *The operational definition of a molecule should be sufficient for a discussion of atomic radii.)*

CRITICAL THINKING Emphasize the key roles electrical attraction and repulsion play within atoms and ions. Review the effects of increasing nuclear charge and changes in the shielding effect of electrons on the size of an atom: nuclear charge increases within groups and across periods; the shielding effect increases within groups, but it is constant across periods. Have students use these effects to describe the trends for atomic size within a period and within groups.

USE AN ANALOGY As an analogy to positions and trends in properties of elements in the periodic table, use seating charts and pricing data from local theaters or sports venues to discover trends. Instruct students to determine patterns that relate the position of a seat to its price. Students should discover that variables such as distance from the stage or field, location relative to the center of the action, and whether the view will be obstructed, all affect price.

Answers

FIGURE 6.15

a. potassium

b. It increases with increasing atomic number.

c. smaller

FIGURE 6.16 the alkali metal

Explore

Ions

PURPOSE To give students practice identifying positive and negative ions

PROCEDURE Give students a list of elements. Ask them to locate each element in the periodic table, and decide whether its atoms are likely to form positive or negative ions. Have students make a list of elements that are likely to form positive ions and another list of elements that are likely to form negative ions.

Misconception Alert

Many students will associate the words "losing" and "gaining" with subtraction and addition, respectively. Make sure they understand that when an atom loses an electron, its charge becomes more positive, rather than more negative. Similarly, when an element gains an electron, it becomes more negative rather than more positive. It may be helpful to remind students that they are adding or subtracting the total *charge* of the electrons gained or lost, rather than the total *number* of electrons. For example, the elemental form has a charge of 0, and it loses a single electron, which has a charge of −1. The charge can be calculated as $0 - (-1) = 0 + 1 = +1$. For chlorine, which gains an electron to become negative, the calculation would be $0 + (-1) = -1$.

Figure 6.17 Cation Formation
When a sodium atom loses an electron, it becomes a positively charged ion.

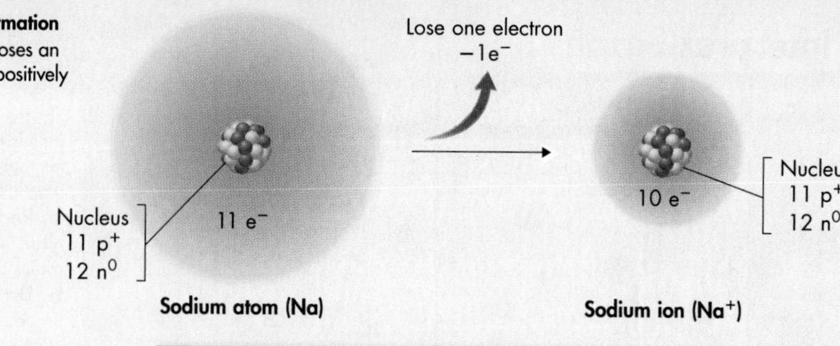

Ions

How do ions form?

Some compounds are composed of particles called ions. An **ion** is an atom or group of atoms that has a positive or negative charge. An atom is electrically neutral because it has equal numbers of protons and electrons. For example, an atom of sodium (Na) has 11 positively charged protons and 11 negatively charged electrons. The net charge on a sodium atom is zero [(+11) + (−11) = 0].

Positive and negative ions form when electrons are transferred between atoms. Atoms of metals, such as sodium, tend to form ions by losing one or more electrons from their highest occupied energy levels. Figure 6.17 compares the atomic structure of a sodium atom and a sodium ion. In the sodium ion, the number of electrons (10) is not equal to the number of protons (11). Because there are more positively charged protons than negatively charged electrons, the sodium ion has a net positive charge. An ion with a positive charge is called a **cation.** The charge for a cation is written as a number followed by a plus sign. If the charge is 1, the number in 1+ is usually omitted from the symbol for the ion. For example, Na^{1+} is written as Na^{+}.

Atoms of nonmetals, such as chlorine, tend to form ions by gaining one or more electrons. Figure 6.18 compares the atomic structure of a chlorine atom and a chloride ion. In a chloride ion, the number of electrons (18) is not equal to the number of protons (17). Because there are more negatively charged electrons than positively charged protons, the chloride ion has a net negative charge. An ion with a negative charge is called an **anion.** The charge for an anion is written as a number followed by a minus sign.

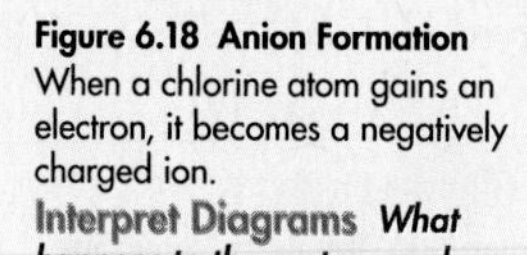

Figure 6.18 Anion Formation
When a chlorine atom gains an electron, it becomes a negatively charged ion.
Interpret Diagrams ***What happens to the protons and neutrons during this change?***

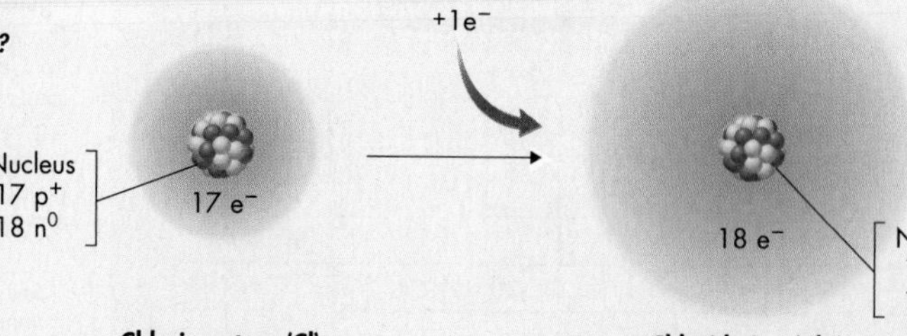

UbD Check for Understanding

How do ions form?

Assess students' knowledge about the formation of ions by asking them the following questions:

a. What occurs when an atom in Group 2 becomes an ion? *(It loses two electrons.)*

b. Is the ion that forms from a Group 2 atom called an anion or a cation? How do you know? *(It is a cation because it has a positive charge, +2.)*

ADJUST INSTRUCTION If students are confused, have them use a copy of the periodic table as a reference as they read the Ions section of the lesson or as they review. Review the information provided in the Misconception Alert with students and show them how to use the group numbers shown in red in Figure 6.9 to help them determine the number of electrons lost or gained in groups 1A–7A.

Trends in Ionization Energy

What are the trends among the elements for first ionization energy?

Recall that electrons can move to higher energy levels when atoms absorb energy. Sometimes the electron has enough energy to overcome the attraction of the protons in the nucleus. The energy required to remove an electron from an atom is called **ionization energy.** This energy is measured when an element is in its gaseous state. The energy required to remove the first electron from an atom is called the first ionization energy. The cation produced has a 1+ charge. **First ionization energy tends to decrease from top to bottom within a group and increase from left to right across a period.**

Ionization energies can help you predict what ions an element will form. Look at the data in Table 6.1 for lithium (Li), sodium (Na), and potassium (K). The increase in energy between the first and second ionization energies is large. It is relatively easy to remove one electron from a Group 1A metal atom, but it is difficult to remove a second electron. This difference indicates that Group 1A metals tend to form ions with a 1+ charge.

Interpret Data

Ionization Energies of First 20 Elements (kJ/mol*)

Symbol	First	Second	Third
H	1312		
He (noble gas)	2372	5247	
Li	520	7297	11,810
Be	899	1757	14,840
B	801	2430	3659
C	1086	2352	4619
N	1402	2857	4577
O	1314	3391	5301
F	1681	3375	6045
Ne (noble gas)	2080	3963	6276
Na	496	4565	6912
Mg	738	1450	7732
Al	578	1816	2744
Si	786	1577	3229
P	1012	1896	2910
S	999	2260	3380
Cl	1256	2297	3850
Ar (noble gas)	1520	2665	3947
K	419	3069	4600
Ca	590	1146	4941

*An amount of matter equal to the atomic mass in grams

Table 6.1 The table compares ionization energies for elements with atomic numbers 1 through 20.

a. Read Tables What are the values for the first, second, and third ionization energies for sodium and aluminum?

b. Compare Is it easier to remove an electron from a sodium (Na) or aluminum (Al) atom? From Na^+ or Al^+? From Na^{2+} or Al^{2+}?

c. Draw Conclusions Which ion is more common—Na^{3+} or Al^{3+}?

Note: The second ionization energy is the energy needed to remove an electron from an ion with a 1+ charge. This produces an ion with a 2+ charge. The third ionization energy is the energy needed to remove an electron from an ion with a 2+ charge. This produces an ion with a 3+ charge.

Differentiated Instruction

L1 LESS PROFICIENT READERS Have students refer to their KWL charts to review their notes on atomic structure. Have them use this information to identify three of the factors that affect ionization energy: nuclear charge, number of energy levels, and shielding.

ELL ENGLISH LANGUAGE LEARNERS Use student volunteers to set up a tug-of-war game to demonstrate why the amount of energy needed to remove successive electrons increases. Assign one team to be "protons" and the other team to be "electrons." Show students how the amount of force exerted by the "protons" on the "electrons" increases as each "electron" is removed from the game, and the "electrons" then have to pull harder against the "protons."

L3 ADVANCED STUDENTS Have students create a three-dimensional tactile or technological model that depicts the energy needed to remove an atom from an electron.

Explain

Trends in Ionization Energy

MAKE A CONNECTION Explain that ionization energy is a measure of the difficulty in removing an electron from the highest occupied energy level. **Ask** Why is the first ionization energy of a nonmetal much higher than that of an alkali metal? *(Because the nuclear charge increases from left to right across a period and the shielding effect stays the same, it is more difficult to remove an electron.)*

Misconception Alert

Some students may be tempted to place a negative sign on the ionization energy value. Students may confuse the phrase "losing an electron" with a negative energy value. Explain that the ionization energy is the energy change associated with removing one electron from a neutral atom. Explain that this energy value is always a positive number since energy must be added to the atom system to remove an electron. *(NOTE: The unit for ionization energy is kJ/mol. The footnote in Table 6.1 supplies an operational definition of mole, which is introduced in Lesson 10.1.)*

Explore

Teacher Demo

PURPOSE To help students understand the concepts of effective nuclear charge and electron shielding

PURPOSE Choose four students to be "protons" and four students to be "electrons." Construct a lithium "nucleus" by having three protons stand together at the front of the room. Note that for purposes of this demo, you are ignoring the neutrons.

- Place two electrons together at a short distance from the nucleus to represent the 1*s* electrons.
- Place the third electron a bit farther away to represent the 2*s* electron.

Point out that there are no other electrons between the 1*s* electrons and the nucleus. Explain that these electrons experience the full impact of the 3+ charge. The third electron's "view" of the nucleus is partially blocked. This means the third electron is shielded somewhat from the full force of the 3+ charge.

Answers

FIGURE 6.18 nothing

TABLE 6.1

a. sodium: 496, 4565, and 6912 kJ/mol; aluminum: 578, 1816, and 2744 kJ/mol

b. Na; Al^+; Al^{2+}

c. Al^{3+}

LESSON 6.3

Explain

Trends in Ionization Energy

USE VISUALS Direct students to Table 6.1, Figure 6.19, and their copy of the periodic table. Show students how to use Table 6.1 and the periodic table to create ordered pairs that are then plotted in the graph in Figure 6.19. Call out various atomic numbers and have students estimate the ionization energy, and vice versa. When students are comfortable reading the graph, direct them to answer the questions.

CRITICAL THINKING Challenge students to explain why the portion of the graph for Periods 4 and 5 is different from the portion of the graph for Periods 2 and 3. *(Periods 4 and 5 include transition metals, whose atoms have electrons in d orbitals.)*

Explore

Teacher Demo

PURPOSE Students observe the relative reactivities of magnesium and calcium and predict relative reactivities for other pairs of elements.

MATERIALS 20 mL 1*M* HCl, two 50-mL beakers, overhead projector, 20 cm magnesium ribbon, 1 g calcium

SAFETY Wear goggles for this demo.

PROCEDURE Pour 20 mL HCl into each beaker. Set the beakers on an overhead projector. Coil the magnesium ribbon and drop it into one beaker. Drop 1 g calcium into the other beaker. Compare the reaction rates in the two beakers. Point out the positions of the two elements in the periodic table, and relate the difference in reactivity to their first and second ionization energies. Ask students to predict the relative reactivities of other pairs of elements in Groups 1A and 2A.

EXPECTED OUTCOME The calcium fizzes in the HCl. The magnesium reacts more slowly with the HCl.

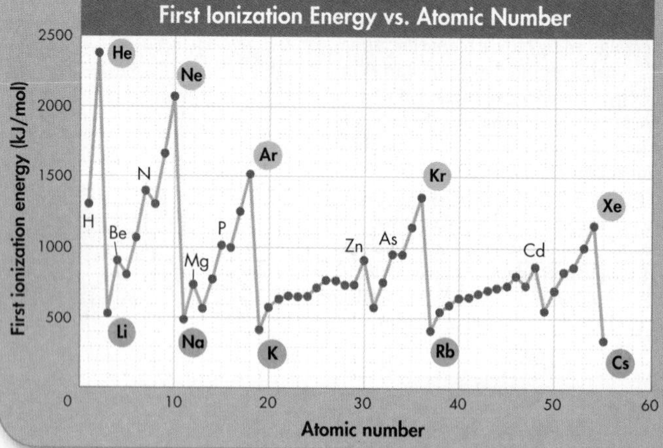

Figure 6.19 This graph reveals group and period trends for ionization energy.
a. Read Graphs Which element in Period 2 has the lowest first ionization energy? In Period 3?
b. Describe What are the group trends for first ionization energy for noble gases and alkali metals?
c. Predict If you drew a graph for second ionization energy, which element would you have to omit? Explain.

Group Trends in Ionization Energy Figure 6.19 is a graph of first ionization energy versus atomic number. Look at the data for the noble gases and the alkali metals. In general, first ionization energy decreases from top to bottom within a group. Recall that the atomic size increases as the atomic number increases within a group. As the size of the atom increases, nuclear charge has a smaller effect on the electrons in the highest occupied energy level. Less energy is required to remove an electron from this energy level, and the first ionization energy is lower.

Period Trends in Ionization Energy In general, the first ionization energy of representative elements tends to increase from left to right across a period. This trend can be explained by the nuclear charge and the shielding effect. The nuclear charge increases across the period, but the shielding effect remains constant. As a result, there is an increase in the attraction of the nucleus for an electron. Thus, it takes more energy to remove an electron from an atom. Figure 6.20 summarizes the group and period trends for first ionization energy.

Figure 6.20
Trends in First Ionization Energy
First ionization energy tends to increase from left to right across a period and decrease from top to bottom within a group.
Predict *Which element would have the larger first ionization energy—an alkali metal in Period 2 or an alkali metal in Period 4?*

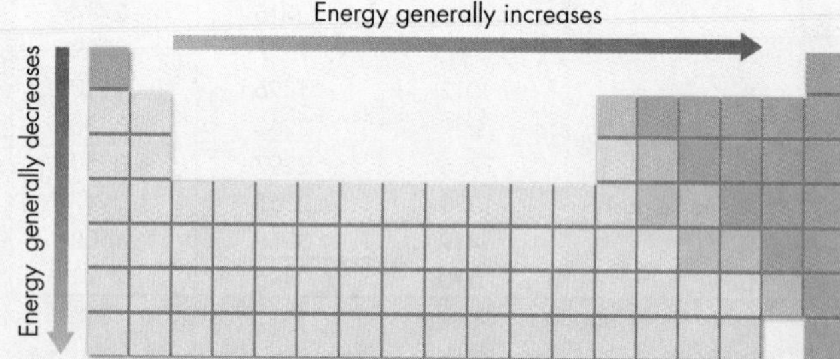

178 Chapter 6 • Lesson 3

UbD Check for Understanding

What are the trends among the elements for first ionization energy and ionic size and electronegativity?

Assess students' knowledge about the trends among the elements for first ionization energy by having students use arm gestures to answer the following questions. **Ask** In which direction on the periodic table does the first ionization energy generally increase in value? *(Accept arm gestures from left to right, and upward.)* **Ask** In which direction on the periodic table does the first ionization energy generally decrease in value? *(Accept arm gestures downward and from right to left.)*

ADJUST INSTRUCTION If students are having trouble answering, have them write some of the values in Table 6.1 on the element squares in a copy of the periodic table and examine their table for ionization patterns.

Trends in Ionic Size

What are the trends among the elements for ionic size?

During reactions between metals and nonmetals, metal atoms tend to lose electrons and nonmetal atoms tend to gain electrons. This transfer of electron has a predictable effect on the size of the ions that form. Cations are always smaller than the atoms from which they form. Anions are always larger than the atoms from which they form. **Ionic size tends to increase from top to bottom within a group. Generally, the size of cations and anions decrease from left to right across a period.**

Group Trends in Ionic Size Figure 6.21 compares the relative sizes of the atoms and ions for three metals in Group 1A—lithium (Li), sodium (Na), and potassium (K). For each of these elements, the ion is much smaller than the atom. For example, the radius of a sodium ion (95 pm) is about half the radius of a sodium atom (191 pm). When a sodium atom loses an electron, the attraction between the remaining electrons and the nucleus is increased. As a result, the electrons are drawn closer to the nucleus. Metals that are representative elements tend to lose all their outermost electrons during ionization. Therefore, the ion has one fewer occupied energy level.

The trend is the opposite for nonmetals, like the halogens in Group 7A. Look at Figure 6.21, and compare the relative sizes of the atoms and ions for fluorine (F), chlorine (Cl), and bromine (Br). For each of these elements, the ion is much larger than the atom. For example, the radius of a fluoride ion (133 pm) is more than twice the radius of a fluorine atom (62 pm). As the number of electrons increases, the attraction of the nucleus for any one electron decreases.

Period Trends in Ionic Size Look ahead at Figure 6.23. From left to right across a period, two trends are visible—a gradual decrease in the size of the positive ions (cations), followed by a gradual decrease in the size of the negative ions (anions). Figure 6.22 summarizes the group and period trends in ionic size.

Figure 6.21
Comparing Atomic and Ionic Sizes
This diagram compares the relative sizes of atoms and ions for selected alkali metals (Group 1A) and halogens (Group 7A). The numbers are measurements of the radii given in picometers (pm).

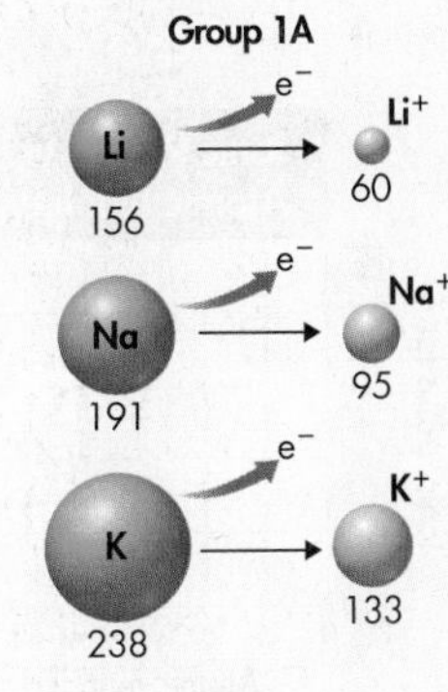

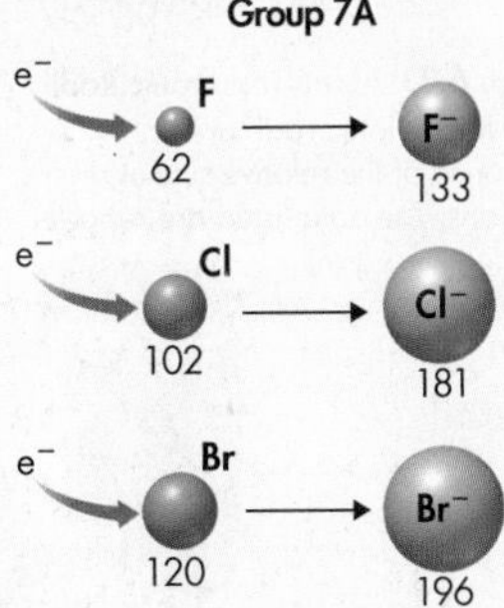

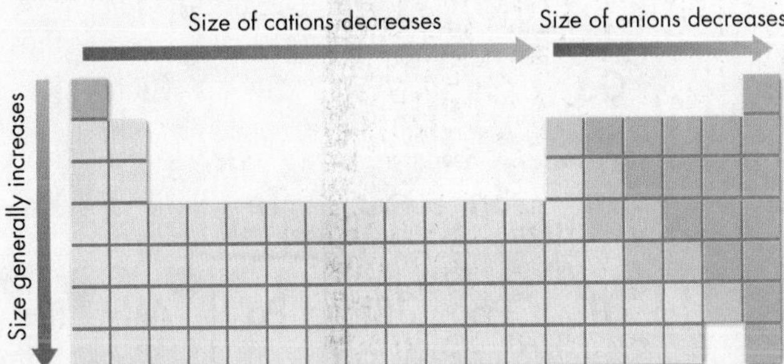

Figure 6.22 Trends in Ionic Size
The ionic radii for cations and anions decrease from left to right across periods and increase from top to bottom within groups.

Differentiated Instruction

L1 LESS PROFICIENT READERS Have students examine Table 6.2. Act out the meaning of electronegativity. Then have students predict if there are any elements other than noble gases that do not have a value for electronegativity. Guide students in researching their prediction. *(Predictions will vary; students will most likely find electronegativity tables lacking values for manmade elements with atomic numbers greater than 102.)* Note these elements rarely, if ever, have been demonstrated to form compounds because their atoms exist only momentarily due to their instability.

ELL SPECIAL NEEDS STUDENTS Provide tactile spherical models of various atoms. Have students arrange the models on a copy of the periodic table to visually convey the trend in ionic size.

L3 ADVANCED STUDENTS Have students research and describe the phenomenon of the lanthanide contraction. Ask them to discuss how the lanthanide contraction accounts for the fact that zirconium and hafnium have virtually the same atomic radius even though hafnium is below zirconium in Group 4B of the periodic table.

Explain

Trends in Ionic Size

MAKING CONNECTIONS Relate the periodic trends in ionic size to those discussed earlier for atomic size. Explain that the effective nuclear charge experienced by an electron in the highest occupied orbital of an atom or ion is equal to the total nuclear charge (the number of protons) minus the shielding effect due to electrons in lower energy levels. Point out that the effective nuclear charge determines the atomic and ionic radii. Explain that as you proceed from left to right in any given period, the principal quantum number, *n*, of the highest occupied energy level remains constant, but the effective nuclear charge increases. Therefore, atomic and ionic radii decrease as you move to the right in a period. Convey that in contrast, within any group, as you proceed from top to bottom, the effective nuclear charge remains nearly constant, but the principal quantum number, *n*, increases. Consequently, point out that atomic and ionic radii increase from top to bottom within a group.

Explore

Teacher Demo

PURPOSE Students observe an analogy for the effect of adding or removing electrons from an atom.

MATERIALS washers or other small circular items, smaller item (such as a button) to represent the nucleus, overhead projector

PROCEDURE On the overhead projector, make a circle of washers to represent an electron cloud in a neutral atom. The washers should be touching. Place the "nucleus" in the center of the circle. Add or subtract washers to mimic ion formation. With each change, adjust the circle so that the washers are still touching. Explain that the change in the diameter of the circle is analogous to the change in the effective attraction of the nuclear charge for electrons.

Answers

FIGURE 6.19

a. lithium; sodium

b. First ionization energy decreases as atomic number increases.

c. Hydrogen; it has only one electron.

FIGURE 6.20 an alkali metal in Period 2

LESSON 6.3

Explore

Trends in Ionic Size

Quick Lab

OBJECTIVE After completing this activity, students will be able to identify periodic trends in ionic size.

SKILLS FOCUS Using tables and graphs, predicting, drawing conclusions

PREP TIME none

CLASS TIME 40 minutes

TEACHING TIPS If time is too limited for students to make the graph, use Figure 6.23 to answer Questions 1, 2, 4, and 5. You may want to reference the radii diagrams in the Elements Handbook on R3, R7, R11, R15, R21, R25, and R29.

EXPECTED OUTCOME Ionic radii increase from top to bottom within a group. The radii of cations and anions decrease from left to right across a period.

ANALYZE AND CONCLUDE

1. Cations are smaller than their atoms; anions are larger than their atoms.
2. Ionic radii increase from top to bottom within a group of metals or within a group of nonmetals.
3. Two portions of the curve slope down from left to right.
4. The trend is similar for the periods.
5. The radii increase within a group because the number of occupied energy levels increases. The radii of cations decrease across a period because the nuclear charge increases, the shielding effect is constant, and the number of electrons decreases. (The effect is smaller with anions because the number of electrons increases.)

FOR ENRICHMENT Have students use the graph on page R37 to describe the periodic trend in atomic size for transition metals. Have students examine how the trend for transition metals compares to the trend for representative elements.

Quick Lab

Periodic Trends in Ionic Radii

Purpose To use a graph to identify period and group trends

Materials
- graph paper
- pencil

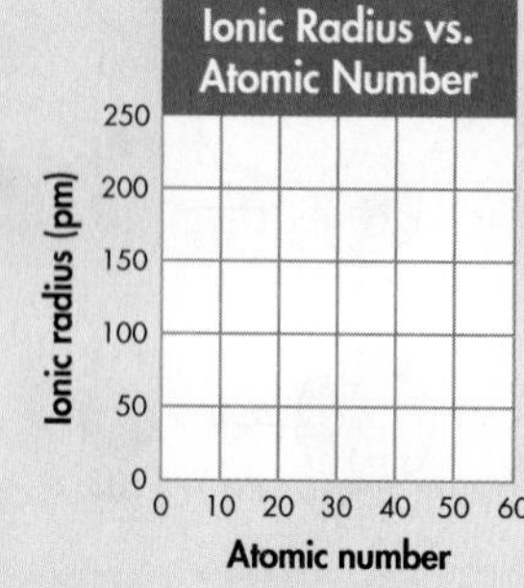

Procedure

Use the data presented in Figure 6.23 to plot ionic radius versus atomic number.

Analyze and Conclude

1. **Compare** How does the size change when an atom forms a cation and when an atom forms an anion?
2. **Describe** How do the ionic radii vary within a group of metals? How do they vary within a group of nonmetals?
3. **Describe** What is the shape of a portion of the graph that corresponds to one period?
4. **Compare and Contrast** Is the trend across a period similar or different for Periods 2, 3, 4, and 5?
5. **Explain** Propose explanations for the trends you have described for ionic radii within groups and across periods.

Figure 6.23 Atomic and Ionic Radii
Atomic and ionic radii are an indication of the relative size of atoms and ions. The data listed are reported in picometers (pm).

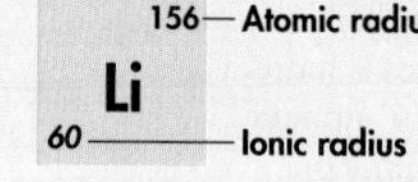

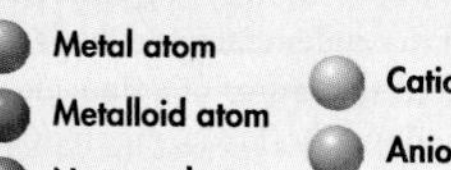

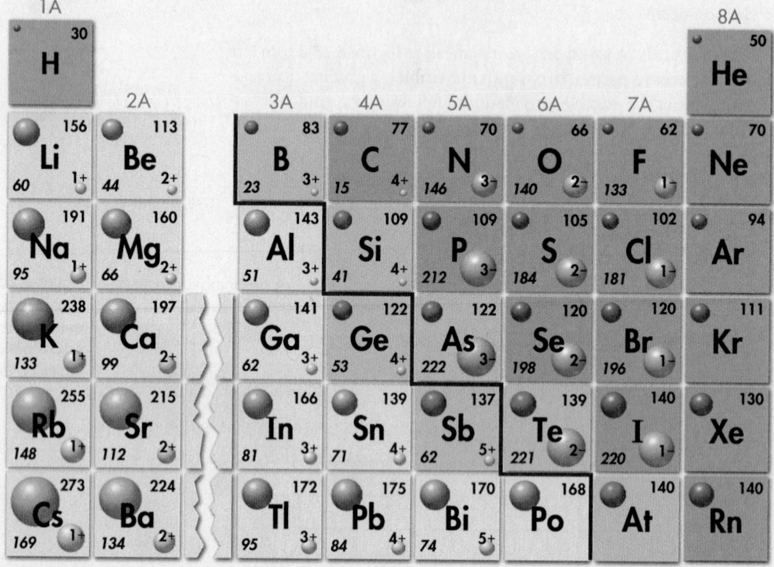

Focus on ELL

4 LANGUAGE PRODUCTION Have students work in groups or pairs to complete the lab. Review plotting data on a coordinate grid with the class. Pair students with higher proficiency in English with students of lower proficiency.

BEGINNING: LOW/HIGH Help students create a set of ordered pairs to plot on the graph. Use gestures to visually show how to plot an ordered pair on a coordinate grid.

INTERMEDIATE: LOW/HIGH Paraphrase the questions in the Analyze and Conclude section. Guide students to answer one question at a time.

ADVANCED: LOW/HIGH Direct students to read the graph aloud, pointing out trends and making predictions prior to answering the questions.

Trends in Electronegativity

What are the trends among the elements for electronegativity?

In Chapters 7 and 8, you will study two types of bonds that can exist in compounds. Electrons are involved in both types of bonds. There is a property that can be used to predict the type of bond that will form during a reaction. This property is called electronegativity. **Electronegativity** is the ability of an atom of an element to attract electrons when the atom is in a compound. Scientists use factors such as ionization energy to calculate values for electronegativity.

Table 6.2 lists electronegativity values for representative elements in Groups 1A through 7A. The elements are arranged in the same order as in the periodic table. The noble gases are omitted because they do not form many compounds. The data in Table 6.2 is expressed in Pauling units. Linus Pauling won a Nobel Prize in Chemistry for his work on chemical bonds. He was the first to define electronegativity.

In general, electronegativity values decrease from top to bottom within a group. For representative elements, the values tend to increase from left to right across a period. Metals at the far left of the periodic table have low values. By contrast, nonmetals at the far right (excluding noble gases) have high values. The electronegativity values among the transition metals are not as regular.

The least electronegative element in the table is cesium, with an electronegativity value of 0.7. It has the least tendency to attract electrons. When it reacts, it tends to lose electrons and form cations. The most electronegative element is fluorine, with a value of 4.0. Because fluorine has such a strong tendency to attract electrons, when it is bonded to any other element it either attracts the shared electrons or forms an anion.

Figure 6.24, on the next page, summarizes several trends that exist among the elements. Refer to this figure as you study the periodic trends presented in this chapter.

Table 6.2

Electronegativity Values for Selected Elements

H 2.1						
Li 1.0	Be 1.5	B 2.0	C 2.5	N 3.0	O 3.5	F 4.0
Na 0.9	Mg 1.2	Al 1.5	Si 1.8	P 2.1	S 2.5	Cl 3.0
K 0.8	Ca 1.0	Ga 1.6	Ge 1.8	As 2.0	Se 2.4	Br 2.8
Rb 0.8	Sr 1.0	In 1.7	Sn 1.8	Sb 1.9	Te 2.1	I 2.5
Cs 0.7	Ba 0.9	Tl 1.8	Pb 1.9	Bi 1.9		

Explore

Trends in Electronegativity

START A CONVERSATION Lead a class discussion on periodic and group trends in electronegativities. Point out that electronegativity values help chemists predict the type of bonding that exists between atoms in compounds. **Ask** Why are the noble gases not included in a discussion on electronegativity? *(because they form very few compounds)* **Ask** Which element represented in Table 6.2 is the most electronegative and which is the least electronegative? *(fluorine; cesium)* Stress that electronegativity is a calculated value rather than a measured quantity.

APPLY CONCEPTS Explain that the values for *electronegativity* are often based on values for ionization energy and electron affinity. Explain that ionization energy is a measure of an atom's ability to lose electrons and electron affinity is a measure of an atom's ability to gain electrons.

Misconception Alert

Students often confuse the meanings of *electronegativity* and *ionization energy*. As a class, brainstorm ideas for how to remember the meanings of each term. Then have students pictorially illustrate both concepts.

Focus on ELL

4 ENABLE LANGUAGE PRODUCTION Have students work in small groups to complete the Small-Scale Lab on page 184. Make sure each group has ELLs of varied language proficiencies, so that more proficient students can help less proficient ones. Have students work according to their proficiency level.

BEGINNING

LOW Model the procedure and have students mimic you. Show students how to read measures from measuring tools.

HIGH Rephrase steps 1 and 3 in the procedure as several single-direction steps.

INTERMEDIATE: LOW/HIGH Restate the Analyze and Conclude questions in simpler terms. Allow students to orally respond to the Analyze and Conclude questions.

ADVANCED: LOW/HIGH Have students paraphrase the questions in Analyze and Conclude and You're the Chemist and read them aloud to students with lower English proficiencies.

Explain

USE VISUALS Direct students' attention to Figure 6.24. Point out that this diagram incorporates information from several diagrams earlier in the chapter. Lead a discussion as to whether students find this or earlier diagrams more helpful, and why.

CHEMISTRY & YOU Trends in atomic size, ionization energy, ionic size, and electronegativity can be identified with the help of the periodic table.

Evaluate

Informal Assessment

Assign each student two elements in the same group and have the student compare the elements in terms of atomic radius, ionic radius, ionization energy, and electronegativity. For successful students, repeat the exercise with a metal and nonmetal from the same period. Have students write general statements to summarize the trends revealed by these comparisons. Then, have students complete the 6.3 Lesson Check.

Reteach

Review the terms used in Figure 6.24. Then, use the periodic table and the terms to play a version of "I'm thinking of . . . ." For example, choose fluorine and say you are thinking of an element that has a very small atomic size and a very high electronegativity. Let students guess, and then discuss the correct answer. Have students continue the game in small groups.

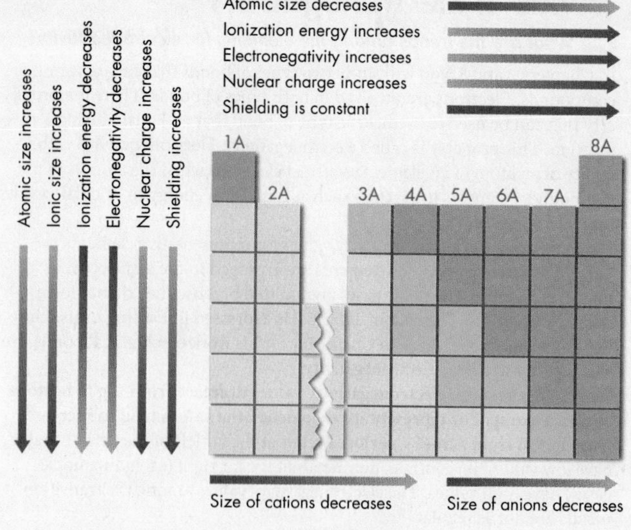

CHEMISTRY & YOU

Q: *You are familiar with using a weather map to identify trends in the weather. For example, certain areas are typically warmer than other areas. What trends in the properties of elements can you identify with the help of the periodic table?*

Figure 6.24 Summary of Periodic Trends
Trends for atomic size, ionization energy, ionic size, and electronegativity vary within groups and across periods. The trends that exist among these properties can be explained by variations in atomic structure. The increase in nuclear charge within groups and across periods explains many trends. Within groups, an increase in the number of occupied energy levels and an increase in shielding both have a significant effect on each trend.
Interpret Diagrams ***Which properties tend to decrease across a period? Which properties tend to decrease down a group?***

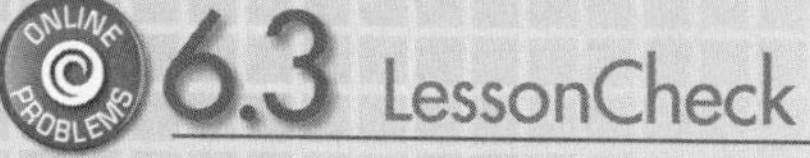

18. **Review** How does atomic size change within groups and across periods?
19. **Explain** When do ions form?
20. **Summarize** How do first ionization energies vary within groups and across periods?
21. **Describe** Compare the size of ions to the size of the atoms from which they form.
22. **Review** How do electronegativity values vary within groups and across periods?
23. **Explain** In general, how can the periodic trends displayed by elements be explained?
24. **Sequence** Arrange these elements in order of decreasing atomic size: sulfur, chlorine, aluminum, and sodium. Does your arrangement demonstrate a periodic trend or a group trend?
25. **Identify** Which element in each pair has the larger first ionization energy?
    a. sodium, potassium
    b. magnesium, phosphorus

Lesson Check Answers

18. Atomic size generally increases within a group and decreases from left to right across a period.
19. Ions form when electrons are transferred between atoms.
20. First ionization energy generally decreases within a group and increases from left to right across a period.
21. Anions are larger and cations are smaller than the atoms from which they form.
22. Electronegativity values generally decrease from top to bottom within a group and increase from left to right across a period.
23. The trends can be explained by variations in atomic structure.
24. sodium, aluminum, sulfur, chlorine; periodic trend
25. **a.** sodium **b.** phosphorus

Elements of Life

Like everything else in the universe, your body is made up of elements. Your body uses these elements for different functions. Roughly 97 percent of the human body consists of just four elements: oxygen, carbon, hydrogen, and nitrogen. The remaining 3 percent contains about 20 other elements that are essential to life.

CIRCULATORY SYSTEM Iron and oxygen are critical to the circulatory system—the system that carries blood throughout the body. Iron, which is contained in red blood cells, helps transport oxygen from the lungs to other cells in your body. Two other elements—copper and cobalt—are necessary for the formation of red blood cells.

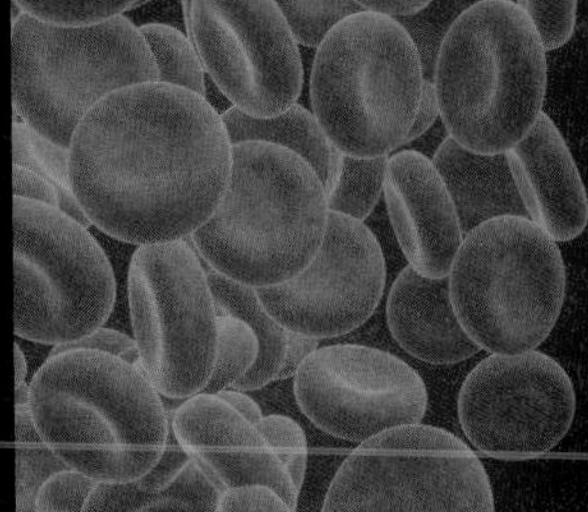

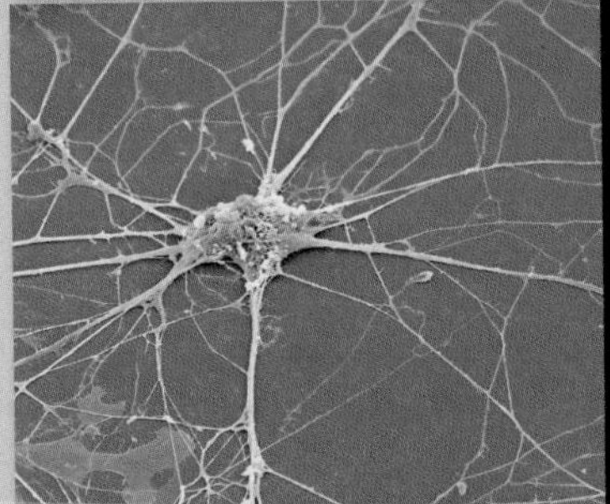

NERVOUS SYSTEM Sodium and potassium are essential to the nervous system, in particular the nerve cells. These elements allow your brain to communicate with other tissues in your body. Other elements that are important for proper nervous system function include calcium, chlorine, zinc, and magnesium.

SKELETAL SYSTEM Your bones and teeth—two components of the skeletal system—are largely comprised of calcium and phosphorus, which give bones and teeth their strength. Fluorine, boron, magnesium, and silicon are also important for bone growth and for maintaining bone strength.

Take It Further

1. Describe Use the information provided on page R1 to estimate the composition of the human body in terms of metals, nonmetals, and metalloids.

2. Predict The elements sodium, magnesium, potassium, and calcium are the most abundant metals in the human body and are present as ions. What is the charge of each of these ions?

3. Sequence Use Figure 6.23 to list the ions in Question 2 from smallest to largest.

Differentiated Instruction

L1 STRUGGLING STUDENTS Help students make a circle graph of the distribution of elements in living organisms as a percentage of body weight. Write these numbers on the board for them to use: oxygen, 61%; carbon, 23%; nitrogen, 2.6%; hydrogen, 10%, and other elements, 3.4%. You may wish to have students use a graphing program to make the graphs, or have them make the graph on the board as a class project.

LPR LESS PROFICIENT READERS Have students read the feature together with a partner. First, have students read a paragraph independently, and then have them discuss what they have read with their partners. Afterwards, discuss the feature together as a class.

L3 ADVANCED STUDENTS Ask students to research and write a report, including tables, that distinguishes between the percentages by *mass* and percentages by *number of atoms* of each of the four main elements in the body.

CHEMISTRY & YOU Have students look at the photos and read about the Elements of Life. Engage students in a conversation of ways the four main elements of life—hydrogen, oxygen, carbon, and nitrogen—are a part of their everyday lives. Students should realize that everything they do, every moment of the day, involves these elements. Pose the following question to students: *How many different encounters have you had with these four elements today?* You may need to assist students in the following ways:

- activities involving eating, drinking, or performing personal hygiene tasks
- activities involved in getting to school
- activities the body carries out to live
- objects encountered in nature

Explain

START A CONVERSATION Explain to students that about 97% of the atoms in the body are either hydrogen, oxygen, carbon, or nitrogen. Encourage them to think about what properties might make these elements so special. Explain that in later chapters they will learn about how atoms combine by forming chemical bonds. Atoms of these four elements are small and light. Also, the arrangement of electrons in the atoms allows them to form bonds in such a way that the atoms can combine into the large, stable molecules necessary to carry out life functions.

Extend

Connect to PHYSIOLOGY

Point out to students that even though hydrogen, oxygen, carbon, and nitrogen are by far the most common atoms in organisms, they are not the *only* atoms that are necessary for life. Have students research the relative percentages of all the different elements present in the human body and to present their findings to the class.

Answers

FIGURE 6.24 sizes of atoms and ions; ionization energy and electronegativity

TAKE IT FURTHER

1. metals: 2%; nonmetals: 98%; metalloids: 0%
2. sodium: 1+; magnesium: 2+; potassium: 1+; calcium: 2+
3. magnesium, sodium, calcium, potassium

SMALL-SCALE LAB

Explore

Small-Scale Lab

OBJECTIVE After completing this activity, students should be able to build concrete models to reinforce periodic trends. They should also be able to apply a procedure to a new variable and design a model on their own.

PREP TIME 10 minutes

CLASS TIME 40 minutes

MATERIALS 96-well spot plates, straws, scissors, metric rulers, permanent fine-line markers

ADVANCE PREPARATION Straws with a 1/4-inch diameter fit snugly in the wells.

TEACHING TIPS Students can use colored straws to color code groups or periods. If you do not have spot plates, press a lump of clay the size of a golf ball flat on a table with a block of wood. Students can mark out a 1-cm square grid and insert the straws in the clay.

EXPECTED OUTCOME Students produce 3-D models for periodic trends.

ANALYZE AND CONCLUDE

1. fluorine
2. Electronegativity generally increases from left to right along a period.
3. Metals, which are on the left side of the table, have lower electronegativity values than nonmetals, which are on the right.
4. Electronegativity generally increases from bottom to top within a group. Except for boron, the rest of Group 3A shows a reverse in this trend.
5. Hydrogen is placed in Group 1A based on its electron configuration, but is classified as a nonmetal.

FOR ENRICHMENT Have students use the data on page R37 to make a 3-D model of trends in atomic size for transition metals.

YOU'RE THE CHEMIST

1. Students divide the values of first ionization energies by 300 and measure the appropriate length of straws.
2. Students must determine their own scale before they begin. Students often use two wells to represent both ionic and atomic radii. Other students cut a straw to a length that represents the larger radius of an atom and mark the straw to show the smaller radius of the corresponding cation.
3. The value for xenon is similar to iodine, which is consistent with the general trend. Based on this value, xenon appears to have the ability to attract electrons and form compounds.

Small-Scale Lab

Periodicity in Three Dimensions

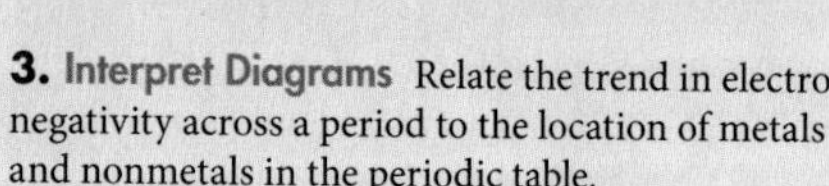

Purpose

To build three-dimensional models for periodic trends

Materials

- 96-well spot plate
- straws
- scissors
- metric ruler
- permanent fine-line marker

Procedure

1. Measure the depth of a well in the spot plate by inserting a straw into a well and holding the straw upright as shown in the photograph. Make a mark on the straw at the point where the straw meets the surface of the plate. Measure the distance from the end of the straw to the mark in centimeters (cm). Record this distance as well depth.
2. Cut the straw to a length that is 4.0 cm plus well depth. The straw will extend exactly 4.0 cm above the surface of the plate.
3. Fluorine has an electronegativity value of 4.0. On a scale of 1.0 cm equals 1.0 unit of electronegativity, the portion of the straw that extends above the surface of the plate represents the electronegativity value for fluorine. Using the same scale, cut straws to represent the electronegativity values for all the elements listed in Table 6.2. Remember to add the well depth to the electronegativity value before cutting a straw. As you cut the straws, mark each straw with the chemical symbol of the element that the straw represents.
4. Arrange the straws in the spot plate in rows and columns to match the locations of the elements in the periodic table.

Analyze and Conclude

1. **Use Models** Which element represented in your model is the most electronegative?
2. **Use Models** Based on your model, what is the general trend in electronegativity from left to right across a period?
3. **Interpret Diagrams** Relate the trend in electronegativity across a period to the location of metals and nonmetals in the periodic table.
4. **Use Models** Based on your model, what is the general trend in electronegativity within a group? Are there any notable exceptions?
5. **Explain** Why do you think that the electronegativity value for hydrogen is so high given its location in the periodic table?

You're the Chemist

1. **Design an Experiment** Construct a similar three-dimensional model for first ionization energies. Use the data in Table 6.1 to construct the model. Use a scale of 1.0 cm equals 300 kJ/mol.
2. **Design an Experiment** Design and construct a three-dimensional model that shows trends in atomic and ionic radii for the elements in Groups 1A and 7A. Devise a way to display both ionic and atomic radii in the same model.
3. **Analyze Data** Xenon has an electronegativity value of 2.6. Cut and place a straw in your first model to represent xenon. Does xenon support the trend for electronegativity across a period? Is xenon likely to form compounds? Explain your answers.

6 Study Guide

BIGIDEA ELECTRONS AND THE STRUCTURE OF ATOMS

Periodic tables may contain each element's name, symbol, atomic number, atomic mass, and number of electrons in each energy level. The electron configuration of an element can be determined based on the location of an element in the periodic table. Atomic size, ionization energy, ionic size, and electronegativity are trends that vary across periods and groups of the periodic table. These trends can be explained by variations in atomic structure. The increase in nuclear charge within groups and across periods explains many trends. Within groups, an increase in electron shielding has a significant effect on these trends.

6.1 Organizing the Elements

Early chemists used the properties of elements to sort them into groups.

Mendeleev arranged the elements in his periodic table in order of increasing atomic mass.

In the modern periodic table, elements are arranged in order of increasing atomic number.

Three classes of elements are metals, nonmetals, and metalloids.

- periodic law (162)
- metal (165)
- nonmetal (165)
- metalloid (166)

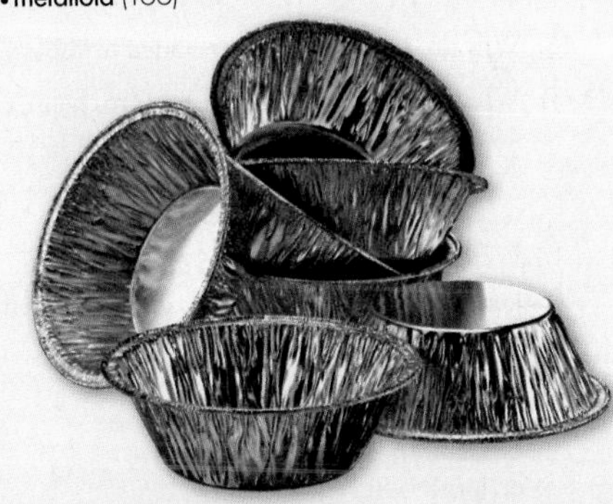

6.2 Classifying the Elements

The periodic table usually displays the symbols and names of elements, along with information about the structure of their atoms.

Elements can be sorted into noble gases, representative elements, transition metals, or inner transition metals based on their electron configurations.

- alkali metal (167)
- alkaline earth metal (167)
- halogen (167)
- noble gas (170)
- representative element (171)
- transition metal (172)
- inner transition metal (172)

6.3 Periodic Trends

In general, atomic size increases from top to bottom within a group and decreases from left to right across a period.

Positive and negative ions form when electrons are transferred between atoms.

First ionization energy tends to decrease from top to bottom within a group and increase from left to right across a period.

Ionic size tends to increase from top to bottom within a group. Generally, the size of cations and anions decrease from left to right across a period.

In general, electronegativity values decrease from top to bottom within a group. For representative elements, the values tend to increase from left to right across a period.

- atomic radius (174)
- ion (176)
- cation (176)
- anion (176)
- ionization energy (177)
- electronegativity (181)

UbD Performance Tasks

EXPOSED TO THE ELEMENTS Have students design a periodic table that contains an eighth period based on their knowledge of how the current table is constructed and their understanding of electron configuration. Students should accompany their table with an explanation of the design. Tables should include space for elements containing a 5*f* sublevel and correctly indicate appropriate atomic numbers for the new elements. Students should also predict the classifications of the new elements based on existing classifications (transition metals, noble gases, alkali metals, and so on).

PERIODIC TRENDS Have students use presentation software to create a multi-slide display about trends in the periodic table. Encourage the use of animations to make their display more appealing and understandable.

Evaluate

Review and Assessment Materials

CHEMISTRY TUTORIAL Encourage students to review electron configurations using the animated step-by-step tutorial for Chapter 6. The tutorial is followed by a similar practice problem.

CHAPTER PROBLEM SET Direct students to additional problems about the periodic table in the Chapter 6 online Problem Set.

DIRECTED VIRTUAL LAB Have students complete the Virtual Lab *Heat Energy and Metals* to help reinforce their understanding of trends exhibited by the elements in the periodic table. In the post-lab exercise, students will analyze the data and draw conclusions based on the results of the experiment.

KINETIC ART Direct students to the Kinetic Art animations *Periodic Table Tour* and *Trends in the Periodic Table* and have them answer the questions that accompany each animation.

Study Tip

REWRITE CLASS NOTES Remind students to take notes during class of the key points that are discussed, and to rewrite the notes so that they are organized and legible. Students can use the textbook to fill in any important points that they missed. They can use the organized notes to review for quizzes and tests.

If your class subscribes to the Interactive Textbook, your students can go online to access an interactive version of the Student Edition and a self-test.

Evaluate

Answers

LESSON 6.1

26. so that undiscovered elements with similar properties could be placed in the same group

27. The close match between the predicted properties and the actual properties of gallium helped gain wider acceptance for Mendeleev's periodic table.

28. a periodic repetition of their physical and chemical properties

29. Yes; both carbon and silicon are in Group 4A, and each has four electrons in the highest occupied energy level.

30. **a.** nonmetal **b.** nonmetal **c.** metal **d.** nonmetal **e.** metal

31. Metalloids have properties that are similar to those of metals and nonmetals. How a metalloid behaves depends on the conditions.

LESSON 6.2

32. Group 1A, Group 2A, Group 7A, and Group 8A, respectively

33. Na, Mg, Cl

34. helium

35. aluminum

36. **a.** $1s^22s^22p^1$
b. $1s^22s^22p^63s^23p^63d^{10}4s^24p^3$
c. $1s^22s^22p^5$
d. $1s^22s^22p^63s^23p^63d^{10}4s^2$
e. $1s^22s^22p^63s^23p^1$

37. **a.** Ar: $1s^22s^22p^63s^23p^6$
b. Si: $1s^22s^22p^63s^23p^2$
c. Mg: $1s^22s^22p^63s^2$

LESSON 6.3

38. **a.** sodium **b.** strontium **c.** germanium **d.** selenium

39. The first ionization energy is the energy needed to remove a first electron from an atom. The second ionization energy is the energy needed to remove a second electron.

40. **a.** boron **b.** magnesium **c.** aluminum

41. **a.** Sr, Mg, Be **b.** Cs, Ba, Bi **c.** Na, Al, S

42. It is relatively easy to remove the first electron from an alkali metal atom; it is much more difficult to remove the second.

43. The ionic radius of a metal cation is smaller than the atomic radius of the metal atom.

44. **a.** Na **b.** S^{2-} **c.** I^- **d.** Al

45. **a.** F **b.** N **c.** Mg **d.** As

46. Noble gases, with rare exception, do not form compounds.

47. **a.** O **b.** F **c.** O **d.** S

48. a and c

6 Assessment

* Solutions appear in Appendix E

Lesson by Lesson

6.1 Organizing the Elements

26. Why did Mendeleev leave spaces in his periodic table?

***27.** What effect did the discovery of gallium have on the acceptance of Mendeleev's table?

28. What pattern is revealed when the elements are arranged in a periodic table in order of increasing atomic number?

29. Based on their locations in the periodic table, would you expect carbon and silicon to have similar properties? Explain your answer.

30. Identify each property below as more characteristic of a metal or a nonmetal.
a. a gas at room temperature
b. brittle
c. malleable
d. poor conductor of electric current
e. shiny

31. In general, how are metalloids different from metals and nonmetals?

6.2 Classifying the Elements

32. Where are the alkali metals, the alkaline earth metals, the halogens, and the noble gases located in the periodic table?

33. Which of the following are symbols for representative elements: Na, Mg, Fe, Ni, Cl?

***34.** Which noble gas does not have eight electrons in its highest occupied energy level?

35. Which of these metals isn't a transition metal?
a. aluminum **c.** iron
b. silver **d.** zirconium

36. Use Figure 6.13 to write the electron configurations of these elements.
a. boron **c.** fluorine **e.** aluminum
b. arsenic **d.** zinc

37. Write the electron configurations of these elements.
a. the noble gas in Period 3
b. the metalloid in Period 3
c. the alkali earth metal in Period 3

6.3 Periodic Trends

***38.** Which element in each pair has atoms with a larger atomic radius?
a. sodium, lithium
b. strontium, magnesium
c. carbon, germanium
d. selenium, oxygen

39. Explain the difference between the first and second ionization energy of an element.

40. Which element in each pair has a greater first ionization energy?
a. lithium, boron
b. magnesium, strontium
c. cesium, aluminum

41. Arrange the following groups of elements in order of increasing ionization energy:
a. Be, Mg, Sr **b.** Bi, Cs, Ba **c.** Na, Al, S

42. Why is there a large increase between the first and second ionization energies of the alkali metals?

***43.** How does the ionic radius of a typical metal compare with its atomic radius?

44. Which particle has the larger radius in each atom/ion pair?
a. Na, Na^+ **c.** I, I^-
b. S, S^{2-} **d.** Al, Al^{3+}

45. Which element in each pair has a higher electronegativity value?
a. Cl, F **c.** Mg, Ne
b. C, N **d.** As, Ca

46. Why are noble gases not included in Table 6.2?

***47.** When the elements in each pair are chemically combined, which element in each pair has a greater attraction for electrons?
a. Ca or O **c.** H or O
b. O or F **d.** K or S

48. For which of these properties does lithium have a larger value than potassium?
a. first ionization energy
b. atomic radius
c. electronegativity
d. ionic radius

Focus on ELL

5 ASSESS UNDERSTANDING Practice for higher-level assessment questions by assigning students Think Critically and Enrichment problems.

BEGINNING: LOW/HIGH Explain the meaning of the boldface words that begin each problem. Explain that these key words help prepare students for the type of thought process needed to solve each problem.

INTERMEDIATE: LOW/HIGH Help students break apart multi-part questions into several simpler questions. Guide students to answer one part of the question at a time. Allow students to answer the questions orally and then help them to write their responses as simple sentences.

ADVANCED: LOW/HIGH Have students read each question, considering the sentence they just read before moving to the next part of the problem. Then, have them write bulleted concepts they need to examine prior to answering the question.

Understand Concepts

49. The bar graph shows how many elements were discovered before 1750 and in each 50-year period between 1750 and 2000.

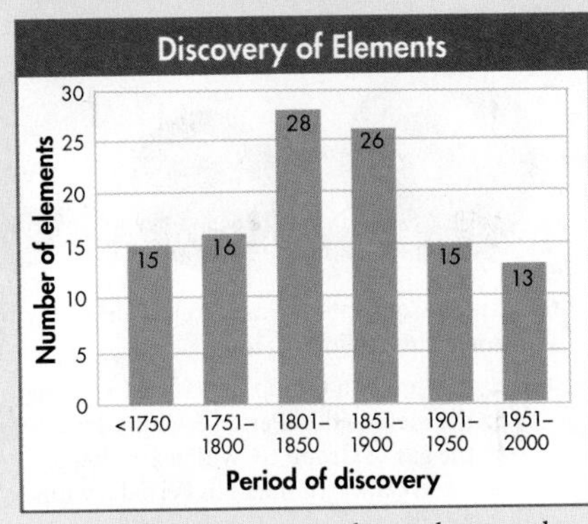

a. In which 50-year period were the most elements discovered?
b. How did Mendeleev's work contribute to the discovery of elements?
c. What percent of these elements were discovered by 1900?

50. Write the symbol of the element or elements that fit each description.

a. a nonmetal in Group 4A
b. the inner transition metal with the lowest atomic number
c. all of the nonmetals for which the atomic number is a multiple of five
d. a metal in Group 5A

***51.** In which pair of elements are the chemical properties of the elements most similar? Explain your reasoning.

a. sodium and chlorine
b. nitrogen and phosphorus
c. boron and oxygen

52. Explain why fluorine has a smaller atomic radius than both oxygen and chlorine.

53. Would you expect metals or nonmetals in the same period to have higher ionization energies? Give a reason for your answer.

54. In each pair, which ion is larger?

a. Ca^{2+}, Mg^{2+} b. Cl^-, P^{3-} c. Cu^+, Cu^{2+}

55. Use the graph in Figure 6.15 to estimate the atomic radius of the indium atom.

***56.** List the symbols for all the elements with electron configurations that end as follows. *Note:* Each *n* represents an energy level.

a. ns^1 b. ns^2np^4 c. ns^2nd^{10}

57. Explain why there should be a connection between an element's electron configuration and its location on the periodic table.

58. Which equation represents the first ionization of an alkali metal atom?

a. $Cl \longrightarrow Cl^+ + e^-$
b. $Ca \longrightarrow Ca^+ + e^-$
c. $K \longrightarrow K^+ + e^-$
d. $H \longrightarrow H^+ + e^-$

59. What trend is demonstrated by the following series of equations?

$$Li + 520\ kJ/mol \longrightarrow Li^+ + e^-$$
$$O + 1314\ kJ/mol \longrightarrow O^+ + e^-$$
$$F + 1681\ kJ/mol \longrightarrow F^+ + e^-$$
$$Ne + 2080\ kJ/mol \longrightarrow Ne^+ + e^-$$

***60.** There is a large jump between the second and third ionization energies of magnesium. There is a large jump between the third and fourth ionization energies of aluminum. Explain these observations.

61. The bar graph shows the relationship between atomic and ionic radii for Group 1A elements.

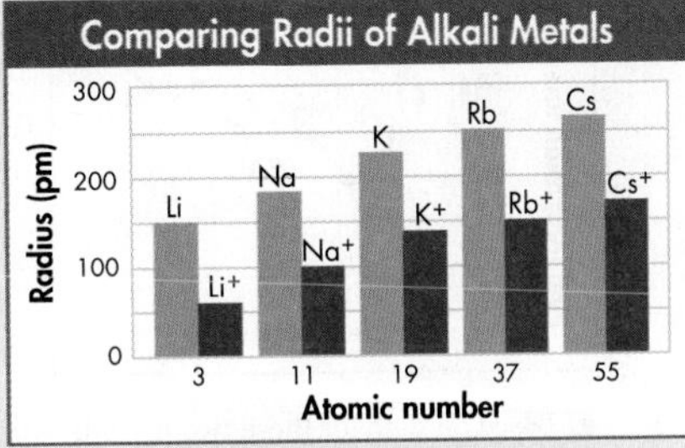

a. Describe and explain the trend in atomic radius within the group.
b. Explain the difference between the size of the atoms and the size of the ions.

62. Locate each of the following elements in the periodic table and decide whether its atoms are likely to form anions or cations.

a. sodium
b. fluorine
c. calcium
d. potassium
e. iodine
f. beryllium
g. oxygen
h. lithium

UNDERSTAND CONCEPTS

49. **a.** 1801–1850
b. Mendeleev's periodic table helped scientists predict the existence of undiscovered elements.
c. 75%

50. **a.** C **b.** La **c.** Ne, P, Br **d.** Bi

51. b; Nitrogen and phosphorus are in the same group (Group 5A).

52. Fluorine has a smaller atomic radius than oxygen because fluorine has a larger nuclear charge. Fluorine has a smaller radius than chlorine because it has eight fewer electrons.

53. Nonmetals; The trend is for ionization energy to increase from left to right across a period.

54. **a.** Ca^{2+} **b.** P^{3-} **c.** Cu^+

55. 170 pm

56. **a.** H, Li, Na, K, Rb, Cs, Fr **b.** O, S, Se, Te, Po
c. Zn, Cd, Hg, Cn

57. An element's electron configuration determines its location (group) in the periodic table.

58. c

59. First ionization energy increases across a period.

60. It is relatively easy to remove two electrons from magnesium; it is much more difficult to remove a third electron. It is relatively easy to remove three electrons from aluminum; it is much more difficult to remove a fourth electron.

61. **a.** The atomic radius increases from top to bottom within the group.
b. Cations are smaller than their corresponding atoms. The attraction between the nucleus and any remaining electron is greater. There is one fewer occupied energy level.

62. **a.** cations
b. anions
c. cations
d. cations
e. anions
f. cations
g. anions
h. cations

ASSESSMENT

Evaluate

Answers

THINK CRITICALLY

63. Yes, 119; 1 through 118 are known.

64. **a.** Electronegativity increases as first ionization energy increases.
b. Both properties depend on the attraction between the nucleus and electrons in the highest occupied energy level. Across a period, the nuclear charge increases, but the shielding effect is constant.

65. **a.** The electrons in calcium are removed from the same energy level. In potassium, the second electron is removed from a lower energy level.
b. Because cesium has a larger atomic radius than lithium, the nuclear charge in a cesium atom has a smaller effect on the electrons in the highest occupied energy level.
c. The third electron removed from a magnesium atom is in a lower energy level.

66. Zinc has a greater nuclear charge (more protons) than calcium.

67. **(a)** is Ca, **(b)** is Ca^{2+}, **(c)** is Mg^{2+}. Cations are always smaller than the corresponding atoms, and ionic size generally increases from top to bottom within a group.

68. **a.** $1s^22s^22p^63s^23p^63d^{10}4s^24p^6$
b. $1s^22s^22p^6$
c. $1s^22s^22p^6$
d. $1s^22s^22p^6$

69. **a.** francium
b. scandium

70. Because the nuclear charge increases from left to right across a period and the shielding effect stays the same, it is more difficult to remove an electron.

71. Density is equal to mass/volume. For this series of elements, the atomic mass increases, and the atomic size (proportional to volume) decreases. So, the density of these elements must increase going from potassium to chromium.

72. Cations are smaller than the corresponding atoms because they have lost one or more electrons and the attraction between the remaining electrons and the nucleus is increased. Anions are larger than the corresponding atoms because they have gained one or more electrons and the attraction for any one electron decreases.

Think Critically

63. Predict Do you think there are more elements left to discover? If so, what is the lowest atomic number a new element could have? Explain.

64. Interpret Graphs The graphs show the relationship between the electronegativities and first ionization energies for Period 2 and Period 3 elements.

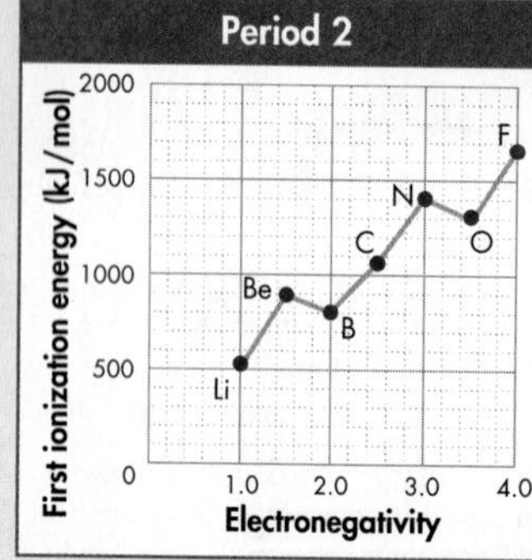

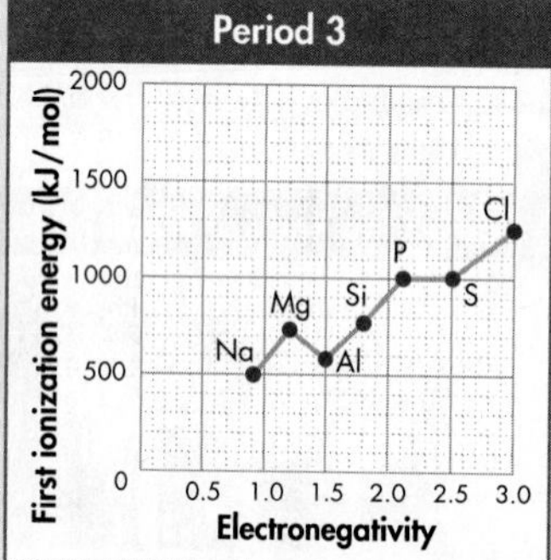

a. Based on data for these two periods, what is the general trend between these two values?
b. Use nuclear charge and shielding effect to explain this trend.

***65. Explain** Give a reason for each of the following comparisons:
a. Calcium has a smaller second ionization energy than does potassium.
b. Lithium has a larger first ionization energy than does cesium.
c. Magnesium has a larger third ionization energy than does aluminum.

***66. Explain** Why does it take more energy to remove a 4*s* electron from zinc than from calcium?

67. Sequence The following spheres represent Ca, Ca^{2+}, and Mg^{2+}. Which one is which? Explain your reasoning.

a.

b.

c.

***68. Apply Concepts** Write the electron configurations of the following ions:
a. the liquid in Group 7A with a 1− charge
b. the metalloid in Period 3 with a 4+ charge
c. the gas in Group 6A with a 2− charge
d. the alkali earth metal in Period 3 with a 2+ charge

69. Interpret Diagrams Use the periodic table and Figure 6.13 to identify the following elements:
a. has its outermost electron in $7s^1$
b. contains only one electron in a *d* orbital

70. Make Generalizations Why is the first ionization energy of a nonmetal much higher than that of an alkali metal?

71. Infer The bar graph shows the densities for the first six elements in Period 4. The density increases across this period from potassium to chromium. Use trends in the periodic table to explain this behavior. *Hint*: What is the equation for determining density?

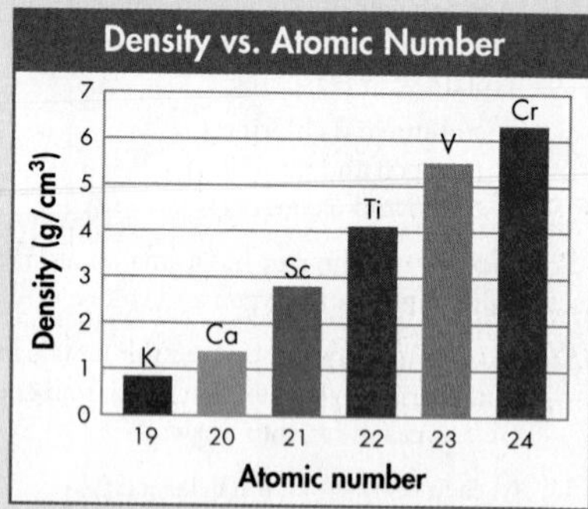

72. Explain Why are cations smaller and anions larger than the corresponding atoms?

Enrichment

73. Analyze Data Make a graph of atomic mass versus atomic number. Choose eleven points (atomic numbers 1, 10, 20, and so forth up to atomic number 100) to make your graph. Use the graph to describe the relationship between atomic mass and atomic number. Is there a 1:1 correlation between atomic mass and atomic number? Explain.

74. Compare and Contrast The Mg^{2+} and Na^{+} ions each have ten electrons. Which ion would you expect to have the smaller radius? Why?

***75. Predict** Electron affinity is a measure of an atom's ability to gain electrons. Predict the trend for electron affinity across a period. Explain your answer.

76. Explain The ions S^{2-}, Cl^{-}, K^{+}, Ca^{2+}, and Sc^{3+} have the same total number of electrons as the noble gas argon. How would you expect the radii of these ions to vary? Would you expect to see the same variation in the series O^{2-}, F^{-}, Na^{+}, Mg^{2+}, and Al^{3+}, in which each ion has the same total number of electrons as the noble gas neon? Explain your answer.

77. Graph The ionization energies for the removal of the first six electrons in carbon are, starting with the first electron, 1086 kJ/mol, 2352 kJ/mol, 4619 kJ/mol, 6220 kJ/mol, 37,820 kJ/mol, and 47,260 kJ/mol.

a. Use these data to construct a graph of ionization energy versus ionization number. *Note*: The ionization number indicates which electron is lost.

b. Between which two ionization numbers does the ionization energy have the largest increase? Why is this behavior predictable?

***78. Infer** Atoms and ions with the same number of electrons are described as *isoelectronic*.

a. Write the symbol for a cation and an anion that are isoelectronic with krypton.

b. Is it possible for a cation to be isoelectronic with an anion in the same period? Explain.

79. Predict Estimate the atomic radius of praseodymium based on the following data for atomic radii of neighboring elements: La (187.9 pm), Ce (183.2 pm), Nd (182.1 pm), and Pm (181.1 pm). Compare your prediction to the value given in a chemistry handbook.

Write About Science

80. Explain Why does the size of an atom tend to increase from top to bottom within a group? Why does the size of an atom tend to decrease from left to right across a period?

81. Connect to the BIGIDEA The ion Zn^{2+} is important in several biological processes. One process depends on Zn^{2+} temporarily binding to a molecule in red blood cells. When Zn^{2+} is absent, Cd^{2+} can bind to the molecule. However, Cd^{2+} binds more strongly and adversely affects the process. The differences in ionic size is one important cause of the difference in biological activity. How would you expect Hg^{2+} to affect this process? Why?

CHEMYSTERY

Made in the USA

Several of the "unnatural" elements were in fact "made in the USA." For example, elements with atomic numbers 94 through 102 were first artificially prepared in California. Three of these elements have names to prove it—americium, berkelium, and californium. Elements such as these are labeled on the periodic table as "Not found in nature" in this book and as "Artificially prepared" in some others.

Most of the artificially prepared elements are actinides or transactinides. Each of these elements has an unstable nucleus. As a result, these elements undergo radioactive decay, which means their nuclei spontaneously break down into smaller parts in the attempt to gain stability.

82. Infer The elements with atomic numbers 99, 101, 104, and 107 were named to honor past influential scientists. Identify the scientist that each element is meant to honor.

83. Connect to the BIGIDEA Many smoke detectors use the artificially prepared element americium. For a challenge, write the electron configuration of americium.

CHEMYSTERY **SUMMARIZE** After students have read through the CHEMystery, call on volunteers to summarize how the "unnatural" elements compare with other elements in the periodic table. Explain that many of these elements exist only momentarily once synthesized because their nuclei are so unstable. **Ask** How could the periodic table help scientists realize that they might be able to synthesize these elements? *(They would know that elements with certain numbers of protons were not yet identified.)*

Explain to students that although some elements are described as not found in nature on Earth, it is likely that many or all of them are found in nature somewhere in the universe. Convey that other elements, unidentified and with higher numbers of protons, may also exist, but conditions on Earth prevent the elements from occurring naturally on this planet.

ENRCHMENT

73. There is not a 1:1 correspondence. Explanations will vary.

74. Mg^{2+} is smaller because it has more protons; its attraction for electrons is greater.

75. Electron affinity increases (becomes more negative) from left to right across a period because the nuclear charge increases and the shielding effect is constant.

76. The ionic radii decrease from S^{2-} to Sc^{3+}. The number of electrons and the shielding effect do not change, but the number of protons increases from left to right in this series. So, the ionic size decreases. The same is true for the series O^{2-} to Al^{3+}.

77. a.

Ionization energy (kJ/mol): 0, 10,000, 20,000, 30,000, 40,000, 50,000
Ionization number: 0, 1, 2, 3, 4, 5, 6

b. The largest increase is between numbers 4 and 5 because carbon easily loses the first four electrons from the second energy level. The fifth electron is removed from the first energy level.

78. a. Possible cations are Rb^{+} and Sr^{2+}; possible anions are Br^{-}, Se^{2-}, and As^{3-}.

b. No, a cation is isoelectronic with the noble gas in the preceding period, and an anion is isoelectronic with the noble gas in the same period.

79. The atomic radius is likely between 182.1 pm (the atomic radius of Nd) and 183.2 pm (the atomic radius of Ce) because praseodymium is between these two elements in the periodic table. The value from the Handbook of Chemistry and Physics is 182.8 pm.

WRITE ABOUT SCIENCE

80. From top to bottom within a group, the effective nuclear charge remains nearly constant and the principal quantum number increases. From left to right in a period, the principal quantum number remains constant, but the effective nuclear charge increases.

81. From their positions in the periodic table, the three ions most likely have ionic sizes that increase going from Zn^{2+} to Cd^{2+} to Hg^{2+}. Since the increase in size from Zn^{2+} to Cd^{2+} adversely affects the process, it follows that a further increase in size could also adversely affect the process. The toxicity of mercury in biological systems is well known.

CHEMYSTERY ANSWERS

82. 99: einsteinium (after Albert Einstein); 101: mendelevium (after Dimitri Mendeleev); 104: rutherfordium (after Ernest Rutherford); 107: bohrium (after Niels Bohr)

83. $1s^22s^22p^63s^23p^63d^{10}4s^24p^64d^{10}4f^{14}5s^25p^6$ $5d^{10}5f^76s^26p^67s^2$

Evaluate

Answers

CUMULATIVE REVIEW

84. Answers will vary but are likely to mention the scientific method.

85. **a.** physical change
b. chemical change
c. physical change
d. chemical change

86. Use a magnet; iron beads are attracted to magnet, copper beads are not. Use tweezers and a magnifying glass; iron beads are silvery black, copper beads are dull red.

87. 4

88. **a.** 3
b. 8, the tenths place

89. The density of the cube is 0.984 g/cm³. The cube will float because its density is less than that of water.

90. **a.** 2.24×10^{-9} m
b. 8.13×10^{-2} m
c. 7.4×10^{-12} m
d. 9.37×10^{-3} m

91. 5.2%

92. 5.2×10^3 g

93. The density of the olive is 1.05 g/cm³. The olive will sink because its density is greater than that of water.

94. 86 round trips

95. The density of sulfur does not vary with mass. The density is constant.

96. 2.57×10^2 mL

97. 4.54 g/cm³

98. **a.** 48
b. 44
c. 114
d. 110

Cumulative Review

84. Explain why science today depends less on chance discoveries than it did in the past.

***85.** Identify each process as a chemical change or a physical change.
a. melting of iron
b. lighting a match
c. grinding corn
d. souring of milk

86. Describe at least two methods to separate a mixture of small copper and iron beads.

87. In the United States, a typical can of cola holds 355 mL. How many 2.00-L bottles could be filled from a 24-can case of cola?

88. The volume of the liquid in the graduated cylinder is reported as 31.8 mL.
a. How many significant figures are there in the measurement?
b. In which digit is there uncertainty?

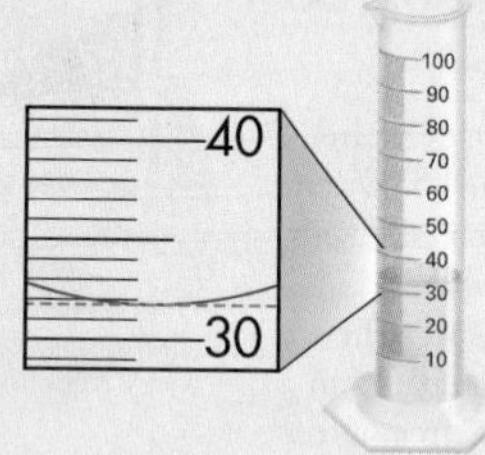

89. A cube of plastic 1.20×10^{-5} km on a side has a mass of 1.70 g. Show by calculation whether this plastic cube will sink or float in pure water.

***90.** Convert the measurements to meters. Express your answers in scientific notation.
a. 2.24 nm
b. 8.13 cm
c. 7.4 pm
d. 9.37 mm

91. An apprentice jeweler determines the density of a sample of pure gold to be 20.3 g/cm³. The accepted value is 19.3 g/cm³. What is the percent error of the jeweler's density measurement?

92. What is the mass of 7.7 L of gasoline at 20°C? Assume the density of gasoline to be 0.68 g/cm³.

***93.** A black olive containing its seed has a mass of 4.5 g and a volume of 4.3 cm³. Will the olive sink or float on water?

94. The distance from the sun to Earth is 1.50×10^8 km. The speed of light is 3.00×10^8 m/s. How many round trips between Earth and the sun could a beam of light make in one day?

95. The table shows how the volume of sulfur varies with mass. How does the density of sulfur vary with mass?

Mass of Sulfur vs. Volume of Sulfur

Mass of sulfur (g)	Volume of sulfur (cm³)
23.5	11.4
60.8	29.2
115	55.5
168	81.1

96. Calculate the volume of acetone with the same mass as 15.0 mL of mercury. The density of mercury is 13.59 g/mL. The density of acetone is 0.792 g/mL.

97. A rectangular container has inside dimensions of 15.2 cm by 22.9 cm and is about 1 meter tall. Water is poured into the container to a height of 55.0 cm. When a jagged rock with a mass of 5.21 kg is placed in the container, it sinks to the bottom. The water level rises to 58.3 cm and completely covers the rock. What is the density of the rock?

***98.** How many neutrons does an atom of each isotope contain?
a. $^{84}_{36}Kr$
b. $^{79}_{35}Br$
c. $^{190}_{76}Os$
d. $^{185}_{75}Re$

99. Name the element and calculate the number of requested subatomic particles in each isotope.
a. neutrons in $^{109}_{47}Ag$
b. protons in $^{118}_{50}Sn$
c. electrons in $^{96}_{42}Mo$
d. electrons in $^{45}_{21}Sc$

***100.** How many filled *p* orbitals do atoms of these elements contain?
a. carbon
b. phosphorus
c. oxygen
d. nitrogen

If You Have Trouble With . . .

Question	84	85	86	87	88	89	90	91	92	93	94	95	96	97	98	99	100
See Chapter	1	2	2	3	3	3	3	3	3	3	3	3	3	3	4	4	5

99. **a.** silver, 62 neutrons
b. tin, 50 protons
c. molybdenum, 42 electrons
d. scandium, 21 electrons

100. **a.** none
b. three, $2p_x$, $2p_y$, $2p_z$
c. one, $2p_x$
d. none

190 Chapter 6 • Assessment

Standardized Test Prep for Indiana

Standard: C.2.5

Select the choice that best answers each question or completes each statement.

1. Which of the following properties increases as you move across a period from left to right?
   I. electronegativity
   II. ionization energy
   III. atomic radius
   (A) I and II only (C) II and III only
   (B) I and III only (D) I, II, and III

2. List the symbols for sodium, sulfur, and cesium in order of increasing atomic radii.
   (A) Na, S, Cs (C) S, Na, Cs
   (B) Cs, Na, S (D) Cs, S, Na

3. The electron configuration for an element in the halogen group should always end with
   (A) ns^2np^6. (C) ns^2np^4.
   (B) ns^2np^5. (D) ns^2np^2.

Use the spheres to answer Questions 4 and 5.

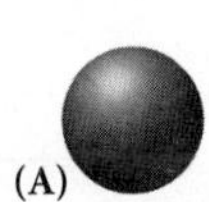
(A)

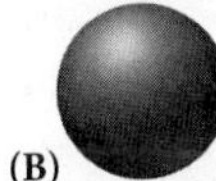
(B)

4. If the spheres represent a potassium atom and a potassium ion, which best represents the ion?

5. If the spheres represent an atom and an anion of the same element, which sphere represents the atom and which represents the anion?

Tips for Success

Interpreting Data Tables Tables can present a large amount of data in a small space. Before you try to answer questions based on a table, look at the table. Read the title, if there is one, and the column headings. Then read the questions. As you read each question, decide which data you will need to use to answer the question.

Use the data table to answer Questions 6–8.

Alkali metal	Atomic radius (pm)	First ionization energy (kJ/mol)	Electronegativity value
Li	152	520	1.0
Na	186	495.8	0.9
K	227	418.8	0.8
Rb	244	250	0.8
Cs	262	210	0.7

6. If you plot atomic radius versus first ionization energy, would the graph reveal a direct or inverse relationship?

7. If you plot atomic radius versus electronegativity, would the graph reveal a direct or inverse relationship?

8. If you plot first ionization energy versus electronegativity, would the graph reveal a direct or inverse relationship?

For each question there are two statements. Decide whether each statement is true or false. Then decide whether Statement II is a correct explanation for Statement I.

	Statement I		Statement II
9.	Electronegativity values are higher for metals than for nonmetals.	BECAUSE	Atoms of nonmetals are among the largest atoms.
10.	A calcium atom is larger than a calcium ion.	BECAUSE	Ions are always larger than the atoms from which they are formed.
11.	The element hydrogen is a metal.	BECAUSE	Hydrogen is on the left in the periodic table.
12.	Among all the elements in a period, the noble gas always has the smallest ionization energy.	BECAUSE	Within any period, atomic radii tend to decrease moving from right to left.

If You Have Trouble With . . .

Question	1	2	3	4	5	6	7	8	9	10	11	12
See Lesson	6.3	6.3	6.2	6.3	6.3	6.3	6.3	6.3	6.3	6.3	6.1	6.3

STP Answers

1. a
2. c
3. b
4. The smaller sphere represents the potassium ion.
5. The smaller sphere represents the atom; the larger sphere represents the anion.
6. inverse
7. inverse
8. direct
9. false, false
10. true, false
11. false, true
12. false, false

7 Ionic and Metallic Bonding Planning Guide

Introducing the BIGIDEA: BONDING AND INTERACTIONS

Forces, called chemical bonds, greatly influence the properties of compounds.

IASS	Lessons and Objectives	Print Resources		
		For the Student	For the Teacher	
C.2.6, C.3.3	**7.1** Ions p 194–199 **7.1.1** Determine the number of valence electrons in an atom of a representative element. **7.1.2** Identify the atoms of elements that tend to lose and tend to gain electrons. **7.1.3** Describe how cations form. **7.1.4** Explain how anions form.	Reading and Study Workbook Lesson 7.1 Lesson Assessment 7.1 p 199 Small-Scale Lab: Analysis of Anions and Cations p 200	Teacher Demo, p 195: Valence Electrons Class Activity, p 197: Forming Cations	
C.3.1, C.3.2, C.3.3	**7.2** Ionic Bonds and Ionic Compounds p 201–207 **7.2.1** Explain the electrical charge of an ionic compound. **7.2.2** Describe three properties of ionic compounds.	Reading and Study Workbook Lesson 7.2 Lesson Assessment 7.2 p 207 Quick Lab: Solutions Containing Ions p 207	Class Activity, p 203: Hardness of Water Class Activity, p 204: Form and Structure of Crystals Class Activity, p 205: Types of Ionic Compounds	
C.3.3	**7.3** Bonding in Metals p 209–212 **7.3.1** Model the valence electrons of metal atoms. **7.3.2** Describe the arrangement of atoms in a metal. **7.3.3** Explain the importance of alloys.	Reading and Study Workbook Lesson 7.3 Lesson Assessment 7.3 p 212	Teacher Demo, p 210: Metal vs. Ionic Compounds Teacher Demo, p 211: Types of Alloys	

Assessing the BIGIDEA: BONDING AND INTERACTIONS

	Essential Questions 1. How do ionic compounds form? 2. How does metallic bonding affect the properties of metals?	Study Guide p 213 STP p 219 Reading and Study Workbook Self-Check and Vocabulary Review Chapter 7		

	PearsonChem.com	
	Editable Worksheets	Online Activities
	Teaching Resources, Lesson 7.1 Review Small-Scale Lab Manual Lab 10: Electronic Configurations of Atoms and Ions	7.1 Lesson Overview
	Teaching Resources, Lesson 7.2 Review Probeware Lab: Quick Lab: Solutions Containing Ions Lab Practical 7-1: Crystal Structures	7.2 Lesson Overview Predicting Formulas of Ionic Compounds Ionic Doubles Partners The Formation of Ionic Compounds Crystal Structures
	Teaching Resources, Lesson 7.3 Review Lab 10: Crystal Structure Lab Practical 7-1: Crystal Structure	7.3 Lesson Overview Properties of Metals Conduction, Ductility, and Malleability
	Core Teaching Resources (includes editable worksheets) • Lesson Reviews • Practice Problems • Interpret Graphs • Vocabulary Review • Chapter Quizzes and Tests • Lab Record Sheets	Chapter 7 Problem Set Percents

Materials List

For the Student

Small-Scale Lab p 200
- reaction surface
- micropipets or droppers
- chemicals: $NaCl$, $MgSO_4$, $AlCl_3$, $FeCl_3$, $CaCl_2$, $NiSO_4$, $CuSO_4$, $ZnCl_2$, $AgNO_3$, $NaOH$

Quick Lab p 207
- 3 D-cell batteries
- masking tape
- 2 30-cm lengths of bell wire with ends scraped bare
- clear plastic cup
- distilled water
- tap water
- vinegar
- table sugar
- table salt
- baking soda
- conductivity probe (optional)

For the Teacher

Teacher Demo p 195
- plastic egg
- 11 marbles

Class Activity p 203
- tap water samples
- 10-mL graduated cylinder
- potassium thiocyanate solution
- dilute ethanoic acid
- sodium oxalate solution
- dropper

Class Activity p 204
- crystals of ionic compounds
- watch glasses
- magnifying glasses

Teacher Demo p 210
- copper metal or alloy
- copper-containing ionic compound

Teacher Demo p 211
- foam balls of different sizes and colors
- toothpicks

Additional Digital Resources

Online Student Edition
Online Teacher's Edition
Exam*View Assessment Suite CD-ROM*
Classroom Resources DVD

What's Online

CHEMISTRY TUTORIAL Students can work step-by-step with the online tutorial to learn how to write formulas of ionic compounds by accessing the Chapter 7 Chemistry Tutorial.

ONLINE PROBLEM SET Additional practice problems about ionic and metallic bonding can be found in the Chapter 7 online Problem Set.

DIRECTED VIRTUAL LAB Encourage students to observe the formation of different ionic compounds using the animated virtual lab for Chapter 7. The animation is followed by analysis questions, where the student is guided through the lab concepts.

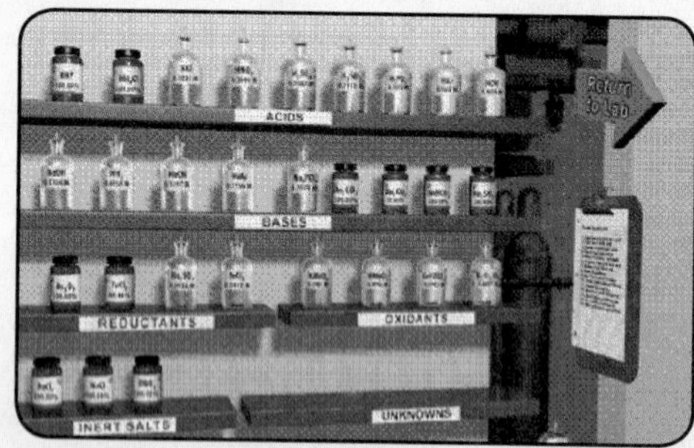

KINETIC ART Have students watch an animation about crystal structures. The animation is followed by analysis questions to enhance understanding of the concepts.

CONCEPTS IN ACTION Students can watch an interactive animation about ionic bonding to see how ions interact like players on a team. The animation is followed by questions to clarify the role of electrons in ionic bonding.

IN ACADEMIC STANDARDS for SCIENCE

Standard 3: Bonding and Molecular Structure

C.3.1 Describe, compare, and contrast the characteristics of the interactions between atoms in ionic and covalent compounds.

C.3.2 Compare and contrast how ionic and covalent compounds form.

C.3.3 Based on the strength of molecular and atomic attractions, explain the physical properties of substances, such as melting and boiling points, electrical and thermal conductivity, and crystal structure.

Also covered C.2.6

Focus on ELL

1 CONTENT AND LANGUAGE Write *ionic, metallic,* and *bonding* on the board. Lead students in discussing the root word for each term.

BEGINNING: LOW/HIGH Visually demonstrate each word. Model an *ion* by drawing an atom and showing it gaining or losing an electron. Identify metal objects in the classroom. Demonstrate a bond as an attractive force by having two people grasp hands.

INTERMEDIATE:

LOW Ask simple questions students can answer with either *ion, metal*, or *bond*. For example, point to a metal object and ask what it is made from.

HIGH Lead students in reviewing ions and metals. Encourage them to use the terms *ion, metal,* and *bond* in the discussion.

ADVANCED: LOW/HIGH Have students read the lesson titles and the essential questions. Ask questions to extend the meanings of *ion, metal*, and *bond*. For example, ask what students think an ionic compound is.

CHAPTER 7

BIGIDEA

BONDING AND INTERACTIONS

Essential Questions:

1. *How do ionic compounds form?*
2. *How does metallic bonding affect the properties of metals?*

CHEMYSTERY

It's Not Easy Being Green

While strolling through Central Park in New York City, you come across this statue of the composer Ludwig van Beethoven. It appears to be made of metal, but its surface is green in color and not very shiny. Is the statue's green complexion due to green paint, or something else?

After doing some research, you learn that the statue is made out of bronze, which is a mixture of metals. The statue was never painted. Instead, the exposed surface of the bronze underwent a chemical change, forming a green film over time. You wonder what the film is made of and how it formed. Are the properties of the film different from the bronze beneath it?

▶ Connect to the BIGIDEA As you read about ionic compounds and metals, think about why the statue changed color. Also, think about how the properties of the green film at the surface differ from the metal beneath it.

INDIANA'S ACADEMIC STANDARDS FOR SCIENCE

Atomic Structure and the Periodic Table C.2.6; **Bonding and Molecular Structure** C.3.1, C.3.2, C.3.3. See lessons for details.

UbD

Understanding by Design

Students are building toward understanding bonding and interactions by learning about ionic and metallic bonding.

PERFORMANCE GOALS At the end of Chapter 7, students will be able to answer the essential questions by applying their knowledge of ionic and metallic bonding.

ESSENTIAL QUESTIONS Read the essential questions aloud. **Ask** How does the charge of an atom change if it gains or loses an electron? *(It will have a positive charge if it loses an electron. It will have a negative charge if it gains an electron.)*

BIGIDEA Use the photo of giant crystals to help students connect to this chapter's concepts. Activate students' prior knowledge by asking if they have looked at small crystals in rocks or seen large crystals at a museum. Discuss the geometric shapes of crystals. Explain that crystals are the result of the orderly arrangement of atoms. Explain that atoms that have gained or lost electrons have negative or positive charges and atoms with opposite charges can form bonds. If the conditions are suitable, the atoms become ordered and bond together.

CHEMYSTERY Have students read over the CHEMystery. Connect the CHEMystery to the Big Idea by pointing out that because the statue is surrounded by air, the metal must be interacting with something in the air. Encourage students to predict what this interaction might be that produces the film on the statue and to consider whether the film would interact with the air in the same way that the metal does.

Introduce the Chapter

CONNECTION TO ELECTRONS IN ATOMS Students have learned that atoms of different elements have different numbers of electrons. Use the activity to review the idea that atoms can gain or lose electrons.

Activity Have each student use modeling clay to make models of a lithium atom and a fluorine atom. Remind them to use the periodic table to determine the correct number of electrons for their atoms. Have students remove one of the lithium atom's electrons. **Ask** Is the lithium atom still neutral? *(No; it has a positive charge.)* Have students give the electron to the fluorine atom. **Ask** Is the fluorine atom still neutral? *(No; it has a negative charge.)* **Ask** What do you think would happen if the charged atoms were near each other? *(They would be attracted to each other.)*

Untamed Science™

Cave Crystals

Diego explores a million-year-old limestone cave to see how different size crystals can produce different looking formations.

LESSON 7.1

Key Objectives

7.1.1 DETERMINE the number of valence electrons in an atom of a representative element.

7.1.2 IDENTIFY the atoms of elements that tend to lose and tend to gain electrons.

7.1.3 DESCRIBE how cations form.

7.1.4 EXPLAIN how anions form.

Additional Resources

Reading and Study Workbook, Lesson 7.1

Available Online or on Digital Media:

- Teaching Resources, Lesson 7.1 Review
- Small-Scale Laboratory Manual, Lab 10

PearsonChem.com

Engage

CHEMISTRY & YOU Have students study the photograph and read the text that opens the lesson. Ask students to suggest reasons for the regular pyrite crystal structure. Do not provide the answer at this point.

Access Prior Knowledge

Have students refer to their copy of the periodic table. **Ask** What does the group number on the periodic table tell you about the electron configuration of a representative element? *(The group number equals the number of electrons in the highest occupied energy level.)* **Ask** *What is formed when electrons are transferred between atoms?* (ions)

IN ACADEMIC STANDARDS for SCIENCE

C.2.6 Use the periodic table and electron configurations to determine an element's number of valence electrons and chemical and physical properties.

C.3.3 Compare and contrast ionic, covalent network, metallic and polar and non-polar molecular crystals with respect to constituent particles, strength of bonds, melting and boiling points, and conductivity; provide examples of each type.

7.1 Ions

C.2.6 Determine valence electrons and properties; **C.3.3** Chemical bonds and intermolecular attractions

CHEMISTRY & YOU

Q: *What is fool's gold?* Pyrite (FeS_2) is often mistaken for gold—hence its nickname, "fool's gold." Pyrite is an example of a crystalline solid. In crystalline solids, the component particles of the substance are arranged in an orderly, repeating fashion. In this chapter, you will learn about crystalline solids, like pyrite, that are composed of ions that are bonded together.

Key Questions

How do you find the number of valence electrons in an atom of a representative element?

How are cations formed?

How are anions formed?

Vocabulary

- valence electron
- electron dot structure
- octet rule
- halide ion

Valence Electrons

How do you find the number of valence electrons in an atom of a representative element?

Mendeleev used similarities in the properties of elements to organize his periodic table. Scientists later learned that all of the elements within each group of the periodic table react in a similar way because they have the same number of valence electrons. **Valence electrons** are the electrons in the highest occupied energy level of an element's atoms. The number of valence electrons largely determines the chemical properties of an element.

Determining the Number of Valence Electrons The number of valence electrons in an atom of an element is related to the element's group number in the periodic table. **To find the number of valence electrons in an atom of a representative element, simply look at its group number.** For example, atoms of the Group 1A elements (hydrogen, lithium, sodium, and so forth) all have one valence electron, corresponding to the 1 in 1A. Carbon and silicon atoms, in Group 4A, have four valence electrons. Figure 7.1 shows some applications of Group 4A elements. Nitrogen and phosphorus atoms, in Group 5A, have five valence electrons, and oxygen and sulfur atoms, in Group 6A, have six. The noble gases (Group 8A) are the only exceptions to the group-number rule: Atoms of helium have two valence electrons, and atoms of all of the other noble gases have eight valence electrons.

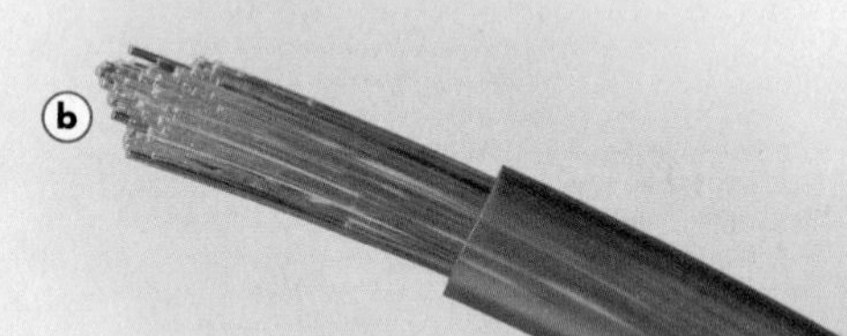

Figure 7.1 Group 4A Elements
Silicon and germanium are Group 4A elements. **a.** Silicon is used in the manufacture of computer chips. **b.** Compounds of germanium are used to make optical fibers.

Focus on ELL

1 CONTENT AND LANGUAGE Read the key questions aloud. Have students write each new term on an index card and its definition on the other side. Review the meaning of *representative element.*

2 FRONTLOAD THE LESSON Review pages 170 to 172 with students. Remind them that elements are organized into groups on the periodic table based on the number of electrons in the highest occupied energy level. Explain that these are the *valence electrons* they will study in this lesson.

3 COMPREHENSIBLE INPUT Review the Pauli exclusion principle and Hund's rule before discussing the diagram of sodium ionization on page 196. Remind students that each box represents an orbital in an atom, and that each orbital can contain a maximum of two electrons spinning in opposite directions, as indicated by the two arrows.

Table 7.1

Electron Dot Structures of Some Group A Elements

	Group							
Period	**1A**	**2A**	**3A**	**4A**	**5A**	**6A**	**7A**	**8A**
1	H·							He:
2	Li·	·Be·	·Ḃ·	·Ċ·	·N̈·	:Ö·	:F̈·	:N̈e:
3	Na·	·Mg·	·Ȧl·	·Ṡi·	·P̈·	:S̈·	:C̈l·	:Är:
4	K·	·Ca·	·Ġa·	·Ġe·	·Äs·	:S̈e·	:B̈r·	:K̈r:

Valence electrons are usually the only electrons involved in chemical bonds. Therefore, as a general rule, only the valence electrons are shown in electron dot structures. **Electron dot structures** are diagrams that show valence electrons in the atoms of an element as dots. Table 7.1 shows electron dot structures for atoms of some Group A elements. Notice that all of the elements within a given group (with the exception of helium) have the same number of electron dots in their structures.

The Octet Rule You learned in Chapter 6 that noble gases, such as neon and argon, are nonreactive in chemical reactions. That is, they are stable. In 1916, chemist Gilbert Lewis used this fact to explain why atoms form certain kinds of ions and molecules. He called his explanation the octet rule. The **octet rule** states that in forming compounds, atoms tend to achieve the electron configuration of a noble gas. An octet is a set of eight. Recall that atoms of each of the noble gases (except helium) have eight electrons in their highest occupied energy levels and the general electron configuration of ns^2np^6. The octet rule takes its name from this fact about noble gases.

Atoms of metals tend to lose their valence electrons, leaving a complete octet in the next-lowest energy level. Atoms of some nonmetals tend to gain electrons or to share electrons with another nonmetal atom or atoms to achieve a complete octet. Although exceptions occur, the octet rule applies to atoms in most compounds.

READING SUPPORT

Build Vocabulary: *Word Origins* *Octet* comes from the Greek word *okto,* meaning "eight." There are eight electrons in the highest occupied energy level of the noble gases, except for helium. ***How do you think the term octet might also be applied to music or poetry?***

Formation of Cations

How are cations formed?

An atom is electrically neutral because it has equal numbers of protons and electrons. An ion forms when an atom or group of atoms loses or gains electrons. **A positively charged ion, or a cation, is produced when an atom loses one or more valence electrons.** Note that for metals, the name of the cation is the same as the name of the element. For example, a sodium atom (Na) forms a sodium cation (Na^+). Likewise, a calcium atom (Ca) forms a calcium cation (Ca^{2+}). Although their names are the same, metals and their cations have many important chemical differences. Sodium metal, for example, reacts explosively with water. By contrast, sodium cations are quite nonreactive. As you may know, sodium cations are a component of table salt, a compound that is very stable in water.

Differentiated Instruction

LPR LESS PROFICIENT READERS Lead a class discussion on writing electron dot structures. **Ask** What is the significance of the electrons that are represented by dots? *(They are valence electrons.)* **Ask** How are they placed? *(symmetrically around the atom according to pairing rules)* **Ask** Why are the non-valence electrons not shown? *(Generally, they are not available for chemical bonding.)*

L1 ENGLISH LANGUAGE LEARNERS If possible, pair beginning English speakers with more proficient English speakers that have the same native language. Write the word *stable* on the board and have each pair of students act out the meaning of the word. Have students identify any other content words that are difficult to understand.

L3 ADVANCED STUDENTS Have students show how the atoms of transition elements become stable with pseudo-noble-gas configurations. (These atoms would have to gain or lose too many electrons to achieve a noble-gas electron configuration.) For example, show how Ag, Zn, and Ga lose 1, 2, and 3 electrons, respectively.

Foundations for Reading

BUILD VOCABULARY Have each student choose a halide ion. Then have students create a graphic organizer that shows how a halide ion relates to the other three vocabulary terms.

READING STRATEGY Have students skim the headings, look at the visuals, and read the boldfaced text to preview the section. Then have students decide what content from previous chapters they should review prior to reading this lesson.

Explain

Valence Electrons

USE VISUALS Direct students to Table 7.1. Have students identify the total number of electrons and the number of valence electrons in selected elements from Table 7.1. Reemphasize that the group number equals the number of valence electrons in an atom of a representative element.

APPLY CONCEPTS Explain that the octet rule is helpful for predicting the ionic charges of most representative elements. Caution students that this rule does not apply to all elements, as some elements have several different ionic charges. Point out examples of stable ions formed by the transition elements, such as Mn^{2+} and Mn^{3+}.

Explore

Teacher Demo

PURPOSE To model valence electrons.

MATERIALS plastic egg, 11 marbles

PROCEDURE Hold up a plastic egg containing 10 marbles. State that the egg represents a sodium atom and that the marbles represent the 10 electrons in the $n = 1$ and $n = 2$ energy levels. Explain that these electrons cannot be "removed" without "breaking" the egg. Now hold up one additional marble next to the egg. State that this marble represents the eleventh electron, and occupies the $n = 3$ energy level. This is the valence electron. Explain that if this electron is lost, the resulting atom has an overall 1+ charge.

EXPECTED OUTCOME Students should be able to distinguish valence electrons from nonvalence electrons.

Answers

READING SUPPORT In music, an octet refers to a group of eight performers. In poetry, an octet refers to a group of eight lines of verse.

LESSON 7.1

Explain

Formation of Cations

USE VISUALS Have students examine Figure 7.2 and have a student volunteer read aloud the caption. Display the diagram adjacent to Figure 7.2 on an overhead projector. Point out that the interaction between atoms that produces bonding involves only the outermost electrons of the atoms. Show students that the inner electrons are locked tightly in filled energy levels and do not participate in bonding. Use a colored pen to circle the outermost electron in the sodium atom. Remind students that the outermost electrons are called valence electrons. Use a different colored pen to circle the octet of electrons in the sodium ion's highest occupied energy level. Circle the corresponding octet of electrons in neon to show the similarity in electron configurations. Have students draw a similar diagram for calcium. *(Check students' diagrams.)*

CRITICAL THINKING Explain that the term *isoelectronic* means having equal numbers of electrons or the same electronic configuration. Then, have students determine the accuracy of this statement: All stable ions of representative elements result in electronic configurations that are isoelectronic with noble gases. Explain that a noble gas configuration is not generally possible with elements that would have to gain or lose many electrons.

Misconception Alert

Students tend to confuse the name of the ion that gets the positive charge with the name of the ion that gets the negative charge. To help students remember, post the following in the classroom:
cation: ca + ion (write the 't' as the + sign)
anion: A Negative ION

Figure 7.2 Sodium Vapor Lamp
The sodium atoms (Na) in a sodium vapor lamp ionize to form sodium cations (Na^+).
Apply Concepts ***How many electrons are in the highest occupied energy level of Na^+?***

Group 1A Cations The most common cations are those produced by the loss of valence electrons from metal atoms. Most of these atoms have one to three valence electrons, which are easily removed. Sodium (atomic number 11) is in Group 1A of the periodic table. Sodium atoms have a total of eleven electrons, including one valence electron. A sodium atom can lose an electron to become a positively charged sodium ion. Sodium atoms become sodium ions in a sodium vapor lamp, which is shown in Figure 7.2. The sodium ion has an electron configuration that is identical to the noble gas neon.

When forming a compound, a sodium atom loses its one valence electron and is left with an octet (eight electrons) in what is now its highest occupied energy level. The number of protons in the sodium nucleus is still eleven, so the loss of one unit of negative charge produces a cation with a charge of 1+. You can represent the loss of the electron, or ionization, of the sodium atom by writing the complete electron configuration of the atom and of the ion formed.

$$\text{Na}\ 1s^2 2s^2 2p^6 3s^1 \xrightarrow{-e^-} \text{Na}^+\ \underbrace{1s^2 2s^2 2p^6}_{\text{octet}}$$

Notice that the electron configuration of the sodium ion ($1s^2 2s^2 2p^6$) is the same as that of a neon atom.

$$\text{Ne}\ \underbrace{1s^2 2s^2 2p^6}_{\text{octet}}$$

The diagrams below help illustrate this point.

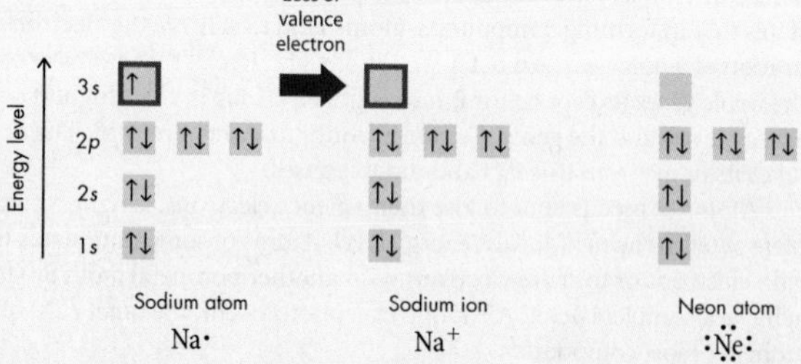

Both the sodium ion and the neon atom have eight electrons in their valence shells (highest occupied energy levels). Using electron dot structures, you can show the ionization more simply.

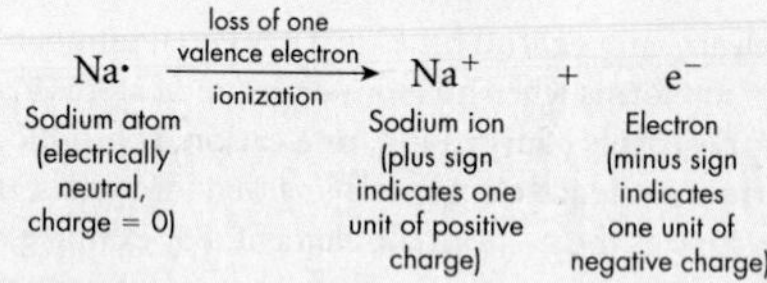

UbD Check for Understanding

How do you find the number of valence electrons in an atom of a representative element?

Call out a representative element. Have students hold up their fingers to show you the number of valence electrons in the atom. Repeat with a few more representative elements. Ask students to give you a thumbs up or a thumbs down to indicate whether they understand that the group number corresponds directly to the number of valence electrons for an atom of a representative element.

ADJUST INSTRUCTION If students are having difficulty answering the question, make sure they are referring to their periodic table. Direct them to reread the text in the Valence Electrons section and reassess their understanding by repeating the activity above.

Group 2A Cations Magnesium (atomic number 12) belongs to Group 2A of the periodic table, so magnesium atoms have two valence electrons. A magnesium atom attains the electron configuration of a neon atom by losing both valence electrons and producing a magnesium cation with a charge of 2+.

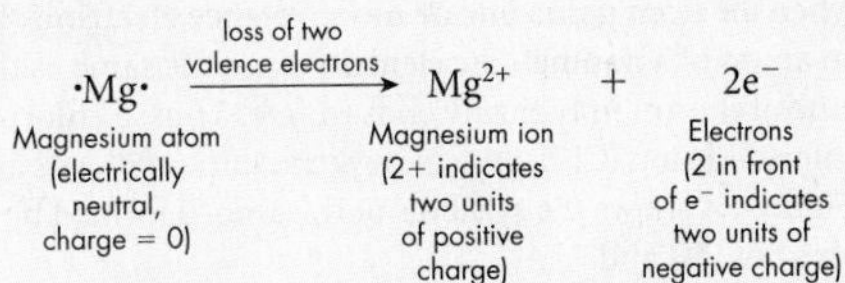

Figure 7.3 lists the symbols of the cations formed by metals in Groups 1A and 2A. Cations of Group 1A elements always have a charge of 1+. Cations of Group 2A elements always have a charge of 2+. This consistency can be explained in terms of the loss of valence electrons by metal atoms: The atoms lose the number of electrons necessary to attain the electron configuration of a noble gas.

Transition Metal Cations The charges of cations of the transition metals may vary. An atom of iron, for example, may lose two valence electrons, forming the Fe^{2+} cation, or three valence electrons, forming the Fe^{3+} cation.

Some ions formed by transition metals do not have noble-gas electron configurations (ns^2np^6) and are therefore exceptions to the octet rule. Silver, with the electron configuration of $1s^22s^22p^63s^23p^63d^{10}4s^24p^64d^{10}5s^1$, is an example. To achieve the structure of krypton, which is the preceding noble gas, a silver atom would have to lose eleven electrons. To acquire the electron configuration of xenon, which is the following noble gas, a silver atom would have to gain seven electrons. Ions with charges of three or greater are uncommon. Thus silver does not achieve a noble-gas configuration. However, if a silver atom loses its $5s^1$ electron, forming a positive ion (Ag^+), the configuration that results ($4s^24p^64d^{10}$), with 18 electrons in the highest occupied energy level and all of the orbitals filled, is relatively favorable. Such a configuration is known as a pseudo noble-gas electron configuration. Other elements that behave similarly to silver are found at the right of the transition metal block of the periodic table. A copper atom loses its lone 4*s* electron to form a copper ion (Cu^+) with a pseudo noble-gas electron configuration as illustrated below.

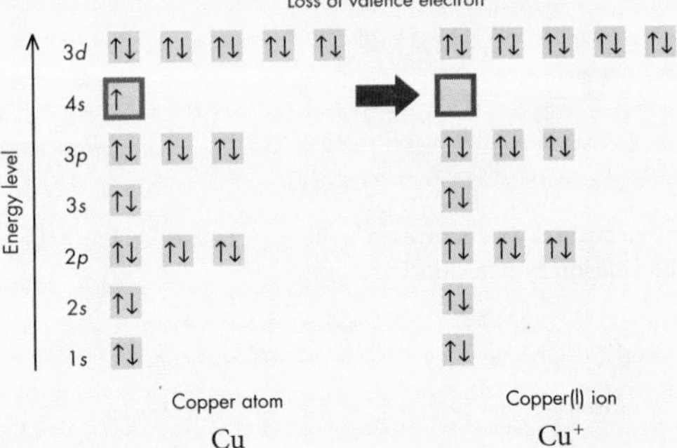

Cations of gold (Au^+), cadmium (Cd^{2+}), and mercury (Hg^{2+}) also have pseudo noble-gas electron configurations.

1A	2A
Li^+	Be^{2+}
Na^+	Mg^{2+}
K^+	Ca^{2+}
Rb^+	Sr^{2+}
Cs^+	Ba^{2+}
Fr^+	Ra^{2+}

Figure 7.3
Group 1A and 2A Cations
Cations of Group 1A elements have a charge of 1+. Cations of Group 2A elements have a charge of 2+.

APPLY CONCEPTS Explain that one method to use to determine the number of valence electrons in an atom is to look at the electron configuration of the atom. Explain that any electron in an atom outside the noble-gas core is called a valence electron. Using diagrams similar to those shown, demonstrate several examples of how various atoms of the representative elements form ions and gain a noble-gas electron configuration. Indicate the noble-gas core and valence electrons in your diagrams.

Explore

Class Activity

PURPOSE Students model the formation of cations using equations.

MATERIALS paper, pencil

PROCEDURE Have students write equations similar to the one shown for Mg in the text, to show the formation of metal cations from metal atoms. Students should show the electron dot structures for the metal atoms and metal cations that are formed. In addition, you may want students to write out the electron configurations for the metal atoms and cations. Check students' work to be sure the correct metal ion is formed.

EXPECTED OUTCOMES Students should be able to use electron dot structures to correctly write equations describing the ionization of metal atoms.

Differentiated Instruction

L1 SPECIAL NEEDS STUDENTS Ask students to work with a partner to practice drawing electron dot structures for Group A elements. Have one student randomly choose an element and write the symbol for that element on a piece of paper. Have another student fill in the electron dots for this element. Students can check their work by referring to Table 7.1. Make sure students understand that the electron dots represent only valence electrons, not the total number of electrons, and that valence electrons are the only electrons involved in chemical reactivity. Also, reinforce the stable octet dot structures for the noble gases and explain how ions of other elements will react to obtain this configuration.

ELL ENGLISH LANGUAGE LEARNERS Explain to students that when one atom forms an ion, that ion is called a monatomic ion. Explain that the prefix *mon-* or *mono-* means "one."

Answers

FIGURE 7.2 8 electrons

LESSON 7.1

Explain

Formation of Anions

INTERPRET DIAGRAMS Direct students' attention to Figure 7.4. Explain that the elements arsenic (As) and tellurium (Te) are metalloids, not nonmetals. However, they form anions that are named according to the same convention as nonmetal anions are named (arsenide, telluride).

START A CONVERSATION Discuss with students how certain elements can form either anions or cations. Have students write the electron configuration of nitrogen ($1s^22s^22p^3$). **Ask** How can a nitrogen atom form a cation that has the electron configuration of a noble gas? *(It can lose five electrons.)* **Ask** How can a nitrogen atom form an anion that has the electron configuration of a noble gas? *(It can gain three electrons.)* Have them repeat this process for carbon. Explain that atoms that have few electrons to gain or lose to achieve an octet are more likely to form ions. Point out that other atoms are not likely to form ions, and the process they undergo will be studied in Chapter 8.

CHEMISTRY & YOU Diagrams should reflect the following electron configuration: $1s^22s^22p^63s^23p^63d^6$.

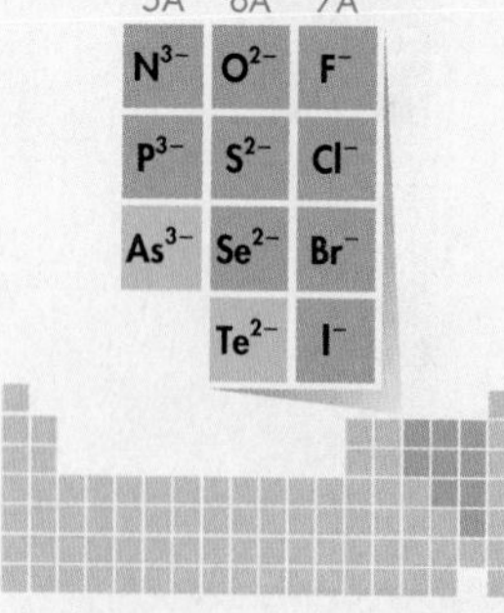

Figure 7.4
Group 5A, 6A, and 7A Anions
Atoms of nonmetals and metalloids form anions by gaining enough valence electrons to attain the electron configuration of the nearest noble gas.
Interpret Diagrams ***To which group of the periodic table do the elements bromine and iodine belong?***

CHEMISTRY & YOU

Q: *Fool's gold is composed of iron(II) cations (Fe^{2+}) and disulfide anions (S_2^{2-}). Write the electron configuration of the Fe^{2+} ion.*

Formation of Anions

How are anions formed?

An anion is an atom or group of atoms with a negative charge. **An anion is produced when an atom gains one or more valence electrons.** Note that the name of an anion of a nonmetallic element is *not* the same as the element name. The name of the anion typically ends in *-ide*. Thus, a chlorine atom (Cl) forms a chloride anion (Cl^-), and an oxygen atom (O) forms an oxide anion (O^{2-}). Figure 7.4 shows the symbols of the anions formed by some elements in Groups 5A, 6A, and 7A.

Atoms of nonmetallic elements attain noble-gas electron configurations more easily by gaining electrons than by losing them because these atoms have relatively full valence shells. For example, chlorine belongs to Group 7A (the halogen family). Atoms of chlorine have seven valence electrons. A gain of one electron gives a chlorine atom an octet and converts a chlorine atom into a chloride ion.

$$\text{Cl}\ \ 1s^22s^22p^63s^23p^5 \xrightarrow{+e^-} \text{Cl}^-\ \ 1s^22s^22p^6\underbrace{3s^23p^6}_{\text{octet}}$$

The chloride ion has a single negative charge. Notice that the electron configuration of the chloride ion ($1s^22s^22p^63s^23p^6$) is the same as that of an argon atom.

$$\text{Ar}\ \ 1s^22s^22p^6\underbrace{3s^23p^6}_{\text{octet}}$$

Chlorine atoms, therefore, need one more valence electron to achieve the electron configuration of the nearest noble gas. The diagrams below illustrate how both the chloride ion and the argon atom have an octet of electrons in their highest occupied energy levels.

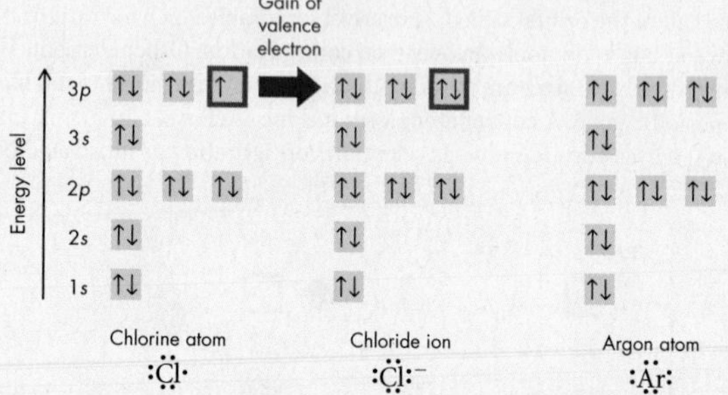

You can use electron dot structures to write an equation showing the formation of a chloride ion from a chlorine atom.

Professional Development Note: Water Purification

Chlorine gas is often used to purify drinking water. It kills a variety of microorganisms, including those that carry diseases. But chlorine also reacts with organic substances in the water to produce chlorinated compounds such as chloroacetonitrile. Chloroacetonitrile has been shown to cause inflammation of the digestive tract in laboratory animals. As an alternative to chlorine, some countries have begun purifying water with ozone. Ozone kills microorganisms even more effectively than chlorine. About one percent of the water supply in the United States is now purified with ozone. It is estimated that it would cost $6 billion to switch completely to ozone for treating all the drinking water supplies.

The ions produced when atoms of chlorine and other halogens gain electrons are called **halide ions.** All halogen atoms have seven valence electrons and need to gain only one electron to achieve the electron configuration of a noble gas. Thus, all halide ions (F^-, Cl^-, Br^-, and I^-) have a charge of 1−. The seawater in Figure 7.5 contains many different ions, but the anions are mostly chloride ions.

Oxygen is in Group 6A, and an oxygen atom has six valence electrons. An oxygen atom attains the electron configuration of neon by gaining two electrons, as shown in the diagrams below.

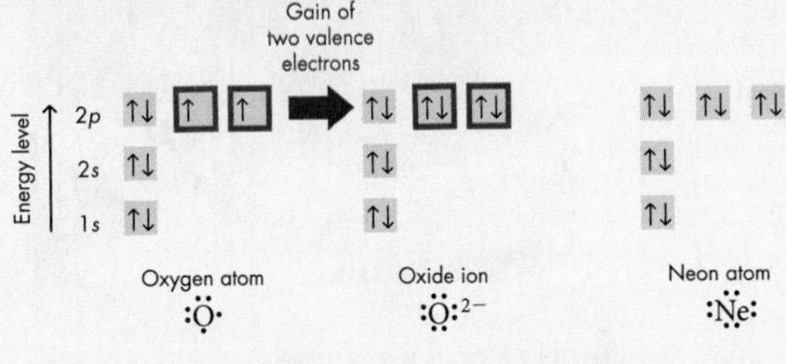

The resulting oxide anion (O^{2-}) has a charge of 2−. You can write the equation for the formation of oxide anions by using electron dot structures.

$$:\ddot{O}\cdot + 2e^- \longrightarrow :\ddot{O}:^{2-}$$

Table 7.2 lists some common anions that you will be learning about in this book.

Figure 7.5 Ions in Seawater Chloride (Cl^-), sodium (Na^+), magnesium (Mg^{2+}), calcium (Ca^{2+}), and potassium (K^+) ions are abundant in seawater.

Table 7.2

Some Common Anions

Name	Symbol	Charge
Fluoride	F^-	1−
Chloride	Cl^-	1−
Bromide	Br^-	1−
Iodide	I^-	1−
Oxide	O^{2-}	2−
Sulfide	S^{2-}	2−
Nitride	N^{3-}	3−
Phosphide	P^{3-}	3−

7.1 LessonCheck

1. **Explain** How can you determine the number of valence electrons in an atom of a representative element?
2. **Describe** How do cations form?
3. **Describe** How do anions form?
4. **Make Generalizations** Atoms of which elements tend to gain electrons? Atoms of which elements tend to lose electrons?
5. **Apply Concepts** How many valence electrons are in each atom?
   a. potassium
   b. carbon
   c. magnesium
   d. oxygen
6. **Use Models** Draw the electron dot structure for each element in Question 5.
7. **Apply Concepts** How many electrons will each element gain or lose in forming an ion?
   a. calcium
   b. fluorine
   c. aluminum
   d. oxygen
8. **Infer** Identify the charge of the ion formed when
   a. a potassium atom loses one electron.
   b. a zinc atom loses two electrons.
   c. a fluorine atom gains one electron.
9. **Describe** Write the electron configuration of Cd^{2+}.

Lesson Check Answers

1. look up the group number of that element
2. An atom loses valence electrons.
3. An atom gains valence electrons.
4. Atoms of nonmetallic elements tend to gain electrons; atoms of metallic elements tend to lose electrons.
5. **a.** 1 **b.** 4 **c.** 2 **d.** 6
6. **a.** K· **b.** ·Ċ· **c.** ·Mg· **d.** :Ö·
7. **a.** lose 2 **b.** gain 1 **c.** lose 3 **d.** gain 2
8. **a.** potassium cation, K^+ **b.** zinc cation, Zn^{2+} **c.** fluoride anion, F^-
9. Cd^{2+}: $1s^22s^22p^63s^23p^63d^{10}4s^24p^64d^{10}$

Evaluate

Informal Assessment

Have students refer to their copy of the periodic table. To determine the students' knowledge about the formation of elemental anions and cations, have them work in small groups to decide whether the following ions are likely to exist, and why:

- H^- *(yes, isoelectronic with He);*
- H^+ *(yes, but without electrons, there is no comparable noble-gas configuration);*
- Sr^{2+} *(yes, isoelectronic with Kr);*
- Al^{3+} *(yes, isoelectronic with Ne);*
- Xe^- *(no, cannot form ions easily due to stable electron configuration);*
- Zn^{6-} *(no, isoelectronic with Kr but formation would require a gain of too many electrons);*
- Zn^{2+} *(yes, not isoelectronic with a noble gas but has pseudo-noble-gas electron configuration with eighteen electrons filling up the outer occupied energy level: $3s^23p^63d^{10}$)*

Then have students complete the 7.1 Lesson Check.

Reteach

Select groups from the periodic table in random order and ask students to predict the common ions that could be formed from elements of each group. Point out that predicting is fairly easy for groups at the far left or far right of the table, but more difficult for groups in the center of the table, which have partially filled *d* and *f* orbitals. Guide students through the information about forming anions and cations, and then repeat the assessment.

Answers

FIGURE 7.4 Group 7A

SMALL-SCALE LAB

Explore

Small-Scale Lab

OBJECTIVE After completing this activity, students will be able to write electron configurations for various transition metals.

SKILLS FOCUS Observing, drawing conclusions, designing experiments

PREP TIME 40 minutes

CLASS TIME 40 minutes

MATERIALS small-scale pipets, solutions

ADVANCE PREPARATION

Solution	Preparation
1.0*M* NaCl	14.6 g in 250 mL
0.2*M* $MgSO_4$	6.0 g in 250 mL
0.2*M* $AlCl_3$	12.1 g in 250 mL
0.1*M* $FeCl_3$	6.8 g in 250 mL 0.1*M* NaCl
0.5*M* $CaCl_2$	14.2 g in 250 mL
0.2*M* $NiSO_4$	2.94 g in 250 mL
0.2*M* $CuSO_4$	12.5 g in 250 mL
0.2*M* $ZnCl_2$	6.8 g in 250 mL
0.05*M* $AgNO_3$	2.1 g in 250 mL
0.5*M* NaOH	20.0 g in 1.0 L
1.0*M* Na_2CO_3	26.5 g in 250 mL
0.1*M* Na_3PO_4	9.5 g in 250 mL

EXPECTED OUTCOME NaCl: colorless, no precipitate; $MgSO_4$, $AlCl_3$, $CaCl_2$, $ZnCl_2$: colorless, white precipitate; $FeCl_3$: yellow, rust-colored precipitate; $NiSO_4$: green, green precipitate; $CuSO_4$: blue, blue precipitate; $AgNO_3$: colorless, brown precipitate

ANALYZE AND CONCLUDE

1. $CuSO_4$, $FeCl_3$, and $NiSO_4$.
2. Cu^{2+}: $1s^22s^22p^63s^23p^63d^9$
   Ag^+: $1s^22s^22p^63s^23p^63d^{10}4s^24p^64d^{10}$
3. It contains a completely full *d* orbital.
   Zn^{2+}: $1s^22s^22p^63s^23p^63d^{10}$
4. Cr^{3+} and V^{2+} are colored because their electron configurations have partially filled *d* orbitals. Hg^{2+} and Cd^{2+} are colorless because their configurations contain completely full *d* orbitals.
5. Yes, except the one produced by silver nitrate.

YOU'RE THE CHEMIST

1. Add one drop of Na_2CO_3 to one drop of each solution. The iron, copper, and nickel salts produce colored precipitates.
2. Add one drop of Na_3PO_4 to one drop of each solution. The iron, copper, and nickel salts produce colored precipitates.

Small-Scale Lab

Electron Configurations of Ions

Purpose

To relate the presence of color in an ionic solution as a characteristic of electron configurations

Materials

- reaction surface
- micropipettes or droppers
- chemicals shown in the grid below
- sodium hydroxide (NaOH) solution

Procedure

1. On separate sheets of paper, draw two grids similar to the one below. Make each square 2 cm on each side.

2. Place a reaction surface over one of the grids, and add one drop of each solution in the indicated places. Record the color of each solution on the other grid.

3. A precipitate is a solid that separates upon mixing solutions. Predict which of the metal cations will form colored precipitates upon the addition of NaOH. Add one drop of NaOH to find out. Record your results.

NaCl	$MgSO_4$	$AlCl_3$
$FeCl_3$	$CaCl_2$	$NiSO_4$
$CuSO_4$	$ZnCl_2$	$AgNO_3$

Analyze and Conclude

1. Draw Conclusions Transition-metal ions with partially filled *d* orbitals usually have color. Based on your observations, which solutions contain transition-metal ions with partially filled *d* orbitals?

2. Analyze Data Write the electron configurations of Cu^{2+} and Ag^+. Is each electron configuration consistent with the color you observed for each cation? Explain.

3. Infer What does the color of the solution containing Zn^{2+} ions suggest about its electron configuration? Write the electron configuration of Zn^{2+}.

4. Predict Which of the following transition-metal ions do you think have color: Cr^{3+}, Cd^{2+}, Hg^{2+}, V^{2+}? Explain your answers.

5. Draw Conclusions Do the colored precipitates all contain transition-metal ions with partially filled *d* orbitals?

You're the Chemist

1. Design an Experiment Predict which of the metal cations in this experiment will form colored precipitates upon the addition of sodium carbonate (Na_2CO_3). Design an experiment to find out.

2. Design an Experiment Design and carry out an experiment to find out which metal ions form precipitates with sodium phosphate (Na_3PO_4). What color are the precipitates?

200 Chapter 7 • Small-Scale Lab

Focus on ELL

4 LANGUAGE PRODUCTION Have students work in pairs to complete the lab. Pair students so that more proficient English-speaking students can help less proficient English-speaking students. Have students work according to their proficiency level.

BEGINNING:

LOW Model this lab and ask students to say the color that each metal produces.

HIGH Provide students with a chart of the solutions and then a column for them to write the color produced.

INTERMEDIATE: LOW/HIGH Have students predict which metals have color. Ask them to verbally compare the actual result with their prediction.

ADVANCED: LOW/HIGH Ask students to complete the Analyze and Conclude questions with written responses.

7.2 Ionic Bonds and Ionic Compounds

C.3.1 Interactions between atoms in ionic and covalent compounds; C.3.2 Formation of ionic and covalent compounds; C.3.3 Chemical bonds and intermolecular attractions

CHEMISTRY & YOU

Q: *Where does table salt come from?* Sodium chloride, or table salt, has been used by people for centuries to add flavor to food and for preserving food. In some countries, salt is obtained by the evaporation of seawater. In other countries, salt is mined from rock deposits deep underground. In this lesson, you will learn how cations and anions combine to form stable compounds such as sodium chloride.

Key Questions

What is the electrical charge of an ionic compound?

What are three properties of ionic compounds?

Vocabulary

- ionic compound
- ionic bond
- chemical formula
- formula unit
- coordination number

Formation of Ionic Compounds

What is the electrical charge of an ionic compound?

Sodium chloride, or table salt, is an ionic compound consisting of sodium cations and chloride anions. An **ionic compound** is a compound composed of cations and anions. **Although they are composed of ions, ionic compounds are electrically neutral.** The total positive charge of the cations equals the total negative charge of the anions.

Ionic Bonds Anions and cations have opposite charges and attract one another by means of electrostatic forces. The electrostatic forces that hold ions together in ionic compounds are called **ionic bonds.**

Sodium chloride provides a simple example of how ionic bonds are formed. Consider the reaction between a sodium atom and a chlorine atom. The sodium atom has a single valence electron that it can easily lose. (If the sodium atom loses its valence electron, it achieves the stable electron configuration of neon.) The chlorine atom has seven valence electrons and can easily gain one electron. (If the chlorine atom gains a valence electron, it achieves the stable electron configuration of argon.) When sodium and chlorine react to form a compound, the sodium atom transfers its one valence electron to the chlorine atom. Thus, sodium and chlorine atoms combine in a one-to-one ratio, and both ions have stable octets.

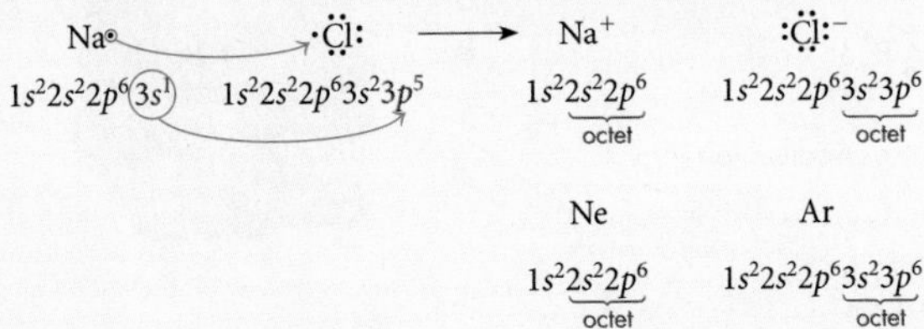

LESSON 7.2

Key Objectives

7.2.1 EXPLAIN the electrical charge of an ionic compound.

7.2.2 DESCRIBE three properties of ionic compounds.

Additional Resources

Reading and Study Workbook, Lesson 7.2

Available Online or on Digital Media:

- Teaching Resources, Lesson 7.2 Review
- Probeware Laboratory Manual, Section 7.2
- Virtual Chemistry Labs, Lab 2

PearsonChem.com

Engage

CHEMISTRY & YOU Have students examine the opening photograph along with Figure 7.7. Discuss how the poisonous elements sodium and chlorine can react to form harmless table salt. **Ask** What characteristics of sodium and chlorine atoms allow them to form a stable compound? *(Sodium atoms can lose an electron easily, and chlorine atoms can accept an electron easily. The resulting ions are attracted to the oppositely charged ions.)* Remind students that NaCl is an example of an ionic compound.

Access Prior Knowledge

Have students review the patterns in Tables 6.1 and 6.2. **Ask** What term is used to describe the ability of an atom of an element to attract electrons? *(electronegativity)*

Focus on ELL

1 CONTENT AND LANGUAGE Write the vocabulary terms on the board along with their phonetic spellings and pronounce them slowly. Discuss the meaning of each word in each term, and then explain the meaning of the combined term. Have students produce a half-page illustration or example for each vocabulary term.

2 FRONTLOAD THE LESSON Have small groups of students read aloud and discuss the lesson headings and key questions. Then have them write at least one sentence for each page summarizing what they will study on that page.

3 COMPREHENSIBLE INPUT Explain to students that they will study the interaction of ions in this lesson. As a visual analogy, show students two magnets with opposite poles attracting each other. Explain that this attraction is similar to the attraction between oppositely charged ions. Be sure to stress that the attraction between the ions is due to electrical charge, not magnetic force.

IN ACADEMIC STANDARDS for SCIENCE

- Develop explanations based on reproducible data and observations gathered during laboratory investigations.

C.3.1 Describe, compare, and contrast the characteristics of the interactions between atoms in ionic and covalent compounds.

C.3.2 Compare and contrast how ionic and covalent compounds form.

Also covered C.3.3.

LESSON 7.2

Foundations for Reading

BUILD VOCABULARY Have students skim the lesson to locate the definition of each vocabulary term. Then have them define each term in their own words.

READING STRATEGY Have students rewrite the headings as how, why, or what questions about ionic compounds. As they read, have them write answers to their questions.

Explain

Formation of Ionic Compounds

APPLY CONCEPTS Explain that the formation of positive ions and negative ions are simultaneous and interdependent processes. Explain that an ionic compound is the result of the transfer of electrons from one set of atoms to another set of atoms. Emphasize that an ionic compound consists entirely of ions. Have students name some substances that they think are ionic compounds. As a class, determine which substances are in fact ionic compounds, and why.

CRITICAL THINKING Direct students' attention to Figure 7.7. **Ask** Why are crystalline ionic compounds generally so rigid and brittle? *(The crystal is rigid because it is held together by a specific three-dimensional array of relatively strong attractive forces between anions and cations, which is accompanied by minimal repulsion of like ions. The crystal is brittle because the attractive interactions are specifically arranged within the crystal structure. If this arrangement is disturbed, as it would be if the crystal were hit with a hammer, repulsion between ions of the same charge can force the crystal to fragment.)*

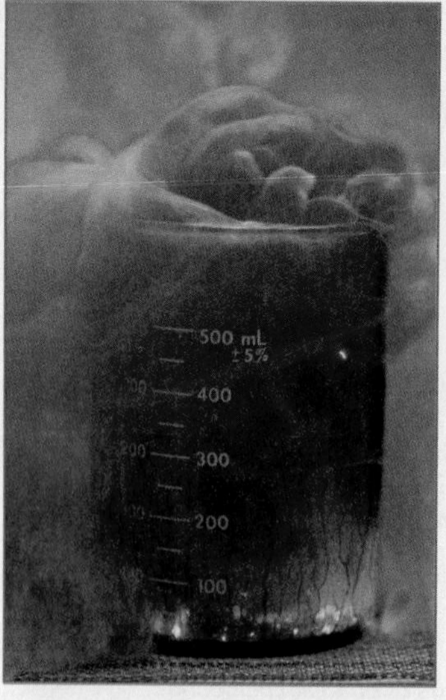

Figure 7.6 Formation of Aluminum Bromide
Aluminum metal and the nonmetal bromine react violently to form the ionic solid aluminum bromide.

Figure 7.6 shows the reaction of aluminum (Al) and bromine (Br_2) to form the ionic compound aluminum bromide ($AlBr_3$). Each aluminum atom has three valence electrons to lose. Each bromine atom has seven valence electrons and readily gains one additional electron. Therefore, when aluminum and bromine react, three bromine atoms combine with each aluminum atom.

Formula Units The ionic compound sodium chloride is composed of equal numbers of sodium cations (Na^+) and chloride anions (Cl^-). Chemists represent the composition of substances by writing chemical formulas. A **chemical formula** shows the numbers of atoms of each element in the smallest representative unit of a substance. For example, NaCl is the chemical formula for sodium chloride.

Note that the formula NaCl does not represent a single physical unit. As shown in Figure 7.7, the ions in solid sodium chloride are arranged in an orderly pattern. Ionic compounds do not exist as single discrete units, but as collections of positively and negatively charged ions arranged in repeating patterns. Therefore, the chemical formula of an ionic compound refers to a ratio known as a formula unit. A **formula unit** is the lowest whole-number ratio of ions in an ionic compound. For sodium chloride, the lowest whole-number ratio of the ions is 1:1 (one Na^+ ion to each Cl^- ion). Thus, the formula unit for sodium chloride is NaCl. Although ionic charges are used to derive the correct formula, they are not shown when you write the formula unit of the compound.

The ionic compound magnesium chloride contains magnesium cations (Mg^{2+}) and chloride anions (Cl^-). In magnesium chloride, the ratio of magnesium cations to chloride anions is 1:2 (one Mg^{2+} ion to two Cl^- ions). Therefore, the formula unit for magnesium chloride is $MgCl_2$. The compound has twice as many chloride anions (each with a 1− charge) as magnesium cations (each with a 2+ charge), so it is electrically neutral. In aluminum bromide, the ratio of aluminum cations to bromide anions is 1:3 (one Al^{3+} ion to three Br^- ions), so the formula unit is $AlBr_3$. Again, the compound is electrically neutral.

Figure 7.7 Formation of Sodium Chloride
Sodium cations and chloride anions form a repeating three-dimensional array in sodium chloride (NaCl).
Infer ***How does the arrangement of ions in a sodium chloride crystal help explain why the compound is so stable?***

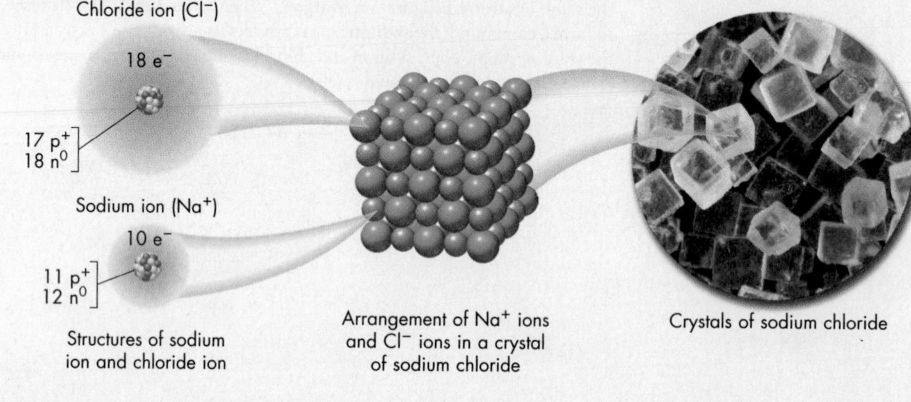

UbD Check for Understanding

The Essential Question How do ionic compounds form?

Assess students' understanding of the process of ionic bond formation by asking them to complete the following cloze prompt:

Ionic bonds form when one or more ________ is transferred from a ________ to a ________ so that both ions formed have a __________. *(electrons, metal, nonmetal, stable octet)*

ADJUST INSTRUCTION If students are having difficulty answering this question, have them reread the last paragraph on page 201 and examine the diagram below it. Then provide students another opportunity to complete the prompt.

SampleProblem 7.1

Predicting Formulas of Ionic Compounds

Use electron dot structures to predict the formulas of the ionic compounds formed from the following elements:

a. potassium and oxygen
b. magnesium and nitrogen

1 Analyze **Identify the relevant concepts.** Atoms of metals lose valence electrons when forming an ionic compound. Atoms of nonmetals gain electrons. Enough atoms of each element must be used in the formula so that electrons lost equals electrons gained.

2 Solve **Apply the concepts to this problem.**

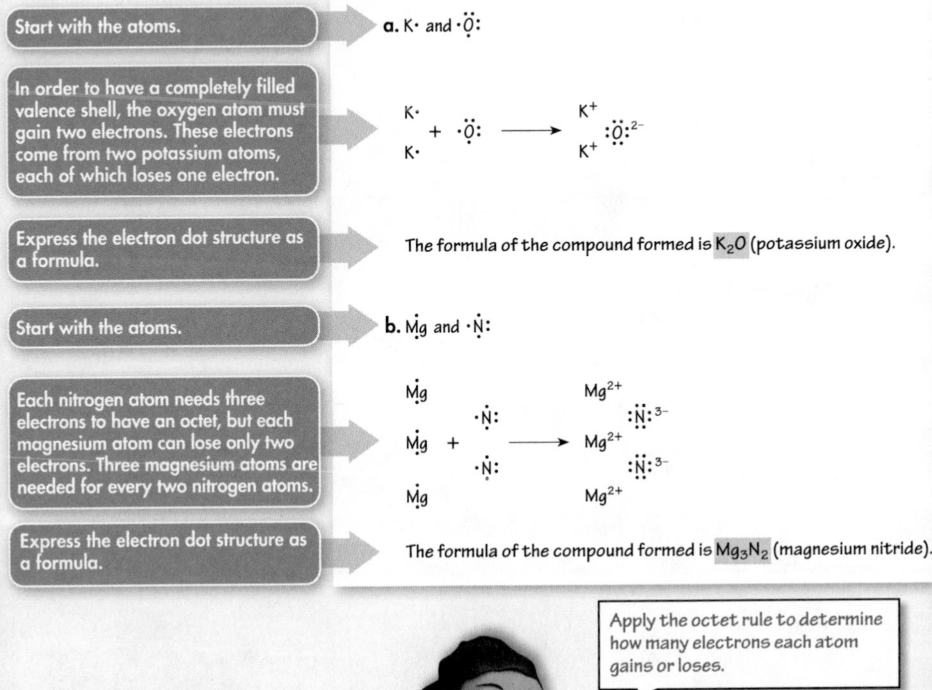

10. Use electron dot structures to determine formulas of the ionic compounds formed when
a. potassium reacts with iodine.
b. aluminum reacts with oxygen.

11. What is the formula of the ionic compound composed of calcium cations and chloride anions?

Sample Practice Problem

Use electron dot structures to determine chemical formulas for the ionic compounds formed when the following elements combine:

a. magnesium and chlorine *($MgCl_2$)*
b. aluminum and sulfur *(Al_2S_3)*
c. calcium and iodine *(CaI_2)*

Explore

Properties of Ionic Compounds

Class Activity

PURPOSE Students detect the presence of ions in water samples.

MATERIALS tap water samples, 10-mL graduated cylinder, potassium thiocyanate solution, dilute ethanoic acid, sodium oxalate solution, dropper

PROCEDURE Have students bring water samples from home to test for ions associated with water hardness. Test 2-mL samples as follows. Add three drops of potassium thiocyanate (KSCN) to the first sample. Add three drops of dilute ethanoic acid, CH_3COOH, and three drops of sodium oxalate, $Na_2C_2O_4$, to the second sample. Mix well.

EXPECTED OUTCOME A red color from the iron(III) thiocyanate ion, $Fe(SCN)^{2+}$, indicates the presence of Fe^{3+} ions. A white precipitate of calcium oxalate, CaC_2O_4, indicates the presence of Ca^{2+} ions.

Professional Development Note: Water Hardness

Water hardness varies with location and source, and is primarily a result of a high calcium and magnesium ion content, although other metal ions contribute to the problem. The United States Geological Survey (USGS) considers water to be hard if it contains greater than 60 mg/L of calcium carbonate. People who live in areas with hard water often see this problem manifested as a build-up of limescale on tubs, sinks, faucets, and other surfaces frequently exposed to water.

Generally, water from groundwater sources is harder than water from surface sources. In the United States, most northeastern, southern, and northwestern states have predominantly soft water. Generally, hard water of varying degrees is found in the southwestern and midwestern states.

Answers

FIGURE 7.7 Each ion is surrounded by ions of opposite charge.

10. **a.** The formula is KI.
b. The formula is Al_2O_3.

11. $CaCl_2$

Explore

Class Activity

PURPOSE Students observe different types of crystals.

MATERIALS crystals of ionic compounds, watch glasses, magnifying glasses

SAFETY Use only nontoxic crystals. Remind students to not touch the crystals.

PROCEDURE Pass around watch glasses with crystals of ionic compounds. Have the students examine the crystals with magnifying glasses and write down their observations. Make a list of these observations and then discuss them in terms of the underlying ionic lattice structures.

EXPECTED OUTCOME Students should observe the geometries of different crystal structures.

CHEMISTRY & YOU *Sample answer:* I would expect NaCl in underground rock deposits to exist as a crystalline solid because NaCl is an ionic compound.

Extend

USE VISUALS Point out the formulas for barite and aragonite in Figure 7.8. Note that those compounds contain anions that are polyatomic ions, which means that they contain more than one type of atom. Tell students that they will learn more about polyatomic ions in Chapter 8. (If students ask about the formulas with multiple cations, explain that those formulas represent complex minerals.)

CHEMISTRY & YOU

Q: *Would you expect to find sodium chloride in underground rock deposits as a solid, liquid, or gas? Explain.*

Properties of Ionic Compounds

What are three properties of ionic compounds?

Figure 7.8 shows the striking beauty of the crystals of some ionic compounds. **Most ionic compounds are crystalline solids at room temperature.** The component ions in such crystals are arranged in repeating three-dimensional patterns. In solid sodium chloride, each sodium ion is surrounded by six chloride ions, and each chloride ion is surrounded by six sodium ions. In this arrangement, each ion is attracted strongly to each of its neighbors, and repulsions are minimized. The large attractive forces result in a very stable structure. This stability is reflected in the fact that NaCl has a melting point of about 800°C. **Ionic compounds generally have high melting points.**

Figure 7.8 Crystalline Solids
The beauty of crystalline solids, such as these, comes from the orderly arrangement of their component ions.

Go online to learn about properties of ionic compounds.

Fluorite (CaF_2)

Grossularite ($Ca_3Al_2(SiO_4)_3$)

Aragonite ($CaCO_3$)

Barite ($BaSO_4$)

Wulfenite ($PbMoO_4$)

Beryl ($Be_3Al_2(SiO_3)_6$)

Hematite (Fe_2O_3)

Cinnabar (HgS)

Differentiated Instruction

ELL ENGLISH LANGUAGE LEARNERS Encourage students to look up and define terms used to describe ionic compounds. Students should define terms such as *crystal* and *formula unit* in English and their native language.

L1 STRUGGLING STUDENTS Students may have difficulty visualizing the ions that represent the coordination number for the example crystals in Figures 7.9 and 7.10. Provide students with copies of both figures, and project the figures on the board. Select a Na^+ ion in NaCl and number the Cl^- ions around it. Then, select a Cl^- ion and number the Na^+ ions around it. Repeat with CsCl and TiO_2, having students direct you as you number each ion.

L3 ADVANCED STUDENTS Have students research how minerals are classified. Suggest that their written reports include information about the physical properties of minerals and how these properties are used in mineral identification. Encourage students to include drawings, photos, or samples of minerals.

The **coordination number** of an ion is the number of ions of opposite charge that surround the ion in a crystal. Figure 7.9a shows the three-dimensional arrangement of ions in NaCl. The coordination number of Na^+ is 6 because each Na^+ ion is surrounded by six Cl^- ions. The coordination number of Cl^- is also 6 because each Cl^- ion is surrounded by six Na^+ ions. Cesium chloride (CsCl) has a formula unit that is similar to that of NaCl. As Figure 7.9b illustrates, both compounds have cubic crystals, but their internal crystal structures are different. Each Cs^+ ion is surrounded by eight Cl^- ions, and each Cl^- ion is surrounded by eight Cs^+ ions. Therefore, each anion and each cation in cesium chloride has a coordination number of 8.

Figure 7.10 shows the crystalline form of titanium dioxide (TiO_2), also known as rutile. In this compound, the coordination number for the cation (Ti^{4+}) is 6. Each Ti^{4+} ion is surrounded by six O^{2-} ions. The coordination number of the anion (O^{2-}) is 3. Each O^{2-} ion is surrounded by three Ti^{4+} ions.

Figure 7.9 Coordination Numbers
Sodium chloride and cesium chloride form cubic crystals. **a.** In NaCl, each ion has a coordination number of 6. **b.** In CsCl, each ion has a coordination number of 8.

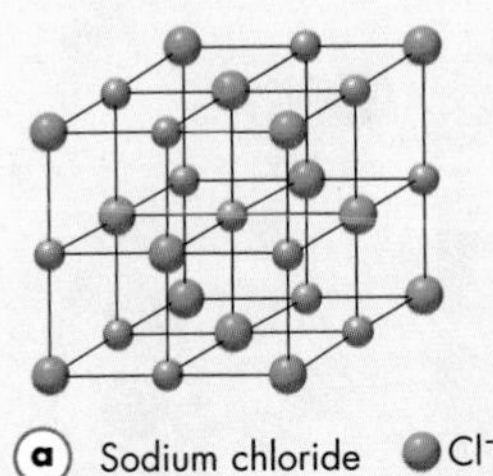

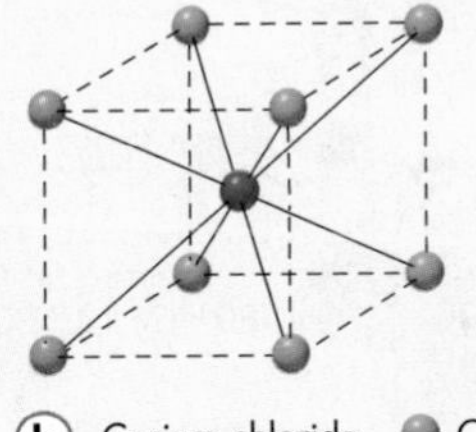

Figure 7.10 Rutile
Titanium dioxide, or rutile, forms tetragonal crystals. In TiO_2, each Ti^{4+} ion has a coordination number of 6, while each O^{2-} ion has a coordination number of 3.

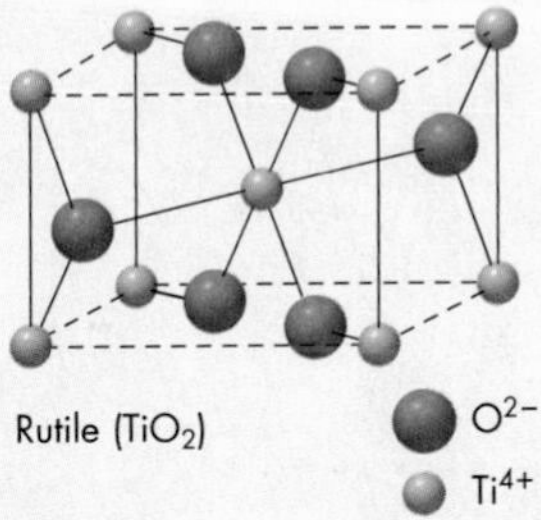

UbD Check for Understanding

BIGIDEA Bonding and Interactions To assess students' knowledge about ionic bonds and ionic compounds, have students write one-minute responses to the following questions:

- Given what you have learned so far, write or illustrate your own definition of an ionic bond. *(an electrical attraction between ions of opposite charge)*
- Explain why crystals of different ionic compounds have different shapes. *(The shapes reflect different geometric arrangements of anions and cations.)*

ADJUST INSTRUCTION If students demonstrate difficulty answering the second question, review Figure 7.10 and discuss how each crystal's shape is determined by the coordination number of an ion. Then, encourage them to go online and watch the Concepts in Action feature about properties of ionic crystals.

Explain

CRITICAL THINKING Point out that the rusting of iron is the production of iron oxide from iron metal and oxygen gas. Point out that Fe^{3+} is a stable cation of Fe. **Ask** What is the stable anion of oxygen? *(O^{2-})* **Ask** What compound is formed from these two ions? *(The compound formed from these ions is iron(III) oxide, or Fe_2O_3.)*

Explore

PURPOSE Students investigate properties of different classes of ionic compounds.

MATERIALS library or Internet access

PROCEDURE Divide the class into groups. Have each group choose a different class of ionic compounds to research and write about. For example, one group could work with oxides while another group works with sulfides. Initially, each student should work alone to discover information such as where the compounds occur in nature, how they are produced, their physical and chemical properties, and any important uses. Finally, students in each group can pool their information to prepare a class display or report.

EXPECTED OUTCOMES Students will discover that different classes of ionic compounds share some properties with other ionic compounds and have some unique properties.

LESSON 7.2

Explore

Quick Lab

OBJECTIVE After completing this activity, students will be able to show that ions in solution conduct an electric current.

Misconception Alert

Students may think that the solutions contain only one ion. Clarify that each solution contains both cations and anions.

SKILLS FOCUS Observing, experimenting, concluding

PREP TIME 20 minutes

CLASS TIME 30 minutes

MATERIALS D-cell batteries, masking tape, 30-cm lengths of bell wire with ends scraped bare, clear plastic cups, distilled water, tap water, vinegar, sucrose, sodium chloride, baking soda, conductivity probe (optional)

SAFETY Students should handle wires with caution. The wires may become hot during the activity.

EXPECTED OUTCOME When ions are present in solution, the solution conducts an electric current.

ANALYZE AND CONCLUDE

1. Solutions of vinegar, sodium chloride, and baking soda (and maybe tap water) produced bubbles. Distilled water and sugar solution (and maybe tap water) do not produce bubbles.
2. The solutions in which bubbles were produced also conduct an electric current. These solutions contain ions.
3. *Sample answer:* A larger number of batteries will increase the current, which will, in turn, cause the rate at which the bubbles appear to increase.

FOR ENRICHMENT Ask What gases form the bubbles you observe? *(hydrogen and oxygen)* **Ask** What is the source of these gases? *(water)* Have students collect the gases produced and check them for the presence of hydrogen and oxygen. Hydrogen will burn when a lit splint is placed in it, and a glowing split will begin to flame when placed in oxygen.

Another characteristic property of ionic compounds relates to conductivity. **Ionic compounds can conduct an electric current when melted or dissolved in water.** When sodium chloride is melted, the orderly crystal structure breaks down. As Figure 7.11a shows, if a voltage is applied across this molten mass, cations migrate freely to one electrode and anions migrate to the other. This movement of ions allows electric current to flow between the electrodes through an external wire. For a similar reason, ionic compounds also conduct electric current if they are dissolved in water. When dissolved, the ions are free to move about in the solution.

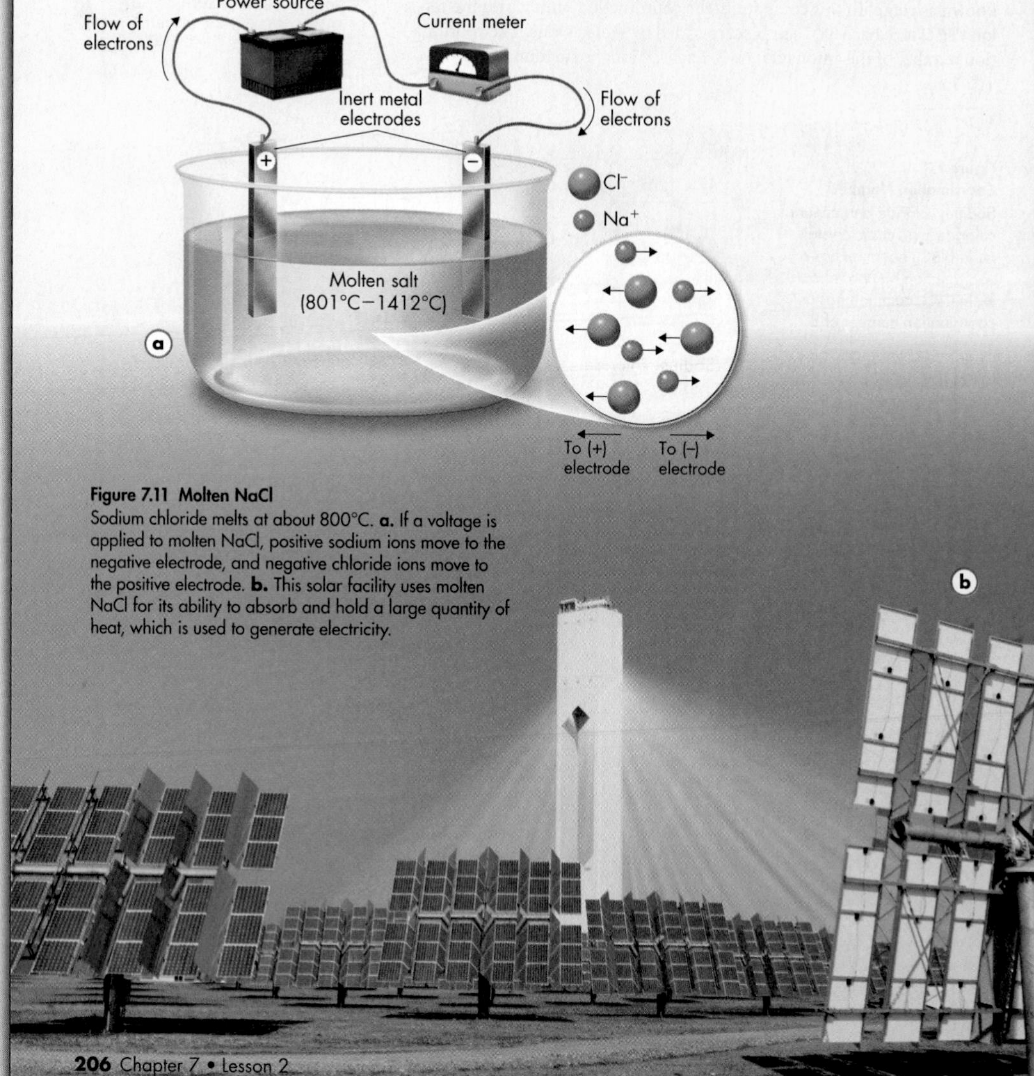

Figure 7.11 Molten NaCl
Sodium chloride melts at about 800°C. **a.** If a voltage is applied to molten NaCl, positive sodium ions move to the negative electrode, and negative chloride ions move to the positive electrode. **b.** This solar facility uses molten NaCl for its ability to absorb and hold a large quantity of heat, which is used to generate electricity.

Focus on ELL

4 LANGUAGE PRODUCTION Have students work in groups of four to complete the Quick Lab. Make sure each group has ELLs of varied language proficiencies, so that more proficient students can help less proficient ones. Have students work according to their proficiency level.

BEGINNING:

LOW Make sure students know what the terminal of a battery is. Discuss each step before groups perform it, using the photo to increase their understanding.

HIGH Recreate the procedure as a numbered list of steps to follow.

INTERMEDIATE: LOW/HIGH Provide students with an organized and labeled chart to complete during the Analyze and Conclude section.

ADVANCED: LOW/HIGH Perform the enrichment activity as a demo and have students write a summary of what happens.

Quick Lab

Purpose To show that ions in solution conduct an electric current

Materials

- 3 D-cell batteries
- masking tape
- 2 30-cm lengths of bell wire with ends scraped bare
- clear plastic cup
- distilled water
- tap water
- vinegar (acetic acid, $C_2H_4O_2$)
- table sugar (sucrose, $C_{12}H_{22}O_{11}$)
- table salt (sodium chloride, NaCl)
- baking soda (sodium hydrogen carbonate, $NaHCO_3$)

Solutions Containing Ions

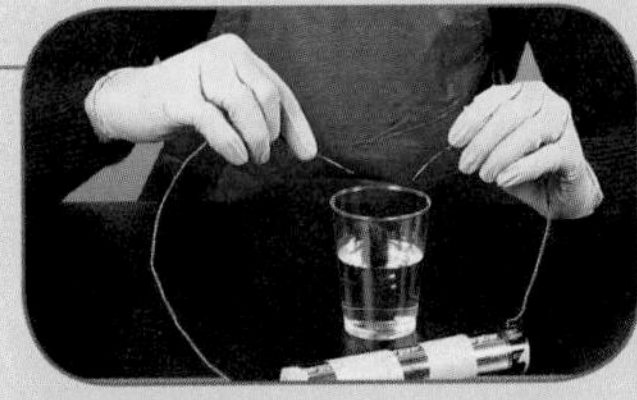

Procedure

1. Tape the batteries together so the positive end of one touches the negative end of another. Tape the bare end of one wire to the positive terminal of the battery assembly and the bare end of the other wire to the negative terminal. **CAUTION** *Bare wire ends can be sharp and scratch skin. Handle with care.*

2. Half fill the cup with distilled water. Hold the bare ends of the wires close together in the water.

3. Look for the production of bubbles. They are a sign that the solution conducts electric current.

4. Repeat Steps 2 and 3 with tap water, vinegar, and concentrated solutions of table sugar, table salt, and baking soda.

Analyze and Conclude

1. Observe Which samples produced bubbles of gas? Which samples did not produce bubbles of gas?

2. Draw Conclusions Which samples conducted an electric current? What do these samples have in common?

3. Predict Would you expect the same results if you used only one battery? If you used six batteries? Explain your answers.

7.2 LessonCheck

ONLINE PROBLEMS

12. **Describe** How can you describe the electrical charge of an ionic compound?

13. **Identify** What properties characterize ionic compounds?

14. **Apply Concepts** Write the correct chemical formula for the compounds formed from each pair of elements.
- **a.** potassium and sulfur
- **b.** calcium and oxygen
- **c.** sodium and oxygen
- **d.** aluminum and nitrogen

15. **Describe** Write formulas for each compound.
- **a.** barium chloride
- **b.** magnesium oxide
- **c.** lithium oxide
- **d.** calcium fluoride

16. **Describe** How can you describe the arrangement of sodium ions and chloride ions in a crystal of sodium chloride?

17. **Relate Cause and Effect** Why do ionic compounds conduct electric current when they are melted or dissolved in water?

18. **Apply Concepts** Read about restoring electrolytes on page R4 of the Elements Handbook. Write electron configurations for the two principal ions found in body fluids.

BIGIDEA BONDING AND INTERACTIONS

19. Which pairs of elements are likely to form ionic compounds? Explain your choices and write the formulas for the compounds that will form.
- **a.** Cl, Br
- **b.** K, He
- **c.** Li, Cl
- **d.** I, Na

Ionic and Metallic Bonding **207**

Lesson Check Answers

12. electrically neutral

13. usually solids at room temperature; have high melting points; conduct an electric current when melted or dissolved in water

14. **a.** K_2S **b.** CaO **c.** Na_2O **d.** AlN

15. **a.** $BaCl_2$ **b.** MgO **c.** Li_2O **d.** CaF_2

16. Acceptable answers should describe a solid containing positive sodium ions and negative chloride ions in an alternating, regular, and repeating three-dimensional pattern.

17. The ions are free to move.

18. Na^+: $1s^22s^22p^6$ K^+: $1s^22s^22p^63s^23p^6$

19. **a.** not likely, both form anions
b. not likely, helium is a noble gas
c. LiCl
d. NaI

LESSON 7.2

Extend

Connect to TECHNOLOGY

Ask for a show of hands of students who are familiar with the term *liquid crystal display* (LCD). Identify any computer monitors or other items in the classroom that have an LCD screen. Explain that LCDs use a type of matter called liquid crystals to display the information or images they see on the LCD screens. Have students use the Internet to discover the properties of liquid crystals and why they have such widespread use. Have students report their findings to the class.

Evaluate

Informal Assessment

Create a game where groups of three students compete to win the Ionic Crystal award. Give each group the name of an ionic compound; have one group member identify the cation, have one group member identify the anion, and have one group member determine the ratio of cations to anions in the given compound. If groups are successful, then have them switch positions in their group and give them the name of another compound. Repeat again until each group member has taken turns identifying a cation, an anion, and the ratio. If the group successfully completes three rounds of the game, award them the "Ionic Crystal." Then have students complete the 7.2 Lesson Check.

Reteach

Emphasize that an ionic solid is a collection of independent ions. Explain that there is no joining of individual particles to form molecules. State that each ion "belongs" as much to one of its nearest neighbors as it belongs to any other. Point out that the arrangement in an ionic crystal is such that each ion is surrounded by ions of opposite charge, which produces a strong bonding force. Work through several examples to be sure students understand this important point.

CHEMISTRY & YOU

CHEMISTRY & YOU Discuss the process of nucleation with students. Then, model nucleation and crystal growth using square blocks or metal cubes. Let one block represent the nucleation crystal. Stack additional blocks on and around the nucleation block to form a crystalline shape. Pose the following question to students: *What interactions do you think occur between the individual crystals to create the larger crystal?* You may need to guide students in the following ways:

- Have students examine the arrangement of sodium and chloride ions in Figure 7.9.
- Ions along the edge of one crystal can interact with ions along the edge of another crystal.
- Have students consider how crystals might share ions along their edges and faces.

Explain

APPLY CONCEPTS After students read the feature, start a discussion of properties of ionic crystals. **Ask** What keeps the ions in their orderly arrangements? *(The electrostatic forces among them.)* **Ask** *Why don't ionic crystals conduct electricity?* (They do not have free electrons or ions.) **Ask** What does "growing a crystal" mean? *(The crystal starts from a small particle and becomes larger as more and more ions attach to the crystal.)*

Extend

Connect to GEOLOGY

In nature, large crystals with clearly defined faces can grow only if there is plenty of free space and the minerals are available from the environment. Molten rock can form crystals if it cools slowly enough that the ions can settle into an orderly arrangement. If the rock cools quickly, such as during a volcanic eruption, the rock is glassy instead of crystalline. Have students look back at the giant crystals on the first page of this chapter. Explain that these crystals in Mexico's Cave of Crystals grew because they were surrounded by very warm water rich in minerals. When the water was pumped out of the cave and the crystals were discovered, the crystals stopped growing.

Answers

THINK ABOUT IT

1. Both crystals have an orderly shape. The sea salt crystals look like little cubes. The Epsom salt crystals are shaped like needles.
2. 2+
3. Possible factors are rate of cooling and purity of the solution.

CHEMISTRY & YOU: EVERYDAY MATTER

Ionic Crystals

What ionic crystal is essential to human life, was found among the funeral offerings of ancient Egyptians, created and destroyed empires, and is now commonly used to season foods? If you said table salt, you'd be right! Table salt, or sodium chloride (NaCl) is an ionic compound composed of sodium cations (Na^+) and chloride anions (Cl^-).

Crystals of ionic compounds, such as sodium chloride, can be grown by a process called nucleation. During nucleation, the ionic compound that is to be crystallized is dissolved in a solvent, such as water. In the dissolution process, the positive and negative ions break away from each other. As the solvent is removed, the ions join together again to form a repeating three-dimensional pattern. Sodium chloride has a cubic crystal structure, but different ionic compounds form crystals with different shapes. Try the On Your Own activity at home and compare the shapes of two different ionic crystals.

On Your Own

1. For this activity you will need **sea salt, Epsom salts, hot tap water, 2 pie plates, a measuring cup, and a stirring rod or spoon.** Epsom salts can be found in the medicine department of most grocery stores.
2. Mix together 1/4 cup of the sea salt and 1/4 cup of hot water in one of the pie plates. Stir to dissolve most of the salt. In the other pie plate, mix together 1/4 cup of the Epsom salts and 1/4 of hot water. Again, stir until most of the salt has dissolved.
3. Place both pie plates in the refrigerator for three hours. Once the crystals have formed, compare the shapes of the crystals made from the two substances. Record your observations.

Think About It

1. **Compare and Contrast** Describe the shapes of the crystals of sea salt and Epsom salts. How are they the same? How are they different?
2. **Identify** Epsom salts are magnesium sulfate ($MgSO_4$) crystals. A magnesium sulfate formula unit consists of a magnesium cation and a sulfate (SO_4^{2-}) anion. What is the charge on the magnesium cation?
3. **Control Variables** What factors do you think affect crystal growth? Identify two possible factors, then repeat the activity to test your hypotheses.

7.3 Bonding in Metals

C.3.3 Chemical bonds and intermolecular attractions

CHEMISTRY & YOU

Q: *What are some properties that are unique to metals?* You have probably seen decorative fences, railings, or weather vanes made of a metal called wrought iron. Wrought iron is a very pure form of iron that contains trace amounts of carbon. It is a tough, malleable, ductile, and corrosion-resistant material that melts at a very high temperature. These properties derive from the way that metal ions form bonds with one another.

Key Questions

How can you model the valence electrons of metal atoms?

Why are alloys important?

Vocabulary

- metallic bond
- alloy

Metallic Bonds and Metallic Properties

How can you model the valence electrons of metal atoms?

Metals consist of closely packed cations and loosely held valence electrons rather than neutral atoms. **The valence electrons of atoms in a pure metal can be modeled as a sea of electrons.** That is, the valence electrons are mobile and can drift freely from one part of the metal to another. **Metallic bonds** are the forces of attraction between the free-floating valence electrons and the positively charged metal ions. These bonds hold metals together.

Properties of Metals The sea-of-electrons model explains many physical properties of metals. Metals are good conductors of electric current because electrons can flow freely in the metal. As electrons enter one end of a bar of metal, an equal number of electrons leaves the other end. Metals are ductile—that is, they can be drawn into wires, as shown in Figure 7.12. Metals are also malleable, which means that they can be hammered or pressed into shapes.

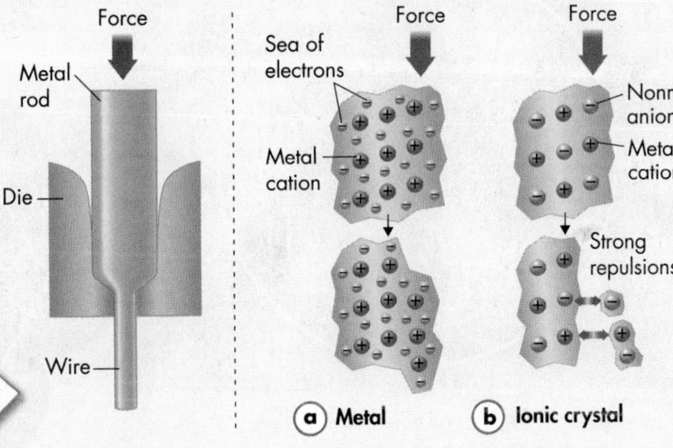

Figure 7.12 Comparing Metals and Ionic Compounds
A metal rod can be forced through a narrow opening to produce wire. **a.** As this occurs, the metal changes shape but remains in one piece. **b.** If an ionic crystal were forced through the opening, it would shatter.
Interpret Diagrams *What causes the ionic crystal to break apart?*

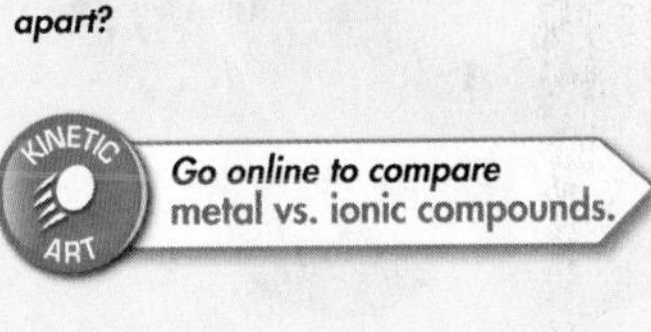

LESSON 7.3

Key Objectives

7.3.1 MODEL the valence electrons of metal atoms.

7.3.2 DESCRIBE the arrangement of atoms in a metal.

7.3.3 EXPLAIN the importance of alloys.

Additional Resources

Reading and Study Workbook, Lesson 7.3

Available Online or on Digital Media:

- Teaching Resources, Lesson 7.3 Review
- Laboratory Manual, Lab 10

PearsonChem.com

Engage

CHEMISTRY & YOU Have students read the text that opens the lesson. Ask students to offer suggestions of how metal ions could form bonds with other metal ions. (Do not provide the correct answer yet.)

Access Prior Knowledge

To assess students' prior knowledge about metals, list the adjectives *tough, malleable, ductile,* and *corrosion-resistant* on the board. **Ask** Which term means able to be hammered? *(malleable)* Which means can be drawn into wires? *(ductile)*

Focus on ELL

1 CONTENT AND LANGUAGE Read the key questions, and write the terms *model, valence electrons, metal,* and *alloy* on the board along with their phonetic spelling. State each word aloud and discuss the meaning of each word. Then help students rephrase the key questions in their own words.

2 FRONTLOAD THE LESSON Hold up a large crystal (preferably one that is long in shape) and a piece of wire or a metal strip. Have a student volunteer to attempt to bend each item. Ask students to use what they know about ionic bonding to explain why the crystal could not be bent. Then, ask students to predict what the difference in bonding in metal crystals is that allows the metal strip to be bent. Tell students that they will explore this concept during the lesson.

3 COMPREHENSIBLE INPUT Have students learn about close packing arrangements by packing table tennis balls or small foam balls into a box.

IN ACADEMIC STANDARDS for SCIENCE

C.3.3 Compare and contrast ionic, covalent network, metallic and polar and non-polar molecular crystals with respect to constituent particles, strength of bonds, melting and boiling points, and conductivity; provide examples of each type.

LESSON 7.3

Foundations for Reading

BUILD VOCABULARY From what they know about the terms *metallic* and *bond,* have students infer the meaning of *metallic bond.*

READING STRATEGY Tell students that before they read, they should write down three things they know about metals. Then, after they read a section, they should explain how the information they already knew about metals helped them to learn something new.

Explore

Teacher Demo

PURPOSE Students compare copper metal and a copper compound.

MATERIALS copper metal or alloy, copper-containing ionic compound

PROCEDURE Show the class a small sample of elemental copper or a copper alloy and a sample of a copper-containing crystalline ionic mineral such as chalcocite (Cu_2S). Wearing safety glasses and standing far from the students, smash both samples with a hammer. Discuss why the two substances respond differently to the stress of the hammer blow.

EXPECTED OUTCOMES The elemental copper will flatten but not break because the cations and electrons are mobile; the crystal will shatter.

Explain

Metallic Bonds and Metallic Properties

START A CONVERSATION Lead a class discussion on the concept of "close packing" of metal cations in pure metals. Use the three different close-packing arrangements shown in Figure 7.14 as a reference. Emphasize that the concept of close packing also relates to more than just metal atoms. Have the students describe other examples of close packing.

CHEMISTRY & YOU Both metals and ionic compounds are held together by strong attractive forces. In ionic solids, the attractions are between ions with opposite charges. In metals, the attractions are between metal ions and mobile valence electrons.

Both the ductility and malleability of metals can be explained in terms of the mobility of valence electrons. A sea of drifting valence electrons insulates the metal cations from one another. When a metal is subjected to pressure, the metal cations easily slide past one another like ball bearings immersed in oil. In contrast, if an ionic crystal is struck with a hammer, the blow tends to push the positive ions close together. The positive ions repel one another, and the crystal shatters.

CHEMISTRY & YOU

Q: *How are metals and ionic compounds different? How are they similar?*

Crystalline Structure of Metals The next time you visit a grocery store, take a look at how the apples or oranges are stacked. More than likely, they will have a close-packed arrangement, as shown in Figure 7.13. This arrangement helps save space while allowing as many oranges as possible to be stacked up high.

Similar close-packed arrangements can be found in the crystalline structures of metals. You may be surprised to learn that metals are crystalline. In fact, metals that contain just one kind of atom are among the simplest forms of all crystalline solids. Metal atoms are arranged in very compact and orderly patterns. For spheres of identical size, such as metal atoms, several closely packed arrangements are possible. Figure 7.14 on the following page shows three such arrangements: body-centered cubic, face-centered cubic, and hexagonal close-packed.

In a body-centered cubic structure, every atom (except those on the surface) has eight neighbors. The metallic elements sodium, potassium, iron, chromium, and tungsten crystallize in a body-centered cubic pattern. In a face-centered cubic arrangement, every atom has twelve neighbors. Among the metals that form a face-centered cubic structure are copper, silver, gold, aluminum, and lead. In a hexagonal close-packed arrangement, every atom also has twelve neighbors. However, because of its hexagonal shape, the pattern is different from the face-centered cubic arrangement. Metals that have a hexagonal close-packed crystal structure include magnesium, zinc, and cadmium.

Figure 7.13
Hexagonal Close Packing
These Thai oranges illustrate a pattern called a hexagonal close-packed arrangement. The same pattern is found in the crystal structures of some metals.

UbD Check for Understanding

The Essential Question How does metallic bonding affect the properties of metals?

Assess students' understanding of this Essential Question by giving them one minute to respond to the following prompt:

How can the conductivity of metals be explained? *(Electrons can flow in and out of a metal because the valence electrons are not fixed, but float freely around the positively charged metal ions.)*

ADJUST INSTRUCTION If students are having difficulty answering, have them consider what they have learned about valence electrons and apply that to the concept of metallic bonding as they examine Figure 7.12.Then ask them to answer the prompt again.

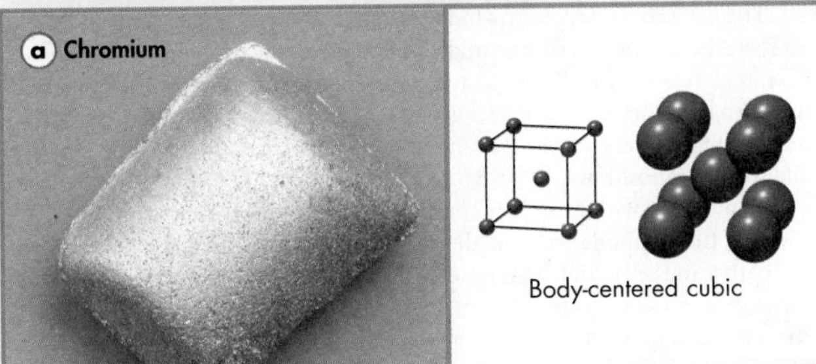

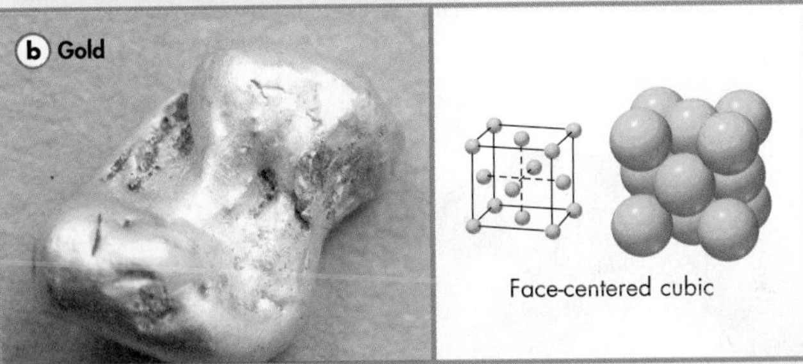

Figure 7.14 Crystalline Structures of Metals
Metal atoms crystallize in characteristic patterns.
a. Chromium atoms have a body-centered cubic arrangement.
b. Gold atoms have a face-centered cubic arrangement.
c. Zinc atoms have a hexagonal close-packed arrangement.
Interpret Diagrams ***Which of these arrangements is the most closely packed?***

Alloys

Why are alloys important?

Every day you use metallic items, such as utensils. However, very few of these objects are made out of a single kind of metal. Instead, most of the metals you encounter are alloys. **Alloys** are mixtures of two or more elements, at least one of which is a metal. Brass, for example, is an alloy of copper and zinc. **Alloys are important because their properties are often superior to those of their component elements.** Sterling silver (92.5 percent silver and 7.5 percent copper) is harder and more durable than pure silver, yet is still soft enough to be made into jewelry and tableware. Bronze is an alloy generally containing seven parts copper to one part tin. Bronze is harder than copper and is easier to cast into molds. Nonferrous (non-iron) alloys, such as bronze, copper-nickel, and aluminum alloys, are commonly used to make coins.

Explore

Alloys

Teacher Demo

PURPOSE Students compare models of interstitial and substitutional alloys.

MATERIALS foam balls of different sizes and colors, toothpicks

PROCEDURE Use foam balls to illustrate the crystal structures of interstitial and substitutional alloys. Use toothpicks to hold the "atoms" together. Point out that brass is a substitutional alloy in which copper atoms are replaced by similarly sized zinc atoms. Steel is an interstitial alloy in which relatively small carbon atoms occupy the interstices between closely packed iron atoms.

EXPECTED OUTCOME Students should be able to distinguish between the two types of alloys.

Differentiated Instruction

L3 ADVANCED STUDENTS Have students develop formulas for calculating the density of a metal given the atomic radius of the metal and the cubic unit cell packing arrangement of the metal atoms. Use sodium, potassium, iron, chromium, and tungsten for body-centered cubic metals. Use copper, silver, gold, aluminum, and lead for face-centered cubic metals. The students may need to make models to see geometric relationships more clearly.

ELL ENGLISH LANGUAGE LEARNERS English language learners may have difficulty understanding the pronunciation of the term *alloy*, despite the pre-lesson exposure to the term. Write the word on the board again, along with its phonetic spelling, and pronounce it slowly. Ask students with higher English proficiencies to translate the term into their native language for lower proficiency students.

Answers

FIGURE 7.12 repulsions between ions of like charge
FIGURE 7.14 face-centered cubic

Extend

Connect to ENGINEERING

Explain that materials for a building must be able to withstand the stresses that a building undergoes. Tensile stress can be observed when a building beam sags. The bottom of a beam undergoes tensile stress as it is slightly stretched. The top of a beam undergoes compression stress, which results when two forces push toward each other through a solid. The top of the beam is compressed as it is slightly shortened. A building undergoes shear stress in a strong wind. In shear stress, forces are applied from different directions, and the building might twist or break. Have students research how steel and iron differ in their response to these types of stresses and report their findings to the class.

Evaluate

Informal Assessment

Have students explain the cause-and-effect relationship between metallic bonding and the properties of metals. Then ask students to explain how this relates to the conductivity of metals. *(Metal ions are surrounded by electrons that can move freely around the metal cations. This freedom of movement allows metallic bonds to remain intact if the metal is stretched or hammered. This freedom of electron movement also allows metals to conduct electricity. As electrons enter the metal as part of an electric current, an equal number leave the opposite end of the metal. The electrons in between those electrons can flow freely toward the opposite end.)* **Ask** If a metal is alloyed with another substance, such as iron and carbon to make steel, how might the properties of the metal be affected? *(Sample response: the resulting alloy may be more corrosion-resistant, stronger, tougher or harder.)* Then have students complete the 7.3 Lesson Check.

Reteach

As a class, create a two-column chart comparing the characteristics of chemical bonding in ionic compounds and chemical bonding in pure metals.

The most important alloys today are steels. The principal elements in most steels, in addition to iron and carbon, are boron, chromium, manganese, molybdenum, nickel, tungsten, and vanadium. As a result, steels have a wide range of useful properties, such as corrosion resistance, ductility, hardness, and toughness. Figure 7.15 shows some items made of common alloys and their compositions.

Alloys can form from their component atoms in different ways. If the atoms of the components in an alloy are about the same size, they can replace each other in the crystal. This type of alloy is called a substitutional alloy. If the atomic sizes are quite different, the smaller atoms can fit into the interstices (spaces) between the larger atoms. Such an alloy is called an interstitial alloy. In the various types of steel, for example, carbon atoms occupy the spaces between the iron atoms. Thus, steels are interstitial alloys.

Figure 7.15 Common Alloys
Alloys are composed of two or more elements. The compositions (by mass) of sterling silver, cast iron, and stainless steel are shown.

7.3 LessonCheck

20. **Describe** How do chemists model the valence electrons of metal atoms?
21. **Explain** Why are alloys more useful than pure metals?
22. **Explain** What is meant by the terms *ductile* and *malleable?*
23. **Relate Cause and Effect** Why is it possible to bend metals but not ionic crystals?
24. **Use Analogies** How is the arrangement of fruit in a stack of oranges similar to the way some metal atoms are arranged in metallic crystals?
25. **Describe** Name two widely used alloys and describe some of their uses.

BIGIDEA BONDING AND INTERACTIONS

26. Describe how the sea-of-electrons model is used to explain the physical properties of metals.

212 Chapter 7 • Lesson 3

Lesson Check Answers

20. metal cations surrounded by a sea of mobile valence electrons
21. The properties of alloys are often superior to their component elements.
22. ductile: can be drawn into wires; malleable: can be hammered into different shapes
23. Under pressure, the cations in a metal slide past each other. The ions in ionic crystals are forced into each other by the rigid structure.
24. The arrangements are orderly and allow the individual items (fruit or cations) to be closely packed.
25. *Sample answer:* Sterling silver used in jewelry is 92.5% silver and 7.5% copper; bronze used in casting is 7 parts copper and 1 part tin.
26. Metal cations are surrounded by free-floating electrons. When metals are hammered or drawn, the cations move past each other.

7 Study Guide

BIG IDEA
BONDING AND INTERACTIONS

Atoms form positive ions (cations) by losing valence electrons and form negative ions (anions) by gaining valence electrons. The electrostatic forces between the oppositely charged ions hold the cations and anions together in an ionic compound. Ionic compounds generally have high melting points and can conduct an electric current in solution and in the molten state. Metals are made up of closely packed cations surrounded by a sea of electrons. The sea-of-electrons model explains why metals are good conductors of electric current and why they are ductile and malleable.

7.1 Ions

- To find the number of valence electrons in an atom of a representative element, simply look at its group number.
- A positively charged ion, or a cation, is produced when an atom loses one or more valence electrons.
- An anion is produced when an atom gains one or more valence electrons.

- valence electron (194)
- electron dot structure (195)
- octet rule (195)
- halide ion (199)

7.2 Ionic Bonds and Ionic Compounds

- Although they are composed of ions, ionic compounds are electrically neutral.
- Most ionic compounds are crystalline solids at room temperature.
- Ionic compounds generally have high melting points.
- Ionic compounds can conduct an electric current when melted or dissolved in water.

- ionic compound (201)
- ionic bond (201)
- chemical formula (202)
- formula unit (202)
- coordination number (205)

7.3 Bonding in Metals

- The valence electrons of atoms in a pure metal can be modeled as a sea of electrons.
- Alloys are important because their properties are often superior to those of their component elements.

- metallic bond (209)
- alloy (211)

UbD Performance Tasks

MODELING IONIC CRYSTALS Have students choose three ionic compounds and do research to find out their crystal structures. Then have students use modeling clay and toothpicks to make three-dimensional models of each crystal structure.

QUENCHING—NOT JUST FOR THIRST Alloys and metals are often quenched in water after being heated and shaped. Have students research and report on the effects of quenching on the crystalline structure of metals and alloys. Also, ask them to compare the resulting crystals to ionic crystals. Reports should include any necessary graphics and explain any changes in metal properties that occur as a result of the quenching process.

Evaluate

Review and Assessment Materials

CHEMISTRY TUTORIAL Encourage students to review solving ionic bonding problems using the animated step-by-step tutorial for Chapter 7. Each tutorial is followed by a similar practice problem, where the student is guided through the problem-solving process.

CHAPTER PROBLEM SET Have students practice more formula writing problems using the Chapter 7 online Problem Set.

DIRECTED VIRTUAL LAB Have students complete the Virtual Lab *The Formation of Ionic Compounds* to help reinforce the concepts of ionic bonding. In the post-lab exercise, students will analyze the data and draw conclusions based on the results of the experiment.

KINETIC ART Direct students to the Kinetic Art animation *Crystal Structures*, and have them answer the questions that accompany the animation.

CONCEPTS IN ACTION If students have trouble understanding ionic bonds, ask them to view the Concepts in Action animation, which presents the material in a familiar context. Have students complete the questions that follow the animation.

Study Tip

Write and answer questions about the content of the chapter. For Chapter 7, you might **ask** the following questions, which require students to summarize the information contained in the chapter.

- What information does the electron dot structure of an element provide? *(the number of valence electrons)*
- In most cases, what is the stable electron configuration of an elemental ion? *(In most cases, a stable configuration is characterized by an octet of valence electrons; the configuration is isoelectronic with that of a noble gas.)*
- Why does NaCl dissolved in water conduct electricity? *(In water, the ionic compound breaks apart into Na^+ and Cl^- ions that are free to move in the solution.)*
- What is the theory of bonding in pure metals? *(metal cations held together by a pool of mobile valence electrons)*

Answers

LESSON 7.1

27. an electron in the highest occupied energy level

28. **a.** Group 5A; 5 valence electrons
b. Group 1A; 1 valence electron
c. Group 5A; 5 valence electrons
d. Group 2A; 2 valence electrons
e. Group 7A; 7 valence electrons
f. Group 4A; 4 valence electrons

29. **a.** :C̈l· **b.** ·S̈· **c.** ·Ȧl· **d.** Li·

30. by gaining or losing electrons

31. **a.** 2 **b.** 3 **c.** 1 **d.** 2

32. **a.** Al^{3+} **b.** Li^{+} **c.** Ba^{2+} **d.** K^{+}
e. Ca^{2+} **f.** Sr^{2+}

33. Most nonmetals gain 1, 2, or 3 electrons to achieve a noble-gas electron configuration.

34. **a.** 3 **b.** 2 **c.** 1 **d.** 3

35. **a.** S^{2-} **b.** Na^{+} **c.** F^{-} **d.** P^{3-}

36. **a.** gain of 1 electron
b. loss of 1 electron
c. gain of 3 electrons
d. loss of 2 electrons
e. loss of 2 electrons
f. loss of 1 electron
g. gain of 1 electron
h. loss of 2 electrons

37. **a.** bromide anion
b. sodium cation
c. arsenide anion
d. barium cation
e. calcium cation
f. copper(I) cation
g. hydride anion
h. copper(II) cation

LESSON 7.2

38. the electrostatic force that holds ions together in ionic compounds

39. The positive charges balance the negative charges.

40. a, c, e

41. a, b, d

42. with the formula that reflects the ratio of ions in the compound

43. **a.** Ca^{2+}, F^{-} **b.** Al^{3+}, Br^{-} **c.** Li^{+}, O^{2-}
d. Al^{3+}, S^{2-} **e.** K^{+}, N^{3-}

44. **a.** K^{+}, Cl^{-} **b.** Ba^{2+}, S^{2-}
c. Mg^{2+}, Br^{-} **d.** Li^{+}, O^{2-}

45. Their network of electrostatic attractions and repulsions forms a rigid structure.

46. Ions are free to move in molten $MgCl_2$.

7 Assessment

* Solutions appear in Appendix E

Lesson by Lesson

7.1 Ions

27. What is a valence electron?

***28.** To which group in the periodic table does each of the following elements belong? How many valence electrons do atoms of each element have?

a. nitrogen
b. lithium
c. phosphorus
d. barium
e. bromine
f. carbon

29. Write electron dot structures for each of the following elements:

a. Cl
b. S
c. Al
d. Li

30. Describe two ways that an ion forms from an atom.

***31.** How many electrons must an atom of each element lose to attain a noble-gas electron configuration?

a. Ca
b. Al
c. Li
d. Ba

32. Write the symbol for the ion formed when each of the following elements loses its valence electrons.

a. aluminum
b. lithium
c. barium
d. potassium
e. calcium
f. strontium

33. Why do nonmetal atoms tend to form anions when they react to form compounds?

***34.** How many electrons must be gained by each of the following atoms to achieve a stable electron configuration?

a. N
b. S
c. Cl
d. P

***35.** What is the formula of the ions formed when atoms of the following elements gain or lose valence electrons and attain noble-gas configurations?

a. sulfur
b. sodium
c. fluorine
d. phosphorus

36. State the number of electrons either lost or gained in forming each ion.

a. Br^{-}
b. Na^{+}
c. As^{3-}
d. Ba^{2+}
e. Ca^{2+}
f. Cu^{+}
g. H^{-}
h. Cu^{2+}

37. Name each ion in Problem 36. Identify each as an anion or a cation.

7.2 Ionic Bonds and Ionic Compounds

38. Define an ionic bond.

39. Explain why ionic compounds are electrically neutral.

***40.** Which of the following pairs of atoms would you expect to combine chemically to form an ionic compound?

a. Li and S
b. O and S
c. Al and O
d. F and Cl
e. I and K
f. H and N

41. Which of the following pairs of elements will not form ionic compounds?

a. sulfur and oxygen
b. sodium and calcium
c. sodium and sulfur
d. oxygen and chlorine

42. How can you represent the composition of an ionic compound?

***43.** Identify the kinds of ions that form each ionic compound.

a. calcium fluoride, CaF_2
b. aluminum bromide, $AlBr_3$
c. lithium oxide, Li_2O
d. aluminum sulfide, Al_2S_3
e. potassium nitride, K_3N

44. Write the formulas for the ions in the following compounds:

a. KCl
b. BaS
c. $MgBr_2$
d. Li_2O

45. Most ionic substances are brittle. Why?

46. Explain why molten $MgCl_2$ does conduct an electric current although crystalline $MgCl_2$ does not.

Focus on ELL

5 ASSESS UNDERSTANDING Engage students in discussing the chapter's key concepts and vocabulary. Encourage them to identify learning strategies they felt were most helpful and those they felt were difficult.

BEGINNING: LOW/HIGH Provide students with a two-column chart which has randomly ordered vocabulary words in one column and definitions in the other. Have students match each word with its definition.

INTERMEDIATE: LOW/HIGH Provide students with a list of one key question from each lesson. Have them write a short answer to each.

ADVANCED: LOW/HIGH Have students write a summary statement for each lesson describing what they have learned.

7.3 Bonding in Metals

*47. How can you describe the arrangement of atoms in metals?

48. Explain briefly why metals are good conductors of electric current.

*49. Name the three crystal arrangements of closely packed metal atoms. Give an example of a metal that crystallizes in each arrangement.

50. Name some alloys that you have used or seen today.

51. Explain why the properties of all steels are not identical.

Understand Concepts

52. Construct a table that shows the relationship among the group number, valence electrons lost or gained, and the formula of the cation or anion produced for the following metallic and nonmetallic elements: Na, Ca, Al, N, S, Br.

53. Write electron dot structures for the following elements.

a. C
b. Be
c. O
d. F
e. Na
f. P

54. Show the relationship between the electron dot structure of an element and the location of the element in the periodic table.

*55. In terms of electrons, why does a cation have a positive charge?

56. Why does an anion have a negative charge?

57. The spheres below represent the relative diameters of atoms or ions. Rearrange the sequences in (a) and (b) so the relative sizes of the particles correspond to the increasing size of the particles as shown in the illustration.

a. oxygen atom, oxide ion, sulfur atom, sulfide ion
b. sodium atom, sodium ion, potassium atom, potassium ion

*58. Write the name and symbol of the ion formed when

a. a sulfur atom gains two electrons.
b. an aluminum atom loses three electrons.
c. a nitrogen atom gains three electrons.
d. a calcium atom loses two electrons.

*59. Write electron configurations for the 2+ cations of these elements.

a. Fe
b. Co
c. Ni

60. Write electron configurations for the 3+ cations of these elements.

a. chromium
b. manganese
c. iron

61. Write the symbol for the ion formed when each element gains electrons and attains a noble-gas electron configuration.

a. Br
b. H
c. As
d. Se

*62. Write electron configurations for the following atoms and ions, and comment on the result.

a. Ar
b. Cl^-
c. S^{2-}
d. P^{3-}

63. Write electron configurations for the following atoms and ions and comment on the result.

a. N^{3-}
b. O^{2-}
c. F^-
d. Ne

64. Name the first four halogens. What group are they in, and how many valence electrons does an atom of each element have?

65. Write complete electron configurations for the following atoms and ions. For each group, comment on the results.

a. Ar, K^+, Ca^{2+}
b. Ne, Na^+, Mg^{2+}, Al^{3+}

66. If ionic compounds are composed of charged particles (ions), why isn't every ionic compound either positively or negatively charged?

*67. Which of the following compounds are most likely not ionic?

a. H_2O
b. Na_2O
c. CO_2
d. CaS
e. SO_2
f. NH_3

Answers

LESSON 7.3

47. *Sample answer:* Atoms in metals are arranged in a compact and orderly manner.

48. They have many mobile valence electrons. Electrons in the current replace the electrons leaving the metal.

49. body-centered cubic: Na, K, Fe, Cr, or W; face-centered cubic: Cu, Ag, Au, Al, or Pb; hexagonal close-packed: Mg, Zn, or Cd

50. Answers could include table-ware, steel in cars and buses, high-speed dental drill bits, solder in stereos and televisions, and structural steel in buildings.

51. The properties of the steel will vary according to its composition. In addition to iron, steel can contain varying amounts of carbon and such metals as chromium, nickel, and molybdenum.

UNDERSTAND CONCEPTS

52.

Group number	Valence electrons lost or gained	Formula of ion
1A	1	Na^+
2A	2	Ca^{2+}
3A	3	Al^{3+}
5A	3	N^{3-}
6A	2	S^{2-}
7A	1	Br^-

53. a. ·Ċ· c. ·Ö· e. Na· b. ·Be· d. :F̤· f. :Ṗ·

54. For the representative elements the number of electrons in the electron dot structure is the group number.

55. It has lost valence electrons.

56. It has gained valence electrons.

57. a. oxygen atom, sulfur atom, oxide ion, sulfide ion
b. sodium ion, potassium ion, sodium atom, potassium atom

58. a. sulfide ion (anion), S^{2-}
b. aluminum ion (cation), Al^{3+}
c. nitride ion (anion), N^{3-}
d. calcium ion (cation), Ca^{2+}

59. a. $1s^22s^22p^63s^23p^63d^6$
b. $1s^22s^22p^63s^23p^63d^7$
c. $1s^22s^22p^63s^23p^63d^8$

60. a. $1s^22s^22p^63s^23p^63d^3$
b. $1s^22s^22p^63s^23p^63d^4$
c. $1s^22s^22p^63s^23p^63d^5$

61. a. Br^- b. H^- c. As^{3-} d. Se^{2-}

62. All have the noble-gas configuration of $1s^22s^22p^63s^23p^6$.

63. All are $1s^22s^22p^6$. All have the same configuration as neon.

64. fluorine, chlorine, bromine, and iodine; Group 7A, 7 valence electrons

65. a. $1s^22s^22p^63s^23p^6$
b. $1s^22s^22p^6$;
Each has a noble-gas electron configuration

66. The positively charged cations exactly balance the negatively charged anions.

67. a, c, e, f

Answers

UNDERSTAND CONCEPTS

68. **a.** $BaBr_2$ **b.** Al_2S_3 **c.** K_3N

69. Their outmost occupied energy levels are filled.

70. **a.** Fe_2O_3
b. PbO_2
c. Li_2O
d. MgO

71. No, the packing of ions in a crystalline structure depends on a number of factors including the relative sizes of the ions. The coordination number of an element can vary from compound to compound.

72. 12

73. Hexagonal close-packed unit cells have twelve neighbors for, every atom or ion. Face-centered cubic unit cells also have twelve neighbors for every atom or ion, with an atom or ion in the center of each face. Body-centered cubic unit cells have eight neighbors for every atom or ion, with an atom or ion at the center of each cube.

74. Brass is a mixture of copper and zinc. The properties of a particular sample of brass will vary with the relative proportions of the two metals.

75. **a.** Cu, Zn **c.** Cu, Sn
b. Ag, Cu **d.** Fe, Cr, Ni, C

THINK CRITICALLY

76. Each dot in the electron dot structure represents a valence electron in the electron configuration diagram.

77. By gaining or losing electrons the atoms of elements achieve a noble-gas electron configuration.

78. No, sodium chloride is composed of equal numbers of sodium ions and chloride ions; the ions are in a 1:1 ratio. Each sodium ion is surrounded by chloride ions, and each chloride is surrounded by sodium ions.

79. **a.** 6A **d.** 6A
b. 7A **e.** 1A
c. 1A **f.** 2A

80. In sodum chloride crystals the sodium and chloride ions vibrate about fixed points; in the molten state, the ions are free to move.

81. Metals are ductile (can be drawn into wires) and malleable (can be hammered into shapes). These changes are possible because a metal consists of metal cations in a sea of valence electrons. When subjected to pressure, the cations easily slide past one another.

82. The spheres are more closely packed in (a); there is less empty space in (a), and a rough count shows 25 spheres in (a) compared with 22 spheres in (b).

***68.** Write the formulas for each ionic compound that can be made by combining each of pair of ions.

a. Ba^{2+} and Br^-
b. Al^{3+} and S^{2-}
c. K^+ and N^{3-}

69. The atoms of the noble gas elements are stable. Explain.

70. What is the simplest formula for the compounds that can form when each of these ions combine with an oxide (O^{2-}) ion?

a. Fe^{3+} **c.** Li^+
b. Pb^{4+} **d.** Mg^{2+}

71. Can you predict the coordination number of an ion from the formula of an ionic compound? Explain.

***72.** Metallic cobalt crystallizes in a hexagonal close-packed structure. How many neighbors will a cobalt atom have?

73. Explain how hexagonal close-packed, face-centered cubic, and body-centered cubic unit cells are different from one another.

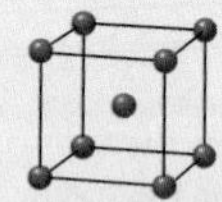
Body-centered cubic

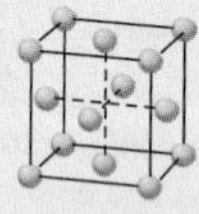
Face-centered cubic

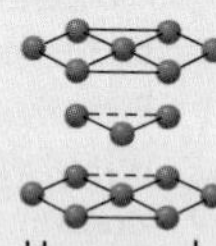
Hexagonal close-packed

74. The properties of all samples of brass are not identical. Explain.

***75.** For each alloy below, list the elements it contains.

a. brass **c.** bronze
b. sterling silver **d.** stainless steel

Think Critically

***76.** **Make Generalizations** What is the relationship between the number of electrons in the valence shells in an electron configuration diagram for an atom and the number of dots in the corresponding electron dot structure?

77. **Relate Cause and Effect** Why are many elements more stable as ions than they are as atoms?

78. **Make Generalizations** Is it accurate to describe sodium chloride (NaCl) as consisting of individual particles, each made up of one Na^+ cation and one Cl^- anion? Explain your answer.

***79.** **Infer** For each ionic formula, identify the A-group number to which element X belongs.

a. CaX **d.** Al_2X_3
b. MgX_2 **e.** XF
c. X_3N **f.** XS

80. **Compare** How do the motions of sodium ions and chloride ions in molten sodium chloride differ from the motions of these ions in sodium chloride crystals?

81. **Relate Cause and Effect** Two physical properties of metals are ductility and malleability. Explain these properties based on what you know about the valence electrons of metal atoms.

82. **Interpret Diagrams** How atoms and ions are arranged in crystals is not just dependent on size. The spheres in each atomic window below are identical in size. The windows have exactly the same area. In which window are the spheres more closely packed? Explain your reasoning.

a.
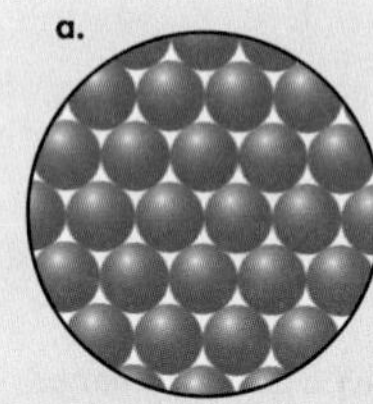

b.
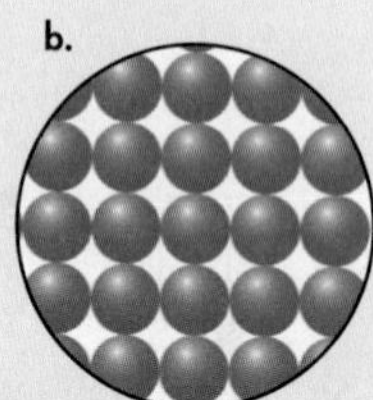

***83.** **Compare and Contrast** Describe the similarities and differences between ionic compounds and metals in terms of their physical and chemical characteristics.

84. **Relate Cause and Effect** How does the octet rule explain the large increase in energy between the first and second ionization energies of Group 1A metals?

85. **Infer** An atom of the element M forms a stable ion in an ionic compound with chlorine having the formula MCl_2. In this compound, the ion of element M has a mass number of 66 and has 28 electrons.

a. What is the identity of the element?
b. How many neutrons does the ion have?

83. Both metals and ionic compounds are composed of ions. Both are held together by electrostatic bonds. Metals always conduct electricity, and ionic compounds conduct only when melted or in water solution. Ionic compounds are composed of cations and anions, but metals are composed of cations and free-floating valence electrons. Metals are ductile, but ionic compounds are brittle.

84. Removal of the first electron results in a very stable electron configuration. Removing a second electron would disrupt that stability.

85. **a.** zinc
b. 38 neutrons

Enrichment

*86. **Apply Concepts** Classify each element in the following list. Will an atom of each element form a cation or an anion, or is the element chemically nonreactive? For the atoms that do form ions during a chemical reaction, write the number of electrons the atom will gain or lose.

a. lithium
b. sodium
c. neon
d. chlorine
e. magnesium

87. **Infer** The chemically similar alkali metal chlorides NaCl and CsCl have different crystal structures, whereas the chemically different NaCl and MnS have the same crystal structures. Why? (*Hint:* Consider periodic and group trends among the properties of the elements.)

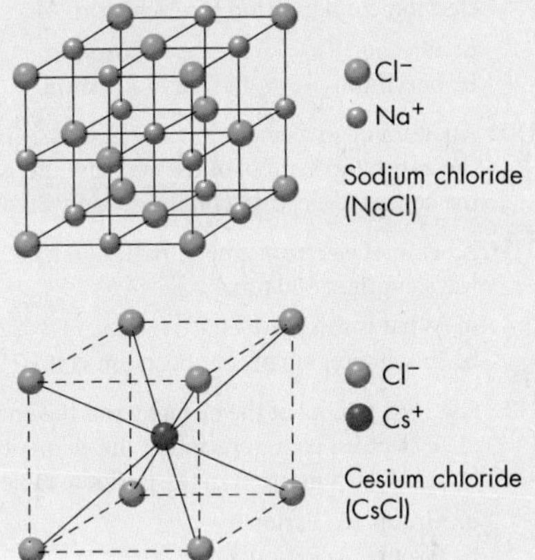

88. **Calculate** Silver crystallizes in a face-centered cubic arrangement. A silver atom is at the edge of each lattice point (the corner of the unit cell). The length of the edge of the unit cell is 0.4086 nm. What is the atomic radius of silver?

*89. **Analyze Data** Consider two ionic compounds, NaCl and CaO.

a. In which compound would you expect the electrostatic forces that hold the compound together to be the strongest? Explain.
b. The melting point of NaCl is 801°C. The melting point of CaO is 2614°C. Does this data support your prediction? Why or why not?

Write About Science

90. **Compare** Describe the formation of a cation that is an exception to the octet rule. In your description, compare the electron configuration of the cation to the electron configurations of the nearest noble gases.

91. **Research** Go online and research X-ray diffraction crystallography. How are the samples prepared? How are the X-rays generated and detected? How is this technique used to study the structure of crystalline substances?

CHEMYSTERY

It's Not Easy Being Green

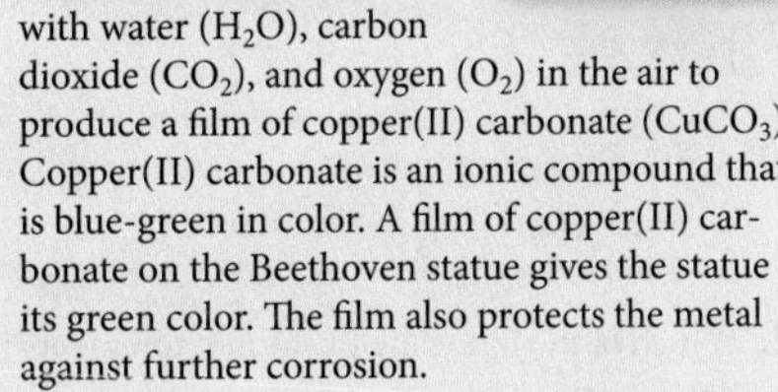

The statue of Ludwig van Beethoven in Central Park is made of bronze. Bronze is an alloy containing copper and tin.

When bronze is exposed to the elements, it reacts with water (H_2O), carbon dioxide (CO_2), and oxygen (O_2) in the air to produce a film of copper(II) carbonate ($CuCO_3$). Copper(II) carbonate is an ionic compound that is blue-green in color. A film of copper(II) carbonate on the Beethoven statue gives the statue its green color. The film also protects the metal against further corrosion.

*92. **Apply Concepts** A copper atom can lose one or two electrons to form a Cu^+ ion or a Cu^{2+} ion, respectively. The charge of the copper ion in $CuCO_3$ is 2+. Write the electron configuration of this cation.

93. **Form an Opinion** Why do you think bronze is often used to create statues?

94. **Connect to the BIGIDEA** How are the properties of the copper(II) carbonate film on the statue different from the properties of the bronze beneath the film? Explain how these properties are a result of the type of bonding present.

CHEMYSTERY **APPLY CONCEPTS** After students have read through the CHEMystery, discuss the principles of ionic and metallic bonding that make the development of the green patina possible. **Ask** What type of bonding occurs? *(The product is an ionic compound composed of copper from the statue and water and carbon dioxide from the air.)*

Ask How is the film able to protect the metal against further corrosion? *(Because the film has a different composition, it doesn't react with air in the same way that the metal does.)*

WRITE ABOUT SCIENCE

91. Remind students to use reliable sources and to document their sources. For such an extensive project, it might make sense for students to work in groups and divide up the tasks.

CHEMYSTERY ANSWERS

92. The configuration is $[Ar]3d^9$.

93. *Sample answer:* Bronze is an attractive metal alloy before and after it forms a patina.

94. The metallically bonded alloy is able to react with elements in its environment to form an ionic compound, which is less reactive.

Answers

86. **a.** cation: lose $1e^-$ **b.** cation: lose $1e^-$ **c.** unreactive **d.** anion: gain $1e^-$ **e.** cation: lose $2e^-$

87. Na^+ and Cs^+ differ greatly in size. Na^+ and Cl^- are similar in size to Mn^{2+} and S^{2-}.

88. 0.1445 nm (You may want to give students this hint: Refer to Fig 7.14b. Draw a right triangle through the centers of three corner atoms. Find the length of the hypotenuse. The length of the hypotenuse is the diameter of the central atom plus the radii of the two corner atoms, so divide by four to find the atomic radius.)

89. **a.** CaO; the electrostatic forces of attraction between Ca^{2+} and O^{2-} are much stronger than the forces between Na^+ and Cl^-.
b. Yes; more energy is needed to overcome the electrostatic attractions in CaO and melt the compound.

WRITE ABOUT SCIENCE

90. Sample answer: An atom of silver has the electron configuration $1s^22s^22p^63s^23p^63d^{10}4s^24p^64d^{10}5s^1$. To achieve the configuration of the preceding noble gas, krypton, a silver atom would have to lose eleven electrons and form Ag^{11+}. To achieve the configuration of the following noble gas xenon, a silver atom must gain seven electrons and form Ag^{7-}. Because ions with such high charges are unlikely, silver does not achieve a noble-gas configuration. However, if a silver atom loses its $5s^1$ electron, the result is an outer electron configuration of eighteen electrons, written as $4s^24p^64d^{10}$. This configuration is favored and stable. It is known as a pseudo-noble-gas configuration.

ASSESSMENT

CUMULATIVE REVIEW

95. Organic chemistry is the study of chemicals containing carbon; inorganic chemistry is the study of chemicals that do not contain carbon.

96. an analytical chemist

97. Conserve energy and develop renewable resources.

98. a, b, and d are chemical changes; c is a physical change

99. b and e; c is not a mixture, it is a pure substance

100. **a.** liquid, vapor
b. vapor
c. liquid, vapor
d. liquid, vapor

101. **a.** 56.6 g
b. 0.0048 m
c. 1.81 L
d. 4.0×10^3 mg

102. b

103. –269°C

104. 27.0 cm^3

105. **a.** $^{64}_{30}Zn$ **c.** $^{3}_{1}H$
b. $^{37}_{17}Cl$ **d.** $^{40}_{20}Ca$

106. 14 amu

107. Each of the isotopes has 8 protons and 8 electrons; oxygen-16 also has 8 neutrons, oxygen-17 has 9 neutrons, and oxygen-18 has 10 neutrons.

108. **a.** 1
b. 3
c. 1
d. 5

109. **a.** N, $1s^22s^22p^3$
b. Be, $1s^22s^2$
c. P, $1s^22s^22p^63s^23p^3$
d. K, $1s^22s^22p^63s^23p^64s^1$

110. chlorine, Cl, $1s^22s^22p^63s^23p^5$

111 **a.** 5×10^{-7} m
b. the visible region, green

112. **a.** K, $1s^22s^22p^63s^23p^64s^1$
b. Al, $1s^22s^22p^63s^23p^1$
c. S, $1s^22s^22p^63s^23p^4$
d. Ba, $1s^22s^22p^63s^23p^63d^{10}4s^24p^64d^{10}5s^25p^66s^2$

113. the electron

114. Possible answers (any two): sodium (Na), cesium (Cs), rubidium (Rb), lithium (Li)

Cumulative Review

95. How is organic chemistry distinguished from inorganic chemistry?

96. What is the name given to a chemist who studies the composition of matter?

97. Explain two ways to meet modern society's need for energy.

***98.** Classify the following actions as chemical or physical changes.
a. Cookies are baked.
b. A firefly emits light.
c. A figure is carved from wood.
d. Caramel is made from sugar.

99. Which of the following substances are not homogeneous mixtures?
a. gold ring
b. spaghetti sauce
c. cane sugar
d. window glass
e. river water
f. bottled water

***100.** What physical state(s) can each of the following substances become as you raise its temperature?
a. silver **c.** ice
b. gasoline **d.** wax

101. Round each measurement to the number of significant figures indicated in parentheses.
a. 56.55 g (3) **c.** 1.8072 L (3)
b. 0.004849 m (2) **d.** 4.007×10^3 mg (2)

***102.** Which of the following linear measurements is the longest?
a. 6×10^4 cm **c.** 0.06 km
b. 6×10^6 mm **d.** 6×10^9 nm

103. Helium has a boiling point of 4 K. This is the lowest boiling point of any liquid. Express this temperature in degrees Celsius.

***104.** The density of silicon is 2.33 g/cm^3. What is the volume of a piece of silicon that has a mass of 62.9 g?

105. Express the composition of each atom in shorthand form.
a. zinc-64 **c.** hydrogen-3
b. chlorine-37 **d.** calcium-40

106. An atom of carbon and an atom of element Z together have a mass of 6 amu less than double the mass of an atom of oxygen. If an atom of oxygen has a mass of 16 amu and the mass of an atom of carbon is 12 amu, what is the mass of an atom of element Z?

107. Determine the number of protons, electrons, and neutrons in each of the three isotopes of oxygen.

***108.** How many orbitals are in the following sublevels?
a. 4*s* sublevel **c.** 3*s* sublevel
b. 2*p* sublevel **d.** 4*d* sublevel

109. Give the symbol for each element and write the electron configuration for each atom.
a. nitrogen **c.** phosphorus
b. beryllium **d.** potassium

110. An atom of an element has 17 electrons. Give the name and symbol of the element and write the complete electron configuration.

***111.** A beam of electromagnetic radiation has a wavelength of 500 nm.
a. What is this wavelength in meters?
b. In what region of the spectrum is this?

***112.** Give the symbol of the element and the complete electron configuration of the element found at each location in the periodic table.
a. Group 1A, Period 4
b. Group 3A, Period 3
c. Group 6A, Period 3
d. Group 2A, Period 6

113. Which subatomic particle plays the most important role in chemistry?

114. Give the name and symbol of two elements that have properties similar to those of potassium.

If You Have Trouble With . . .

Question	95	96	97	98	99	100	101	102	103	104	105	106	107	108	109	110	111	112	113	114
See Chapter	1	1	1	2	2	2	3	3	3	3	4	4	4	5	5	5	5	6	6	6

Standardized Test Prep for Indiana

Standard: C.2.6, C.3.3

Tips for Success

Review All the Answer Choices Even if you find an answer that looks correct, continue reading until you have looked at *every* answer. There may be more than one correct response, or one may be better than another. Also, "all of the above" may be a possible answer. If you stop reading as soon as you find an answer that is correct, you won't notice this option.

Select the choice that best answers each question or completes each statement.

1. Which of these is not an ionic compound?
   (A) KF (B) SiO_2 (C) Na_2SO_4 (D) Na_2O

2. Which statements are correct when barium and oxygen react to form an ionic compound?
   I. Each barium atom loses 2 electrons and forms a cation.
   II. Oxygen atoms form oxide anions (O^{2-}).
   III. The ions are present in a one-to-one ratio in the compound.
   (A) I and II only
   (B) II and III only
   (C) I and III only
   (D) I, II, and III

The lettered choices below refer to Questions 3–6. A lettered choice may be used once, more than once, or not at all.

(A) gains two electrons
(B) loses two electrons
(C) gains three electrons
(D) loses one electron
(E) gains one electron

Which choice describes what likely happens as each of the following elements forms an ion?

3. iodine
4. magnesium
5. cesium
6. phosphorus

7. How many valence electrons does arsenic have?
   (A) 5 (B) 4 (C) 3 (D) 2

8. Which electron configuration represents a nitride ion?
   (A) $1s^2 2s^2 3s^2 4s^2$ (B) $1s^2 2s^2 2p^6$ (C) $1s^2 2s^2 2p^3$ (D) $1s^2$

9. When a bromine atom gains an electron
   (A) a bromide ion is formed.
   (B) the ion formed has a 1− charge.
   (C) the ion formed is an anion.
   (D) all the above are correct.

Use the description and the graph to answer Questions 10–12.

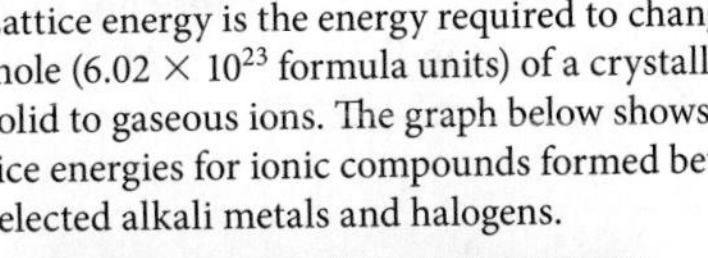

Lattice energy is the energy required to change one mole (6.02×10^{23} formula units) of a crystalline, ionic solid to gaseous ions. The graph below shows the lattice energies for ionic compounds formed between selected alkali metals and halogens.

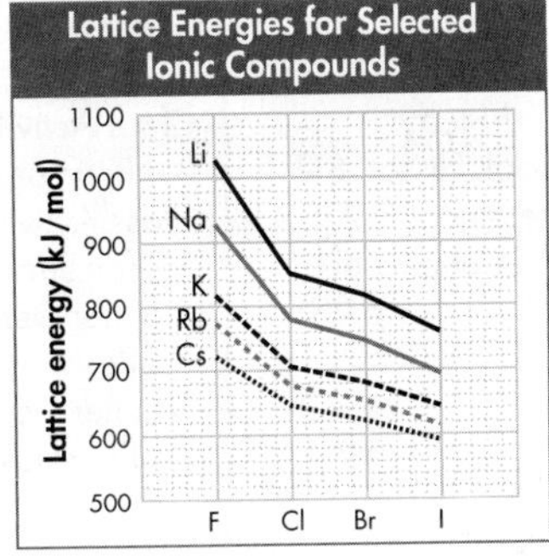

10. For a given alkali metal, what is the trend in lattice energy as the atomic radius of the halogen increases?
11. For a given halogen, what is the trend in lattice energy as the atomic radius of the alkali metal increases?
12. Complete this sentence: "As the atomic radius of either the halogen or the alkali metal increases, the lattice energy __________."

If You Have Trouble With . . .

Question	1	2	3	4	5	6	7	8	9	10	11	12
See Lesson	7.2	7.2	7.1	7.1	7.1	7.1	7.1	7.1	7.1	7.2	7.2	7.2

STP Answers

1. B
2. D
3. E
4. B
5. D
6. C
7. A
8. B
9. D
10. Because the atomic radius increases moving down the halogen group, the lattice energy decreases.
11. Lattice energy decreases as the atomic radius of the alkali metal increases.
12. decreases

8 Covalent Bonding Planning Guide

Introducing the BIGIDEA: BONDING AND INTERACTIONS

Most elements exist in chemical compounds, where elements are held together by chemical bonds.

IASS	Lessons and Objectives	Print Resources: For the Student	Print Resources: For the Teacher	
C.3.1, C.3.2, C.3.3	**8.1** Molecular Compounds p 222–225 **8.1.1** Identify the information a molecular formula provides. **8.1.2** Describe the representative units that define molecular compounds and ionic compounds.	Reading and Study Workbook Lesson 8.1 Lesson Assessment 8.1 p 225	Teacher Demo, p 224: Molecular Structures and Formulas	
C.3.1, C.3.2, C.3.3	**8.2** The Nature of Covalent Bonding p 226–238 **8.2.1** Explain the result of electron sharing in covalent bonds. **8.2.2** Describe how coordinate covalent bonds are different from other covalent bonds. **8.2.3** Identify some exceptions to the octet rule. **8.2.4** Explain how the strength of a covalent bond is related to its bond dissociation energy. **8.2.5** Describe how resonance structures are used.	Reading and Study Workbook Lesson 8.2 Lesson Assessment 8.2 p 238 Quick Lab: Strengths of Covalent Bonds p 238	Class Activity, p 227: Representing Molecules Class Activity, p 229: Bonding for Second Row Elements Teacher Demo, p 234: A Resonance Hybrid Teacher Demo, p 235: Bond Energies	
C.3.1, C.3.2, C.3.3, C.3.4	**8.3** Bonding Theories p 240–246 **8.3.1** Describe the relationship between atomic and molecular orbitals. **8.3.2** Describe how VSEPR theory helps predict the shapes of molecules. **8.3.3** Identify the ways in which orbital hybridization is useful in describing molecules.	Reading and Study Workbook Lesson 8.3 Lesson Assessment 8.3 p 246	Class Activity, p 242: Making Molecular Models	
C.3.1, C.3.2, C.3.3	**8.4** Polar Bonds and Molecules p 247–253 **8.4.1** Describe how electronegativity values determine the charge distribution in a polar molecule. **8.4.2** Evaluate the strengths of intermolecular attractions compared with the strengths of ionic and covalent bonds. **8.4.3** Explain why the properties of covalent compounds are so diverse.	Reading and Study Workbook Lesson 8.4 Lesson Assessment 8.4 p 253 Small-Scale Lab: Paper Chromatography of Food Dyes p 254	Class Activity, p 249: A Magnetic Analogy Teacher Demo, p 250: Observing Evidence of Polarity Teacher Demo, p 252: Evidence of Hydrogen Bonding	

Assessing the BIGIDEA: BONDING AND INTERACTIONS

	Essential Questions 1. How is the bonding in molecular compounds different from the bonding in ionic compounds? 2. How do electrons affect the shape of a molecule? 3. What factors affect molecular properties?	Study Guide p 255 STP p 261 Reading and Study Workbook Self-Check and Vocabulary Review Chapter 8		

	PearsonChem.com	
	Editable Worksheets	Online Activities
	Teaching Resources, Lesson 8.1 Review	8.1 Lesson Overview
	Teaching Resources, Lesson 8.2 Review Probeware Lab: Quick Lab: Strengths of Covalent Bonds	8.2 Lesson Overview Covalent Bonds Drawing Electron Dot Structures The Air Up There
	Teaching Resources, Lesson 8.3 Review Lab 11: Molecular Models Lab Practical 8-1: Model Building	8.3 Lesson Overview
	Small-Scale Lab Manual Lab 11: Paper Chromatography	8.4 Lesson Overview Identifying Bond Type Covalent Bonding Polar Molecules
	Core Teaching Resources (includes editable worksheets) • Lesson Reviews • Practice Problems • Interpret Graphs • Vocabulary Review • Chapter Quizzes and Tests • Lab Record Sheets	Chapter 8 Problem Set Tetrahedral Angles

Materials List

For the Student

Quick Lab p 238
- 170-g (6 oz) can of food
- 2 454-g (16 oz) cans of food
- 3 No. 25 rubber bands
- metric ruler
- coat hanger
- plastic grocery bag
- paper clip
- graph paper
- motion detector (optional)

Small-Scale Lab p 254
- pencil
- ruler
- scissors
- toothpicks
- 4 different colors of food coloring
- plastic cup
- 0.1% NaCl solution
- chromatography paper
- various colored candies
- markers, and powdered drinks
- 70% 2-propanol
- 0.8M CH_3COOH

For the Teacher

Teacher Demo p 224
- molecular model kit or foam balls and toothpicks

Teacher Demo p 235
- 10- to 15-cm piece of clean magnesium ribbon
- a small piece of charcoal
- tongs
- Bunsen burner
- filters for viewing

Class Activity p 249
- magnets
- iron filings
- cardboard

Teacher Demo p 252
- ring stand
- 2 burets
- 2 beakers
- buret clamps
- turpentine
- water
- rubber rod
- glass rod
- fur cloth

Teacher Demo p 234
- small piece of copper metal
- concentrated nitric acid
- evaporating dish
- fume hood

Class Activity p 240
- ball-and-stick molecular model kits or foam balls and toothpicks

Teacher Demo p 252
- four 300-mL stoppered flasks each containing 100 mL of one of these liquids: glycerol, ethylene glycol, water, ethanol

Additional Digital Resources

Online Student Edition
Online Teacher's Edition
Exam*View Assessment Suite CD-ROM*
Classroom Resources DVD

What's Online

CHEMISTRY TUTORIAL Students can work step-by-step with the online tutorial to learn how to draw electron dot structures and identify bond types by accessing the Chapter 8 Chemistry Tutorials.

ONLINE PROBLEM SET Additional practice problems about covalent bonding can be found in the Chapter 8 online Problem Set.

MATH TUTORIAL The online Math Tutorial will help students understand tetrahedral angles.

DIRECTED VIRTUAL LAB Encourage students to observe the properties of polar and nonpolar molecules using the animated virtual lab for Chapter 8. The animation is followed by analysis questions.

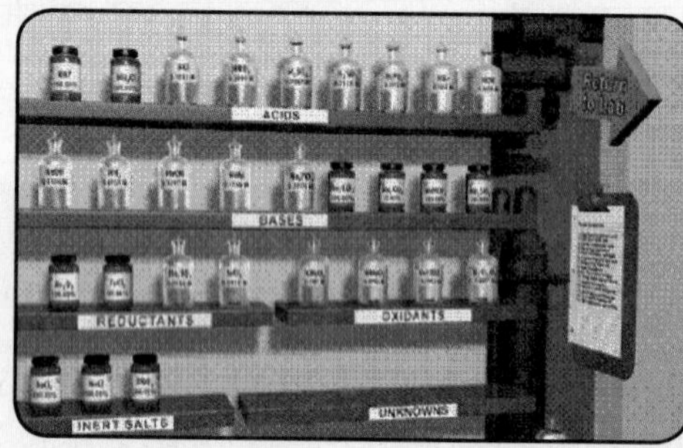

KINETIC ART Have students watch animations about covalent bonds and polar molecules.

CONCEPTS IN ACTION Students can watch an interactive animation that compares ozone, oxygen gas, and oxygen radicals. The animation is followed by questions that highlight the differences between these forms of oxygen.

IN ACADEMIC STANDARDS for SCIENCE

Nature of Science

- Use analogies and models (mathematical and physical) to simplify and represent systems that are difficult to understand or directly experience due to their size, time scale, or complexity, and recognize the limitations of analogies and models.

Standard 3: Bonding and Molecular Structure

C.3.1 Describe, compare, and contrast the characteristics of the interactions between atoms in ionic and covalent compounds.

C.3.2 Compare and contrast how ionic and covalent compounds form.

Also covered C.3.3, C.3.4

8 Covalent Bonding

INSIDE:

- **8.1** Molecular Compounds
- **8.2** The Nature of Covalent Bonding
- **8.3** Bonding Theories
- **8.4** Polar Bonds and Molecules

PearsonChem.com

Water droplets result from attractions between water molecules.

Focus on ELL

1 CONTENT AND LANGUAGE Write the chapter title on the board. Below the title, write *co-valent bonding*. Ask students to predict the meaning of the title based on what they have learned previously about valence electrons and ionic bonding. Explain that the prefix *co-* means "with" or "together" and the verb *valere* means "to be strong." Explain that this chapter is about the strength of the bond between electrons that holds together two atoms.

BEGINNING

LOW Create a picture or symbol to represent the phrase. If literate in the native language, create a native language definition.

HIGH Write the phrase and introduce it to the class while saying it out loud.

INTERMEDIATE: LOW/HIGH Write simplified English definitions of the phrase and present them to the class.

ADVANCED: LOW/HIGH Write the definition of the phrase and present it to the class.

BIGIDEA

BONDING AND INTERACTIONS

Essential Questions:

1. *How is the bonding in molecular compounds different from the bonding in ionic compounds?*
2. *How do electrons affect the shape of a molecule?*
3. *What factors affect molecular properties?*

CHEMYSTERY

What's That Alarm?

A family woke up in the middle of the night to the sound of a piercing alarm. The family thought it must be the fire alarm. They quickly evacuated the house and called 9-1-1. While they waited for the fire department to arrive, they didn't see smoke or any other signs of fire coming from the house.

The fire department inspected the home and told the family that a compound containing carbon and oxygen atoms caused the alarm. Carbon dioxide (CO_2) is made of carbon and oxygen, but the fire department confirmed that it wasn't carbon dioxide. Are other molecules besides carbon dioxide made of carbon and oxygen? What was this mystery substance, and why would it set off an alarm?

▶ Connect to the **BIG**IDEA As you read about covalent bonding, think about how there could be different molecules made of carbon and oxygen atoms.

INDIANA'S ACADEMIC STANDARDS FOR SCIENCE

Bonding and Molecular Structure C.3.1, C.3.2, C.3.3, C.3.4. See lessons for details.

Introduce the Chapter

CONNECTIONS TO IONIC BONDING Students have learned about ionic bonds, in which an electrostatic force between oppositely charged atoms holds particles together. Use this activity to introduce the concept of joining atoms together by sharing electrons.

Activity *You will need 2 egg cartons and 14 table tennis balls or balls of modeling clay.* Cut the top off the egg cartons. Then cut away four cups from each egg carton so that each has only 8 cups (a model octet). Give two students each an egg carton. One should have 7 balls with 1 cup empty. The other should have 1 ball with 7 cups empty. **Ask** How can you model the ionic bonding of sodium and chlorine with the balls? *(A chlorine atom has 7 valence electrons, and a sodium atom has 1. The person with 7 balls can take 1 from the other person so that each atom is stable.)* Next, give two other students the egg cartons. Both should have 7 balls and 1 cup empty to model chlorine atoms. **Ask** How can you model bonding with two chlorine atoms? *(They can each share one of their electrons by nesting two cups from each carton and placing a ball in each of the two nested cups.)* Explain that students have modeled the difference between ionic and covalent bonding.

CHAPTER 8

UbD

Understanding by Design

Students are building toward an understanding of covalent bonding by studying the interaction of particles in compounds.

PERFORMANCE GOALS At the end of Chapter 8, students will be able to answer the essential questions by applying their knowledge of covalent bonding.

ESSENTIAL QUESTIONS Read the essential questions aloud. **Ask** Other than by gaining or losing electrons, how might atoms attain a noble gas configuration? *(sharing electrons)*

BIGIDEA Use the photo of the wet flower to help students connect to this chapter's concepts. Activate prior knowledge by reviewing how electrostatic forces among particles cause particles to attract or repel each other. **Ask** What shape are the water drops? *(round)* **Ask** What forces might hold them in this shape? *(electrostatic forces among particles)* **Ask** How could the particles of water be attracted to each other if all the particles are alike? *(Different parts of the particles have different charges. This causes one water particle to be attracted to another.)*

CHEMYSTERY Connect the CHEMystery to the Big Idea by explaining that bonding can occur among different types of atoms in different ways. For example, a particle of water, H_2O, contains two hydrogen atoms and one oxygen atom. The ratio of hydrogen atoms to oxygen atoms in water is always 2:1. However, hydrogen and oxygen can also combine in other ratios. Have students consider other ratios of carbon and oxygen that might solve the CHEMystery.

UntamedScience™

Painting with Powder?
Margaret visits an engineering facility where chemists are developing powder coating to replace traditional paint coatings.

LESSON 8.1

Key Objectives

8.1.1 **IDENTIFY** the information a molecular formula provides.

8.1.2 **DESCRIBE** the representative units that define molecular compounds and ionic compounds.

Additional Resources

- Reading and study Workbook, Lesson 8.1
- Core Teaching Resources, Lesson 8.1 Review

PearsonChem.com

Engage

CHEMISTRY & YOU Discuss how students used to create objects when playing with logs, blocks, craft sticks, etc. **Ask** If everyone in the room had the same set of toy logs and connector pieces, will everyone in the room create the same model? *(no)* How might the models differ from one another? *(The new models could be different shapes and sizes.)* **Ask** Do you think two atoms can be joined together in more than one way? *(Guide students to predict that it depends on the type of atoms that are joined together.)*

Activate Prior Knowledge

Have students create a three column table. In the first column, instruct students to write a bulleted list of what they learned in Chapter 7 about ionic bonding. In the second column, students should create a bulleted list of what they learned about metallic bonding. As a class, preview this lesson and begin a bulleted list in the third column about covalent bonding.

IN ACADEMIC STANDARDS for SCIENCE

C.3.1 Describe, compare, and contrast the characteristics of the interactions between atoms in ionic and covalent compounds.

C.3.2 Compare and contrast how ionic and covalent compounds form.

C.3.3 Compare and contrast ionic, covalent network, metallic and polar and non-polar molecular crystals with respect to constituent particles, strength of bonds, melting and boiling points, and conductivity; provide examples of each type.

8.1 Molecular Compounds

C.3.1 Interactions between atoms in ionic and covalent compounds; **C.3.2** Formation of ionic and covalent compounds; **C.3.3** Chemical bonds and intermolecular attractions

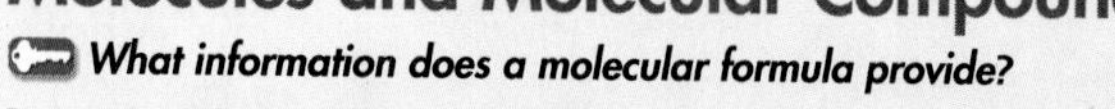

Q: *How are atoms joined together to make compounds with different structures?* This toy model is made from cubes joined together in units by sticks. Although the types of pieces are limited, you can make many different toy models depending on how many pieces you use and how they are arranged. In this lesson, you will learn how atoms are joined together to form units called molecules.

Key Questions

What information does a molecular formula provide?

What representative units define molecular compounds and ionic compounds?

Vocabulary

- covalent bond
- molecule
- diatomic molecule
- molecular compound
- molecular formula

Molecules and Molecular Compounds

What information does a molecular formula provide?

In nature, only the noble gas elements, such as helium and neon, exist as uncombined atoms. They are monatomic; that is they consist of single atoms, as shown in Figure 8.1. But not all elements are monatomic. For example, a key component of the air you breathe is oxygen gas, O_2. As you might guess from the chemical formula, O_2 represents two oxygen atoms that are bonded together.

In Chapter 7, you learned about ionic compounds, which are generally crystalline solids with high melting points. Other compounds, however, have very different properties. For example, water (H_2O) is a liquid at room temperature. Carbon dioxide (CO_2) and nitrous oxide (N_2O) are both gases at room temperature. The attractions that hold together the atoms in O_2, H_2O, CO_2, and N_2O cannot be explained by ionic bonding. These bonds do not involve the transfer of electrons.

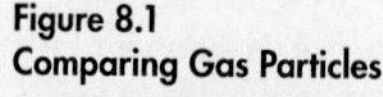

Figure 8.1
Comparing Gas Particles

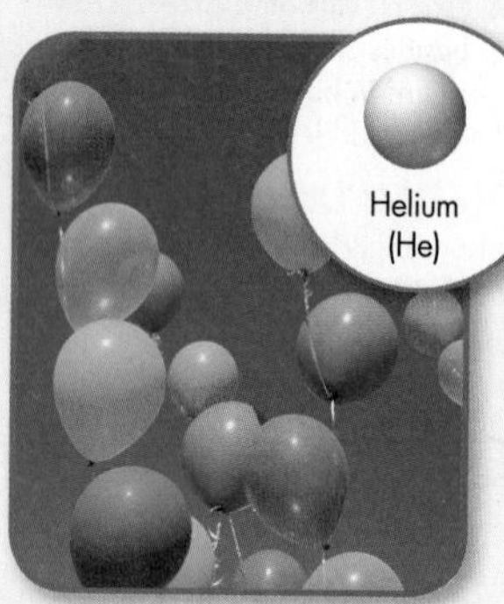

Helium, which is less dense than air, is often used to inflate balloons.

Scuba divers breathe compressed air, a mixture that contains oxygen gas.

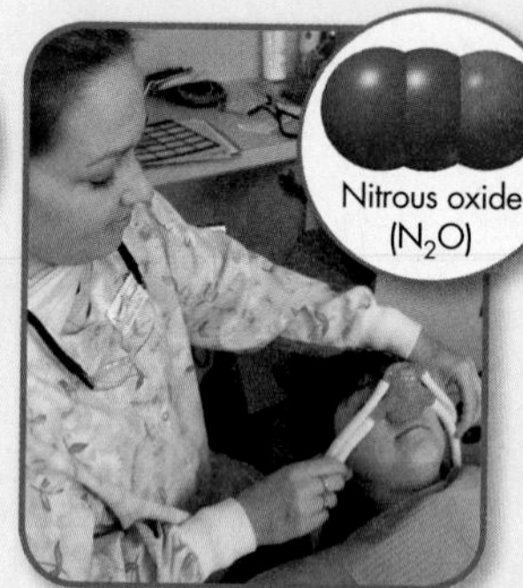

Nitrous oxide (also known as laughing gas) is sometimes used as a mild anesthetic in dental procedures.

Focus on ELL

1 CONTENT AND LANGUAGE As a class, brainstorm the various words that mean the same as *bond*. *(Sample answers: link, union, tie, attachment, connection)* Then have students identify where in this lesson synonyms are used. Then have students identify words that mean the opposite of *bond*. Repeat this exercise with the word *noble*.

2 FRONTLOAD THE LESSON Have students draw a table with column heads labeled *Atoms, Molecules, Diatomic Molecules,* and *Molecular Compounds*. Tell students they will use their tables to organize the names, symbols, and structural drawings of each example illustrated in the lesson.

3 COMPREHENSIBLE INPUT As a class, summarize the paragraph titled Representing Molecules using a two-column table. Under the two column heads labeled *Always True* and *Never True*, construct a list of key points.

Sharing Electrons Recall that ionic bonds form when the combining atoms give up or accept electrons. Another way that atoms can combine is by sharing electrons. Atoms that are held together by sharing electrons are joined by a **covalent bond.** In a covalent bond, a "tug of war" for electrons takes place between the atoms, bonding the atoms together. In Lesson 8.2, you will learn about the different types of covalent bonds.

In Figure 8.1, the representative units shown for oxygen and nitrous oxide are called molecules. A **molecule** is a neutral group of atoms joined together by covalent bonds. Oxygen gas consists of oxygen molecules; each oxygen molecule consists of two covalently bonded oxygen atoms. An oxygen molecule is an example of a **diatomic molecule**—a molecule that contains two atoms. Other elements found in nature in the form of diatomic molecules include hydrogen, nitrogen, and the halogens. Molecules can also be made of atoms of different elements. A compound composed of molecules is called a **molecular compound.** Water is an example of a molecular compound. The molecules in water are all the same; each water molecule is a tightly bound unit of two hydrogen atoms and one oxygen atom.

Figure 8.2 Butane
Butane (C_4H_{10}) is commonly used in lighters and household torches. The butane torch shown here is being used to caramelize sugar on a dessert.

Representing Molecules A **molecular formula** is the chemical formula of a molecular compound. **A molecular formula shows how many atoms of each element a substance contains.** The molecular formula of water is H_2O. Notice that a subscript written after an element's symbol indicates the number of atoms of each element in the molecule. If there is only one atom, the subscript 1 is omitted. The molecular formula of carbon dioxide is CO_2. This formula represents a molecule containing one carbon atom and two oxygen atoms.

Butane, shown in Figure 8.2, is also a molecular compound. The molecular formula for butane is C_4H_{10}. According to this formula, one molecule of butane contains four carbon atoms and ten hydrogen atoms. A molecular formula reflects the actual number of atoms in each molecule. The subscripts are not necessarily lowest whole-number ratios. Note that molecular formulas also describe molecules consisting of atoms of one element. For example, an oxygen molecule consists of two oxygen atoms bonded together; its molecular formula is O_2.

A molecular formula does not tell you about a molecule's structure. In other words, it does not show either the arrangement of the various atoms in space or which atoms are covalently bonded to one another. A variety of diagrams and molecular models, some of them illustrated in Figure 8.3, can be used to show the arrangement of atoms in a molecule. Diagrams and models like these will be used throughout this textbook.

Figure 8.3 Representations of an Ammonia Molecule
The formula NH_3 tells you the composition of an ammonia molecule, but it does not reveal the arrangement of the atoms. Molecular models and structural formulas specify the bonds between atoms and the arrangement of those atoms.

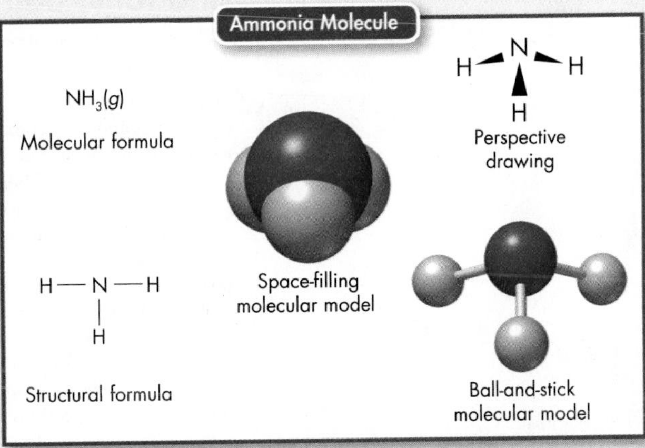

Differentiated Instruction

L1 STRUGGLING STUDENTS Prior to introducing molecular compounds, have students practice identifying elements. To help students interpret a molecular compound, have students assign a colored pencil to each element and rewrite the molecular formula using the colored pencils. Have students write the subscript "1" for single atoms in a compound until they comprehend that it is not necessary.

ELL ENGLISH LANGUAGE LEARNERS Prepare three-dimensional models of ammonia similar to the two models in Figure 8.3. Write the molecular and structural formulas for ammonia on large cards. Hold the models and cards up one by one. Slowly and clearly point out the information that each model or formula provides. Have students make their own drawings indicating the key points.

L3 ADVANCED STUDENTS Have students create a poster or models of the various ways that the bonding in a hydrogen molecule can be represented: molecular formula, structural formula, and electron-dot structure. Then have students research the octet rule and repeat the activity for a chlorine molecule.

Foundations for Reading

BUILD VOCABULARY Explain that the prefix *di-* is a Greek word that means "two." Have students predict the meaning of *diatomic molecule*. *(a molecule containing two atoms)* Have students preview the lesson to verify their prediction. Students should take note of any other words with a common prefix.

READING STRATEGY Encourage students to examine the illustrations carefully as they read. Have students compare the models shown in this lesson with representations of ionic bonding illustrated in Chapter 7. Have students summarize any similarities or differences they notice.

Explain

Molecules and Molecular Compounds

START A CONVERSATION As a class, brainstorm ways in which people join objects together, for instance, attaching papers with a paper clip, attaching a leg of a table to a tabletop with screws, setting a diamond in a ring with prongs, using glue to fix a broken vase, etc. **Ask** What typically guides a person to choose one method for attaching objects over another? *(Sample answers: whether the attachment is permanent or temporary, expensive or inexpensive, etc.)* Guide the discussion toward the concept of stability. Use an analogy involving tape and superglue to help students understand the strength of an ionic or covalent bond. Develop a sense of awareness that the bonding that occurs between two elements varies in strength. Some of the resulting molecules are stronger, meaning more stable, than others. Tell students to keep this in mind as they progress through the chapter.

USE CONTEXT CLUES Have students read the first paragraph. **Ask** What is the meaning of *monatomic*? *(elements that consist of a single atom)* **Ask** Use what you learned about the prefix *di-* from your vocabulary exercise to predict the meaning of the prefix *mono-*. *(one)* Guide students to rely on their language development skills to help them understand difficult or unfamiliar passages.

USING VISUALS Direct students to Figure 8.1. Point out that some other elements are most stable when two atoms are joined together. **Ask** Which of the three examples are monatomic molecules? *(helium)* **Ask** Is H_2 a monatomic or diatomic molecule? *(diatomic molecule)* **Ask** Do you think H_2 could be an ionic compound? *(No, ionic bonds do not form between two atoms of the same element.)* Tell students that H_2 is a molecule held together not by the transfer of an electron but by the sharing of a pair of electrons.

Explain

USE AN ANALOGY Point out that the recipe for a cake provides information about what type and how much of each ingredient is needed. Varying the types and amounts of ingredients leads to cakes that have different tastes and textures. **Ask** How does a molecular formula provide similar information about a compound? *(The subscripts in a molecular formula tell the number of atoms of each element in a compound.)*

Misconception Alert

Students often confuse empirical (simplest) formulas and molecular formulas. Explain that the subscripts in a molecular formula are not necessarily written in the smallest whole-number ratio of the atoms. For example, ethane is correctly represented by the formula C_2H_6 because the molecule actually contains two carbon atoms and six hydrogen atoms. It would be incorrect to represent ethane by its empirical formula, which is CH_3.

CHEMISTRY & YOU Atoms join together by forming covalent bonds. Changing the number, type, and/or order of the atoms changes the compound. Atoms can also be joined by ionic bonds.

Explore

Teacher Demo

PURPOSE Students see the structure associated with molecular formulas.

MATERIALS molecular model kit or foam balls and toothpicks

PROCEDURE Construct models of these pairs of molecules: O_2 and O_3; CO and CO_2; H_2O and H_2O_2. Show students each pair and ask them how many of each type of atom is in each molecule. Write the corresponding chemical formulas on the chalkboard.

EXPECTED OUTCOMES Students will see that the addition of one more oxygen atom to O_2, CO, and H_2O results in new molecules that have very different structures.

Explain

Comparing Molecular and Ionic Compounds

USE VISUALS Direct students to examine Figure 8.5. **Ask** How do the submicroscopic illustrations of NaCl and H_2O differ? *(The ions of sodium and chlorine are separate and arrayed in a regular pattern; two hydrogen atoms and one oxygen atom are bonded together in one molecule.)* Compare the formulas for NaCl and H_2O. *(They look similar, but NaCl represents just one unit in the array of sodium and chloride ions, whereas H_2O represents one separate molecule of water.)*

Figure 8.4
Molecular Formulas and Structures
The formula of a molecular compound indicates the numbers and kinds of atoms in each molecule of the compound.
Use Models ***Which of these molecules has the greatest number of oxygen atoms?***

Hydrogen atom (H)

Oxygen atom (O)

Carbon atom (C)

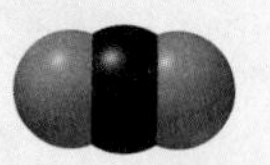
Carbon dioxide (CO_2)
1 molecule of CO_2 contains
2 oxygen atoms
1 carbon atom

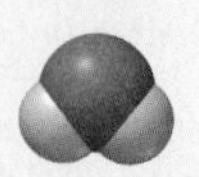
Water (H_2O)
1 molecule of H_2O contains
2 hydrogen atoms
1 oxygen atom

Ethanol (C_2H_6O)
1 molecule of C_2H_6O contains
6 hydrogen atoms
2 carbon atoms
1 oxygen atom

CHEMISTRY & YOU

Q: *Similar to how you can make different types of toy models, there are thousands of different types of molecular structures. How are atoms joined together to make compounds with different structures?*

Figure 8.4 shows the chemical formulas and structures of some other molecular compounds. The arrangement of the atoms within a molecule is called its molecular structure. Carbon dioxide, for example, is a gas produced by the complete burning of carbon. It is found in Earth's atmosphere and is dissolved in seawater. The molecular structure of carbon dioxide shows how the three atoms are arranged in a row. It also shows how the carbon atom in each molecule is in the middle between the two oxygen atoms. The molecular structure of water shows how the oxygen atom is in the middle between the hydrogen atoms. However, the atoms in water are not arranged in a row. Instead, the hydrogen atoms are mainly on one side of the water molecule. The molecular structure of ethanol (C_2H_6O) is more complicated. As you can see in the model, each carbon is bonded to four atoms, each hydrogen is bonded to one atom, and the one oxygen is bonded to two atoms.

Comparing Molecular and Ionic Compounds

What representative units define molecular compounds and ionic compounds?

You have now seen how formulas can be used to describe molecular compounds and ionic compounds. Each type of compound contains atoms of different elements that are combined chemically. However, the formulas describe different representative units. **The representative unit of a molecular compound is a molecule. For an ionic compound, the representative unit is a formula unit.** Recall that a formula unit is the lowest whole-number ratio of ions in an ionic compound. It is important not to confuse formula units with molecules. A molecule is made up of two or more atoms that act as a unit. No such discrete units exist in an ionic compound, which consists of a continuous array of ions. So there is no such thing as a molecule of sodium chloride or a molecule of magnesium chloride. Instead, these compounds exist as collections of positively and negatively charged ions arranged in repeating three-dimensional patterns.

224 Chapter 8 • Lesson 1

UbD Check for Understanding

BIG IDEA BONDING AND INTERACTIONS Give each student an index card. Assess students' knowledge of how bonding in molecular compounds differs from bonding in ionic compounds. Ask students to state the type of compound associated with electron sharing, and the type of compound associated with electron transfer. *(molecular compound; ionic compound)*

ADJUST INSTRUCTION If students are having difficulty answering the question, have them re-read the previous page and then repeat the exercise.

Molecular compounds tend to have relatively lower melting and boiling points than ionic compounds. Many molecular compounds are gases or liquids at room temperature. In contrast to ionic compounds, which are formed from a metal combined with a nonmetal, most molecular compounds are composed of atoms of two or more nonmetals. For example, one atom of carbon can combine with one atom of oxygen to produce one molecule of a compound known as carbon monoxide. Carbon monoxide is a poisonous gas produced by burning gasoline in internal combustion engines or in household gas appliances and furnaces. Figure 8.5 illustrates some differences between molecular and ionic compounds, using water and sodium chloride as examples.

Figure 8.5 Molecular and Ionic Compounds Water, which is a molecular compound, and sodium chloride, which is an ionic compound, are compared here.
Interpret Diagrams *How do molecular compounds differ from ionic compounds?*

8.1 LessonCheck

ONLINE PROBLEMS

1. **Identify** What information does a molecular formula provide?
2. **Compare** How is the representative unit of a molecular compound different from the representative unit of an ionic compound?
3. **Identify** What are the only elements that exist in nature as uncombined atoms? What term is used to describe such elements?
4. **Compare and Contrast** Describe how the molecule whose formula is NO is different from the molecule whose formula is N_2O.
5. **Apply Concepts** Give an example of a diatomic molecule found in Earth's atmosphere.
6. **Identify** What information does a molecular structure give?

Lesson Check Answers

1. Molecular formulas show how many atoms of each element one molecule of a compound contains.
2. The representative unit of a molecular compound is a molecule. The representative unit of an ionic compound is a formula unit.
3. noble gases; monatomic
4. NO has 1 N atom and 1 O atom; N_2O has 2 N atoms and 1 O atom.
5. nitrogen (N_2) or oxygen (O_2)
6. Molecular structures show the arrangement of atoms in a molecule.

CRITICAL THINKING Have students explain why two hydrogen atoms bond by sharing rather than by electron transfer to form a hydrogen molecule. (*Two hydrogen atoms cannot achieve a stable electron configuration by transferring electrons. However, if the hydrogen atoms share their electrons, they can each achieve the stable arrangement of a helium atom.*)

Evaluate

Informal Assessment

Have students number a sheet of paper from 1 to 6. Write the following statements on the board. Have students indicate on their papers if the following statements are true or false:

1. All molecular compounds are composed of atoms of two or more different elements. *(true)*
2. All compounds contain molecules. *(false)*
3. No elements exist as molecules. *(false)*
4. Most molecular compounds are composed of two or more nonmetallic elements. *(true)*
5. Atoms in molecular compounds share electrons. *(true)*
6. Ionic compounds tend to have lower melting and boiling points than molecular compounds. *(false)*

Then have students complete the 8.1 Lesson Check.

Reteach

Use the statements in the Informal Assessment above as a basis for review. Have students locate where in the lesson the concept for each statement is explained. Use the given material to prove each statement as true or provide an example to show that the statement is false.

Answers

FIGURE 8.4 carbon dioxide

FIGURE 8.5 Molecular compounds are made up of molecules and usually have lower melting and boiling points than ionic compounds, which are made up of ions.

LESSON 8.2

Key Objectives

8.2.1 **EXPLAIN** the result of electron sharing in covalent bonds.

8.2.2 **DESCRIBE** how coordinate covalent bonds are different from other covalent bonds.

8.2.3 **IDENTIFY** some exceptions to the octet rule.

8.2.4 **EXPLAIN** how the strength of a covalent bond is related to its bond dissociation energy.

8.2.5 **DESCRIBE** the how resonance structures are used.

Additional Resources

- Reading and Study Workbook, Lesson 8.2
- Core Teaching Resources, Lesson 8.2 Review
- Probeware Laboratory Manual, Quick Lab, p. 238

PearsonChem.com

Engage

CHEMISTRY & YOU Explain that scientists attribute the thinning of the ozone layer, in part, to the action of compounds called chlorofluorocarbons (CFCs), which have been released into the atmosphere. **Ask** Why is ozone important in the atmosphere? *(It filters out radiation that could harm living things on Earth.)*

Activate Prior Knowledge

Prior to beginning this lesson, ask students to recall what they know about electron configuration and the rules that govern it. Have student volunteers demonstrate the proper use of the aufbau principle, the Pauli exclusion principle, and Hund's rule.

IN ACADEMIC STANDARDS for SCIENCE

- Use analogies and models (mathematical and physical) to simplify and represent systems that are difficult to understand or directly experience due to their size, time scale, or complexity, and recognize the limitations of analogies and models.

C.3.1 Describe, compare, and contrast the characteristics of the interactions between atoms in ionic and covalent compounds.

C.3.2 Compare and contrast how ionic and covalent compounds form.

Also covered C.3.3.

8.2 The Nature of Covalent Bonding

C.3.1 Interactions between atoms in ionic and covalent compounds; **C.3.2** Formation of ionic and covalent compounds; **C.3.3** Chemical bonds and intermolecular attractions

CHEMISTRY & YOU

Q: *What is the difference between the oxygen you breathe and the oxygen in ozone in the atmosphere?* Our atmosphere contains two different molecules that are both made of oxygen atoms. One is the oxygen that our cells need to survive. The other molecule containing only oxygen atoms is the ozone that protects us from the sun but also contributes to smog. The colors in this map indicate the concentrations of ozone in various parts of Earth's atmosphere. In this lesson, you will learn how oxygen atoms can join to form the oxygen you breathe and can also join to form ozone.

Key Questions

What is the result of electron sharing in covalent bonds?

How are coordinate covalent bonds different from other covalent bonds?

What are some exceptions to the octet rule?

How is the strength of a covalent bond related to its bond dissociation energy?

How are resonance structures used?

Vocabulary

- single covalent bond
- structural formula
- unshared pair
- double covalent bond
- triple covalent bond
- coordinate covalent bond
- polyatomic ion
- bond dissociation energy
- resonance structure

The Octet Rule in Covalent Bonding

What is the result of electron sharing in covalent bonds?

Recall that when ionic compounds form, electrons tend to be transferred so that each ion acquires a noble gas configuration. A similar rule applies for covalent bonds. **In covalent bonds, electron sharing usually occurs so that atoms attain the electron configurations of noble gases.** For example, a single hydrogen atom has one electron. But a pair of hydrogen atoms shares electrons to form a covalent bond in a diatomic hydrogen molecule. Each hydrogen atom, thus, attains the electron configuration of helium, a noble gas with two electrons. Combinations of atoms of the nonmetals and metalloids in Groups 4A, 5A, 6A, and 7A of the periodic table are likely to form covalent bonds. The combined atoms usually acquire a total of eight electrons, or an octet, by sharing electrons, so that the octet rule applies.

Single Covalent Bonds The hydrogen atoms in a hydrogen molecule are held together mainly by the attraction of the shared electrons to the positive nuclei. Two atoms held together by sharing one pair of electrons are joined by a **single covalent bond.** Hydrogen gas consists of diatomic molecules whose atoms share only one pair of electrons, forming a single covalent bond.

$H\cdot + \cdot H \longrightarrow H{:}H$

Hydrogen atom + Hydrogen atom → Hydrogen molecule (shared pair of electrons)

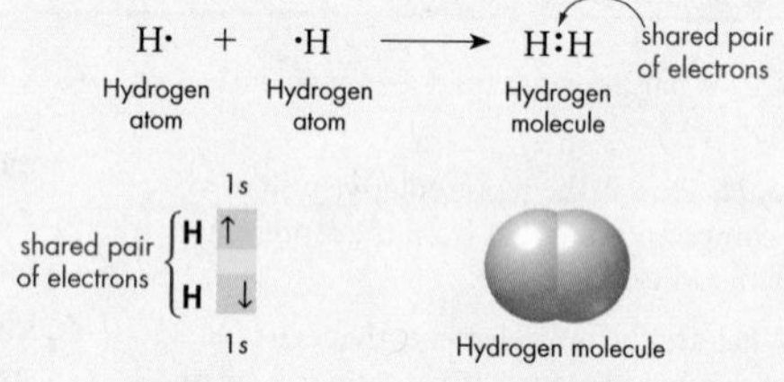

Focus on ELL

1 CONTENT AND LANGUAGE Have each student prepare a set of flash cards for the vocabulary words in this lesson. Each card should list the vocabulary word, its definition, and the page number where it can be found. Cycle through the index cards a few times, reading the words aloud as a class. Point out familiar prefixes, *poly-* and *dis-* and *co-*, and remind students of their meanings. *(many; not; with or together)*

2 FRONTLOAD THE LESSON Divide the class into three groups. Have one group preview all the models in the lesson; another group, the tables, charts, and photos; and the third, the sample problems. Ask each group to answer a series of questions, such as *What is this lesson about? Is there anything I have already learned that will help me learn this material?* Have each group present their answers to the class, as well as any symbols or vocabulary that is unfamiliar.

3 COMPREHENSIBLE INPUT Re-draw the electron dot figures used throughout lesson 8.2, using a separate color for each different element's electrons to help demonstrate the concepts associated with each figure. Have students view the Kinetic Art animation of covalent bonding.

An electron dot structure such as H:H represents the shared pair of electrons of the covalent bond by two dots. The pair of shared electrons forming the covalent bond is also often represented as a dash, as in H—H for hydrogen. A **structural formula** represents the covalent bonds as dashes and shows the arrangement of covalently bonded atoms. In contrast, the molecular formula of hydrogen, H_2, indicates only the number of hydrogen atoms in each molecule.

The halogens also form single covalent bonds in their diatomic molecules. Fluorine is one example. Because a fluorine atom has seven valence electrons, it needs one more to attain the electron configuration of a noble gas. By sharing electrons and forming a single covalent bond, two fluorine atoms each achieve the electron configuration of neon.

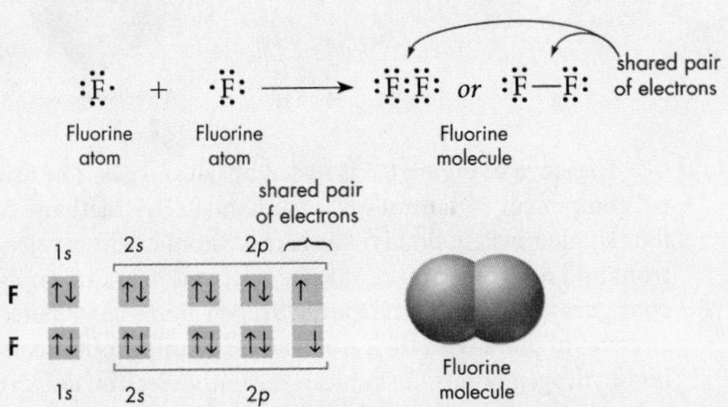

In the F_2 molecule, each fluorine atom contributes one electron to complete the octet. Notice that the two fluorine atoms share only one pair of valence electrons. A pair of valence electrons that is not shared between atoms is called an **unshared pair,** also known as a lone pair or a nonbonding pair. In F_2, each fluorine atom has three unshared pairs of electrons.

You can draw electron dot structures for molecules of compounds in much the same way that you draw them for molecules of diatomic elements. Water (H_2O) is a molecule containing three atoms with two single covalent bonds. Two hydrogen atoms share electrons with one oxygen atom. The hydrogen and oxygen atoms attain noble-gas configurations by sharing electrons. As you can see in the electron dot structures below, the oxygen atom in water has two unshared pairs of valence electrons.

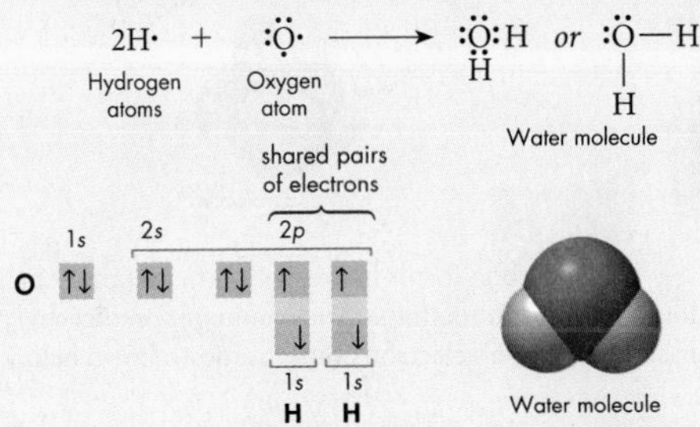

Differentiated Instruction

L1 STRUGGLING STUDENTS Pair each student with a study partner. Have them use the periodic table and quiz each other on writing electron dot structures for single atoms and bonded atoms. Make sure they understand that for representative elements, the group number indicates the number of valence electrons that atom has, and that only valence electrons appear in the electron dot structures.

L1 SPECIAL NEEDS Recreate the electron dot structures on large posters or overheads. With a colored marker, show students how to draw the dots. Point out where on the periodic table the element is located; circle the element and circle the group number. Then explain the reasoning for the placement and the number of dots. Make sure the noble-gas group on their periodic table is highlighted.

Foundations for Reading

BUILD VOCABULARY Explain that the word *structure* comes from the Latin verb *struere*, which means "to build." Convey that a structural formula is one that shows how the atoms of a molecule are joined together by chemical bonds.

READING STRATEGY Identify the main idea in the paragraph titled The Octet Rule in Covalent Bonding. Have students create and complete a KWL chart, with corresponding examples, to support how this rule is applied to a single covalent bond, a double covalent bond, and a triple covalent bond.

Explain

The Octet Rule in Covalent Bonding

APPLY CONCEPTS Write electron configurations for carbon, nitrogen, oxygen, fluorine, and neon on the chalkboard. **Ask** How many electrons would carbon, nitrogen, oxygen, and fluorine need to share in order to achieve the same electron configuration as neon? *(4, 3, 2, and 1, respectively)*

Explore

Class Activity

PURPOSE Students practice different ways to represent molecules.

MATERIALS paper and pencil

PROCEDURE Organize students into groups of three or four. Have them practice drawing molecular diagrams, structural formulas, electron dot structures, and orbital diagrams for molecules such as OF_2, SCl_2, N_2H_4, CCl_4, $CHCl_3$, and C_2H_6.

Explain

INTERPRET DIAGRAMS Direct students to examine the molecular and structural formulas, electron dot structures, and orbital diagrams for the ammonia molecule and compare it to the fluorine and water examples on the previous page. If possible, display physical models. Call attention to the fact that fluorine has one half-filled orbital and forms one bond, oxygen has two half-filled orbitals and forms two bonds, and nitrogen has three and forms three bonds.

MAKE A CONNECTION Direct students to examine the molecular and structural formulas, electron dot structures, and orbital diagrams for the methane molecule. Tell students that carbon has two half-filled orbitals. **Ask** How many covalent bonds do you think carbon forms? *(Students may logically say two.)* Tell students that CH_2 does not represent a stable molecule, but CH_4, methane, is a stable molecule. Explain the concept of electron promotion, which allows carbon to form four single covalent bonds.

CRITICAL THINKING Expand on the concept of electron promotion by showing students that elements in Groups 3A and 4A of the periodic table promote *s* electrons to *p* orbitals, increasing their bonding capacity. Point out that boron's electron configuration is $1s^22s^22p^1$. **Ask** Based on this configuration, how many covalent bonds do you think boron forms? *(Students may logically say one.)* Have students explain how it is possible for the chloride of boron to be BCl_3 rather than BCl. *(The promotion of one 2s electron to the 2p orbital allows for the formation of three bonds.)* **Ask** Since boron does not achieve a noble gas configuration, how does it achieve stability? *(It achieves added stability by forming three bonds rather than one.)*

You can draw the electron dot structure for ammonia (NH_3), a suffocating gas, in a similar way. The ammonia molecule has one unshared pair of electrons.

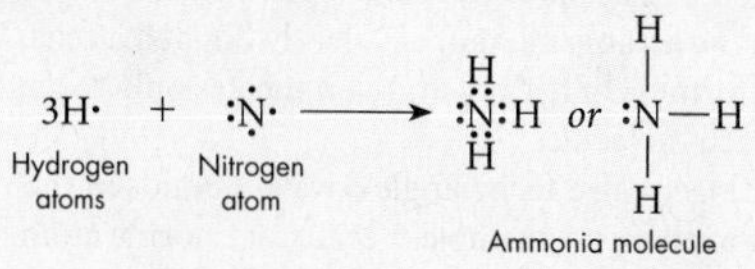

3H· + :N· ⟶ H:N:H (with H) *or* :N—H (with H above and below)

Hydrogen atoms | Nitrogen atom | Ammonia molecule

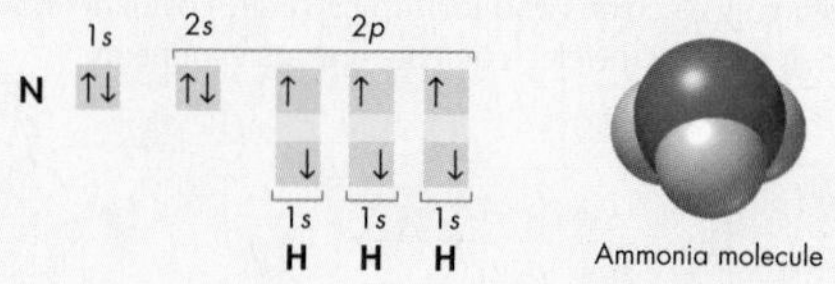

N: 1s ↑↓ | 2s ↑↓ | 2p ↑ ↑ ↑ with ↓ ↓ ↓ from 1s of H, H, H

Ammonia molecule

The stove in Figure 8.6 is fueled by natural gas. The principal component of natural gas is methane (CH_4). Methane contains four single covalent bonds. The carbon atom has four valence electrons and needs four more valence electrons to attain a noble-gas configuration. Each of the four hydrogen atoms contributes one electron to share with the carbon atom, forming four identical carbon-hydrogen bonds. As you can see in the electron dot structure below, methane has no unshared pairs of electrons.

Figure 8.6 Methane
Methane is the principal component of natural gas. Natural gas is commonly used as a fuel for household gas appliances such as gas stoves, water heaters, dryers, and furnaces.

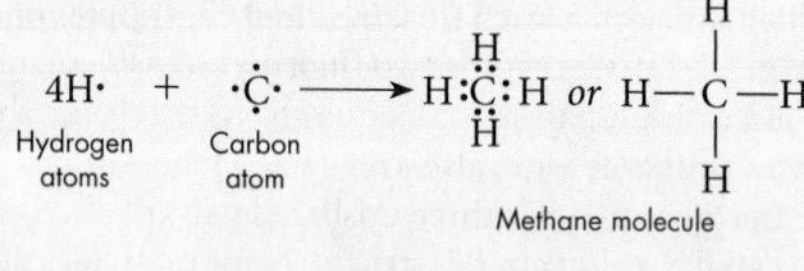

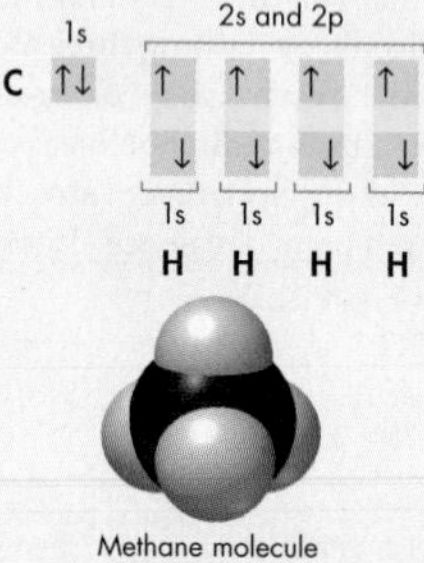

Methane molecule

When carbon forms bonds with other atoms, it usually forms four bonds, as in methane. You would not predict this pattern based on carbon's electron configuration, shown below.

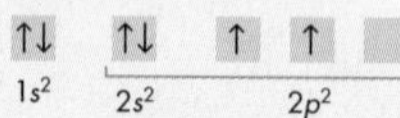

Expanding the Octet

Nonmetals in the third row and beyond, such as phosphorus, sulfur, and iodine, can form more than four bonds because they have empty *d* orbitals. Phosphorus, for example, can unpair a 3*s* electron and promote it to an empty 3*d* orbital. The promotion allows phosphorus to form five bonds. The amount of energy needed to promote an electron is less than the energy released with the formation of an extra bond. Sulfur can promote one 3*s* electron and one 3*p* electron and form two extra bonds.

If you tried to form covalent C—H bonds for methane by combining the two $2p$ electrons of the carbon with two $1s$ electrons of hydrogen atoms, you would incorrectly predict a molecule with the formula CH_2 (instead of CH_4). The formation of four bonds by carbon can be explained by the fact that one of carbon's $2s$ electrons is promoted to the vacant $2p$ orbital to form the following electron configuration:

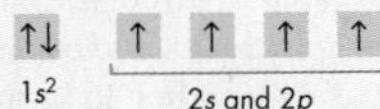

This electron promotion requires only a small amount of energy. The promotion provides four electrons of carbon that are capable of forming covalent bonds with four hydrogen atoms. Methane, the carbon compound formed by electron sharing of carbon with four hydrogen atoms, is much more stable than CH_2. The stability of the resulting methane more than compensates for the small energy cost of the electron promotion. Therefore, formation of methane (CH_4) is more energetically favored than the formation of CH_2.

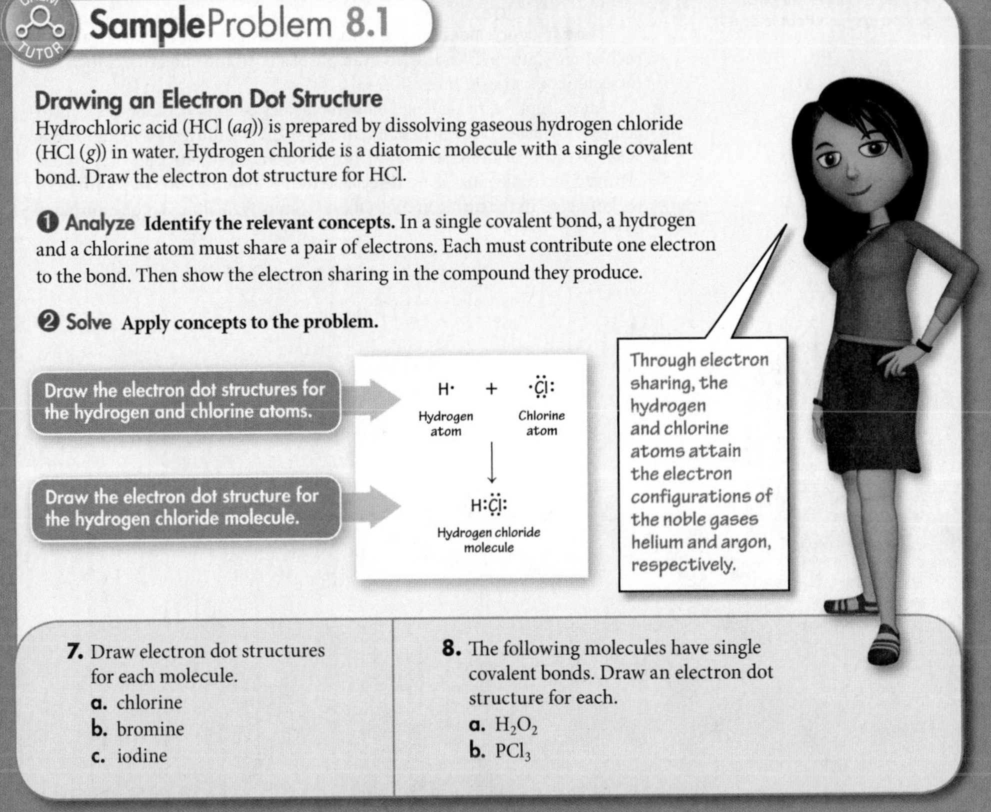

Sample Problem 8.1

Drawing an Electron Dot Structure

Hydrochloric acid (HCl (*aq*)) is prepared by dissolving gaseous hydrogen chloride (HCl (*g*)) in water. Hydrogen chloride is a diatomic molecule with a single covalent bond. Draw the electron dot structure for HCl.

1 Analyze **Identify the relevant concepts.** In a single covalent bond, a hydrogen and a chlorine atom must share a pair of electrons. Each must contribute one electron to the bond. Then show the electron sharing in the compound they produce.

2 Solve **Apply concepts to the problem.**

7. Draw electron dot structures for each molecule.
a. chlorine
b. bromine
c. iodine

8. The following molecules have single covalent bonds. Draw an electron dot structure for each.
a. H_2O_2
b. PCl_3

Check for Understanding

What is the result of electron sharing in covalent bonds?

Assess students' understanding of electron sharing in covalent bonds by asking them to answer the key question in two sentences or less, using their own words. *(Sample answer: Atoms share electrons in order to form an octet. The result is a stable molecule.)*

ADJUST INSTRUCTION If students are having difficulty writing their summaries, review the differences between ionic and covalent compounds. Point out that ions gain or lose electrons to achieve an ionic compound with a noble gas configuration, whereas atoms share electrons to form a molecular compound in which each atom has a noble gas electron configuration.

Explore

Class Activity

PURPOSE Students gain understanding of covalent bonding and distinguish between covalent and ionic bonding.

PROCEDURE Have students draw electron dot structures for each nonmetal in the second row of the periodic table: C, N, O, and F. Then have them answer the following:

- Predict how many bonds each atom must form to attain a noble gas configuration. *(4, 3, 2, 1)*
- Can the metal in Period 2 form a covalent bond and reach stability? *(no)*
- Can each nonmetal reach stability by forming covalent bonds? *(yes)*
- Can fluorine form an ionic bond? *(yes)*
- Are the bonds in nitrogen molecules, N_2, ionic or covalent? *(covalent)*

Sample Practice Problem

The following covalent molecules have only single covalent bonds. Draw an electron dot structure for each.

a. NF_3 :F:N:F: / :F:

b. SBr_2 :Br:S:Br:

Answers

7. a. :Cl:Cl:
b. :Br:Br:
c. :I:I:

8. a. H:O:O:H
b. :Cl:P:Cl: / :Cl:

Explain

START A CONVERSATION To introduce the discussion of multiple covalent bonds, use the electron dot structure for the nitrogen molecule. **Ask** How does the structure of diatomic nitrogen satisfy the octet rule? *(The nitrogen atoms can share six electrons.)* Have students compare the bonding in ammonia with the bonding in nitrogen gas. Then introduce the oxygen molecule. Instruct students to draw a structure that obeys the octet rule. **Ask** Why doesn't oxygen form a triple bond? *(Each oxygen atom needs to share only two electrons to achieve a stable electron configuration.)* Explain that although a double bond in the oxygen molecule fulfills the octet rule, it does not fit with experimental evidence that shows that the oxygen molecule contains two unpaired electrons. Thus the structure of O_2 is an exception to the octet rule.

USE MODELS Help students draw the electron dot structure and orbital diagram for carbon dioxide. **Ask** What type of bonds does carbon form with the two oxygen atoms in CO_2? *(double covalent bonds)* Note that carbon can form single, double, and triple bonds, but a quadruple bond is impossible because of geometric restrictions. Have students draw diagrams for hydrogen cyanide (HCN) and formaldehyde (H_2CO). **Ask** What kind of bonds does carbon form in each of these molecules? *(HCN: one single carbon-to-hydrogen bond and one triple carbon-to-nitrogen bond; H_2CO: two single carbon-to-hydrogen bonds and one double carbon-to-oxygen bond)* As the discussion proceeds, provide models for the molecules discussed, or allow students time to come up with a way to model the examples with materials in the classroom or home.

Figure 8.7 Carbon Dioxide
Carbon dioxide gas is soluble in water and is used to carbonate many beverages. A carbon dioxide molecule has two carbon–oxygen double bonds.

Double and Triple Covalent Bonds Sometimes atoms bond by sharing more than one pair of electrons. **Atoms form double or triple covalent bonds if they can attain a noble gas structure by sharing two pairs or three pairs of electrons.** A **double covalent bond** is a bond that involves two shared pairs of electrons. Similarly, a bond formed by sharing three pairs of electrons is a **triple covalent bond.**

Carbon dioxide (CO_2) is used to carbonate many soft drinks like the one shown in Figure 8.7. The carbon dioxide molecule contains two oxygens, each of which shares two electrons with carbon to form a total of two carbon–oxygen double bonds.

$$\cdot\ddot{\text{O}}\cdot + \cdot\dot{\text{C}}\cdot + \cdot\ddot{\text{O}}\cdot \longrightarrow \ddot{\text{O}}::\text{C}::\ddot{\text{O}} \text{ or } \ddot{\text{O}}=\text{C}=\ddot{\text{O}}$$

Oxygen atom | Carbon atom | Oxygen atom | Carbon dioxide molecule

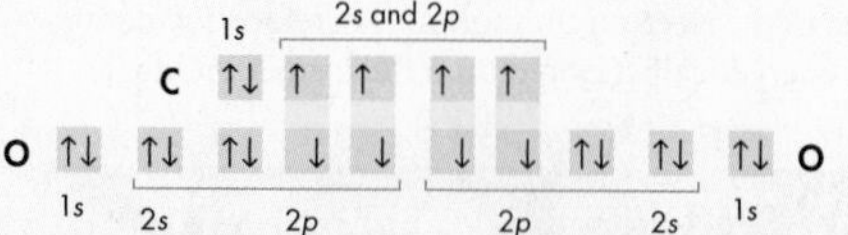

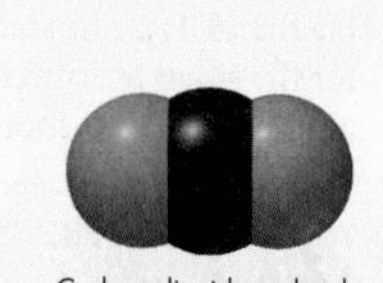

Carbon dioxide molecule

The two double bonds in the carbon dioxide molecule are identical to each other. Carbon dioxide is an example of a triatomic molecule, which is a molecule consisting of three atoms.

An example of an element whose molecules contain triple bonds is nitrogen (N_2), a major component of Earth's atmosphere, illustrated in Figure 8.8. A single nitrogen atom has five valence electrons. Each nitrogen atom in the nitrogen molecule must share three electrons to have the electron configuration of neon. In the nitrogen molecule, each nitrogen atom has one unshared pair of electrons.

$$:\dot{\text{N}}\cdot + \cdot\dot{\text{N}}: \longrightarrow :\text{N}:::\text{N}: \text{ or } :\text{N}{\equiv}\text{N}:$$

Nitrogen atom | Nitrogen atom | Nitrogen molecule

Figure 8.8 Oxygen and Nitrogen
Oxygen and nitrogen are the main components of Earth's atmosphere. The oxygen molecule is an exception to the octet rule. It has two unpaired electrons. Three pairs of electrons are shared in a nitrogen molecule.

Inventing Electron Dot Structures

Gilbert Newton Lewis (1875–1946) was an American chemist who invented electron dot structures, which are often called Lewis structures or diagrams in his honor. These structures supported Lewis's theory of the electron pair in chemical bonding. As a professor of physical chemistry, he expanded the theory of acids and bases by defining an acid as an electron pair acceptor and a base as an electron pair donor. The definitions encompass all Brønsted-Lowry acid-base reactions and include many others not previously categorized as acid-base reactions.

You might think that an oxygen atom, with six valence electrons, would form a double bond by sharing two of its electrons with another oxygen atom.

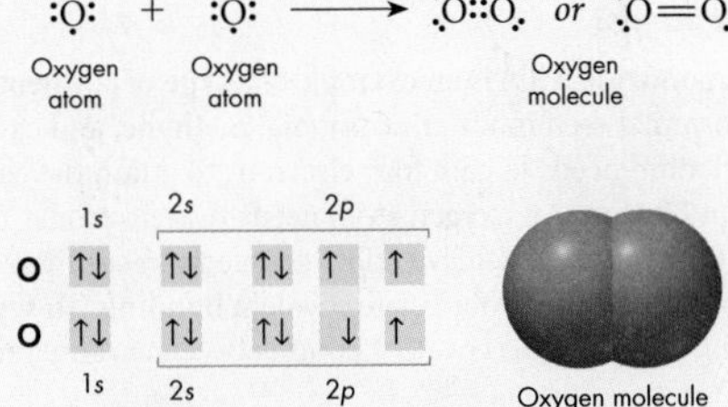

Oxygen molecule

In such an arrangement, all the electrons within the molecule would be paired. Experimental evidence, however, indicates that two of the electrons in O_2 are still unpaired. Thus, the bonding in the oxygen molecule (O_2) does not obey the octet rule. You cannot draw an electron dot structure that adequately describes the bonding in the oxygen molecule.

Nitrogen and oxygen are both diatomic molecules. Table 8.1 lists the properties and uses of these elements and some others that exist as diatomic molecules.

Table 8.1

The Diatomic Elements

Name	Chemical formula	Electron dot structure	Properties and uses
Fluorine	F_2	:F̤̈—F̤̈:	Greenish-yellow reactive toxic gas. Compounds of fluorine, a halogen, are added to drinking water and toothpaste to promote healthy teeth.
Chlorine	Cl_2	:C̤̈l—C̤̈l:	Greenish-yellow reactive toxic gas. Chlorine is a halogen used in household bleaching agents.
Bromine	Br_2	:B̤̈r—B̤̈r:	Dense red-brown liquid with pungent odor. Compounds of bromine, a halogen, are used in the preparation of photographic emulsions.
Iodine	I_2	:Ï̤—Ï̤:	Dense gray-black solid that produces purple vapors; a halogen. A solution of iodine in alcohol (tincture of iodine) is used as an antiseptic.
Hydrogen	H_2	H—H	Colorless, odorless, tasteless gas. Hydrogen is the lightest known element.
Nitrogen	N_2	:N≡N:	Colorless, odorless, tasteless gas. Air is almost 80% nitrogen by volume.
Oxygen	O_2	Inadequate	Colorless, odorless, tasteless gas that is vital for life. Air is about 20% oxygen by volume.

Explain

MAKE A CONNECTION Explain to the class that Earth's atmosphere is approximately 80 percent nitrogen gas. However, surprisingly few nitrogen compounds exist compared with the numerous compounds of oxygen, which constitutes only 20 percent of the atmosphere. Students may correctly surmise that the triple bond in N_2 is harder to break than a double bond and considerably harder to break than a single bond. Thus, N_2 is a stable molecule. Explain that plant and animal life depends on nitrogen, but in order to be usable to living systems, the element must be converted to a compound. Point out that this process is called nitrogen fixation. Tell students that nitrogen fixation occurs naturally when lightning provides the energy for atmospheric nitrogen to react with oxygen to form nitrogen oxides. The oxides dissolve in rain and fall to the ground where they can be utilized by plants. Nitrogen-fixing bacteria in the soil are also able to convert atmospheric nitrogen to usable compounds.

Differentiated Instruction

LPR **LESS PROFICIENT READERS** Divide students into groups of three or four. Have them draw electron dot formulas for oxygen and fluorine and then combine the drawings into the electron dot formula for OF_2. Have them do the same for SCl_2 and CCl_4. When students have mastered these, continue the exercise with $CHCl_3$, N_2H_4, and C_2H_6.

ELL **ENGLISH LANGUAGE LEARNERS** Provide a step-by-step process for creating electron dot structures. Include a step for checking that each element (except hydrogen) has four pairs of electrons around it. Have students draw a circle around each element that has a full octet. Relate the word *octet* to *octopus* and *octagon*, to help students remember that it refers to 8 electrons.

L3 **ADVANCED STUDENTS** Have students research or discover compounds with molecules that do not follow the octet rule. *(Sample answers: NO_2, PF_5)*

Explain

Coordinate Covalent Bonds

APPLY CONCEPTS Explain that a coordinate covalent bond is an exception to the rule that covalent bonding occurs between two half-empty orbitals of two atoms. Show students that in a coordinate covalent bond, one atom has an empty orbital and the other, an orbital filled with an electron pair that is not yet involved in a chemical bond. Note that the bonding still involves only one pair of electrons and one pair of orbitals, but one atom provides both of the shared electrons.

CRITICAL THINKING Ask students to compare the electron dot structure for ammonia, NH_3, and for a hydrogen ion, H^+. After forming the three nitrogen-to-hydrogen bonds, nitrogen has an unshared pair of electrons. The hydrogen ion has no electrons available for bonding. **Ask** How do you explain the existence of the ammonium ion, NH_4^+? *(The empty orbital of the H^+ must overlap the filled orbital in ammonia. The electrons from nitrogen are attracted to both the nitrogen nucleus and the hydrogen nucleus. A bond forms when electrons are simultaneously attracted to two nuclei. This type of bond is called a coordinate covalent bond.)*

Coordinate Covalent Bonds

How are coordinate covalent bonds different from other covalent bonds?

Carbon monoxide (CO) is an example of a type of covalent bonding different from that seen in water, ammonia, methane, and carbon dioxide. A carbon atom needs to gain four electrons to attain the electron configuration of neon. An oxygen atom needs two electrons. Yet it is possible for both atoms to achieve noble-gas electron configurations by a type of bonding called coordinate covalent bonding. To see how, begin by looking at the double covalent bond between carbon and oxygen.

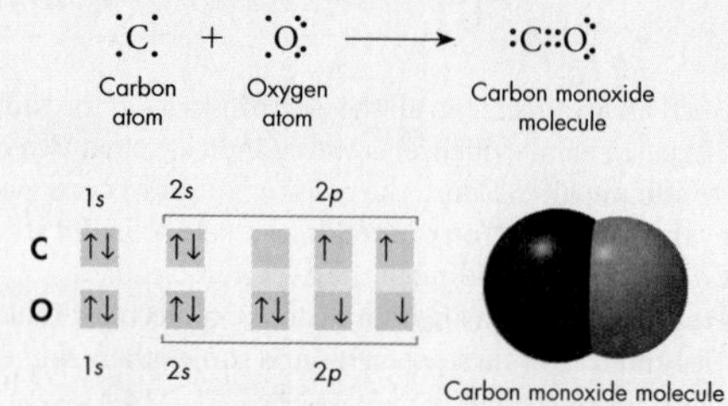

With the double bond in place, the oxygen has a stable configuration, but the carbon does not. As shown below, the dilemma is solved if the oxygen also donates one of its unshared pairs of electrons for bonding.

:C::O: ⟶ :C:::O:

Carbon monoxide molecule

Figure 8.9 Ammonia Fertilizers
Most plants need nitrogen that is already combined in a compound rather than molecular nitrogen (N_2) to grow. The polyatomic ammonium ion (NH_4^+), present in ammonium hydroxide, also called aqua ammonia, is an important component of fertilizer for field crops, home gardens, and potted plants.

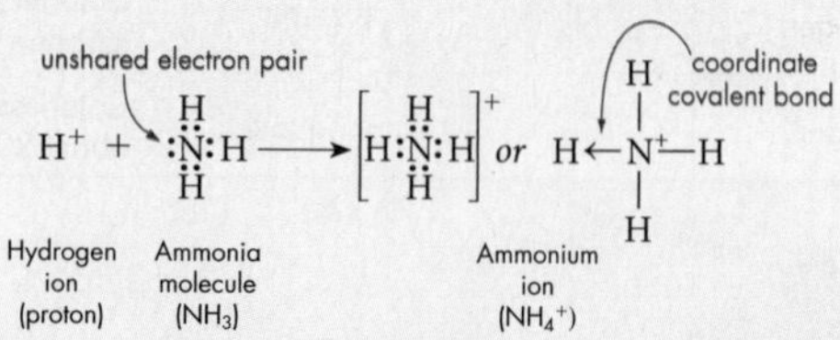

A covalent bond in which one atom contributes both bonding electrons is a **coordinate covalent bond.** In a structural formula, you can show coordinate covalent bonds as arrows that point from the atom donating the pair of electrons to the atom receiving them. The structural formula of carbon monoxide, with two covalent bonds and one coordinate covalent bond, is C⇐O. **In a coordinate covalent bond, the shared electron pair comes from one of the bonding atoms.** Once formed, a coordinate covalent bond is like any other covalent bond.

The ammonium ion (NH_4^+), which is often found in fertilizers like the one in Figure 8.9, consists of atoms joined by covalent bonds, including a coordinate covalent bond. A **polyatomic ion,** such as NH_4^+, is a tightly bound group of atoms that has a positive or negative charge and behaves as a unit. The ammonium ion forms when a positively charged hydrogen ion (H^+) attaches to the unshared electron pair of an ammonia molecule (NH_3).

unshared electron pair; coordinate covalent bond

H^+ + :N̈:H (with H above and below) ⟶ [H:N̈:H]$^+$ (with H above and below) *or* H←N^+—H (with H above and below)

Hydrogen ion (proton) | Ammonia molecule (NH_3) | Ammonium ion (NH_4^+)

UbD Check for Understanding

How are coordinate covalent bonds different from other covalent bonds?

Assess students' understanding of coordinate covalent bonds by having them write a one-minute response that explains the bonding in CO_2 and then explains the bonding in CO using coordinate covalent bonding.

ADJUST INSTRUCTION If students are having difficulty understanding the difference between normal covalent bonding and coordinate covalent bonding, construct a model of each element. Then construct a model of each compound to give students a visual image of the difference in how the electrons are shared.

Most polyatomic cations and anions contain covalent and coordinate covalent bonds. Therefore, compounds containing polyatomic ions include both ionic and covalent bonding. As another example, draw the electron dot structure of the polyatomic ion SO_3^{2-}. First, draw the electron dot structures for the oxygen and sulfur atoms, and the two extra electrons indicated by the charge. Then, join two of the oxygens to sulfur by single covalent bonds.

:Ö· ·S̤: ·Ö: + ·· ⟶ :Ö:S̤: ·Ö:
·Ö: ·Ö:

Next, join the remaining oxygen by a coordinate covalent bond, with sulfur donating one of its unshared pairs to oxygen, and add the two extra electrons. Put brackets around the structure and indicate the 2− charge.

$$:\ddot{O}:\ddot{S}:\ddot{O}: + \cdot\cdot \longrightarrow \left[:\ddot{O}:\ddot{S}:\ddot{O}: \; :\ddot{O}:\right]^{2-}$$

Each of the atoms now has eight valence electrons, satisfying the octet rule. Without the extra electrons, two of the oxygens would be electron-deficient.

Table 8.2 lists electron dot structures of some common compounds with covalent bonds.

Remember, the charge of a negative polyatomic ion is equal to the number of electrons that are in addition to the valence electrons of the atoms present. Since a negatively charged polyatomic ion is part of an ionic compound, the positive charge of the cation of the compound balances the additional electrons.

Sample Problem 8.2

Drawing the Electron Dot Structure of a Polyatomic Ion

The H_3O^+ ion forms when a hydrogen ion is attracted to an unshared electron pair in a water molecule. Draw the electron dot structure of the hydronium ion.

1 Analyze **Identify the relevant concepts.** Each atom must share electrons to satisfy the octet rule.

2 Solve **Apply the concepts to the problem.**

Remember to always include the charge when drawing electron dot structures of polyatomic ions.

Draw the electron dot structure of the water molecule and the hydrogen ion. Then, draw the electron dot structure of the hydronium ion. The oxygen must share a pair of electrons with the added hydrogen ion to form a coordinate covalent bond.

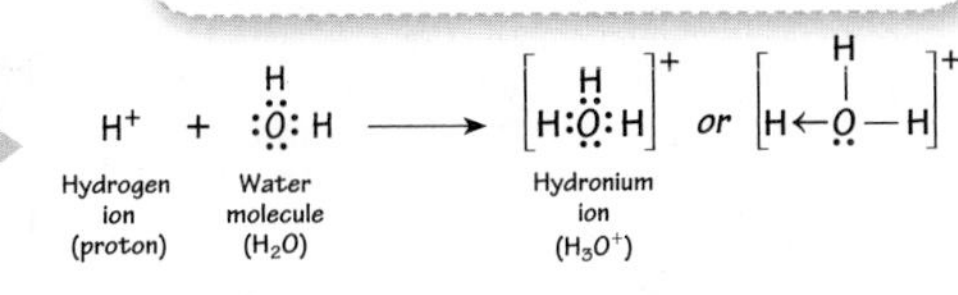

Check that all the atoms have the electrons they need and that the charge is correct.

The oxygen in the hydronium ion has eight valence electrons, and each hydrogen shares two valence electrons, satisfying the octet rule. The water molecule is neutral, and the hydrogen ion has a positive charge, giving the hydronium ion a charge of 1+.

9. Draw the electron dot structure of the hydroxide ion (OH^-).

10. Draw the electron dot structures for sulfate (SO_4^{2-}) and carbonate (CO_3^{2-}). Sulfur and carbon are the central atoms, respectively.

APPLY CONCEPTS Have students write the electron dot structure for SO_2. Emphasize that the structure should satisfy the bonding requirements of all three atoms. Students should find that, to satisfy the octet rule for all the atoms, they must write a structure in which one oxygen atom is double bonded to sulfur. The other oxygen is single bonded by a coordinate covalent bond in which the electrons are donated by sulfur. Point out that experimental evidence indicates that both sulfur-oxygen bonds are identical. Explain that this evidence indicates that the bonding in SO_2 must be some intermediate between a single and double bond. **Ask** How does the formation of a coordinate covalent bond differ from that of a covalent bond? *(In a covalent bond, each atom provides one electon. In a coordinate covalent bond, both electrons are provided by the same atom.)*

Sample Practice Problem

Draw the electron dot structure of the polyatomic chlorate anion, ClO_3^-.

$$\left[:\ddot{O}:\ddot{Cl}:\ddot{O}: \; :\ddot{O}:\right]^-$$

Foundations For Math

PATTERNS Students may start to realize that an element's electron configuration and its location in the periodic table have a useful relationship. Guide students to realize that the number of bonds typically formed by atoms in a covalent compound is equal to the atom's number of valence electrons that the atom must gain in order to achieve an octet.

In **Sample Problem 8.2,** as well as in other examples in this lesson, the number of valence electrons equals the group number for the Group A elements. Have students note this pattern on their periodic table and encourage students to continue to look for these types of patterns as they work through this chapter.

Answers

9. $[H:\ddot{O}:]^-$

10. a. $\left[:\ddot{O}: \; :\ddot{O}:\ddot{S}:\ddot{O}: \; :\ddot{O}:\right]^{2-}$

b. $\left[:\ddot{O}: \; :\ddot{O}:C \; \ddot{O}\right]^{2-}$

LESSON 8.2

LESSON 8.2

Explain

Misconception Alert

Students may mistakenly infer that the atoms that make up polyatomic ions are bonded ionically because polyatomic ions are charged species. However, most polyatomic cations and anions contain covalent and coordinate covalent bonds.

Explore

Teacher Demo

PURPOSE Students observe the formation of nitrogen dioxide and write its resonance structures.

MATERIALS small piece of copper metal, concentrated nitric acid, evaporating dish, fume hood

PROCEDURE Wear safety goggles, gloves, and lab apron. Perform the experiment under an efficient hood.

PROCEDURE Place a small piece of copper in an evaporating dish and cover it with concentrated HNO_3. While students observe the reaction, write the balanced equation for the reaction on the board.

$Cu + 4HNO_3 \rightarrow Cu(NO_3)_2 + 2H_2O + 2NO_2 + \text{energy}$

Tell students that NO_2 is one of the pollutants in automobile exhaust. It gives smog its reddish-brown color and is very reactive and poisonous. Have students try to write the electron dot structure for NO_2. They will be unable to find a way to arrange the 17 electrons around the central nitrogen and the two oxygen atoms so that the octet rule is obeyed. Refer students to the two resonance structures on this page, each with an unpaired electron on the nitrogen atom. Ask them to draw another plausible resonance structure. *(They could draw a structure with the unpaired electron on an oxygen atom.)*

EXPECTED OUTCOMES A reddish-brown gas is produced.

Table 8.2

Some Common Molecular Compounds

Name	Chemical formula	Structure	Properties and uses
Carbon monoxide	CO	:C≡O:	Colorless, highly toxic gas. It is a major air pollutant present in cigarette smoke and exhaust.
Carbon dioxide	CO_2	:Ö=C=Ö:	Colorless unreactive gas. This normal component of the atmosphere is exhaled in the breath of animals and is essential for plant growth.
Hydrogen peroxide	H_2O_2	H—Ö—Ö—H	Colorless, unstable liquid when pure. It is used as rocket fuel. A 3% solution is used as a bleach and antiseptic.
Sulfur dioxide	SO_2	Ö=S̈→Ö:	Oxides of sulfur are produced in the combustion of petroleum products and coal. They are major air pollutants in industrial areas. Oxides of sulfur can lead to respiratory problems.
Sulfur trioxide	SO_3	Ö=S(→Ö:)→Ö:	
Nitric oxide*	NO	Ö=N̈·	Oxides of nitrogen are major air pollutants produced by the combustion of fossil fuels in automobile engines. They irritate the eyes, throat, and lungs. Nitrogen dioxide, a dark-brown gas, readily converts to colorless dinitrogen tetroxide. Dinitrogen tetroxide is used as a rocket fuel.
Nitrogen dioxide*	NO_2	Ö=Ṅ→Ö:	
Dinitrogen tetroxide	N_2O_4	(Ö=)(Ö←)N—N(→Ö)(=Ö)	
Nitrous oxide	N_2O	:Ö←N≡N:	Colorless, sweet-smelling gas. It is used as an anesthetic commonly called laughing gas.
Hydrogen cyanide	HCN	H—C≡N:	Colorless, toxic gas with the smell of almonds.
Hydrogen fluoride	HF	H—F̤̈:	Four hydrogen halides, all extremely soluble in water. Hydrogen chloride, a colorless gas with a pungent odor, readily dissolves in water to give a solution called hydrochloric acid.
Hydrogen chloride	HCl	H—C̤̈l:	
Hydrogen bromide	HBr	H—B̤̈r:	
Hydrogen iodide	HI	H—Ï:	

*Does not obey the octet rule

Free Radical

Nitrogen dioxide has an odd number of valence electrons, 17, so one electron must be unpaired. Molecules with unpaired electrons are called free radicals and tend to be reactive. Two resonance structures for NO_2 appear on page 235. These have the unpaired electron on the nitrogen atom. Two other possible structures place the unpaired electron on oxygen atoms. When free radicals interact, they share their unpaired electrons and create a dimer. Thus, dinitrogen tetroxide, N_2O_4 consists of two nitrogen dioxide molecules joined by a N-N covalent bond. The bond is temperature dependent; at higher temperatures the bond breaks as demonstrated in Figure 18.17 on page 616.

Exceptions to the Octet Rule

What are some exceptions to the octet rule?

The octet rule provides guidance for drawing electron dot structures. For some molecules or ions, however, it is impossible to draw structures that satisfy the octet rule. **The octet rule cannot be satisfied in molecules whose total number of valence electrons is an odd number. There are also molecules in which an atom has less, or more, than a complete octet of valence electrons.** The nitrogen dioxide (NO_2) molecule, for example, contains a total of seventeen, an odd number, of valence electrons. Each oxygen contributes six electrons and the nitrogen contributes five. Two plausible electron dot structures can be drawn for the NO_2 molecule.

```
:Ö=Ṅ—Ö:
:Ö—Ṅ=Ö:
```

Nitrogen dioxide molecule

An unpaired electron is present in each of these structures, both of which fail to follow the octet rule. It is impossible to draw an electron dot structure for NO_2 that satisfies the octet rule for all atoms. Yet, NO_2 does exist as a stable molecule. In fact, it is produced naturally by lightning strikes of the sort shown in Figure 8.10.

A number of other molecules also have an odd number of electrons. In these molecules, as in NO_2, complete pairing of electrons is not possible. It is not possible to draw an electron dot structure that satisfies the octet rule. Examples of such molecules include chlorine dioxide (ClO_2) and nitric oxide (NO).

Several molecules with an even number of valence electrons, such as some compounds of boron, also fail to follow the octet rule. This outcome may occur because an atom acquires less than an octet of eight electrons. The boron atom in boron trifluoride (BF_3), for example, is deficient by two electrons and, therefore, is an exception to the octet rule. Boron trifluoride readily reacts with ammonia to make the compound $BF_3 \cdot NH_3$. In doing so, the boron atom accepts the unshared electron pair from ammonia and completes the octet.

```
   :F:     H              :F: H
    |      |               |  |
:F—B  +  :N—H  ——→  :F—B←N—H
    |      |               |  |
   :F:     H              :F: H
```

Figure 8.10 Nitrogen Dioxide
Lightning is one means by which nitrogen and oxygen in the atmosphere produce nitrogen dioxide.

Covalent Bonding 235

LESSON 8.2

Teacher Demo

PURPOSE Students observe two reactions in which bonds are formed and compare ionic and covalent bond energies.

MATERIALS 10- to 15-cm piece of clean magnesium ribbon, a small piece of charcoal, tongs, Bunsen burner, filters for viewing

SAFETY Wear safety goggles and lab apron. Tie back long hair and loose clothing.

PROCEDURE Using tongs, hold the piece of magnesium in a Bunsen burner flame. CAUTION! Tie back long hair and loose clothing and wear safety goggles. Do not look directly at the flame. Observe through filters. Discuss with students the large amount of heat and light given off in the formation of MgO. Write the balanced equation on the board.

$$2Mg + O_2 \rightarrow 2MgO + \text{energy}$$

Using tongs, place a small piece of charcoal in the Bunsen burner flame and try to ignite it. Write the balanced chemical equation.

$$C + O_2 \rightarrow CO_2 + \text{energy}$$

EXPECTED OUTCOMES Students note that much less energy is given off in forming CO_2 than in forming MgO. **Ask** What kind of bonds are in MgO? *(ionic)* What kind of bonds are in CO_2? *(covalent)* Ionic bond energies are, in general, greater than covalent bond energies.

Paramagnetism

Substances containing unpaired electrons can be identified through a phenomenon called paramagnetism. When molecules with unpaired electrons are placed in a magnetic field, they tend to be drawn into the field. These substances are paramagnetic. In contrast, molecules in which all electrons are paired tend to be pushed from a magnetic field. These substances are diamagnetic.

Paramagnetism differs from ferromagnetism, which is the familiar attraction of metals such as iron, cobalt, and nickel to a magnetic field. The property of paramagnetism is the evidence that shows that the oxygen molecule has unpaired electrons and thus cannot be described exactly by application of the octet rule. However, the oxygen bond does have a double bond character; its bond length and bond energy are similar to those of double bonds in other molecules that conform to the octet rule.

Explain

Bond Dissociation Energies

SUMMARIZE Direct students' attention to Table 8.3. Explain the units used to measure bond energy and bond length. Have students write a summary statement using any patterns they notice between bonds, bond energy, and bond length.

Extend

Have students draw a graph of actual bond dissociation energies versus number of bonds for carbon-carbon single, double, and triple bonds. Have students write a statement that relates this graph to their conclusions in the lab activity.

Connect to PHYSICS

Have students investigate the relationship between potential energy and bond length. Students should be prepared to state the relationship orally in class as well as demonstrate their findings using two magnets. *(Bond length is the measure at which the potential energy of the combined atoms is at its lowest.)*

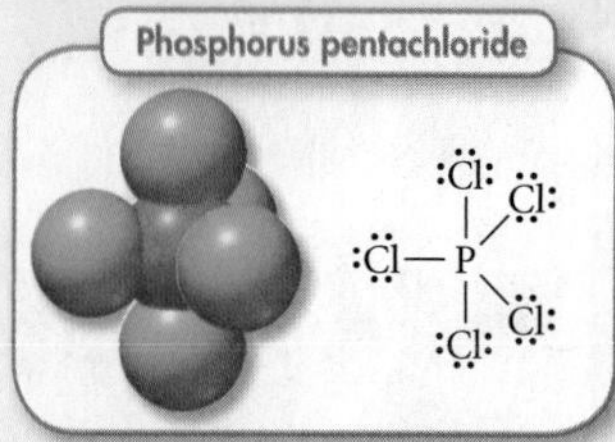

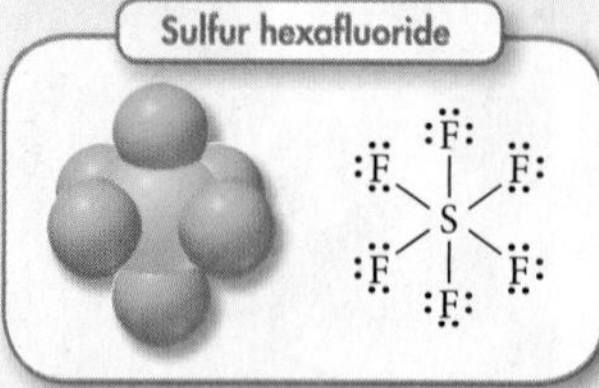

Figure 8.11
Exceptions to the Octet Rule
Phosphorus pentachloride and sulfur hexafluoride, are exceptions to the octet rule.
Interpret Diagrams
How many valence electrons does the sulfur in sulfur hexafluoride (SF_6) have for the structure shown in the figure?

A few atoms, especially phosphorus and sulfur, sometimes expand the octet to ten or twelve electrons. Consider phosphorus trichloride (PCl_3) and phosphorus pentachloride (PCl_5). Both are stable compounds in which all of the chlorine atoms are bonded to a single phosphorus atom. Covalent bonding in PCl_3 follows the octet rule because all the atoms have eight valence electrons. However, as shown in Figure 8.11, the electron dot structure for PCl_5 can be written so that phosphorus has ten valence electrons. The octet is also expanded in sulfur hexafluoride (SF_6). The electron dot structure for SF_6 can be written so that sulfur has twelve valence electrons.

Bond Dissociation Energies

How is the strength of a covalent bond related to its bond dissociation energy?

A large quantity of heat is released when hydrogen atoms combine to form hydrogen molecules. This release of heat suggests that the product is more stable than the reactants. The covalent bond in the hydrogen molecule (H_2) is so strong that it would take 435 kJ of energy to break apart all of the bonds in 1 mole (6.02×10^{23} bonds or about 2 grams) of H_2. (You will study the mole, abbreviated mol, in Chapter 12.) The energy required to break the bond between two covalently bonded atoms is the **bond dissociation energy.** The units for this energy are often given in kJ/mol, which is the energy needed to break one mole of bonds. For example, the bond dissociation energy for the H_2 molecule is 435 kJ/mol.

Table 8.3

Bond Dissociation Energies and Bond Lengths for Covalent Bonds

Bond	Bond dissociation energy (kJ/mol)	Bond length (pm)
H—H	435	74
C—H	393	109
C—O	356	143
C=O	736	121
C≡O	1074	113
C—C	347	154
C=C	657	133
C≡C	908	121
C—N	305	147
Cl—Cl	243	199
N—N	209	140
O—H	464	96
O—O	142	132

UbD Check for Understanding

How is the strength of a covalent bond related to its bond dissociation energy?

Assess students' understanding of covalent bond strength by providing students with a copy of Table 8.3. Have students write down the three strongest covalent bonds in the table and the three weakest. *(Strongest: C triple bond to O; C triple bond to C; C double bond to O; Weakest: O single bond to O; N single bond to N; Cl single bond to Cl)*

ADJUST INSTRUCTION If students are struggling to understand this concept, have students create a new table sorted in order of greatest to least bond energy. As a class, discuss any noticeable patterns.

A large bond dissociation energy corresponds to a strong covalent bond. A typical carbon-carbon single bond has a bond dissociation energy of 347 kJ/mol. Typical carbon-carbon double and triple bonds have bond dissociation energies of 657 kJ/mol and 908 kJ/mol, respectively. Strong carbon-carbon bonds help explain the stability of carbon compounds. Compounds with only C—C and C—H single covalent bonds, such as methane, tend to be quite unreactive. They are unreactive partly because the dissociation energy for each of these bonds is high. Bond dissociation energies of some common bonds are shown in Table 8.3.

Q: *What is the difference between the oxygen you breathe and the oxygen in ozone in the atmosphere?*

Resonance

How are resonance structures used?

Ozone in the upper atmosphere blocks harmful ultraviolet radiation from the sun. At lower elevations, as shown in Figure 8.12, it contributes to smog. The ozone molecule has two possible electron dot structures. Notice that the structure on the left can be converted to the one on the right by shifting electron pairs without changing the positions of the oxygen atoms.

:Ö:Ö::O: ⟷ :O::Ö:Ö:

As drawn, these electron dot structures suggest that the bonding in ozone consists of one single coordinate covalent bond and one double covalent bond. Because earlier chemists imagined that the electron pairs rapidly flip back and forth, or resonate, between the different electron dot structures, they used double-headed arrows to indicate that two or more structures are in resonance.

Double covalent bonds are usually shorter than single covalent bonds, so it was believed that the bond lengths in ozone were unequal. Experimental measurements show, however, that this is not the case. The two bonds in ozone are the same length. This result can be explained if you assume that the actual bonding in the ozone molecule is the average of the two electron dot structures. The electron pairs do not actually resonate back and forth. The actual bonding is a hybrid, or mixture, of the extremes represented by the resonance forms.

The two electron dot structures for ozone are examples of what are still referred to as resonance structures. **Resonance structures** are structures that occur when it is possible to draw two or more valid electron dot structures that have the same number of electron pairs for a molecule or ion. **Chemists use resonance structures to envision the bonding in molecules that cannot be adequately described by a single structural formula.** Although no back-and-forth changes occur, double-headed arrows are used to connect resonance structures.

Learn more about oxygen *in the air* ***online.*** CONCEPTS IN ACTION

Figure 8.12 Ozone Smog
Although ozone high above the ground forms a protective layer that absorbs ultraviolet radiation from the sun, at lower elevations ozone is a pollutant that contributes to smog. The smog shown here in Los Angeles, California, makes it difficult to see the city skyline.

Focus on ELL

4 LANGUAGE PRODUCTION Have students work in groups of four to complete the lab. Make sure each group has ELLs of varied language proficiencies, so that more proficient students can help less proficient ones. Have students work according to their proficiency level.

BEGINNING: LOW/HIGH Model this lab and use student volunteers to help with each step. For example, ask one student to take the measurements, another student to record the measurements, etc.

INTERMEDIATE: LOW/HIGH Add a picture to each step of the lab that will demonstrate the procedure. Provide a chart for recording data and labeled graph paper for the graph.

ADVANCED: LOW/HIGH Ask students to compare and contrast in writing the stretching of rubber bands and the dissociation energy of covalent bonds.

Explain

Resonance

START A CONVERSATION Although the electron dot structures of O_2 on page 229 show that all electrons are paired, experimental evidence reveals that two electrons are unpaired. **Ask** Would you consider oxygen a reactive gas? *(yes)* What could account for oxygen's reactivity? **Ask** *(Unpaired electrons are reactive.)* After a brief discussion of the other exceptions to the octet rule that are presented in this section, remind students that many molecules do follow the octet rule.

SUMMARIZE Have students complete the bulleted list summarizing the types of situations that may have exceptions to the octet rule.

- molecules with an ___ number of electrons *(odd)*
- molecules where one atom has more or less than an ____ *(octet)*

CHEMISTRY & YOU The oxygen you breathe is O_2, while ozone is O_3.

Evaluate

Informal Assessment

Challenge students to write electron dot structures for SBr_2 and C_3H_8. Have students work in pairs to determine whether or not their structures are correct, and if not, how to correct their errors. Then, have students explain coordinate covalent bonding in their own words. *(A coordinate covalent bond is formed when one atom contributes both bonding electrons.)* **Ask** When do resonance structures occur? *(Resonance structures occur when two or more valid electron dot structures can be written for a molecule.)* Then have students complete the 8.2 Lesson Check.

Reteach

As a class, create a concept map interlinking each concept covered in the lesson. Have students use their notes as well as the textbook to add as much detail as possible. Creating a visual relationship of the concepts may help solidify students' understanding.

Answers

INTERPRET DIAGRAMS 12 valence electrons

LESSON 8.2

Explore

Quick Lab

OBJECTIVE After completing this activity, students will understand that the dissociation energy of a covalent bond increases in order from single bond to double bond to triple bond.

SKILLS FOCUS Solving, interpreting

PREP TIME 10 minutes

CLASS TIME 30 minutes

MATERIALS Obtain the necessary materials in advance. You may wish to ask students to help with this task by bringing cans of food, coat hangers, and plastic grocery bags to class.

TEACHING TIP Use tape to secure the hanger in place if necessary. Have groups of students record the data and then create a graph as a class, using an overhead projector.

EXPECTED OUTCOMES As the mass of the load increases, the stretch of the rubber band or bands increases. For a given mass, a single rubber band stretches farther than a double, and a double rubber band stretches farther than a triple.

ANALYZE AND CONCLUDE

1. Triple covalent bonds are stronger than double covalent bonds, which are stronger than single covalent bonds.
2. The change in bond dissociation energies in going from a carbon-carbon single bond to a carbon-carbon double bond to a carbon-carbon triple bond is nearly constant. The change in length of one, two, and three rubber bands, as given by the slopes of the lines, is not constant. It is large going from one to two rubber bands and small going from two to three rubber bands.

Quick Lab

Strengths of Covalent Bonds

Purpose To compare and contrast the stretching of rubber bands and the dissociation energy of covalent bonds

Materials
- 1 170-g (6-oz) can of food
- 2 454-g (16-oz) cans of food
- 3 No. 25 rubber bands
- metric ruler
- coat hanger
- plastic grocery bag
- paper clip
- graph paper
- motion detector (optional)

Procedure

1. Bend the coat hanger to fit over the top of a door. The hook should hang down on one side of the door. Measure the length of the rubber bands (in cm). Hang a rubber band on the hook created by the coat hanger.
2. Place the 170-g can in the plastic bag. Use the paper clip to fasten the bag to the end of the rubber band. Lower the bag gently until it is suspended from the end of the rubber band. Measure and record the length of the stretched rubber band. Using different combinations of food cans, repeat this process three times with the following masses: 454 g, 624 g, and 908 g.
3. Repeat Step 2, first using two rubber bands to connect the hanger and the paper clip, and then using three.
4. Graph the length difference: (stretched rubber band) − (unstretched rubber band) on the y-axis versus mass (kg) on the x-axis for one, two, and three rubber bands. Draw the straight line that you estimate best fits the points for each set of data. (Your graph should have three separate lines.) The x-axis and y-axis intercepts of the lines should pass through zero, and the lines should extend past 1 kg on the x-axis. Determine the slope of each line in cm/kg.

Analyze and Conclude

1. **Analyze Experimental Results** Assuming the rubber bands are models for covalent bonds, what can you conclude about the relative strengths of single, double, and triple bonds?
2. **Evaluate** How does the behavior of the rubber bands differ from that of covalent bonds?

ONLINE PROBLEMS 8.2 LessonCheck

11. **Identify** What electron configurations do atoms usually achieve by sharing electrons to form covalent bonds?
12. **Compare** How is a coordinate covalent bond different from other covalent bonds?
13. **List** List three ways in which the octet rule can sometimes fail to be obeyed.
14. **Explain** How is the strength of a covalent bond related to its bond dissociation energy?
15. **Identify** How are resonance structures used?
16. **Explain** How is an electron dot structure used to represent a covalent bond?
17. **Infer** When are two atoms likely to form a double bond between them? A triple bond?
18. **Identify** What kinds of information does a structural formula reveal about the compound it represents?
19. **Compare** Use the bond dissociation energies of H_2 and of a typical carbon–carbon bond to decide which bond is stronger. Explain your reasoning.
20. **Use Models** Draw electron dot structures for the following molecules, which have only single covalent bonds:
    a. H_2S b. PH_3 c. ClF

Lesson Check Answers

11. the configurations of noble gases
12. The shared electron pair comes from one of the bonding atoms. In other covalent bonds, each bonding atom provides an electron.
13. The octet rule cannot be satisfied in molecules whose total number of valence electrons is an odd number. There are also molecules in which an atom has fewer, or more, than a complete octet of valence electrons.
14. A large bond dissociation energy corresponds to a strong covalent bond.
15. They are used to envision the bonding in molecules that cannot be adequately described by a single structural formula.
16. Two dots represent each covalent bond.
17. when they can attain a noble gas structure by sharing two pairs or three pairs of electrons
18. the arrangement of atoms in a molecule
19. The H-H bond is stronger because it has a greater dissociation energy.
20. a. H:S̈:H
    b. H:P̈:H with H below P
    c. :C̈l:F̈:

Powder Coating

Have you ever admired a new car with its glossy, smooth paint? Car manufacturers use a special process to apply paint to a car. This process is called powder coating or electrostatic spray painting.

In powder coating, a custom-designed spray nozzle wired up to an electric power supply imparts a negative charge to the paint droplets as they exit the spray gun. The negatively charged droplets are attracted to the grounded, positively-charged, metal surface. Painting with attractive forces is very efficient, because almost all the paint is applied to the car body and very little is wasted.

Powder coating isn't just for cars. The process has many different applications, including the painting of motorcycles, outdoor furniture, exercise equipment, office furniture, and metal fencing.

An eye-catching paint finish isn't the only benefit of powder coating, however. This process is also environmentally friendly. Since the paint is actually attracted to its intended surface, the amount of wasted paint is much lower compared to traditional spray painting. Also, the amount of toxic volatile organic compounds (VOCs) released is minimal, if there are any at all.

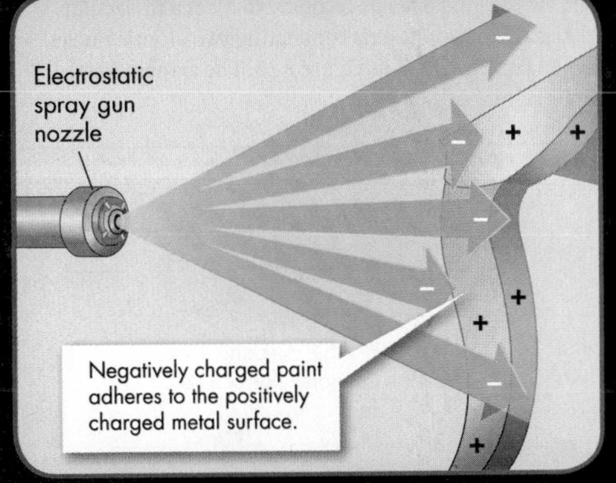

ATTRACTIVE PAINT The paint almost wraps around the metal, sticking to any available charged surface.

APPLYING THE POWDER This worker is using an electrostatic spray gun to apply powder to the metal. Any powder that does not stick to the part can be collected and reused. Once the powder is applied, the part is baked in an oven to cure the paint.

GLOSSY FINISH Powder coating can produce a smooth, glossy paint finish.

Take It Further

1. **Analyze Benefits** Powder coating is being used for more and more applications, partly because of its many benefits. Research other advantages of powder coating that are not mentioned here.
2. **Infer** Powder coating results in a smooth surface, usually without drips and runs. Given what you have learned about attractive forces, why do you think drips and runs are avoided during powder coating?

CHEMISTRY & YOU The automotive industry is an example of an industry that is strongly swayed by consumer concerns.

Pose the following question to students: *What are some likely consumer concerns that have had or could potentially have an effect on the types of technological advancements in the automotive industry, and how did the automotive industry respond to those concerns?*

You may need to guide students in the following ways:

- Safety developments include shoulder restraints, front and side curtain airbags, reinforced frames, anti-lock brakes, enhanced traction systems, in-trunk release latches, xenon headlamps, and LED tail lamps.
- Fuel efficiency developments include computer controlled engines, fuel injection, automated manual and continually variable transmissions, reductions in gross vehicular weight
- Emissions regulation developments include catalytic converters, low carbon fuels (ethanol and CNG), hybrid engines, exhaust gas recirculation, and evaporative emission systems.

Explain

START A CONVERSATION Direct students to read about the powder coating process highlighted in this feature and to examine the photographs. **Ask** What is the basic underlying chemistry-related principle applied in the powder coating process? *(Opposite charges are inherently attracted to each other.)* **Ask** Do you know of other types of coating processes used in other industries? *(Sample answers: electroplating, galvanizing, Parkerizing, bluing, painting, etc.)* **Ask** What do you think the benefits are for using powder coating over one of these other processes? *(Sample answers: durability of the surface, even coating, longevity, avoiding corrosion issues due to weather, etc.)*

21st Century Learning To be successful in the 21st century, students need skills and learning experiences that extend beyond subject matter mastery. The following project helps students build the following 21st Century Skills: *Financial, Economic, Business, and Entrepreneurial Literacy*; *Creativity and Innovation*; *Communication and Collaboration*; *Information, Media, and Technology Literacy*; *Initiative and Self-Direction*; and *Productivity and Accountability*.

CLEANER COATINGS Tell students that due to the social and political viewpoint towards ecological awareness, controlling emissions and waste is an ever increasing concern in all areas of manufacturing. Pose the following challenge to students: *As members of a powder coating company's sales department, it is your job to convince a major plastics manufacturer to replace their liquid coating process with powder coating.* Create "marketing teams" of 4–6 people to develop a multimedia marketing campaign designed to convince the plastics manufacturer of the benefits of switching to powder coating. It should include a comparison of current powder coating and liquid coating processes from the standpoint of environmental impact and worker safety. Record the campaign on DVD for submission.

Answers

1. Answers may vary, but possible benefits include the ability to recover unused or over-sprayed powder, and the production of a strong, scratch-resistant paint finish.
2. Since the paint is attracted to the metal, excess paint that might normally form drips is instead attracted to metal that is not yet covered in paint.

LESSON 8.3

Key Objectives

8.3.1 DESCRIBE the relationship between atomic and molecular orbitals.

8.3.2 DESCRIBE how VSEPR theory helps predict the shapes of molecules.

8.3.3 IDENTIFY the ways in which orbital hybridization is useful in describing molecules.

Additional Resources

- Reading and Study Workbook, Lesson 8.3
- Core Teaching Resources, Lesson 8.3 Review, Interpreting Graphics
- Laboratory Manual, Lab 11
- Small-Scale Chemistry Laboratory Manual, Lab 11

PearsonChem.com

Engage

CHEMISTRY & YOU Have students study the photograph and read the opening text. Explain that topographic maps are usually drawn as a series of lines tracing the elevations of a terrain. The more closely spaced the lines are, the steeper is the terrain. Similarly, scientists can use X-ray diffraction data to map the electron density at any position in a crystal of a particular molecule. Where the lines are closest in the map, the electron density of the molecule is greatest.

Build Background

Have students recall what they know about hybrid organisms or cars. **Ask** What does the term *hybrid* mean to you? *(Sample answer: A hybrid has a combination of characteristics of two or more different organisms or objects, but is different from any of the originals.)* Tell students to keep this in mind as they proceed with the lesson.

IN ACADEMIC STANDARDS for SCIENCE

C.3.1 Describe, compare, and contrast the characteristics of the interactions between atoms in ionic and covalent compounds.

C.3.2 Compare and contrast how ionic and covalent compounds form.

C.3.3 Compare and contrast ionic, covalent network, metallic and polar and non-polar molecular crystals with respect to constituent particles, strength of bonds, melting and boiling points, and conductivity; provide examples of each type.

C.3.4 Draw structural formulas for and name simple molecules.

8.3 Bonding Theories

C.3.1 Interactions between atoms in ionic and covalent compounds; **C.3.2** Formation of ionic and covalent compounds; **C.3.3** Chemical bonds and intermolecular attractions; **C.3.4** Structural formulas

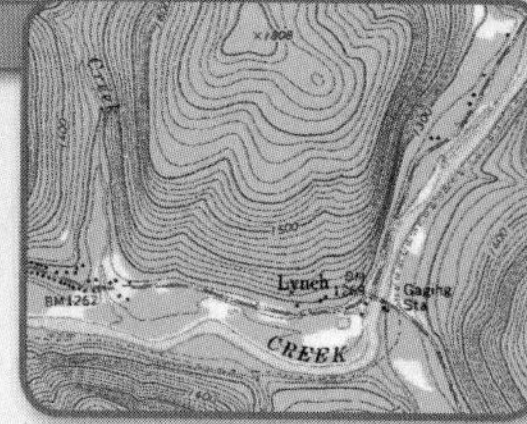

CHEMISTRY & YOU

Q: *How can you predict where an electron is most likely to be found in a molecule?* If you ever go hiking in a hilly area, you might see a topographic map like the one shown here. The lines on a topographic map show you where elevations change. In this lesson, you will learn how to interpret electron "maps" that show where you are most likely to find electrons.

Key Questions

How are atomic and molecular orbitals related?

What do scientists use the VSEPR theory for?

In what ways is orbital hybridization useful in describing molecules?

Vocabulary

- molecular orbital
- bonding orbital
- sigma bond
- pi bond
- tetrahedral angle
- VSEPR theory
- hybridization

Molecular Orbitals

How are atomic and molecular orbitals related?

The model you have been using for covalent bonding assumes the orbitals are those of the individual atoms. There is a quantum mechanical model of bonding, however, that describes the electrons in molecules using orbitals that exist only for groupings of atoms. When two atoms combine, this model assumes that their atomic orbitals overlap to produce **molecular orbitals,** or orbitals that apply to the entire molecule.

In some ways, atomic orbitals and molecular orbitals are similar. **Just as an atomic orbital belongs to a particular atom, a molecular orbital belongs to a molecule as a whole.** Each atomic orbital is filled if it contains two electrons. Similarly, two electrons are required to fill a molecular orbital. A molecular orbital that can be occupied by two electrons of a covalent bond is called a **bonding orbital.**

Sigma Bonds When two atomic orbitals combine to form a molecular orbital that is symmetrical around the axis connecting two atomic nuclei, a **sigma bond** is formed, as illustrated in Figure 8.13. The symbol for this bond is the Greek letter sigma (σ).

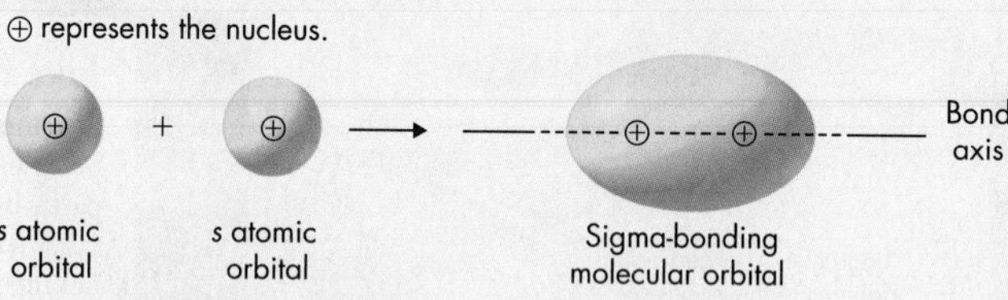

Figure 8.13 *s* Orbital Sigma Bonds
Two *s* atomic orbitals can combine to form a molecular orbital, as in the case of hydrogen (H_2). In a bonding molecular orbital, the electron density between the nuclei is high.

Focus on ELL

1 CONTENT AND LANGUAGE Write the words *axis*, *symmetrical*, and *repulsion* on the board. Define each term and have students add these words to their vocabulary notebooks. Have students identify other terms in the lesson with which they need assistance. Define these terms as needed.

2 FRONTLOAD THE LESSON Blow up and tie off nine balloons. Tie the balloons together with string to form three clusters: two balloons, three balloons, and four balloons. Explain that each balloon represents an electron pair around a central atom. Read aloud the Essential Question, *How do electrons affect the shape of molecule?* Have each group predict the answer based on what they have observed.

3 COMPREHENSIBLE INPUT Show students computerized or ball-and-stick versions of the molecules in Figure 8.18. Rotate the models so students can fully appreciate that the models are three-dimensional.

⊕ represents the nucleus.

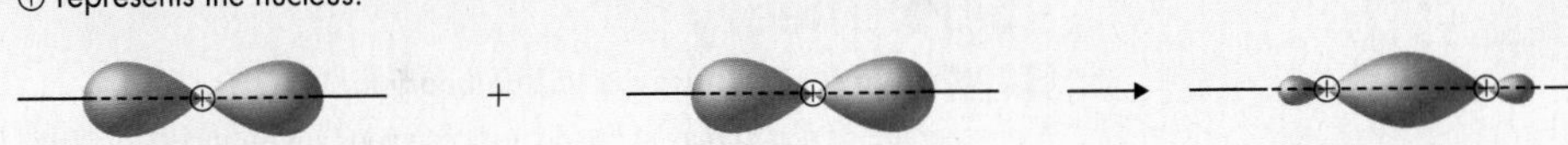

p atomic orbital + p atomic orbital → Sigma-bonding molecular orbital

Figure 8.14 *p* Orbital Sigma Bonds
Two *p* atomic orbitals can combine to form a sigma-bonding molecular orbital, as in the case of fluorine (F_2). Notice that the sigma bond is symmetrical around the bond axis connecting the nuclei.

In general, covalent bonding results from an imbalance between the attractions and repulsions of the nuclei and electrons involved. Because their charges have opposite signs, the nuclei and electrons attract each other. Conversely, nuclei repel other nuclei and electrons repel other electrons because their charges have the same sign. In a hydrogen molecule, the nuclei repel each other, as do the electrons. In a bonding molecular orbital of hydrogen, however, the attractions between the hydrogen nuclei and the electrons are stronger than the repulsions. The balance of all the interactions between the hydrogen atoms is thus tipped in favor of holding the atoms together. The result is a stable diatomic molecule of H_2.

Atomic *p* orbitals can also overlap to form molecular orbitals. A fluorine atom, for example, has a half-filled 2*p* orbital. When two fluorine atoms combine, as shown in Figure 8.14, the *p* orbitals overlap to produce a bonding molecular orbital. There is a high probability of finding a pair of electrons between the positively charged nuclei of the two fluorines. The fluorine nuclei are attracted to this region of high electron density. This attraction holds the atoms together in the fluorine molecule (F_2). The overlap of the 2*p* orbitals produces a bonding molecular orbital that is symmetrical when viewed around the F—F bond axis connecting the nuclei. Therefore, the F—F bond is a sigma bond.

CHEMISTRY & YOU

Q: *Models and drawings are often used to help you visualize where something can be found. How can a drawing show you where an electron is most likely to be found?*

Pi Bonds In the sigma bond of the fluorine molecule, the *p* atomic orbitals overlap end to end. In some molecules, however, orbitals can overlap side by side. As shown in Figure 8.15, the side-by-side overlap of atomic *p* orbitals produces what are called pi molecular orbitals. When a pi molecular orbital is filled with two electrons, a pi bond results. In a **pi bond** (symbolized by the Greek letter π), the bonding electrons are most likely to be found in sausage-shaped regions above and below the bond axis of the bonded atoms. Atomic orbitals in pi bonding overlap less than in sigma bonding. Therefore, pi bonds tend to be weaker than sigma bonds.

⊕ represents the nucleus.

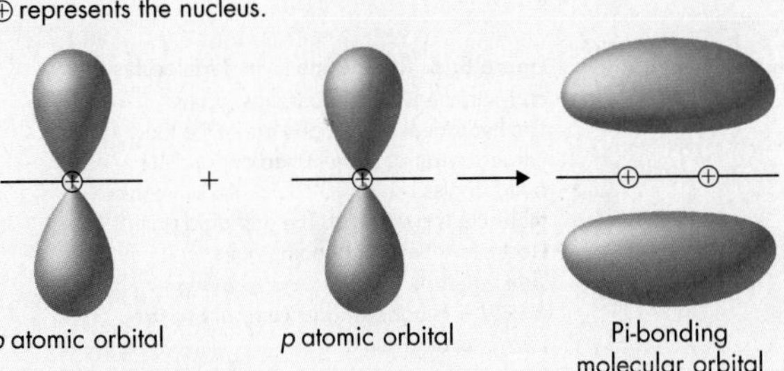

Figure 8.15
***p* Orbital Pi Bonds**
The side-by-side overlap of two *p* atomic orbitals produces a pi-bonding molecular orbital. Together, the two sausage-shaped regions in which the bonding electron pair is most likely to be found constitute one pi-bonding molecular orbital.

UbD Check for Understanding

BIGIDEA BONDING AND INTERACTIONS Have students write one-minute summaries comparing and contrasting sigma and pi bonds.

ADJUST INSTRUCTION If students are having difficulty with their summaries, create and complete a compare and contrast chart on this concept.

Foundations for Reading

BUILD VOCABULARY Explain that the prefix *tetra-* is a Greek word meaning "four," and *hedra-* is a Greek word meaning "face." Explain that a tetrahedron is a four-faced 3-dimensional figure. Prior to introducing the VSEPR theory, remind students that *bi-* means "two," *tri-* means "three," and *octa-* means "eight." Help students to identify the meaning of the acronym VSEPR.

READING STRATEGY As students read about atomic and molecular orbitals, encourage them to make a table in which to list similarities and differences. As you approach the end of the lesson, tell students to expand their table to include hybrid orbitals.

Explain

Molecular Orbitals

USE VISUALS Direct students to examine Figure 8.13. Have them note the positions of the two nuclei in the two hydrogen atomic orbitals. **Ask** Where is the electron most likely to be in the hydrogen atomic orbital? *(In the sphere, close to the nucleus.)* **Ask** Where are the two shared electrons likely to be in the H_2 molecular orbital? *(In the region between the two nuclei.)*

CHEMISTRY & YOU The shape of the drawing represents where electrons are most likely to be found.

APPLY CONCEPTS Tell students to visualize the formation of the covalent bond in H_2 as two separated hydrogen atoms move closer together. Explain that as the distance between the atoms decreases, each nucleus is attracted by both electrons, and the potential energy of the system decreases. Explain that repulsive forces also come into play as the nuclei and the electrons of both atoms begin to repel each other. An interatomic distance is reached at which the potential energy is at a minimum. If the distance decreases beyond this point, the potential energy increases because of increased repulsion. Explain that this results in a chemical bond with a bond length equal to the interatomic distance at which the potential energy is a minimum.

USE VISUALS Have students compare Figures 8.14 and 8.15. **Ask** What is different about the ways that the two *p* orbitals overlap in the two illustrations? *(In Figure 8.14, the* p *orbitals overlap end-to-end, forming an orbital that is symmetrical around the axis joining the nuclei. In Figure 8.15, the* p *orbitals overlap side-by-side and form two areas of electron density, one above and another below the axis joining the nuclei.)* Have students identify the two types of bonds. *(Figure 8.14, sigma bond; Figure 8.15, pi bond)*

Explain

VSEPR Theory

START A CONVERSATION Display the electron configuration for carbon: $1s^2 2s^2 2p^2$. **Ask** How many valence electrons does carbon have? *(4)* **Ask** How many bonds does carbon form? *(4)* **Ask** What do you know about the three *p* orbitals? *(They are dumbbell-shaped orbitals containing two lobes; the three orbitals are perpendicular to each other, oriented on the* x-, y-, *and* z-*axes.)* Convey to students that if carbon forms three of its four bonds using *p* orbitals, then the resulting molecules it forms must be three-dimensional.

Explore

Class Activity

PURPOSE Students achieve an understanding of simple molecular shapes and relate actual three-dimensional geometry to two-dimensional illustrations.

MATERIALS ball-and-stick molecular model kits or foam balls and toothpicks

PROCEDURE Have students make the tetrahedral model of methane. Have them examine the model and compare it with the illustrations in Figure 8.16. Ask them to draw the model and label it *tetrahedral.* Have students remove one ball, leaving the stick in place. The stick can represent an unshared pair of electrons. This is a model for the ammonia molecule. Again, have students compare their models with Figure 8.16 and make their own illustration, labeling it *pyramidal.* Have students remove a second ball. Now the model simulates the water molecule with two sticks (pairs of unshared electrons). Have students compare their models with Figure 8.17 and make their own illustration labeled *bent.*

VSEPR Theory

What do scientists use the VSEPR theory for?

A photograph or sketch may fail to do justice to your appearance. Similarly, electron dot structures fail to reflect the three-dimensional shapes of the molecules illustrated in Figure 8.16. The electron dot structure and structural formula of methane (CH_4), for example, show the molecule as if it were flat and merely two-dimensional.

```
  H              H
H:C:H            |
  H           H—C—H
                 |
                 H
```

Methane (electron dot structure) — Methane (structural formula)

READING SUPPORT

BUILD VOCABULARY: *Word Origins* *Tetrahedral* comes from the Greek *tetra-*, meaning "four," and *hedra*, meaning "face." ***How do these word origins help you understand the shapes of molecules?***

In reality, methane molecules are three-dimensional. As Figure 8.16a shows, the hydrogens in a methane molecule are at the four corners of a geometric solid called a regular tetrahedron. In this arrangement, all of the H—C—H angles are 109.5°, the **tetrahedral angle.**

In order to explain the three-dimensional shape of molecules, scientists use valence-shell electron-pair repulsion theory (VSEPR theory). **VSEPR theory** states that the repulsion between electron pairs causes molecular shapes to adjust so that the valence-electron pairs stay as far apart as possible. The methane molecule has four bonding electron pairs and no unshared pairs. The bonding pairs are farthest apart when the angle between the central carbon and its attached hydrogens is 109.5°. This measurement is the H—C—H bond angle found experimentally.

Unshared pairs of electrons are also important in predicting the shapes of molecules. The nitrogen in ammonia (NH_3) is surrounded by four pairs of valence electrons, so you might predict the tetrahedral angle of 109.5° for the H—N—H bond angle. However, one of the valence-electron pairs shown in Figure 8.16b is an unshared pair. No bonding atom is vying for these unshared electrons. Thus, they are held closer to the nitrogen than are the bonding pairs. The unshared pair strongly repels the bonding pairs, pushing them together. The measured H—N—H bond angle is only 107°.

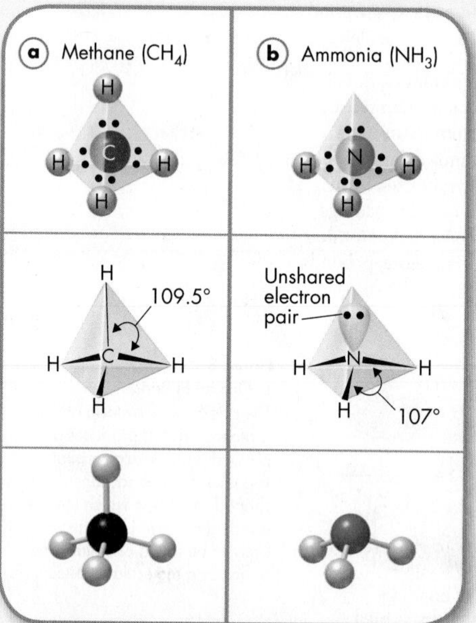

Figure 8.16 Three-Dimensional Molecules **a.** Methane is a tetrahedral molecule. The hydrogens in methane are at the four corners of a regular tetrahedron, and the bond angles are all 109.5°. **b.** An ammonia molecule is pyramidal. The unshared pair of electrons repels the bonding pairs.
Use Models ***How do the resulting H—N—H bond angles compare to the tetrahedral angle?***

Differentiated Instruction

LPR LESS PROFICIENT READERS Give students as many opportunities as possible to see three-dimensional models of molecules. When you write formulas on the chalkboard, try to have a three-dimensional model to hold up or circulate around the class. This will help students interpret the two-dimensional drawings of these three-dimensional figures.

L1 SPECIAL NEEDS Offer students mediums such as modeling clay, colored marshmallows, colored gumdrops, licorice, and straws to mimic the models shown throughout the lesson. Display the models in the classroom and make a point to use these models, giving credit to their creator, when explaining a concept.

L3 ADVANCED STUDENTS Explain that molecular orbital theory is a third way of treating chemical bonding in addition to the valence bond approach and VSEPR theory. Explain that each approach to bonding contributes different yet important ideas. Molecular orbital theory is a quantum mechanical approach involving not just localized bonds but bonds that bind entire molecules. Encourage students to access a college text and learn more about this theory and how it explains the paradox of bonding in the oxygen molecule.

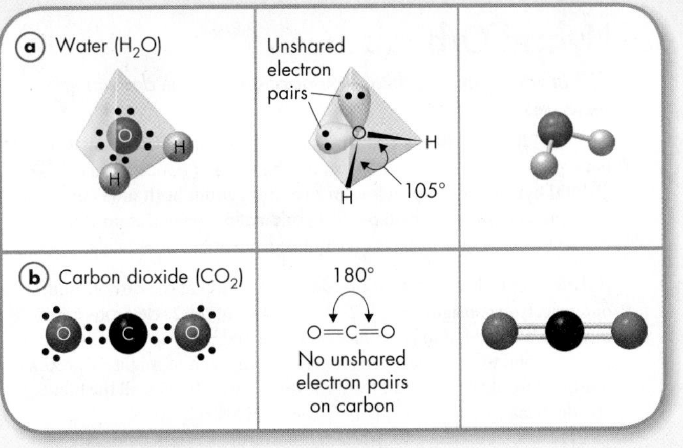

Figure 8.17
Planar and Linear Molecules
This comparison of water and carbon dioxide illustrates how unshared pairs of electrons can affect the shape of a molecule made of three atoms. **a.** The water molecule is bent because the two unshared pairs of electrons on oxygen repel the bonding electrons. **b.** In contrast, the carbon dioxide molecule is linear. The carbon atom has no unshared electron pairs.

In a water molecule, oxygen forms single covalent bonds with two hydrogen atoms. The two bonding pairs and the two unshared pairs of electrons form a tetrahedral arrangement around the central oxygen. Thus, the water molecule is planar (flat) but bent. With two unshared pairs repelling the bonding pairs, the H—O—H bond angle is compressed in comparison with the H—C—H bond angle in methane. The experimentally measured bond angle in water is about 105°, as shown in Figure 8.17a.

In contrast, the carbon in a carbon dioxide molecule has no unshared electron pairs. As illustrated in Figure 8.17b, the double bonds joining the oxygens to the carbon are farthest apart when the O=C=O bond angle is 180°. Thus, CO_2 is a linear molecule. Nine of the possible molecular shapes are shown in Figure 8.18.

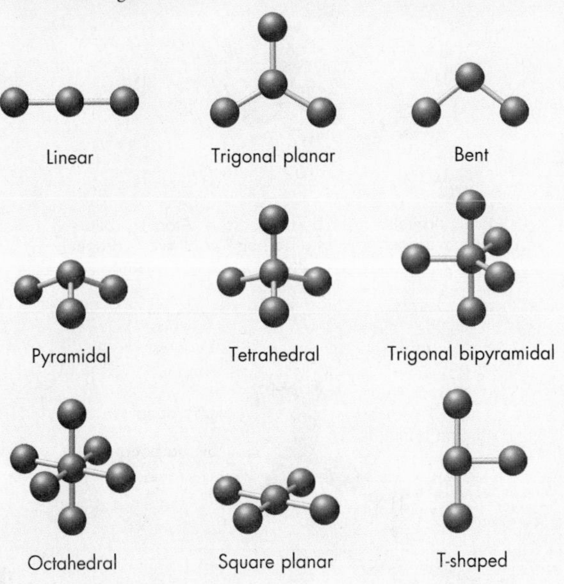

Figure 8.18
Molecular Shapes
Shown here are common molecular shapes.
Infer ***What is the shape of an ammonium ion?***

Explain

START A CONVERSATION Help students understand that VSEPR theory is a model or method used to *predict* molecular shapes based on electron repulsion. In predicting molecular shapes, they may find it helpful to start with an electron dot structure. The electron dot structure shows both the bonding and nonbonding pairs of electrons around the central atom. Point out that when using VSEPR theory to predict molecular shape, double and triple bonds are viewed as single bonds. Each group of electrons in the electron dot structure must be accommodated around the central atom. Because electrons repel each other, they will stay as far apart as possible. When four pairs must be accommodated, the pairs have maximum space when they are arranged tetrahedrally with an angle of 109.5° between them. When three pairs or groups must be accommodated, a trigonal planar arrangement provides the most space with 120° separating them. Finally, when only two groups are present, they will adopt a linear arrangement in which they are separated by 180°. Have students predict the shapes of H_2S and BF_3. *(bent and trigonal planar, respectively)*

UbD Check for Understanding

The Essential Question How do electrons affect the shape of a molecule?

Direct students to Figure 8.18. Have students choose three molecular shapes, then identify and draw a compound that corresponds to each structure. Make sure that students include at least one compound that contains an unshared pair of electrons.

ADJUST INSTRUCTION If students are having difficulty with this concept, review Figures 8.16 and 8.17, making sure to point out the influence exerted over molecular shape by the unshared pairs of electrons associated with nitrogen and oxygen in the illustrated compounds.

Answers

FIGURE 8.16 They are smaller.
FIGURE 8.18 tetrahedral

LESSON 8.3

Explain

Hybrid Orbitals

USE VISUALS Direct students to refer to their periodic table. Verify that students are aware that third row nonmetals such as phosphorus and sulfur, in Groups 5A and 6A, respectively, can form more bonds than the number predicted by the octet rule. For example, show students that phosphorus can form PCl_5 as well as PCl_3, and sulfur can form SF_6 as well as SF_2. Point out that second period elements in the same groups, such as nitrogen and oxygen, do not form compounds that violate the octet rule by expanding the octet. The difference lies in the fact that third period elements can promote electrons to empty 3*d* orbitals. Second period elements have no 2*d* orbitals.

Explain that bonding in phosphorus and sulfur in compounds such as PCl_5 and SF_6 involves hybrid orbitals. For phosphorus, one *s*, three *p*, and one *d* orbitals combine to form five dsp^3 hybrid orbitals. The geometry of a set of five dsp^3 orbitals is trigonal bipyramidal. For sulfur, one *s*, three *p*, and two *d* orbitals combine to form six d^2sp^3 hybrid orbitals. The geometry of six d^2sp^3 orbitals is octahedral.

Hybrid Orbitals

In what ways is orbital hybridization useful in describing molecules?

The VSEPR theory works well when accounting for molecular shapes, but it does not help much in describing the types of bonds formed. **Orbital hybridization provides information about both molecular bonding and molecular shape.** In **hybridization,** several atomic orbitals mix to form the same total number of equivalent hybrid orbitals.

Hybridization Involving Single Bonds Recall that the carbon atom's outer electron configuration is $2s^22p^2$, but one of the 2*s* electrons is promoted to a 2*p* orbital to give one 2*s* electron and three 2*p* electrons, allowing it to bond to four hydrogen atoms in methane. You might suspect that one bond would be different from the other three. In fact, all the bonds are identical. This fact is explained by orbital hybridization.

The one 2*s* orbital and three 2*p* orbitals of a carbon atom mix to form four sp^3 hybrid orbitals. These are at the tetrahedral angle of 109.5°. As you can see in Figure 8.19, the four sp^3 orbitals of carbon overlap with the 1*s* orbitals of the four hydrogen atoms. The sp^3 orbitals extend farther into space than either *s* or *p* orbitals, allowing a great deal of overlap with the hydrogen 1*s* orbitals. The eight available valence electrons fill the molecular orbitals to form four C—H sigma bonds. The extent of overlap results in unusually strong covalent bonds.

Figure 8.19 Methane Molecule
In methane, each of the four sp^3 hybrid orbitals of carbon overlaps with a 1*s* orbital of hydrogen.

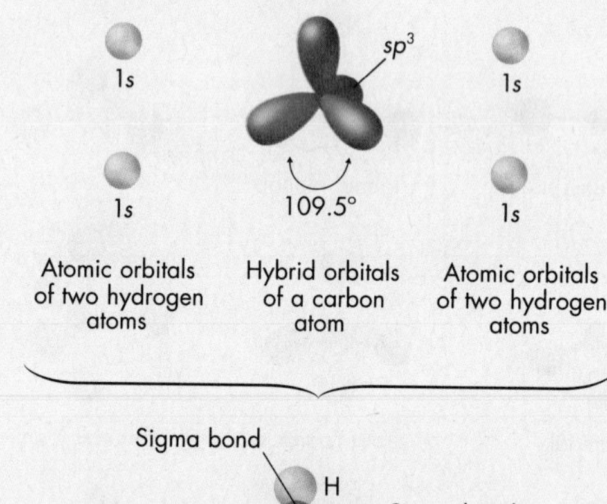

Towering 20th Century Chemist

An esteemed chemist and dedicated humanitarian, Linus Pauling (1916–1994) is the only person to have won two Nobel prizes that were unshared. In 1954 he won the prize for chemistry; in 1962 he won the prize for peace. The peace prize recognized his efforts to bring nuclear weapons under international control and halt nuclear testing. Much of his scientific work involved the structure of simple and complex molecules. His book *The Nature of the Chemical Bond and the Structure of Molecules and Crystals* is one of the most important texts of the century. Pauling introduced the idea of hybrid orbitals to explain why the four bonds around a carbon atom are equivalent. He invented the concept of electronegativity and proposed that most covalent bonds have some ionic character. He also taught that the structures of some compounds must be described by two or more resonance hybrids.

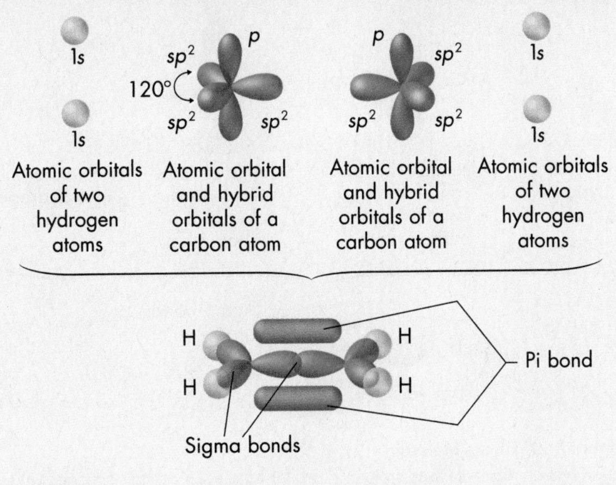

Figure 8.20 Ethene Molecule
In an ethene molecule, two sp^2 hybrid orbitals from each carbon overlap with a 1s orbital of hydrogen to form a sigma bond. The other sp^2 orbitals overlap to form a carbon-carbon sigma bond. The p atomic orbitals overlap to form a pi bond.
Infer ***What region of space does the pi bond occupy relative to the carbon atoms?***

Hybridization Involving Double Bonds Hybridization is also useful in describing double covalent bonds. Ethene is a relatively simple molecule that has one carbon–carbon double bond and four carbon–hydrogen single bonds.

$$\begin{matrix} H & & H \\ & C{=}C & \\ H & & H \end{matrix}$$

Ethene

Experimental evidence indicates that the H—C—H bond angles in ethene are about 120°. In ethene, sp^2 hybrid orbitals form from the combination of one 2*s* and two 2*p* atomic orbitals of carbon. As you can see in Figure 8.20, each hybrid orbital is separated from the other two by 120°. Two sp^2 hybrid orbitals of each carbon form sigma-bonding molecular orbitals with the four available hydrogen 1*s* orbitals. The third sp^2 orbitals of each of the two carbons overlap to form a carbon-carbon sigma-bonding orbital. The nonhybridized 2*p* carbon orbitals overlap side by side to form a pi-bonding orbital. A total of twelve electrons fill six bonding molecular orbitals. Thus, five sigma bonds and one pi bond hold the ethene molecule together. Although they are drawn alike in structural formulas, pi bonds are weaker than sigma bonds. In chemical reactions that involve breaking one bond of a carbon-carbon double bond, the pi bond is more likely to break than the sigma bond.

Hybridization Involving Triple Bonds A third type of covalent bond is a triple bond, which is found in ethyne (C_2H_2), also called acetylene.

$$H{-}C{\equiv}C{-}H$$

As with other molecules, the hybrid orbital description of ethyne is guided by an understanding of the properties of the molecule. Ethyne is a linear molecule. The best hybrid orbital description is obtained if a 2*s* atomic orbital of carbon mixes with only one of the three 2*p* atomic orbitals. The result is two *sp* hybrid orbitals for each carbon.

UbD Check for Understanding

In what ways is orbital hybridization useful in describing molecules?

Write SO_2 on the board. Ask the following question and instruct students to raise their left hand for VSEPR theory and their right hand for orbital hybridization:

What method would you use to determine the bonding between the sulfur and the oxygen atoms and the shape of the molecule?

ADJUST INSTRUCTION If students are struggling with bonding theories, use a comparison chart to show them the information. Acknowledge the limits of each theory for molecules such as SO_2 and H_2Se.

USE VISUALS Direct students to examine Figure 8.20. Call attention to the sigma bonds formed by the two carbon atoms in ethene. Explain that each carbon atom forms three bonds using three sp^2 orbitals, resulting in a planar molecule with the unhybridized *p* orbital forming lobes of electron density above and below the plane.

DRAW A CONCLUSION Have students compare the shapes of the carbon hybrid orbitals in methane (sp^3), ethene (sp^2), and ethyne molecules (*sp*). Guide students to understand that the key concept in hybridization is that when the atomic orbitals combine to form the hybrid orbitals, they lose their identities and entirely new orbitals are formed. These new orbitals have different characteristics than the ones previously. Emphasize that the shape and orientation of the hybrid orbitals cannot be predicted from the shapes of the combining atomic orbitals. Explain that hybrid orbitals of a particular type (sp^3, sp^2, or *sp*) are identical to one another in terms of shape and bonding properties.

Extend

Connect to HORTICULTURE

Ethene, or ethylene, $H_2C = CH_2$, is a gaseous hormone found in most plants. Many fruits produce ethylene as they ripen. Have students research how the produce industry uses this gaseous hormone to its benefit, and present their findings to the class. *(The produce industry can pick and ship unripe fruit and then rely on the production of ethylene to signal synchronized ripening within the shipment. Then the shipment of fruit is ripe by the time it arrives to the market.)*

Answers

FIGURE 8.20 the regions above and below the carbons

LESSON 8.3

Evaluate

Informal Assessment

Have students compare and contrast the characteristics of sigma and pi bonds. *(A sigma bond is formed when two atomic orbitals combine to form a molecular orbital that is symmetrical along the axis connecting the two atomic nuclei. The electrons tend to be found between the two nuclei. A pi bond is formed from the side-to-side overlap of two p atomic orbitals. The electrons in a pi bond are found in sausage-shaped regions above and below the bond axis.)* **Ask** How do unshared pairs of electrons influence the shape of a molecule? *(The unshared pairs take up space in the molecule. They repel the bonding pairs, pushing them closer together and decreasing the bond angle.)* Then have students complete the 8.3 Lesson Check.

Reteach

Have pairs of students determine the shapes of molecules by means of the VSEPR theory. Assign each pair a number from one to five. Write the following information about the central atoms of five molecules on the board and ask the pairs to determine the geometry of the item that corresponds to their number. Have students begin by writing electron dot structures. Allow students to construct the models as part of their response.

1. four bonding pairs + no nonbonding pairs *(tetrahedral)*;
2. three bonding pairs + one nonbonding pair *(pyramidal)*;
3. two bonding pairs + two nonbonding pairs *(bent)*;
4. two bonding pairs + no nonbonding pair *(linear)*;
5. one bonding pair + any number of nonbonding pairs *(linear)*

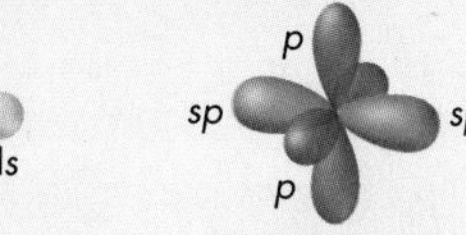

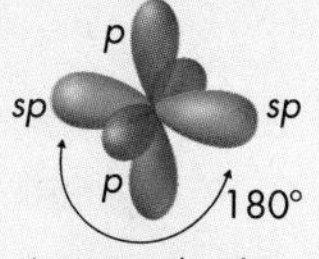

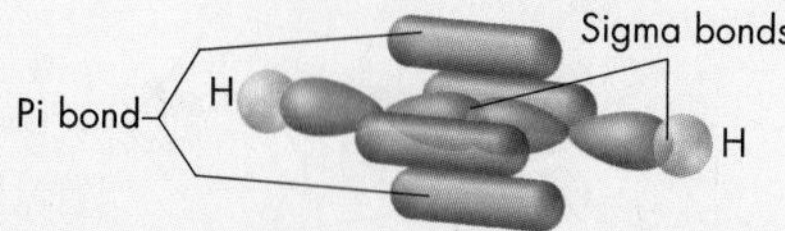

Figure 8.21 Ethyne Molecule
In an ethyne molecule, one *sp* hybrid orbital from each carbon overlaps with a 1*s* orbital of hydrogen to form a sigma bond. The other *sp* hybrid orbital of each carbon overlaps to form a carbon-carbon sigma bond. The two *p* atomic orbitals from each carbon also overlap.
Interpret Diagrams ***How many pi bonds are formed in an ethyne molecule?***

The carbon-carbon sigma-bonding molecular orbital of the ethyne molecule shown in Figure 8.21 forms from the overlap of one *sp* orbital from each carbon. The other *sp* orbital of each carbon overlaps with the 1*s* orbital of each hydrogen, also forming sigma-bonding molecular orbitals. The remaining pair of *p* atomic orbitals on each carbon overlap side by side. They form two pi-bonding molecular orbitals that surround the central carbons. The ten available electrons completely fill five bonding molecular orbitals. The bonding of ethyne consists of three sigma bonds and two pi bonds.

8.3 LessonCheck

21. **Review** How are atomic and molecular orbitals related?
22. **Identify** What do scientists use VSEPR theory for?
23. **Describe** How is orbital hybridization useful in describing molecules?
24. **Classify** What shape would you expect a simple carbon-containing compound to have if the carbon atom has the following hybridizations?
 a. sp^2 **b.** sp^3 **c.** sp
25. **Describe** What is a sigma bond? Describe, with the aid of a diagram, how the overlap of two half-filled 1*s* orbitals produces a sigma bond.
26. **Explain** Use VSEPR theory to predict bond angles in the following covalently bonded molecules. Explain your predictions.
 a. methane **b.** ammonia **c.** water
27. **Identify** How many sigma and how many pi bonds are in an ethyne molecule (C_2H_2)?
28. **Classify** The BF_3 molecule is planar. The attachment of a fluoride ion to the boron in BF_3, through a coordinate covalent bond, creates the BF_4^- ion. What is the geometric shape of this ion?

Lesson Check Answers

21. When two atoms combine, their atomic orbitals overlap to produce molecular orbitals. An atomic orbital belongs to a particular atom, and a molecular orbital belongs to a molecule as a whole.
22. To predict the three-dimensional shapes of molecules
23. Orbital hybridization provides information about both molecular bonding and molecular shape.
24. **a.** trigonal planar
 b. tetrahedral
 c. linear
25. A sigma bond is formed by the overlap of two orbitals along the axis between two nuclei. See Figure 8.13.
26. a. 109.5° (tetrahedral)
 b. 107° (pyramidal)
 c. 105° (bent)
27. 3 sigma bonds and 2 pi bonds
28. tetrahedral

8.4 Polar Bonds and Molecules

C.3.1 Interactions between atoms in ionic and covalent compounds; C.3.2 Formation of ionic and covalent compounds; C.3.3 Chemical bonds and intermolecular attractions

CHEMISTRY & YOU

Q: *How does a snowflake get its shape?* Snow covers approximately 23 percent of Earth's surface. Each individual snowflake is formed from as many as 100 snow crystals. The size and shape of each crystal depends mainly on the air temperature and amount of water vapor in the air at the time the snow crystal forms. In this lesson, you will see how polar covalent bonds in water molecules influence the distinctive geometry of snowflakes.

Key Questions

How do electronegativity values determine the charge distribution in a polar bond?

How do the strengths of intermolecular attractions compare with the strengths of ionic and covalent bonds?

Why are the properties of covalent compounds so diverse?

Vocabulary

- nonpolar covalent bond
- polar covalent bond
- polar bond
- polar molecule
- dipole
- van der Waals forces
- dipole interaction
- dispersion force
- hydrogen bond
- network solid

Bond Polarity

How do electronegativity values determine the charge distribution in a polar bond?

Covalent bonds involve electron sharing between atoms. However, covalent bonds differ in terms of how the bonded atoms share the electrons. The character of the molecule depends on the kind and number of atoms joined together. These features, in turn, determine the molecular properties.

The bonding pairs of electrons in covalent bonds are pulled, as in the tug of war in Figure 8.22, between the nuclei of the atoms sharing the electrons. When the atoms in the bond pull equally (as occurs when identical atoms are bonded), the bonding electrons are shared equally, and each bond formed is a **nonpolar covalent bond.** Molecules of hydrogen (H_2), oxygen (O_2), and nitrogen (N_2) have nonpolar covalent bonds. Diatomic halogen molecules, such as Cl_2, are also nonpolar.

Figure 8.22
Electron Tug of War
The nuclei of atoms pull on the shared electrons, much as the knot in the rope is pulled toward opposing sides in a tug of war.

Focus on ELL

1 CONTENT AND LANGUAGE Write the lesson objectives on the board and read them out loud. Draw students' attention to the word *intermolecular*. Explain that the prefix *inter-* means "between," so the word *intermolecular* means "between molecules." Convey to students that intermolecular forces are different from bonds because these forces do not involve electron donation or sharing.

2 COMPREHENSIBLE INPUT Provide students with a compare and contrast table for them to list important similarities and differences between ionic bonds, covalent bonds, polar covalent bonds, and nonpolar covalent bonds.

Key Objectives

8.4.1 DESCRIBE how electronegativity values determine the charge distribution in a polar molecule.

8.4.2 EVALUATE the strengths of intermolecular attractions compared with the strengths of ionic and covalent bonds.

8.4.3 EXPLAIN why the properties of covalent compounds are so diverse.

Additional Resources

- Reading and Study Workbook, Lesson 8.4
- Core Teaching Resources, Lesson 8.4 Review

Engage

CHEMISTRY & YOU Have students study the photograph and read the text that opens the lesson. **Ask** What kind of bonds are in the water molecule? *(polar)* **Ask** What is the shape of the snowflake? *(hexagon)*

Build Background

Ask students to recall what they know about other topics that involve polarity, such as magnetism and electricity. Remind students that in cases of electrical polarity, the polarity occurs due to the distribution of positive and negative charges. Explain that, in this lesson, they will see that electrical charge is distributed within a molecule, and how this distribution affects molecular properties.

IN ACADEMIC STANDARDS for SCIENCE

C.3.1 Describe, compare, and contrast the characteristics of the interactions between atoms in ionic and covalent compounds.

C.3.2 Compare and contrast how ionic and covalent compounds form.

C.3.3 Compare and contrast ionic, covalent network, metallic and polar and non-polar molecular crystals with respect to constituent particles, strength of bonds, melting and boiling points, and conductivity; provide examples of each type.

Answers

FIGURE 8.21 Two pi bonds are formed.

LESSON 8.4

Foundations for Reading

BUILD VOCABULARY The vocabulary words *polar*, *nonpolar*, and *dipole* all have the same Latin stem *polus,* meaning "pole." A pole is part of a system, large or small, that has opposite electric or magnetic positions. Discuss familiar examples, such as Earth's magnetic north and south poles, and batteries having positive and negative poles or terminals.

READING STRATEGY As students read the lesson, have them focus on the context of the words *polar*, *nonpolar*, and *dipole*. Have students predict what will be taught in each major section of this lesson. Have students check their predictions after reading.

Explain

Bond Polarity

START A CONVERSATION As a class, predict the relationship, if any, that may exist between the electronegativity differences and the type of bond that exists between two elements. **Ask** What does the term *electronegativity* mean? *(It describes the attraction an atom has for shared electrons when the atom is in a compound.)* Remind students that electronegativities are calculated values from 0.0 to 4.0 based on the properties of the elements. Refer students back to Table 6.2. **Ask** Which element is the most electronegative? Which is the least? *(Fluorine has an electronegativity value of 4.0; cesium, the element that has the lowest attraction for electrons, has a value of 0.7.)*

USE VISUALS When students use Table 8.4 to solve Sample Problem 8.3, they may be surprised to discover that bonds previously labeled ionic are now categorized as polar covalent. Explain that the reality is that most bonds are neither totally ionic nor totally covalent. Most fall on a continuum from ionic (complete transfer of electrons) to covalent (equal sharing of electrons). In between are polar-covalent bonds in which electrons are displaced more or less toward one atom or the other. Make sure that students understand the two uses of the Greek letter delta. A lowercase delta (δ) is used to denote partial charges; an uppercase delta (Δ) is used to denote a change in a variable.

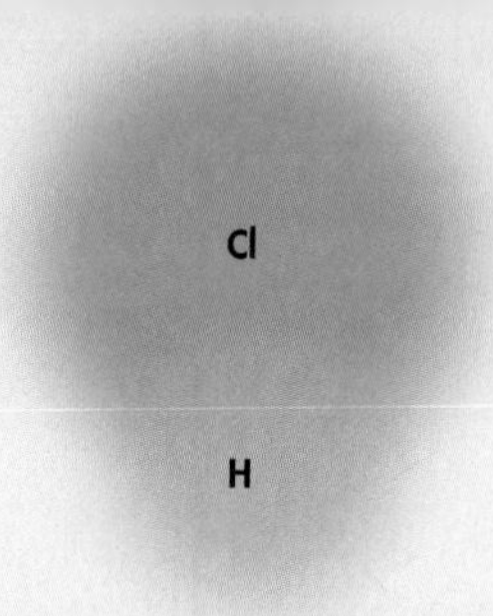

Cl

H

Figure 8.23 Electron Cloud Model of a Polar Bond
This electron-cloud picture of hydrogen chloride shows that the chlorine atom attracts the electron cloud more than the hydrogen atom does.
Infer ***Which atom is more electronegative, a chlorine atom or a hydrogen atom?***

A **polar covalent bond,** known also as a **polar bond,** is a covalent bond between atoms in which the electrons are shared unequally. **The more-electronegative atom attracts electrons more strongly and gains a slightly negative charge. The less-electronegative atom has a slightly positive charge.** Refer back to Table 6.2 in Chapter 6 to see the electronegativities of some common elements. The higher the electronegativity value, the greater the ability of an atom to attract electrons to itself.

Describing Polar Covalent Bonds In the hydrogen chloride molecule (HCl), hydrogen has an electronegativity of 2.1 and chlorine has an electronegativity of 3.0. These values are significantly different, so the covalent bond in hydrogen chloride is polar. The chlorine atom, with its higher electronegativity, acquires a slightly negative charge. The hydrogen atom acquires a slightly positive charge. The lowercase Greek letter delta (δ) denotes that the atoms in the covalent bond acquire only partial charges, less than 1+ or 1−.

$$\overset{\delta+}{H}-\overset{\delta-}{Cl}$$

The minus sign in this notation shows that chlorine has acquired a slightly negative charge. The plus sign shows that hydrogen has acquired a slightly positive charge. These partial charges are shown as clouds of electron density as illustrated in Figure 8.23. The polar nature of the bond may also be represented by an arrow pointing to the more electronegative atom, as shown here:

$$\overset{+\!\!\longrightarrow}{H-Cl}$$

The O—H bonds in a water molecule are also polar. The highly electronegative oxygen partially pulls the bonding electrons away from hydrogen. The oxygen acquires a slightly negative charge. The hydrogen is left with a slightly positive charge.

δ− δ− O, δ+H, Hδ+ *or* O, H, H

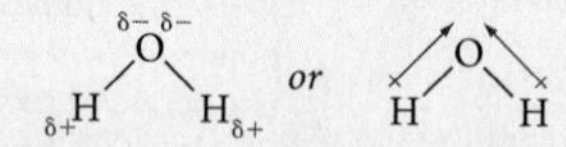

As shown in Table 8.4, the electronegativity difference between two atoms tells you what kind of bond is likely to form. There is no sharp boundary between ionic and covalent bonds. As the electronegativity difference between two atoms increases, the polarity of the bond increases. If the difference is more than 2.0, the electrons will likely be pulled away completely by one of the atoms. In that case, an ionic bond will form.

Table 8.4
Electronegativity Differences and Bond Types

Electronegativity difference range	Most probable type of bond	Example
0.0–0.4	Nonpolar covalent	H—H (0.0)
0.4–1.0	Moderately polar covalent	$\overset{\delta+}{H}-\overset{\delta-}{Cl}$ (0.9)
1.0–2.0	Very polar covalent	$\overset{\delta+}{H}-\overset{\delta-}{F}$ (1.9)
≥2.0	Ionic	Na^+Cl^- (2.1)

UbD Check for Understanding

The Essential Question How is the bonding in molecular compounds different from the bonding in ionic compounds?

Have students write a brief paragraph comparing and contrasting covalent bonding and polar covalent bonding with ionic bonding.

ADJUST INSTRUCTION If students are confused about bonding concepts, use models to explain all three of the bonding types learned thus far.

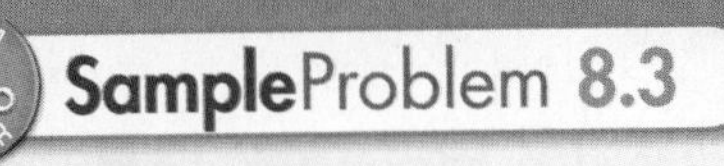

Sample Problem 8.3

Identifying Bond Type

Which type of bond (nonpolar covalent, moderately polar covalent, very polar covalent, or ionic) will form between each of the following pairs of atoms?

a. N and H
b. F and F
c. Ca and Cl
d. Al and Cl

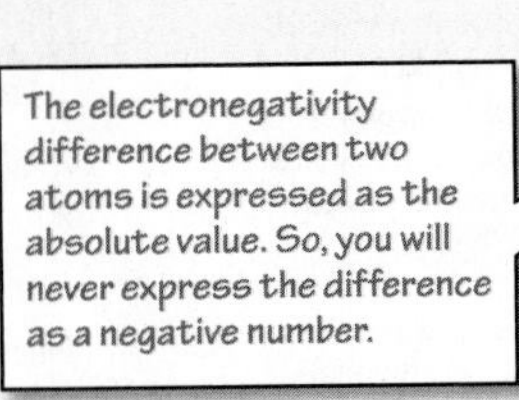

The electronegativity difference between two atoms is expressed as the absolute value. So, you will never express the difference as a negative number.

1 Analyze Identify the relevant concepts. In each case, the pairs of atoms involved in the bonding pair are given. The types of bonds depend on the electronegativity differences between the bonding elements.

2 Solve Apply concepts to this problem.

- Identify the electronegativities of each atom using Table 6.2.
- Calculate the electronegativity difference between the two atoms.
- Based on the electronegativity difference, determine the bond type using Table 8.4.

N (3.0), H (2.1); 0.9; **moderately polar covalent**

F (4.0), F (4.0); 0.0; **nonpolar covalent**

Ca (1.0), Cl (3.0); 2.0; **ionic**

Al (1.5), Cl (3.0); 1.5; **very polar covalent**

29. Identify the bonds between atoms of each pair of elements as nonpolar covalent, moderately polar covalent, very polar covalent, or ionic.

a. H and Br
b. K and Cl
c. C and O
d. Cl and F
e. Li and O
f. Br and Br

30. Place the following covalent bonds in order from least to most polar:

a. H—Cl
b. H—Br
c. H—S
d. H—C

Describing Polar Covalent Molecules The presence of a polar bond in a molecule often makes the entire molecule polar. In a **polar molecule,** one end of the molecule is slightly negative, and the other end is slightly positive. For example, in the hydrogen chloride molecule the partial charges on the hydrogen and chlorine atoms are electrically charged regions, or poles. A molecule that has two poles is called a dipolar molecule, or **dipole.** The hydrogen chloride molecule is a dipole. Look at Figure 8.24. When polar molecules are placed between oppositely charged plates, they tend to become oriented with respect to the positive and negative plates.

Molecular Modeling

The widespread availability of fast computers has greatly benefited chemists interested in the shapes of molecules. Molecular modeling computer programs are becoming an important tool in the pharmaceutical industry for designing drugs. These programs create models of molecules composed of thousands of atoms. Using a special viewing apparatus, chemists can see three-dimensional representations of complex molecules. Models can be rotated, reduced or enlarged, and cut into pieces. Through computer technology called virtual reality, a chemist can virtually get inside a molecular structure to examine it from every possible angle.

Explore

Class Activity

PURPOSE Students experience the attraction of opposite poles of a magnet as an experience similar to the attraction of opposite electric charges.

MATERIALS magnets, iron filings, cardboard

PROCEDURE Place some iron filings on a cardboard. Hold a magnet under the cardboard and move it to create patterns. Allow students to handle magnets and feel the pull when north and south poles are close. Have students write a sentence relating the experience to polar molecules.

Sample Practice Problems

Which type of bond (nonpolar, covalent, moderately covalent, or ionic) will form between each of the following pairs of ions?

A. C and H *(nonpolar covalent)*
B. Be and F *(ionic)*
C. Si and Cl *(very polar covalent)*

Answers

FIGURE 8.23 chlorine

29. **a.** moderately polar covalent
b. ionic
c. moderately to very polar covalent
d. moderately to very polar covalent
e. ionic
f. nonpolar covalent

30. c and d (tie), b, a

Explain

Describing Polar Covalent Bonds

INTERPRET DIAGRAMS Draw diagrams of molecules on the board or overhead projector. Have the class use electronegativity values to determine which element has the partial positive charge and which has the partial negative charge. Explain how the attraction of electrons can be described with a vector. Point out that the shape of the molecule determines the direction of the vector. If the sum of the vectors acting on the central atom in one direction is cancelled by the sum of the vectors acting in the other direction, the molecule has no dipole. If they do not cancel, the molecule is polar.

Explore

Teacher Demo

PURPOSE Students observe what happens to polar molecules when they are placed near positively charged materials.

MATERIALS 2 burets, 2 beakers, ring stand, buret clamps, turpentine, water, rubber rod, glass rod, fur cloth

SAFETY Wear safety goggles and gloves.

PROCEDURE Attach two burets to a ring stand. Fill one with water and the other with turpentine. Place a beaker under each. CAUTION! Turpentine is an irritant. Rub a rubber rod with a fur cloth to charge it negatively. Bring the rod near the open end of the water-containing buret as you slowly open the stopcock. Have students observe the behavior of the water. Repeat the procedure using turpentine. Then repeat using a charged glass rod for both liquids.

EXPECTED OUTCOMES The polar water molecules are attracted to both negatively and positively charged materials, but the nonpolar turpentine molecules are not attracted to either.

Figure 8.24
Polar Molecules in an Electric Field
When polar molecules, such as HCl, are placed in an electric field, the slightly negative ends of the molecules become oriented toward the positively charged plate, and the slightly positive ends of the molecules become oriented toward the negatively charged plate.
Predict *What would happen if, instead, carbon dioxide molecules were placed between the plates? Why?*

See Polar Molecules animated online.

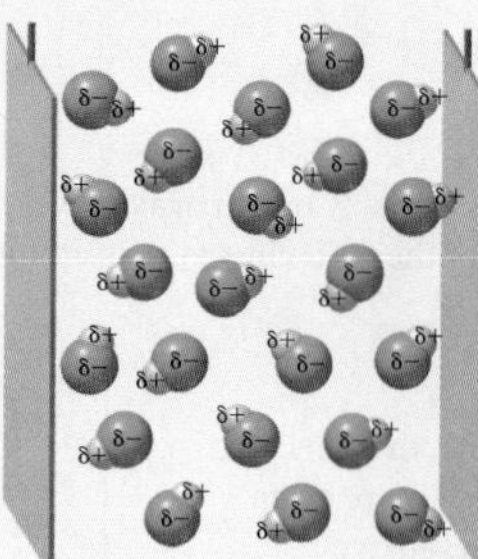
Electric field is absent.
Polar molecules orient randomly.

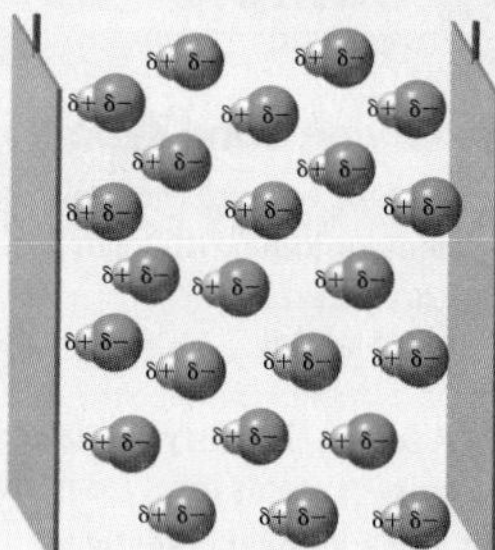
Electric field is on.
Polar molecules line up.

The effect of polar bonds on the polarity of an entire molecule depends on the shape of the molecule and the orientation of the polar bonds. A carbon dioxide molecule, for example, has two polar bonds and is linear.

$$O{=}C{=}O$$

Note that the carbon and oxygens lie along the same axis. Therefore, the bond polarities cancel because they are in opposite directions. Carbon dioxide is thus a nonpolar molecule, despite the presence of two polar bonds.

The water molecule also has two polar bonds. However, the water molecule is bent rather than linear. Therefore, the bond polarities do not cancel and a water molecule is polar.

Attractions Between Molecules

How do the strengths of intermolecular attractions compare with ionic and covalent bonds?

Molecules can be attracted to each other by a variety of different forces. **Intermolecular attractions are weaker than either ionic or covalent bonds.** Nevertheless, you should not underestimate the importance of these forces. Among other things, these attractions are responsible for determining whether a molecular compound is a gas, a liquid, or a solid at a given temperature.

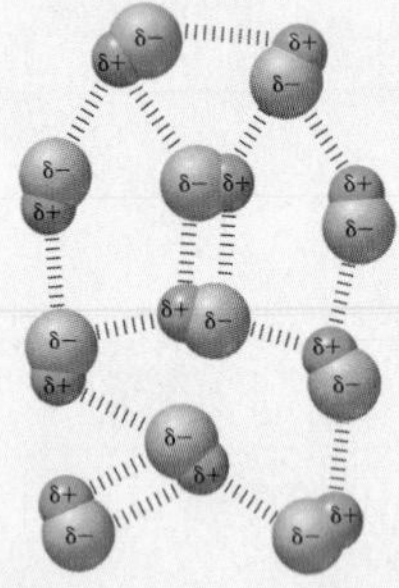
Figure 8.25 Dipole Interactions
Polar molecules are attracted to one another by dipole interactions, a type of van der Waals force.

Van der Waals Forces The two weakest attractions between molecules are collectively called **van der Waals forces,** named after the Dutch chemist Johannes van der Waals (1837–1923). Van der Waals forces consist of dipole interactions and dispersion forces.

Dipole interactions occur when polar molecules are attracted to one another. The electrical attraction involved occurs between the oppositely charged regions of polar molecules, as shown in Figure 8.25. The slightly negative region of a polar molecule is weakly attracted to the slightly positive region of another polar molecule. Dipole interactions are similar to, but much weaker than, ionic bonds.

250 Chapter 8 • Lesson 4

Intermolecular Adjustments

The Dutch physicist Johannes van der Waals (1837–1923) surmised that real gas behavior could not be described exactly by the ideal gas law because the law assumes that gas molecules have no volume and that no attractive forces exist between gas molecules. Van der Waals adjusted the ideal gas equation by introducing two factors: one to take cognizance of the real volume of gas molecules; the other to account for intermolecular forces between molecules. For his work in developing an understanding of intermolecular attractions, he was awarded the Nobel Prize in Physics in 1910, and van der Waals forces, the weak attractive forces he had been studying, were named in his honor.

Dispersion forces, the weakest of all molecular interactions, are caused by the motion of electrons. They occur even between nonpolar molecules. When the moving electrons happen to be momentarily more on the side of a molecule closest to a neighboring molecule, their electric force influences the neighboring molecule's electrons to be momentarily more on the opposite side. This shift causes an attraction between the two molecules similar to, but much weaker than, the force between permanently polar molecules. The strength of dispersion forces generally increases as the number of electrons in a molecule increases. The halogen diatomic molecules, for example, attract each other mainly by means of dispersion forces. Fluorine and chlorine have relatively few electrons and are gases at ordinary room temperature and pressure because of their especially weak dispersion forces. The larger number of electrons in bromine generates larger dispersion forces. Bromine molecules therefore attract each other sufficiently to make bromine a liquid at ordinary room temperature and pressure. Iodine, with a still larger number of electrons, is a solid at ordinary room temperature and pressure.

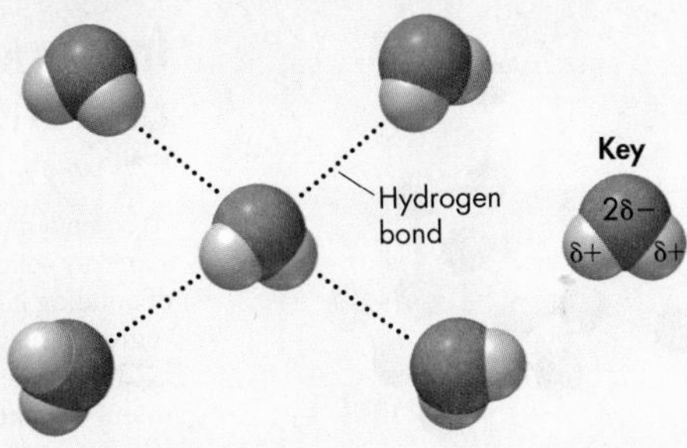

Figure 8.26
Hydrogen Bonds in Water
The strong hydrogen bonding between water molecules accounts for many properties of water, such as the fact that water is a liquid rather than a gas at room temperature.

Hydrogen Bonds The dipole interactions in water produce an attraction between water molecules. Each O—H bond in the water molecule is highly polar, and the oxygen acquires a slightly negative charge because of its greater electronegativity. The hydrogens in water molecules acquire a slightly positive charge. The positive region of one water molecule attracts the negative region of another water molecule, as illustrated in Figure 8.26. This attraction between the hydrogen of one water molecule and the oxygen of another water molecule is strong compared to other dipole interactions. This relatively strong attraction, which is also found in hydrogen-containing molecules other than water, is called a hydrogen bond. Figure 8.26 illustrates hydrogen bonding in water.

Q: *How does a snowflake get its shape?*

Hydrogen bonds are attractive forces in which a hydrogen covalently bonded to a very electronegative atom is also weakly bonded to an unshared electron pair of another electronegative atom. This other atom may be in the same molecule or in a nearby molecule. Hydrogen bonding always involves hydrogen. It is the only chemically reactive element with valence electrons that are not shielded from the nucleus by other electrons.

Remember that for a hydrogen bond to form, a covalent bond must already exist between a hydrogen atom and a highly electronegative atom, such as oxygen, nitrogen, or fluorine. The combination of this strongly polar bond and the lack of shielding effect in a hydrogen atom is responsible for the relative strength of hydrogen bonds. A hydrogen bond has about 5 percent of the strength of an average covalent bond. Hydrogen bonds are the strongest of the intermolecular forces. They are extremely important in determining the properties of water and biological molecules such as proteins. Figure 8.27 shows how the relatively strong attractive forces between water molecules allows the water strider to sit on the surface of the water.

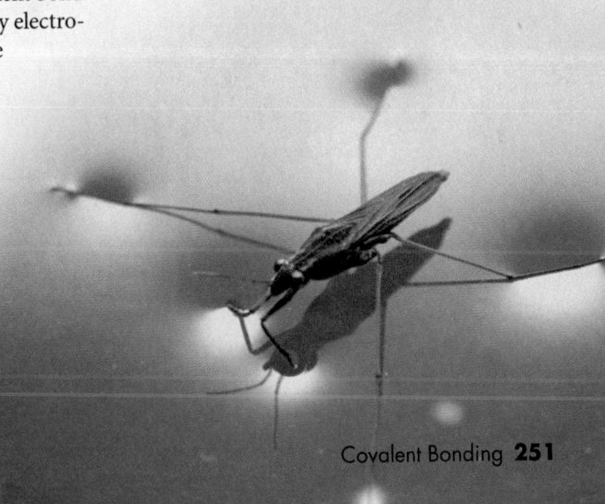

Figure 8.27 Walking on Water
The strong attractions between water molecules allow this water strider to "walk" on water instead of sinking into the water.

Differentiated Instruction

L1 LESS PROFICIENT READERS Use a three-dimensional software program or Internet-based animations to help students predict the polarity and the type of force between molecules.

ELL ENGLISH LANGUAGE LEARNERS Clarify students' understanding of the word *force* as it relates to this lesson. Make sure they are not looking for a quantitative value as they might have learned from physics or mathematics class.

L3 ADVANCED STUDENTS As a preparation for later chapters, have students research the effects of bond polarity on solubility.

Explain

Attractions Between Molecules

APPLY CONCEPTS Discuss how the sharing of electrons binds atoms into molecules. Point out that if no other forces were present, all covalently bonded molecules would be gases at any temperature because no attractive forces would tend to pull them together to form a liquid or solid. Explain, however, that forces such as van der Waals forces and hydrogen bonding do exist between covalently bonded molecules.

CHEMISTRY & YOU A snowflake's shape is influenced by the intermolecular forces of the water molecules that make up the snowflake. Hydrogen bonds cause the water molecules to align themselves so the attractive forces are maximized.

DRAW A DIAGRAM Construct a concept map using the following terms: intermolecular attractions, van der Waals forces, dispersion forces, dipole interactions, and hydrogen bonding. Use the concept map to introduce and discuss each force in detail. Use the halogen family to demonstrate dispersion forces, carbon monoxide to discuss dipole interactions, and water to discuss hydrogen bonding. **Ask** Why does He have a lower boiling point than Rn? *(Rn has more electrons, so its intermolecular forces are stronger than those of He.)* Point out that hydrogen bonds only occur between molecules in which hydrogen is bonded to a strongly electronegative atom such as nitrogen, oxygen, or fluorine. Because living systems contain molecules rich in oxygen and nitrogen, hydrogen bonding plays an important role in the chemistry of living systems. Hydrogen bonding also explains many of the physical properties of water.

Answers

FIGURE 8.24 Carbon dioxide molecules would not change orientation because the CO_2 molecule is nonpolar.

LESSON 8.4

Explore

Intermolecular Attractions and Molecular Properties

Teacher Demo

PURPOSE Students see how hydrogen bonding affects the surface tension and viscosity of fluids.

MATERIALS 4 300-mL stoppered flasks, each containing 100 mL of one of these liquids: glycerol, ethylene glycol, water, ethanol

PROCEDURE Gently rotate one of the flasks so that its contents begin to swirl. Have students measure the time interval between your ceasing to rotate the bottle and the disappearance of the swirling vortex. Repeat the procedure for each fluid. Be sure to rotate each flask the same number of times.

EXPECTED OUTCOMES From longest time to shortest: ethanol, water, ethylene glycol, and glycerol. Point out that hydrogen bonding increases a liquid's surface tension and viscosity. Thus, liquids with a relatively high degree of hydrogen bonding generally stop swirling more quickly. Have students order the fluids according to increasing hydrogen bonding. *(ethanol, water, ethylene glycol, glycerol)* Show the structural formulas of the compounds to demonstrate the origins of the hydrogen bonding.

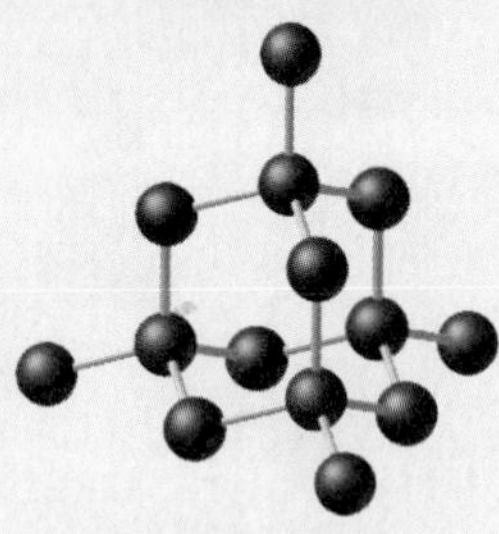

Figure 8.28 Diamond
Diamond is a network-solid form of carbon. Diamond has a three-dimensional structure, with each carbon at the center of a tetrahedron.

Intermolecular Attractions and Molecular Properties

Why are the properties of covalent compounds so diverse?

At room temperature, some compounds are gases, some are liquids, and some are solids. The physical properties of a compound depend on the type of bonding it displays—in particular, on whether it is ionic or covalent. A great range of physical properties occurs among covalent compounds. **The diversity of physical properties among covalent compounds is mainly because of widely varying intermolecular attractions.**

The melting and boiling points of most compounds composed of molecules are low compared with those of ionic compounds. In most solids formed from molecules, only the weak attractions between molecules need to be broken. However, a few solids that consist of molecules do not melt until the temperature reaches 1000°C or higher, or they decompose without melting at all. Most of these very stable substances are **network solids** (or network crystals), solids in which all of the atoms are covalently bonded to each other. Melting a network solid would require breaking covalent bonds throughout the solid.

Diamond is an example of a network solid. As shown in Figure 8.28, each carbon atom in a diamond is covalently bonded to four other carbons, interconnecting carbon atoms throughout the diamond. Cutting a diamond requires breaking a multitude of these bonds. Diamond does not melt; rather, it vaporizes to a gas at 3500°C and above.

Silicon carbide, with the formula SiC and a melting point of about 2700°C, is also a network solid. Silicon carbide is so hard that it is used in grindstones and as an abrasive. It is also used as a coating on materials that are exposed to high temperatures, as in Figure 8.29. The molecular structures of silicon carbide and diamond are similar to each other. You can think of samples of diamond, silicon carbide, and other network solids as single molecules.

Figure 8.29 Silicon Carbide
Surfaces are coated with silicon carbide to make products that are non-adhesive and resistant to extreme temperature, abrasions, and corrosion.

252 Chapter 8 • Lesson 4

UbD Check for Understanding

The Essential Question **What factors affect molecular properties?**

Ask students to summarize why molecular properties among covalent compounds are not all the same. Summaries may be verbal or written.

ADJUST INSTRUCTION Direct students to review Table 8.4 and discuss why and how intermolecular forces allow for such diversity among covalent compounds.

Table 8.5 summarizes some of the characteristic differences between ionic and covalent (molecular) substances. Note that ionic compounds have higher melting points than molecular compounds. Ionic compounds also tend to be soluble in water.

Table 8.5

Characteristics of Ionic and Molecular Compounds

Characteristic	Ionic compound	Molecular compound
Representative unit	Formula unit	Molecule
Bond formation	Transfer of one or more electrons between atoms	Sharing of electron pairs between atoms
Type of elements	Metallic and nonmetallic	Nonmetallic
Physical state	Solid	Solid, liquid, or gas
Melting point	High (usually above 300°C)	Low (usually below 300°C)
Solubility in water	Usually high	High to low
Electrical conductivity of aqueous solution	Good conductor	Poor to nonconducting

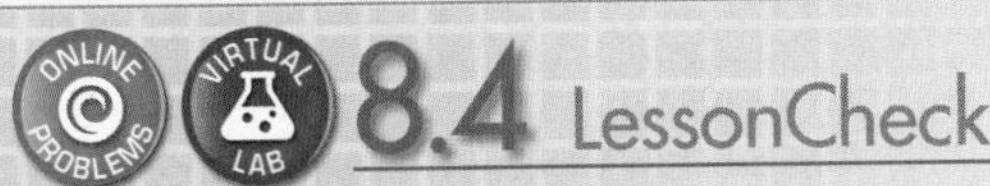

8.4 LessonCheck

31. Explain How do electronegativity values determine the charge distribution in a polar covalent bond?

32. Compare How do the strengths of intermolecular attractions compare to the strengths of ionic bonds and covalent bonds?

33. Explain Why are the properties of covalent compounds so diverse?

34. Explain Explain this statement: Not every molecule with polar bonds is polar. Use CCl_4 as an example.

35. Draw Draw the electron dot structure for each molecule. Identify polar covalent bonds by assigning slightly positive (δ+) and slightly negative (δ−) symbols to the appropriate atoms.

a. HOOH
b. BrCl
c. HBr
d. H_2O

36. Compare How does a network solid differ from most other covalent compounds?

37. What happens when polar molecules are between oppositely charged metal plates?

BIGIDEA
BONDING AND INTERACTIONS

38. Explain how dipole interactions and dispersion forces are related. First, explain what produces the attractions between polar molecules. Then, explain what produces dispersion forces between molecules. Identify what is similar and what is different in the two mechanisms of intermolecular attraction.

Evaluate

Informal Assessment

Have students use their knowledge of bonding to account for the generally low melting points of covalent compounds. *(Intermolecular attractions between covalent compounds are generally not as strong as ionic bonds.)* Then have students complete the 8.4 Lesson Check.

Reteach

Help students list the types of intermolecular attractions that exist between molecules. Give examples of compounds that would fit into each category. Encourage students to continue using their concept maps to organize the chapter material.

Lesson Check Answers

31. The more electronegative atom attracts electrons more strongly and gains a partial negative charge. The less electronegative atom has a partial positive charge.

32. Intermolecular attractions are weaker than either ionic or covalent bonds.

33. Because of widely varying intermolecular attractions

34. The atoms in CCl_4 are oriented so that the bond polarities cancel.

35. a. $\overset{\delta+}{H}:\overset{\delta-}{\ddot{\underset{..}{O}}}:\overset{\delta-}{\ddot{\underset{..}{O}}}:\overset{\delta+}{H}$

b. $:\overset{\delta+}{\ddot{\underset{..}{Br}}}:\overset{\delta-}{\ddot{\underset{..}{Cl}}}:$

c. $\overset{\delta+}{H}:\overset{\delta-}{\ddot{\underset{..}{Br}}}:$

d. $\overset{\delta+}{H}:\overset{\delta-}{\ddot{\underset{..}{O}}}:\overset{\delta+}{H}$

36. The atoms in a network solid are covalently bonded in a large array (or crystal), which can be thought of as a single molecule.

37. Polar molecules tend to become oriented with respect to the positive and negative plates.

38. BIGIDEA In dipole interactions, the slightly positive end of a polar molecule attracts the slightly negative end of another polar molecule. Dispersion forces arise from the movement of electrons in any molecule, polar or nonpolar. If the electrons in one molecule happen to be momentarily more on one side, the electrons in a neighboring molecule can be repulsed. Weak attractions result between the temporary dipoles.

SMALL-SCALE LAB

Explore

Small-Scale Lab

OBJECTIVE After completing this activity, students will be able to infer differences in polarity by observing the chromatographic separation of food dyes.

PREP TIME 15 minutes

CLASS TIME 30 minutes

ADVANCED PREP

- To prepare 0.1% NaCl, dissolve 1.0 g NaCl in 1.0 L water.
- Jelly beans and small candy-coated chocolates are good sources of food dyes. Dissolve a package of unsweetened powdered soft drink in a small amount of water and have students use toothpicks to spot this solution on chromatography paper.
- Coffee filters can substitute for chromatography paper, but caution students not to overload their samples.

TEACHING TIPS Tell students to position the chromatography paper so that the lower end of the paper is in the liquid but the spots of dye are above the surface. A similar, more complete lab on chromatography is found in the Small-Scale Laboratory Manual.

EXPECTED OUTCOME In order of increasing speed of travel: red, yellow, and blue. Green separated into yellow and blue.

YOU'RE THE CHEMIST!

1. Wet candy, blot to remove excess water, press onto chromatography paper, and develop in 0.1% NaCl.
2. Place spots on chromatography paper. Develop in NaCl solution.
3. Use a toothpick to place spots of powdered drink solutions.
4. Rubbing alcohol moves more slowly and gives slightly better separation.
5. With some papers, the positions of Blue No. 1 and Yellow No. 5 are reversed.

ADVANCED LEARNERS Encourage students to investigate how different carrier fluids affect the movement of the four dyes. For example, they might try pure water or ethanol.

ANALYZE

1. Red, yellow, and blue are pure.
2. Green is Yellow No. 5 plus Blue No. 1.
3. Blue No. 1; Red No. 40 (No. 3 if supplies haven't been used up); Yellow No. 5 or Yellow No. 6 if the dye appears orange.
4. See the drawing.
5. Blue is the most polar; red, the least polar.

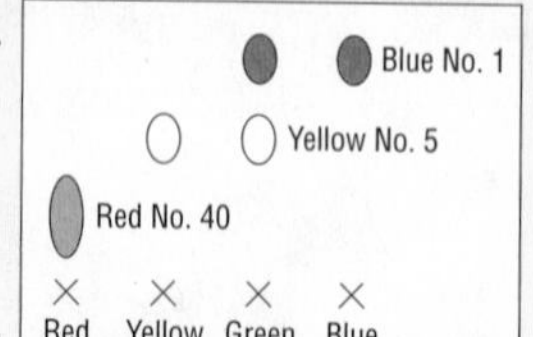

Small-Scale Lab

Paper Chromatography of Food Dyes

Purpose

To use paper chromatography to separate and identify food dyes in various samples

Materials

- pencil
- ruler
- scissors
- toothpicks
- 4 different colors of food coloring
- plastic cup
- 0.1% NaCl solution
- chromatography paper

Procedure

Cut a 5 cm × 10 cm strip of chromatography paper and label it with a pencil, as shown below. Use a different toothpick to place a spot of each of the four food colors on the Xs on your chromatography paper. Allow the spots to dry for a few minutes. Fill the plastic cup so its bottom is just covered with the solvent (0.1% NaCl solution). Wrap the chromatography paper around a pencil. Remove the pencil and place the chromatography paper, color-spot side down, in the solvent. When the solvent reaches the top of the chromatography paper, remove the paper and allow it to dry.

Analysis

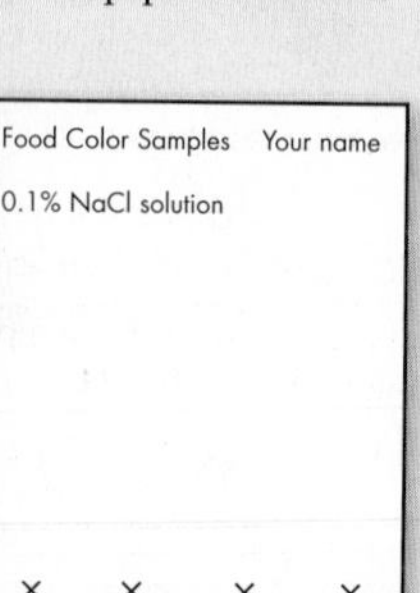

Using your experimental data, record the answers to the following questions below your data table.

1. If a food-color sample yields a single streak or spot, it is usually a pure compound. Which food colors consist of pure compounds?
2. Which food colors are mixtures of compounds?
3. Food colors often consist of a mixture of three colored dyes: Red No. 40, Yellow No. 5, and Blue No. 1. Read the label on the food-color package. Which dyes do your food-color samples contain?
4. Identify each spot or streak on your chromatogram as Red No. 40, Yellow No. 5, or Blue No. 1.
5. Paper chromatography separates polar covalent compounds on the basis of their relative polarities. The dyes that are the most polar migrate the fastest and appear at the top of the paper. Which dye is the most polar? Which dye is the least polar?

You're the Chemist

The following small-scale activities allow you to develop your own procedures and analyze the results.

1. **Design an Experiment** Design and carry out an experiment to identify the dyes in various colored candies.
2. **Design an Experiment** Design and carry out an experiment to identify the dyes in various colored markers using the paper chromatography method.
3. **Design an Experiment** Design and carry out an experiment to identify the dyes in various colored powdered drinks using the paper chromatography method.
4. **Analyze Data** Use different solvents, such as 2-propanol (rubbing alcohol), vinegar, and ammonia, to separate food colors. Does the choice of solvent affect the results?
5. **Analyze Data** Explore the effect of different papers on your results. Try paper towels, notebook paper, and coffee filters. Report your results. Examine the relative positions of Blue No. 1 and Yellow No. 5. What do you observe?

254 Chapter 8 • Small-Scale Lab

Focus on ELL

4 LANGUAGE PRODUCTION Have students work in groups of four to complete the lab. Make sure each group has ELLs of varied language proficiencies, so that more proficient students can help less proficient ones. Have students work according to their proficiency level.

BEGINNING: LOW/HIGH Paraphrase each step of the lab orally and group students with higher language proficiency partners to help with the written answers.

INTERMEDIATE: LOW/HIGH Ask students to read the procedure and then break it down into numbered steps that are easier for them to follow.

ADVANCED: LOW/HIGH Have students help students with lower proficiencies with the written answers and predictions.

8 Study Guide

BIGIDEA
BONDING AND INTERACTIONS

In molecular compounds, bonding occurs when atoms share electrons. In ionic compounds, bonding occurs when electrons are transferred between atoms. Shared electrons and the valence electrons that are not shared affect the shape of a molecular compound, as the valence electrons stay as far apart from each other as possible. The molecular properties of a molecule are affected by intermolecular attractions.

8.1 Molecular Compounds

A molecular formula shows how many atoms of each element a substance contains.

The representative unit of a molecular compound is a molecule. For an ionic compound, the representative unit is a formula unit.

- covalent bond (223)
- molecule (223)
- diatomic molecule (223)
- molecular compound (223)
- molecular formula (223)

8.2 The Nature of Covalent Bonding

In covalent bonds, electron sharing occurs so that atoms attain the configurations of noble gases.

In a coordinate covalent bond, the shared electron pair comes from a single atom.

The octet rule is not satisfied in molecules with an odd number of valence electrons and in molecules in which an atom has less, or more, than a complete octet of valence electrons.

A large bond dissociation energy corresponds to a strong covalent bond.

Chemists use resonance structures to envision the bonding in molecules that cannot be adequately described by a single structural formula.

- single covalent bond (226)
- structural formula (227)
- unshared pair (227)
- double covalent bond (230)
- triple covalent bond (230)
- coordinate covalent bond (232)
- polyatomic ion (232)
- bond dissociation energy (236)
- resonance structure (237)

8.3 Bonding Theories

Just as an atomic orbital belongs to a particular atom, a molecular orbital belongs to a molecule as a whole.

In order to explain the three-dimensional shape of molecules, scientists use the valence-shell electron-pair repulsion theory (VSEPR theory).

Orbital hybridization provides information about both molecular bonding and molecular shape.

- molecular orbital (240)
- bonding orbital (240)
- sigma bond (240)
- pi bond (241)
- tetrahedral angle (242)
- VSEPR theory (242)
- hybridization (244)

8.4 Polar Bonds and Molecules

When different atoms bond, the more-electronegative atom attracts electrons more strongly and acquires a slightly negative charge.

Intermolecular attractions are weaker than either an ionic or covalent bond.

The diversity of physical properties among covalent compounds is mainly because of widely varying intermolecular attractions.

- nonpolar covalent bond (247)
- polar covalent bond (248)
- polar bond (248)
- polar molecule (249)
- dipole (249)
- van der Waals forces (250)
- dipole interaction (250)
- dispersion force (251)
- hydrogen bond (251)
- network solid (252)

Evaluate

Review and Assessment Resources

CHEMISTRY TUTORIAL Encourage students to review drawing electron dot configurations and identifying bond type using the animated step-by-step tutorials for Chapter 8. Each tutorial is followed by a similar practice problem, where the student is guided through the problem-solving process.

CHAPTER PROBLEM SET Have students practice more covalent bonding problems using the Chapter 8 online Problem Set.

DIRECTED VIRTUAL LAB Have students complete the Virtual Lab *Covalent Bonding*. In the post-lab exercise, students will analyze the data and draw conclusions based on the results of the experiment.

KINETIC ART Direct students to the Kinetic Art animations *Covalent Bonds* and *Polar Molecules* and have them answer the questions that accompany each animation.

Study Tip

SWITCH TASKS When you feel yourself losing focus, switch the type of task you are working on, the subject that you are studying, or the environment you are in. Take a break and walk around a bit when you are no longer being productive.

Focus on ELL

5 ASSESS UNDERSTANDING Read the Big Idea paragraph out loud, one sentence at a time. Have students point to the lesson review that corresponds to each sentence. Encourage students to use the Study Guide to prepare for the assessment.

BEGINNING: LOW/HIGH Give each student a copy of the Study Guide. Ask them to make a check mark next to the content they understand and circle the content they do not understand. Create simple matching assessments to assess only the content they circle.

INTERMEDIATE: LOW/HIGH Explain the meaning of common words that appear on multiple-choice questions, such as *likely*, *most*, *probably*, and *best*. Orally pose some questions using these words.

ADVANCED: LOW/HIGH Explain how to interpret multiple-choice questions that use the word *not*. Orally pose some "not" type questions as practice.

UbD Performance Tasks

THREE FOR THREE-D: Have students create three-dimensional models of three different molecular compounds. Each molecule should have a different shape and each part of the molecule should be labeled accurately.

WHAT DID IT? Create a storybook, including pictures, where the beginning of the story is the CHEMystery, "What's that alarm?" Finish the story by explaining what the mystery substance was, how it was created, how the carbon and oxygen were bonded, and how the family prevented the substance from returning into the house.

Evaluate

Answers

LESSON 8.1

39. ionic

40. **a.** 6 C, 8 H, 6 O **b.** 12 C, 22 H, 11 O **c.** 7 C, 5 H, 3 N, 6 O

41. Nitrogen and oxygen achieve stability as diatomic molecules. Argon exists as individual atoms because it has a stable noble-gas electron configuration.

42. Molecular formulas show the number and kind of atoms in the molecule; structural formulas show the covalent bonds in addition to the number and kind of atoms.

43. b, c, and e

LESSON 8.2

44. Neon has an octet of electrons. A chlorine atom achieves an octet by sharing an electron with another chlorine atom.

45. **a.** ionic **b.** ionic **c.** covalent **d.** covalent

46. Ionic bonds depend on electrostatic attraction between ions. Covalent bonds depend on electrostatic attraction between shared electrons and nuclei of combining atoms.

47. A double covalent bond has four shared electrons (two bonding pairs); a triple covalent bond has six shared electrons (three bonding pairs).

48. One atom contributes both electrons to a coordinate covalent bond, as in CO.

49. **a.** :Ï:Ï: **b.** :F̤̈:Ö̤:F̤̈: **c.** H:S̤̈:H **d.** :Ï:N̈:Ï: (with :I̤: below N)

50. An unshared pair of electrons is needed for a coordinate covalent bond. There are no unshared pairs in compounds with only C-H and C-C bonds.

51. [:S̤̈:C::N:]⁻

52. [Ö̤::C:Ö̤:H]⁻ (with :Ö̤: below C)

53. [:Ö̤:N̈::Ö:]⁻ ↔ [:Ö::N̈:Ö̤:]⁻

54. b and c; assuming only single bonds, the P and S atoms each have 10 valence electrons.

55. Bond dissociation energy is defined as the energy needed to break one covalent bond.

56. Increasing bond dissociation energy is linked to lower chemical reactivity.

57. **a.** 4 **b.** 8 **c.** 6 **d.** 2 **e.** 2

58. Check students' electron dot structures.

LESSON 8.3

59. A pi bond is formed by the side-by-side overlap of two half-filled p atomic orbitals to produce a pi molecular orbital. In a pi bond, the bonding electrons are most likely to be found in sausage-shaped regions above and below the bond. See Figure 8.15.

8 Assessment

★ Solutions appear in Appendix E

Lesson by Lesson

8.1 Molecular Compounds

39. The melting point of a compound is 1240°C. Is this compound most likely an ionic compound or a molecular compound?

40. Identify the number and kinds of atoms present in a molecule of each compound.

a. ascorbic acid (vitamin C), $C_6H_8O_6$
b. sucrose (table sugar), $C_{12}H_{22}O_{11}$
c. trinitrotoluene (TNT), $C_7H_5N_3O_6$

41. Which of the following gases in Earth's atmosphere would you expect to find as molecules and which as individual atoms? Explain.

a. nitrogen
b. oxygen
c. argon

42. Describe the differences between molecular formulas and structural formulas for molecular compounds.

43. Identify the phrases that generally apply to molecular compounds.

a. contain metals and nonmetals
b. are often gases or liquids
c. have low melting points
d. contain ionic bonds
e. use covalent bonding

8.2 The Nature of Covalent Bonding

44. Explain why neon is monatomic but chlorine is diatomic.

45. Classify the following compounds as ionic or covalent:

a. $MgCl_2$
b. Na_2S
c. H_2O
d. H_2S

46. Describe the difference between an ionic and a covalent bond.

47. How many electrons do two atoms in a double covalent bond share? How many in a triple covalent bond?

★48. Characterize a coordinate covalent bond and give an example.

49. Draw plausible electron dot structures for the following substances. Each substance contains only single covalent bonds.

a. I_2
b. OF_2
c. H_2S
d. NI_3

★50. Explain why compounds containing C—N and C—O single bonds can form coordinate covalent bonds with H^+ but compounds containing only C—H and C—C single bonds cannot.

51. Draw the electron dot structure of the polyatomic thiocyanate anion (SCN^-).

52. Draw the electron dot structure for the hydrogen carbonate ion (HCO_3^-). Carbon is the central atom, and hydrogen is attached to oxygen in this polyatomic anion.

53. Using electron dot structures, draw at least two resonance structures for the nitrite ion (NO_2^-). The oxygens in NO_2^- are attached to the nitrogen.

★54. Which of these compounds contain elements that do not follow the octet rule? Explain.

a. NF_3
b. PCl_2F_3
c. SF_4
d. SCl_2

55. Explain what is meant by *bond dissociation energy.*

56. What is the relationship between the magnitude of a molecule's bond dissociation energy and its expected chemical reactivity?

57. How many electrons must the atoms of the elements below share with other atoms in covalent bonding to achieve an octet of electrons?

a. S
b. C
c. N
d. Br
e. I

★58. Draw the electron dot structures for each of these molecules.

a. NH_3
b. $BrCl$
c. H_2O_2
d. SiH_4

8.3 Bonding Theories

59. What is a pi bond? Describe, with the aid of a diagram, how the overlap of two half-filled *p* atomic orbitals produces a pi bond.

★60. Use VSEPR theory to predict the shapes of the following compounds:

a. CO_2 **c.** SO_3 **e.** CO
b. $SiCl_4$ **d.** SCl_2 **f.** H_2Se

61. The molecule CO_2 has two carbon–oxygen double bonds. Describe the bonding in the CO_2 molecule, which involves hybridized orbitals for carbon and oxygen.

62. What type of bonding orbital is always formed between hydrogen and another atom in a covalent compound?

★63. What types of hybrid orbitals are involved in the bonding of the carbon atoms in the following molecules?

a. CH_4
b. $H_2C{=}CH_2$
c. $HC{\equiv}CH$
d. $N{\equiv}C{-}C{\equiv}N$

8.4 Polar Bonds and Molecules

64. How must the electronegativities of two atoms compare if a covalent bond between them is to be polar?

★65. The bonds between the following pairs of elements are covalent. Arrange them according to polarity, listing the most polar bond first.

a. H—Cl **c.** H—F **e.** H—H
b. H—C **d.** H—O **f.** S—Cl

66. What is a hydrogen bond?

67. Depict the hydrogen bonding between two ammonia molecules and between one ammonia molecule and one water molecule.

68. Why do compounds with strong intermolecular attractive forces have higher boiling points than compounds with weak intermolecular attractive forces?

★69. Use Table 8.3 to determine how many kilojoules are required to dissociate all the C—H bonds in 1 mol of methane (CH_4).

70. Which of these molecules is least likely to form a hydrogen bond with a water molecule?

a. NH_3 **c.** HF
b. CH_3Cl **d.** H_2O_2

Understand Concepts

★71. Devise a hybridization scheme for PCl_3 and predict the molecular shape based on this scheme.

72. The chlorine and oxygen atoms in thionyl chloride ($SOCl_2$) are bonded directly to the sulfur. Draw an acceptable electron dot structure for thionyl chloride.

73. Explain why each electron dot structure is incorrect. Replace each structure with one that is more acceptable.

a. [:C::N̈:]⁻
b. :F̤̈:P̤::F̤̈: / :F̤: (below P)

74. Use VSEPR theory to predict the geometry of each of the following:

a. $SiCl_4$ **c.** CCl_4
b. CO_3^{2-} **d.** SCl_2

75. The following graph shows how the percent ionic character of a single bond varies according to the difference in electronegativity between the two elements forming the bond. Answer the following questions, using this graph and Table 6.2.

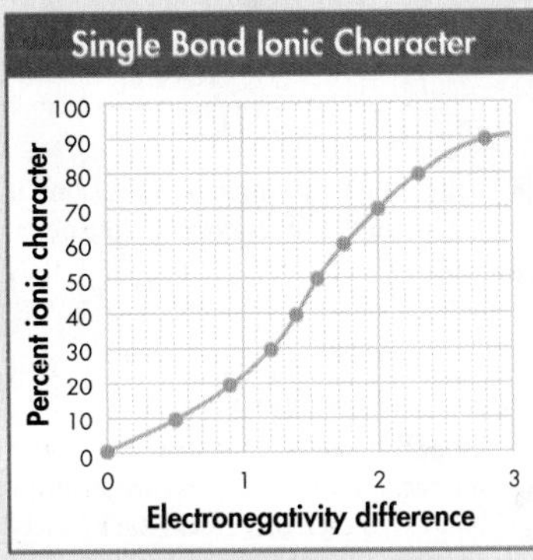

a. What is the relationship between the percent ionic character of single bonds and the electronegativity difference?
b. What electronegativity difference will result in a bond with a 50 percent ionic character?
c. Estimate the percent ionic character of the bonds formed between (1) lithium and oxygen, (2) nitrogen and oxygen, (3) magnesium and chlorine, and (4) nitrogen and fluorine.

Answers

60. **a.** linear **b.** tetrahedral **c.** trigonal planar **d.** bent **e.** linear **f.** bent

61. The 2*s* and the 2*p* orbitals form two *sp* hybrid orbitals on the carbon atom. One *sp* hybrid orbital forms a sigma bond between the carbon atom and each oxygen atom. *Pi* bonds between each oxygen atom and the carbon are formed by the unhybridized 2*p* orbitals.

62. sigma bonding orbital

63. **a.** sp^3 **b.** sp^2 **c.** *sp* **d.** *sp*

LESSON 8.4

64. The electronegativities of the two atoms will differ by about 0.4 to 2.0.

65. c, d, a, f, b, e

66. A hydrogen bond is formed by an electrostatic interaction between a hydrogen atom that is covalently bonded to an electronegative atom, and an unshared electron pair of a nearby atom.

67.
```
 H            H
  \          /
H—N: ····H—N:
  /          \
 H            H

 H            H
  \          /
  :O: ····H—N:
  /          \
 H            H
```

68. More energy is required to separate the molecules.

69. 1572 kJ

70. b. CH_3Cl

UNDERSTAND CONCEPTS

71. The 3*s* and three 3*p* orbitals of phosphorus hybridize to form four sp^3 atomic orbitals. The resulting shape is pyramidal with a bond angle of 107 between the sigma bonds.

72. :C̤̈l:S̤̈:C̤̈l: / :Ö: (below S)

73. **a.** C does not have an octet. [:C:::N:]⁻
b. One F has more than an octet. :F̤̈:P̤̈:F̤̈: / :F̤: (below P)

74. **a.** tetrahedral, 109.5
b. trigonal planar, 120
c. tetrahedral, 109.5
d. bent, 105

75. **a.** The percent ionic character increases as the difference in electronegativities increases.
b. 1.6
c. (1) 85% (2) 10% (3) 62% (4) 23%

ASSESSMENT

Evaluate

Answers

76. **a.** 109.5
b. 120
c. 180

77. **a.** trigonal planar
b. linear
c. pyramidal
d. tetrahedral

78. **a.** Phosphorus in PBr_5 has 10 valence electrons.

79. **a.**

```
  H :Ö:
H:C:C:Ö:H
  H
```

b. No, the molecule contains one carbon-oxygen double bond and one carbon-oxygen single bond.
c. polar bond
d. Yes, it has polar oxygen atoms at one end of the molecule and a nonpolar CH_3– group at the opposite end.

THINK CRITICALLY

80. C, O, H, S, N, F, Cl: These elements are all nonmetals.

81. Answers will vary. Table 8.3 suggests there is no clear difference. The student's argument could be based on chemical properties, such as conductivity of the compound in the liquid state.

82. Ethyl alcohol can form intermolecular hydrogen bonds between its polar –OH groups, but dimethyl ether can not form intermolecular hydrogen bonds.

83. **a.** two covalent bonds to both hydrogens; triple bond between carbons H:C⋮⋮C:H
b. Fluorine and oxygen have only four electrons. :F̤:Ö:H
c. Halogens form one covalent bond, not three. :Ï:C̈l:
d. Nitrogen forms three covalent bonds, not four. H:N̈::N̤:H

84. a. bent
b. tetrahedral
c. pyramidal

85. False. The bond dissociation energies exhibit no particular trend and, in fact, are fairly constant.

76. Give the angles between the orbitals of each hybrid.
a. sp^3 hybrids
b. sp^2 hybrids
c. sp hybrids

77. What is the geometry around the central atom in each of these simple molecules?

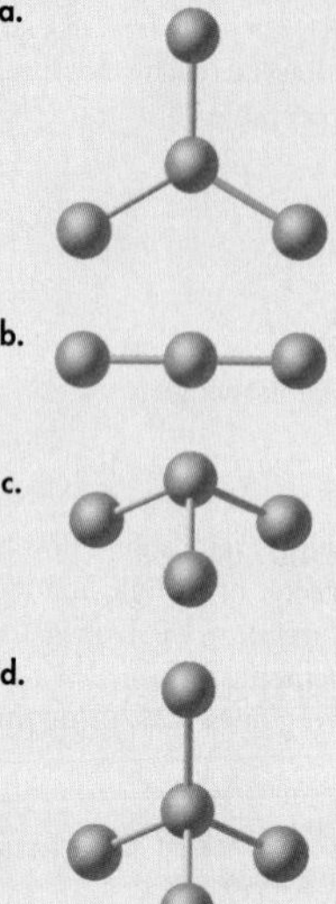

***78.** Which of the following molecules contains a central atom that does not obey the octet rule?
a. PBr_5
b. AlI_3
c. PF_3
d. $SiCl_4$

79. Vinegar contains the compound ethanoic acid, whose molecular formula is CH_3COOH.
a. Draw the electron dot structure of ethanoic acid. (*Hint:* The two carbon atoms are bonded to each other, and the two oxygens are bonded to the same carbon.)
b. Is the bonding between each of the oxygen atoms and the carbon the same?
c. Is the bonding between the carbon atom and each oxygen atom a polar or nonpolar bond?
d. Is ethanoic acid a polar molecule?

Think Critically

80. **Compare** Make a list of the elements in the compounds found in Table 8.2 on page 234. What do the elements that form covalent bonds have in common?

81. **Explain** Is there a clear difference between a very polar covalent bond and an ionic bond? Explain.

***82.** **Explain** Ethyl alcohol (CH_3CH_2OH) and dimethyl ether (CH_3OCH_3) each have the same molecular formula, C_2H_6O. Ethyl alcohol has a much higher boiling point (78°C) than dimethyl ether (–25°C). Propose an explanation for this difference.

83. **Evaluate** Although the relative positions of the atoms are correct in each of these molecules, there are one or more incorrect bonds in each of the electron dot structures. Identify the incorrect bonds. Draw the correct electron dot structure for each molecule.
a. H=C=C=H
b. :F—O—H
c. :I:::Cl:
d. H—N:::N—H

***84.** **Predict** What shape do you expect for a molecule with a central atom and the following pairings?
a. two bonding pairs of electrons and two nonbonding pairs of electrons
b. four bonding pairs and zero nonbonding pairs
c. three bonding pairs and one nonbonding pair

85. **Interpret Tables** Is this statement true or false? "As the electronegativity difference between covalently bonded atoms increases, the strength of the bond increases." Use the table below to justify your answer.

Bond	Electronegativity difference	Bond dissociation energy (kJ/mol)
C—C	2.5 − 2.5 = 0.0	347
C—H	2.5 − 2.1 = 0.4	393
C—N	3.0 − 2.5 = 0.5	305
C—O	3.5 − 2.5 = 1.0	356

Enrichment

86. **Explain** There are some compounds in which one atom has more electrons than the corresponding noble gas. Examples are PCl_5, SF_6, and IF_7. Draw the electron dot structures of P, S, and I atoms and of these compounds. Considering the outer shell configuration of P, S, and I, develop an orbital hybridization scheme to explain the existence of these compounds.

87. **Use Models** Draw the electron dot structure of formic acid, H_2CO_2. The carbon is the central atom, and all the atoms are attached to the carbon except for a hydrogen bonded to an oxygen.

88. **Predict** The electron structure and geometry of the methane molecule (CH_4) can be described by a variety of models, including electron dot structure, simple overlap of atomic orbitals, and orbital hybridization of carbon. Draw the electron dot structure of CH_4. Sketch two molecular orbital pictures of the CH_4 molecule. For your first sketch, assume that one of the paired $2s^2$ electrons of carbon has been promoted to the empty $2p$ orbital. Overlap each half-filled atomic orbital of carbon to a half-filled $2s$ orbital of hydrogen. What is the predicted geometry of the CH_4 molecule, using this simple overlap method? In your second sketch, assume hybridization of the $2s$ and $2p$ orbitals of carbon. Now what geometry would you predict for CH_4? Which picture is preferable based on the facts that all H—C—H bond angles in CH_4 are 109.5° and all C—H bond distances are identical?

89. **Use Models** Oxalic acid, $C_2H_2O_4$, is used in polishes and rust removers. Draw the electron dot structure for oxalic acid given that the two carbons are bonded together but neither of the hydrogen atoms is bonded to a carbon atom.

90. **Use Models** Draw as many resonance structures as you can for HN_3. (*Hint:* The three nitrogen atoms are bonded in a row, and the hydrogen atom is bonded to a nitrogen atom at the end of the row of nitrogens.)

★91. **Explain** Draw an electron dot structure for each molecule and explain why it fails to obey the octet rule.

a. BeF_2 **b.** SiF_6 **c.** ClO_2 **d.** BF_3 **e.** XeF_2

Write About Science

92. **Explain** Describe what a molecular compound is. Explain how a molecular formula is the chemical formula of a molecular compound.

93. **Research a Problem** Research how chemists know that an oxygen molecule has unpaired electrons. Write a brief report on what you find.

CHEMYSTERY

What's That Alarm?

The family realized that the alarm was caused by carbon monoxide (CO). In carbon monoxide, the carbon and oxygen atom are joined by a triple covalent bond. Although carbon monoxide and carbon dioxide are both made of carbon and oxygen atoms, they have very different properties.

Carbon monoxide is an odorless, tasteless gas. When it gets into the bloodstream, it causes the hemoglobin to convert to a form that is unable to transport oxygen. Symptoms of carbon monoxide poisoning include headaches, nausea, vomiting, and mental confusion. Exposure to high levels of carbon monoxide can result in death.

Fuel-burning appliances, such as water heaters, fireplaces, furnaces, and gas stoves, produce carbon monoxide. If the appliance is not functioning properly, it may release unsafe amounts of carbon monoxide. If a home contains one of these appliances, then the homeowners should install carbon monoxide detectors, since the gas cannot be detected by sight or smell.

94. **Use Models** Draw the electron dot structures of carbon monoxide and carbon dioxide. Describe the structural differences between these two molecules.

95. **Connect to the BIGIDEA** How does covalent bonding allow there to be different molecular compounds composed of the same kinds of atoms?

CHEMYSTERY

APPLY CONCEPTS Explain that carbon and oxygen can combine to form two gases, carbon dioxide and carbon monoxide, depending on whether the combustion of carbon is complete or incomplete. Incomplete combustion occurs at high temperatures when there is a limited supply of oxygen. Remind students that CO and CO_2 contain the same elements, but they have different properties because of their molecular structure. As a class, discuss the factors that affect the molecular properties of these two gases. Explain that the effects of CO on human health are due to the structure of CO and how it interacts with hemoglobin molecules in red blood cells.

CHEMYSTERY ANSWERS

94. :C⋮⋮O: Ö::C::Ö Carbon monoxide is one carbon and one oxygen atom joined by a triple bond. Carbon dioxide is one carbon atom joined to two oxygen atoms by double bonds.

95. **BIGIDEA** Atoms joined by covalent compounds can make compounds consisting of the same atoms, but with different numbers of each atom. The type of covalent bond (single, double, or triple) may be different depending on how many atoms are sharing electrons with each other.

Answers

ENRICHMENT

86. [electron dot structures of PCl_5, SF_6, IF_7]

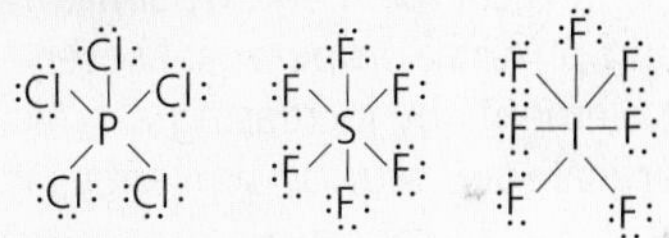

P forms 5 hybrid orbitals (dsp^3), *S* forms 6 hybrid orbitals (d^2sp^3), and I forms 7 hybrid orbitals (d^3sp^3).

87. H:C:O:H with :O: above C

88. [sketches of CH_4: H:C:H electron dot structure; first sketch with 90° angles and "?"; second sketch tetrahedral with 109.5° angles]

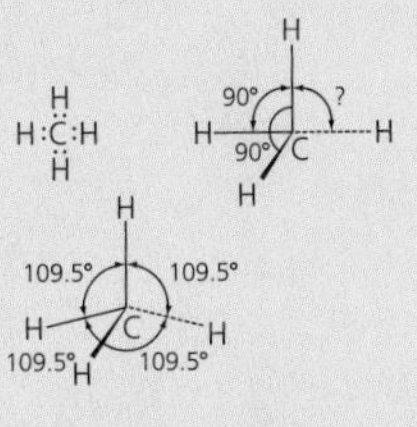

The first sketch shows carbon's three p orbitals oriented at 90° angles, resulting in a pyramidal structure for the carbon atom with three hydrogen atoms. The 4th C-H bond, formed with carbon's 2s orbital and a hydrogen atom's 1s orbital, is at unspecified angles to the other three C-H bonds. The second sketch is tetrahedral. The bond angles in the first sketch are not all the same; some are 90°. The bond angles in the second sketch are all 109.5°. The second sketch is correct. (Note: The wedge-shaped lines come out of the page; the dotted lines recede into the page.)

89. H:Ö:C:C:Ö:H with :O: above each C

90. H:N̈::N::N̈: ⟷ H:N̈:N:::N:

91. **a.** Be has only 4 valence electrons. :F̈—Be—F̈:

b. S has 12 valence electrons. [SF_6 structure]

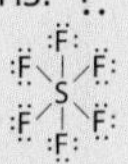

c. Cl has only 7 valence electrons. [ClO_2 structure]

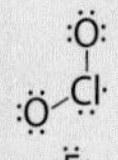

d. B has only 6 valence electrons. [BF_3 structure]

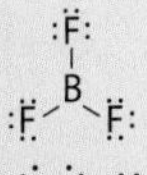

e. Xe has 10 valence electrons. :F̈—Xe—F̈:

WRITE ABOUT SCIENCE

92. Students should write that a molecular compound is a compound composed of molecules. A molecular formula shows how many atoms of each element a molecule contains.

93. Oxygen is paramagnetic, which means that it tends to be attracted to a magnetic field. This behavior is characteristic of substances that have unpaired electrons. Therefore, oxygen must have unpaired electrons.

ASSESSMENT

Evaluate

Answers

CUMULATIVE REVIEW

96. formation of a gas, a change in color or odor, formation of a precipitate

97. **a.** 6.65 x 10^4 micrometers
b. 4 centigrams
c. 5.62 x 10^1 decigram per liter
d. 2.4 x 10^1 meters per second

98. **a.** 2 **b.** 2
c. 4 **d.** 4

99. **a.** 16 **b.** 12
c. 8 **d.** 26

100. Isotopes have the same number of protons and electrons, but different numbers of neutrons.

101. Protons and electrons must be equal.

102. **a.** 6 **b.** 2
c. 5 **d.** 0

103. The wavelength decreases as the frequency increases.

104. The d orbitals related to the third principal energy level contain 5 electrons.

105. **a.** $1s^22s^22p^63s^1$
b. $1s^22s^22p^63s^23p^4$
c. $1s^22s^22p^63s^23p^3$
d. $1s^22s^22p^3$

106. The anion is larger than the corresponding neutral atom.

107. Mendeleev arranged the elements by increasing atomic mass in vertical rows and by similarities in chemical and physical properties. Mosely arranged the elements by increasing atomic number in vertical rows and by similarities in chemical and physical properties.

108. **a.** K, $1s^22s^22p^63s^23p^64s^1$
b. Al, $1s^22s^22p^63s^23p^1$
c. S, $1s^22s^22p^63s^23p^4$
d. Ba, $1s^22s^22p^63s^23p^63d^{10}4s^24p^64d^{10}5s^25p^66s^2$

109. **a.** barium
b. silicon
c. sodium

110. **e.** II and III only

111. All have the same number of electrons as a noble gas.

112. **b.** cesium

113. **a.** 8 **b.** 3
c. 6 **d.** 2

114. **a.** $1s^22s^22p^6$
b. $1s^22s^22p^6$
c. $1s^22s^22p^6$
d. $1s^22s^22p^63s^23p^6$

115. No, an alloy is a homogeneous mixture.

Cumulative Review

96. Name three indicators of chemical change.

*** 97.** Make the following conversions:
a. 66.5 mm to micrometers
b. 4×10^{-2} g to centigrams
c. 5.62 mg/mL to decigrams per liter
d. 85 km/h to meters per second

98. How many significant figures are in each measurement?
a. 0.00052 m **c.** 5.050 mg
b. 9.8×10^4 g **d.** 8.700 mL

99. How many neutrons are in each atom?
a. silicon-30 **c.** nitrogen-15
b. magnesium-24 **d.** chromium-50

100. How do isotopes of an atom differ?

*** 101.** In a neutral atom, the number of which two subatomic particles must always be equal?

102. How many electrons are in the 2*p* sublevel of an atom of each element?
a. aluminum **c.** fluorine
b. carbon **d.** lithium

103. What happens to the wavelength of light as the frequency increases?

104. What does the 5 in $3d^5$ represent?

105. Write correct electron configurations for atoms of the following elements:
a. sodium **c.** phosphorus
b. sulfur **d.** nitrogen

106. How does the ionic radius of a typical anion compare with the radius for the corresponding neutral atom?

107. What criteria did Mendeleev and Moseley use to arrange the elements on the periodic table?

108. Give the electron configuration of the element found at each location in the periodic table.
a. Group 1A, period 4
b. Group 3A, period 3
c. Group 6A, period 3
d. Group 2A, period 6

*** 109.** Identify the larger atom of each pair.
a. calcium and barium
b. silicon and sulfur
c. sodium and nitrogen

110. Which of these statements about the periodic table is correct?
I. Elements are arranged in order of increasing atomic mass.
II. A period is a horizontal row.
III. Nonmetals are located on the right side of the table.
a. I only
b. I and II only
c. I, II, and III
d. I and III only
e. II and III only

*** 111.** Which of the following ions has the same number of electrons as a noble gas?
a. Al^{3+}
b. O^{2-}
c. Br^-
d. N^{3-}

112. Which element is likely to form an ionic compound with chlorine?
a. iodine
b. cesium
c. helium

113. How many valence electrons does each atom have?
a. argon
b. aluminum
c. selenium
d. beryllium

114. Write the electron configuration of each ion.
a. oxide ion
b. magnesium ion
c. nitride ion
d. potassium ion

115. An alloy is composed of two or more elements. Is an alloy a compound? Explain your answer.

If You Have Trouble With . . .

Question	96	97	98	99	100	101	102	103	104	105	106	107	108	109	110	111	112	113	114	115
See Chapter	2	3	3	4	4	4	5	5	5	5	6	6	6	6	6	7	7	7	7	7

Standardized Test Prep for Indiana

Standards: C.3.1, C.3.2, C.3.4

Select the choice that best answers each question or completes each statement.

1. A bond in which two atoms share a pair of electrons is not
   (A) a coordinate covalent bond.
   (B) a polar covalent bond.
   (C) an ionic bond.
   (D) a nonpolar covalent bond.

2. How many valence electrons are in a molecule of phosphoric acid, H_3PO_4?
   (A) 7 (C) 24
   (B) 16 (D) 32

3. Which of these molecules can form a hydrogen bond with a water molecule?
   (A) N_2 (C) O_2
   (B) NH_3 (D) CH_4

4. Which substance contains both covalent and ionic bonds?
   (A) NH_4NO_3 (C) LiF
   (B) CH_3OCH_3 (D) $CaCl_2$

5. Which of these bonds is most polar?
   (A) H—Cl (C) H—F
   (B) H—Br (D) H—I

Use the description and data table below to answer Questions 6–9.

The table relates molecular shape to the number of bonding and nonbonding electron pairs in molecules.

Bonding pairs	Non-bonding pairs	Arrangement of electron pairs	Molecular shape	Example
4	0	tetrahedral	tetrahedral	CH_4
3	1	tetrahedral	pyramidal	NCl_3
2	2	tetrahedral	bent	H_2S
1	3	tetrahedral	linear	HF

6. Draw the electron dot structure for each example molecule.

7. Explain why the arrangement of electron pairs is tetrahedral in each molecule.

8. H_2S has two hydrogen atoms bonded to a sulfur atom. Why isn't the molecule linear?

9. What is the arrangement of electron pairs in PBr_3? Predict the molecular shape of a PBr_3 molecule.

For Questions 10–11, identify the type of intermolecular bonding represented by the dotted lines in the drawings.

10. H_2O

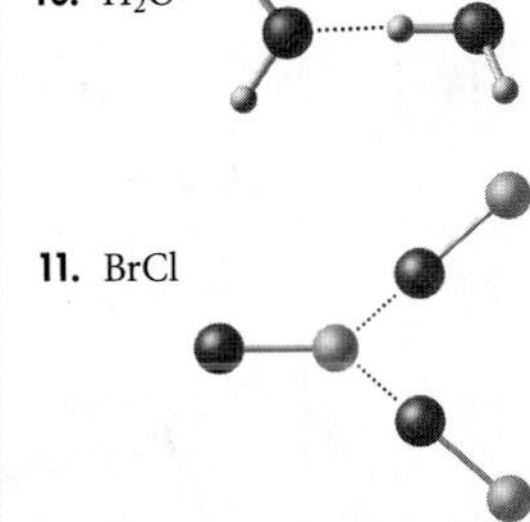

11. BrCl

Tips for Success

Connectors Sometimes two phrases in a true/false question are connected by a word such as *because*. The word implies that one thing caused another thing to happen. Statements that include such words can be false even if both parts of the statement are true by themselves.

In Questions 12–14, a statement is followed by an explanation. Decide if each statement is true, and then decide if the explanation given is correct.

12. A carbon monoxide molecule has a triple covalent bond because carbon and oxygen atoms have an unequal number of valence electrons.

13. Xenon has a lower boiling point than neon because dispersion forces between xenon atoms are stronger than those between neon atoms.

14. The nitrate ion has three resonance structures because the nitrate ion has three single bonds.

If You Have Trouble With . . .

Question	1	2	3	4	5	6	7	8	9	10	11	12	13	14
See Lesson	8.2	8.2	8.4	8.3	8.1	8.3	8.2	8.2	8.2	8.2	8.4	8.4	8.4	8.2

ASSESSMENT

STP Answers

MULTIPLE CHOICE

1. C
2. D
3. B
4. A
5. C
6. H:C:H (with H above and below C) :Cl:N:Cl: (with :Cl: above N) H:S:H H:F:
7. Each central atom has four pairs of electrons that, according to VSEPR theory, assume a tetrahedral shape.
8. The two nonbonding pairs repel the bonding pairs; there are still four pairs of electrons around the sulfur atom.
9. The arrangement of electron pairs is tetrahedral. The electron dot structure shows three bonding electron pairs and one non-bonding electron pair; thus, the predicted molecular shape is pyramidal.
10. hydrogen bonding
11. primarily dispersion forces or dipole interactions
12. true, true but incorrect explanation
13. false, true
14. true, false

Covalent Bonding 261

9 Chemical Names and Formulas Planning Guide

Introducing the BIG IDEA: ELECTRONS AND THE STRUCTURE OF ATOMS; BONDING AND

Elements are composed of particles called atoms, and every atom contains a nucleus and one or more electrons. Most elements exist in

IASS	Lessons and Objectives	Print Resources		
		For the Student	For the Teacher	
C.3.5	**9.1** Naming Ions p 264–269 **9.1.1** Explain how to determine the charges of monatomic ions. **9.1.2** Explain how polyatomic ions differ from and are similar to monatomic ions.	Reading and Study Workbook Lesson 9.1 Lesson Assessment 9.1 p 269	Class Activity, p 265: Symbols for Monatomic Ions Teacher Demo, p 266: Colorful Ions Class Activity, p 268: Shapes of Polyatomic Ions	
C.3.5	**9.2** Naming and Writing Formulas for Ionic Compounds p 271–279 **9.2.1** Apply the rules for naming and writing formulas for binary ionic compounds. **9.2.2** Apply the rules for naming and writing formulas for compounds with polyatomic ions.	Reading and Study Workbook Lesson 9.2 Lesson Assessment 9.2 p 279 Quick Lab: Making Ionic Compounds p 279	Teacher Demo, p 273: Making and Naming an Ionic Compound Class Activity, p 274: Naming Ionic Binary Compounds Class Activity, p 275: Formulas for Binary Ionic Compounds Teacher Demo, p 276: Making and Naming Lead Carbonate	
C.3.4	**9.3** Naming and Writing Formulas for Molecular Compounds p 280–283 **9.3.1** Apply the rules for naming and writing formulas for binary molecular compounds.	Reading and Study Workbook Lesson 9.3 Lesson Assessment 9.3 p 283	Class Activity, p 282: Naming Binary Molecular Compounds	
C.3.4	**9.4** Naming and Writing Formulas for Acids and Bases p 285–288 **9.4.1** Determine the name and formula of an acid. **9.4.2** Determine the name and formula of a base.	Reading and Study Workbook Lesson 9.4 Lesson Assessment 9.5 p 288		
C.3.4	**9.5** The Laws Governing How Compounds Form p 289–294 **9.5.1** Explain how the law of definite proportions is consistent with Dalton's atomic theory. **9.5.2** List the general guidelines that can help you write the name and formula of a chemical compound.	Reading and Study Workbook Lesson 9.5 Lesson Assessment 9.5 p 294 Small-Scale Lab: Names and Formulas for Ionic Compounds p 295		

Assessing the BIG IDEA: ELECTRONS AND THE STRUCTURE OF ATOMS; BONDING AND

	Essential Questions 1. How does the periodic table help you determine the names and formulas of ions and compounds? 2. What is the difference between an ionic and a molecular compound?	Study Guide p 296 Skills Tune-Up p 297 STP p 303 Reading and Study Workbook Self-Check and Vocabulary Review Chapter 9		

INTERACTIONS

chemical compounds, where elements are held together by chemical bonds.

	PearsonChem.com	
	Editable Worksheets	Online Activities
	Teaching Resources, Lesson 9.1 Review	9.1 Lesson Overview
	Teaching Resources, Lesson 9.2 Review Small-Scale Lab Manual Lab 12: Chemical Names and Formulas	9.2 Lesson Overview Writing Formulas for Binary Ionic Compounds Ionic Naming Your Ionic Kitchen (and Bathroom)
	Teaching Resources, Lesson 9.3 Review	9.3 Lesson Overview
	Teaching Resources, Lesson 9.4 Review	9.4 Lesson Overview
	Teaching Resources, Lesson 9.5 Review Lab Practical 9-1: Nomenclature	9.5 Lesson Overview The Law of Multiple Proportions Calculating Mass Ratios Naming Compounds

INTERACTIONS

	Core Teaching Resources (includes editable worksheets) • Lesson Reviews • Practice Problems • Interpret Graphs • Vocabulary Review • Chapter Quizzes and Tests	Chapter 9 Problem Set Unit Ratios

Materials List

For the Student

Quick Lab p 279
- nine small test tubes
- test tube rack
- six solutions in plastic dropper bottles containing 0.05*M* $AgNO_3$, 0.2*M* $Pb(NO_3)_2$, 1.0*M* $NaCO_3$, 0.5*M* NaOH, 0.1*M* $Na_3(PO_4)_2$, and 0.1*M* $FeCl_3$

Small-Scale Lab p 295
- reaction surface
- 0.05*M* $AgNO_3$, 0.2*M* $Pb(NO_3)_2$, 0.5*M* $CaCl_2$, 1.0*M* Na_2CO_3, 0.1*M* Na_3PO_4, 0.5*M* NaOH, 0.2*M* Na_2SO_4, 0.2*M* $CuSO_4$, 0.2*M* $MgSO_4$, 0.1*M* $FeCl_3$

For the Teacher

Class Activity p 265
- blank photocopy of the periodic table for each pair of students

Teacher Demo p 266
- beakers containing solutions of soluble metal salts: $MnCl_2$, $FeCl_2$, $CoCl_2$, $NiCl_2$, and $CuCl_2$

Class Activity p 268
- foam balls
- wooden sticks

Teacher Demo p 273
- 1 g powdered zinc
- 4 g iodine (I_2)
- watch glass
- eyedropper
- 8 mL water

Class Activity p 275
- the list of names of ionic compounds generated in the activity on p 274

Teacher Demo p 276
- 50 mL 0.1*M* lead(II) nitrate
- 50 mL 0.1*M* sodium carbonate
- 100-mL beaker

Additional Digital Resources

Online Student Edition
Online Teacher's Edition
Exam*View Assessment Suite CD-ROM*
Classroom Resources DVD
9.2 Virtual Chem Lab 2: Names and Formulas of Ionic Compounds

CHAPTER 9

What's Online

CHEMISTRY TUTORIAL Students can work step-by-step with the online tutorials to learn how to name and write formulas for binary ionic compounds and to calculate mass ratios by accessing the Chapter 9 Chemistry Tutorials.

ONLINE PROBLEM SET Additional practice problems about chemical names and formulas can be found in the Chapter 9 online Problem Set.

MATH TUTORIAL The online Math Tutorial will help guide students through converting fractions to unit ratios.

DIRECTED VIRTUAL LAB Encourage students to review naming ionic compounds containing transition metals using the animated virtual lab for Chapter 9.

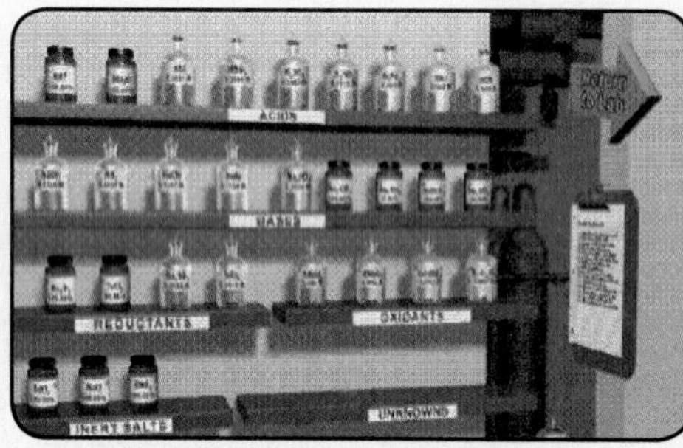

KINETIC ART Have students watch animations about naming compounds and making different compounds with the same elements. Then have students answer the questions that follow each animation.

CONCEPTS IN ACTION Students can watch an interactive animation about ionic compounds that are commonly found in their own houses.

9 Chemical Names and Formulas

INSIDE:

PearsonChem.com

Many transition metals form brightly colored compounds that are used in making artists' paints.

262

IN ACADEMIC STANDARDS for SCIENCE

Nature of Science

- Clearly communicate their ideas and results of investigations verbally and in written form using tables, graphs, diagrams, and photographs.
- Use analogies and models (mathematical and physical) to simplify and represent systems that are difficult to understand or directly experience due to their size, time scale, or complexity, and recognize the limitations of analogies and models.

Standard 3: Bonding and Molecular Structure

C.3.4 Draw structural formulas for and name simple molecules.

C.3.5 Predict chemical formulas for ionic compounds given their name and vice versa.

Focus on ELL

1 CONTENT AND LANGUAGE Write the the chapter title on the board, and underline the words *Names* and *Formulas*. Ask students to predict the difference between a chemical name and a chemical formula.

BEGINNING: LOW/HIGH Rewrite the lesson titles containing a conjugate as two sentences. Read the new lesson titles aloud.

INTERMEDIATE

LOW Present previously learned examples of names and formulas of simple compounds.

HIGH Predict how naming compounds differs from writing formulas for compounds.

ADVANCED: LOW/HIGH Use information from previous chapters to predict what is likely to vary between naming ions, ionic compounds, and molecular compounds. Begin a list of naming rules for each.

CHAPTER 9

BIG IDEAS

- **ELECTRONS AND THE STRUCTURE OF ATOMS**
- **BONDING AND INTERACTIONS**

Essential Questions:

1. *How does the periodic table help you determine the names and formulas of ions and compounds?*
2. *What is the difference between an ionic and a molecular compound?*

CHEMYSTERY

Cucumber Blunder

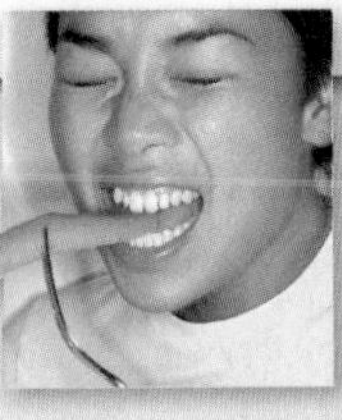

Tara's friend, Ellie, raves about the delicious lime pickles that Tara's grandma makes. As a surprise, Tara decided to make a batch for Ellie. She got the recipe from her grandma. It called for "1 cup of lime for every 3 pounds of cucumbers."

When the pickles were ready, Tara proudly presented the gift to Ellie. She watched Ellie's face as she bit into a pickle. After the first bite, Ellie made an awful face. "What's wrong?" Tara asked. After a short pause, Ellie replied, "They taste . . . funny." Tara tried one and agreed; they tasted nothing like her grandma's pickles. What did Tara do wrong?

► Connect to the **BIGIDEA** As you read about naming compounds, think about what might have caused Tara's cucumber blunder.

INDIANA'S ACADEMIC STANDARDS FOR SCIENCE

Bonding and Molecular Structure C.3.4, C.3.5. See lessons for details.

UbD

Understanding by Design

Students are building toward naming and writing formulas for compounds using their knowledge of atoms, ions, and bonding.

PERFORMANCE GOALS At the end of Chapter 9, students will be able to answer the essential questions by applying their knowledge of ions, ionic compounds, and molecular compounds.

ESSENTIAL QUESTIONS Read the essential questions aloud. **Ask** What is an ion? *(an atom or group of atoms that has a positive or negative charge)*

BIGIDEA Use the photo of an artist's paints to help students connect to this chapter's concepts. Tell students that the artist needs to buy more green paint to finish a picture, but his supplier is out of his favorite brand, and brands use different names for the same hues. **Ask** How could the artist make sure he gets an exact match? *(Answers will vary.)* Explain that all paint pigments have official color index name codes. The artist can use the name code to match the hue.

Explain that, in this chapter, students will discover how chemists make sure that each chemical compound is identified by a unique name and formula.

CHEMYSTERY Connect the CHEMystery to the Big Idea by having students describe what information they get from the chemical formula of a compound. For example, point out that the formula for water, "H_2O," gives information about the composition of water. Ask students to predict why Tara's lime pickles tasted funny. As a hint, ask students how much information they can get from just the name "lime."

Introduce the Chapter

REAL-WORLD CONNECTION Use this activity to draw attention to information students can obtain by looking at the chemical names of ingredients in a variety of household products.

Activity *Display a selection of household products or just their packaging labels. Examples include soap, detergent, toothpaste, shampoo, ibuprofen, bathroom cleaner, windshield wiper fluid, sunscreen, and food. Make sure that at least one chemical name is displayed on the label and that it indicates the elements in the compound.* Have students work with a partner to examine the ingredients listed on each package and record the chemical names with elements they recognize. **Ask** What can you learn from the chemical names? *(the elements in the compounds)* **Ask** Would knowing the chemical formula of an ingredient help you to know more, less, or about the same amount of information? Explain. *(More; the formula would show the proportion of the elements and possibly the type of bonding.)*

Pass Me the Glauber's Salt

Two alchemists who have been lost at sea discover there's a "new" method to name the compounds they use.

LESSON 9.1

Key Objectives

9.1.1 **EXPLAIN** how to determine the charges of monatomic ions.

9.1.2 **EXPLAIN** how polyatomic ions differ from and are similar to monatomic ions.

Additional Resources

Reading and Study Workbook, Lesson 9.1

Available online or DVD-ROM:

- Lesson 9.1 Review
- Interpreting Graphics

PearsonChem.com

Engage

CHEMISTRY & YOU Note that it is difficult to converse with people in a different country unless you are familiar with their language. Emphasize that chemists everywhere use the same chemical language. **Ask** Why is it important that there be only one language of chemistry? *(Chemists need to communicate with one another and having the same names for a compound makes this easier.)*

Activate Prior Knowledge

Have students review the meanings of the terms *anion*, *cation*, and *ion*. Have students locate their working copy of the periodic table and practice determining the charge on elements in the different groups.

9.1 Naming Ions

C.3.5 Write chemical formulas

CHEMISTRY & YOU

Q: *Do you speak "Chemistry"?* Try looking at the ingredient label on a household product—a bottle of shampoo, a tube of toothpaste, a box of detergent. Do the names of the ingredients make sense? To truly understand chemistry, you must learn its language. Part of learning the language of chemistry involves understanding how to name ionic compounds. For this you need to know how to name ions.

Key Questions

How can you determine the charges of monatomic ions?

How do polyatomic ions differ from monatomic ions? How are they similar?

Vocabulary

- monatomic ion

Monatomic Ions

How can you determine the charges of monatomic ions?

Ionic compounds consist of a positive metal ion and a negative nonmetal ion combined in a proportion such that their charges add up to a net charge of zero. For example, the ionic compound sodium chloride (NaCl) consists of one sodium ion (Na^+) and one chloride ion (Cl^-). Probably you are already familiar with the name and formula of sodium chloride, which is common table salt. But it is important, in learning the language of chemistry, to be able to name and write the chemical formulas for all ionic compounds. The first step is to learn about the ions that form ionic compounds. Some ions, called **monatomic ions,** consist of a single atom with a positive or negative charge resulting from the loss or gain of one or more valence electrons, respectively.

Cations Recall that metallic elements tend to lose valence electrons. Lithium, sodium, and potassium in Group 1A lose one electron to form cations. All the Group 1A ions have a 1+ charge (Li^+, Na^+, K^+, Rb^+, Cs^+, and Fr^+). Magnesium and calcium are Group 2A metals. They tend to lose two electrons to form cations with a 2+ charge (Mg^{2+} and Ca^{2+}), as do all the other Group 2A metals. Aluminum is the only common Group 3A metal. As you might expect, aluminum tends to lose three electrons to form a 3+ cation (Al^{3+}). **When the metals in Groups 1A, 2A, and 3A lose electrons, they form cations with positive charges equal to their group number.** Figure 9.1 shows some of the elements whose ionic charges can be obtained from their positions in the periodic table.

The names of the cations of the Group 1A, Group 2A, and Group 3A metals are the same as the name of the metal, followed by the word *ion* or *cation*. Thus Na^+ is the sodium ion (or cation), Ca^{2+} is the calcium ion (or cation), and Al^{3+} is the aluminum ion (or cation).

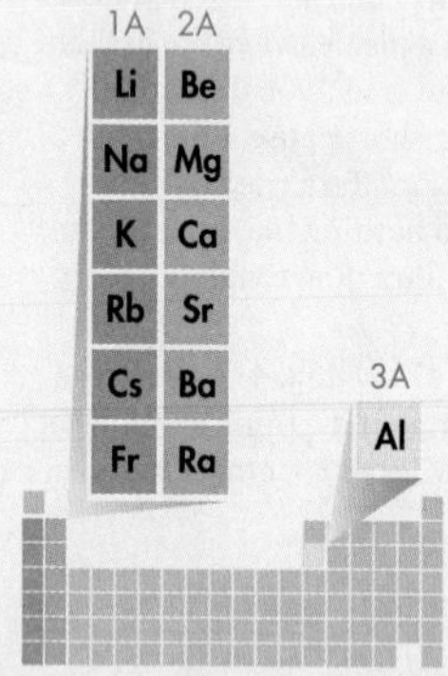

Figure 9.1
Elements That Form Cations
These representative elements form positive ions with charges equal to their group number.

IN ACADEMIC STANDARDS for SCIENCE

- Use analogies and models (mathematical and physical) to simplify and represent systems that are difficult to understand or directly experience due to their size, time scale, or complexity, and recognize the limitations of analogies and models.

C.3.5 Write chemical formulas for ionic compounds given their name and vice versa.

Focus on ELL

1 CONTENT AND LANGUAGE Begin the lesson by drawing students' attention to the word *monatomic* in the vocabulary list. Explain that the word *monatomic* has the prefix *mono-*, which means "single." Have students use this information to predict the meaning of the word *monatomic*.

2 FRONTLOAD THE LESSON Ask students if there is a particular naming system for babies in their home country, such as surname additions or endings. Relate this to the way chemists have a system to name ions.

3 COMPREHENSIBLE INPUT Have students practice think-aloud techniques as they investigate the tables and figures in this lesson. Demonstrate by asking specific questions, such as, *What is the charge on a cesium ion?* or *Which group contains cations with a +3 charge?* Allow students to respond with symbols, by pointing to the answer, or by using gestures to demonstrate their answer.

Table 9.1

Ion Symbols for Some Group A Elements

1A	2A	3A	4A	5A	6A	7A	8A
Li^+	Be^{2+}			N^{3-}	O^{2-}	F^-	
Na^+	Mg^{2+}	Al^{3+}		P^{3-}	S^{2-}	Cl^-	
K^+	Ca^{2+}			As^{3-}	Se^{2-}	Br^-	
Rb^+	Sr^{2+}					I^-	
Cs^+	Ba^{2+}						

Anions Nonmetals tend to gain electrons to form anions, so the charge of a nonmetallic ion is negative. **The charge of any ion of a Group A nonmetal is determined by subtracting 8 from the group number.** The elements in Group 7A form anions with a 1− charge (7 − 8 = −1). The name of an anion is not the same as the element's name. Anion names start with the stem of the element name and end in *-ide*. For example, two elements in Group 7A are fluorine and chlorine. The anions for these nonmetals are the fluor*ide* ion (F^-) and chlor*ide* ion (Cl^-). Anions of nonmetals in Group 6A have a 2− charge (6 − 8 = −2). Group 6A elements, oxygen and sulfur, form the ox*ide* anion (O^{2-}) and the sulf*ide* anion (S^{2-}), respectively. The first three elements in Group 5A, nitrogen, phosphorus, and arsenic, can form anions with a 3− charge (5 − 8 = −3). These anions have the symbols N^{3-}, P^{3-}, and As^{3-} and are called, respectively, nitr*ide* ion, phosph*ide* ion, and arsen*ide* ion. Figure 9.2 shows some common Group A elements that form anions. Table 9.1 summarizes the ionic charges of representative elements that can be obtained from the periodic table. The majority of the elements in Groups 4A and 8A usually do not form ions.

Figure 9.2 Elements That Form Anions
These representative elements form negative ions with charges equal to the group number of the element minus 8.

Metals That Form More Than One Ion The metals of Groups 1A, 2A, and 3A consistently form cations with charges of 1+, 2+, and 3+, respectively. Many of the transition metals (Groups 1B–8B) form more than one cation with different ionic charges. Some of these are shown in Figure 9.3. **The charges of the cations of many transition metal ions must be determined from the number of electrons lost.** For example, the transition metal iron forms two common cations, Fe^{2+} (two electrons lost) and Fe^{3+} (three electrons lost). Cations of tin and lead, the two metals in Group 4A, can also have more than one common ionic charge.

Two methods are used to name these ions. The preferred method is called the Stock system. In the Stock system, you place a Roman numeral in parentheses after the name of the element to indicate the numerical value of the charge. For example, the cation Fe^{2+} is named iron(II) ion. Note that no space is left between the element name and the Roman numeral in parentheses. The name for Fe^{2+} is read "iron two ion." The Fe^{3+} ion is named iron(III) ion and is read "iron three ion."

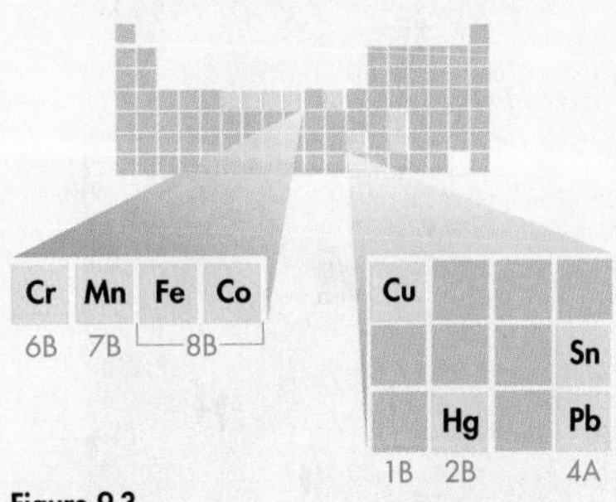

Figure 9.3
Elements That Form More Than One Ion
These metallic elements form more than one positive ion.
Identify ***Which of these elements are transition metals?***

LESSON 9.1

Foundations for Reading

BUILD VOCABULARY As a class, create a list of words that begin with the prefixes *mono-* and *poly-*. *(Students may respond with words such as* monotone, monochrome, monopoly, polynomial, *or* polygon.*)* Discuss the meanings of each word. **Ask** How do you think monatomic ions and polyatomic ions differ? *(A monatomic ion has one atom; a polyatomic ion has more than one atom.)*

READING STRATEGY Before they begin reading the lesson, have students create a chart to compare information about charge, composition, and name endings of various ions.

Explain

Monatomic Ions

USE VISUALS Have students study Figures 9.1 and 9.2. Have students note that in each figure, the periodic table is shown in the background to help them locate the groups being discussed. **Ask** How do the ions formed by metals in Groups 1A, 2A, and 3A differ from the ions formed by nonmetals in Groups 5A, 6A, and 7A? *(Groups 1A, 2A, and 3A form positive ions; Groups 5A, 6A, and 7A form negative ions.)* **Ask** Are the ions formed by the transition metals cations or anions? *(They are cations.)*

Explore

Class Activity

PURPOSE Students will use the periodic table for writing symbols for monatomic ions.

MATERIALS blank photocopy of the periodic table for each pair of students

PROCEDURE Have students work in pairs. Allow them to refer to their textbooks or other resource materials. In the element blocks on the blank periodic table, have them write the symbols and names of the cations and anions of as many elements as possible. Have them use their tables to identify any trends in ionic charges. Encourage them to keep their tables for reference.

Differentiated Instruction

ELL ENGLISH LANGUAGE LEARNERS Have students prepare flash cards with the name of a monatomic or polyatomic ion on one side and its formula on the other side. Pair students so they can use the flash cards to quiz each other.

L1 SPECIAL NEEDS STUDENTS For some students, the charge that results when electrons are transferred is counterintuitive—additions result in a negative charge, subtractions result in a positive charge. Give students different-colored disks to represent protons and electrons in neutral atoms. Remove or add electrons and ask students to find the charge of the resulting "ion."

L3 ADVANCED STUDENTS Ask students to determine the resulting compound that forms between a metal ion and a polyatomic anion, and what type of bonds are involved. *(Ionic compound; the resulting compound contains both ionic bonds and covalent bonds.)* Ask students to determine the type of bonds used to hold atoms together in all polyatomic ions. *(covalent bonds)*

Answers

FIGURE 9.3 Cr, Mn, Fe, Co, Cu, and Hg are transition metals; Sn and Pb are not.

LESSON 9.1

Explain

CHEMISTRY & YOU The periodic table is a very useful tool for naming ions. Cations from Groups 1A–3A are named by stating the element name followed by the word *ion* or *cation*. Anions from Groups 5A–7A are named by adding the suffix *-ide* to the element stem.

Explore

Teacher Demo

PURPOSE Students will see the various colors of solutions of transition metal ions and practice naming the ions.

MATERIALS beakers containing solutions of soluble metal salts, for example: $MnCl_2$, $FeCl_2$, $CoCl_2$, and $CuCl_2$, or ions such as Co^{3+}, Cr^{3+}, Fe^{3+}, and Mn^{2+}

PROCEDURE Write the symbols for the metal ions on the board. Have students describe the colors of the solutions. Ask them to use Table 9.2 to write the Stock and classical names of each cation. Ask students to suggest why chemists prefer to use the Stock system for naming ions.

Extend

Organize students into research teams to gather data about paint pigments from different sources. Assign each group one of the following tasks:

- Examine the pigments available at an art-supply store. Obtain as much information as possible from their labels and then use the Internet to research their composition.
- Create a series of questions and conduct an interview with at least two art teachers about their experience with pigments.
- Research the type of pigments available and popular during earlier eras; guide students to look for information about natural sources of pigments.
- Interview a curator or art restorer from a local museum to talk about using pigments to repair damaged art, authenticate art, or unmask a forgery.

Q: *Suppose you were trying to teach someone how to name ions. Which rules about the "language of chemistry" would you emphasize?*

An older, less useful method for naming these cations uses a root word with different suffixes at the end of the word. The older, or classical, name of the element is used to form the root name for the element. For example, *ferrum* is Latin for iron, so *ferr-* is the root name for iron. The suffix *-ous* is used to name the cation with the lower of the two ionic charges. The suffix *-ic* is used with the higher of the two ionic charges. Using this system, Fe^{2+} is the ferrous ion, and Fe^{3+} is the ferric ion, as shown in Table 9.2. Notice that you can usually identify an element from what may be an unfamiliar classical name by looking for the element's symbol in the name. For example, *ferrous* (Fe) is iron; *cuprous* (Cu) is copper; and *stannous* (Sn) is tin. A major disadvantage of using classical names for ions is that they do not tell you the actual charge of the ion. A classical name tells you only that the cation has either the smaller (*-ous*) or the larger (*-ic*) charge of the pair of possible ions for that element.

A few transition metals have only one ionic charge. The names of these cations do not have a Roman numeral. These exceptions include silver, with cations that have a 1+ charge (Ag^+), as well as cadmium and zinc, with cations that have a 2+ charge (Cd^{2+} and Zn^{2+}). As Figure 9.4 shows, some transition metal ions form colorful solutions.

Figure 9.4
Transition Metal Ions
Compounds of transition metals are often strongly colored. Solutions that contain transition metal ions can also be colored.

Table 9.2

Symbols and Names of Common Metal Ions With More Than One Ionic Charge

Symbol	Stock name	Classical name
Cu^+	Copper(I) ion	Cuprous ion
Cu^{2+}	Copper(II) ion	Cupric ion
Fe^{2+}	Iron(II) ion	Ferrous ion
Fe^{3+}	Iron(III) ion	Ferric ion
$*Hg_2^{2+}$	Mercury(I) ion	Mercurous ion
Hg^{2+}	Mercury(II) ion	Mercuric ion
Pb^{2+}	Lead(II) ion	Plumbous ion
Pb^{4+}	Lead(IV) ion	Plumbic ion
Sn^{2+}	Tin(II) ion	Stannous ion
Sn^{4+}	Tin(IV) ion	Stannic ion
Cr^{2+}	Chromium(II) ion	Chromous ion
Cr^{3+}	Chromium(III) ion	Chromic ion
Mn^{2+}	Manganese(II) ion	Manganous ion
Mn^{3+}	Manganese(III) ion	Manganic ion
Co^{2+}	Cobalt(II) ion	Cobaltous ion
Co^{3+}	Cobalt(III) ion	Cobaltic ion

*A diatomic elemental ion

UbD Check for Understanding

The Essential Question How can you determine the charges of monatomic ions?

Assess students' comprehension by asking them to show you how they use the groups in the periodic table to determine the charge on any given monatomic ion. Students should properly use the terms *cations* and *anions* in their explanation.

ADJUST INSTRUCTION If students are having trouble answering, provide them a clean copy of the periodic table. Instruct them to use colored pencils to color code the groups within the table to help them determine the charges. Direct students to revisit the text following the *Cations* and *Anions* subheads.

SampleProblem 9.1

Naming Cations and Anions

Name the ion formed by each of the following elements:

a. potassium **b.** lead, 4 electrons lost **c.** sulfur

1 Analyze **Identify the relevant concepts.** You can use the periodic table to determine the charge of most Group A elements. Ions with positive charges are cations; ions with negative charges are anions. The names of nonmetallic anions end in *-ide.* Metallic cations take the name of the metal. Some metals, including transition metals, can form more than one cation. Use a Roman numeral in the Stock name or use the classical name with a suffix to name these metals.

2 Solve **Apply the concepts to this problem.**

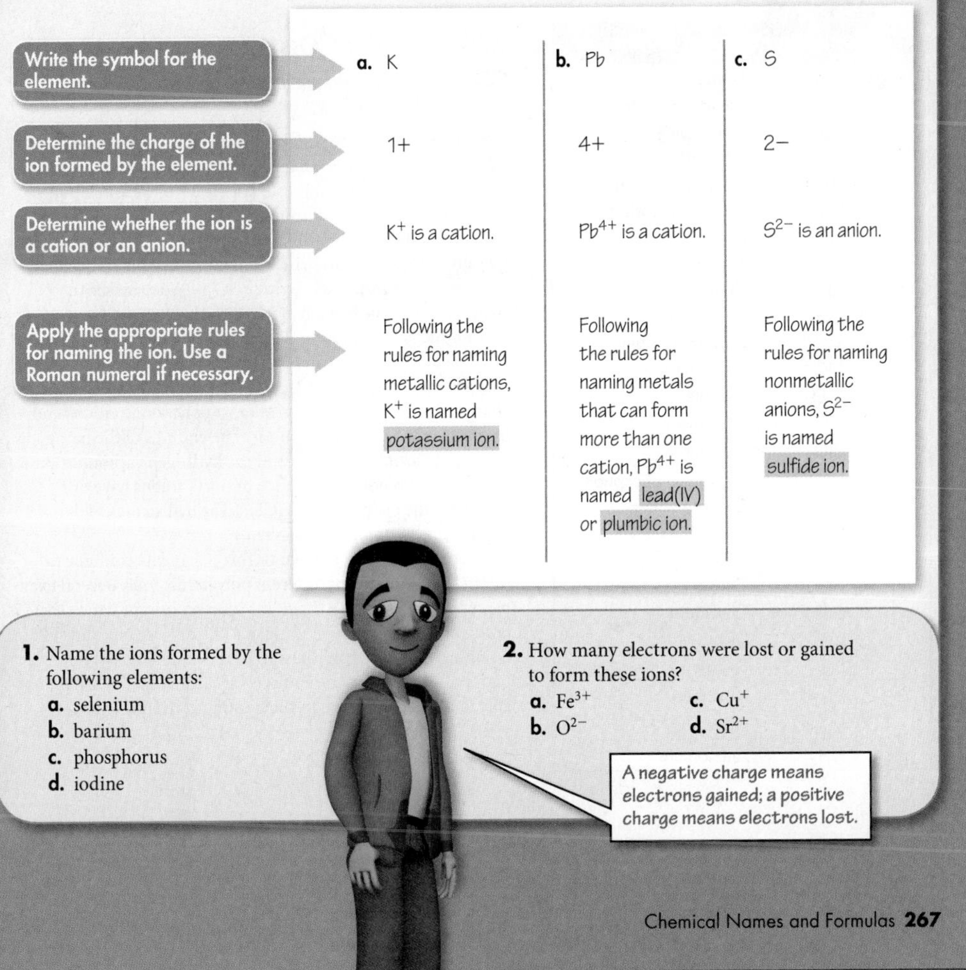

1. Name the ions formed by the following elements:
a. selenium
b. barium
c. phosphorus
d. iodine

2. How many electrons were lost or gained to form these ions?
a. Fe^{3+} **c.** Cu^{+}
b. O^{2-} **d.** Sr^{2+}

A negative charge means electrons gained; a positive charge means electrons lost.

Two Kinds of Bonding

When an ionic compound forms between a metal ion and a polyatomic ion, the resulting compound contains both ionic bonds and covalent bonds. The atoms in all polyatomic ions are held together by covalent bonds.

Explain

Monatomic Ions

Sample Practice Problem

Write the symbol (including charge) for the ion formed by each element. Then classify each as an anion or a cation.

a. arsenic (*As^{3-}, anion*)
b. beryllium (*Be^{2+}, cation*)
c. astatine (*At^{-}, anion*)
d. gallium (*Ga^{3+}, cation*)

Answers

1. **a.** selenide ion
b. barium ion
c. phosphide ion
d. iodide ion

2. **a.** three electrons lost
b. two electrons gained
c. one electron lost
d. two electrons lost

LESSON 9.1

Explain

Polyatomic Ions

START A CONVERSATION Write the following formulas on the board: NaOH, H_2SO_4, NH_4NO_3, and Na_2CO_3. Ask students what these compounds have in common. *(Students may note that they all contain oxygen.)* Tell students that this is an important observation. Explain that all of these compounds contain ions that are made up of more than one type of atom. In ions such as these, oxygen is almost always present.

USE VISUALS. Direct students' attention to Table 9.3. Point out how the table is divided into sections according to charge: more than half of the ions have a charge of 1–, only two ions have charges of 3–, and both of these contain the phosphorus atom. Point out that the ammonium ion is the only common polyatomic ion with a 1+ charge, and it does not contain oxygen. The remaining ions have charges of 2–. Have students write the formulas on the board for all of the *-ate* and *-ite* pairs. **Ask** What do you notice about each of these pairs? *(They have the same charge.)* Emphasize that it is not the charge but rather the number of oxygen atoms in the members of an *-ate* and *-ite* pair that determines the endings of the names.

Explore

Class Activity

PURPOSE Students will visualize the geometry of some polyatomic ions.

MATERIALS foam balls and wooden sticks

PROCEDURE Pairs of students can research the three-dimensional structure of one of the polyatomic ions in Table 9.3. Have them prepare a ball-and-stick model of the ion. They may wish to differentiate atoms by size or color. Make sure that they pay attention to correct geometric placement of atoms.

Ammonium ion (NH_4^+)

Nitrate ion (NO_3^-)

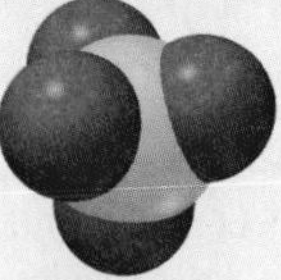
Sulfate ion (SO_4^{2-})

Phosphate ion (PO_4^{3-})

Figure 9.5 Polyatomic Ions
These molecular models show the arrangement of atoms in four common polyatomic ions.
Compare *How does the ammonium ion differ from the other three ions?*

Table 9.3

Common Polyatomic Ions		
Charge	Formula	Name
1–	$H_2PO_4^-$	Dihydrogen phosphate
	$C_2H_3O_2^-$	Ethanoate
	HSO_3^-	Hydrogen sulfite
	HSO_4^-	Hydrogen sulfate
	HCO_3^-	Hydrogen carbonate
	NO_2^-	Nitrite
	NO_3^-	Nitrate
	CN^-	Cyanide
	OH^-	Hydroxide
	MnO_4^-	Permanganate
	ClO^-	Hypochlorite
	ClO_2^-	Chlorite
	ClO_3^-	Chlorate
	ClO_4^-	Perchlorate
2–	HPO_4^{2-}	Hydrogen phosphate
	$C_2O_4^{2-}$	Oxalate
	SO_3^{2-}	Sulfite
	SO_4^{2-}	Sulfate
	CO_3^{2-}	Carbonate
	CrO_4^{2-}	Chromate
	$Cr_2O_7^{2-}$	Dichromate
	SiO_3^{2-}	Silicate
3–	PO_3^{3-}	Phosphite
	PO_4^{3-}	Phosphate
1+	NH_4^+	Ammonium

Polyatomic Ions

How do polyatomic ions differ from monatomic ions? How are they similar?

Some ions, such as the sulfate ion, are called polyatomic ions. **Unlike a monatomic ion, a polyatomic ion is composed of more than one atom. But like a monatomic ion, a polyatomic ion behaves as a unit and carries a charge.** The sulfate anion consists of one sulfur atom and four oxygen atoms. These five atoms together comprise a single anion with an overall 2– charge. The formula is written SO_4^{2-}. You can see the structure of the sulfate ion along with three other common polyatomic ions in Figure 9.5.

The names and formulas of some common polyatomic ions are shown in Table 9.3, grouped according to their charges. Note that the names of most polyatomic anions end in *-ite* or *-ate*. For example, notice the endings of the names of the hypochlor*ite* ion (ClO^-) and the hydrogen carbon*ate* ion (HCO_3^-). Also notice that three important ions have different endings—the cyan*ide* anion (CN^-), the hydrox*ide* anion (OH^-), and the ammon*ium* cation (NH_4^+).

Sometimes the same two or three elements combine in different ratios to form different polyatomic ions. Several examples appear in Table 9.3. Look for pairs of ions for which there is both an *-ite* and an *-ate* ending, for example, sulfite and sulfate. Examine the charge on each ion in the pair. Note the number of oxygen atoms and the endings on each name. You should be able to discern a pattern in the naming convention.

-ite	*-ate*
SO_3^{2-}, sulfite	SO_4^{2-}, sulfate
NO_2^-, nitrite	NO_3^-, nitrate
ClO_2^-, chlorite	ClO_3^-, chlorate

UbD Check for Understanding

BIGIDEA ELECTRONS AND THE STRUCTURE OF ATOMS Assess students' comprehension of the Big Idea by writing on the board the names and formulas of at least ten different polyatomic ions. Purposely mismatch the names so that they do not appear with the correct formulas. Challenge students to see how quickly they can rearrange the names and formulas so that they are paired correctly.

ADJUST INSTRUCTION If students are having trouble answering, have them review Table 9.3 and the supporting text. Then ask them to identify and correct the incorrect pairings.

The charge on each polyatomic ion in a given pair is the same. The *-ite* ending indicates one less oxygen atom than the *-ate* ending. However, the ending does not tell you the actual number of oxygen atoms in the ion. For example, the nitrite ion has two oxygen atoms, and the sulfite ion has three oxygen atoms. All anions with names ending in *-ite* or *-ate* contain oxygen.

When the formula for a polyatomic ion begins with H (hydrogen), you can think of the H as representing a hydrogen ion (H^+) combined with another polyatomic ion. For example, HCO_3^- is a combination of H^+ and CO_3^{2-}. Note that the charge on the new ion is the algebraic sum of the ionic charges of the two component ions.

$$H^+ + CO_3^{2-} \longrightarrow HCO_3^-$$
carbonate → hydrogen carbonate

$$H^+ + PO_4^{3-} \longrightarrow HPO_4^{2-}$$
phosphate → hydrogen phosphate

$$H^+ + HPO_4^{2-} \longrightarrow H_2PO_4^-$$
hydrogen phosphate → dihydrogen phosphate

The hydrogen carbonate anion (HCO_3^-), the hydrogen phosphate anion (HPO_4^{2-}), and the dihydrogen phosphate anion ($H_2PO_4^-$) are essential components of living systems. In contrast, the cyanide ion (CN^-) is extremely poisonous to living systems because it blocks a cell's means of producing energy. Figure 9.6 shows two uses for compounds with hydrogen-containing polyatomic ions.

Figure 9.6 Hydrogen-Containing Polyatomic Ions
Polyatomic ions that contain hydrogen are part of several compounds that affect your daily life.

Antacid Sodium hydrogen carbonate, which contains the HCO_3^- ion, can relieve an upset stomach.

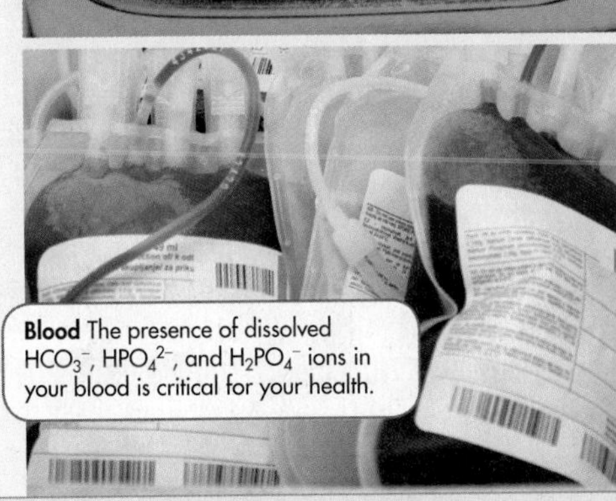

Blood The presence of dissolved HCO_3^-, HPO_4^{2-}, and $H_2PO_4^-$ ions in your blood is critical for your health.

9.1 LessonCheck

3. **Explain** How can you determine the charges of metal cations? Of nonmetal anions? Of transition metal cations?
4. **Review** What are the similarities and differences between polyatomic ions and monatomic ions?
5. **Identify** What are the charges on ions of Group 1A, Group 3A (aluminum), and Group 5A?
6. **Describe** Write the symbol for the ion of each element. Classify the ion as an anion or a cation, and name the ion.
   a. potassium
   b. oxygen
   c. bromine
   d. tin (2 electrons lost)
   e. beryllium
   f. cobalt (3 electrons lost)
7. **Describe** Write the symbol or formula (including charge) for each of the following ions:
   a. ammonium ion
   b. chromium(II) ion
   c. chromate ion
   d. nitrate ion
8. **Compare** How do the differences in the polyatomic ions PO_3^{3-} and PO_4^{3-} help you determine whether each ends in *-ite* or *-ate*?

BIGIDEA ELECTRONS AND THE STRUCTURE OF ATOMS

9. How does the electron configuration of an ion of a Group 1A or Group 7A element compare to that of the nearest noble gas?

Extend

Connect to MEDICINE

Explain to students that there exists a stockpile of a potassium compound for human use in the event of a nuclear accident that results in fallout of radioactive iodine-131. Convey that whether the iodine is radioactive or not, absorbed iodine accumulates in the thyroid gland. Have students find out what purpose potassium iodide would serve in this catastrophic event and why it is useful. *(If a dose of nonradioactive KI is ingested in a timely way after a nuclear accident, the iodide ion is taken up by the thyroid gland, decreasing the chances that the gland will take up iodine-131. Iodine-131 has a relatively short half-life of eight days, but it can cause thyroid cancer, especially in children.)*

Evaluate

Informal Assessment

Write the following symbols on the board: K, NH_3, CN, Mg, O, OH, Co, NH_4, Pb, PO_4, Ar, ClO, H_2O, and Zn. Have students write on an index card which symbol(s) on the board can exist as an ion by simply attaching a charge. *(K, CN, Mg, O, OH, Co, NH_4, Pb, PO_4, ClO, and Zn)* Ask students to identify the charges and name the corresponding ion. Then have students complete the 9.1 Lesson Check.

Reteach

Remind students that one of the most important properties of any compound is that it is electrically neutral. Use the periodic table and Tables 9.1, 9.2, and 9.3, and guide students through the reasoning process for naming and writing formulas of monatomic and polyatomic ions. Make sure students understand that since polyatomic ions carry a charge, they are always combined with other ions in compounds.

Lesson Check Answers

3. Metals in Groups 1A, 2A, and 3A form cations with charges equal to their group numbers. Group A nonmetals form anions with charges equal to their group numbers minus 8. Often, the charge of a transition metal cation must be determined from the number of electrons lost.
4. Both carry a charge. A polyatomic ion is composed of more than one atom; a monatomic ion is a single atom.
5. Group 1A metals, 1+; Group 3A (aluminum), 3+; Group 5A nonmetals, 3−
6. **a.** K^+, cation, potassium ion
   b. O^{2-}, anion, oxide ion
   c. Br^-, anion, bromide ion
   d. Sn^{2+}, cation, tin(II) ion
   e. Be^{2+}, cation, beryllium ion
   f. Co^{3+}, cation, cobalt(III) ion
7. **a.** NH_4^+ **b.** Cr^{2+}
   c. CrO_4^{2-} **d.** NO_3^-
8. One has one more oxygen atom than the other. The one with more oxygen atoms (PO_4^{3-}) gets the *-ate* ending, while the other (PO_3^{3-}) gets the *-ite* ending.
9. **BIGIDEA** The electron configurations for a cation from a Group 1A element and the preceding noble gas are the same. The electron configurations for an anion from a Group 7A element and the following noble gas are the same.

Answers

FIGURE 9.5 The ammonium ion is a cation and does not contain oxygen.

CHEMISTRY & YOU Have students examine the photograph in the background, and ask students to share any experiences they have had with algal blooms. Have them describe whether the algal bloom was near a factory, farmland, or a residential area. Encourage them to think about what the source of phosphates might have been for these algal blooms. Then pose the following question to students: *Consider what you know about algae growth and plant nutrient requirements. Based on this knowledge, how could you explain the response of algae to excess phosphates in their habitat?* You may need to assist students in the following ways:

- Phosphorus is a limiting factor in plant growth.
- High-phosphorus fertilizers are designed to boost plant reproduction by promoting flowering; these fertilizers are often referred to as "bloom boosters" when being marketed for landscaping use.
- Bacterial decay releases phosphate back into the water as part of the phosphorus cycle.

Explain

SUMMARIZE After students read the passage about algal blooms, lead them in a class discussion to summarize the main points and relate them to the chapter. **Ask** What causes freshwater algal blooms? *(excess phosphates in the water)* **Ask** What is the formula for a phosphate ion? *(PO_4^{3-})* **Ask** Are phosphates harmful or useful to algae? *(They are a necessary nutrient.)* **Ask** Why, then, are the phosphates a problem? *(They cause too much algae growth.)* Explain that algae have a short lifespan. After they die, the decay process includes a chemical reaction that uses oxygen. So much oxygen is used in this decay process that fish and other organisms do not have the oxygen they need to survive.

CRITICAL THINKING Have students consider why algal blooms are more common in ponds and lakes than in rivers. As a hint, have them look at characteristics of the lake in the photo. **Ask** What other conditions do you think make an algal bloom more likely to occur? *(warm water, plenty of sunshine, slowly flowing or stagnant water)*

Differentiated Instruction

L1 STRUGGLING STUDENTS Have students identify three things in the photo that can happen when there is an excess of phosphate compounds in a freshwater pond. Make a list on the board as students describe them: excess algae in the water, toxins that can enter drinking water, low oxygen levels that can kill fish and other organisms.

LPR LESS PROFICIENT READERS After students read about algal blooms, have them create a T-chart on paper with the headings *Cause* and *Effect.* Then have them look back through the reading and list items that fit under each heading. If students have difficulty, you may wish to do this as a class project.

L1 ADVANCED STUDENTS Have students research the causes and effects of harmful algal blooms that occur in the ocean, creating so-called "dead zones." Students should also note any detection, prevention, or remediation actions that are being undertaken in response to these algal blooms, and discuss their relative effectiveness.

9.2 Naming and Writing Formulas for Ionic Compounds

C.3.5 Write chemical formulas

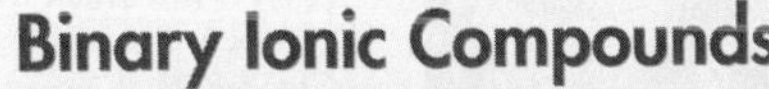

CHEMISTRY & YOU

Q: *What's the name of the secret ingredient?* If this ingredient isn't included in the recipe, the fruit can turn an ugly brown. Think about when you slice an apple at home. The slices don't look that tasty if you let them sit for too long because they begin to change color. But with the recipe and the secret ingredient, your apple slices could keep their color. Chemistry also uses recipes or formulas, but without any secrets. Once you know the rules, you can apply them and name any chemical compound. In this lesson, you will learn how to name ionic compounds.

Key Questions

How do you determine the formula and name of a binary ionic compound?

How do you determine the formula and name of a compound with a polyatomic ion?

Vocabulary

- binary compound

Binary Ionic Compounds

How do you determine the formula and name of a binary ionic compound?

In the days before the science of chemistry developed, the person who discovered a new compound often named it anything he or she wished. It was not uncommon for the name to describe some property of the substance or its source. For example, a common name for potassium carbonate (K_2CO_3) is *potash.* The name evolved because the compound was obtained by boiling wood ashes in iron pots. Baking soda ($NaHCO_3$) is another example. The common name, *baking soda,* describes its use in baking to make baked goods rise. Figure 9.7 shows a compound with the common name of cinnabar. Can you tell what elements are in cinnabar just from looking at the name? Unfortunately, such names do not tell you anything about the chemical composition of the compound.

The French chemist Antoine-Laurent Lavoisier (1743–1794) determined the composition of many compounds in his experiments to show how chemical compounds form. As more and more compounds were identified, Lavoisier recognized that it was becoming impossible to memorize all the unrelated names of the compounds. He worked with other chemists to develop a systematic method for naming chemical compounds. Their work is the basis for naming compounds today.

Figure 9.7 Cinnabar
The red substance that is deposited in this rock is commonly called cinnabar. Cinnabar (HgS) is comprised of mercury(II) ions and sulfide ions.

Key Objectives

9.2.1 APPLY the rules for naming and writing formulas for binary ionic compounds.

9.2.2 APPLY the rules for naming and writing formulas for compounds with polyatomic ions.

Additional Resources

Reading and Study Workbook, Lesson 9.2

Available online or DVD-ROM:

- Lesson 9.2 Review
- Virtual Chem Lab, Lab 2

PearsonChem.com

Engage

CHEMISTRY & YOU Ask students to draw an analogy between a recipe that calls for four cups of sliced apples and one tablespoon of lemon juice and the formula for a chemical compound such as $CaCl_2$. *(Calcium and chlorine correspond to the ingredients sliced apples and lemon juice; the subscripts correspond to the quantities of the ingredients.)*

Build Background

Ask Do you or someone you know have a nickname? Does the nickname tell something about the person? *(Students will have a variety of responses.)* **Ask** Can you use a nickname alone to identify a person? *(no)* Point out that a compound's common name is similar to a nickname. It may be descriptive, but does not necessarily give enough information about the compound to positively determine its chemical composition.

Focus on ELL

1 CONTENT AND LANGUAGE Write the key questions on the board. As a class, rewrite each question as two mini-questions and in words that are easy to understand. Provide a T-chart for students to record the rules for naming ionic compounds and examples of each.

2 FRONTLOAD THE LESSON Note that early clockmakers often used Roman numerals on their clocks, and display a clock face that uses Roman numerals instead of Arabic numerals (select one that uses "IV" rather than "IIII"). Explain that modern chemists also use Roman numerals, but for naming certain compounds rather than keeping time. Have students translate the clock numbers into their Arabic equivalents. If necessary, review with students how the Roman equivalents of 4, 6, 9, 11, and 12 are determined.

3 COMPREHENSIBLE INPUT Have pairs of students create a set of cards where half of the cards have the formula for ionic compounds and the other half has the names of the same ionic compounds. Each pair can then use the cards to play a game of memory, matching the name with the formula.

IN ACADEMIC STANDARDS for SCIENCE

- Develop explanations based on reproducible data and observations gathered during laboratory investigations.

C.3.5 Write chemical formulas for ionic compounds given their name and vice versa.

Answers

1. Na_3PO_4
2. Algal blooms may prevent sunlight from reaching underwater plants. As a result, the plants could die.
3. increase in the water temperature, the amount of sunlight, or other nutrients such as nitrates

Foundations for Reading

BUILD VOCABULARY Explain that the word *compound* comes from the Latin *componere,* which means "to put together." Tell students that parts are put together to make a whole; a compound always contains at least two parts. Explain that binary comes from the Latin word *bini* meaning "two-by-two" or "two-fold." Have students predict what a binary compound is. *(a compound with two parts)*

READING STRATEGY Have students preview the lesson. **Ask** What are the two types of compounds you will be learning about? *(binary ionic compounds and compounds with polyatomic ions)* **Ask** What do you think is the difference between a binary compound and a compound with a polyatomic ion? *(A binary compound contains only two elements. A compound with a polyatomic ion can contain three or more elements because a polyatomic ion can have two or more elements.)*

Explain

Binary Ionic Compounds

USE MODELS Cut index cards to obtain pieces that are one-third and two-thirds of a card. Keep some cards whole. On some whole cards, mark three large positive signs; on others, mark three large negative signs. Mark some of the one-third pieces with single positive signs; others with single negative signs. Mark the two-thirds pieces with two positive signs or two negative signs. On the opposite side of each card, write the formula for an ion having the appropriate charge. Have students match single positive and negative signs. Have them match double positive signs with double negative signs and with two single negative signs. Continue with all possible combinations. Then have students turn the cards over to see the formulas of the ionic compounds they produced.

Figure 9.8 Steelworks
In the process for making steel, iron is extracted from hematite, an ore containing iron(III) oxide.
Apply Concepts ***What is the formula for iron(III) oxide?***

Writing Formulas for Binary Ionic Compounds

A **binary compound** is composed of two elements. Binary compounds can be ionic compounds or molecular compounds. If you know the name of a binary ionic compound, you can write the formula. **To write the formula of a binary ionic compound, first write the symbol of the cation and then the anion. Then add subscripts as needed to balance the charges.** The positive charge of the cation must balance the negative charge of the anion so that the net ionic charge of the formula is zero. The ionic compound potassium chloride is composed of potassium cations (K^+) and chloride anions (Cl^-), so potassium chloride is a binary ionic compound. The charge of each K^+ cation is balanced by the charge of each Cl^- anion. So, in potassium chloride, the potassium and chloride ions combine in a 1:1 ratio. Thus, the formula for potassium chloride is KCl. The net ionic charge of the formula unit is zero.

The binary ionic compound calcium bromide is composed of calcium cations (Ca^{2+}) and bromide anions (Br^-). The two ions do not have equal numerical charges. Thus, each calcium ion with its 2+ charge must combine with (or be balanced by) two bromide ions, each with a 1− charge. That means that the ions must combine in a 1:2 ratio, so the formula for calcium bromide is $CaBr_2$. The net ionic charge of the formula unit is zero.

Figure 9.8 shows one step in the process of making steel from iron ore. Hematite, a common ore of iron, contains iron(III) oxide. What is the formula for this compound? Recall that a Roman numeral in the name of an ion shows the charge of the metal ion. Thus, iron(III) oxide contains Fe^{3+} cations combined with oxide anions (O^{2-}). How can you balance a 3+ charge and a 2− charge? You must find the least common multiple of the charges, which is 6. Iron's three charges taken two times equals six ($3 \times 2 = 6$). Oxygen's two charges taken three times also equals six. Thus, two Fe^{3+} cations (a 6+ charge) will balance three O^{2-} anions (a 6− charge). The balanced formula, then, is Fe_2O_3.

Another approach to writing a balanced formula for a compound is to use the crisscross method. In this method, the numerical value of the charge of each ion is crossed over and becomes the subscript for the other ion. Notice that the signs of the charges are dropped.

$$Fe^{3+} \quad O^{2-}$$

$$Fe_2O_3$$

$$2(3+) + 3(2-) = 0$$

The formula is correct because the overall charge of the formula is zero, and the subscripts are expressed in the lowest whole-number ratio.

READING SUPPORT

Build Reading Skills: *Compare and Contrast* As you read about ionic and molecular compounds, compare the procedures for naming and writing formulas for each. ***How can you tell the difference between an ionic and a molecular compound?***

Differentiated Instruction

LPR LESS PROFICIENT READERS Have students rewrite the answers to the key question as steps in a process. Ask students to identify each step as it is encountered in examples in the text. Allow students to keep these rules handy as they work through the lesson, until the processes become automatic.

L1 STRUGGLING STUDENTS Rewrite the Fe_2O_3 and CaS examples on the board, without the arrows. Guide students through each example step-by-step. Then, write a third pair of ions on the board and have students guide you step-by-step through this example.

L3 ADVANCED STUDENTS Have students research the possible mechanisms for the alleviation of symptoms of mental disorders, such as bipolar disorder, by lithium ions. Challenge students to explain why other Group 1A ions, such as sodium and potassium, which are normally present in substantial concentrations in the body, do not have the same effects as lithium ions even though they have the same charge.

If you use the crisscross method to write the formula for some compounds such as calcium sulfide (Ca^{2+} and S^{2-}), you will obtain the result Ca_2S_2. However, the 2:2 ratio of calcium and sulfide ions is not the lowest whole-number ratio. The correct formula for calcium sulfide is CaS.

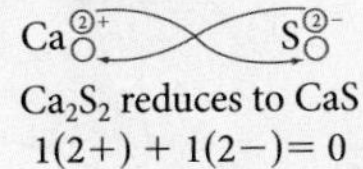

Ca_2S_2 reduces to CaS

$1(2+) + 1(2-) = 0$

Of course, if the magnitudes of the charges of the cation and anion are the same, as they are in this case, the ions combine in a 1:1 ratio, and the charges are balanced.

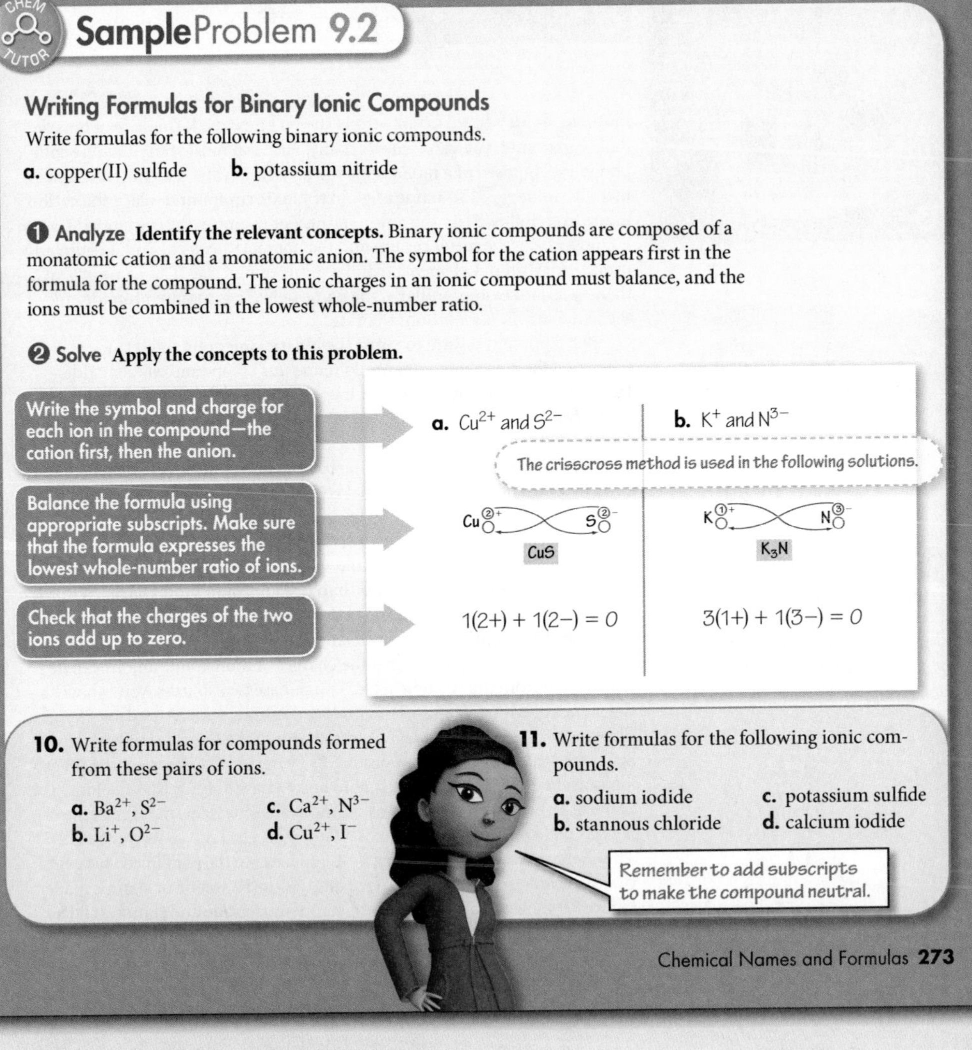

CHEM TUTOR

Sample Problem 9.2

Writing Formulas for Binary Ionic Compounds

Write formulas for the following binary ionic compounds.

a. copper(II) sulfide **b.** potassium nitride

1 Analyze Identify the relevant concepts. Binary ionic compounds are composed of a monatomic cation and a monatomic anion. The symbol for the cation appears first in the formula for the compound. The ionic charges in an ionic compound must balance, and the ions must be combined in the lowest whole-number ratio.

2 Solve Apply the concepts to this problem.

	a.	b.
Write the symbol and charge for each ion in the compound—the cation first, then the anion.	Cu^{2+} and S^{2-}	K^+ and N^{3-}
Balance the formula using appropriate subscripts. Make sure that the formula expresses the lowest whole-number ratio of ions.	CuS	K_3N
Check that the charges of the two ions add up to zero.	$1(2+) + 1(2-) = 0$	$3(1+) + 1(3-) = 0$

The crisscross method is used in the following solutions.

10. Write formulas for compounds formed from these pairs of ions.

a. Ba^{2+}, S^{2-} **c.** Ca^{2+}, N^{3-}
b. Li^+, O^{2-} **d.** Cu^{2+}, I^-

11. Write formulas for the following ionic compounds.

a. sodium iodide **c.** potassium sulfide
b. stannous chloride **d.** calcium iodide

Remember to add subscripts to make the compound neutral.

Foundations for Math

ADDITIVE INVERSES Remind students that the additive inverse of a number is its opposite and that the sum of a number and its additive inverse is 0. For example, the additive inverse of 6 is –6, and 6 + (–6) = 0. When writing the formula for a binary ionic compound, encourage students to use additive inverses to arrive at a net charge of 0.

In **Sample Problem 9.2b,** the additive inverse of 3– is 3+. Since potassium has a charge of 1+ (K^+), the formula for potassium nitride will require three potassium cations, resulting in a total charge of 3(1+) = 3+. If only one nitrogen anion is present, this will yield a net charge of 0 since 3– and 3(1+) = 3+ are additive inverses.

LESSON 9.2

Explore

Making and Naming an Ionic Compound

PURPOSE Students will see a reaction in which an ionic compound is formed and then name the compound and write its formula.

MATERIALS 1 g powdered zinc (Zn), 4 g iodine (I_2), watch glass, eyedropper, 8 mL water

SAFETY Wear goggles and perform the demo in a fume hood. Place the zinc iodide in a plastic or cardboard container and dispose of it in an approved landfill site.

PROCEDURE Wear safety goggles and use a fume hood. Mix 1 g of powdered Zn and 4 g of I_2 on a watch glass. With an eyedropper, carefully add 8 mL of water, one drop at a time. After the reaction is complete, show students the zinc iodide that was formed. Write the formula unit for zinc iodide (ZnI_2) along with its name on the chalkboard. **Ask** What happened to elemental Zn and I_2 during the reaction? *(Zinc and iodine reacted to form an ionic compound composed of Zn^{2+} and I^- ions.)* Use the crisscross method to show how the formula unit for zinc iodide was derived. Why is ZnI_2 named zinc iodide and not zinc(II) iodide? *(Zinc forms only one ion.)*

EXPECTED OUTCOME The masses of Zn and I_2 are the stoichiometic amounts. The reactants should be used up and white, crystalline ZnI_2 should form.

Sample Practice Problem

Write formulas for these compounds.

a. lithium fluoride *(LiF)*
b. aluminum chloride *($AlCl_3$)*
c. sodium nitride *(Na_3N)*
d. ferric oxide *(Fe_2O_3)*

Answers

FIGURE 9.8 Fe_2O_3

READING SUPPORT An ionic compound is composed of ions. A molecular compound is composed of molecules.

10. a. BaS **b.** Li_2O **c.** Ca_3N_2 **d.** CuI_2

11. a. NaI **b.** $SnCl_2$ **c.** K_2S **d.** CaI_2

Explain

Naming Ionic Binary Compounds

No, sodium sulfite is not a binary compound because it is comprised of more than two elements.

Explore

Class Activity

PURPOSE Students will practice naming binary compounds.

MATERIALS chalkboard or whiteboard, chalk or markers

PROCEDURE Divide the class into teams. Have each team write the names of ionic compounds on a sheet of paper as you write formulas on the board. Use chlorides, bromides, phosphides, and oxides of the elements of Groups 1A, 2A, and aluminum in 3A. Mix in sulfides, oxides, chlorides, and bromides of lead, tin, iron, cobalt, copper, and zinc. Have the teams compare their lists of names. When a discrepancy occurs, have them discuss it and come to agreement on the correct name.

Figure 9.9 Tin Compounds
Tin(II) fluoride and tin(IV) oxide have different compositions and uses.
a. Tin(II) fluoride is added to some toothpastes to prevent cavities. **b.** Tin(IV) oxide is used in glazes for pottery.
Identify ***What are the charges of the tin ions in the two compounds?***

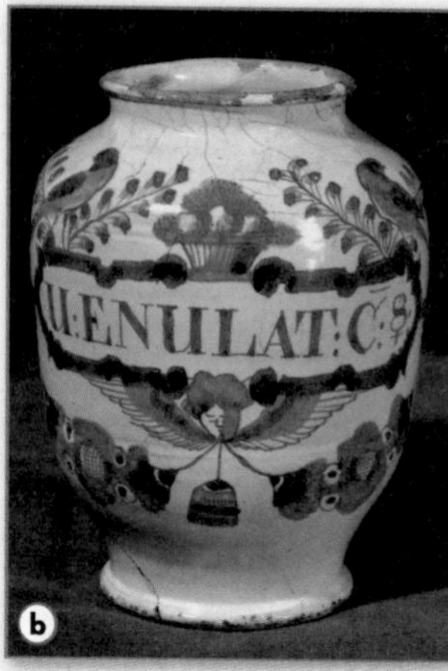

See some everyday chemical names *online.*

Q: *Many companies use sodium sulfite (Na_2SO_3) to keep dried fruit looking delicious. Is Na_2SO_3 a binary compound? Explain.*

Naming Binary Ionic Compounds If you know the formula for a binary ionic compound, you can write its name. First you must verify that the compound is composed of a monatomic metallic cation and a monatomic nonmetallic anion. **To name any binary ionic compound, place the cation name first, followed by the anion name.** For example, the compound Cs_2O is composed of the metal cesium and the nonmetal oxygen. Both cesium and oxygen are Group A elements that have only one charge. The name of Cs_2O, then, is cesium oxide. Similarly, the name of NaBr is sodium bromide, and the name of SrF_2 is strontium fluoride.

But suppose you want to name the binary ionic compound CuO. Following the rule above, you would name this compound copper oxide. However, the name *copper oxide* is incomplete. Recall that copper commonly forms two cations: Cu^+ and Cu^{2+}. The names of these ions are copper(I) ion and copper(II) ion, respectively. How can you tell which of these cations forms the compound CuO? Working backward will help. The formula indicates that the copper cation and the oxide anion combine in a 1:1 ratio. You know that the oxide anion always has a 2− charge. Therefore, the charge of the copper cation must be 2+ in order to balance the 2− charge. The compound CuO must be copper(II) oxide. **If the metallic element in a binary ionic compound has more than one common ionic charge, a Roman numeral must be included in the cation name.**

Table 9.2 lists the symbols and names of the common metals that form more than one cation. Recall that the charges of monatomic anions can be determined from the periodic table. Using these two sources, you can write the names of SnF_2 and SnS_2. Tin (Sn) forms cations with 2+ and 4+ charges. Fluorine is a Group 7A element, so the charge of the fluoride ion is 1−. In SnF_2, the ratio of cations to anions is 1:2. Therefore, the charge of the tin cation must be 2+ to balance the combined 2− charge of two fluoride ions. The name of SnF_2 is tin(II) fluoride or stannous fluoride. However, the name of SnO_2 is not tin(II) oxide. Oxygen is a Group 6A element, so its charge is 2−. The charge of the tin cation must be 4+ to balance the combined charges of two oxide anions. Thus, the name of SnO_2 is tin(IV) oxide or stannic oxide. Figure 9.9 shows examples of uses of stannous fluoride and stannic oxide.

274 Chapter 9 • Lesson 2

Lattice Energy in Ionic Bonding

Ionic compounds tend to be stable—energy in the form of heat or electricity is needed to decompose them. Thus, energy must be released when ionic bonds form. But the released energy is not primarily from the transfer of electrons. Forming a cation is an endothermic process. In addition, both the metal and the nonmetal must be vaporized—another endothermic process—before electron transfer can take place. Some energy (electron affinity) is released when the nonmetal gains one or more electrons, but this is not enough to make the entire process exothermic. It is lattice energy that makes the difference. Lattice energy is the energy released when gaseous cations and anions settle into the orderly crystal structure characteristic of a solid ionic compound.

Sample Problem 9.3

Naming Binary Ionic Compounds

Name the following binary ionic compounds:

a. CoI_2 **b.** Li_2Se

1 Analyze **Identify the relevant concepts.** Confirm that the compound is a binary ionic compound. To name the compound, name the ions in the order written in the formula—the cation name followed by the anion name. The name of a metal ion that has more than one common ionic charge must include a Roman numeral indicating the charge.

2 Solve **Apply the concepts to this problem.**

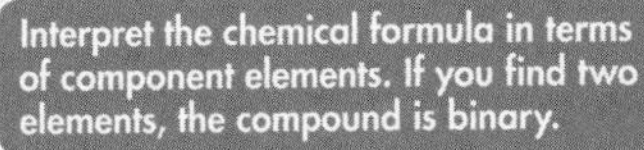

Interpret the chemical formula in terms of component elements. If you find two elements, the compound is binary.

a. CoI_2 contains cobalt cations and iodide anions.

b. Li_2Se contains lithium cations and selenide anions.

Determine whether the metal ion in the compound has more than one common ionic charge.

Cobalt forms two common cations: Co^{2+} and Co^{3+}.

Lithium forms one cation: Li^+.

If the metal ion has more than one ionic charge, use the nonmetal anion to determine which cation is indicated by the formula.

Iodide ion is I^-. The formula CoI_2 specifies two iodide ions, which give a charge of 2−. So the cobalt ion must be Co^{2+} to balance the charge.

This step is not needed for Li_2Se because the lithium ion has only one common charge.

Write the name of the cation, followed by the name of the anion. Include Roman numerals as needed.

cobalt(II) iodide

lithium selenide

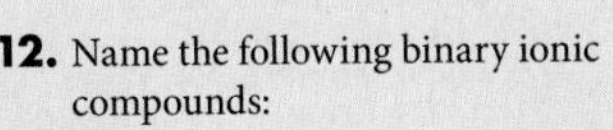

12. Name the following binary ionic compounds:

a. ZnS **b.** KCl **c.** BaO **d.** $CuBr_2$

13. Write the names for these binary ionic compounds.

a. CaO **b.** Cu_2Se **c.** FeS **d.** AlF_3

Check each answer by writing the formula using the ions from the name.

Explain

USE A MODEL Make sure students understand the concept of "electrically neutral" by using a model to indicate cations and anions. Have them use the model each time they write the formula for a binary ionic compound. Models can be a balance, colored chips, stackable blocks, etc. **Ask** Why must the positive charges + the negative charges equal zero? *(Compounds are electrically neutral.)*

Explore

Class Activity

PURPOSE Students become more proficient in writing formulas.

MATERIALS lists of names of ionic compounds generated in the class activity on page 274

PROCEDURE Have teams of students write the formulas for the names on the lists. When the formulas on two or more lists do not agree, have the teams resolve the discrepancy by using the criss-cross method to write the formula on the board.

Sample Practice Problem

Write the names for these binary compounds.

a. Na_2O *(sodium oxide)*

b. Mg_3N_2 *(magnesium nitride)*

c. NaF *(sodium fluoride)*

d. K_3N *(potassium nitride)*

UbD Check for Understanding

How do you determine the formula and name of a binary ionic compound?

Assess students' knowledge about naming and writing binary ionic compounds by asking the following questions. **Ask** When naming a binary compound, what comes first, the anion or the cation? *(cation first, followed by anion)* **Ask** What is the purpose of subscripts in the formula for a binary ionic compound? *(They balance the charges between the cation and the anion.)*

ADJUST INSTRUCTION If students are having trouble answering, have them review Sample Problems 9.2 and 9.3. Then, use the formulas in each problem to walk students through the answers of the two questions.

Answers

FIGURE 9.9 2+ and 4+, respectively

12. **a.** zinc sulfide **b.** potassium chloride **c.** barium oxide **d.** copper(II) bromide

13. **a.** calcium oxide **b.** copper(I) selenide **c.** iron(II) sulfide **d.** aluminum fluoride

LESSON 9.2

Explain

Compounds with Polyatomic Ions

USE VISUALS Direct students to the photographs in Figure 9.10. Tell students that the real-world items shown are only two of many polyatomic ions they encounter in their daily lives. Discuss some other common examples. **Ask** What are the formulas for calcium carbonate, lead(II) sulfate, ammonium sulfate, and ammonium phosphate? *($CaCO_3$, $PbSO_4$, $(NH_4)_2SO_4$, $(NH_4)_3PO_4$)*

Explore

Teacher Demo

PURPOSE Students will watch the formation of an ionic precipitate and postulate possible formulas.

MATERIALS 50 mL 0.1*M* lead(II) nitrate, 50 mL 0.1*M* sodium carbonate, 100-mL beaker

PROCEDURE Wear goggles and gloves. Slowly mix the two solutions. Make sure students see the precipitate of lead carbonate that forms. Write on the board the symbols or formulas for the four ions that are in the two solutions: Pb^{2+}, NO_3^-, Na^+, and CO_3^{2-}. Tell students that the two solutions contained these four ions and that the precipitate is an ionic compound formed by the combination of two of them. **Ask** students to list all possible formula units for the solid and to name each possibility. Identify the solid as lead(II) carbonate ($PbCO_3$).

SAFETY Wear goggles and gloves because lead(II) nitrate is toxic. For disposal, combine the reaction mixture with 10 g of solid NaCl in a beaker. Stir the solution and allow the precipitate to settle. Filter or decant to isolate the precipitate. Place the dry precipitate in a plastic or cardboard container and dispose of it in an approved landfill. Flush the filtrate down the drain with excess water.

Compounds With Polyatomic Ions

How do you determine the formula and name of a compound with a polyatomic ion?

The seashells shown in Figure 9.10 are made of calcium carbonate ($CaCO_3$). Calcium carbonate is obviously not a binary compound because it contains more than two elements. Remember that an *-ate* or *-ite* ending on the name of a compound indicates that the compound contains a polyatomic anion that includes oxygen. Calcium carbonate contains one monatomic ion (Ca^{2+}) and one polyatomic ion (CO_3^{2-}). Figure 9.10 also shows a typical automobile battery called a lead storage battery. The energy-producing reaction inside the battery uses the ionic compound lead(II) sulfate ($PbSO_4$), which consists of the monatomic ion Pb^{2+} and the polyatomic ion SO_4^{2-}.

Figure 9.10
Compounds With Polyatomic Ions
Some examples of ionic compounds that contain polyatomic ions are shown.
Explain *Why is there a Roman numeral in the name lead(II) sulfate?*

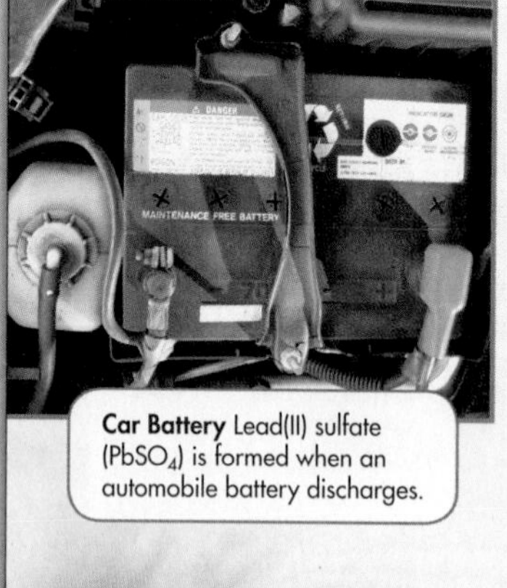

Car Battery Lead(II) sulfate ($PbSO_4$) is formed when an automobile battery discharges.

Writing Formulas for Compounds With Polyatomic Ions How would you write the formula for an ionic compound with a polyatomic ion? For starters, try following the same procedure you used for binary ionic compounds. **To write the formula for a compound with a polyatomic ion, first write the symbol (or formula) for the cation followed by the symbol (or formula) for the anion. Then, add subscripts as needed to balance the charges.** For example, calcium nitrate is composed of a calcium cation (Ca^{2+}) and a polyatomic nitrate anion (NO_3^-). In calcium nitrate, two nitrate anions, each with a 1− charge, are needed to balance the 2+ charge of each calcium cation.

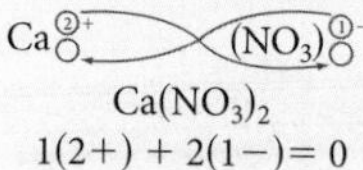

$$Ca(NO_3)_2$$
$$1(2+) + 2(1-) = 0$$

The charge is balanced and the ions are expressed in the lowest whole-number ratio, so the formula is correct. Parentheses are used around the nitrate ion in the formula because more than one nitrate anion is needed. The subscript 2 that follows the parentheses shows that the compound contains two nitrate anions. **Whenever more than one polyatomic ion is needed to balance the charges in an ionic compound, use parentheses to set off the polyatomic ion in the formula.**

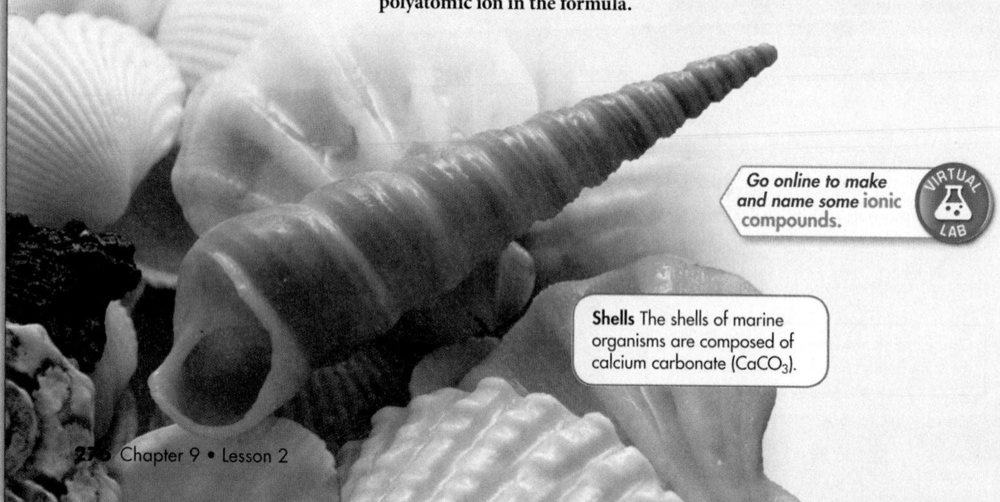

Go online to make and name some ionic compounds. VIRTUAL LAB

Shells The shells of marine organisms are composed of calcium carbonate ($CaCO_3$).

Lithium Carbonate as Medication

When John Cade, a doctor in the Victoria Department of Mental Health in Victoria, Australia, injected guinea pigs with a lithium salt of uric acid in 1949, he found that the usually hyperactive animals became calm and lethargic for a period of time. After further experimentation, Cade injected a severely disturbed patient with lithium carbonate. The results were so astonishingly positive for this patient, and for many others who followed, that within 10 years the treatment had spread throughout Europe and into the United States. Patients who had been confined to mental institutions for virtually their entire lives were able to return to normal living.

Lithium carbonate is a compound that can be prescribed for patients who have mood disorders, such as manic-depressive disorder or bipolar disorder. The compound is composed of lithium cations (Li^{+}) and polyatomic carbonate anions (CO_3^{2-}). In the formula for lithium carbonate, two lithium cations, each with a 1+ charge, are needed to balance the 2− charge of one carbonate anion. Parentheses are not needed to set off the polyatomic carbonate anion.

Li^{1+} $(CO_3)^{2-}$

Li_2CO_3

$2(1+) + 1(2-) = 0$

Strontium sulfate is another example of a compound in which only a single polyatomic ion (SO_4^{2-}) is needed to balance the cation (Sr^{2+}). So, no parentheses are needed when writing the formula: $SrSO_4$.

SampleProblem 9.4

Writing Formulas for Compounds With Polyatomic Ions

What are the formulas for these ionic compounds?

a. magnesium hydroxide **b.** potassium sulfate

1 Analyze Identify the relevant concepts. Write the symbol or formula for each ion in the order listed in the name. Use subscripts to balance the charges. The ions must be combined in the lowest whole-number ratio. If more than one polyatomic ion is needed to balance a formula, place the polyatomic ion formula in parentheses, followed by the appropriate subscript.

2 Solve Apply the concepts to this problem.

	a.	b.
Write the symbol or formula for each ion in the compound—the cation first, then the anion. Include the charge for each ion.	cation: Mg^{2+} anion: OH^{-}	cation: K^{+} anion: SO_4^{2-}
Balance the formula using appropriate subscripts. Make sure that the formula expresses the lowest whole-number ratio of ions.	Mg^{2+} $(OH)^{1-}$ → $Mg(OH)_2$	K^{1+} $(SO_4)^{2-}$ → K_2SO_4
Check that the charges of the two ions add up to zero.	$1(2+) + 2(1-) = 0$	$2(1+) + 1(2-) = 0$

Remember: Only use parentheses if there is more than one polyatomic ion in the balanced formula.

14. Write formulas for compounds formed from these pairs of ions.
a. NH_4^{+}, SO_3^{2-}
b. calcium ion, phosphate ion

15. Write formulas for the following compounds:
a. lithium hydrogen sulfate
b. chromium(III) nitrite

Explain

Misconception Alert

Students may confuse whole numbers and integers. Make sure students understand that the set of whole numbers is the set of natural numbers and zero, whereas the set of integers consist of whole numbers, zero, and their opposites. Point out that a whole number ratio means that the numbers in the ratio will be positive values. Emphasize that decimals and fractions are not considered whole numbers.

MAKE A CONNECTION Explain to students that magnesium hydroxide is a suspension of magnesium hydroxide ($Mg(OH)_2$) in water and is useful as an antacid and laxative. It is an ionic compound with a polyatomic ion. Its common name is milk of magnesia. Explain that in the stomach, the hydroxide ion reacts with excess acid that causes the discomfort of indigestion. In larger amounts, magnesium hydroxide also acts as a muscle relaxant and relieves constipation.

Sample Practice Problem

Write formulas of compounds formed from these pairs of ions:

a. Pb^{2+}, NO_3^{-} (*$Pb(NO_3)_2$*)
b. iron(III) ion and sulfate ion (*$Fe_2(SO_4)_3$*)
c. Cr^{3+}, OH^{-} (*$Cr(OH)_3$*)
d. sodium ion and hydrogen phosphate ion (*Na_2HPO_4*)

Foundations for Math

READING PARENTHESES AND CHARGES When a formula contains parentheses, students may be confused by standard rules of exponents that they have learned in math. For example $(2x)^3$ in math means 2^3x^3—the exponent is applied to both the 2 and the *x* inside the parentheses. Stress that this is not the case when a charge has been assigned to a polyatomic ion. The charge is the total charge for the entire ion and is not distributed to each element in the parentheses.

In **Sample Problem 9.4b,** $(SO_4)^{2-}$ has a total charge of 2−. The 2− is NOT an exponent and is NOT distributed inside the parentheses. A common error would be for students to assign a total charge of 4− to the ion because there are two elements.

Answers

FIGURE 9.10 to specify the charge (Lead can have a 2+ charge or a 4+ charge.)

14. a. $(NH_4)_2SO_3$ **b.** $Ca_3(PO_4)_2$
15. a. $LiHSO_4$ **b.** $Cr(NO_2)_3$

LESSON 9.2

Explain

Sample Practice Problem

Name the following compounds.

a. FeN *(iron(III) nitride)*

b. $Mg(NO_3)_2$ *(magnesium nitrate)*

Extend

Divide the class into groups of three to four students. Have students take turns randomly choosing ions from Tables 9.2 and 9.3 in Lesson 9.1. The other member or the group must write the formula and name the compound. All students in the group should agree on the names and formulas.

Evaluate

Informal Assessment

Write the following cations on one side of the board: NH_4^+, Mg^{2+}, and Pb^{4+}. Write the following anions on the other side of the board: Br^-, S^{2-}, and PO_4^{3-}. Ask students to write formula units for all the possible ionic compounds that these ions could form and name them. Ask them to explain how they arrived at their answers. Then have students complete the 9.2 Lesson Check.

Reteach

Review the lesson with students and help them to organize the information so that the three key items to consider when naming an ionic compound are apparent: (1) the identity of the ions, (2) the order of the names, and (3) the possibility that an element may form cations with more than one charge. Make sure students note that for binary compounds, ions can be identified from their symbols. The suffix *-ide* is used when naming the nonmetal ion, and for compounds with polyatomic ions, the polyatomic ion name can be obtained from Table 9.3 in Lesson 9.1. Students should also note that the name of the cation always precedes the name of the anion. Explain that for elements that form more than one cation, the correct charge is shown by a Roman numeral in parentheses directly after the cation name.

Figure 9.11 Sodium Hypochlorite The compound NaClO is often added to laundry water to bleach (brighten) white fabrics.

Naming Compounds With Polyatomic Ions You have learned to write formulas for compounds containing polyatomic ions when you were given their names. Now, if you were given the formulas for these compounds, could you name them? When naming compounds containing polyatomic ions, you must first identify any polyatomic ions in the formula for the compound. If the polyatomic ion is unfamiliar, find its name in Table 9.3. **To name a compound containing a polyatomic ion, state the cation name first and then the anion name. If the cation is a metallic element that has more than one common ionic charge, include a Roman numeral in the cation name.** Recall that the same rules apply when naming binary ionic compounds.

The compound NaClO is used as a disinfectant for swimming pools and as a bleach, as shown in Figure 9.11. The cation in this compound is sodium ion (Na^+). The other ion, ClO^-, is a polyatomic ion called hypochlorite ion. So, the name for NaClO is sodium hypochlorite.

SampleProblem 9.5

Naming Compounds With Polyatomic Ions

Name the following ionic compounds:

a. $(NH_4)_2C_2O_4$ **b.** $Fe(ClO_3)_3$

1 Analyze **Identify the relevant concepts.** Determine whether there is a polyatomic ion in the formula. To name the compound, list the names of the ions in the order written in the formula—the cation name followed by the anion name. The name of an ion that has more than one common ionic charge must include a Roman numeral indicating the charge.

2 Solve **Apply the concepts to this problem.**

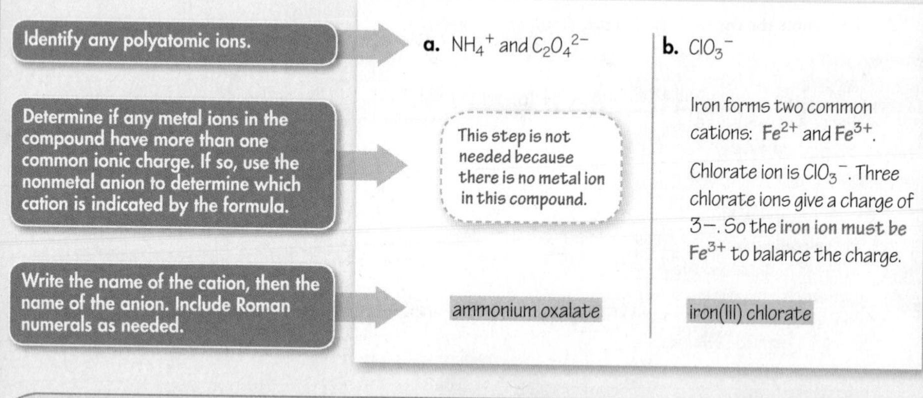

16. Name the following ionic compounds:

a. CaC_2O_4 **b.** KClO **c.** $KMnO_4$ **d.** Li_2SO_3

17. Write the names for these ionic compounds.

a. $Al(OH)_3$ **b.** $NaClO_3$ **c.** $Sn_3(PO_4)_2$ **d.** Na_2CrO_4

Focus on ELL

4 LANGUAGE PRODUCTION Have students work in groups of four to complete the lab. Make sure each group has ELLs of varied language proficiencies, so that more proficient students can help less proficient ones. Have students work according to their proficiency level.

BEGINNING

LOW Have students be in charge of test tube *C*. Students can mimic other students working with test tubes *A* and *B*.

HIGH Have students draw pictures of their observations or use symbols to note their observations.

INTERMEDIATE: LOW/HIGH Provide students with a chart they can use to record their observations. Include a box students can check for "dissolved" or "did not dissolve."

ADVANCED: LOW/HIGH Have students assist students with lower English proficiencies with writing the formulas. Have students read the names and the formulas aloud.

Quick Lab

Purpose To mix solutions containing cations and anions to make ionic compounds

Materials

- 9 small test tubes
- test tube rack
- paper, pencil, ruler
- 6 droppers
- solution A (Fe^{3+} ion)
- solution B (Ag^{+} ion)
- solution C (Pb^{2+} ion)
- solution X (CO_3^{2-} ion)
- solution Y (OH^{-} ion)
- solution Z (PO_4^{3-} ion)

Making Ionic Compounds

Procedure 

1. Label three test tubes A, three test tubes B, and three test tubes C.
2. Add 10 drops (approximately 0.5 mL) of solution A to the test tubes that are labeled A. Add 10 drops of solution B to the test tubes that are labeled B. Repeat this step with solution C.
3. Add 10 drops of solution X to one test tube of A, 10 drops to one test tube of B, and 10 drops to one test tube of C. Observe each test tube for the formation of a solid.
4. Make a 3-by-3 inch grid to record your observations. Label the rows A, B, and C. Label the columns X, Y, and Z. Describe any solid material you observe.
5. Repeat Step 3, adding 10 drops of solution Y to test tubes A, B, and C. Record your observations.
6. Repeat Step 3, adding 10 drops of solution Z to test tubes A, B, and C. Record your observations.

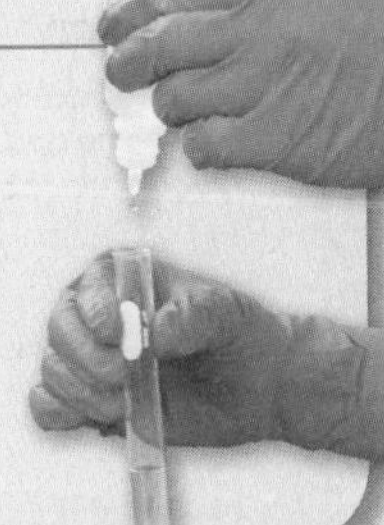

Analyze and Conclude

1. **Infer** Some ionic compounds are insoluble in water. What did you observe? How many of the compounds formed were insoluble?
2. **Describe** Write the formula for each ionic compound formed.
3. **Describe** Name each ionic compound formed.
4. **Draw Conclusions** Will mixing any cation with any anion always lead to the formation of an insoluble ionic compound? Explain your answer.

9.2 LessonCheck

18. **Summarize** Describe the procedures for writing the formulas and names of binary ionic compounds.

19. **Review** How do you write the formulas and the names of compounds with polyatomic ions?

20. **Evaluate** What are the advantages and disadvantages of common names?

21. **Apply Concepts** Write the formula for these binary ionic compounds.
    a. beryllium chloride
    b. cesium sulfide
    c. sodium iodide
    d. strontium oxide

22. **Identify** What condition must be met when writing a formula for an ionic compound?

23. **Apply Concepts** Write the formula for these compounds containing polyatomic ions.
    a. chromium(III) nitrite
    b. sodium perchlorate
    c. magnesium hydrogen carbonate
    d. calcium acetate

24. **Explain** When do you use parentheses in writing a chemical formula?

25. **Describe** Name the following ionic compounds:
    a. LiF
    b. SnS_2
    c. $MnCO_3$
    d. $Sr(H_2PO_4)_2$

26. **Identify** Which of the following formulas are incorrect? Explain your answer.
    a. $Mg_2(SO_4)_3$
    b. $AsRb_3$
    c. $BeCl_3$
    d. NaF

Lesson Check Answers

18. Write the symbol for the cation and then the symbol for the anion. Balance the charges. Write the cation name and then the anion name.
19. Write the symbol or formula for the cation and then for the anion. Balance the charges. Name the cation first followed by the anion.
20. Most common names do not tell you about the chemical composition of a compound. Many common names indicate other physical characteristics of the compound.
21. a. $BeCl_2$ b. Cs_2S c. NaI d. SrO
22. The charges of the ions must balance.
23. a. $Cr(NO_2)_3$ b. $NaClO_4$ c. $Mg(HCO_3)_2$ d. $Ca(C_2H_3O_2)_2$
24. when the compound contains more than one of a particular polyatomic ion
25. a. lithium fluoride
    b. tin(IV) sulfide
    c. manganese(II) carbonate
    d. strontium dihydrogen phosphate
26. a. incorrect; charges are not balanced, $MgSO_4$
    b. incorrect; anion symbol should be second, Rb_3As
    c. incorrect; charges are not balanced, $BeCl_2$

Quick Lab

PURPOSE Students form precipitates and write their formulas and names.

SKILLS FOCUS observing, applying rules

PREP TIME 30 minutes

CLASS TIME 20 minutes

ADVANCED PREPARATION

Solution	Preparation
0.05*M* $AgNO_3$	2.1 g in 250 mL
0.2*M* $Pb(NO_3)_2$	16.6 g in 250 mL
1.0*M* Na_2CO_3	26.5 g in 250 mL
0.5*M* NaOH	5.0 g in 250 mL
0.1*M* Na_3PO_4	9.5 g $Na_3PO_4 \cdot 12H_2O$ in 250 mL
0.1*M* $FeCl_3$	6.8 g $FeCl_3 \cdot 6H_2O$ in 25 mL of 1.0*M* NaCl; dilute to 250 mL

ANALYZE AND CONCLUDE

1. An insoluble compound formed in every tube.
2. $Fe_2(CO_3)_3$, $Fe(OH)_3$, $FePO_4$, Ag_2CO_3, AgOH, Ag_3PO_4, $PbCO_3$, $Pb(OH)_2$, $Pb_3(PO_4)_2$
3. iron(III) carbonate, iron(III) hydroxide, iron(III) phosphate, silver carbonate, silver hydroxide, silver phosphate, lead(II) carbonate, lead(II) hydroxide, lead(II) phosphate
4. No, NaCl, for example, is soluble.

Answers

16. a. calcium oxalate
    b. potassium hypochlorite
    c. potassium permanganate
    d. lithium sulfite
17. a. aluminum hydroxide
    b. sodium chlorate
    c. tin(II) phosphate
    d. sodium chromate

LESSON 9.3

Key Objectives

 **9.3.1 APPLY** the rules for naming and writing formulas for binary molecular compounds.

Additional Resources

Reading and Study Workbook, Lesson 9.3

Available online or DVD-ROM:

- Lesson 9.3 Review

PearsonChem.com

Engage

CHEMISTRY & YOU Have students read the opening text. **Ask** Do you know of any compounds that use prefixes in their names? *(Sample answers: carbon dioxide, carbon monoxide)* Point out that these compounds have prefixes attached to the second component of the compound, but that prefixes can also be attached to the first component as well. Have students speculate as to what information the prefixes provide about the compounds.

Activate Prior Knowledge

Remind students that, earlier in this course, they encountered metric prefixes such as *milli-*, *centi-*, and *kilo-*. Point out that both *milli-* and *centi-* represent fractional amounts, while *kilo-* represents a whole number amount. Explain that, in this lesson, students will encounter prefixes that represent small whole numbers.

IN ACADEMIC STANDARDS for SCIENCE

C.3.4 Draw structural formulas for and name simple molecules.

9.3 Naming and Writing Formulas for Molecular Compounds

C.3.4 Structural formulas and names

CHEMISTRY & YOU

Q: *What numerical prefixes are used in chemistry?* You're already familiar with words containing numerical prefixes. For example, the word *triathlon* contains the prefix *tri-*, indicating the number 3. Athletes sometimes compete in multisport events—pentathlons, heptathlons, decathlons, and so on. The prefix of each tells you how may sports are in the event. Prefixes are used in chemistry, too. In this lesson, you will learn how prefixes in the name of a binary molecular compound tell you its composition.

Key Question

What guidelines are used to write the name and formula of a binary molecular compound?

Q: *How does a triathlon differ from a pentathlon? How does phosphorus trifluoride differ from phosphorus pentafluoride?*

Binary Molecular Compounds

What guidelines are used to write the name and formula of a binary molecular compound?

Recall that binary ionic compounds are composed of the ions of two elements, a metal and a nonmetal. Binary molecular compounds are also composed of two elements, but both elements are nonmetals and are not ions. These differences affect the naming of these compounds and the writing of their formulas. Binary molecular compounds are composed of molecules, not ions, so ionic charges cannot be used to write formulas or to name them.

When two nonmetallic elements combine, they often do so in more than one way. For example, the elements carbon and oxygen combine to form two gaseous compounds, CO and CO_2. How would you name a binary compound formed by the combination of carbon and oxygen atoms? It might seem satisfactory to call it carbon oxide. However, the two carbon oxides, CO and CO_2, are very different compounds. Sitting in a room with small amounts of CO_2 in the air would not present any problems. You exhale CO_2 as a product of your body chemistry, as shown in Figure 9.12. Thus, it is normally present in the air you breathe. On the other hand, if the same amount of CO were in the room, you could die of asphyxiation. The binary compound CO is a poisonous gas that interferes with your blood's ability to carry oxygen to body cells. Obviously, a naming system that distinguishes between these two compounds is needed.

Focus on ELL

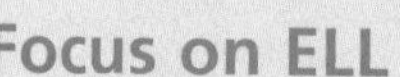

1 CONTENT AND LANGUAGE Write the word *guideline* on the board and pronounce it for the class. Have students locate this word in the text. *(Students should notice that* guideline *is used in the key question and its answer.)* Have students use context clues to define the term. *(A guideline is similar to a rule.)* Add this word to the word wall.

2 FRONTLOAD THE LESSON Preview the prefixes in Table 9.4. Ask student pairs to make a list of familiar words that use these prefixes. Have each pair share their words (and meanings) with the class. Students should add to their list when another pair states a word that is not on their list. Have students keep these examples in mind as they proceed with the lesson.

3 COMPREHENSIBLE INPUT When explaining Sample Problems 9.6 and 9.7, use the same highlighter color to highlight the number 2 and the prefix *di-*, and a different color to highlight the number 1 and the prefix *mono-*. Encourage students to expand on and use this color-coding method during independent practice.

Naming Binary Molecular Compounds Prefixes in the names of binary molecular compounds help distinguish compounds containing different numbers of atoms such as CO and CO_2. Table 9.4 lists the prefixes used to name binary molecular compounds. These prefixes tell how many atoms of an element are present in each molecule of the compound. According to the table, the prefix *mono-* would be used for the single oxygen atom in CO. The prefix *di-* would be used to indicate the presence of the two oxygen atoms in CO_2.

Use the prefixes listed in Table 9.4 along with the following guidelines to name a binary molecular compound. But before you apply these steps, you must confirm that the compound is a binary molecular compound. **To name a binary molecular compound, use the following guidelines:**

1. **Write the names of the elements in the order listed in the formula.**
2. **Use prefixes appropriately to indicate the number of each kind of atom.** If just one atom of the first element is in the formula, omit the prefix *mono-* for that element. Also, the vowel at the end of a prefix is sometimes dropped when the name of the element begins with a vowel.
3. **End the name of the second element with the suffix *-ide*.**

Following these guidelines, CO is named carbon *mon*oxide and CO_2 is named carbon *di*oxide. What about the compound Cl_2O_8? This binary molecular compound consists of two chlorine atoms and eight oxygen atoms. The name is therefore *di*chlorine *oct*oxide.

Table 9.4

Prefixes Used in Naming Binary Molecular Compounds

Prefix	Number
mono-	1
di-	2
tri-	3
tetra-	4
penta-	5
hexa-	6
hepta-	7
octa-	8
nona-	9
deca-	10

Figure 9.12 Carbon Dioxide
When you exhale underwater, bubbles containing CO_2 rise to the surface of the water.

Foundations for Reading

BUILD VOCABULARY Tell students that the word *prefix* comes from the Latin word *praefigere*, meaning "to fasten before." Explain that students will learn to attach a prefix such as *di-*, *tri-*, or *tetra-* to the names of elements to indicate how many atoms of each element are in a formula of a molecular compound.

READING STRATEGY Encourage students to visualize the meaning of the prefixes used in binary molecular compounds. Write some names on the board: for example, carbon tetrachloride. Then, draw one circle representing carbon and four circles representing chlorine. All the circles combined constitute the five atoms in CCl_4. **Ask** How many circles should I draw for fluorine in sulfur hexafluoride? *(six)* How many circles should I draw for sulfur in sulfur hexafluoride? *(one)*

Explain

Binary Molecular Compounds

APPLY CONCEPTS Explain how the order of the elements in the names of most molecular compounds is established. Point out that ordinarily, the less electronegative element appears first in the name. For example, a compound containing carbon and sulfur is carbon disulfide (CS_2). Tell students that since carbon is less electronegative than sulfur, it appears first. Remind students that electronegativity increases from left to right on the periodic table and decreases from top to bottom. Explain that the first element in the name and formula is the element further to the left. **Ask** In the compound disulfur dichloride, S_2Cl_2, why does the sulfur appear first? *(Sulfur is less electronegative than chlorine.)* Explain that for elements in the same column, the element nearer the bottom of the table typically appears first, for example, iodine heptafluoride (IF_7).

CHEMISTRY & YOU A triathlon has three events, and a pentathlon has five events. Likewise, phosphorus trifluoride has three fluorine atoms, and phosphorus pentafluoride has five fluorine atoms.

Differentiated Instruction

L3 ADVANCED STUDENTS Increasingly large quantities of the binary molecular compound CO_2 are released into the atmosphere annually by the burning of fossil fuels. Once in the atmosphere, carbon dioxide traps infrared energy near Earth's surface. As a result, some scientists predict that over time global temperatures will rise. Ask students to learn more about global warming and prepare an oral or written report or a poster that explains the current understanding or perception of this topic.

L1 STRUGGLING STUDENTS Provide students with a copy of Table 9.4, but include an additional column so students can write a couple of examples of compounds for each row. Instruct students to use this table as a study guide.

ELL ENGLISH LANGUAGE LEARNERS Have students write out steps for writing formulas for molecular compounds in their own words. Encourage them to use symbols and examples as needed. Allow beginning speakers to include native language versions as well.

Explore

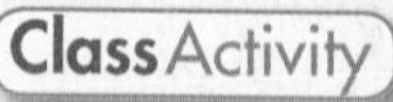

PURPOSE Students will convert the subscripts in a formula to prefixes in the name.

MATERIALS paper and pencil

PROCEDURE Have students make a chart with three columns: *element name, number of atoms,* and *prefix*. Have them fill in the columns for N_2O and then name the compound. For example:

Nitrogen	2	di-
Oxygen	1	mono-

The name is dinitrogen monoxide. Write the molecular formulas for other compounds on the board and have students add them to their charts. Some possibilities are CCl_4, PBr_5, and P_4S_3.

Explain

Binary Molecular Compounds

USE AN ANALOGY Discuss how some families name a male offspring the same name as his father. Explain that this is noted by the person's full name followed by either the word *junior* or by the Roman numeral of his position in the family. For instance, Eric Philip Thomas Junior indicates that he is the second male in his family with the name Eric Philip Thomas, whereas Eric Philip Thomas V indicates that he is the fifth male in his family with this name. Explain that when a person's name is original to the family, such as Wyatt Mitchell Thomas, his name is not followed by a 1st, a Roman numeral I, or the words "the original." When a number or a word does not follow a name, then it is customarily understood that the name is the first occurrence for a male in a family. Use this scenario to help explain why the prefix *mono-* is not used to indicate one atom of the first element in a binary molecular compound. Explain that when a prefix does not appear on the first element in a binary compound, it is customarily understood that there is only one atom. It would be redundant to add the prefix.

Sample Practice Problem

Write the formula for each compound.

a. dinitrogen difluoride *(N_2F_2)*

b. iodine tribromide *(IBr_3)*

Sample Problem 9.6

Naming Binary Molecular Compounds

Name the following binary molecular compounds:

a. N_2O **b.** PCl_3

1 Analyze **Identify the relevant concepts.** Confirm that the compound is a binary molecular compound—a compound composed of two nonmetals. To name the compound, name the elements in the order written in the formula. Use prefixes as necessary to indicate the number of each kind of atom. Use the suffix *-ide* on the name of the second element.

2 Solve **Apply the concepts to this problem.**

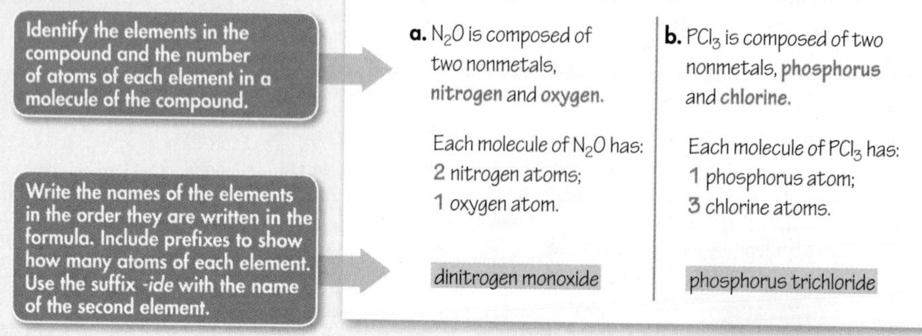

27. Name the following binary molecular compounds:

a. OF_2 **c.** SO_3

b. S_2F_{10} **d.** SF_6

Writing Formulas for Binary Molecular Compounds Suppose you know the name of a molecular compound and want to write the formula. **To write the formula of a binary molecular compound, first use the prefixes in the name to tell you the subscript of each element in the formula. Then, write the correct symbols for the two elements with the appropriate subscripts.** An interesting example is tetraphosphorus trisulfide, which is used in some matches. The name *tetraphosphorus trisulfide* has the prefixes *tetra-* and *tri-*, so the subscripts of phosphorus and sulfur must be 4 and 3, respectively. Thus, the formula for tetraphosphorus trisulfide is P_4S_3.

282 Chapter 9 • Lesson 3

Foundations for Math

PREFIXES Remind students that they probably learned prefixes in algebra or geometry. For example, the names of special polynomials involve prefixes and depend on the number of terms that make up the expression: *mono*mial (1 term), *bi*nomial (2 terms), *tri*nomial (3 terms). **Ask** Which of these prefixes is not used when naming compounds? *(bi-; the prefix when there are 2 atoms of an element is di-)* In geometry, the names of plane shapes are based on the number of sides: *tri*angle (3 sides), *quad*rilateral (4 sides), *penta*gon (5 sides), and so on. **Ask** Which of these prefixes is not used when naming compounds? *(quad-; the prefix when there are 4 atoms of an element is* tetra-*)*

In **Sample Problem 9.6a,** stress that the 2 atoms of nitrogen are read as *dinitrogen*, not *binitrogen*.

SampleProblem 9.7

Writing Formulas for Binary Molecular Compounds

Write formulas for the following binary molecular compounds:

a. nitrogen trifluoride **b.** disulfur dichloride

1 Analyze **Identify the relevant concepts.** The prefixes in the name indicate the subscript of each element in the formula. Write the symbols for the two elements with the appropriate subscripts.

2 Solve **Apply the concepts to this problem.**

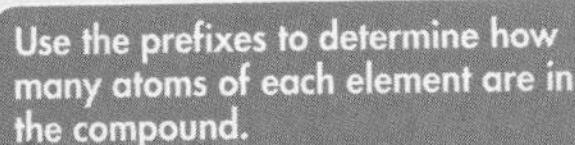
Use the prefixes to determine how many atoms of each element are in the compound.

a. Each molecule of nitrogen trifluoride has: 1 nitrogen atom; 3 fluorine atoms.	**b.** Each molecule of disulfur dichloride has: 2 sulfur atoms; 2 chlorine atoms.
NF_3	S_2Cl_2

Construct the formula using the correct symbols and subscripts.

28. Write formulas for these binary molecular compounds.
a. dinitrogen tetroxide **c.** disulfur decafluoride
b. xenon tetrafluoride **d.** iodine heptafluoride

Note: The number 1 is never used as a subscript in a formula.

9.3 LessonCheck

29. **Review** Explain how to write the name and formula of a binary molecular compound.

30. **Describe** Write the names of these molecular compounds.
a. NCl_3 **b.** BCl_3 **c.** NI_3 **d.** SO_3 **e.** N_2H_4 **f.** N_2O_3

31. **Apply Concepts** Write the formulas for these binary molecular compounds.
a. phosphorus pentachloride
b. iodine heptafluoride
c. chlorine trifluoride
d. iodine dioxide

32. **Describe** Write the formulas or names for these molecular compounds.
a. CS_2 **b.** Cl_2O_7 **c.** carbon tetrabromide **d.** diphosphorus trioxide

33. **Evaluate** The name a student gives for the molecular compound $SiCl_4$ is monosilicon trichloride. Is this name correct? Explain your answer.

34. **Explain** Are the bonds between silicon and chlorine in silicon tetrachloride single bonds? Justify your answer by drawing an electron dot structure of silicon tetrachloride.

35. **Classify** Determine whether each of the following compounds is a molecular compound or an ionic compound. How can you tell?
a. PBr_3 **b.** KBr **c.** iron(III) oxide **d.** carbon tetraiodide

BIGIDEA BONDING AND INTERACTIONS

36. What is the difference between an ionic compound and a molecular compound?

Extend

Connect to GEOLOGY

Remind students that the Mohs scale rates the hardness of various minerals on a scale of 0 to 10. Discuss that silicon carbide is a binary molecular compound that has a Mohs scale rating of 9.3. Have students research this compound and what its rating means in terms of its utility in industry.

Evaluate

Informal Assessment

Write each of the following pairs of compounds on an index card.

a. PBr_3 and $CrBr_3$
b. N_2O and Na_2O
c. CI_4 and PbI_4
d. $P_2O_3$and Fe_2O_3

Ask groups of four students to approach your desk. Have each student select one card and identify what the two compounds have in common. *(Each pair has one element in common.)* Then have each student identify which compound in the pair is a binary molecular compound and which one is a binary ionic compound, and name each compound. Then have students complete the 9.3 Lesson Check.

Reteach

Review the guidelines for naming binary molecular compounds. Then, have students write the names of the following compounds: NO, SiO_2, N_2O_4, and ClF_3. *(nitrogen monoxide, silicon dioxide, dinitrogen tetroxide,* and *chlorine trifluoride)* Have students write the formulas for these compounds: arsenic pentachloride, iodine tribromide, and tetraphosphorus hexoxide. *($AsCl_5$, IBr_3,* and *P_4O_6)*

Lesson Check Answers

29. Name the elements in the order written in the formula, using prefixes as necessary to indicate the number of each kind of atom. Then, use the suffix *-ide* on the name of the second element. Write the symbol for each element with a subscript corresponding to the prefix before each element in the name.

30. **a.** nitrogen trichloride **b.** boron trichloride **c.** nitrogen triiodide **d.** sulfur trioxide **e.** dinitrogen tetrahydride **f.** dinitrogen trioxide

31. **a.** PCl_5 **b.** IF_7 **c.** ClF_3 **d.** IO_2

32. **a.** carbon disulfide **b.** dichlorine heptoxide **c.** CBr_4 **d.** P_2O_3

33. No; the correct name is silicon tetrachloride.

34. Yes; chlorine forms only one bond, and silicon forms four bonds.

35. **a.** molecular; two nonmetals **b.** ionic; metal and nonmetal ions **c.** ionic; metal and nonmetal ions **d.** molecular; two nonmetals

36. **BIGIDEA** An ionic compound is composed of ions. A molecular compound is composed of molecules.

Answers

27. **a.** oxygen difluoride **b.** disulfur decafluoride **c.** sulfur trioxide **d.** sulfur hexafluoride

28. **a.** N_2O_4 **b.** XeF_4 **c.** S_2F_{10} **d.** IF_7

CHEMISTRY & YOU Producers of nutritional supplements are required to list their ingredients in decreasing order by amount on their labels. The ingredients that supply minerals are listed by their chemical names. Because of the cost of redesigning labels, some companies still use common names rather than those based on IUPAC rules. For example, it is not uncommon to see *dibasic calcium phosphate* instead *calcium hydrogen phosphate* listed as a source of calcium and phosphorus. One of the downsides of this continued use of older names is that it is more difficult to write the correct formula for a compound just by looking at its name.

Pose the following question to students: *Why might nutritional supplement companies continue to use non-IUPAC names in their ingredient lists, despite the ease with which IUPAC names can be used to help determine the chemical formula for a compound?* You may need to assist students in the following ways:

- Nutritionists and biochemists tend to be slower to adopt more modern chemical naming conventions than chemists.
- Many older names, such as *dibasic*, convey useful information about a compound's chemical behavior in the body.
- Space on a supplement label is limited. Older name forms are often shorter than those generated using IUPAC naming rules, and thus they take up less space.

Explain

MAKE A CONNECTION Explain that one reason it is important for sports nutrition advisors to understand the "language of chemistry" is that they need to be able to distinguish between good nutritional supplements and supplements that use dangerous ingredients or banned substances. Remind students that many athletic associations ban certain supplement ingredients or other substances because they can give athletes an unfair advantage in a competition. Encourage students who participate in school or extracurricular sporting activities to identify any banned substances in their sport, and explain why the substance has been banned.

Connect to NUTRITION

After reading the passage, instruct students to choose a sport and think about the nutritional requirements of an athlete in that sport. Then have students develop a sample meal plan they feel a nutritionist might recommend to meet those requirements. Encourage students to consider the chemistry of the foods they recommend. For example, students should consider what types of food provide certain vitamins and minerals.

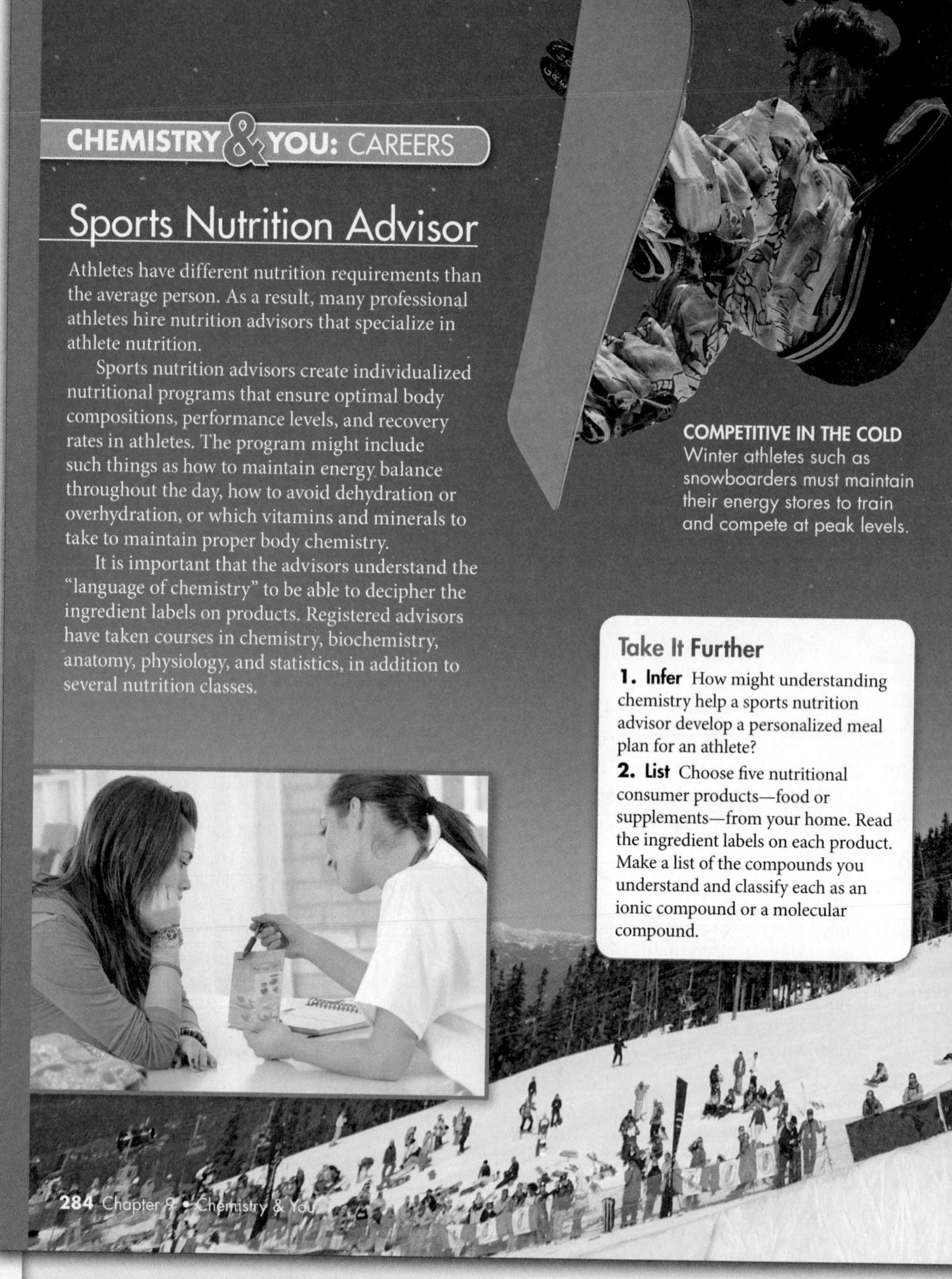

CHEMISTRY & YOU: CAREERS

Sports Nutrition Advisor

Athletes have different nutrition requirements than the average person. As a result, many professional athletes hire nutrition advisors that specialize in athlete nutrition.

Sports nutrition advisors create individualized nutritional programs that ensure optimal body compositions, performance levels, and recovery rates in athletes. The program might include such things as how to maintain energy balance throughout the day, how to avoid dehydration or overhydration, or which vitamins and minerals to take to maintain proper body chemistry.

It is important that the advisors understand the "language of chemistry" to be able to decipher the ingredient labels on products. Registered advisors have taken courses in chemistry, biochemistry, anatomy, physiology, and statistics, in addition to several nutrition classes.

COMPETITIVE IN THE COLD Winter athletes such as snowboarders must maintain their energy stores to train and compete at peak levels.

Take It Further

1. **Infer** How might understanding chemistry help a sports nutrition advisor develop a personalized meal plan for an athlete?
2. **List** Choose five nutritional consumer products—food or supplements—from your home. Read the ingredient labels on each product. Make a list of the compounds you understand and classify each as an ionic compound or a molecular compound.

284 Chapter 9 • Chemistry & You

21st Century Learning To be successful in the 21st century, students need skills and learning experiences that extend beyond subject matter mastery. The following project helps students build the following 21st Century Skills: *Creativity and Innovation, Critical Thinking and Problem Solving, Communication and Collaboration, Information Literacy, Initiative and Self-Direction,* and *Productivity and Accountability.*

CHEMISTRY AND FITNESS Pose the following challenge to your students. *The fitness center you work for has asked you to prepare a brochure about the importance of a balanced mineral intake for sports performance.* The brochure should contain the following information:

- The connection between each mineral and body function.
- Photographic examples of foods that contain each mineral, and the chemical formula of the active form of each mineral in the body.
- An explanation of the relationship between adequate mineral intake and sports performance that notes the consequences of mineral deficiencies (e.g., electrolyte imbalances and muscle cramps).

9.4 Naming and Writing Formulas for Acids and Bases

C.3.4 Structural formulas and names

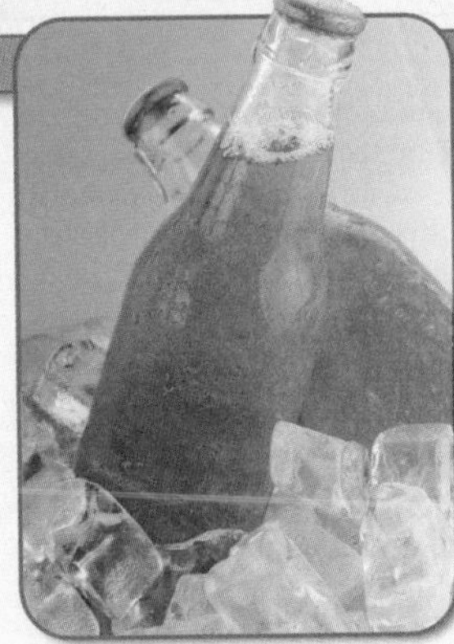

CHEMISTRY & YOU

Q: *What's the name of the acid responsible for the crisp taste in this drink?* There's a certain acid that gives many soft drinks their crisp, enjoyable taste. In this lesson, you will learn the names and formulas of some important acids, including one found in many soft drinks.

Key Questions

How do you determine the name and formula of an acid?

How do you determine the name and formula of a base?

Vocabulary

- acid
- base

Names and Formulas of Acids

How do you determine the name and formula of an acid?

Acids are a group of ionic compounds with unique properties. As you will see in Chapter 19, acids can be defined in several ways. For now, it is enough to know that an **acid** is a compound that contains one or more hydrogen atoms and produces hydrogen ions when dissolved in water. Acids have many uses. For example, sulfuric acid is often used to etch circuit boards like the one shown in Figure 9.13.

When naming an acid, you can consider the acid to consist of an anion combined with as many hydrogen ions as needed to make the molecule electrically neutral. Therefore, the chemical formulas of acids are in the general form H_nX, where X is a monatomic or polyatomic anion and n is a subscript indicating the number of hydrogen ions that are combined with the anion.

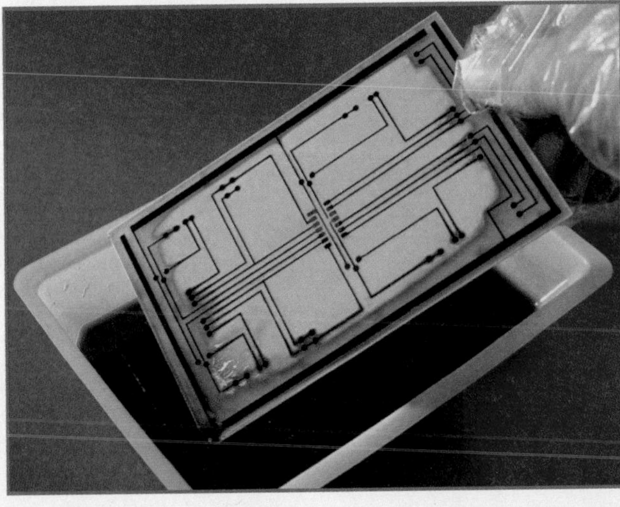

Figure 9.13 Sulfuric Acid
Circuit boards that are used in computers and other electronic devices have grooves (or circuits) that hold the wires for carrying signals. The circuits are often created using a mixture that contains sulfuric acid.

Key Objectives

9.4.1 DETERMINE the name and formula of an acid.

9.4.2 DETERMINE the name and formula of a base.

Additional Resources

Reading and Study Workbook, Lesson 9.4

Available Online or DVD-ROM:

- Lesson 9.4 Review
- Lesson 9.4 Practice Problems

LESSON 9.4

PearsonChem.com

Engage

CHEMISTRY & YOU Have students examine the labels of several soft drinks and juices. Then, direct students to the text and photograph that opens the lesson. **Ask** Which acids did you find listed? *(Sample answers: citric acid, ascorbic acid)* Note that some acids are present for reasons other than flavor, such as preservation or, in the case of ascorbic acid (vitamin C), because they are important nutrients.

Activate Prior Knowledge

As a class, create a list of the reagents you have used thus far in the classroom labs. Have students predict which ones are acids and which ones are bases. Students should notice that acids are more easily identified as such because their names contain the word *acid*. Have students copy the list in their notebooks and add any formulas that are encountered during the lesson.

Focus on ELL

1 CONTENT AND LANGUAGE Clarify that the vocabulary word *base* has both common and scientific meanings. Have students brainstorm other applications of this word. Then have students locate the definition of a chemical base in the text. Allow students with limited English to use their native languages for support.

2 FRONTLOAD THE LESSON Have students bring in a favorite family recipe to share with the class. Have students predict which ingredients are acidic in nature and which are basic.

3 COMPREHENSIBLE INPUT Point out that the rules for naming acids are illustrated in Table 9.5, and associate each rule with its row in the table. As a class, work through the Lesson 9.4 Practice Problems.

IN ACADEMIC STANDARDS for SCIENCE

C.3.4 Draw structural formulas for and name simple molecules.

Answers

TAKE IT FURTHER

1. A sports nutrition advisor must be able to understand the ingredient list on nutrient labels and how ingested compounds behave inside the client's body.
2. Answers will vary.

LESSON 9.4

Foundations for Reading

BUILD VOCABULARY Make a connection between the scientific and common meanings of the word *acid* by pointing out that a sharp or sour taste is a property of acids. This property can be used to recognize foods, such as lemons, that contain acids. Then, ask students to infer what it might mean to say that a person has "an acid tongue."

READING STRATEGY Encourage students to summarize the rules for naming acids and for naming bases in their own words.

Explain

Names and Formulas of Acids

USE VISUALS Direct students' attention to Table 9.5. Remind students that H^+ is the cation in each acid. It is the anion combined with hydrogen that identifies the acid. Call attention to the first column in which the endings of anions are given. Have students name some anions with the endings *-ide, -ite, and -ate. (Sample answers: chloride, bromide, nitrite, sulfite, carbonate, phosphate)* Call attention to the third column. **Ask** When the anion ends in *-ide*, what does the acid name begin with? *(hydro-)* What does the acid name end with? *(-ic)* **Ask** How would you name an acid containing the bromide anion? *(hydrobromic acid)* Continue the questioning and naming through the other two rows of the table and encourage students to use the table until they are comfortable with naming acids. *(NOTE: In Table 9.6, the formula for ethanoic acid is presented in the HA format: $HC_2H_3O_2$. In Chapter 19, students will be introduced to another way of writing the ethanoic acid formula: CH_3COOH.)*

CHEMISTRY & YOU Phosphoric acid is used in many soft drinks. It provides an acidic, crisp taste.

CRITICAL THINKING Explain that Table 9.5 can also be used to help write formulas for acids, if students use it in reverse. Give students an acid name, for example, perchloric acid. **Ask** What does the *-ic* ending indicate? (*There is no* hydro- *in the name, so* -ic *means the anion ends in* -ate.) **Ask** What is the name and formula for the anion? *(perchlorate, ClO_4^-)* **Ask** How many hydrogen atoms will combine with the perchlorate ion? *(one, $HClO_4$)*

Continue the questioning until students are comfortable with writing formulas for acids.

Table 9.5

Naming Common Acids

Anion ending	Example	Acid name	Example
-ide	chloride, Cl^-	*hydro-*(stem)*-ic acid*	*hydrochloric acid*
-ite	sulfite, SO_3^-	(stem)*-ous acid*	sulfur*ous acid*
-ate	nitrate, NO_3^-	(stem)*-ic acid*	nitr*ic acid*

CHEMISTRY & YOU

Q: *An acid that provides the crisp taste in many soft drinks has the formula H_3PO_4. What's the name of this acid?*

Three rules can help you name an acid with the general formula H_nX dissolved in water. Read the rules and the examples carefully. Notice that the naming system depends on the name of the anion (X), in particular the suffix of the anion name. Each rule deals with an anion with a different suffix: *-ide*, *-ite*, and *-ate*.

1. **When the name of the anion ends in *-ide*, the acid name begins with the prefix *hydro-*. The stem of the anion has the suffix *-ic* and is followed by the word *acid*.** Therefore, HCl (X = chloride) is named *hydrochloric acid*.
2. **When the anion name ends in *-ite*, the acid name is the stem of the anion with the suffix *-ous*, followed by the word *acid*.** Thus, H_2SO_3 (X = sulfite) is named sulfur*ous acid*.
3. **When the anion name ends in *-ate*, the acid name is the stem of the anion with the suffix *-ic*, followed by the word *acid*.** Thus, HNO_3 (X = nitrate) is named nitr*ic acid*.

These three rules are summarized in Table 9.5. Use the table to help you write acid names until you become an expert.

Writing Formulas of Acids If you know the name of an acid, you can write its formula. **To write the formula for an acid, use the rule for writing the name of the acid in reverse. Then balance the ionic charges just as you would for any ionic compound.** For example, consider hydrobromic acid. Following Rule 1, hydrobromic acid (*hydro-* prefix and *-ic* suffix) must be a combination of hydrogen ion (H^+) and bromide ion (Br^-). So the formula of hydrobromic acid is HBr. How do you write the formula for phosphorous acid? Using Rule 2, hydrogen ion and phosphite ion (PO_3^{3-}) must be the components of phosphorous acid. You need three hydrogen ions to balance the 3− charge of the phosphite anion. Thus, the formula for phosphorous acid is H_3PO_3. Finally, what is the formula for sulfuric acid? According to Rule 3, sulfuric acid (*-ic* ending) must be a combination of hydrogen ion and sulfate ion (SO_4^{2-}). The formula for sulfuric acid is H_2SO_4 because two hydrogen ions are needed to balance the 2− charge of the sulfate anion.

Many industrial processes, including steel and fertilizer manufacturing, use acids. You should become familiar with the names and formulas of common acids such as those listed in Table 9.6.

Table 9.6

Common Acids

Name	Formula
Hydrochloric acid	HCl
Sulfuric acid	H_2SO_4
Nitric acid	HNO_3
Ethanoic acid	$HC_2H_3O_2$
Phosphoric acid	H_3PO_4
Carbonic acid	H_2CO_3

286 Chapter 9 • Lesson 4

Acid Preservatives Many food labels mention parenthetically that an acid has been added as a preservative. The use of acids as a preservative goes back about 5,000 years, with the discovery that vinegars had a preservative effect on foods. Vinegars, which contain acetic acid, are still used today to preserve cucumbers and other vegetables in a process known as pickling. Other vegetables, and even some livestock feeds, are preserved with the lactic acid from bacterial fermentation within the food itself. Even though the acids differ from one application to the next, they all exert their preservative actions in the same manner—their low pH inhibits the growth of destructive bacteria and fungi in the preserved foods within the anaerobic environment of a sealed jar or silo.

SampleProblem 9.8

Naming Acids

Name the following compounds as acids:

a. HClO **b.** HCN

1 Analyze Identify the relevant concepts. The anion of the acid determines the acid name. (1) If the name of the anion ends in *-ide*, name the acid using the stem of the anion with the prefix *hydro-* and the suffix *-ic*, followed by the word *acid*. (2) If the anion name ends in *-ite*, name the acid using the stem of the anion with the suffix *-ous*, followed by the word *acid*. (3) If the anion name ends in *-ate*, name the acid using the stem of the anion with the suffix *-ic*, followed by the word *acid*.

2 Solve Apply the concepts to this problem.

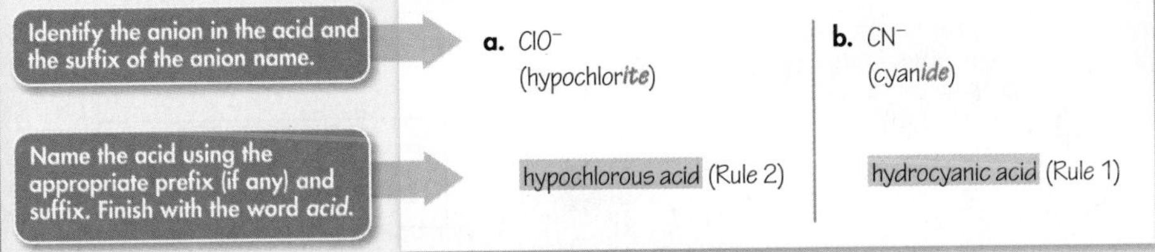

37. Name the following compounds as acids:
a. HF
b. HNO_3
c. H_2SO_3

38. Write formulas for the following acids:
a. perchloric acid
b. hydroiodic acid
c. chlorous acid

Hint: For Problem 38, use the corresponding naming rule in reverse.

Names and Formulas of Bases

How do you determine the name and formula of a base?

A **base** is generally an ionic compound that produces hydroxide ions when dissolved in water. **Bases are named in the same way as other ionic compounds—the name of the cation is followed by the name of the anion.** The common base sodium hydroxide is used in making cleaners, soap, and paper as shown in Figure 9.14. Sodium hydroxide (NaOH) is composed of sodium cations (Na^+) and hydroxide anions (OH^-).

To write the formula for a base, first write the symbol for the metal cation followed by the formula for the hydroxide ion. Then, balance the ionic charges just as you would for any ionic compound. For example, aluminum hydroxide consists of the aluminum cation (Al^{3+}) and the hydroxide anion (OH^-). You need three hydroxide ions to balance the 3+ charge of the aluminum cation. Thus, the formula for aluminum hydroxide is $Al(OH)_3$.

Figure 9.14 Use of Sodium Hydroxide
In the first step of papermaking, manufacturers use NaOH to break down recycled paper and wood to make pulp.

Differentiated Instruction

L1 STRUGGLING STUDENTS Create a set of flash cards by writing the formula for an acid or base on one side of an index card and its name on the other. Alternate showing formulas and names to the class. Have students supply the missing name or formula for each card shown.

ELL ENGLISH LANGUAGE LEARNERS Beginning students may have difficulty distinguishing and pronouncing the different endings for acids, bases, and their associated ions. Write each on the board and pronounce slowly and clearly. Have students practice saying the names to become familiar with their pronunciations.

L3 ADVANCED STUDENTS Have students research the meaning of pH and how it relates to acids and bases. Then have students explain their findings by displaying and classifying common household products such as coffee, antacids, drain cleaner, apple juice, and so on from most acidic to least acidic. Have students explain that least acidic means most basic and define what is meant by neutral on the pH scale.

Explain

USE VISUALS Have students examine the common acids in Table 9.6. Have students match up each of these acids with one of the three rules explained under Table 9.5. Encourage students to recreate the table in their notebooks and add helpful information to it as needed.

MAKE A CONNECTION Refer students to Table 9.3, Common Polyatomic Ions, in Lesson 9.1. Discuss how to combine each polyatomic anion with hydrogen ion(s) to form an acid. Work through the entire table or until students become comfortable with the process.

Sample Practice Problem

Identify which compound is an acid: HOCl, K_2O, or NaOH. Then, name the acid. *(HOCl; hypochlorous acid)*

Names and Formulas of Bases

MAKING CONNECTIONS Direct students to Lesson 9.2 and review the guidelines for naming and writing ionic compounds. Use the flowchart from that lesson, or create a new one for this lesson, for students to follow for naming a base, or for writing a formula for a base. **Ask** What ion is typically seen in the formula for all bases? *(hydroxide ion)* **Ask** What is the charge on this ion? *(1–)*

Answers

37. a. hydrofluoric acid **b.** nitric acid **c.** sulfurous acid

38. a. $HClO_4$ **b.** HI **c.** $HClO_2$

Extend

Connect to APPLIED CHEMISTRY

Explain to the class that sodium hydroxide is a powerful base. Have students research and find at least five different industrial uses of sodium hydroxide. Tell students to explain why this particular base is used for each given application, the concentration of base used, and the appropriate Material Data Safety Sheet (MSDS) for the base at that specific concentration.

Sample Practice Problem

Which of the following is a base? Name the base.
H_2S, CsOH, $FeCl_2$ *(CsOH, cesium hydroxide)*

Evaluate

Informal Assessment

Present only the names of the six common acids in Table 9.6 and a copy of Table 9.5. Have students name the anion and write the formulas for each acid. Ask students to write a statement explaining how to determine whether the compound is a base. Then have students complete the 9.4 Lesson Check.

Reteach

Review with students the rules for writing and naming acids and bases. Create a chart comparing and contrasting the two methods. Then, have students complete the assessment again.

Sample Problem 9.9

Naming Bases

Name the following bases:

a. KOH **b.** $Fe(OH)_2$

1 Analyze **Identify the relevant concepts.** Bases are named like other ionic compounds—the name of the cation is followed by the name of the anion.

2 Solve **Apply the concepts to this problem.**

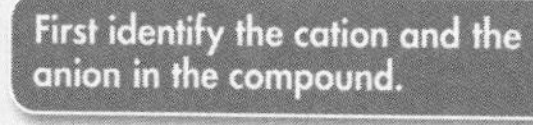

	a.	b.
First identify the cation and the anion in the compound.	cation: K^+ anion: OH^-	cation: Fe^{2+} anion: OH^-
Now write the name of the cation, followed by the name of the anion.	potassium hydroxide	iron(II) hydroxide

39. Name the following bases:
a. $Ba(OH)_2$
b. $Ca(OH)_2$
c. RbOH

40. Write formulas for the following bases:
a. cesium hydroxide
b. beryllium hydroxide
c. manganese(III) hydroxide

9.4 LessonCheck

41. **Review** Explain how to determine the name and formula of an acid.

42. **Review** How are the names and formulas determined for bases?

43. **Identify** Give the names of the following acids:
a. HNO_2 **b.** $HMnO_4$ **c.** HBr **d.** H_2S

44. **Identify** Write the names of these bases.
a. LiOH **b.** $Pb(OH)_2$ **c.** $Mg(OH)_2$ **d.** $Al(OH)_3$

45. **Classify** Identify each compound as an acid or a base. Then name each compound.
a. NH_4OH **b.** $HClO_3$ **c.** $Fe(OH)_3$ **d.** KOH

46. **Describe** Write the formulas for these ionic compounds.
a. carbonic acid **b.** sulfurous acid **c.** iron(III) hydroxide **d.** zinc hydroxide

47. **Compare** What element generally appears in the formula of an acid? What ion generally appears in the formula of a base?

Lesson Check Answers

41. See the rules for naming acids on page 286. Use the rules for naming acids in reverse to write the formula.

42. The name of the cation is followed by the name of the anion. To write the formula, write the symbol for the cation then the formula for the anion. Then, balance the charges.

43. **a.** nitrous acid
b. permanganic acid
c. hydrobromic acid
d. hydrosulfuric acid

44. **a.** lithium hydroxide
b. lead(II) hydroxide
c. magnesium hydroxide
d. aluminum hydroxide

45. **a.** base, ammonium hydroxide
b. acid, chloric acid
c. base, iron(III) hydroxide
d. base, potassium hydroxide

46. **a.** H_2CO_3 **b.** H_2SO_3
c. $Fe(OH)_3$ **d.** $Zn(OH)_2$

47. acid, hydrogen; base, hydroxide ion

9.5 The Laws Governing How Compounds Form

C.3.4 Structural formulas and names

CHEMISTRY & YOU

Q: *Did you know that sand from a beach can be used to make glass?* Sand contains the compound silicon dioxide, which is used in glass making. One molecule of silicon dioxide consists of one silicon atom and two oxygen atoms. In this lesson, you will learn why the ratio of silicon to oxygen atoms in silicon dioxide is always the same.

Key Questions

How is the law of definite proportions consistent with Dalton's atomic theory?

What general guidelines can help you write the name and formula of a chemical compound?

Vocabulary

- law of definite proportions
- law of multiple proportions

The Laws of Definite and Multiple Proportions

How is the law of definite proportions consistent with Dalton's atomic theory?

Consider the compound calcium carbonate ($CaCO_3$), which is commonly found in rocks. Whether you find the compound in a rock in Thailand or New Zealand, it is still calcium carbonate. This statement is true because the three elements—calcium, carbon, and oxygen—are combined in the same proportions in every molecule of $CaCO_3$. Two laws—the law of definite proportions and the law of multiple proportions—describe the proportions in which elements combine to form compounds. The rules for naming and writing formulas for compounds are possible because compounds obey these two rules.

Law of Definite Proportions A chemical formula tells you, by means of subscripts, the ratio of atoms of each element in the compound. Ratios of atoms can also be expressed as ratios of masses. For example, magnesium sulfide (MgS) is composed of magnesium cations and sulfide anions. If you could take 100.00 g of magnesium sulfide and break it down into its elements, you would obtain 43.13 g of magnesium and 56.87 g of sulfur. The Mg:S ratio of these masses is 43.13/56.87 or 0.758:1. This mass ratio does not change regardless of how the magnesium sulfide is formed or the size of the sample. Magnesium sulfide obeys the **law of definite proportions,** which states that in samples of any chemical compound, the masses of the elements are always in the same proportions. This law is consistent with Dalton's atomic theory. **Dalton postulated that atoms combine in simple whole-number ratios. If the ratio of atoms of each element in a compound is fixed, then it follows that the ratio of their masses is also fixed.**

Focus on ELL

1 CONTENT AND LANGUAGE Review mathematical vocabulary such as *ratio*, *proportion*, and *multiple*. Divide students into three small groups. Assign each group one term. Have each group define its term, give an example and a non-example, and present it to the class.

2 FRONTLOAD THE LESSON Discuss the difference between a *law* in science and a *law* in society. Use a Venn diagram to help students see the similarities and the differences.

3 COMPREHENSIBLE INPUT For Figure 9.16, point out the mathematical operators in each graph. Clarify for students that the masses of compounds *X* and *Y* are equal to the sums of the elements *A* and *B* in the two graphs. Make sure that students understand how to use the "yes" and "no" options in Figures 9.18 and 9.19 to navigate through each flowchart.

Key Objectives

9.5.1 EXPLAIN how the law of definite proportions is consistent with Dalton's atomic theory.

9.5.2 LIST the general guidelines that can help you write the name and formula of a chemical compound.

Additional Resources

Reading and Study Workbook, Lesson 9.5

Available Online or DVD-ROM:

- Lesson 9.5 Review

PearsonChem.com

Engage

Activate Prior Knowledge

Ask a student volunteer to explain what a proportion means mathematically. Make sure that students understand that a proportion is an equation stating that two ratios are equal. Review how to use cross products to solve for an unknown quantity in a proportion.

Foundations for Reading

BUILD VOCABULARY A scientific law is a statement of a relationship that is unchanging so far as it is known. Thus, the law of definite proportions and the law of multiple proportions are unchanging, or constant, statements governing the formation of compounds.

READING STRATEGY As you familiarize students with the flowcharts in Figures 9.18 and 9.19, verbalize the routes to a name or formula by asking and answering the questions posed at each intersection.

IN ACADEMIC STANDARDS for SCIENCE

C.3.4 Draw structural formulas for and name simple molecules.

Answers

39. **a.** barium hydroxide
b. calcium hydroxide
c. rubidium hydroxide

40. **a.** CsOH **b.** $Be(OH)_2$ **c.** $Mn(OH)_3$

LESSON 9.5

Explain

The Laws of Definite and Multiple Proportions

USE MODELS Direct students' attention to Figure 9.15. **Ask** How can you tell that the two compounds are different? *(One is green, and the other is blue.)*

APPLY CONCEPTS Students may be able to understand the law of multiple proportions more easily by thinking in terms of atoms. Explain that one atom of carbon in CO_2 has the same mass (12 amu) as one atom of carbon in CO. One atom of oxygen (16 amu) combines with 12 amu of carbon in CO. Two atoms of oxygen (32 amu) combine with 12 amu of carbon in CO_2. The ratio of the masses of oxygen in the two compounds is 32:16 or 2:1. Explain to students that ratios are typically written in one of three ways—as a fraction, as a statement separated by colons, and as a statement separated by the word *to*.

Law of Multiple Proportions In the early 1800s, Dalton and others studied pairs of compounds that contain the same elements but have different physical and chemical properties. Using the results from these studies, Dalton stated the **law of multiple proportions:** Whenever the same two elements form more than one compound, the different masses of one element that combine with the same mass of the other element are in the ratio of small whole numbers. Figure 9.15 shows two compounds—CuCl and $CuCl_2$—that demonstrate the law of multiple proportions.

Figure 9.15 CuCl and $CuCl_2$
a. Copper(I) chloride (CuCl) contains the elements copper and chlorine. This compound is green. **b.** Copper(II) chloride ($CuCl_2$) contains the same two elements as copper(I) chloride—copper and chlorine. But, this compound is blue.

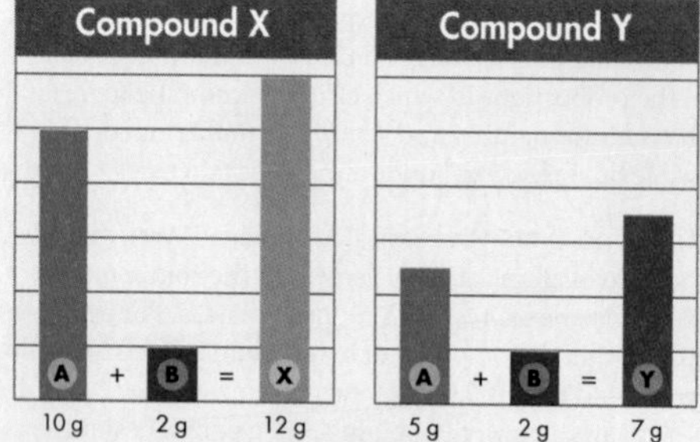

Figure 9.16 Law of Multiple Proportions
Two compounds, X and Y, contain equal masses of element B. The ratio of the masses of A in these compounds is 10:5 or 2:1 (a small whole-number ratio).
Compare ***Would the ratio be different if samples of X and Y contained 3 g of B? Explain.***

Two familiar compounds, water (H_2O) and hydrogen peroxide (H_2O_2), are formed by the same two elements. Although these compounds are formed by the elements hydrogen and oxygen, they have different physical and chemical properties. For example, water does not bleach fabric dyes, but hydrogen peroxide bleaches the dye in most fabrics. Each compound obeys the law of definite proportions. In every sample of hydrogen peroxide, 16.0 g of oxygen are present for each 1.0 g of hydrogen. The mass ratio of oxygen to hydrogen is always 16:1. In every sample of water, the mass ratio of oxygen to hydrogen is always 8:1. If a sample of hydrogen peroxide has the same mass of hydrogen as a sample of water, the ratio of the mass of oxygen in the two compounds is exactly 2:1.

$$\frac{16 \text{ g O (in } H_2O_2 \text{ sample that has 1 g H)}}{8 \text{ g O (in } H_2O \text{ sample that has 1 g H)}} = \frac{16}{8} = \frac{2}{1} = 2{:}1$$

A simple example of the law of multiple proportions is shown in Figure 9.16.

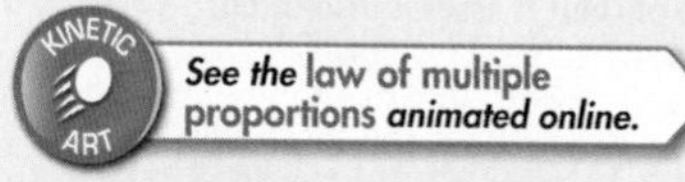

See the law of multiple proportions *animated online.*

Proust's Law

In the early 1800s, many scientists were studying chemical reactions by carefully measuring the masses of reactants and products. The French scientist Joseph Proust (1766–1844) demonstrated that samples of copper carbonate always contain 5.3 parts by mass of copper to 4 parts of oxygen and 1 part of carbon. The principle embodied in this discovery became known as Proust's Law and later as the law of definite proportions. Proust's experiments inspired John Dalton to think in terms of atoms. With atoms in mind, Dalton designed experiments that soon led to the law of multiple proportions and Dalton's atomic theory.

Sample Problem 9.10

Calculating Mass Ratios

Carbon reacts with oxygen to form two compounds. Compound A contains 2.41 g of carbon for each 3.22 g of oxygen. Compound B contains 6.71 g of carbon for each 17.9 g of oxygen. What is the lowest whole-number mass ratio of carbon that combines with a given mass of oxygen?

1 Analyze **List the knowns and the unknown.**
Apply the law of multiple proportions to the two compounds. For each compound, find the grams of carbon that combine with 1.00 g of oxygen. Then find the ratio of the masses of carbon in the two compounds. Confirm that the ratio is the lowest whole-number ratio.

KNOWNS
Compound A = 2.41 g C and 3.22 g O
Compound B = 6.71 g C and 17.9 g O

UNKNOWN
Mass ratio of C per g O in the two compounds = ?

2 Calculate **Solve for the unknown.**

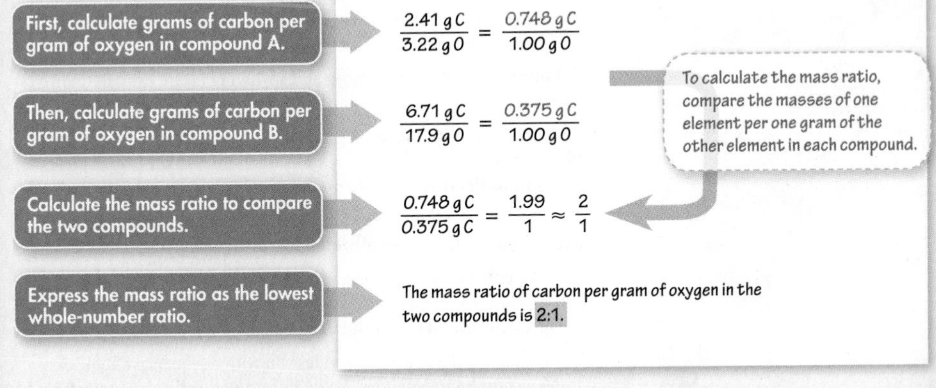

3 Evaluate **Does this result make sense?** The ratio is a low whole-number ratio, as expected. For a given mass of oxygen, compound A contains twice the mass of carbon as compound B.

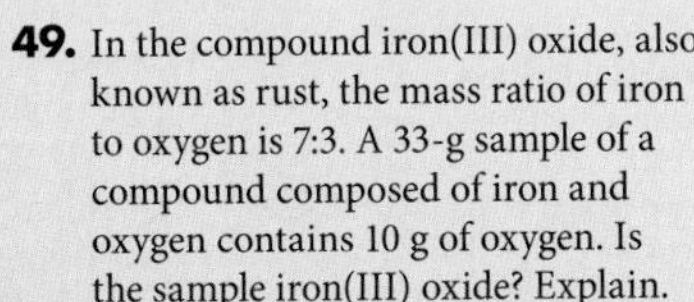

48. Lead forms two compounds with oxygen. One contains 2.98 g of lead and 0.461 g of oxygen. The other contains 9.89 g of lead and 0.763 g of oxygen. For a given mass of oxygen, what is the lowest whole-number mass ratio of lead in the two compounds?

49. In the compound iron(III) oxide, also known as rust, the mass ratio of iron to oxygen is 7:3. A 33-g sample of a compound composed of iron and oxygen contains 10 g of oxygen. Is the sample iron(III) oxide? Explain.

To answer Problem 49, first calculate the mass of iron in the sample.

Foundations for Math

RATIOS In mathematics, a ratio expresses the magnitude of two quantities relative to each other. Ratios can be expressed in multiple forms: , *a* to *b*, or *a*:*b*. When a ratio is in the form , it is clear that it is a quotient and can be simplified using division.

In **Sample Problem 9.10,** the grams of carbon per gram of oxygen in compound A can be found using division: 2.41 ÷ 3.22 = 0.748, which can be written

as . For this calculation, the format is more convenient. However, when

writing the final mass ratio, the format *a*:*b* is typically used because the law of multiple proportions states that the masses will always be in small whole-number ratios.

Explain

USE VISUALS Have students examine Figure 9.16. **Ask** How many grams of *B* are in compound *X*? *(2 g)* **Ask** How many grams of *B* are in compound *Y*? *(2 g)* State that the two compounds have the same masses of *B*. **Ask** What is the ratio between the masses of *A* in these two compounds that have equal masses of *B*? *(10 g:5 g or 2:1)* **Ask** Name the law illustrated by the figure and define it in your own words. *(Sample answer: The law of multiple proportions; When two compounds are formed from the same two elements, and you take a certain mass of one of the elements, the masses of the other element in the two compounds that combined with that certain mass of the first element are in a whole-number ratio).*

Sample Practice Problem

Phosphorus forms two compounds with oxygen. A 3.00 gram sample of Compound *A* contains 1.69 g of phosphorus and 1.31 g of oxygen, while a 5.00 gram sample of Compound *B* contains 2.18 grams phosphorus and 2.82 g of oxygen. For a given mass of oxygen, calculate the lowest whole-number mass ratio of phosphorus in the two compounds.

(1.66:1 ≈ 5:3)

Answers

FIGURE 9.16 No, as long as the two samples have the same mass of *B*.

48. 1:2

49. Yes, because the ratio in the compound is 2.3:1 and the ratio in iron(III) oxide is 2.3:1.

Explain

Practicing Skills: Chemical Names and Formulas

START A CONVERSATION To emphasize the importance of having a consistent method for naming chemical compounds, ask questions that students can relate to in their daily lives. **Ask** Name some situations in which specific names are important for accurate or fast results. *(when you are ordering items by telephone or on the Internet)* **Ask** Why do many bureaucratic agencies use social security or driver's license numbers instead of people's names? *(Many people could have the same name.)*

DRAW CONCLUSIONS Direct students to the opening paragraph on this page. Discuss the importance of naming conventions for compounds. Explain that chemists need to be able to understand papers written by colleagues and to reproduce experimental results. Focus on the problems that could occur if common names in multiple languages were the norm. **Ask** Why would knowing the correct name for a chemical be important to a poison-control center worker? *(The worker needs to know the specific chemical ingested before offering a solution. Administering an antidote for the wrong chemical could cause further harm.)*

Figure 9.17 Petrified Wood
When wood ages, certain compounds from the sediment can replace the dead tissue in the wood. The process is called petrification. Some of these compounds are colored and provide the various colors in the petrified wood.

Practicing Skills: Chemical Names and Formulas

What general guidelines can help you write the name and formula of a chemical compound?

In the average home, you can probably find hundreds of chemicals, including cleaning products, pharmaceuticals, and pesticides. You've probably noticed warning labels on products, which tell about their possible dangers. Most people would not know what to do if a child ingested one of these chemicals. A phone call to a poison control center can provide lifesaving information to victims of such poisonings. But a poison control center can be much more effective if the caller can supply some information about the name or formula of the substance.

Naming Chemical Compounds One of the skills you learned in this chapter is to name chemical compounds. If this is the first time you have tried to master this skill, you may feel a little overwhelmed. For example, you may find it difficult to know when you should or should not use prefixes and Roman numerals in a name. Or you may have trouble determining if a compound's name should end in *-ate, -ide,* or *-ite.*

Here are some guidelines to help you in naming a chemical compound from the chemical formula.

1. **Follow the rules for naming acids when H is the first element in the formula.**
2. **If the compound is binary, generally the name ends with the suffix *-ide.* If the compound is a molecular binary compound, use prefixes to indicate the number of atoms.**
3. **When a polyatomic ion that includes oxygen is in the formula, the compound name generally ends in *-ite* or *-ate.***
4. **If the compound contains a metallic cation that can have different ionic charges, use a Roman numeral to indicate the numerical value of the ionic charge in the compound.**

The flowchart in Figure 9.18 provides you with a sequence of questions for naming a compound when you know its formula. Follow the arrows and answer the questions on the flowchart to write the correct name for a compound. The sequence of questions in the flowchart can help you name compounds you may have in your home as well as the compounds that are responsible for the beautiful colors in the petrified wood shown in Figure 9.17. Apply the general formula Q_xR_y to each compound. Q and R can be atoms, monatomic ions, or polyatomic ions. For example, to name HNO_3, let H = Q and NO_3 = R. Follow the first arrow down to the question "Q = H?" The answer is yes, so the arrow to the right tells you that the compound is an acid. You can then follow the rules for naming acids. HNO_3 is nitric acid.

To name the compound N_2O_3, let N = Q and O = R. The answer to the question "Q = H?" is no, so you follow the arrow down. Does the compound have more than two elements? The answer is no, so you follow the arrow to the left. The compound is binary, and its name ends in *-ide.* Is Q a metal? The answer is no, so you must use prefixes in the name for N_2O_3, which is *di*nitrogen *tri*oxide. Practice naming other compounds using the flowchart. Soon you won't need the flowchart anymore.

Differentiated Instruction

L1 STRUGGLING STUDENTS Students who have difficulty with mathematics may need extra help with the laws of definite proportions and the law of multiple proportions. Assign such students to work with students with strong math skills. Encourage students with strong math skills to develop ways of sharing their problem-solving techniques with others.

LPR LESS PROFICIENT READERS Have students compare Figure 9.18 to the rules in the text on page 292, and Figure 9.19 to the rules on page 293. Give students copies of the two figures and encourage them to write the number of each rule next to the point in the flowchart where it is applied.

L3 ADVANCED STUDENTS Challenge advanced students to use the flowchart as a model for writing a computer program for naming chemical compounds.

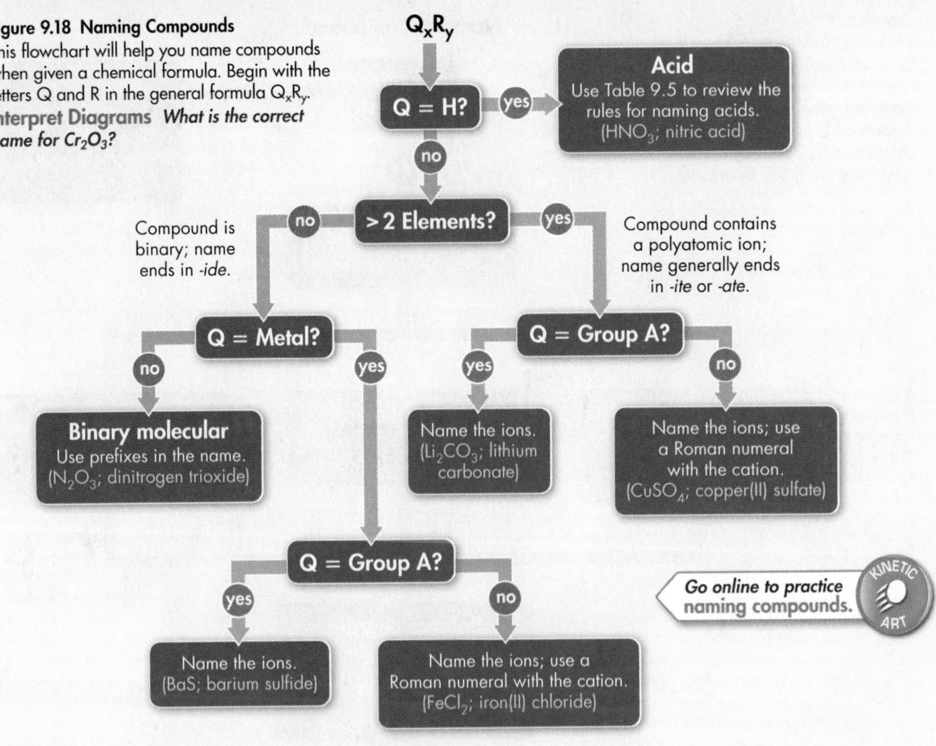

Figure 9.18 Naming Compounds
This flowchart will help you name compounds when given a chemical formula. Begin with the letters Q and R in the general formula Q_xR_y.
Interpret Diagrams ***What is the correct name for Cr_2O_3?***

Go online to practice **naming compounds.** KINETIC ART

Writing Chemical Formulas In writing a chemical formula from a chemical name, it is helpful to remember the following guidelines:

1. **An *-ide* ending generally indicates a binary compound.**
2. **An *-ite* or *-ate* ending means a polyatomic ion that includes oxygen is in the formula.**
3. **Prefixes in a name generally indicate that the compound is molecular.**
4. **A Roman numeral after the name of a cation shows the ionic charge of the cation.**

These guidelines and the questions in the flowchart in Figure 9.19 will help you write the formula for a compound when you know its name. For example, use the flowchart to write the formula for sodium chromate. The name does not contain prefixes, so the compound is ionic. The ions are sodium ion and chromate ion. Sodium is a Group A element, so use the periodic table or Table 9.1 to obtain its ionic charge (1+). Chromate ion is a polyatomic ion, so use Table 9.3 to obtain its charge (2−). Balance the charges to obtain the formula Na_2CrO_4. Use this flowchart to practice writing formulas until you don't need it anymore.

CHEMISTRY & YOU

Q: *Use the flowchart in Figure 9.19 to help you write the formula for silicon dioxide.*

LESSON 9.5

Explain

USE VISUALS Tell students to study Figure 9.18. Point out that the flowchart is like a road map. At every intersection you need to make a decision, and when you make the right decisions you are led to your destination (the correct name for the compound). Lead them through a few examples asking them to respond to the questions at each intersection. Then, have volunteers take the lead to continue the exercise.

MAKE A CONNECTION While students are involved in developing their skills in determining names and writing formulas for ionic and molecular compounds, you may have the opportunity to comment on the characteristics of representative ionic and molecular compounds found in Figure 9.18. For example, call attention to the high melting points of ionic compounds. Iron(II) chloride melts at 677°C; lithium carbonate, at 723°C. In addition, a large amount of energy (heat of fusion) is required to melt these ionic compounds. By contrast, the molecular compound dinitrogen trioxide melts at −101°C, and methane melts at −182°C. Emphasize that the ionic and molecular compounds do not differ just in the way that they are named but fundamentally in the bonding that holds the atoms together in the compound.

CHEMISTRY & YOU The formula for silicon dioxide is SiO_2.

Focus on ELL

4 LANGUAGE PRODUCTION Have students work in groups of four to complete the lab on page 295. Make sure each group has ELLs of varied language proficiencies, so that more proficient students can help less proficient ones. Have students work according to their proficiency level.

BEGINNING: LOW/HIGH Provide the grid for students and model how to complete the reactions using the first square.

INTERMEDIATE: LOW/HIGH Paraphrase the procedure and Analyze and Conclude questions for students.

ADVANCED: LOW/HIGH Have students assist those of lower English proficiencies with the procedure and with answering the questions.

Answers

FIGURE 9.18 chromium(III) oxide

Explain

SUMMARIZE Have students summarize the rules for writing formulas and naming compounds. Point out that naming compounds and writing formulas becomes easier with practice because chemical names and formulas are systematized. Ask students to consider how difficult the task would be if the names were assigned randomly. Because chemists adhere to an established system of naming, it is possible to quickly locate information on physical constants such as melting point, boiling point, and density in chemical handbooks such *Lange's Handbook of Chemistry, The Merck Index,* and the *CRC Handbook of Chemistry and Physics. The Merck Index* is particularly helpful to organic chemists interested in the synthesis and identification of a particular compound. Encourage students to look for these books in their local public library.

Evaluate

Informal Assessment

Ask students to explain what it means when each of the following is included in the name of a compound: an *-ide* ending, an *-ite* or *-ate* ending, a prefix, or a Roman numeral. Then, give students the following names of compounds containing one of these components: hydrogen cyanide, potassium chromate, sulfur tetrafluoride, or copper(I) chloride. Allow students to use Figure 9.19 to determine the corresponding formula. Then have students complete the 9.5 Lesson Check.

Reteach

For students struggling with the assessment, review the four skills for writing names and chemical formulas. Paraphrase each rule so students can clearly understand what is being stated. Work alongside students by stating various compounds and seeing whether students can adequately use Figure 9.18 to determine the names. Then, give them names for different compounds and check whether they can adequately use Figure 9.19 to write the corresponding formulas.

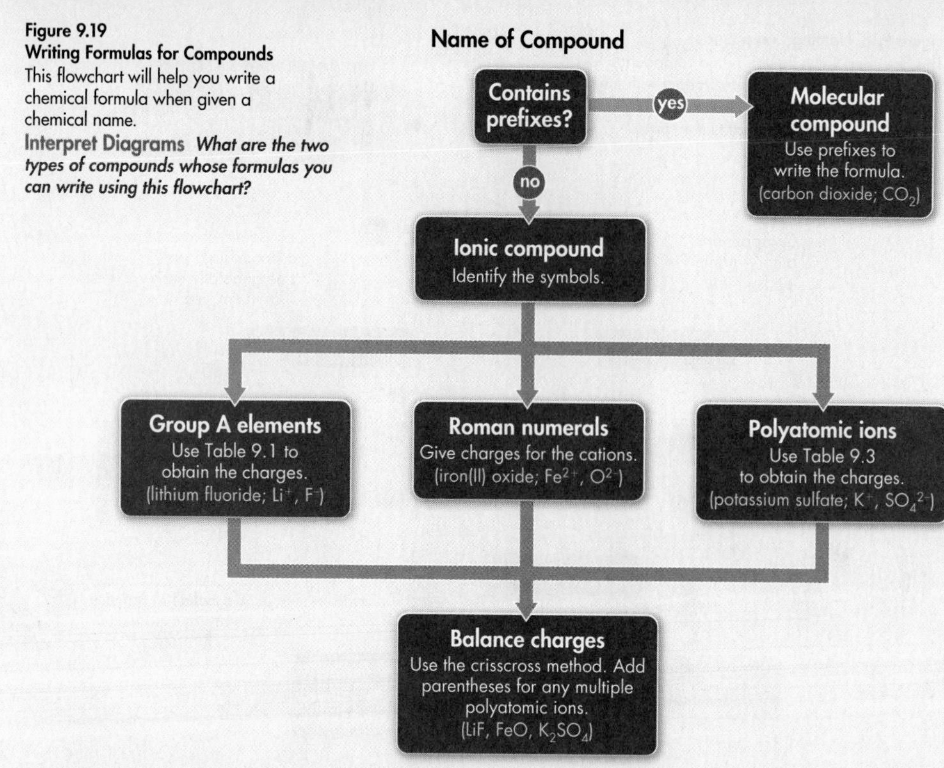

Figure 9.19 **Writing Formulas for Compounds**
This flowchart will help you write a chemical formula when given a chemical name.
Interpret Diagrams *What are the two types of compounds whose formulas you can write using this flowchart?*

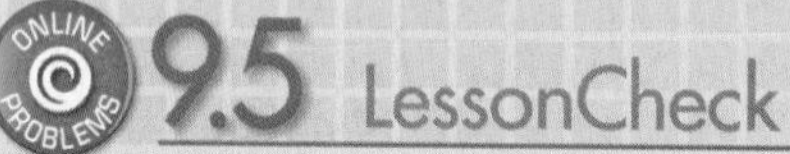

9.5 LessonCheck

50. Review How is the law of definite proportions consistent with Dalton's atomic theory?

51. List What general guidelines can help you write the name and formula of a compound?

52. Compare Two compounds that contain copper and chlorine have the following masses:

Compound A: 32.10 g Cu and 17.90 g Cl
Compound B: 23.64 g Cu and 26.37 g Cl

Are the compounds the same? If not, what is the lowest whole-number mass ratio of copper that combines with a given mass of chlorine?

53. Identify Name the following compounds:
a. $CaCO_3$ **b.** $PbCrO_4$ **c.** $SnCr_2O_7$

54. Describe Write the chemical formulas for each of the following compounds:
a. tin(II) hydroxide **c.** tetraiodide nonoxide
b. barium fluoride **b.** iron(III) oxalate

55. Evaluate Identify any incorrect names or formulas from the following choices. Explain your answer(s).
a. calcium(II) oxide **c.** $Na_2C_2O_4$
b. aluminum oxide **d.** $Mg(NH_4)_2$

BIGIDEA BONDING AND INTERACTIONS

56. Explain why the chemical composition of water (H_2O) is always the same.

Lesson Check Answers

50. Dalton postulated that atoms combine in simple whole-number ratios. If the ratio of atoms of each element in a compound is fixed, then it follows that the ratio of their masses is also fixed.

51. Refer to the rules listed on pages 292 and 293.

52. no; 2:1

53. a. calcium carbonate
b. lead(II) chromate
c. tin(II) dichromate

54. a. $Sn(OH)_2$ **b.** BaF_2 **c.** I_4O_9
d. $Fe_2(C_2O_4)_3$

55. a. incorrect; a Roman numeral is not used with a Group A metal, calcium oxide
d. Incorrect; ammonium ion, a polyatomic cation, will not form a compound with a monatomic cation.

56. BIGIDEA As stated in the law of definite proportions, the masses of the elements in H_2O, hydrogen and oxygen, are always in the same proportion, 2:1.

Small-Scale Lab

Names and Formulas for Ionic Compounds

Purpose

To observe the formation of compounds and to write their names and formulas

Materials

- pencil
- paper
- ruler
- reaction surface
- chemicals shown in Figure A
- chemicals shown in Figure B

Procedure

On separate sheets of paper, draw two grids similar to Figure A. Make each square 2 cm on each side. Draw black X's on one of the grids. Use the other grid as a data table to record your observations. Place a reaction surface over the grid with black X's and add the chemicals as indicated in Figure A.

Analyze and Conclude

1. Observe Describe each precipitate (solid product) that forms. Use terms such as *milky, grainy, cloudy,* or *gelatinous.* Which mixture(s) did not form a precipitate?

2. Describe Write the formulas and names of the chemical compounds produced in the mixings.

You're the Chemist

1. Analyze Data Repeat the experiment, using the chemicals in Figure B. Identify the precipitates, write their formulas, and name them.

2. Explain In ionic equations, the precipitate is written to the right of an arrow, and the ions that produced it are written to the left. Write ionic equations for the precipitates formed from the reactions related to Figure B. For example, the first reaction in Figure B would be written as follows:

$$2Fe^{3+} + 3CO_3^{2-} \longrightarrow Fe_2(CO_3)_3$$

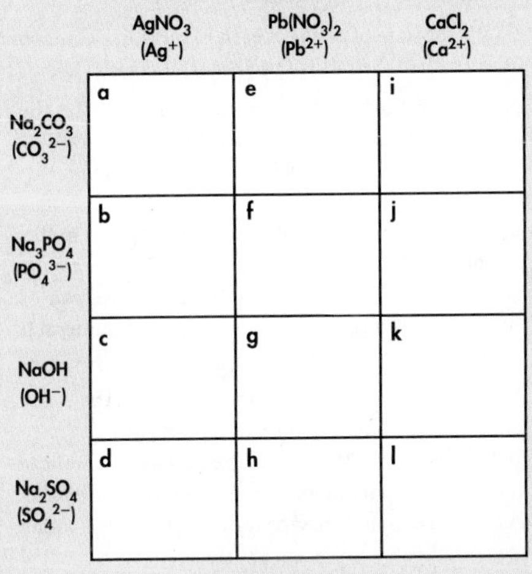

Figure A

	$FeCl_3$ (Fe^{3+})	$MgSO_4$ (Mg^{2+})	$CuSO_4$ (Cu^{2+})
Na_2CO_3 (CO_3^{2-})	a	e	i
Na_3PO_4 (PO_4^{3-})	b	f	j
NaOH (OH^-)	c	g	k
Na_2SO_4 (SO_4^{2-})	d	h	l

Figure B

YOU'RE THE CHEMIST

1. **a.** $Fe_2(CO_3)_3$, iron(III) carbonate
   b. $FePO_4$, iron(III) phosphate
   c. $Fe(OH)_3$, iron(III) hydroxide
   e. $MgCO_3$, magnesium carbonate
   f. $Mg_3(PO_4)_2$, magnesium phosphate
   g. $Mg(OH)_2$, magnesium hydroxide
   i. $CuCO_3$, copper(II) carbonate
   j. $Cu_3(PO_4)_2$, copper(II) phosphate
   k. $Cu(OH)_2$, copper(II) hydroxide

2. **a.** $2Fe^{3+} + 3CO_3^{2-} \rightarrow Fe_2(CO_3)_3(s)$
   b. $Fe^{3+} + PO_4^{3-} \rightarrow FePO_4(s)$
   c. $Fe^{3+} + 3OH^- \rightarrow Fe(OH)_3(s)$
   e. $Mg^{2+} + CO_3^{2-} \rightarrow MgCO_3(s)$
   f. $3Mg^{2+} + 2PO_4^{3-} \rightarrow Mg_3(PO_4)_2(s)$
   g. $Mg^{2+} + 2OH^- \rightarrow Mg(OH)_2(s)$
   i. $Cu^{2+} + CO_3^{2-} \rightarrow CuCO_3(s)$
   j. $3Cu^{2+} + 2PO_4^{3-} \rightarrow Cu_3(PO_4)_2(s)$
   k. $Cu^{2+} + 2OH^- \rightarrow Cu(OH)_2(s)$

FOR ENRICHMENT Have students design an experiment to make a measurable amount of a precipitate, separate it, dry it, and determine its mass.

SMALL-SCALE LAB

Small-Scale Lab

PURPOSE After completing this activity, students will be able to

- recognize and describe precipitates
- write the names and formulas for each

SKILLS FOCUS observing, applying rules

PREP TIME 1 hour

CLASS TIME 30 minutes

ADVANCED PREPARATION

Solution	Preparation
0.05*M* $AgNO_3$	2.1 g in 250 mL
0.2*M* $Pb(NO_3)_2$	16.6 g in 250 mL
0.5*M* $CaCl_2$	13.9 g in 250 mL
1.0*M* Na_2CO_3	26.5 g in 250 mL
0.1*M* Na_3PO_4	9.5 g $Na_3PO_4 \cdot 12H_2O$ in 250 mL
0.5*M* NaOH	20.0 g in 1.0 L
0.2*M* Na_2SO_4	7.1 g in 250 mL
0.2*M* $CuSO_4$	12.5 g $CuSO_4 \cdot 5H_2O$ in 250 mL
0.2*M* $MgSO_4$	6.0 g in 250 mL
0.1*M* $FeCl_3$	6.8 g $FeCl_3 \cdot 6H_2O$ in 25 mL of 1.0*M* NaCl (for stability); dilute to 250 mL

EXPECTED OUTCOME

In Figure A, a, f, g, h, and j: milky white; b and k: cloudy white; e: cloudy tan; i and l: grainy white; c: muddy brown; d: no visible reaction. In Figure B, a, b, and c: orange; e, f, and g: white; i, j, and k: blue; d, h, and l: no visible reaction.

ANALYZE AND CONCLUDE

1. $AgNO_3$ and Na_2SO_4 did not form a precipitate.
2. **a.** Ag_2CO_3, silver carbonate
   b. Ag_3PO_4, silver phosphate
   c. AgOH, silver hydroxide (Note: The product is actually Ag_2O, silver oxide.)
   e. $PbCO_3$, lead(II) carbonate
   f. $Pb_3(PO_4)_2$, lead(II) phosphate
   g. $Pb(OH)_2$, lead(II) hydroxide
   h. $PbSO_4$, lead(II) sulfate
   i. $CaCO_3$, calcium carbonate
   j. $Ca_3(PO_4)_2$, calcium phosphate
   k. $Ca(OH)_2$, calcium hydroxide
   l. $CaSO_4$, calcium sulfate

Answers

FIGURE 9.19 ionic compounds and binary molecular compounds

STUDY GUIDE

Evaluate

Review and Assessment Materials

CHEMISTRY TUTORIAL Encourage students to review writing formulas for binary ionic compounds and calculating mass ratios using the animated step-by-step tutorials for Chapter 9. Each Chemistry Tutorial is followed by a similar practice problem, where the student is guided through the problem-solving process.

CHAPTER PROBLEM SET Have students practice writing formulas, naming compounds, and calculating mass ratios using the Chapter 9 online Problem Set.

MATH TUTORIAL Struggling students can use an online Math Tutorial on unit ratios as a quick review. Ask students to answer the guided practice problem that follows the Math Tutorial.

DIRECTED VIRTUAL LAB Have students complete the Virtual Lab *Ionic Naming* to help reinforce the rules for naming ionic compounds. In the post-lab exercise, students will analyze the data and practice naming new compounds.

KINETIC ART Direct students to the Kinetic Art animations *The Law of Multiple Proportions* and *Naming Compounds,* and have them answer the questions that accompany each animation.

Study Tip

As you review each section, write down the key concepts that are found in bold print in each major division. Then add details and extensions until you have a complete outline of the topic.

9 Study Guide

BIGIDEAS THE STRUCTURE OF ATOMS; BONDING AND INTERACTIONS

An element's position in the periodic table supplies information on ion formation and bonding tendencies, which is used to write the names and formulas of ions and compounds. Ionic and molecular compounds differ in composition—ions form ionic compounds and molecules form molecular compounds.

9.1 Naming Ions

When metals in Groups 1A, 2A, and 3A lose electrons, they form cations with positive charges equal to their group number. The charge of any ion of a Group A nonmetal is determined by subtracting 8 from the group number. The charges of the cations of many transition metals must be determined from the number of electrons lost.

A polyatomic ion is composed of more than one atom that behaves as a unit and carries a charge.

- monatomic ion (264)

9.2 Naming and Writing Formulas for Ionic Compounds

To write the formula of a binary ionic compound, write the symbol of the cation and then the anion. Then balance the charges. The name of a binary ionic compound is the cation name followed by the anion name.

To write formulas for compounds with polyatomic ions, write the symbol for the cation followed by the symbol for the anion. Then balance the charges. To name a compound containing a polyatomic ion, state the cation name followed by the anion name.

- binary compound (272)

9.3 Naming and Writing Formulas for Molecular Compounds

To name a binary molecular compound, write the names of the elements in the order listed in the formula. Use prefixes to indicate the number of each atom. End the name of the second element with *-ide.*

To write the formula of a binary molecular compound, use the prefixes to determine the subscript of each element. Write the symbols for the elements with the subscripts.

9.4 Naming and Writing Formulas for Acids and Bases

If the anion name ends in *-ide,* the acid name begins with the prefix *hydro-*. The stem of the anion has the suffix *-ic* and is followed by the word *acid.* If the anion name ends in *-ite,* the acid name is the stem of the anion with the suffix *-ous,* followed by the word *acid.* If the anion name ends in *-ate,* the acid name is the stem of the anion with the suffix *-ic,* followed by the word *acid.* To write the formula for an acid, use the rule for writing the name of the acid in reverse.

Bases are named like other ionic compounds. To write the formula for a base, write the symbol for the metal cation followed by that of the hydroxide ion. Then, balance the ionic charges.

- acid (285)
- base (287)

9.5 The Laws Governing How Compounds Form

If the ratio of atoms of each element in a compound is fixed, then the ratio of their masses is also fixed.

Follow the rules for naming acids when H is the first element. If the compound is binary, generally the name ends with *-ide.* For a molecular binary compound, use prefixes to indicate the number of atoms. When a polyatomic ion with oxygen is in the formula, the compound name ends in *-ite* or *-ate.* If the compound contains a metallic cation that can have different ionic charges, use a Roman numeral to indicate the ionic charge.

An *-ide* ending usually indicates a binary compound. An *-ite* or *-ate* ending indicates a polyatomic ion with oxygen. Prefixes usually indicate a molecular compound. A Roman numeral after the name of a cation shows the ionic charge of the cation.

- law of definite proportions (289)
- law of multiple proportions (290)

296 Chapter 9 • Study Guide

UbD Performance Tasks

IT'S PROPORTIONAL Have students create a poster explaining the law of multiple proportions using copper(I) oxide and copper(II) oxide as their example compounds. Explanations should include supporting mass ratio equations.

WHAT'S IN A NAME? Have students use the following quote as the opening statement in an article for the school newspaper. *What's in a name?* The article should convey the importance for knowing and understanding the information needed and the information gained from the name and formula of a compound. The article should use as least six different commonly used compounds as examples. Encourage students to include graphics or illustrations as needed to enhance the article.

Skills Tune-Up: Names and Formulas

Problem	❶ Analyze	❷ Solve
Write the name for the binary ionic compound CrI_2. Hint: Refer to Sample Problems 9.2–9.3 if you have trouble identifying binary ionic compounds.	Name the ions in the order written in the formula. Use a Roman numeral if the metal cation in the compound can have more than one common ionic charge.	CrI_2 contains chromium cations and iodide anions. Chromium forms two common cations: Cr^{2+} and Cr^{3+}. The compound CrI_2 is electrically neutral. Iodide ion is I^- and the formula CrI_2 specifies two iodide ions, which give a charge of 2–. So the chromium ion must be Cr^{2+}. The name of the compound is chromium(II) iodide.
Write the name for the binary molecular compound N_2O_5. Hint: Review Sample Problem 9.6 if you need help naming binary molecular compounds.	Name the elements in the order written in the formula. Use prefixes as necessary to indicate the number of each kind of atom. Use the suffix *-ide* on the name of the second element.	N_2O_5 is composed of two nitrogen atoms and five oxygen atoms. The name of the compound is dinitrogen pentoxide.
Write the formula for the ionic compound aluminum sulfate.	Write the symbol or formula for each ion in the order written in the name. Use subscripts to balance the charges. The ions must be combined in the lowest whole-number ratio. Use parentheses if more than one polyatomic ion is needed to balance a formula.	Aluminum sulfate contains Al^{3+} cations and SO_4^{2-} anions. Al 3+ ✕ (SO_4) 2– The formula for aluminum sulfate is $Al_2(SO_4)_3$. Remember: Use parentheses when there is more than one polyatomic ion in the balanced formula.
Write the formula for the binary molecular compound selenium dioxide. Remember: The number 1 is never used as a subscript in a formula.	The prefixes in the name indicate the subscript of each element in the formula. Write the symbols for the two elements with the appropriate subscripts.	Selenium dioxide is composed of one selenium atom and two oxygen atoms. The formula for selenium dioxide is written as SeO_2.

Skills Tune-Up

Example and sample practice problems for writing formulas, naming compounds, and calculating mass ratios can be found on the following pages:

LESSON 9.1 Naming Cations and Anions—page 267

LESSON 9.2 Writing Formulas for Binary Ionic Compounds—page 273; Naming Binary Ionic Compounds—page 275; Writing Formulas for Compounds with Polyatomic Ions—page 277; Naming Compounds with Polyatomic Ions—page 278

LESSON 9.3 Naming Binary Molecular Compounds—page 282; Writing Formulas for Binary Molecular Compounds—page 283

LESSON 9.4 Naming Acids—page 287; Naming Bases—page 288

LESSON 9.5 Calculating Mass Ratios—page 291

SKILLS TUNE-UP

Focus on ELL

5 ASSESS UNDERSTANDING Assign ELL students one of the Performance Tasks as a group for an alternative assessment. Provide guidelines for the assignment and a grading rubric.

BEGINNING

LOW Direct students to create the visual representation for the assignment.

HIGH Direct students to find specific information needed for the assignment from the text.

INTERMEDIATE: LOW/HIGH Direct students to provide the written explanation for the visuals.

ADVANCED: LOW/HIGH Direct students to present the groups' assignment to the class.

Evaluate

Answers

LESSON 9.1

57. **a.** 2– **b.** 1+ **c.** 1– **d.** 3+
58. **a.** 2+ **b.** 2+ **c.** 3+ **d.** 1+
59. **a.** barium ion **b.** iodide ion **c.** silver ion **d.** mercury(II) ion
60. cyanide ion, CN^-, and hydroxide ion, OH^-
61. **a.** hydroxide ion **b.** lead(IV) ion **c.** sulfate ion **d.** oxide ion

LESSON 9.2

62. zero; A compound is electrically neutral.
63. The symbols for the cation and anion are written, and the charges are balanced with subscripts. The name of the cation is followed by the name of the anion.
64. Determine the charge of the anion then work the formula backward to find the charge of the transition metal cation needed to give a net charge of zero for the formula unit.
65. The symbols for the cation and anion are written, and the charges are balanced with subscripts. Parentheses are used around the polyatomic ion if a subscript is needed. The name of the cation is followed by the name of the anion.
66. **a.** NH_4NO_3, ammonium nitrate
b. $Sn(NO_3)_4$, tin(IV) nitrate
c. $Fe(NO_3)_3$, iron(III) nitrate
d. $Mg(NO_3)_2$, magnesium nitrate
e. $(NH_4)_2CO_3$, ammonium carbonate
f. $Sn(CO_3)_2$, tin(IV) carbonate
g. $Fe_2(CO_3)_3$, iron(III) carbonate
h. $MgCO_3$, magnesium carbonate
i. NH_4CN, ammonium cyanide
j. $Sn(CN)_4$, tin(IV) cyanide
k. $Fe(CN)_3$, iron(III) cyanide
l. $Mg(CN)_2$, magnesium cyanide
m. $(NH_4)_3PO_4$, ammonium phosphate
n. $Sn_3(PO_4)_4$, tin(IV) phosphate
o. $FePO_4$, iron(III) phosphate
p. $Mg_3(PO_4)_2$, magnesium phosphate
67. a and c
68. when more than a single polyatomic ion is needed to balance the formula

LESSON 9.3

69. two nonmetals
70. **a.** tri- **b.** mono- **c.** di- **d.** hexa- **e.** penta- **f.** tetra-
71. For formulas, write the correct symbols for each element with a subscript corresponding to the prefix before each element in the name. For naming, name each element in the order given. Use the subscript to determine the prefix before each element in the name. The name ends in *-ide*.
72. **a.** diphosphorus pentoxide
b. carbon tetrachloride
c. BCl_3 **d.** N_2H_4

9 Assessment

★ Solutions appear in Appendix E

Lesson by Lesson

9.1 Naming Ions

57. Give the expected charges on the ions of elements of these groups of the periodic table.
a. Group 6A **b.** Group 1A **c.** Group 7A **d.** Group 3A

★58. Give the expected charge of the cations of the following elements:
a. Sr **b.** Ca **c.** Al **d.** Cs

59. Name these ions, using Table 9.2, if necessary.
a. Ba^{2+} **b.** I^- **c.** Ag^+ **d.** Hg^{2+}

60. Write the names and formulas of the two polyatomic anions in Table 9.3 with names that do not end in *-ite* or *-ate*.

61. Name the following ions:
a. OH^- **b.** Pb^{4+} **c.** SO_4^{2-} **d.** O^{2-}

9.2 Naming and Writing Formulas for Ionic Compounds

62. What is the net charge of every ionic compound? Explain.

63. How are chemical formulas written for binary ionic compounds, given their names? How is the reverse done?

★64. How do you determine the charge of a transition metal cation from the formula of an ionic compound containing that cation?

65. How are formulas written for ionic compounds with polyatomic ions, given their names? How is the reverse done?

66. Complete the table by writing correct formulas for the compounds formed by combining positive and negative ions. Then name each compound.

	NO_3^-	CO_3^{2-}	CN^-	PO_4^{3-}
NH_4^+	a.____	e.____	i.____	m.____
Sn^{4+}	b.____	f.____	j.____	n.____
Fe^{3+}	c.____	g.____	k.____	o.____
Mg^{2+}	d.____	h.____	l.____	p.____

67. Which of the following compounds are binary ionic compounds?
a. KBr
b. K_3PO_4
c. sodium nitride
d. calcium sulfate

68. When must parentheses be used in a formula for a compound?

9.3 Naming and Writing Formulas for Molecular Compounds

69. What are the components of a binary molecular compound?

★70. What prefix indicates each of the following numbers of atoms in the formula of a binary molecular compound?
a. 3 **b.** 1 **c.** 2 **d.** 6 **e.** 5 **f.** 4

71. How are formulas for binary molecular compounds written, given their names? How is the reverse performed, given their formulas?

72. Write the formula or name for the following compounds:
a. P_2O_5
b. CCl_4
c. boron trichloride
d. dinitrogen tetrahydride

9.4 Naming and Writing Formulas for Acids and Bases

73. Give the name or the formula for these acids.
a. HCl **b.** HNO_3 **c.** sulfuric acid **d.** acetic acid

74. Is every compound that contains hydrogen an acid? Explain.

75. Write formulas for these compounds.
a. nitrous acid
b. aluminum hydroxide
c. hydroselenic acid
d. strontium hydroxide
e. phosphoric acid

★76. Write names or formulas for these compounds.
a. $Pb(OH)_2$ **b.** $Co(OH)_2$ **c.** copper(II) hydroxide **d.** iron(II) hydroxide

298 Chapter 9 • Assessment

LESSON 9.4

73. **a.** hydrochloric acid
b. nitric acid
c. H_2SO_4
d. $HC_2H_3O_2$
74. No, to be an acid the compound must produce H^+ ions in an aqueous solution.
75. **a.** HNO_2 **b.** $Al(OH)_3$ **c.** H_2Se **d.** $Sr(OH)_2$ **e.** H_3PO_4
76. **a.** lead(II) hydroxide
b. cobalt(II) hydroxide
c. $Cu(OH)_2$
d. $Fe(OH)_2$

9.5 The Laws Governing How Compounds Form

77. What is the law of definite proportions?

78. Describe the law of multiple proportions.

***79.** Nitrous oxide, laughing gas, is used as an anesthetic in dentistry. The mass ratio of nitrogen to oxygen is 7:4. A 68-g sample of a compound composed of nitrogen and oxygen contains 42 g of nitrogen. Is the sample nitrous oxide?

Understand Concepts

80. Write formulas for these compounds.

a. potassium permanganate
b. calcium hydrogen carbonate
c. dichlorine heptoxide
d. trisilicon tetranitride
e. sodium dihydrogen phosphate
f. phosphorus pentabromide

81. Write formulas for these compounds.

a. magnesium sulfide
b. sodium phosphite
c. barium hydroxide
d. copper(II) nitrite
e. potassium sulfite
f. calcium carbonate
g. sodium bromide
h. ferric sulfate

***82.** Name these compounds.

a. $NaClO_3$
b. Hg_2Br_2
c. K_2CrO_4
d. $HClO_4$
e. SnO_2
f. $Fe(C_2H_3O_2)_3$
g. $KHSO_4$
h. $Ca(OH)_2$
i. BaS

83. Name each substance.

a. $LiClO_4$
b. Cl_2O
c. HgF_2
d. CaO
e. $Ba_3(PO_4)_2$
f. I_2
g. $SrSO_4$
h. $CuC_2H_3O_2$
i. $SiCl_4$

84. Name each compound.

a. $Mg(MnO_4)_2$
b. $Be(NO_3)_2$
c. K_2CO_3
d. N_2H_4
e. $LiOH$
f. BaF_2
g. PI_3
h. ZnO
i. H_3PO_3

85. Write formulas for these compounds.

a. calcium bromide
b. silver chloride
c. aluminum carbide
d. nitrogen dioxide
e. tin(IV) cyanide
f. lithium hydride
g. strontium acetate
h. sodium silicate

***86.** A compound of general formula Q_xR_y contains no hydrogen, and Q and R are both elements. Neither Q nor R is a metal. Is Q_xR_y an acid, a binary ionic compound, or a binary molecular compound?

87. A compound of general formula Q_xR_y contains no hydrogen, Q is the alkali metal of lowest atomic mass, and R contains the elements oxygen and carbon in a 3:1 ratio. Write the name and the formula of the compound.

***88.** Two compounds contain only tin and chlorine. The ratio of the masses of chlorine combined with 1.00 g of tin in the two compounds is 2:1. If one compound has the formula $SnCl_2$, what is the formula for the other compound?

89. Analysis of two compounds shows that they contain only lead and iodine in these amounts:

Compound I: 22.48 g Pb and 27.52 g I
Compound II: 5.80 g Pb and 14.20 g I

a. Determine the ratio of lead contained in the two compounds for every 1 g of iodine.
b. Use your ratio and your knowledge of ionic charges to write the formulas and the names of the two compounds.

90. The U.S. produces thousands of inorganic chemicals. Inorganic chemicals, for the most part, do not contain carbon. The table shows the amounts (in billions of kg) of the top ten inorganic chemicals produced in a recent year.

Chemical name	Amount produced (10^9 kg)
Sulfuric acid	39.4
Nitrogen	26.9
Oxygen	17.7
Ammonia	16.5
Lime	16.3
Phosphoric acid	11.2
Sodium hydroxide	11.0
Chlorine	10.3
Sodium carbonate	9.3
Nitric acid	6.8

a. What percentage of the total production of the top ten is lime (calcium oxide)?
b. Three diatomic gases are on the list. What are their names? What was the combined production of these gases in billions of kilograms?
c. The three acids make up what percentage of the total production of the top ten?
d. Write formulas for the top ten inorganic chemicals.

ASSESSMENT

LESSON 9.5

77. In all samples of the same chemical compound, the masses of the elements are always in the same proportions.

78. Whenever two elements form more than one compound, the different masses of one element that combine with the same mass of the other element are in a ratio of small whole numbers.

79. No; The ratio of nitrogen to oxygen is 42:26, which is not a 7:4 ratio.

UNDERSTAND CONCEPTS

80. **a.** $KMnO_4$ **b.** $Ca(HCO_3)_2$ **c.** Cl_2O_7 **d.** Si_3N_4 **e.** NaH_2PO_4 **f.** PBr_5

81. **a.** MgS **b.** Na_3PO_3 **c.** $Ba(OH)_2$ **d.** $Cu(NO_2)_2$ **e.** K_2SO_3 **f.** $CaCO_3$ **g.** $NaBr$ **h.** $Fe_2(SO_4)_3$

82. **a.** sodium chlorate
b. mercury(I) bromide
c. potassium chromate
d. perchloric acid
e. tin(IV) oxide
f. iron(III) acetate
g. potassium hydrogen sulfate
h. calcium hydroxide
i. barium sulfide

83. **a.** lithium perchlorate
b. dichlorine monoxide
c. mercury(II) fluoride
d. calcium oxide
e. barium phosphate
f. iodine
g. strontium sulfate
h. copper(I) acetate
i. silicon tetrachloride

84. **a.** magnesium permanganate
b. beryllium nitrate
c. potassium carbonate
d. dinitrogen tetrahydride
e. lithium hydroxide
f. barium fluoride
g. phosphorus triiodide
h. zinc oxide
i. phosphorous acid

85. **a.** $CaBr_2$ **b.** $AgCl$ **c.** Al_4C_3 **d.** NO_2 **e.** $Sn(CN)_4$ **f.** LiH **g.** $Sr(C_2H_3O_2)_2$ **h.** Na_2SiO_3

86. binary molecular compound

87. lithium carbonate, Li_2CO_3

88. $SnCl_4$

89. **a.** 2:1 **b.** PbI_2, lead(II) iodide, and PbI_4, lead(IV) iodide

90. **a.** 9.85%
b. nitrogen, oxygen, and chlorine; 54.9 billions of kg
c. 34.7%
d. H_2SO_4, N_2, O_2, NH_3, CaO, H_3PO_4, $NaOH$, Cl_2, Na_2CO_3, HNO_3

Answers

THINK CRITICALLY

91. A molecular formula shows the number of each kind of atom in a molecule of the compound. The formula unit shows the lowest whole-number ratio of ions in a compound.

92. on the right side

93. Common names vary in different languages and are difficult to remember and convert to formulas.

94. The statement is true for the representative metals but not for the transition metals, which often can have different charges.

95. Possible answers include: cations always come before anions; when a cation has more than one ionic charge, the charge is indicated by a Roman numeral; monatomic anions use an *-ide* ending. Each rule has a specific purpose; for example, an ionic charge is necessary information because it determines how many ions are in the formula unit of the compound.

96. a. N_2O, dinitrogen monoxide
b. NO_2, nitrogen dioxide
c. NO, nitrogen monoxide
d. N_2O_4, dinitrogen tetroxide

97. a. Cu_2S, copper(I) sulfide, and CuS, copper(II) sulfide
b. $FeSO_4$, iron(II) sulfate, and $Fe_2(SO_4)_3$, iron(III) sulfate
c. PbO, lead(II) oxide, and PbO_2, lead(IV) oxide
d. MnF_2, manganese(II) fluoride, and MnF_3, manganese(III) fluoride

98. a. The charges do not balance, CsCl.
b. Neon does not form compounds.
c. The charges do not balance, ZnO.
d. The subscripts are not the lowest whole-number ratio, BaS.

99. $NaAl(SO_4)_2$

100. binary ionic compounds: *d* and *g*; binary molecular compounds: *a* and *f*; compounds with polyatomic ions: *b, c, e, h,* and *i*; acids: *b* and *e;* base: *c*

101. a. 3.60 g solid/1.40 g gas = 2.57 and 6.48 g solid/2.52 g gas = 2.57
The compound obeys the law of definite proportions.
b. 3.92 g

102. law of conservation of mass

103. a. $CaCO_3$ and $Mg(OH)_2$
b. NaF and KNO_3
c. TiO_2 and ZnO
d. $FeSO_4$

104. a. PCl_3, phosphorus trichloride, molecular
b. S_2Cl_2, disulfur dichloride, molecular
c. CCl_4, carbon tetrachloride, molecular
d. SCl_2, sulfur dichloride, molecular

Think Critically

91. Compare and Contrast How does the information conveyed by a molecular formula differ from that given by a formula unit of a compound?

92. Make Generalizations Where on the periodic table will you find the two elements in a binary molecular compound?

93. Draw Conclusions Why is it important for chemists to have a system of writing chemical names and formulas?

94. Evaluate Criticize this statement: "The ionic charge of any metal can be determined from the position of the element in the periodic table."

95. Explain Summarize the rules that chemists use for naming ionic compounds. What is the purpose for each rule?

***96. Use Models** Nitrogen and oxygen form a number of stable chemical compounds. In the models below, nitrogen is blue; oxygen is red. Write the molecular formula and name of each.

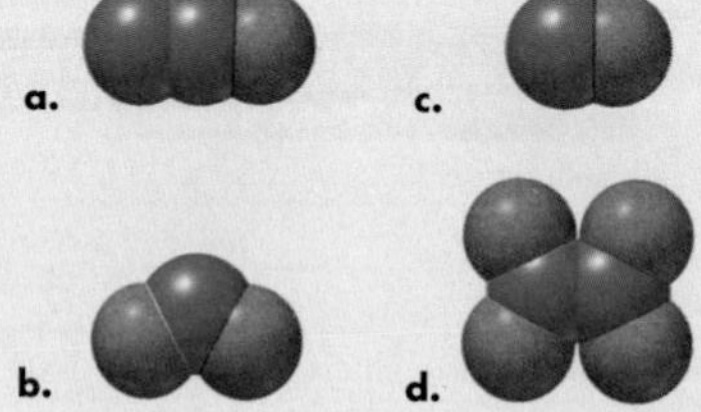

97. Evaluate and Revise Examine the following names for ionic compounds. Show, by writing all possible formulas for the compounds, that the names are incomplete. Then, write each complete name.

a. copper sulfide
b. iron sulfate
c. lead oxide
d. manganese fluoride

***98. Evaluate and Revise** Explain what is wrong with each formula. Write the correct formula.

a. $CsCl_2$
b. LiNe
c. ZnO_2
d. Ba_2S_2

99. Infer Sodium aluminum sulfate is an active ingredient in baking powder. The molecular formula for this ionic compound contains two sulfate ions. Write the complete molecular formula for sodium aluminum sulfate.

***100. Classify** Separate the following compounds into five categories: binary ionic compounds, binary molecular compounds, compounds with polyatomic ions, acids, and bases. Some compounds may fit in more than one category.

a. CBr_4
b. HCN
c. NH_4OH
d. MgS
e. H_2SiO_3
f. ClBr
g. Al_2O_3
h. Na_2HPO_4
i. $KMnO_4$

***101. Calculate** A student heats 5.00 g of a white compound and obtains 3.60 g of a green compound and 1.40 g of a colorless gas. Another student heats a 9.00-g sample of the same compound and obtains 6.48 g of a green compound and 2.52 g of a colorless gas.

a. Show by calculation that the white compound obeys the law of definite proportions.
b. If a third student heats 14.0 g of the white compound, how many grams of colorless gas will be produced?

***102. Draw Conclusions** What other law is illustrated (twice) by the experiment described in the previous question?

103. Apply Concepts Ionic compounds are present in items commonly found in your home. Write formulas for the ionic compounds found in the following common household products:

a. antacid (calcium carbonate and magnesium hydroxide)
b. toothpaste for sensitive teeth (sodium fluoride and potassium nitrate)
c. sunscreen (titanium(IV) oxide and zinc oxide)
d. pasta (ferrous sulfate)

104. Use Models In the models below, chlorine is green; phosphorus is orange; carbon is black; and sulfur is yellow. Write the formula and the name for each compound. Are these compounds ionic or molecular compounds?

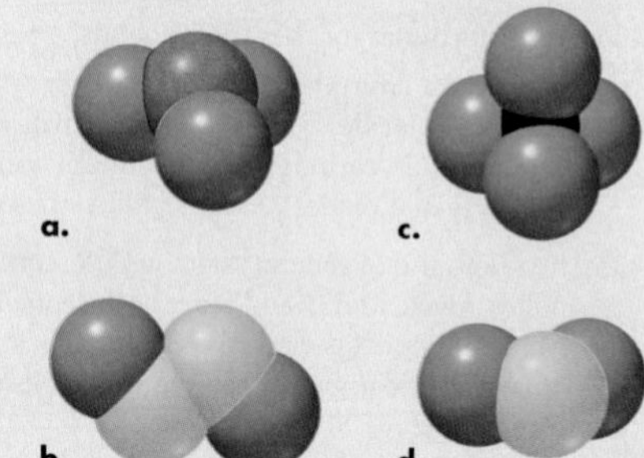

Enrichment

105. Organize Data *CRC Handbook of Chemistry and Physics* is a reference book that contains a wealth of information about elements and compounds. Two sections of this book that you might use are called "Physical Constants of Inorganic Compounds" and "Physical Constants of Organic Compounds." To familiarize yourself with this work, make a table with these headings: Name, Formula, Crystalline Form or Color, Density, Melting Point (°C), Boiling Point (°C), and Solubility in Water.

Name	Formula	Crystalline Form or Color	Density	MP (°C)	BP (°C)

Enter these substances in the body of the table: ammonium chloride, barium, barium sulfate, bromine, calcium carbonate, chlorine, copper(II) sulfate pentahydrate, iodine, iron(II) sulfate pentahydrate, mercury, potassium carbonate, and sulfur. Use the handbook to complete the table.

*106. **Analyze Data** Use the table you prepared for Problem 105 to answer the following questions:

a. You have two unlabeled bottles, each containing a white powder. One of the substances is calcium carbonate, and the other is potassium carbonate. How could you distinguish between these two compounds?

b. How would you distinguish between samples of copper(II) sulfate pentahydrate and iron(II) sulfate pentahydrate?

c. A bottle contains a mixture of ammonium chloride and barium sulfate. How could you separate these two compounds?

d. List the elements in the table in order of increasing density. Identify the elements as metals or nonmetals.

e. List the compounds in the table in order of decreasing density.

f. Calculate the mass of 47.0 cm^3 of mercury.

g. Calculate the volume of 16.6 g of sulfur.

h. How would you distinguish among the Group 7A elements listed in the table?

Write About Science

107. Research a Problem Sodium ions (Na^+) and potassium ions (K^+) are needed for the human body to function. Deficiencies in these ions can have adverse effects on your health. Research where these ions are most likely to be found in the body and the roles they play. Write a brief essay describing your findings.

108. Explain Investigate the role of lithium carbonate in the successful treatment of bipolar disorder. Write a brief report that includes information on bipolar disorder and why lithium carbonate is used to treat it.

109. Connect to the BIGIDEA Choose five personal care products from your home. Read each ingredient label and identify all the compounds that you are able to decipher. Write a short paragraph in which you explain how learning to name chemical compounds has helped you decipher these ingredient labels.

CHEMYSTERY

Cucumber Blunder

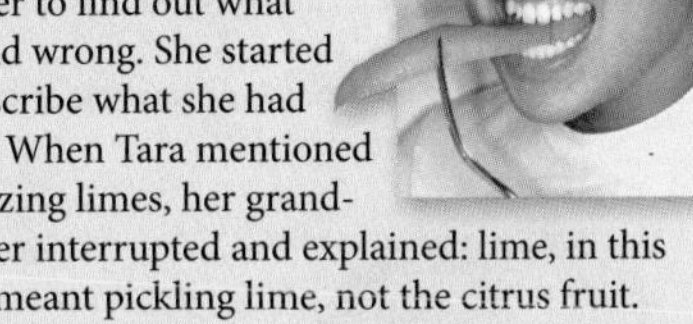

Tara called her grandmother to find out what she did wrong. She started to describe what she had done. When Tara mentioned squeezing limes, her grandmother interrupted and explained: lime, in this case, meant pickling lime, not the citrus fruit. "Pickling lime? I've never heard of that!" Tara exclaimed.

Her grandmother fetched a package of pickling lime from the cupboard to look at the ingredient label. She read aloud "food-grade calcium hydroxide." Her grandmother elaborated, "That's what makes the pickles crunchy like you like 'em."

110. Compare Lime is the common name for calcium hydroxide. What are the advantages and disadvantages of each name?

*111. **Connect to the BIGIDEA** Is calcium hydroxide an ionic or a molecular compound? Write the formula.

CHEMYSTERY

START A CONVERSATION Have students read through the CHEMystery. **Ask** What do you think led Tara to make a mistake when buying her pickle ingredients? *(The recipe simply referred to "lime," and there is more than one type of lime available in the grocery store.)* **Ask** Does the ingredient name *lime* provide useful information about the ingredient's composition? *(no)* **Ask** What can the name *calcium hydroxide* tell you about the composition of the ingredient? *(The compound contains the ions Ca^{2+} and OH^-.)*

APPLY CONCEPTS Have students determine the chemical formula of calcium hydroxide from its name. *($Ca(OH)_2$)* **Ask** Is it an acid or a base? *(base)* Why? *(It is an ionic compound that will produce hydroxide ions when it is dissolved in water.)*

CHEMYSTERY ANSWERS

110. The common name is shorter and easier to remember, but *lime* can also refer to calcium oxide, leading to uncertainty in which compound is intended. When the IUPAC name, calcium hydroxide, is used, the intended compound is clear to those trained in chemistry. For those not trained in chemistry, the IUPAC name may be unfamiliar.

111. BIGIDEA ionic; $Ca(OH)_2$

Answers

ENRICHMENT

105. Spot check student answers.

106. a. Potassium carbonate has greater water solubility than $CaCO_3$.

b. The copper compound is blue; the iron compound is white.

c. Add water to dissolve the NH_4Cl, and then filter out the insoluble $BaSO_4$.

d. chlorine (nonmetal), sulfur (nonmetal), bromine (nonmetal), barium (metal), iodine (nonmetal), mercury (metal)

e. barium sulfate, calcium carbonate, potassium carbonate, copper(II) sulfate pentahydrate, iron(II) sulfate pentahydrate, ammonium chloride

f. 639 g **g.** 8.02 cm^3 **h.** color, density, melting point, and boiling point

WRITE ABOUT SCIENCE

107. Sodium is found in higher concentrations in extra-cellular fluids, such as plasma. Potassium is found in higher concentrations within cells. Both ions are involved in transmission of nerve impulses.

108. Answers should mention that lithium carbonate is used because it is effective in regulating the patient's mental state.

109. Answers will vary.

ASSESSMENT

Answers

CUMULATIVE REVIEW

112. Answers may include: color (physical), solid (physical), magnetic (physical), conducts electricity (physical), burns (chemical).

113. a. 4 **b.** 2 **c.** 2 **d.** 4 **e.** 2 **f.** 1

114. 5.2 cm

115. a. 7.75×10^5 μL **b.** 208 K **c.** 0.832 cg

116. 0.538 g/cm^3

117. Both are in the nucleus and have a mass of about 1 amu. A proton is positively charged; a neutron has no charge.

118. a. 2 **b.** protons **c.** electrons **d.** neutrons

119. a. neon **b.** carbon **c.** boron **d.** helium

120. The metalloids border a line separating the metals from the nonmetals. Their properties are intermediate between those of metals and nonmetals.

121. a. cesium, potassium, sodium, lithium
b. lithium, boron, carbon, fluorine, neon

122. a. Li **b.** I **c.** S **d.** O **e.** N **f.** F

123. When metallic elements of Group 1A and 2A form ions, they lose all their outer shell electrons. This increases the attraction by the nucleus for the fewer remaining electrons and results in ions that are smaller than the neutral atoms. The electron that a Group 7A element gains in forming an ion enters the outer shell, resulting in a decrease in the effective nuclear attraction of the increased number of electrons. The anion is larger than the neutral atom.

124. a. 1 **b.** 6 **c.** 8 **d.** 2 **e.** 7 **f.** 5

125. $1s^22s^22p^6$; Possible answers are N^{3-}, O^{2-}, F^-, Na^+, Mg^{2+}, and Al^{3+}.

126. a. 12 p^+ and 10 e^- **b.** 35 p^+ and 36 e^-
c. 38 p^+ and 36 e^- **d.** 16 p^+ and 18 e^-

127. b and c; molecular compounds formed by two nonmetals have covalent bonds.

128. b, d, and f

129. a. :C̤̈l:C̤̈l: **b.** :C:::O: **c.** :Ö::C::Ö:

d. H:N̤̈:H with H above **e.** :C̤̈l:C̤̈:C̤̈l: with :C̤̈l: above and below

f. H:Ö̤:H **g.** H:C̤̈:H with H above and below

130. A hydrogen bond is an intermolecular force between a hydrogen atom covalently bonded to a very electronegative atom and an unshared pair of electrons on another electronegative atom.

131. ionic bond: electrons are transferred

Na· + ·F̤̈: → Na^+ :F̤̈:$^-$

covalent bond: electrons are shared

H· + ·H → H:H

Cumulative Review

112. List five properties of the chair you are sitting on. Classify each as physical or chemical.

*__113.__ How many significant figures are in the following measurements?

a. 15.05 g
b. 0.31 cm
c. 890 mL
d. 300.0 cm^3
e. 3.0×10^5 kg
f. 0.001 mm

114. Determine the sum of the following measurements to the correct number of significant figures.

1.55 cm + 0.235 cm + 3.4 cm

115. Make the following conversions:

a. 775 mL to microliters (μL)
b. 65°C to K
c. 8.32 mg Ag to centigrams of silver (cg Ag)

116. A student finds that 6.62 g of a substance occupies a volume of 12.3 cm^3. What is the density of the substance?

117. Compare neutrons and protons with respect to their charge, mass, and position in the atom.

*__118.__ The diagrams show two models of the atom.

a. Which model is more accurate?
b. What do the positively charged particles represent?
c. What do the negatively charged particles represent?
d. What major subatomic particle is missing in both of these models?

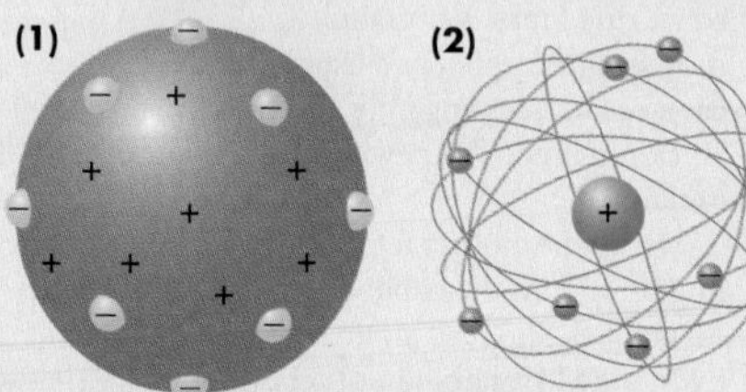

*__119.__ What elements have these electron configurations?

a. $1s^22s^22p^6$
b. $1s^22s^22p^2$
c. $1s^22s^22p^1$
d. $1s^2$

120. Where are the metalloids found on the periodic table? Compare the properties of the metalloids to metals and nonmetals.

121. Arrange the following groups of elements in order of increasing ionization energy.

a. potassium, cesium, lithium, sodium
b. fluorine, boron, lithium, carbon, neon

122. From the positions of the elements in the periodic table, choose the element in each pair with the higher electronegativity.

a. Cs and Li
b. Sr and I
c. S and Mg
d. O and Se
e. Te and N
f. C and F

123. The ions of the elements of Groups 1A and 2A have smaller radii than their neutral atoms, whereas the ions of Group 7A have larger radii than their neutral atoms. Explain.

*__124.__ How many valence electrons do atoms of the following elements have?

a. lithium
b. sulfur
c. neon
d. calcium
e. bromine
f. phosphorus

125. Write the electron configuration for the element neon, then identify three ions that have the same electron configuration.

126. How many protons and electrons are in each ion?

a. magnesium ion
b. bromide ion
c. strontium ion
d. sulfide ion

*__127.__ Which of these compounds would you expect to contain covalent bonds? Why?

a. KCl **b.** PBr_3 **c.** ClBr **d.** NaI

128. Which of these substances would you expect to be polar?

a. Cl_2
b. CO
c. CO_2
d. NH_3
e. CCl_4
f. H_2O
g. CH_4

129. Draw electron dot structures for the substances in Question 128.

130. Explain what a hydrogen bond is and under what conditions a hydrogen bond will form.

131. Explain the difference between an ionic bond and a covalent bond. Use electron dot structures to illustrate your explanation.

If You Have Trouble With . . .

Question	112	113	114	115	116	117	118	119	120	121	122	123	124	125	126	127	128	129	130	131
See Chapter	2	3	3	3	3	4	5	5	6	6	6	6	7	7	7	8	8	8	8	8

Standardized Test Prep for Indiana

Standards: C.3.4, C.3.5

Select the choice that best answers each question or completes each statement.

1. Identify the pair in which the formula does not match the name.
   (A) sulfite, SO_3^{2-} (C) hydroxide, OH^-
   (B) nitrite, NO_3^- (D) dichromate, $Cr_2O_7^{2-}$

2. Which of these compounds are ionic?
   I. $CaSO_4$ II. N_2O_4 III. NH_4NO_3 IV. CaS
   (A) I and II only (C) III and IV only
   (B) II and III only (D) I, III, and IV only

3. What is the name of $AlCl_3$?
   (A) aluminum trichloride
   (B) aluminum(III) chloride
   (C) aluminum chlorite
   (D) aluminum chloride

4. The Roman numeral in manganese(IV) sulfide indicates the
   (A) group number on the periodic table.
   (B) positive charge on the manganese ion.
   (C) number of manganese ions in the formula.
   (D) number of sulfide ions needed in the formula.

Tips for Success

Eliminate Wrong Answers If you don't know which choice is correct, start by eliminating those you know are wrong. If you can rule out some choices, you'll increase your chances of choosing the correct answer.

5. Which of these statements does not describe every binary molecular compound?
   (A) Molecules of binary molecular compounds are composed of two atoms.
   (B) The names of binary molecular compounds contain prefixes.
   (C) The names of binary molecular compounds end in the suffix *-ide*.
   (D) Binary molecular compounds are composed of two nonmetals.

6. What is the formula of ammonium carbonate?
   (A) NH_4CO_3 (C) NH_3CO_4
   (B) $(NH_4)_2CO_3$ (D) NH_4CO_2

The lettered choices below refer to Questions 7–10.
(A) QR (B) QR_2 (C) Q_2R (D) Q_2R_3

Which formula shows the correct ratio of ions in the compound formed by each pair of elements?

	Element Q	Element R
7.	aluminum	sulfur
8.	potassium	oxygen
9.	lithium	chlorine
10.	strontium	bromine

Use the data table to answer Questions 11–12. The table gives formulas for some of the ionic compounds formed when cations (M, N, P) combine with anions (A, B, C, D).

Cation	Anion			
	A	B	C	D
M	MA_2	(1)	(2)	MD
N	(3)	N_2B	(4)	(5)
P	PA_3	(6)	PC	$P_2(D)_3$

11. Use the given formulas to determine the ionic charge of each cation and anion.

12. Write formulas for compounds (1) through (6).

Use the atomic windows to answer Question 13.

13. Classify the contents as elements only, compounds only, or elements and compounds.

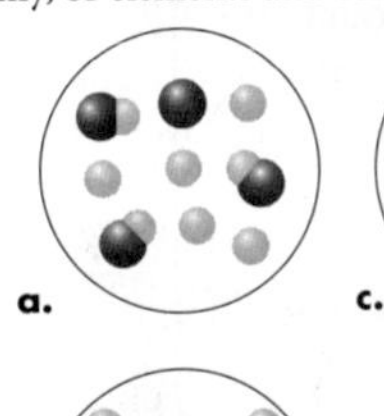
a.

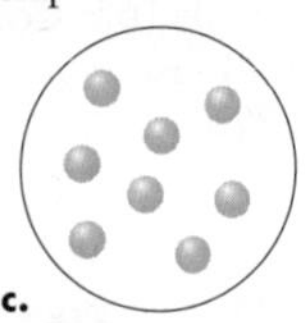
c.

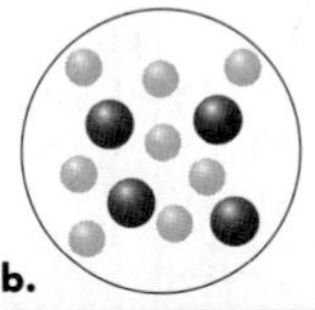
b.

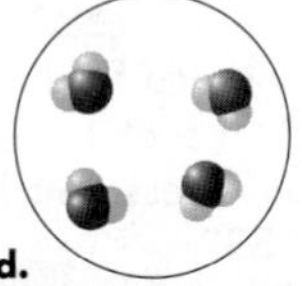
d.

If You Have Trouble With . . .

Question	1	2	3	4	5	6	7	8	9	10	11	12	13
See Lesson	9.1	9.2	9.2	9.2	9.3	9.2	9.2	9.2	9.2	9.2	9.2	9.2	9.2

STP Answers

1. B
2. D
3. D
4. B
5. A
6. B
7. D
8. C
9. A
10. B
11. cations: M^{2+}, N^+, P^{3+}; anions: A^-, B^{2-}, C^{3-}, D^{2-}
12. (1) MB, (2) M_3C_2, (3) NA, (4) N_3C, (5) N_2D, (6) P_2B_3
13. **a** contains elements and compounds, **b** and **c** contain only elements, and **d** contains only a compound.

10 Chemical Quantities Planning Guide

Introducing the BIGIDEA: THE MOLE AND QUANTIFYING MATTER

The mole is the chemist's invaluable unit for specifying the amount of material.

IASS	Lessons and Objectives	Print Resources		
		For the Student	For the Teacher	
C.4.3	**10.1** The Mole: A Measurement of Matter p 306–315 **10.1.1** Convert among the count, mass, and volume of something. **10.1.2** Explain how chemists count the number of atoms, molecules, or formula units in a substance. **10.1.3** Determine the molar mass of an element and of a compound.	Reading and Study Workbook Lesson 10.1 Lesson Assessment 10.1 p 315	Teacher Demo, p 313: Moles and Mass Class Activity, p 314: Calculating Molar Mass	
C.4.3	**10.2** Mole–Mass and Mole–Volume Relationships p 317–323 **10.2.1** Describe how to convert the mass of a substance to the number of moles of a substance, and moles to mass. **10.2.2** Convert the volume of a gas at STP to the number of moles of the gas.	Reading and Study Workbook Lesson 10.2 Lesson Assessment 10.2 p 323 Small-Scale Lab: Counting by Measuring Mass p 324	Teacher Demo, p 321: Molar Volume	
C.4.3, C.4.7	**10.3** Percent Composition and Chemical Formulas p 325–333 **10.3.1** Calculate the percent by mass of an element in a compound. **10.3.2** Calculate the empirical formula of a compound. **10.3.3** Distinguish between empirical and molecular formulas.	Reading and Study Workbook Lesson 10.3 Lesson Assessment 10.3 p 333 Quick Lab: Percent Composition p 328	Class Activity, p 330: Empirical Formulas from Percent Composition	

Assessing the BIGIDEA: THE MOLE AND QUANTIFYING MATTER

	Essential Questions 1. Why is the mole an important measurement in chemistry? 2. How can the molecular formula of a compound be determined experimentally?	Study Guide p 336 Math Tune-Up p 337 STP p 343 Reading and Study Workbook Self-Check and Vocabulary Review Chapter 10		

	PearsonChem.com	
	Editable Worksheets	Online Activities
	Teaching Resources, Lesson 10.1 Review	10.1 Lesson Overview Converting Number of Atoms to Moles and Vice Versa Molar Mass Finding the Molar Mass of a Compound
	Teaching Resources, Lesson 10.2 Review Lab 12: The Masses of Equal Volumes of Gases Small-Scale Lab Manual Lab 13: Measuring Mass: A Means of Counting	10.2 Lesson Overview Converting Moles to Mass and Mass to Moles The Mole Map
	Teaching Resources, Lesson 10.3 Review Lab 13: Empirical Formula Determination Lab Practical 10-1: Empirical Formulas	10.3 Lesson Overview Calculating Percent Composition from Mass and from a Formula Your Daily Percent Composition Determining the Empirical Formula of a Compound
	Core Teaching Resources (includes editable worksheets) • Lesson Reviews • Practice Problems • Interpret Graphs • Vocabulary Review • Chapter Quizzes and Tests • Lab Record Sheets	Chapter 10 Problem Set Unit Conversion Circle Graphs

Materials List

For the Student

Small-Scale Lab p 324
- 1 teaspoon each of water, sodium chloride, and calcium carbonate
- plastic spoon
- weighing paper
- watch glass or small beaker
- balance
- paper
- pencil
- ruler

Quick Lab p 328
- centigram balance
- burner
- 3 medium-sized test tubes
- test tube holder
- test tube rack
- spatula
- 2–3 g each of hydrated salts of copper(II) sulfate, calcium chloride, and sodium sulfate

For the Teacher

Teacher Demo p 313
- 1 mol of each of several different substances
- 1 sealed container per sample

Class Activity p 314
- display from the Teacher Demo on p 313

Teacher Demo p 321
- 50 g dry ice
- towel
- hammer
- large plastic bag
- duct tape
- tongs
- beaker
- balance

Class Activity p 330
- 3 red marbles
- 6 green marbles
- 3 black marbles
- 12 blue marbles

Additional Digital Resources

Online Student Edition
Online Teacher's Edition
Exam*View Assessment Suite CD-ROM*
Classroom Resources DVD
10.2 Virtual Chem Lab 3: Counting by Measuring Mass

What's Online

CHEMISTRY TUTORIAL Students can work step-by-step with the online tutorial to learn how to solve calculations involving moles, calculate percent composition, and determine the empirical formula of a compound, by accessing the Chapter 10 Chemistry Tutorials.

ONLINE PROBLEM SET Additional practice problems about chemical quantities and molar conversions can be found in the Chapter 10 online Problem Set.

MATH TUTORIALS The online Math Tutorials will help guide students through unit conversions and using circle graphs.

DIRECTED VIRTUAL LAB Encourage students to review the relationship between moles and mass using the animated virtual lab for Chapter 10. The animation is followed by analysis questions.

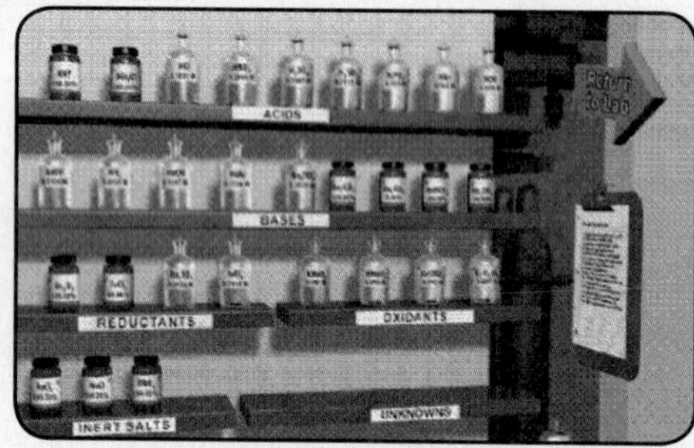

KINETIC ART Have students watch animations about molecular mass and converting between moles, mass, and number of particles. Each animation is followed by analysis questions.

CONCEPTS IN ACTION Students can watch an interactive animation about the percent composition of everyday items and events and then answer questions that apply the concept.

IN ACADEMIC STANDARDS for SCIENCE

Nature of Science

- Apply standard techniques in laboratory investigations to measure physical quantities in appropriate units and convert known quantities to other units as necessary.

Standard 4: Reactions and Stoichiometry

C.4.3 Use the mole concept to determine moles and number of atoms or molecules in samples of elements and compounds, given mass of the sample.

C.4.7 Perform calculations based on percent composition by mass of a compound or mixture when given the formula.

10 Chemical Quantities

INSIDE:

PearsonChem.com

When you shop at the grocery store or farmers' market, you usually buy blueberries by the pint, not by the berry. Similarly, chemists use a unit called the mole to count atoms and molecules.

304

Focus on ELL

1 CONTENT AND LANGUAGE Clarify for students that the word *mole* has various common meanings and scientific meanings. The common meanings have several different origins, but the chemistry meaning is derived from the German word *molekulargewicht,* which means "molecular weight." In chemistry, a mole is a unit of measurement for the amount of a substance.

BEGINNING: LOW/HIGH Make a nonlinguistic representation of the chemistry meaning of *mole* and present it to the class. Have students use a bilingual dictionary to find the definition of *mole*, and write the definition in their notebooks.

INTERMEDIATE: LOW/HIGH Ask students to brainstorm and write a list of other units of SI measures of matter.

ADVANCED: LOW/HIGH Predict the meaning of molar mass. Create an analogy between a grouping unit of measure, such as a pair or dozen, and a chemist's unit of measure, the mole.

BIGIDEA

THE MOLE AND QUANTIFYING MATTER

Essential Questions:

1. *Why is the mole an important measurement in chemistry?*
2. *How can the molecular formula of a compound be determined experimentally?*

CHEMYSTERY

A Formula for Cheating

Anabolic steroids are compounds that are developed to increase muscle size and strength. Stories are often in the news about professional athletes, such as baseball players, cyclists, and track stars, who have used steroids to enhance their performance.

More than 100 different types of anabolic steroids have been developed, and each of these substances is illegal in the United States without a prescription. Steroids have also been banned by many sports organizations because of their dangerous side effects and because they give the user an unfair advantage. Therefore, athletes are often tested for steroid use. So, how can the presence of steroids in the body be detected?

► Connect to the BIGIDEA As you read about the mole and chemical quantities, think about how the molar mass and molecular formula of a compound can be determined and used to identify the presence of steroids in the body.

INDIANA'S ACADEMIC STANDARDS FOR SCIENCE

Reactions and Stoichiometry C.4.3, C.4.7. See lessons for details.

Introduce the Chapter

IDENTIFYING PRECONCEPTIONS Students may not realize that quantifying particles of matter is even possible. Use this activity to introduce the idea of using mass to quantify large numbers of items.

Activity Divide the class into groups and provide each group with a small cup, 15 pennies, 15 dimes, and access to a balance. Alternatively, students can use any two types of small items. Write the mass of a penny (2.500 g) and a dime (2.268 g) on the board. **Ask** How many pennies and dimes would a mass of 7.268 g indicate? *(2 pennies and 1 dime)* Give each group of students a cup with a mixture of pennies and dimes. Have them randomly remove some pennies and dimes and measure the total mass. Then have them count the coins and calculate the mass. Point out that in this chapter they will learn about a similar method, using molecular formulas, to calculate numbers of atoms in a substance.

CHAPTER 10

UbD

Understanding by Design

Students are building toward measuring chemical quantities using the relationships of the mole and quantifying matter.

PERFORMANCE GOALS At the end of Chapter 10, students will be able to answer the essential questions by applying their knowledge of chemical quantities.

ESSENTIAL QUESTIONS Read the essential questions aloud. **Ask** When is it more convenient to count items in groups *(when you are working with large numbers of very small items)* **Ask** What information does a molecular formula tell you? *(the types of atoms and the ratio of each type in a compound)*

BIGIDEA Use the photo of blueberries to help students connect to this chapter's concepts. Activate prior knowledge by asking whether they buy blueberries individually or in groups. Point out that even a small basket holds a large number of blueberries. **Ask** How might grouping the blueberries make it easier to count them? *(grouping makes the number more manageable)*

CHEMYSTERY Connect the CHEMystery to the Big Idea by discussing how a molecular formula helps quantify matter, or describe it using numbers. **Ask** What would a laboratory technician look for in an athlete's urine? *(chemicals that indicated drug use)* As a hint to how the mystery could be solved, encourage students to think about what information the technician would be able to determine once the name of any chemical in the urine is known. *(the chemical formula)*

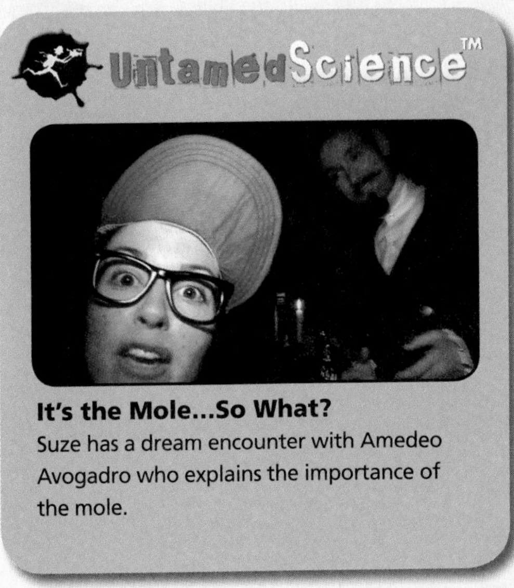

It's the Mole...So What?
Suze has a dream encounter with Amedeo Avogadro who explains the importance of the mole.

LESSON 10.1

Key Objectives

10.1.1 CONVERT among the count, mass, and volume of something.

10.1.2 EXPLAIN how chemists count the number of atoms, molecules, or formula units in a substance.

10.1.3 DETERMINE the molar mass of an element and of a compound.

Additional Resources

Reading and Study Workbook, Lesson 10.1

Available Online or on Digital Media:

- Teaching Resources, Lesson 10.1 Review
- Small-Scale Laboratory Manual, Lab 13

PearsonChem.com

Engage

CHEMISTRY & YOU Have students read the opening paragraph. **Ask** Is it practical to count each grain of sand? *(Students should realize that it is not practical to measure sand by counting individual grains.)* **Ask** How else might you measure, or quantify, the sand? *(Find its mass or volume.)* Lead students to see that just as a small amount of sand contains millions of smaller particles, so also small amounts of chemical substances contain very large numbers of particles.

Activate Prior Knowledge

Remind students that a conversion factor is a ratio equal to one. **Ask** What method is used to change one unit of measure to another unit of measure using conversion factors? *(dimensional analysis)*

IN ACADEMIC STANDARDS for SCIENCE

C.4.3 Use the mole concept to determine the number of moles and number of atoms or molecules in samples of elements and compounds, given mass of the sample.

10.1 The Mole: A Measurement of Matter

C.4.3 The mole concept

CHEMISTRY & YOU

Q: *How can you quantify the amount of sand in a sand sculpture?* Have you ever gone to the beach and created a castle or sculpture out of sand? You could measure the amount of sand in a sculpture by counting the grains of sand. Is there an easier way to measure the amount of sand? Chemists measure the amount of a substance using a unit called the mole.

Key Questions

How can you convert among the count, mass, and volume of something?

How do chemists count the number of atoms, molecules, or formula units in a substance?

How do you determine the molar mass of an element and of a compound?

Vocabulary

- mole
- Avogadro's number
- representative particle
- molar mass

Measuring Matter

How can you convert among the count, mass, and volume of something?

Chemistry is a quantitative science. In your study of chemistry, you will analyze the composition of samples of matter and perform chemical calculations that relate quantities of the reactants in a chemical reaction to quantities of the products. To solve these and other problems, you will have to be able to measure the amount of matter you have.

One way to measure matter is to count how many of something you have. For example, you can count the mp3s in your collection. Another way to measure matter is to determine its mass. You can buy apples by the kilogram or pound, as shown in Figure 10.1. You can also measure matter by volume. For instance, people buy gasoline by the liter or the gallon.

Some of the units used for measuring indicate a specific number of items. For example, a pair always means two. A pair of shoes is two shoes, and a pair of aces is two aces. Similarly, a dozen always means 12. A dozen eggs is 12 eggs, and a dozen pens is 12 pens.

Apples can be measured in three different ways. At a fruit stand, they are often sold by the count. In a supermarket, you usually buy apples by weight or mass. At an orchard, you can buy apples by volume. Each of these different ways to measure apples can be equated to a dozen apples.

By count: 1 dozen apples = 12 apples

For average-sized apples, the following approximations can be used.

By mass: 1 dozen apples = 2.0 kg apples

By volume: 1 dozen apples = 0.20 bushel apples

Figure 10.1 Measuring by Mass
A dozen apples has a mass of about 2.0 kg.

Focus on ELL

1 CONTENT AND LANGUAGE Review common conversion factors like feet to inches, meters to centimeters, hours to minutes, etc.

2 FRONTLOAD THE LESSON Ask students to share unique terms for grouped items from their native culture, and explain the quantity and size of the items in the group that each term represents. Distinguish between groupings in which the number of items varies and those with set numbers of items as a prelude to discussing Avogadro's number.

3 COMPREHENSIBLE INPUT Use a model to introduce the concept of molar mass. Display cartons of small, medium, large, and extra-large eggs as representations of moles of different elements. Tell students to think of the eggs as atoms, with each size being a different element. Explain that one carton of small eggs contains the same number of eggs as a carton of extra-large eggs, but the two cartons of eggs each have different masses because their individual eggs are different sizes (and masses).

Knowing how the count, mass, and volume of an item relate to a common unit allows you to convert among these units. For example, based on the unit relationships given on the previous page, you could calculate the mass of a bushel of apples or the mass of 90 average-sized apples using conversion factors such as the following:

$$\frac{1 \text{ dozen apples}}{12 \text{ apples}} \qquad \frac{2.0 \text{ kg apples}}{1 \text{ dozen apples}} \qquad \frac{1 \text{ dozen apples}}{0.20 \text{ bushel apples}}$$

SampleProblem 10.1

Finding Mass From a Count

What is the mass of 90 average-sized apples if 1 dozen of the apples has a mass of 2.0 kg?

KNOWNS
number of apples = 90 apples
12 apples = 1 dozen apples
1 dozen apples = 2.0 kg apples

UNKNOWN
mass of 90 apples = ? kg

1 Analyze List the knowns and the unknown. Use dimensional analysis to convert the number of apples to the mass of apples.

2 Calculate Solve for the unknown.

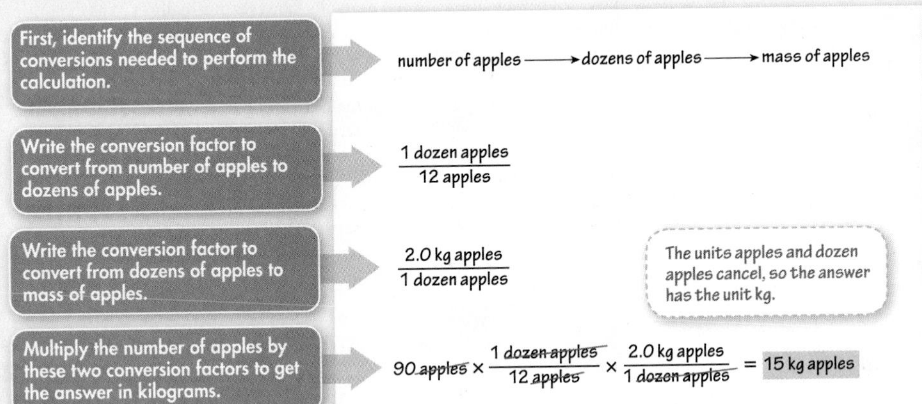

3 Evaluate Does the result make sense? A dozen apples has a mass of 2.0 kg, and 90 apples is less than 10 dozen apples, so the mass should be less than 20 kg of apples (10 dozen × 2.0 kg/dozen).

1. If 0.20 bushel is 1 dozen apples and a dozen apples has a mass of 2.0 kg, what is the mass of 0.50 bushel of apples?

2. Assume 2.0 kg of apples is 1 dozen and that each apple has 8 seeds. How many apple seeds are in 14 kg of apples?

In Problem 1, the desired conversion is bushels of apples → dozens of apples → mass of apples.

In Problem 2, the desired conversion is mass of apples → dozens of apples → number of apples → number of seeds.

Foundations for Math

WRITING CONVERSIONS Work thorough some conversions that students likely encounter in their daily lives. For example: If a recipe calls for six eggs, this quantity can be considered a half-dozen eggs: 6 eggs × (1 dozen/12 eggs) = 0.5 dozen.

As a class, write other familiar relationships, such as 3 feet = 1 yard, 50 cents = ½ dollar, 60 min = 1 h, 30 min = ½ h, etc. Point out that these equalities can be written in either direction, so 60 min = 1 h can also be written as 1 h = 60 min.

In **Sample Problem 10.1,** the mass of a large number of apples is determined by using the mass of a smaller number of apples by means of a conversion factor created from the subset of apples. (Note that, for this example, the mass of one dozen apples is an approximation. In reality, the masses of similar-sized apples vary from one apple to the next.)

Foundations for Reading

BUILD VOCABULARY Have students write two or three sentences, each of which relates a mole to at least one other vocabulary term, such as: a mole contains Avogadro's number of particles; the mass of one mole of a substance is its molar mass, which is found by determining the mass in grams of its representative particles.

READING STRATEGY Students may more easily grasp the meaning of the mole by developing their own mental picture. For instance, suggest visualizing a giant egg carton with 6.02×10^{23} depressions for particles, or a giant sack bulging with Avogadro's number of particles.

Explain

Measuring Matter

USE MODELS Pass around numerous bags, each of which contains a multiple of 12 beans. **Ask** How can you express the quantities of beans in the bags? *(Sample answers: You can count, weigh, or find the volume of the beans.)* Discuss how the word "dozen" can be used as a unit for the quantity of beans in each bag.

START A CONVERSATION Remind students that mass is a measure of the amount of matter that an object contains. **Ask** How are mass and weight related? *(Weight is a force that measures the pull of gravity on a given mass.)*

Sample Practice Problem

Assume 1 dozen oranges has a mass of 1.5 kg and that there are 14 orange slices in each orange. How many slices are in 6 kg of oranges? *(672 slices)*

Answers

1. 0.50 bushel × (1 dozen/0.20 bushel) × (2.0 kg/1 dozen) = 5.0 kg
2. 14 kg × (1 dozen/2.0 kg) × (12 apples/1 dozen) × (8 seeds/1 apple) = 672 seeds

LESSON 10.1

Explain

What Is a Mole?

USE VISUALS Read aloud the caption to Figure 10.2. Then have the class read the text that discusses the number of particles in a mole. Guide students to understand that a mole represents a number of items just as *dozen, gross,* and *ream* all represent a quantity of items. **Ask** If you are counting the number of an extremely small item, should the number of items in a unit be small or large? *(For a small item, a large number per counting unit is more convenient.)* **Ask** How does your answer apply to a mole? *(A mole = 6.02×10^{23} items and is used to measure extremely small objects.)*

CHEMISTRY & YOU You could measure the mass of 1 grain of sand. Then you could measure the mass of the sand castle. If you divide the mass of the sand castle by the mass of 1 grain of sand, you can determine the total number of grains of sand in the castle.

MAKING CONNECTIONS Have students spend 2 minutes writing down what they remember about scientific notation and properties of exponents. As a class, share information and create a summary sheet of the important rules for these two concepts. Ask student volunteers to write several large numbers and small numbers on the board in standard form. Then use the summary sheet to guide students in rewriting the numbers using proper scientific notation. **Ask** When do you typically see measurements written in scientific notation? *(Sample answers: When a quantity is extremely small or extremely large, such as the diameter of a virus or distances in space.)* **Ask** How might scientific notation be useful in calculating the amount of atoms in a given number of moles? *(6.02×10^{23} is an extremely long number when it is not written in scientific notation.)*

Figure 10.2 Grouping Objects
Words other than *mole* are used to describe a number of something—for example, a *dozen* eggs is 12 eggs.

What Is a Mole?

How do chemists count the number of atoms, molecules, or formula units in a substance?

Counting objects as big as apples is a reasonable way to measure how much of the object you have. Picture trying to count the grains of sand in a sand sculpture. It would be an endless job. Recall that matter is composed of atoms, molecules, and ions. These particles are much, much smaller than grains of sand, and an extremely large number of them are in a small sample of a substance. Obviously, counting particles one by one is not practical. However, think about counting eggs. It's easier when the eggs are grouped into dozens, as shown in Figure 10.2. A dozen is a specified number (12) of things.

Counting With Moles Chemists also use a unit that is a specified number of particles. The unit is called the mole. A **mole** (mol) of a substance is 6.02×10^{23} representative particles of that substance and is the SI unit for measuring the amount of a substance. The number of representative particles in a mole, 6.02×10^{23}, is called **Avogadro's number.** It was named in honor of the Italian scientist Amedeo Avogadro di Quaregna (1776–1856), who helped clarify the difference between atoms and molecules.

The term **representative particle** refers to the species present in a substance, usually atoms, molecules, or formula units. The representative particle of most elements is the atom. Iron is composed of iron atoms. Helium is composed of helium atoms. Seven elements, however, normally exist as diatomic molecules (H_2, N_2, O_2, F_2, Cl_2, Br_2, and I_2). The representative particle of these elements and of all molecular compounds is the molecule. The molecular compounds water (H_2O) and sulfur dioxide (SO_2) are composed of H_2O and SO_2 molecules, respectively. For ionic compounds, such as calcium chloride, the representative particle is the formula unit $CaCl_2$. **The mole allows chemists to count the number of representative particles in a substance.** A mole of any substance contains Avogadro's number of representative particles, or 6.02×10^{23} representative particles. Table 10.1 summarizes the relationship between representative particles and moles of substances.

CHEMISTRY & YOU

Q: *What are the different ways you can measure the amount of sand in a sand sculpture?*

Table 10.1

Representative Particles and Moles

Substance	Representative particle	Chemical formula	Representative particles in 1.00 mol
Copper	Atom	Cu	6.02×10^{23}
Atomic nitrogen	Atom	N	6.02×10^{23}
Nitrogen gas	Molecule	N_2	6.02×10^{23}
Water	Molecule	H_2O	6.02×10^{23}
Sucrose	Molecule	$C_{12}H_{22}O_{11}$	6.02×10^{23}
Calcium ion	Ion	Ca^{2+}	6.02×10^{23}
Calcium fluoride	Formula unit	CaF_2	6.02×10^{23}

Differentiated Instruction

ELL ENGLISH LANGUAGE LEARNERS Pair each student with limited English proficiency with a student who has strong mathematical skills. Encourage the English learners to ask their partners for help with any difficulty they are having understanding the terms and how to solve the problems.

L1 STRUGGLING STUDENTS Review the use of parentheses and the fraction bar as ways of grouping symbols to indicate the order of operations. Stress the importance of writing units for each measure in a calculation. Provide additional simple conversion problems for practice before students begin solving problems involving the mole.

L3 ADVANCED STUDENTS Have students create a clever way for their classmates to avoid confusing the concepts of mass and moles.

Converting Between Number of Particles and Moles The relationship, 1 mol = 6.02×10^{23} representative particles, is the basis for the following conversion factors that you can use to convert number of representative particles to moles and moles to number of representative particles.

$$\frac{1 \text{ mol}}{6.02 \times 10^{23} \text{ representative particles}} \quad \text{and} \quad \frac{6.02 \times 10^{23} \text{ representative particles}}{1 \text{ mol}}$$

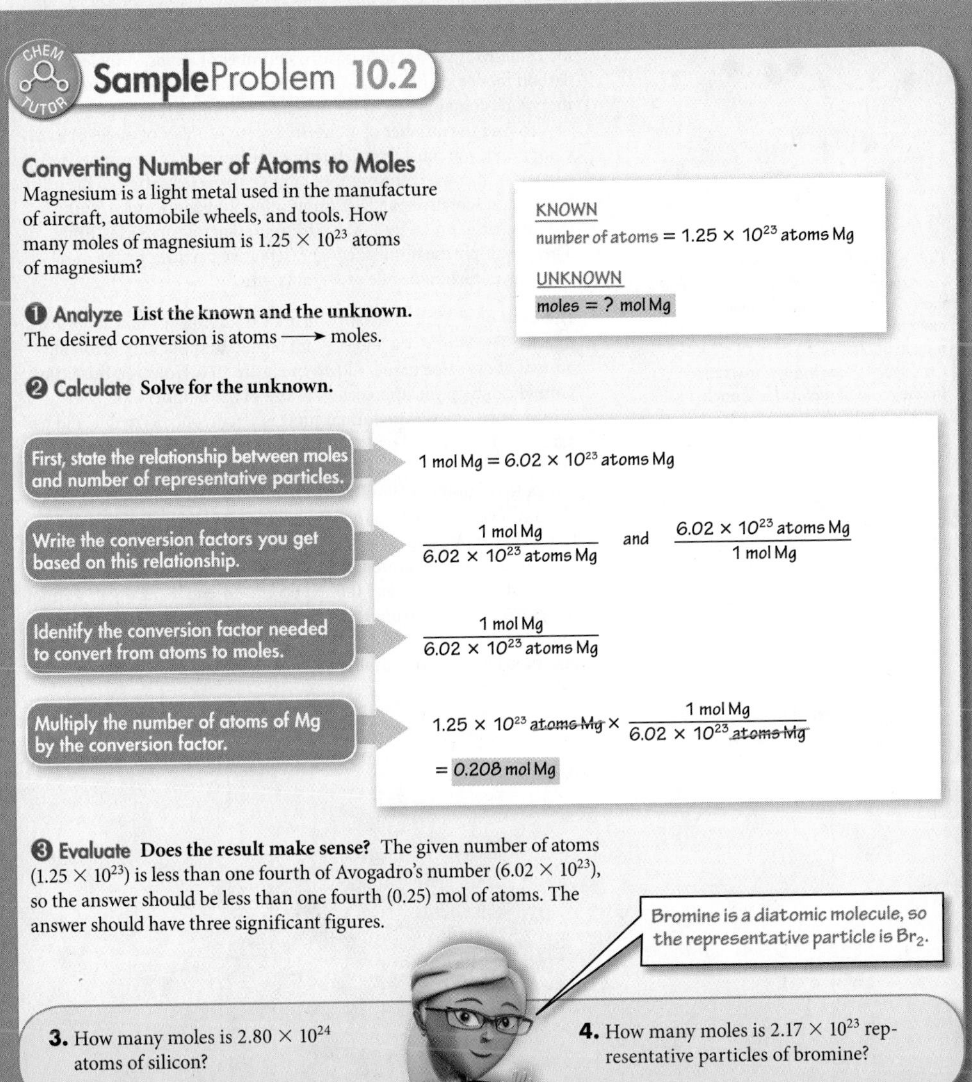

SampleProblem 10.2

CHEM TUTOR

Converting Number of Atoms to Moles

Magnesium is a light metal used in the manufacture of aircraft, automobile wheels, and tools. How many moles of magnesium is 1.25×10^{23} atoms of magnesium?

1 Analyze List the known and the unknown.
The desired conversion is atoms ⟶ moles.

KNOWN
number of atoms = 1.25×10^{23} atoms Mg
UNKNOWN
moles = ? mol Mg

2 Calculate Solve for the unknown.

First, state the relationship between moles and number of representative particles.

$1 \text{ mol Mg} = 6.02 \times 10^{23} \text{ atoms Mg}$

Write the conversion factors you get based on this relationship.

$$\frac{1 \text{ mol Mg}}{6.02 \times 10^{23} \text{ atoms Mg}} \quad \text{and} \quad \frac{6.02 \times 10^{23} \text{ atoms Mg}}{1 \text{ mol Mg}}$$

Identify the conversion factor needed to convert from atoms to moles.

$$\frac{1 \text{ mol Mg}}{6.02 \times 10^{23} \text{ atoms Mg}}$$

Multiply the number of atoms of Mg by the conversion factor.

$$1.25 \times 10^{23} \cancel{\text{atoms Mg}} \times \frac{1 \text{ mol Mg}}{6.02 \times 10^{23} \cancel{\text{atoms Mg}}} = 0.208 \text{ mol Mg}$$

3 Evaluate Does the result make sense? The given number of atoms (1.25×10^{23}) is less than one fourth of Avogadro's number (6.02×10^{23}), so the answer should be less than one fourth (0.25) mol of atoms. The answer should have three significant figures.

3. How many moles is 2.80×10^{24} atoms of silicon?

4. How many moles is 2.17×10^{23} representative particles of bromine?

Explain

APPLY CONCEPTS Explain that the mole is defined as the amount of substance that contains as many representative particles as there are atoms in 12 g of carbon-12 ($^{12}_{6}C$). Tell students that when the mole is used, the representative particles need to be specified as either atoms, molecules, ions, electrons, or formula units. Tell students to pay particular attention to the information given in a problem. Encourage students to underline or highlight the type of representative particle stated in a problem.

Sample Practice Problems

A. How many moles are equal to 3.61×10^{24} representative particles of potassium chloride, KCl? *(6.00 moles)*

B. About how many moles are equal to 5.82×10^{23} atoms of tungsten (W)? *(0.967 moles)*

Extend

Connect to HISTORY Ask students to use the Internet to examine the history of the idea of chemical equivalency. Compare and contrast the ideas of the past with the modern accepted theory. Students should provide a timeline as well as the names of any noted scientists involved with this theory.

Foundations for Math

DIVIDING POWERS OF 10 Tell students that the rules for dividing powers of 10 are the same as those for dividing variables with exponents: as long as the base is the same, you subtract the exponents. For example, . Point out that each term is a power with the same base; caution students not to divide 10 by 10. In **Sample Problem 10.2** have students group the "like" numbers and divide accordingly:

Use a calculator to divide the decimal numbers and use the rules of exponents to divide the powers. In this case, $10^{23-23} = 10^0$, which is equal to 1.

Answers

3. 2.80×10^{24} atoms Si × (1 mol/6.02×10^{23} atoms) = 4.65 mol Si

4. 2.17×10^{23} representative particles × (1 mol/6.02×10^{23} representative particles) = 0.360 mol Br_2

LESSON 10.1

Explain

USE VISUALS Direct students' attention to Figure 10.3. Have students examine the photograph. Note that each cup contains six marbles. **Ask** How much would a dozen cups of marbles hold? *(72 marbles)* **Ask** What are some everyday items that come in a package containing more than one? *(Sample answers: tennis balls, 3; shoes, 2; batteries, 2, 4, 8)* Point out that a dozen packages of tennis balls would be three dozen tennis balls or 36 tennis balls. Point out that this idea can be applied to molecules. For example, a mole of water, H_2O, consists of 2 mol H atoms and 1 mol O atoms. Thus a mole of water contains 3 × 6.02×10^{23} atoms or 1.8×10^{24} atoms.

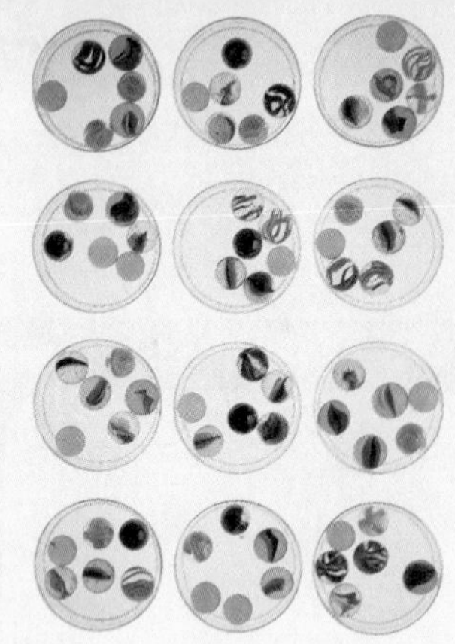

Figure 10.3 Counting Marbles
A dozen cups of marbles contain more than a dozen marbles. Similarly, a mole of molecules contains more than a mole of atoms.
Calculate ***How many atoms are in one mole of molecules if each molecule consists of six atoms?***

Suppose you want to determine how many atoms are in a mole of a compound. To do this, you must know how many atoms are in a representative particle of the compound. This number is determined from the chemical formula. Figure 10.3 illustrates this idea with marbles (atoms) in cups (molecules). The number of marbles in a dozen cups is (6 × 12), or 72 marbles. In the formula for carbon dioxide (CO_2), the subscripts show that one molecule of carbon dioxide is composed of three atoms: one carbon atom and two oxygen atoms. A mole of carbon dioxide contains Avogadro's number of CO_2 molecules. Each molecule contains three atoms, so a mole of carbon dioxide contains three times Avogadro's number of atoms. A molecule of carbon monoxide (CO) consists of two atoms, so a mole of carbon monoxide contains two times Avogadro's number of atoms.

To find the number of atoms in a given number of moles of a compound, you must first determine the number of representative particles. To convert the number of moles of a compound to the number of representative particles (molecules or formula units), multiply the number of moles by 6.02×10^{23} representative particles/1 mol. Then, multiply the number of representative particles by the number of atoms in each molecule or formula unit.

The Size of a Mole Perhaps you are wondering just how large a mole is. The SI unit, the mole, is not related to the small burrowing animal of the same name, shown in Figure 10.4. However, this little animal can help you appreciate the size of the number 6.02×10^{23}. Assume that an average animal-mole is 15 cm long, 5 cm tall, and has a mass of 145 g. Based on this information, the mass of 6.02×10^{23} animal-moles is 8.73×10^{22} kg. That means that the mass of Avogadro's number of animal-moles is equal to more than 60 times the combined mass of Earth's oceans. If spread over the entire surface of Earth, Avogadro's number of animal-moles would form a layer more than 8 million animal-moles thick. What about the length of 6.02×10^{23} animal-moles? If lined up end-to-end, 6.02×10^{23} animal-moles would stretch from Earth to the nearest star, Alpha Centauri, more than two million times. Are you beginning to understand how enormous Avogadro's number is?

Figure 10.4 A Mole of Moles
An average animal-mole has a mass of 145 g. The mass of 6.02×10^{23} animal-moles is 8.73×10^{22} kg.

UbD Check for Understanding

BIGIDEA Assess students' knowledge about the Big Idea of The Mole and Quantifying Matter by projecting Table 10.1 on an overhead. Ask students to briefly state why all the substances in the table have the same number of representative particles per mole. *(One mole of any type of substance is 6.02×10^{23} atoms, molecules, ions, formula units, etc. The type of particle does not affect the number of particles in one mole.)*

ADJUST INSTRUCTION If students are having difficulty with this concept, have them review the text preceding Table 10.1. Then repeat the activity.

SampleProblem 10.3

Converting Moles to Number of Atoms

Propane is a gas used for cooking and heating. How many atoms are in 2.12 mol of propane (C_3H_8)?

1 Analyze **List the knowns and the unknown.**
The desired conversion is moles ⟶ molecules ⟶ atoms.

KNOWNS
number of moles = 2.12 mol C_3H_8
1 mol C_3H_8 = 6.02×10^{23} molecules C_3H_8
1 molecule C_3H_8 = 11 atoms
(3 carbon atoms and 8 hydrogen atoms)

UNKNOWN
number of atoms = ? atoms

2 Calculate **Solve for the unknown.**

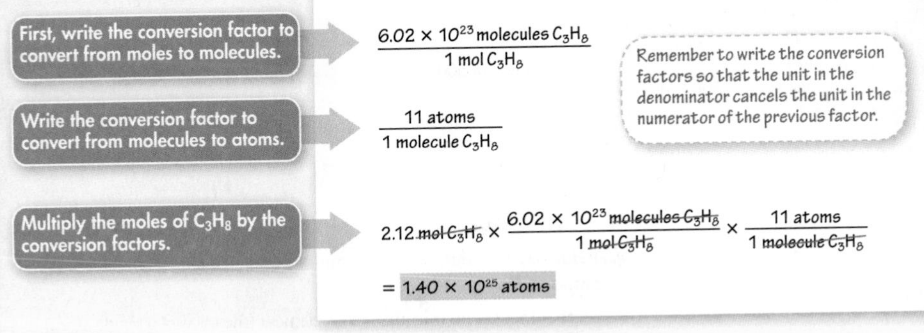

$$\frac{6.02 \times 10^{23}\ \text{molecules } C_3H_8}{1\ \text{mol } C_3H_8}$$

$$\frac{11\ \text{atoms}}{1\ \text{molecule } C_3H_8}$$

$$2.12\ \cancel{\text{mol } C_3H_8} \times \frac{6.02 \times 10^{23}\ \cancel{\text{molecules } C_3H_8}}{1\ \cancel{\text{mol } C_3H_8}} \times \frac{11\ \text{atoms}}{1\ \cancel{\text{molecule } C_3H_8}}$$

$$= 1.40 \times 10^{25}\ \text{atoms}$$

3 Evaluate **Does the result make sense?** There are 11 atoms in each molecule of propane and more than 2 mol of propane, so the answer should be more than 20 times Avogadro's number of propane molecules. The answer has three significant figures based on the three significant figures in the given measurement.

There are 3 atoms of carbon and 8 atoms of hydrogen in 1 molecule of propane.

5. How many atoms are in 1.14 mol of sulfur trioxide (SO_3)?

6. How many carbon atoms are in 2.12 mol of propane? How many hydrogen atoms are in 2.12 mol of propane?

Foundations for Math

CONVERTING PRODUCTS TO SCIENTIFIC NOTATION Point out that if a calculation results in a product that has a power of 10, it may not necessarily be written in proper scientific notation. The coefficient must be a number greater than or equal to 1 and less than 10. Have students write a rule in their own words for converting products in which the coefficient is greater than 10, and in which the coefficient is less than 0. *(Move the decimal to the left and adjust the exponent up; Move the decimal to the right and adjust the exponent down.)*

In **Sample Problem 10.3**, students might opt to multiply 2.12, 6.02, and 11 together first, then multiply the result by 10^{23}. This would give an answer of 140×10^{23}, which is not in proper scientific notation. Point out that an additional conversion must be made by moving the decimal 2 places to the left and adjusting the exponent up by 2 to convert the answer to proper scientific notation.

LESSON 10.1

Explain

CRITICAL THINKING Have students examine the problem solving process for Sample Problems 10.2 and 10.3. **Ask** Why do you divide by Avogadro's number in Sample Problem 10.2, but multiply by it in Sample Problem 10.3? *(In Sample Problem 10.2, the problem asks for the number of moles; in Sample Problem 10.3, the problem asks for the number of atoms.)* Have students write a note to themselves explaining when and how they should use this conversion factor. **Ask** Do the identities of the substances in the problems have any effect on the way you use Avogadro's number in the conversion process? *(no)* Why? *(The number of representative particles in a mole is a constant. The only information supplied by the substance is the number of atoms that makes up one representative particle of the substance. In Sample Problem 10.2, the representative particle is a single atom of Mg. In Sample Problem 10.3, the representative particle contains 11 atoms.)*

Misconception Alert

Some students may think they are finished when the calculator displays the result of the last calculation. Explain that not all calculators automatically convert the final answer to proper scientific notation, or use the proper number of significant figures. Make sure students know how to make the correct conversion to scientific notation with their particular calculator.

Sample Practice Problems

A. How many atoms are in 1.00 mole of glucose, $C_6H_{12}O_6$? *(1.44×10^{25} atoms)*

B. How many atoms of C are in 2.00 moles of $C_6H_{12}O_6$? *(7.22×10^{24} atoms)*

C. How many atoms of H are in 3.00 moles of $C_6H_{12}O_6$? *(2.17×10^{25} atoms)*

D. How many atoms of O are in 1.25 moles of $C_6H_{12}O_6$? *(4.52×10^{24} atoms)*

Answers

FIGURE 10.3 3.61×10^{24} atoms

5. 1.14 mol × (6.02×10^{23} molecules/mol) × (4 atoms/molecule) = 2.75×10^{24} atoms

6. 2.12 mol C_3H_6 × (6.02×10^{23} molecule/mol) × (3 atoms/molecule) = 3.83×10^{24} C atoms
2.12 mol C_3H_6 × (6.02×10^{23} molecule/mol) × (8 atoms/molecule) = 1.02×10^{25} H atoms

Explain

Molar Mass

USE MODELS Provide students with tactile counting objects to model the carbon and hydrogen relationship in Table 10.2. Guide students to use the objects to discover the mass ratio of carbon to hydrogen.

MAKING CONNECTIONS Point out that the mass of a single atom can be expressed in atomic mass units, but it is not realistic to work with single atoms. Explain that chemists work with large numbers of atoms for which the mass can be expressed in grams. In this text, the atomic masses are rounded to one place after the decimal point. Have students solve a given problem multiple times, using a different rounding rule each time, so they can see how rounding the atomic masses affects the answer.

APPLY CONCEPTS Explain that the molar masses of all elements contain the same number of atoms because the atomic masses of the elements are relative values. Present this idea by telling the class that the mass of an atom of element X is twice as great as the mass of an atom of element Y. **Ask** If you have 10 grams of element X and 10 grams of element Y, would you expect both samples to contain the same number of atoms? Why? *(No, because atoms of element X are twice as massive as atoms of element Y. The sample of X would contain only half as many atoms as the sample of Y.)* **Ask** What would you have to do to get the same number of atoms in both samples? *(Double the mass of element X so that it is twice the mass of element Y.)*

Interpret Data

Carbon Atoms		Hydrogen Atoms		Mass Ratio
Number	Mass (amu)	Number	Mass (amu)	$\frac{\text{Mass carbon}}{\text{Mass hydrogen}}$
	12		1	$\frac{12 \text{ amu}}{1 \text{ amu}} = \frac{12}{1}$
	24 (2 × 12)		2 (2 × 1)	$\frac{24 \text{ amu}}{2 \text{ amu}} = \frac{12}{1}$
	120 (10 × 12)		10 (10 × 1)	$\frac{120 \text{ amu}}{10 \text{ amu}} = \frac{12}{1}$
	600 (50 × 12)		50 (50 × 1)	$\frac{600 \text{ amu}}{50 \text{ amu}} = \frac{12}{1}$
Avogadro's number	$(6.02 \times 10^{23}) \times (12)$	Avogadro's number	$(6.02 \times 10^{23}) \times (1)$	$\frac{(6.02 \times 10^{23}) \times (12)}{(6.02 \times 10^{23}) \times (1)} = \frac{12}{1}$

Table 10.2 An average carbon atom is 12 times heavier than an average hydrogen atom.
a. Read Tables What is the mass of 50 carbon atoms? What is the mass of 50 hydrogen atoms?
b. Apply Concepts What is the ratio of the mass of 500 carbon atoms to the mass of 500 hydrogen atoms?
c. Infer Do 36.0 kg of carbon atoms and 3.0 kg of hydrogen atoms contain the same number of atoms? Explain.

Hint: To answer part c, determine the mass ratio of carbon to hydrogen.

Molar Mass

How do you determine the molar mass of an element and of a compound?

Remember that the atomic mass of an element (the mass of a single atom) is expressed in atomic mass units (amu). The atomic masses are relative values based on the mass of the most common isotope of carbon (carbon-12). Table 10.2 shows that an average carbon atom (C) with an atomic mass of 12.0 amu is 12 times heavier than an average hydrogen atom (H) with an atomic mass of 1.0 amu. Therefore, 100 carbon atoms are 12 times heavier than 100 hydrogen atoms. In fact, any number of carbon atoms is 12 times heavier than the same number of hydrogen atoms. So 12.0 g of carbon atoms and 1.0 g of hydrogen atoms must contain the same number of atoms.

If you look at the atomic masses of the elements in the periodic table, you will notice that they are not whole numbers. For example, the atomic mass of carbon is not exactly 12 times the mass of hydrogen. Recall from Chapter 4 that this is because atomic masses are weighted average masses of the isotopes of each element.

312 Chapter 10 • Lesson 1

History of Avogadro's Number

Avogadro's number was not actually developed by Avogadro. In the early 1900s, a French scientist by the name of J. Perrin first used the term "Avogadro's number" to describe the number of particles in a mole. Perrin used Brownian motion to determine the number.

The Mass of a Mole of an Element Quantities measured in grams are convenient for working in the laboratory, so chemists have converted the relative scale of masses of the elements in amu to a relative scale of masses in grams. **The atomic mass of an element expressed in grams is the mass of a mole of the element.** The mass of a mole of an element is its **molar mass.** For carbon, the molar mass is 12.0 g. For atomic hydrogen, the molar mass is 1.0 g. Figure 10.5 shows one mole of carbon, sulfur, and iron. Compare the molar masses in the figure to the atomic masses in your periodic table. Notice that the molar masses are rounded off to one place after the decimal point. All the examples and problems in this text use molar masses that are rounded off in this way. If your teacher uses a different rounding rule for molar masses, your answers to problems may differ slightly from the answers given in the text.

If you were to compare 12.0 g of carbon atoms with 16.0 g of oxygen atoms, you would find they contain the same number of atoms. The molar masses of any two elements must contain the same number of atoms. How many atoms are contained in the molar mass of an element? You already know. The molar mass of any element contains 1 mol or 6.02×10^{23} atoms of that element.

The mole can now be further defined as the amount of substance that contains the same number of representative particles as the number of atoms in 12.0 g of carbon-12. You know that 12.0 g is the molar mass of carbon-12, so 12.0 g of carbon is 1 mol of carbon atoms. The same relationship applies to hydrogen: 1.0 g of hydrogen is 1 mol of hydrogen atoms. Similarly, 24.3 g is the molar mass of magnesium, so 1 mol of magnesium (or 6.02×10^{23} atoms of magnesium) has a mass of 24.3 g. Molar mass is the mass of 1 mol of atoms of any element.

READING SUPPORT

Build Comprehension: *Analogies* You can buy small, medium, and large eggs. The size of the eggs doesn't affect how many eggs are in one dozen. Similarly, the size of the representative particles doesn't affect how many are in one mole. ***Can you think of another analogy to show the relationship between moles and the size of representative particles?***

Figure 10.5 Molar Mass of an Element
One mole of carbon, sulfur, and iron are shown.
Apply Concepts ***How many atoms of each element are present in each beaker?***

Explore

Teacher Demo

PURPOSE Students will observe the difference in the mass and volume of 1 mol of different substances.

MATERIALS 1 mol each of a variety of common substances, 1 sealed container per sample

PROCEDURE Place 1 mol of at least two substances from each of the following categories in containers and seal them: molecular compounds—sucrose, water, paradichlorobenzene; ionic compounds—cobalt(II) chloride, potassium hydroxide, potassium dichromate; elements—sulfur, iron, carbon, mercury. On each container, mark the mass of each substance. Point out that each container holds 1 mol of a substance, no matter whether the representative particles are molecules, formula units, or atoms.

EXPECTED OUTCOME Students observe that a mole of one substance has a different mass from a mole of another substance, even though the same number of representative particles, 6.02×10^{23}, are present.

Differentiated Instruction

L1 SPECIAL NEEDS STUDENTS For the Teacher Demo, mark the levels of the filled containers with tape on the outside of the containers so that sight-impaired students can tell by handling the displays that moles of different substances occupy different volumes and have different masses.

ELL ENGLISH LANGUAGE LEARNERS The differences in the terms *mass*, *atomic mass*, *atomic mass unit*, and *molar mass* may be confusing to English learners. Have students compile a glossary in which they define each term in English and in their native language. Encourage students to then write the word meanings or synonyms on sticky notes to be placed in the text.

L1 STRUGGLING STUDENTS If students are having difficulty with multistep problems, break the problem into simple parts. For instance, for Sample Problem 10.4, insert the following after the first sentence: *Part a: Find the number of grams of hydrogen in 1 mol H_2O_2. Part b: Find the number of grams of oxygen in 1 mol H_2O_2. Part c: Find the molar mass of H_2O_2.*

Answers

INTERPRET DATA

a. mass of 50 carbon atoms = 600 amu, mass of 50 hydrogen atoms = 50 amu

b. 12/1

c. Yes, the mass ratio for carbon to hydrogen is 12 to 1.

READING SUPPORT Answers will vary. The size of the donuts doesn't affect how many are in one dozen.

FIGURE 10.5 6.02×10^{23} atoms

LESSON 10.1

Explore

Class Activity

PURPOSE Students learn to calculate the molar mass of a compound.

MATERIALS the display prepared for the Teacher Demo on p. 313

PROCEDURE Select a compound and show students how to calculate the mass of 1 mol using the atomic masses of the constituent elements from the periodic table. Have them calculate the molar masses of other compounds in the display. Lead students to see they have determined the mass of 6.02×10^{23} representative particles.

Figure 10.6 Molar Mass of a Compound
One mole is shown for each of three molecular compounds.
Infer *How do you know that each sample contains Avogadro's number of molecules?*

KINETIC ART *See the molar masses of compounds animated online.*

The Mass of a Mole of a Compound To find the mass of a mole of a compound, you must know the formula of the compound. The formula of sulfur trioxide is SO_3. A molecule of SO_3 is composed of one atom of sulfur and three atoms of oxygen.

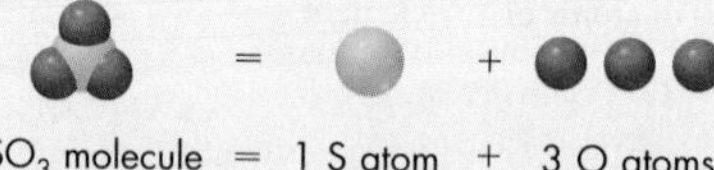

1 SO_3 molecule = 1 S atom + 3 O atoms

You can calculate the mass of a molecule of SO_3 by adding the atomic masses of the atoms making up the molecule. From the periodic table, the atomic mass of sulfur (S) is 32.1 amu. The mass of three atoms of oxygen is three times the atomic mass of a single oxygen atom (O): 3×16.0 amu = 48.0 amu. So, the molecular mass of SO_3 is 32.1 amu + 48.0 amu = 80.1 amu.

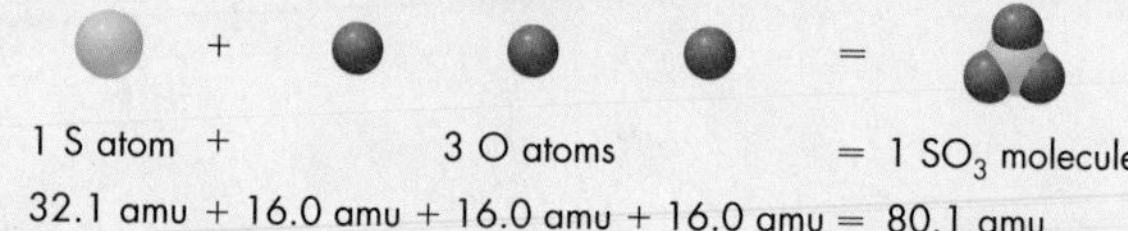

1 S atom + 3 O atoms = 1 SO_3 molecule

32.1 amu + 16.0 amu + 16.0 amu + 16.0 amu = 80.1 amu

Now substitute the unit grams for atomic mass units to find the molar mass of SO_3. The molar mass (g/mol) of any compound is the mass in grams of 1 mol of that compound. Thus, 1 mol of SO_3 has a mass of 80.1 g. This is the mass of 6.02×10^{23} molecules of SO_3.

To calculate the molar mass of a compound, find the number of grams of each element in one mole of the compound. Then add the masses of the elements in the compound. This method for calculating molar mass applies to any compound, molecular or ionic. The molar masses of paradichlorobenzene ($C_6H_4Cl_2$, 147.0 g), water (H_2O, 18.0 g), and glucose ($C_6H_{12}O_6$, 180.0 g) in Figure 10.6 were obtained in this way.

314 Chapter 10 • Lesson 1

UbD Check for Understanding

The Essential Question Why is the mole an important measurement in chemistry?

Assess students' knowledge about the mole by asking them write a one-minute response that answers this Essential Question. *(Sample answer: The mole is an important measurement in chemistry because it lets scientists turn a very large number into one that is small enough to use easily in calculations.)*

ADJUST INSTRUCTION If students are having trouble answering, have them re-read the sections What Is a Mole? and Molar Mass. Then allow them to revise their responses.

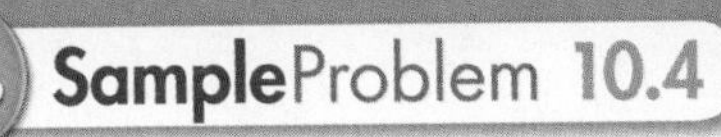

SampleProblem 10.4

Finding the Molar Mass of a Compound

The decomposition of hydrogen peroxide (H_2O_2) provides sufficient energy to launch a rocket. What is the molar mass of hydrogen peroxide?

KNOWNS
molecular formula = H_2O_2
mass of 1 mol H = 1.0 g H
mass of 1 mol O = 16.0 g O

UNKNOWN
molar mass = ? g/mol

1 Analyze List the knowns and the unknown. Convert moles of atoms to grams by using conversion factors (g/mol) based on the molar mass of each element. The sum of the masses of the elements is the molar mass.

2 Calculate Solve for the unknown.

Convert moles of hydrogen and oxygen to grams of hydrogen and oxygen.

$$2\ \text{mol H} \times \frac{1.0\ \text{g H}}{1\ \text{mol H}} = 2.0\ \text{g H}$$

$$2\ \text{mol O} \times \frac{16.0\ \text{g O}}{1\ \text{mol O}} = 32.0\ \text{g O}$$

One mole of H_2O_2 has 2 mol of H atoms and 2 mol of O atoms, so multiply the molar mass of each element by 2.

Add the results.

mass of 1 mol H_2O_2 = 2.0 g H + 32.0 g O = 34.0 g
molar mass of H_2O_2 = 34.0 g/mol

3 Evaluate Does the result make sense? The answer is the sum of two times the molar mass of hydrogen and oxygen (17.0 g/mol). The answer is expressed to the tenths place because the numbers being added are expressed to the tenths place.

One mole of PCl_3 has 1 mol of P atoms and 3 mol of Cl atoms.

7. Find the molar mass of PCl_3.

8. What is the mass of 1.00 mol of sodium hydrogen carbonate?

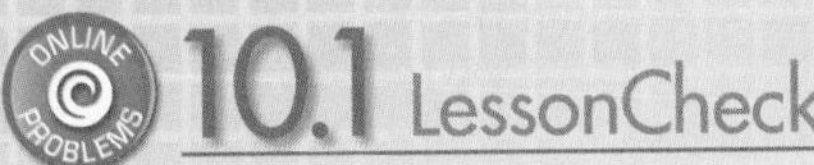

10.1 LessonCheck

9. Review What do you need to know to convert among the count, mass, and volume of something?

10. Describe How do chemists count the number of representative particles in a substance?

11. Explain How do you determine the molar mass of an element? How do you determine the molar mass of a compound?

12. Calculate If a dozen apples has a mass of 2.0 kg and 0.20 bushel is 1 dozen apples, how many bushels of apples are in 1.0 kg of apples?

13. Calculate How many moles is 1.50×10^{23} molecules of NH_3?

14. Calculate How many atoms are in 1.75 mol of $CHCl_3$?

15. Calculate What is the molar mass of $CaSO_4$?

Lesson Check Answers

9. You need a common unit, the mole.

10. Chemists use the mole to count the number of representative particles in a substance.

11. The molar mass of an element is the mass of a mole of the element. To calculate the molar mass of a compound, find the number of grams of each element in one mole of the compound. Then add the masses of the elements in the compound.

12. 0.10 bushel

13. 2.49×10^{-1} mol NH_3

14. 5.27×10^{24} atoms

15. 136.2 g/mol

Explain

Sample Practice Problems

A. Determine the mass of one mole of each of the following compounds: CO_2 *(44.0 g)*, SO_3 *(80.1 g)*, Br_2 *(159.8 g)*, H_2 *(2.0 g)*, N_2 *(28.0 g)*, NaOH *(40.0 g)*, $Al_2(SO_4)_3$ *(342.3 g)*, and $Ba(NO_3)_2$ *(261.3 g)*.

B. What is the mass, in grams, of 1.72 mol $CaCl_2$? *(191 g)*

Evaluate

Informal Assessment

Assess students' understanding of chemical quantities by writing the following three questions on the board and asking students to answer the questions quantitatively or qualitatively.

1. How many particles does one mole of any substance represent? *(6.02×10^{23} representative particles of that substance)*
2. How can you convert the number of atoms of any substance to moles? *(Multiply the number of atoms by the conversion factor 1 mole/6.02×10^{23} representative particles.)*
3. What is the molar mass of $Al_2(CO_3)_3$? *(234.0 g; also accept qualitative description of calculation)*

Then have students complete the 10.1 Lesson Check.

Reteach

Set up cooperative learning groups of three or four students with one student who is proficient in this topic. Have students review each sample problem in the lesson and create a list of parameters to follow for solving each type of question. Then provide the group with problems related to Lesson 10.1.

Answers

7. 1 mol P × (31.0 g P/1 mol P) = 31.0 g P
3 mol Cl × (35.5 g Cl/1 mol Cl) = 106.5 g Cl
31.0 g P + 106.5 g Cl = 137.5 g/mol

8. 1 mol Na × (23.0 g Na/1 mol Na) = 23.0 g Na
1 mol H × (1.0 g H/1 mol H) = 1.0 g H
1 mol C × (12.0 g C/1 mol C) = 12.0 g C
3 mol O × (16.0 g O/1 mol O) = 48.0 g O
23.0 g Na + 1.0 g H + 12.0 g C + 48.0 g O = 84.0 g

CHEMISTRY & YOU Have students read the text and study the figures. Help students understand the nature of a mole by posing the following question: *Why is the statement "a mole of carbon has the same number of particles as a mole of carbon dioxide" true?* You may need to assist students in the following ways:

- The number of particles in a mole is independent of mass.
- A particle of a substance may be a single atom or many atoms joined by covalent bonds, or it may even be a subatomic particle.

Explain

CRITICAL THINKING Students may wonder why a number such as a mole is used. Explain that there are certain numbers in science, called *fundamental constants*, each of which explains some property of nature. Challenge students to name other fundamental constants. Examples include the speed of light in a vacuum, $c = 3.00 \times 10^8$ m/s, and the elementary charge, $e = 1.602 \times 10^{-19}$ coulombs.

How Big Is a Mole?

The mole is an especially useful tool to chemists, because it allows them to express the number of representative particles of a substance in grams. For example, a 1 mol sample of carbon, which contains Avogadro's number of carbon atoms (6.02×10^{23}), has a mass of 12.0 g.

The mole is a huge quantity. Written out, Avogadro's number is 602,000,000,000,000,000,000,000. However, it may be difficult for you to comprehend exactly how big a mole is. Here are some interesting ways to visualize the size of a mole.

TOO BIG TO COUNT If you were able to count at the rate of 1 million numbers per second, it would take almost 20 billion years to count to 6.02×10^{23}.

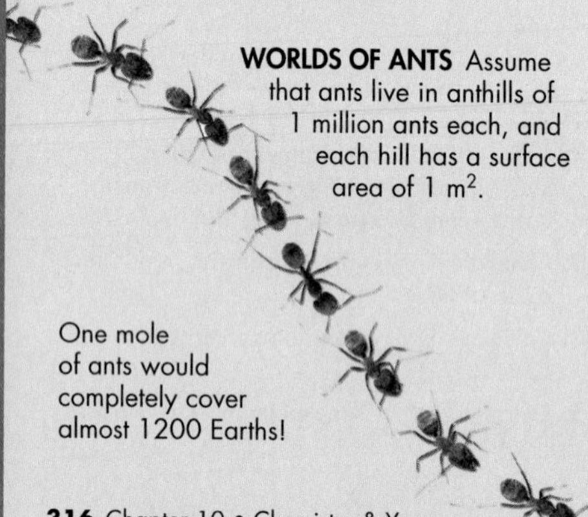

WORLDS OF ANTS Assume that ants live in anthills of 1 million ants each, and each hill has a surface area of 1 m^2.

One mole of ants would completely cover almost 1200 Earths!

Take It Further

1. **Calculate** Show how to calculate the number of years it would take to count to Avogadro's number if you could count at the rate of 1 million numbers per second.
2. **Use Models** Develop your own concept to illustrate the size of Avogadro's number. Show your calculations.
3. **Draw Conclusions** At home, using a food scale, measure out a mole of table sugar (sucrose, $C_{12}H_{22}O_{11}$) or a mole of table salt (sodium chloride, NaCl). What does this measurement tell you about the size of atoms and molecules?

Answers

TAKE IT FURTHER

1. 6.02×10^{23} numbers/(1×10^6 numbers/sec) = 6.02×10^{17} sec, 6.02×10^{17} sec × (1year/3.15×10^7 sec) = 1.91×10^{10} years
2. Answers will vary but students should show the conversion factors in each calculation.
3. The masses of atoms and molecules are not all the same.

21st Century Skills To be successful in the 21st century, students need skills and learning experiences that extend beyond subject matter mastery. The following project helps students build these 21st Century Skills: *Critical Thinking and Problem Solving, Creativity and Innovation, Communication & Collaboration, Information Literacy, Initiative and Self-Direction, and Productivity and Accountability.*

MODELING THE MOLE Divide students into small groups of 4–6 students. Pose the following challenge to students. *An intermediate school teacher has requested your group's assistance in explaining the mole to her eighth grade science class, using a creative and entertaining five-minute animation or a computerized slide show with audio.* The presentation should include multiple examples and demonstrate step-by-step calculations that support the examples. Students should submit their presentations on CD or DVD-ROM.

10.2 Mole–Mass and Mole–Volume Relationships

C.4.3 The mole concept

CHEMISTRY & YOU

Q: *How can you calculate the moles of a substance in a given mass or volume?* Guess how many pennies are in the container and win a prize! You decide to enter the contest, and you win. You estimated the thickness and diameter of a penny to find its approximate volume. Then you estimated the dimensions of the container to obtain its volume. You did the arithmetic and made your guess. In a similar way, chemists use the relationships between the mole and quantities such as mass, volume, and number of particles to solve problems in chemistry.

Key Questions

How do you convert the mass of a substance to the number of moles of the substance?

How do you convert the volume of a gas at STP to the number of moles of the gas?

Vocabulary

- Avogadro's hypothesis
- standard temperature and pressure (STP)
- molar volume

The Mole–Mass Relationship

How do you convert the mass of a substance to the number of moles of the substance?

In the previous lesson, you learned that the molar mass of any substance is the mass in grams of one mole of that substance. This definition applies to all substances—elements, molecular compounds, and ionic compounds. In some situations, however, the term *molar mass* may be unclear. For example, suppose you were asked for the molar mass of oxygen. How you answer this question depends on what you assume to be the representative particle. If you assume the oxygen in the question is molecular oxygen (O_2), then the molar mass is 32.0 g/mol (2 × 16.0 g/mol). If you assume that the question is asking for the mass of a mole of oxygen atoms (O), then the answer is 16.0 g/mol. You can avoid confusion such as this by using the formula of the substance, in this case, O_2 or O.

Suppose you need a given number of moles of a substance for a laboratory experiment. How can you measure this amount? Suppose instead that you obtain a certain mass of a substance in a laboratory experiment. How many moles is this? **Use the molar mass of an element or compound to convert between the mass of a substance and the moles of the substance.** The conversion factors for these calculations are based on the relationship: molar mass = 1 mol.

$$\frac{\text{molar mass}}{1\text{ mol}} \quad \text{and} \quad \frac{1\text{ mol}}{\text{molar mass}}$$

Focus on ELL

1 CONTENT AND LANGUAGE Read aloud the key questions. Explain that the word *convert* in each question is the verb form of *conversion*. Encourage students to write their own definitions for the word *convert* and then to compare their definitions to the definition in a dictionary. Allow beginning speakers to use symbols or their native language for this activity if desired.

2 FRONTLOAD THE LESSON Place three 100 mL beakers on a table. Fill the first beaker with 50 ml of pennies. Repeat with the remaining beakers, using dimes for one and quarters for the other. Ask students to identify the similarities and differences of each beaker. Have them predict how this demonstration might relate to the topics in this lesson.

3 COMPREHENSIBLE INPUT Direct students to Figure 10.8 and read aloud the caption. Have students brainstorm, plan, and construct a model of the mole roadmap. Tell students that their models can help them solve problems.

Key Objectives

10.2.1 DESCRIBE how to convert the mass of a substance to the number of moles of a substance, and moles to mass.

10.2.2 CONVERT the volume of a gas at STP to the number of moles of the gas.

Additional Resources

Reading and Study Workbook, Lesson 10.2

Available Online or on Digital Media:

- Teaching Resources, Lesson 10.2 Review
- Laboratory Manual, Lab 12
- Virtual Chemistry Laboratory Manual, Lab 3

PearsonChem.com

Engage

CHEMISTRY & YOU **Ask** What method, other than estimating the volume of each individual penny, might you use to determine the number of pennies present? *(Sample answer: Divide the total mass of all the pennies by the mass of one penny.)*

Activate Prior Knowledge

Review the mathematical conversions of moles to number of particles and number of particles to moles. Stress that using dimensional analysis in problem solving allows students to perform these calculations without having to memorize the process. Emphasize the use of writing units when solving problems.

IN ACADEMIC STANDARDS for SCIENCE

C.4.3 Use the mole concept to determine the number of moles and number of atoms or molecules in samples of elements and compounds, given mass of the sample.

Foundations for Reading

BUILD VOCABULARY Have students make a graphic organizer for this lesson by creating a KWL chart. In the first column, have students list what they know about each vocabulary term. In the second column, have them list what they want to know about each term. As they progress through the chapter, have them list in the third column what they learn about each term.

READING STRATEGY Instruct students to preview the key concepts, the section heads, the boldfaced items, the sample problems, and the figures. Have them write down a list of three things they expect to learn. After reading, have them state what they learned about each item they listed.

Explain

The Mole–Mass Relationship

MAKING CONNECTIONS Explain that in chemical manufacturing processes, reactants are purchased by mass, and products are sold by mass. However, point out that processes are set up and designed based on the ratio in which moles of reactants combine with each other to form moles of products. **Ask** If one mole of reactant produces one mole of product, predict how you can find the mass of product if you know the mass of the reactant. *(Convert the mass of reactant to number of moles. Then, convert that number of moles to the mass of product.)* Tell students that they will learn more about conversions such as this in Chapter 12.

Sample Practice Problem

Calculate the mass in grams for 0.750 mol of each of the following compounds:

a. cesium iodide (CsI) *(195 g)*

b. aluminum sulfate ($Al_2(SO_4)_3$) *(257 g)*

Converting Moles to Mass

Items made out of aluminum, such as aircraft parts and cookware, are resistant to corrosion because the aluminum reacts with oxygen in the air to form a coating of aluminum oxide (Al_2O_3). This tough, resistant coating prevents any further corrosion. What is the mass, in grams, of 9.45 mol of aluminum oxide?

1 Analyze List the known and the unknown. The mass of the compound is calculated from the known number of moles of the compound. The desired conversion is moles ⟶ mass.

KNOWN
number of moles = 9.45 mol Al_2O_3

UNKNOWN
mass = ? g Al_2O_3

2 Calculate Solve for the unknown.

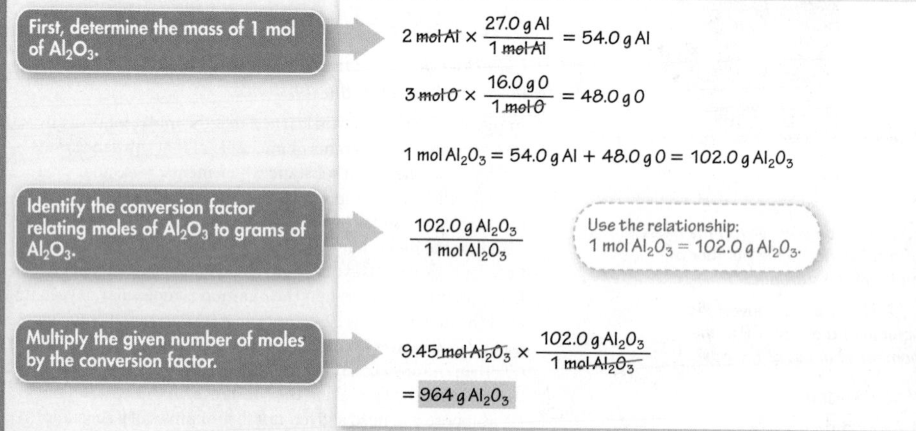

3 Evaluate Does the result make sense? The number of moles of Al_2O_3 is approximately 10, and each has a mass of approximately 100 g. The answer should be close to 1000 g. The answer has been rounded to the correct number of significant figures.

16. Find the mass, in grams, of 4.52×10^{-3} mol $C_{20}H_{42}$.

17. Calculate the mass, in grams, of 2.50 mol of iron(II) hydroxide.

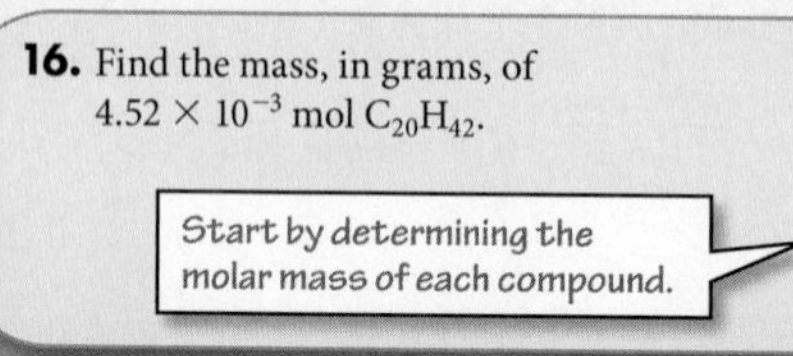

318 Chapter 10 • Lesson 2

Foundations for Math

SIGNIFICANT FIGURES Remind students that answers for calculations involving measurements should be written with the correct number of significant figures. Remind students that the following two rules apply:

1) Adding decimals: perform the operation, and then round the sum to match the measurement with the least number of decimal places.

2) Multiplying or dividing decimals: perform the operation, and then round the answer to match the measurement with the least number of significant figures.

In **Sample Problem 10.5,** the product is $9.45 \times 102.0 = 963.9$. Point out that the final answer is 964. The product is rounded to the nearest ones place because the calculation involved multiplication, and the multiplier with the least number of significant figures is 9.45. If necessary, have students refer to Chapter 3 for additional practice.

SampleProblem 10.6

Converting Mass to Moles

When iron is exposed to air, it corrodes to form red-brown rust. Rust is iron(III) oxide (Fe_2O_3). How many moles of iron(III) oxide are contained in 92.2 g of pure Fe_2O_3?

1 Analyze List the known and the unknown. The number of moles of the compound is calculated from the known mass of the compound. The conversion is mass ⟶ moles.

KNOWN
mass = 92.2 g Fe_2O_3

UNKNOWN
number of moles = ? mol Fe_2O_3

2 Calculate Solve for the unknown.

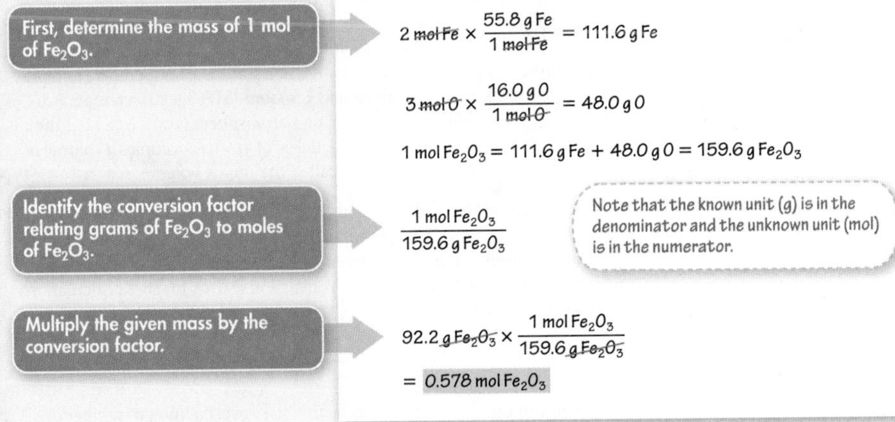

First, determine the mass of 1 mol of Fe_2O_3.

$$2\ \text{mol Fe} \times \frac{55.8\ \text{g Fe}}{1\ \text{mol Fe}} = 111.6\ \text{g Fe}$$

$$3\ \text{mol O} \times \frac{16.0\ \text{g O}}{1\ \text{mol O}} = 48.0\ \text{g O}$$

$$1\ \text{mol } Fe_2O_3 = 111.6\ \text{g Fe} + 48.0\ \text{g O} = 159.6\ \text{g } Fe_2O_3$$

Identify the conversion factor relating grams of Fe_2O_3 to moles of Fe_2O_3.

$$\frac{1\ \text{mol } Fe_2O_3}{159.6\ \text{g } Fe_2O_3}$$

Note that the known unit (g) is in the denominator and the unknown unit (mol) is in the numerator.

Multiply the given mass by the conversion factor.

$$92.2\ \text{g } Fe_2O_3 \times \frac{1\ \text{mol } Fe_2O_3}{159.6\ \text{g } Fe_2O_3} = 0.578\ \text{mol } Fe_2O_3$$

3 Evaluate Does the result make sense? The given mass (about 90 g) is slightly larger than the mass of one-half mole of Fe_2O_3 (about 80 g), so the answer should be slightly larger than one-half (0.5) mol.

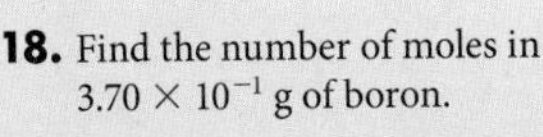

18. Find the number of moles in 3.70×10^{-1} g of boron.

19. Calculate the number of moles in 75.0 g of dinitrogen trioxide.

Again, start by determining the molar mass of each substance.

Explain

Sample Practice Problem

Calculate the number of moles in 4.00×10^2 g of each of the following compounds.

a. cesium iodide *(1.54 mol)*

b. aluminum sulfate *(1.17 mol)*

CRITICAL THINKING Have students examine the problem-solving process for Sample Problems 10.5 and 10.6. Have students create a T-chart to compare and contrast the process for converting between moles to mass and mass to moles. Have students write a "caution" sticky note listing what to look for, what mistakes to avoid, and what process to follow in these types of problems. Tell students to place the sticky note in the text near these sample problems.

Misconception Alert

Students may not understand the difference between moles and mass. Explain that in the types of problems shown in Sample Problems 10.5 and 10.6, mass is typically given in grams. Moles is the unit of measurement for the number of representative particles. Clearly emphasize that mass is not equivalent to moles. Suggest that students create a spider chart to explore the relationship between the mole, mass, volume, and number of particles.

Foundations for Math

ROUNDING IN MULTI-STEP PROBLEMS Explain to students that when a problem involves multiple operations, they should avoid rounding until the final answer. When numbers are rounded in the middle step of a problem and then again at the final step of the problem, the result could be quite different from the correct answer, especially when dealing with larger quantities.

In **Sample Problem 10.6,** the first calculation yields a molar mass of 159.6. Some students may incorrectly assume that since division follows, they should round this mass to 160. Show students that premature rounding affects the accuracy of the final answer: 92.2/160 = 0.576. Explain why the answer of 0.578 is more accurate than 0.576.

Answers

16. 4.52×10^{-3} mol $C_{20}H_{42}$ × (282.0 g $C_{20}H_{42}$/1 mol $C_{20}H_{42}$) = 1.27 g $C_{20}H_{42}$

17. 2.50 mol $Fe(OH)_2$ × (89.8 g $Fe(OH)_2$/1 mol $Fe(OH)_2$) = 225 g $Fe(OH)_2$

18. 3.70×10^{-1} g B × 0 (1 mol B/10.8 g B) = 3.43×10^{-2} mol B

19. 75.0 g N_2O_3 × (1 mol N_2O_3/76.0 g N_2O_3) = 0.987 mol N_2O_3

LESSON 10.2

Explain

The Mole–Volume Relationship

START A CONVERSATION Ask students to name the unit that is used for the mass of a mole. *(grams per mole, g/mol)* **Ask** What unit is used for the volume of a mole? *(Liters per mole, L/mol)* Point out that, unlike solids and liquids, the molar volume of gases is predictable, but it can be affected by temperature and pressure. **Ask** How does temperature affect the volume of a gas? *(When temperature increases, volume increases. When temperature decreases, volume decreases.)* **Ask** How does pressure affect the volume of a gas? *(When pressure increases, volume decreases. A decrease in pressure causes an increase in volume.)* Emphasize that when comparing the molar volumes of gases, it is necessary to have the gases at the same conditions of temperature and pressure.

CHEMISTRY & YOU To calculate the moles of a substance, divide the mass of the substance by the mass of one mole of the substance. To calculate the moles of a gas in a given volume at STP, divide the volume by 22.4 L.

The Mole–Volume Relationship

How do you convert the volume of a gas at STP to the number of moles of the gas?

CHEMISTRY & YOU

Q: *How can you calculate the moles of a substance in a given mass? How can you calculate the moles of a gas in a given volume at STP?*

Look back at Figure 10.6. Notice that the volumes of one mole of different solid and liquid substances are not the same. For example, the volumes of one mole of glucose (blood sugar) and one mole of paradichlorobenzene (moth crystals) are much larger than the volume of one mole of liquid water. What about the volumes of gases? Unlike liquids and solids, the volumes of moles of gases, measured under the same physical conditions, are much more predictable. Why is this?

Avogadro's Hypothesis In 1811, Amedeo Avogadro proposed a groundbreaking explanation. **Avogadro's hypothesis** states that equal volumes of gases at the same temperature and pressure contain equal numbers of particles. The particles that make up different gases are not the same size. However, the particles in all gases are so far apart that a collection of relatively large particles does not require much more space than the same number of relatively small particles. Whether the particles are large or small, large expanses of space exist between individual particles of gas, as shown in Figure 10.7.

The volume of a gas varies with a change in temperature or a change in pressure. Due to these variations with temperature and pressure, the volume of a gas is usually measured at a standard temperature and pressure. **Standard temperature and pressure (STP)** means a temperature of 0°C and a pressure of 101.3 kPa, or 1 atmosphere (atm). At STP, 1 mol, or 6.02×10^{23} representative particles, of any gas occupies a volume of 22.4 L. The quantity, 22.4 L, is called the **molar volume** of a gas.

Calculating the Volume and Moles of a Gas at STP The molar volume of a gas at STP is a useful quantity to chemists. **The molar volume is used to convert between the number of moles of gas and the volume of the gas at STP.** The conversion factors for these calculations are based on the relationship 22.4 L = 1 mol at STP.

$$\frac{22.4\ \text{L}}{1\ \text{mol}} \quad \text{and} \quad \frac{1\ \text{mol}}{22.4\ \text{L}}$$

You can use these conversion factors to convert a known number of moles of gas to the volume of the gas at STP. Similarly, you can convert a known volume of gas at STP to the number of moles of the gas.

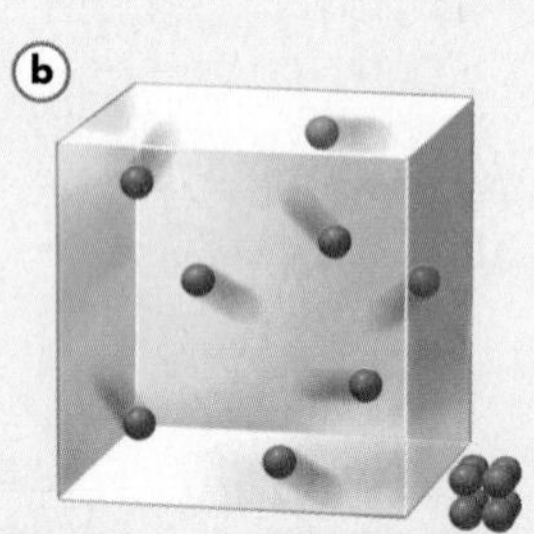

Figure 10.7 Volumes of Gases
In each container, the volume occupied by the gas molecules is small compared with the container's volume. **a.** The molecules in this container are small. **b.** This container can accommodate the same number of larger molecules.
Infer ***If the containers contained liquid molecules, and the molecules in container a were smaller than the molecules in container b, would both containers be able to accommodate the same number of molecules? Explain.***

Differentiated Instruction

LPR LESS PROFICIENT READERS Provide students with several problems and have them identify the known(s) and the unknown(s). Also have them look for key words, such as *total, difference,* and *larger,* that indicate operations or relationships.

L3 ADVANCED STUDENTS Have students apply their problem-solving skills to this question: Salvatore heated a mixture of potassium chlorate and manganese dioxide, producing 0.377 L of oxygen gas at STP. What was the mass of the gas collected? *(0.539 g)*

L1 SPECIAL NEEDS STUDENTS To help students with conversion problems, have them write each of the three pairs of conversion factors in Figure 10.8 on the two sides of an index card. Then have students create individual cards with each of the units of the known and unknown quantities: *grams, moles, liters,* and *representative particles*. Students can use the cards to model a problem as a method of testing their choices of conversion factors before solving the problem.

Sample Problem 10.7

Calculating Gas Quantities at STP

Sulfur dioxide (SO_2) is a gas produced by burning coal. It is an air pollutant and one of the causes of acid rain. Determine the volume, in liters, of 0.60 mol SO_2 gas at STP.

1 Analyze **List the knowns and the unknown.** Since SO_2 is a gas, the volume at STP can be calculated from the known number of moles.

KNOWNS
number of moles = 0.60 mol SO_2
1 mol SO_2 = 22.4 L SO_2 at STP

UNKNOWN
volume = ? L SO_2

2 Calculate **Solve for the unknown.**

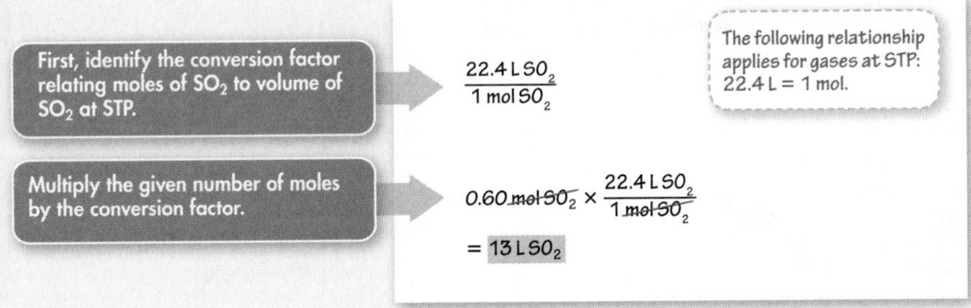

3 Evaluate **Does the result make sense?** One mole of any gas at STP has a volume of 22.4 L, so 0.60 mol should have a volume slightly larger than one half of a mole or 11.2 L. The answer should have two significant figures.

20. What is the volume of these gases at STP?
- **a.** 3.20×10^{-3} mol CO_2
- **b.** 3.70 mol N_2
- **c.** 0.960 mol CH_4

21. At STP, how many moles are in these volumes of gases?
- **a.** 67.2 L SO_2
- **b.** 0.880 L He
- **c.** 1.00×10^3 L C_2H_6

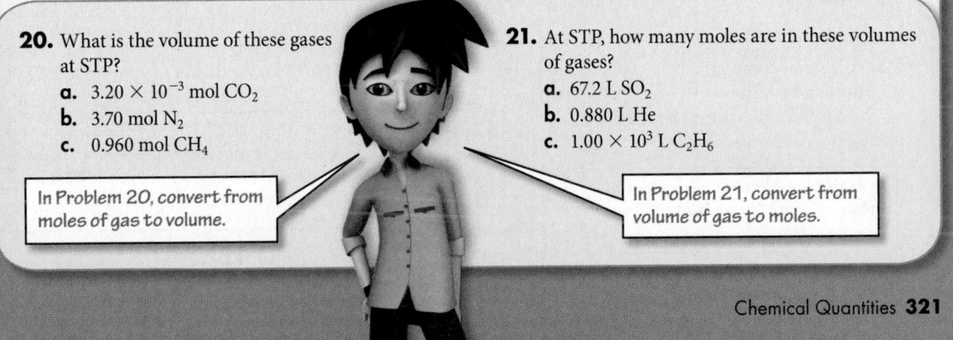

Foundations for Math

DIMENSIONAL ANALYSIS When solving problems using dimensional analysis, students sometimes lose track of their units. Remind students that writing the unknown at the beginning of their equation can serve as an extra checkpoint to verify that the final units are the ones desired before taking the final step to solve the problem.

In **Sample Problem 10.7,** point out that the units of the final element on the right-hand side of the equation are "L SO_2," which match what was asked for, the volume of SO_2.

Explore

Teacher Demo

PURPOSE Students will observe an approximation of molar volume.

MATERIALS Dry ice, towel, hammer, large plastic bag, duct tape, tongs, beaker, balance

PROCEDURE Wrap the dry ice in a towel and hit it with the hammer until it is in small pieces. Place 44 g (1 mol CO_2) of the small pieces in a beaker. Expel any air from the plastic bag, and tape the opening of the bag securely over the top of the beaker. As the dry ice sublimes, the bag will inflate.

SAFETY Wear goggles while crushing the dry ice, and do not allow dry ice to contact skin. Use tongs to handle the dry ice.

EXPECTED OUTCOMES The volume of gas produced will not equal 22.4 L because conditions are not standard. However, the volume will be close to this value.

Sample Practice Problem

At STP, what volume is occupied by each of the following gases?

- **a.** 6.7 mol H_2 *(1.5×10^2 L H_2)*
- **b.** 2.56 mol CH_4 *(5.73×10^1 L CH_4)*

Answers

FIGURE 10.7 No; container (a) would be able to accommodate more molecules than container (b).

20. **a.** 3.20×10^{-3} mol CO_2 × (22.4 L CO_2/1 mol CO_2) = 7.17×10^{-2} L CO_2
b. 3.70 mol N_2 × (22.4 L N_2/1 mol N_2) = 82.9 L N_2
c. 0.960 mol CH_4 × (22.4 L CH_4/1 mol CH_4) = 21.5 L CH_4

21. **a.** 67.2 L SO_2 × (1 mol SO_2/22.4 L SO_2) = 3.00 mol SO_2
b. 0.880 L He × (1 mol He/22.4 L He) = 0.0392 mol He
c. 1.00×10^3 L C_2H_6 × (1 mol C_2H_6/22.4 L C_2H_6) = 44.6 mol C_2H_6

Explain

The Mole-Volume Relationship

APPLY CONCEPTS Review the concept of density as a ratio of mass to volume. Discuss the units of measure that correspond to density. *(g/mL, g/cm^3, or g/L)* **Ask** If you had one mole of gas at STP, how could you calculate the density? *(Divide the number of grams of gas in one mole by 22.4 L.)* Emphasize that if the number of moles is given, the mass of the substance or the volume of a gas can be calculated. This concept will continue to be essential as students study mass–mass and other stoichiometric relationships in Chapter 12.

Sample Practice Problems

A. A gas has a density of 0.791 g/L. What is the molar mass of this gas? *(17.7 g/mol)*

B. What is the density of nitrogen gas at STP? *(1.25 g/L)*

Calculating Molar Mass and Density A gas-filled balloon will either sink or float in the air depending on whether the density of the gas inside the balloon is greater or less than the density of the surrounding air. Different gases have different densities. Usually the density of a gas is measured in grams per liter (g/L) and at a specific temperature. The density of a gas at STP and the molar volume at STP (22.4 L/mol) can be used to calculate the molar mass of the gas. Similarly, the molar mass of a gas and the molar volume at STP can be used to calculate the density of a gas at STP.

You have now examined a mole in terms of particles, mass, and volume of gases at STP. Figure 10.8 summarizes these relationships and illustrates the importance of the mole.

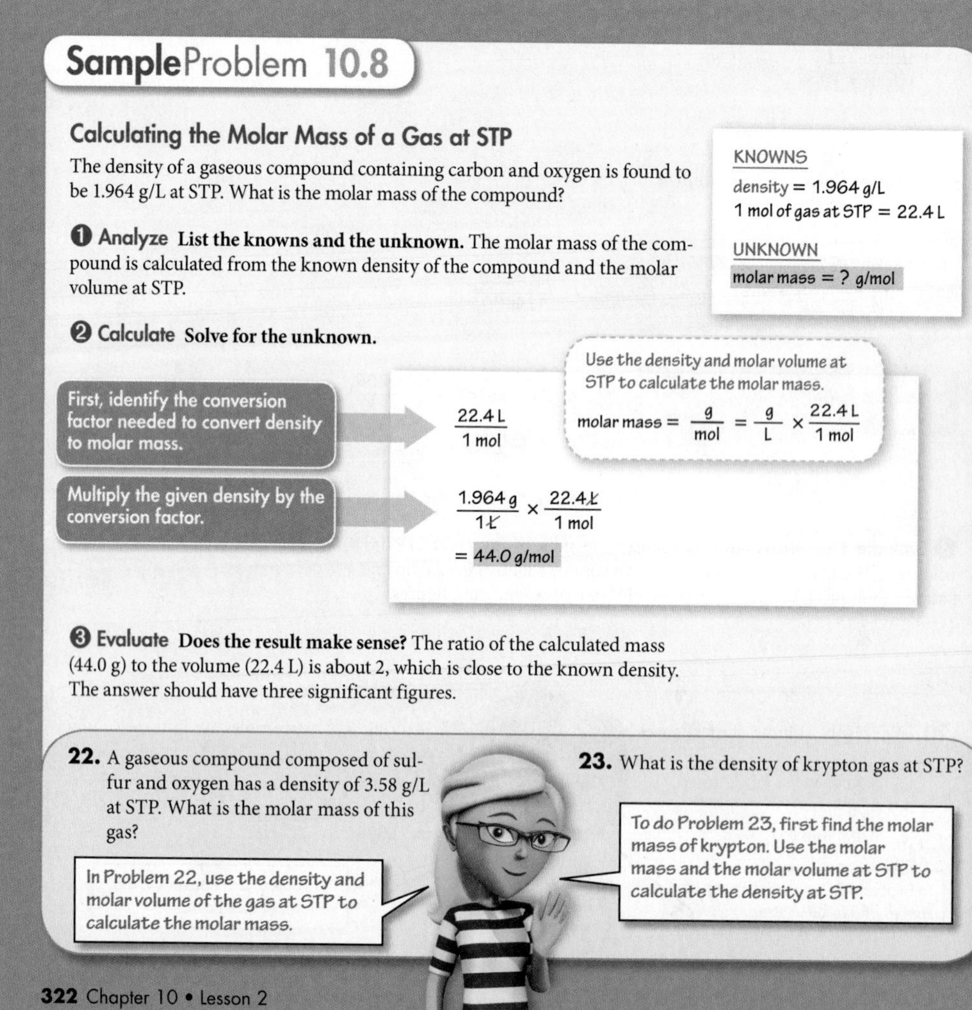

Sample Problem 10.8

Calculating the Molar Mass of a Gas at STP

The density of a gaseous compound containing carbon and oxygen is found to be 1.964 g/L at STP. What is the molar mass of the compound?

KNOWNS
density = 1.964 g/L
1 mol of gas at STP = 22.4 L

UNKNOWN
molar mass = ? g/mol

1 Analyze List the knowns and the unknown. The molar mass of the compound is calculated from the known density of the compound and the molar volume at STP.

2 Calculate Solve for the unknown.

Use the density and molar volume at STP to calculate the molar mass.

$$\text{molar mass} = \frac{g}{mol} = \frac{g}{L} \times \frac{22.4\ L}{1\ mol}$$

First, identify the conversion factor needed to convert density to molar mass.

$$\frac{22.4\ L}{1\ mol}$$

Multiply the given density by the conversion factor.

$$\frac{1.964\ g}{1\ \cancel{L}} \times \frac{22.4\ \cancel{L}}{1\ mol} = 44.0\ g/mol$$

3 Evaluate Does the result make sense? The ratio of the calculated mass (44.0 g) to the volume (22.4 L) is about 2, which is close to the known density. The answer should have three significant figures.

22. A gaseous compound composed of sulfur and oxygen has a density of 3.58 g/L at STP. What is the molar mass of this gas?

23. What is the density of krypton gas at STP?

In Problem 22, use the density and molar volume of the gas at STP to calculate the molar mass.

To do Problem 23, first find the molar mass of krypton. Use the molar mass and the molar volume at STP to calculate the density at STP.

322 Chapter 10 • Lesson 2

Focus on ELL

4 LANGUAGE PRODUCTION Have students work in groups of four to complete the Small-Scale Lab on page 324. Make sure each group has ELLs of varied language proficiencies, so that more proficient students can help less proficient ones. Consider allowing students extra time to complete the Analyze section of the lab.

BEGINNING: LOW/HIGH Provide students with a chart to record their findings. Label the rows with the problem numbers from the Analyze and Conclude section of the lab.

INTERMEDIATE: LOW/HIGH Model the procedure for measuring the mass of sodium chloride for students. Have students mimic you to measure the mass for water and calcium carbonate.

ADVANCED: LOW/HIGH Have students assist students with lower English proficiencies with setting up the calculations and examining the results.

Figure 10.8 Mole Roadmap
The mole is at the center of your chemical calculations. To convert from one unit to another, you must use the mole as an intermediate step. The form of the conversion factor depends on what you know and what you want to calculate.
Interpret Diagrams ***How many conversion factors are needed to convert from the mass of a gas to the volume of a gas at STP?***

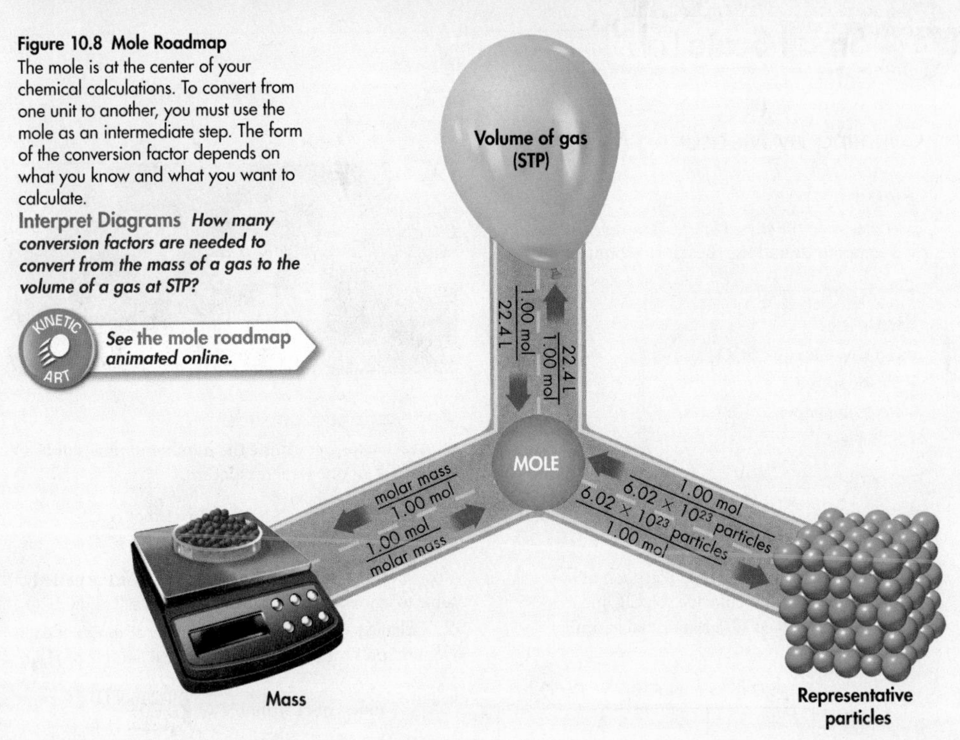

10.2 LessonCheck

ONLINE PROBLEMS

24. **Describe** How do you convert between the mass and the number of moles of a substance?

25. **Describe** How do you convert between the volume of a gas at STP and the number of moles of the gas?

26. **Calculate** How many grams are in 5.66 mol of $CaCO_3$?

27. **Calculate** Find the number of moles in 508 g of ethanol (C_2H_6O).

28. **Calculate** What is the volume, in liters, of 1.50 mol Cl_2 at STP?

29. **Apply Concepts** Three balloons filled with three different gaseous compounds each have a volume of 22.4 L at STP. Do these balloons have the same mass or contain the same number of molecules? Explain.

30. **Calculate** The density of an elemental gas is 1.7824 g/L at STP. What is the molar mass of the element?

31. **Analyze Data** The densities of gases A, B, and C at STP are 1.25 g/L, 2.86 g/L, and 0.714 g/L, respectively. Calculate the molar mass of each substance. Identify each substance as ammonia (NH_3), sulfur dioxide (SO_2), chlorine (Cl_2), nitrogen (N_2), or methane (CH_4).

BIGIDEA
THE MOLE AND QUANTIFYING MATTER

32. A chemist collects 2.94 L of carbon monoxide (CO) gas at STP during an experiment. Explain how she can determine the mass of gas that she collected. Why is the mole important for this calculation?

Lesson Check Answers

24. To convert mass to moles, multiply the given mass by 1 mol/molar mass. To convert moles to mass, multiply the given number of moles by molar mass/1 mol.

25. Divide the volume by 22.4 L.

26. 567 g $CaCO_3$

27. 11.0 mol C_2H_6O

28. 33.6 L Cl_2

29. The balloons have the same number of molecules. Each balloon is filled with one mole of gas, and one mole of any gas has the same number of molecules. The masses of the balloons will probably differ.

30. 39.9 g/mol

31. gas A: 28.0 g, nitrogen; gas B: 64.1 g, sulfur dioxide; gas C: 16.0 g, methane

32. **BIGIDEA** She needs to convert the volume of the gas to moles. Then she can covert moles to mass. She cannot determine mass without knowing moles.

Explain

The Mole Road Map

USE VISUALS Have students study Figure 10.8. Guide them through examples of the various mole conversions. For example, start with 50.0 g of a compound or element and convert it to moles and then to particles. Then, start with a given volume of a gas and convert it to mass or particles. Have students recreate this road map in a visual way that helps them to clearly see the various relationships. Encourage students to use their visual as a reference when attempting homework problems.

Evaluate

Informal Assessment

Have students solve problems in which they use or explain the conversions for converting mass or volume to moles. Ask volunteers to come to the board to show their calculations. Have the class determine whether the calculations are correct. Then have students complete the 10.2 Lesson Check.

Reteach

Review the concept of density as a ratio of mass to volume. **Ask** If you know the molar volume of a gas, how could density help you determine the molar mass? *(Molar mass is the product of density and molar volume.)*

Answers

SAMPLE PROBLEM 10.8

22. 3.58 g/1 L × (22.4 L/1 mol) = 80.2 g/mol

23. 83.8 g/1 mol × (1 mol/22.4 L) = 3.74 g/L

FIGURE 10.8 You need to convert the mass of the gas to the moles of gas. Then convert moles to volume, so you need to use two conversion factors.

SMALL-SCALE LAB

Small-Scale Lab

OBJECTIVE After completing this activity, students will be able to measure masses of chemicals and convert their data to moles and atoms. They will also be able to explore the quantitative chemical compositions of common substances.

SKILLS FOCUS measuring, calculating

PREP TIME 20 minutes

CLASS TIME 40 minutes

TEACHING TIPS Explore with students ways of finding the mass of liquid and solid samples so that the container does not interfere with the measurement. If time allows, have students repeat the procedure and average the data.

EXPECTED OUTCOME See data table.

ANALYZE AND CONCLUDE

Sample calculations using sample data:

1. See data table for answers.
2. See data table for answers.
3. 0.478 mol H × 6.02 × 10^{23} atoms/1 mol H = 2.88 10^{23} atoms H
4. water
5. water

YOU'RE THE CHEMIST

1. Sample answer: Find the mass of 100 drops of water, and then calculate the mass in grams per drop.
2. Sample answer: Find the mass of a piece of chalk. Write your name and find the mass of the chalk again. Convert the mass difference to moles and atoms.

FOR ENRICHMENT Have students use their results from the lab to calculate the volume of one mole of each substance tested. Then, have them use a balance to measure one mole of each substance and a graduated cylinder to find its volume. Have them compare the calculated and experimental values and discuss any discrepancies.

Counting by Measuring Mass

Purpose

To determine the mass of several samples of chemical compounds and use the data to count atoms

Materials

- H_2O, NaCl, and $CaCO_3$
- plastic spoon
- weighing paper
- balance

Procedure 

Measure the mass of one level teaspoon of water (H_2O), sodium chloride (NaCl), and calcium carbonate ($CaCO_3$). Make a table similar to the one below.

	$H_2O(l)$	NaCl(*s*)	$CaCO_3(s)$
Mass (g)			
Molar mass (g/mol)			
Moles of each compound			
Moles of each element			
Atoms of each element			

Analyze and Conclude

1. Calculate Determine the number of moles of H_2O contained in one level teaspoon.

$$\text{moles of } H_2O = \text{g } H_2O \times \frac{1 \text{ mol } H_2O}{18.0 \text{ g } H_2O}$$

Repeat for the remaining compounds. Use the periodic table to calculate the molar masses of NaCl and $CaCO_3$.

2. Calculate Determine the number of moles of each element present in the teaspoon-sized sample of H_2O.

$$\text{moles of H} = \text{mol } H_2O \times \frac{2 \text{ mol H}}{1 \text{ mol } H_2O}$$

Repeat for the other compounds in your table.

3. Calculate Determine the number of atoms of each element present in the teaspoon-sized sample of H_2O.

$$\text{atoms of H} = \text{mol H} \times \frac{6.02 \times 10^{23} \text{ atoms H}}{1 \text{ mol H}}$$

Repeat for the other compounds in your table.

4. Analyze Data Which of the three teaspoon-sized samples contains the greatest number of moles of molecules or formula units?

5. Analyze Data Which of the three compounds contains the greatest number of atoms?

You're the Chemist

1. Design an Experiment Can you count by measuring volume? Design and carry out an experiment to do it.

2. Design an Experiment Design an experiment that will determine the number of atoms of calcium, carbon, and oxygen it takes to write your name on the chalkboard with a piece of chalk. Assume chalk is 100 percent calcium carbonate, $CaCO_3$.

Sample Data

	$H_2O(l)$	NaCl(*s*)	$CaCO_3(s)$
Mass (g)	4.30	5.09	9.68
Molar mass (g/mol)	18.0	58.5	100.1
Moles of compound	0.239	0.0870	0.0967
Moles of elements	0.239 O 0.478 H	0.0870 Na 0.0870 Cl	0.0967 Ca 0.0967 C 0.290 O
Atoms of elements	1.44 × 10^{23} O 2.88 × 10^{23} H	5.24 × 10^{22} Na 5.24 × 10^{22} Cl	5.82 × 10^{22} Ca 5.82 × 10^{22} C 1.75 × 10^{23} O

10.3 Percent Composition and Chemical Formulas

C.4.3 The mole concept; C.4.7 Percent composition, empirical formulas, and molecular formulas

CHEMISTRY & YOU

Q: *What does the percent composition of a compound tell you?* A tag sewn into the seam of a shirt usually tells you what fibers were used to make the cloth and the percent of each. It helps to know the percents of the components in the shirt because they affect how warm the shirt is, whether it will need to be ironed, and how it should be cleaned. Similarly, in chemistry it is important to know the percents of the elements in a compound.

Key Questions

How do you calculate the percent composition of a compound?

How can you calculate the empirical formula of a compound?

How does the molecular formula of a compound compare with the empirical formula?

Vocabulary

- percent composition
- empirical formula

Percent Composition of a Compound

How do you calculate the percent composition of a compound?

In lawn care, the relative amount, or the percent, of each nutrient in fertilizer is important. In spring, you may use a fertilizer that has a high percent of nitrogen to "green" the grass. In fall, you may want to use a fertilizer with a higher percent of potassium to strengthen the root system. Knowing the relative amounts of the components of a mixture or compound is often useful.

The relative amounts of the elements in a compound are expressed as the **percent composition** or the percent by mass of each element in the compound. As shown in Figure 10.9, the percent composition of potassium chromate, K_2CrO_4, is K = 40.3%, Cr = 26.8%, and O = 32.9%. These percents must total 100% (40.3% + 26.8% + 32.9% = 100%). The percent composition of a compound is always the same.

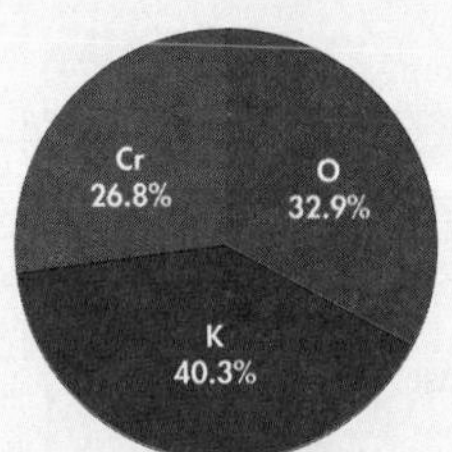

Potassium chromate, K_2CrO_4

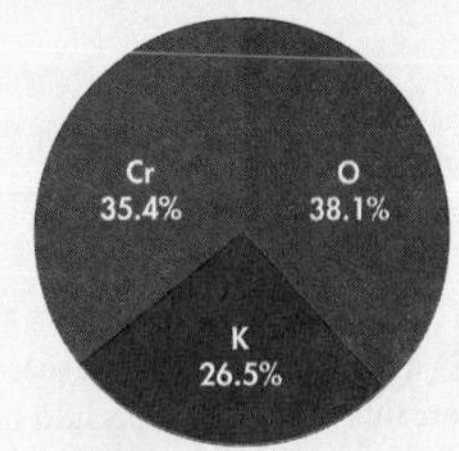

Potassium dichromate, $K_2Cr_2O_7$

Figure 10.9 Percent Composition
Potassium chromate (K_2CrO_4) is composed of 40.3% potassium, 26.8% chromium, and 32.9% oxygen.
Compare *How does this percent composition differ from the percent composition of potassium dichromate ($K_2Cr_2O_7$), a compound composed of the same three elements?*

Key Objectives

10.3.1 CALCULATE the percent composition of a compound.

10.3.2 CALCULATE the empirical formula of a compound.

10.3.3 DISTINGUISH between empirical and molecular formulas.

Additional Resources

Reading and Study Workbook, Lesson 10.3

Available Online or on Digital Media:

- Teaching Resources, Lesson 10.3 Review, Interpreting Graphics
- Laboratory Manual, Lab 13
- Virtual Chemistry Laboratory Manual, Labs 25, 26, and 29

PearsonChem.com

Engage

CHEMISTRY & YOU Have students study the photograph. **Ask** What should be the total percent of materials listed on the label? *(100%)* **Ask** If the shirt were made of 25% polyester, instead of 15%, what percent of the shirt would be made of cotton? *(75%)*

Activate Prior Knowledge

Review the law of definite proportions and the concept of lowest whole-number ratios. Tell students to keep these concepts in mind as they proceed with the lesson.

Focus on ELL

1 CONTENT AND LANGUAGE Ask students if they have heard the word *composition* in their English class. Discuss the similarities and differences between the meaning of this word in English class versus the meaning in chemistry class.

2 FRONTLOAD THE LESSON As a class, determine the percentage of students in the room who are freshmen, sophomores, juniors, and seniors. Remind students that a percentage is determined by writing the part over the total and then multiplying by 100. Explain that they have calculated the percent composition of the class. Ask students to use what they learned from this activity to predict how the percent composition of a chemical compound is calculated.

3 COMPREHENSIBLE INPUT Explain Figure 10.9 and restate information in its caption as needed. Read the questions aloud, and provide students with a double bubble for them to write the similarities and differences between the two compounds.

IN ACADEMIC STANDARDS for SCIENCE

- Develop explanations based on reproducible data and observations gathered during laboratory investigations.
- Clearly communicate their ideas and results of investigations verbally and in written form using tables, graphs, diagrams, and photographs.
- Apply standard techniques in laboratory investigations to measure physical quantities in appropriate units and convert known quantities to other units as necessary.

Also covered C.4.3, C.4.7

LESSON 10.3

Foundations for Reading

BUILD VOCABULARY Have students read the definition of *percent composition* above Figure 10.9 and have them define the term using the definitions of *percent* and *composition* in the dictionary. Have students use the same method to define *empirical formula*.

READING STRATEGY As they read the chapter, students should examine each visual as it is referenced in the text. Have them read each caption and answer any question.

Explain

Percent Composition of a Compound

USE VISUALS Have students study the circle graphs in Figure 10.9 and read the text on percent composition. Point out that the percentages given in the three sectors add up to a total of 100%. **Ask** Which compound is a better source of potassium? Why? *(K_2CrO_4, because it has a greater percentage of potassium than $K_2Cr_2O_7$).*

Sample Practice Problem

What is the percent composition of the compound formed when 2.72 g of potassium react completely with 2.48 grams of chlorine to form 5.20 grams of potassium chloride? *(52.4% K, 47.6% Cl)*

MAKE A CONNECTION Percents are used to show relative parts of mixtures as well as the composition of a compound. But when an extremely small amount of a substance is present in a large amount of another substance, it might not be practical to use percents (parts per one hundred) to show the makeup of the mixture. Instead, concentrations of extremely dilute solutions are sometimes measured in units of parts per million (ppm) or parts per billion (ppb). For example, the composition of a mixture that consists of 1 gram of a substance per 10^6 grams of water (or 1 milligram of substance per liter of water) can be expressed as 1 ppm.

Percent Composition From Mass Data If you know the relative masses of each element in a compound, you can calculate the percent composition of the compound. **The percent by mass of an element in a compound is the number of grams of the element divided by the mass in grams of the compound, multiplied by 100%.**

$$\% \text{ by mass of element} = \frac{\text{mass of element}}{\text{mass of compound}} \times 100\%$$

CHEM TUTOR

Sample Problem 10.9

Calculating Percent Composition From Mass Data

When a 13.60-g sample of a compound containing only magnesium and oxygen is decomposed, 5.40 g of oxygen is obtained. What is the percent composition of this compound?

1 Analyze **List the knowns and the unknowns.** The percent by mass of an element in a compound is the mass of that element divided by the mass of the compound multiplied by 100%.

KNOWNS
mass of compound = 13.60 g
mass of oxygen = 5.40 g O
mass of magnesium = 13.60 g − 5.40 g O = 8.20 g Mg

UNKNOWNS
percent by mass of Mg = ?% Mg
percent by mass of O = ?% O

2 Calculate **Solve for the unknowns.**

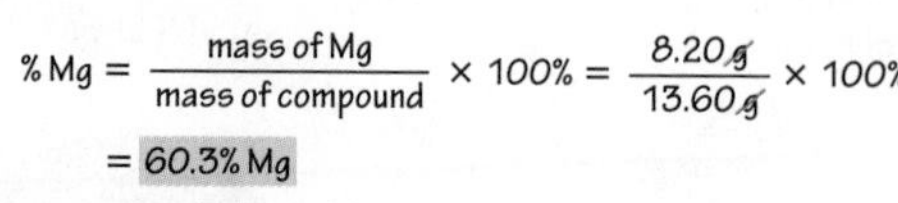

Determine the percent by mass of Mg in the compound.

$$\% \text{Mg} = \frac{\text{mass of Mg}}{\text{mass of compound}} \times 100\% = \frac{8.20 \cancel{g}}{13.60 \cancel{g}} \times 100\%$$

$$= 60.3\% \text{ Mg}$$

Determine the percent by mass of O in the compound.

$$\% \text{O} = \frac{\text{mass of O}}{\text{mass of compound}} \times 100\% = \frac{5.40 \cancel{g}}{13.60 \cancel{g}} \times 100\%$$

$$= 39.7\% \text{ O}$$

3 Evaluate **Does the result make sense?** The percents of the elements add up to 100%.

60.3% + 39.7% = 100%

In Problem 34, calculate the percent by mass of mercury and oxygen in the compound.

33. A compound is formed when 9.03 g Mg combines completely with 3.48 g N. What is the percent composition of this compound?

34. When a 14.2-g sample of mercury(II) oxide is decomposed into its elements by heating, 13.2 g Hg is obtained. What is the percent composition of the compound?

326 Chapter 10 • Lesson 3

Foundations for Math

PERCENTS Remind students that *percent* means "per 100." When you divide a part by a whole, the result is a decimal or fraction. If the result is written as a decimal, it can be converted to a percent by multiplying the decimal by 100. If the result is a fraction, such as ½, divide the numerator by the denominator to get a decimal. Then, multiply the decimal by 100. Note that the decimal is multiplied by 100%, which has a decimal equivalent of 1 (100% = 100/100 = 1).

In **Sample Problem 10.9,** 8.20/13.6 = 0.603 and 5.40/13.6 = 0.397. Convert these decimals to percents by multiplying each decimal by 100: 60.3%, 39.7%. Notice that 60.3% + 39.7% = 100%.

Percent Composition From the Chemical Formula You can also calculate the percent composition of a compound using its chemical formula. The subscripts in the formula are used to calculate the mass of each element in a mole of that compound. Using the individual masses of the elements and the molar mass, you can calculate the percent by mass of each element.

$$\%\text{ by mass of element} = \frac{\text{mass of element in 1 mol compound}}{\text{molar mass of compound}} \times 100\%$$

Sample Problem 10.10

Calculating Percent Composition From a Formula

Propane (C_3H_8), the fuel commonly used in gas grills, is one of the compounds obtained from petroleum. Calculate the percent composition of propane.

1 Analyze **List the knowns and the unknowns.** Calculate the percent by mass of each element by dividing the mass of that element in one mole of the compound by the molar mass of the compound and multiplying by 100%.

KNOWNS

mass of C in 1 mol C_3H_8 = 3 mol × 12.0 g/mol = 36.0 g

mass of H in 1 mol C_3H_8 = 8 mol × 1.0 g/mol = 8.0 g

molar mass of C_3H_8 = 36.0 g/mol + 8.0 g/mol = 44.0 g/mol

UNKNOWNS

percent by mass of C = ?% C

percent by mass of H = ?% H

2 Calculate **Solve for the unknowns.**

Determine the percent by mass of C in C_3H_8.

$$\%C = \frac{\text{mass of C in 1 mol } C_3H_8}{\text{molar mass of } C_3H_8} \times 100\% = \frac{36.0\text{ g}}{44.0\text{ g}} \times 100\%$$

= 81.8% C

Determine the percent by mass of H in C_3H_8.

$$\%H = \frac{\text{mass of H in 1 mol } C_3H_8}{\text{molar mass of } C_3H_8} \times 100\% = \frac{8.0\text{ g}}{44.0\text{ g}} \times 100\%$$

= 18% H

3 Evaluate **Does the result make sense?** The percents of the elements add up to 100% when the answers are expressed to two significant figures (82% + 18% = 100%).

35. Calculate the percent by mass of nitrogen in these fertilizers.
a. NH_3
b. NH_4NO_3

36. Calculate the percent composition of these compounds.
a. ethane (C_2H_6)
b. sodium hydrogen sulfate ($NaHSO_4$)

Foundations for Math

SUM OF THE PARTS Caution students against taking a shortcut when working percentage problems. Since the sum of the percents should be 100%, many students only calculate one of the percents and then simply subtract from 100%. Stress that students should calculate each percent individually and then find the sum to check their answer.

In **Sample Problem 10.10,** the sum of the percents is 100% after rounding both percents to two significant figures.

Sample Practice Problem

Determine the percent composition of the following oxides:

a. Fe_2O_3 *(69.9% Fe, 30.1% O)*
b. HgO *(92.6% Hg, 7.39% O)*
c. Ag_2O *(93.1% Ag, 6.90% O)*
d. Na_2O *(74.2% Na, 25.8% O)*

Extend

Challenge students to develop a spreadsheet that can calculate the percent composition of a chemical compound. Tell students they should consider how to manage the initial inputs, such as the chemical formula and the molar masses of each element in the formula, as well as the formulas needed to complete their final calculations. Encourage students to demonstrate and explain their spreadsheets to interested students.

Answers

FIGURE 10.9 The percent composition of $K_2Cr_2O_7$ is 26.5% K, 35.4% Cr, and 38.1% O.

33. Mass of compound = 9.03 g + 3.48 g = 12.51 g; (9.03 g Mg/12.51 g compound) × 100% = 72.2% Mg; (3.48 g N/12.51 g compound) × 100% = 27.8% N

34. Mass of O = 14.2 g − 13.2 g = 1.0 g O; (1.0 g O/14.2 g) × 100% = 7.0% O; (13.2 g Hg/14.2 g) × 100% = 93.0% Hg

35. **a.** (14.0 g N/17.0 g) × 100% = 82.4% N
b. (28.0 g N/80.0 g) × 100% = 35.0% N

36. **a.** (24.0 g C/30.0 g) × 100% = 80.0% C
(6.00 g H/30.0 g) × 100% = 20.0% H
b. (23.0 g Na/120.1 g) × 100% = 19.2% Na
(1.0 g H/120.1 g) × 100% = 0.83% H
(32.1 g S/120.1 g) × 100% = 26.7% S
(64.0 g O/120.1 g) × 100% = 53.3% O

Explore

Quick Lab

OBJECTIVE After completing this activity, students will be able to determine the percent of water in a hydrate.

SKILLS FOCUS observing, calculating

PREP TIME 20 minutes

CLASS TIME 30 minutes

SAFETY Students should wear safety goggles and tie back loose hair. Caution students that while heating test tubes, they should not aim the opening of the tube toward anyone. Tell them to move the test tube in the flame and not to heat one spot excessively. CAUTION! Be sure that students allow the tubes to cool completely before they touch them. Hot glass looks exactly like cold glass!

TEACHING TIPS For best results, students should do a second heating and cooling of each sample to determine whether all of the water has been driven off.

EXPECTED OUTCOME See the data table at the bottom of the page.

ANALYZE AND CONCLUDE

1–3. See the data table.

4. The hydrated salt of sodium sulfate lost the greatest percent. The hydrated salt of calcium chloride lost the smallest percent.

FOR ENRICHMENT Have students design and conduct a similar experiment to determine the percent of oxygen in potassium chlorate. Tell students that when potassium chlorate is heated, potassium chloride and oxygen are produced: $2KClO_3 \rightarrow 2KCl + 3O_2$. For classroom safety, no more than 5 g of potassium chlorate should be used. Results should show that potassium chlorate is approximately 39% oxygen.

Quick Lab

Percent Composition

Purpose To measure the percent of water in a series of crystalline compounds called hydrates

Materials

- 3 medium-sized test tubes
- balance
- spatula
- hydrated compounds of copper(II) sulfate, calcium chloride, and sodium sulfate
- test tube holder
- gas burner

Procedure

1. Label each test tube with the name of a compound. Measure and record the masses.
2. Add 2–3 g of each compound (a good-sized spatula full) to the appropriately labeled test tube. Measure and record the mass of each test tube and the compound.
3. Using a test tube holder, hold one of the tubes at a 45° angle and gently heat its contents over the burner, slowly passing it in and out of the flame. Note any change in the appearance of the solid compound.
4. As moisture begins to condense in the upper part of the test tube, gently heat the entire length of the tube. Continue heating until all of the moisture is driven from the tube. This process may take 2–3 minutes. Repeat Steps 3 and 4 for the other two tubes.
5. Allow each tube to cool. Then measure and record the mass of each test tube and the heated compound.

Analyze and Conclude

1. **Organize Data** Set up a data table so that you can subtract the mass of the empty tube from the mass of the compound and the test tube, both before and after heating.
2. **Calculate** Find the difference between the mass of each compound before and after heating. This difference represents the amount of water lost by the hydrated compound due to heating.
3. **Calculate** Determine the percent by mass of water lost by each compound.
4. **Analyze Data** Which compound lost the greatest percent by mass of water? Which compound lost the smallest percent by mass of water?

Q: *What information can you get from the percent composition of a compound?*

Percent Composition as a Conversion Factor You can use percent composition to calculate the number of grams of any element in a specific mass of a compound. To do this, multiply the mass of the compound by a conversion factor based on the percent composition of the element in the compound. In Sample Problem 10.10, you found that propane is 81.8 percent carbon and 18 percent hydrogen. That means that in a 100-g sample of propane, you would have 81.8 g of carbon and 18 g of hydrogen. You can use the following conversion factors to solve for the mass of carbon or hydrogen contained in a specific amount of propane.

$$\frac{81.8\text{ g C}}{100\text{ g }C_3H_8} \quad \text{and} \quad \frac{18\text{ g H}}{100\text{ g }C_3H_8}$$

Sample Data

	$CuSO_4 \cdot 5H_2O$	$CaCl_2 \cdot 2H_2O$	$Na_2SO_4 \cdot 10H_2O$
Test tube + hydrate (before heating)	23.88 g	23.60 g	23.92 g
Empty test tube	21.19 g	21.25 g	21.17 g
Mass of hydrate	2.69 g	2.35 g	2.75 g
Test tube + salt (after heating)	22.88 g	23.07 g	22.71 g
Empty test tube	21.19 g	21.25 g	21.17 g
Mass of anhydrous salt	1.69 g	1.82 g	1.54 g
Mass of water lost	1.00 g	0.53 g	1.21 g
Percent water (experimental)	37.2%	22.6%	44.0%
Percent water (theoretical)	36.1%	24.5%	55.9%

Sample Problem 10.11

Calculating the Mass of an Element in a Compound Using Percent Composition

Calculate the mass of carbon and the mass of hydrogen in 82.0 g of propane (C_3H_8).

1 Analyze List the known and the unknowns. Use the conversion factors based on the percent composition of propane to make the following conversions: grams $C_3H_8 \longrightarrow$ grams C and grams $C_3H_8 \longrightarrow$ grams H.

KNOWN	UNKNOWNS
mass of C_3H_8 = 82.0 g	mass of carbon = ? g C
	mass of hydrogen = ? g H

2 Calculate Solve for the unknowns.

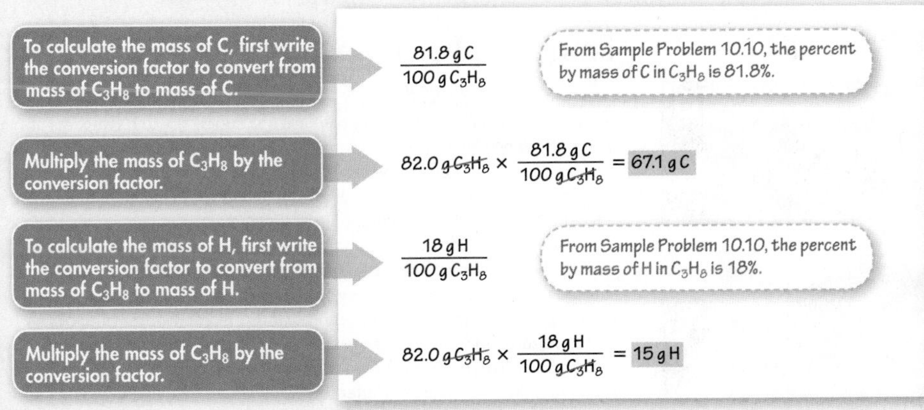

3 Evaluate Does the result make sense? The sum of the two masses equals 82 g, the sample size, to two significant figures (67 g C + 15 g H = 82 g C_3H_8).

37. Calculate the grams of nitrogen in 125 g of each fertilizer.
a. NH_3
b. NH_4NO_3

38. Calculate the mass of hydrogen in each of the following compounds:
a. 350 g ethane (C_2H_6)
b. 20.2 g sodium hydrogen sulfate ($NaHSO_4$)

LESSON 10.3

CHEMISTRY & YOU You can calculate the number of grams of any element in a specific mass of a compound.

Explain

Sample Practice Problem

Calculate the percent composition of H_2SO_4. Then calculate the masses of hydrogen, sulfur, and oxygen in 90.0 g of H_2SO_4. *(2.0% hydrogen, 32.7% sulfur, 65.3% oxygen; 1.8 grams hydrogen, 29.4 grams sulfur, 58.8 grams oxygen)*

Focus on ELL

4 LANGUAGE PRODUCTION Have students work in groups of four to complete the lab. Make sure each group has ELLs of varied language proficiencies, so that more proficient students can model the procedure for less proficient ones. Have students work collaboratively to create anchor charts with challenging words they encounter, along with their meanings

BEGINNING

LOW Model the steps of the procedures for students.

HIGH Paraphrase the steps of the procedure and break long steps into smaller ones.

INTERMEDIATE: LOW/HIGH Provide students with a chart they can use to record their observations and calculations.

ADVANCED: LOW/HIGH Have students assist students with lower English proficiencies with the Analyze and Conclude questions. Make sure students understand the academic vocabulary.

Answers

37. a. 125 g NH_3 × (82.4 g N/100 g NH_3) = 103 g N
b. 125 g NH_4NO_3 × (35.0 g N/100 g NH_4NO_3) = 43.8 g N

38. a. 350 g C_2H_6 × (2.0 × 10^1 g H/100g C_2H_6) = 7.0 × 10^1 g H
b. 20.2 g $NaHSO_4$ × (0.83 g H /100 g $NaHSO_4$) = 0.17 g H

LESSON 10.3

Explore

Empirical Formulas

Class Activity

PURPOSE Students are provided with an analogy that helps clarify the concepts of percent composition and empirical formulas.

MATERIALS 3 red marbles, 6 green marbles, 3 black marbles, and 12 blue marbles

PROCEDURE Provide pairs of students with sets of marbles. Have students express the number of different colored marbles as fractions and percents of the whole collection. **Ask** What percent of the collection do the red marbles represent? *(12.5%)* Show them that the sums of the fractions and percents are equal to 1 and 100%, respectively. **Ask** What is the ratio of red:green:black:blue marbles in lowest terms? *(1:2:1:4)*

This activity can be extended if the different colored marbles are assumed to be atoms of different elements. **Ask** What is the empirical formula of a hypothetical "compound" that consists of 25% red marbles and 75% green marbles? *(The ratio of red marbles to green marbles in the empirical formula would be 1:3.)*

EXPECTED OUTCOME Students express percent composition of the marbles and determine the "empirical formula" of a marble combination.

Figure 10.10 Interpreting Formulas
A formula can be interpreted on a microscopic level in terms of atoms or on a macroscopic level in terms of moles of atoms.

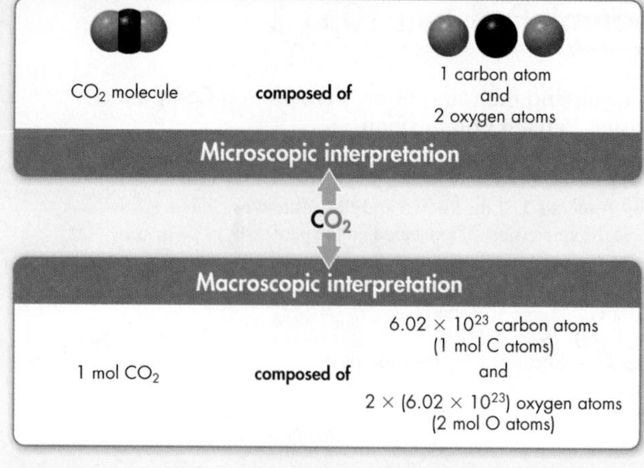

Empirical Formulas

How can you calculate the empirical formula of a compound?

A useful formula for cooking rice is to use one cup of rice and two cups of water. If you needed twice the amount of cooked rice, you would need two cups of rice and four cups of water. The formulas for some compounds also show a basic ratio of elements. Multiplying that ratio by any factor can produce the formulas for other compounds.

The **empirical formula** of a compound gives the lowest whole-number ratio of the atoms or moles of the elements in a compound. Figure 10.10 shows that empirical formulas may be interpreted at the microscopic (atomic) or macroscopic (molar) level. An empirical formula may or may not be the same as a molecular formula. For example, the lowest ratio of hydrogen to oxygen in hydrogen peroxide is 1:1. Thus the empirical formula of hydrogen peroxide is HO. The molecular formula of hydrogen peroxide, H_2O_2, has twice the number of atoms as the empirical formula. Notice that the ratio of hydrogen to oxygen is still the same, 1:1. The molecular formula tells the actual number of each kind of atom present in a molecule of the compound. For carbon dioxide, the empirical and molecular formulas are the same—CO_2. Figure 10.11 shows two compounds of carbon and hydrogen having the same empirical formula (CH) but different molecular formulas.

The percent composition of a compound can be used to calculate the empirical formula of that compound. The percent composition tells the ratio of masses of the elements in a compound. The ratio of masses can be changed to a ratio of moles by using conversion factors based on the molar mass of each element. The mole ratio is then reduced to the lowest whole-number ratio to obtain the empirical formula of the compound.

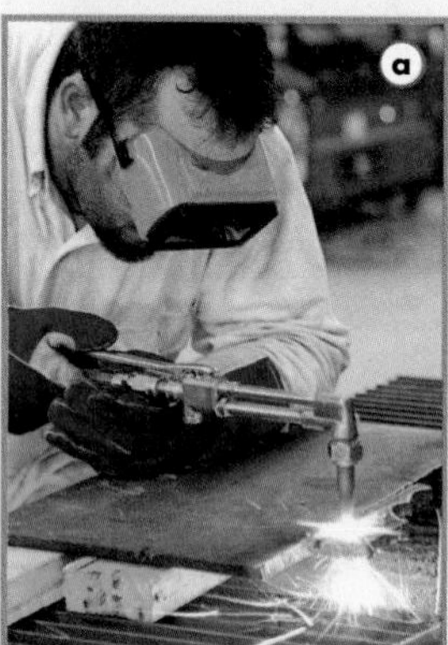

Figure 10.11 Compounds With the Same Empirical Formula
Two different compounds can have the same empirical formula.
a. Ethyne (C_2H_2), also called acetylene, is a gas used in welders' torches. **b.** Styrene (C_8H_8) is used in making polystyrene.
Calculate *What is the empirical formula of ethyne and styrene?*

Differentiated Instruction

LPR LESS PROFICIENT READERS In small groups, have students develop numbered lists of steps to follow for determining empirical formulas from percent composition, determining empirical formulas from mass data, and determining molecular formulas from the empirical formula and molar mass.

L2 SPECIAL NEEDS STUDENTS For the class activity, consider replacing the colored marbles with larger tactile colored objects, larger tactile textured objects, or colored pieces of candy. Issues involving potential swallowing of the materials or sight impairment should be considered when selecting materials for the lab.

L3 ADVANCED STUDENTS Have students research the formulas of the three different oxides of iron. **Ask** Which of the oxides contains a higher percent of iron? *(Of FeO (77.7% Fe), Fe_2O_3 (69.9% Fe), and Fe_3O_4 (72.3% Fe), FeO has the highest percent of iron.)*

Sample Problem 10.12

Determining the Empirical Formula of a Compound

A compound is analyzed and found to contain 25.9% nitrogen and 74.1% oxygen. What is the empirical formula of the compound?

1 Analyze **List the knowns and the unknown.** The percent composition gives the ratio of the mass of nitrogen atoms to the mass of oxygen atoms in the compound. Change the ratio of masses to a ratio of moles and reduce this ratio to the lowest whole-number ratio.

KNOWNS

percent by mass of N = 25.9% N

percent by mass of O = 74.1% O

UNKNOWN

empirical formula = $N_?O_?$

2 Calculate **Solve for the unknown.**

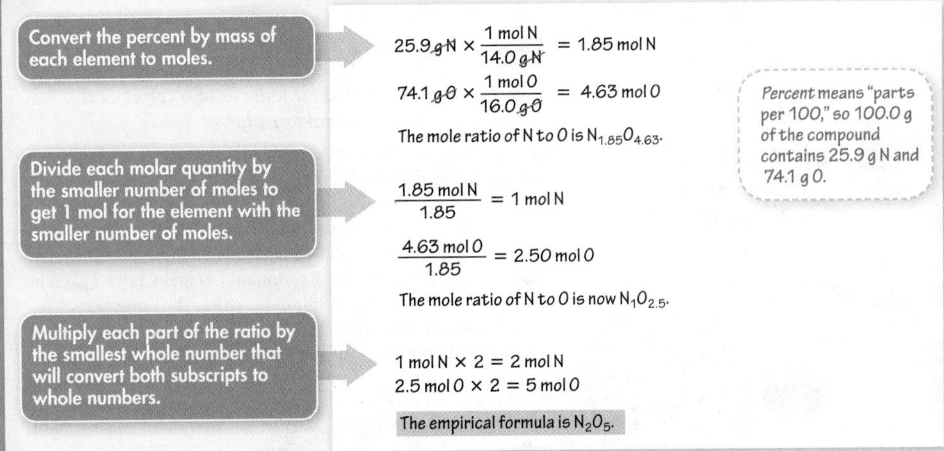

Convert the percent by mass of each element to moles.

$$25.9\ \text{g N} \times \frac{1\ \text{mol N}}{14.0\ \text{g N}} = 1.85\ \text{mol N}$$

$$74.1\ \text{g O} \times \frac{1\ \text{mol O}}{16.0\ \text{g O}} = 4.63\ \text{mol O}$$

The mole ratio of N to O is $N_{1.85}O_{4.63}$.

Percent means "parts per 100," so 100.0 g of the compound contains 25.9 g N and 74.1 g O.

Divide each molar quantity by the smaller number of moles to get 1 mol for the element with the smaller number of moles.

$$\frac{1.85\ \text{mol N}}{1.85} = 1\ \text{mol N}$$

$$\frac{4.63\ \text{mol O}}{1.85} = 2.50\ \text{mol O}$$

The mole ratio of N to O is now $N_1O_{2.5}$.

Multiply each part of the ratio by the smallest whole number that will convert both subscripts to whole numbers.

1 mol N × 2 = 2 mol N

2.5 mol O × 2 = 5 mol O

The empirical formula is N_2O_5.

3 Evaluate **Does the result make sense?** The subscripts are whole numbers, and the percent composition of this empirical formula equals the percents given in the original problem.

39. Calculate the empirical formula of each compound.

a. 94.1% O, 5.9% H

b. 67.6% Hg, 10.8% S, 21.6% O

Start by converting the percent by mass of each element to moles.

40. 1,6-diaminohexane is used to make nylon. What is the empirical formula of this compound if its percent composition is 62.1% C, 13.8% H, and 24.1% N?

Foundations for Math

EQUIVALENT RATIOS The subscripts in empirical formulas are always whole numbers, because compounds cannot have partial atoms. Show students that equivalent ratios can be used to rewrite decimals as whole numbers. To do this, multiply each part of the ratio by the smallest whole number that will convert both subscripts to whole numbers.

In **Sample Problem 10.12,** the ratio of moles of N to moles of O is 1 to 2.5. The smallest whole number to which 2.5 can be converted is 5, which is the product of 2.5 and 2. Because the moles of N and O must remain in the same ratio, the moles of N must also be multiplied by 2. Thus, $N_1O_{2.5}$ is equivalent to N_2O_5.

LESSON 10.3

Explain

APPLY CONCEPTS Direct students to examine Sample Problem 10.12. Explain that when using percent composition to determine an empirical formula, 100.0 g of compound is often arbitrarily chosen because it is easy to use. If an element comprises 28.5% of the mass of a compound, for example, it makes up 28.5 g of a 100.0-g sample. Any other mass of compound can be used but computation will be more difficult.

Sample Practice Problem

What is the empirical formula of each of the following compounds?

a. 36.1% Ca, 63.9% Cl *($CaCl_2$)*

b. 40.0% C, 6.7% H, 53.3% O *(CH_2O)*

c. 3.7% H, 44.4% C, and 51.9% N *(HCN)*

Answers

FIGURE 10.11 CH

39. a. 94.1 g O × (1 mol O/16.0 g O) = 5.88 mol O
5.9 g H × (1 mol H/1.0 g H) = 5.9 mol H
5.88 mol O/5.88 = 1.00 mol O
5.9 mol H/5.88 = 1.0 mol H
Empirical formula = HO

b. 67.6 g Hg × (1 mol Hg/200.6 g Hg) = 0.337 mol Hg
10.8 g S × (1 mol S/32.1 g S) = 0.336 mol S
21.6 g O × (1 mol O/16.0 g O) = 1.35 mol O
0.337 mol Hg/0.336 = 1.00 mol Hg
0.336 mol S/0.336 = 1.00 mol S
1.35 mol O/0.336 = 4.02 mol O
Empirical formula = $HgSO_4$

40. 62.1 g C × (1 mol C/12.0 g C) = 5.18 mol C
13.8 g H × (1 mol H/1.00 g H) = 13.8 mol H
24.1 g N × (1 mol N/14.0 g N) = 1.72 mol N
5.18 mol C/1.72 = 3.01 mol C
13.8 mol H/1.72 = 8.02 mol H
1.72 mol N/1.72 = 1.00 mol N
Empirical formula = C_3H_8N

LESSON 10.3

Explain

Molecular Formulas

CRITICAL THINKING Have students examine the patterns shown in Table 10.3. Then have them add entries for these pairs of compounds: nitrogen dioxide (NO_2) and dinitrogen tetroxide (N_2O_4), diphosphorus pentoxide (P_2O_5) and tetraphosphorus decoxide (P_4O_{10}). Instruct students to determine the empirical formula for each pair.

Extend

Connect to NUTRITION

Tell students that a variety of carbohydrates have the empirical formula CH_2O. Two such carbohydrates are glucose, which is abundant in plants and animals, and fructose, which is found in fruits and honey. Point out that glucose and fructose both share the same molecular formula: $C_6H_{12}O_6$. **Ask** Do you think these carbohydrates have the same structure? Why? *(No, because they have different names.)* **Ask** Do you think there are other carbohydrates with this molecular formula? *(Answers will vary.)* Challenge students to research and draw the structures of glucose and fructose, and list any additional carbohydrates that share the chemical formula $C_6H_{12}O_6$. Tell students they can learn more about these types of monosaccharides in Chapter 24.

Interpret Data

Comparison of Empirical and Molecular Formulas		
Formula (name)	Classification of formula	Molar mass (g/mol)
CH	Empirical	13
C_2H_2 (ethyne)	Molecular	26 (2 × 13)
C_6H_6 (benzene)	Molecular	78 (6 × 13)
CH_2O (methanal)	Empirical and molecular	30
$C_2H_4O_2$ (ethanoic acid)	Molecular	60 (2 × 30)
$C_6H_{12}O_6$ (glucose)	Molecular	180 (6 × 30)

Table 10.3 Different compounds can have the same empirical formula.
a. Read Tables What is the molar mass of benzene, C_6H_6?
b. Interpret Tables Which compounds in the table have the empirical formula CH_2O?
c. Explain Why is the molar mass of glucose ($C_6H_{12}O_6$) equal to six times the molar mass of methanal (CH_2O)?

Hint: How is the formula $C_6H_{12}O_6$ related to the formula CH_2O?

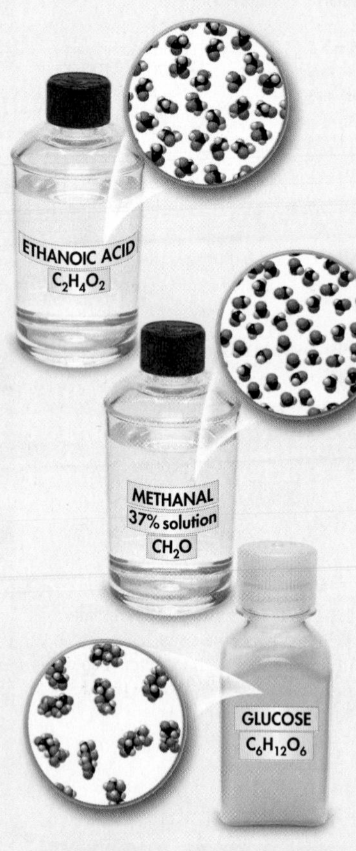

Molecular Formulas

How does the molecular formula of a compound compare with the empirical formula?

Look at the compounds listed in Table 10.3. Ethyne and benzene have the same empirical formula—CH. Methanal, ethanoic acid, and glucose, shown in Figure 10.12, have the same empirical formula—CH_2O. Notice that the molar masses of the compounds in these two groups are simple whole-number multiples of the molar masses of the empirical formulas, CH and CH_2O. **The molecular formula of a compound is either the same as its experimentally determined empirical formula, or it is a simple whole-number multiple of its empirical formula.**

Once you have determined the empirical formula of a compound, you can determine its molecular formula, if you know the compound's molar mass. A chemist often uses an instrument called a mass spectrometer to determine molar mass. The compound is broken into charged fragments (ions) that travel through a magnetic field. The magnetic field deflects the particles from their straight-line paths. The mass of the compound is determined from the amount of deflection experienced by the particles.

You can calculate the empirical formula mass (efm) of a compound from its empirical formula. This is simply the molar mass of the empirical formula. Then you can divide the experimentally determined molar mass by the empirical formula mass. This quotient gives the number of empirical formula units in a molecule of the compound and is the multiplier to convert the empirical formula to the molecular formula.

Figure 10.12 Compounds With the Empirical Formula CH_2O
Methanal (formaldehyde), ethanoic acid (acetic acid), and glucose have the same empirical formula.
Apply Concepts *How could you easily obtain the molar mass of ethanoic acid using the molar mass of methanal?*

UbD Check for Understanding

How does the molecular formula of a compound compare with the empirical formula?

Assess students' understanding of molecular and empirical formulas by asking them a series of true/false questions. Have students give a thumbs-up sign if the statement is true and a thumbs-down sign if the statement is false.

- Empirical and molecular formulas are never the same for any compound. *(false)*
- If you know the empirical formula of the compound, then you know the smallest whole-number ratio of the elements in that compound. *(true)*
- A ratio exists between the empirical formula and the molecular formula of a compound. *(true)*

ADJUST INSTRUCTION Instruct students to reread the text under Empirical Formulas and Molecular Formulas. In groups, have them write a compare and contrast table. Then restate the questions above, and have each group explain why the statement is true or false.

Sample Problem 10.13

Finding the Molecular Formula of a Compound

Calculate the molecular formula of a compound whose molar mass is 60.0 g/mol and empirical formula is CH_4N.

KNOWNS
empirical formula = CH_4N
molar mass = 60.0 g/mol

UNKNOWN
molecular formula = $C_?H_?N_?$

1 Analyze List the knowns and the unknown. Divide the molar mass by the empirical formula mass to obtain a whole number. Multiply the empirical formula subscripts by this value to get the molecular formula.

2 Calculate Solve for the unknown.

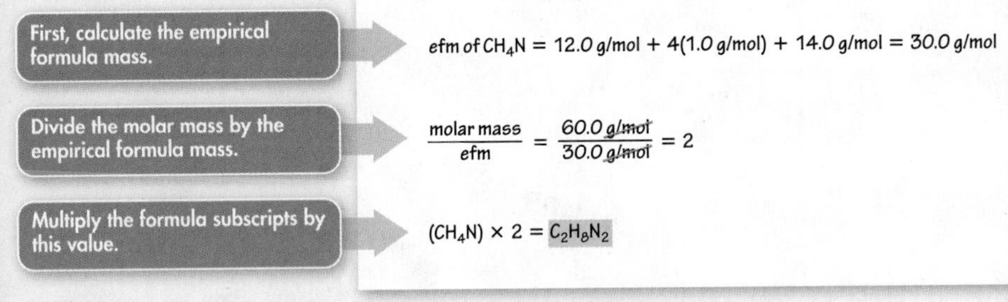

3 Evaluate Does the result make sense? The molecular formula has the molar mass of the compound.

41. What is the molecular formula of a compound with the empirical formula CClN and a molar mass of 184.5 g/mol?

42. Find the molecular formula of ethylene glycol, which is used as antifreeze. The molar mass is 62.0 g/mol, and the empirical formula is CH_3O.

10.3 LessonCheck

43. Review How do you calculate the percent by mass of an element in a compound?

44. Identify What information can you use to calculate the empirical formula of a compound?

45. Explain How is the molecular formula of a compound related to its empirical formula?

46. Calculate Determine the percent composition of the compound that forms when 222.6 g N combines completely with 77.4 g O.

47. Calculate Find the percent composition of calcium acetate, $Ca(C_2H_3O_2)_2$.

48. Calculate Using the results of Problem 47, calculate the grams of hydrogen in 124 g of $Ca(C_2H_3O_2)_2$.

BIGIDEA
THE MOLE AND QUANTIFYING MATTER

49. The compound methyl butanoate smells like apples. Its percent composition is 58.8% C, 9.8% H, and 31.4% O, and its molar mass is 102 g/mol. What is its empirical formula? What is its molecular formula?

Lesson Check Answers

43. Divide the mass of an element in the compound by the mass of the compound; then multiply by 100%.

44. The percent composition of a compound can be used to calculate the empirical formula of a compound.

45. The molecular formula of a compound is a simple whole-number multiple of the empirical formula.

46. 74.2% N, 25.8% O

47. 25.4% Ca, 30.4% C, 3.8% H, 40.5% O

48. 4.7 g

49. **BIGIDEA** $C_5H_{10}O_2$ is both its empirical and molecular formula.

Explain

Molecular Formulas

Sample Practice Problem

A compound is composed of carbon and hydrogen and has a molar mass of 86.0 grams per mole. Its empirical formula is C_3H_7. What is the molecular formula? *(C_6H_{14})*

Evaluate

Informal Assessment

Have students list verbally or in writing the steps they would take to calculate the molecular formula in each of the following situations:

1. The empirical formula and molar mass are known.
2. The percent composition and molar mass are known.

Then have students complete the 10.3 Lesson Check.

Reteach

Point out to students that when they know the percent composition and molar mass of a compound, they must first use the percent composition to calculate the empirical formula. They can then calculate the empirical formula mass and compare it to the molar mass of the molecular compound to determine the molecular formula. Help students to construct a flowchart or a concept map depicting the various scenarios.

Answers

INTERPRET DATA

a. 78 g
b. methanol, ethanoic acid, and glucose
c. The molar mass of glucose is six times greater than the molar mass of methanal.

FIGURE 10.12 Multiply the molar mass of methanal by 2.

41. $C_3Cl_3N_3$

42. molar mass/efm = 62/31 = 2
molecular formula = $2(CH_3O) = C_2H_6O_2$

CHEMISTRY & YOU Lead students in a discussion of Ion Mobility Spectrometry by having them describe experiences they have had going through airport security systems. If any students have gone through a puffer port, encourage them to explain what they saw and felt. Ask if they were concerned about the safety of going through the field or if they thought it was a fun experience. Point out that because of previous airline security alerts, people going through airport security often have to remove their shoes. With the puffer portal, this is unnecessary.

Pose the following question to students: *How do puffer portals differ from metal detectors and x-ray machines in their ability to detect potential weapons?* You may need to assist students in the following ways:

- Metal detectors detect many nonweapon metal objects, such as belt buckles, jewelry, keys, and medical hardware.
- Neither metal detectors nor x-ray machines can detect chemicals of any type on either clothing or other personal belongings.
- Metal detectors and x-ray machines rely heavily on human operators, who must sometimes rescan passengers or luggage and who are prone to fatigue.
- Puffer portals require only seconds to scan and analyze any chemicals detected.

Ion Mobility Spectrometry

In 2001, a terrorist boarded an airline flight with explosives inside his shoes. Since that time, Americans have had to remove their shoes during airport security checks. However, newer airport security devices, known as "puffer portals," allow airport security to scan for minute traces of explosives on a person's body and clothing, without the person having to remove any clothing or shoes.

The puffer portal looks like a standard airport metal detector. There are vents and nozzles on the walls and ceiling of the portal. When a passenger steps inside, the doors close, and the instrument sends sharp bursts of air to dislodge particles from his or her body, hair, and clothing. The air sample is then passed through a chemical analysis system called an ion mobility spectrometer (IMS). The IMS identifies compounds based on the amount of time it takes for ions to pass through an electrified field in a tube filled with a nonreactive gas (drift gas). This "drift time" is then compared to a database of drift times of different compounds. In this way, molecules of known explosive or narcotic materials can be detected and identified. If even a picogram of an explosive is detected, an alarm sounds.

PUFFING OUT EXPLOSIVES Bursts of air dislodge particles from a person's hair, body, and clothes. These particles are then directed to an ion mobility spectrometer (IMS).

334 Chapter 10 • Chemistry & You

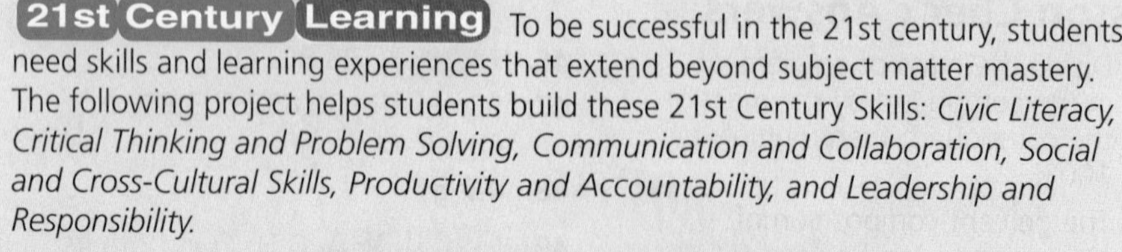

21st Century Learning To be successful in the 21st century, students need skills and learning experiences that extend beyond subject matter mastery. The following project helps students build these 21st Century Skills: *Civic Literacy, Critical Thinking and Problem Solving, Communication and Collaboration, Social and Cross-Cultural Skills, Productivity and Accountability, and Leadership and Responsibility.*

SECURING THE SKIES Pose the following challenge to your students. *Officials at an airport near you have just announced that they are increasing their security procedures. They will continue using metal detectors and luggage checks, but they will also install a puffer portal. Some people have complained that this is an invasion of privacy and unnecessary. Do you agree?* Form debate teams of two students each, and engage in a debate regarding the pros and cons of puffer portals. All teams should research both sides of the economic, security, and privacy issues thoroughly and come prepared to argue either side. Have teams draw straws to determine which position they will argue on debate day.

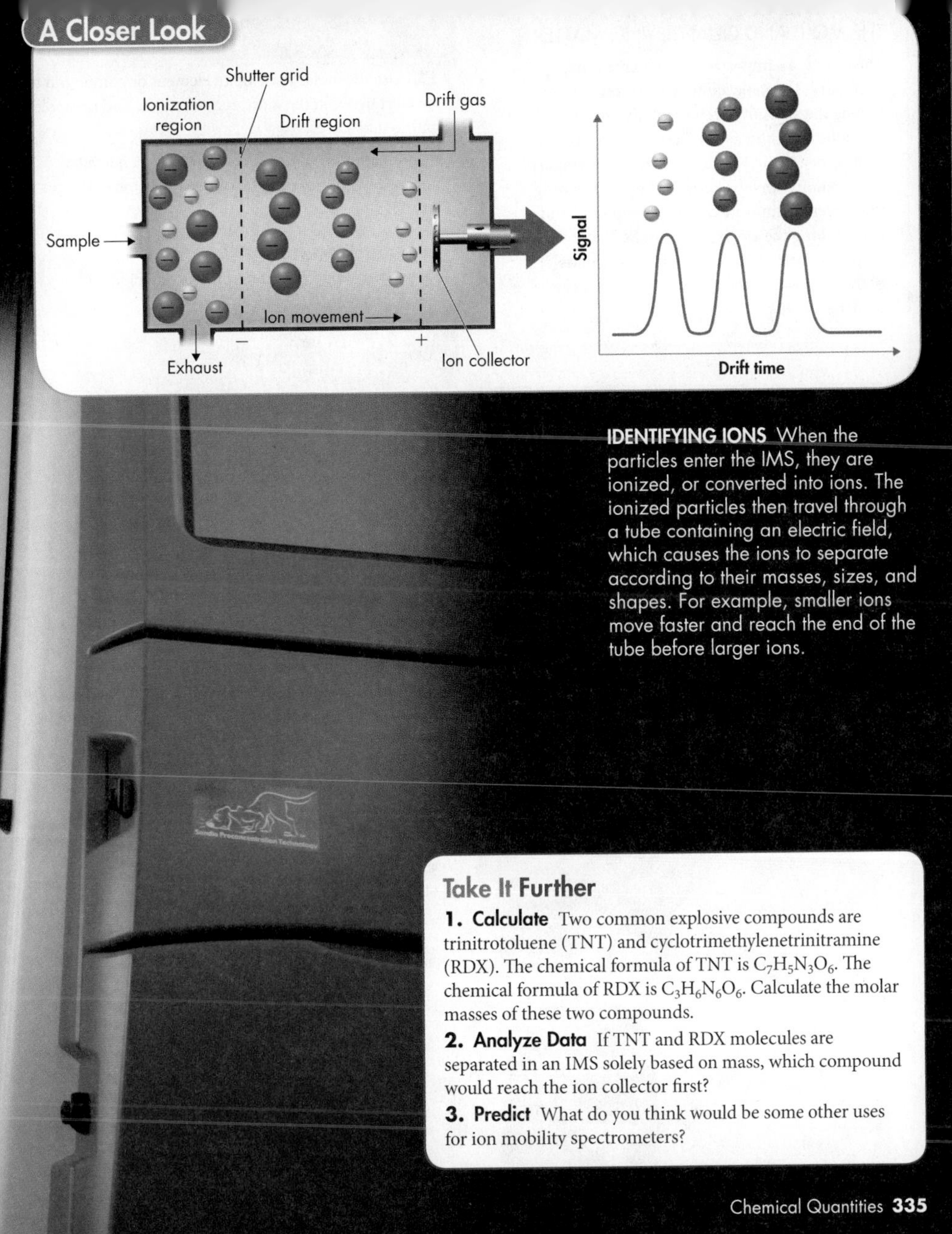

Take It Further

1. **Calculate** Two common explosive compounds are trinitrotoluene (TNT) and cyclotrimethylenetrinitramine (RDX). The chemical formula of TNT is $C_7H_5N_3O_6$. The chemical formula of RDX is $C_3H_6N_6O_6$. Calculate the molar masses of these two compounds.
2. **Analyze Data** If TNT and RDX molecules are separated in an IMS solely based on mass, which compound would reach the ion collector first?
3. **Predict** What do you think would be some other uses for ion mobility spectrometers?

USE VISUALS Have students study the picture of the ion mobility spectrometer. **Ask** Why is it necessary to ionize the particles before they move into the drift region? *(Ions have an electric charge and will respond to the electric field. Uncharged atoms will not.)* **Ask** What can scientists learn about the different masses of the ions? *(The masses indicate the type of element they are.)*

SUMMARIZE Have students write a summary statement explaining what the graph of signal vs. drift time shows. *(The graph shows that particles with different masses arrive at the detector at different times.)*

Extend

Connect to ECONOMICS

Each puffer portal costs well over $100,000. Have students ponder the question, *Is the price really worth the benefit?* Explain to students the concept of cost analysis, in which economists compare the cost of doing something to the potential costs that not doing it might incur. Have students work with a partner to analyze and estimate the costs of installing a puffer portal in an airport (cost of the machine, technician costs, upkeep costs) with the cost of not installing one (possibility of a terrorist threat and the expenses that would involve). *(Although students may have to estimate many of the costs, they should conclude that the cost of not installing the machine at an airport far exceeds the cost of installing it.)*

Differentiated Instruction

L1 STRUGGLING STUDENTS Remind students that an ion is a charged atom. The atom has either gained or lost one or more electrons. As an example, have students draw a diagram of a nitrogen atom, and explain how it can change to an ion.

LPR LESS PROFICIENT READERS Have students work with a partner to identify at least five words in the feature that present them with difficulties. For each word, students should write a definition in their own words and use the word in a sentence.

L3 ADVANCED STUDENTS Have students consider other potential uses for the puffer portals. Examples include other security applications as well as detection of hazards at laboratories. Have students prepare a report describing the uses and possible alternatives to the design used for airport security.

Answers

1. The molar mass of TNT is 227 g. The molar mass of RDX is 222 g.
2. The RDX molecules would reach the ion collector first.
3. Answers will vary but students might state that they could be used at courthouses or bus or train stations.

Evaluate

Review and Assessment Materials

CHEMISTRY TUTORIAL Encourage students to review converting from number of atoms to moles to mass and vice versa, calculating percent composition from mass data and from a formula, and determining the empirical formula of a compound using the animated step-by-step tutorials for Chapter 10. Each Chemistry Tutorial is followed by a similar practice problem, where the student is guided through the problem-solving process.

CHAPTER PROBLEM SET Have students practice more chemical quantities problems using the Chapter 10 online Problem Set.

DIRECTED VIRTUAL LAB Have students complete the Virtual Lab *Counting Atoms by Measuring Mass* to help reinforce the mathematical relationships involved in calculating molar quantities. In the post-lab exercise, students will practice mass to mole conversions for several different compounds.

KINETIC ART Direct students to the Kinetic Art animations *Molar Mass* and *The Mole Map*, and have them answer the questions that accompany each animation.

CONCEPTS IN ACTION If students have trouble understanding percent composition, ask them to view the Concepts in Action animation, which presents the material in a familiar context. Have students complete the questions after the animation.

Study Tip

ORGANIZE NEW INFORMATION Different people learn in different ways.Tell students that if they learn best visually, they should draw concept maps, outlines, and diagrams to help them organize the chapter content. If students learn best by listening, they can take turns reading the chapter aloud with a partner.

10 Study Guide

BIGIDEA
THE MOLE AND QUANTIFYING MATTER

The mole is an important measurement in chemistry. The mole allows you to convert among the amount of representative particles in a substance, the mass of a substance, and the volume of a gas at STP. The molecular formula of a compound can be determined by first finding the percent composition of the compound and determining the empirical formula. Using the empirical formula mass and the molar mass of the compound, the molecular formula can be determined.

10.1 The Mole: A Measurement of Matter

Knowing how the count, mass, and volume of an item relate to a common unit allows you to convert among these units.

The mole allows chemists to count the number of representative particles in a substance.

The atomic mass of an element expressed in grams is the mass of a mole of the element.

To calculate the molar mass of a compound, find the number of grams of each element in one mole of the compound. Then add the masses of the elements in the compound.

- mole (308)
- Avogadro's number (308)
- representative particle (308)
- molar mass (313)

10.2 Mole–Mass and Mole–Volume Relationships

Use the molar mass of an element or compound to convert between the mass of a substance and the moles of the substance.

The molar volume is used to convert between the number of moles of gas and the volume of the gas at STP.

- Avogadro's hypothesis (320)
- standard temperature and pressure (STP) (320)
- molar volume (320)

10.3 Percent Composition and Chemical Formulas

The percent by mass of an element in a compound is the number of grams of the element divided by the mass in grams of the compound, multiplied by 100%.

The percent composition of a compound can be used to calculate the empirical formula of that compound.

The molecular formula of a compound is either the same as its experimentally determined empirical formula, or it is a simple whole-number multiple of its empirical formula.

- percent composition (325)
- empirical formula (330)

Key Equations

$$\text{\% by mass of element} = \frac{\text{mass of element}}{\text{mass of compound}} \times 100\%$$

$$\text{\% by mass of element} = \frac{\text{mass of element in 1 mol compound}}{\text{molar mass of compound}} \times 100\%$$

UbD Performance Tasks

WHAT'S IN A MOLE? Write the following information on the board: Nitrous oxide (N_2O) has a density of 1.977 g/L. Without consulting a periodic table, calculate the following for 7.43 mol of N_2O: molar mass, total mass, number of representative particles, and number of atoms. *(molar mass = 44.3 g/mol; total mass = 3.29×10^2 g; 4.47×10^{24} representative particles; 1.34×10^{25} atoms)*

WHAT PART OF NO... Have students determine the percent composition, empirical formula, and molecular formula for a sample of colorless gas that contains 10.9 g N and 25.1 g O. The molar mass for this compound is 92.0 g/mol. *(percent composition: 30.3% N, 69.7% O; empirical formula: NO_2; molecular formula: N_2O_4)*

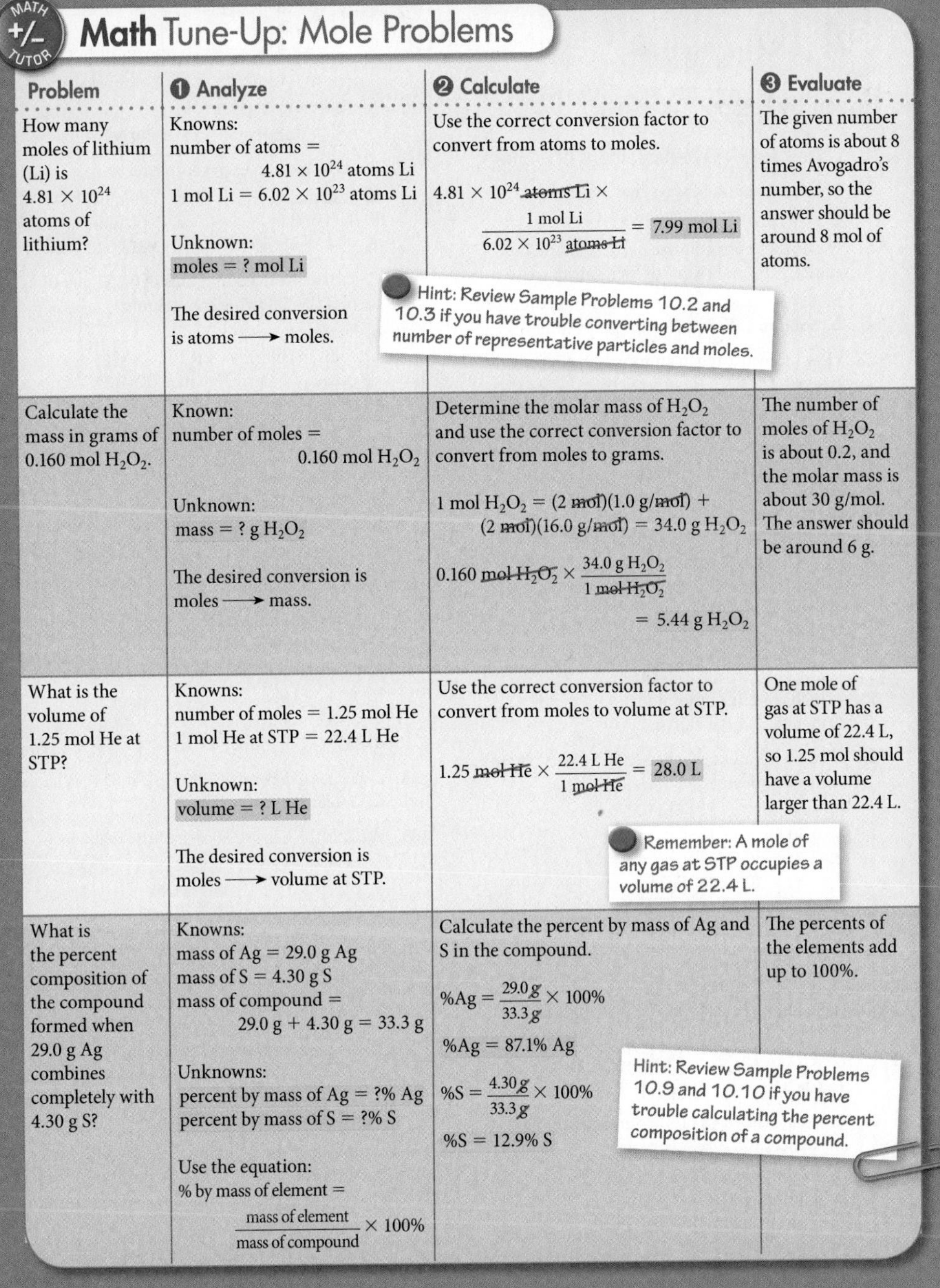

Math Tune-Up: Mole Problems

Problem	❶ Analyze	❷ Calculate	❸ Evaluate
How many moles of lithium (Li) is 4.81×10^{24} atoms of lithium?	Knowns: number of atoms = 4.81×10^{24} atoms Li 1 mol Li = 6.02×10^{23} atoms Li Unknown: moles = ? mol Li The desired conversion is atoms ⟶ moles.	Use the correct conversion factor to convert from atoms to moles. $4.81 \times 10^{24}\ \cancel{\text{atoms Li}} \times \frac{1\ \text{mol Li}}{6.02 \times 10^{23}\ \cancel{\text{atoms Li}}} = 7.99\ \text{mol Li}$	The given number of atoms is about 8 times Avogadro's number, so the answer should be around 8 mol of atoms.
Calculate the mass in grams of 0.160 mol H_2O_2.	Known: number of moles = 0.160 mol H_2O_2 Unknown: mass = ? g H_2O_2 The desired conversion is moles ⟶ mass.	Determine the molar mass of H_2O_2 and use the correct conversion factor to convert from moles to grams. 1 mol H_2O_2 = $(2\ \cancel{\text{mol}})(1.0\ \text{g}/\cancel{\text{mol}}) + (2\ \cancel{\text{mol}})(16.0\ \text{g}/\cancel{\text{mol}}) = 34.0\ \text{g}\ H_2O_2$ $0.160\ \cancel{\text{mol}\ H_2O_2} \times \frac{34.0\ \text{g}\ H_2O_2}{1\ \cancel{\text{mol}\ H_2O_2}} = 5.44\ \text{g}\ H_2O_2$	The number of moles of H_2O_2 is about 0.2, and the molar mass is about 30 g/mol. The answer should be around 6 g.
What is the volume of 1.25 mol He at STP?	Knowns: number of moles = 1.25 mol He 1 mol He at STP = 22.4 L He Unknown: volume = ? L He The desired conversion is moles ⟶ volume at STP.	Use the correct conversion factor to convert from moles to volume at STP. $1.25\ \cancel{\text{mol He}} \times \frac{22.4\ \text{L He}}{1\ \cancel{\text{mol He}}} = 28.0\ \text{L}$	One mole of gas at STP has a volume of 22.4 L, so 1.25 mol should have a volume larger than 22.4 L.
What is the percent composition of the compound formed when 29.0 g Ag combines completely with 4.30 g S?	Knowns: mass of Ag = 29.0 g Ag mass of S = 4.30 g S mass of compound = 29.0 g + 4.30 g = 33.3 g Unknowns: percent by mass of Ag = ?% Ag percent by mass of S = ?% S Use the equation: % by mass of element = $\frac{\text{mass of element}}{\text{mass of compound}} \times 100\%$	Calculate the percent by mass of Ag and S in the compound. $\%Ag = \frac{29.0\ \cancel{g}}{33.3\ \cancel{g}} \times 100\%$ %Ag = 87.1% Ag $\%S = \frac{4.30\ \cancel{g}}{33.3\ \cancel{g}} \times 100\%$ %S = 12.9% S	The percents of the elements add up to 100%.

Hint: Review Sample Problems 10.2 and 10.3 if you have trouble converting between number of representative particles and moles.

Remember: A mole of any gas at STP occupies a volume of 22.4 L.

Hint: Review Sample Problems 10.9 and 10.10 if you have trouble calculating the percent composition of a compound.

Math Tune-Up 337

MATH TUNE-UP

Math Review

Example and sample practice problems for converting chemical quantities can be found on the following pages:

LESSON 10.1 Finding Mass From a Count—page 307; Converting Number of Atoms to Moles—page 309; Converting Moles to Number of Atoms—page 311; Finding the Molar Mass of a Compound—page 315

LESSON 10.2 Converting Moles to Mass—page 318; Converting Mass to Moles—page 319; Calculating Gas Quantities at STP—page 321; Calculating the Molar Mass of a Gas at STP—page 322

LESSON 10.3 Calculating Percent Composition From Mass Data—page 326; Calculating Percent Composition From a Formula—page 327; Calculating the Mass of an Element in a Compound Using Percent Composition—page 329; Determining the Empirical Formula of a Compound—page 331; Finding the Molecular Formula of a Compound—page 333

Focus on ELL

5 ASSESS UNDERSTANDING Use or create a formative assessment for each lesson to determine the students' level of understanding.

BEGINNING

LOW Create an assessment in which students match symbols, pictures, or diagrams to vocabulary words, conversion factors, or concepts.

HIGH Allow students to use their native language glossary as a reference during the assessment.

INTERMEDIATE: LOW/HIGH Assign the Standardized Test Prep. Tell students to cross out wrong answers. Model how to complete the answer sheet.

ADVANCED: LOW/HIGH Assign students a five question assessment involving mass and mole conversion problems and volume and mole conversion problems. Have students indicate the correct answers and provide an analysis for each wrong answer choice.

ASSESSMENT

Evaluate

Answers

LESSON 10.1

50. Number, mass, or volume; examples will vary.

51. **a.** molecule
b. formula unit
c. molecule
d. atom

52. **a.** 3
b. 2
c. 9
d. 10

53. A mole of any substance contains Avogadro's number of representative particles.

54. **a.** 4.00 mol NaCl
b. 15.0 mol Hg
c. 7.72 mol NO_2

55. All contain 6.02×10^{23} molecules

56. 1.00 mol C_2H_6

57. **a.** 1.81×10^{24} atoms Sn
b. 2.41×10^{23} formula units KCl
c. 4.52×10^{24} molecules SO_2
d. 2.89×10^{21} formula units NaI

58. 71.0 g/mol Cl_2

59. Answers will vary but should include
1. Determine the moles of each atom from the formula.
2. Look up the atomic mass of each element.
3. Multiply the number of moles of each atom by its molar mass.
4. Sum these products.

60. **a.** 98.0 g/mol
b. 76.0 g/mol
c. 100.1 g/mol
d. 132.1 g/mol
e. 89.0 g/mol
f. 159.8 g/mol

61. **a.** 60.1 g/mol
b. 28.0 g/mol
c. 106.8 g/mol
d. 63.5 g/mol

LESSON 10.2

62. **a.** 108 g C_5H_{12}
b. 547 g F_2
c. 71.8 g $Ca(CN)_2$
d. 238 g H_2O_2
e. 224 g NaOH
f. 1.88 g Ni

63. **a.** 85.5 g $C_{12}H_{22}O_{11}$
b. 14.6 g NaCl
c. 39.5 g $KMnO_4$

64. **a.** 0.292 mol $C_{12}H_{22}O_{11}$
b. 1.71 mol NaCl
c. 0.633 mol $KMnO_4$

10 Assessment

ONLINE PROBLEMS

$\star$ Solutions appear in Appendix E

Lesson by Lesson

10.1 The Mole: A Measurement of Matter

50. List three common ways that matter is measured. Give examples of each.

$\star$51. Name the representative particle (atom, molecule, or formula unit) of each substance.
a. oxygen gas
b. sodium sulfide
c. sulfur dioxide
d. potassium

$\star$52. How many hydrogen atoms are in a representative particle of each substance?
a. $Al(OH)_3$
b. $H_2C_2O_4$
c. $(NH_4)_2HPO_4$
d. $C_4H_{10}O$

53. Describe the relationship between Avogadro's number and one mole of any substance.

$\star$54. Find the number of moles in each substance.
a. 2.41×10^{24} formula units of NaCl
b. 9.03×10^{24} atoms of Hg
c. 4.65×10^{24} molecules of NO_2

55. Which contains more molecules: 1.00 mol H_2O_2, 1.00 mol C_2H_6, or 1.00 mol CO?

56. Which contains more atoms: 1.00 mol H_2O_2, 1.00 mol C_2H_6, or 1.00 mol CO?

$\star$57. Find the number of representative particles in each substance.
a. 3.00 mol Sn
b. 0.400 mol KCl
c. 7.50 mol SO_2
d. 4.80×10^{-3} mol NaI

58. What is the molar mass of chlorine?

59. List the steps you would take to calculate the molar mass of any compound.

$\star$60. Calculate the molar mass of each substance.
a. H_3PO_4
b. N_2O_3
c. $CaCO_3$
d. $(NH_4)_2SO_4$
e. $C_4H_9O_2$
f. Br_2

61. Calculate the mass of 1.00 mol of each of these substances.
a. silicon dioxide (SiO_2)
b. diatomic nitrogen (N_2)
c. iron(III) hydroxide ($Fe(OH)_3$)
d. copper (Cu)

10.2 Mole–Mass and Mole–Volume Relationships

62. Find the mass of each substance.
a. 1.50 mol C_5H_{12}
b. 14.4 mol F_2
c. 0.780 mol $Ca(CN)_2$
d. 7.00 mol H_2O_2
e. 5.60 mol NaOH
f. 3.21×10^{-2} mol Ni

$\star$63. Calculate the mass in grams of 0.250 mol of each of the following compounds:
a. sucrose ($C_{12}H_{22}O_{11}$)
b. sodium chloride (NaCl)
c. potassium permanganate ($KMnO_4$)

$\star$64. Calculate the number of moles in 1.00×10^2 g of each of the compounds in Problem 63.

65. How many moles is each of the following?
a. 15.5 g SiO_2
b. 0.0688 g AgCl
c. 79.3 g Cl_2
d. 5.96 g KOH
e. 937 g $Ca(C_2H_3O_2)_2$
f. 0.800 g Ca

66. What is the volume of one mole of any gas at STP?

$\star$67. Calculate the volume of each of the following gases at STP.
a. 7.64 mol Ar
b. 1.34 mol SO_2
c. 0.442 mol C_2H_6
d. 2.45×10^{-3} mol H_2S

$\star$68. A gas has a density of 0.902 g/L at STP. What is the molar mass of this gas?

69. What is the density of each of the following gases at STP?
a. C_3H_8
b. O_2
c. Ne
d. NO_2

70. Find each of the following quantities:
a. the volume, in liters, of 835 g SO_3 at STP
b. the mass, in grams, of a molecule of aspirin ($C_9H_8O_4$)
c. the number of atoms in 5.78 mol NH_4NO_3

10.3 Percent Composition and Chemical Formulas

71. What is the percent composition of the compound formed when 2.70 g of aluminum combine with oxygen to form 5.10 g of aluminum oxide?

65. **a.** 0.258 mol SiO_2
b. 4.80×10^{-4} mol AgCl
c. 1.12 mol Cl_2
d. 0.106 mol KOH
e. 5.93 mol $Ca(C_2H_3O_2)_2$
f. 2.00×10^{-2} mol Ca

66. 22.4 L

67. **a.** 171 L Ar
b. 30.0 L SO_2
c. 9.90 L C_2H_6
d. 5.49×10^{-2} L H_2S

68. 20.2 g/mol

69. **a.** 1.96 g/L
b. 1.43 g/L
c. 0.902 g/L
d. 2.05 g/L

70. **a.** 234 L SO_3
b. 2.99×10^{-22} g $C_9H_8O_4$
c. 3.13×10^{25} atoms

LESSON 10.3

71. 52.9% Aluminum and 47.1% Oxygen

*72. Calculate the percent composition when 13.3 g Fe combine completely with 5.7 g O.

*73. Calculate the percent composition of each compound.

a. H_2S
b. $(NH_4)_2C_2O_4$
c. $Mg(OH)_2$
d. Na_3PO_4

*74. Using your answers from Problem 73, calculate the number of grams of these elements.

a. sulfur in 3.54 g H_2S
b. nitrogen in 25.0 g $(NH_4)_2C_2O_4$
c. magnesium in 97.4 g $Mg(OH)_2$
d. phosphorus in 804 g Na_3PO_4

75. Which of the following compounds has the highest percent of iron by mass?

a. $FeCl_2$
b. $Fe(C_2H_3O_2)_3$
c. $Fe(OH)_2$
d. FeO

*76. What is an empirical formula? Which of the following molecular formulas are also empirical formulas?

a. ribose ($C_5H_{10}O_5$)
b. ethyl butyrate ($C_6H_{12}O_2$)
c. chlorophyll ($C_{55}H_{72}MgN_4O_5$)
d. DEET ($C_{12}H_{17}ON$)

77. Which of the following can be classified as an empirical formula?

a. S_2Cl_2
b. $C_6H_{10}O_4$
c. Na_2SO_3

78. Which pair of molecules has the same empirical formula?

a. $C_2H_4O_2$, $C_6H_{12}O_6$
b. $NaCrO_4$, $Na_2Cr_2O_7$

*79. What is the molecular formula for each compound? Each compound's empirical formula and molar mass are given.

a. CH_2O, 90 g/mol
b. HgCl, 472.2 g/mol

Understand Concepts

*80. Table sugar, or sucrose, has the chemical formula $C_{12}H_{22}O_{11}$.

a. How many atoms are in 1.00 mol of sucrose?
b. How many atoms of C are in 2.00 mol of sucrose?
c. How many atoms of H are in 2.00 mol of sucrose?
d. How many atoms of O are in 3.65 mol of sucrose?

81. How can you determine the molar mass of a gaseous compound if you do not know its molecular formula?

*82. A series of compounds has the empirical formula CH_2O. The graph shows the relationship between the molar mass of the compounds and the mass of carbon in each compound.

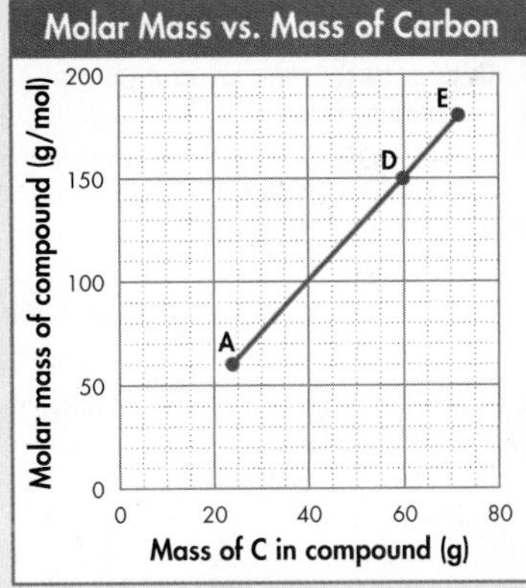

a. What are the molecular formulas for the compounds represented by data points A, D, and E?
b. Find the slope of the line. Is this value consistent with the empirical formula? Explain.
c. Two other valid data points fall on the line between points A and D. What are the *x*, *y* values for these data points?

83. Explain what is wrong with each statement.

a. One mole of any substance contains the same number of atoms.
b. A mole and a molecule of a substance are identical in amount.
c. One molar mass of CO_2 contains Avogadro's number of atoms.

84. Which of the following contains the largest number of atoms?

a. 82.0 g Kr
b. 0.842 mol C_2H_4
c. 36.0 g N_2

*85. Calculate the grams of oxygen in 90.0 g of Cl_2O.

86. What is the total mass of a mixture of 3.50×10^{22} formula units Na_2SO_4, 0.500 mol H_2O, and 7.23 g AgCl?

72. 70% Fe and 30% O

73. **a.** 5.9% H; 94.1%S
b. 22.6% N; 6.5% H; 19.4% C; 51.6% O
c. 41.7% Mg; 54.9% O; 3.4% H
d. 42.1% Na; 18.9% P; 39.0% O

74. **a.** 3.33 g S
b. 5.65 g N
c. 40.6 g Mg
d. 152 g P

75. **d.** 77.7% Fe in FeO

76. An empirical formula has the lowest whole-number ratio of elements.; **a.** molecular; **b.** molecular; **c.** molecular and empirical; **d.** molecular and empirical

77. **a.** molecular **b.** molecular **c.** empirical

78. a

79. **a.** $C_3H_6O_3$
b. Hg_2Cl_2

UNDERSTAND CONCEPTS

80. **a.** 2.71×10^{25} atoms
b. 1.44×10^{25} C atoms
c. 2.65×10^{25} H atoms
d. 2.42×10^{25} O atoms

81. You can measure the mass of 22.4 L of the compound at STP; this is the molar volume of the gas. The mass of the molar volume is the molar mass.

82. **a.** A, $C_2H_4O_2$; D, $C_5H_{10}O_5$; E, $C_6H_{12}O_6$
b. slope = 2.5/1, which is the ratio of the molar mass of the empirical formula to the mass of carbon in the empirical formula: 30/12 = 2.5/1.
c. (36, 90) and (48, 120)

83. **a.** A molecule is composed of two or more atoms.
b. There are 6.02×10^{23} molecules in 1 mol of a molecular substance.
c. A mole of CO_2 has 3 times Avogadro's number of atoms.

84. **b.** 0.842 mol C_2H_4

85. 16.6 g O

86. 24.5 g

ASSESSMENT

Evaluate

Answers

87. 2.73×10^{20} F atoms

88. **a.** CO
b. $C_2O_2NH_5$
c. Cl_2OC

89. **a.** H_2O_2
b. $C_6H_6O_4$
c. K_2CO_3

90. **a.** 27.0 amu
b. aluminum

91. 3×10^{13} km

92. **a.** $CuBr_2$
b. CH_3

93. $C_3H_6O_3$

94. 0.98 g He

95. 3.34×10^{25} molecules H_2O

THINK CRITICALLY

96. C_2H_6O.

97. A molecular formula is a whole-number multiple of its empirical formula.

98. Sulfur atoms have a greater atomic mass. The most abundant sulfur atom has 16 protons, 16 electrons, and 16 neutrons; carbon is composed of 6 protons, 6 electrons, and 6 neutrons. Therefore, 6.02×10^{23} sulfur atoms will have a greater mass than the same number of carbon atoms.

99. Gas molecules are separated by so much empty space their own volumes are insignificant in determining how much space a certain quantity of gas molecules occupies.

100. **a.** $C_9H_{11}O_2N$
b. $C_9H_{11}O_2N$

***87.** The molecular formula of an antibacterial drug is $C_{17}H_{18}FN_3O_3$. How many fluorine atoms are in a 150-mg tablet of this drug?

88. Determine the empirical formulas of compounds with the following percent compositions:

a. 42.9% C and 57.1% O
b. 32.00% C, 42.66% O, 18.67% N, and 6.67% H
c. 71.72% Cl, 16.16% O, and 12.12% C

***89.** Determine the molecular formula for each compound.

a. 94.1% O and 5.9% H; molar mass = 34 g/mol
b. 50.7% C, 4.2% H, and 45.1% O; molar mass = 142 g/mol
c. 56.6% K, 8.7% C, and 34.7% O; molar mass = 138.2 g/mol

***90.** A fictitious "atomic balance" is shown below. Fifteen atoms of boron on the left side of the balance are balanced by six atoms of an unknown element E on the right side.

a. What is the atomic mass of element E?
b. What is the identity of element E?

***91.** A typical virus is 5×10^{-6} cm in diameter. If Avogadro's number of these virus particles were laid in a row, how many kilometers long would the line be?

92. Calculate the empirical formula for each compound.

a. compound consisting of 0.40 mol Cu and 0.80 mol Br
b. compound with 4 atoms of carbon for every 12 atoms of hydrogen

93. Muscle fatigue can result from the buildup of lactic acid. The percent composition of lactic acid is 40.0% C, 6.67% H, and 53.3% O. What is the molecular formula of lactic acid if its molar mass is 90.0 g/mol?

***94.** What mass of helium is needed to inflate a balloon to a volume of 5.50 L at STP?

95. How many water molecules are in a 1.00-L bottle of water? The density of water is 1.00 g/mL.

Think Critically

***96.** **Infer** What is the empirical formula of a compound that has three times as many hydrogen atoms as carbon atoms but only half as many oxygen atoms as carbon atoms?

97. **Apply Concepts** How are the empirical and molecular formulas of a compound related?

98. **Compare** Why does one mole of carbon have a smaller mass than one mole of sulfur? How are the atomic structures of these elements different?

99. **Analyze Data** One mole of any gas at STP equals 22.4 L of that gas. It is also true that different elements have different atomic volumes, or diameters. How can you reconcile these two statements?

***100.** **Interpret Graphs** The graph shows the percent composition of phenylalanine.

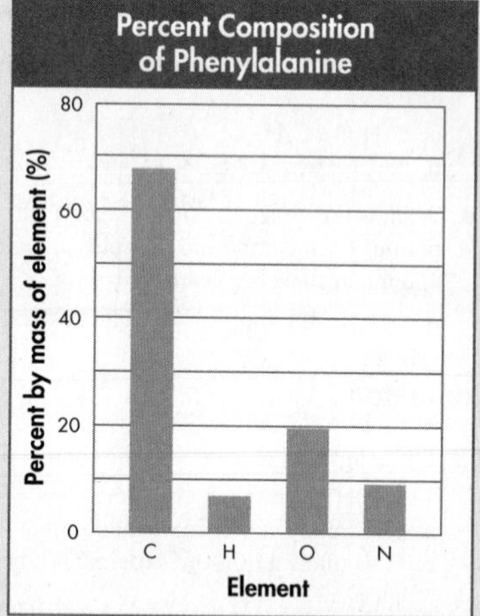

a. What is the empirical formula for phenylalanine?
b. If the molar mass of phenylalanine is 165.2 g/mol, what is its molecular formula?

Enrichment

*101. **Infer** Nitroglycerine contains 60% as many carbon atoms as hydrogen atoms, three times as many oxygen atoms as nitrogen atoms, and the same number of carbon and nitrogen atoms. The number of moles of nitroglycerine in 1.00 g is 0.00441. What is the molecular formula of nitroglycerine?

102. **Calculate** The density of nickel is 8.91 g/cm^3. How large a cube, in cm^3, would contain 2.00×10^{24} atoms of nickel?

*103. **Calculate** Dry air is about 20.95% oxygen by volume. Assuming STP, how many oxygen molecules are in a 75.0-g sample of air? The density of air is 1.19 g/L.

104. **Graph** The table below gives the molar mass and density of seven gases at STP.

Substance	Molar mass (g/mol)	Density (g/L)
Oxygen	32.0	1.43
Carbon dioxide	44.0	1.96
Ethane	30.0	1.34
Hydrogen	2.0	0.089
Sulfur dioxide	64.1	2.86
Ammonia	17.0	0.759
Fluorine	38.0	1.70

a. Plot these data, with density on the *x*-axis.
b. What is the slope of the straight-line plot?
c. What is the molar mass of a gas at STP that has a density of 1.10 g/L?
d. A mole of a gas at STP has a mass of 56.0 g. Use the graph to determine its density.

*105. **Calculate** Avogadro's number has been determined by about 20 different methods. In one approach, the spacing between ions in an ionic substance is determined by using a technique called X-ray diffraction. X-ray diffraction studies of sodium chloride have shown that the distance between adjacent Na^+ and Cl^- ions is 2.819×10^{-8} cm. The density of solid NaCl is 2.165 g/cm^3. By calculating the molar mass to four significant figures, you can determine Avogadro's number. What value do you obtain?

106. **Use Models** In Chapter 3, you learned that the densities of solids and liquids are measured in g/cm^3, but the densities of gases are measured in g/L. Draw atomic diagrams of a solid and of a gas that show why the two different units are practical.

Write About Science

107. **Connect to the BIGIDEA** Research the history of Avogadro's number. What elements other than carbon have been used to define a mole? Write a report that summarizes your findings.

CHEMYSTERY

A Formula for Cheating

Typically, steroids can be detected in an athlete's urine. A urine sample is collected and is first injected into an instrument that separates the chemical compounds in the urine.

The separated compounds are then analyzed using a mass spectrometer. The mass spectrometer provides information such as the molar mass of the compounds present in the urine sample and the molecular structure of these compounds. These structures can be compared against a database of known compounds to identify the presence of steroids in the sample.

*108. **Calculate** Analysis of an athlete's urine found the presence of a compound with a molar mass of 312 g/mol. How many moles of this compound are contained in 30.0 mg? How many molecules of the compound is this?

109. **Connect to the BIGIDEA** The compound found in the athlete's urine, the steroid THG, has a percent composition of 80.8% carbon, 8.97% hydrogen, and 10.3% oxygen. What is the empirical formula of THG? If the molar mass of THG is 312 g/mol, what is the molecular formula?

ASSESSMENT

Explain

CHEMYSTERY

START A CONVERSATION Point out that the mass spectrometer determines the molar masses of all the chemical compounds in the urine, including those that are expected to be present in normal urine. **Ask** How might the laboratory chemist prove that the substance detected by the instrument is not a normal component of urine? *(Analyze a sample of urine that is known to be drug-free, and compare the results of the two samples.)* Note that it is important for the mass spectrometer's data processor to use both structure and molar mass data to identify the drugs, as there may be more than one compound with the same molar mass.

CHEMYSTERY ANSWERS

108. 9.61×10^{-5} mol; 5.79×10^{19} molecules

109. **BIGIDEA** Both the empirical and molecular formulas of THG are $C_{21}H_{28}O_2$.

Answers

ENRICHMENT

101. $C_3H_5O_9N_3$

102. 21.9 cm^3

103. 3.54×10^{23} molcules O_2

104. a.

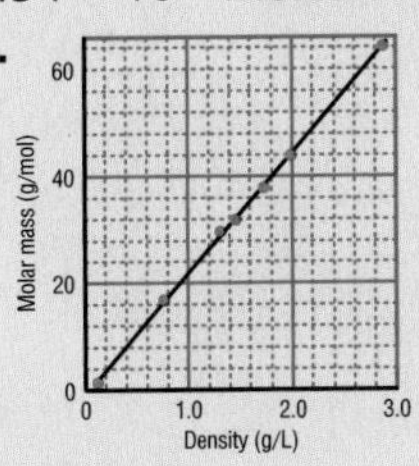

b. 22.4 L/mol

c. 24.6 g/mol

d. 2.5 g/L

105. 6.025×10^{23} formula units/mol

106. Diagrams should show the particles in a solid much more closely packed together than those of a gas. This closely packed arrangement of particles means that solids are much denser than gases. Thus, it makes sense to use a larger unit, g/cm^3, to measure the density of solids.

WRITE ABOUT SCIENCE

107. Students should mention the work of Johann Josef Loschmidt and Jean Perrin in their reports. They should also state why the constant is named after Amedeo Avogadro. Prior to carbon, hydrogen and oxygen were used to define the mole.

Evaluate

Answers

CUMULATIVE REVIEW

110. chemical change: wax burning
physical change: wax melting
physical change: wax vaporizing

111. **a.** physical change
b. chemical change
c. chemical change
d. physical change
e. chemical change
f. physical change

112. b, c

113. No; the student has ignored the units. The density of sugar is 1.59 g/mL; the density of carbon dioxide is much less, 1.83 g/L.

114. The block will float. Its density, 0.848 g/mL, is less than the density of water.

115. **a.** 4.72×10^3 mg
b. 97 km/h
c. 4.4×10^{-2} dm

116. **a.** 40 protons, 40 electrons, 50 neutrons
b. 46 protons, 46 electrons, 62 neutrons
c. 35 protons, 35 electrons, 46 neutrons
d. 51 protons, 51 electrons, 72 neutrons

117. **a.** $1s^22s^22p^5$
b. $1s^22s^1$
c. $1s^22s^22p^63s^23p^63d^{10}4s^24p^65s^1$

118. Magnesium and barium are both in Group 2A and have 2 valence electrons.

119. Cr, Cd, Cu, and Co

120. For Group A elements, the group number equals the number of valence electrons.

121. A molecule is composed of two or more atoms.

122. **a.** pyramidal H:P̤:H (with H below P) **c.** linear :S̤::C::S̤:
b. linear :O::C: **d.** tetrahedral C with four :F: atoms (:F̤:C:F̤: with :F: above and below)

123. For single bond a single line connects the atoms (X—X). Atoms are connected by two lines in a double bond (X=X), and three lines in a triple bond (X≡X). Alternatively, the bonds may be shown as pairs of dots.

124. Answers will vary. **a.** carbon monoxide (CO)
b. ozone (O_3) **c.** nitrogen dioxide (NO_2)

125. Calculate the electronegativity difference between two atoms. If the difference is small (0.0—0.4) the bond is nonpolar covalent. If the difference is greater than 2.0, the bond is most likely ionic. For values between 0.4 and 2.0, the bond is polar covalent.

126. **d.** CaS_2 **f.** BaOH

127. **a.** iron(III) hydroxide
b. ammonium iodide
c. sodium carbonate
d. carbon tetrachloride

128. **a.** KNO_3 **b.** CuO **c.** Mg_3N_2 **d.** AgF

Cumulative Review

★110. Identify at least one chemical change and two physical changes that are occurring in the photo.

★111. Classify each of the following as a physical change or a chemical change.
a. An aspirin tablet is crushed to a powder.
b. A red rose turns brown.
c. Grape juice turns to wine.
d. Fingernail polish remover evaporates.
e. A bean seed sprouts.
f. A piece of copper is beaten into a thin sheet.

112. Which of these statements are true about every solution?
a. Solutions are in the liquid state.
b. Solutions are homogeneous.
c. Solutions are mixtures.
d. Solutions are composed of at least two compounds.

113. A student writes down the density of table sugar as 1.59 and the density of carbon dioxide as 1.83. Can these values be correct? Explain.

★114. A block of wood measuring 2.75 cm × 4.80 cm × 7.50 cm has a mass of 84.0 g. Will the block of wood sink or float in water?

★115. Convert each of the following:
a. 4.72 g to mg
b. 2.7×10^3 cm/s to km/h
c. 4.4 mm to dm

★116. How many protons, electrons, and neutrons are in each isotope?
a. zirconium-90 **c.** bromine-81
b. palladium-108 **d.** antimony-123

★117. Write the complete electron configuration for each atom.
a. fluorine **b.** lithium **c.** rubidium

118. Why do the elements magnesium and barium have similar chemical and physical properties?

★119. Which of the following are transition metals: Cr, Cd, Ca, Cu, Co, Cs, Ce?

120. How can the periodic table be used to infer the number of valence electrons in an atom?

121. How does a molecule differ from an atom?

122. Draw electron dot structures and predict the shapes of the following molecules:
a. PH_3 **b.** CO **c.** CS_2 **d.** CF_4

123. How are single, double, and triple bonds indicated in electron dot structures?

124. Give an example of each of the following:
a. coordinate covalent bonding
b. resonance structures
c. exceptions to the octet rule

125. Explain how you can use electronegativity values to classify a bond as nonpolar covalent, polar covalent, or ionic.

★126. Identify any incorrect formulas among the following:
a. H_2O_2 **d.** CaS_2
b. $NaIO_4$ **e.** $CaHPO_4$
c. SrO **f.** BaOH

★127. Name these compounds.
a. $Fe(OH)_3$ **c.** Na_2CO_3
b. NH_4I **d.** CCl_4

★128. Write formulas for these compounds.
a. potassium nitrate
b. copper(II) oxide
c. magnesium nitride
d. silver fluoride

If You Have Trouble With . . .

Question	110	111	112	113	114	115	116	117	118	119	120	121	122	123	124	125	126	127	128
See Chapter	2	2	2	3	3	3	4	5	6	6	7	8	8	8	8	8	9	9	9

Standardized Test Prep for Indiana

Standards: C.4.3, C.4.7

Tips for Success

Wear a Watch Be aware of how many questions you have to answer and how much time you have to answer them. Look at your watch or a clock frequently to keep track of your progress.

1. Choose the term that best completes the second relationship.
   a. dozen : eggs
   mole : _____
   (A) atoms (B) 6.02×10^{23} (C) size (D) grams
   b. mole : Avogadro's number
   molar volume : _____
   (A) mole (B) water (C) STP (D) 22.4 L

Select the choice that best answers each question or completes each statement.

2. Calculate the molar mass of ammonium phosphate, $(NH_4)_3PO_4$.
   (A) 113.0 g/mol (B) 121.0 g/mol (C) 149.0 g/mol (D) 242.0 g/mol
3. Based on the structural formula below, what is the empirical formula for tartaric acid, a compound found in grape juice?

```
HO—CH—COOH
    |
HO—CH—COOH
```

   (A) $C_2H_3O_3$ (B) $C_4H_6O_6$ (C) CHO (D) $C_1H_{1.5}O_{1.5}$
4. How many hydrogen atoms are in six molecules of ethylene glycol, $C_2H_6O_2$?
   (A) 6 (B) 36 (C) $6 \times (6.02 \times 10^{23})$ (D) $36 \times (6.02 \times 10^{23})$
5. Which of these compounds has the largest percent by mass of nitrogen?
   (A) N_2O (B) NO (C) NO_2 (D) N_2O_3 (E) N_2O_4
6. Which of these statements is true of a balloon filled with 1.00 mol $N_2(g)$ at STP?
   I. The balloon has a volume of 22.4 L.
   II. The contents of the balloon have a mass of 14.0 g.
   III. The balloon contains 6.02×10^{23} molecules.
   (A) I only (B) I and II only (C) I and III only (D) I, II, and III
7. Allicin, $C_6H_{10}S_2O$, is the compound that gives garlic its odor. A sample of allicin contains 3.0×10^{21} atoms of carbon. How many hydrogen atoms does this sample contain?
   (A) 10 (B) 1.0×10^{21} (C) 1.8×10^{21} (D) 5.0×10^{21}

The lettered choices below refer to Questions 8–11. A lettered choice may be used once, more than once, or not at all.

(A) CH (B) CH_2 (C) C_2H_5 (D) CH_3 (E) C_2H_3

Which of the formulas is the empirical formula for each of the following compounds?

8. C_8H_{12}
9. C_6H_6
10. C_2H_6
11. C_4H_{10}

For Questions 12–14, write the molecular formula for each compound whose structural formula is shown. Then calculate the compound's molar mass.

12.
```
   H  OH H
   |  |  |
H—C—C—C—H
   |  |  |
   H  H  H
```

13.
```
      H  O
      |  ||
H2N—C—C—OH
      |
      H
```

14.
```
   H  H        O
   |  |        ||
H—C—C—O—C—H
   |  |
   H  H
```

If You Have Trouble With . . .

Question	1	2	3	4	5	6	7	8	9	10	11	12	13	14
See Lesson	10.2	10.1	10.3	10.1	10.3	10.2	10.1	10.3	10.3	10.3	10.3	10.1	10.1	10.1

ASSESSMENT

STP Answers

MULTIPLE CHOICE

1. a. A b. D
2. C
3. A
4. B
5. A
6. C
7. D
8. E
9. A
10. D
11. C
12. C_3H_8O, 60.0 g/mol
13. $C_2H_5NO_2$, 75.0 g/mol
14. $C_3H_6O_2$, 74.0 g/mol

11 Chemical Reactions Planning Guide

Introducing the BIGIDEA: REACTIONS

Chemical reactions involve processes in which reactants produce products.

IASS	Lessons and Objectives	Print Resources		
		For the Student	For the Teacher	
C.1.4, C.4.1, C.4.2	**11.1** Describing Chemical Reactions p 346–354 **11.1.1** Describe how to write a skeleton equation. **11.1.2** Describe the steps for writing and balancing a chemical equation.	Reading and Study Workbook Lesson 11.1 Lesson Assessment 11.1 p 354 Quick Lab: Removing Silver Tarnish p 354	Teacher Demo, p 350: An Example of a Chemical Change Teacher Demo, p 352: Balance a Chemical Equation	
C.1.4, C.4.1, C.4.5	**11.2** Types of Chemical Reactions p 356–367 **11.2.1** Identify the five general types of reactions.	Reading and Study Workbook Lesson 11.2 Lesson Assessment 11.2 p 367	Teacher Demo, p 360: Single-Replacement Reactions Teacher Demo, p 362: Double-Replacement Reactions Teacher Demo, p 365: Combustion of Iron Teacher Demo, p 366: A Combination Reaction	
C.1.4, C.4.1, C.4.2, C.4.5	**11.3** Reactions in Aqueous Solution p 369–373 **11.3.1** Describe the information found in a net ionic equation. **11.3.2** Predict the formation of a precipitate in a double-replacement reaction.	Reading and Study Workbook Lesson 11.3 Lesson Assessment 11.3 p 373 Small-Scale Lab: Precipitation Reactions: Formation of Solids p 374		

Assessing the BIGIDEA: REACTIONS

	Essential Questions 1. How do chemical reactions obey the law of conservation of mass? 2. How can you predict the products of a chemical reaction?	Study Guide p 375 Skills Tune-Up p 376 STP p 381 Reading and Study Workbook Self-Check and Vocabulary Review Chapter 11		

	PearsonChem.com	
	Editable Worksheets	Online Activities
	Teaching Resources, Lesson 11.1 Review Small-Scale Lab Manual Lab 14: Chemical Equations Small-Scale Lab Manual Lab 21: Synthesis and Qualitative Analysis of Gases	11.1 Lesson Overview Balancing Equations Balancing a Chemical Equation
	Teaching Resources, Lesson 11.2 Review Small-Scale Lab Manual Lab 15: Balancing Chemical Equations Lab 14: Types of Chemical Reactions Lab 15: Reactivity of Metals Lab Practical 11-1: Reactivity of Metals Lab Practical 11-4: Reaction Types	11.2 Lesson Overview Writing Equations for Double-Replacement Reactions A Balanced Equation for Fire Types of Reactions
	Teaching Resources, Lesson 11.3 Review Lab 16: Identification of Anions and Cations in Solution Lab 17: Precipitation Reactions Lab 18: Qualitative Analysis Small-Scale Lab Manual Lab 17: Halogen Ions in Solution Small-Scale Lab Manual Lab 22: Reactions of Aqueous Ionic Compounds Small-Scale Lab Manual Lab 23: Identification of Eight Unknown Solutions Lab Practical 11-2: Identification of Anions and Cations in Solution Lab Practical 11-3: Precipitation Reactions Lab Practical 11-5: Qualitative Analysis	11.3 Lesson Overview Identification of Cations in Solution Writing and Balancing Net Ionic Equations
	Core Teaching Resources (includes editable worksheets) • Lesson Reviews • Practice Problems • Interpret Graphs • Vocabulary Review • Chapter Quizzes and Tests • Lab Record Sheets	Chapter 11 Problem Set Balancing Equations

Materials List

For the Student

Quick Lab p 354
- aluminum foil, 20 cm by 20 cm
- large beaker or glass pan
- tarnished silver fork or spoon
- sodium hydrogen carbonate
- plastic tablespoon
- hot water

Small-Scale Lab p 374
- reaction surface
- 0.05*M* $AgNO_3$, 0.2*M* $Pb(NO_3)_2$, 0.5*M* $CaCl_2$, 1.0*M* Na_2CO_3, 0.1*M* Na_3PO_4, 0.5*M* NaOH, 0.2*M* Na_2SO_4, 1.0*M* NaCl

For the Teacher

Teacher Demo p 350
- 150-mL glass beaker
- table sugar
- fume hood
- concentrated sulfuric acid

Teacher Demo p 352
- 100-mL graduated cylinder
- 0.1*M* copper(II) chloride solution
- 400-mL beaker
- metric ruler
- scissors
- aluminum foil

Teacher Demo p 360
- magnesium ribbon
- 10-mL graduated cylinder
- 1*M* HCl
- sodium
- 2 beakers
- tongs
- water

Teacher Demo p 362
- equimolar solutions of $BaCl_2$, Na_2SO_4, Na_3PO_4, $CaCl_2$, $Pb(NO_3)_2$, and KI

Teacher Demo p 365
- superfine steel wool
- plastic sandwich bag
- balance
- ring stand
- utility clamp
- matches

Teacher Demo p 366
- magnesium ribbon
- large crucible
- lab burner
- cobalt blue glass or exposed photographic film
- crucible tongs
- matches or igniter

Additional Digital Resources

Online Student Edition
Online Teacher's Edition
Exam*View Assessment Suite CD-ROM*
Classroom Resources DVD

11.3 Virtual Chem Lab 22: Precipitation Reactions: Formation of Solids
Virtual Chem Lab 23: Identification of Cations in Solution
Virtual Chem Lab 24: Qualitative Analysis

CHAPTER 11

What's Online

CHEMISTRY TUTORIAL Students can work step-by-step to balance chemical equations, write equations for double-replacement reactions, and write and balance net ionic equations, by accessing the Chapter 11 Chemistry Tutorials.

ONLINE PROBLEM SET Additional practice problems about writing and balancing chemical equations can be found in the Chapter 11 online Problem Set.

MATH TUTORIAL The online Math Tutorial will help guide students through balancing equations.

DIRECTED VIRTUAL LAB Encourage students to review identifying cations in solutions when different ions are combined and react or precipitate, using the animated virtual lab for Chapter 11. The animation is followed by analysis questions.

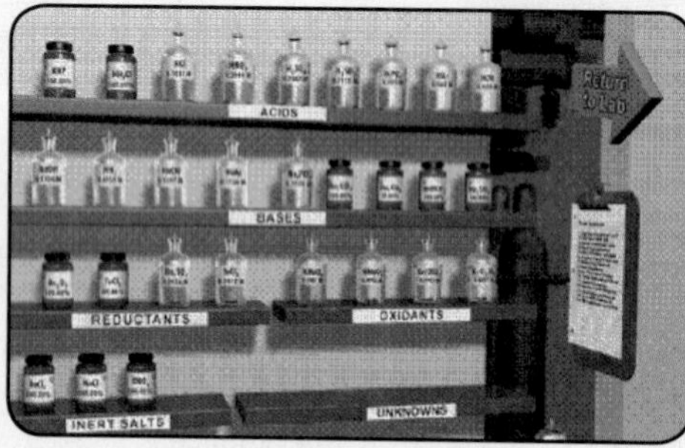

KINETIC ART Have students watch animations about balancing equations and types of reactions. Each animation is followed by analysis questions.

CONCEPTS IN ACTION Students can watch an interactive animation about balancing equations by finding where the products go in a combustion reaction. The animation is followed by questions.

IN ACADEMIC STANDARDS for SCIENCE

Standard 1: Properties and States of Matter

C.1.4 Describe physical and chemical changes at the molecular level.

Standard 4: Reactions and Stoichiometry

C.4.1 Predict products of simple reactions.

C.4.2 Balance chemical equations using the law of conservation of mass and use them to describe chemical reactions.

C.4.5 Recognize and classify types of reactions.

11 Chemical Reactions

INSIDE:

PearsonChem.com

Reef aquariums like this one use an aqueous solution of calcium hydroxide to provide calcium for marine animals such as snails and coral.

Focus on ELL

1 CONTENT AND LANGUAGE Begin the lesson by writing the word *reaction* on the board and defining it as "a response to another event." As a class, create a list of familiar types of reactions on the board, such as emotional, nuclear, light, and reflex.

BEGINNING

LOW Have students create drawings representing reactions they have experienced.

HIGH Have students copy the list created by the class and check off the reactions with which they are familiar.

INTERMEDIATE: LOW/HIGH Using the list created by the class, students should form groups of reactions that have characteristics in common.

ADVANCED: LOW/HIGH Have each student write a journal entry describing a reaction he or she had to an event.

CHAPTER 11

BIG IDEA

REACTIONS

Essential Questions:

1. *How do chemical reactions obey the law of conservation of mass?*
2. *How can you predict the products of a chemical reaction?*

CHEMYSTERY

Order in the Lab

"How does anybody find anything in this lab?" Maria muttered to herself. Her spice cabinet at home was in alphabetical order. If she wanted to find the cinnamon, it was right there between the bay leaves and cumin. She decided to help the teacher by rearranging the chemicals before class.

She found the sodium cyanide and put it next to the sodium sulfide. Then she picked up the sodium hydroxide and looked around until she found the sulfuric acid. Before Maria could put down the sodium hydroxide, her teacher walked in and called out "Maria, leave the chemicals alone! They're arranged the way they are for a reason."

Why did Maria's teacher stop her from rearranging the chemicals?

► Connect to the **BIG**IDEA As you read about chemical reactions, think about how you can predict the outcomes of chemical reactions.

INDIANA'S ACADEMIC STANDARDS FOR SCIENCE

Properties and States of Matter C.1.4; **Reactions and Stoichiometry** C.4.1, C.4.2, C.4.5. See lessons for details.

UbD

Understanding by Design

Students are building toward describing chemical reactions by applying an overall knowledge of reactions.

PERFORMANCE GOALS At the end of Chapter 11, students will be able to answer the essential questions by applying their knowledge of chemical reactions.

ESSENTIAL QUESTIONS Read the essential questions aloud. **Ask** What does *conservation of mass* mean? *(Mass is neither produced nor lost.)* **Ask** What would this mean for a chemical reaction? *(The amount of mass is the same before and after the reaction.)*

BIGIDEA Use the photo of the reef aquarium to help students connect to this chapter's concepts. Activate prior knowledge by asking students if they have ever seen coral. Explain that coral skeletons are made of calcium carbonate. **Ask** Where do you think coral animals get the calcium they need to make the skeletons? *(They use calcium ions in seawater.)* Have students read the photo caption. **Ask** How do you think the calcium hydroxide changes to calcium ions the coral animals can use? *(The calcium hydroxide dissociates into calcium ions and hydroxide ions.)*

CHEMYSTERY Connect the CHEMystery to the Big Idea by explaining that certain chemicals are likely to react with each other and produce new substances. As a result, it is important not to store them side-by-side. The teacher arranged these chemicals to prevent them from coming in contact with each other. Have students predict what might have happened if Maria had placed the sodium hydroxide next to the wrong chemical.

Introduce the Chapter

CONNECTION TO FOOD Students may not realize that they experience chemical reactions frequently each day. Use the activities to demonstrate some chemical reactions.

You will need an effervescent tablet and a glass of water. Drop the tablet into the water so that students can see the bubbles. **Ask** What tells you that a reaction is taking place? (bubbles formed) Show students a photo of a pizza. **Ask** What tells you that a chemical reaction took place? (change in color, odor, and production of gas to raise the dough) You can also show students a cut apple that has turned brown. **Ask** What tells you that a chemical reaction took place? (change in color)

What are some signs that a reaction is taking place? (See above.)

UntamedScience™

Nobel Efforts

The Untamed Science crew surveys a few of the chemists throughout history who have won the Nobel Prize.

LESSON 11.1

Key Objectives

11.1.1 DESCRIBE how to write a skeleton equation.

11.1.2 DESCRIBE the steps for writing and balancing a chemical equation.

Additional Resources

- Reading and Study Workbook, Lesson 11.1

Available Online or on Digital Media:

- Teaching Resources, Lesson 11.1 Review
- Small-Scale Laboratory Manual, Lab 14

PearsonChem.com

Engage

CHEMISTRY & YOU Have students study the photograph and read the text. Discuss why there is an interest in producing hydrogen fuel-cell cars. Write the word equation for the reaction of hydrogen with oxygen to produce water. **Ask** Do you think there is a simpler way to write chemical equations? *(Yes, describe the reaction using chemical symbols of the elements and chemical formulas of compounds.)*

Activate Prior Knowledge

Ask What are the differences between molecular compounds and ionic compounds? *(Molecular compounds are composed of nonmetallic elements. Ionic compounds are composed of a metal cation and an anion.)* **Ask** How are chemical compounds represented? *(By chemical formulas that show the kinds and numbers of atoms in each molecule or formula unit.)*

IN ACADEMIC STANDARDS for SCIENCE

- Develop explanations based on reproducible data and observations gathered during laboratory investigations.
- Clearly communicate their ideas and results of investigations verbally and in written form using tables, graphs, diagrams, and photographs.

C.1.4 Describe physical and chemical changes at the molecular level.

Also covered C.4.1, C.4.2

11.1 Describing Chemical Reactions

C.1.4 Physical and chemical changes; **C.4.1** Predict products; **C.4.2** Balance chemical equations

CHEMISTRY & YOU

Q: *How is a chemical reaction going to change the way you drive?* You've probably heard about hydrogen fuel-cell cars. Fuel cells produce electricity through a chemical reaction without any of the combustion that you find in typical gasoline engines. In this lesson, you'll learn how to write and balance the equations that represent chemical reactions.

Key Questions

How do you write a skeleton equation?

What are the steps for writing and balancing a chemical equation?

Vocabulary

- chemical equation
- skeleton equation
- catalyst
- coefficient
- balanced equation

Introduction to Chemical Equations

How do you write a skeleton equation?

Every minute of the day chemical reactions take place—both inside you and around you. After a meal, a series of chemical reactions take place as your body digests food. Similarly, plants use sunlight to drive the photosynthetic processes needed to produce plant growth. Although the chemical reactions involved in photosynthesis and digestion are different, both chemical reactions are necessary to sustain life. All chemical reactions, whether simple or complex, involve changing substances.

In a chemical reaction, one or more reactants change into one or more products. Cooking food always involves a chemical reaction. In order to bake muffins, you begin with a recipe and ingredients, as shown in Figure 11.1. The recipe tells you which ingredients to mix together and how much of each to use. Chemical reactions take place when the ingredients or reactants are mixed together and heated in the oven. The product, in this case, is a batch of muffins. Chemists use a chemical equation—a quick, shorthand notation—to convey as much information as possible about what happens in a chemical reaction.

Figure 11.1
Reactants and Products
Reactants in the ingredients undergo chemical changes to form the product, the muffins.
Observing ***What evidence shows that chemical changes have occurred?***

Focus on ELL

1 CONTENT AND LANGUAGE Have students create a chart with four columns. Have them copy the key questions and the vocabulary from the beginning of the lesson in the first column. Then, have them title the second column *Answer/Definition*, the third column *Where can I find it?* and the fourth column *Do I know it?* As you proceed through the lesson, encourage students to fill in column two and three of the chart. At the end of the lesson ask them to self-assess their knowledge of each question or definition by filling in the fourth column. You can use this column as a formative assessment to determine if reteaching needs to be done.

2 FRONTLOAD THE LESSON Write the word *skeleton* on the board. Discuss as a class the context in which students are most familiar with the word *skeleton*. Ask students to either discuss with their group or to write in a journal why an equation without specific amounts would be called a *skeleton equation*.

3 COMPREHENSIBLE INPUT Create a one to two word name for each of the rules for writing and balancing equations. Divide the class into 6 groups, assign each group a rule, and have them create a poster describing their rule. Display the posters.

Figure 11.2 Examples of Reactions
Three common chemical reactions are shown below.

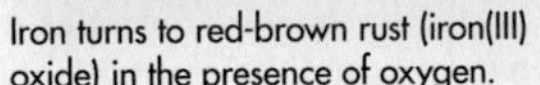
Iron turns to red-brown rust (iron(III) oxide) in the presence of oxygen.

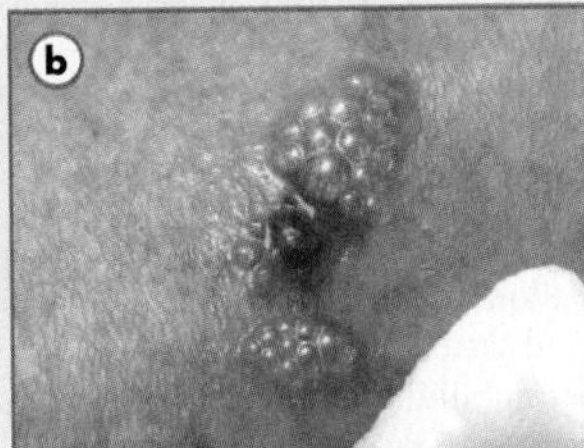
Water and oxygen form when hydrogen peroxide is poured on a cut.

The products of burning methane are carbon dioxide and water.

Word Equations How do you describe what happens in a chemical reaction? Recall from Chapter 2 the shorthand method for writing a description of a chemical reaction. In this method, the reactants were written on the left and the products on the right. An arrow separated them. You read the arrow as *yields, gives,* or *reacts to produce.*

Reactants ⟶ products

How could you describe the rusting of iron shown in Figure 11.2a? You could say: "Iron reacts with oxygen to produce iron(III) oxide (rust)." Although that is a perfectly good description, it's quicker to identify the reactants and product by means of a word equation.

Iron + oxygen ⟶ iron(III) oxide

In a word equation, write the names of the reactants to the left of the arrow, separated by plus signs; write the names of the products to the right of the arrow, also separated by plus signs. Notice that no plus sign is needed on the product side of this equation because iron(III) oxide is the only product.

Have you ever poured the antiseptic hydrogen peroxide on an open cut? Bubbles of oxygen gas form rapidly, as shown in Figure 11.2b. The production of a new substance, a gas, is evidence of a chemical change. Two new substances are produced in this reaction, oxygen gas and liquid water. You could describe this reaction by saying, "Hydrogen peroxide decomposes to form water and oxygen gas." But, you could also write a word equation.

Hydrogen peroxide ⟶ water + oxygen

When you light a burner on your stove, methane gas bursts into flames and produces the energy needed to heat your soup. Methane is the major component of natural gas, a common fuel for heating homes and cooking food. The burning of methane, as shown in Figure 11.2c, is a chemical reaction. How would you write the word equation for this reaction? Burning a substance typically requires oxygen, so methane and oxygen are the reactants. The products are water and carbon dioxide. Thus, the word equation is as follows:

Methane + oxygen ⟶ carbon dioxide + water

Professional Development Note

HYDROGEN PEROXIDE Louis Jacques Thenard (1777–1857), a French chemist, is credited with discovering hydrogen peroxide. Students are likely familiar with the bleaching agent abilities of hydrogen peroxide, because it is used as an ingredient in hair bleach and teeth-whitening products. Students may be surprised to learn that hydrogen peroxide is sometimes used to restore the clarity of old paintings. Old paintings utilized lead-based paints, and these paints tended to darken over time (because PbS forms). Applying hydrogen peroxide to a painting removes sulfur compounds from the surface (converting PbS to $PbSO_4$), resulting in a clearer image.

Foundations for Reading

BUILD VOCABULARY Write the following chemical equation on the board and have students provide synonymous words or phrases for each of the vocabulary words as they describe the equation:

$$2H_2O_2 \xrightarrow{MnO_2} 2H_2O + O_2$$

READING STRATEGY As students read this section, have them write in their own words the sequence of steps that need to be taken to write and balance a chemical equation.

Explain

Introduction to Chemical Equations

MAKING CONNECTIONS Have students study Figure 11.1. Explain that bakers use baking soda and baking powder to make muffin batter rise during baking. Both compounds undergo chemical reactions in the batter when they become wet and are exposed to heat, producing bubbles of CO_2. Explain that the bubbles of trapped CO_2 cause the batter to rise; as the batter bakes, the bubbles leave tiny holes behind. Without the formation of CO_2, the muffins would be flat and heavy, instead of light and fluffy.

SUMMARIZE Have students create a running list in their notebooks of the types of evidence that indicate a chemical change has occurred, with an explanation of why each type of evidence indicates a chemical change.

Answers

FIGURE 11.1 The ingredients have turned into a product with different characteristics such as smell and taste. Bubbles were also formed, as evidenced by the tiny air pockets inside the muffin.

LESSON 11.1

Explain

USE VISUALS Direct students to Table 11.1. Make sure students understand the meaning and role of each symbol. Point out that the phase symbols provide important clues about reactions. Note that the items placed above the reaction arrow represent conditions that must be met before the reaction can take place at a reasonable pace.

USE VISUALS Direct students to Figure 11.3. Point out that although hydrogen peroxide contains the same kinds of elements as water, it has very different properties. Hydrogen peroxide is much less stable than water. Have students write the balanced chemical equation for the reaction described in the caption. Students should indicate the MnO_2 catalyst above the reaction arrow. Write the equation on the board and explain that this is an example of a decomposition reaction—a single compound is broken down into two or more products. Explain that although hydrogen peroxide is an unstable compound that readily decomposes to produce water and oxygen, it does so slowly at room temperature. Therefore, hydrogen peroxide must come in contact with a catalyst in order for the decomposition to proceed rapidly, as shown in Figure 11.3b.

APPLY CONCEPTS Tell students that frequently in nature the products of one reaction become the reactants of a subsequent reaction. As an example, use the combustion of octane (C_8H_{18}). Explain that octane is the primary ingredient of gasoline, which produces carbon dioxide and water vapor upon combustion. Point out that the carbon dioxide can then combine with atmospheric water vapor to produce carbonic acid (H_2CO_3), which makes rain slightly acidic. Ask students to brainstorm other such processes.

Figure 11.3
Speeding Up a Reaction
Hydrogen peroxide decomposes to form water and oxygen gas. **a.** Bubbles of oxygen appear slowly as decomposition proceeds. **b.** With the addition of the catalyst manganese(IV) oxide (MnO_2), decomposition speeds up. The white "smoke" is condensed water vapor.

Chemical Equations Word equations adequately describe chemical reactions, but they are cumbersome. It's easier to use the formulas for the reactants and products to write chemical equations. A **chemical equation** is a representation of a chemical reaction; the formulas of the reactants (on the left) are connected by an arrow with the formulas of the products (on the right). Here is a chemical equation for rusting:

$$Fe + O_2 \longrightarrow Fe_2O_3$$

Equations that show just the formulas of the reactants and products are called skeleton equations. A **skeleton equation** is a chemical equation that does not indicate the relative amounts of the reactants and products. The first step in writing a complete chemical equation is to write the skeleton equation. **To write a skeleton equation, write the chemical formulas for the reactants to the left of the yields sign (arrow) and the formulas for the products to the right.**

To add more information to the equation, you can indicate the physical states of substances by putting a symbol after each formula. Use (*s*) for a solid, (*l*) for a liquid, (*g*) for a gas, and (*aq*) for a substance in aqueous solution (a substance dissolved in water). Here is the unbalanced equation for rusting with symbols for the physical states added:

$$Fe(s) + O_2(g) \longrightarrow Fe_2O_3(s)$$

In many chemical reactions, a catalyst is added to the reaction mixture. A **catalyst** is a substance that speeds up the reaction but is not used up in the reaction. A catalyst is neither a reactant nor a product, so its formula is written above the arrow in a chemical equation. For example, Figure 11.3 shows that the compound manganese(IV) oxide ($MnO_2(s)$) catalyzes the decomposition of an aqueous solution of hydrogen peroxide ($H_2O_2(aq)$) to produce water and oxygen.

$$H_2O_2(aq) \xrightarrow{MnO_2} H_2O(l) + O_2(g)$$

Many of the symbols commonly used in writing chemical equations are listed below.

Table 11.1

Symbols Used in Chemical Equations

Symbol	Explanation
+	Separates two reactants or two products
$\longrightarrow$	"Yields," separates reactants from products
$\rightleftharpoons$	Used in place of $\longrightarrow$ for reversible reactions
(*s*), (*l*), (*g*)	Designates a reactant or product in the solid state, liquid state, and gaseous state; placed after the formula
(*aq*)	Designates an aqueous solution; the substance is dissolved in water; placed after the formula
$\xrightarrow{\Delta}$ $\xrightarrow{heat}$	Indicates that heat is supplied to the reaction
$\xrightarrow{Pt}$	A formula written above or below the yield sign indicates its use as a catalyst (in this example, platinum).

Differentiated Instruction

ELL ENGLISH LANGUAGE LEARNERS Have students copy Table 11.1 into their vocabulary notebooks. Explain that the symbols used to write equations are universal in the scientific community; explain that all scientists use this language of symbols. Next to the symbols in their tables, students should write words from their native languages to help them quickly become familiar with the meaning of the symbols.

L1 STRUGGLING STUDENTS As students learn how to balance equations, have them, working in pairs, take turns writing equations for each other to balance. Two pairs of students may enjoy getting together for a "doubles match" in which each pair works as a team to balance equations that the opposing team creates.

L3 ADVANCED STUDENTS Challenge students to propose a hypothesis about why humans produce an enzyme that helps catalyze the decomposition of hydrogen peroxide. *(to prevent the accumulation in the body of hydrogen peroxide, a toxic byproduct of reactions involving oxygen)*

SampleProblem 11.1

Writing a Skeleton Equation

Hydrochloric acid reacts with solid sodium hydrogen carbonate. The products formed are aqueous sodium chloride, water, and carbon dioxide gas. Write a skeleton equation for this chemical reaction.

1 Analyze **Identify the relevant concepts.**
Write the correct formula for each substance in the reaction. Indicate the state of each substance. Separate the reactants from the products with an arrow. Use plus signs to separate the two reactants and each of the three products.

2 Solve **Apply concepts to this problem.**

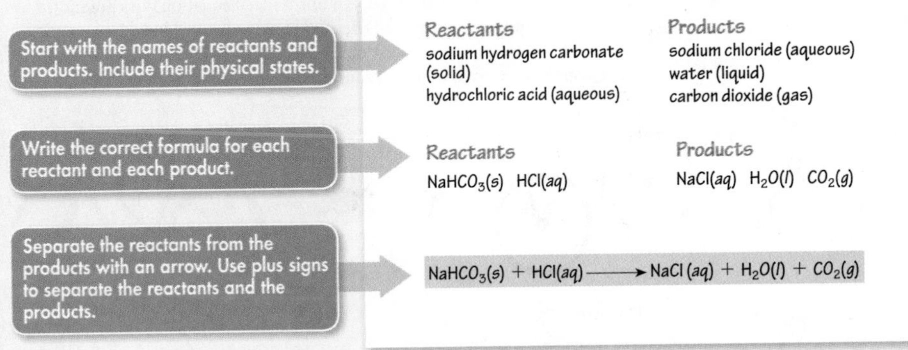

1. Write a sentence that describes this chemical reaction:
$Na(s) + H_2O(l) \longrightarrow NaOH(aq) + H_2(g)$

2. Sulfur burns in oxygen to form sulfur dioxide. Write a skeleton equation for this chemical reaction.

Balancing Chemical Equations

What are the steps for writing and balancing a chemical equation?

How would you write a word equation for the manufacture of bicycles? Simplify your task by limiting yourself to four major components: frames, wheels, handlebars, and pedals. Your word equation for making a bicycle could read like this.

$$\underbrace{\text{Frame} + \text{wheel} + \text{handlebar} + \text{pedal}}_{\text{Reactants}} \longrightarrow \underbrace{\text{bicycle}}_{\text{Product}}$$

Your word equation shows the reactants (the kinds of parts) and the product (a bicycle).

Foundations for Math

TRANSLATING WORDS TO SYMBOLS Students in algebra use key words to translate a word problem into a mathematical expression. For example, *three more than two times a number is at most one-half* translates to $3 + 2x \leq ½$. The word *more* corresponds to +, the word *times* corresponds to placing the coefficient 2 next to the variable *x*, and the phrase *at most* corresponds to ≤. Explain that students should look for key words in a written description of a chemical reaction to help them translate it into a skeleton equation.

In **Sample Problem 11.1**, point out that in the first sentence, the words *reacts with* indicate that the two compounds are the reactants of the equation and should be separated by the + sign, and the word *solid* indicates the phase (*s*) of the particular reactant. Point out that in the next sentence, the words *products formed* indicate the insertion of the reaction arrow to separate the reactants from the products. And the words *aqueous* and *gas* indicate the products' phases (*aq*) and (*g*) respectively.

Explain

Sample Practice Problem

Write a skeleton molecular equation for the following chemical reaction:

Sulfuric acid reacts with solid sulfur to form sulfur dioxide gas and water.

(Sulfuric acid (aq) + Sulfur (s) → Sulfur dioxide (g) + Water (l)
$H_2SO_4(aq) + S(s) \rightarrow SO_2(g) + H_2O(l)$)

Explain

Balancing Chemical Equations

APPLY CONCEPTS Review with students the law of conservation of mass. Emphasize that mass, or matter, cannot be created or destroyed. Point out that in a chemical reaction, atoms are neither created nor destroyed; all reactants must be accounted for in the products. Explain that it is because of this law that a chemical equation is not complete until it is written as a balanced equation.

Misconception Alert

Students often think that they can balance an equation by changing the subscripts in one or more of the formulas. Use the following example to show why this approach is incorrect.

$H_2(g) + O_2(g) \rightarrow H_2O(l)$ could be balanced by changing the formula of the product to H_2O_2. But H_2O_2 (hydrogen peroxide) is not the same substance as water. Therefore, the equation would describe a different reaction.

To help students overcome this misconception, have them draw boxes around the formulas for the reactants and products before they start to balance an equation. Tell them that the boxes are "off-limits" or "out of bounds." Emphasize that they cannot change any number that appears inside a box.

Answers

1. When solid sodium is dropped in water, hydrogen gas and aqueous sodium hydroxide are produced.
2. $S(s) + O_2(g) \rightarrow SO_2(g)$

LESSON 11.1

Explore

An Example of a Chemical Change

Teacher Demo

PURPOSE Students will observe a dramatic example of chemical change.

MATERIALS 150-mL glass beaker, table sugar, hood, concentrated sulfuric acid

PROCEDURE Fill a 150-mL glass beaker to the 30-mL mark with table sugar. Place the beaker in a hood, and add 30 mL of concentrated sulfuric acid, which will dehydrate the sugar.

SAFETY AND DISPOSAL The hood will remove any sulfur dioxide and carbon monoxide, which are possible byproducts of the reaction. Wear goggles, a face shield, a lab apron, and protective gloves. Use a plexiglass shield to separate the students from the reaction. As a further precaution, have students wear goggles. For disposal, wear protective gear and place the beaker and contents into a 500-mL beaker half full of water. Swirl the small beaker in the water and then neutralize using a base with a molarity of less than 1*M*. Decant the liquid and fill the large beaker halfway with water once again. Let stand overnight in a protected location, test for pH, and neutralize. Repeat if necessary. Wrap the carbon in newspaper before placing it in a wastebasket.

EXPECTED OUTCOME The carbon left behind will expand to form a porous foam cylinder as the water vapor escapes.

But if you were responsible for ordering parts to make a bicycle, this word equation would be inadequate because it does not indicate the quantity of each part needed to make one bicycle.

A standard bicycle is composed of one frame (F), two wheels (W), one handlebar (H), and two pedals (P). The formula for a bicycle would be FW_2HP_2. The skeleton equation would be

$$F + 1W + H + 1P \longrightarrow FW_2HP_2$$

This equation is unbalanced. An unbalanced equation does not indicate the quantity of the reactants needed to make the product. A complete description of the reaction must include not only the kinds of parts involved but also the quantities of parts required.

$$F + 2W + H + 2P \longrightarrow FW_2HP_2$$

This equation for making a bicycle is balanced. It tells you that one frame, two wheels, one handlebar, and two pedals produce one bicycle. To balance the equation, the number 2 was placed before wheels and pedals. The number 1 is understood to be in front of *frame, handlebar,* and *bicycle.* These numbers are called **coefficients**—small whole numbers that are placed in front of the formulas in an equation in order to balance it. In this balanced equation, the number of each bicycle part on the reactant side is the same as the number of those parts on the product side. A chemical reaction is also described by a **balanced equation** in which each side of the equation has the same number of atoms of each element and mass is conserved.

Recall that John Dalton's atomic theory states that as reactants are converted to products, the bonds holding the atoms together are broken, and new bonds are formed. The atoms themselves are neither created nor destroyed; they are merely rearranged.

This part of Dalton's theory explains the law of conservation of mass: In any chemical change, mass is conserved. The atoms in the products are the same atoms that were in the reactants—they are just rearranged. Representing a chemical reaction by a balanced chemical equation is a two-step process. **To write a balanced chemical equation, first write the skeleton equation. Then use coefficients to balance the equation so that it obeys the law of conservation of mass.** In every balanced equation, each side of the equation has the same number of atoms of each element.

Differentiated Instruction

L3 ADVANCED STUDENTS Point out that not all chemical equations are easily balanced. Methods other than those currently shown will be used to balance more complex equations, such as those introduced in Chapter 21. Challenge students to balance the following equation: $HCl + HNO_3 \rightarrow HOCl + NO + H_2O$.

($3HCl + 2HNO_3 \rightarrow 3HOCl + 2NO + H_2O$)

LPR LESS PROFICIENT READERS Provide students with molecular model kits or foam balls of different sizes and colors. Using the balls to represent different types of atoms, students should model balancing a simple equation. Provide students with several additional unbalanced chemical equations, and have them balance the equations by using the models. Caution students to keep "atoms" in a compound together so that multiples of the entire unit are used, not multiples of just one atom in the unit.

Sometimes, though, a skeleton equation may already be balanced. For example, carbon burns in the presence of oxygen to produce carbon dioxide.

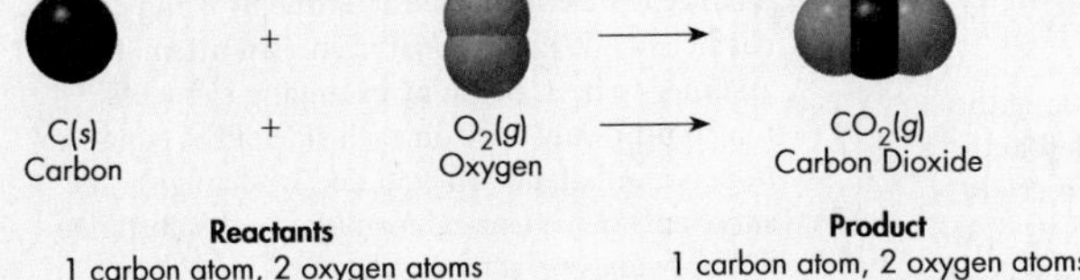

This equation is balanced. One carbon atom and two oxygen atoms are on each side of the equation. You do not need to change the coefficients; they are all understood to be 1.

What about the equation for the reaction of hydrogen and oxygen gas? When hydrogen and oxygen are mixed, the product of the reaction is water. The skeleton equation is as follows:

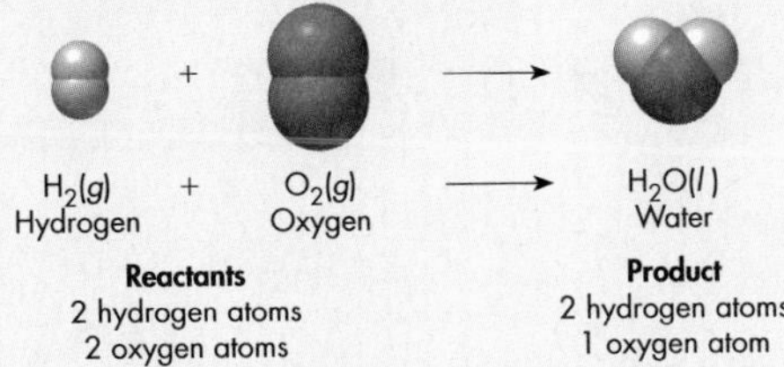

The formulas for all the reactants and the product are correct, but this equation is not balanced. Count the atoms on both sides of the equation. Two oxygen atoms are on the reactant (left) side of the equation and only one oxygen atom is on the product (right) side. As written, the equation does not obey the law of conservation of mass, and so it does not describe quantitatively what really happens. What can you do to balance it?

To balance the equation for the reaction of hydrogen and oxygen, count the number of each kind of atom. Hydrogen is balanced, but oxygen is not. If you put the coefficient 2 in front of H_2O, the oxygen will be balanced. Now twice as many hydrogen atoms are in the product as are in the reactants. To correct this equation, put the coefficient 2 in front of H_2. Four hydrogen atoms and two oxygen atoms are on each side of the chemical equation. The equation is now balanced.

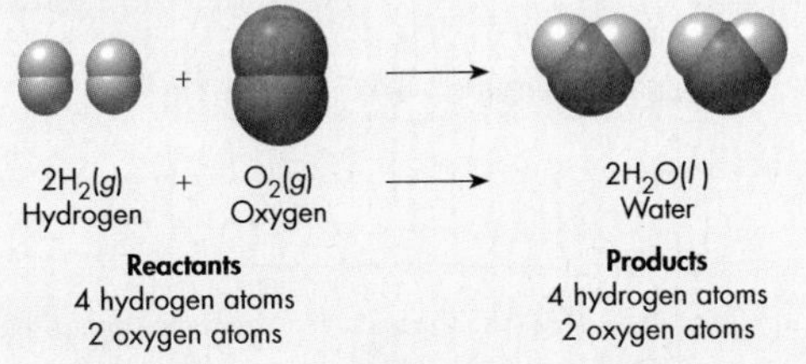

A few guidelines for writing and balancing equations are in the table on the next page.

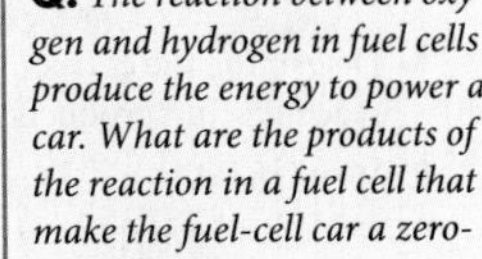

Q: *The reaction between oxygen and hydrogen in fuel cells produce the energy to power a car. What are the products of the reaction in a fuel cell that make the fuel-cell car a zero-emission car?*

UbD Check for Understanding

The Essential Question How do chemical reactions obey the law of conservation of mass?

Assess students' understanding of the law of conservation of mass by asking them to answer the following question:

Why do chemical equations have to be balanced when representing a chemical reaction? *(Atoms cannot be created or destroyed, so they must be conserved. Any atom that is present in a reactant must also be present in a product.)*

ADJUST INSTRUCTION If students are having trouble answering, write the law of conservation of mass on the board, along with a balanced chemical equation. Then ask them to try to answer the question again.

Explain

Misconception Alert

Some students think that the coefficients on both sides of the equation have to be the same in order for the number of atoms of each type to balance. Use the diagram of the formation of CO_2 on the student page to illustrate this concept. The atoms balance, but the coefficients total 2 on the left side of the equation and 1 on the right side. Convey that if the sums of the coefficients on both sides of a balanced chemical equation are the same number, it is coincidental.

CHEMISTRY & YOU Direct students to the reaction of hydrogen and oxygen to yield water. **Ask** What does "zero emission" mean based on the equation? *(Emissions are greenhouse gases and pollutants such as those produced by the combustion of fossil fuels. When hydrogen and oxygen react in a fuel cell, their only product is water.)*

Extend

Connect to TECHNOLOGY

Although fuel-cell technology sounds simple—hydrogen fuel reacts with oxygen from the air to produce water and electricity—explain to the class that in reality, the technology is fairly complicated. Have students research fuel-cell technology and note its challenges and successes in regard to powering automobiles. Instruct students to create a technological presentation explaining the advantages and disadvantages of fuel-cell technology.

LESSON 11.1

Explain

Balancing Chemical Equations

APPLY CONCEPTS Emphasize to students that in order to write a correct balanced chemical equation, they must correctly write the formulas for all reactants and products. Write an unbalanced chemical equation on the board, with all the formulas written correctly. Rewrite the equation with one formula written incorrectly. Identify the incorrect formula so that students consistently recognize that it is not correct. Have students balance the two equations and note the differences. Often an equation with an incorrect formula cannot be balanced.

Sample Practice Problem

Rewrite the following word equation as a balanced chemical equation:

aluminum sulfate + calcium hydroxide → aluminum hydroxide + calcium sulfate

($Al_2(SO_4)_3 + 3Ca(OH)_2 \rightarrow 2Al(OH)_3 + 3CaSO_4$)

Explore

Teacher Demo

PURPOSE Students will observe a chemical reaction and write skeleton and balanced chemical equations for the reaction.

MATERIALS 100-mL graduated cylinder, 0.1*M* copper(II) chloride solution, 400-mL beaker, metric ruler, scissors, aluminum foil

SAFETY $CuCl_2$ is an irritant.

PROCEDURE Add 200 mL of 0.1*M* copper(II) chloride solution to a 400-mL beaker. Cut a 5-cm square piece of aluminum foil. Crumple the piece of aluminum foil into a loose ball and place it in the copper solution. Ask students to note any evidence of chemical change. Write a sentence on the board that describes the chemical reaction. Have students write a skeleton equation for the reaction in their notebooks. Then have students balance the equation. Decant the liquid and flush down the drain with excess water. Dry the copper residue and use it again.

EXPECTED OUTCOME The solution will begin to produce gas, and a red-brown precipitate of copper will fall to the bottom of the beaker. The skeleton equation is the following:

$$CuCl_2(aq) + Al(s) \rightarrow AlCl_3(aq) + Cu(s).$$

The balanced chemical equation is the following:

$$3CuCl_2(aq) + 2Al(s) \rightarrow 2AlCl_3(aq) + 3Cu(s)$$

Rules for Writing and Balancing Equations

1. Determine the correct formulas for all the reactants and products.
2. Write the skeleton equation by placing the formulas for the reactants on the left and the formulas for the products on the right with a yields sign (⟶) in between. If two or more reactants or products are involved, separate their formulas with plus signs.
3. Determine the number of atoms of each element in the reactants and products. Count a polyatomic ion as a single unit if it appears unchanged on both sides of the equation.
4. Balance the elements one at a time by using coefficients. When no coefficient is written, it is assumed to be 1. Begin by balancing elements that appear only once on each side of the equation. Never balance an equation by changing the subscripts in a chemical formula. Each substance has only one correct formula.
5. Check each atom or polyatomic ion to be sure that the number is equal on both sides of the equation.
6. Make sure all the coefficients are in the lowest possible ratio.

CHEM TUTOR

Sample Problem 11.2

Balancing a Chemical Equation

Students suspended copper wire in an aqueous solution of silver nitrate. They noticed a deposit of silver crystals on the copper wire when the copper reacted with the silver nitrate. They recorded the equation for this reaction but didn't balance it. Balance their equation.

$$AgNO_3(aq) + Cu(s) \longrightarrow Cu(NO_3)_2(aq) + Ag(s)$$

1 Analyze **Identify the relevant concepts.**
Apply the rules for balancing equations. Because the nitrate polyatomic ion appears as a reactant and a product, this ion can be balanced as a unit.

2 Solve **Apply concepts to this problem.**

Remember that a coefficient must always go in front of a compound's formula, not in the middle of it.

Balance the nitrate ion. Put a coefficient 2 in front of $AgNO_3(aq)$.

$$2AgNO_3(aq) + Cu(s) \longrightarrow Cu(NO_3)_2(aq) + Ag(s)$$

Balance the silver. Put a coefficient 2 in front of Ag(s).

$$2AgNO_3(aq) + Cu(s) \longrightarrow Cu(NO_3)_2(aq) + 2Ag(s)$$

3. Balance the equation:
$CO + Fe_2O_3 \longrightarrow Fe + CO_2$

4. Write the balanced chemical equation for the reaction of carbon with oxygen to form carbon monoxide.

Foundations for Math

EQUALITY Explain that the equals sign in a mathematical equation is like the midpoint of a balance scale: Each side of the scale must hold the same mass in order for the scale to stay in balance. Likewise, the two sides of a mathematical equation must be equal for the equality to be true.

For instance, $5x^2 + x = 3x^2 + 2x + 2x^2 - x$. Combine terms to show that $5x^2 + x = 5x^2 + x$. Point out that the exponents were not altered; only the coefficients were combined.

In **Sample Problem 11.2**, explain that the yield arrow in a chemical reaction is like the equals sign in a mathematical equation. Convey that only the coefficient in front of a reactant or product can be changed when balancing an equation. Point out that if the subscript of any reactant or product were changed, this would result in a formula change that would alter the entire reaction.

Sample Problem 11.3

Balancing a Chemical Equation

Aluminum is a good choice for outdoor furniture because it reacts with oxygen in the air to form a thin protective coat of aluminum oxide. Balance the equation for this reaction.

$$Al(s) + O_2(g) \longrightarrow Al_2O_3(s)$$

1 Analyze Identify the relevant concepts.
Apply the rules for balancing equations. Notice the odd number of oxygen atoms in the product.

2 Solve Apply concepts to this problem.

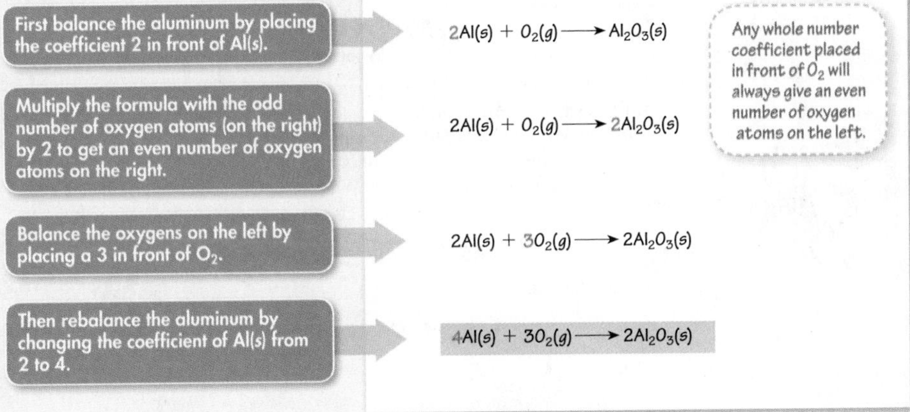

5. Balance each equation.

a. $FeCl_3 + NaOH \longrightarrow Fe(OH)_3 + NaCl$
b. $CS_2 + Cl_2 \longrightarrow CCl_4 + S_2Cl_2$
c. $KI + Pb(NO_3)_2 \longrightarrow PbI_2 + KNO_3$
d. $C_2H_2 + O_2 \longrightarrow CO_2 + H_2O$

6. Write and balance these equations.

a. calcium hydroxide + sulfuric acid ⟶ calcium sulfate + water
b. sodium + water ⟶ sodium hydroxide + hydrogen

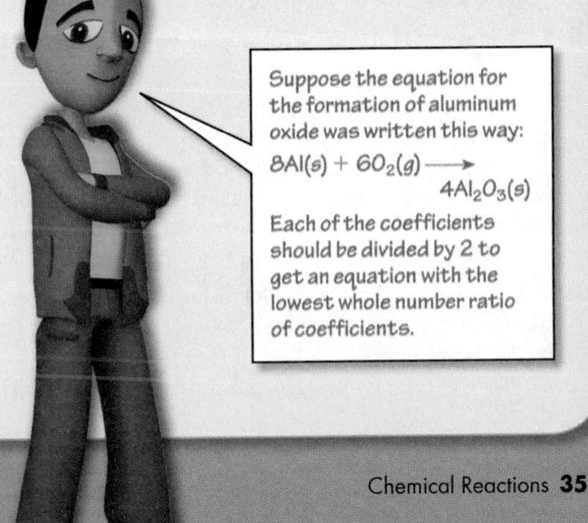

Focus on ELL

4 LANGUAGE PRODUCTION Have students work in groups of four to complete the Quick Lab on page 354. Make sure each group has ELLs of varied language proficiencies so that more proficient students can help less proficient ones. Have students work according to their proficiency level.

BEGINNING: LOW/HIGH Model the procedures. Have students draw a picture of the fork before and after it was placed in the solution.

INTERMEDIATE: LOW/HIGH Restate the procedures in short, simple sentences.

ADVANCED: LOW/HIGH Have these students use the think-aloud strategy when balancing the equation, in order to assist students with lower English proficiencies with this task.

Explain

MAKING CONNECTIONS Point out that balanced chemical equations can help manufacturers who use chemical processes to know how much of each reactant should be purchased and how much of each product will be made. For example, a welder might use oxygen and acetylene (C_2H_2) in a torch. Have students write a balanced chemical equation for the reaction that occurs at the torch. Water and carbon dioxide are produced. *($2C_2H_2 + 5O_2 \rightarrow 4CO_2 + 2H_2O$)* **Ask** From this balanced equation, what would be an effective acetylene:oxygen ratio for the torch? *(2:5)*

Evaluate

Informal Assessment

To evaluate students' understanding of how to write and interpret chemical equations, write two word equations on the board and have students produce skeleton equations. Then, write two unbalanced chemical equations on the board; have students balance them and describe them in words. Finally, write an equation on the board and begin to balance it by changing subscripts. When students object, ask why this approach is incorrect. Then have students complete the 11.1 Lesson Check.

Reteach

Guide students to examine the illustrations of equations throughout the lesson. Have students explain how the illustrations demonstrate the differences between a balanced and an unbalanced chemical equation. Review with students the key steps in writing balanced equations. Have them make a flowchart to describe the best way to subdivide the task.

Answers

3. $3CO + Fe_2O_3 \rightarrow 2Fe + 3CO_2$
4. $2C + O_2 \rightarrow 2CO$
5. **a.** $FeCl_3 + 3NaOH \rightarrow Fe(OH)_3 + 3NaCl$
b. $CS_2 + 3Cl_2 \rightarrow CCl_4 + S_2Cl_2$
c. $2KI + Pb(NO_3)_2 \rightarrow PbI_2 + 2\ KNO_3$
d. $2C_2H_2 + 5O_2 \rightarrow 4CO_2 + 2H_2O$
6. **a.** $Ca(OH)_2 + H_2SO_4 \rightarrow CaSO_4 + 2H_2O$
b. $2Na + 2H_2O \rightarrow 2NaOH + H_2$

Explore

OBJECTIVE After completing this activity, students will be able to describe and write a balanced chemical equation for a single-replacement reaction.

SKILLS FOCUS Observing, inferring, communicating results

PREP TIME 10 minutes

CLASS TIME 20 minutes

EXPECTED OUTCOME Aluminum replaces the silver in the tarnish on the fork or spoon.

ANALYZE AND CONCLUDE

1. After the reaction, the fork looks silvery, and the tarnish is gone.
2. Yes; the aluminum becomes darkened.
3. Al_2S_3
4. $3Ag_2S(s) + 2Al(s) \rightarrow 6Ag(s) + Al_2S_3(s)$

FOR ENRICHMENT Have students repeat the Quick Lab, using a piece of magnesium instead of aluminum foil. Have them compare the results and the speed of the reaction. Students will find that the reaction occurs more quickly with magnesium. After students learn about single-replacement reactions, on page 360, have them explain why the reaction proceeded more rapidly with magnesium. *(Magnesium is more active than aluminum.)*

Extend

Connect to CAREERS

Tell students that materials are classified as hazardous when the chemical reactions they undergo are potentially dangerous to humans or the environment. Explain that a hazardous materials specialist has a constantly changing career because the list of materials that are considered hazardous is constantly changing. Have students research this career and then construct a table that describes the nature of the work, educational and training requirements, employment outlook, working conditions, and other necessary information.

Quick Lab

Removing Silver Tarnish

Materials

- aluminum foil, 20 cm × 20 cm
- large beaker or glass pan
- tarnished silver fork or spoon
- sodium hydrogen carbonate
- plastic tablespoon
- hot water

Procedure

1. Fill the beaker about three-quarters full of hot water and add 2 tablespoons of sodium hydrogen carbonate ($NaHCO_3$).
2. Crush the aluminum foil into a loose ball and place it in the beaker.
3. Write a brief description of the tarnished silver fork; then place it in the beaker so that it is touching the aluminum ball.
4. Allow the beaker to stand undisturbed for 30 minutes.
5. Remove the fork and aluminum ball and rinse them with water.

Analyze and Conclude

1. **Observe** Compare the silver fork with your observations before placing the fork in the water. What changes do you observe?
2. **Explain** Did a chemical reaction occur? How do you know?
3. **Explain** The tarnish on the silver fork is silver sulfide (Ag_2S). Silver becomes tarnished when it is exposed to air, egg yolk, or rubber bands. Each of these substances contains sulfur. Look carefully for a pale yellow precipitate of aluminum sulfide on the bottom of the beaker. Write the formula for aluminum sulfide.
4. **Apply Concepts** The unbalanced equation for the reaction is

$$Ag_2S(s) + Al(s) \longrightarrow Al_2S_3(s) + Ag(s)$$

Balance the equation.

11.1 LessonCheck

ONLINE PROBLEMS

7. **Explain** How do you write a skeleton equation?
8. **Summarize** Describe the steps in writing a balanced chemical equation.
9. **Describe** Write skeleton equations for these reactions.
   a. Heating copper(II) sulfide in the presence of diatomic oxygen produces pure copper and sulfur dioxide gas.
   b. When heated, baking soda (sodium hydrogen carbonate) decomposes to form the products sodium carbonate, carbon dioxide, and water.
10. **Apply Concepts** Balance the following equations:
    a. $SO_2(g) + O_2(g) \longrightarrow SO_3(g)$
    b. $Fe_2O_3(s) + H_2(g) \longrightarrow Fe(s) + H_2O(l)$
    c. $P(s) + O_2(g) \longrightarrow P_4O_{10}(s)$
    d. $Al(s) + N_2(g) \longrightarrow AlN(s)$
11. **Apply Concepts** Write and balance equations for the following reactions:
    a. Iron metal and chlorine gas react to form solid iron(III) chloride.
    b. Solid aluminum carbonate decomposes to form solid aluminum oxide and carbon dioxide gas.
    c. Solid magnesium reacts with aqueous silver nitrate to form solid silver and aqueous magnesium nitrate.

Lesson Check Answers

7. Write the formulas of the reactants to the left of the reaction arrow and the formulas of the products to the right.
8. Write the skeleton equation with the correct formulas of the reactants on the left and the correct formulas of the products on the right. Then balance the equation by using coefficients so that it obeys the law of conservation of mass.
9. a. $CuS(s) + O_2(g) \xrightarrow{heat} Cu(s) + SO_2(g)$
   b. $NaHCO_3(s) \xrightarrow{heat} Na_2CO_3(s) + CO_2(g) + H_2O(l)$
10. a. $2SO_2(g) + O_2(g) \rightarrow 2SO_3(g)$
    b. $Fe_2O_3(s) + 3H_2(g) \rightarrow 2Fe(s) + 3H_2O(l)$
    c. $4P(s) + 5O_2(g) \rightarrow P_4O_{10}(s)$
    d. $2Al(s) + N_2(g) \rightarrow 2AlN(s)$
11. a. $2Fe(s) + 3Cl_2(g) \rightarrow 2FeCl_3(s)$
    b. $Al_2(CO_3)_3(s) \rightarrow Al_2O_3(s) + 3CO_2(g)$
    c. $Mg(s) + 2AgNO_3(aq) \rightarrow 2Ag(s) + Mg(NO_3)_2(aq)$

Kitchen Chemistry

Did you know that your kitchen is a good place to study chemistry? Food preparation generally involves a large number of chemical reactions. Compounds in raw food may combine, decompose, or oxidize to give the finished product. The acids in vinegars, lemon juice, or anything acidic used to marinate meat helps break down the connective tissue of the meat through chemical reactions. If raw food is heated, as in frying and baking, chemical reactions produce many complex compounds. When you eat the cooked food, your body performs another series of chemical reactions that allow the nutrients in the food to nourish your body.

In this lesson, you learned how to describe chemical reactions and write balanced chemical equations. Now you can use ordinary kitchen materials to create your own chemical reactions.

On Your Own

1. For this activity, you'll need a few **paper cups, baking soda, water, vinegar,** and **lemon juice.** You'll also need a **spoon** and an **eyedropper.**
2. Spoon a little bit of baking soda into four cups.
3. Using the eyedropper, add a few drops of water to the first cup. Watch what happens. Then record your observations in a chart similar to the one below.
4. Repeat the process for each substance in the chart.

What Did You See?

Substance	What's the Reaction?
Water	
Vinegar	
Lemon juice	
Your own choice	

Think About It

1. **Draw Conclusions** What clues tell you that a chemical reaction occurred?
2. **Apply Concepts** The skeleton equation for the reaction between baking soda and vinegar is:
$$NaHCO_3 + HC_2H_3O_2 \longrightarrow NaC_2H_3O_2 + H_2O + CO_2$$
Is this equation balanced? Explain.
3. **Connect to the BIGIDEA** How does the law of conservation of mass apply to this experiment?

CHEMISTRY & YOU Draw students' attention to the title of the feature and the photograph of the activity setup. Pose the following question to students: *What are some common chemical reactions that you might encounter in a kitchen?* You may need to assist students in the following ways:

- Ripening in fruits and vegetables involves chemical reactions within their cells.
- Flavors and textures change in some foods as they cook.
- Some reactions occur only when foods are cut.
- Some reactions occur only in the presence of heat, such as with double-acting baking powder.

Explain

APPLY CONCEPTS Explain that chemical reactions can be useful in cooking, but they also can have harmful effects such as food spoilage. Explain that clues that spoilage has occurred include changes in color, the formation of an unpleasant odor, and changes in taste. For example, fats in food can spoil by reacting with oxygen. Encourage students to describe examples they have seen of food spoilage. Help them connect the changes they observed to chemical reactions.

Differentiated Instruction

L1 STRUGGLING STUDENTS Before assigning the On Your Own activity, discuss clues that students should look for that indicate a chemical reaction has occurred. Write the clues on the board for them to use as they make their observations. It may be helpful to have students perform the activity in class rather than at home.

LPR LESS PROFICIENT READERS Before assigning this activity, have students read each step, and then discuss it to be sure students understand the procedure before they perform it.

L3 ADVANCED STUDENTS Have students choose a type of chemical reaction that occurs in the kitchen and prepare a demonstration for the class. Students should write the reaction's chemical equation on the board as part of their demonstration.

Answers

THINK ABOUT IT

1. Bubbles formed.
2. Yes, it's balanced.
3. Atoms are conserved, as shown in the equation. The total mass of reactants should also equal the total mass of the products. (It may not appear to, because the gas produced escapes.)

LESSON 11.2

Key Objective

11.2.1 IDENTIFY the five general types of reactions.

Additional Resources

Reading and Study Workbook, Lesson 11.2

Available Online or on Digital Media:

- Teaching Resources, Lesson 11.2 Review
- Laboratory Manual, Labs 14 and 15
- Small-Scale Laboratory Manual, Lab 15

PearsonChem.com

Engage

CHEMISTRY & YOU Have students study the photograph and read the text. Ask students to predict what happens to the wax as the candle burns.

Activate Prior Knowledge

Review with students the key steps in writing balanced equations. Have them use a flowchart to describe the best way to subdivide the task.

11.2 Types of Chemical Reactions

C.1.4 Physical and chemical changes; **C.4.1** Predict products; **C.4.5** Classify types of reactions

Key Questions

What are the five general types of reactions?

Vocabulary

- combination reaction
- decomposition reaction
- single-replacement reaction
- activity series
- double-replacement reaction
- combustion reaction

CHEMISTRY & YOU

Q: *What happens to the wax when you burn a candle?* You probably have noticed that you have less candle after burning than before, but you may not know that a candle will not burn unless oxygen is present. When you burn a candle, a chemical reaction called combustion takes place. In this lesson, you will learn that if you can recognize the type of reaction, you may be able to predict the products of the reaction.

Classifying Reactions

What are the five general types of reactions?

By classifying chemical reactions, you can more easily predict what products are likely to form. One classification system identifies five general types. **The five general types of reactions include combination, decomposition, single-replacement, double-replacement, and combustion.** Not all chemical reactions fit uniquely into only one category. Occasionally, a reaction may fit equally well into two categories. Nevertheless, recognizing a reaction as a particular type is useful. Patterns of chemical behavior will become apparent and allow you to predict the products of reactions.

Combination Reactions The first type of reaction is the combination, or synthesis, reaction. A **combination reaction** is a chemical change in which two or more substances react to form a single new substance. As shown in Figure 11.4, magnesium metal and oxygen gas combine to form the compound magnesium oxide.

$$2Mg(s) + O_2(g) \longrightarrow 2MgO(s)$$

Notice that in this reaction, as in all combination reactions, the product is a single substance (MgO), which is a compound. The reactants in this combination reaction (Mg and O_2) are two elements, which is often the case. But two compounds may also combine to form a single substance.

When a Group A metal and a nonmetal react, the product is a binary ionic compound.

$$2K(s) + Cl_2(g) \longrightarrow 2KCl(s)$$

When two nonmetals react in a combination reaction, more than one product is often possible.

$$S(s) + O_2(g) \longrightarrow SO_2(g) \text{ sulfur dioxide}$$
$$2S(s) + 3O_2(g) \longrightarrow 2SO_3(g) \text{ sulfur trioxide}$$

IN ACADEMIC STANDARDS for SCIENCE

C.1.4 Describe physical and chemical changes at the molecular level.

C.4.1 Predict products of simple reactions such as synthesis, decomposition, single replacement and double replacement.

C.4.5 Describe, classify and give examples of various kinds of reactions-synthesis (combination), decomposition, single displacement, double displacement and combustion.

Focus on ELL

1 CONTENT AND LANGUAGE Write the lesson objective on the board and read it out loud slowly. Tell students the term *identify* in the objective indicates that they will answer the key question by identifying the five general types of reactions.

2 FRONTLOAD THE LESSON Remind students of the activity they did at the beginning of the chapter when they listed different types of reactions that are familiar to them. Ask a student to read the key question. Draw a diagram that shows the different types of reactions across the top and then under *Chemical Reactions* write the numbers 1–5. Explain to students that they are going to learn the different types of chemical reactions.

3 COMPREHENSIBLE INPUT Assign students to five groups. Assign each group a different reaction and have the students create a poster describing their reaction. Be sure they include pictures and important facts. Display the posters around the room.

More than one product may also result from the combination reaction of a transition metal and a nonmetal.

$$Fe(s) + S(s) \longrightarrow FeS(s) \text{ iron(II) sulfide}$$

$$2Fe(s) + 3S(s) \longrightarrow Fe_2S_3(s) \text{ iron(III) sulfide}$$

Some nonmetal oxides react with water to produce an acid, a compound that produces hydrogen ions in aqueous solution. You will learn about acids in Chapter 19.

$$SO_2(g) + H_2O(l) \longrightarrow H_2SO_3(aq) \text{ sulfurous acid}$$

Some metallic oxides react with water to give a base, or a compound containing hydroxide ions. Again in this case, you can use the ionic charges to derive the formula for the product.

$$CaO(s) + H_2O(l) \longrightarrow Ca(OH)_2(aq) \text{ calcium hydroxide}$$

Figure 11.4 Combination Reaction
When ignited, magnesium ribbon reacts with oxygen in the surrounding air to form magnesium oxide, a white solid. This reaction is a combination reaction.

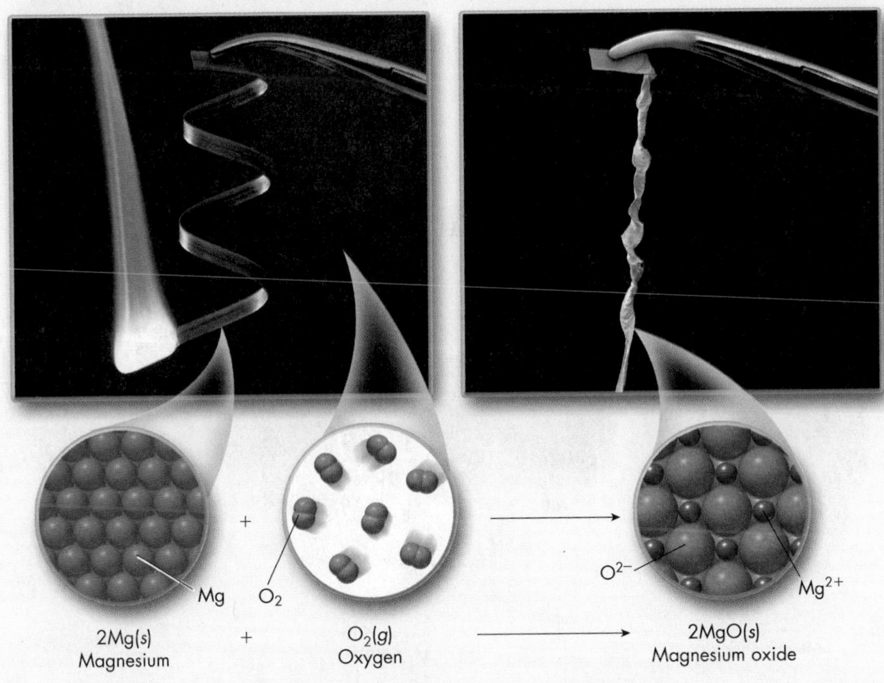

Foundations for Reading

BUILD VOCABULARY Have students draw concept maps titled *Types of Chemical Reactions*. Have them include all the vocabulary terms in the concept map. Ask them to explain how the name of each reaction tells what occurs in the reaction.

READING STRATEGY As students read the lesson, have them visualize the process that occurs in each type of chemical reaction. Suggest that they sketch their visualizations.

Explain

Classifying Reactions

USE VISUALS Direct students' attention to Figure 11.4. Have students study the photographs. Point out that many combination reactions release large amounts of energy. Explain that the reaction of Mg with O_2 was once used in flash photography because it releases energy in the form of a bright, white light. Then, write a number of equations for combination reactions on the board and have students practice balancing them. Rewrite some of the equations with the product on the reactant side and the reactants on the product side. **Ask** Is it possible to reverse the process? *(yes)* **Ask** What is the name of the reverse process? *(a decomposition reaction)* **Ask** Why is energy usually needed for a decomposition reaction to occur? *(Decomposition involves the breaking of bonds, which requires energy.)*

Differentiated Instruction

L3 ADVANCED STUDENTS Combination reactions are sometimes referred to as synthesis reactions. Have students research the meaning of *synthesis.* Point out that synthesis involves making a final product from other substances, but not all synthesis processes—such as those in photosynthesis—are true combination reactions.

ELL ENGLISH LANGUAGE LEARNERS Have students name different elements in their native languages and in English. Then have students list the symbols for several elements in their native languages. Point out that the names might differ, but chemical symbols are the same in any language.

L1 STRUGGLING STUDENTS Have students create a set of clues they can refer to for classifying reactions. Have students write the clues on sticky notes and attach the notes to the appropriate locations in their text.

LESSON 11.2

Explain

Classifying Reactions

USE A VISUAL Help students create a list of criteria for identifying combination and decomposition reactions. For example, decomposition reactions are characterized by a compound on the reactant side and simpler compound or elements on the product side.

Decomposition Reactions Some chemical reactions are the opposite of combination reactions. These kinds of reactions are classified as decomposition reactions. When mercury(II) oxide is heated, it decomposes or breaks down into two simpler substances, as shown in Figure 11.5.

$$2HgO(s) \longrightarrow 2Hg(l) + O_2(g)$$

A **decomposition reaction** is a chemical change in which a single compound breaks down into two or more simpler products. Decomposition reactions involve only one reactant and two or more products. The products can be any combination of elements and compounds. It is usually difficult to predict the products of decomposition reactions. However, when a simple binary compound such as HgO breaks down, you know that the products must be the constituent elements Hg and O_2. Most decomposition reactions require energy in the form of heat, light, or electricity.

Did you know that a decomposition reaction happens when an automobile air bag inflates? A device that can trigger the reaction is placed into the air bag along with sodium azide (NaN_3) pellets. When the device is triggered, the sodium azide pellets decompose and release nitrogen gas, which inflates the air bag quickly.

$$2NaN_3(s) \longrightarrow 2Na(s) + 3N_2(g)$$

Figure 11.5 Decomposition Reaction
When orange-colored mercury(II) oxide is heated, it decomposes into its constituent elements: liquid mercury and gaseous oxygen.
Compare and Contrast *How are the reactions pictured in Figures 11.4 and 11.5 similar? How are they different?*

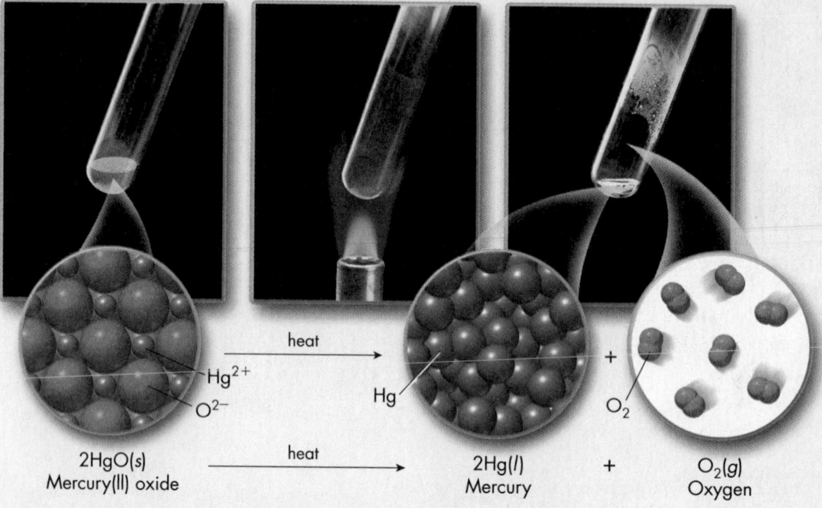

UbD Check for Understanding

BIGIDEA REACTIONS Assess students' knowledge about classifying chemical reactions by directing students to answer the following question with a written explanation or an illustrated example.

How can you tell the difference between a combination reaction and a decomposition reaction? *(A combination reaction forms one product from two or more substances, and a decomposition reaction takes a single compound and breaks it down into simpler compounds or elements.)*

ADJUST INSTRUCTION If students are having trouble answering, draw on the board a compare-contrast table for the two reaction types. Complete the table as a class. Then ask students to identify the key differences between the two reaction types.

SampleProblem 11.4

Writing Equations for Combination and Decomposition Reactions

Write a balanced equation for each of the following reactions.

a. Combination of copper and sulfur:

$Cu(s) + S(s) \longrightarrow$ (two reactions possible)

b. Decomposition of water:

$H_2O(l) \xrightarrow{\text{electricity}}$

❶ Analyze Identify the relevant concepts. Two combination reactions are possible because copper is a transition metal and has more than one common ionic charge (Cu^+ and Cu^{2+}).

❷ Solve Apply the concepts to this problem.

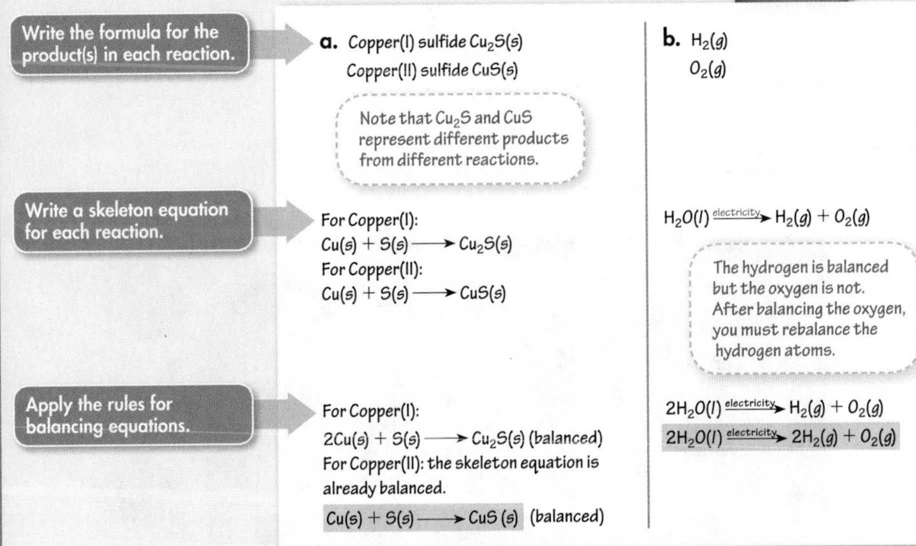

12. Write the formula for the binary compound that decomposes to the products H_2 and Br_2.

13. Complete and balance this decomposition reaction.

$HI \longrightarrow$

14. Write and balance the equation for the formation of magnesium nitride (Mg_3N_2) from its elements.

Explain

Misconception Alert

Emphasize to students that decomposition reactions consist of one substance forming *two or more different substances*. Students tend to limit decomposition reactions to the decomposition of a compound into its component elements. Explain that a compound can also break down into an element and a compound or two or more compounds. Provide students with the following chemical equations as examples of decomposition reactions that result in products other than elements: $2H_2O_2 \rightarrow 2H_2O + O_2$ and $H_2CO_3 \rightarrow H_2O + CO_2$.

Sample Practice Problem

Write and balance an equation for the formation of aluminum chloride ($AlCl_3$) from its elements. *($2Al + 3Cl_2 \rightarrow 2AlCl_3$)*

The Decomposition of Hydrogen Peroxide

Hydrogen peroxide is capable of gently cleansing wounds, but it also has the power to launch small rockets. The secret to its versatility lies in its concentration. Hydrogen peroxide solutions sold in the first-aid section of grocery stores are very dilute (3% v/v). When a dilute solution comes into contact with the broken cells of a wound, an enzyme called catalase catalyzes the decomposition of the hydrogen peroxide into water and oxygen. The free oxygen ions kill bacteria. The resulting bubbles also help loosen dirt and debris and bring them to the surface of the wound.

The decomposition reaction generates more than gas bubbles; it is exothermic as well. While the heat produced by a 3% solution is barely perceptible, the heat generated by highly concentrated solutions can turn water into steam. While this ability to generate heat is useful under certain circumstances, it has sometimes led to disastrous results.

Answers

FIGURE 11.5 Both involve two elements and one compound. In Figure 11.4, a compound is being formed from the elements; in Figure 11.5, a compound is being decomposed into its elements.

12. $HBr(g)$

13. $2HI(g) \rightarrow H_2(g) + I_2(g)$

14. $3Mg(s) + N_2(g) \rightarrow Mg_3N_2(s)$

LESSON 11.2

Explore

Classifying Reactions

Teacher Demo

PURPOSE Students will observe single-replacement reactions.

MATERIALS magnesium ribbon, 10-mL graduated cylinder, 1*M* HCl, sodium, 2 beakers, tongs, water

SAFETY HCl(*aq*) is corrosive and can cause severe burns. The piece of sodium should be no larger than a match head. Wear plastic gloves and use tongs to avoid contact between sodium and your skin.

PROCEDURE Place a 2-cm piece of magnesium ribbon in 10 mL of 1*M* HCl. Using tongs, place a small piece of sodium in 250 mL of cold water.

EXPECTED OUTCOMES In both reactions, the metal replaces hydrogen in a compound. Hydrogen gas is released. The reaction using sodium is much more dramatic and rapid than the reaction involving magnesium.

Single-Replacement Reactions Dropping a small piece of potassium into a beaker of water creates the vigorous reaction shown in Figure 11.6 The reaction produces hydrogen gas and a large quantity of heat. The released hydrogen gas can ignite explosively.

$$2K(s) + 2H_2O(l) \longrightarrow 2KOH(aq) + H_2(g)$$

Similar but less spectacular reactions can occur. For example, if you drop a piece of zinc into a solution of copper nitrate, this reaction occurs:

$$Zn(s) + Cu(NO_3)_2(aq) \longrightarrow Cu(s) + Zn(NO_3)_2(aq)$$

These equations describe two examples of single-replacement reactions. A **single-replacement reaction** is a chemical change in which one element replaces a second element in a compound. You can identify a single-replacement reaction by noting that both the reactants and the products consist of an element and a compound. In the equation above, zinc and copper change places. The reacting element Zn replaces copper in the reactant compound $Cu(NO_3)_2$. The products are the element Cu and the compound $Zn(NO_3)_2$.

Figure 11.6 Single-Replacement Reaction
The alkali metal potassium displaces hydrogen from water and forms a solution of potassium hydroxide in a single-replacement reaction. The heat of the reaction is often sufficient to ignite the hydrogen.
Inferring ***Why are alkali metals stored under mineral oil or kerosene?***

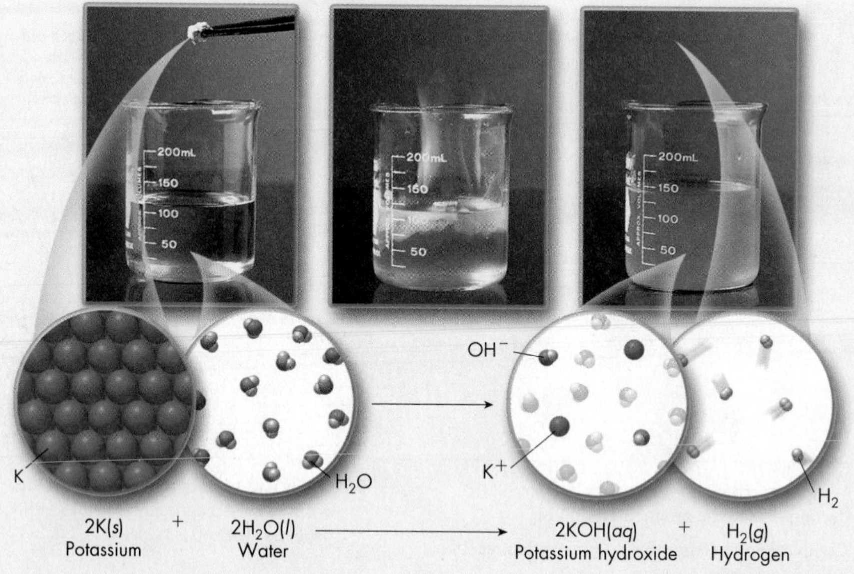

Launching the Space Shuttle

The space shuttle program ran for thirty years. During that time, each mission relied on decomposition and displacement reactions to supply the energy needed for launching the space shuttles and their payloads into space. Have students research what fuels were used to propel the space shuttle orbiters. Have volunteers describe the reactions and how the products provided the thrust necessary to launch the space vehicles.

Sample Problem 11.5

Writing Equations for Single-Replacement Reactions

Write a balanced equation for the single-replacement reaction.

$$Cl_2(aq) + NaBr(aq) \longrightarrow$$

Hint: You're starting with an unequal number of atoms:
- reactants
- 2 chlorine atoms
- 1 sodium atom
- 1 bromine atom
- products
- 1 chlorine atom
- 1 sodium atom
- 2 bromine atoms

1 Analyze **Identify the relevant concepts.** Chlorine is more reactive than bromine and displaces bromine from its compounds.

2 Solve **Apply concepts to this problem.**

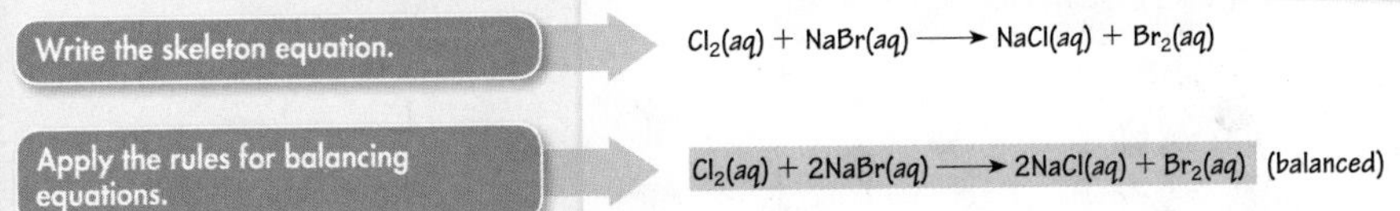

Write the skeleton equation. → $Cl_2(aq) + NaBr(aq) \longrightarrow NaCl(aq) + Br_2(aq)$

Apply the rules for balancing equations. → $Cl_2(aq) + 2NaBr(aq) \longrightarrow 2NaCl(aq) + Br_2(aq)$ (balanced)

15. Complete the equations for these single-replacement reactions in aqueous solution. Balance each equation. Write "no reaction" if a reaction does not occur.

a. $Fe(s) + Pb(NO_3)_2(aq) \longrightarrow$

b. $Cl_2(aq) + NaI(aq) \longrightarrow$

c. $Ca(s) + H_2O(l) \longrightarrow$

d. $Zn(s) + H_2SO_4(aq) \longrightarrow$

Hint: Look at Table 11.2. Zinc displaces hydrogen from an acid and takes its place.

Whether one metal will displace another metal from a compound depends upon the relative reactivities of the two metals. The **activity series** of metals, given in Table 11.2, lists metals in order of decreasing reactivity. A reactive metal will replace any metal listed below it in the activity series. Thus, iron will displace copper from a copper compound in solution, but iron does not similarly displace zinc or calcium.

A halogen can also replace another halogen from a compound. The activity of the halogens decreases as you go down Group 7A of the periodic table—fluorine, chlorine, bromine, and iodine. Bromine is more active than iodine, so this reaction occurs:

$$Br_2(aq) + 2NaI(aq) \longrightarrow 2NaBr(aq) + I_2(aq)$$

But bromine is less active than chlorine, so this reaction does not occur:

$$Br_2(aq) + NaCl(aq) \longrightarrow \text{No reaction}$$

Table 11.2

Activity Series of Metals

	Name	Symbol
Decreasing reactivity ↓	Lithium	Li
	Potassium	K
	Calcium	Ca
	Sodium	Na
	Magnesium	Mg
	Aluminum	Al
	Zinc	Zn
	Iron	Fe
	Lead	Pb
	(Hydrogen)	(H)*
	Copper	Cu
	Mercury	Hg
	Silver	Ag

*Metals from Li to Na will replace H from acids and water; from Mg to Pb they will replace H from acids only.

Foundations for Math

USING TABLES TO SOLVE PROBLEMS Making a table is a problem-solving strategy that is often used in mathematics. A table can help students better visualize relationships in information provided in word problems, and it encourages students to look critically at the information in order to plan and develop a solution.

In **Sample Problem 11.5**, guide students to create and use a table to organize the information presented in the skeleton equation of the single-replacement reaction.

Atom	Reactants	Products
Cl	2	1
Na	1	1
Br	1	2

Point out that sodium atoms are balanced, but the chlorine and the bromine atoms are not yet balanced.

Explain

USE AN ANALOGY Direct students to Table 11.2 and review the definition of a single-replacement reaction. Explain that single-replacement reactions can be compared to cutting in at a dance. A person who is alone approaches a couple and cuts in. The person replaces one partner, who is now left alone. In a chemical reaction, however, only certain substances can replace other substances in a given compound. The activity series of metals shows which metals can replace another metal in a given compound. Review the reactions on this page. Show students how to use Table 11.2 to predict whether a reaction will occur. Direct students' attention to hydrogen's treatment in the table. Remind students that they learned in Chapter 6 that hydrogen is a nonmetal, even though it is located in period 1A of the periodic table. However, it can replace some metals and be replaced by others, so it is included in the table for reference. For this reason, hydrogen is shown enclosed by parentheses.

Sample Practice Problem

Complete the partial equation shown and then balance the equation, which is for a single-replacement reaction that takes place in aqueous solution. If a reaction does not occur, write "no reaction." (Use the activity series.)

$K(s) + H_2SO_4(aq) \rightarrow$

$(2K(s) + H_2SO_4(aq) \rightarrow K_2SO_4(aq) + H_2(g))$

Answers

FIGURE 11.6 to prevent their reaction with water vapor and oxygen in the air

15. **a.** $Fe(s) + Pb(NO_3)_2(aq) \rightarrow Fe(NO_3)_2(aq) + Pb(s)$

b. $Cl_2(aq) + 2NaI(aq) \rightarrow 2NaCl(aq) + I_2(aq)$

c. $Ca(s) + 2H_2O(l) \rightarrow Ca(OH)_2(aq) + H_2(g)$

d. $Zn(s) + H_2SO_4 \rightarrow ZnSO_4 + H_2$

Explain

Classifying Reactions

USE VISUALS Have students examine the reaction in Figure 11.7. **Ask** What characteristics do KI and $Pb(NO_3)_2$ share? *(They are both in aqueous solution, and they are both ionic compounds.)* Explain to students that a double-replacement reaction always involves two ionic compounds in aqueous solution. In addition, one of the products must be a precipitate, a gas, or a molecular compound. **Ask** Does one of the products in this reaction meet the second qualification? Why? *(Yes; PbI_2 is an insoluble compound, so it will precipitate from solution when it forms.)*

Explore

Teacher Demo

PURPOSE Students will observe double-replacement reactions.

MATERIALS 0.1*M* solutions of $BaCl_2$, Na_2SO_4, Na_3PO_4, $CaCl_2$, $Pb(NO_3)_2$, and KI

PROCEDURE Mix equimolar solutions of the following ionic compounds: $BaCl_2$ and Na_2SO_4; Na_3PO_4 and $CaCl_2$; and $Pb(NO_3)_2$ and KI. **Ask** What evidence of a double-replacement reaction do you observe? *(a precipitate)* Write the formula of each precipitate formed in the reactions.

EXPECTED OUTCOME Precipitates of $BaSO_4$, $Ca_3(PO_4)_2$, and PbI_2 form.

Figure 11.7
Double-Replacement Reaction
Aqueous solutions of potassium iodide and lead(II) nitrate react in a double-replacement reaction to form the yellow precipitate lead(II) iodide.

Double-Replacement Reactions Sometimes, when two solutions of ionic compounds are mixed, nothing happens. At other times, the ions in the two solutions react. Figure 11.7 shows that mixing aqueous solutions of potassium iodide and lead(II) nitrate results in a chemical reaction in which a yellow precipitate of solid lead(II) iodide is formed. Potassium nitrate, the other product of the reaction, remains in solution. This reaction is an example of a **double-replacement reaction,** which is a chemical change involving an exchange of positive ions between two compounds. Double-replacement reactions are also referred to as double-displacement reactions. They generally take place in aqueous solution and often produce a precipitate, a gas, or a molecular compound such as water. For a double-replacement reaction to occur, one of the following is usually true:

1. One of the products is only slightly soluble and precipitates from solution. For example, the reaction of aqueous solutions of sodium sulfide and cadmium nitrate produces a yellow precipitate of cadmium sulfide.

$$Na_2S(aq) + Cd(NO_3)_2(aq) \longrightarrow CdS(s) + 2NaNO_3(aq)$$

2. One of the products is a gas. Poisonous hydrogen cyanide gas is produced when aqueous sodium cyanide, also a poison, is mixed with sulfuric acid.

$$2NaCN(aq) + H_2SO_4(aq) \longrightarrow 2HCN(g) + Na_2SO_4(aq)$$

3. One product is a molecular compound such as water. Combining solutions of calcium hydroxide and hydrochloric acid produces water.

$$Ca(OH)_2(aq) + 2HCl(aq) \longrightarrow CaCl_2(aq) + 2H_2O(l)$$

UbD Check for Understanding

The Essential Question How can you predict the products of chemical reactions?

Assess students' understanding of how to predict the products of chemical reactions by having them use the activity series shown in Table 11.2 to answer the following questions:

- What will happen if a piece of iron is placed in a solution of lead(II) nitrate? *(Iron will displace lead.)*
- What will happen if a piece of aluminum is placed in a solution of calcium chloride? *(Nothing will happen because Al will not displace Ca.)*

ADJUST INSTRUCTION If students have difficulty answering, project or write Table 11.2 on the board. Using a think-aloud strategy, demonstrate how to answer the first question by deciding whether iron will displace lead. As a class, work through the answer to the second question. Then provide additional practice using Table 11.2 until students are comfortable predicting reaction products.

Sample Problem 11.6

Writing Equations for Double-Replacement Reactions

A precipitate of barium carbonate is formed when aqueous solutions of barium chloride react with potassium carbonate. Write a balanced chemical equation for the double-replacement reaction.

$$K_2CO_3(aq) + BaCl_2(aq) \longrightarrow$$

1 Analyze **Identify the relevant concepts.**
The driving force behind the reaction is the formation of a precipitate. Write correct formulas of the products using ionic charges. Then balance the equation.

2 Solve **Apply concepts to this problem.**

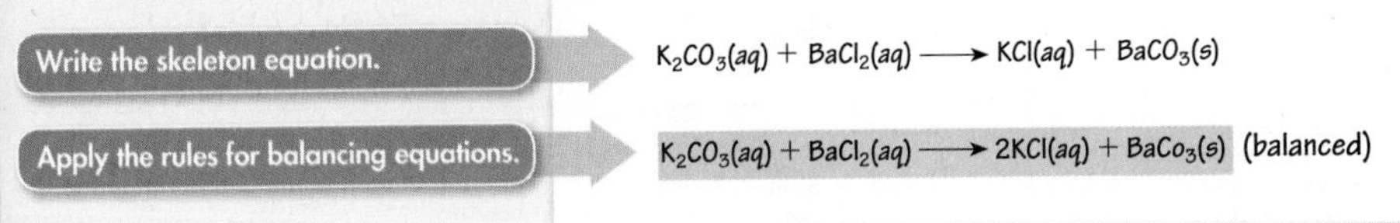

Write the skeleton equation. → $K_2CO_3(aq) + BaCl_2(aq) \longrightarrow KCl(aq) + BaCO_3(s)$

Apply the rules for balancing equations. → $K_2CO_3(aq) + BaCl_2(aq) \longrightarrow 2KCl(aq) + BaCO_3(s)$ (balanced)

16. Write the products of these double-replacement reactions. Then balance each equation.
- **a.** $NaOH(aq) + Fe(NO_3)_3(aq) \longrightarrow$ (Iron(III) hydroxide is a precipitate.)
- **b.** $Ba(NO_3)_2(aq) + H_3PO_4(aq) \longrightarrow$ (Barium phosphate is a precipitate.)
- **c.** $FeS(s) + HCl(aq) \longrightarrow$ (Hydrogen sulfide gas (H_2S) is formed.)

17. Write a balanced equation for each reaction.
- **a.** $KOH(aq) + H_3PO_4(aq) \longrightarrow$ (Water is formed.)
- **b.** $AgNO_3(aq) + NaCl(s) \longrightarrow$ (Silver chloride is a precipitate.)
- **c.** $Ca(OH)_2(aq) + H_3PO_4(aq) \longrightarrow$ (Water is formed.)
- **d.** $KI(aq) + Pb(NO_3)_2(aq) \longrightarrow$ (Lead(II) iodide is a precipitate.)
- **e.** $H_2SO_4(aq) + Al(OH)_3(aq) \longrightarrow$ (Water is formed.)

Hint: Use ionic charges to write the correct formula of the other product.

Combustion Reactions The flames of a campfire, candle, or a gas grill are evidence that a combustion reaction is taking place. A **combustion reaction** is a chemical change in which an element or a compound reacts with oxygen, often producing energy in the form of heat and light. A combustion reaction always involves oxygen as a reactant. Often the other reactant is a hydrocarbon, which is a compound composed of hydrogen and carbon. The complete combustion of a hydrocarbon produces carbon dioxide and water. But if the supply of oxygen is limited during a reaction, the combustion will not be complete. Elemental carbon (soot) and toxic carbon monoxide gas may be additional products.

Foundations for Math

COEFFICIENTS As a class, compare and contrast how coefficients are used in mathematical equations and chemical equations. Point out that a coefficient in front of a variable term such as 3*x* represents the expression "3 times *x*." A coefficient in front of a chemical formula such as 2KCl represents the number of formula units of that compound, in this case 2. In both mathematical and chemical equations, if a coefficient is not present, the number 1 is understood. (Convey that this is also true for subscripts in chemical equations.)

Explain that the balanced chemical equation in **Sample Problem 11.6** is equivalent to the following expanded equation:

$$1K_2C_1O_3(aq) + 1Ba_1Cl_2(aq) \rightarrow 2K_1Cl_1(aq) + 1Ba_1C_1O_3(aq)$$

As students are sure to note, using ones as coefficients and subscripts makes the equation more difficult to read, thus defeating the purpose of using chemical symbols rather than words to express the events in a chemical reaction.

Sample Practice Problem

Write a balanced equation for the double-replacement reaction.

$K_2CrO_4(aq) + Ba(NO_3)_2(aq) \rightarrow$ (Barium chromate is a precipitate.)

($K_2CrO_4(aq) + Ba(NO_3)_2(aq) \rightarrow BaCrO_4(s) + 2KNO_3(aq)$)

Explain

START A CONVERSATION Ask students to consider the three things necessary for a fire to burn. *(fuel, oxygen, and energy to initiate combustion)* **Ask** Why is it not safe to use a single kind of fire extinguisher on all fires? *(A fire extinguisher that controls one type of fire may actually enhance other types of combustion reactions. For example, water is not sprayed on burning magnesium because the intense heat can decompose water, producing flammable hydrogen gas.)*

MAKE A CONNECTION Discuss with students the requirements of a gas that might be used to put out a fire. Explain that the gas cannot itself burn or support the combustion of another material; it also must be heavier than air so that it will settle on a fire, depriving it of oxygen. Discuss some specific gases in terms of these factors. For example, nitrogen and helium are relatively nonreactive, but they are not heavy enough; hydrogen and methane will burn; oxygen supports the burning of the fuel; carbon dioxide will not burn or support combustion and is heavier than air. **Ask** Is a piece of paper combustible? Is it flammable? *(Paper is combustible but not flammable.)* Explain that these two terms are often used synonymously, but *combustible* means that the material will burn, and *flammable* means a material may easily burst into flame.

Answers

16.
- **a.** $3NaOH(aq) + Fe(NO_3)_3(aq) \rightarrow Fe(OH)_3(s) + 3NaNO_3(aq)$
- **b.** $3Ba(NO_3)_2(aq) + 2H_3PO_4(aq) \rightarrow Ba_3(PO_4)_2(s) + 6HNO_3(aq)$
- **c.** $FeS(s) + 2HCl(aq) \rightarrow FeCl_2(s) + H_2S(g)$

17.
- **a.** $3KOH(aq) + H_3PO_4(aq) \rightarrow K_3PO_4(aq) + 3H_2O(l)$
- **b.** $AgNO_3(aq) + NaCl(s) \rightarrow AgCl(s) + NaNO_3(aq)$
- **c.** $3Ca(OH)_2(aq) + 2H_3PO_4(aq) \rightarrow Ca_3(PO_4)_2(s) + 6H_2O(l)$
- **d.** $2KI(aq) + Pb(NO_3)_2(aq) \rightarrow PbI_2(s) + 2KNO_3(aq)$
- **e.** $3H_2SO_4(aq) + 2Al(OH)_3(aq) \rightarrow Al_2(SO_4)_3(aq) + 6H_2O(l)$

Explain

CHEMISTRY & YOU Read the question. Ask students if their prediction about what happens as candles burn was correct.

START A CONVERSATION Point out to students that an important product of most combustion reactions is energy, which is usually in the form of heat or light. Emphasize as well that one of the reactants must be oxygen. Relate this fact to everyday experience by recalling how removing oxygen from a combustion reaction—for example, snuffing out a candle—causes the reaction to stop.

CRITICAL THINKING Ask students to infer why it is important that combustion reactions, such as those used to heat a home or run an automobile, take place in properly ventilated areas. *(Without proper ventilation and enough available oxygen, the combustion may be incomplete, and poisonous carbon monoxide may be produced.)*

USE VISUALS Have students study Figure 11.8. Remind them that the complete combustion of a hydrocarbon, such as methane, always produces water, carbon dioxide, heat, and light. Use a disposable lighter to show students the combustion of butane. Ask students to write the balanced equation for the combustion of butane (C_4H_{10}). Have them use the diagram and chemical equation on this page as an aid.

($2C_4H_{10}(g) + 13O_2(g) \rightarrow 8CO_2(g) + 10H_2O(g)$)

MAKE A CONNECTION Explain that some combustion reactions are also combination reactions in which an element or a compound combines with oxygen to form a single product plus energy. For example: $2Mg(s) + O_2(g) \rightarrow 2MgO(s)$, and $4Fe(s) + 3O_2(g) \rightarrow 2Fe_2O_3(s)$.

Q: *Materials such as candle wax contain hydrogen and carbon. One type of wax has a formula of $C_{25}H_{52}$. The wax reacts with oxygen in the air. So, what happens to the wax as it burns?*

The complete combustion of a hydrocarbon releases a large amount of energy as heat. That's why hydrocarbons such as methane (CH_4), propane (C_3H_8), and butane (C_4H_{10}) are important fuels. The combustion reaction for methane is shown in Figure 11.8. Gasoline is a mixture of hydrocarbons that can be approximately represented by the formula C_8H_{18}. The complete combustion of gasoline in a car engine is shown by this equation.

$$2C_8H_{18}(l) + 25O_2(g) \longrightarrow 16CO_2(g) + 18H_2O(g)$$

The reactions between oxygen and some elements other than carbon are also examples of combustion reactions. For example, both magnesium and sulfur will burn in the presence of oxygen. As you look at these combustion equations, notice that the reactions could also be classified as combination reactions.

$$2Mg(s) + O_2(g) \longrightarrow 2MgO(s)$$

$$S(s) + O_2(g) \longrightarrow SO_2(g)$$

Figure 11.8 Combustion Reaction
Methane gas reacts with oxygen from the surrounding air in a combustion reaction to produce carbon dioxide and water.
Infer ***What else is produced in this reaction?***

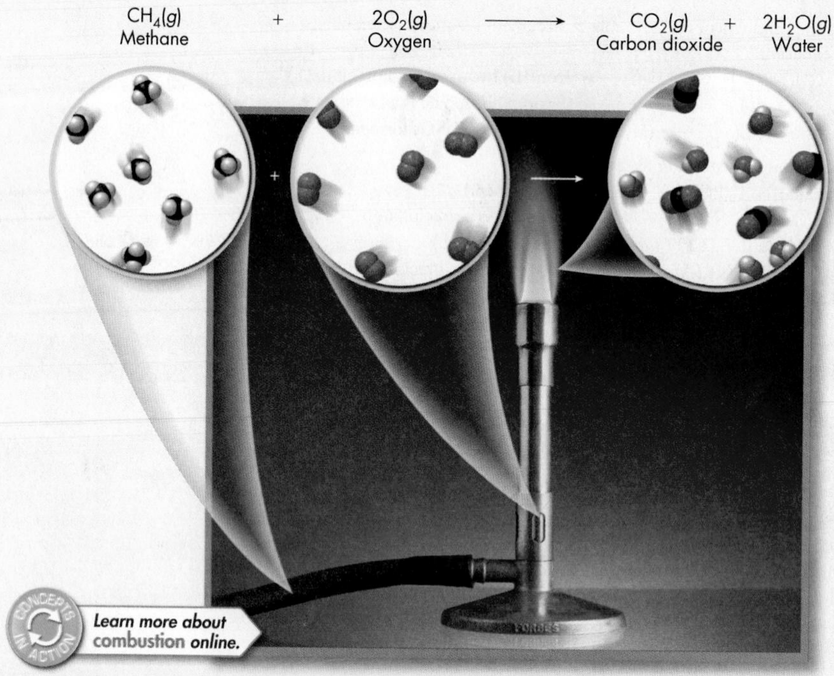

Differentiated Instruction

ELL ENGLISH LANGUAGE LEARNERS Make sure that students understand clearly the five terms that describe general types of chemical reactions: *combination, decomposition, single-replacement, double-replacement*, and *combustion*. Help them to spell and pronounce each term as well as to define it. Encourage them to ask questions about any term they do not fully understand.

L1 STRUGGLING STUDENTS Provide students with general statements that represent each of the first four types of chemical reactions discussed in this lesson, such as X + Y → XY for a combination reaction, XY → X + Y for a decomposition reaction, X + CD → XD + C for a single-displacement reaction, and AB + CD → AD + CB for a double-displacement reaction. Tell students that they can compare actual reactions to these general statements to help them identify the reaction type.

364 Chapter 11 • Lesson 2

Sample Problem 11.7

Writing Equations for Combustion Reactions

An alcohol lamp often uses ethanol as its fuel. Write a balanced equation for the complete combustion of ethanol.

$C_2H_6O(l)$

1 Analyze **Identify the relevant concepts.**
Oxygen is the other reactant in a combustion reaction. The products are CO_2 and H_2O.

2 Solve **Apply concepts to this problem.**

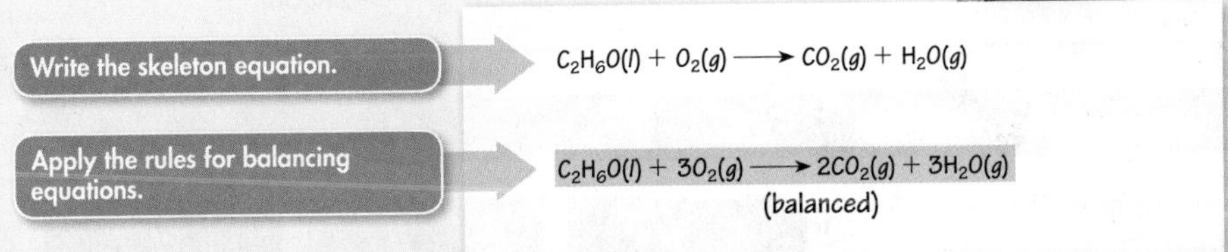

18. Write a balanced equation for the complete combustion of each compound.
a. formaldehyde ($CH_2O(g)$)
b. heptane ($C_7H_{16}(l)$)
c. benzene ($C_6H_6(l)$)

19. Write a balanced equation for the complete combustion of
a. glucose ($C_6H_{12}O_6(s)$)
b. acetone ($C_3H_6O(l)$)
c. pentanol ($C_5H_{12}O(l)$)

Now that you have learned about some of the basic reaction types, you can predict the products of many reactions. The number of elements and/or compounds reacting is a good indicator of possible reaction type and, thus, possible products.

For example, in a combination reaction, two or more reactants (elements or compounds) combine to form a single product. In a decomposition reaction, a single compound is the reactant; two or more substances are the products. An element and a compound are the reactants in a single-replacement reaction. A different element and a new compound are the products. In a double-replacement reaction, two ionic compounds are the reactants; two new compounds are the products. The reactants in a combustion reaction are oxygen and usually a hydrocarbon. The products of most combustion reactions are carbon dioxide and water.

Explore

Teacher Demo

PURPOSE Students will observe the combustion of iron.

MATERIALS superfine steel wool, 250 mL beaker balance, ring stand, utility clamp, matches

SAFETY Steel wool burns readily.

PROCEDURE Place a superfine steel wool pad in a beaker, measure the mass, and write the value on the board. Take the steel wool out of the beaker and unfold it to full length. Clamp it to a ring stand with a utility clamp. Set the steel wool on fire. Ask students to predict what effect the burning will have on the mass of the sample. As soon as the steel wool stops burning, place the remains in the same beaker and measure the mass again. Write the new value on the board, next to the original mass. Discuss why the sample gained mass.

EXPECTED OUTCOMES Two elements, $Fe(s)$ and $O_2(g)$, will combine in a combustion reaction to form a binary compound, Fe_2O_3. The sample will gain mass because initially only the mass of the iron is determined. During the reaction, oxygen from the air will combine with the iron.

Sample Practice Problem

Write a balanced equation for the complete combustion of propane, C_3H_8.
($C_3H_8 + 5O_2 \rightarrow 3CO_2 + 4H_2O$)

Differentiated Instruction

L3 ADVANCED STUDENTS When hydrochloric acid (HCl) is dropped on calcium carbonate ($CaCO_3$), a double-replacement reaction occurs. **Ask** What are the products of this reaction? *($CaCl_2$ and H_2CO_3)* Neither of these products is a gas, but bubbles of gas are released during this reaction. Ask students to infer the identity of the gas. *(CO_2)* **Ask** What type of reaction could produce this gas? *(the decomposition reaction $H_2CO_3 \rightarrow H_2O + CO_2$)* Explain that geologists often make use of this pair of reactions to test rock samples for the presence of limestone, which is formed from the calcium carbonate skeletons of prehistoric corals.

Answers

FIGURE 11.8 energy in the form of heat and light

18. a. $CH_2O + O_2 \rightarrow CO_2 + H_2O$
b. $C_7H_{16} + 11O_2 \rightarrow 7CO_2 + 8H_2O$
c. $2C_6H_6(l) + 15O_2(g) \rightarrow 12CO_2(g) + 6H_2O(g)$

19. a. $C_6H_{12}O_6 + 6O_2 \rightarrow 6CO_2 + 6H_2O$
b. $C_3H_6O(l) + 4O_2(g) \rightarrow 3CO_2(g) + 3H_2O(g)$
c. $2C_5H_{12}O(l) + 15O_2(g) \rightarrow 10CO_2(g) + 12H_2O(l)$

Explore

Classifying Reactions

Teacher Demo

PURPOSE Students will observe a combination reaction.

MATERIALS magnesium ribbon, large crucible, lab burner, cobalt blue glass or exposed photographic film, crucible tongs, matches or igniter

SAFETY Students should not look directly at burning magnesium. Have them observe through pieces of cobalt blue glass or exposed photographic film. Another option is to conduct the reaction inside a large, metal can.

PROCEDURE Explain to students that in a combination reaction, the reactants combine to make one new product. Measure a 5- to 7-cm strip of magnesium ribbon. Light the burner. Hold one end of the magnesium ribbon with a pair of crucible tongs. Ignite the magnesium and hold it above the crucible. Ask students to note any evidence of chemical change. Have students note the condition of the residue compared to the original magnesium. The product may be disposed of in the trash.

EXPECTED OUTCOMES Metallic magnesium and oxygen gas from the air form a white powder, magnesium oxide. Other evidence of chemical change includes release of energy as heat and light. Tell students that small amounts of magnesium nitride also form.

Extend

Extend the demonstration on this page. **Ask** What are the reactants in this combination reaction? *(Mg and O_2)* What is the product of this reaction? *(MgO)* Does this reaction obey the law of conservation of mass? Explain. *(Yes. The sum of the masses of magnesium and oxygen is equal to the mass of the magnesium oxide formed.)* Have students write the balanced equation for the reaction. *($2Mg(s) + O_2(g) \rightarrow 2MgO(s)$)* Challenge students to write the balanced equation for the formation of the small amount of magnesium nitride that forms during this reaction. *($3Mg(s) + N_2(g) \rightarrow Mg_3N_2(s)$)*.

1 Combination Reaction

General Equation: $R + S \longrightarrow RS$

Reactants: Generally two elements, or two compounds (where at least one compound is a molecular compound)

Probable Products: A single compound

Example: Burning magnesium in air

See reactions animated online.

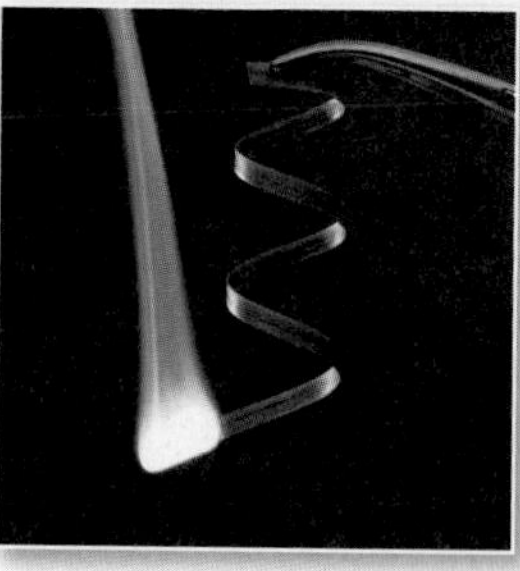

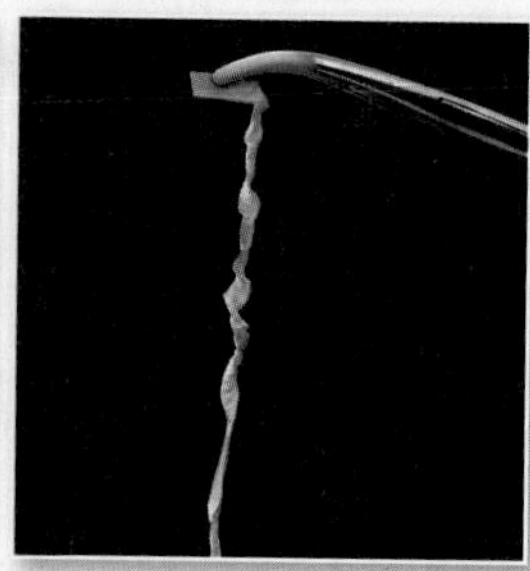

$2Mg(s) + O_2(g) \longrightarrow 2MgO(s)$

2 Decomposition Reaction

General Equation: $RS \longrightarrow R + S$

Reactants: Generally a single binary compound or a compound with a polyatomic ion

Probable Products: Two elements (for a binary compound), or two or more elements and/or compounds (for a compound with a polyatomic ion)

Example: Heating mercury(II) oxide

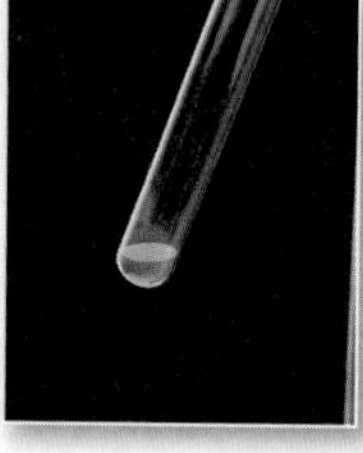

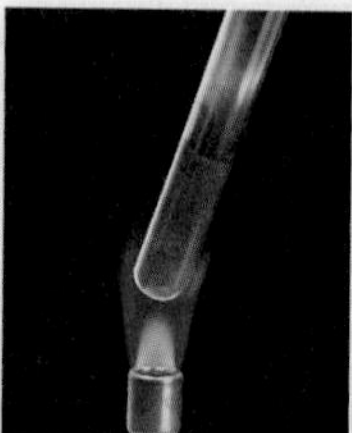

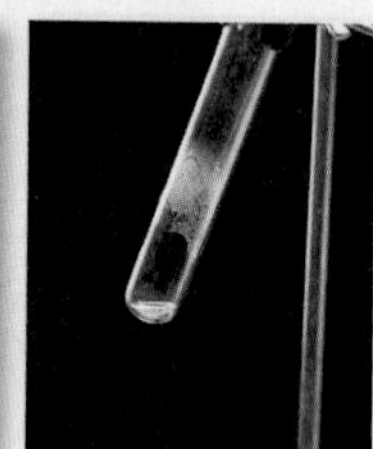

$2HgO(s) \longrightarrow 2Hg(l) + O_2(g)$

3 Single-Replacement Reaction

General Equation: $T + RS \longrightarrow TS + R$

Reactants: An element and a compound

In a single-replacement reaction, an element replaces another element from a compound in aqueous solution. For a single-replacement reaction to occur, the element that is displaced must be less active than the element that is doing the displacing.

Probable Products: A different element and a new compound

Example: Potassium in water

$2K(s) + 2H_2O(l) \longrightarrow 2KOH(aq) + H_2(g)$

Magnesium from Seawater

Magnesium metal is an important component of alloys used to make consumer materials. The main commercial source of Mg(*s*) is seawater. The Mg^{2+} ion is the third most abundant dissolved ion in the oceans. A process for isolating magnesium from seawater depends on the fact that because of a double-replacement reaction, Mg^{2+} will precipitate when OH^- is added. Students can research the commercial process.

4 Double-Replacement Reaction

General Equation: $R^+ S^- + T^+ U^- \longrightarrow R^+ U^- + T^+ S^-$

Reactants: Two ionic compounds

In a double-replacement reaction, two ionic compounds react by exchanging cations to form two different compounds.

Probable Products: Two new compounds

Double-replacement reactions are driven by the formation of a precipitate, a gaseous product, or water.

Example: Reaction of aqueous solutions of potassium iodide and lead(II) nitrate.

$2KI(aq) + Pb(NO_3)_2(aq) \longrightarrow PbI_2(s) + 2KNO_3(aq)$

5 Combustion Reaction

General Equation: $C_xH_y + (x + y/4)\,O_2 \longrightarrow xCO_2 + (y/2)H_2O$

Reactants: Oxygen and a compound of C, H, (O)

When oxygen reacts with an element or compound, combustion may occur.

Probable Products: CO_2 and H_2O

With incomplete combustion, C and CO may also be products.

Example: The combustion of methane gas in air

$CH_4(g) + 2O_2(g) \longrightarrow CO_2(g) + 2H_2O(g)$

11.2 LessonCheck

20. Review What are the five types of chemical reactions?

21. Apply Concepts Classify each reaction and balance the equations.

a. $C_3H_6(g) + O_2(g) \longrightarrow CO_2(g) + H_2O(g)$
b. $Al(OH)_3(s) \longrightarrow Al_2O_3(s) + H_2O(l)$
c. $Li(s) + O_2(g) \longrightarrow Li_2O(s)$
d. $Zn(s) + AgNO_3(aq) \longrightarrow Ag(s) + Zn(NO_3)_2(aq)$

22. Identify Which of the five general types of reaction would most likely occur, given each set of reactants? What are the probable products?

a. an aqueous solution of two ionic compounds
b. a single compound
c. two elements
d. oxygen and a compound of carbon and hydrogen

23. Apply Concepts Complete and balance an equation for each reaction.

a. $CaI_2(aq) + Hg(NO_3)_2(aq) \longrightarrow$ (HgI_2 precipitates.)
b. $Al(s) + Cl_2(g) \longrightarrow$
c. $Ag(s) + HCl(aq) \longrightarrow$
d. $C_2H_2(g) + O_2(g) \longrightarrow$

BIGIDEA REACTIONS

24. After wood burns, the ash weighs much less than the original wood. Explain why the law of conservation of mass is not violated in this situation.

Evaluate

Informal Assessment

Create a matching assessment involving types of reaction in one column and balanced chemical equations in the other column. As a challenge, write on the board only the products of combination, decomposition, and displacement reactions, and challenge students to fill in the reactants. Then have students complete the 11.2 Lesson Check.

Reteach

Help students develop a branched flowchart of chemical reactions that is similar to flowcharts used in qualitative analysis. Start by asking if there is a single reactant. If so, the reaction is a decomposition. If not, proceed to the next step. Ask if oxygen is one of the reactants. If so, the reaction is either combination or combustion. If not, proceed to the next step. Continue through the five general types of reactions.

Lesson Check Answers

20. combination, decomposition, single-replacement, double-replacement, and combustion

21. a. combustion; $2C_3H_6 + 9O_2 \rightarrow 6CO_2 + 6H_2O$
b. decomposition; $2Al(OH)_3 \rightarrow Al_2O_3 + 3H_2O$
c. combination; $4Li + O_2 \rightarrow 2Li_2O$
d. single-replacement; $Zn + 2AgNO_3 \rightarrow 2Ag + Zn(NO_3)_2$

22. a. double replacement; two different compounds
b. decomposition; two or more elements and/or compounds
c. combination; a compound
d. combustion; carbon dioxide and water

23. a. $CaI_2 + Hg(NO_3)_2 \rightarrow HgI_2 + Ca(NO_3)_2$
b. $2Al + 3Cl_2 \rightarrow 2AlCl_3$
c. no reaction
d. $2C_2H_2 + 5O_2 \rightarrow 4CO_2 + 2H_2O$

24. BIGIDEA The combustible compounds in wood react with O_2 to produce CO_2 and H_2O during combustion, leaving the non-combustible components behind as ash. The law is not violated because the total mass of the wood and O_2 before combustion is equal to the total mass of the ash, CO_2 and H_2O produced by the reaction.

CHEMISTRY & YOU

CHEMISTRY & YOU Have students examine the photo of the collapsing building and read the caption. Lead students in discussing the use of decomposition reactions in demolition. Point out that the blasting cap also relies on a decomposition reaction. Pose the following question to students: *Why might an explosion be necessary to detonate dynamite?* You may need to assist students in the following ways:

- Ammonium nitrate replaced nitroglycerin in dynamite because of its greater stability.
- Ammonium nitrate is a common agricultural fertilizer that does not explode despite mixing with air as it is spread on cropland.
- Ammonium nitrate can be melted, at which point additional heat will cause it to decompose into nitrous oxide (N_2O) and water.

Explain

START A CONVERSATION Draw students' attention to the second column of text on the page. **Ask** What are the three reasons mentioned to explain why ammonium nitrate is used today instead of nitroglycerin? *(It makes dynamite stronger, safer, and cheaper.)* Explain that large manufacturing companies often employ chemists who continually research and develop new and better products. Engage students in discussing how a manufacturing company today might use chemistry research to develop an even better explosive than ammonium nitrate

Extend

Connect to PHYSICS

If possible, present a video that shows a controlled demolition of a skyscraper or sports facility and has commentary from the explosives expert taking part in the demolition. Challenge students to identify the factors that allow a relatively small amount of explosive to bring down even the largest sports facility.

CHEMISTRY & YOU: HISTORY

The History of Dynamite

In 1846, Ascanio Sobrero added glycerol to a mixture of concentrated nitric and sulfuric acids. The resulting oily liquid, known as nitroglycerin, turned out to be such a powerful explosive that a small bottle could blow up a building. Unfortunately, it was also extremely unstable, and tended to explode after being handled roughly, or a temperature change.

Alfred Nobel (1833–1896), a Swedish chemist and industrialist, began experimenting with nitroglycerin, looking for a way to make it safe to use. In 1866, Nobel discovered that he could mix nitroglycerine with a fine sand called *kieselguhr* to turn the liquid into paste that could be shaped into rods. The rods were then packed into cylinders made of paper. He named these rods "dynamite."

Originally, Nobel marketed dynamite as Nobel's Blasting Powder.

To safely ignite dynamite and control the timing of a detonation, Nobel also invented blasting caps, which create a small explosion that triggers the larger explosion in the dynamite itself.

Today, ammonium nitrate is used in place of nitroglycerin. This dynamite is stronger, safer, and cheaper than Nobel's original invention.

When he died in 1896, Alfred Nobel left a nine million dollar fortune to be used to fund different fields of study. The Nobel Prize is still awarded to people whose work helps humanity.

BIG BLASTS Nobel began manufacturing the explosive, which was used to blast out dam sites, canal beds, mines, and the foundations for large buildings.

Take It Further

1. **Classify** The equation for the detonation of nitroglycerin is shown below. What kind of reaction is it?
$$4C_3H_5N_3O_9 \longrightarrow 12CO_2 + 10H_2O + 6N_2 + O_2$$
2. **Research** Fireworks are another kind of explosive. How are dynamite explosions different from fireworks explosions?

21st Century Learning To be successful in the 21st century, students need skills and learning experiences that extend beyond subject matter mastery. The following project helps students build the following 21st Century Skills: *Creativity and Innovation; Communication and Collaboration; Initiative and Self-Direction; Productivity and Accountability.*

EXPLOSION! Pose the following challenge to your students. A demolition company has decided to use dynamite to bring down a tall building in a city. Your job is to inform people who work or live in nearby buildings about what to expect. Create an informational flyer that can be posted in the area and passed out to residents and/or businesses in the blast zone. The flyer should describe why dynamite rather than other explosives will be used for the demolition, and it should explain the type of chemical reaction involved in easy-to-understand terms. It should also explain what the explosion will be like at the site of the building and in nearby areas. The flyer should list the precautions that will be taken to protect surrounding properties, and it should include a map of the blast and safety zones.

Answers

TAKE IT FURTHER

1. decomposition reaction
2. Answers will vary.

FIGURE 11.9 nitrate and sodium

11.3 Reactions in Aqueous Solution

C.1.4 Physical and chemical changes; C.4.1 Predict products; C.4.5 Classify types of reactions

CHEMISTRY & YOU

Q: *How did soda straws get into limestone caves?* These "soda straws" are really stalactites in a limestone cave. Soda straws grow on cave ceilings as thin-walled hollow tubes that result from chemical reactions involving water. In this lesson, you will learn to predict the formation of precipitates and write equations to describe the reactions that produce them.

Key Questions

What does a net ionic equation show?

How can you predict the formation of a precipitate in a double-replacement reaction?

Vocabulary

- complete ionic equation
- spectator ion
- net ionic equation

Net Ionic Equations

What does a net ionic equation show?

Your world is water based. More than 70 percent of Earth's surface is covered by water, and about 66 percent of the adult human body is water. It is not surprising, then, that many important chemical reactions take place in water—that is, in aqueous solution. The reaction of aqueous solutions of silver nitrate with sodium chloride to form solid silver chloride and aqueous sodium nitrate is a double-replacement reaction. The reaction is shown in Figure 11.9.

$$AgNO_3(aq) + NaCl(aq) \longrightarrow AgCl(s) + NaNO_3(aq)$$

The equation above reflects the way you have been writing equations involving aqueous solutions of ionic compounds. However, the equation does not show that, like most ionic compounds, the reactants and one of the products dissociate, or separate, into cations and anions when they dissolve in water.

Figure 11.9 Precipitate in a Double Replacement Reaction
A precipitate of silver chloride forms when aqueous solutions of silver nitrate and sodium chloride are mixed.
Inferring *Which ions do not participate in the reaction?*

Key Objectives

11.3.1 DESCRIBE the information found in a net ionic equation.

11.3.2 PREDICT the formation of a precipitate in a double-replacement reaction.

Additional Resources

Reading and Study Workbook, Lesson 11.3

Available Online or on Digital Media:

- Teaching Resources, Lesson 11.3 Review, Interpreting Graphics
- Laboratory Manual, Labs 16, 17, and 18
- Small-Scale Laboratory Manual, Labs 16 and17

LESSON 11.3

PearsonChem.com

Engage

CHEMISTRY & YOU Point out that the formation of stalactites and stalagmites is a precipitation reaction. **Ask** What does the term *precipitation reaction* suggest about the substances that react to create these formations? *(The term precipitation reaction suggests that ions are involved in their creation.)*

Activate Prior Knowledge

Ask In a double-replacement reaction, two ions form a solid precipitate, a molecular compound, or a gas. What happens to the other ions? *(They remain dissolved in the solution.)* **Ask** What would you need to know to predict whether a precipitate forms in a double-replacement reaction? *(the solubility of all compounds that could form)*

Focus on ELL

1 CONTENT AND LANGUAGE Write the word *aqueous* on the board and explain to students that it is the adjective form of the noun *aqua*, which means "water". Remind students that aqueous solutions are solutions made from ionic or molecular compounds dissolved in water.

2 FRONTLOAD THE LESSON Review double-replacement reactions with students. Make sure that students are able to consistently identify monatomic and polyatomic ions and their associated charges by writing a few practice compounds on the board.

3 COMPREHENSIBLE INPUT Preview the equations and sample problems in the lesson. Review the four abbreviations used to indicate the states of matter of reactants and products (*aq*, *s*, *g*, and *l*). Have students list the abbreviations, with definitions, in their notebooks.

IN ACADEMIC STANDARDS for SCIENCE

C.1.4 Describe physical and chemical changes at the molecular level.

C.4.1 Predict products of simple reactions such as synthesis, decomposition, single replacement and double replacement.

C.4.5 Describe, classify and give examples of various kinds of reactions-synthesis (combination), decomposition, single displacement, double displacement and combustion.

LESSON 11.3

Foundations for Reading

BUILD VOCABULARY Help students predict the meaning of the term *net ionic equation* by exploring the meaning of phrases such as *net income* and *net worth.*

READING STRATEGY Stress that it is important to note the physical states of reactants and products in precipitation reactions.

Explain

Net Ionic Equations

APPLY CONCEPTS Explain to students that a net ionic equation differentiates between ions that react to form a solid precipitate, a gas, or water and ions that simply remain in aqueous solution. It is important to note, however, that spectator ions are not completely unaffected by the reaction. They become paired with different anions or cations than they were paired with when the reaction began.

READING SUPPORT

Build Vocabulary: *Word Origins* The word *spectator* comes from the Latin verb *spectare,* meaning "to watch." Thus, a spectator ion can be thought of as only watching a reaction, not participating. ***During a football game, what analogy can you draw to the people in the seats and the football players on the field?***

For example, when sodium chloride dissolves in water, it separates into sodium ions ($Na^+(aq)$) and chloride ions ($Cl^-(aq)$). Similarly, when dissolved in water, silver nitrate dissociates into silver ions ($Ag^+(aq)$) and nitrate ions ($NO_3^-(aq)$). You can use these ions to write a **complete ionic equation,** an equation that shows dissolved ionic compounds as dissociated free ions.

$$Ag^+(aq) + NO_3^-(aq) + Na^+(aq) + Cl^-(aq) \longrightarrow AgCl(s) + Na^+(aq) + NO_3^-(aq)$$

Notice that the nitrate ion and the sodium ion appear unchanged on both sides of the equation. The equation can be simplified by eliminating these ions because they don't participate in the reaction.

$$Ag^+(aq) + \cancel{NO_3^-(aq)} + \cancel{Na^+(aq)} + Cl^-(aq) \longrightarrow AgCl(s) + \cancel{Na^+(aq)} + \cancel{NO_3^-(aq)}$$

An ion that appears on both sides of an equation and is not directly involved in the reaction is called a **spectator ion.** When you rewrite an equation leaving out the spectator ions, you have the net ionic equation. The **net ionic equation** is an equation for a reaction in solution that shows only those particles that are directly involved in the chemical change.

$$Ag^+(aq) + Cl^-(aq) \longrightarrow AgCl(s)$$

In writing balanced net ionic equations, you must make sure that the ionic charge is balanced. For the previous reaction, the net ionic charge on each side of the equation is zero and is therefore balanced. But consider the skeleton equation for the reaction of lead with silver nitrate.

$$Pb(s) + AgNO_3(aq) \longrightarrow Ag(s) + Pb(NO_3)_2(aq)$$

The nitrate ion is the spectator ion in this reaction. The net ionic equation is as follows:

$$Pb(s) + Ag^+(aq) \longrightarrow Ag(s) + Pb^{2+}(aq) \text{ (unbalanced)}$$

Why is this equation unbalanced? Notice that a single unit of positive charge is on the reactant side of the equation. Two units of positive charge are on the product side. Placing the coefficient 2 in front of $Ag^+(aq)$ balances the charge. A coefficient of 2 in front of $Ag(s)$ rebalances the atoms.

$$Pb(s) + 2Ag^+(aq) \longrightarrow 2Ag(s) + Pb^{2+}(aq) \text{ (balanced)}$$

A net ionic equation shows only those particles involved in the reaction and is balanced with respect to both mass and charge. Of the five types of reactions identified in this chapter, both single- and double-replacement reactions can be written as net ionic equations.

Figure 11.10 Silver Halide and Medical X-rays
Small crystals of a silver halide, usually silver bromide, are embedded in the coating on film used to record medical X-rays. The crystals darken when exposed to X-rays that pass through the human body. Dense parts like bones absorb more X-rays; relatively few rays pass through these parts, which appear as light areas on the developed film. More rays pass through soft tissue, which shows up as darker areas on the developed film. Metals also strongly absorb X-rays.

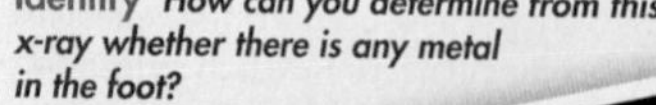

Identify ***How can you determine from this x-ray whether there is any metal in the foot?***

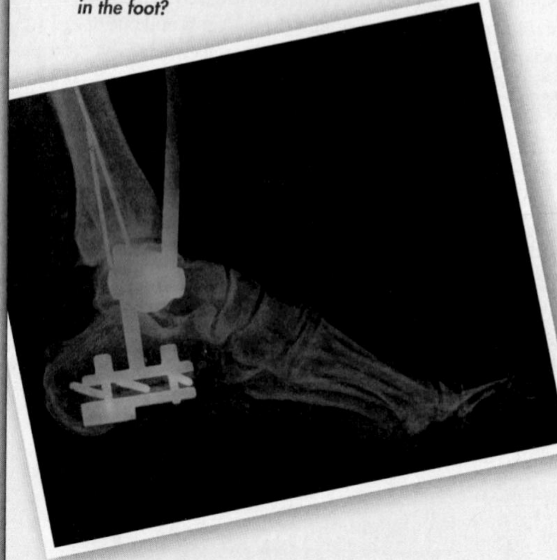

Differentiated Instruction

LPR LESS PROFICIENT READERS Explain that net ionic equations are a direct and simple way to describe chemical change. Ask four student volunteers to represent two people who go to a dance on a double date and then watch while their partners dance together. Explain that the students watching represent spectator ions and the couple dancing represent the ions that form a precipitate.

L3 ADVANCED STUDENTS Explain that precipitation reactions can occur in the human body under certain conditions. Have students research the precipitation reactions involved in the formation of kidney stones and gall stones, and write complete and net ionic equations for each reaction.

L1 STRUGGLING STUDENTS Discuss with students the steps involved in writing and balancing ionic equations. Help students create a checklist they can refer to when writing balanced ionic equations.

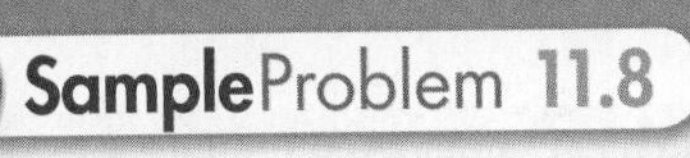

Sample Problem 11.8

Writing and Balancing Net Ionic Equations

Aqueous solutions of iron(III) chloride and potassium hydroxide are mixed. A precipitate of iron(III) hydroxide forms. Identify the spectator ions and write a balanced net ionic equation for the reaction.

1 Analyze Identify the relevant concepts. Write the complete ionic equation. Eliminate aqueous ions that appear in both the reactants and products. Then balance the equation with respect to both mass and charge.

2 Solve Apply the concepts to this problem.

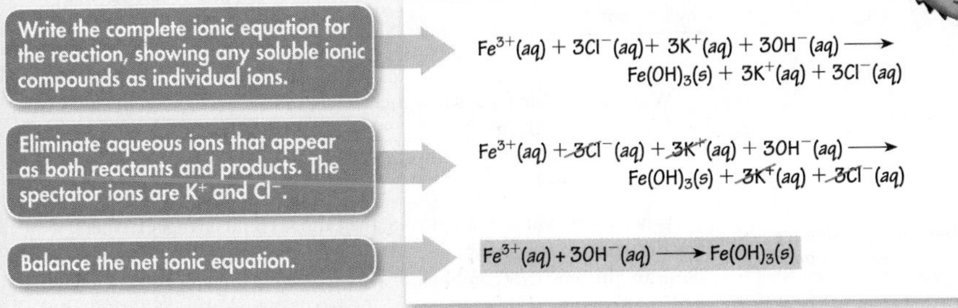

Write the complete ionic equation for the reaction, showing any soluble ionic compounds as individual ions.

$$Fe^{3+}(aq) + 3Cl^-(aq) + 3K^+(aq) + 3OH^-(aq) \longrightarrow Fe(OH)_3(s) + 3K^+(aq) + 3Cl^-(aq)$$

Eliminate aqueous ions that appear as both reactants and products. The spectator ions are K^+ and Cl^-.

$$Fe^{3+}(aq) + \cancel{3Cl^-}(aq) + \cancel{3K^+}(aq) + 3OH^-(aq) \longrightarrow Fe(OH)_3(s) + \cancel{3K^+}(aq) + \cancel{3Cl^-}(aq)$$

Balance the net ionic equation.

$$Fe^{3+}(aq) + 3OH^-(aq) \longrightarrow Fe(OH)_3(s)$$

25. Write the balanced net ionic equation for this reaction.

$$Ca^{2+}(aq) + OH^-(aq) + H^+(aq) + PO_4^{3-}(aq) \longrightarrow Ca^{2+}(aq) + PO_4^{3-}(aq) + H_2O(l)$$

26. Write the complete ionic equation and net ionic equation for the reaction of aqueous calcium hydroxide with phosphoric acid. The products are calcium phosphate and water.

Predicting the Formation of a Precipitate

How can you predict the formation of a precipitate in a double-replacement reaction?

You have seen that mixing solutions of two ionic compounds can sometimes result in the formation of an insoluble salt called a precipitate. Some combinations of solutions produce precipitates, while others do not. Whether or not a precipitate forms depends upon the solubility of the new compounds that form. **By using the general rules for solubility of ionic compounds, you can predict the formation of a precipitate.** These general rules are shown in Table 11.3.

EXPLAIN

Sample Practice Problem

Write a balanced net ionic equation for the reaction between sulfuric acid (H_2SO_4) and sodium hydroxide (NaOH). *($H^+(aq) + OH^-(aq) \rightarrow H_2O(l)$)*

Foundations for Math

SIMPLIFYING EQUATIONS Explain to students that when they simplify a mathematical expression or an equation, they combine like terms and write the answer in simplest form. For example, write the following equation on the board:

$$3x^2 + 4x - 7xy + 2 = 4y + 3x^2 - 7xy$$

Note that crossing out spectator ions from an ionic equation is similar to using inverse operations to solve a mathematical equation. Demonstrate it by crossing out the like terms on the left- and right-hand sides of the equals sign. Note that in the mathematical equation, like terms have the same coefficients, the same variables, and the same mathematical signs in front of them.

In **Sample Problem 11.9**, point out that the spectator ions can be identified by the fact that they are identical in symbol, charge, and coefficient. Remind students that the states of spectator ions will always be aqueous.

Answers

READING SUPPORT The spectators at a football game are present at the game, but they don't participate in it.

FIGURE 11.10 Metal and bone both absorb x-rays. You can examine the lighter areas of the x-ray to see if there are shapes other than bones in the image and determine if those shapes resemble metal objects.

25. $OH^-(aq) + H^+(aq) \rightarrow H_2O(l)$

26. complete ionic equation: $3Ca^{2+}(aq) + 6OH^-(aq) + 6H^+(aq) + 2PO_4^{3-}(aq) \rightarrow Ca_3(PO_4)_2(s) + 6H_2O(l)$
net ionic equation: same as complete ionic equation

LESSON 11.3

Explain

Predicting the Formation of a Precipitate

USE VISUALS Direct students' attention to Table 11.3. Have students practice reading the table and using the solubility rules to decide whether or not some compounds are soluble. Then, have students use the solubility rules to determine whether the following ionic compounds are soluble or not:

a. $PbCl_2$ *(not)*

b. KOH *(soluble)*

c. NaOH *(soluble)*

d. $SrSO_4$ *(not)*

e. Fe_2S_3 *(not)*

APPLY CONCEPTS On the board, write the products only of several double-replacement reactions; challenge students to fill in the reactants. Examples include: $\rightarrow$ AgBr(*s*) + KNO_3(*aq*), $\rightarrow$ $BaSO_4$(*s*) + 2NaCl(*aq*), $\rightarrow$ 2KCl(*aq*) + $CaSO_4$(*s*). *(KBr(aq) + $AgNO_3$(aq), $BaCl_2$(aq) + Na_2SO_4(aq), and $CaCl_2$(aq) + K_2SO_4(aq))* Remind students that in a double-replacement reaction, the positive ions (cations) of one compound trade places with the positive ions of another compound.

CHEMISTRY & YOU Calcium carbonate reacts with carbon dioxide dissolved in water to form soluble calcium hydrogen carbonate. Carbon dioxide then converts calcium hydrogen carbonate back to calcium carbonate, which precipitates and forms the soda straws.

CHEMISTRY & YOU

Q: *How did the soda straws, which are composed of calcium carbonate, get into the cave?*

Table 11.3

Solubility Rules for Ionic Compounds

Compounds	Solubility	Exceptions
Salts of alkali metals and ammonia	Soluble	Some lithium compounds
Nitrate salts and chlorate salts	Soluble	Few exceptions
Sulfate salts	Soluble	Compounds of Pb, Ag, Hg, Ba, Sr, and Ca
Chloride salts	Soluble	Compounds of Ag and some compounds of Hg and Pb
Carbonates, phosphates, chromates, sulfides, and hydroxides	Most are insoluble	Compounds of the alkali metals and of ammonia

Will a precipitate form when aqueous solutions of Na_2CO_3(*aq*) and $Ba(NO_3)_2$(*aq*) are mixed?

$$2Na^+(aq) + CO_3^{2-}(aq) + Ba^{2+}(aq) + 2NO_3^-(aq) \longrightarrow ?$$

VIRTUAL LAB — Learn how to identify ions in solution online.

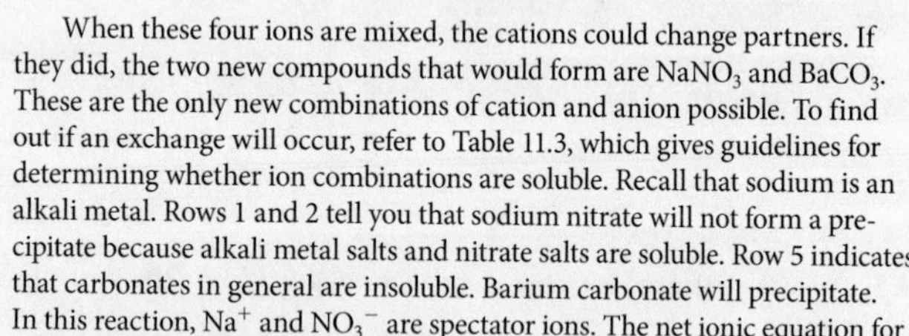

When these four ions are mixed, the cations could change partners. If they did, the two new compounds that would form are $NaNO_3$ and $BaCO_3$. These are the only new combinations of cation and anion possible. To find out if an exchange will occur, refer to Table 11.3, which gives guidelines for determining whether ion combinations are soluble. Recall that sodium is an alkali metal. Rows 1 and 2 tell you that sodium nitrate will not form a precipitate because alkali metal salts and nitrate salts are soluble. Row 5 indicates that carbonates in general are insoluble. Barium carbonate will precipitate. In this reaction, Na^+ and NO_3^- are spectator ions. The net ionic equation for this reaction is as follows:

$$Ba^{2+}(aq) + CO_3^{2-}(aq) \longrightarrow BaCO_3(s)$$

Figure 11.11
Formation of a Precipitate
A precipitate forms when aqueous solutions of sodium sulfate (Na_2SO_4) and barium nitrate ($Ba(NO_3)_2$) are mixed. The net ionic equation for this reaction is as follows:

$$Ba^{2+}(aq) + SO_4^{2-}(aq) \longrightarrow BaSO_4(s)$$

Apply Concepts ***Which ions are present in the final solution but are not part of the net ionic equation?***

Focus on ELL

4 LANGUAGE PRODUCTION Have students work in groups of four to complete the Small-Scale Lab on page 374. Make sure each group has ELLs of varied language proficiencies so that more proficient students can help less proficient ones. Have students work according to their proficiency level.

BEGINNING: LOW/HIGH Paraphrase each procedure and model it for the students. Allow them to draw pictures or diagrams to describe the reaction in each box.

INTERMEDIATE: LOW/HIGH Allow students to use short phrases and simple sentences to answer questions.

ADVANCED: LOW/HIGH Have these students help students with lower English proficiencies to complete their charts and answer the questions.

SampleProblem 11.9

Writing and Balancing Net Ionic Equations

Aqueous potassium carbonate reacts with aqueous strontium nitrate. Identify the precipitate formed and write the net ionic equation for the reaction.

1 Analyze **Identify the relevant concepts.** Write the reactants. Look at possible new pairings of cation and anion that give an insoluble substance. Eliminate the spectator ions.

2 Solve **Apply the concepts to this problem.**

Use the solubility rules in Table 11.3 to identify the precipitate formed.

Write the reactants, showing each as dissociated free ions.	$2K^+(aq) + CO_3^{2-}(aq) + Sr^{2+}(aq) + 2NO_3^-(aq) \longrightarrow ?$
Look at possible new pairings of cation and anion that give an insoluble substance.	Of the two possible combinations, KNO_3 is soluble and $SrCO_3$ is insoluble.
Eliminate the spectator ions and write the net ionic equation.	$CO_3^{2-}(aq) + Sr^{2+}(aq) \longrightarrow SrCO_3(s)$

27. Identify the precipitate formed when solutions of these compounds are mixed. Write the net ionic equation.

$NH_4Cl(aq) + Pb(NO_3)_2(aq) \longrightarrow$

28. Write a complete ionic equation and a net ionic equation for the reaction of aqueous solutions of iron(III) nitrate and sodium hydroxide.

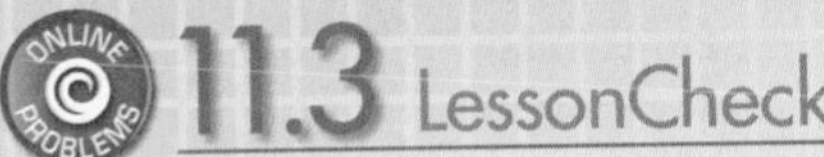

11.3 LessonCheck

29. Review What is a net ionic equation?

30. Explain How can you predict the formation of a precipitate in a double-replacement reaction?

31. Apply Concepts Write a balanced net ionic equation for each reaction.

a. $Pb(NO_3)_2(aq) + H_2SO_4(aq) \longrightarrow PbSO_4(s) + HNO_3(aq)$

b. $Pb(C_2H_3O_2)_2(aq) + HCl(aq) \longrightarrow PbCl_2(s) + HC_2H_3O_2(aq)$

c. $Na_3PO_4(aq) + FeCl_3(aq) \longrightarrow NaCl(aq) + FePO_4(s)$

d. $(NH_4)_2S(aq) + Co(NO_3)_2(aq) \longrightarrow CoS(s) + NH_4NO_3(aq)$

32. Identify List the precipitate formed when solutions of these ionic compounds are mixed.

a. $H_2SO_4 + BaCl_2 \longrightarrow$

b. $Al_2(SO_4)_3 + NH_4OH \longrightarrow$

c. $AgNO_3 + H_2S \longrightarrow$

d. $CaCl_2 + Pb(NO_3)_2 \longrightarrow$

e. $Ca(NO_3)_2 + Na_2CO_3 \longrightarrow$

33. Apply Concepts Hard water contains calcium and magnesium ions. One way to soften water is to add sodium phosphate. Write complete and net ionic equations for the reaction of these two alkaline earth ions with aqueous sodium phosphate.

Lesson Check Answers

29. A net ionic equation shows only those particles involved in the reaction and is balanced with respect to both mass and charge.

30. Use the general rules for solubility of ionic compounds (Table 11.3).

31. a. $Pb^{2+}(aq) + SO_4^{2-}(aq) \rightarrow PbSO_4(s)$

b. $Pb^{2+}(aq) + 2Cl^-(aq) \rightarrow PbCl_2(s)$

c. $Fe^{3+}(aq) + PO_4^{3-}(aq) \rightarrow FePO_4(s)$

d. $Co^{2+}(aq) + S^{2-}(aq) \rightarrow CoS(s)$

32. a. $BaSO_4(s)$ b. $Al(OH)_3(s)$

c. $Ag_2S(s)$ d. $PbCl_2(s)$

e. $CaCO_3(s)$

33. Ca^{2+}, complete: $3Ca^{2+}(aq) + 6Na^+(aq) + 2PO_4^{3-}(aq) \rightarrow 6Na^+(aq) + Ca_3(PO_4)_2(s)$

Ca^{2+}, net: $3Ca^{2+}(aq) + 2PO_4^{3-}(aq) \rightarrow Ca_3(PO_4)_2(s)$

Mg^{2+}, complete: $3Mg^{2+}(aq) + 6Na^+(aq) + 2PO_4^{3-}(aq) \rightarrow 6Na^+(aq) + Mg_3(PO_4)_2(s)$

Mg^{2+}, net: $3Mg^{2+}(aq) + 2PO_4^{3-}(aq) \rightarrow Mg_3(PO_4)_2(s)$

Explain

Sample Practice Problem

Have students write a complete ionic equation and a net ionic equation for the reaction of aqueous solutions of HCl and Na_2S.

(complete: $2H^+(aq) + 2Cl^-(aq) + 2Na^+(aq) + S^{2-}(aq) \rightarrow H_2S(g) + 2Na^+(aq) + 2Cl^-(aq)$; net: $2H^+(aq) + S^{2-}(aq) \rightarrow H_2S(g)$)

Evaluate

Informal Assessment

To evaluate students' understanding of complete ionic equations, net ionic equations, and the formation of precipitates, write the reactants in some precipitation reactions on the board. Include at least one example for which no reaction occurs. Have students write complete and net ionic equations for each set of reactants or indicate that no reaction occurs. Then have students complete the 11.3 Lesson Check.

Reteach

Review with students the writing and balancing of complete and net ionic equations and the use of solubility rules to predict the outcome of double-replacement reactions. Stress that it is important to note the physical states of reactants and products in precipitation reactions. Remind students that in a double-replacement reaction, the positive ions (cations) of one compound trade places with the positive ions of another compound.

Answers

FIGURE 11.11 sodium and nitrate

27. $2Cl^-(aq) + Pb^+(aq) \rightarrow PbCl_2(s)$

28. $Fe^{3+}(aq) + 3NO_3^-(aq) + 3Na^+(aq) + 3OH^-(aq) \rightarrow Fe(OH)_3(s) + 3Na^+(aq) + 3NO_3^-(aq)$

$Fe^{3+}(aq) + 3OH^-(aq) \rightarrow Fe(OH)_3(s)$

SMALL-SCALE LAB

Explore

Small-Scale Lab

OBJECTIVE Students observe, identify, and write balanced equations for precipitation reactions.

PREP TIME 30 minutes

CLASS TIME 40 minutes

ADVANCE PREPARATION

Solution	Preparation
0.05*M* $AgNO_3$	2.1 g in 250 mL
0.2*M* $Pb(NO_3)_2$	16.6 g in 250 mL
0.5*M* $CaCl_2$	13.9 g in 250 mL
1.0*M* Na_2CO_3	26.5 g in 250 mL
0.1*M* Na_3PO_4	9.5 g $Na_3PO_4{\bullet}12H_2O$ in 250 mL
0.5*M* NaOH	20.0 g in 1.0 L
0.2*M* Na_2SO_4	7.1 g in 250 mL
1.0*M* NaCl	14.6 g in 250 mL

DISPOSAL Dispose of precipitates in the trash and flush the remaining solutions down the drain with excess water.

ANALYZE

1. a. $Na_2CO_3 + 2AgNO_3 \rightarrow 2NaNO_3 + Ag_2CO_3(s)$
   b. $2Na_3PO_4 + 3Pb(NO_3)_2 \rightarrow 6NaNO_3 + Pb_3(PO_4)_2(s)$
2. Sodium hydroxide and calcium chloride form sodium chloride and solid calcium hydroxide.
3. Mixings d, n, and o do not react.
4. $Na_3PO_4 + 3AgNO_3 \rightarrow 3NaNO_3 + Ag_3PO_4(s)$
   $NaOH + AgNO_3 \rightarrow NaNO_3 + AgOH(s)$
   (actual products: $NaNO_3 + Ag_2O(s) + H_2O$)
   $NaCl + AgNO_3 \rightarrow NaNO_3 + AgCl(s)$
   $Na_2CO_3 + Pb(NO_3)_2 \rightarrow 2NaNO_3 + PbCO_3(s)$
   $2NaOH + Pb(NO_3)_2 \rightarrow 2NaNO_3 + Pb(OH)_2(s)$
   $Na_2SO_4 + Pb(NO_3)_2 \rightarrow 2NaNO_3 + PbSO_4(s)$
   $2NaCl + Pb(NO_3)_2 \rightarrow 2NaNO_3 + PbCl_2(s)$
   $Na_2CO_3 + CaCl_2 \rightarrow 2NaCl + CaCO_3(s)$
   $2Na_3PO_4 + 3CaCl_2 \rightarrow 6NaCl + Ca_3(PO_4)_2(s)$
   $2NaOH + CaCl_2 \rightarrow 2NaCl + Ca(OH)_2(s)$
5. $3Ag^+ + PO_4^{3-} \rightarrow Ag_3PO_4(s)$
   $Ag^+ + OH^- \rightarrow AgOH(s)$
   $Ag^+ + Cl^- \rightarrow AgCl(s)$
   $Pb^{2+} + CO_3^{2-} \rightarrow PbCO_3(s)$
   $Pb^{2+} + 2OH^- \rightarrow Pb(OH)_2(s)$
   $Pb^{2+} + SO_4^{2-} \rightarrow PbSO_4(s)$
   $Pb^{2+} + 2Cl^- \rightarrow PbCl_2(s)$
   $Ca^{2+} + CO_3^{2-} \rightarrow CaCO_3(s)$
   $3Ca^{2+} + 2PO_4^{3-} \rightarrow Ca_3(PO_4)_2(s)$
   $Ca^{2+} + 2OH^- \rightarrow Ca(OH)_2(s)$

Small-Scale Lab

Precipitation Reactions: Formation of Solids

Purpose

To observe, identify, and write balanced equations for precipitation reactions

Materials

- pencil
- paper
- ruler
- reaction surface
- chemicals shown in the grid to the right

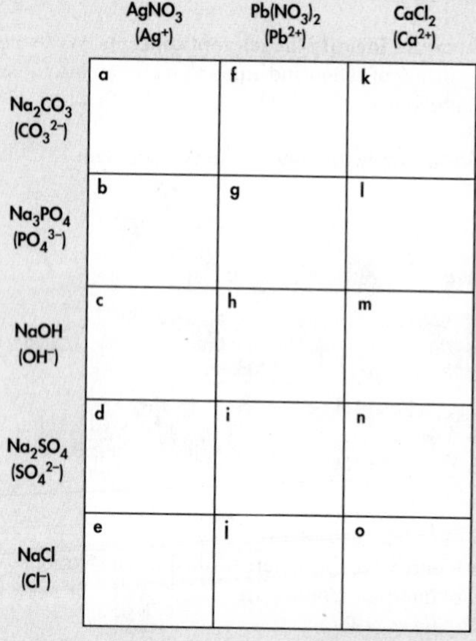

Procedure

1. Copy the grid on two sheets of paper.
2. Make each square 2 cm on each side.
3. Draw large black Xs on one of the grids.
4. Place a reaction surface over the grid with black Xs and add the chemicals as shown. Use the other grid as a data table to record your observations for each solution.

Analyze

Using your experimental data, record your answers to the following in the space below your data table.

1. **Explain** Translate the following word equations into balanced chemical equations and explain how the equations represent what happens in grid spaces *a* and *g*.
   a. In grid space *a*, sodium carbonate reacts with silver nitrate to produce sodium nitrate and solid silver carbonate.
   b. In grid space *g*, sodium phosphate reacts with lead(II) nitrate to produce sodium nitrate and solid lead(II) phosphate.
2. **Describe** Write a word equation to represent what happens in grid space *m*.
3. **Explain** What happens in grid space *d*? Which other mixings gave similar results? Is it necessary to write an equation when no reaction occurs?
4. **Describe** Write balanced equations for the other precipitation reactions you observed.
5. **Describe** Write balanced net ionic equations for the other precipitation reactions you observed.

You're the Chemist

The following small-scale activities allow you to develop your own procedures and analyze the results.

1. **Explain** Mix a solution of potassium iodide (KI) with silver nitrate. Then mix potassium iodide solution with lead(II) nitrate. Describe your results. Write balanced equations and net ionic equations for each reaction.
2. **Design an Experiment** Table salt is mostly sodium chloride. Design and carry out an experiment to find out if table salt will form a precipitate with either lead(II) nitrate or silver nitrate. Interpret your results.
3. **Design an Experiment** Design and carry out an experiment to show that iodized table salt contains potassium iodide.

YOU'RE THE CHEMIST

1. $KI + AgNO_3 \rightarrow KNO_3 + AgI(s)$
   $Ag^+ + I^- \rightarrow AgI(s)$
   $2KI + Pb(NO_3)_2 \rightarrow 2KNO_3 + PbI_2(s)$
   $Pb^{2+} + 2I^- \rightarrow PbI_2(s)$
2. Add a drop of $Pb(NO_3)_2$ or $AgNO_3$ to a few grains of salt. Look for white crystals.
3. Place a drop of $Pb(NO_3)_2$ on a small pile of dry table salt. Keep part of the pile dry. Look for yellow lead iodide.

11 Study Guide

BIGIDEA REACTIONS

The law of conservation of mass states that mass is neither created nor destroyed. In order to show that mass is conserved during a reaction, a chemical equation must be balanced. You can predict the products of most chemical reactions by identifying the reaction type. To determine the reaction type, consider the number of reacting elements and compounds.

11.1 Describing Chemical Reactions

To write a skeleton equation, write the formulas for the reactants to the left of the yields sign and the formulas for the products to the right.

After writing the skeleton equation, use coefficients to balance the equation so that it obeys the law of conservation of mass.

- chemical equation (348)
- skeleton equation (348)
- catalyst (348)
- coefficient (350)
- balanced equation (350)

11.2 Types of Chemical Reactions

The five general types of reactions are combination, decomposition, single-replacement, double-replacement, and combustion.

The number of elements and/or compounds reacting is a good indicator of possible reaction type and, thus, possible products.

In a combination reaction, there is always a single product.

A decomposition reaction involves the breakdown of a single compound into two or more simpler substances.

In a single-replacement reaction, both the reactants and the products are an element and a compound.

A double-replacement reaction generally takes place between two ionic compounds in aqueous solution.

A combustion reaction always involves oxygen as a reactant.

- combination reaction (356)
- decomposition reaction (358)
- single-replacement reaction (360)
- activity series (361)
- double-replacement reaction (362)
- combustion reaction (363)

11.3 Reactions in Aqueous Solution

A net ionic equation shows only those particles involved in the reaction and is balanced with respect to mass and charge.

By using the general rules for solubility of ionic compounds, you can predict the formation of a precipitate.

- complete ionic equation (370)
- spectator ion (370)
- net ionic equation (370)

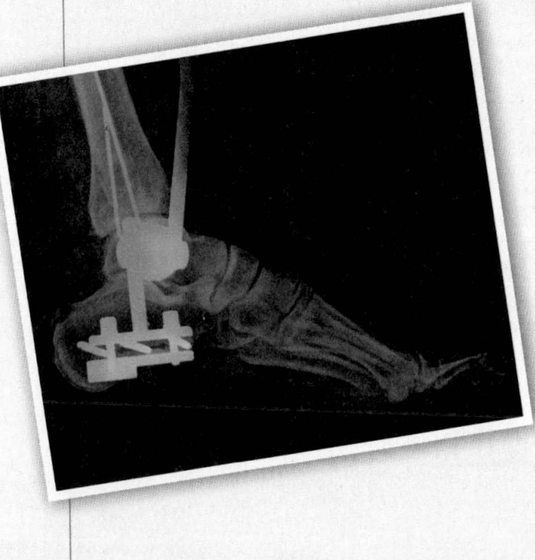

UbD Performance Tasks

IDENTIFYING REACTIONS Provide students with the reactants only of a series of reactions. Have students write balanced equations to complete each reaction and identify the type of reaction. Students should explain the reasoning behind each classification.

CORAL REEFS Explain that limestone is sedimentary rock formed from the exoskeletons of marine corals. Have students research the formation of these exoskeletons and write a report on the chemical processes involved. Reports should include balanced complete and net ionic equations representing the precipitation reactions responsible.

Evaluate

Review and Assessment Materials

CHEMISTRY TUTORIAL Encourage students to review balancing chemical equations, writing equations for double-replacement reactions, and writing and balancing net ionic equations using the animated step-by-step tutorials for Chapter 11. Each Chemistry Tutorial is followed by a similar practice problem, where the student is guided through the problem-solving process.

CHAPTER PROBLEM SET Have students practice more chemical reaction problems using the Chapter 11 online Problem Set.

MATH TUTORIAL Struggling students can use a tutorial on balancing equations as a quick review. Ask students to answer the guided practice problem that follows the Math Tutorial.

DIRECTED VIRTUAL LAB Have students complete the Virtual Lab *Identifying Cations in Solution* to help reinforce the concepts involved in identifying ionic precipitates in reactions between solutions. In the post-lab exercise, students will practice using solubility data to design lab procedures for identifying cations in solutions.

KINETIC ART Direct students to the Kinetic Art animations *Balancing Equations* and *Types of Reactions*, and have them answer the questions that accompany each animation.

Study Tip

Encourage students to use physical models to clarify concepts. For example, for balancing chemical equations or classifying a reaction according to its type, physical models of atoms can aid visual learners. Molecular model kits, foam balls, gumdrops, or modeling clay can be used to represent different types of atoms.

Evaluate

Skills Review

Example and sample practice problems related to balancing chemical equations can be found on the following pages:

LESSON 11.1 Writing a Skeleton Equation—page 349; Balancing a Chemical Equation—pages 352 and 353

LESSON 11.2 Writing Equations for Combination and Decomposition Reactions—page 359; Writing the Equation for Single-Replacement Reactions— page 361; Writing Equations for Double-Replacement Reactions—page 363; Writing Equations for Combustion Reactions—page 365

LESSON 11.3 Writing and Balancing Net Ionic Equations—pages 371 and 373

Skills Tune-Up: Balancing Chemical Equations

Problem	❶ Analyze	❷ Solve
Write the balanced equation for the following reaction: $C_2H_4 + O_2 \longrightarrow$	The reactants are a hydrocarbon and oxygen. The hydrocarbon tells you that the products must be CO_2 and H_2O. The oxygen tells you that this is a combustion reaction.	First write a skeleton equation. $C_2H_4 + O_2 \longrightarrow CO_2 + H_2O$ (unbalanced) Balance the C atoms and the H atoms first. $C_2H_4 + O_2 \longrightarrow 2CO_2 + 2H_2O$ (unbalanced) Balance the O atoms next. $C_2H_4 + 3O_2 \longrightarrow 2CO_2 + 2H_2O$ (balanced)
Write the balanced equation for the following reaction: $Al + Cu(NO_3)_2 \longrightarrow$	$Cu(NO_3)_2$ is an ionic compound, and Al is an element. This is a single-replacement reaction. Check Table 11.2 to be sure a reaction will take place.	First write a skeleton equation. $Al + Cu(NO_3)_2 \longrightarrow Al(NO_3)_3 + Cu$ (unbalanced) Balance the equation. $2Al + 3Cu(NO_3)_2 \longrightarrow 2Al(NO_3)_3 + 3Cu$ (balanced)
Write the balanced equation for the following reaction: $Na(OH)(aq) + Ba(NO_3)_2(aq) \longrightarrow$	Both reactants are ionic compounds, so this is a double-replacement reaction. In a double-replacement reaction, two compounds exchange positive ions. They often produce a gas, a precipitate, or another molecular compound such as water.	Write the reactants, showing each as dissociated free ions. $Na^+(aq) + OH^-(aq) + Ba^{2+}(aq) + 2NO_3^-(aq) \longrightarrow$ Look at the possible new pairings of cation and anion that give an insoluble substance. Of the two possible combinations, $Na(NO)_3$ is soluble and $Ba(OH)_2$ is insoluble. Balance the equation. $2NaOH(aq) + Ba(NO_3)_2(aq) \longrightarrow 2NaNO_3(aq) + Ba(OH)_2(s)$ (balanced)

A subscript in a polyatomic ion moves with the ion. So the 3 in NO_3 stays with the ion. But the subscript 2 is there only to balance the charges. It's not part of the ion and doesn't move with it.

Use the solubility rules in Table 11.3 to identify the precipitate formed.

Focus on ELL

5 ASSESS UNDERSTANDING Use a variety of informal assessments to help monitor students' progress in this chapter. Include various types of learning assignments, based on each student's language proficiency.

BEGINNING: LOW/HIGH Have students use diagrams to answer questions or describe reactions.

INTERMEDIATE

LOW Ask students to create a song or poem that demonstrates their knowledge of the five different types of chemical reactions.

HIGH Use the students' participation in the labs to assess their ability to demonstrate understanding.

ADVANCED: LOW/HIGH Include questions from previous lessons as part of assessments so that students can retain information.

11 Assessment

* Solutions appear in Appendix E

Lesson by Lesson

11.1 Describing Chemical Reactions

34. Identify the reactants and products in each chemical reaction.

a. Hydrogen gas and sodium hydroxide are formed when sodium is dropped into water.
b. In photosynthesis, carbon dioxide and water react to form oxygen gas and glucose.

35. Write sentences that completely describe each of the chemical reactions shown in these skeleton equations.

a. $NH_3(g) + O_2(g) \xrightarrow{Pt} NO(g) + H_2O(g)$
b. $H_2SO_4(aq) + BaCl_2(aq) \longrightarrow BaSO_4(s) + HCl(aq)$
c. $N_2O_3(g) + H_2O(l) \longrightarrow HNO_2(aq)$

36. The equation for the formation of water from its elements, $H_2(g) + O_2(g) \longrightarrow H_2O(l)$, can be "balanced" by changing the formula of the product to H_2O_2. Explain why this is incorrect.

*** 37.** Balance the following equations:

a. $PbO_2(s) \longrightarrow PbO(s) + O_2(g)$
b. $Fe(OH)_3(s) \longrightarrow Fe_2O_3(s) + H_2O(s)$
c. $(NH_4)_2CO_3(s) \longrightarrow NH_3(g) + H_2O(g) + CO_2(g)$
d. $CaCl_2(aq) + H_2SO_4(aq) \longrightarrow CaSO_4(s) + HCl(aq)$

11.2 Types of Chemical Reactions

***38.** Write balanced chemical equations for the following combination reactions:

a. $Mg(s) + O_2(g) \longrightarrow$
b. $P(s) + O_2(g) \longrightarrow$ diphosphorus pentoxide
c. $Ca(s) + S(s) \longrightarrow$

39. Write a balanced chemical equation for each decomposition reaction.

a. $Ag_2O(s) \xrightarrow{\Delta}$
b. ammonium nitrate $\xrightarrow{\Delta}$ dinitrogen monoxide + water

40. Use the activity series of metals to write a balanced chemical equation for each single-replacement reaction.

a. $Au(s) + KNO_3(aq) \longrightarrow$
b. $Zn(s) + AgNO_3(aq) \longrightarrow$
c. $Al(s) + H_2SO_4(aq) \longrightarrow$

41. Write a balanced equation for each of the following double-replacement reactions:

a. $H_2C_2O_4(aq) + KOH(aq) \longrightarrow$
b. $CdBr_2(aq) + Na_2S(aq) \longrightarrow$ (Cadmium sulfide is a precipitate.)

42. Write a balanced equation for the complete combustion of each compound.

a. butene (C_4H_8) **b.** propanal (C_3H_6O)

43. Balance each equation and identify its type.

a. $Hf(s) + N_2(g) \longrightarrow Hf_3N_4(s)$
b. $Mg(s) + H_2SO_4(aq) \longrightarrow MgSO_4(aq) + H_2(g)$
c. $C_2H_6(g) + O_2(g) \longrightarrow CO_2(g) + H_2O(g)$
d. $Pb(NO_3)_2(aq) + NaI(aq) \longrightarrow PbI_2(s) + NaNO_3(aq)$

44. What is a distinguishing feature of every decomposition reaction?

11.3 Reactions in Aqueous Solution

45. What is a spectator ion?

***46.** Write a balanced net ionic equation for the following reactions:

a. $HCl(aq) + Ca(OH)_2(aq) \longrightarrow$
b. $AgNO_3(aq) + AlCl_3(aq) \longrightarrow$ (Silver chloride is a precipitate.)

47. Complete each equation and then write a net ionic equation.

a. $Al(s) + H_2SO_4(aq) \longrightarrow$
b. $HCl(aq) + Ba(OH)_2(aq) \longrightarrow$
c. $Au(s) + HCl(aq) \longrightarrow$

Understand Concepts

48. Write a balanced chemical equation for each reaction. Use the necessary symbols from Table 11.1 to describe the reaction completely.

a. Bubbling chlorine gas through a solution of potassium iodide gives elemental iodine and a solution of potassium chloride.
b. Bubbles of hydrogen gas and aqueous iron(III) chloride are produced when metallic iron is dropped into hydrochloric acid.
c. Solid tetraphosphorus decaoxide reacts with water to produce phosphoric acid.

Evaluate

Answers

LESSON 11.1

34. **a.** reactants: sodium and water; products: hydrogen and sodium hydroxide
b. reactants: carbon dioxide and water; products: oxygen and glucose

35. **a.** Gaseous ammonia and oxygen react in the presence of a platinum catalyst to produce nitrogen monoxide gas and water vapor.
b. Aqueous solutions of sulfuric acid and barium chloride are mixed to produce a precipitate of barium sulfate and aqueous hydrochloric acid.
c. The gas dinitrogen trioxide reacts with water to produce an aqueous solution of nitrous acid.

36. A formula is a unique identifier of a substance. A different formula would indicate a different substance, not the one that is taking part in the reaction you are trying to balance.

37. **a.** $2PbO_2(s) \rightarrow 2PbO(s) + O_2(g)$
b. $2Fe(OH)_3(s) \rightarrow Fe_2O_3(s) + 3H_2O(l)$
c. $(NH_4)_2CO_3(s) \rightarrow 2NH_3(g) + H_2O(g) + CO_2(g)$
d. $CaCl_2(aq) + H_2SO_4(aq) \rightarrow CaSO_4(s) + 2HCl(aq)$

LESSON 11.2

38. **a.** $2Mg(s) + O_2(g) \rightarrow 2MgO(g)$
b. $4P(s) + 5O_2(g) \rightarrow 2P_2O_5(s)$
c. $Ca(s) + S(s) \rightarrow CaS(s)$

39. **a.** $2Ag_2O \xrightarrow{\Delta} 4Ag + O_2$
b. $NH_4NO_3 \xrightarrow{\Delta} N_2O + 2H_2O$

40. **a.** no reaction
b. $Zn(s) + 2AgNO_3(aq) \rightarrow Zn(NO_3)_2(aq) + 2Ag(s)$
c. $2Al(s) + 3H_2SO_4(aq) \rightarrow Al_2(SO_4)_3(aq) + 3H_2(g)$

41. **a.** $H_2C_2O_4(aq) + 2KOH(aq) \rightarrow K_2C_2O_4(aq) + 2H_2O(l)$
b. $CdBr_2(aq) + Na_2S(aq) \rightarrow CdS(s) + 2NaBr(aq)$

42. **a.** $C_4H_8(g) + 6O_2(g) \rightarrow 4CO_2(g) + 4H_2O(g)$
b. $C_3H_6O(l) + 4O_2(g) \rightarrow 3CO_2(g) + 3H_2O(g)$

43. **a.** $3Hf(s) + 2N_2(g) \rightarrow Hf_3N_4(s)$; combination
b. $Mg(s) + H_2SO_4(aq) \rightarrow MgSO_4(aq) + H_2(g)$; single replacement
c. $2C_2H_6(g) + 7O_2(g) \rightarrow 4CO_2(g) + 6H_2O(g)$; combustion
d. $Pb(NO_3)_2(aq) + 2NaI(aq) \rightarrow PbI_2(s) + 2NaNO_3(aq)$; double replacement

44. a single reactant

LESSON 11.3

45. an ion that does not participate in the reaction

46. **a.** $H^+(aq) + OH^-(aq) \rightarrow H_2O(l)$
b. $Ag^+(aq) + Cl^-(aq) \rightarrow AgCl(s)$

47. **a.** $2Al(s) + 6H^+(aq) \rightarrow 2Al^{3+}(aq) + 3H_2(g)$
b. $H^+(aq) + OH^-(aq) \rightarrow H_2O(l)$
c. no reaction

UNDERSTAND CONCEPTS

48. **a.** $Cl_2(g) + 2KI(aq) \rightarrow I_2(aq) + 2KCl(aq)$
b. $2Fe(s) + 6HCl(aq) \rightarrow 2FeCl_3(aq) + 3H_2(g)$
c. $P_4O_{10}(s) + 6H_2O(l) \rightarrow 4H_3PO_4(aq)$

Evaluate

Answers

49. a. $Cl_2 + 2NaI \rightarrow 2NaCl + I_2$
b. $2NH_3 \rightarrow N_2 + 3H_2$
c. $4Na + O_2 \rightarrow 2Na_2O$

50. a. $ZnS(aq) + H_2SO_4(aq) \rightarrow H_2S(g) + ZnSO_4(aq)$
b. $NaOH(aq) + HNO_3(aq) \rightarrow H_2O(l) + NaNO_3(aq)$
c. $2KF(aq) + Ca(NO_3)_2(aq) \rightarrow CaF_2(s) + 2KNO_3(aq)$

51. a. $Na_2O(s) + H_2O(l) \rightarrow 2NaOH(aq)$
b. $H_2(g) + Br_2(g) \rightarrow 2HBr(g)$
c. $Cl_2O_7(l) + H_2O(l) \rightarrow 2HClO_4(aq)$

52. a. $Fe(s) + H_2SO_4(aq) \rightarrow FeSO_4(aq) + H_2(g)$
b. no reaction
c. $Br_2(l) + BaI_2(aq) \rightarrow BaBr_2(aq) + I_2(aq)$

53. a. tube A
b. $2Na(s) + 2H_2O(l) \rightarrow 2NaOH(aq) + H_2(g)$
single-replacement

54. a. $2C_8H_{18} + 25O_2 \rightarrow 16CO_2 + 18H_2O$
b. $C_6H_{12}O_6 + 6O_2 \rightarrow 6CO_2 + 6H_2O$
c. $HC_2H_3O_2 + 2O_2 \rightarrow 2CO_2 + 2H_2O$

55. a. $2Al_2O_3 \xrightarrow{\text{energy}} 4Al + 3O_2$
b. $Sn(OH)_4 \xrightarrow{\Delta} SnO_2 + 2H_2O$
c. $Ag_2CO_3 \xrightarrow{\Delta} Ag_2O + CO_2$

56. a. $H^+(aq) + OH^-(aq) \rightarrow H_2O(l)$
b. $S^{2-}(aq) + 2H^+(aq) \rightarrow H_2S(g)$
c. $3OH^-(aq) + Fe^{3+}(aq) \rightarrow Fe(OH)_3(s)$

57. a. $CdS(s)$
b. $Na^+(aq)$ and $NO_3^-(aq)$
c. $Cd^{2+}(aq) + S^{2-}(aq) \rightarrow CdS(s)$

THINK CRITICALLY

58. a. (1) combination
(2) single-replacement
(3) combustion
(4) double-replacement
b. (1) $2Al(s) + 3Br_2(l) \rightarrow 2AlBr_3(s)$
(2) $Cu(s) + 2AgNO_3(aq) \rightarrow Cu(NO_3)_2(aq) + 2Ag(s)$
(3) $C_3H_8(g) + 5O_2(g) \rightarrow 3CO_2(g) + 4H_2O(g)$
(4) $Pb(NO_3)_2(aq) + 2KI(aq) \rightarrow PbI_2(s) + 2KNO_3(aq)$

---

49. Each equation is incorrect. Find the errors, then rewrite and balance each equation.

a. $Cl_2 + NaI \longrightarrow NaCl_2 + I$
b. $NH_3 \longrightarrow N + H_3$
c. $Na + O_2 \longrightarrow NaO_2$

50. Write balanced chemical equations for these double-replacement reactions that occur in aqueous solution.

a. Zinc sulfide is added to sulfuric acid.
b. Sodium hydroxide reacts with nitric acid.
c. Solutions of potassium fluoride and calcium nitrate are mixed.

***51.** Write a balanced chemical equation for each combination reaction.

a. sodium oxide + water
b. hydrogen + bromine
c. dichlorine heptoxide + water

52. Write a balanced chemical equation for each single-replacement reaction that takes place in aqueous solution. Write "no reaction" if a reaction does not occur.

a. Steel wool (iron) is placed in sulfuric acid.
b. Mercury is poured into an aqueous solution of zinc nitrate.
c. Bromine reacts with aqueous barium iodide.

***53.** Pieces of sodium and magnesium are dropped into separate water-filled test tubes (A and B). There is vigorous bubbling in Tube A but not in Tube B.

a. Which tube contains the sodium metal?
b. Write an equation for the reaction in the tube containing the sodium metal. What type of reaction is occurring in this tube?

54. Write a balanced equation for the complete combustion of each compound. Assume that the products are carbon dioxide and water.

a. octane (C_8H_{18})
b. glucose ($C_6H_{12}O_6$)
c. ethanoic acid ($HC_2H_3O_2$)

55. Write balanced chemical equations for these decomposition reactions.

a. Aluminum is obtained from aluminum oxide with the addition of a large amount of electrical energy.
b. Heating tin(IV) hydroxide gives tin(IV) oxide and water.
c. Silver carbonate decomposes into silver oxide and carbon dioxide when it is heated.

56. Write a balanced net ionic equation for each reaction. The product that is not ionized is given.

a. $H_2C_2O_4 + KOH \longrightarrow [H_2O]$
b. $Na_2S + HCl \longrightarrow [H_2S]$
c. $NaOH + Fe(NO_3)_3 \longrightarrow [Fe(OH)_3]$

***57.** A yellow precipitate formed when aqueous solutions of sodium sulfide and cadmium nitrate were mixed in a beaker.

a. Write the formula of the yellow precipitate.
b. Identify the spectator ions in the solution.
c. Write the net ionic equation for the reaction.

Think Critically

58. Interpret Photos The photos show various types of reactions.

(1)
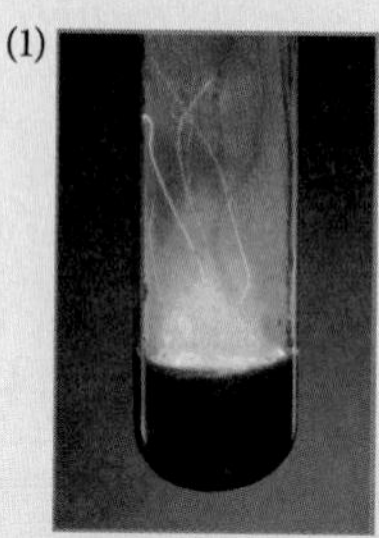
(2)
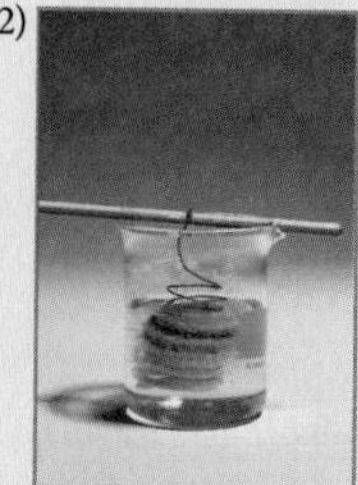
(3)

(4)

(1) Aluminum reacting with bromine

(2) The reaction of copper with aqueous silver nitrate

(3) Propane (C_3H_8) reacting with oxygen

(4) The reaction of lead(II) nitrate with potassium iodide

a. Identify each type of reaction.
b. Write the equation for each type of reaction.

378 Chapter 11 • Assessment

59. **Apply Concepts** Write a balanced chemical equation for each reaction. Classify each by type.

a. Sodium iodide reacts with phosphoric acid.
b. Potassium oxide reacts with water.
c. Heating sulfuric acid produces water, oxygen, and sulfur dioxide.
d. Aluminum reacts with sulfuric acid.
e. Pentane (C_5H_{12}) reacts with oxygen.

*60. **Draw Conclusions** When pale yellow chlorine gas is bubbled through a clear, colorless solution of sodium iodide, the solution turns brown.

a. What type of reaction is taking place?
b. Write the net ionic equation.

Enrichment

61. **Interpret Graphs** Alkanes are hydrocarbon molecules that have the general formula C_nH_{2n+2}. The graph shows the number of oxygen, carbon dioxide, and water molecules needed to balance the equations for the complete combustion of every alkane having from one to ten carbon atoms.

C_nH_{2n+2} + ___ O_2 ⟶ ___ CO_2 + ___ H_2O

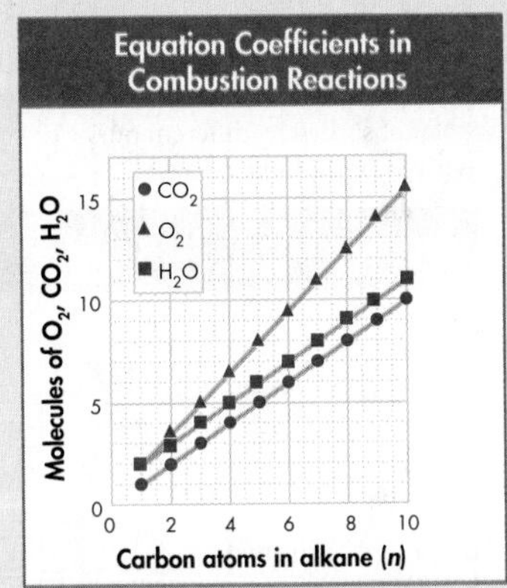

a. Use the graph to write balanced equations for the combustion of C_5H_{12} and C_9H_{20}.
b. Extrapolate the graph and write balanced equations for the combustion of $C_{12}H_{26}$ and $C_{17}H_{36}$.
c. The coefficient for O_2 in the general equation is as follows:

$$n + \frac{n+1}{2}$$

What are the coefficients for CO_2 and H_2O?

*62. **Apply Concepts** Fill in the missing reactant, and then balance each equation.

a. $K(s)$ + ___ ⟶ $KOH(aq) + H_2(g)$
b. $C_2H_5OH(l)$ + ___ ⟶ $CO_2(g) + H_2O(g)$
c. $Bi(NO_3)_3(aq)$ + ___ ⟶ $Bi_2S_3(s) + HNO_3(aq)$
d. $Al(s)$ + ___ ⟶ $AlBr_3(s)$

Write About Science

63. **Explain** Research organisms such as fireflies and jellyfish that use bioluminescence, including information on the discovery of green fluorescent protein (GFP). In a pamphlet or poster, explain how bioluminescence works and how each organism uses it.

64. **Observe** Make a list of five chemical reactions that happen in your kitchen. Describe and name each reaction on your list.

*65. **Relate Cause and Effect** Why is smoking not permitted near an oxygen source? What would happen if a match were struck in a room filled with oxygen?

CHEMYSTERY

Order in the Lab

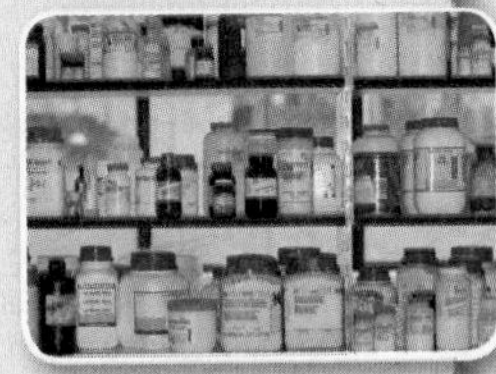

Chemicals should not be stored in alphabetical order because some chemicals that will react if mixed could end up next to each other. For example, acids should not be stored near cyanides, sulfides and other chemicals that produce toxic gases when combined. Acids should also not be stored near bases or active metals. Reactions between acids and bases produce heat. Acids and active metals react to produce gases and heat. Acids and flammables should have separate, dedicated storage areas.

66. **Connect to the BIGIDEA** Should sulfuric acid be stored next to sodium hydroxide? Explain your answer. If they should not be stored next to each other, write a balanced chemical equation to support your answer.

ASSESSMENT

CHEMYSTERY **APPLY CONCEPTS** After students have read through the CHEMystery, call on volunteers to write chemical equations for reactions involving the chemicals that Maria rearranged. **Ask** What gaseous product would form by reacting sulfuric acid with sodium cyanide? *(hydrogen cyanide, HCN)* **Ask** What gaseous product would form from the reaction of sulfuric acid and sodium sulfide? *(hydrogen sulfide, H_2S)* Explain that both HCN and H_2S are gases that can be lethal if inhaled. Because sulfuric acid is a liquid, a very serious situation would have occurred in the lab if a spill occurred while it was stored next to either Na_2S or NaCN.

CHEMYSTERY ANSWERS

66. **BIGIDEA** No; NaOH is a base, and bases should not be stored next to acids because reactions between these two classes of chemicals generate heat.
$H_2SO_4(aq) + 2NaOH(s) \rightarrow Na_2SO_4(aq) + 2H_2O(l)$

Answers

59. a. $3NaI + H_3PO_4 \rightarrow 3HI + Na_3PO_4$; double-replacement
b. $K_2O + H_2O \rightarrow 2KOH$; combination
c. $2H_2SO_4 \xrightarrow{\Delta} 2H_2O + O_2 + 2SO_2$; decomposition
d. $2Al + 3H_2SO_4 \rightarrow 3H_2 + Al_2(SO_4)_3$; single-replacement
e. $C_5H_{12} + 8O_2 \rightarrow 5CO_2 + 6H_2O$; combustion

60. a. single-replacement
b. $Cl_2(g) + 2I^-(aq) \rightarrow I_2(aq) + 2Cl^-(aq)$

ENRICHMENT

61. a. $C_5H_{12} + 8O_2 \rightarrow 5CO_2 + 6H_2O$
$C_9H_{20} + 14O_2 \rightarrow 9CO_2 + 10H_2O$
b. $2C_{12}H_{26} + 37O_2 \rightarrow 24CO_2 + 26H_2O$
$C_{17}H_{36} + 26O_2 \rightarrow 17CO_2 + 18H_2O$
c. $n = CO_2$; $(n + 1) = H_2O$

62. a. $2K(s) + 2H_2O(l) \rightarrow 2KOH(aq) + H_2(g)$
b. $C_2H_5OH(l) + 3O_2(g) \rightarrow 2CO_2(g) + 3H_2O(g)$
c. $2Bi(NO_3)_3(aq) + 3H_2S(g) \rightarrow Bi_2S_3(s) + 6HNO_3(aq)$
d. $2Al(s) + 3Br_2(l) \rightarrow 2AlBr_3(s)$

WRITE ABOUT SCIENCE

63. Answers will vary, but should indicate an understanding that bioluminescence is a combination reaction that requires a catalyst, and produces an oxidized protein and light.

64. Answers will vary, but should include examples and descriptions of combination, decomposition, single-replacement, double-replacement, and combustion reactions.

65. Answers will vary, but should indicate an understanding that as the oxygen concentration in a combustion reaction increases, the more rapidly the reaction will proceed and the more likely combustion be complete. A fire in an oxygen-filled room will spread much more rapidly and cause significantly more damage than in a room with the concentration of oxygen found in air.

Cumulative Review

67. **a.** water
b. water vapor in the air
c. physical change

68. Element: gold; compounds: sodium chloride, ice with water; homogeneous mixtures, salt water, air; heterogeneous mixture: salt and sand; substance: sodium chloride, gold, water with ice

69. 36.6 kg

70. 22 protons, 28 neutrons, and 22 electrons

71. **a.** $1s^22s^22p^63s^23p^64s^23d^{10}4p^6$
b. $1s^22s^22p^63s^23p^6$
c. $1s^22s^22p^63s^23p^63d^{10}$
d. $1s^22s^22p^63s^23p^63d^{10}$

72. Electronegativity is the tendency for an atom to attract bonded electrons to itself. Electronegativity values increase from left to right in a period.

73. **a.** incorrect; KBr
b. correct
c. incorrect; Ca_3N_2
d. correct

74. **a.** K_2CrO_4
b. $NaHSO_3$
c. permanganic acid
d. potassium oxalate

75. **a.** 2.41 mol
b. 6.91×10^{-2} mol
c. 0.934 mol
d. 7.09 mol

76. **a.** compound 1: $FeSO_4$, compound 2: $FeSO_3$
b. $FeSO_4$, iron(II) sulfate; $FeSO_3$, iron(II) sulfite

77. $C_8H_{10}O_2N_4$

78. **a.**

$CaCl_2$ (mol)	H_2O (mol)
a. 0.156	e. 0.312
b. 0.439	f. 0.878
c. 1.12	g. 2.24
d. 3.03	h. 6.06

b.

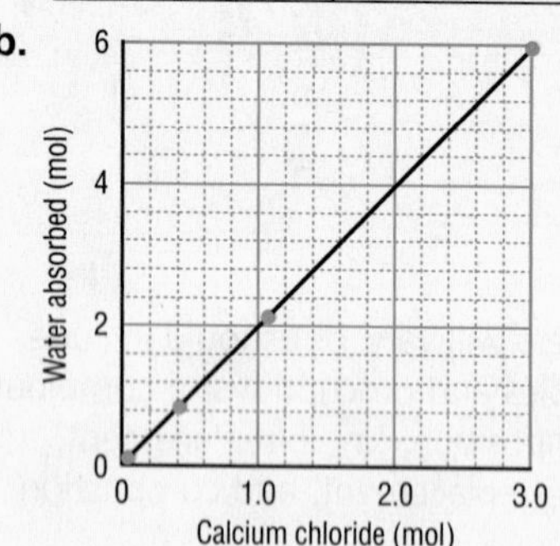

c. Two molecules of water.

Cumulative Review

67. When you take a glass of cold liquid outside on a warm, humid day, drops of liquid soon form on the outside of the glass.
a. What is the liquid?
b. Where did the liquid come from?
c. Did a chemical or physical change occur?

68. Classify each of the following as an element, a compound, a homogeneous mixture, a heterogeneous mixture, or a substance. Some may fit in more than one category.
a. salt water
b. sodium chloride
c. air
d. salt and sand
e. gold
f. water with ice

69. A block of ice measures 25.0 cm × 42.0 cm × 38.0 cm. What is the mass of the ice in kilograms? The density of ice is 0.917 g/cm³.

***70.** List the number of protons, neutrons, and electrons in this isotope of titanium: $^{50}_{22}Ti$.

71. Write electron configurations for the following ions.
a. Sr^{2+} **b.** S^{2-} **c.** Ga^{3+} **d.** Cu^+

72. Explain what is meant by *electronegativity.* How do electronegativity values change across a row of representative elements?

73. Are any of the following formulas for ionic compounds incorrect? If so, write the correct formulas.
a. K_2Br **b.** Na_2S **c.** CaN_2 **d.** Al_2O_3

***74.** Give the name or formula for the following compounds:
a. potassium chromate
b. sodium hydrogen sulfite
c. $HMnO_4$
d. $K_2C_2O_4$

75. Calculate the number of moles in each substance.
a. 54.0 L of nitrogen dioxide (at STP)
b. 1.68 g of magnesium ions
c. 69.6 g of sodium hypochlorite
d. 4.27×10^{24} molecules of carbon monoxide

76. The graph shows the percent composition of two different compounds formed by the elements iron, oxygen, and sulfur.

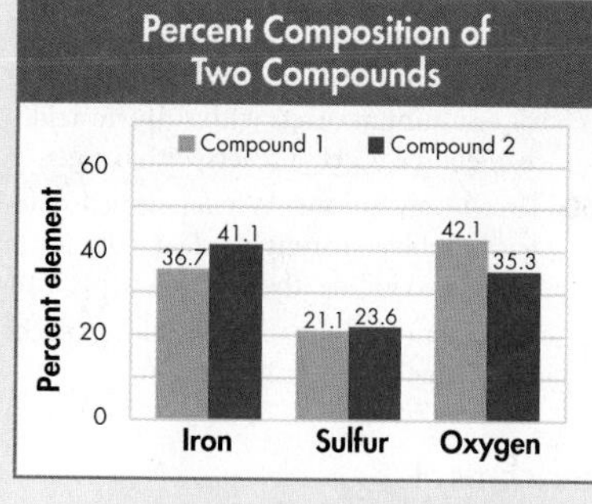

a. Using the data on the graphs, calculate the empirical formula of each compound.
b. Name each compound.

***77.** Many coffees and colas contain the stimulant caffeine. The percent composition of caffeine is 49.5% C, 5.20% H, 16.5% O, and 28.9% N. What is the molecular formula of caffeine if its molar mass is 194.1 g/mol?

78. Calcium chloride ($CaCl_2$) is a white solid used as a drying agent. The maximum amount of water absorbed by different quantities of $CaCl_2$ is given in the table below.

$CaCl_2$ (g)	$CaCl_2$ (mol)	H_2O (g)	H_2O (mol)
17.3	a.______	5.62	e.______
48.8	b.______	15.8	f.______
124	c.______	40.3	g.______
337	d.______	109	h.______

a. Complete the table.
b. Plot the moles of water absorbed (*y*-axis) versus the moles of $CaCl_2$.
c. Based on your graph, how many molecules of water does each formula unit of $CaCl_2$ absorb?

If You Have Trouble With . . .

Question	67	68	69	70	71	72	73	74	75	76	77	78
See Chapter	2	2	3	4	5	6	7	9	10	10	10	10

Standardized Test Prep for Indiana

Standards: C.4.1, C.4.2, C.4.5

Select the choice that best answers each question or completes each statement.

1. When the equation $Fe_2O_3 + H_2 \longrightarrow Fe + H_2O$ is balanced using whole-number coefficients, what is the coefficient of H_2?
   (A) 6 (B) 3 (C) 2 (D) 1

2. Identify the spectator ion in this reaction.
   $Ba(OH)_2(aq) + H_2SO_4(aq) \longrightarrow BaSO_4(s) + H_2O(l)$
   (A) Ba^{2+}
   (B) SO_4^{2-}
   (C) OH^-
   (D) H^+
   (E) There is no spectator ion.

3. Magnesium ribbon reacts with an aqueous solution of copper(II) chloride in a single-replacement reaction. Which are the products of the balanced net ionic equation for the reaction?
   (A) $Mg^{2+}(aq) + 2Cl^-(aq) + Cu(s)$
   (B) $Mg^+(aq) + Cl^-(aq) + Cu^+(aq)$
   (C) $Mg^{2+}(aq) + Cu(s)$
   (D) $Cu(s) + 2Cl^-(aq)$

Use the following description and data table to answer Questions 4–6.

Dropper bottles labeled P, Q, and R contain one of three aqueous solutions: potassium carbonate, K_2CO_3; hydrochloric acid, HCl; and calcium nitrate, $Ca(NO_3)_2$. The table shows what happens when pairs of solutions are mixed.

Solution	P	Q	R
P	—	Precipitate	No reaction
Q	Precipitate	—	Gas forms.
R	No reaction	Gas forms.	—

4. Identify the contents of each dropper bottle.
5. Write the net ionic equation for the formation of the precipitate.
6. Write the complete ionic equation for the formation of the gas.

7. Which are the expected products of the decomposition reaction of potassium oxide, K_2O?
   (A) $K^+(s)$ and $O^{2-}(g)$
   (B) $K^+(s)$ and $O_2(g)$
   (C) $K(s)$ and $O_2^{2-}(g)$
   (D) $K(s)$ and $O_2(g)$

Tips for Success

Interpreting Diagrams Before you answer questions about a diagram, study the diagram carefully. Read all captions and labels. Look at all the information in the diagram and think about how it all interrelates.

Use the diagram to answer Questions 8–11.

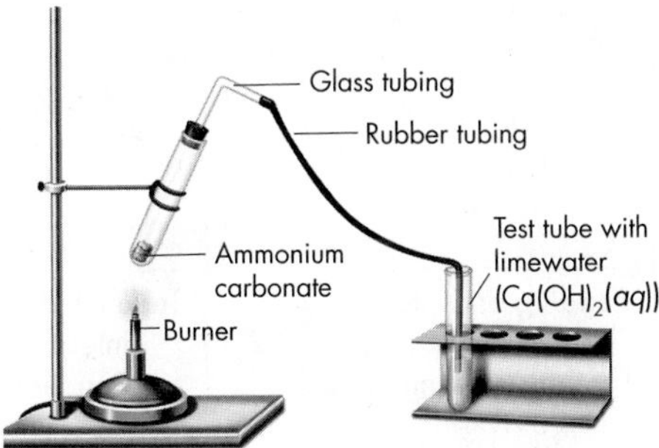

8. When ammonium carbonate is heated, water, ammonia, and carbon dioxide are produced. What type of chemical reaction is occurring?
9. Write formulas for the reaction products.
10. Write a balanced equation for the reaction. Include states for reactants and products.
11. Limewater is used to test for the presence of carbon dioxide gas. The products of the reaction of $Ca(OH)_2$ with CO_2 are calcium carbonate and water. Write a balanced equation for the reaction.

If You Have Trouble With . . .

Question	1	2	3	4	5	6	7	8	9	10	11
See Lesson	11.1	11.3	11.2	11.3	11.3	11.3	11.2	11.2	11.1	11.2	11.2

STP Answers

1. B
2. E
3. C
4. P is calcium nitrate, Q is potassium carbonate, and R is hydrochloric acid.
5. $Ca^{2+}(aq) + CO_3^{2-}(aq) \rightarrow CaCO_3(s)$
6. $2K^+(aq) + CO_3^{2-}(aq) + 2H^+(aq) + 2Cl^-(aq) \rightarrow 2K^+(aq) + 2Cl^-(aq) + H_2O(l) + CO_2(g)$
7. D
8. decomposition reaction
9. H_2O, NH_3, and CO_2
10. $(NH_4)_2CO_3(s) \rightarrow 2NH_3(g) + CO_2(g) + H_2O(g)$
11. $Ca(OH)_2 + CO_2 \rightarrow CaCO_3 + H_2O$

12 Stoichiometry Planning Guide

Introducing the BIGIDEA: THE MOLE; REACTIONS

Chemists use the mole to make sure that they measure the right amount of reacting material.

IASS	Lessons and Objectives	Print Resources		
		For the Student	For the Teacher	
C.1.6, C.4.2, C.4.3	**12.1** The Arithmetic of Equations p 384–389 **12.1.1** Describe how chemists use balanced chemical equations. **12.1.2** Describe the quantities you can use to interpret a balanced chemical equation.	Reading and Study Workbook Lesson 12.1 Lesson Assessment 12.1 p 389	Teacher Demo, p 387: Interpreting a Chemical Equation	
C.1.6, C.4.2, C.4.3, C.4.4, C.5.3	**12.2** Chemical Calculations p 390–398 **12.2.1** Explain how mole ratios are used in chemical calculations. **12.2.2** Explain the general procedure for solving a stoichiometric problem.	Reading and Study Workbook Lesson 12.2 Lesson Assessment 12.2 p 398 Small-Scale Lab: Analysis of Baking Soda p 399	Teacher Demo, p 392: Interpreting a Chemical Equation Class Activity, p 396: Stoichiometric Flash Cards	
C.4.2, C.4.3, C.4.4, C.5.3	**12.3** Limiting Reagent and Percent Yield p 400–408 **12.3.1** Explain how the amount of product in a reaction is affected by an insufficient quantity of any of the reactants. **12.3.2** Explain what the percent yield of a reaction measures.	Reading and Study Workbook Lesson 12.3 Lesson Assessment 12.3 p 408 Quick Lab: Limiting Reagents p 404	Teacher Demo, p 401: Limiting Factor Class Activity, p 403: Molecular Models and Limiting Reagent Class Activity, p 405: Actual Yield and Heat	

Assessing the BIGIDEA: THE MOLE; REACTIONS

	Essential Questions 1. How are balanced chemical equations used in stoichiometric calculations? 2. How can you calculate amounts of reactants and products in a chemical reaction?	Study Guide p 409 Math Tune-Up p 410 STP p 417 Reading and Study Workbook Self-Check and Vocabulary Review Chapter 12		

	PearsonChem.com	
	Editable Worksheets	Online Activities
	Teaching Resources, Lesson 12.1 Review	12.1 Lesson Overview Using a Balanced Equation as a Recipe Skateboard Stoichiometry Interpreting a Balanced Chemical Equation
	Teaching Resources, Lesson 12.2 Review Lab 19: Quantitative Analysis Lab Practical 12-1: Stoichiometry in a Reaction Small-Scale Lab Manual Lab 16: Titration of Bleach Probeware Lab: Small-Scale Lab: Analysis of Baking Soda	12.2 Lesson Overview Analysis of Baking Soda Calculating the Mass of a Product
	Teaching Resources, Lesson 12.3 Review Lab 20: Balanced Chemical Equations Lab Practical 12-2: Limiting Reagent	12.3 Lesson Overview Limiting Reagents Determining the Limiting Reagent in a Reaction Calculating the Theoretical and Percent Yield of a Reaction
	Core Teaching Resources (includes editable worksheets) • Lesson Reviews • Practice Problems • Interpret Graphs • Vocabulary Review • Chapter Quizzes and Tests • Lab Record Sheets	Chapter 12 Problem Set Percents Limiting Factors

Materials List

For the Student

Small-Scale Lab p 399
- baking soda
- 3 plastic cups
- soda straw
- balance
- pipets of HCl, NaOH, and thymol blue
- pH sensor (optional)

Quick Lab p 404
- graduated cylinder
- balance
- 3 250-mL Erlenmeyer flasks
- 3 rubber balloons
- 4.2 g magnesium ribbon
- 300 mL 1.0*M* hydrochloric acid

For the Teacher

Teacher Demo p 387
- 2.5–3.5-cm strip of magnesium
- 50 mL 1*M* HCl
- 100-mL beaker
- baking soda

Teacher Demo p 392
- 0.1*M* KI
- 0.1*M* $Pb(NO_3)_2$
- 250-mL beakers

Class Activity p 396
- 8 index cards
- 1 colored index card
- paper punch
- 2 brass paper fasteners

Teacher Demo p 401
- 15 plastic bottles
- 30 plastic caps to fit the bottles
- 6 containers to hold 5 caps each

Teacher Demo p 403
- 20 metal paper clips (symbol M)
- 20 same-colored vinyl-coated paper clips (symbol C)
- plastic sandwich bag

Class Activity p 405
- 3 foam cups
- a thermometer
- 100 mL of 1.0*M* HCl
- approximately 200 mL of 1.0*M* NaOH

Additional Digital Resources

Online Student Edition
Online Teacher's Edition
Exam*View Assessment Suite CD-ROM*
Classroom Resources DVD
12.2 Virtual Chem Lab 28: Analysis of Baking Soda

What's Online

CHEMISTRY TUTORIAL Students can work step-by-step with the online tutorial to learn how to use balanced equations to predict components of a reaction and to solve stoichiometry problems by accessing the Chapter 12 Chemistry Tutorials.

ONLINE PROBLEM SET Additional practice problems about stoichiometry can be found in the Chapter 12 online Problem Set.

MATH TUTORIALS The online Math Tutorials will help guide students through converting ratios to percents and identifying limiting factors in problems.

DIRECTED VIRTUAL LAB Encourage students to review the mathematical concepts used in stoichiometric calculations of percent purity using the virtual lab for Chapter 12. The animation is followed by questions.

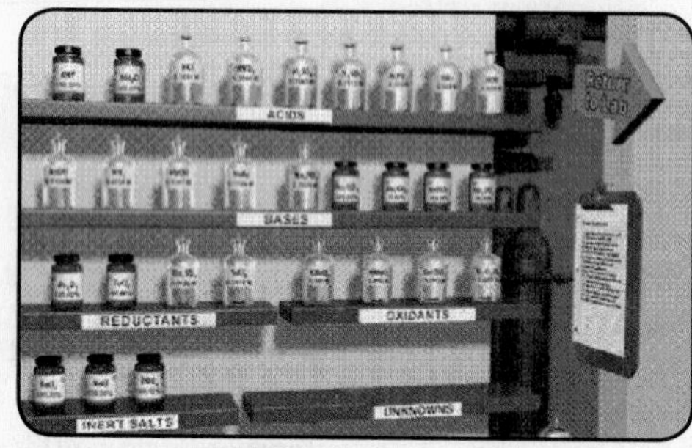

KINETIC ART Have students watch animations about interpreting a balanced equation and limiting reagents in reactions. Each animation is followed by questions.

CONCEPTS IN ACTION Students can watch an interactive animation about balancing chemical equations. The animation is followed by questions.

IN ACADEMIC STANDARDS for SCIENCE

Nature of Science

- Use analogies and models (mathematical and physical) to simplify and represent systems that are difficult to understand or directly experience due to their size, time scale, or complexity, and recognize the limitations of analogies and models.

Standard 4: Reactions and Stoichiometry

C.4.3 Use the mole concept to determine moles and number of atoms or molecules in samples of elements and compounds, given mass of the sample.

C.4.4 Using a balanced chemical equation, calculate the amounts of reactants needed or products made in a chemical reaction that goes to completion.

Also covered C.1.6, C.4.2, C.5.3.

12 Stoichiometry

INSIDE:

PearsonChem.com

Like a chemical equation, a recipe tells you the amount of each ingredient (your reactants) needed to make your product, in this case, bread.

Focus on ELL

1 CONTENT AND LANGUAGE Present the chapter title by writing *stoichiometry* on the board and the phonetic spelling, stoy kee AHM uh tree. Model pronunciation and have students repeat after you. Direct students' attention to the photograph. Ask students what a recipe has to do with stoichiometry. Help students see that a recipe and stoichiometry both measure reactants and products and predict results.

BEGINNING LOW Have students write the word *stoichiometry* on one side of a note card, and on the other side, have them copy the phonetic spelling of it.

HIGH Have students determine the quantities of ingredients they would use to make a ham and cheese sandwich and write an equation.

INTERMEDIATE: LOW/HIGH Paraphrase the Essential Questions. Summarize the CHEMystery story by writing the main idea of the story.

ADVANCED: LOW/HIGH Develop a list of words that have the suffix *-metry*, which means "to measure." Present the words to the class, along with their meanings.

BIGIDEAS

- **THE MOLE**
- **REACTIONS**

Essential Questions:

1. *How are balanced chemical equations used in stoichiometric calculations?*
2. *How can you calculate amounts of reactants and products in a chemical reaction?*

CHEMYSTERY

Cookie Crumbles

For the school bake sale, Jack wanted to make cookies to sell. He looked in cookbooks to find a good recipe. The recipe he chose called for specific amounts of butter, flour, sugar, eggs, vanilla, and baking soda. Jack wanted to make sure that his cookies were delicious and sweet. He didn't think there was enough sugar in the recipe, so he added twice as much sugar as the recipe called for.

Jack mixed the ingredients, put balls of the dough on a cookie sheet, and placed them in the oven to bake. When the bake time was up, Jack was very disappointed in his cookies. Instead of sweet, delicious cookies, his cookies were brown, hard, and crumbly. What happened? He checked the oven temperature and the amount of time that the cookies were in the oven. The time and temperature matched the directions in the recipe. Why didn't Jack's cookies turn out as he expected?

▶ Connect to the **BIGIDEA** As you read about quantifying chemical reactions, think about what could have happened to Jack's cookies.

INDIANA'S ACADEMIC STANDARDS FOR SCIENCE

Properties and States of Matter C.1.6; **Reactions and Stoichiometry** C.4.2, C.4.3, C.4.4; **Behavior of Gases** C.5.3. See lessons for details.

Introduce the Chapter

IDENTIFY PRECONCEPTIONS Students often enter into the study of stoichiometry with preconceptions. Use the activity to show them that all reactants may not be used up during a reaction.

Activity *Place 5 chocolate pieces, 8 graham crackers, and 9 marshmallows in a bowl.* **Ask** What's the ratio of reactants if you need one chocolate piece, two graham crackers and two mini-marshmallows to make one s'more? *(1:2:2 yields 1.)* Have students predict the number of s'mores you will be able to make with your reactants. *(Answers will vary.)* Then, make as many s'mores as you can with the reactants that you have. **Ask** How many s'mores did I make? *(four)* **Ask** Did I use up all my reactants? *(no)* **Ask** What prevented me from being able to make more s'mores? *(I ran out of graham crackers.)*

CHAPTER 12

UbD

Understanding by Design

Students are building toward measuring the efficiency of a chemical reaction by using the relationships of the mole and quantifying matter.

PERFORMANCE GOALS At the end of Chapter 12, students will be able to answer the essential questions by applying their knowledge of stoichiometry.

ESSENTIAL QUESTIONS Read the essential questions aloud. **Ask** What is conserved in a balanced chemical equation? *(atoms and mass)* **Ask** What information do you need to calculate the amount of a reactant or product? *(the number of moles of the substance and the number of grams or liters of the substance in each mole)*

BIGIDEA Use the chapter opener photo to help students connect with the Big Idea. Invite students to share their experiences with recipes and ingredients. **Ask** How would you adjust the recipe to double it? Guide students to an understanding of the importance of maintaining the correct ratios of ingredients when adjusting a recipe.

CHEMYSTERY Connect the CHEMystery to the Big Idea by explaining that the amounts of reactants and products of a reaction can be predicted using a balanced chemical equation. Mole ratios from a balanced chemical equation can be used to relate the moles of reactant to the moles of product. Ask students to predict how the additional sugar affected the cookies. As a hint, tell students that when they adjust a recipe, the ratios of ingredients must be the same in order to achieve the desired product.

UntamedScience™

Avogadro's Cookies

Amedeo Avogadro uses stoichiometry to figure out how much baking soda is needed in a cookie recipe.

LESSON 12.1

Key Objectives

12.1.1 DESCRIBE how chemists use balanced chemical equations.

12.1.2 DESCRIBE the quantities you can use to interpret a balanced chemical equation.

12.1.3 IDENTIFY the quantities that are always conserved in chemical reactions.

Additional Resources

- Reading and Study Workbook, Lesson 12.1
- Probeware Laboratory Manual, Lab 19
- Teaching Resources, Lesson 12.1 Review

PearsonChem.com

Engage

CHEMISTRY & YOU Have students study the photograph and read the text that opens the lesson. **Ask** How many wheels would be needed to make 200 bikes? *(two per bike, or 400)* **Ask** How would you calculate the number of frames? *(one per bike, or 200)*

Activate Prior Knowledge

Tell students that the word *stoichiometry* comes from the Greek words *stoicheion,* meaning "element," and *metron*, meaning "measure." Draw a two-column chart on the board, with *Reactants* on one side and *Products* on the other. Invite students to brainstorm a list of measurements that a chemist might use in relation to products and reactants, such as mass and moles. Ask students to compare the list to measurements studied in chemical reactions that relate to reactants and products.

IN ACADEMIC STANDARDS for SCIENCE

- Apply standard techniques in laboratory investigations to measure physical quantities in appropriate units and convert known quantities to other units as necessary.
- Use analogies and models (mathematical and physical) to simplify and represent systems that are difficult to understand or directly experience due to their size, time scale, or complexity, and recognize the limitations of analogies and models.

C.4.2 Balance chemical equations using the law of conservation of mass and use them to describe chemical reactions.

Also covered C.1.6, C.4.3

12.1 The Arithmetic of Equations

C.1.6 Conservation of mass: chemical changes; **C.4.2** Balance chemical equations; **C.4.3** The mole concept

CHEMISTRY & YOU

Q: *How do you figure out how much starting material you need to make a finished product?* Whenever you make something, you need to have the ingredients or the parts that make up the desired product. When making bikes, you need parts such as wheels, handlebars, pedals, and frames. If a factory needs to make 200 bikes, then the workers would need to calculate how many of each part they need to produce the 200 bikes. In this lesson, you will learn about how chemists determine how much of each reactant is needed to make a certain amount of product.

Key Questions

How do chemists use balanced chemical equations?

In terms of what quantities can you interpret a balanced chemical equation?

Vocabulary

- stoichiometry

Using Equations

How do chemists use balanced chemical equations?

One example of something that you might make is food. When you make cookies, for instance, you probably use a recipe. A cookie recipe tells you the precise amounts of ingredients to mix to make a certain number of cookies. If you need a larger number of cookies than the recipe provides for, you can double or triple the amounts of all the ingredients. In a way, a cookie recipe provides the same kind of information that a balanced chemical equation provides. In a cookie recipe, you can think of the ingredients as the reactants and the cookies as the products.

Everyday Equations The making of tricycles, like bikes and cookies, is a job that requires quantitative information to create the final product. Let's say you are in charge of manufacturing for the Travel Time Tricycle Company. The business plan for Travel Time requires the production of 640 custom-made tricycles each week. One of your responsibilities is to make sure there are enough parts available at the start of each workweek to make these tricycles. How can you determine the number of parts you need per week?

To simplify this discussion, assume that the major components of the tricycle are the frame (F), the seat (S), the wheels (W), the handlebars (H), and the pedals (P)—in other words, the reactants. The figure below illustrates how an equation can represent the manufacturing of a single tricycle.

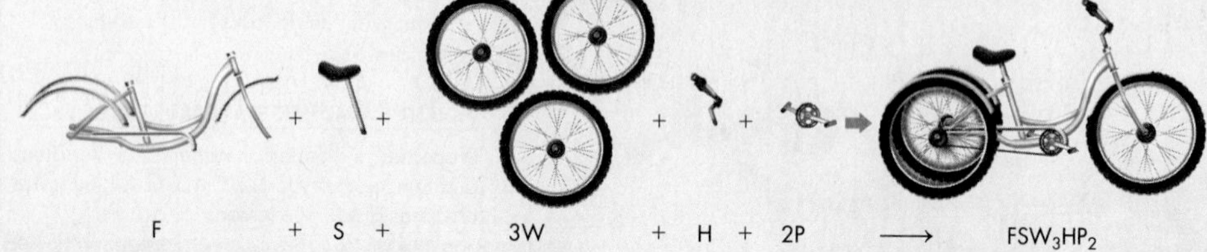

$$F + S + 3W + H + 2P \longrightarrow FSW_3HP_2$$

Focus on ELL

1 CONTENT AND LANGUAGE Begin the lesson by reviewing the meaning of the words *interpret, balanced, conservation,* and *arithmetic.* Review the key symbols used in chemical equations as well as the rules for writing and balancing a chemical equation.

2 FRONTLOAD THE LESSON Have students bring in their favorite recipes from their native countries, and guide students to understand how their recipes are similar to balanced chemical equations.

3 COMPREHENSIBLE INPUT Have students model Figure 12.2 using colored paper clips. Use one color to represent a nitrogen atom and one color to represent a hydrogen atom. Explain that 2 links of the same color represent either a nitrogen molecule or a hydrogen molecule.

The finished tricycle, your product, has a "formula" of FSW_3HP_2. The balanced equation for making a single tricycle is

$$F + S + 3W + H + 2P \longrightarrow FSW_3HP_2$$

This balanced equation is a "recipe" to make a single tricycle: Making a tricycle requires assembling one frame, one seat, three wheels, one handlebar, and two pedals. Now look at Sample Problem 12.1. It shows you how to use the balanced equation to calculate the number of parts needed to manufacture a given number of tricycles.

SampleProblem 12.1

Using a Balanced Equation as a Recipe

In a five-day workweek, Travel Time is scheduled to make 640 tricycles. How many wheels should be in the plant on Monday morning to make these tricycles?

1 Analyze List the knowns and the unknown. Use the balanced equation to identify a conversion factor that will allow you to calculate the unknown. The conversion you need to make is from tricycles (FSW_3HP_2) to wheels (W).

KNOWNS

number of tricycles = 640 tricycles = 640 FSW_3HP_2

$F + S + 3W + H + 2P \longrightarrow FSW_3HP_2$

UNKNOWN

number of wheels = ? W

2 Calculate Solve for the unknown.

Identify a conversion factor that relates wheels to tricycles. You can write two conversion factors relating wheels to tricycles.

$$\frac{3\ W}{1\ FSW_3HP_2} \quad \text{and} \quad \frac{1\ FSW_3HP_2}{3\ W}$$

The desired unit is W; so use the conversion factor on the left. Multiply the number of tricycles by the conversion factor.

$$640\ \cancel{FSW_3HP_2} \times \frac{3\ W}{1\ \cancel{FSW_3HP_2}} = 1920\ W$$

When using conversion factors, remember to cancel like units when they are in both the numerator and denominator. This tells you that you are using the correct conversion factor.

3 Evaluate Does the result make sense? If three wheels are required for each tricycle and more than 600 tricycles are being made, then a number of wheels in excess of 1800 is a logical answer. The unit of the known (FSW_3HP_2) cancels, and the answer has the correct unit (W).

1. Travel Time has decided to make 288 tricycles each day. How many tricycle seats, wheels, and pedals are needed for each day?

2. Write an equation that gives your own "recipe" for making a skateboard.

Foundations for Math

RATIOS AND PROPORTIONS Explain that a ratio represents the relationship between two quantities. For example, the ratio 1:3 can be used to represent the relationship 1 handlebar to 3 wheels in **Sample Problem 12.1,** or 1 molecule of nitrogen to 3 molecules of hydrogen in the balanced equation for ammonia. Explain that the ratio may not be a true representation if the equations are not balanced. Set two ratios equal to each other to show students how a proportion can be used to determine an unknown quantity. Solve the proportion by using cross products, and point out how the units cancel, leaving you with the desired unit of the quantity you were trying to find.

Foundations for Reading

BUILD VOCABULARY Have students use an analogy of ingredients and product of a recipe to explain their interpretation of stoichiometry.

READING STRATEGY Have students construct tables similar to Figure 12.2 to help them understand the quantities involved with any balanced chemical equation they encounter in this chapter.

Explain

Using Equations

START A CONVERSATION Direct students to look at the visual representation of a tricycle. Then write this statement on the board "A frame, a seat, wheels, a handlebar, and pedals are needed to assemble a complete tricycle." **Ask** How can this statement be revised so that it more accurately describes the process of assembling a tricycle? *(One frame, one seat, three wheels, one handlebar, and two pedals are needed to assemble a complete tricycle.)*

MAKE A CONNECTION Write the equation on the board.

$$F + S + 3W + H + 2P \rightarrow FSW_3HP_2$$

Point out that the balanced equation contains information that not only relates reactants to product but relates one reactant to another.

Ask Describe a method you can use to determine how many handlebars are needed to complete tricycles if you know how many pedals are available. *(Set up and solve a proportion.)*

Sample Practice Problems

A. Set up and solve a proportion to find how many handlebars are needed to complete tricycles if you know there are 24 pedals available. (*12*)

B. How many handlebars are needed to complete tricycles if 24 seats are available? (*24*)

C. How many handlebars are needed to complete tricycles if 24 wheels are available? (*8*)

Answers

1. 288 seats, 864 wheels, 576 pedals
2. Answers will vary but should include the correct number of "parts" to make the product.

Explain

Chemical Equations

START A CONVERSATION Write the following two chemical reactions on the board:

A. $CuO(s) + NH_3(aq) \rightarrow Cu(s) + H_2O(l) + N_2(g)$
B. $4NH_3(g) + 5O_2(g) \rightarrow 4NO(g) + 6H_2O(g)$

Ask Which equation can be used in its current form to determine how much product is formed by the reactants? *(equation B)*

Ask If you used equation A. in its current form to make stoichiometric calculations, why would your answers not be accurate? *(The equation is not balanced.)*

Misconception Alert

Make sure students understand that they cannot balance an equation by changing the subscripts in a formula. A change to the subscripts causes the chemical identity of the substance to change.

CHEMISTRY & YOU Use a balanced chemical equation.

Extend

Connect to ENVIRONMENTAL SCIENCE

Tell students that Earth's atmosphere contains 0.01 parts per million of ammonia, and small amounts of ammonia occur in volcanic gases. Most ammonia cycles through the living world without returning to the atmosphere. Ammonia plays a role in several stages of the nitrogen cycle. Nitrogen-fixing bacteria form nodules, or swellings, on the roots of plants in the legume family, such as beans and clover plants. These bacteria change atmospheric nitrogen into ammonia molecules or ammonium ions. Other bacteria break down the nitrogenous material in dead plants and animals into ammonia molecules. Certain soil bacteria oxidize these molecules into nitrate ions, the form readily absorbed by plant roots. When a plant dies, this cycle begins again.

Q: *How can you determine the amount of each reactant you need to make a product?*

Learn more about stoichiometry *online.*

READING SUPPORT

Build Vocabulary: *Word Origins* *Stoichiometry* comes from the combination of the Greek words *stoikheioin,* meaning "element," and *metron,* meaning "to measure." Stoichiometry is the calculation of amounts of substances involved in chemical reactions. ***What do you first need to know about a chemical reaction before doing stoichiometry calculations?***

Balanced Chemical Equations Nearly everything you use is manufactured from chemicals—soaps, shampoos and conditioners, CDs, cosmetics, medicines, and clothes. When manufacturing such items, the cost of making them cannot be greater than the price at which they are sold. Otherwise, the manufacturer will not make a profit. Therefore, the chemical processes used in manufacturing must be carried out economically. A situation like this is where balanced equations help.

A balanced chemical equation tells you what amounts of reactants to mix and what amount of a product to expect. **Chemists use balanced chemical equations as a basis to calculate how much reactant is needed or how much product will be formed in a reaction.** When you know the quantity of one substance in a reaction, you can calculate the quantity of any other substance consumed or created in the reaction. Quantity usually means the amount of a substance expressed in grams or moles. However, quantity could just as well be in liters, tons, or molecules.

The calculation of quantities in chemical reactions is a subject of chemistry called **stoichiometry.** Calculations using balanced equations are called stoichiometric calculations. For chemists, stoichiometry is a form of bookkeeping. For example, accountants can track income, expenditures, and profits for a small business by tallying each in dollars and cents. Chemists can track reactants and products in a reaction by stoichiometry. It allows chemists to tally the amounts of reactants and products using ratios of moles or representative particles derived from chemical equations.

Chemical Equations

In terms of what quantities can you interpret a balanced chemical equation?

In gardens such as the one shown in Figure 12.1, fertilizers are often used to improve the growth of flowers. Ammonia is widely used as a fertilizer. Ammonia is produced industrially by the reaction of nitrogen with hydrogen.

$$N_2(g) + 3H_2(g) \longrightarrow 2NH_3(g)$$

The balanced chemical equation tells you the relative amounts of reactants and product in the reaction. However, your interpretation of the equation depends on how you quantify the reactants and products. **A balanced chemical equation can be interpreted in terms of different quantities, including numbers of atoms, molecules, or moles; mass; and volume.** As you study stoichiometry, you will learn how to interpret a chemical equation in terms of any of these quantities.

Number of Atoms At the atomic level, a balanced equation indicates the number and types of atoms that are rearranged to make the product or products. Remember, both the number and types of atoms are not changed in a reaction. In the synthesis of ammonia, the reactants are composed of two atoms of nitrogen and six atoms of hydrogen. These eight atoms are recombined in the product.

$$\text{2 atoms N} + \text{6 atoms H} \longrightarrow \text{2 atoms N and 6 atoms H}$$
$$\text{8 atoms} \longrightarrow \text{8 atoms}$$

Differentiated Instruction

L1 LESS PROFICIENT READERS Have students create a two-column table with the following row headings: *Number of Atoms, Number of Molecules, Moles, Mass, and Volume*. Then have students complete their tables by writing an example of the formation of ammonia, $N_2(g) + 3H_2(g) \rightarrow 2NH_3(g)$, and how the corresponding quantity is interpreted from the balanced equation.

ELL ENGLISH LANGUAGE LEARNERS Divide students into groups of five. Have each group draw or model the formation of ammonia, $N_2(g) + 3H_2(g) \rightarrow 2NH_3(g)$. Then have each student in the group read one of the paragraphs in the Chemical Equations section. Students should use the model to explain the concept in their paragraph.

L3 ADVANCED STUDENTS Ask students to explain to the class the meaning of STP, the values of STP, the molar volume of any gas at STP, and how many particles it contains. *(standard temperature and pressure; 0°C and 101.3 kPa; 22.4 L/mol; 22.4 L of any ideal gas at STP contains 6.02×10^{23} particles of that gas.)*

Number of Molecules The balanced equation indicates that one molecule of nitrogen reacts with three molecules of hydrogen. Nitrogen and hydrogen will always react to form ammonia in a 1:3:2 ratio of molecules. If you could make 10 molecules of nitrogen react with 30 molecules of hydrogen, you would expect to get 20 molecules of ammonia. Of course, it is not practical to count such small numbers of molecules and allow them to react. You could, however, take Avogadro's number of nitrogen molecules and make them react with three times Avogadro's number of hydrogen molecules. This value would be the same 1:3 ratio of molecules of reactants. The reaction would form two times Avogadro's number of ammonia molecules.

$$1 \times \begin{matrix} 6.02 \times 10^{23} \\ \text{molecules } N_2 \end{matrix} + 3 \times \begin{matrix} 6.02 \times 10^{23} \\ \text{molecules } H_2 \end{matrix} \longrightarrow 2 \times \begin{matrix} 6.02 \times 10^{23} \\ \text{molecules } NH_3 \end{matrix}$$

Moles You know that Avogrado's number of representative particles is equal to one mole of a substance. Therefore, since a balanced chemical equation tells you the number of representative particles, it also tells you the number of moles. The coefficients of a balanced chemical equation indicate the relative numbers of moles of reactants and products in a chemical reaction. These numbers are the most important pieces of information that a balanced chemical equation provides. Using this information, you can calculate the amounts of reactants and products. In the synthesis of ammonia, one mole of nitrogen molecules reacts with three moles of hydrogen molecules to form two moles of ammonia molecules. As you can see from this reaction, the total number of moles of reactants does not equal the total number of moles of product.

$$1 \text{ mol } N_2 + 3 \text{ mol } H_2 \longrightarrow 2 \text{ mol } NH_3$$

Mass A balanced chemical equation obeys the law of conservation of mass. This law states that mass can be neither created nor destroyed in an ordinary chemical or physical process. As you recall, the number and type of atoms does not change in a chemical reaction. Therefore, the total mass of the atoms in the reaction does not change. Using the mole relationship, you can relate mass to the number of atoms in the chemical equation. The mass of 1 mol of N_2 (28.0 g) plus the mass of 3 mol of H_2 (6.0 g) equals the mass of 2 mol of NH_3 (34.0 g). Although the number of moles of reactants does not equal the number of moles of product, the total number of grams of reactants does equal the total number of grams of product.

$$28.0 \text{ g } N_2 + (3 \times 2.0 \text{ g } H_2) \longrightarrow (2 \times 17.0 \text{ g } NH_3)$$
$$34.0 \text{ g} \longrightarrow 34.0 \text{ g}$$

Volume If you assume standard temperature and pressure, the equation also tells you about the volumes of gases. Recall that 1 mol of any gas at STP occupies a volume of 22.4 L. The equation indicates that 22.4 L of N_2 reacts with 67.2 L (3 × 22.4 L) of H_2. This reaction forms 44.8 L (2 × 22.4 L) of NH_3.

$$22.4 \text{ L } N_2 + 67.2 \text{ L } H_2 \longrightarrow 44.8 \text{ L } NH_3$$

Figure 12.1 Use of Ammonia
Gardeners use ammonium salts as fertilizer. The nitrogen in these salts is essential to plant growth.

LESSON 12.1

Explore

Teacher Demo

PURPOSE Students interpret a balanced equation of the reaction of magnesium and hydrochloric acid.

MATERIALS 2.5–3.5-cm strip of magnesium, 50 mL 1*M* HCl(*aq*) in a 100-mL beaker, baking soda

SAFETY Wear safety glasses and an apron. Neutralize remaining HCl(*aq*) with baking soda before flushing down the drain.

PROCEDURE Identify the two reactants as magnesium and hydrochloric acid. Have students observe the reaction as you carefully add the magnesium strip to the acid. Ask students to write a balanced chemical equation for the reaction of magnesium and hydrochloric acid. ($Mg(s) + 2HCl(aq) \rightarrow MgCl_2(aq) + H_2(g)$) Have students interpret the equation in terms of particles, moles, and molar masses.

EXPECTED OUTCOME Students should express the balanced equation at the particle level as one atom of magnesium reacts with two molecules of hydrogen chloride to produce one formula unit of magnesium chloride and one molecule of hydrogen gas. Similarly, one mole of magnesium reacts with two moles of hydrogen chloride to produce one mole of magnesium chloride and one mole of hydrogen gas. Finally, 24.31 g Mg + 72.92 g HCl produces 95.21 g $MgCl_2$ + 2.02 g H_2.

Check for Understanding

The Essential Question What is the purpose of balanced chemical equations in stoichiometric calculations?

Assess students' knowledge of stoichiometry by asking them to interpret the following equation in terms of atoms, molecules, moles, mass, and/or volume.

$$2KClO_3(s) \rightarrow 2KCl(s) + 3O_2(g)$$

Elicit oral responses from students to determine if they understand why chemical equations must be balanced.

Explain

CRITICAL THINKING Have students try to find a counterexample to the statement "The conservation of moles exists for every chemical reaction," and have them try to find a counterexample to the statement "The conservation of mass exists for every chemical reaction" to prove that the law of conservation exists for mass and moles of atoms, but not necessarily for moles of compound. *(Answers may vary.)*

SampleProblem 12.2

Interpreting a Balanced Chemical Equation

Hydrogen sulfide, which smells like rotten eggs, is found in volcanic gases. The balanced equation for the burning of hydrogen sulfide is

$$2H_2S(g) + 3O_2(g) \longrightarrow 2SO_2(g) + 2H_2O(g)$$

Interpret this equation in terms of

a. numbers of representative particles and moles.

b. masses of reactants and products.

Remember that atoms and molecules are both representative particles. In this equation, all the reactants and products are molecules; so all the representative particles are molecules.

1 Analyze **Identify the relevant concepts.** The coefficients in the balanced equation give the relative number of representative particles and moles of reactants and products. A balanced chemical equation obeys the law of conservation of mass.

2 Solve **Apply concepts to this situation.**

Use the coefficients in the balanced equation to identify the number of representative particles and moles.

a. 2 molecules H_2S + 3 molecules $O_2 \longrightarrow$ 2 molecules SO_2 + 2 molecules H_2O
2 mol H_2S + 3 mol $O_2 \longrightarrow$ 2 mol SO_2 + 2 mol H_2O

Use the periodic table to calculate the molar mass of each reactant and product.

b. 1 mol H_2S = 34.1 g H_2S
1 mol O_2 = 32.0 g O_2
1 mol SO_2 = 64.1 g SO_2
1 mol H_2O = 18.0 g H_2O

Multiply the number of moles of each reactant and product by its molar mass.

2 mol H_2S + 3 mol $O_2 \longrightarrow$ 2 mol SO_2 + 2 mol H_2O.

$$\left(2\,\cancel{mol} \times 34.1\,\frac{g}{\cancel{mol}}\right) + \left(3\,\cancel{mol} \times 32.0\,\frac{g}{\cancel{mol}}\right) \longrightarrow \left(2\,\cancel{mol} \times 64.1\,\frac{g}{\cancel{mol}}\right) + \left(2\,\cancel{mol} \times 18.0\,\frac{g}{\cancel{mol}}\right)$$

68.2 g H_2S + 96.0 g $O_2 \longrightarrow$ 128.2 g SO_2 + 36.0 g H_2O

164.2 g = 164.2 g

3. Interpret the equation for the formation of water from its elements in terms of numbers of molecules and moles, and volumes of gases at STP.

$$2H_2(g) + O_2(g) \longrightarrow 2H_2O(g)$$

4. Balance the following equation:

$$C_2H_4(g) + O_2(g) \longrightarrow CO_2(g) + H_2O(g)$$

Interpret the balanced equation in terms of relative numbers of moles, volumes of gas at STP, and masses of reactants and products.

Foundations for Math

ORDER OF OPERATIONS Explain that students should always follow the order of operations to correctly determine the answer to a mathematical problem. Remind students of PEMDAS, a mnemonic commonly used to help students remember the order of operations:

> **P**lease **E**xcuse **M**y **D**ear **A**unt **S**ally: **P**arentheses first, **E**xponents next, **M**ultiplication and **D**ivision from left to right, **A**ddition and **S**ubtraction from left to right

In **Sample Problem 12.2 part b,** point out that to correctly determine the total mass of the reactants and the total mass of the products, multiplication is computed prior to addition.

Figure 12.2 summarizes the information derived from the balanced chemical equation for the formation of ammonia. As you can see, the mass of the reactants equals the mass of the products. In addition, the number of atoms of each type in the reactants equals the number of atoms of each type in the product. Mass and atoms are conserved in every chemical reaction. However, molecules, formula units, moles, and volumes are not necessarily conserved—although they may be. Consider, for example, the formation of hydrogen iodide.

$$H_2(g) + I_2(g) \longrightarrow 2HI(g)$$

In this reaction, molecules, moles, and volume are all conserved. But in the majority of chemical reactions, they are not.

$N_2(g)$	+	$3H_2(g)$	→	$2NH_3(g)$
	+		→	
2 atoms N	+	6 atoms H	→	2 atoms N and 6 atoms H
1 molecule N_2	+	3 molecules H_2	→	2 molecules NH_3
10 molecules N_2	+	30 molecules H_2	→	20 molecules NH_3
$1 \times$ 6.02×10^{23} molecules N_2	+	$3 \times$ 6.02×10^{23} molecules H_2	→	$2 \times$ 6.02×10^{23} molecules NH_3
1 mol N_2	+	3 mol H_2	→	2 mol NH_3
28.0 g N_2	+	3×2.0 g H_2	→	2×17.0 g NH_3
		34.0 g reactants	→	34.0 g products
Assume STP: 22.4 L	+	22.4 L 22.4 L 22.4 L	→	22.4 L 22.4 L
22.4 L N_2		67.2 L H_2		44.8 L NH_3

Figure 12.2 Interpreting a Balanced Chemical Equation
The balanced chemical equation for the formation of ammonia can be interpreted in several ways.
Predict ***How many molecules of NH_3 could be made from 5 molecules of N_2 and 15 molecules of H_2?***

See balancing chemical equations animated online.

12.1 LessonCheck

5. Explain How do chemists use balanced equations?

6. Identify Chemical reactions can be described in terms of what quantities?

7. Explain How is a balanced equation similar to a recipe?

8. Identify What quantities are always conserved in chemical reactions?

9. Apply Concepts Interpret the given equation in terms of relative numbers of representative particles, numbers of moles, and masses of reactants and products.

$$2K(s) + 2H_2O(l) \longrightarrow 2KOH(aq) + H_2(g)$$

10. Apply Concepts Balance this equation:

$$C_2H_5OH(l) + O_2(g) \longrightarrow CO_2(g) + H_2O(g)$$

Show that the balanced equation obeys the law of conservation of mass.

USE VISUALS Direct students to Figure 12.2. Remind students that the term *STP* represents "standard temperature and pressure." **Ask** What are the values of STP? *(0°C and 101.3 kPa)* **Ask** Why is the volume of a gas usually measured at STP? *(because its volume varies with temperature and pressure)* **Ask** What is the molar volume of any gas at STP? *(22.4L/mol of any ideal gas at STP)* How many particles does it contain? *(22.4 L of any gas contains 6.02×10^{23} particles.)*

Evaluate

Informal Assessment

Divide the class into three teams. Have Team 1 write five balanced chemical equations on separate note cards and place them in a box. Also have them create four note cards with one word per card: *atoms, molecules, mass, moles*. Have Team 2 draw an equation from the box, and have Team 3 randomly select one of the four cards. Team 3 tells Team 2 which quantitative relationship of their selected equation they have to determine within a one minute time period. Rotate teams after each round. Students can be assessed on participation, correctness of answers, and creativity of balanced equations.

Reteach

Construct a "gizmo" from a large flask (F) fitted with a one-hole stopper (S) and glass tube (T). Show students the gizmo and write its balanced equation on the board as the following:

__F + __S + __T = __FST.

Explain that any counted quantity such as 2 dozen or 12 ten-packs could be used to fill any set of four blanks in the equation. However, a physical quantity such as 10 kg would not balance the equation. Have students explain why.

Lesson Check Answers

5. as a basis to calculate how much reactant is needed or product is formed in a reaction
6. numbers of atoms, molecules, or moles; mass; and volumes
7. Both a balanced equation and a recipe give quantitative information about the starting and end materials.
8. mass and atoms
9. 2 atoms K + 2 molecules H_2O → 2 formula units KOH + 1 molecule H_2
   2 mol K + 2 mol H_2O → 2 mol KOH + 1 mol H_2
   78.2 g K + 36.0 g H_2O → 112.2 g KOH + 2.0 g H_2
10. $C_2H_5OH + 3O_2 \rightarrow 2CO_2 + 3H_2O$
    46.0 g C_2H_5OH + 96.0 g O_2 → 88.0 g CO_2 + 54.0 g H_2O
    142.0 g reactants → 142.0 g products

Answers

FIGURE 12.2 10 molecules NH_3

3. 2 molecules H_2 + 1 molecule O_2 → 2 molecules H_2O
   2 mol H_2 + 1 mol O_2 → 2 mol H_2O
   44.8 L H_2 + 22.4 L O_2 → 44.8 L H_2O
4. 1 mol C_2H_4 + 3 mol O_2 → 2 mol CO_2 + 2 mol H_2O
   22.4 L C_2H_4 + 67.2 L O_2 → 44.8 L CO_2 + 44.8 L H_2O
   124 g reactants → 124 g products

LESSON 12.2

Key Objectives

12.2.1 Explain how mole ratios are used in chemical calculations.

12.2.2 Explain the general procedure for solving a stoichiometric problem.

Additional Resources

- Reading and Study Workbook, Lesson 12.2
- Probeware Laboratory Manual, Lab 19
- Virtual Chemistry Labs, Analysis of Baking Soda
- Core Teaching Resources, Lesson 12.2 Review

PearsonChem.com

Engage

CHEMISTRY & YOU Have students study the photograph and read the text. Ask students what ammonia is used for. *(Ammonia is used in cleaning products, in fertilizers, and in the manufacture of other chemicals.)* Tell students to consider how stoichiometry might be used to calculate the amount of product as they read through the lesson.

Activate Prior Knowledge

Engage students in a review of moles and molar mass. **Ask** What is a *mole*? *(A mole is equivalent to 6.02×10^{23} particles of substance.)* **Ask** How can you determine the number of moles of a substance in a chemical equation? *(The number of moles is represented by the substance's coefficient.)* **Ask** What is *molar mass*? *(Molar mass is the mass of one mole of a substance.)*

IN ACADEMIC STANDARDS for SCIENCE

- Apply standard techniques in laboratory investigations to measure physical quantities in appropriate units and convert known quantities to other units as necessary.

C.4.3 Use the mole concept to determine the number of moles and number of atoms or molecules in samples of elements and compounds, given mass of the sample.

C.4.4 Using a balanced chemical equation, calculate the quantities of reactants needed and products made in a chemical reaction that goes to completion.

Also covered C.1.6, C.5.3.

12.2 Chemical Calculations

C.1.6 Conservation of mass: chemical changes; **C.4.3** The mole concept; **C.4.4** Calculate reactants and products; **C.5.3** Reactant and product calculations: gases

CHEMISTRY & YOU

Q: *How do manufacturers know how to make enough of their desired product?* Chemical plants produce ammonia by combining nitrogen with hydrogen. If too much ammonia is produced, then it might be wasted. But if too little is produced, then there might not be enough for all their customers. In this lesson, you will learn how to use a balanced chemical equation to calculate the amount of product formed in a chemical reaction.

Key Questions

How are mole ratios used in chemical calculations?

What is the general procedure for solving a stoichiometric problem?

Vocabulary

- mole ratio

Writing and Using Mole Ratios

How are mole ratios used in chemical calculations?

As you learned in the previous lesson, a balanced chemical equation provides a great deal of quantitative information. It relates particles (atoms, molecules, formula units), moles of substances, and masses. A balanced chemical equation also is essential for all calculations involving amounts of reactants and products. For example, suppose you know the number of moles of one substance. The balanced chemical equation allows you to determine the number of moles of all other substances in the reaction.

Look again at the balanced equation for the production of ammonia.

$$N_2(g) + 3H_2(g) \longrightarrow 2NH_3(g)$$

The most important interpretation of this equation is that 1 mol of nitrogen reacts with 3 mol of hydrogen to form 2 mol of ammonia. Based on this interpretation, you can write ratios that relate moles of reactants to moles of product. A **mole ratio** is a conversion factor derived from the coefficients of a balanced chemical equation interpreted in terms of moles. **In chemical calculations, mole ratios are used to convert between a given number of moles of a reactant or product to moles of a different reactant or product.** Three mole ratios derived from the balanced equation above are

$$\frac{1\ \text{mol}\ N_2}{3\ \text{mol}\ H_2} \quad \frac{2\ \text{mol}\ NH_3}{1\ \text{mol}\ N_2} \quad \frac{3\ \text{mol}\ H_2}{2\ \text{mol}\ NH_3}$$

Mole-Mole Calculations In the mole ratio below, *W* is the unknown, wanted, quantity and *G* is the given quantity. The values of *a* and *b* are the coefficients from the balanced equation. Thus, a general solution for a mole-mole problem, such as Sample Problem 12.3, is given by

$$\underset{\text{Given}}{x\ \text{mol}\ G} \times \underset{\text{Mole ratio}}{\frac{b\ \text{mol}\ W}{a\ \text{mol}\ G}} = \underset{\text{Calculated}}{\frac{xb}{a}\ \text{mol}\ W}$$

Focus on ELL

1 CONTENT AND LANGUAGE Present academic vocabulary that appears in the lesson, such as *instantaneously*, *conversion*, and *excess*. Pair students and have them find these words in the lesson and use context clues to predict their meanings. Have students share their interpretations with the class.

2 FRONTLOAD THE LESSON Have students discuss their experience with ratios from mathematics class. To check understanding, have students determine the ratio of males to females in the class and the ratio of each gender to total students.

3 COMPREHENSIBLE INPUT Explain **Sample Problem 12.3** using a diagram. Write each reactant using a different color. Continue to use the corresponding color throughout the calculation. Speak clearly and slowly, and use block arrows to show the steps of the process.

SampleProblem 12.3

Calculating Moles of a Product

How many moles of NH_3 are produced when 0.60 mol of nitrogen reacts with hydrogen?

1 Analyze List the known and the unknown. The conversion is mol $N_2 \longrightarrow$ mol NH_3. According to the balanced equation, 1 mol N_2 combines with 3 mol H_2 to produce 2 mol NH_3. To determine the number of moles of NH_3, the given quantity of N_2 is multiplied by the form of the mole ratio from the balanced equation that allows the given unit to cancel.

KNOWN

moles of nitrogen = 0.60 mol N_2

UNKNOWN

moles of ammonia = ? mol NH_3

2 Calculate Solve for the unknown.

Write the mole ratio that will allow you to convert from moles N_2 to moles NH_3.

$$\frac{2 \text{ mol } NH_3}{1 \text{ mol } N_2}$$

Multiply the given quantity of N_2 by the mole ratio in order to find the moles of NH_3.

$$0.60 \text{ mol } N_2 \times \frac{2 \text{ mol } NH_3}{1 \text{ mol } N_2} = 1.2 \text{ mol } NH_3$$

3 Evaluate Does the result make sense? The ratio of 1.2 mol NH_3 to 0.60 mol N_2 is 2:1, as predicted by the balanced equation.

Remember that the mole ratio must have N_2 on the bottom so that the mol N_2 in the mol ratio will cancel with mol N_2 in the known.

11. This equation shows the formation of aluminum oxide, which is found on the surface of aluminum objects exposed to the air.

$$4Al(s) + 3O_2(g) \longrightarrow 2Al_2O_3(s)$$

a. Write the six mole ratios that can be derived from this equation.

b. How many moles of aluminum are needed to form 3.7 mol Al_2O_3?

12. According to the equation in Problem 11,

a. How many moles of oxygen are required to react completely with 14.8 mol Al?

b. How many moles of Al_2O_3 are formed when 0.78 mol O_2 reacts with aluminum?

Mass-Mass Calculations No laboratory balance can measure substances directly in moles. Instead, the amount of a substance is usually determined by measuring its mass in grams. From the mass of a reactant or product, the mass of any other reactant or product in a given chemical equation can be calculated. The mole interpretation of a balanced equation is the basis for this conversion. If the given sample is measured in grams, then the mass can be converted to moles by using the molar mass. Then the mole ratio from the balanced equation can be used to calculate the number of moles of the unknown. If it is the mass of the unknown that needs to be determined, the number of moles of the unknown can be multiplied by the molar mass. As in mole-mole calculations, the unknown can be either a reactant or a product.

Foundations for Math

DIMENSIONAL ANALYSIS Remind students that to solve a problem by using dimensional analysis, they need to first find the conversion factor. A conversion factor is a ratio of equivalent measurements. **Sample problem 12.3** asks students to find the moles of ammonia produced, so the answer needs to be in units of "moles of ammonia."

Remind students that they can multiply any quantity by anything equal to 1 without changing the value of the quantity.

Since the number of moles of nitrogen is given, the ratio of moles of nitrogen to moles of ammonia can be used as the conversion factor.

Foundations for Reading

BUILD VOCABULARY Have students paraphrase the meaning of **mole ratio** using words or symbols. Guide students to use what they have already learned about balanced chemical equations from the previous lesson and about ratios from mathematics class to form their definitions. Have students read their definitions to the class.

READING STRATEGY As students read the section about Mass-Mass Calculations, have them identify and list the main ideas presented by the text.

Explain

Writing and Using Mole Ratios

START A CONVERSATION Discuss with students what they know about the relationship of the total mass of the reactants and the total mass of the products in a chemical reaction. **Ask** Why do you think this relationship is important when trying to determine quantitative information about a chemical reaction? *(Answers will vary.)*

Be sure students understand the difference between moles and mole ratio. Point out to students that moles are always involved in stoichiometry problems. Mole ratios are used to convert from the amount of one substance in the balanced equation to the amount of another substance.

Sample Practice Problem

Iron(III) oxide reacts with carbon monoxide to yield iron and carbon dioxide in the following reaction:

$$Fe_2O_3(s) + 3CO(g) \rightarrow 2Fe(s) + 3CO_2(g)$$

A. How many mole ratios can be derived from this equation? What are they? *(12;*

and their reciprocals.

B. How many moles of Fe are produced from 1.8 mol of Fe_2O_3? *(3.6 mol)*

Answers

11. a. $\frac{4 \text{ mol Al}}{3 \text{ mol } O_2}$ $\frac{3 \text{ mol } O_2}{4 \text{ mol Al}}$ $\frac{4 \text{ mol Al}}{2 \text{ mol } Al_2O_3}$ $\frac{2 \text{ mol } Al_2O_3}{4 \text{ mol Al}}$ $\frac{3 \text{ mol } O_2}{2 \text{ mol } Al_2O_3}$ $\frac{2 \text{ mol } Al_2O_3}{3 \text{ mol } O_2}$

b. 7.4 mol

12. a. 11.1 mol

b. 0.52 mol

LESSON 12.2

Explore

Teacher Demo

PURPOSE Students interpret a balanced equation in terms of moles and mass.

MATERIALS Prior to the demonstration, prepare 0.1*M* solutions of potassium iodide and lead(II) nitrate. Measure 50.0 mL of $Pb(NO_3)_2$ and 150 mL of KI into separate 250-mL beakers.

SAFETY Wear safety glasses and an apron.

PROCEDURE Tell students that you are going to mix 0.005 mole of lead(II) nitrate with excess potassium iodide. Have students observe as you combine both solutions in the 250-mL beaker. Have students write a balanced chemical equation for the observed reaction.

$$[2KI(aq) + Pb(NO_3)_2(aq) \rightarrow 2KNO_3(aq) + PbI_2(s)]$$

Have students predict the number of moles of product produced. *(0.005 mole PbI_2 assuming the reaction was complete)* Note that, in an actual reaction, the amounts of reactants often are not present in the mole ratios predicted by the coefficients in a balanced equation. Explain the importance of the mole ratios in an equation for calculating relative quantities. **Ask** What is the mass of lead(II) nitrate reacted and the mass of lead(II) iodide produced? *(1.66 g $Pb(NO_3)_2$ and 2.30 g PbI_2)*

EXPECTED OUTCOME A bright yellow precipitate will form.

Misconception Alert

Students sometimes try to do mass-mass conversions by incorrectly using the mole ratio as a mass ratio. That is, they use grams instead of moles as the units in the mole ratio and then skip the mass-mole conversion step. Stress that because the number of grams in one mole of a substance varies with its molar mass, a mass-mole conversion is a necessary intermediate step in mass-mass stoichiometric problems.

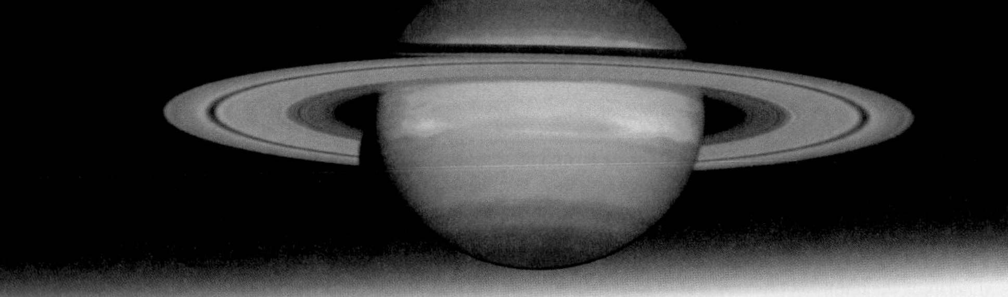

Figure 12.3 Ammonia in Space
In this Hubble Space Telescope image, clouds of condensed ammonia are visible covering the surface of Saturn.

Steps for Solving a Mass-Mass Problem Mass-mass problems are solved in basically the same way as mole-mole problems. The steps for the mass-mass conversion of any given mass (*G*) to any wanted mass (*W*) are outlined below.

1. Change the mass of *G* to moles of *G* (mass $G \longrightarrow$ mol *G*) by using the molar mass of *G*.

$$\text{mass } G \times \frac{1 \text{ mol } G}{\text{molar mass } G} = \text{mol } G$$

2. Change the moles of *G* to moles of *W* (mol $G \longrightarrow$ mol *W*) by using the mole ratio from the balanced equation.

$$\text{mol } G \times \frac{b \text{ mol } W}{a \text{ mol } G} = \text{mol } W$$

3. Change the moles of *W* to grams of *W* (mol $W \longrightarrow$ mass *W*) by using the molar mass of *W*.

$$\text{mol } W \times \frac{\text{molar mass } W}{1 \text{ mol } W} = \text{mass } W$$

Figure 12.4 shows another way to represent the steps for doing mole-mass and mass-mole stoichiometric calculations. For a mole-mass problem, the first conversion (from mass to moles) is skipped. For a mass-mole problem, the last conversion (from moles to mass) is skipped. You can use parts of the three-step process shown in Figure 12.4 as they are appropriate to the problem you are solving.

Figure 12.4 Mass-Mass Conversion Steps
This general solution diagram indicates the steps necessary to solve a mass-mass stoichiometry problem: Convert mass to moles, use the mole ratio, and then convert moles to mass.
Infer *Is the given always a reactant?*

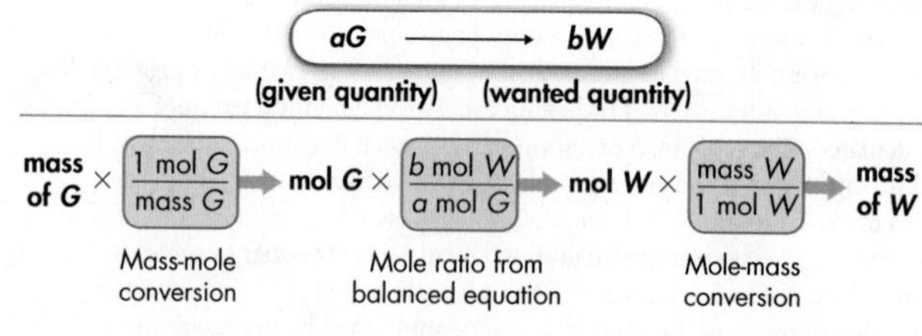

Atmospheric Ammonia

Ammonia is found in trace amounts in the atmospheres of three Jovian planets—Jupiter, Saturn, and Uranus. In Jupiter's atmosphere, the clouds of ammonia consist of frozen ammonia droplets changing to liquid ammonia droplets nearer the planet's surface. Because of colder temperatures, the ammonia clouds in the atmosphere of Saturn and Uranus consist of frozen ammonia droplets.

Sample Problem 12.4

Calculating the Mass of a Product

Ammonia (NH_3) clouds are present around some planets, as in Figure 12.3. Calculate the number of grams of NH_3 produced by the reaction of 5.40 g of hydrogen with an excess of nitrogen. The balanced equation is

$$N_2(g) + 3H_2(g) \longrightarrow 2NH_3(g)$$

1 Analyze **List the knowns and the unknown.** The mass of hydrogen will be used to find the mass of ammonia: g $H_2 \longrightarrow$ g NH_3. The coefficients of the balanced equation show that 3 mol H_2 reacts with 1 mol N_2 to produce 2 mol NH_3. The following steps are necessary to determine the mass of ammonia:

$$\text{g } H_2 \longrightarrow \text{mol } H_2 \longrightarrow \text{mol } NH_3 \longrightarrow \text{g } NH_3$$

KNOWNS

mass of hydrogen = 5.40 g H_2

2 mol NH_3/3 mol H_2 (from balanced equation)

1 mol H_2 = 2.0 g H_2 (molar mass)

1 mol NH_3 = 17.0 g NH_3 (molar mass)

UNKNOWN

mass of ammonia = ? g NH_3

2 Calculate **Solve for the unknown.**

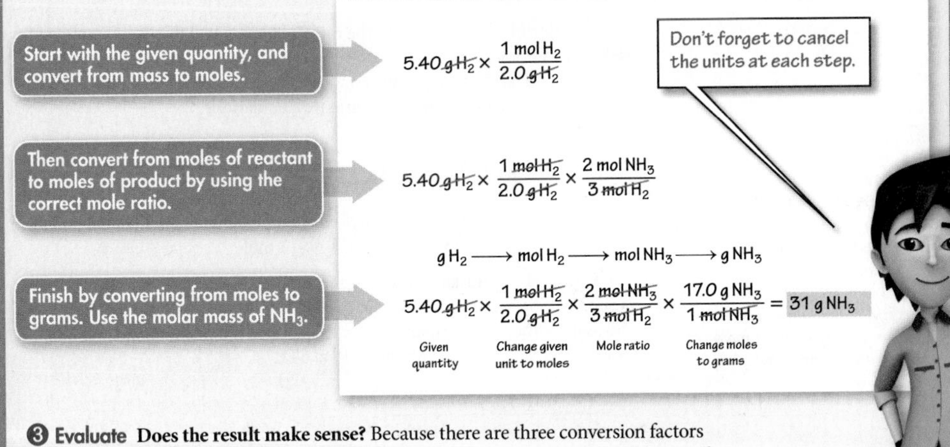

3 Evaluate **Does the result make sense?** Because there are three conversion factors involved in this solution, it is more difficult to estimate an answer. However, because the molar mass of NH_3 is substantially greater than the molar mass of H_2, the answer should have a larger mass than the given mass. The answer should have two significant figures.

13. Acetylene gas (C_2H_2) is produced by adding water to calcium carbide (CaC_2).

$$CaC_2(s) + 2H_2O(l) \longrightarrow C_2H_2(g) + Ca(OH)_2(aq)$$

How many grams of acetylene are produced by adding water to 5.00 g CaC_2?

14. Use the equation in Question 13 to determine how many moles of CaC_2 are needed to react completely with 49.0 g H_2O.

Explain

START A CONVERSATION Take a quick vote in class to see who believes there is a "law of conservation of moles." Discuss the reasons why students believe this law exists. **Ask** In the following reaction, $2H_2O(l) \rightarrow 2H_2(g) + O_2(g)$, is the number of moles of the products greater than, less than, or equal to the number of moles of the reactant? *(greater than)* **Ask** In the following reaction, $2Mg(s) + O_2(g) \rightarrow 2MgO(s)$, is the number of moles of product greater than, less than, or equal to the number of moles of the reactants? *(less than)*

Guide students to understand that there are many types of reactions where reactants can combine or decompose to produce fewer or more moles of product. Explain that although the total mass of reactants and products is constant, and although no atoms are gained or lost, the number of moles of particles can increase or decrease depending on the final grouping of atoms.

Sample Practice Problem

Rust (Fe_2O_3) is produced when iron (Fe) reacts with oxygen (O_2): $4Fe(s) + 3O_2(g) \rightarrow 2Fe_2O_3(s)$.

How many grams of Fe_2O_3 are produced when 12.0 g of iron rusts? *(17.2 g)*

Foundations for Math

SIGNIFICANT FIGURES The significant figures in a measurement are all the digits known with certainty plus one estimated digit.

The number of significant figures of a product or quotient should equal the least number of significant figures of any measurement in the calculation. For example, 5.5 cm × 2.24 cm = 12 cm²; notice that the rounded product has 2 significant figures, the same as 5.5 cm.

The number of decimal places of a sum or difference should equal the least number of decimal places in any of the measurements being added or subtracted. For example, 5.5 cm + 2.24 cm = 7.7 cm; notice that the rounded sum has 2 significant figures, the same as 5.5 cm.

Answers

FIGURE 12.4 No; the given could be a product.

13. 2.03 g C_2H_2

14. 1.36 mol CaC_2

LESSON 12.2

Explain

Other Stoichiometric Calculations

START A CONVERSATION On the board, write equations for reactions in which the reactants are both gases or are a gas and a solid. Ask students how the reactants and products in each reaction would most likely be measured. Have students relate these measurements to the concept of a mole.

CHEMISTRY & YOU Manufacturers use balanced chemical equations to calculate the amount of product formed from a given quantity of reactants.

USE VISUALS Direct students to Figure 12.5, and walk them through it carefully. Although it looks complicated, it's really three steps: Take what you're given, and find a way to change it to moles. Then use a mole ratio from the balanced equation to get moles of the second substance. Finally find a way to convert the moles into the units that you need for the final answer.

Other Stoichiometric Calculations

What is the general procedure for solving a stoichiometric problem?

Q: *How do you think air bag manufacturers know how to get the right amount of air in an inflated air bag?*

As you already know, you can obtain mole ratios from a balanced chemical equation. From the mole ratios, you can calculate any measurement unit that is related to the mole. The given quantity can be expressed in numbers of representative particles, units of mass, or volumes of gases at STP. The problems can include mass-volume, particle-mass, and volume-volume calculations. For example, you can use stoichiometry to relate volumes of reactants and products in the reaction shown in Figure 12.5. **In a typical stoichiometric problem, the given quantity is first converted to moles. Then, the mole ratio from the balanced equation is used to calculate the number of moles of the wanted substance. Finally, the moles are converted to any other unit of measurement related to the unit mole, as the problem requires.**

Thus far, you have learned how to use the relationship between moles and mass (1 mol = molar mass) in solving mass-mass, mass-mole, and mole-mass stoichiometric problems. The mole-mass relationship gives you two conversion factors.

$$\frac{1 \text{ mol}}{\text{molar mass}} \text{ and } \frac{\text{molar mass}}{1 \text{ mol}}$$

Recall from Chapter 10 that the mole can be related to other quantities as well. For example, 1 mol = 6.02×10^{23} representative particles, and 1 mol of a gas = 22.4 L at STP. These two relationships provide four more conversion factors that you can use in stoichiometric calculations.

$$\frac{1 \text{ mol}}{6.02 \times 10^{23} \text{ particles}} \text{ and } \frac{6.02 \times 10^{23} \text{ particles}}{1 \text{ mol}}$$

$$\frac{1 \text{ mol}}{22.4 \text{ L}} \text{ and } \frac{22.4 \text{ L}}{1 \text{ mol}}$$

Figure 12.5 summarizes the steps for a typical stoichiometric problem. Notice that the units of the given quantity will not necessarily be the same as the units of the wanted quantity. For example, given the mass of G, you might be asked to calculate the volume of W at STP.

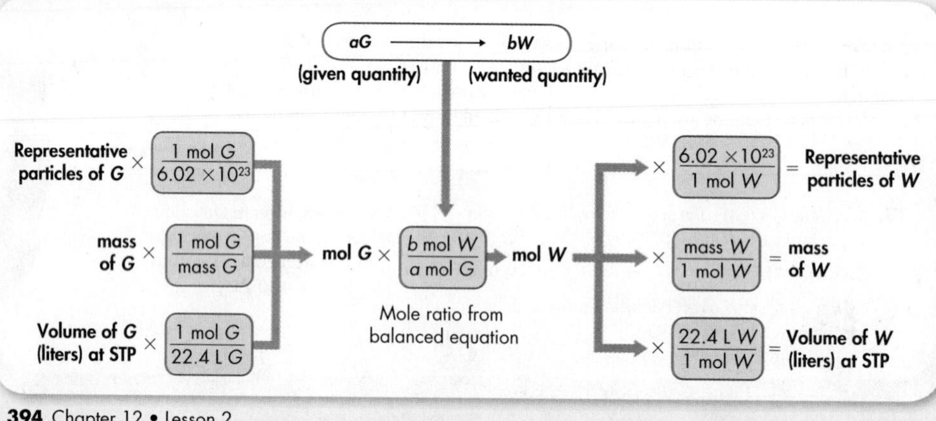

Figure 12.5
Solving Stoichiometric Problems
With your knowledge of conversion factors and this problem-solving approach, you can solve a variety of stoichiometric problems.
Identify ***What conversion factor is used to convert moles to representative particles?***

UbD Check for Understanding

The Essential Question How do you calculate amounts of reactants and products in a chemical reaction?

Assess students' understanding of stoichiometry by showing students various equations for reactions in which the reactants are both gases or are a gas and a solid, such as $2Mg(s) + O_2(g) \rightarrow 2MgO(s)$. Have students create three note cards, one with the words *mass-volume*, one with the words *particle-mass*, and one with the words *volume-volume*. As you show students an equation, have them raise the card that names how they think the reactants and products in each reaction would most likely be measured.

ADJUST INSTRUCTION If students are still having difficulty deciding which ratios to use, review Figures 12.4 and 12.5 with them.

SampleProblem 12.5

Calculating Molecules of a Product

How many molecules of oxygen are produced when 29.2 g of water is decomposed by electrolysis according to this balanced equation?

$$2H_2O(l) \xrightarrow{\text{electricity}} 2H_2(g) + O_2(g)$$

❶ Analyze List the knowns and the unknown. The following calculations need to be performed:

$$\text{g } H_2O \longrightarrow \text{mol } H_2O \longrightarrow \text{mol } O_2 \longrightarrow \text{molecules } O_2$$

The appropriate mole ratio relating mol O_2 to mol H_2O from the balanced equation is 1 mole O_2/2 mol H_2O.

KNOWNS

mass of water = 29.2 g H_2O

1 mol O_2/2 mol H_2O (from balanced equation)

1 mol H_2O = 18.0 g H_2O (molar mass)

1 mol O_2 = 6.02×10^{23} molecules O_2

UNKNOWN

molecules of oxygen = ? moleciules O_2

❷ Calculate Solve for the unknown.

Start with the given quantity, and convert from mass to moles.

$$29.2 \text{ g } H_2O \times \frac{1 \text{ mol } H_2}{18.0 \text{ g } H_2O}$$

Remember to start your calculations with the given quantity, even if the given quantity is a product in the reaction.

Then, convert from moles of reactant to moles of product.

$$29.2 \text{ g } H_2O \times \frac{1 \text{ mol } H_2}{18.0 \text{ g } H_2O} \times \frac{1 \text{ mol } O_2}{2 \text{ mol } H_2O}$$

Finish by converting from moles to molecules.

$$\underbrace{29.2 \text{ g } H_2O}_{\text{Given quantity}} \times \underbrace{\frac{1 \text{ mol } H_2}{18.0 \text{ g } H_2O}}_{\text{Change to moles}} \times \underbrace{\frac{1 \text{ mol } O_2}{2 \text{ mol } H_2O}}_{\text{Mole ratio}} \times \underbrace{\frac{6.02 \times 10^{23} \text{ molecules } O_2}{1 \text{ mol } O_2}}_{\text{Change to molecules}}$$

$$= 4.88 \times 10^{23} \text{ molecules } O_2$$

❸ Evaluate Does the result make sense? The given mass of water should produce a little less than 1 mol of oxygen, or a little less than Avogadro's number of molecules. The answer should have three significant figures.

15. How many molecules of oxygen are produced by the decomposition of 6.54 g of potassium chlorate ($KClO_3$)?

$$2KClO_3(s) \longrightarrow 2KCl(s) + 3O_2(g)$$

16. The last step in the production of nitric acid is the reaction of nitrogen dioxide with water.

$$3NO_2(g) + H_2O(l) \longrightarrow 2HNO_3(aq) + NO(g)$$

How many grams of nitrogen dioxide must react with water to produce 5.00×10^{22} molecules of nitrogen monoxide?

DRAW A DIAGRAM On the board or overhead projector, draw a diagram showing the relationships that are useful for solving stoichiometry problems. One simple model reaction is A → B. Use double-headed arrows to connect these terms: *Particles of A, Moles of A, Grams of A, Moles of B, Particles of B*, and *Grams of B*. Above the appropriate arrows, write *Avogadro's number, Coefficients*, and *Molar mass*. Explain that the only "transitions" are allowed between quantities connected by arrows. Point out that the required conversion factor to make a "transition" is written above each arrow. Encourage students to refer to the diagram when working practice problems.

Foundations for Math

CONVERSION FACTORS Students often struggle with the proper use of conversion factors. Conversion factors should always be oriented so that like units cancel, leaving the desired unit in the numerator. Explain to students that planning the steps of the solution can make the task simpler.

In **Sample Problem 12.5,** the goal is to find the number of molecules of O_2 produced from 29.2 g of H_2O. The solution should end with molecules of O_2 in the numerator of the ratio. With this in mind, students should first look for a molar ratio that is a direct comparison of H_2O and O_2. In this case, the ratio is 2 mol H_2O to 1 mol O_2. Now they know that the first step in their solution is to determine the number of moles in 29.2 g H_2O. Then they can determine the number of moles of O_2 that will be produced and finally the number of molecules of O_2.

Answers

FIGURE 12.5 6.02×10^{23} representative particles/1 mol

15. 4.82×10^{22} molecules O_2

16. 11.5 g NO_2

Explore

Class Activity

PURPOSE Students practice sequencing the steps in solving stoichiometric problems.

MATERIALS 8 white index cards, 1 colored index card, paper punch, 2 brass paper fasteners

PROCEDURE Distribute the white cards to the students. Have them divide the cards into two piles of four cards each. On the first card of the first pile, have them write *Converting a given measured quantity to moles.* On each of the three remaining cards, have students write the conversion factors for converting mass to moles, representative particles to moles, and volume to moles, respectively. For the second set of cards, have students label the first card *Changing moles of wanted substances to a measured quantity.* On each of the remaining cards, have them write the appropriate conversion factor. On the colored card, have students write *Converting moles of given to moles of wanted using mole ratio from balanced chemical equation b mol W/a mol G.* Have the students use the paper punch to punch each of the two sets of cards. Then have them fasten each set with a brass paper fastener. Allow students to practice using the cards to solve the Practice Problems.

EXPECTED OUTCOME The cards should aid in sequencing the steps in solving stoichiometric problems.

Sample Practice Problem

Ammonia (NH_3) reacts with oxygen (O_2) to produce nitrogen monoxide (NO) and water.

$$4NH_3(g) + 5O_2(g) \rightarrow 4NO(g) + 6H_2O(l)$$

How many liters of NO are produced when 1.40 L of oxygen reacts with ammonia? *(1.12 L)*

Sample Problem 12.6

Volume-Volume Stoichiometric Calculations

Nitrogen monoxide and oxygen gas combine to form the brown gas nitrogen dioxide, which contributes to photochemical smog. How many liters of nitrogen dioxide are produced when 34 L of oxygen react with an excess of nitrogen monoxide? Assume conditions are at STP.

$$2NO(g) + O_2(g) \longrightarrow 2NO_2(g)$$

1 Analyze List the knowns and the unknown. The following calculations need to be performed:

$$L\ O_2 \longrightarrow mol\ O_2 \longrightarrow mol\ NO_2 \longrightarrow L\ NO_2$$

For gaseous reactants and products at STP, 1 mol of a gas has a volume of 22.4 L.

KNOWNS

volume of oxygen = 34 L O_2

2 mol NO_2/1 mol O_2 (from balanced equation)

1 mol O_2 = 22.4 L O_2 (at STP)

1 mol NO_2 = 22.4 L NO_2 (at STP)

UNKNOWN

volume of nitrogen dioxide = ? L NO_2

2 Calculate Solve for the unknown.

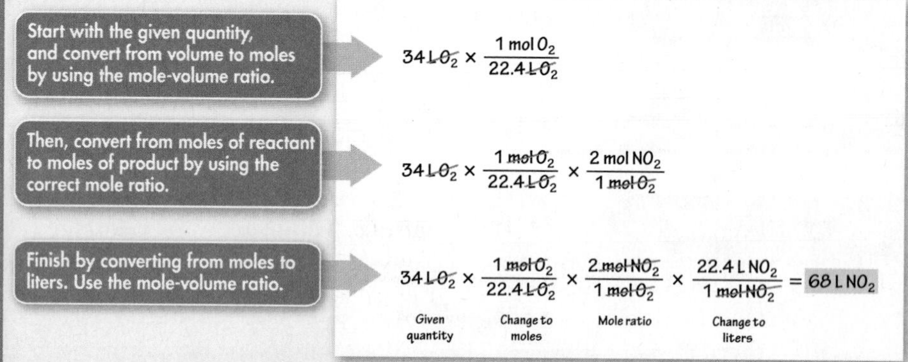

3 Evaluate Does the result make sense? Because 2 mol NO_2 are produced for each 1 mol O_2 that reacts, the volume of NO_2 should be twice the given volume of O_2. The answer should have two significant figures.

17. The equation for the combustion of carbon monoxide is

$$2CO(g) + O_2(g) \longrightarrow 2CO_2(g)$$

How many liters of oxygen are required to burn 3.86 L of carbon monoxide?

18. Phosphorus and hydrogen can be combined to form phosphine (PH_3).

$$P_4(s) + 6H_2(g) \longrightarrow 4PH_3(g)$$

How many liters of phosphine are formed when 0.42 L of hydrogen reacts with phosphorus?

Differentiated Instruction

L1 STRUGGLING STUDENTS Encourage students to find a method of problem solving that capitalizes on their strengths, such as drawing pictures of reactants and products. Make molecular model kits available to help students visualize reactions.

LPR LESS PROFICIENT READERS Direct students' attention to the key questions and answers. Rewrite the answers on the board, and then revise them by writing simplified sentences or bulleted lists.

L3 ADVANCED STUDENTS Have computer-literate students use the calculations in the sample problem as the basis for a general algorithm in a spreadsheet or computer program to solve stoichiometric problems. Have students demonstrate and explain their programs to interested classmates.

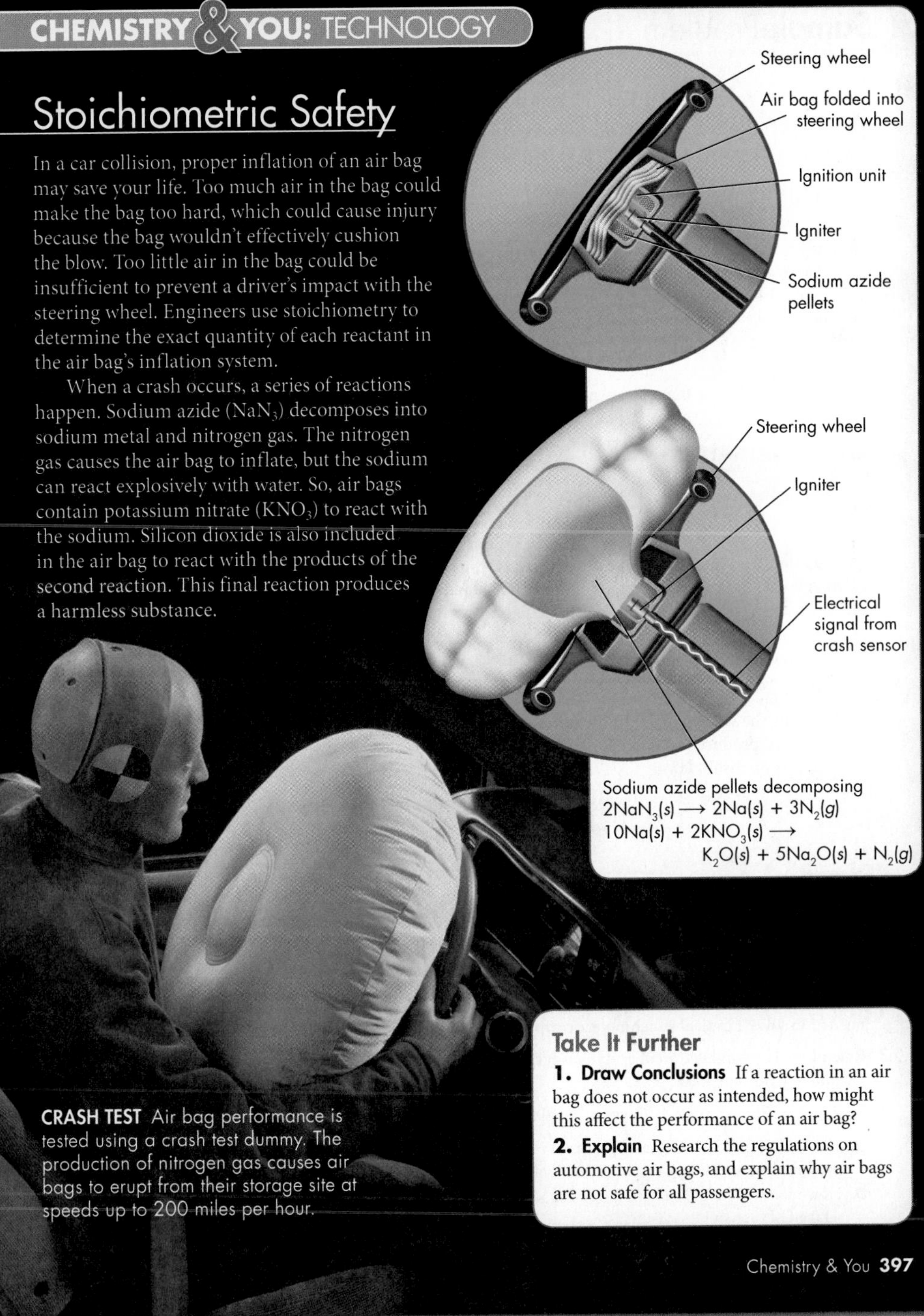

CHEMISTRY & YOU: TECHNOLOGY

Stoichiometric Safety

In a car collision, proper inflation of an air bag may save your life. Too much air in the bag could make the bag too hard, which could cause injury because the bag wouldn't effectively cushion the blow. Too little air in the bag could be insufficient to prevent a driver's impact with the steering wheel. Engineers use stoichiometry to determine the exact quantity of each reactant in the air bag's inflation system.

When a crash occurs, a series of reactions happen. Sodium azide (NaN_3) decomposes into sodium metal and nitrogen gas. The nitrogen gas causes the air bag to inflate, but the sodium can react explosively with water. So, air bags contain potassium nitrate (KNO_3) to react with the sodium. Silicon dioxide is also included in the air bag to react with the products of the second reaction. This final reaction produces a harmless substance.

CRASH TEST Air bag performance is tested using a crash test dummy. The production of nitrogen gas causes air bags to erupt from their storage site at speeds up to 200 miles per hour.

Take It Further

1. Draw Conclusions If a reaction in an air bag does not occur as intended, how might this affect the performance of an air bag?

2. Explain Research the regulations on automotive air bags, and explain why air bags are not safe for all passengers.

Chemistry & You 397

21st Century Learning To be successful in the 21st century, students need skills and learning experiences that extend beyond subject matter mastery. The following project helps students build 21st Century Skills.

BROADCASTING FOR SAFETY Pose the following challenge to students: *A major broadcasting corporation is creating a new automotive-themed podcast and has hired you as its host. Form groups of four to five students to do the following:*

- Write, produce, and record a 35-40 minute podcast, with one student serving as the host, one as an expert on air bag restraint systems, and two to three students as callers to the show.
- The show host will conduct a 20-minute interview with the expert, eliciting important information about current air bag technologies and those in development for future car models.
- The callers will use the final 15-20 minutes of the show to ask questions about safety and environmental concerns. The podcast will be submitted as an MP3 file either via e-mail or on a CD or DVD.

CHEMISTRY & YOU Write the following reactions on the board.

Reaction I $2NaN_3(s) \rightarrow 2Na(s) + 3N_2(g)$

Reaction II $10Na(s) + 2KNO_3(s) \rightarrow K_2O(s) + 5Na_2O(s) + N_2(g) + \text{heat}$

Point out that the proper inflation of the air bag requires two reactions. Explain that an electrical current produced by the igniter causes the decomposition of sodium azide into sodium metal and nitrogen gas. Note that the sodium metal produced is dangerously reactive. In a second reaction, potassium nitrate reacts with the elemental sodium and forms potassium oxide, sodium oxide, and additional nitrogen gas. The heat causes all the solid products to fuse with SiO_2, powdered sand, which is also part of the reaction mixture. The fused product is a safe, unreactive glass.

Ask How many moles of potassium nitrate must be included in the reaction mixture to consume the sodium produced by the decomposition of one mole of sodium azide? *(0.2 mol KNO_3)* **Ask** How many liters of N_2 are produced at STP if 1.0 mole of sodium azide and 0.20 mole of potassium nitrate react? *(36 L)*

Have students speculate how the pressure of the gas inside the air bag depends on the number of moles of nitrogen produced and the temperature inside the air bag. *(Acceptable answers should indicate that because gas pressure depends on the number of gas particles present, pressure depends on the number of moles of gas particles present. The heat released by this reaction raises the temperature of the gaseous products, helping the bag inflate even faster.)*

Extend

Connect to LANGUAGE ARTS

Explain to students that the use of air bag restraint systems (seatbelt/air bags) reduces the risk of fatalities in accidents by about 70%. However, not all passengers in a vehicle benefit from air bags, which can deliver a significant blow to a passenger when activated. Have students research the injuries that can occur as the result of air bag deployment, then have students write an argumentative essay that advocates the use of air bags despite the potential for injury. Essays should include current recommendations and restrictions for the use of air bags with small children and infants.

Answers

17. 1.93 L O_2

18. 0.28 L PH_3

TAKE IT FURTHER

1. Answers will vary. The air bag could under inflate or inflate too slowly.
2. Answers will vary. Some students may find that passenger-side air bags are not safe for young children.

Evaluate

Informal Assessment

Write a balanced equation on the board, such as $H_2(g) + I_2(g) \rightarrow 2HI(g)$. Have students orally state all the different mole ratios for the reaction; write all of the given ratios on the board. Then pose a problem, such as, *How many moles of hydrogen iodide are formed when 0.75 mol I_2 gas is reacted with excess hydrogen gas?* Have students go to the board and place their initials next to the correct mole ratio for the problem. Repeat with various types of reactions. Then have students complete the 12.2 Lesson Check.

Reteach

Use molecular models to review the importance of mole ratios. Illustrate how the mole ratios from the balanced chemical equation are related to the individual atoms, formula units, and molecules of the reactants and products as described by the equation.

Sample Problem 12.7

Finding the Volume of a Gas Needed for a Reaction

Assuming STP, how many milliliters of oxygen are needed to produce 20.4 mL SO_3 according to this balanced equation?

$$2SO_2(g) + O_2(g) \longrightarrow 2SO_3(g)$$

1 Analyze **List the knowns and the unknown.** For a reaction involving gaseous reactants or products, the coefficients also indicate relative amounts of each gas. So, you can use volume ratios in the same way you have used mole ratios.

KNOWNS

volume of sulfur trioxide = 20.4 mL

1 ml O_2/2 ml SO_3 (from balanced equation)

UNKNOWN

volume of oxygen = ? mL O_2

2 Calculate **Solve for the unknown.**

Multiply the given volume by the appropriate volume ratio.

$$20.4\ \text{mL } SO_3 \times \frac{1\ \text{mL } O_2}{2\ \text{mL } SO_3} = 10.2\ \text{mL } O_2$$

The volume ratio can be written using milliliters as the units instead of liters.

3 Evaluate **Does the result make sense?** Because the volume ratio is 2 volumes SO_3 to 1 volume O_2, the volume of O_2 should be half the volume of SO_3. The answer should have three significant figures.

Use the following chemical equation to answer Problems 19 and 20.

$$CS_2(l) + 3O_2(g) \longrightarrow CO_2(g) + 2SO_2(g)$$

19. Calculate the volume of sulfur dioxide, in milliliters, produced when 27.9 mL O_2 reacts with carbon disulfide.

20. How many deciliters of carbon dioxide are produced when 0.38 L SO_2 is formed?

12.2 LessonCheck

21. Explain How are mole ratios used in chemical calculations?

22. Sequence Outline the sequence of steps needed to solve a typical stoichiometric problem.

23. Calculate The combustion of acetylene gas is represented by this equation:

$$2C_2H_2(g) + 5O_2(g) \longrightarrow 4CO_2(g) + 2H_2O(g)$$

a. How many grams of CO_2 and grams of H_2O are produced when 52.0 g C_2H_2 burn in oxygen?

b. How many moles of H_2O are produced when 64.0 g C_2H_2 burn in oxygen?

24. Apply Concepts Write the 12 mole ratios that can be derived from the equation for the combustion of isopropyl alcohol.

$$2C_3H_7OH(l) + 9O_2(g) \longrightarrow 6CO_2(g) + 8H_2O(g)$$

BIGIDEA THE MOLE AND QUANTIFYING MATTER

25. Use what you have learned about stoichiometric calculations to explain the following statement: Stoichiometric calculations are not possible without a balanced chemical equation.

Lesson Check Answers

21. Mole ratios are written using the coefficients from a balanced chemical equation. They are used to relate moles of reactants and products in stoichiometric calculations.

22. Convert the given quantity to moles; use the mole ratio from the equations to find the moles of the wanted; convert moles of wanted to the desired unit.

23. **a.** 176 g CO_2, 36.0 g H_2O **b.** 2.46 mol H_2O

24.

25. A chemical reaction's mole ratios are derived from the relationships between coefficients in a balanced chemical equation.

Small-Scale Lab

Analysis of Baking Soda

Purpose

To determine the mass of sodium hydrogen carbonate in a sample of baking soda, using stoichiometry

Materials

- baking soda
- 3 plastic cups
- soda straw
- balance
- pipets of HCl, NaOH, and thymol blue

Procedure

A. Measure the mass of a clean, dry plastic cup.

B. Using the straw as a scoop, fill one end with baking soda to a depth of about 1 cm. Add the sample to the cup and measure its mass again.

C. Place two HCl pipets that are about 3/4 full into a clean cup and measure the mass of the system.

D. Transfer the contents of both HCl pipets to the cup containing baking soda. Swirl until the fizzing stops. Wait 5–10 minutes to be sure the reaction is complete. Measure the mass of the two empty HCl pipets in their cup again.

E. Add 5 drops of thymol blue to the plastic cup.

F. Place two full NaOH pipets in a clean cup and measure the mass of the system.

G. Add NaOH slowly to the baking soda/HCl mixture until the pink color just disappears. Measure the mass of the NaOH pipets in their cup again.

Analyze

Using your experimental data, record the answers to the following questions below your data table.

1. Evaluate Write a balanced equation for the reaction between baking soda ($NaHCO_3$) and HCl.

2. Calculate Calculate the mass in grams of the baking soda.

(Step B − Step A)

3. Calculate Calculate the total mmol of 1*M* HCl. *Note:* Every gram of HCl contains 1 mmol.

(Step C − Step D) × 1.00 mmol/g

4. Calculate Calculate the total mmol of 0.5*M* NaOH. *Note:* Every gram of NaOH contains 0.5 mmol.

(Step F − Step G) × 0.500 mmol/g

5. Calculate Calculate the mmol of HCl that reacted with the baking soda. *Note:* The NaOH measures the amount of HCl that did not react.

(Step 3 − Step 4)

6. Calculate Calculate the mass of the baking soda from the reaction data.

(0.084 g/mmol × Step 5)

7. Calculate Calculate the percent error of the experiment.

$$\frac{(\text{Step 2} - \text{Step 6})}{\text{Step 2}} \times 100\%$$

You're the Chemist

The following small-scale activities allow you to develop your own procedures and analyze the results.

1. Analyze Data For each calculation you did, substitute each quantity (number and unit) into the equation and cancel the units to explain why each step gives the quantity desired.

2. Design an Experiment Baking powder consists of a mixture of baking soda, sodium hydrogen carbonate, and a solid acid, usually calcium dihydrogen phosphate ($Ca(H_2PO_4)_2$). Design and carry out an experiment to determine the percentage of baking soda in baking powder.

ANALYZE

1. $HCl(aq) + NaHCO_3(s) \rightarrow CO_2(g) + H_2O(l) + NaCl(aq)$
2. 3.28 g – 2.83 g = 0.45 g
3. (10.70 – 4.29) g × 1.00 mmol/g = 6.41 mmol
4. (10.53 – 8.78) g × 0.500 mmol/g = 0.875 mmol (0.875 mmol HCl unreacted)
5. 6.41 mmol total – 0.875 mmol unreacted = 5.54 mmol (5.54 mmol $NaHCO_3$)
6. (0.0840 g/mmol) × 5.53 mmol = 0.46 g
7. (0.46 – 0.45) g/0.45 g × 100% = 2% error (assuming baking soda is one hundred percent sodium hydrogen carbonate)

YOU'RE THE CHEMIST

1. See Steps 2–7.
2. Repeat Steps A–G and 1–7 using baking powder instead of baking soda. The percent error is the percent of baking soda in baking powder, assuming no other errors.

FOR ENRICHMENT Ask students to predict how much baking soda and 1 *M* HCl are needed to produce enough CO_2 to fill a 1-L plastic bag. Have them write a procedure and then carry out the experiment.

Explore

Small-Scale Lab

OBJECTIVE Students calculate the mass of $NaHCO_3$ in a sample using stoichiometry.

PREP TIME 1 hour

CLASS TIME 30 minutes

MATERIALS Baking soda; plastic cups; soda straws; mass balances; pipets of HCl, NaOH, and thymol blue

ADVANCE PREP

Solution	Preparation
0.5*M* NaOH	20.0 g in 1.0 L
1.0*M* HCl	82 mL of 12*M* in 1.0 L **Caution** *Always add acid to water carefully and slowly.*
0.04% TB	100 mg in 21.5 mL of 0.01*M* NaOH; dilute to 250 mL

SAFETY Have students wear safety glasses and follow the standard safety procedures.

TEACHING TIPS

- Stress that the procedure measures the amount of excess HCl that is not reacted with the baking soda (Step 4). Because this excess HCl reacts with the NaOH in a 1:1 mole ratio, the moles of NaOH equal the moles of HCl in excess. Subtracting the excess moles of HCl from the total moles used in the experiment (Step 5) yields the moles reacted with the baking soda, which is 100% $NaHCO_3$.
- If the mixture does not turn red when thymol blue is added, the student should find the mass of a third pipet and add just enough HCl to turn the mixture cherry red. Then the student should find the mass of the half-empty pipet so the mass of HCl added can be calculated and added to the total mass used.

EXPECTED OUTCOME Sample data: Step A. 2.83 g, B. 3.28 g, C. 10.70 g, D. 4.29 g, F. 10.53 g, G. 8.78 g

Answers

19. 18.6 mL SO_2

20. 1.9 dL CO_2

LESSON 12.3

Key Objectives

12.3.1 EXPLAIN how the amount of product in a reaction is affected by an insufficient quantity of any of the reactants.

12.3.2 EXPLAIN what the percent yield of a reaction measures.

Additional Resources

- Reading and Study Workbook, Lesson 12.3
- Teaching Resources, Lesson 12.3
- Laboratory Manual, Lab 20

PearsonChem.com

Engage

CHEMISTRY & YOU Direct students to the photograph at the beginning of the lesson. **Ask** If you have two tabletops and seven table legs and you want to construct built-in tables using two legs for support of each table, what part limits the number of tables you can construct? *(tabletops)* **Ask** How many unused parts would remain? *(3 legs)*

Build Background

Have students think about the yield information in their favorite cookie recipes. **Ask** Do you always get the number of cookies your recipe supposedly yields when you bake a batch? *(no)* **Ask** What affects the number of cookies in a batch? *(Sample answers: inaccurate measurements, size of cookies, an ingredient left out)* Tell students to keep this example in mind as they work through the lesson.

IN ACADEMIC STANDARDS for SCIENCE

- Apply standard techniques in laboratory investigations to measure physical quantities in appropriate units and convert known quantities to other units as necessary.

C.4.2 Balance chemical equations using the law of conservation of mass and use them to describe chemical reactions.

C.5.3 Given the equation for a chemical reaction involving one or more gases as reactants and/or products, calculate the volumes of gas assuming the reaction goes to completion and the ideal gas law holds.

Also covered C.4.3, C.4.4

12.3 Limiting Reagent and Percent Yield

C.4.3 The mole concept; **C.4.4** Calculate reactants and products; **C.5.3** Reactant and product calculations: gases

CHEMISTRY & YOU

Q: *What determines how much product you can make?* If a carpenter had two tabletops and seven table legs, he would have difficulty building more than one functional four-legged table. The first table would require four of the legs, leaving just three legs for the second table. In this case, the number of table legs is limiting the construction of four-legged tables. In this lesson you will learn how the amount of product is limited in a chemical reaction.

Key Questions

How is the amount of product in a reaction affected by an insufficient quantity of any of the reactants?

What does the percent yield of a reaction measure?

Vocabulary

- limiting reagent
- excess reagent
- theoretical yield
- actual yield
- percent yield

Limiting and Excess Reagents

How is the amount of product in a reaction affected by an insufficient quantity of any of the reactants?

Many cooks follow a recipe when making a new dish. They know that sufficient quantities of all the ingredients must be available in order to follow the recipe. Suppose, for example, that you are preparing to make tacos like the ones in Figure 12.6. You would have more than enough meat, cheese, lettuce, tomatoes, sour cream, salsa, and seasoning on hand. However, you have only two taco shells. The quantity of taco shells you have will limit the number of tacos you can make. Thus, the taco shells are the limiting ingredient in this cooking venture. A chemist often faces a similar situation. **In a chemical reaction, an insufficient quantity of any of the reactants will limit the amount of product that forms.**

Figure 12.6 Limiting Ingredients
The amount of product is determined by the quantity of the limiting reagent. In this example, the taco shells are the limiting reagent. No matter how much of the other ingredients you have, with two taco shells you can make only two tacos.

Focus on ELL

1 CONTENT AND LANGUAGE Direct students' attention to the word *reagent* in the lesson title, and tell students that it is derived from the Latin word for "react." Explain that reagents are substances that chemically react with other substances to form new substances, so a reagent is a reactant in a chemical equation.

2 FRONTLOAD THE LESSON Ask students to think about items that need to be assembled by the purchaser, such as furniture or models sold as kits. Point out that even though materials lists in these kits state that each kit contains enough parts to complete the item, some kits may be missing parts and some may have extra parts. Ask volunteers to share their experiences with such items.

3 COMPREHENSIBLE INPUT Use ball-and-stick models to demonstrate the concepts illustrated in Figure 12.7. Say and write instructions when demonstrating the sample problems.

Chemical Equations						
	$N_2(g)$	+	$3H_2(g)$	$\longrightarrow$	$2NH_3(g)$
Microscopic recipe:	1 molecule N_2	+	3 molecules H_2	$\longrightarrow$	2 molecules NH_3
Macroscopic recipe:	1 mol N_2	+	3 mol H_2	$\longrightarrow$	2 mol NH_3

Experimental Conditions	Reactants		Products
Before reaction	2 molecules N_2	3 molecules H_2	0 molecules NH_3
After reaction	1 molecule N_2	0 molecules H_2	2 molecules NH_3

Figure 12.7 Limiting Reagent The "recipe" calls for three molecules of H_2 for every one molecule of N_2.

In this particular experiment, H_2 is the limiting reagent and N_2 is in excess.
Infer ***How would the amount of products formed change if you started with four molecules of N_2 and three molecules of H_2?***

As you know, a balanced chemical equation is a chemist's recipe. You can interpret the recipe on a microscopic scale (interacting particles) or on a macroscopic scale (interacting moles). The coefficients used to write the balanced equation give both the ratio of representative particles and the mole ratio. Recall the equation for the preparation of ammonia:

$$N_2(g) + 3H_2(g) \longrightarrow 2NH_3(g)$$

When one molecule (mole) of N_2 reacts with three molecules (moles) of H_2, two molecules (moles) of NH_3 are produced. What would happen if two molecules (moles) of N_2 reacted with three molecules (moles) of H_2? Would more than two molecules (moles) of NH_3 be formed? Figure 12.7 shows both the particle and the mole interpretations of this problem.

Before the reaction takes place, nitrogen and hydrogen are present in a 2:3 molecule (mole) ratio. The reaction takes place according to the balanced equation. One molecule (mole) of N_2 reacts with three molecules (moles) of H_2 to produce two molecules (moles) of NH_3. At this point, all the hydrogen has been used up, and the reaction stops. One molecule (mole) of unreacted nitrogen is left in addition to the two molecules (moles) of NH_3 that have been produced by the reaction.

In this reaction, only the hydrogen is completely used up. This reactant is the **limiting reagent,** or the reactant that determines the amount of product that can be formed by a reaction. The reaction occurs only until the limiting reagent is used up. By contrast, the reactant that is not completely used up in a reaction is called the **excess reagent.** In this example, nitrogen is the excess reagent because some nitrogen remains unreacted.

Sometimes in stoichiometric problems, the given quantities of reactants are expressed in units other than moles. In such cases, the first step in the solution is to convert the quantity of each reactant to moles. Then the limiting reagent can be identified. The amount of product formed in a reaction can be determined from the given amount of limiting reagent.

See limiting reagents animated online.

CHEMISTRY & YOU

Q: *What determines how much product you can make in a chemical reaction?*

Differentiated Instruction

LPR LESS PROFICIENT READERS Help students understand the concept of a limiting reagent by comparing it to everyday situations. For example, suppose a person is planning a dinner party. He has 6 glasses, 8 plates, and 4 chairs. How many people can he invite? Or, you might ask, how many party invitations can you send if you have 24 invitations, 12 envelopes, and 10 stamps?

ELL ENGLISH LANGUAGE LEARNERS Discuss with students the meaning of the word *reagent*. Ask them to define *reagent* in their own words. Then ask students to define the words *limit* and *excess* in their own words. Then ask students to state the meanings of *limiting reagent* and *excess reagent* in their own words.

READING STRATEGY Have students quickly read through the lesson and use context clues to help define the meaning of *limiting reagent* and *excess reagent*. Discuss real-world meanings of the terms *limiting* and *excess*.

Explain

Limiting and Excess Reagents

USING VISUALS Direct students' attention to Figure 12.6. **Ask** Where else do you see examples of processes or activities being "limited" on an everyday basis? *(Sample answers: the number of people you can fit into a car and the number of biscuits that can be made with one cup of flour)*

Explore

Teacher Demo

PURPOSE Students model the concept of a limiting reagent.

MATERIALS 15 plastic bottles, 30 plastic caps to fit the bottles, 6 containers to hold 5 caps each

PROCEDURE Before the demonstration, place five caps in each container and put them out of the sight of the students. Set out all 15 plastic bottles. Bring out one container of caps. **Ask** How many closed plastic bottles can we have? *(5)* **Ask** What limits the number? *(caps)* Bring out another container of caps. Again, ask the students what will happen. Continue bringing out caps and matching them to the bottles.

EXPECTED OUTCOME With two containers of caps, students should indicate that the caps are the limiting factor. With three containers, students should realize that there is a one-to-one correspondence and therefore no limiting factor. Students may also argue that both the bottles and caps are equally limiting factors. With the remaining containers, the caps are in excess and students should indicate that the bottles are the limiting factor.

the limiting reagent

Answers

FIGURE 12.7: The amount would remain the same; two molecules of NH_3 would form.

Explain

APPLY CONCEPTS Write the balanced chemical equation for the following reaction on the board: $2Na(s) + Cl_2(g) \rightarrow 2NaCl(s)$. Ask a series of questions that require students to predict the amount of product formed, based on molar quantities of reactant supplied by you. Be sure to use small whole numbers that students can easily manipulate without calculators. As students solve the problems, record the molar quantities under each reactant and product in the equation. Point out how these values are related for each set of conditions. In each case, ask students to state which reactant is the limiting reagent and circle the value in the list to designate it as the limiting quantity.

Misconception Alert

Caution students not to make quick judgments when finding the limiting reagent in a reaction. Often, students make a decision based on mass, or they do not completely finish the calculation and answer prematurely. Encourage students to read a problem slowly, formulate a plan, and write each calculation in steps until a final answer is obtained. Emphasize the importance of checking the reasonableness of an answer.

Sample Practice Problem

Phosphoric acid reacts with sodium hydroxide according to the following equation:

$$H_3PO_4(aq) + 3NaOH(aq) \rightarrow Na_3PO_4(aq) + 3H_2O(l)$$

If 1.75 mol H_3PO_4 is made to react with 5.00 mol NaOH, what is the limiting reagent? *(5.25 mol NaOH is required; NaOH is the limiting reagent.)*

Sample Problem 12.8

Determining the Limiting Reagent in a Reaction

Copper reacts with sulfur to form copper(I) sulfide according to the following balanced equation:

$$2Cu(s) + S(s) \longrightarrow Cu_2S(s)$$

What is the limiting reagent when 80.0 g Cu reacts with 25.0 g S?

KNOWNS
mass of copper = 80.0 g Cu
mass of sulfur = 25.0 g S
1 mol S / 2 mol Cu

UNKNOWN
limiting reagent = ?

1 Analyze List the knowns and the unknown. The number of moles of each reactant must first be found. The balanced equation is used to calculate the number of moles of one reactant needed to react with the given amount of the other reactant.

2 Calculate Solve for the unknown.

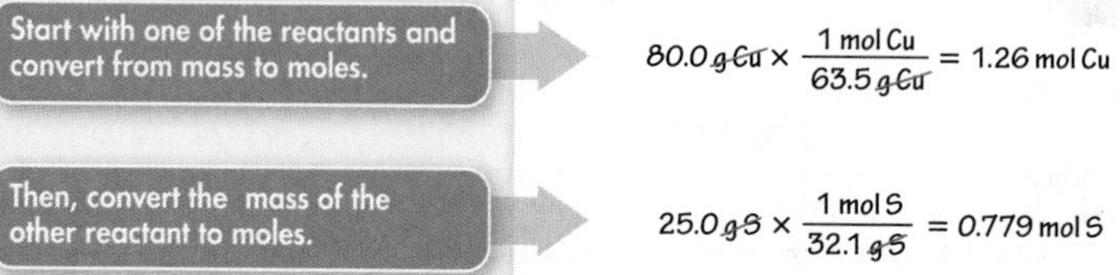

Start with one of the reactants and convert from mass to moles.

$$80.0\text{ g Cu} \times \frac{1\text{ mol Cu}}{63.5\text{ g Cu}} = 1.26\text{ mol Cu}$$

Then, convert the mass of the other reactant to moles.

$$25.0\text{ g S} \times \frac{1\text{ mol S}}{32.1\text{ g S}} = 0.779\text{ mol S}$$

Now convert moles of Cu to moles of S needed to react with 1.26 moles of Cu.

$$1.26\text{ mol Cu} \times \frac{1\text{ mol S}}{2\text{ mol Cu}} = 0.630\text{ mol S}$$

Given quantity | Mole ratio | Needed amount

Compare the amount of sulfur needed with the given amount of sulfur.

0.630 mol S (amount needed to react) < 0.779 mol S (given amount)
Sulfur is in excess, so copper is the limiting reagent.

It doesn't matter which reactant you use. If you used the actual number of moles of S to find the amount of copper needed, then you would still identify copper as the limiting reagent.

3 Evaluate Do the results make sense? Since the ratio of the given mol Cu to mol S was less than the ratio (2:1) from the balanced equation, copper should be the limiting reagent.

26. The equation for the complete combustion of ethene (C_2H_4) is

$$C_2H_4(g) + 3O_2(g) \longrightarrow 2CO_2(g) + 2H_2O(g)$$

If 2.70 mol C_2H_4 reacts with 6.30 mol O_2, identify the limiting reagent.

27. Hydrogen gas can be produced by the reaction of magnesium metal with hydrochloric acid.

$$Mg(s) + 2HCl(aq) \longrightarrow MgCl_2(aq) + H_2(g)$$

Identify the limiting reagent when 6.00 g HCl reacts with 5.00 g Mg.

Differentiated Instruction

L1 SPECIAL NEEDS Modify the sample problems so you can explain the concepts of limiting reagents and excess reagents by using manipulatives, such as coins, chips, or food items. These visual representations will engage students' senses and allow them to experience the concepts.

ELL ENGLISH LANGUAGE LEARNERS Label a number line from 0 to 1 by increments of 0.1. Using the overhead projector, plot the number of moles calculated for sulfur in **Sample Problem 12.8** on the number line to show students that 0.779 > 0.630. Encourage students to use number lines when determining which number is greater.

In Sample Problem 12.8, you may have noticed that even though the mass of copper used in the reaction is greater than the mass of sulfur, copper is the limiting reagent. The reactant that is present in the smaller amount by mass or volume is not necessarily the limiting reagent.

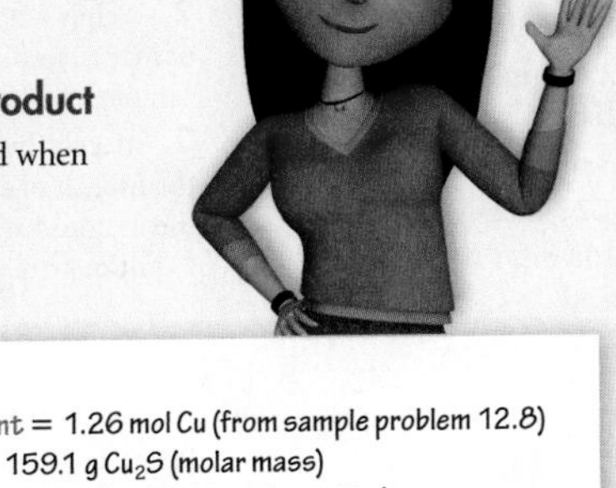

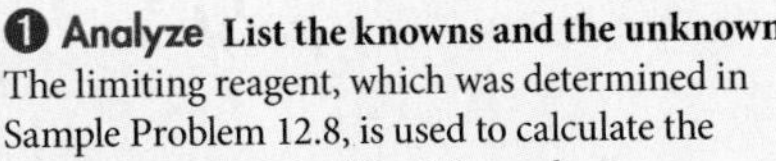

Sample Problem 12.9

Using a Limiting Reagent to Find the Quantity of a Product

What is the maximum number of grams of Cu_2S that can be formed when 80.0 g Cu reacts with 25.0 g S?

$$2Cu(s) + S(s) \longrightarrow Cu_2S(s)$$

1 Analyze **List the knowns and the unknown.** The limiting reagent, which was determined in Sample Problem 12.8, is used to calculate the maximum amount of Cu_2S formed.

$$\text{mol Cu} \longrightarrow \text{mol } Cu_2S \longrightarrow \text{g } Cu_2S$$

KNOWNS

limiting reagent = 1.26 mol Cu (from sample problem 12.8)

1 mol Cu_2S = 159.1 g Cu_2S (molar mass)

1 mol Cu_2S / 2 mol Cu (from balanced equation)

UNKNOWN

yield = ? g Cu_2S

2 Calculate **Solve for the unknown.**

Start with the moles of the limiting reagent and convert to moles of the product. Use the mole ratio from the balanced equation.

$$1.26 \cancel{\text{mol Cu}} \times \frac{1 \text{ mol } Cu_2S}{2 \cancel{\text{mol Cu}}}$$

Finish the calculation by converting from moles to mass of product.

$$1.26 \cancel{\text{mol Cu}} \times \frac{1 \cancel{\text{mol } Cu_2S}}{2 \cancel{\text{mol Cu}}} \times \frac{159.1 \text{ g } Cu_2S}{1 \cancel{\text{mol } Cu_2S}} = 1.00 \times 10^2 \text{ g } Cu_2S$$

3 Evaluate **Do the results make sense?** Copper is the limiting reagent in this reaction. The maximum number of grams of Cu_2S produced should be more than the amount of copper that initially reacted because copper is combining with sulfur. However, the mass of Cu_2S produced should be less than the total mass of the reactants (105.0 g) because sulfur was in excess.

28. The equation below shows the incomplete combustion of ethene.

$$C_2H_4(g) + 2O_2(g) \longrightarrow 2CO(g) + 2H_2O(g)$$

If 2.70 mol C_2H_4 is reacted with 6.30 mol O_2,

a. identify the limiting reagent.

b. calculate the moles of water produced.

29. The heat from an acetylene torch is produced by burning acetylene (C_2H_2) in oxygen.

$$2C_2H_2(g) + 5O_2(g) \longrightarrow 4CO_2(g) + 2H_2O(g)$$

How many grams of water can be produced by the reaction of 2.40 mol C_2H_2 with 7.40 mol O_2?

Foundations for Math

SCIENTIFIC NOTATION Explain that it is typical for very large and very small numbers to be written in scientific notation, simply for the ease of writing or stating the number. For example, it is cumbersome for a scientist to write the number 0.0000000000082 and much simpler to write 8.2×10^{-12}.

The number 8.2×10^{-12} is basically the product of a decimal coefficient, 8.2, and an exponential term, 10^{-12}. The decimal coefficient is ≥ 1 and < 10. It also indicates the number of significant figures in a measurement. The exponent indicates the placement of the decimal point. A **positive** exponent indicates to move the decimal point that number of places to the right. A **negative** exponent indicates to move the decimal point that number of places to the left.

The measure 3,440,000 cm can be written as 3.44×10^6 cm. This measure has 3 significant figures, whereas 0.00047 cm = 4.7×10^{-4} cm has only 2 significant figures.

Explore

Class Activity

PURPOSE Students will use simple molecular models to demonstrate the concept of a limiting reagent in a chemical reaction.

PREP TIME 5 minutes

CLASS TIME 15 minutes

MATERIALS 20 metal paper clips (symbol M), 20 same-colored vinyl-coated paper clips (symbol C), plastic sandwich bag

PROCEDURE

1. Link two paper clips of the same type several times to form models representing 10 diatomic molecules of each reactant. Place these molecules in the plastic bag.
2. Without looking, remove 15 molecules from the plastic bag.
3. Line up the M_2 and C_2 molecules in two adjacent vertical rows.
4. Pair reactant molecules in the 1:3 M_2-to-C_2 ratio as shown in the equation $M_2 + 3C_2 \rightarrow 2MC_3$.
5. Make the molecules "react" by taking pairs of clips apart and forming two molecules of the product.
6. Continue making M_2 and C_2 react in a 1:3 ratio until you run out of one of the reactants.

THINK ABOUT IT

1. List the number of each type of reactant molecule that you removed from the bag. *(Answers may vary.)*
2. How many molecules of the product did you form? *(Answers may vary: It depends on n_{M_2} and n_{C_2}.)*
3. Which reactant molecule did you run out of first? *(Answers may vary: It depends on n_{M_2} and n_{C_2}.)*
4. How many molecules of each reactant remained at the completion of the reaction? *(Answers may vary: It depends on n_{M_2} and n_{C_2}.)*
5. Repeat the experiment and compare the results. *(Sample answer: The number of MC_3 molecules formed and the number and type of reactant molecules remaining depend on the initial number of M_2 and C_2 molecules and the ratio of M to C atoms in the molecular formula.)*

Answers

26. O_2 is the limiting reagent.

27. HCl is the limiting reagent.

28. **a.** 5.40 mol O_2 is required; C_2H_4 is the limiting reactant.

b. 5.40 mol H_2O

29. 43.2 g H_2O

LESSON 12.3

Explore

Quick Lab

OBJECTIVE Students observe the effect that changing the amount of limiting reagent in a reaction has on the amounts of products.

SKILLS FOCUS Observing, measuring, calculating

PREP TIME 15 minutes

CLASS TIME 20 minutes

MATERIALS graduated cylinders, mass balances, 250-mL Erlenmeyer flasks, rubber balloons, magnesium ribbons, 1.0*M* hydrochloric acid

SAFETY The balloons contain hydrogen gas and should be kept away from heat and open flames during and after the experiment. Caution students about the corrosive nature of HCl. Students should wear safety glasses and aprons. Once the reaction in a flask is complete, carefully remove the balloon. Disperse the gas in a fume hood or outside. Neutralize remaining HCl(*aq*) with baking soda before flushing down the drain.

TEACHING TIPS This is a discrepant event lab. Allow students to make their predictions and do the lab. As they are analyzing the data, remind them to check for a limiting reagent.

EXPECTED OUTCOME The approximate volumes of the balloons are 0.60 g Mg, 0.5 L H_2; 1.2 g Mg, 1 L H_2; 2.4 g Mg, 1 L H_2.

ANALYZE AND CONCLUDE

1. Those students who failed to balance the equation, obtain the mole ratio, and compare the molar amounts of reactants likely predicted a doubling of the volume with each doubling of the mass of magnesium.
2. $Mg(s) + 2HCl(aq) \rightarrow MgCl_2(aq) + H_2(g)$
3. A mass of 0.60 g Mg is 0.025 mol Mg. Because Mg and HCl react in a 1:2 mol ratio, 0.10 mol HCl is in excess and Mg is limiting. According to the balanced equation, 0.025 mol Mg should produce 0.025 mol H_2. A mass of 1.2 g Mg is 0.050 mol Mg. According to the balanced equation, 0.050 mol Mg will react with 0.1 mol HCl to produce 0.05 mol H_2.
4. A mass of 2.4 g Mg is 0.10 mol Mg. Because there is only 0.1 mol HCl in the flask and, according to the equation, 0.20 mol HCl is needed to react with 0.10 mol Mg, HCl is limiting. Therefore, only 0.05 mol H_2 is produced.

FOR ENRICHMENT Ask What other clue could be used to determine if HCl was the limiting reagent? *(If HCl were the limiting reagent, remaining magnesium metal would be present after the reaction.)*

Quick Lab

Limiting Reagents

Purpose To illustrate the concept of a limiting reagent in a chemical reaction

Materials

- graduated cylinder
- balance
- 3 250-mL Erlenmeyer flasks
- 3 rubber balloons
- 4.2 g magnesium ribbon
- 300 mL 1.0*M* hydrochloric acid

Procedure

1. Add 100 mL of the hydrochloric acid solution to each flask.
2. Weigh out 0.6 g, 1.2 g, and 2.4 g of magnesium ribbon, and place each sample into its own balloon.
3. Stretch the end of each balloon over the mouth of each flask. Do not allow the magnesium ribbon in the balloon to fall into the flask.
4. Magnesium reacts with hydrochloric acid to form hydrogen gas. When you mix the magnesium with the hydrochloric acid in the next step, you will generate a certain volume of hydrogen gas. How do you think the volume of hydrogen produced in each flask will compare?
5. Lift up on each balloon and shake the magnesium into each flask. Observe the volume of gas produced until the reaction in each flask is completed. Record your observations.

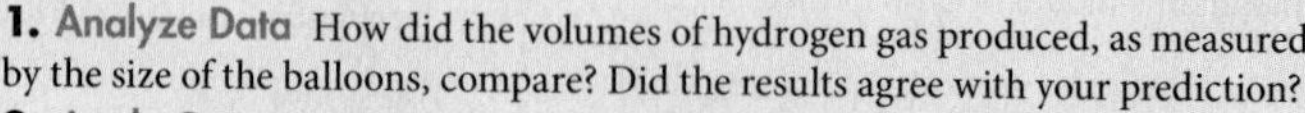

Analyze and Conclude

1. **Analyze Data** How did the volumes of hydrogen gas produced, as measured by the size of the balloons, compare? Did the results agree with your prediction?
2. **Apply Concepts** Write a balanced equation for the reaction you observed.
3. **Calculate** The 100 mL of hydrochloric acid contained 0.10 mol HCl. Show by calculation why the balloon with 1.2 g Mg inflated to about twice the size of the balloon with 0.60 g Mg.
4. **Calculate** Show by calculation why the balloons with 1.2 g and 2.4 g Mg inflated to approximately the same volume. What was the limiting reagent when 2.4 g Mg was added to the acid?

Percent Yield

What does the percent yield of a reaction measure?

When a teacher gives an exam to the class, every student could get a grade of 100 percent. However, this outcome generally does not occur. Instead, the performance of the class is usually spread over a range of grades. Your exam grade, expressed as a percentage, is a ratio of two items. The first item is the number of questions you answered correctly. The second is the total number of questions. The grade compares how well you performed with how well you could have performed if you had answered all the questions correctly. Chemists perform similar calculations in the laboratory when the product from a chemical reaction is less than expected, based on the balanced chemical equation.

Figure 12.8 Batting Average
A batting average is actually a percent yield. A batting average is calculated by dividing the number of hits a batter has had (actual yield) by the number of at-bats (theoretical yield).

404 Chapter 12 • Lesson 3

Focus on ELL

4 LANGUAGE PRODUCTION Have students work in groups to complete the lab. Make sure each group has ELLs of varied language proficiency so that the more proficient students can help the less proficient ones. Have students work according to their proficiency level.

BEGINNING: LOW/HIGH Allow students to use drawings, single words, or short phrases. Demonstrate Steps 1–3 while reading each step out loud.

INTERMEDIATE:

LOW Write simplified procedures on the board or overhead projector.

HIGH Have students draft sentences to answer the questions that require writing.

ADVANCED: LOW/HIGH Have students assist classmates with lower language proficiency to set up and balance equations and answer questions.

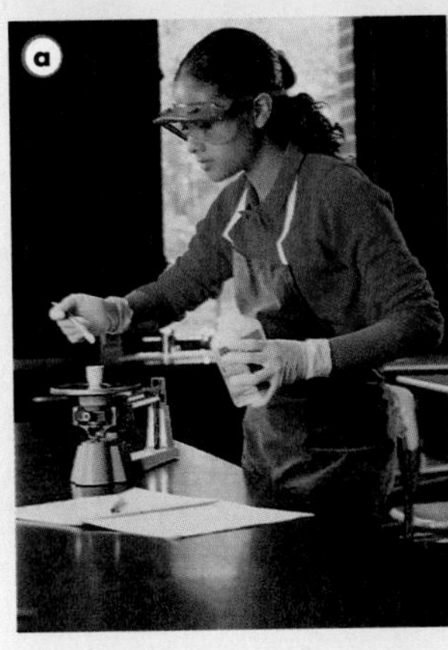

Figure 12.9 Determining Percent Yield
Sodium hydrogen carbonate ($NaHCO_3$) will decompose when heated. **a.** The mass of $NaHCO_3$, the reactant, is measured. **b.** The reactant is heated. **c.** The mass of one of the products, sodium carbonate (Na_2CO_3), the actual yield, is measured. The percent yield is calculated once the actual yield is determined.
Predict ***What are the other products of this reaction?***

When a balanced chemical equation is used to calculate the amount of product that will form during a reaction, the calculated value represents the theoretical yield. The **theoretical yield** is the maximum amount of product that could be formed from given amounts of reactants. In contrast, the amount of product that actually forms when the reaction is carried out in the laboratory is called the **actual yield**. The **percent yield** is the ratio of the actual yield to the theoretical yield expressed as a percent.

$$\text{Percent yield} = \frac{\text{actual yield}}{\text{theoretical yield}} \times 100\%$$

Because the actual yield of a chemical reaction is often less than the theoretical yield, the percent yield is often less than 100 percent. **The percent yield is a measure of the efficiency of a reaction carried out in the laboratory.** This yield is similar to an exam score measuring your efficiency of learning or a batting average measuring your efficiency of hitting a baseball, as in Figure 12.8.

Stoichiometry and conservation of mass dictate that yields of greater than 100 percent are not possible. However, errors and lack of knowledge in a process can cause a reaction to apear to have a yield that is more than 100 percent. For example, if air or water leaks into a system, then more product may be formed than expected.

Many factors can cause percent yields to be less than 100 percent. Reactions do not always go to completion; when a reaction is incomplete, less than the calculated amount of product is formed. Impure reactants and competing side reactions may cause unwanted products to form. Actual yield can also be lower than the theoretical yield due to a loss of product during filtration or in transferring between containers. Moreover, if reactants or products have not been carefully measured, a percent yield of 100 percent is unlikely.

An actual yield is an experimental value. Figure 12.9 shows a typical laboratory procedure for determining the actual yield of a product of a decomposition reaction. For reactions in which percent yields have been determined, you can calculate and, therefore, predict an actual yield if the reaction conditions remain the same.

Comparative Torch Temperatures

- Acetylene welding torches produce flame temperatures of 3300–3400°C.
- Propane torches reach flame temperatures of about 1400°C.
- The flame of a match has a temperature range of 600–800°C.

Percent Yield

APPLY CONCEPTS For industrial chemists or chemical engineers, the goal is to find cost-effective methods for converting reactants into products. Industrial chemists and chemical engineers want to achieve the maximum product yield at the lowest cost. One way to control costs is to minimize the amount of excess reagent by calculating stoichiometric quantities of the reactants.

Class Activity

PURPOSE Students assess actual yield of a reaction, based on temperature changes.

MATERIALS 3 foam cups, a thermometer, 100 mL of 1.0*M* HCl, and approximately 200 mL of 1.0*M* NaOH

SAFETY Students should wear safety glasses and aprons. HCl and NaOH are corrosive. Caution students to avoid skin contact with these chemicals. Neutralized solutions can be flushed down the drain with excess water.

PROCEDURE Write the balanced chemical equation on the board for students to use as a reference: $HCl(aq) + NaOH(aq) \rightarrow H_2O(l) + NaCl(aq) + \text{heat}$. Have students interpret the balanced equation as 1 mole of HCl reacts with 1 mole of NaOH to form 1 mole of NaCl and 1 mole of water. Have students transfer 30 mL of the HCl to a foam cup and measure the temperature. Next, have them add 10 mL of NaOH and gently stir the contents with the thermometer. Students should record the highest temperature reached as the reaction proceeds. Using a new cup each time, they should repeat the measurement with 30 mL of HCl and 30 mL of NaOH, and with 30 mL of HCl and 60 mL of NaOH.

EXPECTED OUTCOMES The second trial, in which reactants are present in a 1:1 mole ratio, produces the greatest temperature change. A slightly lower temperature change for the third trial can be accounted for by the greater mass of reagents. **Ask** Which trial had the highest yield? *(the one that produced the greatest temperature change)*

Foundations for Reading

BUILD VOCABULARY Have students use the LINCS strategy for the terms *theoretical yield, actual yield,* and *percent yield.* Students should **L**ist the parts of a term they know; **I**magine a picture of the term; **N**ote a "sound-alike word" for the term; **C**onnect the terms; and **S**elf-test the terms.

Explain

Percent Yield

CRITICAL THINKING Explain that some chemical processes are extremely costly, and low yields can have adverse effects on a company's profits. Have students research a pharmaceutical manufacturing company and learn about one of its processes. **Ask** What are some of the reasons for the expensive cost of the process? *(Sample answers: raw materials, materials of construction, high government standards of cleanliness)* **Ask** Why might this company *not* want to find cheaper raw materials for its product? *(Sample answer: It could affect the quality and effectiveness of the pharmaceutical product.)*

APPLY CONCEPTS Point out to students that solving stoichiometric problems requires a synthesis, or putting together, of chemical concepts, terms, and factual information. Write several problems on the board and have groups of students solve different problems, identifying the concepts, terms, and information necessary to do the problem.

MAKE A CONNECTION Relate percent yield to the mathematical concept of probability. Remind students that the experimental probability of an event is determined by the following:

$$\text{Probability} = \frac{\text{number of favorable outcomes}}{\text{total number of possible outcomes}}$$

Show students how the percent yield of a reaction is similar to probability; both concepts divide the actual result by the total possible.

$$\text{Percent Yield} = \frac{\text{actual amount of outcome}}{\text{total possible amount of outcome}}$$

Calculating the Theoretical Yield of a Reaction

Calcium carbonate, which is found in seashells, is decomposed by heating. The balanced equation for this reaction is

$$CaCO_3(s) \xrightarrow{\Delta} CaO(s) + CO_2(g)$$

What is the theoretical yield of CaO if 24.8 g $CaCO_3$ is heated?

KNOWNS
mass of calcium carbonate = 24.8 g $CaCO_3$
1 mol $CaCO_3$ = 1 mol CaO

UNKNOWN
theoretical yield = ? g CaO

1 Analyze List the knowns and the unknown. Calculate the theoretical yield using the mass of the reactant:
g $CaCO_3$ ⟶ mol $CaCO_3$ ⟶ mol CaO ⟶ g CaO

2 Calculate Solve for the unknown.

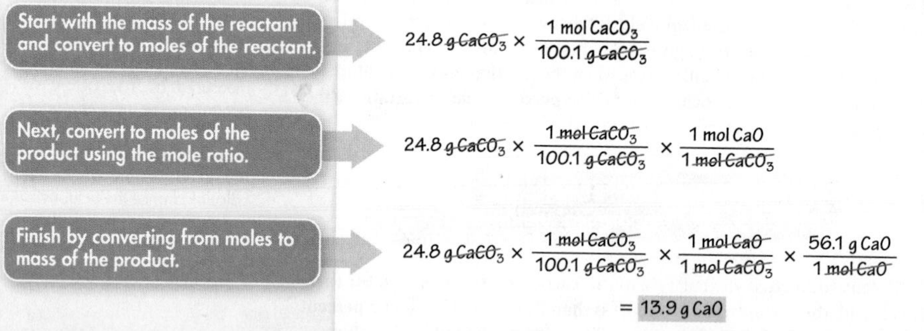

3 Evaluate Does the result make sense? The mole ratio of CaO to $CaCO_3$ is 1:1. The ratio of their masses in the reaction should be the same as the ratio of their molar masses, which is slightly greater than 1:2. The result of the calculations shows that the mass of CaO is slightly greater than half the mass of $CaCO_3$.

30. When 84.8 g of iron(III) oxide reacts with an excess of carbon monoxide, iron is produced.

$$Fe_2O_3(s) + 3CO(g) \longrightarrow 2Fe(s) + 3CO_2(g)$$

What is the theoretical yield of iron?

If there is an excess of a reactant, then there is more than enough of that reactant and it will not limit the yield of the reaction.

31. When 5.00 g of copper reacts with excess silver nitrate, silver metal and copper(II) nitrate are produced. What is the theoretical yield of silver in this reaction?

Foundations for Math

PERCENTS Caution students that the percent yield of a chemical reaction will not be greater than 100%. Point out that if the value they calculate is greater than 100%, they likely switched the placement of the numerator and denominator.

For struggling students, restate the percent yield formula out loud as "The percent yield is equal to the amount the reaction actually produced divided by the total possible amount the reaction could have produced." Encourage students to write the formula in their own words.

(Sample answer: $\text{Percent Yield} = \frac{\textit{what I got}}{\textit{what was possible to get}}$*)*

Success Stats

You may not use the term "percent yield" outside of chemistry class, but there are many examples of percent yield in our lives. In chemical reactions, percent yield refers to the amount of product formed in a reaction compared to how much product was possible. In school, percent yield could refer to the graduation rate or a score on a test. In sports, percent yield could refer to the percent of shots that make it into a goal. The actual performance of a product compared to its advertised performance is also an example of percent yield.

Whether it's in the chemistry lab or anywhere else, percent yield is a way to measure how successfully something or someone has performed. The next time you calculate the percent yield of a chemical reaction, think about how this skill could be used in other situations outside of chemistry class.

85 PERCENT PERFORMANCE
Actual Yield: 156 minutes during which a drink stayed hot in thermos
Theoretical Yield: 180 minutes, as advertised by the thermos manufacturer

38 PERCENT SHOT-CONVERSION RATE
Actual Yield: 8 goals scored
Theoretical Yield: 21 shots on goal

Take It Further

1. **Calculate** Sara's car is advertised to get 43 miles per gallon. Sara calculated her gas mileage over the last month and found that it was 39 miles per gallon. What is the percent yield of Sara's gas mileage?
2. **Identify** The percent yield of a reaction may be different each time the reaction occurs. Similarly, the performance of an athlete may vary. What are some factors that might affect percent yield?

95 PERCENT GRADUATION RATE
Actual Yield: 305 students graduating
Theoretical Yield: 321 students in the senior class

CHEMISTRY & YOU Draw students' attention to the photo of the insulated container and to its caption. Point out that products like this container are manufactured to meet a certain set of standards specified by the manufacturer. However, quality control inspectors are allowed to pass products that perform within a certain narrow range on either side of the given standards. In the case of the insulated container in the photo, the advertised number of hours that the container will keep a drink hot is most likely based on the properties of the container's materials and the properties of a standard beverage.

Pose the following question to students: *What factors determine how long the container keeps its contents hot? Assume that during manufacturing, the properties of the materials used to make all the containers are consistent. Also assume that all the containers meet the manufacturer's specifications.*

You may need to assist students in the following ways:

- The contents are most likely solutions (beverages) or mixtures of solids and liquids (soups, stews, etc.).
- The wide variety of environmental conditions to which the container will be exposed during normal use cannot be accurately duplicated in a quality control laboratory.
- The container's owner may not adhere to the manufacturer's care instructions for the container.

Differentiated Instruction

L1 STRUGGLING STUDENTS If students are having difficulty with Question 1, have them review **Sample Problem 12.11** in Lesson 12.3. If students are having difficulty with Question 2, have them review the Percent Yield section of Lesson 12.3.

ELL ENGLISH LANGUAGE LEARNERS Have students work in groups of mixed proficiencies to answer questions.

L3 ADVANCED STUDENTS Have students research and compare analytical balances of different sensitivities, and have students show mathematically how using a balance of too little or too great a sensitivity can affect theoretical and percent yields.

Answers

30. 59.3 g Fe
31. 17.0 g Ag

TAKE IT FURTHER

1. 91%
2. Answers may vary. Environmental conditions of the reaction and the quality of the reactants may affect the percent yield.

CHEMISTRY & YOU

START A CONVERSATION Ask students if they have ever observed a campfire or fireplace the next day after the fire has gone out. In most cases, there will be some unburned or partially burned pieces of wood left. Similarly, during a chemical reaction, some portions of the reactants may remain unreacted.

Evaluate

Informal Assessment

Following the safety precautions listed in the Class Activity on page 405, place a small piece of zinc in a large beaker of dilute hydrochloric acid. Explain the reaction to the students.

$$Zn(s) + 2HCl(aq) \rightarrow ZnCl_2(aq) + H_2(g)$$

Ask Which of the two reactants is the limiting reagent? *(If the zinc completely disappears after some time, students may conclude that the zinc is used up first, and is, therefore, the limiting reagent.)* **Ask** How can you test your hypothesis? *(Sample answers: Measure the volume of H_2 produced for various amounts of zinc added to a fixed volume of HCl; determine the stoichiometric quantity of zinc, the amount that reacts to give the greatest volume of H_2.)*

SampleProblem 12.11

Calculating the Percent Yield of a Reaction

What is the percent yield if 13.1 g CaO is actually produced when 24.8 g $CaCO_3$ is heated?

$$CaCO_3(s) \xrightarrow{\Delta} CaO(s) + CO_2(g)$$

1 Analyze List the knowns and the unknown. Use the equation for percent yield. The theoretical yield for this problem was calculated in Sample Problem 12.10.

KNOWNS
actual yield = 13.1 g CaO
theoretical yield = 13.9 g CaO (from Sample Problem 12.10)

UNKNOWN
percent yield = ? %

2 Calculate Solve for the unknown.

$$\text{percent yield} = \frac{\text{actual yield}}{\text{theoretical yield}} \times 100\%$$

Substitute the values for actual yield and theoretical yield into the equation for percent yield.

$$\text{percent yield} = \frac{13.1 \cancel{g\,CaO}}{13.9 \cancel{g\,CaO}} \times 100\% = 94.2\%$$

3 Evaluate Does the result make sense? In this example, the actual yield is slightly less than the theoretical yield. Therefore, the percent yield should be slightly less than 100 percent.

32. If 50.0 g of silicon dioxide is heated with an excess of carbon, 27.9 g of silicon carbide is produced.

$$SiO_2(s) + 3C(s) \xrightarrow{\Delta} SiC(s) + 2CO(g)$$

What is the percent yield of this reaction?

33. If 15.0 g of nitrogen reacts with 15.0 g of hydrogen, 10.5 g of ammonia is produced. What is the percent yield of this reaction?

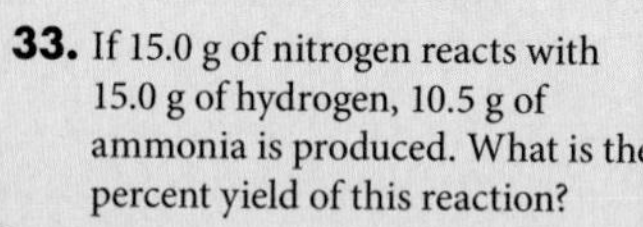

12.3 LessonCheck

34. **Relate Cause and Effect** In a chemical reaction, how does an insufficient quantity of a reactant affect the amount of product formed?

35. **Explain** How can you gauge the efficiency of a reaction carried out in the laboratory?

36. **Define** What is a limiting reagent? An excess reagent?

37. **Calculate** How many grams of SO_3 are produced when 20.0 g FeS_2 reacts with 16.0 g O_2 according to this balanced equation?

$$4FeS_2(s) + 15O_2(g) \longrightarrow 2Fe_2O_3(s) + 8SO_3(g)$$

38. **Calculate** What is the percent yield if 4.65 g of copper is produced when 1.87 g of aluminum reacts with an excess of copper(II) sulfate?

$$2Al(s) + 3CuSO_4(aq) \longrightarrow Al_2(SO_4)_3(aq) + 3Cu(s)$$

Lesson Check Answers

34. In a chemical reaction, an insufficient quantity of any of the reactants will limit the amount of product that forms.

35. The efficiency of a reaction carried out in a laboratory can be measured by calculating the percent yield.

36. A limiting reagent is a reagent that determines how much product can be formed in a reaction. An excess reagent is a reactant that is not completely used up in a reaction.

37. 21.4 g SO_3

38. 70.5%

12 Study Guide

BIGIDEAS
- THE MOLE AND QUANTIFYING MATTER
- REACTIONS

Balanced chemical equations are the basis for stoichiometric calculations. The coefficients of a balanced equation indicate the number of particles, moles, or volumes of gas in the reaction. Mole ratios from the balanced equation are used to calculate the amount of a reactant or product in a chemical reaction from a given amount of one of the reactants or products.

12.1 The Arithmetic of Equations

Chemists use balanced chemical equations as a basis to calculate how much reactant is needed or product is formed in a reaction.

A balanced chemical equation can be interpreted in terms of different quantities, including numbers of atoms, molecules, or moles; mass; and volume.

- stoichiometry (386)

12.2 Chemical Calculations

In chemical calculations, mole ratios are used to convert between a given number of moles of a reactant or product to moles of a different reactant or product.

In a typical stoichiometric problem, the given quantity is first converted to moles. Then, the mole ratio from the balanced equation is used to calculate the moles of the wanted substance. Finally, the moles are converted to any other unit of measurement related to the unit mole.

- mole ratio (390)

Key Equation

mole-mole relationship for $aG \rightarrow bW$:

$$x \text{ mol } G \times \frac{b \text{ mol } W}{a \text{ mol } G} = \frac{xb}{a} \text{ mol } W$$

12.3 Limiting Reagent and Percent Yield

In a chemical reaction, an insufficient quantity of any of the reactants will limit the amount of product that forms.

The percent yield is a measure of the efficiency of a reaction performed in the laboratory.

- limiting reagent (401)
- excess reagent (401)
- theoretical yield (405)
- actual yield (405)
- percent yield (405)

Key Equation

$$\text{percent yield} = \frac{\text{actual yield}}{\text{theoretical yield}} \times 100\%$$

UbD Performance Tasks

CHEMICAL MANUFACTURING Have students write a paragraph describing a hypothetical chemical product that they are manufacturing. Students should describe their product, explain why their process has excess or limiting reagents, and show how they determine the actual, theoretical, and percent yields of their product.

STOICHIOMETRIC SHOWDOWN Write a set of unbalanced chemical equations on the board. Have students balance each equation. Have students determine the number of representative particles and moles, the molar mass of each reactant and product, and the mass of each reactant and product. Then, have students identify all of the mole ratios in each equation.

What's Online

Review and Assessment Resources

CHEMISTRY TUTORIAL Encourage students to use the animated, step-by-step tutorials for Chapter 12. These tutorials cover using balanced equations, calculating the mass of a product, determining the limiting reagent, and calculating the theoretical and percent yields. Each Chemistry Tutorial is followed by a similar practice problem, where the student is guided through the problem-solving process.

CHAPTER PROBLEM SET Have students practice more stoichiometry problems using the Chapter 12 Online Problem Set.

KINETIC ART Direct students to the Kinetic Art animations *Interpreting a Balanced Equation* and *Limiting Reagents*, and have them answer the questions that accompany each animation.

CONCEPTS IN ACTION If students have trouble understanding why all of the elements in an equation must balance, ask them to view the Concepts in Action animation, which presents the material in a familiar context. Have students complete the questions after the animation.

Study Tip

PRIORITIZE Schedule your time realistically. Stick to your deadlines.

Answers

TAKE IT FURTHER

32. 83.5%
33. 57.7%

MATH TUNE-UP

Math Review

Example and sample practice problems for stoichiometry problems can be found on the following pages:

LESSON 12.1 Using a Balanced Equation as a Recipe—page 385; Interpreting a Balanced Chemical Equation—page 388

LESSON 12.2 Calculating Moles of a Product—page 391; Calculating the Mass of a Product—page 393; Calculating Molecules of a Product—page 395; Volume-Volume Stoichiometric Calculations—page 396; Finding the Volume of a Gas Needed for a Reaction—page 398

LESSON 12.3 Determining the Limiting Reagent in a Reaction—page 402; Using a Limiting Reagent to Find the Quantity of a Product—page 403; Calculating the Theoretical Yield of a Reaction—page 406; Calculating the Percent Yield of a Reaction—page 408

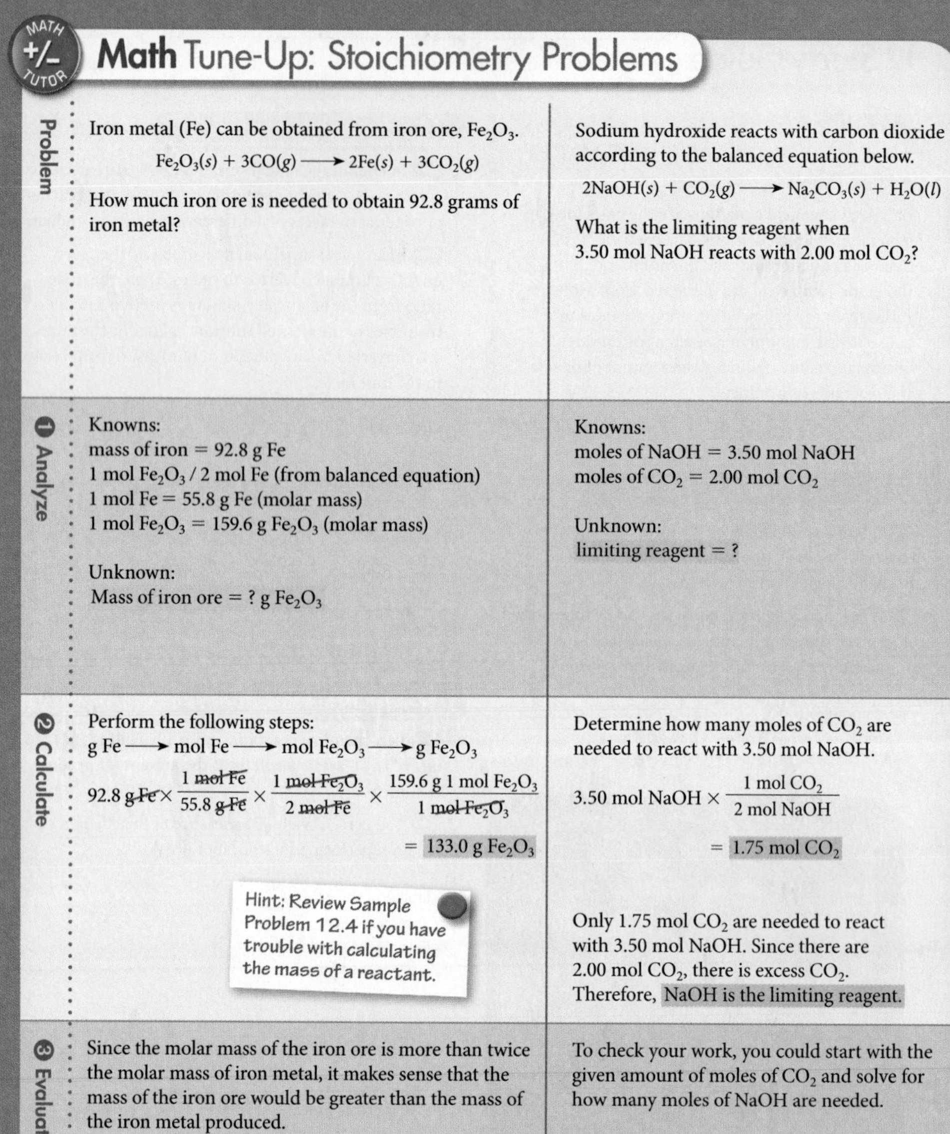

MATH +/– TUTOR

Math Tune-Up: Stoichiometry Problems

Problem	Iron metal (Fe) can be obtained from iron ore, Fe_2O_3. $Fe_2O_3(s) + 3CO(g) \longrightarrow 2Fe(s) + 3CO_2(g)$ How much iron ore is needed to obtain 92.8 grams of iron metal?	Sodium hydroxide reacts with carbon dioxide according to the balanced equation below. $2NaOH(s) + CO_2(g) \longrightarrow Na_2CO_3(s) + H_2O(l)$ What is the limiting reagent when 3.50 mol NaOH reacts with 2.00 mol CO_2?
1 Analyze	Knowns: mass of iron = 92.8 g Fe 1 mol Fe_2O_3 / 2 mol Fe (from balanced equation) 1 mol Fe = 55.8 g Fe (molar mass) 1 mol Fe_2O_3 = 159.6 g Fe_2O_3 (molar mass) Unknown: Mass of iron ore = ? g Fe_2O_3	Knowns: moles of NaOH = 3.50 mol NaOH moles of CO_2 = 2.00 mol CO_2 Unknown: limiting reagent = ?
2 Calculate	Perform the following steps: g Fe $\longrightarrow$ mol Fe $\longrightarrow$ mol Fe_2O_3 $\longrightarrow$ g Fe_2O_3 $92.8\ \cancel{g\ Fe} \times \frac{1\ \cancel{mol\ Fe}}{55.8\ \cancel{g\ Fe}} \times \frac{1\ \cancel{mol\ Fe_2O_3}}{2\ \cancel{mol\ Fe}} \times \frac{159.6\ g\ 1\ mol\ Fe_2O_3}{1\ \cancel{mol\ Fe_2O_3}}$ = 133.0 g Fe_2O_3 Hint: Review Sample Problem 12.4 if you have trouble with calculating the mass of a reactant.	Determine how many moles of CO_2 are needed to react with 3.50 mol NaOH. $3.50\ mol\ NaOH \times \frac{1\ mol\ CO_2}{2\ mol\ NaOH}$ = 1.75 mol CO_2 Only 1.75 mol CO_2 are needed to react with 3.50 mol NaOH. Since there are 2.00 mol CO_2, there is excess CO_2. Therefore, NaOH is the limiting reagent.
3 Evaluate	Since the molar mass of the iron ore is more than twice the molar mass of iron metal, it makes sense that the mass of the iron ore would be greater than the mass of the iron metal produced.	To check your work, you could start with the given amount of moles of CO_2 and solve for how many moles of NaOH are needed. Hint: Review Sample Problem 12.8 if you have trouble identifying the limiting reagent.

Focus on ELL

5 ASSESS UNDERSTANDING Place students in groups of three for an alternative assessment assignment. Provide each group of students with a grading rubric, and explain that they are to depict the *question* and *answer* to **one** key question from **each** lesson in Chapter 12. Suggest the following ideas to the appropriate groups:

BEGINNING

LOW Create cartoon-type drawings.

HIGH Draw pictures with simple labels or captions.

INTERMEDIATE

LOW Design and act out a role-playing scenario.

HIGH Use models or create a diagram and provide simple descriptions.

ADVANCED: LOW/HIGH Create a poster and give an oral presentation to the class.

12 Assessment

★ Solutions appear in Appendix E

Lesson by Lesson

12.1 The Arithmetic of Equations

39. Interpret each chemical equation in terms of interacting particles.

a. $2KClO_3(s) \longrightarrow 2KCl(s) + 3O_2(g)$
b. $4NH_3(g) + 6NO(g) \longrightarrow 5N_2(g) + 6H_2O(g)$
c. $4K(s) + O_2(g) \longrightarrow 2K_2O(s)$

40. Interpret each equation in Problem 39 in terms of interacting numbers of moles of reactants and products.

41. Calculate and compare the mass of the reactants with the mass of the products for each equation in Problem 39. Show that each balanced equation obeys the law of conservation of mass.

42. Balance the following equation:

$$C_5H_{12}(g) + O_2(g) \longrightarrow CO_2(g) + H_2O(g)$$

Interpret the balanced equation in terms of relative number of moles, volumes of gas at STP, and masses of reactants and products.

12.2 Chemical Calculations

43. Explain the term *mole ratio* in your own words. When would you use this term?

44. What ratio is used to carry out each conversion?

a. mol CH_4 to g CH_4
b. L $CH_4(g)$ to mol $CH_4(g)$ (at STP)
c. molecules CH_4 to mol CH_4

★45. Carbon disulfide is an important industrial solvent. It is prepared by the reaction of coke with sulfur dioxide.

$$5C(s) + 2SO_2(g) \longrightarrow CS_2(l) + 4CO(g)$$

a. How many moles of CS_2 form when 2.7 mol C reacts?
b. How many moles of carbon are needed to react with 5.44 mol SO_2?
c. How many moles of carbon monoxide form at the same time that 0.246 mol CS_2 forms?
d. How many mol SO_2 are required to make 118 mol CS_2?

★46. Methanol (CH_3OH) is used in the production of many chemicals. Methanol is made by reacting carbon monoxide and hydrogen at high temperature and pressure.

$$CO(g) + 2H_2(g) \longrightarrow CH_3OH(g)$$

a. How many moles of each reactant are needed to produce 3.60×10^2 g CH_3OH?
b. Calculate the number of grams of each reactant needed to produce 4.00 mol CH_3OH.
c. How many grams of hydrogen are necessary to react with 2.85 mol CO?

47. The reaction of fluorine with ammonia produces dinitrogen tetrafluoride and hydrogen fluoride.

$$5F_2(g) + 2NH_3(g) \longrightarrow N_2F_4(g) + 6HF(g)$$

a. If you have 66.6 g NH_3, how many grams of F_2 are required for a complete reaction?
b. How many grams of NH_3 are required to produce 4.65 g HF?
c. How many grams of N_2F_4 can be produced from 225 g F_2?

48. What information about a chemical reaction is derived from the coefficients in a balanced equation?

49. Rust is produced when iron reacts with oxygen.

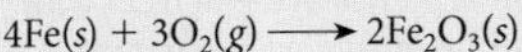

$$4Fe(s) + 3O_2(g) \longrightarrow 2Fe_2O_3(s)$$

How many grams of Fe_2O_3 are produced when 12.0 g of iron rusts?

★50. Lithium nitride reacts with water to form ammonia and aqueous lithium hydroxide.

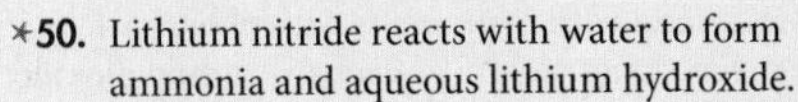

$$Li_3N(s) + 3H_2O(l) \longrightarrow NH_3(g) + 3LiOH(aq)$$

a. What mass of water is needed to react with 32.9 g Li_3N?
b. When the above reaction takes place, how many molecules of NH_3 are produced?
c. Calculate the number of grams of Li_3N that must be added to an excess of water to produce 15.0 L NH_3 (at STP).

Evaluate

ASSESSMENT

Answers

LESSON 12.1

39. a. Two formula units $KClO_3$ decompose to form two formula units KCl and three molecules O_2.
b. Four molecules NH_3 react with six molecules NO to form five molecules N_2 and six molecules H_2O.
c. Four atoms K react with one molecule O_2 to form two formula units K_2O.

40. a. Two mol $KClO_3$ decompose to form two mol KCl and three mol O_2.
b. Four mol NH_3 react with six mol NO to form five mol N_2 and six mol H_2O.
c. Four mol K react with one mol O_2 to form two mol K_2O.

41. a. 245.2 g **b.** 248.0 g **c.** 188.4 g
All obey the law of conservation of mass.

42. 1 mol $C_5H_{12}(g)$ + 8 mol $O_2(g) \rightarrow$ 5 mol CO_2 (g) + 6 mol $H_2O(g)$; 22.4 L $C_5H_{12}(g)$ + 179 L $O_2(g) \rightarrow$ 112 L $CO_2(g)$ + 134 L $H_2O(g)$; 328 g reactants $\rightarrow$ 328 g products

LESSON 12.2

43. Acceptable answers include the idea of writing a ratio using the coefficients of two substances from a balanced equation as the number of moles of each substance reacting or being formed.

44. a. 16.0 g CH_4/1 mol CH_4
b. 1 mol CH_4/22.4 L CH_4
c. 1 mol CH_4/6.02×10^{23} molecules CH_4

45. a. 0.54 mol **b.** 13.6 mol **c.** 0.984 mol
d. 236 mol

46. a. 11.3 mol CO, 22.5 mol H_2
b. 112 g CO, 16.0 g H_2
c. 11.4 g H_2

47. a. 372 g F_2
b. 1.32 g NH_3
c. 123 g N_2F_4

48. The coefficients indicate the relative numbers of moles (or particles) of reactants and products.

49 17.2 g

50. a. 51.2 g H_2O
b. 5.71×10^{23} molecules NH_3
c. 23.2 g Li_3N

ASSESSMENT

Evaluate

Answers

LESSON 12.3

51. The amount of the limiting reagent determines the maximum amount of product that can be formed. The excess reagent is only partially consumed in the reaction.

52. To identify the limiting reagent, express quantities of reactants as moles; compare to the mole ratios from the balanced equation.

53. **a.** Al
b. 3.0 mol $AlCl_3$
c. 0.8 mol Cl_2

54. 91.1%

55. 5.25 mol NaOH required; NaOH is the limiting reagent

UNDERSTAND CONCEPTS

56. **a.** 2.36 g H_3PO_4
b. 1.89 g CO_2

57. **a.** 5.70×10^{21} atoms Zn
b. 95.2 g Zn

58. 87.0%

59. **a.** Initially, the amount of NaCl formed increases as the amount of Na used increases. For this part of the curve, sodium is the limiting reagent. Beyond a mass of about 2.5 g of Na, the amount of product formed remains constant because chlorine is now the limiting reagent.
b. Chlorine becomes the limiting reagent when the mass of sodium exceeds 2.5 g. This corresponds to a mass of about 3.9 g chlorine.

60 **a.** 7.0×10^2 L N_2
b. The excess remaining reagent is 0.2 kg O_2.

12.3 Limiting Reagent and Percent Yield

51. What is the significance of the limiting reagent in a reaction? What happens to the amount of any reagent that is present in an excess?

52. How would you identify a limiting reagent in a chemical reaction?

***53.** In a reaction chamber, 3.0 mol of aluminum is mixed with 5.3 mol Cl_2 and reacts. The following balanced chemical equation describes the reaction:

$$2Al(s) + 3Cl_2(g) \longrightarrow 2AlCl_3(s)$$

a. Identify the limiting reagent for the reaction.
b. Calculate the number of moles of product formed.
c. Calculate the number of moles of excess reagent remaining after the reaction.

***54.** Heating an ore of antimony (Sb_2S_3) in the presence of iron gives the element antimony and iron(II) sulfide.

$$Sb_2S_3(s) + 3Fe(s) \longrightarrow 2Sb(s) + 3FeS(s)$$

When 15.0 g Sb_2S_3 reacts with an excess of Fe, 9.84 g Sb is produced. What is the percent yield of this reaction?

55. Phosphoric acid reacts with sodium hydroxide according to the equation:

$$H_3PO_4(aq) + 3NaOH(aq) \longrightarrow Na_3PO_4(aq) + 3H_2O(l)$$

If 1.75 mol H_3PO_4 is made to react with 5.00 mol NaOH, identify the limiting reagent.

Understand Concepts

56. Calcium carbonate reacts with phosphoric acid to produce calcium phosphate, carbon dioxide, and water.

$$3CaCO_3(s) + 2H_3PO_4(aq) \longrightarrow Ca_3(PO_4)_2(aq) + 3CO_2(g) + 3H_2O(l)$$

a. How many grams of phosphoric acid react with excess calcium carbonate to produce 3.74 g $Ca_3(PO_4)_2$?
b. Calculate the number of grams of CO_2 formed when 0.773 g H_2O is produced.

***57.** Nitric acid and zinc react to form zinc nitrate, ammonium nitrate, and water.

$$4Zn(s) + 10HNO_3(aq) \longrightarrow 4Zn(NO_3)_2(aq) + NH_4NO_3(aq) + 3H_2O(l)$$

a. How many atoms of zinc react with 1.49 g HNO_3?
b. Calculate the number of grams of zinc that must react with an excess of HNO_3 to form 29.1 g NH_4NO_3.

58. If 75.0 g of silderite ore ($FeCO_3$) is heated with an excess of oxygen, 45.0 g of ferric oxide (Fe_2O_3) is produced.

$$4FeCO_3(s) + O_2(g) \longrightarrow 2Fe_2O_3(s) + 4CO_2(g)$$

What is the percent yield of this reaction?

59. In an experiment, varying masses of sodium metal are reacted with a fixed initial mass of chlorine gas. The following graph shows the amounts of sodium used and the amounts of sodium chloride formed.

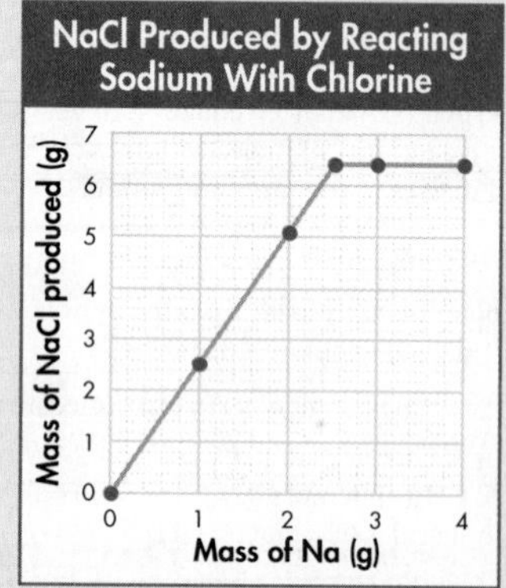

a. Explain the general shape of the graph.
b. Estimate the amount of chlorine gas used in this experiment at the point where the curve becomes horizontal.

***60.** Hydrazine (N_2H_4) is used as rocket fuel. It reacts with oxygen to form nitrogen and water.

$$N_2H_4(l) + O_2(g) \longrightarrow N_2(g) + 2H_2O(g)$$

a. How many liters of N_2 (at STP) form when 1.0 kg N_2H_4 reacts with 1.2 kg O_2?
b. How many grams of the excess reagent remain after the reaction?

61. When 50.0 g of silicon dioxide is heated with an excess of carbon, 32.2 g of silicon carbide is produced.

$$SiO_2(s) + 3C(s) \longrightarrow SiC(s) + 2CO(g)$$

a. What is the percent yield of this reaction?
b. How many grams of CO gas are made?

62. If the reaction below proceeds with a 96.8% yield, how many kilograms of $CaSO_4$ are formed when 5.24 kg SO_2 reacts with an excess of $CaCO_3$ and O_2?

$$2CaCO_3(s) + 2SO_2(g) + O_2(g) \longrightarrow 2CaSO_4(s) + 2CO_2(g)$$

63. Ammonium nitrate will decompose explosively at high temperatures to form nitrogen, oxygen, and water vapor.

$$2NH_4NO_3(s) \longrightarrow 2N_2(g) + 4H_2O(g) + O_2(g)$$

What is the total number of liters of gas formed when 228 g NH_4NO_3 is decomposed? (Assume STP.)

***64.** Hydrogen gas can be made by reacting methane (CH_4) with high-temperature steam:

$$CH_4(g) + H_2O(g) \longrightarrow CO(g) + 3H_2(g)$$

How many hydrogen molecules are produced when 158 g of methane reacts with steam?

65. Suppose hydrogen gas and iodine vapor react to give gaseous hydrogen iodide.

a. Write the balanced equation for the reaction.
b. In the atomic window below, which reactant is the limiting reagent?
c. How many molecules of the reagent in excess remain at the completion of the reaction?
d. How many molecules of the limiting reagent need to be added to the atomic window so that all the reactants will react to form products?

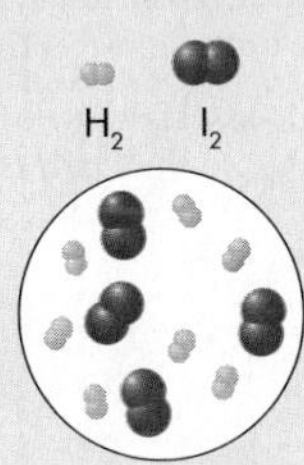

66. The following reaction occurs when an automobile battery is charged.

$$PbSO_4(s) + H_2O(l) \longrightarrow PbO_2(s) + Pb(s) + H_2SO_4(aq)$$

a. Balance the equation.
b. How many grams of sulfuric acid are produced when 68.1 g of lead(II) sulfate react?

***67.** Liquid sulfur difluoride reacts with fluorine gas to form gaseous sulfur hexafluoride.

a. Write the balanced equation for the reaction.
b. How many fluorine molecules are required to react with 5.00 mg of sulfur difluoride?
c. What volume of fluorine gas at STP is required to react completely with 6.66 g of sulfur difluoride?

68. Ammonia (NH_3) reacts with oxygen (O_2) to produce nitrogen monoxide (NO) and water.

$$4NH_3(g) + 5O_2(g) \longrightarrow 4NO(g) + 6H_2O(l)$$

How many liters of NO are produced when 1.40 L of oxygen reacts with ammonia?

***69.** The manufacture of compound F requires five separate chemical reactions. The initial reactant, compound A, is converted to compound B, compound B is converted to compound C, and so on. The diagram below summarizes the steps in the manufacture of compound F, including the percent yield for each step. Provide the missing quantities or missing percent yields. Assume that the reactant and product in each step react in a one-to-one mole ratio.

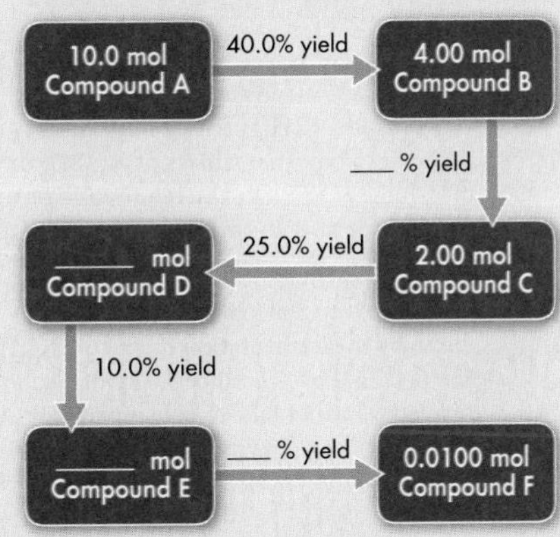

61. a. 96.4%
b. 44.9 g
62. 10.7 kg $CaSO_4$
63. 223 L gas
64. 1.78×10^{25} molecules H_2
65. a. $H_2(g) + I_2(g) \rightarrow 2HI(g)$
b. I_2
c. 3 molecules H_2
d. 3 molecules of I_2
66. a. $2PbSO_4(s) + 2H_2O(l) \rightarrow PbO_2(s) + Pb(s) + 2H_2SO_4(aq)$
b. 22.1 g H_2SO_4
67. a. $SF_2(l) + 2F_2(g) \rightarrow SF_6(g)$
b. 8.59×10^{19} molecules
c. 4.26 L F_2
68. 1.12 L NO
67. a. Initially, the amount of NaCl formed increases as the amount of Na used increases. For this part of the curve, sodium is the limiting reagent. Beyond a mass of about 2.5 g of Na, the amount of product formed remains constant because chlorine is now the limiting reagent.
b. Chlorine becomes the limiting reagent when the mass of sodium exceeds 2.5 g. This corresponds to a mass of about 3.9 g chlorine.
69. 50.0% yield; 0.500 mol; 0.0500 mol; 20.0% yield

Stoichiometry 413

ASSESSMENT

THINK CRITICALLY

70. The percent yield is 115%; such a yield could be attributed to experimenter error, to unreacted starting material, or to outside materials contaminating the product.

71. Yes, a net ionic equation is balanced and thus obeys the law of conservation of mass.

72. **a.** $2Ca_3(PO_4)_2 + 6SiO_2 \rightarrow P_4O_{10} + 6CaSiO_3$
$P_4O_{10} + 10C \rightarrow P_4 + 10CO$
b. SiO_2
c. 7.9×10^4 g P_4
d. 7.7×10^4 g C

73. **a.** $H_2SO_4 + Ca(OH)_2 \rightarrow CaSO_4 + 2H_2O$
b. 2.26 g H_2SO_4

74. 1.1×10^6 L air

75. 13 days

76. **a.** 29 frames
b. 58 wheels
c. 174 pedals
d. 87 seats

ENRICHMENT

77. 1.86 g/L

78. 87.4% $CaCO_3$

79. KOH is the limiting reagent.

Think Critically

70. Evaluate Given a certain quantity of reactant, you calculate that a particular reaction should produce 55 g of a product. When you perform the reaction, you find that you have produced 63 g of product. What is your percent yield? What could have caused a percent yield greater than 100 percent?

71. Explain Would the law of conservation of mass hold in a net ionic equation? Explain.

***72. Calculate** The element phosphorus is manufactured from a mixture of phosphate rock ($Ca_3(PO_4)_2$), sand (SiO_2), and coke (C) in an electric furnace. The chemistry is complex but is summarized by these two equations.

$$Ca_3(PO_4)_2 + SiO_2 \longrightarrow P_4O_{10} + CaSiO_3$$

$$P_4O_{10} + C \longrightarrow P_4 + CO$$

An excess of coke is reacted with 5.5×10^5 g of calcium phosphate and 2.3×10^5 g of sand.

a. Balance each of the equations.
b. What is the limiting reagent?
c. How many grams of phosphorus are produced?
d. How many grams of carbon are consumed?

73. Calculate Sulfuric acid reacts with calcium hydroxide to form calcium sulfate and water.

a. Write the balanced equation for the reaction.
b. Find the mass of unreacted starting material when 75.0 g sulfuric acid reacts with 55.0 g calcium hydroxide.

74. Apply Concepts A car gets 9.2 kilometers to a liter of gasoline. Assuming that gasoline is 100% octane (C_8H_{18}), which has a density of 0.69 g/cm^3, how many liters of air (21% oxygen by volume at STP) will be required to burn the gasoline for a 1250-km trip? Assume complete combustion.

***75. Calculate** Ethyl alcohol (C_2H_5OH) can be produced by the fermentation of glucose ($C_6H_{12}O_6$). If it takes 5.0 h to produce 8.0 kg of alcohol, how many days will it take to consume 1.0×10^3 kg of glucose? (An enzyme is used as a catalyst.)

$$C_6H_{12}O_6 \xrightarrow{\text{enzyme}} 2C_2H_5OH + 2CO_2$$

76. Calculate A bicycle built for three has a frame, two wheels, six pedals, and three seats. The balanced equation for this bicycle is

$$F + 2W + 6P + 3S \longrightarrow FW_2P_6S_3$$

How many of each part are needed to make 29 bicycles built for three?

a. frames
b. wheels
c. pedals
d. seats

Enrichment

77. Calculate A 1004.0-g sample of $CaCO_3$ that is 95.0% pure gives 225 L CO_2 at STP when reacted with an excess of hydrochloric acid.

$$CaCO_3 + 2HCl \longrightarrow CaCl_2 + CO_2 + H_2O$$

What is the density (in g/L) of the CO_2?

***78. Calculate** The white limestone cliffs of Dover, England, contain a large percentage of calcium carbonate ($CaCO_3$). A sample of limestone with a mass of 84.4 g reacts with an excess of hydrochloric acid to form calcium chloride.

$$CaCO_3 + 2HCl \longrightarrow CaCl_2 + H_2O + CO_2$$

The mass of calcium chloride formed is 81.8 g. What is the percentage of calcium carbonate in the limestone?

79. Calculate For the reaction below there are 100.0 g of each reactant available. Which reactant is the limiting reagent?

$$2MnO_2 + 4KOH + O_2 + Cl_2 \longrightarrow 2KMnO_4 + 2KCl + 2H_2O$$

414 Chapter 12 • Assessment

80. Calculate The equation for one of the reactions in the process of reducing iron ore to the metal is

$$Fe_2O_3(s) + 3\,CO(g) \longrightarrow 2\,Fe(s) + 3\,CO_2(g)$$

a. What is the maximum mass of iron, in grams, that can be obtained from 454 g (1.00 lb) of iron(III) oxide?

b. What mass of CO is required to reduce the iron(III) oxide to iron metal?

81. Calculate Esters are a class of compounds that impart a characteristic odor to some fruits. The ester pentyl acetate, composed of carbon, hydrogen, and oxygen, has the odor of bananas. When 7.44 g of this compound undergoes complete combustion, 17.6 g CO_2 and 7.21 g H_2O are produced.

a. What is the empirical formula of pentyl acetate? (*Hint:* All the carbon ends up in the CO_2; all the hydrogen ends up in the H_2O.)

b. The molar mass of pentyl acetate is 130.0 g. What is the molecular formula of this compound?

c. Write the equation for the complete combustion of pentyl acetate.

d. Check your work by using your equation from part c to calculate the grams of CO_2 and H_2O produced by the complete combustion of 7.44 g of pentyl acetate.

*__82. Calculate__ Nitric acid, HNO_3, is produced in a complex three-step process summarized by these unbalanced equations.

Step 1: $NH_3 + O_2 \longrightarrow NO + H_2O$

Step 2: $NO + O_2 \longrightarrow NO_2$

Step 3: $NO_2 + H_2O \longrightarrow HNO_3 + NO$

Notice that the nitric oxide, NO, produced in Step 3 is recycled into Step 2.

a. Balance each of the equations.

b. Assuming all the nitrogen from the ammonia will eventually be incorporated into the nitric acid, calculate the mass of nitric acid obtained from 88.0 g NH_3.

c. The concentrated nitric acid used in the lab is a 70.0% by mass solution of HNO_3 in water. Using your answer from part b, calculate the mass of ammonia needed to prepare 1.00 kg of concentrated nitric acid.

83. Calculate SO_3 can be produced in the following two-step process:

$$FeS_2 + O_2 \longrightarrow Fe_2O_3 + SO_2$$

$$SO_2 + O_2 \longrightarrow SO_3$$

Assuming that all the FeS_2 reacts, how many grams of SO_3 are produced when 20.0 g of the FeS_2 reacts with 16.0 g of O_2?

Write About Science

84. Explain Explain this statement: "Mass and atoms are conserved in every chemical reaction, but moles are not necessarily conserved."

85. Explain Review the "mole road map" at the end of Lesson 10.2. Explain how this road map ties into the summary of steps for stoichiometric problems shown in Figure 12.5.

CHEMYSTERY

Cookie Crumbles

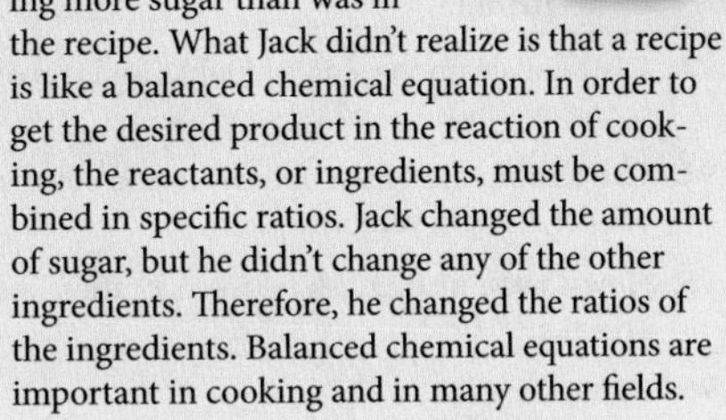

Jack tried to make cookies that were extra sweet by adding more sugar than was in the recipe. What Jack didn't realize is that a recipe is like a balanced chemical equation. In order to get the desired product in the reaction of cooking, the reactants, or ingredients, must be combined in specific ratios. Jack changed the amount of sugar, but he didn't change any of the other ingredients. Therefore, he changed the ratios of the ingredients. Balanced chemical equations are important in cooking and in many other fields.

86. Infer If Jack's recipe calls for 2.5 cups of flour and 2 eggs, and Jack wants to scale up the recipe by 50 percent, then how much flour and eggs will he need?

87. Connect to the BIGIDEA How does Jack's baking experience illustrate the concept of a limiting reagent?

80. a. 347 g Fe
b. 239 g CO

81. a. $C_7H_{14}O_2$
b. $C_7H_{14}O_2$
c. $2\,C_2H_{14}O_2 + 9O_2 \rightarrow 4CO_2 + 14H_2O$
d. 17.6 g CO_2 and 7.21 g H_2O are produced

82. a. $4NH_3 + 5O_2 \rightarrow 4NO + 6H_2O$
$2NO + O_2 \rightarrow 2NO_2$
$3NO_2 + H_2O \rightarrow 2HNO_3 + NO$
b. 326 g HNO_3
c. 189 g NH_3

83. 6.51 g SO_3

WRITE ABOUT SCIENCE

84. Moles of reactants might combine, recombine in different ratios, or decompose into components. Therefore, a mole of reactant does not necessarily translate into a mole of product.

85. Figure 10.8, the "mole road map," describes how the mass, volume, and number of representative particles of a substance can be converted into moles (and vice versa). These relationships apply to six of the calculations shown in Figure 12.5; the seventh is a mole-mole conversion, based on a mole ratio.

CHEMYSTERY **START A CONVERSATION** Ask students if anyone watches cooking shows on television. Ascertain if they have seen a chef on one of these shows cook a dish without measuring ingredients. *(Sample answers: chefs often use approximate measurements, "eyeball" ingredients, or add pinches of seasonings to meats, vegetables, and other savory dishes.)* **Ask** Do these same chefs use any approximate measurements when baking? If so, which ingredients do the chefs always make sure to measure? *(They always make sure to measure butter and other fats, sugar, flour, eggs, and leavenings such as baking soda and baking powder. They may use approximate measurements for salt and flavorings.)* Point out that the ingredients which are measured are those that must be present in specific ratios in order for the finished product to turn out successfully. For example, cookies need to have butter and sugar present in a different ratio than a cake does. Explain that when a chef creates a new recipe for a baked good, he or she uses science and math to determine the amounts of the basic ingredients in the new creation. Then, he or she is free to use creativity to develop a distinctive new flavor. Knowing baking ratios allows television chefs who compete in cook-offs to confidently create a cake, a custard, cookies, a pastry, or a bread without referring to a recipe.

CHEMYSTERY ANSWERS

86. 3.75 cups flour 3 eggs

87. **BIGIDEA** When Jack added more sugar than the recipe called for, the sugar became an excess reagent and all the other ingredients were limiting reagents.

Evaluate

Answers

88. a. 22, 22, 25
b. 50, 50, 70
c. 8, 8, 10
d. 12, 12, 14

89. a. ultraviolet
b. ultraviolet
c. ultraviolet

90. a. sodium
b. arsenic
c. cesium

91. a. Cs·
b. :Br·
c. ·Ca·
d. ·P·

92. c and d

93. single bond, one pair of shared electrons; double bond, two pairs of shared electrons; triple bond, three pairs of shared electrons

94. Yes, an ionic compound with at least one polyatomic ion has covalent bonds.

95. A cation has a positive charge, and an anion has a negative charge.

96. a. phosphate ion
b. aluminum ion
c. selenide ion
d. ammonium ion

97. a. silicon dioxide
b. potassium sulfate
c. carbonic acid
d. magnesium sulfide

98. a. $Al_2(CO_3)_3$
b. NO_2
c. K_2S
d. $MnCrO_4$
e. NaBr

99. 7.38 g Be

100. 1.30×10^{-22} g

101. $C_2H_2O_4$

102. a. 0.473 mol KNO_3
b. 9.91×10^{-2} mol SO_2
c. 3.74×10^{-2} mol PCl_3

103. a. $2Pb(NO_3)_2 \rightarrow 2PbO + 4NO_2 + O_2$
b. $2C_3H_7OH + 9O_2 \rightarrow 6CO_2 + 8H_2O$
c. $2Al + 3FeO \rightarrow 3Fe + Al_2O_3$

104. a. 1, 1, 1, 2
b. 1, 3, 3, 1
c. 1, 1, 1, 2

105. a. $Ba^{2+}(aq) + SO_4^{2-}(aq) \rightarrow BaSO_4(s)$
b. $Ag^+(aq) + Cl^-(aq) \rightarrow AgCl(s)$
c. $H^+(aq) + OH^-(aq) \rightarrow H_2O(l)$

106. a. sodium ion and nitrate ion
b. aluminum ion and nitrate ion
c. magnesium ion and sulfate ion

107. $C_5H_{10}O_5 + 5O_2 \rightarrow 5CO_2 + 5H_2O$

Cumulative Review

*88. How many electrons, protons, and neutrons are in an atom of each isotope?
a. titanium-47
b. tin-120
c. oxygen-18
d. magnesium-26

89. When comparing ultraviolet and visible electromagnetic radiation, which has
a. a higher frequency?
b. a higher energy?
c. a shorter wavelength?

90. Identify the larger atom of each pair.
a. sodium and chlorine
b. arsenic and nitrogen
c. fluorine and cesium

91. Write electron dot formulas for the following atoms:
a. Cs
b. Br
c. Ca
d. P

92. Which of these elements form ions with a 2+ charge?
a. potassium
b. sulfur
c. barium
d. magnesium

93. Distinguish among single, double, and triple covalent bonds.

94. Can a compound have both ionic and covalent bonds? Explain your answer.

95. How do you distinguish between a cation and an anion?

96. Name these ions.
a. PO_4^{3-}
b. Al^{3+}
c. Se^{2-}
d. NH_4^+

97. Name each substance.
a. SiO_2
b. K_2SO_4
c. H_2CO_3
d. MgS

98. Write the formula for each compound.
a. aluminum carbonate
b. nitrogen dioxide
c. potassium sulfide
d. manganese(II) chromate
e. sodium bromide

*99. How many grams of beryllium are in 147 g of the mineral beryl ($Be_3Al_2Si_6O_{18}$)?

100. What is the mass, in grams, of a molecule of benzene (C_6H_6)?

*101. What is the molecular formula of oxalic acid, molar mass 90 g/mol? Its percent composition is 26.7% C, 2.2% H, and 71.1% O.

102. How many moles is each of the following?
a. 47.8 g KNO_3
b. 2.22 L SO_2 (at STP)
c. 2.25×10^{22} molecules PCl_3

103. Write a balanced chemical equation for each reaction.
a. When heated, lead(II) nitrate decomposes to form lead(II) oxide, nitrogen dioxide, and molecular oxygen.
b. The complete combustion of isopropyl alcohol (C_3H_7OH) produces carbon dioxide and water vapor.
c. When a mixture of aluminum and iron(II) oxide is heated, metallic iron and aluminum oxide are produced.

104. Balance each equation.
a. $Ba(NO_3)_2(aq) + Na_2SO_4(aq) \longrightarrow BaSO_4(s) + NaNO_3(aq)$
b. $AlCl_3(aq) + AgNO_3(aq) \longrightarrow AgCl(s) + Al(NO_3)_3(aq)$
c. $H_2SO_4(aq) + Mg(OH)_2(aq) \longrightarrow MgSO_4(aq) + H_2O(l)$

105. Write a net ionic equation for each reaction in Problem 104.

106. Identify the spectator ions in each reaction in Problem 104.

107. Write a balanced chemical equation for the complete combustion of ribose, $C_5H_{10}O_5$.

If You Have Trouble With . . .

Question	88	89	90	91	92	93	94	95	96	97	98	99	100	101	102	103	104	105	106	107
See Chapter	4	5	6	7	7	8	8	9	9	9	9	10	10	10	10	11	11	11	11	11

Standardized Test Prep for Indiana

Standards: C.1.6, C.4.2, C.4.3, C.4.4

Tips for Success

Anticipate the answer. Use what you know to predict what you think the answer should be. Then look to see if your answer, or one much like it, is given as an option.

Select the choice that best answers each question or completes each statement.

1. Nitric acid is formed by the reaction of nitrogen dioxide with water.

   $$3NO_2(g) + H_2O(l) \longrightarrow NO(g) + 2HNO_3(aq)$$

   How many moles of water are needed to react with 8.4 mol NO_2?

   (A) 2.8 mol **(C)** 8.4 mol
   (B) 3.0 mol **(D)** 25 mol

2. Phosphorus trifluoride is formed from its elements.

   $$P_4(s) + 6F_2(g) \longrightarrow 4PF_3(g)$$

   How many grams of fluorine are needed to react with 6.20 g of phosphorus?

   (A) 2.85 g **(C)** 11.4 g
   (B) 5.70 g **(D)** 37.2 g

3. Magnesium nitride is formed in the reaction of magnesium metal with nitrogen gas.

   $$3Mg(s) + N_2(g) \longrightarrow Mg_3N_2(s)$$

   The reaction of 4.0 mol of nitrogen with 6.0 mol of magnesium produces

   (A) 2.0 mol of Mg_3N_2 and no excess N_2.
   (B) 2.0 mol of Mg_3N_2 and 2.0 mol of excess N_2.
   (C) 4.0 mol of Mg_3N_2 and 1.0 mol of excess Mg.
   (D) 6.0 mol of Mg_3N_2 and 3.0 mol of excess N_2.

Use the reaction below to answer Questions 4 and 5.

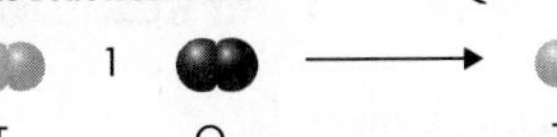

4. Write a balanced equation for the reaction between element T and element Q.

5. Based on the atomic windows below, identify the limiting reagent.

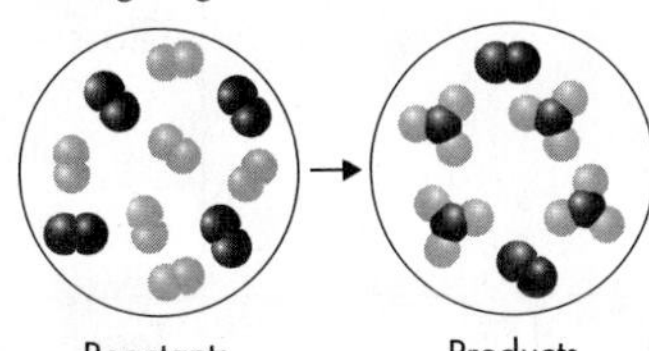

For each question, there are two statements. Decide whether each statement is true or false. Then decide whether Statement II is a correct explanation for Statement I.

	Statement I		Statement II
6.	Every stoichiometry calculation uses a balanced equation.	BECAUSE	Every chemical reaction obeys the law of conservation of mass.
7.	A percent yield is always greater than 0% and less than 100%.	BECAUSE	The actual yield in a reaction is never more than the theoretical yield.
8.	The amount of the limiting reagent left after a reaction is zero.	BECAUSE	The limiting reagent is completely used up in a reaction.
9.	The coefficients in a balanced equation represent the relative masses of the reactants and products.	BECAUSE	The mass of the reactants must equal the mass of the products in a chemical reaction.
10.	A mole ratio is always written with the larger number in the numerator.	BECAUSE	A mole ratio will always be greater than 1.

If You Have Trouble With . . .

Question	1	2	3	4	5	6	7	8	9	10
See Lesson	12.2	12.2	12.3	12.1	12.3	12.1	12.3	12.3	12.1	12.2

STP Answers

1. A
2. C
3. B
4. $3T_2 + Q_2 \rightarrow 2T_3Q$
5. T_2 is the limiting reagent.
6. True, True, correct explanation
7. False, True
8. True, True, correct explanation
9. False, True
10. False, False

13 States of Matter Planning Guide

Introducing the BIGIDEA: KINETIC THEORY

The ways in which particle motions vary with changes in temperature and pressure determine whether a substance will be a solid, liquid, or gas.

IASS	Lessons and Objectives	Print Resources		
		For the Student	For the Teacher	
C.1.5, C.6.1, C.6.2	**13.1** The Nature of Gases p 420–424 **13.1.1** Describe the three assumptions of the kinetic theory as it applies to gases. **13.1.2** Interpret gas pressure in terms of kinetic theory. **13.1.3** Define the relationship between the temperature in kelvins and the average kinetic energy of particles.	Reading and Study Workbook Lesson 13.1 Lesson Assessment 13.1 p 424	Teacher Demo, p 421: Elastic Collisions Teacher Demo, p 422: Air Pressure Class Activity, p 423: Kinetic Energy and Frequency of Collision	
C.1.5, C.3.3, C.6.1, C.6.2	**13.2** The Nature of Liquids p 425–430 **13.2.1** Identify the factors that determine physical properties of a liquid. **13.2.2** Define evaporation in terms of kinetic energy. **13.2.3** Define the conditions under which a dynamic equilibrium can exist between a liquid and its vapor. **13.2.4** Identify the conditions under which boiling occurs.	Reading and Study Workbook Lesson 13.2 Lesson Assessment 13.2 p 430	Teacher Demo, p 427: Vapor Pressure Teacher Demo, p 428: Comparing Vapor Pressures of Water and Alcohol Class Activity, p 429: Temperature and Boiling	
C.1.5, C.3.3, C.6.1	**13.3** The Nature of Solids p 431–434 **13.3.1** Describe how the structure and properties of solids are related. **13.3.2** Identify the factors that determine the shape of a crystal.	Reading and Study Workbook Lesson 13.3 Lesson Assessment 13.3 p 434 Small-Scale Lab: The Behavior of Liquids and Solids p 435	Class Activity, p 432: Wallpaper Lattices Teacher Demo, p 433: Crystalline Solid Model	
C.1.5	**13.4** Changes of State p 436–439 **13.4.1** Identify the conditions necessary for sublimation. **13.4.2** Determine how the conditions at which phases are in equilibrium are represented on a phase diagram.	Reading and Study Workbook Lesson 13.4 Lesson Assessment 13.4 p 439 Quick Lab: Sublimation p 437		

Assessing the BIGIDEA: KINETIC THEORY

	Essential Questions 1. What factors determine the physical state of a substance? 2. What are the characteristics that distinguish gases, liquids, and solids? 3. How do substances change from one state to another?	Study Guide p 442 STP p 447 Reading and Study Workbook Self-Check and Vocabulary Review Chapter 13		

	PearsonChem.com	
	Editable Worksheets	Online Activities
	Teaching Resources, Lesson 13.1 Review	13.1 Lesson Overview Converting Between Units of Pressure
	Teaching Resources, Lesson 13.2 Review	13.2 Lesson Overview Effects of Ionic Compounds on the Boiling and Freezing Points of Water Evaporation
	Teaching Resources, Lesson 13.3 Review Lab 21: Allotropic Forms of Sulfur	13.3 Lesson Overview Brick and Mortar Crystal Palaces
	Teaching Resources, Lesson 13.4 Review Lab 22: Changes of Physical State Lab Practical 13-1: Changes of Physical State	13.4 Lesson Overview Change in State
	Core Teaching Resources (includes editable worksheets) • Lesson Reviews • Practice Problems • Interpret Graphs • Vocabulary Review • Chapter Quizzes and Tests • Lab Record Sheets	Chapter 13 Problem Set Reading Graphs

Materials List

For the Student

Small-Scale Lab p 435
- plastic Petri dish
- water
- ice
- rubbing alcohol
- calcium chloride
- graph paper (1-cm cycle)
- bromothymol blue solution
- vinegar
- aqueous ammonia

Quick Lab p 437
- small pieces of solid air freshener
- small shallow container
- 2 clear plastic cups (8 oz each)
- hot tap water
- ice
- 3 thick cardboard strips

For the Teacher

Teacher Demo p 421
- Newtonian cradle

Teacher Demo p 422
- empty aluminum beverage can
- water
- hot plate
- foil or other material for sealing opening of can

Class Activity p 423
- cardboard containers (such as half-pint milk cartons)
- ball bearings

Teacher Demo p 427
- manometer
- 1-L flask with 2-hole stopper
- separatory funnel
- short piece of glass tubing bent at a 90° angle
- rubber hose
- 50 mL acetone
- ice water

Teacher Demo p 428
- 2 cotton balls
- water
- rubbing alcohol

Class Activity p 429
- thermometer
- beaker
- water
- ring stand
- thermometer clamp
- hot plate

Class Activity p 432
- samples of wallpaper with repeating patterns

Teacher Demo p 433
- dishwashing detergent
- watch glass
- overhead projector
- drinking straw

Additional Digital Resources

Online Student Edition
Online Teacher's Edition
Exam*View Assessment Suite CD-ROM*
Classroom Resources DVD

CHAPTER 13

What's Online

CHEMISTRY TUTORIAL Students can work step-by-step with the online tutorial to learn how to convert between different units of pressure by accessing the Chapter 13 Chemistry Tutorial.

ONLINE PROBLEM SET Additional practice problems about states of matter can be found in the Chapter 13 online Problem Set.

MATH TUTORIAL The online Math Tutorial will help guide students through reading graphs.

DIRECTED VIRTUAL LAB Encourage students to review the effect of dissolving ionic compounds in water to change boiling and freezing points using the animated virtual lab for Chapter 13. The animation is followed by analysis questions.

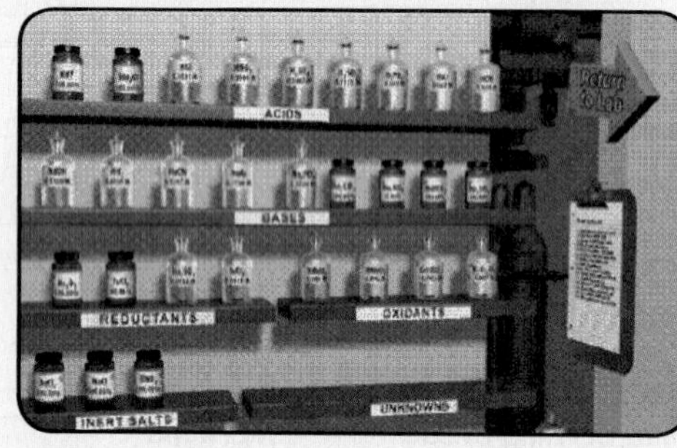

KINETIC ART Have students watch animations about evaporation and changes in state. Each animation is followed by analysis questions.

CONCEPTS IN ACTION Students can watch an interactive animation about how bricks are like unit cells of ionic crystals. The animation is followed by questions to clarify how unit cells affect the overall crystal structure.

IN ACADEMIC STANDARDS for SCIENCE

Standard 1: Properties and States of Matter

C.1.5 Describe the characteristics of solids, liquids, and gases and state changes at the molecular level.

Standard 6: Thermochemistry

C.6.1 Explain that atoms and molecules that make up matter are in constant motion and that this motion results in collisions between molecules and subsequent energy transfer between them.

Also covered C.6.2.

13 States of Matter

INSIDE:

PearsonChem.com

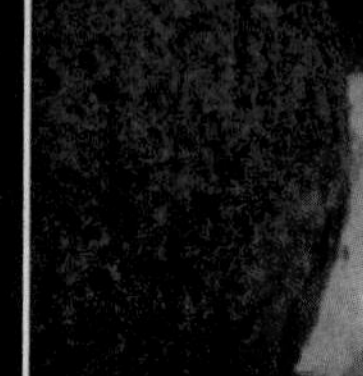

Hot, liquid lava flows from a volcano. When it cools to a solid, new rock will be formed.

Focus on ELL

1 CONTENT AND LANGUAGE Have students follow along as you read the lesson titles and Essential Questions out loud. Then write the word *nature* on the board and explain that it has many meanings. Have students use the lesson titles and Essential Questions to formulate their own definition of *nature* as it is used in the lesson titles.

BEGINNING: LOW/HIGH If literate in students' native language, translate the lesson titles and Essential Questions. Provide students with a simplified English definition of the term *nature*.

INTERMEDIATE: LOW/HIGH Have students use context clues in the lesson titles and Essential Questions to develop their own definition for the term *nature*.

ADVANCED: LOW/HIGH Have students look up the definitions of the term *nature* in the dictionary. Have them use context clues from the lesson titles and Essential Questions to determine which definition is best suited for the term in this context.

BIGIDEA

KINETIC THEORY

Essential Questions:

1. *What factors determine the physical state of a substance?*
2. *What are the characteristics that distinguish gases, liquids, and solids?*
3. *How do substances change from one state to another?*

CHEMYSTERY

Foggy Car Windows

It's a cold, rainy day in September, and you and a friend are heading out to a movie. When you first get into your mom's car, you can clearly see nearby trees swaying in the wind. But shortly after your mom starts the car, the glass fogs up, making it almost impossible to see outside. Your mom sighs, and turns on the heat, which only makes the foggy windows worse. Then she turns on the air conditioner, and the fog is gone in seconds. Why do car windows fog up when it is cold or raining outside? Why does the fog go away when you turn on the air conditioner?

▶ Connect to the BIGIDEA As you read about states of matter, think about what might cause car windows to fog.

INDIANA'S ACADEMIC STANDARDS FOR SCIENCE

Properties and States of Matter C.1.5; **Thermochemistry** C.6.1, C.6.2. See lessons for details.

Introduce the Chapter

IDENTIFYING PRECONCEPTIONS Students may have not have considered that substances can change from one state to another at temperatures other than their boiling points and melting points and that these points can vary. Use the activities to introduce these concepts.

Activity 1 *You will need pea-sized bits of dry ice, a beaker with water, insulated gloves, and tongs.* As students watch, use tongs to drop bits of dry ice into the water. Students will see the bits swirl as the dry ice becomes warmer and forms carbon dioxide gas. Explain that the solid is changing to a gas by sublimation.

Activity 2 Ask At what temperature does liquid water change to water vapor? *(Students may say 100°C.)* Point out that water also changes to a gas by evaporation at any temperature. Have students slightly dampen a sheet of paper and observe it after a day. The paper will be dry because the water evaporated.

CHAPTER 13

UbD

Understanding By Design

Students are building toward understanding states of matter by relating them to the kinetic theory of matter.

PERFORMANCE GOALS At the end of Chapter 13, students will be able to answer the essential questions by applying their knowledge of states of matter.

ESSENTIAL QUESTIONS Read the essential questions aloud. **Ask** Why are solids, liquids, and gases called physical states and not chemical states? *(If a substance changes from one state to another, only its physical properties change.)* **Ask** Is the temperature at which a substance changes from one state to another always the same? Explain. *(No. The temperature at which a substance changes state depends on both temperature and pressure.)*

BIGIDEA Use the photo of a lava flow to help students connect to this chapter's concepts. Activate prior knowledge by discussing the properties of the lava. **Ask** What two states of matter can you see in the photo? *(solid and liquid)* **Ask** How are the properties of the solid and liquid lava different? *(The solid is hard, cannot flow, and has a fixed shape. The liquid is not hard and can flow. Its shape changes and it is hotter than the solid.)*

CHEMYSTERY Connect the CHEMystery to the Big Idea by explaining that the state of matter depends on how the particles of a substance move. Ask students to predict why the fog on car windows goes away when the air conditioner is turned on. As a hint, suggest that students think about how much water the air can hold at different temperatures.

UntamedScience™

Allotrope Applications

Hazen visits a company that is working on building stronger touch screens using carbon nanotube technology.

LESSON 13.1

Key Objectives

13.1.1 **DESCRIBE** the three assumptions of the kinetic theory as it applies to gases.

13.1.2 **INTERPRET** gas pressure in terms of kinetic theory.

13.1.3 **DEFINE** the relationship between the temperature in kelvins and the average kinetic energy of particles.

Additional Resources

Reading and Study Workbook, Lesson 13.1

Available Online or on Digital Media:

- Teaching Resources, Lesson 13.1 Review

PearsonChem.com

Engage

CHEMISTRY & YOU Display a weather map from your local paper and explain what the symbols represent. Ask student volunteers to interpret the map and predict the weather for the week. **Ask** What type of weather can you expect if the map shows a dry, cold air mass moving into your area during the summer? *(cool temperatures, dry air, and clear skies)* Explain that winds often blow from areas of high pressure towards areas of low pressure. **Ask** What causes pockets of air in Earth's atmosphere to have different pressures? *(uneven heating of Earth's surface)*

IN ACADEMIC STANDARDS for SCIENCE

C.1.5 Describe the characteristics of solids, liquids, and gases and state changes at the molecular level.

C.6.1 Explain that atoms and molecules that make up matter are in constant motion and that this motion increases as thermal energy increases.

Also covered C.6.2.

13.1 The Nature of Gases

C.1.5 States of matter; **C.6.1** Explain that atoms and molecules are in constant motion; **C.6.2** Distinguish between temperature and heat flow

CHEMISTRY & YOU

Q: *What factors most strongly affect the weather?* The atmosphere is a gas, and the factors that determine the behavior of gases—temperature and pressure—affect the weather in the atmosphere. That is why weather maps show temperature readings and areas of high and low pressure. In this lesson, you will learn how temperature and pressure affect the particles of a gas.

Key Questions

What are the three assumptions of the kinetic theory as it applies to gases?

How does kinetic theory explain gas pressure?

What is the relationship between the temperature in kelvins and the average kinetic energy of particles?

Vocabulary

- kinetic energy
- kinetic theory
- gas pressure
- vacuum
- atmospheric pressure
- barometer
- pascal (Pa)
- standard atmosphere (atm)

Kinetic Theory and a Model for Gases

What are the three assumptions of the kinetic theory as it applies to gases?

The word *kinetic* refers to motion. The energy an object has because of its motion is called **kinetic energy.** According to the **kinetic theory,** all matter consists of tiny particles that are in constant motion. The particles in a gas are usually molecules or atoms. The kinetic theory as it applies to gases includes the following fundamental assumptions about gases.

The particles in a gas are considered to be small, hard spheres with an insignificant volume. Within a gas, the particles are relatively far apart compared with the distance between particles in a liquid or solid. Between the particles, there is empty space. No attractive or repulsive forces exist between the particles.

The motion of the particles in a gas is rapid, constant, and random. As a result, gases fill their containers regardless of the shape and volume of the containers. An uncontained gas can spread out into space without limit. The particles travel in straight-line paths until they collide with another particle, or another object, such as the wall of their container. The particles change direction only when they rebound from collisions with one another or with other objects.

Measurements indicate that the average speed of oxygen molecules in air at 20°C is an amazing 1700 km/h! At these high speeds, the odor from a hot cheese pizza in Washington, D.C., should reach Mexico City in about 115 minutes. That does not happen, however, because the molecules responsible for the odor are constantly striking molecules in air and rebounding in other directions. Their path of uninterrupted travel in a straight line is very short. The aimless path the molecules take is called a random walk.

All collisions between particles in a gas are perfectly elastic. During an elastic collision, kinetic energy is transferred without loss from one particle to another, and the total kinetic energy remains constant. The diagrams in Figure 13.1 illustrate the assumptions of kinetic theory as applied to gases.

Focus on ELL

1 CONTENT AND LANGUAGE Assign small groups one content vocabulary term. Groups should prepare a learning tool for the class. This tool should show how to pronounce the assigned vocabulary term and provide a meaning for it. Encourage groups to use word-analysis strategies, pronunciation spelling, vowel patterning, sounds-like clues, and illustrations. Groups should present their learning tool to the class and address common misinterpretations and mispronunciations of their assigned word

2 FRONTLOAD THE LESSON Engage students in a discussion of what they already know about gases. Discuss why the intensity of odors vary depending on proximity to the source. Discuss the danger of odorless gases, such as carbon monoxide. Finally, point out that gasoline used in cars is often referred to as *gas,* but it is actually a liquid.

3 COMPREHENSIBLE INPUT Use Figure 13.1 to introduce properties of gases. Explain that by considering the small scale action of individual gas particles students can get a sense of the large scale action of a gas.

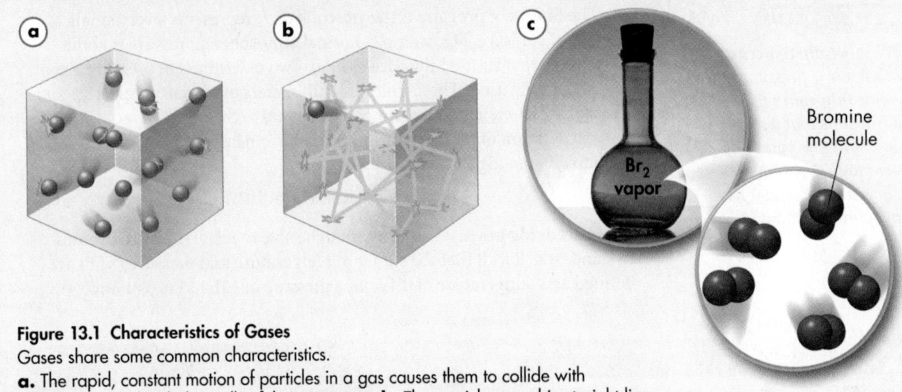

Figure 13.1 Characteristics of Gases
Gases share some common characteristics.
a. The rapid, constant motion of particles in a gas causes them to collide with one another and with the walls of their container. **b.** The particles travel in straight-line paths between collisions. **c.** A gas fills all the available space in its container.
Relate Cause and Effect ***Why don't the particles in a gas eventually slow down and stop?***

Gas Pressure

How does kinetic theory explain gas pressure?

A balloon filled with helium or hot air maintains its shape because of the pressure of the gas within it. **Gas pressure** results from the force exerted by a gas per unit surface area of an object. What causes this force? Moving bodies exert a force when they collide with other bodies. Although a single particle in a gas is a moving body, the force it exerts is extremely small. Yet it is not hard to imagine that simultaneous collisions involving many particles would produce a measurable force on an object. **Gas pressure is the result of billions of rapidly moving particles in a gas simultaneously colliding with an object.** If no particles are present, no collisions can occur. Consequently, there is no pressure. An empty space with no particles and no pressure is called a **vacuum.**

You are already familiar with a gas pressure caused by a mixture of gases—air. Air exerts pressure on Earth because gravity holds the particles in air within Earth's atmosphere. The collisions of atoms and molecules in air with objects results in **atmospheric pressure.** Atmospheric pressure decreases as you climb a mountain because the density of Earth's atmosphere decreases as the elevation increases.

A **barometer** is a device that is used to measure atmospheric pressure. Figure 13.2 shows an early type of mercury barometer. The height of the mercury column in the tube depends on the pressure exerted by particles in air colliding with the surface of the mercury in the dish. Atmospheric pressure depends on weather and on altitude. In fair weather at sea level, the atmospheric pressure is sufficient to support a mercury column 760 mm high.

Figure 13.2 Atmospheric Pressure
At sea level, air exerts enough pressure to support a 760-mm column of mercury. On top of Mount Everest, at 9000 m, the air exerts only enough pressure to support a 253-mm column of mercury.

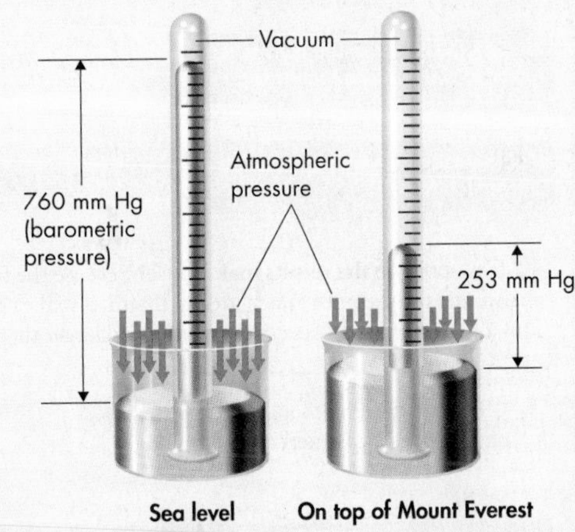

Differentiated Instruction

ELL ENGLISH LANGUAGE LEARNERS Have available the weather section of the local newspaper, and the weather reports from students' native cities. Help students determine the day's pressure, in all locations, in units of mm Hg, kPa, and atm, and to compare data. Students may wish to track the daily pressure and display the changes on a graph.

L1 STRUGGLING STUDENTS To help convey the meaning of the word *kinetic*, show a video of kinetic sculptures created by sculptor Theo Jansen. Challenge students to create a kinetic sculpture that reflects the concepts in this lesson.

L3 ADVANCED STUDENTS Unlike traditional barometers, aneroid barometers do not balance atmospheric pressure against a liquid of known density. In fact, they contain no liquid and are independent of gravity. If possible, bring one to class and explain its components. Have interested students research the advantages of aneroid barometers. *(They are compact, self enclosed, and easily transported. They can operate under conditions not suited to mercury barometers.)*

Explain

Kinetic Theory and a Model for Gases

USE VISUALS Direct students to Figure 13.1. Be sure students understand that the space between the gas particles is not filled with air or any other substance. Help students visualize that the pressure of a gas is caused by collisions of the particles with the walls of the container; point out that the magnitude of the pressure is related to how often and how hard the particles strike the walls.

Explore

Gas Pressure

Teacher Demo

PURPOSE Students will observe and differentiate elastic collisions from perfectly elastic collisions.

MATERIALS Newtonian cradle (a device in which small steel balls are suspended by thin nylon tethers to horizontal wooden sticks)

PROCEDURE Explain that in an elastic collision, kinetic energy is transferred between the objects that collide. In a perfectly elastic collision, the total amount of kinetic energy remains constant. Then pull back one of the balls and let it fall into the other balls. Let the balls come to rest. **Ask** Were the collisions between the balls elastic? *(Yes, because kinetic energy was transferred with each collision.)* **Ask** Why did the balls eventually stop moving? *(The collisions were not perfectly elastic; some kinetic energy was lost as heat during each collision.)* Remind students that the kinetic theory assumes that all collisions between particles in a gas are perfectly elastic. (Note: In Chapter 14, students will learn that the assumptions of the kinetic theory do not hold true for real gases under all conditions.)

Explain

USE VISUALS Direct students to Figure 13.2. Point out the parts of the barometer and explain how the barometer is used to measure atmospheric pressure. **Ask** What is the decrease in pressure from sea level to the top of Mount Everest in kPa? *(507 mm Hg)*

Answers

FIGURE 13.1 The particles in a gas collide elastically meaning that the total amount of kinetic energy remains constant, and the particles do not slow down and stop.

LESSON 13.1

Explore

Gas Pressure

CHEMISTRY & YOU The height of the mercury column in a barometer decreases as a storm approaches because the atmospheric pressure decreases

Sample Practice Problem

A. What pressure, in mm Hg and atm, does a sample of neon gas exert at 75.0 kPa? *(563 mm Hg, 0.740 atm)*

B. What pressure, in mm Hg and kPa, does a sample of argon gas exert at 1.561 atm? *(1186 mm Hg, 158.1 kPa)*

Explore

Teacher Demo

PURPOSE Students will become aware of the tremendous pressure exerted by Earth's atmosphere.

MATERIALS empty aluminum beverage can, water, hot plate, tongs, foil or other material for sealing opening of can

PROCEDURE Explain to students that they live at the bottom of a very deep "ocean" of air. Fill an empty aluminum beverage container with water to a depth of about 1 to 2 cm. Set the can on a hot plate and bring the water to a boil. Allow the water to boil vigorously for a minute. Using tongs, remove the can from the heat source, seal the opening, and allow the can to cool at room temperature or invert the can in a pan of cool water.

EXPECTED OUTCOME Once the water vapor inside the can condenses, the atmospheric pressure will crush the can. Ask students to interpret their observations.

Q: *When weather forecasters state that a low-pressure system is moving into your region, it usually means that a storm is coming. What do you think happens to the column of mercury in a barometer as a storm approaches? Why?*

The SI unit of pressure is the **pascal (Pa).** It represents a very small amount of pressure. For example, normal atmospheric pressure is about 100,000 Pa, that is, 100 kilopascals (kPa). Two older units of pressure are still commonly used. These units are millimeters of mercury (mm Hg) and atmospheres. One **standard atmosphere (atm)** is the pressure required to support 760 mm of mercury in a mercury barometer at 25°C. The numerical relationship among the three units is

$$1 \text{ atm} = 760 \text{ mm Hg} = 101.3 \text{ kPa}$$

When studying gases, it is important to be able to relate measured values to standards. Recall that the standard temperature and pressure (STP) are defined as a temperature of 0°C and a pressure of 101.3 kPa, or 1 atm.

Sample Problem 13.1

Converting Between Units of Pressure

A pressure gauge records a pressure of 450 kPa. Convert this measurement to

a. atmospheres. b. millimeters of mercury.

KNOWNS	UNKNOWNS
pressure = 450 kPa	pressure = ? atm
1 atm = 101.3 kPa	pressure = ? mm Hg
1 atm = 760 mm Hg	

1 Analyze **List the knowns and the unknowns.** The given pressure is converted into the desired unit by multiplying by the proper conversion factor.

2 Calculate **Solve for the unknowns.**

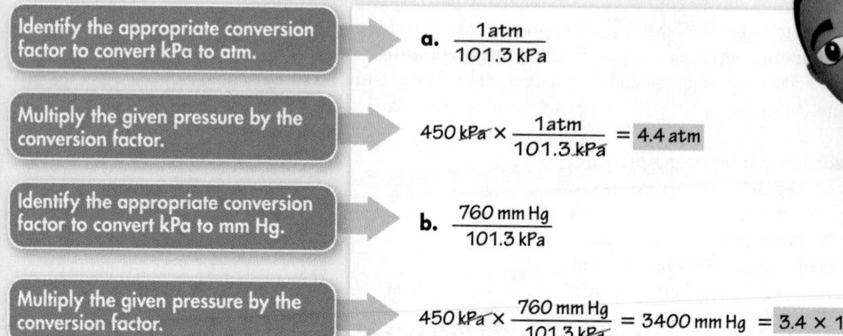

Identify the appropriate conversion factor to convert kPa to atm.

a. $\dfrac{1 \text{ atm}}{101.3 \text{ kPa}}$

Multiply the given pressure by the conversion factor.

$$450 \text{ kPa} \times \frac{1 \text{ atm}}{101.3 \text{ kPa}} = 4.4 \text{ atm}$$

Identify the appropriate conversion factor to convert kPa to mm Hg.

b. $\dfrac{760 \text{ mm Hg}}{101.3 \text{ kPa}}$

Multiply the given pressure by the conversion factor.

$$450 \text{ kPa} \times \frac{760 \text{ mm Hg}}{101.3 \text{ kPa}} = 3400 \text{ mm Hg} = 3.4 \times 10^3 \text{ mm Hg}$$

3 Evaluate **Do the results make sense?** Because the first conversion factor is much less than 1 and the second is much greater than 1, it makes sense that the values expressed in atm and mm Hg are respectively smaller and larger than the value expressed in kPa.

1. What pressure, in kilopascals and in atmospheres, does a gas exert at 385 mm Hg?

2. The pressure at the top of Mount Everest is 33.7 kPa. Is that pressure greater or less than 0.25 atm?

422 Chapter 13 • Lesson 1

Foundations for Math

CHECKING ANSWERS Remind students that scientists check their answers when making conversions to be sure they are reasonable. This is an especially important step when working with measurements that can be converted between more than one acceptable unit of measurement.

In **Sample Problem 13.1,** answers are evaluated in relation to the conversion factor used for each calculation. For part **a**, 450 kPa is being divided by a value of just over 100, so it is reasonable to expect a value approximately equal to 4.5 atm. In part **b**, 450 kPa is being multiplied by a factor of approximately 7.6. Because 7.6 rounds to 8, it is reasonable to expect an answer of approximately 3600 mm Hg. The actual answer, 3.4×10^3 mm Hg, is reasonable because 7.6 is less than 8.

Kinetic Energy and Temperature

What is the relationship between the temperature in kelvins and the average kinetic energy of particles?

As a substance is heated, its particles absorb energy, some of which is stored within the particles. This stored portion of the energy, or potential energy, does not raise the temperature of the substance. The remaining absorbed energy does speed up the particles—that is, increases their kinetic energy. This increase in kinetic energy results in an increase in temperature.

Average Kinetic Energy The particles in any collection of atoms or molecules at a given temperature have a wide range of kinetic energies. Most of the particles have kinetic energies somewhere in the middle of this range. Therefore, we use average kinetic energy when discussing the kinetic energy of a collection of particles in a substance. At any given temperature, the particles of all substances, regardless of physical state, have the same average kinetic energy. For example, the ions in table salt, the molecules in water, and the atoms in helium all have the same average kinetic energy at room temperature, even though the three substances are in different physical states.

Figure 13.3 shows the distribution of kinetic energies of water molecules at two different temperatures. The green curve shows the distribution of kinetic energy among the water molecules in cold water. The purple curve shows the distribution of kinetic energy among the water molecules in hot water. In both cases, most of the molecules have intermediate kinetic energies, which are close to the average value. Notice that molecules at the higher temperature have a wider range of kinetic energies.

The average kinetic energy of the particles in a substance is directly related to the substance's temperature. An increase in the average kinetic energy of the particles causes the temperature of a substance to rise. As a substance cools, the particles tend to move more slowly, and their average kinetic energy decreases.

READING SUPPORT

Build Vocabulary:
Word Origins Kinetic comes from the greek word *kinetos*, meaning "to move." Kinetic energy is the energy an object has because of its motion. ***Some sculptures are kinetic. What characteristic do they share?***

Interpret Graphs

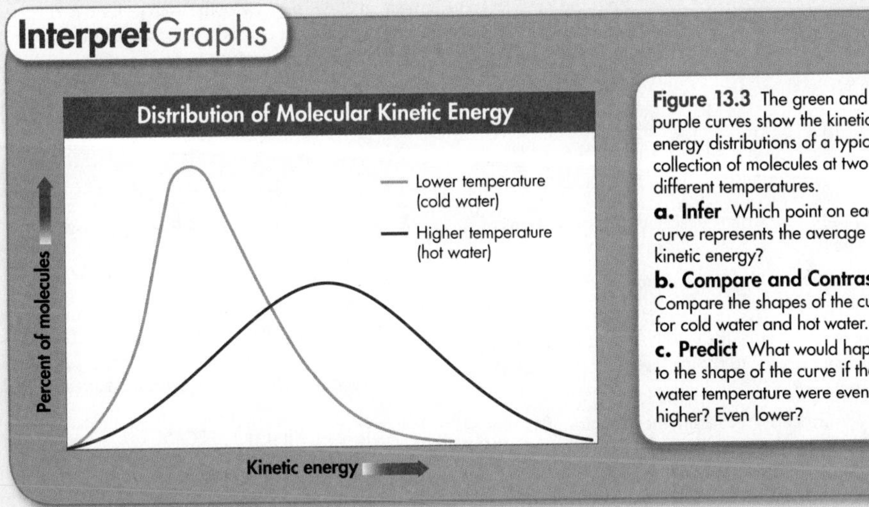

Figure 13.3 The green and purple curves show the kinetic energy distributions of a typical collection of molecules at two different temperatures.
a. Infer Which point on each curve represents the average kinetic energy?
b. Compare and Contrast Compare the shapes of the curves for cold water and hot water.
c. Predict What would happen to the shape of the curve if the water temperature were even higher? Even lower?

UbD Check for Understanding

How does kinetic theory explain gas pressure?

To assess students' understanding of the kinetic theory, ask them to work in pairs and to create a scenario that is analogous to the kinetic theory. Have each pair illustrate in a diagram what happens to "gas particles" in their scenario when conditions such as pressure and temperature change.
(check students' diagram)

ADJUST INSTRUCTION If students have difficulty creating or diagramming their scenarios, have them create a bulleted list of facts about the kinetic theory as it relates to gas pressure and temperature. Then provide students with additional practice where they can reference their list to describe the behavior of gas particles under conditions of varying pressure and temperature such as those described in the text.

Explain

Kinetic Energy and Temperature

CRITICAL THINKING Direct students to Figure 13.3. **Ask** What does the intersection point of the green and purple curves indicate? *(The particles have the same kinetic energy.)*

Explore

Class Activity

PURPOSE Students will be able to visualize what is happening in a gas as the average kinetic energy increases.

MATERIALS cardboard containers (e.g., half-pint milk cartons), ball bearings

PROCEDURE Give students cardboard boxes containing the same number of ball bearings. Have students shake the boxes slowly at first. Then have them gradually increase the rate of shaking and observe what happens. Tell students that the shaking is analogous to adding energy to the particles in a gas. The greater the rate of the shaking, the more energy is added. Then ask how an increase in energy affects the frequency of collisions.

EXPECTED OUTCOME Student should hear and feel the increase in number of collisions as the rate of shaking (or the amount of energy added) increases.

Answers

1. 51.3 kPa, 0.507 atm
2. 33.7 kPa is greater than 0.25 atm

READING SUPPORT A kinetic sculpture has moving parts. Some are free-hanging mobiles; some are stabiles with motors.

FIGURE 13.3

a. a point near the peak of the curve
b. The curves have the same overall shape, but the curve for hot water is wider with a lower peak.
c. At an even higher temperature, the graph would be wider than the purple curve with a lower peak; at an even lower temperature, the graph would be narrower than the green curve with a higher peak.

LESSON 13.1

Extend

Connect to ASTRONOMY

Astronomers classify nebulae into two broad categories—bright and dark. Bright nebulae are close enough to stars that they glow. Dark nebulae are illuminated only if something bright, like a star cluster, is in the background. Dark nebulae are the sites of star birth while bright nebulae are the sites of star death. But nebulae do not stay in one category or the other forever. For example, once a star is born, a dark nebula changes into a bright nebula. There are several types of dark and bright nebulae. These nebulae are named for their physical appearance and/or chemical makeup. The boomerang nebula shown in Figure 13.4 is a type of bright nebula called a planetary nebula. Planetary nebulae are named for their round shape and resemblance to planets. An emission nebula is another type of bright nebula. Have interested students research why an emission nebula is so named. Then have them compare and contrast emission nebulae to planetary nebulae.

Evaluate

Informal Assessment

To assess students' comprehension of kinetic theory as it applies to gases, pose the following questions. **Ask** What is kinetic energy? *(energy due to the motion of an object)* **Ask** How is the average kinetic energy of a collection of particles related to temperature? *(Average kinetic energy is directly proportional to the Kelvin temperature. Higher temperatures reflect a greater average kinetic energy.)*

Then direct students to complete 13.1 Lesson Check

Reteach

Have students look at Figure 13.3 again. Point out that the distribution of kinetic energy in cold water is represented by the green curve. Point out that the distribution of kinetic energy in hot water is represented by the purple curve. **Ask** In which sample is the average kinetic energy of the particles higher? *(hot water)* Explain how this graph applies to gases.

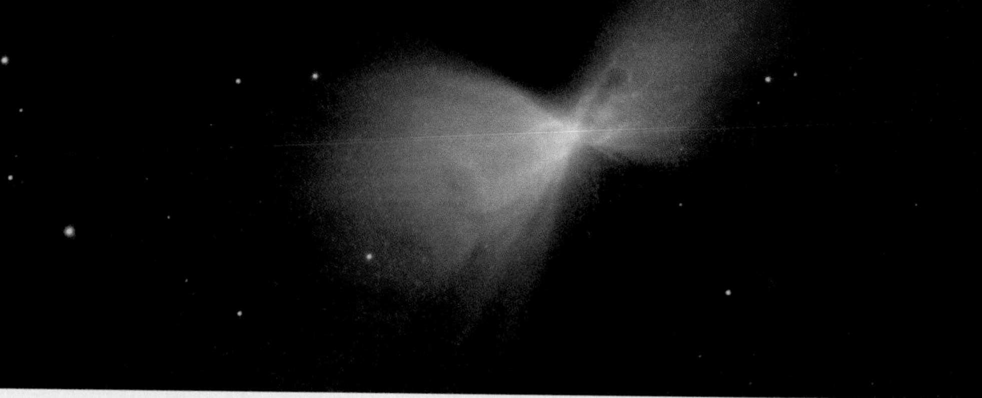

Figure 13.4
The Coldest Place in the Universe
The boomerang nebula is the coldest known region of space. A nebula is a large cloud of gas and dust spread out in an immense volume. Gases are moving rapidly away from a dying star at the center of this nebula. The rapid expansion of these gases is the reason why this nebula is so cold.

You could reasonably expect the particles of all substances to stop moving at some very low temperature. The particles would have no kinetic energy at that temperature because they would have no motion. Absolute zero (0 K, or −273.15°C) is the temperature at which the motion of particles theoretically ceases. No temperature can be lower than absolute zero. Absolute zero has never been produced in the laboratory. However, a near-zero temperature of about 0.000 000 000 1 K (0.1×10^{-9} K), which is 0.1 nanokelvin, has been achieved. The coldest temperatures recorded outside of the laboratory are from space. In 1995, astronomers used a radio telescope to measure the temperature of the boomerang nebula shown in Figure 13.4. At a temperature of about 1 K, it is the coldest known region of space.

Average Kinetic Energy and Kelvin Temperature The Kelvin temperature scale reflects the relationship between temperature and average kinetic energy. **The Kelvin temperature of a substance is directly proportional to the average kinetic energy of the particles of the substance.** For example, the particles in helium gas at 200 K have twice the average kinetic energy as the particles in helium gas at 100 K. The effects of temperature on particle motion in liquids and solids are more complex than in gases.

ONLINE PROBLEMS 13.1 LessonCheck

3. **Describe** Briefly describe the assumptions of kinetic theory as applied to gases.
4. **Explain** Use kinetic theory to explain what causes gas pressure.
5. **Explain** How is the Kelvin temperature of a substance related to the average kinetic energy of its particles?
6. **Apply Concepts** Describe the behavior of an oxygen molecule in a sealed container of air. Include what happens when the molecule collides with another molecule or the container walls.
7. **Calculate** Convert the following pressures to kilopascals.
   a. 0.95 atm
   b. 45 mm Hg
8. **Predict** A cylinder of oxygen gas is cooled from 300 K (27°C) to 150 K (−123°C). By what factor does the average kinetic energy of the oxygen molecules in the cylinder decrease?

BIGIDEA KINETIC THEORY

9. Why does a gas take the shape and volume of its container?

Lesson Check Answers

3. A gas is composed of tiny particles that are in constant motion. This motion is rapid and random. Collisions between particles are perfectly elastic.
4. Gas pressure is the result of simultaneous collisions of billions of rapidly moving particles with an object.
5. The Kelvin temperature is directly proportional to the average kinetic energy of the particles.
6. The oxygen molecule moves in straight-line random motion until it collides with another molecule or with the side of the container. After a collision, the direction of the motion changes.
7. a. 96 kPa
   b. 6.0 kPa
8. by one-half
9. **BIGIDEA** As gas particles move, they spread apart filling all available space.

13.2 The Nature of Liquids

C.1.5 States of matter; **C.3.3** Chemical bonds and intermolecular attractions; **C.6.1** Explain that atoms and molecules are in constant motion; **C.6.2** Distinguish between temperature and heat flow

CHEMISTRY & YOU

Q: *How hot should water be when you make coffee?* Ground coffee beans contain many different oils, which contribute to the flavor and aroma of the brew. If the water used for brewing the coffee is too hot, some of these oils vaporize and escape into the air, leaving the coffee less rich and flavorful. For this reason, you should let boiling water cool a little before using it to brew your coffee. In this section, you will learn why different liquids have different boiling points.

Key Questions

What factors determine the physical properties of a liquid?

What is the relationship between evaporation and kinetic energy?

When can a dynamic equilibrium exist between a liquid and its vapor?

Under what conditions does boiling occur?

Vocabulary

- vaporization
- evaporation
- vapor pressure
- boiling point
- normal boiling point

A Model for Liquids

What factors determine the physical properties of a liquid?

According to the kinetic theory, both the particles in gases and the particles in liquids have kinetic energy. This energy allows the particles in gases and liquids to flow past one another, as shown in Figure 13.5. Substances that can flow are referred to as fluids. The ability of gases and liquids to flow allows them to conform to the shape of their containers.

Gases and liquids have a key difference between them. According to the kinetic theory, there are no attractions between the particles in a gas. However, the particles in a liquid are attracted to each other. These intermolecular attractions keep the particles in a liquid close together, which is why liquids have a definite volume. **The interplay between the disruptive motions of particles in a liquid and the attractions among the particles determines the physical properties of liquids.**

Intermolecular attractions reduce the amount of space between the particles in a liquid. Thus liquids are much more dense than gases. Increasing the pressure on a liquid has hardly any effect on its volume. The same is true for solids. Therefore, liquids and solids are known as condensed states of matter.

Figure 13.5
Comparing Liquids and Gases
Both liquids and gases can flow. The liquid on the left is colored water. The gas on the right is bromine vapor. If a gas is denser than air, it can be poured from one container into another. These pictures were taken in a fume hood because bromine is both toxic and corrosive.
Predict *Over time, what will happen to the gas in the uncovered beaker? Explain.*

Key Objectives

13.2.1 IDENTIFY the factors that determine physical properties of a liquid.

13.2.2 DEFINE evaporation in terms of kinetic energy.

13.2.3 DEFINE the conditions under which a dynamic equilibrium can exist between a liquid and its vapor.

13.2.4 IDENTIFY the conditions under which boiling occurs.

Additional Resources

Reading and Study Workbook, Lesson 13.2

Available Online or on Digital Media:

- Teaching Resources, Lesson 13.2 Review
- Small-Scale Laboratory Manual, Lab 20

PearsonChem.com

Engage

CHEMISTRY & YOU Explain that in the coffee brewing process, water acts as a solvent. The water passes through the coffee grounds, dissolving soluble solids as it goes, and into the brew. The non-soluble solids remain in the filter. As a class brainstorm factors that may affect coffee brewing results. *(Sample answers: the ratio of water-to-coffee; the particle size of the coffee; the temperature of the water; the amount of time water and coffee are in contact with each other; etc.)* **Ask** Why do you think coffee is brewed with hot water? *(water is a better solvent at hot, or near-boiling temperatures)*

IN ACADEMIC STANDARDS for SCIENCE

C.1.5 Describe the characteristics of solids, liquids, and gases and state changes at the molecular level.

C.6.1 Explain that atoms and molecules that make up matter are in constant motion and that this motion increases as thermal energy increases.

Also covered C.3.3, C.6.2.

Focus on ELL

1 CONTENT AND LANGUAGE Read the lesson title and each key question aloud several times, slowly. Have students signal to you when they hear a word they do not understand. Write these words along with the words *factor, property, relationship, dynamic, equilibrium,* and *condition* on the board. Discuss the academic meaning and the contextual meaning of each word. Then help students paraphrase the title and the questions in a way that is meaningful to them.

2 FRONTLOAD THE LESSON Briefly review kinetic theory as it relates to gases. Then, preview Figure 13.5 to bridge the content students studied about gases in Lesson 13.1 with the discussion of liquids in this lesson.

3 COMPREHENSIBLE INPUT Assign several students the role of gas particle, and assign several others the role of liquid particle. Have them act out various properties of liquids and gases for the class.

Answers

FIGURE 13.5 It will disperse in the surrounding air.

LESSON 13.2

Activate Prior Knowledge

Have students create a two column table, one column titled *Gases* and the other *Liquids*. Have students list everything they know about gases, including properties, characteristics, behavior regarding pressure and temperature changes, units of measure, and kinetic theory. Then have students predict how liquids will compare to gases in each category. Have students periodically check the accuracy of their predictions as they proceed with the lesson.

Foundations for Reading

BUILD VOCABULARY After students read the Vapor Pressure section, have them create a concept map using the following terms: *evaporation, condensation, dynamic equilibrium,* and *vapor pressure.*

READING STRATEGY As students read a particular section, have them write two or three real-world scenarios that correspond to each section on sticky notes. Tell students to place the sticky notes at each section head.

Explain

A Model for Liquids

USE VISUALS Have students study Figure 13.5. **Ask** How are the colored water and the bromine vapor similar? *(They can both flow and take the shape of their beakers.)* **Ask** How are the water and the bromine vapor different? *(Water has a definite volume. The vapor doesn't have a definite volume.)* **Ask** According to kinetic theory, why doesn't the water spontaneously flow out of the beaker? *(Attractions among water molecules hold the molecules together.)*

Evaporation

USE VISUALS Use the photographs and drawings in Figure 13.6 to compare the evaporation of water in open and closed systems. Make sure students understand the difference between an open and closed system. In an open system, some of the molecules that evaporate from the surface collide with particles in the air and return to the liquid. But because the system is open, all the water molecules will eventually evaporate. In a closed system, the vapor accumulates above the liquid. The amount of vapor will increase until the rates of evaporation and condensation are equal.

Figure 13.6 Open vs. Closed Systems
The process of evaporation has a different outcome in an open system, such as a lake or an open container, than in a closed system, such as a terrarium or a sealed container.
a. In an open system, molecules that evaporate can escape from the system.
b. In a closed system, the molecules cannot escape. They collect as a vapor above the liquid. Some molecules condense back into a liquid.
Predict ***Does the water level change over time in either the open container or the sealed container? Why?***

Evaporation

What is the relationship between evaporation and kinetic energy?

As you probably know, water in an open container, like the one in Figure 13.6a, eventually escapes into the air as water vapor. The conversion of a liquid to a gas or vapor is called **vaporization.** When this conversion occurs at the surface of a liquid that is not boiling, the process is called **evaporation.** Most of the molecules in a liquid don't have enough kinetic energy to overcome the attractive forces and escape into the gaseous state. **During evaporation, only those molecules with a certain minimum kinetic energy can escape from the surface of the liquid.** Even some of the particles that do escape collide with molecules in the air and rebound back into the liquid.

You may have noticed that a liquid evaporates faster when heated. This occurs because heating the liquid increases the average kinetic energy of its particles. The added energy enables more particles to overcome the attractive forces keeping them in the liquid state. As evaporation occurs, the particles with the highest kinetic energy tend to escape first. The particles left in the liquid have a lower average kinetic energy than the particles that have escaped. The process is similar to removing the fastest runner from a race. The remaining runners have a lower average speed. As evaporation takes place, the liquid's temperature decreases. Therefore, evaporation is a cooling process.

You can observe the effects of evaporative cooling on hot days. When you perspire, water molecules in your perspiration absorb heat from your body and evaporate from the skin's surface. This evaporation leaves the remaining perspiration cooler. The perspiration that remains cools you further by absorbing more body heat.

Differentiated Instruction

L3 ADVANCED STUDENTS Have students follow the scientific method to examine whether the surface area of a liquid has an effect on evaporation rate. Tell students to submit a lab report of their hypothesis, procedure, data, and conclusion. Students should conclude that surface area does have an effect on evaporation rate.

ELL ENGLISH LANGUAGE LEARNERS Write the word *evaporate* on the board. Pronounce it slowly and have students repeat. Explain that evaporate comes from the Latin *evaporare,* a term that was formed by combining the Latin prefix *e-,* meaning "out", with the verb form *vaporare,* meaning "to emit vapor". Note that when a liquid evaporates it is emitting a vapor (gas).

L1 SPECIAL NEEDS STUDENTS The amount of visuals used in the lesson could be overwhelming to some students. Consider making individual copies of each visual in the lesson and introducing the concepts one at a time. Modify font size, adjust colors, and rephrase captions if necessary.

Interpret Data

Vapor Pressure (in kPa) of Three Substances at Different Temperatures						
Substance	0°C	20°C	40°C	60°C	80°C	100°C
Water	0.61	2.33	7.37	19.92	47.34	101.33
Ethanol	1.63	5.85	18.04	47.02	108.34	225.75
Diethyl ether	24.70	58.96	122.80	230.65	399.11	647.87

Table 13.1 The table compares vapor pressure values for water, ethanol, and diethyl ether at six temperatures.

a. Infer At a given temperature, ethanol has a higher vapor pressure than water. What does that say about the relative strength of attraction between particles of each substance?

b. Draw Conclusions How does an increase in temperature affect a compound's ability to evaporate?

Vapor Pressure

When can a dynamic equilibrium exist between a liquid and its vapor?

The evaporation of a liquid in a closed system differs from evaporation in an open system. No particles of liquid can escape into the outside air from the sealed container in Figure 13.6b. When a partially filled container of liquid is sealed, some of the particles at the surface of the liquid vaporize. These particles collide with the walls of the sealed container, producing pressure. A measure of the force exerted by a gas above a liquid is called **vapor pressure.** Over time, the number of particles entering the vapor increases and some of the particles condense and return to the liquid state. The following equation summarizes the process.

$$\text{Liquid} \underset{\text{condensation}}{\overset{\text{evaporation}}{\rightleftharpoons}} \text{Vapor (gas)}$$

Eventually, the number of particles condensing will equal the number of particles vaporizing. The vapor pressure will then remain constant. **In a system at constant vapor pressure, a dynamic equilibrium exists between the vapor and the liquid. The system is in equilibrium because the rate of evaporation of liquid equals the rate of condensation of vapor.**

At equilibrium, the particles in the system continue to evaporate and condense, but no net change occurs in the number of particles in the liquid or vapor. The sealed terrarium in Figure 13.6b is an example of a closed system at equilibrium. The moisture on the inner walls of the terrarium is a sign that equilibrium has been established. Particles that once evaporated are condensing, but other particles are evaporating to take their place.

Vapor Pressure and Temperature Change An increase in the temperature of a contained liquid increases the vapor pressure. This happens because the particles in the warmed liquid have increased kinetic energy. As a result, more of the particles will reach the minimum kinetic energy necessary to escape the surface of the liquid. The particles escape the liquid and collide with the walls of the container at a greater frequency. Table 13.1 gives the vapor pressures of some common liquids at various temperatures. The vapor pressure data indicates how volatile a given liquid is, or how easily it evaporates. Of the three liquids shown, diethyl ether is the most volatile and water is the least volatile.

CHEMISTRY & YOU

Q: *To make the best tasting coffee, many people grind the coffee beans just prior to brewing the coffee. Also, they are careful not to grind the coffee beans too much. Explain how both of these methods help prevent the natural oils in coffee beans from vaporizing.*

Differentiated Instruction

L1 STRUGGLING STUDENTS If students are having difficulty grasping the concept of a dynamic equilibrium, try this analogy. There are three students on one side of a room and three students on the other side. One group has 30 foam balls. The other group has none. The balls represent water molecules. Students in the first group begin to toss the balls toward the other group. This action represents evaporation. At first the only process occurring is evaporation. But soon students in the second group begin to toss the balls back. This action represents condensation. When the number of balls being tossed in each direction at any given moment is equal, a dynamic equilibrium exists.

L3 ADVANCED STUDENTS Ask students to use the data in Table 13.1 to construct a graph of vapor pressure versus temperature. They will need to plot the natural log (ln) of P versus $1/T$, where P is the vapor pressure in mm Hg and T is the temperature in kelvins.

Explore

Vapor Pressure

Teacher Demo

PURPOSE Students will observe how the vapor pressure of a liquid changes with temperature.

MATERIALS manometer, 1-L flask with 2-hole stopper, separatory funnel, short piece of glass tubing bent at a 90 angle, rubber hose, 50 mL acetone, ice water

SAFETY To insert the glass tubing in the stopper, use glycerol. Wear gloves or hold the tubing in a towel to protect your hands. Be sure there are no open flames in the room because acetone is flammable.

PROCEDURE You will need to obtain or construct a manometer and assemble the following apparatus. Insert a two-hole stopper into a 1-L flask. Into one hole of the stopper, insert a separatory funnel capable of containing 50 mL of acetone. Insert the bent glass tubing into the other hole. Use a rubber hose to connect the glass tubing to one end of the manometer. Be sure the hose fits both connections tightly. Use the manometer to measure the initial pressure. Then open the valve on the separatory funnel, and let the acetone drain into the flask. Have students observe the pressure in the flask. The maximum pressure observed is the vapor pressure of acetone at room temperature. Next, place the flask in ice water, and have students observe the change in pressure.

EXPECTED OUTCOMES The pressure rises immediately as the acetone flows into the flask. The pressure falls when the flask is cooled.

CHEMISTRY & YOU Grinding the coffee beans increases the surface area of the coffee. A greater surface area increases the rate at which the natural oils in coffee will evaporate. Overgriding the beans will heat the oils also causing them to evaporate.

Answers

FIGURE 13.6 The water level will decrease over time in the open container because the evaporating water molecules can escape from the container.

INTERPRET DATA

a. The strength of attractions between the particles in water is stronger than in ethanol.

b. An increase in temperature makes it easier for a substance to evaporate.

Explore

Vapor Pressure

Teacher Demo

PURPOSE Students will infer the relative vapor pressures of two liquids.

MATERIALS 2 cotton balls, water, rubbing alcohol

PROCEDURE Use cotton balls to simultaneously dab spots of water and rubbing alcohol onto the chalkboard. Have the class observe what happens to the spots. Ask students to infer which liquid has the greater vapor pressure at room temperature and explain their reasoning.

Explain

Boiling Point

APPLY CONCEPTS Explain that atmospheric pressure opposes the formation of bubbles in a liquid; bubbles form only when the vapor pressure inside a bubble is equal to atmospheric pressure.

Misconception Alert

Many students use the terms boiling and evaporation interchangeably. Explain that boiling is different from evaporation because evaporation can occur at any temperature, whereas boiling occurs at specific temperatures and pressures.

Vapor Pressure Measurements The vapor pressure of a liquid can be determined with a device called a manometer. Figure 13.7 shows how a simple manometer works. One end of a U-shaped glass tube containing mercury is attached to a closed container. The other end of the tube is open to the atmosphere. When there is only air in the container, the pressure is the same on both sides of the tube and the mercury level is the same in each arm of the tube. When a liquid is added to the container, the pressure in the container increases due to the vapor pressure of the liquid. The vapor pressure of the liquid pushes the mercury down on the container side of the U-tube. The levels of mercury in the U-tube are no longer the same. You can determine the vapor pressure in mm of Hg by measuring the difference between the two levels of mercury. As the vapor pressure increases, so does the difference between the two levels.

Boiling Point

Under what conditions does boiling occur?

The rate of evaporation of a liquid from an open container increases as the liquid is heated. Heating allows a greater number of particles at the liquid's surface to overcome the attractive forces that keep them in the liquid state. The remaining particles in the liquid move faster and faster as they absorb the added energy. Thus, the average kinetic energy of the particles in the liquid increases and the temperature of the liquid rises. **When a liquid is heated to a temperature at which particles throughout the liquid have enough kinetic energy to vaporize, the liquid begins to boil.** Bubbles of vapor form throughout the liquid, rise to the surface, and escape into the air. The **boiling point** (bp) is the temperature at which the vapor pressure of the liquid is just equal to the external pressure on the liquid.

Figure 13.7 Manometer
The vapor pressure of a contained liquid can be measured in a manometer. The vapor pressure is equal to the difference in height of the mercury in the two arms of the U-tube.
Calculate *What is the difference in vapor pressure between ethanol at 0°C and ethanol at 20°C?*

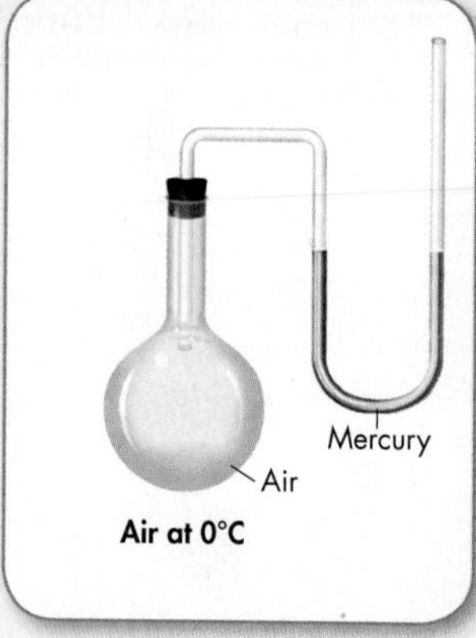

Air at 0°C

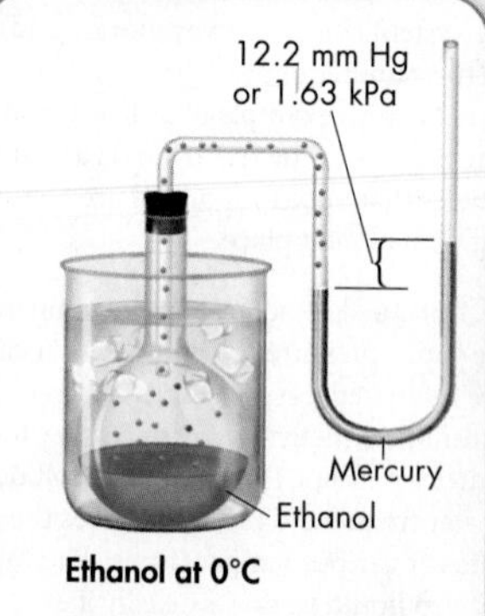

Ethanol at 0°C

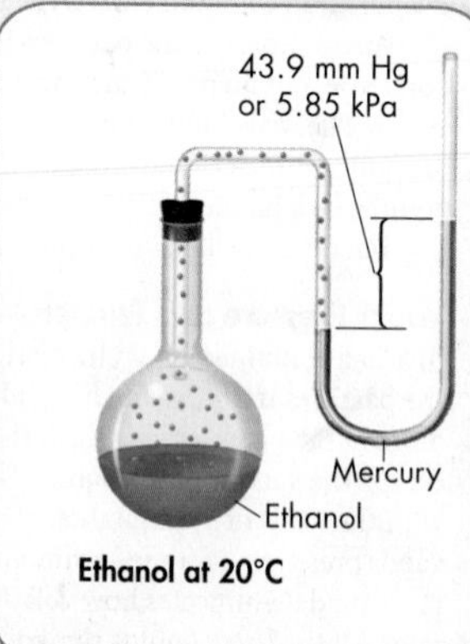

Ethanol at 20°C

UbD Check for Understanding

BIGIDEA KINETIC THEORY To assess students' understanding of the liquid processes explained in this lesson, have them write a short paragraph or create a two-column table, comparing and contrasting evaporation and boiling.

ADJUST INSTRUCTION If students struggle to explain the relationship between evaporation and boiling, have them reread the Evaporation and Boiling Point sections in the lesson. Then discuss various real-world examples to help students better connect these concepts.

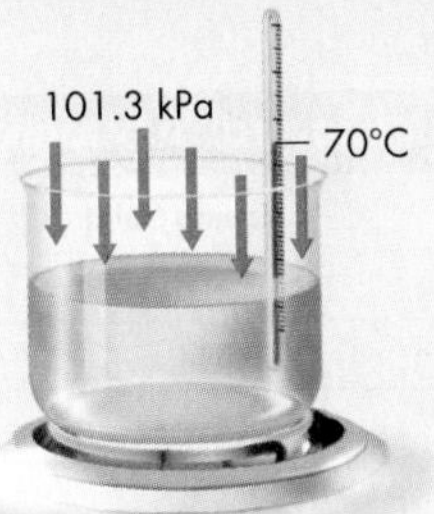

Sea Level
Atmospheric pressure at the surface of water at 70°C is greater than its vapor pressure. Bubbles of vapor cannot form in the water, and it does not boil.

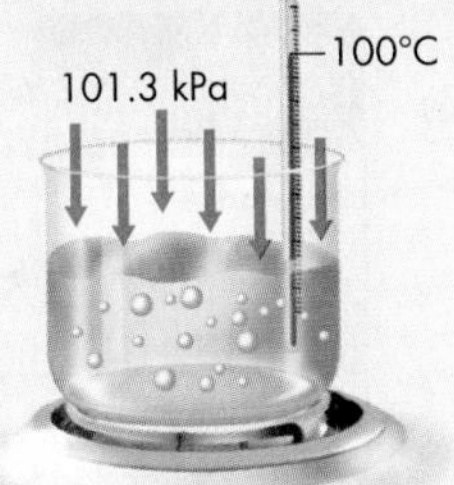

Sea Level
At the boiling point, the vapor pressure is equal to atmospheric pressure. Bubbles of vapor form in the water, and it boils.

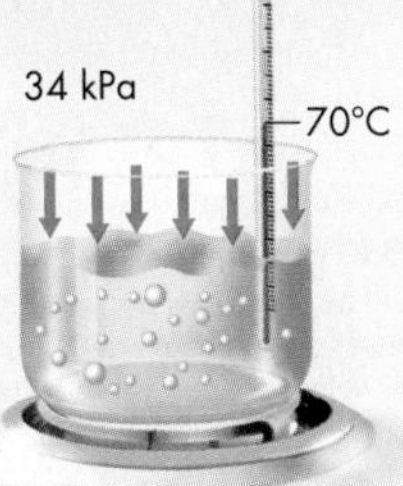

Atop Mount Everest
At higher altitudes, the atmospheric pressure is lower than it is at sea level. Thus the water boils at a lower temperature.

Figure 13.8
Boiling Point and Altitude
A liquid boils when the vapor pressure of particles within the liquid equals the atmospheric pressure. The boiling point varies with altitude.

Boiling Point and Pressure Changes Because a liquid boils when its vapor pressure is equal to the external pressure, liquids don't always boil at the same temperature. Figure 13.8 shows how a change in altitude affects the boiling point of water. Because atmospheric pressure is lower at higher altitudes, boiling points decrease at higher altitudes. For example, in Denver, which is 1600 m above sea level, the average atmospheric pressure is 85.3 kPa. So water boils at about 95°C. In a pressure cooker, the vapor cannot escape and the vapor pressure increases. So water boils at a temperature above 100°C and food can cook more quickly.

Look at the vapor pressure versus temperature graph in Figure 13.9. You can use the graph to show how the boiling point of a liquid is related to vapor pressure. At a lower external pressure, the boiling point decreases. The particles in the liquid need less kinetic energy to escape from the liquid. At a higher external pressure, the boiling point increases. The particles in the liquid need more kinetic energy to escape from the liquid.

InterpretGraphs

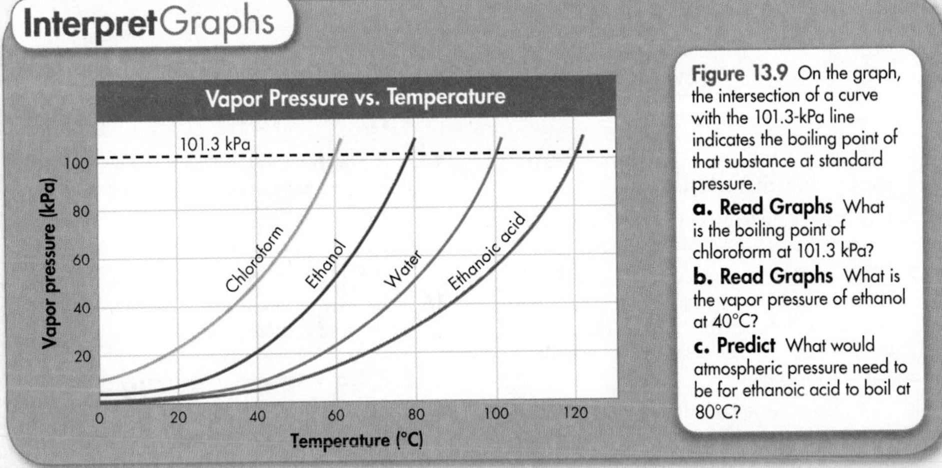

Figure 13.9 On the graph, the intersection of a curve with the 101.3-kPa line indicates the boiling point of that substance at standard pressure.

a. Read Graphs What is the boiling point of chloroform at 101.3 kPa?

b. Read Graphs What is the vapor pressure of ethanol at 40°C?

c. Predict What would atmospheric pressure need to be for ethanoic acid to boil at 80°C?

Explore

PURPOSE To dispel the misconception that the temperature of a liquid increases as it boils

MATERIALS thermometer, beaker, ring stand, thermometer clamp, water, hot plate

SAFETY Students must wear safety goggles, an apron, and mitts to protect them from the boiling water and steam. **CAUTION** *Steam can cause a more severe burn than boiling water can.*

PROCEDURE Have students measure the temperature every 30 seconds for several minutes as water is heated and then boiled. They should insert the bulb of the thermometer into the water, but take care not to let it touch the beaker. Have students use their data to construct a graph of temperature versus time. Ask students to write a short paragraph explaining their results.

MAKE A CONNECTION Discuss the effect of altitude on cooking time. **Ask** Why does it take longer to cook food in water at high altitudes? *(Because atmospheric pressure decreases with altitude, water boils at a lower temperature and the food doesn't cook as quickly.)* Why does food take less time to cook in a pressure cooker? *(In a pressure cooker the pressure can be greater than atmospheric pressure. Thus, the boiling point of water can be higher, and the food takes less time to cook.)*

Explain

USE VISUALS Direct students' attention to Figure 13.9. **Ask** Why doesn't the temperature/vapor pressure relationship show a straight line for the four substances shown? *(The relationship between temperature and vapor pressure is not linear because there is not a constant change in vapor pressure with a constant change in temperature.)*

UbD Check for Understanding

What is the relationship between evaporation and kinetic energy?

Place a beaker of water over a flame or on a hot plate and let it come to a boil. Have students write a one-sentence answer to the following question.

Once the water begins to boil, why doesn't all of the water immediately evaporate? *(There are so many molecules that need to enter the gas phase that it takes some time, depending on how much water is in the pot.)*

ADJUST INSTRUCTION If students have difficulty answering the question, have them reread the text in the Evaporation section. Have students consider how the mass of the water affects the amount of kinetic energy it takes to evaporate the water.

Answers

FIGURE 13.7 4.22 kPa or 31.7 mm Hg

INTERPRET GRAPHS

a. 60°C
b. about 20 kPa
c. about 30 kPa

LESSON 13.2

Evaluate

Informal Assessment

Ask students to summarize the general characteristics that distinguish a liquid from a gas. *(Gases can expand to fill their containers. Gases are easily compressed. Liquids conform to the shape of their containers, but do not expand to fill them. Liquids are not as easy to compress as gases because the spaces between particles in a liquid are much smaller than those in a gas.)*

Then direct students to complete the 13.2 Lesson Check.

Reteach

Illustrate for the students that the temperature at which the vapor pressure of a liquid is equal to the atmospheric pressure is the boiling point of the liquid. Depict that the exact temperature at which boiling occurs depends on the strength of the intermolecular attractions between molecules in the liquid and the magnitude of the atmospheric pressure. Point out that when the atmospheric pressure is reduced below the vapor pressure of water, water will boil at room temperature.

Figure 13.10 Boiling Water
You should be careful when working with boiling water as you could get a burn from either the water or the invisible water vapor, or steam, in the air just above the water.

Table 13.2

Normal Boiling Points of Several Substances

Substance	Boiling Point (°C)
Carbon disulfide (CS_2)	46.0
Chloroform ($CHCl_3$)	61.7
Methanol (CH_4O)	64.7
Carbon tetrachloride (CCl_4)	76.8
Ethanol (C_2H_6O)	78.5
Water (H_2O)	100.0

Boiling is a cooling process, similar to evaporation. During boiling, the particles with the highest kinetic energy escape first when the liquid is at the boiling point. Turning off the source of external heat drops the liquid's temperature below its boiling point. Supplying more heat allows more particles to acquire enough kinetic energy to escape. However, the temperature of the boiling liquid never rises above its boiling point. If heat is supplied at a greater rate, the liquid only boils faster. The vapor produced is at the same temperature as that of the boiling liquid. Although the vapor has the same average kinetic energy as the liquid, its potential energy (or stored energy) is much higher. Thus, a burn from steam is more severe than one from an equal mass of boiling water, even though they are both at the same temperature. This is one of the reasons why you need to be careful when straining boiling water from pasta or vegetables, as shown in Figure 13.10.

Normal Boiling Point A liquid can have various boiling points depending on pressure. The **normal boiling point** is defined as the boiling point of a liquid at a pressure of 101.3 kPa. Table 13.2 lists the normal boiling points of six molecular compounds.

ONLINE PROBLEMS 13.2 LessonCheck

10. **Identify** What factors help determine the physical properties of liquids?
11. **Explain** In terms of kinetic energy, explain how a molecule in a liquid evaporates.
12. **Describe** A liquid is in a closed container and has a constant vapor pressure. What is the relationship between the rate of evaporation of the liquid and the rate of condensation of the vapor in the container?
13. **Relate Cause and Effect** What conditions must exist for a liquid to boil?
14. **Interpret Graphs** Use Figure 13.9 to determine the boiling point of ethanoic acid at 27 kPa and chloroform at 80 kPa.
15. **Explain** Why does the boiling point of a liquid vary with atmospheric pressure?
16. **Infer** Explain how evaporation lowers the temperature of a liquid.

BIGIDEA KINETIC THEORY

17. Why does a liquid take the shape but not the volume of its container?

Lesson Check Answers

10. the interplay between the disruptive motions of particles in a liquid and the attractions among the particles
11. A molecule with a certain minimum kinetic energy can escape from the surface of the liquid and vaporize.
12. Rate of evaporation equals the rate of condensation.
13. Particles throughout the liquid must have enough kinetic energy to vaporize.
14. about 76°C; about 52°C
15. Boiling occurs when the vapor pressure of a liquid equals the external pressure. If the atmospheric pressure changes, the boiling point will change.
16. When the molecules with the highest kinetic energy escape from the liquid, the average kinetic energy of the remaining particles is lower and the temperature decreases.
17. **BIGIDEA** A liquid has a definite volume because of the attractive forces between the particles. A liquid conforms to the shape of its container because the kinetic energy of the particles allows them to move past one another.

13.3 The Nature of Solids

C.1.5 States of matter; C.3.3 Chemical bonds and intermolecular attractions; C.6.1 Explain that atoms and molecules are in constant motion

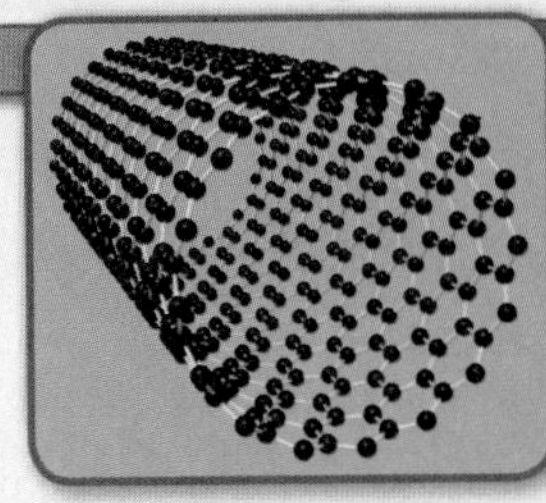

CHEMISTRY & YOU

Q: *What is the strongest material in the world?* It's not steel or any synthetic plastic, but a form of pure carbon known as fullerene nanotubes. These cylindrical structures, formed from carbon atoms linked together in hexagonal patterns, are over 300 times stronger than steel. They can be made with a diameter of only a few nanometers (hence the name) but several millimeters in length. Researchers in many different fields are finding new applications and possibilities for these structures, part of the new area of research known as nanotechnology. You will find out about fullerenes in this section.

Key Questions

How are the structure and properties of solids related?

What determines the shape of a crystal?

Vocabulary

- melting point
- freezing point
- crystal
- unit cell
- allotropes
- amorphous solid
- glass

A Model for Solids

How are the structure and properties of solids related?

The particles in liquids are relatively free to move. The particles in solids, however, are not. **The general properties of solids reflect the orderly arrangement of their particles and the fixed locations of their particles.** In most solids, the atoms, ions, or molecules are packed tightly together. These solids are dense and not easy to compress. Because the particles in solids tend to vibrate about fixed points, solids do not flow.

When you heat a solid, its particles vibrate more rapidly as their kinetic energy increases. The organization of particles within the solid breaks down, and eventually the solid melts. The **melting point** (mp) is the temperature at which a solid changes into a liquid. At this temperature, the disruptive vibrations of the particles are strong enough to overcome the attractions that hold them in fixed positions. The **freezing point** (fp) is the temperature at which a liquid changes into a solid. The melting and freezing points of a substance are at the same temperature. At that temperature, the liquid and solid phases are in equilibrium.

$$\text{Solid} \underset{\leftarrow \text{freezing}}{\overset{\text{melting} \rightarrow}{\rightleftharpoons}} \text{Liquid}$$

In general, ionic solids have high melting points because relatively strong forces hold them together. Sodium chloride, an ionic compound, has a rather high melting point of 801°C. By contrast, molecular solids have relatively low melting points. For example, hydrogen chloride, a molecular compound, melts at −112° C. Not all solids melt, however. Wood and cane sugar, for example, decompose when heated.

Key Objectives

13.3.1 DESCRIBE how the structure and properties of solids are related.

13.3.2 IDENTIFY the factors that determine the shape of a crystal.

Additional Resources

Reading and Study Workbook, Lesson 13.3

Available Online or on Digital Media:

- Teaching Resources, Lesson 13.3 Review
- Laboratory Manual, Lab 21

PearsonChem.com

Engage

CHEMISTRY & YOU Have students read the opening paragraph and examine the illustration. **Ask** What does the number of bonds and bond angles between carbons in fullerene suggest about its structure? *(Some of the carbons are connected by double bonds.)* Explain that the double bonds partially explain the strength of the nanotubes. Tell students that they will discover the rest of the explanation as they proceed with the lesson.

Activate Prior Knowledge

Assess students' knowledge of types of compounds by asking them to compare the structures of molecular and ionic compounds. *(Ionic compounds are composed of ions and are usually formed from a metal and a nonmetal. Molecular compounds are composed of molecules formed from two or more nonmetallic elements.)*

Focus on ELL

1 CONTENT AND LANGUAGE Read aloud each of the vocabulary words, and guide students to use prior knowledge to predict their meanings. Help students pronounce the words *allotrope* and *amorphous*.

2 FRONTLOAD THE LESSON Explain that this is the third state of matter students will study. Review the key questions and answers in Lessons 13.1 and 13.2. Then have students read the key questions for this lesson and predict the answers based on what they now understand about gases and liquids

3 COMPREHENSIBLE INPUT Briefly review what students learned about crystalline structure with respect to coordinate numbers in ionic crystals and atom arrangement in metallic crystals (see Lessons 7.2 and 7.3). Then preview Figures 13.11 through 13.13, pointing out the connections to the concepts from Chapter 7.

IN ACADEMIC STANDARDS for SCIENCE

C.1.5 Describe the characteristics of solids, liquids, and gases and state changes at the molecular level.

C.3.3 Compare and contrast ionic, covalent network, metallic and polar and non-polar molecular crystals with respect to constituent particles, strength of bonds, melting and boiling points, and conductivity; provide examples of each type.

C.6.1 Explain that atoms and molecules that make up matter are in constant motion and that this motion increases as thermal energy increases.

LESSON 13.3

Foundations for Reading

BUILD VOCABULARY Explain that the word *amorphous* comes from the Greek prefix *a-*, meaning "not," and the Greek root *morph*, meaning "form." Have students use this information to predict the meaning of the term *amorphous*. Have students use a similar word strategy to help them learn the meaning of allotrope. (*Allo-* means "different" and *-trope* means "turns toward.")

READING STRATEGY Have students make a table summarizing the characteristics of the seven crystal systems described in this section.

Explore

Crystal Structure and Unit Cells

PURPOSE Students will explore an analogy for the crystal lattices in solids.

MATERIALS samples of wallpaper with repeating patterns

PROCEDURE Explain that wallpaper patterns often consist of a small group of images that are repeated to obtain the overall effect. The group of images is like a unit cell in a crystal lattice. Ask students to identify the "unit cells" in wallpaper samples. Show students how to obtain a set of "lattice points" by choosing the same point in each unit of the repeating pattern. The collection of lattice points shows the fundamental arrangement of the units in the pattern. Ask students if they could arrange the components of each unit differently to produce a different "unit cell."

EXPECTED OUTCOME Students observe two-dimensional arrangements at the macroscopic level that are analogous to the three-dimensional arrangements in crystal lattices.

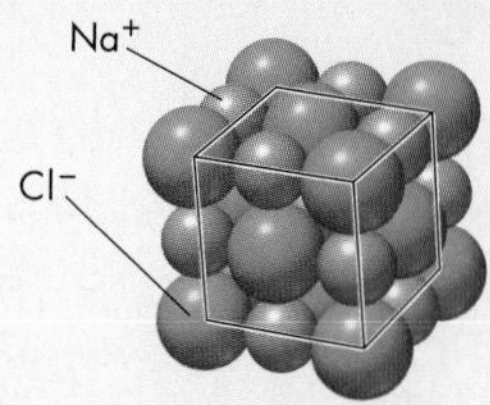

Figure 13.11 Ionic Crystal
In sodium chloride (NaCl), sodium ions (Na^+) and chloride ions (Cl^-) are closely packed in a regular array. The ions vibrate about fixed points on the crystal.

Crystal Structure and Unit Cells

What determines the shape of a crystal?

Most solid substances are crystalline. In a **crystal,** the particles are arranged in an orderly, repeating, three-dimensional pattern called a crystal lattice. Figure 13.11 shows part of the crystal lattice in sodium chloride. **The shape of a crystal reflects the arrangement of the particles within the solid.**

Crystal Systems A crystal has sides, or faces. The angles at which the faces of a crystal intersect are always the same for a given substance and are characteristic of that substance. Crystals are classified into seven groups, or crystal systems, which have the characteristic shapes shown in Figure 13.12. The edges are labeled a, b, and c. The angles are labeled α, β, and γ. The seven crystal systems differ in terms of the angles between the faces and in the number of edges of equal length on each face.

The shape of a crystal depends on the arrangement of the particles within it. The smallest group of particles within a crystal that retains the geometric shape of the crystal is known as a **unit cell.** A crystal lattice is a repeating array of any one of fourteen kinds of unit cells. Each crystal system can be composed of from one to four types of unit cells. Figure 13.13 shows the three kinds of unit cells that can make up a cubic crystal system.

Figure 13.12 Crystal Systems
Crystals are classified into seven crystal systems.
Classify ***In which of the systems are all three angles equal to 90°?***

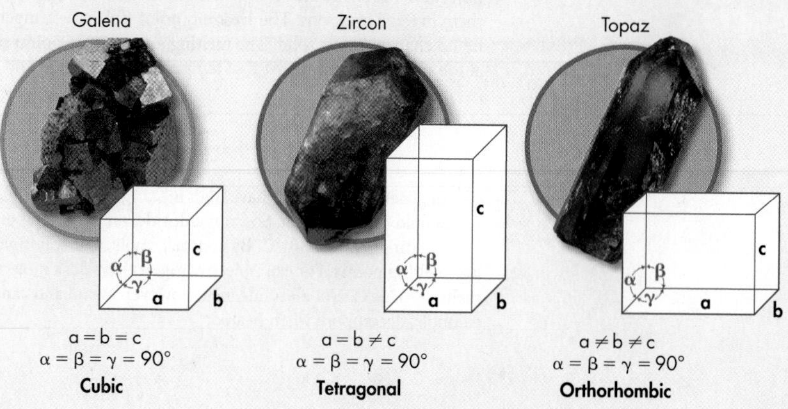

Differentiated Instruction

L3 ADVANCED STUDENTS Have students do research on the crystalline structure and formation of natural, synthetic, and simulated diamonds. Have students write a short paper comparing the three kinds of crystalline structures.

LPR LESS PROFICIENT READERS Have students create a concept map centered around the term *solid.* Their maps should highlight the relationship between solids and crystals and amorphous solids. Students should determine where unit cells, allotropes, and amorphous solids fit into their concept map.

ELL SPECIAL NEEDS STUDENTS Explain that morphing is a special effect used in films or videos to change the shape or form of an object. Have students go online to find an example of morphing. Have them compare their example of morphing to an amorphous solid. Allow class time for them to share their findings.

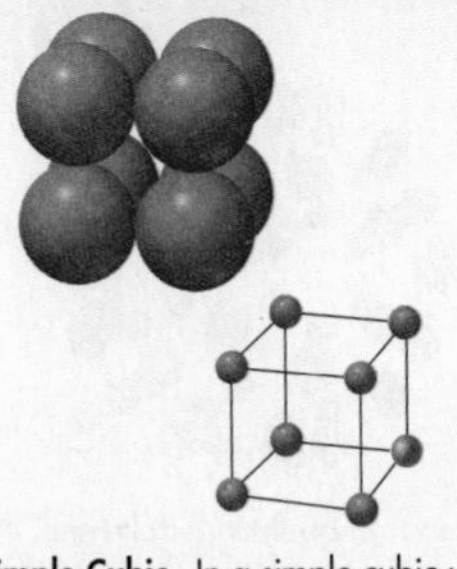

Simple Cubic In a simple cubic unit cell, the atoms or ions are arranged at the corners of an imaginary cube.

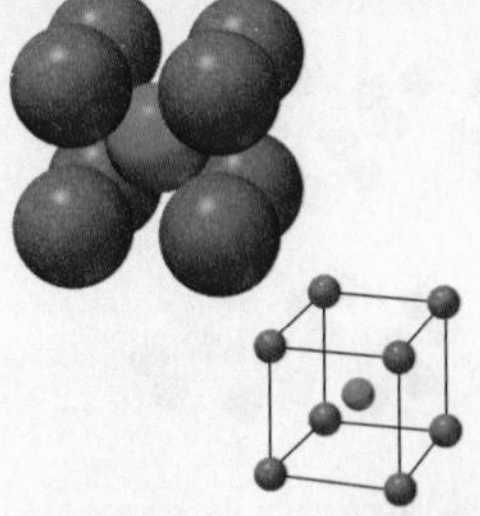

Body-Centered In a body-centered cubic unit cell, the atoms or ions are at the corners and in the center of an imaginary cube.

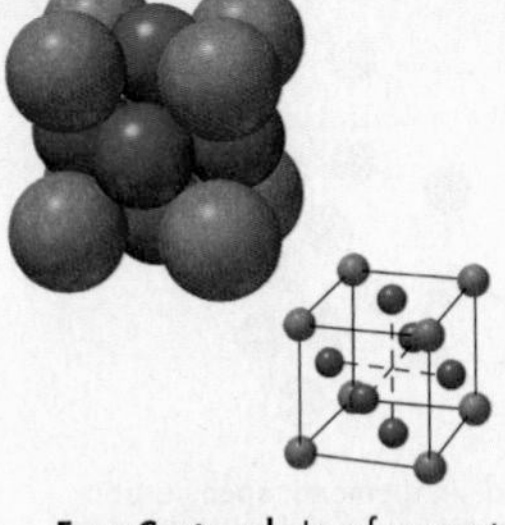

Face-Centered In a face-centered cubic unit cell, there are atoms or ions at the corners and in the center of each face of the imaginary cube.

Figure 13.13 Unit Cells
The unit cell in a cubic crystal system may be simple cubic, body-centered cubic, or face-centered cubic. In the space-filling models and line drawings, the spheres represent atoms or ions.

Allotropes Some solid substances can exist in more than one form. A good example is the element carbon. Diamond is one crystalline form of carbon. It forms when carbon crystallizes under tremendous pressure (thousands of atmospheres). A different crystalline form of carbon is graphite. The lead in a pencil is not the element lead; it is graphite. In graphite, the carbon atoms are packed in sheets rather than in the extended three-dimensional array that is characteristic of diamond.

In 1985, a third crystalline form of carbon was discovered in ordinary soot. This form of carbon is called buckminsterfullerene. The carbon atoms in buckminsterfullerene are bonded together in groups of 60 atoms to form a hollow sphere, or cage, known as a buckyball. The atoms are arranged in a pattern of hexagons and pentagons on the surface of the cage, similar to the pattern on the surface of a soccer ball. Since 1985, other molecules of carbon with hollow cages have been discovered. The one with 70 carbon atoms is shaped like a football. As a group, these forms of carbon are called fullerenes.

Learn more about crystal structures of solids *online.*

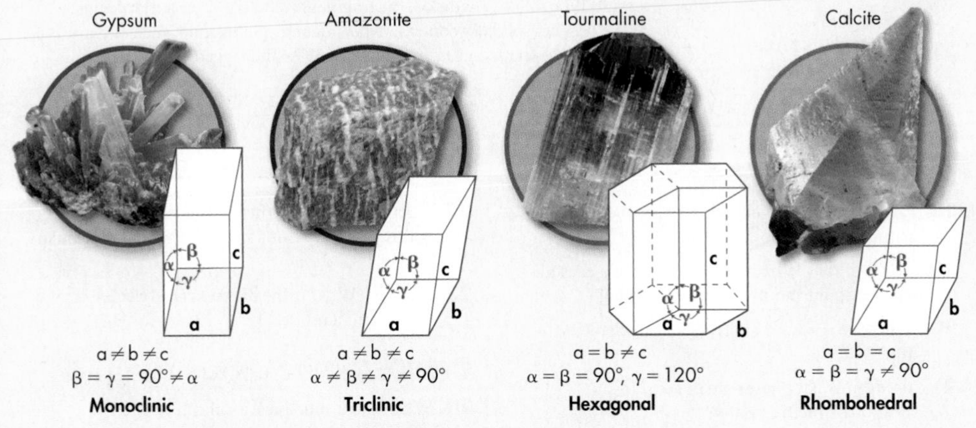

Explore

Teacher Demo

MATERIALS dishwashing detergent, watch glass, overhead projector, drinking straw

PROCEDURE You can use an aqueous solution of detergent to make a model of a crystalline solid. Place the solution in a watch glass on an overhead projector and use a straw to blow bubbles into the solution. The bubbles form a regular, repeating pattern that is analogous to the arrangement of unit cells in a crystalline solid.

Explain

USE VISUALS Review basic geometric concepts of three-dimensional solids prior to introducing Figure 13.12. Explain that the properties of the crystals shown in the diagram are based on their geometric construction.

MAKING CONNECTIONS Have students describe how a higher degree of organization and stronger intermolecular attractions between particles distinguish solids from gases and liquids.

Natural Glass

The most familiar examples of glass are made by humans: plate glass, blown glass, and pressed glass. However, glass also forms naturally as the result of geologic and meteorologic processes. Volcanic glass is called obsidian, and forms when molten silicon dioxide (silica) cools so rapidly that mineral crystals of any significant size are unable to form. Because obsidian develops the characteristic razor-sharp edges associated with broken and chipped glass, it was highly prized by Stone Age toolmakers. Later civilizations used it for ceremonial weaponry and other ornaments.

Meteorological events that generate significantly high temperatures are also capable of forming glass. Lightning strikes in sandy areas form structures called *fulgurites*, which are essentially hollow glass tubes. Meteor impacts form rounded shapes called *tektites*.

Answers

FIGURE 13.12 cubic, tetragonal, and orthorhombic

Explain

CRITICAL THINKING Ask students to determine if any elements besides carbon exist in different allotropic forms. *(Elemental oxygen occurs as O_2 and O_3, both of which are gases; sulfur can exist as monoclinic or orthorhombic crystals; phosphorus can exist as red or white phosphorus, which are solids at room temperature.)*

CHEMISTRY & YOU As Figure 13.14 shows, the arrangement of the atoms give fullerene its hollowed-out shape. Its shape is the reason why fullerene is so strong.

Extend

Connect to TECHNOLOGY

Explain that scientists can manipulate the crystal structure of various materials to improve certain physical properties. Note that one such method is the production of alloys such as steel, as discussed in Lesson 7.4. Another method is the production of semiconductors, which led to the development of many important advances in electronics. Have students research the history, structure and uses of key semiconductors in today's electronics, and report their findings to the class.

Evaluate

Informal Assessment

Ask students to describe orally, or in paragraph form the distinguishing characteristics of crystalline solids and amorphous solids.

(Crystalline solids are characterized by an orderly, repeating three-dimensional arrangement of atoms, ions, or molecules. All crystals have a regular shape that reflects the arrangement of particles in the solid. Amorphous solids lack a well-defined arrangement of particles.)

Reteach

Describe how a higher degree of organization and stronger intermolecular attractions between particles distinguish solids from gases and liquids. Have students consider how these factors are related to the ability of solids to retain their shape without a container.

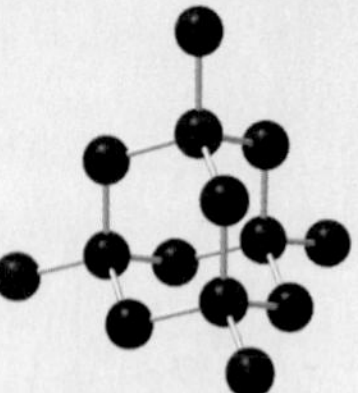

Diamond In diamond, each carbon atom in the interior of the diamond is strongly bonded to four others. The array is rigid and compact.

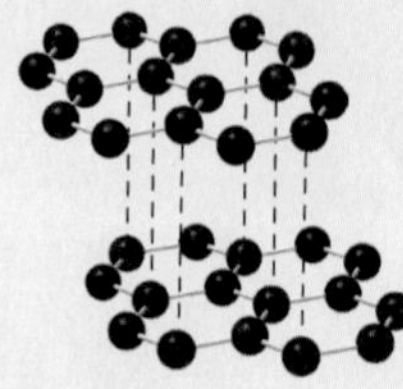

Graphite In graphite, the carbon atoms are linked in widely spaced layers of hexagonal (six-sided) arrays.

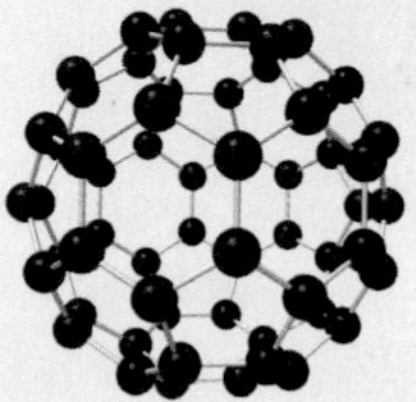

Fullerene In buckminsterfullerene, 60 carbon atoms form a hollow sphere. The carbons are arranged in pentagons and hexagons.

Figure 13.14 Allotropes of Carbon
Diamond, graphite, and fullerenes are allotropes of carbon.
Compare ***Based on the arrangements of their atoms, explain why the properties of fullerenes are closer to those of diamond than of graphite.***

The physical properties of diamond, graphite, and fullerenes are quite different. Diamond has a high density and is very hard. Graphite has a relatively low density and is soft and slippery. The hollow cages in fullerenes give them great strength and rigidity. Diamond, graphite, and fullerenes are crystalline allotropes of carbon. **Allotropes** are two or more different molecular forms of the same element in the same physical state. Although allotropes are composed of atoms of the same element, they have different properties because their structures are different. Figure 13.14 compares the structures of carbon allotropes. Only a few elements have allotropes. In addition to carbon, these elements include phosphorus, sulfur, oxygen, boron, and antimony.

CHEMISTRY & YOU

Q: *What structural properties make fullerene nanotubes the strongest material in the world? Refer to Figure 13.14 in your answer.*

Non-Crystalline Solids Not all solids are crystalline in form; some solids are amorphous. An **amorphous solid** lacks an ordered internal structure. Rubber, plastic, and asphalt are amorphous solids. Their atoms are randomly arranged. Other examples of amorphous solids are glasses. A **glass** is a transparent fusion product of inorganic substances that have cooled to a rigid state without crystallizing. Glasses are sometimes called supercooled liquids. The irregular internal structures of glasses are intermediate between those of a crystalline solid and those of a free-flowing liquid. Glasses do not melt at a definite temperature. Instead, they gradually soften when heated. When a crystalline solid is shattered, the fragments have the same surface angles as the original solid. By contrast, when an amorphous solid, such as glass, is shattered, the fragments have irregular angles and jagged edges.

13.3 LessonCheck

ONLINE PROBLEMS

18. **Describe** In general, how are the particles arranged in solids?
19. **Explain** What does the shape of a crystal tell you about the structure of a crystal?
20. **Compare** How do allotropes of an element differ?
21. **Identify** What phases are in equilibrium at a substance's melting point?
22. **Compare** How do the melting points of ionic solids generally compare with those of molecular solids?
23. **Explain** What is the difference between a crystal lattice and a unit cell?

BIGIDEA KINETIC THEORY

24. Why does a solid have a definite shape and a definite volume?

Lesson Check Answers

18. Particles in solids are packed tightly together in an orderly arrangement. The locations of the particles are fixed.
19. The shape of a crystal reflects the arrangement of the particles within the solid.
20. Allotropes are different molecular forms of the same element in the same physical state.
21. The liquid and solid states are in equilibrium.
22. Ionic solids generally have higher melting points than do molecular solids.
23. A crystal lattice is a repeating array of unit cells.
24. **BIGIDEA** The particles in a solid are packed tightly together and are not free to move. This arrangement of particles gives solids its definite shape and volume.

Small-Scale Lab

The Behavior of Liquids and Solids

Purpose

To explore and explain some behaviors of liquids and solids

Materials

- plastic Petri dish
- water
- ice
- rubbing alcohol
- graph paper, 1-cm
- calcium chloride

Procedure

1. In your notebook, make a copy of the table shown below. Add a column for your observations. In the experiments, you will place substances labeled A and B inside the Petri dish and substances labeled C on top of the dish.

2. For Experiment 1, place one drop of water in the Petri dish. Replace the cover and place a small piece of ice on top of the cover.

3. After a few minutes, observe the interior surface of the Petri dish cover and the contents of the dish. Record your observations. Clean and dry the Petri dish and its cover.

4. Repeat Steps 2 and 3 for Experiments 2–5, using the materials listed in the table. For Experiment 4, place the Petri dish on the graph paper so that you can place the water and the calcium chloride about 3 cm apart.

Experiment	Substance A	Substance B	Substance C
1	drop of water	–	ice cube
2	drop of water	–	drop of water
3	drop of rubbing alcohol	–	drop of water
4	drop of water	piece of $CaCl_2$	–
5	–	several pieces of $CaCl_2$	ice cube

Analyze and Conclude

1. Draw Conclusions Explain your observations in Experiment 1 in terms of the behavior of liquids.

2. Explain Why is ice not needed for cloud formation in Experiment 2?

3. Compare What differences do you observe about the behavior of rubbing alcohol in Experiment 3 and the behavior of water in the previous experiments? Explain.

4. Predict What happens to solid calcium chloride in a humid environment?

5. Draw Conclusions Propose an explanation for no cloud formation in Experiment 5.

You're the Chemist

1. Analyze Data Place a drop of water and a drop of rubbing alcohol about 3 cm apart in a Petri dish. Cover the dish and place it on a piece of graph paper. Be careful not to mix the contents. Observe what happens to the size of the water drops over time. Provide an explanation for what you observe.

2. Observe Add a drop of bromothymol blue (BTB) to a drop of vinegar. What happens?

3. Design an Experiment Vinegar is a solution of water and ethanoic acid, $C_2H_4O_2$. Design and carry out an experiment using what you learned from the previous exercise to see if ethanoic acid will evaporate from a drop of vinegar.

4. Design an Experiment Design and carry out an experiment to see if ammonia will evaporate from a drop of aqueous ammonia.

Focus on ELL

4 LANGUAGE PRODUCTION Have students work in groups of four to complete the lab. Make sure each group has ELLs of varied language proficiencies, so that more proficient students can help less proficient ones. Have students work according to their proficiency level.

BEGINNING: LOW/HIGH Restate each step in simple terms. Demonstrate each step before students perform it.

INTERMEDIATE: LOW/HIGH Paraphrase the procedures before students begin. Allow them to answer the Analyze and Conclude questions with a partner.

ADVANCED: LOW/HIGH Discuss the procedures and the Analyze and Conclude questions before students begin. Discuss the meanings of words that may be unfamiliar, such as *formation* and *aqueous*.

Small-Scale Lab

OBJECTIVE After completing this activity, students will be able to explain some behaviors of liquids and solids.

SKILLS FOCUS Observing, inferring, drawing conclusions

PREP TIME 20 minutes

CLASS TIME 40 minutes

MATERIALS plastic Petri dish, water, ice, rubbing alcohol, calcium chloride, graph paper (1-cm), bromothymol blue solution (BTB), vinegar, aqueous ammonia

ADVANCE PREPARATION Purchase vinegar, aqueous ammonia, and rubbing alcohol. If you purchase BTB as a powder, prepare a 0.04% solution by dissolving 100 mg BTB powder in 16.0 mL 0.01*M* NaOH. Dilute to 250 mL.

SAFETY Be sure students wear goggles and aprons. BTB can stain clothing and skin. The room should be well ventilated and there should be no flames.

EXPECTED OUTCOMES The answers to the Analyze questions include a description of the expected outcome for each experiment.

ANALYZE AND CONCLUDE

1. Water in the dish evaporates and condenses when it contacts the cold surface under the ice.
2. The drop of water on top of the dish provides enough cooling to cause condensation.
3. Water beads up and alcohol spreads because the intermolecular attractions in water are stronger.
4. Calcium chloride absorbs water from the environment in the dish.
5. The many pieces of calcium chloride effectively dry the atmosphere, leaving no water vapor in the dish.

YOU'RE THE CHEMIST

1. The water drop increases in diameter over time as the alcohol evaporates and is "captured" by the water drop. The attractions in the resulting mixture are weaker overall.
2. The BTB turns from green to yellow in the presence of vinegar.
3. Place a drop of vinegar and a drop of BTB about 3 cm apart in a Petri dish. Cover and observe. The BTB slowly changes from green to yellow. Ethanoic acid that evaporates is "captured" by the BTB.
4. Place a drop of ammonia and a drop of BTB about 3 cm apart in a Petri dish. Cover and observe. The BTB slowly changes from green to blue. Ammonia that evaporates is "captured" by the BTB.

Answers

FIGURE 13.14 The hollow cages have a structure that is closer to diamond's rigid structure than to the widely spaced layers in graphite.

LESSON 13.4

Key Objectives

13.4.1 IDENTIFY the conditions necessary for sublimation.

13.4.2 DETERMINE how the conditions at which phases are in equilibrium are represented on a phase diagram.

Additional Resources

Reading and Study Workbook, Lesson 13.4

Available Online or on Digital Media:

- Teaching Resources, Lesson 13.4 Review, Interpreting Graphics
- Laboratory Manual, Lab 22

PearsonChem.com

Engage

CHEMISTRY & YOU Explain that rain is the condensation product of water vapor in the atmosphere. **Ask** Describe the water cycle in terms of phase changes. *(Answers will vary, but responses should focus on evaporation, condensation, and precipitation.)* Use the discussion to introduce sublimation.

Activate Prior Knowledge

Display in the room a large poster of the water cycle. Use colored sticky notes to represent solids, liquids, and gases. Have student volunteers place the sticky notes in the correct location on the water cycle poster. Then discuss the relationship between temperature and the three phases of water as an introduction to the concepts addressed in Figure 13.18.

IN ACADEMIC STANDARDS for SCIENCE

- Develop explanations based on reproducible data and observations gathered during laboratory investigations.

C.1.5 Describe the characteristics of solids, liquids, and gases and state changes at the molecular level.

13.4 Changes of State

C.1.5 States of matter

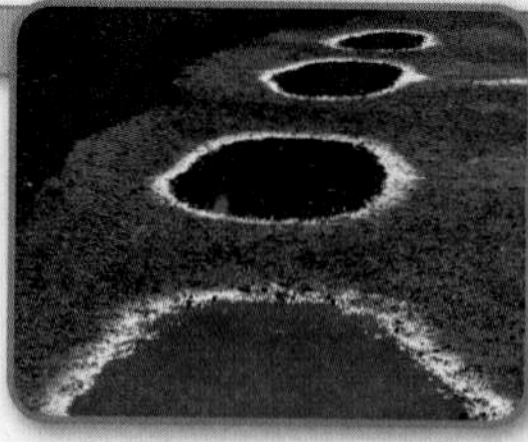

CHEMISTRY & YOU

Q: *Where does rain water go when a puddle dries up?* Water cycles through much of the matter on Earth. It falls as liquid rain or solid snow, collects in rivers, oceans, or glaciers and returns to the air as a gas through evaporation. All living organisms use water and release it back into the air. Other elements, such as carbon and nitrogen, also cycle through Earth as solids, liquids, or gases. In this section, you will learn what conditions can control the state of a substance.

Key Questions

When can sublimation occur?

How are the conditions at which phases are in equilibrium represented on a phase diagram?

Vocabulary

- sublimation
- phase diagram
- triple point

Sublimation

When can sublimation occur?

If you hang wet laundry on a clothesline on a very cold day, the water in the clothes quickly freezes to ice. Eventually, however, if the day is sunny, the clothes become dry although the ice never thaws. The ice changes directly to water vapor without melting and passing through the liquid state. The change of a substance from a solid to a vapor without passing through the liquid state is called **sublimation.** Sublimation can occur because solids, like liquids, have a vapor pressure. **Sublimation occurs in solids with vapor pressures that exceed atmospheric pressure at or near room temperature.**

$$\text{Solid} \underset{\leftarrow \text{deposition}}{\overset{\text{sublimation} \rightarrow}{\rightleftharpoons}} \text{Vapor}$$

Iodine is another example of a substance that undergoes sublimation. This violet-black solid ordinarily changes into a purple vapor without passing through a liquid state. Notice in Figure 13.15 how dark crystals of iodine deposit on the outside of a test tube placed inside another test tube. The outer test tube contains solid iodine that is being gently heated. The inner test tube contains liquid water and ice. The iodine crystals at the bottom of the outer test tube change directly to iodine vapor. When the vapor reaches the cool surface of the inner test tube, it goes directly from the gaseous to the solid state.

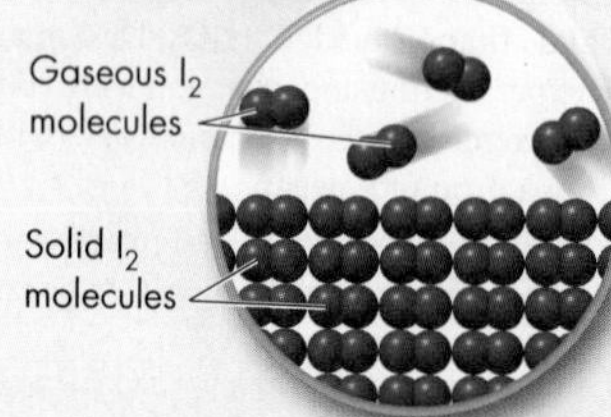

Figure 13.15 Sublimation
When solid iodine is heated, the crystals sublime, going directly from the solid to the gaseous state. When the vapor cools, it goes directly from the gaseous to the solid state.

Focus on ELL

1 CONTENT AND LANGUAGE Students who know that the prefix *sub-* means "below or under" may be confused by the definition of sublimation. Explain that this word comes from the Latin word *sublimare* which is related to the word *sublime* and means "to raise or elevate."

2 FRONTLOAD THE LESSON Ask students to predict whether a solid can change to a gas without first becoming a liquid. *(Answers will vary.)* Then, display a piece of dry ice in a beaker placed in a well-ventilated area, and have students examine the beaker for the presence of liquid (CAUTION: do not handle dry ice without appropriate safety precautions and gear, as it can cause serious skin damage on contact.). Students should notice that there is no liquid in the beaker, and that the solid CO_2 enters the gas phase directly. Explain that this type of phase change is not limited to dry ice, and that they will explore why such a phase change is possible.

Sublimation has many useful applications. If freshly brewed coffee is frozen and the water vapor is removed with a vacuum pump (a sublimation process), the result is freeze-dried coffee. The solid carbon dioxide (dry ice) shown in Figure 13.16 is often used as a coolant for goods, such as ice cream, that must remain frozen during shipment. Dry ice has a low temperature of −78°C. Because it sublimes, it does not produce a liquid, as ordinary ice does when it melts. Solid air fresheners contain a variety of substances that sublime at room temperature. Sublimation is also useful for separating substances. Organic chemists use sublimation to separate mixtures and to purify compounds.

Figure 13.16 Dry Ice Solid carbon dioxide, or dry ice, sublimes at normal atmospheric pressures. As it changes state, dry ice absorbs heat, keeping materials nearby cool and dry.
Interpret Photos *Why does fog form in the air around the dry ice?*

Quick Lab

Sublimation

Purpose
To observe the sublimation of air freshener

Materials
- small pieces of solid air freshener
- small shallow container
- 2 clear 8-oz plastic cups
- hot tap water
- ice
- 3 thick cardboard strips

Procedure

1. Place a few pieces of air freshener in one of the cups. **CAUTION** *Work in a well-ventilated room.*
2. Bend the cardboard strips and place them over the rim of the cup that has the air freshener pieces.
3. Place the second cup inside the first. The base of the second cup should not touch the air freshener. Adjust the cardboard as necessary. This assembly is your sublimator.
4. Fill the top cup with ice. Do not get any ice or water in the bottom cup.
5. Fill the shallow container about one-third full with hot tap water.

6. Carefully place your sublimator in the hot water. Observe what happens.

Analyze and Conclude

1. **Define** What is sublimation?
2. **Predict** What do you think would happen if the water in the shallow container were at room temperature? If it were boiling?
3. **Explain** Why is it possible to separate the substances in some mixtures by sublimation?

Focus on ELL

4 LANGUAGE PRODUCTION Have students work in groups of four to complete the lab. Make sure each group has ELLs of varied language proficiencies, so that more proficient students can help less proficient ones. Have students work according to their proficiency level.

BEGINNING: LOW/HIGH Rephrase each Procedure step in simple terms. Have students follow along as you demonstrate each step.

INTERMEDIATE: LOW/HIGH Slowly read each Procedure step aloud. Answer the Analyze and Conclude questions as a class. Have students copy the answers.

ADVANCED: LOW/HIGH Have more capable students help others with a lower language proficiency by posing the Analyze and Conclude questions as true/false questions.

Foundations for Reading

BUILD VOCABULARY Explain that the word *sublimation* comes from the Latin word *sublimare*, meaning "to elevate." Convey that the meaning of *sublimare* was expanded by the Middle Ages to include "to refine" or purify. The process of sublimation can be used to remove impurities from a substance.

READING STRATEGY Have students note key points about sublimation and evaporation as they read this lesson. Then ask them to summarize their key points in a paragraph comparing the two processes.

Explore

Sublimation

Quick Lab

OBJECTIVE After completing this activity, students will be able to describe the process of sublimation.

SKILLS FOCUS Observing, inferring, predicting, communicating

PREP TIME 15 minutes

CLASS TIME 20 minutes

SAFETY Perform this activity in a well-ventilated room. Some people are allergic to substances in air fresheners.

TEACHING TIPS Students should not fill the shallow containers to the point where the sublimator floats, to prevent it from tipping over. If a colored air freshener is used, the condensed solid will be white.

EXPECTED OUTCOME The heat from the hot water causes the air freshener to sublime. When the vapor reaches the cold surface of the upper cup, it condenses back into a solid.

ANALYZE AND CONCLUDE

1. The change of a substance from a solid to a gas without passing through the liquid state.
2. At room temperature, there would be less sublimation; at the boiling point, there would be more sublimation.
3. When a substance sublimes in a mixture, other components remain as residue.

FOR ENRICHMENT

If you used a colored air freshener, ask students to explain why the condensed solid is white. *(The dye does not sublime and remains behind at the bottom of the sublimator.)*

Explain

Phase Diagrams

USE VISUALS Draw the phase diagram of water in Figure 13.18 on the board. Have volunteers label the states of matter. Using simple words, ask students to demonstrate melting, and have a volunteer draw an arrow on the diagram from the solid to the liquid area. Repeat this for other changes of state.

Explain

USE VISUALS Display Figure 13.18 on an overhead projector. Explain that a phase diagram is a convenient way to summarize the conditions of temperature and pressure at which the solid, liquid, and gas phases of a substance are most stable. Point out that each line in the diagram represents the set of possible temperature-pressure values at which different phases are in dynamic equilibrium with one another. For example, in a closed system at 100°C and 101.3 kPa, liquid water is in dynamic equilibrium with water vapor. (Note that the *x*- and *y*-axes are not drawn to scale so that the triple point and normal boiling point can fit on the graph.) **Ask** At what temperature and pressure are the liquid, solid, and vapor phases of water in dynamic equilibrium? *(0.016°C, 0.61 kPa)*

Extend

By this point in the course students should have a strong understanding that reactions can occur between different states of matter as well as between liquids and between gases. However, they may not realize that reactions can also occur between solids. Explain to students that these types of reactions produce a number of useful products, including semi-conductors and light-sensitive photopolymers. A number of chemists specialize in solid-state chemistry. Have students research current areas of experimentation in solid state chemistry either on the Internet or by interviewing a solid-state chemist at a local college or university, and prepare a poster showing examples of applications of this research that are found in everyday objects.

Figure 13.17 Triple Point
This flask is at the triple point, where ice, liquid water, and water vapor exist in equilibrium. Freezing, melting, boiling, and condensation are all occuring at the same time in the flask.

Phase Diagrams

How are the conditions at which phases are in equilibrium represented on a phase diagram?

The relationships among the solid, liquid, and vapor states (or phases) of a substance in a sealed container can be represented in a single graph. The graph is called a phase diagram. A **phase diagram** gives the conditions of temperature and pressure at which a substance exists as solid, liquid, or gas (vapor). **The conditions of pressure and temperature at which two phases exist in equilibrium are indicated on a phase diagram by a line separating the two regions representing the phases.**

Figure 13.18 shows the phase diagram for water. In each of the colored regions of the phase diagram, water is in a single phase. The curving line that separates water's vapor phase from its liquid phase describes the equilibrium conditions for liquid and vapor. Under these conditions, water can change its state from liquid to vapor or from vapor to liquid. The same line also illustrates how the vapor pressure of water varies with temperature. The other two lines describe the conditions for equilibrium between liquid water and ice and between water vapor and ice. The point on the diagram at which all three lines meet is called the triple point. The **triple point** describes the only set of conditions at which all three phases can exist in equilibrium with one another. For water, the triple point is a temperature of 0.016°C and a pressure of 0.61 kPa (0.0060 atm). Figure 13.17 shows water at its triple point.

Interpret Graphs

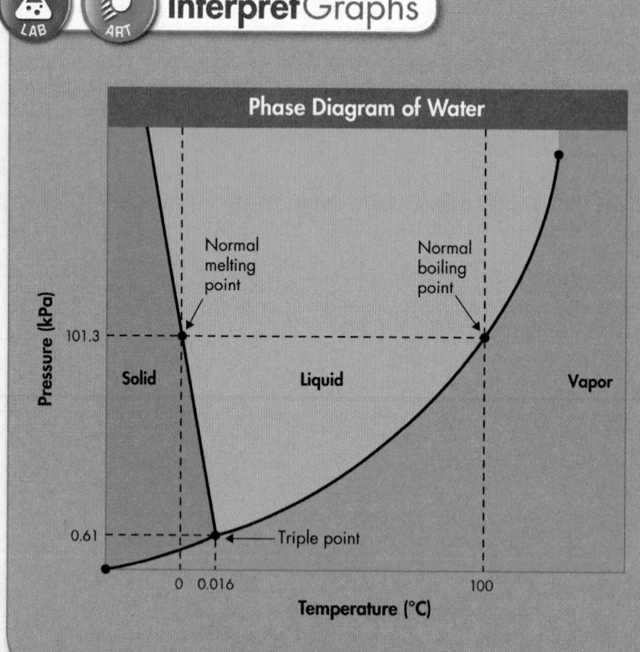

Figure 13.18 The phase diagram of water shows the relationship among pressure, temperature, and the physical states of water. Note that the scale on the axes is not linear.

a. Read Graphs At the triple point of water, what are the values of temperature and pressure?

b. Identify What states of matter are present at the triple point of water?

c. Analyze Assuming standard pressure, at what temperature is there an equilibrium between water vapor and liquid water? Between liquid water and ice?

Hint: Each line represents the set of possible temperature-pressure values at which the phases are in dynamic equilibrium.

Professional Development Note: Freeze-Dried Foods

Astronauts and backpackers use freeze-dried foods because they are lightweight and easily reconstituted by adding water. Freeze-dried foods don't require refrigeration because bacteria cannot multiply in the absence of moisture. The sublimation of ice at low pressure makes the process of freeze-drying possible. Food is frozen and placed in a chamber attached to a vacuum pump. Then the pressure in the chamber is lowered until the ice crystals sublime. Freeze-dried food tends to maintain its flavor because the substances in the food that impart flavor remain in the food as the ice sublimes.

By referring to Figure 13.18, you can determine what happens if you melt ice or boil water at pressures less than 101.3 kPa. A decrease in pressure lowers the boiling point and raises the melting point. An increase in pressure will raise the boiling point and lower the melting point.

Look at Figure 13.18. Follow the equilibrium line between liquid water and water vapor to the triple point. Below the triple point, the vapor and liquid cannot exist in equilibrium. Increasing the pressure won't change the vapor to a liquid. The solid and the vapor are in equilibrium at temperatures below 0.016°C. With an increase in pressure, the vapor begins to behave more like a solid. For example, it is no longer easily compressed.

Q: *Describe how water might move from Earth's surface to Earth's atmosphere and back again as part of the water cycle. Be sure to include any phase changes that occur in your description.*

Figure 13.18 also illustrates how an increase in pressure affects the melting point of ice. For years, the accepted hypothesis for how ice skaters move along the ice was the following. The blades of the skates exert pressure, which lowers the melting point of the ice. The ice melts and a film of water forms under the blades of the skates. This film acts as a lubricant, enabling the skaters to glide gracefully over the ice as shown in Figure 13.19. This hypothesis fails to explain why skiers also glide along very nicely on another solid form of water—snow. Wide skis exert much less pressure per unit area of snow than narrow skate blades exert on ice. Recent research shows that the surface of ice has a slippery, water-like surface layer that exists well below ice's melting point. Even ice that is at −129°C has this layer. A new hypothesis proposes that the liquid-like surface layer provides the lubrication needed for smooth skating and skiing.

Figure 13.19 Water on Ice
The surface of ice has a thin layer of water above it.
Apply Concepts ***Research shows that the colder the ice is, the thinner the layer of water above the ice is. In which sports—ice hockey, speed skating or figure skating—do you think that it would benefit the athletes more to keep the ice colder?***

13.4 LessonCheck

25. Identify What properties must a solid have to undergo sublimation?

26. Explain What do the curved lines on a phase diagram represent?

27. Apply Concepts Describe one practical use of sublimation.

28. Interpret Graphs Using Figure 13.18, estimate the boiling point of water at a pressure of 50 kPa.

29. Explain What does the triple point on a phase diagram describe?

30. Classify Would you expect a substance that sublimates at or near room temperature to be a molecular substance or an ionic substance? Use what you learned about bonding in Chapters 7 and 8 to answer the question.

CHEMISTRY & YOU The water in a pond evaporates. The water vapor in the atmosphere condenses to form clouds. When the water droplets in clouds get heavy enough, they fall to the ground as rain.

Evaluate

Informal Assessment

List a series of temperature–pressure values on the board. Have students state or point to which phase of water is most stable under each set of conditions and explain why.

Reteach

On the board, draw a stoppered, sidearm flask containing water and a chunk of ice. Note that the sidearm is connected to a vacuum pump. Tell students that the diagram illustrates ice, liquid water, and water vapor in equilibrium. Have groups of students approach the diagram and one at a time have a student add an arrow and label to the diagram to represent a phase change. Continue until there are six arrows on the diagram.

Lesson Check Answers

25. Sublimation occurs in solids that have vapor pressures that exceed atmospheric pressure at or near room temperature.
26. The lines show the conditions of temperature and pressure at which two phases exist in equilibrium.
27. Sample answers: freeze-dried coffee, dry ice as a coolant, air fresheners, separating mixtures, and purifying substances
28. below 100°C
29. The triple point describes the only set of conditions at which three phases can exist in equilibrium.
30. The substance is likely molecular because the bonds between ions are typically stronger than the attractions between molecules.

Answers

INTERPRET GRAPHS

a. 0.016°C and 0.61 kPa
b. solid, liquid, and vapor
c. 100°C; 0°C

FIGURE 13.19 A thick layer of water means that there is more friction, which can slow the skaters down. In ice hockey and speed skating, where speed is crucial, it is better to skate on colder ice. Also, the warmer the ice is the softer the surface of the ice is. In figure skating, the layer of water provides a cushion for the skaters if they fall.

CHEMISTRY & YOU

CHEMISTRY & YOU Hazardous wastes such as chemicals and medical items (biological fluids and tissues, syringe needles, and so on) are often buried in landfills or incinerated in traditional high-temperature incinerators. Both methods are effective, but have some serious disadvantages. For example, landfills rely on bacterial decomposition to destroy hazards. However, this requires time, and some chemicals do not undergo biological decomposition. Landfills also can leak, contaminating groundwater. High-temperature incinerators rely on combustion to destroy hazardous wastes, but must consistently reach temperatures greater than 1200°C to break chemical bonds. At lower temperatures, incomplete combustion or the formation of hazardous by-products may occur. Incomplete combustion results in soot formation. It may also take several hours for solid wastes to be completely incinerated.

Pose the following question to students: *Based on the description given in the feature, what are some benefits of using a plasma waste converter instead of burning waste material?*

You may need to guide students in the following ways:

- The temperature in a plasma furnace can be hotter than the surface of the Sun, which is 5497°C.
- Plasma furnaces are enclosed, unlike incinerators where waste is burned.
- Some high-temperature incinerators produce liquid waste that must be treated before it can be discharged into the sewer system.

Explain

START A CONVERSATION Engage students in discussing how a plasma is different from other states of matter. They should understand that, unlike other states of matter, a plasma is not made of neutral atoms. The particles of a plasma have so much energy that the electrons are stripped away, leaving positively charged ions and negatively charged electrons. Point out that, although other states of matter are more common on Earth, plasma is the most common state of matter in the universe. Stars, including our Sun, are made of plasma.

CHEMISTRY & YOU: GREEN CHEMISTRY

Plasma Waste Converter

You are very familiar with solids, liquids, and gases, but did you know that there's a fourth state of matter called plasma? A plasma is a gaseous mixture of freely moving positive ions and electrons. Plasmas exist naturally in the sun and other stars, and in lightning. Plasmas can also be artificially created and used in a variety of technologies. One new application of plasma technology is plasma waste converters, which treat most waste material, including many hazardous wastes.

Plasma waste converters produce plasmas with temperatures as high as 7,000°C. The energy within the plasma breaks the waste material's molecular bonds so that the material separates into its elemental components. The end products are a hydrogen-rich gas known as syngas and a solid known as slag.

Plasma waste converters have many benefits over traditional waste treatments. They do not require the use of dumps or landfills, and they emit much less greenhouse gases and other pollutants than incinerators. Despite these benefits there are currently only two plasma waste treatment plants that process municipal solid waste. This is mostly due to the cost since each plant must be custom-built. Standardizing production should result in the construction of more plants.

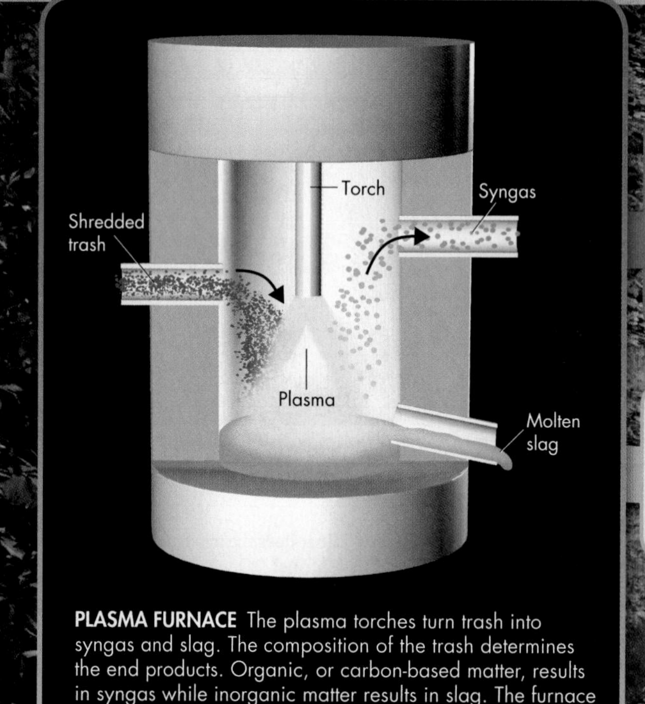

PLASMA FURNACE The plasma torches turn trash into syngas and slag. The composition of the trash determines the end products. Organic, or carbon-based matter, results in syngas while inorganic matter results in slag. The furnace is airtight, so that none of the syngas can escape.

SYNGAS
The gases pass from the furnace into a chamber where any remaining organic matter or hazardous wastes are removed. The gases can then be used as a fuel source or to make electricity.

SLAG
Any hazardous materials in the molten slag are inert and will not dissolve out of the slag. The slag takes different forms depending on how it is cooled.

21st Century Learning To be successful in the 21st century, students need skills and learning experiences that extend beyond subject matter mastery. The following project helps students build the following 21st Century Skills: *Creativity and Innovation; Communication and Collaboration; Information Literacy; Initiative and Self-Direction.*

USING PLASMA Pose the following challenge to your students. *Suppose you are the safety supervisor for a welding company. It is your responsibility to train new welders about workplace safety in regards to plasma welding equipment. Create a multimedia presentation on workplace safety with regards to handling and using plasma welding equipment that includes a 15-20 minute computerized slide show and a take-home brochure.*

The presentation and brochure should explain the chemical and physical properties of plasma, why it is used for welding, its advantages and disadvantages compared to other types of welding, and include all safety considerations.

Take It Further

1. **Infer** Describe how a gas might become a plasma.
2. **Compare and Contrast** Is it misleading to refer to a plasma waste converter as a plasma incinerator? Research how incinerators work and compare them to how plasma waste converters work.
3. **Evaluate the Impact on Society** How might the use of plasma waste converters change society?

USE VISUALS Have students study the diagram of a plasma furnace and think about its different parts. **Ask** Why do you think the trash is shredded? *(to make breaking it down easier)* **Ask** How is the syngas (synthetic gas, pronounced sin gas) separated from the slag? *(The less-dense gas moves upward, and the denser slag flows out.)* **Ask** Why do you think the slag in the photo forms a glass rather than crystals? *(Like volcanic rocks, the slag cools so quickly that the particles do not have time to arrange themselves into a repeating pattern before the slag hardens.)*

Extend

Connect to ECONOMICS

Encourage students to consider whether a plasma waste converter ultimately has a higher or lower cost for society. Point out that although a plasma waste converter is expensive to build, the costs of not building it might be high also. For example, have students consider the costs to society of dealing with polluted air and storing toxic wastes such as PCBs.

Differentiated Instruction

L1 STRUGGLING STUDENTS Have students study the feature with a partner. Instruct students to read each paragraph and discuss it before going to the next paragraph. Check with pairs frequently to provide guidance and answer questions.

ELL ENGLISH LANGUAGE LEARNERS List idiomatic and unfamiliar words in the feature on the board. For example, you might list *municipal, the grid, sellable, molten, fill 'er up, concrete,* and *asphalt.* Discuss their meanings before students attempt to read the feature.

L3 ADVANCED STUDENTS Have students research a municipal plasma waste conversion facility that exists today, or one that is being planned or built. Instruct students to provide a short summary for the class about the facility.

Answers

TAKE IT FURTHER

1. Heat a gas until electrons begin to strip off the atoms.
2. Yes, it is misleading to refer to a plasma waste converter as a plasma incinerator because a plasma waste converter does not burn waste. Rather it breaks the waste down into its elemental components. Furthermore, a byproduct of incineration is ash. Ash must still be disposed of in landfills.
3. Less green space would have to be set aside for landfills. There may not be the need to recycle. Also, plasma waste converters could reduce society's dependence on fossil fuels.

STUDY GUIDE

Evaluate

Review and Assessment Materials

CHEMISTRY TUTORIAL Encourage students to review converting between units of pressure using the animated step-by-step tutorial for Chapter 13. Each Chemistry Tutorial is followed by a similar practice problem, where the student is guided through the problem-solving process.

CHAPTER PROBLEM SET Have students practice more kinetic theory and states of matter problems using the Chapter 13 online Problem Set.

MATH TUTORIAL Struggling students can use the tutorial on reading graphs as a quick review. Ask students to answer the guided practice problem that follows the Math Tutorial.

KINETIC ART Direct students to the Kinetic Art animations *Evaporation* and *Changes in State*, and have them answer the questions that accompany each animation.

Study Tip

One excellent way to evaluate your knowledge of a key concept is to try to explain the concept to another person. The questions the other person asks will help you assess the depth of your knowledge.

Understanding by Design

Performance Tasks

COMIC STRIP Have students create and illustrate a comic strip that explains kinetic theory as it relates to different states of matter. Students may opt to generate their strips using an online comic strip creator or computer art software. The strips should utilize a storyline involving characters and a situation that demonstrates students' ability to successfully apply the kinetic theory to one or more substances.

TEACH A LESSON Divide the class into groups of four students. Assign each student within the group a topic: solid, liquid, gas, or change of state. Instruct students to prepare and teach their topic to others in their group. Encourage students to use multi-media or tactile examples as part of their instruction.

13 Study Guide

BIGIDEA KINETIC THEORY

The state of a substance is determined by conditions of pressure and temperature. Low pressures and high temperatures favor gases, in which particles move randomly in constant, high-speed motion. At higher pressures and lower temperatures, the particles slow down and exert attractive forces on one another, producing a liquid. At still higher pressures and lower temperatures, the particles become fixed in orderly arrangements, producing solids. A substance changes state when there is a change in the balance between the random motion of its particles and the attractions among those particles.

13.1 The Nature of Gases

Particles in a gas are considered to be small, hard spheres with an insignificant volume. The motion of the particles in a gas is rapid, constant, and random. All collisions between particles in a gas are perfectly elastic.

Gas pressure is the result of billions of rapidly moving particles in a gas simultaneously colliding with an object.

The Kelvin temperature of a substance is directly proportional to the average kinetic energy of the particles of the substance.

- kinetic energy (420)
- kinetic theory (420)
- gas pressure (421)
- vacuum (421)
- atmospheric pressure (421)
- barometer (421)
- pascal (Pa) (422)
- standard atmosphere (atm) (422)

13.2 The Nature of Liquids

The interplay between the disruptive motions of particles in a liquid and the attractions among the particles determines the physical properties of liquids.

During evaporation, only those molecules with a certain minimum kinetic energy can escape from the surface of the liquid.

In a system at constant vapor pressure, a dynamic equilibrium exists between the vapor and the liquid. The rates of evaporation and condensation are equal.

At a temperature at which particles throughout a liquid have enough kinetic energy to vaporize, the liquid begins to boil.

- vaporization (426)
- evaporation (426)
- vapor pressure (427)
- boiling point (428)
- normal boiling point (430)

13.3 The Nature of Solids

The general properties of solids reflect the orderly arrangement and the fixed locations of their particles.

The shape of a crystal reflects the arrangement of the particles within the solid.

- melting point (431)
- freezing point (431)
- crystal (432)
- unit cell (432)
- allotropes (434)
- amorphous solid (434)
- glass (434)

13.4 Changes of State

Sublimation occurs in solids with vapor pressures that exceed atmospheric pressure at or near room temperature.

The conditions of pressure and temperature at which two phases exist in equilibrium are indicated on a phase diagram by a line separating the two regions representing the phases.

- sublimation (436)
- phase diagram (438)
- triple point (438)

Focus on ELL

5 ASSESS UNDERSTANDING Use the phase diagram in Figure 13.18 as an alternative oral assessment tool.

BEGINNING: LOW/HIGH Point to the *solid, liquid,* and *gas* areas. Have students use simple phrases to explain two properties of each state of matter.

INTERMEDIATE: LOW/HIGH Ask simple questions about the diagram, such as *What is sublimation?* Allow students to answer orally, using the diagram as a visual aid during their response.

ADVANCED: LOW/HIGH Have students use complete sentences to describe the different states of matter and explain changes of state using the diagram.

13 Assessment

* Solutions appear in Appendix E

Lesson by Lesson

13.1 The Nature of Gases

*31. What is meant by an elastic collision?

32. Which of these statements are characteristic of matter in the gaseous state?
    a. Gases fill their containers completely.
    b. Gases exert pressure.
    c. Gases have mass.
    d. The pressure of a gas is independent of the temperature.
    e. Gases are compressible.

33. List the various units used to measure pressure, and identify the SI unit.

34. Change 1656 kPa to atm.

*35. Change 190 mm Hg to kPa and atm.

36. Explain the relationship between the Kelvin temperature of a substance and the kinetic energy of its particles.

37. How is the average kinetic energy of water molecules affected when you pour hot water into cups at the same temperature as the water?

38. What does the abbreviation STP represent?

39. What is significant about the temperature absolute zero?

*40. By what factor does the average kinetic energy of the molecules of gas in an aerosol container increase when the temperature is raised from 27°C (300 K) to 627°C (900 K)?

13.2 The Nature of Liquids

41. Explain why liquids and gases differ in density and the ability to be compressed.

42. Compare evaporation of a liquid in a closed container with that of liquid in an open container.

*43. Describe what is happening at the molecular level when a dynamic equilibrium occurs.

44. Explain why increasing the temperature of a liquid increases its rate of evaporation.

45. Would you expect a dynamic equilibrium in a liquid in an open container? Explain.

46. Describe the effect that increasing temperature has on the vapor pressure of a liquid.

47. Distinguish between the boiling point and the normal boiling point of a liquid.

48. Use the graph to answer each question.

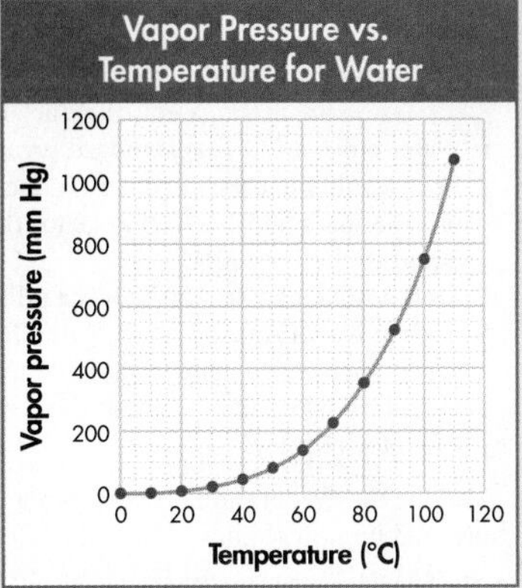

a. What is the vapor pressure of water at 40°C?
b. At what temperature is the vapor pressure of water 600 mm Hg?
c. What is the significance of the vapor pressure of water at 100°C?

*49. Explain how boiling is a cooling process.

13.3 The Nature of Solids

50. Name at least one physical property that would permit you to distinguish a molecular solid from an ionic solid.

51. Describe what happens when a solid is heated to its melting point.

*52. Explain why molecular solids usually have lower melting points than ionic solids.

13.4 Changes of State

*53. When you remove the lid from a food container that has been left in a freezer for several months, you discover a large collection of ice crystals on the underside of the lid. Explain what has happened.

54. Explain why a liquid stays at a constant temperature while it is boiling.

Evaluate

Answers

LESSON 13.1

31. In an elastic collision, energy is transferred between particles.

32. a, b, c, and e

33. pascal (Pa), SI; millimeter of mercury (mm Hg); atmosphere (atm)

34. 16.35 atm

35. 25 kPa and 0.25 atm

36. The Kelvin temperature is directly proportional to the average kinetic energy.

37. If the temperature does not change, the average kinetic energy is not affected.

38. STP stands for standard temperature (0°C) and pressure (101.3 kPa or 1 atm).

39. At absolute zero, the motion of particles would theoretically cease.

40. The average kinetic energy triples.

LESSON 13.2

41. The particles in a gas are relatively far apart compared to the particles in a liquid. Because of the extra space between particles, a gas is less dense and easier to compress.

42. In both cases, particles with sufficient kinetic energy move from the liquid to the vapor phase. In a closed container, a dynamic equilibrium is set up between the contained liquid and its vapor.

43. Two opposing processes are occurring at identical rates.

44. More molecules have enough energy to escape the attractions within the liquid.

45. No. In an open container, most of the particles that escape from the surface of the liquid as vapor do not condense back to liquid.

46. The average kinetic energy increases, which allows more vapor to form above the liquid.

47. The boiling point is the temperature at which the vapor pressure equals the external pressure. At the normal boiling point, the external pressure is 101.3 kPa.

48. **a.** about 50 mm Hg **b.** about 94°C
c. 760 mm Hg is standard pressure

49. Escaping molecules have more kinetic energy than the average. Thus, the average kinetic energy and temperature of the remaining molecules are lower.

LESSON 13.3

50. Ionic compounds generally have higher melting points than molecular solids do.

51. The particles have sufficient kinetic energy to overcome the attractive forces holding them in place.

52. The intermolecular attractions between molecules are weaker than the attractions between ions.

LESSON 13.4

53. Water from the food sublimed and then condensed on the lid.

54. The temperature remains constant while the liquid boils because the energy that is added is used to vaporize the molecules.

ASSESSMENT

Evaluate

UNDERSTAND CONCEPTS

55. The average kinetic energy of the molecules is greater because, by definition, a fever is a state of increased body temperature.

56. **a.** 121°C **b.** chloroform **c.** chloroform

d. The external pressure on ethanol would have to increase; the external pressure on ethanoic acid would have to decrease.

e. about 77°C

57. Evaporation is the conversion of a liquid to a gas or vapor when the liquid is below its boiling point. The vapor pressure is the force per unit area exerted by the vaporized particles on the walls of a sealed container. The boiling point is the temperature at which the vapor pressure of the liquid is equal to the external pressure.

58. Although the net amounts of vapor and liquid remain constant, some molecules are evaporating while an equal number of molecules are condensing.

59. **a.** 80°C **b.** 100°C

60. decrease; As the attractions become stronger, it becomes more difficult for molecules to overcome the attractions and vaporize.

61. **a.** liquid **b.** vapor **c.** liquid **d.** vapor **e.** solid **f.** solid

62. Atmospheric pressure results from the collisions of particles in air with objects. There are fewer particles in a given volume of air at the top of a mountain than at sea level.

63. The Kelvin temperature is directly proportional to the average kinetic energy. As the temperature rises, the air particles speed up and increase in kinetic energy, which causes the raft to expand. As the temperature drops, the air particles slow down resulting in a decrease in kinetic energy, which causes the raft to not be fully inflated.

THINK CRITICALLY

64. When you draw on a straw, the pressure inside the straw is less than the pressure (atmospheric) on the liquid in the container. So liquid is pushed up the straw.

65. Possible answer: Since the beaker is an open container, the water should boil at 100°C at or close to sea level. Your partner probably misread the thermometer and should recheck the value.

66. Possible answers: Odors will travel through a room; ink will spread throughout a beaker of water.

67. The average kinetic energy is the same, because the temperature is the same.

68. No. Collisions between large objects involve some loss of kinetic energy to heat.

69. Since molecules require kinetic energy to evaporate, the evaporation removes energy from your skin, cooling it down.

Understand Concepts

***55.** What happens to the average kinetic energy of the water molecules in your body when you have a fever?

56. Refer back to Figure 13.9 to answer the following questions.

a. What is the normal boiling point of ethanoic acid?

b. Which liquid has the highest vapor pressure at 40°C?

c. At standard atmospheric pressure, which of the substances are in the gaseous state at 70°C?

d. Water boils at 100°C at standard pressure. How would the pressure on ethanol and on ethanoic acid have to change for these liquids to boil at 100°C?

e. Mount McKinley in Alaska is the tallest peak in North America at 6194 m. The atmospheric pressure at its peak is 44 kPa. What is the boiling point of water at the peak of Mount McKinley?

57. Describe the evaporation process, vapor pressure, and boiling point.

58. Why is the equilibrium that exists between a liquid and its vapor in a closed container called a dynamic equilibrium?

59. The table gives the vapor pressure of isopropyl alcohol at various temperatures. Graph the data. Use a smooth curve to connect the data points.

Temperature (°C)	Vapor pressure (kPa)
0	1.11
25	6.02
50	23.9
75	75.3
100	198
125	452

a. What is the estimated normal boiling point of isopropyl alcohol?

b. What is the boiling point of isopropyl alcohol when the external pressure is increased to twice standard pressure?

***60.** In a series of liquids, as the intermolecular forces of attraction strengthen, would you expect the vapor pressure to increase or decrease? Explain.

61. Predict the physical state of each substance at the specified temperature. Use the melting point and boiling point data from the table below.

a. phenol at 99°C

b. ammonia at −25°C

c. methanol in an ice-water bath

d. methanol in a boiling-water bath

e. ammonia at −100°C

f. phenol at 25°C

Substance	Melting Point (°C)	Boiling Point (°C)
ammonia	−77.7	−33.4
methanol	−97.7	64.7
water	0	100
phenol	40.9	181.9

62. Why is atmospheric pressure much lower on the top of a mountain than it is at sea level?

***63.** A fully inflated raft is left outside overnight. The next morning the raft is not fully inflated. As the air temperature rises during the day, the raft expands to the size it was the night before. Assuming that the amount of air inside the raft did not change, use kinetic theory to explain why the size of the raft changes.

Think Critically

64. Relate Cause and Effect What role does atmospheric pressure play when someone is drinking a liquid through a straw?

***65. Analyze Data** Your lab partner measures the boiling point of water in an open beaker as 108.2°C. You know that water can be made to boil at this temperature, but you still ask your partner to repeat the measurement. Explain.

66. Infer What everyday evidence suggests that all matter is in constant motion?

67. Explain Is the average kinetic energy of the particles in a block of ice at 0°C the same as or different from the average kinetic energy of the particles in a gas-filled weather balloon at 0°C? Explain.

68. Infer Can objects that are large enough for you to see collide elastically? Explain.

69. Apply Concepts How does perspiration help cool your body on a hot day?

444 Chapter 13 • Assessment

70. **Relate Cause and Effect** Why do different liquids have different normal boiling points?

71. **Explain** A liquid-vapor equilibrium exists in a container. Explain why the equilibrium vapor pressure in the container is not affected when the volume of the container is changed.

*72. **Analyze Data** A teacher wants to demonstrate that unheated water can boil at room temperature in a beaker within a bell jar connected to a vacuum pump. However, the vacuum pump is faulty and can reduce pressures only to 15 kPa. Can the teacher use this pump to perform the demonstration successfully? Explain your answer.

73. **Compare** You have two sealed jars of water at the same temperature. In the first jar, there is a large amount of water. In the second jar, there is a small amount of water. Explain how the vapor pressure can be the same in both jars.

74. **Draw Conclusions** Why are pressure cookers recommended for cooking at high altitude?

75. **Propose a Solution** A mixture of gases contains oxygen, nitrogen, and water vapor. What physical process could you use to remove the water vapor from the sample?

Enrichment

*76. **Apply Concepts** Relative humidity is defined by the following equation

$$\text{Relative humidity} = \frac{(a)}{(b)} \times 100\%$$

where (a) is the pressure of water vapor in the air and (b) is the equilibrium vapor pressure of water in the air at the same temperature. Can the relative humidity exceed 100%? Explain.

77. **Use Models** The ions in sodium chloride are arranged in a face-centered cubic pattern. Draw a layer of ions in a sodium chloride crystal.

78. **Use Models** Using Figure 13.12, identify the crystal systems described below.

a. three unequal edges meet at right angles
b. three equal edges with three equal angles that are not right angles
c. two equal edges and one unequal edge meet at right angles
d. three unequal edges do not meet at right angles
e. three equal edges meet at right angles

*79. **Use Models** Use this drawing to answer the questions.

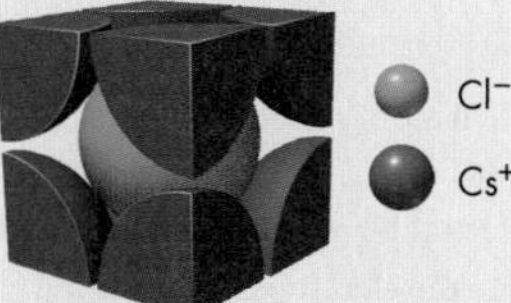

a. What type of unit cell is in a lattice of cesium chloride?
b. What is the coordination number of Cs^+? (*Hint:* Refer back to Section 7.2 for the definition of coordination number.)
c. Based on the diagram, what is the formula of cesium chloride? Explain your answer.

80. **Predict** The solid-liquid equilibrium line in the phase diagram of a given substance slants to the right. How is the substance's freezing point affected by increased pressure?

CHEMYSTERY

Foggy Car Windows

Car windows fog up because of condensation. When warm, moist air comes into contact with a cold surface, the water vapor in the air condenses into liquid drops on the cold surface. When it's cold outside, your body and breath warm the air inside the car. This warm, moist air comes in contact with the cold surface of the car window, causing it to fog up. Using a car heater increases the amount of moisture in the air, since warm air can hold more water vapor than cold air. The amount of moisture in the air will also be higher on a rainy day because some of the water on the passengers evaporates. An air conditioner cools and removes moisture from the air.

81. **Explain** Describe why opening a window can also help defog a car windshield.

82. **Connect to the BIGIDEA** How does the average motion of the water molecules change as water vapor condenses on a car window? Why does the motion change?

ASSESSMENT

Explain

CHEMYSTERY

APPLY CONCEPTS After students have read through the CHEMystery, call on volunteers to summarize why car windows fog up when it is cold or when it is raining, and why the fog goes away when you turn on the air conditioner. **Ask** On a cold day, is condensation more likely to form if you have cold air or warm air inside a car? Why? *(Condensation is more likely to form if you have warm air inside the car. When the water vapor inside the car touches the cold window, it cools and changes to a liquid.)* **Ask** Why does warm air hold more water vapor than cold air? *(The higher the air temperature is the higher the average kinetic energy of the water molecules is. As a result, more water molecules will have enough kinetic energy to overcome the forces of attraction and evaporate.)*

CHEMYSTERY ANSWERS

81. When you open a car window, fresh air will enter the car replacing some of the moist air inside of the car.

82. The average kinetic energy of the water molecules decreases as the water vapor condenses on the car window. The motion of the water molecules decreases as a result.

Answers

70. The intermolecular attractions in some compounds are stronger than in others.

71. The vapor pressure depends only on the kinetic energy of the escaping molecules.

72. No. At 15 kPa, water would boil at a temperature of about 50°C, which is much higher than room temperature.

73. The kinetic energy of the molecules in the vapor is the same in both cases; so the vapor pressure is the same.

74. At high altitude, the boiling point of water is less than 100°C because the atmospheric pressure is lower. The increased pressure in the pressure cooker increases the temperature at which water boils.

75. condensation of the water vapor on a cold surface

ENRICHMENT

76. No; if (a) > (b) then water vapor will condense at a greater rate than the liquid evaporates.

77. $Na^+Cl^-Na^+Cl^-$
$Cl^-Na^+Cl^-Na^+$
$Na^+Cl^-Na^+Cl^-$
$Cl^-Na^+Cl^-Na^+$

78. **a.** orthorhombic **b.** rhombohedral **c.** tetragonal **d.** triclinic **e.** cubic

79. **a.** body-centered cubic **b.** 8
c. CsCl (one Cl^- ion and 8 × 1/8 equals one Cs^+ ion)

80. The freezing point increases.

ASSESSMENT

Cumulative Review

83. inversely

84. a. S **b.** K

85. a. $1s^22s^22p^63s^23p^6$ **b.** $1s^22s^22p^63s^23p^6$ **c.** $1s^2$

86. a. 2 **b.** 2

87. dispersion forces, dipole interactions, hydrogen bonds

88. a. Cu_2SO_3 **b.** HNO_2

89. a. CO **b.** PBr_3

90. a. Fe^{3+} **b.** Cd^{2+}

91. a. 53.7% Fe **b.** 34.6% Al

92. a. 13.9 mol SO_2 **b.** 0.0472 mol NH_3

93. a. 51.2 g Cl_2O_7 **b.** 30.6 mL H_2O

94. a. 0.38 mol CO_2 **b.** 3.73 mol NH_3

95. $H_2S(aq) + Cd(NO_3)_2(aq) \rightarrow 2HNO_3(aq) + CdS(s)$

96. a. $V_2O_5 + 2H_2 \rightarrow V_2O_3 + 2H_2O$

b. $(NH_4)_2Cr_2O_7 \rightarrow Cr_2O_3 + N_2 + 4H_2O$

97. a. Mg **b.** Li

98. a. combination **b.** combustion

99. a. 198 g H_2O **b.** 23 mol **c.** 144 g C

100. C_2H_4

101. 39.4 g FeS (0.448 mol)

102. 41.9%

Cumulative Review

83. How are the frequency and wavelength of light waves related?

84. Which atom in each pair has the larger atomic radius?

a. O and S **b.** K and Br

***85.** Write the electron configuration of each ion.

a. Ca^{2+} **b.** S^{2-} **c.** Li^+

86. How many unshared pairs of electrons are in each molecule?

a. H_2O **b.** CO

87. List the intermolecular attractions between molecules in order of increasing strength.

a. dispersion forces
b. hydrogen bonds
c. dipole interactions

88. Write a correct formula for each compound.

a. copper(I) sulfite **b.** nitrous acid

89. Identify the binary molecular compound in each pair of substances.

a. NaCl or CO **b.** PBr_3 or LiOH

90. Write formulas for these ions.

a. iron(III) ion **b.** cadmium ion

91. Calculate the percent by mass of the metal in each compound.

a. Fe_2S_3 **b.** $Al(OH)_3$

92. How many moles are there in each sample?

a. 888 g of sulfur dioxide
b. 2.84×10^{22} molecules of ammonia

***93.** Perchloric acid forms by the reaction of water with dichlorine heptoxide.

$$Cl_2O_7 + H_2O \longrightarrow 2HClO_4$$

a. How many grams of Cl_2O_7 react with an excess of H_2O to form 56.2 g of $HClO_4$?
b. How many mL of water are needed to form 3.40 mol $HClO_4$?

94. How many moles are there in each sample?

a. 8.6 L CO_2 (at STP) **b.** 63.4 g NH_3

***95.** When hydrogen sulfide gas is bubbled into a solution of cadmium nitrate in water, the products are nitric acid and a precipitate of cadmium sulfide. Write a balanced equation for the reaction. Include physical states for all reactants and products. (Hydrogen sulfide gas is soluble in water.)

96. Balance these equations.

a. $V_2O_5 + H_2 \longrightarrow V_2O_3 + H_2O$
b. $(NH_4)_2Cr_2O_7 \longrightarrow Cr_2O_3 + N_2 + H_2O$

97. List the metal that ranks higher in the activity series of metals.

a. magnesium or mercury
b. potassium or lithium

98. Classify each reaction as a combination, decomposition, single-replacement, double-replacement, or combustion.

a. $2Li(s) + Br_2(l) \longrightarrow 2LiBr(s)$
b. $2C_2H_6(g) + 7O_2(g) \longrightarrow 4CO_2(g) + 6H_2O(g)$

***99.** The complete decomposition of sucrose (table sugar) caused by strong heating may be represented by this equation.

$$C_{12}H_{22}O_{11}(s) \longrightarrow 11H_2O(l) + 12C(s)$$

For the decomposition of 1.00 mol of sucrose:

a. How many grams of H_2O are produced?
b. What is the total number of moles of products produced?
c. How many grams of C are produced?

***100.** Hydrogen reacts with ethene (C_2H_4) to form ethane (C_2H_6).

$$C_2H_4 + H_2 \longrightarrow C_2H_6$$

What is the limiting reagent when 40.0 g C_2H_4 reacts with 3.0 g H_2?

101. Iron(II) sulfide is produced when iron is heated with sulfur.

$$Fe(s) + S(s) \xrightarrow{\Delta} FeS(s)$$

What is the theoretical yield of FeS if 25.0 g Fe is heated with 32.0 g S?

102. What is the percent yield in Question 101 if 16.5 g of FeS is produced?

If You Have Trouble With . . .

Question	83	84	85	86	87	88	89	90	91	92	93	94	95	96	97	98	99	100	101	102
See Chapter	5	6	7	8	8	9	9	9	10	10	10	10	11	11	11	11	12	12	12	12

Standardized Test Prep for Indiana

Tips for Success

Interpreting Graphs A line graph helps you see the relationship between two variables. Before you answer a question about a graph, identify the variables and the general relationship between the variables based on the shape of the curve.

Use the graph to answer Questions 1 and 2.

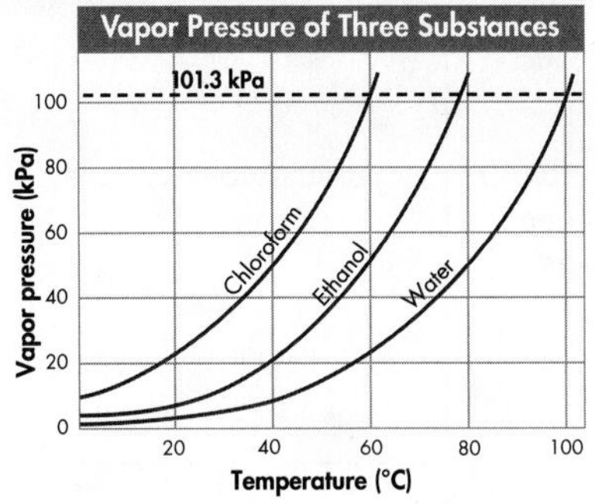

1. What is the normal boiling point of ethanol?
2. Can chloroform be heated to 90°C in an open container?

3. Which sequence has the states of CH_3OH correctly ordered in terms of increasing average kinetic energy?
   (A) $CH_3OH(s)$, $CH_3OH(g)$, $CH_3OH(l)$
   (B) $CH_3OH(g)$, $CH_3OH(l)$, $CH_3OH(s)$
   (C) $CH_3OH(l)$, $CH_3OH(g)$, $CH_3OH(s)$
   (D) $CH_3OH(s)$, $CH_3OH(l)$, $CH_3OH(g)$

Use the drawing to answer Questions 4–6. The same liquid is in each flask.

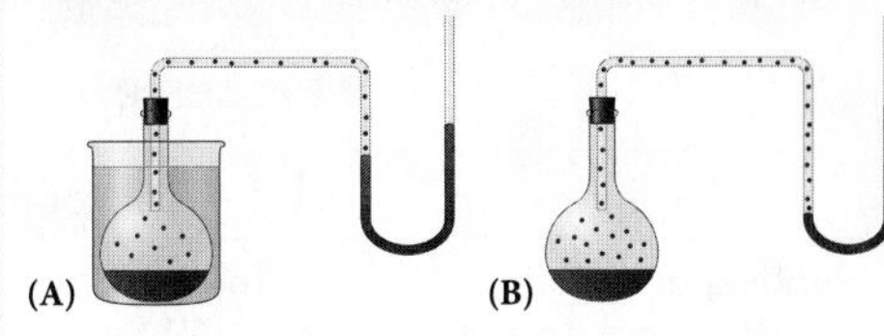

4. In which flask is the vapor pressure lower? Give a reason for your answer.
5. In which flask is the liquid at the higher temperature? Explain your answer.
6. How can the vapor pressure in each flask be determined?

For each question below, there are two statements. Decide whether each statement is true or false. Then decide whether Statement II is a correct explanation for Statement I.

	Statement I		Statement II
7.	In an open container, the rate of evaporation of a liquid always equals the rate of condensation.	BECAUSE	A dynamic equilibrium exists between the liquid and its vapor in an open container.
8.	Water boils at a temperature below 100°C on top of a mountain.	BECAUSE	Atmospheric pressure decreases with an increase in altitude.
9.	The temperature of a substance always increases as heat is added to the substance.	BECAUSE	The average kinetic energy of the particles in a substance increases with an increase in temperature.
10.	Solids have a fixed volume.	BECAUSE	Particles in a solid cannot move.
11.	Gases are more compressible than liquids.	BECAUSE	There is more space between particles in a gas than between particles in a liquid.

If You Have Trouble With . . .

Question	1	2	3	4	5	6	7	8	9	10	11
See Lesson	13.2	13.2	13.4	13.2	13.2	13.2	13.2	13.2	13.1	13.3	13.1

STP Answers

1. about 78°C
2. no
3. D
4. Vapor pressure is lower in flask A because the level of mercury in the left arm of AB is higher, which indicates less pressure on the mercury.
5. Temperature is higher in flask B because vapor increases with temperature.
6. by measuring the difference in heights between the two arms
7. False, False
8. True, True, correct explanation
9. False, True
10. True, False
11. True, True, correct explanation

14 The Behavior of Gases Planning Guide

Introducing the BIGIDEA: KINETIC THEORY

Kinetic theory states that all matter consists of tiny particles that are in constant motion.

IASS	Lessons and Objectives	Print Resources		
		For the Student	For the Teacher	
C.5.1	**14.1** Properties of Gases p 450–454 **14.1.1** Explain why gases are easier to compress than solids or liquids. **14.1.2** Describe the three factors that affect gas pressure.	Reading and Study Workbook Lesson 14.1 Lesson Assessment 14.1 p 463	Teacher Demo, p 452: Effect of Particle Size on Gas Pressure	
C.5.1, C.6.2	**14.2** The Gas Laws p 456–463 **14.2.1** Describe the relationship among the pressure, volume, and temperature of a gas.	Reading and Study Workbook Lesson 14.2 Lesson Assessment 14.2 p 463	Teacher Demo, p 458: Pressure and Volume Class Activity, p 460: Observing the Effects of Pressure on Temperature	
C.5.1, C.5.2, C.5.3	**14.3** Ideal Gases p 464–468 **14.3.1** Calculate the amount of a contained gas when the pressure, volume, and temperature are specified. **14.3.2** Define the conditions when real gases are most likely to differ from ideal gases.	Quick Lab: Carbon Dioxide from Antacid Tablets p 467 Reading and Study Lesson 14.3 Lesson Assessment 14.3 p 468		
C.4.2	**14.4** Gases: Mixtures and Movements p 469–474 **14.4.1** Relate the total pressure of a mixture of gases to the partial pressures of the component gases. **14.4.2** Explain how the molar mass of a gas affects the rate at which gas diffuses and effuses.	Reading and Study Workbook Lesson 14.4 Small-Scale Lab: Diffusion p 475 Lesson Assessment 14.4 p 474	Class Activity, p 470: Model Partial Pressure Teacher Demo, p 472: Effusion	

Assessing the BIGIDEA: KINETIC THEORY

	Essential Questions 1. How do gases respond to changes in pressure, volume, and temperature? 2. Why is the ideal gas law useful even though ideal gases do not exist?	Study Guide p 478 Math Tune-Up p 479 STP p 485 Reading and Study Workbook Self-Check and Vocabulary Review Chapter 14		

	PearsonChem.com	
	Editable Worksheets	Online Activities
	Teaching Resources, Lesson 14.1 Review	14.1 Lesson Overview Triathalons Are a Gas!
	Teaching Resources, Lesson 14.2 Review Lab 23: Pressure-Volume Relationships for Gases Lab 24: Temperature-Volume Relationships for Gases Lab Practical 14-1: Boyle's Law Lab Practical 14-2: Charles's Law	14.2 Lesson Overview Using the Gas Laws Using the Combined Gas Laws Charles's Law
	Teaching Resources, Lesson 14.3 Review Probeware Lab: Quick Lab: Carbon Dioxide from Antacid Tablets	14.3 Lesson Overview Using the Ideal Gas Law Comparing Real and Ideal Gases
	Teaching Resources, Lesson 14.4 Review Lab 25: Diffusion of Gases Small-Scale Lab Manual Lab 21: Synthesis and Qualitative Analysis of Gases	14.4 Lesson Overview Partial Pressure of Gases Using Dalton's Law of Partial Pressures
	Core Teaching Resources (includes editable worksheets) • Lesson Reviews • Practice Problems • Interpret Graphs • Vocabulary Review • Chapter Quizzes and Tests • Lab Record Sheets	Chapter 14 Problem Set Multistep Equations Direct and Inverse Relationships

Materials List

For the Student

Quick Lab p 467
- 6 effervescent antacid tablets
- 3 rubber balloons
- plastic medicine dropper
- water
- watch
- metric tape measure
- graph paper

Small-Scale Lab p 475
- clear plastic cup
- dropper bottles with bromothymol blue
- hydrochloric acid and sodium hydrogen sulfite solutions
- ruler
- cotton swab
- NaOH, NH_4Cl, KI, and $NaNO_2$ (optional)

For the Teacher

Teacher Demo p 452
- 2 identical glass containers
- 2 different sizes of beads

Teacher Demo p 458
- vacuum pump
- bell jar
- marshmallows

Class Activity p 460
- 2 large vats
- ice water
- hot water
- inflated bicycle tire

Class Activity p 470
- balance
- 4 marbles
- 2 buttons
- 3 pennies
- 4 nickels

Teacher Demo p 472
- 2 round identical balloons
- helium
- metric tape measure

Additional Digital Resources

Online Student Edition
Online Teacher's Edition
Exam*View Assessment Suite CD-ROM*
Classroom Resources DVD
14.1 Virtual Chem Lab 15: Investigation of Gas Pressure and Mass
14.2 Virtual Chem Lab 11: Pressure-Volume Relationship for Gases
14.2 Virtual Chem Lab 12: Temperature-Volume Relationship for Gases
14.3 Virtual Chem Lab 13: Derivation of the Ideal Gas Law
14.3 Virtual Chem Lab 14: Ideal vs. Real Gases

CHAPTER 14

What's Online

CHEMISTRY TUTORIAL Students can work step-by-step with the online tutorial to learn how to use the gas laws, the combined gas law, the Ideal Gas Law, and Dalton's law of partial pressures by accessing the Chapter 14 Chemistry Tutorials.

ONLINE PROBLEM SET Additional practice problems about the gas laws can be found in the Chapter 14 online Problem Set.

MATH TUTORIALS The online Math Tutorials will help guide students through solving equations and recognizing direct and inverse relationships.

DIRECTED VIRTUAL LAB Encourage students to review the pressure and volume properties of real and ideal gases using the animated virtual lab for Chapter 14. The animation is followed by analysis questions.

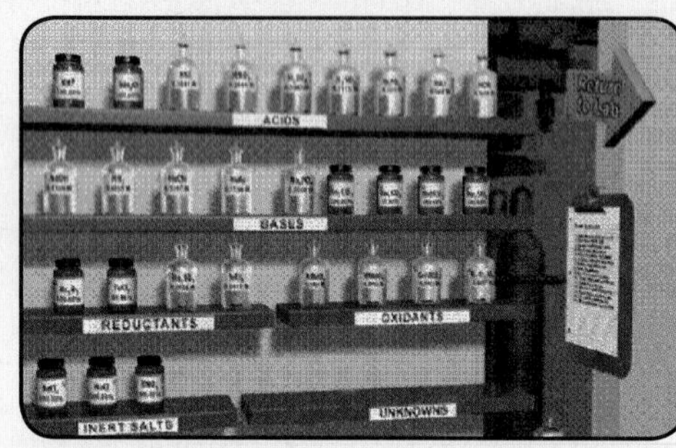

KINETIC ART Have students watch animations about Charles's Law and about comparing real and ideal gases.

CONCEPTS IN ACTION Students can watch an interactive animation about how the gas laws show up in everyday life. The animation is followed by questions.

14 The Behavior of Gases

INSIDE:

PearsonChem.com

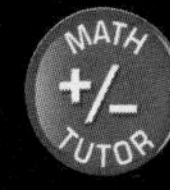

Aviators known as cluster balloonists rise above the clouds by harnessing themselves to balloons filled with helium gas.

IN ACADEMIC STANDARDS for SCIENCE

Nature of Science

- Explain how scientific knowledge can be used to guide decisions on environmental and social issues.

Standard 5: Behavior of Gases

C.5.1 Use kinetic molecular theory to explain changes in gas volumes, pressure, moles, and temperature.

C.5.2 Using the ideal gas equation of state, $pv = nRT$, calculate the change in one variable when another variable is changed and the others are held constant. Distinguish among isothermal, isobaric, isometric, and adiabatic processes.

Also covered C.4.2, C.5.3, C.6.2

Focus on ELL

1 CONTENT AND LANGUAGE Clarify for students that the words *property(ies)*, *gas(es)*, and *law(s)* each have both a common and a scientific meaning. Explain how to create the plural form for each. Assign students to three groups. Have each group complete a Frayer Model diagram for one word—with a definition, characteristics, an example, and a nonexample. Have each group teach the word to the class.

BEGINNING

LOW Create a symbol or drawing to represent each word. If you can, create a native language definition.

HIGH Write each word and introduce it to the class by pointing and saying it aloud.

INTERMEDIATE: LOW/HIGH Brainstorm examples and nonexamples and present them to the class.

ADVANCED: LOW/HIGH Write each definition and present it to the class. Explain to the class how the examples represent the word.

KINETIC THEORY

Essential Questions:

1. *How do gases respond to changes in pressure, volume, and temperature?*
2. *Why is the ideal gas law useful even though ideal gases do not exist?*

CHEMYSTERY

Under Pressure

Just after 2 P.M., Becki completes her eighth scuba dive over a four-day period off the coast of Belize. After the dive, she feels fine.

A few hours later at dinner, Becki feels tired. She thinks that her fatigue is probably due to the many hours she had spent swimming during her vacation. But she also begins to feel itchy and notices a blotchy rash on her skin. Did she get stung by a sea creature during her last dive? Becki decides to go back to her hotel room to get some rest. As she is walking, she begins to feel severe pains in the joints of her arms and legs and feels achy all over her body. Becki feels like she is coming down with the flu, but she realizes that her symptoms are related to her dives. What is wrong with Becki?

▶ Connect to the **BIG**IDEA As you read about the behavior of gases, think about what may have caused Becki's symptoms.

INDIANA'S ACADEMIC STANDARDS FOR SCIENCE

Reactions and Stoichiometry C.4.2; **Behavior of Gases** C.5.1, C.5.2, C.5.3; **Thermochemistry** C.6.2. See lessons for details.

CHAPTER 14

UbD

Understanding by Design

Students are building toward understanding gas behavior by using the kinetic theory of gases.

PERFORMANCE GOALS At the end of Chapter 14, students will be able to answer the essential questions by applying their knowledge of gas laws.

ESSENTIAL QUESTIONS Read the essential questions aloud. **Ask** Compared with solids and liquids, what's unique about gases? *(Gas particles are much farther apart from each other than particles in solids or liquids.)* **Ask** Why do you suppose we study ideal gases, which don't exist?

BIGIDEA Use the photo of the cluster balloonist to help students connect to this chapter's concepts. Activate prior knowledge by reviewing kinetic theory. **Ask** What makes helium ideal for use by these balloonists? *(Helium is less dense than any other gas except hydrogen and is less dense than air.)*

CHEMYSTERY Connect the CHEMystery to the Big Idea by explaining that the behavior of gases can be predicted. The gas laws describe how gases respond to changes in pressure, volume, and temperature. Ask students to predict what happens to the pressure of the gases that Becki breathes in as she dives and then ascends back to the surface. As a hint, tell students that the pressure of the gases that Becki breathes in is the same as the pressure outside her body. As Becki dives deeper, the pressure outside her body increases. As she rises to the surface, the pressure decreases.

Introduce the Chapter

IDENTIFY PRECONCEPTIONS Students often enter into the study of gases with preconceptions. Use the following activities to show them that gases such as air—even though invisible—are matter and are made up of atoms and molecules that take up space.

Activity 1 *You'll need a bottle and a balloon.* **Ask** What's inside the bottle? *(Most will say "nothing.")* Hold the top of the balloon and push the bottom into the bottle. Stretch the top of the balloon over the mouth of the bottle and try to inflate the balloon by blowing into it. The result is that the balloon expands only slightly because the bottle is filled with air. **Ask** Why doesn't the balloon expand completely? *(because the bottle is filled with air)*

Activity 2 Take the balloon out of the bottle, inflate it, and tie it off. Then try to compress it. **Ask** Why does an inflated balloon resist being compressed? *(because the balloon is filled with air)*

A Fresh Look at Fresh Food

Danni visits a hypobaric research lab to learn how chemists are developing new methods to store fresh food.

LESSON 14.1

Key Objectives

14.1.1 EXPLAIN why gases are easier to compress than solids or liquids are.

14.1.2 DESCRIBE the three factors that affect gas pressure.

Additional Resources

- Reading and Study Workbook, Lesson 14.1
- Core Teaching Resources, Lesson 14.1 Review

PearsonChem.com

Engage

CHEMISTRY & YOU Ask students to look at the photograph that opens the section. **Ask** What factors do you think affect the pressure of the air inside the ball? *(the temperature of the air in the ball; the volume of the ball)*

Activate Prior Knowledge

Point out that in this chapter, students will build upon their knowledge of kinetic theory to describe the behavior of gases. **Ask** What is kinetic energy? *(energy of motion)* How are temperature and kinetic energy related? *(Temperature is a measure of average kinetic energy.)*

14.1 Properties of Gases

C.5.1 Kinetic molecular theory

CHEMISTRY & YOU

Q: *Why is there a recommended pressure range for the air inside a soccer ball?* In organized soccer, there are rules about the equipment used in a game. For example, in international competitions, the ball's mass must not be more than 450 grams and not less than 410 grams. The pressure of the air inside the ball must be no lower than 0.6 atmospheres and no higher than 1.1 atmospheres at sea level.

In this lesson, you will study variables that affect the pressure of a gas. As you will discover, gas pressure is useful in a number of different objects, including auto air bags, inflatable rafts, aerosol sprays, and, yes, soccer balls.

Key Questions

Why are gases easier to compress than solids or liquids?

What are the three factors that affect gas pressure?

Vocabulary

- compressibility

Compressibility

Why are gases easier to compress than solids or liquids?

Recall from Chapter 13 that a gas can expand to fill its container, unlike a solid or liquid. The reverse is also true. Gases are easily compressed, or squeezed into a smaller volume. **Compressibility** is a measure of how much the volume of matter decreases under pressure.

The compressibility of a gas plays an important role in auto safety. When a car comes to a sudden stop, the people in the car will continue to move forward unless they are restrained. The driver and any passengers are more likely to survive a collision if they are wearing seat belts to restrict their forward movement. Cars also contain air bags as a second line of defense. A sudden reduction in speed triggers a chemical reaction inside an air bag. One product of the reaction is nitrogen gas, which causes the bag to inflate. An inflated air bag keeps the driver from colliding with the steering wheel. On the front passenger side of the car, an inflated air bag keeps a passenger from colliding with the dashboard or windshield.

Why does a collision with an inflated air bag cause much less damage than a collision with a steering wheel or dashboard? When a person collides with an inflated air bag, as shown in Figure 14.1, the impact forces the molecules of nitrogen gas in the bag closer together. The compression of the gas absorbs the energy of the impact.

IN ACADEMIC STANDARDS for SCIENCE

C.5.1 Use kinetic molecular theory to explain changes in gas volumes, pressure, moles, and temperature.

Focus on ELL

1 CONTENT AND LANGUAGE Ask students to state the key questions in their own words. Have pairs make and use flashcards for key vocabulary. They should write the word on one side and a definition, visual, and sentence with the word on the other. Allow students with limited English to use their native languages for support.

2 FRONTLOAD THE LESSON Have students discuss games they've played that involve inflatable balls. Ask them what differences they've noticed, if any, when balls are underinflated. Direct them to the information in Chemistry & You.

3 COMPREHENSIBLE INPUT Explain Figure 14.5 and restate information in its caption as needed. Read the question aloud, and draw a cause-and-effect chart on the board. As you discuss the question, have students fill in the chart.

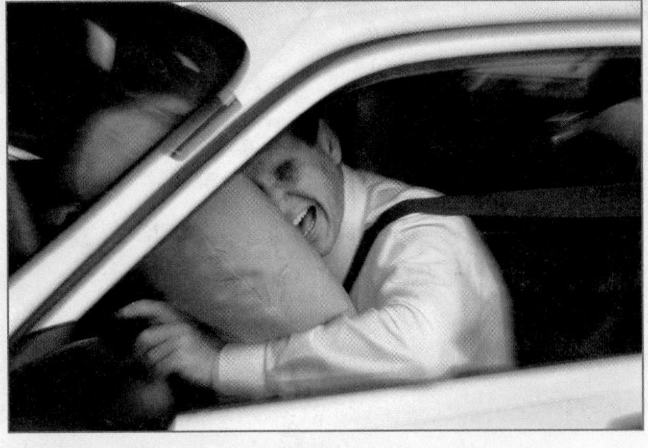

Figure 14.1 Compression of a Gas
Because gases can be compressed, the air bag absorbs some of the energy from the impact of a collision. Air bags work best when combined with seat belts.
Describe ***What happens to the gas molecules inside an air bag when a driver collides with the bag?***

Kinetic theory can explain why gases are compressed more easily than liquids or solids. **Gases are easily compressed because of the space between the particles in a gas.** Remember that the volume of the particles in a gas is small compared to the overall volume of the gas. So the distance between particles in a gas is much greater than the distance between particles in a liquid or solid. Under increased pressure, the particles in a gas are forced closer together, or compressed.

Figure 14.2 is a model of identical air samples in two different containers. Only oxygen and nitrogen—the two main gases in air—are represented. Each container has 8 nitrogen molecules and 2 oxygen molecules. In the larger container, the molecules are farther apart. In the smaller container, the air sample is compressed, and the molecules are closer together. Note that at STP, the distance between particles in an enclosed gas is about 10 times the diameter of a particle. However, it isn't practical to represent the actual distances between particles in all the molecular drawings of gases in this book. In order for the drawings to fit easily on a page, the particles are drawn closer together.

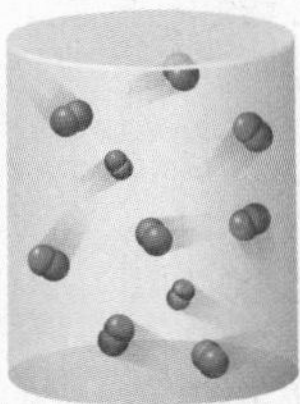

Figure 14.2 Modeling Air at Two Different Pressures
Air is primarily a mixture of two gases, nitrogen (N_2) and oxygen (O_2). A sample of air contains about 4 nitrogen molecules for every oxygen molecule.

Factors Affecting Gas Pressure

What are the three factors that affect gas pressure?

Kinetic theory can help explain other properties of gases, such as their ability to expand and take the shape and volume of their containers. Recall these assumptions about the particles in a gas. The particles move along straight-line paths until they collide with other particles or the walls of their container. The motion of the particles is constant and random. Because kinetic theory assumes there are no significant forces of attraction or repulsion among particles in a gas, particles in a gas can move freely.

Four variables are generally used to describe a gas. The variables and their common units are pressure (P) in kilopascals, volume (V) in liters, temperature (T) in kelvins, and the number of moles (n). **The amount of gas, the volume, and the temperature are factors that affect gas pressure.**

Foundations for Reading

BUILD VOCABULARY Have students look up the meanings of the prefix *com-* *("together")*, the word *press* *("to act on with a steady force or weight")*, and the suffix *-ibility* *("ability or tendency")*. Have them use these meanings to define *compressibility*.

READING STRATEGY For each paragraph in this section, have students identify the main idea. They can later use their list of main ideas as a study guide.

LESSON 14.1

Explain

Compressibility

USE VISUALS Figure14.2 illustrates the amount of space in a gas under ordinary conditions. Ask students to estimate the amount of space occupied by the molecules in the figure. *(Students are likely to say less than 5 percent.)* Explain that for practical reasons, in most drawings the particles will be too close together for their depicted size. Remind students how small the actual particles are and how many particles there are in a small volume of gas.

Factors Affecting Gas Pressure

APPLY CONCEPTS Help students review their knowledge of kinetic theory. **Ask** How would you describe gas pressure? *(Gas pressure is the result of collisions of particles with their container.)* **Ask** How does increasing the number of particles in a gas affect the pressure? *(The more particles colliding with the walls of a container, the higher the pressure.)*

MAKE A CONNECTION Lead the class in a discussion of the four variables—volume, temperature, pressure, and number of moles—used to describe a gas. Discuss how the properties of gases lead to some unique uses of them. For example, because the particles in a gas are relatively far apart, gases are poor thermal conductors and are used for insulation between the glass panes in double-paned windows.

Gas and Momentum

The particles in a gas have momentum, the product of their mass and velocity (mv). The pressure exerted by a gas in a given volume is proportional to the average kinetic energy of its molecules ($1/2\ mv^2$). The mass of the particles is constant, so any change in momentum is due to a change in velocity. A rise in temperature reflects a rise in average kinetic energy, corresponding to an increase in momentum and pressure.

Answers

FIGURE 14.1 The gas molecules inside the air bag are forced closer together as a result of the impact. (The air bag eventually deflates when the compressed gas escapes through small holes in the bag.)

LESSON 14.1

Explain

USE VISUALS Point out Figures 14.3 and 14.4 to students. Make sure they understand the difference between a container with a volume that can vary (within limits), such as a balloon or raft, and a rigid container with a constant volume.

Explore

Teacher Demo

PURPOSE Students use an analogy to infer the effect of particle size on gas pressure.

MATERIALS 2 identical glass containers, 2 different sizes of beads

PROCEDURE Add small beads to a glass container until it is approximately one-eighth full. Add the same number of larger beads to an identical container. Have the students assume that the beads represent particles of gases in sealed containers at the same temperature. Ask students to compare the pressures in the two containers.

EXPECTED OUTCOME The pressures will be identical because both gases contain the same number of particles with the same average kinetic energy in the same volume.

Figure 14.3 Gas in a Flexible Container
The volume of an air-filled raft is much larger than its volume before it was inflated. Using a pump to force air into a raft increases the pressure of the air inside the raft.
Compare *How does an underinflated raft compare with a fully inflated raft? Why do you think an underinflated raft might be dangerous to ride in?*

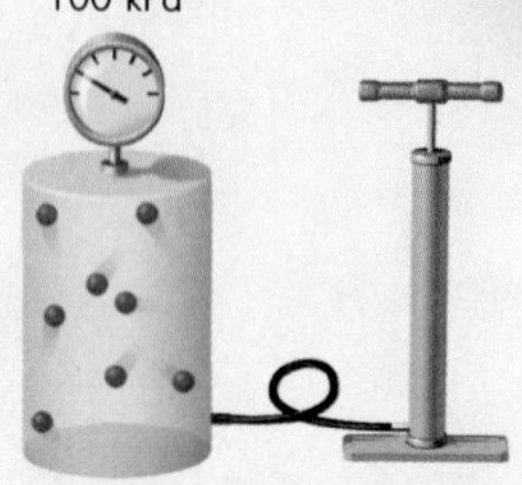

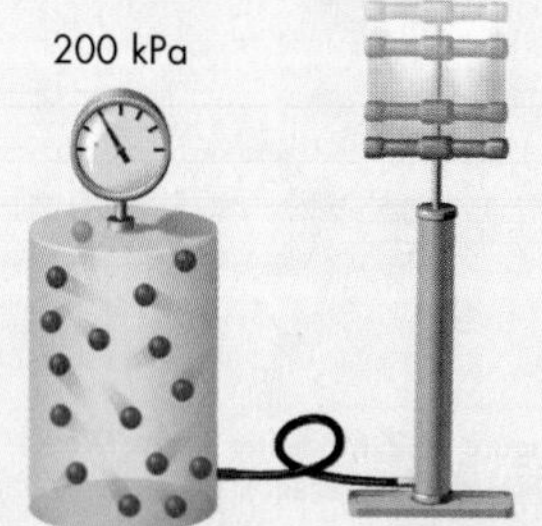

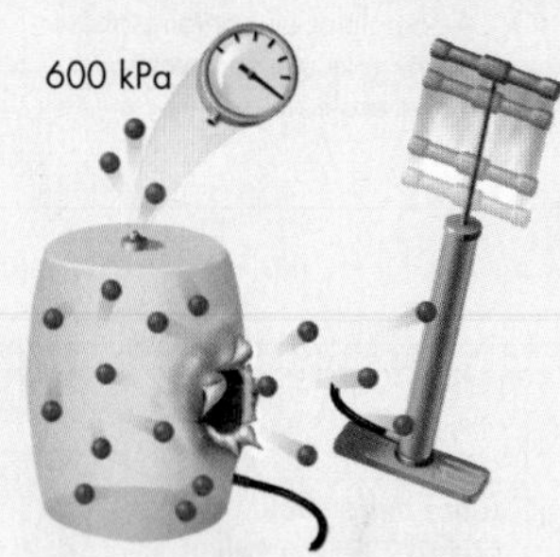

Figure 14.4 Gas in a Rigid Container When a gas is pumped into a closed rigid container, the pressure increases as more particles are added. If the number of particles is doubled, the pressure will double.

Amount of Gas An air-filled raft blasts through a narrow opening between rocks and plummets over a short waterfall into churning white water below. The raft bends and twists, absorbing some of the pounding energy of the river. The strength and flexibility of the raft rely on the pressure of the gas inside the raft. The raft must be made of a material that is strong enough to withstand the pressure of the air inside the raft. The material must also keep air from leaking out of the raft. The volume of the inflated raft in Figure 14.3 is dramatically larger than the volume of the raft before it is inflated. As air is added, the raft expands to its intended volume. The pressure of the air inside the raft keeps the raft inflated.

You can use kinetic theory to predict and explain how gases will respond to a change of conditions. If you inflate an air raft, for example, the pressure inside the raft will increase. Collisions of gas particles with the inside walls of the raft result in the pressure that is exerted by the enclosed gas. By adding gas, you increase the number of particles. Increasing the number of particles increases the number of collisions, which explains why the gas pressure increases.

Figure 14.4 shows what happens when gas is added to an enclosed, rigid container. Because the container is rigid, the volume of the gas is constant. Assume also that the temperature of the gas does not change. Under these conditions, doubling the number of particles of gas doubles the pressure. Tripling the number of particles triples the pressure, and so on. With a powerful pump and a strong container, you can generate very high pressures by adding more and more gas. However, once the pressure exceeds the strength of the container, the container will burst. Removing gas from a rigid container has the opposite effect. As the amount of gas is reduced, the pressure inside the container is reduced. If the number of particles in the container were cut in half, the pressure would drop by half.

Differentiated Instruction

L1 STRUGGLING STUDENTS Have students draw a balloon and use circles to represent gas particles. Then ask students to draw a smaller balloon with the same number of particles to illustrate what happens to the gas particles in the balloon when it's compressed.

ELL ENGLISH LANGUAGE LEARNERS Assign students to groups of three. Have each student silently read one of the paragraphs in the sections titled Amount of Gas, Volume, and Temperature. Ask students to cover the page and restate to the group one fact they read.

L3 ADVANCED STUDENTS Ask students to explain why a fully inflated polyester film balloon appears underinflated after moving from a warm place to a cool one.

If the pressure of the gas in a sealed container is lower than the outside air pressure, air will rush into the container when the container is opened. This movement causes the whoosh you hear when you open a vacuum-packed container. When the pressure of a gas in a sealed container is higher than the outside air pressure, the gas will flow out of the container when the container is unsealed.

The operation of an aerosol can depends on the movement of a gas from a region of high pressure to a region of lower pressure. Aerosol cans may contain whipped cream, hair mousse, or spray paint. Figure 14.5 shows how a can of spray paint works. The can contains a gas stored at high pressure. The air outside the can is at a lower pressure. Pushing the spray button creates an opening between the inside of the can and the air outside. The gas flows through the opening to the lower pressure region outside. The movement of the gas propels, or forces, the paint out of the can. As the gas is depleted, the pressure inside the can decreases until the gas can no longer propel paint from the can.

Volume You can raise the pressure exerted by a contained gas by reducing its volume. The more the gas is compressed, the more pressure the gas exerts inside the container. When gas is in a cylinder, as in an automobile engine, a piston can be used to reduce its volume. The snug-fitting piston keeps gas from escaping as the cylinder moves down and up.

Figure 14.6 shows a cylinder of gas under two different conditions. When the cylinder has a volume of 1 L, the gas exerts a pressure of 100 kPa. When the volume is halved to 0.5 L, the pressure is doubled to 200 kPa. Increasing the volume of the contained gas has the opposite effect. If the volume is doubled, the particles can expand into a volume that is twice the original volume. With the same number of particles in twice the volume, the pressure of the gas is cut in half.

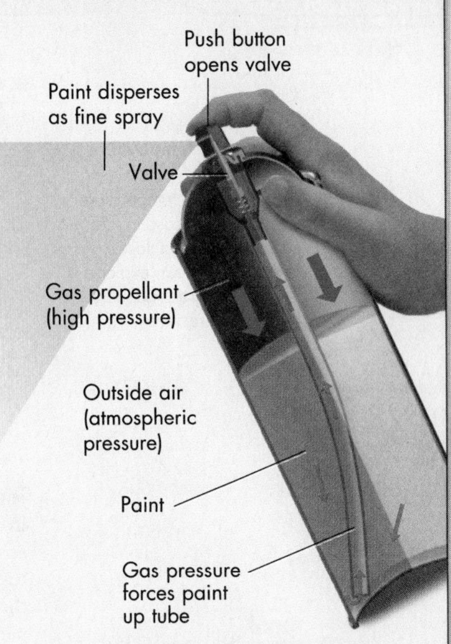

Figure 14.5 Aerosol Can
The pressure of the gas inside a new can of spray paint is greater than the air pressure outside the can. When gas rushes though an opening in the top of the can, it propels, or forces, paint out of the can. As the can is used, the pressure of the propellant decreases.
Relate Cause and Effect ***What happens when the pressure of the propellant equals the air pressure outside the can?***

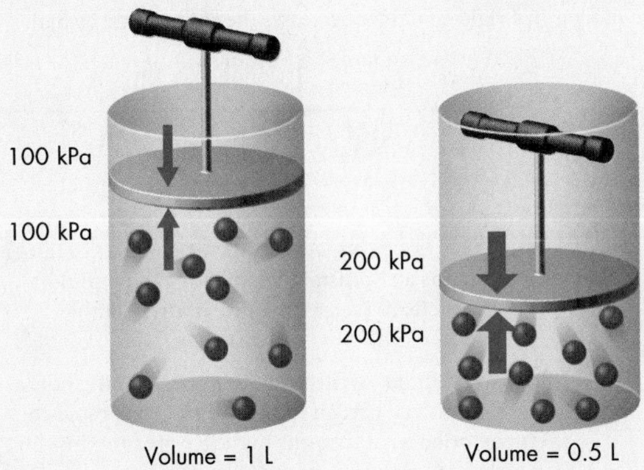

Figure 14.6 Pressure and Volume
A piston can be used to force a gas in a cylinder into a smaller volume. When the volume is decreased, the pressure the gas exerts is increased.
Interpret Diagrams ***What happens to the gas pressure when the volume is reduced from 1 L to 0.5 L?***

Explain

START A CONVERSATION Point out Figure 14.5. Ask students to think about why the instructions on an aerosol can of paint say to hold it upright while spraying. **Ask** If the can is not held upright, what material will come out of the can? *(the propellant)* **Ask** How would the loss of propellant affect the use of the can when it is later held upright and sprayed? *(The amount of propellant would be reduced, so there would be less pressure to force the paint out of the can.)*

USE VISUALS Figure 14.6 is the first of the drawings that show a piston and cylinder. If possible, show students an actual piston and cylinder. Explain how the tight fit of the piston in the cylinder keeps the gas contained and, thus, able to be compressed. **Ask** What will happen to the pressure in the cylinder as the volume is decreased? *(Pressure will increase.)* Point out that the number of particles in both diagrams is the same. **Ask** Why does the pressure of a contained gas double when the volume is reduced by one half? *(Reducing the volume by one half doubles the number of simultaneous collisions of particles with the walls of the piston, doubling the pressure.)*

UbD Check for Understanding

The Essential Question How do gases respond to changes in pressure, volume, and temperature?

Assess students' understanding of gas pressure. **Ask** What would be the effect on pressure if the number of particles of a gas in a closed container were tripled? *(The pressure would triple.)* **Ask** What effect would doubling the volume of an enclosed gas have on the pressure? *(The pressure would decrease by one half.)* **Ask** How does the pressure of an enclosed gas change with increasing temperature? *(The number and force of collisions increase with temperature, and the pressure increases.)*

ADJUST INSTRUCTION If students are having difficulty answering, refer them to Figures 14.4, 14.6, and 14.7 for review, and then ask the questions again.

Answers

FIGURE 14.3 An underinflated raft has a smaller volume and is less rigid than a fully inflated raft. An underinflated raft would be more susceptible to folding or collapsing during the ride.

FIGURE 14.5 The gas can no longer propel the product out of the can.

FIGURE 14.6 The pressure is doubled from 100kPa to 200 kPa.

LESSON 14.1

Explain

CHEMISTRY & YOU A ball that is properly inflated will rebound faster and travel farther than a ball that is underinflated. If the pressure is too high, the ball may burst when it is kicked.

Extend

Connect to PHYSICS

BALLS IN SPORTS Have students research the relationship between gas compressibility and the transfer of kinetic energy in inflatable balls such as soccer balls and basketballs. Ask them to write a brief report comparing and contrasting energy transfer in inflatable balls to energy transfer in solid core balls such as baseballs.

Evaluate

Informal Assessment

Divide the class into two teams. Have each team write five to eight questions about factors that affect gas pressure. After all the questions are put into a box, have Team 1 and then Team 2 draw a question and attempt to answer it in one minute. Continue for several rounds. The team with the most correct answers wins. Then have students complete the 14.1 Lesson Check.

Reteach

Help students make a table to summarize how gas pressure changes as variables increase or decrease. Label the rows *Volume*, *Temperature*, and *Number of Particles*. Label the columns *Increase* and *Decrease*. Have students use up or down arrows to complete the table.

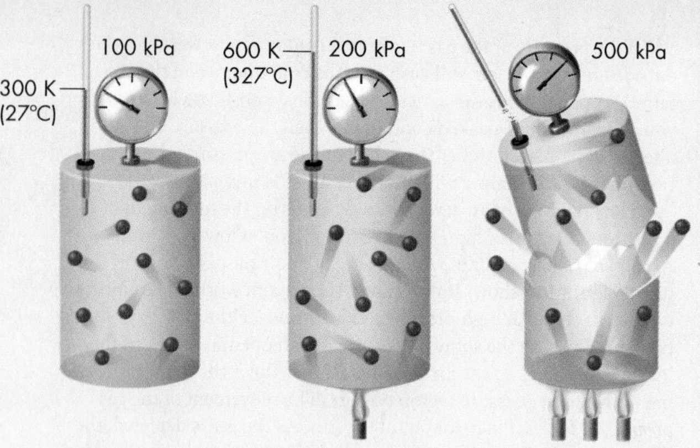

Figure 14.7 Temperature and Pressure
An increase in temperature causes an increase in the pressure of an enclosed gas. The container can explode if there is too great an increase in the gas pressure.

Temperature A sealed bag of potato chips bulges at the seams when placed in a sunny location. The bag bulges because an increase in the temperature of an enclosed gas causes an increase in its pressure. You can use kinetic theory to explain what happens. As a gas is heated, the temperature increases and the average kinetic energy of the particles in the gas increases. Faster-moving particles strike the walls of their container with more energy.

Look at Figure 14.7. The volume of the container and the amount of gas is constant. When the Kelvin temperature of the enclosed gas doubles from 300 K to 600 K, the pressure of the enclosed gas doubles from 100 kPa to 200 kPa. A gas in a sealed container may generate enormous pressure when heated. For that reason, an aerosol can, even an "empty" one, may explode if thrown onto a fire.

By contrast, as the temperature of an enclosed gas decreases, the pressure decreases. The particles, on average, move more slowly and have less kinetic energy. They strike the container walls with less force. Halving the Kelvin temperature of a gas in a rigid container decreases the gas pressure by half.

CHEMISTRY & YOU

Q: *Which do you think would travel farther if kicked with the same amount of force: a properly inflated soccer ball or an under-inflated soccer ball? What might happen to an overinflated soccer ball if you kicked it too hard?*

14.1 LessonCheck

ONLINE PROBLEMS

1. **Review** Why is a gas easy to compress?
2. **Identify** List three factors that can affect gas pressure.
3. **Compare and Contrast** Why does a collision with an air bag cause less damage than a collision with a steering wheel?
4. **Explain** How does a decrease in temperature affect the pressure of a contained gas?
5. **Apply Concepts** If the temperature is constant, what change in volume would cause the pressure of an enclosed gas to be reduced to one quarter of its original value?
6. **Apply Concepts** Assuming the gas in a container remains at a constant temperature, how could you increase the gas pressure in the container a hundredfold?
7. **Summarize** Write a paragraph explaining how a pressurized garden sprayer works. Make sure to describe what happens to the air pressure inside the sprayer as it is pumped by hand.

BIGIDEA KINETIC THEORY

8. Use the kinetic theory of gases to explain why a gas can be easily squeezed into a smaller volume.

Lesson Check Answers

1. Gases are easily compressed because of the space between gas particles.
2. temperature, volume, and amount of gas
3. Because the gas in the inflated air bag can be compressed, the bag absorbs some of the energy from the impact of a collision. The solid steering wheel cannot do this.
4. If the temperature decreases, the pressure will also decrease.
5. The volume would need to increase by a factor of four.
6. Increase the amount of gas in the container a hundredfold.
7. Sample answer: The pressure of the gas inside the sprayer is greater than atmospheric pressure. As the gas propels the contents out, the number of gas particles in the sprayer decreases and the gas pressure decreases.
8. **BIGIDEA** According to kinetic theory, the particles in a gas have insignificant volume, and the distance between particles in a gas is much greater than the distance between particles in a liquid or solid. Under pressure, the particles in a gas are forced closer together, resulting in a smaller overall volume.

Atmospheric Chemist

Earth's atmosphere is a mixture of many gases, including oxygen, nitrogen, water vapor, carbon dioxide, methane, and ozone. Each one has an impact on life on Earth. The study of the chemical composition of the atmosphere is called atmospheric chemistry. Atmospheric chemists analyze the concentrations of atmospheric gases and determine how these gases chemically interact.

An important part of atmospheric research involves developing models that can predict the effects of fossil fuel emissions and other pollutants on air quality, climate, and the biosphere. Some atmospheric chemists study volcanic plumes, which are mixtures of hot gases and dust given off by volcanoes. Atmospheric chemistry is not limited to Earth's atmosphere. With the aid of telescopes, atmospheric chemists can study the composition of atmospheres of distant planets.

Atmospheric research is often a collaboration among scientists from different disciplines, including chemistry, physics, climatology, and oceanography. Atmospheric chemists typically have a bachelor's degree in chemistry or atmospheric science. Many also have a graduate degree in a specific field of research.

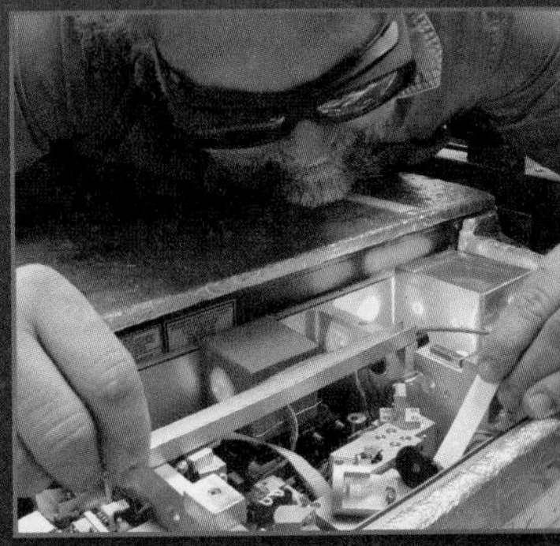

TOOLS AND TECHNOLOGY An atmospheric chemist adjusts a device used to analyze the motion and composition of air in the atmosphere.

AIR QUALITY Smog is a form of air pollution caused by tailpipe and smokestack emissions. The work of atmospheric chemists can help communities better understand how human activity impacts local air quality.

Take It Further

1. **Infer** What kinds of data do you think atmospheric chemists collect to study gases in the atmosphere?
2. **Research a Problem** Ozone (O_3) is one of many gases that atmospheric chemists study. Research the ozone layer and describe how atmospheric ozone levels have changed over time.

CHEMISTRY & YOU Point out to students that many weather reports now contain updates on air quality. Special alerts are issued when the amounts of certain pollutants that adversely affect human health are predicted to be above accepted maximum levels. Have students investigate the models used to predict these levels, and then have them write a summary of the pollutants monitored and the effects each has on health.

Connect to CIVICS

Point out to students that regulations to control greenhouse gas emissions begin in the houses of the United States Congress as bills, before being signed into law and inserted into the federal code. Assign students to teams acting as members of Congress who write a bill to regulate greenhouse gas emissions. Students should write the bill in the formal language and style used by Congress.

Differentiated Instruction

LPR LESS PROFICIENT READERS Have students identify and list unfamiliar words in the text. Assist students in using context clues and word roots to write a definition of each word. Check definitions for accuracy.

L1 STRUGGLING STUDENTS Review the chemical structures and formulas of methane and ozone. Explain how each substance is formed. Refer students to page R27 of the Elements Handbook for additional information on the ozone layer.

ELL ENGLISH LANGUAGE LEARNERS Clarify for students the roles of the Environmental Protection Agency (EPA) and the National Oceanographic and Atmospheric Administration (NOAA) in identifying, researching, and regulating greenhouse gas emissions. Ask students to share their knowledge of their native countries' government agencies responsible for these tasks, and use this information to draw parallels with the EPA and NOAA.

Answers

TAKE IT FURTHER

1. Sample answers: pressure, temperature, mass, concentration, altitude, geographic coordinates, time/date.
2. Due to ozone-depleting pollutants such as chlorofluorocarbons (CFCs), the amount of ozone in the ozone layer decreased sharply by the mid-1980s. Scientists estimate that it will take decades for the ozone layer to recover.

LESSON 14.2

Key Objectives

14.2.1 DESCRIBE the relationships among the temperature, pressure, and volume of a gas.

Additional Resources

- Reading and Study Workbook, Lesson 14.2
- Core Teaching Resources, Lesson14.2 Review
- Laboratory Manual, Labs 23–24
- Laboratory Practicals 14–1, 14–2
- Virtual ChemLab, Labs 11, 12

Engage

CHEMISTRY & YOU Have students study the photograph and read the text that opens the section. **Ask** What is the effect of heating a gas at constant pressure? *(The density of the gas decreases. As the density of the gas inside the balloon is lowered relative to the density of the gas outside the balloon, the balloon rises.)*

Activate Prior Knowledge

Review compressibility and the factors that affect gas pressure. **Ask** If the amount of gas in a cylinder remains constant, what happens to the volume if the gas is compressed? *(It decreases.)* What happens to the pressure? *(It increases.)* **Ask** How does increasing the temperature of a gas affect its pressure? *(The pressure increases.)* Its volume? *(The volume increases.)*

Explain that these basic relationships help form the basis of the gas laws.

IN ACADEMIC STANDARDS for SCIENCE

C.5.1 Use kinetic molecular theory to explain changes in gas volumes, pressure, moles, and temperature.

C.6.2 Distinguish between the concepts of temperature and heat flow in macroscopic and microscopic terms.

14.2 The Gas Laws

C.5.1 Kinetic molecular theory; **C.6.2** Distinguish between temperature and heat flow

CHEMISTRY & YOU

Q: *How do you fill up a hot air balloon?* A hot air balloon works on the principle that warm air is less dense than cooler air. To make a hot air balloon rise, the pilot heats the air inside the balloon. To make the balloon descend, the pilot releases hot air through a vent in the top of the balloon. In this section, you'll study the laws that allow you to predict gas behavior.

Key Question

How are the pressure, volume, and temperature of a gas related?

Vocabulary

- Boyle's law
- Charles's law
- Gay-Lussac's law
- combined gas law

Boyle's Law

How are the pressure and volume of a gas related?

Kinetic theory tells you that there is empty space between the particles in a gas. Imagine how an increase in pressure would affect the volume of a contained gas. **If the temperature is constant, as the pressure of a gas increases, the volume decreases.** In turn, as the pressure decreases, the volume increases. Robert Boyle was the first person to study this pressure-volume relationship in a systematic way. In 1662, Boyle proposed a law to describe the relationship. **Boyle's law** states that for a given mass of gas at constant temperature, the volume of the gas varies inversely with pressure.

Look at Figure 14.8. A gas with a volume of 1.0 L (V_1) is at a pressure of 100 kPa (P_1). As the volume increases to 2.0 L (V_2), the pressure decreases to 50 kPa (P_2). The product $P_1 \times V_1$ (100 kPa × 1.0 L = 100 kPa·L) is the same as the product $P_2 \times V_2$ (50 kPa × 2.0 L = 100 kPa·L). As the volume decreases to 0.5 L (V_3), the pressure increases to 200 kPa (P_3). Again, the product of the pressure and the volume equals 100 kPa·L.

Interpret Graphs

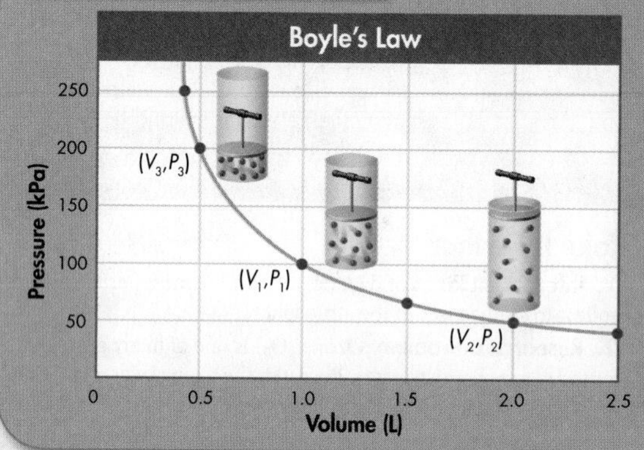

Figure 14.8 The pressure of a gas changes as the volume changes.
a. Read Graphs When the volume is 2.0 L, what is the pressure?
b. Predict What would the pressure be if the volume were increased to 3.0 L?
c. Draw Conclusions Based on the shape of the graph, describe the general pressure-volume relationship.

Focus on ELL

1 CONTENT AND LANGUAGE Ask students to state the key questions in their own words. Have pairs make and use flashcards with key equations. Have the pairs write the name of each gas law on one side and the equation on the other side. Allow students with limited English to use their native languages for support.

2 FRONTLOAD THE LESSON Encourage students to think about what happens when a partially filled balloon is squeezed. Ask them if they would notice a difference in the air pressure and volume. Direct them to the information in Chemistry & You.

3 COMPREHENSIBLE INPUT Explain the graphs of Boyle's and Charles's laws. Be sure to point out the different labels on the *x* and *y* axis of each graph and the letter for each variable. Discuss with students the meanings of *pressure, temperature,* and *volume.*

In an inverse relationship, the product of the two variable quantities is constant. So the product of pressure and volume at any two sets of pressure and volume conditions is always constant at a given temperature. The mathematical expression of Boyle's law is as follows.

$$P_1 \times V_1 = P_2 \times V_2$$

The graph of an inverse relationship is always a curve, as in Figure 14.8.

Sample Problem 14.1

CHEM TUTOR

Using Boyle's Law

A balloon contains 30.0 L of helium gas at 103 kPa. What is the volume of the helium when the balloon rises to an altitude where the pressure is only 25.0 kPa? (Assume that the temperature remains constant.)

1 Analyze **List the knowns and the unknown.** Use Boyle's law ($P_1 \times V_1 = P_2 \times V_2$) to calculate the unknown volume (V_2).

KNOWNS	UNKNOWN
$P_1 = 103$ kPa	$V_2 = ?$ L
$V_1 = 30.0$ L	
$P_2 = 25.0$ kPa	

2 Calculate **Solve for the unknown.**

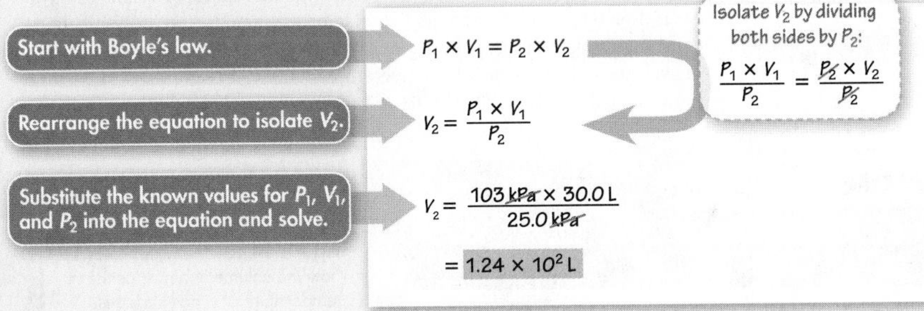

3 Evaluate **Does the result make sense?** A decrease in pressure at constant temperature must correspond to a proportional increase in volume. The calculated result agrees with both kinetic theory and the pressure-volume relationship. The units have canceled correctly.

9. Nitrous oxide (N_2O) is used as an anesthetic. The pressure on 2.50 L of N_2O changes from 105 kPa to 40.5 kPa. If the temperature does not change, what will the new volume be?

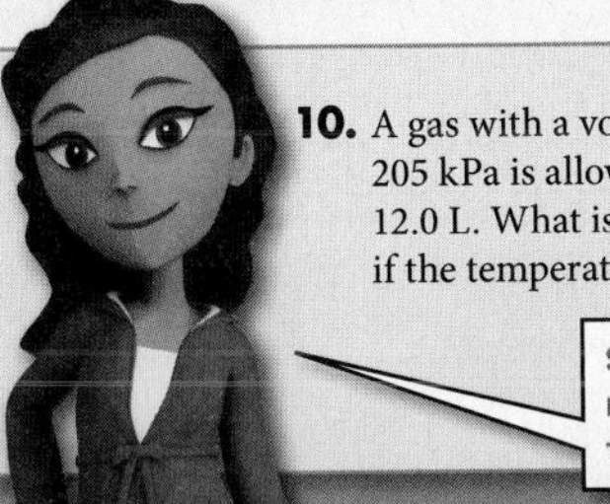

10. A gas with a volume of 4.00 L at a pressure of 205 kPa is allowed to expand to a volume of 12.0 L. What is the pressure in the container if the temperature remains constant?

Solve Problem 10 by rearranging Boyle's law to isolate P_2.

Foundations for Reading

BUILD VOCABULARY Have students make a compare-contrast table for the gas laws. Have students include variables, constants, and the classification "direct or inverse."

READING STRATEGY Have students look at Figures 14.8 and 14.10 and use the graphs to write preliminary versions of Boyle's law and Charles's law.

Explain

Boyle's Law

APPLY CONCEPTS Have students consider what will happen when a helium-filled balloon is released into the sky. Assume the temperature remains constant. Remind students that as elevation increases, atmospheric pressure decreases. **Ask** If the balloon contains 30 L of gas at 100 kPa, what would its volume be at 25 kPa? *(120 L)*

Sample Practice Problem

The volume of a gas at 99.6 kPa and 24°C is 4.23 L. What volume will it occupy at 93.3 kPa and 24°C? *(4.52 L)*

Foundations for Math

CONVERSION FACTORS Before beginning to solve a problem, students should determine if the units of measurement in the problem match the units needed in the answer. If the units do not match, students should determine what conversion factors are needed. For temperature, the possible units are degrees Farenheit (°F), degrees Celsius (°C), and kelvins (K).

Look at the first part of Step 2 in **Sample Problem 14.2.** In this step, temperature is converted from degrees Celsius to kelvins by adding 273 to the values of T_1 and T_2.

Answers

INTERPRET GRAPHS

a. 50 kPa

b. about 33 kPa

c. As either pressure or volume increases, the other variable decreases.

9. 105 kPa × 2.50 L = 40.5 kPa × V_2
V_2 = 105 kPa × 2.50 L/40.5 kPa = 6.48 L

10. 205 kPa × 4.00 L = P_2 × 12.0 L
P_2 = 205 kPa × 4.00 L/12.0 L = 68.3 kPa

Extend

Teacher Demo

PURPOSE Students observe the effect that changing pressure has on the volume of a gas.

MATERIALS vacuum pump, bell jar, marshmallows

PROCEDURE Explain that marshmallows contain trapped air. Place several marshmallows in the bell jar, and then pull a vacuum in the jar. Ask students to explain why the marshmallows increase in size.

EXPECTED OUTCOME The removal of air surrounding the marshmallows reduces the pressure on them. Air trapped inside the marshmallows can expand to a greater volume.

Explain

Charles's Law

MAKE A CONNECTION Point out that when the pressure and amount of a gas are unchanged, the ratio of the volume of the gas to the absolute temperature of the gas is a constant. Mathematically, this constant can be expressed as $V_1/T_1 = V_2/T_2$ and is known as Charles's law.

USE VISUALS Direct students to the line on the graph in Figure 14.10. Point out the ΔV and ΔT labels in the graph. **Ask** What does the ratio $\Delta V/\Delta T$ represent? *(slope)*

APPLY CONCEPTS While solving Charles's law problems, students should remember that the new volume of a gas is equal to its original volume times a quotient. The value of the quotient indicates whether the gas is heated or cooled. If the gas is heated ($T_2 > T_1$), the new volume is greater because the gas expands. So, the quotient (T_2/T_1) has to be greater than 1. If the gas is cooled ($T_2 < T_1$), the new volume is smaller because the gas contracts. So, the quotient must be less than 1.

Figure 14.9 Cooling Balloons in Liquid Nitrogen
When the gas in a balloon is cooled at constant pressure, the volume of the gas decreases.
Predict *What would happen if you removed the balloons from the beaker and allowed them to warm back up to room temperature?*

Charles's Law

How are the temperature and volume of a gas related?

Figure 14.9 shows inflated balloons being dipped into a beaker of liquid nitrogen. For each balloon, the amount of air and the pressure are constant. As the air inside rapidly cools, the balloon shrinks. In fact, the gas volume decreases so much that all the cooled balloons can easily fit inside the beaker.

In 1787, the French physicist Jacques Charles studied the effect of temperature on the volume of a gas at constant pressure. When he graphed his data, Charles observed that a graph of gas volume versus temperature (in °C) is a straight line for any gas. **As the temperature of an enclosed gas increases, the volume increases if the pressure is constant.** When Charles extrapolated, or extended, the line to zero volume ($V = 0$), the line always intersected the temperature axis at −273.15°C. This value is equal to 0 on the Kelvin temperature scale. The observations that Charles made are summarized in Charles's law. **Charles's law** states that the volume of a fixed mass of gas is directly proportional to its Kelvin temperature if the pressure is kept constant. Look at the graph in Figure 14.10. When the temperature is 300 K, the volume is 1.0 L. When the temperature is 900 K, the volume is 3.0 L. In both cases, the ratio of V to T is 0.0033.

Interpret Graphs

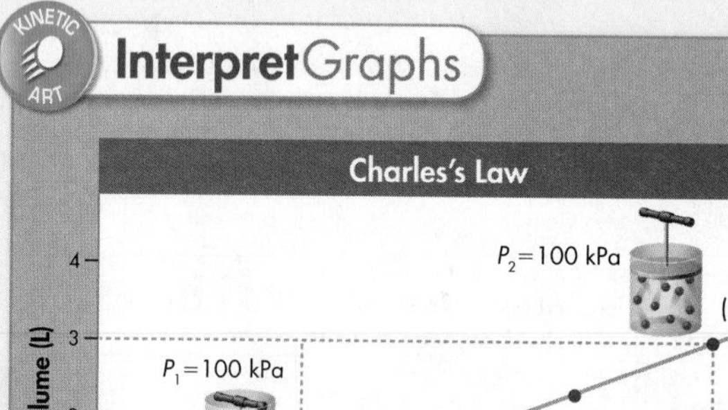

Figure 14.10 The graph shows how the volume changes as the temperature of a gas changes. View the Kinetic Art to see an online simulation of Charles's law.

a. Read Graphs In what unit is the temperature data expressed?

b. Draw Conclusions What happens to the volume as the temperature rises?

c. Predict If the temperature of a gas were 0 K, what would the volume of the gas be?

Hint: ΔV is the change in gas volume resulting from temperature change ΔT.

Differentiated Instruction

L3 GIFTED AND TALENTED Explain to students that Charles summarized his observations of the relationship between the volume and temperature of a gas in the following equation: $V = V_0(1 + aT)$ where V_0 is the volume of the gas at 0°C, T is its temperature expressed in °C, and a is a constant for all gases. Have students show that the numerical value of a is approximately 1/273.

ELL ENGLISH LANGUAGE LEARNERS Have English learners make a list of terms in this section that they don't understand. Pair them with other students who can use paraphrasing to explain the terms.

L1 SPECIAL NEEDS Consider having students work in pairs to solve the practice problems in this chapter. Match up students who have mastery of algebraic equations with students who need more practice solving for an unknown.

The ratio V_1/T_1 is equal to the ratio V_2/T_2. Because this ratio is constant at all conditions of temperature and volume, when the pressure is constant, you can write Charles's law as follows.

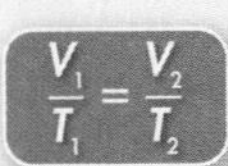

$$\frac{V_1}{T_1} = \frac{V_2}{T_2}$$

The ratio of the variables is always a constant in a direct relationship, and the graph is always a straight line. It is not a direct relationship if the temperatures are expressed in degrees Celsius. So when you solve gas law problems, the temperature must always be expressed in kelvins.

Q: *A hot air balloon contains a propane burner onboard to heat the air inside the balloon. What happens to the volume of the balloon as the air is heated?*

Sample Problem 14.2

Using Charles's Law

A balloon inflated in a room at 24°C has a volume of 4.00 L. The balloon is then heated to a temperature of 58°C. What is the new volume if the pressure remains constant?

❶ Analyze **List the knowns and the unknown.** Use Charles's law ($V_1/T_1 = V_2/T_2$) to calculate the unknown volume (V_2).

KNOWNS	UNKNOWN
$V_1 = 4.00\ L$	$V_2 = ?\ L$
$T_1 = 24°C$	
$T_2 = 58°C$	

❷ Calculate **Solve for the unknown.**

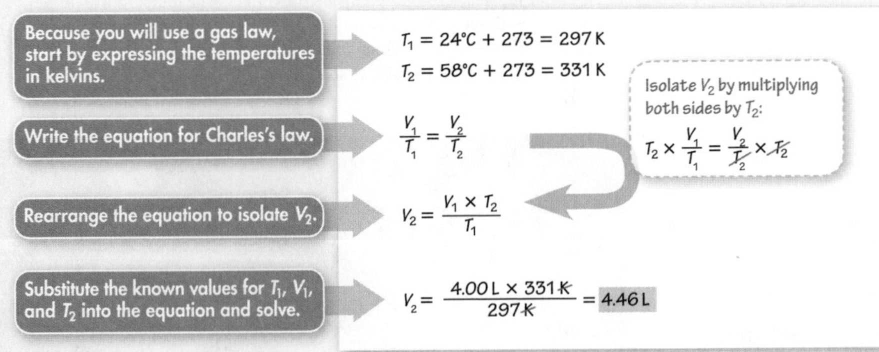

❸ Evaluate **Does the result make sense?** The volume increases as the temperature increases. This result agrees with both the kinetic theory and Charles's law.

11. If a sample of gas occupies 6.80 L at 325°C, what will its volume be at 25°C if the pressure does not change?

12. Exactly 5.00 L of air at −50.0°C is warmed to 100.0°C. What is the new volume if the pressure remains constant?

Foundations for Math

ISOLATING A VARIABLE The common transposition method is to do the same thing (mathematically) to both sides of an equation in order to get like terms together and isolate the variable, or unknown. In a shortcut method, students isolate the variable by moving like terms to one side of the equation. To maintain the equality of the two sides, there are two rules for the shortcut method: *When moving a term to the other side of the equal sign:* ***Rule 1*** Change the sign (for positive or negative values); ***Rule 2*** Invert the coefficient.

Point out that in **Sample Problem 14.2** Rule 1 is unnecessary because are there are no additions or subtractions in the equation for Charles's law. The only action taken was to invert the coefficient, $1/T_2$.

CHEMISTRY & YOU As the air inside a hot air balloon is heated, it expands at constant pressure. Charles's law states that the volume of a fixed amount of gas is directly proportional to its Kelvin temperature if the pressure is kept constant.

CRITICAL THINKING Explain to students that home cooks and food manufacturers rely on Charles's law to ensure a proper seal on glass jars with screw caps during the canning process. Lidded jars are placed in a water bath and heated for a specified time before being removed to cool. During the cooling process, the centers of the rubber-lined metal caps depress with an audible click as the vacuum seals form. Have students use Charles's law to explain why this process results in a vacuum seal. *(During heating, the air temperature in the jar increases, causing the air's volume to increase and forcing some molecules out of the jar. As the jar begins to cool, the temperature of the air decreases, causing its volume to decrease. This causes a seal to form between the lid and the lip of the glass jar. The click is caused by the vacuum pulling the center of the jar lid inward until it flexes.)*

Sample Practice Problem

The volume of a gas is 0.80 L at 101.3 kPa and 0°C. **Ask** What volume will it occupy at 101.3 kPa and 24°C? *(0.87 L)*

Answers

FIGURE 14.9 The balloons would expand back to their original size.

INTERPRETING GRAPHS

a. kelvins

b. The volume increases

c. The volume would be 0.

11. $T_1 = 325\ C + 273 = 598\ K$
$T_2 = 25\ C + 273 = 298\ K$
$6.80\ L/598\ K = V_2/298\ K$
$V_2 = 298\ K \times 6.80\ L/598\ K = 3.39\ L$

12. $T_1 = -50.0\ C + 273 = 223\ K$
$T_2 = 100.0\ C + 273 = 373\ K$
$5.00\ L/223\ K = V_2/373\ K$
$V_2 = 373\ K \times 5.00\ L/223\ K = 8.36\ L$

Explain

Gay-Lussac's Law

MAKE A CONNECTION Use tire pressure to discuss the relationship between gas pressure and Kelvin temperature. **Ask** Why do auto tire manufacturers recommend checking for proper inflation before driving the car more than a mile? *(The tires get warm as the car moves, increasing the pressure inside the tires.)*

Explore

Student Activity

PURPOSE Students observe the relationship between the pressure and temperature of a gas.

MATERIALS 2 large vats, ice water, hot water, inflated bicycle tire

PROCEDURE Fill a large vat with ice water and a second vat with hot water. First, have students squeeze an inflated bicycle tire to assess its firmness. Next, immerse the tire in ice water and have them feel its firmness. Finally, immerse the tire in hot water and have them assess the firmness. Have the students describe the relationship between pressure and temperature at constant volume.

EXPECTED OUTCOME A direct relationship exists between pressure and temperature at a constant volume.

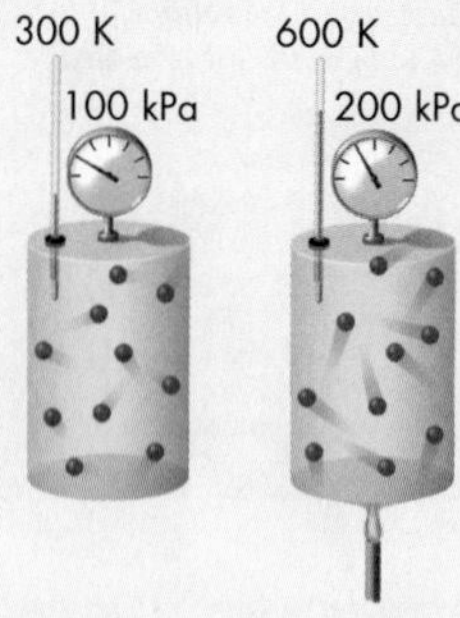

Figure 14.11 Gay-Lussac's Law
When a gas is heated at constant volume, the pressure increases.
Interpret Diagrams ***How can you tell from the drawings that there is a fixed amount of gas in the cylinders?***

Gay-Lussac's Law

How are the pressure and temperature of a gas related?

When tires are not inflated to the recommended pressure, fuel efficiency and traction decrease. Treads can wear down faster. Most importantly, improper inflation can lead to tire failure. A driver should not check tire pressure after driving a long distance because the air in a tire heats up during a drive. **As the temperature of an enclosed gas increases, the pressure increases if the volume is constant.**

Joseph Gay-Lussac (1778–1850), a French chemist, discovered the relationship between the pressure and temperature of a gas in 1802. The gas law that describes the relationship bears his name. **Gay-Lussac's law** states that the pressure of a gas is directly proportional to the Kelvin temperature if the volume remains constant. Look at Figure 14.11. When the temperature is 300 K, the pressure is 100 kPa. When the temperature is doubled to 600 K, the pressure doubles to 200 kPa. Because Gay-Lussac's law involves direct proportions, the ratios P_1/T_1 and P_2/T_2 are equal at constant volume. You can write Gay-Lussac's law as follows:

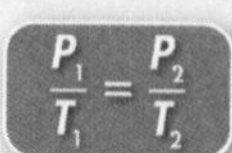

$$\frac{P_1}{T_1} = \frac{P_2}{T_2}$$

Gay-Lussac's law can be applied to reduce the time it takes to cook food. One cooking method involves placing food above a layer of water and heating the water. The water vapor, or steam, that is produced cooks the food. Steam that escapes from the pot is at a temperature of about 100°C when the pressure is near one atmosphere. In a pressure cooker, like the one shown in Figure 14.12, steam is trapped inside the cooker. The temperature of the steam reaches about 120°C. The food cooks faster at this higher temperature, but the pressure rises, which increases the risk of an explosion. A pressure cooker has a valve that allows some vapor to escape when the pressure exceeds the set value.

Figure 14.12 Pressure Cooker
A pressure cooker is a gas-tight container in which pressurized steam is used to cook food. With the lid locked, the volume of steam and the number of water molecules are constant. So any increase in temperature causes an increase in pressure.

Differentiated Instruction

L3 GIFTED AND TALENTED Gases such as oxygen, nitrogen, helium, and hydrogen can be liquefied by lowering the temperature and increasing the pressure. But different gases must be cooled to different temperatures before they condense, no matter how high the pressure is raised. This temperature is called the critical temperature. Have students look up the critical temperatures for oxygen, nitrogen, helium, and hydrogen. **Ask** What do these temperatures indicate about the relative strengths of the intermolecular attractions in these gases? *(The lower the critical temperature, the lower the intermolecular attractive forces.)*

460 Chapter 14 • Lesson 2

Using Gay-Lussac's Law

Aerosol cans carry labels warning not to incinerate (burn) the cans or store them above a certain temperature. This problem will show why it is dangerous to dispose of aerosol cans in a fire. The gas in a used aerosol can is at a pressure of 103 kPa at 25°C. If the can is thrown onto a fire, what will the pressure be when the temperature reaches 928°C?

1 Analyze List the knowns and the unknown. Use Gay-Lussac's law ($P_1/T_1 = P_2/T_2$) to calculate the unknown pressure (P_2). Remember, because this problem involves temperatures and a gas law, the temperatures must be expressed in kelvins.

KNOWNS	UNKNOWN
P_1 = 103 kPa	P_2 = ? kPa
T_1 = 25°C	
T_2 = 928°C	

2 Calculate Solve for the unknown.

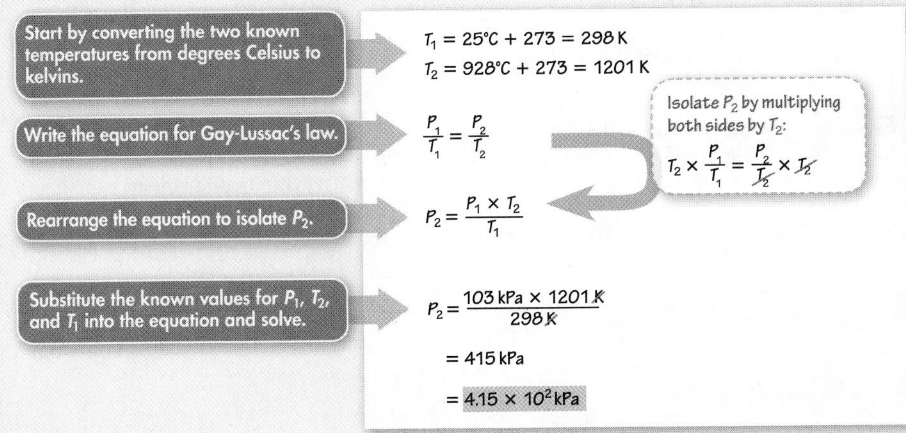

3 Evaluate Does the result make sense? From the kinetic theory, one would expect the increase in temperature of a gas to produce an increase in pressure if the volume remains constant. The calculated value does show such an increase.

13. The pressure in a sealed plastic container is 108 kPa at 41°C. What is the pressure when the temperature drops to 22°C? Assume that the volume has not changed.

14. The pressure in a car tire is 198 kPa at 27°C. After a long drive, the pressure is 225 kPa. What is the temperature of the air in the tire? Assume that the volume is constant.

To solve Problem 14, rearrange Gay-Lussac's law to isolate T_2.

LESSON 14.2

Explain

START A CONVERSATION The direct relationship between pressure and temperature at constant volume can be expressed as $P_1/T_1 = P_2/T_2$. In Gay-Lussac's law problems, the new pressure of the gas is equal to its original pressure times a quotient. The value of the quotient indicates whether the gas is heated or cooled. If the gas is heated ($T_2 > T_1$), the new pressure is greater. So, the quotient (T_2/T_1) has to be greater than 1. If the gas is cooled ($T_2 < T_1$), the new pressure is less. So, the quotient must be less than 1.

Sample Practice Problem

A sample of nitrogen gas has a pressure of 6.58 kPa at 539 K. If the volume does not change, what will the pressure be at 211K? *(2.58 kPa)*

Foundations for Math

CHECK ANSWERS Whenever students have solved a problem, they should check the answer by substituting it back into the original equation. This lets them determine whether or not the value on each side of the equal sign is the same. At the same time, they should make sure that the units of the answer are appropriate for the unknown.

In **Sample Problem 14.3**, substituting 4.15×10^2 kPa for P_2 in the Gay-Lussac's law equation gives a value of 0.346 kPa/K. Since 103 kPa / 298 K = 0.346 kPa/K as well, the answer is correct. Note also that kPa is the appropriate unit of measurement for pressure.

Answers

FIGURE 14.11 The number of particles in each container is the same.

13. T_1 = 41 C + 273 = 314 K
T_2 = 22 C + 273 = 295 K
108 kPa/314 K = P_2/295 K
P_2 = 295 K × 108 kPa/314 K = 101 kPa

14. T_1 = 27 C + 273 = 300 K
198 kPa/300 K = 225 kPa/T_2
T_2 = 225 kPa × 300 K/198 kPa = 341 K *(68 C)*

LESSON 14.2

Explain

The Combined Gas Law

USE VISUALS Write the equation for the combined gas law on the board. **Ask** What variable that is used to describe a gas is missing from this equation? *(n, the number of moles of gas)*

USE VISUALS Direct students to the weather balloon in Figure 14.13 on the following page. Remind students that in many situations, a sample of gas is simultaneously subjected to pressure and temperature changes that have opposite effects on volume. Consider a weather balloon rising through the atmosphere. The higher it rises, the colder the temperature and the lower the volume. At the same time, atmospheric pressure decreases, allowing the gas to expand. The combined gas law allows students to determine which variable has the greater effect on the volume.

Sample Practice Problem

The volume of a gas at 26°C and 75 kPa is 10.5 L. The pressure is increased to 116 kPa. What final temperature would be required to reduce the volume to 9.5 L? *(418 K, or 145°C)*

Extend

Connect to BIOLOGY

Ask students to inhale, hold, and exhale. Ask students to think about Boyle's law and how it applies to breathing. **Ask** How does Boyle's law explain why air enters your lungs when you inhale and leaves when you exhale? *(When you inhale, lung volume increases. Pressure decreases and air moves in. When you exhale, lung volume decreases. Pressure increases and air moves out.)* **Ask** Why do you think scuba divers are taught never to hold their breath as they ascend from deep water? *(As a diver ascends, pressure decreases. A decrease in pressure means an increase in volume. If a diver holds her breath while ascending, air volume in the lungs will increase.)*

The Combined Gas Law

How are the pressure, volume, and temperature of a gas related?

There is a single expression, called the **combined gas law**, that combines Boyle's law, Charles's law, and Gay-Lussac's law.

$$\frac{P_1 \times V_1}{T_1} = \frac{P_2 \times V_2}{T_2}$$

When only the amount of gas is constant, the combined gas law describes the relationship among pressure, volume, and temperature.

Sample Problem 14.4

Using the Combined Gas Law

The volume of a gas-filled balloon is 30.0 L at 313 K and 153 kPa pressure. What would the volume be at standard temperature and pressure (STP)?

1 Analyze List the knowns and the unknown. Use the combined gas law ($P_1V_1/T_1 = P_2V_2/T_2$) to calculate the unknown volume (V_2).

KNOWNS	UNKNOWN
$V_1 = 30.0$ L	$V_2 = ?$ L
$T_1 = 313$ K	
$P_1 = 153$ kPa	
$T_2 = 273$ K (standard temperature)	
$P_2 = 101.3$ kPa (standard pressure)	

2 Calculate Solve for the unknown.

State the combined gas law.

$$\frac{P_1 \times V_1}{T_1} = \frac{P_2 \times V_2}{T_2}$$

Isolate V_2 by multiplying both sides by T_2 and dividing both sides by P_2:

$$\frac{T_2}{P_2} \times \frac{P_1 \times V_1}{T_1} = \frac{P_2 \times V_2}{T_2} \times \frac{T_2}{P_2}$$

Rearrange the equation to isolate V_2.

$$V_2 = \frac{P_1 \times V_1 \times T_2}{P_2 \times T_1}$$

Substitute the known quantities into the equation and solve.

$$V_2 = \frac{153\ \text{kPa} \times 30.0\ \text{L} \times 273\ \text{K}}{101.3\ \text{kPa} \times 313\ \text{K}} = 39.5\ \text{L}$$

3 Evaluate Does the result make sense? A decrease in temperature and a decrease in pressure have opposite effects on the volume. To evaluate the increase in volume, multiply V_1 (30.0 L) by the ratio of P_1 to P_2 (1.51) and the ratio of T_2 to T_1 (0.872). The result is 39.5 L.

15. A gas at 155 kPa and 25°C has an initial volume of 1.00 L. The pressure of the gas increases to 605 kPa as the temperature is raised to 125°C. What is the new volume?

16. A 5.00-L air sample has a pressure of 107 kPa at a temperature of −50.0°C. If the temperature is raised to 102°C and the volume expands to 7.00 L, what will the new pressure be?

Foundations for Math

IDENTIFYING APPROPRIATE EQUATIONS An important step in solving gas law problems is determining which equation to use, since many gas laws use similar variables. Remind students that to identify the appropriate equation, they must first identify the known and unknown variables. Then they should ask themselves which equations use those variables. If more than one equation uses the same set of variables, they should reread the text of the question. They can then use the information in the question to identify the appropriate equation.

Look at **Sample Problem 14.4.** The question provides the known initial values for *P*, *V*, and *T* and asks students to calculate a final volume at STP. Since the combined gas law is the only equation that utilizes pressure, volume, and temperature, it is the appropriate equation for this calculation.

Weather balloons, like the one in Figure 14.13, carry a package of data-gathering instruments up into the atmosphere. At an altitude of about 27,000 meters, the balloon bursts. The combined gas law can help to explain this situation. Both outside temperature and pressure drop as the balloon rises. These changes have opposite effects on the volume of the weather balloon. A drop in temperature causes the volume of an enclosed gas to decrease. A drop in outside pressure causes the volume to increase. Given that the balloon bursts, the drop in pressure must affect the volume more than the drop in temperature does.

The combined gas law can also help you solve gas problems when only two variables are changing. It may seem challenging to remember four different expressions for the gas laws. But you actually only need to remember one expression—the combined gas law. You can derive the other laws from the combined gas law by holding one variable constant.

To illustrate, suppose you hold the temperature constant ($T_1 = T_2$). Rearrange the combined gas law so that the two temperature terms are on the same side of the equation. Because $T_1 = T_2$, the ratio of T_1 to T_2 is equal to one. Multiplying by 1 does not change a value in an equation. So when the temperature is constant, you can delete the temperature ratio from the rearranged combined gas law. What you are left with is the equation for Boyle's law.

$$P_1 \times V_1 = P_2 \times V_2 \times \frac{T_1}{T_2}$$

$$P_1 \times V_1 = P_2 \times V_2$$

A similar process yields Charles's law when pressure remains constant and Gay-Lussac's law when volume remains constant.

Figure 14.13 Weather Balloon
Meteorologists use weather balloons to gather data about Earth's atmosphere.
Infer ***Why is helium more likely to be used in weather balloons than air?***

14.2 LessonCheck

ONLINE PROBLEMS

17. Review How are the pressure and volume of a gas related at constant temperature?

18. Review If pressure is constant, how does a change in temperature affect the volume of a gas?

19. Review What is the relationship between the temperature and pressure of a contained gas at constant volume?

20. Describe In what situations is the combined gas law useful?

21. Define Write the mathematical equation for Boyle's law and explain the symbols.

22. Calculate A given mass of air has a volume of 6.00 L at 101 kPa. What volume will it occupy at 25.0 kPa if the temperature does not change?

23. Explain How can Charles's law be derived from the combined gas law?

24. Apply Concepts The volume of a weather balloon increases as the balloon rises in the atmosphere. Why doesn't the drop in temperature at higher altitudes cause the volume to decrease?

BIGIDEA KINETIC THEORY

25. Why do you think scientists cannot collect temperature and volume data for an enclosed gas at temperatures near absolute zero?

Lesson Check Answers

17. At constant temperature, volume decreases as pressure increases.

18. At constant pressure, volume increases as temperature increases.

19. At constant volume, pressure increases as temperature increases.

20. The combined gas law allows you to do calculations when the only constant is the amount of gas.

21. $P_1 \times V_1 = P_2 \times V_2$; P_1 = initial pressure; V_1 = initial volume; P_2 = final pressure; V_2 = final volume

22. 24.2 L

23. When the pressure is constant, $P_1 = P_2$, so the pressure terms cancel, leaving an equation for Charles's law.

24. The outside pressure decreases, causing a greater increase in the balloon's volume.

25. **BIGIDEA** As temperatures decrease toward absolute zero, particles in the gas slow down, and attractions between particles increase. A gas would liquefy and then solidify as it cooled to temperatures near absolute zero.

Evaluate

Informal Assessment

Sketch two balloons (one twice as large as the other) on the board. Label the smaller balloon V_1 and the other V_2, and indicate that $V_2 = 2V_1$. Copy the following table onto the chalkboard, listing only those values in boldface type. Explain that this is the same balloon under the same atmospheric pressure but at different temperatures. Have students complete the table.

T_1	T_2
(256 K)	**512 K**
40°C	*(353 K)*
T_1	$2T_1$
(–136°C)	**1°C**

Then have students complete the 14.2 Lesson Check.

Reteach

Write the combined gas law on the board. Remind students that the law is a combination of Boyle's, Charles's, and Gay-Lussac's laws. Have students examine the equation and rewrite the combined gas law, using a different color for the portion of the equation representing Boyle's law. Have them repeat this process for Charles's law and Gay-Lussac's law.

Answers

15. T_1 = 25°C + 273 = 298 K
T_2 = 125°C + 273 = 398 K
155 kPa × 1.00 L/298 K = 605 kPa × V_2/398 K
V_2 = 398 K × 155 kPa × 1.00 L/(298 K × 605 kPa) = 0.342 L

16. T_1 = –50°C + 273 = 223 K
T_2 = 102°C + 273 = 375 K
107 kPa × 5.00 L/223 K = P_2 × 7.00 L/375 K
P_2 = 375 K × 107 kPa × 5.00 L/(223 K × 7.00 L) = 1.29×10^2 kPa

FIGURE 14.13 Helium is less dense than air.

LESSON 14.3

Key Objectives

14.3.1 CALCULATE the amount of a contained gas when the pressure, volume, and temperature are specified.

14.3.2 DETERMINE conditions under which real gases are most likely to differ from ideal gases.

Additional Resources

- Reading and Study Workbook, Lesson 14.3
- Probeware Laboratory Manual, 14.3 Quick Lab
- Virtual ChemLab, Labs 13, 14

PearsonChem.com

Engage

CHEMISTRY & YOU Have students study the photograph and read the text that opens the lesson. Ask students to predict an answer, and write it on the board.

Activate Prior Knowledge

Have students draw a KWL chart in their notebooks as you draw one on the board. Together, fill in the K column with all that students can remember about the gas laws. Then in groups, have students fill in the W column. As students read the lesson, have them fill in the L column.

Foundations for Reading

BUILD VOCABULARY Have students look up the meanings of *ideal* and *real*. Have them write definitions that include the terms *abstract* and *actual*.

READING STRATEGY Have students use their understanding of the terms *real* and *ideal* to infer the differences between real and ideal gases. As the students read the section, have them evaluate and revise their inferences.

IN ACADEMIC STANDARDS for SCIENCE

- Develop explanations based on reproducible data and observations gathered during laboratory investigations.

C.5.2 Using the ideal gas equation of state, $pv = nRT$, calculate the change in one variable when another variable is changed and the others are held constant.

C.5.3 Given the equation for a chemical reaction involving one or more gases as reactants and/or products, calculate the volumes of gas assuming the reaction goes to completion and the ideal gas law holds.

Also covered C.5.1.

14.3 Ideal Gases

C.5.1 Kinetic molecular theory; C.5.2 Ideal gas equation; C.5.3 Reactant and product calculations: gases

CHEMISTRY & YOU

Q: *How can you make fog indoors?* Carbon dioxide freezes at −78.5°C, which is much colder than the ice in your freezer. Solid carbon dioxide, or dry ice, can be used to make stage fog. Dry ice doesn't melt—it sublimes. As solid carbon dioxide changes to gas, water vapor in the air condenses and forms a white fog. Dry ice can exist because gases don't obey the assumptions of kinetic theory at all conditions. In this section, you will learn how real gases differ from the ideal gases on which the gas laws are based.

Key Questions

How can you calculate the amount of a contained gas when the pressure, volume, and temperature are specified?

Under what conditions are real gases most likely to differ from ideal gases?

Vocabulary

- ideal gas constant
- ideal gas law

Ideal Gas Law

How can you calculate the amount of a contained gas when the pressure, volume, and temperature are specified?

Up to this point, you have worked with three variables that describe a gas: pressure, volume, and temperature. There is a fourth variable still to be considered: the amount of gas in the system, expressed in terms of the number of moles.

Suppose you want to calculate the number of moles (n) of a gas in a fixed volume at a known temperature and pressure. By modifying the combined gas law, you can solve for n. First, you must recognize that the volume occupied by a gas at a specified temperature and pressure depends on the number of particles. The number of moles of gas is directly proportional to the number of particles. So moles must be directly proportional to volume as well. You can now introduce moles into the combined gas law by dividing each side of the equation by n.

$$\frac{P_1 \times V_1}{T_1 \times n_1} = \frac{P_2 \times V_2}{T_2 \times n_2}$$

This equation shows that $(P \times V)/(T \times n)$ is a constant. This constant holds for what are called ideal gases—gases that conform to the gas laws.

If you know the values for P, V, T, and n for one set of conditions, you can calculate a value for the constant. Recall that 1 mol of every gas occupies 22.4 L at STP (101.3 kPa and 273 K). You can use these values to find the value of the constant, which has the symbol R and is called the ideal gas constant. Insert the values of P, V, T, and n into $(P \times V)/(T \times n)$.

$$R = \frac{P \times V}{T \times n} = \frac{101.3\ \text{kPa} \times 22.4\ \text{L}}{273\ \text{K} \times 1\ \text{mol}} = 8.31\ (\text{L} \cdot \text{kPa})/(\text{K} \cdot \text{mol})$$

Focus on ELL

1 CONTENT AND LANGUAGE Direct students' attention to the vocabulary terms on page 464 and the heading on page 467. Clarify that the words *ideal, constant,* and *real* have both common and scientific meanings. Have pairs make and use flashcards with these terms. They should write a term on one side and the common and scientific definitions on the other. Have students practice writing sentences using each term in both its common and scientific context. Allow students with limited English to use their native languages for support.

2 FRONTLOAD THE LESSON Have students share their experiences with stage fog and/or dry ice sublimation, and ask them under what temperatures or seasons they encountered these events. Direct them to the information in Chemistry & You.

3 COMPREHENSIBLE INPUT Explain Figure 14.14. to students. Point out that scientists make use of the extremely cold temperature of liquid nitrogen to preserve biological samples.

The **ideal gas constant** (R) has the value 8.31 (L·kPa)/(K·mol). The gas law that includes all four variables—P, V, T, and n—is called the **ideal gas law.** It is usually written as follows.

$$P \times V = n \times R \times T \text{ or } PV = nRT$$

When the pressure, volume, and temperature of a contained gas are known, you can use the ideal gas law to calculate the number of moles of the gas. The amount of helium in a balloon, the amount of air in a scuba tank or a bicycle tire—each of these quantities can be calculated using the ideal gas law as long as you know the values for P, V, and T in each case.

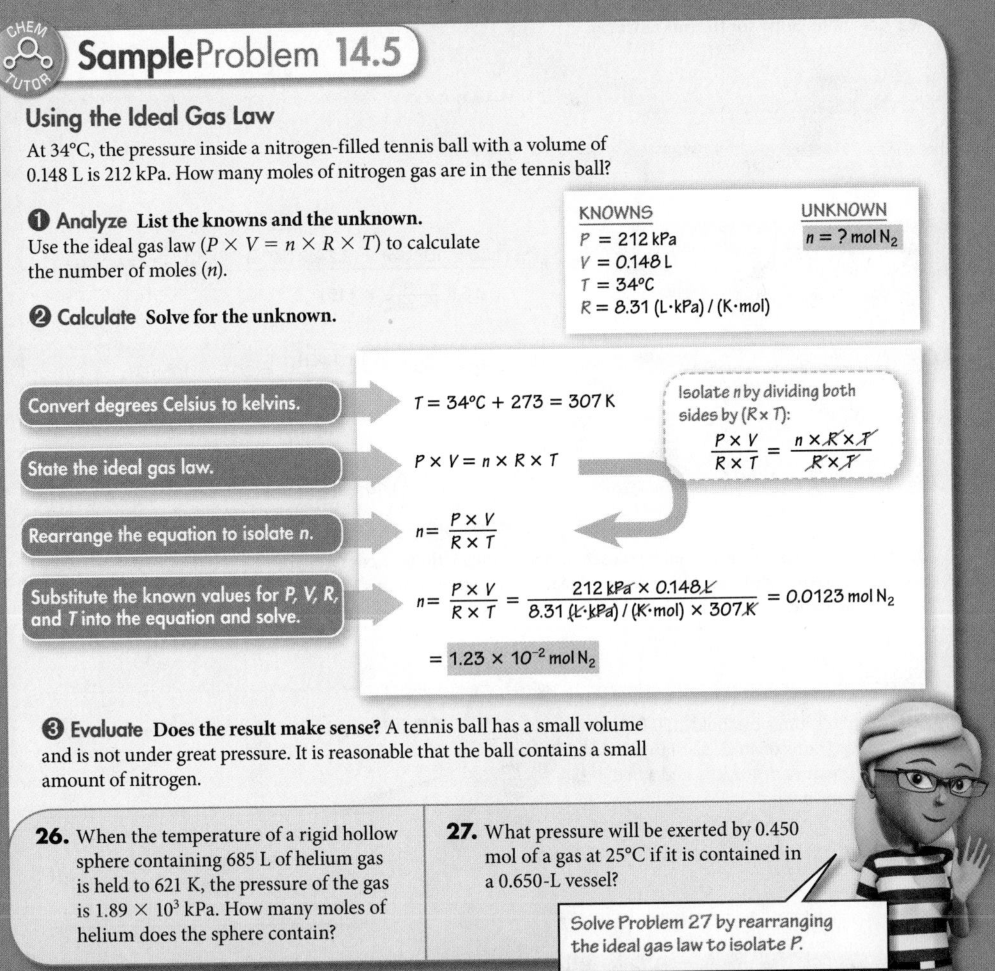

Sample Problem 14.5

Using the Ideal Gas Law

At 34°C, the pressure inside a nitrogen-filled tennis ball with a volume of 0.148 L is 212 kPa. How many moles of nitrogen gas are in the tennis ball?

1 Analyze **List the knowns and the unknown.** Use the ideal gas law ($P \times V = n \times R \times T$) to calculate the number of moles (n).

KNOWNS	UNKNOWN
P = 212 kPa	n = ? mol N_2
V = 0.148 L	
T = 34°C	
R = 8.31 (L·kPa) / (K·mol)	

2 Calculate **Solve for the unknown.**

Convert degrees Celsius to kelvins.

$$T = 34°C + 273 = 307 \text{ K}$$

State the ideal gas law.

$$P \times V = n \times R \times T$$

Isolate n by dividing both sides by ($R \times T$):

$$\frac{P \times V}{R \times T} = \frac{n \times \cancel{R} \times \cancel{T}}{\cancel{R} \times \cancel{T}}$$

Rearrange the equation to isolate n.

$$n = \frac{P \times V}{R \times T}$$

Substitute the known values for P, V, R, and T into the equation and solve.

$$n = \frac{P \times V}{R \times T} = \frac{212 \text{ kPa} \times 0.148 \text{ L}}{8.31 (\text{L} \cdot \text{kPa}) / (\text{K} \cdot \text{mol}) \times 307 \text{ K}} = 0.0123 \text{ mol } N_2$$

$$= 1.23 \times 10^{-2} \text{ mol } N_2$$

3 Evaluate **Does the result make sense?** A tennis ball has a small volume and is not under great pressure. It is reasonable that the ball contains a small amount of nitrogen.

26. When the temperature of a rigid hollow sphere containing 685 L of helium gas is held to 621 K, the pressure of the gas is 1.89×10^3 kPa. How many moles of helium does the sphere contain?

27. What pressure will be exerted by 0.450 mol of a gas at 25°C if it is contained in a 0.650-L vessel?

Solve Problem 27 by rearranging the ideal gas law to isolate P.

The Behavior of Gases 465

Explain

Ideal Gas Law

USE VISUALS Sketch two containers with identical volumes on the board. Tell students one container is filled with neon gas and the other with helium at the same temperature and pressure. Ask students to use the ideal gas law to find the equation for the number of moles in each container. Students should determine that $n_{He} = n_{Ne}$.

START A CONVERSATION Explain that Avogadro's hypothesis makes it possible to relate the molar quantity of a gas to its temperature, volume, and pressure. The hypothesis assumes that as long as the particles are not tightly packed, equal volumes of gases at the same temperature and pressure contain equal numbers of particles. Emphasize that Avogadro's hypothesis works only for gases because a large portion of their volume is empty space. At low pressure, the volumes of individual molecules are negligible compared to the volume of the container holding the gas. The volume of a gas depends on the number of particles present, not their size. **Ask** How would you determine the mass of the helium in a balloon, at STP, without making any mass measurements? *(According to Avogadro's hypothesis, one mole of a gas has a volume of 22.4 L at STP. Calculate the volume of the helium in the balloon, in liters at STP. Then, convert from liters to moles. To find the mass, multiply the number of moles by the molar mass of He.)*

Sample Practice Problem

A helium-filled balloon contains 0.16 mol He at 101 kPa and a temperature of 23°C. What is the volume of the gas in the balloon? *(3.9 L)*

Foundations for Math

SCIENTIFIC NOTATION Gas law calculations often result in very large or very small numbers. Converting these numbers to scientific notation makes them more manageable and easier to read. Before converting to scientific notation, students should identify the direction in which the decimal point needs to move. This determines the sign of the exponent. The direction of decimal point movement is determined by whether the number is greater than or less than one.

In **Sample Problem 14.5**, the answer is less than one, so the decimal point needs to move to the right in order to convert the number to scientific notation. Moving the decimal point to the right gives a negative exponent. In the sample problem, the decimal point moves two places to the right, so the exponent for 10 will be -2.

Answers

26. 2.51×10^2 mol

27. 1.71×10^3 kPa

Explain

Sample Practice Problems

A. When the temperature of a rigid cylinder containing 235 L of argon gas is held to 365 K, the pressure of the gas is 768 kPa. How many moles of argon does the cylinder contain? *(59.5 mol Ar)*

B. What is the volume of a vessel that contains 0.500 mol of a gas, if the gas exerts 3.75×10^3 kPa of pressure at 25°C? *(0.330 L)*

Using the Ideal Gas Law

A deep underground cavern contains 2.24×10^6 L of methane gas (CH_4) at a pressure of 1.50×10^3 kPa and a temperature of 315 K. How many kilograms of CH_4 does the cavern contain?

KNOWNS	UNKNOWN
$P = 1.50 \times 10^3$ kPa	$m = ?$ kg CH_4
$V = 2.24 \times 10^6$ L	
$T = 315$ K	
$R = 8.31$ (L·kPa) / (K·mol)	
molar mass$_{CH_4}$ = 16.0g	

❶ Analyze List the knowns and the unknown. Calculate the number of moles (n) using the ideal gas law. Use the molar mass of methane to convert moles to grams. Then convert grams to kilograms.

❷ Calculate Solve for the unknown.

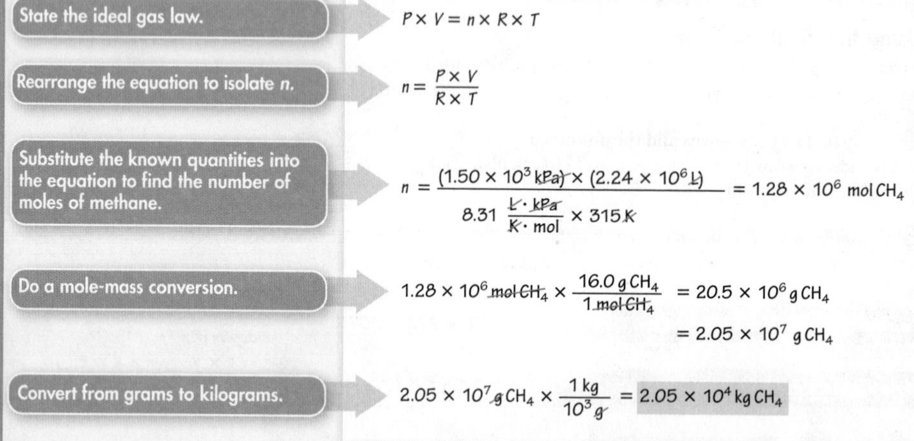

❸ Evaluate Does the result make sense? Although the methane is compressed, its volume is still very large. So it is reasonable that the cavern contains a large mass of methane.

28. A child's lungs can hold 2.20 L. How many grams of air do her lungs hold at a pressure of 102 kPa and a body temperature of 37°C? Use a molar mass of 29 g for air, which is about 20% O_2 (32 g/mol) and 80% N_2 (28 g/mol).

29. What volume will 12.0 g of oxygen gas (O_2) occupy at 25°C and a pressure of 52.7 kPa?

In Problems 28 and 29, make sure to express the temperature in kelvins before substituting for T in the ideal gas law equation.

Foundations for Math

RECIPROCALS When dividing fractions, you have to find the reciprocal. Two fractions are reciprocals of each other if their product is 1. For example, $\frac{2}{3}$ and $\frac{3}{2}$ are reciprocals since $\frac{2}{3} \times \frac{3}{2} = 1$.

This mathematical property holds true when converting units. Look at the third part of Step 2 in **Sample Problem 14.6**. When all the units are canceled, you are left with 1/1/mol. Keeping reciprocals in mind, you know that 1/1/mol is the same as 1 x mol/1, so the units are moles.

Ideal Gases and Real Gases

Under what conditions are real gases most likely to differ from ideal gases?

An ideal gas is one that follows the gas laws at all conditions of pressure and temperature. Such a gas would have to conform precisely to the assumptions of kinetic theory. Its particles could have no volume, and there could be no attraction between particles in the gas. As you probably suspect, there is no gas for which these assumptions are true. So an ideal gas does not exist. Nevertheless, at many conditions of temperature and pressure, a real gas behaves very much like an ideal gas.

The particles in a real gas do have volume, and there are attractions between the particles. Because of these attractions, a gas can condense, or even solidify, when it is compressed or cooled. For example, if water vapor is cooled below 100°C at standard atmospheric pressure, it condenses to a liquid. The behavior of other real gases is similar, although lower temperatures and greater pressures may be required. Such conditions are required to produce the liquid nitrogen in Figure 14.14. **Real gases differ most from an ideal gas at low temperatures and high pressures.**

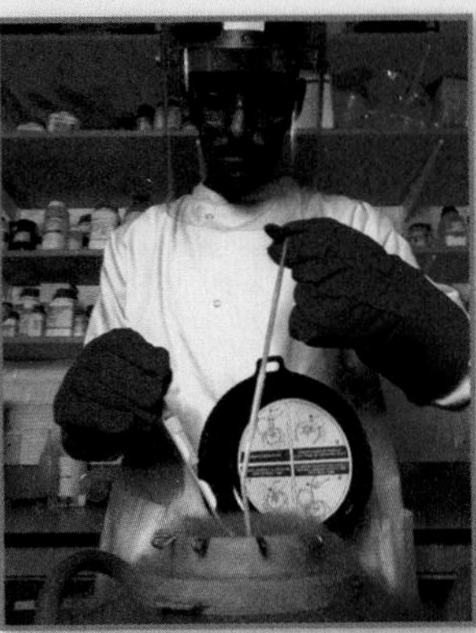

Figure 14.14 Liquid Nitrogen
A lab technician places a cell sample into an insulated tank containing liquid nitrogen. Nitrogen boils at −196°C.

Quick Lab

Purpose To measure the amount of carbon dioxide gas given off when antacid tablets dissolve in water

Materials
- 6 effervescent antacid tablets
- 3 rubber balloons (spherical)
- plastic medicine dropper
- water
- clock or watch
- metric tape measure
- graph paper
- water
- pressure sensor (optional)

Carbon Dioxide From Antacid Tablets

Procedure 

1. Break six antacid tablets into small pieces. Keep the pieces from each tablet in a separate pile. Put the pieces from one tablet into the first balloon. Put the pieces from two tablets into a second balloon. Put the pieces from three tablets into a third balloon. **CAUTION** *If you are allergic to latex, do not handle the balloons.*

2. After you use the medicine dropper to squirt about 5 mL of cold water into each balloon, immediately tie off each balloon.

3. Shake the balloons to mix the contents. Allow the contents to warm to room temperature.

4. Measure and record the circumference of each balloon several times during the next 20 minutes.

5. Use the maximum circumference of each balloon to calculate its volume. (*Hint:* For the volume of a sphere, use $V = \frac{4}{3}\pi r^3$ and $r = \text{circumference}/2\pi$.)

Analyze and Conclude

1. Graph Make a graph of volume versus number of tablets. Use your graph to describe the relationship between the number of tablets used and the volume of the balloon.

2. Calculate Assume that the balloon is filled with carbon dioxide gas at 20°C and standard pressure. Calculate the mass and the number of moles of CO_2 in each balloon at maximum inflation.

3. Analyze Data If a typical antacid tablet contains 2.0 g of sodium hydrogen carbonate, how many moles of CO_2 should one tablet yield? Compare this theoretical value with your results.

Focus on ELL

4 LANGUAGE PRODUCTION Have students work in groups to complete the lab. Make sure each group has ELLs of varied language proficiencies so that more proficient students can help less proficient ones. Have students work according to their proficiency level.

BEGINNING: LOW/HIGH Allow students to use drawings, single words, or short phrases. Have them copy the written work of more proficient students for future reference.

INTERMEDIATE: LOW/HIGH Have students draft sentences to answer the questions that require writing.

ADVANCED: LOW/HIGH Have students assist and/or edit the work of classmates with lower language proficiency.

Explain

Ideal Gases and Real Gases

APPLY CONCEPTS Have students apply the ideal gas law to a sample of gas in a balloon. Ask them to explain why in a balloon *n* and *R* are constants and *P*, *V*, and *T* are variables.

Explore

Quick Lab

PURPOSE After completing this activity, students will be able to measure the amount of carbon dioxide gas given off when antacid tablets dissolve in water.

SKILLS FOCUS observing, calculating, measuring

PREP TIME 10 minutes

CLASS TIME 40 minutes

SAFETY If you use latex balloons, check to see if any students are allergic to latex.

EXPECTED OUTCOME The volumes of CO_2 produced will reflect the amount of antacid each balloon contains.

ANALYZE AND CONCLUDE

1. The volume of the balloon is directly proportional to the number of tablets.
2. Answers will vary, but the masses and numbers of moles should be in ratios of 1:2:3 for the three balloons.
3. Possible answer: 2.0 g of $NaHCO_3$ (molar mass = 84.01 g) should yield about 1.2×10^{-2} mol of CO_2.

EXTENSION Have students use a similar procedure to compare different brands of effervescent antacids instead of different amounts of the same antacid.

Answers

28. 2.5 g air
29. 17.6 L

Explain

INTERPRET GRAPHS Investigate how real gases are different from ideal gases. For an ideal gas, by definition $PV = nRT$, and the ratio of PV/nRT is 1. **Ask** When the volume of a gas is greater than expected, what is the value of the ratio of PV to nRT? *(PV to nRT tends to be greater than 1.)* **Ask** When the volume is less than expected, what is the value of the ratio of PV to nRT? *(PV to nRT tends to be less than 1.)*

CHEMISTRY & YOU Sublimation (when the solid carbon dioxide changes to gas) and condensation (when water vapor in the air changes to liquid).

Evaluate

Informal Assessment

Have students apply the ideal gas law to a sample of gas in a balloon. Ask them to explain why in a balloon n and R are constants and P, V, and T are variables. Then have students complete the 14.3 Lesson Check.

Reteach

If students are having trouble understanding Boyle's law, have them illustrate Boyle's law using dots to represent gas particles. They should begin with an illustration that shows the particles in a starting condition. The other illustrations should show the particles' positions after changes in pressure.

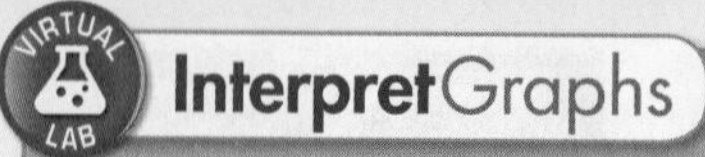

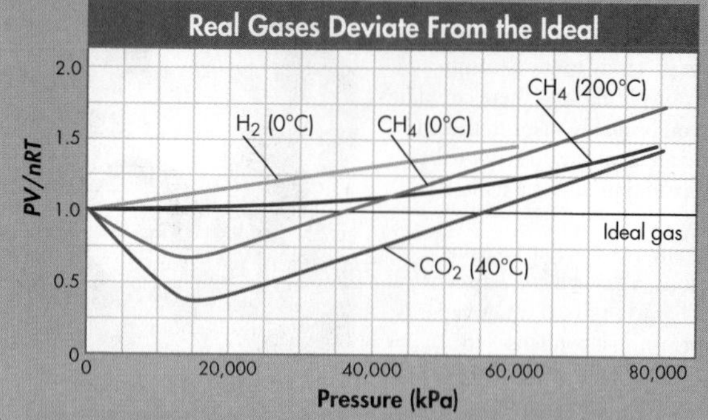

Figure 14.15 This graph shows how real gases deviate from the ideal gas law at high pressures.
a. Read Graphs What are the values of (PV/nRT) for an ideal gas at 20,000 and 60,000 kPa?
b. Identify What variable is responsible for the differences between the two methane (CH_4) curves?
c. Make Generalizations How does an increase in pressure affect the value of (PV/nRT) for real gases?

CHEMISTRY & YOU

Q: *Certain types of fog machines use dry ice and water to create stage fog. Chunks of dry ice are added to hot water, which causes the dry ice to sublime. The cold carbon dioxide gas causes water vapor in the surrounding air to condense into small droplets, resulting in a smoky fog. What phase changes occur when stage fog is made?*

Figure 14.15 shows how the value of the ratio (PV/nRT) changes as pressure increases. For an ideal gas, the result is a horizontal line because the ratio is always equal to 1. For real gases at high pressure, the ratio may deviate, or depart, from the ideal. When the ratio is greater than 1, the curve rises above the ideal gas line. When the ratio is less than 1, the curve drops below the line. The deviations can be explained by two factors. As attractive forces reduce the distance between particles, a gas occupies less volume than expected, causing the ratio to be less than 1. But the actual volume of the molecules causes the ratio to be greater than 1.

In portions of the curves below the line, intermolecular attractions dominate. In portions of the curves above the line, molecular volume dominates. Look at the curves for methane (CH_4) at 0°C and at 200°C. At 200°C, the molecules have more kinetic energy to overcome intermolecular attractions. So the curve for CH_4 at 200°C never drops below the line.

14.3 LessonCheck

ONLINE PROBLEMS

30. **Review** How can you determine the number of moles of a contained gas when the pressure, volume, and temperature are known values?

31. **Identify** Under what conditions do real gases deviate most from ideal behavior?

32. **Calculate** Determine the volume occupied by 0.582 mol of a gas at 15°C if the pressure is 81.8 kPa.

33. **Calculate** You fill a rigid steel cylinder that has a volume of 20.0 L with nitrogen gas to a final pressure of 2.00×10^4 kPa at 28°C. How many kilograms of N_2 does the cylinder contain?

34. **Compare** What is the difference between a real gas and an ideal gas?

35. **Analyze Data** At standard pressure, ammonia condenses at −33.3°C but nitrogen does not condense until −195.79°C. Use what you know about bond polarity to explain this difference.

BIG IDEA KINETIC THEORY

36. Use the kinetic theory of gases to explain this statement: No gas exhibits ideal behavior at all temperatures and pressures.

Lesson Check Answers

30. by using the equation for the ideal gas law *($PV = nRT$)*

31. Real gases deviate from ideal behavior at low temperatures and high pressures.

32. 17.0 L

33. 4.48 kg

34. An ideal gas obeys the assumptions of kinetic theory. A real gas deviates from ideal behavior except within a small range of conditions.

35. The nitrogen molecule is nonpolar and the ammonia molecule is polar. So there are stronger intermolecular attractions in ammonia.

36. **BIGIDEA** In real gases, there are attractions between molecules, and the molecules have volume. At low temperatures, attractions between molecules pull them together and reduce the volume. At high pressures, the volume occupied by the molecules is a significant part of the total volume.

14.4 Gases: Mixtures and Movements

CHEMISTRY & YOU

Q: *Why do balloons filled with helium deflate faster than balloons filled with air?* You have probably seen party balloons inflated with air or helium. The surface of a latex balloon has tiny pores through which gas particles can pass, causing the balloon to deflate over time. The rate at which the balloon deflates depends on the gas it contains.

Key Questions

How is the total pressure of a gas mixture related to the partial pressures of the component gases?

How does the molar mass of a gas affect the rate at which the gas diffuses or effuses?

Vocabulary

- partial pressure
- Dalton's law of partial pressures
- diffusion • effusion
- Graham's law of effusion

Dalton's Law

How is the total pressure of a gas mixture related to the partial pressures of the component gases?

Gas pressure results from collisions of particles in a gas with an object. If the number of particles increases in a given volume, more collisions occur. If the average kinetic energy of the particles increases, more collisions occur. In both cases, the pressure increases. Gas pressure depends only on the number of particles in a given volume and on their average kinetic energy. Particles in a mixture of gases at the same temperature have the same average kinetic energy. So the kind of particle is not important.

Table 14.1 shows the composition of dry air, or air that does not contain any water vapor. The contribution each gas in a mixture makes to the total pressure is called the **partial pressure** exerted by that gas. In dry air, the partial pressure of nitrogen is 79.11 kPa. **In a mixture of gases, the total pressure is the sum of the partial pressures of the gases.**

Interpret Data

Composition of Dry Air

Component	Volume (%)	Partial pressure (kPa)
Nitrogen	78.08	79.11
Oxygen	20.95	21.22
Carbon dioxide	0.04	0.04
Argon and others	0.93	0.95
Total	**100.00**	**101.32**

Table 14.1 The total pressure of dry air is the sum of the partial pressures of its component gases.
a. Read Tables What is the partial pressure of oxygen in dry air?
b. Predict As altitude increases, atmospheric pressure decreases. What do you think happens to the partial pressure of oxygen in the air as altitude increases?

LESSON 14.4

Key Objectives

14.4.1 RELATE the total pressure of a mixture of gases to the partial pressures of the component gases.

14.4.2 EXPLAIN how the molar mass of a gas affects the rate at which the gas diffuses and effuses.

Additional Resources

- Core Teaching Resources, Lesson 14.4 Review
- Reading and Study Workbook, Lesson 14.4
- Laboratory Manual, Lab 25
- Small-Scale Laboratory Manual, Lab 21

PearsonChem.com

Engage

CHEMISTRY & YOU Have students study the photograph and read the opening text. Ask students to predict an answer, and write it on the board.

Build Background

Have students speculate on how smelling salts work. Explain that most smelling salts contain ammonia (NH_3). If possible bring smelling salts to class and demonstrate. *(Ammonia gas diffuses into the air and irritates the membranes of the nose, triggering an inhalation reflex.)*

Foundations for Reading

BUILD VOCABULARY Students may have trouble with the scientific definitions of *effusion* and *diffusion* because the common meanings often overlap. Ask students what strategy they will use to help them remember that particles move apart during *diffusion* but through a tiny hole during *effusion*.

Focus on ELL

1 CONTENT AND LANGUAGE Ask students to restate the key questions in their own words. Have pairs make and use flashcards with key vocabulary. They should write the term on one side and a definition, visual, and sentence with the term on the other. Allow students with limited English to use their native languages for support.

2 FRONTLOAD THE LESSON Have students discuss their experiences with helium balloons, and ask if they know why these balloons float. Direct them to the information in Chemistry & You.

3 COMPREHENSIBLE INPUT Explain Figure 14.17, restating information in its captions as needed. Model solving Sample Problem 14.7 by explaining your thought process aloud while writing on the board.

Answers

INTERPRET GRAPHS

a. 1.0 in both cases
b. temperature
c. The ratio increases in the long run. (For some gases it decreases first.)

INTERPRET DATA

a. 21.22 kPa
b. The partial pressure decreases.

LESSON 14.4

Explain

Dalton's Law

START A CONVERSATION Assist students in understanding that the relative partial pressure exerted by a gas in a mixture of gases does not vary with temperature, pressure, or volume of the mixture. The ratio of n_1/n (the relative amount of the gas in the mixture) remains the same. Because the value of n_1/n remains constant, the relative partial pressure (P_1/P) exerted by the gas remains the same.

KINETIC ART See Dalton's Law animated online.

The data in Table 14.1 illustrate a law proposed by the chemist John Dalton. **Dalton's law of partial pressures** states that, at constant volume and temperature, the total pressure exerted by a mixture of gases is equal to the sum of the partial pressures of the component gases. You can express Dalton's law mathematically as follows:

$$P_{total} = P_1 + P_2 + P_3 + \ldots$$

Dalton's law holds true because each component gas exerts its own pressure independent of the pressure exerted by the other gases. Look at Figure 14.16a. The container is filled with heliox, a helium-oxygen gas mixture used in deep-sea scuba diving. The helium component of this mixture is shown in Figure 14.16b at the same volume and temperature. The oxygen component is shown in Figure 14.16c, also at the same volume and temperature. Each gas in the mixture exerts the pressure it exerted before the gases were mixed to make heliox. So the pressure in the container of heliox (500 kPa) is the sum of the pressures in the containers of helium and oxygen (400 kPa + 100 kPa).

If the percent composition of a mixture of gases does not change, the fraction of the pressure exerted by a gas does not change as the total pressure changes. This fact is important for people who must operate at high altitudes. For example, at the top of Mount Everest, the total atmospheric pressure is 33.73 kPa. This pressure is about one third of its value at sea level. The partial pressure of oxygen is also reduced by one third, to 7.06 kPa. But in order to support respiration in humans, the partial pressure of oxygen must be 10.67 kPa or higher. So climbers of Mount Everest need an oxygen mask and a cylinder of compressed oxygen to survive the ascent.

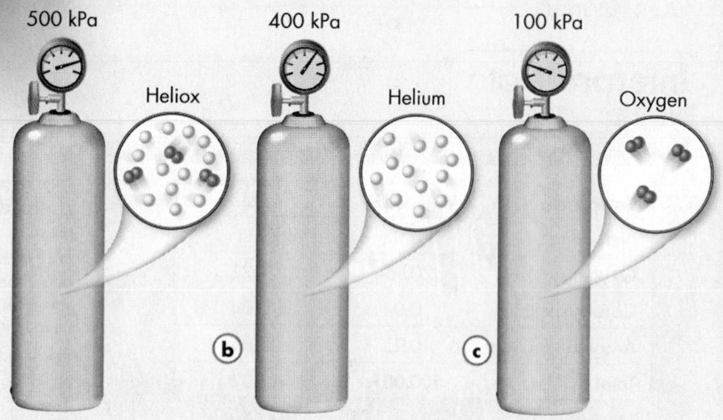

500 kPa
Heliox
a
400 kPa
Helium
b
100 kPa
Oxygen
c

Figure 14.16 Dalton's Law
Heliox is a mixture of helium and oxygen gas. The pressure exerted by the helium in the mixture is independent of the pressure exerted by the oxygen. **Compare** ***How does the ratio of helium atoms to oxygen molecules in heliox compare to the ratio of the partial pressures?***

Decompression Sickness

The invention of compressed air in the 1840s allowed people to work underwater, with one major drawback. Decompression sickness was initially called caisson disease because workers constructing the Brooklyn Bridge worked in watertight containers called caissons. The pressure of the air in the caissons needed to be greater than atmospheric pressure to withstand the pressure of the surrounding water. The disease the workers experienced was also called "the bends." In 1878, Paul Bert stated that workers could avoid the bends if they ascended gradually to the surface. Bert referred to the work of Robert Boyle. In 1667, Boyle observed a bubble form in the eye of a viper that was placed in a compressed atmosphere and then removed. (Boyle reported that the viper appeared distressed by the experience.)

SampleProblem 14.7

Using Dalton's Law of Partial Pressures

Air contains oxygen, nitrogen, carbon dioxide, and trace amounts of other gases. What is the partial pressure of oxygen (P_{O_2}) at 101.30 kPa of total pressure if the partial pressures of nitrogen, carbon dioxide, and other gases are 79.10 kPa, 0.040 kPa, and 0.94 kPa, respectively?

1 Analyze **List the knowns and the unknown.** Use the equation for Dalton's law of partial pressures ($P_{total} = P_{O_2} + P_{N_2} + P_{CO_2} + P_{others}$) to calculate the unknown value (P_{O_2}).

KNOWNS	UNKNOWN
$P_{N_2} = 79.10$ kPa	$P_{O_2} = ?$ kPa
$P_{CO_2} = 0.040$ kPa	
$P_{others} = 0.94$ kPa	
$P_{total} = 101.30$ kPa	

2 Calculate **Solve for the unknown.**

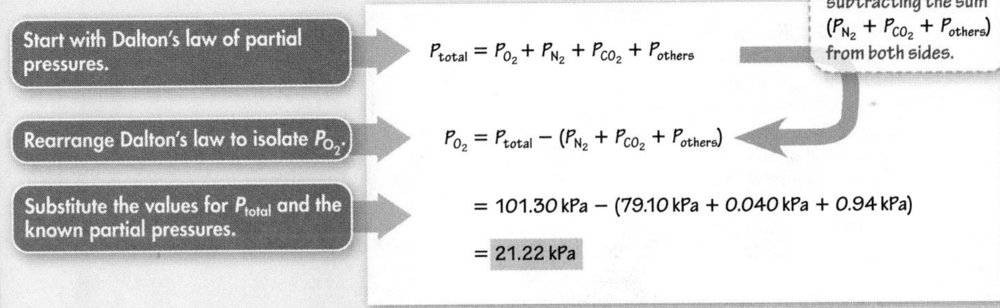

3 Evaluate **Does this result make sense?** The partial pressure of oxygen must be smaller than that of nitrogen because P_{total} is only 101.30 kPa. The other partial pressures are small, so the calculated answer of 21.22 kPa seems reasonable.

37. A gas mixture containing oxygen, nitrogen, and carbon dioxide has a total pressure of 32.9 kPa. If $P_{O_2} = 6.6$ kPa and $P_{N_2} = 23.0$ kPa, what is P_{CO_2}?

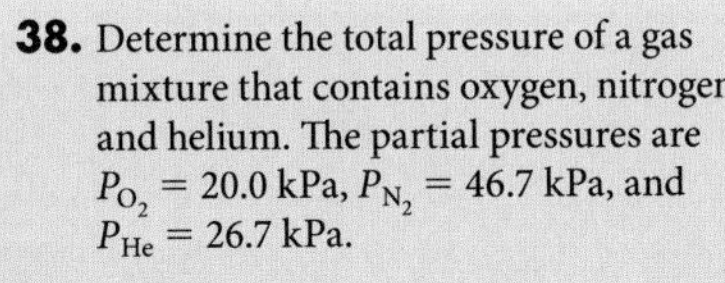

38. Determine the total pressure of a gas mixture that contains oxygen, nitrogen, and helium. The partial pressures are $P_{O_2} = 20.0$ kPa, $P_{N_2} = 46.7$ kPa, and $P_{He} = 26.7$ kPa.

In Problem 38, the unknown is P_{total}, so you can solve without having to rearrange Dalton's law.

Foundations for Math

ISOLATING A VARIABLE Remind students that there are two shortcut methods for maintaining the equality of the two sides of an equation when isolating a variable: ***Rule 1*** Change the sign (for positive or negative values); ***Rule 2*** Invert the coefficient.

Point out that in **Sample Problem 14.7**, Rule 1 is used to isolate P_{O_2}. The sign change is accomplished by subtracting the sum of partial pressures of the remaining gases in air from both sides of the equation to yield the second step of the equation:

$$P_{total} - (P_{N_2} + P_{CO_2} + P_{others}) = P_{O_2} + P_{N_2} + P_{CO_2} + P_{others} - (P_{N_2} + P_{CO_2} + P_{others})$$

Students should recognize that subtracting the sum of the partial pressures is the same as subtracting each partial pressure from both sides separately.

Explain

MAKE A CONNECTION Explain to students that Dalton's law of partial pressures has important applications for pilots and mountain climbers. People who work at high altitudes must have a source of supplementary oxygen to ensure that the pressure of oxygen in their blood is high enough for them to survive. Adding oxygen to a mixture results in a higher percentage of oxygen and thus a higher partial pressure. Oxygen masks, for example, provide a higher partial pressure of oxygen. Ask students for examples of where an oxygen mask is used. *(in hospitals where people have trouble breathing; at fire scenes where people have inhaled smoke; at sporting events where athletes cannot get enough oxygen)*

USE AN ANALOGY Use the following analogy to help students with the concept of partial pressures. If possible, try to bring a bag of chocolates to demonstrate.

You are given a 2.00-kg bag of chocolates and in the bag you find there are four kinds of chocolates. You can separate each kind into a pile and find the mass of each pile. To find the percentage mass of each kind of chocolate, you divide each partial mass by the total and multiply by 100.

Kind of chocolate	Mass	Percentage of total
red	0.85 kg	42.5%
yellow	0.32 kg	16%
green	0.66 kg	33%
brown	0.17 kg	8.5%
Total	**2.00 kg**	**100%**

Sample Practice Problem

In a gas mixture of H_2, He, and Ar the total pressure is 101.3 kPa. The partial pressure of Ar is 50.7kPa, and the partial pressure of He is 10.1 kPa. What is the pressure of H_2? *(40.5 kPa)*

Answers

FIGURE 14.16 The ratios are equal.

37. 32.9 kPa – 6.6 kPa – 23.0 kPa = 3.3 kPa

38. 20.0 kPa + 46.7 kPa + 26.7 kPa = 93.4 kPa, or 9.34×10^1 kPa

LESSON 14.4

Explain

Graham's Law

USE VISUALS Have students study the photographs in Figure 14.17. Explain that diffusion is a relatively slow process whose rate is inversely proportional to the square root of the molar mass of the gas.

Explore

Teacher Demo

PURPOSE Students compare effusion rates of helium and air.

MATERIALS round, identical balloons; helium; metric tape measure.

SAFETY Wear safety goggles when handling balloons.

PROCEDURE Fill two identical, round nonlatex balloons with equal volumes of helium and air. Have volunteers measure the circumferences of the balloons. One day later, they should measure the circumferences again.

EXPECTED OUTCOME Helium effused from the balloon at a faster rate.

Misconception Alert

Be sure to tell students that effusion requires that a gas effuses into a vacuum or that two gases effuse into each other with no other molecules present. It's unlikely that students have seen a demonstration of true effusion.

Graham's Law

How does the molar mass of a gas affect the rate at which the gas diffuses or effuses?

READING SUPPORT

Build Vocabulary: *Prefixes* *Diffusion* and *effusion* come from the Latin *fundere* meaning "to pour." They differ only in their prefixes. The prefix *dif-* means "apart." The prefix *ex-* means "out." ***How do these prefixes help to contrast what happens to a gas during diffusion and effusion?***

Suppose you open a perfume bottle in one corner of a room. At some point, a person standing in the opposite corner will be able to smell the perfume. Molecules in the perfume evaporate and diffuse, or spread out, through the air in the room. **Diffusion** is the tendency of molecules to move toward areas of lower concentration until the concentration is uniform throughout.

The photo sequence in Figure 14.17 illustrates the diffusion process for bromine vapor. In Figure 14.17a, a glass cylinder containing air is inverted and sealed onto a cylinder containing bromine vapor. Figure 14.17b shows the bromine vapor diffusing through the air. The bromine vapor in the bottom cylinder has started to move upward into the top cylinder, where there is a lower concentration of bromine. In Figure 14.17c, the bromine has diffused to the top of the column formed by the combined cylinders. The concentration of bromine is now the same throughout the column.

Figure 14.17 Diffusion
The diffusion of one substance through another is a relatively slow process.
Describe ***How does the concentration of bromine in the bottom part of the column change during this sequence?***

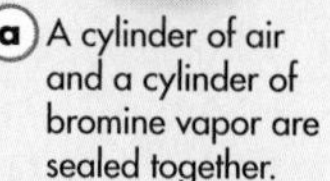

(a) A cylinder of air and a cylinder of bromine vapor are sealed together.

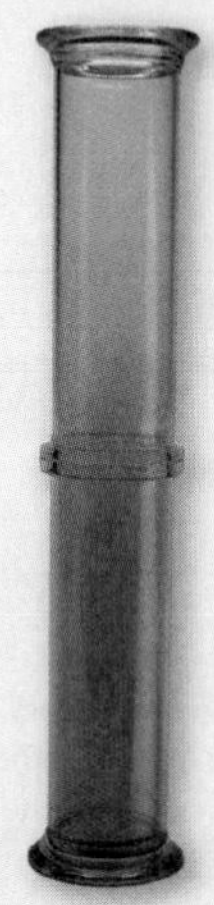

(b) Bromine vapor diffuses upward through the air.

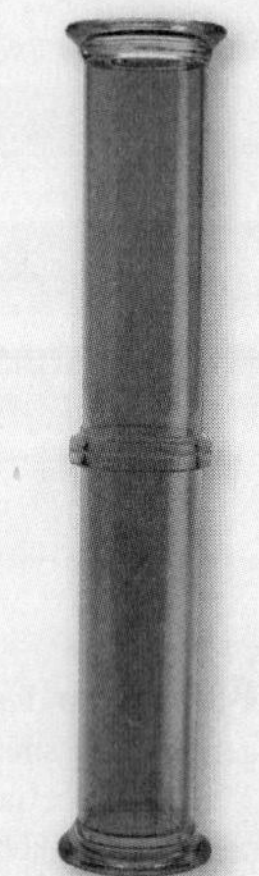

(c) After several hours, bromine vapors reach the top of the column.

There is another process that involves the movement of molecules in a gas. This process is called effusion. During **effusion,** a gas escapes through a tiny hole in its container. With effusion and diffusion, the type of particle is important. **Gases of lower molar mass diffuse and effuse faster than gases of higher molar mass.**

Thomas Graham's Contribution The Scottish chemist Thomas Graham studied rates of effusion during the 1840s. From his observations, he proposed a law. **Graham's law of effusion** states that the rate of effusion of a gas is inversely proportional to the square root of the gas's molar mass. This law can also be applied to the diffusion of gases.

Differentiated Instruction

L2 STRUGGLING STUDENTS Have students work in groups to rank the relative rates of diffusion at a constant temperature for carbon dioxide, helium, and nitrogen. *(helium > nitrogen > carbon dioxide)*

LPR LESS PROFICIENT READERS If students are having difficulty remembering the difference between effusion and diffusion, have them draw and label a model of each, indicating the movement of the molecules.

L3 ADVANCED STUDENTS Have students compare the use of compressed gas at high altitudes and the use of compressed gas during deep water dives.

Figure 14.18 Blimps
The cigar-shaped part of a blimp, called an envelope, is a sealed container of helium gas. **Infer** ***What properties do you think are desirable for the materials used to make the envelope of a blimp?***

Graham's law makes sense if you know how the mass, velocity, and kinetic energy of a moving object are related. The expression that relates the mass (m) and the velocity (v) of an object to its kinetic energy (KE) is $\frac{1}{2}mv^2$. For the kinetic energy to be constant, any increase in mass must be balanced by a decrease in velocity. For example, a ball with a mass of 2 g must travel at 5 m/s to have the same kinetic energy as a ball with a mass of 1 g traveling at 7 m/s. There is an important principle here. If two objects with different masses have the same kinetic energy, the lighter object must move faster.

CHEMISTRY & YOU

Q: *Why do balloons filled with helium deflate faster than balloons filled with air? Use Graham's law of effusion to explain your answer.*

Comparing Effusion Rates The blimp shown in Figure 14.18 is inflated with helium, which is less dense than air. One of the challenges in maintaining blimps is to keep the helium from seeping out. You may have noticed that party balloons filled with either helium or air gradually deflate over time. Both helium atoms and the molecules in air can pass through the tiny pores in a latex balloon. But a helium-filled balloon will deflate faster than an air-filled balloon. Kinetic theory can explain the difference.

Suppose you have two balloons, one filled with helium and the other filled with air. If the balloons are at the same temperature, the particles in each balloon have the same average kinetic energy. But helium atoms are less massive than oxygen or nitrogen molecules. So the molecules in air move more slowly than helium atoms with the same kinetic energy. Because the rate of effusion is related only to a particle's speed, Graham's law can be written as follows for two gases, A and B.

$$\frac{\text{Rate}_A}{\text{Rate}_B} = \sqrt{\frac{\text{molar mass}_B}{\text{molar mass}_A}}$$

In other words, the rates of effusion of two gases are inversely proportional to the square roots of their molar masses. Sample Problem 14.8 on the following page compares the effusion rates of helium and nitrogen.

Foundations for Math

SQUARE ROOTS Students should remember that when dividing one square root by another, they can simplify the problem by dividing before taking the square root. The order of operations is interchangeable because taking a square root is a form of division.

Have students study **Sample Problem 14.8** on the following page. Point out that the square root of 7.0 is 2.6 (rounded to 2 significant figures). They can find the same answer by dividing the square root of 28.0 by the square root of 4.0.

Explain

CHEMISTRY & YOU Helium has a smaller molar mass than the nitrogen and oxygen molecules that make up air. So, according to Graham's law of effusion, helium atoms will effuse faster through the pores in rubber balloon.

CRITICAL THINKING Ask a student volunteer to read the second paragraph of the Comparing Effusion Rates section out loud. Have students consider the statement *But helium atoms are less massive than oxygen or nitrogen molecules*. Ask them to use their knowledge of atomic and molecular structure of the three gases and their knowledge of kinetic energy to explain why this statement is true. *(Helium is a monatomic gas with a low atomic mass while oxygen and nitrogen are more massive and diatomic. Thus, the masses of oxygen and nitrogen molecules are much greater than the mass of helium atoms. Since KE = ½mv², the much smaller mass of the helium atoms gives them a much greater velocity compared to the oxygen and nitrogen molecules.)*

Challenge students to use the kinetic energy equation to show mathematically that the velocity of helium atoms is 2.8 times the velocity of oxygen molecules and 2.6 times the velocity of nitrogen molecules.

Answers

READING SUPPORT During diffusion, the particles move apart. During effusion, the particles move out of a container through a tiny hole. (Students may have trouble with the scientific definitions of *effusion* and *diffusion* because of the multiple, often overlapping, meanings of these terms in common usage. For example, in many dictionaries, one definition of *effuse* is "to spread out; diffuse; radiate.")

FIGURE 14.17 The bromine concentration in the bottom part of the column gradually decreases.

FIGURE 14.18 Sample answer: The envelope material should be strong, flexible, lightweight (so as not to make the blimp too heavy), and gastight (to prevent seepage of helium).

LESSON 14.4

Explain

Sample Practice Problem

How many times as fast does helium (He) effuse compared to oxygen (O_2)? *(2.8 times as fast)*

Evaluate

Informal Assessment

Tell students that the partial pressures of oxygen and hydrogen gases in a container are both 100 kPa. **Ask** In which sample are there more molecules? In which sample do the molecules have greater average kinetic energy? *(Both gases have the same number of molecules and same average kinetic energy.)* About how many times as fast does helium diffuse compared to oxygen? *(Helium diffuses almost three times as fast as oxygen.)* Then have students complete the 14.4 Lesson Check.

Reteach

Remind students that *diffusion* is a general term that applies to molecules moving away from a region of high concentration. Effusion is a specific example of diffusion and involves molecules passing through a narrow opening. Graham's law applies to both.

SampleProblem 14.8

Comparing Effusion Rates

How much faster does helium (He) effuse than nitrogen (N_2) at the same temperature?

1 Analyze List the knowns and the unknown. Use Graham's law and the molar masses of the two gases to calculate the ratio of effusion rates.

KNOWNS
molar $mass_{He}$ = 4.0 g
molar $mass_{N_2}$ = 28.0 g

UNKNOWN
ratio of effusion rates = ?

2 Calculate Solve for the unknown. Helium effuses nearly three times faster than nitrogen at the same temperature.

Start with the equation for Graham's law of effusion.

$$\frac{Rate_{He}}{Rate_{N_2}} = \frac{\sqrt{\text{molar mass}_{N_2}}}{\sqrt{\text{molar mass}_{He}}}$$

Substitute the molar masses of nitrogen and helium into the equation.

$$\frac{Rate_{He}}{Rate_{N_2}} = \sqrt{\frac{28.0\,\cancel{g}}{4.0\,\cancel{g}}} = \sqrt{7.0} = 2.7$$

3 Evaluate Does the result make sense? Helium atoms are less massive than nitrogen molecules, so it makes sense that helium effuses faster than nitrogen.

39. Calculate the ratio of the velocity of hydrogen molecules to the velocity of carbon dioxide molecules at the same temperature.

Use what you know about chemical formulas and the mole to write the molar mass of each gas.

14.4 LessonCheck

40. Review In a mixture of gases, how is the total pressure determined?

41. Review What is the effect of molar mass on rates of diffusion and effusion?

42. Explain How is the partial pressure of a gas in a mixture calculated?

43. Calculate The pressure in an automobile tire filled with air is 245.0 kPa. If P_{O_2} = 51.3 kPa, P_{CO_2} = 0.10 kPa, and P_{others} = 2.3 kPa, what is P_{N_2}?

44. Compare What distinguishes effusion from diffusion? How are these processes similar?

45. Relate Cause and Effect Explain why the rates of diffusion of nitrogen gas and carbon monoxide are almost identical at the same temperature.

46. Analyze Data Both Table 14.1 on page 469 and the Elements in the Atmosphere table on page R1 list data on the composition of air. Look at the data included in each table. Identify two ways in which the tables are similar. Describe at least three differences.

Lesson Check Answers

40. Total pressure is equal to the sum of the partial pressures of the components.

41. Gases with lower molar masses diffuse and effuse faster than gases with higher molar masses.

42. Subtract the partial pressures of the other gases from the total pressure.

43. 191.3 kPa

44. During effusion, a gas escapes through a tiny hole in its container. In both cases, a gas becomes more spread out and the rate depends on the molar mass.

45. Carbon monoxide and nitrogen have identical molar masses when the masses are rounded to two significant figures (28 g).

46. Both tables report data for dry air as mainly nitrogen and oxygen. Table 14.1 uses percents and includes a compound. The table on R1 uses ppm and provides specific data for elements.

Small-Scale Lab

Diffusion

Purpose

To infer diffusion of a gas by observing color changes during chemical reactions

Materials

- clear plastic cup or petri dish
- reaction surface
- dropper bottles containing bromothymol blue, hydrochloric acid, and sodium hydrogen sulfite
- ruler
- cotton swab
- NaOH, NH_4Cl (optional)

Procedure

1. Use the plastic cup or petri dish to draw the large circle shown below on a sheet of paper.

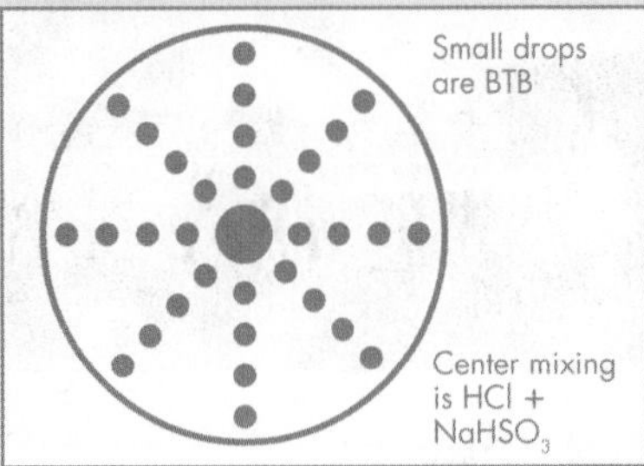

2. Place a reaction surface over the grid and add small drops of bromothymol blue (BTB) in the pattern shown by the small circles. Make sure the drops do not touch one another.

3. Mix one drop each of hydrochloric acid (HCl) and sodium hydrogen sulfite ($NaHSO_3$) in the center of the pattern.

4. Place the cup or petri dish over the grid and observe what happens.

5. If you plan to do Activity 1 in the You're the Chemist section, don't dispose of your materials yet.

Analyze

1. Observe Describe in detail the changes you observed in the drops of BTB over time. Draw pictures to illustrate the changes.

2. Describe Draw a series of pictures showing how one of the BTB drops might look over time if you could view the drop from the side.

3. Explain The BTB changed even though you added nothing to it. If the mixture in the center circle produced a gas, would this explain the change in the drops of BTB? Use kinetic theory to explain your answer.

4. Describe Translate the following word equation into a balanced chemical equation: Sodium hydrogen sulfite reacts with hydrochloric acid to produce sulfur dioxide gas, water, and sodium chloride.

You're the Chemist

The following activities allow you to develop your own procedures and analyze the results.

1. Analyze Data Carefully absorb the center mixture of the original experiment onto a cotton swab and replace it with one drop of NaOH and one drop of NH_4Cl. Describe what happens and explain in terms of kinetic theory. Ammonium chloride reacts with sodium hydroxide to produce ammonia gas, water, and sodium chloride. Write and balance a chemical equation to describe this reaction.

2. Design an Experiment Design an experiment to observe the effect of the size of the BTB drops on the rate at which they change. Explain your results in terms of kinetic theory.

Focus on ELL

4 LANGUAGE PRODUCTION Have students work in groups of four to complete the lab. Make sure each group has ELLs of varied language proficiencies so that more proficient students can help less proficient ones. Have students work according to their proficiency level.

BEGINNING: LOW/HIGH Restate each step in simple sentences. Demonstrate each step, and then have students perform it before proceeding to the next step.

INTERMEDIATE: LOW/HIGH Paraphrase the procedure and Analyze and Conclude questions for students.

ADVANCED: LOW/HIGH Have students assist classmates with lower language proficiency in understanding procedures and drafting answers.

Explore

Small-Scale Lab

OBJECTIVE After completing this activity, students will be able to infer diffusion of a gas by observing color changes during reaction.

SKILLS FOCUS inferring, drawing conclusions

PREP TIME 30 minutes

CLASS TIME 40 minutes

ADVANCE PREPARATION

Solution	Preparation
1.0*M* HCl	82 mL of 12*M* in 1.0 L
0.04% BTB	100 mg in 16.0 mL 0.01*M* NaOH, dilute to 250 mL
0.5*M* NaOH	20.0 g in 1.0 L
1.0*M* NH_4Cl	13.4 g in 250 mL
1.0*M* $NaHSO_3$	26 g in 250 mL

SAFETY Always slowly add acid to water.

TEACHING TIPS Before class, add a drop of BTB to the paper to be sure the paper is not acidic. The BTB should remain blue.

EXPECTED OUTCOME The drops of BTB turn yellow, starting with those closest to the center.

ANALYZE AND CONCLUDE

1. The drops near the center change immediately. As the gas diffuses, all the drops change color. The color change begins at the outer edge of each drop.
2. Students' drawings will vary.
3. The particles of gas produced are in motion. As the particles diffuse from the center, they collide and react with molecules of BTB.
4. $NaHSO_3 + HCl \rightarrow SO_2 + H_2O + NaCl$

YOU'RE THE CHEMIST

1. As the ammonia diffuses, BTB changes from yellow to blue.
$$NH_4Cl + NaOH \rightarrow NH_3 + H_2O + NaCl$$
2. Vary the size of the BTB drops from "pinheads" to "puddles." Tiny drops are better able to detect small quantities of gas.

FOR ENRICHMENT Have students design an experiment that shows the effect of temperature on rate of diffusion. (The drops change color more quickly at higher temperatures.)

Answers

39. 4.7/1

CHEMISTRY & YOU

CHEMISTRY & YOU Draw students' attention to the photos of the CNG fueling station and the pressurized NGV fuel tank. Emphasize that CNG is a gas—unlike gasoline, which is a liquid. Call on volunteers to identify the local price of gasoline. Some students' answers may not specify volumetric units. Explain that the price is in fact a ratio expressed as dollars ($) per gallon.

Pose the following question to the students: *Knowing that CNG is a gas and gasoline is a liquid, how would environmental factors play a role in determining the price of CNG?* You may need to guide the students in the following ways:

- Temperature and pressure do not significantly affect the volume of liquid gasoline.
- Price assumes that the volume is constant.
- Gases are easily compressed into smaller volumes.
- A gallon of natural gas at a high pressure contains more representative particles than a gallon of natural gas at a low temperature.

CHEMISTRY & YOU: TECHNOLOGY

Natural Gas Vehicles

Most of the cars you see on the road run on gasoline, better known as gas—although the fuel itself is a liquid, not a gas. But in some U.S. cities, as well as parts of South America and Asia, vehicles that run on gaseous fuel are becoming increasingly common.

A natural gas vehicle, or NGV, runs on compressed natural gas (CNG), which is natural gas that has been compressed to less than 1 percent of its volume at standard pressure. The fuel is stored in a pressurized cylinder. Like other vehicles that burn fossil fuels, NGVs emit CO_2. However, natural gas burns cleaner than gasoline or diesel fuel. Many gasoline-powered cars can be retrofitted with NGV technology so that the driver can choose to run the car on either gasoline or CNG.

Pros & Cons

Advantages of NGVs	Disadvantages of NGVs
✔ **Less pollution** NGVs produce much less carbon monoxide, nitrogen oxides, and toxins than gasoline-powered vehicles.	✘ **More expensive** NGVs tend to cost more than comparable gasoline-powered cars.
✔ **Less maintenance** Because natural gas burns cleaner than gasoline, the engines of NGVs require less servicing than those of gasoline-powered vehicles.	✘ **Less roomy** Due to the CNG tank, NGVs have less trunk space.
✔ **Cheaper fuel** Natural gas costs less than gasoline.	✘ **Limited travel range** On a single tank of gas, NGVs can travel only about 60 percent as far as gasoline-powered cars before needing more fuel.
✔ **Safety** The fuel tanks in NGVs are stronger and safer than gasoline storage tanks.	✘ **Hard to refuel** CNG refueling stations are currently few and far between.
	✘ **Still fossil-fueled** Like oil, natural gas is a nonrenewable resource.

21st Century Learning To be successful in the 21st century, students need skills and learning experiences that extend beyond subject matter mastery. The following project helps students build the following 21st Century Skills: Financial, *Economic, Business, and Entrepreneurial Literacy; Creativity and Innovation; Communication and Collaboration; Information, Media, and Technology Literacy; Initiative and Self-Direction; Social Skills; and Productivity and Accountability.*

SELL THAT CAR! Pose the following challenge to your students. *A major automobile manufacturer has just introduced an NGV to its lineup of cars. The manufacturer is in need of a marketing firm to create an ad campaign promoting the new model. Form a "firm" of 3–5 people to do the following:* Create a TV or magazine ad campaign that will be presented to the class.

- The ad should be based on facts about NGVs currently on the market.
- The ad should describe how these types of cars are different from gasoline-powered vehicles.

ON THE ROAD Most of the NGVs found in the United States are buses. But don't be surprised if you start to see passenger cars sporting the "NGV" logo.

FILL 'ER UP Any building with a natural gas line can be equipped with a pressurized fueling device that delivers CNG. But NGV drivers must be patient—an empty CNG tank takes much longer to fill than an empty gasoline tank.

A REAL GAS IN THE GAS TANK A mid-sized NGV sedan has a fuel tank in the back with a volume of 8 gallons, or 30 L. At ambient temperatures, the pressure inside a full tank of CNG is about 3600 pounds per square inch (psi), or 25,000 kPa.

Take It Further

1. Calculate The natural gas in a 30-L NGV fuel tank has a pressure of 2.05×10^4 kPa at a temperature of 297 K. How many kilograms of fuel are in the tank? (Use a molar mass of 19 g/mol for natural gas.)

2. Calculate Natural gas is 89% methane (CH_4), 5% ethane (C_2H_6), 5% butane (C_4H_{10}), and 1% propane (C_3H_8). Use the data from Question 1 to determine the partial pressures of each component gas in the fuel tank.

Differentiated Instruction

L1 STRUGGLING STUDENTS If your students have trouble with Question 1, have them review Practice Problem 14.5 (on the ideal gas law) in Lesson 14.3. If students have trouble with Question 2, have them review Practice Problem 14.7 (on Dalton's law of partial pressures) in Lesson 14.4.

LPR LESS PROFICIENT READERS

L3 ADVANCED STUDENTS Assign the lab simulation Real vs. Ideal Gases in **Virtual ChemLab.** Natural gas contains mostly methane (CH_4). Ask students to investigate the conditions under which methane deviates from ideal gas behavior.

Explain

START A CONVERSATION After students read the feature, begin a discussion of NGVs. Ask students to compare NGVs and gasoline-powered cars. **Ask** Why do you think NGVs have become common in some densely populated cities, such as Bangkok and Rio de Janeiro? *(Because NGVs have a more limited travel range than gasoline-powered cars and because CNG refueling stations are not yet as common as (liquid) gas stations, NGVs are better suited to city driving than highway driving.)*

Extend

NATURAL GAS VEHICLES Have students do research on electric cars and fuel-cell cars. Ask them to create a pros-and-cons table for each type of car (similar to the one for NGVs). Ask them to suggest how the technology of NGVs, electric cars, and fuel-cell cars might be improved to meet the needs of everyday drivers.

Connect to MATH

The standard cubic foot (scf) is a unit used to measure the quantity of a gas. 1 scf equals the amount of gas filling 1 ft^3 at 14.73 psi and 60°F (101.6 kPa and 289 K). To compare mileages of gasoline-powered cars with those of NGVs, 1 gallon of gasoline is considered equivalent to 125 scf of natural gas.

Answers

TAKE IT FURTHER

1. 4.7 kg
2. $P_{methane} = 1.82 \times 10^4$ kPa
   $P_{ethane} = P_{butane} = 1.03 \times 10^3$ kPa
   $P_{propane} = 2.05 \times 10^2$ kPa

Evaluate

Review and Assessment Resources

CHEMISTRY TUTORIAL Encourage students to review using the gas laws, combined gas laws, the ideal gas law, and Dalton's law of partial pressures, using the animated step-by-step tutorials for Chapter 14. Each Chemistry Tutorial is followed by a similar practice problem, where the student is guided through the problem-solving process.

CHAPTER PROBLEM SET Have students practice more gas law problems using the Chapter 14 online Problem Set.

DIRECTED VIRTUAL LAB Have students complete the Virtual Lab *Comparing Real and Ideal Gas Laws* to help reinforce the mathematical relationships described by the gas laws. In the post-lab exercise, students will analyze the data and draw conclusions based on the results of the experiment.

KINETIC ART Direct students to the Kinetic Art animations *Charles's Law* and *Partial Pressures of Gases*, and have them answer the questions that accompany each animation.

14 Study Guide

BIGIDEA KINETIC THEORY

Ideal gases conform to the assumptions of kinetic theory. The behavior of ideal gases can be predicted by the gas laws. With the ideal gas law, the number of moles of a gas in a fixed volume at a known temperature and pressure can be calculated. Although an ideal gas does not exist, real gases behave ideally under a variety of temperature and pressure conditions.

14.1 Properties of Gases

Gases are easily compressed because of the space between the particles in a gas.

The amount of gas (n), volume (V), and temperature (T) are factors that affect gas pressure (P).

- compressibility (450)

14.2 The Gas Laws

If the temperature is constant, as the pressure of a gas increases, the volume decreases.

As the temperature of an enclosed gas increases, the volume increases if the pressure is constant.

As the temperature of an enclosed gas increases, the pressure increases if the volume is constant.

When only the amount of gas is constant, the combined gas law describes the relationship among pressure, volume, and temperature.

- Boyle's law (456)
- Charles's law (458)
- Gay-Lussac's law (460)
- combined gas law (462)

Key Equations

Boyle's law: $P_1 \times V_1 = P_2 \times V_2$

Charles's law: $\frac{V_1}{T_1} = \frac{V_2}{T_2}$

Gay-Lussac's law: $\frac{P_1}{T_1} = \frac{P_2}{T_2}$

combined gas law: $\frac{P_1 \times V_1}{T_1} = \frac{P_2 \times V_2}{T_2}$

14.3 Ideal Gases

When the pressure, volume, and temperature of a contained gas are known, you can use the ideal gas law to calculate the number of moles of the gas.

Real gases differ most from an ideal gas at low temperatures and high pressures.

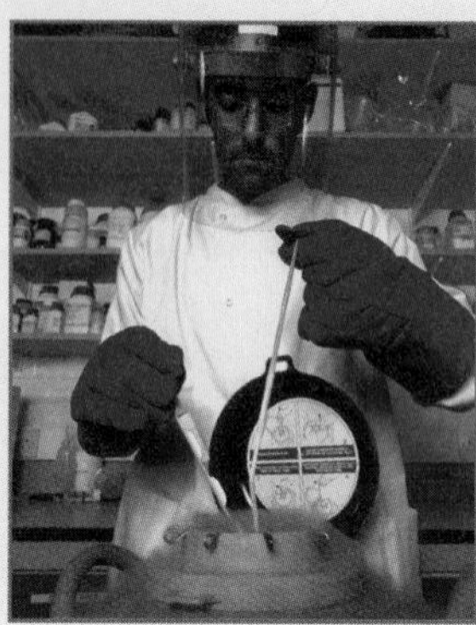

- ideal gas constant (465)
- ideal gas law (465)

Key Equation

ideal gas law: $P \times V = n \times R \times T$ or $PV = nRT$

14.4 Gases: Mixtures and Movements

In a mixture of gases, the total pressure is the sum of the partial pressures of the gases.

Gases of lower molar mass diffuse and effuse faster than gases of higher molar mass.

- partial pressure (469)
- Dalton's law of partial pressure (470)
- diffusion (472)
- effusion (472)
- Graham's law of effusion (472)

Key Equations

Dalton's law: $P_{total} = P_1 + P_2 + P_3 + ...$

Graham's law: $\frac{Rate_A}{Rate_B} = \sqrt{\frac{molar\ mass_B}{molar\ mass_A}}$

UbD Performance Tasks

MARINE ANIMALS Have students research how the tremendous pressure of the water affects the blood gases of marine animals when they dive. Have the students present their findings to the class, using photographs or charts.

UP, UP, AND AWAY Have students research what happens to helium-filled rubber and plastic film balloons when they are released into the atmosphere. Have students write a report detailing the events in terms of the applicable gas laws.

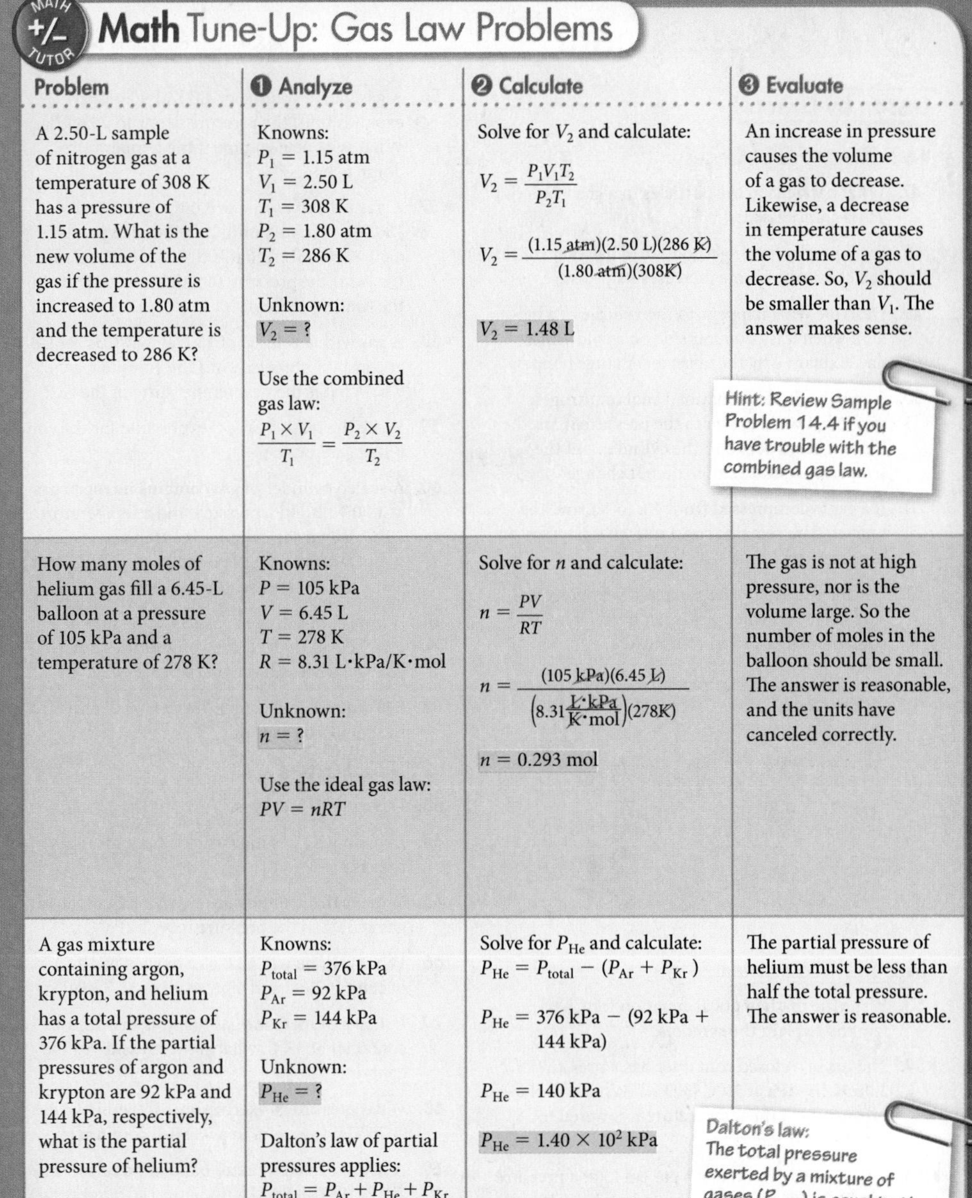

Math Tune-Up: Gas Law Problems

Problem	❶ Analyze	❷ Calculate	❸ Evaluate
A 2.50-L sample of nitrogen gas at a temperature of 308 K has a pressure of 1.15 atm. What is the new volume of the gas if the pressure is increased to 1.80 atm and the temperature is decreased to 286 K?	Knowns: $P_1 = 1.15$ atm $V_1 = 2.50$ L $T_1 = 308$ K $P_2 = 1.80$ atm $T_2 = 286$ K Unknown: $V_2 = ?$ Use the combined gas law: $\frac{P_1 \times V_1}{T_1} = \frac{P_2 \times V_2}{T_2}$	Solve for V_2 and calculate: $V_2 = \frac{P_1V_1T_2}{P_2T_1}$ $V_2 = \frac{(1.15\ \text{atm})(2.50\ \text{L})(286\ \text{K})}{(1.80\ \text{atm})(308\ \text{K})}$ $V_2 = 1.48$ L	An increase in pressure causes the volume of a gas to decrease. Likewise, a decrease in temperature causes the volume of a gas to decrease. So, V_2 should be smaller than V_1. The answer makes sense. Hint: Review Sample Problem 14.4 if you have trouble with the combined gas law.
How many moles of helium gas fill a 6.45-L balloon at a pressure of 105 kPa and a temperature of 278 K?	Knowns: $P = 105$ kPa $V = 6.45$ L $T = 278$ K $R = 8.31$ L·kPa/K·mol Unknown: $n = ?$ Use the ideal gas law: $PV = nRT$	Solve for n and calculate: $n = \frac{PV}{RT}$ $n = \frac{(105\ \text{kPa})(6.45\ \text{L})}{\left(8.31 \frac{\text{L}\cdot\text{kPa}}{\text{K}\cdot\text{mol}}\right)(278\ \text{K})}$ $n = 0.293$ mol	The gas is not at high pressure, nor is the volume large. So the number of moles in the balloon should be small. The answer is reasonable, and the units have canceled correctly.
A gas mixture containing argon, krypton, and helium has a total pressure of 376 kPa. If the partial pressures of argon and krypton are 92 kPa and 144 kPa, respectively, what is the partial pressure of helium?	Knowns: $P_{total} = 376$ kPa $P_{Ar} = 92$ kPa $P_{Kr} = 144$ kPa Unknown: $P_{He} = ?$ Dalton's law of partial pressures applies: $P_{total} = P_{Ar} + P_{He} + P_{Kr}$	Solve for P_{He} and calculate: $P_{He} = P_{total} - (P_{Ar} + P_{Kr})$ $P_{He} = 376\ \text{kPa} - (92\ \text{kPa} + 144\ \text{kPa})$ $P_{He} = 140$ kPa $P_{He} = 1.40 \times 10^2$ kPa	The partial pressure of helium must be less than half the total pressure. The answer is reasonable. Dalton's law: The total pressure exerted by a mixture of gases (P_{total}) is equal to the sum of the partial pressures of the component gases.

Math Tune-Up 479

Math Review

Example and sample practice problems for the various gas laws can be found on the following pages:

LESSON 14.2 Boyle's law—page 421; Charles's law—page 423; Gay-Lussac's law—page 425; the combined gas law—page 426

LESSON 14.3 The ideal gas law—pages 429 and 430

LESSON 14.4 Dalton's law of partial pressures—435; Graham's law of effusion—page 438

Focus on ELL

5 ASSESS UNDERSTANDING Have ELLs complete one of the Performance Tasks as a group for an alternative assessment. Provide guidelines on the information that it must cover, and provide a grading rubric.

BEGINNING

LOW Have students create visuals for the presentation.

HIGH Have students provide simple labels or captions as needed.

INTERMEDIATE: LOW/HIGH Have students provide more extended written descriptions for the visuals.

ADVANCED: LOW/HIGH Have students give the oral presentation to the class.

ASSESSMENT

Evaluate

Answers

LESSON 14.1

47. The space between particles is reduced.

48. The increased kinetic energy of the particles causes collisions to occur more frequently and with more force.

49. The volume decreases. The molecules have less kinetic energy and cause less pressure on the inside of the balloon.

50. The pressure doubles.

51. The pressure quadruples.

52. As particles are removed from the container, there is a decrease in the number of collisions.

LESSON 14.2

53. $V_1/T_1 = V_2/T_2$
V_1 and V_2 are the initial and final volumes; T_1 and T_2 are the initial and final temperatures.

54. 1.00×10^2 kPa

55. 1.80 L

56. 18 L

57. 846 K (573°C)

58. 200 K (-73°C)

59. $(P_1V_1)/T_1 = (P_2V_2)/T_2$

60. 1.10×10^3 kPa

61. 2.58 kPa

62. The volume is constant, $V_1 = V_2$, so the volume terms cancel, leaving the equation for Gay-Lussac's law.

LESSON 14.3

63. Its particles have no volume, there are no attractions between them, and collisions are elastic. An ideal gas follows the gas laws at all temperatures and pressures.

64. Particles in a real gas have a finite volume and are attracted to one another at low temperatures and high pressures.

65. 33.0 L

66. 17.6 L

67. 3.60×10^2 kPa

68. 1.71×10^3 kPa

69. 42 g

14 Assessment

* Solutions appear in Appendix E

Lesson by Lesson

14.1 Properties of Gases

47. What happens to the particles in a gas when the gas is compressed?

48. Explain why heating a contained gas that is held at a constant volume increases its pressure.

49. Describe what happens to the volume of a balloon when it is taken outside on a cold winter day. Explain why the observed change happens.

50. A metal cylinder contains 1 mol of nitrogen gas. What will happen to the pressure if another mole of gas is added to the cylinder, but the temperature and volume do not change?

51. If a gas is compressed from 4 L to 1 L and the temperature remains constant, what happens to the pressure?

52. Use the drawing to help explain why gas pressure decreases when gas is removed from a container with a fixed volume.

200 kPa

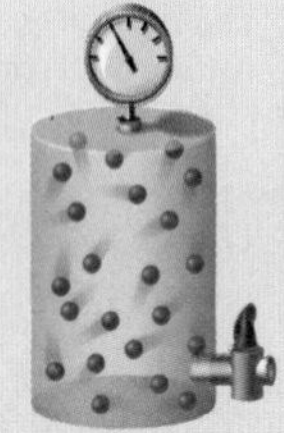

Decreasing Pressure

14.2 The Gas Laws

53. Write the mathematical equation for Charles's law and explain the symbols.

***54.** The gas in a closed container has a pressure of 3.00×10^2 kPa at 30°C (303 K). What will the pressure be if the temperature is lowered to −172°C (101 K)?

55. Calculate the volume of a gas (in L) at a pressure of 1.00×10^2 kPa if its volume at 1.20×10^2 kPa is 1.50×10^3 mL.

56. A gas with a volume of 4.0 L at 90.0 kPa expands until the pressure drops to 20.0 kPa. What is its new volume if the temperature doesn't change?

***57.** A gas with a volume of 3.00×10^2 mL at 150.0°C is heated until its volume expands to 6.00×10^2 mL. What is the new temperature of the gas if the pressure remains constant during the heating process?

***58.** A gas with a volume of 15 L at 327°C is cooled at constant pressure until the volume reaches 5 L. What is the new temperature of the gas?

59. Write the mathematical expression for the combined gas law.

60. A sealed cylinder of gas contains nitrogen gas at 1.00×10^3 kPa pressure and a temperature of 20°C. When the cylinder is left in the sun, the temperature of the gas increases to 50°C. What is the new pressure in the cylinder?

***61.** A sample of nitrogen gas has a pressure of 6.58 kPa at 539 K. If the volume does not change, what will the pressure be at 211 K?

62. Show how Gay-Lussac's law can be derived from the combined gas law.

14.3 Ideal Gases

63. Describe an ideal gas.

64. Explain why it is impossible for an ideal gas to exist.

***65.** What is the volume occupied by 1.24 mol of a gas at 35°C if the pressure is 96.2 kPa?

66. What volume will 12.0 g of oxygen gas (O_2) occupy at 25°C and a pressure of 52.7 kPa?

***67.** If 4.50 g of methane gas (CH_4) is in a 2.00-L container at 35°C, what is the pressure in the container?

68. What pressure is exerted by 0.450 mol of a gas at 25°C if the gas is in a 0.650-L container?

***69.** A helium-filled weather balloon has a volume of 2.4×10^2 L at 99 kPa pressure and a temperature of 0°C. What is the mass of the helium in the balloon?

14.4 Gases: Mixtures and Movements

70. In your own words, state Dalton's law of partial pressure.

71. Which gas effuses faster: hydrogen or chlorine? How much faster?

72. Which gas effuses faster at the same temperature: molecular oxygen or atomic argon?

*73. Calculate the ratio of the velocity of helium atoms to the velocity of neon atoms at the same temperature.

74. Calculate the ratio of the velocity of helium atoms to the velocity of fluorine molecules at the same temperature.

Understand Concepts

75. How does kinetic theory explain the compressibility of gases?

76. A teacher adds enough water to cover the bottom of an empty metal can with a screw cap. Using a stove, the teacher heats the can with the cap off until the water boils, and then screws on the cap tightly. When the sealed can is dunked in cold water, the sides of the can immediately collapse inward as though crushed in a trash compactor.

a. Use kinetic theory to explain why the can collapsed inward.

b. If the experiment were done with a dry can, would the results be similar? Explain.

77. Explain how the compressed gas in an aerosol can forces paint out of the can. Make sure to describe how the gas pressure inside the can changes as the paint is sprayed. (Refer to Figure 14.5 in Lesson 14.1.)

78. Why do aerosol containers display the warning, "Do not incinerate"?

79. The manufacturer of an aerosol deodorant packaged in a 150-mL container plans to produce a container of the same size that will hold twice as much gas. How will the pressure of the gas in the new product compare with that of the gas in the original container?

80. Why must Kelvin temperatures be used in calculations that involve gases?

81. Explain how using a pressure cooker reduces the time required to cook food.

82. The ratio of two variables is always a constant. What can you conclude about the relationship between the two variables?

*83. A 3.50-L gas sample at 20°C and a pressure of 86.7 kPa expands to a volume of 8.00 L. The final pressure of the gas is 56.7 kPa. What is the final temperature of the gas in degrees Celsius?

84. Explain the reasons why real gases deviate from ideal behavior.

85. How would the number of particles of two gases compare if their partial pressures in a container were identical?

86. Why does a balloon filled with helium deflate more quickly than a balloon filled with air?

*87. A certain gas effuses four times as fast as oxygen (O_2). What is the molar mass of the gas?

*88. During an effusion experiment, a certain number of moles of an unknown gas passed through a tiny hole in 75 seconds. Under the same conditions, the same number of moles of oxygen gas passed through the hole in 30 seconds. What is the molar mass of the unknown gas?

89. The photograph shows a tube with cotton balls at each end. The cotton ball at the left was soaked with hydrochloric acid. The cotton ball on the right was soaked with a solution of ammonia. When these compounds react, they form a white solid, ammonium chloride. Based on the location of the ammonium chloride in the tube, which gas diffuses at a faster rate, hydrogen chloride or ammonia? Explain.

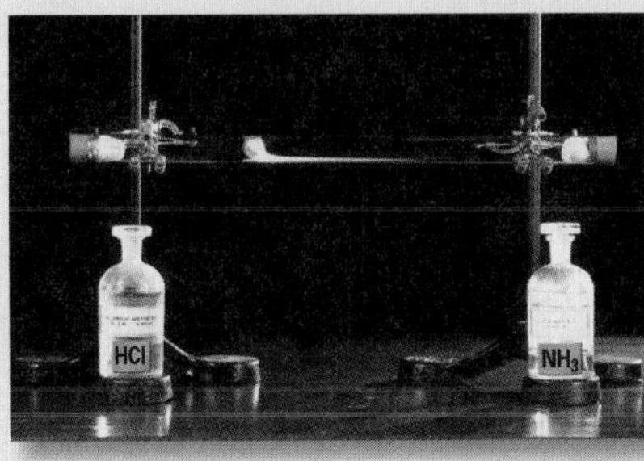

Answers

LESSON 14.4

70. The total pressure of a gaseous mixture is equal to the sum of the individual pressures of each gas.

71. Hydrogen gas diffuses faster than chlorine gas by an approximate factor of six.

72. molecular oxygen

73. 2.25:1

74. 3.08:1

UNDERSTAND CONCEPTS

75. Gases are easily compressed because there is a lot of space between the particles.

76. **a.** Boiling the water fills the can with steam. When the can is plunged into cold water, the steam is trapped and rapidly condenses, reducing gas pressure inside the can. The walls of the can are not strong enough to withstand the comparatively high atmospheric pressure, which crushes the can. **b.** The results would be much less dramatic. The change in volume (and in internal pressure) from heated air to cold air is much less dramatic than when steam (a vapor) condenses to water (a liquid).

77. The gas propellant inside the can is at a higher pressure than the air outside the can. When the valve at the top of the can is opened, gas rushes out of the can, propelling the paint in the process. The pressure of the propellant decreases as the can is used. When there is no longer a pressure difference between the propellant and the outside air, the can no longer works.

78. High temperatures increase the pressure of the contents of the container and may cause it to explode.

79. The pressure will double.

80. Temperatures measured on the Kelvin scale are directly proportional to the average kinetic energy of the particles. Celsius temperatures are not.

81. The water boils at a higher temperature, which speeds the cooking process.

82. The variables are directly proportional.

83. 165°C

84. The particles in a real gas have a finite volume and are attracted to one another.

85. The number of particles would be equal.

86. Helium atoms have a smaller molar mass than oxygen and nitrogen molecules and effuse faster through pores in the balloon.

87. 2.0 g

88. 2.0×10^2 g

89. Ammonia diffuses faster than hydrogen chloride. Based on the location of the reaction product, the ammonia molecules travel about twice the distance of the hydrogen chloride molecules, in the same time.

ASSESSMENT

Evaluate

Answers

THINK CRITICALLY

90. A vacuum contains no matter to allow the transfer of kinetic energy between molecules.

91. The gases that make up the atmosphere, just like any other form of matter, are held near Earth by the force of gravity.

92. Helium gas is composed of small atoms with little attraction for each other.

93. The partial pressure of oxygen would be greater than at sea level, because the atmospheric pressure would be greater than at sea level.

94. **a.** 1.63×10^2 kPa
b. 4.49×10^2 kPa

95. There could be different amounts of each gas, or the gases could be at different pressures.

96. **a.**

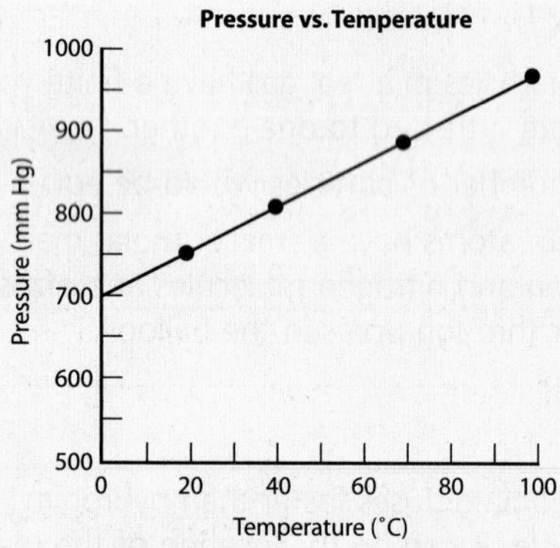

b. 700 mm Hg
c. directly proportional
d. The pressure rises 2.64 mm Hg for every 1°C increase in temperature.
e. $P = (2.64 \text{ mm Hg/°C})T + 700 \text{ mm Hg}$
f. Gay-Lussac's law

97. **a.** methane at 200°C
b. Attractions between the particles and the physical volume of the particles. At 0°C, the methane molecules are moving relatively slowly. The attractions between the molecules are sufficiently strong so that, at low pressures, the curve is below the $PV/nRT = 1$ line. At higher pressures, the space between molecules is reduced. The actual physical volume of the methane molecules now becomes more important, and the curve is above the $PV/nRT = 1$ line.

Think Critically

90. Infer Figure 14.14 in Lesson 14.3 shows an insulated tank used to store liquid nitrogen. How does the vacuum between the walls of the tank prevent heat transfer?

91. Infer Gases will diffuse from a region of higher concentration to a region of lower concentration. Why don't the gases in Earth's atmosphere escape into the near-vacuum of space?

92. Apply Concepts What real gas comes closest to having the characteristics of an ideal gas? Explain your answer.

93. Predict Death Valley in California is at 86 m below sea level. Will the partial pressure of oxygen in Death Valley be the same, lower, or higher than the partial pressure of oxygen at sea level? Give a reason for your answer.

***94. Calculate** The following reaction takes place in a sealed 40.0-L container at a temperature of 120°C:

$$4NH_3(g) + 5O_2(g) \longrightarrow 4NO(g) + 6H_2O(g)$$

a. When 34.0 g of NH_3 reacts with 96.0 g of O_2, what is the partial pressure of NO in the sealed container?

b. What is the total pressure in the container?

95. Interpret Graphs The graph shows the direct relationship between volume and temperature for three different gas samples. Offer at least one explanation for why the graphs are not identical for the three samples. (*Hint:* What variables other than temperature and volume can be used to describe a gas?)

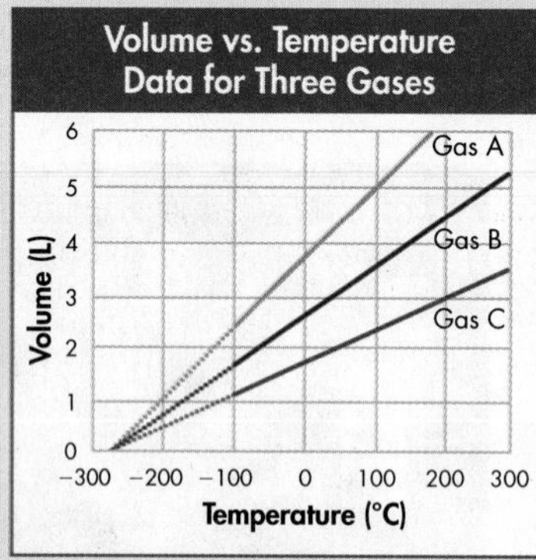

***96. Analyze Data** A student collected the following data for a fixed volume of gas.

Temperature (°C)	Pressure (mm Hg)
10	726
20	750
40	800
70	880
100	960

a. Graph the data, using pressure as the dependent variable.

b. What is the pressure of the gas at 0°C?

c. Is the relationship between the variables directly or inversely proportional?

d. How does the pressure of the gas change with each degree Celsius change in the temperature?

e. Write an equation relating the pressure and temperature of the gas.

f. Which gas law is illustrated by the data? Select two data points on your graph to confirm your answer.

97. Interpret Graphs The graph shows how the ratio (PV/nRT) changes with increasing pressure for methane (CH_4) at 0°C and 200°C.

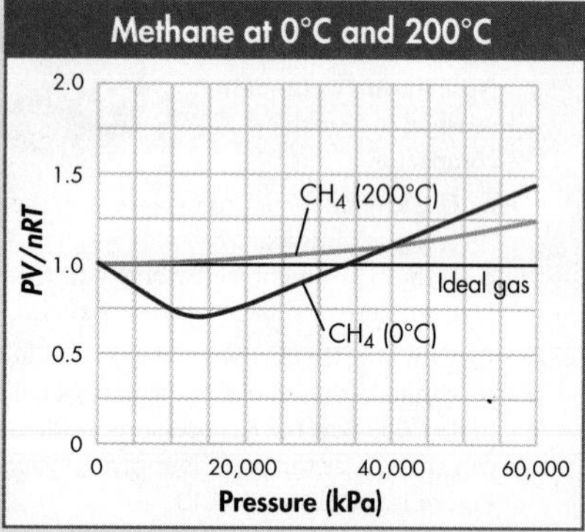

a. At lower pressures, which gas behaves more like an ideal gas: methane at 0°C or methane at 200°C.

b. The curve for methane at 0°C shows that the ratio PV/nRT is less than 1 at lower pressures and greater than 1 at higher pressures. What characteristics of real gases can explain these deviations?

482 Chapter 14 • Assessment

Enrichment

98. Analyze Data Oxygen is produced in the laboratory by heating potassium nitrate (KNO_3). The data table below gives the volume of oxygen produced at STP from different quantities of KNO_3. Use the data to determine the mole ratio by which KNO_3 and O_2 react.

Mass of KNO_3 (g)	Volume of O_2 (cL)
0.84	9.3
1.36	15.1
2.77	30.7
4.82	53.5
6.96	77.3

***99. Calculate** A mixture of ethyne gas (C_2H_2) and methane gas (CH_4) occupied a certain volume at a total pressure of 16.8 kPa. When the sample burned, the products were CO_2 gas and H_2O vapor. The CO_2 was collected and its pressure found to be 25.2 kPa in the same volume and at the same temperature as the original mixture. What percentage of the original mixture was methane?

***100. Calculate** A 0.10-L container holds 3.0×10^{20} molecules of H_2 at 100 kPa and 0°C.

a. If the volume of a hydrogen molecule is 6.7×10^{-24} mL, what percentage of the volume of the gas is occupied by its molecules?

b. If the pressure is increased to 100,000 kPa, the volume of the gas is 1×10^{-4} L. What fraction of the total volume do the hydrogen molecules now occupy?

101. Draw Conclusions Many gases that have small molecules, such as N_2 and O_2, have the expected molar volume of 22.41 L at STP. However, other gases behave in a very non-ideal manner, even if extreme pressures and temperatures are not involved. The molar volumes of CH_4, CO_2, and NH_3 at STP are 22.37 L, 22.26 L, and 22.06 L, respectively. Explain the reasons for these large departures from the ideal.

Write About Science

102. Explain Why does a tennis ball bounce higher in the summer than it does in the winter? Use what you know about gas behavior to explain your answer.

103. Research a Problem Cars that run on natural gas or hydrogen require different fuel tanks and different refueling stations than cars that run on gasoline, which is a liquid at STP. How would you design a fuel tank for storing a gas? How would you design a gas pump that pumped a gas instead of a liquid? Research a kind of vehicle that runs on gaseous fuel and explain how these questions have been addressed.

CHEMYSTERY

Under Pressure

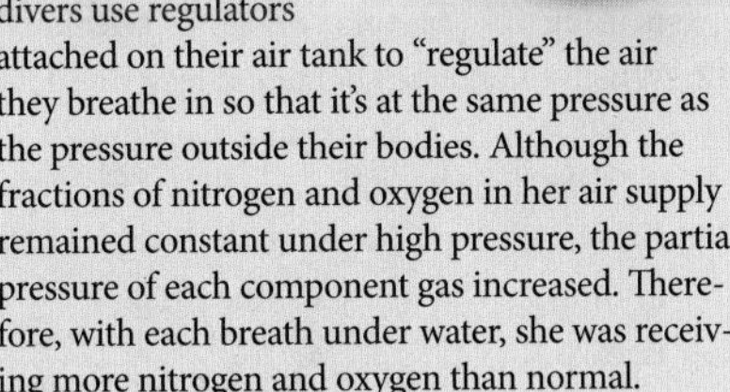

Becki realized that she had decompression sickness, also known as the bends. Recreational divers use regulators attached on their air tank to "regulate" the air they breathe in so that it's at the same pressure as the pressure outside their bodies. Although the fractions of nitrogen and oxygen in her air supply remained constant under high pressure, the partial pressure of each component gas increased. Therefore, with each breath under water, she was receiving more nitrogen and oxygen than normal.

As Becki ascended and the pressure on her body decreased, the excess nitrogen formed bubbles in her blood and tissues, causing pain and other symptoms. Serious cases of the bends require treatment in a high-pressure chamber. The pressure is reduced gradually so that the excess nitrogen can leave the body harmlessly.

104. Infer How could Becki have prevented getting the bends?

105. Connect to the BIGIDEA What would have happened if Becki held her breath while ascending from a dive? Use the gas laws to explain.

ASSESSMENT

CHEMYSTERY After students have read through the CHEMystery, call on volunteers to summarize how the behavior of gases helps explain why Becki fell ill after her dives.

Ask Why does Becki breathe in more nitrogen and oxygen during her dives? *(As Becki dives, the pressure of the air she breathes increases. Although the composition of the gases in the air doesn't change with increasing pressure, their partial pressures do. Therefore, with each breath, she breathes in more nitrogen and oxygen underwater than she does at the surface.)* **Ask** What caused the pain that Becki experienced after her dives? *(As Becki ascended from her dives, the pressure decreased and the nitrogen formed bubbles in her blood and tissues, which caused pain.)*

CHEMYSTERY ANSWERS

104. Becki could have prevented getting the bends by ascending slowly. This would have allowed the excess nitrogen to leave her body harmlessly.

105. BIGIDEA Holding her breath while ascending would cause the volume of the gases in Becki's lungs to increase. This could cause her lungs to rupture if she does not exhale.

Answers

ENRICHMENT

98. 2 mol KNO_3 for each 1 mol O_2

99. 50% CH_4

100. a. 2.0×10^{-3}%

b. 2.0%

101. Because attractions between molecules in gases such as nitrogen and oxygen are insignificant, these gases have the molar volume of an ideal gas, 22.4 L at STP. Based on their molar volumes at STP, there are attractions between molecules of CH_4, CO_2, and NH_3. These attractions increase in strength from methane to carbon dioxide to ammonia.

WRITE ABOUT SCIENCE

102. The higher the air pressure inside the ball, the higher the bounce. Assuming constant volume, the air pressure is directly proportional to the Kelvin temperature. Average temperature is higher in summer than in winter, which means the air pressure (and the height of the bounce) will be higher during the summer.

103. Answers will vary. The gas cylinders in CNG-fueled vehicles are designed to withstand high pressures. CNG dispensers must be gastight and capable of high compression.

Evaluate

Answers

CUMULATIVE REVIEW

106. K = °C + 273

107. 10.6 g/cm^3

108. 82 protons, 82 electrons, 124 neutrons

109. nickel, Ni

110. b. tungsten

111. c. selenium

112. b. SO_2

113. a. tin(II) bromide
b. barium sulfate
c. magnesium hydroxide
d. iodine pentafluoride

114. 206 g

115. a. 158.1 g
b. 98.0 g
c. 342.0 g
d. 331.2 g

116. It is the volume occupied by 1 mol of a gas at STP.

117. a. $C_4H_8O_2$
b. C_8H_8
c. $C_2H_2O_4$

118. 60.0% C, 13.3% H, 26.7% O

119. a. single-replacement
b. decomposition

120. a. $P_4O_{10} + 6H_2O \rightarrow 4H_3PO_4$
b. $Al_2S_3 + 6H_2O \rightarrow 2Al(OH)_3 + 3H_2S$

121. a. $4Al + 3O_2 \rightarrow 2Al_2O_3$
b. 3.09×10^2 g Al; 2.75×10^2 g O_2

122. The motion of particles in a gas is constant, random, and rapid.

123. Gas pressure is the result of the collisions of gas particles with the container wall.

Cumulative Review

106. What is the mathematical relationship between the Kelvin and Celsius temperature scales?

***107.** A metal sample has a mass of 9.92 g and measures 4.50 cm × 1.30 cm × 1.60 mm. What is the density of the metal?

108. How many electrons, protons, and neutrons are there in an atom of lead-206?

109. Which element has the following electron configuration?

1s	2s	2p	3s	3p	4s	3d
↑↓	↑↓	↑↓	↑↓	↑↓	↑↓	↑↓
		↑↓		↑↓		↑↓
		↑↓		↑↓		↑↓
						↑
						↑

110. Which of these elements are metals?
a. arsenic **b.** tungsten **c.** xenon

111. Which element is most likely to form a compound with strontium?
a. neon **b.** tin **c.** selenium

112. Which compound contains at least one double bond?
a. H_2Se **b.** SO_2 **c.** PCl_3

***113.** Name each compound.
a. $SnBr_2$
b. $BaSO_4$
c. $Mg(OH)_2$
d. IF_5

114. An atom of lead has a mass 17.16 times greater than the mass of an atom of carbon-12. What is the molar mass of this isotope of lead?

***115.** Calculate the molar mass of each substance.
a. $Ca(CH_3CO_2)_2$
b. H_3PO_4
c. $C_{12}H_{22}O_{11}$
d. $Pb(NO_3)_2$

116. What is the significance of the volume 22.4 L?

***117.** Calculate the molecular formula of each of the following compounds.
a. The empirical formula is C_2H_4O, and the molar mass is 88 g.
b. The empirical formula is CH, and the molar mass is 104 g.
c. The molar mass is 90 g. The percent composition is 26.7% C, 71.1% O, and 2.2% H.

118. Calculate the percent composition of 2-propanol (C_3H_7OH).

119. What type of reaction is each of the following?
a. Calcium reacts with water to form calcium hydroxide and hydrogen gas.
b. Mercury and oxygen are prepared by heating mercury(II) oxide.

120. Write balanced equations for the following chemical reactions.
a. Tetraphosphorus decoxide reacts with water to form phosphoric acid.
b. Aluminum hydroxide and hydrogen sulfide form when aluminum sulfide reacts with water.

***121.** Aluminum oxide is formed from its elements.

$$Al(s) + O_2(g) \longrightarrow Al_2O_3(s)$$

a. Balance the equation.
b. How many grams of each reactant are needed to form 583 g $Al_2O_3(s)$?

122. Explain why a gas expands until it takes the shape and volume of its container.

123. Use the drawings to explain how gas pressure is produced.

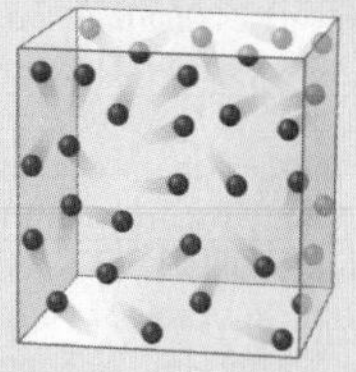

If You Have Trouble With . . .

Question	106	107	108	109	110	111	112	113	114	115	116	117	118	119	120	121	122	123
See Chapter	3	3	4	5	6	7	8	9	10	10	10	10	10	11	11	12	13	13

Standardized Test Prep for Indiana

Standard: C.5.1

Select the choice that best answers each question or completes each statement.

1. A gas in a balloon at constant pressure has a volume of 120.0 mL at −123°C. What is its volume at 27.0°C?
   (A) 60.0 mL (C) 26.5 mL
   (B) 240.0 mL (D) 546 mL

2. If the Kelvin temperature of a gas is tripled and the volume is doubled, the new pressure will be
   (A) 1/6 the original pressure.
   (B) 2/3 the original pressure.
   (C) 3/2 the original pressure.
   (D) 5 times the original pressure.

3. Which of these gases effuses fastest?
   (A) Cl_2 (C) NH_3
   (B) NO_2 (D) N_2

4. All the oxygen gas from a 10.0-L container at a pressure of 202 kPa is added to a 20.0-L container of hydrogen at a pressure of 505 kPa. After the transfer, what are the partial pressures of oxygen and hydrogen?
   (A) Oxygen is 101 kPa; hydrogen is 505 kPa.
   (B) Oxygen is 202 kPa; hydrogen is 505 kPa.
   (C) Oxygen is 101 kPa; hydrogen is 253 kPa.
   (D) Oxygen is 202 kPa; hydrogen is 253 kPa.

5. Which of the following changes would increase the pressure of a gas in a closed container?
   I. Part of the gas is removed.
   II. The container size is decreased.
   III. Temperature is increased.
   (A) I and II only
   (B) II and III only
   (C) I and III only
   (D) I, II, and III

6. A real gas behaves most nearly like an ideal gas
   (A) at high pressure and low temperature.
   (B) at low pressure and high temperature.
   (C) at low pressure and low temperature.
   (D) at high pressure and high temperature.

Use the graphs to answer Questions 7–10. A graph may be used once, more than once, or not at all.

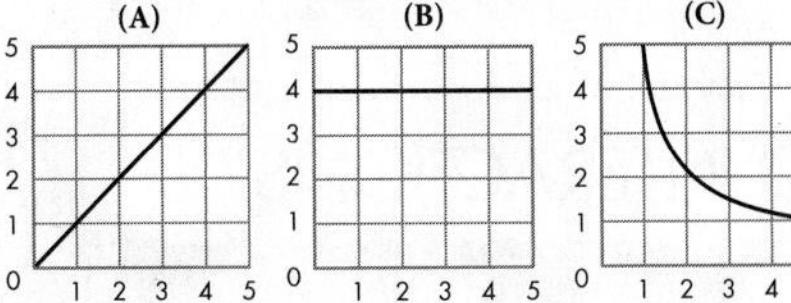

Which graph shows each of the following?

7. directly proportional relationship
8. graph with slope = 0
9. inversely proportional relationship
10. graph with a constant slope

Use the drawing to answer Questions 11 and 12.

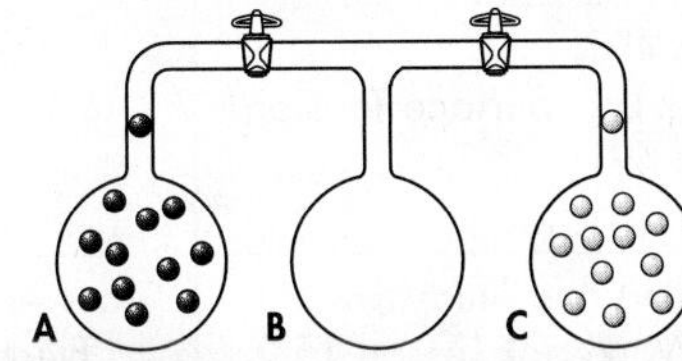

11. Bulb A and bulb C contain different gases. Bulb B contains no gas. If the valves between the bulbs are opened, how will the particles of gas be distributed when the system reaches equilibrium? Assume none of the particles are in the tubes that connect the bulbs.

Tips for Success

Constructing a Diagram If you are asked to draw a diagram, sketch lightly at first (so you can erase easily), or do a sketch on a separate piece of paper. Once you are sure of your answer, draw the final diagram.

12. Make a three-bulb drawing with 6 blue spheres in bulb A, 9 green spheres in bulb B, and 12 red spheres in bulb C. Then draw the setup to represent the distribution of gases after the valves are opened and the system reaches equilibrium.

If You Have Trouble With . . .

Question	1	2	3	4	5	6	7	8	9	10	11	12
See Lesson	14.2	14.2	14.4	14.3	14.1	14.3	14.2	14.2	14.2	14.2	14.4	14.4

ASSESSMENT

STP Answers

MULTIPLE CHOICE

1. B
2. C
3. C
4. A
5. B
6. B
7. A
8. B
9. C
10. A and B

OPEN RESPONSE

11. There will be four black spheres and four white spheres in each bulb.
12. Each bulb will contain two blue, three green, and four red spheres.

15 Water and Aqueous Systems Planning Guide

Introducing the BIGIDEA: BONDING AND INTERACTIONS

Chemical and physical properties of materials can be explained by the structure and arrangement of atoms, ions, and molecules and the

IASS	Lessons and Objectives	Print Resources: For the Student	Print Resources: For the Teacher	
C.3.3, C.7.1	**15.1** Water and Its Properties p 488–493 **15.1.1** Identify the factor that causes the high surface tension, low vapor pressure, and high boiling point of water. **15.1.2** Describe the structure of ice.	Reading and Study Workbook Lesson 15.1 Lesson Assessment 15.1 p 493 Quick Lab: Surface Tension p 491		
C.7.1	**15.2** Homogeneous Aqueous Systems p 494–501 **15.2.1** Identify the types of substances that dissolve most readily in water. **15.2.2** Explain why all ionic compounds are electrolytes. **15.2.3** Explain why hydrates easily lose and regain water.	Reading and Study Workbook Lesson 15.2 Lesson Assessment 15.2 p 501	Class Activity, p 497: Electrolytes Teacher Demo, p 498: Magic Writing	
C.7.1	**15.3** Heterogeneous Aqueous Systems p 504–507 **15.3.1** Distinguish between a suspension and a solution. **15.3.2** Identify how to distinguish a colloid from a suspension and a solution.	Reading and Study Workbook Lesson 15.3 Lesson Assessment 15.3 p 507 Small-Scale Lab: Electrolytes p 508	Teacher Demo, p 506: Tyndall Effect	

Assessing the BIGIDEA: BONDING AND INTERACTIONS

	Essential Questions 1. How do the interactions between water molecules account for the unique properties of water? 2. How do aqueous solutions form?	Study Guide p 509 STP p 515 Reading and Study Workbook Self-Check and Vocabulary Review Chapter 15		

forces between them.

	PearsonChem.com	
	Editable Worksheets	Online Activities
	Teaching Resources, Lesson 15.1 Review	15.1 Lesson Overview Hydrogen Bonding
	Teaching Resources, Lesson 15.2 Review Lab 27: The Solvent Properties of Water Lab 28: Water of Hydration Lab 29: Electrolytes and Nonelectrolytes Small-Scale Lab Manual Lab 24: Electrolytes Small-Scale Lab Manual Lab 25: Hard and Soft Water Lab Practical 8-2: Empirical Formula of a Hydrated Compound Lab Practical 15-1: Electrolytes and Nonelectrolytes Probeware Lab: Small-Scale Lab: Electrolytes	15.2 Lesson Overview Solvation of Ionic Compounds The Conductivity of Electrolyte Solutions Finding the Percent by Mass of Water in a Hydrate
	Teaching Resources, Lesson 15.3 Review	15.3 Lesson Overview An Afternoon at Home, Emulsified
	Core Teaching Resources (includes editable worksheets) • Lesson Reviews • Practice Problems • Interpret Graphs • Vocabulary Review • Chapter Quizzes and Tests • Lab Record Sheets	Chapter 15 Problem Set

Materials List

For the Student

Quick Lab p 491
- shallow dish or Petri dish
- water
- paper clip
- rubber band (approximately 5 cm in diameter)
- 2 micropipets or droppers
- vegetable oil
- liquid dish detergent

Small-Scale Lab p 508
- paper
- pencil
- ruler
- reaction surface
- water
- micropipet or dropper
- $NaCl(s)$; $Na_2CO_3(s)$; $NaHCO_3(s)$; $KCl(s)$; $MgSO_4(s)$
- table sugar
- cornstarch
- $KI(s)$
- aqueous solutions of HCl, H_2SO_4, HNO_3, CH_3COOH, NH_3, and $NaOH$
- rubbing alcohol
- distilled water
- various liquid foods

For the Teacher

Class Activity p 497
- light bulb in a porcelain socket
- 9 V or lantern battery
- 2 copper metal strips
- lamp cord
- alligator clips
- 0.1*M* solutions of glucose, alanine, glycine, ascorbic acid, malonic acid, citric acid, acetic acid, hydrochloric acid
- 8 beakers

Teacher Demo p 498
- $CoCl_2\ 6H_2O$ solution
- cotton swab
- white paper
- hot plate
- misting bottle
- water

Teacher Demo p 506
- whole milk
- water
- beaker
- stirring rod
- projector or laser pointer

Additional Digital Resources

Online Student Edition
Online Teacher's Edition
Exam*View Assessment Suite CD-ROM*
Classroom Resources DVD
15.2 Virtual Chem Lab 21: Electrolytes

CHAPTER 15

What's Online

CHEMISTRY TUTORIAL Students can work step-by-step with the online tutorial to learn how to find the percent by mass of water in a hydrate by accessing the Chapter 15 Chemistry Tutorial.

ONLINE PROBLEM SET Additional practice problems about aqueous systems can be found in the Chapter 15 online Problem Set.

DIRECTED VIRTUAL LAB Encourage students to review how different electrolytes change the conductivity of solutions using the animated virtual lab for Chapter 15. The animation is followed by analysis questions, where the student is guided through the lab concepts.

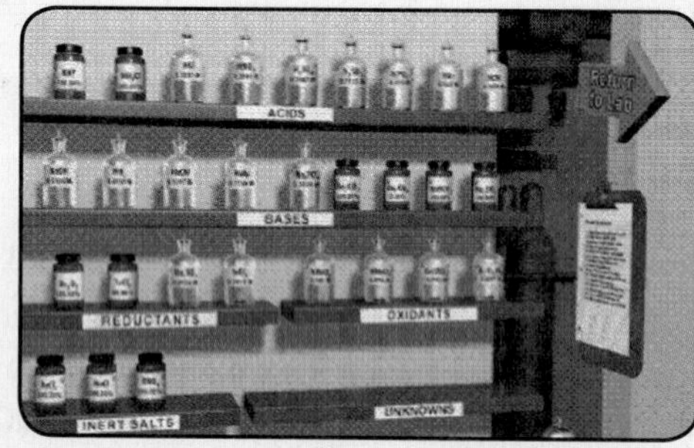

KINETIC ART Have students watch animations about hydrogen bonding and the solvation of ionic compounds. Each animation is followed by analysis questions to guide further understanding of the concepts.

CONCEPTS IN ACTION Students can watch an interactive animation about emulsions and emulsifying agents. The animation is followed by questions that help reinforce understanding of these concepts.

IN ACADEMIC STANDARDS for SCIENCE

Nature of Science

- Develop explanations based on reproducible data and observations gathered during laboratory investigations.

Standard 7: Solutions

C.7.1 Describe the composition and properties of types of solutions.

15 Water and Aqueous Systems

Water has many unique properties. In this chapter, you will learn about the interactions between water molecules.

INSIDE:

PearsonChem.com

486

Focus on ELL

1 CONTENT AND LANGUAGE Write the words *homogeneous*, *heterogeneous*, and *aqueous* on the board. Assign students to three groups. Assign each group of students one of the words, and ask them to brainstorm the meaning of the prefix of their word. Then have each group complete a Frayer model diagram for the word, with a definition, characteristics, an example, and a nonexample. Have each group teach the word to the class.

BEGINNING

LOW Draw a symbol to represent the meaning of the prefix.

HIGH Use a bilingual dictionary or glossary to find any cognates for the prefix.

INTERMEDIATE: LOW/HIGH List other words that use the same prefix.

ADVANCED: LOW/HIGH After the class presentations, have students predict the meaning of the lesson titles *Homogeneous Aqueous Systems* and *Heterogeneous Aqueous Systems*.

BIGIDEA

BONDING AND INTERACTIONS

Essential Questions:

1. *How do the interactions between water molecules account for the unique properties of water?*
2. *How do aqueous solutions form?*

CHEMYSTERY

Coming Clean

On a beautiful Saturday afternoon, Wes decided to take his bicycle out for a ride. He set off on a long ride through the trails at a nearby park.

When Wes returned from his bike ride, he found that his socks had quite a few stains on them. There was dirt from the trail as well as grease from the bicycle chain. Wes figured he could just clean the socks in the sink. He tried soaking the socks in water, but neither the dirt, nor the grease would rinse off. Isn't water supposed to clean everything? Apparently not. If Wes had known more about the chemistry of water, he might have tried something different to clean those dirty socks.

▶ Connect to the BIGIDEA As you read about water and aqueous systems, think about how Wes could remove the dirt and grease from his socks.

INDIANA'S ACADEMIC STANDARDS FOR SCIENCE

Solutions C.7.1. See lessons for details.

UbD

Understanding by Design

Students are building toward the ability to predict how particles bond and interact by understanding how the properties of water and aqueous solutions are affected by interactions among water molecules.

PERFORMANCE GOALS At the end of Chapter 15, students will be able to answer the essential questions by applying their knowledge of water and aqueous solutions.

ESSENTIAL QUESTIONS Read the essential questions aloud. **Ask** What are some properties of water? *(Sample answers: It forms spherical drops. It can dissolve many substances. It has three states of matter.)* **Ask** How do you think a water molecule's structure relates to water's ability to form spherical drops?

BIGIDEA Use the photo of water drops to help students connect to this chapter's concepts. Activate prior knowledge by reviewing the covalent bonding in a water molecule. **Ask** Why do you think people use water to wash hands, dishes, and laundry? *(Sample answer: Water often cleans things well.)*

CHEMYSTERY Connect the CHEMystery to the Big Idea by asking students to think about how the bonding in a water molecule might affect the ability of water to remove the dirt and grease from Wes's socks. Challenge students to predict what Wes could use to clean his dirty socks. As a hint, suggest that students think about what they use at home to wash things.

CHAPTER 15

Introduce the Chapter

IDENTIFYING MISCONCEPTIONS Students may think a solution is always tiny bits of a solute floating in a solvent that is continuous; that is, the solvent is not made of individual particles. As a result, they may think any solute will dissolve in water. Use the following activity to help them realize that water will not dissolve all substances.

Activity *You will need glass beakers, stirring rods, water, and various substances that are or are not soluble in water.* **Ask** Do you think all these substances will dissolve in water? *(Answers will vary.)* Have students predict whether each substance will dissolve in water at room temperature. Then have student volunteers try to dissolve the substances in water. Conclude by stating that water's properties cause it to dissolve some substances but not others. Explain that students will learn the difference as they study the chapter.

UntamedScience™

S'mores the Merrier

Margaret dives into the thick of understanding colloids at a marshmallow factory.

LESSON 15.1

Key Objectives

15.1.1 IDENTIFY the factor that causes the high surface tension, low vapor pressure, and high boiling point of water.

15.1.2 DESCRIBE the structure of ice.

Additional Resources

- Reading and Study Workbook, Lesson 15.1
- Probeware Laboratory Manual, Core Teaching Resources, Lesson 15.1

PearsonChem.com

Engage

CHEMISTRY & YOU Have students study the photograph and read the text. **Ask** What are the white swirls in the photo? *(clouds)* **Ask** How are the white swirls relevant to the subject matter of this chapter? *(Clouds form when water vapor in the atmosphere condenses into tiny liquid droplets.)*

Activate Prior Knowledge

Before students begin to read this chapter, review the concepts of polarity and hydrogen bonding. Remind students that water is a liquid rather than a gas at room temperature because of the strong hydrogen bonding between water molecules.

15.1 Water and Its Properties

C.3.3 Chemical bonds and intermolecular attractions; C.7.1 Composition and properties of solutions

CHEMISTRY & YOU

Q: *What properties of water make it essential to life on Earth?* When the *Apollo 8* astronauts first saw their home planet from a distance of thousands of kilometers, they called it the big blue marble. Water covers about three quarters of Earth's surface. In addition to making up Earth's oceans, water forms the polar ice caps and cycles through the atmosphere. All known life forms, including the penguin in Figure 15.1, are made mostly of water.

Key Questions

What factor causes the high surface tension, low vapor pressure, and high boiling point of water?

How can you describe the structure of ice?

Vocabulary

- surface tension
- surfactant

Water in the Liquid State

What factor causes the high surface tension, low vapor pressure, and high boiling point of water?

You couldn't live without water, nor could all the plants and animals that share space on the "big blue marble." Besides the water visible on Earth's surface, immense reserves of water exist deep underground. Water in the form of ice and snow dominates the polar regions of Earth. Water vapor from the evaporation of surface water and from steam spouted from geysers and volcanoes is always present in Earth's atmosphere.

Recall that water, H_2O, is a simple molecule consisting of three atoms. The oxygen atom forms a covalent bond with each of the hydrogen atoms. Oxygen has a greater electronegativity than hydrogen, so the oxygen atom attracts the electron pair of the covalent O—H bond to a greater extent than the hydrogen atom. Thus, the O—H bond is highly polar. As a result, the oxygen atom acquires a partial negative charge ($\delta-$). The less electronegative hydrogen atoms acquire partial positive charges ($\delta+$).

Figure 15.1 Water Is Vital to Life
The oceans supply penguins with an abundant supply of food.

IN ACADEMIC STANDARDS for SCIENCE

C.3.3 Compare and contrast ionic, covalent network, metallic and polar and non-polar molecular crystals with respect to constituent particles, strength of bonds, melting and boiling points, and conductivity; provide examples of each type.

C.7.1 Describe the composition and properties of types of solutions.

Focus on ELL

1 CONTENT AND LANGUAGE Create a KWL chart with the students, beginning with the things they know about water in all of its forms. Ask students to describe personal experiences they have had with liquid water or ice such as swimming, ice-skating, and seeing fog.

2 FRONTLOAD THE LESSON Display pictures of the planets of our solar system, and encourage students to share what they know about each. Point out that only Earth has the characteristic blue hue that indicates the presence of surface water. Explain that the presence of water on Earth is one of the primary reasons that Earth can support life. Tell students to make note of each of water's life-sustaining properties that they will read about in the lesson.

3 COMPREHENSIBLE INPUT Have students view the Kinetic Art animation on hydrogen bonding. Use a role-playing activity to model the effect of adding a surfactant to water to demonstrate how surfactant molecules limit the ability of water molecules to hydrogen bond.

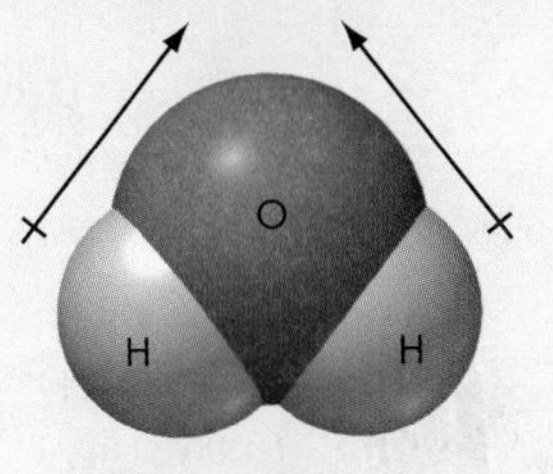

Polar bonds

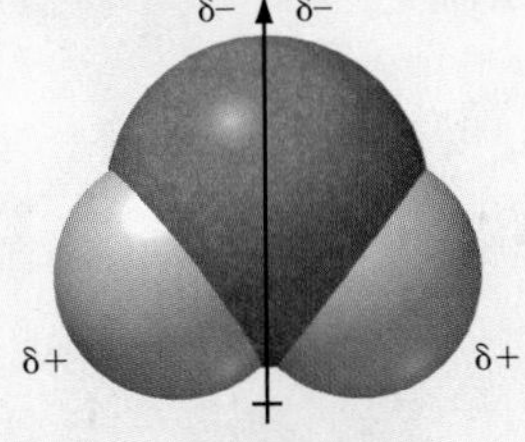

Molecule has net polarity

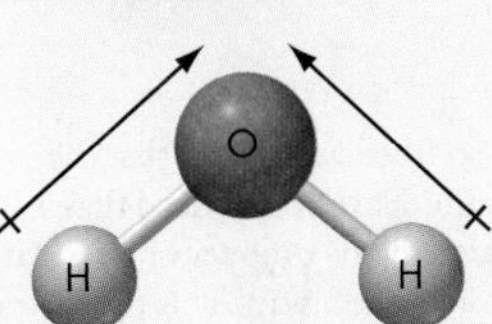

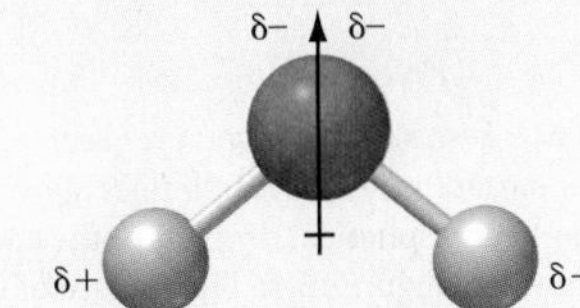

Figure 15.2 Polarity of H_2O
In a water molecule, the bond polarities are equal, but the two poles do not cancel each other because a water molecule is bent. The molecule as a whole is polar.
Apply Concepts ***Which element in water has the higher electronegativity?***

How do the polarities of the two O—H bonds affect the polarity of the molecule? The shape of the molecule is the determining factor. The bond angle of the water molecule is approximately 105°, which gives the molecule a bent shape. The two O—H bond polarities do not cancel, so the water molecule as a whole is polar. The net polarity of the water molecule is illustrated in Figure 15.2.

In general, polar molecules are attracted to one another by dipole interactions. The negative end of one molecule attracts the positive end of another molecule. However, in water, this attraction results in hydrogen bonding, as illustrated in Figure 15.3. Recall that hydrogen bonds are attractive forces that arise when a hydrogen atom is covalently bonded to a very electronegative atom and also weakly bonded to an unshared electron pair of another electronegative atom. Hydrogen bonds are not as strong as covalent bonds, but they are stronger than other intermolecular forces. **Many unique and important properties of water—including its high surface tension, low vapor pressure, and high boiling point—result from hydrogen bonding.**

Figure 15.3 Hydrogen Bonding in Water
The polarity of the water molecule results in hydrogen bonding. **a.** The oxygen atom has a partial negative charge. Each hydrogen atom has a partial positive charge. **b.** Hydrogen bonds form between the hydrogen atom of one water molecule and the oxygen atom of an adjacent water molecule.
Infer ***To form a hydrogen bond, what must be true about hydrogen and the element to which it is hydrogen bonded?***

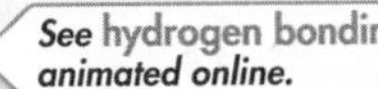

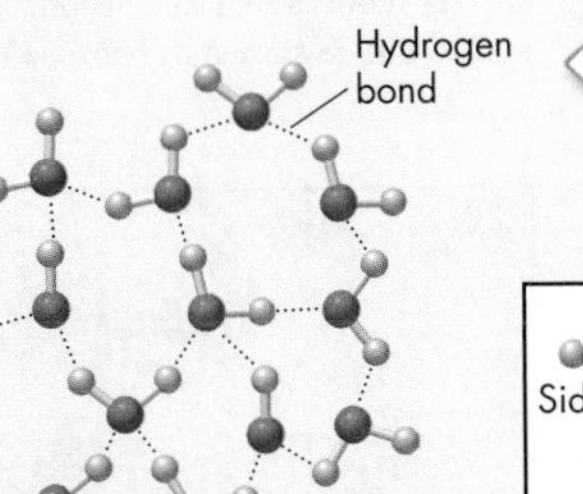

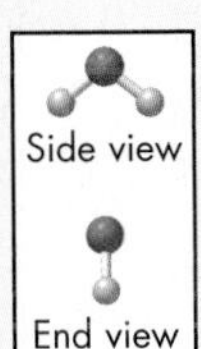

Professional Development Note

Water Worth Drinking

The year 2009 marked the 35th year of public health protection under the Safe Drinking Water Act. The Act, passed in 1974 and amended in 1986 and 1996, gives the Environmental Protection Agency (EPA) the authority to set drinking water standards. Drinking water standards apply to public water systems.

Standards for safety are constantly being examined by the EPA and change as new data become available. These standards set upper limits on potentially harmful substances that could be present in water. These substances include pesticides, household chemicals, and various microorganisms. Some microorganisms occur naturally in drinking water sources, but others indicate contamination by sewage or animal wastes. Two microorganisms are routinely monitored: coliform bacteria and the *Cryptosporidium* parasite. Both organisms can cause gastrointestinal disorders ranging from mild to severe. Cryptosporidiosis is potentially fatal for patients with compromised immune systems.

Foundations for Reading

BUILD VOCABULARY The word *tension* comes from the Latin *tendere*, which means "to stretch." Thus, surface tension causes the surface of a liquid to appear stretched and taut. Have students use the derivation of *tension* to explain the expression *nervous tension*. *(a condition of stress or tautness in an individual, brought about by some external cause)*

READING STRATEGY Have students preview the figures and read the related captions. Encourage students to use the caption information to predict the answers to any questions associated with the figures.

Explain

Water in the Liquid State

START A CONVERSATION Tell students that, although three-quarters of Earth's surface is water, only a small fraction of it can be used for drinking. The majority of Earth's surface water is in the oceans and contains too high a concentration of salts for the human body to tolerate.

USE VISUALS Have students study Figures 15.2 and 15.3. **Ask** What is meant by the term *polarity*? *(Polarity refers to the net molecular dipole resulting from electronegativity differences between covalently bonded atoms.)* **Ask** What element in a water molecule has a partial negative charge? *(the oxygen atom)* **Ask** What element has a partial positive charge? *(the hydrogen atoms)* **Ask** Predict the type of substances you would expect to dissolve in water. *(polar substances, ionic compounds)*

APPLY CONCEPTS Write the structural formula of water on the board, or hold up a model of a water molecule. Have students use Table 6.2 on page 181 to find and compare the electronegativity of oxygen and hydrogen. Use VSEPR theory and the model to show students how the oxygen's non-bonding electron pairs reduce the bond angle in a water molecule to 105°. Point out how the net molecular dipole and hydrogen bonding properties of water are due to the large electronegativity difference between hydrogen and oxygen. Note that hydrogen bonding contributes to water's unusually low vapor pressure and high surface tension.

Answers

FIGURE 15.2 oxygen

FIGURE 15.3 The hydrogen atom must be covalently bonded to a very electronegative atom with one or more lone pairs of electrons.

Explain

USE VISUALS Direct students' attention to Figure 15.4. To help explain why liquid water assumes a spherical shape on many surfaces, explain that nature tends to find the path of least resistance. Guide students to understand that a spherical shape provides the minimum surface area for a given volume.

USE AN ANALOGY Convey that the *skin-like* qualities of water are due to an exceptionally high surface tension that is created by an extensive network of hydrogen bonds. As an analogy, describe the following scene. A crowd of autograph seekers surrounds a celebrity. As each person approaches as closely as possible, an impenetrable circular barrier forms. This barrier will remain intact as long as there is a net attraction toward the circle's center. The process is dynamic. As one person wiggles closer to the celebrity, another is forced to retreat, but the overall shape does not change. Similarly, within a drop of water, attractions between individual molecules may shift, but the overall shape remains constant as molecules continue to be drawn toward a central focal point.

Figure 15.4
Surface Tension of Water
Water forms nearly spherical drops on a leaf. Water molecules at the surface of the water drop cannot form hydrogen bonds with molecules in the air, so they are drawn into the body of the liquid.

Surface Tension Have you ever seen a glass so filled with water that the water surface is not flat but bulges above the rim? Have you noticed that water forms nearly spherical droplets at the end of a medicine dropper or on a leaf, as shown in Figure 15.4? The surface of water acts like a skin. This property of water's surface is explained by the ability of water molecules to form hydrogen bonds. The water molecules within the body of the liquid form hydrogen bonds with other molecules that surround them on all sides. The attractive forces on each of these molecules are balanced. However, water molecules at the surface of the liquid experience an unbalanced attraction. You can see in Figure 15.4 that the water molecules are hydrogen bonded on only the inside of the drop. As a result, water molecules at the surface tend to be drawn inward. The inward force, or pull, that tends to minimize the surface area of a liquid is called **surface tension.**

All liquids have a surface tension, but water's surface tension is higher than most. This is why, on some surfaces, water tends to bead up rather than spread out. The surface tension of water tends to hold a drop of liquid in a spherical shape. For example, you may notice that water tends to form beads on the surface of a newly waxed car. The wax molecules are nonpolar, so there is little or no attraction between the wax molecules and the polar water molecules. The drops are not perfect spheres because the force of gravity tends to pull them down, causing them to flatten.

It is possible to decrease the surface tension of water by adding a surfactant. A **surfactant** is any substance that interferes with the hydrogen bonding between water molecules and thereby reduces surface tension. Soaps and detergents are surfactants. Adding a detergent to beads of water on a greasy surface reduces surface tension, causing the beads of water to collapse and spread out, as shown in Figure 15.5.

Figure 15.5 Effect of a Surfactant
Water drops bead up on some surfaces. When detergent, a surfactant, is added to water, the drop spreads out.
Compare *Which liquid drop has a higher surface tension?*

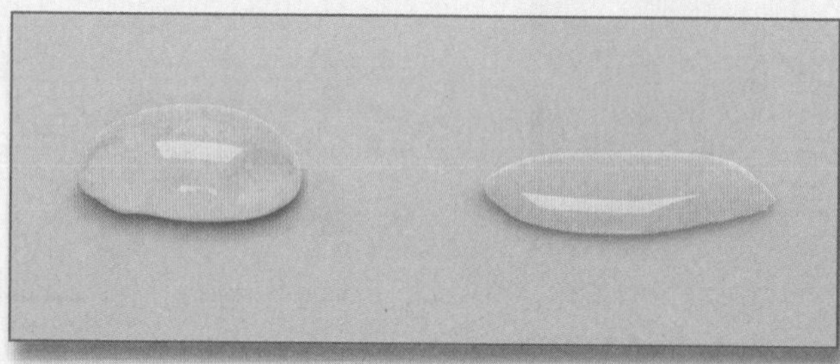

Differentiated Instruction

L3 ADVANCED STUDENTS Explain that the structure of ice was determined using X-ray diffraction, a method that was also used by Rosalind Franklin in her work with DNA. This work was critical to the discovery of the helical structure of DNA by James Watson and Francis Crick. Have students research how X-ray diffraction is used to determine the structure of a substance.

L1 STRUGGLING STUDENTS Have students use prior knowledge and create a visual of the water cycle. Ask them to explain changes of state. Have students refer to their diagram as you teach the more unfamiliar properties of water.

LPR LESS PROFICIENT READERS Allow students to use food coloring, cotton swabs, and a protractor to visually grasp the concept of polarity and bond angle in water molecules. Students can use the food coloring to stain one end of the cotton swab green and the other end red to represent negative and positive charge and then correctly orient their cotton swabs to form a 105° angle.

Vapor Pressure Hydrogen bonding between water molecules also explains water's unusually low vapor pressure. Remember that the vapor pressure of a liquid is the result of molecules escaping from the surface of the liquid and entering the vapor phase. An extensive network of hydrogen bonds holds the molecules in liquid water to one another. These hydrogen bonds must be broken before water changes from the liquid to the vapor state, so the tendency of these molecules to escape is low and evaporation is slow. Imagine what would happen if it were not. All the lakes and oceans, with their large surface areas, would rapidly evaporate!

Boiling Point Molecular compounds of low molar mass are usually gases or liquids with low boiling points at normal atmospheric pressure. Ammonia (NH_3), a molecular compound, has a molar mass of 17.0 g/mol and boils at about −33°C. Water has a molar mass of 18.0 g/mol, but it has a boiling point of 100°C. The difference between the boiling points of these two compounds is due to hydrogen bonding, which is more extensive in water than in ammonia. It takes much more heat to disrupt the attractions between water molecules than those between ammonia molecules. If the hydrogen bonding in water were as weak as it is in ammonia, water would be a gas at the usual temperatures found on Earth.

READING SUPPORT

Build Reading Skills: ***Inference*** If water molecules did not form hydrogen bonds with each other, water would have a much lower boiling point. ***What do you think this would mean for life as we know it?***

Quick Lab

Purpose To observe an unusual surface property of water that results from hydrogen bonding

Materials

- shallow dish or petri dish
- water
- paper clip
- rubber band, approximately 5 cm in diameter
- micropipets or droppers (2)
- vegetable oil
- liquid dish detergent

Surface Tension

Procedure

1. Thoroughly clean and dry the dish.
2. Fill the dish almost full with water. Dry your hands.
3. Being careful not to break the surface, gently place the paper clip on the water. Observe what happens.
4. Repeat Steps 1 and 2.
5. Gently place the open rubber band on the water.
6. Slowly add the vegetable oil drop by drop onto the water encircled by the rubber band until that water is covered with a layer of oil. Observe for 15 seconds.
7. Allow one drop of dish detergent to fall onto the center of the oil layer. Observe the system for 15 seconds.

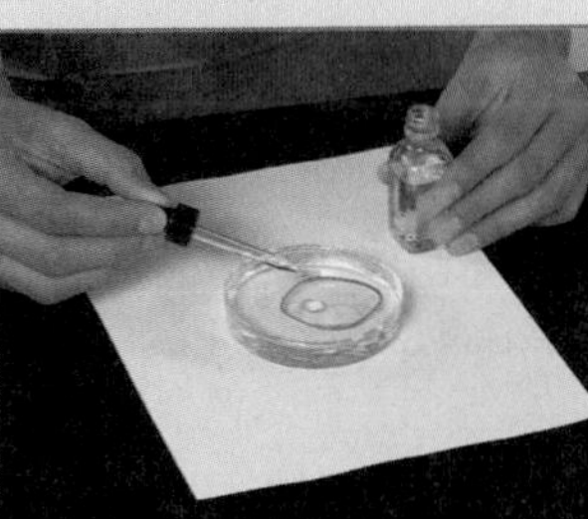

Analyze and Conclude

1. **Observe** What happened to the paper clip in Step 3? Why?
2. **Predict** If a paper clip becomes wet, does it float? Explain your answer.
3. **Observe** What shape did the rubber band take when the water inside it was covered with oil? Why did it take the observed shape?
4. **Describe** What happened when a drop of dish detergent was placed onto the layer of oil?

Focus on ELL

4 LANGUAGE PRODUCTION Have students work in groups of four to complete the lab. Make sure each group has ELLs of varied language proficiencies so that more proficient students can help less proficient ones. Have students work according to their proficiency level.

BEGINNING: LOW/HIGH Allow students to draw a picture of their observations.

INTERMEDIATE: LOW/HIGH Paraphrase the procedure and the Analyze and Conclude questions for students.

ADVANCED: LOW/HIGH Have these students assist those with lower English proficiencies to complete the procedure and answer the questions.

Misconception Alert

Students may believe that the surface of a volume of water contains a different kind of water molecule than its interior. Remind students that surface tension still exists even after you mix the water sample to redistribute the molecules.

Explore

Quick Lab

OBJECTIVE After completing this activity, students will be able to identify a property of water that results from hydrogen bonding and will be able to describe the effect of a surfactant on surface tension.

SKILLS FOCUS Observing, inferring, drawing conclusions

PREP TIME 10 minutes

CLASS TIME 20 minutes

ADVANCE PREPARATION Gather paper clips, rubber bands, detergent, and vegetable oil.

TEACHING TIPS Have students dry their hands before touching the paper clips.

EXPECTED OUTCOME The paper clip floats. The rubber band takes on a circular shape when filled with oil. Detergent destroys the surface tension and causes the rubber band to return to its original, noncircular shape.

ANALYZE AND CONCLUDE

1. The paper clip floated. The surface tension of the water supported the paper clip.
2. No. The surface tension is broken, and the paper clip sinks because it is made of a metal that has a greater density than water.
3. The rubber band took a circular shape. The oil spreads equally in all directions, pushing the rubber band into a circular shape.
4. The oil layer was broken, and the rubber band returned to its original shape.

FOR ENRICHMENT Explain that some insects, such as a water strider, depend on the surface tension of water for moving around and capturing food. Tell students that detergents are increasingly found as pollutants in bodies of water in nature. Have students design an experiment to determine the concentration of detergent at which water can no longer support the weight of a water strider.

Answers

FIGURE 15.5 The water droplet on the left has a higher surface tension.

READING SUPPORT Sample Answer: There would be less water in the liquid state for animals to drink and other organisms to use.

LESSON 15.1

Explain

Water in the Solid State

USE VISUALS Direct students' attention to Table 15.1. **Ask** What relation do you notice between temperature and density? *(The density of liquid water increases over most of the range of decreasing temperature.)* **Ask** Why does this trend not continue below 4°C? *(Below 4°C, the kinetic energy of the molecules is insufficient to overcome hydrogen bonding, which holds the water molecules in fixed positions.)* If students have difficulty discerning the pattern, use the overhead and plot the data in the table on a coordinate grid; place temperature on the *x*-axis and density on the *y*-axis.

CRITICAL THINKING Have students explain why ice, which has a smaller mass per volume ratio than liquid water, results in a lower relative density. *(Density is based on mass per volume; the substance with the smaller mass per volume ratio has the lower density.)*

Misconception Alert

Make sure students understand that hydrogen bonding always involves hydrogen, but it is not a covalent or ionic bond. Rather, it explains a type of intermolecular attractive force. Explain that polarity leads to intermolecular forces—attractions between molecules. Covalent bonding—the bond between hydrogen and oxygen in a water molecule—is a type of intramolecular force.

Water in the Solid State

How can you describe the structure of ice?

You have seen that water in the liquid state exhibits some unique properties. The same is true for water in the solid state. For example, ice cubes float in your glass of iced tea because solid water has a lower density than liquid water. This situation is not usual for liquids. As a typical liquid cools, it begins to contract and its density increases gradually. The density increases because the molecules of the liquid move closer together so that a given volume of the liquid contains more molecules and thus more mass. If the cooling continues, the liquid eventually solidifies with a density greater than the density of the liquid. A typical solid sinks in its own liquid because the density of the solid is greater than that of the corresponding liquid.

As water begins to cool, it behaves initially like a typical liquid. It contracts slightly and its density gradually increases, as shown in Table 15.1. Notice that at 4°C, the density of water is at its maximum of 1.0000 g/cm^3. When the temperature of the water falls below 4°C, the density of water actually starts to decrease. Below 4°C, water no longer behaves like a typical liquid. Ice, which forms at 0°C, has about a 10 percent lower density than liquid water at 0°C. You may have noticed that ice begins to form at the surface of a pond when the temperature reaches 0°C, but the ice does not sink. It floats at the surface, making ice skating and ice fishing possible. Ice is one of only a few solids that floats in its own liquid.

Why is ice less dense than liquid water? As you can see in Figure 15.6, hydrogen bonds hold the water molecules in place in the solid phase. **The structure of ice is a regular open framework of water molecules in a hexagonal arrangement.** When ice melts, the framework collapses. Looking back at Figure 15.3, you can see that the water molecules pack closer together in liquid water, making it more dense than ice.

Table 15.1
Density of Liquid Water and Ice

Temperature (°C)	Density (g/cm^3)
100 (liquid water)	0.9584
50	0.9881
25	0.9971
10	0.9997
4	1.0000
0 (liquid water)	0.9998
0 (ice)	0.9168

Figure 15.6 Structure of Ice
The unique properties of ice are a result of hydrogen bonding. **a.** Extensive hydrogen bonding in ice holds the water molecules farther apart in a more ordered arrangement than in liquid water. **b.** The hexagonal symmetry of a snowflake reflects the structure of the ice crystal.
Compare and Contrast ***How are the structures of liquid water (shown in Figure 15.3) and ice similar? How are they different?***

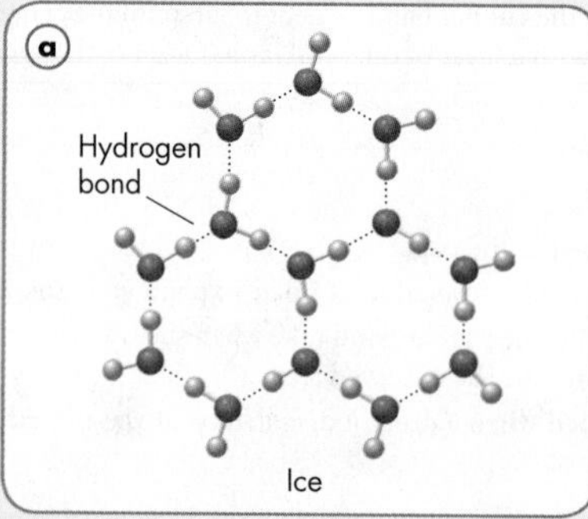

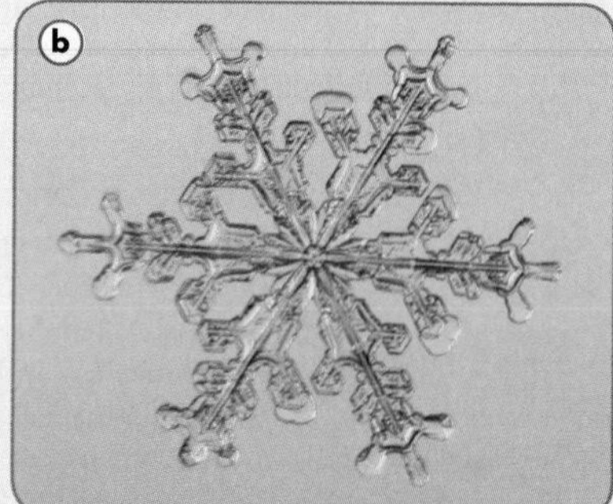

UbD Check for Understanding

The Essential Question How do the interactions between water molecules account for the unique properties of water?

Assess students' knowledge about the importance of hydrogen bonding in water by asking them to write a paragraph that explains how hydrogen bonding is responsible for one of the following properties of water: high surface tension, high boiling point, low vapor pressure, or high melting point.

ADJUST INSTRUCTION If students are having difficulty with this exercise, have them reread the portion of this lesson associated with their chosen property. Then provide students with the opportunity to revise their paragraphs.

The fact that ice floats has important consequences for organisms. A layer of ice on the top of a body of water, such as the one shown in Figure 15.7, acts as an insulator for the water beneath, preventing the water from freezing solid except under extreme conditions. The liquid water at the bottom of an otherwise frozen body of water is warmer than 0°C, so fish and other aquatic life are better able to survive. If ice were denser than liquid water, bodies of water would tend to freeze solid during the winter months, destroying many types of organisms.

Ice melts at 0°C, which is a high melting temperature for a molecule with such a low molar mass. A considerable amount of energy is required to return water molecules in the solid state to the liquid state. The heat absorbed when 1 g of water at 0°C changes from a solid to a liquid is 334 J. This heat is the same amount of energy that is needed to raise the temperature of 1 g of liquid water from 0°C to 80°C.

Q: *What properties of water that result from hydrogen bonding make it essential to life on Earth?*

Figure 15.7 Ice Floats in Liquid Water
Many organisms that live in water would not survive if ice were more dense than liquid water.

1. **Review** What causes the high surface tension, low vapor pressure, and high boiling point of water?
2. **Describe** How are water molecules arranged in ice?
3. **Explain** Why does water form spherical drops on some surfaces?
4. **Relate Cause and Effect** What effect does a surfactant have on the surface tension of water?
5. **Infer** Water (H_2O) and methane (CH_4) have similar molar masses. Methane changes from a liquid to a gas at −161°C. Water becomes a gas at 100°C. What could account for the difference?
6. **Apply Concepts** What causes water pipes to break in freezing weather?

BIGIDEA BONDING AND INTERACTIONS

7. Describe how hydrogen bonding accounts for the properties of water.

Lesson Check Answers

1. Water molecules are hydrogen bonded to each other but not to air molecules. Net attraction is inward, minimizing the water surface area. Hydrogen bonding makes it more difficult for water molecules to escape from the liquid phase to the vapor phase.
2. Ice has a honeycomb-like structure of water molecules.
3. The surface tension of a liquid tends to hold a drop of liquid in a spherical shape.
4. Surfactants lower the surface tension of water by interfering with hydrogen bonding.
5. Water has hydrogen bonding between its molecules; methane does not.
6. Water expands as it freezes.
7. Water molecules at the surface of a water drop are drawn into the body of the liquid producing water's unusually high surface tension. Hydrogen bonds also hold water molecules to one another resulting in water's unusually low vapor pressure.

CHEMISTRY & YOU Ice is less dense than liquid water because of hydrogen bonding. As a result, ice floats on the surface of water providing a layer of insulation for the water beneath. Many organisms are able to survive underwater in the winter because the water is not frozen throughout.

Extend

Connect to ▶ SOCIAL STUDIES

Have students compare areas of population density and the distribution of water. Provide students with two different maps. One map should show the locations of cities and towns. The other should be a regional contour map showing water courses, rivers, lakes, and reservoirs. Have students determine why most large cities or historical sites are located near water sources, and have them explain the effects when such places are not near water. Ask students to examine the vast irrigation networks and water channels that crisscross the nation's agricultural areas. Have students determine the ways in which water is transported to arid regions and to population centers with insufficient fresh water.

Evaluate

Informal Assessment

Ask Why is the surface tension of water so high compared to that of other liquids? *(Water molecules form a large number of hydrogen bonds, in addition to dipole-dipole forces between molecules.)*

Reteach

Have students explain the property of water that allows ice to float in its liquid phase. *(Ice has a lower density than liquid water. Because there is more space between water molecules in the solid state than the liquid state, there are fewer molecules in any given volume. Less mass per volume results in a lower relative density of ice as compared to water.)*

Answers

FIGURE 15.6 There are two hydrogen atoms attached to one oxygen atom in both structures. The molecules in liquid water are further apart and have a less ordered arrangement than ice.

LESSON 15.2

Key Objectives

15.2.1 **IDENTIFY** the types of substances that dissolve most readily in water.

15.2.2 **EXPLAIN** why all ionic compounds are electrolytes.

15.2.3 **EXPLAIN** why hydrates easily lose and regain water.

Additional Resources

- Reading and Study Workbook, 15.2
- Probeware Laboratory Manual, Lesson 15.2
- Core Teaching Resources, Lesson 15.2
- Laboratory Manual, Labs 26, 27, 28, 29
- Small-Scale Chemistry Laboratory Manual, Labs 22, 23, 24
- Laboratory Practicals, 15.1

PearsonChem.com

Engage

CHEMISTRY & YOU Have students study the photograph and read the text. **Ask** What is it about the pickle that allows it to produce light? *(It conducts electricity.)* **Ask** What are some other materials that conduct electricity? *(metals and ionic solutions)*

Access Prior Knowledge

Explain that in this lesson students will rely on their understanding of polarity, ionic compounds, covalent compounds, and molar mass to understand the concepts of aqueous systems.

IN ACADEMIC STANDARDS for SCIENCE

C.7.1 Describe the composition and properties of types of solutions.

15.2 Homogeneous Aqueous Systems

C.7.1 Composition and properties of solutions

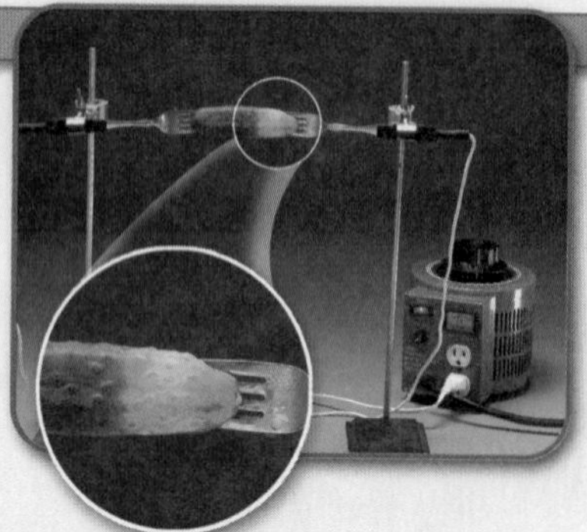

CHEMISTRY & YOU

Q: *How can you make a pickle glow?* Is it possible to read by the light of a glowing pickle? Although it sounds absurd, an ordinary dill pickle from the deli can be a source of light! Metal forks are inserted into the ends of the pickle and connected to a source of alternating electric current. After a time, during which the pickle becomes hot and produces steam, the pickle begins to glow. The mechanism by which the light is generated is not fully understood, but it is clear that conduction of electric current by the pickle is an important factor.

Key Questions

What types of substances dissolve most readily in water?

Why are all ionic compounds electrolytes?

Why do hydrates easily lose and regain water?

Vocabulary

- aqueous solution • solvent
- solute • solvation
- electrolyte • nonelectrolyte
- strong electrolyte
- weak electrolyte
- water of hydration • hydrate
- anhydrous • effloresce
- hygroscopic • desiccant
- deliquescent

Solutions

What types of substances dissolve most readily in water?

Water dissolves so many of the substances that it comes in contact with that you won't find chemically pure water in nature. Even the tap water you drink is a solution that contains varying amounts of dissolved minerals and gases. An **aqueous solution** is water that contains dissolved substances.

Solvents and Solutes In a solution, the dissolving medium is the **solvent.** The dissolved particles in a solution are the **solute.** A solvent dissolves the solute, and the solute becomes dispersed in the solvent. Solvents and solutes may be gases, liquids, or solids.

Recall that solutions are homogeneous mixtures. They are also stable mixtures. For example, sodium chloride (NaCl) does not settle out when its solutions are allowed to stand, provided other conditions, such as temperature, remain constant. Solute particles can be atoms, ions, or molecules, and their average diameters are usually less than 1 nm (10^{-9} m). Therefore, if you filter a solution through filter paper, both the solute and the solvent pass through the filter.

Substances that dissolve most readily in water include ionic compounds and polar covalent compounds. Nonpolar covalent compounds, such as methane, and compounds found in oil, grease, and gasoline, do not dissolve in water. However, oil and grease will dissolve in gasoline. To understand this difference, you must know more about the structures of the solvent and the solute and what attractions exist between them.

Focus on ELL

1 CONTENT AND LANGUAGE Have small groups of students create flashcards with the vocabulary terms by writing each term on one side of a card and a definition and visual on the other. Allow students with limited English to use their native languages for support.

2 FRONTLOAD THE LESSON Have students discuss aqueous systems they have made or observed at home. Ask students to state whether the items in these aqueous systems are solids, liquids, and/or gases.

3 COMPREHENSIBLE INPUT Have students bring in a grocery item that they believe to be an aqueous system. Write the definition for *homogeneous aqueous system* on the board, and have students place their items on a specified desk if they can defend that their items satisfy the definition.

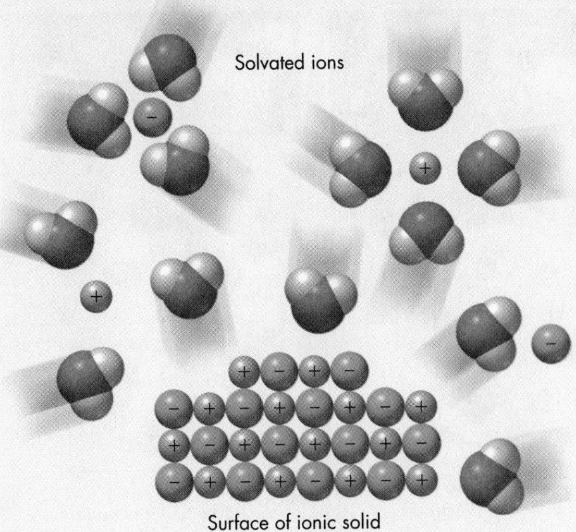

Figure 15.8 Solvation of an Ionic Solid
When an ionic solid dissolves, the ions become solvated, or surrounded by solvent molecules.
Infer ***Why do the water molecules orient themselves differently around the anions and the cations?***

See solvation animated online.

The Solution Process Water molecules are in continuous motion because they have kinetic energy. When a crystal of sodium chloride is placed in water, the water molecules collide with the crystal. Remember that a water molecule is polar, with a partial negative charge on the oxygen atom and partial positive charges on the hydrogen atoms. The polar solvent molecules (H_2O) attract the solute ions (Na^+, Cl^-). As individual solute ions break away from the crystal, the negatively and positively charged ions become surrounded by solvent molecules and the ionic crystal dissolves. The process by which the positive and negative ions of an ionic solid become surrounded by solvent molecules is called **solvation.** Figure 15.8 shows a model of the solvation of an ionic solid such as sodium chloride.

In some ionic compounds, the attractions among the ions in the crystals are stronger than the attractions exerted by water. These compounds cannot be solvated to any significant extent and are therefore nearly insoluble. Barium sulfate ($BaSO_4$) and calcium carbonate ($CaCO_3$) are examples of nearly insoluble ionic compounds.

Figure 15.9 shows that oil and water do not mix. What about oil in gasoline? Both oil and gasoline are composed of nonpolar molecules. The attractive forces that hold two molecules in oil together are similar in magnitude to the forces that hold two molecules in gasoline together. Molecules in oil can easily separate and replace molecules in gasoline to form a solution. As a rule, polar solvents such as water dissolve ionic compounds and polar compounds; nonpolar solvents such as gasoline dissolve nonpolar compounds. This relationship can be summed up in the expression "like dissolves like."

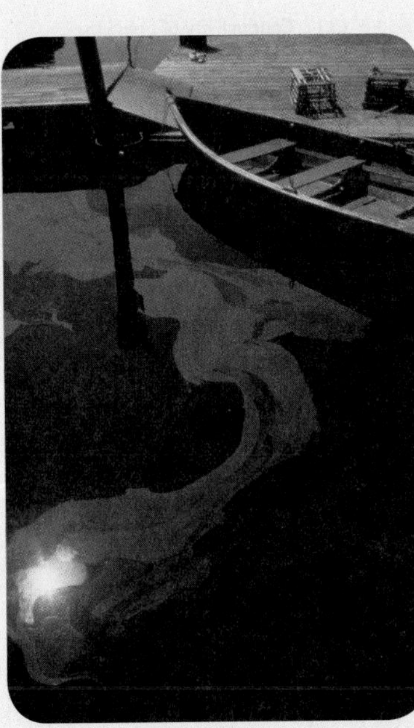

Figure 15.9 Oil and Water
Oil and water do not mix. Oil is less dense than water, so it floats on top. The colors result from the bending of light rays by the thin film of oil.

Differentiated Instruction

LPR LESS PROFICIENT READERS Have students preview the section by looking for vocabulary and other unfamiliar terms. Encourage students to write the terms, with phonetic respellings and definitions, in their notebooks.

L1 SPECIAL NEEDS STUDENTS Have students explore the phrase "like dissolves like." Provide student pairs with pairs of liquids that are polar, pairs that are nonpolar, and pairs that contain one polar and one nonpolar liquid. Have students mix the two liquids in each pair and decide whether the liquids are like or unlike.

L3 ADVANCED STUDENTS Ask students to determine a method for proving if a bottled water product is pure water or if it is an aqueous solution. Have students present bottled water products with misleading labels.

Foundations for Reading

BUILD VOCABULARY Have students make a concept map using the following words: *solvation, solute, aqueous solution, solvent.*

READING STRATEGY For each section in this lesson, have students write a short summary identifying the main idea of the passage. Students can refer to these summaries when preparing for an assessment.

Explain

Solutions

START A CONVERSATION Remind students that an anion is any negatively charged atom or group of atoms. A cation is any positively charged atom or group of atoms. Have students suggest a memory aid or a symbol to help them remember these terms. *(Sample answer: The first two letters of* anion, a *and* n, *could represent "a negative", since anions are negative ions.)*

USE VISUALS Have students study Figure 15.8. Stress the idea that when a solid dissolves in water, it breaks into particles. **Ask** What must happen for an ionic solid to dissolve? *(The molecules of the solvent must be able to overcome the attractive forces acting between ions and holding the solid together.)* **Ask** What part of a water molecule is attracted to a negatively charged solute ion? *(the hydrogen atoms)*

Extend

Connect to PHYSIOLOGY

Remind students that solutions are homogeneous mixtures containing a solvent and one or more solutes. Explain that the solvent is typically defined as the component in the system that is present in the greatest amount. Point out that a water-soluble solute can be a solid, liquid, or gas. Have students identify some of the solutes found in blood. *(Solutes in blood are typically ions such as sodium, potassium, calcium, chloride, hydrogen carbonate, and phosphate. Glucose, a covalent compound, also is dissolved in blood. Dissolved gases such as oxygen and carbon dioxide are also present.)*

Answers

FIGURE 15.8 The positive hydrogen end of the water molecule orients itself toward the anion. The negative oxygen end of the water molecule orients itself toward the cation.

Extend

Electrolytes and Nonelectrolytes

START A CONVERSATION Display to the class the warning labels on some small electrical appliances, such as a hair dryer or an electric razor. **Ask** Why do these appliances warn you not to use them near water? *(because there are electrolytes in tap water and therefore it conducts electricity)* **Ask** Why does tap water conduct electricity? *(Tap water comes from rivers, streams, and wells and contains dissolved ionic particles from rocks and river banks. These act as electrolytes.)*

SUMMARIZE Have students summarize in a chart the types of compounds that strongly conduct an electrical current, weakly conduct an electrical current, and do not conduct an electrical current.

Figure 15.10 Conductivity of Solutions
A solution conducts an electric current if it contains ions. **a.** Sodium chloride, a strong electrolyte, is nearly 100 percent dissociated into ions in water. **b.** Mercury(II) chloride, a weak electrolyte, is only partially dissociated in water. **c.** Glucose, a nonelectrolyte, does not dissociate in water.
Predict *Silver chloride dissolves only slightly in water. If the electrodes were immersed in a solution of silver chloride, would the bulb glow brightly, dimly, or not at all?*

Electrolytes and Nonelectrolytes

Why are all ionic compounds electrolytes?

Remember the glowing pickle? The pickle contained an electrolyte. An **electrolyte** is a compound that conducts an electric current when it is in an aqueous solution or in the molten state. Conduction of an electric current requires ions that are mobile and, thus, able to carry charges through a liquid. **All ionic compounds are electrolytes because they dissociate into ions.** Sodium chloride, copper(II) sulfate, and sodium hydroxide are typical water-soluble electrolytes. Barium sulfate is an ionic compound that cannot conduct an electric current in aqueous solution because it is insoluble, but it can conduct in the molten state.

The experimental setup in Figure 15.10 can be used to determine whether a solution contains an electrolyte. In order for the bulb to light, an electric current must flow between the two electrodes that are immersed in the solution. If ions are present in the solution, they carry electrical charge from one electrode to the other, completing the electrical circuit.

A **nonelectrolyte** is a compound that does not conduct an electric current in either an aqueous solution or the molten state. Many molecular compounds are nonelectrolytes because they are not composed of ions. Most compounds of carbon, such as table sugar (sucrose) and the alcohol in rubbing alcohol (2-propanol), are nonelectrolytes.

UbD Check for Understanding

The Essential Question How do aqueous solutions form?

Assess students' understanding of the difference between how electrolyte and nonelectrolyte solutions form by asking students to give a thumbs up if they agree with the following statement: *Electrolyte solutions are characterized by charged ions separated by molecules of water. Nonelectrolyte solutions are characterized by uncharged molecules separated by water molecules.*

ADJUST INSTRUCTION If students have difficulty with this concept, review Figure 15.10 with students, and then ask them to evaluate the statement again.

Some polar molecular compounds are nonelectrolytes in the pure state but become electrolytes when they dissolve in water. This change occurs because such compounds ionize in solution. For example, neither ammonia ($NH_3(g)$) nor hydrogen chloride ($HCl(g)$) is an electrolyte in the pure state. Yet an aqueous solution of ammonia conducts an electric current because ammonium ions (NH_4^+) and hydroxide ions (OH^-) form when ammonia dissolves in water.

$$NH_3(g) + H_2O(l) \longrightarrow NH_4^+(aq) + OH^-(aq)$$

Similarly, in an aqueous solution, hydrogen chloride produces hydronium ions (H_3O^+) and chloride ions (Cl^-). An aqueous solution of hydrogen chloride conducts an electric current and is therefore an electrolyte.

$$HCl(g) + H_2O(l) \longrightarrow H_3O^+(aq) + Cl^-(aq)$$

Not all electrolytes conduct an electric current to the same degree. In Figure 15.10, the bulb glows brightly when the electrodes are immersed in a sodium chloride solution. The bright glow shows that sodium chloride is a strong electrolyte because nearly all the dissolved sodium chloride exists as separate Na^+ and Cl^- ions. In a solution that contains a **strong electrolyte,** all or nearly all of the solute exists as ions. The ions move in solution and conduct an electric current. Most soluble salts, inorganic acids, and inorganic bases are strong electrolytes.

The bulb glows dimly when the electrodes are immersed in a mercury(II) chloride solution because mercury(II) chloride is a weak electrolyte. A **weak electrolyte** conducts an electric current poorly because only a fraction of the solute in the solution exists as ions. Organic acids and bases are also examples of weak electrolytes. In a solution of glucose, the bulb does not glow. Glucose ($C_6H_{12}O_6$) is a molecular compound. It does not form ions, so it is a nonelectrolyte.

Electrolytes are essential to all metabolic processes. Your cells use electrolytes, such as sodium and potassium ions, to carry electrical impulses internally and to other cells. These impulses are crucial to nerve and muscle function. The kidneys help to maintain balanced electrolyte concentrations in the blood. However, an electrolyte imbalance can occur if you become dehydrated. For example, when you exercise, you can lose water and electrolytes from your body through perspiration. The athlete in Figure 15.11 understands that it is important to replenish these electrolytes by eating salty foods or by drinking sports drinks.

CHEMISTRY & YOU

Q: *Pickles contain table salt. Why can electric current flow through a pickle, causing it to glow?*

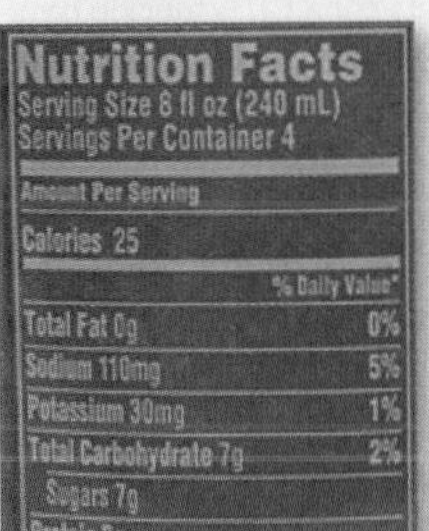

Figure 15.11 Sports Drinks
It is important to replenish electrolytes when exercising or sweating. Sports drinks often contain sodium and potassium.

Metabolic Process

Electrolytes are essential to all metabolic processes. Sodium and potassium ions control nerve impulse transmission and muscle contraction. If renal function is impaired or malabsorption from the gut disturbs optimum sodium and potassium levels, then serious nervous-system problems arise. Loss of consciousness or difficulty in maintaining muscle coordination could result. Any condition that causes prolonged bouts of diarrhea can be life threatening because of the dramatic loss of electrolytes. Electrolytes are excreted through the skin via sweat, and they must be replenished or cramps and heat stroke may occur. Sports drinks are a good source of electrolytes; they contain Na^+, K^+, and Ca^{2+}.

Table salt, or sodium chloride, is a strong electrolyte.

Explore

Class Activity

PURPOSE Students will compare electrolytes by using a conductivity tester.

MATERIALS light bulb in a porcelain socket; 9V or lantern battery; 2 copper metal strips; lamp cord; alligator clips; 0.1*M* solutions of glucose ($C_6H_{12}O_6$), alanine ($HC_3H_6O_2N$), glycine ($HC_2H_4O_2N$), ascorbic acid ($HC_6H_7O_6$), malonic acid ($H_2C_3H_2O_4$), citric acid ($H_3C_6H_5O_7$), acetic acid ($HC_2H_3O_2$), and hydrochloric acid (HCl); 8 beakers labeled 1–8; marking pen

Set up a light bulb (in a porcelain socket), a 9V or lantern battery, and two copper metal strips to be immersed in the test solution. Make connections to the battery and light bulb socket by using the lamp cord and alligator clips. If possible, prepare eight such testers so that students can directly compare the brightnesses of the light bulbs.

PROCEDURE Write the names and numbers of the beakers on the board. Pour an equal volume of a different solution into each labeled beaker. Remind students that all the solutions have the same concentration (0.1*M*). Immerse a pair of electrodes in each solution and have students compare the brightnesses of the light bulb. Direct students to list the substances in order from strongest to weakest electrolytes. Remind students that a strong electrolyte is a substance that exists in solution almost entirely as ions. A weak electrolyte is a substance that only partly dissociates into ions in solution.

EXPECTED OUTCOME Students infer that the different substances are dissociated, or ionized, to different extents.

Answers

FIGURE 15.10 The bulb will glow dimly.

Explain

Hydrates

USE VISUALS Have students examine Figure 15.12. **Ask** What is a hydrate? *(a compound that contains weakly bonded water molecules as part of its crystal structure)* **Ask** What does the solid dot between the $CuSO_4$ and the $5H_2O$ in the compound formula mean? *(It indicates that five water molecules are bonded with each formula unit of the compound.)* **Ask** How can the presence of water in the structure of the crystal lattice be inferred? *(by the change in the appearance of the hydrated substance when the water is driven off)*

Explore

Teacher Demo

PURPOSE Students will observe the color change when a hydrate gains and loses water.

MATERIALS $CoCl_2 \cdot 6H_2O$ solution, cotton swab, white paper, hot plate, misting bottle, water.

PROCEDURE Use a cotton swab to draw a picture or write a message on a piece of white paper by using a solution of hydrated cobalt(II) chloride. Place the paper on a warm (not hot) hot plate until the writing dries. Use a misting bottle to mist water onto the paper. The paper can also be used to indicate the presence of humidity in the air.

SAFETY AND DISPOSAL Wear gloves. Cobalt chloride is toxic if ingested. Collect the cobalt chloride solution, evaporate it to dryness, place the cooled residue in a bag, and label the bag for disposal.

EXPECTED OUTCOMES The picture or message will appear blue when the paper dries and pink when the paper is misted with water.

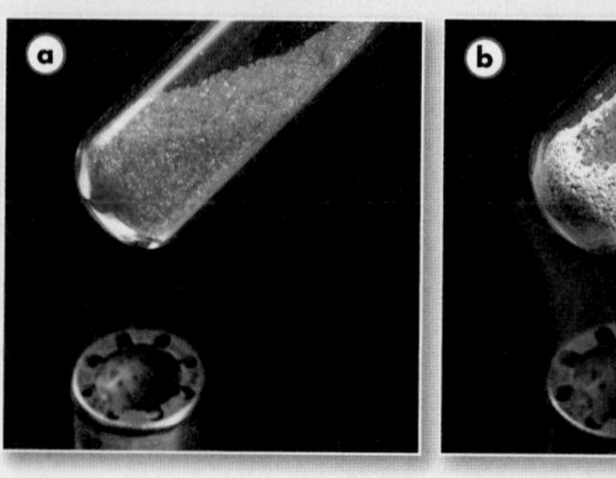

Figure 15.12 Heating a Hydrate Water can be driven from a hydrate by heating it. **a.** Heating of a sample of blue $CuSO_4 \cdot 5H_2O$ begins. **b.** After a time, much of the blue hydrate has been converted to white anhydrous $CuSO_4$.

Hydrates

Why do hydrates easily lose and regain water?

When an aqueous solution of copper(II) sulfate is allowed to evaporate, deep-blue crystals of copper(II) sulfate pentahydrate are deposited. The chemical formula for this compound is $CuSO_4 \cdot 5H_2O$. Water molecules are an integral part of the crystal structure of copper(II) sulfate pentahydrate and many other substances. The water contained in a crystal is called the **water of hydration** or water of crystallization. A compound that contains water of hydration is called a **hydrate.** In writing the formula of a hydrate, use a dot to connect the formula of the compound and the number of water molecules per formula unit. Crystals of copper(II) sulfate pentahydrate always contain five molecules of water for each copper and sulfate ion pair. The deep-blue crystals are dry to the touch. They are unchanged in composition or appearance in normally moist air. However, when the crystals are heated above 100°C, they lose their water of hydration. **The forces holding the water molecules in hydrates are not very strong, so the water is easily lost and regained.** Figure 15.12 shows how the blue crystals of $CuSO_4 \cdot 5H_2O$ crumble to a white anhydrous powder that has the formula $CuSO_4$. A substance that is **anhydrous** does not contain water. If anhydrous copper(II) sulfate is treated with water, the blue pentahydrate is regenerated.

$$CuSO_4 \cdot 5H_2O(s) \underset{-\text{heat}}{\overset{+\text{heat}}{\rightleftharpoons}} CuSO_4(s) + 5H_2O(g)$$

Another compound that changes color in the presence of moisture is cobalt(II) chloride. A piece of filter paper that has been dipped in an aqueous solution of cobalt(II) chloride and then dried is blue in color (anhydrous $CoCl_2$). As you can see in Figure 15.13, when the paper is exposed to moist air, it turns pink because of the formation of the hydrate cobalt(II) chloride hexahydrate ($CoCl_2 \cdot 6H_2O$). The blue paper could be used to test for the presence of water.

Figure 15.13 Exposing a Hydrate to Moist Air Paper dipped in an aqueous cobalt(II) chloride solution and then dried is blue. In the presence of moisture, the paper turns pink. **Infer** *How could you change the pink paper back to blue?*

Some familiar hydrates are listed in Table 15.2. Each one contains a fixed quantity of water and has a definite composition. To determine what percent by mass of a hydrate is water, first determine the mass of water in one mole of hydrate. Then determine the molar mass of the hydrate. The percent by mass of water can be calculated using the following equation:

$$\text{Percent by mass } H_2O = \frac{\text{mass of water}}{\text{mass of hydrate}} \times 100\%$$

UbD Check for Understanding

Why are all ionic compounds electrolytes?

Assess students' knowledge about electrolytes by asking them to signal true or false to the following statements. Have students signal true with a thumbs up and false with a thumbs down.

- All solutions are conductors of electricity. *(false)*
- Degree of electrical conductivity is directly related to kinetic energy. *(false)*
- Ionic compounds conduct electricity because they contain ions. *(true)*
- Molecular compounds conduct electricity because they do not contain ions. *(false)*

ADJUST INSTRUCTION If students are having difficulty answering, direct them to reexamine Figure 15.10 and reread the text that follows. Then ask them to respond to the statements again.

Table 15.2

Some Common Hydrates		
Formula	**Chemical name**	**Common name**
$MgSO_4 \cdot 7H_2O$	Magnesium sulfate heptahydrate	Epsom salt
$Ba(OH)_2 \cdot 8H_2O$	Barium hydroxide octahydrate	
$CaCl_2 \cdot 2H_2O$	Calcium chloride dihydrate	
$CuSO_4 \cdot 5H_2O$	Copper(II) sulfate pentahydrate	Blue vitriol
$Na_2SO_4 \cdot 10H_2O$	Sodium sulfate decahydrate	Glauber's salt
$KAl(SO_4)_2 \cdot 12H_2O$	Potassium aluminum sulfate dodecahydrate	Alum
$Na_2B_4O_7 \cdot 10H_2O$	Sodium tetraborate decahydrate	Borax
$FeSO_4 \cdot 7H_2O$	Iron(II) sulfate heptahydrate	Green vitriol
$H_2SO_4 \cdot H_2O$	Sulfuric acid monohydrate (mp 8.6°C)	

Efflorescent Hydrates The water molecules in hydrates are held by weak forces, so hydrates often have an appreciable vapor pressure. If a hydrate has a vapor pressure higher than the pressure of water vapor in the air, the hydrate will lose its water of hydration or **effloresce.** For example, copper(II) sulfate pentahydrate has a vapor pressure of about 1.0 kPa at room temperature. The average pressure of water vapor at room temperature is about 1.3 kPa. Copper(II) sulfate pentahydrate is stable until the humidity decreases. When the vapor pressure drops below 1.0 kPa, the hydrate effloresces. Washing soda, or sodium carbonate decahydrate ($Na_2CO_3 \cdot 10H_2O$), is efflorescent. As the crystals lose water of hydration, they effloresce and become coated with a white powder of anhydrous sodium carbonate (Na_2CO_3).

Figure 15.14 Desiccator
A desiccator may contain calcium chloride. Substances that must be kept dry are stored inside.

Hygroscopic Hydrates Hydrated ionic compounds that have a low vapor pressure remove water from moist air to form higher hydrates. These hydrates and other compounds that remove moisture from air are called **hygroscopic.** For example, calcium chloride monohydrate spontaneously absorbs a second molecule of water when exposed to moist air.

$$CaCl_2 \cdot H_2O(s) \xrightarrow{\text{moist air}} CaCl_2 \cdot 2H_2O(s)$$

Calcium chloride is used as a desiccant in the laboratory. A **desiccant** is a substance used to absorb moisture from the air and create a dry atmosphere. Anhydrous calcium chloride can be placed in the bottom of a tightly sealed container called a desiccator, which is shown in Figure 15.14. Substances that must be kept dry are stored inside. A solid desiccant such as calcium sulfate ($CaSO_4$) can also be added to a liquid solvent, such as ethanol, to keep it dry. The calcium sulfate does not dissolve appreciably in the solvent but absorbs water from the ethanol. When a desiccant has absorbed all the water it can hold, the compound can be returned to its anhydrous state by heating.

Desiccants

Cameras and electronic equipment made in Japan and other Asian countries are usually shipped to the United States in the holds of ships. Because ocean air contains so much moisture, the packaging for this equipment usually contains small packets of a desiccant. The desiccant absorbs the water vapor from the air, preventing it from condensing on the delicate wiring and circuit boards of the equipment, possibly causing corrosion and short circuits.

START A CONVERSATION Many of the terms used to describe the properties of hydrates may be unfamiliar to students. Review the definitions and pronunciations of the terms *effloresce, hygroscopic, desiccant,* and *deliquescent.* Have students write the definition for each term in their vocabulary notebooks.

Demonstrate some of these properties of hydrates by setting out watch glasses containing fresh sodium hydroxide pellets, anhydrous calcium chloride, and anhydrous copper sulfate. Advise students not to touch the compounds. Point out the compounds that are hygroscopic and those that are deliquescent. Many students may have seen rice in salt shakers at restaurants or at home. **Ask** What purpose does the rice serve? (*Rice acts as a desiccant.*) Explain why hygroscopic agents are preferred over deliquescent agents for use as desiccants, or drying agents. Emphasize the reversible properties of hydration. Point out that desiccants can be recycled by heating to drive off absorbed water. If possible, show students a desiccator, and explain why desiccators are used in the laboratory to store hygroscopic chemicals.

USE VISUALS Display Table 15.2 on an overhead projector. Review the nomenclature used to name salts and the use of prefixes to indicate the number of water molecules in the crystalline hydrate. **Ask** Do you know some of the familiar uses for the compounds in the table? *(Epsom salt is used as a laxative; Glauber's salt is used as a laxative or diuretic; alum is used as an astringent, as an emetic, and in the manufacture of baking powder, dyes, and paper; borax is used in the manufacture of glass, enamel, artificial gems, soaps, and antiseptics.)*

Answers

FIGURE 15.13 Heat it gently to drive off the water.

LESSON 15.2

Explain

Sample Practice Problem

Determine the percent by mass of water in calcium chloride dihydrate to the nearest tenth: $CaCl_2 \cdot 2H_2O$. *(24.5%)*

Extend

After students have studied Sample Problem 15.1, have them work in groups to find the percent by mass of water in the compounds listed in Table 15.2. *($MgSO_4 \cdot 7H_2O$, 51.1%; $Ba(OH)_2 \cdot 8H_2O$, 45.7%; $CaCl_2 \cdot 2H_2O$, 24.5%; $CuSO_4 \cdot 5H_2O$, 36.1%; $Na_2SO_4 \cdot 10H_2O$, 55.9%; $KAl(SO_4)_2 \cdot 12H_2O$, 45.5%; $Na_2B_4O_7 \cdot 10H_2O$, 47.2%; $FeSO_4 \cdot 7H_2O$, 45.3%; $H_2SO_4 \cdot H_2O$, 15.5%)*

Sample Problem 15.1

Finding the Percent by Mass of Water in a Hydrate

Calculate the percent by mass of water in washing soda, sodium carbonate decahydrate ($Na_2CO_3 \cdot 10H_2O$).

1 Analyze List the known and the unknown. To determine the percent by mass, divide the mass of water in the one mole of the hydrate by the molar mass of the hydrate and multiply by 100 percent.

KNOWN
formula of hydrate = $Na_2CO_3 \cdot 10H_2O$

UNKNOWN
percent by mass H_2O = ?%

2 Calculate Solve for the unknown.

For every 1 mol of $Na_2CO_3 \cdot 10H_2O$, there are 10 mol of H_2O.

Determine the mass of 10 mol of water.

$$\text{mass of 10 mol } H_2O = 10[(2 \times 1.0\,g) + 16.0\,g] = 180.0\,g$$

Determine the mass of 1 mol of the hydrated compound.

$$\text{mass of 1 mol } Na_2CO_3 \cdot 10H_2O = (2 \times 23.0\,g) + 12.0\,g + (3 \times 16.0\,g) + 180.0\,g = 286.0\,g$$

Calculate the percent by mass of water in the hydrate.

$$\text{percent by mass } H_2O = \frac{\text{mass of water}}{\text{mass of hydrate}} \times 100\% = \frac{180.0\,g}{286.0\,g} \times 100\% = 62.94\%$$

3 Evaluate Does the result make sense? The mass of the water accounts for more than half the molar mass of the compound, so a percentage greater than 50 percent is expected.

8. What is the percent by mass of water in $CuSO_4 \cdot 5H_2O$?

In Problem 8, start by determining the mass of 5 mol of water and 1 mol of the hydrate.

9. If you need 5.00 g of anhydrous Na_2CO_3 for your reaction, how many grams of $Na_2CO_3 \cdot 10H_2O$ could you use instead?

You know from the Sample Problem above that 62.94% of the hydrate is water, so 37.06 g out of every 100 g of the hydrate is Na_2CO_3.

Foundations for Math

ORDER OF OPERATIONS AND THE CALCULATOR Point out that finding the percent by mass of water in hydrates involves several uses of the order of operations. Students must use the correct order of operations each time they find the molar mass. If students use a calculator, encourage them to use grouping symbols when they enter the values in the calculator.

In **Sample Problem 15.1,** students should enter the calculation for the molar mass of Na_2CO_3 as (2 × 23) + 12 + (3 × 16) to ensure that the calculator follows the correct order of operations.

Deliquescent Compounds Have you ever noticed the small packets of silica gel that are often packaged with electronic equipment and leather goods? Although the structure of silica gel is not the same as a hydrated salt, it is a hygroscopic substance used to absorb moisture from the surrounding air to prevent damage to sensitive equipment and materials. Some compounds are so hygroscopic that they become wet when exposed to normally moist air. These compounds are **deliquescent,** which means that they remove sufficient water from the air to dissolve completely and form solutions. Figure 15.15 shows that pellets of sodium hydroxide are deliquescent. For this reason, containers of sodium hydroxide and other deliquescent chemicals should always be tightly stoppered and the chemicals should never come in contact with your skin. The solution formed by a deliquescent substance has a lower vapor pressure than that of the water in the air.

Figure 15.15 Sodium Hydroxide
Deliquescent substances, such as sodium hydroxide, can remove water from the air. **a.** Sodium hydroxide pellets absorb moisture from the air. **b.** Eventually a solution is formed.
Classify ***What is the solvent? What is the solute?***

(a)

(b)

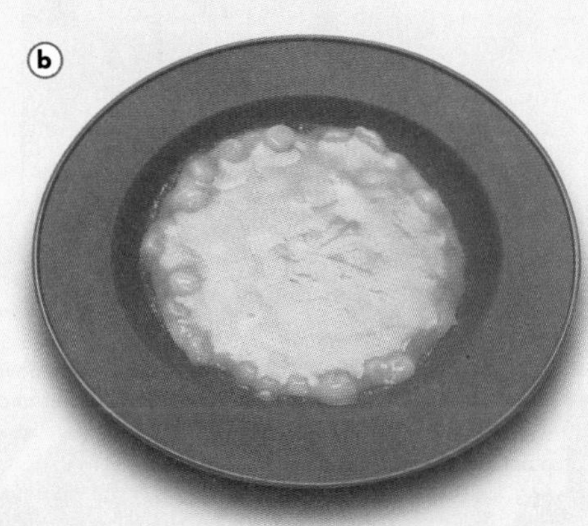

15.2 LessonCheck

ONLINE PROBLEMS

10. Identify What types of substances dissolve most readily in water?

11. Review What property of all ionic compounds make them electrolytes?

12. Explain Why do hydrates easily lose water when heated and regain water when exposed to moisture?

13. Classify Identify the solvent and the solute in vinegar, a dilute aqueous solution of acetic acid.

14. Calculate What is the percent by mass of water in iron(II) sulfate heptahydrate ($FeSO_4 \cdot 7H_2O$)?

15. Compare Is the percent by mass of copper in $CuSO_4 \cdot 5H_2O$ the same as in $CuSO_4$? Explain.

16. Compare and Contrast Distinguish between efflorescent and hygroscopic substances.

BIGIDEA BONDING AND INTERACTIONS

17. Which of the following substances dissolve to a significant extent in water? Explain your answer in terms of the interactions between the solvent and solute.

a. CH_4
b. KCl
c. I_2
d. $MgSO_4$
e. sucrose ($C_{12}H_{22}O_{11}$)
f. $NaHCO_3$

Lesson Check Answers

10. Ionic compounds and polar covalent compounds

11. They dissociate into ions.

12. The forces holding the water molecules in hydrates are not very strong, so the water is easily lost and regained.

13. The solute is acetic acid. The solvent is water.

14. 45.3%

15. The percents of copper in copper sulfate and copper sulfate pentahydrate are different because the molar masses of the two compounds are different.

16. Efflorescent compounds such as certain hydrates lose water to the air. Hygroscopic compounds remove moisture from air, sometimes forming hydrates.

17. **a.** insoluble, nonpolar
b. soluble, ionic
c. insoluble, nonpolar
d. soluble, ionic
e. soluble, polar
f. soluble, ionic

Evaluate

Informal Assessment

To assess students' understanding of the properties of aqueous solutions, show students the following activity and ask them to answer each question.

Dissolve a small amount of nickel(II) chloride in 500 mL of distilled water.

1. What is the solute and what is the solvent? *(Water is the solvent and nickel(II) chloride is the solute.)*

2. What type of electrolyte is nickel(II) chloride? How do you know? *(Nickel(II) chloride is an ionic compound and is soluble in water; therefore, it is a strong electrolyte.)*

3. What types of substances will dissolve in water to form aqueous solutions? *(ionic compounds and polar covalent molecules)*

Then have students complete the 15.2 Lesson Check.

Reteach

Project models of water molecules, cations, and anions on an overhead projector to show how water molecules orient their dipoles to solvate ions. Point out that cations and anions are attracted to different ends of the water molecule. Stress that the classification of water-soluble substances as strong electrolytes, weak electrolytes, or nonelectrolytes is determined by the relative number of ions in solution. Explain that nonelectrolytes do not produce ions.

Answers

8. 36.1%

9. 13.5 g $Na_2CO_3 \cdot 10H_2O$

FIGURE 15.15 The solvent is water and the solute is sodium hydroxide.

CHEMISTRY & YOU

CHEMISTRY & YOU Students should be familiar with the process of osmosis in biological systems and understand that water flows across a membrane *toward* an area of high concentration. In contrast, reverse osmosis requires water to flow *away* from an area of high concentration. Just as in a biological system, the process of moving water away from a highly concentrated solution toward a more dilute one requires an input of energy. In this case, a high-pressure differential is used to force the water through a semi-permeable membrane. It is this requirement of high pressure that creates a problem for reverse osmosis: the higher the efficiency of the process, the higher the pressure provided must be, thus increasing the energy costs.

Pose the following question to students: *Considering both its benefits and problems, why is reverse osmosis a highly desirable solution for water filtration?* You may need to assist students in the following ways:

- Reverse osmosis can be used to filter any water, not just saltwater.
- A reasonably high efficiency of reverse osmosis can be achieved only in large filtration plants, which can provide the high pressure needed.
- Reverse osmosis can be performed on a small scale, even in faucet and aquarium filters.
- Membranes used for reverse osmosis must be frequently replaced.

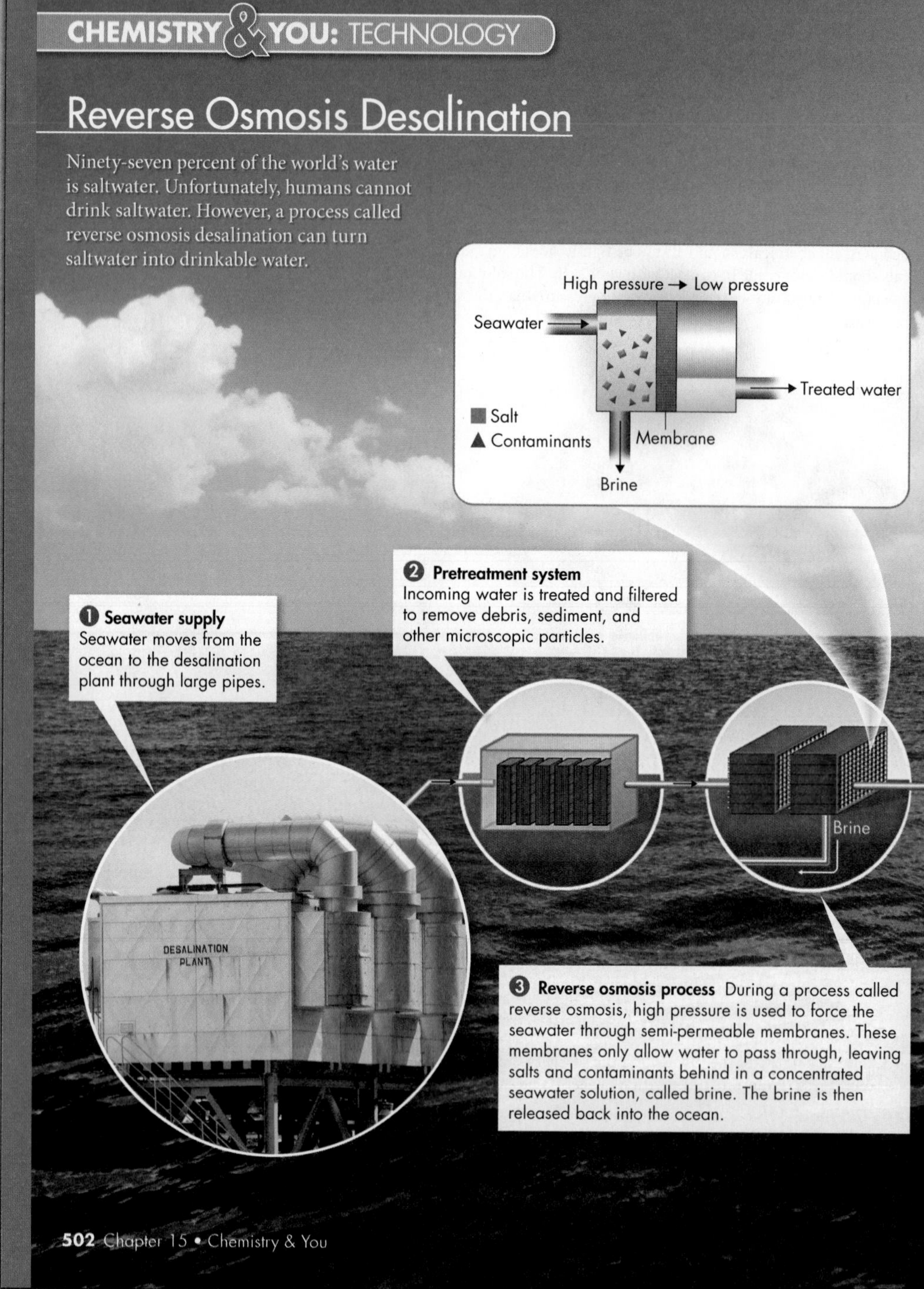

21st Century Learning To be successful in the 21st century, students need skills and learning experiences that extend beyond subject matter mastery. The following project helps students build the following 21st Century Skills: *Financial, Economic, Business, and Entrepreneurial Literacy; Creativity and Innovation; Critical Thinking and Problem Solving; Communication and Collaboration; Information Literacy;* and *Initiative and Self-Direction.*

Pose the following challenge to students: *A local coastal government is trying to decide between distillation and reverse osmosis for a city's new water filtration plant. Form a team of 3–5 students to represent a local citizens' group that favors reverse osmosis over distillation, and prepare an advocacy report supporting your views on this technology.* Presentations should be prepared and submitted electronically and should include a suggested site for the plant. Students should include information on the pros and cons of the technology, its potential environmental impact, and its economic feasibility.

Pros & Cons

Advantages of Reverse Osmosis Desalination	Disadvantages of Reverse Osmosis Desalination
✓ **Expands drinking water resources** Some areas of the world do not have abundant sources of drinking water. Reverse osmosis desalination provides drinking water in cases of drought, water shortages, or national disasters.	✗ **High energy consumption** The cost of running a reverse osmosis desalination plant is high compared to other methods for obtaining drinking water.
✓ **Low environmental impact** The conversion of saltwater to drinking water via reverse osmosis desalination uses minimal chemicals.	✗ **Low efficiency** The volume of drinking water produced is low compared to the volume of seawater treated.
✓ **Low start-up costs** The cost to install a reverse osmosis desalination facility is relatively low.	✗ **Potentially harmful to sea life** The process of removing seawater and returning brine to the ocean can disturb marine environments.

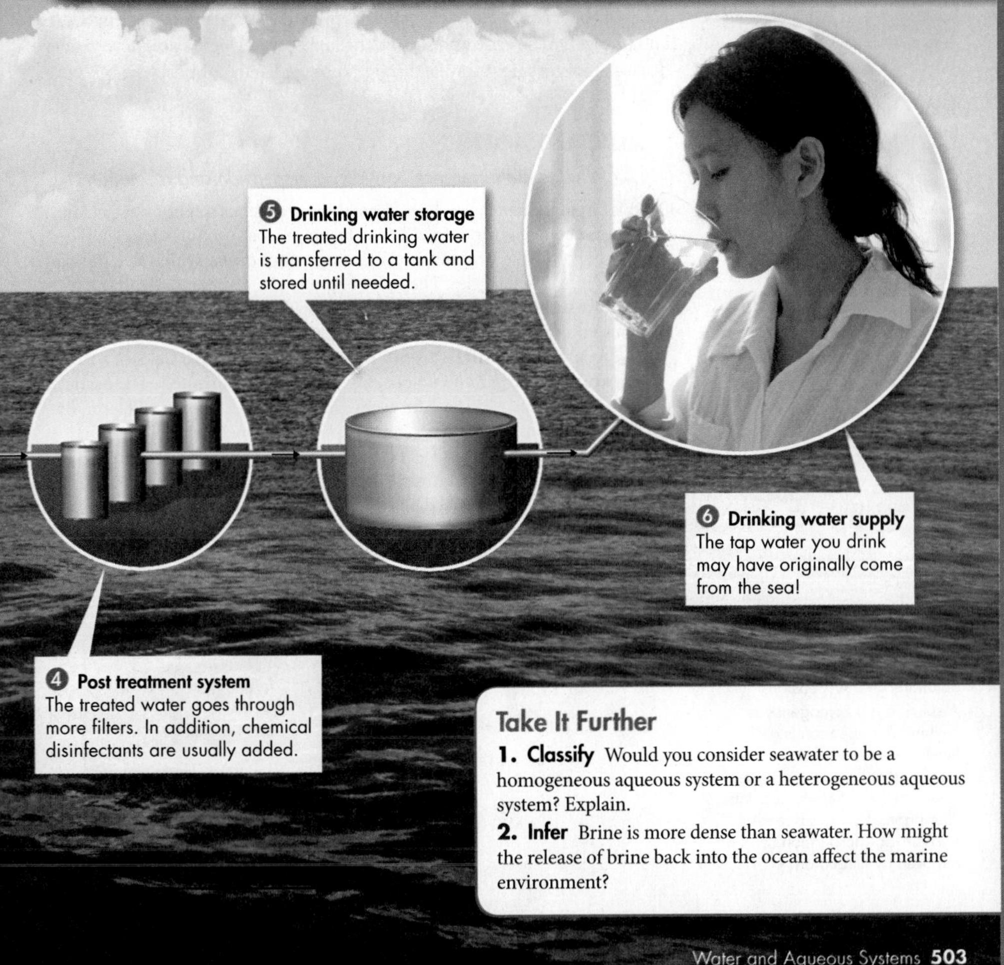

5 Drinking water storage The treated drinking water is transferred to a tank and stored until needed.

6 Drinking water supply The tap water you drink may have originally come from the sea!

4 Post treatment system The treated water goes through more filters. In addition, chemical disinfectants are usually added.

Take It Further

1. Classify Would you consider seawater to be a homogeneous aqueous system or a heterogeneous aqueous system? Explain.

2. Infer Brine is more dense than seawater. How might the release of brine back into the ocean affect the marine environment?

Explain

MAKE A CONNECTION One of the key difficulties in setting up reverse osmosis facilities is the high pressure required to efficiently process saltwater. Remind students that the principle of osmosis is ultimately the reason for the requirement of a high-pressure differential. In a situation where the solute concentrations are drastically different on the two sides of a membrane, there is naturally a relatively high pressure exerted toward the higher concentration. This causes difficulty during the process of reverse osmosis because not only must the water be forced through the membrane but the system must also overcome the natural pressure of osmosis.

Extend

Connect to ENVIRONMENTAL SCIENCE

The principal environmental concern from reverse osmosis desalination is the fact that an inevitable byproduct of the process is a concentrated mix of salts and impurities known as brine. The brine has a much higher salinity than ocean water and is therefore poisonous to marine wildlife. It is especially harmful to bottom-dwelling animals because brine sinks due to its higher solute concentration. Have students brainstorm potential ways in which the environmental impact due to the release of brine could be minimized.

Differentiated Instruction

L1 STRUGGLING STUDENTS Have students work in small groups to produce a poster demonstrating the process by which a desalination plant filters saltwater. Posters should include the saltwater intake, flow through the semi-permeable membrane, and water and brine outflow.

LPR LESS PROFICIENT READERS Help students understand the term *desalination* by explaining the connection between the term *saline* and saltwater. Point out that *desalination* can be referred to as the process of "de-salting" water.

L3 ADVANCED STUDENTS Divide students into pairs. Have the pairs research how desalination has evolved over the years and what advantages reverse osmosis desalination has over earlier methods. Have students present their findings to the class. Encourage creativity in the presentation medium.

Answers

1. Seawater is a homogeneous aqueous system because it is an aqueous solution.
2. Since brine is more dense than seawater, it will sink to the bottom of the ocean. The brine could affect marine life that live on the ocean floor.

Key Objectives

15.3.1 DISTINGUISH between a suspension and a solution.

15.3.2 IDENTIFY how to distinguish a colloid from a suspension and a solution.

Additional Resources

- Reading and Study Workbook, Lesson 15.3
- Core Teaching Resources, Lesson 15.3 Review
- Small-Scale Chemistry Laboratory Manual, Lab 24

PearsonChem.com

Engage

CHEMISTRY & YOU Have students study the photograph and read the text. Comment that the atmosphere is referred to as a homogeneous mixture called a colloid. **Ask** What does that tell you about the composition of the atmosphere? *(Its composition is not uniform throughout.)* **Ask** If a colloid such as the atmosphere differs from a solution with respect to the size of its particles, would you expect a colloid to have smaller or larger particles than a solution? *(larger)*

Activate Prior Knowledge

Review solutions and the solution process. Ask students to explain why the particles in a homogeneous aqueous system can pass through a filter, and have them predict whether the same can happen when filtering heterogeneous systems.

Foundations for Reading

BUILD VOCABULARY Have students make a Venn diagram using the following vocabulary terms: *solution, suspension, colloid, Tyndall effect, Brownian motion.*

IN ACADEMIC STANDARDS for SCIENCE

C.7.1 Describe the composition and properties of types of solutions.

15.3 Heterogeneous Aqueous Systems

C.7.1 Composition and properties of solutions

CHEMISTRY & YOU

Q: *Why are some sunsets red?* Have you ever wondered what causes the red glow of the evening sky? The atmosphere contains particles of water and dust. As sunlight passes through the particles, it is scattered. However, not all wavelengths are scattered to the same extent. The shorter wavelengths of visible light (blue and green) are scattered more than the longer wavelengths (red and orange). At sunrise and sunset, the longer wavelengths are more visible because the sun's light travels through more of Earth's atmosphere.

Key Questions

What is the difference between a suspension and a solution?

What distinguishes a colloid from a suspension and a solution?

Vocabulary

- suspension
- colloid
- Tyndall effect
- Brownian motion
- emulsion

Suspensions

What is the difference between a suspension and a solution?

So far in this chapter, you have learned about aqueous solutions, which are homogeneous mixtures. In contrast, heterogeneous mixtures are not solutions. If you shake a container containing a piece of clay with water, the clay breaks into fine particles. The water becomes cloudy because the clay particles are suspended in the water. If you stop shaking, the particles settle out. A **suspension** is a mixture from which particles settle out upon standing. **A suspension differs from a solution because the particles of a suspension are much larger and do not stay suspended indefinitely.** The particles in a typical suspension have an average diameter greater than 1000 nm. By contrast, the particle size in a solution is usually about 1 nm.

Suspensions are heterogeneous because at least two substances can be clearly identified. In the example of clay particles mixed with water, you can clearly see the dispersed phase (clay) in the dispersion medium (water). Figure 15.16 shows how the difference between a solution and suspension is easily seen when each type of mixture is filtered.

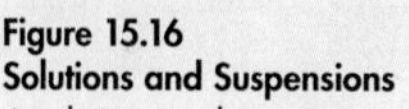

Figure 15.16
Solutions and Suspensions
A solution is a homogeneous mixture. A suspension is a heterogeneous mixture. **a.** The small size of the solute particles in a solution allows them to pass through filter paper. **b.** The suspended particles of a suspension can be removed by filtration.

Focus on ELL

1 CONTENT AND LANGUAGE Have students use a dictionary to look up the various uses of the word *disperse* or *dispersion*. Ask students to explain how the seeds of a plant being dispersed by the wind are related to the dispersed phase and dispersion medium of a colloid.

2 FRONTLOAD THE LESSON Provide students with a partially completed compare-contrast table for heterogeneous solutions, and ask them to complete the table as they read the lesson. Students should then use the self-questioning strategy to note the similarities and differences between the solutions.

3 COMPREHENSIBLE INPUT Provide small groups of students samples of saltwater, oil-and-vinegar salad dressing, mayonnaise, and grape jelly to compare macroscopically and microscopically. Assign students the online Concepts in Action activity on emulsions.

504 Chapter 15 • Lesson 3

Colloids

What distinguishes a colloid from a suspension and a solution?

Gelatin is an example of a type of mixture called a colloid. A **colloid** is a heterogeneous mixture containing particles that range in size from 1 nm to 1000 nm. The particles are spread, or dispersed, throughout the dispersion medium, which can be a solid, liquid, or a gas. The first substances to be identified as colloids were glues. Other colloids include such mixtures as paint, aerosol sprays, and smoke. Table 15.3 lists some common colloidal systems and gives examples of familiar colloids.

How do the properties of colloids differ from those of suspensions and solutions? Many colloids are cloudy or milky in appearance, like suspensions, when they are concentrated. Colloids may look clear or almost clear, like solutions, when they are dilute. The important difference between colloids and solutions and suspensions is in the size of the particles. **Colloids have particles smaller than those in suspensions and larger than those in solutions.** These intermediate-sized particles cannot be retained by filter paper as are the larger particles of a suspension, and they do not settle out with time. Colloids can be distinguished by a phenomenon called the Tyndall effect and by the observation of Brownian motion. They are also subject to coagulation or clumping together, and they can be emulsified or made stable.

Table 15.3

Some Colloidal Systems

System			
Dispersed phase	Dispersion medium	Type	Example
Gas	Liquid	Foam	Whipped cream
Gas	Solid	Foam	Marshmallow
Liquid	Liquid	Emulsion	Milk, mayonnaise
Liquid	Gas	Aerosol	Fog, aerosol
Solid	Gas	Smoke	Dust in air
Solid	Liquid	Sols, gels	Egg white, jelly, paint, blood, starch in water, gelatin

UbD Understanding by Design

What is the difference between a suspension and a solution?

Assess students' knowledge about heterogeneous aqueous systems by asking them to explain how to tell if a sample of an aqueous system is homogeneous or heterogeneous.

ADJUST INSTRUCTION If students are having difficulty distinguishing between the two systems, have them review the Solutions section in Lesson 15.2 and the Suspensions section in Lesson 15.3. Then have students revise their answers.

READING STRATEGY Before students begin to read this lesson, have them preview the pages and predict how the properties of suspensions and colloids differ from those of solutions. Then, as they read, have them correct any misconceptions they may have.

Explain

Suspensions

APPLY CONCEPTS Explain to students that municipal wastewater treatment plants take advantage of the properties of suspensions during the water treatment process. Point out that sewage is a suspension rather than a solution. This allows treatment facilities to use sedimentation to remove larger particles. Then, lime and alum are added to help coagulate smaller particles before the remaining liquid is filtered through sandy loams (diatomaceous earth). This process removes the remaining suspended particles. The diatomaceous earth may also absorb colloidal material.

Colloids

USE VISUALS Display Table 15.3 on an overhead projector. Explain that colloids are characterized by the physical state of the *dispersed phase* and of the *continuous phase*—also called the *dispersion medium*. **Ask** What is the dispersion medium in fog? *(air)* **Ask** What is dispersed in fog? *(water droplets)* Help students understand this two-part system by comparing it to a solution, which, instead of a dispersed phase and a dispersion medium, consists of a solute and a solvent.

Extend

Explain that homogenized milk is a colloid of protein and fat. Curdling occurs when bacteria produce enough lactic acid to cause dispersed protein and fat particles to coagulate into larger particles, which then separate from the rest of the mixture. Have students write an analogy to explain how this process is comparable to the clotting of blood.

Explain

START A CONVERSATION Point out that the main difference between solutions, suspensions, and colloids is particle size. Solution particles are typically less than 1 nm in diameter. Colloid particles are between 1 nm and 1000 nm in diameter. Suspension particles are typically larger than 1000 nm. Smaller particles are less susceptible to the effects of gravity and are influenced more by the effects of Brownian motion. The collisions of molecules with extremely small colloidal particles are sufficiently energetic to move colloidal particles in a random fashion that prevents their settling to the bottom.

Explore

The Tyndall Effect

Teacher Demo

PURPOSE Students will observe the Tyndall effect.

MATERIALS whole milk, water, beaker, stirring rod, projector or laser pointer

PROCEDURE Add a small amount of whole milk to some water in a beaker, and stir to mix. In a darkened room, shine a light from a projector or a laser pointer through the beaker.

SAFETY If a laser pointer is used, warn students not to look directly at it.

EXPECTED OUTCOME When viewed from the side, the path of the light beam is observed. This is the Tyndall effect. Students also can observe Brownian motion of airborne dust particles in the light beam.

CHEMISTRY & YOU You will have a better chance of seeing a red sunset if the air contains a lot of water and dust particles. That way more of the shorter wavelengths of light will be scattered from the sunlight leaving the the sky with a reddish appearance.

Figure 15.17 Scattering of Light
The path of light is visible only when the light is scattered by particles. **a.** Fog or mist is a colloid and thus exhibits the Tyndall effect. **b.** Solutions do not scatter light. Particles in colloids and suspensions reflect or scatter light in all directions.
Explain ***Why is it easier to see the light beam of an automobile on a foggy night than on a clear night?***

The Tyndall Effect Ordinarily you cannot see a beam of sunlight unless the light passes through particles of water (mist) or dust in the air. These particles scatter the sunlight. Similarly, a beam of light is visible as it passes through a colloid. The scattering of visible light by colloidal particles is called the **Tyndall effect.** Suspensions also exhibit the Tyndall effect. Solutions do not exhibit the Tyndall effect. The particles in solutions are too small to scatter light. Figure 15.17 shows how the Tyndall effect can differentiate solutions from colloids and suspensions.

Brownian Motion Flashes of light, or scintillations, are seen when colloids are studied under a microscope. Colloids scintillate because the particles reflecting and scattering the light move erratically. The chaotic movement of colloidal particles, which was first observed by the Scottish botanist Robert Brown (1773–1858), is called **Brownian motion.** Brownian motion is caused by collisions of the molecules of the dispersion medium with the small, dispersed colloidal particles. These collisions help prevent the colloidal particles from settling.

CHEMISTRY & YOU

Q: *What would be the ideal conditions to see a red sunset?*

Coagulation Colloidal particles also tend to stay suspended because they become charged by adsorbing ions from the dispersing medium onto their surface. *Adsorption* means to adhere to a surface. Some colloidal particles become positively charged by adsorbing positively charged ions. Other colloidal particles become negatively charged by adsorbing negatively charged ions. All the colloidal particles in a particular colloidal system will have the same charge, although the colloidal system is neutral. The repulsion between the like-charged particles prevents the particles from forming heavier aggregates that would have a greater tendency to settle out. Thus, a colloidal system can be destroyed or coagulated by the addition of electrolytes. The added ions neutralize the charged colloidal particles. The particles can clump together to form heavier aggregates and settle out from the dispersion.

Paint

Students may be interested in knowing why paint must be stirred before it is used. Explain that latex paint is a complex mixture of binders, pigments, and drying agents. Some ingredients are water soluble; others are present as dispersed particles. For uniform color and composition, the paint must be stirred well before being used.

Emulsions Mayonnaise is an example of a colloidal system called an emulsion. An **emulsion** is a colloidal dispersion of a liquid in a liquid. An emulsifying agent is essential for the formation of an emulsion and for maintaining the emulsion's stability. For example, oils and greases are not soluble in water. However, oils and greases readily form a colloidal dispersion if soap or detergent is added to the water. Soaps and detergents are emulsifying agents. One end of a large soap or detergent molecule is polar and is attracted to water molecules. The other end of the soap or detergent molecule is nonpolar and is soluble in oil or grease. Soaps and other emulsifying agents thus allow the formation of colloidal dispersions between liquids that do not ordinarily mix. Mayonnaise is a heterogeneous mixture of oil and vinegar. Such a mixture would quickly separate without the presence of egg yolk, which is the emulsifying agent. Other foods such as milk, margarine, and butter are also emulsions. Cosmetics, shampoos, and lotions are formulated with emulsifiers to maintain consistent quality. Table 15.4 summarizes the properties of solutions, colloids, and suspensions.

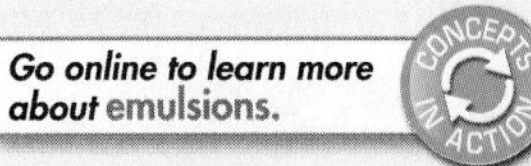

Table 15.4

Properties of Solutions, Colloids, and Suspensions

	System		
Property	**Solution**	**Colloid**	**Suspension**
Particle type	Ions, atoms, small molecules	Large molecules or particles	Large particles or aggregates
Particle size	0.1–1 nm	1–1000 nm	1000 nm and larger
Effect of light	No scattering	Exhibits Tyndall effect	Exhibits Tyndall effect
Effect of gravity	Stable, does not separate	Stable, does not separate	Unstable, sediment forms
Filtration	Particles not retained on filter	Particles not retained on filter	Particles retained on filter
Uniformity	Homogeneous	Heterogeneous	Heterogeneous

15.3 LessonCheck

ONLINE PROBLEMS

18. Describe How does a suspension differ from a solution?

19. Explain What distinguishes a colloid from a suspension and a solution?

20. Apply Concepts How can you determine through observation that a mixture is a suspension?

21. Explain Could you separate a colloid by filtering? Explain.

22. Infer How can the Tyndall effect be used to distinguish between a colloid and a solution?

23. Relate Cause and Effect Can the presence of Brownian motion distinguish between a solution and a colloid? Explain.

Lesson Check Answers

18. The particles of a suspension are much larger and do not stay suspended indefinitely.

19. Colloids have particles smaller than those in suspensions and larger than those in solutions.

20. The particles in a suspension will settle out over time.

21. Particles in a colloid such as gelatin are smaller than the holes in filter paper and cannot be removed by filtering.

22. A beam of light is visible as it passes through a colloid; a beam of light is invisible as it passes through a solution.

23. Flashes of light, or scintillations, are seen when colloids are studied under a microscope. Colloids scintillate because the particles move erratically. The particles in a solution are too small to be seen under a microscope and do not cause scintillations.

Evaluate

Informal Assessment

Ask In what way are colloids similar to solutions? *(In both, dispersed particles are small enough to pass through standard filter paper and to withstand the pull of gravity.)* **Ask** In what way are colloids similar to suspensions? *(Both types of mixtures produce the Tyndall effect.)* Then have students complete the 15.3 Lesson Check.

Reteach

On the board, draw the relative sizes of solute, colloid, and suspension particles by making an analogy to golf balls, baseballs, and basketballs. Ask students to identify which "particles" would get caught in a sieve with holes a bit smaller than a basketball hoop and which particles would pass through. Remind students that solution particles are typically less than 1 nm in diameter, whereas colloid particles are ten to one thousand times as large and those of suspensions exceed the largest colloid particles.

Answers

FIGURE 5.17 Fog is a colloid that produces the Tyndall effect. Bright lights produce a higher degree of light scattering in all directions, including straight back into the driver's eyes.

SMALL-SCALE LAB

Small-Scale Lab

OBJECTIVE After completing this activity, students will be able to classify compounds as electrolytes or nonelectrolytes by testing them for conductivity.

ADDRESS MISCONCEPTIONS Students often think that a substance will conduct electricity if it dissolves in water. Tell students that some covalent compounds dissolve in water but do not conduct electricity.

PREP TIME 45 minutes

CLASS TIME 20 minutes

ADVANCE PREPARATION Make up the reagents:

Solution	Preparation
1.0*M* HCl	82 mL of 12*M* in 1.0 L
1.0*M* HNO_3	63 mL of 15.8*M* in 1.0 L
1.0*M* H_2SO_4	56 mL of 18*M* in 1.0 L

CAUTION: *Always add acid to water.*

1.0*M* NH_3	67 mL of 15*M* in 1.0 L

CAUTION: *Using proper ventilation, add ammonia to water.*

CH_3COOH	Use white vinegar.
0.5*M* NaOH	20.0 g in 1.0 L

Put out table sugar, cornstarch, and various liquids such as rubbing alcohol, distilled water, soft drinks, orange juice, pickle juice, and coffee.

SAFETY Be sure students wear goggles and aprons.

TEACHING TIP Ask students to bring in small samples of orange juice and other liquid foods.

EXPECTED OUTCOME None of the compounds conduct electricity when in the solid state. Aqueous solutions of NaCl, $MgSO_4$, Na_2CO_3, $NaHCO_3$, KCl, and KI conduct electricity and are thus electrolytes. Cornstarch and table sugar do not conduct electricity and are nonelectrolytes.

ANALYZE AND CONCLUDE

1. Electrolytes: NaCl, $MgSO_4$, Na_2CO_3, $NaHCO_3$, KCl, KI; nonelectrolytes: table sugar, cornstarch
2. no, because the ions are locked in a crystal lattice and cannot move
3. Table sugar and cornstarch are covalent compounds. NaCl, $MgSO_4$, Na_2CO_3, $NaHCO_3$, KCl, and KI are ionic compounds. An electrolyte is a compound that dissociates into ions in solution.
4. It must conduct electricity.

YOU'RE THE CHEMIST

1. $NaHCO_3(s) \rightarrow Na^+(aq) + HCO_3^-(aq)$
   $KCl(s) \rightarrow K^+(aq) + Cl^-(aq)$
   $MgSO_4(s) \rightarrow Mg^{2+}(aq) + SO_4^{2-}(aq)$
   $KI(s) \rightarrow K^+(aq) + I^-(aq)$
   Diagrams should be similar to Figure 15.10.
2. Test each solution. Strong electrolytes: HCl, H_2SO_4, HNO_3, NaOH; Weak electrolytes: CH_3COOH, NH_3; Nonelectrolytes: rubbing alcohol, distilled water
3. Strong electrolytes: soft drinks, pickle juice Weak electrolytes: orange juice, coffee

Electrolytes

Purpose

To classify compounds as electrolytes by testing their conductivity in aqueous solution

Materials

- reaction surface
- chemicals shown in the grid below
- conductivity tester
- micropipet or dropper
- water

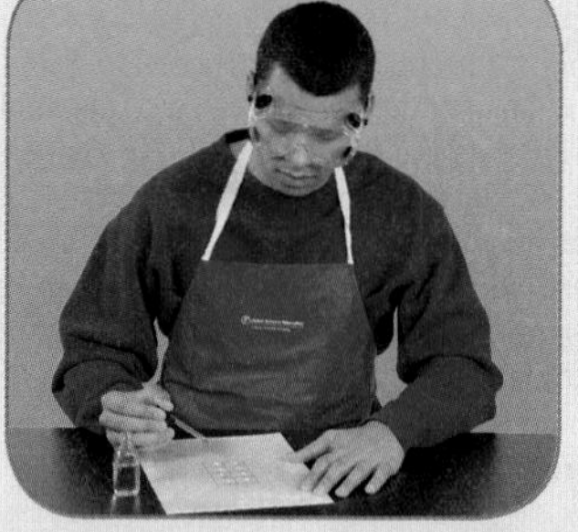

Procedure

1. On separate sheets of paper, draw two grids similar to the one below. Make each square 2 cm on each side.
2. Place a reaction surface over one of the grids and place a few grains of each solid in the indicated places.
3. Test each solid for conductivity.
4. Add 1 drop of water to each solid and test the wet mixture for conductivity. Be sure to clean and dry the conductivity leads between each test.

NaCl(s)	$MgSO_4(s)$
$Na_2CO_3(s)$	Table sugar $C_{12}H_{22}O_{11}$
$NaHCO_3(s)$	Cornstarch $(C_6H_{10}O_5)n$
KCl(s)	KI(s)

Analyze and Conclude

1. **Infer** Which compounds in your table are electrolytes? Which are not electrolytes?
2. **Observe** Do any of these electrolytes conduct electric current in the solid form? Explain.
3. **Classify** Identify each compound in the grid as ionic or covalent.
4. **Draw Conclusions** For a compound to be an electrolyte, what must happen when it dissolves in water?

You're the Chemist

1. **Analyze Data** When an ionic solid dissolves in water, water molecules attract the ions, causing them to come apart, or dissociate. The resulting dissolved ions are electrically charged particles that allow the solution to conduct electric current. The following chemical equations represent this phenomenon.

$$NaCl(s) \xrightarrow{H_2O} Na^+(aq) + Cl^-(aq)$$

$$Na_2CO_3(s) \xrightarrow{H_2O} 2Na^+(aq) + CO_3^{2-}(aq)$$

Write a similar chemical equation for each electrolyte you tested. Draw diagrams to explain how the ions conduct electric current.

2. **Design an Experiment** Obtain the following aqueous solutions: HCl, H_2SO_4, HNO_3, $C_2H_4O_2$, NH_3, NaOH, rubbing alcohol, and distilled water. Design and carry out an experiment to test their conductivities. Use your data to classify each substance as a strong electrolyte, weak electrolyte, or nonelectrolyte.
3. **Design an Experiment** Test various liquids for conductivity. Try soft drinks, orange juice, pickle juice, and coffee. Which liquids are electrolytes?

Focus on ELL

4 LANGUAGE PRODUCTION Have students work in groups of four to complete the lab. Make sure each group has ELLs of varied language proficiencies so that more proficient students can help less proficient ones. Have students work according to their proficiency level.

BEGINNING: LOW/HIGH Provide students with a checklist for each solid. For example, if the solid is conductive, they should check yes.

INTERMEDIATE: LOW/HIGH Break down each step in the procedure into smaller steps with fewer words.

ADVANCED: LOW/HIGH Have students help classmates of lower English proficiencies with following the procedures and drawing conclusions.

15 Study Guide

BIGIDEA
BONDING AND INTERACTIONS

Water molecules are held together through hydrogen bonds. The hydrogen bonding interactions between water molecules accounts for the unique properties of water including its high surface tension, low vapor pressure, and high boiling point. Hydrogen bonding also accounts for the fact that ice is less dense than liquid water. Ionic compounds and polar covalent compounds dissolve most readily in water to form aqueous solutions. Ionic compounds dissolve in water when the polar water molecules attract the ions of the solute, causing the individual solute ions to break away from the ionic crystal.

15.1 Water and Its Properties

Many unique and important properties of water—including its high surface tension, low vapor pressure, and high boiling point—result from hydrogen bonding.

The structure of ice is a regular open framework of water molecules in a hexagonal arrangement.

- surface tension (490)
- surfactant (490)

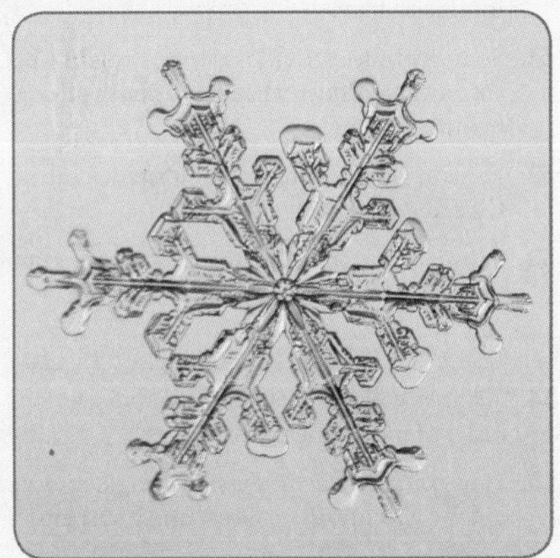

15.2 Homogeneous Aqueous Systems

Substances that dissolve most readily in water include ionic compounds and polar covalent compounds.

All ionic compounds are electrolytes because they dissociate into ions.

The forces holding the water molecules in hydrates are not very strong, so the water is easily lost and regained.

- aqueous solution (494)
- solvent (494)
- solute (494)
- solvation (495)
- electrolyte (496)
- nonelectrolyte (496)
- strong electrolyte (497)
- weak electrolyte (497)
- water of hydration (498)
- hydrate (498)
- anhydrous (498)
- effloresce (499)
- hygroscopic (499)
- desiccant (499)
- deliquescent (501)

Key Equation

$$\text{Percent by mass } H_2O = \frac{\text{mass of water}}{\text{mass of hydrate}} \times 100\%$$

15.3 Heterogeneous Aqueous Systems

A suspension differs from a solution because the particles of a suspension are much larger and do not stay suspended indefinitely.

Colloids have particles smaller than those in suspensions and larger than those in solutions.

- suspension (504)
- colloid (505)
- Tyndall effect (506)
- Brownian motion (506)
- emulsion (507)

Focus on ELL

5 ASSESS UNDERSTANDING After students complete the Standardized Test Prep, have them review the topics they scored poorly on in the test.

BEGINNING: LOW/HIGH Have students underline the high-frequency words they understand and use context clues to eliminate distractors.

INTERMEDIATE: LOW/HIGH Explain to students that they can skip a question that they do not understand immediately. Show them how to mark it so they can come back to it. Remind them to also skip that answer space on the answer sheet.

ADVANCED: LOW/HIGH Ask students to write out each question they answered incorrectly and to explain in writing why they missed the question and where to find the correct information.

Evaluate

Review and Assessment Materials

CHEMISTRY TUTORIAL Encourage students to review finding the percent of water in a hydrate using the animated tutorial for Chapter 15. Each Chemistry Tutorial is followed by a practice problem.

CHAPTER PROBLEM SET Have students practice more problems about the properties of water using the Chapter 15 online Problem Set.

DIRECTED VIRTUAL LAB Have students complete the Virtual Lab *The Conductivity of Electrolyte Solutions* to help reinforce the concepts of ionic compounds and electrolytes. In the post-lab exercise, students will analyze the data and draw conclusions.

KINETIC ART Direct students to the Kinetic Art animations *Hydrogen Bonding* and *Solvation of Ionic Compounds*, and have them answer the questions that accompany each animation.

CONCEPTS IN ACTION If students have trouble understanding properties of emulsions, ask them to view the Concepts in Action animation, which presents the material in a familiar context.

Study Tip

ORGANIZE NEW INFORMATION Suggest that students review the chapter content by using each of the vocabulary terms correctly in a sentence. Check their sentences for errors. Then, have students rewrite their sentences as fill-in-the-blank questions by leaving out the vocabulary terms. Students can use the questions to quiz classmates.

UbD

Understanding by Design

Performance Tasks

OH, THE SUSPENSE! Have each student hypothesize whether plain yogurt is a colloid, a suspension, or a solution. Have them develop a method to test the hypothesis and then write an illustrated report detailing the hypothesis, materials, methods, results, and conclusions.

WALKING ON WATER Have students research which animals have the ability to walk on the surface of water. Tell them to explain why those animals have such ability and humans do not. Students should refer to surface tension in their answers and should present their findings to the class by using a slide-show presentation.

ASSESSMENT

Evaluate

Answers

LESSON 15.1

24. The hydrogen atoms have a slightly positive charge. The oxygen atom has a slightly negative charge.

25. Surface molecules are attracted to the liquid molecules below but not to the air. Molecules inside the liquid are attracted in all directions.

26. Strong hydrogen bonding causes an inward pull that tends to minimize the surface area.

27. Drops are spherical; objects denser than a liquid can float on its surface.

28. A surfactant is a wetting agent such as soap or detergent. A surfactant interferes with hydrogen bonding between water molecules and reduces surface tension.

29. Hydrogen bonds hold water molecules to each other, so the tendency for them to escape the solution is low.

30. Water has a relatively low vapor pressure.

31. Water has low vapor pressure.

32. Ice is less dense than liquid water. The molecules in ice are farther apart in ice than in liquid water.

33. Hydrogen bonds between water molecules hold the molecules in a regular, open structure.

34. Ice is a regular open framework of hydrogen-bonded water molecules arranged like a honeycomb. This structure collapses in liquid water.

35. Bodies of water would freeze from the bottom up. This would kill many forms of aquatic life.

LESSON 15.2

36. An aqueous solution is a solution that has water as the solvent.

37. The solvent is the part of the solution present in the largest amount. It is the dissolving medium. The solute is the dissolved particles.

38. Solvent: water; solute: sugar

39. No; the molecules and ions are smaller than the pores of the filter, and would therefore pass through the filter.

40. Solvent molecules surround positively charged and negatively charged ions.

41. Polar water molecules electrostatically attract ions and polar covalent molecules, but nonpolar compounds are unaffected because they have no charges.

42. Water is polar, and gasoline is nonpolar.

43. **a.** HCl (polar) dissolves.
b. K_2SO_4 (ionic) dissolves
c. NaI (ionic) dissolves.
d. C_2H_6 (nonpolar) will not dissolve.
e. NH_3 (polar) dissolves.
f. $CaCO_3$ (strong ionic forces) will not dissolve.

44. ions

45. Its ions are free to move toward positively and negatively charged electrodes.

46. A strong electrolyte is almost totally ionized.

47. water in a crystal structure of a substance

48. **a.** $Na_2SO_4 \cdot 10H_2O$ **b.** $CaCl_2 \cdot 2H_2O$ **c.** $Ba(OH)_2 \cdot 8H_2O$

49. **a.** tin(IV) chloride pentahydrate
b. iron(II) sulfate heptahydrate
c. barium bromide tetrahydrate
d. iron(III) phosphate tetrahydrate

50. $MgSO_4 \cdot 7H_2O \rightarrow MgSO_4 \cdot H_2O + 6H_2O$

51. Efflorescence is the loss of water of hydration that occurs when the hydrate has a higher vapor pressure than that of the water vapor in air.

52. Hygroscopic substances absorb water vapor from the air and create a dry environment in a sealed container.

53. They absorb water vapor from the air.

LESSON 15.3

54. solutions, colloids, suspensions

55. Colloids and suspensions exhibit the Tyndall effect but solutions do not. The particles in a suspension will settle out over time.

56. scattering of visible light by colloids and suspensions

15 Assessment

★ Solutions appear in Appendix E

Lesson by Lesson

15.1 Water and Its Properties

24. Explain why water molecules are polar.

★**25.** Why do the particles at the surface of a liquid behave differently from those in the bulk of the liquid?

26. Why does water have a high surface tension?

27. Describe some observable effects that are produced by the surface tension of a liquid.

★**28.** What is a surfactant? Explain how it works.

29. How can the unusually low vapor pressure of water be explained?

30. Explain why water has a relatively high boiling point.

★**31.** Explain why bodies of water with large surface areas such as lakes and oceans do not evaporate rapidly.

32. What characteristic of ice distinguishes it from other solid substances?

33. Explain the role of hydrogen bonds in ice.

34. How does the structure of ice differ from the structure of water?

★**35.** What would be some of the consequences if ice were denser than water?

15.2 Homogeneous Aqueous Systems

36. Distinguish between a solution in general and an aqueous solution.

37. In the formation of a solution, how does the solvent differ from the solute?

★**38.** Identify the solvent and the solute in a solution of table sugar in water.

★**39.** Suppose an aqueous solution contains both table sugar and table salt. Can you separate either of these solutes from the water by filtration? Explain your reasoning.

★**40.** Describe the process of solvation.

41. Why is water an excellent solvent for most ionic and polar covalent compounds but not for nonpolar compounds?

42. Explain why gasoline does not dissolve in water.

★**43.** Which of the following substances dissolve appreciably in water? Give reasons for your choices.

a. HCl
b. K_2SO_4
c. NaI
d. C_2H_6
e. NH_3
f. $CaCO_3$

44. What particles must be present in a solution if it is to conduct an electric current?

★**45.** Why does molten sodium chloride conduct an electric current?

46. What is the main distinction between an aqueous solution of a strong electrolyte and an aqueous solution of a weak electrolyte?

47. What is meant by a substance's water of hydration?

★**48.** Write formulas for these hydrates.

a. sodium sulfate decahydrate
b. calcium chloride dihydrate
c. barium hydroxide octahydrate

★**49.** Name each hydrate.

a. $SnCl_4 \cdot 5H_2O$
b. $FeSO_4 \cdot 7H_2O$
c. $BaBr_2 \cdot 4H_2O$
d. $FePO_4 \cdot 4H_2O$

★**50.** Epsom salt ($MgSO_4 \cdot 7H_2O$) changes to the monohydrate form at 150°C. Write an equation for this change.

51. Some hydrates are efflorescent. Explain what that means. Under what conditions will a hydrate effloresce?

★**52.** Explain why a hygroscopic substance can be used as a desiccant.

53. Why is it important to keep some hygroscopic substances in tightly sealed containers?

15.3 Heterogeneous Aqueous Systems

★**54.** Arrange colloids, suspensions, and solutions in order of increasing particle size.

★**55.** How could you distinguish through observation among a solution, a colloid, and a suspension?

56. What is the Tyndall effect?

57. Why don't solutions demonstrate the Tyndall effect?

58. What causes Brownian motion?

*__59.__ What are two circumstances that help keep colloidal particles in suspension?

60. How can a colloid be destroyed?

61. What makes a colloidal dispersion stable?

Understand Concepts

*__62.__ From your knowledge of intermolecular forces, arrange these liquids in order of increasing surface tension: water (H_2O), hexane (C_6H_{14}), ethanol (C_2H_6O).

*__63.__ The graph below shows the density of water over the temperature range 0°C to 20°C.

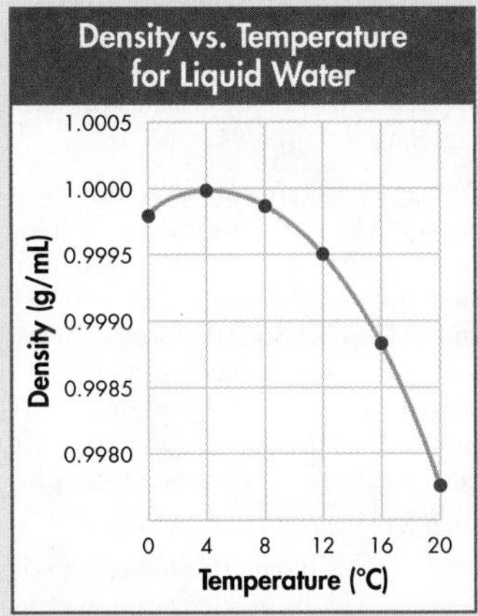

a. What is the maximum density of water?
b. At what temperature does the maximum density of water occur?
c. Would it be meaningful to expand the smooth curve of the graph to the left to temperatures below 0°C?

64. Explain why ions become solvated in aqueous solution.

*__65.__ Methanol (CH_4O) and hydrobromic acid (HBr) are both molecular compounds. However, an aqueous solution of methanol does not conduct an electric current, but an aqueous solution of hydrobromic acid does. Account for this difference.

*__66.__ Explain which properties of water are responsible for these occurrences.

a. Water in tiny cracks in rocks helps break up the rocks when it freezes.
b. Water beads up on a newly waxed car.
c. A longer time is needed for a teaspoon of water to evaporate than a teaspoon of alcohol.

67. Water has its maximum density at 4°C. Discuss the consequences of this fact.

*__68.__ Water is a polar solvent; gasoline is a nonpolar solvent. Decide which compounds are more likely to dissolve in water and which are more likely to dissolve in gasoline.

a. CCl_4
b. CH_4
c. Na_2SO_4
d. KCl

*__69.__ You have a solution containing either table sugar or table salt dissolved in water.

a. Can you tell which it is by visual inspection? Explain.
b. Give two ways by which you could easily tell which it is.

70. Explain why ethanol (C_2H_6O) will dissolve in both gasoline and water.

71. Are all liquids soluble in each other? Explain.

*__72.__ Write equations to show how these substances ionize or dissociate in water.

a. NH_4Cl
b. CH_3COOH
c. $Cu(NO_3)_2$
d. $HgCl_2$

*__73.__ Name these hydrates and determine the percent by mass of water in each.

a. $Na_2CO_3 \cdot H_2O$
b. $MgSO_4 \cdot 7H_2O$

74. Calculate the percent by mass of water in calcium chloride hexahydrate ($CaCl_2 \cdot 6H_2O$).

*__75.__ Barium hydroxide forms an octahydrate.

a. Write the equation for the formation of this hydrate from the anhydrous salt.
b. Calculate the percent by mass of water in barium hydroxide octahydrate.

*__76.__ How many grams of copper(II) sulfate pentahydrate would you need to measure in order to have 10.0 g of anhydrous copper(II) sulfate?

77. Explain the structural difference between ice and liquid water that accounts for the lower density of ice.

Evaluate

Answers

57. The molecules or ions are too small to have reflective surfaces.

58. collisions of the molecules of the dispersing medium with the small, dispersed colloidal particles

59. Brownian motion and repulsion between like-charged ions adsorbed on the surfaces of colloidal particles.

60. Add ions with a charge opposite that of the colloidal particles.

61. the addition of an emulsifier

UNDERSTAND CONCEPTS

62. hexane, ethanol, water

63. **a.** 1.0000 g/mL
b. 4°C
c. No; there would be a break in the curve at 0°C as liquid water at 0°C changes to ice at 0°C.

64. Anions are attracted to the hydrogen atoms of the water molecule because they have a partial positive charge. Cations are attracted to the oxygen atom in the water molecule because it has a partial negative charge.

65. Hydrobromic acid disassociates into hydrogen and bromide ions when dissolved in water, but methanol does not.

66. **a.** Water expands when it freezes to ice.
b. Water is polar and wax is nonpolar, and water has a higher surface tension.
c. Water has a lower vapor pressure than alcohol.

67. Ice has a lower density than liquid water, so it floats. Warmer water is less dense so it tends to layer over colder water.

68. **a.** gasoline **b.** gasoline **c.** water **d.** water

69. **a.** No, both form clear, colorless solutions.
b. Evaporate the water to examine the crystals; test for electrical conductivity; do a flame test

70. Ethanol has both a polar hydroxyl end (–OH) that dissolves in water and a nonpolar hydrocarbon end (C_2H_5–) that dissolves in gasoline.

71. No; nonpolar molecules do not dissolve in polar solvents.

72. **a.** $NH_4Cl(s) \rightarrow NH_4^+(aq) + Cl^-(aq)$
b. $CH_3COOH(s) \rightarrow H^+(aq) + CH_3COO^-(aq)$
c. $Cu(NO_3)_2(s) \rightarrow Cu^{2+}(aq) + 2NO_3^-(aq)$
d. $HgCl_2(s) \rightarrow Hg^{2+}(aq) + 2Cl^-(aq)$

73. **a.** sodium carbonate monohydrate, 14.5% H_2O
b. magnesium sulfate heptahydrate, 51.1% H_2O

74. 49.3% H_2O

75. **a.** $Ba(OH)_2(s) + 8H_2O(l) \rightarrow Ba(OH)_2 \cdot 8H_2O(s)$
b. 45.7% H_2O

76. 15.6 g $CuSO_4 \cdot 5H_2O$

77. The molecules in ice are further apart than the molecules in liquid water.

Evaluate

Answers

78. **a.** (1), (3), (6), (8) **b.** (1), (2), (5), (6), (7), (9) **c.** (2), (4), (5)

79. The Tyndall Effect

80. **a.** nonelectrolyte **b.** weak electrolyte **c.** strong electrolyte

THINK CRITICALLY

81. The container would break because water expands as it freezes.

82.

```
H— Ö:           :Ö —H
    |            |
    H            H
      \        /
         :O:
        H   H
      /       \
H— Ö:           :Ö —H
    |            |
    H            H
```

83. The surface tension of water keeps the strider from sinking. The surfactant would reduce the surface tension and the strider would shrink.

84. Ice would form on top and sink to the bottom. The pond would freeze more quickly because no surface ice would insulate the water below from the freezing air.

85. The hydrogen-bonded structures in liquid water are disrupted when ethyl alcohol is added because the alcohol competes for hydrogen bonds with water molecules and the water structure collapses. Thus, mixtures of water and ethyl alcohol have less volume than the sum of the volumes of the components. Mixing two liquids could result in a volume greater than the sum of the volumes of the components if the structural ordering in the mixture is greater than in the separated components.

86. Most of the important chemical reactions of life take place in aqueous solutions inside cells.

87. The liquid would form a layer on top of the water if its density was less than that of water and on the bottom if its density was greater. The mixture would form a temporary emulsion.

88. Solid A is potassium nitrate. Solid B is barium sulfate. Solid C is fructose.

89. **a.** pink **b.** pink **c.** blue **d.** 45.4 % H_2O **e.** water or water vapor

***78.** Match each term with the following descriptions. A description may apply to more than one term.

a. solution **b.** colloid **c.** suspension

(1) does not settle out on standing
(2) heterogeneous mixture
(3) particle size less than 1 nm
(4) particles can be filtered out
(5) demonstrates the Tyndall effect
(6) particles are invisible to the unaided eye
(7) homogenized milk
(8) saltwater
(9) jelly

79. A student standing in front of a murky aquarium sees a sharply focused flashlight beam shining through the aquarium as only a broad, diffuse light. What phenomenon is the student observing?

***80.** The diagrams below represent aqueous solutions of three different substances. Identify each substance as a strong electrolyte, weak electrolyte, or nonelectrolyte.

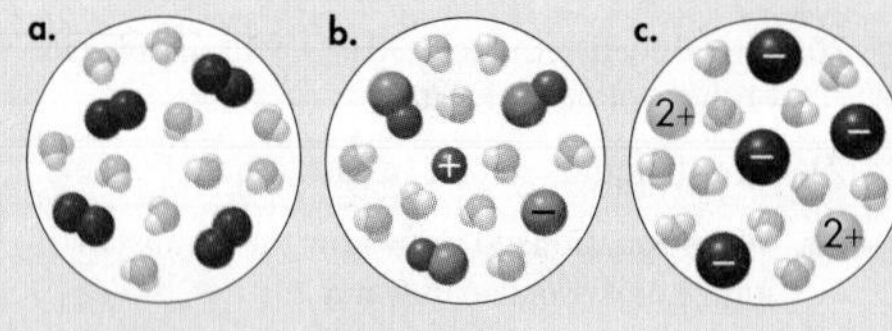

Think Critically

***81. Predict** Describe what might happen if you put a sealed glass container full of water into a freezer.

82. Use Models Make a drawing to show how one molecule of water can be connected to as many as four other molecules of water by hydrogen bonds. Write an explanation of your drawing.

***83. Relate Cause and Effect** A water strider is an insect with elongated legs that can easily walk on the surface of water. What would you expect to happen to the insect if a small amount of a surfactant were added to the water? Explain the reasoning behind your answer.

84. Predict Describe what would happen to a pond at 0°C if the density of ice were greater than the density of water. Do you think the pond would freeze more quickly? Explain.

***85. Infer** When ethanol (C_2H_6O) dissolves in water, the volume of the final solution is less than the separate volumes of the water and alcohol added together. Can you explain this result? Do you think that it might be possible to mix two different liquids and get a mixture volume that is larger than the sum of the volumes of the two components? Explain.

***86. Draw Conclusions** When the humidity is low and the temperature high, humans must take in large quantities of water or face serious dehydration. Why do you think water is so important for the proper functioning of your body?

87. Apply Concepts Describe as specifically as possible what would happen if a nonpolar molecular liquid were added to water. What would form if you shook this mixture vigorously?

88. Analyze Data You are given three white solids: A, B, and C. You know that one of the solids must be fructose ($C_6H_{12}O_6$), one must be potassium nitrate (KNO_3), and one must be barium sulfate ($BaSO_4$). Solid A dissolves in water and the resulting solution conducts an electric current. Solid B is insoluble. Solid C dissolves in water and the resulting solution does not conduct an electric current. Identify the solids A, B, and C.

***89. Infer** Cobalt(II) chloride test paper is blue. This paper is made by soaking strips of paper in an aqueous solution of $CoCl_2 \cdot 6H_2O$. The paper strips are then dried in an oven.

$$\underset{\text{pink}}{CoCl_2 \cdot 6H_2O} \xrightarrow{\text{heat}} \underset{\text{blue}}{CoCl_2} + 6H_2O$$

a. When cobalt(II) chloride hexahydrate is dissolved in water, what is the color of the solution?
b. What is the color of wet cobalt(II) chloride paper?
c. What is the color of dry cobalt(II) chloride paper?
d. What is the percent by mass of water in the hexahydrate?
e. What does cobalt(II) chloride test paper test for?

Enrichment

90. Compare How do the volumes of the following compare?

a. 1 g of ice at 0°C and 1 g of liquid water at 0°C

b. 1 g of liquid water at 100°C and 1 g of steam at 100°C

***91. Relate Cause and Effect** When spring comes, ice melting at the surface of a pond begins a beneficial process that stirs up the water of the pond. Explain why the pond water begins to mix when the ice melts. (*Hint:* Consider the changes in density.)

***92. Infer** A problem for firefighters is that much of the water they spray on a fire doesn't soak in but runs off carrying debris and pollution into the environment. Explain how the addition of a surfactant to water used to fight fires could help put out the fire more rapidly and protect the environment.

***93. Relate Cause and Effect** After a winter of alternate periods of freezing and thawing, some roads have broken pavement and potholes. Using what you know of the properties of water, explain why potholes form.

94. Interpret Diagrams Detergents consist of molecules with a polar or charged head and a long, oil-like tail. In water, detergent molecules can aggregate into organized structures called micelles. As shown below, the polar heads face the water molecules and the oily chains are inside the micelle. Suggest an explanation for the formation of micelles.

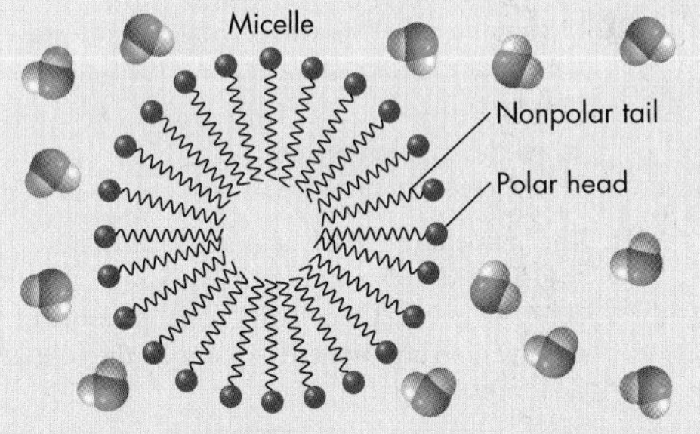

Write About Science

95. Explain Research the importance of electrolytes in the body. Write a paragraph explaining why the concentration of these ions may decline and how they can be restored.

96. Research Clothes dryer sheets are often added to tumble dryers to prevent clothes from wrinkling. Do research to learn what compounds are used in dryer sheets and how they work. Write a brief report of your findings.

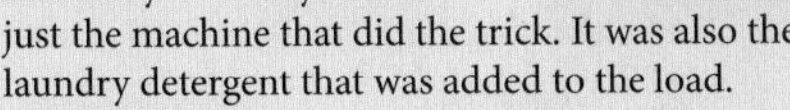

Coming Clean

Wes's dirty socks finally did become clean, but only after being thrown into the washing machine with a load of his family's laundry. It's not just the machine that did the trick. It was also the laundry detergent that was added to the load.

Water alone cannot remove many common stains. The particles of dirt and grease on Wes's socks were trapped in the cloth fibers and could not be dissolved by water. Also, because of its high surface tension, the water could not penetrate the fibers. Laundry detergents typically contain one or more surfactants. The surfactants reduce the surface tension of water, so that it can effectively wet and penetrate the fibers. The surfactants also act as emulsifiers. One end of the surfactant molecule is nonpolar and can dissolve the molecules in dirt and grease. The other end of the surfactant molecule is polar and can dissolve in water. Agitation provided by the washing machine helps pull the stain free from the cloth fibers.

*** 97. Infer** What happens to the molecules in the dirt and grease once they are lifted from the cloth fibers?

98. Connect to the BIGIDEA Explain how detergents remove stains from clothing by describing the interactions among the molecules in the detergent, the molecules in the stain, and the water molecules.

CHEMYSTERY

START A CONVERSATION After students have read through the CHEMystery, engage students in a discussion of the properties of water, dirt, and grease that prevented Wes from being able to clean his socks with just water. **Ask** Why wasn't Wes able to remove the dirt and grease by using plain water? *(Dirt and grease are nonpolar. Water is polar. Water will easily dissolve only polar compounds.)* **Ask** Would the surfactant be more or less effective in removing the dirt and grease if both ends of it were polar? Why? *(It would be less effective. One end needs to be nonpolar to dissolve the nonpolar dirt and grease.)*

CHEMYSTERY ANSWERS

97. The molecules in the dirt and grease are removed with the soapy water.

98. BIGIDEA Dirt and grease contain nonpolar molecules. One end of the surfactant molecules in detergent is nonpolar and can dissolve the dirt and grease. The other end of the surfactant molecules is polar and can dissolve in the polar molecules of water.

Answers

ENRICHMENT

90. **a.** The volume of 1 g of ice at 0°C is greater.
b. The volume of 1 g of liquid water at 100°C is less.

91. In spring, when the ice melts and the temperature of the surface water increases to 4°C, it becomes heavier than the water below it and sinks. The downward movement of surface water forces water in the deeper parts of the lake upward, where it is warmed.

92. A surfactant helps to wet the burning material, so less water is needed to put out the fire. Thus, less water carries pollutants into the environment.

93. Water enters cracks in pavement and expands when it freezes, creating larger cracks. Continuous freeze-thaw cycles cause pavement to break up and form potholes.

94. The polar heads are attracted to the water molecules so they line up on the outside of the micelle. The nonpolar tails are not attracted to the water molecules, so they shift to the inside of the micelle.

WRITE ABOUT SCIENCE

95. Electrolytes are essential to all metabolic processes. Electrolytes are lost through diarrhea, vomiting and sweating. Sport drinks are a good source of electrolytes.

96. Fabrics build up electrostatic charges from rubbing against one another and then cling to each other and wrinkle. Dryer sheets are saturated with a liquid surfactant, a polar molecule that bonds to the surface of negatively charged fabrics. The other side of the surfactant molecule is a long chain fatty acid or alcohol. The fatty material leaves the surface of the fabric slightly oily preventing the buildup of electrostatic charges.

CUMULATIVE REVIEW

99. **a.** 195 g **b.** 1.95×10^5 mg **c.** 0.195 kg

100. **a.** 5 **b.** 2 **c.** 2 **d.** 4

101. $1s^22s^22p^6$, neon

102. $H^+ + H:\ddot{O}:H \longrightarrow [H:\ddot{O}:H]^+$ (with H bonded below O in the product)

103. **a.** 22.4 L.
b. mass He: 4.00 g; mass CH_4: 16.0 g; mass O_2: 32.0 g
c. density He: 0.179 g/L; density CH_4: 0.714 g/L; density O_2: 1.43 g/L
d. Balloons containing He and CH_4 will rise; O_2 balloon will sink.

104. **a.** $6CO_2(g) + 6H_2O(l) \rightarrow C_6H_{12}O_6(s) + 6O_2(g)$
b. $2Na(s) + 2H_2O(l) \rightarrow 2Na^+(aq) + 2OH^-(aq) + H_2(g)$

105. 9.0 g H_2; 72.0 g O_2

106. 0.0360 g H_2O, 0.0224 L O_2

107. 636 g C_2H_4O

108. **a.** 0.300 mol O_2 **b.** 6.72 L O_2

109. **a.** hydrogen **b.** 0.049 g H_2O **c.** oxygen **d.** 0.0090 L

110. **a.** raises the boiling point
b. lowers the boiling point

111. 1.27 atm

Cumulative Review

99. A cylindrical vessel, 28.0 cm in height and 3.00 cm in diameter, is filled with water at 50°C. The density of water is 0.988 g/cm^3 at this temperature. Express the mass of water in the vessel in the following units.

a. grams **b.** milligrams **c.** kilograms

★**100.** How many significant figures are in each measurement?

a. 56.003 g
b. 750 mL
c. 0.0056 cm
d. 0.4005 dg

101. Write the correct electron configuration for the oxide ion. Which noble gas has the same electron configuration?

★**102.** When a proton is attracted to the unshared electron pair of a water molecule, the polyatomic hydronium ion (H_3O^+) is formed. Draw electron dot structures to show the formation of this ion.

103. The balloons contain 1 mol of He, CH_4, and O_2 at STP.

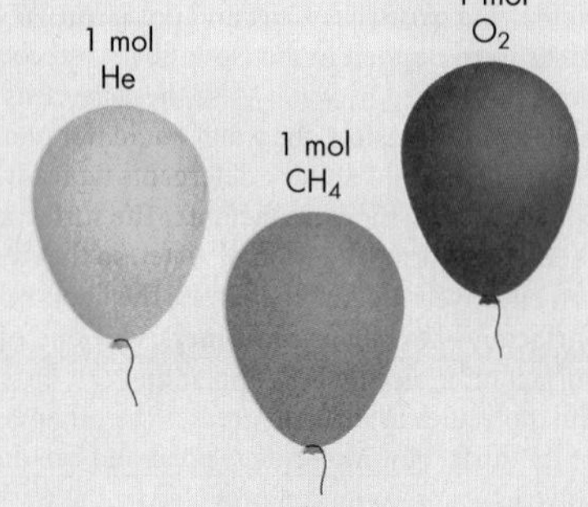

a. What is the volume of each balloon?
b. What is the mass of the gas in each balloon?
c. Calculate the density of the gas in each balloon.
d. The density of air at room temperature is about 1.2 g/mL. Predict whether each balloon will rise or sink when released.

★**104.** Balance the following equations.

a. $CO_2(g) + H_2O(l) \longrightarrow C_6H_{12}O_6(s) + O_2(g)$
b. $Na(s) + H_2O(l) \longrightarrow Na^+(aq) + OH^-(aq) + H_2(g)$

105. How many grams each of hydrogen gas and oxygen gas are required to produce 4.50 mol of water?

★**106.** The decomposition of hydrogen peroxide is given by this equation.

$$2H_2O_2(l) \longrightarrow 2H_2O(l) + O_2(g)$$

Calculate the mass of water and the volume of oxygen at STP formed when 2.00×10^{-3} mol of hydrogen peroxide is decomposed.

★**107.** Acetaldehyde (C_2H_4O) is produced commercially by the reaction of acetylene (C_2H_2) with water, as shown by this equation.

$$C_2H_2(g) + H_2O(l) \longrightarrow C_2H_4O(l)$$

How many grams of C_2H_4O can be produced from 2.60×10^2 g H_2O, assuming sufficient C_2H_2 is present?

108. Hydrogen reacts with oxygen to form water.

$$2H_2(g) + O_2(g) \longrightarrow 2H_2O(l)$$

a. How many moles of oxygen are required to produce 10.8 g H_2O?
b. How many liters of oxygen is this at STP?

★**109.** A mixture of 40 cm^3 of oxygen gas and 60 cm^3 of hydrogen gas at STP is ignited.

a. Which gas is the limiting reagent?
b. What is the mass of water produced?
c. Which gas remains after reaction?
d. What is the volume, at STP, of the remaining gas?

110. Explain how the following changes in the pressure on the surface of water affect the water's boiling point.

a. an increase in pressure
b. a decrease in pressure

★**111.** The temperature of 1 L of steam at constant volume and 1.00 atm pressure is increased from 100°C to 200°C. Calculate the final pressure of the steam in atmospheres, assuming the volume does not change.

If You Have Trouble With . . .

Question	99	100	101	102	103	104	105	106	107	108	109	110	111
See Chapter	3	3	7	8	10	11	12	12	12	12	12	13	14

514 Chapter 15 • Assessment

Standardized Test Prep for Indiana

Standard: C.7.1

Select the choice that best answers each question or completes each statement.

1. When a sugar cube completely dissolves in a glass of water, it forms
   (A) a colloid. (C) an emulsion.
   (B) a suspension. (D) a solution.

2. How many water molecules are tied up per formula unit of a compound that is an octahydrate?
   (A) nine (B) eight (C) seven (D) six

3. Which property is characteristic of water?
   (A) relatively high surface tension
   (B) relatively high vapor pressure
   (C) relatively low solvent ability
   (D) relatively low polarity

Use the atomic windows to answer Question 4.

4. Atomic window (A) represents solute particles in a given volume of solution. Which window represents the solute particles in the same volume of solution when the amount of solvent is doubled?

(A)

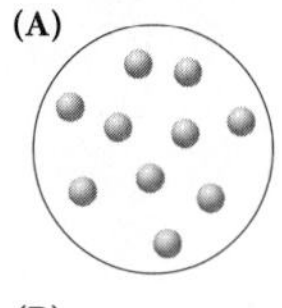

(C)

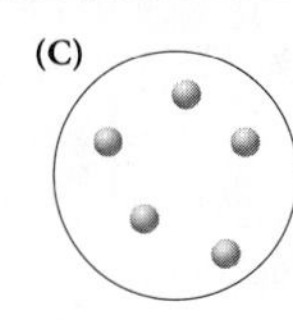

(B)

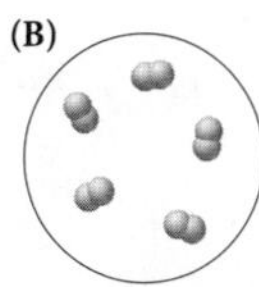

(D)

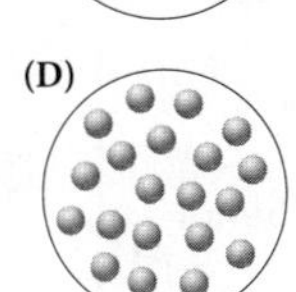

Tips for Success

Constructed Response You will probably answer most constructed response questions by writing a sentence or a paragraph. Put as much information into your answer as possible, but avoid unnecessary words. Be sure to address all the points asked for in the question.

Use the description and the data table to answer Questions 5–7.

A student measured the conductivity of six aqueous solutions. Each solution had equal concentrations of solute. The magnitude of the conductivity value is proportional to the number of ions in the solution. The SI conductivity unit is the microsiemens/cm (μS/cm). The table gives the student's results.

Solution	Conductivity (μS/cm)
Potassium chloride, KCl	2050
Aluminum chloride, $AlCl_3$	4500
Calcium chloride, $CaCl_2$	3540
Sodium hydroxide, $NaOH$	2180
Ethanol, C_2H_6O	0
Magnesium bromide, $MgBr_2$	3490

5. Why does the ethanol solution have zero conductivity?
6. Explain why two pairs of conducting solutions have similar conductivities.
7. The $AlCl_3$ solution has a conductivity that is about twice that of the KCl solution. Explain.

For each question there are two statements. Decide whether each statement is true or false. Then decide whether Statement II is a correct explanation for Statement I.

	Statement I		Statement II
8.	Water has a relatively high surface tension.	BECAUSE	Water molecules form strong hydrogen bonds with other water molecules.
9.	Particles in a colloid settle out faster than particles in a solution.	BECAUSE	Particles in a colloid are larger than particles in a solution.
10.	Water molecules are polar.	BECAUSE	The bond between hydrogen and oxygen atoms in a water molecule is polar.

If You Have Trouble With . . .

Question	1	2	3	4	5	6	7	8	9	10
See Lesson	15.3	15.2	15.1	15.2	15.2	15.2	15.2	15.1	15.3	15.1

STP Answers

1. D
2. B
3. A
4. C
5. Ethanol is a molecular compound that does not ionize in water.
6. KCl and NaOH have similar conductivities because they each provide two moles of ions per mole of compound. $CaCl_2$ and $MgBr_2$ have similar conductivities because they each provide three moles of ions per mole of compound.
7. Aluminum chloride provides twice as many ions per mole as potassium chloride does.
8. True, True, correct explanation
9. False, True
10. True, True, correct explanation

16 Solutions Planning Guide

Introducing the BIGIDEA: THE MOLE AND QUANTIFYING MATTER

Chemists use the mole to measure how concentrated a solution is.

IASS	Lessons and Objectives	Print Resources	
		For the Student	For the Teacher
C.7.1, C.7.2	**16.1** Properties of Solutions p 518–524 **16.1.1** Identify the factors that affect how fast a substance dissolves. **16.1.2** Describe the equilibrium in a saturated solution. **16.1.3** Describe the factors that affect the solubility of a substance.	Reading and Study Workbook Lesson 16.1 Lesson Assessment 16.1 p 524 Quick Lab: Solutions and Colloids p 519	Teacher Demo, p 522: Solubility of Gases
C.7.1, C.7.4	**16.2** Concentrations of Solutions p. 525–531 **16.2.1** Calculate the molarity of a solution. **16.2.2** Describe the effect of dilution on the total moles of solute in solution. **16.2.3** Express solution concentration as a percent by volume or percent by mass.	Reading and Study Workbook Lesson 16.2 Lesson Assessment 16.2 p 531	Class Activity, p 527: Preparing Solutions Class Activity, p 528: Solution Calculations Teacher Demo, p 530: Serial Dilutions
C.7.1	**16.3** Colligative Properties of Solutions p 534–537 **16.3.1** Explain how colligative properties can be explained on a particle basis.	Reading and Study Workbook Lesson 16.3 Lesson Assessment 16.3 p 537	Class Activity, p 536: Freezing Point Demonstration
C.7.1, C.7.3, C.7.4	**16.4** Calculations Involving Colligative Properties p 538–544 **16.4.1** Identify the two ways of expressing the ratio of solute to solvent in a solution. **16.4.2** Describe how the freezing-point depression and boiling-point elevation are related to molality.	Reading and Study Workbook Lesson 16.4 Lesson Assessment 16.4 p 544 Small-Scale Lab: Making a Solution p 545	Class Activity, p 540: Diagramming Methods of Concentration Calculation

Assessing the BIGIDEA: THE MOLE AND QUANTIFYING MATTER

	Essential Questions 1. What properties are used to describe the nature of solutions? 2. In what ways can you quantify the concentration of a solution?	Study Guide p 546 Math Tune-Up p 547 STP p 553 Reading and Study Workbook Self-Check and Vocabulary Review Chapter 16	

PearsonChem.com	
Editable Worksheets	Online Activities
Teaching Resources, Lesson 16.1 Review Small-Scale Lab Manual Lab 26: Solubility Rules Lab 30: Factors Affecting Solution Formation Lab 31: Supersaturation Lab Practical 16-1: Chromatography	16.1 Lesson Overview Dynamic Equilibrium of Saturated Solutions Using Henry's Law
Teaching Resources, Lesson 16.2 Review	16.2 Lesson Overview Calculating Molarity Preparing a Dilute Solution
Teaching Resources, Lesson 16.3 Review Lab 33: Freezing Point	16.3 Lesson Overview Vapor Pressure Sorbet Solutions Solutions
Teaching Resources, Lesson 16.4 Review	16.4 Lesson Overview Calculating the Freezing-Point Depression of a Solution
Core Teaching Resources (includes editable worksheets) • Lesson Reviews • Practice Problems • Interpret Graphs • Vocabulary Review • Chapter Quizzes and Tests • Lab Record Sheets	Chapter 16 Problem Set Converting Units Solving Equations

Materials List

For the Student

Quick Lab p 519
- sodium hydrogen carbonate
- cornstarch
- stirring rod
- distilled water
- flashlight
- masking tape
- 3 jars with parallel sides
- teaspoon cup

Small-Scale Lab p 545
- NaCl (lab grade)
- water
- 50-mL volumetric flask
- balance

For the Teacher

Teacher Demo p 522
- bottle of warm soda
- bottle of cold soda
- 2 other bottles of soda

Class Activity p 527
- water, NaCl, sucrose
- 50-mL, 100-mL, and 1-L volumetric flasks
- balance
- graph paper

Class Activity p 528
- calculator

Teacher Demo p 530
- four 20-mL test tubes
- 10 mL of 0.5*M* sucrose
- 15 mL water
- 10-mL pipet

Class Activity p 536
- ice
- water
- rock salt
- thermometer
- foam cup

Additional Digital Resources

Online Student Edition
Online Teacher's Edition
Exam*View Assessment Suite CD-ROM*
Classroom Resources DVD

CHAPTER 16

What's Online

CHEMISTRY TUTORIAL Students can work step-by-step with the online tutorial to learn how to use Henry's Law, calculate molarity, and calculate the freezing-point depression of a solution by accessing the Chapter 16 Chemistry Tutorials.

ONLINE PROBLEM SET Additional practice problems about solutions can be found in the Chapter 16 online Problem Set.

MATH TUTORIALS The online Math Tutorials will help guide students through converting units and solving equations.

DIRECTED VIRTUAL LAB Encourage students to review the colligative properties of solutions including the effect of different ionic compounds on boiling point using the animated virtual lab for Chapter 16. The animation is followed by analysis questions.

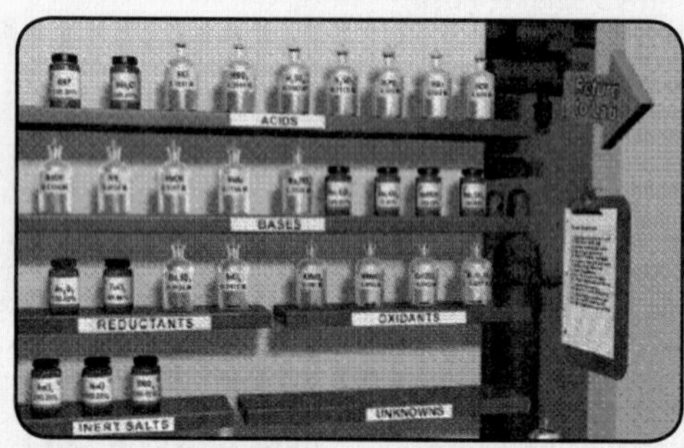

KINETIC ART Have students watch animations about the dynamic equilibrium of saturated solutions and vapor pressure. Each animation is followed by questions.

CONCEPTS IN ACTION Students can watch an interactive animation about how making sorbet is based on having the right solutions at the right concentrations.

IN ACADEMIC STANDARDS for SCIENCE

Nature of Science

- Recognize that their explanations must be based both on their data and other known information from investigations of others.
- Apply standard techniques in laboratory investigations to measure physical quantities in appropriate units and convert known quantities to other units as necessary.

Solutions

C.7.1 Describe the composition and properties of types of solutions.

Also covered C.7.2, C.7.3, C.7.4

16 Solutions

INSIDE:

- **16.1** Properties of Solutions
- **16.2** Concentrations of Solutions
- **16.3** Colligative Properties of Solutions
- **16.4** Calculations Involving Colligative Properties

PearsonChem.com

River water contains many dissolved ions, including sodium, calcium, magnesium, chloride, and sulfate ions.

516

Focus on ELL

1 CONTENT AND LANGUAGE Begin the class by focusing students' attention on the words *solutions, concentrations,* and *colligative properties* in the lesson titles. Write the words on the board along with their phonetic spellings and pronounce them slowly. Group students in threes and have them create a KWL chart for each word. Have them list details for the first two columns. Instruct groups to complete the last column after they read the lesson.

BEGINNING: LOW/HIGH Ask students to volunteer to explain aloud their KWL chart, in English or in the students' native language.

INTERMEDIATE: LOW/HIGH Verbally brainstorm questions for the *W* column based on the chapter opener photo and CHEMystery.

ADVANCED: LOW/HIGH Have students teach the words to less proficient students, using a creative and appropriate medium to show how the words are related.

BIGIDEA

THE MOLE AND QUANTIFYING MATTER

Essential Questions:

1. *What properties are used to describe the nature of solutions?*
2. *In what ways can you quantify the concentration of a solution?*

CHEMYSTERY

That Sinking Feeling

It was there one minute and gone the next—a stretch of road that looked like it had been swallowed up by the ground beneath it. The pavement had fallen into a sinkhole caused by moving groundwater. Groundwater consists of rain and melted snow that has soaked into the ground. Over time, groundwater can dissolve huge amounts of rock, gradually forming beautiful limestone caves, as well as destructive sinkholes. One recorded sinkhole swallowed a house, several other buildings, five cars, and a swimming pool!

What factors do you think contribute to the formation of sinkholes? How do you think sinkhole accidents can be prevented?

▶ Connect to the **BIG**IDEA As you read about solutions, try to identify the factors that cause sinkholes to form.

INDIANA'S ACADEMIC STANDARDS FOR SCIENCE

Solutions C.7.1, C.7.2, C.7.3, C.7.4. See lessons for details.

CHAPTER 16

UbD

Understanding by Design

Students are building toward an understanding of solutions, by applying the mole concept and dimensional-analysis skills.

PERFORMANCE GOALS At the end of Chapter 16, students will be able to answer the essential questions by applying their knowledge of solutions.

ESSENTIAL QUESTIONS Read the essential questions aloud. **Ask** In a solution of salt water, what is the solute and what is the solvent? *(Salt is the solute. Water is the solvent.)* **Ask** Are the parts of a solution chemically or physically combined? *(physically)*

BIGIDEA Use the photo of the canyon to help students connect to this chapter's concepts. Activate prior knowledge by having students describe similar places. **Ask** Why are the sides of the canyon vertical? *(The water has worn them down.)* **Ask** What do you think happened to the minerals in the rock? *(Some dissolved in the water; others broke loose into small fragments. Then, the moving water transported the minerals downstream.)* Point out that you could describe the concentration of minerals in the water if you knew how much of each mineral was in a certain quantity of river water.

CHEMYSTERY Connect the CHEMystery to the Big Idea by explaining that various factors may affect the ability of one substance to dissolve in another. Point out that, just as a river can dissolve minerals in rock, groundwater can dissolve minerals underground. Then, as a hint to solving the CHEMystery, have students think about what would happen if soluble minerals such as salt were in the underground rock.

Introduce the Chapter

IDENTIFYING PRECONCEPTIONS Students may believe that dissolution is an all-or-none process. Use the activity to introduce the idea that the ability of one substance to dissolve in another depends on various factors.

Activity *Each pair of students will need 2 beakers (about 100 mL), 2 plastic spoons, warm and cold water, and sugar.* Have one student in a group half-fill a beaker with warm water and the other half-fill a beaker with cold water. Have the students stir sugar, one spoonful at a time, into the water. Students should compare how fast it dissolves in the cold and warm water, and the total amount that dissolves. Repeat the activity, having students use water at the same temperature, but stirring at different rates.

The Truth About Solutions

Rob is the truth sleuth who is hired to investigate various solutions found around the house.

LESSON 16.1

Key Objectives

16.1.1 **IDENTIFY** the factors that affect how fast a substance dissolves.

16.1.2 **DESCRIBE** the equilibrium in a saturated solution.

16.1.3 **DESCRIBE** the factors that affect the solubility of a substance.

Additional Resources

Reading and Study Workbook, Lesson 16.1

Available Online or on Digital Media:

- Teaching Resources, Lesson 16.1 Review
- Kinetic Art, Vapor Pressure
- Chem Tutor: Using Henry's Law
- Laboratory Manual, Labs 30 and 31
- Small-Scall Lab Manual, Lab 26

PearsonChem.com

Engage

CHEMISTRY & YOU Instruct students to read the opening text and to examine the photo. Encourage students to share experiences with crystal-growing kits or making rock candy. **Ask** What do these processes have in common with the one shown in the photo? *(They each start with a heated solution.)*

Activate Prior Knowledge

Engage students in a discussion about the process of making instant hot chocolate. Elicit that the hot water is poured over the particles of hot chocolate mix, and then stirred to aid in the dissolving of the mix. Explain that students will use their knowledge of creating mixtures such as hot chocolate or lemonade, to learn about the properties of chemical solutions.

IN ACADEMIC STANDARDS for SCIENCE

- Develop explanations based on reproducible data and observations gathered during laboratory investigations.

C.7.1 Describe the composition and properties of types of solutions.

C.7.2 Explain how temperature, pressure and polarity of the solvent affect the solubility of a solute.

16.1 Properties of Solutions

C.7.1 Composition and properties of solutions; **C.7.2** Factors affecting solubility

CHEMISTRY & YOU

Q: *How can you grow a tree made out of crystals?* You're already familiar with the concept of liquids freezing. But what about crystals growing from a solution? The crystallization of a solute from solution is a physical change that is different from freezing. The crystal tree shown here began its "life" as an ordinary aqueous solution. The tree trunk, made of absorbent paper, soaks up the liquid. As water evaporates from the solution, the solutes crystallize onto the paper, forming delicate "leaves." Not all solutions will crystallize as this one did. The rate of crystallization depends on the nature of the solute and solvent, as well as on the temperature and humidity of the surroundings.

Key Questions

What factors affect how fast a substance dissolves?

How can you describe the equilibrium in a saturated solution?

What factors affect the solubility of a substance?

Vocabulary

- saturated solution
- solubility
- unsaturated solution
- miscible
- immiscible
- supersaturated solution
- Henry's law

Solution Formation

What factors affect how fast a substance dissolves?

Have you noticed, when making tea, that granulated sugar dissolves faster than sugar cubes, and that both granulated sugar and sugar cubes dissolve faster in hot tea or when you stir the tea? Figure 16.1 illustrates these observations. You will be able to explain these observations once you have gained an understanding of the properties of solutions.

Recall that solutions are homogeneous mixtures that may be solid, liquid, or gaseous. The compositions of the solvent and the solute determine whether or not a substance will dissolve. **Factors that affect how fast a substance dissolves include agitation, temperature, and the particle size of the solute.** Each of these factors involves the contact of the solute with the solvent.

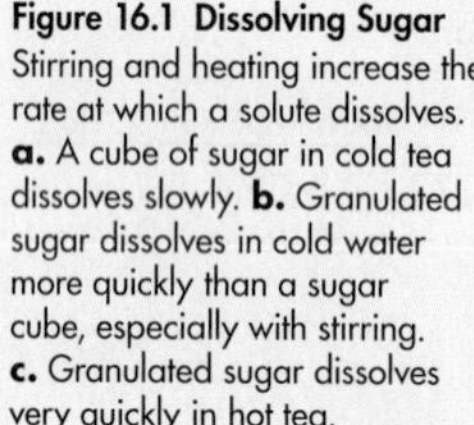

Figure 16.1 Dissolving Sugar
Stirring and heating increase the rate at which a solute dissolves. **a.** A cube of sugar in cold tea dissolves slowly. **b.** Granulated sugar dissolves in cold water more quickly than a sugar cube, especially with stirring. **c.** Granulated sugar dissolves very quickly in hot tea.

Focus on ELL

1 CONTENT AND LANGUAGE Divide the class into seven groups; assign one vocabulary word to each group. Instruct each group to pronounce, define, create a visual example, and create a picture or symbol of their word. Have each group teach their word to the class.

2 FRONTLOAD THE LESSON Ask students to describe the places in their native countries where crystals form naturally, such as shorelines and caves. Allow time for students to share information about their country that explains why the formation of crystals does or does not occur, such as particular climatic or geologic conditions.

3 COMPREHENSIBLE INPUT Physically demonstrate the examples of dissolution described in the Solution Formation section. Use the formation of rock candy on a string to supplement the Kinetic Art animation of a saturated solution.

Agitation If a teaspoon of granulated sugar (sucrose) is placed in a glass of tea, the crystals dissolve slowly. If the contents of the glass are stirred, however, the crystals dissolve more quickly. The dissolving process occurs at the surface of the sugar crystals. Stirring speeds up the process because fresh solvent (the water in tea) is continually brought into contact with the surface of the solute (sugar). It's important to realize, however, that agitation (stirring or shaking) affects only the rate at which a solid solute dissolves. It does not influence the amount of solute that will dissolve. An insoluble substance remains undissolved regardless of how vigorously or for how long the solvent/solute system is agitated.

Temperature Temperature also influences the rate at which a solute dissolves. Sugar dissolves much more rapidly in hot tea than in iced tea. At higher temperatures, the kinetic energy of water molecules is greater than at lower temperatures, so the molecules move faster. The more rapid motion of the solvent molecules leads to an increase in the frequency and the force of the collisions between water molecules and the surfaces of the sugar crystals.

Particle Size of the Solute The rate at which a solute dissolves also depends upon the size of the solute particles. A spoonful of granulated sugar dissolves more quickly than a sugar cube because the smaller particles in granulated sugar expose a much greater surface area to the colliding water molecules. Remember, the dissolving process is a surface phenomenon. The more surface area of the solute that is exposed, the faster the rate of dissolving.

Quick Lab

Purpose To classify mixtures as solutions or colloids using the Tyndall effect

Materials

- sodium hydrogen carbonate
- cornstarch
- stirring rod
- distilled water (or tap water)
- flashlight
- masking tape
- 3 jars with parallel sides
- teaspoon
- cup

Solutions and Colloids

Procedure 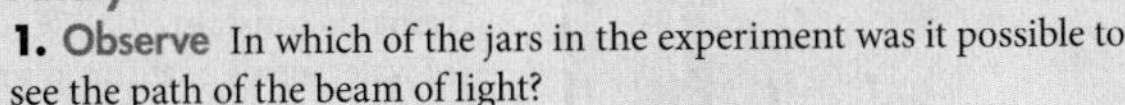

1. In a cup, make a paste: Mix one-half teaspoon of cornstarch with 4 teaspoons of water.
2. Fill one jar with water. Add one-half teaspoon of sodium hydrogen carbonate to a second jar and fill with water. Stir to mix. Add the cornstarch paste to the third jar and fill with water. Stir to mix.
3. Turn out the lights in the room. Shine the beam of light from the flashlight at each of the jars and record your observations.

Analyze and Conclude

1. **Observe** In which of the jars in the experiment was it possible to see the path of the beam of light?
2. **Infer** What made the light beam visible?
3. **Explain** If a system that made the light beam visible were filtered, would the light beam be visible in the filtrate? Explain your answer.
4. **Predict** What would you observe if you were to replace the sodium hydrogen carbonate with sucrose (cane sugar) or sodium chloride (table salt)? If you were to replace the cornstarch with flour or diluted milk?

Focus on ELL

4 LANGUAGE PRODUCTION Have students work in pairs to complete the Quick lab. Make sure students are paired so that more proficient students can help less proficient ones.

BEGINNING

LOW Model the procedure and have students repeat your actions.

HIGH Have students draw a picture or use symbols to note their observations.

INTERMEDIATE: LOW/HIGH Ask students to answer the questions in Analyze and Conclude aloud.

ADVANCED: LOW/HIGH Have students create a pictorial procedure that other students can follow to test the predictions made in question 4 of the Quick Lab.

Foundations for Reading

BUILD VOCABULARY Have students create a compare/contrast table to compare the following solution types: *saturated*, *unsaturated*, and *supersaturated*. Have them include definitions and examples of each solution type. Encourage students to draw a sketch to depict each word.

READING STRATEGY For each section, have students identify three things they learned, two things they found particularly interesting, and one question they have after reading the text.

Explain

Solution Formation

MAKE A CONNECTION Prepare a bulletin board display using pictures that illustrate the importance of solution processes in nature. Examples include a farmer applying fertilizer to a field, a person eating food, fish swimming, or a volcano erupting. Encourage students to contribute to the display by posting pictures they find in periodicals and newspapers, or with pictures they have taken themselves.

Explore

Quick Lab

OBJECTIVE After completing this activity, students will be able to classify mixtures as solutions or colloids using the Tyndall effect.

SKILLS FOCUS observing, inferring, predicting

PREP TIME 30 minutes

CLASS TIME 20 minutes

EXPECTED OUTCOME The beam of light is visible in the cornstarch/water mixture but not in water or sodium hydrogen carbonate solution.

ANALYZE AND CONCLUDE

1. undissolved cornstarch
2. Cornstarch particles reflect light.
3. yes for a colloid; no for a suspension
4. Sucrose and sodium chloride are both soluble in water, so the beam of light would not be visible in either of the resulting solutions. Flour and milk do not dissolve in water (flour and water form a suspension; milk and water form a colloid), so the beam of light would be visible in either mixture.

FOR ENRICHMENT Have students test and classify various household mixtures using the Tyndall effect.

LESSON 16.1

Explain

Solubility

USE AN ANALOGY Use the concept of population density to help convey the meaning of solubility. Both population density and solubility are quantities that describe concentration using derived units. Population density typically means the number of individuals per unit area—for example, persons per square kilometer, birds per square mile (of land), or fish per hectare (of water surface). Provide pictorial representations to visually convey the population density of a species for a given area or territory. There is a maximum population that a given amount of land or water can sustain (this is known as carrying capacity). This is analogous to saturation of a solution. Overcrowding is analogous to supersaturation; both lead to instability.

USE AN ANALOGY Equilibrium may be a difficult concept for some students to comprehend. Emphasize the dynamic nature of saturation. Point out that initially, salt readily dissolves in water because the ions can be solvated by available water molecules. Explain that the number of solvated ion-complexes that can inhabit a given volume of water is limited, however. If more ions are to join in the solvation, some ions must give up their places. Use substitutions in a basketball game as an analogy for the dynamic nature of saturation. Explain that in basketball only five players from each team can be on the court at one time. A player on the bench cannot enter the game until a player from the same team comes off the court. Convey that the players on the bench are analogous to the ions in a crystal; and those on the court are analogous to the dissolved ions.

USE VISUALS Have students examine Figure 16.4. **Ask** In terms of polarity and bonds, why are oil and water insoluble in one another? *(For a nonpolar liquid, such as oil, to mix with water, hydrogen bonds must be broken and replaced by much weaker forces between water and the nonpolar compound.)*

Figure 16.2 Saturated Solution
In a saturated solution, a state of dynamic equilibrium exists between the solution and the excess solute. The rate of solvation (dissolving) equals the rate of crystallization, so the total amount of dissolved solute remains constant.
Predict *What would happen if you added more solute to this saturated solution?*

See a saturated solution animated online.

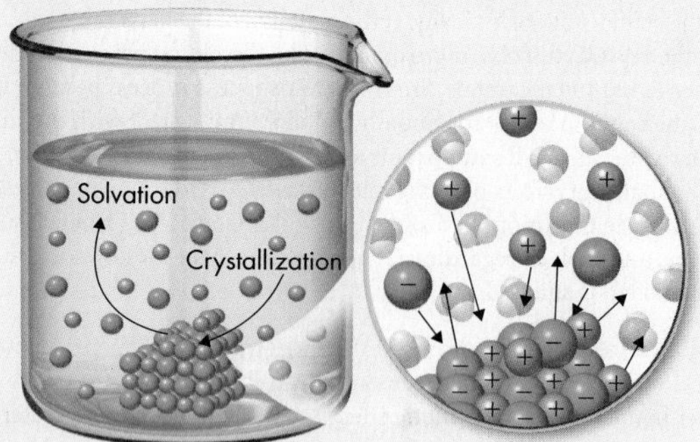

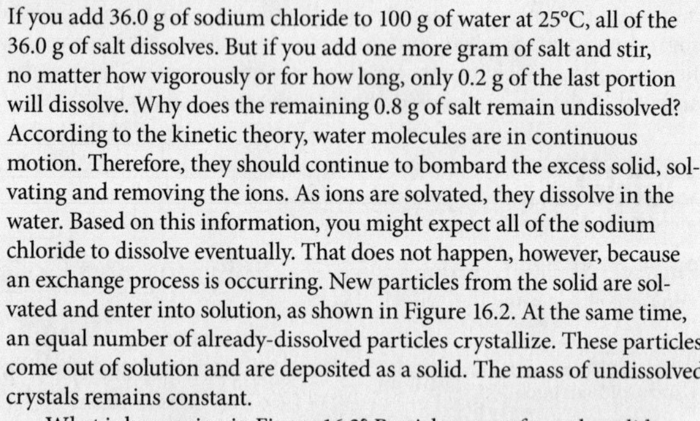

Solubility

How can you describe the equilibrium in a saturated solution?

If you add 36.0 g of sodium chloride to 100 g of water at 25°C, all of the 36.0 g of salt dissolves. But if you add one more gram of salt and stir, no matter how vigorously or for how long, only 0.2 g of the last portion will dissolve. Why does the remaining 0.8 g of salt remain undissolved? According to the kinetic theory, water molecules are in continuous motion. Therefore, they should continue to bombard the excess solid, solvating and removing the ions. As ions are solvated, they dissolve in the water. Based on this information, you might expect all of the sodium chloride to dissolve eventually. That does not happen, however, because an exchange process is occurring. New particles from the solid are solvated and enter into solution, as shown in Figure 16.2. At the same time, an equal number of already-dissolved particles crystallize. These particles come out of solution and are deposited as a solid. The mass of undissolved crystals remains constant.

What is happening in Figure 16.2? Particles move from the solid into the solution. Some dissolved particles move from the solution back to the solid. Because these two processes occur at the same rate, no net change occurs in the overall system. Such a solution is said to be saturated. A **saturated solution** contains the maximum amount of solute for a given quantity of solvent at a constant temperature and pressure. **In a saturated solution, a state of dynamic equilibrium exists between the solution and any undissolved solute, provided that the temperature remains constant.** At 25°C, 36.2 g of sodium chloride dissolved in 100 g of water forms a saturated solution. If additional solute is added to this solution, it will not dissolve.

The **solubility** of a substance is the amount of solute that dissolves in a given quantity of a solvent at a specified temperature and pressure to produce a saturated solution. Solubility is often expressed in grams of solute per 100 g of solvent (g/100 g H_2O). Sometimes the solubility of a gas is expressed in grams per liter of solution (g/L). A solution that contains less solute than a saturated solution at a given temperature and pressure is an **unsaturated solution.** If additional solute is added to an unsaturated solution, the solute will dissolve until the solution is saturated.

Figure 16.3 Hot Spring
The water in this hot spring in Yellowstone National Park is saturated with minerals. As the water cools near the edges of the spring, some of the minerals crystallize because they are less soluble at the lower temperature.

Check for Understanding

How can you describe the equilibrium in a saturated solution?

Have students create their own analogy to demonstrate their understanding of equilibrium in a saturated solution. Ask volunteers to share their analogies with the class.

ADJUST INSTRUCTION If students are having trouble answering, have them re-read the section on Solubility. Then re-state the basketball player substitution analogy, emphasizing the parts of the analogy most suited to clarifying any continuing misconceptions.

Some liquids—for example, water and ethanol—are infinitely soluble in each other. Any amount of ethanol will dissolve in a given volume of water, and vice versa. Similarly, ethylene glycol and water mix in all proportions. Pairs of liquids such as these are said to be completely miscible. Two liquids are **miscible** if they dissolve in each other in all proportions. In such a solution, the liquid that is present in the larger amount is usually considered the solvent. Liquids that are slightly soluble in each other—for example, water and diethyl ether—are partially miscible. Liquids that are insoluble in one another are **immiscible.** As you can see in Figure 16.4, oil and water are examples of immiscible liquids.

Figure 16.4 Oil and Water Vegetable oil is not soluble in water. Liquids that are insoluble in one another are immiscible.

Factors Affecting Solubility

What factors affect the solubility of a substance?

You have read that solubility is defined as the mass of solute that dissolves in a given mass of a solvent at a specified temperature. **Temperature affects the solubility of solid, liquid, and gaseous solutes in a solvent; both temperature and pressure affect the solubility of gaseous solutes.**

Temperature The solubility of most solid substances increases as the temperature of the solvent increases. For sodium chloride (NaCl), the increase in solubility is small—from 36.2 g per 100 g of water at 25°C to 39.2 g per 100 g of water at 100°C. Figure 16.5 shows how the solubility of several substances varies with temperature.

For a few substances, solubility decreases with temperature. For example, the solubility of ytterbium sulfate ($Yb_2(SO_4)_3$) in water drops from 44.2 g per 100 g of water at 0°C to 5.8 g per 100 g of water at 90°C. Table 16.1 on the next page lists the solubilities of some common substances in water at various temperatures.

Interpret Graphs

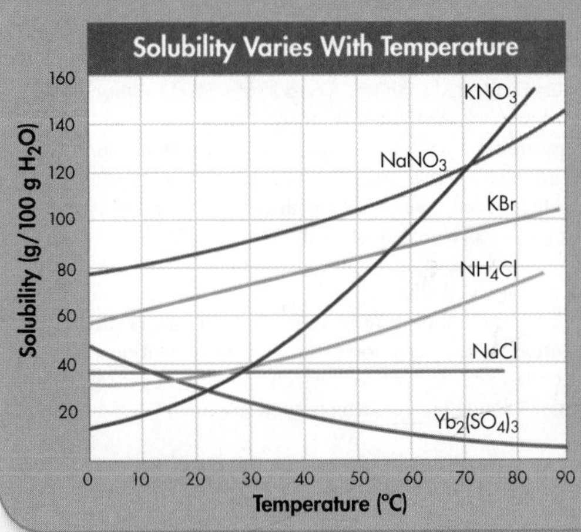

Figure 16.5 Changing the temperature usually affects the solubility of a substance.

a. Read Graphs What happens to the solubility of KNO_3 as the temperature increases?

b. Identify Which substance exhibits the least change in solubility as temperature increases?

c. Predict Suppose you added some solid NaCl to a saturated solution of NaCl at 20°C and warmed the mixture to 40°C. What would happen to the added NaCl?

d. Infer The mineral deposits around the hot spring in Figure 16.3 include NaCl and KCl. How do you think the solubility of KCl changes as the temperature decreases? Explain your answer.

Explain

Factors Affecting Solubility

INTERPRET DIAGRAMS Direct students' attention to Figure 16.5. Explain that the graph is used to show the way temperature affects the solubilities of six different ionic compounds. If students struggle to interpret the graph, consider showing the coordinate grid on an overhead with just the *x*- and *y*-labels and the title of the graph. Provide students with a table of data for each substance and, as a class, plot the data points in the table and then draw smooth curves through the points. Work with the data of only one substance at a time. Use interpolation and extrapolation techniques to make sure students understand how to interpret the graph. When the class is ready, direct them to answer the questions that correspond to Figure 16.5.

Extend

Direct students to review the compounds in Figure 16.5. **Ask** Predict the compounds you think have negative molar heats of solution. *(Answers will vary.)* Have students look up the ΔH_{soln} for each compound in the graph in a resource book such as *CRC Handbook of Chemistry and Physics*. **Ask** Summarize how the solubility curve for a compound is related to its ΔH_{soln}. *(Students should find that with the exception of $Yb_2(SO_4)_3$, for which data is not given, all of the compounds in the graph have positive ΔH_{soln} values at 25°C. Thus, heat is absorbed in the solution process for these compounds. Students may predict, then, that providing more heat by raising the temperature of a solution will increase solubility. Students may also note that NaCl, which shows the least increase in solubility with temperature, has the lowest positive ΔH_{soln} among the compounds discussed.)*

Differentiated Instruction

L1 SPECIAL NEED STUDENTS Adapt demonstrations so that students who have difficulties in managing visual or auditory information can experience the concepts. For example, fill a large bowl with warm water and guide students to agitate the water with one hand as you gradually add Epsom salts to the water. Ask students to describe what they feel at various points of the demonstration.

LPR LESS PROFICIENT READERS Have students draw and label a diagram illustrating the dynamic equilibrium of a saturated solution, with arrows describing the motion of particles between the crystalline and solvated states.

ELL ENGLISH LANGUAGE LEARNERS Convey that the word *miscible* comes from the Latin word *miscere*, meaning "to mix." Explain that the prefix *im–* means not. Point out that *immense* means "not capable of normal measurement" and *immobile* means "not capable of being moved." Help students to deduce that two liquids that are not soluble in each other are called *immiscible*.

Answers

FIGURE 16.2 It would not dissolve.

INTERPRETING GRAPHS

a. The solubility of KNO_3 increases as the temperature increases.

b. NaCl

c. Only a negligible amount of NaCl would go into solution, if any.

d. The solubility of KCl decreases as the temperature decreases. As the saturated hot spring water cools, it can no longer maintain the same concentration of KCl. As a result, some of the KCl crystallizes out of solution, forming the solid KCl deposits.

LESSON 16.1

Explore

Factors Affecting Solubility

Teacher Demo

PURPOSE Students will observe the effects of pressure and agitation on the solubility of gases.

MATERIALS bottle of warm soda, bottle of chilled soda, 2 other bottles of soda, 2 balloons

PROCEDURE Open a bottle of warm soda and a bottle of chilled soda. **Ask** What do you observe? *(Carbon dioxide gas comes out of solution much more quickly in the warm soda.)* Explain that the rush of bubbles in a warm bottle of soda can be a model for the "bends," a malady that afflicts divers when dissolved nitrogen forms bubbles in the bloodstream. Show students that the model also shows how the escape of carbon dioxide from rising magma can trigger a volcanic eruption. Point out that in each case, high pressure keeps a gas in solution; reduced pressure allows it to escape. With two open bottles at the same temperature, demonstrate that agitation increases the rate at which gases come out of solution. Put a balloon over the mouth of each one and then agitate one of the bottles. Set the bottles aside for a while and observe.

SAFETY Be sure to open the shaken bottle in a sink or other partially enclosed area, to avoid spilling.

CHEMISTRY & YOU Crystal growing kits work by the evaporation of solvent (usually water) from a saturated or supersaturated solution. As the solvent evaporates, the solution can no longer contain the same amount of dissolved solute, and the solute gradually precipitates and forms solid crystals.

Table 16.1

Solubilities of Substances in Water at Various Temperatures

Substance		Solubility (g/100 g H_2O)			
Name	Formula	0°C	20°C	50°C	100°C
Barium hydroxide	$Ba(OH)_2$	1.67	31.89	—	—
Barium sulfate	$BaSO_4$	0.00019	0.00025	0.00034	—
Calcium hydroxide	$Ca(OH)_2$	0.189	0.173	—	0.07
Lead(II) chloride	$PbCl_2$	0.60	0.99	1.70	—
Lithium carbonate	Li_2CO_3	1.5	1.3	1.1	0.70
Potassium chlorate	$KClO_3$	4.0	7.4	19.3	56.0
Potassium chloride	KCl	27.6	34.0	42.6	57.6
Sodium chloride	NaCl	35.7	36.0	37.0	39.2
Sodium nitrate	$NaNO_3$	74	88.0	114.0	182
Aluminum chloride	$AlCl_3$	30.84	31.03	31.60	33.32
Silver nitrate	$AgNO_3$	122	222.0	455.0	733
Lithium bromide	LiBr	143.0	166	203	266.0
Sucrose (table sugar)	$C_{12}H_{22}O_{11}$	179	230.9	260.4	487
Hydrogen*	H_2	0.00019	0.00016	0.00013	0.0
Oxygen*	O_2	0.0070	0.0043	0.0026	0.0
Carbon dioxide*	CO_2	0.335	0.169	0.076	0.0

*Gas at 101 kPa (1 atm) total pressure

Q: *How do you think crystal growing kits work? Use what you know about solubility and saturated solutions to explain your answer.*

Suppose you make a saturated solution of sodium ethanoate (sodium acetate) at 30°C and let the solution stand undisturbed as it cools to 25°C. Because the solubility of this compound is greater at 30°C than at 25°C, you expect that solid sodium ethanoate will crystallize from the solution as the temperature drops. But no crystals form. You have made a supersaturated solution. A **supersaturated solution** contains more solute than it can theoretically hold at a given temperature. The crystallization of a supersaturated solution can be initiated if a very small crystal, called a seed crystal, of the solute is added. The rate at which excess solute deposits upon the surface of a seed crystal can be very rapid, as shown in Figure 16.6. Crystallization can also occur if the inside of the container is scratched.

Another example of crystallization in a supersaturated solution is the production of rock candy. A solution is supersaturated with sugar. Seed crystals cause the sugar to crystallize out of solution onto a string for you to enjoy.

Fertilizer Runoff

Most farmers use fertilizers that contain salts of one or more of three elements essential to plant growth: potassium, nitrogen, and phosphorus. When more fertilizer is applied than the soil can absorb, rain washes off the excess salts. Not only is this economically wasteful, it is hazardous to the environment. The water containing these dissolved salts flows into streams and rivers, where the salts contribute to the eutrophication of the water. Eutrophied waters are rich in nutrients but deficient in dissolved oxygen due to overgrowth of algae and oxygen-consuming bacteria. Signs of eutrophication include overgrowth of water plants, algal blooms, and bad odors resulting from the growth of bacteria that do not need oxygen. In recent years, the problem of fertilizer runoff has lessened with the introduction of slow-release nitrogenous fertilizers. Many farmers have also adopted new methods to reduce runoff.

Figure 16.6 Supersaturated Solution
A supersaturated solution of sodium ethanoate ($NaC_2H_3O_2(aq)$) crystallizes rapidly when disturbed.
a. The solution is clear before a seed crystal is added. **b.** Crystals begin to form in the solution immediately after the addition of a seed crystal. **c–d.** The excess solute crystallizes rapidly.
Infer ***When the crystallization has ceased, will the solution be saturated or unsaturated?***

The effect of temperature on the solubility of gases in liquid solvents is opposite that of solids. The solubilities of most gases are greater in cold water than in hot. For example, Table 16.1 shows that the most important component of air for living beings—oxygen—becomes less soluble in water as the temperature of the solution rises. This fact has some important consequences. When an industrial plant takes water from a lake to use for cooling and then dumps the resulting heated water back into the lake, the temperature of the entire lake increases. Such a change in temperature is known as thermal pollution. Aquatic animal and plant life can be severely affected because the increase in temperature lowers the concentration of dissolved oxygen in the lake water.

Pressure Changes in pressure have little effect on the solubility of solids and liquids, but pressure strongly influences the solubility of gases. Gas solubility increases as the partial pressure of the gas above the solution increases. Carbonated beverages are a good example. These drinks contain large amounts of carbon dioxide (CO_2) dissolved in water. Dissolved CO_2 makes the liquid fizz and your mouth tingle. The drinks are bottled under a high pressure of CO_2 gas, which forces large amounts of the gas into solution. When a carbonated-beverage container is opened, the partial pressure of CO_2 above the liquid decreases. Immediately, bubbles of CO_2 form in the liquid and escape from the open bottle, as shown in Figure 16.7. As a result, the concentration of dissolved CO_2 decreases. If the bottle is left open, the drink becomes "flat" as the solution loses most of its CO_2.

How is the partial pressure of carbon dioxide gas related to the solubility of CO_2 in a carbonated beverage? The relationship is described by **Henry's law,** which states that at a given temperature, the solubility (S) of a gas in a liquid is directly proportional to the pressure (P) of the gas above the liquid. In other words, as the pressure of the gas above the liquid increases, the solubility of the gas increases. Similarly, as the pressure of the gas decreases, the solubility of the gas decreases. You can write the relationship in the form of an equation.

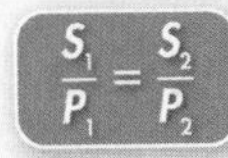

$$\frac{S_1}{P_1} = \frac{S_2}{P_2}$$

S_1 is the solubility of a gas at one pressure, P_1; S_2 is the solubility at another pressure, P_2.

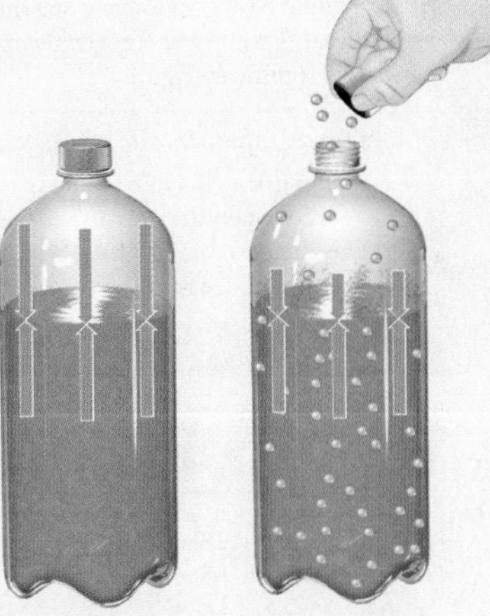

Figure 16.7 CO_2 in Solution
When a carbonated-beverage bottle is sealed, the pressure of CO_2 above the liquid is high, and the concentration of CO_2 in the liquid is also high. When the cap is removed, the pressure of CO_2 gas above the liquid decreases, and carbon dioxide bubbles out of the liquid.

Explain

USE AN ANALOGY Explain that the action of a gas pressing down on the surface of a liquid, thus increasing the solubility of the gas in the liquid, is like the action of a hammer driving a nail into a piece of wood. Explain or demonstrate that the harder a nail is hit, the farther the nail is driven into the wood. Convey that the hammer is analogous to pressure, the nail to solute gas, and the wood to solvent liquid.

Foundations for Math

PROPORTIONAL RELATIONSHIPS Explain that Henry's law is a proportion since it expresses two ratios that are equal. To find an unknown value in a proportion, students can use cross products and inverse operations.

In **Sample Problem 16.1,** cross products and inverse operations are used to isolate S_2 in order to identify the solubility of the gas at 1.0 atm.

$\frac{S_1}{P_1} = \frac{S_2}{P_2}$ Start with Henry's Law equaton.

$S_1P_2 = S_2P_1$ Use cross products.

$\frac{S_1P_2}{P_1} = S_2$ Use the inverse property of multiplication.

Point out that any variable in Henry's Law can be determined using the same method.

Answers

FIGURE 16.6 saturated

LESSON 16.1

Explain

Sample Practice Problem

The solubility of a gas is 1.04 g/L at 98.0 kPa and 25°C. What is the solubility of the same gas at 125 kPa and 25°C? *(1.33 g/L)*

Extend

Connect to GEOLOGY

Explain that caves form as the result of different processes, one of which is solution formation. Solutional caves form from limestone and gypsum. Encourage students to research the processes by which solutional caves, such as the Carlsbad Caverns, form. Ask students to share the results of their research with the class.

Evaluate

Informal Assessment

Assess students' understanding of the properties of solutions by writing the following words on the board: agitation, temperature, and particle size. **Ask** Which of these factors affects the solubility of a substance? *(temperature only)* **Ask** How is the solubility of a gas affected by the pressure of the gas above the liquid? *(The solubility of the gas increases with increasing pressure.)* Then have students complete the 16.1 Lesson Check.

Reteach

For students who struggle with the assessment, review the factors that affect the aqueous solubilities of solid and gaseous substances—temperature and pressure. Make sure that students are able to distinguish between the factors that affect the solubility and those that affect the rate of solution, such as agitation and particle size. Tell students that solubility is the amount of a substance that will dissolve in a given amount of solvent, whereas rate of solution is the amount of solute entering a solution per unit time. Convey that temperature is the only factor discussed that affects both the rate of solution and the solubility.

Using Henry's Law

If the solubility of a gas in water is 0.77 g/L at 3.5 atm of pressure, what is its solubility (in g/L) at 1.0 atm of pressure? (The temperature is held constant at 25°C.)

1 Analyze List the knowns and the unknown. Use Henry's law to solve for the unknown solubility.

KNOWNS	UNKNOWN
$P_1 = 3.5$ atm	$S_2 = ?$ g/L
$S_1 = 0.77$ g/L	
$P_2 = 1.0$ atm	

2 Calculate Solve for the unknown.

State the equation for Henry's law.

$$\frac{S_1}{P_1} = \frac{S_2}{P_2}$$

Isolate S_2 by multiplying both sides by P_2:

$$P_2 \times \frac{S_1}{P_1} = \frac{S_2}{P_2} \times P_2$$

Solve Henry's law for S_2. Substitute the known values and calculate.

$$S_2 = \frac{S_1 \times P_2}{P_1} = \frac{0.77 \text{ g/L} \times 1.0 \text{ atm}}{3.5 \text{ atm}} = 0.22 \text{ g/L}$$

3 Evaluate Does the result make sense? The new pressure is approximately one third of the original pressure, so the new solubility should be approximately one third of the original. The answer is correctly expressed to two significant figures.

In Problem 1, you're solving Henry's law for an unknown solubility. In Problem 2, you're solving for an unknown pressure.

1. The solubility of a gas in water is 0.16 g/L at 104 kPa. What is the solubility when the pressure of the gas is increased to 288 kPa? Assume the temperature remains constant.

2. A gas has a solubility in water at 0°C of 3.6 g/L at a pressure of 1.0 atm. What pressure is needed to produce an aqueous solution containing 9.5 g/L of the same gas at 0°C?

ONLINE PROBLEMS 16.1 LessonCheck

3. Review What determines how fast a substance will dissolve?

4. Describe How can you describe the state of equilibrium in a saturated solution that contains undissolved solute?

5. Describe What condition(s) determine the solubilities of solid, liquid, and gaseous solutes in a solvent?

6. Identify Name a unit used to express solubility.

7. Describe What determines whether or not a substance will dissolve?

8. Explain What would you do to change
- **a.** a saturated solid/liquid solution to an unsaturated solution?
- **b.** a saturated gas/liquid solution to an unsaturated solution?

9. Calculate The solubility of a gas is 0.58 g/L at a pressure of 104 kPa. What is its solubility if the pressure increases to 250 kPa at the same temperature?

Lesson Check Answers

3. agitation, temperature, particle size of the solute
4. A dynamic equilibrium exists between the solution and the undissolved solute, provided that the temperature remains constant. As new particles from the solid are solvated and enter into solution, an equal number of already-dissolved particles crystallize.
5. temperature (if the solute is a solid or liquid); temperature and pressure (if the solute is a gas)
6. Sample answer: grams of solute per 100 g of solvent
7. chemical composition of the solute and solvent
8. **a.** Sample answers: add solvent, increase the temperature
   b. Sample answers: increase the pressure, decrease the temperature, agitate it
9. 1.4 g/L

16.2 Concentrations of Solutions

C.7.1 Composition and properties of solutions; **C.7.4** Prepare a solution

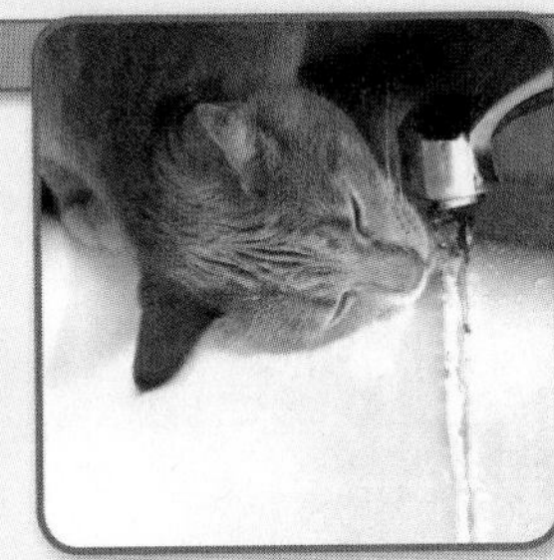

CHEMISTRY & YOU

Q: *How can you describe the concentration of a solution?* Clean drinking water is important for all communities. What constitutes clean water? Your federal and state governments set standards limiting the amount of contaminants allowed in drinking water. These contaminants include metals, pesticides, and bacteria. Water must be tested continually to ensure that the concentrations of these contaminants do not exceed established limits.

Key Questions

How do you calculate the molarity of a solution?

What effect does dilution have on the amount of solute?

How do percent by volume and percent by mass differ?

Vocabulary

- concentration
- dilute solution
- concentrated solution
- molarity (M)

Molarity

How do you calculate the molarity of a solution?

You have learned that a substance can dissolve to some extent in a particular solvent to form a solution. This lesson focuses on ways to express the actual extent of dissolving. The **concentration** of a solution is a measure of the amount of solute that is dissolved in a given quantity of solvent. A solution that contains a relatively small amount of solute is a **dilute solution.** By contrast, a **concentrated solution** contains a large amount of solute. An aqueous solution of sodium chloride containing 1 g NaCl per 100 g H_2O might be described as dilute when compared with a sodium chloride solution containing 30 g NaCl per 100 g H_2O. But the same solution might be described as concentrated when compared with a solution containing only 0.01 g NaCl per 100 g H_2O. You can see that the terms *concentrated* and *dilute* are only qualitative descriptions of the amount of a solute in solution.

How can concentration be expressed quantitatively? In chemistry, the most important unit of concentration is molarity. **Molarity (M)** is the number of moles of solute dissolved in one liter of solution. Molarity is also known as molar concentration. When the symbol M is accompanied by a numerical value, it is read as "molar." Figure 16.8 illustrates the procedure for making a 0.5M, or 0.5-molar, solution. Note that the volume involved is the total volume of the resulting solution, not the volume of the solvent alone.

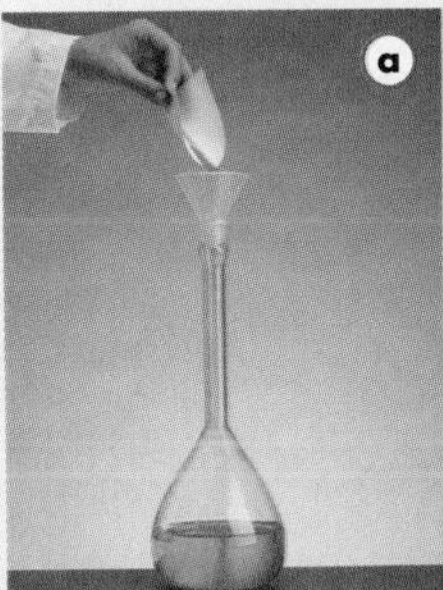

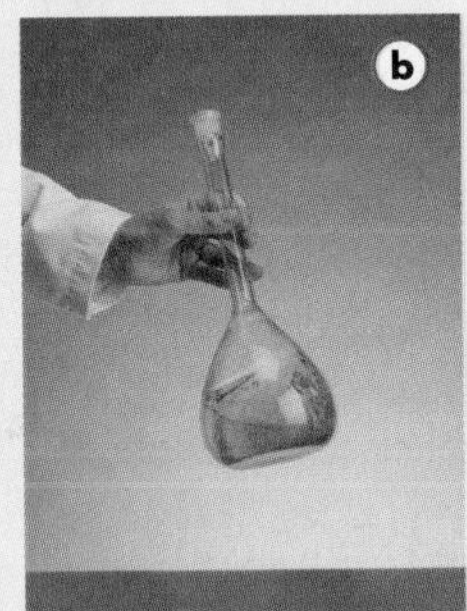

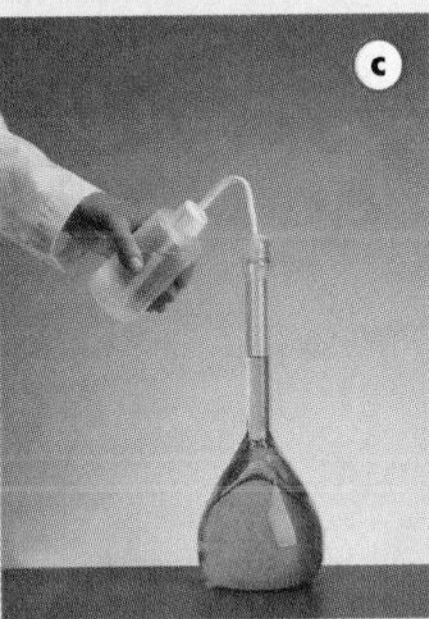

Figure 16.8 How to Make a 0.5M Solution
a. Add 0.5 mol of solute to a 1-L volumetric flask that is half filled with distilled water.
b. Swirl the flask carefully to dissolve the solute.
c. Fill the flask with water exactly to the 1-L mark.

Key Objectives

16.2.1 CALCULATE the molarity of a solution.

16.2.2 DESCRIBE the effect of dilution on the total moles of solute in solution.

16.2.3 EXPRESS solution concentration as a percent by volume or percent by mass.

Additional Resources

- Reading and Study Workbook, Lesson 16.2

Available Online or on Digital Media:

- Teaching Resources, Lesson 16.2 Review
- Laboratory Manual, Lab 32
- Laboratory Practicals, 16-1
- Chem Tutor: Calculating Molarity; Preparing a Dilute Solution

PearsonChem.com

Engage

CHEMISTRY & YOU Have students study the photograph and read the text. **Ask** What substances may be contaminants in drinking water? *(metals, pesticides, bacteria, water treatment by-products)* **Ask** How is the quality of drinking water assessed? *(Federal and state governments set standards that limit the concentration of contaminants allowed in drinking water.)*

Activate Prior Knowledge

Review the concepts of chemical quantities that were introduced in Chapter 10. **Ask** How many moles are in 12.0 g of NaCl? *(0.205 mol)* **Ask** How many grams are in 1.50 mol of NaOH? *(60.0 g)* **Ask** Which contains more molecules: 1.00 mol SO_2 or 1.00 mol SO_3? *(Numbers are equal.)* Which has more mass? *(1.00 mol SO_3)*

IN ACADEMIC STANDARDS for SCIENCE

C.7.1 Describe the composition and properties of types of solutions.

C.7.4 Prepare a specific volume of a solution of a given molarity when provided with a known solute.

Focus on ELL

1 CONTENT AND LANGUAGE Prepare students for the mathematics in this lesson by helping them to create a useful reference sheet. Have students preview the lesson and create a listing of key words and their corresponding math symbols. Tell students to also include on their sheet the units of measure and their corresponding chemical quantity.

2 FRONTLOAD THE LESSON Ask students to share their experiences with canned soup. Point out that soups which have the term *condensed* on their labels require the addition of more liquid before consumption. When they add water or another liquid to a condensed soup, they are making a dilution. Tell students to keep this in mind as they read the lesson.

3 COMPREHENSIBLE INPUT Walk students through Sample Problems 16.2 and 16.4. Then, physically demonstrate the processes of making the solutions that result from the calculations as a visual support for the concepts of molarity and dilution.

Answers

1. $S_2 = (S_1 \times P_2)/P_1 = (0.16\ \text{g/L} \times 288\ \text{kPa})/104\ \text{kPa} = 4.4 \times 10^{-1}\ \text{g/L}$
2. $P_2 = (P_1 \times S_2)/S_1 = (1.0\ \text{atm} \times 9.5\ \text{g/L})/3.6\ \text{g/L} = 2.6\ \text{atm}$

Foundations for Reading

BUILD VOCABULARY Have students make a web diagram showing how *dilute solution, concentrated solution,* and *molarity* are related to *concentration*.

READING STRATEGY Have students identify the main ideas of this section by listing the key concepts under each heading.

Explain

Molarity

USE VISUALS Direct students' attention to Figure 16.8 on page 525. Students may think that solutions of given molarity can be correctly prepared by addition of the solute to a premeasured volume of solvent. They often forget that the solute will take up some of the available space in the volumetric flask. Explain that the sequence of steps is important when making up standard solutions because the solute plus the solvent may take up more volume than the solvent alone. **Ask** Why is it important that the solute be added to some of the solvent and dissolved, and then solvent be added up to the 1 L mark on the volumetric flask? *(The total volume of the mixture is likely to exceed the desired volume if the solute is directly added to a full liter of solvent and the concentration will be different.)*

Sample Practice Problem

A solution of KCl contains 34.75 g of solute in 250 mL of solution. Determine the molarity of the solution. *(1.86M KCl)*

Extend

Connect to MEDICINE

Tell students that a patient in a hospital is often administered an intravenous (IV) drip containing an aqueous solution. The patient may need the solution to prevent dehydration or to administer nutrients or medicines. Explain that great care must be taken when fluids are introduced into the bloodstream in order to keep the balance of dissolved electrolytes inside and outside of the tissue cells from being altered. Have students research the effects of an electrolyte imbalance on humans as a result of dehydration.

To calculate the molarity of a solution, divide the number of moles of solute by the volume of the solution in liters.

$$\text{Molarity } (M) = \frac{\text{moles of solute}}{\text{liters of solution}}$$

For example, suppose 2 mol of glucose are dissolved in 5 L of solution. You would calculate the molarity of the solution as follows:

$$\frac{2 \text{ mol glucose}}{5 \text{ L solution}} = 0.4 \text{ mol/L} = 0.4M$$

If the amount of solute in a solution is expressed in mass units instead of moles, you can calculate molarity by using the appropriate conversion factors, as shown in Sample Problem 16.2.

Sample Problem 16.2

CHEM TUTOR

Calculating Molarity

Intravenous (IV) saline solutions are often administered to patients in the hospital. One saline solution contains 0.90 g NaCl in exactly 100 mL of solution. What is the molarity of the solution?

KNOWNS
solution concentration = 0.90 g NaCl/100 mL
molar mass of NaCl = 58.5 g/mol

UNKNOWN
solution concentration = ?M

1 Analyze List the knowns and the unknown. Convert the concentration from g/100 mL to mol/L. The sequence is g/100 mL ⟶ mol/100 mL ⟶ mol/L.

2 Calculate Solve for the unknown.

Use the molar mass to convert g NaCl/100 mL to mol NaCl/100 mL. Then convert the volume units so that your answer is expressed in mol/L.

$$\text{Solution Concentration} = \frac{0.90 \text{ g NaCl}}{100 \text{ mL}} \times \frac{1 \text{ mol NaCl}}{58.5 \text{ g NaCl}} \times \frac{1000 \text{ mL}}{1 \text{ L}}$$

$$= 0.15 \text{ mol/L}$$

$$= 0.15M$$

The relationship 1 L = 1000 mL gives you the conversion factor 1000 mL/1 L.

3 Evaluate Does the result make sense? The answer should be less than 1*M* because a concentration of 0.90 g/100 mL is the same as 9.0 g/1000 mL (9.0 g/1 L), and 9.0 g is less than 1 mol of NaCl. The answer is correctly expressed to two significant figures.

10. A solution has a volume of 2.0 L and contains 36.0 g of glucose ($C_6H_{12}O_6$). If the molar mass of glucose is 180 g/mol, what is the molarity of the solution?

11. A solution has a volume of 250 mL and contains 0.70 mol NaCl. What is its molarity?

Foundations for Math

CONVERSION FACTORS Tell students if they identify the unit of measure of the desired answer first, they can work backwards to determine which conversion factors are needed in the entire problem. Show students a simple example converting days to seconds. Point out that this same strategy can be used to organize molarity calculations.

In **Sample Problem 16.2,** the desired units are mol/L, but the starting units are mass (in grams) and volume (in mL). Point out that to arrive at an answer in moles per liter, students will need two conversion factors: one for converting grams to moles and another that converts milliliters to liters.

In some cases, you may need to determine the number of moles of solute dissolved in a given volume of solution. You can do this if the molarity of the solution is known. For example, how many moles are in 2.00 L of 2.5*M* lithium chloride (LiCl)? First, rearrange the formula for molarity to solve for the number of moles. Then, substitute the known values for molarity and volume.

$$\text{Molarity } (M) = \frac{\text{moles of solute}}{\text{liters of solution } (V)}$$

$$\text{Moles of solute} = \text{molarity } (M) \times \text{liters of solution } (V)$$

$$= 2.5M \times 2.00\ \text{L} = \left(\frac{2.5\ \text{mol}}{1\ \cancel{\text{L}}}\right) \times 2.00\ \cancel{\text{L}}$$

$$= 5.0\ \text{mol}$$

Thus, 2.00 L of 2.5*M* lithium chloride solution contains 5.0 mol of LiCl.

SampleProblem 16.3

Calculating the Moles of Solute in a Solution

Household laundry bleach is a dilute aqueous solution of sodium hypochlorite (NaClO). How many moles of solute are present in 1.5 L of 0.70*M* NaClO?

1 Analyze List the knowns and the unknown. The conversion is volume of solution ⟶ moles of solute. Molarity has the units mol/L, so you can use it as a conversion factor between moles of solute and volume of solution.

KNOWNS

volume of solution = 1.5 L

solution concentration = 0.70*M* NaClO

UNKNOWN

moles solute = ? mol

2 Calculate Solve for the unknown.

Multiply the given volume by the molarity expressed in mol/L.

$$1.5\ \cancel{\text{L}} \times \frac{0.70\ \text{mol NaClO}}{1\ \cancel{\text{L}}} = 1.1\ \text{mol NaClO}$$

3 Evaluate Does the result make sense? The answer should be greater than 1 mol but less than 1.5 mol, because the solution concentration is less than 0.75 mol/L and the volume is less than 2 L. The answer is correctly expressed to two significant figures.

12. How many moles of ammonium nitrate are in 335 mL of 0.425*M* NH_4NO_3?

13. How many moles of solute are in 250 mL of 2.0*M* $CaCl_2$? How many grams of $CaCl_2$ is this?

Make sure that your volume units cancel when you do these problems. If they don't, then you're probably missing a conversion factor in your calculations.

Foundations for Math

UNITS OF MEASUREMENT As discussed in Sample Problem 16.2, an important step in making calculations of chemical quanties is identifying the necessary conversion factors. Another useful step is writing a skeleton equation. This allows students the opportunity to test their choice of conversion factors to insure the proper units once the problem is solved.

For **Sample Problem 16.3**, the following skeleton equation can be written to make sure that the final units of measurement will be correct:

$$\text{L} \times \frac{\text{mol}}{\text{L}} = \text{mol}$$

Because the desired units is the number of moles, the conversion factor is correct and the problem can be solved once the proper quantities are added to the units.

Explore

Class Activity

PURPOSE Students will learn how to prepare molar solutions.

MATERIALS water; NaCl; sucrose; 50-mL, 100-mL, and 1-L volumetric flasks; balance; graph paper

PROCEDURE Have students determine the mass of a 50-mL, 100-mL, and a 1-L volumetric flask. Review the definition of 1 *M* and how to calculate the mass of 1 mol of NaCl (58.5 g). Add 500 mL of water to the 1-L volumetric flask and add 1 mol of NaCl. Swirl the flask until the NaCl dissolves. Add water up to the 1-L mark.

Ask Does the flask contain a liter of water? *(No, because the dissolved NaCl takes up some volume.)* Have students prepare 0.0625*M*, 0.125*M*, 0.250*M*, and 0.500*M* solutions of sucrose using the 100-mL and 50-mL volumetric flasks. **Ask** What is the mass of each solution? **Ask** What is the density of each solution? *(Density = (mass of flask + solution – mass of flask)/mL of solution.)* Have students make a graph of density versus concentration; include the density of pure water (1 g/mL).

EXPECTED OUTCOME The graph shows that density is directly proportional to concentration.

Sample Practice Problem

How many grams of solute are in 2.40 L of 0.650*M* $HClO_2$? *(107 g)*

Answers

10. (36.0 g/2.0 L solution) × (1 mol/180 g) = $1.0 \times 10^{-1}M$

11. (0.70 mol/250 mL) × (1000 mL/ 1 L) = 2.8*M*

12. 335 mL × (1L/1000 mL) × (0.425 mol/1L) = 1.42×10^{-1} mol NH_4NO_3

13. 250 mL × (1L/1000 mL) × 2.0 mol/L = 5.0×10^{-1} mol $CaCl_2$; 5.6×10^{1} g $CaCl_2$

LESSON 16.2

Explain

Making Dilutions

APPLY CONCEPTS Remind students that when substances react, they do so as particles and that an equal number of moles of different substances contain the same number of representative particles. Because many reactions occur in solution, a unit of concentration based on moles, such as molarity, is needed to provide information about the number of reactant particles in solution.

USE VISUALS Have students examine Figure 16.10. Remind students that when making dilute solutions, the accuracy of the resulting solution's concentration is dependent upon the measuring devices used. Show students that using a graduated cylinder to measure is much less accurate than using a volumetric pipette or buret.

Explore

Class Activity

PURPOSE Students practice calculating the variables in the definition of molarity.

MATERIALS calculator, paper, pencil

PROCEDURE Divide students into groups of three. One student should name a solute; the other two should each specify one of three quantities: *solute mass, solution volume,* or *molarity.* The unspecified quantity should then be calculated with input from each member. For example, the first student may say "sodium chloride," the second may say "12 grams," and the third may say "2.5 liters." The group would then have to calculate the molarity of the resulting solution. Each group should carry out at least one calculation involving each of the possible unknown quantities.

Figure 16.9 Dilution
Adding solvent to a concentrated solution lowers the concentration, but the total number of moles of solute present remains the same.

Making Dilutions

What effect does dilution have on the amount of solute?

Both of the solutions in Figure 16.9 contain the same amount of solute. You can tell by the color of solution (a) that it is more concentrated than the solution (b); that is, solution (a) has the greater molarity. The more dilute solution (b) was made from solution (a) by adding more solvent. **Diluting a solution reduces the number of moles of solute per unit volume, but the total number of moles of solute in solution does not change.** You can also express this concept by writing an equation.

$$\text{Moles of solute before dilution} = \text{moles of solute after dilution}$$

Now recall the definition of molarity and how it can be rearranged to solve for moles of solute.

$$\text{Molarity } (M) = \frac{\text{moles of solute}}{\text{liters of solution } (V)}$$

$$\text{Moles of solute} = \text{molarity } (M) \times \text{liters of solution } (V)$$

The total number of moles of solute remains unchanged upon dilution, so you can now write this equation:

$$\text{Moles of solute} = M_1 \times V_1 = M_2 \times V_2$$

M_1 and V_1 are the molarity and volume of the initial solution, and M_2 and V_2 are the molarity and volume of the diluted solution. Volumes can be in liters or milliliters, as long as the same units are used for both V_1 and V_2. Figure 16.10 illustrates the procedure used for making a dilution in the lab.

Figure 16.10 Making a Dilution
The student is preparing 100 mL of 0.40*M* $MgSO_4$ from a stock solution of 2.0*M* $MgSO_4$. **a.** She measures 20 mL of the stock solution with a 20-mL pipette. **b.** She transfers the 20 mL to a 100-mL volumetric flask. **c.** She carefully adds water to the mark to make 100 mL of solution.
Compare *How many moles of $MgSO_4$ are in 20 mL of the stock solution? In 100 mL of the dilute solution?*

UbD Check for Understanding

What effect does dilution have on the amount of solute?

Assess students' understanding of dilution by asking them to complete the following cloze prompt:

When a dilution is made, the ________ decreases, but the ________ remains the same. *(molarity; moles of solute or amount of solute)*

ADJUST INSTRUCTION If students are having difficulty answering the question, have them review the third equation shown in the text: moles of solute = molarity (*M*) × liters of solution (*V*). Remind students that for moles of solute to remain the same, molarity must decrease if volume increases.

Sample Problem 16.4

Preparing a Dilute Solution

How many milliliters of aqueous $2.00M$ $MgSO_4$ solution must be diluted with water to prepare 100.0 mL of aqueous $0.400M$ $MgSO_4$?

KNOWNS

$M_1 = 2.00M$ $MgSO_4$
$M_2 = 0.400M$ $MgSO_4$
$V_2 = 100.0$ mL of $0.400M$ $MgSO_4$

UNKNOWN

$V_1 = ?$ mL of $2.00M$ $MgSO_4$

1 Analyze **List the knowns and the unknown.** Use the equation $M_1 \times V_1 = M_2 \times V_2$ to solve for the unknown initial volume of solution (V_1) that is diluted with water.

2 Calculate **Solve for the unknown.**

Solve for V_1 and substitute the known values into the equation.

$$V_1 = \frac{M_2 \times V_2}{M_1} = \frac{0.400\cancel{M} \times 100.0\text{ mL}}{2.00\cancel{M}} = 20.0\text{ mL}$$

Thus 20.0 mL of the initial solution must be diluted by adding enough water to increase the volume to 100.0 mL.

3 Evaluate **Does the result make sense?** The initial concentration is five times larger than the dilute concentration. Because the number of moles of solute does not change, the initial volume of solution should be one fifth the final volume of the diluted solution.

14. How many milliliters of a solution of $4.00M$ KI are needed to prepare 250.0 mL of $0.760M$ KI?

15. How could you prepare 250 mL of $0.20M$ NaCl using only a solution of $1.0M$ NaCl and water?

What kind of volume-measuring device would you use to make the dilution in Sample Problem 16.4? The dilution requires a molarity with three significant figures, so you would need to measure 20.0 mL of the $2.00M$ $MgSO_4$ solution with a 20-mL volumetric pipette or a burette. (A graduated cylinder would not provide enough precision.) You would transfer the solution to a 100-mL volumetric flask and add distilled water to the flask exactly up to the etched line. The contents would then be 100.0 mL of $0.400M$ $MgSO_4$.

Percent Solutions

How do percent by volume and percent by mass differ?

If both the solute and the solvent are liquids, a convenient way to make a solution is to measure the volumes of the solute and the solution. The concentration of the solute is then expressed as a percent of the solution by volume. **Percent by volume of a solution is the ratio of the volume of solute to the volume of solution.** For example, isopropyl alcohol (2-propanol) is sold as a 91 percent solution by volume. You could prepare such a solution by diluting 91 mL of pure isopropyl alcohol with enough water to make 100 mL of solution. The concentration is written as 91 percent by volume, 91 percent (volume/volume), or 91% (v/v).

$$\text{Percent by volume (\%(v/v))} = \frac{\text{volume of solute}}{\text{volume of solution}} \times 100\%$$

Foundations for Math

SUBSTITUTION If students are not comfortable manipulating a literal equation for a desired variable, explain that they can arrive at the same solution if they substitute given values first and then solve for the unknown.

Solve **Sample Problem 16.4** by first substituting the given values into the formula. Then use inverse operations to solve for the unknown volume.

$$M_1 \times V_1 = M_2 \times V_2$$

$$2.00M\ MgSO_4 \times V_1 = 0.400M\ MgSO_4 \times 100.0\text{ mL}$$

$$\frac{\cancel{2.00M\ MgSO_4} \times V_1}{\cancel{2.00M\ MgSO_4}} = \frac{0.400M\ \cancel{MgSO_4} \times 100.0\text{ mL}}{2.00M\ \cancel{MgSO_4}}$$

$$V_1 = \frac{0.400 \times 100.0\text{ mL}}{2.0} = 20.0\text{ mL}$$

Explain

Dilutions

MAKE A CONNECTION If students are having difficulty with the concepts of molarity and dilution, have them consider diluting a container of juice concentrate (frozen or liquid). Convey that no matter how much water is added to the concentrate, the initial quantity of concentrate does not change. Explain that what changes is the amount of concentrate in a given volume of the diluted solution. Have students consider using only a fraction of the original concentrate. For example, a gallon of juice prepared from half a can will contain half as much concentrate as a gallon prepared from the entire can. Students should be able to deduce that, in this case, the concentration would be reduced by one-half.

USE MODELS Have students work in groups of three to explore dilution. Have each group bring in a shoe box and put 20 red beads into it. Tell students that these red beads represent solute molecules. Then have students add 40 green beads to the box. Tell students that these beads represent solvent molecules. Students should see that the box contains twice as many green beads as red beads; thus, the concentration is 0.5 red/green. Now add another 40 green beads. Explain that this model is analogous to a dilution from 0.5 to 0.25 red/green. Point out that although the concentration of red beads has been halved, the total number of red beads has not changed. Show students that if each red bead represents a mole of solute, 20 moles of solute would still be in solution.

Sample Practice Problem

How many milliliters of a $6.00M$ stock solution of $NH_3(aq)$ would you need to prepare 100.0 mL of $0.30M$ $NH_3(aq)$? *(5.0 mL)*

Answers

FIGURE 16.10 0.040 mol $MgSO_4$(both before and after the dilution)

14. $V_1 = (M_2 \times V_2)/M_1 = (0.760M \times 250.0\text{ mL})/4.00M = 4.75 \times 10^1$ mL

15. Use a volumetric pipette to transfer 5.0×10^1 mL of the $1.0M$ solution to a 250-ml volumetric flask. Then add distilled water up to the mark.

LESSON 16.2

Explain

Percent Solutions

MAKE A CONNECTION Display various fruit-flavored drinks. **Ask** What method is used to determine the percent of juice in the drink, percent by volume, or percent by mass? *(Answers will vary.)* Have students consider whether one or the other method might present the product in a more favorable light. **Ask** Do you think the name "fruit drink" is misleading to consumers? *(Answers will vary.)*

Sample Practice Problem

Ethanol is mixed with gasoline to make gasohol. What is the percent by volume of ethanol in gasohol when 95 mL of ethanol is added to sufficient gasoline to make 1.0 L of gasohol? *(9.5% (v/v))*

CHEMISTRY & YOU Molarity = moles of solute/liters of solution; percent by volume = volume of solute/volume of solution × 100%; percent by mass = mass of solute/mass of solution × 100%

Explore

Teacher Demo

PURPOSE Students will learn how to prepare standard solutions efficiently.

MATERIALS four 20-mL test tubes, 10 mL of 0.5*M* sucrose, 15 mL water, 10-mL pipet

PROCEDURE Remind students of the activity in which they prepared molar solutions of sucrose. Convey that each solution was prepared independently. **Ask** Is there a more efficient way to prepare this same set of molar solutions? *(Yes; make serial dilutions.)* Place 10 mL of 0.5*M* sucrose in one test tube and 5 mL water in each of three others. Using a 10-mL pipette, transfer 5 mL of the 0.5*M* solution to the second test tube; mix by pipeting up and down. Repeat this process for the third and fourth test tubes. Have students calculate the molarities of each of the resulting solutions. Tell students that serial dilutions are used to prepare extremely dilute solutions, such as 0.000001*M* sucrose. The procedure eliminates the error that would arise from having to measure out very small volumes of stock solutions.

Sample Problem 16.5

Calculating Percent by Volume

What is the percent by volume of ethanol (C_2H_6O, or ethyl alcohol) in the final solution when 85 mL of ethanol is diluted to a volume of 250 mL with water?

KNOWNS
volume of solute = 85 mL ethanol
volume of solution = 250 mL

UNKNOWN
Percent by volume = ? % ethanol (v/v)

1 Analyze **List the knowns and the unknown.** Use the known values for the volume of solute and volume of solution to calculate percent by volume.

2 Calculate **Solve for the unknown.**

State the equation for percent by volume.

$$\text{Percent by volume (\% (v/v))} = \frac{\text{volume of solute}}{\text{volume of solution}} \times 100\%$$

Substitute the known values into the equation and solve.

$$\text{\% (v/v)} = \frac{85 \text{ mL ethanol}}{250 \text{ mL}} \times 100\%$$
$$= 34\% \text{ ethanol (v/v)}$$

3 Evaluate **Does the result make sense?** The volume of the solute is about one third the volume of the solution, so the answer is reasonable. The answer is correctly expressed to two significant figures.

16. If 10 mL of propanone (C_3H_6O, or acetone) is diluted with water to a total solution volume of 200 mL, what is the percent by volume of propanone in the solution?

17. A bottle of the antiseptic hydrogen peroxide (H_2O_2) is labeled 3.0% (v/v). How many mL H_2O_2 are in a 400.0-mL bottle of this solution?

Another way to express the concentration of a solution is as a percent by mass, or percent (mass/mass). **Percent by mass of a solution is the ratio of the mass of the solute to the mass of the solution.**

$$\text{Percent by mass (\%(m/m))} = \frac{\text{mass of solute}}{\text{mass of solution}} \times 100\%$$

CHEMISTRY & YOU

Q: *What are three ways to calculate the concentration of a solution?*

You can also define percent by mass as the number of grams of solute per 100 g of solution. Percent by mass is sometimes a convenient measure of concentration when the solute is a solid. For example, a solution containing 7 g of sodium chloride in 100 grams of solution has a concentration of 7 percent by mass—also written as 7 percent (mass/mass) or 7% (m/m).

You have probably seen information on food labels expressed as a percent composition. For example, the label on a fruit-flavored drink often indicates the "percent juice" contained in the product. Such information can be misleading unless the units are given. When you describe percent solutions, be sure to specify whether the concentration is % (v/v) or % (m/m).

Differentiated Instruction

L1 STRUGGLING STUDENTS Write the dilution equation on the board: $M_1 \times V_1 = M_2 \times V_2$. Show students that when mol/*V* is substituted for *M*, the volumes cancel out, leaving moles. Emphasize that, when diluting a solution, the initial number of moles is the same as the final number of moles. Work through several examples using the dilution equation.

ELL ENGLISH LANGUAGE LEARNERS Restate the steps for solving the sample problems in simplified language. Use drawings when possible to promote comprehension by beginning and intermediate speakers. Allow more advanced speakers to translate the sample problems into their native language for their classmates when appropriate.

Sample Problem 16.6

Using Percent by Mass as a Conversion Factor

How many grams of glucose ($C_6H_{12}O_6$) are needed to make 2000 g of a 2.8% glucose (m/m) solution?

1 Analyze List the knowns and the unknown. The conversion is mass of solution ⟶ mass of solute. In a 2.8% $C_6H_{12}O_6$ (m/m) solution, each 100 g of solution contains 2.8 g of glucose. Used as a conversion factor, the concentration allows you to convert g of solution to g of $C_6H_{12}O_6$.

KNOWNS
mass of solution = 2000 g
percent by mass = 2.8% $C_6H_{12}O_6$ (m/m)

UNKNOWN
mass of solute = ? g $C_6H_{12}O_6$

2 Calculate Solve for the unknown.

Write the percent by mass as a conversion factor with g $C_6H_{12}O_6$ in the numerator.

$$\frac{2.8\ g\ C_6H_{12}O_6}{100\ g\ solution}$$

Multiply the mass of the solution by the conversion factor.

$$2000\ \cancel{g\ solution} \times \frac{2.8\ g\ C_6H_{12}O_6}{100\ \cancel{g\ solution}} = 56\ g\ C_6H_{12}O_6$$

3 Evaluate Does the result make sense? The prepared mass of the solution is 20×100 g. Since a 100-g sample of 2.8% (m/m) solution contains 2.8 g of solute, you need $20 \times 2.8\ g = 56$ g of solute. To make the solution, mix 56 g of $C_6H_{12}O_6$ with 1944 g of solvent. (56 g of solute + 1944 g of solvent = 2000 g of solution)

18. Calculate the grams of solute required to make 250 g of 0.10% $MgSO_4$ (m/m).

You can solve this problem by using either dimensional analysis or algebra.

16.2 LessonCheck

19. **Review** How do you calculate the molarity of a solution?

20. **Compare** How does the number of moles of solute before a dilution compare with the number of moles of solute after the dilution?

21. **Identify** What are two ways of expressing the concentration of a solution as a percent?

22. **Calculate** What is the molarity of a solution containing 400 g $CuSO_4$ in 4.00 L of solution?

23. **Calculate** How many milliliters of a stock solution of 2.00*M* KNO_3 would you need to prepare 100.0 mL of 0.150*M* KNO_3?

24. **Calculate** How many moles of solute are present in 50.0 mL of 0.20*M* KNO_3?

25. **Calculate** What is the concentration, in percent (v/v), of a solution containing 50 mL of diethyl ether ($C_4H_{10}O$) in 2.5 L of solution?

26. **Calculate** What mass of K_2SO_4 would you need to prepare 1500 g of 5.0% K_2SO_4 (m/m) solution?

BIGIDEA
THE MOLE AND QUANTIFYING MATTER

27. What information would you need in order to convert molarity to percent by volume?

Lesson Check Answers

19. If the number of moles and the volume of a solution is known, its molarity is determined by dividing the moles of solute by the volume of the solution.

20. Diluting a solution reduces the number of moles of solute per unit volume, but the total number of moles of solute in solution does not change.

21. Express the concentration as the ratio of the volume of the solute to the volume of the solution (v/v) or as the ratio of the mass of the solute to the mass of the solution (m/m).

22. $6.27 \times 10^{-1}M$ $CuSO_4$

23. 7.50 mL

24. 1.0×10^{-2} mol KNO_3

25. 2.0% (v/v) diethyl ether

26. 7.5×10^1g K_2SO_4

27. **BIGIDEA** *M* = mol of solute/L of solution; %(v/v) = volume of solute/volume of solution. You would need to know the molar mass and density of the solute in order to convert moles of solute to volume of solute.

Sample Practice Problem

How many grams of KI would you need to prepare 0.50 g of 7.0% KI (m/m) solution? *(35 grams)*

Evaluate

Informal Assessment

Have students compare percent by mass and percent by volume with molarity as ways of expressing the concentration of a solution. Have students write out a numbered list of steps they would follow to prepare a 1*M* aqueous solution of KCl. Then have students complete the 16.2 Lesson Check.

Reteach

Review the process of making dilutions. Stress to students that M_1 will always be greater than M_2, and V_1 will always be less than V_2. In other words, when making a dilution, molarity always decreases and volume always increases. Remind students that the only value that remains constant during a dilution is the moles of solute.

Answers

16. %(v/v) = 10 mL propanone/200 mL × 100% = 5.0% v/v

17. mL H_2O_2 = (400.0 mL × 3) / 100 = 1.2×10^1 mL

18. 250 g solution × (0.10 g $MgSO_4$/100 g solution) = 0.25 g $MgSO_4$

CHEMISTRY & YOU

CHEMISTRY & YOU Have students read the title of the feature, and draw their attention to the many types of foods shown in the photos that can be pickled. Encourage students to share their own experiences with either preparing or eating pickled foods. If possible, provide a variety of acid and brine pickles for students to examine.

Pose the following question to students: *Why do you think pickling is referred to as an* art *in the title of this feature?* You may need to guide students in the following ways:

- Consider how pickling relates to what you have studied so far in this chapter about solutions.
- The method of pickling varies for different types of food.
- Spices and herbs are added for flavor, but they have no effect on the pickling process itself. The amount of spice added is determined by personal preference.
- It is important to maintain the right concentration in a pickling solution in order to produce a pleasant-tasting food, achieve the proper texture, and avoid the development of harmful bacteria.

CHEMISTRY & YOU: EVERYDAY MATTER

Art of the Pickle

Every culture has its own version of the pickle, and the art of making pickles dates back to ancient history. The earliest known pickles were produced more than 4000 years ago using cucumbers native to India. In Korea, kimchi (pickled cabbage) has been produced for more than 3000 years. Ancient Egyptians and Greeks wrote about the nutritive value and healing power of pickles.

Nowadays, you can often count on the savory taste of a pickle when you order food at a restaurant or from a street vendor. Pickles come in many varieties. You might already know the taste of the dill pickles that accompany sandwiches, or the relish found at a hot dog stand. Other kinds of pickles include Japanese pickled ginger (often served with sushi), European pickled herring, and the hot pickled peppers found in some Mexican salsas.

Pickling is a way of preserving food using a solution of salt, acid (usually vinegar), spices, and/or sugar. Soaking vegetables and meats in the pickling solution prevents the growth of harmful bacteria and imparts a tangy, savory flavor to the food.

Main Types of Pickling

Pickling in Acid Solution	Pickling in Brine Solution
Process This type of pickling typically involves immersing and/or cooking vegetables for a few hours or days in a solution containing vinegar and spices. Vinegar is a 5% (v/v) aqueous solution of acetic acid ($C_2H_4O_2$).	**Process** In brine-based pickling, food is soaked in a salt solution called brine for 4 to 6 weeks. The brine encourages the growth of acid-producing bacteria. The concentration of the brine is usually 5–10% NaCl (m/m).
Examples Acid solutions are used to prepare pickled ginger (above), pickled beets, bread-and-butter pickles, pickled herring, and hot dog relish.	**Examples** Common brine pickles include dill cucumber pickles, sauerkraut (pickled fermented cabbage), Middle Eastern pickled turnips, and Indian mango pickles and chutneys.

21st Century Learning To be successful in the 21st century, students need skills and learning experiences that extend beyond subject matter mastery. The following project helps students build the following 21st Century Skills: *Creativity and Innovation, Communication and Collaboration, Information Literacy, Flexibility and Adaptability, Social and Cross-Cultural Skills, Productivity and Accountability, and Leadership and Responsibility.*

Pickled Peppers Pose the following challenge to students. *You've been hired by a well-known canning supply company to author an illustrated how-to booklet on the preparation of pickled peppers. Booklets should include the following:*

- A brief history and explanation of the use of pickling as a method of preserving food.
- Step-by-step instructions for pickling peppers, with metric measurements. For each step, write a sentence or two explaining why it is important.
- An explanation of the choice of pickling solution (vinegar versus salt).
- Recipe credit and appropriate references.

SALTY SOLUTIONS As pickles soak, the salt in the solution encourages the growth of acid-producing bacteria, which prevent the growth of bacteria that cause spoilage. The salt concentration needs to be as exact as possible. If it's too high, the acid producers can't flourish; if it's too low, then not enough acid producers will grow, and your pickles may spoil.

Take It Further

1. Calculate A pickle recipe calls for 1.5 cups of pickling salt (NaCl) to be mixed with 1 gallon of water. If 1 cup of pickling salt has a mass of 220 grams, what is the percent by mass of NaCl in the pickling solution? (1 gal = 3785 cm^3)

2. Evaluate Is enough information given in the previous question to calculate the percent by volume of NaCl in the pickling solution? Explain your answer.

3. Calculate How many grams of pickling salt would you need to prepare 3500 g of 5.0% NaCl (m/m) solution for a batch of cucumber pickles?

Explain

MAKE A CONNECTION Help students connect the properties of solutions, principles of molarity, and concentration of solutions to the process of pickling. Have them read the Main Types of Pickling feature. Point out that brine pickling uses a 5–10% (m/m) NaCl solution. **Ask** If the concentration of a brine solution used for pickling is 5.0% NaCl (m/m), how many moles of NaCl are in 4.0 kg of solution? *(3.4 mol)* **Ask** Do you have enough information to calculate the molarity of the solution? Explain your answer. *(No; although you know how many moles of solute the solution contains, you do not know the volume of the solution. Only the mass of the solution is known. Molarity is moles of solute per unit volume of solution—not per unit mass of solution.)*

Extend

Connect to FOOD SCIENCE

One version of brine pickling involves soaking meats such as beef brisket in salt water, a process known as *corning*. During this process, salts move into the cells of the meat to equalize sodium chloride concentrations. The cells then absorb water to keep the overall solution concentrations the same. Draw a diagram on the board of a cell and the surrounding solution. Show students how salt and water migrate through the cell wall.

Differentiated Instruction

L1 STRUGGLING STUDENTS Work through this sample calculation before having students answer the Take It Further problem. Suppose the recipe calls for a 10% (m/m) brine solution using 350 g of salt. How many grams of solution does this recipe yield? *(3500 g)* How many grams of water are needed? *(3150 g)*

LPR LESS PROFICIENT READERS Work with students in small groups. Have them read each paragraph independently. Then lead them in discussing the main idea of the paragraph. Afterwards, have them write a sentence describing the paragraph. Continue with the remaining paragraphs.

L3 ADVANCED STUDENTS Commercial production of pickles requires that the pickling solutions be monitored very carefully to guarantee quality. Have students split into groups and research the large-scale production of a type of pickle. Have each group write a report detailing which physical properties of the pickling solution are controlled and how.

Answers

TAKE IT FURTHER

1. mass of solute = 1.5 cup NaCl × (220 g NaCl/1 cup) = 330 g NaCl; mass of solvent = 1 gal H_2O × (3785 cm^3/1 gal) × (1.0 g H_2O/1 cm^3) = 3785 g H_2O; mass of solution = 330 g NaCl + 3785 g H_2O = 4115 g; %(m/m) = 330 g NaCl/4115 g = 8.0% NaCl (m/m)
2. No; the volume of the solution is unknown.
3. 3500 g solution × (5.0 g NaCl/100 g solution) = 1.8×10^2 g NaCl

LESSON 16.3

Key Objectives

16.3.1 EXPLAIN how colligative properties can be explained on a particle basis.

Additional Resources

Reading and Study Workbook, Lesson 16.3

Available Online or on Digital Media:

- Teaching Resources, Lesson 16.3 Review
- Kinetic Art: Vapor Pressure of Solutions

PearsonChem.com

Engage

CHEMISTRY & YOU Have students read the opening text. **Ask** What do you think would happen if you forgot to add rock salt to the ice in an ice-cream machine? *(The ice cream mixture would not get cold enough to freeze.)* **Ask** What do you think would happen if you added too much salt to the ice in an ice cream machine? *(The ice cream mixture could freeze too quickly, making the texture of the resulting ice cream hard or crusty.)*

Activate Prior Knowledge

Have students write the chemical equations for the following examples to refresh their understanding of aqueous solutions.

- hydrochloric acid in water
  ($HCl(g) \rightarrow H^{+}(aq)^{+} Cl^{-}(aq)$)
- magnesium chloride in water
  ($MgCl_2(s) \rightarrow Mg^{2+}(aq) + 2Cl^{-}(aq)$)
- glucose in water
  ($C_6H_{12}O_6(s) \rightarrow C_6H_{12}O_6(aq)$)

IN ACADEMIC STANDARDS for SCIENCE

C.7.1 Describe the composition and properties of types of solutions.

16.3 Colligative Properties of Solutions

C.7.1 Composition and properties of solutions

CHEMISTRY & YOU

Q: *Why do you need salt to make ice cream?* Here's a hint—it's *not* because ice cream is supposed to taste salty. Temperatures below 0°C are needed to make ice cream. Ice-cream makers know that if you add rock salt to ice, the mixture freezes at a few degrees below 0°C. In this lesson, you will discover how a solute can change the freezing point of a solution.

Key Question

What are three colligative properties of solutions?

Vocabulary

- colligative property
- freezing-point depression
- boiling-point elevation

Describing Colligative Properties

What are three colligative properties of solutions?

You already know that the physical properties of a solution differ from those of the pure solvent used to make the solution. After all, tea is not the same as pure water. But it might surprise you to learn that some of these differences in properties have little to do with the specific identity of the solute. Instead, they depend upon the mere presence of solute particles in the solution.

A **colligative property** is a property of solutions that depends only upon the number of solute particles, not upon their identity. **Three important colligative properties of solutions are vapor-pressure lowering, freezing-point depression, and boiling-point elevation.**

Vapor-Pressure Lowering Recall that vapor pressure is the pressure exerted by a vapor that is in dynamic equilibrium with its liquid in a closed system. A solution that contains a solute that is nonvolatile (not easily vaporized) always has a lower vapor pressure than the pure solvent, as shown in Figure 16.11. Glucose, a molecular compound, and sodium chloride, an ionic compound, are examples of nonvolatile solutes. When glucose or sodium chloride is dissolved in a solvent, the vapor pressure of the solution is lower than the vapor pressure of the pure solvent. Why is this true?

KINETIC ART See vapor pressure animated online.

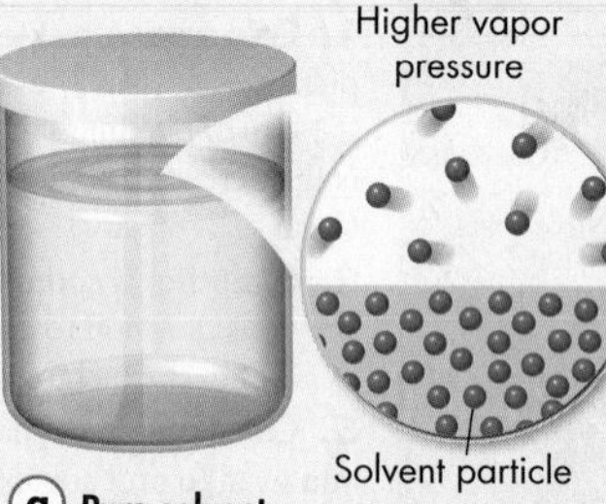

Figure 16.11 Vapor Pressure The vapor pressure of a solution of a nonvolatile solute is less than the vapor pressure of a pure solvent. **a.** Equilibrium is established between the liquid and vapor in a pure solvent. **b.** In a solution, solute particles reduce the number of solvent particles able to escape the liquid. Equilibrium is established at a lower vapor pressure.

Focus on ELL

1 CONTENT AND LANGUAGE Write the words *property*, *depression*, and *elevation* on the board. Explain that these three words have both common and scientific meanings. Divide the class into three groups, and assign one word to each group. Have groups look up the definitions for their word, then explain the similarities and differences between the common and scientific meanings to the rest of the class.

2 FRONTLOAD THE LESSON Have students share their experiences with antifreeze, car de-icers, and road brine treatments. Tell students that each of these products work because of their colligative properties in solution.

3 COMPREHENSIBLE INPUT Use phase diagrams to help illustrate the effects of solutes on vapor pressure, boiling point, and freezing point of a solution. Have students investigate the Kinetic Art animation and perform the ice slurry/rock salt activity on page 536 as adjuncts to classroom discussion of vapor pressure and freezing-point depression, respectively.

Figure 16.12 Molecular vs. Ionic Solutes
Particle concentrations differ for dissolved molecular and ionic compounds in water.
Compare ***Which solution has the lowest vapor pressure? The highest?***

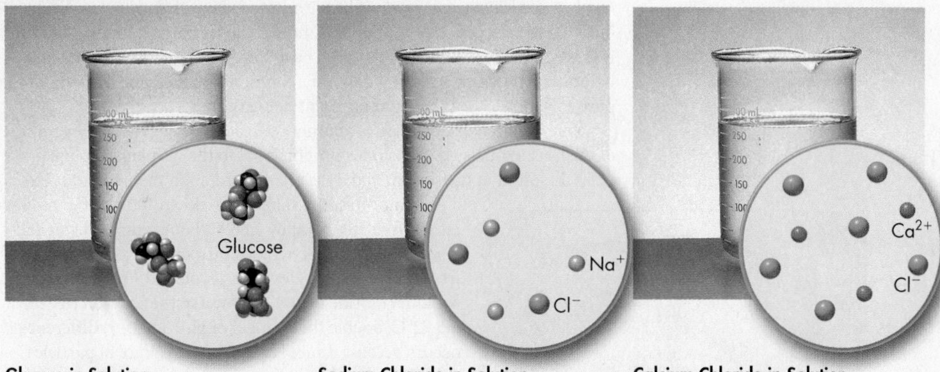

Glucose in Solution
Three moles of glucose dissolved in water produce 3 mol of particles because glucose does not dissociate.

Sodium Chloride in Solution
Three moles of sodium chloride dissolved in water produce 6 mol of particles because each formula unit of NaCl dissociates into two ions.

Calcium Chloride in Solution
Three moles of calcium chloride dissolved in water produce 9 mol of particles because each formula unit of $CaCl_2$ dissociates into three ions.

In an aqueous solution of sodium chloride, sodium ions and chloride ions are dispersed throughout the liquid water. Both within the liquid and at the surface, the ions are surrounded by layers of associated water molecules, or shells of water of solvation. The formation of these shells of water of solvation reduces the number of solvent molecules that have enough kinetic energy to escape as vapor. Thus, the solution has a lower vapor pressure than the pure solvent (water) would have at the same temperature.

Ionic solutes that dissociate, such as sodium chloride and calcium chloride, have greater effects on the vapor pressure than does a nondissociating solute such as glucose. Recall that each formula unit of the sodium chloride (NaCl) produces two particles in solution, a sodium ion and a chloride ion.

$$NaCl(s) \xrightarrow{H_2O} Na^+(aq) + Cl^-(aq)$$

Each formula unit of calcium chloride ($CaCl_2$) produces three particles, a calcium ion and two chloride ions.

$$CaCl_2(s) \xrightarrow{H_2O} Ca^{2+}(aq) + 2Cl^-(aq)$$

When glucose dissolves, the molecules do not dissociate.

$$C_6H_{12}O_6(s) \xrightarrow{H_2O} C_6H_{12}O_6(aq)$$

Figure 16.12 compares the number of particles in three solutions of the same concentration. The decrease in a solution's vapor pressure is proportional to the number of particles the solute makes in solution. For example, the vapor-pressure lowering caused by 0.1 mol of sodium chloride in 1000 g of water is twice that caused by 0.1 mol of glucose in the same quantity of water. In the same way, 0.1 mol of $CaCl_2$ in 1000 g of water produces three times the vapor-pressure lowering as 0.1 mol of glucose in the same quantity of water.

READING SUPPORT

Build Study Skills: *Concept Map*
As you read, construct a concept map that organizes the major ideas of this lesson. ***What factor determines the magnitude of colligative properties of solutions?***

Osmotic Pressure

Another colligative property, which is important in many biological processes, is osmotic pressure. Osmotic pressure is the tendency for solvent to flow through a semipermeable membrane from a region of low solute concentration to a region of higher solute concentration. This process helps regulate the movement of fluids across cell membranes. Cells remain healthy as long as they are continually bathed in an isotonic medium, which maintains the correct osmotic balance between the inside and outside of the cell membrane. Physicians make use of the osmotic effect to treat patients with kidney failure. Afflicted individuals undergo dialysis, a process in which the blood is circulated through a machine equipped with a semipermeable membrane. In the machine, osmotic pressure is used to separate waste materials from the blood.

Foundations for Reading

BUILD VOCABULARY Explain that the adjective *colligative* comes from the Latin *colligare,* meaning "to tie." Explain that a colligative property is one in which particles are seemingly tied together in their actions and act as a group rather than as individuals.

READING STRATEGY Have students create a main ideas and details chart to organize their notes on this lesson. Encourage students to add drawings to their charts to clarify important concepts.

Explain

Describing Colligative Properties

MAKING CONNECTIONS Remind students that ionic compounds and certain molecular compounds, such as HCl, produce two or more particles when they dissolve in water. Most molecular compounds, such as glucose, do not dissociate when they dissolve in water. For each formula unit of $MgCl_2$ that dissolves, three particles are formed in solution. **Ask** How many particles are formed when $FeCl_3$ dissolves in water? *(4)*

USE VISUALS Direct students to Figure 16.12. Emphasize that colligative properties do not depend on the type of particles; rather, they depend on the concentration. For colligative properties, a mole of one kind of particle has the same effect as a mole of any other kind of particle. **Ask** Which produces a greater change in colligative properties—an ionic solid or a molecular solid? *(An ionic solid produces a greater change because it will produce two or more moles of ions for every mole of solid that dissolves.)*

Answers

FIGURE 16.12 The glucose solution has the highest vapor pressure; the calcium chloride solution has the lowest vapor pressure.

LESSON 16.3

Explore

Class Activity

PURPOSE To observe the freezing point depression of ice by the addition of rock salt (NaCl) to a slurry of ice-water

MATERIALS thermometer, foam cup, rock salt, water, ice.

PROCEDURE Have students make an ice-water slurry in their cups. Ask them to measure the initial temperature of the ice-water slurry and the lowest temperature reached after the addition of the rock salt. Explain that the observed difference in temperature is the freezing point depression for this solution of NaCl in water.

Freezing-Point Depression When a substance freezes, the particles of the solid take on an orderly pattern. The presence of a solute in water disrupts the formation of this pattern because of the shells of water of solvation. As a result, more kinetic energy must be withdrawn from a solution than from the pure solvent to cause the solution to solidify. The freezing point of a solution is lower than the freezing point of the pure solvent. The difference in temperature between the freezing point of a solution and the freezing point of the pure solvent is called the **freezing-point depression.**

Freezing-point depression is another colligative property. The magnitude of the freezing-point depression is proportional to the number of solute particles dissolved in the solvent and does not depend upon their identity. The addition of 1 mol of solute particles to 1000 g of water lowers the freezing point by 1.86°C. For example, if you add 1 mol (180 g) of glucose to 1000 g of water, the solution freezes at −1.86°C. However, if you add 1 mol (58.5 g) of sodium chloride to 1000 g of water, the solution freezes at −3.72°C, double the change for glucose. This difference occurs because 1 mol NaCl produces 2 mol of particles and, thus, doubles the freezing-point depression.

The freezing-point depression of aqueous solutions plays an important role in helping to keep travelers safe in cold, icy weather. The truck in Figure 16.13 spreads a layer of salt on the icy road to make the ice melt. The melted ice forms a solution with a lower freezing point than that of pure water. Similarly, ethylene glycol ($C_2H_6O_2$, antifreeze) is added to the water in automobile cooling systems to depress the freezing point of the water below 0°C. Automobiles can thus withstand subfreezing temperatures without freezing up.

Figure 16.13 De-icing Measures
Roads can be free of ice even at temperatures below 0°C if salt is applied. A common de-icer used on aircraft is a mixture of water and propylene glycol.
Infer *Why do you think $CaCl_2$ is a more effective road de-icer than NaCl?*

UbD Check for Understanding

BIGIDEA THE MOLE AND QUANTIFYING MATTER Provide students with a handout copy of Figure 16.12, without the explanatory text. Have students write a one-minute summary on the handout that explains why the three solutions have different freezing points and boiling points, despite the fact that their molar concentrations are equal.

ADJUST INSTRUCTION If students are having difficulty writing their summaries, review the equations in the text on page 535. Have students compare the number of representative particles that result when each solid is dissolved in water. Then, allow students to revise their summaries as needed.

Figure 16.14 Antifreeze The fluid circulating through a car's cooling system is a solution of water and ethylene glycol, or antifreeze. The resulting mixture freezes below 0°C and boils above 100°C.

Boiling-Point Elevation The boiling point of a substance is the temperature at which the vapor pressure of the liquid phase equals atmospheric pressure. As you just learned, adding a nonvolatile solute to a liquid solvent decreases the vapor pressure of the solvent. Because of the decrease in vapor pressure, additional kinetic energy must be added to raise the vapor pressure of the liquid phase of the solution to atmospheric pressure and initiate boiling. Thus, the boiling point of a solution is higher than the boiling point of the pure solvent. The difference in temperature between the boiling point of a solution and the boiling point of the pure solvent is the **boiling-point elevation.**

Figure 16.14 shows antifreeze being poured into a car's coolant tank. The antifreeze doesn't just lower the freezing point of the water in the cooling system. It also elevates the boiling point, which helps protect the engine from overheating in the summer.

Boiling-point elevation is a colligative property; it depends on the concentration of particles, not on their identity. Therefore, you can think about boiling-point elevation in terms of particles. It takes additional kinetic energy for the solvent particles to overcome the attractive forces that keep them in the liquid. Thus, the presence of a solute elevates the boiling point of the solvent. The magnitude of the boiling-point elevation is proportional to the number of solute particles dissolved in the solvent. The boiling point of water increases by 0.512°C for every mole of particles that the solute forms when dissolved in 1000 g of water.

CHEMISTRY & YOU

Q: *Solutes other than NaCl could be used to produce the same freezing-point depression in an ice-cream machine. What factors do you think make NaCl a good choice?*

16.3 LessonCheck

ONLINE PROBLEMS

28. Identify Name three colligative properties of solutions.

29. Explain Why does a solution have a lower vapor pressure than the pure solvent of that solution?

30. Explain Why does a solution have a depressed freezing point and an elevated boiling point compared with the pure solvent?

31. Compare Would a dilute or a concentrated sodium fluoride solution have a higher boiling point? Explain.

32. Compare An equal number of moles of KI and MgI_2 are dissolved in equal volumes of water. Which solution has the higher

a. boiling point?

b. vapor pressure?

c. freezing point?

33. Apply Concepts Review what you learned in Lesson 13.2 about the relationship between the vapor pressure of liquids and their boiling points. Explain why only nonvolatile solutes cause the elevation of the solvent's boiling point.

Lesson Check Answers

28. vapor-pressure lowering, boiling-point elevation, and freezing-point depression

29. Formation of solvation shells around solute particles reduces the number of water molecules with sufficient kinetic energy to escape the solution. Therefore, the vapor pressure is lower relative to pure solvent.

30. Because vapor pressure has been reduced, more kinetic energy is needed to reach the boiling point. For a solution to freeze, it must lose more kinetic energy than the pure solvent does.

31. concentrated sodium fluoride, because the magnitude of the boiling-point elevation is proportional to the number of solute particles dissolved in the solvent

32. **a.** MgI_2 solution

b. KI solution

c. KI solution

33. Volatile solutes quickly evaporate at higher temperatures and so would not be present to cause the elevation of the solvent's boiling point.

Explain

CRITICAL THINKING Ask a student volunteer to explain how a car's cooling system works. Include visuals if time allows. Lead a discussion comparing the consequences of cooling a car engine with pure water to those of using antifreeze in the cooling system. Point out that, when water freezes, it must form a crystalline lattice. Encourage students to think about what effect adding a solute has on the ability of water to form that lattice readily at normal freezing temperatures.

CHEMISTRY & YOU Compared to other ionic solids, NaCl is inexpensive and also food-safe (in case it accidentally comes into contact with the ice-cream mixture).

Evaluate

Informal Assessment

Have students explain, using words or diagrams, how the addition of solute particles to a solvent causes the vapor pressure and freezing point of the solution to decrease relative to the pure solvent and the boiling point to increase. Then, have students complete the 16.3 Lesson Check.

Reteach

Guide students to understand that colligative properties are those physical properties of solutions that depend only on the number of particles of solute in solution and not on the chemical composition of the particles. Explain that the magnitudes of the observed physical changes are proportional to the quantity of solute particles in solution.

Answers

FIGURE 16.13 $CaCl_2$ produces three ions in solution; NaCl produces two.

LESSON 16.4

Key Objectives

16.4.1 DESCRIBE a solution in terms of molality and mole fraction.

16.4.2 EXPLAIN how the freezing-point depression and boiling-point elevation are related to molality.

Additional Resources

Reading and Study Workbook, Lesson 16.4

Available Online or on Digital Media:

- Teaching Resources, Lesson 16.4 Review
- Laboratory Manual, Lab 33
- Virtual Lab, Boiling and Freezing Points of Solutions
- Chem Tutor: Calculating Freezing-Point Depression

PearsonChem.com

Engage

CHEMISTRY & YOU **Ask** *What effect does adding salt have on the cooking process? (Adding salt increases the boiling point of the cooking water.)* **Ask** Do you think the resulting boiling point increase would be enough to significantly shorten the time required for cooking? *(Students are likely to predict, correctly, that the change would be negligible due to the low concentration of salt.)*

Build Background

Ask How does the movement of molecules in ice compare to the movement of molecules in water? *(Molecules move faster in water than in ice.)* Discuss why salt or other de-icing chemicals are spread on roads, driveways, and parking lots during wet weather in temperatures less than 32°F. **Ask** Why is salt useful in these conditions? *(It lowers the freezing point of water so the molecules do not slow down to the point of solidifying into ice.)*

IN ACADEMIC STANDARDS for SCIENCE

C.7.1 Describe the composition and properties of types of solutions.

C.7.3 Describe the concentration of solutes in solution in terms of molarity. Perform calculations using molarity, mass, and volume.

16.4 Calculations Involving Colligative Properties

C.7.1 Composition and properties of solutions; **C.7.3** Molarity

CHEMISTRY & YOU

Q: *How hot is a pot of boiling pasta?* Cooking instructions for a wide variety of foods, from dried pasta to fresh vegetables, often call for the addition of a small amount of salt to the cooking water. Most people like the flavor of food cooked with salt.

But adding salt can have another effect on the cooking process. Recall that dissolved salt elevates the boiling point of water. Suppose you added a teaspoon of salt to two liters of water. A teaspoon of salt has a mass of about 20 g. Would the resulting boiling-point increase be enough to shorten the time required for cooking? In this lesson, you will learn how to calculate the amount the boiling point of the cooking water would rise.

Key Questions

What are two ways of expressing the ratio of solute to solvent in a solution?

How are freezing-point depression and boiling-point elevation related to molality?

Vocabulary

- molality (m)
- mole fraction
- molal freezing-point depression constant (K_f)
- molal boiling-point elevation constant (K_b)

Molality and Mole Fraction

What are two ways of expressing the ratio of solute to solvent in a solution?

Recall that colligative properties of solutions depend only on the number of solute particles dissolved in a given amount of solvent. **Chemists use two ways to express the ratio of solute particles to solvent particles: in molality and in mole fractions.**

Molality (m) is the number of moles of solute dissolved in 1 kilogram (1000 grams) of solvent. Molality is also known as molal concentration.

$$\text{Molality } (m) = \frac{\text{moles of solute}}{\text{kilogram of solvent}}$$

Note that molality is not the same as molarity. Molality refers to moles of solute per kilogram of solvent rather than moles of solute per liter of solution. In the case of water as the solvent, 1 kg or 1000 g equals a volume of 1000 mL, or 1 L.

You can prepare a solution that is 1.00 molal (1m) in glucose, for example, by adding 1.00 mol (180 g) of glucose to 1000 g of water. A 0.500 molal (0.500m) sodium chloride solution is prepared by dissolving 0.500 mol (29.3 g) of NaCl in 1.000 kg (1000 g) of water.

Focus on ELL

1 CONTENT AND LANGUAGE Provide students with a spider diagram. Have them write the word *molality* in the center circle. As students progress through the lesson, have them add any new concepts to their diagram. Stop often and help students pronounce the words on their diagrams. Take this time to verify that students understand how the concepts relate to *molality*.

2 FRONTLOAD THE LESSON Explain to students that this lesson involves the term *molality*, which sounds and looks the very similar to another term that they recently learned, *molarity*. Point out that both terms apply to solutions. Work with students to develop a visual or verbal cue to verify which term is being used in order to avoid confusion due to pronunciation similarities.

3 COMPREHENSIBLE INPUT Explain Figure 16.16 and restate information in its caption as needed. Read the question aloud, and provide students with their own copy of the graph. As you discuss the questions, have students write notes on the graph. Then, have students access the figure's associated Virtual Lab.

Sample Problem 16.7

Using Molality

How many grams of potassium iodide must be dissolved in 500.0 g of water to produce a 0.060 molal KI solution?

❶ Analyze **List the knowns and the unknown.** According to the definition of molality, the final solution must contain 0.060 mol KI per 1000 g H_2O. Use the molality as a conversion factor to convert from mass of the solvent (H_2O) to moles of the solute (KI). Then use the molar mass of KI to convert from mol KI to g KI. The steps are as follows: mass of $H_2O \longrightarrow$ mol KI $\longrightarrow$ g KI.

KNOWNS
mass of water = 500.0 g = 0.5000 kg
solution concentration = 0.060*m*
molar mass of KI = 166.0 g/mol

UNKNOWN
mass of solute = ? g KI

❷ Calculate **Solve for the unknown.**

Identify the conversion factor based on 0.060*m* that allows you to convert from g H_2O to mol KI.

$$\frac{0.060\ \text{mol KI}}{1.000\ \text{kg}\ H_2O}$$

Identify the conversion factor based on the molar mass of KI that allows you to convert from mol KI to g KI.

$$\frac{166.0\ \text{g KI}}{1\ \text{mol KI}}$$

Multiply the known solvent volume by the conversion factors.

$$0.5000\ \cancel{\text{kg}\ H_2O} \times \frac{0.060\ \cancel{\text{mol KI}}}{1.000\ \cancel{\text{kg}\ H_2O}} \times \frac{166.0\ \text{g KI}}{1\ \cancel{\text{mol KI}}} = 5.0\ \text{g KI}$$

To make the 0.060-molal KI solution, you would dissolve 5.0 g of KI in 500.0 g of water.

❸ Evaluate **Does this result make sense?** A 1-molal KI solution is one molar mass of KI (166.0 g) dissolved in 1000 g of water. The desired molal concentration (0.060*m*) is about $\frac{1}{20}$ of that value, so the mass of KI should be much less than the molar mass. The answer is correctly expressed to two significant figures.

34. How many grams of sodium fluoride are needed to prepare a 0.400*m* NaF solution that contains 750 g of water?

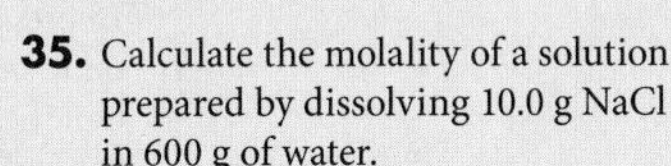

35. Calculate the molality of a solution prepared by dissolving 10.0 g NaCl in 600 g of water.

Remember: Molality equals moles of solute dissolved per kilogram of solvent.

Foundations for Reading

BUILD VOCABULARY Have students create a three-column chart for each vocabulary term. Students can use the chart to organize the terms, definitions, and associated mathematical formulas.

READING STRATEGY Prior to the lesson, have students write down what they know about fractions, moles, freezing points, and boiling points. Then, have students preview the chapter. Ask students to circle the items on their list that will be useful for this lesson. Then, have students write down any concerns they may have about concepts that seem familiar, but are in fact different.

Explain

Molality and Mole Fraction

APPLY CONCEPTS Write the expressions defining molarity and molality on the board. Compare the chemical quantities in each expression. Point out that molarity is denoted by *M* and molality by *m*. Explain that the molality of a solution does not vary with temperature because the mass of the solvent does not change. In contrast, the molarity of a solution does vary with temperature because the liquid can expand and contract. Tell students that when studying colligative properties such as boiling-point elevation and freezing-point depression, it is preferable to use a concentration that does not depend on temperature.

Sample Practice Problem

How many grams of potassium bromide must be dissolved in 0.750 kg of water to produce a 0.150 molal KBr solution? *(13.4 g)*

Foundations for Math

METRIC CONVERSIONS It may make more sense to some students to immediately convert all given quantities to the desired unit in a problem.

In **Sample Problem 16.7,** the mass of the water is given in grams, molar mass is measured in grams/mol, and the desired unit of KI is grams. The only quantity not given in grams is the molality, which is always given in moles per kilogram. Tell students they will arrive at the same answer if they begin by converting molality to moles per *x* grams. **Ask** How many grams are in 1 kilogram? *(1,000)* **Ask** How can you write molality in terms of grams? *(mol/kg = mol/1000 g)* Solve by writing molality in terms of grams:

500 g H_2O × (0.060 mol KI/1000 g H_2O) × (166.0 g KI/1 mol KI).

Show each of the units cancelling to arrive at a final solution of 4.98 g KI, which rounds to 5.0 g KI.

Answers

34. 750 g water × (0.400 mol NaF/1000 g water) × (42.0 g NaF/1 mol NaF) = 1.26×10^1 g NaF

35. (10.0 g NaCl/600 g water) × (1 mol NaCl/58.5 g NaCl) × (1000 g water/1 kg water) = 2.85×10^{-1}*m* NaCl

Explain

MAKE A CONNECTION Guide students to understand that the mole fraction compares the number of moles of a solute to the total number of moles in the solution. Explain that organic chemists, who frequently work with non-aqueous solvent systems, often use this method of expressing concentration. Tell students that the mole fraction is also used when calculating the vapor pressure of a solution.

Explore

Class Activity

PURPOSE Students will organize various methods used to calculate the concentration of a solution.

PROCEDURE Have students design a diagram that includes a definition of each type of concentration unit. More than one design is possible, but each should include the terms *concentration of a solution, percent by mass, percent by volume, molarity, molality, and mole fraction.* Students may wish to include additional detail by showing interconversions between mass of solute and moles of solute. Underneath each concentration unit, students should provide examples of when the unit would be used in the laboratory and in everyday life.

The concentration of a solution also can be expressed as a mole fraction. The **mole fraction** of a solute in a solution is the ratio of the moles of that solute to the total number of moles of solvent and solute. In a solution containing n_A mol of solute A and n_B mol of solvent B, the mole fraction of solute A (X_A) and the mole fraction of solvent B (X_B) can be expressed as follows:

$$X_A = \frac{n_A}{n_A + n_B} \qquad X_B = \frac{n_B}{n_A + n_B}$$

Note that mole fraction is a dimensionless quantity. The sum of the mole fractions of all the components in a solution equals unity, or one.

Figure 16.15 below and Sample Problem 16.8 on the next page illustrate how to calculate the mole fractions of the solute and solvent for a solution of ethylene glycol (EG) in water.

Figure 16.15 Mole Fraction
Ethylene glycol (EG) is added to water as antifreeze in the proportions shown. A mole fraction is the ratio of the number of moles of one substance to the total number of moles of all substances in the solution.
Infer *What is the sum of all mole fractions in a solution?*

Total moles = Ⓐ+Ⓑ = 5.25 mol

Mole fraction EG $= \frac{Ⓐ}{Ⓐ+Ⓑ} = \frac{1.25}{5.25}$

Mole fraction $H_2O = \frac{Ⓑ}{Ⓐ+Ⓑ} = \frac{4.00}{5.25}$

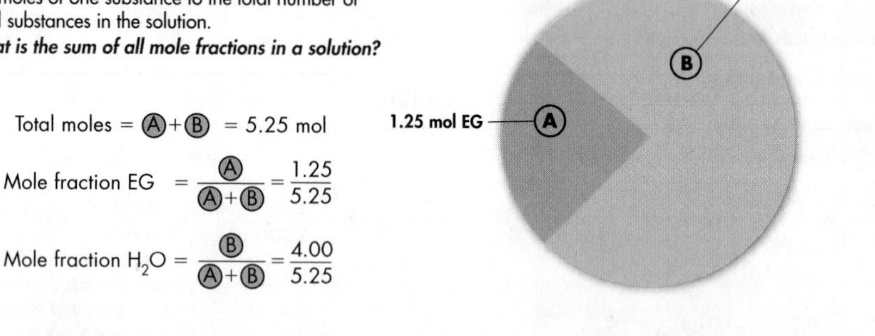

Freezing-Point Depression and Boiling-Point Elevation

How are freezing-point depression and boiling-point elevation related to molality?

Depressions of freezing points and elevations of boiling points are usually quite small. For example, if you add a teaspoon of salt to a pot of water and boil the resulting solution, you will have a hard time detecting any change in the boiling point using a cooking thermometer. It turns out that the elevation is just a small fraction of a degree Celsius. To measure colligative properties accurately, you would need a thermometer that can measure temperatures to the nearest 0.001°C.

Another way to determine the magnitudes of colligative properties is by calculating them. You can do this if you know the molality of the solution and some reference data about the solvent.

Differentiated Instruction

LPR LESS PROFICIENT READERS Have students write sentences using the words *molarity* and *molality*, and have them circle the letter in each word that makes them distinct.

L1 SPECIAL NEEDS STUDENTS Use fraction models or pie models when introducing mole fractions. Use only whole numbers in simple examples until students understand the relationship.

L3 ADVANCED STUDENTS The wood frog is an amphibian that can survive being frozen. Scientists believe that the frog's cells contain a substance that acts as a natural antifreeze. Have students research wood frogs and other animals with this ability, and report back on current theories on the identity of the substance or substances responsible for protecting the animals' cells.

Sample Problem 16.8

Calculating Mole Fractions

Ethylene glycol (EG, or $C_2H_6O_2$) is added to automobile cooling systems to protect against cold weather. What is the mole fraction of each component in a solution containing 1.25 mol of ethylene glycol and 4.00 mol of water?

1 Analyze **List the knowns and the unknowns.** The given quantities of solute (EG) and solvent (water) are expressed in moles. Use the equations for mole fraction of a solute and mole fraction of a solvent to solve this problem. (The pie graph in Figure 16.15 gives you a visual representation of the mole fraction of each component.)

KNOWNS

moles of ethylene glycol (n_{EG}) = 1.25 mol EG

moles of water (n_{H_2O}) = 4.00 mol H_2O

UNKNOWNS

mole fraction EG (X_{EG}) = ?

mole fraction H_2O (X_{H_2O}) = ?

2 Calculate **Solve for the unknowns.**

Write the equation for the mole fraction of ethylene glycol (X_{EG}) in the solution.

$$X_{EG} = \frac{n_{EG}}{n_{EG} + n_{H_2O}}$$

Note that the denominator for each mole fraction is the same: the total number of moles of solvent and solute in the solution.

Write the equation for the mole fraction of water (X_{H_2O}) in the solution.

$$X_{H_2O} = \frac{n_{H_2O}}{n_{EG} + n_{H_2O}}$$

Substitute the known values into each equation.

$$X_{EG} = \frac{n_{EG}}{n_{EG} + n_{H_2O}} = \frac{1.25 \text{ mol}}{1.25 \text{ mol} + 4.00 \text{ mol}} = 0.238$$

$$X_{H_2O} = \frac{n_{H_2O}}{n_{EG} + n_{H_2O}} = \frac{4.00 \text{ mol}}{1.25 \text{ mol} + 4.00 \text{ mol}} = 0.762$$

3 Evaluate **Does the result make sense?** The sum of the mole fractions of all the components in the solution equals 1 ($X_{EG} + X_{H_2O} = 1.000$). Each answer is correctly expressed to three significant figures.

36. What is the mole fraction of each component in a solution made by mixing 300 g of ethanol (C_2H_6O) and 500 g of water?

37. A solution contains 50.0 g of carbon tetrachloride (CCl_4) and 50.0 g of chloroform ($CHCl_3$). Calculate the mole fraction of each component in the solution.

Remember: Mole fraction is a mole ratio, not a mass ratio. If the given quantities are masses, you must first convert each mass to moles using the molar mass of the substance.

Sample Practice Problem

Calculate the mole fraction of each component in a solution of 42 g CH_3OH, 35 g C_2H_5OH, and 50 g C_3H_7OH. *($X_{CH_3OH} = 0.45$; $X_{C_2H_5OH} = 0.26$; $X_{C_3H_7OH} = 0.29$)*

Foundations for Math

FRACTIONS Point out that a mole fraction is simply a ratio that compares a part to a whole. Explain that calculating mole fractions is the same as determining the parts in any other whole, such as slices in a pizza. Therefore, the sum will always be unity, or 1.

In **Sample Problem 16.8,** stress that the formula may look intimidating, but that it simply represents the ratio of parts to the whole, given that the quantities are in moles. Point out that the mole fraction for water is the ratio of water to the total quantity of solution, and the mole fraction for ethylene glycol is the ratio of ethylene glycol to the total quantity of solution.

Answers

FIGURE 16.15 1 (unity)

36. $X_{C_2H_5OH} = 0.190; X_{H_2O} = 0.810$

37. $X_{CCl_4} = 0.437; X_{CHCl_3} = 0.563$

Explain

Freezing-Point Depression and Boiling-Point Elevation

USE VISUALS Display Figure 16.16 on an overhead projector. Choose an arbitrary concentration for an aqueous solution of NaCl or ethylene glycol. Calculate the boiling-point elevation and freezing-point depression. Then, write the temperatures on the horizontal axis. Show students how to read the data for the boiling points of the pure solvent and of the solution. Then, have students read the corresponding values for the freezing points. **Ask** What are ΔT_b and ΔT_f for this solution? (*Point out that the solute affects both the freezing point and boiling point of a liquid.*)

CRITICAL THINKING Have students consider the usefulness of boiling-point elevation and freezing-point depression calculations in formulating automobile antifreeze. **Ask** Why would knowing the boiling-point elevation and freezing-point depression be important when choosing antifreeze for car radiators? *(Despite its name, antifreeze protects against both freezing and overheating.)*

MISCONCEPTION ALERT Some students may think that freezing points and boiling points can be depressed or elevated without end. Explain that as the concentration of a solute, such as ethylene glycol, increases, there comes a point when the quantity of solute exceeds the quantity of solvent. Ethylene glycol becomes the solvent and water becomes the solute; the trends in colligative properties begin to reflect ethylene glycol instead of water. Explain that if the solute is a solid, such as NaCl, eventually the solution becomes saturated. But point out that even before this saturation point is reached, the magnitude of certain colligative properties may reach a maximum.

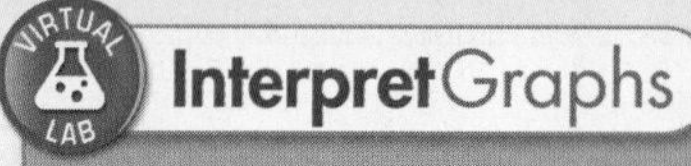

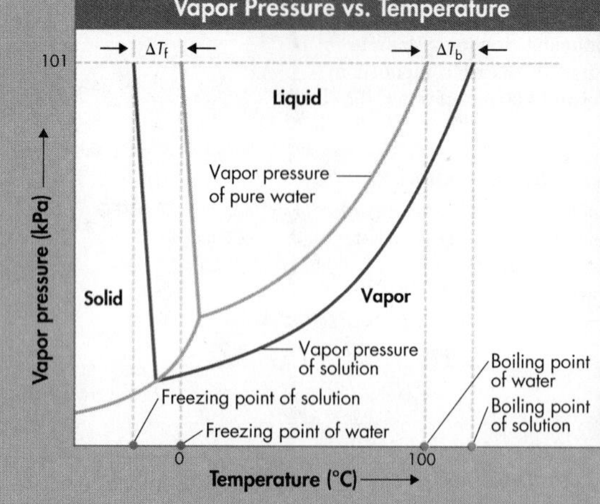

Figure 16.16 The graph shows the relationship between vapor pressure and temperature for pure water and aqueous solutions.

a. Read Graphs What is the freezing point of water? What is the boiling point?

b. Compare How do the freezing and boiling points of the solution compare to those of pure water?

c. Draw Conclusions Does adding a solute to water allow it to remain as a liquid over a longer or shorter temperature range? Explain.

The graph in Figure 16.16 shows that the freezing point of a solvent is lowered, and its boiling point is raised, by the addition of a nonvolatile solute. **The magnitudes of the freezing-point depression (ΔT_f) and the boiling-point elevation (ΔT_b) of a solution are directly proportional to the molal concentration (m), assuming the solute is molecular, not ionic.**

$$\Delta T_f \propto m$$
$$\Delta T_b \propto m$$

Table 16.2

K_f and K_b Values for Some Common Solvents

Solvent	K_f (°C/m)	K_b (°C/m)
Acetic acid	3.90	3.07
Benzene	5.12	2.53
Camphor	37.7	5.95
Cyclohexane	20.2	2.79
Ethanol	1.99	1.19
Nitrobenzene	7.00	5.24
Phenol	7.40	3.56
Water	1.86	0.512

The change in the freezing temperature (T_f) is the difference between the freezing point of the solution and the freezing point of the pure solvent. Similarly, the change in the boiling temperature (T_b) is the difference between the boiling point of the solution and the boiling point of the pure solvent. The term m is the molal concentration of the solution.

With the addition of a constant, the proportionality between the freezing point depression (ΔT_f) and the molality m can be expressed as an equation.

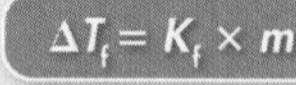

$$\Delta T_f = K_f \times m$$

The constant, K_f, is the **molal freezing-point depression constant,** which is equal to the change in freezing point for a 1-molal solution of a nonvolatile molecular solute. The value of K_f depends upon the solvent. Its units are °C/m. Table 16.2 lists the K_f values for water and some other solvents.

542 Chapter 16 • Lesson 4

UbD Check for Understanding

BIGIDEA THE MOLE AND QUANTIFYING MATTER Have students write one-minute summaries comparing and contrasting molarity and molality.

ADJUST INSTRUCTION If students are having difficulty with their summaries, write the formulas for each quantity on the board, highlighting the differences in a different color. Then, review the appropriate times to use each quantity.

Calculating the Freezing-Point Depression of a Solution

Antifreeze protects a car from freezing. It also protects it from overheating. Calculate the freezing-point depression of a solution containing exactly 100 g of ethylene glycol ($C_2H_6O_2$) antifreeze in 0.500 kg of water.

1 Analyze **List the knowns and the unknown.** Calculate the number of moles of $C_2H_6O_2$ and the molality of the solution. Then calculate the freezing-point depression using $\Delta T_f = K_f \times m$.

KNOWNS	UNKNOWN
mass of $C_2H_6O_2$ = 100 g	ΔT_f = ?°C
mass of water = 0.500 kg	
K_f for H_2O = 1.86°C/m	
molar mass of $C_2H_6O_2$ = 62.0 g/mol	

2 Calculate **Solve for the unknown.**

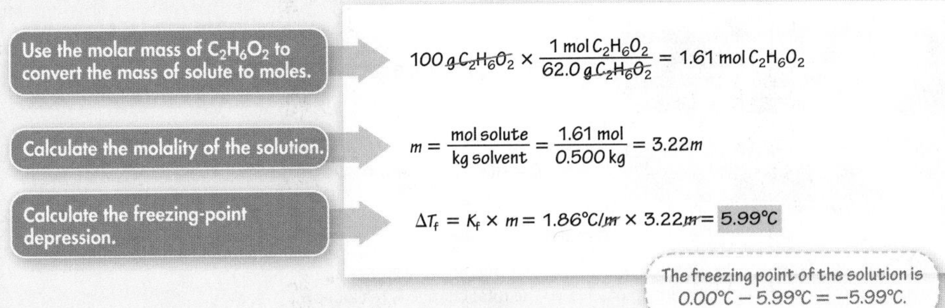

The freezing point of the solution is 0.00°C − 5.99°C = −5.99°C.

3 Evaluate **Does the result make sense?** A 1-molal solution reduces the freezing temperature by 1.86°C, so a decrease of 5.99°C for an approximately 3-molal solution is reasonable.

38. What is the freezing-point depression of an aqueous solution of 10.0 g of glucose ($C_6H_{12}O_6$) in 50.0 g H_2O?

39. Calculate the freezing-point depression of a benzene solution containing 400 g of benzene and 200 g of the molecular compound acetone (C_3H_6O). K_f for benzene is 5.12°C/m.

As you might expect, the boiling-point elevation of a solution can also be expressed as an equation. In this case, the proportionality constant is K_b, the **molal boiling-point elevation constant,** which is equal to the change in boiling point for a 1-molal solution of a nonvolatile molecular solute.

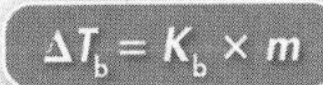

Table 16.2 lists the K_b values for some solvents. Like K_f, K_b has units of °C/m.

Sample Problem 16.9 above described how to determine ΔT_f if the solute is a molecular compound. But for ionic compounds, both ΔT_f and ΔT_b depend upon the number of ions produced by each formula unit. This number is used to calculate an effective molality, as you'll see in Sample Problem 16.10.

Q: *Does pasta cook at 100°C? After you read Sample Problem 16.10, calculate the boiling-point elevation for the solution described at the beginning of the lesson on page 538.*

Focus on ELL

4 LANGUAGE PRODUCTION Have students work in groups of four to complete the Small-Scale Lab on page 545. Make sure each group has ELLs of varied language proficiencies, so that more proficient students can help less proficient ones. Have students work according to their proficiency level.

BEGINNING: LOW/HIGH Recreate the procedure as a numbered list of steps to follow. Model the procedure for students.

INTERMEDIATE: LOW/HIGH Provide students with an organized and labeled chart to complete during the Analyze and Conclude section.

ADVANCED: LOW/HIGH Have students create their own data chart to record the answers to their Analyze and Conclude calculations.

Extend

Have students consider the following question: *Which aqueous solution would have the larger boiling-point elevation and freezing-point depression: 1m KCl or 1m $CaBr_2$?* Challenge students to answer the question, and use their knowledge of molality, boiling-point elevation, and freezing-point depression to justify their choice. *($CaBr_2$; because both solutes are ionic, their effects on freezing-point depression and boiling-point elevation are determined by the number of particles each places into solution. Even though both solutions have a molality of 1, the solution $CaBr_2$ contains 3 mol of ions per kilogram of water, and the KCl solution contains 2 mol of ions per kilogram of water. Theoretically, the $CaBr_2$ solution will exhibit a freezing-point depression and boiling-point elevation one and one-half times as great as those exhibited by the KCl solution.)*

ΔT_b = 0.09°C

Sample Practice Problem

What is the freezing point of a 0.800m aqueous solution of glucose, given that K_f for H_2O = 1.86°C/m? *(−1.49°C)*

Answers

INTERPRET GRAPHS

- **a.** 0°C; 100°C
- **b.** The freezing point of the solution is lower than that of pure water and the boiling point is higher than that of pure water.
- **c.** Adding a solute to water allows it to remain as a liquid over a longer temperature range because the solution changes to a solid at a lower temperature and changes to a vapor at a higher temperature.

38. mol $C_6H_{12}O_6$ = 10.0 g glucose × (1 mol/180.2 g) = 0.0556 mol; m = (0.0556 mol/50.0 g H_2O) × (1000 g/1 kg) = 1.11m; $\Delta T_f = K_f \times m$ = 1.86°C/m × 1.11m = 2.06°C

39. mol C_3H_6O = 200 g C_3H_6O × (1 mol/58.0 g C_3H_6O) = 3.45 mol C_3H_6O; m = (3.45 mol C_3H_6O/400 g) × (1000 g/1 kg) = 8.63m; $K_f \times m$ = 5.12°C/m × 8.63m = 44.2°C

LESSON 16.4

Explain

Sample Practice Problem

What is the boiling point of a solution containing 96.7 g of sucrose ($C_{12}H_{22}O_{11}$) in 250.0 g water at 1 atm? *(100.579°C)*

Evaluate

Informal Assessment

Assess students' understanding of colligative properties by posing the following questions.

Ask Which solution has a higher boiling point, 1 mol of $Al(NO_3)_3$ in 1000 g of water or 1.5 mol of KCl in 1000 g of water? Have students explain their answers. *(The solution of $Al(NO_3)_3$ has a higher boiling point because $Al(NO_3)_3$ dissociates into a larger number of particles.)* **Ask** Does volatility affect the colligative properties of a compound? Explain. *(Yes; Volatile solutes quickly evaporate at higher temperatures, which would change the molal concentration of the solution.)*

Then, have students complete the 16.4 Lesson Check.

Reteach

Remind students that when they do calculations involving ionic solids, they must find the molality of the solution in terms of the total number of particles. Work through several examples to be sure they understand this important point.

Sample Problem 16.10

Calculating the Boiling Point of a Solution

What is the boiling point of a 1.50*m* NaCl solution?

1 Analyze List the knowns and the unknown. Each formula unit of NaCl dissociates into two particles, according to the equation $NaCl(s) \longrightarrow Na^+(aq) + Cl^-(aq)$. Based on the total number of dissociated particles, the effective molality is $2 \times 1.50m = 3.00m$. Calculate the boiling-point elevation (using the equation $\Delta T_b = K_b \times m$), and then add it to 100°C.

KNOWNS

solution concentration = 1.50*m* NaCl

K_b for H_2O = 0.512°C/*m*

UNKNOWN

boiling point = ?°C

2 Calculate Solve for the unknown.

Calculate the boiling-point elevation, making sure to use the molality of total dissociated particles in solution.

$\Delta T_b = K_b \times m = 0.512°C/m \times 3.00m = 1.54°C$

Calculate the boiling point of the solution.

$T_b = 100°C + 1.54°C = 101.54°C$

3 Evaluate Does the result make sense? The boiling point increases about 0.5°C for each mole of solute particles, so the total change is reasonable. Because the boiling point of water is defined as exactly 100°C, this value does not limit the number of significant figures in the solution of the problem.

40. What is the boiling point of a solution that contains 1.25 mol $CaCl_2$ in 1400 g of water?

41. What mass of NaCl would have to be dissolved in 1.000 kg of water to raise the boiling point by 2.00°C?

16.4 LessonCheck

ONLINE PROBLEMS

42. List What are two ways of expressing the ratio of solute particles to solvent particles?

43. Explain How are freezing-point depression and boiling-point elevation related to molality?

44. Calculate How many grams of sodium bromide must be dissolved in 400.0 g of water to produce a 0.500 molal solution?

45. Calculate Calculate the mole fraction of each component in a solution of 2.50 mol ethanoic acid ($C_2H_4O_2$) in 10.00 mol of water.

46. Predict What is the freezing point of a solution of 12.0 g of CCl_4 dissolved in 750.0 g of benzene? The freezing point of benzene is 5.48°C; K_f is 5.12°C/*m*.

47. Make Generalizations Look at the table on page R1 of the Elements Handbook showing the distribution of elements in the oceans. What generalization can you make about the temperature at which ocean water will freeze? What effect does the presence of dissolved elements in the ocean have on the rate of evaporation of ocean water?

544 Chapter 16 • Lesson 4

Answers

40. $m = (1.25 \text{ mol } CaCl_2/1400 \text{ g}) \times (1000 \text{ g}/1 \text{ kg}) = 0.893m\ CaCl_2$; each formula unit of $CaCl_2$ dissociates into three particles; $0.893m\ CaCl_2 \times 3 = 2.68m$; $\Delta T_b = K_b \times m = 0.512°C/m \times 2.68m = 1.37°C$; 100°C + 1.37°C = 101.37°C

41. 2.00°C/0.512°C/*m* = 3.91*m*; Each formula unit of NaCl dissociates into two particles; 3.91*m*/2 = 1.96*m*; in 1 kg of water, 1.96 mol NaCl × (58.5 g/1 mol) = 115 g NaCl

Lesson Check Answers

42. Molality and mole fractions are two convenient ways of expressing the ratio of solute particles to solvent particles.

43. The magnitudes of the freezing-point depression (ΔT_f) and the boiling-point elevation (ΔT_b) of a solution are directly proportional to the molal concentration (*m*) when the solute is molecular, not ionic.

44. 20.6 g NaBr

45. $X_{CH_3COOH} = 0.200$; $X_{H_2O} = 0.800$

46. 4.95°C

47. Given the list of dissolved elements and their concentrations in ppm, students should infer that the freezing point of ocean water is less than 0°C at 1 atm and that the rate of evaporation of ocean water is slower because of lowered vapor pressure.

Small-Scale Lab

Making a Solution

Purpose

To make a solution and use carefully measured data to calculate the solution's concentration

Materials

- solid NaCl
- water
- 50-mL volumetric flask
- balance

Procedure 

Measure the mass of a clean, dry, volumetric flask. Add enough solid NaCl to fill approximately one tenth of the volume of the flask. Measure the mass of the flask again. Half fill the flask with water and shake it gently until all the NaCl dissolves. Fill the flask with water to the 50-mL mark and measure the mass again.

Analyze and Conclude

Answer the following questions based on your data.

1. Percent by mass tells how many grams of solute are present in 100 g of solution.

$$\% \text{ by mass} = \frac{\text{mass of solute}}{\text{mass of solute} + \text{solvent}} \times 100\%$$

a. Calculate the mass of the solute (NaCl).
b. Calculate the mass of the solvent (water).
c. Calculate the percent by mass of NaCl in the solution.

2. Mole fraction tells how many moles of solute are present for every 1 mol of total solution.

$$\text{Mole fraction} = \frac{\text{mol NaCl}}{\text{mol NaCl} + \text{mol } H_2O}$$

a. Calculate the moles of NaCl solute.
Molar mass of NaCl = 58.5 g/mol
b. Calculate the moles of water.
Molar mass of H_2O = 18.0 g/mol
c. Calculate the mole fraction of your solution.

3. Molality (m) tells how many moles of solute are present in 1 kg of solvent.

$$m = \frac{\text{mol NaCl}}{\text{kg } H_2O}$$

Calculate the molality of your solution.

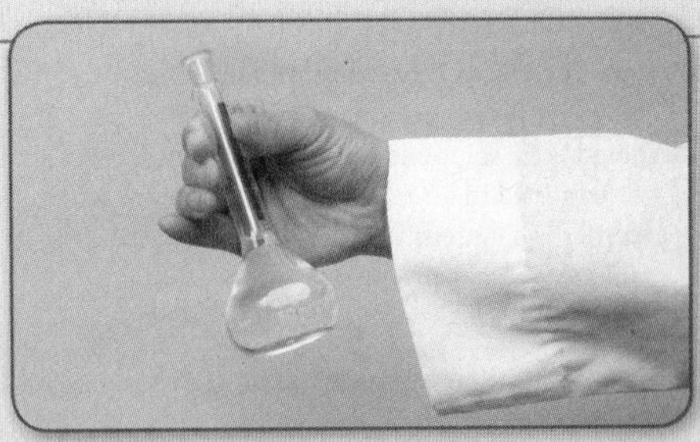

4. Molarity (M) tells how many moles of solute are dissolved in 1 L of solution.

$$M = \frac{\text{mol NaCl}}{\text{L solution}}$$

a. Calculate the liters of solution.
b. Calculate the molarity of the NaCl solution.

5. Density tells how many grams of solution are present in 1 mL of solution.

$$\text{Density} = \frac{\text{g solution}}{\text{mL solution}}$$

Calculate the density of the solution.

You're the Chemist

The following small-scale activities allow you to develop your own procedures and analyze the results.

1. Analyze Data Measure the mass of an empty volumetric flask. Use a small-scale pipette to extract a sample of your NaCl solution and deliver it to the flask. Measure the mass of the flask again and fill it with water to the 50-mL line. Measure the mass of the flask again. Calculate the concentration of this dilute solution using the same units you used to calculate the concentration of the NaCl solution. Are the results you obtained reasonable?

2. Design an Experiment Design and carry out an experiment to make a solution of table sugar quantitatively. Calculate the concentration of the table sugar solution using the same units you used to calculate the concentration of the NaCl solution. Is the effective molality of the table sugar solution the same as the effective molality of a sodium chloride solution of the same concentration? Recall that effective molality is the concentration value used to calculate boiling-point elevation and freezing-point depression.

Small-Scale Lab

Making a Solution

OBJECTIVE After completing this activity, students will be able to make a solution and to calculate the solution's molarity, molality, percent by mass, and mole fraction.

PREP TIME 20 minutes

CLASS TIME 20 minutes

SAFETY Wear safety goggles and a lab apron.

TEACHING TIPS Household table salt contains a small amount of desiccant, such as sodium silicate, which gives the salt solution a hazy appearance. Use lab grade NaCl instead. Remind students about significant figures. Point out that the balance is probably more accurate than the volumetric flask. Calculations based only on measurements using the balance will have a larger number of significant figures.

EXPECTED OUTCOME Students make a solution and calculate the concentration in various units.

ANALYZE AND CONCLUDE

Sample data:
dry bottle = 15.98 g
flask + NaCl = 22.88 g
flask + NaCl + water = 69.09 g

1. **a.** 6.90 g **b.** 46.21 g **c.** 13.0%
2. **a.** 0.118 mol NaCl **b.** 2.57 mol H_2O **c.** 0.0439
3. 2.55m
4. **a.** 0.050 L **b.** 2.4M
5. 1.1 g/mL

YOU'RE THE CHEMIST!

1. Sample data:
dry flask = 15.98 g
flask + NaCl solution = 22.88 g
flask + NaCl solution + water = 69.09 g
mass of NaCl solution = 2.09 g
mass of water = 47.77 g
mass of NaCl = 0.271 g
percent mass of NaCl = 0.544%
moles NaCl = 4.64×10^{-3} mol
mass of water = 49.59 g
moles of water = 2.76 mol
mole fraction = 1.68×10^{-3}
molality = 0.0936m
density = 1.0 g mL

2. Sample data:
mass of dry flask = 16.72 g
mass of flask + sugar = 20.85 g
mass of flask + sugar + water = 69.53 g
mass of sugar = 4.13 g
mass of solvent = 48.68 g
percent mass of sugar = 7.82%
moles of sugar = 0.0121 mol
moles of water = 2.70 mol
mole fraction = 4.46×10^{-3}
molality = 0.249m
molarity = 0.242M
density = 1.1 g/mL

No, the effective molality of the NaCl solution is twice the effective molality of the sugar solution.

FOR ENRICHMENT

Students could compare the freezing point depressions of the sugar and salt solutions. (One way to do this is to partially freeze the solutions and determine the temperature of the mixtures of liquid and solid.)

Evaluate

Review and Assessment Materials

CHEMISTRY TUTORIAL Encourage students to review solving Henry's Law, molarity, and freezing-point depression problems using the animated step-by-step tutorials for Chapter 16. Each Chemistry Tutorial is followed by a similar practice problem, where the student is guided through the problem-solving process.

CHAPTER PROBLEM SET Have students practice more molarity and chemical quantity problems using the Chapter 16 Online Problem Set.

PROBLEM SETS Have students practice more chemical quantity problems by using the Chapter 16 Online Problem Set.

DIRECTED VIRTUAL LAB Have students complete the Virtual Lab *Solutions* to help reinforce the concept of boiling point elevation. In the post-lab exercise, students will analyze the data and draw conclusions based on the results of the experiment.

KINETIC ART Direct students to the Kinetic Art animations *Dynamic Equilibrium of Saturated Solutions* and *Vapor Pressure*, and have them answer the questions that accompany each animation.

CONCEPTS IN ACTION If students have trouble understanding freezing-point depression, ask them to view the Concepts in Action animation, which presents the material in a familiar context. Have students complete the questions after the animation.

Study Tip

Encourage students to create flashcards to help them visualize relationships and formulas. Tell them to try to integrate what they're studying with what they already know.

16 Study Guide

BIGIDEA

THE MOLE AND QUANTIFYING MATTER

Solubility, miscibility, concentration, and colligative properties are used to describe and characterize solutions. Solution concentration can be quantified in terms of molarity (moles of solute per liter of solution), molality (moles of solute per kilogram of solvent), percent by volume, and percent by mass.

16.1 Properties of Solutions

Factors that determine how fast a substance dissolves are stirring, temperature, and surface area.

In a saturated solution, a state of dynamic equilibrium exists between the solution and any undissolved solute, provided that the temperature remains constant.

Temperature affects the solubility of solid, liquid, and gaseous solutes in a solvent; both temperature and pressure affect the solubility of gaseous solutes.

- saturated solution (520)
- solubility (520)
- unsaturated solution (520)
- miscible (521)
- immiscible (521)
- supersaturated solution (522)
- Henry's law (523)

Key Equation

$$\text{Henry's law: } \frac{S_1}{P_1} = \frac{S_2}{P_2}$$

16.2 Concentrations of Solutions

To calculate the molarity of a solution, divide the moles of solute by the volume of the solution in liters.

Diluting a solution reduces the number of moles of solute per unit volume, but the total number of moles of solute in solution does not change.

Percent by volume is the ratio of the volume of solute to the volume of solution. Percent by mass is the ratio of the mass of the solute to the mass of the solution.

- concentration (525)
- dilute solution (525)
- concentrated solution (525)
- molarity (M) (525)

Key Equations

$$\text{Molarity } (M) = \frac{\text{moles of solute}}{\text{liters of solution}}$$

$$M_1 \times V_1 = M_2 \times V_2$$

$$\text{Percent by volume} = \frac{\text{volume of solute}}{\text{volume of solution}} \times 100\%$$

$$\text{Percent by mass} = \frac{\text{mass of solute}}{\text{mass of solution}} \times 100\%$$

16.3 Colligative Properties of Solutions

Colligative properties of solutions include vapor-pressure lowering, freezing-point depression, and boiling-point elevation.

- colligative property (534)
- freezing-point depression (536)
- boiling-point elevation (537)

16.4 Calculations Involving Colligative Properties

Chemists use two ways to express the ratio of solute to solvent: in molality and in mole fractions.

The magnitudes of freezing-point depression and boiling-point elevation are proportional to molality.

- molality (m) (538)
- mole fraction (540)
- molal freezing-point depression constant (K_f) (542)
- molal boiling-point elevation constant (K_b) (543)

Key Equations

$$\text{Molality } (m) = \frac{\text{moles of solute}}{\text{kilogram of solvent}}$$

$$\text{Mole fractions: } X_A = \frac{n_A}{n_A + n_B} \qquad X_B = \frac{n_B}{n_A + n_B}$$

$$\Delta T_f = K_f \times m$$

$$\Delta T_b = K_b \times m$$

UbD Performance Tasks

STANDARDIZED ROAD PREP Have students research the most commonly used brine and solid road de-icers used to maintain roads during winter weather events. Students should identify the chemicals used, discover the freezing-point depression provided by each chemical, and use this value to calculate the molality of the solution formed by the combination of melted precipitation and de-icer.

SOLUBILITY 911 Have students determine the impact a recent well-known chemical spill has had on the environment. They should explain how emergency clean-up specialists used their knowledge of solubility to aid in the cleanup. Challenge students to create a list of emergency procedures that cleanup workers used for the chemical spill, and use solubility concepts to explain the rationale for each step.

Math Tune-Up: Solution Concentration Problems

Problem	❶ Analyze	❷ Calculate	❸ Evaluate
What volume of 12.00M sulfuric acid is required to prepare 1.00 L of 0.400M sulfuric acid?	Knowns: $M_1 = 12.00M\ H_2SO_4$ $M_2 = 0.400M\ H_2SO_4$ $V_2 = 1.00\text{ L of }0.400M\ H_2SO_4$ Unknown: $V_1 = ?\text{ L of }12.00M\ H_2SO_4$ Use the following equation to solve for the unknown initial volume of solution that is diluted: $M_1 \times V_1 = M_2 \times V_2$	Solve the equation for V_1 and substitute. $V_1 = \frac{M_2 \times V_2}{M_1}$ $V_1 = \frac{0.400\cancel{M} \times 1.00\text{ L}}{12.00\cancel{M}}$ $V_1 = 0.0333\text{ L}$	The concentration of the initial solution (12.00M) is 30 times larger than the concentration of the diluted solution (0.400M). So, the volume of the solution to be diluted should be one thirtieth the final volume of the diluted solution.
Ethanol is mixed with gasoline to make a solution called gasohol. What is the percent by volume of ethanol in gasohol when 95 mL of ethanol is added to sufficient gasoline to make 1.0 L of gasohol?	Knowns: volume of ethanol = 95 mL volume of solution = 1.0 L Unknown: solution concentration = ? % (v/v) Use the equation for percent by volume: $\%\text{ (v/v)} = \frac{\text{volume of solute}}{\text{volume of solution}} \times 100\%$	Make sure the known volumes are expressed in the same units. Then calculate percent by volume of ethanol. $\%\text{ (v/v)} = \frac{0.095\ \cancel{L}}{1.00\ \cancel{L}} \times 100\%$ $= 9.5\%\text{ (v/v)}$	The volume of the solute is about one tenth the volume of the solution, so the answer is reasonable. The answer is correctly expressed to two significant figures.
Calculate the molality of a solution prepared by mixing 5.40 g LiBr with 444 g of water.	Knowns: mass of solute = 5.40 g mass of water = 444 g molar mass of LiBr = 86.8 g Unknown: solution concentration = ?m Use the equation for molal concentration: $\text{Molality} = \frac{\text{mol of solute}}{\text{kg of solvent}}$	Convert the mass of the solute to moles of solute. $5.40\ \cancel{\text{g LiBr}} \times \frac{1\text{ mol LiBr}}{86.8\ \cancel{\text{g LiBr}}} =$ 0.0622 mol LiBr Calculate molality. $\text{Molality} = \frac{0.0622\text{ mol LiBr}}{0.444\text{ kg }H_2O}$ $= 0.140m$	The answer has the correct units (mol of solute per kg of solvent) and is correctly expressed to three significant figures.

Remember: Molality is mol of solute per kg of solvent. Make sure you have the correct mass units in the denominator.

MATH TUNE-UP

Math Review

Example and sample practice problems involving solutions can be found on the following pages:

LESSON 16.1 Using Henry's Law—page 524

LESSON 16.2 Calculating Molarity—page 526; Calculating the Moles of Solute in a Solution—page 527; Preparing a Dilute Solution—page 529; Calculating Percent by Volume—page 530; Using Percent by Mass as a Conversion Factor—page 531

LESSON 16.4 Use Molality—page 539; Calculating Mole Fractions—page 541; Calculating the Freezing-Point Depression of a Solution—page 543; Calculating the Boiling Point of a Solution—page 544

Focus on ELL

5 ASSESS UNDERSTANDING Have students complete the Standardized Test Prep for this chapter. After the results of the assessment are given, have students review the topics they do not understand.

BEGINNING: LOW/HIGH Have students underline the high-frequency words they understand and use context clues to eliminate unlikely distracters. Make sure students know how to correctly use an answer sheet.

INTERMEDIATE: LOW/HIGH Explain to students that they can skip a question that they do not understand immediately. Have students mark the question so that they remember to come back to it. Remind them to skip that answer on the answer sheet.

ADVANCED: LOW/HIGH Ask students to write down the questions they missed. Then, have them identify and explain why their answer was incorrect.

Math Tune-Up 547

Evaluate

Answers

LESSON 16.1

48. The solvent is the substance in which the solute is dissolved.

49. Random collisions of the solvent molecules with the solute particles provide enough force to overcome gravity.

50. solubility: the amount of a substance that dissolves in a given quantity of solvent at specified conditions of temperature and pressure to produce a saturated solution; saturated solution: a solution containing the maximum amount of solute for a given amount of solvent at a constant temperature and pressure; unsaturated solution: a solution that contains less solute than a saturated solution at a given temperature and pressure; miscible: describes liquids that dissolve in each other; immiscible: describes liquids that are insoluble in each other.

51. Particles of solute crystallize.

52. No; if there were undissolved solute, the excess solute would come out of a supersaturated solution.

53. 5.6×10^2 g $AgNO_3$

54. Solubility increases with pressure.

55. **a.** 1.6×10^{-2} g/L **b.** 4.7×10^{-2} g/L

LESSON 16.2

56. *Dilute* and *concentrated* are relative terms and are not quantitative. Molarity provides the exact number of moles of solute per liter of solution.

57. Molarity is the number of moles of solute dissolved in one liter of solution.
a. 1.3*M* KCl **b.** 3.3×10^{-1}*M* $MgCl_2$

58. 2.00×10^1 mL

59. 0.83*M*

60. **a.** 5.0×10^{-1} mol NaCl, 29 g NaCl
b. 1.0 mol KNO_3, 1.0×10^2 g KNO_3 **c.** 2.5×10^{-2} mol $CaCl_2$, 2.8 g $CaCl_2$ **d.** 0.60 mol Na_2SO_4, 85 g Na_2SO_4

61. **a.** 2.3×10^1 g NaCl
b. 2.0 g $MgCl_2$

62. **a.** 5.5% (m/m) **b.** 3.0% (m/m) **c.** 15% (m/m)

63. **a.** 17% (v/v) ethanol
b. 63.6% (v/v) isopropyl alcohol

LESSON 16.3

64. Colligative properties are properties of a solution that depend only on the number of solute particles; boiling-point elevation, freezing-point depression, and vapor-pressure lowering. Boiling points are elevated because solvent molecules surround solute particles, reducing the number of solvent molecules that have sufficient energy to escape the solution; relative to the pure solvent, the amount of energy required to cause vaporization or boiling increases. Solutes disrupt the ordering of the solvent structure, so more kinetic energy must be withdrawn from a solution for it to solidify. This lowers the freezing point of the solution.

65. **a.** sea water **b.** 1.5*M* KNO_3
c. 0.100*M* $MgCl_2$

66. The effective molality of the $Ca(NO_3)_2$ solution is 3*m*. The effective molality of the $NaNO_3$ solution is 2*m*.

67. When vapor pressure is lowered relative to pure solvent, more energy must be supplied to reach the boiling point; thus the boiling point is increased relative to pure solvent.

68. The added glucose lowers the freezing point of the solution within the frog's cells.

16 Assessment

* Solutions appear in Appendix E

Lesson by Lesson

16.1 Properties of Solutions

48. Name and distinguish between the two components of a solution.

49. Explain why the dissolved component does not settle out of a solution.

50. Define the following terms: *solubility, saturated solution, unsaturated solution, miscible,* and *immiscible.*

51. If a saturated solution of sodium nitrate is cooled, what change might you observe?

52. Can a solution with undissolved solute be supersaturated? Explain.

53. What mass of $AgNO_3$ can be dissolved in 250 g of water at 20°C? Use Table 16.1.

54. What is the effect of pressure on the solubility of gases in liquids?

***55.** The solubility of methane, the major component of natural gas, in water at 20°C and 1.00 atm pressure is 0.026 g/L. If the temperature remains constant, what will be the solubility of this gas at the following pressures?

a. 0.60 atm
b. 1.80 atm

16.2 Concentrations of Solutions

56. Knowing the molarity of a solution is more meaningful than knowing whether a solution is dilute or concentrated. Explain.

57. Define *molarity,* and then calculate the molarity of each solution.

a. 1.0 mol KCl in 750 mL of solution
b. 0.50 mol $MgCl_2$ in 1.5 L of solution

***58.** How many milliliters of 0.500*M* KCl solution would you need to dilute to make 100.0 mL of 0.100*M* KCl?

***59.** Calculate the molarity of a solution that contains 0.50 g of NaCl dissolved in 100 mL of solution.

60. Calculate the moles and grams of solute in each solution.

a. 1.0 L of 0.50*M* NaCl
b. 5.0×10^2 mL of 2.0*M* KNO_3
c. 250 mL of 0.10*M* $CaCl_2$
d. 2.0 L of 0.30*M* Na_2SO_4

***61.** Calculate the grams of solute required to make the following solutions:

a. 2500 g of saline solution (0.90% NaCl (m/m))
b. 0.050 kg of 4.0% (m/m) $MgCl_2$

62. What is the percent by mass of sodium chloride in each of the following solutions?

a. 44 g NaCl dissolved in 756 g H_2O
b. 15 g NaCl dissolved in 485 g H_2O
c. 135 g NaCl dissolved in 765 g H_2O

***63.** What is the concentration (in % (v/v)) of the following solutions?

a. 25 mL of ethanol (C_2H_6O) is diluted to a volume of 150 mL with water.
b. 175 mL of isopropyl alcohol (C_3H_8O) is diluted with water to a total volume of 275 mL.

16.3 Colligative Properties of Solutions

64. What are colligative properties? Identify three colligative properties and explain why each occurs.

65. Which has the higher boiling point:

a. seawater or distilled water?
b. 1.0*M* KNO_3 or 1.5*M* KNO_3?
c. 0.100*M* KCl or 0.100*M* $MgCl_2$?

66. Why does a 1*m* solution of calcium nitrate have a lower freezing point than a 1*m* solution of sodium nitrate?

67. Explain how a decrease in the vapor pressure of a solution results in an increase in its boiling point.

68. When the water inside a living cell freezes, the ice crystals damage the cell. The wood frog is a unique creature that can survive being frozen. In extremely cold conditions, the frog's liver produces large amounts of glucose ($C_6H_{12}O_6$), which becomes concentrated in the frog's cells. How does the glucose help prevent ice from forming in the frog's cells?

16.4 Calculations Involving Colligative Properties

69. Distinguish between a 1*M* solution and a 1*m* solution.

70. Describe how you would make an aqueous solution of methanol (CH_4O) in which the mole fraction of methanol is 0.40.

71. What is the boiling point of each solution?
   a. 0.50 mol glucose in 1000 g H_2O
   b. 1.50 mol NaCl in 1000 g H_2O

*72. What is the freezing point of each solution?
   a. 1.40 mol Na_2SO_4 in 1750 g H_2O
   b. 0.060 mol $MgSO_4$ in 100 g H_2O

73. Determine the freezing points of each 0.20*m* aqueous solution.
   a. K_2SO_4
   b. $CsNO_3$
   c. $Al(NO_3)_3$

Understand Concepts

74. Different numbers of moles of two different solutes, A and B, were added to identical quantities of water. The graph shows the freezing point of each of the solutions formed.

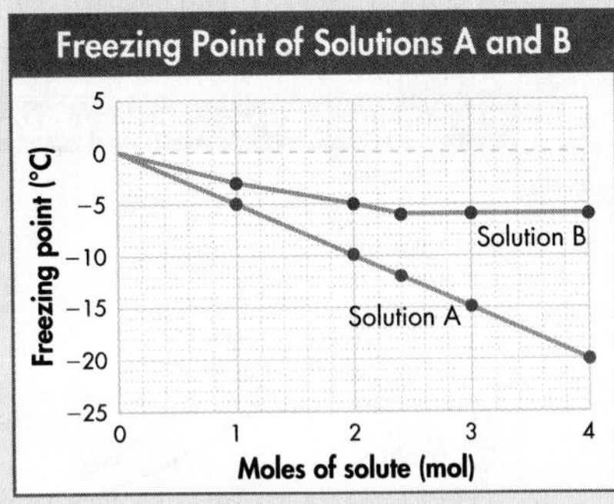

   a. Explain the relative slopes of the two lines between 0 and 2 mol of solute added.
   b. Why does the freezing point for solution A not continue to drop as amounts of solute A are added beyond 2.4 mol?

75. A mixture of ethylene glycol (EG) and water is used as antifreeze in automobile engines. The freezing point and density of the mixture vary with the percent by mass of (EG) in the mixture. On the following graph, point A represents 20% (EG) by mass; point B, 40%; and point C, 60%.

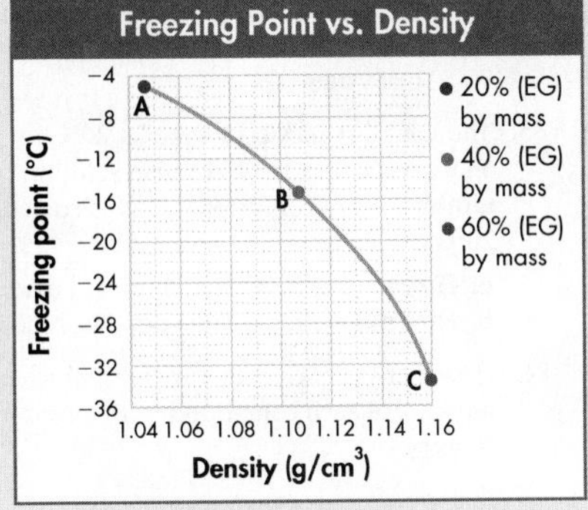

   a. What is the density of the antifreeze mixture that freezes at −25°C?
   b. What is the freezing point of a mixture that has a density of 1.06?
   c. Estimate the freezing point of a mixture that is 30% by mass (EG).

*76. Calculate the freezing- and boiling-point changes for a solution containing 12.0 g of naphthalene ($C_{10}H_8$) in 50.0 g of benzene.

77. Describe how you would prepare an aqueous solution of acetone (C_3H_6O) in which the mole fraction of acetone is 0.25.

78. The solubility of sodium hydrogen carbonate ($NaHCO_3$) in water at 20°C is 9.6 g/100 g H_2O. What is the mole fraction of $NaHCO_3$ in a saturated solution? What is the molality of the solution?

79. A solution is labeled 0.150*m* NaCl. What are the mole fractions of the solute and solvent in this solution?

80. You are given a clear aqueous solution containing KNO_3. How would you determine experimentally if the solution is unsaturated, saturated, or supersaturated?

81. Plot a graph of solubility versus temperature for the three gases listed in Table 16.1.

LESSON 16.4

69. 1*M* solution: 1 mol of solute in 1 L of solution; 1*m* solution: 1 mol of solute in 1000 g of solvent

70. Add 27.0 g H_2O to 32.0 g CH_4O.

71. **a.** 100.26°C **b.** 101.54°C

72. **a.** −4.46°C **b.** −2.2°C

73. **a.** −1.1°C **b.** −0.74°C **c.** −1.5°C

UNDERSTAND CONCEPTS

74. **a.** The freezing-point depression is twice as great for solute A; solute A must provide twice as many particles in solution.
b. Solute B probably forms a saturated solution.

75. **a.** about 1.14 g/cm³
b. about −7.2°C
c. about −9.5°C

76. $\Delta T_f = 9.63°C$; $\Delta T_b = 4.76°C$

77. Each gram of acetone requires 0.93 g of water.

78. The mole fraction of $NaHCO_3$ is 0.020; of water is 0.98. The solution is 1.1*m*.

79. The mole fraction of NaCl is 2.69×10^{-3}; the mole fraction of H_2O is 9.97×10^{-1}.

80. Add one crystal of KNO_3. If the solution is supersaturated, crystallization occurs. If it is saturated, the crystal does not dissolve; if unsaturated, the crystal dissolves.

81.

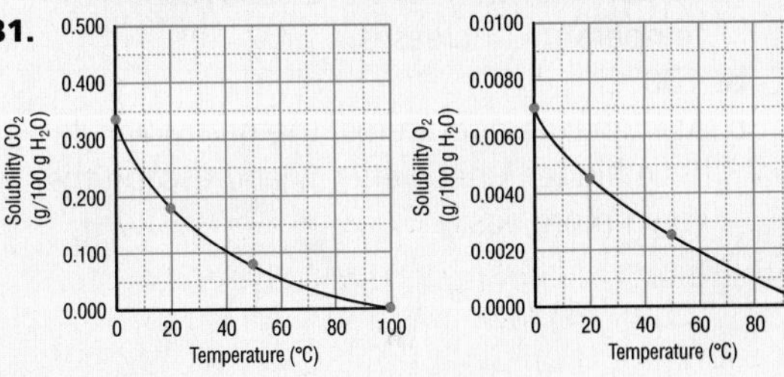

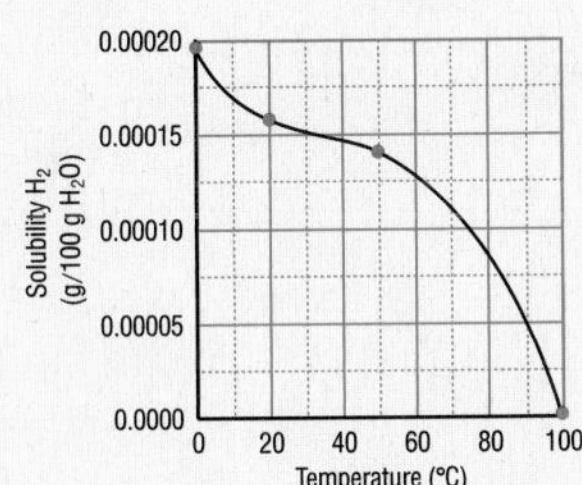

Solutions 549

Evaluate

Answers

82. fp = −1.86°C; bp = 100.512°C
83. $X_{C_2H_6O}$ = 0.20; X_{H_2O} = 0.80
84. −0.413°C
85. **a.** 44.2 g KCl **b.** 5.8 g KCl
86. **a.** 0.30 mol **b.** 0.40 mol **c.** 0.50 mol **d.** 0.20 mol

THINK CRITICALLY

87. unsaturated
88. 100.680°C
89. **a.** 7.5 g H_2O_2 **b.** 8.8×10^{-1}M
90. 5.2×10^1 g $NaNO_3$
91. 8.55×10^1 g/mol
92. X_{H_2O} = 0.972; $X_{C_{12}H_{22}O_{11}}$ = 0.028
93. $CaCl_2$ produces three particles upon dissolving; NaCl produces only two particles. Freezing-point depression depends on the number of solute particles in the solvent.
94. chloride: 96.9 g; sodium: 54.1 g; magnesium: 6.44 g; calcium: 2.07 g; potassium: 1.99 g
95. The solution with the higher concentration of ions will have the greater boiling point elevation; 6.00 g $Ca(NO_3)_2$ in 30 g of water.
96. **a.** Solubility of each gas decreases as temperature increases.
b. 100°C
c. At higher temperatures, the gas molecules have higher kinetic energy and escape the liquid more easily.

82. Calculate the freezing point and the boiling point of a solution that contains 15.0 g of urea (CH_4N_2O) in 250 g of water. Urea is a covalently bonded compound.

83. Calculate the mole fractions in a solution that is 25.0 g of ethanol (C_2H_6O) and 40.0 g of water.

84. Estimate the freezing point of an aqueous solution of 20.0 g of glucose ($C_6H_{12}O_6$) dissolved in 500.0 g of water.

*** 85.** The solubility of KCl in water at 20°C is 34.0 g KCl/100 g H_2O. A warm solution containing 50.0 g KCl in 130 g H_2O is cooled to 20°C.

a. How many grams of KCl remain dissolved?
b. How many grams came out of solution?

86. How many moles of ions are present when 0.10 mol of each compound is dissolved in water?

a. K_2SO_4
b. $Fe(NO_3)_3$
c. $Al_2(SO_4)_3$
d. $NiSO_4$

Think Critically

*** 87. Analyze Data** A solution contains 26.5 g NaCl in 75.0 g H_2O at 20°C. Determine if the solution is unsaturated, saturated, or supersaturated. (The solubility of NaCl at 20°C is 36.0 g/100 g H_2O.)

88. Infer An aqueous solution freezes at −2.47°C. What is its boiling point?

89. Calculate Percent (mass/volume), or % (m/v), is the number of grams of solute per 100 mL of solution. Hydrogen peroxide is often sold commercially as a 3.0% (m/v) aqueous solution.

a. If you buy a 250-mL bottle of 3.0% H_2O_2 (m/v), how many grams of hydrogen peroxide have you purchased?
b. What is the molarity of this solution?

90. Calculate How many grams of $NaNO_3$ will precipitate if a saturated solution of $NaNO_3$ in 200 g H_2O at 50°C is cooled to 20°C?

*** 91. Calculate** What is the molar mass of a nondissociating compound if 5.76 g of the compound in 750 g of benzene gives a freezing-point depression of 0.460°C?

*** 92. Calculate** The molality of an aqueous solution of sugar ($C_{12}H_{22}O_{11}$) is 1.62*m*. Calculate the mole fractions of sugar and water.

93. Apply Concepts Why might calcium chloride spread on icy roads be more effective at melting ice than an equal amount of sodium chloride?

94. Calculate The following table lists the molar concentrations of the most abundant monatomic ions in seawater. Calculate the mass in grams of each ion contained in 5.00 L of seawater. The density of seawater is 1.024 g/mL.

Ion	Molarity (*M*)
Chloride	0.546
Sodium	0.470
Magnesium	0.053
Calcium	0.0103
Potassium	0.0102

95. Compare and Contrast Which will have a greater boiling point elevation: 3.00 g $Ca(NO_3)_2$ in 60.0 g of water or 6.00 g $Ca(NO_3)_2$ in 30.0 g of water?

96. Interpret Graphs The graph shows the effect of temperature on the solubilities of oxygen gas (O_2), nitrogen gas (N_2), and nitrogen monoxide (NO) in water.

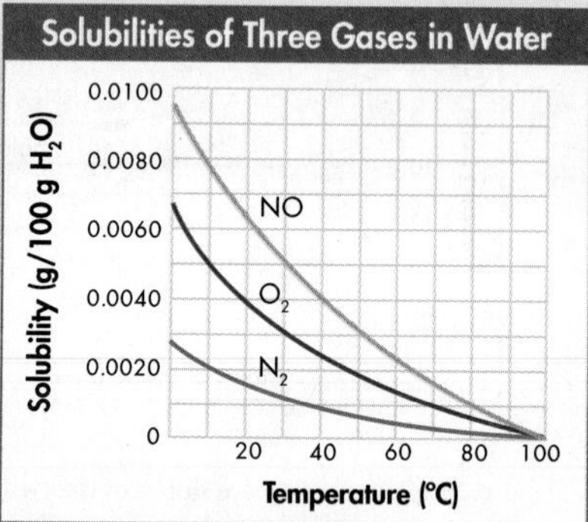

a. How does an increase in temperature affect the solubility in water of each gas?
b. At what temperature do the gases become virtually insoluble?
c. Use kinetic theory to explain the solubility behavior shown in the graph.

550 Chapter 16 • Assessment

Enrichment

97. Calculate When an excess of zinc is added to 800 mL of a hydrochloric acid solution, the solution evolves 1.21 L of hydrogen gas measured over water at 21°C and 747.5 mm Hg. What was the molarity of the acid? The vapor pressure of water at 21°C is 18.6 mm Hg.

98. Calculate How many milliliters of 1.50*M* HNO_3 contain enough nitric acid to dissolve an old copper penny with a mass of 3.94 g?

$$3Cu + 8HNO_3 \longrightarrow 3Cu(NO_3)_2 + 2NO + 4H_2O$$

99. Graph One way to express the solubility of a compound is in terms of moles of compound that will dissolve in 1 kg of water. Solubility depends on temperature. Plot a graph of the solubility of potassium nitrate (KNO_3) from the following data:

Temperature (°C)	Solubility (mol/kg)
0	1.61
20	2.80
40	5.78
60	11.20
80	16.76
100	24.50

Using your graph, estimate

a. the solubility of KNO_3 at 76°C and at 33°C.

b. the temperature at which its solubility is 17.6 mol/kg of water.

c. the temperature at which the solubility is 4.24 mol/kg of water.

★**100. Calculate** A 250-mL sample of Na_2SO_4 is reacted with an excess of $BaCl_2$. If 5.28 g $BaSO_4$ is precipitated, what is the molarity of the Na_2SO_4 solution?

101. Design an Experiment Suppose you have an unknown compound and want to identify it by means of its molar mass. Design an experiment that uses the concept of freezing-point depression to obtain the molar mass. What laboratory measurements would you need to make? What calculations would be needed?

Write About Science

102. Describe Find a recipe for rock candy online or in a cookbook. Write a short paragraph describing how the recipe applies key concepts that you have learned about solutions. Use the terms *solute, solvent, solubility, crystallization,* and *supersaturated solution* in your paragraph.

★**103. Sequence** Write a stepwise procedure for preparing 100 mL of 0.50*M* KCl, starting with a stock solution that is 2.0*M* KCl.

CHEMYSTERY

That Sinking Feeling

Although you can't see it happening, the groundwater beneath your feet is very slowly dissolving away rocks and minerals below ground. Eventually, enough of these mineral solutes dissolve to hollow out underground cavities or caverns. A sinkhole occurs when the roof of an underground cavern depresses or collapses.

104. Explain Why do you think areas underlain by salt beds are prone to sinkholes?

105. Infer As you read earlier in this chapter, agitation can speed up the rate at which a solid dissolves in liquid. What forces might contribute to agitating groundwater as it dissolves minerals underground?

106. Connect to the BIGIDEA Limestone, which contains mostly calcium carbonate ($CaCO_3$), is insoluble in water. Yet areas underlain by limestone are prone to sinkholes. Read the article on limestone caves on page R9. How does the "dissolving" of limestone differ from solvation?

ASSESSMENT

CHEMYSTERY

START A CONVERSATION After students have read through the CHEMystery, call on volunteers to describe factors that might contribute to the formation of sinkholes and identify ways that sinkhole development might be prevented. Point out sinkholes can form for a variety of reasons. For example, sinkholes can form when the level of groundwater decreases, causing soft soil to collapse. Sinkholes might also form when fast-moving groundwater washes away underlying soil. The scenario described in the CHEMystery has another cause: groundwater dissolves minerals in the rock, leaving empty spaces that eventually collapse. **Ask** How might the rate at which groundwater moves affect the rate at which sinkholes develop? *(If sinkholes develop due to dissolving minerals, they might develop faster if the water is moving faster.)*

CHEMYSTERY ANSWERS

104. Salt beds consist of rock salt (NaCl), which is soluble in water.

105. Sample answers: gravity; earthquakes or volcanic activity; heavy precipitation or meltwater.

106. BIGIDEA Unlike solvation, which is a physical change involving a solute and solvent, the dissolving of calcium carbonate by carbonic acid is a chemical reaction.

Answers

ENRICHMENT

97. $1.2 \times 10^{-1}M$ HCl

98. 1.10×10^2 mL HNO_3

99. a. 76°C: 15 mol/kg; 33°C: 5 mol/kg
b. 82°C **c.** 30°C

100. $9.0 \times 10^{-2}M$ Na_2SO_4

101. Determine the freezing point of a suitable solvent. Dissolve a known mass of the unknown molecular compound in a known mass of the solvent. Determine the freezing point of the solution. This gives the freezing point depression. Use $\Delta T_f = K_f \times m$ to find the molality of the solution. Use the molality to find the moles of solute. Use the moles of solute and the measured mass of solute to calculate the molar mass.

$$\frac{\text{g solute}}{\text{g solvent}} \times \frac{1000\text{ g solvent}}{1\text{ kg solvent}} \times \frac{1\text{ kg solvent}}{1\text{ mol solute}} = \frac{\text{g solute}}{1\text{ mol solute}}$$

The molar mass obtained is valid only for an undissociating molecular solute.

WRITE ABOUT SCIENCE

102. Answers will vary.

103. Transfer 25 mL of the stock solution to a 100-mL volumetric flask. Then add water to make 100 mL of solution.

ASSESSMENT

Cumulative Review

107. a. 3.47×10^{-1} kg
b. 7.3 10^{-5} kg
c. 9.43×10^{-6} kg
d. 8.77×10^{-4} kg
108. 85.468 amu
109. Rutherford's model contains a nucleus.
110. a. manganese, Mn **b.** indium, In
c. francium, Fr **d.** polonium, Po
111. Calcium permanganate is $Ca(MnO_4)_2$. Four formula units contain 4 Ca atoms, 8 Mn atoms, and 32 O atoms.
112. a. :I· **b.** ·Te· **c.** :Sb· **d.** ·Sr·
113. $C_8H_6O_4$
114. a. 5.58×10^1 g Fe, 6.35×10^1 g Cu, 2.01×10^2 g Hg, 3.21×10^1 g S
b. Each sample contains 6.02×10^{23} atoms.
c. 4.48×10^{-1} mol Fe, 3.94×10^{-1} mol Cu, 1.25×10^{-1} mol Hg, 7.79×10^{-1} mol S
115. 1.7×10^4 L
116. a. combination **b.** decomposition
c. single-replacement **d.** combustion
e. single-replacement **f.** double-replacement
117. $2H^+(aq) + S^{2-}(aq) \rightarrow H_2S(g)$
118. a. $NH_4Cl(s) \rightarrow NH_4^+(aq) + Cl^-(aq)$
b. $Cu(NO_3)_2(s) \rightarrow Cu^{2+}(aq) + 2NO_3^-(aq)$
c. $HNO_3(aq) \rightarrow H^+(aq) + NO_3^-(aq)$
d. $HC_2H_3O_2(l) \rightarrow H^+(aq) + C_2H_3O_2^-(aq)$
e. $Na_2SO_4(s) \rightarrow 2Na^+(aq) + SO_4^{2-}(aq)$
f. $HgCl_2(s) \rightarrow Hg^{2+}(aq) + 2Cl^-(aq)$
119. 36.8 L O_2
120. 1.08×10^2 kPa
121. The particles in an ideal gas have no volume and there are no attractions between particles.
122. Unbalanced intermolecular attractions between molecules at the surface of the liquid and those below the surface create an inward pull, or force, that minimizes the surface area to create surface tension.
123. The very polar hydrogen chloride molecule produces hydronium ions (H_3O^+) and chloride ions (Cl^-) that are stabilized by becoming surrounded by water molecules in aqueous solution. Hydrogen chloride does not dissociate in nonpolar benzene. Polar compounds generally have low solubility in nonpolar solvents.
124. a solution, if the soap mixture is very dilute; for example, soap bubble solutions are clear. More concentrated mixtures of soap in water form colloids, as evidenced by their milky, cloudy appearance.

Cumulative Review

107. Convert each of the following mass measurements to its equivalent in kilograms.
a. 347 g
b. 73 mg
c. 9.43 mg
d. 877 mg

***108.** Rubidium has two naturally occurring isotopes. Rubidium-85 (72.165%) has a mass of 84.912 amu. Rubidium-87 (27.835%) has a mass of 86.909 amu. Calculate the average atomic mass of rubidium.

109. What is the most significant difference between the Thomson model of the atom and the Rutherford model?

110. Name and give the symbol for the element in the following positions in the periodic table:
a. Group 7B, Period 4
b. Group 3A, Period 5
c. Group 1A, Period 7
d. Group 6A, Period 6

111. How many atoms of each element are present in four formula units of calcium permanganate?

112. Draw electron dot structures for the following atoms:
a. I **b.** Te **c.** Sb **d.** Sr

113. Terephthalic acid is an organic compound used in the synthesis of polyesters. Terephthalic acid contains 57.8 percent C, 3.64 percent H, and 38.5 percent O. The molar mass is approximately 166 g/mol. What is the molecular formula of terephthalic acid?

***114.** The photograph shows one mole each of iron, copper, mercury, and sulfur.
a. What is the mass of each element?
b. How many atoms are in each sample?
c. How many moles is 25.0 g of each element?

***115.** What is the volume occupied by 1500 g of hydrogen gas (H_2) at STP?

116. Identify the type of chemical reaction.
a. $H_2(g) + Cl_2(g) \longrightarrow 2HCl(g)$
b. $2H_2O(l) \longrightarrow O_2(g) + 2H_2(g)$
c. $2K(s) + 2H_2O(l) \longrightarrow 2KOH(aq) + H_2(g)$
d. $C_2H_6O(l) + 3O_2(g) \longrightarrow 2CO_2(g) + 3H_2O(l)$
e. $Cl_2(aq) + 2KBr(aq) \longrightarrow 2KCl(aq) + Br_2(aq)$
f. $Pb(NO_3)_2(aq) + 2NaCl(aq) \longrightarrow PbCl_2(s) + 2NaNO_3(aq)$

117. Write the net ionic equation for the following reaction:

$$2HI(aq) + Na_2S(aq) \longrightarrow H_2S(g) + 2NaI(aq)$$

118. Indicate by simple equations how the following substances ionize or dissociate in water:
a. NH_4Cl
b. $Cu(NO_3)_2$
c. HNO_3
d. $HC_2H_3O_2$
e. Na_2SO_4
f. $HgCl_2$

119. The equation for the combustion of methanol (CH_4O) is the following:

$$2CH_4O(l) + 3O_2(g) \longrightarrow 2CO_2(g) + 4H_2O(l)$$

What volume of oxygen, measured at STP, is required to completely burn 35.0 g of methanol?

***120.** A cylinder of nitrogen gas at 25°C and 101.3 kPa is heated to 45°C. What is the new pressure of the gas?

121. Why does an ideal gas not exist?

122. What relationship exists between surface tension and intermolecular attractions in a liquid?

123. The solubility of hydrogen chloride gas in the polar solvent water is much greater than its solubility in the nonpolar solvent benzene. Why?

124. When soap is shaken with water, which is formed: a solution, a suspension, or a colloid? Explain.

If You Have Trouble With . . .

Question	107	108	109	110	111	112	113	114	115	116	117	118	119	120	121	122	123	124
See Chapter	3	4	5	6	7	7	10	10	10	11	11	11	12	14	14	15	15	15

Standardized Test Prep for Indiana

Standard: C.7.2

Select the choice that best answers each question or completes each statement.

1. An aqueous solution is 65% (v/v) rubbing alcohol. How many milliliters of water are in a 95-mL sample of this solution?
   (A) 62 mL (C) 33 mL
   (B) 1.5 mL (D) 30 mL

2. When 2.0 mol of methanol is dissolved in 45 g of water, the mole fraction of methanol is
   (A) 0.44. (C) 2.25.
   (B) 0.043. (D) 0.55.

The lettered choices below refer to Questions 3–6. A lettered choice may be used once, more than once, or not at all.

(A) moles/liter of solution
(B) grams/mole
(C) moles/kilogram of solvent
(D) °C/molal
(E) no units

Which of the above units is appropriate for each measurement?

3. molality
4. mole fraction
5. molar mass
6. molarity

Use the atomic windows to answer Questions 7–9. The windows show water and two aqueous solutions with different concentrations. Black spheres represent solute particles; gray spheres represent water.

(A) 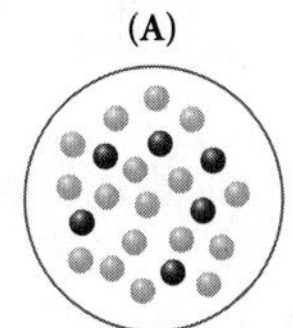
(B) 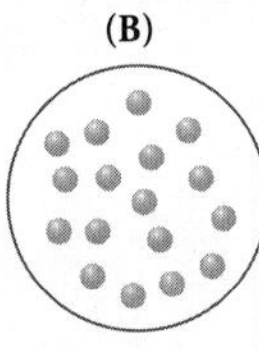
(C) 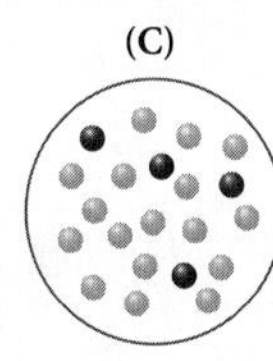

7. Which solution has the highest vapor pressure?
8. Which solution has the lowest vapor pressure?
9. Which solution has the lowest boiling point?

10. Which of these actions will cause more sugar to dissolve in a saturated sugar water solution?
    I. Add more sugar while stirring.
    II. Add more sugar and heat the solution.
    III. Grind the sugar to a powder; then add while stirring.
    (A) I only
    (B) II only
    (C) III only
    (D) I and II only
    (E) II and III only

Tips for Success

Reading Data Tables Data tables are used to summarize data. When reading a table, try to figure out the relationships between the different columns and rows of information.

Use the description and the data table to answer Questions 11–14.

A student measured the freezing points of three different aqueous solutions at five different concentrations. The data are shown below.

Molarity (M)	Freezing Point Depression (°C)		
	NaCl	$CaCl_2$	C_2H_6O
0.5	1.7	2.6	0.95
1.0	3.5	5.6	2.0
1.5	5.3	8.3	3.0
2.0	7.2	11.2	4.1
2.5	9.4	14.0	5.3

11. Graph the data for all three solutes on the same graph, using molarity as the independent variable.
12. Summarize the relationship between molarity and freezing-point depression.
13. Compare the slopes of the three lines and explain any difference.
14. If you collected similar data for KOH and added a fourth line to your graph, which existing line would the new line approximate?

If You Have Trouble With . . .

Question	1	2	3	4	5	6	7	8	9	10	11	12	13	14
See Lesson	16.2	16.4	16.4	16.4	16.2	16.2	16.3	16.3	16.3	16.1	16.3	16.4	16.3	16.3

ASSESSMENT

STP Answers

MULTIPLE CHOICE

1. C
2. A
3. C
4. E
5. B
6. A
7. B
8. A
9. B
10. B

OPEN RESPONSE

11. 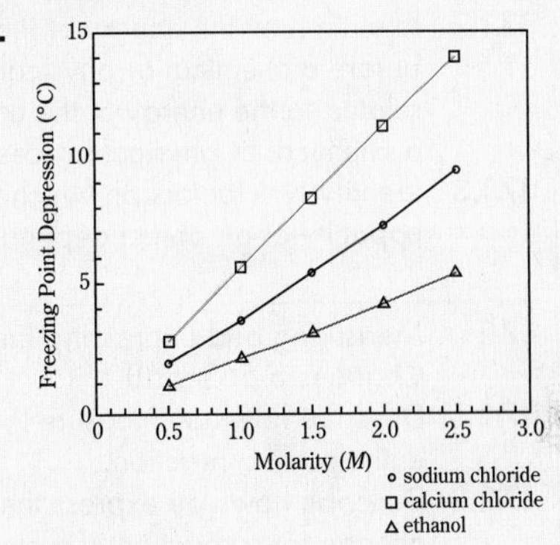

12. As the molarity increases, the freezing-point depression increases.
13. The slopes are an approximate 1:2:3 ratio that reflects the relative number of particles per mole of each solute in solution.
14. NaCl

17 Thermochemistry Planning Guide

Introducing the BIGIDEA: MATTER AND ENERGY

Every chemical process uses or produces energy, often in the form of heat.

IASS	Lessons and Objectives	Print Resources		
		For the Student	For the Teacher	
C.6.3	**17.1** The Flow of Energy p 556–561 **17.1.1** Explain the ways in which energy changes can occur. **17.1.2** Explain how the energy of the universe before a chemical or physical process is related to the energy of the universe after a chemical or physical process. **17.1.3** Identify two factors on which the heat capacity of an object depends.	Reading and Study Workbook Lesson 17.1 Lesson Assessment 17.1 p 561	Teacher Demo, p 557: An Endothermic Reaction Class Activity, p 559: Heat Transfer	
C.6.3, C.6.4	**17.2** Measuring and Expressing Enthalpy Changes p 562–568 **17.2.1** Describe how you measure the change in enthalpy of a reaction. **17.2.2** Describe how you express the enthalpy change for a reaction in a chemical equation.	Reading and Study Workbook Lesson 17.2 Lesson Assessment 17.2 p 568	Class Activity, p 563: Calorimetry Measurements Teacher Demo, p 566: An Exothermic Reaction	
C.6.3, C.6.4	**17.3** Heat in Changes of State p 569–575 **17.3.1** Compare the quantity of heat absorbed by a melting solid to the quantity of heat released when the liquid solidifies. **17.3.2** Compare the quantity of heat absorbed by a vaporizing liquid to the quantity of heat released when the vapor condenses. **17.3.3** Describe thermochemical changes that occur when a solution forms.	Reading and Study Workbook Lesson 17.3 Lesson Assessment 17.3 p 575 Quick Lab: Heat of Fusion of Ice p 571	Class Activity, p 570: Melting and Boiling Points Teacher Demo, p 572: Exothermic Solidification Class Activity, p 574: Hot and Cold Packs	
C.6.3	**17.4** Calculating Heats of Reaction p 578–582 **17.4.1** Identify two ways that you can determine the heat of reaction when it cannot be directly measured.	Reading and Study Workbook Lesson 17.4 Lesson Assessment 17.4 p 582 Small-Scale Lab: Heat of Combustion of a Candle p 583		

Assessing the BIGIDEA: MATTER AND ENERGY

	Essential Questions 1. How is energy conserved in a chemical or physical process? 2. How can you determine the amount of energy absorbed or released in a chemical or physical process?	Study Guide p 584 Math Tune-Up p 585 STP p 591 Reading and Study Workbook Self-Check and Vocabulary Review Chapter 17		

	PearsonChem.com	
	Editable Worksheets	Online Activities
	Teaching Resources, Lesson 17.1 Review Lab 34: The Specific Heat of a Metal Lab Practical 17-1: Specific Heat	17.1 Lesson Overview A Tremendous Capacity for Heat Calculating the Specific Heat of a Substance
	Teaching Resources, Lesson 17.2 Review Lab 34: The Specific Heat of a Metal Lab 35: Heats of Reaction Lab Practical 17-1: Specific Heat	17.2 Lesson Overview Bomb Calorimeter Heat of Combustion Enthalpy Change in a Calorimetry Experiment
	Teaching Resources, Lesson 17.3 Review Small-Scale Lab Manual Lab 27: Heat of Fusion of Ice Probeware Lab: Quick Lab: Heat of Fusion of Ice	17.3 Lesson Overview Using the Heat of Fusion and the Heat of Vaporization in Phase-Change Calculations
	Teaching Resources, Lesson 17.4 Review Lab 35: Heats of Reaction Probeware Lab: Small-Scale Lab: Heat of Combustion of a Candle Lab Practical 17-2: Heat of Reaction	17.4 Lesson Overview Hess's Law Calculating the Standard Heat of Reaction
	Core Teaching Resources (includes editable worksheets) • Lesson Reviews • Practice Problems • Interpret Graphs • Vocabulary Review • Chapter Quizzes and Tests	Chapter 17 Problem Set Solving Multistep Literal Equations

Materials List

For the Student

Quick Lab p 571
- 100-mL graduated cylinder
- hot water
- foam cup
- thermometer
- ice

Small-Scale Lab p 583
- candle
- metric ruler
- aluminum foil
- balance
- safety matches
- clock or watch
- temperature probe (optional)

For the Teacher

Teacher Demo p 557
- barium hydroxide octahydrate
- ammonium chloride
- 250-mL Erlenmeyer flask with stopper

Class Activity p 559
- sheet metal (13 cm × 18 cm)
- foam (13 cm × 18 cm)
- wood (40 cm × 20 cm)
- glue

Class Activity p 563
- 1M HCl
- foam cup
- thermometer
- mossy zinc
- magnesium turnings

Teacher Demo p 566
- CaO
- metal tray
- thin aluminum pan
- water
- egg (raw)

Class Activity p 570
- beaker
- water
- boiling chips
- hot plate
- thermometer
- ice cubes
- foam cup

Teacher Demo p 572
- 3 wide-mouth test tubes
- sodium thiosulfate pentahydrate
- hot water bath
- 30°C water bath
- thermometers
- stirring rods

Class Activity p 574
- hot and cold packs

Additional Digital Resources

Online Student Edition
Online Teacher's Edition
Exam*View Assessment Suite CD-ROM*
Classroom Resources DVD
17.1 Virtual Chem Lab 16: The Specific Heat of a Metal
17.2 Virtual Chem Lab 19: Heat of Combustion
17.3 Virtual Chem Lab 17: Heat of Fusion of Water
17.4 Virtual Chem Lab 18: Heats of Reaction

CHAPTER 17

What's Online

CHEMISTRY TUTORIAL Students can work step-by-step with the online tutorial to calculate the specific heat of a substance, determine enthalpy change, and calculate the standard heat of a reaction by accessing the Chapter 17 Chemistry Tutorials.

ONLINE PROBLEM SET Additional practice problems about thermochemistry can be found in the Chapter 17 online Problem Set.

MATH TUTORIAL The online Math Tutorial will help guide students through solving multistep word problems.

DIRECTED VIRTUAL LAB Encourage students to review the heat of combustion of several common compounds using the animated virtual lab for Chapter 17. The animation is followed by analysis questions.

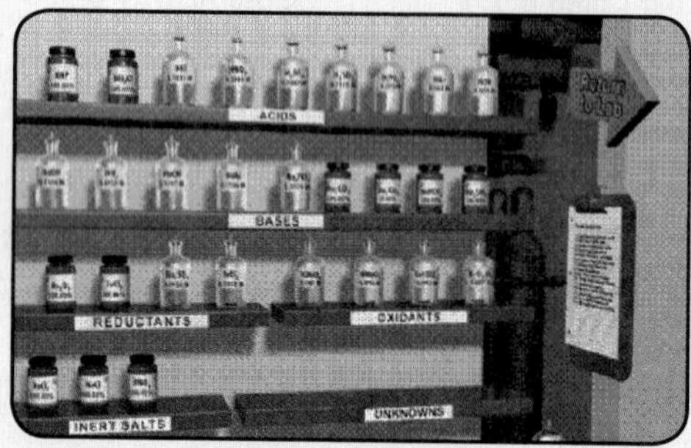

KINETIC ART Have students watch animations about a bomb calorimeter and Hess's Law. Each animation is followed by analysis questions.

CONCEPTS IN ACTION Students can watch an interactive animation about the heat capacity of several objects. The animation is followed by questions.

IN ACADEMIC STANDARDS for SCIENCE

Nature of Science

- Apply standard techniques in laboratory investigations to measure physical quantities in appropriate units and convert known quantities to other units as necessary.

Thermochemistry

C.6.3 Solve problems involving heat flow and temperature changes, using known values of specific heat and/or phase change constants (latent heats).

C.6.4 Classify chemical reactions and phase changes as exothermic or endothermic.

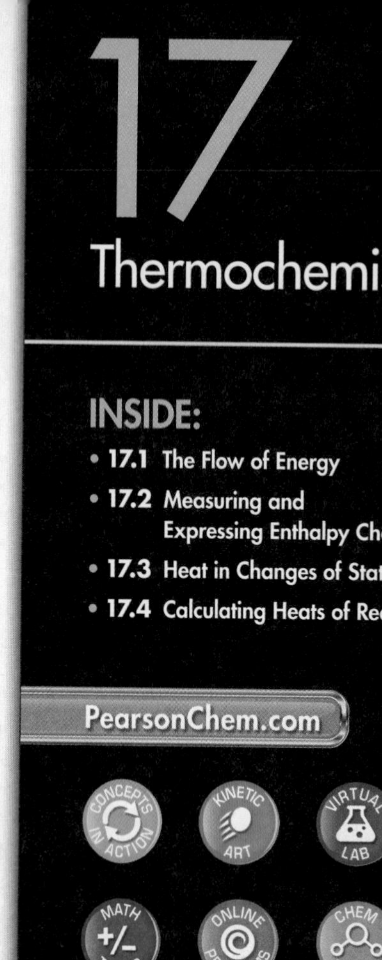

17 Thermochemistry

INSIDE:

PearsonChem.com

This solar furnace in Font Romeu, France converts light from the sun into heat.

Focus on ELL

1 CONTENT AND LANGUAGE Write the word *thermochemistry* on the board. Slowly enunciate the word and have students repeat it. Point out that the prefix *thermo-* means "heat." Have students use this information to predict the meaning of *thermochemistry*.

BEGINNING

LOW Have students write any unfamiliar words from the lesson titles on cards and practice saying each word.

HIGH Have students use a bilingual dictionary or glossary to find any cognates to any unfamiliar words from the lesson titles.

INTERMEDIATE: LOW/HIGH Have students use an English dictionary or the glossary to define any unfamiliar words from the lesson titles.

ADVANCED: LOW/HIGH Have students use the lesson titles to predict the purpose or main idea of the chapter.

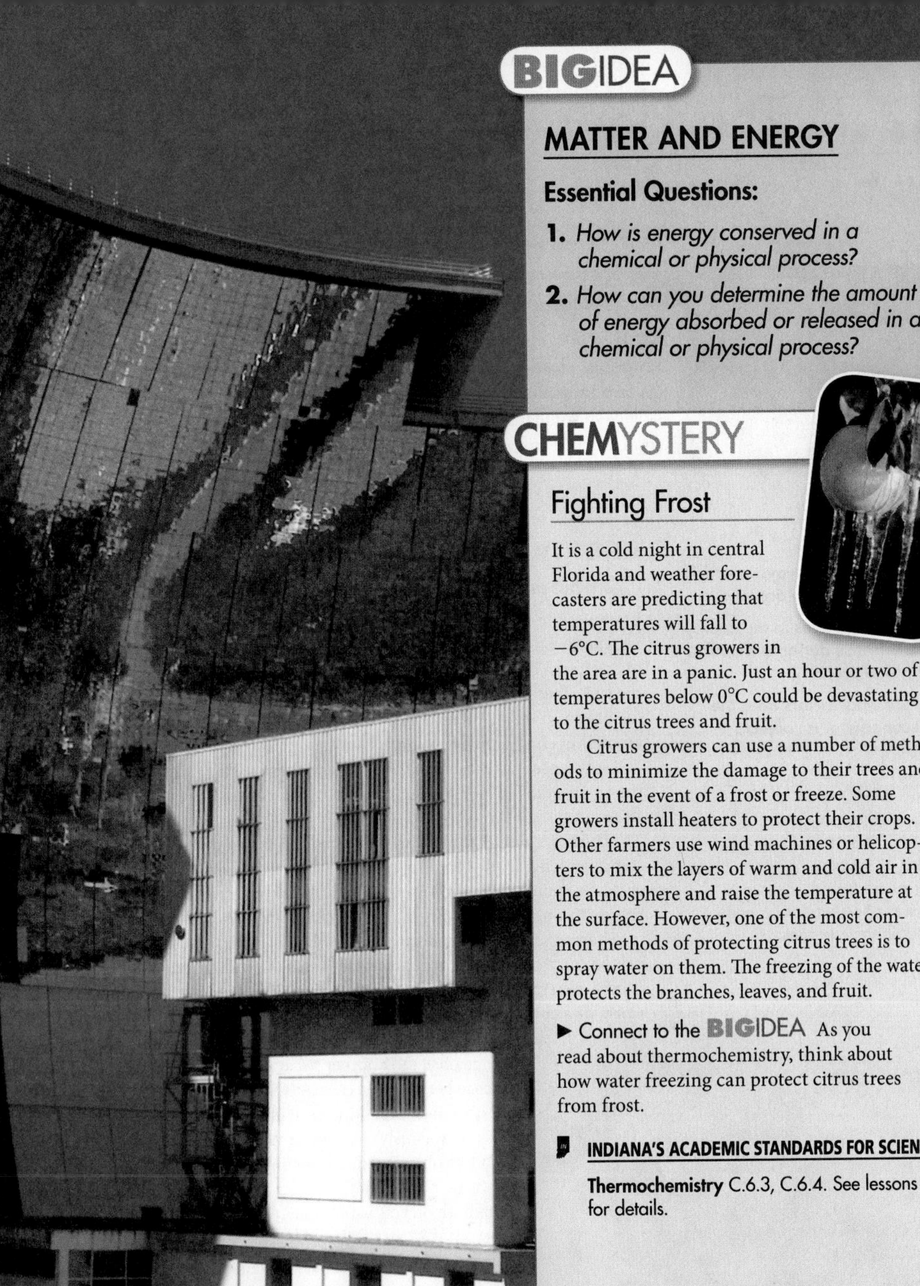

BIGIDEA

MATTER AND ENERGY

Essential Questions:

1. *How is energy conserved in a chemical or physical process?*
2. *How can you determine the amount of energy absorbed or released in a chemical or physical process?*

CHEMYSTERY

Fighting Frost

It is a cold night in central Florida and weather forecasters are predicting that temperatures will fall to −6°C. The citrus growers in the area are in a panic. Just an hour or two of temperatures below 0°C could be devastating to the citrus trees and fruit.

Citrus growers can use a number of methods to minimize the damage to their trees and fruit in the event of a frost or freeze. Some growers install heaters to protect their crops. Other farmers use wind machines or helicopters to mix the layers of warm and cold air in the atmosphere and raise the temperature at the surface. However, one of the most common methods of protecting citrus trees is to spray water on them. The freezing of the water protects the branches, leaves, and fruit.

▶ Connect to the **BIG**IDEA As you read about thermochemistry, think about how water freezing can protect citrus trees from frost.

INDIANA'S ACADEMIC STANDARDS FOR SCIENCE

Thermochemistry C.6.3, C.6.4. See lessons for details.

Introduce the Chapter

IDENTIFYING MISCONCEPTIONS Since students may know that some reactions release heat, they may assume that other reactions release cold, rather than absorbing heat. Use the activity to identify students who may have this misconception.

Activity *You'll need two or more chemical heat packs or hand warmers and two or more chemical cold packs.* **Ask** How does a chemical reaction affect the total energy of the reactants? *(Answers will vary. Students may say that the reactants lose energy.)* Follow the directions for activating the heat packs, and pass them around the room. **Ask** Why does the heat pack feel warm? *(Sample answer: heat is flowing from the pack to my hand.)* Follow the directions for activating the cold packs, and pass them around the room. **Ask** Why does the cold pack feel cold? *(Sample answer: cold is flowing from the pack to my hand.)* Explain that cold is not flowing from the pack to the students' hands, but that heat is flowing from their hands to the pack. Sensors in their skin detect the change in temperature, and their brains interpret this as a feeling of cold. Tell students to keep this activity in mind as they read the chapter.

UbD

Understanding by Design

Students are building toward describing energy changes during chemical or physical processes by using what they know about matter and energy.

PERFORMANCE GOALS At the end of Chapter 17, students will be able to answer the essential questions by applying their knowledge of thermochemistry.

ESSENTIAL QUESTIONS Read the essential questions aloud. **Ask** What is meant by "energy is conserved" in the first question? *(Energy cannot be created or destroyed during any chemical or physical process.)* **Ask** How is it possible, then, for energy to be absorbed or released in a reaction? *(Energy can change to other forms, or it can move from one substance to another.)*

BIGIDEA Use the photo of the solar thermal power plant to help students connect to this chapter's concepts. Activate prior knowledge by reviewing methods of energy transfer. Point out that mirrors direct the Sun's rays toward the side of the building. The rays then reflect onto collectors in the nearby tower. **Ask** What type of energy is carried by the Sun's rays? *(radiant energy)* **Ask** What happens to this energy? *(It is used as a source of heat.)*

CHEMYSTERY Connect the CHEMystery to the Big Idea by explaining that freezing occurs because of the transfer of heat energy. Have students use this fact to predict why the growers might protect the citrus trees by spraying them with water. As a hint, remind students that a change in the state of matter involves a gain or loss of heat.

UntamedScience™

Turning Up the Heat

Diego and Danni find some common, everyday examples of exothermic and endothermic chemical reactions.

LESSON 17.1

Key Objectives

17.1.1 **EXPLAIN** the ways in which energy changes can occur.

17.1.2 **EXPLAIN** how the energy of the universe before a chemical or physical process is related to the energy of the universe after a chemical or physical process.

17.1.3 **IDENTIFY** two factors on which the heat capacity of an object depends.

Additional Resources

Reading and Study Workbook, Lesson 17.1

Available Online or on Digital Media:

- Teaching Resources, Lesson 17.1 Review
- Laboratory Manual, Lab 34
- Virtual Chemistry Labs, Lab 16

PearsonChem.com

Engage

CHEMISTRY & YOU **Ask** What happens to the heat in lava once it flows out of a volcano? *(The heat is released.)* **Ask** Why does lava cool more quickly in water than on land? *(Sample answer: Water has a greater capacity to absorb heat than does air; the temperature difference between the lava and water may be greater than between the lava and the air.)*

Activate Prior Knowledge

To assess whether students know the difference between temperature and energy, **Ask** What does a thermometer measure? *(average kinetic energy)* **Ask** Does a thermometer measure heat? *(no)*

IN ACADEMIC STANDARDS for SCIENCE

C.6.3 Solve problems involving heat flow and temperature changes, using known values of specific heat and/or phase change constants (latent heats).

17.1 The Flow of Energy

C.6.3 Heat flow and temperature change problems

CHEMISTRY & YOU

Q: *Why does lava cool faster in water than in air?* Lava flowing out of an erupting volcano is very hot. Its temperature ranges from 550°C to 1400°C. As lava flows down the side of a volcano, it loses heat and begins to cool slowly. In some instances, the lava may flow into the ocean, where it cools more rapidly. In this lesson, you will learn about heat flow.

Key Questions

What are the ways in which energy changes can occur?

What happens to the energy of the universe during a chemical or physical process?

On what factors does the heat capacity of an object depend?

Vocabulary

- thermochemistry
- chemical potential energy
- heat • system
- surroundings
- law of conservation of energy
- endothermic process
- exothermic process
- heat capacity
- specific heat

Energy Transformations

What are the ways in which energy changes can occur?

Energy is the capacity for doing work or supplying heat. Unlike matter, energy has neither mass nor volume. Energy is detected only because of its effects. For example, a car moves because of the energy supplied by the fuel. **Thermochemistry** is the study of energy changes that occur during chemical reactions and changes in state. Every substance has a certain amount of energy stored inside it. The energy stored in the chemical bonds of a substance is called **chemical potential energy.** The kinds of atoms and the arrangement of the atoms in a substance determine the amount of energy stored in the substance.

During a chemical reaction, a substance is transformed into another substance with a different amount of chemical potential energy. When you buy gasoline, shown in Figure 17.1, you are actually buying the stored potential energy it contains. The controlled explosions of the gasoline in a car's engine transform the potential energy into useful work, which can be used to propel the car. At the same time, however, heat is also produced, making the car's engine extremely hot. **Energy changes occur as either heat transfer or work, or a combination of both.**

Heat, represented by q, is energy that transfers from one object to another because of a temperature difference between the objects. One of the effects of adding heat to an object is an increase in its temperature. Heat flows spontaneously from a warmer object to a cooler object. If two objects remain in contact, heat will flow from the warmer object to the cooler object until the temperature of both objects is the same.

Figure 17.1 Chemical Potential Energy
Chemical potential energy is stored within the bonds of the molecules in gasoline.

Focus on ELL

1 CONTENT AND LANGUAGE Divide the class into five groups and assign each group two key vocabulary words from the lesson. Have each group pronounce the words, define the words, give a visual example of the words, and create a non-linguistic representation of the words. Have each group teach their words to the class.

2 FRONTLOAD THE LESSON Write the word *calorie* on the board. Discuss as a class the context in which students are most familiar with this word. Ask students to share three things that they have heard about calories. Explain how food calories are related to the unit of measure, calorie. Clarify any misconceptions between these two concepts.

3 COMPREHENSIBLE INPUT When discussing direction of heat flow with respect to the system, sketch the images from the various photos in the lesson on the board. Label the system and surroundings in each and draw arrows to indicate the direction of heat flow with respect to the system. Have students classify each process as endothermic or exothermic.

Endothermic and Exothermic Processes

What happens to the energy of the universe during a chemical or physical process?

Chemical reactions and changes in physical state generally involve either the absorption or the release of heat. In studying energy changes, you can define a **system** as the part of the universe on which you focus your attention. Everything else in the universe makes up the **surroundings.** In thermochemical experiments, you can consider the region in the immediate vicinity of the system as the surroundings. Together, the system and its surroundings make up the universe.

A major goal of thermochemistry is to examine the flow of heat between the system and its surroundings. The **law of conservation of energy** states that in any chemical or physical process, energy is neither created nor destroyed. **During any chemical or physical process, the energy of the universe remains unchanged.** If the energy of the system increases during that process, the energy of the surroundings must decrease by the same amount. Likewise, if the energy of the system decreases during that process, the energy of the surroundings must increase by the same amount.

Direction of Heat Flow In thermochemical calculations, the direction of heat flow is given from the point of view of the system. Heat is absorbed from the surroundings in an **endothermic process.** In an endothermic process, the system gains heat as the surroundings lose heat. In Figure 17.2a, the system (the body) gains heat from its surroundings (the fire). Heat flowing into a system from its surroundings is defined as positive; *q* has a positive value. An **exothermic process** is one that releases heat to its surroundings. In an exothermic process, the system loses heat as the surroundings gain heat. In Figure 17.2b, the system (the body) loses heat to the surroundings (the perspiration on the skin, and the air). Heat flowing out of a system into its surroundings is defined as negative; *q* has a negative value.

Figure 17.2 Heat Flow
Heat flow is defined from the point of view of the system. **a.** In an endothermic process, heat flows into the system from the surroundings. **b.** In an exothermic process, heat flows from the system to the surroundings. In both cases, energy is conserved.
Apply Concepts *In which process does q have a negative value?*

Differentiated Instruction

LPR LESS PROFICIENT READERS Have students write definitions for each of the key terms in this chapter. Special attention should be given to those terms that are defined by mathematical relationships. Example calculations showing how to apply the formulas should be included in their notes.

L1 SPECIAL NEEDS STUDENTS Provide tactile examples of endothermic and exothermic reactions tailored to the main senses that the student relies on for learning.

ELL ENGLISH LANGUAGE LEARNERS Have students compare and contrast the prefixes *endo-* and *exo-*. Guide students to create symbols they can use to mark a reaction as heat absorbing or heat releasing.

Explain

Energy Transformations

USE VISUALS Direct students' attention to Figure 17.1. Have students consider the use of fuel in automobiles. **Ask** What work is done as the chemical potential energy of the fuel is released? *(The automobile moves along the road.)* **Ask** In addition to work, what other type of energy change occurs as the fuel is burned? *(heat is released)*

Exothermic and Endothermic Processes

MAKE A CONNECTION Discuss with students the distinction between kinetic and potential energy. Kinetic energy is the energy associated with an object because of its motion. Potential energy is the energy associated with an object because of its position in a field of force or due to its particular chemical composition. Explain that the potential energy of a reactant or a product in a chemical reaction is determined by the strengths of the attractive and repulsive forces between atoms. In a chemical reaction, atoms are rearranged into new groupings that have different relative potential energies. The change in potential energy is either the result of absorption of energy from the surroundings (endothermic reaction) or the release of energy to the surroundings (exothermic reaction).

Explore

Teacher Demo

PURPOSE Students will observe an endothermic reaction between two solids.

MATERIALS barium hydroxide octahydrate, ammonium chloride, 250-mL Erlenmeyer flask with stopper

SAFETY Barium salts are poisonous if ingested. Inhalation of concentrated ammonia vapors can be dangerous. Perform this demo in a fume hood or a well-ventilated room. Disposal: Flush the solution down the drain with excess water.

PROCEDURE Mix 32 g of barium hydroxide octahydrate ($Ba(OH)_2 \cdot 8H_2O$) with 11 g of ammonium chloride (NH_4Cl) in a 250-mL Erlenmeyer flask fitted with a stopper. Swirl to mix.

EXPECTED OUTCOME The flask becomes extremely cold. Place the flask on a wet surface and the flask will freeze to it.

Answers

FIGURE 17.2b In Figure 17.2b, The system loses heat to its surroundings, so *q* is negative.

LESSON 17.1

Explain

Endothermic and Exothermic Processes

MAKING CONNECTIONS As a class, brainstorm everyday processes students experience that absorb or release heat. Create a list on the board and randomly select students to write *–heat* or *+heat* next to each process to indicate the direction of heat flow.

Sample Practice Problem

During photosynthesis, plants use the energy from the sun to convert carbon dioxide and water into glucose and oxygen. Is photosynthesis an endothermic or exothermic process? (*endothermic*)

Misconception Alert

Make sure students understand the terms introduced in this section and how to convert from one to another. Students may have difficulty with the distinction between *calorie* and *Calorie* because everyday usage of the terms is often not accurate. The distinction between *heat capacity* and *specific heat* can be hard to remember because the terms are so closely related and sound so similar. Tell students to remember that *specific* heats of different materials can be compared because the quantity of matter involved (1 g) is *specified*.

SampleProblem 17.1

Recognizing Endothermic and Exothermic Processes

On a sunny winter day, the snow on a rooftop begins to melt. As the melted water drips from the roof, it refreezes into icicles. Describe the direction of heat flow as the water freezes. Is this process endothermic or exothermic?

1 Analyze **Identify the relevant concepts.** Heat flows from a warmer object to a cooler object. An endothermic process absorbs heat from the surroundings. An exothermic process releases heat to the surroundings.

2 Solve **Apply the concepts to this problem.**

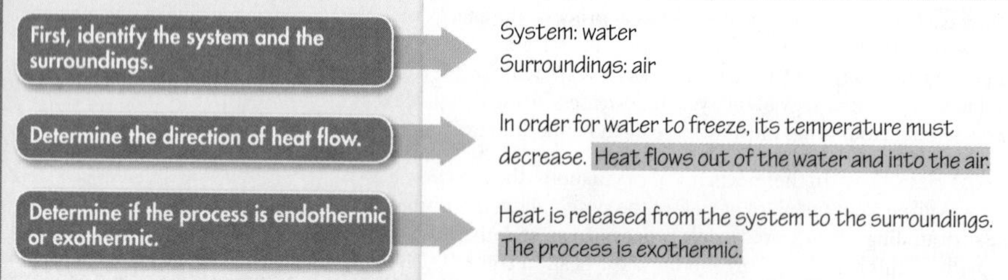

1. A container of melted wax stands at room temperature. What is the direction of heat flow as the liquid wax solidifies? Is the process endothermic or exothermic?

2. When barium hydroxide octahydrate, $Ba(OH)_2 \cdot 8H_2O$ is mixed in a beaker with ammonium thiocyanate, NH_4SCN, a reaction occurs. The beaker becomes very cold. Is the reaction endothermic or exothermic?

First, identify the system and surroundings in each situation. Then, determine the direction of heat flow.

Units for Measuring Heat Flow Describing the amount of heat flow requires units different than those used to describe temperature. Heat flow is measured in two common units, the calorie and the joule. You have probably heard of someone exercising to "burn calories." During exercise, your body breaks down sugars and fats in a process that releases heat. Although there is not an actual fire burning the sugars and fats within your body, chemical reactions accomplish the same result. For example, in breaking down 10 g of sugar, your body releases the same amount of heat that would be released if 10 g of sugar were completely burned in a fire.

A calorie (cal) is defined as the quantity of heat needed to raise the temperature of 1 g of pure water 1°C. The word *calorie* is written with a small *c* except when referring to the energy contained in food. The dietary Calorie, written with a capital *C*, always refers to the energy in food. One dietary Calorie is equal to one kilocalorie, or 1000 calories.

$$1 \text{ Calorie} = 1 \text{ kilocalorie} = 1000 \text{ calories}$$

Dietary Calories

Your proper caloric intake depends on your level of physical activity. In an eight-hour day at a desk, you burn about 800 Calories. This is about the number of Calories in two helpings of spaghetti. When exercising, however, you become a relative biochemical blast furnace. In vigorous activities such as running and jumping, you expend 7–10 Calories per minute, or 420–600 Calories per hour. At these rates, a runner who covers a 26-mile marathon course in 3 hours might expend 1800 Calories, or the equivalent of 4.5 helpings of spaghetti.

The joule (J) is the SI unit of energy. One joule of heat raises the temperature of 1 g of pure water 0.2390°C. You can convert between calories and joules using the following relationships:

$$1 \text{ J} = 0.2390 \text{ cal} \qquad 4.184 \text{ J} = 1 \text{ cal}$$

Heat Capacity and Specific Heat

On what factors does the heat capacity of an object depend?

The amount of heat needed to increase the temperature of an object exactly 1°C is the **heat capacity** of that object. **The heat capacity of an object depends on both its mass and its chemical composition.** The greater the mass of the object, the greater its heat capacity. One of the massive steel cables on the bridge in Figure 17.3, for example, requires much more heat to raise its temperature 1°C than a small steel nail does.

Different substances with the same mass may have different heat capacities. On a sunny day, a 20-kg puddle of water may be cool, while a nearby 20-kg iron sewer cover may be too hot to touch. This situation illustrates how different heat capacities affect the temperature of objects. Assuming that both the water and the iron absorb the same amount of radiant energy from the sun, the temperature of the water changes less than the temperature of the iron in the same amount of time because the specific heat capacity of water is larger than the specific heat capacity of iron.

The specific heat capacity, or simply the **specific heat,** of a substance is the amount of heat it takes to raise the temperature of 1 g of the substance 1°C. Table 17.1 gives specific heats for some common substances. Water has a very high specific heat compared with the other substances in the table. Metals generally have low specific heats. The same amount of heat affects the temperature of objects of the same mass with a high specific heat much less than the temperature of those with a low specific heat.

Figure 17.3 Heat Capacity
A massive steel cable has a higher heat capacity than a steel nail.
Compare ***Which has a greater heat capacity: a cup of water or a drop of water?***

InterpretData

Specific Heats of Some Common Substances

Substance	Specific heat	
	J/(g·°C)	cal/(g·°C)
Liquid water	4.18	1.00
Ethanol	2.4	0.58
Ice	2.1	0.50
Steam	1.9	0.45
Chloroform	0.96	0.23
Aluminum	0.90	0.21
Iron	0.46	0.11
Silver	0.24	0.057
Mercury	0.14	0.033

Table 17.1 The specific heat of a substance can be expressed in J/(g·°C) or cal/(g·°C).
a. Read Tables What is the specific heat of chloroform in cal/(g·°C)?
b. Compare Which metal in the table has the highest specific heat?
c. Calculate Show how to convert the specific heat of liquid water from J/(g·°C) to cal/(g·°C).

Hint: For part c, use the relationship 1 J = 0.2390 cal to write the appropriate conversion factor.

Explain

Heat Capacity and Specific Heat

USE VISUALS Have students study Table 17.1. **Ask** Which has a higher specific heat capacity—water or steam? *(water)* **Ask** What factors do you think affect the specific heat of each substance? *(Specific heat is affected by the amount of heat and the change in temperature.)*

Explore

Class Activity

PURPOSE Students will compare heat transfer of different materials.

MATERIALS sheet metal (13 cm × 18 cm), foam (13 cm × 18 cm), wood (40 cm × 20 cm), glue

PROCEDURE Glue a 13 cm × 18 cm piece of sheet metal and a 13 cm × 18 cm piece of foam to a piece of wood that is about 40 cm × 20 cm. Pass the wood around the classroom, and ask students to put their hands on each of the three surfaces and describe the temperature of each.

EXPECTED OUTCOME Usually students indicate that the metal feels the coldest and the foam feels the same temperature as their hands. Explain that all of the surfaces are colder than their hand, and that heat is being transferred from their hand to those surfaces. Convey that metals conduct heat away from the hand more efficiently than foam, so they feel colder than foam.

UbD Check for Understanding

What happens to the energy of the universe during a chemical or physical process?

Access students' understanding of energy flow by using the following activity. Have each student take an ice cube out of a cooler. Ask them to identify the surroundings, if the ice cube is the system. Then, have them explain the direction of energy flow that occurs as it relates to the ice cube. *(The hand and air are the surroundings. Energy flows from the skin to the ice cube, giving a sense that it is cold.)* **Ask** Is the process of ice melting endothermic or exothermic?

ADJUST INSTRUCTION If students are having difficulty identifying the surroundings, have students reread the first paragraph of the Endothermic and Exothermic Processes section. Then, ask students where the heat transfer occurs between the hand and the ice cube. *(The transfer is occurring between the surface molecules of the skin and the surface molecules of the ice.)* Explain that the process is endothermic because the ice gains heat from the hand.

Answers

1. Heat flows from the system (wax) to the surroundings (air). The process is exothermic.
2. Since the beaker becomes cold, heat is absorbed by the system (chemicals within the beaker) from the surroundings (beaker and surrounding air). The process is endothermic.

FIGURE 17.3 a cup of water

INTERPRET DATA

a. 0.23 cal/(g·°C)

b. aluminum

c. $4.18 \text{ J/(g}\cdot\text{°C)} \times \dfrac{0.2390 \text{ cal}}{1 \text{ J}} = 1.00 \text{ cal/(g}\cdot\text{°C)}$

Explain

CRITICAL THINKING Direct students to Figure 17.4 and discuss the difference in temperature between a coastal city such as San Francisco and an inland city such as Fresno. State the following remark once made by Mark Twain: "The coldest winter I ever spent was the summer I spent in San Francisco." **Ask** How does this quote relate to the temperature difference between a coastal city and an inland city? *(The relatively high heat capacity of the large bodies of water near San Francisco help to moderate local temperatures. During summer, the ocean water warms slowly as it absorbs large amounts of heat from warm air masses. Onshore breezes off the cool water keep coastal land temperatures from rising to match those of inland areas. Condensation of water vapor in the air produces a cool fog, which probably enhanced Mark Twain's impression of the cool of San Francisco summers.)*

CHEMISTRY & YOU A given mass of water can absorb more heat from the lava than an equal mass of air.

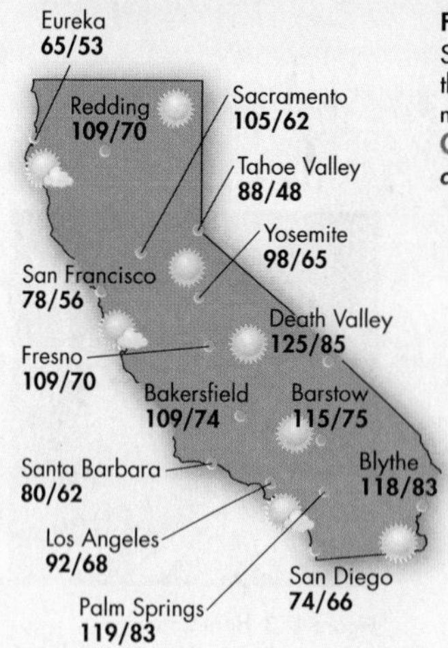

Figure 17.4 Temperature Moderation
San Francisco is located on the Pacific coast. The high specific heat of the water in the ocean helps keep the temperature in San Francisco much more moderate than that of the towns and cities farther inland.
Compare ***Describe how the ocean affects the temperature of coastal areas in the summer and in the winter.***

CHEMISTRY & YOU

Q: *Heat will flow from the lava to the surroundings until the lava and surroundings are at the same temperature. Air has a smaller specific heat than water. Why would lava then cool more quickly in water than in air?*

Specific Heat of Water Just as it takes a lot of heat to raise the temperature of water, water also releases a lot of heat as it cools. Water in lakes and oceans absorbs heat from the air on hot days and releases it back into the air on cool days. As illustrated in Figure 17.4, this property of water is responsible for moderate climates in coastal areas. Citrus farmers often spray their trees with water to protect the fruit from frost damage during icy weather. As the water freezes, it releases heat, which helps prevent the fruit from freezing. When a freshly baked apple pie, such as the one shown in Figure 17.5, comes out of the oven, both the filling and crust are at the same temperature. However, the filling, which is mostly water, has a higher specific heat than the crust. In order to cool down, the filling must give off a lot of heat. This release of heat is why you have to be careful not to burn your tongue when eating hot apple pie.

Calculating Specific Heat To calculate the specific heat (C) of a substance, you divide the heat input by the mass of the substance times the temperature change.

$$C = \frac{q}{m \times \Delta T} = \frac{\text{heat (J or cal)}}{\text{mass (g)} \times \text{change in temperature (°C)}}$$

In the equation above, q is heat and m is mass. The symbol ΔT (read "delta T") represents the change in temperature. ΔT is calculated from the equation $\Delta T = T_f - T_i$, where T_f is the final temperature and T_i is the initial temperature. As you can see from the equation and in Table 17.1 on the previous page, heat may be expressed in terms of joules or calories. Therefore, the units of specific heat are either J/(g·°C) or cal/(g·°C).

Figure 17.5 Cooling of Water
The filling of a hot apple pie is mostly water, so it is much more likely to burn your tongue than the crust.

Foundations for Math

SYMBOLS Students often struggle with calculations involving the delta (Δ) symbol, as some calculations require the subtraction of the smallest value from the largest, while in others the initial value is subtracted from the final value. Explain that in specific heat calculations, the presence of Δ in the denominator of the equation indicates that the initial temperature should be subtracted from the final temperature.

In **Sample Problem 17.2,** the initial temperature is 25°C and the final temperature is 48.0°C. Explain to students that they can verify that their order is correct by looking for words that indicate the direction of change, such as *increase* and *decrease*. In this problem, the phrase *increases from 25.0°C* provides additional evidence that the lower value is the initial value.

Sample Problem 17.2

Calculating the Specific Heat of a Substance

The temperature of a 95.4-g piece of copper increases from 25.0°C to 48.0°C when the copper absorbs 849 J of heat. What is the specific heat of copper?

KNOWNS

$m_{Cu} = 95.4\ g$

$\Delta T = 48.0°C - 25.0°C = 23.0°C$

$q = 849\ J$

UNKNOWN

$C_{Cu} = ?\ J/(g \cdot °C)$

1 Analyze **List the knowns and the unknown.** Use the known values and the definition of specific heat.

2 Calculate **Solve for the unknown.**

Start with the equation for specific heat.

$$C_{Cu} = \frac{q}{m_{Cu} \times \Delta T}$$

Substitute the known quantities into the equation to calculate the unknown value C_{Cu}.

$$C_{Cu} = \frac{849\ J}{95.4\ g \times 23.0°C} = 0.387\ J/(g \cdot °C)$$

3 Evaluate **Does the result make sense?** Remember that liquid water has a specific heat of 4.18 J/(g·°C). Metals have specific heats lower than water. Thus the calculated value of 0.387 J/(g·°C) seems reasonable.

3. When 435 J of heat is added to 3.4 g of olive oil at 21°C, the temperature increases to 85°C. What is the specific heat of the olive oil?

4. How much heat is required to raise the temperature of 250.0 g of mercury 52°C?

You can find the specific heat of mercury on Table 17.1.

17.1 LessonCheck

5. Review What are the ways that energy conversion can occur?

6. Describe What happens to the energy of the universe during a physical or chemical process?

7. List On what two factors does the heat capacity of an object depend?

8. Classify On a cold night you use an electric blanket to warm your body. Describe the direction of heat flow. Is this process endothermic or exothermic?

9. Calculate A chunk of silver has a heat capacity of 42.8 J/°C and a mass of 181 g. Calculate the specific heat of silver.

10. Calculate Using calories, calculate how much heat 32.0 g of water absorbs when it is heated from 25.0°C to 80.0°C. How many joules is this?

BIGIDEA MATTER AND ENERGY

11. How is the energy of the universe conserved during the combustion of gasoline in a car engine?

Lesson Check Answers

5. Energy changes occur as either heat transfer or work, or a combination of both.
6. The energy of the universe remains unchanged.
7. mass and chemical composition
8. Heat flows from the blanket to your body. If your body is the system, the process is endothermic.
9. 2.36×10^{-1} J/(g · °C)
10. 1.76×10^{3} cal (1.76 kcal); 7.36×10^{3} J (7.36 kJ)
11. Chemical energy in the gasoline is transformed into energy to move the car, and heat. None of the energy is lost in the process.

Sample Practice Problem

How much heat is required to raise the temperature of 3.5 g of grain alcohol (ethanol) by 15°C? *(1.26×10^{2} J)*

Evaluate

Informal Assessment

Place an ice cube in a beaker of cool water. Have students discuss the flow of heat when the ice and water are the system and surroundings, respectively, and when the ice/water mixture and a 37°C room are the system and surroundings, respectively. Then have students complete the 17.1 Lesson Check.

Reteach

Emphasize that substances vary in their response to an input of heat; a given amount of heat raises the temperature of some substances (such as metals) far more than others (such as nonmetals). Point out that specific heat is a property of a substance; it is a measure of the ability of a substance to store heat.

Answers

3. 2.0 J/(g · °C)
4. 1.8 kJ

LESSON 17.2

Key Objectives

17.2.1 DESCRIBE how to measure the change in enthalpy of a reaction.

17.2.2 DESCRIBE how to express the enthalpy change for a reaction in a chemical equation.

Additional Resources

Reading and Study Workbook, Lesson 17.2

Available Online or on Digital Media:

- Teaching Resources, Lesson 17.2 Review
- Laboratory Manual, Labs 33, 34, and 35
- Virtual Chemistry Labs, 17 and 19

PearsonChem.com

Engage

CHEMISTRY & YOU Tell students that a match will not ignite unless you strike it and add heat to it. **Ask** Where does the heat that is added to the match come from? *(friction)* **Ask** Is the burning of a match an endothermic reaction? Explain. *(No; the reaction releases more energy, in the form of heat and light, than the amount it absorbs to start.)*

Build Background

Ask students to predict whether there is a way to measure how much heat is released from a burning match. *(Answers will vary.)* Explain that it is possible, but only indirectly. If the reaction were confined, then any temperature changes in the surroundings could be attributed to heat transfer from the reaction. Tell students to keep this in mind as they proceed with the lesson.

IN ACADEMIC STANDARDS for SCIENCE

C.6.3 Solve problems involving heat flow and temperature changes, using known values of specific heat and/or phase change constants (latent heats).

C.6.4 Classify chemical reactions and phase changes as exothermic or endothermic.

17.2 Measuring and Expressing Enthalpy Changes

C.6.3 Heat flow and temperature change problems, **C.6.4** Exothermic and endothermic reactions

CHEMISTRY & YOU

Q: *How can you measure the amount of heat released when a match burns?* When you strike a match, heat is released to the surroundings. In addition to describing the direction of heat flow, you may also want to determine the quantity of heat that is transferred. The concept of specific heat allows you to measure heat flow in chemical and physical processes.

Key Questions

How can you measure the change in enthalpy of a reaction?

How can you express the enthalpy change for a reaction in a chemical equation?

Vocabulary

- calorimetry
- calorimeter
- enthalpy
- thermochemical equation
- heat of reaction
- heat of combustion

Calorimetry

How can you measure the change in enthalpy of a reaction?

Heat that is absorbed or released during many chemical reactions can be measured by a technique called calorimetry. **Calorimetry** is the measurement of the heat flow into or out of a system for chemical and physical processes. In a calorimetry experiment involving an endothermic process, the heat absorbed by the system is equal to the heat released by its surroundings. In an exothermic process, the heat released by a system is equal to the heat absorbed by its surroundings. The insulated device used to measure the absorption or release of heat in chemical or physical processes is called a **calorimeter.**

Constant-Pressure Calorimeters Foam cups can be used as simple calorimeters because they do not let much heat in or out. The heat flows for many chemical reactions can be measured in a constant-pressure calorimeter similar to the one shown in Figure 17.6. Most chemical reactions and physical changes carried out in the laboratory are open to the atmosphere and, thus, occur at constant pressure. The **enthalpy** (*H*) of a system accounts for the heat flow of the system at constant pressure.

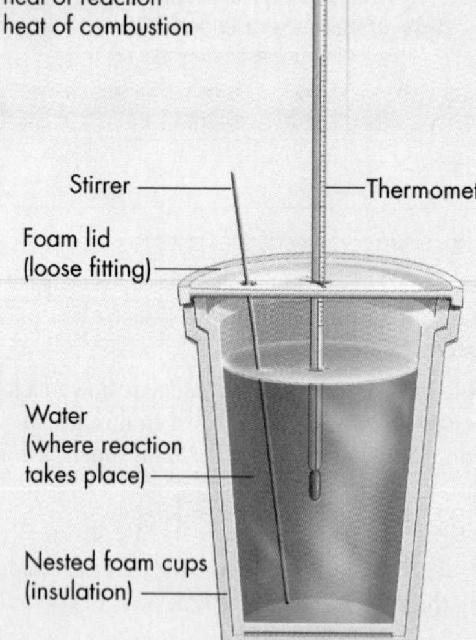

Figure 17.6 Constant-Pressure Calorimeter
In a simple constant-pressure calorimeter, a thermometer records the temperature change as chemicals react in water. The reacting substances constitute the system. The water constitutes the surroundings.
Relate Cause and Effect ***What happens to the temperature of the water if heat is released by the reaction in the calorimeter?***

Focus on ELL

1 CONTENT AND LANGUAGE Discuss the meaning of the suffixes *–metry* and *–meter* and how to use suffixes to determine the meaning of words. Then ask students to predict the meaning of the words *calorimetry* and *calorimeter*.

2 FRONTLOAD THE LESSON Have students brainstorm enthalpy changes that occur in their daily lives. Ask them to consider how these changes can be measured. *(Sample answer: a change in temperature)* **Ask** How could you make sure you are measuring and identifying only the enthalpy change in the system you are studying? Explain that, in this lesson, they will learn the answer to this question.

3 COMPREHENSIBLE INPUT When explaining the signs of enthalpy changes in endothermic and exothermic reactions, make the connection between a + sign, addition of energy to the system, and the meaning of the prefix *endo-*. Make a similar connection between a – sign, release (subtraction) of energy from the system, and the meaning of the prefix *exo-*.

The heat absorbed or released by a reaction at constant pressure is the same as the change in enthalpy, symbolized as ΔH. **The value of ΔH of a reaction can be determined by measuring the heat flow of the reaction at constant pressure.** In this textbook, the terms *heat* and *enthalpy change* are used interchangeably because the reactions presented occur at constant pressure. In other words, $q = \Delta H$.

To measure the enthalpy change for a reaction in aqueous solution in a foam cup calorimeter, dissolve the reacting chemicals (the system) in known volumes of water (the surroundings). Measure the initial temperature of each solution, and mix the solutions in the foam cup. After the reaction is complete, measure the final temperature of the mixed solutions. You can calculate the heat absorbed or released by the surroundings (q_{surr}) using the formula for specific heat, the initial and final temperatures, and the heat capacity of water.

$$q_{surr} = m \times C \times \Delta T$$

In this expression, m is the mass of the water, C is the specific heat of water, and $\Delta T = T_f - T_i$. The heat absorbed by the surroundings is equal to, but has the opposite sign of, the heat released by the system. Conversely, the heat released by the surroundings is equal to, but has the opposite sign of, the heat absorbed by the system. Therefore, the enthalpy change for the reaction (ΔH) can be written as follows:

$$q_{sys} = \Delta H = -q_{surr} = -m \times C \times \Delta T$$

The sign of ΔH is positive for an endothermic reaction and negative for an exothermic reaction.

Constant-Volume Calorimeters Calorimetry experiments can also be performed at constant volume using a device called a bomb calorimeter. In a bomb calorimeter, which is shown in Figure 17.7, a sample of a compound is burned in a constant-volume chamber in the presence of oxygen at high pressure. The heat that is released warms the water surrounding the chamber. By measuring the temperature increase of the water, it is possible to calculate the quantity of heat released during the combustion reaction.

CHEMISTRY & YOU

Q: *What type of calorimeter would you use to measure the heat released when a match burns? Describe the experiment and how you would calculate the heat released.*

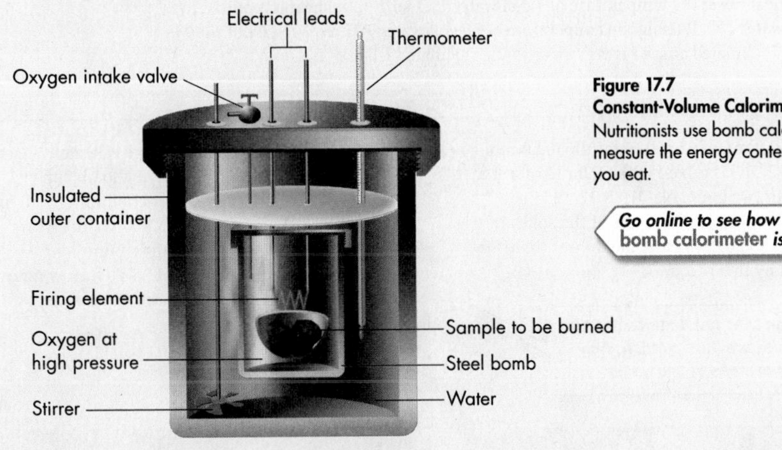

Figure 17.7
Constant-Volume Calorimeter
Nutritionists use bomb calorimeters to measure the energy content of the foods you eat.

KINETIC ART *Go online to see how a* **bomb calorimeter** *is used.*

UbD Check for Understanding

How is energy conserved in a chemical or physical process?

Assess students' understanding of this Essential Question by having them write a one-minute explanation of how the relationship between q_{sys} and q_{surr} in a calorimetric experiment proves the law of conservation of energy.

ADJUST INSTRUCTION If students have difficulty writing an explanation, have students reread the text in the Constant-Pressure Calorimeters section. Then, review the specific heat equations given in the text with students. Allow students an opportunity to revise their explanations following this review.

Explain

Calorimetry

CHEMISTRY & YOU You would use a constant-volume calorimeter. You would burn the match in a steel bomb. Then you would measure the temperature increase of the surrounding water. Finally, you multiply the mass of water by the specific heat of water and by the temperature change of water and you get the heat released by the match.

Explore

Class Activity

PURPOSE Students will make calorimetry measurements and calculations. Two separate but related oxidation-reduction reactions will be studied by different groups of students. The results can then be pooled and compared to provide insight into how calorimetry can be used as a tool to investigate the physical and chemical properties of substances.

MATERIALS 1*M* HCl, foam cup, thermometer, mossy zinc, magnesium turnings

SAFETY Students must wear protective goggles and aprons. Remind them of the safety precautions needed for handling HCl. Use a fume hood if possible. Students should not be allowed to dispense concentrated HCl themselves.

PROCEDURE Pour 100 mL of 1*M* HCl into a foam cup and determine the temperature to the nearest tenth of a degree Celsius. Add 0.5 g of mossy zinc or 0.5 g of magnesium turnings.

EXPECTED OUTCOME For calculations, ignore the mass of the solids added. Assume that HCl is primarily water and has the same specific heat and density as water.

$\Delta H = 100 \text{ g} \times 4.18 \text{ J/(g} \cdot °\text{C)} \times \Delta T$

ΔH_{rxn} Zn = –150 kJ/mol

ΔH_{rxn} Mg = –460 kJ/mol

Discuss sources of error such as heat lost to surroundings.

Answers

FIGURE 17.6 The temperature of the water increases.

LESSON 17.2

Explain

Calorimeters

MAKING CONNECTIONS Display photographs or actual examples of the types of calorimeter(s) used in laboratories today. Discuss the use of calorimetry as an analytical tool; for example, in investigating the caloric content of foods.

Misconception Alert

Make sure students understand all the terms in the equation $q_{sys} = \Delta H = -m \times C \times \Delta T$. Emphasize that the delta notation (Δ) represents an algebraic operation involving two separate measurements. Explain that when determining the algebraic value of ΔT, the algebraic signs of the *final* and *initial* temperatures must be included in the operation.

Sample Practice Problem

A copper coin is heated and placed in a foam cup calorimeter containing 32.0 mL of water at 22.0°C. The water reaches a temperature of 24.0°C. How many joules of heat were released by the coin? *(268 J)*

Extend

Connect to ENVIRONMENTAL SCIENCE

Tell students that a large percentage of the electricity used in the United States is produced by the burning of fossil fuels. Explain that power plants generate electricity by transferring heat to water during the combustion of fossil fuels; the heated water turns to steam, which drives turbines that generate electricity. Explain that in order to condense and recycle the steam, a source of coolant water is needed; consequently, power plants (including nuclear power plants) are often built near a source of water, such as an ocean, lake, or river. However, when the coolant water is returned to its source, it is usually warmer. Explain that this increase in temperature is called thermal pollution. Have students research this topic and then conduct a class debate to determine if this label is justified.

Enthalpy Change in a Calorimetry Experiment

When 25.0 mL of water containing 0.025 mol HCl at 25.0°C is added to 25.0 mL of water containing 0.025 mol NaOH at 25.0°C in a foam cup calorimeter, a reaction occurs. Calculate the enthalpy change (in kJ) during this reaction if the highest temperature observed is 32.0°C. Assume the densities of the solutions are 1.00 g/mL and that the volume of the final solution is equal to the sum of the volumes of the reacting solutions.

1 Analyze List the knowns and the unknown. Use dimensional analysis to determine the mass of the water. You must also calculate ΔT. Use $\Delta H = -q_{surr} = -m \times C \times \Delta T$ to solve for ΔH.

KNOWNS	UNKNOWN
$C_{water} = 4.18\ J/(g \cdot °C)$	$\Delta H = ?\ kJ$
$V_{final} = V_{HCl} + V_{NaOH}$	
$= 25.0\ mL + 25.0\ mL = 50.0\ mL$	
$T_i = 25.0°C$	
$T_f = 32.0°C$	
$density_{solution} = 1.00\ g/mL$	

2 Calculate Solve for the unknown.

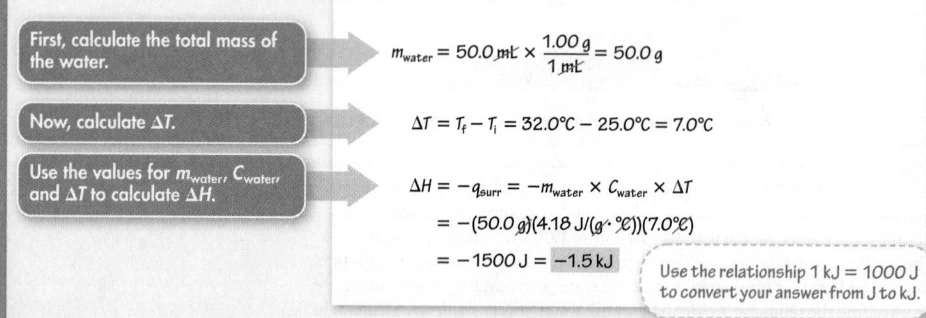

3 Evaluate Does the result make sense? The temperature of the solution increases, which means that the reaction is exothermic, and thus the sign of ΔH should be negative. About 4 J of heat raises the temperature of 1 g of water 1°C, so 200 J of heat is required to raise 50 g of water 1°C. Raising the temperature of 50 g of water 7°C requires about 1400 J, or 1.4 kJ. This estimated answer is very close to the calculated value of ΔH.

12. When 50.0 mL of water containing 0.50 mol HCl at 22.5°C is mixed with 50.0 mL of water containing 0.50 mol NaOH at 22.5°C in a calorimeter, the temperature of the solution increases to 26.0°C. How much heat (in kJ) is released by this reaction?

Assume that the densities of the solutions are 1.00 g/mL to find the total mass of the water.

13. A small pebble is heated and placed in a foam cup calorimeter containing 25.0 mL of water at 25.0°C. The water reaches a maximum temperature of 26.4°C. How many joules of heat are released by the pebble?

Foundations for Math

MULTI-STEP PROBLEMS Although the equation for calculating ΔH is straightforward, more than one step may be necessary before this calculation can be successfully completed. This occurs when the quantity of substance is given as a volume or a molar quantity, or when the value of ΔT must be calculated.

In **Sample Problem 17.3,** the system is comprised of equal volumes of HCl and NaOH solutions, which must be converted to mass in grams. Since the reaction is a neutralization reaction between a strong acid and a strong base, you can assume that the density of the solution is approximately equal to the density of water. Once the mass of the solution and the temperature change are determined, the values can be plugged into the equation as the final step.

Thermochemical Equations

How can you express the enthalpy change for a reaction in a chemical equation?

If you mix calcium oxide with water, the water in the mixture becomes warm. This exothermic reaction occurs when cement, which contains calcium oxide, is mixed with water to make concrete. When 1 mol of calcium oxide reacts with 1 mol of water, 1 mol of calcium hydroxide forms and 65.2 kJ of heat is released. **In a chemical equation, the enthalpy change for the reaction can be written as either a reactant or a product.** In the equation describing the exothermic reaction of calcium oxide and water, the enthalpy change can be considered a product.

$$CaO(s) + H_2O(l) \longrightarrow Ca(OH)_2(s) + 65.2\text{ kJ}$$

This equation is presented visually in Figure 17.8. A chemical equation that includes the enthalpy change is called a **thermochemical equation.**

Heats of Reaction The **heat of reaction** is the enthalpy change for the chemical equation exactly as it is written. You will usually see heats of reaction reported as ΔH, which is equal to the heat flow at constant pressure. The physical state of the reactants and products must also be given. The standard conditions are that the reaction is carried out at 101.3 kPa (1 atm) and that the reactants and products are in their usual physical states at 25°C. The heat of reaction, or ΔH, in the above example is −65.2 kJ. Each mole of calcium oxide and water that reacts to form calcium hydroxide produces 65.2 kJ of heat.

$$CaO(s) + H_2O(l) \longrightarrow Ca(OH)_2(s) \qquad \Delta H = -65.2\text{ kJ}$$

In this and other exothermic processes, the chemical potential energy of the reactants is higher than the chemical potential energy of the products.

READING SUPPORT

Build Comprehension: *Use Prior Knowledge* Thermochemical equations are just like other balanced equations. If heat is absorbed in the reaction, it is written as a reactant. If heat is released, it is written as a product. ***Recall chemical reactions from Chapter 11. For the combustion of methane, on which side of the reaction arrow would you write the heat absorbed or released by the reaction?***

Figure 17.8 Exothermic Process
Calcium oxide is one of the components of cement. The reaction of calcium oxide and water is an exothermic process.

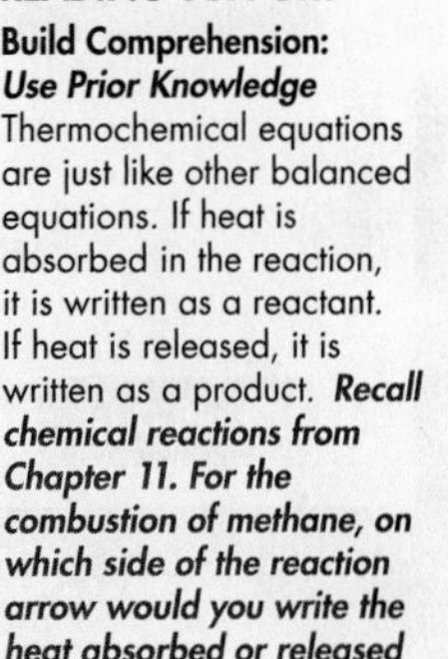

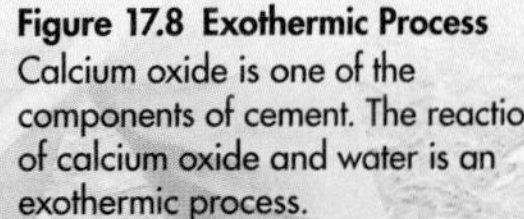

Differentiated Instruction

L1 SPECIAL NEEDS STUDENTS Iron wool burns rapidly in high concentrations of oxygen. It oxidizes slowly at the concentrations found in air. Rusting of iron is essentially a slow combustion reaction. Demonstrate that the reaction is exothermic by wrapping moist iron wool around the bulb of a thermometer. Have students work in small groups to quantify the thermodynamic parameters of the reaction by using a foam cup calorimeter.

LPR LESS PROFICIENT READERS Special attention should be given to vocabulary terms in the chapter that are defined by mathematical relationships. Have students write definitions of these terms. Sample calculations showing how to apply the formulas should be included in students' notes.

Explain

Thermochemical Equations

APPLY CONCEPTS To illustrate writing a thermochemical equation, refer to two familiar reactions. Show students that the combustion of methane in a Bunsen burner adds energy to the surroundings, thus, the burning of methane has a negative ΔH.

$$CH_4(g) + 2O_2(g) \rightarrow CO_2(g) + 2H_2O(g) + 890\text{ kJ}$$
$$\Delta H = -890\text{ kJ/mol}$$

Explain that during the electrolysis of water, electrical energy is used to decompose the water into hydrogen gas and oxygen gas. Point out that energy is added to the system from the surroundings. Thus, the electrolysis of water has a positive ΔH.

$$241.8\text{ kJ} + 2H_2O(l) \rightarrow 2H_2(g) + O_2(g)$$
$$\Delta H = +241.8\text{ kJ/mol}$$

Misconception Alert

Remind students that the sign of ΔH indicates whether the reaction is endothermic (+) or exothermic (–). Although an exothermic reaction has a negative ΔH by convention, heat is expressed as an absolute value when it is shown as a "product" in an equation.

USE VISUALS Have students examine Figure 17.8. Instruct students to draw similar diagrams for the burning of CH_4 and the electrolysis of H_2O. For each reaction, **Ask** Is the potential energy of the reactants higher or lower than the products? *(higher for CH_4, lower for H_2O)* **Ask** In what direction does heat move in the CaO reaction? *(into the surroundings)*

Answers

12. 1.46 kJ
13. 150 J

READING SUPPORT Heat is released so you write it on the right side of the equation.

LESSON 17.2

Explain

MAKE A CONNECTION Tell interested students that the decomposition of sodium bicarbonate to form carbon dioxide gas is essential for successful baking. Sodium bicarbonate, also called baking soda, is the source of the carbon dioxide that causes cakes, muffins, and quick breads to rise in the oven. **Ask** What is the result if you forget to add this ingredient to a cookie recipe? *(Without baking soda, the result is a flat and heavy, and not very tasty cookie.)*

Explore

An Exothermic Reaction

Teacher Demo

PURPOSE Students will observe an exothermic reaction and write a thermochemical equation.

MATERIALS CaO, metal tray, thin aluminum pan, water, egg (raw)

SAFETY CaO is corrosive.

PROCEDURE Place 100 g of CaO in a metal tray. Place a thin aluminum pan or pie tin in the tray so that its bottom is completely in contact with the calcium oxide. Pour just enough water into the tray to cover the calcium oxide. Then break a fresh egg into the aluminum pan. **Ask** Is this an exothermic or endothermic reaction? *(Exothermic; 65.2 kJ of heat is released per mole of CaO.)* Have students write a thermochemical equation to describe the reaction. Check to see that students express the enthalpy of reaction as a negative value. Have students use the thermochemical equation to answer stoichiometric questions such as: How much heat is released when 100.0 g of calcium oxide reacts with excess water? *(116 kJ)*

EXPECTED OUTCOME Students should observe that enough heat is liberated by the reaction to cook the egg.

Figure 17.9 Endothermic Process
Muffin batter often contains baking soda, also known as sodium bicarbonate. The decomposition of sodium bicarbonate is an endothermic process.

Other reactions absorb heat from the surroundings. For example, baking soda (sodium bicarbonate) decomposes when it is heated. The carbon dioxide released in the reaction causes muffins to rise while baking. This process is endothermic.

$$2NaHCO_3(s) + 85\text{ kJ} \longrightarrow Na_2CO_3(s) + H_2O(l) + CO_2(g)$$

Remember that ΔH is positive for endothermic reactions. Therefore, you can write the reaction as follows:

$$2NaHCO_3(s) \longrightarrow Na_2CO_3(s) + H_2O(l) + CO_2(g) \qquad \Delta H = 85\text{ kJ}$$

Figure 17.9 shows the enthalpy diagram for this reaction.

Chemistry problems involving enthalpy changes are similar to stoichiometry problems. The amount of heat released or absorbed during a reaction depends on the number of moles of the reactants involved. The decomposition of 2 mol of sodium bicarbonate, for example, requires 85 kJ of heat. Therefore, the decomposition of 4 mol of the same substance would require twice as much heat, or 170 kJ. In this and other endothermic processes, the chemical potential energy of the products is higher than the chemical potential energy of the reactants.

To see why the physical state of the reactants and products in a thermochemical reaction must be stated, compare the following two equations for the decomposition of 1 mol H_2O:

$$H_2O(l) \longrightarrow H_2(g) + \tfrac{1}{2}O_2(g) \qquad \Delta H = 285.8\text{ kJ}$$
$$H_2O(g) \longrightarrow H_2(g) + \tfrac{1}{2}O_2(g) \qquad \Delta H = 241.8\text{ kJ}$$
$$\text{difference} = 44.0\text{ kJ}$$

Although the two equations are very similar, the different physical states of H_2O result in different ΔH values. In one case, the reactant is a liquid; in the other case, the reactant is a gas. The vaporization of 1 mol of liquid water to water vapor at 25°C requires 44.0 kJ of heat.

$$H_2O(l) \longrightarrow H_2O(g) \qquad \Delta H = 44.0\text{ kJ}$$

566 Chapter 17 • Lesson 2

Keeping Warm

How do animals such as polar bears and seals survive the cold land and water temperatures where they live? A good coat of fur helps, but it is not enough to keep them warm. These animals also have unique fat cells that help generate heat. These fat cells are in tissue called brown fat. The cells of brown fat are unlike other fat cells in these animals' bodies. Most other cells store chemical energy released during the breakdown of carbohydrates and fatty acids in the bonds of adenosine triphosphate (ATP). ATP acts as the central source of energy and growth of all organisms. Heat is mostly a waste product in ATP-producing cells. The heat generated by the brown fat tissue, however, helps the animal keep relatively comfortable even at subzero temperatures.

Sample Problem 17.4

Using the Heat of Reaction to Calculate Enthalpy Change

Calculate the amount of heat (in kJ) required to decompose 2.24 mol $NaHCO_3(s)$.

$$2NaHCO_3(s) + 85 \text{ kJ} \longrightarrow Na_2CO_3(s) + H_2O(l) + CO_2(g)$$

1 Analyze **List the knowns and the unknown.** Use the thermochemical equation above to write a conversion factor relating kilojoules of heat and moles of $NaHCO_3$. Then use the conversion factor to determine ΔH for 2.24 mol $NaHCO_3$.

KNOWNS

amount of $NaHCO_3(s)$ that decomposes = 2.24 mol

ΔH = 85 kJ for 2 mol $NaHCO_3$

UNKNOWN

ΔH = ? kJ for 2.24 mol $NaHCO_3$

2 Calculate **Solve for the unknown.**

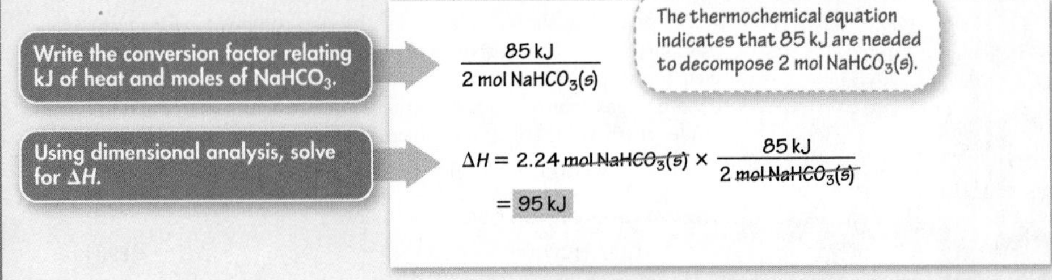

Write the conversion factor relating kJ of heat and moles of $NaHCO_3$.

$$\frac{85 \text{ kJ}}{2 \text{ mol } NaHCO_3(s)}$$

The thermochemical equation indicates that 85 kJ are needed to decompose 2 mol $NaHCO_3(s)$.

Using dimensional analysis, solve for ΔH.

$$\Delta H = 2.24 \text{ mol } NaHCO_3(s) \times \frac{85 \text{ kJ}}{2 \text{ mol } NaHCO_3(s)}$$

$$= 95 \text{ kJ}$$

3 Evaluate **Does the result make sense?** The 85 kJ in the thermochemical equation refers to the decomposition of 2 mol $NaHCO_3(s)$. Therefore, the decomposition of 2.24 mol should absorb more heat than 85 kJ. The answer of 95 kJ is consistent with this estimate.

To do Problem 15, first convert from mass of CS_2 to moles of CS_2.

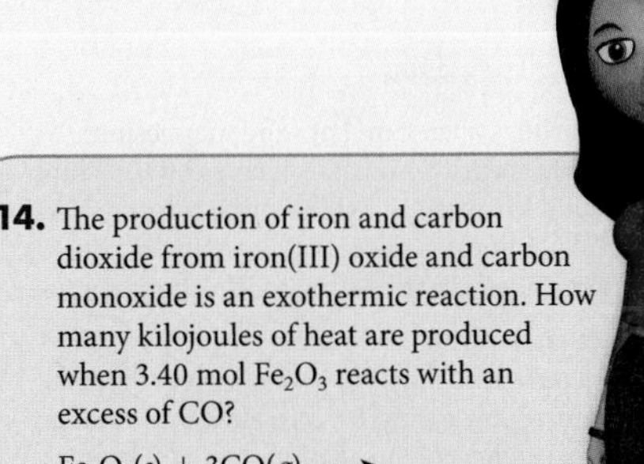

14. The production of iron and carbon dioxide from iron(III) oxide and carbon monoxide is an exothermic reaction. How many kilojoules of heat are produced when 3.40 mol Fe_2O_3 reacts with an excess of CO?

$$Fe_2O_3(s) + 3CO(g) \longrightarrow 2Fe(s) + 3CO_2(g) + 26.3 \text{ kJ}$$

15. When carbon disulfide is formed from its elements, heat is absorbed. Calculate the amount of heat (in kJ) absorbed when 5.66 g of carbon disulfide is formed.

$$C(s) + 2S(s) \longrightarrow CS_2(l)$$

$$\Delta H = 89.3 \text{ kJ}$$

Foundations for Math

CONVERSION FACTORS Stress to students that it is essential that they include units throughout a problem, not only for the given quantities but also for the conversion factors. Point out that this is true for two reasons: 1) cancelling the units helps demonstrate that the conversion factors are oriented correctly; and 2) the final answer will contain the correct units.

In **Sample Problem 17.4,** a conversion factor relating kilojoules of heat and moles of sodium bicarbonate is needed. By including the units, you can see that the conversion factor should be written with mol $NaHCO_3$ in the denominator so that they will cancel. This means that kJ should be in the numerator, and the final answer will have the desired units for heat.

Explain

Sample Practice Problem

Calculate the amount of heat (in kJ) required to decompose 2.0 mol hydrogen peroxide in the following decomposition reaction.

$$2H_2O_2(l) \rightarrow 2H_2O(l) + O_2(g)$$

$$\Delta H = -98.2 \text{ kJ (for 1 mol } H_2O_2)$$

(–196.0 kJ)

Extend

CRITICAL THINKING Tell students that many metals react with oxygen to produce metal oxides. Have students research the combustion reactions of a series of selected metals (e.g., Mg, Ca, Fe). Students should write thermochemical equations for each reaction and compare the relative amounts of heat released in each case. Ask students to describe any trend(s) they observe.

Answers

14. 89.4 kJ
15. 6.63 kJ

LESSON 17.2

Extend

Thermochemical Equations

Connect to TECHNOLOGY

Have students research and write a report on a topic related to solar power that draws on their newly gained knowledge of enthalpy changes. Possible topics include:

- hybrid solar power plants that combine solar energy collection with fossil fuel combustion
- mirror technologies used in solar power plants
- heat-transfer fluids used in CSP technologies (e.g., air, steam, molten salt, oil)
- houses and other buildings that have solar-powered heating and hot-water systems
- photovoltaic technologies used in solar power plants, solar cars, and solar houses

Evaluate

Informal Assessment

To determine students' knowledge about measuring and expressing heat changes, pose the following questions:

A. Explain the difference between H and ΔH. *(H is enthalpy or heat content; ΔH represents a change in heat content.)*

B. Use the data in Table 17.2 to write the thermochemical equation for the combustion of propane.

($C_3H_8(g) + 5O_2(g) \rightarrow 3CO_2(g) + 4H_2O(l)$; $\Delta H = -2220$ kJ)

Then have students complete the 17.2 Lesson Check.

Reteach

Select a substance from Table 17.2, and write the thermochemical equation on the chalkboard. Discuss the ΔH for the reaction, why it is negative or positive, and draw the enthalpy diagram for the reaction. Remind students that heat is the energy that flows into or out of a thermodynamic system because of differences in temperature between the system and its surroundings. Remind students that the total energy, potential and kinetic, of the system and the surroundings must remain the same in any chemical or physical process.

Figure 17.10 Combustion
The combustion of natural gas is an exothermic reaction. As bonds in methane (the main component of natural gas) and oxygen are broken and bonds in carbon dioxide and water are formed, large amounts of energy are released.

Table 17.2

Heats of Combustion at 25°C

Substance	Formula	ΔH (kJ/mol)
Hydrogen	$H_2(g)$	−286
Carbon	C(*s*, graphite)	−394
Methane	$CH_4(g)$	−890
Acetylene	$C_2H_2(g)$	−1300
Ethanol	$C_2H_6O(l)$	−1368
Propane	$C_3H_8(g)$	−2220
Glucose	$C_6H_{12}O_6(s)$	−2808
Octane	$C_8H_{18}(l)$	−5471
Sucrose	$C_{12}H_{22}O_{11}(s)$	−5645

Heats of Combustion Table 17.2 lists heats of combustion for some common substances. The **heat of combustion** is the heat of reaction for the complete burning of one mole of a substance. Figure 17.10 shows the combustion of natural gas, which is mostly methane. Small amounts of natural gas within crude oil are burned off at oil refineries. This is an exothermic reaction.

$$CH_4(g) + 2O_2(g) \longrightarrow CO_2(g) + 2H_2O(l) + 890 \text{ kJ}$$

You can also write this equation as follows:

$$CH_4(g) + 2O_2(g) \longrightarrow CO_2(g) + 2H_2O(l) \qquad \Delta H = -890 \text{ kJ}$$

Burning 1 mol of methane releases 890 kJ of heat. The heat of combustion (ΔH) for this reaction is −890 kJ per mole of methane burned.

Like other heats of reaction, heats of combustion are reported as the enthalpy changes when the reactions are carried out at 101.3 kPa of pressure and the reactants and products are in their physical states at 25°C.

ONLINE PROBLEMS 17.2 LessonCheck

16. **Describe** How can you determine the value of ΔH of a reaction?

17. **Review** How are enthalpy changes expressed in chemical equations?

18. **Calculate** A lead mass is heated and placed in a foam cup calorimeter containing 40.0 mL of water at 17.0°C. The water reaches a temperature of 20.0°C. How many joules of heat were released by the lead?

19. **Explain** What does the term *heat of combustion* refer to?

20. **Describe** When 2 mol of solid magnesium (Mg) combines with 1 mol of oxygen gas (O_2), 2 mol of solid magnesium oxide (MgO) is formed and 1204 kJ of heat is released. Write the thermochemical equation for this combustion reaction.

21. **Calculate** Gasohol contains ethanol, $C_2H_6O(l)$. When ethanol burns, it reacts with $O_2(g)$ to produce $CO_2(g)$ and $H_2O(l)$. How much heat is released when 12.5 g of ethanol burns?

$$C_2H_6O(l) + 3O_2(g) \longrightarrow 2CO_2(g) + 3H_2O(l)$$
$$\Delta H = -1368 \text{ kJ}$$

Lesson Check Answers

16. The value of ΔH of a reaction can be determined by measuring the heat flow of the reaction at constant pressure.

17. The enthalpy change in a chemical reaction can be written as either a reactant or a product.

18. 502 J

19. Heat of combustion is the heat of reaction for the complete burning of one mole of a substance.

20. $2Mg(s) + O_2(g) \rightarrow 2MgO(s) + 1204$ kJ, or $2Mg(s) + O_2(g) \rightarrow 2MgO(s)$ $\Delta H = -1204$ kJ

21. 3.72×10^2 kJ

17.3 Heat in Changes of State

C.6.3 Heat flow and temperature change problems, C.6.4 Exothermic and endothermic reactions

CHEMISTRY & YOU

Q: *Why does sweating help cool you off?* An athlete can burn a lot of calories during a game. These calories are either used to do work or are released as heat. When your body heats up, you start to sweat. The evaporation of sweat is your body's way of cooling itself to a normal temperature.

Key Questions

What is the relationship between molar heat of fusion and molar heat of solidification?

What is the relationship between molar heat of vaporization and molar heat of condensation?

What thermochemical changes can occur when a solution forms?

Vocabulary

- molar heat of fusion
- molar heat of solidification
- molar heat of vaporization
- molar heat of condensation
- molar heat of solution

Heats of Fusion and Solidification

What is the relationship between molar heat of fusion and molar heat of solidification?

What happens if you place an ice cube on a table in a warm room? The ice cube is the system, and the table and air around it are the surroundings. The ice absorbs heat from its surroundings and begins to melt. The temperature of the ice and the liquid water produced remains at 0°C until all of the ice has melted.

Like ice cubes, all solids absorb heat as they melt to become liquids. The gain of heat causes a change of state instead of a change in temperature. Whenever a change of state occurs by a gain or loss of heat, the temperature of the substance undergoing the change remains constant. The heat absorbed by one mole of a solid substance as it melts to a liquid at a constant temperature is the **molar heat of fusion** (ΔH_{fus}). The **molar heat of solidification** (ΔH_{solid}) is the heat lost when one mole of a liquid substance solidifies at a constant temperature. **The quantity of heat absorbed by a melting solid is exactly the same as the quantity of heat released when the liquid solidifies; that is, $\Delta H_{fus} = -\Delta H_{solid}$.**

The melting of 1 mol of ice at 0°C to 1 mol of liquid water at 0°C requires the absorption of 6.01 kJ of heat. This quantity of heat is the molar heat of fusion of water. Likewise, the conversion of 1 mol of liquid water at 0°C to 1 mol of ice at 0°C releases 6.01 kJ of heat. This quantity of heat is the molar heat of solidification of water.

$$H_2O(s) \longrightarrow H_2O(l) \qquad \Delta H_{fus} = 6.01 \text{ kJ/mol}$$

$$H_2O(l) \longrightarrow H_2O(s) \qquad \Delta H_{solid} = -6.01 \text{ kJ/mol}$$

Key Objectives

17.3.1 COMPARE the quantity of heat absorbed by a melting solid to the quantity of heat released when the liquid solidifies.

17.3.2 COMPARE the quantity of heat absorbed by a vaporizing liquid to the quantity of heat released when the vapor condenses.

17.3.3 DESCRIBE thermochemical changes that occur when a solution forms.

Additional Resources

Reading and Study Workbook, Lesson 17.3

Available Online or on Digital Media:

- Teaching Resources, Lesson 17.3 Review
- Small-Scale Laboratory Manual, Lab 27

PearsonChem.com

LESSON 17.3

Engage

CHEMISTRY & YOU Have students read the opening text and study the photograph. Remind them that heat is generated during chemical reactions in muscles, but the movement of heat from the point of reaction to the surface of the skin is a series of transfers between body tissues and fluids. Some heat is dissipated through radiation, convection and conduction. The rest is dissipated during the evaporation of sweat. **Ask** What type of process is evaporation? *(a change in phase from a liquid to a gas)*

Focus on ELL

1 CONTENT AND LANGUAGE Divide the class into 5 groups and assign each group one vocabulary term. Encourage each group to create a picture or word equation to illustrate the meaning of their word. Have groups present their illustrations to the class, allowing time for students to ask questions for clarification. Provide students with the accepted definitions for each term and allow them to make corrections to their illustrations where necessary.

2 FRONTLOAD THE LESSON Review phase changes and phase diagrams with students, and verify that students have a clear understanding of the difference between endothermic and exothermic processes. Demonstrate the use of chemical equations to represent phase changes.

3 COMPREHENSIBLE INPUT Prior to beginning the Quick Lab, review Figure 17.11. Ask students to identify which arrow represents ice melting. Point out that this figure can be used as a reference to identify whether a process is endothermic or exothermic and to determine which mathematical sign should be associated with a given enthalpy change.

IN ACADEMIC STANDARDS for SCIENCE

- Develop explanations based on reproducible data and observations gathered during laboratory investigations.
- Clearly communicate their ideas and results of investigations verbally and in written form using tables, graphs, diagrams, and photographs.
- Apply standard techniques in laboratory investigations to measure physical quantities in appropriate units and convert known quantities to other units as necessary.

Also covered C.6.3, C.6.4.

LESSON 17.3

Activate Prior Knowledge

Remind students that evaporation is a physical change from a liquid state to a gas (vapor).
Ask On a molecular level, what is the difference between liquid water and water vapor? *(In the vapor phase, the water molecules are farther apart and not as strongly attracted to each other.)* **Ask** What must happen for a liquid to change into a gas? *(Energy must be added to overcome the attraction.)* **Ask** Is evaporation endothermic or exothermic? *(endothermic)*

Foundations for Reading

BUILD VOCABULARY Point out that each vocabulary word has the same first three words, "molar heat of...". Have students predict the meaning of each vocabulary term and then confirm or modify their predictions once they read the appropriate section.

READING STRATEGY After students read about the enthalpy change that occurs during a particular phase change, have them examine Figure 17.11 and find the arrow that applies to that phase change. Tell students to look for other useful visual aids in the lesson that will help them understand their reading, or to create their own as needed.

Explore

Heats of Fusion and Solidification

Class Activity

PURPOSE Students will measure temperatures during melting of ice and boiling of water.

MATERIALS beaker, water, boiling chips, hot plate, thermometer, ice cubes, foam cup

PROCEDURE Have students record the temperature of gently boiling water in a beaker (use boiling chips) and of melting ice in a foam cup. Students should record the temperatures every 30 seconds.

EXPECTED OUTCOMES Even if the thermometers do not record exactly 100°C or 0°C, students should note that the temperature does not change during boiling and melting.

Sample Practice Problem

How much heat must be removed to freeze a container of water if the water has a mass of 1.50×10^2 g? *(50.0 kJ)*

Sample Problem 17.5

Using the Heat of Fusion in Phase-Change Calculations

How many grams of ice at 0°C will melt if 2.25 kJ of heat are added?

1 Analyze **List the knowns and the unknown.** Find the number of moles of ice that can be melted by the addition of 2.25 kJ of heat. Convert moles of ice to grams of ice.

KNOWNS
Initial and final temperatures are 0°C
$\Delta H_{fus} = 6.01$ kJ/mol
$\Delta H = 2.25$ kJ

UNKNOWN
$m_{ice} = ?$ g

2 Calculate **Solve for the unknown.**

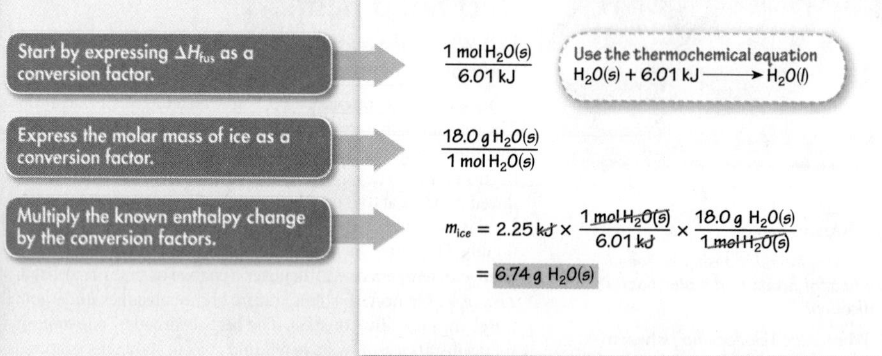

Start by expressing ΔH_{fus} as a conversion factor.

$$\frac{1 \text{ mol } H_2O(s)}{6.01 \text{ kJ}}$$

Use the thermochemical equation $H_2O(s) + 6.01 \text{ kJ} \longrightarrow H_2O(l)$

Express the molar mass of ice as a conversion factor.

$$\frac{18.0 \text{ g } H_2O(s)}{1 \text{ mol } H_2O(s)}$$

Multiply the known enthalpy change by the conversion factors.

$$m_{ice} = 2.25 \text{ kJ} \times \frac{1 \text{ mol } H_2O(s)}{6.01 \text{ kJ}} \times \frac{18.0 \text{ g } H_2O(s)}{1 \text{ mol } H_2O(s)} = 6.74 \text{ g } H_2O(s)$$

3 Evaluate **Does the result make sense?** To melt 1 mol of ice, 6.01 kJ of energy is required. Only about one third of this amount of heat (roughly 2 kJ) is available, so only about one-third mol of ice, or 18.0 g/3 = 6 g, should melt. This estimate is close to the calculated answer.

22. How many grams of ice at 0°C could be melted by the addition of 0.400 kJ of heat?

23. How many kilojoules of heat are required to melt a 50.0-g popsicle at 0°C? Assume the popsicle has the same molar mass and heat of fusion as water.

To do Problem 23, first convert from mass to moles. Then, express ΔH_{fus} as a conversion factor to convert from moles of ice to kJ of heat.

Focus on ELL

4 LANGUAGE PRODUCTION Have students work in pairs to complete the lab. Make sure each ELL student is paired with a stronger English proficient student. Allow time for students to silently read and translate the lab into their own language if necessary.

BEGINNING: LOW Use the think-aloud strategy while modeling the procedure.

HIGH Rewrite the procedure using smaller steps and fewer words.

INTERMEDIATE: LOW Provide students with a chart to record their observations and measurements.

HIGH Demonstrate how to use the change in volume of H_2O and its density to calculate the mass of the ice.

ADVANCED: LOW/HIGH Have students assist lower proficiency students with their calculations.

Quick Lab

Purpose To estimate the heat of fusion of ice

Materials

- 100-ml graduated cylinder
- hot tap water
- foam cup
- thermometer
- ice

Heat of Fusion of Ice

Procedure

1. Fill the graduated cylinder with hot tap water and let stand for 1 minute. Pour the water into the sink.
2. Measure 70 mL of hot water. Pour the water into the foam cup. Measure the temperature of the water.
3. Add an ice cube to the cup of water. Gently swirl the cup. Measure the temperature of the water as soon as the ice cube has completely melted.
4. Pour the water into the graduated cylinder and measure the volume.

Analyze and Conclude

1. **Calculate** Determine the mass of the ice. (*Hint:* Use the increase in the volume of water and the density of water.) Convert this mass into moles.
2. **Calculate** Determine the heat transferred from the water to the ice using the mass of the hot water, the specific heat of liquid water, and the change in temperature.
3. **Calculate** Determine ΔH_{fus} of ice (kJ/mol) by dividing the heat transferred from the water by the moles of ice melted.
4. **Perform Error Analysis** Compare your experimental value of ΔH_{fus} of ice with the accepted value of 6.01 kJ/mol. Account for any error.

Heats of Vaporization and Condensation

What is the relationship between molar heat of vaporization and molar heat of condensation?

A liquid that absorbs heat at its boiling point becomes a vapor. The amount of heat required to vaporize one mole of a given liquid at a constant temperature is called its **molar heat of vaporization** (ΔH_{vap}). Table 17.3 lists the molar heats of vaporization for several substances at their normal boiling points.

The molar heat of vaporization of water is 40.7 kJ/mol. This means that it takes 40.7 kJ of energy to convert 1 mol of liquid water to 1 mol of water vapor at the normal boiling point of water (100°C at 101.3 kPa).

$$H_2O(l) \longrightarrow H_2O(g) \qquad \Delta H_{vap} = 40.7 \text{ kJ/mol}$$

Diethyl ether ($C_4H_{10}O$) has a boiling point of 34.6°C and a molar heat of vaporization (ΔH_{vap}) of 26.5 kJ/mol. If liquid diethyl ether is poured into a beaker on a warm, humid day, the ether will absorb heat from the beaker walls and evaporate rapidly. If the beaker loses enough heat, the water vapor in the air may condense and freeze on the beaker walls, forming a coating of frost on the outside of the beaker.

$$C_4H_{10}O(l) \longrightarrow C_4H_{10}O(g) \qquad \Delta H_{vap} = 26.5 \text{ kJ/mol}$$

Table 17.3

Heats of Physical Change

Substance	ΔH_{fus} (kJ/mol)	ΔH_{vap} (kJ/mol)
Ammonia (NH_3)	5.66	23.3
Ethanol (C_2H_6O)	4.93	38.6
Hydrogen (H_2)	0.12	0.90
Methanol (CH_4O)	3.22	35.2
Oxygen (O_2)	0.44	6.82
Water (H_2O)	6.01	40.7

Quick Lab

OBJECTIVE Students will melt an ice cube and calculate the heat of fusion of ice.

SKILLS FOCUS observing, measuring, calculating, designing experiments

PREP TIME 10 minutes

CLASS TIME 20 minutes

MATERIALS 100-mL graduated cylinder, hot water, foam cup, thermometer, temperature probe (optional), ice

ADVANCE PREPARATION Provide students with a source of water between 50°C and 60°C.

SAFETY Students should wear safety goggles. Warn students not to use the thermometer as a stirring rod and discuss what students should do if the thermometer breaks.

TEACHING TIPS Remind students to make all temperature measurements to the nearest tenth of a degree and all volume measurements to the nearest milliliter. Small ice cubes or crushed ice work best.

EXPECTED OUTCOME Students should calculate values of approximately 6.01 kJ/mol for the heat of fusion of ice.

ANALYZE AND CONCLUDE

1. The mass of ice will be equal to the final volume of water minus the initial volume of water, expressed in grams.
2. Answers will depend on what the students measure for the change in temperature. Students will need to multiply the volume of water by the density of water to get the mass of water. Then they can solve for the heat transferred from the water to the ice.
3. See Expected Outcome.
4. See Expected Outcome. If the experiment is not done quickly, loss of heat to the classroom may cause ΔT to be too large, resulting in a larger ΔH_{fus}. Careful reading of the thermometer and graduated cylinder is crucial. Some water may adhere to the thermometer and cup; this may affect volume measurements.

FOR ENRICHMENT Have students suggest how to revise the procedures to achieve more accurate results. Then, have them design a new experiment that incorporates their suggestions for revising the procedure. If students perform the new experiment, they should compare the results of the two experiments and calculate the respective percent errors.

Answers

22. 1.20 g
23. 16.7 kJ

LESSON 17.3

Explain

Heats of Vaporization and Condensation

CHEMISTRY & YOU When the sweat evaporates, it absorbs heat from your body leaving you feeling cooler.

CRITICAL THINKING Direct students' attention to Figure 17.12. **Ask** How would the rate of heating affect the heating curve? *(The faster the rate of heating is, the steeper the slope of the curve will be.)*

Explore

Teacher Demo

PURPOSE Students will observe the release of heat when a liquid solidifies.

MATERIALS 3 wide-mouth test tubes, sodium thiosulfate pentahydrate, hot water bath, 30°C water bath, thermometers, stirring rods

PROCEDURE Set three wide-mouth test tubes each containing about 20 grams of sodium thiosulfate pentahydrate ($Na_2S_2O_3 \cdot 5H_2O$) in a hot water bath until the chemical is completely melted. Transfer the tubes to a cool water bath at about 30°C. Let the test tubes sit undisturbed until they are close to 30°C and feel lukewarm to touch. Give the tubes to three student volunteers at different locations in the classroom so that many students can see and feel what happens. Give each volunteer a thermometer and a seed crystal. Ask them to quickly take the temperature of the liquid, drop in the seed crystal, and stir gently.

EXPECTED OUTCOMES The temperature of the test tube will rise. Students will observe the change by touching the test tube and reading the thermometer.

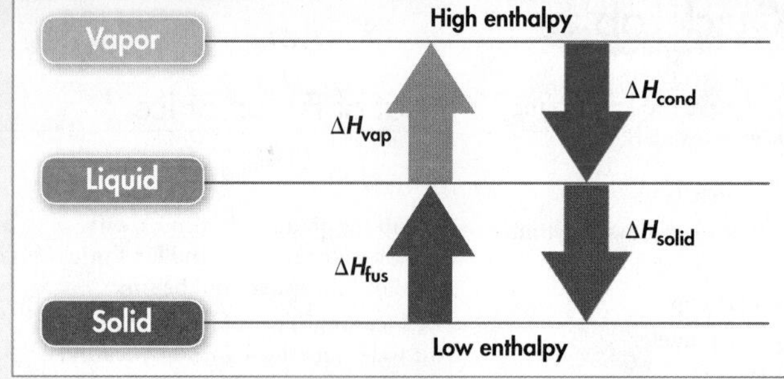

Figure 17.11 Changes in State
Enthalpy changes accompany changes in state. Fusion and vaporization are endothermic processes. Solidification and condensation are exothermic processes.
Interpret Diagrams ***Which arrows represent processes that release heat to the surroundings?***

Condensation is the exact opposite of vaporization. When a vapor condenses, heat is released. The **molar heat of condensation** (ΔH_{cond}) is the amount of heat released when one mole of a vapor condenses at its normal boiling point. **The quantity of heat absorbed by a vaporizing liquid is exactly the same as the quantity of heat released when the vapor condenses; that is, $\Delta H_{vap} = -\Delta H_{cond}$.** Figure 17.11 shows the relationships between the molar heat of fusion and molar heat of solidification and between the molar heat of vaporization and molar heat of condensation.

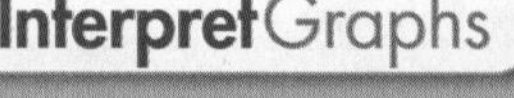

Q: *Explain why the evaporation of sweat off your body helps to cool you off.*

Figure 17.12 summarizes the enthalpy changes that occur as ice is heated to a liquid and then to a vapor. You should be able to identify certain trends regarding the temperature during changes of state and the energy requirements that accompany these changes from the graph. The large values for ΔH_{vap} and ΔH_{cond} are the reason hot vapors such as steam can be very dangerous. You can receive a scalding burn from steam when the heat of condensation is released as the steam touches your skin.

$$H_2O(g) \longrightarrow H_2O(l) \qquad \Delta H_{cond} = -40.7 \text{ kJ/mol}$$

Interpret Graphs

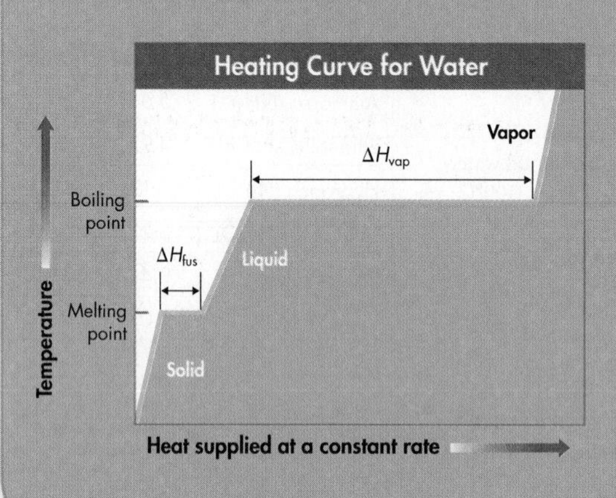

Figure 17.12 A heating curve graphically describes the enthalpy changes that take place during phase changes.
a. Identify In which region(s) of the graph is temperature constant?
b. Compare How does the amount of energy required to melt a given mass of ice compare to the energy required to vaporize the same mass of liquid water? Explain.
c. Apply Concepts Which region of the graph represents the coexistence of solid and liquid? Liquid and vapor?

Remember: The temperature of a substance remains constant during a change in state.

UbD Check for Understanding

BIGIDEA How can you determine the amount of energy absorbed or released in a chemical or physical process?

Give students 60 seconds to explain, orally, pictorially, or in writing, how to use molar heats to determine the quantity of heat absorbed by a melting solid and the quantity of heat released when a liquid solidifies. *(The molar heat of fusion should be used to determine the energy absorbed by a melting solid and the molar heat of solidification should be used to determine the energy released by a solidifying liquid.)*

ADJUST INSTRUCTION If students struggle to answer, have them reread the Heats of Fusion and Solidification section, then review the two equations at the end of the section and Sample Problem 17.5. Clarify that, except for mathematical sign, a substance's molar heat of fusion is exactly the same as its molar heat of solidification.

SampleProblem 17.6

Using the Heat of Vaporization in Phase-Change Calculations

How much heat (in kJ) is absorbed when 24.8 g $H_2O(l)$ at 100°C and 101.3 kPa is converted to $H_2O(g)$ at 100°C?

1 Analyze List the knowns and the unknown. First, convert grams of water to moles of water. Then, find the amount of heat that is absorbed when the liquid water is converted to steam.

KNOWNS

Initial and final conditions are 100°C and 101.3 kPa

mass of liquid water converted to steam = 24.8 g

ΔH_{vap} = 40.7 kJ/mol

UNKNOWN

ΔH = ? kJ

2 Calculate Solve for the unknown.

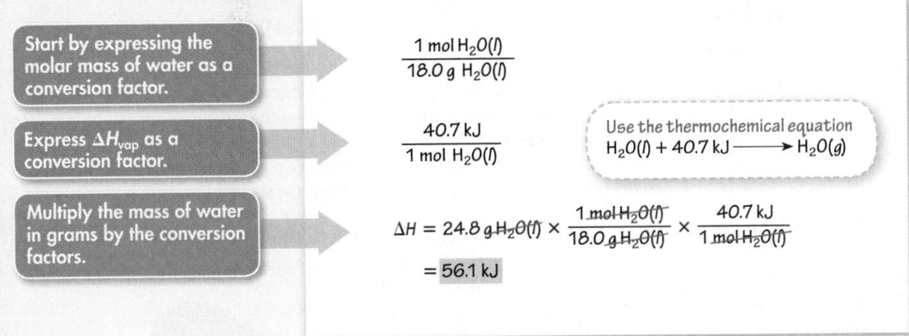

Start by expressing the molar mass of water as a conversion factor.

$$\frac{1 \text{ mol } H_2O(l)}{18.0 \text{ g } H_2O(l)}$$

Express ΔH_{vap} as a conversion factor.

$$\frac{40.7 \text{ kJ}}{1 \text{ mol } H_2O(l)}$$

Use the thermochemical equation $H_2O(l) + 40.7 \text{ kJ} \longrightarrow H_2O(g)$

Multiply the mass of water in grams by the conversion factors.

$$\Delta H = 24.8 \text{ g } H_2O(l) \times \frac{1 \text{ mol } H_2O(l)}{18.0 \text{ g } H_2O(l)} \times \frac{40.7 \text{ kJ}}{1 \text{ mol } H_2O(l)}$$

$$= 56.1 \text{ kJ}$$

3 Evaluate Does the result make sense? Knowing the molar mass of water is 18.0 g/mol, 24.8 g $H_2O(l)$ can be estimated to be somewhat less than 1.5 mol H_2O. The calculated enthalpy change should be a little less than 1.5 mol × 40 kJ/mol = 60 kJ, and it is.

24. How much heat is absorbed when 63.7 g $H_2O(l)$ at 100°C and 101.3 kPa is converted to $H_2O(g)$ at 100°C? Express your answer in kJ.

25. How many kilojoules of heat are absorbed when 0.46 g of chloroethane (C_2H_5Cl, bp 12.3°C) vaporizes at its normal boiling point? The molar heat of vaporization of chloroethane is 24.7 kJ/mol.

For Problem 25, start by writing the thermochemical equation for the vaporization of chloroethane.

Explain

SUMMARIZE Have students look up the molar heat of fusion, the molar heat of vaporization, and the specific heat of ethanol (grain alcohol) in Tables 17.1 and 17.3. Tell students that ethanol freezes at 158.7 K and boils at 351.5 K. Have students summarize the enthalpy changes that occur as solid ethanol is heated to a liquid and then to a gas by constructing a heating curve like the one shown in Figure 17.12 for water. Ask students to find the amount of heat necessary to convert 1 mole of solid ethanol to ethanol vapor.

Sample Practice Problem

The molar heat of vaporization of water is 40.7 kJ/mol. How many kilojoules of heat are required to vaporize 60.5 g of water at its boiling point of 100°C? *(137 kJ)*

Differentiated Instruction

LPR LESS PROFICIENT READERS Provide at least two brands of cold or hot packs. Have students activate the packs at exactly the same time and then measure the maximum or minimum temperature achieved. Have students determine how long the packs remain effective. Ask students to decide, based on the data and the cost of each pack, which brand provided the best value.

ELL ENGLISH LANGUAGE LEARNERS Have students create a flow chart to summarize the enthalpy changes involved when a substance undergoes a phase change. Guide students to study fusion and solidification and vaporization and condensation in pairs. Allow students to create their charts using symbols, colors, and their own words so the diagrams are useful to them for studying.

L3 ADVANCED STUDENTS Have students devise an experiment to test whether initial water temperature affects the rate at which water freezes.

Answers

INTERPRET GRAPHS

a. at the melting point and the boiling point

b. It takes much less energy to melt a given mass of ice than to vaporize the same mass of water. The ΔH_{fus} is much smaller than ΔH_{vap}.

c. the plateau of the curve between solid and liquid; the plateau between liquid and vapor

FIGURE 17.11 condensation and solidification

24. 144 KJ

25. 0.18 KJ

LESSON 17.3

Explain

Heat of Solution

USE VISUALS Direct students' attention to Figure 17.13. **Ask** Is the process that occurs in a cold pack endothermic or exothermic? How do you know? *(Endothermic; heat is absorbed and the temperature of the surroundings decreases.)* **Ask** In which direction is heat flowing? *(from the surroundings to the system)* **Ask** Is the process in a hot pack endothermic or exothermic? How do you know? *(Exothermic; heat is released and the temperature of the surroundings increases.)* **Ask** In which direction is heat flowing? *(from the system to the surroundings)*

Explore

Class Activity

Purchase a number of hot and cold packs used by athletes and show the class how they work. Most instant cold and hot packs operate by utilizing the heat released or absorbed when certain substances dissolve in water. Hot packs usually contain calcium chloride, $CaCl_2(s)$, while cold packs usually contain ammonium nitrate, $NH_4NO_3(s)$. If the labels specify the amount of salt contained in each package, have students use the molar heats of solution provided in the text to calculate the amount of heat absorbed and released in each case.

Heat of Solution

What thermochemical changes can occur when a solution forms?

If you've ever used a hot pack or a cold pack, then you have felt the enthalpy changes that occur when a solute dissolves in a solvent. **During the formation of a solution, heat is either released or absorbed.** The enthalpy change caused by the dissolution of one mole of substance is the **molar heat of solution** (ΔH_{soln}). For example, when 1 mol of sodium hydroxide, $NaOH(s)$, is dissolved in water, the solution can become so hot that it steams. The heat from this process is released as the sodium ions and the hydroxide ions interact with the water. The temperature of the solution increases, releasing 44.5 kJ of heat as the molar heat of solution.

$$NaOH(s) \longrightarrow Na^+(aq) + OH^-(aq)$$
$$\Delta H_{soln} = -44.5 \text{ kJ/mol}$$

A practical application of an exothermic dissolution process is a hot pack. In a hot pack, calcium chloride, $CaCl_2(s)$, mixes with water, producing heat.

$$CaCl_2(s) \longrightarrow Ca^{2+}(aq) + 2Cl^-(aq)$$
$$\Delta H_{soln} = -82.8 \text{ kJ/mol}$$

The dissolution of ammonium nitrate, $NH_4NO_3(s)$, is an example of an endothermic process. When ammonium nitrate dissolves in water, the solution becomes so cold that frost may form on the outside of the container. The cold pack in Figure 17.13 contains solid ammonium nitrate crystals and water. Once the solute dissolves in the solvent, the pack becomes cold. In this case, the solution process absorbs energy from the surroundings.

$$NH_4NO_3(s) \longrightarrow NH_4^+(aq) + NO_3^-(aq)$$
$$\Delta H_{soln} = 25.7 \text{ kJ/mol}$$

Figure 17.13 Cold Pack
The cold pack shown has two sealed plastic bags, one inside the other. The outer bag contains ammonium nitrate crystals. The inner bag contains liquid water. When the pack is squeezed, the inner bag breaks, allowing the ammonium nitrate and water to mix.
Infer ***How would you define the system and the surroundings in this process?***

Foundations for Math

MULTIPLYING SIGNIFICANT FIGURES Remind students that when multiplying measured quantities, the final result should have the same number of significant figures as the measurement with the least number of significant figures. For a thorough review of significant figures, have students revisit section 3.1 of the text.

In **Sample Problem 17.7,** multiplying 2.50 by −44.5 yields −111.25. Some students may think that they should report the answer as −111.2, because both multipliers have digits to the right of the decimal. Stress that the presence of a decimal point in a multiplier is relevant only in the respect that its presence makes a 0 significant, as it does in 2.50. The presence of a decimal in both multipliers is not a guarantee of the presence of a decimal in the reported answer. However, if the answer were to be reported in scientific notation, then a decimal *would* be necessary in order to insure that the answer contained three significant figures.

Sample Problem 17.7

Calculating the Enthalpy Change in Solution Formation

How much heat (in kJ) is released when 2.50 mol NaOH(s) is dissolved in water?

1 Analyze List the knowns and the unknown. Use the heat of solution for the dissolution of NaOH(s) in water to solve for the amount of heat released (ΔH).

KNOWNS
$\Delta H_{soln} = -44.5$ kJ/mol
amount of NaOH(s) dissolved = 2.50 mol

UNKNOWN
$\Delta H = ?$ kJ

2 Calculate Solve for the unknown.

Start by expressing ΔH_{soln} as a conversion factor.

$$\frac{-44.5\ \text{kJ}}{1\ \text{mol NaOH}(s)}$$

Use the thermochemical equation
$NaOH(s) \longrightarrow Na^+(aq) + OH^-(aq) + 44.5$ kJ/mol

Multiply the number of moles by the conversion factor.

$$\Delta H = 2.50\ \cancel{\text{mol NaOH}(s)} \times \frac{-44.5\ \text{kJ}}{1\ \cancel{\text{mol NaOH}(s)}} = -111\ \text{kJ}$$

3 Evaluate Does the result make sense? ΔH is 2.5 times greater than ΔH_{soln}, as it should be. Also, ΔH should be negative, as the dissolution of NaOH(s) in water is exothermic.

26. How much heat (in kJ) is released when 0.677 mol NaOH(s) is dissolved in water?

27. How many moles of $NH_4NO_3(s)$ must be dissolved in water so that 88.0 kJ of heat is absorbed from the water?

ΔH_{soln} for the dissolution of $NH_4NO_3(s)$ in water is 25.7 kJ/mol.

17.3 LessonCheck

28. Describe How does the molar heat of fusion of a substance compare to its molar heat of solidification?

29. Describe How does the molar heat of vaporization of a substance compare to its molar heat of condensation?

30. Identify What enthalpy changes occur when a solute dissolves in a solvent?

31. Calculate How much heat must be removed to freeze a tray of ice cubes at 0°C if the water has a mass of 225 g?

32. Calculate How many kilojoules of heat are required to vaporize 50.0 g of ethanol, C_2H_6O? The boiling point of ethanol is 78.3°C. Its molar heat of vaporization is 38.6 kJ/mol.

33. Calculate How many kilojoules of heat are released when 25.0 g of NaOH(s) is dissolved in water?

BIGIDEA MATTER AND ENERGY

34. Use what you know about hydrogen bonding to explain why water has such a large heat of vaporization.

Lesson Check Answers

28. Molar heat of fusion and molar heat of solidification have an identical numerical value but are of opposite sign.

29. Molar heat of vaporization and molar heat of condensation have an identical numerical value but are of opposite sign.

30. Heat is either released or absorbed in the formation of a solution.

31. 75.1 kJ

32. 42.0 kJ

33. –27.9 kJ

34. Water molecules are very polar, and many hydrogen bonds form among the molecules. It takes a great deal of energy to break the hydrogen bonds and pull the water molecules apart. SC.912.L.18.12

Sample Practice Problem

How many kilojoules of heat are released when 46.0 g of $CaCl_2$ are dissolved in water? *(–34.3 kJ)*

Evaluate

Informal Assessment

Ask students to sketch a heating curve for 1 mol of ice being heated from –10°C to 110°C. Have students write brief explanations of their graphs. Ask students to calculate how much energy would be required to change 54.0 g of ice at –10.0°C to water vapor at 110°C. *(165 kJ)* Then have students complete the 17.3 Lesson Check.

Reteach

Display Figure 17.12 on an overhead projector. Remind students that during endothermic phase changes, melting or evaporation, energy absorbed does not increase the temperature because the energy is being used to overcome attractions between particles. The energy can be released back to the surroundings during the opposing exothermic phase change, condensation or freezing.

Answers

FIGURE 17.13 The system is the cold pack; the athlete and the air around the cold pack constitute the surroundings.

26. –30.1 kJ

27. 3.42 mol $NH_4NO_3(s)$

CHEMISTRY & YOU Engage students in a discussion of the photo of the geothermal energy plant and the swimming pool. Explain that the people are swimming in the Blue Lagoon, Iceland. The saltwater output from the nearby geothermal energy plant produces a pool with a water temperature of about 40°C.

Pose the following question to students: *If geothermal energy is a virtually unlimited supply of heat and power, what might be some reasons that it is not more widely used?*

You may need to guide students by asking the following questions:

- What would be the heat source for the power plant?
- How would this heat source be accessed?
- Where are similar heat sources located in relation to population centers?

Explain

START A CONVERSATION One type of geothermal power plant, called a *dry steam design,* has two nearby wells. An *injection well* pumps water down onto hot underground rock where it changes to steam. A nearby *production well* pumps this dry steam up from hot underground rock and uses it to spin a turbine. The turbine then turns a generator for the production of electricity. Enlist students' help to draw a diagram on the board of this type of system. Discuss how energy is taken into the plant or released during the process.

Geothermal Energy

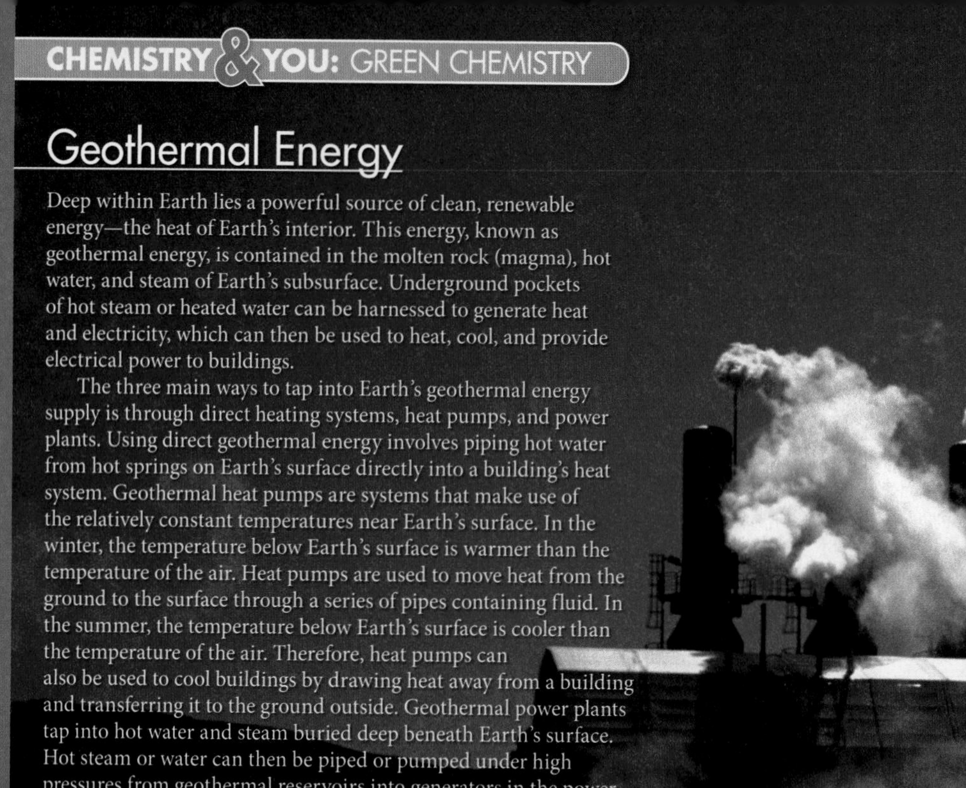

Deep within Earth lies a powerful source of clean, renewable energy—the heat of Earth's interior. This energy, known as geothermal energy, is contained in the molten rock (magma), hot water, and steam of Earth's subsurface. Underground pockets of hot steam or heated water can be harnessed to generate heat and electricity, which can then be used to heat, cool, and provide electrical power to buildings.

The three main ways to tap into Earth's geothermal energy supply is through direct heating systems, heat pumps, and power plants. Using direct geothermal energy involves piping hot water from hot springs on Earth's surface directly into a building's heat system. Geothermal heat pumps are systems that make use of the relatively constant temperatures near Earth's surface. In the winter, the temperature below Earth's surface is warmer than the temperature of the air. Heat pumps are used to move heat from the ground to the surface through a series of pipes containing fluid. In the summer, the temperature below Earth's surface is cooler than the temperature of the air. Therefore, heat pumps can also be used to cool buildings by drawing heat away from a building and transferring it to the ground outside. Geothermal power plants tap into hot water and steam buried deep beneath Earth's surface. Hot steam or water can then be piped or pumped under high pressures from geothermal reservoirs into generators in the power plants on the surface. These power plants provide electricity to homes and businesses.

Pros & Cons

Advantages of Using Geothermal Energy	Disadvantages of Using Geothermal Energy
✔ **Low operating costs** Once the initial cost of constructing a geothermal energy facility has been paid for, there are no additional fuel costs.	✘ **Large initial expense** The initial cost of drilling wells to reach geothermal reservoirs and installing geothermal power plants is millions of dollars.
✔ **Clean Energy** Direct geothermal heating systems, geothermal heat pumps, and geothermal power plants emit little or no greenhouse gases.	✘ **Large space requirements** Geothermal heat pumps and power plants require large expanses of land for pipes and wells.
✔ **Sustainable Energy Source** Unlike fossil fuels, heat from beneath Earth's surface is a renewable resource.	✘ **Disruptive to the environment** Deep drilling can cause small earthquakes and hot water and steam can bring contaminants to Earth's surface.

21st Century Learning To be successful in the 21st century, students need skills and learning experiences that extend beyond subject matter mastery. The following project helps students build the following 21st Century Skills: *Critical Thinking and Problem Solving; Creativity and Innovation; Communication and Collaboration; Information, Media, and Technology Literacy; Productivity and Accountability; Global Awareness.*

POWER PLANT Divide students into groups of 3-5. Pose the following challenge. *A power company wants to develop a new geothermal plant in the United States. The type of plant and its location are undecided. Your consulting group is responsible for determining the best type of plant and the most appropriate location.* Prepare a written, illustrated report that identifies three possible locations for the plant that can serve areas with high population densities; identifies and explains the most appropriate plant type for the location, and summarizes the environmental impact on the ecosystem at each location.

THE POWER OF STEAM This power plant in Iceland taps into a geothermal reservoir 2000 meters below the surface. The steam, which can reach 240°C, is used to generate electricity. The condensed steam (liquid water) is then directed to a lagoon, where people bathe in it for healing and wellness purposes.

NATURAL HEAT Geysers and hot springs on Earth's surface indicate the presence of subsurface heat sources. These monkeys are keeping warm in one of Japan's hot springs.

Take It Further

1. Classify Describe the direction of heat flow when a geothermal heat pump is used to heat a building. What is the system and what are the surroundings? Is this process endothermic or exothermic?

2. Calculate How much energy is absorbed or released when: 535 kg of steam at 154°C cools to 100°C, the steam condenses at 100°C, and the liquid water cools from 100°C to 37°C?

3. Compare and Contrast Compare geothermal energy to one or more other sources of energy. Assess the costs and benefits of each.

Explain

CRITICAL THINKING Explain to students that a common use of geothermal energy is heat pumps. Unlike geothermal power plants, which use extreme underground temperatures, geothermal heat pumps operate at very shallow depths, sometimes as little as several meters below Earth's surface. The ground temperature here may be a cool 10°C. In winter, the heat pump transfers heat from the ground. In summer, it transfers heat into the ground. Write the following steps for a geothermal heat pump, and have students apply what they have learned about heat transfer in this chapter:

- A compressor increases the pressure of a gas, which increases its temperature.
- The hot gas moves to a condenser where it changes back to a liquid. **Ask** Does this step take in or release energy? Why? *(It releases energy. This is the molar heat of condensation.)*
- The cool liquid moves back to the evaporator and takes in geothermal energy. **Ask** What effect does the increase in energy have on the liquid? Why? *(It changes the liquid back into a gas. This is the molar heat of vaporization.)*

Extend

Connect to EARTH SCIENCE

Geothermal power plants are often considered to be a more environmentally-friendly source of electricity. However, there is some concern about how much of an environmental impact geothermal plants have. Divide students into small groups and assign them the task of researching the potential environmental effects of a modern geothermal plant versus a comparable fossil fuel plant. Afterwards, lead students in discussing the issue. Provide students with the opportunity to express their opinions.

Differentiated Instruction

L1 STRUGGLING STUDENTS Point out to students that in a geothermal power plant, steam is pumped from underground to power a turbine. It is then pumped back underground as water. Discuss with students whether this involves a state change or a chemical reaction *(state change),* and whether it is an endothermic or exothermic reaction. *(Vaporization is endothermic. Condensation is exothermic.)*

LPR LESS PROFICIENT READERS Provide students with copies of an illustration of a geothermal plant. Engage them in a discussion of how heat flows through the system and where it might be lost.

L3 ADVANCED STUDENTS One version of a geothermal plant, called a binary cycle plant, uses lower temperature water to heat a second fluid with a low boiling point such as butane which actually drives the turbines. Have students research how binary cycle plants operate and what thermodynamic principles they demonstrate during operation. Allow time for students to present their findings to the class.

Answers

TAKE IT FURTHER

1. The system is the building. The surroundings are the hot water and steam below Earth's surface. Heat flows from the surroundings into the system so the process is endothermic.
2. 1.4×10^6 kJ of energy are released.
3. Students should compare the costs and benefits of geothermal energy to another source of energy. Possible sources of energy include nuclear fusion, fossil fuels, solar and wind.

LESSON 17.4

Key Objectives

17.4.1 IDENTIFY two ways that the heat of reaction can be determined when it cannot be directly measured.

Additional Resources

Reading and Study Workbook, Lesson 17.4

Available Online or on Digital Media:

- Teaching Resources, Lesson 17.4 Review
- Laboratory Manual, Lab 35
- Virtual Chemistry Lab, Lab 18

PearsonChem.com

Engage

CHEMISTRY & YOU Have students examine the photograph of the diamond. **Ask** Given that a diamond is too valuable to be destroyed, is there a way to determine the heat of reaction without actually performing the reaction? *(Sample answer: The heat of reaction might be determined indirectly by studying the enthalpy changes for related reactions.)*

Activate Prior Knowledge

Have students recall how the overall enthalpy change for converting one mole of ice to water vapor can be calculated by summing the enthalpy changes for each step in the process. A similar strategy can be used for determining the enthalpy change for a chemical reaction.

IN ACADEMIC STANDARDS for SCIENCE

C.6.3 Solve problems involving heat flow and temperature changes, using known values of specific heat and/or phase change constants (latent heats).

17.4 Calculating Heats of Reaction

C.6.3 Heat flow and temperature change problems

CHEMISTRY & YOU

Q: *How much heat is released when a diamond changes into graphite?* Diamonds are gemstones composed of carbon. Over a time period of millions and millions of years, diamond will break down into graphite, which is another form of carbon. How then can you determine the enthalpy change for the reaction?

Key Question

How can you calculate the heat of reaction when it cannot be directly measured?

Vocabulary

- Hess's law of heat summation
- standard heat of formation

Hess's Law

How can you calculate the heat of reaction when it cannot be directly measured?

Sometimes it is hard to measure the enthalpy change for a reaction. The reaction might take place too slowly to measure the enthalpy change, or the reaction might be an intermediate step in a series of reactions. Fortunately, it is possible to determine a heat of reaction indirectly using Hess's law of heat summation. **Hess's law of heat summation** states that if you add two or more thermochemical equations to give a final equation, then you can also add the heats of reaction to give the final heat of reaction. **Hess's law allows you to determine the heat of reaction indirectly by using the known heats of reaction of two or more thermochemical equations.**

Consider the conversion of diamond to graphite, discussed above.

$$C(s, \text{diamond}) \longrightarrow C(s, \text{graphite})$$

Although the enthalpy change for this reaction cannot be measured directly, you can use Hess's law to find the enthalpy change for the conversion of diamond to graphite by using the following combustion reactions and Figure 17.14:

a. $C(s, \text{graphite}) + O_2(g) \longrightarrow CO_2(g) \quad \Delta H = -393.5 \text{ kJ}$

b. $C(s, \text{diamond}) + O_2(g) \longrightarrow CO_2(g) \quad \Delta H = -395.4 \text{ kJ}$

Write equation **a** in reverse to give:

c. $CO_2(g) \longrightarrow C(s, \text{graphite}) + O_2(g) \quad \Delta H = 393.5 \text{ kJ}$

When you write a reverse reaction, you must also change the sign of ΔH. If you add equations **b** and **c**, you get the equation for the conversion of diamond to graphite. The $CO_2(g)$ and $O_2(g)$ terms on both sides of the summed equations cancel. If you also add the values of ΔH for equations **b** and **c**, you get the heat of reaction for this conversion.

$$C(s, \text{diamond}) + \cancel{O_2(g)} \longrightarrow \cancel{CO_2(g)} \quad \Delta H = -395.4 \text{ kJ}$$

$$\cancel{CO_2(g)} \longrightarrow C(s, \text{graphite}) + \cancel{O_2(g)} \quad \Delta H = 393.5 \text{ kJ}$$

$$C(s, \text{diamond}) \longrightarrow C(s, \text{graphite}) \quad \Delta H = -1.9 \text{ kJ}$$

Focus on ELL

1 CONTENT AND LANGUAGE Provide students with a double bubble diagram. Have students write how the two methods of determining the heat of reaction are different on the lines on the outside of each bubble, and write how they are similar on the lines between the bubbles.

2 FRONTLOAD THE LESSON Ask each student to draw a map from their house to the school using two different routes. Have each student present the two routes and explain the pros and cons of each route. Relate this to the two methods of determining the heats of reaction in the lesson.

3 COMPREHENSIBLE INPUT Explain Figures 17.14–17.15 and read aloud the captions. Provide students with their own copy of the diagrams. As you discuss the questions in the captions, encourage students to take notes directly on the diagram. Assign students to compare the diagrams and to create a summary paragraph explaining what they have learned.

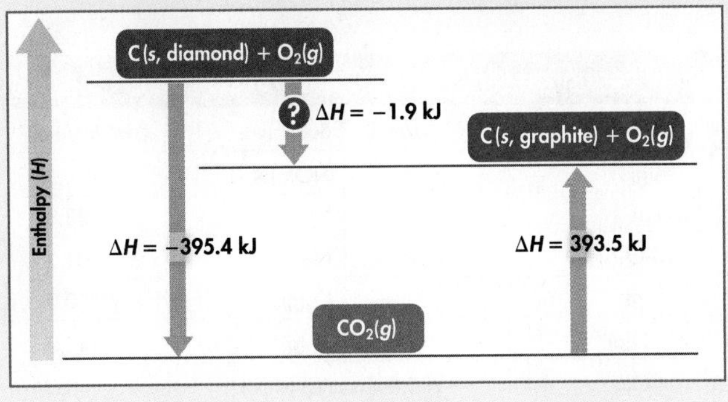

Figure 17.14 Conversion of Diamond to Graphite
Hess's law is used to determine the enthalpy change for the conversion of diamond to graphite.

Go online to see how Hess's law is used.

Another case where Hess's law is useful is when reactions yield products in addition to the product of interest. Suppose you want to determine the enthalpy change for the formation of carbon monoxide from its elements. You can write the following equation for this reaction.

$$C(s, \text{graphite}) + \tfrac{1}{2}O_2(g) \longrightarrow CO(g) \qquad \Delta H = ?$$

Although it is easy to write the equation, carrying out the reaction in the laboratory as written is virtually impossible. Carbon dioxide is produced along with carbon monoxide. Therefore, any measured heat of reaction is related to the formation of both $CO(g)$ and $CO_2(g)$, and not $CO(g)$ alone. However, you can calculate the desired enthalpy change by using Hess's law and the following two reactions that can be carried out in the laboratory:

a. $C(s, \text{graphite}) + O_2(g) \longrightarrow CO_2(g) \qquad \Delta H = -393.5\text{ kJ}$
b. $CO(g) + \tfrac{1}{2}O_2(g) \longrightarrow CO_2(g) \qquad \Delta H = -283.0\text{ kJ}$

Writing the reverse of equation **b** yields equation **c.**

c. $CO_2(g) \longrightarrow CO(g) + \tfrac{1}{2}O_2(g) \qquad \Delta H = 283.0\text{ kJ}$

Adding equations **a** and **c** gives the expression for the formation of $CO(g)$ from its elements. The enthalpy diagram for this heat summation is shown in Figure 17.15. Notice that only $\tfrac{1}{2}O_2(g)$ cancels from each equation.

$$C(s, \text{graphite}) + O_2(g) \longrightarrow \cancel{CO_2(g)} \qquad \Delta H = -393.5\text{ kJ}$$
$$\cancel{CO_2(g)} \longrightarrow CO(g) + \tfrac{1}{2}O_2(g) \qquad \Delta H = 283.0\text{ kJ}$$
$$C(s, \text{graphite}) + \tfrac{1}{2}O_2(g) \longrightarrow CO(g) \qquad \Delta H = -110.5\text{ kJ}$$

Q: *How can you determine ΔH for the conversion of diamond to graphite without performing the reaction?*

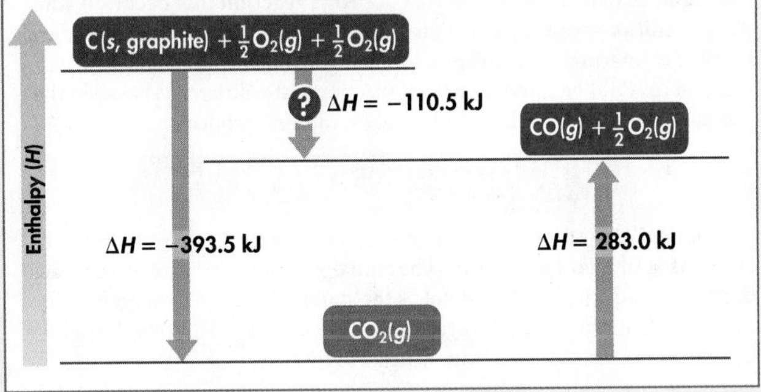

Figure 17.15 Formation of CO(g) From Its Elements
Hess's law is used to determine the enthalpy change for the formation of CO(g) from its elements.
Interpret Diagrams ***How does the diagram represent endothermic and exothermic reactions differently?***

UbD Check for Understanding

How can you determine the amount of energy absorbed or released in a chemical or physical process?

Write the following reactions related to the dissociation of sulfuric acid on the board.

$H_2SO_4(l) \rightarrow SO_3(g) + H_2O(g)$

a. $H_2S(g) + 2O_2(g) \rightarrow H_2SO_4(l) \qquad \Delta H = -235.5\text{ kJ}$
b. $H_2S(g) + 2O_2(g) \rightarrow SO_3(g) + H_2O(l) \qquad \Delta H = -207\text{ kJ}$
c. $H_2O(l) \rightarrow H_2O(g) \qquad \Delta H = 44\text{ kJ}$

Assess students' understanding of this Essential Question by asking them to use Hess's law to calculate the enthalpy change for this dissociation. *(Write equation (a) in reverse and change the sign on ΔH to positive. Then add ΔH for the three reactions: 235.5 kJ + (–207 kJ) + 44 kJ = 73kJ)*

Explain

Hess's Law

USE VISUALS Have students examine Figure 17.14. **Ask** What is the enthalpy change of the reverse reaction—from graphite to diamond? Why? *(ΔH = 1.9 kJ, because the enthalpy change for the reverse reaction is equal to the negative ΔH of the forward reaction. The negative of –1.9kJ is 1.9 kJ.)*

USE AN ANALOGY Provide the following analogy for enthalpy change:

A group of hikers is attempting to reach the summit of a mountain. Two routes are possible: one is a short, steep path that goes directly to the top; the other is a long path that is less steep but involves a number of changes in direction. Although the distances of the two routes are different, the final altitude is the same.

Explain that the change in altitude is analogous to the enthalpy change for a reaction. The distances traveled by the hikers on each path is analogous to the number of steps required to form a given set of products. No matter how one goes from a given set of reactants to a given set of products, in one step or in several, the overall enthalpy change is the same.

CHEMISTRY & YOU You can use Hess's Law.

Explain

USE VISUALS Have students study Figure 17.15. **Ask** How can you tell from the diagram that the formation of CO is exothermic? *(The arrow points downward and ΔH is negative.)*

Extend

Connect to GEOLOGY

Explain that diamonds are made almost entirely of carbon, and are the world's oldest substance. At 3.3 billion years of age, they are almost as old as Earth itself. Guide students to understand that diamonds only form under extremely intense temperature and pressure. Explain that the only place on Earth these conditions are found naturally is 100 miles underground. **Ask** Why are diamonds discovered after a volcanic eruption? *(The eruption lifts diamonds and other minerals to the surface.)*

Answers

FIGURE 17.15 The arrow points downward for an exothermic reaction and upward for an endothermic reaction.

LESSON 17.4

Explain

Standard Heats of Formation

USE VISUALS Display Table 17.4 on an overhead projector. Define *standard heat of formation* as the change in enthalpy that accompanies the formation of 1 mole of a compound from its elements with all substances in their standard states at 25°C and 101.3 kPa. Point out that the standard heat of formation of a free element in its standard state is defined as 0. Demonstrate how to use standard heats of formation to calculate heats of reaction at standard conditions.

USE VISUALS Have students study Figure 17.16. **Ask** Does water have a lower or higher enthalpy than the elements from which it is formed? *(lower)* **Ask** On what other basis can you account for your answer? (*It takes energy to decompose water.*)

Table 17.4

Standard Heats of Formation ($\Delta H_f°$) at 25°C and 101.3 kPa

Substance	$\Delta H_f°$ (kJ/mol)	Substance	$\Delta H_f°$ (kJ/mol)	Substance	$\Delta H_f°$ (kJ/mol)
$Al_2O_3(s)$	−1676.0	$F_2(g)$	0.0	NO(g)	90.37
$Br_2(g)$	30.91	Fe(s)	0.0	$NO_2(g)$	33.85
$Br_2(l)$	0.0	$Fe_2O_3(s)$	−822.1	NaCl(s)	−411.2
C(s, diamond)	1.9	$H_2(g)$	0.0	$O_2(g)$	0.0
C(s, graphite)	0.0	$H_2O(g)$	−241.8	$O_3(g)$	142.0
$CH_4(g)$	−74.86	$H_2O(l)$	−285.8	P(s, white)	0.0
CO(g)	−110.5	$H_2O_2(l)$	−187.8	P(s, red)	−18.4
$CO_2(g)$	−393.5	$I_2(g)$	62.4	S(s, rhombic)	0.0
$CaCO_3(s)$	−1207.0	$I_2(s)$	0.0	S(s, monoclinic)	0.30
CaO(s)	−635.1	$N_2(g)$	0.0	$SO_2(g)$	−296.8
$Cl_2(g)$	0.0	$NH_3(g)$	−46.19	$SO_3(g)$	−395.7

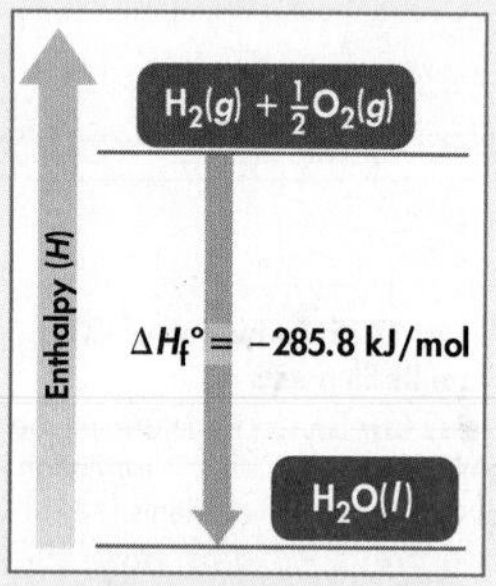

Standard Heats of Formation

How can you calculate the heat of reaction when it cannot be directly measured?

Enthalpy changes generally depend on conditions of the process. To compare enthalpy changes, scientists specify a common set of conditions. These conditions, called the standard state, refer to the stable form of a substance at 25°C and 101.3 kPa. The **standard heat of formation** ($\Delta H_f°$) of a compound is the change in enthalpy that accompanies the formation of one mole of a compound from its elements with all substances in their standard states. The $\Delta H_f°$ of a free element in its standard state is arbitrarily set at zero. Thus, $\Delta H_f° = 0$ kJ/mol for the diatomic molecules $H_2(g)$, $N_2(g)$, $O_2(g)$, $F_2(g)$, $Cl_2(g)$, $Br_2(l)$, and $I_2(s)$. Similarly, $\Delta H_f° = 0$ kJ/mol for the graphite form of carbon, C(*s*, graphite). Table 17.4 lists $\Delta H_f°$ values for some common substances.

Standard heats of formation provide an alternative to Hess's law in determining heats of reaction indirectly. **For a reaction that occurs at standard conditions, you can calculate the heat of reaction by using standard heats of formation.** Such an enthalpy change is called the standard heat of reaction ($\Delta H°$). The standard heat of reaction is the difference between the standard heats of formation of all the reactants and products.

$$\Delta H° = \Delta H_f°\text{(products)} - \Delta H_f°\text{(reactants)}$$

Figure 17.16 Standard Heat of Formation of Water
This enthalpy diagram shows the standard heat of formation of water.
Classify *Is the reaction endothermic or exothermic?*

Figure 17.16 is an enthalpy diagram for the formation of water from its elements at standard conditions. The enthalpy difference between the reactants and products, −285.8 kJ/mol, is the standard heat of formation of liquid water from the gases hydrogen and oxygen. Notice that water has a lower enthalpy than the elements from which it is formed.

Differentiated Instruction

L1 STRUGGLING STUDENTS If students are having trouble visualizing the application of Hess's law, give them the following simplified equations.

$1 + 2 \rightarrow 3$

$3 + 4 \rightarrow 5 + 2$

Show students how 2 and 3 can be cancelled from both sides of the equations. The two equations can then be added together to produce $1 + 4 \rightarrow 5$.

L1 SPECIAL NEEDS STUDENTS When assessing students, have them give a thumbs up or a nod when they want to keep an enthalpy value the same, and a thumbs down or a head shake to indicate that they want to change the sign on the enthalpy value.

L3 ADVANCED STUDENTS Tell students to prepare a 10 minute lesson on Hess's Law directed to struggling students. Tell students to use an analogy, such as moving a ball down a football field or soccer field, to visually explain the concept.

Sample Problem 17.8

Calculating the Standard Heat of Reaction

What is the standard heat of reaction ($\Delta H°$) for the reaction of $CO(g)$ with $O_2(g)$ to form $CO_2(g)$?

1 Analyze List the knowns and the unknown. Balance the equation of the reaction of $CO(g)$ with $O_2(g)$ to form $CO_2(g)$. Then determine $\Delta H°$ using the standard heats of formation of the reactants and products.

KNOWNS
(from Table 17.4)
$\Delta H_f°CO(g) = -110.5\ kJ/mol$
$\Delta H_f°O_2(g) = 0\ kJ/mol$ (free element)
$\Delta H_f°CO_2(g) = -393.5\ kJ/mol$

UNKNOWN
$\Delta H° = ?\ kJ$

2 Calculate Solve for the unknown.

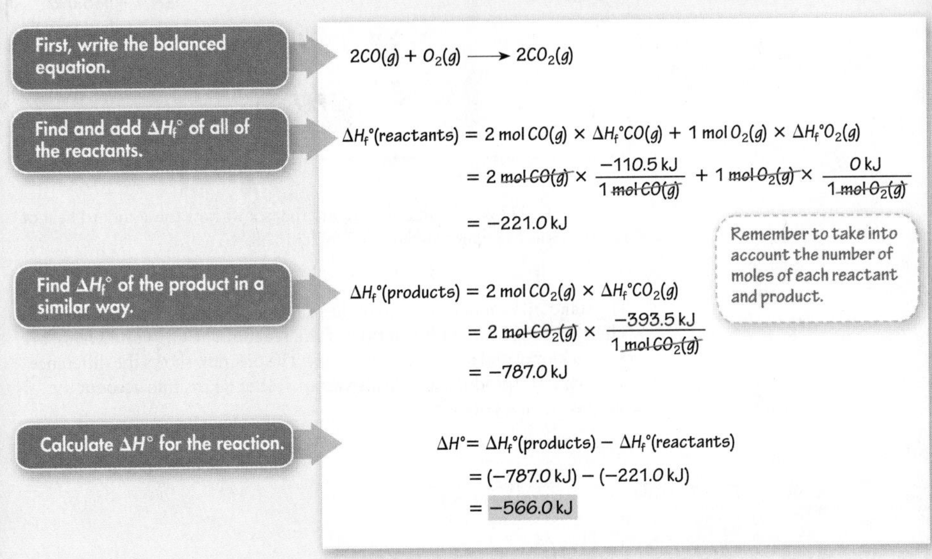

3 Evaluate Does the result make sense? The $\Delta H°$ is negative, so the reaction is exothermic. This outcome makes sense because combustion reactions always release heat.

Foundations for Math

COEFFICIENTS When calculating standard heats of reaction, the number of moles of each reactant and product must be taken into account. Remind students that standard heats of formation have units of kJ/mol, but that the heat of reaction has units of kJ. Thus, they must account for the number of moles of each reactant and each product in their calculations in order to obtain answers with units of kJ.

In **Sample Problem 17.8,** there are 2 mol CO, 1 mol O_2, and 2 mol CO_2. Point out that the standard heat of formation for each compound is multiplied by the number of moles of the compound to determine the total heat of formation for that compound in the reaction.

Extend

Connect to ENGINEERING

In this thermite reaction, iron(III) oxide is reduced to metallic iron by aluminum.

$$Fe_2O_3(s) + 2Al(s) \rightarrow 2Fe(s) + Al_2O_3(s)$$
$$\Delta H = -853.9\ kJ$$

The reaction reaches a temperature of about 3000°C. **Ask** students to research the resulting products of this reaction and describe the condition of each. *(The products are iron and aluminum oxide, and are molten.)*

Sample Practice Problem

Calculate the standard heat of reaction for the following reaction:

$$2SO_2(g) + O_2(g) \rightarrow 2SO_3(g)$$
(–197.8 kJ)

Answers

FIGURE 17.16 exothermic

35. -3.091×10^1 kJ

36. -1.130×10^2 kJ

LESSON 17.4

Evaluate

Informal Assessment

Have students explain two ways that scientists can determine the heat of a reaction indirectly. Encourage students to use examples, symbols, and/or diagrams in their explanations. *(1. Add two or more thermochemical equations to give the desired equation. 2. Subtract the standard heats of formation for all reactants from the standard heats of formation for all products in the reaction.)* Then direct students to complete 17.4 Lesson Check.

Reteach

Write an overall reaction on the board, using compounds found in Table 17.4. Challenge students to identify the reaction steps, if any, needed to account for the enthalpy change. Then, have students use Hess's law of heat summation and the standard heats of formation to indirectly determine the enthalpy change for the reaction.

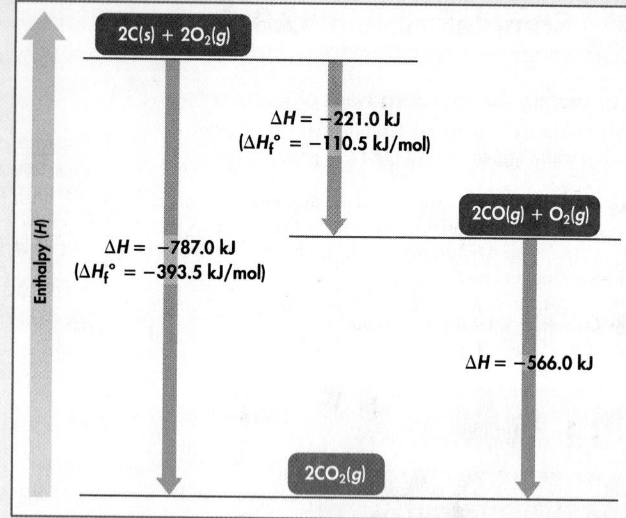

Figure 17.17 Reaction of Carbon Monoxide and Oxygen
Standard heats of formation are used to calculate the enthalpy change for the reaction of carbon monoxide and oxygen.
Explain ***How does this diagram also demonstrate Hess's law?***

Figure 17.17 is an enthalpy diagram that shows how the standard heat of reaction was calculated in Sample Problem 17.8.

$$2CO(g) + O_2(g) \longrightarrow 2CO_2(g)$$

The standard heat of formation of the product, $CO_2(g)$, is −393.5 kJ/mol. The standard heats of formation of the reactants, $CO(g)$ and $O_2(g)$, are −110.5 kJ/mol and 0 kJ/mol, respectively. The diagram shows the difference between $\Delta H_f°$(product) and $\Delta H_f°$(reactants) after taking into account the number of moles of each.

17.4 LessonCheck

ONLINE PROBLEMS

37. Describe What are two ways the heat of reaction can be determined when it cannot be directly measured?

38. Calculate What is the enthalpy change (ΔH) in kJ for the following reaction?

$$2Al(s) + Fe_2O_3(s) \longrightarrow 2Fe(s) + Al_2O_3(s)$$

Use the enthalpy changes for the combustion of aluminum and iron:

$$2Al(s) + \tfrac{3}{2}O_2(g) \longrightarrow Al_2O_3(s) \quad \Delta H = -1676.0 \text{ kJ}$$
$$2Fe(s) + \tfrac{3}{2}O_2(g) \longrightarrow Fe_2O_3(s) \quad \Delta H = -822.1 \text{ kJ}$$

39. Explain How can you calculate the standard heat of reaction?

40. Calculate What is the standard heat of reaction ($\Delta H°$) for the decomposition of hydrogen peroxide?

$$2H_2O_2(l) \longrightarrow 2H_2O(l) + O_2(g)$$

BIGIDEA MATTER AND ENERGY

41. Use Hess's law and two thermochemical equations on page R34 to calculate ΔH for the following reaction:

$$2H_2O(g) + CH_4(g) \longrightarrow CO_2(g) + 4H_2(g)$$

Lesson Check Answers

37. Use Hess's law of heat summation or use standard heats of formation.
38. −8.539 x 10^2 kJ
39. $\Delta H° = \Delta H_f°$(products) − $\Delta H_f°$(reactants)
40. −1.960 × 10^2 kJ
41. 165 KJ

Answers

FIGURE 17.17 It shows the summation of thermochemical equations to give a final equation. It also shows the enthalpy change for each step.

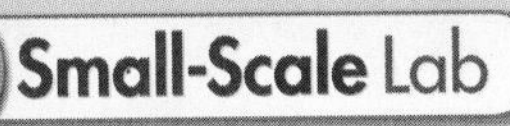

Small-Scale Lab

Heat of Combustion of a Candle

Purpose

To observe a burning candle and calculate the heat associated with the combustion reaction

Materials

- ruler • candle • aluminum foil
- balance • safety matches

Procedure

1. Measure and record the length of a candle in centimeters.

2. Place the candle on a small piece of aluminum foil and measure the mass of the foil-candle system.

3. Note the time as you light the candle. Let the candle burn for about five minutes. **CAUTION** *Keep clothing away from the flame.* While you wait, begin answering the Analyze and Conclude questions.

4. Extinguish the candle and record the time.

5. Measure the mass of the foil-candle system again. **DO NOT** try to measure the mass while the candle is burning.

Analyze and Conclude

1. Observe While the candle is burning, draw a picture of what you see.

2. Observe Examine the flame closely. Is it the wax or the wick that burns?

3. Infer If you said the wax, how does the wax burn without touching the flame? If you said the wick, what is the function of the wax?

4. Analyze Data If you could measure the temperature near the flame, you would find that the air is much hotter above the flame than it is beside it. Why?

5. Draw Conclusions How much length and mass did the candle lose? Are these data more consistent with the wax or the wick burning?

6. Infer Keeping in mind that *wick* is also a verb, explain how a candle works.

7. Describe The formula for candle wax can be approximated as $C_{20}H_{42}$. Write a balanced equation for the complete combustion of candle wax.

8. Calculate Determine the number of moles of candle wax burned in the experiment.

9. Calculate What is the heat of combustion of candle wax in kJ/mol? The standard heat of formation of candle wax ($C_{20}H_{42}$) is −2230 kJ/mol. The standard heats of formation of carbon dioxide gas and liquid water are −394 kJ/mol and −286 kJ/mol, respectively.

10. Calculate Determine the amount of heat (in kJ) released in your reaction. (*Hint:* Multiply the number of moles of candle wax burned in the experiment by the heat of combustion of candle wax.)

You're the Chemist

1. Design an Experiment Design an experiment to show that the candle wax does not burn with complete combustion.

2. Design an Experiment Design an experiment to show that water is a product of the combustion of a candle.

Focus on ELL

4 LANGUAGE PRODUCTION Have students work in groups of four to complete the Lab. Make sure each group has ELLs of varied language proficiencies, so that more proficient students can help less proficient ones. Have students work according to their proficiency level.

BEGINNING: LOW/HIGH Break the procedure down into simpler bulleted steps.

INTERMEDIATE: LOW/HIGH Simplify the wording of the Analyze and Conclude questions. Convert multi-part questions into multiple questions.

ADVANCED: LOW/HIGH Pair students with writing partners to write their lab reports. Have partners evaluate each other's reports and provide feedback for revisions.

Small-Scale Lab

OBJECTIVE After completing this activity, students will be able to explore the way a candle burns. They will also be able to apply basic principles of thermochemistry to the combustion reaction and calculate the heat of combustion.

MISCONCEPTION ALERT Students have many misconceptions about the way candles burn. Questions 1 and 2 may prompt students to wrongly conclude that the wick burns and the wax just slows the rate of burning. Questions 5 and 6 focus students on the key experiment, the determination of the mass loss of the candle, and allow them to revise their opinions.

PREP TIME 10 minutes

CLASS TIME 20 minutes

MATERIALS candle, metric ruler, aluminum foil, balance, safety matches, clock or watch, temperature probe (optional)

ADVANCE PREPARATION Purchase or collect candles and a roll of aluminum foil.

SAFETY Remind students to wear safety glasses and to keep the flame away from any combustible material.

TEACHING TIPS For best results and greater safety, use short, stout votive candles. The day before the lab, remind students to tie long hair back and to avoid wearing loose-sleeved clothes.

EXPECTED OUTCOME The candle loses mass and length as it burns.

ANALYZE AND CONCLUDE

1. Drawings should show the wax melting.
2. Many students incorrectly choose the wick.
3. The wick draws melted wax to the flame. Those who think the wick burns may suggest that the wax slows the rate of burning.
4. The hot gases expand and rise.
5. A few millimeters and a few tenths of a gram are reasonable answers. The mass loss is consistent with burning wax.
6. Heat from the combustion melts wax, which is drawn up by the wick, evaporated, and burned.
7. $C_{20}H_{42} + \frac{61}{2}O_2 \longrightarrow 20CO_2 + 21H_2O$
8. Sample answer: 0.0018 mol
9. Sample Answer: −11,700 kJ/mol
10. Sample Answer: 21 kJ

YOU'RE THE CHEMIST

1. Soot will appear on a glass Petri dish held over the flame.
2. Water will condense on the bottom of an ice-filled, glass Petri dish held over the flame.

FOR ENRICHMENT Have students design an experiment to determine the heat of combustion of table sugar. This is a reaction that you might wish to demonstrate for the class.

Evaluate

Review and Assessment Materials

CHEMISTRY TUTORIAL Encourage students to review solving specific heat, enthalpy, heat of fusion, vaporization, and standard heat of reaction problems using the animated step-by-step tutorials for Chapter 17. Each Chemistry Tutorial is followed by a similar practice problem, where the student is guided through the problem-solving process.

CHAPTER PROBLEM SET Have students practice more thermochemistry problems using the Chapter 17 online Problem Set.

DIRECTED VIRTUAL LAB Have students complete the Virtual Lab *Heat of Combustion* to help reinforce the mathematical relationships involved in calculating the enthalpy changes in different reactions. In the post-lab exercise, students will analyze the data and draw conclusions based on the results of the experiment.

KINETIC ART Direct students to the Kinetic Art animations *Bomb Calorimeter* and *Hess's Law*, and have them answer the questions that accompany each animation.

CONCEPTS IN ACTION If students have trouble understanding difference between heat and temperature, ask them to view the Concepts in Action animation, which presents the material in a familiar context. Have students complete the questions after the animation.

Study Tip

STUDY WITH FRIENDS Divide the class into small groups, and ask each group to write a review question for each Key Concept and each vocabulary term in one of the sections of the chapter. When students have finished writing their questions, have the groups place their lists in a central location. Then, ask each group to pick up a list of questions for one of the other three sections. After groups have had a chance to collaborate on answering the questions, repeat the process, with groups answering the questions for a third and fourth section.

17 Study Guide

BIGIDEA MATTER AND ENERGY

During a chemical or physical process, the energy of the universe is conserved. If energy is absorbed by the system in a chemical or physical process, the same amount of energy is released by the surroundings. Conversely, if energy is released by the system, the same amount of energy is absorbed by the surroundings. The heat of reaction or process can be determined experimentally through calorimetry. The heat of reaction can also be calculated by using the known heats of reaction of two or more thermochemical equations or by using standard heats of formation.

17.1 The Flow of Energy

- Energy changes occur as either heat transfer or work, or a combination of both.
- During any chemical or physical process, the energy of the universe remains unchanged.
- The heat capacity of an object depends on both its mass and its chemical composition.

- thermochemistry (556)
- chemical potential energy (556) • heat (556)
- system (557) • surroundings (557)
- law of conservation of energy (557)
- endothermic process (557)
- exothermic process (557)
- heat capacity (559) • specific heat (559)

Key Equation

$$C = \frac{q}{m \times \Delta T}$$

17.2 Measuring and Expressing Enthalpy Changes

- The value of ΔH of a reaction can be determined by measuring the heat flow of the reaction at constant pressure.
- In a chemical equation, the enthalpy change for the reaction can be written as either a reactant or a product.

- calorimetry (562)
- calorimeter (562)
- enthalpy (562)
- thermochemical equation (565)
- heat of reaction (565)
- heat of combustion (568)

Key Equation

$$q_{sys} = \Delta H = -q_{surr} = -m \times C \times \Delta T$$

17.3 Heat in Changes of State

- The quantity of heat absorbed by a melting solid is exactly the same as the quantity of heat released when the liquid solidifies; that is, $\Delta H_{fus} = -\Delta H_{solid}$.
- The quantity of heat absorbed by a vaporizing liquid is exactly the same as the quantity of heat released when the vapor condenses; that is, $\Delta H_{vap} = -\Delta H_{cond}$.
- During the formation of a solution, heat is either released or absorbed.

- molar heat of fusion (569)
- molar heat of solidification (569)
- molar heat of vaporization (571)
- molar heat of condensation (572)
- molar heat of solution (574)

17.4 Calculating Heats of Reaction

- Hess's law allows you to determine the heat of reaction indirectly by using the known heats of reaction of two or more thermochemical equations.
- For a reaction that occurs at standard conditions, you can calculate the heat of reaction by using standard heats of formation.

- Hess's law of heat summation (578)
- standard heat of formation (580)

Key Equation

$$\Delta H° = \Delta H_f°(\text{products}) - \Delta H_f°(\text{reactants})$$

UbD Understanding by Design

Performance Tasks

CHILLING OUT Research the mechanism used in refrigerators and freezers to keep food cold. Create a flowchart that illustrates and explains the heat transfer processes and direction of heat flow that occurs between the refrigerator and its contents. Include an explanation of how frost-free units prevent a build-up of frost in the freezer compartment. The flowchart should document all relevant enthalpy changes.

HOT AND COLD PACKS Have students design and perform an experiment that measures the temperature change in a hot or cold pack. Consider changing the solute to a compound not mentioned in the Lesson 17.3 (e.g., sodium acetate) and have students conclude whether the solution process is endothermic or exothermic.

Math Tune-Up: Calculating Enthalpy Changes

Problem	❶ Analyze	❷ Calculate	❸ Evaluate
When 75.0 mL of water containing 0.100 mol HCl at 21.0°C is added to 75.0 mL of water containing 0.100 mol NaOH at 21.0°C in a foam cup calorimeter, the temperature of the solution increases to 29.6°C. Calculate the enthalpy change (in kJ) during this reaction.	Knowns: $C_{water} = 4.18\ J/(g{\cdot}°C)$ $V_{final} = V_{HCl} + V_{NaOH}$ $= 75.0\ mL + 75.0\ mL$ $= 150.0\ mL$ $\Delta T = T_f - T_i$ $= 29.6°C - 21.0°C$ $= 8.6°C$ $density_{solution} = 1.00\ g/mL$ Unknown: $\Delta H = ?\ kJ$ Use $\Delta H = -q_{surr}$ $= -m \times C \times \Delta T$	Calculate the mass of water. $m_{water} = 150.0\ \cancel{mL} \times \frac{1.00\ g}{1\ \cancel{mL}}$ $= 150.0\ g$ Use the values for m_{water}, C_{water}, and ΔT to calculate ΔH. $\Delta H = -(150.0\ \cancel{g})(4.18\ J/(\cancel{g}{\cdot}\cancel{°C}))$ $\times (8.6\cancel{°C})$ $= -5400\ J$ $\Delta H = -5.4\ kJ$	About 4 J of heat raises the temperature of 1 g of water 1°C, so 600 J of heat is required to raise the temperature of 150 g of water 1°C. To heat 150 g of water 9°C requires about 5400 J, or 5.4 kJ.
How much heat is absorbed when 54.9 g $H_2O(l)$ at 100°C and 101.3 kPa is converted to $H_2O(g)$ at 100°C?	Knowns: Initial and final conditions are 100°C and 101.3 kPa mass of liquid water converted to steam = 54.9 g $\Delta H_{vap} = 40.7\ kJ/mol$ Unknown: $\Delta H = ?\ kJ$ Refer to the thermochemical equation $H_2O(l) + 40.7\ kJ \longrightarrow H_2O(g)$	The required conversion factors come from the molar mass of water and ΔH_{vap}. $\frac{1\ mol\ H_2O(l)}{18.0\ g\ H_2O(l)}$ and $\frac{40.7\ kJ}{1\ mol\ H_2O(l)}$ Multiply the mass of water by the conversion factors. $\Delta H = 54.9\ \cancel{g\ H_2O(l)} \times \frac{1\ \cancel{mol\ H_2O(l)}}{18.0\ \cancel{g\ H_2O(l)}}$ $\times \frac{40.7\ kJ}{1\ \cancel{mol\ H_2O(l)}}$ $\Delta H = 124\ kJ$	Knowing the molar mass of water is 18.0 g/mol, 54.9 g $H_2O(l)$ is about 3 mol H_2O. The calculated enthalpy change should be about 3 $\cancel{mol}$ × 40 kJ/$\cancel{mol}$ = 120 kJ, and it is.
What is the standard heat of reaction ($\Delta H°$) for the reaction of $SO_2(g)$ with $O_2(g)$ to form $SO_3(g)$?	Knowns: (from Table 17.4) $\Delta H_f° SO_2(g) = -296.8\ kJ/mol$ $\Delta H_f° O_2(g) = 0\ kJ/mol$ $\Delta H_f° SO_3(g) = -395.7\ kJ/mol$ Unknown: $\Delta H° = ?\ kJ$ Use the standard heats of formation for the reactants and products to calculate $\Delta H°$. $\Delta H° = \Delta H_f°(products) - \Delta H_f°(reactants)$	Write the balanced equation. $2SO_2(g) + O_2(g) \longrightarrow 2SO_3(g)$ Find $\Delta H_f°$ of the reactants. $\Delta H_f°(reactants) = 2\ \cancel{mol\ SO_2(g)} \times \frac{-296.8\ kJ}{1\ \cancel{mol\ SO_2(g)}} + 0\ kJ = -593.6\ kJ$ Find $\Delta H_f°$ of the product. $\Delta H_f°(product) = 2\ \cancel{mol\ SO_3(g)} \times \frac{-395.7\ kJ}{1\ \cancel{mol\ SO_3(g)}} = -791.4\ kJ$ Calculate $\Delta H°$ for the reaction. $\Delta H° = (-791.4\ kJ) - (-593.6\ kJ)$ $\Delta H° = -197.8\ kJ$	The $\Delta H°$ is negative, so the reaction is exothermic. This makes sense because combustion reactions always release heat.

Note: For reactions in aqueous solutions, you can assume that the densities of the solutions are 1.00 g/mL.

Remember: The $\Delta H_f°$ of a free element in its standard state is 0.

MATH TUNE-UP

Math Review

Example and sample practice problems related to calculating enthalpy changes can be found on the following pages:

LESSON 17.1 Calculating the Specific Heat of a Metal—page 561

LESSON 17.2 Enthalpy Change in a Calorimetry Experiment—page 564; Using the Heat of Reaction to Calculate Enthalpy Change—page 567

LESSON 17.3 Using the Heat of Fusion in Phase-Change Calculations—page 570; Using the Heat of Vaporization in Phase-Change Calculations—page 573; Calculating the Enthalpy Change in Solution Formation—page 575

LESSON 17.4 Calculating the Standard Heat of Reaction—page 581

Focus on ELL

5 ASSESS UNDERSTANDING Assess students using a portfolio for this chapter. Include various types of learning assignments based on the student's language proficiency.

BEGINNING: LOW Include models or pictures of models, lists, or diagrams created by the student.

HIGH Include copies of the graphic organizers used to take notes or show how concepts are related.

INTERMEDIATE: LOW Include any predictions made on the topic of lessons or pre-reading and the result of those predictions.

HIGH Include any lab write-ups that were assigned during the chapter.

ADVANCED: LOW/HIGH Include an essay that explains why freezing can protect a citrus tree.

ASSESSMENT

Evaluate

Answers

LESSON 17.1

42. Chemical potential energy is energy stored in a substance because of its chemical composition.

43. Heat flows from the object at the higher temperature to the object at the lower temperature. For example, holding a cold soft-drink can makes your hand cold.

44. Answers will vary, but should mention that thermochemistry measures heat flow across the boundary between the system and the surroundings.

45. Answers will vary, but should include the idea that energy is conserved in every physical and chemical process.

46. In an endothermic process, the system absorbs heat from the surroundings. In an exothermic process, the system releases heat to the surroundings.

47. **a.** exothermic
b. The immediate surroundings are the glass beaker and the air. If one or more of the substances is in water, the water is also considered part of the surroundings.

48. **a.** exothermic **b.** endothermic
c. exothermic **d.** endothermic

49. A negative sign is given to heat flow from the system to the surroundings. A positive sign is given to heat flow to the system from the surroundings.

50. 1 Cal = 1000 cal = 1 kcal

51. **a.** 0.85 Cal
b. 1.86×10^3 J
c. 1.8×10^3 J
d. 1.1×10^2 cal

52. the chemical composition of the substance and its mass

53. 4.3×10^3 J

LESSON 17.2

54. The heat released by a system is equal to the heat absorbed by its surroundings. The heat absorbed by a system is equal to the heat released by its surroundings.

55. A calorimeter is an instrument used to measure enthalpy changes in physical or chemical processes.

56. enthalpy

57. bomb calorimeter

58. amount of heat released or absorbed in a chemical change at constant pressure

59. -4.5×10^2 kJ

60. one atmosphere pressure (101.3 kPa); all reactants and products in their normal physical state at 25°C.

LESSON 17.3

61. Heat is being used to melt the ice.

62. **a.** –21.0 kJ
b. –18 kJ
c. –55.6 kJ
d. 5.8 kJ

63. increase

LESSON 17.4

64. Hess's law allows the calculation of the enthalpy change of a reaction from the known enthalpy changes of two or more other reactions.

65. **a.** 1676 kJ **b.** ΔH is positive so the reaction is endothermic.

17 Assessment

★ Solutions appear in Appendix E

Lesson by Lesson

17.1 The Flow of Energy

42. Define *chemical potential energy.*

43. What always happens when two objects of different temperatures come in contact? Give an example from your own experience.

44. Why do you think it is important to define the system and the surroundings?

45. Explain in your own words the law of conservation of energy.

46. How do endothermic processes differ from exothermic processes?

★**47.** Two substances in a glass beaker chemically react, and the beaker becomes too hot to touch.
a. Is the reaction endothermic or exothermic?
b. If the two substances are defined as the system, what constitutes the surroundings?

★**48.** Classify these processes as endothermic or exothermic.
a. condensing steam
b. evaporating alcohol
c. burning alcohol
d. baking a potato

49. Describe the sign convention that is used when describing heat flow in a system.

50. What is the relationship between a calorie and a Calorie?

★**51.** Make the following conversions.
a. 8.50×10^2 cal to Calories
b. 444 cal to joules
c. 1.8 kJ to joules
d. 4.5×10^{-1} kJ to calories

52. What factors determine the heat capacity of an object?

★**53.** How much heat is required to raise the temperature of 400.0 g of silver 45°C?

17.2 Measuring and Expressing Enthalpy Changes

54. Calorimetry is based on what basic concepts?

55. What is the function of a calorimeter?

56. What is the property that describes heat change at constant pressure?

57. What device would you use to measure the heat released at constant volume?

58. What information is given in a thermochemical equation?

★**59.** The burning of magnesium is a highly exothermic reaction.

$$2Mg(s) + O_2(g) \longrightarrow 2MgO(s) + 1204\ kJ$$

How many kilojoules of heat are released when 0.75 mol of Mg burn in an excess of O_2?

60. Give the standard conditions for heat of combustion.

17.3 Heat in Changes of State

61. Explain why ice melts at 0°C without an increase of temperature, even though heat flows from the surroundings to the system (the ice).

★**62.** Calculate the quantity of heat gained or lost in the following changes:
a. 3.50 mol of water freezes at 0°C
b. 0.44 mol of steam condenses at 100°C
c. 1.25 mol $NaOH(s)$ dissolves in water
d. 0.15 mol $C_2H_6O(l)$ vaporizes at 78.3°C

63. Sodium ethanoate dissolves readily in water according to the following equation:

$$NaC_2H_3O_2(s) \longrightarrow NaC_2H_3O_2(aq)$$
$$\Delta H = -17.3\ kJ/mol$$

Would this process increase or decrease the temperature of the water?

17.4 Calculating Heats of Reaction

64. Explain Hess's law of heat summation.

★**65.** A considerable amount of heat is required for the decomposition of aluminum oxide.

$$2Al_2O_3(s) \longrightarrow 4Al(s) + 3O_2(g)$$
$$\Delta H = 3352\ kJ$$

a. What is the enthalpy change for the formation of 1 mol of aluminum oxide from its elements?
b. Is the reaction endothermic or exothermic?

*66. Calculate the enthalpy change for the formation of lead(IV) chloride by the reaction of lead(II) chloride with chlorine.

$$PbCl_2(s) + Cl_2(g) \longrightarrow PbCl_4(l)$$

$$\Delta H = ?$$

Use the following thermochemical equations:

$$Pb(s) + 2Cl_2(g) \longrightarrow PbCl_4(l)$$

$$\Delta H = -329.2 \text{ kJ}$$

$$Pb(s) + Cl_2(g) \longrightarrow PbCl_2(s)$$

$$\Delta H = -359.4 \text{ kJ}$$

67. What is the standard heat of formation of a compound?

68. What is the standard heat of formation of a free element in its standard state?

Understand Concepts

69. How many kilojoules of heat are absorbed when 1.00 L of water is heated from 18°C to 85°C?

*70. Equal masses of two substances absorb the same amount of heat. The temperature of substance A increases twice as much as the temperature of substance B. Which substance has the higher specific heat? Explain.

71. Identify each enthalpy change by name and classify each change as endothermic or exothermic.
   - a. 1 mol $C_3H_8(l) \longrightarrow$ 1 mol $C_3H_8(g)$
   - b. 1 mol $Hg(l) \longrightarrow$ 1 mol $Hg(s)$
   - c. 1 mol $NH_3(g) \longrightarrow$ 1 mol $NH_3(l)$
   - d. 1 mol $NaCl(s)$ + 3.88 kJ $\longrightarrow$ 1 mol $NaCl(aq)$
   - e. 1 mol $NaCl(s) \longrightarrow$ 1 mol $NaCl(l)$

72. Name at least three sources of error in experiments that use foam cups as calorimeters.

73. Calculate the enthalpy change in calories when 45.2 g of steam at 100°C condenses to water at the same temperature. What is the enthalpy change in joules?

*74. A 1.55-g piece of stainless steel absorbs 141 J of heat when its temperature increases by 178°C. What is the specific heat of the stainless steel?

75. With one exception, the standard heats of formation of Na(s), $O_2(g)$, $Br_2(l)$, CO(g), Fe(s), and He(g) are identical. What is the exception?

*76. Calculate the change in enthalpy (in kJ) for the following reactions using standard heats of formation ($\Delta H_f°$):
   - a. $CH_4(g) + \frac{3}{2}O_2(g) \longrightarrow CO(g) + 2H_2O(l)$
   - b. $2CO(g) + O_2(g) \longrightarrow 2CO_2(g)$

77. The amounts of heat required to change different quantities of carbon tetrachloride, $CCl_4(l)$, into vapor are given in the table.

Mass of CCl_4 (g)	Heat (J)	Heat (cal)
2.90	652	156
7.50	1689	404
17.0	3825	915
26.2	5894	1410
39.8	8945	2140
51.0	11453	2740

   - a. Graph the data, using heat as the dependent variable.
   - b. What is the slope of the line?
   - c. The heat of vaporization of $CCl_4(l)$ is 53.8 cal/g. How does this value compare with the slope of the line?

*78. Find the enthalpy change for the formation of phosphorus pentachloride from its elements.

$$2P(s) + 5Cl_2(g) \longrightarrow 2PCl_5(s)$$

Use the following thermochemical equations:

$$PCl_5(s) \longrightarrow PCl_3(g) + Cl_2(g)$$

$$\Delta H = 156.5 \text{ kJ}$$

$$2P(s) + 3Cl_2(g) \longrightarrow 2PCl_3(g)$$

$$\Delta H = -574.0 \text{ kJ}$$

79. Use standard heats of formation ($\Delta H_f°$) to calculate the change in enthalpy for these reactions.
   - a. $2C(s, \text{graphite}) + O_2(g) \longrightarrow 2CO(g)$
   - b. $2H_2O_2(l) \longrightarrow 2H_2O(l) + O_2(g)$
   - c. $4NH_3(g) + 5O_2(g) \longrightarrow 4NO(g) + 6H_2O(g)$
   - d. $CaCO_3(s) \longrightarrow CaO(s) + CO_2(g)$

80. The molar heat of vaporization of ethanol, $C_2H_6O(l)$, is 38.6 kJ/mol. Calculate the heat required to vaporize 25.0 g of ethanol at its boiling point.

ASSESSMENT

66. ΔH = 30.2 kJ
67. the change in enthalpy that accompanies the formation of 1 mole of a compound from its elements with all substances in their standard states at 25°C and 101.3 kPa.
68. zero

UNDERSTAND CONCEPTS

69. 281 kJ
70. Substance B; For equal masses, the substance with the greater heat capacity undergoes the smaller temperature change.
71. **a.** molar heat of vaporization; endothermic
    b. molar heat of solidification; exothermic
    c. molar heat of condensation; exothermic
    d. molar heat of solution; endothermic
    e. molar heat of fusion; endothermic
72. The foam cup will absorb heat. Some heat will be lost to the air. If the reactants are not completely mixed, temperature measurements will not be accurate.
73. 2.44×10^4 cal; 1.02×10^5 J
74. 0.511 J/(g · °C)
75. CO(g)
76. **a.** –607.2 kJ
    b. –566.0 kJ
77. **a.** Graphs should show the mass on the *x*-axis and heat (in calories) on the *y*-axis.
    b. about 54 cal/g
    c. The two values are essentially the same.
78. ΔH = –750 kJ
79. **a.** -2.21×10^2 kJ
    b. -1.96×10^2 kJ
    c. -9.046×10^2 kJ
    d. 178.4 kJ
80. 21.0 kJ

ASSESSMENT

Evaluate

Answers

81. 1.41×10^3 g
82. 2.4×10^3 kJ
83. 1.8×10^2 kJ
84. 67.1 kJ
85. 40 grams of liquid water produced; 9.6×10^2 g of ice remain

THINK CRITICALLY

86. Your fingers feel cold because heat always passes from a warmer object (fingers) to a cooler object (ice).
87. a. 1.5×10^2 kJ
b. The refrigerator absorbs 1.5×10^2 kJ of heat.
c. assumes the mineral water has the same specific heat as chemically pure water, that no heat is lost by the refrigerator, and the density of mineral water is 1.0 g/mL
88. The statement is true, since stability implies lower energy. The greater the release of heat, the more stable is the compound relative to its elements (all of which have $\Delta H_f^\circ = 0$)
89. a. –184.76 kJ
b. –138.6 kJ
c. –46.19 kJ
90. When a solid reaches its melting point, additional heat must be absorbed to convert it to a liquid. Therefore, fusion of a solid is endothermic. This heat of fusion is released when a liquid freezes, so freezing is exothermic.
91. a. 1.34×10^4 J, 3.21×10^3 cal, 3.21 kcal
b. 128 g H_2O
92. When the vapor condenses to a liquid, the heat of vaporization is released. Therefore, the energy content of the vapor must have been higher than that of the liquid. The statement should state, "The energy content of a substance is higher in the vapor phase than in the liquid phase at the same temperature."
93. –1207 kJ
94. 2820 kJ/mol

*81. An orange contains 106 Calories. What mass of water could this same amount of energy raise from 25.0°C to the boiling point?

82. The combustion of ethene (C_2H_4) is an exothermic reaction.

$$C_2H_4(g) + 3O_2(g) \longrightarrow 2CO_2(g) + 2H_2O(l)$$
$$\Delta H = -1.40 \times 10^3 \text{ kJ}$$

Calculate the amount of heat liberated when 4.79 g C_2H_4 reacts with excess oxygen.

83. Calculate the enthalpy change (ΔH) for the formation of nitrogen monoxide from its elements.

$$N_2(g) + O_2(g) \longrightarrow 2NO(g)$$

Use the following thermochemical equations:

$$4NH_3(g) + 3O_2(g) \longrightarrow 2N_2(g) + 6H_2O(l)$$
$$\Delta H = -1.53 \times 10^3 \text{ kJ}$$
$$4NH_3(g) + 5O_2(g) \longrightarrow 4NO(g) + 6H_2O(l)$$
$$\Delta H = -1.17 \times 10^3 \text{ kJ}$$

84. How much heat must be removed from a 45.0-g sample of liquid naphthalene ($C_{10}H_8$) at its freezing point to bring about solidification? The heat of fusion of naphthalene is 19.1 kJ/mol.

*85. If 3.20 kcal of heat is added to 1.00 kg of ice at 0°C, how much liquid water at 0°C is produced, and how much ice remains?

Think Critically

86. **Relate Cause and Effect** Your fingers quickly begin to feel cold when you touch an ice cube. What important thermochemical principle does this change illustrate?

87. **Calculate** You place a bottle containing 2.0 L of mineral water at 25°C into a refrigerator to cool to 7°C.
a. How many kJ of heat are lost by the water?
b. How many kJ of heat are absorbed by the refrigerator?
c. What assumptions did you make in your calculations?

88. **Evaluate** Consider the statement, "the more negative the value of ΔH_f°, the more stable the compound." Is this statement true or false? Explain.

*89. **Calculate** When 1.000 mol of $N_2(g)$ reacts completely with 3.000 mol of $H_2(g)$, 2.000 mol of $NH_3(g)$ and 92.38 kJ of heat are produced.

$$N_2(g) + 3H_2(g) \longrightarrow 2NH_3(g) + 92.38 \text{ kJ}$$

Use this thermochemical equation to calculate ΔH for the following reactions:
a. $2N_2(g) + 6H_2(g) \longrightarrow 4NH_3(g)$
b. $\frac{3}{2}N_2(g) + \frac{9}{2}H_2(g) \longrightarrow 3NH_3(g)$
c. $\frac{1}{2}N_2(g) + \frac{3}{2}H_2(g) \longrightarrow NH_3(g)$

90. **Explain** Why is fusion an endothermic process, but freezing is an exothermic process?

*91. **Calculate** An ice cube with a mass of 40.0 g melts in water originally at 25.0°C.
a. How much heat does the ice cube absorb from the water when it melts? Report your answer in calories, kilocalories, and joules.
b. Calculate the number of grams of water that can be cooled to 0°C by the melting ice cube.

92. **Evaluate and Revise** Evaluate this statement: "The energy content of a substance is higher in the liquid phase than in the vapor phase at the same temperature." If the statement is incorrect, restate it so it is correct.

93. **Apply Concepts** Using the following equations,

$$Ca(s) + 2C(s) \longrightarrow CaC_2(s)$$
$$\Delta H = -62.8 \text{ kJ}$$
$$CO_2(g) \longrightarrow C(s) + O_2(g)$$
$$\Delta H = 393.5 \text{ kJ}$$
$$CaCO_3(s) + CO_2(g) \longrightarrow CaC_2(s) + \tfrac{5}{2}O_2(g)$$
$$\Delta H = 1538 \text{ kJ}$$

determine the heat of reaction (in kJ) for

$$Ca(s) + C(s) + \tfrac{3}{2}O_2(g) \longrightarrow CaCO_3(s)$$

*94. **Calculate** The sugar glucose ($C_6H_{12}O_6$) is an important nutrient for living organisms to meet their energy needs. The standard heat of formation (ΔH_f°) of glucose is −1260 kJ/mol. Calculate how much heat (in kJ/mol) is released at standard conditions if 1 mol of glucose undergoes the following reaction:

$$C_6H_{12}O_6(s) + 6O_2(g) \longrightarrow 6CO_2(g) + 6H_2O(l)$$

588 Chapter 17 • Assessment

Enrichment

95. Calculate Ethane, $C_2H_6(g)$, can be formed by the reaction of ethene, $C_2H_4(g)$, with hydrogen gas.

$$C_2H_4(g) + H_2(g) \longrightarrow C_2H_6(g)$$

Use the heats of combustion for the following reactions to calculate the heat change for the formation of ethane from ethene and hydrogen.

$$2H_2(g) + O_2(g) \longrightarrow 2H_2O(l)$$
$$\Delta H = -5.72 \times 10^2 \text{ kJ}$$
$$C_2H_4(g) + 3O_2(g) \longrightarrow 2CO_2(g) + 2H_2O(l)$$
$$\Delta H = -1.401 \times 10^3 \text{ kJ}$$
$$2C_2H_6(g) + 7O_2(g) \longrightarrow 4CO_2(g) + 6H_2O(l)$$
$$\Delta H = -3.100 \times 10^3 \text{ kJ}$$

96. Infer An ice cube at 0°C was dropped into 30.0 g of water in a cup at 45.0°C. At the instant that all of the ice was melted, the temperature of the water in the cup was 19.5°C. What was the mass of the ice cube?

★97. Calculate A 41.0-g piece of glass at 95°C is placed in 175 g of water at 21°C in an insulated container. They are allowed to come to the same temperature. What is the final temperature of the glass-water mixture? The specific heat of glass is 2.1 cal/(g·°C).

98. Interpret Graphs The molar heat of vaporization of water at various temperatures is given in the graph. Estimate the amount of heat required to convert 1 L of water to steam on the summit of Mount Everest (8850 m), where the boiling temperature of water is 70°C.

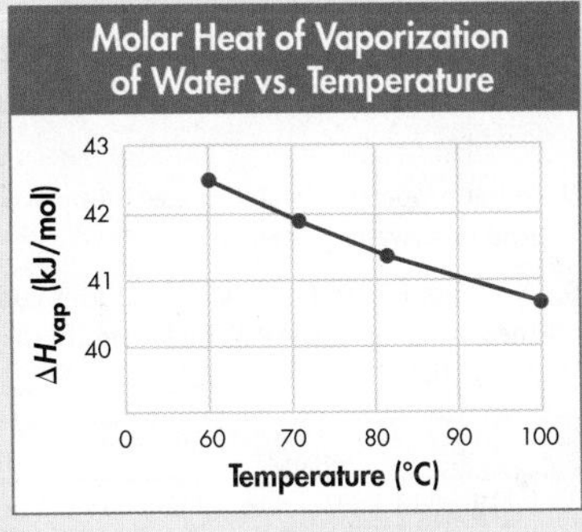

Write About Science

99. Explain Use the concept of heat capacity to explain why on a sunny day, the concrete deck around an outdoor swimming pool becomes hot, while the water stays cool.

100. Compare Why is a burn from steam potentially far more serious than a burn from very hot water?

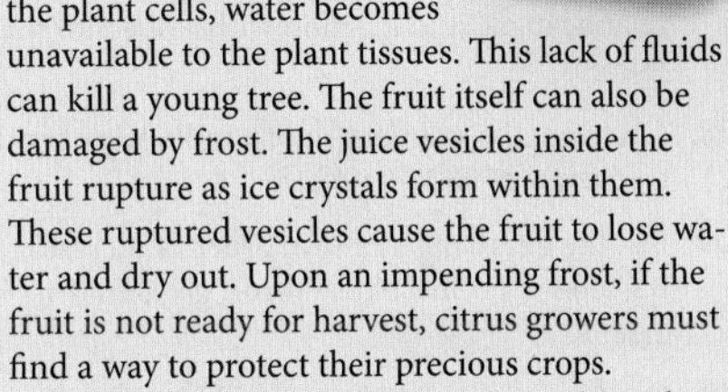

CHEMYSTERY

Fighting Frost

If the temperature of the branches, leaves, and fruit of a citrus tree falls below 0°C, severe damage can occur. When ice crystals form in the plant cells, water becomes unavailable to the plant tissues. This lack of fluids can kill a young tree. The fruit itself can also be damaged by frost. The juice vesicles inside the fruit rupture as ice crystals form within them. These ruptured vesicles cause the fruit to lose water and dry out. Upon an impending frost, if the fruit is not ready for harvest, citrus growers must find a way to protect their precious crops.

Spraying the trees with water throughout the duration of a frost is an effective way to prevent the trees and fruit from freezing. The water freezes directly on the branches, leaves, and fruit. Freezing is an exothermic process. As the water freezes, it releases heat and prevents the plant cells from reaching freezing temperatures.

101. Apply Concepts Identify the system and the surroundings when water freezes on a citrus fruit.

★102. Predict Evaporation of the water on a plant surface can occur under dry and windy conditions. How would this affect the citrus tree and fruit?

103. Connect to the BIGIDEA Explain, in terms of the law of conservation of energy, why the freezing of water on a citrus tree can cause the temperature of the tree to increase.

CHEMYSTERY APPLY CONCEPTS After students have read through the CHEMystery, call on volunteers to summarize why spraying trees and fruit with water and letting it freeze is an effective way to protect the plants during freezing temperatures. **Ask** In terms of heat flow, what would most likely happen if the growers did *not* spray the plants with water during freezing temperatures? *(Heat would flow from the plant directly to the surrounding air, and water in the plant cells would freeze. Because freezing water expands, this process causes cell membranes and walls to rupture, leading to cell death.)* **Ask** Why must water be continually sprayed on the plants while the air temperatures are freezing? *(Although the ice provides a buffer that slows the loss of heat from the plant to the colder air, it is the freezing process that protects the plant by losing heat to the environment.)*

Point out that while most fruit trees tend to survive frost and freeze events, the freezing weather results in economic losses due to the destruction of existing fruit and flower buds. Annual and perennial vegetable and fruit plants are more severely affected by freezing because they do not have woody stems and have more delicate leaves. As a result, producers lose the plants along with the produce.

CHEMYSTERY ANSWERS

101. The system is the branches, leaves, and fruit. The surroundings are the water.

102. Evaporation is an endothermic process. As the water evaporates, it absorbs heat from the trees and fruit, causing their temperature to decrease. This can lead to the branches, leaves and fruit freezing.

103. BIGIDEA As the water freezes, it releases a certain amount of energy. In accordance with the law of conservation of energy, the tree gains the same amount of energy. The temperature of the tree increases as a result. No energy is created or destroyed during this process.

Answers

ENRICHMENT

95. -1.37×10^2 kJ

96. 9.6 g

97. 45.6°C

98. ΔH_{vap} for water at 70°C is approximately 42 kJ/mol. 1 L of water (1000 mL) has a mass of 1000 g and contains 55.6 mol water. Therefore, the amount of heat required is 42 kJ/mol × 55.6 mol = 2.34×10^3 kJ

WRITE ABOUT SCIENCE

99. Concrete has a lower specific heat capacity than water. As a result, it takes more heat to increase a certain amount of water of by 1°C than it does to increase the same amount of concrete by 1°C. Assuming that the water and concrete absorb the same amount of radiant energy from the sun, the temperature of the concrete will rise faster.

100. Steam at 100°C contains more energy than liquid water at 100°C. If steam touches your skin, it will condense releasing the heat of condensation.

ASSESSMENT

Cumulative Review

104. The independent variable is the variable you change during an experiment. The dependent variable is the variable you observe during an experiment.

105. **a.** Cr **b.** Cu **c.** C **d.** Ca **e.** Cs

106. **a.** 6.99 m^2 **b.** 10.68 g **c.** 3.6×10^2 m/s **d.** 4.44 °C

107.

Particle	Charge	Relative Mass	Location
proton	1+	1	nucleus
neutron	0	1	nucleus
electron	1–	1/1840	outside nucleus

108. 3.22 m

109. phosphorus, arsenic, germanium

110. **a.** 2
b. 2
c. 3
d. 1

111. A covalent bond is nonpolar when the electrons of the bonding atoms are shared equally. A covalent bond is polar when the electrons of the bonding atoms are shared unequally. The bond between oxygen atoms in the oxygen molecule (O_2) is nonpolar. The bond between a carbon atom and an oxygen atom in the carbon monoxide molecule is polar.

112. **a.** K_3N **b.** Al_2S_3 **c.** $Ca(NO_3)_2$ **d.** $CaSO_4$

113. 1.20×10^{24} H_2 molecules

114. $Ag^+(aq) + Cl^-(aq) \rightarrow AgCl(s)$

115. $N_2(g) + O_2(g) \rightarrow 2NO(g)$
$2NO(g) + O_2(g) \rightarrow 2NO_2(g)$

116. 11.8 g O_2

117. 0.40

118. 11.1 L

119. Water is a very good solvent and dissolves at least a small part of most of what it comes in contact with in nature.

120. solutions; suspensions

Cumulative Review

104. Explain the difference between an independent variable and a dependent variable.

105. Write the correct chemical symbol for each element.

a. chromium
b. copper
c. carbon
d. calcium
e. cesium

***106.** Express the results of the following calculations with the correct number of significant figures.

a. 6.723 m × 1.04 m
b. 8.934 g + 0.2005 g + 1.55 g
c. 864 m ÷ 2.4 s
d. 9.258°C − 4.82°C

107. List three kinds of subatomic particles in an atom. Describe each kind in terms of charge, relative mass, and location with respect to the nucleus.

***108.** Calculate the wavelength of a radio wave with a frequency of $93.1 \times 10^6\ s^{-1}$.

109. List the following atoms in order of increasing atomic radius: phosphorus, germanium, arsenic.

110. How many chloride ions would be required to react with these cations to make an electrically neutral particle?

a. strontium cation
b. calcium cation
c. aluminum cation
d. lithium cation

111. How does a polar covalent bond differ from a nonpolar covalent bond? Which type of bond is found in molecular oxygen (O_2)? In carbon monoxide (CO)?

***112.** Write formulas for the following compounds:

a. potassium nitride
b. aluminum sulfide
c. calcium nitrate
d. calcium sulfate

***113.** How many hydrogen molecules are in 44.8 L $H_2(g)$ at STP?

114. Write the net ionic equation for the reaction of aqueous solutions of sodium chloride and silver acetate.

***115.** When lightning flashes, nitrogen and oxygen combine to form nitrogen monoxide. The nitrogen monoxide reacts with oxygen to form nitrogen dioxide. Write equations for these two reactions.

116. How many grams of oxygen are formed by the decomposition of 25.0 g of hydrogen peroxide?

$$2H_2O_2(l) \longrightarrow 2H_2O(l) + O_2(g)$$

117. What fraction of the average kinetic energy of hydrogen gas at 100 K does hydrogen gas have at 40 K?

***118.** A gas has a volume of 8.57 L at 273 K. What will be the volume at 355 K if its pressure does not change?

119. What property of water makes it impossible to find pure water in nature?

120. Do colloids, suspensions, or solutions contain the smallest particles? Which contain the largest particles?

If You Have Trouble With . . .

Question	104	105	106	107	108	109	110	111	112	113	114	115	116	117	118	119	120
See Chapter	1	2	3	4	5	6	7	8	9	10	11	11	12	13	14	15	15

Standardized Test Prep for Indiana

Standard: C.6.3

Select the choice that best answers each question or completes each statement.

Tips for Success

Eliminate Wrong Answers If you don't know which response is correct, start by eliminating those you know are wrong. If you can rule out some choices, you'll have fewer left to consider and you'll increase your chances of choosing the correct answer.

1. The ΔH_{fus} of ethanol (C_2H_6O) is 4.93 kJ/mol. How many kilojoules are required to melt 24.5 g of ethanol at its freezing point?
   (A) 2.63 kJ (C) 9.27 kJ
   (B) 4.97 kJ (D) 263 kJ

2. How much heat, in kilojoules, must be added to 178 g of liquid water to increase the temperature of the water by 5.0°C ?
   (A) 890 kJ (C) 3.7 kJ
   (B) 36 kJ (D) 0.093 kJ

3. The standard heat of formation of a free element in its standard state is always
   (A) zero.
   (B) positive.
   (C) negative.
   (D) higher for solids than for gases.

4. If ΔH for the reaction $2HgO(s) \longrightarrow 2Hg(l) + O_2(g)$ is 181.66 kJ, then ΔH for the reaction $Hg(l) + \frac{1}{2}O_2(g) \longrightarrow HgO(s)$ is
   (A) 90.83 kJ. (C) −90.83 kJ.
   (B) 181.66 kJ. (D) −181.66 kJ.

5. The specific heat capacity of ethanol is ten times larger than the specific heat capacity of silver. A hot bar of silver with a mass of 55 g is dropped into an equal mass of cool alcohol. If the temperature of the silver bar drops 45°C, the temperature of the alcohol
   (A) increases 45°C.
   (B) decreases 4.5°C.
   (C) increases 4.5°C.
   (D) decreases 45°C.

6. Hydrogen gas and fluorine gas react to form hydrogen fluoride, HF. Calculate the enthalpy change (in kJ) for the conversion of 15.0 g of $H_2(g)$ to $HF(g)$ at constant pressure.

$$H_2(g) + F_2(g) \longrightarrow 2HF(g)$$
$$\Delta H = -536 \text{ kJ}$$

Use the graph and table to answer Questions 7–10. Assume 1.00 mol of each substance.

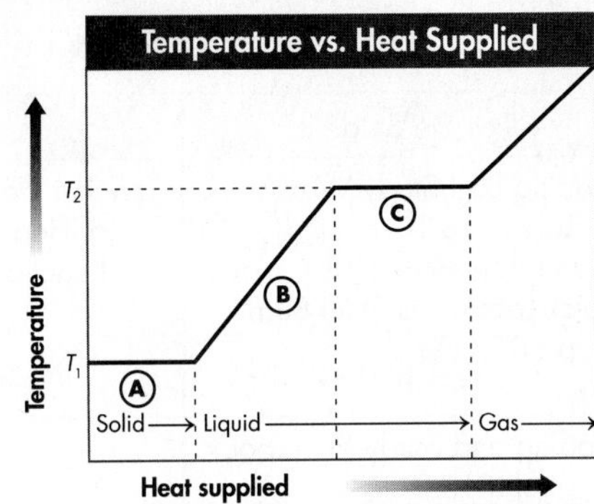

Substance	Freezing point (K)	ΔH_{fus} (kJ/mol)	Boiling point (K)	ΔH_{vap} (kJ/mol)
Ammonia	195.3	5.66	239.7	23.3
Benzene	278.7	9.87	353.3	30.8
Methanol	175.5	3.22	337.2	35.2
Neon	24.5	0.33	27.1	1.76

7. Calculate heat absorbed in region A for neon.
8. Calculate heat absorbed in region C for ammonia.
9. Calculate heat absorbed in regions B and C for methanol. [specific heat = 2.53 J/(g·°C)]
10. Calculate heat absorbed in regions A, B, and C for benzene. [specific heat = 1.74 J/(g·°C)]

If You Have Trouble With . . .

Question	1	2	3	4	5	6	7	8	9	10
See Lesson	17.3	17.1	17.4	17.4	17.1	17.2	17.3	17.3	17.3	17.3

ASSESSMENT

STP Answers

1. A
2. C
3. A
4. B
5. C
6. −3.98 × 10³ kJ
7. 0.33 kJ
8. 23.3 kJ
9. region B: 13.1 kJ; region C: 35.2 kJ
10. region A: 5.65 kJ; region B: 26.5 kJ; region C: 23.4 kJ

18 Reaction Rates and Equilibrium Planning Guide

Introducing the BIGIDEA: REACTIONS; MATTER AND ENERGY

Chemical reactions are important to the chemistry of living and nonliving things. The rate of a chemical reaction can be controlled by

IASS	Lessons and Objectives	Print Resources		
		For the Student	For the Teacher	
C.7.5	**18.1** Rates of Reaction p 594–601 **18.1.1** Describe how to express the rate of a chemical reaction. **18.1.2** Identify four factors that influence the rate of a chemical reaction.	Reading and Study Workbook Lesson 18.1 Lesson Assessment 18.1 p 601 Quick Lab: Does Steel Burn? p 600	Teacher Demo, p 599: Use of Heat and Catalyst in a Reaction	
C.7.5	**18.2** The Progress of Chemical Reactions p 604–608 **18.2.1** Describe the relationship between the value of the specific rate constant and the speed of a chemical reaction. **18.2.2** Describe how most reactions progress from start to finish.	Reading and Study Workbook Lesson 18.2 Lesson Assessment 18.2 p 608		
C.7.6	**18.3** Reversible Reactions and Equilibrium p 609–620 **18.3.1** Describe what happens at the molecular level in a chemical system at equilibrium. **18.3.2** Identify the three stresses that can cause a change in the equilibrium position of a chemical system. **18.3.3** Describe what the size of an equilibrium constant indicates about a system at equilibrium.	Reading and Study Workbook Lesson 18.3 Lesson Assessment 18.3 p 620	Teacher Demo, p 612: Banana Oil Demonstration Teacher Demo, p 615: Temperature and Equilibrium	
C.7.2	**18.4** Solubility Equilibrium p 621–626 **18.4.1** Describe the relationship between the solubility product constant and the solubility of a compound. **18.4.2** Predict whether a precipitation will occur when two solutions are mixed.	Reading and Study Workbook Lesson 18.4 Lesson Assessment 18.4 p 626	Class Activity, p 623: Solubility Tables Class Activity, p 625: Common Ion Effect	
	18.5 Free Energy and Entropy p 627–634 **18.5.1** Identify the two characteristics of spontaneous reactions. **18.5.2** Identify the part entropy plays in a chemical reaction. **18.5.3** Identify the two factors that determine whether a reaction is spontaneous. **18.5.4** Describe how the value of ΔG is related to the spontaneity of a reaction.	Reading and Study Workbook Lesson 18.5 Lesson Assessment 18.5 p 634 Small-Scale Lab: Enthalpy and Entropy p 635	Teacher Demo, p 631: The Entropy of Water Teacher Demo, p 633: Observing Spontaneous Reaction	

Assessing the BIGIDEA: REACTIONS; MATTER AND ENERGY

	Essential Questions 1. How can the rate of a chemical reaction be controlled? 2. What is the role of energy in chemical reactions? 3. Why do some reactions occur naturally?	Study Guide p 636 Math Tune-Up p 637 STP p 643 Reading and Study Workbook Self-Check and Vocabulary Review		

adjusting temperature, concentration, or particle size.

PearsonChem.com	
Editable Worksheets	**Online Activities**
Teaching Resources, Lesson 18.1 Review Lab 36: Factors Affecting Reaction Rates Lab 37: The Clock Reaction Lab Practical 18-1: Reaction Rate Small-Scale Lab Manual Lab 28: Factors Affecting the Rate of a Chemical Reaction	18.1 Lesson Overview Collision Theory
Teaching Resources, Lesson 18.2 Review	18.2 Lesson Overview
Teaching Resources, Lesson 18.3 Review Lab 38: Disturbing Equilibrium Small-Scale Lab Manual Lab 29: Le Châtelier's Principle and Chemical Equation	18.3 Lesson Overview Reversible Reactions Expressing and Calculating K_{eq}
Teaching Resources, Lesson 18.4 Review Lab 39: A Solubility Product Constant	18.4 Lesson Overview Finding the Ion Concentrations in a Saturated Solution Finding Equilibrium Ion Concentrations in the Presence of a Common Ion
	18.5 Lesson Overview Enthalpy and Entropy Entropy—Gone to the Dogs
Core Teaching Resources (includes editable worksheets) • Lesson Reviews • Practice Problems • Interpret Graphs • Vocabulary Review • Chapter Quizzes and Tests • Lab Record Sheets	Chapter 18 Problem Set Slope of a Curve

Materials List

For the Student

Quick Lab p 600
- #0000 steel-wool pad
- tissue paper
- tongs
- Bunsen burner
- heat-resistant pad

Small-Scale Lab p 635
- $NaCl$, NH_4Cl, $CaCl_2$
- thermometer
- 4 1-oz plastic cups
- plastic spoon
- crushed ice
- water
- stirring rod
- Optional solids: KCl, $NaHCO_3$, Na_3PO_4

For the Teacher

Teacher Demo p 599
- 10 mL ethanoic acid
- 10 mL isoamyl alcohol
- 2 test tubes
- anhydrous calcium sulfate
- sulfuric acid, water

Teacher Demo p 612
- 10 mL ethanoic acid
- 10 mL isoamyl alcohol
- 2 test tubes
- anhydrous calcium sulfate
- sulfuric acid
- water

Teacher Demo p 615
- 3 g copper turnings
- 3 heat-resistant test tubes
- 10 mL of 6*M* nitric acid
- one-hole stopper
- 2 solid stoppers
- glass delivery tube
- rubber tubing
- warm water bath
- ice water bath

Class Activity p 623
- calculator
- paper
- pencil
- colored markers

Class Activity p 625
- saturated solutions of $AgCl$ and $AgNO_3$
- 2 test tubes
- dropper

Teacher Demo p 631
- block of ice
- 2 beakers
- water • hot plate

Teacher Demo p 633
- 0.1*M* solution of $KMnO_4$ in 0.1*M* HCl
- graduated cylinder
- 1000-mL beaker
- thermometer
- 20 mL of 30% H_2O_2
- watch glass
- splint • matches

Additional Digital Resources

Online Student Edition
Online Teacher's Edition
Exam*View Assessment Suite CD-ROM*
Classroom Resources DVD
18.5 Virtual Chem Lab 20: Enthalpy and Entropy

CHAPTER 18

What's Online

CHEMISTRY TUTORIAL Students can work step-by-step with the online tutorial to learn how to calculate equilibrium constants and find equilibrium ion concentrations by accessing the Chapter 18 Chemistry Tutorials.

ONLINE PROBLEM SET Additional practice problems about reaction rates and equilibrium can be found in the Chapter 18 online Problem Set.

MATH TUTORIAL The online Math Tutorial will help guide students through finding the slope of a curve.

DIRECTED VIRTUAL LAB Encourage students to review the concepts of entropy and enthalpy by using the virtual lab for Chapter 18. The animation is followed by questions.

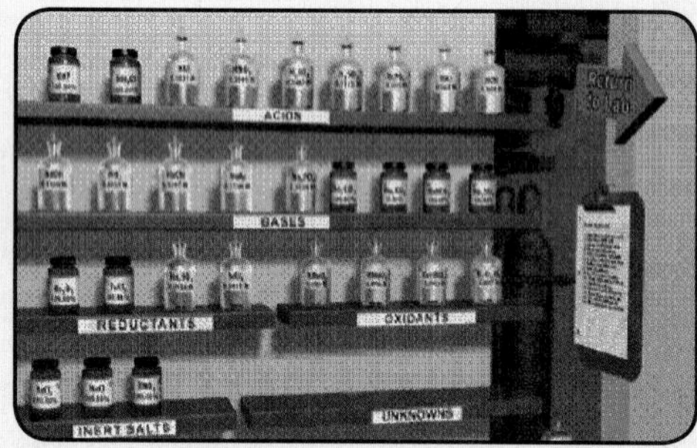

KINETIC ART Have students watch animations about collision theory and reversible reactions. Each animation is followed by analysis questions.

CONCEPTS IN ACTION Students can watch an interactive animation about how entropy can change. The animation is followed by questions.

18 Reaction Rates and Equilibrium

INSIDE:

PearsonChem.com

If the rate at which vehicles enter the city equals the rate at which vehicles exit the city, the number of vehicles in the city is constant. This situation represents a system in a state of balance. Similarly, chemical reactions can also reach a state of balance.

592

Focus on ELL

1 CONTENT AND LANGUAGE Write the title of the chapter and lesson titles on the board. Underline the word *equilibrium* in the appropriate titles. Explain that *equilibrium* is from the Latin *aequi-*, meaning "equal," and *libra*, meaning "balance." Have students use this information to predict what a chemical reaction at *equilibrium* means.

BEGINNING

LOW Have students use pictures or symbols to represent unfamiliar words from the titles. Rewrite the titles using these symbols in place of the words.

HIGH Have students use a bilingual dictionary or glossary to find any cognates to any unfamiliar words from the titles.

INTERMEDIATE: LOW/HIGH Have students write simplified English definitions of the titles and present them to the class.

ADVANCED: LOW/HIGH Have students write the definition of each word in each title and surmise to the class the purpose of the chapter and the main ideas of each lesson.

BIGIDEAS

- **REACTIONS**
- **MATTER AND ENERGY**

Essential Questions:

1. *How can the rate of a chemical reaction be controlled?*
2. *What is the role of energy in chemical reactions?*
3. *Why do some reactions occur naturally, and others do not?*

CHEMYSTERY

Explosive Sugar

At a sugar refinery, sugar is processed until it is pure enough to sell to consumers. Conveyor belts connect large storage silos to the packaging warehouse. In February 2008, workers at a refinery in Georgia heard a small explosion. The sound came from the location where the conveyor belts entered the warehouse.

Five minutes later, a much larger explosion occurred, which destroyed the warehouse. The fire spread to the storage silos. Thirteen workers died, and dozens more were injured. It took four days for 232 firefighters to put out the fire in the warehouse. The fire in the silos took a few days longer to extinguish.

How did the first small explosion lead to the second, more destructive explosion?

▶ Connect to the **BIGIDEA** As you read about reaction rates, think about what could cause an explosion at a sugar refinery.

INDIANA'S ACADEMIC STANDARDS FOR SCIENCE

Solutions C.7.2, C.7.5, C.7.6. See lessons for details.

Introduce the Chapter

IDENTIFY MISCONCEPTIONS Some students may not realize that some chemical reactions are reversible. Use the activity to introduce the concepts of reversible reactions and equilibrium.

ACTIVITY *You will need one or more light-sensitive or temperature-sensitive products, such as eyeglasses or toys.* Allow time for students to handle the products in different lighting or water baths of different temperatures to observe how the color changes in response to lighting conditions or temperature. Point out how the color gradually changes and adjusts to the level of light or temperature. **Ask** Why do you think the color changes? *(The ultraviolet light or heat causes a chemical change in the material.)* What causes the product to change back to its original color? *(A different chemical reaction occurs in the absence of heat or light.)* Point out that in these reactions, the degree of color change depends on a balance between the reaction that changes the color and the reverse reaction that changes the color back.

CHAPTER 18

UbD

Understanding by Design

Students are building toward understanding how matter and energy behave in chemical reactions.

PERFORMANCE GOALS At the end of Chapter 18, students will be able to answer the essential questions by applying their knowledge of reaction rates and equilibrium.

ESSENTIAL QUESTIONS Read the essential questions aloud. **Ask** How does raising the temperature of a substance affect particle movement? *(The particles move faster.)* How would this affect the rate of a chemical reaction? *(It would most likely increase the rate.)*

BIGIDEA Use the photo of vehicles to help students connect to this chapter's concepts. Activate prior knowledge by discussing traffic within a city. **Ask** Identify similar scenarios in which people move into and out of an area. *(Sample answer: students moving into and out of a lunchroom.)* **Ask** What would have to happen for the number of people in the area to remain unchanged? *(The number moving in must equal the number moving out.)*

CHEMYSTERY Connect the CHEMystery to the Big Idea by having students think about how the concentration of substances might relate to the explosions discussed in the CHEMystery. Explain that sugar dust is highly combustible. Point out that an explosion is the result of a combustion reaction, which is exothermic, and the energy released from an exothermic reaction can take more than one form. For example, explosions generate a shock wave that causes objects to vibrate. If these objects are covered in a layer of dust, the dust is released into the air. Have students predict how the release of a large cloud of sugar dust might make a second explosion more likely.

The Chemistry of Diabetes

Danni and Suze talk with a researcher at a clinic to learn how applying chemistry can help control diabetes.

LESSON 18.1

Key Objectives

DESCRIBE how to express the rate of a chemical reaction.

IDENTIFY four factors that influence the rate of a chemical reaction.

Additional Resources

- Reading and Study Workbook, Lesson 18.1
- Probeware Laboratory Manual
- Virtual Chemistry Labs

PearsonChem.com

Engage

CHEMISTRY & YOU Explain that all chemical reactions occur at rates specific to the nature of the reactants. **Ask** How can the rate of the corrosion of iron-magnesium alloy be increased? *(by adding salt water)* **Ask** What do you think can be added to the iron-magnesium alloy to increase the speed at which rusting occurs? *(add salt water)*

Build Background

Brainstorm reasons for the length of time it takes for a person to undergo a chemical hair treatment. **Ask** Do you know why there is a waiting period? *(It takes time for the chemical color or other treatment to affect the hair.)* **Ask** Why do you sometimes sit under a hair dryer? *(To speed up the process.)* Explain to the class that a hair coloring process involves a chemical reaction. The rate of the reaction depends on the initial characteristics of the client's hair as well as a variety of factors.

IN ACADEMIC STANDARDS for SCIENCE

- Develop explanations based on reproducible data and observations gathered during laboratory investigations.

C.7.5 Explain how the rate of a reaction is qualitatively affected by changes in concentration, temperature, surface area, and the use of a catalyst.

18.1 Rates of Reaction

C.7.5 Factors affecting rate of reaction

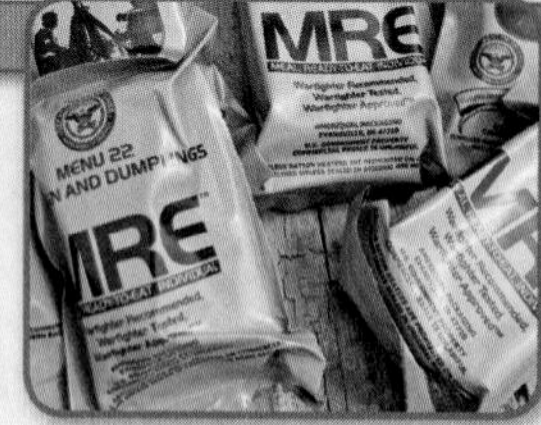

CHEMISTRY & YOU

Q: *How can rusting be used to cook a meal?* Sometimes a soldier or hiker wants a hot meal but has no place to cook it. Normally, rusting takes place at such a slow rate that the heat released could not be used to cook a meal. However, there are products that use the rusting of an iron-magnesium alloy to heat packaged food. These products are known as Meals Ready to Eat, or MREs for short.

Key Questions

How is the rate of a chemical reaction expressed?

What four factors influence the rate of a chemical reaction?

Vocabulary

- rate
- collision theory
- activation energy
- activated complex
- inhibitor

Describing Reaction Rates

How is the rate of a chemical reaction expressed?

When you strike a match, it erupts into flame almost instantly and burns quickly. Other reactions occur more slowly. For example, millions of years were required for plants buried beneath Earth's surface to be converted to coal. These examples show that the speed of chemical reactions can vary from very fast to extremely slow.

The concept of speed is a familiar one. In a race, speed determines the winner. The swimmers in Figure 18.1 are competing in a 100-meter race. (The pool is 25 meters long, so they must swim four lengths of the pool to finish the race.) A world-class swimmer might finish the race in 54.5 seconds. His average speed can be calculated by using the following equation:

$$\text{Average speed (m/s)} = \frac{\text{Distance (m)}}{\text{Time (s)}}$$

The swimmer's average speed, which is expressed in meters per second, is 100 m/54.5 s = 1.83 m/s. A slower swimmer might take 60.0 seconds to cover the same distance. He swims at an average speed of 1.67 m/s. Both 1.83 m/s and 1.67 m/s express rates of travel.

Figure 18.1 Speed
Speed is a change in distance over a given interval of time. A swimmer's average speed can be expressed in units of m/s.

Focus on ELL

1 CONTENT AND LANGUAGE Write the words *rate* and *speed* on the board. Explain that these words are sometimes used interchangeably in everyday conversation but do not have the same meaning in science. Direct students' attention to the paragraph that explains the term *rate*. Emphasize that in science, referring to the *speed* of a reaction would be inappropriate, because speed is a change in distance over time.

2 FRONTLOAD THE LESSON Preview the figures in the lesson. Point out the use of the words *rate, progress,* and *change* in the titles, labels, and captions to reinforce the concept that this lesson is about change over time. As a class, brainstorm a list of other familiar rates and write them on the board.

3 COMPREHENSIBLE INPUT When discussing collision theory, demonstrate the clay ball analogy described in the text.

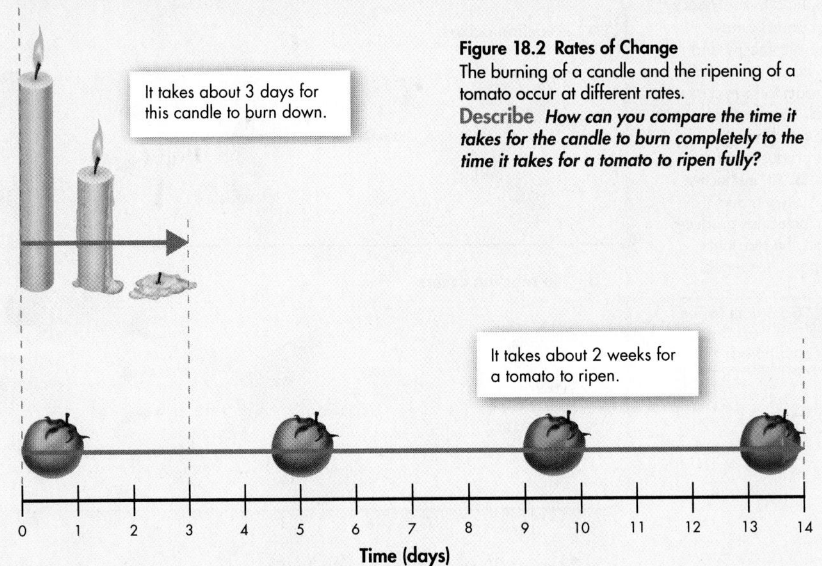

Figure 18.2 Rates of Change
The burning of a candle and the ripening of a tomato occur at different rates.
Describe ***How can you compare the time it takes for the candle to burn completely to the time it takes for a tomato to ripen fully?***

Rates of Change Speed is an example of a rate. A **rate** is a measure of how much something changes within a specified amount of time. The interval of time can range from less than a second to centuries. Figure 18.2 compares the rate of two changes—the burning of a candle and the ripening of a tomato.

In chemistry, the rate of a chemical reaction, or the reaction rate, is usually expressed as the change in the amount of reactant or product per unit time. The amount of a reactant is usually expressed in moles. For example, if one half of a 1-mole piece of iron turns to rust in one year, the rate at which iron rusts might be expressed as 0.5 mol/yr. This number is an average rate. Figure 18.3 illustrates the progress of a typical reaction. Over time, the amount of reactant decreases and the amount of product increases.

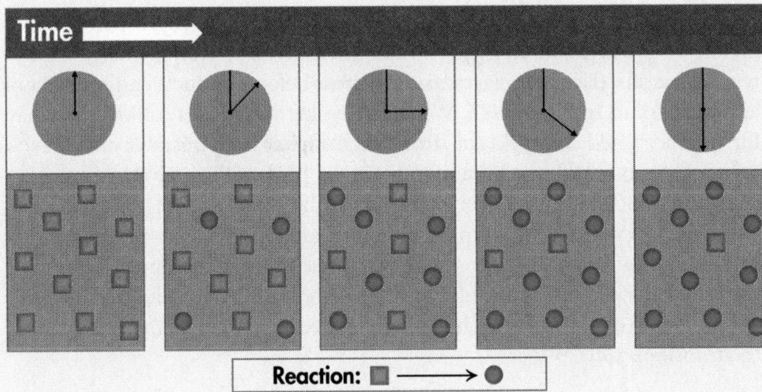

Figure 18.3 Reaction Progress
During a reaction, reactants are converted into products. The red squares represent the reactants. The blue circles represent the products. Assume that each box represents the same time interval.
Interpret Diagrams ***Is the rate of conversion of reactant to product constant throughout the reaction? How can you tell?***

UbD Check for Understanding

How is the rate of a chemical reaction expressed?

Assess students' understanding of the expression of reaction rate as a change in the amount of reactant per unit time by having students give a thumbs-up sign if they understand the concept and a thumbs-down if they do not.

ADJUST INSTRUCTION If students indicate that they do not understand the concept, direct their attention to Figure 18.3. Demonstrate mathematically that if the time elapsed for the reaction is 0.5 minute and each red square represents 1 mole reactant, then the reactant is decreasing at an average rate of 10 mol/0.5 minute, or 0.33 mol/s.

Foundations for Reading

BUILD VOCABULARY Have students create a concept map around the term *collision theory*. They should complete the map with remaining vocabulary terms as they read the lesson.

READING STRATEGY Have students create a flow chart for each factor that affects reaction rates. The chart should show the factor, its direct effect, and its effect on the rate of reaction.

Explain

Describing Reaction Rates

START A CONVERSATION Point out that the time required for a chemical change to take place may be less than 10^{-6} seconds or more than 10^6 years. Draw three vertical columns on the board and label the headers of each column "very fast," "average," and "very slow." Have the class brainstorm scenarios of chemical changes that fit into each category. If students are having trouble, guide students to think of common changes that occur within their own homes, for instance, how quickly bleach removes mold from a shower stall or how quickly it removes coloring from an article of clothing, or how quickly milk spoils if kept in the refrigerator versus being left outside on a warm day, or how long it takes for grass clippings to decompose in a compost pile.

USE VISUALS Direct students to study Figure 18.3. **Ask** As you move from the diagram on the far left to the diagram on the far right, how does the concentration of the reactant change? *(It decreases.)* Explain that the rates of chemical reactions are often measured by the decrease in the concentration of one of the reactants or the increase in the concentration of one of the products. **Ask:** What happens to the reaction rate as the concentration of reactants decreases? *(It slows down.)*

Answers

FIGURE 18.2 You can compare the two rates by dividing one rate by the other rate expressed in the same units: 14/3 = 4.7 (The candle burns about 4.7 times as fast as the tomato ripens.) or 3/14 = 0.21

FIGURE 18.3 No; The number of additional blue circles—the products—decreases in each successive time interval.

LESSON 18.1

Explain

USE VISUALS Direct students to Figure 18.4. Point out that in both (a) and (b) the particles collide. But only in (a) do the particles actually stick together. Stress that for particles to react and form bonds, three events must occur:

1. The particles must collide.
2. The particles must collide at the proper orientation.
3. The particles must have enough kinetic energy to cross the activation energy barrier.

Misconception Alert

Students may have difficulty accepting collision theory because they tend to visualize atoms and molecules wedged together with no spaces in between them, as shown in space-filling models. Remind students that in Chapter 13 they learned that, according to kinetic theory, all matter is made of tiny particles that are in constant motion, and that between the particles there is empty space.

USE AN ANALOGY Relate the activation energy barrier to a hill covered in snow. Tell the students to imagine they are rolling a large snowball up the hill. Once they get to the top of the hill, they will push the snowball over the hill. Explain that when they first begin to move the snowball, very few students are needed to move it because they are moving over flat ground, but as they begin moving up the hill, more students must join in to continue moving the snowball forward. **Ask** What would happen if enough people are not available to push the snowball all the way to the top of the hill? *(The snowball would not go over the hill.)* **Ask** How is the least number of students needed to push the snowball over the top of the hill analogous to the activation energy of a reaction? *(The activation energy is the minimum amount of energy needed for the reactants to form an activation complex.)*

Figure 18.4 Collision Theory
If colliding particles have enough kinetic energy and collide at the right orientation, they can react to form a new product. **a.** An effective collision of oxygen and hydrogen molecules produces water molecules. **b.** An ineffective collision of oxygen and hydrogen molecules produces no reaction; the reactants bounce apart unchanged.

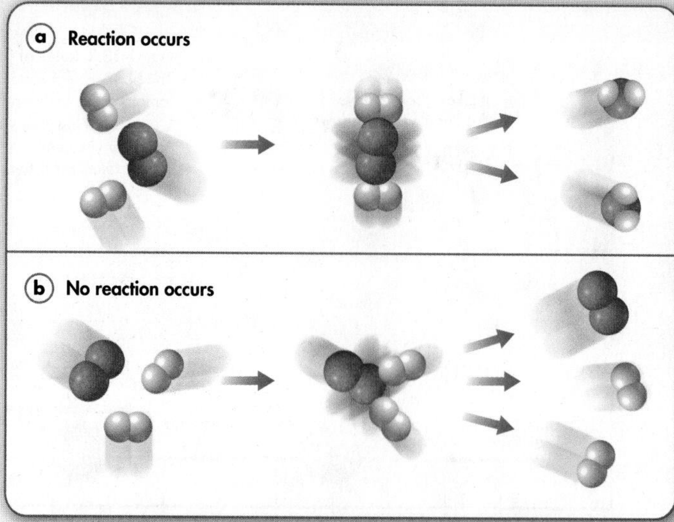

Collision Theory A model called collision theory is used to relate the properties of particles to the rates of reactions. According to **collision theory,** atoms, ions, and molecules can react to form products when they collide if the particles have enough kinetic energy. Particles that do not have enough energy to react bounce apart unchanged when they collide. Figure 18.4 shows what can happen when hydrogen and oxygen molecules collide.

You can use two balls of soft modeling clay to illustrate collision theory. If you throw the balls of clay together gently, they don't stick to one another. This situation is analogous to colliding particles of low energy that fail to react. If the same balls of clay are thrown together with great force, they will stick tightly to each other. This situation is analogous to the collision of two high-energy particles that results in the formation of a product.

You can also use modeling clay to illustrate another point about chemical reactions. Roll the clay into a rope and begin to shake one end more and more vigorously. At some point, the clay rope will break. Similarly, if enough energy is applied to a molecule, the bonds holding the molecule together can break, forming different products.

The minimum energy that colliding particles must have in order to react is called the **activation energy.** You can think of the activation energy for a reaction as a barrier that reactants must cross before products can form. Look at the diagrams in Figure 18.5. When two reactant particles collide, they may form an activated complex. An **activated complex** is an unstable arrangement of atoms that forms for a moment at the peak of the activation-energy barrier. The activated complex forms only if the colliding particles have enough energy and if the atoms are oriented properly. The lifetime of an activated complex is typically about 10^{-13} seconds. Its brief existence ends with the reformation of the reactants or with the formation of products. The two outcomes are equally likely. Thus, the activated complex is sometimes called the *transition state.*

596 Chapter 18 • Lesson 1

Differentiated Instruction

ELL ENGLISH LANGUAGE LEARNERS Verify students' understanding of the terminology in the questions in Figure 18.5 by having students restate the questions in their own words.

L1 SPECIAL NEEDS Create a model of colliding molecules with fuzzy, lightweight balls and a hook and loop backboard. Tell students that if they throw the balls hard enough, the "molecules" will stick to the backboard or *react*. If they miss or throw with too little kinetic energy, the molecules will not *react*.

L3 ADVANCED STUDENTS Tell students that collision theory uses two-particle collisions to explain reaction mechanisms because the probability of three or more particles colliding simultaneously is unlikely. Have students research typical reactions that require the collision of three or more particles and summarize their findings in a report that they present to the class.

Collision theory explains why some reactions are extremely slow at room temperature. For example, carbon and oxygen react when charcoal burns, but the reaction has a high activation energy. The O—O bonds and C—C bonds must be broken to form the activated complex. At room temperature, the collisions of oxygen and carbon molecules are not energetic enough to break the bonds. Thus, the reaction rate of carbon with oxygen at room temperature is essentially zero.

Interpret Graphs

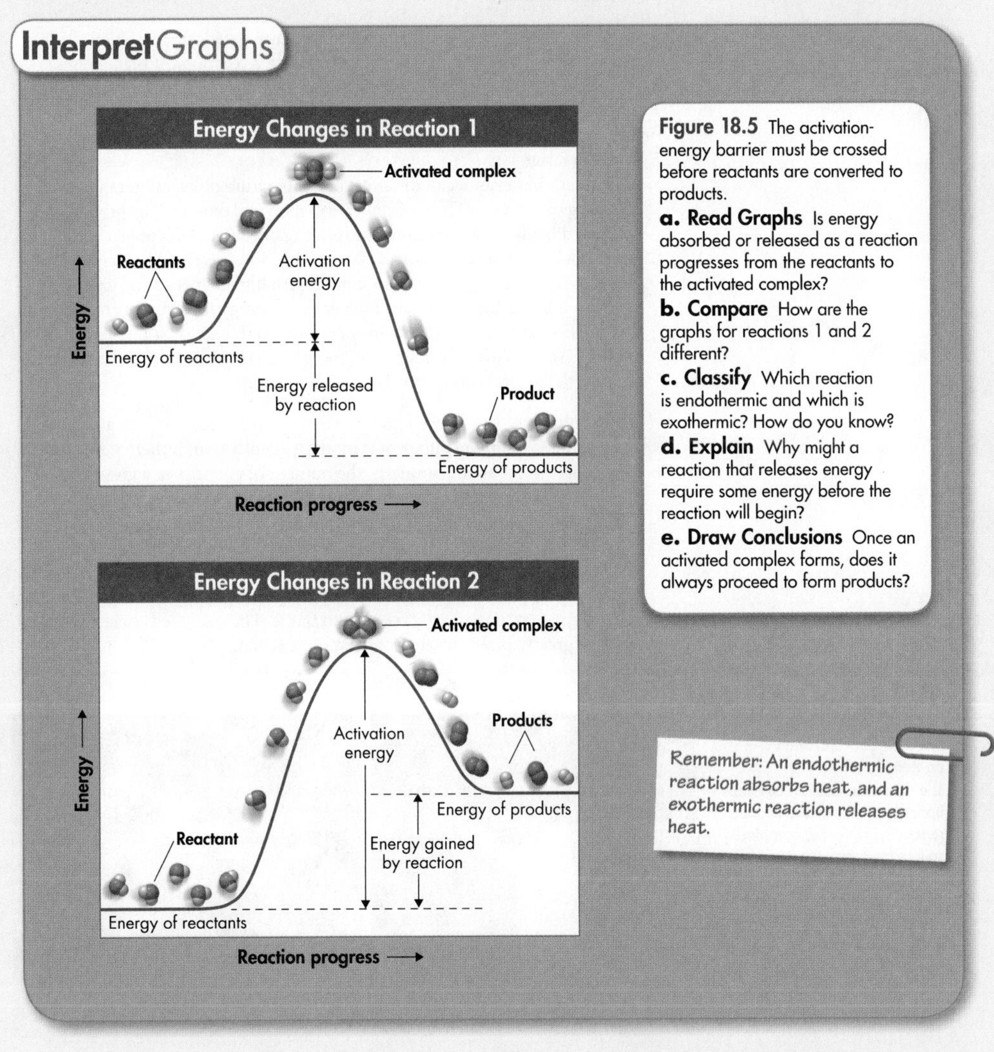

Figure 18.5 The activation-energy barrier must be crossed before reactants are converted to products.

a. Read Graphs Is energy absorbed or released as a reaction progresses from the reactants to the activated complex?

b. Compare How are the graphs for reactions 1 and 2 different?

c. Classify Which reaction is endothermic and which is exothermic? How do you know?

d. Explain Why might a reaction that releases energy require some energy before the reaction will begin?

e. Draw Conclusions Once an activated complex forms, does it always proceed to form products?

USE VISUALS Direct students' attention to Figure 18.5. Point out that the diagram graphically depicts the energy changes that occur during a chemical reaction. Explain that the activation-energy hill must be crossed before reactants can be converted to products. **Ask** How does a large activation energy affect reaction time? *(It causes reaction time to increase, slowing the reaction.)* Explain that when the activation energy is great, fewer molecules are colliding with enough energy to form an activated complex at any one time, so it takes longer for the reaction to proceed to completion.

CRITICAL THINKING Have students consider whether changing the temperature of the reactants will affect the size of the activation energy barrier. *(No; an increase in temperature will only increase the number of reactants having the activation energy needed to react.)* Tell students that they will explore this concept in more detail shortly.

UbD Check for Understanding

BIGIDEA MATTER AND ENERGY Have each student write a one-minute summary explaining the role of activation energy in a chemical reaction. Tell students they can include a graph or an analogy to help them convey the information.

ADJUST INSTRUCTION If students are having difficulty with this concept, guide them in a step-by-step review of Figure 18.5.

Answers

INTERPRET GRAPHS

a. absorbed
b. The graph for reaction 2 is the opposite of the graph for reaction 1.
c. Reaction 1 is exothermic because energy (in the form of heat) is released. Reaction 2 is endothermic because energy (in the form of heat) is gained.
d. An activated complex, which has a higher energy, needs to be formed before the reaction will proceed.
e. No; it could also revert to the reactants.

Reaction Rates and Equilibrium 597

Explain

Factors Affecting Reaction Rates

CRITICAL THINKING Point out that an increase in temperature causes an increase in average kinetic energy. Explain that the number of collisions increases because the molecules are moving faster. More important, a greater number of the collisions are effective. **Ask** How do refrigerators prevent food from spoiling? *(The kinetic energy of reactants is reduced, so fewer particles possess sufficient energy to react. As a result, reactions that lead to spoilage take longer to occur.)*

Extend

Connect to FOOD CHEMISTRY

One of the key reasons that food spoils, other than microbial activity, is that many foods contain compounds that are highly reactive with oxygen. When these compounds react with oxygen, they oxidize. This oxidation causes the food to become rancid and turn brown. Even though refrigeration slows these reactions, food chemists also rely on a class of compounds called *antioxidants* to further slow food spoilage. Some fresh foods are natural sources of antioxidants, which gives them a slightly longer shelf life than those which are low in these compounds. Many processed foods have these substances added during the manufacturing process.

Have students research antioxidants and create posters that explain what antioxidants are, how they work, and that identify natural and processed foods that contain antioxidants.

Factors Affecting Reaction Rates

What four factors influence the rate of a chemical reaction?

Every chemical reaction proceeds at its own rate. Some reactions are naturally fast, and some are naturally slow under the same conditions. However, by varying the conditions, you can modify the rate of almost any reaction. **Factors that can affect the rate of a chemical reaction are temperature, concentration, particle size, and the use of a catalyst.** Collision theory helps explain why changing one or more of these factors may affect the rate of a chemical reaction.

Temperature Usually, raising the temperature speeds up a reaction. Lowering the temperature usually slows down a reaction. At higher temperatures, particles move faster. The frequency of collisions increases along with the percentage of particles that have enough kinetic energy to slip over the activation-energy barrier. Thus, an increase in temperature causes products to form faster.

A familiar example of the effect of temperature on reaction rate is the burning of charcoal. The reactants are carbon and oxygen. The product is carbon dioxide. At room temperature, a bag of charcoal in contact with air does not burn. However, when a starter flame touches the charcoal, carbon atoms and oxygen molecules collide with higher energy and greater frequency. Some collisions are at a high enough energy to form the product. The heat released by the reaction then supplies enough energy to get more carbon and oxygen over the activation-energy barrier. When the starter flame is removed, the reaction continues.

Concentration In a crowded room where people are moving about, you may find yourself bumping into people more frequently than if there were only a few people in the room. Similarly, the number of particles in a given volume affects the rate at which reactions occur. Cramming more particles into a fixed volume increases the concentration of reactants and, thus, the frequency of collision. Increased collision frequency leads to a higher reaction rate.

The lighted splint in Figure 18.6 glows in air and soon dies out because air is only 20 percent oxygen. But when the glowing splint is plunged into pure oxygen, it immediately bursts into flame. The increased concentration of oxygen greatly speeds up the combustion reaction.

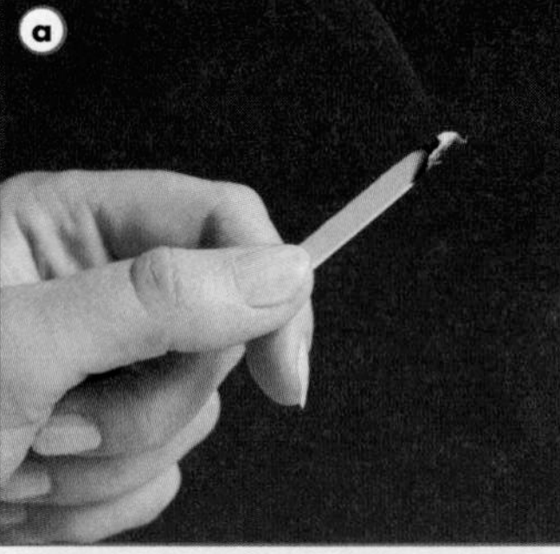

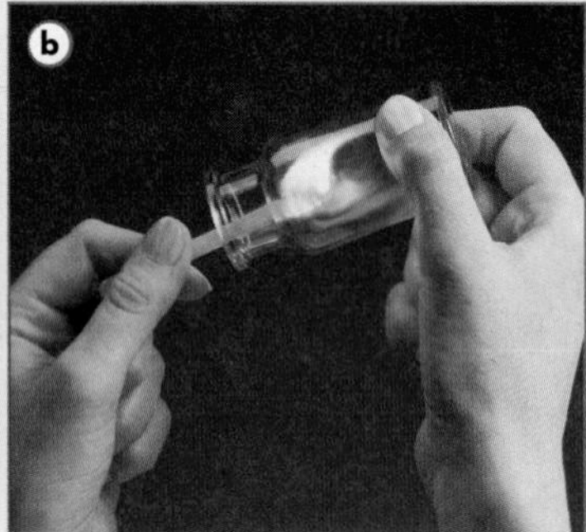

Figure 18.6
Effect of Concentration on Reaction Rate
The rate of a reaction depends upon the concentrations of the reactants. **a.** In air, a lighted splint glows and soon goes out. **b.** When placed in a vial containing pure oxygen, the splint bursts into flame.
Infer ***In areas where oxygen tanks are used, why do warning signs say "No smoking. No open flames"?***

Northern Lights

The term *northern lights* is commonly used to refer to the aurora borealis, an astronomical spectacle that happens as a result of electrically charged particles reacting with gases in the upper atmosphere. The particles that help to produce this colorful display of light originate from the massive nuclear reactions that occur in the sun. Highly energized fragments of atoms released from these reactions travel at extreme rates of speed and bombard the upper atmosphere, causing some of the nitrogen molecules to dissociate. When two excited nitrogen atoms recombine and produce the fluorescence of molecular nitrogen, the green, blue, or purple lights of the aurora borealis are the result. Scientists have tried with no success to recreate these colored lights in the laboratory. So far, there is no amount of thermal energy available under laboratory conditions equal to the energy of the bombarding charged particles released by the sun.

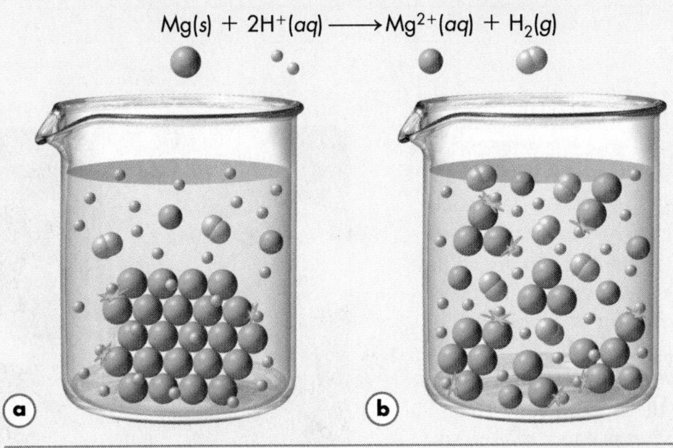

Figure 18.7 Effect of Particle Size on Reaction Rate
When a piece of magnesium is placed in a dilute acid, hydrogen ions can collide with magnesium atoms. **a.** Only atoms at the surface of the metal are available for reaction. **b.** Dividing the metal into smaller pieces increases the surface area and the number of particle collisions.
Explain ***How does increasing the number of collisions increase the reaction rate?***

Particle Size If you put a bundle of sticks on a fire, they burn quickly. A log with the same mass burns more slowly. The small pieces of wood have more surface area than the log, and surface area plays an important role in determining the rate of the combustion reaction. The same is true for other chemical reactions because collisions occur at the surface of particles.

The total surface area of a solid or liquid reactant affects the rate of a reaction. The smaller the particle size, the greater the surface area is for a given mass of particles. Figure 18.7 shows how a decrease in particle size affects the amount of a reactant exposed for reaction. When a piece of metal is divided into several smaller pieces, the total surface area increases. The result of an increase in surface area is an increase in the frequency of collisions and the reaction rate.

Another way to increase the surface area of solids is to dissolve them. In a solution, particles are separated and more accessible to other reactants. You can also increase the surface area of a solid by grinding it into a fine powder. Small dustlike particles, however, can be dangerous when suspended in air. As coal miners know, coal dust mixed with air is an explosive hazard because of the large surface area of the coal dust particles. The same risk exists in flour mills, grain elevators, and sugar refineries. The photograph in Figure 18.8 shows a sugar refinery in Georgia after such an explosion.

Figure 18.8 Dust Explosion
An explosion destroyed this sugar refinery. The tiny size of the reactant particles (sugar dust) caused the reaction of sugar with oxygen in the air to be explosive.

Kinetics

In 1903, Svante Arrhenius, a Swedish chemist associated with two concepts present in most kinetic theories, *activation energy* and the *Arrhenius equation*, became the first Swede to be awarded the Nobel Prize in chemistry. The Arrhenius equation,

$$E_a = -RT \ln\left(\frac{k}{A}\right)$$

quantifies the relationship between the temperature dependence on the rate constant k and the activation energy E_a for any given chemical reaction. Arrhenius suggested that when a reaction occurs at an absolute temperature T, a proportional relationship exists between the fraction of molecules that have a kinetic energy greater than E_a and the probability distribution of molecular speed, otherwise known as the Maxwell-Boltzmann distribution of statistical mechanics.

Explore

Teacher Demo

PURPOSE Students observe the effect of heat and a catalyst on the rate of a chemical reaction.

MATERIALS 10 mL ethanoic acid, 10 mL isoamyl alcohol, 2 test tubes, anhydrous calcium sulfate, concentrated sulfuric acid, water

PROCEDURE Mix 5 mL ethanoic acid *(acetic acid)* and 5 mL isoamyl alcohol in each of two test tubes. Add a few granules of anhydrous calcium sulfate to each test tube. To one test tube, add three drops of concentrated sulfuric acid and heat this test tube in a boiling water bath for a few minutes. Write the general chemical equation for the formation of an ester on the board.

(organic acid + alcohol → ester + water)

Inform students that esters are responsible for the characteristic odors of many fruits and may be prepared by reacting an organic acid with an alcohol. Have students smell the odor emanating from both test tubes by carefully wafting the odor towards their noses. **Ask** What do you observe? *(The heated solution containing sulfuric acid smells like bananas while the other solution has a strong odor of vinegar (ethanoic acid).)* Tell students that a change in odor is a sign of a reaction. **Ask** What purpose does the boiling water bath serve in this reaction? *(It increases the temperature of the reactants, which speeds up the reaction rate.)* **Ask** What is the purpose of the sulfuric acid in this reaction? *(The sulfuric acid is a catalyst for the reaction; it speeds up the rate of reaction.) You can allow students to smell some diluted sulfuric acid to demonstrate that it is not responsible for the odor.*

EXPECTED OUTCOME

The solution containing sulfuric acid smells like bananas after being heated, whereas the other solution smells like vinegar, or ethanoic acid.

Answers

FIGURE 18.6 The concentration of the reactant, oxygen, is increased.

FIGURE 18.7 More frequent collisions mean more frequent reactions between colliding particles. The percentage of particles that have enough energy to react remains constant, but the number of successful collisions in a given time interval is greater.

LESSON 18.1

Explore

Quick Lab

OBJECTIVE After completing this activity, students will be able to relate particle size or surface area to the rate of a reaction and write a balanced equation for the reaction.

SKILLS FOCUS observing, writing equations, inferring

PREP TIME 20 minutes

CLASS TIME 20 minutes

SAFETY Steel wool burns readily. Use proper safety precautions.

EXPECTED OUTCOME

Tight ball: except for the glowing of a few loose steel fibers on the surface, there are no apparent changes; loose ball: glows; loose fibers: burn readily.

ANALYZE AND CONCLUDE

1. Except for the glowing of a few loose steel fibers on the surface, no apparent changes occurred when the tight ball was heated. The loose ball glowed. The loose fibers burst into flame almost immediately.
2. The ball acts like a single large particle of steel that is in contact with atmospheric oxygen only at its relatively small surface. The loose fibers act like many small particles, with larger total surface area, and each small particle is exposed to atmospheric oxygen.
3. $4Fe(s) + 3O_2(g) \rightarrow 2Fe_2O_3(s)$
4. The combustion of the steel wool is faster because the high temperature provides the energy for the iron and oxygen to overcome the activation-energy barrier. The small particle size exposes more of the iron surface to atmospheric oxygen.
5. The steel wool may ignite.

Quick Lab

Purpose To determine whether steel will burn

Materials

- #0000 steel wool pad
- tongs
- gas burner
- heat-resistant pad
- tweezers

Does Steel Burn?

Procedure

1. Roll a small piece of steel wool into a very tight, pea-sized ball.
2. Holding the ball with tongs, heat the steel wool in the blue-tip flame of the burner for no longer than 10 seconds. **CAUTION** *Observe all precautions for working with flames.*
3. Place the heated steel wool on the heat-resistant pad to cool. Record your observations.
4. Gently roll a second small piece of steel wool into a loose ball. Repeat Steps 2 and 3.

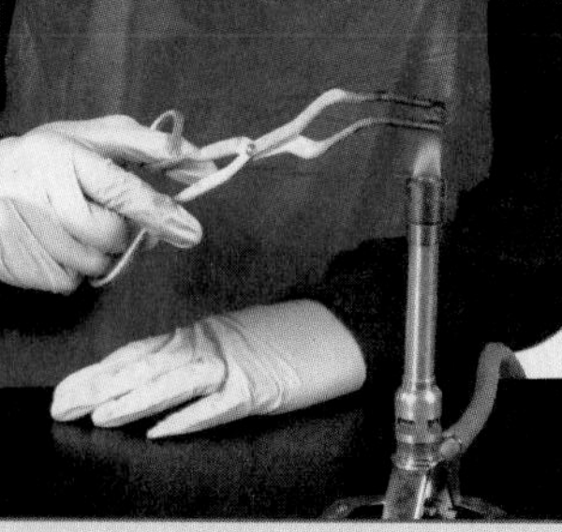

5. Use tweezers to pull a few individual fibers of steel wool from the pad. Hold one end of the loose fibers with the tongs. Repeat Steps 2 and 3.

Analyze and Conclude

1. **Observe** What differences did you observe when the tight ball, the loose ball, and the loose fibers were heated in the flame?
2. **Relate Cause and Effect** Give a reason for any differences you observed.
3. **Describe** Write the balanced equation for any chemical reaction you may have observed. Assume that the steel wool is composed mainly of iron.
4. **Compare and Contrast** How do your results differ from those observed in the rusting of an automobile body?
5. **Apply Concepts** Explain why steel wool is a hazard in shops where there are hot plates, open flames, or sparking motors.

CHEMISTRY & YOU

Q: *When salt water is added to the metal alloy in an MRE, the rate of the rusting reaction increases, and heat is produced rapidly. Which factor that can affect reaction rates is being applied in this situation?*

Catalysts Increasing the temperature is not always the best way to increase the rate of a reaction. A catalyst is often better. Recall that a catalyst is a substance that increases the rate of a reaction without being used up during the reaction. Catalysts permit reactions to proceed along a lower energy path. Figure 18.9 shows that the activation-energy barrier for the catalyzed reaction is lower than that of the uncatalyzed reaction. When the barrier is lower, a greater fraction of reactants have the energy to form products within a given time. For instance, the rate of the reaction of hydrogen and oxygen at room temperature is negligible. But with a small amount of platinum (Pt) as a catalyst, the reaction is rapid.

$$2H_2(g) + O_2(g) \xrightarrow{Pt} 2H_2O(l)$$

A catalyst is not consumed during a reaction. Therefore, it does not appear as a reactant in the chemical equation. Instead, the catalyst is often written above the yield arrow, as in the equation above.

Focus on ELL

4 LANGUAGE PRODUCTION Have students work in groups of four to complete the lab. Make sure each group has ELLs of varied language proficiencies, so that more proficient students can help less proficient ones. Have students work according to their proficiency level.

BEGINNING: LOW/HIGH Accept drawings, single words, or short phrases as answers to the questions. Have students copy the written work completed by more proficient students for future reference.

INTERMEDIATE: LOW/HIGH Create and distribute a chart with three columns, titled "tight," "loose," and "loose fibers" for students to record their observations.

ADVANCED: LOW/HIGH Have students assist and/or edit the work of classmates with lower language proficiency.

InterpretGraphs

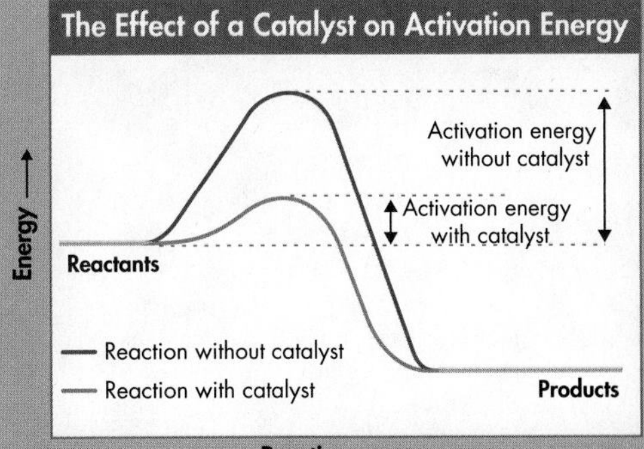

Figure 18.9 A catalyst lowers the activation-energy barrier for a reaction. With a lower barrier, more particles have enough energy for a successful collision.
a. Read Graphs How does using a catalyst change the amount of energy needed for products to form?
b. Make Generalizations Does the catalyst change the amount of energy released in the reaction?
c. Compare Along which of the two reaction paths are reactants converted more rapidly to products?

Hint: The reactant particles must have an energy equal to or greater than the activation energy to be converted to products.

Normal body temperature is only about 37°C. Your body needs to maintain this temperature to avoid damage to cells. At 37°C, reactions in the body would be too slow without catalysts. The catalysts that increase the rates of biological reactions are called enzymes. When you eat a meal containing protein, enzymes in your digestive tract help break down the protein molecules in a few hours. Without enzymes, the digestion of protein at 37°C would take many years!

An **inhibitor** is a substance that interferes with the action of a catalyst. Some inhibitors work by reacting with, or "poisoning," the catalyst itself. Thus, the inhibitor reduces the amount of catalyst available for a reaction. Reactions slow or even stop when a catalyst is poisoned.

18.1 LessonCheck

1. **Explain** How can you express the rate of a chemical reaction?
2. **Identify** List the four factors that can influence the rate of a chemical reaction.
3. **Describe** A thin, 0.2-mol sheet of zinc is completely converted to zinc oxide (ZnO) in one month. How would you express the rate of conversion of the zinc?
4. **Summarize** Does every collision between reacting particles lead to products? Explain.
5. **Relate Cause and Effect** Food stored in a refrigerator can stay fresh for long periods. Why does the same food stored at room temperature quickly spoil?

BIGIDEA MATTER AND ENERGY

6. Make a Venn diagram with two circles. Label one Matter and the other Energy. Choose a location in the diagram for each factor that can influence the rate of a reaction. Write a paragraph explaining your choices.

Lesson Check Answers

1. The rate of a chemical reaction is the amount of reactant changing per unit of time.
2. temperature, concentration, particle size, use of a catalyst
3. Rate = 0.2 mol Zn/month
4. No, the collision must have sufficient energy to break and form bonds.
5. Chemical reactions involved in food spoilage occur faster at higher temperatures because more energy is available.
6. **BIGIDEA** Concentration and particle size should be in the circle labeled *Matter* because each of these factors increase the reactant itself that is available for collisions. Temperature and catalysts should be in the circle labeled *Energy* because each of these factors manipulate the energy of the reactant or the energy of the activated complex, respectively.

Explain

USE VISUALS Direct students to Figure 18.9. **Ask** How would adding an inhibitor affect both graphs? *(The activation energy would be higher than the one shown in the graph for the reaction with a catalyst, but the inhibitor would have no effect on the reaction without a catalyst.)*

APPLY CONCEPTS Explain that home cooks often use inhibitors to poison catalysts on purpose. Have students consider what happens to the color of an apple after it is sliced and left exposed to air. *(The cut surfaces turn brown.)* Explain that the color change is the result of oxidation. The cut cells of the apple release an enzyme that increases the rate of oxidation. In order to prevent an apple from oxidizing, you would want to apply an inhibitor, such as ascorbic or citric acid, to the exposed surface.

Evaluate

Informal Assessment

Direct students' attention to Figure 18.5. Challenge students to use the activated complex theory to explain why hydrogen and oxygen do not react at room temperature but do combine explosively in the presence of a spark or flame. *(The explosive reaction indicates that the product—water—has less energy than the reactants. The reaction does not occur at room temperature because the reaction has a very high activation energy.)* Then have students complete the 18.1 Lesson Check.

Reteach

Have students revisit the section on Factors Affecting Reaction Rates. Remind students that chemical changes occur as a result of individual reactions between particles. Therefore, factors that affect collisions between particles are the factors that influence the rate of reaction. Help students to create and complete a table showing the relationship between increasing and decreasing each factor and the effect on the number of collisions. For visual learners, sketch example graphs of the activation complex for each of the four factors.

Answers

INTERPRET GRAPHS

a. It lowers the activation energy.
b. no
c. the line for reaction with a catalyst

CHEMISTRY & YOU

CHEMISTRY & YOU Draw students' attention to the diagrams and their labels. Pose the following question to students: *Why do you think catalytic converters work better in gasoline engines than diesel engines for reducing nitrogen oxides?* You may need to guide students in the following ways:

- Gasoline exhaust has a higher temperature than diesel exhaust.
- Diesel fuel contains sulfur, which can poison the catalyst in the converter.
- Catalytic converters need high temperatures to properly reduce nitrogen oxides.

CHEMISTRY & YOU: TECHNOLOGY

Catalytic Converters

Vehicles that run on fossil fuels are a major source of air pollution. One technology that significantly reduces air pollution caused by vehicles is called a catalytic converter. The device keeps pollutants from being released into the air by converting them into less-harmful emissions. A typical catalytic converter is capable of eliminating about 98 percent of a vehicle's pollution—unreacted hydrocarbons, carbon monoxide, and nitrogen oxides. Precious metals such as platinum, palladium, and rhodium are catalysts for the reactions that occur inside the converter.

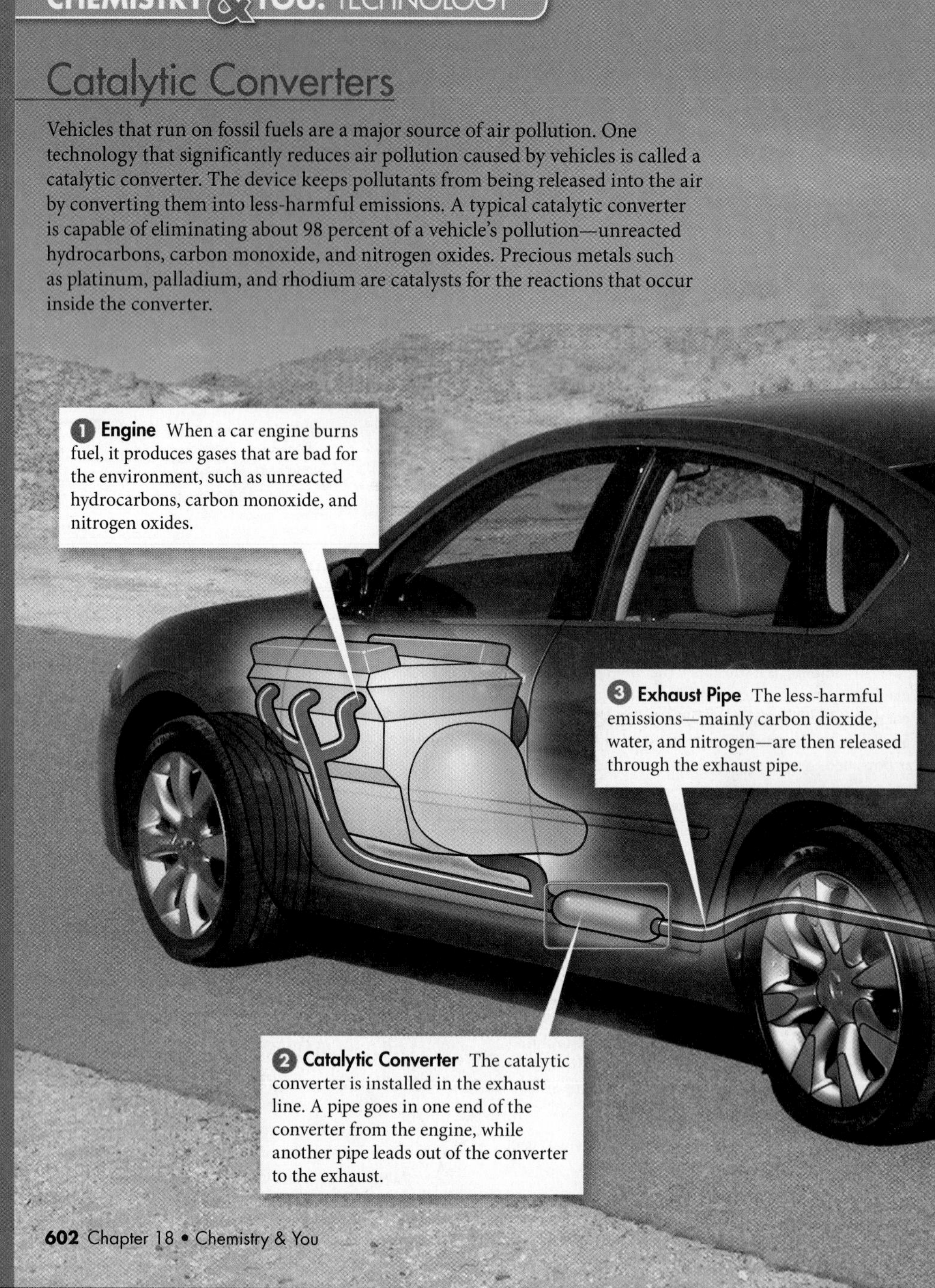

1 Engine When a car engine burns fuel, it produces gases that are bad for the environment, such as unreacted hydrocarbons, carbon monoxide, and nitrogen oxides.

2 Catalytic Converter The catalytic converter is installed in the exhaust line. A pipe goes in one end of the converter from the engine, while another pipe leads out of the converter to the exhaust.

3 Exhaust Pipe The less-harmful emissions—mainly carbon dioxide, water, and nitrogen—are then released through the exhaust pipe.

21st Century Learning To be successful in the 21st century, students need skills and learning experiences that extend beyond subject matter mastery. The following project helps students build the following 21st Century Skills: *Global Awareness*; *Civic Literacy*; *Critical Thinking and Problem Solving*; *Communication and Collaboration*; *Information, Media, and Technology Literacy*; *Initiative and Self-Direction*; and *Productivity and Accountability*.

ENVIRONMENTAL DEFENSE States have laws that require catalytic converters in all vehicles, but the details and penalties of the laws vary. Pose the following challenge to your students. *You have been asked to serve as an expert on a taskforce convened by the EPA to revise and toughen rules regarding installation of aftermarket catalytic converters. Create a slide presentation for the taskforce that does the following:*

- explains the history of the development of the first automotive catalytic converter
- provides a clear description of how catalytic converters work
- explains the difference between the three general types of converters and the dangers of installing the wrong converter on a vehicle
- summarizes the environmental impact of catalytic converter use and disposal.

Honeycomb Core Inside the catalytic converter, gases from the engine flow through a ceramic honeycomb-like structure that is coated with metal catalysts. The honeycomb design increases the surface area available to support the metal catalysts.

Take It Further

1. Infer Why does maximizing the surface area of the core increase the effectiveness of a catalytic converter?

2. Describe In a catalytic converter, platinum catalyzes a combination reaction between oxygen and carbon monoxide to form carbon dioxide. Write a balanced chemical equation for this catalyzed reaction.

3. Research a Problem Cars that are equipped with catalytic converters are designed to run on unleaded gasoline. A single tank of leaded gasoline can seriously impair the performance of the converter. Conduct an Internet search to determine the reason. Write a short paragraph to summarize what you learn.

Differentiated Instruction

L1 STRUGGLING STUDENTS Work together to draw a flow chart on the board to depict the steps involved in the production and conversion of exhaust from a vehicle. Have each student name a step in the process and explain why it is important.

LPR LESS PROFICIENT READERS Have students draw a T-chart in their vocabulary notebooks. In the left column, have them list five unfamiliar words in the feature. Have students use a dictionary to define the words and write each definition in the right column of the table, along with a symbol or drawing for additional clarity.

ELL ENGLISH LANGUAGE LEARNERS Write the words *catalytic* and *catalyst* on the board. Underline the similar parts of the words. Explain that the word *catalyst* is a noun, and the word *catalytic* is an adjective, which is a word that describes a noun. Lead students in discussing how the meanings of the words are related.

Explain

USE VISUALS Have students study the picture of the catalytic converter and its honeycomb core. **Ask** What do you think are some reasons for using a honeycomb design? *(It provides a strong surface through which the exhaust easily can flow. The design provides a way to expose a large amount of the catalysts to the exhaust.)* **Ask** Do you think the precious metals get used up so that the catalytic converter has to be replaced? Why or why not? *(No; Because they are catalysts, the precious metals participate in the chemical reactions that convert the exhaust, but they are neither reactants nor products of the reactions. They do not get used up.)*

Extend

Connect to ENVIRONMENTAL SCIENCE

Catalytic converters greatly reduce the hydrocarbon, carbon monoxide, and nitrogen oxide emissions from a vehicle. However, they increase the emission of carbon dioxide, which may contribute to global warming. Global warming is the gradual increase in Earth's average temperature. Excess carbon dioxide, along with water vapor, methane, ozone, and other greenhouse gases in the atmosphere, contribute to global warming by trapping some of the sun's radiation that would otherwise escape into space. Consequences include melting glaciers, severe droughts, rising sea levels, an increase in the intensity of hurricanes, and destruction of coral reefs. The addition of catalytic converters significantly reduces the air pollution from vehicles, but other technologies are needed to avoid the effect on global warming. Ask students to research current statistics on global numbers of gasoline- and diesel-powered vehicles on the world's roads, and the percentage of vehicular carbon dioxide emissions attributed to the five most populous countries.

Answers

TAKE IT FURTHER

1. It increases the frequency of particle collisions.
2. $\frac{1}{2}O_2(g) + CO(g) \rightarrow CO_2(g)$
3. The lead binds to the catalytic particles and inhibits their activity.

Key Objectives

DESCRIBE the relationship between the value of the specific rate constant and the speed of a reaction.

DESCRIBE how most reactions progress from start to finish.

Additional Resources

- Reading and Study Workbook, Lesson 18.2
- Probeware Laboratory Manual
- Virtual Chemistry Labs

CHEMISTRY & YOU Have students read the opening text. Display a picture of the Pyrenees mountain range. Explain that the Tour's route is broken into a series of stages, many of which travel through the Pyrenees mountains. **Ask** How is the Tour de France bicycle race like the activation energy diagram for a complex chemical reaction? *(A series of chemical reactions may require more activation energy at some points and may release energy at other points.)*

Activate Prior Knowledge

Have students recall the four factors that can influence the rate of a chemical reaction. *(temperature, concentration, particle size, and the presence of a catalyst)* **Ask** How does the concentration of reactants affect the rate of a chemical reaction? *(The higher the concentration, the faster the reaction rate, and vice versa.)* Have students keep this in mind as they read this lesson.

Foundations for Reading

BUILD VOCABULARY Write the word *intermediate* on the board. Have students look up the prefix *inter- (between)* and the stem *mediate (middle)* and write a definition for the word in their own terms. Have students compare their definitions to the dictionary definition.

ACADEMIC STANDARDS for SCIENCE

C.7.5 Explain how the rate of a reaction is qualitatively affected by changes in concentration, temperature, surface area, and the use of a catalyst.

18.2 The Progress of Chemical Reactions

C.7.5 Factors affecting rate of reaction

CHEMISTRY & YOU

Q: *How is a bicycle race like a chemical reaction?* The Tour de France is one of the most famous bicycle races in the world. It is held from mid-July to early August every year. During the race, cyclists travel almost 4000 kilometers. At one stage of the race, riders must cross steep mountains with heights of 1900 meters or more. The riders need extra energy to ride through these steep mountains.

Key Questions

What is the relationship between the value of the specific rate constant, k, and the speed of a chemical reaction?

How do most reactions progress from start to finish?

Vocabulary

- rate law
- specific rate constant
- first-order reaction
- elementary reaction
- reaction mechanism
- intermediate

Rate Laws

What is the relationship between the value of the specific rate constant, k, and the speed of a chemical reaction?

The rate of a reaction depends in part on the concentrations of the reactants. Suppose there were a reaction with only one reactant and one product. You could write a simple equation for this reaction.

$$A \longrightarrow B$$

The rate at which A forms B can be expressed as the change in A (ΔA) with time, where $[A_1]$ is the initial molar concentration of A at time t_1 and $[A_2]$ is the molar concentration of A at a later time, t_2.

$$\text{Rate} = \frac{\Delta A}{\Delta t} = \frac{[A_2] - [A_1]}{t_2 - t_1}$$

The rate of disappearance of A is proportional to the concentration of A.

$$\frac{\Delta A}{\Delta t} \propto [A]$$

The proportionality can be expressed as the concentration of A, [A], multiplied by a constant, k.

$$\text{Rate} = \frac{\Delta A}{\Delta t} = k \times [A]$$

This equation is a **rate law**, an expression for the rate of a reaction in terms of the concentration of reactants. The **specific rate constant** (k) for a reaction is a proportionality constant relating the concentrations of reactants to the rate of the reaction. The value of the specific rate constant depends on the conditions of the reaction and is determined through experiments. **The value of the specific rate constant, k, in a rate law is large if the products form quickly; the value is small if the products form slowly.**

604 Chapter 18 • Lesson 2

Focus on ELL

1 CONTENT AND LANGUAGE Write the words *order*, *elementary*, and *intermediate* on the board. Pronounce each word and ask students to repeat. Explain that these words have both common and scientific meanings. Show students how the common meaning of each word is related to its scientific meaning in relation to this lesson.

2 FRONTLOAD THE LESSON Provide a quick review of mathematical terminology such as *proportional, interpret, constant, slope, tangent, known,* and *unknowns* prior to explaining the rate laws. Explain that this lesson will use mathematics to quantify concepts learned in Lesson 18.1.

3 COMPREHENSIBLE INPUT During the evaluation step of Sample Problem 18.1, demonstrate mathematically that the concentration of Trial 2 is double that of Trial 1, and that the initial concentration of Trial 3 is four times that of Trial 1. When discussing reaction mechanisms, write the equations for the decomposition of nitrous oxide on the board and demonstrate step-by-step how to add the two equations together to attain the overall equation.

InterpretGraphs

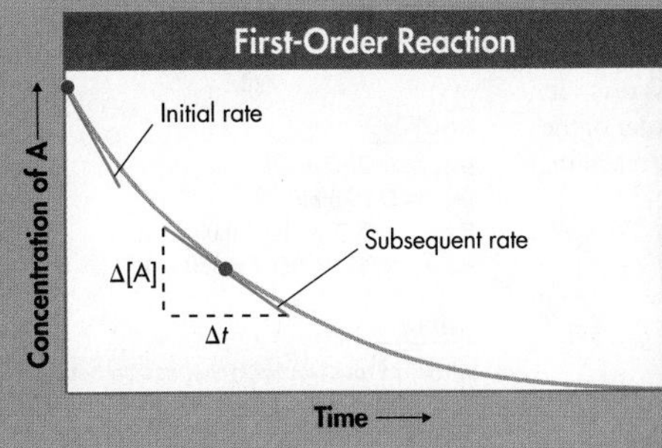

Figure 18.10 The graph shows how the concentration of a reactant changes during a first-order reaction.
a. Identify What are the independent and dependent variables?
b. Describe What happens to the concentration of reactant A over time?
c. Compare The short red lines (tangents) illustrate the reaction rates at two different points in the reaction. Which rate is faster—the *initial rate* or the *subsequent rate*?

Hint: The slope of the tangent to the curve at any point is equal to the rate at that point.

First-Order Reactions The order of a reaction is the power to which the concentration of a reactant must be raised to match the experimental data on concentration and rate. In a **first-order reaction,** the rate is directly proportional to the concentration of only one reactant. Assume the reaction $A \longrightarrow B$ is an example of a first-order reaction. The reaction rate is proportional to the concentration of A raised to the first power: $[A]^1 = [A]$.

The graph in Figure 18.10 shows the progress of a first-order reaction. Over time, the rate of reaction decreases because the concentration of the reactant is decreasing. For a first-order reaction, if [A] is reduced by one half, the reaction rate is reduced by one half. The rate ($\Delta A/\Delta t$) at any point on the graph equals the slope of the tangent to the curve at that point.

Higher-Order Reactions In some reactions, two substances react to give products. One example is a double-replacement reaction. The general equation for a double-replacement reaction can be written as follows. The coefficients are represented by lowercase letters.

$$aA + bB \longrightarrow cC + dD$$

For the reaction of A with B, the rate of reaction is dependent on the concentrations of both A and B.

$$\text{Rate} = k[A]^x[B]^y$$

When each exponent in the rate law equals 1 (that is, $x = y = 1$), the reaction is said to be first order in A and first order in B. The overall order of a reaction is the sum of the exponents for the individual reactants. A reaction that is first order in A and first order in B is thus second order overall.

You might assume that the coefficients in a chemical equation and the exponents in the rate law for that reaction would always be the same. If your assumption were true, you could use the coefficients of the reactants to find the order of a reaction. For most reactions, however, the exponents in the rate law and the coefficients in the equation do not correspond. Most reactions are more complex than the reactions used in the examples. Thus, the actual order of a reaction must be determined by experiment.

UbD Check for Understanding

What is the relationship between the value of the specific rate constant *k* and the speed of a chemical reaction?

Assess students' understanding by asking them to consider what can be implied if the magnitude of the rate constants from two different chemical reactions is known? *(Sample answer: Whether the products formed quickly or slowly.)*

ADJUST INSTRUCTION If students are having trouble answering, have them re-read the section on Rate Laws. Then, provide students with specific rate constants for two reactions, such as a reaction with and without a catalyst added, for comparison.

READING STRATEGY Have students create a main ideas and details chart with the red and blue headings. Have them complete their charts as they read the lesson.

Explain

Rate Laws

APPLY CONCEPTS Point out that the rate law contains a rate constant, *k,* which is specific to a reaction. Demonstrate how to set up the rate equation for a chemical reaction such as the formation of water from hydrogen and oxygen. Emphasize that this constant only applies to the concentration of the reactants over time, not the products.

USE MODELS Review linear functions with the class. **Ask** What ratio do you use for the slope of a line? *(the slope of a line is the ratio of the rise* (Δy) *to the run* (Δx).*)* Demonstrate how to calculate the slope for several lines on a coordinate grid using the concept of rise over run between two points of a given line.

USE VISUALS Ask students to examine the title and the labels on Figure 18.10. **Ask** As time increases, what happens to the concentration of reactant A? *(it decreases)* Discuss the effect of concentration on reaction rate. Explain how this relationship can be expressed quantitatively. Show students how they can use the relationship between concentration and time to predict the actual reaction rate as well as to describe the step-by-step mechanism by which the reaction occurs. Make sure students understand that the the rate *must* be calculated; it cannot be "read" directly from the graph.

INTERPRET DIAGRAMS Direct students again to Figure 18.10. **Ask** Is the graph a linear function? Explain. *(No; the graph is not a straight line.)* If the relationship in the graph formed a straight line, how would you find the slope? *(rise over run, or the ratio of* ΔA *to* Δt*)* Show students that to find the rate of reaction at a specific instant during the reaction, they can draw a line tangent to the curve at the point on the line associated with that instant. Then they can find the slope of that line to determine the rate at that specific time during the reaction. Demonstrate this concept by showing students how to calculate the rates of reaction for the two points shown in Figure 18.10.

Answers

INTERPRET GRAPHS

a. independent: time; dependent: relative concentration of A
b. It decreases.
c. initial rate

LESSON 18.2

Explain

START A CONVERSATION Point out that most reactions are more complex than the one-step reactions used as examples in the text. In most cases, reaction order cannot be predicted solely on the basis of a balanced equation; it must be determined experimentally. This is because the rate depends primarily on the slowest step in the reaction mechanism.

Sample Practice Problem

A first-order reaction has an initial rate of 0.6 mol/L·s. What is the rate when one-third of the starting material remains? *(0.2 mol/L·s)* When one-eighth of the starting material remains? *(0.075 mol/L·s)*

SampleProblem 18.1

Finding the Order of a Reaction From Experimental Data

Consider the reaction $a\text{A} \longrightarrow \text{B}$. The rate law for this reaction is Rate = $k[\text{A}]^x$. From the data in the table, find the order of the reaction with respect to A and the overall order of the reaction.

Trial	Initial concentration of A (mol/L)	Initial rate (mol/(L·s))
1	0.050	3.0×10^{-4}
2	0.10	12×10^{-4}
3	0.20	48×10^{-4}

KNOWNS
$[\text{A}]_1 = 0.050$ mol/L
$[\text{A}]_2 = 0.10$ mol/L
$\text{Rate}_1 = 3.0 \times 10^{-4}$ mol/(L·s)
$\text{Rate}_2 = 12 \times 10^{-4}$ mol/(L·s)

UNKNOWNS
Order of reaction with respect to A = ?
Overall order of the reaction = ?

1 Analyze **List the knowns and the unknowns.** Use the first two trials to calculate the order and the third to evaluate your answer.

2 Calculate **Solve for the unknowns.**

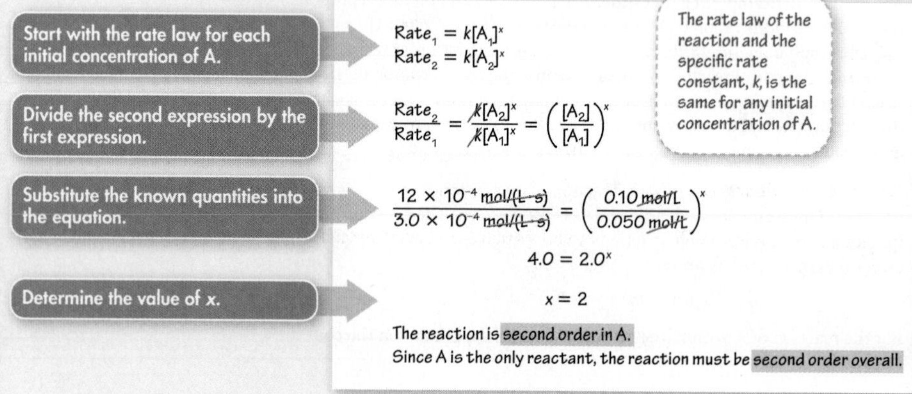

3 Evaluate **Does this result make sense?** If the reaction was first order in A, doubling the concentration would double the rate. However, Rate_2 is *four* times Rate_1. So the reaction is second order for A and second order overall because A is the only reactant. As a further check, look at what happens to the rate when the concentration doubles again from 0.10 mol/L to 0.20 mol/L.

7. Suppose the initial rate for a first-order reaction is 0.5 mol/(L·s). What is the rate when half the reactant remains? When one fourth of the reactant remains?

8. Show that the unit of k for a first-order reaction is a reciprocal unit of time, such as a reciprocal second (s^{-1}). Begin with the expression Rate = $k[\text{A}]$.

To do Problem 8, isolate k, and then substitute the units for rate and concentration.

Foundations for Math

ISOLATING A VARIABLE Tell students that solving rate problems often involves solving for a variable in an exponent. Explain that to isolate a variable that is an exponent, they must use natural logarithms (natural log, or ln).

To solve for *a* in **Sample Problem 18.1,** begin by taking the natural log of both sides of the equation $4.0 = 2^a$:

$$\ln(4.0) = \ln(2^a)$$

Use the properties of logs to pull *a* out of the exponent by rewriting the equation as:

$$a \ln 2 = \ln 4.0$$

Divide both sides of the equation by ln 2 to isolate *a*:

$$\frac{a \ln 2}{\ln 2} = \frac{\ln 4.0}{\ln 2}$$

Solving for *a* gives $a = \ln 4.0/\ln 2 = 1.38629/0.69314 = 2.00001$, which rounds to 2. So $a = 2$. Remind students that natural log is not the same as antilog (inverse log). Help them locate the natural log key on their calculators.

Reaction Mechanisms

How do most reactions progress from start to finish?

Balanced equations are extremely useful. They tell you what reactants are present at the start of a reaction and what products are present at the end. What a balanced equation does not tell you is *how* the reaction occurred. For example, plants use photosynthesis to capture and store light energy. The process can be summarized by stating that carbon dioxide and water yield simple sugars and oxygen. However, the process of photosynthesis is not as simple as this summary implies.

One-Step and Multistep Reactions If you had enough data, you could graph all the energy changes that occur as reactants are converted to products in a chemical reaction. Such a graph is called a reaction progress curve, or reaction profile. For an elementary reaction, you would get a graph like the ones in Figure 18.5. An **elementary reaction** is a reaction in which reactants are converted to products in a single step. This type of reaction has only one activation-energy peak and one activated complex.

Most chemical reactions consist of two or more elementary reactions. The series of elementary reactions or steps that take place during the course of a complex reaction is called a **reaction mechanism.** Figure 18.11 shows a reaction progress curve for a complex chemical reaction. The graph has a peak for each activated complex and a valley for each intermediate.

An **intermediate** is a product of one step in a reaction mechanism and a reactant in the next step. An intermediate has a more stable structure and a longer lifetime than an activated complex. Yet, an intermediate is reactive enough to take part in the next step. Intermediates do not appear in the overall chemical equation for a reaction.

CHEMISTRY & YOU

Q: *In the mountain stage of the Tour de France, a rider encounters a series of peaks and valleys. In terms of energy, how does the trip through the mountains compare to what happens during a multistep reaction?*

InterpretGraphs

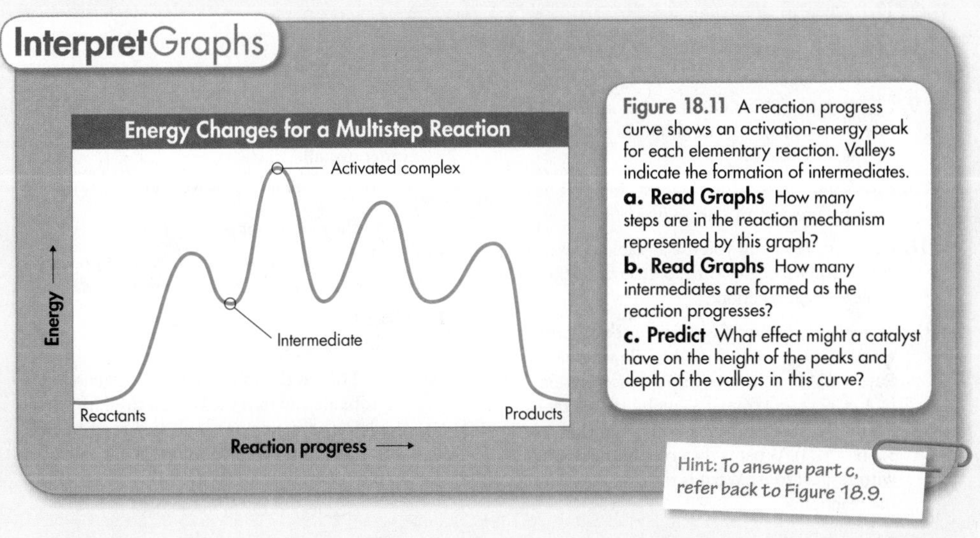

Figure 18.11 A reaction progress curve shows an activation-energy peak for each elementary reaction. Valleys indicate the formation of intermediates.

a. Read Graphs How many steps are in the reaction mechanism represented by this graph?

b. Read Graphs How many intermediates are formed as the reaction progresses?

c. Predict What effect might a catalyst have on the height of the peaks and depth of the valleys in this curve?

Reaction Mechanisms

APPLY CONCEPTS Explain that even complex reactions occur as a series of simple steps. Each step usually involves the collision of no more than two different particles. Use the reaction between hydrogen and bromine as an example. Show students that the overall reaction, $H_2 + Br_2 \rightarrow 2HBr$, occurs as a sequence of three elementary steps:

$Br_2 \rightarrow 2Br$

$Br + H_2 \rightarrow HBr + H$

$H + Br \rightarrow HBr$

INTERPRET DIAGRAMS Direct students' attention to Figure 18.11 and explain that the activation energy of a reaction, determined experimentally, is plotted versus time or reaction progress. Remind students that the independent variable is shown on the *x*-axis and the dependent variable is shown on the *y*-axis. **Ask** Why does this graph have more hills and valleys than the graphs in Lesson 18.1? *(This graph is a multi-step reaction, but the graphs in Lesson 18.1 were only one-step graphs.)* Explain that for each step in a reaction there must be enough activation energy for the reaction to proceed. **Ask** Is the activation energy for each step in the multi-step reaction shown in the graph equal to all the other activation energies in the reaction? Explain. *(No; the peaks of the hills all correspond to different y-values.)* **Ask** Does each step in the reaction take the same amount of time? How do you know? *(No; the change in the x-values is not constant.)*

MAKING CONNECTIONS Explain that displaying the activation energy data in a graph provides a more complete picture of the reaction process than a table of data. Relate this concept to piece-wise or step functions learned in algebra.

Differentiated Instruction

ELL ENGLISH LANGUAGE LEARNERS Speak clearly and slowly when reading the symbols used in the equations aloud. Be sure to phrase the relationships using precise mathematical verbiage. Define the math words and symbols as they appear. Add these terms to the word wall.

L1 STRUGGLING STUDENTS Point out that the units of concentration of a reactant or product are often given in moles per unit volume. However, clarify that concentration is not dependent on the entire volume of the reaction.

L3 ADVANCED STUDENTS Challenge students to compare and contrast the concepts of average rate and instantaneous rate. Ask students to describe which rate is best suited to depict the progress of a chemical reaction and why.

Answers

7. 0.250 mol/ (L · s); 0.125 mol/ (L · s)
8. Rate = k[A]; rate is expressed in units of mol/(L · s). [A] is expressed in units of mol/L.
   k = Rate/[A]
   $k = 1/s = s^{-1}$

FIGURE 18.11

a. four
b. three
c. It would lower the activation energy for the slowest step in the reaction mechanism.

LESSON 18.2

Extend

Explain to students that intermediate products often react with reactants during a multi-step reaction to produce other intermediates or final products. Tell students that one reaction in which this occurs is the thermal decomposition of nitryl chloride (NO_2Cl) to NO_2 and Cl_2. This process is believed to occur in two steps. Challenge students to suggest a possible reaction mechanism for this reaction.

($NO_2Cl \rightarrow NO_2 + Cl$
$NO_2Cl + Cl \rightarrow NO_2 + Cl_2$)

Evaluate

Informal Assessment

Write on the board the two-step reaction mechanism for the production of NO_2F from NO_2 and F_2.

$NO_2 + F_2 \rightarrow NO_2F + F$ (slow)
$NO_2 + F \rightarrow NO_2F$ (fast)

$2\,NO_2 + F_2 \rightarrow 2NO_2F$

Ask students to construct the reaction rate equation for the reaction mechanism shown. *(The first step of the reaction is the slowest step so it is the rate-determining step. The rate law is rate = $k[NO_2][F_2]$.)* Then have students complete the 18.2 Lesson Check.

Reteach

Scaffold learning by showing students how to predict the reaction rate with a simple one-step reaction with one reactant, then two reactants, and then a multi-step reaction.

Rate-Determining Steps In a multistep chemical reaction, the steps do not all progress at the same rate. One step will be slower than the others. The slowest step will determine, or limit, the rate of the overall reaction. As an analogy, think about shopping in a supermarket. First, you collect the items you want to buy. Then, you go through the checkout line and pay for your purchases. If you are buying many items, the first step is likely to take longer than the second. If you are buying only one or two items, getting through the checkout line may be the slower step.

Consider the reaction mechanism for the decomposition of nitrous oxide (N_2O). Experiments have shown that the mechanism consists of the two steps shown below.

Step 1: $N_2O(g) \longrightarrow N_2(g) + O(g)$ (slow)
Step 2: $N_2O(g) + O(g) \longrightarrow N_2(g) + O_2(g)$ (fast)
Overall: $2N_2O(g) \longrightarrow 2N_2(g) + O_2(g)$

In the first step, nitrous oxide decomposes into nitrogen gas and oxygen atoms. The oxygen atoms are an intermediate. They react with nitrous oxide in the second step to produce nitrogen molecules and oxygen molecules. The reaction in the second, faster step cannot occur until the first, slower step occurs. Therefore, the rate of the overall reaction depends on the rate of the first step. To increase the rate of the overall reaction, you would need to increase the rate of the first step. Thus, for the decomposition of nitrous oxide, the first step is called the *rate-determining step.*

When the equations for the two steps are summed, the oxygen atoms disappear. This example illustrates why you cannot use the overall chemical equation for a complex reaction to determine the reaction mechanism. Instead, the mechanism must be determined through experiments.

ONLINE PROBLEMS

18.2 LessonCheck

9. Explain What does the size of the specific rate constant, k, indicate about the rate of a chemical reaction?

10. Summarize How do most chemical reactions progress from start to finish?

11. Describe How can you use a graph of reactant concentration versus time to determine the rate of a chemical reaction?

12. Draw Conclusions Consider the one-step reaction $aA \longrightarrow B$, with the following rate law: Rate = $k[A]^x$. When the initial concentration of A is increased from 0.35 mol/L to 0.70 mol/L, the initial rate increases from 1.6 mol/(L·s) to 3.2 mol/(L·s). What is the order of the reaction with respect to A? What is the overall order of the reaction?

13. Apply Concepts The rate law for the following reaction is first order in NO and O_3, and second order overall.

$$NO(g) + O_3(g) \longrightarrow NO_2(g) + O_2(g)$$

Write the rate law for this reaction.

14. Explain What is an elementary reaction, and how is it related to a reaction mechanism?

15. Classify Is an intermediate a reactant or a product? Explain.

16. Infer Look at the equation for the reaction of nitric oxide and oxygen. Do you think this is a single-step reaction or a multistep reaction? Use what you know about collision theory to explain your answer.

$$2NO(g) + O_2(g) \longrightarrow 2NO_2(g)$$

Lesson Check Answers

9. The larger the specific rate constant, k, the faster the reaction.
10. Most reactions consist of two or more elementary reactions, each resulting in the formation of an intermediate.
11. The rate at any given point on the graph is determined by calculating the slope of the tangent to the curve at that point.
12. first order in A; first order overall
13. Rate = $k[NO][O_3]$
14. An elementary reaction is a reaction in which reactants are converted to products in a single step. A reaction mechanism is a series of elementary reactions that comprise a complex chemical reaction.
15. Both, it is a product in one step in a reaction mechanism and a reactant in the next step.
16. multistep; To proceed in a single-step reaction, three molecules would have to collide in the correct orientation, which is unlikely.

18.3 Reversible Reactions and Equilibrium

C.7.6 Reversible reactions

CHEMISTRY & YOU

Q: *How did chemists help farmers produce more food?* Fertilizers can increase the amount of a crop per unit of land. Most fertilizers contain ammonia or nitrogen compounds made from ammonia. For years, scientists tried and failed to develop an efficient way to produce ammonia from nitrogen and hydrogen. In the early 1900s, two German chemists, Fritz Haber and Karl Bosch, found the solution.

Key Questions

What happens at the molecular level in a chemical system at equilibrium?

What three stresses can cause a change in the equilibrium position of a chemical system?

What does the size of an equilibrium constant indicate about a system at equilibrium?

Vocabulary

- reversible reaction
- chemical equilibrium
- equilibrium position
- Le Châtelier's principle
- equilibrium constant

Reversible Reactions

What happens at the molecular level in a chemical system at equilibrium?

Based on the chemical equations you have seen, you may have inferred that chemical reactions always progress in one direction. This inference is not true. Some reactions are reversible. A **reversible reaction** is one in which the conversion of reactants to products and the conversion of products to reactants occur at the same time. Here's an example of a reversible reaction.

$$\text{Forward reaction: } 2SO_2(g) + O_2(g) \longrightarrow 2SO_3(g)$$

$$\text{Reverse reaction: } 2SO_2(g) + O_2(g) \longleftarrow 2SO_3(g)$$

In the first reaction, which is read from left to right, sulfur dioxide and oxygen produce sulfur trioxide. In the second reaction, which is read from right to left, sulfur trioxide decomposes into oxygen and sulfur dioxide. The first reaction is called the forward reaction. The second is called the reverse reaction. The two equations can be combined into one using a double arrow. The double arrow tells you that the reaction is reversible.

$$\underset{\text{Sulfur dioxide}}{2SO_2(g)} + \underset{\text{Oxygen}}{O_2(g)} \rightleftharpoons \underset{\text{Sulfur trioxide}}{2SO_3(g)}$$

Figure 18.12 models what is happening at the molecular level.

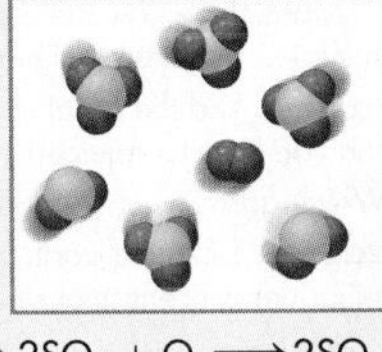

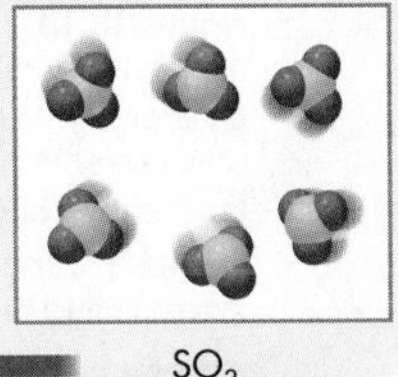

SO_2 and O_2 ⟶ $2SO_2 + O_2 \rightleftharpoons 2SO_3$ ⟵ SO_3

Figure 18.12 Opposite Reactions Molecules of SO_2 and O_2 react to give SO_3. Molecules of SO_3 decompose to give SO_2 and O_2.

See reversible reactions animated online.

LESSON 18.3

Key Objectives

DESCRIBE what happens at the molecular level in a chemical system at equilibrium.

IDENTIFY the three stresses that can cause a change in the equilibrium position of a chemical system.

DESCRIBE what the size of an equilibrium constant indicates about a system at equilibrium.

Additional Resources

- Reading and Study Workbook, Lesson 18.3
- Probeware Laboratory Manual,
- Virtual Chemistry Labs,

PearsonChem.com

Engage

CHEMISTRY & YOU Point out that the two reactants involved in production of ammonia are gases. **Ask:** Based on what you have previously learned about gases and reaction rates, what conditions involving the reaction of gaseous nitrogen and hydrogen, if altered, might affect the reaction? *(altering temperature and pressure)*

Access Prior Knowledge

Briefly review the ideal gas law and ask students to explain how real gases differ from ideal gases. Then, ask students how a reaction rate is determined and the factors that can affect the reaction rate.

IN ACADEMIC STANDARDS for SCIENCE

C.7.6 Write equilibrium expressions for reversible reactions.

Focus on ELL

1 CONTENT AND LANGUAGE Write the word *equilibrium* on the board. Display a picture of a pan balance, and explain that the Latin root of *equilibrium*, *libra*, means "balance." Have students use word-analysis strategies to determine the meaning of the prefix *equi-*, and write their own definitions of *equilibrium*.

2 FRONTLOAD THE LESSON Display a reversible jacket, or ask a student with one to show how it can be worn with either side facing out. Then ask students to think about how a car can go forward, but also can go in reverse. Tell students that they will explore how reactions can also be reversible as they work through this lesson.

3 COMPREHENSIBLE INPUT Show students a recorded gymnastics competition or pictures of gymnasts in various poses displaying the concept of equilibrium. Point out the concept of forward/reversible with the forward/backward motion of a gymnast.

LESSON 18.3

Foundations for Reading

BUILD VOCABULARY Have students use the LINCS strategy to learn what Le Châtelier's principle means. In LINCS exercises, students list what they know about each term, imagine a picture that describes the term, note a reminding "sound-alike" word, connect the terms to the sound-alike word by making up a short story, and then perform a brief self-test.

READING STRATEGY Ask students to rewrite the headings of each section as *how, why,* and *what* questions. Tell students to consider these questions as they read the corresponding information.

Explain

Reversible Reactions

USE VISUALS Have students examine Figure 18.13. Point out that this visual shows the reaction of the reactants and the decomposition of the products. **Ask** Does the reaction continue when the products and reactants are at equilibrium? *(Yes; reactants change into products, and products reform into reactants, but their net amounts stay the same.)*

CRITICAL THINKING Have students study Figure 18.13 again. **Ask** If you changed the volume, pressure, or temperature of the reaction, would the graphs change? Explain. *(The concentrations of both the reactants and products would change because changing the volume of a gas's container, its pressure, and its temperature can affect concentration. Changing the temperature also affects equilibrium in other ways.)*

Establishing Equilibrium What actually happens when sulfur dioxide and oxygen gases are mixed in a sealed container? The forward reaction begins at a given rate. Because no sulfur trioxide is present at the start, the initial rate of the reverse reaction is zero. As sulfur trioxide forms, however, the decomposition of sulfur trioxide begins. The rate of the reverse reaction is slow at first. Its rate increases as the concentration of sulfur trioxide increases. At the same time, the rate of the forward reaction decreases because sulfur dioxide and oxygen are being used up. Eventually, sulfur trioxide is decomposing as fast as sulfur dioxide and oxygen are combining. When the rates of the forward and reverse reactions are equal, the reaction has reached a state of balance called **chemical equilibrium.**

Look at the graphs in Figure 18.13. The graph on the left shows the progress of a reaction that starts with initial concentrations of SO_2 and O_2, but with zero SO_3. The graph on the right shows the progress of a reaction that begins with an initial concentration of SO_3 and zero concentrations for SO_2 and O_2. Notice that after a certain time, equilibrium is achieved and all the concentrations remain constant. The amount of SO_3 in the equilibrium mixture is the maximum amount that can be produced by this reaction under the conditions of the reaction.

Conditions at Equilibrium The unchanging amounts of SO_2, O_2, and SO_3 in the reaction mixture at equilibrium might cause you to think that both reactions have stopped. This is not the case. Chemical equilibrium is a dynamic state. **At chemical equilibrium, both the forward and reverse reactions continue, but because their rates are equal, no net change occurs in the concentrations of the reaction components.** Figure 18.14 provides an analogy for how an equilibrium is established and maintained.

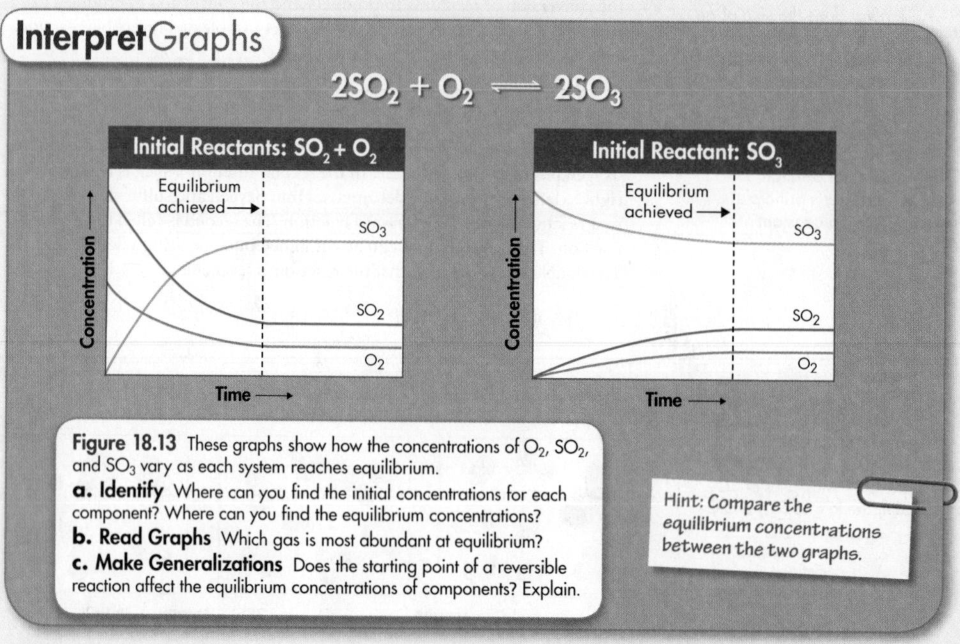

Figure 18.13 These graphs show how the concentrations of O_2, SO_2, and SO_3 vary as each system reaches equilibrium.
a. Identify Where can you find the initial concentrations for each component? Where can you find the equilibrium concentrations?
b. Read Graphs Which gas is most abundant at equilibrium?
c. Make Generalizations Does the starting point of a reversible reaction affect the equilibrium concentrations of components? Explain.

UbD Check for Understanding

What happens at the molecular level in a chemical system at equilibrium?

Instruct students to write a brief statement explaining the information they know about a reaction if the reaction is reversible *and* at chemical equilibrium.

ADJUST INSTRUCTION If students are having difficulty writing their statements, ask them to read the text about conditions at equilibrium and organize their thoughts with a bulleted list. Then have students revise their statements.

Figure 18.14 Equilibrium Positions
The up and down escalators in a store could represent a forward reaction and its reverse reaction. **a.** When the store opens, only the forward reaction occurs as shoppers head to the second floor. **b.** Equilibrium is reached when the rate at which shoppers move from the first floor to the second is equal to the rate at which shoppers move from the second floor to the first.
Use Analogies ***Do the number of people on each floor have to be equal at equilibrium? Explain.***

Concentrations at Equilibrium Although the rates of the forward and reverse reactions are equal at equilibrium, the concentrations of the components usually are not. Look again at Figure 18.13. At equilibrium, the concentration of SO_3 is much greater than the concentrations of SO_2 and O_2. The relative concentrations of the reactants and products at equilibrium mark the **equilibrium position** of a reaction. This position tells you whether the forward or reverse reaction is more likely to happen. Suppose a single reactant, A, forms a single product, B. If the equilibrium mixture contains 1 percent A and 99 percent B, then the formation of B is said to be favored.

$$\underset{1\%}{A} \rightleftharpoons \underset{99\%}{B}$$

If the mixture contains 99 percent A and 1 percent B at equilibrium, then the formation of A is favored.

$$\underset{99\%}{A} \rightleftharpoons \underset{1\%}{B}$$

Notice that the equilibrium arrows are not of equal length. The longer of the two arrows indicates which reaction is favored and whether reactants or products will be more common at equilibrium.

In principle, almost all reactions are reversible to some extent under the right conditions. In practice, one set of components is often so favored at equilibrium that the other set cannot be detected. When no reactants can be detected, you can say that the reaction has gone to completion, or is irreversible. When no products can be detected, you can say that no reaction has taken place. Reversible reactions occupy a middle ground between the extremes of irreversibility and no reaction.

APPLY CONCEPTS Point out that, in many cases, the concept of reactants turning into products, as described by a chemical equation, is an oversimplification. Explain that many reactions reach an equilibrium state in which the reaction mixture contains both reactant and product particles. The percent of reactants converted to products varies considerably. Show students that when hydrogen and oxygen combine to form water, for example, nearly all of the molecules react, as opposed to the reaction of hydrogen and nitrogen to form ammonia, in which the percent of reactants converted to product is much smaller. Stress that equilibrium refers to the rates of the forward or reverse reactions, not to the quantities of reactants and products.

CRITICAL THINKING As a class, discuss how the concepts of endothermic and exothermic reactions may relate to the concept of reversible reactions.

Extend

Connect to PHYSICS Ask students to research Newton's Third Law—the Law of Reciprocal Actions—and create a representation to show how this law helps to explain the concept of chemical equilibrium.

Differentiated Instruction

L1 LESS PROFICIENT READERS Have students identify and list words in the lesson with which they are having difficulty. Assist students in using context clues to determine the meaning of the listed words.

ELL ENGLISH LANGUAGE LEARNERS Avoid using the terms *unbalanced* and *balanced* to convey disequilibrium and equilibrium, as students may confuse the meaning with balanced reactions. Act out the concept of equilibrium by asking students to stand with one foot extended in front of them, and arms extended to the side for as long as possible. Point out those students who are steady and not moving as students at equilibrium.

L3 ADVANCED STUDENTS Ask students to determine what is meant by a static reaction and a dynamic reaction. Have students explain to the class why a reaction at equilibrium is said to be a dynamic system.

Answers

INTERPRET GRAPHS

a. Initial concentrations are at the far left, and equilibrium concentrations are at the far right.
b. SO_3
c. No, as long as the reaction conditions are the same, the equilibrium concentrations are the same.

FIGURE 18.14 No, the numbers can be different, but the number of shoppers moving between floors must be the same.

Explain

Factors Affecting Equilibrium: Le Châtelier's Principle

MAKE A CONNECTION Ask students to recall the formation of banana oil demonstration in Lesson 18.1. Write the general word equation for the formation of an ester on the board (organic acid + alcohol → ester + water). Add a reverse arrow to show the possibility of the opposite reaction. Remind students that the reaction was effective in the test tube to which anhydrous calcium sulfate had been added. Calcium sulfate reacted with water to form a hydrate, which limited the reverse reaction by removing a product.

Explore

Teacher Demo

PURPOSE Students observe the reaction with the greatest yield.

MATERIALS 10 mL ethanoic acid, 10 mL isoamyl alcohol, 2 test tubes, anhydrous calcium sulfate, concentrated sulfuric acid, water

PROCEDURE Mix 5 mL ethanoic acid (acetic acid) and 5 mL isoamyl alcohol in each of two test tubes. To one test tube, add a few granules of anhydrous calcium sulfate. Then add 3 drops of concentrated sulfuric acid to each test tube. Heat the test tubes in a boiling water bath for a few minutes. **Ask** Predict which reaction you think will yield the most product. *(Answers will vary.)* Then compare the results qualitatively by strength of odor or quantitatively if you have access to gas phase chromatography (GC) equipment.

EXPECTED OUTCOME

The tube containing anhydrous calcium sulfate yields the larger amount of product.

Factors Affecting Equilibrium: Le Châtelier's Principle

What three stresses can cause a change in the equilibrium position of a chemical system?

READING SUPPORT

Build Reading Skills: ***Cause and Effect*** As you read about Le Châtelier's principle, identify the effect that each type of stress (cause, in this case) has on the equilibrium position. ***How do changes in concentration affect the equilibrium position?***

The balance that exists in a chemical system at equilibrium is a delicate one. It can be disrupted when one or more of the conditions of the reaction are changed. When the equilibrium of a system is disturbed, the system makes adjustments to restore the equilibrium. However, the equilibrium position of the restored equilibrium will not be the same as the original equilibrium position. That is, the amount of reactants may have increased or the amount of products may have increased. Such a change is called a shift in the equilibrium position.

The French chemist Henri Le Châtelier (1850–1936) studied how the equilibrium position shifts as a result of changing conditions. He proposed what has come to be called **Le Châtelier's principle:** If a stress is applied to a system in dynamic equilibrium, the system changes in a way that relieves the stress. **Stresses that upset the equilibrium of a chemical system include changes in concentration of reactants or products, changes in temperature, and changes in pressure.**

In the examples of Le Châtelier's principle presented in this lesson, reactants will be to the left of the double arrow and products will be to the right. Blue arrows show the shifts when something is added to or removed from the system. The blue arrows always point in the direction of the shift in the equilibrium position—that is, toward the favored side.

Concentration Changing the amount, or concentration, of any reactant or product in a system at equilibrium disturbs the equilibrium. The system will adjust to minimize the effects of the change. Consider the decomposition of carbonic acid (H_2CO_3) in aqueous solution. The products are carbon dioxide and water. The system has reached equilibrium. The amount of carbonic acid is less than 1 percent.

$$\underset{<1\%}{H_2CO_3(aq)} \underset{\text{Remove } CO_2 \text{; Direction of shift} \rightarrow}{\overset{\text{Add } CO_2 \text{; } \leftarrow \text{Direction of shift}}{\rightleftharpoons}} \underset{>99\%}{CO_2(aq) + H_2O(l)}$$

Suppose carbon dioxide is added to the system. This increase in the concentration of CO_2 causes the rate of the reverse reaction to increase. As more reactant (H_2CO_3) is formed, the rate of the forward reaction also begins to increase. In time, the rates of the forward and reverse reactions again become equal. A new equilibrium is established with a higher concentration of reactant (H_2CO_3). Adding a product to a reaction at equilibrium pushes a reversible reaction in the direction of the reactants.

Suppose, on the other hand, carbon dioxide is removed. This decrease in the concentration of CO_2 causes the rate of the reverse reaction to decrease. As less reactant (H_2CO_3) is formed, the rate of the forward reaction also begins to decrease. In time, the rates of the forward and reverse reactions again become equal. The equilibrium is restored but at a different equilibrium position. Removing a product always pulls a reversible reaction in the direction of the products.

612 Chapter 18 • Lesson 3

Environmental Issues

The ammonia produced by the Haber-Bosch process reacts with nitric acid to create ammonium nitrate, a product typically used as a fertilizer for enhanced crop growth. There are environmental disadvantages to using nitrogen fertilizer to generate a greater crop yield. Excess fertilizer can flow into streams and rivers after a large rainstorm and adversely affect various habitats due to the chain reaction caused by the contamination. The fertilizer causes a rapid increase in algae, which exceeds the ability of its consumers to control its population. The abundance of dead algae leads to an massive increase in the population of bacterial decomposers in the pond. The bacteria consume large quantities of dissolved oxygen, leading to the death of a large number of fish. Also, high levels of nitrates in drinking water have been associated with methemoglobinemia in humans. Methemoglobin is a form of hemoglobin that cannot bind oxygen.

Farmers use the removal of a product to increase the yield of eggs laid by hens. Hens lay eggs and then proceed to hatch them. If the eggs are removed after they are laid, the hen will lay more eggs. Similarly as products are removed from a reaction mixture, the system attempts to restore equilibrium by producing more products. However, the reaction can never reestablish equilibrium because the products are constantly being removed. The reaction continues to produce products until the reactants are used up.

Your body provides another example of the effect of removing a product. Carbon dioxide is a product of the reactions that provide your body with energy. Carbon dioxide produced in the cells is carried through blood vessels to the lungs. Some of the carbon dioxide diffuses from the blood into the lungs and is exhaled into the air. An equilibrium between carbonic acid, carbon dioxide, and water exists in your blood.

The athletes in Figure 18.15 use more energy than a person at rest. During exercise, the concentration of CO_2 in the blood increases. This increase in CO_2 shifts the equilibrium in the direction of carbonic acid. The level of H_2CO_3 in the blood needs to remain fairly constant. Fortunately, the increase in the level of CO_2 also triggers an increase in the rate of breathing. With more breaths per minute, more CO_2 is removed through the lungs. The removal of CO_2 causes the equilibrium to shift toward the products, which reduces the amount of H_2CO_3.

The same principle applies to adding or removing reactants. When a reactant is added to a system at equilibrium, the reaction shifts in the direction of the formation of products. When a reactant is removed, the reaction shifts in the direction of formation of reactants.

Figure 18.15 Effect of Concentration on Equilibrium
Carbon dioxide is a product of reactions in cells. Rapid breathing during and after vigorous exercise helps reduce the level of CO_2 in the blood and, thus, helps control the level of H_2CO_3.
Predict ***Is the concentration of CO_2 greater in inhaled air or in exhaled air? Explain.***

613

Explain

USE VISUALS Direct students' attention to the inset diagram in Figure 18.15. Discuss the effects of adding carbon dioxide to and removing carbon dioxide from the system. Point out that gaseous carbon dioxide in the lungs is in equilibrium with aqueous, or dissolved, carbon dioxide in the blood. Explain that the removal of carbon dioxide from the lungs by exhalation allows more carbon dioxide to diffuse from the blood into the lungs.

MAKE A CONNECTION Ask a volunteer to explain a time when he or she was exerting so much energy during an athletic contest that he or she could not catch his or her breath. **Ask** What changes do you feel happening in your body when you are exercising? *(Sample answers: faster breathing rate, faster heart rate, sweating)* Point out that during and after vigorous exercise, more CO_2 than usual needs to be removed to maintain the body's equilibrium. During exercise, the increased breathing rate also is needed to supply muscles with extra oxygen.

APPLY CONCEPTS Write the following equation on the board. $2H_2(g) + O_2(g) + \text{heat} \rightleftharpoons 2H_2O(g)$
Prior to showing students how stresses on a reaction affect the reaction, pair students and have them follow the Think-Pair-Share strategy to predict what happens with the reaction if:

- concentration of hydrogen increases *(The reaction will shift to the right to produce more product.)*
- concentration of hydrogen decreases *(The reaction will shift to the left.)*
- temperature is increased *(Reaction will shift to the right.)*
- pressure is increased *(The reaction will shift to the right.)*
- a catalyst is added *(The reaction will occur faster.)*

Continue teaching this section and have each pair of students revisit their predictions to clarify any misunderstandings.

Carbon Monoxide

Molecules such as carbon monoxide (CO) that are structurally similar to the oxygen molecule can bind to the hemoglobin molecule. Therefore, they compete with oxygen for the same binding locations. This competition is the cause of carbon monoxide poisoning. Because their affinity for the heme group is greater, carbon monoxide molecules can displace oxygen molecules from hemoglobin.

$$HbO_2 + CO \rightleftharpoons HbCO + O_2$$

The equilibrium in this reaction lies far to the right because the affinity of carbon monoxide for heme is approximately 150 times stronger than that of oxygen.

Answers

READING SUPPORT An increase in concentration of a reactant or product will cause a shift in the equilibrium position in the reverse direction.

FIGURE 18.15 exhaled air

LESSON 18.3

Explain

USE VISUALS Direct students to Figure 18.16. **Ask** How does the equilibrium in cylinder *a* relate to the equilibrium in cylinder c? *(Increasing the pressure on the system in cylinder a caused the reaction to shift to the right, producing more molecules of product in cylinder* c.*)* **Ask** What state(s) of matter can a change in pressure affect? *(gas)*

MAKE A CONNECTION Point out that the industrial production of ammonia is an example of a practical application of Le Châtelier's principle. Write the equilibrium equation on the board. **Ask** Based on the equilibrium equation, select ideal temperature and pressure conditions for the highest possible yield of ammonia. *(high pressure and low temperature)* Tell students that the industrial process that was devised by Carl Bosch actually uses both high temperature and pressure plus an iron-based catalyst. **Ask** Why doesn't Bosch's process use the conditions predicted by Le Châtelier's principle? *(A higher-than-ideal temperature—plus the catalyst—is needed to increase the reaction rate.)* Explain that an efficient industrial process must often strike a balance between multiple goals, in this case, the need for high yield and fast reaction time.

Sample Practice Problem

The gases dinitrogen tetroxide and nitrogen dioxide exist in equilibrium according to the equation

$$N_2O_4(g) + 58\ kJ \rightleftharpoons 2NO_2(g)$$

What effect do the following changes have on the equilibrium position?

a. addition of heat *(favors products)*

b. decrease in pressure *(favors products)*

c. addition of NO_2 *(favors reactants)*

d. removal of N_2O_4 *(favors reactants)*

CHEMISTRY & YOU

Q: *Fritz Haber and Karl Bosch figured out how to increase the yield of ammonia when nitrogen and hydrogen react. Their success came from controlling the temperature and pressure. In which direction did they adjust each factor, and why?*

Temperature Increasing the temperature causes the equilibrium position of a reaction to shift in the direction that absorbs heat. In other words, it will shift in the direction that reduces the stress. For example, the reaction that occurs when ammonia (NH_3) is produced from N_2 and H_2 is exothermic.

$$N_2(g) + 3H_2(g) \underset{\text{Remove heat (cool). Direction of shift}\rightarrow}{\overset{\text{Add heat. }\leftarrow\text{Direction of shift}}{\rightleftharpoons}} 2NH_3(g) + \text{heat}$$

Heat can be considered to be a product, just like NH_3. Heating the reaction mixture at equilibrium pushes the equilibrium position to the left, which favors the reactants. As a result, the product yield decreases. Cooling, or removing heat, pulls the equilibrium position to the right, and the product yield increases.

Pressure Equilibrium systems in which some reactants and products are gases can be affected by a change in pressure. A shift will occur only if there are an unequal number of moles of gas on each side of the equation. The reaction in which ammonia forms is a useful example. Figure 18.16 shows how a change in pressure affects the system. The three gases are in a cylinder with a piston attached to a plunger. When the plunger is pushed down, the volume decreases and the pressure increases. You can predict which way the equilibrium position will shift by comparing the number of molecules of reactants and products. When two molecules of ammonia form, four molecules of reactants are used up (three of hydrogen and one of nitrogen). A shift toward ammonia (the product) will reduce the number of molecules. This shift will decrease the pressure but not to the original pressure.

$$N_2(g) + 3H_2(g) \underset{\text{Reduce pressure. }\leftarrow\text{Direction of shift}}{\overset{\text{Increase pressure. Direction of shift}\rightarrow}{\rightleftharpoons}} 2NH_3(g)$$

A change in pressure can be used to favor the reactants in the ammonia reaction. Pulling up on the plunger in Figure 18.16 will increase the volume that the gases occupy. This increase in volume causes a decrease in pressure. Adding molecules to the container can help to restore the pressure. Thus, the decomposition of ammonia, which increases the number of molecules, is favored. A new equilibrium is established at a pressure that is higher than when the system was disturbed but not as high as the original pressure.

Figure 18.16
Effect of Pressure on Equilibrium
Pressure affects a mixture of nitrogen, hydrogen, and ammonia.
a. The system is at equilibrium.
b. Equilibrium is disturbed by an increase in pressure. **c.** A new equilibrium position is established with fewer molecules.
Explain ***How does the ideal gas law (PV = nRT) explain this shift in the equilibrium position?***

Ammonia molecule (NH_3)
Hydrogen molecule (H_2)
Nitrogen molecule (N_2)

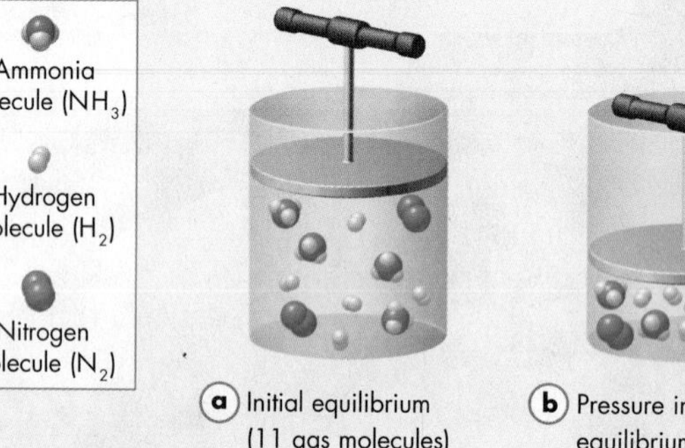
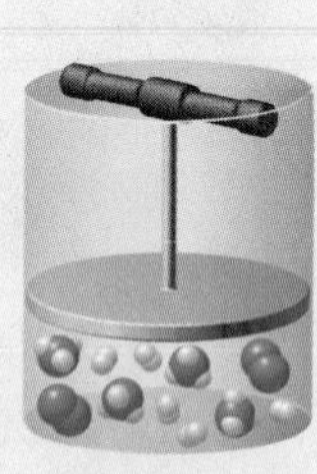
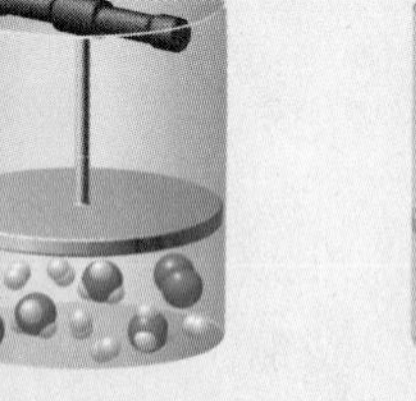
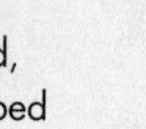
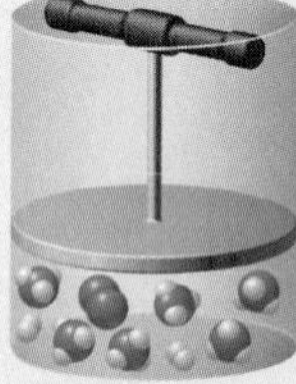

(a) Initial equilibrium (11 gas molecules)
(b) Pressure increased, equilibrium disturbed
(c) New equilibrium (9 gas molecules)

UbD Check for Understanding

What three stresses can cause a change in the equilibrium position of a chemical system?

Ask students to indicate with a thumbs-up or thumbs-down whether they understand that changes in temperature, concentration, or pressure are stresses that can affect or disturb a chemical reaction at equilibrium.

ADJUST INSTRUCTION If students are having difficulty answering the question, have them review the Conditions at Equilibrium section as well as the discussion on Le Châtelier's principle to clarify their understanding.

Catalysts and Equilibrium Catalysts decrease the time it takes to establish equilibrium. However, they do not affect the amounts of reactants and products present at equilibrium. The energy path for a reverse reaction is the exact opposite of the energy path for the forward reaction. So adding a catalyst lowers the energy path by the same amount for both reactions.

SampleProblem 18.2

Applying Le Châtelier's Principle

What effect will each of the following changes have on the equilibrium position for this reversible reaction?

$$PCl_5(g) + \text{heat} \rightleftharpoons PCl_3(g) + Cl_2(g)$$

a. Cl_2 is added.
b. Pressure is increased.
c. Heat is removed.
d. PCl_3 is removed as it forms.

1 Analyze **Identify the relevant concepts.** According to Le Châtelier's principle, the equilibrium position will shift in a direction that minimizes the imposed stress.

2 Solve **Apply the concepts to this problem.**

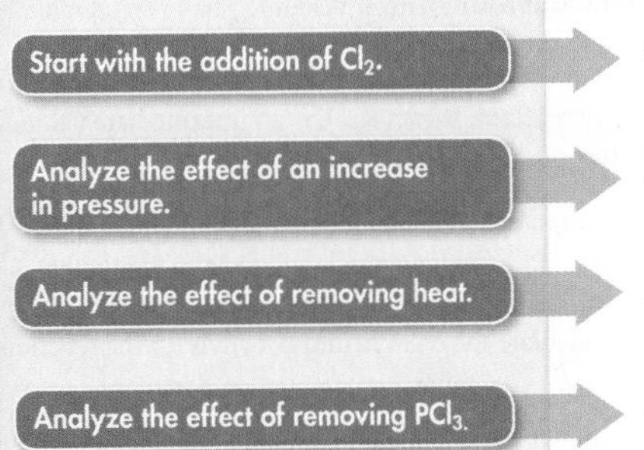

a. Cl_2 is a product. Increasing the concentration of a product shifts the equilibrium to the left.

b. Reducing the number of gas molecules decreases the pressure. The equilibrium shifts to the left.

c. The reverse reaction produces heat. The removal of heat causes the equilibrium to shift to the left.

d. PCl_3 is a product. Removal of a product as it forms causes the equilibrium to shift to the right.

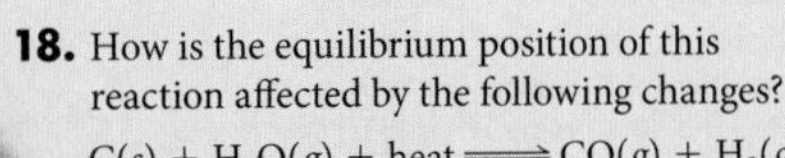

17. How is the equilibrium position of this reaction affected by the following changes?

$$2SO_2(g) + O_2(g) \rightleftharpoons 2SO_3(g) + \text{heat}$$

a. lowering the temperature
b. decreasing the pressure
c. removing oxygen
d. adding sulfur trioxide (SO_3)

18. How is the equilibrium position of this reaction affected by the following changes?

$$C(s) + H_2O(g) + \text{heat} \rightleftharpoons CO(g) + H_2(g)$$

a. lowering the temperature
b. increasing the pressure
c. removing hydrogen
d. adding water vapor

For a change in pressure, compare the number of gas molecules on both sides of the equation.

Differentiated Instruction

L1 LESS PROFICIENT READERS Allow students to visually discover the effect of decreasing pressure on a substance by showing them how and why a carbonated beverage goes flat.

L1 SPECIAL NEEDS Engage students in **Sample Problem 18.2** by having them stand up and spread their arms wide. Tell students that their arms represent the dual arrows indicating the reversible reaction. Have students lean one way or the other to indicate how the reaction is affected by each stress.

L3 ADVANCED STUDENTS Have students find an online applet that simulates Le Châtelier's principle. Have students create a graphical representation of what they discovered.

Explore

Factors Affecting Equilibrium: Le Châtelier's Principle

Teacher Demo

PURPOSE Students observe the effect of temperature changes on the equilibrium position of an exothermic reaction.

MATERIALS 3 g copper turnings, 3 heat-resistant test tubes, 10 mL 6*M* nitric acid, one-hole stopper, 2 solid stoppers, glass delivery tube, rubber tubing, warm water bath, ice water bath

PROCEDURE Place a small amount (3 g or less) of copper turnings in a test tube, and add 10 mL of 6*M* nitric acid (HNO_3). Insert a one-hole stopper, fitted with a glass delivery tube and rubber tubing, into the test tube. Fill two test tubes with the brown gas that forms. When the color is the same as in the reacting test tube, stopper the tubes. Once sealed, the tubes can be saved indefinitely for repeated use, or discharged in a fume hood.

SAFETY Use a fume hood. Nitric acid is corrosive, and NO_2 is a toxic gas.

PROCEDURE Set up an ice bath and a warm water bath (70–80°C). Hold up the tubes for students to see and tell them that each tube contains NO_2 gas. Place one tube in warm water and the other in cold water. After a few minutes, remove the tubes and have students compare their colors. Now reverse the positions of the tubes in the water baths and after a few minutes, have students again observe the colors.

Write the reaction equation on the board:

$$2NO_2(g) \rightleftharpoons N_2O_4(g) + \text{heat}$$

Explain that nitrogen dioxide is brown and dinitrogen tetroxide is colorless. Ask students to explain the effect of changing temperature. *(Adding heat shifts the equilibrium in the direction that absorbs heat, to the left. Removing heat shifts the equilibrium in the direction that produces heat, to the right.)*

Answers

FIGURE 18.16 Increasing the pressure causes a decrease in volume, which shifts the equilibrium in the direction of fewer gas molecules.

17. **a.** favors products
b. favors reactants
c. favors reactants
d. favors reactants

18. **a.** favors reactants
b. favors reactants
c. favors products
d. favors products

LESSON 18.3

Explain

Equilibrium Constants

START A CONVERSATION Remind students of the specific reaction rate constant they learned in Lesson 18.1 and the way in which they learned to determine the order of a reaction. Then explain that the fixed numerical relationship between the reactants and products of a chemical system at equilibrium is called the equilibrium constant, typically symbolized by K_{eq}. **Ask** Based on what you have learned about equilibrium so far, what predictions do you think you can make given the value of an equilibrium constant of a reaction? *(the equilibrium position of the reaction)*

INTERPRET A DIAGRAM Direct students to Figure 18.17. Tell students to count the number of circles and the number of squares in each flask. **Ask** What did you notice about the numbers of circles and squares in each flask? *(The number of circles in the warm flask is equal to the number of squares in the cool flask; and vice versa.)* Explain that each diagram represents a system at equilibrium. Point out that whenever a chemical system reaches equilibrium, the reactants and products have a fixed numerical relationship. Once students show understanding, guide them to Sample Problem 18.3, in which they can calculate the constant for the reaction represented by this diagram.

Extend

Have students write the equilibrium expression for the production of ammonia from nitrogen and hydrogen.

$$\left[K_{eq} = \frac{[NH_3]^2}{[N_2][H_2]^3}\right]$$

Equilibrium Constants

What does the size of an equilibrium constant indicate about a system at equilibrium?

Chemists express the equilibrium position as a numerical value. This value relates the amounts of reactants to products at equilibrium. Look at the following general reaction in which two reactants form two products. The coefficients *a*, *b*, *c*, and *d* represent the number of moles.

$$aA + bB \rightleftharpoons cC + dD$$

The **equilibrium constant** (K_{eq}) is the ratio of product concentrations to reactant concentrations at equilibrium. Each concentration is raised to a power equal to the number of moles of that substance in the balanced chemical equation. Here is the expression for the equilibrium constant for the general reaction stated above.

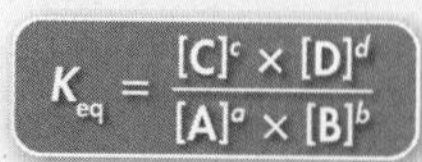

$$K_{eq} = \frac{[C]^c \times [D]^d}{[A]^a \times [B]^b}$$

The exponents in the equilibrium-constant expression are the coefficients in the balanced chemical equation. The concentrations of substances are in moles per liter (mol/L). The value of K_{eq} depends on the temperature of the reaction. If the temperature changes, the value of K_{eq} also changes.

The size of the equilibrium constant indicates whether reactants or products are more common at equilibrium. When K_{eq} has a large value, such as 3.1×10^{11}, the reaction mixture at equilibrium will consist mainly of product. When K_{eq} has a small value, such as 3.1×10^{-11}, the mixture at equilibrium will consist mainly of reactant. When K_{eq} has an intermediate value, such as 0.15 or 50, the mixture will have significant amounts of both reactant and product.

When the value of an equilibrium constant is calculated, the cancellation of units may or may not lead to a unit for the constant. As a result, chemists have agreed to report equilibrium constants without a stated unit. Sample Problem 18.3 shows how to calculate the equilibrium constant for the reaction illustrated in Figure 18.17.

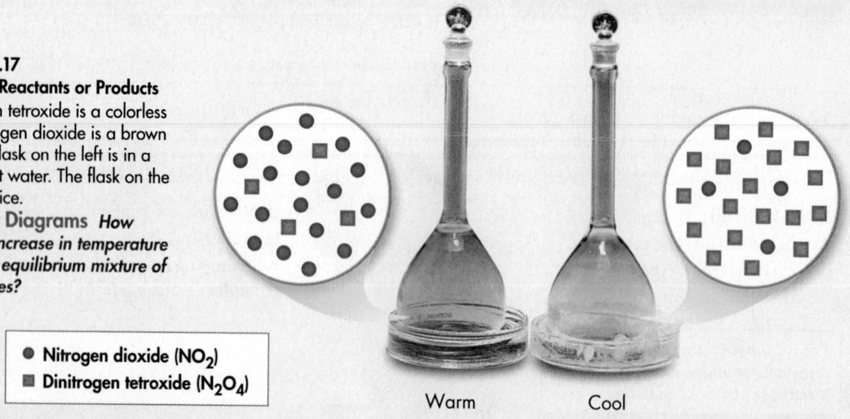

Figure 18.17
Favoring Reactants or Products
Dinitrogen tetroxide is a colorless gas. Nitrogen dioxide is a brown gas. The flask on the left is in a dish of hot water. The flask on the right is in ice.
Interpret Diagrams ***How does an increase in temperature affect the equilibrium mixture of these gases?***

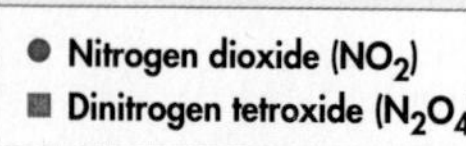

Warm

Cool

UbD Check for Understanding

What does the size of an equilibrium constant indicate about a system at equilibrium?

Assess students' understanding about an equilibrium system by asking them to explain mathematically how an equilibrium constant can be used to make a prediction about the relative concentrations of reactants and products at equilibrium. *(Sample answer: K_{eq} = [products]/[reactants]; If you divide a small number by a large number, the quotient is a very small number. So, if K_{eq} is a small value, then the concentration of reactants has to be much larger than the concentration of products, and vice versa.)*

ADJUST INSTRUCTION If students are having trouble answering, show them a simplified version of a ratio representing the equilibrium constant. Have students use large and small and similar values for the concentrations and calculate the constant. Tell students to look for a pattern they can use to make predictions.

Sample Problem 18.3

Expressing and Calculating K_{eq}

The colorless gas dinitrogen tetroxide (N_2O_4) and the brown gas nitrogen dioxide (NO_2) exist in equilibrium with each other.

$$N_2O_4(g) \rightleftharpoons 2NO_2(g)$$

A liter of the gas mixture at equilibrium contains 0.0045 mol of N_2O_4 and 0.030 mol of NO_2 at 10°C. Write the expression for the equilibrium constant (K_{eq}) and calculate the value of the constant for the reaction.

1 Analyze List the knowns and the unknowns. Modify the general expression for the equilibrium constant and substitute the known concentrations to calculate K_{eq}.

KNOWNS	UNKNOWNS
$[N_2O_4] = 0.0045$ mol/L	K_{eq} (algebraic expression) = ?
$[NO_2] = 0.030$ mol/L	K_{eq} (numerical value) = ?

2 Calculate Solve for the unknowns.

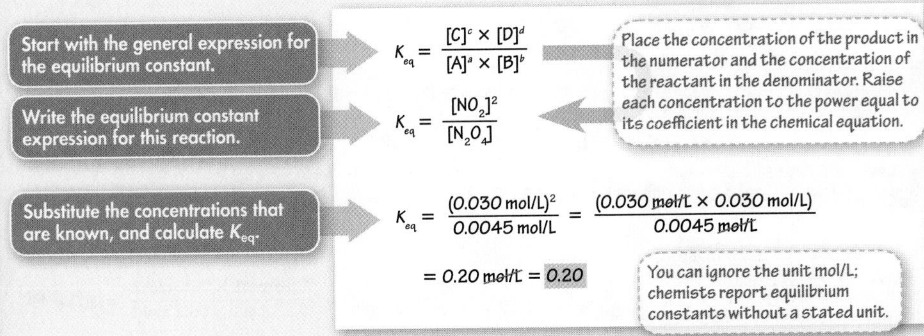

3 Evaluate Does the result make sense? Each concentration is raised to the correct power. The numerical value of the constant is correctly expressed to two significant figures. The value for the K_{eq} is appropriate for an equilibrium mixture that contains significant amounts of both gases.

Here's a hint for Problem 20: Calculate $1/K_{eq}$ for the forward reaction.

19. The reaction in which ammonia is formed is $N_2(g) + 3H_2(g) \rightleftharpoons 2NH_3(g)$. At equilibrium, a 1-L flask contains 0.15 mol H_2, 0.25 mol N_2, and 0.10 mol NH_3. Calculate K_{eq} for this reaction.

20. Using the equilibrium conditions described in Problem 19, Calculate K_{eq} for $2NH_3(g) \rightleftharpoons N_2(g) + 3H_2(g)$. How is the K_{eq} for a forward reaction related to the K_{eq} for a reverse reaction?

Foundations for Math

WORKING WITH EXPONENTS Remind students to apply the properties of powers when simplifying quantities and units. In Sample Problem 18.3, the Power of Product rule, $(a^x)^y = a^{xy}$, and the Power of a Quotient rule, $\frac{a^x}{a^y} = a^{x-y}$, when $a \neq 0$ can be used to simplify Step 2.

Remind students that anything inside a set of parentheses that is raised to a given power is affected by the exponent, including units of measurements. This is demonstrated in the third equation of step 2 above, where $(0.030\ \text{mol/L})^2$ is rewritten as $(0.030\ \text{mol/L}) \times (0.030\ \text{mol/L})$, rather than (0.030×0.030) mol/L.

Explain

APPLY CONCEPTS Show students how to set up and solve problems involving equilibrium constants. Stress the following:

- the only quantities that are used to determine the value of the equilibrium constant expression are concentrations, or quantities which are related to concentrations, like partial pressures
- the units for concentration are often given as moles per liter
- an equilibrium constant does not have units; it is a unitless constant
- be sure to subtract the number of moles that react from the number of moles originally present in order to determine the number remaining at equilibrium.

CRITICAL THINKING Have students examine the similarities and differences in the expressions used to determine an equilibrium constant with the way to determine the specific reaction rate constant in Lesson 18.2

Sample Practice Problem

Analysis of an equilibrium mixture of hydrogen, iodine, and hydrogen iodide contained in a 10.0-L flask at a certain temperature gives the following results: hydrogen, 0.15 mol; iodine, 0.15 mol; hydrogen iodide, 0.87 mol. Calculate K_{eq} for the reaction:

$H_2(g) + I_2(g) \rightleftharpoons 2HI(g)$

(K_{eq} = 34)

Answers

FIGURE 18.17 An increase in temperature causes the equilibrium to shift toward the formation of more NO_2.

19. $K_{eq} = 12$

20. $K_{eq} = 8.4 \times 10^{-2}$; One is the inverse of the other.

Explain

CRITICAL THINKING Students often think they need to use all of the information provided in a problem statement. Help students to understand that some of the values that are given in an equilibrium problem involve the conditions of the reaction. When determining the equilibrium constant, these values are not needed in the calculation. Students may, however, be able to draw conclusions about the factors that are driving the equilibrium in one direction or another. Ask students to infer which of the three factors that affect an equilibrium is providing stress to the system in Sample Problem 18.4. *(high temperature)*

Sample Practice Problems

A. Write the equilibrium constant for the following reaction.

$$H_2(g) + Br_2(g) \rightleftharpoons 2HBr(g)$$

$$\left(K_{eq} = \frac{[HBr]^2}{[H_2][Br_2]}\right)$$

B. Determine the value of K_{eq} for the reaction in problem A for the following concentrations at equilibrium in a 2.0-L container: 0.70 mol of H_2, 0.50 mol of Br_2, and 0.80 mol of HBr.
$(K_{eq} = 1.8)$

Sample Problem 18.4

Finding the Equilibrium Constant

One mole of colorless hydrogen gas and one mole of violet iodine vapor are sealed in a 1-L flask and allowed to react at 450°C. At equilibrium, 1.56 mol of colorless hydrogen iodide is present, together with some of the reactant gases. Calculate K_{eq} for the reaction.

$$H_2(g) + I_2(g) \rightleftharpoons 2HI(g)$$

1 Analyze List the knowns and the unknowns. Find the concentrations of the reactants at equilibrium. Then substitute the equilibrium concentrations in the expression for the equilibrium constant for this reaction.

KNOWNS	UNKNOWN
$[H_2]$ (initial) = 1.00 mol/L	K_{eq} = ?
$[I_2]$ (initial) = 1.00 mol/L	
[HI] (equilibrium) = 1.56 mol/L	

2 Calculate Solve for the unknown.

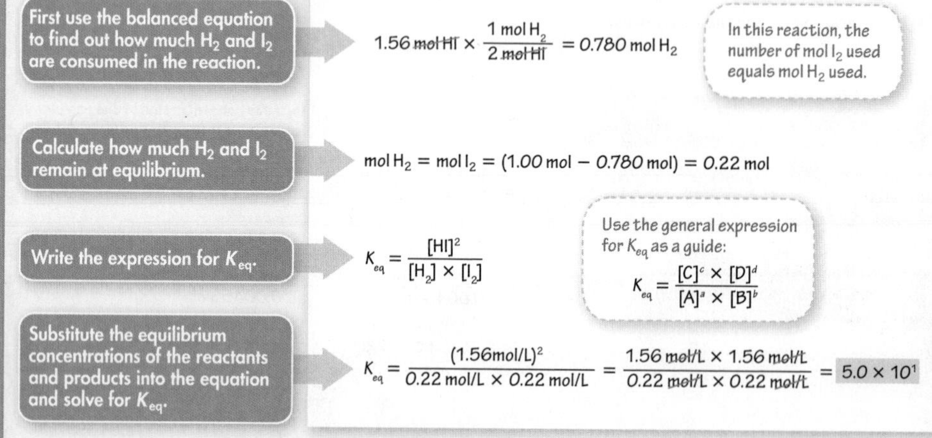

3 Evaluate Does the result make sense? Each concentration is raised to the correct power. The value of the constant reflects the presence of significant amounts of reactants and product in the equilibrium mixture.

In Problem 22, you must first find the equilibrium concentrations of H_2 and CO_2.

21. Suppose the following system reaches equilibrium:
$N_2(g) + O_2(g) \rightleftharpoons 2NO(g)$
Analysis of the equilibrium mixture in a 1-L flask gives the following results: 0.50 mol of N_2, 0.50 mol of O_2, and 0.020 mol of NO. Calculate K_{eq} for the reaction.

22. At 750°C, the following reaction reaches equilibrium in a 1-L flask:
$H_2(g) + CO_2(g) \rightleftharpoons H_2O(g) + CO(g)$
The reaction begins with 0.10 mol H_2 and 0.10 mol CO_2. At equilibrium, there is 0.047 mol H_2O and 0.047 mol CO. Calculate K_{eq} for the reaction.

Foundations for Math

COMBINING LIKE TERMS Remind students that the first step in solving multi-step equations often involves combining like terms. Explain that to add and subtract terms in an expression, they must have the same variable raised to the same power.

In Sample Problem 18.4, since the moles of each reactant are equal, x can be used to represent the number of moles of H_2 used as well as the number of moles of I_2 used. Thus, the total moles of reactants used is $x + x$, or $1x + 1x$. Point out that the coefficient and the exponent are both understood to be 1. Since the base of each term in the expression is the same, x, and the exponent is the same, 1, then to combine the terms, you add coefficients and keep the exponent the same. So, $1x + 1x = 2x$.

Sample Problem 18.5

Finding Concentrations at Equilibrium

Bromine chloride (BrCl) decomposes to form bromine and chlorine.

$$2BrCl(g) \rightleftharpoons Br_2(g) + Cl_2(g)$$

At a certain temperature, the equilibrium constant for the reaction is 11.1. A sample of pure BrCl is placed in a 1-L container and allowed to decompose. At equilibrium, the reaction mixture contains 4.00 mol Cl_2. What are the equilibrium concentrations of Br_2 and BrCl?

1 Analyze **List the knowns and the unknowns.** Use the balanced equation, equilibrium constant, and the equilibrium constant expression to find the unknown concentrations. According to the balanced equation, when BrCl decomposes, equal numbers of moles of Br_2 and Cl_2 are formed.

KNOWNS	UNKNOWNS
$[Cl_2]$ (equilibrium) = 4.00 mol/L	$[Br_2]$ (equilibrium) = ? mol/L
K_{eq} = 11.1	[BrCl] (equilibrium) = ? mol/L

2 Calculate **Solve for the unknowns.**

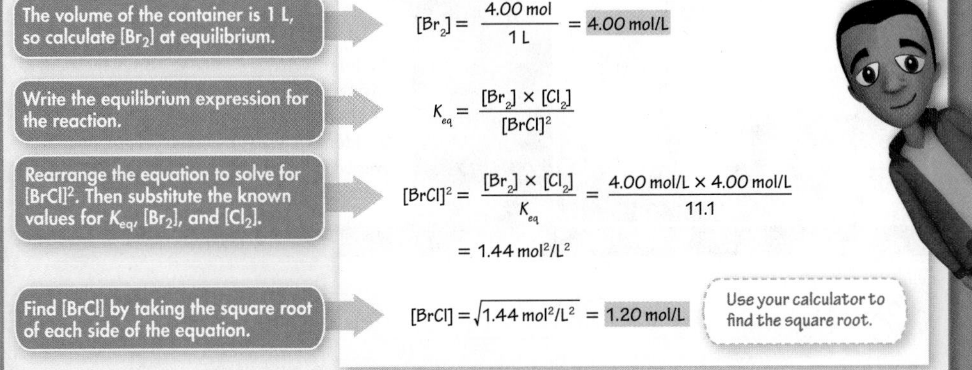

The volume of the container is 1 L, so calculate $[Br_2]$ at equilibrium.

$$[Br_2] = \frac{4.00\ \text{mol}}{1\ \text{L}} = 4.00\ \text{mol/L}$$

Write the equilibrium expression for the reaction.

$$K_{eq} = \frac{[Br_2] \times [Cl_2]}{[BrCl]^2}$$

Rearrange the equation to solve for $[BrCl]^2$. Then substitute the known values for K_{eq}, $[Br_2]$, and $[Cl_2]$.

$$[BrCl]^2 = \frac{[Br_2] \times [Cl_2]}{K_{eq}} = \frac{4.00\ \text{mol/L} \times 4.00\ \text{mol/L}}{11.1}$$

$$= 1.44\ \text{mol}^2/\text{L}^2$$

Find [BrCl] by taking the square root of each side of the equation.

$$[BrCl] = \sqrt{1.44\ \text{mol}^2/\text{L}^2} = 1.20\ \text{mol/L}$$

Use your calculator to find the square root.

3 Evaluate **Does the result make sense?** It makes sense that the equilibrium concentration of the reactant and the products are both present in significant amounts because K_{eq} has an intermediate value.

23. At a certain temperature, the equilibrium constant for the reaction in which nitrogen dioxide forms dinitrogen tetroxide is 5.6.

$$2NO_2(g) \rightleftharpoons N_2O_4(g)$$

In a 1-L container, the amount of N_2O_4 at equilibrium is 0.66 mol. What is the equilibrium concentration of NO_2?

24. Hydrogen iodide decomposes to form hydrogen and iodine.

$$2HI(g) \rightleftharpoons H_2(g) + I_2(g)$$

In a 1-L container at 450°C, the equilibrium mixture contains 0.50 mol of hydrogen. What are the equilibrium concentrations of hydrogen and hydrogen iodide? (K_{eq} = 0.020)

Explain

APPLY CONCEPTS Point out that a reversible reaction can be written in either order, reactants followed by products, or products followed by reactants. **Ask** Does a reversible reaction have the same equilibrium constant regardless of the way it is written? Why? *(No; the equilibrium constant for the reverse reaction is the reciprocal of the equilibrium constant of the forward reaction.)*

MAKING CONNECTIONS Relate the procedure for finding the unknown in Sample Problem 18.5 to solving a multi-step equation in algebra. Have students identify the key words to use in order to correctly set up the problem.

Sample Practice Problem

At elevated temperatures, molecular hydrogen and molecular bromine react to form hydrogen bromide:

$$H_2(g) + Br_2(g) \rightleftharpoons 2HBr(g)$$

A mixture of 0.77 mol of H_2 and 0.57 mol of Br_2 is combined in a 1-L vessel. At equilibrium, at a high temperature, there are 0.49 mol of Br_2 present. How many moles of H_2 are there at equilibrium? *(0.69 mol)*

Foundations for Math

SOLVING EQUATIONS WITH SQUARE ROOTS Remind students that when solving an equation involving a variable with an exponent, such as x^2, students should isolate this variable term and then take the square root of both sides in order to find the value for x. Point out that $x^2 = a$, so $x = \sqrt{a}$

In Sample Problem 18.5, let x = [BrCl] and show students that $[BrCl]^2 = 1.44\ \text{mol}^2/\text{L}^2$ can be represented algebraically as $x^2 = 1.44\ \text{mol}^2/\text{L}^2$.

$$x^2 = 1.44\ \text{mol}^2/\text{L}^2$$

$$\sqrt{x^2} = \sqrt{1.44\ \text{mol}^2\ \text{L}^2}$$

$$\sqrt{x^2} = \sqrt{1.44} \cdot \sqrt{\frac{\text{mol}^2}{\text{L}^2}}$$

$$x = 1.20\ \text{mol/L}$$

Answers

21. $K_{eq} = 1.6 \times 10^{-3}$

22 $K_{eq} = 0.79$

23. $[NO_2] = 0.34M$

24. $[H_2] = [I_2] = 0.50M$; $[HI] = 3.5M$

CHEMISTRY & YOU

CHEMISTRY & YOU Chemical reaction kinetics, coupled with courses in materials science and thermodynamics, equip chemical engineers with the necessary tools to design, improve, create, and investigate substances found in sundry disciplines. Although the coursework to become a chemical engineer is heavy in chemistry courses, chemical engineers are also required to have a strong understand of physics and higher-level mathematics in order to understand how the three disciplines are related.

Extend

Ask students to research and find the properties of wax that allow for it to be a useful product for surfers. Have students generate a list of parameters that were likely involved in determining the type of substance needed for this surfing application.

Evaluate

Informal Assessment

Write several reversible chemical equations on the board. Have students use Le Châtelier's principle to predict how they could increase the yield of a reaction product.

Then write the general reversible chemical reaction shown below on the board.

$$aA + bB \rightleftharpoons cC + dD$$

Ask students write the equation for calculating the equilibrium constant for this reaction.

$$K_{eq} = \left[\frac{[C]^c[D]^d}{[A]^a[B]^b}\right]$$

Reteach

If students have trouble understanding Le Châtelier's principle and the equilibrium constant, list the steps for writing the correct equilibrium constant expression from a balanced equation. Then guide them through Sample Problems 18.2–18.5 again. Encourage students to summarize their understanding of each concept in a bulleted list.

CHEMISTRY & YOU: CAREERS

Chemical Engineer

You may think that surfing is about as far away from chemistry as you can get. But you'd be wrong! Surfboards, wet suits, and even the synthetic waxes used to help keep your feet on the board were created using products developed by chemical engineers.

Chemical engineers are primarily concerned with the large-scale manufacture of chemical products, such as plastics and other petroleum products, pharmaceuticals, and foods. Chemical engineers may also be involved in research, the design and development of processing plants, the evaluation of operating processes, and the extraction and processing of raw materials.

Take It Further

1. **Infer** What are some reasons why chemical engineers might need a wide range of science and engineering knowledge in their work?
2. **Identify** What are some products you use every day that could have been developed by a chemical engineer?

18.3 LessonCheck

25. **Review** What happens to the amounts of reactants and products after a reaction has reached chemical equilibrium?
26. **List** What are the three stresses that can upset the equilibrium of a chemical system?
27. **Explain** What does the value of the equilibrium constant tell you about the amounts of reactants and products present at equilibrium?
28. **Relate Cause and Effect** Can a change in pressure shift the equilibrium position in every reversible reaction? Explain.
29. **Describe** How can you use a balanced chemical equation to write an equilibrium-constant expression?
30. **Apply Concepts** Which of the equilibrium constants indicates a reaction in which the amount of product is much larger than the amount of reactant at equilibrium? Explain.
    a. $K_{eq} = 1 \times 10^8$
    b. $K_{eq} = 3 \times 10^{-6}$
31. **Calculate** The equilibrium mixture for the reaction $2HI(g) \rightleftharpoons H_2(g) + I_2(g)$ contains 0.050 mol H_2. How many moles of I_2 and HI are present at equilibrium ($K_{eq} = 0.018$)?

BIGIDEA REACTIONS

32. Review the concept of percent yield from Lesson 12.3. How can understanding Le Châtelier's principle help chemists increase the percent yield of a reversible chemical reaction?

Lesson Check Answers

25. There is not a net change in the amounts of reactants or products.
26. changes in concentrations of reactants and products, changes in temperature, and changes in pressure
27. For a large K_{eq}, products dominate the reaction mixture; for a small K_{eq}, reactants dominate the reaction mixture.
28. no; only in reversible reactions in which the mole ratios of gaseous reactants and products are unequal
29. An equilibrium constant expression is a ratio. In the numerator, product concentrations are multiplied. In the denominator, reactant concentrations are multiplied. Each concentration is raised to a power equal to the coefficient for that species in the balanced equation.
30. a; At large K_{eq} values, products dominate the reaction mixture at equilibrium.
31. 0.050 mol I_2 and 0.37 mol HI
32. **BIGIDEA** They will know how to manipulate concentrations, temperature, and pressure to increase the percent yield of the reaction.

620 Chapter 18 • Lesson 3

18.4 Solubility Equilibrium

C.7.2 Factors affecting solubility

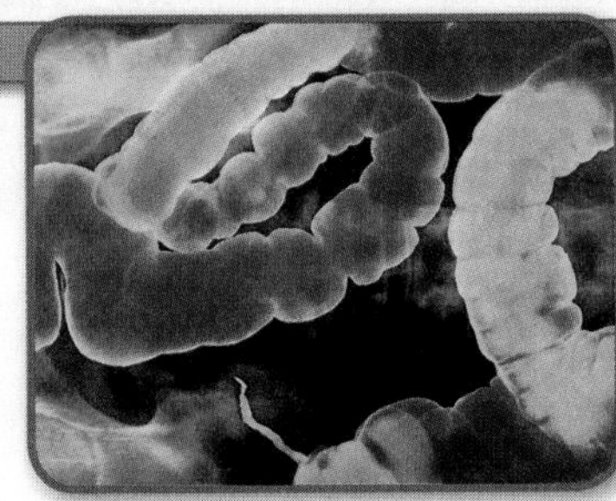

CHEMISTRY & YOU

Q: *How is it possible to ingest a poison without being harmed?* Chemical substances are needed to make organs other than bones visible on X-ray images. In one test, a patient drinks a thick liquid mixture that contains barium sulfate, which is a poison. Because barium sulfate can absorb X-rays, tissues coated with the liquid will appear as light areas in the X-ray images. This lesson will help you understand how a poison like barium sulfate can be safely used for this test.

Key Questions

What is the relationship between the solubility product constant and the solubility of a compound?

How can you predict whether precipitation will occur when two solutions are mixed?

Vocabulary

- solubility product constant
- common ion
- common ion effect

The Solubility Product Constant

What is the relationship between the solubility product constant and the solubility of a compound?

Most ionic compounds containing alkali metals are soluble in water. For example, more than 35 g of sodium chloride will dissolve in only 100 g of water. By contrast, some ionic compounds are insoluble in water. For example, compounds that contain phosphate, sulfite, or carbonate ions tend not to dissolve in water. Exceptions to this rule are compounds in which these ions are combined with ammonium ions or alkali metal ions. Table 18.1 provides some general rules for the solubility of ionic compounds in water.

Table 18.1

Solubility of Ionic Compounds in Water

Compounds	Solubility	Exceptions
Salts of Group 1A metals and ammonia	Soluble	Some lithium compounds
Ethanoates, nitrates, chlorates, and perchlorates	Soluble	Few exceptions
Sulfates	Soluble	Compounds of Pb, Ag, Hg, Ba, Sr, and Ca
Chlorides, bromides, and iodides	Soluble	Compounds of Ag and some compounds of Hg and Pb
Sulfides and hydroxides	Most are insoluble	Alkali metal sulfides and hydroxides are soluble. Compounds of Ba, Sr, and Ca are slightly soluble.
Carbonates, phosphates, and sulfites	Insoluble	Compounds of the alkali metals and of ammonium ions

LESSON 18.4

Key Objectives

DESCRIBE the relationship between the solubility product constant and the solubility of a compound.

PREDICT whether precipitation will occur when two solutions are mixed.

Additional Resources

- Reading and Study Workbook, Lesson 18.4
- Probeware Laboratory Manual
- Virtual Chemistry Labs

Engage

CHEMISTRY & YOU Direct students to the photograph and the question at the lesson opener. **Ask** Why is $BaSO_4$ ingested by a patient before having an X-ray taken of the stomach? *(Soft tissues don't show up well in X-rays.)* **Ask** Why is the digestive system well suited to this type of image enhancement? *(Chemicals are easily brought into the system and easily removed.)*

Access Prior Knowledge

Ask students to identify the types of reactants and products studied in Lesson 18.3. *(Both reactants and products were gases.)* Explain that in this lesson, students will continue to learn about equilibrium; however, the reactants and products will be liquids and solids. Ask students to draw on their prior knowledge to predict the form of the equilibrium constant expression for the dissolving of an ionic compound in water.

IN ACADEMIC STANDARDS for SCIENCE

C.7.2 Explain how temperature, pressure and polarity of the solvent affect the solubility of a solute.

Focus on ELL

1 CONTENT AND LANGUAGE Present the key questions slowly, then paraphrase the questions or use visual aids to help explain the intent of each question. Write word pairs such as *precipitation* and *precipitates*, *dissociate* and *associate*, and *common* and *uncommon* on the word wall. As a class, discuss the meaning of these words in a chemistry course.

2 FRONTLOAD THE LESSON Have students establish a purpose for reading. Students should create a list of three to four active questions they can answer as they read the text, the sample problems, and the tables.

3 COMPREHENSIBLE INPUT Introduce the concept of limited solubility by demonstrating the solubilities of equal amounts of sodium chloride and silver chloride in equal volumes of water. During the discussion of common ion effect, perform the teacher demonstration with silver chloride and silver nitrate.

Take It Further

1. Sample answers: to know how the manufacturing process affects the environment (environmental science); to know how to design a properly functioning reactor (physics, math, engineering)
2. Answers will vary.

LESSON 18.4

Foundations for Reading

BUILD VOCABULARY Make sure students understand the difference between the terms *solubility* and *solubility product constant*. Have them write the definitions in their own words.

READING STRATEGY After reading this lesson, ask students to summarize what they learned about solubility constants and the common ion effect. Suggest that they use an equation to illustrate their summary.

Explain

The Solubility Product Constant

START A CONVERSATION Explain that any ionic solid placed in water, if it does not dissolve completely, establishes an equilibrium between its dissociated ions in solution and the solid that is undissolved. **Ask** Do you think all solids dissolve entirely in water, partially in water, or are insoluble in water? *(Answers will vary.)* As a class, brainstorm examples of solids that always dissolve in water, never dissolve in water, or partially dissolve in water. Explain that the solubility product constant is an equilibrium constant that describes the equilibrium between a solid and its ions in solution.

USE MODELS Explain that the solubility product constant is closely related to the equilibrium constant that students learned about in Lesson 18.2. Write the equilibrium equation on the board and compare it to the solubility product equation. Point out similarities in the equations for these two quantities. Write the sample problems on the board and carry out the solutions for the solubility product, K_{sp}. Focus on how to set up the algebraic expression.

USE VISUALS Have students examine Table 18.2. Call out a few of the compounds in Table 18.2 and have students identify the solubility product constant for each compound. Explain that chemists often utilize data tables such as this to look up critical information about a specific chemical. Be sure to explain that students will not have to memorize the information in a table such as Table 18.2 but that they will be expected to extract the necessary information from the table in order to answer a question.

Misconception Alert

Students often forget to balance a chemical equation prior to determining the solubility product constant. Make sure students understand that the chemical equation should be written with the solid as the reactant and the ions as the products.

Figure 18.18 Silver Chloride
Some ionic compounds, such as silver chloride, are slightly soluble in water.
Predict ***Would adding solid silver chloride to this test tube increase the concentrations of silver ions and chloride ions?***

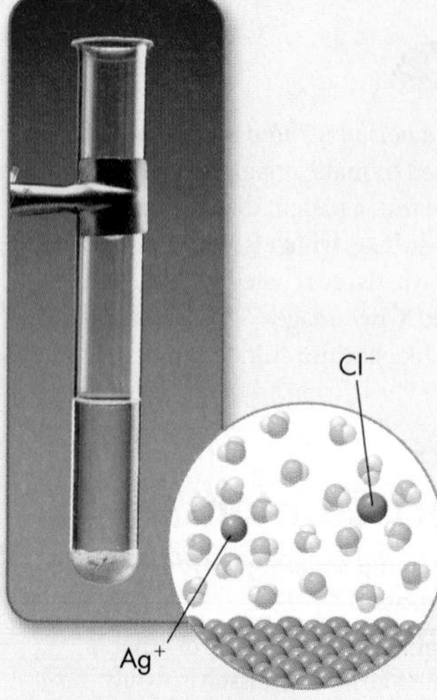

Most insoluble ionic compounds will actually dissolve to some extent in water. These compounds are said to be slightly soluble in water. Figure 18.18 models what happens when the "insoluble" compound silver chloride is mixed with water. A very small amount of silver chloride dissolves in the water. An equilibrium is established between the solid and the dissolved ions in the saturated solution.

$$AgCl(s) \rightleftharpoons Ag^+(aq) + Cl^-(aq)$$

You can write an equilibrium-constant expression for this process.

$$K_{eq} = \frac{[Ag^+] \times [Cl^-]}{[AgCl]}$$

The expression includes the dissolved ions, which are in solution, and the solid AgCl, which is not. To compare the solubility of salts, it is useful to have a constant that reflects only the concentrations of the dissolved ions. This constant is called the **solubility product constant** (K_{sp}), which is equal to the product of the concentrations of the ions each raised to a power equal to the coefficient of the ion in the dissociation equation.

$$K_{sp} = [A]^a \times [B]^b$$

The coefficients for the dissociation of silver chloride are each 1, so the K_{sp} expression for the dissociation is written as follows:

$$K_{sp} = [Ag^+] \times [Cl^-]$$

What does the size of the solubility product constant tell you about the solubility of a compound? **The smaller the value of the solubility product constant, the lower the solubility of the compound.** Table 18.2 lists the K_{sp} values for some ionic compounds that are slightly soluble in water.

Table 18.2

Solubility Product Constants (K_{sp}) at 25°C

Ionic compound	K_{sp}	Ionic compound	K_{sp}	Ionic compound	K_{sp}
Halides		**Sulfates**		**Hydroxides**	
AgCl	1.8×10^{-10}	$PbSO_4$	6.3×10^{-7}	$Al(OH)_3$	3.0×10^{-34}
AgBr	5.0×10^{-13}	$BaSO_4$	1.1×10^{-10}	$Zn(OH)_2$	3.0×10^{-16}
AgI	8.3×10^{-17}	$CaSO_4$	2.4×10^{-5}	$Ca(OH)_2$	6.5×10^{-6}
$PbCl_2$	1.7×10^{-5}	**Sulfides**		$Mg(OH)_2$	7.1×10^{-12}
$PbBr_2$	2.1×10^{-6}	NiS	4.0×10^{-20}	$Fe(OH)_2$	7.9×10^{-16}
PbI_2	7.9×10^{-9}	CuS	8.0×10^{-37}	**Carbonates**	
PbF_2	3.6×10^{-8}	Ag_2S	8.0×10^{-51}	$CaCO_3$	4.5×10^{-9}
CaF_2	3.9×10^{-11}	ZnS	3.0×10^{-23}	$SrCO_3$	9.3×10^{-10}
Chromates		FeS	8.0×10^{-19}	$ZnCO_3$	1.0×10^{-10}
$PbCrO_4$	1.8×10^{-14}	CdS	1.0×10^{-27}	Ag_2CO_3	8.1×10^{-12}
Ag_2CrO_4	1.2×10^{-12}	PbS	3.0×10^{-28}	$BaCO_3$	5.0×10^{-9}

UbD Check for Understanding

What is the relationship between the solubility product constant and the solubility of a compound?

Assess students' understanding of solubility equilibrium by asking them to complete the following statement:

The more soluble a compound is, the ________ its solubility product constant. *(greater)*

ADJUST INSTRUCTION If students are misinterpreting the solubility product constant, have students read the paragraph directly above Table 18.2. Have students think about how small the number of moles of ions in solution must be to give a product as small as the values in Table 18.2.

Sample Problem 18.6

Finding the Ion Concentrations in a Saturated Solution

What is the concentration of lead ions and chromate ions in a saturated solution of lead(II) chromate at 25°C? ($K_{sp} = 1.8 \times 10^{-14}$)

KNOWNS	UNKNOWNS
$K_{sp} = 1.8 \times 10^{-14}$	$[Pb^{2+}]$ = ? M
$PbCrO_4(s) \rightleftharpoons Pb^{2+}(aq) + CrO_4^{2-}(aq)$	$[CrO_4^{2-}]$ = ? M

1 Analyze **List the knowns and the unknowns.** Write the expression for K_{sp}. Then modify it so that there is a single unknown.

2 Calculate **Solve for the unknowns.**

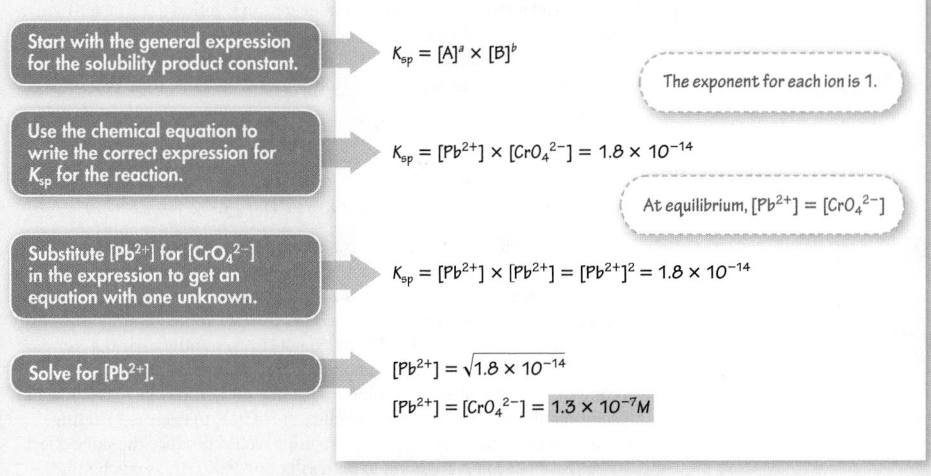

3 Evaluate **Does the result make sense?** Calculate $[Pb^{2+}] \times [CrO_4^{2-}]$ to evaluate the answer. The result is 1.7×10^{-14}, which is close to the value for K_{sp}. The result varies slightly from the actual value because the answers were rounded to two significant figures.

33. Lead(II) sulfide (PbS) has a K_{sp} value of 3.0×10^{-28}. What is the concentration of lead(II) ions in a saturated solution of PbS at 25°C?

34. What is the concentration of calcium ions in a saturated calcium carbonate solution at 25°C? Use the K_{sp} value for calcium carbonate from Table 18.2.

Explore

Class Activity

PURPOSE Students illustrate the solubilities of salts containing various anions.

MATERIALS calculator, paper, pencil, colored markers or highlighters

PROCEDURE Divide the class into groups of four students. Ask the groups to create color-coded tables illustrating the solubilities of salts containing various anions. For example, students can design separate tables for nitrates, sulfates, chlorides, sulfides, hydroxides, and carbonates. Have them use different colors in each table to show which elements form soluble compounds containing a given anion and which form insoluble compounds. Students should divide the tasks of looking up solubilities, designing the tables, and drawing the final tables. Remind students that the compounds listed in Table 18.2 are only a small sampling of possible compounds these anions can form.

Sample Practice Problem

What is the concentration of barium and sulfate ions in a saturated barium sulfate solution at 25°C? $K_{sp} = 1.1 \times 10^{-10}$ *($1.0 \times 10^{-5} M$)*

Foundations for Math

SUBSTITUTION Tell students that when solving solubility product equations, there are a lot of letters, subscripts, and superscripts involved and that the process for solving the equation may be easier if they use a variable to represent the components of the reaction. In **Sample Problem 18.6**, let A represent $[Pb^{2+}]$ and let B represent $[CrO_4^{2-}]$. Then, the equation can be rewritten as $A \times B = 1.8 \times 10^{-14}$. Since $[Pb^{2+}] = [CrO_4^{2-}]$ at equilibrium, then $A = B$, and the equation becomes $A \times A = 1.8 \times 10^{-14}$. Now use algebra to solve for A.

$$A \times A = 1.8 \times 10^{-14}$$
$$A^2 = 1.8 \times 10^{-14}$$
$$\sqrt{A^2} = \sqrt{1.8 \times 10^{-14}}$$
$$A = 1.3 \times 10^{-7}$$

Use the value of the variable to answer the question:

$$[A] = [Pb^{2+}] = [CrO_4^{2-}] = 1.3 \times 10^{-7}$$

Answers

FIGURE 18.18 No, because the concentration of AgCl is constant.

33. $1.7 \times 10^{-14} M$

34. $6.7 \times 10^{-5} M$

LESSON 18.4

Explain

The Common Ion Effect

CHEMISTRY & YOU Only a tiny amount of the barium sulfate dissolves; most of it simply passes through the digestive system.

APPLY CONCEPTS Point out that the common ion effect is not a new concept, but an example of Le Châtelier's principle at work. Explain that adding an ion to an equilibrium system containing that ion will cause the equilibrium to shift in the direction that uses up the ion.

DRAW CONCLUSIONS Have students hypothesize how the common ion effect could be used to identify ions in solution. Suggest that they use Table 18.1 to find a common property of salts listed in Table 18.2. *(If a precipitate forms when a few drops of an unknown solution is added to a known saturated solution, the unknown solution may have an ion in common with the known solution, or ions in the two solutions may have combined to form an insoluble product.)* Point out that in either case, additional tests would need to be performed to determine the identity of the precipitate.

The Common Ion Effect

How can you predict whether precipitation will occur when two solutions are mixed?

In a saturated solution of lead(II) chromate, an equilibrium is established between the solid lead(II) chromate and its ions in solution.

$$PbCrO_4(s) \rightleftharpoons Pb^{2+}(aq) + CrO_4^{2-}(aq) \quad K_{sp} = 1.8 \times 10^{-14}$$

What would happen if you added some lead nitrate to this solution? Lead(II) nitrate, $Pb(NO_3)_2$, is soluble in water. So adding $Pb(NO_3)_2$ causes the concentration of lead ion to increase. The product of $[Pb^{2+}]$ and $[CrO_4^{2-}]$ would be greater than the K_{sp} for lead(II) chromate. The addition of lead ions is a stress on the equilibrium. Applying Le Châtelier's principle, the stress can be relieved if the reaction shifts to the left. Figure 18.19 shows the result. The excess lead ions combine with chromate ions in solution to form additional solid $PbCrO_4$. Lead(II) chromate continues to precipitate from the solution until the product of $[Pb^{2+}]$ and $[CrO_4^{2-}]$ once again equals 1.8×10^{-14}. The difference is that now the lead ions in solution come from two sources, $PbCrO_4$ and $Pb(NO_3)_2$.

$$K_{sp} \text{ for the original solution: } [Pb^{2+}][CrO_4^{2-}] = 1.8 \times 10^{-14}$$

$$K_{sp} \text{ after addition of } Pb(NO_3)_2\text{: } \Big[Pb^{2+}\Big][CrO_4^{2-}] = 1.8 \times 10^{-14}$$

CHEMISTRY & YOU

Q: *The K_{sp} of $BaSO_4$ is 1.1×10^{-10}. How can patients ingest the toxic $BaSO_4$ without being harmed?*

In this example, the lead ion is a common ion. A **common ion** is an ion that is found in both ionic compounds in a solution. The lowering of the solubility of an ionic compound as a result of the addition of a common ion is called the **common ion effect.** Chemists can use an ionic compound with a common ion to cause a solid to precipitate from solution. For the common ion effect to work, the added compound must be more soluble than the compound that is already in solution. Adding lead(II) nitrate to a saturated solution of $PbCrO_4$ causes the solubility of $PbCrO_4$ to decrease. Adding sodium chromate, Na_2CrO_4, to the solution would produce the same effect. The chromate ion is common to both salts, and Na_2CrO_4 is much more soluble in water than is $PbCrO_4$.

Figure 18.19 Lead(II) Chromate Lead(II) chromate, $PbCrO_4$, is slightly soluble in water. **a.** The yellow solid in the test tube is $PbCrO_4$. It cannot dissolve because the solution is saturated with Pb^{2+} and CrO_4^{2-} ions. **b.** Adding some lead(II) nitrate, $Pb(NO_3)_2$, to the solution causes $PbCrO_4$ to precipitate from solution, which increases the amount of solid $PbCrO_4$.
Relate Cause and Effect ***How does adding lead nitrate disturb the $PbCrO_4$ equilibrium?***

Differentiated Instruction

L1 STRUGGLING STUDENTS Pair students who have trouble with word problems with those who are proficient at algebra and problem solving. Have each pair work through the practice problems, taking turns being the lead solver. The lead student should explain the process being used to arrive at the solution.

ELL ENGLISH LANGUAGE LEARNERS Have students draw and label a diagram demonstrating the common ion effect with silver chloride and silver nitrate, using different colors to represent the silver ions from each compound. Have students use arrows to indicate the combination of the silver ions from silver nitrate with the chloride ions.

L3 ADVANCED STUDENTS Have students research and explain why photographic film is covered with a light-sensitive emulsion containing silver bromide.

Sample Problem 18.7

Finding Equilibrium Ion Concentrations in the Presence of a Common Ion

Small amounts of silver bromide can be added to the lenses used for eyeglasses. The silver bromide causes the lenses to darken in the presence of large amounts of UV light. The K_{sp} of silver bromide is 5.0×10^{-13}. What is the concentration of bromide ion in a 1.00-L saturated solution of AgBr to which 0.020 mol of $AgNO_3$ is added?

1 Analyze List the knowns and the unknown. Use one unknown to express both $[Ag^+]$ and $[Br^-]$. Let x be the equilibrium concentration of bromide ion and $x + 0.020$ be the equilibrium concentration of silver ion.

KNOWNS	UNKNOWN
$K_{sp} = 5.0 \times 10^{-13}$	$[Br^-] = ?\ M$
moles of $AgNO_3$ added = 0.020 mol	
$AgBr(s) \rightleftharpoons Ag^+(aq) + Br^-(aq)$	

2 Calculate Solve for the unknown.

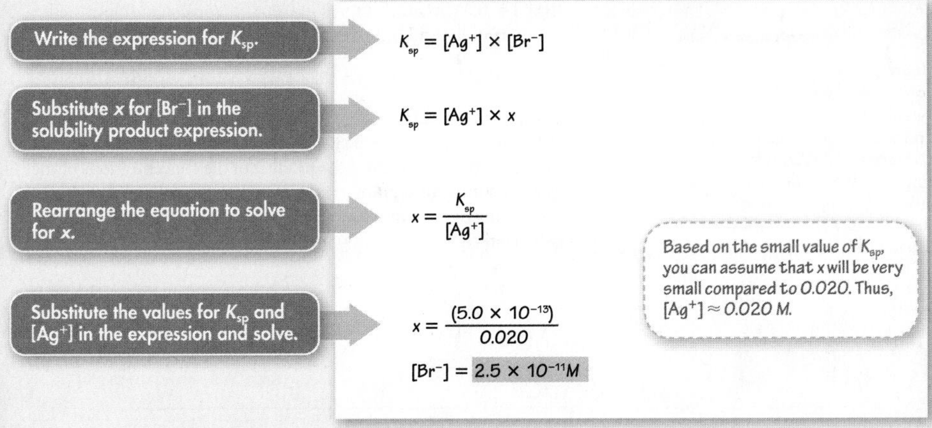

Based on the small value of K_{sp}, you can assume that x will be very small compared to 0.020. Thus, $[Ag^+] \approx 0.020\ M$.

3 Evaluate Does the result make sense? The concentration of Br^- in a saturated solution of AgBr is $7.0 \times 10^{-7} M$ (the square root of the K_{sp}). It makes sense that the addition of $AgNO_3$ would lower the concentration of Br^- because the presence of a common ion, Ag^+, causes AgBr to precipitate from solution.

35. What is the concentration of sulfide ion in a 1.0-L solution of iron(II) sulfide to which 0.04 mol of iron(II) nitrate is added? The K_{sp} of FeS is 8×10^{-19}.

36. The K_{sp} of $SrSO_4$ is 3.2×10^{-7}. What is the equilibrium concentration of sulfate ion in a 1.0-L solution of strontium sulfate to which 0.10 mol of $Sr(CH_3CO_2)_2$ is added?

You can usually assume that the amount of the slightly soluble compound that dissociates is small compared to the amount of the common ion that is added.

Foundations for Math

CHECK YOUR ANSWER Remind students that checking their answers is a valuable way to know if their answer is correct. Remind students of the inverse properties of multiplication and division so they can quickly check their results.

In **Sample Problem 18.7** the answer can be substituted into the K_{sp} expression. To check that the answer is correct, multiply $[Br^-]$ by $[Ag^+]$. The product should equal K_{sp}.

$(0.020) \times (2.5 \times 10^{-11}) = 5.0 \times 10^{-13}$

Since 5.0×10^{-13} is the value of K_{sp}, the answer checks.

Explore

Class Activity

PURPOSE Students hypothesize the result of mixing silver nitrate and silver chloride.

MATERIALS silver chloride, silver nitrate, 2 test tubes, dropper

PROCEDURE Prepare saturated solutions of silver chloride and silver nitrate in two test tubes. Write the solubility product constant expression for silver chloride on the board.

$$K_{sp} = [Ag^+] \times [Cl^-] = 1.8 \times 10^{-10}$$

Have students hypothesize what would occur if a few drops of the silver nitrate solution were added to the test tube containing the silver chloride. Add the drops of silver nitrate solution and ask students what they observe. *(A precipitate, silver chloride, comes out of solution.)* Explain that the value of $[Ag^+]$ increases in the K_{sp} expression as the silver nitrate is added, so that silver chloride must precipitate until the solubility product again equals K_{sp}.

Sample Practice Problem

What is the concentration of chloride ion in a 1.0-L solution of silver chloride to which 0.030 mol of silver nitrate has been added? The K_{sp} of AgCl is 1.8×10^{-10}. *(6.0×10^{-9})*

Answers

FIGURE 18.19 An increase in $Pb^{2+}(aq)$ shifts the equilibrium to the left, creating more $PbCrO_4(s)$.

35. $2 \times 10^{-17} M$

36. $3.2 \times 10^{-6} M$

LESSON 18.4

Evaluate

Informal Assessment

Have students classify the salts in Table 18.2 on the basis of solubility. **Ask** What is the relationship between the solubility product constant and the relative solubility of a salt? *(The larger the solubility product constant, the more soluble the salt.)*

Reteach

Project a transparency of Table 18.1. Point out some salient features, for example, water is the solvent, and the last column is for exceptions to the first column. Ask students to look in Table 18.2 for examples of ions described as being slightly soluble in Table 18.1. Help students sketch a graph of the relationship between solubility product constants and solubility.

Figure 18.20
Formation of a Precipitate
As solutions of barium nitrate and sodium sulfate are mixed, a precipitate of $BaSO_4$ is formed.
Explain ***Why is barium sulfate the only compound that precipitates when the solutions are mixed?***

You can use the solubility product constant to predict whether a precipitate will form when two solutions are mixed. **A precipitate will form if the product of the concentrations of two ions in the mixture is greater than the K_{sp} value for the compound formed from the ions.** After the precipitate forms, the solution is saturated for that compound. If the product of the concentrations is less than the K_{sp} value, no precipitate will form and the solution is unsaturated.

Suppose you mix two solutions, 0.50 L of 0.002*M* $Ba(NO_3)_2$ and 0.50 L of 0.008*M* Na_2SO_4. The mixture will have a volume of one liter. The compound that might form a precipitate is barium sulfate ($BaSO_4$), which has a K_{sp} value of 1.1×10^{-10}. Precipitation will occur if the product of the concentrations of Ba^{2+} and SO_4^{2-} is greater than the K_{sp} value.

To predict whether a precipitate will form, you need to know the concentration of the ions after the solutions were mixed. During mixing, each solution is diluted with an equal volume of the other solution. So the concentrations of both Ba^{2+} and SO_4^{2-} will be half of their original concentrations. Therefore, in the combined solution, $[Ba^{2+}] = 0.001M$ and $[SO_4^{2-}] = 0.004M$. You can multiply these concentrations and compare the result with the K_{sp} value.

$$[Ba^{2+}] \times [SO_4^{2-}] = (0.001M) \times (0.004M) = 4 \times 10^{-6}$$

The calculated result is larger than the K_{sp} value for barium sulfate (1.1×10^{-10}). Therefore, $BaSO_4$ will precipitate from the solution. The process will continue until the product of the concentrations of the ions remaining in solution is equal to 1.1×10^{-10}. Figure 18.20 shows the mixing of the two solutions and the formation of the barium sulfate precipitate.

ONLINE PROBLEMS

18.4 LessonCheck

37. **Summarize** What does the solubility product constant, K_{sp}, tell you about the solubility of a compound?
38. **Identify** What two values should you compare to predict whether a precipitate will form when two solutions are mixed?
39. **Apply Concepts** Write the solubility product expression for Ag_2CO_3.
40. **Compare** Which compound has the higher solubility: FeS ($K_{sp} = 8.0 \times 10^{-19}$) or CuS ($K_{sp} = 8.0 \times 10^{-37}$)?
41. **Calculate** What is the K_{sp} of nickel(II) sulfide if the equilibrium concentrations of Ni^{2+} and S^{2-} in a saturated solution of NiS are each $2 \times 10^{-10}M$?
42. **Calculate** What is the concentration of lead ions and sulfide ions in a saturated solution of lead sulfide (PbS) solution at 25°C? ($K_{sp} = 3.0 \times 10^{-28}$)
43. **Calculate** The K_{sp} value for barium sulfate is 1.1×10^{-10}. What is the sulfate-ion concentration of a 1.00 L saturated solution of $BaSO_4$ to which 0.015 mol of $Ba(NO_3)_2$ is added?
44. **Predict** Will a precipitate form when 500 mL of a 0.02*M* solution of $AgNO_3$ is mixed with 500 mL of a 0.001*M* solution of NaCl? Explain.

BIGIDEA REACTIONS

45. Explain how the common ion effect illustrates Le Châtelier's principle.

Lesson Check Answers

37. The smaller the K_{sp}, the lower the solubility of the compound.
38. If the product of the concentrations of two ions in a mixture is greater than the K_{sp} of the compound formed from the ions, a precipitate will form.
39. $[Ag^+]^2 \times [CO_3^{2-}] = K_{sp}$
40. FeS
41. 4×10^{-20}
42. $2 \times 10^{-14}M$
43. $7.3 \times 10^{-9}M$
44. Yes; The product $[Ag^+] \times [Cl^-] = 0.01M \times 0.0005M = 5 \times 10^{-6}$ is greater than the K_{sp} of AgCl (1.8×10^{-10})
45. **BIGIDEA** Adding a common ion (a product in the dissolution expression) disturbs the equilibrium and causes a shift toward the reactant side of the expression.

18.5 Free Energy and Entropy

CHEMISTRY & YOU

Q: *How can a fire start on its own?* Sometimes a fire can occur without an external source of ignition, such as a match or an electrical spark. The fuel might be a pile of oily rags or a stack of hay that has not been thoroughly dried. *Spontaneous combustion* is the term used to describe these fires. In this lesson, you will learn about the conditions that can cause such a fire.

Key Questions

What are two characteristics of spontaneous reactions?

What part does entropy play in a reaction?

What two factors determine whether a reaction is spontaneous?

How is the value of ΔG related to the spontaneity of a reaction?

Vocabulary

- free energy
- spontaneous reaction
- nonspontaneous reaction
- entropy
- law of disorder

Free Energy and Spontaneous Reactions

What are two characteristics of spontaneous reactions?

Some of the energy released in a chemical reaction can be harnessed to do work, such as pushing the pistons in an internal-combustion engine. The energy that is available to do work is called **free energy.** Just because energy is available to do work, however, does not mean that it can be used efficiently. For example, an internal-combustion engine in a car is only about 30 percent efficient. That is, only about 30 percent of the free energy released as gasoline burns is used to move the car. The remaining 70 percent is lost as friction and waste heat. No process can be made 100 percent efficient. Even in living things, which are among the most efficient users of free energy, processes are seldom more than 70 percent efficient.

Spontaneous Versus Nonspontaneous Reactions Energy can be obtained from a reaction only if the reaction actually occurs. In other words, you can write a balanced equation for a chemical reaction, but the reaction may not actually take place. For example, you can write an equation for the decomposition of carbon dioxide to carbon and oxygen.

$$CO_2(g) \longrightarrow C(s) + O_2(g)$$

This equation, which represents the reverse of combustion, is balanced. However, experience tells you that this reaction does not tend to occur. Carbon and oxygen react to form carbon dioxide, not the reverse. The world of balanced chemical equations is really divided into two groups. One group contains equations representing reactions that actually occur. The other contains equations representing reactions that do not tend to occur, or at least not efficiently.

Key Objectives

IDENTIFY the two characteristics of spontaneous reactions.

IDENTIFY the part entropy plays in a chemical reaction.

IDENTIFY the two factors that determine whether a reaction is spontaneous.

DESCRIBE how the value of ΔG is related to the spontaneity of a reaction.

Additional Resources

- Reading and Study Workbook, Lesson 18.5
- Probeware Laboratory Manual
- Virtual Chemistry Labs

LESSON 18.5

Engage

CHEMISTRY & YOU Explain when a substance starts to combust without an external energy source, it is called spontaneous combustion. **Ask** Considering what you have learned about reactions so far, what factors do you think spontaneous chemical reactions have in common? *(Sample answers: reactions that have low activation energy; reactions that involve active elements such as alkali metals; combustion reactions above the kindling temperature; reactions that produce a precipitate.)*

Access Prior Knowledge

Begin a cluster diagram on an overhead relating the word *energy* to exothermic reactions, endothermic reactions, and the requirements for a chemical reaction to occur. Ask students to recall what they have learned about these three concepts in regard to energy.

Focus on ELL

1 CONTENT AND LANGUAGE Present the following pairs of words in a T-chart: *endothermic* and *exothermic*, *spontaneous* and *nonspontaneous, order* and *disorder*, *forward* and *reverse*, *increase* and *decrease*, *absorb* and *release*, and *favorable* and *unfavorable*. Explain that the pairs are *antonyms*, word pairs with opposite meanings.

2 FRONTLOAD THE LESSON Have students describe their first or favorite fireworks display. Note that fireworks are examples of combustion reactions. Ask students if they recall any fireworks that might not have behaved as expected during the display. Explain that in this lesson they will learn why a firework may or may not explode.

3 COMPREHENSIBLE INPUT When discussing Figure 18.26, be sure that students understand that the length of the colored arrows represents the sum of their associated black arrows. Point out that black arrows will be shorter for reactions in which the colored arrows point in opposite directions.

Answers

FIGURE 18.20 Sodium nitrate is a soluble compound.

LESSON 18.5

Foundations for Reading

BUILD VOCABULARY Have students write definitions for *spontaneous* and *nonspontaneous* and *order* and *disorder* in their own words. Guide students to consider the meaning of these words as they relate to human behavior.

READING STRATEGY Have students create a compare/contrast table to track similarities and differences between spontaneous and nonspontaneous reactions.

Explain

Free Energy and Spontaneous Reactions

MAKE A CONNECTION Convey to students that the words *force* and *work* have both common and scientific meanings. Remind students that, in science, a *force* can be defined as a push or a pull, and is measured in newtons (N). *Work* is done on an object when a force is applied through a distance and the force is in the same direction as the distance moved. Work is measured in joules.

Point out that work is done on a gas when force is applied to reduce its volume. Rapidly expanding gases, such as those produced in an explosion, also do work. **Ask** Is work done on molecules when the temperature is raised or when chemical bonds are broken? *(yes, in both cases)* **Ask** How is energy related to work? *(Work and energy have the same units.)*

USE VISUALS Direct students to Figure 18.21. Explain that the highly spontaneous reactions in fireworks require very high initiation temperatures to produce the fast reaction rate observed. They also release a large quantity of free energy. Clarify, however, that spontaneous reactions are not directly related to reaction rate.

Misconception Alert

Students may think that a nonspontaneous reaction is one in which no products are formed. Explain that the difference between spontaneous and nonspontaneous reactions is the amount of products at equilibrium. Remind students that a reversible reaction under a given set of conditions has a both a spontaneous reaction and a nonspontaneous reaction.

Figure 18.21 Spontaneous Reaction Fireworks displays are the result of highly favored spontaneous reactions. A large quantity of free energy is released.

Some chemical reactions are spontaneous. A **spontaneous reaction** occurs naturally and favors the formation of products at the stated conditions. Figure 18.21 shows an example of a spontaneous reaction. **Spontaneous reactions produce large amounts of products and release free energy.** A chemical reaction that does not favor the formation of products at the stated conditions is called a **nonspontaneous reaction.** Such reactions produce little, if any, product.

Reversible Reactions In nearly all reversible reactions, one reaction is favored over the other. Consider the decomposition of carbonic acid in water.

$$\underset{<1\%}{H_2CO_3(aq)} \rightleftharpoons \underset{>99\%}{CO_2(g) + H_2O(l)}$$

In the forward reaction, carbonic acid is the reactant. Suppose you could start with pure carbonic acid in water and let the system come to equilibrium. More than 99 percent of the reactant would be converted to the products carbon dioxide and water. These products are highly favored at equilibrium. The natural tendency is for carbonic acid to decompose to carbon dioxide and water. Thus, the forward reaction is spontaneous and releases free energy. In the reverse reaction, carbon dioxide and water are the reactants, and carbonic acid is the product. Suppose you allow a solution of carbon dioxide in water to come to equilibrium. Less than 1 percent of the reactants will combine to form carbonic acid. The reactants show little natural tendency to go to products. Thus, the combination of carbon dioxide and water to form carbonic acid is a nonspontaneous reaction.

Figure 18.22 shows another example of a reversible reaction. When solutions of cadmium nitrate and sodium sulfide are mixed, the products are aqueous sodium nitrate and solid yellow cadmium sulfide. Cadmium sulfide is highly favored. Thus, the forward reaction is spontaneous. The reverse reaction, the production of cadmium nitrate and sodium sulfide from cadmium sulfide and sodium nitrate, is nonspontaneous.

$$Cd(NO_3)_2(aq) + Na_2S(aq) \rightleftharpoons CdS(s) + 2NaNO_3(aq)$$

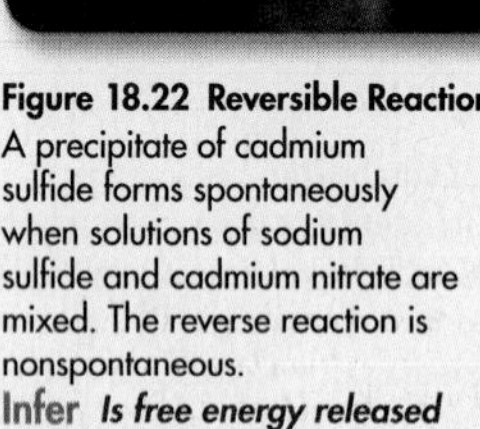

Figure 18.22 Reversible Reaction A precipitate of cadmium sulfide forms spontaneously when solutions of sodium sulfide and cadmium nitrate are mixed. The reverse reaction is nonspontaneous.

Infer *Is free energy released in this reaction?*

Differentiated Instruction

L1 LESS PROFICIENT READERS Clarify the concepts of free energy and energy efficiency. Steer students away from relating the words *free* and *efficient*. Explain that energy efficiency is a measure of the amount of free energy that is used to do work.

ELL ENGLISH LANGUAGE LEARNERS Provide students with a key for each symbol presented in this lesson. Include the various arrow symbols as well as delta (Δ). Encourage students to use these symbols when answering questions in the labs and in the assessments.

L3 ADVANCED STUDENTS As a preparation for biochemistry, have students write down a coupled reaction involving adenosine triphosphate (ATP). Have them research and explain why ATP has the ability to enter into various types of coupled reactions.

Figure 18.23 Photosynthesis
Outside of plants, carbon dioxide and water do not normally combine to produce glucose ($C_6H_{12}O_6$) and oxygen. Photosynthesis, which is a series of reactions, does occur in plants.
Infer ***What happens to the energy that is stored in glucose?***

The Rate of Spontaneous Reactions It is important to note that the terms *spontaneous* and *nonspontaneous* do not refer to the rate of a reaction. Some spontaneous reactions are so slow that they appear to be nonspontaneous. The reaction of table sugar and oxygen is an example of such a reaction.

$$C_{12}H_{22}O_{11} + 12O_2 \longrightarrow 12CO_2 + 11H_2O$$

Nothing appears to be happening in a bowl of sugar sitting on a table. You might assume that the reaction is nonspontaneous. In fact, the reaction is highly spontaneous, but at room temperature the reaction is so slow that it would take thousands of years for the reaction to reach completion. When you supply energy in the form of heat, the reaction is fast. Then it is obvious that the formation of carbon dioxide and water is highly favored.

Changing the conditions of a chemical reaction can affect more than the reaction rate. It can also affect whether a reaction will occur. A reaction that is nonspontaneous in one set of conditions may be spontaneous in other conditions. Consider the photosynthesis reaction, which is summarized in Figure 18.23. This multistep reaction takes place in plant leaves. This complex process could not happen without the energy supplied by sunlight. Other requirements include plant pigments, such as chlorophyll, that absorb the light energy.

Coupled Reactions Sometimes a nonspontaneous reaction can be made to occur if it is coupled to a spontaneous reaction. One reaction releases energy that is used by the other reaction. Coupled reactions are common in the complex biological processes that take place in living organisms. Within cells, a series of spontaneous reactions release the energy stored in glucose. Molecules in cells capture and transfer the free energy to nonspontaneous reactions, such as the formation of proteins. Chapter 24 describes these processes.

CHEMISTRY & YOU

Q: *Decomposition reactions that occur inside a pile of oily rags or a damp stack of hay cause heat to build up. If the heat cannot escape, the temperature within the pile or stack will rise. How can a rise in temperature cause a fire to start on its own? Hint: Think about the reaction of table sugar and oxygen.*

UbD Check for Understanding

The Essential Question Why do some reactions occur naturally, and others do not?

Assess students' knowledge about reactions by asking them to show you a thumbs-up for *true* and a thumbs-down for *false* for the following questions.

- Chemical reactions that occur naturally only occur in nature. *(false)*
- Spontaneous reactions occur without energy input. *(true)*
- A reaction that occurs naturally almost always releases free energy. *(true)*
- Any reaction that occurs immediately after the temperature is increased is a spontaneous reaction. *(false)*

ADJUST INSTRUCTION If students struggle to answer the questions, have them review the section on Free Energy and Spontaneous Reactions. As a class complete a T-chart listing characteristics of natural/unnatural reactions.

Explain

Misconception Alert

Students may think that any reaction that does not involve human involvement may occur spontaneously. Caution students that just because a process occurs in nature does not necessarily mean that the reaction occurs spontaneously.

CRITICAL THINKING Have students consider the combustion of glucose to produce carbon dioxide and water. Point out that the reverse of this reaction is the overall reaction for photosynthesis, the process by which plants make glucose using carbon dioxide and water. Ask students to consider the magnitude of the free-energy value for the photosynthesis reaction, +2868 kJ/mol. **Ask** Under ordinary conditions, do you think photosynthesis is a spontaneous reaction or a nonspontaneous reaction? Why? *(nonspontaneous; because it requires an energy source)*

Answers

FIGURE 18.22 Free energy is released because the reaction is spontaneous.

FIGURE 18.23 It is used later during cellular respiration.

LESSON 18.5

Explain

Entropy

START A CONVERSATION Ask students to raise their hands and state a word that is a synonym of order. *(orderliness, tidiness, cleanliness, pattern, organization, distribution, regulation)* Neatly write each word on the board. Next repeat this activity but tell the class they have 30 seconds to shout out words that are synonyms of disorder. *(chaos, confusion, messiness, turmoil, untidiness, unruliness, disarray, clutter)* Quickly and haphazardly write the words on the board.

Ask Which activity did you enjoy more, the orderly brainstorming session or the disorderly brainstorming session? Why? *(Sample answers: disorderly; it was more fun, we could be loud, we could speak whenever we wanted)* Explain that the way in which students enjoyed the chaos of the brainstorming session is similar to the way in which a chemical reaction proceeds. Convey to the class that molecules have a natural tendency to move away from order toward disorder. State that molecules, like students, would rather be able to act unruly too.

INTERPRET DIAGRAMS Have students study Figure 18.24. Before students read the caption, **Ask** Do you think dogs would rather be on a leash or would rather be running freely on their own? *(freely on their own)* **Ask** What do you think the purpose is for each picture? *(to show order and disorder)*

USE AN ANALOGY Ask students to describe other human-made or natural environments that are analogous to the concept of entropy. *(Sample answers: weeds in a garden, books dumped haphazardly on library shelves, children at recess, fans entering a stadium, a table of sweaters on a sale day, volcanoes, the contents of a house after a flood or earthquake)*

Figure 18.24 Order and Disorder
A dog walker with several dogs could represent relative order and disorder. **a.** All of the dogs are on leashes and are strolling orderly along the path. **b.** The dogs are no longer wearing leashes and are running freely. This situation represents disorder.

Entropy

What part does entropy play in a reaction?

Recall that changes in the heat content, or enthalpy, of a system occur with most chemical and physical processes. These changes help determine whether a process is spontaneous. For example, the combustion of carbon (graphite) is exothermic. The reaction releases 393.5 kJ for each mole of carbon burned. The reaction is spontaneous.

$$C(s, \text{graphite}) + O_2(g) \longrightarrow CO_2(g) + 393.5 \text{ kJ/mol}$$

You might expect that only exothermic reactions are spontaneous. Some processes, however, are spontaneous even though they absorb heat. Consider what happens as ice melts. As it changes from a solid to a liquid, 1 mol of ice at 25°C absorbs 6.0 kJ of heat from its surroundings. The liquid water has a higher energy than the solid ice.

$$H_2O(s) + 6.0 \text{ kJ/mol} \longrightarrow H_2O(l)$$

If you consider only enthalpy changes, it is difficult to explain why the ice melts. In spontaneous processes, the rule seems to be that the direction of energy change is from higher energy to lower energy. Yet the ice does melt. Some factor other than the enthalpy change must help determine whether a physical or chemical process is spontaneous.

The other factor is related to order. You are probably familiar with everyday ideas about order and disorder. For example, a handful of marbles is relatively ordered in the sense that all the marbles are collected in one place. If the marbles are dropped, it is not probable that they will end up in the same neat arrangement. Instead, the marbles scatter on the ground. They become disordered. Scattered marbles have a higher entropy than a handful of marbles. **Entropy** is a measure of the disorder of a system. Figure 18.24 provides another comparison of relative order and disorder.

The **law of disorder** states that the natural tendency is for systems to move in the direction of increasing disorder or randomness. You are probably familiar with this tendency. For example, you start the school year with an empty locker that you gradually fill with items. For a few weeks, it is easy to find items in your locker. Over time, your locker may tend to become more disordered unless you put energy (work) into maintaining the order.

The law of disorder also applies at the level of atoms and molecules. So entropy can affect the direction of a reaction. **Reactions in which entropy increases as reactants form products tend to be favored.** Figure 18.25 provides some general rules to help you predict the effect of entropy on a reaction.

Origin of Entropy

The term *entropy* was coined by the German physicist Rudolf Clausius in 1865. He used the term to describe the portion of free energy that is unavoidably lost as waste heat. Clausius pointed out that in any process involving the flow of energy, there is always some loss, so the entropy of the universe is always increasing.

Figure 18.25 Entropy
Here are four examples of changes that can increase the entropy of a system.

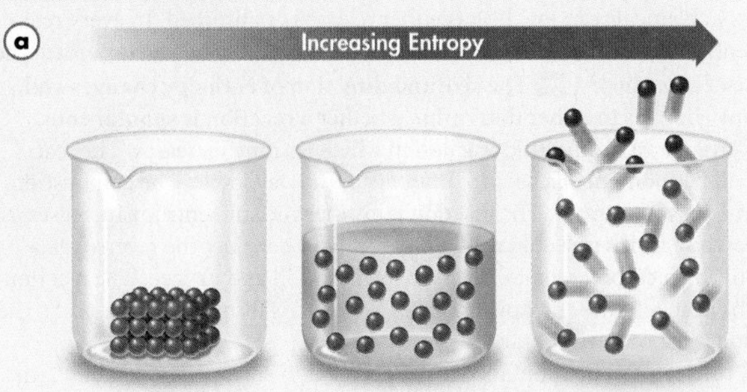

For a given substance, the entropy of the gas is greater than the entropy of the liquid or the solid. Similarly, the entropy of the liquid is greater than that of the solid. Thus, entropy increases in reactions in which solid reactants form liquid or gaseous products. Entropy also increases when liquid reactants form gaseous products.

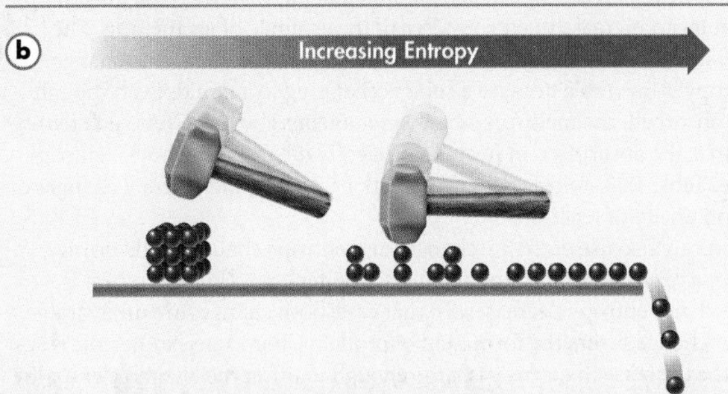

Entropy increases when a substance is divided into parts. For instance, entropy increases when an ionic compound, such as sodium chloride, dissolves in water. The sodium ions and chloride ions are less ordered in solution than they are in the solid crystal.

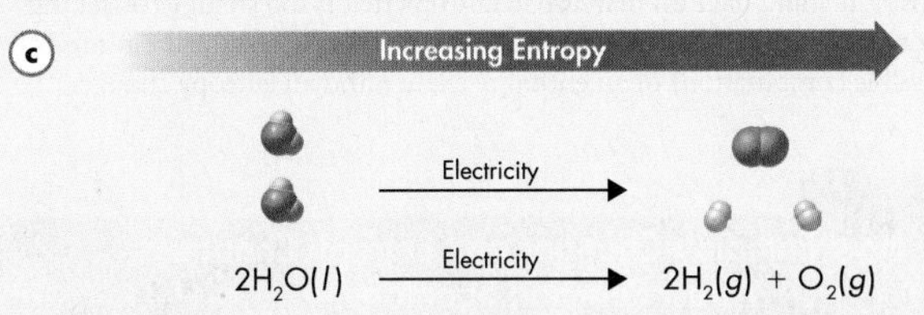

Entropy tends to increase in chemical reactions in which the total number of product molecules is greater than the total number of reactant molecules.

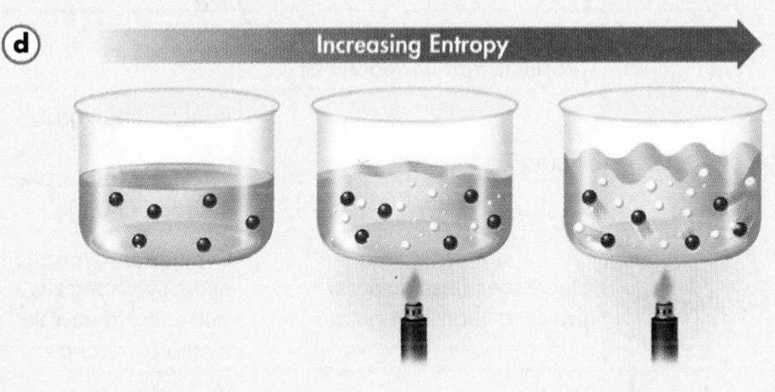

Entropy tends to increase when the temperature increases. As the temperature rises, the molecules move faster and faster, which increases the disorder.

Differentiated Instruction

ELL ENGLISH LANGUAGE LEARNERS Terms such as *entropy* and *enthalpy* may be confusing to English language learners due to their similar sounds. When discussing these topics, pair each term with a descriptive phrase such as "molecular disorder" for entropy and "heat content" for enthalpy to help them distinguish which term is being used.

L1 LESS PROFICIENT READERS Prove the concept that a system moves more toward disorder than order by giving each student a packet of salt or pepper. Ask students to open their packet and pour the contents into something else without losing one particle. Have students describe the difficulty in doing this, and guide them to understand how the particles naturally moved toward disorder rather than order.

L3 ADVANCED STUDENTS State that entropy decreases in the reaction of iron, air, and water to form rust. Have students explain why this reaction is spontaneous. *(An ordered solid (Fe) and a less-ordered gas (O_2) release enough energy to balance the production of a more-ordered solid (Fe_2O_3) in the presence of a catalyst (H_2O).)*

Explore

Teacher Demo

PURPOSE Students infer the relative magnitude of the entropy of water in different states.

MATERIALS block of ice, beaker of water, rapidly boiling water, hot plate

PROCEDURE Place a block of ice, a beaker of water, and a beaker of rapidly boiling water in front of the class. Explain that in standard conditions, the natural tendency of the solid form of water is to turn to liquid and the natural tendency of the liquid form of water is to turn to vapor. **Ask** Which state of matter has the least entropy? *(the block of ice)* The greatest entropy? *(water vapor)* Why? *(In solid ice, the water molecules are highly ordered in a lattice. In liquid water, despite the hydrogen bonding between molecules, the molecules are relatively free to move around. In the water vapor produced by boiling water, randomness is greatly increased because the water molecules are able to move freely.)*

Extend

Connect to METEOROLOGY

Ask for a show of hands of students who have heard of the *butterfly effect*. Meteorologists once joked that the fluttering of a single butterfly's wings on one day in one part of the world creates a tiny disturbance in the air that ultimately results in completely altering weather patterns several months into the future. Explain that the butterfly effect has become a meteorological metaphor for chaos theory, a branch of mathematics devoted to the study of changes in systems with sensitive initial conditions. The butterfly effect is used to explain why weather can be so accurately predicted 24 hours into the future, but forecasts for a week to 10 days later often bear little resemblance to the actual weather events on those days. Have students demonstrate the butterfly effect by making a record of the day's weather forecast, making note of the 24-hour, 2-day, and 7- or 10-day forecast. Have students track the actual weather events that occur on each day and compare them back to the original forecast, then present their findings to the class.

Interested students may also enjoy reading the book *Chaos: Making a New Science* by James Gleick. The book is a non-technical, entertaining account of how chaos theory came about.

LESSON 18.5

Explain

Enthalpy and Entropy

APPLY CONCEPTS Explain that evidence suggests that Earth is being warmed because of the enhanced greenhouse effect, which is largely due to the release into the atmosphere of CO_2 produced by combustion. Another aspect of the greenhouse effect that relates to the topic of spontaneity is the increase in the quantity of atmospheric CO_2 gas that would occur if the oceans became warmer.

Remind students that the dissolution of CO_2 in bodies of water is an important part of the carbon cycle. However, CO_2, like all other gases, is less soluble in warm ocean water than in cool ocean water. There is concern that if the oceans become warmer, more CO_2 would be released into the atmosphere. In warmer oceans, the process whereby CO_2 dissolves in ocean water would become nonspontaneous, and the reverse process, the release of CO_2 gas, would become spontaneous. Guide students to understand that if this occurred, then the increased temperature would cause even more CO_2 release, which would further increase temperature, and the cycle would continue on.

Connect to CHEMICAL ENGINEERING

Explain that thermodynamics is the study of the relationship between heat, work, and energy. Although a complex science, it can generally be summarized by the three laws of thermodynamics. The British scientist and author C.P. Snow used a common sense approach for remembering the three laws of thermodynamics. Have students research the actual laws of thermodynamics and provide a statement explaining each of C.P. Snow's analogies:

- **Law 1:** You cannot win. *(The total energy of a system remains constant, even if it is converted from one form to another.)*
- **Law 2:** You cannot break even. *(Heat naturally flows in the direction of a hotter region to a cooler region, but not the other way around; You cannot return to the same energy state, because the disorder of a system tends to increase.)*
- **Law 3:** You cannot get out of the game. *(Since molecules are never completely at rest, absolute zero is impossible to reach since there will always be some type of disorder.)*

Enthalpy and Entropy

What two factors determine whether a reaction is spontaneous?

In every chemical reaction, heat is either released or absorbed. In every reaction, entropy either increases or decreases. How do these two factors affect the course of a reaction? **The size and direction of enthalpy changes and entropy changes together determine whether a reaction is spontaneous.**

Consider an exothermic reaction in which entropy increases. The reaction will be spontaneous because both factors are favorable. The combustion of carbon is an example. The reaction is exothermic, and entropy increases as solid carbon forms gaseous carbon dioxide. Now consider the reverse reaction in which carbon dioxide reacts to form carbon and oxygen. The reaction absorbs heat, and the entropy decreases as a solid is formed from gases. So the reaction must be nonspontaneous.

A reaction can be spontaneous if a decrease in entropy is offset by a large release of heat. A reaction can also be spontaneous if an increase in enthalpy is offset by an increase in entropy. Recall the example of ice melting. The change in enthalpy is not favorable because heat is absorbed. The change in entropy is favorable because a solid is changing to a liquid. Even though heat is absorbed, the melting of ice is spontaneous above 0°C. At such temperatures, the absorption of heat is sufficiently offset by a favorable entropy change. Table 18.3 summarizes the effect of enthalpy and entropy changes on the spontaneity of reactions.

You can also use enthalpy changes and entropy changes to determine whether a reaction is nonspontaneous. Consider a reaction when heat is absorbed and entropy decreases. In that case, both changes are unfavorable. Neither change favors the formation of products. For some exothermic reactions, the decrease in entropy is large enough to offset the favorable enthalpy change. In that case, the reaction is not spontaneous. Finally, a reaction that absorbs heat could have an increase in entropy that is too small to offset the unfavorable enthalpy change. Figure 18.26 uses visuals to show the results of six possible combinations of an enthalpy change and an entropy change.

Table 18.3

How Enthalpy Changes and Entropy Changes Affect Reaction Spontaneity

Enthalpy change	Entropy change	Is the reaction spontaneous?
Decreases (exothermic)	Increases (more disorder in products than in reactants)	Yes
Increases (endothermic)	Increases	Only if unfavorable enthalpy change is offset by favorable entropy change
Decreases (exothermic)	Decreases (less disorder in products than in reactants)	Only if unfavorable entropy change is offset by favorable enthalpy change
Increases (endothermic)	Decreases	No

The Evolution of Thermodynamics

Throughout the nineteenth century many scientists studied machinery and efficiency. The hope was to find a way to efficiently convert all of the heat produced by a machine into work. This would have resulted in a machine that could be powered by its own recycled heat, producing a boundless cycle of work, or perpetual motion. Unfortunately what scientists discovered did not lead to a perpetual-motion machine. Instead their findings led to a solid understanding of the relationship between heat, energy, and work in a system, hence the laws of thermodynamics.

The first law states that energy can neither be created nor destroyed, and is also known as the Law of Conservation of Energy. The second law states that in an isolated system, natural processes are spontaneous when they lead to an increase in disorder. The third law states that as temperature approaches absolute zero, the entropy of a system approaches a constant minimum.

Figure 18.26 Enthalpy and Entropy
The combination of the enthalpy change and the change in entropy for a reaction determines whether or not the reaction is spontaneous.
Summarize ***Explain why a spontaneous reaction can occur when one factor is unfavorable but not when both factors are unfavorable.***

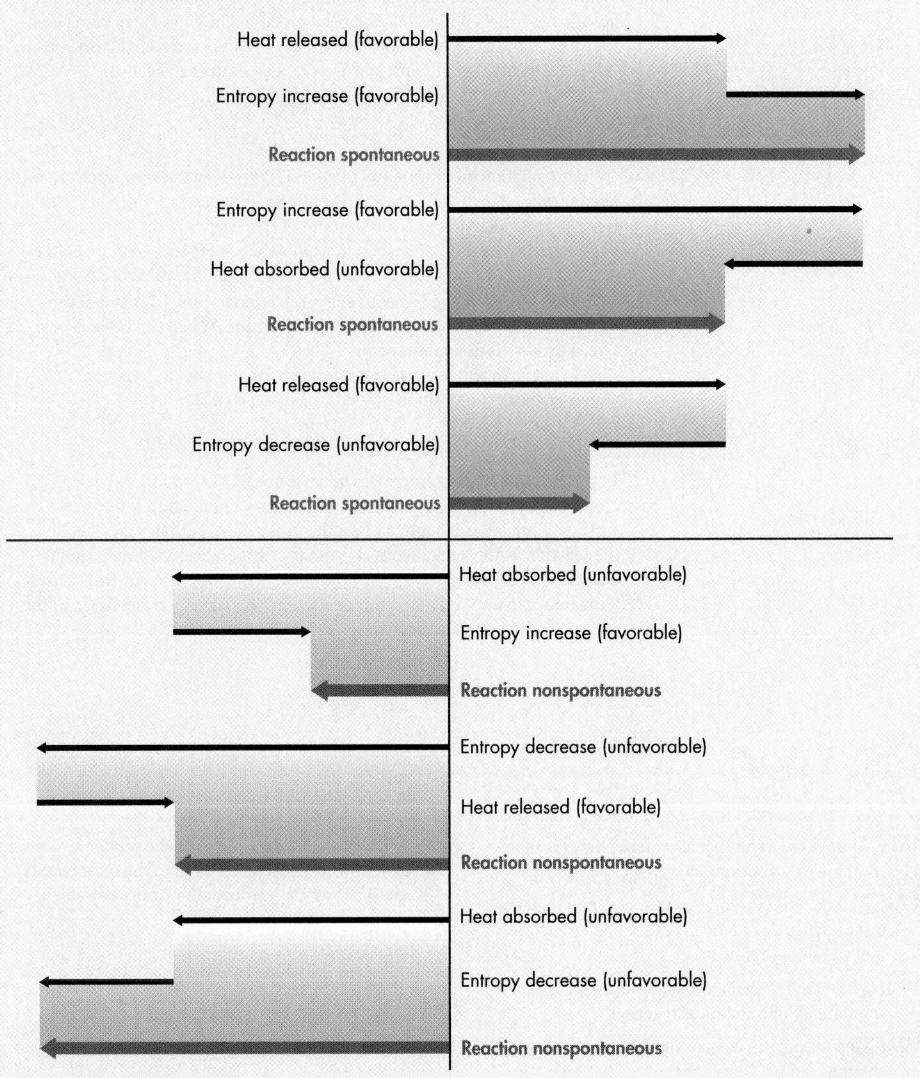

Explore

Teacher Demo

PURPOSE Students observe a spontaneous reaction with a negative ΔH and a positive ΔS.

MATERIALS 0.1M solution of $KMnO_4$ in 0.1M HCl, graduated cylinder, 1000-mL beaker, thermometer, 20 mL of 30% H_2O_2, watch glass, splint, matches

SAFETY Wear gloves and safety goggles. Potassium permanganate is a powerful oxidizing agent and a strong skin irritant. The manganese ion is considered hazardous waste and must be disposed of in compliance with local regulations.

PROCEDURE Transfer 100 mL of the $KMnO_4$ solution to a 1000-mL beaker. Record its temperature. With a graduated cylinder, measure 20 mL of 30% H_2O_2. (Use a more dilute solution for a less vigorous reaction.) Record its temperature. Add the peroxide to the permanganate and cover the beaker with a watch glass. Record the temperature when the most vigorous reaction has subsided. Test the gas produced with a glowing splint.

ANALYZE AND CONCLUDE Write on the board the chemical reaction that was demonstrated.

$$2MnO_4^-(aq) + 5H_2O_2(l) + 6H^+(aq) \rightarrow 2Mn^{2+}(aq) + 8H_2O(l) + 5O_2(g).$$

Ask Was the reaction spontaneous? *(yes)* **Ask** Is the change in free energy positive or negative? Why? *(ΔG is negative since spontaneous reactions release free energy.)* **Ask** Is the entropy change positive or negative? How do you know? *(positive, because a gas is produced.)* **Ask** Considering the free-energy equation, $\Delta G = \Delta H - T\Delta S$, and the temperature change, what can you conclude about the change in enthalpy at room temperature? *(ΔH is negative at room temperature.)*

UbD Check for Understanding

The Essential Question What is the role of energy in chemical reactions?

Ask students whether each reaction described in the statements below is spontaneous or nonspontaneous

- An endothermic reaction occurs that has a large entropy increase that offsets the enthalpy change. *(spontaneous)*
- A reaction is exothermic with an increase in entropy. *(spontaneous)*
- An endothermic reaction results in a decrease in entropy. *(nonspontaneous)*

ADJUST INSTRUCTION Have students review Table 18.3 and Figure 18.26. Then have them revisit their incorrect answers.

Answers

FIGURE 18.26 Both factors (changes in enthalpy and entropy) together determine the spontaneity of the reaction. If both factors are unfavorable, the reaction is not spontaneous. But, if one of the factors is favorable, it can offset the unfavorable factor.

LESSON 18.5

Explain

Free Energy Change

DRAW CONCLUSIONS Point out that free energy refers to more than just the enthalpy change that occurs in a chemical reaction; it combines the change in enthalpy with the change in entropy. Explain that it is usually assumed that ΔH and ΔS are constant for the same reactions at different temperatures, so the contribution to the $T\Delta S$ term in the free energy equation comes mostly from the temperature itself.

Evaluate

Informal Assessment

Have students pair up to review what they have learned. One student should explain the significance of positive and negative changes in entropy. The pair then should evaluate the explanation for defects. The second student should then explain how the values of ΔG, ΔH, and ΔS determine whether a reaction is spontaneous. Again, the students should evaluate the explanation. You may wish to write the following questions on the board as a guide for discussion:

- What energy changes occur during a reaction that favors spontaneity? *(changes in which heat is released and entropy increases)*
- Under what conditions is a reaction least likely to be spontaneous? *(when heat is absorbed and entropy decreases)*
- What happens when one change favors spontaneity and one change does not? *(The relative strengths of the opposing changes determine the outcome.)*

Then have students complete the 18.5 Lesson Check.

Reteach

Rewrite Table 18.3 on the board, inserting an extra column labeled *Temperature change* between the second and third columns, and relabeling the fourth column *Free energy change*. Add the appropriate conditions to the temperature column necessary to make the fourth column true.

Free Energy Change

How is the value of ΔG related to the spontaneity of a reaction?

Free energy is often expressed as Gibbs free energy. This term is named for Josiah Gibbs, the scientist who defined this thermodynamic property. The symbol for Gibbs free energy is G. Free energy can either be released or absorbed during a physical or chemical process. The equation below is used to calculate the change in Gibbs free energy (ΔG). ΔS is the change in entropy, ΔH is the change in enthalpy, and T is the temperature in kelvins.

$$\Delta G = \Delta H - T\Delta S$$

The equation provides a way to quantify the effect of enthalpy changes and entropy changes on a process. Recall that spontaneous processes release free energy, which decreases the amount of free energy in the system. In contrast, work must be done for a nonspontaneous process to occur. So the amount of free energy in the system increases. Thus, the value of ΔG indicates whether a process is spontaneous at a given temperature. **When the value of ΔG is negative, a process is spontaneous. When the value is positive, a process is nonspontaneous.**

A change in temperature can affect the spontaneity of a reaction. Consider the decomposition of solid calcium carbonate.

$$CaCO_3(s) + 178\text{ kJ} \longrightarrow CaO(s) + CO_2(g)$$

In this reaction, entropy increases because one of the products is a gas. However, the reaction absorbs heat. At ordinary temperatures, the increase in entropy is not large enough to offset the unfavorable enthalpy change. Thus, the reaction is nonspontaneous at ordinary temperatures. However, in the equation, ΔS is multiplied by T. So the effect of entropy is magnified as the temperature rises. At temperatures above 850°C, $T\Delta S$ can offset ΔH, and the reaction becomes spontaneous.

ONLINE PROBLEMS 18.5 LessonCheck

46. Review What two characteristics do all spontaneous reactions share?

47. Explain How can you use entropy to determine whether a chemical reaction is more or less likely to be spontaneous?

48. Review What two factors determine the spontaneity of a chemical reaction?

49. Summarize When the Gibbs free-energy change for a process is negative, what does this tell you about the process?

50. Define What is free energy?

51. Draw Conclusions Suppose the products of a reaction are more ordered than the reactants. Is the entropy change favorable or unfavorable?

52. Explain How can a reaction be spontaneous at some conditions but nonspontaneous at other conditions?

53. Infer Nitroglycerine decomposes as follows:

$$4C_3H_5(NO_3)_3(l) \longrightarrow 6N_2(g) + O_2(g) + 12CO_2(g) + 10H_2O(g)$$

The ΔH value for this reaction is -1427 kJ/mol. Use what you know about enthalpy and entropy to explain why this reaction is so explosive.

Lesson Check Answers

46. They produce substantial amounts of products at equilibrium and release free energy.
47. An increase in entropy favors a spontaneous reaction; a decrease favors a nonspontaneous reaction.
48. The size and direction of enthalpy and entropy changes together determine whether a reaction is spontaneous.
49. The reaction is spontaneous.
50. the energy available to do work
51. unfavorable
52. A change in state of a reactant, due to change in temperature or pressure, may change the balance between ΔS and ΔH.
53. The reaction of nitroglycerine releases 1427 kJ of heat, and in the process, converts 4 mol of a liquid into 29 mol of four different gases. Thus, both enthalpy and entropy changes favor spontaneity.

Small-Scale Lab

Enthalpy and Entropy

Purpose
To observe and measure energy changes during the formation of a solution and to describe and explain these changes in terms of enthalpy and entropy

Materials
- alcohol thermometer
- four 1-oz plastic cups
- plastic spoon
- sodium chloride
- ammonium chloride
- calcium chloride
- water
- crushed ice
- stirring rod

Procedure
1. Make a data table like the one below on a separate sheet of paper.
2. Place two level spoonfuls of water in one plastic cup. Measure and record the water temperature (T_1).
3. Dry the spoon. Then, add one level spoonful of NaCl to the cup. Stir gently with the thermometer.
4. Record the highest or lowest temperature that results (T_2).
5. Rinse the thermometer and spoon.
6. Repeat Steps 2–5, using NH_4Cl and $CaCl_2$.

Mixture	T_1	T_2	ΔT
$NaCl(s) + H_2O(l)$			
$NH_4Cl(s) + H_2O(l)$			
$CaCl_2(s) + H_2O(l)$			

Analyze and Conclude
1. **Calculate** Determine ΔT for each mixture using the following equation: $\Delta T = T_2 - T_1$
2. **Observe** An exothermic process releases heat. An endothermic process absorbs heat. In which mixture was the process of dissolving endothermic? In which mixture was the process exothermic? In which mixture, was heat neither released nor absorbed? Which solution(s) had little or no change in temperature?

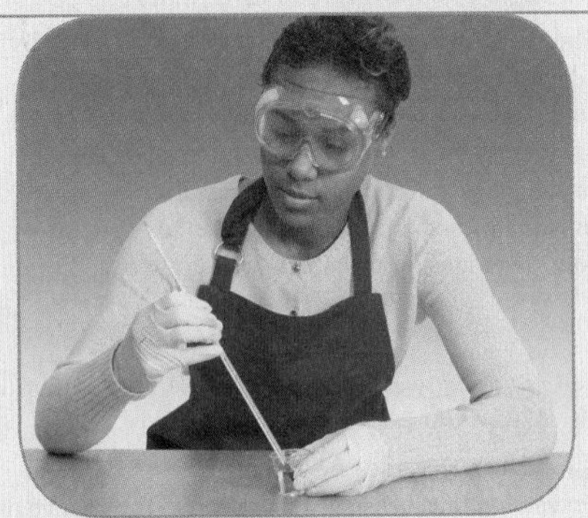

3. **Describe** This is the equation for the dissociation of NaCl in water:

$$NaCl(s) \longrightarrow Na^+(aq) + Cl^-(aq)$$

Write similar ionic equations to show how NH_4Cl and $CaCl_2$ dissociate in water. Include heat as a reactant or a product in each equation.
4. **Draw Conclusions** When a solid dissolves in water, does entropy increase or decrease? Explain your reasoning.
5. **Relate Cause and Effect** Both the entropy change and the enthalpy change were favorable for a spontaneous change in which mixture? In which mixture was one change favorable and the other unfavorable for a spontaneous change? Explain why the change still happened?

You're the Chemist
1. **Analyze Data** Mix a tablespoon of crushed ice with a tablespoon of NaCl. Stir gently with a stirring rod. Then measure and record the lowest temperature reached. Compare the temperature change for this mixture with the results for the mixture of NaCl and liquid water.
2. **Analyze Data** Explain what you observed in the first activity. *Hint:* Is the process of melting ice exothermic or endothermic?
3. **Design an Experiment** Will the results be the same when an ionic compound other than NaCl is mixed with crushed ice? Try doing Activity 1 with NH_4Cl and $CaCl_2$.

Focus on ELL

4 ENABLE LANGUAGE PRODUCTION Have students work in groups of varied language proficiencies, so that more proficient students can help less proficient ones. Have students work according to their proficiency level.

BEGINNING

LOW Clarify the procedures by demonstrating that the same process is to be followed 3 times; only the solid compound will change.

HIGH Provide students with a data table similar to the one in the lab but add an additional column labeled *Heat*. Tell them to write + for endothermic, – for exothermic, and 0 for neither in the *Heat* column.

INTERMEDIATE: LOW/HIGH Have students predict which equation will give off heat, absorb heat, or neither. Ask them to compare the actual result with their prediction. Allow students to orally present their results.

ADVANCED: LOW/HIGH Provide groups with reformatted Analyze and Conclude questions. Reformat the multi-part questions so each question is on a separate line.

Explore

Small-Scale Lab

OBJECTIVE After completing this lab, students will be able to:
- investigate the energy associated with chemical reactions;
- relate changes to exothermic and endothermic reactions.

MATERIALS NaCl, NH_4Cl, $CaCl_2$, alcohol thermometer, 4 1-oz. plastic cups, plastic spoon, crushed ice, water, stirring rod. Optional solids: KCl, $NaHCO_3$, Na_3PO_4

CLASS TIME 40 minutes

SAFETY Use small alcohol thermometers. If students are careful, they can use the thermometers to gently stir the contents of the plastic cup. Remind students of the hazard posed by broken glass. Never use mercury thermometers due to their poisonous contents. Look for a research lab that will trade alcohol thermometers for mercury thermometers.

Analyze and Conclude
1. Sample data for ΔT: **a.** 0°C; **b.** –16°C; **c.** +32°C
2. $NH_4Cl + H_2O$ is endothermic. ΔH is positive. $CaCl_2 + H_2O$ is exothermic. ΔH is negative. $NaCl + H_2O$ is neither endothermic nor exothermic. ΔH is close to 0.
3. heat + $NH_4Cl(s) \rightarrow NH_4^+(aq) + Cl^-(aq)$
   $CaCl_2(s) \rightarrow Ca^{2+}(aq) + 2Cl^-(aq)$ + heat
4. Entropy usually increases during solvation. ΔS is positive in each case.
5. $NH_4Cl + H_2O$; $CaCl_2 + H_2O$; The change still happened because the favorable change in entropy outweighed the unfavorable enthalpy change.

You're the Chemist
1. The ice mixture has a larger change in temperature. The temperature can drop as much as 15°C.
2. The melting of ice is endothermic. Endothermic processes absorb heat and cool the environment.
3. Since they are soluble, they should all lower the temperature of the ice-water mixture.

ENRICHMENT Have students use and then dissect a chemical heat pack that uses sodium acetate and stainless steel. **Ask:** What process occurs that makes the heat pack an exothermic process? *(crystallization of sodium acetate)* What is the purpose of the steel? *(acts as a catalyst)*

STUDY GUIDE

Evaluate

Review and Assessment Resources

CHEMISTRY TUTORIAL Encourage students to review solving equilibrium constant and equilibrium ion concentration problems using the animated step-by-step tutorials for Chapter 18. Each Chemistry Tutorial is followed by a similar practice problem, where the student is guided through the problem-solving process.

CHAPTER PROBLEM SET Have students practice more reaction rate and equilibrium problems using the Chapter 18 online Problem Set.

DIRECTED VIRTUAL LAB Have students complete the Virtual Lab *Enthalpy and Entropy* to help reinforce the concepts of enthalpy, entropy, and spontaneous reactions. In the post-lab exercise, students will analyze the data and draw conclusions based on the results of the experiment.

KINETIC ART Direct students to the Kinetic Art animations *Collision Theory* and *Reversible Reactions*, and have them answer the questions that accompany each animation.

Study Tip

SWITCH TASKS When you feel yourself losing focus, switch the type of task you are working on, the subject that you are studying, or the environment that you are in. Take a break and walk around a bit. Stop studying when you are no longer being productive.

18 Study Guide

BIGIDEAS
REACTIONS, MATTER AND ENERGY

The rate of a chemical reaction can be controlled by adjusting temperature, concentration, or particle size. Adding a catalyst speeds up a reaction by lowering the activation energy. Energy is released in some reactions and absorbed in others. Changes in enthalpy and entropy can be used to explain why some reactions occur naturally and others do not.

18.1 Rates of Reaction

In chemistry, the rate of a chemical reaction, or the reaction rate, is usually expressed as the change in the amount of reactant or product per unit time.

Factors that can affect the rate of a chemical reaction are temperature, concentration, particle size, and the use of a catalyst.

- rate (595)
- collision theory (596)
- activation energy (596)
- activated complex (596)
- inhibitor (601)

18.2 The Progress of Chemical Reactions

The value of the specific rate constant, *k*, in a rate law is large if products form quickly; the value is small if the products form slowly.

Most chemical reactions consist of two or more elementary reactions.

- rate law (604)
- specific rate constant (604)
- first-order reaction (605)
- elementary reaction (607)
- reaction mechanism (607)
- intermediate (607)

18.3 Reversible Reactions and Equilibrium

At chemical equilibrium, both the forward and reverse reactions continue, but because their rates are equal, no net change occurs in the concentrations of the reaction components.

Stresses that upset the equilibrium of a chemical system include changes in concentration of reactants or products, changes in temperature, and changes in pressure.

The size of the equilibrium constant indicates whether reactants or products are more common at equilibrium.

- reversible reaction (609)
- chemical equilibrium (610)
- equilibrium position (611)
- Le Châtelier's principle (612)
- equilibrium constant (616)

18.4 Solubility Equilibrium

The smaller the value of the solubility product constant, the lower the solubility of the compound.

A precipitate will form if the product of the concentrations of two ions in the mixture is greater than the K_{sp} value of the compound formed from the ions.

- solubility product constant (622)
- common ion (624)
- common ion effect (624)

18.5 Free Energy and Entropy

Spontaneous reactions produce large amounts of products and release free energy.

Reactions in which entropy increases as reactants form products tend to be favored.

The size and direction of enthalpy changes and entropy changes together determine whether a reaction is spontaneous.

When the value of ΔG is negative, a process is spontaneous. When the value is positive, a process is nonspontaneous.

- free energy (627)
- spontaneous reaction (628)
- nonspontaneous reaction (628)
- entropy (630)
- law of disorder (630)

Key Equations

$$\text{Rate} = \frac{\Delta A}{\Delta t} = k \times [A] \qquad \text{Rate} = k[A]^x[B]^y$$

$$K_{eq} = \frac{[C]^c \times [D]^d}{[A]^a \times [B]^b}$$

$$K_{sp} = [A]^a \times [B]^b$$

$$\Delta G = \Delta H - T\Delta S$$

UbD Performance Tasks

DIRECT ORDER: Create or download a computer program for your graphing calculator that can be used to determine the order of a chemical reaction. Create a mini-manual detailing the instructions for how to operate the program. The manual should include a cover page, a table of contents, an index defining concept vocabulary, a listing of the information needed about a chemical reaction, and step-by-step keystroke guidelines for operating the program. The manual should also include an example problem, as well as two to three practice problems.

NATURALLY IMPULSIVE: Research and find a reaction that occurs spontaneously in nature. List the reactants and the products of the reaction. Describe how the reaction takes place and the rate of the reaction. Include a graph of the reaction as well as the rate law. Prove that the reaction is spontaneous and describe any factors that increase the reaction's entropy. Organize and present the information in a computer slide-show presentation. Use graphics to create interest in your reaction.

Math Tune-Up: Equilibrium Problems

Problem	❶ Analyze	❷ Calculate	❸ Evaluate
Gaseous sulfur dioxide reacts with oxygen in a 1.0-L container at 600°C to form gaseous sulfur trioxide. $2SO_2(g) + O_2(g) \rightleftharpoons 2SO_3(g)$ At equilibrium, a mixture of these gases contains 1.5 mol SO_2, 1.2 mol O_2, and 3.4 mol SO_3. What is the equilibrium constant for this reaction?	Knowns: $[SO_2] = 1.5$ mol/L $[O_2] = 1.2$ mol/L $[SO_3] = 3.4$ mol/L Unknown: $K_{eq} = ?$ Use the expression for K_{eq}: $K_{eq} = \frac{[C]^c \times [D]^d}{[A]^a \times [B]^b}$	Write the expression for K_{eq} for this equation and substitute the equilibrium concentrations in the expression. $K_{eq} = \frac{[SO_3]^2}{[SO_2]^2 \times [O_2]}$ $= \frac{(3.4 \text{ mol/L})^2}{(1.5 \text{ mol/L})^2 \times (1.2 \text{ mol/L})}$ $K_{eq} = 4.3$	Each concentration is raised to the correct power. The value for K_{eq} is appropriate for an equilibrium mixture that contains a larger number of moles of products than of reactants.
What is the concentration of silver ions and chloride ions in a saturated solution of silver chloride at 25°C? ($K_{sp} = 1.8 \times 10^{-10}$)	Knowns: $K_{sp} = 1.8 \times 10^{-10}$ $AgCl(s) \rightleftharpoons Ag^+(aq) + Cl^-(aq)$ Unknowns: $[Ag^+] = ?M$ $[Cl^-] = ?M$ Use the expression for K_{sp}: $K_{sp} = [A]^a \times [B]^b$	Write the expression for K_{sp}, modify it so there is a single unknown, and solve for that unknown. $K_{sp} = [Ag^+] \times [Cl^-]$ $1.8 \times 10^{-10} = [Ag^+]^2$ $[Ag^+] = \sqrt{1.8 \times 10^{-10}}$ $[Ag^+] = 1.3 \times 10^{-5}M$ $[Cl^-] = 1.3 \times 10^{-5}M$	If you multiply $[Ag^+] \times [Cl^-]$, the result is 1.7×10^{-10}, which is close to the value of K_{sp}. The value is not exactly the same because the answers are rounded to two significant figures. Hint: At equilibrium, $[Ag^+] = [Cl^-]$.
Predict whether a precipitate will form when 0.50 L of 0.001M $Ca(NO_3)_2$ is mixed with 0.50 L of 0.0008M Na_2CO_3 to form one liter of solution. The K_{sp} of $CaCO_3$ is 4.5×10^{-9}.	Knowns: 0.50 L of 0.001M $Ca(NO_3)_2$ 0.50 L of 0.0008M Na_2CO_3 K_{sp} of $CaCO_3 = 4.5 \times 10^{-9}$ Unknown: $[Ca^{2+}] \times [CO_3^{2-}] > K_{sp}$? Precipitation will occur if the product of the concentrations of the two ions exceeds the K_{sp} value of $CaCO_3$.	Divide the initial concentrations in half because the volume of the solution has doubled. Multiply the concentrations together as a trial product and compare with K_{sp}. $[Ca^{2+}]$ (final) $= 0.0005M$ $[CO_3^{2-}]$ (final) $= 0.0004M$ $[Ca^{2+}] \times [CO_3^{2-}] = (0.0005M) \times (0.0004M) = 2 \times 10^{-7}$ $2 \times 10^{-7} > K_{sp}$, so a precipitate will form.	A precipitate will form in this case because 2.0×10^{-7} is greater than 4.5×10^{-9}. Remember: Equilibrium constants are not reported with units.

Math Review

Example and sample practice problems for the various reaction rate and equilibrium problems can be found on the following pages:

LESSON 18.2 Finding the Order of a Reaction From Experimental Data—page 606

LESSON 18.3 Applying Le Châtelier's Principle—page 615; Expressing and Calculating K_{eq}—page 617; Finding the Equilibrium Constant—page 618; Finding Concentrations at Equilibrium—page 619

LESSON 18.4 Finding the Ion Concentrations in a Saturated Solution—page 623; Finding Equilibrium Ion Concentrations in the Presence of a Common Ion—page 625

Focus on ELL

5 ASSESS UNDERSTANDING Practice for standardized tests by assigning students to complete the Standardized Test Prep for this chapter. Make sure students know how to correctly use the state-specific answer sheet. If students struggle with questions, suggest they review the material with a partner.

BEGINNING: LOW/HIGH Have students underline the high-frequency words they understand and use context clues to eliminate unlikely distracters.

INTERMEDIATE: LOW/HIGH Explain to students that they can skip a question they do not understand immediately. Tell students to mark that test question with a symbol so they remember to come back to it. Tell students to also skip the corresponding item number on the answer sheet.

ADVANCED: LOW/HIGH Have students write out each question they answered incorrectly because they did not understand the concept. Tell students to find the information they need and then try the question again.

Evaluate

Answers

LESSON 18.1

54. Atoms, ions, or molecules can react to form products, or they can bounce apart unchanged.

55. the minimum energy that colliding particles must have in order to react

56. c

57. Gas molecules and oxygen molecules do not have enough energy to react at room temperature. The flame raises the temperature and the energy of collisions, so the reaction begins. The heat released by the reaction maintains the high temperature, and the reaction continues spontaneously.

58. A catalyst increases the rate of reactions by providing an alternative reaction mechanism with a lower activation energy.

LESSON 18.2

59. an expression of the rate of reaction in terms of concentrations of reactants

60. A first-order reaction has a reaction rate that is directly proportional to concentration of only one reactant. A higher-order reaction has a reaction rate that is dependent on the concentrations of more than one reactant.

61.

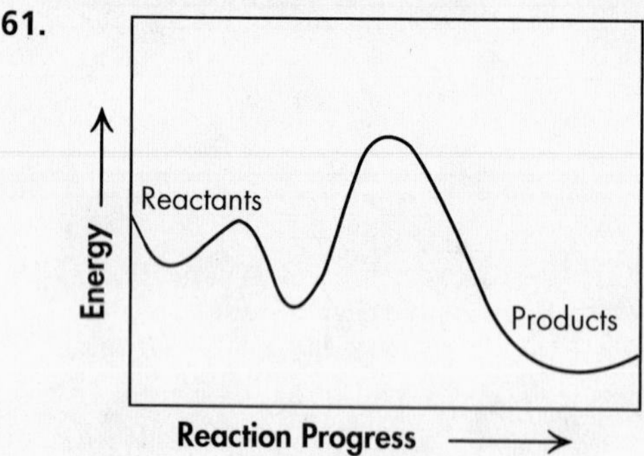

The intermediate is $N_2O_2(g)$.

62. $2NO(g) + O_2(g) \rightarrow 2NO_2(g)$

LESSON 18.3

63. In a reversible reaction, reactants are continually forming products and products are continually forming reactants.

64. They are equal.

65. It states that a system in dynamic equilibrium changes to relieve a stress applied to it.

66. **a.** $K_{eq} = \dfrac{[H_2S]^2 \times [CH_4]}{[H_2]^4 \times [CS_2]}$ **b.** $K_{eq} = \dfrac{[PCl_3] \times [Cl_2]}{[PCl_5]}$

67. **a.** highly favorable **b.** favorable **c.** highly unfavorable

68. c; If the pressure is increased, the equilibrium position will shift to left (reactant side) to relieve the stress by reducing the number of gas molecules.

18 Assessment

ONLINE PROBLEMS

★ Solutions appear in Appendix E

Lesson by Lesson

18.1 Rates of Reaction

★54. According to collision theory, what are two things that can happen when atoms, ions, or molecules collide?

55. What is activation energy?

56. Which of these statements is true?

a. Chemical reactions tend to slow down when the temperature rises.

b. Once a chemical reaction starts, the reacting particles no longer have to collide for products to form.

c. Increasing the total surface area of solid or liquid reactants increases the rate of the reaction.

57. When the gas to a stove is turned on, the gas does not burn until it is lit by a flame. Explain this observation in terms of the effect of temperature on reaction rate.

★58. Explain how a catalyst is able to change the rate of a reaction.

18.2 The Progress of Chemical Reactions

59. What is a rate law?

60. How is a first-order reaction different from higher-order reactions?

★61. Sketch a reaction progress curve for the overall reaction with the following mechanism:

$$2NO(g) \longrightarrow N_2O_2(g) \text{ (fast)}$$

$$N_2O_2(g) + O_2(g) \longrightarrow 2NO_2(g) \text{ (slow)}$$

What is the intermediate in this reaction?

62. Write the balanced equation for the overall reaction described in Question 61.

18.3 Reversible Reactions and Equilibrium

63. In your own words, define a reversible reaction.

64. Compare the rates of the forward and reverse reactions when a chemical equilibrium has been established.

65. What is Le Châtelier's principle?

★66. Write the expression for the equilibrium constant for each reaction.

a. $4H_2(g) + CS_2(g) \rightleftharpoons CH_4(g) + 2H_2S(g)$

b. $PCl_5(g) \rightleftharpoons PCl_3(g) + Cl_2(g)$

67. For each reaction, decide whether products or reactants will be more common at equilibrium.

a. $H_2(g) + F_2(g) \rightleftharpoons 2HF(g); K_{eq} = 1 \times 10^{13}$

b. $SO_2(g) + NO_2(g) \rightleftharpoons NO(g) + SO_3(g); K_{eq} = 1 \times 10^2$

c. $2H_2O(g) \rightleftharpoons 2H_2(g) + O_2(g); K_{eq} = 6 \times 10^{-28}$

68. For which reaction will a change in pressure affect the equilibrium position? How will the position change, and why?

a. $H_2(g) + F_2(g) \longrightarrow 2HF(g)$

b. $SO_2(g) + NO_2(g) \longrightarrow NO(g) + SO_3(g)$

c. $2H_2O(g) \longrightarrow 2H_2(g) + O_2(g)$

18.4 Solubility Equilibrium

69. What does the solubility product constant (K_{sp}) represent?

70. Write the expression for K_{sp} for each of the following compounds:

a. NiS **b.** $BaCO_3$

★71. Use Table 18.2 to rank these compounds from most soluble to least soluble.

a. CuS **b.** $BaSO_4$ **c.** $SrCO_3$ **d.** AgI

72. What happens when a common ion is added to a saturated solution of an ionic compound?

18.5 Free Energy and Entropy

73. Explain what it means to say that a process is 50 percent efficient.

74. Compare the amount of product at equilibrium in a spontaneous reaction to the amount of product at equilibrium in a nonspontaneous reaction.

75. State the law of disorder in your own words.

76. Which system in each example below has the lower entropy?

a. 50 mL of liquid water or 50 mL of ice

b. 10 g of sodium chloride crystals or a solution containing 10 g of sodium chloride

LESSON 18.4

69. the product of the ion concentrations raised to the power of their coefficients

70. **a.** $[Ni^{2+}][S^{2-}]$ **b.** $[Ba^{2+}][CO_3^{2-}]$

71. c, b, d, a

72. It lowers the solubility of the ionic compound.

LESSON 18.5

73. Only 50 percent of the free energy is actually used to do work.

74. The amount of product at equilibrium in a spontaneous reaction is much greater.

75. Systems naturally move toward maximum disorder.

76. **a.** ice **b.** sodium chloride crystals

*77. Does entropy increase or decrease in each of the following reactions:
a. $CaCO_3(s) \longrightarrow CaO(s) + CO_2(g)$
b. $NH_3(g) + HCl(g) \longrightarrow NH_4Cl(s)$

78. Is it true that all spontaneous processes are exothermic? Explain your answer.

79. Explain why steam condenses to liquid water at normal atmospheric pressure even though the entropy change is unfavorable. *Hint:* Is condensation exothermic or endothermic?

80. How can a change in Gibbs free energy be used to predict whether a reaction will be spontaneous?

Understand Concepts

*81. Which of the following statements are always true for a reaction that is spontaneous?
a. The reaction is exothermic.
b. Entropy is increased in the reaction.
c. Free energy is released in the reaction.

82. Use Le Châtelier's principle to explain why carbonated drinks go flat when their containers are left open.

83. Consider the decomposition of N_2O_5 in carbon tetrachloride (CCl_4) at 45°C.

$$2N_2O_5(soln) \longrightarrow 4NO_2(g) + O_2(g)$$

The reaction is first order in N_2O_5, with the specific rate constant $6.08 \times 10^{-4}\ s^{-1}$. Calculate the reaction rate at these conditions.
a. $[N_2O_5] = 0.200$ mol/L
b. $[N_2O_5] = 0.319$ mol/L

84. Consider the following reversible reaction:

$$2NO_2(g) \rightleftharpoons N_2O_4(g)$$

What will happen to the reaction rate if the concentration of NO_2 is changed from 0.020 mol/L to 0.030 mol/L? What will happen to the equilibrium position?

*85. For the reaction $A + B \rightleftharpoons C$, the activation energy of the forward reaction is 5 kJ, and the total energy change is −20 kJ. What is the activation energy of the reverse reaction?

86. Sketch a reaction progress curve for a reaction that has an activation energy of 22 kJ and a total energy change of −103 kJ.

87. A large box is divided into two compartments with a door between them. In (a), equal quantities of two different monatomic gases are placed in the two compartments. In (b), the door between the compartments is opened, and the gases start to mix. Why would it be highly unlikely for the system in (b) to progress to the system in (c)?

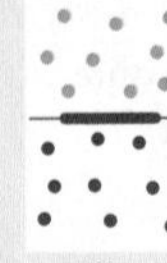
a.

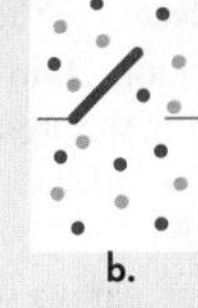
b.

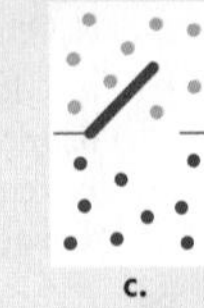
c.

88. Would you expect the entropy to increase in each of the following reactions? Explain your reasoning.
a. $C(s) + O_2(g) \longrightarrow CO_2(g)$
b. $2Al_2O_3(s) \longrightarrow 4Al(s) + 3O_2(g)$
c. $2N(g) \longrightarrow N_2(g)$
d. $N_2(g) \longrightarrow 2N(g)$

*89. What would be the effect on the equilibrium position if the volume is decreased in the following reaction?

$$4HCl(g) + O_2(g) \rightleftharpoons 2Cl_2(g) + 2H_2O(g)$$

90. Write the equilibrium-constant expression for this reaction.

$$2SO_2(g) + O_2(g) \rightleftharpoons 2SO_3(g)$$

*91. A mixture at equilibrium at 827°C contains 0.552 mol CO_2, 0.552 mol H_2, 0.448 mol CO, and 0.448 mol H_2O. The balanced equation is shown below.

$$CO_2(g) + H_2(g) \rightleftharpoons CO(g) + H_2O(g)$$

What is the value of K_{eq}?

*92. What must be true about the concentration of two ions if precipitation occurs when solutions of the two ions are mixed?

93. What is the concentration of carbonate ions in a saturated solution of $SrCO_3$? ($K_{sp} = 9.3 \times 10^{-10}$)

94. Predict what will happen to the rate of a slow reversible reaction if a catalyst is added. What will happen to the equilibrium position?

95. Make a list of five things you did today that resulted in an increase in entropy.

77 **a.** Entropy increases. **b.** Entropy decreases.

78. No, some endothermic processes are spontaneous because of their favorable change in entropy.

79. The favorable exothermic change of the condensation process offsets the unfavorable entropy change.

80. A spontaneous reaction has a negative free energy.

UNDERSTAND CONCEPTS

81. c

82. Carbonated drinks in closed containers are in a state of dynamic equilibrium between the CO_2 in the liquid and above the liquid. When the container is opened, CO_2 gas above the liquid escapes. The gas bubbles out of the liquid in an attempt to reestablish equilibrium.

83. a. $1.22\ 10^{-4}$ mol/(L · s) **b.** $1.94\ 10^{-4}$ mol/(L · s)

84. The rate will increase. The equilibrium position will shift to the right.

85. Since the total energy change for the forward reaction (–20 kJ) is negative, energy is released, and the reaction is exothermic. The reverse reaction is endothermic (energy is absorbed). The activation energy of the reverse reaction is the total height of the peak (20 kJ + 5 kJ = 25 kJ).

86.

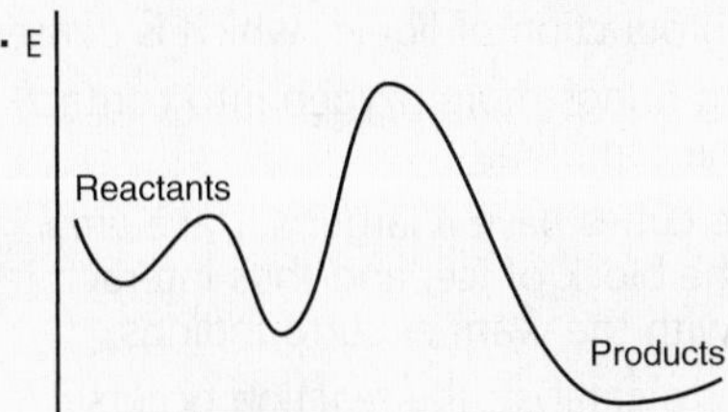

87. The change from **a** to **b** is spontaneous, favored by an increase in entropy. The change from **b** to **c** will not occur, because it would result in a decrease in entropy, causing the process to be nonspontaneous.

88. a. yes; a solid becomes a gas.
b. yes; a single solid becomes two products, one of them a gas.
c. no; two particles become one.
d. yes; one particle becomes two.

89. increase in products

90. $K_{eq} = \dfrac{[SO_3]^2}{[SO_2]^2 \times [O_2]}$

91. $K_{eq} = 6.59 \times 10^{-1}$

92. The product of the concentrations of the ions must be greater than the ion-product constant (K_{sp}).

93. $3.0 \times 10^{-5} M$

94. A catalyst will establish equilibrium more quickly, but it does not change the equilibrium position.

95. Possible answers: use a blow dryer, flush a toilet, mow the lawn, cook breakfast, drive a car, and breathing

Evaluate

Answers

96. 2.0 X $10^{-8}M$

97. a. favors products

b. no effect

c. favors reactants

d. no effect

98. Yes, the product of the concentration of the two ions (3.8 x 10^{-4}) is greater than the K_{sp} value for the compound (1.7 x 10^{-5}).

THINK CRITICALLY

99. A catalyst increases the efficiency of the collisions; more collisions result in the formation of the product.

100. a. IO^-

b. two

c. the slow reaction

d. $2H_2O_2 \rightarrow 2H_2O + O_2$

e. No, the I^- is changed in the reaction. A catalyst would not appear in the reaction as a reactant, an intermediate, or a product.

101. Increasing pressure tends to reduce volume and increase density, so the system responds by favoring production of liquid, which is denser.

102. a. Fanning brings more oxygen into contact with the campfire.

b. The ice cubes have a larger surface area than the block of ice, and thus interact more with the warmer surroundings.

c. MnO_2 is a catalyst; the reaction occurs explosively.

103. first order in NO_2^-, first order in NH_4^+, second order overall

104. The purpose of a speed bump is to hinder a driver from speeding in a particular area. The activation energy hinders reactants from being converted to products.

105. Increase pressure, cool the reaction mixture, and remove NH_3

106. The coefficients in the equation seldom correspond with the exponents in the rate law.

107. Lowering the temperature of the jam slows down the rates of the reactions that cause the jam to spoil.

96. What is the equilibrium concentration of barium ion in a 1.0-L saturated solution of barium carbonate to which 0.25 mol K_2CO_3 is added?

***97.** Suppose equilibrium is established for the following reaction at 425 K:

$$Fe_3O_4(s) + 4H_2(g) \rightleftharpoons 3Fe(s) + 4H_2O(g)$$

How would the equilibrium concentration of H_2O be affected by these actions?

a. adding more H_2 to the mixture
b. increasing the pressure
c. removing $H_2(g)$
d. adding a catalyst

98. A student prepares a solution by combining 0.025 mol $CaCl_2$ and 0.015 mol $Pb(NO_3)_2$ and adding water to make 1.0 L of solution. Will a precipitate of $PbCl_2$ form in this solution?

Think Critically

99. Relate Cause and Effect An increase in temperature raises the energy of the collisions between reactant molecules. An increase in the concentration of reactants increases the number of collisions. What is the effect of a catalyst on the collisions between molecules?

***100. Draw Conclusions** The mechanism for the decomposition of hydrogen peroxide is thought to be as follows:

$$H_2O_2(aq) + I^-(aq) \longrightarrow H_2O(l) + IO^-(aq) \text{ (slow)}$$

$$IO^-(aq) + H_2O_2(aq) \longrightarrow H_2O(l) + O_2(g) + I^-(aq) \text{ (fast)}$$

a. What is the intermediate?
b. What is the minimum number of activated complexes needed to describe the reaction?
c. Which of the two reactions has the smaller specific rate constant?
d. Write the overall balanced chemical equation for the reaction.
e. Does I^- qualify as a catalyst? Explain.

101. Relate Cause and Effect The freezing of liquid water at 0°C can be represented as follows.

$$H_2O(l) \rightleftharpoons H_2O(s)$$

The density of the liquid water is 1.00 g/cm^3. The density of the ice is 0.92 g/cm^3. Explain why applying pressure causes ice to melt.

102. Apply Concepts Explain what is happening in each of the following situations.

a. A campfire is "fanned" to help get it going.
b. Ice cubes melt faster than a block of ice with the same mass.
c. A pinch of powdered manganese dioxide causes hydrogen peroxide to explode even though the manganese dioxide is not changed.

***103. Analyze Data** Ammonium ions and nitrite ions react in water to form nitrogen gas.

$$NO_2^-(aq) + NH_4^+(aq) \longrightarrow N_2(g) + 2H_2O(l)$$

From the following data, decide the order of the reaction with respect to NH_4^+ and NO_2^- and the overall order of the reaction.

Initial $[NO_2^-]$ (mol/L)	Initial $[NH_4^+]$ (mol/L)	Initial rate (mol/(L·s))
0.0100	0.200	5.4×10^{-7}
0.0200	0.200	10.8×10^{-7}
0.0400	0.200	21.5×10^{-7}
0.0600	0.200	32.3×10^{-7}
0.200	0.0202	10.8×10^{-7}
0.200	0.0404	21.6×10^{-7}
0.200	0.0606	32.4×10^{-7}
0.200	0.0808	43.3×10^{-7}

104. Use Analogies A speed bump is a raised ridge that runs across the surface of a street, a parking lot, or driveway. What is the purpose of a speed bump, and how is it similar to activation energy?

***105. Apply Concepts** In the reversible reaction for the formation of ammonia, the enthalpy change and the entropy change both favor the reverse reaction.

$$N_2(g) + 3H_2(g) \rightleftharpoons 2NH_3(g) + \text{heat}$$

You are the chemical engineer in charge of the large-scale production of ammonia. What are three things you can do to increase the percent yield of your product?

106. Compare and Contrast When you calculate the K_{eq} for a reaction, you use the coefficient from the balanced equation. Why are the coefficients not used to find the order of a reaction?

107. Infer Jars of jam are stored on grocery shelves at room temperature. Why does the instruction "Refrigerate after opening" appear on the label?

Enrichment

108. Graph The following data were collected for the decomposition of compound AB into its elements. The reaction is first order in AB. Answer the following questions.

[AB] (mol/L)	Time (s)
0.300	0
0.246	50
0.201	100
0.165	150
0.135	200
0.111	250
0.090	300
0.075	350

a. Make a graph of concentration (y-axis) versus time (x-axis).

b. Determine the rate of the reaction at $t = 100$ seconds and $t = 250$ seconds.

***109. Interpret Graphs** When table sugar, or sucrose, is dissolved in an acid, the sucrose slowly decomposes into two simpler sugars: fructose and glucose. Use the graph to answer the following questions:

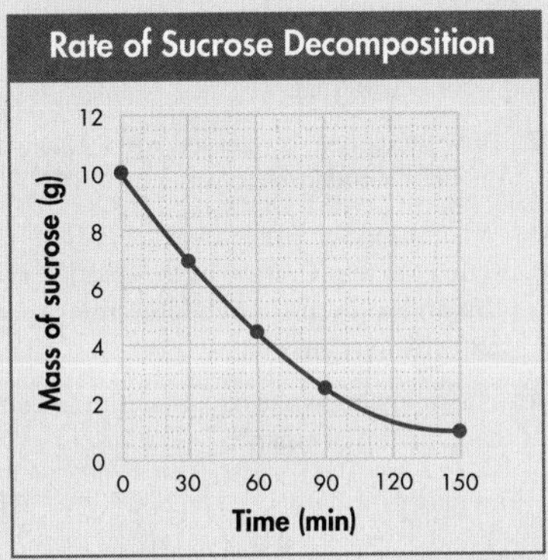

a. How many grams of sucrose decompose in the first 30 minutes?

b. How many grams of sucrose decompose in the interval between 90 minutes and 120 minutes?

c. In general, what happens to the rate of decomposition with time?

Write About Science

110. Explain Compost is a mixture that gardeners add to soil to improve the soil. Many gardeners make their own compost by recycling yard and kitchen waste. What types of reactions take place in a compost pile? Why is it important that the compost pile be turned regularly?

111. Connect to the BIGIDEA A discussion of reaction rates often focuses on ways to speed up reactions. However, sometimes it is important to know how to slow down a reaction, such as combustion. Use Le Châtelier's principle to explain why some of the methods that are used to fight fires work.

CHEMYSTERY

Explosive Sugar

The U.S. Chemical Safety Board (CSB) investigates chemical accidents in the workplace. The agency uses what it learns to recommend ways to improve safety. In 2009, the CSB issued a report on the explosion at the sugar refinery in Georgia.

A year before the blast, a steel cover was added to the conveyor belt where the first explosion took place. The cover was added to keep the sugar clean. But the cover also allowed dangerous concentrations of sugar dust to build up around the belt. It was difficult for the CSB to pinpoint the exact cause of the first explosion because of the amount of damage. However, they suspected that an overheated part of the belt ignited the dust.

Vibrations from the small explosion shook loose sugar dust that had collected on many surfaces in the warehouse. Suddenly the air inside the warehouse was filled with tiny particles of sugar.

112. Relate Cause and Effect Explain how the first explosion increased the likelihood that a second, more violent explosion would occur?

113. Connect to the BIGIDEA Suppose you worked for the CSB. What recommendations would you make to prevent similar fires in the future?

ASSESSMENT

CHEMYSTERY SUMMARIZE Have students read over the CHEMystery. Then have volunteers produce a flow chart on the board to summarize the events at the sugar refinery and relate this to what they have learned about concentration and chemical reactions. **Ask** What would happen to the concentration of sugar dust in the air when the first small explosion shook the warehouse? *(The concentration would increase.)* How do you think the increase in concentration would affect the likelihood of a larger explosion? Why? *(The likelihood of a larger explosion would increase because the particles of sugar are more likely to touch an overheated surface in the warehouse and ignite.)*

CHEMYSTERY ANSWERS

112. The first explosion produced vibrations that shook surfaces in the warehouse. Sugar dust on these surfaces was thrown upward, resulting in a high concentration of sugar dust in the air. The high concentration of sugar dust increased the likelihood that particles of sugar would touch an overheated surface and ignite.

113. BIGIDEA Sample answer: To reduce the chance of an explosion, the concentration of sugar dust in the air must be kept low. I would recommend that the sugar refinery frequently clean the sugar dust from all surfaces.

Answers

ENRICHMENT

108. a.

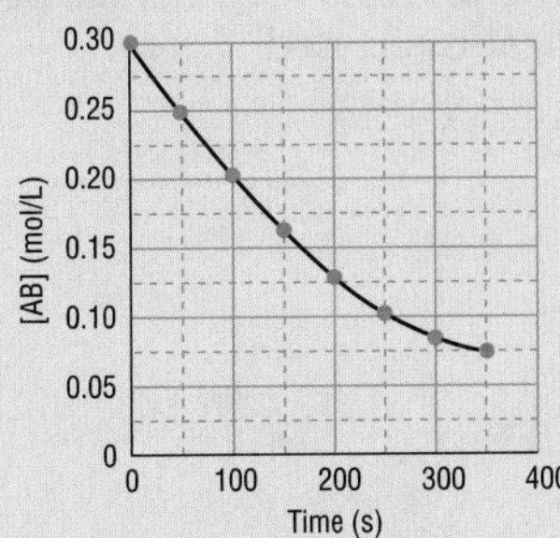

b. The rate when $t = 100$ seconds is 8×10^{-4} mol/(L · s). The rate when $t = 250$ seconds is 4×10^{-4} mol/(L · s).

109. a. about 3 g

b. about 1.3 g

c. The rate of decomposition slows down over time.

WRITE ABOUT SCIENCE

110. decomposition reactions; Most decomposition reactions require energy, in this case, in the forms of heat and light.

111. BIGIDEA The main products of a combustion reaction are water and carbon dioxide. If one of these products is supplied to the fire, the combustion reaction will shift toward the reactants and cause the fire to slow or stop. (Some fire extinguishers contain pressurized carbon dioxide gas.)

ASSESSMENT

Evaluate

Answers

CUMULATIVE REVIEW

114. a. $1s^22s^22p^63s^23p^63d^{10}4s^24p^2$ ·Ġe·

b. $1s^22s^22p^63s^23p^64s^2$ ·Ca·

c. $1s^22s^22p^4$:Ȯ·

d. $1s^22s^22p^63s^23p^6$:Är:

e. $1s^22s^22p^63s^23p^5$:Cl·

f. $1s^22s^22p^63s^23p^3$:Ṗ·

115. Potassium chloride is an ionic compound, not a molecular compound.

116. a. fluoride anion

b. copper(II) cation

c. phosphide anion

d. hydrogen cation

e. sodium cation

f. iodide anion

g. oxide anion

h. magnesium cation

117. a. sodium perchlorate, 1–

b. potassium permanganate, 1–

c. calcium phosphate, 3–

d. magnesium carbonate, 2–

e. sodium sulfate, 2–

f. potassium dichromate, 2–

118. positive: b, e–h; negative: a, c, d

119. a. 2.51×10^2 g

b. 4.60×10^1 g

c. 2.99×10^{-22} g

d. 9.57 g

120. a. $Ag^+(aq) + NO_3^-(aq) + K^+(aq) + I^-(aq) \rightarrow AgI(s) + K^+(aq) + NO_3^-(aq)$

b. potassium ions and nitrate ions

c. $Ag^+(aq) + I^-(aq) \rightarrow AgI(s)$

121. a. $2KClO_3(s) + \text{heat} \rightarrow 2KCl(s) + 3O_2(g)$

b. 1.91 g O_2

122. Pascal (Pa), atmosphere (atm), millimeters of mercury (mm Hg)

123. No; the boiling point is the temperature at which the vapor pressure of the liquid equals the atmospheric pressure; it changes if the atmospheric pressure changes.

124. a. increases

b. decreases

c. increases

125. 1.71×10^1 L

126. a, c, d

127. 19.5%

128. b

129. 1.6 mol KCl

130. 1.25 L

131. a. The solute is ethanol; the solvent is water.

b. below

132. 2.80×10^1 kJ

133. The process is exothermic because heat is released.

134. 1.65×10^3 kJ

Cumulative Review

114. Write electron configurations and draw electron dot structures for the following elements:

a. Ge c. O e. Cl
b. Ca d. Ar f. P

***115.** Why is it wrong to say that solid potassium chloride is composed of KCl molecules?

116. Name each ion and then identify it as an anion or a cation.

a. F^- c. P^{3-} e. Na^+ g. O^{2-}
b. Cu^{2+} d. H^+ f. I^- h. Mg^{2+}

***117.** Name the following compounds and give the charge on the anion for each.

a. $NaClO_4$ c. $Ca_3(PO_4)_2$ e. Na_2SO_4
b. $KMnO_4$ d. $MgCO_3$ f. $K_2Cr_2O_7$

118. Which atoms from the following list would you expect to form positive ions, and which would you expect to form negative ions?

a. Cl c. P e. Cu g. K
b. Ca d. Se f. Sn h. Fe

119. Find the mass in grams of each quantity.

a. 4.50 mol Fe
b. 36.8 L CO (at STP)
c. 1 molecule of glucose, $C_6H_{12}O_6$
d. 0.0642 mol ammonium phosphate

120. Aqueous silver nitrate reacts with aqueous potassium iodide to form the precipitate silver iodide.

a. Write the complete ionic equation.
b. What are the spectator ions?
c. Write the net ionic equation.

***121.** When heated, potassium chlorate decomposes into potassium chloride and oxygen gas.

a. Write the balanced equation for this chemical reaction.
b. How many grams of oxygen are formed when 4.88 g $KClO_3$ decompose?

122. Give the names and abbreviations of three units of pressure.

123. Is the boiling point of a liquid substance a constant? Explain.

124. What happens to the pressure of a contained gas in each instance?

a. More gas particles are added.
b. The temperature of the gas is decreased.
c. The volume of the container is reduced.

125. What volume will 24.5 g of carbon dioxide gas occupy at 55°C and a pressure of 88.8 kPa?

126. Which of these compounds should readily dissolve in water?

a. KI(s) c. $NH_4Cl(s)$
b. $C_2H_6(g)$ d. $Na_3PO_4(s)$

127. Calculate the percent by mass of water in barium bromide tetrahydrate.

128. For which of these substances would the solubility in water be most likely to decrease with an increase in temperature?

a. $NH_4NO_3(s)$ c. KI(s)
b. $NH_3(g)$ d. NaCl(s)

***129.** How many moles of solute are in 2.40 L of 0.66*M* KCl?

130. How many liters of a stock solution of 6.00*M* HCl would you need to prepare 15.0 L of 0.500*M* HCl?

131. A small amount of ethanol (C_2H_5OH) is dissolved in a large beaker of water.

a. Identify the solute and the solvent.
b. Is the freezing point of the solution above or below 0°C?

132. How much heat is released when 12.4 g of steam at 100°C condenses to water at 100°C?

***133.** When solid sodium hydroxide is dissolved in water, the temperature of the solution rises. Is this an exothermic or endothermic process? Explain your answer.

134. The following is the equation for the complete combustion of ethene:

$$C_2H_4(g) + 3O_2(g) \longrightarrow 2CO_2(g) + 2H_2O(g) \quad \Delta H = -1411 \text{ kJ}$$

How many kilojoules of heat are released when 32.8 g of ethene are burned?

If You Have Trouble With . . .

Question	114	115	116	117	118	119	120	121	122	123	124	125	126	127	128	129	130	131	132	133	134
See Chapter	5	7	9	9	9	10	11	12	13	13	14	14	15	15	16	16	16	16	17	17	17

Standardized Test Prep for Indiana

Select the choice that best answers each question or completes each statement.

1. Which reaction is represented by the following expression for an equilibrium constant?

$$K_{eq} = \frac{[CO]^2 \times [O_2]}{[CO_2]^2}$$

(A) $2CO_2 \rightleftharpoons O_2 + 2CO$
(B) $CO_2{}^2 \rightleftharpoons O_2 + 2CO^2$
(C) $O_2 + 2CO \rightleftharpoons 2CO_2$
(D) $O_2 + CO_2 \rightleftharpoons CO_2{}^2$

2. At 25°C, zinc sulfide has a K_{sp} of 3.0×10^{-23}, zinc carbonate has a K_{sp} of 1.0×10^{-10}, and silver iodide has a K_{sp} of 8.3×10^{-17}. Order these salts from most soluble to least soluble.
(A) zinc carbonate, zinc sulfide, silver iodide
(B) silver iodide, zinc carbonate, zinc sulfide
(C) zinc carbonate, silver iodide, zinc sulfide
(D) zinc sulfide, silver iodide, zinc carbonate

Tips for Success

Multiple Parts Sometimes two phrases in a true/false question are connected with a word such as *because, therefore,* or *so.* These words imply a cause-and-effect relationship between the two phrases. Be aware that the overall statement can be false even if each phrase is true on its own.

Use the table to answer Questions 3 and 4.

ΔS	ΔH	ΔG	Spontaneous?
+	–	(a)	Yes
+	(b)	+ or –	At high T
(c)	+	+	No
–	–	(d)	At low T

3. The value of ΔG depends on the enthalpy (ΔH) and entropy (ΔS) terms for a reaction. The value of ΔG also varies as a function of temperature. Use the data in the table to identify the missing entries (a), (b), (c), and (d).

4. Which of these reactions would you expect to be spontaneous at relatively low temperatures? At relatively high temperatures?
(A) $H_2O(l) \longrightarrow H_2O(g)$
(B) $H_2O(g) \longrightarrow H_2O(l)$
(C) $H_2O(s) \longrightarrow H_2O(l)$

5. The atomic windows below represent different degrees of entropy. Arrange the windows in order of increasing entropy.

a. 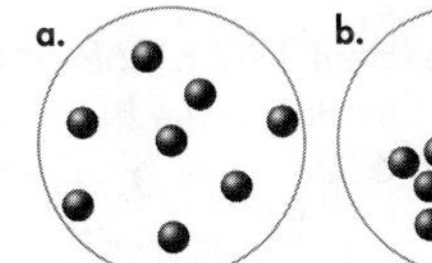

b. 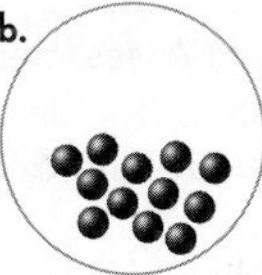

c. 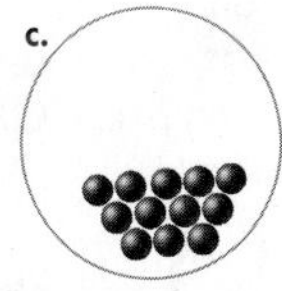

For each question below there are two statements. Decide whether each statement is true or false. Then decide whether Statement II is a correct explanation for Statement I.

	Statement I		Statement II
6.	A catalyst lowers the activation energy for a chemical reaction.	BECAUSE	A catalyst makes a reaction more exothermic.
7.	The entropy of ice is greater than the entropy of steam.	BECAUSE	The density of ice is greater than the density of steam.
8.	The rate of a chemical reaction is affected by a change in temperature.	BECAUSE	The kinetic energy of particles is related to the temperature.
9.	A large value for an equilibrium constant indicates that products are favored at equilibrium.	BECAUSE	The ratio of products to reactants at equilibrium is always > 1.

If You Have Trouble With . . .

Question	1	2	3	4	5	6	7	8	9
See Lesson	18.3	18.4	18.5	18.5	18.5	18.1	18.5	18.1	18.3

ASSESSMENT

STP Answers

1. A
2. C
3. (a) –
(b) +
(c) –
(d) + or –
4. a and b at low temperatures; c at high temperatures
5. c, b, a
6. true, false
7. false, true
8. true, true, correct explanation
9. true, false

19 Acids, Bases, and Salts Planning Guide

Introducing the BIGIDEA: REACTIONS

Chemical reactions involve processes in which reactants produce products.

IASS	Lessons and Objectives	Print Resources: For the Student	Print Resources: For the Teacher
C.8.1, C.8.2	**19.1** Acid-Base Theories p 646–652 **19.1.1** Define an acid and a base according to Arrhenius. **19.1.2** Distinguish an acid from a base in the Brønsted-Lowry theory. **19.1.3** Define an acid and a base according to Lewis.	Reading and Study Workbook Lesson 19.1 Lesson Assessment 19.1 p 652	Teacher Demo, p 647: Reactive Acids
C.8.2, C.8.4	**19.2** Hydrogen Ions and Acidity p 653–662 **19.2.1** Describe how [H^+] and [OH^+] are related in an aqueous solution. **19.2.2** Classify a solution as neutral, acidic, or basic using pH. **19.2.3** Identify two methods that are used to measure pH.	Reading and Study Workbook Lesson 19.2 Lesson Assessment 19.2 p 662 Quick Lab: Indicators From Natural Sources p 662	Class Activity, p 659: Using a pH Meter Teacher Demo, p 660: pH Indicators
C.8.2, C.8.3	**19.3** Strengths of Acids and Bases p 664–669 **19.3.1** Describe how acids and bases are classified as either strong or weak.	Reading and Study Workbook Lesson 19.3 Lesson Assessment 19.3 p 669 Small-Scale Lab: Dissociation Constants of Weak Acids p 670	Teacher Demo, p 658: Observing pH Change
C.8.2, C.8.5	**19.4** Neutralization Reactions p 672–675 **19.4.1** Identify the products that form when an acid and a base react. **19.4.2** Identify the point in a titration when neutralization will occur.	Reading and Study Workbook Lesson 19.4 Lesson Assessment 19.5 p 675	Teacher Demo, p 674: Titration Using Indicators Teacher Demo, p 674: Titration Using a pH Meter
C.8.2	**19.5** Salts in Solution p 676–680 **19.5.1** Describe when a solution of a salt is acidic or basic. **19.5.2** Identify the components of a buffer.	Reading and Study Workbook Lesson 19.5 Lesson Assessment 19.5 p 680	Teacher Demo, p 677: Predicting pH of Solutions Teacher Demo, p 678: Comparing Commercial Buffers

Assessing the BIGIDEA: REACTIONS

	Essential Questions	For the Student	For the Teacher
	Essential Questions 1. What are the different ways chemists define acids and bases? 2. What does the pH of a solution mean? 3. How do chemists use acid-base reactions?	Study Guide p 682 Math Tune-Up p 683 STP p 689 Reading and Study Workbook Self-Check and Vocabulary Review Chapter 19	

	PearsonChem.com	
	Editable Worksheets	Online Activities
	Teaching Resources, Lesson 19.1 Review	19.1 Lesson Overview Dissociation of Acids and Bases
	Teaching Resources, Lesson 19.2 Review Lab 40: Estimation of pH Small-Scale Lab Manual Lab 30: A Small-Scale Colorimetric pH Meter Lab Practical 19-1: Determining pH	19.2 Lesson Overview Calculating pH from $[H^+]$ and $[OH^-]$
	Teaching Resources, Lesson 19.3 Review Lab Practical 19-2: Acids and Bases	19.3 Lesson Overview Calculating a Dissociation Constant
	Teaching Resources, Lesson 19.4 Review Lab 42: Neutralization Reactions Lab 43: Acid-Base Titrations Small-Scale Lab Manual Lab 14: Chemical Equations Small-Scale Lab Manual Lab 15: Balancing Chemical Equations Small-Scale Lab Manual Lab 16: Titration of Bleach Small-Scale Lab Manual Lab 18: Titration: Determining How Much Acid Is in a Solution Small-Scale Lab Manual Lab 19: Mass Titrations: Measuring Molar Concentrations Small-Scale Lab Manual Lab 31: Titration Curves Lab 32: Strong and Weak Acids and Bases Lab Practical 19-3: Titration Probeware Lab Quick Lab: The Neutralizing Power of Antacids Probeware Lab Small-Scale Lab: Small-Scale Titrations	19.4 Lesson Overview Acid Clues to a Marble Mystery Titrations Acids and Bases Determining Concentration by Titration
	Teaching Resources, Lesson 19.5 Review Lab 44: Salt Hydrolysis Lab 45: Buffers Small-Scale Lab Manual Lab 33: Buffers	19.5 Lesson Overview Using Equations to Illustrate the Action of a Buffer
	Core Teaching Resources (includes editable worksheets) • Lesson Reviews • Practice Problems • Interpret Graphs • Vocabulary Review • Chapter Quizzes and Tests	Chapter 19 Problem Set Logarithmic Scale

Materials List

For the Student

Quick Lab p 662
- See SE p 662.

Small-Scale Lab p 670
- reaction surface
- metric ruler
- pH buffer
- 0.04% solutions of BCG, BTB, BPB, MCP, and TB
- 0.05% MO, 0.02% AYR, 0.1% phenolphthalein

For the Teacher

Teacher Demo p 647
- 5–10 mL of 1*M* HCl
- test tube
- small pieces of zinc
- safety match

Teacher Demo p 658
- lemon juice
- tea
- antacid tablets
- glass of 10% grape juice
- household ammonia

Class Activity p 659
- several test tubes
- 0.1*M* HCl, 0.1*M* NaOH
- household products such as lemon juice, vinegar, shampoo, liquid detergent
- pH meter

Teacher Demo p 660
- 0.1% solutions of thymol blue
- methyl red, bromothymol blue, phenolphthalein
- aqueous buffers spanning pH 4 to pH 10
- test tubes
- pH meter

Teacher Demo p 674
- equimolar solutions of HCl and NaOH
- phenolphthalein
- electric stirrer with a magnetic stir bar
- beaker
- table salt

Teacher Demo p 677
- 1*M* solutions of NH_4NO_3, KCl, $NaHCO_3$, Na_2SO_4
- pH meter or universal pH paper

Teacher Demo p 678
- antacid tablets
- flasks and burets
- 0.5*M* HCl and NaOH
- distilled water
- methyl red
- bromothymol blue

Additional Digital Resources

Online Student Edition
Online Teacher's Edition
Exam*View Assessment Suite CD-ROM*
Classroom Resources DVD
19.3 Virtual Chem Lab 27: Ionization Constants of Weak Acids
19.5 Virtual Chem Lab 25: Study of Acid-Base Titrations
Virtual Chem Lab 26: Acid-Base Titrations
Virtual Chem Lab 29: Molecular Weight Determination by Acid-Base Titration

CHAPTER 19

What's Online

CHEMISTRY TUTORIAL Students can work step-by-step with the online tutorial to learn how to calculate pH and dissociation constants. Students will also learn how to use titrations to determine concentration by accessing the Chapter 19 Chemistry Tutorials.

MATH TUTORIAL The online Math Tutorial will help guide students through understanding logarithmic scales.

DIRECTED VIRTUAL LAB Encourage students to review the principles of neutralization and titration, using the equivalence point in calculations using the animated virtual lab for Chapter 19. The animation is followed by analysis questions, where the student is guided through the lab concepts.

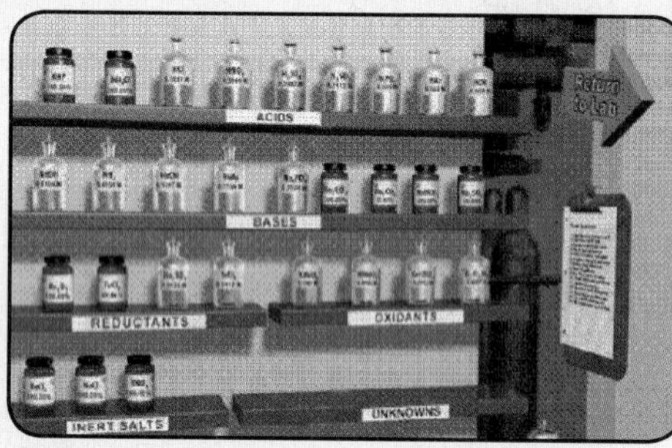

KINETIC ART Have students watch an animation about the dissociation of acids and bases. The animation is followed by analysis questions to guide further understanding of the concepts.

CONCEPTS IN ACTION Students can watch an interactive animation about how acid rain can lead to harmful acid base reactions. The animation is followed by questions to explore the factors that affect the rates and products of acid base reactions.

IN ACADEMIC STANDARDS for SCIENCE

Nature of Science

- Develop explanations based on reproducible data and observations gathered during laboratory investigations.

Standard 8: Acids and Bases

C.8.1 Use Arrhenius and Bronsted-Lowry models to classify substances as acids or bases.

C.8.2 Describe the properties of acids and bases.

Also covered C.8.3, C.8.4, C.8.5

19 Acids, Bases, and Salts

INSIDE:

PearsonChem.com

Artists often use hydrofluoric acid to etch designs on glass.

Focus on ELL

1 CONTENT AND LANGUAGE Make a three-column chart on the board, with the following headings: *Acids*, *Bases*, *Salts*. Have students, as a class, list what they know about each word. Have students identify any unfamiliar words in the lesson titles. Add these words to the classroom word wall.

BEGINNING:

LOW Make a nonlinguistic representation of each unfamiliar word.

HIGH Ask student volunteers to present *acid*, *base*, and *salt*. Have each one write a word on the board and introduce it to the class by pointing at it and saying it aloud.

INTERMEDIATE: LOW/HIGH Have students brainstorm and write examples of *acids, bases, and salts.*

ADVANCED:

LOW Write the definition of the each word and present it to the class.

HIGH Have students use context clues to predict the meanings of unfamiliar words.

CHAPTER 19

BIGIDEA

REACTIONS

Essential Questions:

1. *What are the different ways chemists define acids and bases?*
2. *What does the pH of a solution mean?*
3. *How do chemists use acid-base reactions?*

CHEMYSTERY

Paper Trail

The invention of the printing press in 1440 increased the number of books that could be printed and the need for paper on which to print these books. The main ingredient in paper was cotton, which is almost pure cellulose. This type of paper is referred to as rag-based paper. When books printed on high-quality rag-based paper are stored correctly, they can last for hundreds of years.

By 1880, the demand for paper was so great that printers switched from rag-based paper to wood-based paper. Over time, the wood-based paper tended to yellow and crack. Sometimes the paper was so brittle that it crumbled when touched.

Why did a change in the content of paper cause such a dramatic change in the properties of the paper?

▶ Connect to the BIGIDEA As you read about acids and bases, think about what could cause paper to crumble.

INDIANA'S ACADEMIC STANDARDS FOR SCIENCE

Acids and Bases C.8.1, C.8.2, C.8.3, C.8.4, C.8.5. See lessons for details.

UbD

Understanding by Design

Students are building toward understanding chemical reactions by using knowledge of acids, bases, and salts.

PERFORMANCE GOALS At the end of Chapter 19, students will be able to answer the essential questions by applying their knowledge of acids, bases, and salts.

ESSENTIAL QUESTIONS Read the essential questions aloud. **Ask** What are some properties of acids? *(Sample answer: They can burn or corrode materials.)* **Ask** What do you think it means when an advertisement claims that an antacid tablet can "neutralize" stomach acid? *(Sample answer: The tablet can change the acid so that it is no longer harmful.)*

BIGIDEA Use the photo of etched glass to help students connect to this chapter's concepts. Activate prior knowledge by explaining that pH is a measure of acidity and having students name ways they have heard about acids or pH in everyday life. *(Sample answers: acid rain, aquarium or pool water pH)* Draw students' attention to the photo. **Ask** What caused parts of the glass to change from clear to frosted? *(Acid reacted with the glass.)* Engage students in a discussion of how acids react in the examples they gave.

CHEMYSTERY Connect the CHEMystery to the Big Idea by pointing out that even though both cotton- and wood-based papers contain cellulose, the cotton-based paper does not break down as easily as the wood-based paper does. Have them consider what differences in the paper types might cause the wood-based paper to deteriorate more rapidly.

Introduce the Chapter

IDENTIFY PRECONCEPTIONS Students may believe that all acids are strong and highly caustic. Use the following activity to help them understand that some familiar foods contain acids that are not harmful.

Activity *You will need baking soda, cola, vinegar, lemon juice, small paper cups, and plastic spoons.* **Ask** What happens if you touch an acid? *(Students may say that acids will burn you.)* Explain that baking soda reacts with acids. Have students place a spoonful of baking soda into a cup. Then have them pour a few spoonfuls of cola into the cup and watch the reaction. Repeat with vinegar and lemon juice. Explain that strong acids are harmful, but cola, vinegar, and lemon juice all contain acids that are weak enough to ingest.

CO_2 and Thinning Shells
Suze dives into how atmospheric CO_2 is increasing the acidity of Earth's oceans.

LESSON 19.1

Key Objectives

19.1.1 **DEFINE** an acid and a base according to Arrhenius.

19.1.2 **DISTINGUISH** an acid from a base in the Brønsted-Lowry theory.

19.1.3 **DEFINE** an acid and a base according to Lewis.

Additional Resources

Reading and Study Workbook, Lesson 19.1

Available Online or on Digital Media:

- Teaching Resources, Lesson 19.1 Review

PearsonChem.com

Engage

CHEMISTRY & YOU Have students study the photograph and read the text that opens the lesson. **Ask** Why do visitors to Bracken Cave in Texas need to wear protective goggles and respirators? *(Students will likely say that ammonia has a strong, pungent odor, and high levels of ammonia are dangerous.)*

Activate Prior Knowledge

Have students review what they know about the characteristics of hydrogen ions, hydronium ions, and hydroxide ions. Then review how each is formed. Explain that each of these ions plays an important role in one or more of the acid-base theories discussed in this lesson.

IN ACADEMIC STANDARDS for SCIENCE

C.8.1 Use Arrhenius and Bronsted-Lowry definitions to classify substances as acids or bases.

C.8.2 Describe the characteristic properties of acids and bases.

19.1 Acid-Base Theories

C.8.1 Arrhenius and Bronsted-Lowry; **C.8.2** Properties of acids and bases.

CHEMISTRY & YOU

Q: *Why are high levels of ammonia harmful to you?* Bracken Cave, near San Antonio, Texas, is home to millions of Mexican free-tailed bats. Nitrogen compounds in bat urine can decompose and release ammonia into the air. Visitors to the cave must wear protective goggles and respirators. They need this protection because of what happens when ammonia reacts with water. Ammonia is an example of a base. In this lesson, you will learn about some of the properties of acids and bases.

Key Questions

How did Arrhenius define an acid and a base?

What distinguishes an acid from a base in the Brønsted-Lowry theory?

How did Lewis define an acid and a base?

Vocabulary

- hydronium ion (H_3O^+)
- conjugate acid
- conjugate base
- conjugate acid-base pair
- amphoteric
- Lewis acid
- Lewis base

Arrhenius Acids and Bases

How did Arrhenius define an acid and a base?

Acids and bases have distinctive properties. Many of the foods you eat, including those shown in Figure 19.1a, contain acids. Acids give foods a tart or sour taste. Lemons, which taste sour enough to make your mouth pucker, contain citric acid. Aqueous solutions of acids are strong or weak electrolytes. Recall that an electrolyte can conduct electricity. The electrolyte in a car battery is an acid. Acids cause certain chemical dyes, called indicators, to change color. Many metals, such as zinc and magnesium, react with aqueous solutions of acids to produce hydrogen gas.

The soap in Figure 19.1b is a familiar material that has the properties of a base. If you have accidentally tasted soap, you know that it has a bitter taste. This bitter taste is a general property of bases, but one that is dangerous to test. The slippery feel of soap is another property of bases. Like acids, bases will cause an indicator to change color. Bases also form aqueous solutions that are strong or weak electrolytes.

Chemists had known the properties of acids and bases for many years. Yet they were not able to propose a theory to explain this behavior. Then, in 1887, the Swedish chemist Svante Arrhenius proposed a new way of defining and thinking about acids and bases. **According to Arrhenius, acids are hydrogen-containing compounds that ionize to yield hydrogen ions (H^+) in aqueous solution. Bases are compounds that ionize to yield hydroxide ions (OH^-) in aqueous solution.**

Figure 19.1 Acids and Bases
Many items contain acids or bases, or produce acids and bases when dissolved in water. **a.** Citrus fruits contain citric acid ($HC_6H_7O_7$). **b.** Many soaps are made using the common base sodium hydroxide (NaOH).

Focus on ELL

1 CONTENT AND LANGUAGE Read aloud each key question, making sure to enunciate each word. Have students repeat the three theory names. Define the term *distinguish*, and then ask students to rely on this definition to orally paraphrase the key questions.

2 FRONTLOAD THE LESSON Write the term *battery acid* on the board, and ask volunteers to explain why acids are used to make batteries. *(They conduct electricity.)* Remind students that chemical compounds that conduct electricity when in aqueous solution are called *electrolytes*. Have students review what they know about the behavior of electrolytes in solution.

3 COMPREHENSIBLE INPUT Copy Table 19.1 on the board. Add a column labeled *Equations*. Write chemical equations to show the behavior of acids in solution and demonstrate the multiple dissociations possible for polyprotic acids.

Arrhenius Acids Table 19.1 lists six common acids. They vary in the number of hydrogens they contain that can form hydrogen ions. A hydrogen atom that can form a hydrogen ion is described as *ionizable*. Nitric acid (HNO_3) has one ionizable hydrogen, so nitric acid is classified as a *monoprotic* acid. The prefix *mono-* means "one," and the stem *protic* reflects the fact that a hydrogen ion is a proton. Acids that contain two ionizable hydrogens, such as sulfuric acid (H_2SO_4), are called *diprotic* acids. Acids that contain three ionizable hydrogens, such as phosphoric acid (H_3PO_4), are called *triprotic* acids.

Not all compounds that contain hydrogen are acids. Also, some hydrogens in an acid may not form hydrogen ions. Only a hydrogen that is bonded to a very electronegative element can be released as an ion. Recall that such bonds are highly polar. When a compound that contains such bonds dissolves in water, it releases hydrogen ions. An example is the hydrogen chloride molecule, shown below.

$$\underset{\text{Hydrogen chloride}}{\overset{\delta+}{H}-\overset{\delta-}{Cl}(g)} \xrightarrow{H_2O} \underset{\text{Hydrogen ion}}{H^+(aq)} + \underset{\text{Chloride ion}}{Cl^-(aq)}$$

(hydrochloric acid)

However, in an aqueous solution, hydrogen ions are not present. Instead, the hydrogen ions are joined to water molecules as hydronium ions. A **hydronium ion (H_3O^+)** is the ion that forms when a water molecule gains a hydrogen ion. As seen in Figure 19.2, hydrogen chloride ionizes to form an aqueous solution of hydronium ions and chloride ions.

Table 19.1

Some Common Acids

Name	Formula
Hydrochloric acid	HCl
Nitric acid	HNO_3
Sulfuric acid	H_2SO_4
Phosphoric acid	H_3PO_4
Ethanoic acid	CH_3COOH
Carbonic acid	H_2CO_3

HCl Hydrogen chloride + H_2O Water ⟶ H_3O^+ Hydronium ion + Cl^- Chloride ion

Figure 19.2 Hydrochloric Acid
Hydrochloric acid is actually an aqueous solution of hydrogen chloride. Hydrogen chloride forms hydronium ions, making this compound an acid.
Explain ***Why does hydrogen chloride release a hydrogen ion when dissolved in water?***

See hydrochloric acid animated online.

In contrast to hydrogen chloride, methane (CH_4) is an example of a hydrogen-containing compound that is not an acid. The four hydrogen atoms in methane are attached to the central carbon atom by weakly polar C—H bonds. Thus, methane has no ionizable hydrogens and is not an acid. Ethanoic acid (CH_3COOH), which is commonly called acetic acid, is an example of a molecule that contains both hydrogens that do not ionize and a hydrogen that does ionize. Although its molecules contain four hydrogens, ethanoic acid is a monoprotic acid. The structural formula shows why.

$$\begin{array}{ccccccc} & & H & & O & & \\ & & | & & \| & & \\ H & - & C & - & C & - & O-H \\ & & | & & & & \\ & & H & & & & \end{array}$$

Ethanoic acid
(CH_3COOH)

The three hydrogens attached to a carbon atom are in weakly polar bonds. They do not ionize. Only the hydrogen bonded to the highly electronegative oxygen can be ionized. For complex acids, you need to look at the structural formula to recognize which hydrogens can be ionized.

Acetic Acid

Acetic acid is the common name for ethanoic acid, the acid found in vinegar. The word *acetic* comes from the Latin word *acetum*, meaning "vinegar." This word origin is fitting because early vinegars were products of the souring of wine. Wines sour due to the activity of *Acetobacter* bacteria, which convert ethanol into ethanoic acid. This genus of bacteria is especially resistant to high concentrations of both ethanol and ethanoic acid, making it unusual in the bacterial world. Early wine vinegars developed as the result of natural contamination by these bacteria, which are plentiful in water and soil as well as on crops and fruit. Ethanoic acid intended for human consumption is still made using bacterial fermentation, but industrial ethanoic acid is usually produced by reacting carbon monoxide with methanol in the presence of a catalyst such as hydrogen iodide.

Foundations for Reading

BUILD VOCABULARY Explain that one definition of the word *conjugate* is "having features in common but opposite or inverse in some particular." Have students keep this definition in mind as they read the lesson.

READING STRATEGY Have students preview the lesson. Then have them create a compare- and-contrast table to differentiate between the three acid-base theories.

Explain

Arrhenius Acids and Bases

USE AN ANALOGY Explain that in this lesson students will be exposed to three different ways in which an acid or base is defined. Help students to understand that the definitions presented are all correct, but some definitions include more possible substances. Explain that where one definition lacks, another fills in. Relate the lesson content to a coaching strategy or a gaming strategy, in which there may be more than one possible correct play or solution that leads to victory, even though some plays or solutions may provide more avenues to success than others.

USE VISUALS Direct students' attention to Table 19.1. Note that the acids in the table are listed in order of decreasing tendency to yield hydrogen ions. **Ask** What element do all the acids in this table have in common? *(hydrogen)*

Explore

Properties of Acids and Bases

PURPOSE Demonstrate that many metals and acids react to produce hydrogen gas.

MATERIALS 5–10 mL of 1*M* HCl, test tube, small pieces of zinc, match

SAFETY Wear safety goggles, gloves, and a lab apron while performing the demonstration.

PROCEDURE Add 5–10 mL 1*M* HCl to a test tube and add a few small pieces of zinc metal.

EXPECTED OUTCOME Gas bubbles are liberated from the solution. Explain that many metals react with aqueous acids to produce hydrogen gas. Remind students that hydrogen is a highly flammable gas. Demonstrate the liberation of the gas by placing a lit match at the mouth of the test tube to ignite the hydrogen.

LESSON 19.1

Explain

USE VISUALS Have students study Table 19.2. Have them note that the bases containing Group 1 elements appear to be highly soluble in water, while those containing Group 2 elements are not. **Ask** Based on this generalization, is LiOH likely or not likely to be soluble in water? *(LiOH is likely to be soluble in water.)* Explain that hydroxides of Group 2 elements increase in solubility going down the periodic table column. **Ask** Do you think $Be(OH)_2$ is soluble in water? *(insoluble)* $Sr(OH)_2$? *(quite soluble)*

Extend

Explain that acid–base concepts have been a part of chemistry for more than 300 years. Convey that over time, explanations were developed to explain the observed behavior of acids and bases. Point out that Arrhenius's explanation addresses compounds whose formulas contain obvious clues to their acidic or basic nature. Guide a student-led class discussion about the pros and cons of Arrhenius's explanation.

CHEMISTRY & YOU *Sample answer*: Ammonia can react with water in the eyes and air passages to form a corrosive solution.

Figure 19.3 Clogged Drains
Sometimes water backs up in a sink because the drain is clogged. A plumber can take apart the pipes to remove a clog, or a drain cleaner containing sodium hydroxide can be used to eat away the clog.

CHEMISTRY & YOU

Q: *Visitors to Bracken Cave wear protective gear to keep ammonia gas out of their eyes and respiratory tracts. Think about the properties of bases. Why are high levels of ammonia harmful?*

Table 19.2

Some Common Bases

Name	Formula	Solubility in water
Sodium hydroxide	NaOH	High
Potassium hydroxide	KOH	High
Calcium hydroxide	$Ca(OH)_2$	Very low
Magnesium hydroxide	$Mg(OH)_2$	Very low

Arrhenius Bases Table 19.2 lists four common bases. You may be familiar with the base sodium hydroxide (NaOH), which is also known as lye. Sodium hydroxide is an ionic solid. It dissociates into sodium ions and hydroxide ions in aqueous solution.

$$\underset{\text{Sodium hydroxide}}{NaOH(s)} \xrightarrow{H_2O} \underset{\text{Sodium ion}}{Na^+(aq)} + \underset{\text{Hydroxide ion}}{OH^-(aq)}$$

Sodium hydroxide is extremely caustic. A caustic substance can burn or eat away materials with which it comes in contact. This property is the reason that sodium hydroxide is a major component of products that are used to clean clogged drains. Figure 19.3 shows a drain cleaner that contains sodium hydroxide.

Potassium hydroxide (KOH) is another ionic solid. It dissociates to produce potassium ions and hydroxide ions in aqueous solution.

$$\underset{\text{Potassium hydroxide}}{KOH(s)} \xrightarrow{H_2O} \underset{\text{Potassium ion}}{K^+(aq)} + \underset{\text{Hydroxide ion}}{OH^-(aq)}$$

Sodium and potassium are Group 1A elements. Elements in Group 1A, the alkali metals, react violently with water. The products of these reactions are aqueous solutions of a hydroxide and hydrogen gas. The following equation summarizes the reaction of sodium with water.

$$\underset{\text{Sodium metal}}{2Na(s)} + \underset{\text{Water}}{2H_2O(l)} \longrightarrow \underset{\text{Sodium hydroxide}}{2NaOH(aq)} + \underset{\text{Hydrogen}}{H_2(g)}$$

Sodium hydroxide and potassium hydroxide are very soluble in water. Thus, making concentrated solutions of these compounds is easy. The solutions would have the typically bitter taste and slippery feel of a base. However, these are not properties that you would want to confirm. The solutions are extremely caustic to the skin. They can cause deep, painful, slow-healing wounds if not immediately washed off.

Differentiated Instruction

L3 ADVANCED STUDENTS Have students research the effect of electronegativity on the ease with which an acid releases its hydrogen ions. Then, have students use this information to demonstrate numerically why the acids at the top of Table 19.1 give up their hydrogen ions more easily than do those at the bottom of the table.

ELL ENGLISH LANGUAGE LEARNERS Have students create a list of all the words used to describe ion transfer, such as *donate*, *donor*, *accept*, *acceptor*, *loses*, *gains*. Have students look up the definition of each word and write the words and definitions in their vocabulary notebooks. Add the words to the word wall.

L1 STRUGGLING STUDENTS Have students copy all the tables in the lesson on one page in their notebooks. Have students examine the tables, looking for similarities and differences. Then have them re-create tables on a separate page that will help them to remember the important concepts.

Calcium hydroxide, $Ca(OH)_2$, and magnesium hydroxide, $Mg(OH)_2$, are compounds of Group 2A metals. These compounds are not very soluble in water. Their solutions are always very dilute, even when saturated. A saturated solution of calcium hydroxide has only 0.165 g $Ca(OH)_2$ per 100 g of water. Magnesium hydroxide is even less soluble than calcium hydroxide. A saturated solution has only 0.0009 g $Mg(OH)_2$ per 100 g of water. Figure 19.4 shows a suspension of magnesium hydroxide in water. Some people use this suspension as an antacid and as a mild laxative.

Figure 19.4 Milk of Magnesia
This product is a suspension of magnesium hydroxide in water. Most bases are too caustic to be swallowed. But, the low solubility of magnesium hydroxide makes the suspension safe to consume.

Brønsted-Lowry Acids and Bases

What distinguishes an acid from a base in the Brønsted-Lowry theory?

The Arrhenius definition of acids and bases is not a very broad one. It excludes some substances that have acidic or basic properties, on their own or in solution. For example, sodium carbonate (Na_2CO_3) and ammonia (NH_3) act as bases when they form aqueous solutions. Yet neither of these compounds is a hydroxide-containing compound, so neither compound would be classified as a base by the Arrhenius definition.

In 1923, the Danish chemist Johannes Brønsted and the English chemist Thomas Lowry were working independently. Each chemist proposed the same definition of acids and bases. **According to the Brønsted-Lowry theory, an acid is a hydrogen-ion donor and a base is a hydrogen-ion acceptor.** This theory includes all the acids and bases that Arrhenius defined. It also includes some compounds that Arrhenius did not classify as bases.

You can use the Brønsted-Lowry theory to understand why ammonia is a base. Ammonia gas is very soluble in water. When ammonia dissolves in water, hydrogen ions are transferred from water to ammonia to form ammonium ions and hydroxide ions.

$$NH_3(aq) + H_2O(l) \rightleftharpoons NH_4^+(aq) + OH^-(aq)$$

Figure 19.5 illustrates how each water molecule donates a hydrogen ion to ammonia. Ammonia is a Brønsted-Lowry base because it accepts hydrogen ions. Water is a Brønsted-Lowry acid because it donates hydrogen ions.

KINETIC ART *See ammonia in water animated online.*

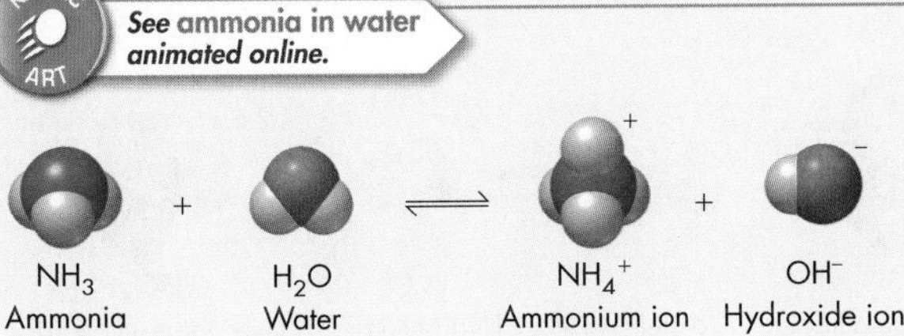

Figure 19.5
Ammonia in Water
When ammonia and water react, water molecules donate hydrogen ions to ammonia molecules. The reaction produces ammonium ions and hydroxide ions.
Explain ***Why is ammonia not classified as an Arrhenius base?***

Check for Understanding

What properties are used to classify acids and bases?

Write the following list of properties on the board and have students classify each as a property of an acid, a base, or both: bitter taste, reacts with metals to form hydrogen gas, forms electrolytes in aqueous solution, sour taste, slippery feel. *(bitter taste, base; reacts with metals to form hydrogen gas, acid; forms electrolytes in aqueous solution, both; sour taste, acid; slippery feel, base)*

ADJUST INSTRUCTION If students are having difficulty answering, have them reread the first two paragraphs of the Arrhenius Acids and Bases section. Then allow them to revise their answers as appropriate.

Explain

Brønsted-Lowry Acids and Bases

APPLY CONCEPTS Explain that Brønsted arrived at his theory of acids and bases through his work in kinetics and thermodynamics. Point out that Brønsted noticed that a great many compounds that would not be classified as acids and bases by the Arrhenius definition behaved like acids and bases in his experiments. He found that these compounds were capable of hydrogen-ion transfer reactions. This led him to define acids as hydrogen-ion donors and bases as hydrogen-ion acceptors.

Misconception Alert

Some students may think the conjugate base of an acid in a given reaction appears on the same side of the equation as the acid. Explain that an acid does not react with its conjugate base but instead produces it or is produced by it in a reaction.

Answers

FIGURE 19.5 Ammonia is not a hydroxide.

LESSON 19.1

Explain

Brønsted-Lowry Acids and Bases

USE VISUALS Direct students' attention to Table 19.3. Have students study the Brønsted-Lowry acid-base conjugate pairs. **Ask** How are hydrogen ions related to Brønsted-Lowry acids? *(Acids lose hydrogen ions.)* **Ask** How are hydrogen ions related to Brønsted-Lowry bases? *(Bases gain hydrogen ions.)*. Point out that a hydrogen ion, H^+, is essentially just a proton. This is why acids are sometimes called "proton donors" and bases are sometimes called "proton acceptors." Clarify that this is still a chemical reaction and does not involve the nucleus.

Extend

Have students locate the three substances in Table 19.3 that are amphoteric. *(HSO_4^-, H_2O, and HCO_3^-)* Then, have them write reactions demonstrating each substance's role as an acid and as a base.

Conjugate Acids and Bases All gases become less soluble in water as the temperature rises. Thus, when the temperature of an aqueous solution of ammonia is increased, ammonia gas is released. This release acts as a stress on the system. In response to this stress, NH_4^+ reacts with OH^- to form more NH_3 and H_2O. In the reverse reaction, ammonium ions donate hydrogen ions to hydroxide ions. Thus, NH_4^+ (the donor) acts as a Brønsted-Lowry acid, and OH^- (the acceptor) acts as a Brønsted-Lowry base. In essence, the reversible reaction of ammonia and water has two acids and two bases.

$$\underset{\text{Base}}{NH_3(aq)} + \underset{\text{Acid}}{H_2O(l)} \rightleftharpoons \underset{\text{Conjugate acid}}{NH_4^+(aq)} + \underset{\text{Conjugate base}}{OH^-(aq)}$$

In the equation, the products of the forward reaction are distinguished from the reactants by the use of the adjective *conjugate*. This term comes from a Latin word meaning "to join together." A **conjugate acid** is the ion or molecule formed when a base gains a hydrogen ion. In the reaction above, NH_4^+ is the conjugate acid of the base NH_3. A **conjugate base** is the ion or molecule that remains after an acid loses a hydrogen ion. In the reaction above, OH^- is the conjugate base of the acid H_2O.

Conjugate acids are always paired with a base, and conjugate bases are always paired with an acid. A **conjugate acid-base pair** consists of two ions or molecules related by the loss or gain of one hydrogen ion. The ammonia molecule and the ammonium ion are a conjugate acid-base pair. The water molecule and the hydroxide ion are also a conjugate acid-base pair.

$$\underset{\text{Base}}{NH_3(aq)} + \underset{\text{Acid}}{H_2O(l)} \rightleftharpoons \underset{\text{Conjugate acid}}{NH_4^+(aq)} + \underset{\text{Conjugate base}}{OH^-(aq)}$$

The dissociation of hydrogen chloride in water provides another example of conjugate acids and bases.

$$\underset{\text{Acid}}{HCl(g)} + \underset{\text{Base}}{H_2O(l)} \rightleftharpoons \underset{\text{Conjugate acid}}{H_3O^+(aq)} + \underset{\text{Conjugate base}}{Cl^-(aq)}$$

In this reaction, hydrogen chloride is the hydrogen-ion donor. Thus, it is by definition a Brønsted-Lowry acid. Water is the hydrogen-ion acceptor and a Brønsted-Lowry base. The chloride ion is the conjugate base of the acid HCl. The hydronium ion is the conjugate acid of the base water.

Figure 19.6 shows the reaction that takes place when sulfuric acid dissolves in water. The products of this reaction are hydronium ions and hydrogen sulfate ions. Use the figure to identify the two conjugate acid-base pairs.

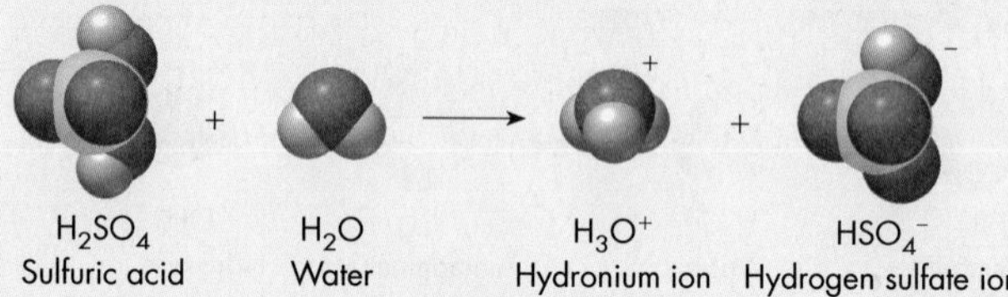

Figure 19.6 Sulfuric Acid
When sulfuric acid and water react, they form hydronium ions and hydrogen sulfate ions.
Identify ***Which product is the conjugate acid, and which is the conjugate base?***

Differentiated Instruction

L1 STRUGGLING STUDENTS Help students create a concept map that connects the three acid-base theories. Assist students in choosing connecting words and phrases. Encourage students to use both names and chemical symbols when referring to ions, and electron dot structures to illustrate Lewis acids and bases.

ELL ENGLISH LANGUAGE LEARNERS To help students visualize words with the prefix *amph-*, show students a photo of an amphibian, such as a frog or salamander, and explain that it is an animal with aquatic larvae that have gills and grow into air-breathing adults with lungs Then show students a photo of an amphibious airplane and direct them to answer the reading support question.

L3 ADVANCED STUDENTS Have students study a college text to find support for the following statement: *The formation of amphoteric hydroxides is a characteristic of metalloids.* Then have them write a brief paragraph explaining why metalloids have this characteristic.

Amphoteric Substances Look at Table 19.3. Note that water appears in both the list of acids and the list of bases. Sometimes water accepts a hydrogen ion. At other times, it donates a hydrogen ion. How water behaves depends on the other reactant. A substance that can act as either an acid or a base is said to be **amphoteric.** Water is amphoteric. In the reaction with hydrochloric acid, water accepts a proton and is therefore a base. In the reaction with ammonia, water donates a proton and is therefore an acid. Look for two other substances in Table 19.3 that are amphoteric.

Table 19.3

Some Conjugate Acid-Base Pairs	
Acid	**Base**
HCl	Cl^-
H_2SO_4	HSO_4^-
H_3O^+	H_2O
HSO_4^-	SO_4^{2-}
CH_3COOH	CH_3COO^-
H_2CO_3	HCO_3^-
HCO_3^-	CO_3^{2-}
NH_4^+	NH_3
H_2O	OH^-

Lewis Acids and Bases

How did Lewis define an acid and a base?

The work that Gilbert Lewis (1875–1946) did on bonding led to a new concept of acids and bases. **According to Lewis, an acid accepts a pair of electrons and a base donates a pair of electrons during a reaction.** This definition is more general than those offered by Arrhenius or by Brønsted and Lowry. A **Lewis acid** is a substance that can accept a pair of electrons to form a covalent bond. Similarly, a **Lewis base** is a substance that can donate a pair of electrons to form a covalent bond.

The Lewis definitions include all the Brønsted-Lowry acids and bases. Consider the reaction of H^+ and OH^-. The hydrogen ion donates itself to the hydroxide ion. Therefore, H^+ is a Brønsted-Lowry acid and OH^- is a Brønsted-Lowry base. The hydroxide ion can bond to the hydrogen ion because it has an unshared pair of electrons. Thus, OH^- is also a Lewis base, and H^+, which accepts the pair of electrons, is a Lewis acid.

$$\underset{\text{Lewis acid}}{H^+} + \underset{\text{Lewis base}}{{}^-\!:\!\ddot{\text{O}}\!-\!H} \longrightarrow H\!-\!\ddot{\text{O}}\!-\!H$$

READING SUPPORT

Building Vocabulary: ***Prefixes*** The prefix *amphi-* is from a Greek word meaning "of both kinds." An amphibian is an animal that is capable of living both on land and in the water. ***What does it mean to describe an airplane as amphibious?***

A second example of a reaction between a Lewis acid and a Lewis base is what happens when ammonia dissolves in water. Hydrogen ions from the dissociation of water are the electron-pair acceptor and the Lewis acid. Ammonia is the electron-pair donor and the Lewis base.

Table 19.4 compares the definitions of acids and bases. The Lewis definition is the broadest. It extends to compounds that the Brønsted-Lowry theory does not classify as acids and bases. Sample Problem 19.1 provides some examples of those compounds.

Table 19.4

Acid-Base Definitions		
Type	**Acid**	**Base**
Arrhenius	H^+ producer	OH^- producer
Brønsted-Lowry	H^+ donor	H^+ acceptor
Lewis	electron-pair acceptor	electron-pair donor

UbD Check for Understanding

What distinguishes an acid from a base in the Brønsted-Lowry theory?

Assess students' knowledge about the Brønsted-Lowry theory by asking them to fill in the blank of the following prompt:

Ammonia is a base because it _______ a hydrogen ion to become a hydronium ion. *(accepts)*

ADJUST INSTRUCTION If students are having trouble answering, have them write the chemical formulas for ammonia and the ammonium ion. Then have students decide whether a hydrogen ion was donated or accepted by ammonia.

Explore

Lewis Acids and Bases

APPLY CONCEPTS Point out that Lewis was the first scientist to discuss the significance of electron pairs in bonding, providing the basis for Lewis electron-dot diagrams. Explain that his theory of acids and bases was an extension of his concept of electron pairs. **Ask** A Lewis acid accepts a pair of electrons to form what type of bond? *(covalent bond)* **Ask** Does a Lewis base accept or donate a pair of electrons to form a covalent bond? *(donates)*

Extend

Explain that many cleaning supplies contain acids or bases. Tell students to go to local grocery, hardware, or home supply stores and study the labels on various cleaning products, such as those that clean lime deposits from tile and porcelain or mortar stains from brick and tile. Have students record the name of each cleaning product, the purpose of the product, and the specific name of the acid or base in the product. Explain that some of the chemical names may be familiar, but others, such as *muriatic acid*, may not be. Have students determine whether the unfamiliar names are older common names, and have them identify these chemicals by both their common and official names.

Answers

FIGURE 19.6 hydronium ion; hydrogen sulfate ion

READING SUPPORT An amphibious airplane is designed to take off from and land on either land or water.

LESSON 19.1

Explain

Sample Practice Problem

Identify each reactant as an acid or a base.

$NaHCO_3 + HCl \rightarrow H_2CO_3 + NaCl$

($NaHCO_3$ is the base; HCl is the acid.)

Evaluate

Informal Assessment

Have students use hydrochloric acid and ammonia as examples to illustrate the differences between the three methods for classifying acids and bases. Then direct students to complete the 19.1 Lesson Check.

Reteach

Review the concept of conjugate acids and bases by first writing on the board the equilibrium equation for the conversion of ammonia and water to ammonium ions and hydroxide ions. Label the acid, base, conjugate acid, and conjugate base. Beneath that, write the reverse reaction. Stress that for the purpose of identifying conjugate acid-base pairs, the reactants of the forward reaction are considered to be the acid and the base, and the products of the reaction are their conjugates. Next, label the reactants and products of the reverse reaction. Point out that because the reaction is proceeding in the reverse direction, ammonium ions and hydroxide ions are now considered the acid and base, and ammonia and water are the conjugates.

Answers

1. a. H^+ is the Lewis acid; H_2O is the Lewis base.
   b. $AlCl_3$ is the Lewis acid; Cl^- is the Lewis base.
2. Lewis base; it has a nonbonding pair of electrons that it can donate.

SampleProblem 19.1

Identifying Lewis Acids and Bases

Identify the Lewis acid and the Lewis base in this reaction between ammonia and boron trifluoride.

$$NH_3 + BF_3 \longrightarrow NH_3BF_3$$

❶ **Analyze** **Identify the relevant concepts.** When a Lewis acid reacts with a Lewis base, the base donates a pair of electrons and the acid accepts the donated pair.

❷ **Solve** **Apply the concepts to this problem.**

Draw electron dot structures to identify which reactant has an unshared pair of electrons.

Identify the reactant with the unshared pair of electrons and the reactant that can accept the pair of electrons.

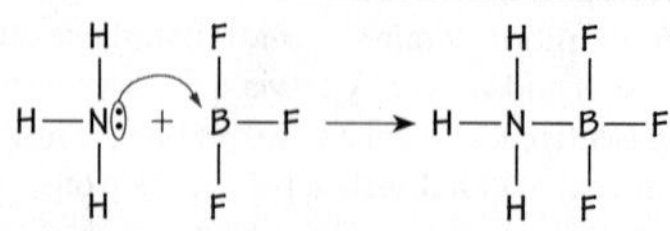

Ammonia has an unshared pair of electrons to donate. The boron atom can accept the donated electrons.

Classify the reactants based on their behavior.

Lewis bases donate a pair of electrons, so ammonia is the Lewis base. Lewis acids accept a pair of electrons, so boron trifluoride is the Lewis acid.

1. Identify the Lewis acid and Lewis base in each reaction.
   a. $H^+ + H_2O \longrightarrow H_3O^+$
   b. $AlCl_3 + Cl^- \longrightarrow AlCl_4^-$

2. Predict whether PCl_3 would be a Lewis acid or a Lewis base in typical chemical reactions. Explain your prediction.

19.1 LessonCheck

ONLINE PROBLEMS

3. **Review** What is the Arrhenius definition of an acid and a base?

4. **Describe** How are acids and bases defined by the Brønsted-Lowry theory?

5. **Explain** How did Lewis broaden the definition of acids and bases?

6. **Compare and Contrast** How are the properties of acids and bases similar? How are they different?

7. **Classify** Determine whether the following acids are monoprotic, diprotic, or triprotic:
a. H_2CO_3 **b.** H_3PO_4 **c.** HCl **d.** H_2SO_4

8. **Apply Concepts** Write a chemical equation for the ionization of HNO_3 in water and for the reaction of CO_3^{2-} with water. Identify the hydrogen-ion donor and the hydrogen-ion acceptor in each equation. Then, label each conjugate acid-base pair in the two equations.

BIGIDEA REACTIONS

9. Some household drain cleaners contain both sodium hydroxide and small particles of aluminum or zinc. Research how adding these metals can increase the effectiveness of the product.

652 Chapter 19 • Lesson 1

Lesson Check Answers

3. An acid gives hydrogen ions. A base gives hydroxide ions.
4. Acids are hydrogen-ion donors and bases are hydrogen-ion acceptors.
5. A Lewis acid is an electron-pair acceptor; a Lewis base is an electron-pair donor.
6. Both are electrolytes and change the color of an acid-base indicator. Acids have a sour taste; bases taste bitter.
7. a. diprotic
   b. triprotic
   c. monoprotic
   d. diprotic
8. $HNO_3 + H_2O \rightarrow H_3O^+ + NO_3^-$; HNO_3 is the hydrogen-ion donor; its conjugate base is NO_3^-. H_2O is the hydrogen-ion acceptor; its conjugate acid is H_3O^+.
   $CO_3^{2-} + H_2O \rightleftharpoons HCO_3^- + OH^-$; CO_3^{2-} is the hydrogen ion acceptor; its conjugate acid is HCO_3^-. H_2O is the hydrogen-ion donor; its conjugate base is OH^-.
9. **BIGIDEA** The reaction of NaOH with aluminum generates heat, which softens greases and oils and hydrogen, which agitates the mixture.

19.2 Hydrogen Ions and Acidity

C.8.2 Properties of acids and bases; **C.8.4** Calculate pH and hydronium ion concentration.

CHEMISTRY & YOU

Q: *What factors do you need to control so a fish has healthy water to live in?* Goldfish can live for 20 years or more in an aquarium if the conditions are right. The water in the aquarium must be cleaned regularly. You must also control the temperature of the water. In this lesson, you will study another factor that affects the ability of fish to survive.

Key Questions

How are $[H^+]$ and $[OH^-]$ related in an aqueous solution?

How is pH used to classify a solution as neutral, acidic, or basic?

What are two methods that are used to measure pH?

Vocabulary

- self-ionization
- neutral solution
- ion-product constant for water (K_w)
- acidic solution
- basic solution
- pH

Hydrogen Ions From Water

How are $[H^+]$ and $[OH^-]$ related in an aqueous solution?

Water molecules are highly polar and are in constant motion, even at room temperature. On occasion, the collisions between water molecules are energetic enough for a reaction to occur. When this happens, a hydrogen ion is transferred from one water molecule to another, as illustrated below. A water molecule that gains a hydrogen ion becomes a hydronium ion (H_3O^+). A water molecule that loses a hydrogen ion becomes a hydroxide ion (OH^-).

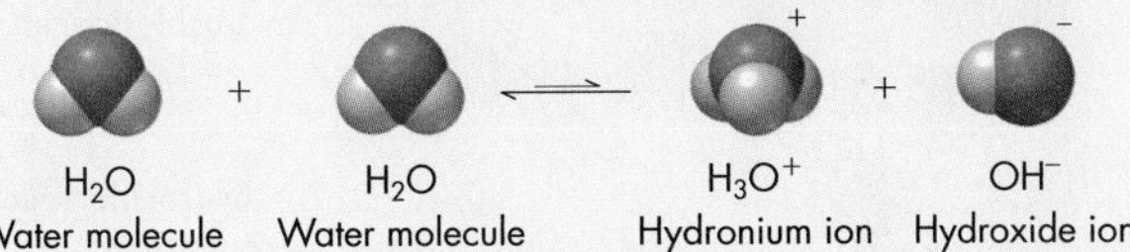

Self-Ionization of Water The reaction in which water molecules produce ions is called the **self-ionization** of water. This reaction can be written as a simple dissociation.

$$H_2O(l) \rightleftharpoons \underset{\text{Hydrogen ion}}{H^+(aq)} + \underset{\text{Hydroxide ion}}{OH^-(aq)}$$

In water or in an aqueous solution, hydrogen ions are always joined to water molecules as hydronium ions. Yet chemists may still refer to these ions as hydrogen ions or even protons. In this textbook, either H^+ or H_3O^+ is used to represent hydrogen ions in aqueous solution.

The self-ionization of water occurs to a very small extent. In pure water at 25°C, the concentration of hydrogen ions is only $1 \times 10^{-7}M$. The concentration of OH^- is also $1 \times 10^{-7}M$ because the numbers of H^+ and OH^- ions are equal in pure water. Any aqueous solution in which $[H^+]$ and $[OH^-]$ are equal is a **neutral solution.**

LESSON 19.2

Key Objectives

19.2.1 DESCRIBE how $[H^+]$ and $[OH^-]$ are related in an aqueous solution.

19.2.2 CLASSIFY a solution as neutral, acidic, or basic by using pH.

19.2.3 IDENTIFY two methods that are used to measure pH.

Additional Resources

Reading and Study Workbook, Lesson 19.2

Available Online or on Digital Media:

- Teaching Resources, Lesson 19.2 Review
- Laboratory Manual, Lab 40
- Small-Scale Chemistry Laboratory Manual, Lab 30

PearsonChem.com

Engage

CHEMISTRY & YOU Have students read the text that opens the lesson. Ask students to suggest other factors that might affect the ability of fish to survive, but do not provide the answer at this point.

Activate Prior Knowledge

Ask students to define the term *amphoteric*. *(A substance that can act as both an acid and a base is amphoteric.)* Then, ask them to explain why water is considered amphoteric. *(It can gain a proton to form a hydronium ion or lose a proton to form a hydroxide ion.)*

IN ACADEMIC STANDARDS for SCIENCE

- Develop explanations based on reproducible data and observations gathered during laboratory investigations.

C.8.2 Describe the characteristic properties of acids and bases.

C.8.4 Given the hydronium (H3O+) ion concentration in a solution, calculate the pH, and vice versa. Explain the meaning of these values.

Focus on ELL

1 CONTENT AND LANGUAGE Write the key questions on the board. As a class, work to paraphrase these questions so they are easier to understand. Read aloud each new paraphrased question and discuss what students have learned previously that will help them in this lesson.

2 FRONTLOAD THE LESSON Ask a student volunteer to explain the process for maintaining the water in a swimming pool or a hot tub. Guide the volunteer to share how he or she measures the conditions of the water and then the type of chemicals used to adjust the water conditions.

3 COMPREHENSIBLE INPUT Remind students that square brackets around a chemical formula indicate a molar concentration (moles per liter). Write the equation for K_w on the board as shown on page 654. Then rewrite the equation, substituting $1 \times 10^{-7}M$ for $[H^+]$ and $[OH^-]$ to show how the value $1 \times 10^{-14}M$ is determined.

Foundations for Reading

BUILD VOCABULARY Guide students to use a compare-and-contrast table to define and then to compare and contrast the properties of a *neutral solution*, an *acid solution*, and a *basic solution*.

READING STRATEGY Have students write at least two questions that challenge them to investigate the relationships between two or more vocabulary words or concepts. Then, have students answer their questions as they read the lesson.

Explain

Hydrogen Ions From Water

USE MODELS Direct students to examine the illustration of the self-ionization of water. On the board, show how two water molecules can react to yield a hydronium ion and a hydroxide ion. **Ask** Which element is donating a pair of electrons? *(oxygen)* **Ask** Which element is accepting a pair of electrons? *(hydrogen)*

MAKING CONNECTIONS Have students look at Figure 19.7. Remind students of the hydrochloric acid–zinc metal demonstration they observed in Lesson 19.1. Point out that pitting of the metal guitar parts resulted from a similar reaction between hydrochloric acid and the metal.

Figure 19.7 Aged by Acid Sometimes guitar players want a new guitar to look like it is old or "vintage." The guitarist can remove the shiny new metal parts of the guitar and expose them to hydrochloric acid. The acid will make the metal parts look dull. Both of the guitars in the photo below are new, but the bottom one has been aged with acid.

Ion-Product Constant for Water The ionization of water is a reversible reaction, so Le Châtelier's principle applies. Adding either hydrogen ions or hydroxide ions to an aqueous solution is a stress to the system. In response, the equilibrium will shift toward the formation of water. The concentration of the other ion will decrease. In any aqueous solution, when $[H^+]$ increases, $[OH^-]$ decreases. Likewise, when $[H^+]$ decreases, $[OH^-]$ increases.

$$H^+(aq) + OH^-(aq) \rightleftharpoons H_2O(l)$$

For aqueous solutions, the product of the hydrogen-ion concentration and the hydroxide-ion concentration equals 1.0×10^{-14}.

$$[H^+] \times [OH^-] = 1.0 \times 10^{-14}$$

This equation is true for all dilute aqueous solutions at 25°C. When substances are added to water, the concentrations of H^+ and OH^- may change. However, the product of $[H^+]$ and $[OH^-]$ does not change. The product of the concentrations of the hydrogen ions and hydroxide ions in water is called the **ion-product constant for water (K_w)**.

$$K_w = [H^+] \times [OH^-] = 1.0 \times 10^{-14}$$

Acidic Solutions Not all solutions are neutral. When some substances dissolve in water, they release hydrogen ions. For example, when hydrogen chloride dissolves in water, it forms hydrochloric acid.

$$HCl(aq) \longrightarrow H^+(aq) + Cl^-(aq)$$

In hydrochloric acid, the hydrogen-ion concentration is greater than the hydroxide-ion concentration. (The hydroxide ions come from the self-ionization of water.) A solution in which $[H^+]$ is greater than $[OH^-]$ is an **acidic solution.** In acidic solutions, the $[H^+]$ is greater than $1 \times 10^{-7}M$. Figure 19.7 shows a guitar that was artificially aged by using hydrochloric acid.

Basic Solutions When sodium hydroxide dissolves in water, it forms hydroxide ions in solution.

$$NaOH(aq) \longrightarrow Na^+(aq) + OH^-(aq)$$

In such a solution, the hydrogen-ion concentration is less than the hydroxide-ion concentration. Remember, the hydrogen ions are present from the self-ionization of water. A **basic solution** is one in which $[H^+]$ is less than $[OH^-]$. The $[H^+]$ of a basic solution is less than $1 \times 10^{-7}M$. Basic solutions are also known as alkaline solutions.

Differentiated Instruction

L1 STRUGGLING STUDENTS Students without a solid foundation in scientific notation may have trouble understanding the relationship between hydrogen- and hydroxide-ion concentrations. Draw diagrams to help students visualize the shift in equilibrium that occurs when one or the other type of ion is added or removed.

ELL ENGLISH LANGUAGE LEARNERS Discuss the meaning of *neutral* in the following sentences: *Switzerland was neutral in World War II. The umpire should be neutral during a baseball game. A neutral particle carries no overall electric charge. Then discuss its meaning in terms of pH.*

L1 LESS PROFICIENT READERS Have students create flashcards to help remember the symbols or abbreviations used in this lesson. For example, square brackets represent *molar concentration*, double arrows represent a reversible reaction, and *(aq)* means *aqueous*.

Sample Problem 19.2

Using the Ion-Product Constant for Water

If the $[H^+]$ in a solution is $1.0 \times 10^{-5}M$, is the solution acidic, basic, or neutral? What is the $[OH^-]$ of this solution?

KNOWNS
$[H^+] = 1.0 \times 10^{-5}M$
$K_w = 1 \times 10^{-14}$

UNKNOWNS
Is the solution acidic, basic, or neutral?
$[OH^-] = ?M$

1 Analyze List the knowns and the unknowns. Use the expression for the ion-product constant for water and the known concentration of hydrogen ions to find the concentration of hydroxide ions.

2 Calculate Solve for the unknowns.

Use $[H^+]$ to determine whether the solution is acidic, basic, or neutral.

$[H^+]$ is $1.0 \times 10^{-5}M$, which is greater than $1.0 \times 10^{-7}M$. Thus, the solution is acidic.

Rearrange the expression for the ion-product constant to solve for $[OH^-]$.

$$K_w = [H^+] \times [OH^-]$$

$$[OH^-] = \frac{K_w}{[H^+]}$$

Substitute the known values of $[H^+]$ and K_w. Then, solve for $[OH^-]$.

$$[OH^-] = \frac{1.0 \times 10^{-14}}{1.0 \times 10^{-5}}$$

$$= 1.0 \times 10^{-9}M$$

When you divide numbers written in scientific notation, subtract the exponent in the denominator from the exponent in the numerator.

3 Evaluate Does the result make sense? If $[H^+]$ is greater than $1.0 \times 10^{-7}M$, then $[OH^-]$ must be less than $1.0 \times 10^{-7}M$. $1 \times 10^{-9}M$ is less than $1 \times 10^{-7}M$. To check your calculation, multiply the values for $[H^+]$ and $[OH^-]$ to make sure the result equals 1×10^{-14}.

For Problem 11, rearrange the expression for the ion-product constant to solve for $[H^+]$.

10. Classify each solution as acidic, basic, or neutral.
a. $[H^+] = 6.0 \times 10^{-10}M$
b. $[OH^-] = 3.0 \times 10^{-2}M$
c. $[H^+] = 2.0 \times 10^{-7}M$
d. $[OH^-] = 1.0 \times 10^{-7}M$

11. If the hydroxide-ion concentration of an aqueous solution is $1 \times 10^{-3}M$, what is the $[H^+]$ in the solution? Is the solution acidic, basic, or neutral?

Foundations for Math

SUBTRACTING INTEGERS When explaining the power of a quotient, make sure students remember how to subtract integers. Tell students that to subtract an integer, you simply add the opposite. Use a number line to prove why this method works.

In **Sample Problem 19.2,** show students that the exponent for the quotient can be determined by subtracting the integer exponents. Point out that to find the difference of –14 – (–5), *you add the opposite*. Change the subtraction sign to a positive and change the sign on the subtrahend to its opposite—in this case, to a +. So, –14 – (–5) = –14 + 5 = –9.

Extend

Tell students that pure water (i.e., distilled, deionized water) self-ionizes to form hydrogen and hydroxide ions, yet it does not conduct electric current well. **Ask** Why does this occur? *(Pure water is a poor conductor because the concentrations of the ions are low.)* **Ask** Why do tap water and water from natural sources conduct electricity? *Water from these sources contains dissolved ions in large enough concentrations to conduct an electrical current.)*

USE CONTEXT CLUES Draw students' attention to the final sentence on the preceding page. Explain that sometimes words that mean the same thing may be interchanged in the sample problems or the text, as in the case of *alkaline* and *basic*. **Ask** What are some other terms that can be used to describe hydrogen ions? *(H^+, H_3O^+, hydronium ions, and protons)* Alert students to use context clues to help them understand an unfamiliar term, because the term might be just another way of stating something they already know.

Sample Practice Problem

The $[OH^-]$ of a grapefruit juice is $3.4 \times 10^{-11}M$. What is the $[H^+]$ of this solution? Is the solution acidic or basic? *(2.9×10^{-4}M; acidic)*

Answers

10. a. basic **b.** basic **c.** acidic **d.** neutral
11. $1 \times 10^{-11}M$; basic

LESSON 19.2

Explain

The pH Concept

START A CONVERSATION Introduce pH as a simpler way to express hydrogen-ion concentration. State the following sentences aloud.

- The pH of a solution is 3.00.
- The hydrogen ion concentration is equal to 1.0×10^{-3} moles per liter.

Point out that the two sentences express exactly the same information about the hydrogen ion concentration of a solution, but the first sentence expresses the information in a more concise manner.

APPLY CONCEPTS For students who have little or no experience with logarithms, explain that pH is found by taking the negative of the power (exponent) of 10 that expresses the hydrogen ion concentration. Show students this example: $[H^+] = 0.00010 = 1.0 \times 10^{-4}M$; $pH = -\log_{10}(1.0 \times 10^{-4}) = 4.00$. Explain that most pH values are positive, but negative values are also possible. Assuming 100% ionization, for example, the pH of a 10*M* HCl solution is −1.00. Challenge students to prove mathematically why this is true. *(10M = 1×10^1M; $pH = -\log_{10}(1.0 \times 10^1) = -1.00$).*

CHEMISTRY & YOU *Sample answer*: The difference in the ideal pH range most likely reflects the typical pH ranges in the natural environments of freshwater and saltwater fish.

Q: *In an aquarium, the pH of water is another factor that affects the ability of fish to survive. Most freshwater fish need a slightly acidic or neutral pH. For a saltwater tank, the ideal pH is slightly basic. What might explain this difference in the ideal pH range?*

The pH Concept

How is pH used to classify a solution as neutral, acidic, or basic?

Expressing hydrogen-ion concentration in molarity is not practical. A more widely used system for expressing $[H^+]$ is the pH scale, proposed in 1909 by the Danish scientist Søren Sørensen. The pH scale ranges from 0 to 14.

Hydrogen Ions and pH The **pH** of a solution is the negative logarithm of the hydrogen-ion concentration. The pH may be represented mathematically using the following equation:

$$pH = -\log[H^+]$$

In pure water or a neutral solution, the $[H^+] = 1 \times 10^{-7}M$, and the pH is 7.

$$\begin{aligned} pH &= -\log(1 \times 10^{-7}) \\ &= -(\log 1 + \log 10^{-7}) \\ &= -(0.0 + (-7.0)) = 7.0 \end{aligned}$$

If the $[H^+]$ of a solution is greater than $1 \times 10^{-7}M$, the pH is less than 7.0. If the $[H^+]$ of the solution is less than $1 \times 10^{-7}M$, the pH is greater than 7.0. **A solution with a pH less than 7.0 is acidic. A solution with a pH of 7.0 is neutral. A solution with a pH greater than 7.0 is basic.** Table 19.5 summarizes the relationship among $[H^+]$, $[OH^-]$, and pH. It also indicates the pH values of some common aqueous systems, including milk and blood.

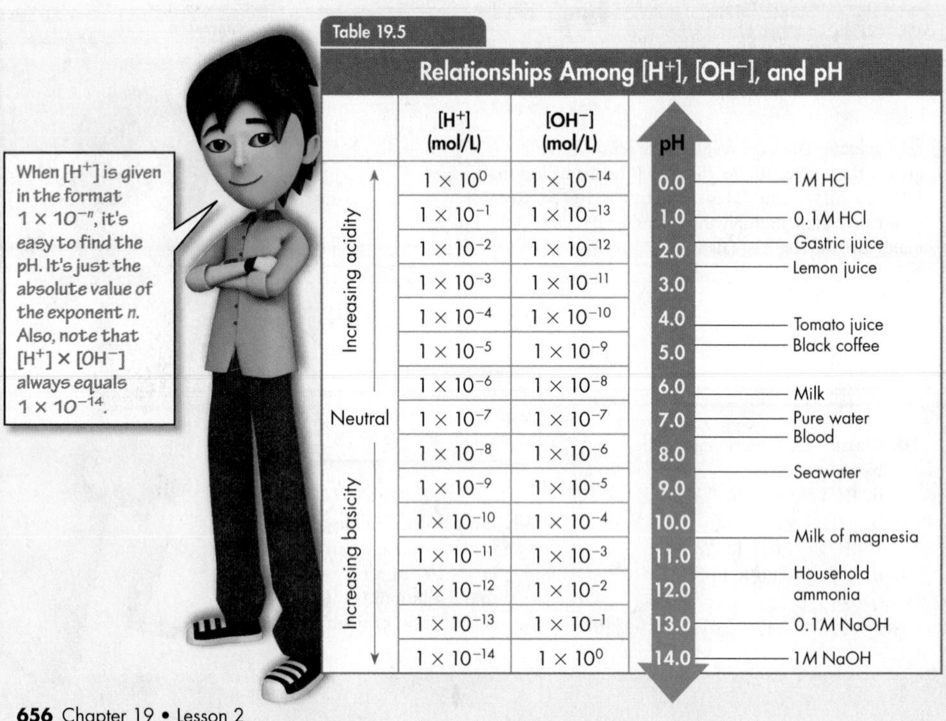

Table 19.5

Relationships Among $[H^+]$, $[OH^-]$, and pH

	$[H^+]$ (mol/L)	$[OH^-]$ (mol/L)	pH	
Increasing acidity	1×10^{0}	1×10^{-14}	0.0	1*M* HCl
	1×10^{-1}	1×10^{-13}	1.0	0.1*M* HCl
	1×10^{-2}	1×10^{-12}	2.0	Gastric juice; Lemon juice
	1×10^{-3}	1×10^{-11}	3.0	
	1×10^{-4}	1×10^{-10}	4.0	Tomato juice
	1×10^{-5}	1×10^{-9}	5.0	Black coffee
	1×10^{-6}	1×10^{-8}	6.0	Milk
Neutral	1×10^{-7}	1×10^{-7}	7.0	Pure water; Blood
Increasing basicity	1×10^{-8}	1×10^{-6}	8.0	Seawater
	1×10^{-9}	1×10^{-5}	9.0	
	1×10^{-10}	1×10^{-4}	10.0	Milk of magnesia
	1×10^{-11}	1×10^{-3}	11.0	Household ammonia
	1×10^{-12}	1×10^{-2}	12.0	
	1×10^{-13}	1×10^{-1}	13.0	0.1*M* NaOH
	1×10^{-14}	1×10^{0}	14.0	1*M* NaOH

Ions in Water

All aqueous systems contain both hydrogen and hydroxide ions due to self-ionization of water. At 25°C, about 1 molecule of water out of 550,000,000 will dissociate. This amounts to 1 g of hydrogen ions and 17 g of hydroxide ions in 10,000,000 L of water. No matter how small the concentrations, both ions are always present in a water solution. The relationship between the concentrations of H^+ and OH^- in aqueous solutions at constant temperature is similar to the relationship between the pressure and volume of a gas at constant temperature. In both cases, the relationship is inverse. The product of the two quantities is a constant—as one quantity increases, the other decreases.

Calculating pH From [H⁺] Expressing [H⁺] in scientific notation can make it easier to calculate pH. For example, you would rewrite $0.0010M$ as $1.0 \times 10^{-3}M$. The coefficient 1.0 has two significant figures. The pH for a solution with this concentration is 3.00. The two numbers to the right of the decimal point represent the two significant figures in the concentration.

It is easy to find the pH for solutions when the coefficient is 1.0. The pH of the solution equals the exponent, with the sign changed from minus to plus. For example, a solution with $[H^+] = 1 \times 10^{-2}M$ has a pH of 2.0. When the coefficient is a number other than 1, you will need to use a calculator with a log function key to calculate pH.

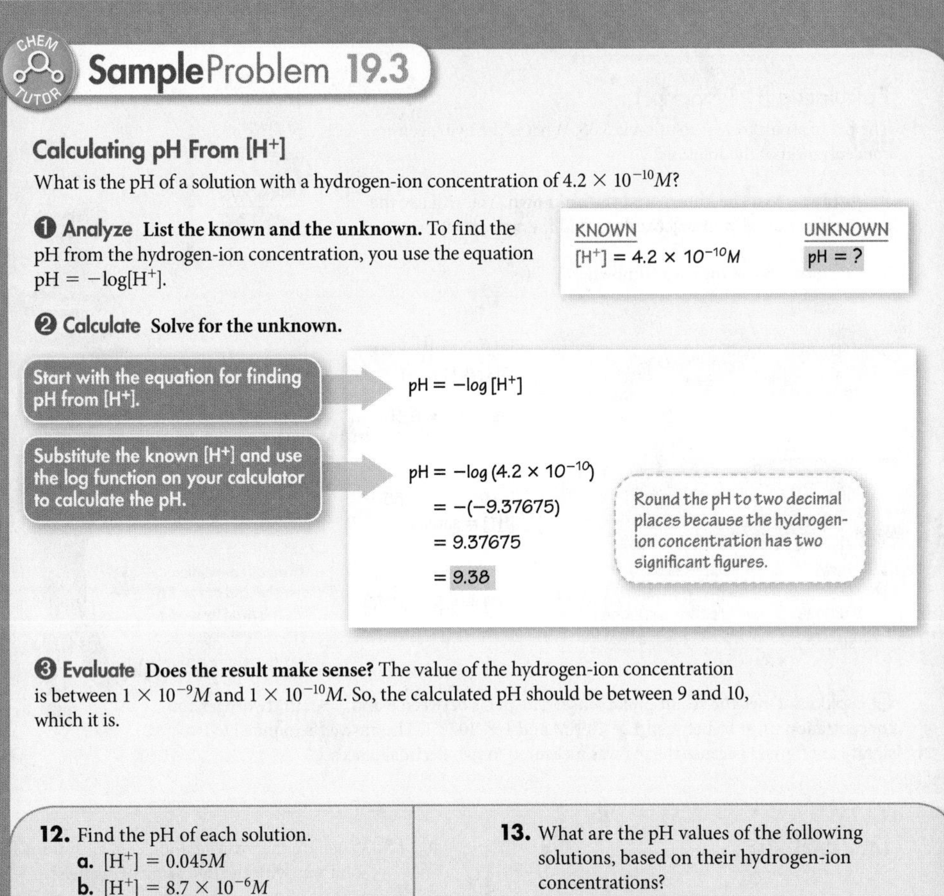

SampleProblem 19.3

Calculating pH From [H⁺]

What is the pH of a solution with a hydrogen-ion concentration of $4.2 \times 10^{-10}M$?

1 Analyze **List the knowns and the unknown.** To find the pH from the hydrogen-ion concentration, you use the equation $pH = -\log[H^+]$.

KNOWN	UNKNOWN
$[H^+] = 4.2 \times 10^{-10}M$	pH = ?

2 Calculate **Solve for the unknown.**

Start with the equation for finding pH from [H⁺].

$$pH = -\log[H^+]$$

Substitute the known [H⁺] and use the log function on your calculator to calculate the pH.

$$pH = -\log(4.2 \times 10^{-10})$$
$$= -(-9.37675)$$
$$= 9.37675$$
$$= 9.38$$

Round the pH to two decimal places because the hydrogen-ion concentration has two significant figures.

3 Evaluate **Does the result make sense?** The value of the hydrogen-ion concentration is between $1 \times 10^{-9}M$ and $1 \times 10^{-10}M$. So, the calculated pH should be between 9 and 10, which it is.

12. Find the pH of each solution.
- **a.** $[H^+] = 0.045M$
- **b.** $[H^+] = 8.7 \times 10^{-6}M$
- **c.** $[H^+] = 0.0015M$
- **d.** $[H^+] = 1.2 \times 10^{-3}M$

13. What are the pH values of the following solutions, based on their hydrogen-ion concentrations?
- **a.** $[H^+] = 1.0 \times 10^{-12}M$
- **b.** $[H^+] = 1 \times 10^{-4}M$

Foundations for Math

PROPERTIES OF LOGARITHMS Show students that the distributive property holds true for logarithms. Students who are not proficient with a scientific calculator, students who want to check their answers for sense, and even mathematically proficient students may benefit from the following kind of derivation.

In **Sample Problem 19.3,** show students that in Step 2 the value of $\log(4.2 \times 10^{-10})$ can be determined by breaking the problem into simpler parts. Point out that by the distributive property, the expression $\log(4.2 \times 10^{-10})$ is equivalent to $\log(4.2) + \log(10^{-10})$. By definition of logarithms, show students that $\log(10^{-10}) = -10$. Now the expression can be simplified as $\log(4.2) + (-10)$, or $\log(4.2) - 10$.

Explain

MAKE CONNECTIONS Have students examine Table 19.5. Point out that as the concentration of hydrogen ions in an aqueous solution increases, the concentration of hydroxide ions in the same solution decreases so that the product of the two concentrations always equals $1 \times 10^{-14}M$ at 25°C. The pH scale in the third column shows the relationship between the concentrations of the two ions and the resulting pH of the solution.

Sample Practice Problem

What are the pH values of the following three solutions, based on their hydrogen ion concentrations?

- **a.** $[H^+] = 1.0 \times 10^{-5}M$ *(5.00)*
- **b.** $[H^+] = 1.0 \times 10^{-9}M$ *(9.00)*
- **c.** $[H^+] = 0.001M$ *(3.0)*

Extend

Connect to GEOLOGY

Point out to students that the pH scale can be compared to the Richter scale, which measures the magnitude of earthquakes and is also based on a logarithmic scale. Explain that on each scale, a change of one unit represents a tenfold change in the value being measured. As a comparison, show students that on the Richter scale, a tremor measuring 4.0 is ten times as strong as one measuring 3.0 and on the pH scale, the hydrogen-ion concentration of a solution with a pH of 3.0 is ten times the concentration of a solution with a pH of 4.0. Have students research five of the most recent earthquakes. Instruct them to graph and compare the magnitudes.

Answers

12. **a.** pH = 1.35 **b.** pH = 5.06 **c.** pH = 2.82 **d.** pH = 2.92

13. **a.** $pH = -\log[H^+] = -\log(1.0 \times 10^{-12}) = 12$
b. $pH = -\log[H^+] = -\log(1 \times 10^{-4}) = 4$

Explain

Sample Practice Problem

A beverage that is of 88% apple juice has a pH of 2.7. What is the hydrogen-ion concentration of the beverage? *(2×10^{-3}M)*

CRITICAL THINKING Challenge students to hypothesize what the letters *p* and *H* represent in the abbreviation *pH*. As a hint, remind them that pH is an expression of the hydronium ion concentration in an aqueous solution. *(Students should correctly deduce that the H represents something having to do with hydronium ions.)* Write the equation pH = $-\log[H_3O^+]$ on the board, using a contrasting color to write *p* and *–log*. Give students an opportunity to suggest that *p* is another way of representing the negative logarithm.

Explore

Teacher Demo

PURPOSE Students will observe a color change resulting from a change in pH.

MATERIALS lemon juice, tea, antacid tablet, glass of 10% grape juice, household ammonia

PROCEDURE Without explanation, add lemon juice to tea, add an antacid tablet to grape juice, and add a few drops of ammonia to grape juice. Have the students try to explain the changes they observe.

EXPECTED OUTCOME Color changes occur when lemon juice, an antacid tablet, or ammonia is added. The pigments in tea and grape juice are weak acids that change color with varying pH.

Calculating [H⁺] From pH You can calculate the hydrogen-ion concentration of a solution if you know the pH. If the pH is an integer, it is easy to find the value of $[H^+]$. For a pH of 9.0, $[H^+] = 1 \times 10^{-9}M$. For a pH of 4.0, $[H^+]$ is $1 \times 10^{-4}M$.

However, most pH values are not whole numbers. For example, milk of magnesia has a pH of 10.50. The $[H^+]$ must be less than $1 \times 10^{-10}M$ (pH 10.0) but greater than $1 \times 10^{-11}M$ (pH 11.0). The hydrogen-ion concentration is $3.2 \times 10^{-11}M$. When the pH value is not a whole number, you will need a calculator with an antilog (10^x) function to get an accurate value for the hydrogen-ion concentration.

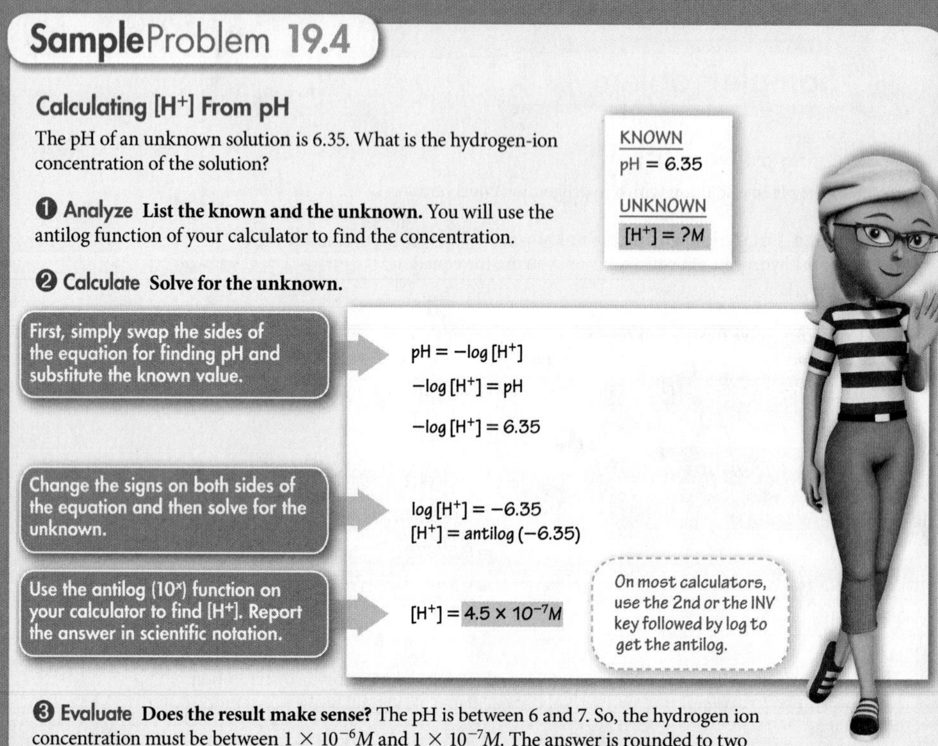

Sample Problem 19.4

Calculating [H⁺] From pH

The pH of an unknown solution is 6.35. What is the hydrogen-ion concentration of the solution?

KNOWN
pH = 6.35

UNKNOWN
$[H^+]$ = ?*M*

1 Analyze **List the known and the unknown.** You will use the antilog function of your calculator to find the concentration.

2 Calculate **Solve for the unknown.**

First, simply swap the sides of the equation for finding pH and substitute the known value.

$\text{pH} = -\log[H^+]$

$-\log[H^+] = \text{pH}$

$-\log[H^+] = 6.35$

Change the signs on both sides of the equation and then solve for the unknown.

$\log[H^+] = -6.35$

$[H^+] = \text{antilog}(-6.35)$

Use the antilog (10^x) function on your calculator to find $[H^+]$. Report the answer in scientific notation.

$[H^+] = 4.5 \times 10^{-7}M$

On most calculators, use the 2nd or the INV key followed by log to get the antilog.

3 Evaluate **Does the result make sense?** The pH is between 6 and 7. So, the hydrogen ion concentration must be between $1 \times 10^{-6}M$ and $1 \times 10^{-7}M$. The answer is rounded to two significant figures because the pH was measured to two decimal places.

14. Calculate $[H^+]$ for each solution.
a. pH = 5.00
b. pH = 12.83

15. What are the hydrogen-ion concentrations for solutions with the following pH values?
a. 4.00
b. 11.55

Foundations for Math

LOGS AND ANTILOGS Explain that pH is based on common, or base 10, logarithms. A common logarithm is an exponent of the number 10, and 10 raised to that exponent is its antilog. For example, log (10,000) = log (10^4) = 4. In this example, 4 is the log and its antilog is 10^4. Have students consider the equation $-\log[H^+] = \text{pH}$. Rewriting to isolate the log gives $\log[H^+] = -\text{pH}$. The log is –pH and its antilog is $10^{-\text{pH}}$. So $[H^+] = 10^{-\text{pH}}$.

For **Sample Problem 19.4**, $[H^+]$ = antilog (–6.35) can be rewritten as $[H^+] = 10^{-6.35}$. This relationship can help students remember that they need to use the 10^x function on their calculators to determine $[H^+]$ and to include the minus sign when making their antilog calculations. Suggest to students that they write $[H^+] = 10^{-\text{pH}}$ in their notebooks along with their notes on how to use the antilog function of their calculator.

Calculating pH From $[OH^-]$ If you know the $[OH^-]$ of a solution, you can find its pH. Recall that the ion-product constant for water defines the relationship between $[H^+]$ and $[OH^-]$. Therefore, you can use the ion-product constant for water to determine $[H^+]$ for a known $[OH^-]$. Then, you use $[H^+]$ to calculate the pH. For practice, try doing Sample Problem 19.5.

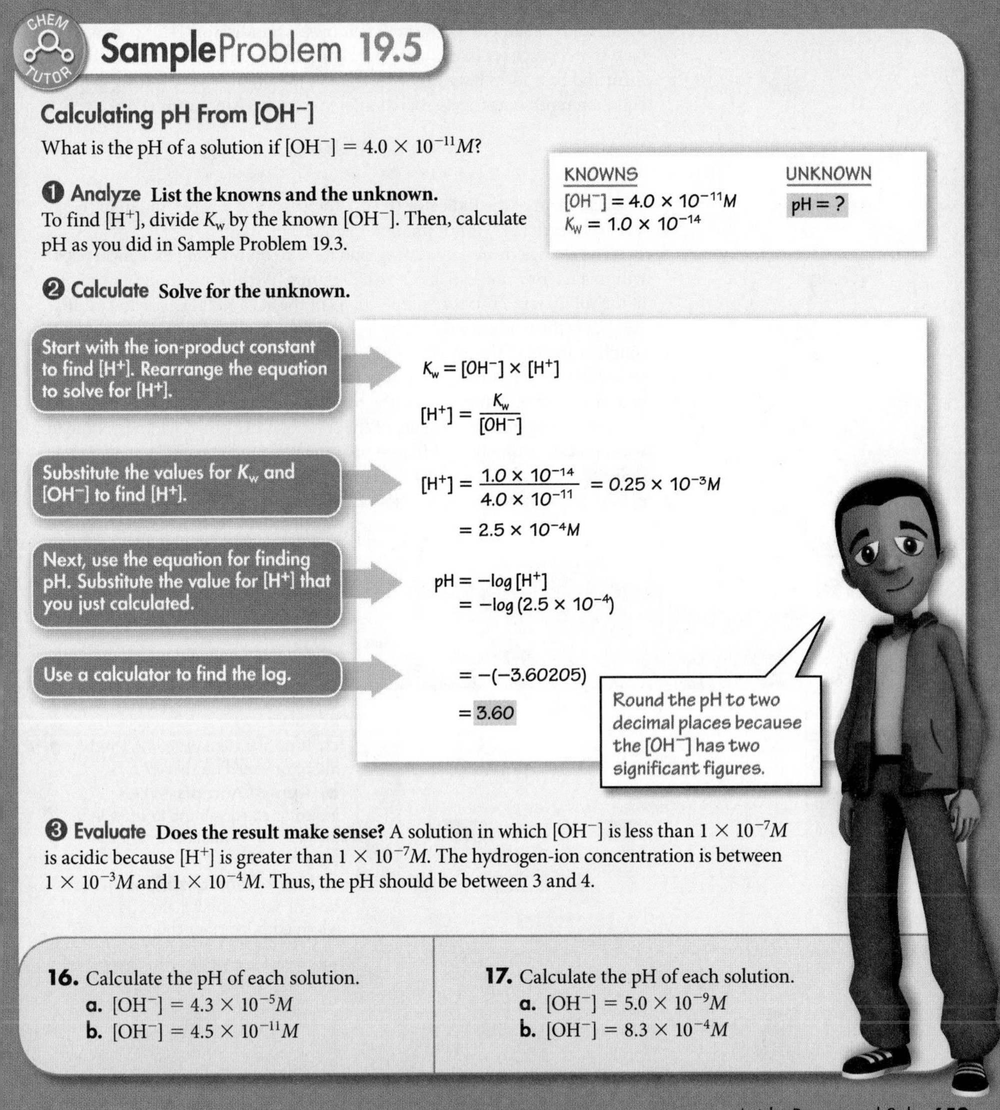

SampleProblem 19.5

Calculating pH From $[OH^-]$

What is the pH of a solution if $[OH^-] = 4.0 \times 10^{-11}M$?

1 Analyze **List the knowns and the unknown.**
To find $[H^+]$, divide K_w by the known $[OH^-]$. Then, calculate pH as you did in Sample Problem 19.3.

KNOWNS	UNKNOWN
$[OH^-] = 4.0 \times 10^{-11}M$	pH = ?
$K_w = 1.0 \times 10^{-14}$	

2 Calculate **Solve for the unknown.**

Start with the ion-product constant to find $[H^+]$. Rearrange the equation to solve for $[H^+]$.

$$K_w = [OH^-] \times [H^+]$$

$$[H^+] = \frac{K_w}{[OH^-]}$$

Substitute the values for K_w and $[OH^-]$ to find $[H^+]$.

$$[H^+] = \frac{1.0 \times 10^{-14}}{4.0 \times 10^{-11}} = 0.25 \times 10^{-3}M$$

$$= 2.5 \times 10^{-4}M$$

Next, use the equation for finding pH. Substitute the value for $[H^+]$ that you just calculated.

$$pH = -\log[H^+]$$

$$= -\log(2.5 \times 10^{-4})$$

Use a calculator to find the log.

$$= -(-3.60205)$$

$$= 3.60$$

3 Evaluate **Does the result make sense?** A solution in which $[OH^-]$ is less than $1 \times 10^{-7}M$ is acidic because $[H^+]$ is greater than $1 \times 10^{-7}M$. The hydrogen-ion concentration is between $1 \times 10^{-3}M$ and $1 \times 10^{-4}M$. Thus, the pH should be between 3 and 4.

16. Calculate the pH of each solution.
a. $[OH^-] = 4.3 \times 10^{-5}M$
b. $[OH^-] = 4.5 \times 10^{-11}M$

17. Calculate the pH of each solution.
a. $[OH^-] = 5.0 \times 10^{-9}M$
b. $[OH^-] = 8.3 \times 10^{-4}M$

Foundations for Math

NEGATIVE MULTIPLIERS Explain that when a negative sign is located just outside of a grouping symbol, the negative really represents the value –1. It is common in mathematics not to show the 1; explain that the 1 is "understood." Point out that since the negative is next to the grouping symbol, everything within the grouping symbol should be multiplied by –1.

In **Sample Problem 19.5,** students may attempt to calculate the log of -2.5×10^{-4}. Remind students that the log function serves as a grouping symbol, so the log should be calculated *before* multiplying by –1.

Explore

Class Activity

PURPOSE Students observe how a pH meter measures the pH of solutions.

MATERIALS several test tubes; 0.1*M* HCl; 0.1*M* NaOH; household products such as lemon juice, vinegar, shampoo, and liquid detergent; pH meter

PROCEDURE Set up test tubes with acids and bases of varying strength. In addition to HCl and NaOH, include household products such as lemon juice, vinegar, shampoo, and liquid detergent. Ask the students to predict the order from least acidic to most acidic and list their predictions in a table on the board. Use a pH meter to measure the pH of each solution. Explain that the reading on the pH meter depends upon the concentration of hydrogen ions in the solutions. Write the results on the board. Discuss any differences between students' predictions and the results of the experiment.

Explain

Sample Practice Problem

Calculate the pH of each solution.

a. $[OH^-] = 1.0 \times 10^{-11}$ (*3.00*)

b. $[H^+] = 7.7 \times 10^{-4}$ (*3.11*)

Answers

SAMPLE PROBLEMS 19.4 AND 19.5

14. a. $[H^+] = 1.0 \times 10^{-5}M$
b. $[H^+] = 1.5 \times 10^{-13}M$

15. a. $[H^+] = 1.0 \times 10^{-4}M$
b. $[H^+] = 2.8 \times 10^{-12}M$

16. a. $[H^+] = 9.63$
b. $[H^+] = 3.65$

17. a. pH = 5.70
b. pH = 10.92

Explore

Measuring pH

Teacher Demo

PURPOSE Students observe how pH indicators react to the acidity of their environment.

MATERIALS 0.1% solutions of thymol blue, methyl red, bromothymol blue, and phenolphthalein; test tubes; aqueous buffers spanning pH 4 to pH 10 (Either create a range of buffers spanning pH 4 to pH 10 or higher, or purchase standard buffers with different pH values.)

PROCEDURE Explain that many natural and synthetic pigments are weak acids that change color with varying pH. Transfer a small volume of each buffer to a test tube. Label the tubes with their respective pH values. Add 5 drops of indicator to each test tube. Set up a separate rack of tubes for each indicator and ask students to infer the pH range for which each indicator is most suitable. Students should note the colors and pH values of the solutions in tubes. In addition, add 5 drops of the indicators to two or three tubes containing a solution with an unknown pH and have students estimate its pH value using the reference tubes. **Ask** Why would an investigator prefer to use a pH meter to measure pH? *(Individual indicators are responsive only to pH changes in narrow ranges. Many different indicators are needed to span the entire pH range. A pH meter allows an investigator to collect quantitative values on a continuous basis throughout the entire pH range. The precision and accuracy of pH meters are superior to those of standard indicators. pH meters can be calibrated and used reliably at different temperatures.)*

Explain

USE VISUALS Have students examine Figure 19.8. **Ask** Which indicator(s) would best show a change in pH from 4.6 to 4.9? *(Using both bromocresol green and methyl red would give the best results.)*

Measuring pH

What are two methods that are used to measure pH?

In many situations, knowing the pH is useful. A custodian might need to maintain the correct acid-base balance in a swimming pool. A gardener may want to know if a certain plant will thrive in a yard. A doctor might be trying to diagnose a medical condition. **Either acid-base indicators or pH meters can be used to measure pH.**

Acid-Base Indicators An indicator is often used for initial pH measurements and for samples with small volumes. An indicator (HIn) is an acid or a base that dissociates in a known pH range. Indicators work because their acid form and base form have different colors in solution. The following general equation represents the dissociation of an acid-base indicator (HIn).

$$\underset{\text{Acid form}}{\mathrm{HIn}(aq)} \rightleftharpoons \mathrm{H^+}(aq) + \underset{\text{Base form}}{\mathrm{In^-}(aq)}$$

The acid form of the indicator (HIn) is dominant at low pH and high $[H^+]$. The base form (In^-) is dominant at high pH and high $[OH^-]$.

The change from dominating acid form to dominating base form occurs within a narrow range of about two pH units. Within this range, the color of the solution is a mixture of the colors of the acid and the base forms. If you know the pH range over which this color change occurs, you can make a rough estimate of the pH of a solution. At all pH values below this range, you would see only the color of the acid form. At all pH values above this range, you would see only the color of the base form.

For a more precise estimate of the solution's pH, you could repeat the test with indicators that have different pH ranges for their color change. Many indicators are needed to span the entire pH spectrum. Figure 19.8 shows the pH ranges of some common acid-base indicators.

Interpret Graphs

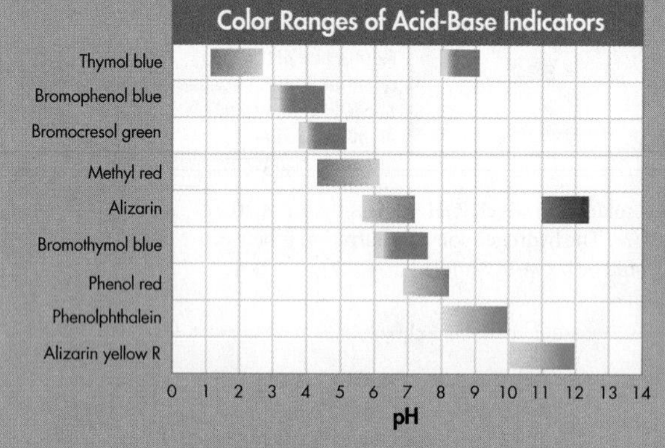

Figure 19.8 Each indicator is useful for a specific range of pH values.
a. Identify At a pH of 12, which indicator would be yellow?
b. Apply Concepts Which indicator could you use to show that the pH of a solution has changed from 3 to 5?
c. Make Generalizations What do you notice about the range over which each indicator changes color?

Differentiated Instruction

L3 ADVANCED LEARNERS Have students identify occupations in which people need to measure pH. Interested students may select a local person in an occupation from the list and ask to interview that person about his or her use of pH measurement. Have students write up their interviews as if they were preparing them for a blog.

L1 SPECIAL NEEDS STUDENTS Students with colorblindness will be unable to distinguish many of the color changes that are shown in Figure 19.8 and that occur during demonstrations, class activities, and labs involving color indicators. Pair these individuals with a lab partner who has normal vision, and provide these students with alternative assessment items when necessary.

LPR LESS PROFICIENT READERS Provide students with a copy of Figure 19.8, Have students write *acidic, neutral*, and *basic* below the pH values on the *x*-axis, using brackets to encompass pH ranges where appropriate. Have students refer to Table 19.5 for help. Encourage students to refer to the graph during the teacher demo.

Figure 19.9 Effects of Soil Acidity
Soil pH can affect how plants develop. **a.** In acidic soils, hydrangeas produce blue flowers. **b.** In basic soils, hydrangeas produce pink flowers.

Indicators have certain properties that limit their usefulness. The pH values of indicators are usually given for 25°C. At other temperatures, an indicator may change color at a different pH. If the solution being tested is not colorless, the color of the indicator may be misleading. Dissolved salts in a solution may affect an indicator's dissociation. Using indicator strips can help overcome these problems. An indicator strip is a piece of paper or plastic that has been soaked in an indicator, and then dried. The paper is dipped into an unknown solution. The color that results is compared with a color chart to measure the pH. Some indicator paper has absorbed multiple indicators. The colors that result will cover a wide range of pH values. Before planting the shrub shown in Figure 19.9, you might want to test the pH of your soil.

pH Meters Your chemistry laboratory probably has a pH meter. A pH meter is used to make rapid, continuous measurements of pH. The measurements of pH obtained with a pH meter are typically accurate to within 0.01 pH unit of the true pH. If the pH meter is connected to a computer or chart recorder, the user will have a record of the pH changes.

A pH meter can be easier to use than liquid indicators or indicator strips. As shown in Figure 19.10, the pH reading is visible in a display window on the meter. Hospitals use pH meters to find small but meaningful changes in the pH of blood and other body fluids. Sewage, industrial wastes, and soil pH are also easily monitored with a pH meter. The color and cloudiness of the solution do not affect the accuracy of the pH value obtained.

Figure 19.10 pH Meter
A pH meter provides a quick and accurate way to measure the pH of a system.

Focus on ELL

4 LANGUAGE PRODUCTION Have students work in groups of four to complete the Quick Lab on the following page. Make sure each group has ELLs of varied language proficiencies so that more proficient students can help less proficient ones. Have students work according to their proficiency level.

BEGINNING: LOW/HIGH Have students use colored markers correlated to a color key to record their observations.

INTERMEDIATE: LOW/HIGH Model Step 3 of the procedure–creating the pH scale. Then have students follow your example.

ADVANCED: LOW/HIGH Ask students to create a sorting method for less proficient students to use for help in answering Analyze and Conclude questions 3 and 4.

Evaluate

Informal Assessment

Have students use equations to describe the relationship between the concentrations of hydrogen and hydroxide ions in pure water and to show what happens to the equilibrium when HCl is added to water. Repeat the exercise with NaOH. *(At 25°C, $[H^+] \times [OH^-] = 1.0 \times 10^{-14}$. When acid is added, the $[H^+]$ increases and the $[OH^-]$ decreases. When base is added, the $[OH^-]$ increases and the $[H^+]$ decreases.)* Then direct students to complete the 19.2 Lesson Check.

Reteach

Review the key steps in calculating pH and the relationship between K_w, $[H^+]$, and $[OH^-]$. Then provide students with additional practice in calculating pH, $[H^+]$, and $[OH^-]$.

Answers

INTERPRET GRAPHS

a. methyl red
b. bromophenol blue or bromocresol green
c. The indicators change color over a very limited range of pH. This allows for a more accurate reading of pH change.

LESSON 19.2

Explore

Quick Lab

OBJECTIVE Students will measure the pH of various household materials by using a natural indicator and an indicator chart.

PREP TIME 20 minutes

CLASS TIME If the period allotted to lab work is short, spread the work over two days.

SAFETY Perform this lab in a well-ventilated room.

EXPECTED OUTCOME Students will use their natural indicators and pH scales to determine if tested materials are acidic, basic, or neutral.

ANALYZE AND CONCLUDE

1. The initial color is purple. In an acidic solution, the indicator is red; in a neutral solution, it is blue-purple; and in a basic solution, it is green.
2. Changes in the relative number of H^+ and OH^- ions present are responsible for the color changes.
3. Answers will vary depending on the materials chosen.
4. Personal hygiene items mainly test neutral; cleaning materials, such as soap, mainly test basic.

Quick Lab

Indicators From Natural Sources

Purpose To measure the pH of household materials using a natural indicator

Materials

- red cabbage leaves
- 1-cup measure
- hot water
- 2 jars
- spoon
- cheesecloth
- 3 sheets of plain white paper
- transparent tape
- metric ruler
- pencil
- 10 small clear plastic cups
- permanent marker
- white vinegar (CH_3COOH)
- baking soda ($NaHCO_3$)
- spatula
- household ammonia
- dropper
- assorted household materials

Procedure

1. Put one-half cup of finely chopped red cabbage leaves in a jar and add one-half cup of hot water. Stir and crush the leaves with a spoon. Continue this process until the water has a distinct color.

2. Strain the mixture through a piece of clean cheesecloth into a clean jar. The liquid that collects in the jar is your indicator.

3. Tape three sheets of paper end to end. Draw a line along the center of the taped sheets. Label the line at 5-cm intervals with the numbers 1 to 14. This labeled line is your pH scale.

4. Use the permanent marker to label three cups vinegar, baking soda, and ammonia. Pour indicator into each cup to a depth of about 1 cm.

5. Add several drops of vinegar to the first cup. Use a spatula to add a pinch of baking soda to the second cup. Add several drops of ammonia to the third cup. The pH values for the solutions of vinegar, baking soda, and household ammonia are about 3, 9, and 11, respectively. Record the colors you observe at the correct locations on your pH scale.

6. Repeat the procedure for household items such as table salt, milk, lemon juice, laundry detergent, milk of magnesia, toothpaste, shampoo, and carbonated beverages.

Analyze and Conclude

1. Observe What color is the indicator in acidic, neutral, and basic solutions?

2. Relate Cause and Effect What caused the color of the indicator to change when a material was added to a cup?

3. Classify Divide the household materials you tested into three groups—acidic, basic, and neutral.

4. Analyze Data Which group contains items used for cleaning? Which group contains items used for personal hygiene?

19.2 LessonCheck

ONLINE PROBLEMS

18. Review How are the concentrations of hydrogen ions and hydroxide ions related in an aqueous solution?

19. Identify What is the range of pH values in the following solutions?
a. basic **b.** acidic **c.** neutral

20. List What methods can you use to measure the pH of a solution?

21. Relate Cause and Effect What happens to the $[H^+]$ as the pH of a solution increases?

22. Calculate Determine the pH of each solution.
a. $[H^+] = 1 \times 10^{-6}M$
b. $[H^+] = 0.00010M$
c. $[OH^-] = 1 \times 10^{-2}M$
d. $[OH^-] = 1 \times 10^{-11}M$

23. Compare In terms of ion concentrations, how do basic solutions differ from acidic solutions?

24. Calculate Find the hydroxide-ion concentrations for solutions with the following pH values:
a. 6.00 **b.** 9.00 **c.** 12.00

Lesson Check Answers

18. $[H^+] \times [OH^-] = 1.0 \times 10^{-14}$; when $[H^+]$ in a solution increases, the $[OH^-]$ decreases.

19. basic: greater than 7; acidic: less than 7; neutral: 7

20. acid-base indicator or a pH meter

21. $[H^+]$ decreases as pH increases.

22. a. 6.0 **b.** 4.00 **c.** 12.0 **d.** 3.0

23. In basic solutions the $[OH^-]$ is greater than $[H^+]$; in acidic solutions $[H^+]$ is greater than $[OH^-]$.

24. a. $1.0 \times 10^{-8}M$ **b.** $1.0 \times 10^{-5}M$ **c.** $1.0 \times 10^{-2}M$

Agronomist

Do you like the idea of working with plants but find chemistry more interesting than biology? If so, you might want to consider a career in agronomy. Agronomy is a branch of agriculture that deals with the interactions between plants, soils, and the environment. Agronomists use their knowledge of chemistry to help produce healthy crops and increase yields, while preserving the environment.

The opportunities for agronomists extend beyond laboratories and greenhouses. Agronomists also work for business firms, government agencies, conservation groups, philanthropic organizations, and universities. Agronomists can also use their knowledge of water and land management to address such issues as urban area beautification and highway landscaping.

SOIL ACIDITY The pH of the soil is among the most important factors in crop production. Agronomists can help farmers obtain the right soil pH for a specific crop.

FIELD CHEMISTRY Many agronomists have the opportunity to work with local communities. This agronomist is advising a Kenyan farming group on how to improve their crops using science.

Take It Further

1. **Apply Concepts** The ideal soil pH for corn is around 6.0. If $[H^+]$ equals $2.14 \times 10^{-5}M$, is the soil too acidic or too basic for growing corn?
2. **Infer** What are two nonfood items that an agronomist might help produce?

21st Century Learning To be successful in the 21st century, students need skills and learning experiences that extend beyond subject matter mastery. This project helps students build the following 21st Century Skills: *Creativity and Innovation; Critical Thinking and Problem Solving; Information Literacy; Information, Communication, and Technology Literacy; Initiative and Self-Direction.*

PLAN A GARDEN Pose the following challenge to your students. *You have been hired to plan a garden for a large company. After you complete the following tasks, prepare a three-page report summarizing your plan.*

- Research to find out the typical soil acidity for your area. Which plants will do well in this acidity? Which should you avoid?
- Include a variety of plants. Explain why each plant might do well.
- Include instructions for maintaining the soil pH. What tests should be done? What should be the response to the tests?

START A CONVERSATION Draw students' attention to the photo at the top of the page. Point out that the agronomist is gathering a soil sample to take back to a lab for testing. Pose the following questions. How could the agronomist determine the pH of the soil? What changes might the agronomist recommend if the pH were too high or too low? You may need to guide students in the following ways:

- Would using an acid-base indicator or a pH meter be more convenient for testing the soil's acidity?
- What could be done to change the soil if it is too acidic? What could be done if the soil is too basic?

Explain

CRITICAL THINKING Lead students in discussing factors that might affect the pH of soil. **Ask** What do you think might contribute to high soil acidity? *(Sample answers: acid rain, the rocks from which the soil formed, fertilizers added to the soil)*

Extend

Connect to BIOLOGY

No single soil pH is best for all plants. Many plants prefer slightly acidic soil, although some prefer slightly basic soil. Explain that different types of nutrients tend to be available to plants in soils with different pH values.

Answers

TAKE IT FURTHER

1. The pH is 4.67, which is probably too acidic for growing corn.
2. Possible answers are cotton, corn that is used to produce ethanol, sod, and pharmaceutical crops, which are harvested for their proteins.

LESSON 19.3

Key Objectives

19.3.1 IDENTIFY the property used to classify acids and bases as strong or weak.

Additional Resources

Reading and Study Workbook, Lesson 19.3

Available Online or on Media:

- Teaching Resources, Lesson 19.3 Review
- Virtual Chemistry Laboratory Manual, Lab 27

PearsonChem.com

Engage

CHEMISTRY & YOU Have students study the photograph and read the text that opens the lesson. **Ask** Do you think citric acid is a strong acid or weak acid? *(weak)* **Ask** Do you think sulfuric acid is a strong acid or a weak acid? *(strong)*

Activate Prior Knowledge

Write the ion-product constant (K_w) on the board. **Ask** What does this constant represent? *(the product of the concentrations of hydronium ions and hydroxide ions in water)* **Ask** Why does water contain these ions? *(It undergoes self-ionization.)* Explain that another term used for the process of ionization is *dissociation*. Tell students to keep this term in mind as they read this lesson.

19.3 Strengths of Acids and Bases

C.8.2 Properties of acids and bases; **C.8.3** Dissociation and strength of acids and bases.

CHEMISTRY & YOU

Q: *What makes one acid safer than another?* Lemon juice, which contains citric acid, has a pH of about 2.3. Yet, you consume lemon juice. When you cut a lemon, you usually don't wear gloves or safety goggles. But some acids do require such precautions. This lesson will explain the difference between a "weak" acid such as citric acid and a "strong" acid such as sulfuric acid.

Key Question

How are acids and bases classified as either strong or weak?

Vocabulary

- strong acid
- weak acid
- acid dissociation constant (K_a)
- strong base
- weak base
- base dissociation constant (K_b)

Strong and Weak Acids and Bases

How are acids and bases classified as either strong or weak?

Table 19.6 compares the strengths of some acids and bases. **Acids and bases are classified as strong or weak based on the degree to which they ionize in water.** Hydrochloric acid and sulfuric acid are examples of strong acids. In general, a **strong acid** is completely ionized in aqueous solution.

$$HCl(g) + H_2O(l) \longrightarrow H_3O^+(aq) + Cl^-(aq)$$

100%

A **weak acid** ionizes only slightly in aqueous solution. The ionization of ethanoic acid (CH_3COOH), a typical weak acid, is not complete.

$$CH_3COOH(aq) + H_2O(l) \rightleftharpoons H_3O^+(aq) + CH_3COO^-(aq)$$

< 1%

Table 19.6

Relative Strengths of Common Acids and Bases

Substance	Formula	Relative strength
Hydrochloric acid	HCl	Strong acids
Nitric acid	HNO_3	
Sulfuric acid	H_2SO_4	
Phosphoric acid	H_3PO_4	Increasing strength of acid ↑
Ethanoic acid	CH_3COOH	
Carbonic acid	H_2CO_3	
Hypochlorous acid	$HClO$	
		Neutral solution
Ammonia	NH_3	Increasing strength of base ↓
Sodium silicate	Na_2SiO_3	
Calcium hydroxide	$Ca(OH)_2$	Strong bases
Sodium hydroxide	$NaOH$	
Potassium hydroxide	KOH	

IN ACADEMIC STANDARDS for SCIENCE

C.8.2 Describe the characteristic properties of acids and bases.

C.8.3 Compare and contrast the dissociation and strength of acids and bases in solution.

Focus on ELL

1 CONTENT AND LANGUAGE Assign students to four groups. Assign each group one of the words *strong*, *weak*, *concentrated*, or *diluted*. Have each group complete a Frayer model diagram for the word, with a definition, characteristics, an example and a nonexample. Have each group teach the word to the class.

2 FRONTLOAD THE LESSON Ask students to work with a partner to draw a picture or to cut out magazine pictures that represent the words *strong* and *weak*. Display the results. As students view the display, have them try to deduce what is meant by these two terms in relation to acids and bases.

3 COMPREHENSIBLE INPUT Review the process for writing an equilibrium-constant expression before introducing K_a. Then, demonstrate mathematically that the equation for the dissociation constant (K_a) is achieved by multiplying both sides of the equation by $[H_2O]$. Explain how the $[H_2O]$ values on the right-hand side of the equation cancel each other.

Interpret Graphs

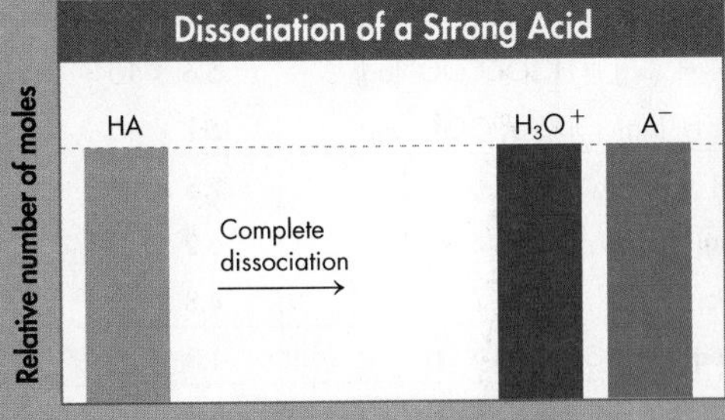

$HA(aq) + H_2O(l) \longrightarrow H_3O^+(aq) + A^-(aq)$

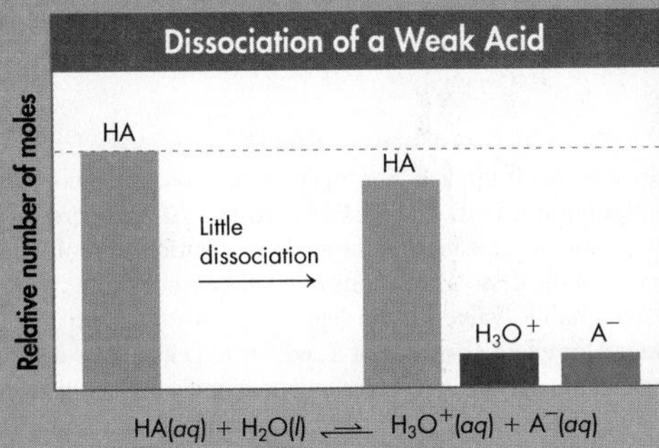

$HA(aq) + H_2O(l) \rightleftharpoons H_3O^+(aq) + A^-(aq)$

Figure 19.11 Dissociation of an acid (HA) in water yields H_3O^+ and an anion, A^-. The bar graphs compare the extent of dissociation of a strong acid and a weak acid.

a. Explain Why is there only one bar for HA in the graph for the strong acid, but two bars for HA in the graph for the weak acid?

b. Apply Concepts In the graph for the strong acid, why do the bars for H_3O^+ and A^- have the same height as the bar for HA?

c. Infer In the graph for the weak acid, why is the height of the bar for H_3O^+ equal to the distance from the top of the second HA bar to the dotted line?

Hint: The bars represent the relative amounts of the acid and the ions it forms in solution.

Acid Dissociation Constant Figure 19.11 compares the extent of dissociation of strong and weak acids. A strong acid, such as hydrochloric acid, completely dissociates in water. As a result, $[H_3O^+]$ is high in an aqueous solution of a strong acid. By contrast, weak acids remain largely undissociated. For example, in an aqueous solution of ethanoic acid, less than 1 percent of the molecules are ionized at any time.

You can use a balanced equation to write the equilibrium-constant expression for a reaction. The equilibrium-constant expression shown below is for ethanoic acid.

$$K_{eq} = \frac{[H_3O^+] \times [CH_3COO^-]}{[CH_3COOH] \times [H_2O]}$$

For dilute aqueous solutions, the concentration of water is a constant. This constant can be combined with K_{eq} to give an acid dissociation constant. An **acid dissociation constant (K_a)** is the ratio of the concentration of the dissociated form of an acid to the concentration of the undissociated form. The dissociated form includes both the H_3O^+ and the anion.

$$K_{eq} \times [H_2O] = K_a = \frac{[H_3O^+] \times [CH_3COO^-]}{[CH_3COOH]}$$

Differentiated Instruction

LPR LESS PROFICIENT READERS Some students may find the tables and figures overwhelming. Present the visuals on an overhead. During lecture, show only the parts of the visuals necessary for conveying the concept and cover up the rest. Use Socratic questioning to engage students in determining the information provided by each visual.

ELL ENGLISH LANGUAGE LEARNERS At the beginning of the discussion on dissociation constants, provide each student with a sticky note containing the word *dissociation*. During class, clearly pronounce *dissociation (dih so see* AY *shun)*, and have the class repeat slowly. Then brainstorm a list of synonyms for the word and elicit from students that the antonym of dissociation is *association*.

L1 SPECIAL NEEDS STUDENTS Explain the graph in Figure 19.11 by using a visual and tactile activity. Provide students with stackable colored blocks and help them to create structures similar in height to bars in Figure 19.11. Use Socratic questioning to help students understand how a strong acid differs from a weak acid.

Foundations for Reading

BUILD VOCABULARY Encourage students to create a two-column table, labeling the columns *Acid* and *Base*. Under each column, students should write the related vocabulary words and their definitions.

READING STRATEGY Have students read the key question and then quickly preview the lesson by skimming the headings, visuals, and boldfaced text. Then have students predict the answer to the key question.

Explain

Strong and Weak Acids and Bases

USE VISUALS Have students study Table 19.6. **Ask** Which is the weakest acid in the table? *(hypochlorous acid)* **Ask** Which is the weakest base in the table? *(ammonia)*

CRITICAL THINKING Point out the two equations above Table 19.6. **Ask** What do you notice about the arrows in each equation? *(One equation has a single arrow, and the other equation has a double arrow.)* **Ask** What do these arrows tell you about the degree to which an acid ionizes in water? *(The single arrow pointing to the right indicates that the acid completely ionizes; the double arrow indicates that the acid is not completely ionized.)* Point out the differing lengths of the two arrows. Explain that the shorter length of the top arrow indicates that ionization is occurring but that the preferred direction of the reaction is to the left. **Ask** If there was a greater degree of dissociation occurring, which arrow would be longer—the one for the forward reaction or the one for the reverse reaction? *(The arrow for the forward reaction would be longer.)*

USE VISUALS Direct students to Figure 19.11. Explain to students that the bar graphs compare the extent of ionization or dissociation of strong and weak acids. **Ask** What is the primary difference between strong and weak acids? *(Strong acids are nearly 100% dissociated, whereas weak acids are only partially dissociated.)*

Answers

INTERPRET GRAPHS

a. For a strong acid, there is virtually no HA left after dissociation. For a weak acid, a considerable amount of HA remains after dissociation.

b. The strong acid dissociates completely, forming equal amounts of ions.

c. The amount of acid that dissociates is equal to the amount of each resulting ion.

LESSON 19.3

Explain

Strong and Weak Acids and Bases

APPLY CONCEPTS Remind students that in solution, the negative ion of an acid is a base. Explain that strong acids, such as nitric acid, dissociate almost completely because the negative ion is a very weak base. That is, its tendency to combine with a hydrogen ion is slight. In contrast, the dissociation of a weak acid, such as carbonic acid, is limited, because its negative ion is a relatively strong base. Explain that these ions combine with most of the available hydrogen ions to form the undissociated form of the acid again.

Misconception Alert

The terms *strong* and *weak*, as applied to acids, are often confused with the concept of concentration. Stress that the dissociation of an acid or base into ions involves the establishment of equilibrium. The terms *strong* and *weak* refer to the position of the equilibrium. Explain that when a strong acid or base dissolves, the equilibrium favors the products. However, when a weak acid or base dissolves, the equilibrium favors the reactants. Show students that the extent to which products or reactants are favored can be determined from the values of K_a or K_b. Discuss how values of K_a and K_b can be used to compare the strengths of acids and bases.

USE VISUALS Have students examine Table 19.7. Write out the equation for calculating the first K_a of oxalic acid to reinforce the concept that the K_a for each acid is equal to the product of the concentrations of the products divided by the concentration of the non-ionized acid. Also explain that the K_a value is a quantitative indication of acid strength. A lower K_a value means the reactants are favored in the equilibrium. Thus, the acid is weaker.

CRITICAL THINKING Have students study the dissociation constants for phosphoric acid in Table 19.7. **Ask** What pattern do you observe in the values for K_a? *(Each subsequent K_a is smaller than the previous value.)* Explain that as the charge on an acid becomes more negative, its attraction for the remaining protons increases, decreasing the likelihood that additional dissociations will occur to any great degree.

Table 19.7

Dissociation Constants of Weak Acids

Acid	Chemical equation for dissociation	K_a (25°C)
Oxalic acid	$HOOCCOOH(aq) \rightleftharpoons H^+(aq) + HOOCCOO^-(aq)$	5.6×10^{-2}
	$HOOCCOO^-(aq) \rightleftharpoons H^+(aq) + OOCCOO^{2-}(aq)$	5.1×10^{-5}
Phosphoric acid	$H_3PO_4(aq) \rightleftharpoons H^+(aq) + H_2PO_4^-(aq)$	7.5×10^{-3}
	$H_2PO_4^-(aq) \rightleftharpoons H^+(aq) + HPO_4^{2-}(aq)$	6.2×10^{-8}
	$HPO_4^{2-}(aq) \rightleftharpoons H^+(aq) + PO_4^{3-}(aq)$	4.8×10^{-13}
Methanoic acid	$HCOOH(aq) \rightleftharpoons H^+(aq) + HCOO^-(aq)$	1.8×10^{-4}
Benzoic acid	$C_6H_5COOH(aq) \rightleftharpoons H^+(aq) + C_6H_5COO^-(aq)$	6.3×10^{-5}
Ethanoic acid	$CH_3COOH(aq) \rightleftharpoons H^+(aq) + CH_3COO^-(aq)$	1.8×10^{-5}
Carbonic acid	$H_2CO_3(aq) \rightleftharpoons H^+(aq) + HCO_3^-(aq)$	4.3×10^{-7}
	$HCO_3^-(aq) \rightleftharpoons H^+(aq) + CO_3^{2-}(aq)$	4.8×10^{-11}

The acid dissociation constant (K_a) reflects the fraction of an acid that is ionized. For this reason, dissociation constants are sometimes called ionization constants. If the degree of dissociation or ionization of the acid in a solution is small, the value of the dissociation constant will be small. Weak acids have small K_a values. If the degree of ionization of an acid is more complete, the value of K_a will be larger. The stronger an acid is, the larger its K_a value will be. For example, nitrous acid (HNO_2) has a K_a of 4.4×10^{-4}, but ethanoic acid (CH_3COOH) has a K_a of 1.8×10^{-5}. This means that nitrous acid is more ionized in solution than ethanoic acid. Therefore, nitrous acid is a stronger acid than ethanoic acid.

Table 19.7 shows the ionization equations and dissociation constants of a few weak acids. Some of the acids have more than one dissociation constant because they have more than one ionizable hydrogen. Oxalic acid, for example, is a diprotic acid. It loses two hydrogens, one at a time. Therefore, it has two dissociation constants. Oxalic acid is found naturally in certain herbs and vegetables, such as those pictured in Figure 19.12.

The acids in Table 19.7 are ranked by the value of the first dissociation constant. Observe what happens to the K_a with each ionization. The K_a decreases from the first ionization to the second. It decreases again from the second ionization to the third.

Figure 19.12 Oxalic Acid
Chives and parsley have relatively high amounts of oxalic acid compared to other fruits and vegetables.

UbD Check for Understanding

What property is used to classify acids and bases as strong or weak?

Assess students' knowledge about acids and bases by saying that substance X is reported to be a stronger acid than substance Y. **Ask** What information could you use to determine if this statement is true or false? Explain your answer.

(I would compare the dissociation constants for the two acids. A greater K_a for substance X than for substance Y would show that it is a stronger acid because K_a increases as the tendency for an acid to ionize in aqueous solution increases. The stronger the acid, the greater its tendency to ionize in solution.)

ADJUST INSTRUCTION If students are having trouble answering, have them reread the text following Table 19.7. Then pose the question again.

Calculating Dissociation Constants To calculate the acid dissociation constant (K_a) of a weak acid, you need to know the initial molar concentration of the acid and the $[H^+]$ (or alternatively, the pH) of the solution at equilibrium. You can use these data to find the equilibrium concentrations of the acid and the ions. These values are then substituted into the expression for K_a.

In general, you can find the K_a of an acid in water by substituting the equilibrium concentrations of the acid, [HA], the anion from the dissociation of the acid, $[A^-]$, and the hydrogen ion, $[H^+]$, into the equation below.

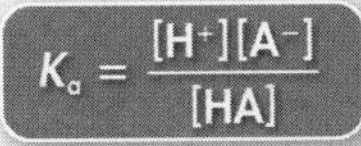

$$K_a = \frac{[H^+][A^-]}{[HA]}$$

CHEM TUTOR

SampleProblem 19.6

Calculating a Dissociation Constant

In a $0.1000M$ solution of ethanoic acid, $[H^+] = 1.34 \times 10^{-3}M$. Calculate K_a of this acid. Refer to Table 19.7 for the ionization equation for ethanoic acid.

1 Analyze **List the knowns and the unknown.**

KNOWNS

[ethanoic acid] = $0.1000M$

$[H^+] = 1.34 \times 10^{-3}M$

UNKNOWN

$K_a = ?$

2 Calculate **Solve for the unknown.**

Start by determining the equilibrium concentration of the ions.

$[H^+] = [CH_3COO^-] = 1.34 \times 10^{-3}M$

Each molecule of CH_3COOH that ionizes gives an H^+ ion and a CH_3COO^- ion.

Then determine the equilibrium concentrations of each component.

$(0.1000 - 0.00134)M = 0.0987M$

Concentration	$[CH_3COOH]$	$[H^+]$	$[CH_3COO^-]$
Initial	0.1000	0	0
Change	-1.34×10^{-3}	1.34×10^{-3}	1.34×10^{-3}
Equilibrium	0.0987	1.34×10^{-3}	1.34×10^{-3}

Substitute the equilibrium values into the expression for K_a.

$$K_a = \frac{[H^+] \times [CH_3COO^-]}{[CH_3COOH]} = \frac{(1.34 \times 10^{-3}) \times (1.34 \times 10^{-3})}{0.0987}$$

$$= 1.82 \times 10^{-5}$$

3 Evaluate **Does the result make sense?** The calculated value of K_a is consistent with that of a weak acid.

25. In a $0.1000M$ solution of methanoic acid, $[H^+] = 4.2 \times 10^{-3}M$. Calculate the K_a of this acid.

26. In a $0.2000M$ solution of a monoprotic weak acid, $[H^+] = 9.86 \times 10^{-4}M$. What is the K_a for this acid?

Foundations for Math

ASSOCIATIVE PROPERTY AND SCIENTIFIC NOTATION Explain that in multiplication of numbers written in scientific notation, numbers can be regrouped using the Associative Property of Multiplication: $(xy)(xz) = (xx)(yz)$.

In the last step of **Sample Problem 19.6,** show students how to regroup the two factors written in scientific notation in the numerator of the fraction:

$$(1.34 \times 10^{-3})(1.34 \times 10^{-3}) = (1.34 \times 1.34)(10^{-3} \times 10^{-3})$$

Next, group the decimals and group the powers:

$$= 1.34^{1+1} \times 10^{-3+(-3)}$$

Then have students use exponent rules to find the product:

$$= 1.34^2 \times 10^{-6} = 1.80 \times 10^{-5}$$

Explain

APPLY CONCEPTS Explain how the dissociation constant of an acid or base can be determined experimentally from the concentration of the solution and its pH. Tell students that the calculation is based on the assumption that the acid molecule forms an equal number of hydrogen ions and negative ions when it dissociates. Explain that this is true only if there is no additional source of either hydrogen ions or the negative ions.

Tell students that the process of calculating experimental dissociation constants is more complicated in the case of polyprotic acids, because there are multiple equilibrium reactions taking place simultaneously and they each influence the others. With polyprotic acids, all of the reactions will produce H_3O^+ but in differing amounts. The scientist must make assumptions about the relative amounts of H_3O^+ contributed by each reaction. Also, the products of some reactions are reactants in others, so the net change in their concentrations must be taken into account. Explain that for this course, students will only be expected to calculate dissociation constants for monoprotic acids.

Sample Practice Problem

A solution of a weak acid, which is exactly $0.725M$, has a $[H^+] = 5.23 \times 10^{-6}M$.

a. What is the pH of this solution? *(5.28)*

b. What is the value of K_a for this acid? *(3.77×10^{-11})*

Answers

SAMPLE PROBLEM 19.6

25. $K_a = 1.8 \times 10^{-4}$

26. $K_a = 4.89 \times 10^{-6}$

LESSON 19.3

Extend

Ask for a show of hands of students who have heard the phrase *pH-balanced shampoo*. Explain that healthy hair has a pH range of 4.5 to 5.5 and that alkaline shampoos can damage the cuticle of hairs. Divide the class into small groups to research the types of shampoo ingredients and their uses. Then, provide a selection of shampoos for normal, dry, and oily hair for students to test. Have students conduct a survey of ingredients and pH levels for the different shampoos and present their results. Supply each group with a 10-mL graduated cylinder, universal indicator paper, and beakers. Suggest that students prepare standard 1% solutions of shampoo for testing pH. Have groups present their findings to the class.

Connect to ENVIRONMENTAL SCIENCE

Have students do research about the sources and causes of acid rain. *(Acid rain is thought to be due principally to the release of sulfur oxides and nitrogen oxides into the atmosphere.)* **Ask** How do weather patterns affect the distribution of environmental damage due to acid rain? *(Prevailing winds generally carry pollutants from west to east.)* **Ask** What types of vegetation, if any, are most resistant to acid rain or tend to thrive in acidic soils? *(Students may wish to speak to staff at nurseries for help in answering this question.)* **Ask** What types of conservation efforts are used to adjust the pH of lakes and rivers that become too acidic? *(A special slurry of $CaCO_3$, or limestone, is sometimes added to lakes and rivers to control pH.)*

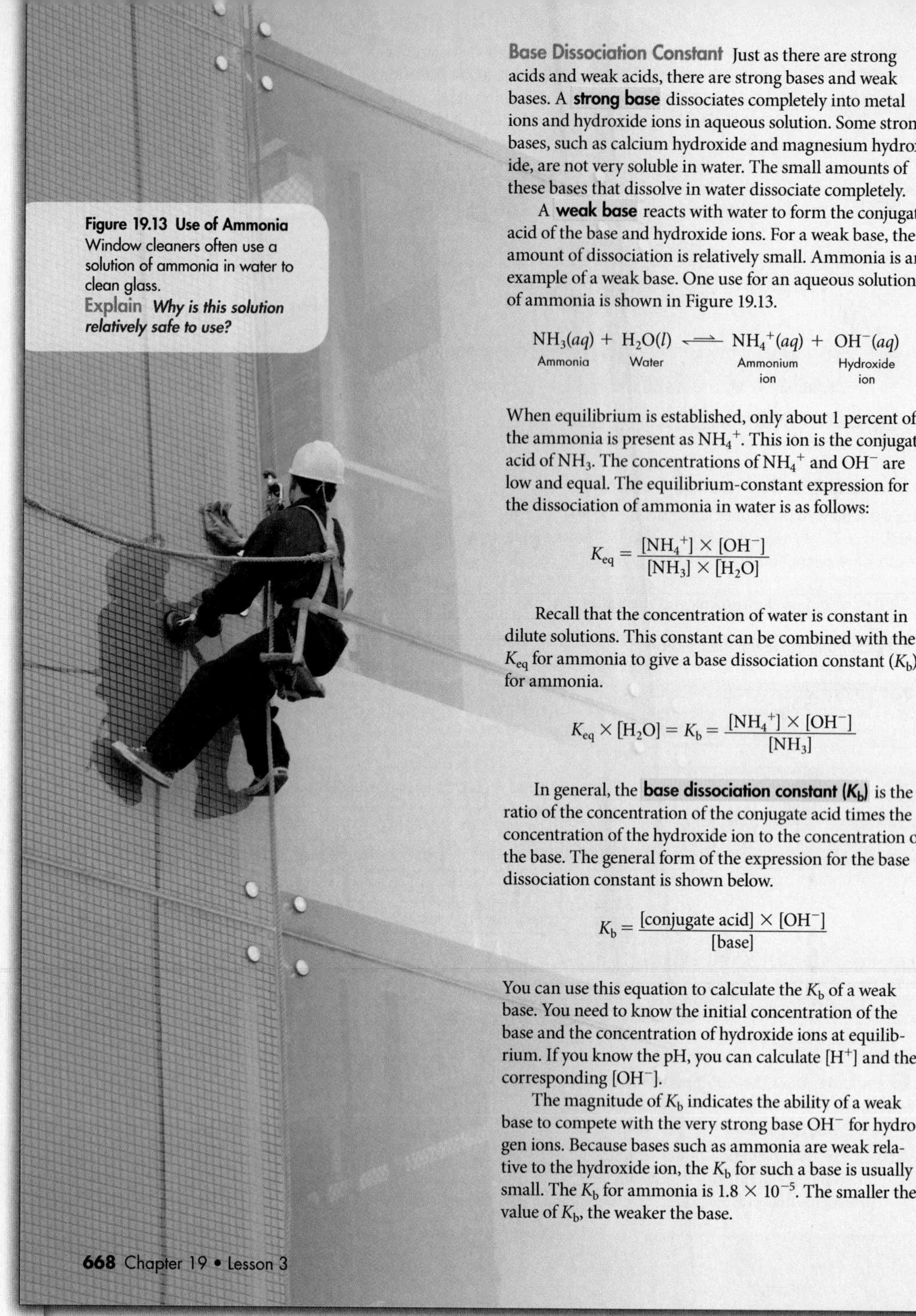

Figure 19.13 Use of Ammonia
Window cleaners often use a solution of ammonia in water to clean glass.
Explain *Why is this solution relatively safe to use?*

Base Dissociation Constant Just as there are strong acids and weak acids, there are strong bases and weak bases. A **strong base** dissociates completely into metal ions and hydroxide ions in aqueous solution. Some strong bases, such as calcium hydroxide and magnesium hydroxide, are not very soluble in water. The small amounts of these bases that dissolve in water dissociate completely.

A **weak base** reacts with water to form the conjugate acid of the base and hydroxide ions. For a weak base, the amount of dissociation is relatively small. Ammonia is an example of a weak base. One use for an aqueous solution of ammonia is shown in Figure 19.13.

$$\underset{\text{Ammonia}}{NH_3(aq)} + \underset{\text{Water}}{H_2O(l)} \rightleftharpoons \underset{\text{Ammonium ion}}{NH_4^+(aq)} + \underset{\text{Hydroxide ion}}{OH^-(aq)}$$

When equilibrium is established, only about 1 percent of the ammonia is present as NH_4^+. This ion is the conjugate acid of NH_3. The concentrations of NH_4^+ and OH^- are low and equal. The equilibrium-constant expression for the dissociation of ammonia in water is as follows:

$$K_{eq} = \frac{[NH_4^+] \times [OH^-]}{[NH_3] \times [H_2O]}$$

Recall that the concentration of water is constant in dilute solutions. This constant can be combined with the K_{eq} for ammonia to give a base dissociation constant (K_b) for ammonia.

$$K_{eq} \times [H_2O] = K_b = \frac{[NH_4^+] \times [OH^-]}{[NH_3]}$$

In general, the **base dissociation constant (K_b)** is the ratio of the concentration of the conjugate acid times the concentration of the hydroxide ion to the concentration of the base. The general form of the expression for the base dissociation constant is shown below.

$$K_b = \frac{[\text{conjugate acid}] \times [OH^-]}{[\text{base}]}$$

You can use this equation to calculate the K_b of a weak base. You need to know the initial concentration of the base and the concentration of hydroxide ions at equilibrium. If you know the pH, you can calculate $[H^+]$ and the corresponding $[OH^-]$.

The magnitude of K_b indicates the ability of a weak base to compete with the very strong base OH^- for hydrogen ions. Because bases such as ammonia are weak relative to the hydroxide ion, the K_b for such a base is usually small. The K_b for ammonia is 1.8×10^{-5}. The smaller the value of K_b, the weaker the base.

668 Chapter 19 • Lesson 3

Focus on ELL

4 LANGUAGE PRODUCTION Have students work in groups of four to complete the Small-Scale Lab on page 670. Make sure each group has ELLs of varied language proficiencies so that more proficient students can help less proficient ones. Have students work according to their proficiency level.

BEGINNING: LOW/HIGH Have students use colored pencils or crayons to draw pictures of their observations.

INTERMEDIATE: LOW/HIGH Provide students with a chart they can use to record their observations.

ADVANCED: LOW/HIGH Have students assist students with lower English proficiencies with the calculations and answering the questions.

Table 19.8

Comparing Concentration and Strength of Acids

Acidic solution	Concentration: Quantitative (or Molar)	Concentration: Relative	Strength
Hydrochloric acid	12*M* HCl	Concentrated	Strong
Gastric juice	0.08*M* HCl	Dilute	Strong
Ethanoic acid	17*M* CH_3COOH	Concentrated	Weak
Vinegar	0.2*M* CH_3COOH	Dilute	Weak

Q: *Despite its relatively low pH, lemon juice is safe to consume because citric acid is a weak acid. Citric acid has three K_a values. What does this information tell you about citric acid?*

Concentration Versus Strength Sometimes people confuse the concepts of concentration and strength. The words *concentrated* and *dilute* indicate how much of an acid or base is dissolved in solution. These terms refer to the number of moles of the acid or base in a given volume. The words *strong* and *weak* refer to the extent of ionization or dissociation of an acid or base.

Table 19.8 shows four possible combinations of concentration and strength for acids. Hydrochloric acid, HCl(*aq*), is a strong acid because it completely dissociates into ions. The gastric juice in your stomach is a dilute solution of HCl. The relatively small number of HCl molecules in a given volume of gastric juice are all dissociated into ions. To summarize, even when concentrated hydrochloric acid is diluted with water, it is still a strong acid. Conversely, ethanoic acid (acetic acid) is a weak acid because it ionizes only slightly in solution. Vinegar is a dilute solution of ethanoic acid. Even at a high concentration, ethanoic acid is still a weak acid.

The same concepts apply to bases. A solution of ammonia can be either dilute or concentrated. However, in any solution of ammonia, the relative amount of ionization will be small. Thus, ammonia is a weak base at any concentration. Likewise, sodium hydroxide is a strong base at any concentration.

19.3 LessonCheck

ONLINE PROBLEMS

27. Review What factor determines whether an acid or base is strong or weak?

28. Compare How do acid dissociation constants vary between strong acids and weak acids?

29. Draw Conclusions Which of the acids in Table 19.6 would you expect to have the lowest dissociation constant?

30. Describe How do you determine the K_a of a weak acid or the K_b of a weak base?

31. Predict Acid HX has a very small K_a. How will the relative amounts of H^+ and HX compare at equilibrium?

32. Calculate A 0.500*M* solution of an acid has a hydronium-ion concentration of $5.77 \times 10^{-6}M$. Calculate the K_a of this acid.

33. Describe Write a chemical equation for the dissociation of each of the following acids and bases in water.

a. nitric acid

b. ethanoic acid

c. ammonia

d. magnesium hydroxide

34. Classify A 15*M* solution of an acid has a K_a of 7.5×10^{-3}. Explain how you would classify this solution in terms of concentration and strength.

Lesson Check Answers

27. the degree to which they ionize in water

28. A strong acid is completely ionized in aqueous solution and has a large K_a. A weak acid ionizes only slightly in aqueous solution and has a small K_a.

29. hypochlorous acid

30 Substitute the measured concentrations of all the substances present at equilibrium into the expressions for K_a or K_b.

31. [HX] will be much greater than [H+] at equilibrium.

32. $K_a = 6.66 \times 10^{-11}$.

33. a. $HNO_3 + H_2O \rightarrow H_3O^+ + NO_3^-$

b. $CH_3COOH + H_2O \rightleftharpoons CH_3COO^- + H_3O^+$

c. $NH_3 + H_2O \rightleftharpoons NH_4^+ + OH^-$

d. $Mg(OH)_2 \rightarrow Mg^{2+} + 2OH^-$

34. It is concentrated based on its molarity, but weak based on its K_a.

The fact that citric acid has three acid dissociation constants indicates that citric acid is a triprotic acid.

Evaluate

Informal Assessment

Direct students to write or state the two conditions needed to calculate the dissociation constant of a weak acid. *(You must know the initial molar concentration of the acid and the pH or [H+] of the solution at equilibrium.)* Then direct students to complete the 19.3 Lesson Check.

Reteach

Place students in small groups. Have each group prepare a numbered list of steps for calculating dissociation constants. Have groups share their lists with the class. Create an overall class list of steps to follow for calculating dissociation constants. Provide a problem similar to Sample Problem 19.6, and have students use the class-generated procedure to work through the problem.

Answers

FIGURE 19.13 Ammonia is a weak base.

Explore

Small-Scale Lab

OBJECTIVE Students measure ionization constants of weak acids.

PREP TIME 1 hour

CLASS TIME 30 minutes

ADVANCE PREPARATION SAFETY Always add acid to water carefully and slowly!

Solution	Preparation
*0.04% BTB	100 mg in 16.0 mL 0.01*M* NaOH, dilute to 250 mL
*0.04% BCG	100 mg in 14.3 mL 0.01*M* NaOH, dilute to 250 mL
*0.04% BPB	100 mg in 14.9 mL 0.01*M* NaOH, dilute to 250 mL
*0.04% MCP	100 mg in 26.2 mL 0.01*M* NaOH, dilute to 250 mL
*0.04% TB	100 mg in 21.5 mL 0.01*M* NaOH, dilute to 250 mL
0.1% phenolphthalein	250 mg in 250 mL of 70% 2-propanol
0.05% MO	125 mg in 250 mL
0.02% AYR	50 mg in 250 mL
Solution A	14.9 g boric acid and 12.6 g citric acid in 1200 mL
Solution B	45.6 g $Na_3PO_4 \cdot 12H_2O$ in 1200 mL
1.0*M* H_3PO_4	14 mL of 15*M* H_3PO_4 in 200 mL

*Can dissolve the sodium salt of this indicator directly in 250 mL of distilled water.

pH*	Solution A	Solution B
2	195 mL	5 mL
3	176 mL	24 mL
4	155 mL	45 mL
5	134 mL	66 mL
6	118 mL	82 mL
7	99 mL	101 mL
8	85 mL	115 mL
9	69 mL	131 mL
10	54 mL	146 mL
11	44 mL	156 mL
12	17 mL	183 mL

*Use 1.0*M* H_3PO_4 for pH 1.

Small-Scale Lab

Dissociation Constants of Weak Acids

Purpose

To measure dissociation constants of weak acids

Materials

- paper, pencil, and ruler
- reaction surface
- 12 solutions with different pH values
- bromocresol green
- solutions of other acid-base indicators

Procedure

1. On two separate sheets of paper, draw two grids similar to the one below. Make each square 2 cm on each side.

2. Place a reaction surface over one of the grids and place one drop of bromocresol green in each square.

3. Add one drop of the solution with a pH of 1 to the square labeled 1. Add one drop of the solution with a pH of 2 to the square labeled 2. Continue adding drops in this manner until you have added a drop to each square.

4. Use the second grid as a data table to record your observations for each square.

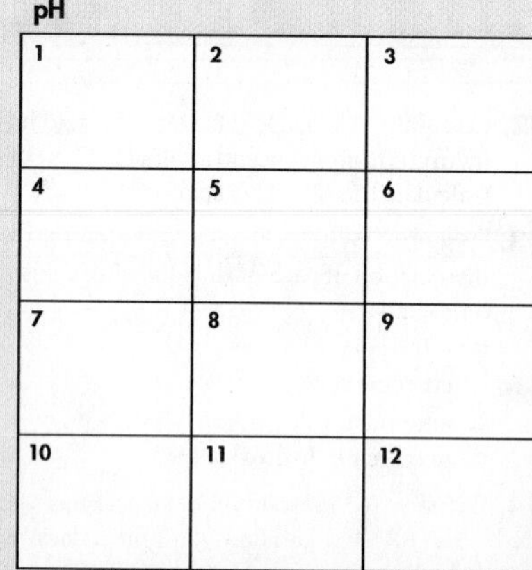

Analyze and Conclude

1. Observe What colors are the solutions with the lowest pH and the highest pH?

2. Observe At which pH does the bromocresol green change from one color to the other?

3. Infer Acid-base indicators, such as bromocresol green, are usually weak acids. Because bromocresol green has a fairly complex formula, it is convenient to represent its formula as HBCG. HBCG dissociates in water according to the following equation. HBCG and BCG^- are a conjugate acid-base pair.

$$\underset{\text{(yellow)}}{HBCG(aq)} + H_2O(l) \rightleftharpoons \underset{\text{(blue)}}{BCG^-(aq)} + H_3O^+(aq)$$

The K_a expression is

$$K_a = \frac{[BCG^-] \times [H_3O^+]}{[HBCG]}$$

When $[BCG^-] = [HBCG]$, $K_a = [H_3O^+]$.

What color is the conjugate base of HBCG? What color is the conjugate acid of BCG^-?

4. Draw Conclusions At what pH is there an equal amount of the conjugate acid and conjugate base? How can you tell?

5. Calculate What is the K_a for the solution described in Question 4?

You're the Chemist

1. Design an Experiment Design and carry out an experiment to measure the dissociation constants of some other acid-base indicators. Record the color of each conjugate acid and conjugate base. Calculate the K_a for each acid.

2. Explain How can you measure the dissociation constant of an acid-base indicator? Describe what to do and how to interpret the results.

670 Chapter 19 • Small-Scale Lab

ANALYZE AND CONCLUDE

1. THe lowest pH solutions are yellow and the highest pH solutions are blue.
2. pH 4
3. The conjugate acid, HBCG, is yellow. The conjugate base, BCG^-, is blue.
4. at pH 4, because the indicator is green
5. 1.0×10^{-4}

YOU'RE THE CHEMIST

1. Results will vary depending on the indicators chosen.
2. To measure the K_a of a colored acid-base indicator, mix one drop of the indicator solution with one drop of each pH 1–12 buffer solution. Look for the pH of the color change. The $[H^+]$ at this pH is close to the K_a of the acid.

Stone Erosion

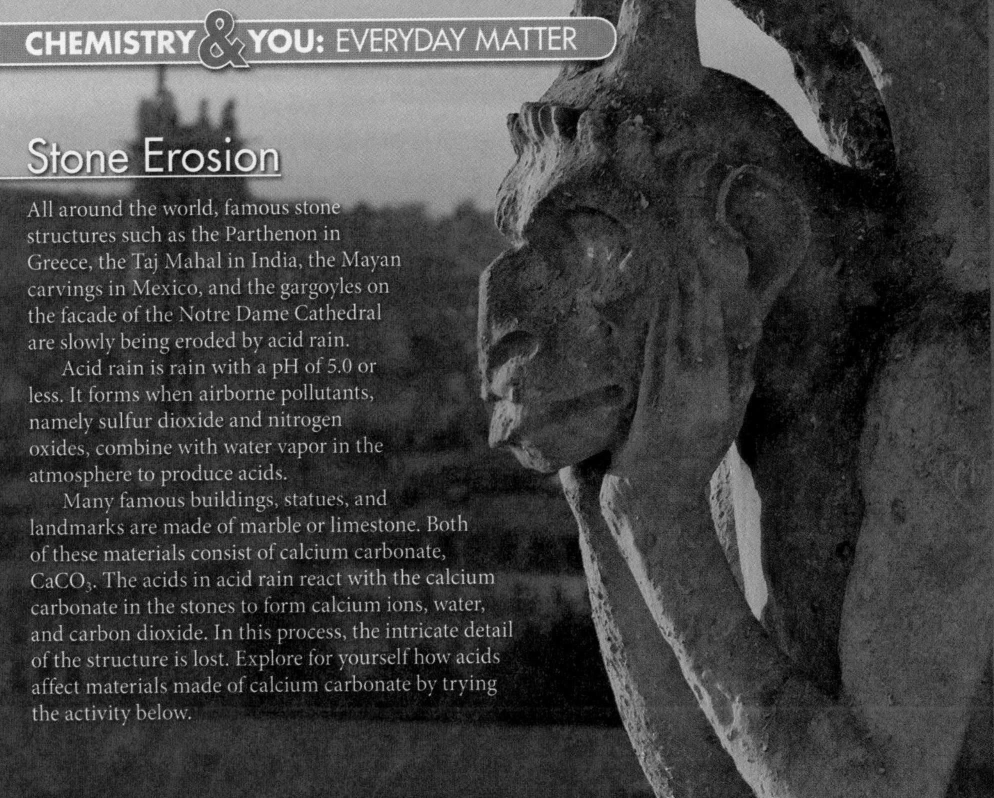

All around the world, famous stone structures such as the Parthenon in Greece, the Taj Mahal in India, the Mayan carvings in Mexico, and the gargoyles on the facade of the Notre Dame Cathedral are slowly being eroded by acid rain.

Acid rain is rain with a pH of 5.0 or less. It forms when airborne pollutants, namely sulfur dioxide and nitrogen oxides, combine with water vapor in the atmosphere to produce acids.

Many famous buildings, statues, and landmarks are made of marble or limestone. Both of these materials consist of calcium carbonate, $CaCO_3$. The acids in acid rain react with the calcium carbonate in the stones to form calcium ions, water, and carbon dioxide. In this process, the intricate detail of the structure is lost. Explore for yourself how acids affect materials made of calcium carbonate by trying the activity below.

On Your Own

1. For this activity you will need **2 bowls, 2 same-size pieces of chalk made of calcium carbonate, white vinegar, tap water, masking tape, a permanent marker,** and **a paper towel.** Put vinegar in the first bowl. Label the bowl with a piece of masking tape and a marker. Put the same amount of water in the second bowl and label it. *Optional*: If you'd like, you can scratch a design (with a needle, nail, or thumb tack) into each piece of chalk before going to Step 2.

2. Place a piece of chalk in each bowl. Observe what happens to the chalk.

3. After about 5 minutes, remove the chalk pieces from the bowls and place the pieces on a paper towel. Compare the pieces of chalk.

Think About It

1. Compare How did the vinegar affect the chalk compared to the water?

2. Infer What causes the bubbles you see when the chalk is placed in vinegar?

3. Describe Write a balanced equation to explain what happens to the chalk when it is placed in vinegar. *Note*: Let $H^+(aq)$ represent the acid.

4. Draw Conclusions Why does acid rain result in a loss of detail in the gargoyle above? What would be the effect of acid rain on a statue over a long period of time?

CHEMISTRY & YOU Draw students' attention to the statue in the picture. Point out the pitted stone and the discoloration on the statue's face and neck. Ask students if they have seem similar damage to stone structures in your community and, if so, to describe the damage.

Explain

CRITICAL THINKING Oxides of sulfur and nitrogen in the air can produce acid rain. **Ask** How would the concentration of these compounds in air affect the pH of rain? *(Because these compounds can combine with water vapor to form acids, an increase in concentration should cause a decrease in pH.)*

Differentiated Instruction

L1 STRUGGLING STUDENTS Have students perform the activity with a partner. Discuss the questions as a class.

LPR LESS PROFICIENT READERS Have students make a list of difficult words in the feature, and discuss their meanings. Lead students in summarizing the feature before they perform the activity.

L3 ADVANCED STUDENTS Have students test several rocks for the presence of calcium carbonate. They should research the procedure and the types of rocks to test.

Answers

THINK ABOUT IT

1. The vinegar had a much more noticeable effect on the chalk.
2. The reaction between the acid in vinegar and the calcium carbonate in chalk produced carbon dioxide.
3. $2H^+ + CaCO_3 \rightarrow Ca^{2+} + H_2O + CO_2$
4. *Sample answer*: The acid reacts with and erodes those parts of the stone that project out from the surface. With time fewer and fewer of the original details remain.

LESSON 19.4

Key Objectives

19.4.1 **DEFINE** the products that form when an acid and a base react.

19.4.2 **IDENTIFY** the point in a titration when neutralization will occur.

Additional Resources

Reading and Study Workbook, Lesson 19.4

Available Online or on Digital Media:

- Teaching Resources, Lesson 19.4 Review
- Laboratory Manual, Labs 41-43
- Small-Scale Laboratory Manual, Lab 31

PearsonChem.com

Engage

CHEMISTRY & YOU Have students read the opening text and study the photograph. Tell students that adding iron to soil with a high pH will not reverse chlorosis. Ask students to hypothesize what role soil pH might play in the availability of iron to plants.

Activate Prior Knowledge

Have students review what they know about acid-base indicators. Direct students to examine Figures 19.8 and 19.9 as a prelude to discussing titration.

Foundations for Reading

BUILD VOCABULARY Have students write *standard solution, titration,* and *end point* in the first column of a three-column chart. In the second column, they should write each term's definition. Have students use the third column to explain how each term is related to the others.

IN ACADEMIC STANDARDS for SCIENCE

C.8.2 Describe the characteristic properties of acids and bases.

C.8.5 For an acid-base titration, calculate the concentration of an unknown solution.

19.4 Neutralization Reactions

C.8.2 Properties of acids and bases; **C.8.5** Titration calculations.

CHEMISTRY & YOU

Q: *What could cause leaves to turn yellow during the growing season?* You may have noticed yellow leaves like these during a season when the leaves should still be green. This condition is called *chlorosis* because the plant lacks a pigment called chlorophyll. To produce chlorophyll, plants need to absorb nutrients, such as iron, from the soil. Sometimes there is plenty of iron, but it is not taken up by the roots of the plant because the pH of the soil is too high.

Key Questions

What products form when an acid and a base react?

At what point in a titration does neutralization occur?

Vocabulary

- neutralization reaction
- titration
- standard solution
- equivalence point
- end point

Acid-Base Reactions

What products form when an acid and a base react?

Suppose you mix a solution of a strong acid, such as HCl, with a solution of a strong base, such as NaOH. The products are sodium chloride and water.

$$HCl(aq) + NaOH(aq) \longrightarrow NaCl(aq) + H_2O(l)$$

In general, acids and bases react to produce a salt and water. The complete reaction of a strong acid and a strong base produces a neutral solution. Thus, this type of reaction is called a **neutralization reaction.**

When you hear the word *salt,* you may think of the substance that is used to flavor food. Table salt (NaCl) is only one example of a salt. Salts are ionic compounds consisting of an anion from an acid and a cation from a base.

A reaction between an acid and base will go to completion when the solutions contain equal numbers of hydrogen ions and hydroxide ions. The balanced equation provides the correct ratio of acid to base. For hydrochloric acid and sodium hydroxide, the mole ratio is 1:1.

$$\underset{1\text{ mol}}{HCl(aq)} + \underset{1\text{ mol}}{NaOH(aq)} \longrightarrow \underset{1\text{ mol}}{NaCl(aq)} + \underset{1\text{ mol}}{H_2O(l)}$$

For sulfuric acid and sodium hydroxide, the ratio is 1:2. Two moles of the base are required to neutralize one mole of the acid.

$$\underset{1\text{ mol}}{H_2SO_4(aq)} + \underset{2\text{ mol}}{2NaOH(aq)} \longrightarrow \underset{1\text{ mol}}{Na_2SO_4(aq)} + \underset{2\text{ mol}}{2H_2O(l)}$$

Similarly, hydrochloric acid and calcium hydroxide react in a 2:1 ratio.

$$\underset{2\text{ mol}}{2HCl(aq)} + \underset{1\text{ mol}}{Ca(OH)_2(aq)} \longrightarrow \underset{1\text{ mol}}{CaCl_2(aq)} + \underset{2\text{ mol}}{2H_2O(l)}$$

Learn more about acid-base reactions online.

Focus on ELL

1 CONTENT AND LANGUAGE Begin the lesson by writing the words *neutralization* and *titration* on the board. Pronounce each word slowly and have students repeat. Write various forms of each word. As a class, discuss other contexts where students have seen or studied the term *neutral*. Explain that the word *titration* stems from the French word *titre*, meaning "to rank." Have students predict how *neutralization* and *titration* are related in the study of acids and bases.

2 FRONTLOAD THE LESSON Have students participate in a Think-Pair-Share exercise to share personal or academic experiences with neutralization and titration.

3 COMPREHENSIBLE INPUT During the teacher demonstrations, use the think-aloud strategy by stopping intermittently during the demonstrations and explaining your thought process. During the titration, point out that a momentary color change at the point of drop impact signals that the end point of the titration is near.

Sample Problem 19.7

Finding the Moles Needed for Neutralization

The term *neutralization* is used to describe both the reaction and the point at which a neutralization reaction is complete. How many moles of sulfuric acid are required to neutralize 0.50 mol of sodium hydroxide? The equation for the reaction is

$$H_2SO_4(aq) + 2NaOH(aq) \longrightarrow Na_2SO_4(aq) + 2H_2O(l)$$

1 Analyze List the knowns and the unknown. To determine the number of moles of acid, you need to know the number of moles of base and the mole ratio of acid to base.

KNOWNS	UNKNOWN
mol NaOH = 0.50 mol	mol H_2SO_4 = ? mol
1 mol H_2SO_4/2 mol NaOH (from balanced equation)	

2 Calculate Solve for the unknown.

Use the mole ratio of acid to base to determine the number of moles of acid.

$$0.50 \text{ mol NaOH} \times \frac{1 \text{ mol } H_2SO_4}{2 \text{ mol NaOH}} = 0.25 \text{ mol } H_2SO_4$$

3 Evaluate Does the result make sense? Because the mole ratio of H_2SO_4 to NaOH is 1:2, the number of moles of H_2SO_4 should be half the number of moles of NaOH.

35. How many moles of potassium hydroxide are needed to neutralize 1.56 mol of phosphoric acid?

To solve each problem, begin by writing a balanced equation.

36. How many moles of sodium hydroxide are required to neutralize 0.20 mol of nitric acid?

Titration

At what point in a titration does neutralization occur?

You can use a neutralization reaction to determine the concentration of an acid or base. The process of adding a measured amount of a solution of known concentration to a solution of unknown concentration is called a **titration.** The steps in an acid-base titration are as follows.

1. A measured volume of an acid solution of unknown concentration is added to a flask.
2. Several drops of an indicator are added to the solution while the flask is gently swirled.
3. Measured volumes of a base of known concentration are mixed into the acid until the indicator just barely changes color.

Foundations for Math

DECIMALS Explain that students can convert between equivalent decimals and fractions when solving problems. Either number form will lead them to the correct answer as long as they complete the dimensional analysis correctly.

In **Sample Problem 19.7,** show students that the moles of sodium hydroxide, 0.50 mol, is equivalent to 50 one-hundredths, or 50/100. Then show the conversion problem as follows. Have students complete the problem without the use of a calculator.

$$\frac{50 \text{ mol NaOH}}{100} \times \frac{1 \text{ mol } H_2SO_4}{2 \text{ mol NaOH}} = \frac{50 \text{ mol } H_2SO_4}{200} = \frac{1}{4} \text{ mol } H_2SO_4 = 0.25 \text{ mol } H_2SO_4.$$

READING STRATEGY Have students skim through the text to identify and list unfamiliar words before reading for content. Help students use context clues to deduce the meaning of these words. Have students write the definition of each of these words on a sticky note and place it at point of use for reference.

Explain

Acid-Base Reactions

MAKING CONNECTIONS Ask students if they have witnessed or read a news report about a chemical spill on a highway or at a factory. **Ask** If a spill involves an acid, what type of treatment is possible? *(A weak base in powder form, such as sodium carbonate, can be spread over the spill to absorb, react with, and neutralize the acid.)*

Sample Practice Problem

How many moles of aluminum hydroxide are needed to neutralize 2.30 mol of sulfuric acid? *(1.53 mol)*

APPLY CONCEPTS Point out that neutralization is a process that occurs whenever an acid reacts with a base in the mole ratios specified by the balanced equation. Explain that not all neutralization reactions produce neutral solutions. The relative strengths of the reactants determine whether the solution will be acidic, basic, or neutral.

Extend

Explain that curling hair with a permanent wave process involves an acid-base neutralization reaction. Pass around pictures or use a student volunteer to convey what a permanent wave looks like. Explain that a permanent wave process involves rolling the hair with curlers, applying a waving solution, rinsing with water, then finally applying a neutralizing solution. Have students research the chemistry behind this hair treatment process. Have students report back to the class on their findings. *(Students should discover that the waving solution is basic with a pH ≈ 9.2, and the neutralizer is acidic with a pH range of 2 to 6. The typical active ingredient in the waving solution is an ammonium compound, and prolonged exposure to this compound before neutralization can result in damaged hair.)*

Answers

SAMPLE PROBLEM 19.7

35. 4.68 mol KOH

36. 0.20 mol NaOH

LESSON 19.4

Explore

Titration

Teacher Demo

PURPOSE Students will observe the process of titration using indicators.

MATERIALS equimolar solutions of HCl and NaOH, phenolphthalein, electric stirrer with a magnetic stir bar (if available), beaker, table salt

SAFETY Wear goggles and laboratory apron.

PROCEDURE Write the following equation on the board: $HCl(aq) + NaOH(aq) \rightarrow NaCl(aq) + H_2O(l)$

Prepare equimolar solutions of HCl and NaOH. Add a few drops of phenolphthalein indicator to the HCl solution. Slowly add the NaOH to the HCl while carefully stirring. If possible, use an electric stirrer with a magnetic stir bar. **Ask** What happens to the color of the solution as the base is added? *(A spot of pink is observed, which quickly fades. When the pH of the solution reaches 8, the solution remains pink.)* Remind students that phenolphthalein is an acid-base indicator that is colorless at pH lower than 8. Have a few volunteers come up and gently touch the outside of the reaction beaker. Have students describe their observations to the class. **Ask** Is the neutralization reaction endothermic or exothermic? *(exothermic)*

Display a quantity of table salt equivalent to the amount of salt that could be retrieved if the neutral solution were evaporated.

EXTENSION Repeat the demo, placing the base in a beaker and the acid in a buret. Use a pH meter to monitor the titration. Construct a graph similar to the one in Figure 19.15. Compared with Figure 19.15, the graph will be reversed because an acidic solution is being added to a basic solution. **Ask** What happens to the graph as the titration nears its equivalence point? *(The slope becomes very steep.)* **Ask** What would happen to the pH if a small amount of base were added to the beaker after the equivalence point is reached? *(The pH would increase dramatically.)*

Add some base to the beaker and observe the pH. **Ask** What can be done to regain the equivalence point? *(Add acid from the buret.)*

Figure 19.14 Titration
The photographs show steps in an acid-base titration.
a. A flask with a known volume of acid (and some phenolphthalein indicator) is placed beneath a buret that is filled with a base of known concentration.
b. Base is slowly added from the buret to the acid while the flask is gently swirled. **c.** A change in the color of the solution is the signal that neutralization has occurred.

CHEMISTRY & YOU

Q: *Iron compounds need to dissociate before the iron can be absorbed by plants. However, these compounds become less soluble as the pH rises. For most plants, a pH between 5.0 and 6.5 will provide enough usable iron. How could you change the pH of soil?*

The solution of known concentration is the **standard solution.** The steps in the titration of an acid of unknown concentration with a standard base are shown in Figure 19.14. You can use a similar procedure to find the concentration of a base using a standard acid.

Neutralization occurs when the number of moles of hydrogen ions is equal to the number of moles of hydroxide ions. Two things that are equal in value are said to be *equivalent.* Thus, the point at which neutralization occurs is called the **equivalence point.** The indicator that is chosen for a titration must change color at or near the pH of the equivalence point. The point at which the indicator changes color is the **end point** of the titration.

Figure 19.15 shows how the pH of a solution changes during the titration of a strong acid (HCl) with a strong base (NaOH). The initial acid solution has a low pH (about 1). As NaOH is added, the pH increases because some of the acid reacts with the base. The equivalence point for this reaction occurs at a pH of 7. As the titration nears the equivalence point, the pH rises dramatically because hydrogen ions are being used up. Extending the titration beyond the point of neutralization produces a further increase of pH. If the titration of HCl and NaOH could be stopped right at the equivalence point, the solution in the beaker would consist of only H_2O and NaCl, plus a small amount of indicator.

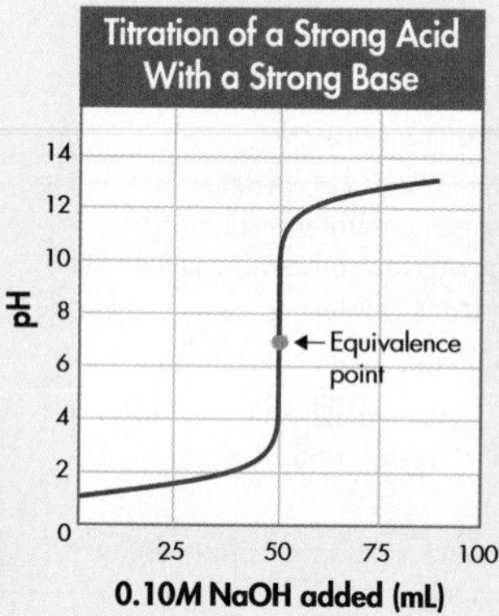

Figure 19.15 Titration Curve
In this titration, 0.10*M* NaOH is slowly added to 50.0 mL of 0.10*M* HCl. The pH of the solution is measured and recorded periodically to construct a titration curve. The equivalence point is located at the midpoint of the vertical part of the curve. Neutralization occurs when 50.0 mL of NaOH have been added to the flask.
Compare *How are $[H^+]$ and $[OH^-]$ related at the equivalence point?*

UbD Check for Understanding

The Essential Question What happens when acids and bases react?

Assess students' understanding of the events that occur when an acid and base react by asking students to briefly explain what occurs during a neutralization reaction. *(In a neutralization reaction, an acid and a base react to form a salt and water, and the reaction is considered to have gone to completion when there are equal numbers of hydrogen ions and hydroxide ions in solution)*

ADJUST INSTRUCTION If students have difficulty answering, review the Titration section. Be sure students recognize that achieving an equal number of hydrogen ions and hydroxide ions at neutralization is not synonymous with a 1:1 mole ratio between acid and base, as is demonstrated by the mole ratio between hydrochloric acid and calcium hydroxide.

Sample Problem 19.8

Determining Concentration by Titration

A 25-mL solution of H_2SO_4 is neutralized by 18 mL of 1.0*M* NaOH. What is the concentration of the H_2SO_4 solution? The equation for the reaction is

$$H_2SO_4(aq) + 2NaOH(aq) \longrightarrow Na_2SO_4(aq) + 2H_2O(l)$$

KNOWNS
[NaOH] = 1.0*M*
V_{NaOH} = 18 mL = 0.018 L
$V_{H_2SO_4}$ = 25 mL = 0.025 L

UNKNOWN
$[H_2SO_4]$ = ?*M*

1 Analyze **List the knowns and the unknown.** The conversion steps are as follows: L NaOH ⟶ mol NaOH ⟶ mol H_2SO_4 ⟶ *M* H_2SO_4.

2 Calculate **Solve for the unknown.**

Convert volumes to liters because molarity is in moles per liter.

Use the molarity to convert the volume of base to moles of base.

$$0.018\ \text{L NaOH} \times \frac{1.0\ \text{mol NaOH}}{1\ \text{L NaOH}} = 0.018\ \text{mol NaOH}$$

Use the mole ratio to find the moles of acid.

$$0.018\ \text{mol NaOH} \times \frac{1\ \text{mol } H_2SO_4}{2\ \text{mol NaOH}} = 0.0090\ \text{mol } H_2SO_4$$

Calculate the molarity by dividing moles of acid by liters of solution.

$$\text{molarity} = \frac{\text{mol of solute}}{\text{L of solution}} = \frac{0.0090\ \text{mol}}{0.025\ \text{L}} = 0.36M\ H_2SO_4$$

3 Evaluate **Does the result make sense?** If the acid had the same molarity as the base (1.0*M*), 50 mL of base would neutralize 25 mL of acid. Because the volume of base is much less than 50 mL, the molarity of the acid must be much less than 1.0*M*.

37. How many milliliters of 0.45*M* HCl will neutralize 25.0 mL of 1.00*M* KOH?

38. What is the molarity of a solution of H_3PO_4 if 15.0 mL is neutralized by 38.5 mL of 0.150*M* NaOH?

19.4 LessonCheck

39. Review What are the products of a reaction between an acid and a base?

40. Explain Why is the point in the titration when neutralization occurs called the equivalence point?

41. Calculate How many moles of HCl are required to neutralize aqueous solutions of these bases?
- **a.** 0.03 mol KOH
- **b.** 2 mol NH_3
- **c.** 0.1 mol $Ca(OH)_2$

42. Describe Write complete balanced equations for the following acid-base reactions:
- **a.** $H_2SO_4(aq) + KOH(aq) \longrightarrow$
- **b.** $H_3PO_4(aq) + Ca(OH)_2(aq) \longrightarrow$
- **c.** $HNO_3(aq) + Mg(OH)_2(aq) \longrightarrow$

BIGIDEA REACTIONS

43. Review the information on types of chemical reactions in Chapter 11. Which types of reactions are neutralization reactions? Explain your answer.

Lesson Check Answers

39. a salt and water

40. At that point, the number of moles of hydrogen ions is equal to the number of moles of hydroxide ions.

41. **a.** 0.03 mol
b. 2 mol
c. 0.2 mol

42. **a.** $H_2SO_4 + 2KOH \rightarrow 2H_2O + K_2SO_4$
b. $2H_3PO_4 + 3Ca(OH)_2 \rightarrow 6H_2O + Ca_3(PO_4)_2$
c. $2HNO_3 + Mg(OH)_2 \rightarrow 2H_2O + Mg(NO_3)_2$

43. **BIGIDEA** Acid-base neutralizations are double displacement reactions. The positive ions are exchanged between the acid and base reactants; the products are a salt and water.

Explain

CHEMISTRY & YOU Depending on the pH, you could add an acidic substance or an alkaline substance.

Sample Practice Problem

What is the molarity of sulfuric acid if 25.0 mL of the solution is neutralized by 25.5 mL of 0.50*M* KOH? *(0.26M)*

Evaluate

Informal Assessment

Ask students to summarize the titration process in their own words. *(During a titration, equivalent volumes of an acid and of a base are determined. One is a standard solution with a known concentration; the concentration of the other solution is unknown.)* Then direct students to complete the 19.4 Lesson Check.

Reteach

Project a stacked image of Figure 19.14 and the graph from Figure 19.15 on a whiteboard. Draw arrows connecting the events shown in Figure 19.14 to the graph in Figure 19.15, making sure that students understand why Figure 19.14c represents the equivalence point and not the end of the graphed curve.

Answers

FIGURE 19.15 The concentrations are equal.
SAMPLE PROBLEM 19.8
37. 56 mL HCl
38. 0.128*M* H_3PO_4

LESSON 19.5

Key Objectives

19.5.1 **DESCRIBE** when a solution of a salt is acidic or basic.

19.5.2 **IDENTIFY** the components of a buffer.

Additional Resources

Reading and Study Workbook, Lesson 19.5

Available Online or on Digital Media:

- Teaching Resources, Lesson 19.5 Review, Interpreting Graphics
- Laboratory Manual, Labs 44-45
- Small-Scale Laboratory Manual, Labs 32-33
- Virtual Chemistry Laboratory Manual, Labs 25, 26, and 29

PearsonChem.com

Engage

CHEMISTRY & YOU Have students read the opening paragraph. **Ask** What is the normal pH of human blood? *(7.4)* **Ask** What is the safe pH range for human blood? *(between 6.8 and 7.8)* As they proceed with the lesson, have students consider how the body might use acid-base reactions to maintain a safe blood pH

Activate Prior Knowledge

Briefly review acid-base neutralization reactions. **Ask** What types of acids and bases were involved in the neutralization reactions in Lesson 19.4? *(strong acids and strong bases.)* **Ask** What were the products of these reactions? *(a salt and water)* **Ask** What is the pH at the equivalence point when a strong acid and strong base neutralize each other? *(7)*

IN ACADEMIC STANDARDS for SCIENCE

C.8.2 Describe the characteristic properties of acids and bases.

19.5 Salts in Solution

C.8.2 Properties of acids and bases.

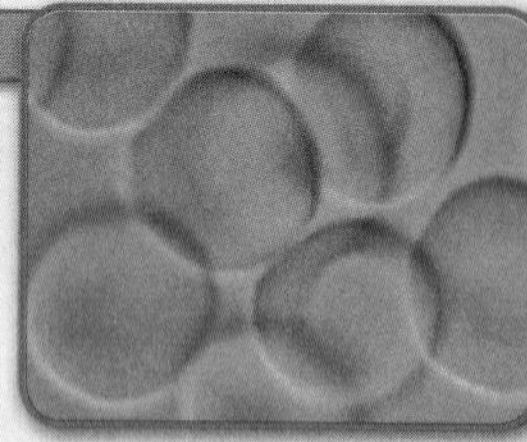

CHEMISTRY & YOU

Q: *How is the pH of blood controlled in the human body?* Chemical reactions in cells are very sensitive to slight changes in pH. For example, the pH of human blood needs to be kept close to 7.4. A person cannot survive for more than a few minutes if the pH of blood drops below 6.8 or rises above 7.8. This lesson will explain the process that prevents such a life-threatening event.

Key Questions

When is the solution of a salt acidic or basic?

What are the components of a buffer?

Vocabulary

- salt hydrolysis
- buffer
- buffer capacity

Salt Hydrolysis

When is the solution of a salt acidic or basic?

Recall that a salt is one of the products of a neutralization reaction. A salt consists of an anion from an acid and a cation from a base. The solutions of many salts are neutral. Salts that form neutral solutions include sodium chloride and potassium sulfate. Some salts form acidic or basic solutions, as shown in Figure 19.16. The indicator used for Figure 19.16 is called universal indicator because it can be used for a wide range of pH values.

Figure 19.17 shows two titration curves. One curve is for the addition of sodium hydroxide, a strong base, to ethanoic acid, a weak acid. An aqueous solution of sodium ethanoate exists at the equivalence point.

$$\underset{\text{Ethanoic acid}}{CH_3COOH(aq)} + \underset{\text{Sodium hydroxide}}{NaOH(aq)} \longrightarrow \underset{\text{Sodium ethanoate}}{CH_3COONa(aq)} + \underset{\text{Water}}{H_2O(l)}$$

The second titration curve is for the reaction between hydrochloric acid, which is a strong acid, and sodium hydroxide. This second curve should look familiar. It first appeared in Lesson 19.4 in the section on titrations.

Figure 19.16
The pH of Salt Solutions
Universal indicator was added to 0.10*M* aqueous salt solutions. Based on the indicator color, the solutions can be classified as follows: **a.** Ammonium chloride, $NH_4Cl(aq)$, is acidic (pH of about 5.3). **b.** Sodium chloride, $NaCl(aq)$, is neutral (pH of 7). **c.** Sodium ethanoate, $CH_3COONa(aq)$, is basic (pH of about 8.7).

Focus on ELL

1 CONTENT AND LANGUAGE Write the word *buffer* on the board and say it out loud. Explain that this word has both common and scientific meanings. One of the common meanings is "something that provides a protective barrier." Explain that a chemical buffer performs a similar function.

2 FRONTLOAD THE LESSON Divide the class into four groups of varying English speaking capabilities. Assign each group a different feature of the lesson to preview: Vocabulary, Figures and Tables, Sample Problem, and Lesson Check. Have groups present what they discovered to the class.

3 COMPREHENSIBLE INPUT Place two colors of plastic blocks (snapped together) in a large beaker to model that the sole source of ethanoate ions resulting from salt hydrolysis is ethanoic acid. Then add free blocks of a single color to show that both ethanoic acid and its acid salt are potential sources of ethanoate ions in an ethanoic acid/ethanoate buffer system.

KINETIC ART VIRTUAL LAB **Interpret Graphs**

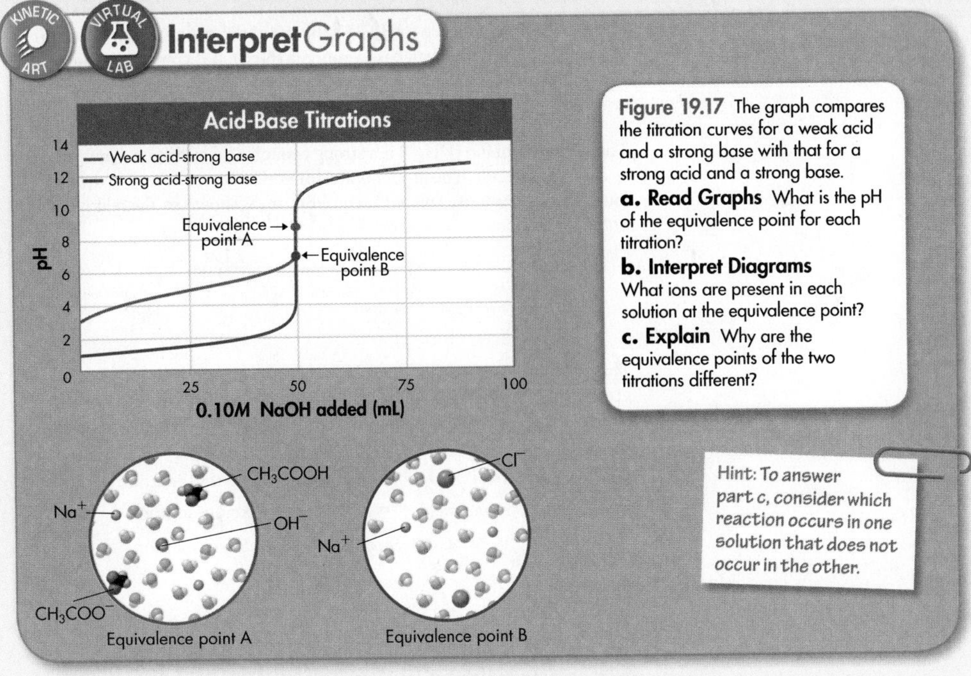

Figure 19.17 The graph compares the titration curves for a weak acid and a strong base with that for a strong acid and a strong base.

a. Read Graphs What is the pH of the equivalence point for each titration?

b. Interpret Diagrams What ions are present in each solution at the equivalence point?

c. Explain Why are the equivalence points of the two titrations different?

Hint: To answer part c, consider which reaction occurs in one solution that does not occur in the other.

The pH at the equivalence point for the weak acid-strong base titration is basic. For a strong acid-strong base titration, the pH at the equivalence point is neutral. This difference in pH exists because hydrolysis occurs with some salts in solution. In **salt hydrolysis,** the cations or anions of a dissociated salt remove hydrogen ions from, or donate hydrogen ions to, water. **Salts that produce acidic solutions have positive ions that release hydrogen ions to water. Salts that produce basic solutions have negative ions that attract hydrogen ions from water.**

Sodium ethanoate (CH_3COONa) is the salt of a weak acid and a strong base. In solution, the salt is completely ionized.

$$\underset{\text{Sodium ethanoate}}{CH_3COONa(aq)} \longrightarrow \underset{\text{Ethanoate ion}}{CH_3COO^-(aq)} + \underset{\text{Sodium ion}}{Na^+(aq)}$$

The ethanoate ion is a Brønsted-Lowry base, which means it is a hydrogen-ion acceptor. It reacts with water to form ethanoic acid and hydroxide ions. At equilibrium, the reactants are favored.

$$\underset{\substack{\text{H}^+\text{ acceptor,}\\\text{Brønsted-Lowry base}}}{CH_3COO^-(aq)} + \underset{\substack{\text{H}^+\text{ donor,}\\\text{Brønsted-Lowry acid}}}{H_2O(l)} \rightleftharpoons CH_3COOH(aq) + \underset{\substack{\text{(makes the}\\\text{solution basic)}}}{OH^-(aq)}$$

This process is called hydrolysis because a hydrogen ion is split off a water molecule. The suffix *-lysis* comes from a Greek word meaning to "separate" or "loosen." In the solution, the hydroxide-ion concentration is greater than the hydrogen-ion concentration. Thus, the solution is basic.

UbD Check for Understanding

When is the solution of a salt acidic or basic?

Assess students' understanding of the Key Question by asking them to complete the following cloze prompt:

At the equivalence point of the titration of a _______ acid with a _______ base, the salt hydrolyzes to yield negative ions that attract hydrogen ions from water. *(weak; strong)*

ADJUST INSTRUCTION If students are having difficulty answering, have them examine the equation at the bottom of the page. **Ask** What type of ions are present in the solution at equilibrium? *(negative)* **Ask** What is the conjugate acid of ethanoate? *(ethanoic acid)* Point out that the products of the salt hydrolysis are the original reactants in the titration. Have students identify the strength of each reactant and then answer the cloze prompt again.

Foundations for Reading

BUILD VOCABULARY Have students look up the definitions of *capacity* and determine which definition is most appropriate for this lesson. Then have them define *buffer capacity* in their own words.

READING STRATEGY Have students construct a KWL chart about buffers, with the three column titles *What I Know About Buffers; What I Want to Know About Buffers; and What I Learned About Buffers.* Have students complete the appropriate columns before, during, and after reading this lesson.

Explain

Salt Hydrolysis

USE VISUALS Have students compare Figures 19.16 and 19.17. Note that the equivalence point is not always at pH 7. Explain that the salts may hydrolyze, resulting in a reversible reaction that produces a higher concentration of hydrogen or hydroxide ions. **Ask** How do neutralization reactions that produce a neutral solution at the equivalence point differ from those that produce acidic or basic solutions? *(In the strengths of the reactants; when a strong acid and a strong base react, the solution is neutral; when a strong base and a weak acid react, the solution is basic; when a strong acid and a weak base react, the solution is acidic.)*

Explore

Teacher Demo

PURPOSE Students predict the pH of different salt solutions and give reasons for their predictions.

MATERIALS 1*M* solutions of NH_4NO_3, KCl, $NaHCO_3$, Na_2SO_4; pH meter or universal pH paper

PROCEDURE Have students predict the pH of these solutions and explain their reasoning. After some discussion, check the pH of the solutions. Then ask students to explain the data. *(NH_4NO_3 is acidic because the salt results from a strong acid and a weak base; KCl is neutral because the salt results from a strong acid and a strong base; $NaHCO_3$ is basic because the salt results from a strong base and a weak acid; Na_2SO_4 is neutral because the salt results from a strong acid and a strong base.)*

Answers

INTERPRET GRAPHS

a. A: 8.7, B: 7

b. A. Na^+, OH^-, CH3COO^-; B: Na^+, Cl^-

c. CH_3COO^- can react with water to produce OH^-.

LESSON 19.5

MAKE A CONNECTION Point out that the salts that give non-neutral solutions in the pH demo are classified as Brønsted-Lowry acids and bases because the cations and anions of these salts accept hydrogen ions from or donate hydrogen ions to water (hydrolysis). Point out that the following relationships are a guide to predicting the type of solution expected from each combination of acid and base, and suggest that students commit them to memory:

Strong acid + Strong base → Neutral solution

Strong acid + Weak base → Acidic solution

Weak acid + Strong base → Basic solution

Explain

Buffers

USE VISUALS Direct students' attention to Figure 19.18. Explain that buffers are solutions in which the pH stays relatively constant when small amounts of either acid or base are added. Point out that in the buffered solution on the left, there is little change in pH.

Explore

Teacher Demo

PURPOSE To demonstrate that the neutralizing capacities of commercial buffers differ, students compare antacid tablets, buffered aspirin, and aspirin.

MATERIALS antacid tablets, aspirin, buffered aspirin, flasks and burets, 0.5*M* HCl and NaOH, distilled water, methyl red, bromothymol blue

PROCEDURE Set up pairs of flasks containing 100 mL of distilled water to test each product. Dissolve equal masses of the tablets in each flask. Add a few drops of methyl red to one of the paired flasks and titrate with 0.5*M* NaOH. To the other flask, add a few drops of bromothymol blue and titrate with 0.5*M* HCl. Titrate until the end point is reached. Explain that the buffering capacity is directly related to the amount of acid or base required to reach the end point. Have students compare the buffering capacities of the tablets tested.

Ammonium chloride (NH_4Cl) is the salt of the strong acid hydrochloric acid (HCl) and the weak base ammonia (NH_3). It is completely ionized in solution.

$$NH_4Cl(aq) \longrightarrow NH_4^+(aq) + Cl^-(aq)$$

The ammonium ion (NH_4^+) is a strong enough acid to donate a hydrogen ion to a water molecule. The products are ammonia molecules and hydronium ions. The reactants are favored at equilibrium, as shown by the relative sizes of the forward and reverse arrows.

$$\underset{\text{H}^+\text{ donor, Brønsted-Lowry acid}}{NH_4^+(aq)} + \underset{\text{H}^+\text{ acceptor, Brønsted-Lowry base}}{H_2O(l)} \rightleftharpoons NH_3(aq) + \underset{\text{(makes the solution acidic)}}{H_3O^+(aq)}$$

This process is another example of hydrolysis. At equilibrium the $[H_3O^+]$ is greater than the $[OH^-]$. Thus, a solution of ammonium chloride is acidic. To determine if a salt will form an acidic or basic solution, remember the following rules:

Strong acid + **Strong base** ⟶ Neutral solution

Strong acid + Weak base ⟶ **Acidic** solution

Weak acid + **Strong base** ⟶ **Basic** solution

Buffers

What are the components of a buffer?

Suppose you add 10 mL of 0.10*M* sodium hydroxide to 1 L of pure water. The pH will increase about 4 pH units—from 7.0 to about 11.0. This change is a relatively large increase in pH. Now consider a solution containing 0.20*M* each of ethanoic acid and sodium ethanoate. This solution has a pH of 4.76. If you add 10 mL of 0.10*M* sodium hydroxide to 1 L of this solution, the pH increases 0.01 pH units—from 4.76 to 4.77. This is a relatively small change in pH. If 10 mL of acid had been added instead of the base, the amount of change in pH would also have been small.

The solution of ethanoic acid and sodium ethanoate is an example of a buffer. A **buffer** is a solution in which the pH remains fairly constant when small amounts of acid or base are added. **A buffer is a solution of a weak acid and one of its salts or a solution of a weak base and one of its salts.** Figure 19.18 compares what happens when 1.0 mL of 0.01*M* HCl solution is added to an unbuffered solution and to a solution with a buffer.

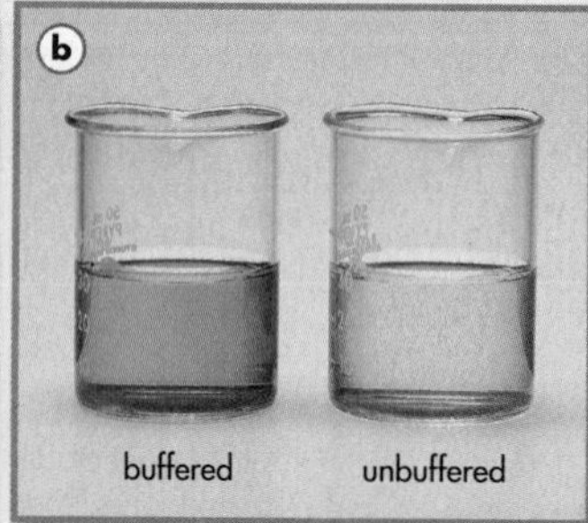

Figure 19.18 Effect of a Buffer In a buffer solution, the pH does not shift dramatically. **a.** The indicator shows that both solutions are basic (pH of about 8). **b.** HCl is added to each solution. The indicator shows no visible pH change in the buffered solution. The color change in the unbuffered solution indicates a change in pH from 8 to about 3.
Predict ***How would the original solutions respond if NaOH were added?***

Differentiated Instruction

L1 STRUGGLING STUDENTS For students who may benefit from a closer examination of the teacher demos, allow time before class for them to come in and observe the lab on a closer level. Allow students to participate in the lab, and guide student understanding by using the thinking aloud technique.

ELL ENGLISH LANGUAGE LEARNERS Students with limited English proficiency may be confused by the terms *equivalents* and *equivalence*, which sound alike and have similar spellings. Have them write out both words and underline or circle the differences in the endings. Help them write out meaningful definitions for both words.

L3 ADVANCED STUDENTS Have students design and implement a lab to determine the difference between a buffered aspirin and a regular aspirin.

How Buffers Work A buffer solution is better able to resist drastic changes in pH than is pure water. The reason is fairly simple. A buffer solution contains one component that can react with hydrogen ions (a hydrogen-ion acceptor) and another component that can react with hydroxide ions (a hydrogen-ion donor). These components act as reservoirs of neutralizing power that can be tapped when either hydrogen ions or hydroxide ions are added to the solution.

The ethanoic acid–ethanoate ion buffer can be used to show how a buffer works. When an acid is added to the buffer, the ethanoate ions (CH_3COO^-) act as a hydrogen-ion "sponge." As the ethanoate ions react with the hydrogen ions, they form ethanoic acid. This weak acid does not ionize extensively in water, so the change in pH is very slight.

$$\underset{\text{Ethanoate ion}}{CH_3COO^-(aq)} + \underset{\text{Hydrogen ion}}{H^+(aq)} \rightleftharpoons \underset{\text{Ethanoic acid}}{CH_3COOH(aq)}$$

When hydroxide ions are added to the buffer, the ethanoic acid and the hydroxide ions react to produce water and the ethanoate ion.

$$\underset{\text{Ethanoic acid}}{CH_3COOH(aq)} + \underset{\text{Hydroxide ion}}{OH^-(aq)} \rightleftharpoons \underset{\text{Ethanoate ion}}{CH_3COO^-(aq)} + \underset{\text{Water}}{H_2O(l)}$$

The ethanoate ion is not a strong enough base to accept hydrogen ions from water to a great extent. Therefore, the reverse reaction is minimal and the change in pH is very slight.

The Capacity of a Buffer Buffer solutions have their limits. As acid is added to an ethanoate buffer, eventually no more ethanoate ions will be present to accept the hydrogen ions. At that point, the buffer can no longer control the pH. The ethanoate buffer also becomes ineffective when too much base is added. In that case, no more ethanoic acid molecules are present to donate hydrogen ions. Adding too much acid or base will exceed the buffer capacity of a solution. The **buffer capacity** is the amount of acid or base that can be added to a buffer solution before a significant change in pH occurs.

Table 19.9 lists some common buffer systems. Two of these buffer systems help maintain optimal human blood pH. One is the carbonic acid–hydrogen carbonate buffer system. The other is the dihydrogen phosphate–hydrogen phosphate buffer system.

CHEMISTRY & YOU

Q: *The equilibrium between carbonic acid (H_2CO_3) and hydrogen carbonate ions (HCO_3^-) helps keep the pH of blood within a narrow range (7.35–7.45). If the pH rises, molecules of carbonic acid donate hydrogen ions. What can happen if the pH drops, that is, if the $[H^+]$ increases?*

Table 19.9

Important Buffer Systems

Buffer name	Formulas	Buffer pH*
Ethanoic acid–ethanoate ion	CH_3COOH / CH_3COO^-	4.76
Dihydrogen phosphate ion–hydrogen phosphate ion	$H_2PO_4^- / HPO_4^{2-}$	7.20
Carbonic acid–hydrogen carbonate ion (solution saturated with CO_2)	H_2CO_3 / HCO_3^-	6.46
Ammonium ion–ammonia	NH_4^+ / NH_3	9.25

*Components have concentrations of 0.1*M*.

Blood Buffers

Your body functions properly only when the pH of your blood lies between 7.35 and 7.45. A blood pH too far outside these narrow limits can be life threatening. To keep your blood at the proper pH, your blood contains buffers. The most important blood buffer consists of carbonic acid and hydrogen carbonate ions. Hydrogen ions produced by chemical reactions in your body are the main threat to your blood's pH. This buffer maintains the desired blood pH by removing excess hydrogen ions.

$$HCO_3^-(aq) + H^+(aq) \rightleftharpoons H_2CO_3(aq)$$

As long as there are hydrogen carbonate ions available, the excess hydrogen ions are removed, and the pH of the blood changes very little.

Explain

APPLY CONCEPTS Explain that buffers present a practical application of concepts in this chapter. Point out that the limited dissociation of weak acids and bases gives these substances the ability to act as buffers. This ability is greatly expanded if the salt of the acid or base is added to the solution. Explain that buffer action utilizes the equilibrium established when a weak acid or base is combined with its salt. Convey that when a base is added to a buffered solution, the acidic form removes hydroxide ions from the solution. When an acid is added to a buffered solution, the basic form removes hydrogen ions from the solution.

CHEMISTRY & YOU If the pH drops, carbonate ions react with the excess hydrogen ions and form carbonic acid.

Extend

Connect to BIOCHEMISTRY

Explain that the choice of buffering system is determined by its effective pH range, which can be determined by calculating the pK_a of the system's acid. **Ask** How do you think that a pK_a is calculated? *(Students should notice that pK_a looks similar to pH and should suggest that it can be calculated by taking the negative logarithm of the acid's dissociation constant.)* Explain that the effective pH range is approximately equal to the pK_a +/– 1. **Ask** Why do you think many biochemical experiments use buffers with pK_a values in the 7 to 8 range? *(Many biochemical experiments involve chemicals found in the body, which has a pH of 7.4, which is referred to as physiological pH. Enzymes in particular are sensitive to pH. Buffers with pK_a values in the 7 to 8 range are used because they help maintain the pH of the experimental system at physiological pH.)*

Answers

FIGURE 19.18 There would be no visible change in the buffered solution. There would be a color change in the unbuffered solution indicating an increase in pH.

LESSON 19.5

Evaluate

Informal Assessment

Have students explain in their own words what determines whether a solution containing a hydrolyzing salt is acidic or basic. *(Salts formed from a strong acid and strong base are neutral. Those formed from a strong base and weak acid are basic; those formed from a strong acid and weak base are acidic.)* **Ask** How can ethanoate ions effectively remove hydrogen ions from solution? *(The ethanoate ions are a Brønsted–Lowry base that accept the hydrogen ions to form ethanoic acid.)* Then have students complete the 19.5 Lesson Check.

Reteach

Point out that a hydrolysis reaction always uses one part of the water molecule and leaves behind the other. Use equations to show students how the salt of a weak acid always reacts with water to produce hydroxide ions and the salt of a weak base always reacts to produce hydrogen ions. Remind students that chemical buffers are equilibrium systems that tend to resist change due to external influences.

Sample Practice Problem

Contact lens solution often contains boric acid as a buffer. Write the equation for the reaction that shows what happens when an acid is added to a buffer system containing boric acid (H_3BO_3) and dihydrogen borate ions ($H_2BO_3^-$). *($H_2BO_3^- + H^+ \rightarrow H_3BO_3$)*

Sample Problem 19.9

Describing Buffer Systems

Write balanced chemical equations to show how the carbonic acid–hydrogen carbonate buffer can "mop up" added hydroxide ions and hydrogen ions.

1 Analyze **Identify the relevant concepts.** A buffer contains two components: a hydrogen-ion acceptor (which can react with H^+) and a hydrogen-ion donor (which can react with OH^-).

2 Solve **Apply the concepts to this problem.**

Identify the hydrogen-ion acceptor and the hydrogen-ion donor. → H_2CO_3, a weak acid, can release hydrogen ions. HCO_3^- is the conjugate base, which can accept hydrogen ions.

Write the equation for the reaction that occurs when a base is added to the buffer. → When a base is added, the hydroxide ions react with H_2CO_3.

$$H_2CO_3(aq) + (OH^-)(aq) \rightleftharpoons HCO_3^-(aq) + H_2O(l)$$

Write the equation for the reaction that occurs when an acid is added to the buffer. → When an acid is added, the hydrogen ions react with HCO_3^-.

$$HCO_3^-(aq) + (H^+)(aq) \rightleftharpoons H_2CO_3(aq)$$

44. Write equations to show what happens in the following situations:
- **a.** Acid is added to a solution that contains HPO_4^{2-} ions.
- **b.** Base is added to a solution that contains $H_2PO_4^-$ ions.

45. A buffer consists of methanoic acid (HCOOH) and methanoate ion ($HCOO^-$). Write an equation to show what happens when an acid is added to this buffer.

19.5 LessonCheck

46. **Review** What type of salt produces an acidic solution? What type of salt produces a basic solution?

47. **Describe** What types of substances can be combined to make a buffer solution?

48. **Classify** Which of these salts would hydrolyze to produce an acidic aqueous solution, and why?
- **a.** $KC_2H_3O_2$
- **b.** LiCl
- **c.** $NaHCO_3$
- **d.** $(NH_4)_2SO_4$

49. **Identify** Which of the following pairs can form a buffer solution? Explain.
- **a.** NH_3 and HCO_3^-
- **b.** C_6H_5COOH and $C_6H_5COO^-$

50. **Describe** Write a balanced chemical equation to show what happens when an acid is added to an ammonium ion–ammonia buffer. Write an equation to show what happens when a base is added.

51. **Relate Cause and Effect** Use Le Châtelier's principle to explain how a buffer system maintains the pH of a solution.

Lesson Check Answers

46. acidic solution: salt of a strong acid and weak base; basic solution: salt of a weak acid and a strong base

47. a weak acid and one of its salts or a weak base and one of its salts

48. d; it is the product of a strong acid, H_2SO_4, and a weak base, NH_3.

49. Pair (b) because it consists of a weak acid and one of its salts.

50. $NH_3 + H^+ \rightarrow NH_4^+$
$NH_4^+ + OH^- \rightarrow NH_3 + H_2O$

51. According to Le Châtelier, when a system in equilibrium is disturbed, the response is an attempt to restore the equilibrium. When the stress is the addition of an acid or a base, a buffer system responds by partially restoring equilibrium.

Answers

SAMPLE PROBLEM 19.9

44. **a.** $HPO_4^{2-} + H^+ \rightarrow H_2PO_4^-$
b. $H_2PO_4^- + OH^- \rightarrow HPO_4^{2-} + H_2O$

45. $HCOO^- + H^+ \rightarrow HCOOH$

Ocean Buffers

As atmospheric carbon dioxide (CO_2) levels rise due to increased burning of fossil fuels, global warming is not the only potential environmental problem that Earth faces. Additionally, ocean water becomes more acidic, which can disrupt the ocean ecosystem.

The oceans naturally absorb CO_2 from the atmosphere. Some of the absorbed CO_2 is converted to carbonic acid (H_2CO_3), which can lower the pH of ocean water. Fortunately, the oceans have an excellent natural buffer system that helps maintain the optimal pH for supporting ocean life—about 8.2. The ocean's buffer system is largely based on the hydrogen carbonate–carbonate ion buffer system. However, the buffer capacity of the buffers in ocean water is limited, and current human activities are pushing these limits.

Try the following activity at home to gain a better understanding of the effects of CO_2 and sea salt on the pH of ocean water.

On Your Own

1. For this activity, you will need the following materials: **4 drinking glasses, masking tape, a permanent marker, distilled water, carbonated water, sea salt, a measuring cup, ¼ teaspoon measuring spoon,** and **4 pH test strips.** (You can ask your teacher for pH test strips if you do not have them at home.) Use the masking tape and marker to label the glasses 1, 2, 3, and 4.

2. Add ½ cup of distilled water to glasses 1 and 2. Add ½ cup of carbonated water to glasses 3 and 4.

3. Add ¾ teaspoon of sea salt to containers 2 and 4. Stir until the salt is dissolved. (This ratio of sea salt to water is similar to that found in the ocean.)

4. Measure the pH of each solution and record it in a table similar to the one shown to the right.

What Did You Find?

Glass	1	2	3	4
Contents				
pH				

Think About It

1. Compare Determine whether each solution is acidic, basic, or neutral. How do the four solutions differ?

2. Explain How does a solution containing hydrogen carbonate ions (HCO_3^-) and carbonate ions (CO_3^{2-}) act as a buffer? Use chemical equations to support your explanation.

3. Draw Conclusions What does this experiment demonstrate about the effects of dissolved CO_2 (in the carbonated water) and sea salt on the pH of ocean water?

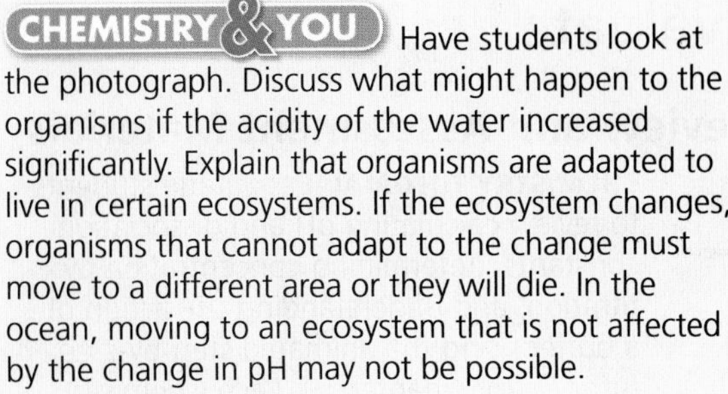

CHEMISTRY & YOU Have students look at the photograph. Discuss what might happen to the organisms if the acidity of the water increased significantly. Explain that organisms are adapted to live in certain ecosystems. If the ecosystem changes, organisms that cannot adapt to the change must move to a different area or they will die. In the ocean, moving to an ecosystem that is not affected by the change in pH may not be possible.

Explain

Draw a Diagram To help students understand the purpose and importance of ocean buffers, have a student volunteer make a flow chart on the board, with the help of the class. Starting with factors that increase carbon dioxide levels in the atmosphere, go to the absorption of carbon dioxide by ocean water, and then to the action of buffers in the ocean. Encourage students to add information they already know to the facts provided in the feature. After the flowchart is complete, discuss causes and results of each step. For example, burning fossils fuel can increase the carbon dioxide levels in the atmosphere. Water can absorb carbon dioxide through the following reaction.

$$CO_2(g) + H_2O + Ca^{2+}(aq) \rightleftharpoons CaCO_3(s) + 2H^+(aq)$$

Extend

Connect to BIOLOGY

Explain that coral reefs are made from the skeletons of corals, and these skeletons are made of calcium carbonate. Coral reefs are home to a vast number of organisms. **Ask** Why could even a slight increase in ocean acidity be devastating to a coral reef ecosystem? *(The acidic water could dissolve the calcium carbonate, and destroy the reef.)*

Differentiated Instruction

L1 STRUGGLING STUDENTS Arrange for students to work on the activity in pairs. Discuss each step with them beforehand. Ask questions to be sure they understand what to do.

LPR LESS PROFICIENT READERS Help students make a checklist that includes gathering the materials for the activity and following each step. They should rewrite the instructions for each step in simple words.

L3 ADVANCED STUDENTS Have students gather the data from the entire class and prepare a class chart or graph. Students should display their chart or graph in the classroom and explain it.

Answers

THINK ABOUT IT

1. Cup 1 is neutral, cup 2 is basic, cups 3 and 4 are acidic, but cup 4 is less acidic.
2. The sea salt added hydroxide ions to the water. The hydrogen carbonate ions reacted with the hydroxide ions, producing carbonate ions and water.
3. Sea salt increase the pH of water. The dissolved carbon dioxide establishes a buffer system that helps maintain the pH at an optimum level.

STUDY GUIDE

Evaluate

Review and Assessment Materials

CHEMISTRY TUTORIAL Encourage students to review calculating pH and dissociation constants, determining concentration by titration, and understanding the action of a buffer using the animated step-by-step tutorials for Chapter 19. Each Chemistry Tutorial is followed by a similar practice problem, where the student is guided through the problem-solving process.

CHAPTER PROBLEM SET Have students practice more problems about the properties of acids and bases and neutralization using the Chapter 19 online Problem Set.

DIRECTED VIRTUAL LAB Have students complete the Virtual Lab *Acids and Bases* to help reinforce the concepts of acid-base titrations. In the post-lab exercise, students will analyze the data and draw conclusions based on the results of the experiment.

KINETIC ART Direct students to the Kinetic Art animation *Dissociation of Acids and Bases*, and have them answer the questions that accompany the animation.

CONCEPTS IN ACTION If students have trouble understanding neutralization reactions, ask them to view the Concepts in Action animation, which presents the material in a familiar context. Have students complete the questions after the animation.

Study Tip

GET INTO THE HABIT OF ACTIVELY STUDYING EACH DAY Offer students the following advice: Take time each night to review the day's work. Make notes of difficult concepts to ask your teacher about or to discuss with a study friend.

19 Study Guide

BIGIDEA REACTIONS

Chemists define acids and bases according to the ions they yield in aqueous solution. Chemists also define acids and bases based on whether they accept or donate hydrogen ions, and whether they are electron-pair donors or acceptors. The pH of a solution reflects the hydrogen-ion concentration. Chemists use acid-base reactions to determine the concentration of an acid or a base in solution.

19.1 Acid-Base Theories

According to Arrhenius, acids are hydrogen-containing compounds that ionize to yield hydrogen ions in aqueous solution. Bases are compounds that ionize to yield hydroxide ions in aqueous solution.

According to the Brønsted-Lowry theory, an acid is a hydrogen-ion donor and a base is a hydrogen-ion acceptor.

According to Lewis, an acid accepts a pair of electrons and a base donates a pair of electrons.

- hydronium ion (H_3O^+) (647)
- conjugate acid (650)
- conjugate base (650)
- conjugate acid-base pair (650)
- amphoteric (651)
- Lewis acid (651)
- Lewis base (651)

19.2 Hydrogen Ions and Acidity

For aqueous solutions, the product of the hydrogen-ion concentration and the hydroxide-ion concentration equals 1×10^{-14}.

A solution with a pH less than 7.0 is acidic. A solution with a pH of 7 is neutral. A solution with a pH greater than 7.0 is basic.

Either acid-base indicators or pH meters can be used to measure pH.

- self-ionization (653)
- neutral solution (653)
- ion-product constant for water (K_w) (654)
- acidic solution (654)
- basic solution (654)
- pH (656)

Key Equations

$$K_w = [H^+] \times [OH^-] = 1.0 \times 10^{-14}$$

$$pH = -\log[H^+]$$

19.3 Strengths of Acids and Bases

Acids and bases are classified as strong or weak based on the degree to which they ionize in water.

- strong acid (664)
- weak acid (664)
- acid dissociation constant (K_a) (665)
- strong base (668)
- weak base (668)
- base dissociation constant (K_b) (668)

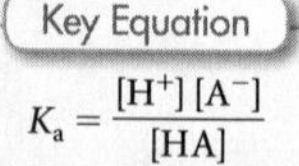

Key Equation

$$K_a = \frac{[H^+][A^-]}{[HA]}$$

19.4 Neutralization Reactions

In general, acids and bases react to produce a salt and water.

Neutralization occurs when the number of moles of hydrogen ions is equal to the number of moles of hydroxide ions.

- neutralization reaction (672)
- titration (673)
- standard solution (674)
- equivalence point (674)
- end point (674)

19.5 Salts in Solution

Salts that produce acidic solutions have positive ions that release hydrogen ions to water. Salts that produce basic solutions have negative ions that attract hydrogen ions from water.

A buffer is a solution of a weak acid and one of its salts or a weak base and one of its salts.

- salt hydrolysis (677)
- buffer (678)
- buffer capacity (679)

UbD Performance Tasks

KITCHEN CHEMISTRY Have students randomly select six solutions from their kitchens to investigate. Tell them to create a table noting the common name, chemical name, and purpose for each substance. Review the lists for safety before students determine whether the solutions are acidic, basic, or neutral. Tell students to create a plot that visually shows the items from least basic to most basic and to present their findings to the class.

PHUN IN CHEMISTRY Tell students to create a computer slide show that compares the two methods of measuring pH and explains how, when, and why you use each method. Have them determine five different household items that can each be used as an acid/base indicator if pH paper or a pH meter were not available. Tell them to explain how they know these items can be used as indicators and to describe the method for preparing and using each of them.

Math Tune-Up: Acid-Base Problems

Problem	1 Analyze	2 Calculate	3 Evaluate
The pH of an unknown solution is 3.70. What is the hydrogen-ion concentration?	Known: pH = 3.70 Unknown: $[H^+] = ?M$ Use the following equation: $pH = -\log[H^+]$	Rearrange the expression for pH and substitute the known pH value to solve for the unknown: $-\log[H^+] = pH$ $-\log[H^+] = 3.70$ $\log[H^+] = -3.70$ The antilog of −3.70 is 2.0×10^{-4}. Thus, $[H^+] = 2.0 \times 10^{-4}M$.	The pH is between 3 and 4. So, the hydrogen ion concentration must be between $1 \times 10^{-3}M$ and $1 \times 10^{-4}M$.
In a 0.500*M* solution of a weak acid (HA), the $[H^+]$ is 4.02×10^{-3} at equilibrium. Find the K_a for this acid. The acid dissociates as follows: $HA \rightleftharpoons H^+ + A^-$	Knowns: $[HA] = 0.500M$ $[H^+] = 4.02 \times 10^{-3}$ Unknown: $K_a = ?$ Use the general expression for K_a: $K_a = \frac{[H^+] \times [A^-]}{[HA]}$	At equilibrium, $[H^+]$ is equal to $[A^-]$: $[H^+] = [A^-] = 4.02 \times 10^{-3}$ Calculate [HA] at equilibrium: $0.500M - 0.00402M = 0.496M$ Substitute the equilibrium concentrations into the equation for K_a and solve: $K_a = \frac{(4.02 \times 10^{-3}) \times (4.02 \times 10^{-3})}{0.496}$ $K_a = 3.26 \times 10^{-5}$	The value of K_a is consistent with that of a weak acid.
How many moles of KOH are needed to neutralize 0.25 mol of H_2SO_4? The equation for the reaction is $2KOH(aq) + H_2SO_4(aq) \longrightarrow K_2SO_4(aq) + 2H_2O(l)$	Known: mol H_2SO_4 = 0.25 mol Unknown: mol KOH = ? mol	Use the mole ratio of base to acid (2 mol KOH to 1 mol H_2SO_4) to determine the number of moles of base: $0.25 \text{ mol } H_2SO_4 \times \frac{2 \text{ mol KOH}}{1 \text{ mol } H_2SO_4}$ = 0.50 mol KOH	The mole ratio of KOH to H_2SO_4 is 2:1. So, the number of moles of KOH should be twice the number of moles of H_2SO_4.

Note: To determine the antilog on most calculators, press the 2nd or INV key then the log key.

Hint: Review Sample Problem 19.6 for help on finding equilibrium concentrations.

Focus on ELL

5 ASSESS UNDERSTANDING Assess students' understanding by using a portfolio. Include various learning assignments, based on each student's language proficiency.

BEGINNING:

LOW Include models or pictures of models created by the student.

HIGH Include copies of the graphic organizers used to take notes or any cause-and- effect charts.

INTERMEDIATE:

LOW Include any prereading predictions about lesson topics, as well as the results of those predictions.

HIGH Include any lab write-ups that were assigned during the chapter.

ADVANCED: LOW/HIGH Include an essay that requires the student to compare and contrast different acid-base theories.

Math Review

Example and sample practice problems for Chapter 19 can be found on the following pages:

LESSON 19.1 Identifying Lewis Acids and Bases—page 652

LESSON 19.2 Using the Ion-Product Constant for Water—page 655; Calculating pH from $[H^+]$—page 657; Calculating $[H^+]$ from pH—page 658; Calculating pH from $[OH^-]$—page 659

LESSON 19.3 Calculating a Dissociation Constant—page 667

LESSON 19.4 Finding the Moles Needed for Neutralization—page 673; Determining Concentration by Titration—page 675

LESSON 19.5 Describing Buffer Systems—page 680

Evaluate

Answers

LESSON 19.1

52. Acids ionize to give hydrogen ions in aqeous solution. Bases ionize to give hydroxide ions in aqueous solution.

53. **a.** base **b.** acid **c.** acid
d. base **e.** acid **f.** acid

54. **a.** $KOH \rightarrow K^+ + OH^-$
b. $Mg(OH)_2 \rightarrow Mg^{2+} + 2OH^-$

55. **a.** $2Li + 2H_2O \rightarrow 2LiOH + H_2$
b. $Ba + 2H_2O \rightarrow Ba(OH)_2 + H_2$

56. **a.** HNO_3, acid; H_2O, base
b. CH_3COOH, acid; H_2O, base
c. H_2O, acid; NH_3, base
d. H_2O, acid, CH_3COO^-, base

57. **a.** HNO_3 with NO_3^-, H_2O with H_3O^+
b. CH_3COOH with CH_3COO^-, H_2O with H_3O^+
c. H_2O with OH^-, NH_3 with NH_4^+
d. H_2O with OH^-, CH_3COO^-) with CH_3COOH

58. A Lewis acid accepts a pair of electrons to form a covalent bond. A Lewis base donates a pair of electrons to form a covalent bond.

LESSON 19.2

59. $H_2O \rightleftharpoons H^+ + OH^-$

60. 1.0×10^{-7} *M* for both H^+ and OH^- at 25°C

61. the negative logarithm of the $[H^+]$

62. The hydrogen ion concentration in pure water at 25°C is $1 \times 10^{-7}M$. The negative logarithm or pH of this concentration is 7.0.

63. **a.** 12; basic **b.** 2; acidic

64. **a.** $1.0 \times 10^{-10}M$ **b.** $1.0 \times 10^{-6}M$
c. $1.0 \times 10^{-2}M$

65. **a.** 5.62 **b.** $6.3 \times 10^{-14}M$

LESSON 19.3

66. **a.** strong base **b.** weak base
c. strong acid **d.** strong acid

67. A strong acid is completely disassociated; K_a must be large

68. They have high K_b values. Their concentration in a saturated solution is low because of their low solubility.

69. **a.** $K_a = \dfrac{[H^+][F^-]}{[HF]}$ **b.** $K_a = \dfrac{[H^+][HCO_3^-]}{[H_2CO_3]}$

LESSON 19.4

70. Acid + base → a salt + water

71. **a.** $HNO_3 + KOH \rightarrow KNO_3 + H_2O$
b. $2HCl + Ca(OH)_2 \rightarrow CaCl_2 + 2H_2O$
c. $H_2SO_4 + 2NaOH \rightarrow Na_2SO_4 + 2H_2O$

72. The indicator changes color.

73. **a.** 1.40*M* **b.** 2.61*M*

ONLINE PROBLEMS

19 Assessment

* Solutions appear in Appendix E

Lesson by Lesson

19.1 Acid-Base Theories

*52. How did Arrhenius describe acids and bases?

53. Classify each compound as an Arrhenius acid or an Arrhenius base.
a. $Ca(OH)_2$ **b.** C_2H_5COOH **c.** HNO_3 **d.** KOH **e.** HBr **f.** H_2SO_4

54. Write an equation for the dissociation of each compound in water.
a. KOH **b.** $Mg(OH)_2$

*55. Write balanced equations for the reaction of each metal with water.
a. lithium **b.** barium

56. Identify each reactant in the following equations as a hydrogen-ion donor (acid) or a hydrogen-ion acceptor (base). All the reactions take place in aqueous solution.
a. $HNO_3 + H_2O \longrightarrow H_3O^+ + NO_3^-$
b. $CH_3COOH + H_2O \rightleftharpoons H_3O^+ + CH_3COO^-$
c. $NH_3 + H_2O \rightleftharpoons NH_4^+ + OH^-$
d. $H_2O + CH_3COO^- \rightleftharpoons CH_3COOH + OH^-$

57. Label the conjugate acid-base pairs for each equation in Question 56.

58. What is a Lewis acid? What is a Lewis base?

19.2 Hydrogen Ions and Acidity

59. Write an equation showing the self-ionization of water.

60. What are the concentrations of H^+ and OH^- in pure water at 25°C?

61. How is the pH of a solution calculated?

62. Why is the pH of pure water at 25°C equal to 7.0?

*63. Calculate the pH for the following solutions and indicate whether each solution is acidic or basic.
a. $[OH^-] = 1 \times 10^{-2}M$ **b.** $[H^+] = 1 \times 10^{-2}M$

64. What are the hydroxide-ion concentrations for solutions with the following pH values?
a. 4.00 **b.** 8.00 **c.** 12.00

65. Calculate the pH or $[H^+]$ for each solution.
a. $[H^+] = 2.4 \times 10^{-6}M$ **b.** pH = 13.20

19.3 Strengths of Acids and Bases

*66. Identify each compound as a strong or weak acid or base.
a. NaOH **b.** NH_3 **c.** H_2SO_4 **d.** HCl

67. Would a strong acid have a large or a small K_a? Explain your answer.

68. Why are $Mg(OH)_2$ and $Ca(OH)_2$ classified as strong bases even though their saturated solutions are only mildly basic?

69. Write the expression for K_a for each acid. Assume only one hydrogen is ionized.
a. HF **b.** H_2CO_3

19.4 Neutralization Reactions

70. Write a general word equation for a neutralization reaction.

71. Identify the products and write balanced equations for each neutralization reaction.
a. $HNO_3(aq) + KOH(aq) \longrightarrow$
b. $HCl(aq) + Ca(OH)_2(aq) \longrightarrow$
c. $H_2SO_4(aq) + NaOH(aq) \longrightarrow$

72. How is it possible to recognize the end point of a titration?

*73. What is the molarity of sodium hydroxide if 20.0 mL of the solution is neutralized by each of the following 1.00*M* solutions?
a. 28.0 mL of HCl
b. 17.4 mL of H_3PO_4

19.5 Salts in Solution

74. What kinds of salts hydrolyze water?

*75. Write an equation showing why an aqueous solution of sodium hydrogen carbonate is basic.

76. Explain why solutions of salts that hydrolyze water do not have a pH of 7.

77. Predict whether an aqueous solution of each salt will be acidic, basic, or neutral.
a. $NaHCO_3$ **b.** NH_4NO_3 **c.** KCl **d.** Na_2CO_3 **e.** Na_2SO_4 **f.** NH_4Cl

78. Explain why a buffered solution cannot absorb an unlimited amount of acid or base.

684 Chapter 19 • Assessment

LESSON 19.5

74. salts with a cation from a weak base and an anion from a strong acid, or with a cation from a strong base and an anion from a weak acid.

75. $HCO_3^-\ (aq) + H_2O\ (l) \rightarrow H_2CO_3(aq) + OH^-\ (aq)$

76. Weak-acid anions accept protons from water, increasing the pH of the solution. Weak-base cations donate protons to water, decreasing the pH.

77. **a.** basic **b.** acidic **c.** neutral
d. basic **e.** neutral **f.** acidic.

78. Eventually the buffer capacity of the buffer is exceeded and the pH will change significantly with the addition of a strong acid or base.

Understand Concepts

79. Explain how the Lewis theory is a more general classification system than either the Arrhenius descriptions or the Brønsted-Lowry theory.

80. Is it possible to have a concentrated weak acid? Explain.

★81. Write equations showing that the hydrogen phosphate ion (HPO_4^{2-}) is amphoteric.

82. The pH of a 0.5000*M* HNO_2 solution is 1.83. What is the K_a of this acid?

83. How do the $[H^+]$ and the $[OH^-]$ compare in each type of solution?

a. neutral solution
b. basic solution
c. acidic solution

84. Write the formula and name of the conjugate base of each Brønsted-Lowry acid.

a. HCO_3^- b. NH_4^+ c. HI d. H_2SO_3

★85. Write the formula and name of the conjugate acid of each Brønsted-Lowry base.

a. ClO_2^- b. H_2O c. $H_2PO_4^-$ d. NH_3

86. Calculate the $[OH^-]$ or pH of each solution.

a. pH = 4.60
b. pH = 9.30
c. $[OH^-] = 1.8 \times 10^{-2}M$
d. $[OH^-] = 7.3 \times 10^{-9}M$

87. Write the three equations for the stepwise ionization of phosphoric acid.

88. Use the Brønsted-Lowry and Lewis definitions of acids and bases to identify each reactant as an acid or a base.

a. $KOH(aq) + HBr(aq) \longrightarrow KBr(aq) + H_2O(l)$
b. $HCl(aq) + H_2O(l) \longrightarrow Cl^-(aq) + H_3O^+(aq)$

89. Write the formula for the conjugate base of each of the following acids:

a. H_2SO_4 b. CH_3COOH c. H_2O

90. Use the phosphate buffer ($H_2PO_4^-/HPO_4^{2-}$) to illustrate how a buffer system works. Use equations to show how the pH of a solution can be kept almost constant when small amounts of acid or base are added.

★91. Write an equation for the reaction of each antacid with hydrochloric acid.

a. magnesium hydroxide
b. calcium carbonate
c. aluminum hydroxide

92. How would the addition of each substance affect the equilibrium between hypochlorous acid and the hypochlorite ion?

$$HOCl(aq) + OH^-(aq) \rightleftharpoons OCl^-(aq) + H_2O(l)$$

a. HCl b. NaOH

93. The following data were collected from a titration of 50.00 mL of ethanoic acid (CH_3COOH) of unknown concentration with 0.100*M* NaOH. Plot these data to obtain a titration curve. Place pH on the *y*-axis.

Volume of NaOH (mL)	pH	Volume of NaOH (mL)	pH
0	3.18	50.00	8.73
10.00	4.15	50.01	8.89
25.00	4.76	51.00	11.00
40.00	5.36	60.00	11.96
49.00	6.45	75.00	12.30
49.99	8.55	100.00	12.52

a. What is the pH at the end point of this titration?
b. Use Figure 19.8 to identify one or more acid-base indicators that could be used to determine the end point in this titration.

94. Write an equation to show that an aqueous solution of sodium ethanoate will be basic.

★95. Arrange the following solutions in order of decreasing acidity:

a. 0.1*M* NaOH
b. 0.1*M* HCl
c. 0.1*M* NH_4Cl
d. 0.1*M* CH_3COONa

96. Vapors of the strong acid HCl(*aq*) and the weak base $NH_3(aq)$ combine to form a white salt.

a. What is the name and formula of the salt?
b. What kind of solution will this salt form when it dissolves in water?

ASSESSMENT

Answers

UNDERSTAND CONCEPTS

79. The Lewis theory explains the behavior of compounds that act like bases without accepting hydrogen ions or that act like acids without donating hydrogen.

80. Yes, acids like acetic acid dissolve well but ionize poorly.

81. $HPO_4^{2-} \rightarrow H^+ + PO_4^{3-}$ (acting as an acid); $HPO_4^{2-} + H^+ \rightarrow H_2PO_4^-$ (acting as a base)

82. 4.6×10^{-4}

83. a. The concentrations are equal.
b. $[OH^-]$ is greater than $[H^+]$
c. $[H^+]$ is greater than $[OH^-]$

84. a. CO_3^{2-}, carbonate ion
b. NH_3, ammonia
c. I^-, iodide ion
d. HSO_3^-, hydrogen sulfite ion.

85. a. $HClO_2$, chlorous acid
b. H_3O^+, hydronium ion
c. H_3PO_4, phosphoric acid
d. NH_4^+, ammonium ion.

86. a. $[OH^-] = 4.0 \times 10^{-10}M$
b. $[OH^-] = 2.0 \times 10^{-5}M$
c. pH = 12.26
d. pH = 5.86

87. $H_3PO_4 \rightleftharpoons H^+ + H_2PO_4^-$;
$H_2PO_4^- \rightleftharpoons H^+ + HPO_4^{2-}$;
$HPO_4^{2-} \rightleftharpoons H^+ + PO_4^{3-}$

88. a. KOH is the base; HBr is the acid.
b. HCl is the acid; H_2O is the base.

89. a. HSO_4^-
b. CH_3COO^-
c. OH^-

90. $H_2PO_4^- + OH^- \rightleftharpoons H_2O + HPO_4^{2-}$
$HPO_4^{2-} + H^+ \rightleftharpoons H_2PO_4^-$

91. a. $2HCl + Mg(OH)_2 \rightarrow MgCl_2 + 2H_2O$
b. $2HCl + CaCO_3 \rightarrow H_2O + CO_2 + CaCl_2$
c. $Al(OH)_3 + 3HCl \rightarrow AlCl_3 + 3H_2O$

92. a. shift to the left
b. shift to the right

93. a. 8.73
b. phenolphthalein or phenol red

94. $NaC_2H_3O_2 + H_2O \rightarrow Na^+ + HC_2H_3O_2 + OH^-$

95. b, c, d, a

96. a. NH_4Cl, ammonium chloride
b. acidic

ASSESSMENT

Evaluate

Answers

THINK CRITICALLY

97. a. Students are likely to consider the Arrhenius explanation the easiest to understand, and the Lewis theory the best because it is the most general.

b. All three explanations provide definitions and describe accepted behavior of a certain group of compounds. The Brønsted-Lowry theory includes a greater number of compounds than the Arrhenius explanation because it is more general, and the Lewis theory includes the greatest number of compounds because it is the most general.

98. CO_2 concentration is higher in pure water. Less CO_2 becomes carbonate because pure water does not have OH^- ions needed to reduce H^+ concentration.

99. a. False; an indicator determines a range of pH values.

b. False; an Arrhenius base dissociates to give hydroxide ions in aqueous solution. Ammonia does not do this.

c. False; strength is a measure of dissociation or ionization, not concentration.

100. $HOCN + OH^- \rightleftharpoons H_2O + OCN^-$
$OCN^- + H^+ \rightleftharpoons HOCN$

101. The *y*-axis might correspond to $[H^+]$ because HCl is a strong acid.

102. 3.88×10^{-4}

103. a. $NaOH + HCl \rightarrow NaCl + H_2O$

b. 0.72*M* NaOH

104. 50.0 mL; The pH is 7 when $[H^+] = [OH^-]$. Because HCl is a strong acid that supplies one hydrogen ion per formula unit and NaOH is a strong base that supplies one hydroxide ion per formula unit, $[H^+] = [OH^-]$ when equal volumes of solutions of the same molarity are combined.

105. a. neutral

b. neutral

c. basic

d. acidic

e. neutral

106. $pOH = -\log [OH^-]$

107. pOH = 7.3

108. Because $\log (a \times b) = \log a + \log b$, and $K_w = [H^+] \times [OH^-] = 1.0 \times 10^{-14}$, taking the log of each concentration term gives: $\log [H^+] + \log [OH^-] = -14$; $-\log [H^+] - \log [OH^-] = 14$; $pH + pOH = 14$.

Think Critically

97. Compare Arrhenius, Brønsted-Lowry, and Lewis all offered explanations for the behavior of acids and bases.

a. Which explanation is easiest for you to understand?

b. How is it possible for all three explanations to be accepted by chemists?

98. Predict The solubility of carbon dioxide in water depends on four different reversible reactions.

$$CO_2(g) \rightleftharpoons CO_2(aq)$$
$$CO_2(aq) + H_2O(l) \rightleftharpoons H_2CO_3(aq)$$
$$H_2CO_3(aq) \rightleftharpoons H^+(aq) + HCO_3^-(aq)$$
$$HCO_3^-(aq) \rightleftharpoons H^+(aq) + CO_3^{2-}(aq)$$

If seawater is slightly alkaline, would you expect the concentration of dissolved CO_2 to be higher or lower than in pure water? Explain your answer.

99. Evaluate Critique the accuracy of each of these statements.

a. Indicators such as methyl red provide accurate and precise measurements of pH.

b. According to the Arrhenius definition of acids and bases, ammonia qualifies as a base.

c. The strength of an acid or base changes as its concentration changes.

★100. Relate Cause and Effect Use the cyanate buffer HOCN/OCN^- to explain how a buffer system works. Use equations to show how the pH of a solution can be kept almost constant when small amounts of acid or base are added.

★101. Identify Which quantity might correspond to the *y*-axis on this graph: $[H^+]$, pH, or $[OH^-]$? Explain your answer.

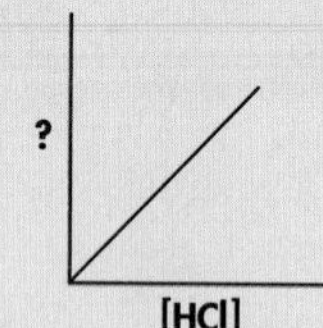

102. Calculate The sugar substitute saccharin ($HNC_7H_4SO_3$) has one acidic hydrogen. A 1.000*M* aqueous solution of saccharin has a pH of 1.71. Calculate the K_a of saccharin.

103. Interpret Graphs The graph shows the number of millimoles (mmol) of water formed as drops of 1.0*M* HCl are added to a 25.0-mL sample of NaOH of unknown concentration.

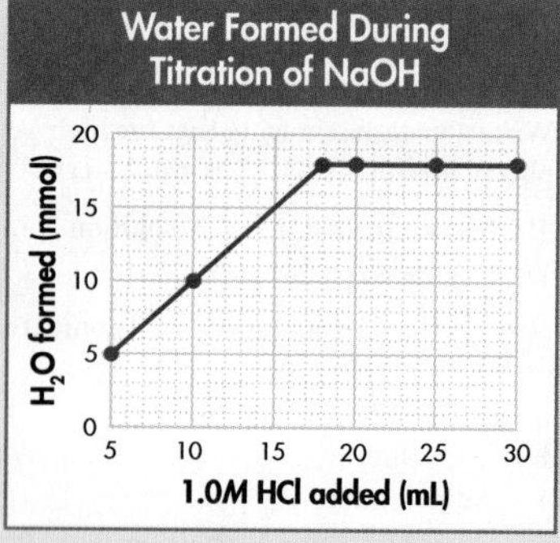

a. Write an equation for the reaction.

b. Estimate the concentration of the NaOH.

104. Calculate Suppose you slowly add 0.1*M* NaOH to 50.0 mL of 0.1*M* HCl. What volume of NaOH must you add before neutralization will occur? Explain your reasoning.

105. Predict Will the resulting solutions be neutral, acidic, or basic at the equivalence point for each of the following titrations? Explain.

a. HCl titrated with NaOH

b. NaOH titrated with HCl

c. CH_3COOH titrated with NaOH

d. NH_3 titrated with HCl

e. CH_3COOH titrated with NH_3

★106. Use Models You can use the following expression to find the pH of a solution:

$$pH = -\log[H^+]$$

What expression could you use to find the analogous quantity, the pOH of a solution?

107. Apply Concepts Milk, an aqueous emulsion, has a pH of about 6.7. Calculate the pOH of milk using the equation you derived in Question 106. If the equation you derived is correct, the sum of the values for pH and pOH will equal 14.

108. Apply Concepts Use the expression for K_w to demonstrate the following relationship:

$$pH + pOH = 14$$

Enrichment

109. Calculate What is the pH of a 0.010*M* solution of NaCN ($K_b = 2.1 \times 10^{-5}$)?

★**110. Make Generalizations** Show that for any conjugate acid-base pair, $K_a \times K_b = K_w$.

111. Interpret Data The K_w of water varies with temperature, as shown in the table.

Temperature (°C)	K_w	pH
0	1.137×10^{-15}	a. ______
10	2.917×10^{-15}	b. ______
20	6.807×10^{-15}	c. ______
30	1.469×10^{-14}	d. ______
40	2.917×10^{-14}	e. ______
50	5.470×10^{-14}	f. ______

a. Calculate the pH of water for each temperature in the table.

b. Use the data to prepare a graph of pH versus temperature. Use the graph to estimate the pH of water at 5°C.

c. At what temperature is the pH of water approximately 6.85?

112. Calculate What is the molarity of an H_2SO_4 solution if 80.0 mL of the solution reacts with 0.424 g Na_2CO_3?

$$H_2SO_4(aq) + Na_2CO_3(aq) \longrightarrow H_2O(l) + CO_2(aq) + Na_2SO_4(aq)$$

★**113. Apply Concepts** The hydrogen carbonate ion–carbonic acid buffer system is an important buffer system in the blood. This system is represented by the following equations:

$$H_2O(l) + CO_2(g) \rightleftharpoons H_2CO_3(aq)$$
$$H_2CO_3(aq) \rightleftharpoons H^+(aq) + HCO_3^-(aq)$$

Reactions in cells produce carbon dioxide. Excess carbon dioxide is released through the lungs. How could rapid breathing lead to an abnormally high blood pH (alkalosis)? How could slow breathing lead to an abnormally low blood pH (acidosis)?

114. Calculate Household bleach is a solution of sodium hypochlorite. What is the $[OH^-]$ in an aqueous solution that is 5.0% NaClO by mass? What is the pH of the solution? (The density of the solution is 1.0 g/mL, and $K_a = 3.5 \times 10^{-8}$.)

Write About Science

115. Research The main cause of tooth decay is the weak acid lactic acid (C_2H_5OCOOH). Lactic acid forms when bacteria, such as *Streptococcus mutans*, feed on sugar. In the mouth, sugars are present in the sticky plaque on tooth surfaces. Starting with the information on page R30, research current efforts to fight tooth decay. Write a report summarizing your findings.

116. Connect to the BIGIDEA Hypochlorite salts are used to disinfect swimming pools. On page R30 in the Elements Handbook, read about what happens when chlorine compounds are added to pool water. Use hydrolysis reactions to explain how the pH of the water affects the concentration of hypochlorous acid (HOCl).

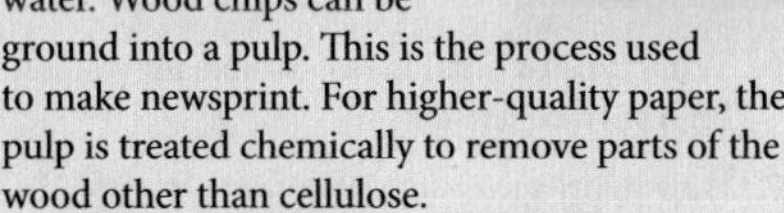

Paper Trail

The wood pulp used to make paper is a suspension of cellulose fibers in water. Wood chips can be ground into a pulp. This is the process used to make newsprint. For higher-quality paper, the pulp is treated chemically to remove parts of the wood other than cellulose.

The paper is often coated with a chemical such as aluminum sulfate to keep it from absorbing too much ink. The chemicals that were used for this purpose often left a residue of acid in the paper. Over time, the acid caused the cellulose fibers to decay.

117. Infer The process of treating the paper is called deacidification. The first step in a popular deacidification method is to immerse the paper in a dilute solution of calcium hydroxide. Write a chemical equation to describe what occurs in this deacidification step.

★**118. Connect to the BIGIDEA** What type of reaction is performed in the process of deacidification? Would you expect the pH of the paper to be raised or lowered in the process of deacidification?

Answers

ENRICHMENT

109. pH = 10.66

110.

$$K_a \times K_b = \frac{[H^+][A^-]}{[HA]} \times \frac{[HA][OH^-]}{[A^-]} = [H^+][OH^-] = K_w$$

111. a. 7.4721, 7.2675, 7.0835, 6.9165, 6.7675, 6.6310

b. 7.37

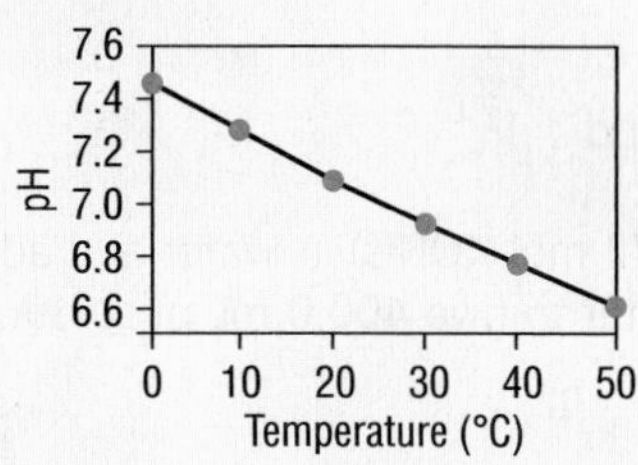

c. 35°C

112. $5.00 \times 10^{-2}M$;

113. Hyperventilation releases CO_2. The equilibrium shift causes H_2CO_3 and H^+ concentrations to decrease. The loss of H^+ increases the basicity of the blood and alkalosis results. Hypoventilation does not release enough CO_2. The CO_2 build up increases the H_2CO_3 and H^+ concentrations. The H^+ concentration increases the acidity of the blood and acidosis results.

114. $[OH^-] = 4.4 \times 10^{-4}M$; pH = 10.66

WRITE ABOUT SCIENCE

115. To fight tooth decay, reduce sugar intake; brush and floss regularly; use a fluoride toothpaste or rinse; and have regular dental checkups.

116. Students should note that the pH affects the relative concentrations of ClO^- and HClO in pool water. If the pH of the water is too high, solid sodium hydrogen sulfate can be used to lower the pH. If the pH is too low, sodium carbonate can be used to neutralize some of the acid.

CHEMYSTERY

START A CONVERSATION Have students reread CHEMystery. Call on volunteers to summarize why the wood-based paper tended to yellow and crack over time. **Ask** How did chemical reactions cause the paper to become crumbly? *(An acid used in the production of the paper reacted with the cellulose, causing the fibers to decay.)*
Ask Why don't the books made of rag-based paper become crumbly? Rag-based paper is acid-free.

CHEMYSTERY ANSWERS

117. $H_2SO_4 + Ca(OH)_2 \rightarrow CaSO_4 + 2H_2O$

118. BIGIDEA Deacidificaiton involves a double-replacement (neutralization) reaction, during which the pH increases.

Cumulative Review

119. a. $K_2O(s)$
b. $CaS(s)$
c. $AlF_3(s)$
120. 131 g O_2
121. liquid
122. The total pressure in a mixture of gases is equal to the sum of the partial pressures of each gas in the mixture.
123. b. Boyle's law
124. suspension
125. c. $SiO_2(s)$
126. hydrogen bond
127. 2.25 g KCl
128. Dissolve 0.272 mol KOH(s) in water and add sufficient water to give 400.0 mL of solution.
129. 0.47 L
130. b, c, and d
131. 1.8×10^4 J
132. a. 144 J
b. 1.0×10^3 kJ
c. 82.9 cal
133. 2.8 kJ
134. The product of the concentrations of two ions must be greater than the solubility product.
135. a. $K_{eq} = \dfrac{[CO]^2 \times [O_2]}{[CO_2]^2}$
b. $K_{eq} = \dfrac{[NH_3]^2}{[N_2] \times [H_2]^3}$
136. 2.0×10^{-8} *M*
137. a. NaCl(*aq*)
b. $CO_2(g)$
c. hot water
138. a. shift right
b. no change
c. shift right
d. shift right
e. shift rght
139. A: first order
B: second order
C: zero order; overall: third order

Cumulative Review

119. Write the product of each of these combination reactions.
a. $K(s) + O_2(g) \longrightarrow$
b. $Ca(s) + S(s) \longrightarrow$
c. $F_2(g) + Al(s) \longrightarrow$

120. How many grams of oxygen are needed to completely burn 87.4 g of sulfur to form sulfur trioxide?

$$S(s) + O_2(g) \longrightarrow SO_3(g)$$

***121.** Which state of matter is not part of the process of sublimation?

122. State Dalton's law of partial pressures.

123. Which of these laws describes an inverse relationship?
a. Charles's law
b. Boyle's law
c. Gay-Lussac's law

124. Which has the largest particles, a solution, a colloid, or a suspension?

125. Which of these is not an electrolyte?
a. NaCl(*l*)
b. $KNO_2(aq)$
c. $SiO_2(s)$
d. NaCl(*aq*)

***126.** What type of bond is responsible for water's high surface tension?

127. How many grams of potassium chloride are in 45.0 mL of a 5.00% (by mass) solution?

128. How would you prepare 400.0 mL of a 0.680*M* KOH solution?

***129.** How many liters of 8.0*M* HCl are needed to prepare 1.50 L of 2.5*M* HCl?

130. Which of these is an endothermic process? Provide an explanation.
a. burning wax
b. evaporating water
c. melting wax
d. roasting a marshmallow

131. How many joules of heat are required to melt a 55.0-g ice cube at 0°C?

***132.** Make the following conversions:
a. 34.5 cal to joules
b. 250 Cal to kilojoules
c. 0.347 kJ to calories

133. The specific heat capacity of iron is 0.46 J/(g·°C). How many kilojoules of energy are needed to raise the temperature of a 432-g iron bar 14°C?

134. What must be true about the concentration of two ions if precipitation occurs when solutions of the two ions are mixed?

135. Write an equilibrium-constant expression for each equation.
a. $2CO_2(g) \rightleftharpoons 2CO(g) + O_2(g)$
b. $N_2(g) + 3H_2(g) \rightleftharpoons 2NH_3(g)$

136. What is the equilibrium concentration of barium ion in a 1.0-L saturated solution of $BaCO_3$ to which 0.25 mol K_2CO_3 has been added?

***137.** In each pair, which has the higher entropy?
a. NaCl(*s*) or NaCl(*aq*)
b. $CO_2(s)$ or $CO_2(g)$
c. hot water or cold water

138. How would each change affect the position of equilibrium of this reaction?

$$2H_2(g) + O_2(g) \rightleftharpoons 2H_2O(g) + \text{heat}$$

a. increasing the pressure
b. adding a catalyst
c. increasing the concentration of $H_2(g)$
d. cooling the reaction mixture
e. removing water vapor from the container

139. For the reaction $A(g) + B(g) + C(g) \longrightarrow D(g)$, the following data were obtained at a constant temperature. From the data, determine the order of reaction with respect to A, B, and C, and the overall order of reaction.

Initial [A] (mol/L)	Initial [B] (mol/L)	Initial [C] (mol/L)	Initial rate (mol/(L·min))
0.0500	0.0500	0.0100	6.25×10^{-3}
0.1000	0.0500	0.0100	1.25×10^{-2}
0.1000	0.1000	0.0100	5.00×10^{-2}
0.0500	0.0500	0.0200	6.25×10^{-3}

If You Have Trouble With . . .

Question	119	120	121	122	123	124	125	126	127	128	129	130	131	132	133	134	135	136	137	138	139
See Chapter	11	12	13	14	14	15	15	15	16	16	16	17	17	17	17	18	18	18	18	18	18

Standardized Test Prep for Indiana

Standards: C.8.4, C.8.5

Select the choice that best answers each question or completes each statement.

1. If an acid has a measured K_a of 3×10^{-6},
   (A) the acid is a strong acid.
   (B) an aqueous solution of the acid would have a pH < 7.
   (C) the acid is a strong electrolyte.
   (D) All of the above are correct.
2. The pH of a sample of orange juice is 3.5. A sample of tomato juice has a pH of 4.5. Compared to the $[H^+]$ of orange juice, the $[H^+]$ of tomato juice is
   (A) 1.0 times higher. (C) 10 times higher.
   (B) 10 times lower. (D) 1.0 times lower.

Tips for Success

Eliminate Wrong Answers If you don't know which choice is correct, eliminate those you know are wrong. If you can rule out some choices, you'll increase your chances of choosing the correct answer.

3. Which ion or molecule is the conjugate base of the ammonium ion, NH_4^+?
   (A) H_2O (B) OH^- (C) NH_3 (D) H_3O^+
4. How many moles of NaOH are required to neutralize 2.4 mol H_2SO_4?
   (A) 1.2 mol (B) 2.4 mol (C) 3.6 mol (D) 4.8 mol
5. A solution with a hydrogen-ion concentration of $2.3 \times 10^{-8}M$ has a pH between
   (A) 2 and 3. (B) 3 and 4. (C) 7 and 8. (D) 8 and 9.
6. The net ionic equation for the neutralization reaction between solutions of potassium hydroxide and hydrochloric acid is
   (A) $H^+(aq) + OH^-(aq) \longrightarrow H_2O(l)$
   (B) $KOH(aq) + HCl(aq) \longrightarrow H_2O(l) + KCl(aq)$
   (C) $K^+(aq) + Cl^-(aq) \longrightarrow KCl(aq)$
   (D) $K^+(aq) + OH^-(aq) + H^+(aq) + Cl^-(aq) \longrightarrow KCl(aq) + H_2O(l)$
7. Calculate the molarity of an HCl solution if 25.0 mL of the solution is neutralized by 15.5 mL of 0.800*M* NaOH.
   (A) 0.248*M* (B) 0.496*M* (C) 1.29*M* (D) 0.645*M*
8. Which combination of compound and ion would not make a useful buffer solution?
   (A) ammonium ion and ammonia
   (B) hydrogen carbonate ion and carbonic acid
   (C) sulfate ion and sulfuric acid
   (D) ethanoate ion and ethanoic acid

The lettered choices below refer to Questions 9–11. In each formula, P is the cation, and Q is the anion.

(A) PQ (B) P_2Q_3 (C) PQ_3 (D) P_3Q

Which of the choices is the general formula for the salt formed in each of the following neutralization reactions?

9. $H_3PO_4 + NaOH \longrightarrow$
10. $H_2SO_4 + Mg(OH)_2 \longrightarrow$
11. $HNO_3 + Al(OH)_3 \longrightarrow$

Use the drawings below to answer Questions 12 and 13. Water molecules have been omitted from the solution windows.

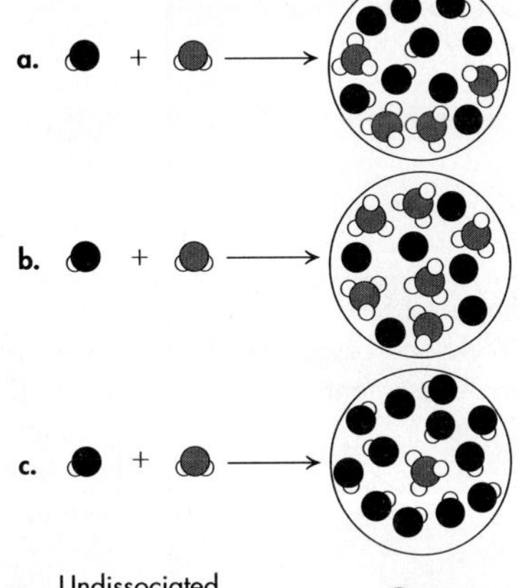

12. Rank the acids in order of increasing strength.
13. How many of the acids are strong acids?

If You Have Trouble With . . .

Question	1	2	3	4	5	6	7	8	9	10	11	12	13
See Lesson	19.3	19.2	19.1	19.4	19.2	19.4	19.4	19.5	19.4	19.4	19.4	19.3	19.3

ASSESSMENT

STP Answers

1. B
2. B
3. C
4. D
5. C
6. A
7. B
8. C
9. D
10. A
11. C
12. c, a, b
13. one (b)

20 Oxidation-Reduction Reactions Planning Guide

Introducing the BIGIDEA: REACTIONS

In oxidation-reduction reactions, electrons are transferred from one reacting species to another.

IASS	Lessons and Objectives	Print Resources		
		For the Student	For the Teacher	
C.1.3, C.4.6	**20.1** The Meaning of Oxidation and Reduction p 692–699 **20.1.1** Describe what happens to a substance that undergoes oxidation and a substance that undergoes reduction. **20.1.2** Explain how the presence of salts and acids accelerates the corrosion of metals.	Reading and Study Workbook Lesson 20.1 Lesson Assessment 20.1 p 699 Quick Lab: Bleach It! Oxidize the Color Away p 699	Teacher Demo, p 693: Oxidation and Reduction Class Activity, p 697: Metal Corrosion	
C.4.6	**20.2** Oxidation Numbers p 701–706 **20.2.1** State the general rule for assigning oxidation numbers. **20.2.2** Define oxidation and reduction in terms of a change in oxidation number.	Reading and Study Workbook Lesson 20.2 Lesson Assessment 20.2 p 706	Class Activity, p 702: Assigning Oxidation Numbers Teacher Demo, p 703: Redox Reactions Class Activity, p 705: Identifying Redox Reactions	
C.4.6	**20.3** Describing Redox Equations p 707–715 **20.3.1** Identify the two classes of chemical reactions. **20.3.2** Describe two different methods for balancing a redox equation.	Reading and Study Workbook Lesson 20.3 Lesson Assessment 20.3 p 715 Small-Scale Lab: Half-Reactions p 717	Teacher Demo, p 710: Color Changes in Redox Reactions Teacher Demo, p 713: Preventing Oxidation	

Assessing the BIGIDEA: REACTIONS

	Essential Questions 1. What happens during oxidation and reduction? 2. How can you balance redox equations?	Study Guide p 718 Skills Tune-Up p 719 STP p 725 Reading and Study Workbook Self-Check and Vocabulary Review Chapter 20		

	PearsonChem.com	
	Editable Worksheets	Online Activities
	Teaching Resources, Lesson 20.1 Review Lab 47: Corrosion Probeware Lab: Quick Lab: Bleach It! Oxidize the Color Away	20.1 Lesson Overview Oxidation of Methane Identifying Oxidized and Reduced Reactants Used Cars 101 Lesson 20: Be Wary of Where Your Car Used to Live Corrosion
	Teaching Resources, Lesson 20.2 Review	20.2 Lesson Overview Assigning Oxidation Numbers to Atoms
	Teaching Resources, Lesson 20.3 Review Lab 41: Reactions of Acids Lab 46: Oxidation-Reduction Reactions Small-Scale Lab Manual Lab 34: Determination of an Activity Series Small-Scale Lab Manual Lab 35: Oxidation-Reduction Reactions Lab Practical 20-1: Reduction of Metals	20.3 Lesson Overview Redox Titrations: Determining the Purity of a Sample Balancing Redox Equations by Oxidation-Number Change
	Core Teaching Resources (includes editable worksheets) • Lesson Reviews • Practice Problems • Interpret Graphs • Vocabulary Review • Chapter Quizzes and Tests • Lab Record Sheets	Chapter 20 Problem Set Balanced Equations

Materials List

For the Student

Quick Lab p 699
- spot plates
- medicine droppers
- liquid and powder bleach
- 3% H_2O_2
- 1% oxalic acid
- 0.2*M* $Na_2S_2O_3$, I_2 in $KI(aq)$
- $KMnO_4(aq)$
- grape juice
- rusty water
- colored fabric
- grass-stained fabric
- flower petals

Small-Scale Lab p 717
- reaction surfaces
- 1.0*M* HCl, HNO_3, and H_2SO_4
- pieces of Zn, Mg, Cu, and Fe

For the Teacher

Teacher Demo p 693
- 3 g copper(II) oxide (CuO)
- 3 g charcoal (C)
- crucible with cover
- Bunsen burner
- tongs
- glass plate

Teacher Demo p 703
- 100 mL of 0.1*M* lead(II) acetate $[Pb(CH_3COO)_2]$
- 150-mL beaker
- 1-cm × 3-cm zinc strip
- sodium sulfide
- 3*M* NaOH
- 1*M* iron(III) chloride
- stirring rod
- plastic container

Class Activity p 705
- 3 small glass test tubes
- zinc
- copper
- magnesium
- 0.1*M* HCl
- medicine dropper or pipet

Teacher Demo p 710
- 6 crystals of potassium permanganate ($KMnO_4$)
- 540 mL of water, large beaker
- 1 g sodium hydrogen sulfite ($NaHSO_3$)
- two 50-mL beakers
- 1 g barium chloride dihydrate ($BaCl_2 \cdot 2H_2O$)
- 3*M* H_2SO_4
- stirring rod
- plastic container

Class Activity p 713
- potatoes
- apples
- beakers
- water
- boiled water
- lemon juice
- carbonated beverage
- sugar
- salt
- vinegar solutions
- paper towels

Additional Digital Resources

Online Student Edition
Online Teacher's Edition
Exam*View Assessment Suite CD-ROM*
Classroom Resources DVD
20.3 Virtual Chem Lab 30: Redox Titrations: Determination of Iron

CHAPTER 20

What's Online

CHEMISTRY TUTORIAL Students can work step-by-step with the online tutorial to identify oxidized and reduced reactants, assign oxidation numbers to atoms, and balance redox equations by accessing the Chapter 20 Chemistry Tutorials.

ONLINE PROBLEM SET Additional practice problems about redox reactions can be found in the Chapter 20 online Problem Set.

MATH TUTORIAL The Online Math Tutorial will help guide students through balancing equations.

DIRECTED VIRTUAL LAB Encourage students to review how redox reactions can be tracked with titrations, to find the purity of a sample using the animated virtual lab for Chapter 20. The animation is followed by analysis questions.

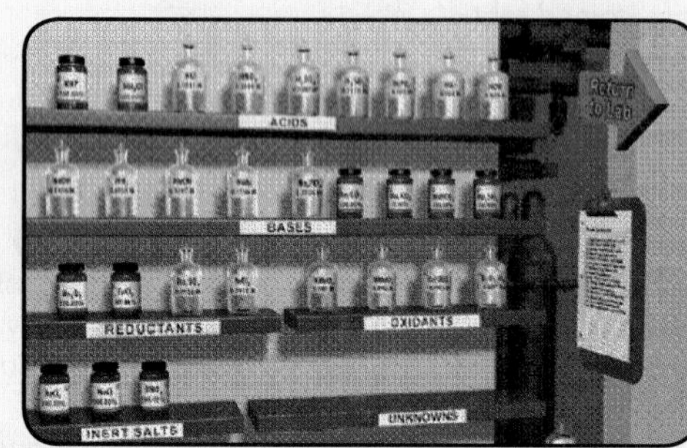

KINETIC ART Have students watch animations about the oxidation of methane and corrosion. Each animation is followed by analysis questions.

CONCEPTS IN ACTION Students can watch an interactive animation about corrosion. The animation is followed by questions.

IN ACADEMIC STANDARDS for SCIENCE

Nature of Science

- Develop explanations based on reproducible data and observations gathered during laboratory investigations.
- Clearly communicate their ideas and results of investigations verbally and in written form using tables, graphs, diagrams, and photographs.

Standard 4: Reactions and Stoichiometry

C.4.6 Determine oxidation states and identify the substances gaining and losing electrons in redox reactions.

20 Oxidation-Reduction Reactions

INSIDE:

- **20.1** The Meaning of Oxidation and Reduction
- **20.2** Oxidation Numbers
- **20.3** Describing Redox Equations

PearsonChem.com

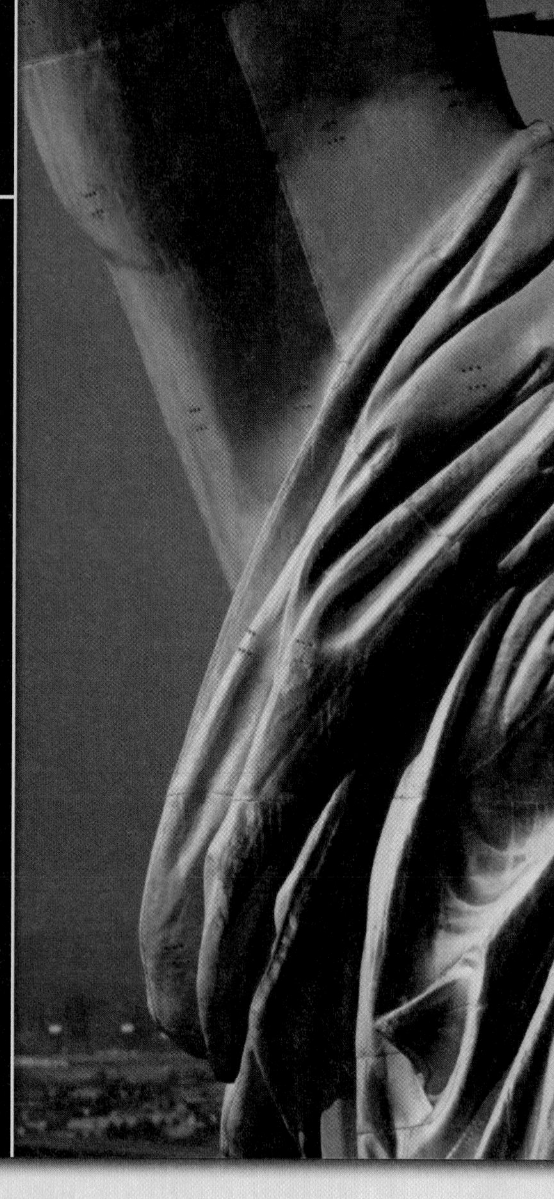

The color of the Statue of Liberty is the result of the copper in her outer shell reacting with water, carbon dioxide, and other compounds in the air.

Focus on ELL

1 CONTENT AND LANGUAGE Write the chapter title on the board. Below the title, write the word *redox* from the title of Lesson 20.3. Slowly read the words aloud, taking care to enunciate the prefixes in *oxidation* and *reduction*. Convey that the word *redox* is a simpler way to say *oxidation-reduction*.

BEGINNING: LOW/HIGH Use colored markers to combine the words *oxidation-reduction* to form the word *redox*. Have students share simple words in their native language that are the combination of two larger words.

INTERMEDIATE: LOW/HIGH Write simplified English definitions of the words *oxidation* and *reduction* and present them to the class.

ADVANCED: LOW/HIGH Have students look up *oxidation* and *reduction* in a dictionary and use the definitions to predict what they will learn in each lesson.

BIGIDEA

REACTIONS

Essential Questions:

1. *What happens during oxidation and reduction?*
2. *How can you balance redox equations?*

CHEMYSTERY

Tarnished Treasure

When a ship sinks to the bottom of the ocean, the ship and everything on it is exposed to the harsh elements of the sea. The shipwreck slowly starts to fall apart. Shipwrecks of Spanish galleons are no different.

Many of the Spanish galleons contained different types of coins. Silver coins discovered on shipwrecks look like they might be just junk, as they are often black and corroded. However, gold coins still retain their gold finish. Why would a gold coin withstand the elements of the sea better than a silver coin?

▶ Connect to the **BIG**IDEA As you read about oxidation and reduction chemical reactions, think about why different metals might react differently even if they are exposed to the same conditions.

INDIANA'S ACADEMIC STANDARDS FOR SCIENCE

Reactions and Stoichiometry C.4.6. See lessons for details.

CHAPTER 20

UbD

Understanding by Design

Students are building toward describing what happens during oxidation and reduction, using their knowledge of chemical reactions.

PERFORMANCE GOALS At the end of Chapter 20, students will be able to answer the essential questions by applying their knowledge of oxidation and reduction.

ESSENTIAL QUESTIONS Read the essential questions aloud. **Ask** What do you think *oxidation* and *reduction* mean? *(Students may say oxidation is related to oxygen and reduction is related to decreasing.)* Tell students that sometimes names have historical meanings that are no longer accurate.

BIGIDEA Use the photo of the Statue of Liberty to help students connect to this chapter's concepts. Activate prior knowledge by asking students to describe the statue's color. Point out that the statue is made of copper, which naturally has a reddish-brown color. A chemical reaction changed the color of the statue's outer layer to the green patina. **Ask** What are other examples of chemical reactions changing a metal object? *(Sample answers: A bike left outside rusts; silver tarnishes.)* Discuss conditions students have noticed when these changes occur.

CHEMYSTERY Connect the CHEMystery to the Big Idea by reminding students that whether a reaction occurs depends on the types of elements involved and on the environment of the elements. Although silver and gold are in the same group on the periodic table, their tendencies to react are different. Have students predict how this might cause silver to corrode but gold to retain its finish.

Introduce the Chapter

IDENTIFY PRECONCEPTIONS Students may incorrectly interpret redox reactions in terms of unrelated everyday experiences. Use the activity below to encourage them to consider chemical explanations for changes they see in their everyday life. (Note: this activity requires the treated metal to sit overnight for the results to develop.)

Activity *You will need some corroded and pitted nickels and pennies, some shiny strips of nickel and copper metal, and some clear nail polish.* Allow students to observe the coins. **Ask** What do you think caused the differences between the coins and the metal strips? *(Students may say the corroded coins are just old, have been damaged by misuse, or have something on them.)* Have students coat half of each shiny metal strip with nail polish. After the polish dries, have students place the metal strips in water and then set them on a paper towel to dry overnight. The next day, students will observe that only the uncoated portions of the strips corroded.

Building Better Bridges

Rob investigates a rusting bridge and visits a chemist who is developing methods to slow down the rusting process.

LESSON 20.1

Key Objectives

20.1.1 DESCRIBE what happens to a substance that undergoes oxidation and a substance that undergoes reduction.

20.1.2 EXPLAIN how the presence of salts and acids accelerates the corrosion of metals.

Additional Resources

Reading and Study Workbook, Lesson 20.1

Available Online or on Digital Media:

- Teaching Resources, Lesson 20.1 Review
- Laboratory Manual, Labs 46 and 47
- Small-Scale Chemistry Laboratory Manual, Lab 34

PearsonChem.com

Engage

CHEMISTRY & YOU Have students read the opening paragraph. **Ask** What are some ways that you can protect a car against this type of corrosion? *(Put a protective undercoat on the car or use plastic car parts.)*

Activate Prior Knowledge

Review the concept of valence electrons. Then assess students' knowledge about electron transfer. **Ask** What happens to magnesium and oxygen when they react to form magnesium oxide, MgO? *(Mg loses electrons to form Mg^{2+}, and oxygen gains electrons to form O^{2-}.)*

IN ACADEMIC STANDARDS for SCIENCE

- Develop explanations based on reproducible data and observations gathered during laboratory investigations.

C.4.6 Determine oxidation states and identify the substances apparently gaining and losing electrons in redox reactions.

20.1 The Meaning of Oxidation and Reduction

C.4.6 Redox reactions

CHEMISTRY & YOU

Q: *Why do you need to wash the bottom of your car?* A clean car may look better than a dirty car, but looks aren't the only reason to wash your car. During winter in cold climates, salt is often spread on roads to lower the freezing point of water and thereby prevent the buildup of slippery ice. Salt may make driving safer, but the salt that clings to the metallic parts of cars can cause damage to your car. In this lesson, you will learn about oxidation-reduction reactions, including those that damage salty cars by causing them to corrode or rust relatively quickly.

Key Questions

What happens to a substance that undergoes oxidation? What happens to a substance that undergoes reduction?

How does the presence of salts and acids accelerate the corrosion of metals?

Vocabulary

- oxidation-reduction reactions
- oxidation
- reduction
- reducing agent
- oxidizing agent

What Are Oxidation and Reduction?

What happens to a substance that undergoes oxidation? What happens to a substance that undergoes reduction?

The combustion of gasoline in an automobile engine and the burning of wood in a fireplace are reactions that require oxygen as they release energy. The reactions that break down food in your body and release energy use oxygen from the air you breathe.

Oxygen and Redox Early chemists saw oxidation only as the combining of an element with oxygen to produce an oxide. The burning of a fuel is also an oxidation reaction that uses oxygen. For example, when methane (CH_4), the main component of natural gas, burns in air, it oxidizes and forms oxides of carbon and hydrogen, as shown in Figure 20.1. One oxide of carbon is carbon dioxide, CO_2.

KINETIC ART *See Oxidation animated online.*

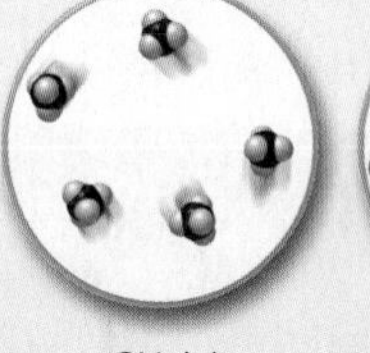

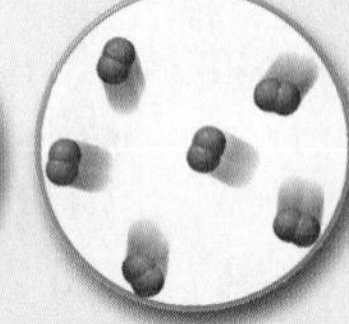

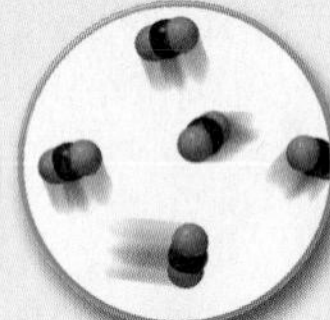

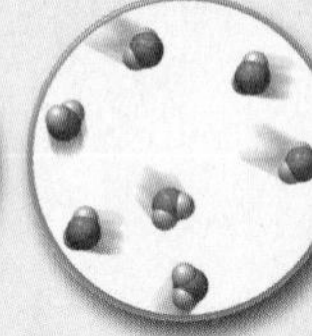

$CH_4(g)$ + $2O_2(g)$ ⟶ $CO_2(g)$ + $2H_2O(g)$

Figure 20.1
Oxidation of Methane
In a Bunsen burner, oxygen from the air is mixed with methane from the gas line. The methane is oxidized, and carbon dioxide and water are produced.

Focus on ELL

1 CONTENT AND LANGUAGE Write the word *reduction* on the board and pronounce it slowly. Explain that this term comes from the Latin word *reducere*, meaning "to lead back." Have students brainstorm phrases that refer to reductions, and help them connect the concept of reduction with a decrease in something, such as mass or volume.

2 FRONTLOAD THE LESSON Ask students if they have ever worn jewelry that turned their skin green or had jewelry that turned dark. Explain that these color changes are the result of oxidation-reduction reactions, and both reactions were related to the metals' ability to lose electrons—also known as their reactivities.

3 COMPREHENSIBLE INPUT On the board, rewrite the redox reaction from page 695 by using electron dot structures with different colors for the Mg and S electrons. Then draw the arrows showing the transfer of electrons. Use the Kinetic Art activity to reinforce the discussion on corrosion.

Not all oxidation processes that use oxygen involve burning. For example, when elemental iron turns to rust as in Figure 20.2, it slowly oxidizes to compounds such as iron(III) oxide (Fe_2O_3). Bleaching stains in fabrics is another example of oxidation that does not involve burning. Common liquid household bleach contains sodium hypochlorite (NaClO), a substance that releases oxygen, which oxidizes stains to a colorless form. Powder bleaches may contain calcium hypochlorite ($Ca(ClO)_2$), sodium perborate ($NaBO_3$), or sodium percarbonate ($2Na_2CO_3 \cdot 3H_2O_2$). Hydrogen peroxide (H_2O_2) also releases oxygen when it decomposes. It is both a bleach and a mild antiseptic that kills bacteria by oxidizing them.

A process called reduction is the opposite of oxidation. Originally, reduction meant the loss of oxygen from a compound. The reduction of iron ore to metallic iron involves the removal of oxygen from iron(III) oxide. The reduction is accomplished by heating the ore with carbon, usually in the form of coke. The equation for the reduction of iron ore is shown below.

$$\underset{\text{Iron(III) oxide}}{2Fe_2O_3(s)} + \underset{\text{Carbon}}{3C(s)} \longrightarrow \underset{\text{Iron}}{4Fe(s)} + \underset{\text{Carbon dioxide}}{3CO_2(g)}$$

The reduction of iron also includes an oxidation process. As iron(III) oxide is reduced to iron by losing oxygen, carbon oxidizes to carbon dioxide by gaining oxygen. Oxidation and reduction always occur simultaneously. **A substance that undergoes oxidation gains oxygen. A substance that undergoes reduction loses oxygen.** No oxidation occurs without reduction, and no reduction occurs without oxidation. Reactions that involve the processes of oxidation and reduction are therefore called **oxidation-reduction reactions.** Oxidation-reduction reactions are also known as *redox reactions.*

Figure 20.2 Oxidation of Iron
When items made of iron are exposed to moist air, the Fe atoms react with O_2 molecules. The iron rusts; it is oxidized to compounds such as iron(III) oxide (Fe_2O_3).

Differentiated Instruction

L3 ADVANCED STUDENTS Have advanced students do research on redox reactions involving oxygen that are important in nature. Examples are the reduction of carbon in carbon dioxide to form carbohydrates during photosynthesis and the oxidation of glucose during cell respiration to release energy. Have students prepare reports or posters based on their research.

L1 SPECIAL NEEDS STUDENTS Give each student two index cards, one with the word GAIN and the other with the word LOSE. To assess students' understanding, tell them to answer reduction questions by holding up the GAIN card to represent a gain in electrons and to answer oxidation questions by holding up the LOSE card to represent a loss of electrons.

Foundations for Reading

BUILD VOCABULARY Use a web diagram to connect the vocabulary terms. Place the terms *oxidation-reduction reactions* and *redox reactions* in the center, and then place the four remaining terms around the center, along with an explanation of how the terms relate.

READING STRATEGY Have students preview the lesson content by creating an outline using the orange and blue section headings.

Explain

What Are Oxidation and Reduction?

START A CONVERSATION Recall the electron structure of an oxygen atom and how oxygen atoms bond with other atoms. Be sure to identify the source of the electrons an oxygen atom gains during bonding. **Ask** When a metal combines with oxygen, does the metal gain or lose electrons? *(It loses electrons.)* **Ask** When oxygen is removed from the oxide of a metal, does the metal gain or lose electrons? *(It gains electrons.)* Explain to students that this information is what led to the broader definition of *oxidation and reduction* as an exchange of electrons.

Explore

Teacher Demo

PURPOSE Students will observe how pure metals are obtained from their oxide ores by reduction.

MATERIALS 3 g copper(II) oxide (CuO), 3 g charcoal (C), crucible with cover, Bunsen burner, tongs, glass plate

PROCEDURE Mix 3 g copper(II) oxide (CuO) and 3 g charcoal (C) in a crucible. Cover the crucible. Then heat thoroughly and evenly over a Bunsen burner for 10 minutes. Cool the crucible and remove the copper with tongs. Pass the copper around on a glass plate for the students to examine. Pass around a bit of the original CuO as well so that the students can see the change in appearance. On the board, write the reaction that occurred: $2CuO(s) + C(s) \rightarrow 2Cu(s) + CO_2(g)$. **Ask** Based on loss or gain of oxygen, what species were oxidized and reduced? *(C was oxidized and CuO was reduced.)*

EXPECTED OUTCOME Copper metal is formed.

LESSON 20.1

Explore

USE AN ANALOGY To reinforce the dual nature of redox reactions, compare redox reactions to acid-base reactions. Explain that the transfer of electrons from a reducing agent to an oxidizing agent is analogous to the transfer of protons from acids to bases. In neither case can the donor act independently from an acceptor.

Extend

Connect to PHYSICAL CHEMISTRY

Show students how to construct Born-Haber cycles for the redox reactions of sodium and magnesium with oxygen by finding tabulated energy values for the heats of formation of MgO and Na_2O; heats of vaporization of Mg(*s*) and Na(*s*); heat of forming O atoms from O_2; ionization energies of gaseous Mg and Na; and the electron affinity of O. *(From their diagrams, the students should be able to determine the lattice energies of MgO and Na_2O. Large lattice energies indicate that these redox reactions are highly exothermic.)*

Electron Shift in Redox Reactions The modern concepts of oxidation and reduction have been extended to include many reactions that do not even involve oxygen. You learned in Chapter 6 that, with the exception of fluorine, oxygen is the most electronegative element. As a result, when oxygen bonds with an atom of a different element (other than fluorine), electrons from that atom shift toward oxygen. Redox reactions are currently understood to involve any shift of electrons between reactants. **Oxidation** is now defined to mean complete or partial loss of electrons or gain of oxygen. **Reduction** is now defined to mean complete or partial gain of electrons or loss of oxygen.

Oxidation	Reduction
Loss of electrons Gain of oxygen	Gain of electrons Loss of oxygen

Redox Reactions That Form Ions During a reaction between a metal and a nonmetal, electrons are transferred from atoms of the metal to atoms of the nonmetal. For example, when magnesium metal is heated with the nonmetal sulfur, the ionic compound magnesium sulfide is produced as shown in Figure 20.3. Two electrons are transferred from a magnesium atom to a sulfur atom. The magnesium atoms are made more stable by the loss of electrons. The sulfur atoms become more stable due to the gain of electrons.

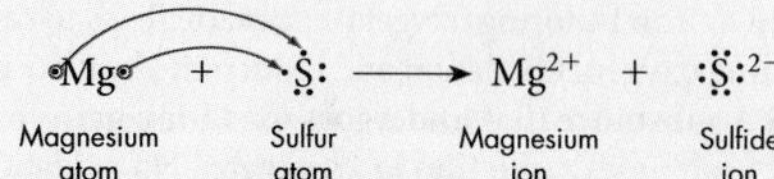

Mg + S → Mg^{2+} + S^{2-}

Magnesium atom, Sulfur atom, Magnesium ion, Sulfide ion

Because it loses electrons, the magnesium atom is said to be oxidized to a magnesium ion. Simultaneously, the sulfur atom gains two electrons and is reduced to a sulfide ion. The overall process is represented as the two component processes below.

Oxidation: $\cdot Mg \cdot \longrightarrow Mg^{2+} + 2e^-$ (loss of electrons)

Reduction: $S + 2e^- \longrightarrow S^{2-}$ (gain of electrons)

Figure 20.3
Synthesis of an Ionic Compound
When magnesium and sulfur are heated together, they undergo an oxidation-reduction reaction to form magnesium sulfide.

Professional Development Note

The History of the Word *Copper*

Reduction processes were used to produce some metals more than 5000 years ago. Copper oxide and copper sulfide ores were roasted with charcoal to form copper, carbon dioxide, and sulfur dioxide. Large amounts of copper ore were mined on the island of Cyprus, off the coast of the area now known as Turkey. The Romans, who eventually imported this valuable ore, called it *aes cyprium*, or *ore of Cyprus*. The name was later shortened to *cyprium* and then corrupted to *cuprum*. The English word *copper* derives from this corruption of the Latin word for Cyprus.

A substance that undergoes oxidation loses electrons. A substance that undergoes reduction gains electrons. The substance that loses electrons is the **reducing agent.** By losing electrons to sulfur, magnesium reduces the sulfur. Magnesium is thus the reducing agent. The substance that accepts electrons is the **oxidizing agent.** By accepting electrons from magnesium, sulfur oxidizes the magnesium. Sulfur is the oxidizing agent. Another way to identify oxidizing and reducing agents is to remember that the species that is reduced is the oxidizing agent and the species oxidized is the reducing agent.

READING SUPPORT

Build Study Skills: *Use Mnemonics* To help you remember the definitions of *oxidation* and *reduction,* use the phrase *"LEO the lion goes GER."* LEO stands for Losing Electrons is Oxidation; GER stands for Gaining Electrons is Reduction.

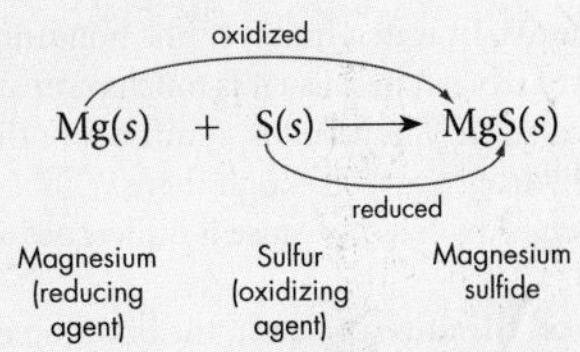

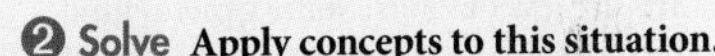

SampleProblem 20.1

Identifying Oxidized and Reduced Reactants

Silver nitrate reacts with copper to form copper nitrate and silver. From the equation below, determine what is oxidized and what is reduced. Identify the oxidizing agent and the reducing agent.

$$2AgNO_3(aq) + Cu(s) \longrightarrow Cu(NO_3)_2(aq) + 2Ag(s)$$

1 Analyze Identify the relevant concepts. Identify the ions in the reaction and then trace how the electrons were transferred.

2 Solve Apply concepts to this situation.

Rewrite the equation in ionic form so it will be easier to analyze the reaction.

$$2Ag^+ + 2NO_3^- + Cu \longrightarrow Cu^{2+} + 2NO_3^- + 2Ag$$

In this reaction, two electrons are lost from a copper atom (Cu) when it becomes a Cu^{2+} ion. These electrons are gained by two silver ions (Ag^+), which become neutral silver atoms.

The species that loses electrons is oxidized and is the reducing agent. The species that gains electrons is reduced and is the oxidizing agent.

Oxidation: $Cu \longrightarrow Cu^{2+} + 2e^-$ (loss of electrons)

Reduction: $2Ag^+ + 2e^- \longrightarrow 2Ag$ (gain of electrons)

The Cu is the reducing agent. The Ag^+ is the oxidizing agent.

1. Determine what is oxidized and what is reduced in each reaction. Identify the oxidizing agent and reducing agent in each case.
a. $2Na(s) + S(s) \longrightarrow Na_2S(s)$
b. $4Al(s) + 3O_2(g) \longrightarrow 2Al_2O_3(s)$

2. Identify these processes as either oxidation or reduction.
a. $Li \longrightarrow Li^+ + e^-$
b. $2I^- \longrightarrow I_2 + 2e^-$
c. $Zn^{2+} + 2e^- \longrightarrow Zn$
d. $Br_2 + 2e^- \longrightarrow 2Br^-$

Explore

PURPOSE Students will write a reaction equation and identify the oxidizing and reducing agents.

MATERIALS paper, pencil, periodic table

PROCEDURE Assign students to groups of two or three. Have one student in each group select an active metal and a nonmetal from the periodic table. Have this student ask the other student(s) to identify the ionic compound that forms when electron transfer occurs between the metal and the nonmetal. Have students write the equation for the reaction and identify the oxidizing and reducing agents.

Sample Practice Problem

Determine what is oxidized and what is reduced in the following reaction. Also identify the oxidizing agent and reducing agent.

$6Li(s) + N_2(g) \rightarrow 2Li_3N(s)$

(Li is oxidized and is the reducing agent; N is reduced and is the oxidizing agent.)

UbD Check for Understanding

What happens during oxidation and reduction?

To assess students' understanding of electron transfer during a redox reaction with ionic compounds, ask students to use a thumbs up to indicate *true* or a thumbs down to indicate *false* in response to the following statement:

A substance that undergoes oxidation becomes more negative, and a substance that undergoes reduction becomes more positive. *(false)*

ADJUST INSTRUCTION If students are having difficulty with this concept, have them review the oxidation and reduction reactions for Mg and S on the previous page. Point out that the charge on Mg becomes more positive when it loses two electrons during oxidation, while the charge on S becomes more negative. It may be helpful to plot the charges on a number line so that students can visually trace the direction of change.

Answers

1. **a.** Na: oxidized (reducing agent)
   S: reduced (oxidizing agent)
   b. Al: oxidized (reducing agent)
   O_2: reduced (oxidizing agent)
2. **a.** oxidation
   b. oxidation
   c. reduction
   d. reduction

LESSON 20.1

Explain

APPLY CONCEPTS As a class, generate supporting examples of redox reactions based on the processes listed in Table 20.1. Then, have students identify the oxidizing and reducing agents. Emphasize that for any given reaction, one or more of the processes may be in effect. Point out any such examples generated by the class. Tell students that the issue of oxidation numbers will be addressed in Lesson 20.2.

USE VISUALS Have students examine the photograph in Figure 20.4. Have a volunteer slowly read the caption aloud. On the board, write the balanced redox reaction for hydrogen burning in oxygen. Identify the reducing and oxidizing agents, and explain the identities on the basis of partial loss or gain of electrons. State that another gas, acetylene, C_2H_2, is also used in torches because it produces a large amount of heat when it burns. **Ask** What is the balanced chemical equation for this redox reaction? *($2C_2H_2(g) + 5O_2(g) \rightarrow 4CO_2(g) + 2H_2O(g)$)* Then, have students underline the oxidizing agents and circle the reducing agents. As a class, discuss the reasoning behind students' answers.

Figure 20.4 Welding
This welder is using a torch fueled with a mixture of H_2 and O_2 called oxyhydrogen to cut and weld steel. When hydrogen burns in oxygen, the redox reaction generates temperatures of about 2600°C.
Infer ***What is the product of this redox reaction?***

Redox With Covalent Compounds When a metal and a nonmetal react and form ions, it is easy to identify complete transfers of electrons. But some reactions involve covalent compounds, that is, compounds in which complete electron transfer does not occur. One example is the reaction of hydrogen with oxygen.

$$2H_2(g) + O_2(g) \longrightarrow 2H_2O(l)$$

Consider what happens to the bonding electrons in the formation of a water molecule. In each reactant hydrogen molecule, the bonding electrons are shared equally between the hydrogen atoms. In water, however, the bonding electrons are pulled toward oxygen because it is much more electronegative than hydrogen. The result is a shift of bonding electrons away from hydrogen, even though there is not a complete transfer. Hydrogen is oxidized because it undergoes a partial loss of electrons.

In oxygen, the other reactant, the bonding electrons are shared equally between oxygen atoms in the reactant oxygen molecule. However, when oxygen bonds to hydrogen in the water molecule, there is a shift of electrons toward oxygen. Oxygen is thus reduced because it undergoes a partial gain of electrons.

H—H (electrons shared equally) O—O (electrons shared equally) H—O—H (shift of bonding electrons away from hydrogen and towards oxygen)

In the reaction of hydrogen and oxygen to produce water, hydrogen is the reducing agent because it is oxidized. Oxygen is the oxidizing agent because it is reduced. This redox reaction is highly exothermic—that is, it releases a great deal of energy, as shown in Figure 20.4.

In some reactions involving covalent reactants or products, the partial electron shifts are less obvious. Some general guidelines are helpful. For example, for carbon compounds, the addition of oxygen or the removal of hydrogen is always oxidation. Table 20.1 below lists processes that constitute oxidation and reduction. The last entry in the table refers to oxidation numbers, which are introduced in Lesson 20.2.

Table 20.1

Processes Leading to Oxidation and Reduction

Oxidation	Reduction
Complete loss of electrons (ionic reactions)	Complete gain of electrons (ionic reactions)
Shift of electrons *away* from an atom in a covalent bond	Shift of electrons *toward* an atom in a covalent bond
Gain of oxygen	Loss of oxygen
Loss of hydrogen by a covalent compound	Gain of hydrogen by a covalent compound
Increase in oxidation number	Decrease in oxidation number

Developing Film

Black-and-white photography involves oxidation-reduction reactions. Exposing black-and-white film to light activates very fine grains of silver bromide in the film. The film is developed by placing it in a developing solution that is actually a reducing agent. The developer, usually an organic chemical, such as hydroquinone, $C_6H_4(OH)_2$, reduces the activated silver bromide to finely divided, black metallic silver. Any silver bromide that remains unactivated is removed from the film by using a solvent called a fixer. Sodium thiosulfate, commonly called hypo, is used for this purpose. On the film, the areas exposed to the most light appear darkest because they have the highest concentration of metallic silver. The reversed image, called a negative, is used to produce a positive print, the black-and-white photograph, of the image.

Corrosion

How does the presence of salts and acids accelerate the corrosion of metals?

Billions of dollars are spent yearly to prevent and to repair damage caused by the corrosion of metals. Iron, a common construction metal often used in the form of the alloy steel, corrodes by being oxidized by oxygen to ions of iron. Water in the environment accelerates the rate of corrosion. Oxygen, the oxidizing agent, is reduced to oxide ions (in compounds such as Fe_2O_3) or to hydroxide ions. The following equations describe the corrosion of iron to iron hydroxides in moist conditions.

$$2Fe(s) + O_2(g) + 2H_2O(l) \longrightarrow 2Fe(OH)_2(s)$$

$$4Fe(OH)_2(s) + O_2(g) + 2H_2O(l) \longrightarrow 4Fe(OH)_3(s)$$

Corrosion occurs more rapidly in the presence of salts and acids. **The presence of salts and acids accelerates corrosion by producing conductive solutions that make electron transfer easier.** The corrosion of some metals can be a desirable feature, as Figure 20.5 shows.

Resistance to Corrosion Not all metals corrode easily. Gold and platinum are called noble metals because they are very resistant to losing their electrons by corrosion. Other metals lose electrons easily but are protected from extensive corrosion by the oxide coating formed on their surface. For example, aluminum oxidizes quickly in air to form a coating of very tightly packed aluminum oxide particles. This coating protects the aluminum object from further corrosion, as shown in Figure 20.6. Iron also forms a coating when it corrodes, but the coating of iron oxide that forms is not tightly packed. Water and air can penetrate the coating and attack the iron metal beneath it. The corrosion continues until the iron object becomes only a pile of rust.

Figure 20.5 Corrosion
Oxidation-reduction reactions cause corrosion. The copper on this building reacted with water vapor, carbon dioxide, and other substances in the air to form a patina. This patina consists of a pale-green film of basic copper(II) carbonate. Patinas enhance the surface appearance of copper objects.

KINETIC ART See corrosion animated online.

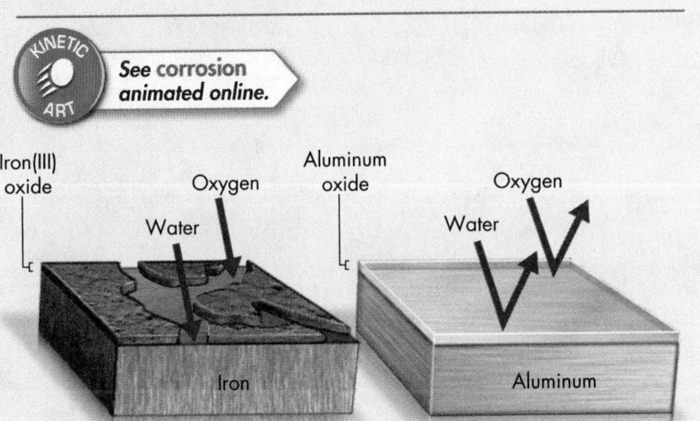

Figure 20.6 Comparing Oxides
Oxidation causes the complete corrosion of some metals. **a.** Iron reacts with water and oxygen to form iron(III) oxide, or rust. **b.** Aluminum, however, resists such corrosion because it forms a protective coating of aluminum oxide.
Apply Concepts ***How does the aluminum oxide on aluminum differ from the iron(III) oxide formed on corroding iron?***

CHEMISTRY & YOU

Q: *If your car is exposed to salt on the roads in the winter, why is it important to wash the salt off your car?*

UbD Check for Understanding

What happens during oxidation and reduction?

Assess students' knowledge of the difference between redox reactions involving ionic compounds and redox reactions involving covalent compounds by having students write a one-minute comparison of electron transfer in the two types of compounds. *(In ionic compounds, electrons are transferred between ions. In covalent compounds, there is a shift of electrons toward one of the atoms in the compound, but no electrons are actually transferred between the atoms.)*

ADJUST INSTRUCTION If students are having difficulty answering the question, have them review Table 20.1 and the preceding text, and then allow them the opportunity to revise their answers. It may be helpful to compare electron transfer between ions to passengers switching seats from one side of a bus to the other when going around a curve. Then compare passengers leaning into the curve but staying in the same seat to electrons shifting toward one atom in covalent compounds.

Explain

Corrosion

START A CONVERSATION Discuss with the class what happens if they leave a bike, skateboard, or wagon outside for a period of time. **Ask** Have you ever noticed any changes to the metal parts on these objects after they have been exposed to the weather? *(Sample answer: Yes; the parts rust.)* **Ask** Do you think all metals react the same way to long-term weather exposure? *(Answers will vary.)* Explain that rusting is a type of corrosion. Convey that many metals oxidize when exposed to the oxygen in air. **Ask** Can you identify the common chemical characteristic of all metal corrosion? *(the transfer of electrons from metals to oxidizing agents)* **Ask** What are some factors that increase the rate of corrosion? *(Sample answers: temperature, concentration of salts and acids, exposure.)* **Ask** What are some ways in which the rate of corrosion can be decreased or eliminated? *(Sample answers: Protect the objects from the weather; coat the metal parts; replace metal parts with plastic.)*

Explore

Class Activity

PURPOSE Students will order metals based on their ease of corrosion.

PROCEDURE Have students try to prepare a list of metals ordered by their ease of corrosion. Have students base their lists on known behaviors of specific metals. As a clue to metals' behavior, suggest that students consider how different metals are used. After students compile their lists, refer them to Table 11.2, the activity series for metals, on page 361. Have students write a paragraph discussing how closely their lists correspond to the actual activity series. **Ask** Why might there be discrepancies between your lists and the activity series? *(Reasons might include unavailability of data for some metals or difficulty in rating metals whose behaviors and uses are similar.)*

CHEMISTRY & YOU Salt causes metal car parts to corrode more quickly.

Answers

FIGURE 20.4 H_2O

FIGURE 20.6 Unlike iron(III) oxide, the tough aluminum oxide coating on the surface of the metal actually protects the aluminum object from further corrosion.

Extend

Students may incorrectly assume that corrosion is always undesirable, because people spend a lot of time, money, and energy trying to prevent or inhibit the redox reaction that causes corrosion. Point out that corrosion is not always a bad thing. Explain that because the oxidation of metal is generally an exothermic reaction, it can be used to generate useful heat—as long as the oxidation takes place quickly enough to release the heat energy in a short span of time. Have students research ways in which metal oxidation can be useful to humans. Then, have them identify the reaction involved and write an equation for it. Students should also identify which reactant was oxidized and which one was reduced.

Evaluate

Informal Assessment

On the board, write several equations for redox reactions. Ask students to identify the oxidizing and reducing agents. Be sure to assess the logic students use to arrive at their answers. Then have students complete the 20.1 Lesson Check.

Reteach

Use the illustrations in this lesson as a basis for review. Starting with Figure 20.1 and continuing to Figure 20.7, question students about what is being illustrated and what is significant about it. Have students use information supplied by the captions of Figures 20.5, 20.6, and 20.7 to write redox equations for the reactions taking place in the images.

Controlling Corrosion The corrosion of objects such as shovels or knives is a common problem but not usually a serious one. In contrast, the corrosion of a steel support pillar of a bridge or the steel hull of an oil tanker is much more serious and costly. To prevent corrosion in such cases, the metal surface may be coated with oil, paint, plastic, or another metal, as shown in Figure 20.7. These coatings exclude air and water from the surface, thus preventing corrosion. If the coating is scratched or worn away, however, the exposed metal will begin to corrode.

In another method of corrosion control, one metal is "sacrificed," or allowed to corrode, to save a second metal. For example, to protect an iron object, a piece of magnesium (or another active metal) may be placed in electrical contact with the iron. When oxygen and water attack the iron object, the iron atoms lose electrons as the iron begins to be oxidized. However, because magnesium is a better reducing agent than iron, the magnesium immediately transfers electrons to the iron atoms, preventing their oxidation.

Sacrificial zinc and magnesium blocks are sometimes attached to piers and ship hulls to prevent corrosion damage in areas submerged in water. Figure 20.7b shows zinc blocks attached to the steel hull of a ship. The blocks corrode instead of the iron. Underground pipelines and storage tanks may be connected to magnesium blocks for protection. It is easier and cheaper to replace a block of magnesium or zinc than to replace a bridge or a pipeline.

Figure 20.7 Corrosion Control
Painting a surface, like this bridge, protects it from the effects of the environment. **a.** Chromium metal also serves as a protective coating and imparts an attractive, mirrorlike finish. Like aluminum, chromium forms a corrosion-resistant oxide film on its surface. **b.** Zinc blocks are attached to the steel (iron) hull of this ship. The zinc blocks oxidize (corrode) instead of the iron, preventing the hull from corroding.

CONCEPTS IN ACTION *Learn more about corrosion online.*

Focus on ELL

4 LANGUAGE PRODUCTION Have students work in groups of four to complete the Quick Lab on the following page. Make sure each group has ELLs of varied language proficiencies so that more proficient students can help less proficient ones. Have students work according to their proficiency level.

BEGINNING: LOW/HIGH Provide the data grid for students and model how to record the observations. Allow students to use single words or symbols in their answers.

INTERMEDIATE: LOW/HIGH Paraphrase the procedure and Analyze and Conclude questions for students. Have students generate their own questions to help them compare and contrast the oxidizing agents.

ADVANCED: LOW Pair students with writing partners for peer feedback.

HIGH Ask students to complete the Analyze and Conclude questions with written responses.

Quick Lab

Purpose To test the effect of oxidizing agents on stains and dyes

Materials

- spot plate
- dropper
- water
- colorimeter (optional)

Oxidizing Agents

- liquid chlorine bleach (5% (m/v) sodium hypochlorite)
- powder bleach
- oxalic acid solution (1% (m/v))
- sodium thiosulfate solution (hypo) (0.2M $Na_2S_2O_3$)
- hydrogen peroxide (3% (v/v) H_2O_2)

Samples

- iodine solution (1% I_2 in 2% (m/v) KI)
- potassium permanganate solution (0.05M $KMnO_4$)
- grape juice
- rusty water
- piece of colored fabric
- colored flower petals
- grass stain on piece of white fabric

Bleach It! Oxidize the Color Away

Procedure

1. Place samples on a spot plate. Use 4 drops of each liquid or a small piece of each solid.
2. Describe the color and appearance of each sample in Step 1.
3. Add a few drops of the first oxidizing agent to each sample.
4. Describe any immediate change in appearance and any further change after 15 minutes.
5. Repeat Steps 1–4 with each oxidizing agent, using a new sample for each oxidizing agent.

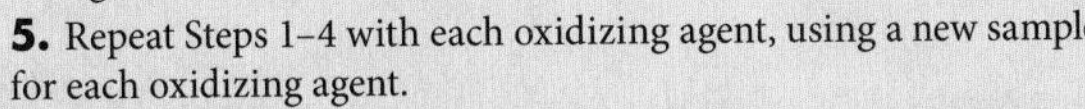

Analyze and Conclude

1. **Organize Data** Make a grid and record your observations.
2. **Compare and Contrast** Compare the oxidizing power of the oxidizing agents.
3. **Explain** How do you know that chemical changes have occurred?

20.1 LessonCheck

ONLINE PROBLEMS

3. **Define** Define oxidation and reduction in terms of the gain or loss of oxygen.
4. **Define** Define oxidation and reduction in terms of the gain or loss of electrons.
5. **Explain** How does the presence of salts and acids accelerate the corrosion of metals?
6. **Explain** How do you identify the oxidizing agent and the reducing agent in a redox reaction?
7. **Identify** Which of the following would most likely be oxidizing agents and which would most likely be reducing agents? (*Hint:* Think in terms of tendencies to lose or gain electrons.)
   a. Cl_2 **b.** K **c.** Ag^+
8. **Apply Concepts** Use electron transfer or electron shift to identify what is oxidized and what is reduced in each reaction. Use the electronegativity values in Table 6.2 in Chapter 6, for covalent compounds.
   a. $2Na(s) + Br_2(l) \longrightarrow 2NaBr(s)$
   b. $H_2(g) + Cl_2(g) \longrightarrow 2HCl(g)$
   c. $2Li(s) + F_2(g) \longrightarrow 2LiF(s)$
   d. $S(s) + Cl_2(g) \longrightarrow SCl_2(g)$
   e. $N_2(g) + 2O_2(g) \longrightarrow 2NO_2(g)$
   f. $Mg(s) + Cu(NO_3)_2(aq) \longrightarrow Mg(NO_3)_2(aq) + Cu(s)$
9. **Identify** For each reaction in Problem 8, identify the reducing agent and the oxidizing agent.

Lesson Check Answers

3. Oxidation is the gain of oxygen. Reduction is the loss of oxygen. SC.912.P.8.10
4. Oxidation is the loss of electrons. Reduction is the gain of electrons.
5. Salts and acids produce conductive solutions that make electron transfer easier.
6. The species reduced is the oxidizing agent. The species oxidized is the reducing agent.
7. **a.** oxidizing agent **b.** reducing agent **c.** both
8. **a.** Na oxidized; Br_2 reduced **b.** H_2 oxidized; Cl_2 reduced **c.** Li oxidized; F_2 reduced **d.** S oxidized; Cl_2 reduced **e.** N_2 oxidized; O_2 reduced **f.** Mg oxidized; Cu reduced
9. **a.** reducing agent, Na; oxidizing agent, Br_2
   b. reducing agent, H_2; oxidizing agent, Cl_2
   c. reducing agent, Li; oxidizing agent, F_2
   d. reducing agent, S; oxidizing agent, Cl_2
   e. reducing agent, N_2; oxidizing agent, O_2
   f. reducing agent, Mg; oxidizing agent, Cu^{2+}

Explore

Quick Lab

OBJECTIVE After completing this activity, students will be able to test the effect of oxidizing agents on stains and dyes.

SKILLS FOCUS observing, comparing, drawing conclusions

PREP TIME 20 minutes

MATERIALS spot plates, medicine droppers, liquid and powder bleach, 3% H_2O_2, 1% oxalic acid, 0.2M $Na_2S_2O_3$, I_2 in KI(aq), $KMnO_4(aq)$, grape juice, rusty water, colored fabric, grass-stained fabric, flower petals

ADVANCE PREPARATION Prepare the oxidizing agents and other solutions before class. Arrange to bring items such as grape juice and flower petals to class on the day of the lab.

SAFETY The oxidizing agents in this lab pose a slight contact hazard and should be handled carefully. Safety goggles should be worn at all times. Students should also be careful to avoid splashing and spilling because the oxidizing solutions can bleach or destroy clothing; lab coats are recommended.

EXPECTED OUTCOME Various color changes occur.

ANALYZE AND CONCLUDE

1. Grids contain student observations.
2. Rankings depend on observations.
3. Color changes occur.

FOR ENRICHMENT Have students extend this lab by bringing in stained or old clothing from home. Have them test the effect of the oxidizing agents on these materials.

CHEMISTRY & YOU Fireworks are probably one of the most spectacular demonstrations of oxidation-reduction reactions that students will encounter in their daily lives. Have students look at the different colors produced by the fireworks display in the photo. Point out that the differences in the display show that different elements were involved in the chemical reactions that produced the explosions. Encourage students to describe fireworks displays they have seen. Emphasize that each of these displays depended on redox reactions.

Explain

APPLY CONCEPTS Explain that a fireworks display might include the reaction of potassium nitrate (KNO_3) and sulfur (S) to produce potassium oxide (K_2O), nitrogen gas (N_2), and sulfur dioxide (SO_2). Have students write an equation for this reaction, and then use oxidation numbers to balance it. *($4KNO_3 + 5S \rightarrow 2K_2O + 2N_2 + 5SO_2$)* **Ask** What is the oxidizing agent in this reaction? *(potassium nitrate)* **Ask** What is the reducing agent? *(sulfur)*

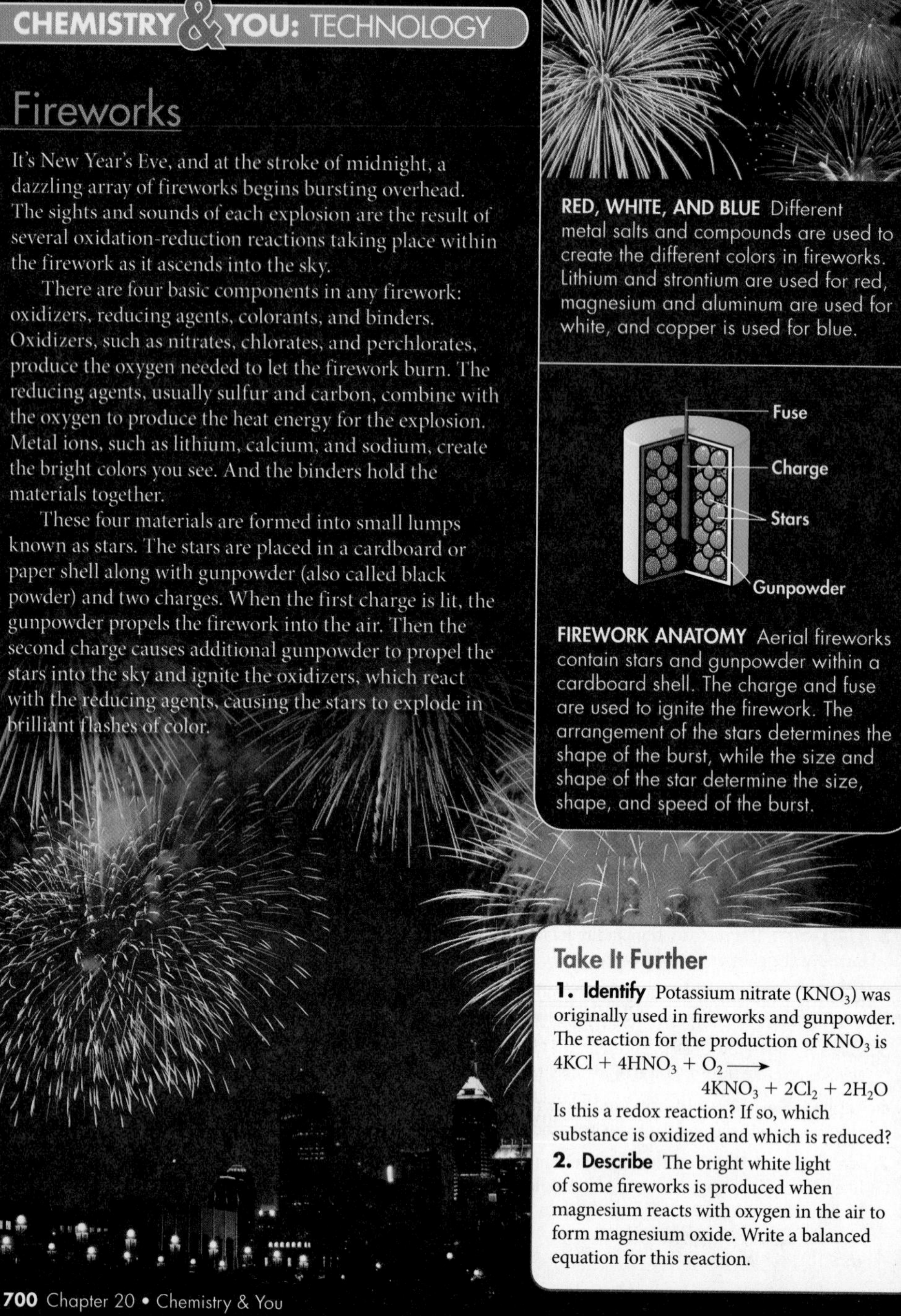

CHEMISTRY & YOU: TECHNOLOGY

Fireworks

It's New Year's Eve, and at the stroke of midnight, a dazzling array of fireworks begins bursting overhead. The sights and sounds of each explosion are the result of several oxidation-reduction reactions taking place within the firework as it ascends into the sky.

There are four basic components in any firework: oxidizers, reducing agents, colorants, and binders. Oxidizers, such as nitrates, chlorates, and perchlorates, produce the oxygen needed to let the firework burn. The reducing agents, usually sulfur and carbon, combine with the oxygen to produce the heat energy for the explosion. Metal ions, such as lithium, calcium, and sodium, create the bright colors you see. And the binders hold the materials together.

These four materials are formed into small lumps known as stars. The stars are placed in a cardboard or paper shell along with gunpowder (also called black powder) and two charges. When the first charge is lit, the gunpowder propels the firework into the air. Then the second charge causes additional gunpowder to propel the stars into the sky and ignite the oxidizers, which react with the reducing agents, causing the stars to explode in brilliant flashes of color.

RED, WHITE, AND BLUE Different metal salts and compounds are used to create the different colors in fireworks. Lithium and strontium are used for red, magnesium and aluminum are used for white, and copper is used for blue.

FIREWORK ANATOMY Aerial fireworks contain stars and gunpowder within a cardboard shell. The charge and fuse are used to ignite the firework. The arrangement of the stars determines the shape of the burst, while the size and shape of the star determine the size, shape, and speed of the burst.

Take It Further

1. Identify Potassium nitrate (KNO_3) was originally used in fireworks and gunpowder. The reaction for the production of KNO_3 is
$$4KCl + 4HNO_3 + O_2 \longrightarrow 4KNO_3 + 2Cl_2 + 2H_2O$$
Is this a redox reaction? If so, which substance is oxidized and which is reduced?

2. Describe The bright white light of some fireworks is produced when magnesium reacts with oxygen in the air to form magnesium oxide. Write a balanced equation for this reaction.

700 Chapter 20 • Chemistry & You

21st Century Learning To be successful in the 21st century, students need skills and learning experiences that extend beyond subject matter mastery. The following project helps students build the following 21st Century Skills: *Creativity and Innovation; Critial Thinking and Problem Solving; Communication and Collaboration; Information Literacy; Media Literacy; Information, Communications and Technology Literacy; Initiative and Self-Direction; Productivity and Accountability.*

FIREWORKS CHEMISTRY Pose the following challenge to your students. *Suppose you work for a fireworks company, and it is your job to choose the oxidizing and reducing chemicals to use for one of their fireworks. What will you use?*

- Research various chemicals used to produce colors in fireworks.
- Choose one color, and produce a creative computer display that includes the redox reaction as a balanced equation.
- Identify the color of the fireworks produced, the oxidizing agent, and the reducing agent.

Answers

1. yes; the chlorine ion in KCl is oxidized and O_2 is reduced
2. $2Mg + O_2 \rightarrow 2MgO$

20.2 Oxidation Numbers

C.4.6 Redox reactions

CHEMISTRY & YOU

Q: *Why does a sparkler have such a bright light?* If you have ever seen or held a sparkler, then you know that sparklers give off very bright light. They are like handheld fireworks. Sparklers contain powdered metals. As the sparklers burn, a chemical reaction changes the metals. In this lesson, you will learn about how oxidation and reduction in reactions like these are defined in terms of a change in oxidation number.

Key Questions

What is the general rule for assigning oxidation numbers?

How are oxidation and reduction defined in terms of a change in oxidation number?

Vocabulary

- oxidation number

Assigning Oxidation Numbers

What is the general rule for assigning oxidation numbers?

An **oxidation number** is a positive or negative number assigned to an atom to indicate its degree of oxidation or reduction. **As a general rule, a bonded atom's oxidation number is the charge that it would have if the electrons in the bond were assigned to the atom of the more electronegative element.** In the next lesson, you will learn that equations of complex redox reactions can be balanced by the use of oxidation-number changes. The set of rules on the next page should help you determine oxidation numbers.

In binary ionic compounds, such as NaCl and $CaCl_2$, the oxidation numbers of the atoms equal their ionic charges (Rule 1). The compound sodium chloride is composed of sodium ions (Na^{1+}) and chloride ions (Cl^{1-}). Thus the oxidation number of sodium is +1, and that of chlorine is −1. In $CaCl_2$, the oxidation number of calcium is +2, and that of chlorine is −1. Notice that the sign is put before the oxidation number.

Because water is a molecular compound, no ionic charges are associated with its atoms. However, oxygen is reduced in the formation of water. Oxygen is more electronegative than hydrogen. So, in water, the two shared electrons in the H—O bond are shifted toward oxygen and away from hydrogen. Picture the electrons contributed by the two hydrogen atoms as being completely transferred to the oxygen. The charges that would result from this transfer are the oxidation numbers of the bonded elements. The oxidation number of oxygen is −2, and the oxidation number of each hydrogen is +1 (Rules 2 and 3). Oxidation numbers are often written above the chemical symbols in a formula. For example, water can be represented as

$$\overset{+1}{H_2}\overset{-2}{O}$$

Focus on ELL

1 CONTENT AND LANGUAGE Direct students' attention to the chart called Rules for Assigning Oxidation Numbers, on the following page. Guide students in creating a visual aid using colors, symbols, and examples of the rules. Allow them to refer to this visual throughout the chapter.

2 FRONTLOAD THE LESSON Review the concepts of oxidation and reduction by asking students to explain how each process affects an atom's oxidation number. Then, write a sample reaction on the board. Ask students to write an oxidation reaction and a reduction reaction to help them identify the oxidizing and reducing agents in the reaction.

3 COMPREHENSIBLE INPUT Prepare a two-column chart of the rules for assigning oxidation numbers. Write a simplified form of each rule in the first column, and place examples in the second column. When discussing the sample problems, identify any polyatomic ions, and make sure that students understand the charge distribution within them before proceeding with the solutions.

Key Objectives

20.2.1 STATE the general rule for assigning oxidation numbers.

20.2.2 DEFINE oxidation and reduction in terms of a change in oxidation number.

Additional Resources

Reading and Study Workbook, Lesson 20.2

Available Online or on Digital Media:

- Teaching Resources, Lesson 20.2

PearsonChem.com

Engage

CHEMISTRY & YOU Have students read the opening paragraph. Explain that sparklers burn when lit with a match because the reaction of metal with oxygen is a combustion reaction. **Ask** Does a metal gain or lose electrons when it reacts with oxygen? *(It loses electrons.)* Remind students that oxidation is the complete or partial loss of electrons or gain of oxygen.

Activate Prior Knowledge

To prepare students for learning the rules for assigning oxidation numbers, have them define and provide two examples of the following: monoatomic ion, metal hydride, peroxide, element, neutral compound, polyatomic ion.

Foundations for Reading

BUILD VOCABULARY Have students relate the vocabulary term, *oxidation number*, with other, similar terms used in the chapter, such as *oxidize* and *oxide*.

IN ACADEMIC STANDARDS for SCIENCE

C.4.6 Determine oxidation states and identify the substances apparently gaining and losing electrons in redox reactions.

READING STRATEGY Have students preview the lesson's content by making index cards, one for each of the six rules for assigning oxidation numbers. As they read the lesson, students should try to find an example for each rule and write it on the back of the corresponding card.

Explain

Assigning Oxidation Numbers

START A CONVERSATION Explain that determining oxidation numbers of elements in compounds is a way for chemists to keep track of electron transfer during redox reactions. Point out that oxidation numbers are an accounting system for electrons. Then review the chart called Rules for Assigning Oxidation Numbers with the class, going over many examples of each rule until students are comfortable with these numbers.

Misconception Alert

Students may tend to assign an element the same oxidation number in every compound. Explain that the same element can have several oxidation numbers, depending on the compound. **Ask** What are the oxidation numbers of iodine in HIO_4, HIO_3, HIO, I_2, and HI? *(+7, +5, +1, 0, −1, respectively)*

Extend

Connect to ENVIRONMENTAL SCIENCE

Have students examine why carbon monoxide is extremely poisonous to living systems whereas carbon dioxide is relatively harmless. Have students write a paragraph using what they have learned about oxidation numbers to support their findings. *(Carbon in carbon monoxide has an oxidation number that is lower than the oxidation number for carbon in carbon dioxide. Carbon monoxide is a better reducing agent than carbon dioxide.)*

Explore

Class Activity

PURPOSE Students will work cooperatively to assign oxidation numbers to various elements.

PROCEDURE Write formulas for ionic and molecular compounds on the board. Divide the class into groups of two and have students work together to assign oxidation numbers to all of the elements in each compound. Students can use the problem-solving strategy outlined in Sample Problem 20.2.

Many elements can have several different oxidation numbers. Use rules 5 and 6 to determine the oxidation number of atoms of these elements, plus other elements not covered in the first four rules. All the substances shown in Figure 20.8 contain chromium, but chromium has a different oxidation number in its uncombined state and in each compound.

Rules for Assigning Oxidation Numbers

1. The oxidation number of a monatomic ion is equal in magnitude and sign to its ionic charge. For example, the oxidation number of the bromide ion (Br^{1-}) is −1; that of the Fe^{3+} ion is +3.
2. The oxidation number of hydrogen in a compound is +1, except in metal hydrides, such as NaH, where it is −1.
3. The oxidation number of oxygen in a compound is −2, except in peroxides, such as H_2O_2, where it is −1, and in compounds with the more electronegative fluorine, where it is positive.
4. The oxidation number of an atom in uncombined (elemental) form is 0. For example, the oxidation number of the potassium atoms in potassium metal (K) or of the nitrogen atoms in nitrogen gas (N_2) is 0.
5. For any neutral compound, the sum of the oxidation numbers of the atoms in the compound must equal 0.
6. For a polyatomic ion, the sum of the oxidation numbers must equal the ionic charge of the ion.

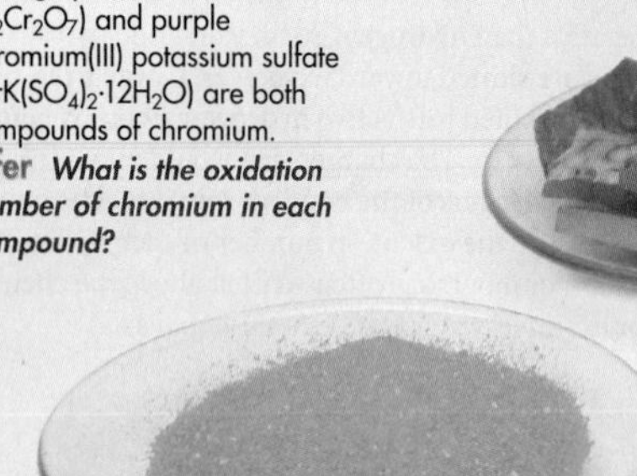

Figure 20.8
Chromium in its uncombined state is a dull silvery color. Orange potassium dichromate ($K_2Cr_2O_7$) and purple chromium(III) potassium sulfate ($CrK(SO_4)_2 \cdot 12H_2O$) are both compounds of chromium.
Infer *What is the oxidation number of chromium in each compound?*

Differentiated Instruction

L3 ADVANCED STUDENTS Tell students that a breath analyzer is used to estimate the level of alcohol in a person's bloodstream. The chemical potassium dichromate is used in this process. Ask students to research the color change and the change in oxidation number of chromium if any ethanol is present in a person's breath when breathing into the apparatus. *(The Cr^{6+} in the potassium dichromate (orange) is reduced to one of its lower states, such as Cr^{3+} (green). The color change can be related to the amount of alcohol in the bloodstream.)*

ELL ENGLISH LANGUAGE LEARNERS Students may have difficulty distinguishing the word *assign* from the word *sign* when spoken. Write the word *assign* on the board and pronounce it slowly, being sure to clearly enunciate the first syllable. To avoid confusion, be sure to use the term *mathematical sign* instead of *sign* when referring to the charge on an oxidation number.

SampleProblem 20.2

Assigning Oxidation Numbers to Atoms

What is the oxidation number of each kind of atom in the following ions and compounds?

a. SO_2 **c.** Na_2SO_4
b. CO_3^{2-} **d.** $(NH_4)_2S$

1 Analyze Identify the relevant concepts. Use the set of rules you just learned to assign and calculate oxidation numbers.

2 Solve Apply concepts to this situation.

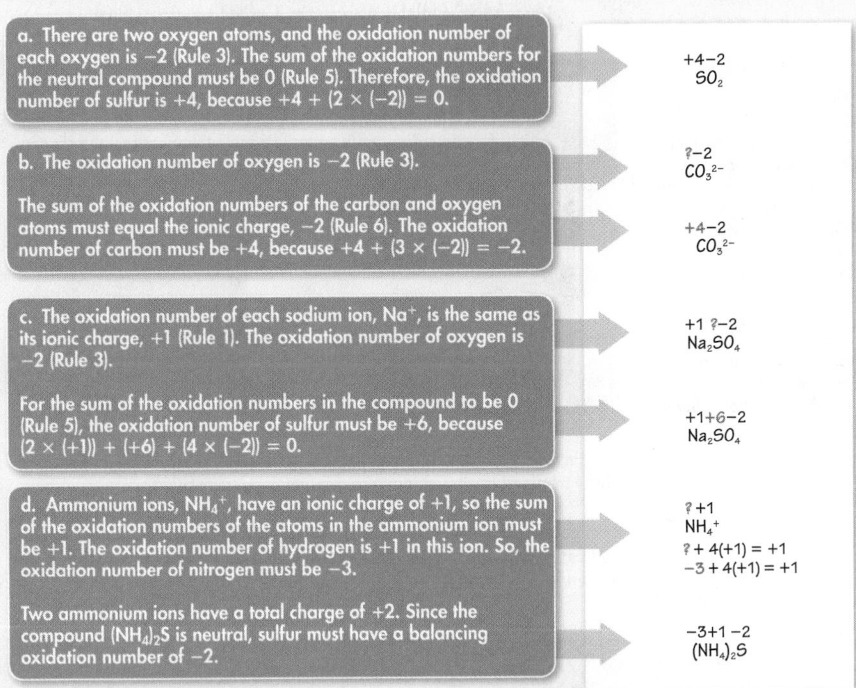

3 Evaluate Do the results make sense? The results are consistent with the rules for determining oxidation numbers. Also, addition of the oxidation numbers correctly gives the final overall charge for the ion and the three neutral compounds.

10. Determine the oxidation number of each element in the following:
a. S_2O_3 **c.** P_2O_5
b. Na_2O_2 **d.** NO_3^-

11. Determine the oxidation number of chlorine in each of the following substances:
a. $KClO_3$ **c.** $Ca(ClO_4)_2$
b. Cl_2 **d.** Cl_2O

Foundations for Math

WRITING EQUATIONS Show students that they can set up and solve an equation to assign an oxidation number to an ion or a compound.

In **Sample Problem 20.2c,** use an equation to find the oxidation number for Na_2SO_4. Let x equal the oxidation number for sulfur; set the sum of the charges equal to zero since the compound is neutral.

Follow the order of operations and use an inverse operation to isolate x.

$(+1)(2) + x + (-2)(4) = 0$ Sum of the charges = 0.
$2 + x - 8 = 0$ Multiply from left to right.
$x - 6 = 0$ Add.
$x = 6$ Add −6 to both sides.

The charge on sulfur is +6.

Sample Practice Problem

Determine the oxidation number for the carbon atom in each of the following compounds:

CH_4, CH_3OH, H_2CO

(CH_4 (−4); CH_3OH (−2); H_2CO (0))

Explore

Teacher Demo

PURPOSE Students will observe a redox reaction and write a balanced equation describing it.

MATERIALS 100 mL of 0.1*M* lead(II) acetate [$Pb(CH_3COO)_2$], 150-mL beaker, 1-cm × 3-cm zinc strip, sodium sulfide, 3*M* NaOH, 1*M* iron(III) chloride, stirring rod, plastic container

SAFETY Lead(II) acetate is a suspected carcinogen and mutagen. Perform the activity in a well-ventilated area. Wear safety goggles, an apron, and disposable gloves.

PROCEDURE Pour 100 mL of 0.1*M* lead(II) acetate [$Pb(CH_3COO)_2$] into a 150-mL beaker and add a 1-cm × 3-cm zinc strip.

EXPECTED OUTCOME A "tree" of lead will form over the course of a day. Ask students to write a balanced equation for the redox reaction that occurred. Have them assign oxidation numbers for Zn and Pb. Based on changes in oxidation number, have students identify the elements that were oxidized and reduced.

ANSWERS $Zn(s) + Pb(CH_3COO)_2(aq) \rightarrow Pb(s) + Zn(CH_3COO)_2(aq)$

Reactants: Zn^0, Pb^{2+}; Products: Zn^{2+}, Pb^0
Zn was oxidized. Pb was reduced.

DISPOSAL Dispose of the reaction mixture by adding a threefold molar excess of sodium sulfide to the contents of the beaker and stirring occasionally for one hour. Adjust the pH to neutral with 3*M* NaOH. Filter or decant the mixture. Allow the residue to dry. Add a threefold excess of 1*M* iron(III) chloride to the filtrate, stirring slowly. Allow the precipitate to settle, and then filter or decant. Put the dry precipitate and filtered residue in a plastic container and bury the container at an approved landfill site. Flush the neutral filtrate down the drain with excess water.

Answers

FIGURE 20.8 +6, +3

10. **a.** S, +3; O, −2
b. Na, +1; O, −1
c. P, +5; O, −2
d. N, +5; O, −2

11. **a.** +5 **b.** 0 **c.** +7 **d.** +1

LESSON 20.2

Explain

Oxidation-Number Changes in Chemical Reactions

CHEMISTRY & YOU The oxidation numbers of the metals change.

USE VISUALS Have students examine Figure 20.10. **Ask** What would happen if a copper nail were added to a solution of iron(II) sulfate instead? Explain. *(Nothing would happen because copper is below iron in the activity series of metals.)*

Q: *What happens to the oxidation numbers of metals as they burn in a sparkler?*

Oxidation-Number Changes in Chemical Reactions

How are oxidation and reduction defined in terms of a change in oxidation number?

Figure 20.9 shows what happens when copper wire is placed in a solution of silver nitrate. In this reaction, the oxidation number of silver decreases from +1 to 0 as each silver ion (Ag^{1+}) gains an electron and is reduced to silver metal (Ag^0). Copper's oxidation number increases from 0 to +2 as each atom of copper metal (Cu^0) loses two electrons and is oxidized to a copper(II) ion (Cu^{2+}). Here is the equation with oxidation numbers added:

$$\overset{+1\ +5-2}{2AgNO_3}(aq) + \overset{0}{Cu}(s) \longrightarrow \overset{+2\ +5-2}{Cu(NO_3)_2}(aq) + 2\overset{0}{Ag}(s)$$

Figure 20.9
Oxidation of Copper
Copper reacts with silver nitrate. **a.** A copper wire is placed in a silver nitrate solution. **b.** Crystals of silver coat the wire, and the solution slowly turns blue as a result of the formation of copper(II) nitrate.
Draw Conclusions ***What change occurs in the oxidation number of silver? How does the oxidation number of copper change?***

Figure 20.10 illustrates a redox reaction that shows what occurs when a shiny iron nail is dipped into a solution of copper(II) sulfate.

You can define oxidation and reduction in terms of a change in oxidation number. **An increase in the oxidation number of an atom or ion indicates oxidation. A decrease in the oxidation number of an atom or ion indicates reduction.**

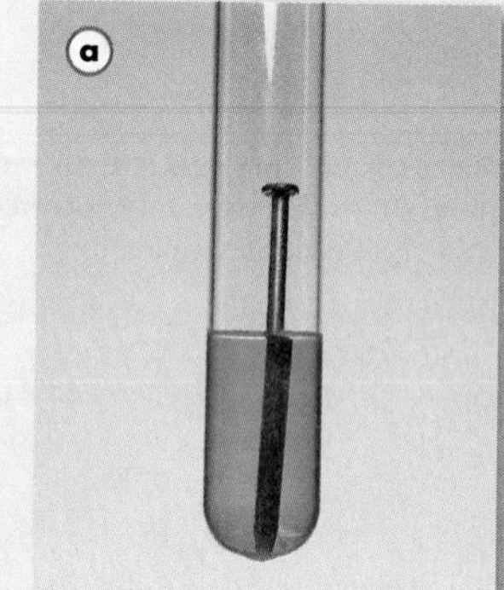

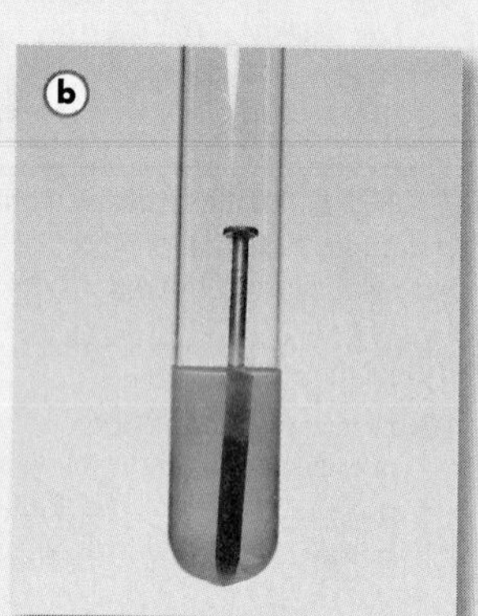

Figure 20.10
Reduction of Copper
A redox reaction occurs between iron and copper. **a.** An iron nail is placed in a copper(II) sulfate solution. **b.** The iron reduces Cu^{2+} ions in solution and is simultaneously oxidized to Fe^{2+}. The iron becomes coated with metallic copper.

UbD Check for Understanding

How are oxidation and reduction defined in terms of a change in oxidation number?

Assess students' understanding of this concept by having them write a balanced equation for the reaction shown in Figure 20.10 and assign an oxidation number to each atom or ion. Have them then show mathematically the changes in oxidation number that indicate reduction and oxidation. *[Fe(s) + $CuSO_4$(aq) → Cu(s) + $FeSO_4$(aq); Fe: 0, $CuSO_4$: +2,+6, −2, Cu: 0, $FeSO_4$: +2,+6, −2; Fe: 0 to +2 (oxidation), Cu: +2 to 0 (reduction)]*

Ask students to write a one-minute answer to this question, using the reaction between $AgNO_3$ and Cu (above) as an example.

ADJUST INSTRUCTION If students are having difficulty answering this question, have them reread the first paragraph on this page and the bold text following the key icon in the last paragraph. Then walk them through the process of answering the question.

Sample Problem 20.3

Identifying Oxidized and Reduced Atoms

Use changes in oxidation number to identify which atoms are oxidized and which are reduced in the following reactions. Also identify the oxidizing agent and the reducing agent.

a. $Cl_2(g) + 2HBr(aq) \longrightarrow 2HCl(aq) + Br_2(l)$
b. $C(s) + O_2(g) \longrightarrow CO_2(g)$

1 Analyze Identify the relevant concepts. An increase in oxidation number indicates oxidation. A decrease in oxidation number indicates reduction. The substance that is oxidized in a redox reaction is the reducing agent. The substance that is reduced is the oxidizing agent.

2 Solve Apply concepts to this situation.

The oxidation number of each chlorine in Cl_2 is 0 because of Rule 4.

a. Use the rules to assign oxidation numbers to each atom in the equation.

$$\overset{0}{Cl_2}(g) + 2\overset{+1-1}{HBr}(aq) \longrightarrow 2\overset{+1-1}{HCl}(aq) + \overset{0}{Br_2}(l)$$

Then, use the changes in oxidation numbers to identify which atoms are oxidized and which are reduced.

The element chlorine is reduced because its oxidation number decreases (0 to −1). The bromide ion from HBr(*aq*) is oxidized because its oxidation number increases (−1 to 0). Chlorine is reduced, so Cl_2 is the oxidizing agent.

Finally, identify the oxidizing and reducing agent.

The bromide ion from HBr(*aq*) is oxidized, so Br^- is the reducing agent.

b. Use the rules to assign oxidation numbers to each atom in the equation.

$$\overset{0}{C}(s) + \overset{0}{O_2}(g) \longrightarrow \overset{+4-2}{CO_2}(g)$$

Then, use the changes in oxidation numbers to identify which atoms are oxidized and which are reduced.

The element carbon is oxidized because its oxidation number increases (0 to +4). The element oxygen in reduced because its oxidation decreases (0 to −2).

Finally, identify the oxidizing agent and reducing agent.

Carbon is oxidized, so C is the reducing agent. Oxygen is reduced, so O_2 is the oxidizing agent.

3 Evaluate Do the results make sense? It makes sense that what is oxidized in a chemical reaction is the reducing agent because it loses electrons—it becomes the agent by which the atom that is reduced gains electrons. Conversely, it makes sense that what is reduced in a chemical reaction is the oxidizing agent because it gains electrons—it is the agent by which the atom that is oxidized loses electrons.

12. Use the changes in oxidation numbers to identify which atoms are oxidized and which are reduced in each reaction.
a. $2H_2(g) + O_2(g) \longrightarrow 2H_2O(l)$
b. $2KNO_3(s) \longrightarrow 2KNO_2(s) + O_2(g)$

13. Identify the oxidizing agent and the reducing agent in each equation in Problem 12.

Explain

Sample Practice Problem

For the following equation, identify which atoms are oxidized and which atoms are reduced. Then identify the oxidizing and reducing agents.

$$2HgO \rightarrow O_2 + 2Hg$$

(Mercury is reduced; it is the oxidizing agent. Oxygen is oxidized; it is the reducing agent.)

Explore

Class Activity

PURPOSE Students will identify and write equations for redox reactions.

MATERIALS 3 small glass test tubes, zinc, copper, magnesium, 0.1*M* HCl, medicine dropper or pipet

PROCEDURE Have students work in pairs. Give each pair a set of three small glass test tubes containing a small portion of zinc, copper, and magnesium. Have students add approximately 10 drops of 0.1*M* HCl to the metals and note signs of reaction. They should see bubbles form in the tubes containing zinc and magnesium but not in the tube containing copper. **Ask** Can you guess the identity of the gas in the tubes containing zinc and magnesium? *($H_2(g)$)* Have students write equations for the reactions of zinc and magnesium with HCl. Have students explain why these reactions are redox reactions.

Foundations for Math

COMPARING INTEGERS Explain that the charges of the atoms are integer values. To determine which atoms in a redox equation are oxidized and which are reduced, students need to simply compare integers.

In **Sample Problem 20.3a** have students use the symbols >, <, and = to compare the charges in the order in which they appear in the reaction.

For chlorine: $0 > -1$

The charge goes from a greater value to a lesser value, so chlorine is reduced.

For hydrogen: $+1 = +1$

The charges are equal, so hydrogen is neither oxidized nor reduced.

For bromine: $-1 < 0$

The charge goes from a lesser value to a greater value, so bromine is oxidized.

For struggling students, use an integer number line to plot and compare the values.

Answers

FIGURE 20.9 The oxidation number of silver changes from +1 to 0; the oxidation number of copper changes from 0 to +2.

12. a. H_2 oxidized, O_2 reduced
b. N reduced, O oxidized

13. a. H_2 reducing agent, O_2 oxidizing agent
b. N oxidizing agent, O reducing agent

LESSON 20.2

Explain

Sample Practice Problem

For the following equation, identify which atoms are oxidized and which atoms are reduced. Then identify the oxidizing and reducing agents.

$$2B_2O_3(s) + 6Cl_2(g) \rightarrow 4BCl_3(l) + 3O_2(g)$$

(Chlorine is reduced; it is the oxidizing agent. Oxygen is oxidized; it is the reducing agent.)

Evaluate

Informal Assessment

To assess students' understanding of oxidation numbers, ask students to determine the oxidation numbers for the following compounds: KNO_3 *(K +1, N +5, O −2)*; Na_2SO_3 *(Na +1, S +4, O −2)*; P_2O_5 *(P +5, O −2)*; MgH_2 *(Mg +2, H −1)*; ClF_3 *(Cl +3, F −1)*; HIO *(H +1, I +1, O −2)*. **Ask** What do the sign and magnitude of the oxidation number of an element in a compound indicate? *(The sign indicates loss (+) or gain (−) of electrons compared to the isolated element; the magnitude indicates the degree of the gain or loss.)* **Ask** If fluorine is converted to F^- from F_2, what is the change in oxidation number? Is this oxidation or reduction? *(Change is 0 to −1; reduction.)* Then have students complete the 20.2 Lesson Check.

Reteach

Stress that oxidation numbers refer to the combining capacity of single atoms. Thus, a change in oxidation number must be calculated on a per-atom basis. Review the rules for assigning oxidation numbers and use examples such as Mn_2O_3 and $Cr_2O_7^{2-}$ to explain to students that they must consider the number of atoms involved. Then have students repeat using two different formulas.

Answers

14. **a.** N in NH_4^+ is oxidized (−3 to 0); H is unchanged; N in NO_2^- is reduced (+3 to 0); O is unchanged.
b. Pb is reduced (+4 to +2); O is unchanged; H is unchanged; I is oxidized (−1 in HI to 0 in I_2).

15. **a.** N in NH_4^+ is the reducing agent; N in NO_2^- is the oxidizing agent.
b. Pb is the oxidizing agent; I^- is the reducing agent.

FIGURE 20.11 The oxidation number of potassium changes from 0 to +1, so this is a redox reaction. The oxidation number of zinc changes from 0 to +2, so this is a redox reaction.

SampleProblem 20.4

Identifying Oxidized and Reduced Atoms

Use changes in oxidation number to identify which atoms are oxidized and which are reduced in the following reactions. Also identify the oxidizing agent and the reducing agent.

$$Zn(s) + 2MnO_2(s) + 2NH_4Cl(aq) \longrightarrow ZnCl_2(aq) + Mn_2O_3(s) + 2NH_3(g) + H_2O(l)$$

1 Analyze Identify the relevant concepts.

2 Solve Apply concepts to this situation.

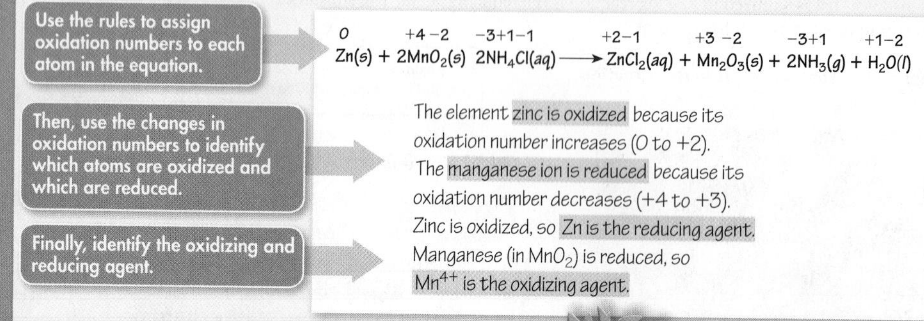

14. Identify which atoms are oxidized and which are reduced in each reaction.
a. $NH_4NO_2(s) \longrightarrow N_2(g) + 2H_2O(g)$
b. $PbO_2(aq) + 4HI(aq) \longrightarrow I_2(aq) + PbI_2(s) + 2H_2O(l)$

15. Identify the oxidizing agent and the reducing agent in each equation in Problem 14.

16. **Explain** What is the general rule for assigning oxidation numbers?

17. **Explain** How is a change in oxidation number related to the process of oxidation and reduction?

18. **Explain** How is charge used to assign oxidation numbers to the elements in a polyatomic ion?

19. **Identify** Use the changes in oxidation numbers to identify which atoms are oxidized and which are reduced in each reaction.
a. $2Na(s) + Cl_2(g) \longrightarrow 2NaCl(s)$
b. $2HNO_3(aq) + 6HI(aq) \longrightarrow 2NO(g) + 3I_2(s) + 4H_2O(l)$
c. $3H_2S(g) + 2HNO_3(aq) \longrightarrow 3S(s) + 2NO(g) + 4H_2O(l)$
d. $2PbSO_4(s) + 2H_2O(l) \longrightarrow Pb(s) + PbO_2(s) + 2H_2SO_4(aq)$

20. **Identify** Identify the oxidizing agent and the reducing agent in each reaction in Problem 19.

Lesson Check Answers

16. The oxidation number is the charge a bonded atom would have if the electrons in the bond were assigned to the more electronegative element.

17. An increase in oxidation number indicates oxidation; a decrease in oxidation number indicates reduction.

18. For a polyatomic ion, the sum of the oxidation numbers must equal the ionic charge of the ion.

19. **a.** Na, oxidized; Cl_2, reduced
b. I, oxidized; N, reduced
c. S, oxidized; N, reduced
d. Pb, oxidized and reduced

20. **a.** Na reducing agent, Cl_2 oxidizing agent
b. N in HNO_3 oxidizing agent, I in HI reducing agent
c. N in HNO_3, oxidizing agent; S in H_2S, reducing agent
d. Pb in $PbSO_4$, oxidizing-reducing agent

20.3 Describing Redox Equations

C.4.6 Redox reactions

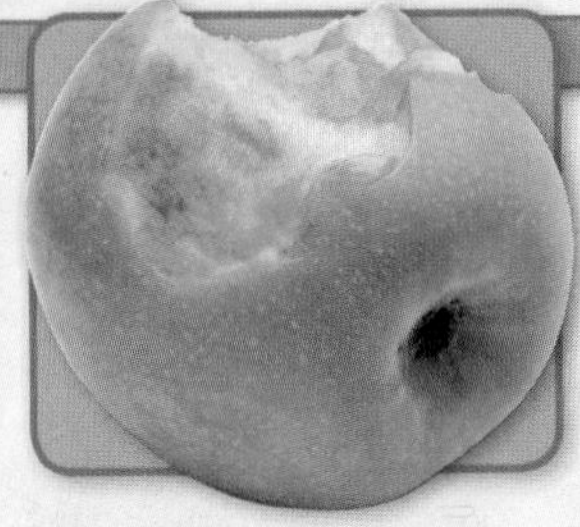

CHEMISTRY & YOU

Q: *Why does cut fruit turn brown?* If you have eaten an apple, then you have probably noticed that the flesh of the apple turns brown after you remove the skin. The apple is still safe to eat; it just doesn't look as appetizing. As you may have guessed, the browning is a chemical reaction. In this lesson, you will learn more about identifying certain types of chemical reactions and how to write and balance chemical equations for redox reactions.

Key Questions

What are the two classes of chemical reactions?

What are two different methods for balancing a redox equation?

Vocabulary

- oxidation-number-change method
- half-reaction
- half-reaction method

Identifying Redox Reactions

What are the two classes of chemical reactions?

In general, all chemical reactions can be assigned to one of two classes. **One class of chemical reactions is oxidation-reduction (redox) reactions, in which electrons are transferred from one reacting species to another. The other class includes all other reactions, in which no electron transfer occurs.** Many single-replacement reactions, combination reactions, decomposition reactions, and combustion reactions are redox reactions. Two examples of redox reactions are shown in Figure 20.11. Figure 20.11a shows what happens when potassium metal reacts with water. Figure 20.11b shows the reaction of zinc with hydrochloric acid. Examples of reactions that are not redox reactions include double-replacement reactions and acid-base reactions.

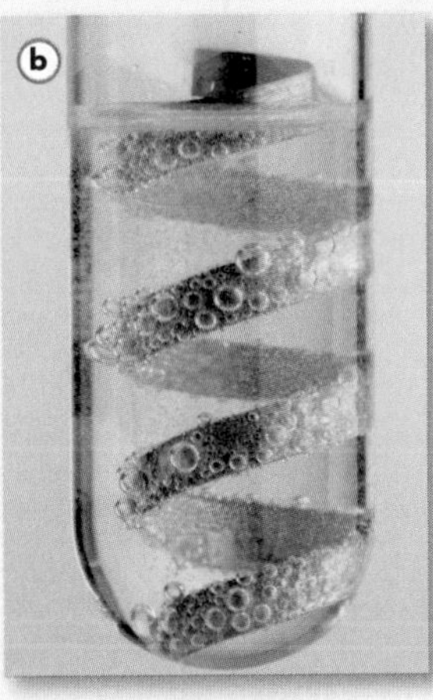

Figure 20.11 Single-Replacement Redox Reactions
Single-replacement reactions are redox reactions. **a.** Potassium metal reacts violently with water to produce hydrogen gas (which ignites) and potassium hydroxide. **b.** Zinc metal reacts vigorously with hydrochloric acid to produce hydrogen gas and zinc chloride.
Apply Concepts *Explain why each reaction is a redox reaction.*

LESSON 20.3

Key Objectives

20.3.1 IDENTIFY the two classes of chemical reactions.

20.3.2 BALANCE a redox equation by using the oxidation-number-change method.

20.3.3 BALANCE a redox equation using half-reactions.

Additional Resources

Reading and Study Workbook, Lesson 20.3

Available Online or on Digital Media:

- Small-Scale Laboratory Manual, Lab 35
- Virtual Chemistry Lab, Lab 30

PearsonChem.com

Engage

CHEMISTRY & YOU Direct students to look at the photograph of the apple and to read the corresponding text. Explain that the chemicals involved in the browning of fruit are called phenols. **Ask** Do you think the phenols in the apple are oxidized or reduced during the browning process? Why? *(The chemicals are oxidized because they are reacting with oxygen.)*

Activate Prior Knowledge

Ask students to recall the types of chemical reactions they studied in previous chapters. Write an example of each kind of equation on the board or overhead projector. Ask students to predict which equations represent redox reactions.

Focus on ELL

1 CONTENT AND LANGUAGE Direct students' attention to the Key Questions. Read each question aloud. Write the words *classes*, *balance*, and *procedure* on the board. Have students use a dictionary to look up the various uses of these words. Ask students to retell, in their own words, what the questions are asking before they begin reading.

2 FRONTLOAD THE LESSON Review the process of assigning oxidation numbers and have students practice determining oxidation numbers for various polyatomic ions. Then review the concept of ionic equations and have students practice writing complete and net ionic equations in preparation for balancing redox reactions by using the half-reaction method.

3 COMPREHENSIBLE INPUT Create large visuals to display the steps to follow for using oxidation-number changes and for using the half-reaction method to balance redox equations. Simplify the steps and use symbols below words when appropriate.

IN ACADEMIC STANDARDS for SCIENCE

C.4.6 Determine oxidation states and identify the substances apparently gaining and losing electrons in redox reactions.

Foundations for Reading

BUILD VOCABULARY Explain the meaning of the new terms by using examples. Then have students create a short summary in their own words to convey their understanding of each term.

READING STRATEGY Tell students to consider previous concepts, such as the law of conservation of mass, when reading about balancing redox equations.

Explain

Identifying Redox Reactions

APPLY CONCEPTS Remind students of the concepts that underlie the need to balance chemical equations. Explain that for all equations, mass must be conserved by balancing the number of atoms and electrical neutrality must be maintained by balancing charges. **Ask** How does the statement *Electrons cannot be created or destroyed* relate to the balancing of redox reactions? *(The balancing of redox reactions must also reflect the electrons gained and lost by oxidizing and reducing agents.)*

USE AN ANALOGY To convey the concepts involved with balancing chemical equations, use an analogy of a checking account. Explain that to maintain sufficient funds to cover checks written, money withdrawn needs to be balanced by deposits.

CHEMISTRY & YOU A redox reaction is occurring on the surface of the fruit.

Figure 20.12 Redox in Lightning When lightning bolts heat the air to extreme temperatures, the redox reaction that takes place between nitrogen and oxygen forms nitrogen monoxide.

During an electrical storm, as shown in Figure 20.12, oxygen molecules and nitrogen molecules in the air react to form nitrogen monoxide. This reaction is an example of a combination reaction. The equation for the reaction is shown below.

$$N_2(g) + O_2(g) \longrightarrow 2NO(g)$$

How can you tell if this is a redox reaction? If the oxidation number of an element in a reacting species changes, then that element has undergone either oxidation or reduction. Therefore, the reaction as a whole must be a redox reaction. In the example above, the oxidation number of nitrogen increases from 0 to +2, while the oxidation number of oxygen decreases from 0 to −2. Therefore, the reaction between nitrogen and oxygen to form nitrogen monoxide is a redox reaction.

CHEMISTRY & YOU

Q: *Some fruits, including apples, turn brown when you cut them. What do you think is happening on the surface of the fruit that causes it to turn brown?*

Many reactions in which color changes occur are redox reactions. One example is the apple on the first page of this lesson. Another example is shown in Figure 20.13. Written in ionic form, the unbalanced equation for this reaction is

$$MnO_4^-(aq) + Br^-(aq) \longrightarrow Mn^{2+}(aq) + Br_2(aq)$$

$MnO_4^-(aq)$	$Br^-(aq)$	$Mn^{2+}(aq)$	$Br_2(aq)$
Permanganate ion (purple)	Bromide ion (colorless)	Manganese(II) ion (colorless)	Bromine (brown)

Figure 20.13 Color Clues A color change can signal a redox reaction. When a colorless solution containing bromide ions (Br^-) is added to a solution containing permanganate ions (MnO_4^-), the distinctive purple color of the permanganate ion is replaced by the pale brown color of bromine.

Differentiated Instruction

LPR LESS PROFICIENT READERS Have students create a two-column table with labels for the two classes of chemical reactions as the column headings. As you discuss a particular reaction, have students note the reaction in the appropriate column.

L3 STRUGGLING STUDENTS Use a lab balance and colored blocks to visually demonstrate whether a reaction is a redox reaction. Use a specific color of block for each element in the reaction, determine the charge for each element or compound, and place the number of corresponding blocks on the reactant and the product sides of the balance. Have students observe the balance to determine whether or not it is an example of a redox reaction.

Sample Problem 20.5

Identifying Redox Reactions

Use the change in oxidation number to identify whether each reaction is a redox reaction or a reaction of some other type. If a reaction is a redox reaction, identify the element reduced, the element oxidized, the reducing agent, and the oxidizing agent.

a. $Cl_2(g) + 2NaBr(aq) \longrightarrow 2NaCl(aq) + Br_2(aq)$
b. $2NaOH(aq) + H_2SO_4(aq) \longrightarrow Na_2SO_4(aq) + 2H_2O(l)$

1 Analyze **Identify the relevant concepts.**
If changes in oxidation number occur, the reaction is a redox reaction. The element whose oxidation number increases is oxidized and is the reducing agent. The element whose oxidation number decreases is reduced and is the oxidizing agent.

2 Solve **Apply concepts to this situation.**

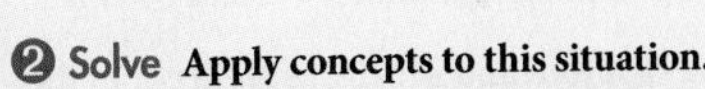

a. Assign oxidation numbers.

$\overset{0}{Cl_2}(g) + 2\overset{+1}{Na}\overset{-1}{Br}(aq) \longrightarrow 2\overset{+1}{Na}\overset{-1}{Cl}(aq) + \overset{0}{Br_2}(aq)$

Interpret the change (or lack of change) in oxidation numbers to identify if the reaction is a redox reaction.

This is a redox reaction. The chlorine is reduced. The bromide ion is oxidized. Chlorine is the oxidizing agent; the bromide ion is the reducing agent.

b. Assign oxidation numbers.

$2\overset{+1}{Na}\overset{-2}{O}\overset{+1}{H}(aq) + \overset{+1}{H_2}\overset{+6}{S}\overset{-2}{O_4}(aq) \longrightarrow \overset{+1}{Na_2}\overset{+6}{S}\overset{-2}{O_4}(aq) + 2\overset{+1}{H_2}\overset{-2}{O}(l)$

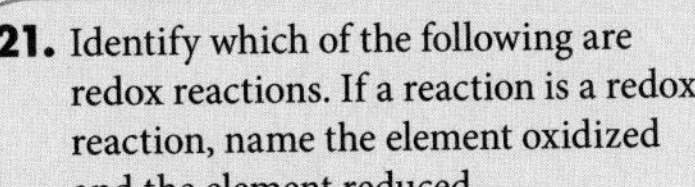

Interpret the change (or lack of change) in oxidation numbers to identify if the reaction is a redox reaction.

None of the elements change in oxidation number. This is not a redox reaction.

This is an acid-base (neutralization) reaction.

21. Identify which of the following are redox reactions. If a reaction is a redox reaction, name the element oxidized and the element reduced.
a. $Mg(s) + Br_2(l) \longrightarrow MgBr_2(s)$
b. $H_2CO_3(aq) \longrightarrow H_2O(l) + CO_2(g)$

22. Identify which of the following are oxidation-reduction reactions. If a reaction is a redox reaction, name the element oxidized and the element reduced.
a. $CaCO_3(s) + 2HCl(aq) \longrightarrow CaCl_2(aq) + H_2O(l) + CO_2(g)$
b. $CuO(s) + H_2(g) \longrightarrow Cu(s) + H_2O(l)$

Balancing Redox Equations

What are two different methods for balancing a redox equation?

Many oxidation-reduction reactions are too complex to be balanced by trial and error. Fortunately, two systematic methods are available. **Two different methods for balancing redox equations are the oxidation-number-change method and the half-reaction method.** These two methods are based on the fact that the total number of electrons gained in reduction must equal the total number of electrons lost in oxidation. One method uses oxidation-number changes, and the other uses half-reactions.

Sample Practice Problem

Have students determine if the following reaction is a redox reaction. Then have them determine which element is oxidized and which element is reduced.

$$2KMnO_4 + 3KCN + H_2O \rightarrow 2MnO_2 + 2KOH + 3K(OCN)$$

(It is a redox reaction; C is oxidized, and Mn is reduced.)

CRITICAL THINKING As a class, discuss examples in which "balancing" is required to achieve conservation. **Ask** How does the first law of thermodynamics, which states that energy must be conserved, relate to balancing? *(To conserve energy, any energy change in a system must be balanced by an equal and opposite energy change in the surroundings.)*

Foundations for Math

USE A TABLE Some students may benefit from using a table to organize the information about a reaction. Explain that a table is a good visual tool that can be helpful in keeping track of all of the elements and corresponding charges in a reaction.

In **Sample Problem 20.5a** show students how to create a table and organize the information about the reaction. Then point out how easy it is to see if the reaction is a redox reaction.

Element	Reactant Charge	Product Charge	Compare Charges	
Cl	0	−1	decreased	→ oxidizing agent
Na	1	1	neutral	→ neutral
Br	−1	0	increased	→ reducing agent

Answers

21. a. Redox reaction: Mg is oxidized, and Br_2 is reduced.
b. not a redox reaction

22. a. not a redox reaction
b. Redox reaction: H_2 is oxidized, and Cu is reduced.

Explore

Teacher Demo

PURPOSE Students will observe that redox reactions can sometimes be identified as a result of color changes.

MATERIALS 6 crystals of potassium permanganate ($KMnO_4$), 540 mL of water, large beaker, 1 g sodium hydrogen sulfite ($NaHSO_3$), two 50-mL beakers, 1 g barium chloride dihydrate ($BaCl_2{\cdot}2H_2O$), 3*M* H_2SO_4, stirring rod, plastic container

CAUTION Potassium permanganate is an irritant and a suspected mutagen. Barium chloride is toxic.

PROCEDURE Dissolve 6 crystals of potassium permanganate ($KMnO_4$) in 500 mL of water in a large beaker. Dissolve 1 g sodium hydrogen sulfite ($NaHSO_3$) in 20 mL of water in a 50-mL beaker. Add this solution to the $KMnO_4$ solution and stir. Dissolve 1 g barium chloride dihydrate ($BaCl_2 \cdot 2H_2O$) in 20 mL of water in another 50-mL beaker. Add this solution to the large beaker and stir.

EXPECTED OUTCOMES In the first reaction, the purple solution becomes colorless. When the $BaCl_2$ solution is added, the reaction mixture turns cloudy-white. The first reaction involves reduction of MnO_4^- to colorless Mn^{2+} and oxidation of HSO_3^- to HSO_4^-. The second reaction is not a redox reaction; it involves the formation of insoluble $BaSO_4$.

DISCUSS Ask students to use the oxidation-number-change method to balance this redox equation:

$$Cu(s) + HNO_3(aq) \rightarrow Cu(NO_3)_2(aq) + NO(g) + H_2O(l)$$

[3Cu(s) + 8HNO₃(aq) →
3Cu(NO₃)₂(aq) + 2NO(g) + 4H₂O(l)]

$$[3Cu(s) + 8HNO_3(aq) \rightarrow 3Cu(NO_3)_2(aq) + 2NO(g) + 4H_2O(l)]$$

DISPOSAL Dispose of the waste solution by adding 3*M* H_2SO_4 until precipitation of $BaSO_4$ appears to be complete. Let stand overnight. Filter or decant the liquid. Dry the residue, place it in a plastic container, and bury the container at an approved landfill site. Neutralize the filtrate, and flush it down the drain with excess water.

Figure 20.14
Reduction of Iron Ions
In a blast furnace like this one, air is blown through a combination of iron ore and coke. The carbon monoxide produced from the oxidation of coke reduces the Fe^{3+} ions to metallic iron.

Using Oxidation-Number Changes You can use oxidation numbers to keep track of electron transfers. In the **oxidation-number-change method,** you balance a redox equation by comparing the increases and decreases in oxidation numbers. To use this method, start with the skeleton equation for the redox reaction. As an example, look at the process used to obtain metallic iron from iron ore in a blast furnace, shown in Figure 20.14.

$$Fe_2O_3(s) + CO(g) \longrightarrow Fe(s) + CO_2(g) \text{ (unbalanced)}$$

Step 1: Assign oxidation numbers to all the atoms in the equation. Write the numbers above the atoms.

$$\overset{+3\,-2}{Fe_2O_3}(s) + \overset{+2\,-2}{CO}(g) \longrightarrow \overset{0}{Fe}(s) + \overset{+4\,-2}{CO_2}(g)$$

The oxidation number is stated per atom. So although the total positive charge of Fe ions in Fe_2O_3 is 6+, the oxidation number of each Fe ion is +3.

Step 2: Identify which atoms are oxidized and which are reduced. In this reaction, iron decreases in oxidation number from +3 to 0, a change of −3. Therefore, iron is reduced. Carbon increases in oxidation number from +2 to +4, a change of +2. Thus, carbon is oxidized.

Step 3: Use one bracketing line to connect the atoms that undergo oxidation and another such line to connect those that undergo reduction. Write the oxidation-number change at the midpoint of each line.

+2 (oxidation)

$$\overset{+3\,-2}{Fe_2O_3}(s) + \overset{+2\,-2}{CO}(g) \longrightarrow \overset{0}{Fe}(s) + \overset{+4\,-2}{CO_2}(g)$$

−3 (reduction)

Remember that a change in oxidation number represents the number of electrons transferred. Each carbon atom in CO loses 2 electrons in oxidation, and each iron atom in Fe_2O_3 accepts 3 electrons in reduction. As the equation is written, the number of electrons transferred in oxidation does not equal the number of electrons transferred in reduction. Step 4 will make the oxidation-number changes equal.

Step 4: Make the total increase in oxidation number equal to the total decrease in oxidation number by using appropriate coefficients. In this example, the oxidation-number increase should be multiplied by 3 and the oxidation-number decrease should be multiplied by 2, which gives an increase of +6 and a decrease of −6. This equalization can be achieved in the equation by placing the coefficient 2 in front of Fe on the right side and the coefficient 3 in front of both CO and CO_2. The formula Fe_2O_3 does not need a coefficient because the formula already indicates 2 Fe.

$3 \times (+2) = +6$

$$Fe_2O_3(s) + 3CO(g) \longrightarrow 2Fe(s) + 3CO_2(g)$$

$2 \times (-3) = -6$

Step 5: Finally, make sure the equation is balanced for both atoms and charge. If necessary, finish balancing the equation by inspection.

$$Fe_2O_3(s) + 3CO(g) \longrightarrow 2Fe(s) + 3CO_2(g)$$

710 Chapter 20 • Lesson 3

Redox Reactions and Rockets

Evidence exists that the use of rockets—and of redox reactions to power them—began in the thirteenth century when Chinese armies used "arrows of flying fire" to defend their cities. These rockets were leather cylinders filled with gunpowder and mounted on a stick. It is generally thought that the gunpowder mixture originated in ninth-century China.

The essential components of a rocket propellant are a source of energy (the fuel) and an oxidizing agent to react with the fuel. When the mixture is ignited, the fuel is oxidized rapidly and the oxidizing agent is reduced. Gunpowder is a mixture of charcoal (carbon), sulfur, and potassium nitrate. It reacts explosively to produce potassium sulfide, gaseous carbon dioxide, and nitrogen. In the redox reaction, carbon is the fuel and other components are oxidizing agents.

CHEM TUTOR

SampleProblem 20.6

Balancing Redox Equations by Oxidation-Number Change

Balance this redox equation by using the oxidation-number-change method.

$$K_2Cr_2O_7(aq) + H_2O(l) + S(s) \longrightarrow KOH(aq) + Cr_2O_3(s) + SO_2(g)$$

1 Analyze Identify the relevant concepts. You can balance redox equations by determining changes in oxidation numbers and applying the five steps.

2 Solve Apply the concepts to this problem.

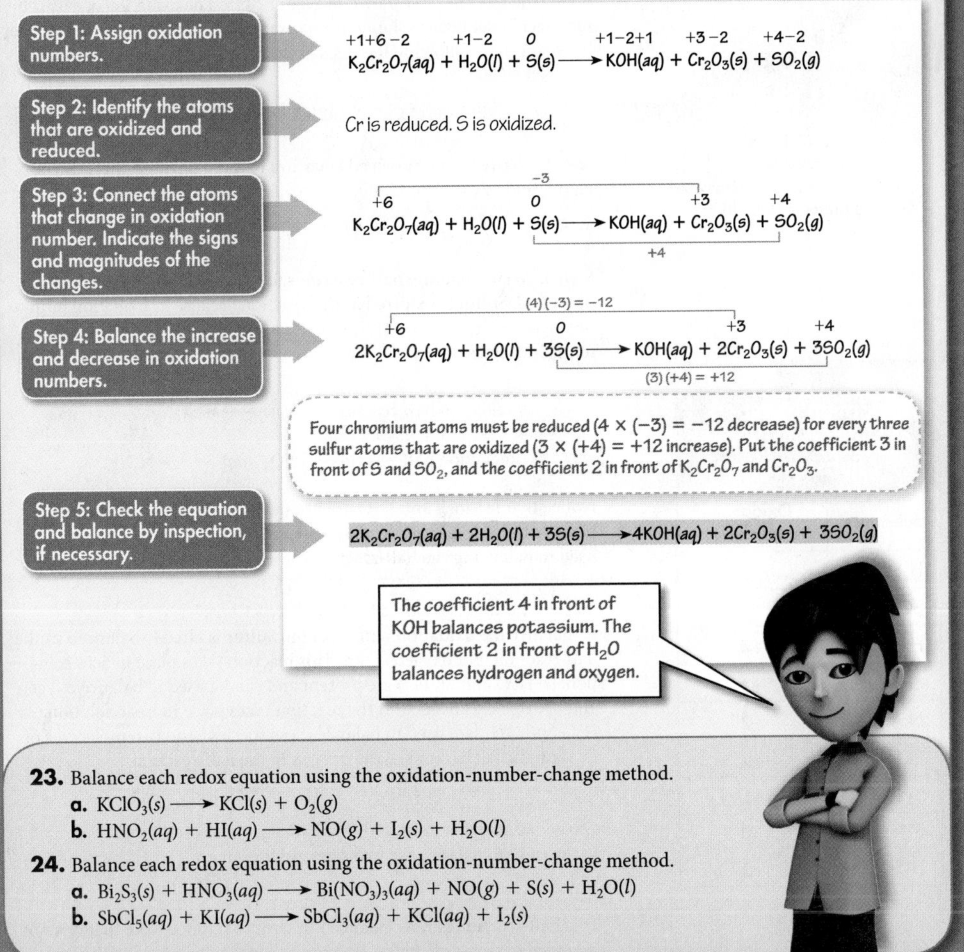

23. Balance each redox equation using the oxidation-number-change method.

a. $KClO_3(s) \longrightarrow KCl(s) + O_2(g)$

b. $HNO_2(aq) + HI(aq) \longrightarrow NO(g) + I_2(s) + H_2O(l)$

24. Balance each redox equation using the oxidation-number-change method.

a. $Bi_2S_3(s) + HNO_3(aq) \longrightarrow Bi(NO_3)_3(aq) + NO(g) + S(s) + H_2O(l)$

b. $SbCl_5(aq) + KI(aq) \longrightarrow SbCl_3(aq) + KCl(aq) + I_2(s)$

Explain

Sample Practice Problem

Write chemical equations for the following redox reactions. Balance each equation by using the oxidation-number-change method.

A. Solid barium chlorate decomposes when heated, yielding solid barium chloride and gaseous oxygen. *($Ba(ClO_3)_2(s) \rightarrow BaCl_2(s) + 3O_2(g)$)*

B. Solid lead(II) sulfide reacts with gaseous oxygen to produce solid lead(II) oxide and gaseous sulfur dioxide. *($2PbS(s) + 3O_2(g) \rightarrow 2PbO(s) + 2SO_2(g)$)*

Foundations for Math

LEAST COMMON MULTIPLE (LCM) Explain to students that they can use the LCM to balance redox equations. They should find the LCM of the absolute values of the changes in oxidation number of the reducing agent and the oxidizing agent. Remind students that the LCM is the smallest number that is a multiple of two or more numbers.

In **Sample Problem 20.6** the changes in oxidation number are –3 and +4. Show students how to find the LCM of 3 and 4.

Multiples of 3: 3, 6, 9, **12**, 15, 18

Multiples of 4: 4, 8, **12**, 16, 20

Point out that the smallest number that is a multiple of 3 and 4 is 12.

Answers

23. a. 2, 2, 3
b. 2, 2, 2, 1, 2

24. a. 1, 8, 2, 2, 3, 4
b. 1, 2, 1, 2, 1

LESSON 20.3

Explain

Balancing Redox Equations

USE VISUALS Have students study Table 20.2. **Ask** What are the oxidation numbers of all the elements in H_2SO_4? *(H is +1, S is +6, and O is –2.)* What are the oxidation numbers of all the elements in $Na_2S_2O_3$? *(Na is +1, S is +2, and O is –2.)*

APPLY CONCEPTS Tell students that the underlying assumption when using half-reactions is that any redox reaction can always be considered as two separate processes occurring simultaneously. Therefore, the "whole" reaction can be thought of as the sum of two halves—an oxidation reaction and a reduction reaction. Go through the half-reaction method with a specific example, clearly explaining each step. Emphasize the following points:

- Oxidation half-reactions are always written with the electrons on the right-hand side of the equation.
- Reduction half-reactions are always written with the electrons on the left-hand side of the equation.

CRITICAL THINKING Point out that in Step 3, H_2O and either H^+ or OH^- can be added to either side of an equation to balance O and H atoms. **Ask** Why is it acceptable that these ions can be added to either side of the equation? *(The reaction is taking place in an aqueous solution where H_2O, H^+, and OH^- are all abundant.)*

Figure 20.15
Sulfur Dioxide
Sulfur dioxide is often used to preserve dried fruits.

Using Half-Reactions The second method for balancing redox equations involves the use of half-reactions. A **half-reaction** is an equation showing just the oxidation or just the reduction that takes place in a redox reaction. In the **half-reaction method,** you write and balance the oxidation and reduction half-reactions separately before combining them into a balanced redox equation. The procedure is different, but the outcome is the same as with the oxidation-number-change method.

Sulfur is an element that can have several different oxidation numbers, as you can see in Table 20.2. Sulfur dioxide can be used to preserve dried fruit like the fruit shown in Figure 20.15. The oxidation of sulfur by nitric acid in aqueous solution is one example of a redox reaction that can be balanced by following the steps of the half-reaction method, which is outlined below.

$$S(s) + HNO_3(aq) \longrightarrow SO_2(g) + NO(g) + H_2O(l) \text{ (unbalanced)}$$

Step 1: Write the unbalanced equation in ionic form. In this case, only HNO_3 is ionized. The products are covalent compounds.

$$S(s) + H^+(aq) + NO_3^-(aq) \longrightarrow SO_2(g) + NO(g) + H_2O(l)$$

Step 2: Write separate half-reactions for the oxidation and reduction processes. Sulfur is oxidized in this reaction because its oxidation number increases from 0 to +4. Nitrogen is reduced because its oxidation number decreases from +5 to +2.

$$\text{Oxidation half-reaction: } \overset{0}{S}(s) \longrightarrow \overset{+4}{S}O_2(g)$$

$$\text{Reduction half-reaction: } \overset{+5}{N}O_3^-(aq) \longrightarrow \overset{+2}{N}O(g)$$

Notice that H^+ ions and H_2O are not included in the half-reactions because they are neither oxidized nor reduced. However, they will be used in balancing the half-reactions.

Step 3: Balance the atoms in the half-reactions.
a. Balance the oxidation half-reaction. Sulfur is already balanced in the half-reaction, but oxygen is not. This reaction takes place in acid solution, so H_2O and $H^+(aq)$ are present and can be used to balance oxygen and hydrogen as needed. If the reaction takes place in basic solution, H_2O and OH^- are used to balance these species. Add two molecules of H_2O on the left to balance the oxygen in the half-reaction.

$$2H_2O(l) + S(s) \longrightarrow SO_2(g)$$

Oxygen is now balanced, but four hydrogen ions ($4H^+$) must be added to the right to balance the hydrogen on the left.

$$2H_2O(l) + S(s) \longrightarrow SO_2(g) + 4H^+(aq)$$

This half-reaction is now balanced in terms of atoms. Note that it is not balanced in terms of charge. The charges will be balanced in Step 4.

Table 20.2

Oxidation Numbers of Sulfur in Different Substances

Substance	Oxidation number
H_2SO_4	+6
SO_3	+6
H_2SO_3	+4
SO_2	+4
$Na_2S_2O_3$	+2
SCl_2	+2
S_2Cl_2	+1
S	0
H_2S	–2

Differentiated Instruction

LPR **LESS PROFICIENT READERS** Balancing redox reactions by the half-reaction method is a lengthy process that requires good organizational skills and attention to detail. Students who are impatient or easily frustrated may have difficulty with these problems. Working in small groups can help them organize their efforts. Supervise the work closely to be sure each student is contributing.

L1 **STRUGGLING STUDENTS** Hang a large sheet of butcher paper on the wall to teach example problems involving the half-reaction method. Use the think-aloud method and ask questions such as "What is the charge of this element?" and "Should I add water and hydrogen ion?" Have a student volunteer use sticky notes to place the answers to these questions in the proper locations on the reactions.

b. Balance the reduction half-reaction. Nitrogen is already balanced. Add two molecules of H_2O on the right to balance the oxygen.

$$NO_3^-(aq) \longrightarrow NO(g) + 2H_2O(l)$$

Oxygen is balanced, but four hydrogen ions ($4H^+$) must be added to the left to balance hydrogen.

$$4H^+(aq) + NO_3^-(aq) \longrightarrow NO(g) + 2H_2O(l)$$

This half-reaction is now balanced in terms of atoms.

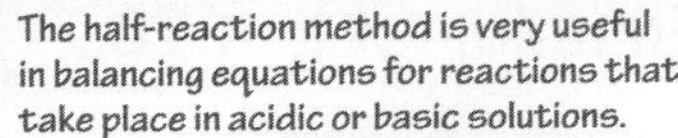

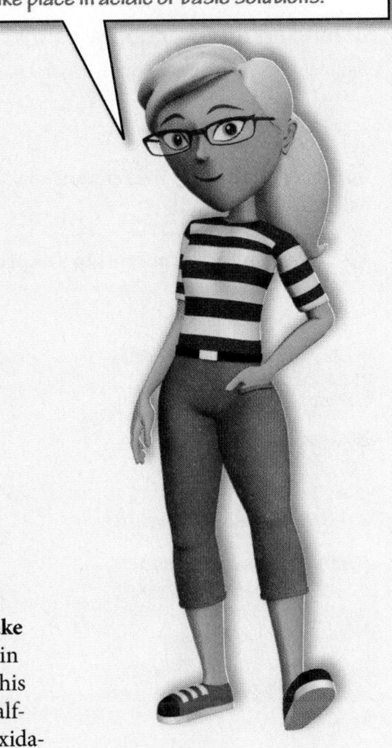

Step 4: Add enough electrons to one side of each half-reaction to balance the charges. Note that neither half-reaction is balanced for charge. Four electrons are needed on the right side in the oxidation half-reaction.

$$\text{Oxidation: } 2H_2O(l) + S(s) \longrightarrow SO_2(g) + 4H^+(aq) + 4e^-$$

Three electrons are needed on the left side in the reduction half-reaction.

$$\text{Reduction: } 4H^+(aq) + NO_3^-(aq) + 3e^- \longrightarrow NO(g) + 2H_2O(l)$$

Each half-reaction is now balanced with respect to both atoms and charge.

Step 5: Multiply each half-reaction by an appropriate number to make the numbers of electrons equal in both. The number of electrons lost in oxidation must equal the number of electrons gained in reduction. In this case, the oxidation half-reaction is multiplied by 3 and the reduction half-reaction is multiplied by 4. Therefore, the number of electrons lost in oxidation and the number of electrons gained in reduction both equal 12.

$$\text{Oxidation: } 6H_2O(l) + 3S(s) \longrightarrow 3SO_2(g) + 12H^+(aq) + 12e^-$$

$$\text{Reduction: } 16H^+(aq) + 4NO_3^-(aq) + 12e^- \longrightarrow 4NO(g) + 8H_2O(l)$$

Step 6: Add the balanced half-reactions to show an overall equation.

$$6H_2O(l) + 3S(s) + 16H^+(aq) + 4NO_3^-(aq) + 12e^- \longrightarrow 3SO_2(g) + 12H^+(aq) + 12e^- + 4NO(g) + 8H_2O(l)$$

Then, subtract terms that appear on both sides of the equation.

$$3S(s) + 4H^+(aq) + 4NO_3^-(aq) \longrightarrow 3SO_2(g) + 4NO(g) + 2H_2O(l)$$

Step 7: Add the spectator ions and balance the equation. Recall that spectator ions are present but do not participate in or change during a reaction. Because none of the ions in the reactants appear in the products, there are no spectator ions in this particular example. The balanced equation is correct. However, it can be written to show the HNO_3 as not ionized.

$$3S(s) + 4HNO_3(aq) \longrightarrow 3SO_2(g) + 4NO(g) + 2H_2O(l)$$

UbD Check for Understanding

BIGIDEA How can you balance redox equations?

Assess students' understanding of balancing redox equations by asking them to write a compare-and-contrast statement comparing the oxidation-number-change method to the half-reaction method. Answers may be in the form of a written summary or a two-column table.

ADJUST INSTRUCTION If students are having difficulty with this exercise, have them read the first paragraph of the Balancing Redox Equations section and review the listed steps for each method.

LESSON 20.3

Explore

Teacher Demo

PURPOSE Students will observe the oxidation process in everyday foods.

MATERIALS potatoes; apples; beakers; water; boiled water; lemon juice; carbonated beverage; sugar, salt and vinegar solutions; paper towels

PROCEDURE Students should be familiar with the "browning" of sliced potatoes or apples upon exposure to oxygen. The oxidation is catalyzed by enzymes in food. Explain that oxidation can be prevented or slowed down. The outer skin provides a natural physical barrier to oxygen. Chemical protection can be provided by antioxidants, which act as reducing agents, or by inhibiting enzymes. Liquids such as lemon juice contain antioxidants.

Have students form groups of two or three, and give each group a potato and an apple. Provide beakers filled with tap water, water that has been boiled to exclude most dissolved oxygen, lemon juice, a carbonated beverage, plus sugar, salt, and vinegar solutions. Have students quickly cut the potato and apple into slices and immediately place a slice in each beaker. Have them label areas on the paper towels to match the labels on the beakers. Have them remove the slices after they have soaked for several minutes and place them on the paper towels. Have them make periodic observations of the slices to determine the degree of "browning." On the basis of their observations, students should decide which liquids offer the best protection. Have them try to explain why the liquids are good or bad protectants. Ask students to present their findings in the form of a poster, a computerized slide show presentation, or a written or oral report.

Explain

Balancing Redox Equations

APPLY CONCEPTS When showing students various examples using the half-reaction method for balancing redox equations, students may question why hydrogen ions, hydroxide ions, or water molecules can be added to an equation during the process. Remind them that the aqueous solutions in which these reactions take place contain an abundance of these species. **Ask** Why is there an abundance of these species? *(Water exists in a state of equilibrium. Water molecules are constantly dissociating into hydrogen ions and hydroxide ions, making them available to take part in the half-reactions.)*

CRITICAL THINKING A reaction in which the same compound is both oxidized and reduced is called *disproportional*. When NO_2 is dissolved in water, it disproportionates to form nitric acid (HNO_3) and nitric oxide (NO). Have students balance the equation for this reaction.

($3NO_2(g) + H_2O(l) \rightarrow 2H^+(aq) + 2NO_3^-(aq) + NO(g)$)

Extend

Connect to BIOLOGY

Explain to the class that some organisms produce light by means of oxidation-reductions in a process known as bioluminescence. Have students examine this process to determine what is involved. *(Light is given off when one of a class of compounds called luciferins is oxidized by combining with oxygen. A product of these reactions is energy in the form of light.)* Ask students to determine the function bioluminescence serves for fireflies and squids. *(Fireflies use bioluminescence to attract mates; squids release a glowing cloud of "ink" to confuse and escape from their predators.)*

SampleProblem 20.7

Balancing Redox Equations by Half-Reactions

Balance this redox equation using the half-reaction method.

$$KMnO_4(aq) + HCl(aq) \longrightarrow MnCl_2(aq) + Cl_2(g) + H_2O(l) + KCl(aq)$$

1 Analyze **Identify the relevant concepts.** You can use the seven steps of the half-reaction method.

2 Solve **Apply the concepts to this problem.**

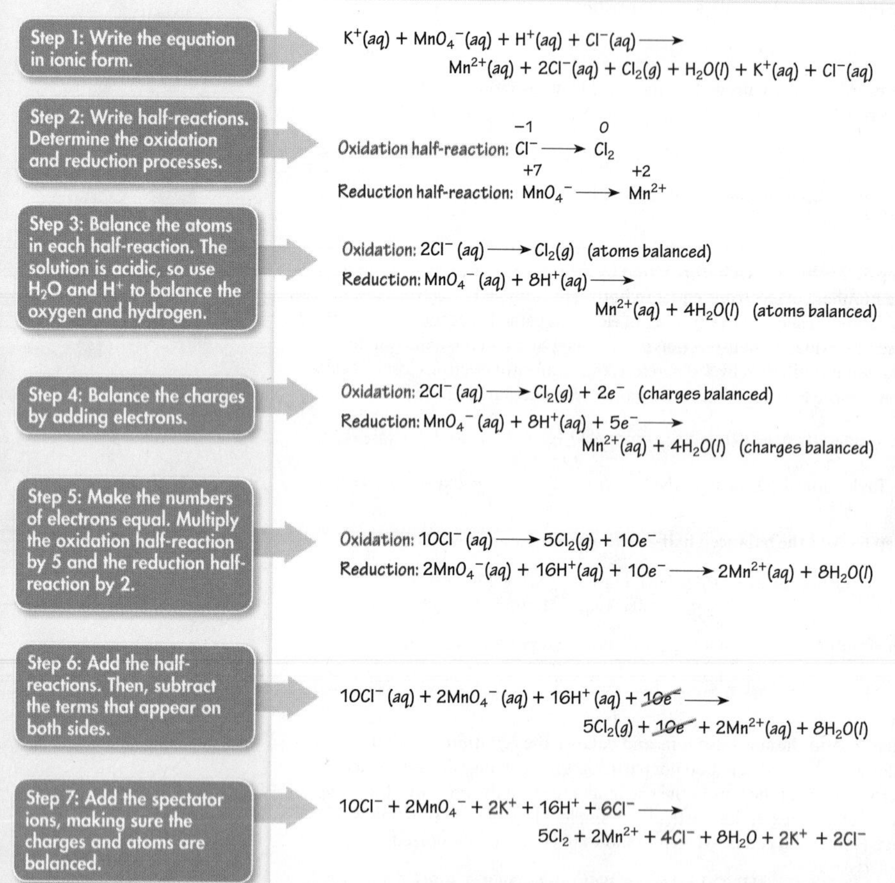

Step 1: Write the equation in ionic form.

$K^+(aq) + MnO_4^-(aq) + H^+(aq) + Cl^-(aq) \longrightarrow Mn^{2+}(aq) + 2Cl^-(aq) + Cl_2(g) + H_2O(l) + K^+(aq) + Cl^-(aq)$

Step 2: Write half-reactions. Determine the oxidation and reduction processes.

Oxidation half-reaction: $\overset{-1}{Cl^-} \longrightarrow \overset{0}{Cl_2}$

Reduction half-reaction: $\overset{+7}{MnO_4^-} \longrightarrow \overset{+2}{Mn^{2+}}$

Step 3: Balance the atoms in each half-reaction. The solution is acidic, so use H_2O and H^+ to balance the oxygen and hydrogen.

Oxidation: $2Cl^-(aq) \longrightarrow Cl_2(g)$ (atoms balanced)

Reduction: $MnO_4^-(aq) + 8H^+(aq) \longrightarrow Mn^{2+}(aq) + 4H_2O(l)$ (atoms balanced)

Step 4: Balance the charges by adding electrons.

Oxidation: $2Cl^-(aq) \longrightarrow Cl_2(g) + 2e^-$ (charges balanced)

Reduction: $MnO_4^-(aq) + 8H^+(aq) + 5e^- \longrightarrow Mn^{2+}(aq) + 4H_2O(l)$ (charges balanced)

Step 5: Make the numbers of electrons equal. Multiply the oxidation half-reaction by 5 and the reduction half-reaction by 2.

Oxidation: $10Cl^-(aq) \longrightarrow 5Cl_2(g) + 10e^-$

Reduction: $2MnO_4^-(aq) + 16H^+(aq) + 10e^- \longrightarrow 2Mn^{2+}(aq) + 8H_2O(l)$

Step 6: Add the half-reactions. Then, subtract the terms that appear on both sides.

$10Cl^-(aq) + 2MnO_4^-(aq) + 16H^+(aq) + \cancel{10e^-} \longrightarrow 5Cl_2(g) + \cancel{10e^-} + 2Mn^{2+}(aq) + 8H_2O(l)$

Step 7: Add the spectator ions, making sure the charges and atoms are balanced.

$10Cl^- + 2MnO_4^- + 2K^+ + 16H^+ + 6Cl^- \longrightarrow 5Cl_2 + 2Mn^{2+} + 4Cl^- + 8H_2O + 2K^+ + 2Cl^-$

714 Chapter 20 • Lesson 3

Foundations for Math

LIKE TERMS Remind students that in algebra they learned that they can add or subtract terms in an expression or equation only if the terms are like terms. Explain that like terms have the same variable raised to the same power, but the coefficients may differ.

- $3x$ and $5x$ are like terms.
- $3x$ and $4x^2$ are not like terms.

In **Sample Problem 20.7**, point out that the step following Step 7 is similar to combining like terms. Point out that among the reactants, $10Cl^-$ and $6Cl^-$ are like terms and can be combined to get a sum of $16Cl^-$. Point out that among the products, only $4Cl^-$ and $2Cl^-$ can be combined to get a sum of $6Cl^-$; $5Cl_2$ is not a like term.

Combine the spectator and nonspectator Cl^- on each side.

$16Cl^-(aq) + 2MnO_4^-(aq) + 2K^+(aq) + 16H^+(aq) \longrightarrow$
$5Cl_2(g) + 2Mn^{2+}(aq) + 6Cl^-(aq) + 8H_2O(l) + 2K^+(aq)$

Show the balanced equation for the substances given in the question (rather than for ions).

$2KMnO_4(aq) + 16HCl(aq) \longrightarrow 2MnCl_2(aq) + 5Cl_2(g) + 8H_2O(l) + 2KCl(aq)$

25. The following reaction takes place in an acidic solution. Balance the equation using the half-reaction method.
$Sn^{2+}(aq) + Cr_2O_7^{2-}(aq) \longrightarrow Sn^{4+}(aq) + Cr^{3+}(aq)$

26. The following reaction takes place in basic solution. Balance the equation using the half-reaction method.
$Zn(s) + NO_3^-(aq) \longrightarrow NH_3(aq) + Zn(OH)_4^{2-}(aq)$

For a solution that is basic, use H_2O and OH^- in Step 3 to balance the atoms.

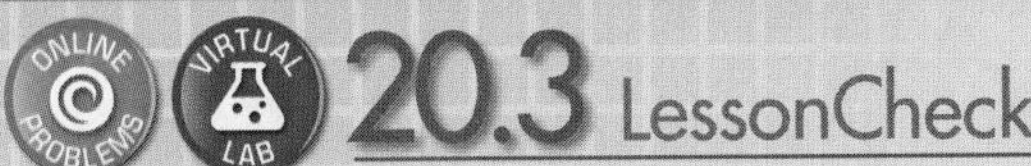

20.3 LessonCheck

27. **Identify** What are the two classes of chemical reactions?

28. **Compare** What are two different methods for balancing a redox equation?

29. **Calculate** Balance each redox equation, using the oxidation-number-change method.

a. $ClO_3^-(aq) + I^-(aq) \longrightarrow Cl^-(aq) + I_2(aq)$ [acidic solution]

b. $C_2O_4^{2-}(aq) + MnO_4^-(aq) \longrightarrow Mn^{2+}(aq) + CO_2(g)$ [acidic solution]

c. $Br_2(l) + SO_2(g) \longrightarrow Br^-(aq) + SO_4^{2-}(aq)$ [acidic solution]

30. **Calculate** Use the half-reaction method to write a balanced ionic equation for each reaction.

a. $Zn(s) + As_2O_3(aq) \longrightarrow AsH_3(aq) + Zn^{2+}(aq)$ [basic solution]

b. $NiO_2(s) + S_2O_3^{2-}(aq) \longrightarrow Ni(OH)_2(s) + SO_3^{2-}(aq)$ [basic solution]

31. **Identify** Review the equations for the production of sodium, bromine, and iodine found on pages R2 and R28 of the Elements Handbook. Identify the oxidizing agent and the reducing agent in each reaction.

Lesson Check Answers

27. Reactions in which electrons are transferred from one reacting species to another (redox reactions) and reactions that do not involve a transfer of electrons.

28. The oxidation-number-change method and the half-reaction method.

29. **a.** $ClO_3^-(aq) + 6I^-(aq) + 6H^+(aq) \rightarrow Cl^-(aq) + 3I_2(aq) + 3H_2O(l)$

b. $5C_2O_4^{2-}(aq) + 2MnO_4^-(aq) + 16H^+(aq) \rightarrow 2Mn^{2+}(aq) + 10CO_2(g) + 8H_2O(l)$

c. $Br_2(l) + SO_2(g) + 2H_2O(l) \rightarrow 2Br^-(aq) + SO_4^{2-}(aq) + 4H^+(aq)$

30. **a.** $6Zn(s) + As_2O_3(aq) + 9H_2O(l) \rightarrow 6Zn^{2+}(aq) + 2AsH_3(aq) + 12OH^-(aq)$

b. $2NiO_2(s) + S_2O_3^{2-}(aq) + H_2O(l) + 2OH^-(aq) \rightarrow 2SO_3^{2-}(aq) + 2Ni(OH)_2(s)$

31. oxidizing agent, Na^+; reducing agent, Cl^-

oxidizing agent, Cl_2; reducing agent, Br^-

oxidizing agent, IO_3^- or I^{5+}; reducing agent, HSO_3^- or S^{4+}

LESSON 20.3

Evaluate

Informal Assessment

In order to assess students' knowledge of redox reactions, pose the following questions. **Ask** Copper(II) hydroxide ($Cu(OH)_2$) is converted to copper(II) oxide (CuO) and water upon heating. Is this a redox reaction? Why or why not? *(No, it is not; none of the elements undergo a change in oxidation number.)* **Ask** Copper(II) oxide (CuO) is converted to copper metal upon heating in the presence of carbon. Is this a redox reaction? Why or why not? *(Yes, copper is reduced; its oxidation number decreases from 2 to 0. Carbon is oxidized; its oxidation number increases from 0 to 4 when it is converted to CO_2.)*

Then have students complete the 20.3 Lesson Check.

Reteach

Point out that the distinguishing feature of redox reactions is the change in oxidation numbers that occurs. To decide whether a given reaction is a redox reaction, students need only compare the oxidation numbers of the reactants to the oxidation numbers of the products. If a change has occurred, the reaction is a redox reaction.

Answers

25. $3Sn^{2+} + Cr_2O_7^{2-} + 14H^+ \rightarrow 3Sn^{4+} + 2Cr^{3+} + 7H_2O$

26. $4Zn + NO_3^- + 6H_2O + 7OH^- \rightarrow 4Zn(OH)_4^{2-} + NH_3$

CHEMISTRY & YOU Have students consider the relationship between color and light. Point out that color, or lack of color, in a gemstone is related to its ability to transmit light. Colorless stones transmit all wavelengths of visible light. Impurities, such as the ions listed in the text, absorb certain wavelengths and transmit others. Aquamarine is blue because its impurities transmit the wavelengths responsible for creating the characteristic blue color of that stone, and absorb all other wavelengths. Each type of stone in the photographs has its own absorption spectrum.

Point out that while heat can be used to change the color of some stones, the color of other stones can be changed just by changing the type of light used to illuminate them. In these stones, the oxidation state of the impurity is not changed. Alexandrite, another stone that contains Cr^{3+}, usually appears purple in sunlight. However, in candlelight it appears red and in fluorescent light it appears blue. Pose the following question to students: *What is the most likely explanation for the different colors observed in alexandrite?* You may need to assist students in the following ways:

- The chromium is not oxidized or reduced.
- Sunlight contains all wavelengths of visible light.
- Candlelight is made up mostly of red wavelengths.
- Fluorescent light is made up mostly of blue wavelengths.
- Purple light can be transmitted by a stone that has the ability to transmit both red and blue wavelengths.

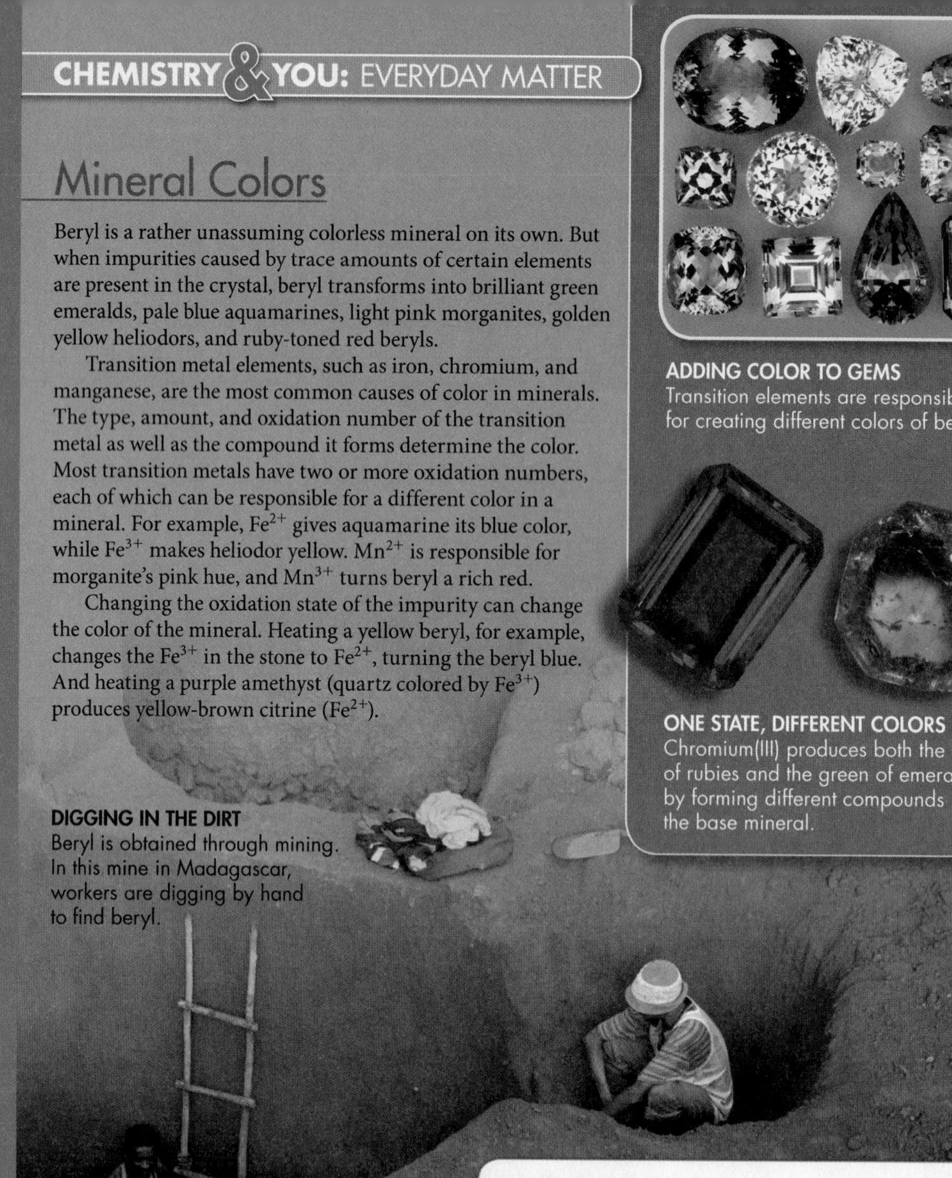

CHEMISTRY & YOU: EVERYDAY MATTER

Mineral Colors

Beryl is a rather unassuming colorless mineral on its own. But when impurities caused by trace amounts of certain elements are present in the crystal, beryl transforms into brilliant green emeralds, pale blue aquamarines, light pink morganites, golden yellow heliodors, and ruby-toned red beryls.

Transition metal elements, such as iron, chromium, and manganese, are the most common causes of color in minerals. The type, amount, and oxidation number of the transition metal as well as the compound it forms determine the color. Most transition metals have two or more oxidation numbers, each of which can be responsible for a different color in a mineral. For example, Fe^{2+} gives aquamarine its blue color, while Fe^{3+} makes heliodor yellow. Mn^{2+} is responsible for morganite's pink hue, and Mn^{3+} turns beryl a rich red.

Changing the oxidation state of the impurity can change the color of the mineral. Heating a yellow beryl, for example, changes the Fe^{3+} in the stone to Fe^{2+}, turning the beryl blue. And heating a purple amethyst (quartz colored by Fe^{3+}) produces yellow-brown citrine (Fe^{2+}).

ADDING COLOR TO GEMS Transition elements are responsible for creating different colors of beryl.

ONE STATE, DIFFERENT COLORS Chromium(III) produces both the red of rubies and the green of emeralds by forming different compounds in the base mineral.

DIGGING IN THE DIRT Beryl is obtained through mining. In this mine in Madagascar, workers are digging by hand to find beryl.

Take It Further

1. **Identify** Determine the oxidation number for each of the following transition metal ions: Fe^{2+}, Fe^{3+}, Mn^{2+}, Mn^{3+}.
2. **Apply Concepts** When amethyst is heated to produce citrine, is the iron in the mineral oxidized or reduced? Explain your answer.

716 Chapter 20 • Chemistry & You

Differentiated Instruction

L3 ADVANCED STUDENTS Explain that synthetic gemstones have the same chemical makeup and physical properties as their natural counterparts. Challenge students to research the processes involved in creating synthetic versions of the gemstones discussed in this feature and share their results with the class.

L1 LESS PROFICIENT READERS Have students organize the minerals, their impurities and their colors in a three-column table. Have students refer to their tables during discussion to help them identify the colors that result from changes in oxidation state.

Answers

1. +2, +3, +2, +3
2. Reduced; the oxidation number changes from +3 to +2.

Small-Scale Lab

Half-Reactions

Purpose

To observe redox reactions and to write half-reactions that describe them

Materials

- ruler
- reaction surface
- chemicals listed in the grid

Procedure

1. On separate sheets of paper, draw two grids similar to the one below.

	HCl	HNO_3	H_2SO_4
Zn			
Mg			
Cu			
Fe			

2. Make each square 2 cm on each side. Place a reaction surface over one of the grids and add one drop of each acid solution to one piece of each metal, as shown above. Use the second grid as a data table to record your observations for each solution.

Analyze

Using your data, answer the following questions.

1. Explain Which metal is the most reactive? How do you know? Which metal did not react with any of the acids?

2. Infer List the metals in order of decreasing reactivity.

3. Infer What is the chemical formula of the gas produced in each reaction?

4. Calculate An active metal reacts with an acid to produce hydrogen gas and a salt. Write chemical equations and net ionic equations to describe the reactions you observed. Are all of these redox reactions? Explain.

5. Calculate The half-reaction for the oxidation of zinc is shown below.

$$Zn(s) \longrightarrow Zn^{2+}(aq) + 2e^-$$

Write the oxidation half-reaction for the other metals that react.

6. Calculate The half-reaction for the reduction of hydrogen from the acid is shown below.

$$2H^+ + 2e^- \longrightarrow H_2(g)$$

Notice that this half-reaction is the same for all the acids. Demonstrate how adding this half-reaction to each oxidation half-reaction results in the overall net ionic equations.

You're the Chemist

The following small-scale activities allow you to develop your own procedures and analyze the results.

1. Analyze Data Pennies minted after 1982 are made of zinc with a thin copper coating. Use a penny that has been damaged so that a portion of the zinc shows through to compare the reactivity of the zinc and the copper toward various acids.

2. Design an Experiment Many household products, such as toilet-bowl cleaners and vinegar, contain acids. Design and carry out experiments to find out if these products also react with metals.

Focus on ELL

4 LANGUAGE PRODUCTION Have students work in groups of two to complete the lab. Be sure to pair ELLs with writing/reading partners so that more proficient students can help less proficient ones. Have students work according to their proficiency level.

BEGINNING: LOW/HIGH As you place the chemicals on the grid, state each chemical name aloud. Students can practice stating the name of each chemical. Allow students to respond orally to the Analyze questions.

INTERMEDIATE: LOW/HIGH Allow students to present their results by using symbols as needed. Orally question students about their findings.

ADVANCED

LOW Have students write draft responses to the Calculate questions from the lab and then revise their answers after consultation with you.

HIGH Ask students to complete the You're the Chemist activities with clearly planned oral responses to the class. Allow time for classmates to ask questions.

Small-Scale Lab

OBJECTIVE After completing this activity, students will be able to write half-reactions to describe observed redox reactions between acids and metals.

PREP TIME 45 minutes

CLASS TIME 45 minutes

MATERIALS zinc, magnesium, copper, and iron (in granular form or small strips)

ADVANCE PREPARATION Prepare the following solutions.

Solution	Preparation
1.0*M* HCl	82 mL of 12*M* in 1.0 L
1.0*M* HNO_3	63 mL of 15.8*M* in 1.0 L
1.0*M* H_2SO_4	56 mL of 18*M* in 1.0 L

SAFETY Always add acid to water carefully and slowly.

TEACHING TIPS Emphasize that a half-reaction shows what happens to electrons in redox reactions. Use magnesium turnings, or cut magnesium ribbon into 0.5-cm pieces. Use galvanized nails for zinc, staples from a stapler for iron, and pennies for copper. Recycle all the metals except the staples.

ANALYZE AND CONCLUDE

1. Mg is most reactive because it bubbles most vigorously. Cu does not react.
2. The order of reactivity is Mg > Zn > Fe > Cu.
3. $H_2(g)$ is produced in all the reactions.
4. $Mg(s) + 2HCl(aq) \rightarrow H_2(g) + MgCl_2(aq)$
   $Mg(s) + 2H^+(aq) \rightarrow H_2(g) + Mg^{2+}(aq)$
   Reactions of HCl with Zn and Fe are similar.
   All are redox reactions because the oxidation numbers of reactants change.
5. $Mg(s) \rightarrow Mg^{2+}(aq) + 2e^-$ $Cu(s) \rightarrow Cu^{2+}(aq) + 2e^-$
   $Fe(s) \rightarrow Fe^{2+}(aq) + 2e^-$
6. $2H^+ + 2e^- \rightarrow H_2(g)$
   $Mg(s) \rightarrow Mg^{2+}(aq) + 2e^-$
   $2H^+ + Mg(s) \rightarrow H_2(g) + Mg^{2+}(aq)$

YOU'RE THE CHEMIST

1. Add a drop of any acid to the damaged part of the penny and notice that only the zinc interior reacts.
2. Vinegar and many toilet-bowl cleaners dissolve metals. Keep products containing acids away from metal pipes and fixtures.

FOR ENRICHMENT Have students research the redox reactions that take place in the black-and-white photography process. Have them write half-reactions and complete redox equations for the reactions.

STUDY GUIDE

Evaluate

Review and Assessment Materials

CHEMISTRY TUTORIAL Encourage students to review identifying oxidizing and reducing reagents, assigning oxidation numbers to numbers, and balancing redox equations using the animated step-by-step tutorials for Chapter 20. Each Chemistry Tutorial is followed by a similar practice problem, where the student is guided through the problem-solving process.

CHAPTER PROBLEM SET Have students practice more redox problems using the Chapter 20 online Problem Set.

KINETIC ART Direct students to the Kinetic Art animations *Oxidation of Methane* and *Corrosion* and have them answer the questions that accompany each animation.

CONCEPTS IN ACTION If students have trouble understanding corrosion, ask them to view the Concepts in Action animation, which presents the material in a familiar context. Have students complete the questions after the animation.

Study Tip

Tell students that organizing information will help them remember it more easily. Suggest that students write key information from the chapter on index cards or large self-stick notes and use them to create large concept maps.

20 Study Guide

BIGIDEA REACTIONS

Oxidation-reduction reactions always occur simultaneously in redox reactions. Losing electrons is oxidation. Gaining electrons is reduction. If oxygen is involved in the reaction, then the substance gaining oxygen is oxidized, while the substance losing oxygen is reduced. The species that is reduced is the oxidizing agent, while the oxidized species is the reducing agent. Redox reactions are identified by changes in oxidation number. Redox equations can be balanced by two methods, the oxidation-number-change method and by balancing the oxidation and reduction half-reactions.

20.1 The Meaning of Oxidation and Reduction

A substance that undergoes oxidation gains oxygen or loses electrons, while a substance that undergoes reduction loses oxygen or gains electrons.

The presence of salts and acids accelerates corrosion by producing conductive solutions that make electron transfer easier.

- oxidation-reduction reaction (693)
- oxidation (694)
- reduction (694)
- reducing agent (695)
- oxidizing agent (695)

20.2 Oxidation Numbers

As a general rule, a bonded atom's oxidation number is the charge that it would have if the electrons in the bond were assigned to the atom of the more electronegative element.

An increase in the oxidation number of an atom or ion indicates oxidation. A decrease in the oxidation number of an atom or ion indicates reduction.

- oxidation number (701)

20.3 Describing Redox Equations

One class of chemical reactions is oxidation-reduction (redox) reactions, in which electrons are transferred from one reacting species to another. The other class includes all other reactions, in which no electron transfer occurs.

To balance a redox equation using the oxidation-number change method, the total increase in oxidation number of the species oxidized must be balanced by the total decrease in the oxidation number of the species reduced.

To balance a redox reaction using half-reactions, write separate half-reactions for the oxidation and the reduction. After you balance atoms in each half-reaction, balance electrons gained in the reduction with electrons lost in the oxidation.

- oxidation-number-change method (710)
- half-reaction (712)
- half-reaction method (712)

UbD Performance Tasks

PLAN AND PROVE! Have students create a pros and cons assessment of balancing redox equations by using changes in oxidation numbers versus balancing equations by using the half-reaction method. Students should provide examples supporting their opinions.

RESEARCHING REDOX Have students investigate and report via a computerized slide show presentation two different technological applications that utilize redox reactions. Students should include the equations and the corresponding half-reactions used in these applications. Students should summarize with a statement generalizing the ways in which their lives would differ if redox reactions did not exist.

Skills Tune-Up: Redox Reactions

Problem	Use the changes in oxidation number to identify which atoms are oxidized and which are reduced in each reaction. $2ZnS(s) + 3O_2(g) \longrightarrow 2ZnO(s) + 2SO_2(g)$	The following reaction takes place in a basic solution. Use the half-reaction method to write a balanced ionic equation. $MnO_4^-(aq) + I^-(aq) \longrightarrow MnO_2(s) + I_2(s)$
1 Analyze	Assign an oxidation number to each atom on both sides of the equation. A decrease in oxidation number indicates reduction. An increase in oxidation number indicates oxidation.	Follow the steps of the half-reaction method to balance the equation. Write separate half-reactions for the oxidation and the reduction. After you balance atoms in each half-reaction, balance electrons gained in the reduction with electrons lost in the oxidation.
2 Solve	Use the Rules for Assigning Oxidation Numbers to identify the oxidation number of each atom. $2\overset{+2}{Zn}\overset{-2}{S}(s) + 3\overset{0}{O_2}(g) \longrightarrow 2\overset{+2}{Zn}\overset{-2}{O}(s) + 2\overset{+4}{S}\overset{-2}{O_2}(g)$ The oxidation number of sulfur increases from +2 to +4, so **sulfur is oxidized.** The oxidation number of oxygen decreases from 0 to −2, so **oxygen is reduced.** *Remember that the sum of the oxidation numbers of the atoms in a neutral compound must equal zero.*	Perform the steps below. **Step 1:** $MnO_4^-(aq) + I^-(aq) \longrightarrow MnO_2(s) + I_2(s)$ **Step 2:** Oxidation half-reaction: $I^-(aq) \longrightarrow I_2(s)$ Reduction half-reaction: $MnO_4^-(aq) \longrightarrow MnO_2(s)$ **Step 3:** Balance the oxidation half-reaction. $I^-(aq) \longrightarrow I_2(s)$ $2I^-(aq) \longrightarrow I_2(s)$ *Hint: First, balance the equation as though it is an acidic solution, using H^+ ions and H_2O. Then, add as many OH^- ions to each side of the equation as there are H^+ ions on one side.* Balance the reduction half-reaction. $MnO_4^-(aq) \longrightarrow MnO_2(s) + 2H_2O$ $4H^+ + MnO_4^-(aq) \longrightarrow MnO_2(s) + 2H_2O$ $4OH^- + 4H^+ + MnO_4^-(aq) \longrightarrow MnO_2(s) + 2H_2O + 4OH^-$ $4H_2O + MnO_4^-(aq) \longrightarrow MnO_2(s) + 2H_2O + 4OH^-$ $2H_2O + MnO_4^-(aq) \longrightarrow MnO_2(s) + 4OH^-$ **Step 4:** Oxidation: $2I^-(aq) \longrightarrow I_2(s) + 2e^-$ Reduction: $3e^- + 2H_2O + MnO_4^-(aq) \longrightarrow MnO_2(s) + 4OH^-$ **Step 5:** Oxidation: $6I^-(aq) \longrightarrow 3I_2(s) + 6e^-$ Reduction: $6e^- + 4H_2O + 2MnO_4^-(aq) \longrightarrow 2MnO_2(s) + 8OH^-$ **Step 6:** $6I^-(aq) + \cancel{6e^-} + 4H_2O + 2MnO_4^-(aq) \longrightarrow 3I_2(s) + \cancel{6e^-} + 2MnO_2(s) + 8OH^-$ **$6I^-(aq) + 4H_2O + 2MnO_4^-(aq) \longrightarrow 3I_2(s) + 2MnO_2(s) + 8OH^-$**

SKILLS TUNE-UP

Skills Review

Example and sample practice problems related to redox reactions can be found on the following pages:

LESSON 20.1 Identifying Oxidized and Reduced Reactants—page 695

LESSON 20.2 Assigning Oxidation Numbers to Atoms—page 703; Identifying Oxidized and Reduced Atoms—pages 705 and 706

LESSON 20.3 Identifying Redox Reactions—page 709; Balancing Redox Equations by Oxidation-Number Change—page 711; Balancing Redox Equations by Half-Reactions—page 714

Focus on ELL

5 ASSESS UNDERSTANDING Preview the questions in the chapter assessment. Have students underline the academic or vocabulary words they do not understand or have trouble pronouncing. Review these words by stating them aloud and referring students to the word wall, a bilingual glossary, or a dictionary.

BEGINNING: LOW/HIGH Rephrase the directions in simplified English. Repeat the directions and emphasize key words.

INTERMEDIATE

LOW Allow students to use a bilingual glossary during the assessment.

HIGH Help students with questions that require an explanation by showing them how to outline their answers prior to writing.

ADVANCED: LOW/HIGH Guide students in using the Skills Tune-Up for this chapter to help with application problems.

ASSESSMENT

Evaluate

Answers

LESSON 20.1

32. oxidation

33. The oxidizing agent is reduced.

34. **a.** $2Ba(s) + O_2(g) \rightarrow 2BaO(s)$; barium is oxidized
b. $CuO(s) + H_2(g) \rightarrow Cu(s) + H_2O(l)$; copper is reduced
c. $C_2H_4(g) + 3O_2(g) \rightarrow 2CO_2(g) + 2H_2O(l)$; carbon is oxidized
d. $3CaO(s) + 2Al(s) \rightarrow Al_2O_3(s) + 3Ca(s)$; calcium is reduced

35. **a.** oxidation
b. oxidation
c. oxidation
d. oxidation

36. **a.** oxidizing agent
b. reducing agent
c. oxidizing agent
d. oxidizing agent

37. **a.** H_2 is oxidized; S is reduced.
b. N_2 is reduced; H_2 is oxidized.
c. S is oxidized; O_2 is reduced.
d. H_2 is oxidized; O_2 is reduced.

38. **a.** H_2 is the reducing agent; S is the oxidizing agent.
b. N_2 is the oxidizing agent; H_2 is the reducing agent.
c. S is the reducing agent; O_2 is the oxidizing agent.
d. H_2 is the reducing agent; O_2 is the oxidizing agent.

LESSON 20.2

39. An oxidation number is the charge an atom would have if the electrons in each bond were assigned to the atoms of the more electronegative element.

40. c is false.

41. **a.** +2
b. +3
c. Na, +1; Cr, +6
d. +5
e. +7

42. **a.** O, –2; H, +1
b. P, +5; O, –2
c. I, +5; O, –2
d. H, +1; P, +5; O, –2
e. H, +1; S, +6; O, –2

LESSON 20.3

43. **a.** Al is oxidized; Mn is reduced.
b. K is oxidized; H is reduced.
c. Hg is reduced; O is oxidized.
d. P is oxidized; O is reduced.

44. **a.** $2Al(s) + 3Cl_2(g) \rightarrow 2AlCl_3(s)$
b. $2Al(s) + Fe_2O_3(s) \rightarrow Al_2O_3(s) + 2Fe(s)$
c. $3Cl_2(g) + 6KOH(aq) \rightarrow KClO_3(aq) + 5KCl(aq) + 3H_2O(l)$
d. $2HNO_3(aq) + 3H_2S(aq) \rightarrow 3S(s) + 2NO(g) + 4H_2O(l)$
e. $KIO_4(aq) + 7KI(aq) + 8HCl(aq) \rightarrow 8KCl(aq) + 4I_2(s) + 4H_2O(l)$

45. redox: a, b, c, d, e

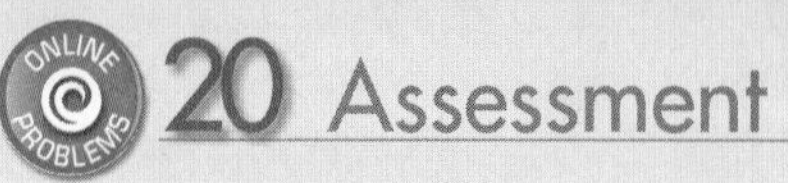

20 Assessment

* Solutions appear in Appendix E

Lesson by Lesson

20.1 The Meaning of Oxidation and Reduction

32. What chemical process must always accompany a reduction process?

33. What happens to an oxidizing agent during a redox reaction?

***34.** Balance each redox equation and identify whether the first substance in each equation was oxidized or reduced.
a. $Ba(s) + O_2(g) \longrightarrow BaO(s)$
b. $CuO(s) + H_2(g) \longrightarrow Cu(s) + H_2O(l)$
c. $C_2H_4(g) + O_2(g) \longrightarrow CO_2(g) + H_2O(l)$
d. $CaO(s) + Al(s) \longrightarrow Al_2O_3(s) + Ca(s)$

35. Identify each process as either oxidation or reduction.
a. $Al \longrightarrow Al^{3+} + 3e^-$
b. $2Cl^- \longrightarrow Cl_2 + 2e^-$
c. $S^{2-} \longrightarrow S + 2e^-$
d. $Sr \longrightarrow Sr^{2+} + 2e^-$

***36.** Which of the following would most likely be oxidizing agents, and which would most likely be reducing agents? (*Hint:* Think in terms of tendencies to lose or gain electrons.)
a. Cl_2
b. K
c. Ag^+
d. Zn^{2+}

***37.** Refer to the electronegativity values in Table 6.2 to determine which reactant is oxidized and which reactant is reduced in each reaction.
a. $H_2(g) + S(s) \longrightarrow H_2S(g)$
b. $N_2(g) + 3H_2(g) \longrightarrow 2NH_3(g)$
c. $S(s) + O_2(g) \longrightarrow SO_2(g)$
d. $2H_2(g) + O_2(g) \longrightarrow 2H_2O(l)$

38. Identify the oxidizing agent and the reducing agent for each of the reactions in Problem 37.

20.2 Oxidation Numbers

39. In your own words, what is an oxidation number?

40. Which of these statements is false?
a. The oxidation number of an uncombined element is zero.
b. The sum of the oxidation numbers of the atoms in a polyatomic ion must equal the charge of the ion.
c. Every element has a single oxidation number.
d. The oxidation number of oxygen in a compound or a polyatomic ion is almost always –2.

41. Determine the oxidation number of each metal atom.
a. Ca^{2+} **b.** Al_2S_3 **c.** Na_2CrO_4 **d.** V_2O_5 **e.** MnO_4^-

***42.** Assign oxidation numbers to the atoms in the following ions:
a. OH^- **b.** PO_4^{3-} **c.** IO_3^- **d.** $H_2PO_4^-$ **e.** HSO_4^-

20.3 Describing Redox Equations

43. Use the changes in oxidation numbers to identify which atoms are oxidized and which are reduced in each reaction.
a. $Al(s) + MnO_2(s) \longrightarrow Al_2O_3(s) + Mn(s)$
b. $K(s) + H_2O(l) \longrightarrow KOH(aq) + H_2(g)$
c. $HgO(s) \longrightarrow Hg(l) + O_2(g)$
d. $P_4(s) + O_2(g) \longrightarrow P_4O_{10}(s)$

***44.** Balance each redox equation.
a. $Al(s) + Cl_2(g) \longrightarrow AlCl_3(s)$
b. $Al(s) + Fe_2O_3(s) \longrightarrow Al_2O_3(s) + Fe(s)$
c. $Cl_2(g) + KOH(aq) \longrightarrow KClO_3(aq) + KCl(aq) + H_2O(l)$
d. $HNO_3(aq) + H_2S(g) \longrightarrow S(s) + NO(g) + H_2O(l)$
e. $KIO_4(aq) + KI(aq) + HCl(aq) \longrightarrow KCl(aq) + I_2(s) + H_2O(l)$

***45.** Identify which of these unbalanced equations represent redox reactions.
a. $Li(s) + H_2O(l) \longrightarrow LiOH(aq) + H_2(g)$
b. $K_2Cr_2O_7(aq) + HCl(aq) \longrightarrow KCl(aq) + CrCl_3(aq) + H_2O(l) + Cl_2(g)$
c. $Al(s) + HCl(aq) \longrightarrow AlCl_3(aq) + H_2(g)$
d. $Cl_2(g) + H_2O(l) \longrightarrow HCl(aq) + HClO(aq)$
e. $I_2O_5(s) + CO(g) \longrightarrow I_2(s) + CO_2(g)$
f. $H_2O(l) + SO_3(g) \longrightarrow H_2SO_4(aq)$

*46. Use the half-reaction method to write a balanced ionic equation for each reaction. All occur in acidic solutions.

a. $CuS(s) + NO_3^-(aq) \longrightarrow Cu(NO_3)_2(aq) + NO_2(g) + SO_2(g)$

b. $I^-(aq) + NO_3^-(aq) \longrightarrow I_2(s) + NO(g)$

*47. Use the half-reaction method to write a balanced ionic equation for each reaction. All occur in basic solutions.

a. $MnO_4^-(aq) + ClO_2^-(aq) \longrightarrow MnO_2(s) + ClO_4^-(aq)$

b. $Cr^{3+}(aq) + ClO^-(aq) \longrightarrow CrO_4^{2-}(aq) + Cl^-(aq)$

c. $Mn^{3+}(aq) + I^-(aq) \longrightarrow Mn^{2+}(aq) + IO_3^-(aq)$

Understand Concepts

48. Balance the equations in Problem 43 by an appropriate method.

49. Balance the equations in Problem 45 by an appropriate method.

*50. Determine the oxidation number of phosphorus in each substance.

a. P_4O_8
b. PO_4^{3-}
c. P_2O_5
d. P_4O_6
e. $H_2PO_4^-$
f. PO_3^{3-}

51. What is the oxidation number for chromium in each of these compounds?

a. K_2CrO_4
b. Cr_2O_3

52. Identify the element oxidized, the element reduced, the oxidizing agent, and the reducing agent in each unbalanced redox equation.

a. $MnO_2(s) + HCl(aq) \longrightarrow MnCl_2(aq) + Cl_2(g) + H_2O(l)$

b. $Cu(s) + HNO_3(aq) \longrightarrow Cu(NO_3)_2(aq) + NO_2(g) + H_2O(l)$

c. $P(s) + HNO_3(aq) + H_2O(l) \longrightarrow NO(g) + H_3PO_4(aq)$

d. $Bi(OH)_3(s) + Na_2SnO_2(aq) \longrightarrow Bi(s) + Na_2SnO_3(aq) + H_2O(l)$

53. Balance each redox equation in Problem 52 by using the oxidation-number-change method.

*54. An alcohol-content measuring device is used to test a person's breath for the alcohol ethanol, C_2H_5OH. In this test, ethanol reacts with an acidic solution of orange dichromate ion to form green chromium(III) ion.

$$Cr_2O_7^{2-}(aq) + C_2H_5OH(aq) \longrightarrow Cr^{3+}(aq) + CO_2(g)$$

The amount of color change is proportional to the amount of ethanol in the exhaled breath.

a. Balance this equation by the half-reaction method.

b. Is dichromate ion an oxidizing agent or a reducing agent?

55. The metallic element tungsten, used as a filament in incandescent light bulbs, is obtained by heating tungsten(VI) oxide with hydrogen.

$$WO_3(s) + H_2(g) \longrightarrow W(s) + H_2O(g)$$

a. Balance the equation.

b. What is the reducing agent in this reaction?

c. Which element undergoes an increase in oxidation number?

56. Silver tarnishes when it reacts with hydrogen sulfide in the air.

$$Ag(s) + H_2S(g) \longrightarrow Ag_2S(s) + H_2(g)$$

a. Is silver oxidized or reduced in this reaction?

b. Identify the oxidizing agent and the reducing agent.

c. Balance the equation.

57. The following equation represents an oxidation-reduction reaction that uses oxygen. Show how this reaction can also be defined as an oxidation-reduction reaction in terms of electron transfer.

$$Pb(s) + O_2(g) \longrightarrow PbO(s)$$

58. Does each of the following equations represent a redox reaction? Explain how you know.

a. $Bi_2O_3(s) + 3C(s) \longrightarrow 2Bi(s) + 3CO(g)$

b. $Cr_2O_3(s) + 3H_2S(g) \longrightarrow Cr_2S_3(s) + 3H_2O(l)$

c. $BCl_3(g) + 3H_2O(l) \longrightarrow H_3BO_3(s) + 3HCl(g)$

ASSESSMENT

Evaluate

Answers

46. a. $CuS(s) + 8NO_3^-(aq) + 8H^+(aq) \rightarrow Cu(NO_3)_2(aq) + SO_2(g) + 6NO_2(g) + 4H_2O(l)$

b. $6I^-(aq) + 2NO_3^-(aq) + 8H^+(aq) \rightarrow 3I_2(s) + 2NO(g) + 4H_2O(l)$

47. a. $4MnO_4^-(aq) + 3ClO_2^-(aq) + 2H_2O(l) \rightarrow 4MnO_2(s) + 3ClO_4^-(aq) + 4OH^-(aq)$

b. $2Cr^{3+}(aq) + 3ClO^-(aq) + 10OH^-(aq) \rightarrow 2CrO_4^{2-}(aq) + 3Cl^-(aq) + 5H_2O(l)$

c. $6Mn^{3+}(aq) + I^-(aq) + 6OH^-(aq) \rightarrow 6Mn^{2+}(aq) + IO_3^-(aq) + 3H_2O(l)$

UNDERSTAND CONCEPTS

48. a. $4Al(s) + 3MnO_2(s) \rightarrow 2Al_2O_3(s) + 3Mn(s)$

b. $2K(s) + 2H_2O(l) \rightarrow 2KOH(aq) + H_2(g)$

c. $2HgO(s) \rightarrow 2Hg(l) + O_2(g)$

d. $P_4(s) + 5O_2(g) \rightarrow P_4O_{10}(s)$

49. a. $2Li(s) + 2H_2O(l) \rightarrow 2LiOH(aq) + H_2(g)$

b. $K_2Cr_2O_7(aq) + 14HCl(aq) \rightarrow 2KCl(aq) + 2CrCl_3(aq) + 7H_2O(l) + 3Cl_2(g)$

c. $2Al(s) + 6HCl(aq) \rightarrow 2AlCl_3(aq) + 3H_2(g)$

d. $Cl_2(g) + H_2O(l) \rightarrow HCl(aq) + HClO(aq)$

e. $I_2O_5(s) + 5CO(g) \rightarrow I_2(s) + 5CO_2(g)$

f. $H_2O(l) + SO_3(g) \rightarrow H_2SO_4(aq)$

50. a. +4
b. +5
c. +5
d. +3
e. +5
f. +3

51. a. +6
b. +3

52. a. Cl oxidized, Mn reduced, Mn oxidizing agent, Cl reducing agent

b. Cu oxidized, N reduced, N oxidizing agent, Cu reducing agent

c. P oxidized, N reduced, N oxidizing agent, P reducing agent

d. Sn oxidized, Bi reduced, Bi oxidizing agent, Sn reducing agent

53. a. $MnO_2(s) + 4HCl(aq) \rightarrow MnCl_2(aq) + Cl_2(g) + 2H_2O(l)$

b. $Cu(s) + 4HNO_3(aq) \rightarrow Cu(NO_3)_2(aq) + 2NO_2(g) + 2H_2O(l)$

c. $3P(s) + 5HNO_3(aq) + 2H_2O(l) \rightarrow 5NO(g) + 3H_3PO_4(aq)$

d. $2Bi(OH)_3(s) + 3Na_2SnO_2(aq) \rightarrow 2Bi(s) + 3Na_2SnO_3(aq) + 3H_2O(l)$

54. a. $16H^+(aq) + 2Cr_2O_7^{2-}(aq) + C_2H_5OH(aq) \rightarrow 4Cr^{3+}(aq) + 2CO_2(g) + 11H_2O(l)$

b. oxidizing agent

55. a. $WO_3(s) + 3H_2(g) \rightarrow W(s) + 3H_2O(g)$

b. H_2

c. H

56. a. oxidized

b. H is the oxidizing agent; Ag is the reducing agent.

c. $2Ag(s) + H_2S(s) \rightarrow Ag_2S(s) + H_2(g)$

57. Lead atoms are oxidized by losing 2 electrons to form Pb^{2+} ions. Oxygen atoms are reduced by gaining 2 electrons to form O^{2-} ions.

58. a. Yes, the oxidation number of bismuth changes from +3 to zero; the oxidation number of carbon changes from zero to +2.

b. No, there is no change in oxidation number of any of the atoms in this reaction.

c. No, there is no change in oxidation number of any of the atoms in this reaction.

ASSESSMENT

Evaluate

Answers

59. **a.** $Ba(ClO_3)_2(s) \rightarrow BaCl_2(s) + 3O_2(g)$
b. $2PbS(s) + 3O_2(g) \rightarrow 2PbO(s) + 2SO_2(g)$

60. $N_2O_4(l) + 2N_2H_4(l) \rightarrow 3N_2(g) + 4H_2O(g)$
Based on oxidation number changes, nitrogen atoms from N_2O_4 gain electrons in forming N_2 and nitrogen atoms from N_2H_4 lose electrons in forming N_2.

61. **a.** reactant, 0; product, +3
b. reactant, –2; product, –2
c. X
d. H

THINK CRITICALLY

62. When one reactant loses electrons another reactant must gain them.

63. $1s^22s^22p^63s^23p^5$ A chlorine atom can "lose" its seven valence electrons or it can gain one electron to fill the last 3*p* orbital.

64. A sodium atom achieves a stable electron configuration by losing its one valence electron, making it a reducing agent. A sodium ion has a stable electron configuration.

65. Double-replacement reactions never involve the transfer of electrons; instead they involve the transfer of ions in aqueous solution.

66. In every redox reaction one species loses one or more electrons and is a reducing agent. Another substance gains one or more electrons and is an oxidizing agent.

67. Iron is easily oxidized, forming rust. Over thousands of years, an iron artifact would most likely have oxidized and crumbled, thereby destroying the artifact.

68. **a.** SO_4^{2-} **b.** H_2O_2 **c.** NO_3^- **d.** $Cr_2O_7^{2-}$ **e.** H_2O

69. Group 1A metals; a reducing agent must give up electrons. Group 1A metals achieve stability by giving up electrons and forming positively charged ions.

70. **a.** $Rb(s) + I_2(s) \rightarrow RbI_2(s)$; oxidizing agent is I
b. $Ba(s) + 2H_2O(l) \rightarrow Ba(OH)_2(aq) + H_2(g)$; oxidizing agent is H
c. $2Al(s) + 3FeSO_4(aq) \rightarrow Al_2(SO_4)_3(aq) + 3Fe(s)$; oxidizing agent is Fe
d. $C_4H_8(g) + 6O_2(g) \rightarrow 4CO_2(g) + 4H_2O(l)$; oxidizing agent is O
e. $Zn(s) + 2HBr(aq) \rightarrow ZnBr_2(aq) + H_2(g)$; oxidizing agent is H
f. $Mg(s) + Br_2(l) \rightarrow MgBr_2(s)$, oxidizing agent is Br

71. Rhenium is oxidized and selenium is reduced, because the bonded electrons would shift toward the more electronegative element, selenium.

72. MnO_4^-, because the manganese is at its highest oxidation state.

***59.** Write chemical equations for the following redox reactions. Balance each equation using the oxidation-number-change method.

a. Solid barium chlorate decomposes when heated, yielding solid barium chloride and gaseous oxygen.

b. Solid lead(II) sulfide reacts with gaseous oxygen to produce solid lead(II) oxide and gaseous sulfur dioxide.

60. The following unbalanced equation represents a reaction that can occur when dinitrogen tetroxide, N_2O_4, is combined with hydrazine, N_2H_4.

$$N_2O_4(l) + N_2H_4(l) \longrightarrow N_2(g) + H_2O(g)$$

Balance the equation and describe in words the electron transfer that takes place.

61. Examine the following hypothetical redox equation.

$$2X + 3H_2Y \longrightarrow X_2Y_3 + 3H_2$$

a. What is the oxidation number of element X on each side of the equation?

b. What is the oxidation number of element Y on each side of the equation?

c. What is oxidized in this equation?

d. What is reduced in this equation?

Think Critically

62. Explain Why must the number of electrons lost equal the number of electrons gained in every redox reaction.

***63. Explain** The highest possible oxidation number that chlorine exhibits in any compound is +7, whereas its most negative oxidation number is −1. Write the electron configuration of chlorine, and explain why these are the limiting oxidation numbers for chlorine.

64. Explain Why is a sodium atom a reducing agent but a sodium ion is not?

65. Explain Many decomposition, single-replacement, combination, and combustion reactions are also redox reactions. Why is a double-replacement reaction never a redox reaction?

***66. Make Generalizations** Why must every redox reaction have a reducing agent and an oxidizing agent?

67. Infer Humankind began to make and use iron tools more than 3000 years ago, but few iron artifacts from ancient times have survived. Explain.

68. Identify Which substance in each pair is more likely to be an oxidizing agent?

a. S^{2-} or SO_4^{2-}

b. H_2O or H_2O_2

c. NO_2^- or NO_3^-

d. $Cr_2O_7^{2-}$ or Cr^{3+}

e. H_2 or H_2O

69. Identify Which is more likely to be a strong reducing agent, a group 1A metal or a group 7A nonmetal? Explain.

***70. Predict** Predict the product(s) and write the balanced equation for each of these redox reactions. Identify the oxidizing agent in each reaction.

a. rubidium + iodine $\longrightarrow$

b. barium + water $\longrightarrow$

c. aluminum + iron(II) sulfate $\longrightarrow$

d. butene (C_4H_8) + oxygen $\longrightarrow$

e. zinc + hydrobromic acid $\longrightarrow$

f. magnesium + bromine $\longrightarrow$

71. Explain The electronegativity of rhenium, Re, is 1.9, and the electronegativity of selenium is 2.4. If rhenium were to react with selenium to form a compound, which element would be oxidized and which element would be reduced? Explain.

72. Explain Which of the following ions is most likely to be an oxidizing agent? Explain your choice.

$$MnO_4^-, MnO_4^{2-}, Mn^{+2}$$

Enrichment

***73. Calculate** How many grams of copper are needed to reduce completely the silver ions in 85.0 mL of 0.150*M* $AgNO_3(aq)$ solution?

74. Calculate How many milliliters of 0.280*M* $K_2Cr_2O_7(aq)$ solution are needed to oxidize 1.40 g of sulfur? First, balance the equation.

$$K_2Cr_2O_7(aq) + H_2O(l) + S(s) \longrightarrow SO_2(g) + KOH(aq) + Cr_2O_3(s)$$

722 Chapter 20 • Assessment

ENRICHMENT

73. 0.405 g Cu

74. 104 mL $K_2Cr_2O_7$

75. Calculate Carbon monoxide can be removed from the air by passing it over solid diiodine pentoxide.

$$CO(g) + I_2O_5(s) \longrightarrow I_2(s) + CO_2(g)$$

a. Balance the equation.
b. Identify the element being oxidized and the element being reduced.
c. How many grams of carbon monoxide can be removed from the air by 0.55 g of diiodine pentoxide (I_2O_5)?

76. Calculate What is the oxidation number of nitrogen in each of these species?

a. HNO_3 c. N_2O_3 e. N_2O g. NO
b. NH_3 d. NO_2^- f. NH_4Cl h. NO_2

77. Calculate The elements fluorine and oxygen can react to form fluorine monoxide, F_2O. Write the balanced chemical equation for this reaction. Check electronegativity values, and then identify the elements oxidized and reduced.

78. Explain The oxidation number of nitrogen can range from a minimum of −3 to a maximum of +5. Use this information to explain why the nitride ion, N^{3-}, can act only as a reducing agent, and why the nitrate ion, NO_3^-, can act only as an oxidizing agent.

79. Calculate Oxidation-reduction reactions form the basis for chemical analysis by redox titration. Potassium permanganate, a good oxidizing agent, is sometimes used as a titrant because it undergoes a color change as it is reduced.

$$\underset{\text{(purple)}}{MnO_4^-(aq)} \longrightarrow \underset{\text{(colorless)}}{Mn^{2+}(aq)}$$

Write a balanced redox equation for the oxidation of stannous ion to stannic ion in acidic solution using permanganate as the oxidizing agent.

***80. Describe** There are a number of anions that are commonly used as oxidizing or reducing agents in the laboratory. Balance each of these half reactions. Identify each of the anions as an oxidizing or reducing agent.

a. $Cr_2O_7^{2-}(aq) \rightarrow Cr^{3+}(aq)$ [acidic solution]
b. $S_2O_3^{2-}(aq) \rightarrow SO_4^{2-}(aq)$ [acidic solution]
c. $CrO_4^{2-}(aq) \rightarrow Cr(OH)_3(aq)$ [basic solution]
d. $MnO_4^-(aq) \rightarrow Mn^{2+}(aq)$ [acidic solution]
e. $C_2O_4^{2-}(aq) \rightarrow CO_2(g)$ [acidic solution]
f. $MnO_4^-(aq) \rightarrow MnO_2(s)$ [basic solution]

81. Calculate Combine each of the following pairs of half-reactions into a complete, balanced, ionic redox equation.

a. $Hg^{2+} + 2e^- \longrightarrow Hg$
$Al \longrightarrow Al^{3+} + 3e^-$
b. $MnO_2 + 4H^+ + 2e^- \longrightarrow Mn^{2+} + 2H_2O$
$Fe \longrightarrow Fe^{2+} + 2e^-$
c. $Fe^{3+} + e^- \longrightarrow Fe^{2+}$
$Cd \longrightarrow Cd^{2+} + 2e^-$

Write About Science

82. Explain Silver recovered from shipwrecks may have layers of black tarnish. Research and write a report on how the thick layers of tarnish are removed from the silver artifacts.

83. Connect to the BIGIDEA Write a paragraph about five different real-world examples of oxidation-reduction reactions.

CHEMYSTERY

Tarnished Treasure

Gold and silver coins recovered from the same shipwrecked Spanish galleons were exposed to the same elements of the sea. However, the silver coins corrode faster than gold coins because silver is more easily oxidized than gold. Oxidation reactions cause silver coins to corrode, eventually causing their surface to become black and crumbly. Since gold is not as easily oxidized, the gold coins retain their gold color and do not show as many signs of oxidation.

84. Infer Suppose there were two types of coins, one made of magnesium and one made of iron. Which coin would you expect to tarnish faster if they were both exposed to the same conditions?

85. Connect to the BIGIDEA Why would it be helpful for shipbuilders to have a knowledge of redox reactions?

Evaluate

CHEMYSTERY **SUMMARIZE** After students have read through the CHEMystery, call on volunteers to summarize why a gold coin would withstand the elements of the sea better than a silver coin. **Ask** Does the oxidation number for silver atoms in the coins increase or decrease while the coins are underwater? *(Because the silver is oxidized, the oxidation number increases.)* **Ask** How could you use gold's and silver's places in the activity series of metals to predict whether they would be easily oxidized? *(Gold is at the bottom of the activity series, which suggests it is very difficult to oxidize. Silver is near the bottom of the list, which suggests it is also hard to oxidize, but easier than gold.)*

Ask Is the black coating on the coins also silver? *(No; it is a silver compound.)* Explain that the black material that forms is silver sulfide, Ag_2S. It forms when silver reacts with hydrogen sulfide, H_2S, releasing hydrogen gas. Have students write a balanced equation for this reaction. *($2Ag + H_2S \rightarrow Ag_2S + H_2$)* **Ask** Based on this equation, how does the oxidation number of silver change during the reaction? *(It changes from 0 to +1.)*

CHEMYSTERY ANSWERS

84. the magnesium coin

85. BIGIDEA Since ships are usually exposed to water and salty conditions, shipbuilders should be aware of which metals are less reactive and therefore less likely to corrode.

Answers

75. **a.** $5CO + I_2O_5 \rightarrow I_2 + 5CO_2$
b. C is oxidized. I is reduced.
c. 0.22 g CO

76. **a.** +5 **b.** −3 **c.** +3
d. +3 **e.** +1 **f.** −3
g. +2 **h.** +4

77. $2F_2 + O_2 \rightarrow 2F_2O$; fluorine is reduced; oxygen is oxidized

78. Nitride ion has the minimum oxidation number of -3, therefore it cannot gain additional electrons and be an oxidizing agent. It can lose electrons, however, and be a reducing agent. Nitrogen in the nitrate ion has the maximum oxidation number of +5, therefore it cannot lose additional electrons and enable NO_3^- to be a reducing agent. It can gain electrons, however, and be an oxidizing agent.

79. $2MnO_4^-(aq) + 5Sn^{2+}(aq) + 16H^+(aq) \rightarrow 2Mn^{2+}(aq) + 5Sn^{4+}(aq) + 8H_2O(l)$

80. **a.** $Cr_2O_7^{2-}(aq) + 14H^+(aq) + 6e^- \rightarrow 2Cr^{3+}(aq) + 7H_2O(l)$
oxidizing agent
b. $S_2O_3^{2-}(aq) + 5H_2O(l) \rightarrow 2SO_4^{2-}(aq) + 10H^+(aq) + 8e^-$
reducing agent
c. $CrO_4^{2-}(aq) + 4H_2O(l) + 3e^- \rightarrow Cr(OH)_3(aq) + 5OH^-(aq)$
oxidizing agent
d. $MnO_4^-(aq) + 8H^+(aq) + 5e^- \rightarrow Mn^{2+}(aq) + 4H_2O(l)$
oxidizing agent
e. $C_2O_4^{2-}(aq) \rightarrow 2CO_2(g) + 2e^-$
reducing agent
f. $MnO_4^-(aq) + 2H_2O(l) + 3e^- \rightarrow MnO_2(s) + 4OH^-(aq)$
oxidizing agent

81. **a.** $3Hg^{2+} + 2Al \rightarrow 3Hg + 2Al^{3+}$
b. $MnO_2 + 4H^+ + Fe \rightarrow Mn^{2+} + 2H_2O + Fe^{2+}$
c. $2Fe^{3+} + Cd \rightarrow 2Fe^{2+} + Cd^{2+}$

WRITE ABOUT SCIENCE

82. Silver tarnish is mainly silver sulfide. Tarnish can be removed if the silver is placed in a dilute NaOH bath and weak current is passed through the solution.

83. Answers may vary.

ASSESSMENT

CUMULATIVE REVIEW

86. a.

Alkane burned	O_2 used (mol)	CO_2 produced (mol)	H_2O produced (mol)
CH_4	2	1	2
C_2H_6	3.5	2	3
C_3H_8	5	3	4
C_4H_{10}	6.5	4	5
C_5H_{12}	8	5	6
C_6H_{14}	9.5	6	7

b. $C_xH_y + [x + (y/4)]O_2 \rightarrow xCO_2 + (y/2)H_2O$

87. sublimation

88. 1.8×10^2 kPa

89. suspension; particles settle out of a suspension; not a solution

90. a, c, and d

91. Dilute 110 mL of 6.0*M* HCl to 440 mL total volume.

92. $Ca(NO_3)_2$; boiling point elevation is a colligative property that depends on the number of particles in solution. $Ca(NO_3)_2$ gives three particles per formula unit; LiF gives two particles per formula unit.

93. 0.379*M* H_3PO_4

94. Solubility $PbBr_2 = 8.1 \times 10^{-3}M$

95. large

96. Test tube B has the NaCl added to it. Due to the common ion effect, the addition of either sulfate ion or barium ion to a saturated solution of $BaSO_4$ will cause the solubility product of $BaSO_4$ to be exceeded and barium sulfate will precipitate as shown in test tubes A and C.

97. a. $1.0 \times 10^{-2}M$
b. $1.0 \times 10^{-11}M$
c. $1.6 \times 10^{-9}M$

98. 56.3 mL KOH

99. a. NH_4^+ and NH_3; H_2O and H_3O^+
b. H_2SO_3 and HSO_3^-; NH_2^- and NH_3
c. HNO_3 and NO_3^-; I^- and HI

100. a. 5.00; acidic
b. 10.00; basic
c. 13.00; basic
d. 6.52; acidic

Cumulative Review

***86.** The complete combustion of hydrocarbons involves the oxidation of both carbon and hydrogen atoms by oxygen. The following table lists the moles of O_2 used, and the moles of CO_2 and moles of H_2O produced, when a series of hydrocarbons called alkanes are burned.

Alkane burned	O_2 used (mol)	CO_2 produced (mol)	H_2O produced (mol)
CH_4	2	1	2
C_2H_6	3.5	2	3
C_3H_8	5	3	4
C_4H_{10}	a. ____	d. ____	g. ____
C_5H_{12}	b. ____	e. ____	h. ____
C_6H_{14}	c. ____	f. ____	i. ____

a. Complete the table.
b. Based on the data, write a balanced generalized equation for the complete oxidation of any alkane. Use the following form, and write the coefficients in terms of *x* and *y*:

$$C_xH_y + \_\_ O_2 \longrightarrow \_\_ CO_2 + \_\_ H_2O$$

87. Name a change of state that does not involve a liquid.

***88.** A gas cylinder has a volume of 6.8 L and is filled with 13.8 g of N_2. Calculate the pressure of N_2 at 25°C.

89. A particular paint must be stirred before using. Is the stirred paint a solution or a suspension? Explain.

90. Which of these are nonelectrolytes?
a. $S(s)$
b. $NH_4Cl(aq)$
c. $SiO_2(s)$
d. $F_2(g)$

***91.** How would you make 440 mL of 1.5*M* HCl solution from a stock solution of 6.0*M* HCl?

92. One mole of LiF and $Ca(NO_3)_2$ are each dissolved in 1.0 L of water. Which solution has the higher boiling point? Explain.

93. What is the molarity of the solution prepared by dissolving 46.4 g H_3PO_4 in enough water to make 1.25 L of solution?

***94.** The K_{sp} of lead(II) bromide ($PbBr_2$) at 25°C is 2.1×10^{-6}. What is the solubility of $PbBr_2$ (in mol/L) at this temperature?

95. A reaction goes essentially to completion. Do you expect the value of *K* to be large or small?

96. Bottles containing 0.1*M* solutions of Na_2SO_4, $BaCl_2$, and NaCl have had their labels accidentally switched. To discover which bottle contains the NaCl, you place a clear saturated solution of $BaSO_4$ ($K_{sp} = 1.1 \times 10^{-10}$) into three test tubes. To each test tube you add a few drops of each mislabeled solution. The results are shown below. To which tube was NaCl added? Explain.

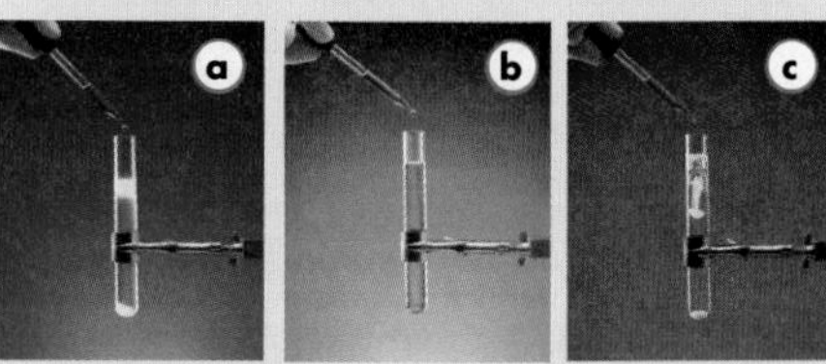

97. What is the hydrogen-ion concentration of solutions with the following pH?
a. 2.00
b. 11.00
c. 8.80

98. How many milliliters of a 4.00*M* KOH solution are needed to neutralize 45.0 mL of 2.50*M* H_2SO_4 solution?

99. Identify the conjugate acid-base pairs in each equation.
a. $NH_4^+(aq) + H_2O(l) \longrightarrow NH_3(aq) + H_3O^+(aq)$
b. $H_2SO_3(aq) + NH_2^-(aq) \longrightarrow HSO_3^-(aq) + NH_3(aq)$
c. $HNO_3(aq) + I^-(aq) \longrightarrow HI(aq) + NO_3^-(aq)$

***100.** Calculate the pH of solutions with the following hydrogen-ion or hydroxide-ion concentrations. Classify each as acidic, basic, or neutral.
a. $[H^+] = 0.000\ 010M$
b. $[OH^-] = 1.0 \times 10^{-4}M$
c. $[OH^-] = 1.0 \times 10^{-1}M$
d. $[H^+] = 3.0 \times 10^{-7}M$

If You Have Trouble With . . .

Question	86	87	88	89	90	91	92	93	94	95	96	97	98	99	100
See Chapter	12	13	14	15	15	16	16	16	18	18	19	19	19	19	19

Standardized Test Prep for Indiana

Standard: C.4.6

Select the choice that best answers each question or completes each statement.

1. Which of these processes is not an oxidation?
   (A) a decrease in oxidation number
   (B) a complete loss of electrons
   (C) a gain of oxygen
   (D) a loss of hydrogen by a covalent molecule

2. In which of these pairs of nitrogen-containing ions and compounds is the oxidation number of nitrogen in the ion higher than in the nitrogen compound?
   I. N_2H_4 and NH_4^+
   II. NO_3^- and N_2O_4
   III. N_2O and NO_2^-
   (A) I only
   (B) I and II only
   (C) I and III only
   (D) II and III only
   (E) I, II, and III

3. Identify the elements oxidized and reduced in this reaction.
   $$2ClO^- + H_2 + 2e^- \longrightarrow 2Cl^- + 2OH^-$$
   (A) Cl is oxidized; H is reduced
   (B) H is oxidized; Cl is reduced
   (C) Cl is oxidized; O is reduced
   (D) O is oxidized; Cl is reduced

4. Which of these half-reactions represents a reduction?
   I. $Fe^{2+} \longrightarrow Fe^{3+}$
   II. $Cr_2O_7^{2-} \longrightarrow Cr^{3+}$
   III. $MnO_4^- \longrightarrow Mn^{2-}$
   (A) I and II only
   (B) II and III only
   (C) I and III only
   (D) I, II, and III

5. Which of these general types of reactions is not a redox reaction?
   (A) single replacement
   (B) double replacement
   (C) combustion
   (D) combination

6. What is the reducing agent in this reaction?
   $$MnO_4^- + SO_2 \longrightarrow Mn^{2+} + SO_4^{2-}$$
   (A) SO_2
   (B) SO_4^{2-}
   (C) Mn^{2+}
   (D) MnO_4^-

Tips for Success

Interpreting Data Tables To interpret the content in a table, start by reading the title (if there is one). Then, read the headings. Try to figure out the relationship between the different columns and rows of information. Ask yourself: *What information is related in the table? How are the relationships represented?*

Use the table to answer Questions 7–9.

Group	Metal	Metal ion
3	K	K^+
3	Ca	Ca^{2+}
3	Na	Na^+
4	Mg	Mg^{2+}
4	Fe	Fe^{2+}
4	Sn	Sn^{2+}
4	Pb	Pb^{2+}
5	Cu	Cu^{2+}
5	Ag	Ag^+

(Arrow 1 points upward along the Metal column; arrow 2 points downward along the Metal ion column.)

7. Which arrow indicates increasing ease of oxidation? Of reduction?
8. Which numbered group of metals are the strongest reducing agents? Which numbered group of metals are the most difficult to oxidize?
9. Which is a stronger oxidizing agent, Na or Fe?

Use this diagram to answer Questions 10 and 11. It shows the formation of an ion from an atom.

10. Does the diagram represent oxidation or reduction? Does the oxidation number increase or decrease when the ion forms?
11. Draw a diagram showing the formation of a sulfide ion from a sulfur atom. Make the relative sizes of the atom and ion realistic. Does your drawing represent an oxidation or a reduction?

If You Have Trouble With . . .

Question	1	2	3	4	5	6	7	8	9	10	11
See Lesson	20.1	20.2	20.2	20.2	20.3	20.3	20.2	20.2	20.2	20.2	20.1

ASSESSMENT

STP Answers

1. a
2. d
3. b
4. b
5. b
6. a
7. arrow 1; arrow 2
8. group 3, group 5
9. Fe
10. Oxidation; oxidation number increases.
11. Sulfide ion should be larger than sulfur atom; reduction.

21 Electrochemistry Planning Guide

Introducing the BIGIDEA: MATTER AND ENERGY

Every chemical process uses or produces energy.

IASS	Lessons and Objectives	Print Resources		
		For the Student	For the Teacher	
	21.1 Electrochemical Cells p 728–736 **21.1.1** Identify the type of chemical reaction that is involved in all electrochemical processes. **21.1.2** Describe how a voltaic cell produces electrical energy. **21.1.3** Identify the current applications that use electrochemical processes to produce electrical energy.	Reading and Study Workbook Lesson 21.1 Lesson Assessment 21.1 p 736	Teacher Demo, p 730: A Redox Reaction Teacher Demo, p 732: Inside a Dry Cell Teacher Demo, p 733: Making a Lead Cell	
	21.2 Half-Cells and Cell Potentials p 737–743 **21.2.1** Identify what causes the electrical potential of an electrochemical cell. **21.2.2** Determine the standard reduction potential of a half-cell. **21.2.3** Determine if a redox reaction is spontaneous or nonspontaneous.	Reading and Study Workbook Lesson 21.2 Lesson Assessment 21.2 p 743	Teacher Demo, p 739: The Corrosion of Iron Class Activity, p 741: Combining Half-Reactions	
	21.3 Electrolytic Cells p 745–751 **21.3.1** Distinguish between electrolytic and voltaic cells. **21.3.2** Describe some applications that use electrolytic cells.	Reading and Study Workbook Lesson 21.3 Lesson Assessment 21.3 p 751 Quick Lab: Electrochemical Analysis of Metals p 750 Small-Scale Lab: Electrolysis of Water p 752	Teacher Demo, p 747: The Electrolysis of Water	

Assessing the BIGIDEA: MATTER AND ENERGY

	Essential Questions 1. How is energy produced in an electrochemical process? 2. How can energy be used to drive an electrochemical process?	Study Guide p 753 STP p 759 Reading and Study Workbook Self-Check and Vocabulary Review Chapter 21		

	PearsonChem.com	
	Editable Worksheets	Online Activities
	Teaching Resources, Lesson 21.1 Review Small-Scale Lab Manual Lab 36: Small-Scale Voltaic Cells Lab 47: Corrosion	21.1 Lesson Overview Voltaic Cell Chemistry Lessons from Lasagna and Lemons
	Teaching Resources, Lesson 21.2 Review	21.2 Lesson Overview Writing the Cell Reaction Calculating the Standard Cell Potential Electrochemistry
	Teaching Resources, Lesson 21.3 Review Lab 48: Electrochemistry Lab Practical 21-1: Electroplating	21.3 Lesson Overview Voltaic and Electrolytic Cells
	Core Teaching Resources (includes editable worksheets) • Lesson Reviews • Practice Problems • Interpret Graphs • Vocabulary Review • Chapter Quizzes and Tests • Lab Record Sheets	Chapter 21 Problem Set Systems of Equations

Materials List

For the Student

Quick Lab p 750
- 9-V battery, sodium sulfate solution
- copper penny
- nickel coin
- iron nail
- filter paper
- aluminum foil
- reaction surface

Small-Scale Lab p 752
- reaction surface
- electrolysis device
- droppers
- 0.2*M* Na_2SO_4
- 0.1*M* KI
- 0.2*M* $CuSO_4$
- 0.04% BTB
- 1.0*M* NaCl
- 0.2*M* KBr
- liquid starch

For the Teacher

Teacher Demo p 730
- 200 mL of 0.1*M* silver nitrate
- 250-mL beaker
- glass stirring rod
- strip of polished copper metal

Teacher Demo p 733
- 2 strips of lead
- wooden rod
- 2 connecting wires
- 250-mL beaker
- dilute sulfuric acid
- 6-V DC power supply
- doorbell

Teacher Demo p 732
- dry cell
- voltmeter
- 2 wire leads
- zinc electrode
- carbon electrode
- manganese dioxide/ammonium chloride paste

Teacher Demo p 739
- 2 pennies
- metal wires from twist ties
- table salt
- paper towel

Teacher Demo p 747
- Hoffman apparatus or apparatus constructed using iron nails as electrodes
- 2 test tubes
- 6*M* NaOH
- 250- or 500-mL beaker
- 2 insulated wire leads

Additional Digital Resources

Online Student Edition
Online Teacher's Edition
Exam*View Assessment Suite CD-ROM*
Classroom Resources DVD
21.2 Virtual Chem Lab 30: Redox Titrations: Determination of Iron

What's Online

CHEMISTRY TUTORIAL Students can work step-by-step with the online tutorial to learn how to write cell reactions and calculate the standard cell potential by accessing the Chapter 21 Chemistry Tutorials.

ONLINE PROBLEM SET Additional practice problems about electrochemistry can be found in the Chapter 21 online Problem Set.

MATH TUTORIAL The Online Math Tutorial will help guide students through solving systems of equations.

DIRECTED VIRTUAL LAB Encourage students to review half-cell voltage potentials in several chemical reactions using the animated virtual lab for Chapter 21. The animation is followed by analysis questions, where the student is guided through the lab concepts.

KINETIC ART Have students watch animations about voltaic and electrolytic cells. Each animation is followed by analysis questions to guide further understanding of the concepts.

CONCEPTS IN ACTION Students can watch an interactive animation about how batteries can be made out of everyday materials. The animation is followed by questions to explore the characteristics of the materials that can be used to make a battery.

21 Electrochemistry

INSIDE:

- **21.1** Electrochemical Cells
- **21.2** Half-Cells and Cell Potentials
- **21.3** Electrolytic Cells

PearsonChem.com

An electrochemical process was used to produce the shiny chrome finish on this car.

726

IN ACADEMIC STANDARDS for SCIENCE

Nature of Science

- Develop explanations based on reproducible data and observations gathered during laboratory investigations.

Standard 7: Solutions

C.7.2 Explain how changes in temperature, surface area, and pressure affect solubility.

C.7.6 Explain how the rate of a reaction is qualitatively affected by changes in concentration, temperature, surface area, and the use of a catalyst.

C.7.7 Describe dynamic equilibrium in terms of changes in amounts of reactants and products.

C.7.8 Write equilibrium expressions for reversible reactions.

Focus on ELL

1 CONTENT AND LANGUAGE Begin the class by writing *Electrochemistry* on the board and pronouncing it slowly. Have students determine the meaning of the word from its prefix. Then, present a complete list of the vocabulary terms in this chapter. Have students examine how these terms are related to the chapter content.

BEGINNING: LOW/HIGH Create illustrations to represent the terms. If literate in the student's native language, create a native language definition.

INTERMEDIATE: LOW/HIGH Provide students with sentences in which key words have been replaced with blanks. Have students complete the blanks with words from a given word bank as they read through the chapter.

ADVANCED: LOW/HIGH Challenge students to create their own sentences in which key words have been replaced with blanks. Have students trade papers with a partner to help them review lesson terminology.

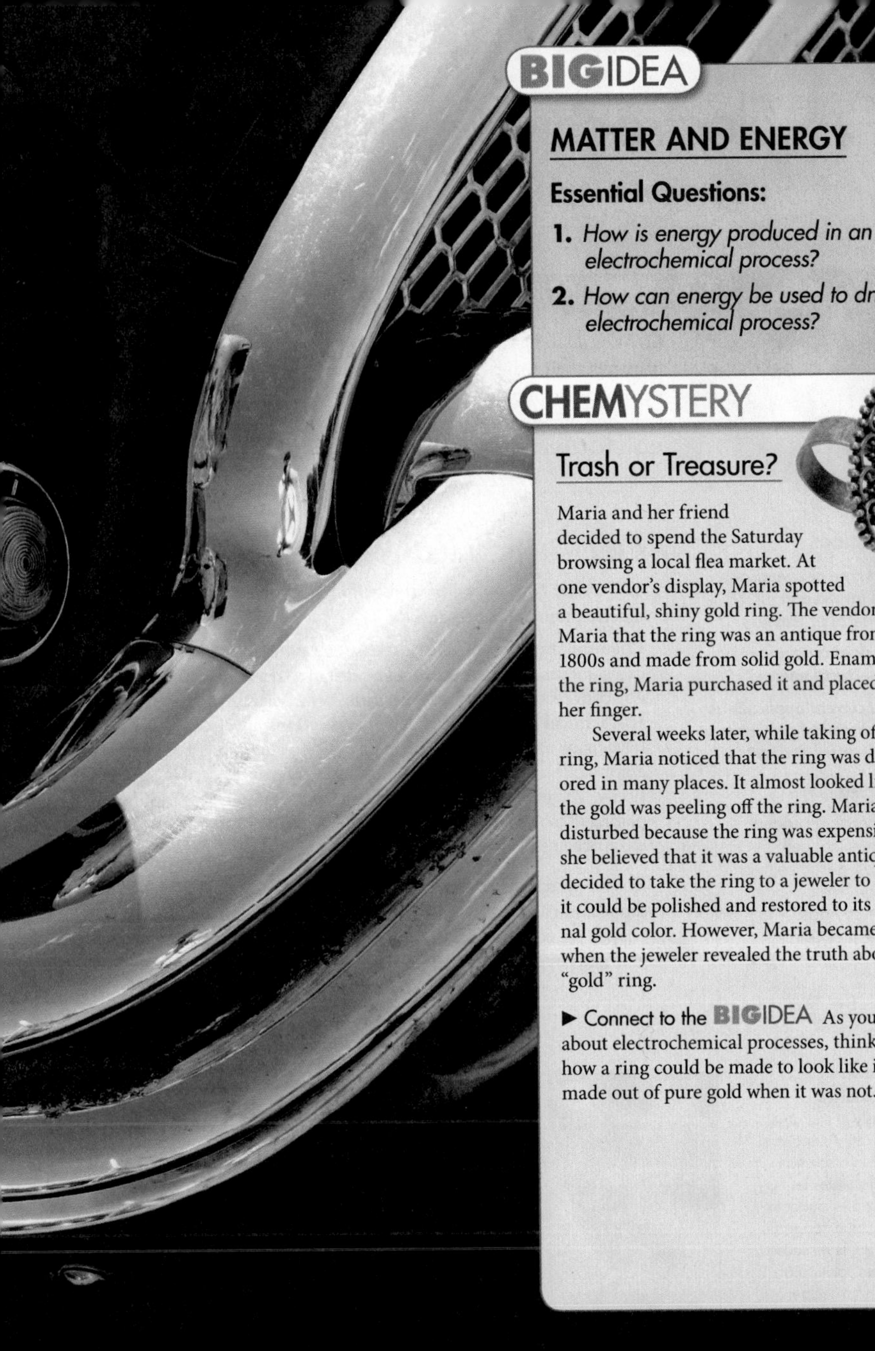

BIGIDEA

MATTER AND ENERGY

Essential Questions:

1. *How is energy produced in an electrochemical process?*
2. *How can energy be used to drive an electrochemical process?*

CHEMYSTERY

Trash or Treasure?

Maria and her friend decided to spend the Saturday browsing a local flea market. At one vendor's display, Maria spotted a beautiful, shiny gold ring. The vendor told Maria that the ring was an antique from the 1800s and made from solid gold. Enamored by the ring, Maria purchased it and placed it on her finger.

Several weeks later, while taking off the ring, Maria noticed that the ring was discolored in many places. It almost looked like the gold was peeling off the ring. Maria was disturbed because the ring was expensive, and she believed that it was a valuable antique. She decided to take the ring to a jeweler to see if it could be polished and restored to its original gold color. However, Maria became upset when the jeweler revealed the truth about her "gold" ring.

► Connect to the **BIG**IDEA As you read about electrochemical processes, think about how a ring could be made to look like it was made out of pure gold when it was not.

UbD

Understanding by Design

Students are building toward understanding electrochemistry using the relationship between matter and energy.

PERFORMANCE GOALS At the end of Chapter 21, students will be able to answer the essential questions by applying their knowledge of electrochemistry.

ESSENTIAL QUESTIONS Read the essential questions aloud. **Ask** What two main types of energy do you think are involved in an electrochemical process? *(electrical and chemical)* **Ask** Why would electrical energy be involved in redox reactions? *(It involves the flow of electrons.)*

BIGIDEA Use the photo of the chrome on a car to help students connect to this chapter's concepts. Activate prior knowledge by asking students to name chrome items familiar to them. **Ask** Is chrome an element? *(No. It is an alloy, or mixture of metals, containing chromium that is deposited onto a surface.)* **Ask** How do you think they get the chrome onto the surface? *(using a redox reaction)*

CHEMYSTERY Have students read over the CHEMystery. Connect the CHEMystery to the Big Idea of Matter and Energy by reminding students that matter can undergo both physical changes and chemical changes. Encourage students to think about whether a physical change or chemical change might have been used to make Maria's ring look like solid gold. Then, have them predict what the change was. As a hint, suggest that they consider how electrical energy could have been used to change the ring.

Introduce the Chapter

CONNECTIONS TO REDOX REACTIONS In Chapter 20, students learned how oxidation-reduction reactions can change materials. Use these activities to introduce students to the connection between redox reactions and electrochemistry.

Activity 1 *You will need a simple circuit composed of wire, a dry cell battery, and a bulb.* Have a student connect the wire to complete the circuit. Point out that an electrical current is needed to light the bulb. Explain that charge moves through the battery by a redox reaction. Tell students that they will learn about electrical applications of redox reactions in this chapter.

Activity 2 *You will need rusted nails, a nut or bolt electroplated with zinc, and a shiny chrome material which has scratches with rust.* Pass the materials around for students to observe. Point out that the coated metal did not rust. Ask How do you think the coating was applied to the metals? *(Students may think it was painted on.)* Explain that electrical energy was used to cause a redox reaction that coated the metal.

Better Batteries Through Chemistry

Haley visits a chemical engineer at MIT to learn how his research is helping to develop batteries with much shorter recharge times.

Key Objectives

21.1.1 IDENTIFY the type of chemical reaction that is involved in all electrochemical processes.

21.1.2 DESCRIBE how a voltaic cell produces electrical energy.

21.1.3 IDENTIFY the current applications that use electrochemical processes to produce electrical energy.

Additional Resources

Reading and Study Workbook, Lesson 21.1

Available Online or on Digital Media:

- Teaching Resources, Lesson 21.1 Review
- Laboratory Manual, Lab 47
- Small-Scale Chemistry Laboratory Manual, Lab 36

PearsonChem.com

Engage

CHEMISTRY & YOU Have students read the opening paragraph. **Ask** What have you learned that shows that energy can be released in the form of light? *(When electrons drop from higher to lower energy levels in an atom they emit energy in the form of light.)*

Activate Prior Knowledge

Review the concepts of reversible reactions and spontaneous and nonspontaneous reactions that were introduced in Chapter 18. **Ask** If a reaction releases free energy, is the reaction spontaneous or nonspontaneous? *(spontaneous)*

21.1 Electrochemical Cells

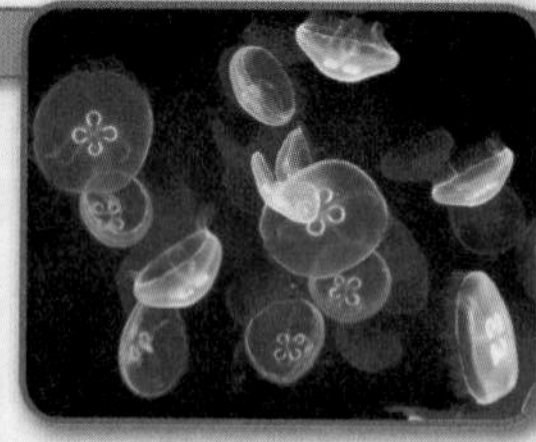

CHEMISTRY & YOU

Q: *Why do some kinds of jellyfish glow?* On a summer evening, fireflies glow to attract their mates. In the ocean depths, angelfish emit light to attract prey. Luminous shrimp, squid, jellyfish, and even bacteria also exist. These organisms, and others, are able to give off energy in the form of light as a result of redox reactions.

Key Questions

What type of chemical reaction is involved in all electrochemical processes?

How does a voltaic cell produce electrical energy?

What current applications use electrochemical processes to produce electrical energy?

Vocabulary

- electrochemical process
- electrochemical cell
- voltaic cell • half-cell
- salt bridge • electrode
- anode • cathode • dry cell
- battery • fuel cell

Electrochemical Processes

What type of chemical reaction is involved in all electrochemical processes?

Chemical processes can either release energy or absorb energy. The energy can sometimes be in the form of electricity. An **electrochemical process** is any conversion between chemical energy and electrical energy. **All electrochemical processes involve redox reactions.** Electrochemical processes have many applications in the home as well as in industry. Flashlight and automobile batteries are familiar examples of devices used to generate electricity. The manufacture of sodium and aluminum metals and the silver-plating of tableware involve the use of electricity. Biological systems also use electrochemistry to carry nerve impulses.

Redox Reactions and the Activity Series When a strip of zinc metal is dipped into an aqueous solution of blue copper(II) sulfate, the zinc becomes copper-plated, as shown in Figure 21.1. The net ionic equation involves only zinc and copper.

$$Zn(s) + Cu^{2+}(aq) \longrightarrow Zn^{2+}(aq) + Cu(s)$$

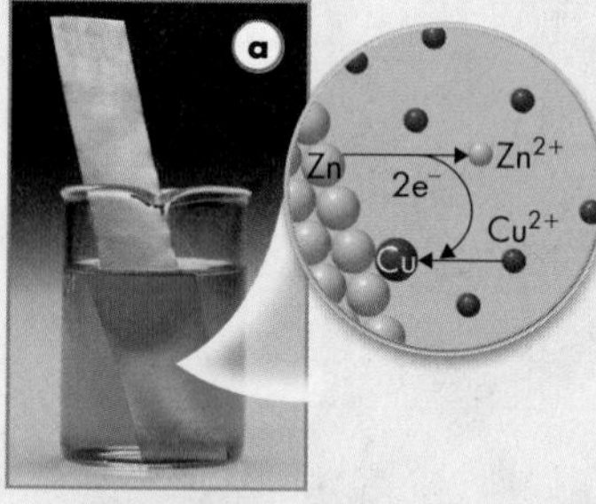

Figure 21.1
Redox Reaction
Zinc metal oxidizes spontaneously in a copper-ion solution. **a.** A zinc strip is immersed in a solution of copper(II) sulfate. **b.** As the copper plates out onto the zinc, the blue copper(II) sulfate solution is replaced by a colorless solution of zinc sulfate. The copper appears black because it is in a finely divided state.

Focus on ELL

1 CONTENT AND LANGUAGE Draw students' attention to the words *anode* and *cathode*. Explain that the word cathode comes from the Greek *kathodos* meaning "way down" and that the word anode comes from the Greek *anodos* meaning "way up." Thus the two words, cathode and anode, are opposites. Write the symbols – and + next to words.

2 FRONTLOAD THE LESSON Ask students to recall what they know about electricity, electrical current, and circuits. Then, have students examine Figure 21.3 and identify features that the voltaic cell shares with an electrical circuit. Explain that, in this lesson, they will learn how an electrical current can be generated using chemicals.

3 COMPREHENSIBLE INPUT Help students visualize a dry cell as a classic aqueous voltaic cell by drawing Figure 21.3 on the board. Have students use a think-aloud strategy to determine which of the dry cell components in Figure 21.4 are equivalent to the aqueous cell components as you write the labels on the drawing.

Interpret Data

Activity Series of Metals

	Element	Oxidation half-reaction
Most active and most easily oxidized	Lithium	$Li(s) \longrightarrow Li^+(aq) + e^-$
	Potassium	$K(s) \longrightarrow K^+(aq) + e^-$
	Barium	$Ba(s) \longrightarrow Ba^{2+}(aq) + 2e^-$
	Calcium	$Ca(s) \longrightarrow Ca^{2+}(aq) + 2e^-$
	Sodium	$Na(s) \longrightarrow Na^+(aq) + e^-$
	Magnesium	$Mg(s) \longrightarrow Mg^{2+}(aq) + 2e^-$
	Aluminum	$Al(s) \longrightarrow Al^{3+}(aq) + 3e^-$
Decreasing activity ↓	Zinc	$Zn(s) \longrightarrow Zn^{2+}(aq) + 2e^-$
	Iron	$Fe(s) \longrightarrow Fe^{2+}(aq) + 2e^-$
	Nickel	$Ni(s) \longrightarrow Ni^{2+}(aq) + 2e^-$
	Tin	$Sn(s) \longrightarrow Sn^{2+}(aq) + 2e^-$
	Lead	$Pb(s) \longrightarrow Pb^{2+}(aq) + 2e^-$
	Hydrogen*	$H_2(g) \longrightarrow 2H^+(aq) + 2e^-$
	Copper	$Cu(s) \longrightarrow Cu^{2+}(aq) + 2e^-$
	Silver	$Ag(s) \longrightarrow Ag^+(aq) + e^-$
	Mercury	$Hg(s) \longrightarrow Hg^{2+}(aq) + 2e^-$
Least active and least easily oxidized	Gold	$Au(s) \longrightarrow Au^{3+}(aq) + 3e^-$

*Hydrogen is included for reference purposes.

Table 21.1 The half-reaction for the oxidation of each metal is shown.

a. Read Tables What is the half-reaction for the oxidation of nickel?

b. Compare Which metal is more readily oxidized, lead or magnesium?

c. Relate Cause and Effect What will happen if a strip of copper is dipped in a solution of silver nitrate? If a reaction occurs, write the half-reactions.

Remember: The more active metal will be oxidized; the less active metal will be reduced.

Electrons are transferred from zinc atoms to copper ions. This is a spontaneous redox reaction. Zinc atoms lose electrons as they are oxidized to zinc ions, while copper ions in solution gain the electrons lost by the zinc. The copper ions are reduced to copper atoms and are deposited as metallic copper. As the copper ions in solution are gradually replaced by zinc ions, the blue color of the solution fades. The balanced half-reactions for this redox reaction can be written as follows:

Oxidation: $Zn(s) \longrightarrow Zn^{2+}(aq) + 2e^-$

Reduction: $Cu^{2+}(aq) + 2e^- \longrightarrow Cu(s)$

In the activity series of metals in Table 21.1, zinc is above copper on the list. For any two metals in an activity series, the more active metal is the more readily oxidized. Zinc is more readily oxidized than copper. When zinc is dipped into a copper(II) sulfate solution, zinc becomes plated with copper. In contrast, when a copper strip is dipped into a solution of zinc sulfate, the copper does not spontaneously become plated with zinc. This is because copper metal is not oxidized by zinc ions.

Differentiated Instruction

L3 ADVANCED STUDENTS Challenge students to research the electrochemical nature of the nervous system. Students may want to focus on different aspects of the system and combine their findings into a visual display and/or oral report. Possible areas to be addressed include the role of sodium and potassium ions, how signals are transmitted across synapses, what an EEG measures, and the value of squid for nervous system research.

L1 STRUGGLING STUDENTS Have students think about the half-reactions in Table 21.1 in terms of where the element is located in the periodic table. Remind them of what they learned in Chapter 6 about periodic trends in chemical properties. Point out that the elements on the left side of the periodic table are more easily oxidized (lose electrons) than the elements on the right side of the periodic table. Explain that Group 1A and 2A metals, listed at the top of the activity series, are some of the most reactive elements in nature.

Foundations for Reading

BUILD VOCABULARY Have students create a concept map from the vocabulary terms for the lesson. Encourage students to include drawings or symbols to enhance the descriptions used to explain the connections made in the map.

READING STRATEGY Show students how to make an outline as they read this section. Have them use the red headings as their main entries and the blue headings as secondary entries. Ask them to include, under each heading, a sentence summarizing the important idea.

Explain

Electrochemical Processes

USE VISUALS Direct students to Figure 21.1. **Ask** What is the color of the copper (II) sulfate solution in Figure 21.1a? *(blue)* Is the color different in Figure 21.1b? *(It's still blue but lighter in color.)* What is the significance of the lighter color? *(Blue copper (II) ions have been removed from the solution and changed into the copper atoms on the zinc.)* Next, direct students' attention to Table 21.1. **Ask** In regard to what you have learned previously, what does *most active* mean? *(The ionization energy needed for an atom to lose an electron is less, therefore it is more likely to occur.)* Have students locate Zn and Cu in the table. **Ask** Look back at the net ionic equation for the copper-plating reaction. Which species was oxidized in that reaction? *(Zn)* **Ask** If a gold bar was placed in the $CuSO_4$ solution, would you expect it to become copper-plated? Why? *(No, it is lower on the activity series than copper, so it would be more likely to be reduced than oxidized.)*

APPLY CONCEPTS Explain that redox reactions make it possible for energy interconversion between electrical energy and chemical energy. Review oxidation and reduction half-cell reactions and stress that the electrons must be balanced. Show how these reactions are combined to form the net ionic equation for an electrochemical process. Point out that the net equation summarizes the transfer of electrons from the species being oxidized to the species being reduced.

Answers

INTERPRET DATA

a. $Ni(s) \rightarrow Ni^{2+}(aq) + 2e^-$

b. magnesium

c. The copper becomes plated with silver.
$Cu(s) \rightarrow Cu^{2+}(aq) + 2e^-$
$Ag^+(aq) + e^- \rightarrow Ag(s)$

LESSON 21.1

Explore

Teacher Demo

PURPOSE Students will observe a redox reaction.

MATERIALS 200 mL 0.1*M* silver nitrate ($AgNO_3$) in a 250 mL beaker, glass stirring rod, strip of polished copper metal

SAFETY Wear gloves, apron, and safety goggles.

PROCEDURE Put a glass stirring rod across the top of the beaker and suspend a polished copper strip from the rod so that the strip dips into the $AgNO_3$ solution. The metal surface will darken and appear "fuzzy" as silver metal is deposited. Over time, a layer of silver will form on the copper. **Ask** What is the significance of the blue color of the solution? *(Blue copper(II) ions are being produced.)* Write on the board the net reaction for the electrochemical process:

$$2Ag^{+}(aq) + Cu(s) \rightarrow 2Ag(s) + Cu^{2+}(aq)$$

Point out that the reaction involves a transfer of electrons from copper atoms to silver ions. **Ask** What was oxidized? What was reduced? *(Copper was oxidized; silver ions were reduced.)* Combine liquid wastes and add a 50% molar excess of NaCl. Filter or decant and dry the AgCl residue. Put in a plastic container and bury in an approved landfill. Flush the filtrate down the drain with excess water.

EXPECTED OUTCOME Silver crystals form on the copper wire. The solution turns blue because of dissolved copper(II) ions.

Electrochemical Cells When a zinc strip is dipped into a copper(II) sulfate solution, electrons are transferred from zinc atoms to copper ions. This flow of electrons is an electric current. If a redox reaction is to be used as a source of electrical energy, the two half-reactions must be physically separated. In the case of the zinc-metal–copper-ion reaction, the electrons released by the zinc atoms must pass through an external circuit to reach the copper ions if useful electrical energy is to be produced. In that situation, the system serves as an electrochemical cell. Alternatively, an electric current can be used to produce a chemical change. That system also serves as an electrochemical cell. Any device that converts chemical energy into electrical energy or electrical energy into chemical energy is an **electrochemical cell.** Redox reactions occur in all electrochemical cells.

Q: *Jellyfish and other creatures that glow contain compounds that undergo redox reactions. What do these reactions have in common with redox reactions that occur in electrochemical cells?*

Voltaic Cells

How does a voltaic cell produce electrical energy?

In 1800, the Italian physicist Alessandro Volta built the first electrochemical cell that could be used to generate a direct electric current (DC). Named after its inventor, a **voltaic cell** is an electrochemical cell used to convert chemical energy into electrical energy. **Electrical energy is produced in a voltaic cell by a spontaneous redox reaction within the cell.** You can find voltaic cells everywhere. They power your flashlight and your mp3 player, as shown in Figure 21.2.

Figure 21.2 Voltaic Cell
A voltaic cell powers this mp3 player.
Predict ***What other items contain voltaic cells?***

Constructing a Voltaic Cell A voltaic cell consists of two half-cells. A **half-cell** is one part of a voltaic cell in which either oxidation or reduction occurs. A typical half-cell consists of a piece of metal immersed in a solution of its ions. Figure 21.3 on the following page shows a voltaic cell that makes use of the zinc–copper reaction. In this cell, one half-cell is a zinc strip immersed in a solution of zinc sulfate. The other half-cell is a copper strip immersed in a solution of copper(II) sulfate.

The half-cells are connected by a **salt bridge,** which is a tube containing a strong electrolyte, often potassium sulfate (K_2SO_4). Salt bridges also contain agar, a gelatinous substance. A porous plate may be used instead of a salt bridge. The salt bridge or porous plate allows ions to pass from one half-cell to the other but prevents the solutions from mixing completely. A wire carries the electrons in the external circuit from the zinc strip to the copper strip. A voltmeter or light bulb can be connected in the circuit. The driving force of such a voltaic cell is the spontaneous redox reaction between zinc metal and copper ions in solution.

The zinc and copper strips in this voltaic cell serve as the electrodes. An **electrode** is a conductor in a circuit that carries electrons to or from a substance other than a metal. The reaction at the electrode determines whether the electrode is labeled as an anode or a cathode. The electrode at which oxidation occurs is called the **anode.** Electrons are produced at the anode. Therefore, the anode is labeled the negative electrode in a voltaic cell. The electrode at which reduction occurs is called the **cathode.** Electrons are consumed at the cathode in a voltaic cell. As a result, the cathode is labeled the positive electrode. Neither electrode is actually charged, however. All parts of the voltaic cell remain balanced in terms of charge at all times. The moving electrons balance any charge that might build up as oxidation and reduction occur.

UbD Check for Understanding

What type of chemical reaction is involved in all electrochemical processes?

Assess students' understanding that electrochemical processes are redox reactions by asking the following questions. **Ask** What is transferred between ions in a redox reaction? *(electrons)* **Ask** What is transferred between ions in an electrochemical reaction? *(electrons)* **Ask** What happens to the ion that gives up electrons in an electrochemical reaction? *(It is oxidized.)* **Ask** What happens to the ion that receives electrons in an electrochemical reaction? *(It is reduced.)*

ADJUST INSTRUCTION If students are confusing oxidation and reduction, have them reread Lesson 20.1 on the Meaning of Oxidation-Reduction as well as the text following the section on Electrochemical Processes in Lesson 21.1. Then, write the chemical reaction shown at the bottom of the previous page on the board and diagram the movement of electrons between the reactants and products to reinforce the concept that this electrochemical reaction is an oxidation-reduction reaction.

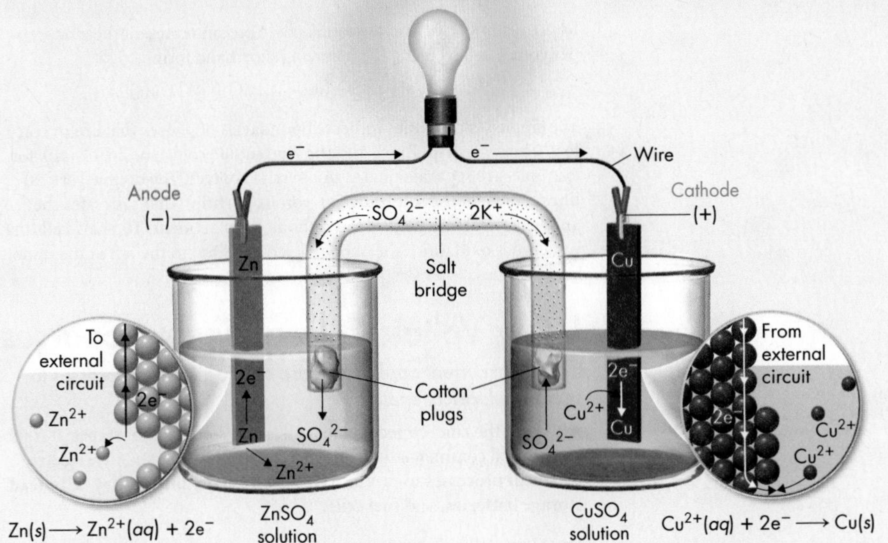

How a Voltaic Cell Works The electrochemical process that occurs in a zinc–copper voltaic cell can best be described in a number of steps. These steps actually occur at the same time.

Step 1 Electrons are produced at the zinc strip according to the oxidation half-reaction:

$$Zn(s) \longrightarrow Zn^{2+}(aq) + 2e^-$$

Zinc is oxidized at the zinc strip, so the zinc strip is the anode, or negative electrode, in the voltaic cell.

Step 2 The electrons leave the zinc anode and pass through the external circuit to the copper strip. (If a bulb is in the circuit, the electron flow will cause it to light. If a voltmeter is present, it will indicate a voltage.)

Step 3 Electrons enter the copper strip and interact with copper ions in solution. There, the following reduction half-reaction occurs:

$$Cu^{2+}(aq) + 2e^- \longrightarrow Cu(s)$$

Copper ions are reduced at the copper strip, so the copper strip is the cathode, or positive electrode, in the voltaic cell.

Step 4 To complete the circuit, both positive and negative ions move through the aqueous solutions via the salt bridge. The two half-reactions can be summed to show the overall cell reaction. Note that the electrons must cancel.

$$\begin{array}{l} Zn(s) \longrightarrow Zn^{2+}(aq) + \cancel{2e^-} \\ Cu^{2+}(aq) + \cancel{2e^-} \longrightarrow Cu(s) \\ \hline Zn(s) + Cu^{2+}(aq) \longrightarrow Zn^{2+}(aq) + Cu(s) \end{array}$$

Figure 21.3 Zinc–Copper Voltaic Cell
In this voltaic cell, the electrons generated from the oxidation of Zn to Zn^{2+} flow through the external circuit (the wire) into the copper strip. These electrons reduce the surrounding Cu^{2+} to Cu. To maintain neutrality in the electrolytes, anions flow through the salt bridge.
Explain ***What is the purpose of the salt bridge?***

See voltaic cells animated online.

LESSON 21.1

Explain

Voltaic Cells

USE VISUALS Direct students' attention to Figure 21.3. Explain that voltaic cells can be used as sources of electrical energy because the two half-reactions are physically separated. The reaction in the illustration could take place in a single beaker, but it would not be possible to produce a stream of electrons. **Ask** At which electrode does oxidation (loss of electrons) take place? *(Oxidation occurs at the anode (negative electrode).)* **Ask** Where does reduction (gain of electrons) take place? *(Reduction occurs at the cathode (positive electrode).)* **Ask** What path do the electrons given up by zinc follow? *(They go through the wire and the electric light to the copper electrode.)* **Ask** What happens to the electrons at the copper electrode? *(They reduce copper ions to copper.)*

APPLY CONCEPTS In discussing the need for a salt bridge in the Zn/Cu voltaic cell, explain that as zinc is oxidized at the anode, Zn^{2+} ions enter the solution. Explain that they have no negative ions to balance their charges, so a positive charge tends to build up around the anode. Similarly, point out that at the cathode, Cu^{2+} ions are reduced to Cu and taken out of the solution, leaving behind unbalanced negative ions. Thus, a negative charge tends to develop around the cathode. Convey that the salt bridge allows negative ions, such as SO_4^{2-}, to be drawn to the anode compartment to balance the growing positive charge. Likewise, positive ions, such as K^+, are drawn from the salt bridge to balance the growing negative charge at the cathode.

Differentiated Instruction

L1 LESS PROFICIENT READERS Students may have difficulty remembering that oxidation occurs at the anode of a voltaic cell and reduction occurs at the cathode. Tell them that consonants go together and vowels go together. Both cathode and reduction begin with consonants and anode and oxidation start with vowels.

ELL ENGLISH LANGUAGE LEARNERS Clarify students' understanding of the word *cell* as it relates to this lesson. Make sure they are not looking for a living cell as they might have learned from biology class. Stress that an electrochemical cell is a chemical system that generates electrical energy.

L1 SPECIAL NEEDS STUDENTS Provide simple DC devices, such as flashlights, cell phones, walkie-talkies, etc. Have students determine how the battery must be inserted in order for the object to power up. Tell students to pay particular attention to the markings on the batteries and the battery compartments.

Answers

FIGURE 21.2 Most handheld electronic devices are powered by voltaic cells, including calculators and flashlights.

FIGURE 21.3 to prevent significant build-up of charges at the electrodes

LESSON 21.1

Explore

Using Voltaic Cells as Energy Sources

Teacher Demo

PURPOSE Students will observe the construction of a dry cell.

MATERIALS dry cell, voltmeter, 2 wire leads, zinc electrode, carbon electrode, manganese dioxide/ammonium chloride paste (Prepare the paste by adding saturated $NH_4Cl(aq)$ to powdered MnO_2 until thick.)

PROCEDURE Before class, cut through a dry cell with a hacksaw from the top down on the right side of the central carbon electrode. In class, have students identify the three parts of the cell. *(The graphite rod is the cathode, the zinc container is the anode, and the manganese dioxide/ammonium chloride paste is the electrolyte.)* Construct a dry cell by placing a zinc electrode and a carbon electrode into a manganese dioxide/ammonium chloride paste. Connect the cell to a voltmeter and measure the potential.

EXPECTED OUTCOME Students should be able to identify the anode, cathode, and electrolyte in each cell.

Explain

APPLY CONCEPTS Explain to students that, by itself, a dry cell does not provide a complete circuit, that is, electrons cannot flow from the anode to the cathode. When devices using dry cells are turned on, an external circuit is completed, allowing the flow of electrons from the anode to the cathode.

USE VISUALS Direct students to Figure 21.4. Ask students to describe the difference between the common dry cell and the alkaline battery. *(In the alkaline dry cell, the electrolyte is a basic KOH paste, thus the distinction of this dry cell as alkaline.)*

Representing Electrochemical Cells You can represent the zinc–copper voltaic cell by using the following shorthand form.

$$Zn(s) \mid ZnSO_4(aq) \parallel CuSO_4(aq) \mid Cu(s)$$

The single vertical lines indicate boundaries of phases that are in contact. The zinc strip, $Zn(s)$, and the zinc sulfate solution, $ZnSO_4(aq)$, for example, are separate phases in physical contact. The double vertical lines represent the salt bridge or porous partition that separates the anode compartment from the cathode compartment. The half-cell that undergoes oxidation (the anode) is written first, to the left of the double vertical lines.

Using Voltaic Cells as Energy Sources

What current applications use electrochemical processes to produce electrical energy?

Although the zinc–copper voltaic cell is of historical importance, it is no longer used commercially. **Current applications that use electrochemical processes to produce electrical energy include dry cells, lead storage batteries, and fuel cells.**

Dry Cells When a compact, portable electrical energy source is required, a dry cell is usually chosen. A **dry cell** is a voltaic cell in which the electrolyte is a paste. In one type of dry cell, a zinc container is filled with a thick, moist electrolyte paste of manganese(IV) oxide (MnO_2), zinc chloride ($ZnCl_2$), ammonium chloride (NH_4Cl), and water (H_2O). As shown in Figure 21.4a, a graphite rod is embedded in the paste. The zinc container is the anode, and the graphite rod is the cathode. The thick paste and its surrounding paper liner prevent the contents of the cell from freely mixing, so a salt bridge is not needed. The half-reactions for this cell are shown below.

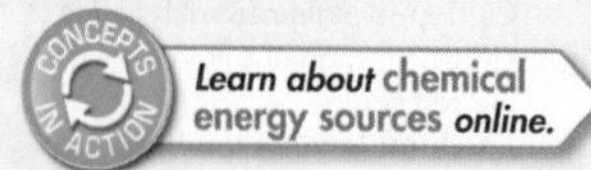

Oxidation: $Zn(s) \longrightarrow Zn^{2+}(aq) + 2e^-$ (at anode)

Reduction: $2MnO_2(s) + 2NH_4^+(aq) + 2e^- \longrightarrow Mn_2O_3(s) + 2NH_3(aq) + H_2O(l)$ (at cathode)

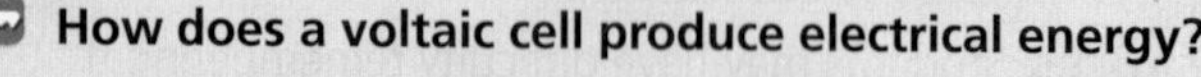

Figure 21.4 Dry Cells Both dry cells and alkaline batteries are single electrochemical cells that produce about 1.5 V. **a.** The dry cell is inexpensive, has a short shelf life, and suffers from voltage drop when in use. **b.** The alkaline battery has a longer shelf life and does not suffer from voltage drop.
Apply Concepts *What is oxidized in these cells and what is reduced?*

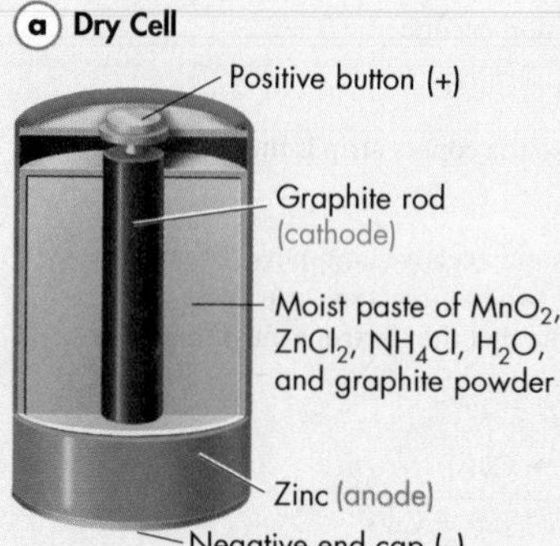

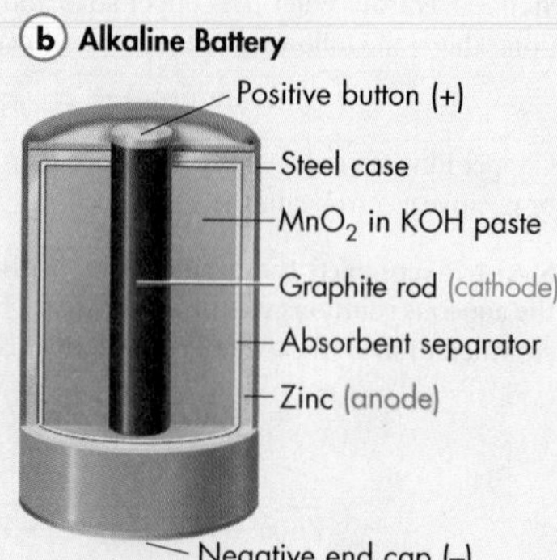

UbD Check for Understanding

How does a voltaic cell produce electrical energy?

Assess students' understanding of how a voltaic cell produces electrical energy by listing the following vocabulary terms on the board: *voltaic cell, half-cell, salt bridge, electrode, anode*, and *cathode*. Then, have students write a brief paragraph using the vocabulary terms that answers the key question.

ADJUST INSTRUCTION If students are having difficulty with this concept, have students reread the Voltaic Cells section. Then, draw or project Figure 21.3 on the board. Read each numbered step in the section How A Voltaic Cell Works, and have students indicate where on the figure that step takes place. Provide students with an opportunity to revise their paragraphs.

In an ordinary dry cell, the graphite rod serves only as a conductor and does not undergo reduction, even though it is the cathode. The manganese in MnO_2 is the species that is actually reduced. The electrical potential of this cell starts out at 1.5 V but decreases steadily during use to about 0.8 V. Dry cells of this type are not rechargeable because the cathode reaction is not reversible.

The alkaline battery, shown in Figure 21.4b on the previous page, is an improved dry cell. In the alkaline battery, the reactions are similar to those in the common dry cell, but the electrolyte is a basic KOH paste. This change in design eliminates the buildup of ammonia gas and maintains the zinc electrode, which corrodes more slowly under basic, or alkaline, conditions.

READING SUPPORT

Build Reading Skills: ***Compare and Contrast*** An alkaline battery is actually a dry cell, whereas a lead storage battery is considered to be a true battery. ***How are alkaline batteries and lead storage batteries similar? How are they different?***

Lead Storage Batteries People depend on lead storage batteries to start their cars. A **battery** is a group of voltaic cells connected together. A 12-V car battery consists of six voltaic cells connected together. Each cell produces about 2 V and consists of lead grids, as shown in Figure 21.5. One set of grids, the anode, is packed with spongy lead. The other set, the cathode, is packed with lead(IV) oxide (PbO_2). The electrolyte for both half-cells in a lead storage battery is sulfuric acid. Using the same electrolyte for both half-cells allows the cell to operate without a salt bridge or porous separator. The half-reactions are as follows:

$$\text{Oxidation: } Pb(s) + SO_4^{2-}(aq) \longrightarrow PbSO_4(s) + 2e^-$$

$$\text{Reduction: } PbO_2(s) + 4H^+(aq) + SO_4^{2-}(aq) + 2e^- \longrightarrow PbSO_4(s) + 2H_2O(l)$$

When a lead storage battery discharges, it produces the electrical energy needed to start a car. The overall spontaneous redox reaction that occurs is the sum of the oxidation and reduction half-reactions.

$$Pb(s) + PbO_2(s) + 2H_2SO_4(aq) \longrightarrow 2PbSO_4(s) + 2H_2O(l)$$

This equation shows that lead(II) sulfate forms during discharge. The sulfate slowly builds up on the plates, and the concentration of the sulfuric acid electrolyte decreases.

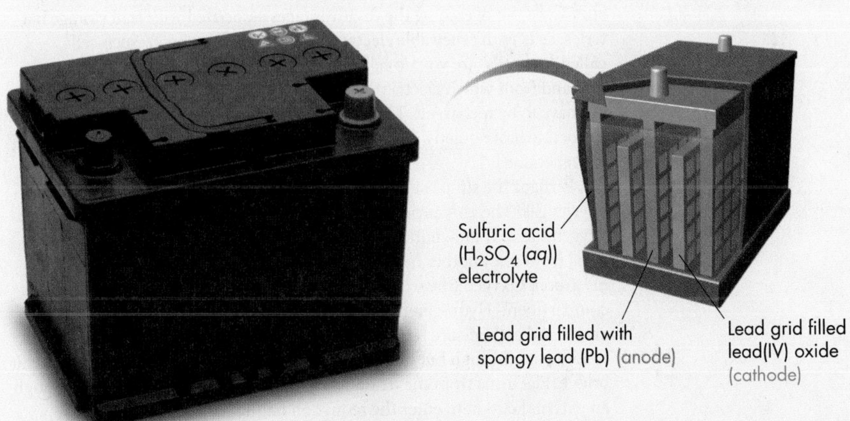

Figure 21.5 Lead Storage Battery
One cell of a 12-V lead storage battery is illustrated here. Current is produced when lead at the anode and lead(IV) oxide at the cathode are both converted to lead(II) sulfate. These processes decrease the sulfuric acid concentration in the battery. Reversing the reaction recharges the battery.

Car Battery

The forerunner of today's car battery was built by the French physicist Gaston Planté in 1859. Planté's first battery consisted of a single cell containing two sheets of lead separated by rubber strips and immersed in a 10% sulfuric acid solution. A year later, he presented another lead-acid battery consisting of nine cells enclosed in a box with outside terminals, to the Académie des Sciences (French Academy of Sciences). Planté's invention was significant because it provided the first means of storing electrical energy.

Explore

Teacher Demo

PURPOSE Students will observe a simple version of the lead storage battery.

MATERIALS 2 strips of lead, wooden rod, 2 connecting wires, 250-mL beaker, dilute sulfuric acid (H_2SO_4), 6-V DC power supply, doorbell

SAFETY Sulfuric acid is corrosive. Wear safety goggles, gloves, and a lab apron.

PROCEDURE Attach two lead strips to a wooden rod so that the strips hang vertically in a 250-mL beaker. Place the strips 4 cm apart. Pour sufficient dilute sulfuric acid into the beaker to cover two-thirds of each strip. Connect a 6-V DC power supply to the strips with wires, and charge the battery for a few minutes. Then, connect the cell to a doorbell.

EXPECTED OUTCOME Students observe that after charging the battery, the bell rings. Discuss the half-cell reactions that occur. Ask students to write the shorthand notation for the electrochemical cell.

[$Pb(s) \mid PbSO_4(s) \parallel PbO_2(s) \mid PbSO_4(s)$]

Explain

START A CONVERSATION Direct students' attention to Figure 21.5. Call attention to the grids, two containing Pb and two containing PbO_2. **Ask** What are the oxidation numbers of the metal lead (Pb) and the lead in lead(IV) oxide (PbO_2)? *(oxidation number of elemental Pb is 0; oxidation number of Pb in PbO_2 is +4.)* **Ask** Which is the anode and which is the cathode? *(Pb is the anode and PbO_2 is the cathode.)* How do you know? *(Pb can only lose electrons and be oxidized, and Pb^{4+} can gain electrons and be reduced.)* Write on the board the overall reaction for the battery when it is discharging. Call attention to the products ($PbSO_4$ and H_2O) and note that Pb is oxidized to form $PbSO_4$ and Pb^{4+} (in PbO_2) is reduced to also form $PbSO_4$. **Ask** What is the oxidation number of the lead in $PbSO_4$? *(+2; Pb is oxidized to Pb^{2+} and Pb^{4+} is reduced to Pb^{2+}.)*

Answers

FIGURE 21.4 In both the alkaline battery and the dry cell, zinc is oxidized and MnO_2 is reduced.

Explain

Using Voltaic Cells as Energy Sources

MAKING CONNECTIONS Point out that fuel cells were developed for space travel where lightweight, reliable power systems were needed. Explain that fuel cells differ from lead storage batteries in that they are not self-contained. Their operation depends on a steady flow of fuel and oxygen into the cell—where combustion takes place—and the flow of the combustion product out of the cell.

Explain that in the case of the hydrogen fuel cell, the product is pure water. Convey that both the electricity generated and the water produced is consumed in space flights. Tell students that fuel cells convert 75% of the available energy into electricity, in contrast with a conventional electric power plant that converts about 35 to 40% of the energy of coal to electricity.

Point out fuels other than hydrogen can also be used, for example, ammonia (NH_3), hydrazine (N_2H_4), and methane (CH_4). Students may be interested in the equations for the reactions that take place in these cells.

$$4NH_3(g) + 3O_2(g) \rightarrow 2N_2(g) + 6H_2O(g)$$

$$2N_2H_4(g) + 2O_2(g) \rightarrow 2N_2(g) + 4H_2O(g)$$

$$CH_4(g) + 2O_2(g) \rightarrow CO_2(g) + 2H_2O(g)$$

Explain that in each case, the products are gases and water vapor which are normally found in Earth's atmosphere.

Extend

Connect to TECHNOLOGY

Divide the class into groups of three or four students. Have them research the use of fuel cells in the space shuttle and by utility companies. Ask them to prepare posters showing different types of fuel cells and their applications. Where possible, have students include the half-cell and overall reactions for each fuel cell. Have them describe the advantages of using fuel cells rather than more conventional sources of electrical energy.

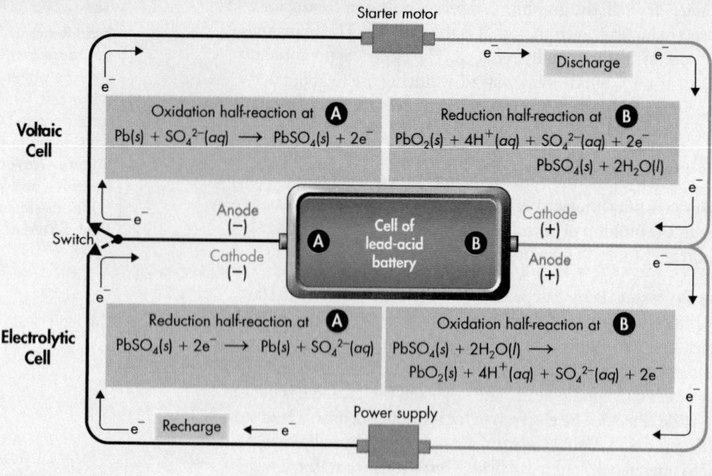

Figure 21.6 Discharge and Recharge of a Lead-Acid Battery
The lead-acid battery in an automobile acts as a voltaic cell (top) when it supplies current to start the engine. Some of the power from the running engine is used to recharge the battery, which then acts as an electrolytic cell (bottom). You will learn more about electrolytic cells in Lesson 21.3.

The reverse reaction occurs when a lead storage battery is recharged. This reaction occurs whenever the car's generator is working properly.

$$2PbSO_4(s) + 2H_2O(l) \longrightarrow Pb(s) + PbO_2(s) + 2H_2SO_4(aq)$$

This is not a spontaneous reaction. To make the reaction proceed as written, a direct current must pass through the cell in a direction opposite that of the current flow during discharge. The processes that occur during the discharge and recharge of a lead-acid battery are summarized in Figure 21.6. In theory, a lead storage battery can be discharged and recharged indefinitely, but in practice its lifespan is limited. Small amounts of lead(II) sulfate fall from the electrodes and collect on the bottom of the cell. Eventually, the electrodes lose so much lead(II) sulfate that the recharging process is ineffective or the cell is shorted out. The battery must then be replaced.

Fuel Cells To overcome the disadvantages associated with lead storage batteries, cells with renewable electrodes have been developed. Such cells, called **fuel cells,** are voltaic cells in which a fuel substance undergoes oxidation and from which electrical energy is continuously obtained. Fuel cells do not have to be recharged. They can be designed to emit no air pollutants and to operate more quietly and more cost-effectively than a conventional electrical generator.

Perhaps the simplest fuel cell involves the reaction of hydrogen gas and oxygen gas. The only product of the reaction is liquid water. In the hydrogen–oxygen fuel cell shown in Figure 21.7a, there are three compartments separated from one another by two electrodes. The electrodes are usually made of carbon. Oxygen (the oxidizing agent) from the air flows into the cathode compartment. Hydrogen (the fuel) flows into the anode compartment. The anode and cathode are separated by a thin membrane that allows hydrogen ions to pass through but not electrons. The membrane therefore acts as a salt bridge. Electrons from the oxidation half-reaction at the anode pass through an external circuit to enter the reduction half-reaction at the cathode.

UbD Check for Understanding

The Essential Question How is energy produced in an electrochemical process?

To assess students' understanding of voltaic cells, give students one to two minutes to explain, (orally, written, or visually) how a voltaic cell produces energy.

ADJUST INSTRUCTION If students' are struggling in their response, direct them to Figure 21.3. Project the figure on the board or a screen, and use various colors of highlighters to visually show each step of how a voltaic cell works.

The half-reactions in this type of hydrogen–oxygen fuel cell are as follows:

Oxidation: $2H_2(g) \longrightarrow 4H^+(aq) + 4e^-$ (at anode)

Reduction: $O_2(g) + 4H^+(aq) + 4e^- \longrightarrow 2H_2O(g)$ (at cathode)

The overall reaction is the oxidation of hydrogen to form water.

$$2H_2(g) + O_2(g) \longrightarrow 2H_2O(g)$$

Other fuels, such as methane (CH_4) and ammonia (NH_3), can be used in place of hydrogen. Other oxidizing agents, such as chlorine (Cl_2) and ozone (O_3), can be used in place of oxygen.

Since the 1960s, astronauts have used fuel cells as an energy source aboard spacecraft. Hydrogen–oxygen fuel cells with a mass of approximately 100 kg each were used in the Apollo spacecraft missions. Fuel cells are well suited for extended space missions because they offer a continuous energy source that releases no pollutants. On space shuttle missions, for example, astronauts drink the water produced by onboard hydrogen–oxygen fuel cells.

The use of fuel cells is no longer limited to space travel. Scientists and engineers have developed fuel-cell cars. These vehicles, such as the one shown in Figure 21.7b, are propelled by electric motors, which are powered by fuel cells. Fuel-cell vehicles can be fueled with pure hydrogen gas, which is stored in high-pressure tanks. However, more research and development is needed before fuel-cell vehicles will predominate the roadways. Currently, fuel cells are expensive to make, and it is difficult to store hydrogen. Nevertheless, you may soon be seeing fuel-cell cars, buses, and bicycles. You may even one day own a cellphone or laptop that is powered by a miniature fuel cell.

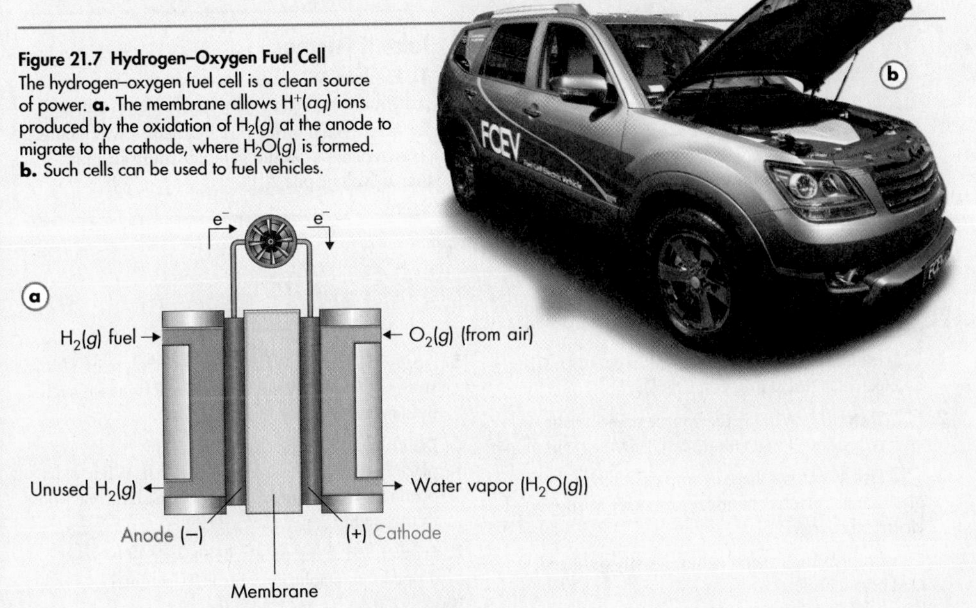

Figure 21.7 Hydrogen–Oxygen Fuel Cell
The hydrogen–oxygen fuel cell is a clean source of power. **a.** The membrane allows $H^+(aq)$ ions produced by the oxidation of $H_2(g)$ at the anode to migrate to the cathode, where $H_2O(g)$ is formed. **b.** Such cells can be used to fuel vehicles.

Explain

Connect to ENGINEERING

Tell students that electric cars are quiet and nonpolluting—no noisy engines spew noxious gases into the atmosphere. So why have car makers not built, promoted, and sold many electric vehicles? Have interested students research the current status of electric cars in the marketplace. Have them find out what obstacles stand in the way of mass usage of these cars. Students could choose to write a report or prepare an oral presentation. *(One obstacle that students may discover is the need for inexpensive, lightweight batteries for efficient storing of electrical energy to lengthen the distance of travel without recharging.)*

Evaluate

Informal Assessment

Have students make a sketch of a tin/lead voltaic cell (*Sn* | *SnSO*$_4$ || *PbSO*$_4$ | *Pb*). Students should label the cathode and anode, and indicate the direction of electron flow.

(Tin is the anode; lead is the cathode. The electrons flow from tin to lead.) Ask students to then write the equations for the half-reactions.

$$[Sn(s) \rightarrow Sn^{2+}(aq) + 2e^-$$
$$Pb^{2+}(aq) + 2e^- \rightarrow Pb(s)]$$

Then, have students complete the 21.1 Lesson Check.

Reteach

Emphasize that a chemical reaction can produce a flow of electrons or a flow of electrons can cause a chemical reaction to occur. Note that reduction always occurs at the cathode, and oxidation always occurs at the anode. Discuss the half-reactions for the charging process in a lead cell and compare them to the half-reactions when the cell is producing electric current.

Electroplating

Electroplating is the use of electrolysis to deposit a thin coating of metal on an object, often to enhance its appearance or value. Materials often used for electroplating include gold, silver, copper, nickel, and chromium. In an electrolytic cell, used for plating, the object to be plated is the cathode and the anode is the plating metal. The cathode in an electroplating apparatus must be able to conduct electricity. Thus, one might think that nonmetallic substances such as wood, plastic, and leather could not be electroplated. However, chromium-plated plastic automobile parts enhance the appearance of most cars. Interested students may want to research the pretreatment of plastics that makes electroplating possible, or how a base metal can be electroplated with an alloy, such as brass or bronze.

Electrochemistry 735

CHEMISTRY & YOU

Tell students that Italian physicist Alessandro Volta had already been experimenting with static electricity when, in 1780, his friend Luigi Galvani demonstrated that a current of electricity could be generated if two different metals, copper and zinc, were placed in contact with the muscle of a frog. At that time, scientists believed that nerves acted as water pipes. Galvani was convinced that he'd discovered a new form of electricity—one that was generated by animal tissues and transmitted through the nerves as an electrical fluid. Volta later showed that animal tissue was not necessary—current flowed with only the two dissimilar metals. Galvani's research led to the eventual discovery that nerves transmit electrical impulses. Pose the following challenge to students. *Compare and contrast the electrical current generated by a nerve cell and the current generated by a voltaic cell.* You may need to assist students in the following ways:

- Nerve cells contain sodium and potassium ions.
- Nerve impulses are generated when sodium ions flow into the cell and potassium ions flow out of the cell in response to a stimulus, creating an imbalance in charge across the membrane.

Answers

TAKE IT FURTHER

1. to conduct current between the two metals
2. The "crown of cups" has a sufficient amount of electrolyte solution, whereas the "voltaic pile" can easily dry out.

CHEMISTRY & YOU: HISTORY

Alessandro Volta

In the late 1770s, Italian physicist Alessandro Volta (1745–1827) discovered that contact between two different metals could produce electricity. Using this knowledge, Volta began experimenting with ways to produce a steady electric current. In 1799, he built a stack of alternating zinc and copper discs, separated by pasteboard soaked in saltwater. When he connected a wire to both ends of the pile, a steady current flowed. This device, called the "voltaic pile," was the first battery.

Volta found that different types of metals could change the amount of current produced and that he could increase the current by adding disks to the stack. Later, he improved on the pile by creating a "crown of cups"—separate cups of salt solution linked by metal straps. In 1810, Napoleon Bonaparte gave Volta the title of Count in honor of his work, and in 1881 the unit of electrical potential was named the "volt" in Volta's honor.

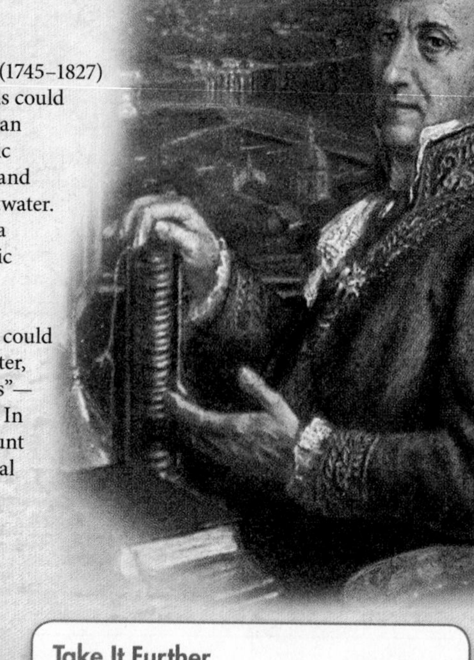

THE BATTERY REINVENTED Modern batteries look much different from Volta's alternating zinc and copper discs.

Take It Further

1. **Explain** What was the function of the pasteboard soaked in saltwater in Volta's battery?
2. **Infer** What is one possible reason why Volta's "crown of cups" would generate more current than a "voltaic pile"?

ONLINE PROBLEMS 21.1 LessonCheck

1. **Identify** What type of reaction occurs during an electrochemical process?
2. **Describe** What is the source of electrical energy produced in a voltaic cell?
3. **List** What are three examples of technologies that use electrochemical processes to supply electrical energy?
4. **Compare** Which metal is more easily oxidized, lead or calcium?
5. **Apply Concepts** What is the electrolyte in a lead storage battery? Write the half-reactions for such a battery.
6. **Describe** Write the overall reaction that takes place in a hydrogen–oxygen fuel cell. What product(s) are formed? Describe the half-reactions in this cell.
7. **Predict** What happens when a strip of copper is dipped into a solution of iron(II) sulfate?

Lesson Check Answers

1. a redox reaction
2. spontaneous redox reactions within the cell
3. fuel cells, lead storage batteries, and dry cells
4. calcium
5. concentrated sulfuric acid;
   Anode:
   $Pb(s) + SO_4^{2-}(aq) \rightarrow PbSO_4(s) + 2e^-$
   Cathode:
   $PbO_2(s) + 4H^+(aq) + SO_4^{2-}(aq) + 2e^- \rightarrow PbSO_4(s) + 2H_2O(l)$
6. $2H_2(g) + O_2(g) \rightarrow 2H_2O(l)$; water is the product. H_2 is oxidized at the anode; O_2 is reduced at the cathode.
7. no reaction

21.2 Half-Cells and Cell Potentials

CHEMISTRY & YOU

Q: *How can you calculate the electrical potential of a cell in a laptop battery?* Batteries provide current to power lights and many kinds of electronic devices—such as the laptop shown here. The electrical potential between the negative and positive terminals of a lithium laptop battery is 3.7 V. In this lesson, you will learn how to calculate electrical potential.

Key Questions

What causes the electrical potential of an electrochemical cell?

How can you determine the standard reduction potential of a half-cell?

How can you determine if a redox reaction is spontaneous?

Vocabulary

- electrical potential
- reduction potential
- cell potential
- standard cell potential
- standard hydrogen electrode

Electrical Potential

What causes the electrical potential of an electrochemical cell?

The **electrical potential** of a voltaic cell is a measure of the cell's ability to produce an electric current. Electrical potential is usually measured in volts (V). The potential of an isolated half-cell cannot be measured. For example, you cannot measure the electrical potential of a zinc half-cell or of a copper half-cell separately. When these two half-cells are connected to form a voltaic cell, however, the difference in potential can be measured. **The electrical potential of a cell results from a competition for electrons between two half-cells.** The half-cell that has a greater tendency to acquire electrons is the one in which reduction occurs. Oxidation occurs in the other half-cell. The tendency of a given half-reaction to occur as a reduction is called the **reduction potential.** The half-cell in which reduction occurs has a greater reduction potential than the half-cell in which oxidation occurs. The difference between the reduction potentials of the two half-cells is called the **cell potential.**

$$\text{cell potential} = \begin{pmatrix}\text{reduction potential}\\ \text{of half-cell in which}\\ \text{reduction occurs}\end{pmatrix} - \begin{pmatrix}\text{reduction potential}\\ \text{of half-cell in which}\\ \text{oxidation occurs}\end{pmatrix}$$

$$\text{or } E_{cell} = E_{red} - E_{oxid}$$

The **standard cell potential** (E°_{cell}) is the measured cell potential when the ion concentrations in the half-cells are 1*M*, any gases are at a pressure of 101 kPa, and the temperature is 25°C. The symbols E°_{red} and E°_{oxid} represent the standard reduction potentials for the reduction and oxidation half-cells, respectively. The relationship between these values follows the general relationship for cell potential above.

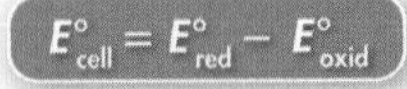

$$E^\circ_{cell} = E^\circ_{red} - E^\circ_{oxid}$$

Focus on ELL

1 CONTENT AND LANGUAGE Begin the lesson by writing the word *potential* on the board and pronouncing it slowly. Explain that the word potential comes from the Latin *potens* meaning "power." Explain that electrical potential is the ability of a cell to produce an electric current.

2 FRONTLOAD THE LESSON Read the first page of the text aloud slowly, and with the participation of the class, summarize the contents in a sentence or two. Ask students to write the summary in their notebooks. Continue this process of summarizing the principal ideas for each heading.

3 COMPREHENSIBLE INPUT Create a voltaic cell and have students assist you in calculating its cell potential. Then, connect a voltmeter to the electrodes and allow students to observe the display to compare the calculated results to the actual potential.

Key Objectives

21.2.1 IDENTIFY what causes the electrical potential of an electrochemical cell.

21.2.2 DETERMINE the standard reduction potential of a half-cell.

21.2.3 DETERMINE if a redox reaction is spontaneous or nonspontaneous.

Additional Resources

Reading and Study Workbook, Lesson 21.2

Available Online or on Digital Media:

- Teaching Resources, Lesson 21.2 Review
- Laboratory Manual, Lab 47

PearsonChem.com

Engage

CHEMISTRY & YOU Have students read the opening paragraph. Point out that many rechargeable batteries contain lithium. **Ask** What characteristic makes lithium a good choice for a battery terminal? *(It is the most easily oxidized metal element.)* Have students keep this characteristic in mind as they proceed with the lesson.

Activate Prior Knowledge

Prepare students for the content in this lesson by asking a few questions about electrochemical and voltaic cells. **Ask** What process goes on at the anode? *(oxidation)* **Ask** What process goes on the cathode? *(reduction)* **Ask** In which process are electrons lost? *(oxidation)*

LESSON 21.2

Foundations for Reading

BUILD VOCABULARY Clarify that the word *cell* as used in this lesson refers to electrochemical cells that convert chemical energy into electrical energy. Have students use their understanding of electrochemical cells and of potential energy to predict the meaning of the vocabulary words. Then, have students check and correct their predictions.

READING STRATEGY Have students monitor their understanding by identifying what they don't yet understand after reading a section. Have students attempt to rewrite the section using a bulleted list, or encourage them to ask for help. Then, have students write a statement showing a clarified understanding of the section.

Explain

Electrical Potential

USE VISUALS Have students study the diagram in Figure 21.8. **Ask** What is the significance of the standard hydrogen electrode and how is it different from other electrodes? *(It has been assigned a standard reduction potential of 0.00 V at 25°C.)* **Ask** What does the standard hydrogen half-cell consist of? *(A platinum electrode with hydrogen gas at atmospheric pressure bubbling over its surface; the electrode is immersed in 1.00M acid solution at 25°C.)*

USE AN ANALOGY Students may be confused by the arbitrary assignment of 0.00 V to the hydrogen half-cell. Use a diagram of the Celsius temperature scale as an analogy. Explain that when the Celsius temperature scale was invented, 0°C was arbitrarily assigned to the temperature at which water freezes; all measured temperatures are relative to that assigned value. Explain that temperatures higher than the freezing point of water are positive values of temperature and temperatures lower than 0°C are negative values of temperature.

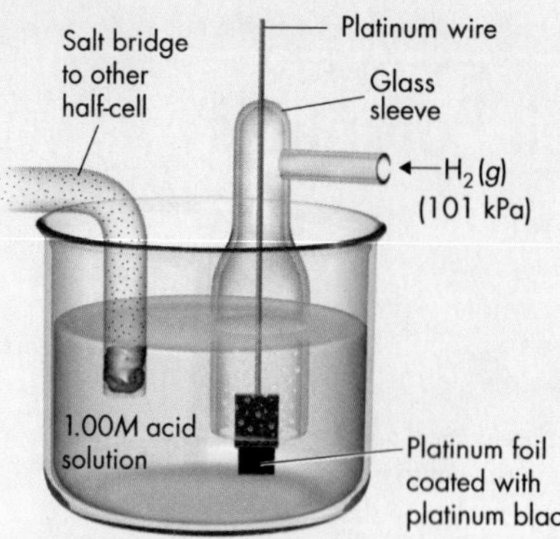

Figure 21.8 Standard Hydrogen Electrode
The standard hydrogen electrode is arbitrarily assigned a standard reduction potential of 0.00 V at 25°C.

Half-cell potentials cannot be measured directly, so scientists have chosen an arbitrary electrode to serve as a reference. The **standard hydrogen electrode** is used with other electrodes, so the reduction potentials of the other cells can be measured. The standard reduction potential of the hydrogen electrode has been assigned a value of 0.00 V. The standard hydrogen electrode, which is illustrated in Figure 21.8, consists of a platinum electrode immersed in a solution with a hydrogen-ion concentration of 1*M*. The solution is at 25°C. The electrode itself is a small square of platinum foil coated with finely divided platinum, known as platinum black. Hydrogen gas at a pressure of 101 kPa is bubbled around the platinum electrode. The half-cell reaction that occurs at the platinum black surface is as follows:

$$2H^+(aq, 1M) + 2e^- \rightleftharpoons H_2(g, 101\ kPa) \quad E^\circ_{H^+} = 0.00\ V$$

The double arrows in the equation indicate that the reaction is reversible. The symbol $E^\circ_{H^+}$ represents the standard reduction potential of H^+. The standard reduction potential of H^+ is the tendency of H^+ ions to acquire electrons and be reduced to $H_2(g)$. Whether this half-cell reaction occurs as a reduction or as an oxidation is determined by the reduction potential of the half-cell to which the standard hydrogen electrode is connected.

Standard Reduction Potentials

How can you determine the standard reduction potential of a half-cell?

A voltaic cell can be made by connecting a standard hydrogen half-cell to a standard zinc half-cell, as shown in Figure 21.9. To determine the overall reaction for this cell, first identify the half-cell in which reduction takes place. In all electrochemical cells, reduction takes place at the cathode, and oxidation takes place at the anode. A voltmeter gives a reading of +0.76 V when the zinc electrode is connected to the negative terminal and the hydrogen electrode is connected to the positive terminal. The zinc is oxidized, which means that it is the anode. Hydrogen ions are reduced, which means that the hydrogen electrode is the cathode. You can now write the half-reactions and the overall cell reaction.

Oxidation:	$Zn(s) \longrightarrow Zn^{2+}(aq) + 2e^-$ (at anode)
Reduction:	$2H^+(aq) + 2e^- \longrightarrow H_2(g)$ (at cathode)
Cell reaction:	$Zn(s) + 2H^+(aq) \longrightarrow Zn^{2+}(aq) + H_2(g)$

You can determine the standard reduction potential of a half-cell by using a standard hydrogen electrode and the equation for standard cell potential. In the zinc–hydrogen cell, zinc is oxidized and hydrogen ions are reduced. Let $E^\circ_{red} = E^\circ_{H^+}$ and $E^\circ_{oxid} = E^\circ_{Zn^{2+}}$ in the standard cell potential equation.

$$E^\circ_{cell} = E^\circ_{red} - E^\circ_{oxid}$$
$$E^\circ_{cell} = E^\circ_{H^+} - E^\circ_{Zn^{2+}}$$

Differentiated Instruction

L1 SPECIAL NEEDS STUDENTS Engage students in a role playing activity to reinforce the direction of electron flow in a voltaic cell. Give students small cards with a large e^- printed on each one. Divide students into pairs. Let one student role play being an anode and the other a cathode. The anode should hand one or more cards to the cathode.

LPR LESS PROFICIENT READERS Have pairs of students look at the diagram of the voltaic cell in Figure 21.9. Have them trace the movement of the electrons from one electrode to the other. They should be able to explain that the electrons travel up from the zinc anode into the external wire, through the voltmeter, and down through the wire into the hydrogen/platinum cathode. Have them identify where oxidation is taking place and where reduction is taking place. Refer them to Table 21.2 on page 740 to find the appropriate half-cell reactions and have them write the overall reaction for the cell.

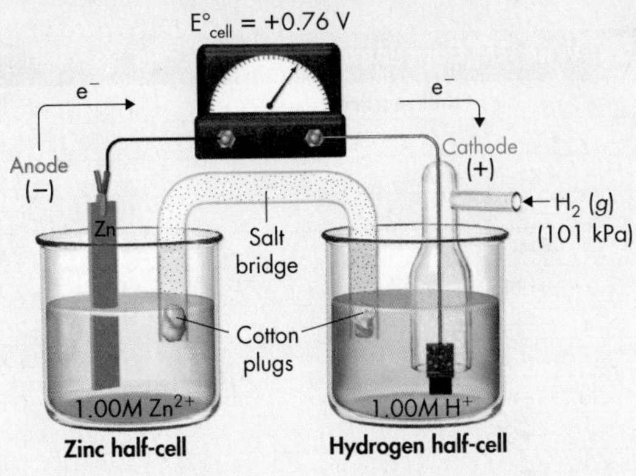

Figure 21.9 Zinc–Hydrogen Cell
This voltaic cell consists of zinc and hydrogen half-cells.
Interpret Diagrams ***Where does reduction occur? What species is reduced in this cell?***

The cell potential (E°_{cell}) is measured at $+0.76$ V. The reduction potential of the hydrogen half-cell is a defined standard: $E^\circ_{H^+}$ always equals 0.00 V. Substituting these values into the preceding equation will give the standard reduction potential for the zinc half-cell.

$$+0.76 \text{ V} = 0.00 \text{ V} - E^\circ_{Zn^{2+}}$$
$$E^\circ_{Zn^{2+}} = -0.76 \text{ V}$$

The standard reduction potential for the zinc half-cell is -0.76 V. The value is negative because the tendency of zinc ions to be reduced to zinc metal in this cell is less than the tendency of hydrogen ions to be reduced to hydrogen gas.

Many different half-cells can be paired with the hydrogen half-cell in a similar manner. Using this method, the standard reduction potential for each half-cell can be obtained. For a standard copper half-cell, for example, the measured standard cell potential is $+0.34$ V when the copper electrode is connected to the positive electrode and the hydrogen electrode is connected to the negative terminal. Copper is the cathode, and copper ions are reduced to copper metal when the cell operates. The hydrogen half-cell is the anode, and hydrogen gas is oxidized to hydrogen ions. You can calculate the standard reduction potential for copper as follows:

$$E^\circ_{cell} = E^\circ_{red} - E^\circ_{oxid}$$
$$E^\circ_{cell} = E^\circ_{Cu^{2+}} - E^\circ_{H^+}$$
$$+0.34 \text{ V} = E^\circ_{Cu^{2+}} - 0.00 \text{ V}$$
$$E^\circ_{Cu^{2+}} = +0.34 \text{ V}$$

This value is positive because the tendency for copper ions to be reduced in the cell is greater than the tendency of hydrogen ions to be reduced.

Table 21.2 on the following page lists some standard reduction potentials at 25°C. The half-reactions are arranged in increasing order of their tendency to occur in the forward direction—that is, as a reduction. Thus, the half-reactions at the top of the table have the least tendency to occur as reductions. The half-reactions at the bottom of the table have the greatest tendency to occur as reductions.

Q: *What do you need to know to calculate the electrical potential of a cell in a laptop battery?*

UbD Check for Understanding

How can you determine the standard reduction potential of a half-cell?

Assess students' understanding of the role of the standard hydrogen electrode in determining standard reduction potentials by having them write a one-minute summary explaining how a standard hydrogen electrode and the standard cell potential equation are used to determine the standard reduction potential of a metal. *(A voltaic cell is created using the metal as one half-cell and a standard hydrogen electrode as the other half-cell. The cell potential (E°_{cell}) is measured and plugged into the standard cell potential equation along with the standard reduction potential of the hydrogen electrode. Once the cathode and anode in the reaction are identified, the equation can be solved for the standard reduction potential of the metal.)*

LESSON 21.2

Explore

Standard Reduction Potentials

Teacher Demo

PURPOSE Students will observe the galvanic corrosion of iron compared to copper.

MATERIALS 2 pennies, metal wires from twist ties, table salt, paper towel

PROCEDURE Thoroughly clean the pennies and the wires. Moisten a paper towel with tap water and sprinkle it with salt. Wrap one wire around a penny and place it on the paper towel. Place an unwrapped penny and a coil of wire on the paper towel as a control. Observe what happens after at least 24 hours.

EXPECTED OUTCOMES Iron corroded; copper did not. Ask students to explain the outcomes. *(In galvanic corrosion, the corrosion rate (oxidation) of one metal accelerates while that of the other is decreased. Because iron is more active than copper, its corrosion was accelerated.)*

Answers

FIGURE 21.9 Reduction occurs at the cathode. In this cell, hydrogen ions are reduced to hydrogen gas.

Explain

APPLY CONCEPTS Explain that the two half-cells of a voltaic cell are competing for electrons and that oxidation or reduction could occur in either cell. Point out that the half-cell with the more positive reduction potential will win the competition and undergo reduction; oxidation will occur in the other half-cell. Explain that the potential produced by the electrochemical cell is the difference in the reduction potentials of the two half-cell reactions. Explain that the quantitative value of any half-cell potential is obtained by measuring it against the standard hydrogen electrode. Point out that Table 21.2 contains tabulated values for some half-cells. These can be used to calculate cell potentials.

Table 21.2

Reduction Potentials at 25°C With 1*M* Concentrations of Aqueous Species

	Electrode	Half-reaction	$E°$ (V)
Least tendency to occur as a reduction	Li^+/Li	$Li^+ + e^- \longrightarrow Li$	−3.05
	K^+/K	$K^+ + e^- \longrightarrow K$	−2.93
	Ba^{2+}/Ba	$Ba^{2+} + 2e^- \longrightarrow Ba$	−2.90
	Ca^{2+}/Ca	$Ca^{2+} + 2e^- \longrightarrow Ca$	−2.87
	Na^+/Na	$Na^+ + e^- \longrightarrow Na$	−2.71
Increasing tendency to occur as a reduction (stronger oxidizing agent) ↓	Mg^{2+}/Mg	$Mg^{2+} + 2e^- \longrightarrow Mg$	−2.37
	Al^{3+}/Al	$Al^{3+} + 3e^- \longrightarrow Al$	−1.66
	H_2O/H_2	$2H_2O + 2e^- \longrightarrow H_2 + 2OH^-$	−0.83
	Zn^{2+}/Zn	$Zn^{2+} + 2e^- \longrightarrow Zn$	−0.76
	Cr^{3+}/Cr	$Cr^{3+} + 3e^- \longrightarrow Cr$	−0.74
	Fe^{2+}/Fe	$Fe^{2+} + 2e^- \longrightarrow Fe$	−0.44
	H_2O/H_2 (pH 7)	$2H_2O + 2e^- \longrightarrow H_2 + 2OH^-$	−0.42
	Cd^{2+}/Cd	$Cd^{2+} + 2e^- \longrightarrow Cd$	−0.40
	$PbSO_4/Pb$	$PbSO_4 + 2e^- \longrightarrow Pb + SO_4^{2-}$	−0.36
	Co^{2+}/Co	$Co^{2+} + 2e^- \longrightarrow Co$	−0.28
	Ni^{2+}/Ni	$Ni^{2+} + 2e^- \longrightarrow Ni$	−0.25
	Sn^{2+}/Sn	$Sn^{2+} + 2e^- \longrightarrow Sn$	−0.14
	Pb^{2+}/Pb	$Pb^{2+} + 2e^- \longrightarrow Pb$	−0.13
	Fe^{3+}/Fe	$Fe^{3+} + 3e^- \longrightarrow Fe$	−0.036
	H^+/H_2	$2H^+ + 2e^- \longrightarrow H_2$	0.000
	$AgCl/Ag$	$AgCl + e^- \longrightarrow Ag + Cl^-$	+0.22
	Hg_2Cl_2/Hg	$Hg_2Cl_2 + 2e^- \longrightarrow 2Hg + 2Cl^-$	+0.27
	Cu^{2+}/Cu	$Cu^{2+} + 2e^- \longrightarrow Cu$	+0.34
	O_2/OH^-	$O_2 + 2H_2O + 4e^- \longrightarrow 4OH^-$	+0.40
	Cu^+/Cu	$Cu^+ + e^- \longrightarrow Cu$	+0.52
	I_2/I^-	$I_2 + 2e^- \longrightarrow 2I^-$	+0.54
	Fe^{3+}/Fe^{2+}	$Fe^{3+} + e^- \longrightarrow Fe^{2+}$	+0.77
	Hg_2^{2+}/Hg	$Hg_2^{2+} + 2e^- \longrightarrow 2Hg$	+0.79
	Ag^+/Ag	$Ag^+ + e^- \longrightarrow Ag$	+0.80
	O_2/H_2O (pH 7)	$O_2 + 4H^+ + 4e^- \longrightarrow 2H_2O$	+0.82
	Hg^{2+}/Hg	$Hg^{2+} + 2e^- \longrightarrow Hg$	+0.85
	Br_2/Br^-	$Br_2 + 2e^- \longrightarrow 2Br^-$	+1.07
	O_2/H_2O	$O_2 + 4H^+ + 4e^- \longrightarrow 2H_2O$	+1.23
	MnO_2/Mn^{2+}	$MnO_2 + 4H^+ + 2e^- \longrightarrow Mn^{2+} + 2H_2O$	+1.28
	$Cr_2O_7^{2-}/Cr^{3+}$	$Cr_2O_7^{2-} + 14H^+ + 6e^- \longrightarrow 2Cr^{3+} + 7H_2O$	+1.33
	Cl_2/Cl^-	$Cl_2 + 2e^- \longrightarrow 2Cl^-$	+1.36
	PbO_2/Pb^{2+}	$PbO_2 + 4H^+ + 2e^- \longrightarrow Pb^{2+} + 2H_2O$	+1.46
Greatest tendency to occur as a reduction	MnO_4^-/Mn^{2+}	$MnO_4^- + 8H^+ + 5e^- \longrightarrow Mn^{2+} + 4H_2O$	+1.51
	$PbO_2/PbSO_4$	$PbO_2 + 4H^+ + SO_4^{2-} + 2e^- \longrightarrow PbSO_4 + 2H_2O$	+1.69
	F_2/F^-	$F_2 + 2e^- \longrightarrow 2F^-$	+2.87

Differentiated Instruction

L1 STRUGGLING STUDENTS Some students may have trouble remembering which reactions in Table 21.2 have the greatest tendency to occur as reductions. Have them compare the table with the activity series of metals in Table 21.1 in Lesson 21.1. Point out that the most active metals are at the top of Table 21.1; ions of these same metals are at the top of Table 21.2. Explain that because active metals lose electrons easily, they are most easily oxidized. By the same reasoning, point out that their ions are least likely to be reduced.

L3 ADVANCED STUDENTS Have students investigate the practice of attaching large pieces of magnesium, called hull anodes, to the hulls of ships having steel (iron) hulls. Ask them to explain the reason for this practice and provide any supporting figures they can. *(The more active metal, magnesium, is oxidized rather than the iron. The reduction potential for magnesium (Mg^{2+}/Mg) is –2.37 V and the reduction potential for iron (Fe^{3+}/Fe) is –0.036 V).*

Calculating Standard Cell Potentials

How can you determine if a redox reaction is spontaneous?

In an electrochemical cell, the half-cell reaction having the more positive (or less negative) reduction potential occurs as a reduction in the cell. You can use the known standard reduction potentials for the half-cells (from Table 21.2) to predict the half-cells in which reduction and oxidation will occur and to find the E°_{cell} value without having to actually assemble the cell. **If the cell potential for a given redox reaction is positive, then the reaction is spontaneous as written. If the cell potential is negative, then the reaction is nonspontaneous.** This latter reaction will be spontaneous in the reverse direction, and the cell potential will then have a numerically equal but positive value.

Sample Problem 21.1

Determining Reaction Spontaneity

Show that the following redox reaction between zinc metal and silver ions is spontaneous.

$$Zn(s) + 2Ag^+(aq) \longrightarrow Zn^{2+}(aq) + 2Ag(s)$$

1 Analyze **List the knowns and the unknown.** Identify the half-reactions, and calculate the standard cell potential ($E^\circ_{cell} = E^\circ_{red} - E^\circ_{oxid}$). If E°_{cell} is positive, the reaction is spontaneous.

KNOWNS

cell reaction: $Zn(s) + 2Ag^+(aq) \longrightarrow Zn^{2+}(aq) + 2Ag(s)$

UNKNOWN

Is the reaction spontaneous?

2 Calculate **Solve for the unknown.**

First, identify the half-reactions.

Oxidation: $Zn(s) \longrightarrow Zn^{2+}(aq) + 2e^-$

Reduction: $Ag^+(aq) + e^- \longrightarrow Ag(s)$

Write both half-cells as reductions with their standard reduction potentials.

$Zn^{2+}(aq) + 2e^- \longrightarrow Zn(s) \quad E^\circ_{Zn^{2+}} = -0.76\,V$

$Ag^+(aq) + e^- \longrightarrow Ag(s) \quad E^\circ_{Ag^+} = +0.80\,V$

Calculate the standard cell potential.

$E^\circ_{cell} = E^\circ_{red} - E^\circ_{oxid} = E^\circ_{Ag^+} - E^\circ_{Zn^{2+}}$

$= +0.80\,V - (-0.76\,V) = +1.56\,V$

$E^\circ_{cell} > 0$, so the reaction is spontaneous.

Use Table 21.2 to look up the standard reduction potentials for the half-cells.

3 Evaluate **Does the result make sense?** Zinc is above silver in the activity series for metals. It makes sense that zinc is oxidized in the presence of silver ions.

8. Determine whether the following redox reaction will occur spontaneously:

$3Zn^{2+}(aq) + 2Cr(s) \longrightarrow 3Zn(s) + 2Cr^{3+}(aq)$

9. Is this redox reaction spontaneous as written?

$Co^{2+}(aq) + Fe(s) \longrightarrow Fe^{2+}(aq) + Co(s)$

Foundations for Math

ADDING AND SUBTRACTING INTEGERS Remind students that to add integers with different signs, find the absolute value of each number. Then, subtract the lesser absolute value from the greater absolute value. The sum has the sign of the integer with the greater absolute value. Subtracting an integer is the same as adding its opposite. For example: 2 – 3 is the same as 2 + (–3). Have students follow the rules for adding integers with different signs to get a sum of –1.

In **Sample Problem 21.1,** explain that the rules for adding and subtracting integers also apply to real numbers. The standard cell potential is calculated as follows:

$E^\circ = 0.80\text{ V} - (-0.76\text{ V})$

$E^\circ = 0.80\text{ V} + (0.76\text{ V})$ Add the opposite.

$E^\circ = 1.56\text{ V}$ Find the sum.

Explore

Calculating Standard Cell Potentials

Class Activity

PURPOSE Students become familiar with combining reduction half-reactions.

PROCEDURE Divide the class into groups of three or four. Have them use Table 21.2 to combine half-reactions and create cells having potentials that fall in the following ranges: +3.0 V to +2.5 V, +2.4 V to +1.4 V, +1.3 V to +0.5 V, and +0.4 V to +0.1 V.

EXPECTED OUTCOMES Several cells could be created in each category. Typical cells might be Na^+/Na with Sn^{2+}/Sn, +2.57 V; Mg^{2+}/Mg with Pb^{2+}/Pb, +2.24 V; Ca^{2+}/Ca with Al^{3+}/Al, 1.21 V; $PbSO_4/Pb$ with Ni^{2+}/Ni, 0.11 V.

Sample Practice Problem

Determine the half-reactions of the following reaction. Then, determine if the reaction will occur spontaneously. Explain how you determined your response.

$Fe(s) + Cl_2(g) \rightarrow Fe^{2+}(aq) + 2Cl^-(aq)$

[$Cl_2 + 2e^- \rightarrow 2Cl^-$ $\quad E^\circ_{Cl^-} = +1.36V$ (red.)

$Fe^{2+} + 2e^- \rightarrow Fe$ $\quad E^\circ_{Fe^{2+}} = -0.44V$ (ox.)

The reaction is spontaneous since the standard cell potential is a positive value: $E^\circ = +1.80$ V]

Answers

8. $E^\circ_{cell} = -0.02$ V; $E^\circ_{cell} < 0$, so the reaction is not spontaneous.

9. $E^\circ_{cell} = +0.16$ V ; $E^\circ_{cell} > 0$, so the reaction is spontaneous.

Explain

Misconception Alert

As students work Sample Problems 21.1 and 21.2, they may think that when a half-cell reaction must be multiplied by a factor such as 2 or 3, the $E°$ value must also be multiplied by that factor. Point out that even though two or three times as many electrons are present, the tendency for the electrons to flow is not two or three times greater. The tendency, which is measured by the $E°$ value, remains the same.

MAKING CONNECTIONS Point out that, until now, it has been assumed that reactions in standard half-cells always go in the spontaneous direction. For example, a Ag/Cu cell can be constructed in such a way that the only substance available for oxidation is silver metal and the only substance available for reduction is copper ion. In this case, the reaction is nonspontaneous and the cell potential is negative. Thus, the sign of the cell potential is an indicator of reaction spontaneity. Students will learn in the following lesson that nonspontaneous reactions can be driven by an outside source of energy.

Sample Practice Problem

A. Explain which half-reaction is more likely to be a reduction reaction, a half-reaction with a standard cell potential of –0.12 V or a half reaction with a standard cell potential of +0.12 V. *(The positive standard cell value is more likely to be the reduction reaction since the positive value aligns more to the cathode in a cell.)*

B. A voltaic cell is constructed using the following half-reactions:

$Li^{+}(aq) + e^{-} \rightarrow Li(s)$ $E° = -3.05$ V

$Mg^{2+}(aq) + 2e^{-} \rightarrow Mg(s)$ $E° = -2.37$ V

Determine the standard cell reaction.

[$2Li(s) + Mg^{2+}(aq) \rightarrow Mg(s) + 2Li^{+}(aq)$]

Sample Problem 21.2

Writing the Cell Reaction

Determine the cell reaction for a voltaic cell composed of the following half-cells:

$$Fe^{3+}(aq) + e^{-} \longrightarrow Fe^{2+}(aq) \quad E°_{Fe^{3+}} = +0.77\text{ V}$$
$$Ni^{2+}(aq) + 2e^{-} \longrightarrow Ni(s) \quad E°_{Ni^{2+}} = -0.25\text{ V}$$

1 Analyze Identify the relevant concepts. The half-cell with the more positive reduction potential is the one in which reduction occurs (the cathode). The oxidation reaction occurs at the anode. Add the half-reactions, making certain that the number of electrons lost equals the number of electrons gained.

2 Solve Apply the concepts to this problem.

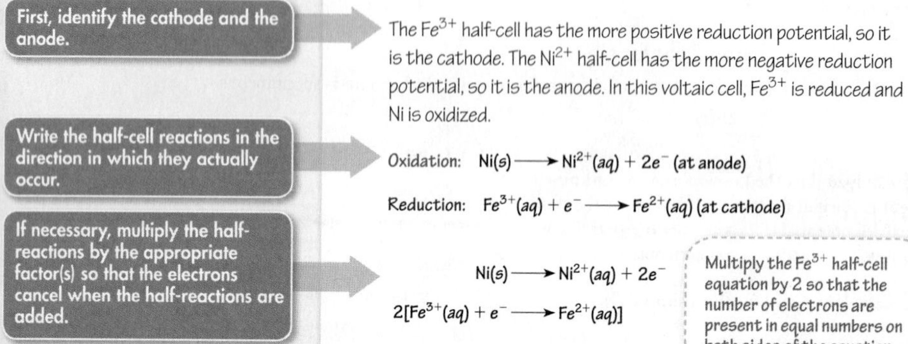

First, identify the cathode and the anode.

The Fe^{3+} half-cell has the more positive reduction potential, so it is the cathode. The Ni^{2+} half-cell has the more negative reduction potential, so it is the anode. In this voltaic cell, Fe^{3+} is reduced and Ni is oxidized.

Write the half-cell reactions in the direction in which they actually occur.

Oxidation: $Ni(s) \longrightarrow Ni^{2+}(aq) + 2e^{-}$ (at anode)

Reduction: $Fe^{3+}(aq) + e^{-} \longrightarrow Fe^{2+}(aq)$ (at cathode)

If necessary, multiply the half-reactions by the appropriate factor(s) so that the electrons cancel when the half-reactions are added.

$Ni(s) \longrightarrow Ni^{2+}(aq) + 2e^{-}$

$2[Fe^{3+}(aq) + e^{-} \longrightarrow Fe^{2+}(aq)]$

Multiply the Fe^{3+} half-cell equation by 2 so that the number of electrons are present in equal numbers on both sides of the equation.

Add the half-reactions.

$Ni(s) \longrightarrow Ni^{2+}(aq) + 2e^{-}$

$2Fe^{3+}(aq) + 2e^{-} \longrightarrow 2Fe^{2+}(aq)$

$Ni(s) + 2Fe^{3+}(aq) \longrightarrow Ni^{2+}(aq) + 2Fe^{2+}(aq)$

The electrons lost by the species that is oxidized must be equal to the electrons gained by the species that is reduced.

The half-cell with the more positive reduction potential is the cathode.

10. A voltaic cell is constructed using the following half-reactions:

$Cu^{2+}(aq) + 2e^{-} \longrightarrow Cu(s)$

$E°_{Cu^{2+}} = +0.34$ V

$Al^{3+}(aq) + 3e^{-} \longrightarrow Al(s)$

$E°_{Al^{3+}} = -1.66$ V

Determine the cell reaction.

11. A voltaic cell is constructed using the following half-reactions:

$Ag^{+}(aq) + e^{-} \longrightarrow Ag(s)$

$E°_{Ag^{+}} = +0.80$ V

$Cu^{2+}(aq) + 2e^{-} \longrightarrow Cu(s)$

$E°_{Cu^{2+}} = +0.34$ V

Determine the cell reaction.

742 Chapter 21 • Lesson 2

Foundations for Math

SUBTRACTING REAL NUMBERS Remind students that scientific calculators are programmed to abide by the rules for order of operations. Emphasize that students should know what keys are available on their calculators and how to use them. Point out that on most scientific calculators, the key marked (+/–) changes the sign of the following number (On some scientific calculators, it changes the sign of the previous number.). The key marked – by the + key *subtracts* the following number from the number before it. If you confuse the two keys, you will either get a syntax error, or you will get the wrong value.

In **Sample Problem 21.2 and 21.3,** show students the appropriate keystrokes and values needed to find the standard cell potential. The sample problem shows:

$$E° = +0.77\text{ V} - (-0.25)\text{ V}$$

The following keystrokes should be used to solve this problem (brackets represent function keys):

.77 [–] [+/–] .25 [=]

Sample Problem 21.3

Calculating the Standard Cell Potential

Calculate the standard cell potential for the voltaic cell described in Sample Problem 21.2. The half-reactions are as follows:

$$Fe^{3+}(aq) + e^- \longrightarrow Fe^{2+}(aq) \quad E^\circ_{Fe^{3+}} = +0.77\text{ V}$$

$$Ni^{2+}(aq) + 2e^- \longrightarrow Ni(s) \quad E^\circ_{Ni^{2+}} = -0.25\text{ V}$$

1 Analyze List the knowns and the unknown. Use the equation $E^\circ_{cell} = E^\circ_{red} - E^\circ_{oxid}$ to calculate the standard cell potential.

KNOWNS	UNKNOWN
$E^\circ_{Fe^{3+}} = +0.77\text{ V}$	$E^\circ_{cell} = ?$
$E^\circ_{Ni^{2+}} = -0.25\text{ V}$	
anode: Ni^{2+} half-cell	
cathode: Fe^{3+} half-cell	

2 Calculate Solve for the unknown.

First, write the equation for the standard cell potential.

$$E^\circ_{cell} = E^\circ_{red} - E^\circ_{oxid} = E^\circ_{Fe^{3+}} - E^\circ_{Ni^{2+}}$$

Substitute the values for the standard reduction potentials and solve the equation.

$$E^\circ_{cell} = +0.77\text{ V} - (-0.25\text{ V}) = +1.02\text{ V}$$

3 Evaluate Does the result make sense? The reduction potential of the reduction is positive, and the reduction potential of the oxidation is negative. Therefore, E°_{cell} must be positive.

12. Calculate the standard cell potential of a voltaic cell constructed using the half-reactions described in Problem 10.

13. Calculate the standard cell potential of a voltaic cell constructed using the half-reactions described in Problem 11.

If you did Problem 10, you know what the anode and cathode are in this cell.

If you did Problem 11, you know what the anode and cathode are in this cell.

21.2 LessonCheck

14. Explain What causes the electrical potential of a cell?

15. Describe How can you find the standard reduction potential of a half-cell?

16. Review What cell potential values indicate a spontaneous reaction? What cell potential values indicate a nonspontaneous reaction?

17. Calculate Determine whether the following redox reaction will occur spontaneously:

$$2Al^{3+}(aq) + 3Mg(s) \longrightarrow 2Al(s) + 3Mg^{2+}(aq)$$

18. Calculate Determine the cell reaction and standard cell potential for a voltaic cell composed of the following half-cells:

$$Li^+(aq) + e^- \longrightarrow Li(s) \quad E^\circ_{Li^+} = -3.05\text{ V}$$

$$Mg^{2+}(aq) + 2e^- \longrightarrow Mg(s) \quad E^\circ_{Mg^{2+}} = -2.37\text{ V}$$

BIGIDEA MATTER AND ENERGY

19. Why does a lead-acid battery produce energy? Calculate the standard cell potential of one voltaic cell in a lead-acid battery.

Lesson Check Answers

14. Electrical potential of cell results from competition for electrons between two half-cells.

15. by connecting it to a standard hydrogen electrode and measuring the cell potential, or looking it up in a table

16. If the cell potential for a redox reaction is positive, the reaction is spontaneous; if it is negative, the reaction is nonspontaneous.

17. spontaneous; $E^\circ_{cell} = +0.71\text{ V}$

18. $2Li(s) + Mg^{2+}(aq) \rightarrow Mg(s) + 2Li^+(aq)$; $E^\circ_{cell} = +0.68$

19. There is a positive electrical potential between the negative and positive terminals of the battery. Cell reaction: $Pb + PbO_2 + 2H_2SO_4 \rightarrow 2PbSO_4 + 2H_2O$; $E^\circ_{cell} = +2.05\text{ V}$

Sample Practice Problem

Calculate the standard cell potential of the following reaction to determine whether the reaction will occur spontaneously.

$Pb^{2+}(aq) + 2e^- \rightarrow Pb(s)$ $\quad E^\circ_{Pb^{2+}} = -0.13$

$I_2(s) + 2e^- \rightarrow 2I^-(aq)$ $\quad E^\circ_{I^-} = +0.54$

(spontaneous; $E^\circ_{cell} = +0.67$)

Evaluate

Informal Assessment

Ask students to refer to Table 21.2. Tell students to use the table to create an electrochemical cell that will operate spontaneously. Instruct students to submit the following:

- calculations showing the cell potential
- the equations for the two half-reactions
- the overall cell reaction

Reteach

Use the think-aloud process to help students understand the thought process that is involved in determine the cell potential of a nonspontaneous reaction. On the overhead, write the following equation:

$$Cd^{2+}(aq) + Ag(s) \rightarrow Cd(s) + Ag^+(aq)$$

Ask What element is being reduced? *(cadmium)* Have students direct you in writing the half-reaction for the reduction. *[$Cd^{2+}(aq) + 2e^- \rightarrow Cd(s)$]*

Ask What element is being oxidized? *(silver)* Write the half-reaction with student input.

$$[Ag(s) \rightarrow Ag^+(aq) + e^-]$$

Guide students through the process of calculating the cell potential. *(–1.20 V)* **Ask** Does the cell operate spontaneously? How do you know? *(No, it has a negative potential.)*

Answers

10. $2Al(s) + 3Cu^{2+}(aq) \rightarrow 2Al^{3+}(aq) + 3Cu(s)$

11. $Cu(s) + 2Ag^+(aq) \rightarrow Cu^{2+}(aq) + 2Ag(s)$

12. $E^\circ_{cell} = +2.00\text{ V}$

13. $E^\circ_{cell} = +0.46\text{ V}$

CHEMISTRY & YOU Have students examine the photo of the lemon battery. Then, ask students to recall the discussion of Luigi Galvani's generation of an electric current through a frog's leg in the Chemistry and You feature at the end of Lesson 21.1. Point out that, unlike a frog's leg, a lemon does not have nerves, and yet it still can conduct electricity. Pose the following question to students: *What characteristics do lemon cells share with frog nerve cells that can aid in transmitting an electrical current through their tissues?* You may need to assist students in the following ways:

- Both types of cells contain water.
- Both types of cells have semipermeable membranes.
- Both types of cells have the ability to actively transport ions across their membranes.

Explain

USE VISUALS Draw a diagram on the board of the voltaic cell and circuit created in the activity. Have student volunteers label the different parts. Also have them draw arrows on the diagram to indicate the direction of ion movement to reinforce the concepts they have learned in the chapter.

START A CONVERSATION Encourage students to think of various alterations they could try if they have difficulty getting their lemon battery to work. Possible alterations include using more than one lemon with the nails connected by a wire, using an LED bulb, and using a thinner wire.

CHEMISTRY & YOU: EVERYDAY MATTER

A Lemon Battery

Batteries provide power to many of the electronic devices you may use in your daily life, including laptops, cellphones, and mp3 players. Batteries come in all different shapes and sizes, from the large 12-V battery in cars, to the AAA batteries in calculators, to the tiny batteries in watches.

A battery is essentially a container that contains an anode, where oxidation occurs, a cathode, where reduction occurs, and an electrolyte. Electrons flow from the anode to the cathode through an external circuit, which is usually the device to which the battery is supplying energy. Positive and negative ions move through the electrolyte to complete the circuit.

Did you know that you can make a battery at home with only nails or wires and fruit? Try this activity at home and see how much of a voltage you can generate.

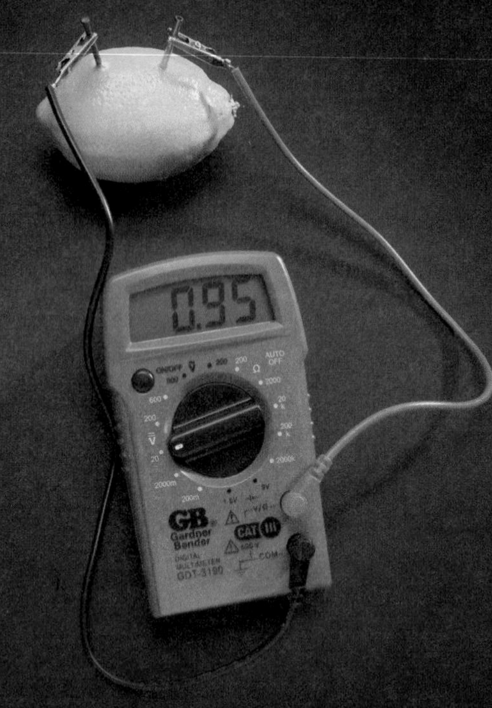

On Your Own

1. For this activity, you will need a **copper nail, screw, or wire** (about 5 cm long) and a **zinc or galvanized nail or screw** (about 5 cm long). These will serve as your electrodes. You will also need a **whole lemon** and a **multimeter** set to measure voltage (if you have one). Clean the electrodes with hot, soapy water, rinse, and dry them. Soften the lemon by rolling it on a table. Be careful not to break the skin.

2. Insert the zinc and copper electrodes into the lemon. The electrodes should not touch and should be about 2–3 cm apart.

3. Briefly touch both electrodes at the same time with your tongue and observe what happens.

4. Measure the voltage of your lemon battery with the multimeter. Attach the positive (red) lead to the copper electrode and the negative (black) lead to the zinc electrode.

Think About It

1. Observe Describe the sensation you felt on your tongue.

2. Measure What was the voltage of your battery? What would happen if you attached the positive lead of the multimeter to the zinc electrode and the negative lead to the copper electrode?

3. Predict Could you generate enough voltage to start a car if you connected 12 lemon batteries together in series? Explain.

21st Century Learning The following project helps students build the following 21st Century Skills: *Global Awareness; Financial, Economic, Business and Entrepreneurial Literacy; Civic Literacy; Critical Thinking and Problem Solving; Information, Communications and Technology Literacy.*

GREEN BATTERIES Pose the following challenge to your students. *A company you work for has decided to invest in developing a line of batteries that are better for the environment. Your job is to design an environmentally-friendly 1.5-V battery and develop a presentation promoting your design to the company president.* Designs should include information on materials used for construction, equations for half-cell reactions, calculations of reduction potentials showing that the choice of materials for the electrodes will provide an $E^{\circ}_{cell} = 1.5$ V, and an estimate of the expected life of the battery. An environmental impact statement should be included that compares the potential impact of the battery to that of current batteries.

21.3 Electrolytic Cells

CHEMISTRY & YOU

Q: *How can an aluminum bottle be colored?* You probably know that aluminum is a metal that is silver in color. How then can objects made from aluminum be colored? If an aluminum object has a color other than silver, it was anodized before it was dyed. In the anodizing process, the aluminum object is the anode in an electrolytic cell. The electrolyte is a dilute acid. When an electric current flows through the cell, aluminum oxide forms on the surface of the aluminum. In this lesson, you will learn about electrochemical processes that require electrical energy.

Key Questions

How do voltaic and electrolytic cells differ?

What are some applications that use electrolytic cells?

Vocabulary

- electrolysis
- electrolytic cell

Electrolytic vs. Voltaic Cells

How do voltaic and electrolytic cells differ?

In Lesson 21.1, you learned how a spontaneous chemical reaction can be used to generate a flow of electrons (an electric current). In this lesson, you will learn how an electric current can be used to make a nonspontaneous redox reaction occur. The process in which electrical energy is used to bring about such a chemical change is called **electrolysis.** Although you may not have realized it, you are already familiar with some results of electrolysis, such as gold-plated jewelry and chrome-plated automobile parts. Figure 21.10 shows some silver-plated dishes.

The apparatus in which electrolysis is carried out is an electrolytic cell. An **electrolytic cell** is an electrochemical cell used to cause a chemical change through the application of electrical energy. An electrolytic cell uses electrical energy (direct current) to make a nonspontaneous redox reaction proceed to completion.

Figure 21.10
Products of Electrolysis
Pure silver can be plated onto steel using electrolysis.

Key Objectives

21.3.1 DISTINGUISH between electrolytic and voltaic cells.

21.3.2 DESCRIBE some applications that use electrolytic cells.

Additional Resources

Reading and Study Workbook, Lesson 21.3

Available on Digital Media:

- Laboratory Manual, Labs 17 and 18
- Virtual Chemistry Labs, Labs 22, 23, and 24
- Small-Scale Labs, Lab 17

LESSON 21.3

PearsonChem.com

Engage

CHEMISTRY & YOU Have students study the photograph and read the opening text. If possible, display a colored aluminum sport bottle. **Ask** Based on what you just read, do you think anodization is a spontaneous reaction? Why? *(No, it requires an input of electrical energy.)* **Ask** Do you think anodization takes place in a voltaic cell? Why? *(No, because reactions in voltaic cells are spontaneous.)* Tell students to keep this in mind as they proceed with the lesson.

Activate Prior Knowledge

Review the characteristics of voltaic and electrochemical cells with students. Then, create a compare/contrast table for voltaic and electrolytic cells, and fill in the information for voltaic cells. Have students complete the table during the lesson.

Focus on ELL

1 CONTENT AND LANGUAGE Write the following terms on the board: *apparatus, brine, circuit,* and *net result*. Pronounce each word and ask students to repeat. Clarify the part of speech and discuss any cognates. Have students look up the definitions and add them to their notebooks.

2 FRONTLOAD THE LESSON Display a variety of metal items, which have been plated using either a voltaic or an electrolytic process. Provide information about the base metal and the plating metal of each item. Allow students to examine the objects and predict which ones were plated spontaneously and which ones required energy for plating to occur. Encourage students to refer back to Table 21.1 as they make their predictions.

3 COMPREHENSIBLE INPUT Project Figure 21.11 on the board, and circle the similarities between the two cells as students identify them. Use a different color to circle the differences. Then, have students view the Kinetic Art animation of this figure.

Answers

1. a tingling sensation
2. about 1 V; A voltage would not be measured.
3. Twelve lemons in series do create a potential difference of 12 volts. However, lemon batteries do not supply enough power to start a car.

Foundations for Reading

BUILD VOCABULARY *Electr-* and *electro-* are combining forms from the New Latin word *electricus*. Ask students what words they know outside of chemistry that contain these combining forms. *(electricity, electrocardiogram, electronic, electromagnet)*

READING STRATEGY Before beginning this section, have students preview it by leafing through the pages and reading the headings. Then, have students read the boldface sentences under each heading. Ask them if they now have an idea of what this section is about.

Explain

Electrolytic vs. Voltaic Cells

MAKE A CONNECTION The spontaneous cell reaction in a lead-storage battery provides immediate energy to start a car. **Ask** What reaction is occurring in a car battery that makes it function? *(Lead and lead(IV) oxide undergo a redox reaction in which lead(II) sulfate is formed.)* Students may have experienced a dead battery. Often, recharging can revive a dead battery. **Ask** What reaction occurs during recharging? *(Lead(II) ions in lead(II) sulfate are oxidized to lead(IV) and reduced to lead.)* Point out that until now it has been assumed that reactions in standard half-cells always go in the spontaneous direction. Explain that recharging the car battery is a nonspontaneous reaction that requires energy from another source.

APPLY CONCEPTS Tell students that electrolytic processes are used to separate active metals such as aluminum, magnesium, and sodium from their salts. Explain that this same process is used to recover metals from ores. As a class, compare electrolytic cells to voltaic cells. Point out that in both, oxidation occurs at the anode and reduction at the cathode. Show students that in the voltaic cell, a spontaneous redox reaction occurs, whereas in the electrolytic cell, external electrical energy makes a nonspontaneous redox reaction occur.

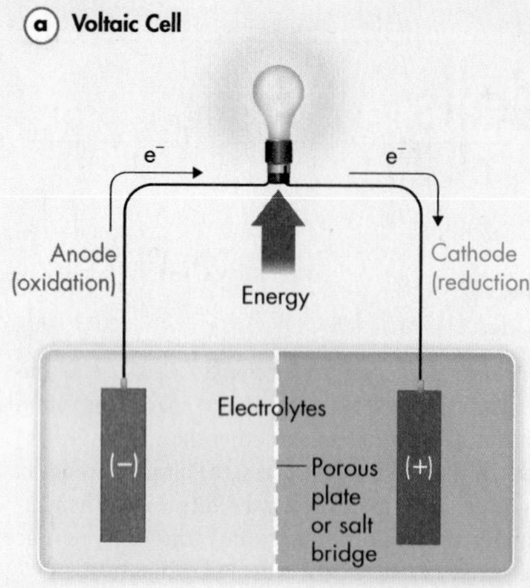

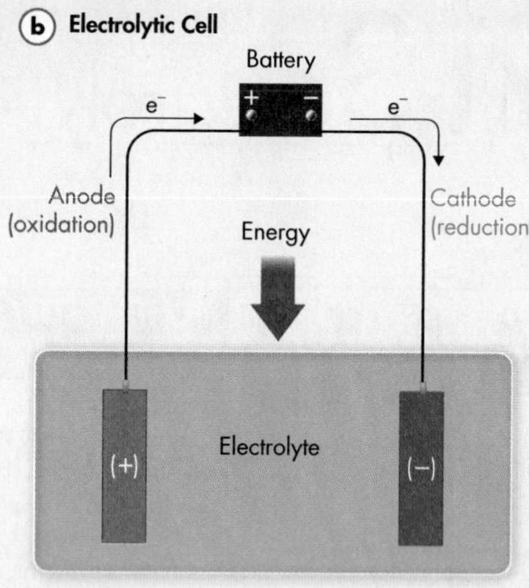

Figure 21.11 Voltaic and Electrolytic Cells
Electrochemical cells can be classified as voltaic or electrolytic. **a.** In a voltaic cell, energy is released from a spontaneous redox reaction. The system (cell) does work on the surroundings (light bulb). **b.** In an electrolytic cell, energy is absorbed to drive a nonspontaneous reaction. The surroundings (battery or power supply) do work on the system (cell).

In both voltaic and electrolytic cells, electrons flow from the anode to the cathode in the external circuit. As shown in Figure 21.11, for both types of cells, the electrode at which reduction occurs is the cathode. The electrode at which oxidation occurs is the anode.

The key difference between voltaic and electrolytic cells is that in a voltaic cell, the flow of electrons is the result of a spontaneous redox reaction, whereas in an electrolytic cell, electrons are caused to flow by an outside power source, such as a battery. The redox process in the voltaic cell is spontaneous; in the electrolytic cell, the redox process is nonspontaneous. Electrolytic and voltaic cells also differ in the assignment of charge to the electrodes. In a voltaic cell, the anode is the negative electrode and the cathode is the positive electrode. In an electrolytic cell, the cathode is considered to be the negative electrode because it is connected to the negative electrode of the battery. The anode in the electrolytic cell is considered to be the positive electrode because it is connected to the positive electrode of the battery. It is important to remember these conventions about the two kinds of cells.

Driving Nonspontaneous Processes

What are some applications that use electrolytic cells?

Electrolysis can be used to drive otherwise nonspontaneous reactions that are of commercial importance. **Electrolysis of a solution or of a melted, or molten, ionic compound can result in the separation of elements from compounds. Electrolytic cells are also commonly used in the plating, purifying, and refining of metals.**

Anodizing

Anodizing is an electrolytic process for applying a corrosion-resistant oxide coating to aluminum and certain other metals. When an electric current flows through an electrolytic cell containing an aluminum anode, a thin coat of aluminum oxide forms on the surface of the aluminum. Immersion of the aluminum object in pure boiling water permanently seals the film against corrosion. Dyeing of anodized aluminum can produce metallic red, metallic blue, and a host of other metallic colors on the metal surface. Also, the high porosity of the oxide coating permits it to absorb pigments and lubricants.

Electrolysis of Water When a current is applied to two electrodes immersed in pure water, nothing happens. There is no current flow and no electrolysis. However, when an electrolyte such as H_2SO_4 or KNO_3 in low concentration is added to the pure water, the solution conducts an electric current and electrolysis occurs. This process is illustrated in Figure 21.12. The products of the electrolysis of water are hydrogen gas and oxygen gas. This process is sometimes used to produce hydrogen gas for fuel cells.

Water is oxidized at the anode according to the following oxidation half-reaction:

$$\text{Oxidation: } 2H_2O(l) \longrightarrow O_2(g) + 4H^+(aq) + 4e^- \text{ (at anode)}$$

Water is reduced to hydrogen at the cathode according to the following reduction half-reaction:

$$\text{Reduction: } 2H_2O(l) + 2e^- \longrightarrow H_2(g) + 2OH^-(aq) \text{ (at cathode)}$$

The region around the anode turns acidic due to an increase in H^+ ions. The region around the cathode turns basic due to the production of OH^- ions. The overall cell reaction is obtained by adding the half-reactions (after doubling the reduction half-reaction equation to balance electrons).

$$\begin{aligned} \text{Oxidation: } & 2H_2O(l) \longrightarrow O_2(g) + 4H^+(aq) + 4e^- \\ \text{Reduction: } & 2[2H_2O(l) + 2e^- \longrightarrow H_2(g) + 2OH^-(aq)] \\ \hline \text{Overall cell reaction: } & 6H_2O(l) \longrightarrow 2H_2(g) + O_2(g) + \\ & \qquad 4H^+(aq) + 4OH^-(aq) \end{aligned}$$

The ions produced tend to recombine to form water.

$$4H^+(aq) + 4OH^-(aq) \longrightarrow 4H_2O(l)$$

Therefore, the net reaction can be written as follows:

$$2H_2O(l) \longrightarrow 2H_2(g) + O_2(g)$$

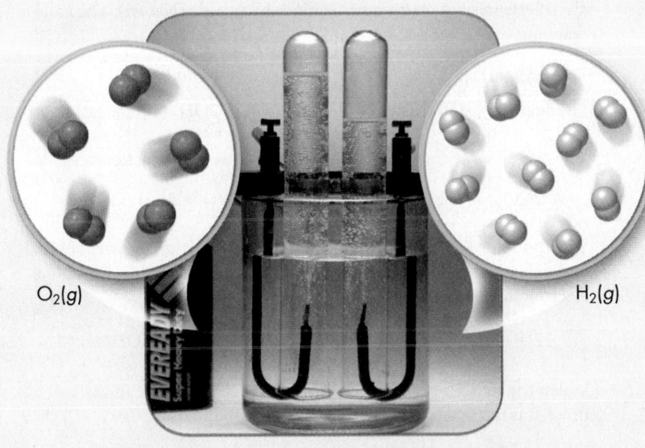

Figure 21.12 Electrolysis of Water
When an electric current is passed through water, the water decomposes into oxygen gas and hydrogen gas.
Interpret Photos *Which electrode in the photograph is the cathode? Which is the anode?*

UbD Check for Understanding

How do voltaic and electrolytic cells differ?

Have students answer the key question by briefly summarizing the similarities and differences between the two cells with a compare/contrast statement.

ADJUST INSTRUCTION If students are having difficulty writing their statements, have them examine Figure 21.11 and identify the similarities and differences. Then, reinforce the concept that voltaic cell reactions are spontaneous, while those that take place in electrolytic cells do so only when an input of energy occurs.

Explain

Driving Nonspontaneous Processes

USE VISUALS Direct students' attention to Figure 21.12. Call attention to the volumes of hydrogen and oxygen that have been collected in the photograph. **Ask** Why is one volume larger than the other? *(When water is decomposed, two molecules of hydrogen are formed for every one molecule of oxygen.)* For emphasis, write the equation for the decomposition of water on the board. **Ask** Why is it necessary to add an acid to the water in the cell? *(Water is a molecular compound; it does not form a significant number of ions. The liquid in the cell must be an electrolyte, that is, it must contain ions.)*

Explore

Teacher Demo

PURPOSE Students will observe the electrolysis of water and note that energy is required for the process.

MATERIALS Hoffman apparatus or construct a substitute using iron nails as electrodes, 2 test tubes, 6*M* NaOH, 250- or 500-mL beaker, 2 insulated wires

SAFETY Wear gloves, apron and safety goggles. NaOH is corrosive.

PROCEDURE If you have sufficient class time, assemble the apparatus in front of the class. Otherwise, prepare the apparatus in advance. Insert the iron electrodes into the lower ends of the test tubes. Use insulated wires to connect the electrodes to a power supply to avoid contact with the electrolyte. Fill the test tubes with the NaOH solution and invert them in a beaker that is one-third filled with the solution. When all is ready, turn on the power supply.

EXPECTED OUTCOMES Hydrogen gas is collected at the negative electrode (cathode); oxygen gas is collected at the positive electrode (anode). Demonstrate with a lit match (H_2) and a glowing splint (O_2) that one test tube contains hydrogen and the other oxygen.

Answers

Figure 21.12 The electrode on the right is the cathode; the electrode on the left is the anode.

LESSON 21.3

Explain

USE VISUALS Have students study the diagram in Figure 21.13. **Ask** What is the purpose of the battery? *(The battery is a voltaic cell that supplies the electrical energy needed to drive the nonspontaneous redox reaction in the electrolytic cell.)* **Ask** What substance is being oxidized in the electrolytic cell? *(Cl^-)* **Ask** What substance is being reduced? *(H_2O)*

START A CONVERSATION Explain that AC current alternates, meaning that it flows in both directions around a circuit (unless the circuit has diodes), and that in DC current, electric current flows in one direction only. Refer students to Figure 21.13. **Ask** Why is DC current, and not AC current, used to run the reaction? *(DC devices maintain a constant potential difference. The voltage potential from an AC generator alternates in sign.)* Write the overall equation for the reaction on the board and ask students to use the standard reduction potentials in Table 21.2 to show why this reaction is nonspontaneous. [*$Cl_2(g) + 2e^- \rightarrow 2Cl^-(aq)$; $E° = +1.36\ V$*

$$2H_2O(l) + 2e^- \rightarrow H_2(g) + 2OH^-(aq);\ E° = -0.42\ V$$

$$E°_{cell} = E°_{red} - E°_{oxid} = -0.42\ V - 1.36\ V = -1.78\ V$$

The potential is negative; therefore, energy is needed to make this reaction occur.]

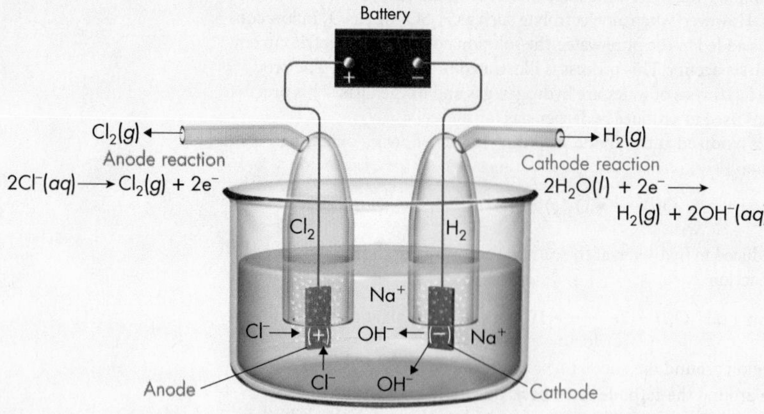

Figure 21.13 Electrolysis of Brine
Chlorine gas, hydrogen gas, and sodium hydroxide are produced when an electric current is passed through an aqueous solution of sodium chloride.
Interpret Diagrams ***Which substances are produced by oxidation? Which substances are produced by reduction?***

Electrolysis of Brine If the electrolyte in an aqueous solution is more easily oxidized or reduced than water, then the products of electrolysis will be substances other than hydrogen and oxygen. An example is the electrolysis of brine, a concentrated aqueous solution of sodium chloride. This process simultaneously produces three important industrial chemicals: chlorine gas, hydrogen gas, and sodium hydroxide. The electrolytic cell for the electrolysis of brine is shown in Figure 21.13.

During electrolysis of brine, chloride ions are oxidized to produce chlorine gas at the anode. Water is reduced to produce hydrogen gas at the cathode. Sodium ions are not reduced to sodium metal in the process because water molecules are more easily reduced than are sodium ions. The reduction of water also produces hydroxide ions. Thus, the electrolyte in solution becomes sodium hydroxide (NaOH). The half-reactions are as follows:

Oxidation: $2Cl^-(aq) \longrightarrow Cl_2(g) + 2e^-$ (at anode)

Reduction: $2H_2O(l) + 2e^- \longrightarrow H_2(g) + 2OH^-(aq)$ (at cathode)

The overall ionic equation is the sum of the two half-reactions.

$$2Cl^-(aq) + 2H_2O(l) \longrightarrow Cl_2(g) + H_2(g) + 2OH^-(aq)$$

The spectator ion Na^+ can be included in the equation (as part of NaCl and of NaOH) to show the formation of sodium hydroxide during the electrolytic process.

$$2NaCl(aq) + 2H_2O(l) \longrightarrow Cl_2(g) + H_2(g) + 2NaOH(aq)$$

When the sodium hydroxide solution is about 10 percent (mass/volume), it is removed from the cell and processed further.

Differentiated Instruction

L3 ADVANCED STUDENTS Explain that electrolytic cells are used in the recovery of low-grade ores in mining. Have students select a metal and research how it is recovered and purified by electrolysis.

L1 LESS PROFICIENT READERS Make a sketch of the cell in Figure 21.13 on the board. Use different colors to label the cathode (1) and anode (2) on the battery. Use gestures to show the flow of electrons from the anode (negative terminal) of the battery to the cathode (negative electrode) of the electrolytic cell, and from the anode (positive electrode) of the electrolytic cell to the cathode (positive terminal) of the battery. Emphasize that in both the voltaic and electrolytic cells, the definitions of a cathode and an anode are the same: reduction always takes place at the cathode and oxidation always takes place at the anode.

ELL ENGLISH LANGUAGE LEARNERS Explain that AC current is an abbreviation for alternating current, and that DC current is an abbreviation for direct current.

Electrolysis of Molten Sodium Chloride Both sodium and chlorine are commercially important. Sodium is used in sodium vapor lamps and as the coolant in some nuclear reactors. Chlorine, a greenish-yellow gas, is used to sterilize drinking water and is important in the manufacture of polyvinyl chloride (PVC) and various pesticides. These two elements are produced through the electrolysis of pure molten sodium chloride, rather than an aqueous solution of NaCl. Chlorine gas is produced at the anode, and molten sodium collects at the cathode. The half-reactions for the electrolysis of molten sodium chloride are as follows:

$$\text{Oxidation: } 2Cl^-(l) \longrightarrow Cl_2(g) + 2e^- \text{ (at anode)}$$

$$\text{Reduction: } 2Na^+(l) + 2e^- \longrightarrow 2Na(l) \text{ (at cathode)}$$

The overall equation is the sum of the two half-reactions.

$$2NaCl(l) \longrightarrow 2Na(l) + Cl_2(g)$$

The electrolytic cell in which this commercial process is carried out is called the Downs cell and is shown in Figure 21.14. The cell operates at a temperature of 801°C so that the sodium chloride is maintained in the molten state. A perforated iron screen separates the circular cathode from the graphite anode. The sodium, with a melting point of 97.8°C, remains in liquid form. The liquid sodium floats on the more dense molten sodium chloride and is drawn off as it is formed. The chlorine gas is collected after it bubbles up and out of the molten salt. The design of the Downs cell allows fresh sodium chloride to be added as required. The design also separates the products so they will not recombine to re-form sodium chloride.

Figure 21.14 Downs Cell
The Downs cell produces sodium metal and chlorine gas from the electrolysis of molten sodium chloride.

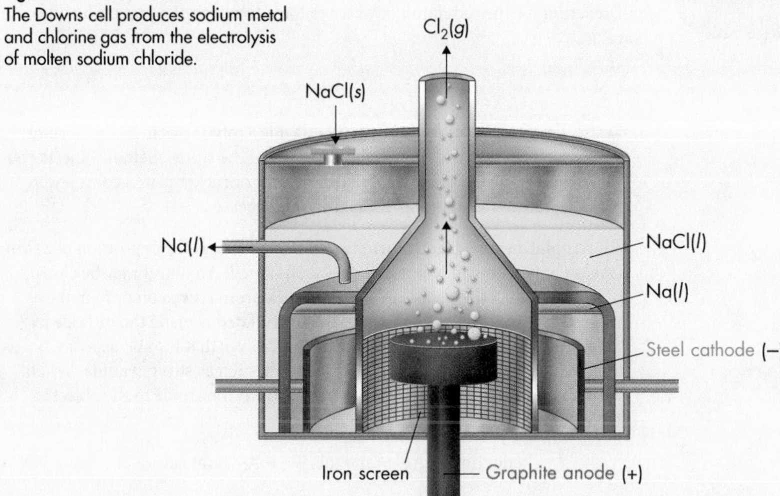

CRITICAL THINKING Note that metallic sodium has a much lower melting point than sodium chloride. **Ask** How are the two substances different? *(Metallic sodium is an element in its pure form, while sodium chloride is an ionic compound.)* **Ask** What type of intermolecular forces hold the ions in NaCl crystals together? *(the attraction between the positively charged sodium ions and the negatively charged chloride ions)* Point out that this is an electrostatic attraction. Explain that the atoms in a metal are also held in place in their crystal lattice by electrostatic attractions. Because metals so readily give up electrons to form ions, the positively charged sodium ions are surrounded by a so-called "sea" composed of the electrons that were given up by the atoms in forming the sodium ions. These electrons form electrostatic attractions with the sodium ions in the lattice. **Ask** How can the melting points of metallic sodium and NaCl be so different when their crystals are both held together by electrostatic attractions? *(The electrostatic attraction between sodium and chloride ions must be greater than the electrostatic attraction between sodium ions and individual electrons, so a greater input of energy is required to disrupt the attraction between the sodium and chloride ions.)*

Extend

Connect to ENVIRONMENTAL SCIENCE

Explain that the need for metallic sodium has decreased since gasoline engines changed to unleaded fuel. Have students determine the relationship between metallic sodium and unleaded gasoline and explain the environmental concerns associated with its use.

Focus on ELL

4 LANGUAGE PRODUCTION Have students work in small groups to complete the Small-Scale Lab on page 752. Make sure each pairing has ELLs of varied language proficiencies, so that a more proficient student can help a less proficient one. Have students work according to their proficiency level.

BEGINNING: LOW/HIGH Model this lab for these students. Use think-aloud and gestures during the demonstrations. Pause and have students mimic you as you move through the procedure.

INTERMEDIATE: LOW/HIGH Restate the steps in simpler terms. Allow students to present their results orally.

ADVANCED: LOW/HIGH Ask students to present to the class their analysis of the experiment and how they used what they learned to design the experiment described in #3 of You're the Chemist.

Answers

FIGURE 21.13 Oxidation produces chlorine gas. Reduction produces hydrogen gas and hydroxide ions.

LESSON 21.3

Explore

Quick Lab

OBJECTIVE Students electrochemically oxidize metals and identify the products.

SKILLS FOCUS observing, inferring

PREP TIME 15 minutes

CLASS TIME 20 minutes

MATERIALS 9-volt battery, copper penny, nickel coin, iron nail, filter paper, aluminum foil, reaction surface, Na_2SO_4

ADVANCE PREPARATION

Solution	Preparation
0.2*M* Na_2SO_4	7.1 g in 250 mL water

EXPECTED OUTCOMES In Step 3, a light blue image of the penny appears on the filter paper. In Step 4, a blue-green image of the nickel appears. In Step 5, a brownish-red image of the nail appears.

ANALYZE AND CONCLUDE

1. See Expected Outcomes.
2. Each oxidation occurs at the anode:
   $Cu(s) \rightarrow Cu^{2+}(aq) + 2e^-$
   $Ni(s) \rightarrow Ni^{2+}(aq) + 2e^-$
   $Fe(s) \rightarrow Fe^{2+}(aq) + 2e^-$
3. The battery oxidized each metal, and the resulting colored metal ion migrated to the filter paper.
4. $2H_2O(l) + 2e^- \rightarrow H_2(g) + 2OH^-(aq)$
5. $$2H_2O(l) + 2e^- \rightarrow H_2(g) + 2OH^-(aq)$$
   $$Cu(s) \rightarrow Cu^{2+}(aq) + 2e^-$$
   $$2H_2O(l) + Cu(s) \rightarrow H_2(g) + 2OH^-(aq) + Cu^{2+}(aq)$$

FOR ENRICHMENT Have students repeat Steps 1–3 but replace $Na_2SO_4(aq)$ with one drop of 0.1*M* KI and one drop of 20% (v/v) starch solution. Explain that the battery oxidized the iodide ions to form iodine. Ask students to write a half-reaction for the oxidation and explain the color change. *($2I^- \rightarrow I_2 + 2e^-$; I^- ions are oxidized to I_2, which reacts with starch to form a blue-black substance.)*

Quick Lab

Purpose To electrochemically oxidize metals and identify the products

Materials

- reaction surface
- aluminum foil
- filter paper
- micropipette or dropper
- sodium sulfate (Na_2SO_4) solution
- penny
- 9-volt battery
- nickel coin
- iron nail

Electrochemical Analysis of Metals

Procedure

1. Stack the following in order on a reaction surface: a 3-cm square of aluminum foil, a 2-cm square of filter paper, 1 drop of Na_2SO_4 solution, and a penny. The penny should be roughly centered on the filter paper, which should be roughly centered on the foil.
2. Apply the negative (−) terminal of the 9-volt battery to the aluminum foil and the positive (+) terminal to the penny for no more than three seconds.
3. Remove the penny and observe the filter paper.
4. Repeat Steps 1–3, replacing the penny with the nickel coin.
5. Repeat Steps 1–3, replacing the penny with the iron nail.

Analyze and Conclude

1. **Observe** What colors formed on the filter paper for each object?
2. **Describe** For each metal object you tested, the battery oxidized the metal atoms to form metal cations with a 2+ charge. Write a half-reaction for each metal oxidation you observed. Did these reactions take place at the anode or the cathode?
3. **Relate Cause and Effect** Explain in your own words why the colors formed on the filter paper.
4. **Describe** The aluminum foil serves as the cathode, where the reduction of water takes place. Write the half-reaction for the reduction of water.
5. **Describe** Combine the half-reaction for the oxidation of copper with the half-reaction for the reduction of water to form the overall equation for the cell reaction.

Using Electrolysis in Metal Processing Electrolysis has many important applications in the field of metallurgy. Many of the shiny, metallic objects you see every day—such as chrome-plated fixtures or nickel-plated coins—were manufactured with the help of electrolytic processes.

Electroplating and Electroforming Electroplating is the deposition of a thin layer of a metal on an object in an electrolytic cell. An object may be electroplated to protect the surface of the base metal from corrosion or to make it more attractive. An object that is to be silver-plated is made the cathode in an electrolytic cell. The anode is the metallic silver that is to be deposited, and the electrolyte is a solution of a silver salt, such as silver cyanide. When a direct current is applied, silver ions move from the anode to the object to be plated.

$$\text{Reduction: } Ag^+(aq) + e^- \longrightarrow Ag(s) \text{ (at cathode)}$$

Focus on ELL

4 LANGUAGE PRODUCTION Have students work in groups of four to complete the lab. Make sure each group has ELLs of varied language proficiencies, so that more proficient students can help less proficient ones. Have students work according to their proficiency level.

BEGINNING: LOW/HIGH Paraphrase orally the procedural steps and group students with higher language proficiency partners.

INTERMEDIATE: LOW/HIGH Have students brainstorm what they think they may observe, prior to performing each step. Allow students to orally present their results.

ADVANCED: LOW/HIGH Have students model the procedures for group members and then assist the group as they repeat the procedure using the nickel coin and the iron.

The net result is that silver transfers from the silver anode to the object being plated. Figure 21.15 shows statuettes that were electroplated with copper, nickel, and 24-carat gold. Many factors contribute to the quality of the metal coating that forms. In the plating solution, the concentration of the cations to be reduced must be carefully controlled. The solution must also contain compounds to control the acidity and to increase the conductivity. Other compounds may be used to make the metal coating brighter or smoother.

Electroforming is a process in which an object is reproduced by making a metal mold of it at the cathode of a cell. For example, a mold of an object can be coated with metal so it will conduct a current. It is then electroplated with a thick coating of metal. This coating can then be stripped off of the mold. Electroforming is used to make jewelry and tubing for chemical instrumentation, such as gas chromatographs.

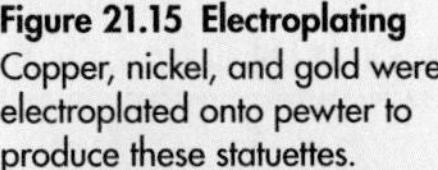

Figure 21.15 Electroplating Copper, nickel, and gold were electroplated onto pewter to produce these statuettes.

Electrowinning and Electrorefining In a process called electrowinning, impure metals can be purified in electrolytic cells. The cations of molten salts or aqueous solutions are reduced at the cathode to give very pure metals. A common use of electrowinning is in the extraction of aluminum from its ore, bauxite. Bauxite is impure alumina (Al_2O_3). In a method known as the Hall-Heroult process, purified alumina is dissolved in molten cryolite (Na_3AlF_6), and heated to above 1000°C in a carbon-lined tank. The carbon lining, connected to a direct current, serves as the cathode. The anode consists of carbon rods dipped into the tank. At the cathode, Al^{3+} ions are reduced, forming molten aluminum. At the anode, carbon is oxidized, forming carbon dioxide gas. The overall reaction is as follows:

$$2Al_2O_3(l) + 3C(s) \longrightarrow 4Al(l) + 3CO_2(g)$$

In the process of electrorefining, a piece of impure metal is made the anode of the cell. It is oxidized to the cation and then reduced to the pure metal at the cathode. This technique is used to obtain ultrapure silver, lead, and copper.

Other Processes Other electrolytic processes are centered on the anode rather than the cathode. In electropolishing, for example, the surface of an object at the anode is dissolved selectively to give it a high polish. In electromachining, a piece of metal at the anode is partially dissolved until the remaining portion is an exact copy of the object at the cathode.

CHEMISTRY & YOU

Q: *In the process of anodizing aluminum, the aluminum object serves as the anode. To which electrode of the power source is the anode connected?*

21.3 LessonCheck

ONLINE PROBLEMS

20. **Describe** What is the difference between an electrolytic cell and a voltaic cell?
21. **List** What are some applications of electrolytic cells?
22. **Compare** What is the charge on the anode of an electrolytic cell? What is the charge on the anode of a voltaic cell?
23. **Apply Concepts** Which process, oxidation or reduction, always occurs at the cathode of an electrolytic cell?
24. **Explain** Read about the Hall-Heroult process on page R13. Write the half-reactions for the electrolysis that takes place, and explain the function of cryolite in the process.

BIGIDEA MATTER AND ENERGY

25. Does the redox reaction in a voltaic cell release free energy? If no electrical energy is supplied, does the redox reaction in an electrolytic cell release free energy? Explain.

Lesson Check Answers

20. A voltaic cell uses an electrochemical reaction to produce electrical energy; an electrolytic cell uses electrical energy to bring about a chemical change.
21. to separate elements from compounds and to plate, purify, and refine metals
22. electrolytic cell anode (+); voltaic cell anode (–)
23. Reduction always occurs at the cathode.
24. Anode: $C(s) + 2O^{2-}(l) \rightarrow CO_2(g) + 4e^-$
    Cathode: $3e^- + Al^{3+}(l) \rightarrow Al(l)$
    Adding cryolite to aluminum oxide results in a mixture that melts at a much lower temperature (1012°C) than pure aluminum oxide (2045°C).
25. Yes, the electrical energy produced in a voltaic cell can do work. An electrolytic cell does not produce free energy; it requires energy.

Extend

Connect to METALLURGY

Explain that gold and silver are the most common metals used for plating jewelry. Have students research how electrolysis is used in plating jewelry and share their findings with the class. *(A thin layer of the precious metal is deposited over much less expensive base metals such as copper. Nonmetallic objects may be used as the base material as long as their surface is first coated with a conductor such as graphite. Plating solutions contain a suitable electrolyte (an acid or base) to adjust pH. Small amounts of other substances may be added to vary the color of gold jewelry. Nickel, silver, copper, cobalt, and iridium may be deposited along with the gold.)*

Evaluate

Informal Assessment

Have students draw and label a simple electrolytic cell for the hydrolysis of water. Students should include labels for the cathode, anode, and the positive and negative terminals of the battery, as well as indicate the paths of electrons, hydroxide ions, and hydrogen ions between the anode and the cathode.

Then, direct students to complete 21.3 Lesson Check.

Reteach

Use the diagrams in this lesson as a basis for review. Starting with Figure 21.11 and continuing to Figure 21.14, question students about what is being illustrated and what is significant about it.

Then, direct students to study the voltaic and electrolytic cells compare/contrast chart they created. Have students add to this chart or clear up any misconceptions.

Small-Scale Lab

OBJECTIVE Students electrolyze solutions and interpret their observations.

SKILLS FOCUS observing, drawing conclusions

MATERIALS reaction surface, electrolysis device, reagents listed below

PREP TIME 30 minutes

CLASS TIME 30 minutes

ADVANCE PREPARATION

Solution	Preparation
0.2*M* Na_2SO_4	7.1 g in 250 mL
0.1*M* KI	4.2 g in 250 mL
0.2*M* $CuSO_4$	12.5 g $CuSO_4 \cdot 5H_2O$ in 250 mL
Starch	50 mL liquid starch in 200 mL
0.04% BTB	100 mg in 16.0 mL of 0.01M NaOH; dilute to 250 mL
1.0*M* NaCl	14.6 g in 250 mL
0.2*M* KBr	6.0 g in 250 mL

EXPECTED OUTCOMES

Figure A *H_2O:* no visible reaction; *Na_2SO_4:* bubbles at anode and cathode; *Na_2SO_4 + BTB:* anode yellow, cathode blue.

Figure B *KI:* bubbles at cathode, yellow solution at anode; *KI + starch:* bubbles at cathode, black solution at anode; *KI + BTB:* bubbles and blue solution at cathode, yellow solution at anode.

ANALYZE AND CONCLUDE

1. Pure water has too few ions to carry an electric current.
2. In solution, sodium sulfate dissociates into ions, which carry an electric current.
3. The bubbles are $H_2(g)$, and the blue BTB solution indicates the presence of OH^- ions.
4. The bubbles are $O_2(g)$, and H^+ ions in solution impart the yellow color to the BTB solution.
5. $H_2O(l) \rightarrow H_2(g) + \frac{1}{2}O_2(g)$

YOU'RE THE CHEMIST

1. Bubbles are $H_2(g)$; blue solution indicates presence of OH^- ions; yellow solution is $I_2(aq)$, which is black in the presence of starch.
2. I^- is more likely to oxidize than H_2O because it has a more positive $E°$ value:
   $H_2O(l) \rightarrow \frac{1}{2}O_2(g) + 2H^+ + 2e^-$; $E° = -0.82$ V
   $2I^-(aq) \rightarrow I_2(aq) + 2e^-$; $E° = -0.54$ V
3. **NaCl** Cathode: $2H_2O(l) + 2e^- \rightarrow H_2(g) + 2OH^-(aq)$ (bubbles, blue BTB); Anode: $2Cl^-(aq) \rightarrow Cl_2(aq) + 2e^-$ (yellow solution)
   KBr Cathode: $2H_2O(l) + 2e^- \rightarrow H_2(g) + 2OH^-(aq)$ (bubbles, blue BTB); Anode: $2Br^-(aq) \rightarrow Br_2(aq) + 2e^-$ (yellow solution)
   $CuSO_4$ Cathode: $Cu^{2+} + 2e^- \rightarrow Cu$ (copper plates out);
   Anode: $H_2O(l) \rightarrow \frac{1}{2}O_2(g) + 2H^+(g) + 2e^-$ (bubbles, yellow BTB)

Small-Scale Lab

Electrolysis of Water

Purpose

To electrolyze solutions and interpret your observations in terms of chemical reactions and equations

Materials

- reaction surface
- micropipettes or droppers
- electrolysis device
- water
- sodium sulfate (Na_2SO_4) solution
- bromothymol blue (BTB) solution

Procedure

1. On separate sheets of paper, draw two grids similar to Figure A. Make each square 2 cm on each side.
2. Place a reaction surface over one of the grids and add one drop of each solution shown in Figure A.
3. Apply the leads of the electrolysis device to each solution. Be sure to clean the leads between each experiment. Look carefully at the cathode (negative lead) and the anode (positive lead).
4. Use the second grid as a data table to record your observations for each solution.

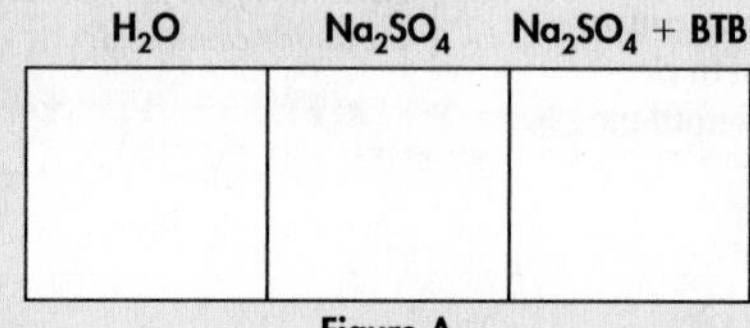

Figure A

Analyze and Conclude

1. **Relate Cause and Effect** Explain why pure water does not conduct an electric current and does not undergo electrolysis.
2. **Relate Cause and Effect** Explain why water with sodium sulfate conducts an electric current and undergoes electrolysis.
3. **Analyze Data** The cathode provides electrons to water, and the following half-reaction occurs:
   $$2H_2O(l) + 2e^- \longrightarrow H_2(g) + 2OH^-(aq)$$
   Explain how your observations correspond to the products shown in this reaction.
4. **Analyze Data** The anode removes electrons from water, and the following half-reaction occurs:
   $$H_2O(l) \longrightarrow \tfrac{1}{2}O_2(g) + 2H^+(aq) + 2e^-$$
   Explain how your observations correspond to the products shown in this reaction.
5. **Summarize** Add the two half-reactions to obtain the overall reaction for the electrolysis of water. Simplify the result by adding the OH^- and H^+ ions to get H_2O, and then canceling anything that appears on both sides of the equation.

You're the Chemist

1. **Analyze Data** Perform the above experiment using the solutions shown in Figure B. Record your results. The cathode and anode reactions are
   $$2H_2O(l) + 2e^- \longrightarrow H_2(g) + 2OH^-(aq) \text{ (at cathode)}$$
   $$2I^-(aq) \longrightarrow I_2(aq) + 2e^- \text{ (at anode)}$$
   Explain how your observations correspond to the products shown in these half-reactions.

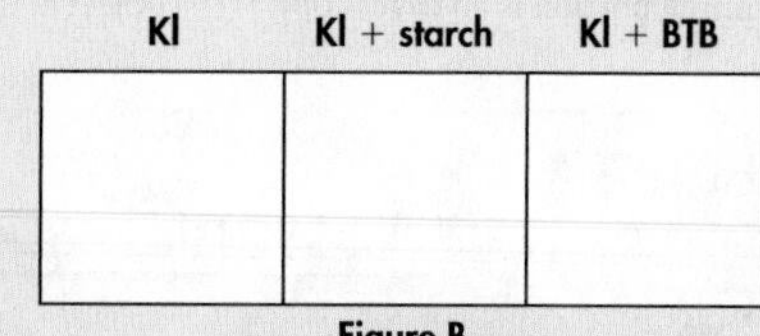

Figure B

2. **Analyze Data** For each half-reaction listed above, look up the $E°$ values in Table 21.2. Show that the $E°$ values are consistent with what you observe.
3. **Design an Experiment** Design an experiment to explore what happens when you electrolyze NaCl, KBr, and $CuSO_4$ with and without BTB. Write half-reactions to predict your results.

FOR ENRICHMENT In 1985, treasure hunter Mel Fisher salvaged tons of silver coins from the shipwrecked galleon *Atocha*, which sank in 1622. Have students research the electrochemistry Fisher used to rescue the treasure from corrosion.

21 Study Guide

BIGIDEA MATTER AND ENERGY

The two types of electrochemical cells are voltaic cells and electrolytic cells. In a voltaic cell, electric current is produced by a spontaneous redox reaction. Voltaic cells are used in batteries and fuel cells. In an electrolytic cell, a nonspontaneous redox reaction is driven by the application of electrical energy. Electrolytic cells are used to produce commercially important chemicals and to plate, purify, and refine metals.

21.1 Electrochemical Cells

All electrochemical processes involve redox reactions.

Electrical energy is produced in a voltaic cell by a spontaneous redox reaction within the cell.

Current applications that use electrochemical processes to produce electrical energy include dry cells, lead storage batteries, and fuel cells.

- electrochemical process (728)
- electrochemical cell (730)
- voltaic cell (730)
- half-cell (730)
- salt bridge (730)
- electrode (730)
- anode (730)
- cathode (730)
- dry cell (732)
- battery (733)
- fuel cell (734)

21.2 Half-Cells and Cell Potentials

The electrical potential of a cell results from a competition for electrons between two half-cells.

You can determine the standard reduction potential of a half-cell by using a standard hydrogen electrode and the equation for standard cell potential.

If the cell potential for a given redox reaction is positive, then the reaction is spontaneous as written. If the cell potential is negative, then the reaction is nonspontaneous.

- electrical potential (737)
- reduction potential (737)
- cell potential (737)
- standard cell potential (737)
- standard hydrogen electrode (738)

Key Equation

$$E^\circ_{cell} = E^\circ_{red} - E^\circ_{oxid}$$

21.3 Electrolytic Cells

The key difference between voltaic and electrolytic cells is that in a voltaic cell, the flow of electrons is the result of a spontaneous redox reaction, whereas in an electrolytic cell, electrons are caused to flow by an outside power source, such as a battery.

Electrolysis of a solution or of a melted, or molten, ionic compound can result in the separation of elements from compounds. Electrolytic cells are also commonly used in the plating, purifying, and refining of metals.

- electrolysis (745)
- electrolytic cell (745)

Evaluate

Review and Assessment Materials

CHEMISTRY TUTORIAL Encourage students to review writing cell reactions and calculating standard cell potentials using the animated step-by-step tutorials for Chapter 21. Each Chemistry Tutorial is followed by a similar practice problem, where the student is guided through the problem-solving process.

CHAPTER PROBLEM SET Have students practice more electrochemistry problems using the Chapter 21 online Problem Set.

KINETIC ART Direct students to the Kinetic Art animations *Voltaic Cell* and *Voltaic and Electrolytic Cells*, and have them answer the questions that accompany each animation.

CONCEPTS IN ACTION If students have trouble understanding the components of voltaic cells, ask them to view the Concepts in Action animation, which presents the material in a familiar context. Have students complete the questions after the animation.

Study Tip

Study for short, focused blocks of time. Take a break when you feel tired. When you are no longer being productive, take a short walk or stretch.

Focus on ELL

5 ASSESS UNDERSTANDING Have students read the Big Idea paragraph aloud to themselves. Have students create a list of words that they cannot pronounce or do not understand. Have students use the lesson review to locate the words in the text so they can define them.

BEGINNING: LOW/HIGH Give each student a copy of the Study Guide. Ask them to make a check mark next to the content they understand and circle the content they do not understand. Create assessments with illustrations and symbols to assess only the content they circle.

INTERMEDIATE: LOW/HIGH Explain the meaning of common words or phrases that appear on multiple choice questions, such as *Which is the most/least likely, or Which best explains, etc*. Orally pose some questions using these phrases.

Understanding by Design

Performance Tasks

ELECTROPLATING Have students work with a partner to research, design, and perform an electroplating demonstration. Students should prepare a report that explains the electrochemical process in their demonstration. The report should include a sketch that identifies the cathode and anode and describe the reactions that occur in each half-cell with the appropriate chemical reactions

EVERYDAY ELECTROCHEMISTRY Have students create a slide-show presentation about everyday examples of electrochemistry. The presentations should explain voltaic cells and electrolytic cells, and it should describe at least two examples of each. Encourage students to include photographs and illustrations in their presentations.

Evaluate

Answers

LESSON 21.1

26. the metal with the higher activity

27. Oxidation: $Al(s) \rightarrow Al^{3+}(aq) + 3e^-$
Reduction: $Cu^{2+}(aq) + 2e^- \rightarrow Cu(s)$

28. **a.** Cu **b.** Ca **c.** Mg **d.** Sn **e.** Zn **f.** Al

29. nothing

30. A half-cell is one part of a voltaic cell in which either oxidation or reduction occurs.

31. The salt bridge allows ions to pass from one half-cell to the other but prevents the solutions from mixing.

32. cathode

33. A zinc container (anode) filled with electrolyte paste; the cathode is a graphite rod embedded in the paste. Zn is oxidized; MnO_2 is reduced.

34. Dry cells and alkaline batteries are both single electrochemical cells that produce about 1.5 V. The two cells share a similar structure (a graphite cathode embedded in an electrolyte paste, with a zinc container that serves as the anode) but use different electrolyte mixtures. Compared to a dry cell, an alkaline battery has a longer shelf life and produces a steadier voltage.

35. Water is produced by the redox reaction and sulfuric acid is used up; water has a lower density than sulfuric acid.

36. $Pb(s)|PbSO_4(aq)||PbO_2(s)|PbSO_4(s)$

37. A fuel cell needs no recharging; it does not produce toxic wastes; and if the fuel is hydrogen gas, it is lighter.

38. Fuel cells cannot generate electricity as economically as more conventional forms of electrical generation.

LESSON 21.2

39. the ability of a voltaic cell to produce a current

40. The standard reduction potential of a half-cell is a measure of the tendency of a given half-reaction to occur as a reduction under standard conditions. The difference between the standard reduction potentials of the two half-cells is the standard cell potential.

41. 0.00 V; the value was set arbitrarily.

42. Connect the aluminum half-cell to a standard hydrogen half-cell and use a voltmeter. The indicated voltage is the standard reduction potential for the aluminum half-cell. Alternatively, look up the standard reduction potential in a reference table.

43. The relative order is the same because both tables rank the elements according to their tendency to undergo oxidation/reduction.

44. **a.** nonspontaneous, –0.34 V
b. nonspontaneous, –1.24 V

45. **a.** +1.10 V **b.** +1.61 V **c.** +0.94 V

21 Assessment

* Solutions appear in Appendix E

Lesson by Lesson

21.1 Electrochemical Cells

26. If the relative activities of two metals are known, which metal is more easily oxidized?

27. Write the half-reactions that occur when a strip of aluminum is dipped into a solution of copper(II) sulfate.

***28.** For each pair of metals listed below, decide which metal is more readily oxidized.

a. Hg, Cu
b. Ca, Al
c. Ni, Mg
d. Sn, Ag
e. Pb, Zn
f. Cu, Al

29. What would you expect to happen when a strip of lead is placed in an aqueous solution of magnesium nitrate?

30. What is meant by the term *half-cell*?

31. Explain the function of the salt bridge in a voltaic cell.

32. At which electrode in a voltaic cell does reduction always occur?

***33.** Describe the structure of a dry cell. What substance is oxidized? What substance is reduced?

34. How are dry cells and alkaline batteries similar? How are they different?

35. Explain why the density of the electrolyte in a lead storage battery decreases during the discharge process.

***36.** Use the shorthand method to represent the electrochemical reaction in a lead storage battery.

37. List the advantages of a fuel cell over a lead storage battery.

38. Fuel cells can be designed to generate electrical energy while emitting no air pollutants, yet they are not widely used. Explain.

21.2 Half-Cells and Cell Potentials

39. What is the electrical potential of a voltaic cell?

***40.** What is the difference between standard cell potential and standard reduction potential?

41. What is the electrical potential of a standard hydrogen electrode? How was it determined?

42. Explain how to determine the standard reduction potential for the aluminum half-cell.

43. How does the order of the metals in Table 21.1 compare with the order in Table 21.2? Explain.

***44.** Determine whether these redox reactions will occur spontaneously. Calculate the standard cell potential in each case.

a. $Cu(s) + 2H^+(aq) \longrightarrow Cu^{2+}(aq) + H_2(g)$
b. $2Ag(s) + Fe^{2+}(aq) \longrightarrow 2Ag^+(aq) + Fe(s)$

45. Use the information in Table 21.2 to calculate standard cell potentials for these voltaic cells.

a. $Zn \mid Zn^{2+} \parallel Cu^{2+} \mid Cu$
b. $Ni \mid Ni^{2+} \parallel Cl_2 \mid Cl^-$
c. $Sn \mid Sn^{2+} \parallel Ag^+ \mid Ag$

21.3 Electrolytic Cells

***46.** Distinguish between voltaic and electrolytic cells.

Use the diagram to answer Problems 47–49.

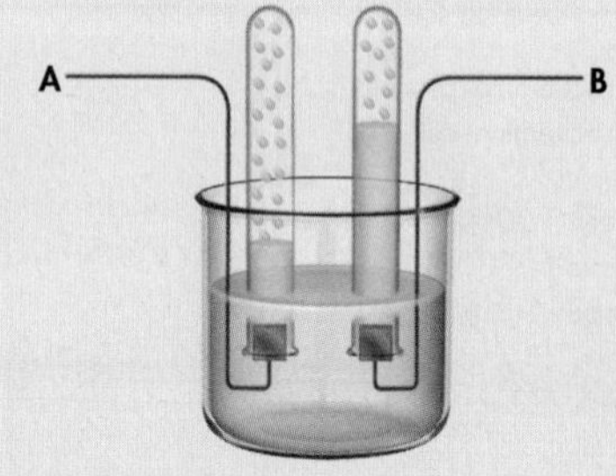

47. Write the equation for the decomposition of water by electrolysis.

48. At which electrode, A or B, is hydrogen produced?

49. The equation for the electrolysis of brine is

$$2NaCl(aq) + 2H_2O(l) \longrightarrow Cl_2(g) + H_2(g) + 2NaOH(aq)$$

How would you modify the electrolysis diagram to make it quantitatively represent the formation of hydrogen and chlorine?

***50.** What chemical changes occur during the electrolysis of brine?

LESSON 21.3

46. Voltaic cells convert chemical energy into electrical energy. Electrolytic cells use electrical energy to cause a chemical reaction.

47. $2H_2O(l) \rightarrow O_2(g) + 2H_2(g)$

48. electrode A

49. The small spheres representing molecules of H_2 and Cl_2 would be in 1:1 ratio; the volumes of the gases would be equal.

50. Chloride ions are oxidized to produce chlorine gas, and water is reduced to produce hydrogen gas.

51. What are the products of the electrolysis of molten sodium chloride?

52. What are some applications of electrolysis in the field of metallurgy?

53. Describe briefly how you would electroplate a teaspoon with silver.

Understand Concepts

54. Describe the composition of the anode, cathode, and electrolyte in the following:
   a. a fully charged lead storage battery
   b. a fully discharged lead storage battery

*55. Predict what will happen, if anything, when an iron nail is dipped into a solution of copper(II) sulfate. Write the oxidation and reduction half-reactions for this process and the balanced equation for the overall reaction.

56. The standard reduction potential for a cadmium half-cell is −0.40 V. What does this mean?

*57. Calculate E°_{cell} and write the overall cell reaction for these cells.
   a. $Sn \mid Sn^{2+} \parallel Pb^{2+} \mid Pb$
   b. $H_2 \mid H^+ \parallel Br_2 \mid Br^-$

58. Why is it not possible to measure the potential of an isolated half-cell?

59. Complete the data table below for the electrolysis of water.

H_2O used	H_2 formed	O_2 formed
a. 2.0 mol	____ mol	____ mol
b. ____ g	____ g	16.0 g
c. ____ mL	10.0 g	____ g
d. 44.4 g	____ g	____ g
e. ____ g	8.80 L (STP)	____ L (STP)
f. 66.0 mL	____ g	____ L (STP)

*60. What is the relationship between the voltage produced by a redox reaction and the spontaneity of the reaction?

61. The reactions that take place in voltaic cells produce electric current, and the reactions in electrolytic cells can be made to take place when an electric current is applied. What common feature do these redox reactions share?

*62. In one process used to produce aluminum metal, ore containing aluminum oxide is converted to aluminum chloride. Aluminum metal and chlorine gas are then produced by the electrolysis of molten aluminum chloride ($AlCl_3$).
   a. Write the half-reactions that take place at the cathode and the anode.
   b. Write the equation for the overall cell reaction.
   c. Identify the products produced at the cathode and the anode.

63. Use the information in Table 21.2 to determine which of the following cell reactions will proceed spontaneously.
   a. $Zn + Pb^{2+} \longrightarrow ?$
   b. $Cu + Fe^{2+} \longrightarrow ?$
   c. $Ag + Cu^{2+} \longrightarrow ?$
   d. $H_2 + Cu \longrightarrow ?$
   e. $Fe + Pb^{2+} \longrightarrow ?$
   f. $Na + Cl_2 \longrightarrow ?$

64. For each spontaneous reaction in Problem 63, write the half-reaction that takes place at the anode. Write the half-reaction that takes place at the cathode.

65. Write the overall balanced equation for each spontaneous reaction in Problem 63.

66. Determine the standard cell potential for each spontaneous reaction in Problem 63.

67. In certain cases, more than one reaction is possible at an electrode. How can you determine which reaction actually takes place?

*68. Answer the following questions for the electrolysis of brine (concentrated sodium chloride solution).
   a. Write the equations for the two possible reactions that can take place at the anode.
   b. Write the equations for the two possible reactions that can take place at the cathode.
   c. Which reaction actually takes place at the anode? Explain why this reaction takes place in preference to the other possible reaction.
   d. Which reaction actually takes place at the cathode? Explain why this reaction takes place in preference to the other possible reaction.

*69. Identify the stronger oxidizing agent in each of the following pairs:
   a. Li^+, Ca^{2+}
   b. Fe^{3+}, Hg_2^{2+}
   c. Cu^{2+}, Cu^+
   d. Hg^{2+}, I_2

Evaluate

Answers

51. sodium metal and chlorine gas
52. Sample answers: silver-plated jewelry, electrorefining impure lead or copper.
53. Set up an electrolytic cell with a silver rod as the anode, the object to be silver-plated as the cathode, and silver cyanide solution as the electrolyte.
54. a. The anode grid is packed with spongy lead, and the cathode grid is packed with Pb. The electrolyte is concentrated sulfuric acid.
   b. The anode and cathode grids are both packed with $PbSO_4$. The electrolyte is very dilute sulfuric acid.
55. Some of the iron dissolves and the nail becomes coated with copper.
   Oxidation: $Fe \rightarrow Fe^{2+} + 2e^-$
   Reduction: $Cu^{2+} + 2e^- \rightarrow Cu$
   $Fe(s) + CuSO_4(aq) \rightarrow FeSO_4(aq) + Cu(s)$
56. The cadmium half-cell has a tendency to undergo oxidation.
57. a. $Sn(s) + Pb^{2+}(aq) \rightarrow Sn^{2+}(aq) + Pb(s)$
   $E^\circ_{cell} = +0.01$ V
   b. $H_2(g) + Br_2(l) \rightarrow 2H^+(aq) + 2Br^-(aq)$
   $E^\circ_{cell} = +1.07$ V
58. Two half-cells are needed because oxidation or reduction cannot occur in isolation. One half-cell gains electrons and one loses them, producing an electric current.
59. a. 2.0, 1.0
   b. 18.0, 2.0
   c. 90, 80
   d. 4.9, 39.5
   e. 7.07, 4.40
   f. 7.3, 41.1
60. If the cell potential for a redox reaction is positive, the reaction is spontaneous as written.
61. In each type of cell, oxidation occurs at the anode and reduction occurs at the cathode.
62. a. Oxidation: $6Cl^-(l) \rightarrow 3Cl_2(g) + 6e^-$ (anode)
   Reduction: $2Al^{3+}(l) + 6e^- \rightarrow 2Al(l)$ (cathode)
   b. $2AlCl_3(l) \rightarrow 2Al(l) + 3Cl_2(g)$
   c. Chlorine gas is produced at anode; liquid aluminum is produced at cathode.
63. a, e, and f are spontaneous; b, c, and d are nonspontaneous.
64. a. $Zn \rightarrow Zn^{2+} + 2e^-$ (anode)
   $Pb^{2+} + 2e^- \rightarrow Pb$ (cathode)
   e. $Fe \rightarrow Fe^{2+} + 2e^-$ (anode)
   $Pb^{2+} + 2e^- \rightarrow Pb$ (cathode)
   f. $Na \rightarrow Na^+ + e^-$ (anode)
   $Cl_2 + 2e^- \rightarrow 2Cl^-$ (cathode)
65. a. $Zn + Pb^{2+} \rightarrow Zn^{2+} + Pb$
   e. $Fe + Pb^{2+} \rightarrow Fe^{2+} + Pb$
   f. $2Na + Cl_2 \rightarrow 2Na^+ + 2Cl^-$
66. a. +0.63 V e. +0.21 V f. +4.07 V
67. It will be the reaction that has the larger cell potential. In some cases, an indicator can be used to identify the product.
68. a. possible oxidation reactions at anode:
   (i) $2Cl^-(aq) \rightarrow Cl_2(g) + 2e^-$
   (ii) $2H_2O(l) \rightarrow O_2(g) + 4H^+(aq) + 4e^-$
   b. possible reduction reactions at cathode:
   (i) $Na^+(aq) + e^- \rightarrow Na(s)$
   (ii) $2H_2O(l) + 2e^- \rightarrow H_2(g) + 2OH^-(aq)$
   c. (i) Chloride ions are more readily oxidized to chlorine gas than water molecules are oxidized to oxygen.
   d. (ii) Water molecules are more easily reduced than sodium ions.
69. a. Ca^{2+} b. Hg_2^{2+} c. Cu^+ d. Hg^{2+}

ASSESSMENT

Evaluate

Answers

70. Gold belongs near the bottom, below silver, because it is one of the least active metals.

71. A direct current flows in one direction only.

72. a. $Cr(s) + 3Ag^{+}(aq) \rightarrow Cr^{3+}(aq) + 3Ag(s)$
$E^{\circ}_{cell} = +1.54\ V$

b. $2Al(s) + 3Cd^{2+}(aq) \rightarrow 2Al^{3+}(aq) + 3Cd(s)$
$E^{\circ}_{cell} = +1.26\ V$

THINK CRITICALLY

73. The paste in a dry cell allows for the movement of electrons but not ions.

74. The chemists' definition focuses on the electrons that are produced by oxidation at the anode of a voltaic cell; the dictionary definition is probably based on an electrolytic cell, whose electrodes are defined by the battery terminals to which they are attached.

75. The spoon is being plated with silver.

76. As electrons flow from the anode to the cathode in the external circuit, anions must flow from the cathode compartment to the anode compartment to maintain neutrality in the electrolytes. Anions cannot flow through wire made of copper or any other metal. The cell will not function if the salt bridge is replaced with a metal wire.

77. $PbSO_4$ and PbO_2 are both solids.

78. a. negative electrode (cathode)
b. positive electrode (anode)
c. negative electrode (cathode)
d. negative electrode (cathode)

79. d; the voltage falls steadily.

70. Gold is not included in Table 21.2. Where in the table does gold belong?

★71. Why is direct current, not alternating current, used in the electroplating of metals?

72. Write the overall cell reactions and calculate E°_{cell} for voltaic cells composed of the following sets of half-reactions.

a. $Ag^{+}(aq) + e^{-} \longrightarrow Ag(s)$
$Cr^{3+}(aq) + 3e^{-} \longrightarrow Cr(s)$

b. $Al^{3+}(aq) + 3e^{-} \longrightarrow Al(s)$
$Cd^{2+}(aq) + 2e^{-} \longrightarrow Cd(s)$

Think Critically

★73. Relate Cause and Effect Lead storage batteries can be recharged. Why are dry cells not rechargeable?

74. Infer For any voltaic cell, chemists consider the electrode that produces electrons to be negative, and they call it the anode. Most dictionaries, however, define the anode as the positively charged electrode. Explain.

★75. Interpret Diagrams Describe the process that is occurring in the illustration below.

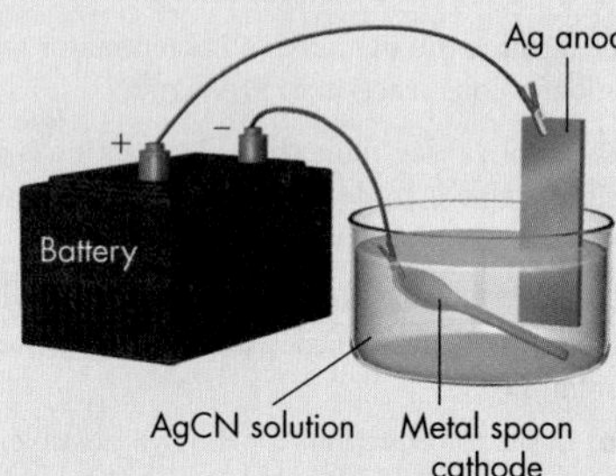

76. Relate Cause and Effect In most voltaic cells, the half-cells are connected by a salt bridge or a porous barrier instead of a piece of wire made of copper or some other metal. Why is a metal wire not suitable for connecting the half-cells of a voltaic cell? Explain your answer.

77. Infer What property do lead(II) sulfate and lead(IV) oxide have that makes salt bridges unnecessary in a lead storage battery?

78. Interpret Diagrams An electrolytic cell is shown below. The substance MX is an ionic compound.

a. At which electrode does reduction occur?
b. At which electrode are electrons being released by ions?
c. At which electrode are electrons entering the cell?
d. At which electrode does elemental M form?

★79. Interpret Graphs Which plot is characteristic of a dry cell? Explain your answer.

a.
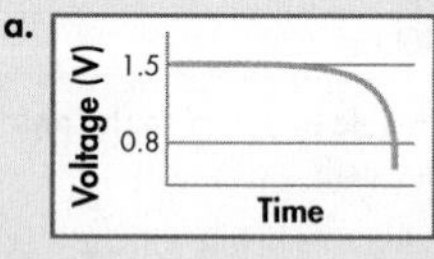

b.
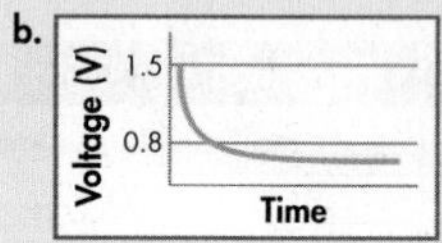

c.
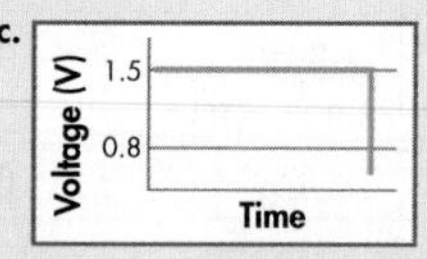

d.
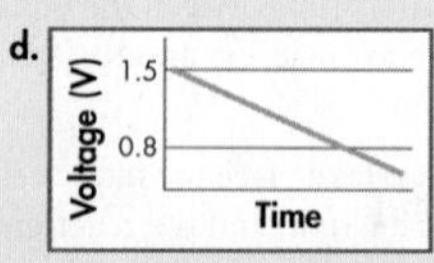

756 Chapter 21 • Assessment

Enrichment

*80. **Calculate** Write the overall cell reactions and calculate E°_{cell} for voltaic cells composed of the following sets of half-reactions.

a. $AgCl(s) + e^- \longrightarrow Ag(s) + Cl^-(aq)$
$Ni^{2+}(aq) + 2e^- \longrightarrow Ni(s)$

b. $Al^{3+}(aq) + 3e^- \longrightarrow Al(s)$
$Cl_2(g) + 2e^- \longrightarrow 2Cl^-(aq)$

81. **Evaluate** An engineer has proposed a new battery design that uses silver as the electrode in both half-cells. As a chemist, what would you tell the engineer about the proposal?

82. **Apply Concepts** Impure copper is purified in an electrolytic cell. Design an electrolytic cell, with H_2SO_4 as the electrolyte, that will allow you to carry out this process. Give the oxidation and reduction half-reactions and a balanced equation for the overall reaction.

83. **Apply Concepts** This spontaneous redox reaction occurs in the voltaic cell illustrated below.

$$Ni^{2+}(aq) + Fe(s) \longrightarrow Ni(s) + Fe^{2+}(aq)$$

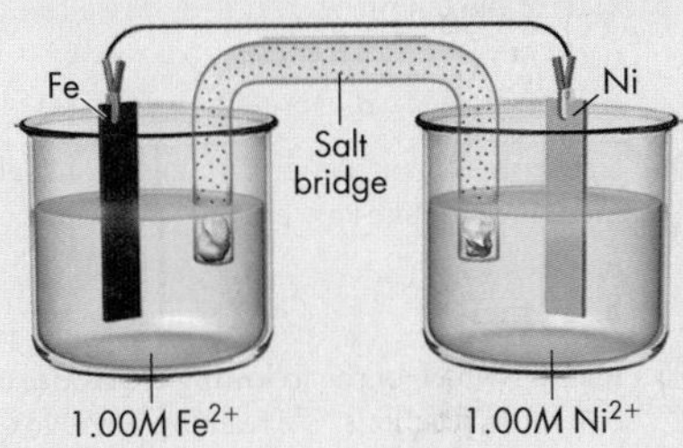

a. Identify the anode and the cathode.
b. Assign charges to the electrodes.
c. Write the half-reactions.
d. Calculate the standard cell potential when the half-cells are at standard conditions.

*84. **Relate Cause and Effect** The cells in an automobile battery are separated from each other. However, the electrolyte is the same in all the cells. What would be the consequence of placing all the cells into a single container of electrolyte solution rather than into separate compartments?

Write About Science

85. **Explain** Write a paragraph explaining how a zinc–copper voltaic cell works. Make sure to mention the half-reactions and the overall reaction in your explanation. (*Hint:* Use Figure 21.3 as a reference.)

86. **Compare and Contrast** What do voltaic cells and electrolytic cells have in common? How do they differ?

CHEMYSTERY

Trash or Treasure?

The jeweler informed Maria that she did not purchase a pure gold antique ring but an inexpensive piece of costume jewelry. The ring was in fact made of stainless steel that was electroplated with gold. As Maria wore the ring, the thin layer of gold began to wear off, revealing the stainless steel underneath. However, the jeweler offered to re-plate the inexpensive ring with more gold to restore its original color.

The process by which gold is plated onto another metal is similar to the process by which silver is plated onto an object. The object that is to be gold-plated is the cathode in the electrolytic cell. The anode is metallic gold, which supplies the gold to be deposited onto the object. The electrolyte is a solution of a gold compound. Electroplating of the object occurs when a direct current is applied.

87. **Describe** Draw the electrolytic cell that is used to plate gold onto an object. Label the anode and the cathode and assign charges to the electrodes. Indicate the direction of electron flow and write the half-reactions.

*88. **Infer** How could you remove the gold from a gold-plated object using an electrolytic cell?

89. **Connect to the BIGIDEA** Will the half-reactions in an electrochemical cell used to plate gold onto an object occur if no electric current is supplied to the cell? Explain.

Evaluate

CHEMYSTERY

APPLY CONCEPTS After students have read through the CHEMystery, call on volunteers to summarize how Maria's ring could have been made to look like it was pure gold when it was not. **Ask** Was the redox process used to coat the ring spontaneous or nonspontaneous reaction? *(nonspontaneous)* **Ask** What type of energy was used to drive this reaction? *(electrical energy)* **Ask** During the process of electroplating the ring with gold, does oxidation or reduction occur at the surface of the ring? How do you know? *(Reduction occurs at the surface of the ring because it is the cathode.)*

CHEMYSTERY ANSWERS

87. Students' diagrams should look like 21.11b on page 746. The anode (positive electrode) is gold; the cathode (negative electrode) is the object to be plated. The electrons flow from the anode to the cathode. The half-reactions are as follows:
Anode: $Au(s) \rightarrow Au^+(aq) + e^-$
Cathode: $Au^+(aq) + e^- \rightarrow Au(s)$

88. Connect the positive electrode of the power source to the gold object and the negative electrode to another metal. The gold object serves as the anode and the gold will be oxidized.

89. **BIGIDEA** No, the process is not spontaneous (E°_{cell} = 0.00 V).

ASSESSMENT

Answers

ENRICHMENT

80. a. $2AgCl + Ni \rightarrow 2Ag + NiCl_2$
$E^\circ_{cell} = +0.47$ V
b. $3Cl_2 + 2Al \rightarrow 2AlCl_3$
$E^\circ_{cell} = +3.02$ V

81. The battery would not produce a voltage (E°_{cell} = 0.00 V).

82. Oxidation:
$2Cu(impure) + 2H_2SO_4 \rightarrow 2Cu^{2+} + 2H_2 + 2SO_4^{2-}$
Reduction:
$2Cu^{2+} + 2SO_4^{2-} + 2H_2O \rightarrow 2Cu(pure) + 2H_2SO_4 + O_2$
Overall reaction:
$2Cu(impure) + 2H_2O \rightarrow 2Cu(pure) + 2H_2 + O_2$

83. a. The iron electrode is the anode; the nickel electrode is the cathode.
b. The anode is negative; the cathode is positive.
c. Oxidation: $Fe(s) \rightarrow Fe^{2+}(aq) + 2e^-$ (anode)
Reduction: $Ni^{2+}(aq) + 2e^- \rightarrow Ni(s)$ (cathode)
d. $E^\circ_{cell} = +0.19$ V

84. The battery output would not be 12 V.

WRITE ABOUT SCIENCE

85. In a zinc–copper voltaic cell, zinc is oxidized and copper ions are reduced.
Anode: $Zn(s) \rightarrow Zn^{2+}(aq) + 2e^-$
Cathode: $Cu^{2+}(aq) + 2e^- \rightarrow Cu(s)$
A zinc rod serves as the anode; a copper rod serves as the cathode. The half-cells are connected by a salt bridge, through which both positive and negative ions migrate. The overall reaction is as follows:
$Zn(s) + Cu^{2+}(aq) \rightarrow Zn^{2+}(aq) + Cu(s)$

86. In both cells, electrons flow from the anode to the cathode through an external circuit. Reduction occurs at the anode; oxidation occurs at the cathode. In a voltaic cell, the flow of electrons is spontaneous; an outside source of power is needed to drive the reaction in an electrolytic cell. The negative electrode is the anode in a voltaic cell; it is the cathode in an electrolytic cell.

ASSESSMENT

Cumulative Review

90. **a.** $3H_2S(g) + 2HNO_3(aq) \rightarrow 3S(s) + 2NO(g) + 4H_2O(l)$
b. $2AgNO_3(aq) + Pb(s) \rightarrow Pb(NO_3)_2(aq) + 2Ag(s)$
c. $3Cl_2(g) + 6NaOH(aq) \rightarrow 5NaCl(aq) + NaClO_3(aq) + 3H_2O(l)$

91. 467 mL

92. **a.** $SnCl_4 \cdot 5H_2O$
b. $MgSO_4 \cdot 7H_2O$
c. $FePO_4 \cdot 4H_2O$
d. $CaCl_2 \cdot 2H_2O$

93. **a.** 2.3 g NaCl
b. 101 g KNO_3

94. Dilute 31 mL 16*M* HNO_3 to 500 mL with water.

95. **a.** 4.32×10^2 kJ
b. 2.55×10^5 cal
c. 2.70×10^3 J

96. **a.** -2.78×10^1 kJ **b.** 7.03 kJ
c. -12 kJ **d.** -27 kJ

97. -267 kJ

98. a and c; $K_{eq} < 1$

99. $\dfrac{[N_2] \times [H_2]^3}{[NH_3]^2}$

100. **a.** pH = 8.00
b. pH = 5.00
c. pH = 10.00
d. pH = 5.00

101. **a.** $[OH^-] = 1 \times 10^{-7}M$
b. $[OH^-] = 1 \times 10^{-10}M$
c. $[OH^-] = 1 \times 10^{-5}M$

102. **a.** $2Na(s) + 2H_2O(l) \rightarrow 2NaOH(aq) + H_2(g)$
b. $Ca(s) + 2H_2O(l) \rightarrow Ca(OH)_2(aq) + H_2(g)$

103. **a.** +6 **b.** −2 **c.** +4
d. +2 **e.** 0 **f.** +4

104. **a.** Ca, +2; Cr, +6; O, −2
b. K, +1; Mn, +7; O, −2
c. Ca, +2; N, +5; O, −2
d. Al, +3; O, −2; H, +1

105. b; Ca is oxidized to Ca^{2+}, and Cl_2 is reduced to Cl^-.
c; Ca is oxidized to Ca^{2+}, and H^+ is reduced to H_2.

106. **a.** $3Br_2(g) + 6NaOH(aq) \rightarrow NaBrO_3(aq) + 5NaBr(aq) + 3H_2O(l)$
b. $Fe_2O_3(s) + 3H_2(g) \rightarrow 2Fe(s) + 3H_2O(l)$

Cumulative Review

90. Balance each equation.

a. $H_2S(g) + HNO_3(aq) \longrightarrow S(s) + NO(g) + H_2O(l)$

b. $AgNO_3(aq) + Pb(s) \longrightarrow Pb(NO_3)_2(aq) + Ag(s)$

c. $Cl_2(g) + NaOH(aq) \longrightarrow NaCl(aq) + NaClO_3(aq) + H_2O(l)$

***91.** A sample of oxygen gas has a volume of 425 mL at 30°C. What is the new volume of the gas if the temperature is raised to 60°C while the pressure is kept constant?

92. Write formulas for these hydrates.

a. tin(IV) chloride pentahydrate
b. magnesium sulfate heptahydrate
c. iron(III) phosphate tetrahydrate
d. calcium chloride dihydrate

***93.** Calculate the grams of solute required to make the following solutions:

a. 250 g of 0.90% NaCl (m/m)
b. 500 mL of 2.0*M* KNO_3

94. Concentrated nitric acid is 16*M*. How would you prepare 500 mL of 1.0*M* HNO_3 from the concentrated acid?

95. Convert the following:

a. 4.32×10^5 joules to kilojoules
b. 255 Calories to calories
c. 645 calories to joules

96. Calculate the quantity of heat lost or gained in the following changes:

a. 0.625 mol NaOH(*s*) dissolves in water
b. 1.17 mol water freezes at 0°C
c. 0.30 mol $C_2H_6O(l)$ vaporizes
d. 0.66 mol of steam condenses at 100°C

***97.** The combustion of natural gas, methane (CH_4), is an exothermic reaction.

$$CH_4(g) + 2O_2(g) \longrightarrow CO_2(g) + 2H_2O(l) \quad \Delta H = -890 \text{ kJ}$$

Calculate the amount of heat liberated when 4.80 g CH_4 reacts with an excess of oxygen.

***98.** Four reactions have the following equilibrium constants. Identify in which of these reactions the reactants are favored over products. Why?

a. $K_{eq} = 0.006$ **c.** $K_{eq} = 8 \times 10^{-4}$
b. $K_{eq} = 5.3$ **d.** $K_{eq} = 2 \times 10^3$

99. Give the equilibrium-constant expression for the decomposition of ammonia to nitrogen and hydrogen.

$$2NH_3(g) \rightleftharpoons N_2(g) + 3H_2(g)$$

***100.** Determine the pH for each solution.

a. $[H^+] = 1.0 \times 10^{-8}M$
b. $[H^+] = 0.000010M$
c. $[OH^-] = 1.0 \times 10^{-4}M$
d. $[OH^-] = 1.0 \times 10^{-9}M$

101. Three solutions have the following pH values. What are the hydroxide-ion concentrations of these solutions?

a. pH = 7.0 **b.** pH = 4.0 **c.** pH = 9.0

102. Write a balanced equation for the reaction of each of the following metals with water:

a. sodium **b.** calcium

***103.** Determine the oxidation number of sulfur in each of the following:

a. H_2SO_4 **c.** SO_2 **e.** S
b. H_2S **d.** $Na_2S_2O_3$ **f.** SO_3^{2-}

104. Determine the oxidation number of each element in these substances.

a. $CaCr_2O_7$ **c.** $Ca(NO_3)_2$
b. $KMnO_4$ **d.** $Al(OH)_3$

***105.** Identify which of the following are oxidation-reduction reactions. If a reaction is a redox reaction, name the element oxidized and the element reduced.

a. $CaCO_3(s) \longrightarrow CaO(s) + CO_2(g)$
b. $Ca(s) + Cl_2(g) \longrightarrow CaCl_2(s)$
c. $Ca(s) + 2H_2O(l) \longrightarrow Ca(OH)_2(aq) + H_2(g)$

106. Balance each redox equation.

a. $Br_2(g) + NaOH(aq) \longrightarrow NaBrO_3(aq) + NaBr(aq) + H_2O(l)$
b. $Fe_2O_3(s) + H_2(g) \longrightarrow Fe(s) + H_2O(l)$

If You Have Trouble With . . .

Question	90	91	92	93	94	95	96	97	98	99	100	101	102	103	104	105	106
See Chapter	11	14	15	16	16	17	17	17	18	18	19	19	19	20	20	20	20

758 Chapter 21 • Assessment

Standardized Test Prep for Indiana

Select the choice that best answers each question or completes each statement.

Tips for Success

Eliminate Wrong Answers If you don't know which response is correct, start by eliminating those you know are wrong. If you can rule out some choices, you'll have fewer left to consider and you'll increase your chances of choosing the correct answer.

1. Which statement describes electrolysis?
(A) Reduction occurs at the anode.
(B) Energy is produced.
(C) Oxidation occurs at the cathode.
(D) Positive ions move to the cathode.

2. A voltaic cell is constructed using the following half-reactions:

$$Cd^{2+}(aq) + 2e^- \longrightarrow Cd(s) \quad E^\circ_{Cd^{2+}} = -0.40\ V$$
$$Sn^{2+}(aq) + 2e^- \longrightarrow Sn(s) \quad E^\circ_{Sn^{2+}} = -0.14\ V$$

What is the standard cell potential for this voltaic cell?
(A) −0.54 V (C) +0.26 V
(B) −0.26 V (D) +0.54 V

3. Which of the following is a product of the reaction that occurs in a hydrogen–oxygen fuel cell?
(A) $CO(g)$ (C) $H_2O(g)$
(B) $CO_2(g)$ (D) $H_2O_2(l)$

4. Magnesium metal is prepared by the electrolysis of molten $MgCl_2$. One half-reaction is

$$Mg^{2+}(l) + 2e^- \longrightarrow Mg(l)$$

Which of the following statements is true?
(A) This half-reaction occurs at the cathode.
(B) Magnesium ions are oxidized.
(C) Chloride ions are reduced at the anode.
(D) Chloride ions gain electrons during this process.

5. If the cell potential for a redox reaction is positive,
(A) the redox reaction is spontaneous.
(B) the redox reaction is not spontaneous.
(C) the reaction only occurs during electrolysis.
(D) More than one statement is correct.

Use the data table to answer Questions 6–12. Hydrogen is included as a reference point for the metals.

Activity Series of Selected Metals

Element	Oxidation half-reaction
Lithium	$Li(s) \longrightarrow Li^+(aq) + e^-$
Potassium	$K(s) \longrightarrow K^+(aq) + e^-$
Sodium	$Na(s) \longrightarrow Na^+(aq) + e^-$
Aluminum	$Al(s) \longrightarrow Al^{3+}(aq) + 3e^-$
Zinc	$Zn(s) \longrightarrow Zn^{2+}(aq) + 2e^-$
Iron	$Fe(s) \longrightarrow Fe^{2+}(aq) + 2e^-$
Hydrogen	$H_2(g) \longrightarrow 2H^+(aq) + 2e^-$
Copper	$Cu(s) \longrightarrow Cu^{2+}(aq) + 2e^-$

6. Which metal will more easily lose an electron, sodium or potassium?

7. Which metal is more easily oxidized, copper or aluminum?

8. What is the relationship between ease of oxidation and the activity of a metal?

9. Describe what would happen if you placed a clean strip of aluminum in a solution of copper(II) sulfate. Explain your answer.

10. Would a copper strip placed in a solution containing zinc ions react spontaneously with the zinc ions? Explain your reasoning.

11. Based on the positions of zinc and iron in the table, explain how attaching zinc blocks to a steel ship hull protects the steel from corrosion.

12. Write the half-reaction for the reduction of aluminum ions.

13. An electrolytic cell is shown at the right. Draw this cell on a separate piece of paper and label the anode, cathode, and direction of electron flow.

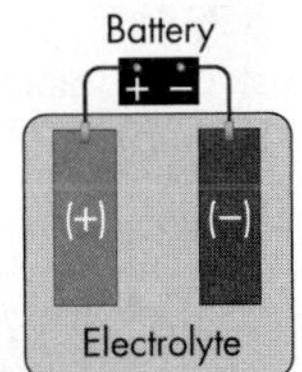

If You Have Trouble With . . .

Question	1	2	3	4	5	6	7	8	9	10	11	12	13
See Lesson	21.3	21.2	21.1	21.3	21.2	21.1	21.1	21.1	21.1	21.1	21.1	21.1	21.3

ASSESSMENT

STP Answers

1. D
2. C
3. C
4. A
5. A
6. potassium
7. aluminum
8. The more active the metal, the more easily it is oxidized.
9. Aluminum would be oxidized and copper ions would be reduced. Copper would plate out on the aluminum.
10. No, because zinc is more easily oxidized than copper, copper cannot reduce the zinc ions in solution.
11. Zinc is more readily oxidized than iron. If iron atoms lose electrons through oxidation, zinc can act as a reducing agent to replace those lost electrons.
12. $Al^{3+}(aq) + 3e^+ \rightarrow Al(s)$
13. Answers should look like the labels on Figure 21.11b on page 746.

22 Hydrocarbon Compounds Planning Guide

Introducing the BIGIDEA: CARBON CHEMISTRY

There are many different hydrocarbon compounds. Some are important in your life.

IASS	Lessons and Objectives	Print Resources		
		For the Student	For the Teacher	
C.2.2, C.3.4, C.9.1, C.9.2	**22.1** Hydrocarbons p 762–771 **22.1.1** Explain why a carbon atom forms four covalent bonds. **22.1.2** Identify two possible arrangements of carbon atoms in an alkane.	Reading and Study Workbook Lesson 22.1 Lesson Assessment 22.1 p 771	Teacher Demo, p 763: Methane Shape Class Activity, p 764: Model of Ethane Teacher Demo, p 765: Alkane Structures	
C.3.4, C.9.1, C.9.2	**22.2** Unsaturated Hydrocarbons p 772–773 **22.2.1** Describe the structural characteristics of alkenes. **22.2.2** Identify the structural characteristics of alkynes.	Reading and Study Workbook Lesson 22.2 Lesson Assessment 22.2 p 773		
C.3.4, C.9.1, C.9.2	**22.3** Isomers p 775–778 **22.3.1** Explain how the properties of constitutional isomers differ. **22.3.2** Identify two types of stereoisomers.	Reading and Study Workbook Lesson 22.3 Lesson Assessment 22.3 p 778 Quick Lab: Isomers of Heptane p 778	Class Activity, p 776: Modeling Isomers Teacher Demo, p 777: Enantiomers	
	22.4 Hydrocarbon Rings p 779–781 **22.4.1** Identify the general structure of a cyclic hydrocarbon. **22.4.2** Describe bonding in a benzene ring.	Reading and Study Workbook Lesson 22.4 Lesson Assessment 22.5 p 781		
	22.5 Hydrocarbons From Earth's Crust p 782–786 **22.5.1** Identify the hydrocarbons in natural gas. **22.5.2** Describe the first step in the refining of petroleum. **22.5.3** Classify coal.	Reading and Study Workbook Lesson 22.5 Lesson Assessment 22.5 p 786 Small-Scale Lab: Hydrocarbon Isomers: p 787		

Assessing the BIGIDEA: CARBON CHEMISTRY

	Essential Questions 1. How are hydrocarbons named? 2. What are the general properties of hydrocarbons?	Study Guide p 788 Skills Tune-Up p 789 STP p 795 Reading and Study Workbook Self-Check and Vocabulary Review Chapter 22		

	PearsonChem.com	
	Editable Worksheets	Online Activities
	Teaching Resources, Lesson 22.1 Review	22.1 Lesson Overview Hydrocarbon Inferno Naming Branched-Chain Alkanes
	Teaching Resources, Lesson 22.2 Review Lab Practical 22-1: Model Building with Organic Compounds	22.2 Lesson Overview Saturated and Unsaturated
	Teaching Resources, Lesson 22.3 Review Small-Scale Lab Manual Lab 37: Molecular Structure of Hydrocarbons	22.3 Lesson Overview Asymmetric Carbons Identifying Asymmetric Carbon Atoms
	Teaching Resources, Lesson 22.4 Review Lab 49: Hydrocarbons: A Structural Study	22.4 Lesson Overview
	Teaching Resources, Lesson 22.5 Review	22.5 Lesson Overview Fractional Distillation of Hydrocarbons
	Core Teaching Resources (includes editable worksheets) • Lesson Reviews • Practice Problems • Interpret Graphs • Vocabulary Review • Chapter Quizzes and Tests	Chapter 22 Problem Set Graphing Inequalities

Materials List

For the Student

Quick Lab p 778
- ball-and-stick molecular model kits

Small-Scale Lab p 787
- toothpicks
- modeling clay

For the Teacher

Teacher Demo p 763
- colored foam balls
- applicator sticks

Class Activity p 764
- foam balls and applicator sticks or molecular model sets

Teacher Demo p 765
- foam balls and applicator sticks or molecular model sets

Class Activity p 769
- different colored markers

Class Activity p 776
- molecular model sets

Teacher Demo p 777
- two ball-and-stick models of CHFClBr

Additional Digital Resources

Online Student Edition
Online Teacher's Edition
Exam*View Assessment Suite CD-ROM*
Classroom Resources DVD

CHAPTER 22

What's Online

CHEMISTRY TUTORIAL Students can work step-by-step with the online tutorial to learn how to name branched-chain alkanes and identify asymmetric carbon atom structures by accessing the Chapter 22 Chemistry Tutorials.

ONLINE PROBLEM SET Additional practice problems about hydrocarbons can be found in the Chapter 22 online Problem Set.

MATH TUTORIAL The Online Math Tutorial will help guide students through graphing inequalities.

DIRECTED VIRTUAL LAB Encourage students to review the characteristics of saturated and unsaturated hydrocarbons using the animated virtual lab for Chapter 22. The animation is followed by analysis questions, where the student is guided through the lab concepts.

KINETIC ART Have students watch an animation about fractional distillation of hydrocarbons. Each animation is followed by analysis questions to guide further understanding of the concepts.

CONCEPTS IN ACTION Students can watch an interactive animation about how different hydrocarbons burn. The animation is followed by questions to uncover the trends in flash points and heat of combustion of hydrocarbons.

22 Hydrocarbon Compounds

INSIDE:

PearsonChem.com

Turkmenistan has some of the world's largest deposits of natural gas. This natural gas deposit in the city of Darvaza has been burning for over three decades!

760

IN ACADEMIC STANDARDS for SCIENCE

Nature of Science

- Develop explanations based on reproducible data and observations gathered during laboratory investigations.
- Use analogies and models (mathematical and physical) to simplify and represent systems that are difficult to understand or directly experience due to their size, time scale, or complexity, and recognize the limitations of analogies and models.

Standard 9: Organic Chemistry and Biochemistry

C.9.2 Identify the variety of large molecules formed by the covalent bonding of carbon atoms and describe the functions of these molecules.

Also covered C.2.2, C.3.4, C.9.1

Focus on ELL

1 CONTENT AND LANGUAGE Begin the lesson by drawing students' attention to the word *hydrocarbon* or *hydrocarbons* in the chapter title and lesson titles. Explain that the prefix *hydro-* means "containing hydrogen or water." Have students use this information to predict the meaning of the word *hydrocarbon*.

BEGINNING

LOW Write any unfamiliar words from the lesson titles on cards, and have students practice saying each word.

HIGH Use a bilingual dictionary or glossary to find any cognates to any unfamiliar words from the titles.

INTERMEDIATE: LOW/HIGH Use an English dictionary or the glossary to define any unfamiliar words from the titles.

ADVANCED: LOW/HIGH Using the titles only, explain to the group the purpose or main idea of the chapter.

CARBON CHEMISTRY

Essential Questions:

1. *How are hydrocarbons named?*
2. *What are the general properties of hydrocarbons?*

CHEMYSTERY

Nose for Hire

Walking home from school one day, Anthony spotted a poster soliciting participants for a smell test. The poster offered participants "$50 for less than an hour of your time." So he decided to go for it.

When Anthony reported for the study, after filling out some paperwork, a researcher asked him to smell two chemical samples. The first one smelled like a freshly cut orange. The second had an odor that reminded him of pine trees.

Anthony was curious. "What am I smelling?" he asked. "Limonene," the researcher answered. "What about the second one?" Anthony asked. The researcher gave the same reply: "Limonene." Anthony was perplexed. How could two substances, both with the name limonene, smell so different?

▶ Connect to the **BIG**IDEA As you read about hydrocarbons, think about what could make this phenomenon possible.

INDIANA'S ACADEMIC STANDARDS FOR SCIENCE

Atomic Structure and the Periodic Table C.2.2; **Bonding and Molecular Structure** C.3.4; **Organic Chemistry and Biochemistry** C.9.1, C.9.2. See lessons for details.

UbD

Understanding by Design

Students are building toward understanding carbon chemistry by studying hydrocarbon compounds.

PERFORMANCE GOALS At the end of Chapter 22, students will be able to answer the essential questions by applying their knowledge of hydrocarbon compounds.

ESSENTIAL QUESTIONS Read the essential questions aloud. **Ask** What are factors that often determine the names of compounds? *(structure and properties)* **Ask** If hydrocarbons contain just hydrogen and carbon atoms, how is it possible for there to be different types of hydrocarbons? *(The number and arrangement of the two types of atoms must be different.)*

BIGIDEA Use the photo of the burning natural gas deposit to help students connect to this chapter's concepts. Activate prior knowledge by reminding students that natural gas is a fossil fuel. Explain that the main component of natural gas is the hydrocarbon methane, CH_4. **Ask** What property does the photo show methane has? *(It is combustible.)*

CHEMYSTERY Connect the CHEMystery to the Big Idea by pointing out that the name of the compound Anthony smelled was the same each time. **Ask** If the name is the same, then what must be true about the molecular formulas? *(They must be the same.)* As a hint, have students consider what must be different about compounds that have the same formula but different properties. Students should conclude that the compounds might have different molecular structures.

CHAPTER 22

Introduce the Chapter

IDENTIFYING MISCONCEPTIONS Students may assume that molecules can have only one arrangement. Use these activities to help them realize that compounds can have the same molecular formula but different structures and properties.

Activity 1 *You will need toothpicks and two colors of modeling clay.* Have students make 4 model carbon atoms and 10 model hydrogen atoms. Draw the structure of butane, and have students model it. **Ask** What is its formula? *(C_4H_{10})* Repeat for 2-methylpropane. Explain that the formulas are the same, but the structures and properties differ.

Activity 2 *You will need toothpicks and five colors of modeling clay.* Have students make one model atom each of carbon, hydrogen, fluorine, chlorine, and bromine. Then have them connect the atoms of fluorine and hydrogen to either side of the carbon atom. **Ask** Where should you connect the chlorine and bromine atoms? *(There are two possibilities, depending on the side where each is placed.)* Explain that the two ways represent two compounds with the same molecular formula but different structures and properties.

UntamedScience™

Cracking the Crude

Danni visits a state-of-the-art oil refinery to learn the role that chemical properties play in turning crude oil into gasoline.

LESSON 22.1

Key Objectives

EXPLAIN why a carbon atom forms four covalent bonds.

IDENTIFY two possible arrangements of carbon atoms in an alkane.

Additional Resources

Reading and Study Workbook, Lesson 22.1

PearsonChem.com

Engage

CHEMISTRY & YOU Have students read the introduction. **Ask** What do many fuels have in common? *(They are mixtures of compounds called hydrocarbons.)* **Ask** Based on its name, predict the composition of a hydrocarbon. *(Hydrocarbons are compounds that contain only hydrogen and carbon.)*

Activate Prior Knowledge

Briefly review bonding concepts with students. **Ask** What is the difference between an ionic bond and a covalent bond? *(Electrons are transferred between atoms in ionic bonds and shared in covalent bonds.)* **Ask** What type of bonding occurs in carbon? *(covalent bonding)* **Ask** What are valence electrons? *(electrons in the highest occupied energy level of an atom)*

Foundations for Reading

BUILD VOCABULARY Have students use simpler words or phrases to rewrite the definition of *homologous series.*

IN ACADEMIC STANDARDS for SCIENCE

C.3.4 Draw structural formulas for and name simple molecules.

C.9.1 Use structural formulas to illustrate carbon atoms' ability to bond covalently to form many different substances.

C.9.2 Identify the variety of molecular types formed by the covalent bonding of carbon atoms and describe the typical properties of these molecular types.

Also covered C.2.2

22.1 Hydrocarbons

C.2.2 Subatomic particles and atomic structure; **C.3.4** Structural formulas and names; **C.9.1** Use structural formulas: carbon atoms; **C.9.2** Identify the variety of carbon molecules.

CHEMISTRY & YOU

Q: *Why are some fossil fuels gases, some liquids, and some solids?* The gasoline used to fuel this motorcycle is a liquid at STP. So are the diesel fuel used in trucks and buses, and the kerosene used in lanterns. Other fuels are gases or solids. For example, the fuel used in a furnace might be natural gas or a solid such as coal. All these fuels contain mixtures of compounds called hydrocarbons. In this lesson, you will learn about the structure and properties of hydrocarbons.

Key Questions

Why does a carbon atom form four covalent bonds?

What are two possible arrangements of carbon atoms in an alkane?

Vocabulary

- hydrocarbon
- alkane
- straight-chain alkane
- homologous series
- condensed structural formula
- substituent
- alkyl group
- branched-chain alkane

Organic Chemistry and Hydrocarbons

Why does a carbon atom form four covalent bonds?

Fewer than 200 years ago, it was thought that only living organisms could synthesize the carbon compounds found in their cells. So these compounds were classified as organic compounds, and the study of these compounds was known as organic chemistry. Many people thought that a mysterious vital force directed the formation of carbon compounds. A German chemist, Friedrich Wöhler (1800–1882), refuted this idea in 1828. He was able to use inorganic substances to synthesize urea—an organic compound found in urine. Today, organic chemistry includes the chemistry of almost all carbon compounds, regardless of their origin.

Introduction to Hydrocarbons In a reference book that lists properties of common compounds, the list of organic compounds is much longer than the list of inorganic compounds. In fact, new organic compounds are synthesized every day. The simplest organic compounds are called hydrocarbons. By definition, a compound contains at least two elements. A **hydrocarbon** is an organic compound that contains only carbon and hydrogen. The two simplest hydrocarbons are methane and ethane.

Methane (CH_4) is the major component of natural gas. It is sometimes called marsh gas because it is formed by the action of bacteria on decaying plants in swamps and other marshy areas. Livestock and termites also emit substantial quantities of methane as a product of digestion.

Focus on ELL

1 CONTENT AND LANGUAGE Write the words *primary*, *secondary*, *tertiary*, and *quaternary* on the board and pronounce them for the students. Explain that these words have both common and scientific meanings. Provide the common definitions, and have students write those definitions in their vocabulary notebooks, leaving room to add the scientific definitions as they are encountered in the lesson.

2 FRONTLOAD THE LESSON Review bonding concepts by asking students to explain the difference between ionic and covalent bonds. **Ask** What type of bonding occurs in carbon? *(covalent)* Remind students that only valence electrons participate in bonding.

3 COMPREHENSIBLE INPUT As students create their physical models throughout the lesson, guide them to think aloud. Students should verbalize their thought process by asking themselves questions such as, *How many bonds do I need to show? What type of shape does this molecule represent? If I attach this hydrogen atom here, then where should I place the next hydrogen atom?*, and so on.

Recall that a carbon atom has four valence electrons and a hydrogen atom has one valence electron. So one carbon atom can form a single covalent bond with four hydrogen atoms as shown below.

$$\cdot\ddot{C}\cdot + 4H\cdot \longrightarrow H\!:\!\underset{H}{\overset{H}{\ddot{C}}}\!:\!H$$

Carbon atom — Hydrogen atoms — Methane molecule

Because carbon has four valence electrons, a carbon atom always forms four covalent bonds. Remembering this principle will help you to write correct structures for organic compounds.

Methane is not typical of the vast majority of organic compounds because there isn't a bond between carbon atoms in a methane molecule. But there is a carbon-carbon bond in ethane. In an ethane molecule, two carbon atoms share a pair of electrons. The remaining six valence electrons form bonding pairs with the electrons from six hydrogen atoms.

$$2\cdot\ddot{C}\cdot + 6H\cdot \longrightarrow \cdot\ddot{C}\!:\!\ddot{C}\cdot + 6H\cdot \longrightarrow H\!:\!\underset{H\,H}{\overset{H\,H}{\ddot{C}\!:\!\ddot{C}}}\!:\!H$$

Carbon atoms — Hydrogen atoms — Ethane molecule

The ability of carbon to form stable carbon-carbon bonds is one reason that carbon can form so many different compounds.

Representing Hydrocarbons Table 22.1 shows the structural formulas, ball-and-stick models, and space-filling models for methane and ethane. Structural formulas are a convenient way to show the arrangement of atoms in a molecule. But two-dimensional structural formulas do not provide accurate information about how the atoms in a molecule are arranged in space. Three-dimensional molecular models represent the shapes of molecules more accurately. Throughout this chapter and the next, ball-and-stick models and space-filling models will be used along with structural formulas to represent organic molecules. Hybrid orbital theory and VSEPR theory are used to predict the molecular shapes.

READING SUPPORT

Build Study Skills: *Cluster Diagram* As you read through the chapter, construct a cluster diagram organizing the different types of hydrocarbons. Use color-coding to distinguish among the groups. ***How will this diagram help you learn about hydrocarbons?***

Table 22.1

Different Ways of Representing Hydrocarbons

Name	Structural formula	Ball-and-stick model	Space-filling model
Methane	$\text{H—}\underset{\text{H}}{\overset{\text{H}}{\text{C}}}\text{—H}$		
Ethane	$\text{H—}\underset{\text{H}}{\overset{\text{H}}{\text{C}}}\text{—}\underset{\text{H}}{\overset{\text{H}}{\text{C}}}\text{—H}$		

UbD Check for Understanding

Why does a carbon atom form four covalent bonds?

Provide each student with enough modeling supplies to create a model that can be used to explain why a carbon atom is able to form four covalent bonds. Ask students to answer the key question with a one-minute response, using the model as a visual aid.

ADJUST INSTRUCTION If students are struggling with this concept, refer them to Lesson 8.3 and ask them to revisit VSEPR theory. Encourage students to revise their models.

READING STRATEGY Preview the visuals, having students pay particular attention to the terms associated with the chemical structures. Have students make flashcards to assist them in recalling important parts of hydrocarbon structures, such as methylene groups and the four types of carbons.

Explain

Organic Chemistry and Hydrocarbons

USE VISUALS Draw students' attention to the electron dot structures representing methane and ethane in the equations at the top of the page. Then, have students study the structural formulas and ball-and-stick models in Table 22.1. Explain that in these types of structures, lines and sticks are used to represent the electron pairs represented by dots in the dot structures. **Ask** How many lines or sticks are needed to represent one pair of electrons in a covalent bond? *(one line or stick)* Tell students to keep this in mind as they proceed through the lesson.

Explore

Teacher Demo

PURPOSE Students observe the three-dimensional shape of the simplest hydrocarbon, methane.

MATERIALS colored foam balls, applicator sticks

PROCEDURE Use colored foam balls and applicator sticks to construct models that are large enough for all the students to see. Review the tetravalency of carbon as well as the octet rule. To assess students' knowledge about molecular shape, **Ask** Why does methane have a tetrahedral shape? *(VSEPR theory states that because electron pairs repel each other, electron pairs are as far apart as possible in molecules.)* Explain that one of the key ideas of organic chemistry is that structure dictates function; the arrangement of atoms is as important as the number and type of atoms.

Extend

Connect to GEOLOGY

Explain that because the carbon bond is incredibly strong and can occur with so many nonmetals, including carbon itself, millions of carbon compounds exist. Through the study of biochemistry, scientists have learned that these carbon-based, organic compounds are fundamental to life. In a similar manner, silicon, also a Group 4 element, is fundamental to most inorganic substances on Earth. Have students create a visual representation that demonstrates the analogous relationship between carbon–organic–life and silicon–inorganic–Earth.

LESSON 22.1

Explore

Class Activity

PURPOSE Students construct a ball-and-stick model of ethane.

MATERIALS colored foam balls, applicator sticks

PROCEDURE Have students construct a ball-and-stick model of ethane. Check the models for tetrahedral geometry at each carbon. **Ask** How many covalent bonds are there in ethane? *(7)* **Ask** Why are three-dimensional models a more accurate representation of organic molecules than are two-dimensional models? *(Three-dimensional models convey the spatial relationships among the atoms.)*

Extend

Connect to PHYSIOLOGY Prepare a large test tube with equal volumes of water and either mineral oil or baby oil. A small amount of food coloring can be added to the water to more clearly demarcate the polar and nonpolar phases. Shake the test tube vigorously and allow the phases to separate. Tell students to pay attention to the behavior of the oil as it separates from the water. Remind students that an enzyme molecule can catalyze the breakdown of an organic molecule only if it can attach to the molecule. Point out that in the small intestine, the digestive matter, called *chyme*, contains both fats and aqueous materials. **Ask** How does the digestive system reduce the tendency of fats and water to separate? *(The liver secretes bile to emulsify the fat.)* Have students create a poster illustrating how bile works to emulsify fat and how an inability of the liver to produce bile affects nutrient absorption.

Explain

Alkanes

USE VISUALS Direct students' attention to the ball-and-stick structures in Figure 22.2. Have students compare these structures to the one they created for ethane. Point out that, although all three structures are called straight-chain alkanes, in reality chains of three or more carbons are not arranged in a perfectly straight line because the bond angles of a carbon atom are not equal to 180°.

Figure 22.1 Oil and Water
The nonpolar molecules in this cooking oil are not attracted to the polar water molecules, so they do not mix.
Interpret Photos *What evidence of insolubility do you see in the photo?*

Properties of Hydrocarbons The electron pair in a carbon-hydrogen bond or a carbon-carbon bond is shared almost equally by the nuclei of the atoms forming the bond. Thus, hydrocarbons are nonpolar molecules. The attractions between nonpolar molecules are weak van der Waals forces. So hydrocarbons with low molar masses tend to be gases or liquids that boil at a low temperature.

Recall the general rule "like dissolves like." Two nonpolar compounds will form a solution, as will two polar compounds. But a nonpolar compound and a polar compound will not form a solution. For example, because oil is a mixture of hydrocarbons, oil and water do not mix. So oil will float on top of water, as shown in Figure 22.1.

Alkanes

What are two possible arrangements of carbon atoms in an alkane?

Methane and ethane are examples of alkanes. An **alkane** is a hydrocarbon in which there are only single covalent bonds. In any alkane, all the carbon-carbon bonds are single covalent bonds, and all the other bonds are carbon-hydrogen bonds. **The carbon atoms in an alkane can be arranged in a straight chain or in a chain that has branches.**

Straight-Chain Alkanes Ethane is the simplest **straight-chain alkane,** which is an alkane that contains any number of carbon atoms, one after the other, in a chain. Propane (C_3H_8) has three carbon atoms bonded in a chain with eight electrons shared with eight hydrogen atoms. Butane (C_4H_{10}) has a chain of four carbons and ten hydrogens. Figure 22.2 shows how propane and butane can be used.

Figure 22.2 Hydrocarbon Fuels
Hydrocarbons are commonly used as fuels. **a.** Pressurized tanks of propane are used to fuel the burners in hot-air balloons. **b.** Butane serves as the fuel for many kitchen torches.

Differentiated Instruction

LPR LESS PROFICIENT READERS Have students list the lesson objectives in their notebooks, leaving space after each objective. As they read the section, they should list terms and phrases that provide the information requested for each objective. Check their entries for accuracy.

ELL ENGLISH LANGUAGE LEARNERS Encourage students to locate the vocabulary terms in the lesson. Have students compile an illustrated glossary in which they define each term in English and in their native language.

L3 ADVANCED STUDENTS Have students research sources of methane and methane's relationship to the greenhouse effect and global warming. Questions to be answered could include *What is the greenhouse effect, and how does methane contribute to it?*, *What are the major sources of methane in the Earth's atmosphere?*, and *Is the methane content of Earth's atmosphere increasing or decreasing?*

Interpret Graphs

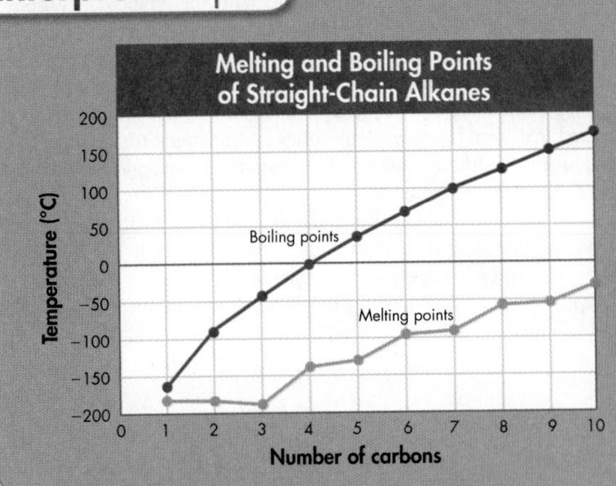

Figure 22.3 The graph illustrates how the melting and boiling points vary with the number of carbons in straight-chain alkanes.

a. Describe Determine whether each alkane is a solid, liquid, or gas at room temperature.

b. Read Graphs How do the added carbons affect the boiling points of the straight-chain alkanes?

c. Predict Estimate a boiling point for undecane, the straight-chain alkane with eleven carbons.

Figure 22.3 shows the melting and boiling points for straight-chain alkanes containing up to ten carbons. The straight-chain alkanes are an example of a homologous series. A group of compounds forms a **homologous series** if there is a constant increment of change in molecular structure from one compound in the series to the next. A CH_2 group is the increment of change in straight-chain alkanes. This change is summarized below for the initial part of the homologous series.

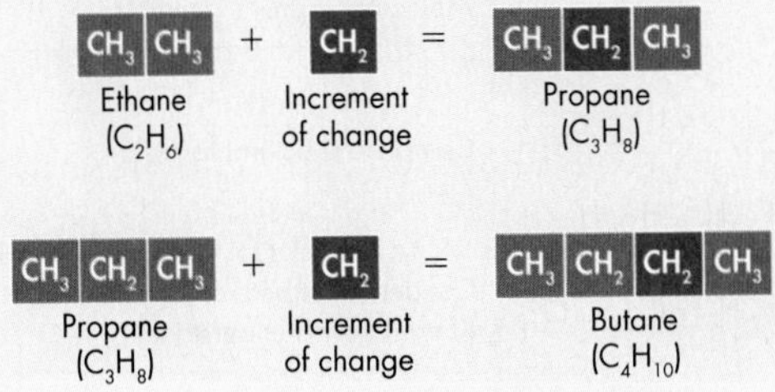

Notice in Figure 22.3 that the boiling points of the straight-chain alkanes increase as the number of carbons in the chain increase. The melting points increase in a similar way.

Naming Straight-Chain Alkanes The names of straight-chain alkanes follow rules established by the International Union of Pure and Applied Chemistry (IUPAC). Every alkane has a name that ends with the suffix *-ane*. For the straight-chain alkanes with one to four carbon atoms, the official names and the common names are the same. They are methane, ethane, propane, and butane, respectively. A mixture of Latin and Greek prefixes are used to name the hydrocarbons having straight chains longer than four carbon atoms. The prefixes are *pent-* for 5, *hex-* for 6, *hept-* for 7, and so on. Use Table 22.2 to memorize the names of the first ten straight-chain alkanes.

Table 22.2

Straight-Chain Alkanes

Name	Formula
Methane	CH_4
Ethane	C_2H_6
Propane	C_3H_8
Butane	C_4H_{10}
Pentane	C_5H_{12}
Hexane	C_6H_{14}
Heptane	C_7H_{16}
Octane	C_8H_{18}
Nonane	C_9H_{20}
Decane	$C_{10}H_{22}$

Organic Discoveries

An organic chemist uses scientific theories and scientific methods to design and build a molecule with a specific set of new properties. To confirm the identity of new molecules, organic chemists use infrared (IR) spectroscopy and nuclear magnetic resonance (NMR) spectrometry to analyze the structures. IR spectroscopy determines the frequencies of infrared radiation absorbed by a molecule. The resulting spectrum is a "fingerprint" of the compound. With NMR spectrometry, many atoms can be identified by the characteristic radio frequencies that they absorb when placed in a magnetic field.

Explore

Teacher Demo

PURPOSE Students compare molecular models and structural formulas of alkanes.

MATERIALS foam balls and applicator sticks or molecular model sets

PROCEDURE Write the complete structural formulas for methane, ethane, propane, and butane on the board. Ask students how the structures are similar and how they are different. Then show students a molecular model for each of the compounds. Pull apart an ethane model and show how a —CH_2— (methylene) group can be inserted between the —CH_3 (methyl) groups to make larger alkanes. Use the models of propane and butane to show how the tetrahedral arrangement of the atoms around each carbon results in a "zigzag" shape for the carbon backbone. Illustrate the rotational freedom of the atoms in a saturated hydrocarbon by twisting the models of propane and butane about the carbon-carbon axis. Once students understand how the atoms are joined in the larger alkanes, write the condensed structural formulas under the complete structural formulas for ethane, propane, and butane. Explain that the condensed formulas are a convenient shorthand notation.

DRAW A CONCLUSION If possible, bring samples of methane, hexane, and paraffin wax to class. On the board, create a table containing the condensed formulas and boiling points of methane (CH_4, −161.0°C) and hexane (C_6H_{14}, 68.7°C). Add paraffin wax to the table, below hexane. Explain to students that paraffin wax is a mixture of alkanes and so does not have a specific condensed chemical formula. Point out that it has a melting point of approximately 56°C, at which point it becomes an oil. **Ask** Based on the melting point of the solid, how do you think the boiling point of paraffin oil compares to those of methane and hexane? *(It is higher.)* Tell students that the boiling point of paraffin oil is significantly higher—at least 260°C. **Ask** Based on this information, what inference can you make about the chain lengths of the alkanes in paraffin? *(They are greater in length than hexane.)* **Ask** What can you conclude about the attraction between molecules as the number of carbon atoms increases? *(The intermolecular attractions increase.)*

Answers

FIGURE 22.1 The two phases remain separate.

INTERPRET GRAPHS

a. Methane through butane are gases; pentane through decane are liquids.

b. As carbon number increases, the boiling point increases.

c. A good estimate is around 200°C. The actual value is 196°C.

Explain

CHEMISTRY & YOU Some hydrocarbon fuels, such as natural gas, consist of lighter hydrocarbons that have boiling points below 20°C. These fuels are gases. Other hydrocarbon fuels, such as gasoline or diesel fuel, are comprised of heavier hydrocarbons, which have comparatively higher boiling points and thus are liquids.

USE AN ANALOGY Introduce Table 22.3 by asking students to write down every name and nickname they are called or referred to by their friends, families, teachers, strangers, and so on. Ask volunteers to share their lists and to share how and why the various names came to be. **Ask** Who has the greatest number of names? **Ask** Does the name used to address you change who you are as a person? *(no)* State that these various references to a person are like the various ways in which a chemical formula may be written. Just as the name that is used to address you does not change the essence of you, the formula used to describe butane or any other carbon compound does not change the properties and characteristics of the compound.

APPLY CONCEPTS Suggest that students create flashcards of the names and structures of the first ten straight-chain alkanes and commit them to memory. **Ask** Why are some alkanes called straight-chain alkanes? *(The carbon atoms form an unbroken line. Each carbon atom within the chain is bonded to no more than two other carbon atoms.)* **Ask** What do you think is meant by the term *branched-chain alkane*? *(Sample answer: The carbons branch off from the carbon chain.)*

Q: *Hydrocarbons such as alkanes can be used as fuels. You just learned that the boiling and melting points of straight-chain alkanes increase as the number of carbons increase. Now explain why some hydrocarbon fuels are gases and others are liquids.*

Drawing Structural Formulas for Straight-Chain Alkanes To draw the structural formula for a straight-chain alkane, write the symbol for carbon as many times as necessary to get the proper chain length. Then complete the formula with hydrogens and lines representing covalent bonds. Complete structural formulas show all the atoms and bonds in a molecule. Sometimes, however, shorthand or condensed structural formulas work just as well. In a **condensed structural formula,** some bonds and/or atoms are left out of the structural formula. Although the bonds and atoms do not appear, you must understand that they are there.

A condensed structural formula in the form of $CH_3(CH_2)_nCH_3$ uses a different kind of shorthand. The CH_2 unit in parentheses is called a methylene group. The subscript *n* to the right of the parentheses indicates the number of methylene groups that are linked together. This shorthand method applies to butane as follows:

$$CH_3(CH_2)_2CH_3$$

Methylene unit — Subscript

The subscript after the parentheses in the condensed structural formula for butane is 2. This means two methylene groups are linked together in the structure. Table 22.3 shows several ways to draw condensed structural formulas, using butane as an example.

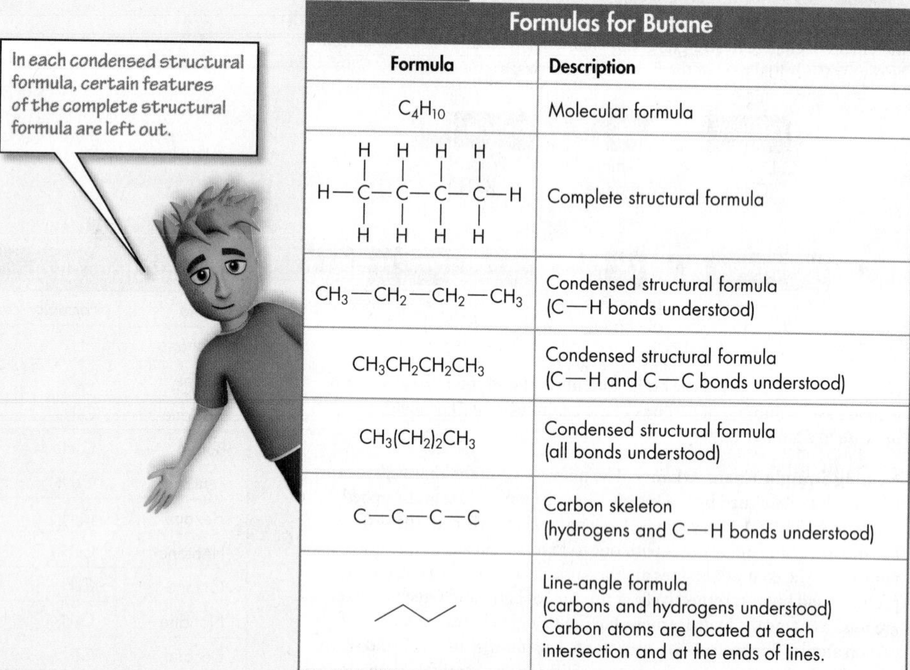

Table 22.3

Formulas for Butane

Formula	Description
C_4H_{10}	Molecular formula
H H H H H—C—C—C—C—H H H H H	Complete structural formula
$CH_3—CH_2—CH_2—CH_3$	Condensed structural formula (C—H bonds understood)
$CH_3CH_2CH_2CH_3$	Condensed structural formula (C—H and C—C bonds understood)
$CH_3(CH_2)_2CH_3$	Condensed structural formula (all bonds understood)
C—C—C—C	Carbon skeleton (hydrogens and C—H bonds understood)
	Line-angle formula (carbons and hydrogens understood) Carbon atoms are located at each intersection and at the ends of lines.

Carbocations

George Olah won the 1994 Nobel Prize in Chemistry for his work with hydrocarbons. Olah showed that some chemical reactions involving hydrocarbons proceed through short-lived, highly reactive intermediate compounds, which he called carbocations. Olah was able to produce, isolate, and preserve carbocations so that they could be studied in detail.

The name *carbocations* means "carbon cations." A carbocation has a positive charge on carbon, thus only three shared pairs of electrons around the element, instead of four. Because the octet is not complete, carbocations are unstable and reactive. When used as an intermediate in an organic reaction, a carbocation is typically considered to be the rate-determining step for the reaction.

SampleProblem 22.1

Drawing Structural Formulas for Alkanes

Draw complete structural formulas for the straight-chain alkanes that have

a. three carbon atoms. **b.** four carbon atoms.

1 Analyze **Identify the relevant concepts.** In an alkane, each carbon atom forms four covalent bonds to hydrogen or to other carbon atoms. Because these are straight-chain alkanes, write the appropriate number of carbons in a straight line, connected to each other by single bonds. Then add the appropriate number of hydrogen atoms.

2 Solve **Apply the concepts to this problem.**

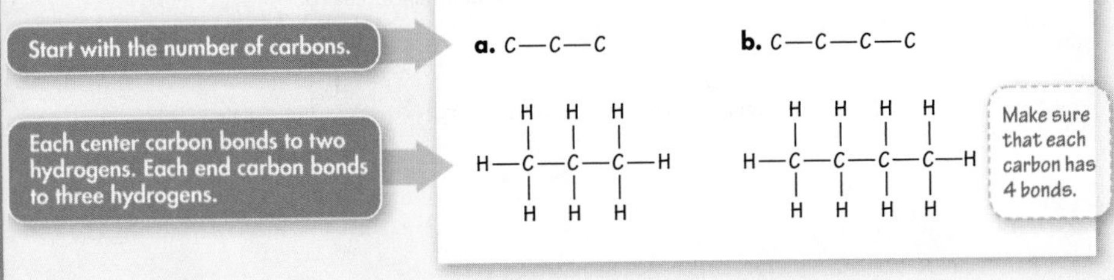

1. Draw complete structural formulas for the straight-chain alkanes with five and six carbon atoms.

2. How many single bonds are there in a propane molecule?

Branched-Chain Alkanes Alkanes and other hydrocarbons do not always have carbon atoms bonded in straight chains. Because a carbon atom forms four covalent bonds, it can bond not only to one or two other carbon atoms, but also to three or even four other carbons, resulting in branched chains. In organic chemistry, branches on a hydrocarbon chain are discussed as if they were substituted for a hydrogen atom on the chain. An atom or group of atoms that can take the place of a hydrogen atom on a parent hydrocarbon molecule is called a **substituent.** Look at the diagrams below. The longest continuous carbon chain of a branched-chain hydrocarbon is called the parent alkane. All other carbon atoms or groups of carbon atoms are regarded as substituents. In Chapter 23, you will study compounds in which atoms such as halogens, oxygen, and nitrogen can take the place of a hydrogen atom on the carbon chain.

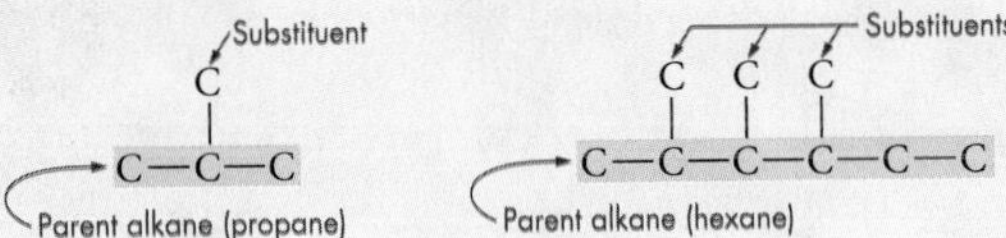

UbD Check for Understanding

To assess students' understanding of the lesson vocabulary, ask students to create a crossword puzzle utilizing the concepts in the lesson, the vocabulary, the labs, and so on. Students should provide a clue for each item. Encourage students to use the Internet to find a free crossword puzzle creator. Have pairs of students exchange crosswords and complete and evaluate each other's creations.

ADJUST INSTRUCTION If students are having difficulty understanding the vocabulary, review the explanations in the text and help students write definitions using their own words. Verify that students' definitions are accurate.

Extend

Connect to PHYSICAL CONSTANTS

Have students organize several of the first C_nH_{2n+2} compounds, where $n \le 40$, into groups with similar characteristics. For instance, students might group compounds according to their states of matter at STP. Ask students to research paraffin and how it relates to hydrocarbons. *(Long-chain alkanes are called paraffins. In their solid state, they are known as paraffin wax. In their liquid state, they are known as either paraffin oil or mineral oil.)*

Sample Practice Problem

A. How many single bonds are there in a straight-chain pentane molecule? *(16)*

B. How many single bonds are there in a butane molecule? *(13)*

Answers

1.

```
  H H H H H
  | | | | |
H-C-C-C-C-C-H
  | | | | |
  H H H H H

  H H H H H H
  | | | | | |
H-C-C-C-C-C-C-H
  | | | | | |
  H H H H H H
```

2. 10 single bonds

LESSON 22.1

Explain

MAKE A CONNECTION Draw the structural formulas for the following alkyl groups in a line down the board: methyl, ethyl, and propyl. To the right of each structure, draw the structures for the following alkanes, using a different color for the last hydrogen: methane, ethane, and propane. Point out to the students that they can use the shared word root to determine the number of carbons in an unfamiliar alkyl group or to identify an alkyl group on the basis of its structure.

APPLY CONCEPTS Direct students to the photograph in Figure 22.4. State that gas stations typically offer three different fuels: regular, mid-grade, and premium. **Ask** What is the basis for your decision in choosing the type of gasoline you (or your family or friends) pump into your car? *(Sample answers: octane rating, cost per gallon, type of car, type of engine, performance, speed, manufacturer's recommendation)* **Ask** Why do you think one type of gasoline costs more than another type? *(Sample answers: higher octane rating, preservatives added, better engine performance)* Explain that prior to pumping gasoline into a car, the owner should check the car manual to determine the fuel specifications needed for that specific car. Explain that the fuel choices at the pump—regular, mid-grade, and premium—relate to the octane rating of the fuel. The octane level required by an engine is determined by its compression ratio—cars with higher-compression engines require higher octane, or premium, fuel.

Misconception Alert

Students may have heard that higher octane fuel will make a car go faster. Explain that this is not true; the octane rating simply rates the fuel's resistance to knocking.

A hydrocarbon substituent that is derived from an alkane is called an **alkyl group.** You can think of an alkyl group as just an alkane with one of the hydrogens removed. An alkyl group can be one carbon or several carbons long. Alkyl groups are named by removing the *-ane* ending from the parent hydrocarbon name and adding *-yl.* The three smallest alkyl groups are the methyl group ($—CH_3$), the ethyl group ($—CH_2CH_3$), and the propyl group ($—CH_2CH_2CH_3$).

When a substituent alkyl group is attached to a straight-chain hydrocarbon, branches are formed. An alkane with one or more alkyl groups is called a **branched-chain alkane.** Each carbon in an organic molecule can be categorized as a primary, secondary, tertiary, or quaternary carbon. If the carbon in question has only one carbon attached to it, then the carbon is a primary carbon. If two carbons are attached to the carbon in question, the carbon is a secondary carbon; if three carbons, a tertiary carbon; and if four carbons, a quaternary carbon. Examples of primary, secondary, tertiary, and quaternary carbons are labeled in the structural formulas below.

$CH_3—CH_3$ (Primary carbons) $CH_3—CH_2—CH_3$ (Secondary carbon) $CH_3—CH(CH_3)—CH_3$ (Tertiary carbon) $CH_3—C(CH_3)_2—CH_3$ (Quaternary carbon)

Isooctane is a hydrocarbon that contains each of these types of carbons. You have probably seen labels for octane ratings in gasoline as shown in Figure 22.4. Isooctane is the standard for determining octane ratings of the mixtures of hydrocarbons that make up gasoline. A gasoline's octane rating is a measure of its ability to prevent engine knock, which is the sound an internal combustion engine makes when gasoline ignites too soon. Knocking reduces a vehicle's performance and can eventually lead to engine damage.

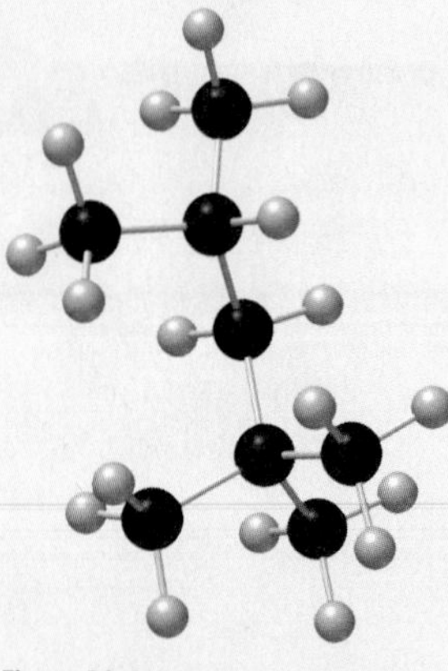

Figure 22.4 Isooctane
Octane ratings of gasoline are relative to that of isooctane, a good anti-knock fuel, which is assigned a value of 100.
Interpret Diagrams ***Find each type of carbon—primary, secondary, tertiary, and quaternary—in the model of isooctane.***

Octane Ratings

The octane system was developed as a way of determining how likely a fuel is to cause engine knocking. The fuels are tested against two standard fuels: isooctane and heptane. Isooctane is very resistant to knocking and has an arbitrary octane rating of 100. Heptane is very prone to knocking and has an arbitrary rating of 0. Every fuel tested is assigned an octane number relative to the controls, based on the fuel's performance in a special test engine. A fuel with octane rating of 89 performs the same as a mixture that is 89 parts isooctane and 11 parts heptane.

Fuels are tested under two sets of driving conditions. The motor (M) octane rating test simulates high-speed driving with heavy loads. The research (R) octane rating test simulates normal driving conditions. The (R + M)/2 octane rating at the gas pump is an average of the two tests' results.

Naming Branched-Chain Alkanes The IUPAC rules for naming branched-chain alkanes are quite straightforward. The name of a branched-chain alkane is based on the name of the longest continuous carbon chain. Each substituent is named according to the length of its chain and numbered according to its position on the main chain. The compound with the structural formula shown to the right can be used as an example.

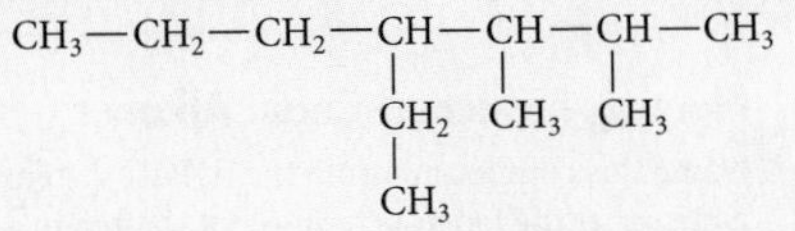

1. Find the longest continuous chain of carbons in the molecule. This chain is considered the parent hydrocarbon.

The longest chain is highlighted in the example. It contains seven carbon atoms. So, the parent hydrocarbon is heptane.

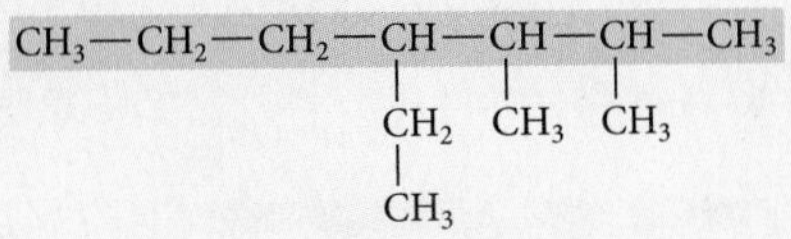

2. Number the carbons in the main chain in sequence. To do this, start at the end that will give the substituent groups attached to the chain the smallest numbers.

Numbering the chain from right to left gives the substituents the lowest numbers (2, 3, and 4). Numbering the chain the other way violates the rule.

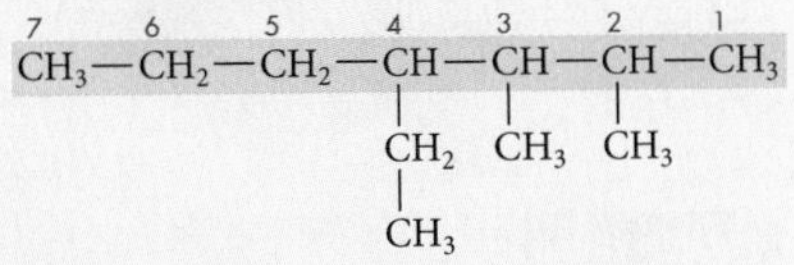

3. Add numbers to the names of the substituent groups to identify their positions on the chain. These numbers become prefixes to the name of the substituent group.

The substituents and positions are 2-methyl, 3-methyl, and 4-ethyl.

4. Use prefixes to indicate the appearance of the same group more than once in the structural formula. Common prefixes are *di-* (twice), *tri-* (three times), and *tetra-* (four times).

The two methyl groups are combined as 2,3-dimethyl.

5. List the names of alkyl substituents in alphabetical order. For purposes of alphabetizing, ignore the prefixes *di-*, *tri-*, and so on.

The 4-ethyl group is listed first, followed by 2,3-dimethyl.

6. Combine all the parts and use proper punctuation. Write the entire name without any spaces. Use commas to separate numbers and use hyphens to separate numbers and words.

The correct name of the compound is 4-ethyl-2,3-dimethylheptane. It is incorrect to write the name as 4-ethyl-2,3-dimethyl heptane.

LESSON 22.1

Explain

USE A MODEL Discuss how branched-chain alkanes differ from straight-chain alkanes. Remind students of the activity they performed at the beginning of the chapter, using clay and toothpicks to build butane. Use ball-and-stick models to remind students how straight-chain butane (C_4H_{10}) can be rearranged as a branched-chain structure (2-methylpropane) with the same molecular formula. Stress that each arrangement of atoms represents a different compound with different physical and chemical properties. Repeat the demonstration, this time beginning with a model of pentane, and rearranging it into other branched-chain alkanes. Display the IUPAC naming rules using an overhead or LCD projector. Have students take turns using the rules to name the structures you create.

Explore

Class Activity

PURPOSE Students distinguish between the parent structures and the substituents in a branched-chain hydrocarbon.

MATERIALS different-colored markers

PROCEDURE Write structural formulas for branched-chain hydrocarbons on the board in order from simplest to most complex. Include formulas in which the longest chain is not drawn in a straight line. Have students use one color to circle the carbons in the parent structure, another to circle the methyl groups, and so on.

Differentiated Instruction

L1 LESS PROFICIENT READERS Provide students with a copy of the IUPAC rules for naming branched-chain alkanes. Have students organize the rules into a two-column chart, one column for the rule, the other for a description and an example.

ELL ENGLISH LANGUAGE LEARNERS Prior to introducing the IUPAC rules, write a few names of alkanes on the board and read them aloud for the class. Have students practice stating the numbers and punctuation associated with the names. Provide students with a separate reference sheet that includes common prefixes as well as rules for when and how to use commas and hyphens.

L3 ADVANCED STUDENTS Have students create a set of index cards with large colorful hyphens, numbers, prefixes, names of substituents, and the symbols C, CH, CH_2, CH_3, and CH_4 on separate cards. Then, direct students to use their cards to guide others through the IUPAC rules. Have students use the symbol and hyphen cards to create a representation of a branched-chain alkane. Then have students use the prefix, hyphen, number, and substituent cards to "build" the name of the compound, and vice-versa.

Answers

FIGURE 22.4 Check students' diagrams.

Explain

Misconception Alert

Students may shy away from the longer, more confusing structures, thinking that the naming procedure is too difficult to follow. Guide students to simply follow one rule at a time. Encourage them to cover up the bulk of the structure and focus directly on the simple alkane that is its root.

Sample Practice Problems

Name the following compounds according to IUPAC system.

A.
```
CH3—CH—CH2—CH2—CH3
    |
    CH3
```
(2-methylpentane)

B.
```
CH2—CH2—CH2—CH—CH2
|            |   |
CH3          CH2 CH3
             |
             CH3
```
(3-ethylheptane)

SampleProblem 22.2

Naming Branched-Chain Alkanes

Name this compound using the IUPAC system. Notice that the longest chain is not written in a straight line.

```
           CH3
           |
CH3—CH2—C—CH3
           |
           CH2
           |
           CH2
           |
           CH3
```

1 Analyze Identify the relevant concepts. The parent structure is the longest chain of carbons. All other groups are substituents. Number the carbons to give the first substituent the lowest possible number. These location numbers become part of the name as prefixes. List the names of the substituents in alphabetical order with numbers separated by commas, and numbers and words separated by hyphens.

2 Solve Apply the concepts to this problem.

Step	Explanation
Identify the longest carbon chain in the molecule.	The longest chain has six carbons, so the name ends with hexane.
Identify the substituents and their positions on the parent hydrocarbon.	There are two methyl substituents on carbon 3, so the prefix is 3,3-dimethyl.
Put everything together. You can skip the alphabetizing step because there is only one type of substituent.	The correct IUPAC name is 3,3-dimethylhexane.

3. Name these compound according to the IUPAC system.

a.
```
CH2—CH2—CH—CH2—CH3
|        |
CH3      CH2
         |
         CH3
```

b.
```
CH3—CH2—CH—CH3
         |
         CH3
```

4. Name the following compound according to the IUPAC system.

```
CH3—CH2—CH2—CH—CH2—CH3
             |
             CH2
             |
       CH3—CH
             |
             CH3
```

Drawing Structural Formulas for Branched-Chain Alkanes With the name of a branched-chain alkane and knowledge of the IUPAC rules, it is easy to reconstruct the structural formula. First, find the root word (ending in *-ane*) in the hydrocarbon name. Then, draw the longest carbon chain to create the parent hydrocarbon, and number the carbons on the chain. Next, identify the substituent groups in the hydrocarbon name. Attach the substituents to the numbered parent chain at the proper positions. Complete the structural formula by adding hydrogens as needed.

Foundations for Math

NUMBER VARIATION Explain that the names of chemical compounds, as well as ions, molecules, and so on, rely on a numbering system that requires an understanding of the English and Roman numbering system as well as Greek and Latin numerical prefixes.

In Sample Problem 22.2, the name of the structure is *3,3-dimethylhexane*. The *3,3-di* indicates that the third carbon in the chain has two substituents attached to it. In **Sample Problem 22.3**, the name of the given compound begins with *2,2,4-tri*. This indicates that there are a total of three substituents, two attached to the second carbon in the chain and one attached to the fourth carbon in the chain.

SampleProblem 22.3

Drawing Structural Formulas for Branched-Chain Alkanes

Draw the structural formula for 2,2,4-trimethylpentane, or isooctane.

1 Analyze **Identify the relevant concepts.** The part of the name that ends in *-ane* indicates the parent structure. Prefixes indicate the types of substituents, the number of times each appears, and their locations on the parent chain. Hydrogens are added as needed.

2 Solve **Apply the concepts to this problem.**

Draw the number of carbons needed to represent the parent structure indicated in the name, and number the carbons on the chain.

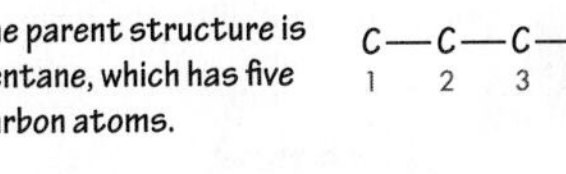

The parent structure is pentane, which has five carbon atoms.

```
C—C—C—C—C
1  2  3  4  5
```

Attach each substituent as indicated in the prefix.

There are two methyl groups on carbon 2 and one on carbon 4.

```
      CH3       CH3
      |         |
C—C—C—C—C
1  |2  3  4  5
      CH3
```

Finish by adding hydrogens where needed in the formula.

A total of nine hydrogens need to be added to complete the structure.

```
        CH3          CH3
        |            |
CH3—C—CH2—CH—CH3
        |
        CH3
```

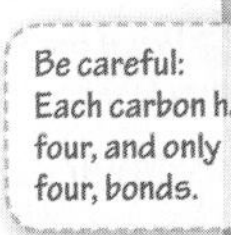

Be careful: Each carbon has four, and only four, bonds.

5. Draw the structural formula for 2,3-dimethylhexane.

6. Draw the structural formula for 4-ethyl-2,3,4-trimethyloctane.

22.1 LessonCheck

7. Review Why do carbon atoms form four covalent bonds?

8. Identify What are two ways that carbon atoms can be arranged in an alkane?

9. Infer Explain why mineral oil, which is a mixture of hydrocarbons, is not soluble in water.

10. Describe Draw complete structural formulas for the following alkanes:

a. octane

b. 3-ethylhexane

11. Describe Write a condensed structural formula for 2,2-dimethylbutane.

12. Identify Name the following alkanes using the IUPAC system.

a.
```
    H   H   H
    |   |   |
H—C—C—C—H
    |   |   |
    H   H   H
```

b.
```
CH3—CH—CH—CH2—CH3
        |     |
        CH3  CH3
```

BIGIDEA CARBON CHEMISTRY

13. How would you characterize alkanes in terms of bond polarity?

Lesson Check Answers

7. because a carbon atom contains four valence electrons
8. in a straight chain or in a chain with branches
9. The nonpolar hydrocarbons in mineral oil will not form a solution with a polar compound such as water.
10. a.
```
    H   H   H   H   H   H   H   H
    |   |   |   |   |   |   |   |
H—C—C—C—C—C—C—C—C—H
    |   |   |   |   |   |   |   |
    H   H   H   H   H   H   H   H
```
b.
```
    H   H      H      H   H   H
    |   |      |      |   |   |
H—C—C — C — C—C—C—H
    |   |      |      |   |   |
    H   H H—C—H  H   H   H
               |
            H—C—H
               |
               H
```
11.
```
        CH3
        |
CH3—C—CH2—CH3
        |
        CH3
```
12. a. propane
    b. 2,3-dimethylpentane
13. **BIGIDEA** Alkanes are nonpolar.

MAKE A CONNECTION Tell students that the branched-chain alkane in Sample Problem 22.3 is also known as isooctane. This isomer of octane defines the 100 point maximum on the octane rating scale for fuel.

Sample Practice Problems

Draw the structural formulas for these compounds.

A. 3-methylhexane
```
CH3—CH—CH2—CH2
        |           |
        CH2        CH3
        |
        CH3
```

B. 3-ethyl-3,4-dimethyloctane
```
              CH2CH3
              |
CH3—CH2—C—CH—CH2—CH2—CH2—CH3
              |    |
              CH3 CH3
```

Evaluate

Informal Assessment

Assess students' knowledge of alkanes by asking them to explain why straight-chain alkanes are considered a homologous series. *(There is a constant increment of change in molecular structure from one compound in the series to the next.)* **Ask** How do straight-chain and branched-chain alkanes differ? *(In branched-chain alkanes, substituents are attached to carbons on the parent chain.)* Then, have students complete the 22.1 Lesson Check.

Reteach

Quiz students orally on the first ten prefixes used in hydrocarbon nomenclature. Then, write the following names on the board: *3-ethylpentane, 2,3,4-trimethylhexane,* and *2,2-dimethylbutane.* Have students write structural formulas for these compounds. Remind students to (1) write the alkane carbon skeleton, (2) attach the substituent groups, and (3) add hydrogen atoms to all the unfilled bonding sites.

Answers

3. a. 3-ethylhexane b. 2-methylbutane
4. 4-ethyl-2-methylheptane
5.
```
       CH3
       |
CH3—CH—CH—CH2—CH2—CH3
             |
             CH3
```
6.
```
       CH3   CH3   CH3
       |     |     |
CH3—CH—CH—C—CH2—CH2—CH2—CH3
                   |
                   CH2
                   |
                   CH3
```

Key Objectives

DESCRIBE the structural characteristics of alkenes.

DESCRIBE the structural characteristics of alkynes.

Additional Resources

Reading and Study Workbook, Lesson 22.2

Engage

CHEMISTRY & YOU Have students study the photo and read the opening text. Ask students to share what they know about the effects of consuming saturated and unsaturated fats on health. *(Consumption of saturated fats is associated with heart disease, atherosclerosis, and high blood cholesterol levels. Unsaturated fats lower these risks.)*

Activate Prior Knowledge

Assess students' prior knowledge of multiple covalent bonds. **Ask** What kind of bonding occurs between carbon and oxygen in a molecule of CO_2 and between nitrogen atoms in a molecule of N_2? *(a double bond; a triple bond)*

Foundations for Reading

BUILD VOCABULARY Draw students' attention to the terms *alkene* and *alkyne*. Explain that the root that they share with *alkane*, *alk-*, is derived from the word *alkyl*. Point out that this root word indicates that alkenes and alkynes have many of the structural characteristics exhibited by alkanes.

READING STRATEGY Have students make a compare/contrast table for saturated and unsaturated hydrocarbons.

IN ACADEMIC STANDARDS for SCIENCE

C.3.4 Draw structural formulas for and name simple molecules.

C.9.1 Use structural formulas to illustrate carbon atoms' ability to bond covalently to form many different substances.

C.9.2 Identify the variety of molecular types formed by the covalent bonding of carbon atoms and describe the typical properties of these molecular types.

22.2 Unsaturated Hydrocarbons

C.3.4 Structural formulas and names; **C.9.1** Use structural formulas: carbon atoms; **C.9.2** Identify the variety of carbon molecules.

CHEMISTRY & YOU

Q: *What does it mean if a fat is unsaturated?* You've probably read nutrition labels listing saturated and unsaturated fat content. For example, olives are high in unsaturated fat but low in saturated fat. In this lesson, you'll find out what the terms *saturated* and *unsaturated* mean.

Key Questions

What are the structural characteristics of alkenes?

What are the structural characteristics of alkynes?

Vocabulary

- saturated compound
- unsaturated compound
- alkene • alkyne

Alkenes

What are the structural characteristics of alkenes?

An organic compound that contains the maximum number of hydrogen atoms per carbon atom is called a **saturated compound.** Alkanes are saturated compounds because the only bonds in alkanes are single covalent bonds. An organic compound that contains double or triple carbon-carbon bonds is called an **unsaturated compound.** The ratio of hydrogen atoms to carbon atoms is lower in an unsaturated compound than in a saturated compound. An **alkene** is a hydrocarbon that contains one or more carbon-carbon double covalent bonds. A carbon-carbon double bond is shown in structural formulas as two parallel lines. **At least one carbon-carbon bond in an alkene is a double covalent bond. Other bonds may be single carbon-carbon bonds and carbon-hydrogen bonds.**

Ethene (C_2H_4) is the simplest alkene. It is often called by the common name ethylene. Figure 22.5 shows the ball-and-stick model of ethene. To name an alkene by the IUPAC system, find the longest chain in the molecule that contains the double bond. This chain is the parent alkene. It has the root name of the alkane with the same number of carbons plus the ending *-ene*. The chain is numbered so that the carbon atoms of the double bond have the lowest possible numbers. Substituents on the chain are named and numbered in the same way they are for alkanes. Some examples of the structures and IUPAC names of simple alkenes are shown below.

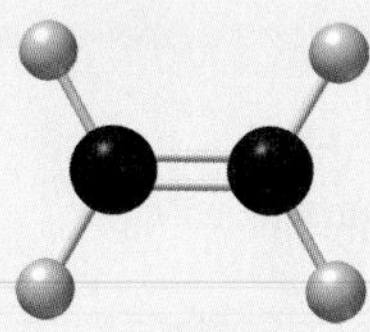

Figure 22.5 Ethene
Because rotation is restricted around the double bond, atoms in ethene lie in one plane.

$CH_2{=}CH_2$
Ethene

$CH_3{-}C(H){=}C(H){-}H$
Propene

$CH_2{=}CH{-}CH_2{-}CH_3$
1-butene

$CH_3{-}C(H){=}C(H){-}CH_3$
2-butene

$CH_3{-}CH(CH_3){-}C(H){=}C(H){-}CH_3$
4-methyl-2-pentene

CHEMISTRY & YOU

Q: *You just read about saturated and unsaturated compounds. Use what you have learned to describe how saturated and unsaturated fats differ structurally.*

Focus on ELL

1 CONTENT AND LANGUAGE Remind students that an *alkane* is a hydrocarbon in which there are only single covalent bonds. Write the words *alkane, alkene,* and *alkyne* on the board. Underline the root of each word and circle each suffix. Write the words *single bond, double bond,* and *triple bond.* Draw arrows to the appropriate term. Guide student to see that the suffix in the names of these compounds tells you the type of bond.

2 FRONTLOAD THE LESSON Have students list the first ten straight-chain alkanes on the board, and review the IUPAC naming rules. Then, have students preview the chemical structures discussed throughout the lesson, noting the similarities between the names of the unsaturated hydrocarbons and their related alkanes.

3 COMPREHENSIBLE INPUT Have students form human models of the hydrocarbons in this lesson. Divide students into three groups, each representing a different class of hydrocarbon: alkane, alkene, alkyne. Provide short lengths of rope to represent the bonds between the atoms and "C" and "H" signs to indicate which students are carbon and hydrogen atoms.

InterpretGraphs

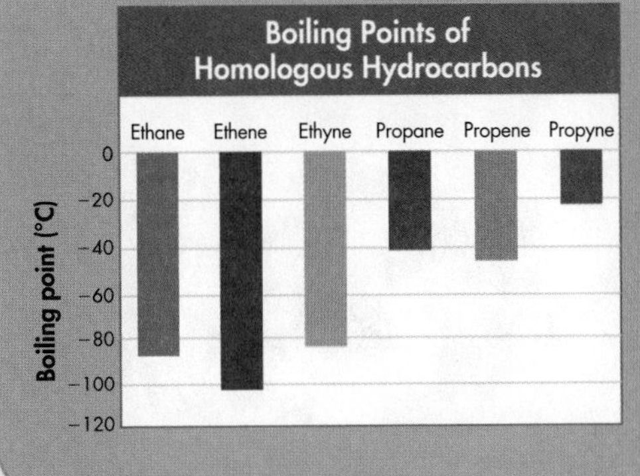

Figure 22.6 The graph shows the relationship between the boiling points and the degree of saturation for some hydrocarbons.

a. Read Graphs Determine the boiling point for each hydrocarbon in the graph.

b. Identify Draw a condensed structural formula for each hydrocarbon.

c. Make Generalizations How does the degree of saturation affect the boiling points of hydrocarbons with the same number of carbon atoms?

Note: The *degree of saturation* refers to the ratio of hydrogen atoms to carbon atoms.

Alkynes

What are the structural characteristics of alkynes?

A hydrocarbon that contains one or more carbon-carbon triple covalent bonds is called an **alkyne.** A carbon-carbon triple bond is shown in structural formulas as three parallel lines. **At least one carbon-carbon bond in an alkyne is a triple covalent bond. Other bonds may be single or double carbon-carbon bonds and carbon-hydrogen bonds.** Like alkenes, alkynes are unsaturated compounds.

Alkynes are not plentiful in nature. The simplest alkyne is ethyne (C_2H_2), which has the common name acetylene. Acetylene is the fuel burned in oxy-acetylene torches used for welding. Figure 22.7 shows that the single bonds that extend from the carbons in the carbon-carbon triple bond of ethyne are separated by an angle of 180°, which makes ethyne a linear molecule.

Like alkanes, the major attractions between alkenes and alkynes are weak van der Waals forces. As a result, the introduction of a double or triple bond into a hydrocarbon does not have a dramatic effect on physical properties such as boiling point. Compare the boiling points for alkanes, alkenes, and alkynes with two and three carbons in Figure 22.6.

Figure 22.7 Ethyne
The triple bond restricts rotation in an ethyne molecule, which has a linear shape.
Identify ***What is the ratio of hydrogen to carbon in ethyne?***

22.2 LessonCheck

14. Review Describe the bonding between atoms in an alkene.

15. Identify What types of bonds are present in an alkyne?

16. Explain What is the difference between saturated and unsaturated hydrocarbons?

17. Make Generalizations How do the boiling points of alkenes and alkynes compare to those of alkanes?

18. Apply Concepts Draw electron dot structures for ethene and ethyne. Use your knowledge of bonding theories to describe the shape of each molecule.

Lesson Check Answers

14. At least one carbon-carbon bond in an alkene is a double bond. Other bonds may be single C–C bonds or C–H bonds.

15. An alkyne contains at least one carbon-carbon triple bond. Other bonds may be single or double C–C bonds or C–H bonds.

16. Saturated hydrocarbons have the maximum number of hydrogen atoms per carbon. Unsaturated hydrocarbons have fewer hydrogen atoms per carbon

17. The boiling points are similar for hydrocarbons with the same number of carbon atoms.

18.

```
 H H
H:C::C:H        H:C:::C:H
 ethene           ethyne
```

Ethene is planar. Ethyne is linear.

Explain

Alkenes

USE MODELS Direct students to Figures 22.5 and 22.7. Have students compare the models of ethene and ethyne. *(Both molecules have symmetrical carbon atoms that can be superimposed upon one another. In ethene, the carbon atoms are joined by a double bond; in ethyne, by a triple bond. Ethyne has two fewer hydrogen atoms and is linear.)* Then, model the rigidity of the double bond for students. Touch the first fingers of both hands together to represent a single bond, then rotate your hands in opposite directions. Repeat with two fingers of each hand touching, demonstrating that rotation is impossible around a double bond.

CHEMISTRY & YOU Unsaturated fats have double carbon-carbon bonds. All the carbon-carbon bonds in saturated fats are single bonds.

Evaluate

Informal Assessment

Have students explain why hydrocarbons with double or triple bonds are called unsaturated hydrocarbons. Then, have students complete the 22.2 Lesson Check.

Reteach

Have students construct ball-and-stick models of ethane, ethene, and ethyne. Ask students to write and compare the complete structural formulas for the molecules and to describe the molecular geometry of each.

Answers

INTERPRET GRAPHS

a. Answers will be approximate. Actual boiling points are: ethane, –89; ethene, –104; ethyne, –82; propane, –42; propene, –47; propyne, –23

b. CH_3CH_3, $CH_2{=}CH_2$, $CH{\equiv}CH$, $CH_3CH_2CH_3$, $CH_2{=}CHCH_3$, $CH{\equiv}CCH_3$

c. For hydrocarbons with the same number of carbons, the boiling points are similar.

FIGURE 22.7 1:1

Explain

CHEMISTRY & YOU Have students read the feature and examine the photographs.

An arson investigator can often tell by the burn markings at a fire scene if an accelerant was used. The use of an accelerant is typical in arson-related fires because it makes a fire burn hotter and longer than normal, producing the most damage in a short period of time. Most accelerants used by an arsonist are organic compounds containing mixtures of hydrocarbons. Pose the following challenge to students: *Imagine that you are investigating a suspected arson fire in a wood-framed building. Use your knowledge of hydrocarbons to search for signs that an accelerant was used.* You may need to guide students in the following ways:

- Liquids tend to seek the lowest level in a structure.
- Hydrocarbons burn faster and hotter than wood.
- Hydrocarbons cannot ignite spontaneously.
- Hydrocarbons are not miscible in water.
- Hydrocarbon vapors are heavier than air and flow downward.
- Many hydrocarbons are powerful solvents.

Extend

Connect to TECHNOLOGY

Explain that as accelerants evaporate, the hydrocarbons vaporize into the air space above the remaining ash. Explain that one of the most advanced techniques for detecting accelerants is gas chromatography, a technique used to separate volatile organic compounds into specific hydrocarbons. Tell students that once the likely ignition point of the fire has been located, samples of solid debris are collected from the area and placed in an airtight container to prevent evaporation. The container is then heated, releasing any hydrocarbons into the trapped space above the sample. A needle is inserted through the top of the container to remove a sample of the hydrocarbons and then injected into the chromatography equipment for separation and analysis. The result identifies each specific compound of the gas mixture, and if the compounds are hydrocarbons, the determination of the crime of arson can be confirmed.

Answers

TAKE IT FURTHER

1. $$\begin{array}{c} \quad\;\; CH_3 \\ \quad\;\; | \\ CH_3CH_2CHCH_2CH_2CH_3 \end{array}$$
2. Gasoline contains a wider range of compounds, with a majority of lighter compounds. Lighter fluid consists of a narrower range of compounds, which are in the heavier range.

CHEMISTRY & YOU: CAREERS

Arson Investigator

Arson is a dangerous crime in which a person starts a fire with the intent to cause property damage or to harm another person. The person responsible, an arsonist, often uses an accelerant to initiate the fire. Gasoline and lighter fluid—both ignitable fluids composed of hydrocarbons—are two of the most commonly used accelerants. If arson is suspected, an arson investigator carefully examines the fire scene and looks for evidence of a crime.

Arson investigators often collect debris from the scene to take back to the laboratory to analyze for accelerants. Investigators frequently analyze the samples with an instrument called a gas chromatograph. The output of the analysis is sometimes called a "fingerprint" because, like a human fingerprint, each accelerant shows a characteristic pattern. With the knowledge of the molecules in each accelerant, the investigator can identify the accelerant from the fingerprint. Knowing which accelerant was used in the crime can help police narrow the search for the arsonist.

WHO'S YOUR PARTNER? Trained dogs with keen noses are often used to guide the evidence collection.

Gas Chromatographic Analysis

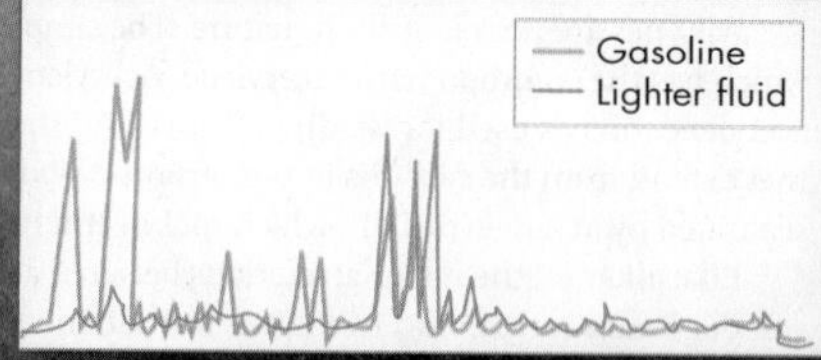

ACCELERANTS Gasoline and lighter fluid are complex mixtures of hydrocarbons. Each peak in the chromatograph represents a different chemical compound.

Take It Further

1. **Describe** One of the hydrocarbons found in gasoline is the branched-chain alkane 3-methylhexane. Write the condensed structural formula for this compound.
2. **Analyze Data** Use the data in the gas chromatographic analysis above to compare the number of hydrocarbon compounds in gasoline and lighter fluid.

Differentiated Instruction

ELL ENGLISH LANGUAGE LEARNERS Relate the word *accelerant* to *accelerate*. Act them out to visually show that both words imply to go faster, to hasten, to move quickly, to speed up, and then act out the opposite meaning, slowing down.

LPR LESS PROFICIENT READERS Have students hypothesize why there is a dog in the photograph of a fire scene. Then, ask these students to perform an Internet search on arson dogs and to orally explain what they learned.

L3 ADVANCED STUDENTS Have students compile a list of hydrocarbons typically found in liquid fuel and determine the properties of these compounds that allow for the separation and identification with the gas chromatography equipment.

22.3 Isomers

C.3.4 Structural formulas and names; C.9.1 Use structural formulas: carbon atoms; C.9.2 Identify the variety of carbon molecules.

CHEMISTRY & YOU

Q: *How do isomers help you see?* Like many other biological processes, vision involves chemistry. Light entering the eye causes a change in the three-dimensional structure of retinal molecules in your eye. The overall structure of retinal molecules appears bent before the light strikes; afterward the structure appears relatively straight. This structural change results in vision.

Key Questions

How do the properties of constitutional isomers differ?

What are two types of stereoisomers?

Vocabulary

- isomer
- constitutional isomer
- stereoisomer
- *cis-trans* isomer
- *cis* configuration
- *trans* configuration
- asymmetric carbon
- enantiomer

Constitutional Isomers

How do the properties of constitutional isomers differ?

You may have noticed that the structures of some hydrocarbons differ only in the positions of substituents or of multiple bonds. Look at the structural formulas for butane and 2-methylpropane and at the models in Figure 22.8.

$$CH_3—CH_2—CH_2—CH_3$$

Butane (C_4H_{10}) (bp −0.5°C)

$$\begin{array}{c} CH_3 \\ | \\ CH_3—CH—CH_3 \end{array}$$

2-methylpropane (C_4H_{10}) (bp −11.7°C)

Even though both compounds have the formula C_4H_{10}, their boiling points and other properties differ. Because their structures are different, they are different substances. Compounds that have the same molecular formula but different molecular structures are called **isomers.**

Butane and 2-methylpropane represent a category of isomers called constitutional isomers, or structural isomers. **Constitutional isomers** are compounds that have the same molecular formula, but the atoms are joined together differently. **Constitutional isomers differ in physical properties such as boiling point and melting point. They also have different chemical reactivities.** In general, the more highly branched the hydrocarbon structure is, the lower the boiling point of the isomer will be compared with less branched isomers.

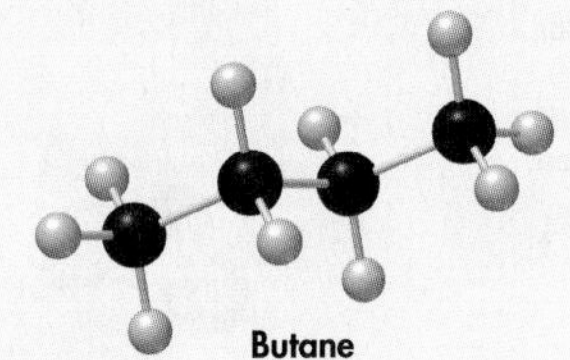

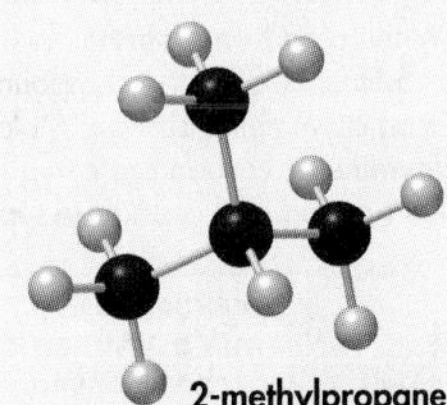

Figure 22.8 Constitutional Isomers
Both butane and 2-methylpropane have the molecular formula C_4H_{10}. But the atoms in each compound are arranged differently.

Key Objectives

EXPLAIN how the properties of constitutional isomers differ.

IDENTIFY two types of stereoisomers.

Additional Resources

- Reading and Study Workbook, Lesson 22.3

PearsonChem.com

Engage

CHEMISTRY & YOU Have students read the introduction. Explain that retinal is a molecule in the rod and cone cells of the retina. It straightens when it is excited by light energy. **Ask** What information does the three-dimensional structure of a molecule reveal? *(how its atoms are arranged in space)*

Build Backround

Bring a folded folding chair to class and show it to students. **Ask** What will happen to the structure of this chair if I open it? *(It will change.)* **Ask** Will opening the chair change the kinds of atoms in the chair? *(no)* Point out that opening or folding the chair is an analogy for *cis-trans* isomerization. (To model constitutional isomers, you would reassemble the chair in a different configuration.)

Foundations for Reading

BUILD VOCABULARY Explain that *isomers*, is Greek for "equal parts." **Ask** What is another chemical term that begins with the prefix *iso-*? What does it mean? (isotopes; *atoms of the same elements with different masses)*

Focus on ELL

1 CONTENT AND LANGUAGE Write the term *asymmetric carbon* on the board, and pronounce it for the class. Remind students that in biology, animals are considered symmetric if the two halves of their bodies are mirror images of each other. Explain that the prefix *a-* means "without." A carbon with four different atoms or groups attached to it is asymmetrical.

2 FRONTLOAD THE LESSON Demonstrate that one of your hands is the mirror image of the other when placed with palms together. Then, have students place both hands palm down and overlap them. Point out that in this position their hands are not *superimposable* on each other. Tell students to keep this activity in mind as they discuss asymmetric carbons.

3 COMPREHENSIBLE INPUT Use ball-and-stick models to demonstrate the isomers in Figures 22.8, 22.9, and 22.10. Provide students with enough balls and sticks to create butane and *trans*-2-pentene, and have them practice rearranging the structures to form each molecule's isomers.

IN ACADEMIC STANDARDS for SCIENCE

- Use analogies and models (mathematical and physical) to simplify and represent systems that are difficult to understand or directly experience due to their size, time scale, or complexity, and recognize the limitations of analogies and models.

C.9.1 Use structural formulas to illustrate carbon atoms' ability to bond covalently to form many different substances.

Also covered C.3.4, C.9.2.

LESSON 22.3

Explain

Constitutional Isomers

START A CONVERSATION Point out that the number of possible constitutional isomers for an alkane increases dramatically with increasing numbers of carbon atoms. Explain that with more carbon atoms, there are more ways to arrange the atoms in space. Remind students that each isomer has a unique set of physical and chemical properties. Point out that the existence of isomers is a major reason that so many different organic compounds exist.

CHEMISTRY & YOU *Cis-trans* isomerism aids in vision. Before light strikes it, the retinal structure around the red double bond is in the *cis* configuration and appears bent. After light strikes, the retinal structure around the red double bond is in the *trans* configuration, and the structure appears straight.

cis configuration —Light→ *trans* configuration

Explore

Class Activity

PURPOSE Students use models to identify isomers.

PROCEDURE Have students use model kits to construct isomers of pentane (C_5H_{12}) and hexane (C_6H_{14}). Have them draw the structural formula for each isomer. **Ask** How does the number of pentane isomers compare with the number of hexane isomers? (There are more hexane isomers.)

Explain

Stereoisomers

USE VISUALS Point out that the number of possible isomers for an alkene can be greater than for an alkane. Draw the complete structural formulas of 1-butene, *cis*-2-butene, and *trans*-2-butene on the board to show the three possible isomers for butene (C_4H_8). Introduce the term cis-trans *isomers* and explain how *cis*-2-butene and *trans*-2-butene are different molecules due to the lack of free rotation about the carbon-carbon double bond.

Cis configuration

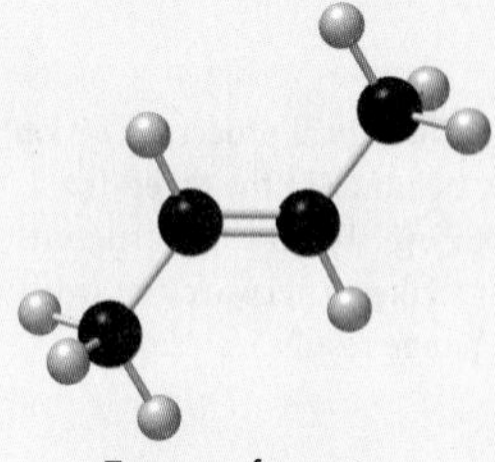

Trans configuration

Figure 22.9 *Cis-Trans* Isomers
The properties of *cis-trans* isomers are different. The boiling point of *cis*-2-butene is 3.7°C, and the boiling point of *trans*-2-butene is 0.8°C.
Compare ***How are the cis and trans configurations different?***

Stereoisomers

What are two types of stereoisomers?

Remember that molecules are three-dimensional structures. So molecules with the same molecular formula and with atoms joined in exactly the same order may still be isomers. **Stereoisomers** are molecules in which the atoms are joined in the same order, but the positions of the atoms in space are different. **Two types of stereoisomers are *cis-trans* isomers and enantiomers.**

Cis-Trans Isomers A double bond between two carbon atoms prevents other atoms in the molecule from rotating, or spinning, with respect to each other. Because of this lack of rotation, groups on either side of the double bond can have different orientations in space. ***Cis-trans* isomers,** also known as geometric isomers, have atoms joined in the same order, but the spatial orientation of the groups differs. Although *cis-trans* isomerism is possible in other molecules, the most common example occurs in molecules with double bonds. Look at the models of 2-butene in Figure 22.9. Two arrangements are possible for the methyl groups and hydrogen atoms with respect to the rigid double bond. In the ***cis* configuration,** similar groups are on the same side of the double bond. But, when similar groups extend from opposite sides of the double bond, the isomer is in the ***trans* configuration.** *Cis-trans* isomers have different physical and chemical properties.

You should be able to identify *cis-trans* isomers of alkenes when each carbon of the double bond has one substituent and one hydrogen. Notice that the substituent groups attached to the carbons of the double bond do not need to be the same as illustrated in the structures below.

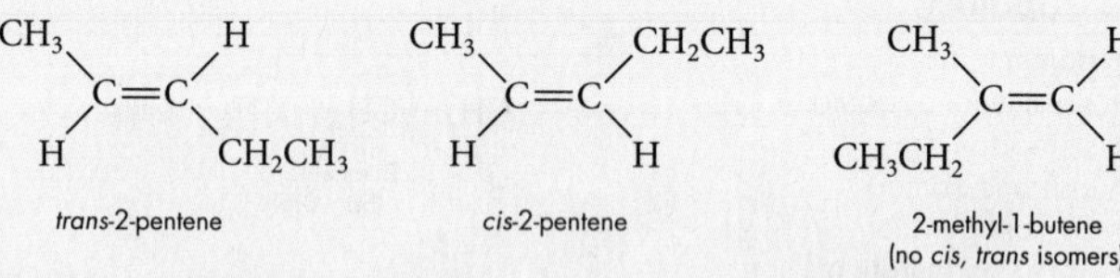

Enantiomers The second category of stereoisomerism occurs whenever a central atom has four different atoms or groups attached. Most commonly the central atom is carbon. A carbon with four different atoms or groups attached is an **asymmetric carbon.** Look at the molecular models in Figure 22.10. Because H, F, Cl, and Br atoms are attached to a single carbon atom, the carbon is an asymmetric carbon. The relationship between the two molecules is similar to the relationship between right and left hands. Sometimes the terms *right-handed* and *left-handed* are used to describe compounds with an asymmetric carbon.

Figure 22.10 Asymmetric Carbons
When a carbon atom has four different substituents, as in the compound CHFClBr, it is an asymmetric carbon. Molecules with an asymmetric carbon are either right-handed or left-handed and cannot be superimposed.

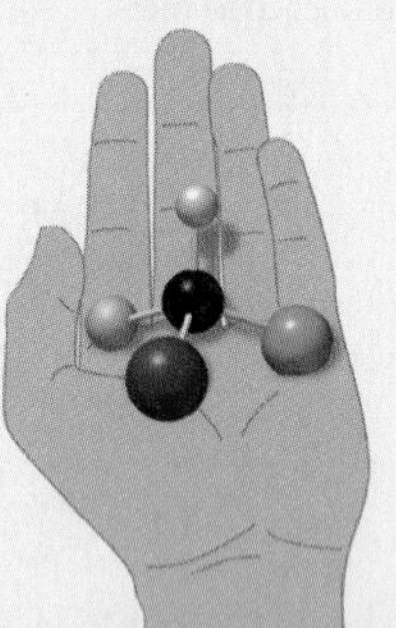

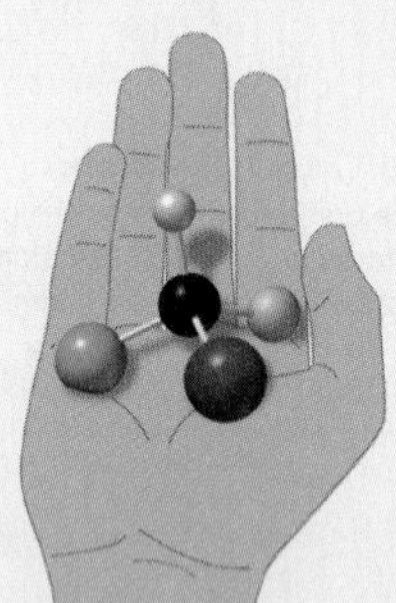

Differentiated Instruction

L1 STRUGGLING STUDENTS Create ball-and-stick models of various constitutional isomers and *cis-trans* isomers. Have students practice identifying the type of isomer and in the case of a *cis-trans* isomer, the configuration.

ELL ENGLISH LANGUAGE LEARNERS Have students create flashcards for the vocabulary terms. On the front side, students should write the term. On the back, students should write the definition and draw an example structure.

L3 ADVANCED STUDENTS Have students research the origin of the International Union of Pure and Applied Chemistry (IUPAC). Tell them to find out when, where, and why the first international congress was held. Findings should be presented in a written report.

To understand the stereoisomerism that involves asymmetric carbons, you need to visualize the relationship between an object and its mirror image. If the object is symmetrical, like a ball, then its mirror image can be superimposed. That is, the appearance of the ball and its reflection are indistinguishable. By contrast, a pair of hands is distinguishable even though the hands have identical parts. The right hand reflects as a left hand, and the left hand reflects as a right hand. When you try to stack your hands on top of one another, the thumb of one hand lines up with the little finger of the other hand. No matter how you turn your hands, you can't get them to look exactly alike.

Pairs of molecules that are mirror images and not superimposable are called **enantiomers,** or optical isomers. The molecules shown in Figure 22.10 are examples of enantiomers. Unlike other isomers, enantiomers have identical physical properties such as boiling points and densities. Enantiomers do, however, behave differently when they interact with other molecules that have asymmetric carbons. In Chapter 24, you will learn that many molecules in your body have asymmetric carbons. As a result, each enantiomer can have a different effect on the body.

CHEMISTRY & YOU

Q: *What kind of isomerism aids in vision? Look at Figure 22.9. Explain how the overall structure of a retinal molecule would appear relatively straight or bent depending on which cis-trans isomer it is.*

Sample Problem 22.4

Identifying Asymmetric Carbon Atoms

Which compound has an asymmetric carbon?

a. CH_3CHCH_3 with OH on the central carbon   **b.** $CH_3CHCH_2CH_3$ with OH on the second carbon

The central carbon in compound (a) has two CH_3 groups attached, so the carbon is not asymmetric.

1 Analyze Identify the relevant concepts.
An asymmetric carbon has four different substituents attached.

2 Solve Apply the concepts to this problem.

Draw the structure in a way that makes it easier to compare the four different groups attached to the central carbon.

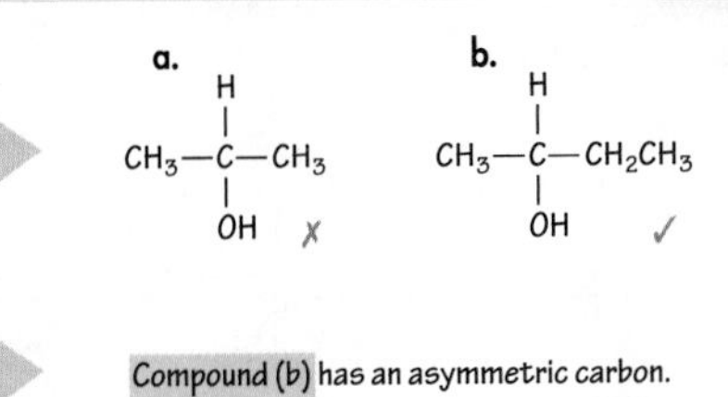

Compare the groups. If all four groups are unique, the central carbon is asymmetric. If any two are the same, the central carbon is not asymmetric.

Compound (b) has an asymmetric carbon.

19. Identify the asymmetric carbon, if any, in each of the following compounds:

a. CH_3CHCHO with Cl on the second carbon   **b.** CH_3CHOH with CH_3 on the second carbon

20. Identify any asymmetric carbons in the following compounds:

a. CH_3CH_2—C—Br with CH_3 above and F below the carbon   **b.** CH_2Cl_2

Focus on ELL

4 LANGUAGE PRODUCTION Have students work in groups of two to complete the lab. Be sure to pair ELLs with writing/reading partners so that more proficient students can help less proficient ones. Have students work according to their proficiency level.

BEGINNING: LOW/HIGH Create drawings of the structural formulas of the nine different constitutional isomers of heptane, utilizing color to emphasize the differences.

INTERMEDIATE: LOW/HIGH Orally question students about their findings. Use the terms *right-handed* and *left-handed* when asking questions about asymmetric carbon.

ADVANCED

LOW Have students write draft explanations and answers to the questions from the lab and then revise their answers after consultation with you.

HIGH Ask students to complete the Analyze and Conclude questions with clearly displayed, written responses. Post their responses for other students' refererence.

Explore

Teacher Demo

PURPOSE Students observe three-dimensional models of enantiomers.

MATERIALS two ball-and-stick models of CHFClBr

PROCEDURE Demonstrate how it is not possible to superimpose one model on the other.

Explain

MAKE A CONNECTION Tell students that mirror images exist in nature. For example, 99 percent of the shells of some snail species are right-handed spirals and 1 percent are left-handed spirals. The right- and left-handed spirals are mirror images. Other species have shells that are normally left-handed.

Sample Practice Problem

Determine if 2-chloropropane has an asymmetric carbon. *(no)*

Answers

FIGURE 22.9 The groups are on opposite sides of the double bond in the *trans* configuration; the groups are on the same side of the bond in the *cis* configuration.

19. a. CH_3—*C—CHO, with H above and Cl below the asymmetric carbon (*)

b. no asymmetric carbon

20. a. CH_3CH_2—*C—Br, with CH_3 above and F below the asymmetric carbon (*)

b. no asymmetric carbon

LESSON 22.3

Explore

Isomers

Quick Lab

PURPOSE After completing this activity, students will be able to build ball-and-stick models of heptane isomers and name them.

SKILLS FOCUS using models, observing, analyzing

EXPECTED OUTCOMES Students will model and name nine constitutional isomers of heptane.

PREP TIME 5 minutes

CLASS TIME 25 minutes

ANALYZE AND CONCLUDE

1. heptane, 2-methylhexane, 3-methylhexane, 2,3-dimethylpentane, 2,4-dimethylpentane, 2,2-dimethylpentane, 3,3-dimethylpentane, 3-ethylpentane, 2,2,3-trimethylbutane
2. 4
3. Each is a unique compound.
4. 3-methylhexane and 2,3-dimethylpentane

ENRICHMENT Students can model, draw, and name the 18 constitutional isomers of octane.

Evaluate

Informal Assessment

Have students verbally explain what the term *isomer* means and then describe the differences among constitutional isomers, *cis-trans* isomers, and *enantiomers* in 60 seconds or less. Ask students to draw an example of each type of isomer. Then have students complete the 22.3 Lesson Check.

Reteach

Use ball-and-stick models to visually explain the requirements for classifying two molecules with the same molecular formula as constitutional isomers, *cis-trans* isomers, or enantiomers.

Answer

27.

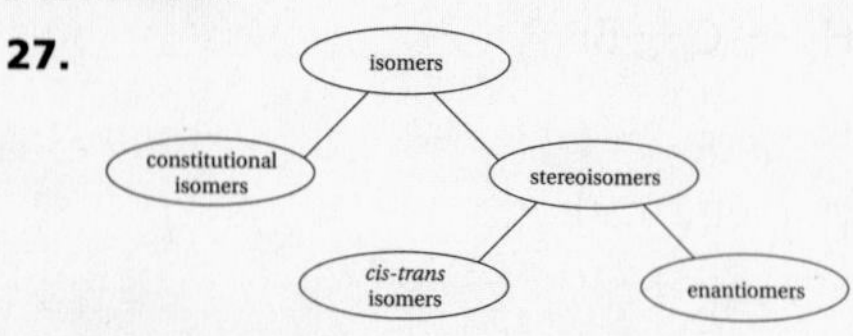

Quick Lab

Isomers of Heptane

Purpose To build ball-and-stick models and name the nine constitutional isomers of heptane (C_7H_{16})

Materials

- **ball-and-stick molecular model kit** (Colors used to represent elements in the kit may not match colors used to represent elements in this book.)
- **pencil and paper**

Procedure

1. Build a model for the straight-chain isomer of C_7H_{16}. Draw the structural formula for this isomer.

2. Remove one carbon atom from the end of the chain and reattach it as a methyl substituent to form a branched-chain alkane. Draw the structural formula for this isomer.

3. Move the methyl group to a new position on the chain. Then draw this third isomer. Is there another position that the methyl group can be placed on the chain of six carbons to form yet another isomer?

4. Make other constitutional isomers by shortening the longest straight chain and using the removed carbons as substituents. Draw the structural formulas for each isomer.

Heptane

Analyze and Conclude

1. List What are the names of the nine constitutional isomers of C_7H_{16}?

2. Identify What is the shortest possible straight carbon chain in the group of heptane isomers?

3. Explain Why does each constitutional isomer have its own unique name?

4. Use Models Look carefully at the structural formulas you drew for the nine constitutional isomers. Identify any that have an asymmetric carbon.

ONLINE PROBLEMS 22.3 LessonCheck

21. Explain Why would you expect two constitutional isomers to have different properties such as boiling points?

22. Review Name two types of stereoisomers.

23. Apply Concepts Draw structural formulas for the following alkenes. If a compound has *cis-trans* isomers, draw both the *cis* and *trans* forms.

a. 1-pentene
b. 2-hexene
c. 2-methyl-1-butene
d. 2,5-dimethyl-3-hexene

24. Compare How are constitutional isomers and stereoisomers similar? How are they different?

25. Explain How can you identify an asymmetric carbon?

26. Use Analogies Think of an analogy to describe the relationship between two molecules that are enantiomers.

27. Summarize Draw a concept map to show how the following vocabulary words are related: isomers, constitutional isomers, stereoisomers, *cis-trans* isomers, and enantiomers.

Lesson Check Answers

21. Constitutional isomers are different compounds with different physical properties.
22. *cis-trans* isomers and enantiomers
23. a. $(H)(H)C{=}C(H)(CH_2CH_2CH_3)$
    1-pentene

    b. $(CH_3)(H)C{=}C(H)(CH_2CH_2CH_3)$
    trans-2-hexene

    $(CH_3)(H)C{=}C(CH_2CH_2CH_3)(H)$
    cis-2-hexene

    c. $(H)(H)C{=}C(CH_3)(CH_2CH_3)$
    2–methyl–1–butene

    d. $((CH_3)_2CH)(H)C{=}C(CH(CH_3)_2)(H)$
    cis-2,5-dimethyl-3-hexene

    $(H)((CH_3)_2CH)C{=}C(CH(CH_3)_2)(H)$
    trans-2,5-dimethyl-3-hexene
24. Both types of isomers have the same molecular formula but different molecular structures. Constitutional isomers differ in the order the atoms are joined; stereoisomers differ in the way the atoms are arranged in space.
25. It has four different substituents.
26. Answers will vary.

22.4 Hydrocarbon Rings

C.3.4 Structural formulas and names; C.9.1 Use structural formulas: carbon atoms; C.9.2 Identify the variety of carbon molecules.

CHEMISTRY & YOU

Q: *Does a compound have to be smelly in order to be classified as aromatic?* When you hear the word *aromatic*, you may think of perfume or flowers. But in organic chemistry, *aromatic* means something else.

Key Questions

What is the general structure of a cyclic hydrocarbon?

What is the most accurate description of bonding in benzene?

Vocabulary

- cyclic hydrocarbon
- cycloalkane
- aromatic compound
- aliphatic compound

Cyclic Hydrocarbons

What is the general structure of a cyclic hydrocarbon?

Not all hydrocarbons are straight chains or branched chains. **In some hydrocarbon compounds, the carbon chain is in the form of a ring.** A compound that contains a hydrocarbon ring is called a **cyclic hydrocarbon.** Figure 22.11 shows the structures of some examples. Many molecules found in nature contain cyclic hydrocarbons. Rings with five and six carbons are the most abundant.

Just as straight-chain and branched-chain alkanes can be either saturated or unsaturated, so can cyclic hydrocarbons. A cyclic hydrocarbon that contains only single bonds, and is therefore saturated, is called a **cycloalkane.** To determine the IUPAC name of a cycloalkane, first count the number of carbons in the ring and assign the corresponding alkane name. Then simply add the prefix *cyclo-* to the alkane name. For example, the three-carbon ring in Figure 22.11 is named cyclopropane.

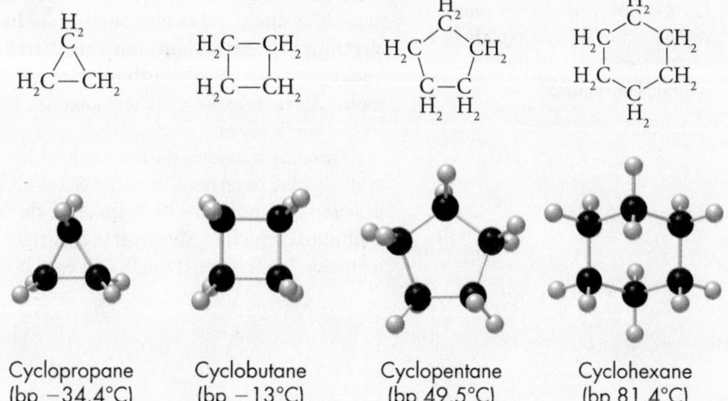

Figure 22.11 Cycloalkanes These illustrations show the first four members of the homologous series of cycloalkanes.
Predict *How would you expect the boiling point of cycloheptane to compare to the boiling points of these cycloalkanes?*

Focus on ELL

1 CONTENT AND LANGUAGE Begin the lesson by writing the root word *cycle* on the board. Assign students to groups and ask them to create a list of words using this root word. *(Sample answers:* bicycle, cyclical, cycle, motorcycle, tricycle, cyclone*)* Explain that *cycle* means "circle, wheel, or circular." Have students predict the meaning of the vocabulary terms *cyclic hydrocarbon* and *cycloalkane*.

2 FRONTLOAD THE LESSON Have students preview the models and structural formulas of the featured hydrocarbons. Discuss how these visuals differ from the other representations of hydrocarbons learned in prior lessons.

3 COMPREHENSIBLE INPUT Show students digital three-dimensional models of the structures in this lesson. Then have students create physical models of benzene by using foam balls and pipe cleaners, or pretzel sticks and marshmallows.

LESSON 22.4

Key Objectives

IDENTIFY the general structure of a cyclic hydrocarbon.

DESCRIBE bonding in a benzene ring.

Additional Resources

- Reading and Study Workbook, Lesson 22.4

PearsonChem.com

Engage

CHEMISTRY & YOU Explain that *aromatic* comes from a Greek word meaning "sweet spice." Tell students that aromatherapy is the use of aromatic oils for their supposed therapeutic effects when applied to the skin or inhaled. Predict why scientists used the name *aromatic* to describe a particular type of hydrocarbon.

Build Background

Allow students to handle models representing 6 carbon atoms and 12 hydrogen atoms as you begin the lesson. **Ask** How could you form a ring or a circle with these atoms, rather than a chain? *(The two carbons at the end of the chain could bond to one another.)*

Foundations for Reading

BUILD VOCABULARY Explain that the prefixes *ortho-*, *meta-*, and *para-* used to name some aromatic compounds are all derived from Greek, meaning "straight or correct," "beyond or after," and "beside or alongside of," respectively.

READING STRATEGY Have students use a Venn diagram to compare and contrast aliphatic cyclic hydrocarbons and aromatic cyclic hydrocarbons.

IN ACADEMIC STANDARDS for SCIENCE

C.3.4 Draw structural formulas for and name simple molecules.

C.9.2 Identify the variety of molecular types formed by the covalent bonding of carbon atoms and describe the typical properties of these molecular types.

Also covered C.9.1

Answers

FIGURE 22.11 It should be higher (118°C)

Explain

Cyclic Hydrocarbons

USE VISUALS Have students examine Figure 22.11. **Ask** Why does cyclohexane have two fewer hydrogen atoms than hexane? *(When carbon atoms at the end of a chain bond to form a ring, the atoms can form only two additional bonds instead of three.)*

APPLY CONCEPTS Point out that the formation of a ring causes bond strain in all cycloalkanes with rings that contain fewer than six carbons. The bond angle in these cycloalkanes tends to be less than the normal carbon-carbon bond angle of 109.5°. Explain that bond strain explains the increased chemical reactivity of small-ring cycloalkanes.

Aromatic Hydrocarbons

CHEMISTRY & YOU Aromatic compounds contain benzene or another ring that has bonding like that of benzene. Bonding in benzene and other aromatic compounds is special in that the bonding electrons between ring members are shared evenly around the ring.

Extend

The concept of benzene as a hybrid of two conventional structures was a major breakthrough in chemistry. Using a CAD program or an actual three-dimensional model, show students that the benzene ring structure consists of σ bonds formed from overlap of sp^2 hybrid orbitals on the neighboring carbon atoms. Point out how the unhybridized carbon p_z orbitals project above and below the plane of the ring. Describe how, in theory, the lobes of the orbitals join to form the circular rings of electron density above and below the plane. These rings make up the other part of the carbon-carbon double bond in benzene. The two forms of this ring represent the total resonance of the system. Have students investigate Kekulé structures and describe and draw how these structures relate to the concepts in this lesson.

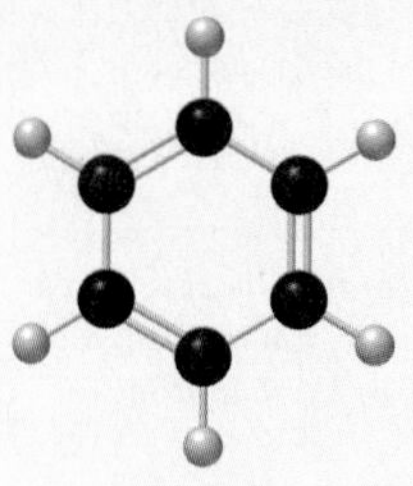

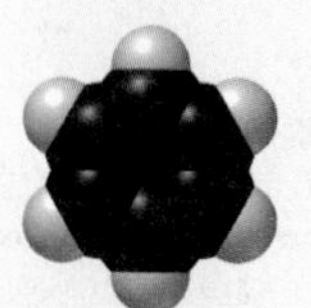

Figure 22.12 Benzene
All the atoms in the aromatic compound benzene lie in a single plane.
Use Models *What is the molecular formula for benzene?*

Aromatic Hydrocarbons

What is the most accurate description of bonding in benzene?

There is a class of unsaturated cyclic hydrocarbons that are responsible for the aromas of spices such as vanilla, cinnamon, cloves, and ginger. These compounds were originally called aromatic compounds because they have distinct, pleasant aromas. However, not all compounds classified as aromatic have pleasant odors. In fact, many do not have an odor at all.

Benzene is the simplest aromatic compound. An **aromatic compound,** or arene, is now defined as an organic compound that contains a benzene ring or other ring in which the bonding is like that of benzene. Any compound not classified as an aromatic compound is an **aliphatic compound.** The compounds you studied earlier in this chapter—alkanes, alkenes, alkynes, and cycloalkanes—are aliphatic compounds. The properties of aromatic compounds are quite different from those of aliphatic compounds.

The Structure of Benzene Friedrich Kekulé (1829–1896) made a major contribution to chemistry. He was the first to describe the structure of a benzene molecule. Look at the models of benzene shown in Figure 22.12. The benzene molecule is a six-membered carbon ring with one hydrogen atom attached to each carbon. This arrangement leaves one electron from each carbon free to participate in a double bond. Two different structures with alternating double bonds can be written for benzene.

[Structural formulas of the two benzene resonance structures, labeled "Single bond" and "Double bond"]

These structural formulas show only the extremes in electron sharing between any two adjacent carbons in benzene. One extreme is a normal single bond. The other extreme is a normal double bond. Recall that when two or more equally valid structures can be drawn for a molecule, resonance occurs. The actual bonding in a benzene ring doesn't alternate between the two extreme resonance structures. Rather, all the bonds in the ring are identical hybrids of single and double bonds. **In a benzene molecule, the bonding electrons between carbon atoms are shared evenly around the ring.** Benzene and other molecules that exhibit resonance are more stable than similar molecules that do not exhibit resonance. Thus, benzene is not as reactive as six-carbon alkenes.

Drawing a solid or dashed circle inside a hexagon is a good way to represent benzene in terms of how the electrons are distributed. However, such a drawing does not show the number of electrons involved. For this reason, the traditional structure, shown at the far right in the series below, is used in this textbook. Remember, though, that each bond in the ring is identical.

  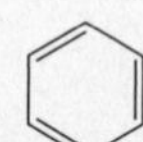

CHEMISTRY & YOU

Q: *You just read about aromatic compounds. Can you explain to a classmate what it means if a compound is aromatic?*

Differentiated Instruction

L1 STRUGGLING STUDENTS Have students summarize the lesson in a main-ideas-and-details chart. Check their charts for accuracy and suggest corrections where needed.

LPR LESS PROFICIENT READERS Have students restate the definitions of *aromatic compound* and *aliphatic compound* in their own words. Encourage students to list an example of each type of compound and explain why the example fits into its classification.

L3 ADVANCED STUDENTS Challenge students to draw the electron dot structure for benzene. Completed structures should include both resonance forms of benzene.

Substituted Aromatic Compounds Many dyes used to produce the intense colors of your clothing, such as the blue shown in Figure 22.13 are substituted aromatic compounds. Compounds containing substituents attached to a benzene ring are named using benzene as the parent hydrocarbon. When the benzene ring is a substituent, the C_6H_5 group is called a phenyl group.

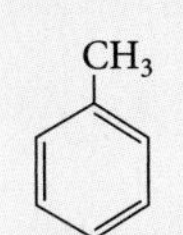

Methylbenzene (toluene) | Ethylbenzene | 3-phenylhexane

Some derivatives of benzene have two substituents. These derivatives are called disubstituted benzenes. Dimethylbenzene, also called xylene, is an example of a disubstituted benzene. There are three constitutional isomers for dimethylbenzene ($C_6H_4(CH_3)_2$). The boiling points of the three compounds are a reminder that constitutional isomers have different physical properties.

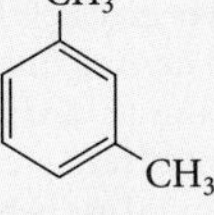
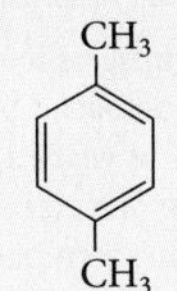

1,2-dimethylbenzene (*o*-xylene) (bp 144°C) | 1,3-dimethylbenzene (*m*-xylene) (bp 139°C) | 1,4-dimethylbenzene (*p*-xylene) (bp 138°C)

In the IUPAC naming system, the possible positions of two substituents in disubstituted benzene are designated as 1,2; 1,3; or 1,4. Common names for disubstituted benzenes use the terms *ortho*, *meta*, and *para* (abbreviated as *o*, *m*, and *p*) in place of numbers.

Figure 22.13 Dyes
Many dye molecules have phenyl groups. For example, indigo, which is used to color blue jeans, has two phenyl groups.

22.4 LessonCheck

28. Define What is a cyclic hydrocarbon?

29. Review Describe the bonding between carbon atoms in benzene.

30. Identify Name the following compounds using the IUPAC system.

a. CH_2CH_3 (on benzene ring)

c. CH_2CH_3, $CH_2CH_2CH_3$ (on benzene ring)

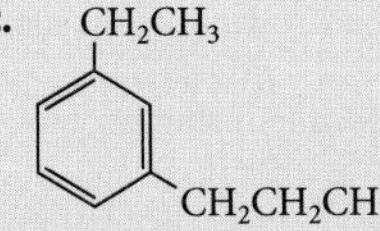

b. (two linked benzene rings)

31. Evaluate The alternate name for an aromatic compound is arene. Do you think arene is a good option? Consider what the suffix *-ene* means when used with aliphatic compounds.

BIGIDEA CARBON CHEMISTRY

32. Hexane, 1-hexene, cyclohexane, and benzene each have six carbon atoms. What is the difference between these hydrocarbons? Use the words *aliphatic, aromatic, saturated,* and *unsaturated* to explain your answer.

Lesson Check Answers

28. In a cyclic hydrocarbon, the carbon chain is in the form of a ring.
29. In benzene, the bonding electrons between carbon atoms are shared evenly around the ring.
30. **a.** ethylbenzene
    b. 1-ethyl-3-propylbenzene
    c. phenylbenzene
31. The suffix *-ene* indicates the presence of a double bond, but the bonds in the rings in aromatic compounds are hybrid bonds, not alternating single and double bonds.
32. **BIGIDEA** Hexane is a straight-chain, aliphatic, saturated hydrocarbon. 1-hexene is a straight-chain, aliphatic, unsaturated hydrocarbon. Cyclohexane is a cyclic, aliphatic, saturated hydrocarbon. Benzene is an aromatic hydrocarbon, which is unsaturated, by definition.

Explain

START A CONVERSATION Compare the structure of benzene to that of a hexagon. Remind students that the prefix *hex-* indicates six and *hexagon* means "six-sided polygon." Show students that the benzene ring consists of six carbon atoms bonded in a flat hexagonal ring. Point out that each carbon is bonded to only one hydrogen atom because of the three alternating double bonds. Ask students to describe the molecular geometry at each carbon atom. *(It is trigonal planar.)*

Extend

Have students draw and present to the class six examples of disubstituted benzenes. Students should present the names of the compounds as well as a summary list explaining how to name them.

Evaluate

Informal Assessment

To asses students' understanding of the lesson, ask them to create a compare-and-contrast table and complete it for each set of terms in relation to hydrocarbons.

1. *aromatic* and *aliphatic*
2. *chain* and *cyclic*
3. *saturated* and *unsaturated*

Then, have students complete the 22.4 Lesson Check.

Reteach

As a class, create a concept map to help clarify relationships and characteristics of the various types of hydrocarbons learned in this lesson. If students are still struggling, include on the concept map the relationships and characteristics of all the hydrocarbons learned in this chapter.

Answers

FIGURE 22.12 C_6H_6

LESSON 22.5

Key Objectives

IDENTIFY hydrocarbons found in natural gas.

DESCRIBE the first step in the refining of petroleum.

CLASSIFY coal.

Additional Resources

- Reading and Study Workbook, Lesson 22.5

PearsonChem.com

Engage

CHEMISTRY & YOU Have students read the introduction. Have students share what they know about gasoline. *(Sample answers: mixture of alkanes, flammable, volatile)* **Ask** Do you think gasoline occurs naturally or do you think it is a manufactured product? *(Answers will vary.)* Explain that gasoline is not manufactured, but it must go through a refining process before it can be used as a vehicular fuel.

Build Background

Relate the chemical process of cracking in a petroleum refinery to the physical process when an object cracks or breaks apart.

Foundations for Reading

BUILD VOCABULARY Explain that the word *petroleum* was created by combining the Latin word for oil, *oleum*, with the Greek prefix *petr-*, meaning "stone."

22.5 Hydrocarbons From Earth's Crust

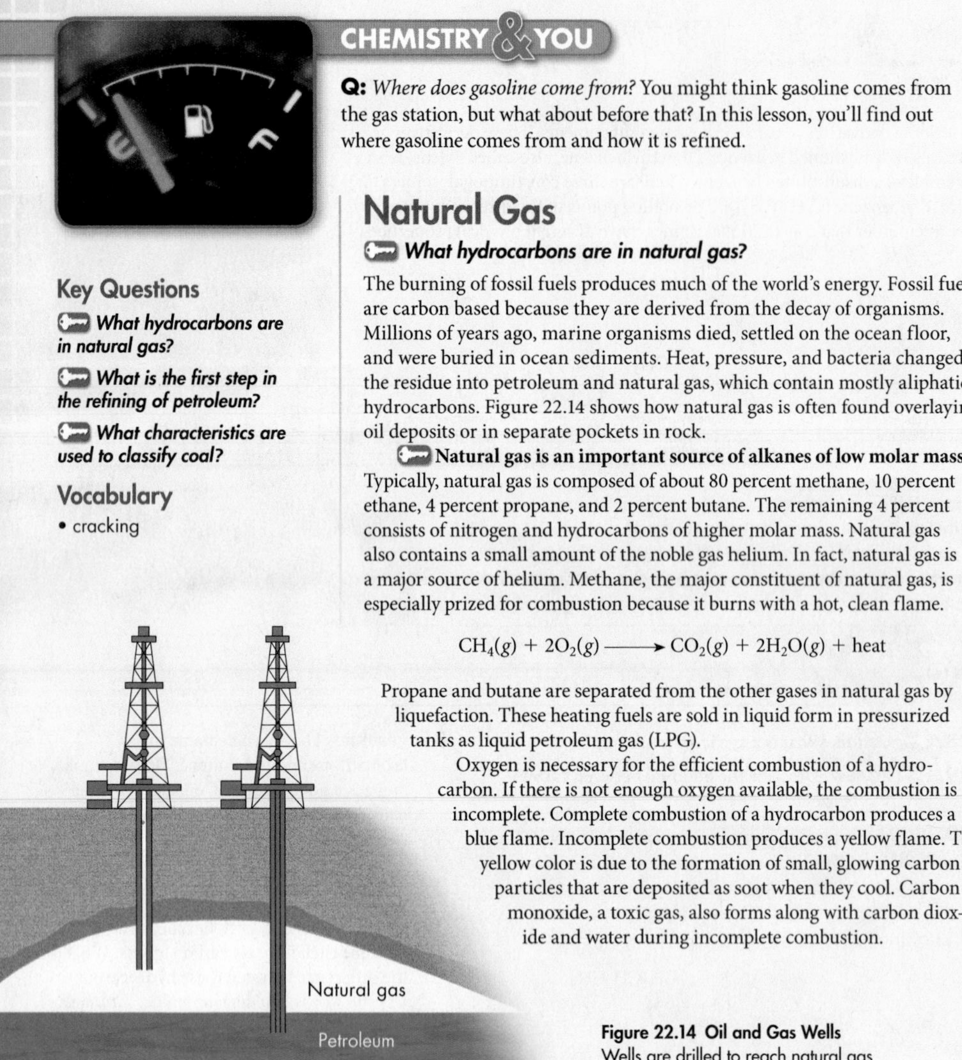

CHEMISTRY & YOU

Q: *Where does gasoline come from?* You might think gasoline comes from the gas station, but what about before that? In this lesson, you'll find out where gasoline comes from and how it is refined.

Key Questions

What hydrocarbons are in natural gas?

What is the first step in the refining of petroleum?

What characteristics are used to classify coal?

Vocabulary

- cracking

Natural Gas

What hydrocarbons are in natural gas?

The burning of fossil fuels produces much of the world's energy. Fossil fuels are carbon based because they are derived from the decay of organisms. Millions of years ago, marine organisms died, settled on the ocean floor, and were buried in ocean sediments. Heat, pressure, and bacteria changed the residue into petroleum and natural gas, which contain mostly aliphatic hydrocarbons. Figure 22.14 shows how natural gas is often found overlaying oil deposits or in separate pockets in rock.

Natural gas is an important source of alkanes of low molar mass. Typically, natural gas is composed of about 80 percent methane, 10 percent ethane, 4 percent propane, and 2 percent butane. The remaining 4 percent consists of nitrogen and hydrocarbons of higher molar mass. Natural gas also contains a small amount of the noble gas helium. In fact, natural gas is a major source of helium. Methane, the major constituent of natural gas, is especially prized for combustion because it burns with a hot, clean flame.

$$CH_4(g) + 2O_2(g) \longrightarrow CO_2(g) + 2H_2O(g) + \text{heat}$$

Propane and butane are separated from the other gases in natural gas by liquefaction. These heating fuels are sold in liquid form in pressurized tanks as liquid petroleum gas (LPG).

Oxygen is necessary for the efficient combustion of a hydrocarbon. If there is not enough oxygen available, the combustion is incomplete. Complete combustion of a hydrocarbon produces a blue flame. Incomplete combustion produces a yellow flame. The yellow color is due to the formation of small, glowing carbon particles that are deposited as soot when they cool. Carbon monoxide, a toxic gas, also forms along with carbon dioxide and water during incomplete combustion.

Figure 22.14 Oil and Gas Wells Wells are drilled to reach natural gas and petroleum deposits.

Focus on ELL

1 CONTENT AND LANGUAGE Have students read the lesson title and each section heading. Have a discussion about what students already know about these subjects. Record their statements in a KWL chart. Provide students with a copy of the chart to complete as you proceed with the lesson.

2 FRONTLOAD THE LESSON Review the carbon cycle with students. Ask students to identify the processes of the carbon cycle that are associated with natural gas, petroleum, and coal formation.

3 COMPREHENSIBLE INPUT Use a lit Bunsen burner to demonstrate the difference in flame color between incomplete and complete combustion. Assign the Kinetic Art activity to accompany the discussion on refining petroleum.

Petroleum

What is the first step in the refining of petroleum?

The organic compounds found in petroleum, or crude oil, are more complex than those in natural gas. Most of the hydrocarbons in petroleum are straight-chain and branched-chain alkanes. But petroleum also contains small amounts of aromatic compounds and sulfur-, oxygen-, and nitrogen-containing organic compounds.

Humans have known about petroleum for centuries; ancient peoples found it seeping from the ground in certain areas. In the late 1850s, a vast deposit of petroleum was discovered in Pennsylvania when a well was drilled to obtain petroleum for use as a fuel. Within decades, petroleum deposits had also been found in the Middle East, Europe, and the East Indies. Petroleum has since been found in other parts of the world as well.

Crude oil is a mixture of hydrocarbons having anywhere from 1 carbon atom to more than 40 carbon atoms. Without further treatment, crude oil is not very useful. The mixture must be separated, or refined, into parts called fractions, which have many commercial uses. **Petroleum refining begins with the distillation of crude oil into fractions according to boiling point.** Figure 22.15 shows a schematic of a petroleum refining distillation tower. Each distillation fraction contains several different hydrocarbons.

Note that the gasoline fraction makes up just 40 percent of the crude oil mixture. However, gasoline is by far the most commonly used petroleum product. To make the supply meet the demand, other processes such as cracking are used. **Cracking** is a controlled process by which hydrocarbons are broken down or rearranged into smaller, more useful molecules. For example, fractions containing compounds of higher molar mass are "cracked" to produce the more useful short-chain components of gasoline and kerosene. Hydrocarbons are cracked with the aid of a catalyst and with heat. This process also produces low-molar-mass alkanes, which are used to manufacture paints and plastics. Other catalytic processes besides cracking are used to increase the amounts of components that improve the performance of gasoline.

Figure 22.15 Fractional Distillation of Crude Oil
The crude oil is heated so that it vaporizes and rises through the fractionating column. Compounds with the highest boiling points condense near the bottom. Compounds with the lowest boiling points condense near the top.
Infer ***In which fraction would you expect to find decane?***

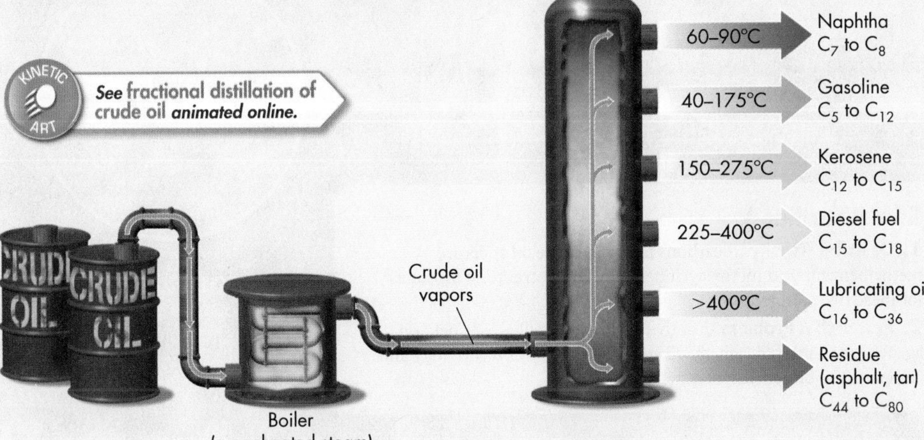

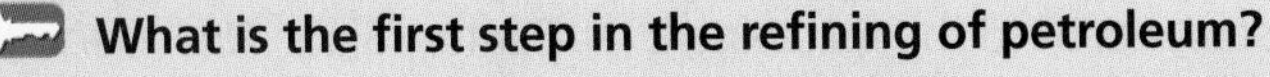

UbD Check for Understanding

What is the first step in the refining of petroleum?

Ask students to give a one-minute response that describes the difference between distillation and cracking. Answers may be oral or written.

ADJUST INSTRUCTION If students are having difficulty with this exercise, have them reread the last two paragraphs of the Petroleum section and review Figure 22.15. Clarify that cracking is a process that takes place after fractional distillation and is performed to reduce higher molecular mass fractions into more useful short-chain hydrocarbons.

Explain

Petroleum

START A CONVERSATION Explain to the class that it was not until the invention of the kerosene lamp that people valued petroleum. By the time electric lights reduced the need for kerosene, automobiles required gasoline, which previously had been discarded as a useless byproduct of kerosene production. Explain that before the petroleum era began, fats from domesticated animals, whale oil, wood, and coal were used as energy sources.

USE VISUALS Direct students' attention to Figure 22.15. Explain that the crude oil is pumped through the boiler, where it vaporizes. The petroleum vapor enters the fractionating column. The column is divided into chambers with openings in the floor of each chamber to allow the passage of vapors into the chamber. Each chamber is kept at a temperature just below the boiling point of a specific fraction. **Ask** Why do you think the temperature is kept below the boiling point of each fraction? *(The temperature is kept below the boiling point of the given fraction so that those particular vapors will condense back to a liquid, while allowing the other vapors to rise into the next chamber.)*

Extend

Divide the class into five groups. Have each group research a different fraction obtained from crude oil. Challenge each group of students to create a pie graph of the percents of each fraction, highlighting the slice that represents their fraction. The display should also include information on the chemical composition of the fraction and the uses of the products derived from that fraction. Encourage students to illustrate their displays with original art or magazine photographs.

Answers

FIGURE 22.15 Decane has ten carbons, so it should be in the gasoline fraction.

CHEMISTRY & YOU

CHEMISTRY & YOU Draw students' attention to the background image and the inset of the remediation worker. Pose the following question to students: *How do the challenges associated with the containment and remediation of an oil spill on the open sea differ from those associated with a land-based spill?* You may need to guide the students in the following ways:

- Computer modeling can be used to predict the spread and movement of ocean spills.
- Shorelines have different uses and characteristics, as well as different sensitivities to oil spills because of the flora and fauna at shorelines.
- Successful bioremediation requires high concentrations of microbes to be in close proximity to the oil.
- Land-based spills pose the risk of contaminants entering the groundwater.

Explain

USE AN ANALOGY Explain that bioremediation is analogous to photosynthesis in plants. Photosynthesis, like bioremediation, uses a process that turns a harmful byproduct of human respiration, carbon dioxide, into oxygen.

Extend

Connect to ENVIRONMENTAL ENGINEERING

Have students research a product that is currently available for bioremediation and write a short paragraph detailing how the product optimizes bacterial consumption of oil.

CHEMISTRY & YOU: TECHNOLOGY

Bioremediation

Oil and water don't mix. You may have witnessed this fact watching footage of an oil spill in the news. Oil spills can lead to the deaths of seabirds and marine mammals and can contaminate soil and drinking water.

One tool being used to clean up spilled oil is a relatively new technology called bioremediation. The technology uses "oil-eating" microbes—particularly bacteria—to remedy the spill. For these microbes, the hydrocarbons in crude oil are not a contaminant but a food source. During the digestion process, harmful hydrocarbons are converted to less harmful products—mainly carbon dioxide and water. Bioremediation is a safe, simple, and relatively inexpensive method of dealing with spilled oil. However, the process takes time to work. In addition, it is usually only effective on residual oil on shorelines, after part of the spill has been removed by other means.

Take It Further

1. Describe Two hydrocarbons found in crude oil spills are methylbenzene and methylcyclopentane. Draw structural formulas for these two compounds.

2. Research a Problem Another technology called a dispersion agent is often used to remedy oil spills. Research this technology and compare it to bioremediation.

HIGH-TECH . . . AND ALIVE The degradation of petroleum in the marine environment is carried out by diverse microorganisms, including the *Pseudomonas* species shown here.

784 Chapter 22 • Chemistry & You

21st Century Learning To be successful in the 21st century, students need skills and learning experiences that extend beyond subject matter mastery. The following project helps students build the following 21st Century Skills: *Financial, Economic, Business, and Entrepreneurial Literacy*; *Creativity and Innovation*; *Communication and Collaboration*; *Information, Media, and Technology Literacy*; *Initiative and Self-Direction*; and *Productivity and Accountability*.

CALL TO ACTION Tell students that a prominent wildlife protection society is concerned about the ongoing environmental problems at the site of an old oil spill. The society has need of an Internet-savvy public relations agency to create an awareness campaign about the site. Pose the following challenge to students: *Form an "agency" of four to six students to create a Web site designed to showcase the history of the spill and its environmental impact, remediation progress, and cumulative costs.* Base the campaign on an actual oil spill that affected a marine ecosystem, an estuary/wetland ecosystem, or a human community. Include an appeal to the public to support safer, improved oil transportation/storage methods and remediation technologies.

Coal

What characteristics are used to classify coal?

Geologists think that coal had its origin some 300 million years ago when huge tree ferns and mosses grew abundantly in swampy tropical regions. When the plants died, they formed thick layers of decaying vegetation. Layer after layer of soil and rock eventually covered the decaying vegetation, which caused a buildup of intense pressure. This pressure, together with heat from Earth's interior, slowly turned the plant remains into coal.

Coal Formation The first stage in the formation of coal is an intermediate material known as peat. Peat, shown in Figure 22.16, is a soft, brown, spongy, fibrous material. When first dug out of a bog, peat has a very high water content. After it has been allowed to dry, it produces a low-cost but smoky fuel. If peat is left in the ground, it continues to change. After a long period of time, peat loses most of its fibrous texture and becomes lignite, or brown coal. **Coal is classified by its hardness and carbon content.** For example, lignite is much harder than peat and has a higher carbon content (about 30 percent). The water content, however, is still high. Continued pressure and heat slowly change lignite into bituminous coal, or soft coal, which is harder than lignite. Bituminous coal has a lower water content and higher carbon content (35 percent to 85 percent) than lignite. In some regions of Earth's crust, even greater pressures have been exerted. In those places, such as the earth beneath eastern Pennsylvania, soft coal has been changed into anthracite, or hard coal. Anthracite has a carbon content that exceeds 85 percent, making it an excellent fuel source.

Coal, which is usually found in seams from 1 to 3 meters thick, is obtained from both underground and surface mines. In North America, coal mines are usually less than 100 meters underground. Much of the coal is so close to the surface that it is strip mined. By contrast, many coal mines in Europe and Asia extend 1000 to 1500 meters below Earth's surface.

Q: *Where does gasoline come from? Where does coal come from? Use Figures 22.14, 22.15, and 22.16 to explain your answers.*

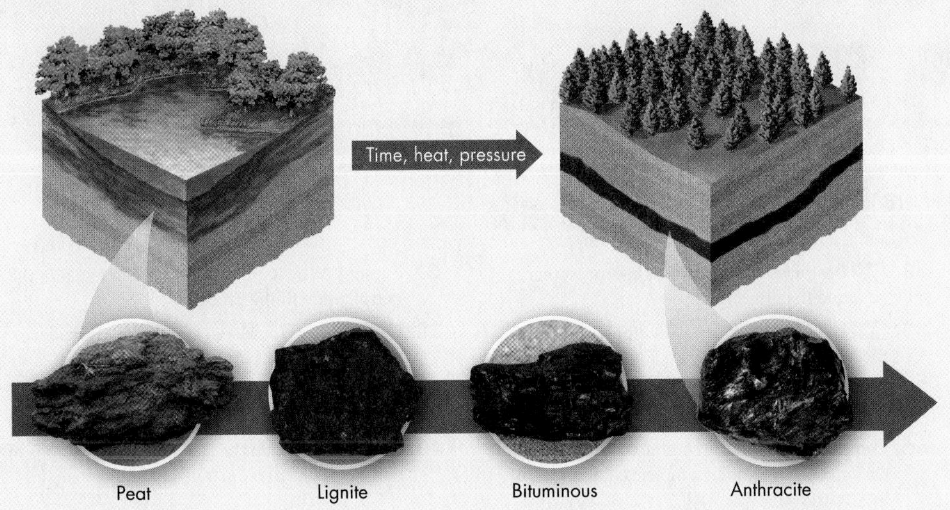

Figure 22.16 Coal Formation
When tree ferns and mosses die, their decomposing remains build up. Over millions of years, layers of decaying organic material pile up and form peat. Continued pressure and heat transform peat into lignite, bituminous coal, and anthracite coal.
Identify ***What three variables contribute to coal formation?***

Differentiated Instruction

LPR LESS PROFICIENT READERS Present the information about coal formation, coal composition, and coal production in a bulleted summary list. Have students use the list to create a flowchart.

ELL ENGLISH LANGUAGE LEARNERS Relate water wells to oil and gas wells. Use pictures and diagrams to demonstrate the similarities and differences between water wells, oil wells, and natural gas wells.

L3 ADVANCED STUDENTS Have students research the energy sources used in the United States. Have them summarize their findings in a bar chart showing to what extent the United States depends on each type of energy source. The bar chart should profile the energy sources for the years 1850, 1900, 1950, 1990, 2000, and 2010.

Explain

Coal

USE A DIAGRAM Have students study Figure 22.16 and read the caption. If possible, show the students actual samples of peat, lignite, bituminous coal, and anthracite coal. Explain that coal can be converted to methane by a process called coal gasification, in which coal and hydrogen react under high temperature and pressure. Synthetic crude oil can be made by a process called coal liquefaction, in which coal is dissolved in a solvent at high pressure and temperature. Point out that the product, called solvent-refined coal, is then purified and heated. The hard, brittle residue that remains can be melted and used as a liquid fuel.

CRITICAL THINKING Both coal gasification and coal liquefaction produce products that burn more cleanly and efficiently than coal and can be used in place of coal and petroleum products to power steam generators in electrical power plants. Make sure students understand that, like conventional coal and petroleum products, the products of coal gasification and liquefaction produce greenhouse gases when combusted. However, the carbon dioxide produced while burning the coal gas can more easily be captured, or *sequestered*, before it escapes into the atmosphere. Challenge students to explain how the use of coal-based products in place of coal or petroleum might benefit both humans and the environment. *(fewer emissions due to burning more cleanly than unprocessed coal; lower cost to electrical customers due to increased efficiency; less carbon dioxide entering the atmosphere if sequestered; potential for less dependence on foreign oil resources)*

Answers

TAKE IT FURTHER

1.

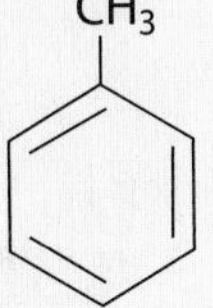

Methylbenzene

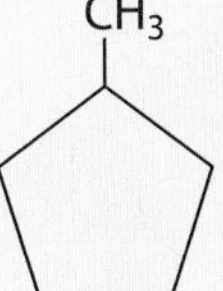

Methylcyclopentane

2. The goal when using a dispersion agent is simply to mix the oil with the water, whereas bioremediation aims to rid the water of the harmful hydrocarbons of the oil spill.

FIGURE 22.16 time, heat, and pressure

LESSON 22.5

Extend

Scientists have tried for years to determine whether hydrocarbons could have been created deep in the Earth and formed without organic matter. Have students research this issue and report back the current status of findings. *(As of July 2009, scientists have found that ethane and heavier hydrocarbons can be formed under the pressure-temperature conditions of the upper mantle of the Earth.)*

Evaluate

Informal Assessment

Have students create a three-column chart to compare and contrast the formation and composition of coal, natural gas, and petroleum. *(Petroleum and natural gas are formed from marine organisms buried under ocean sediments; coal is formed from buried wetland plants. Natural gas and petroleum contain mostly aliphatic compounds; coal contains a large percentage of aromatic compounds.)* Then have students complete the 22.5 Lesson Check.

Reteach

Display Figure 22.15 on an overhead projector. If possible, bring a sample of crude oil to class so that students can observe its color and odor. Explain that petroleum is a complex mixture of hydrocarbons whose molecules range in size from 1 to more than 40 carbon atoms. To be used effectively, the crude oil must be refined by fractional distillation, which separates compounds on the basis of their boiling points. Explain that in general, low molar mass compounds have low boiling points and high molar mass compounds have high boiling points.

Composition of Coal Coal consists largely of condensed aromatic compounds of extremely high molar mass. These compounds have a high proportion of carbon compared with hydrogen. Due to the high proportion of aromatic compounds, coal leaves more soot when burned than the more aliphatic fuels obtained from petroleum. Coal also contains a small amount of sulfur. As coal burns, the sulfur oxidizes to form SO_2 and SO_3, two major air pollutants that contribute to acid rain and the smog shown in Figure 22.17. In order to reduce air pollution, the majority of sulfur is removed before the coal is burned. Any sulfur oxides present after combustion are generally captured before the emissions leave the smokestack.

Coal may be distilled to obtain a variety of products: coal gas, coal tar, ammonia, and coke. Coke is the solid material left after coal distillation. It is used as a fuel in many industrial processes and is the crucial reducing agent in the smelting of iron ore. Coal gas consists mainly of hydrogen, methane, and carbon monoxide. Coal tar can be distilled further into benzene, toluene, naphthalene, phenol, and pitch. The ammonia from distilled coal is converted to ammonium sulfate for use as a fertilizer.

Figure 22.17 Smog in Beijing
Sulfur oxides from emissions of coal-fired plants in China often contribute to dense smog.

22.5 LessonCheck

33. **Describe** Which type of hydrocarbons can be found in natural gas?
34. **Review** Describe the first process used in the refining of petroleum.
35. **Identify** What are the two characteristics used to classify coal?
36. **Compare** How do the combustion products differ for complete and incomplete combustion of hydrocarbons?
37. **Explain** Why is cracking a necessary step in the petroleum refining process?
38. **List** What are some common products made from natural gas, petroleum, and coal?
39. **Summarize** How did each of the three major fossil fuels form?
40. **Compare** How do the chemical compositions of natural gas, petroleum, and coal differ?

Lesson Check Answers

33. alkanes of low molar mass
34. The refining process starts with the distillation of petroleum into fractions according to boiling point.
35. hardness and carbon content
36. For complete combustion, the products are carbon dioxide and water. For incomplete combustion, carbon monoxide and soot form in addition to carbon dioxide and water.
37. Amounts of products obtained by fractional distillation don't match demands. Cracking breaks down hydrocarbons into smaller, more useful components.
38. sample answers: petroleum: paints and plastics; coal: ammonia fertilizer
39. Heat, pressure, and bacteria changed marine organisms buried in ocean sediments into petroleum and natural gas. Heat and pressure changed buried layers of vegetation into coal.
40. Natural gas largely consists of lighter alkanes such as methane, ethane, propane, and butane. Coal consists largely of aromatic compounds of extremely high molar mass. Petroleum consists largely of hydrocarbons that fall between those of natural gas and coal.

786 Chapter 22 • Lesson 5

Small-Scale Lab

Hydrocarbon Isomers

Purpose

To draw line-angle formulas and name some of the isomers in gasoline

Materials

- toothpicks
- modeling clay
- pencil
- paper

Procedure

Gasoline is a complex mixture of hydrocarbon molecules. Generally, each molecule contains between five and twelve carbon atoms. Many of the components of gasoline are isomers with the same molecular formula. These components include the isomers of pentane. Study the formulas and names of the constitutional isomers of C_5H_{12} in the table below. Make a model of each isomer using toothpicks and modeling clay, using the space-filling models as a guide. Compare the models for each isomer.

Isomers of C_5H_{12}

Condensed formula	Line-angle formula	Space-filling model
$CH_3CH_2CH_2CH_2CH_3$ pentane		
$CH_3CHCH_2CH_3$ with CH_3 on the second carbon; 2-methylbutane		
CH_3CCH_3 with CH_3 above and below the central carbon; 2,2-dimethylpropane		

Analyze and Conclude

1. Describe Draw the complete structural formula for each isomer of C_5H_{12} in the table.

2. Infer In a line-angle formula, each line represents a carbon-carbon bond. Each end of a line, as well as the intersection of lines, represents a carbon atom. Knowing that carbon always forms four covalent bonds, explain how to determine the number of hydrogen atoms bonded to each carbon in a line-angle formula.

3. Describe Because butane can vaporize readily, it is used in the formulations of gasolines in cold climates during winter. Draw condensed structural formulas and line-angle formulas for the two isomers of butane (C_4H_{10}).

You're the Chemist

1. Analyze Data Gasoline contains isomers of hexane, too. Draw the line-angle formulas and name the five isomers of C_6H_{14}. Make a model of each isomer.

2. Design an Experiment Gasoline also contains small amounts of the six isomers of pentene. Two of the isomers are *cis* and *trans* configurations of the same constitutional isomer. Experiment with your models to make the six isomers. Use two toothpicks to represent a double bond. Draw line-angle formulas for each isomer. Name each compound.

Focus on ELL

4 LANGUAGE PRODUCTION Have students work in groups of four to complete the lab. Make sure each group has ELLs of varied language proficiencies so that more proficient students can help less proficient ones. Have students work according to their proficiency level.

BEGINNING: LOW/HIGH Use the table of isomers to explain the purpose and the procedure of the lab. Assist students in creating their clay models. Encourage students to make a sketch of what they want each model to look like.

INTERMEDIATE: LOW/HIGH Paraphrase the procedure for students. Read out loud the Analyze and Conclude questions.

ADVANCED: LOW/HIGH Have students develop their own questions that could be answered by another student working on the same lab.

Explore

Small-Scale Lab

OBJECTIVE After completing this activity, students will be able to write line-angle formulas for hydrocarbons.

SKILLS FOCUS analyze

PREP TIME 5 minutes

CLASS TIME 20 minutes

ANALYZE AND CONCLUDE

1. Students should draw structural formulas for pentane, 2-methylbutane, and 2, 2-dimethylpropane.
2. Subtract from 4 the number of lines drawn to any point.
3. $CH_3(CH_2)_2CH_3$ butane; $(CH_3)_2CHCH_3$ methylpropane

YOU'RE THE CHEMIST

1. 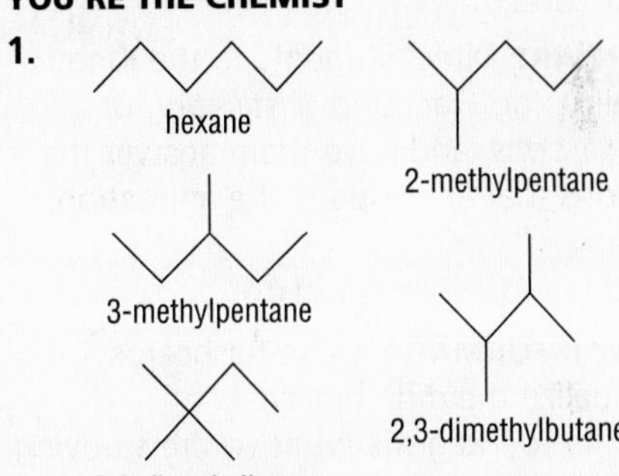

2. 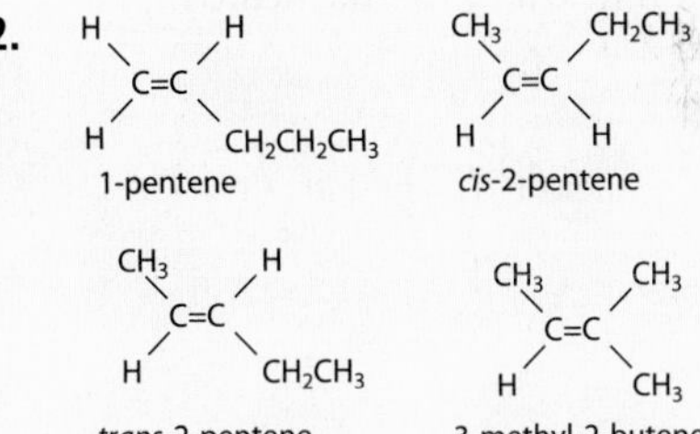

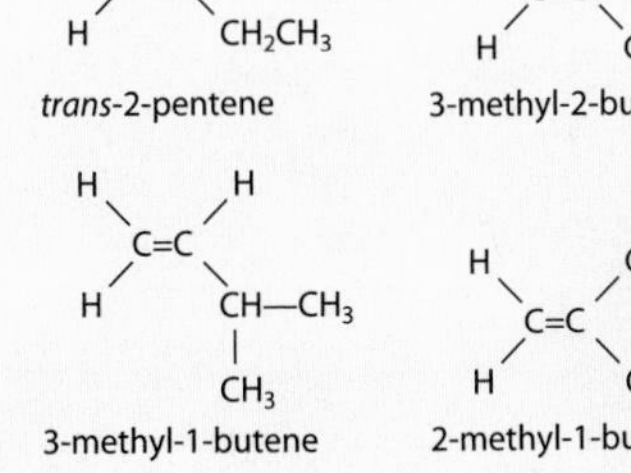

Evaluate

Review and Assessment Resources

CHEMISTRY TUTORIAL Encourage students to review naming branched-chain alkanes and identifying asymmetric carbon atom structures using the animated step-by-step tutorials for Chapter 22. Each Chemistry Tutorial is followed by a similar practice problem, where the student is guided through the problem-solving process.

CHAPTER PROBLEM SET Have students continue practicing naming hydrocarbons using the Chapter 22 online Problem Set.

DIRECTED VIRTUAL LAB Have students complete the Virtual Lab *Saturated and Unsaturated* to help reinforce the type and strength of bonds in saturated and unsaturated hydrocarbons. In the post-lab exercise, students will analyze the data and draw conclusions based on the results of the experiment.

KINETIC ART Direct students to the Kinetic Art animation *Fractional Distillation of Hydrocarbons*, and have them answer the questions that accompany the animation.

Study Tip

ORGANIZE NEW INFORMATION Use flashcards to help you visualize the different types of hydrocarbons. Try to integrate what you're studying with what you already know. Find a study partner and quiz one another using the flashcards.

22 Study Guide

BIGIDEA CARBON CHEMISTRY

Hydrocarbons are named using the IUPAC system, although sometimes common names are used for certain familiar compounds. All hydrocarbons are nonpolar molecules. In general, the fewer carbon atoms in a hydrocarbon, the lower its melting and boiling points. Hydrocarbons with the same molecular formula but different molecular structures can have different properties.

22.1 Hydrocarbons

Because carbon has four valence electrons, a carbon atom always forms four covalent bonds.

The carbon atoms in an alkane can be arranged in a straight chain or in a chain that has branches.

- hydrocarbon (762)
- alkane (764)
- straight-chain alkane (764)
- homologous series (765)
- condensed structural formula (766)
- substituent (767)
- alkyl group (768)
- branched-chain alkane (768)

22.2 Unsaturated Hydrocarbons

At least one carbon-carbon bond in an alkene is a double covalent bond. Other bonds may be single carbon-carbon bonds and carbon-hydrogen bonds.

At least one carbon-carbon bond in an alkyne is a triple covalent bond. Other bonds may be single or double carbon-carbon bonds and carbon-hydrogen bonds.

- saturated compound (772)
- unsaturated compound (772)
- alkene (772)
- alkyne (773)

22.3 Isomers

Constitutional isomers differ in physical properties such as boiling point and melting point. They also have different chemical reactivities.

Two types of stereoisomers are *cis-trans* isomers and enantiomers.

- isomer (775)
- constitutional isomer (775)
- stereoisomer (776)
- *cis-trans* isomer (776)
- *cis* configuration (776)
- *trans* configuration (776)
- asymmetric carbon (776)
- enantiomer (777)

22.4 Hydrocarbon Rings

In some hydrocarbon compounds, the carbon chain is in the form of a ring.

In a benzene molecule, the bonding electrons between carbon atoms are shared evenly around the ring.

- cyclic hydrocarbon (779)
- cycloalkane (779)
- aromatic compound (780)
- aliphatic compound (780)

22.5 Hydrocarbons From Earth's Crust

Natural gas is an important source of alkanes of low molar mass.

Petroleum refining begins with the distillation of crude oil into fractions according to boiling point.

Coal is classified by its hardness and carbon content.

- cracking (783)

788 Chapter 22 • Study Guide

UbD Performance Tasks

ROLE REVERSAL Swap roles with the students, and have them create a computerized presentation and give a lecture on naming hydrocarbons. Presentations should include visual representations as well as step-by-step procedures. Explain that the lecture should include various examples, give tips to avoid common mistakes, and conclude with a brief set of practice problems.

SAY IT WITH GRAPHS Have students create large, colorful graphs of the first ten alkanes to convey the ratio of composition of carbon/hydrogen by percent weight. Students should arrange their graphs so that any pattern they notice between the carbon atoms and the percentage of carbon in the alkanes is obvious. Ask students to include an overall statement summarizing their findings.

Skills Tune-Up: Structural Formulas

Problem	❶ Analyze	❷ Solve
Name this compound using the IUPAC system. $CH_3-CH_2-CH_2-CH_2-CH(CH_2-CH_3)-CH(CH_3)-CH(CH_3)-CH_3$ (vertical chain, top to bottom)	• The carbons on the parent hydrocarbon should be numbered to give the substituents the lowest possible numbers. • The location numbers for the substituents are included in the prefix. If there are multiple substituents of the same type, include that information in the prefix as well, for example, *di-* or *tri-* for two or three substituents, respectively. • The names of the substituents should be listed in alphabetical order with correct punctuation. Hint: Review Sample Problem 22.2 if you have trouble naming branched-chain alkanes.	Parent hydrocarbon: eight carbon atoms (octane) Substituents and their locations: • methyl group on carbon 2 • methyl group on carbon 3 • ethyl group on carbon 4 8 CH_3 – 7 CH_2 – 6 CH_2 – 5 CH_2 – 4 CH–[CH_2-CH_3] – 3 CH–[CH_3] – 2 CH–[CH_3] – 1 CH_3 (Substituents boxed) Prefix: 4-ethyl-2,3-dimethyl The IUPAC name is 4-ethyl-2,3-dimethyloctane.
Draw a condensed structural formula for 4-ethyl-2-methylheptane. Hint: Review Sample Problem 22.3 if you have trouble drawing structural formulas for branched-chain alkanes.	• The part of the name that ends in *-ane* identifies the parent hydrocarbon. • Prefixes identify the substituents, how many times they appear, and the location of each on the parent hydrocarbon. • Each carbon atom must have four covalent bonds. Hydrogens should be added to meet this need.	Parent hydrocarbon: heptane (seven carbon atoms) Substituents and their locations: • ethyl group on carbon 4 • methyl group on carbon 2 The structural formula for 4-ethyl-2-methylheptane is $CH_3-CH(CH_3)-CH_2-CH(CH_2-CH_3)-CH_2-CH_2-CH_3$ (vertical chain, top to bottom)

Focus on ELL

5 ASSESS UNDERSTANDING Assess students' understanding by using a portfolio for this chapter. Include various types of learning assignments, based on each student's language proficiency.

BEGINNING

LOW Include models or pictures of models created by the student.

HIGH Include copies of the graphic organizers used to take notes or any double-bubble diagrams that show similarities and differences between the different types of hydrocarbons.

INTERMEDIATE

LOW Include any prereading predictions made about lesson topics, as well as the results of those predictions.

HIGH Include any lab write-ups that were assigned during the chapter.

ADVANCED: LOW/HIGH Include an essay that asks the student to compare and contrast different types of hydrocarbons.

Skills Review

Example and sample practice problems for the various hydrocarbon problems can be found on the following pages:

LESSON 22.1 Drawing Structural Formulas for Alkanes—page 767; Naming Branched-Chain Alkanes—page 770; Drawing Structural Formulas for Branched-Chain Alkanes—page 771

LESSON 22.3 Identifying Asymmetric Carbon Atoms—page 777

ASSESSMENT

Answers

LESSON 22.1

41. The carbon-carbon bonds are nonpolar and the carbon-hydrogen bonds are very weakly polar.

42. pentane: $CH_3CH_2CH_2CH_2CH_3$
hexane: $CH_3CH_2CH_2CH_2CH_2CH_3$

43. **a.** propane
b. octane
c. pentane

44.

```
   H         H  H          H  H  H
   |         |  |          |  |  |
H—C—     H—C—C—      H—C—C—C—
   |         |  |          |  |  |
   H         H  H          H  H  H
 methyl     ethyl         propyl
```

45. **a.** 2-methylbutane
b. 2,3-dimethylbutane
c. 3-methylhexane

LESSON 22.2

46. **a.** propene
b. *trans*-2-pentene
c. 4-methyl-1-pentene
d. 3-ethyl-2-methyl-2-pentene

47. **a.** unsaturated, because there is a double carbon-carbon bond
b. saturated, because the compound contains the maximum number of hydrogen atoms per carbon atom

LESSON 22.3

48.

```
C—C—C—C—C—C              C
    hexane               |
                      C—C—C—C
C                        |
|                        C
C—C—C—C—C          2, 2-dimethylbutane
2-methylpentane

      C                  C  C
      |                  |  |
C—C—C—C—C             C—C—C—C
3-methylpentane     2, 3-dimethylbutane
```

49. **a.** Accept any isomer with 5 carbons and 12 hydrogens.
b. Accept any isomer with 7 carbons and 16 hydrogens.

50. **a.**

```
CH3      H         CH3      CH2CH3
   \    /             \    /
    C=C                C=C
   /    \             /    \
  H      CH2CH3      H      H
 trans-2-pentene     cis-2-pentene
```

b.

```
CH3      H
   \    /
    C=C
   /    \
CH3      CH2CH3
2-methyl-2-pentene
```

c.

```
CH3      CH2CH3
   \    /
    C=C
   /    \
  H      CH2CH3
3-ethyl-2-pentene
```

51. No, hexane contains no substituted groups.

52. No, only molecules with at least one asymmetric carbon can have enantiomers.

22 Assessment

ONLINE PROBLEMS

★ Solutions appear in Appendix E

Lesson by Lesson

22.1 Hydrocarbons

41. Why are alkane molecules nonpolar?

★**42.** Draw condensed structural formulas for pentane and hexane. Assume that the C—H and C—C bonds are understood.

43. Name the alkanes that have the following molecular or structural formulas.
a. $CH_3CH_2CH_3$
b. $CH_3(CH_2)_6CH_3$
c.

```
   H  H  H  H  H
   |  |  |  |  |
H—C—C—C—C—C—H
   |  |  |  |  |
   H  H  H  H  H
```

44. Draw structures for the alkyl groups derived from methane, ethane, and propane.

★**45.** Give the IUPAC name for each compound.

a.

```
CH3—CH—CH2
     |    |
    CH3  CH3
```

b.

```
CH3—CH—CH—CH3
     |   |
    CH3 CH3
```

c.

```
CH3—CH—CH2—CH2
     |        |
    CH2      CH3
     |
    CH3
```

22.2 Unsaturated Hydrocarbons

46. Give the IUPAC name for these alkenes.
a. $CH_3CH{=}CH_2$
b.

```
CH3      H
   \    /
    C=C
   /    \
  H      CH2CH3
```

c.

```
CH3CHCH2CH=CH2
   |
  CH3
```

d.

```
CH3      CH2CH3
   \    /
    C=C
   /    \
CH3      CH2CH3
```

★**47.** Classify each of the following compounds as saturated or unsaturated. Explain.
a. $CH_3CH{=}CHCH_2CH_3$
b. $CH_3CH_2CH_3$

22.3 Isomers

48. Draw and name all the constitutional isomers with the molecular formula C_6H_{14}.

49. Draw one constitutional isomer of each compound.

a.

```
      CH3
      |
CH3—C—CH3
      |
      CH3
```

b.

```
       CH3
       |
CH3—CH—CH—CH3
            |
            CH2
            |
            CH3
```

50. Draw a structural formula or carbon skeleton for each of the following alkenes. If *cis* and *trans* forms are present, include both forms.
a. 2-pentene
b. 2-methyl-2-pentene
c. 3-ethyl-2-pentene

51. Can you draw a constitutional isomer of hexane that has an asymmetric carbon? Explain.

★**52.** Do all molecules have enantiomers? Explain.

22.4 Hydrocarbon Rings

53. Give the IUPAC name for the following cyclic hydrocarbons.

a. 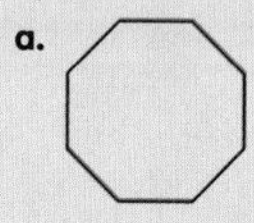

b. 

54. Explain why both of these structures represent 1,2-diethylbenzene.

(two benzene rings, each with CH_2CH_3 substituents on adjacent carbons)

★**55.** Draw a structural formula for each compound.
a. 1,4-diethylbenzene
b. 2-methyl-3-phenylpentane
c. 1,3-dimethylbenzene

LESSON 22.4

53. **a.** cyclooctane
b. cyclobutane

54. Two different structural formulas are possible because a benzene ring exhibits resonance.

55. **a.** benzene ring with CH_2CH_3 at top and CH_2CH_3 at bottom (para)

b.

```
      CH3
      |
CH3—CH—CH—CH2—CH3
          |
       (benzene ring)
```

c. benzene ring with CH_3 and CH_3 in meta positions

22.5 Hydrocarbons From Earth's Crust

56. Rank these materials in order of increasing carbon content: bituminous coal, peat, lignite, and anthracite coal.

57. How are catalysts used in petroleum refining?

***58.** What happens to the sulfur when coal burns?

Understand Concepts

59. Why are the following names incorrect? What are the correct names?

a. 2-dimethylpentane
b. 1,3-dimethylpropane
c. 3-methylbutane
d. 3,4-dimethylbutane

***60.** For each hydrocarbon shown, identify the type of covalent bonds and name the compound.

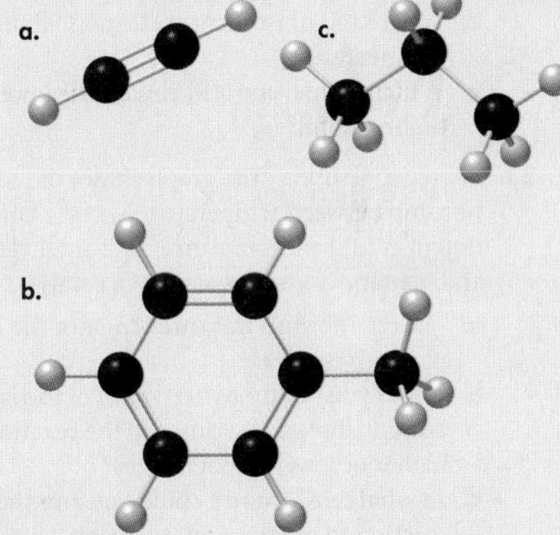

61. Write structural formulas for these compounds.

a. propyne
b. cyclohexane
c. 2-phenylpropane
d. 2,2,4-trimethylpentane

***62.** After ethane, what are the next three members of the homologous series of alkanes?

63. Compare *cis-trans* isomers and enantiomers.

64. Draw electron dot structures for each compound.

a. propene
b. propane
c. propyne
d. cyclobutane

65. Write an equation for the combustion of octane.

***66.** Compare these three molecular structures. Which would you expect to be most stable? Explain your answer.

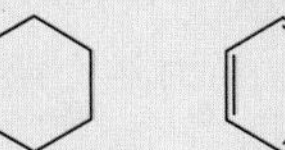 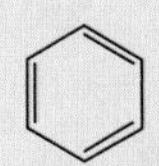 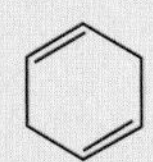

67. The four hydrocarbons produced in the largest amounts in the United States in a recent year are listed in the table below. Answer the following questions based on the data given.

Chemical	Amount produced (billions of kg)
Ethene	15.9
Propene	8.4
Benzene	5.3
Ethylbenzene	4.3

a. How many billion kilograms of aromatic compounds were produced?
b. Of the total mass of all four compounds produced, what percent by mass was made up of aliphatic compounds?

***68.** Are these two structures *cis-trans* isomers? Explain your answer.

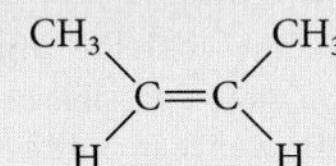

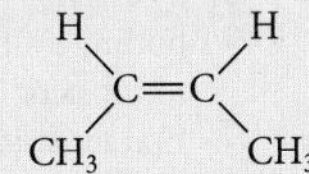

69. Use the labeled features in the molecular structure below to answer the following questions.

(Structure: benzene ring with substituents ❶ C_6H_5; ❷ $CH_2CH_2CH_3$; H; ❸ $CH{=}CH_2$; ❺ CH_3; and $Cl{-}C(F){-}Br$ with ❹ pointing to the carbon.)

a. Which label identifies a double bond?
b. Which label identifies a phenyl group?
c. Which label identifies a methyl group?
d. Which label identifies an asymmetric carbon?
e. Which label identifies a propyl group?

70. Does ethylcyclohexane have an asymmetric carbon? Explain.

ASSESSMENT

LESSON 22.5

56. peat, lignite, bituminous coal, anthracite coal

57. During cracking, catalysts are used to produce more short-chain components.

58. The combustion of sulfur in coal produces the air pollutants SO_2 and SO_3.

UNDERSTAND CONCEPTS

59. a. The *di-* indicates two methyl groups, but only one location is given; 2,2-dimethlypentane.
b. Alkyl groups on end carbons are part of the chain; pentane.
c. The chain wasn't numbered so the substituent has the lowest possible number; 2-methylbutane.
d. The methyl group on carbon 4 is part of the chain; 3-methylpentane

60. a. Ethyne (acetylene) has one triple carbon-carbon bond and two single carbon-hydrogen bonds.
b. In methylbenzene, there are hybrid bonds within the ring. All other bonds are single bonds.
c. All the bonds in propane are single bonds.

61. a. $CH{\equiv}C{-}CH_3$

b.

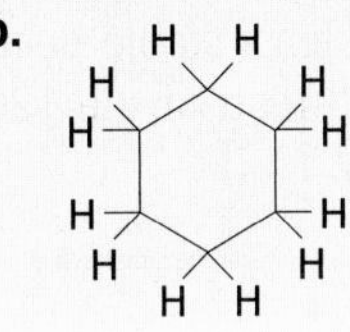

c. $CH_3{-}CH{-}CH_3$ (with a phenyl ring attached to the central CH)

d.

```
     CH3      CH3
      |        |
CH3—C—CH2—CH—CH3
      |
     CH3
```

62. propane, butane, pentane

63. *Cis-trans* isomers and enantiomers are both types of stereoisomers. *Cis-trans* isomers differ in the arrangement of substituent groups attached to each carbon in a double bond. In enantiomers, there is at least one asymmetric carbon.

64.

```
a.  H H H              c.        H
    C::C:C:H              H:C:::C:C:H
    H   H                        H

b.   H H H             d.   H H
   H:C:C:C:H              H:C:C:H
     H H H                H:C:C:H
                            H H
```

65. $2C_8H_{18}(l) + 25O_2(g) \rightarrow 16CO_2(g) + 18H_2O(l)$

66. The middle structure is most stable due to resonance within the ring.

67. a. 9.6 billion kg
b. 72%

68. No, the structures are identical; one has been flipped over.

69. a. 3 **b.** 1
c. 5 **d.** 4
e. 2

70. No, two of the R groups (those forming the ring) are identical.

ASSESSMENT

71. a. 4
b. 2
c. 3
d. 1

THINK CRITICALLY

72. The amount of heat per carbon is higher for methane (–890 kJ/mol of carbon burned) than for benzene (–545 kJ/mol of carbon burned). Burning aromatic compounds produces more soot.

73. *Meth-* implies one carbon atom; *ene-* implies a double bond, which requires two carbon atoms.

74. The methyl and ethyl groups can be on the same side of the bond or opposite sides.

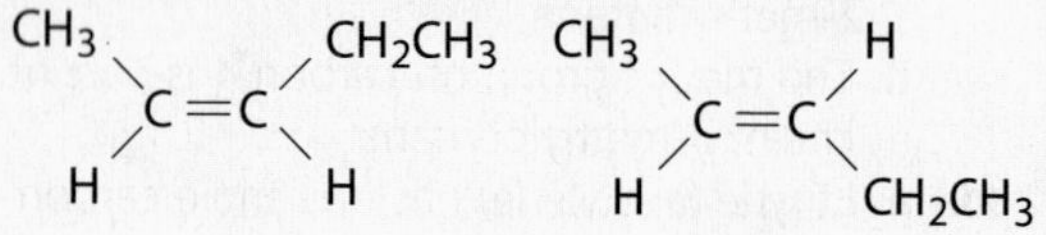

trans-2-pentene   *cis*-2-pentene

75. Students may infer that the ring structures in cyclic hydrocarbons produce stronger van der Waals attractions, which increase the energy required for vaporization.

76. $CH_2=C=CH_2$

77. The cycloalkene would be least stable because the bond angles are 90° instead of the 120° predicted by VSEPR theory.

$CH\equiv CCH_2CH_3$ (alkyne)

(cycloalkene)

$CH_2=CHCH=CH_2$ (alkadiene)

78. a. $CH_3CH_2C(CH_3)=C(CH_3)CH_2CH_3$

b.

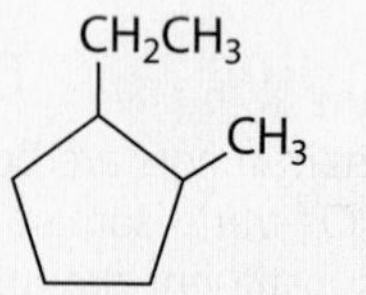

c. $CH_3(CH_2)_3C(CH_2CH_2CH_3)_2(CH_2)_4CH_3$

79. Alkanes contain only single bonds. Alkenes contain at least one double bond. Aromatic hydrocarbons contain a benzene ring or a similar ring. Cycloalkanes contain aliphatic chains that are arranged in a ring.

80. b.

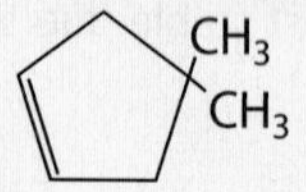

c. $CH_3-C\equiv C-CH_2-CH_3$

d. $CH_2=CH-CH_2-CH_3$

81. The alkane is ethane.

$2CH_3CH_3 + 7O_2 \rightarrow 4CO_2 + 6H_2O$

***71.** Use the labeled features in the molecular structure below to answer the following questions.

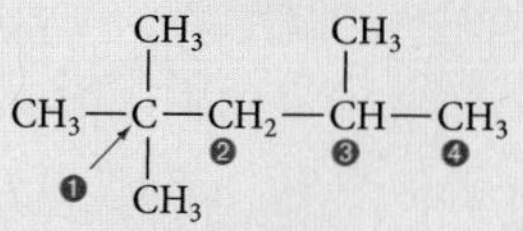

a. Which label identifies a primary carbon?
b. Which label identifies a secondary carbon?
c. Which label identifies a tertiary carbon?
d. Which label identifies a quaternary carbon?

Think Critically

***72. Infer** Methane (CH_4), a widely used fuel, has a heat of combustion (ΔH) of −890 kJ/mol. The ΔH for benzene (C_6H_6) is much higher, −3268 kJ/mol, yet benzene alone is never used as a fuel. Suggest some reasons why benzene is a less desirable fuel than methane.

73. Analyze Explain why you cannot draw a structural formula for methene.

74. Use Models Use the isomers of 2-pentene to show how lack of rotation about a carbon-carbon double bond leads to *cis-trans* isomers.

75. Infer Most cyclic hydrocarbons have higher boiling points than alkanes with the same number of carbons. Suggest a possible explanation for this general difference in boiling points.

***76. Apply Concepts** Alkadienes are hydrocarbons with two double bonds. Draw the structural formula of the alkadiene with the molecular formula C_3H_4.

77. Predict The molecular formula C_4H_6 could represent an alkyne, a cycloalkene, or a hydrocarbon with two double bonds. Write a condensed structural formula for each. Which compound do you think is the least stable? Why?

***78. Apply Concepts** Draw structural formulas for the following compounds:

a. 3,4-dimethyl-3-hexene
b. 1-ethyl-2-methylcyclopentane
c. 5,5-dipropyldecane

79. Compare What structural feature is associated with each of these hydrocarbons: an alkane, an alkene, an aromatic hydrocarbon, and a cycloalkane?

80. Evaluate and Revise Draw a correct structure for any of the choices below that are incorrect.

a. $CH_3-CH=CH-CH_2-CH_3$

b.

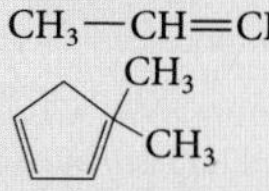

c. $CH_3-C\equiv CH-CH_2-CH_3$

d. $CH_3=CH-CH_2-CH_3$

81. Apply Concepts Upon complete combustion, a gaseous alkane gives a volume of carbon dioxide that is twice the volume of the starting alkane at the same temperature and pressure. Identify the alkane and write the balanced equation for its combustion.

***82. Apply Concepts** The alkanes 2-methylbutane and pentane are readily interconverted, or isomerized, in the presence of a catalyst.

a. Write a balanced chemical equation for this isomerization reaction.
b. What kind of isomers are 2-methylbutane and pentane?
c. Which isomer most likely has the lowest boiling point?

83. Interpret Graphs The graph shows the relationship between temperature and the composition of equilibrium mixtures of 2-methylbutane and pentane in the presence of a catalyst.

a. The equilibrium mixtures contain the most of which isomer?
b. Could you obtain better yields of the isomer 2-methylbutane by running the reaction at higher or lower temperatures?
c. At what temperature could you run the reaction to obtain a mixture that contains about 25 percent pentane and 75 percent 2-methylbutane?

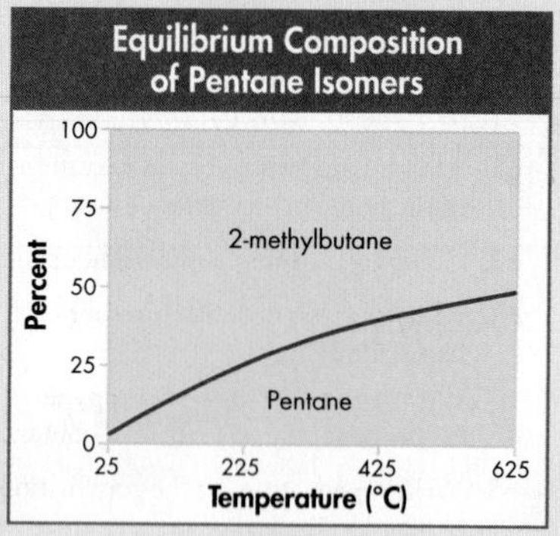

82. a. $CH_3CH_2CH_2CH_2CH_3 \rightleftharpoons CH_3CH(CH_3)CH_2CH_3$

b. constitutional isomers
c. 2-methylbutane

83. a. 2-methylbutane
b. lower temperatures
c. about 225°C

Enrichment

84. **Predict** Use the graph in Figure 22.3 to predict the boiling point of dodecane, the straight-chain alkane containing twelve carbons. Then use a chemistry handbook to find the actual boiling point of dodecane. Compare the actual boiling point with your prediction.

85. **Apply Concepts** Correct each of the following names and draw the correct structural formulas.
    a. 4-methylhexane
    b. 1,4-diethyl cyclopentane
    c. 3,3methyl-4-ethyloctane
    d. 4,4-dimethylpentane
    e. 2-ethylheptane
    f. phenylmethane

*86. **Interpret Graphs** The graph shows the number of constitutional isomers for alkanes with three to ten carbon atoms.
    a. How many constitutional isomers are there for the C_6, C_7, C_8, C_9, and C_{10} alkanes?
    b. The difference between the number of isomers for C_7 and C_8 is 9. The difference between the number of isomers for C_9 and C_{10} is 40. In each case, one additional carbon atom is added to the molecule. Why is the change in the number of isomers so different?

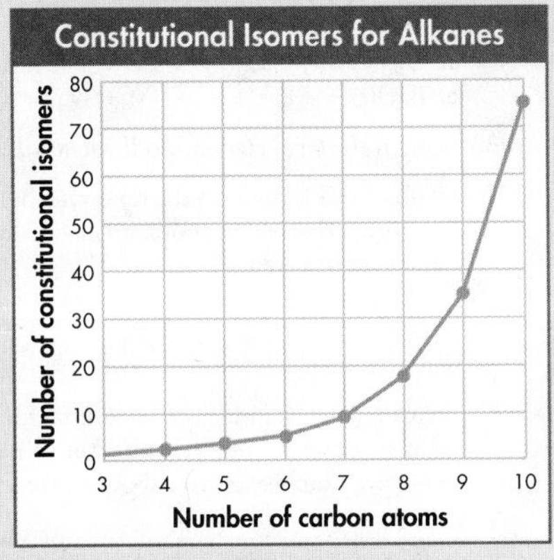

87. **Evaluate** Fossil fuels such as oil and natural gas are the raw materials for many consumer products. Should this information affect the decision to develop energy sources other than fossil fuels? Explain.

Write About Science

88. **Research a Problem** When fossil fuels burn in an internal combustion engine, the exhaust contains more than carbon dioxide gas and water vapor. Research how catalytic converters work, and write a paragraph explaining what happens to pollutants in a catalytic converter.

89. **Connect to the BIGIDEA** You've probably seen the instruction "Shake well before using" on a salad dressing bottle. These instructions usually appear on dressings that contain olive oil and water as two of the main ingredients. The molecules in olive oil have long hydrocarbon tails. Explain why you need to shake these types of salad dressings before pouring them onto your salad.

CHEMYSTERY

Nose for Hire

After finishing up with the smell test, Anthony immediately went home and jumped on the Internet to research limonene. He discovered that the two limonene compounds were enantiomers. The right-handed enantiomer smells like oranges, and the left-handed one smells like pine trees. He read on to learn that a nose, like all human tissue, is composed of molecules with asymmetric carbons. As a result, the molecules in your nose that are responsible for detecting smell react differently to the two enantiomers of limonene.

This phenomenon is common in compounds with asymmetric carbons. Another example is the two enantiomers of the compound carvone. One enantiomer smells like spearmint and the other enantiomer smells like the spice caraway.

*90. **Infer** Do physical properties such as color, density, and boiling point differ for the two enantiomers of limonene?

91. **Connect to the BIGIDEA** Many medications have asymmetric carbons. There are cases in which one enantiomer helps, while the other causes severe harm. How is this occurrence possible?

CHEMYSTERY

MAKE A CONNECTION Direct students to read the CHEMystery. Explain that humans have an area in the back of their nasal cavities that is layered with chemically sensitive nerve cells. These nerve cells react with the emitted molecules and send a signal off to the brain, triggering the involuntary smell response.

Since humans are not alike, it makes sense that each person's response to a smell is not always the same as other people's responses. Guide interested students to research the work of British scientist John Amoore, his conclusions about people's perception of smell, and a condition called anosmia. Amoore's findings account for why some people have a stronger sense of smell than other people have.

CHEMYSTERY ANSWERS

90. No, enantiomers have identical physical properties.

91. **BIGIDEA** Medications must often interact with a particular protein, which also has asymmetric carbons. Sometimes, one enantiomer of the medication is metabolized differently or interacts with a completely different protein, causing adverse effects.

Answers

ENRICHMENT

84. The estimated boiling point should be greater than 200°C. The actual boiling point is 216°C.

85. a.
```
              CH3
              |
CH3CH2CHCH2CH2CH3
```
3-methylhexane

b.
```
  CH2CH3
  (cyclopentane ring)
        CH2CH3
```
1,3-diethylcyclopentane

c.
```
         CH3 CH2CH3
          |   |
CH3CH2—C—CHCH2CH2CH2CH3
          |
         CH3
```
3,3-dimethyl-4-ethyloctane

d.
```
           CH3
            |
CH3CH2CH2CCH3
            |
           CH3
```
2, 2-dimethylpentane

e.
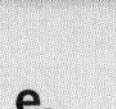
```
   CH2CH3
   |
CH3CHCH2CH2CH2CH2CH3
```
3-methyloctane

f.
```
 CH3
  |
(benzene ring)
```
Methylbenzene

86. a. C_6 = 5, C_7 = 9, C_8 = 18, C_9 = 35, C_{10} = 75

b. As the molecule gets larger, the number of different ways that the carbon atoms can be bonded together to form constitutional isomers increases dramatically.

87. Answers will vary.

WRITE ABOUT SCIENCE

88. Answers should mention that a catalytic converter uses a catalyst to help oxidize some of the toxic by-products of combustion and turn them into less toxic compounds.

89. The long hydrocarbon tails on the molecules in olive oil are nonpolar and do not mix with the polar water molecules.

ASSESSMENT

CUMULATIVE REVIEW

92. a. 13.9 L
b. 1 L
c. 20 kPa

93. 1.13 mol KNO_3; 114 g KNO_3

94. 3.03×10^4 calories

95. 1.67×10^2 J

96. 1 cal = 4.184 J; 1 kcal = 4.184×10^3 J

97. a. Decreasing particle size speeds up the reaction.
b. Increasing the temperature usually speeds up the reaction.

98. a. favors reactants **b.** favors products

99. a. $K_{eq} = \dfrac{[ICl]^2}{[I_2][Cl_2]}$

b. $K_{eq} = \dfrac{[H_2][Br_2]}{[HBr]^2}$

c. $K_{eq} = \dfrac{[HCl]^4[S]^3[SO_2]}{[S_2Cl_2]^2[H_2O]^2}$

d. $K_{eq} = \dfrac{[NH_3]^2}{[N_2][H_2]^3}$

100. a. 10.00 **b.** 7.59
c. 12.00 **d.** 11.70

101. a. ~8 **b.** Use a pH meter.

102. a. H_3PO_4 **b.** CsOH **c.** H_2CO_3 **d.** $Be(OH)_2$

103. a. $NaOH(aq) \rightarrow Na^+(aq) + OH^-(aq)$
b. $Ba(OH)_2(aq) \rightarrow Ba^{2+}(aq) + 2OH^-(aq)$

104. a. Ca, +2; C, +4; O, −2
b. Cl, 0
c. Li, +1; I, +5; O, −2
d. Na, +1; S, +4; O, −2

105. a. reduction **b.** reduction **c.** reduction
d. oxidation

106. a. +4 **b.** +5
c. +4 **d.** −3
e. −3 **f.** +2

107. a. The coefficients are 2, 9, 6, and 8.
b. The coefficients are 3, 2, 1, and 3.

108. It is the cell potential when the ion concentrations in the half-cells are 1*M*, the temperature is 25°C, and the pressure of any gas present is 101.3 kPa.

109. Reduction occurs in the half-cell with the more positive, or less negative, reduction potential. Ni^{2+} is reduced, and Al is oxidized.
Cell reaction: $3Ni^{2+}(aq) + 2Al(s) \rightarrow 3Ni(s) + 2Al^{3+}(aq)$; $E°_{cell} = +1.41$ V

110. The reaction is nonspontaneous.

111. reduction; reduction

Cumulative Review

***92.** Calculate the following quantities:

a. The number of liters occupied at STP by 6.20×10^{-1} mol $Cl_2(g)$
b. The volume of a gas at 3 kPa of pressure if the same gas has a volume of 6 L at 0.5 kPa and the temperature is constant
c. The partial pressure of gas X (P_X) in a mixture of three gases, X, Y, and Z, if the total pressure (P_{total}) is 50 kPa and the sum of the partial pressures of Y and Z is 30 kPa

***93.** How many moles of solute are in 750 mL of 1.50*M* KNO_3? How many grams of KNO_3 is this?

94. How many calories are absorbed when 56.0 g of liquid water at 100°C is vaporized to steam?

95. A silver dollar is heated and placed in a foam cup calorimeter containing 50.0 mL of water at 26.5°C. The water reached a maximum temperature of 27.3°C. How many joules of heat were released by the silver dollar?

96. What is the relationship between a calorie and a joule? How many joules is 1 kcal?

97. How does (a) particle size and (b) temperature affect the rate of a chemical reaction?

***98.** Explain how the equilibrium position of this reaction is affected by (a) decreasing the temperature and (b) removing CO_2.

$CaCO_3(s) + \text{heat} \rightleftharpoons CaO(s) + CO_2(g)$

99. Write equilibrium constant expressions for the following reactions:

a. $Cl_2(g) + I_2(g) \rightleftharpoons 2ICl(g)$
b. $2HBr(g) \rightleftharpoons H_2(g) + Br_2(g)$
c. $2S_2Cl_2(g) + 2H_2O(g) \rightleftharpoons 4HCl(g) + 3S(g) + SO_2(g)$
d. $N_2(g) + 3H_2(g) \rightleftharpoons 2NH_3(g)$

***100.** What are the pH values for aqueous solutions containing each of the following hydroxide-ion concentrations?

a. $1.0 \times 10^{-4}M$ **c.** $0.010M$
b. $3.9 \times 10^{-7}M$ **d.** $0.0050M$

101. A colorless solution of unknown pH turns blue when tested with the acid-base indicator bromothymol blue. It remains colorless when tested with phenolphthalein.

a. What is the approximate pH of the solution?
b. How could you determine the pH more accurately?

***102.** Write the formula for each acid or base.

a. phosphoric acid **c.** carbonic acid
b. cesium hydroxide **d.** barium hydroxide

103. Write the reaction for the dissociation of each of the following compounds in water.

a. sodium hydroxide **b.** barium hydroxide

***104.** Give the oxidation number of each element in the following substances:

a. $CaCO_3$ **c.** $LiIO_3$
b. Cl_2 **d.** Na_2SO_3

105. Are these processes oxidation or reduction?

a. $Fe^{3+} + e^- \rightarrow Fe^{2+}$
b. $Cl_2 + 2e^- \rightarrow 2Cl^-$
c. $Fe^{3+} + 3e^- \rightarrow Fe$
d. $Zn \rightarrow Zn^{2+} + 2e^-$

106. Determine the oxidation number of nitrogen in the following compounds and ions:

a. N_2O_4 **c.** NO_2 **e.** NH_3
b. NO_3^- **d.** NH_4^+ **f.** NO

107. Balance these redox equations.

a. $C_3H_7OH(l) + O_2(g) \rightarrow CO_2(g) + H_2O(l)$
b. $BaO(s) + Al(s) \rightarrow Al_2O_3(s) + Ba(s)$

108. Explain the term *standard cell potential.*

109. A voltaic cell is made of the following half-cells. Determine the cell reaction and calculate the standard cell potential.

$Al^{3+}(aq) + 3e^- \rightarrow Al(s)$ $E°_{Al^{3+}} = -1.66$V
$Ni^{2+}(aq) + 2e^- \rightarrow Ni(s)$ $E°_{Ni^{2+}} = -0.25$V

***110.** The calculated standard cell potential for a redox reaction is a negative number. What does a negative number tell you about the reaction?

111. What process always occurs at the cathode of an electrolytic cell? At the cathode of a voltaic cell?

If You Have Trouble With . . .

Question	92	93	94	95	96	97	98	99	100	101	102	103	104	105	106	107	108	109	110	111
See Chapter	14	16	17	17	17	18	18	18	19	19	19	19	20	20	20	20	21	21	21	21

794 Chapter 22 • Assessment

Standardized Test Prep for Indiana

Standard: C.3.4

Select the choice that best answers each question or completes each statement.

1. What is the name of the compound with the following structural formula?

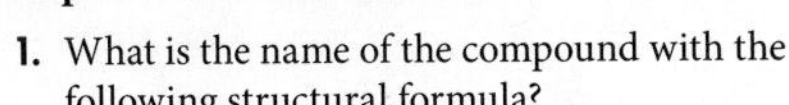

```
  CH3 H  CH3
   |  |   |
CH3—C—C—C—CH3
   |  |   |
   H  H   H
```

(A) 1,2,3,3-tetramethylpropane
(B) heptane
(C) 2,4-dimethylpentane
(D) 1,5-dimethylbutane

2. Which of these are characteristic of all alkenes?
   I. unsaturated
   II. carbon-carbon double bond
   III. enantiomers

(A) I and II only
(B) II and III only
(C) I and III only
(D) I, II, and III

3. How many carbon atoms are in a molecule of 4,5-diethyloctane?
(A) 10 (C) 14
(B) 12 (D) 16

4. *Cis-trans* isomerism is possible in
(A) 2-pentene. (C) propyne.
(B) 2-butane. (D) benzene.

Tips for Success

Eliminate Wrong Answers If you don't know which response to a question is correct, start by eliminating those you know are wrong. If you can rule out some choices, you'll have fewer choices left to consider, which will increase your chances of choosing the correct answer.

5. A constitutional isomer of heptane is
(A) methylbenzene.
(B) 3,3-dimethylpentane.
(C) cycloheptane.
(D) 3-methylhexene.

6. Which molecule can have enantiomers?
(A) CH_4 (C) CFClBrI
(B) CF_2H_2 (D) CF_2ClH

7. Draw structural formulas for three constitutional isomers of pentane, C_5H_{12}. Name each isomer.

8. Write structural formulas for the four constitutional isomers of cyclopentane. Name each of the isomers.

The lettered choices below refer to Questions 9–12. A lettered choice may be used once, more than once, or not at all.

(A) alkene
(B) arene
(C) alkyne
(D) alkane

To which of the above classes of hydrocarbons does each of the following compounds belong?

9. C_7H_{16}
10. C_5H_8
11. C_6H_6
12. C_8H_{16}

Use the molecular structures below to answer Questions 13–16. A molecular structure may be used once, more than once, or not at all.

(A) 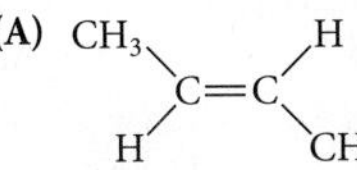

(C) 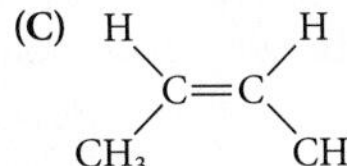

(B) $CH_3CH_2CH_2CH_3$

(D) 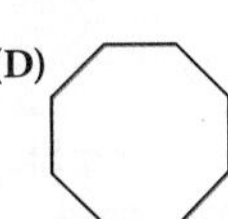

13. Which structure is a cycloalkane?
14. Which structure is a saturated hydrocarbon?
15. Which structure is a *cis*-isomer?
16. Which structure is a *trans*-isomer?

If You Have Trouble With . . .

Question	1	2	3	4	5	6	7	8	9	10	11	12	13	14	15	16
See Lesson	22.1	22.2	22.1	22.3	22.3	22.3	22.3	22.3	22.1	22.2	22.4	22.2	22.4	22.2	22.3	22.3

STP Answers

1. C
2. A
3. B
4. A
5. B
6. C
7.

```
  H H H H H
  | | | | |
H—C—C—C—C—C—H
  | | | | |
  H H H H H
```

pentane

```
  H H H H
  | | | |
H—C—C—C—C—H
  | | | |
  H | H H
  H—C—H
    |
    H
```

2-methylbutane

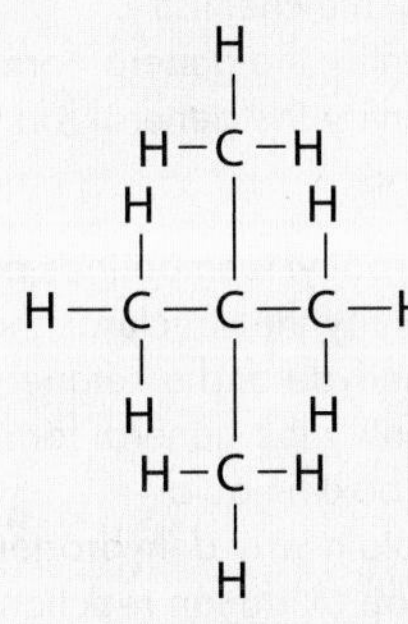

2,2-dimethylpropane

8.

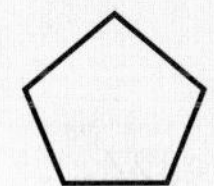

cyclopentane

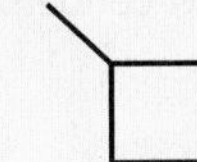

methylcyclobutane

1,1-dimethylcyclopropane

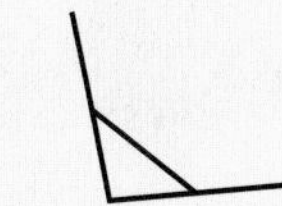

1,2-dimethylcyclopropane

9. D
10. C
11. B
12. A
13. D
14. B, D
15. C
16. A

23 Functional Groups Planning Guide

Introducing the BIGIDEA: CARBON CHEMISTRY; REACTIONS

Carbon compounds can have functional groups, which have an impact on the properties of the compound, including its chemical reactivity.

IASS	Lessons and Objectives	Print Resources		
		For the Student	For the Teacher	
C.4.2, C.4.5, C.9.2	**23.1** Introduction to Functional Groups p 798–802 **23.1.1** Classify organic compounds. **23.1.2** Identify the general formula of a halocarbon. **23.1.3** Describe how substitution reactions are used in organic chemistry.	Reading and Study Workbook Lesson 23.1 Lesson Assessment 23.1 p 802	Class Activity, p 799: Functional Groups Class Activity, p 800: Halocarbon Structures and Names Teacher Demo, p 801: Testing for Halocarbons	
C.4.5	**23.2** Alcohols, Ethers, and Amines p 804–811 **23.2.1** Identify the general formula of an alcohol. **23.2.2** Explain how addition reactions are used in organic chemistry. **23.2.3** Identify the general formula of an ether. **23.2.4** Identify the general formula of an amine.	Reading and Study Workbook Lesson 23.2 Lesson Assessment 23.2 p 811	Teacher Demo, p 806: Are Alcohols Basic? Class Activity, p 807: Household Products with Alcohols Teacher Demo, p 808: Test for Unsaturation	
C.4.5	**23.3** Carbonyl Compounds p 812–820 **23.3.1** Identify the structural characteristic that an aldehyde and a ketone share. **23.3.2** Identify the general formula of a carboxylic acid. **23.3.3** Explain why dehydrogenation is classified as an oxidation reaction. **23.3.4** Identify the general formula of an ester.	Reading and Study Workbook Lesson 23.3 Lesson Assessment 23.3 p 820 Quick Lab: Testing for an Aldehyde p 818	Class Activity, p 813: Odors of Functional Groups Teacher Demo, p 817: Oxidation of Alcohols Teacher Demo, p 819: Making an Ester	
	23.4 Polymers p 822–827 **23.4.1** Describe how addition polymers are formed. **23.4.2** Describe how condensation polymers are formed.	Reading and Study Workbook Lesson 23.4 Lesson Assessment 23.4 p 827 Small-Scale Lab: Polymers p 828	Teacher Demo, p 824: Building a Polymer Model Teacher Demo, p 826: Making Nylon	

Assessing the BIGIDEA: CARBON CHEMISTRY; REACTIONS

	Essential Questions 1. What effect does a functional group have on an organic compound? 2. How are chemical reactions used in organic chemistry?	Study Guide p 829 STP p 835 Reading and Study Workbook Self-Check and Vocabulary Review Chapter 23		

	PearsonChem.com	
	Editable Worksheets	Online Activities
	Teaching Resources, Lesson 23.1 Review	23.1 Lesson Overview By Organic Compound, We Don't Mean a Campground Full of People Growing Veggies Without Pesticides
	Teaching Resources, Lesson 23.2 Review	23.2 Lesson Overview Hydration and Hydrogenation
	Teaching Resources, Lesson 23.3 Review Lab 50: Esters of Carboxylic Acids Small-Scale Lab Manual Lab 38: Vitamin C in Tablets Small-Scale Lab Manual Lab 39: Vitamin C in Drinks Lab Practical 23-1: Esters of Carboxylic Acids	23.3 Lesson Overview Properties of Organic Molecules Redox Reactions
	Teaching Resources, Lesson 23.4 Review	23.4 Lesson Overview
	Core Teaching Resources (includes editable worksheets) • Lesson Reviews • Practice Problems • Interpret Graphs • Vocabulary Review • Chapter Quizzes and Tests	Chapter 23 Problem Set

Materials List

For the Student

Quick Lab p 818
- 1M NaOH
- 5% silver nitrate
- 6M aqueous ammonia
- small test tubes
- test tube rack
- droppers
- glucose solution
- ethanol

Small-Scale Lab p 828
- 3.5-oz cup
- soda straw
- powdered guar gum
- 4% borax solution
- pipet

For the Teacher

Class Activity p 799
- 2-propanol
- propanoic or butanoic acid
- butyl amine
- vanillin or cinnamaldehyde
- octyl acetate or ethyl butyrate

Teacher Demo p 801
- copper wire
- Bunsen burner
- forceps
- bromobenzene
- benzoic acid

Teacher Demo p 806
- several alcohols
- litmus paper

Class Activity p 807
- common household products that contain alcohols

Teacher Demo p 817
- CrO_3
- concentrated sulfuric acid
- water
- 3 test tubes
- acetone
- ethanol
- isopropanol
- *t*-butyl alcohol

Teacher Demo p 807
- 10 or 15 test tubes with stoppers
- distilled water
- ethanol
- 1-butanol
- 1-hexanol
- 1-octanol
- a hexanediol

Teacher Demo p 808
- $KMnO_4$
- NaOH
- distilled water
- cyclohexane
- cyclohexene
- 2 test tubes with stoppers

Class Activity p 813
- benzaldehyde
- isopentyl acetate
- vanillin

Teacher Demo p 819
- water bath
- hot plate
- test tubes
- several carboxylic acids
- several corresponding alcohols
- concentrated sulfuric acid
- filter paper
- watch glass

Additional Digital Resources

Online Student Edition
Online Teacher's Edition
Exam*View Assessment Suite CD-ROM*
Classroom Resources DVD

CHAPTER 23

What's Online

ONLINE PROBLEM SET Additional practice problems about functional groups can be found in the Chapter 23 online Problem Set.

DIRECTED VIRTUAL LAB Encourage students to review the effects of different functional groups on the boiling points of organic compounds using the animated virtual lab for Chapter 23. The animation is followed by analysis questions, where the student is guided through the lab concepts.

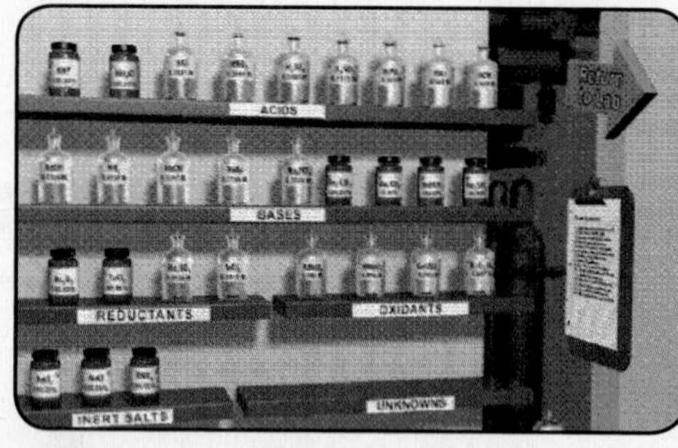

KINETIC ART Have students watch animations about hydration and hydrogenation and redox reactions that occur with organic compounds. Each animation is followed by analysis questions to guide further understanding of the concepts.

CONCEPTS IN ACTION Students can watch an interactive animation about how chemists classify compounds as organic, as compared to the common understanding of the term. The animation is followed by questions to explore characteristics and misunderstandings about organic compounds.

IN ACADEMIC STANDARDS for SCIENCE

Nature of Science

- Develop explanations based on reproducible data and observations gathered during laboratory investigations.
- Clearly communicate their ideas and results of investigations verbally and in written form using tables, graphs, diagrams, and photographs.

Standard 4: Reactions and Stoichiometry

C.4.5 Recognize and classify types of reactions.

Also covered C.4.2, C.9.2

23 Functional Groups

INSIDE:

PearsonChem.com

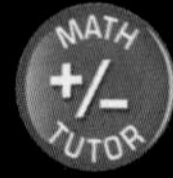

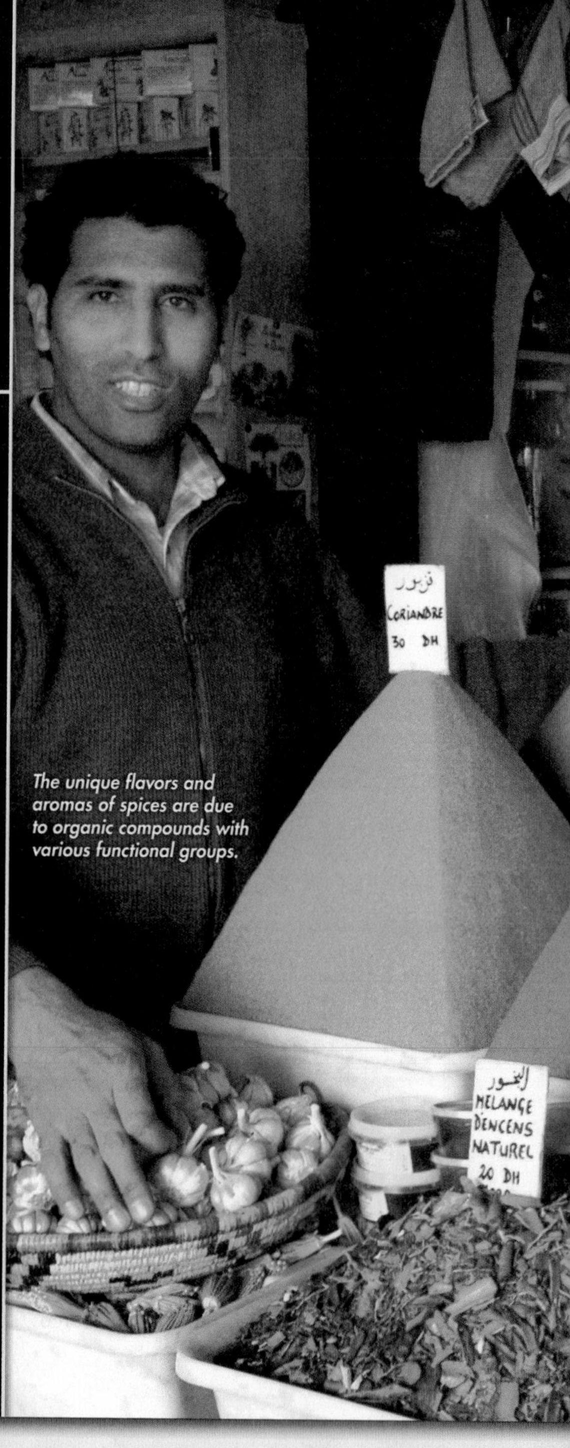

The unique flavors and aromas of spices are due to organic compounds with various functional groups.

796

Focus on ELL

1 CONTENT AND LANGUAGE Read aloud Essential Question 1, emphasizing the word *effect*. Have students each prepare a cause and effect diagram to organize their answer to this question as they read the chapter. Read aloud Essential Question 2, emphasizing the word *used*. Have students each prepare a cluster diagram to organize their answer to this question as they read the chapter.

BEGINNING

LOW Make a non-linguistic representation of each essential question.

HIGH Allow students to complete the charts in their native language.

INTERMEDIATE: LOW/HIGH Ask students to complete their charts by verbalizing predictions based on their knowledge from Chapter 22.

ADVANCED: LOW/HIGH Have students preview lesson titles, key questions, and vocabulary for information they can use to fill in their charts.

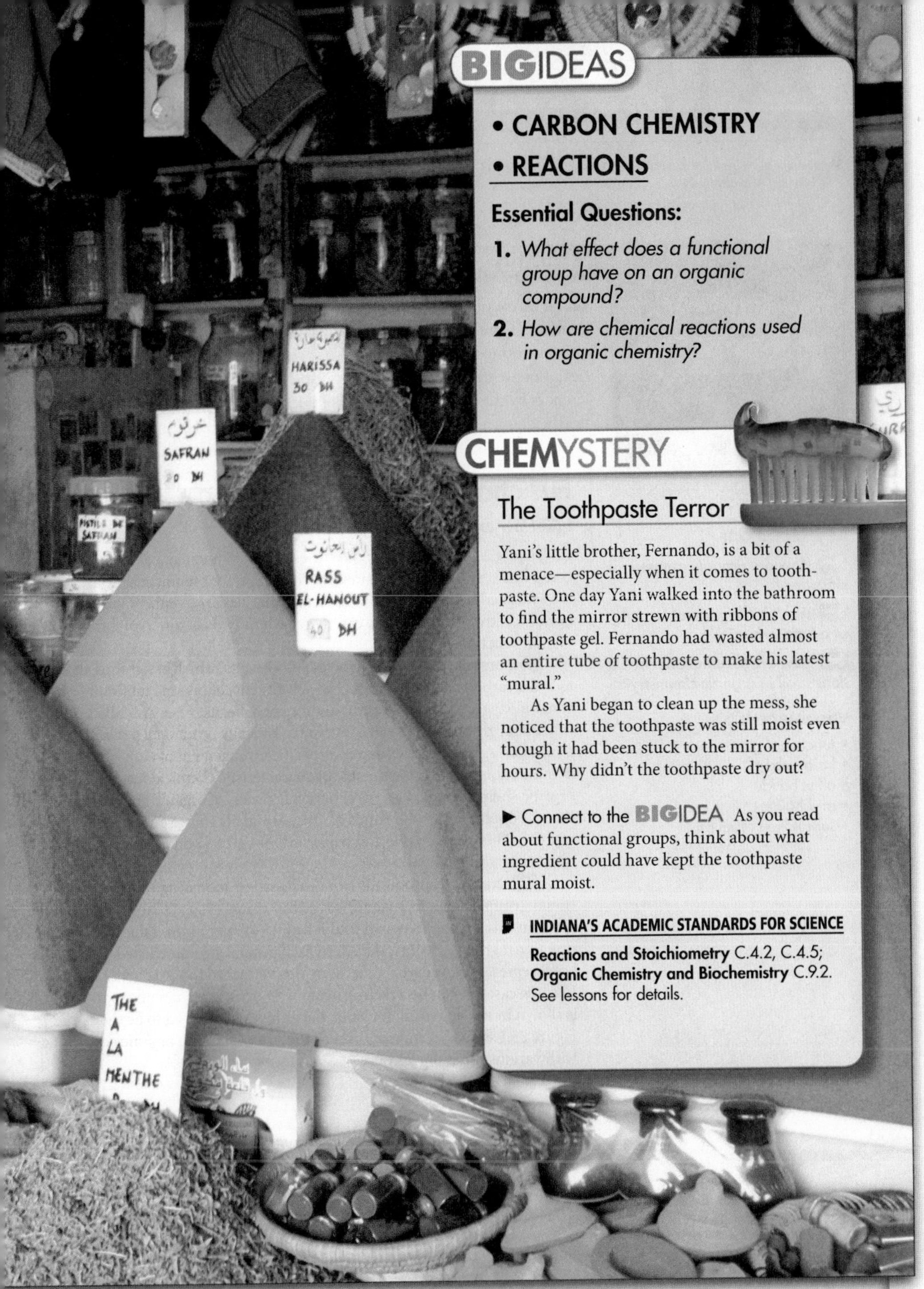

BIGIDEAS

- **CARBON CHEMISTRY**
- **REACTIONS**

Essential Questions:

1. *What effect does a functional group have on an organic compound?*
2. *How are chemical reactions used in organic chemistry?*

CHEMYSTERY

The Toothpaste Terror

Yani's little brother, Fernando, is a bit of a menace—especially when it comes to toothpaste. One day Yani walked into the bathroom to find the mirror strewn with ribbons of toothpaste gel. Fernando had wasted almost an entire tube of toothpaste to make his latest "mural."

As Yani began to clean up the mess, she noticed that the toothpaste was still moist even though it had been stuck to the mirror for hours. Why didn't the toothpaste dry out?

▶ Connect to the **BIGIDEA** As you read about functional groups, think about what ingredient could have kept the toothpaste mural moist.

INDIANA'S ACADEMIC STANDARDS FOR SCIENCE

Reactions and Stoichiometry C.4.2, C.4.5; **Organic Chemistry and Biochemistry** C.9.2. See lessons for details.

CHAPTER 23

UbD

Understanding by Design

Students are building toward understanding chemical reactions in carbon chemistry using the relationships between organic compounds and their functional groups.

PERFORMANCE GOALS At the end of Chapter 23, students will be able to answer the essential questions by applying their knowledge of functional groups.

ESSENTIAL QUESTIONS Read the essential questions aloud. **Ask** How do you think an alkane (with only single bonds) might be affected if you replaced one of its hydrogen atoms with a halogen atom? *(The product would have greater mass and a higher boiling point.)* **Ask** How do you think adding a group of atoms to an alkene, which has double bonds, might be able to change it to an alkane? *(An added atom might bond with one of the double-bonded carbon atoms, reducing the carbon-carbon double bond to a single bond.)*

BIGIDEA Use the photo of the spices to help students connect to this lesson's concepts. Activate prior knowledge by asking which spices students like in various foods. **Ask** What do you think makes them smell and taste different? *(They must have some different atoms or groups of atoms.)*

CHEMYSTERY Connect the CHEMystery to the Big Idea by reminding students that chemical reactions create new products with different properties from the reactants. Note that sometimes this change can be achieved by replacing a single atom or by joining whole molecules together. Ask students to predict why the toothpaste did not dry out.

Introduce the Chapter

REAL-WORLD CONNECTIONS Students may not realize that many products they frequently use contain compounds with functional groups. Use the activities to relate the chapter to their everyday lives.

Activity 1 *Display a variety of common products that contain alcohols. Examples include hand lotion with propylene glycol, lilac perfume with cinnamic alcohol, toothpaste with glycerol, and shaving lotion with menthol.) Allow students to smell the various products.* **Ask** What do you think all of these products have in common? *(Students may offer various opinions.)* Explain that each of them contains a hydrocarbon with a hydroxyl group, —OH.

Activity 2 *You will need small bottles filled with various artificial flavors with the labels covered.* Allow time for students to smell the items and try to guess the flavors. Tell students that all of these contain a hydrocarbon with a group called an ester. Draw an ester group on the board. Explain that each flavor depends on the organic compound to which the ester group is attached.

UntamedScience™

Protection From the Flames

Erica visits a polymer chemist who helps protect firefighters while they put out fires.

LESSON 23.1

Key Objectives

23.1.1 EXPLAIN how organic compounds are classified.

23.1.2 IDENTIFY the general formula of a halocarbon.

23.1.3 DESCRIBE how substitution reactions are used in organic chemistry.

Additional Resources

- Reading and Study Workbook, Lesson 23.1
- Probeware Laboratory Manual
- Virtual Chemistry Labs

PearsonChem.com

Engage

CHEMISTRY & YOU Direct students to the photograph and the text that opens the lesson. **Ask** Using the context of the band, predict how the functional groups determine the character of organic compounds. *(Each functional group has certain chemical properties that affect the behavior of any compound that contains the functional group.)*

Activate Prior Knowledge

Have students review what they know about hydrocarbons and substituents from Chapter 22. Explain that in this lesson, students will explore non-carbon substituents and the reactions that are used to add them to hydrocarbons.

Foundations for Reading

BUILD VOCABULARY Use set notation or a Venn diagram to convey the interrelationship among the vocabulary words.

IN ACADEMIC STANDARDS for SCIENCE

C.4.2 Balance chemical equations using the law of conservation of mass and use them to describe chemical reactions.

C.4.5 Describe, classify and give examples of various kinds of reactions-synthesis (combination), decomposition, single displacement, double displacement and combustion.

C.9.2 Identify the variety of molecular types formed by the covalent bonding of carbon atoms and describe the typical properties of these molecular types.

23.1 Introduction to Functional Groups

C.4.5 Classify types of reactions; **C.9.2** Identify the variety of carbon molecules **Also covered** C.4.2.

CHEMISTRY & YOU

Q: *How can you tell the difference between one organic compound and another?* If you've ever seen a marching band, you know that all the members wear the same uniform. From a distance, all the musicians look nearly identical. But are they? If you look more closely, you can distinguish their differences by the instruments they hold. In a similar way, one hydrocarbon is nearly identical to another until it is distinguished by a functional group.

Key Questions

How are organic compounds classified?

What is the general formula of a halocarbon?

How are substitution reactions used in organic chemistry?

Vocabulary

- functional group
- halocarbon
- alkyl halide
- aryl halide
- substitution reaction

Functional Groups

How are organic compounds classified?

In Chapter 22, you learned about the essential components of every organic compound: hydrocarbon chains and rings. In most organic reactions, the saturated hydrocarbon skeletons of molecules are chemically inert, or nonreactive. So how, then, can there be hundreds of different kinds of organic reactions?

Most organic chemistry involves substituents, which are groups attached to hydrocarbon chains or rings. The substituents of organic molecules often contain oxygen, nitrogen, sulfur, and/or phosphorus. They are called functional groups because they are the chemically functional parts of the molecules. A **functional group** is a specific arrangement of atoms in an organic compound that is capable of characteristic chemical reactions. Most organic chemistry involves the functional groups of organic molecules. Note that the double and triple bonds of alkenes and alkynes are chemically reactive. Therefore, double and triple carbon-carbon bonds are considered functional groups.

Organic compounds can be classified according to their functional groups. Table 23.1 identifies the functional groups that you will learn about in this chapter. You will find it helpful to refer to this table as new functional groups are introduced. In each general structure listed, the symbol R represents any carbon chains or rings attached to the functional group. In some cases, R can be a hydrogen atom. When more than one R group is shown in the structural formula, the groups do not need to be the same. Figure 23.1 shows several consumer products containing organic compounds with various functional groups.

Focus on ELL

1 CONTENT AND LANGUAGE Rephrase the key questions on the board using familiar synonyms for *classified* and *general*. Explain how you substituted one word in the question for another to help explain the meaning of substitution.

2 FRONTLOAD THE LESSON Ask the class to preview Table 23.1. Make sure students understand that the letter "R" in each the structure in the middle column represents a hydrocarbon segment or chain. Ask students to find the sentence or sentences in the lesson that refer to Table 23.1. Discuss the meaning of the sentences.

3 COMPREHENSIBLE INPUT When discussing the Substitution Reactions section, rewrite the example equations by expanding the condensed formulas into structural formulas. Use different colors for each reactant and carry these colors into the products.

Table 23.1

Organic Compounds Classified by Functional Group

Compound type	General structure	Functional group
Halocarbon	R—X (X = F, Cl, Br, or I)	Halogen
Alcohol	R—OH	Hydroxy
Ether	R—O—R	Ether
Amine	$R—NH_2$	Amino
Aldehyde	R—C(=O)—H	Carbonyl
Ketone	R—C(=O)—R	Carbonyl
Carboxylic acid	R—C(=O)—OH	Carboxyl
Ester	R—C(=O)—O—R	Ester
Amide	R—C(=O)—N(R)—R	Amide

Figure 23.1 Consumer Products Many items contain hydrocarbon derivatives. The hydrocarbon skeletons in these products are chemically similar. Functional groups give each product unique properties and uses.

UbD Check for Understanding

How are organic compounds classified?

Assess students' understanding about the classification of organic compounds by drawing various functional groups on the board, and ascertaining via a thumbs-up/thumbs-down test whether students know the identity of each one.

ADJUST INSTRUCTION If students are indicating a significant number of thumbs-down responses, have them review Table 23.1. Repeat the exercise until all students consistently recognize functional groups by their names and structures.

Explain

Functional Groups

START A CONVERSATION Stress the importance of structure in the chemistry of organic molecules. Explain that molecular shape is dictated by functional groups. **Ask** How does the structure of a molecule relate to its taste or smell? *(Both senses rely on the ability of chemoreceptor molecules on cell surfaces, which recognize a molecule by its shape.)*

USE VISUALS Have students examine Table 23.1 for patterns and relationships. Convey the meaning of the symbols *R* and *X*, and encourage students to memorize the general compound structure for each functional group. List specific examples of compounds for each functional group to help students become more familiar with these groups in larger molecules. **Ask** Why are the carbon–carbon double and triple bonds considered to be functional groups? *(They are chemically reactive.)* **Ask** What do most of the functional groups in the table have in common? *(Most functional groups contain oxygen.)*

Explore

Class Activity

PURPOSE Students will learn that compounds containing the same functional group have similar chemical and physical properties.

MATERIALS 2-propanol, propanoic or butanoic acid, butyl amine, vanillin or cinnamaldehyde, octyl acetate or ethyl butyrate

SAFETY Have students carefully waft vapors of the compounds in order to detect their odors. Remind students that normally it is not safe to smell chemicals. An exception is being made for these known compounds.

PROCEDURE Bring samples of organic compounds to class. Allow students to observe the characteristic odors associated with alcohols (2-propanol), acids (propanoic or butanoic acid), amines (butyl amine), aldehydes (vanillin or cinnamaldehyde), and esters (the orange odor of octyl acetate and the pineapple odor of ethyl butyrate). Have students record their observations, and then make a summary table on the board. Ask students to consider what other properties molecules with the same functional groups might have in common.

LESSON 23.1

Explain

Halocarbons

USE VISUALS Direct students' attention to Table 23.2. Display the table using a transparency and overhead projector and discuss the rules for naming alkyl halides. Create large examples of primary, secondary, tertiary, vinyl, and aryl halides on sheets of paper and display them around the classroom. Help students to recognize the alkyl portion of the molecule by circling it in a different color of chalk. Write the structure of the original parent hydrocarbon next to the alkyl halide. Allow pairs of students to go on a gallery walk to review these examples up close. At each example station, students should actively be adding to their notes.

USE CONTEXT CLUES After students have mastered the meaning of the terms *primary, secondary,* and *tertiary,* show them the derivation of the common names: *sec*-butylbromide and *tert*-butylbromide.

Explore

Class Activity

Divide the class into groups of two or three students. Have each group practice naming halocarbons from structural formulas and writing structural formulas from their names. Students in each group should take turns devising questions and giving answers.

Extend

Use examples from the discussion on halogen substitution to show how a chemist can start with a simple alkane, change it to an alkyl halide, then change the halide to an alcohol. Point out that this is one way organic chemists synthesize molecules that are not found in nature. Have students research one such reaction and write a chemical equation to represent the reaction.

Figure 23.2 Halocarbons A halocarbon is a carbon-containing compound with a halogen substituent.
Classify *Are these halocarbons alkyl halides or aryl halides?*

Chloromethane

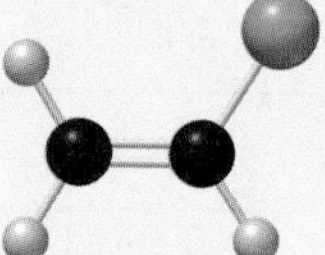
Chloroethene

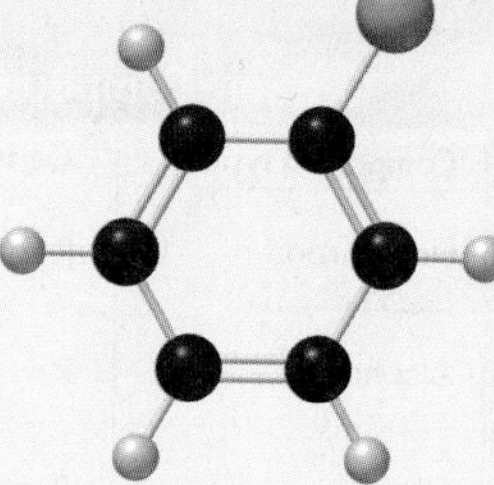
Chlorobenzene

Halocarbons

What is the general formula of a halocarbon?

A **halocarbon** is an organic compound that contains at least one covalently bonded fluorine, chlorine, bromine, or iodine atom. **The general formula of a halocarbon is RX, where X is a halogen substituent.** The IUPAC rules for naming halocarbons are based on the name of the parent hydrocarbon. The halogen groups are named as substituents. Examples of IUPAC names for some simple halocarbons are shown below. The common names are given in parentheses. Figure 23.2 shows the ball-and-stick models for these halocarbons.

CH_3-Cl — Chloromethane (methyl chloride)

$(H)(H)C=C(H)(Cl)$ — Chloroethene (vinyl chloride)

C_6H_5-Cl — Chlorobenzene (phenyl chloride)

Common names of halocarbons consist of two parts. The first part names the hydrocarbon portion of the molecule as an alkyl group, such as *methyl-* or *ethyl-*. The second part names the halogen with an *-ide* ending. On the basis of their common names, a halocarbon in which a halogen is attached to a carbon of an aliphatic chain is called an **alkyl halide.** A halocarbon in which a halogen is attached to a carbon of an arene ring is called an **aryl halide.** Table 23.2 lists the names of some substituent groups other than methyl, ethyl, and propyl.

The attractions between halocarbon molecules are primarily the result of the weak van der Waals interactions. Generally, these attractions increase with the degree of halogen substitution. This means that more highly halogenated organic compounds have higher boiling points, as illustrated in Table 23.3.

Very few halocarbons are found in nature, but they can be readily prepared and used for many purposes. For example, hydrofluorocarbons (HFCs) are used as refrigerants in automobile air-conditioning systems. Halocarbons are also used as solvents and as ingredients of stretchable polymers.

Table 23.2

Some Common Substituent Groups

Name	Group structure
Isopropyl	$CH_3-\underset{H}{\overset{CH_3}{C}}-$
Isobutyl	$CH_3-\overset{CH_3}{CH}-CH_2-$
Secondary butyl (*sec*-butyl)	$CH_3-CH_2-\underset{\vert}{CH}-CH_3$
Tertiary butyl (*tert*-butyl)	$CH_3-\underset{CH_3}{\overset{CH_3}{C}}-$
Vinyl	$(H)(H)C=C(H)-$
Phenyl	C_6H_5-

Differentiated Instruction

ELL ENGLISH LANGUAGE LEARNERS Encourage students to practice writing and then naming the structures of 1-, 2-, and 3-carbon alkyl halides. Have them build a model of one of these structures and discuss how it is produced from the corresponding alkane.

L1 STRUGGLING STUDENTS Explain that students used the concept of substitution when they learned to evaluate an expression in algebra; i.e., evaluate $3xy + 2x$ for $x = 3$ and $y = 4$. Show students how this concept is analogous to substitution reactions.

L3 ADVANCED STUDENTS Have students research examples in which the structure of a chemical plays an important role in a biological process. Suggestions for research include neurotransmitters, such as acetylcholine, and hormones, such as estradiol and testosterone. Have students research how environmental pollutants can interfere with the action of these natural chemicals.

Table 23.3

Comparing Methane and Chloromethanes

Molecular formula	Name	Molar mass (g)	Boiling point (°C)
CH_4	Methane	16.0	−161
CH_3Cl	Chloromethane (methyl chloride)	50.5	−24
CH_2Cl_2	Dichloromethane (methylene chloride)	85.0	40
$CHCl_3$	Trichloromethane (chloroform)	119.5	61
CCl_4	Tetrachloromethane (carbon tetrachloride)	154.0	74

Substitution Reactions

How are substitution reactions used in organic chemistry?

Organic reactions often proceed more slowly than inorganic reactions because organic reactions commonly involve the breaking of relatively strong covalent bonds. They often require catalysts. Many organic reactions are complex, and they usually give a mixture of products. The desired product must then be separated by distillation, crystallization, or other means. A common type of organic reaction is a **substitution reaction,** in which an atom, or a group of atoms, replaces another atom or group of atoms.

Substitution reactions are an important method of introducing new functional groups to organic molecules. For example, a halogen atom can replace a hydrogen atom on an alkane to produce a halocarbon. The symbol X stands for a halogen in this generalized equation.

$$\underset{\text{Alkane}}{R{-}H} + \underset{\text{Halogen}}{X_2} \longrightarrow \underset{\text{Halocarbon}}{R{-}X} + \underset{\text{Hydrogen halide}}{HX}$$

From the generalized equation, you can write a specific one. This type of reaction is also called a halogenation reaction because the reaction introduces a halogen atom into the molecule. Sunlight or another source of ultraviolet radiation usually serves as a catalyst.

$$\underset{\text{Methane}}{CH_4} + \underset{\text{Chlorine}}{Cl_2} \xrightarrow{\text{UV light}} \underset{\text{Chloromethane}}{CH_3Cl} + \underset{\text{Hydrogen chloride}}{HCl}$$

Even under controlled conditions, this simple substitution reaction produces a mixture of mono-, di-, tri-, and tetrachloromethanes.

Halogenation of benzene in the presence of a catalyst causes the substitution of a hydrogen atom on the ring. Iron compounds are often used as catalysts for substitution reactions in aromatic compounds.

$$\underset{\text{Benzene}}{C_6H_6} + \underset{\text{Bromine}}{Br_2} \xrightarrow{\text{catalyst}} \underset{\text{Bromobenzene}}{C_6H_5Br} + \underset{\text{Hydrogen bromide}}{HBr}$$

Tetrachloroethylene

Tetrachloroethylene (formerly called perchloroethylene, perc, or PCE) is a manufactured solvent used in about 95 percent of dry cleaning businesses in the United States. For years it has been the subject of close government and public scrutiny as a possible carcinogen. Recently, the Department of Health and Human Services determined that tetrachloroethylene may reasonably be considered to be a carcinogen. The National Institute for Occupational Safety and Health (NIOSH) recommends that tetrachloroethylene be handled as a possible carcinogen and recommends that exposure levels at the workplace be as low as possible. The Occupational Safety and Health Administration (OSHA) has a limit of 100 ppm for an 8-hour workday over a 40-hour workweek in order to protect those people having to work in this environment.

Explain

Substitution Reactions

USE VISUALS Direct students to Table 23.3. Point out that the subscript on the chlorine molecule in each molecular formula increases from 0 to 4. Have students look for a trend in boiling-point temperatures. **Ask** What can you infer about the cause of this pattern in boiling-point temperatures? *(The more halogen substituents a molecule has, the greater will be the attraction of the molecule to other halocarbon molecules.)*

Explore

Teacher Demo

PURPOSE Students will observe the Beilstein test for determining the presence of halogen functionalities in organic compounds. Point out that organic chemists often perform qualitative chemical tests to identify unknown organic compounds.

SAFETY Wear safety goggles and a laboratory apron while performing the demo.

MATERIALS copper wire, Bunsen burner, forceps, bromobenzene, benzoic acid

PROCEDURE Prepare a piece of copper wire with a small loop in one end. Heat the loop in a Bunsen burner flame and allow the wire to cool. Next, dip the loop into a sample of aryl halide (bromobenzene). Heat the wire in the flame again. Repeat the test with a negative control, benzoic acid.

EXPECTED OUTCOMES A green flame will be produced with bromobenzene, indicating the presence of a halogen. The flame will not be green when testing benzoic acid.

Answers

FIGURE 23.2 Chloromethane and chloroethene are alkyl halides. Chlorobenzene is an aryl halide.

LESSON 23.1

Explore

Substitution Reactions

MAKE A CONNECTION Explain that when triglycerides are combined with aqueous sodium hydroxide, the products are glycerol and fatty acid salts. This process is called *saponification*. Encourage interested students to look ahead in Chapter 24 to discover how the concepts learned in this lesson translate to the study of lipids.

Extend

Ask students to prepare a computer slideshow presentation explaining how it is beneficial to be able to deduce the properties of halogens from what we know about them as a functional group. Presentations should show several real-world applications that have benefited society. Presentations should convey the various uses of halocarbons: anesthetics, cleaning agents, non-stick surfaces, pipes and other plastic items, and refrigerants.

Evaluate

Informal Assessment

Write the structural formulas for ethene, chloroethane, and ethanol on the board. Ask students to name the compounds and identify their functional groups. *(double bond, halogen, hydroxyl group)* Ask students to choose two of the compounds to illustrate a substitution reaction. *(Hydroxide ions can replace the chlorine in ethyl chloride to form ethanol.)*

Reteach

Work with students to construct a concept map showing the connection between alkanes and their derivatives: alkenes, alkynes, halocarbons, and alcohols. Use propane as a specific example and start with the heading *alkanes*. Write the structural formula for the propyl derivative in each case.

CHEMISTRY & YOU

Q: *Earlier you learned that adding a halogen substituent to a hydrocarbon affects certain physical properties. How can you distinguish between hydrocarbons and halocarbons in terms of structures and properties?*

Halocarbons can be converted to other types of compounds by substitution reactions. For example, hydroxide ions can displace halogen atoms on carbon chains to form an alcohol. (Fluorine is one exception. Because fluoro groups are not easily displaced, they are seldom used to prepare alcohols.) The general reaction for the formation of an alcohol from a halocarbon is as follows:

$$\underset{\text{Halocarbon}}{R{-}X} + \underset{\text{Hydroxide ion}}{OH^-} \xrightarrow[100°C]{H_2O} \underset{\text{Alcohol}}{R{-}OH} + \underset{\text{Halide ion}}{X^-}$$

Chemists usually use aqueous solutions of sodium hydroxide or potassium hydroxide as the source of hydroxide ions. The chemical equation for two specific examples are shown below.

$$\underset{\text{Iodomethane}}{CH_3{-}I(l)} + \underset{\text{Potassium hydroxide}}{KOH(aq)} \xrightarrow{100°C} \underset{\text{Methanol}}{CH_3{-}OH(l)} + \underset{\text{Potassium iodide}}{KI(aq)}$$

$$\underset{\text{Bromoethane}}{CH_3CH_2Br(l)} + \underset{\text{Sodium hydroxide}}{NaOH(aq)} \xrightarrow{100°C} \underset{\text{Ethanol}}{CH_3CH_2OH(l)} + \underset{\text{Sodium Bromide}}{NaBr(aq)}$$

Halocarbons can also be converted to other halocarbons, amines, or ethers by similar substitution reactions.

ONLINE PROBLEMS

23.1 LessonCheck

1. **Review** How are organic compounds classified?
2. **Identify** What is the general formula of a halocarbon?
3. **Explain** Why are substitution reactions useful in organic chemistry?
4. **Classify** Identify the functional group in each structure. Then classify the compound according to its functional group.
   a. $CH_3{-}OH$
   b. $CH_3{-}CH_2{-}NH_2$
   c. $C_6H_5{-}\overset{O}{\overset{\|}{C}}{-}OH$ (benzene ring bonded to C(=O)—OH)
   d. $CH_3{-}CH_2{-}CH_2{-}Br$
   e. $CH_3{-}CH_2{-}O{-}CH_2{-}CH_3$
   f. $CH_3{-}CH_2{-}I$
5. **Describe** Draw a structural formula for each compound listed below.
   a. isopropyl chloride
   b. 1-iodo-2,2-dimethylpentane
   c. *p*-bromotoluene
   d. bromoethene
   e. 2-bromo-2-chloro-1,1,1-trifluoroethane
   f. vinyl iodide
6. **Compare** How would you expect the boiling points of ethane and chloroethane to compare? Explain your answer.
7. **Apply Concepts** Draw the structural formulas and write the IUPAC names of all possible dichloropropanes that could result from a substitution reaction involving chlorine and propane.
8. **Describe** Write a balanced chemical equation for the preparation of 2-propanol from potassium hydroxide and 2-bromopropane.

802 Chapter 23 • Lesson 1

Lesson Check Answers

1. according to their functional groups
2. RX, where X is a covalently bonded halogen substituent such as fluorine, chlorine, bromine, or iodine
3. They are used to introduce new functional groups to organic molecules.
4. a. —OH, hydroxyl group; alcohol
   b. $—NH_2$, amino group; amine
   c. —COOH, carboxyl group; carboxylic acid
   d. —Br, halogen; halocarbon
   e. —C—O—C—, ether group; ether
   f. —I, halogen; halocarbon
5. a. $CH_3{-}\underset{H}{\overset{CH_3}{C}}{-}Cl$
   b. $I{-}CH_2\underset{CH_3}{\overset{CH_3}{C}}CH_2CH_2CH_3$
   c. benzene ring with CH_3 at the top and Br at the bottom (para)
   d. $\underset{H}{\overset{H}{}}C{=}C\underset{H}{\overset{Br}{}}$
   e. $F{-}\underset{F}{\overset{F}{C}}{-}\underset{Cl}{\overset{Br}{C}}{-}H$
   f. $\underset{H}{\overset{H}{}}C{=}C\underset{H}{\overset{I}{}}$
6. Chloroethane would have a higher boiling point because of the increased strength of the interactions between halocarbon molecules compared to those between hydrocarbons.
7. $CHCl_2CH_2CH_3$, 1,1-dichloropropane
   $CH_2ClCHClCH_3$, 1,2-dichloropropane
   $CH_3CCl_2CH_3$, 2,2-dichloropropane
   $CH_2ClCH_2CH_2Cl$, 1,3-dichloropropane
8. $CH_3CHBrCH_3 + KOH \rightarrow CH_3CHOHCH_3 + KBr$

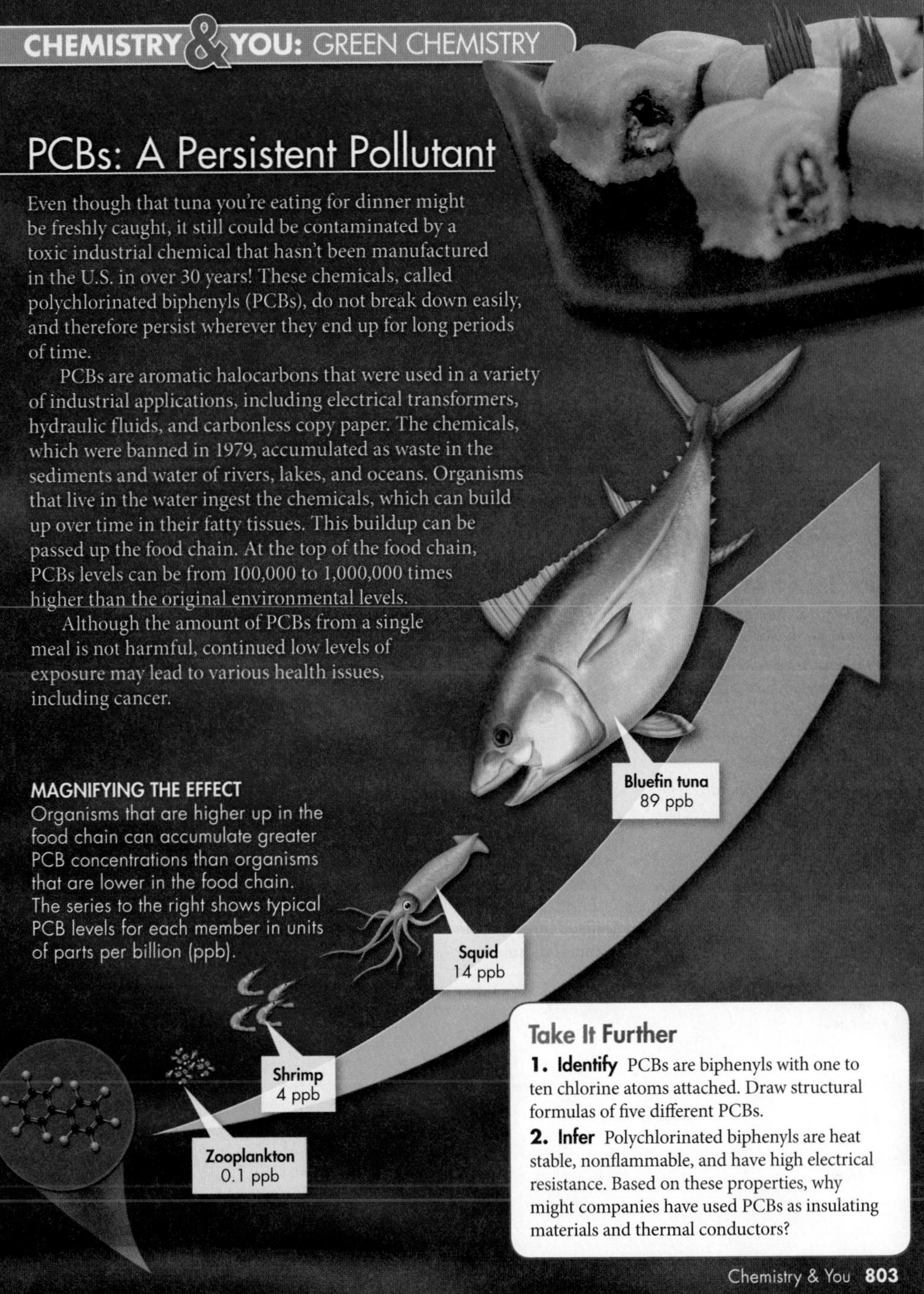

CHEMISTRY & YOU: GREEN CHEMISTRY

PCBs: A Persistent Pollutant

Even though that tuna you're eating for dinner might be freshly caught, it still could be contaminated by a toxic industrial chemical that hasn't been manufactured in the U.S. in over 30 years! These chemicals, called polychlorinated biphenyls (PCBs), do not break down easily, and therefore persist wherever they end up for long periods of time.

PCBs are aromatic halocarbons that were used in a variety of industrial applications, including electrical transformers, hydraulic fluids, and carbonless copy paper. The chemicals, which were banned in 1979, accumulated as waste in the sediments and water of rivers, lakes, and oceans. Organisms that live in the water ingest the chemicals, which can build up over time in their fatty tissues. This buildup can be passed up the food chain. At the top of the food chain, PCBs levels can be from 100,000 to 1,000,000 times higher than the original environmental levels.

Although the amount of PCBs from a single meal is not harmful, continued low levels of exposure may lead to various health issues, including cancer.

MAGNIFYING THE EFFECT
Organisms that are higher up in the food chain can accumulate greater PCB concentrations than organisms that are lower in the food chain. The series to the right shows typical PCB levels for each member in units of parts per billion (ppb).

Take It Further

1. **Identify** PCBs are biphenyls with one to ten chlorine atoms attached. Draw structural formulas of five different PCBs.
2. **Infer** Polychlorinated biphenyls are heat stable, nonflammable, and have high electrical resistance. Based on these properties, why might companies have used PCBs as insulating materials and thermal conductors?

Chemistry & You 803

CHEMISTRY & YOU Discuss with students the effects of PCBs on various ecosystems or food chains. Direct students' attention to the diagram of the food chain and have them read the accompanying text. Explain that biologists refer to the accumulation of toxins in each successive trophic level of an ecosystem as *biological magnification*.

Pose the following question to students. ***If humans are at the top of this food chain, how are they affected by the biological magnification of PCBs in aquatic food webs?*** You may need to assist students in the following ways:

- The greatest levels of PCBs are found in fresh and saltwater sportfish species, such as the bluefin tuna shown in the diagram.
- Human consumption of PCBs can lead to skin and liver disorders, cancer, low infant birth weights, and developmental delays in children.
- One part per billion (ppb) is another way of representing one microgram per kilogram (μg/kg).

MAKE A CONNECTION Have students research the concept of half life and find out what the approximate half life is for PCBs in marine animals. Have students use the concept of half life to explain why PCBs pose a risk to humans for a very long time.

Connect to GOVERNMENT

Have students research and identify the two federal agencies that have responsibility in protecting humans from exposure to harmful levels of PCBs. Students should determine the regulations set in regard to PCBs. *(FDA—the allowable level of PCBs is currently 2.0 ppm for fish and shellfish. EPA—the limit for all PCBs in fish tissue is 0.01 ppm.)* Have students also determine other examples of industrial contamination for which these agencies have had to implement standards in order to protect human health. *(Sample answers: lead, mercury, and other heavy metals)*

21st Century Learning To be successful in the 21st century, students need skills and learning experiences that extend beyond subject matter mastery. The following project helps students build the following 21st Century Skills: *Financial, Economic, Business, and Entrepreneurial Literacy*; *Creativity and Innovation*; *Communication and Collaboration*; *Information, Media, and Technology Literacy*; *Initiative and Self-Direction*; and *Productivity and Accountability*.

YOU ARE WHAT YOU EAT! Pose the following challenge to your students. *The EPA wants to convey that a clean-up effort in one of the Great Lakes has resulted in a dramatic reduction of PCBs in its fish population. They have hired you to create a video-taped public service announcement to explain the circumstances leading to the agency's removal of this lake's fish from its watch list.* The announcement should demonstrate an awareness of the environment as it relates to PCBs and how they affect an aquatic ecosystem.

Answers

TAKE IT FURTHER

1. Check students' structural formulas.
2. It is desirable that the items do not melt or catch fire when exposed to heat or a nearby flame or spark.

LESSON 23.2

Key Objectives

23.2.1 IDENTIFY the general formula of an alcohol.

23.2.2 DESCRIBE how addition reactions are used in organic chemistry.

23.2.3 IDENTIFY the general formula of an ether.

23.2.4 IDENTIFY the general formula of an amine.

Additional Resources

- Reading and Study Workbook, Lesson 23.2
- Probeware Laboratory Manual
- Virtual Chemistry Labs

PearsonChem.com

Engage

CHEMISTRY & YOU Pass out mints for students to smell, taste, and observe, and then direct them to the opening text and photograph. Ask students to compare the sensations of aftershaves or muscle rubs on their skin to that from the candy in their mouths.

Access Prior Knowledge

Prior to beginning the lesson, have students create a KWL chart about what they already know about organic chemistry and functional groups. Have students continue their charts as they proceed with the lesson.

Foundations for Reading

BUILD VOCABULARY To help students relate vocabulary to concepts discussed in this section, have them distinguish between hydration and hydrogenation. Although both terms come from the same root word, hydration is the addition of water, whereas hydrogenation is the addition of hydrogen.

IN ACADEMIC STANDARDS for SCIENCE

C.4.5 Describe, classify and give examples of various kinds of reactions-synthesis (combination), decomposition, single displacement, double displacement and combustion.

23.2 Alcohols, Ethers, and Amines

C.4.5 Classify types of reactions

CHEMISTRY & YOU

Q: *How can organic chemistry cool you down?* Aftershave lotion contains a certain organic compound that imparts a cooling sensation on the skin. The same ingredient can be found in mint-flavored candy, cough drops, and muscle rubs. In this lesson, you'll read about three classes of organic compounds that have a wide variety of uses.

Key Questions

What is the general formula of an alcohol?

How are addition reactions used in organic chemistry?

What is the general formula of an ether?

What is the general formula of an amine?

Vocabulary

- alcohol
- hydroxy group
- fermentation
- addition reaction
- hydration reaction
- hydrogenation reaction
- ether
- amine

Alcohols

What is the general formula of an alcohol?

What do mouthwash, perfume, and hairspray have in common? They all contain an alcohol of some type. An **alcohol** is an organic compound with an —OH group. **The general formula of an alcohol is ROH.**

R—Ö—H (bent)

Alcohol

The —OH functional group in alcohols is called a **hydroxy group,** sometimes called a hydroxyl group. The oxygen atom in the —OH group has two pairs of nonbonding electrons, which compress the R—O—H bond angle. As a result, an alcohol functional group has a bent shape.

Aliphatic alcohols can be classified into structural categories according to the number of carbons attached to the carbon with the hydroxy group. If only one carbon (or no carbon) is attached to C—OH, the latter carbon is considered a primary carbon, and the alcohol is a primary alcohol. If two carbons are attached, the carbon is a secondary carbon, which gives a secondary alcohol; if three carbons, a tertiary carbon, and a tertiary alcohol. This nomenclature is summarized below using R to represent any carbon chains or rings.

Primary alcohol	$R—CH_2—OH$	Only one R group is attached to C—OH of a primary (abbreviated 1°) alcohol.
Secondary alcohol	R—CH(R)—OH	Two R groups are attached to C—OH of a secondary (2°) alcohol.
Tertiary alcohol	R—C(R)(R)—OH	Three R groups are attached to C—OH of a tertiary (3°) alcohol.

Focus on ELL

1 CONTENT AND LANGUAGE Write the suffixes *-tion* and *-ation* on the board. Explain that these two suffixes both mean "action or process." Have students identify the vocabulary terms containing these suffixes and use their definition to define the vocabulary terms.

2 FRONTLOAD THE LESSON Review the vocabulary and key questions from Lesson 23.2. Ask students to recall and draw the functional groups for alcohols, ethers, and amines.

3 COMPREHENSIBLE INPUT Use the Kinetic Art feature to show students animated examples of addition reactions. Help students understand the process of naming ethers by highlighting the two R groups with different colors, using the same color to represent the smaller group each time.

Naming Alcohols Both IUPAC and common names are used for alcohols. To name aliphatic alcohols using the IUPAC system, drop the *-e* ending of the parent hydrocarbon name and add the ending *-ol.* The parent hydrocarbon is the longest continuous chain of carbons that includes the carbon attached to the hydroxy group. In numbering the parent hydrocarbon, the position of the hydroxy group is given the lowest possible number. Some alcohols have more than one hydroxy group. To name these alcohols using the IUPAC system, simply add the ending *-diol* or *-triol,* to the parent hydrocarbon name if the alcohol has two or three hydroxy groups, respectively.

The common names of aliphatic alcohols are written in the same way as those of halocarbons. The alkyl group ethyl, for example, is named and followed by the word *alcohol,* as in *ethyl alcohol.* The common name for an alcohol with two hydroxy groups is a glycol.

When the hydroxy group is attached directly to an aromatic ring, the compound is called a phenol. To assign the IUPAC name, phenol is used as the parent hydrocarbon. For example, a phenol with a methyl group attached is called a methylphenol. Cresol is the common name for the *o, m,* and *p* constitutional isomers of methylphenol. Table 23.4 lists the names, structural formulas, and ball-and-stick models of some simple alcohols and phenols.

Table 23.4

Some Common Alcohols and Phenols

IUPAC name (common name)	Structural formula	Ball-and-stick model
Ethanol (ethyl alcohol)	$CH_3—CH_2—OH$	
2-propanol (isopropyl alcohol)	$CH_3—CH(OH)—CH_3$	
1,2-ethanediol (ethylene glycol)	$CH_2(OH)—CH_2(OH)$	
1,2,3-propanetriol (glycerol)	$CH_2(OH)—CH(OH)—CH_2(OH)$	
2-methylphenol (*o*-cresol)	benzene ring with CH_3 and OH on adjacent carbons	

READING STRATEGY Have students write a short paragraph comparing structural formulas and ball-and-stick models. Students should explain what information each representation best shows about a compound and what each representation does not show.

Explain

Alcohols

APPLY CONCEPTS Explain that many biochemically active molecules contain carbon atoms bonded to oxygen atoms. Point out that the presence of a single oxygen atom significantly alters the chemical and physical properties of a hydrocarbon. Explain that alcohols and ethers are two important classes of organic compounds that contain oxygen.

MAKE A CONNECTION Discuss the IUPAC conventions for naming aliphatic alcohols. Write the condensed structural formulas of several primary, secondary, and tertiary alcohols on the board, and help students to name them. Remind students how to distinguish primary, secondary, and tertiary carbons. Write the structures of methanol, ethanol, and isopropyl alcohol on the board and point out that students already know a lot about the uses and properties of these alcohols. Help students name the structures shown. **Ask** What are some everyday examples where these alcohols are used? *(Methanol is a good fuel for internal combustion engines; it is sometimes used for race cars. Methanol is also an excellent solvent, and is used for paints, shellacs, and varnishes. Ethanol is found in alcoholic beverages and is sometimes used as a gasoline additive to produce cleaner burning fuels. Isopropyl alcohol is used as an antiseptic—rubbing alcohol.)* **Ask** How does a hydroxyl group differ from a hydroxide ion? *(The bonding between carbon and the hydroxyl group in alcohols is covalent. The bonding between a metal and the hydroxide ion in bases is ionic.)*

Antifreeze Poisoning

Every year, about 10,000 dogs and cats are victims of accidental poisoning by automobile antifreeze. A pet does not have to drink a lot of antifreeze to be poisoned. Most brands of commercial antifreeze consist of 95 percent ethylene glycol, an extremely toxic chemical. Even a few licks of this sweet-tasting liquid can be fatal to a cat or dog. For a medium-sized dog, ingestion of about 3 to 4 tablespoons of antifreeze is toxic. For cats, as little as 1 to 2 teaspoons can be lethal. Antifreeze poisoning commonly occurs in spring and fall when car owners replace the old antifreeze in their car radiators with fresh antifreeze. However, poisoning can happen anytime, particularly when a car boils over or when a hose leaks, releasing the antifreeze.

Explain

Alcohols

USE MODELS Write the structure of water and the general structures of alcohols and ethers on the board. Point out that alcohols and ethers can be viewed as derivatives of water. Allow students to handle ball-and-stick and space-filling models of water, ethanol, and diethyl ether. Write the condensed structural formula for each on the board. **Ask** What are the structural similarities between all three compounds? *(Students should identify the bent geometry around the oxygen atom.)* **Ask** What significance does this bent shape have for the overall polarity of the molecules? *(Each O—H and O—C bond is polar. Because the molecules are bent, the bond polarities do not cancel, and the molecules as a whole are polar. The region around the oxygen atom has a slight negative charge.)*

CRITICAL THINKING Point out that the polar character of alcohols and ethers is tempered by the alkyl groups, which are nonpolar. Challenge students to name the important intermolecular forces holding molecules together in alcohols and ethers. **Ask** How are water, ethanol, and diethyl ether ranked in order of increasing boiling point? *(Ether, ethanol, water; ether molecules are attracted to one another by intermolecular dipolar attractions, while ethanol and water molecules have strong intermolecular hydrogen bonding.)*

Explore

Teacher Demo

PURPOSE Students will observe that, although alcohols contain —OH groups, they are not hydroxides.

MATERIALS several alcohols, litmus paper

PROCEDURE Test several alcohols with litmus paper to show that they have no basic properties.

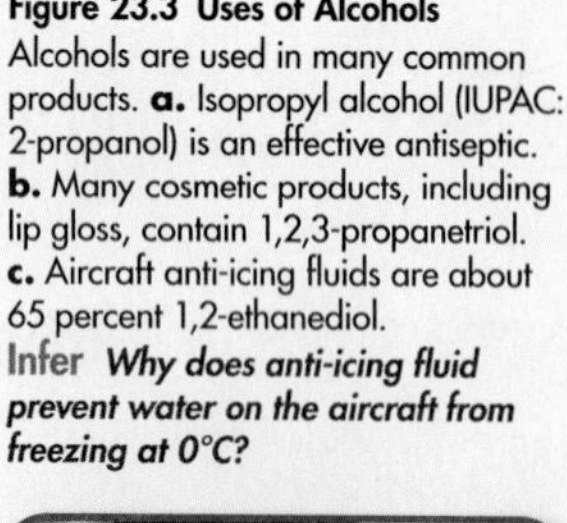

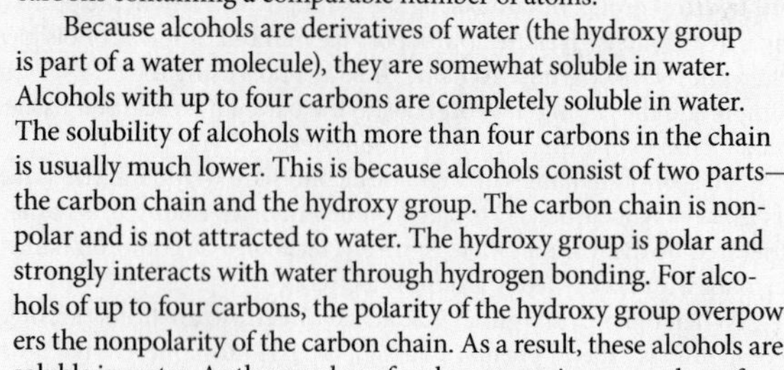

Properties of Alcohols Alcohols are capable of intermolecular hydrogen bonding, which is reflected in their physical properties. For example, alcohols boil at higher temperatures than alkanes and halocarbons containing a comparable number of atoms.

Because alcohols are derivatives of water (the hydroxy group is part of a water molecule), they are somewhat soluble in water. Alcohols with up to four carbons are completely soluble in water. The solubility of alcohols with more than four carbons in the chain is usually much lower. This is because alcohols consist of two parts—the carbon chain and the hydroxy group. The carbon chain is nonpolar and is not attracted to water. The hydroxy group is polar and strongly interacts with water through hydrogen bonding. For alcohols of up to four carbons, the polarity of the hydroxy group overpowers the nonpolarity of the carbon chain. As a result, these alcohols are soluble in water. As the number of carbon atoms increases above four, however, the nonpolarity of the carbon chain dominates, and the solubility of the alcohol decreases.

Uses of Alcohols Figure 23.3 illustrates some common uses of alcohols. For example, 2-propanol, which is more familiarly known as rubbing alcohol, is often used as an antiseptic. It is also used as a base for perfumes, creams, lotions, and other cosmetics. Another alcohol, 1,2,3-propanetriol, is highly soluble in water because it has three hydroxy groups. It also has a tendency to absorb water from its surroundings. This quality makes 1,2,3-propanetriol a valuable moistening agent in cosmetics, foods, and pharmaceuticals.

Some antifreezes use 1,2-ethanediol as the main ingredient. This alcohol has a high boiling point, 197°C, which helps prevent vehicle engines from overheating. Its advantages over other liquids with high boiling points are its solubility in water and its low freezing point, −17.4°C. When water is added to 1,2-ethanediol, the mixture freezes at even lower temperatures. For example, a 50% (v/v) aqueous solution of 1,2-ethanediol freezes at −36°C. This property also makes 1,2-ethanediol an effective anti-icing agent.

Figure 23.3 Uses of Alcohols
Alcohols are used in many common products. **a.** Isopropyl alcohol (IUPAC: 2-propanol) is an effective antiseptic. **b.** Many cosmetic products, including lip gloss, contain 1,2,3-propanetriol. **c.** Aircraft anti-icing fluids are about 65 percent 1,2-ethanediol.
Infer ***Why does anti-icing fluid prevent water on the aircraft from freezing at 0°C?***

Differentiated Instruction

L1 STRUGGLING STUDENTS Students may not understand why breaking one bond in a double bond opens up two bonding sites. Have them count the number of bonds each carbon atom can form after the bond has been broken. Remind them that each carbon atom can form single bonds with four other atoms, allowing one atom to be added to each carbon.

ELL ENGLISH LANGUAGE LEARNERS Suggest that beginning and intermediate speakers copy the examples of each reaction type (addition, hydration, halogenation, and hydrogenation) and write a brief native-language explanation beside each type. Encourage advanced speakers to assist students of lower proficiencies with writing their explanations.

L3 ADVANCED STUDENTS Have students research fermentation as a cellular process in animal and microbial cells. Have students design a simple experiment to show that fermentation can take place when the starting ingredient is table sugar (sucrose) rather than glucose.

Figure 23.4 Fermentation in Bread
Yeasts, which are part of a bread recipe, break down sugars in the dough mixture. The carbon dioxide gas that is produced diffuses and causes the dough to rise. Tiny holes remain in the baked bread as evidence of the carbon dioxide gas.

Ethanol is an important industrial chemical. Most ethanol is produced by yeast fermentation of sugar. **Fermentation** is the production of ethanol from sugars by the action of yeasts or bacteria. The enzymes of the yeasts or bacteria serve as catalysts for the transformation. The bread in Figure 23.4 rises because of fermentation. Glucose molecules in the dough are broken down by the following fermentation reaction.

$$\underset{\text{Glucose}}{C_6H_{12}O_6(aq)} \longrightarrow \underset{\text{Ethanol}}{2CH_3CH_2OH(aq)} + \underset{\text{Carbon dioxide}}{2CO_2(g)}$$

The ethanol in alcoholic beverages is generally produced by fermentation. Ethanol is an intoxicating substance; it is a depressant that can be fatal if taken in large doses at once.

The ethanol used in many industrial applications is denatured. Denatured alcohol is ethanol that is unsuitable for consumption due to the presence of an additive. That added substance, or denaturant, is often methanol. Methanol is extremely toxic. As little as 10 mL has been reported to cause blindness, and as little as 30 mL has been known to cause death.

READING SUPPORT
Build Study Skills: *Compare/Contrast Table* As you read about addition reactions, use a compare/contrast table to organize the information. In total, three kinds of addition reactions are discussed. Include a column in your table for each kind. ***What do the reactions have in common?***

Addition Reactions

How are addition reactions used in organic chemistry?

The carbon–carbon single bonds in alkanes are not easy to break. In an alkene, however, one of the bonds in the double bond is somewhat weaker. This bond is easier to break than a carbon–carbon single bond. So it is sometimes possible for a compound of general structure X—Y to add to a double bond. In an **addition reaction,** a substance is added at the double or triple bond of an alkene or alkyne. **Addition reactions are an important method of introducing new functional groups to organic molecules. They are also used to convert alkenes to alkanes.** In the general addition reaction shown below, X and Y represent the two parts of the reagent that are added to the alkene.

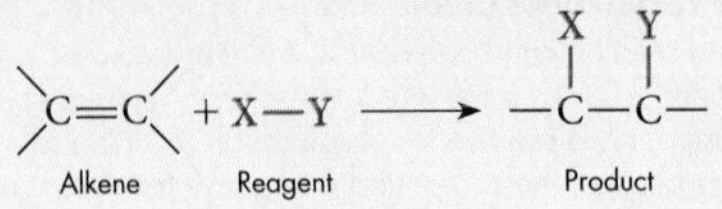

$$\underset{\text{Alkene}}{>C{=}C<} + \underset{\text{Reagent}}{X{-}Y} \longrightarrow \underset{\text{Product}}{-\overset{X}{\overset{|}{C}}-\overset{Y}{\overset{|}{C}}-}$$

UbD Check for Understanding

The Essential Question What effect does a functional group have on an organic compound?

Have students summarize their understanding of alcohols by writing or verbalizing a one-minute response to the Essential Question as it pertains to alcohols.

ADJUST INSTRUCTION If students are having difficulty with their summaries, have them reread the section Properties of Alcohols, and then allow them to revise their summaries.

LESSON 23.2

Explore

Class Activity

Bring household products that contain alcohols to class. Read the name of the alcohol or alcohols given on each label, and help students to write the structure for each alcohol. **Ask** Why do you think alcohols are used in so many commercial products? *(Because alcohols have both polar and nonpolar characteristics, and are capable of forming hydrogen bonds, they make excellent solvents for a wide variety of polar and nonpolar compounds.)*

Teacher Demo

PURPOSE Students will observe the aqueous solubilities of alcohols and glycols.

MATERIALS 10 or 15 test tubes with stoppers, distilled water, ethanol, 1-butanol, 1-hexanol, 1-octanol, a hexanediol

PROCEDURE Prepare two or three sets of 5 test tubes, labeled 1 to 5, each containing a small volume of distilled water and an equal volume of ethanol, 1-butanol, 1-hexanol, 1-octanol, and a hexanediol. Stopper the tubes securely. Instruct students to mix the contents of each tube and allow the mixture to settle. Have students record their observations and describe any trends. Students should generalize about the aqueous solubility of straight-chain alcohols and about the role that —OH groups play in this solubility.

EXPECTED OUTCOMES Students should observe that the solubilities of the alcohols decrease as the alkyl chains increase in length.

Answers

FIGURE 23.3 A nonvolatile solute depresses water's freezing point.

READING SUPPORT Each reaction results in a product that is more saturated than the reactant.

LESSON 23.2

Explain

Addition Reactions

USE VISUALS Direct students' attention to Figure 23.5. **Ask** If the orange color remains, what would you conclude about the sample? *(The sample is saturated.)* Explain that a double or triple bond is more reactive than a single covalent bond. Write the equation for the addition of Br_2 to 2-butene on the board. Explain that when a molecule containing a double bond reacts, one of the bonds in the double bond is broken. This reaction opens up two bonding sites, one on each of the carbon atoms that previously shared the double bond. Explain that the addition of molecules to alkenes makes it possible for chemists to synthesize a large variety of compounds, including polymers.

Explore

Teacher Demo

PURPOSE Students will observe the reactions of alkenes and alkynes with oxidizing agents such as potassium permanganate. This reaction can be used to test for unsaturated compounds.

MATERIALS $KMnO_4$, NaOH, distilled water, cyclohexane, cyclohexene, 2 test tubes with stoppers

PROCEDURE Prepare an alkaline solution of potassium permanganate by dissolving 0.1 g $KMnO_4$ and 15 g NaOH in 100 mL distilled water. Put 12 mL cyclohexane in one test tube and 12 mL cyclohexene in a second test tube. Add 1 mL of the alkaline permanganate solution to each. Stopper and shake.

EXPECTED OUTCOMES The decolorization of cyclohexene indicates the presence of a double or triple bond. Cyclohexane does not decolorize. The general equation for the reaction of permanganate with an unsaturated hydrocarbon in the presence of water is

$$RCH{=}CHR + MnO_4^- \rightarrow RCHOHCHOHR + MnO_2$$

Hydration Reactions The addition of water to an alkene is a **hydration reaction.** A hydration reaction results in the formation of an alcohol. Hydration reactions usually occur when the alkene and water are heated to about 100°C in the presence of a small amount of a strong acid. The acid, usually hydrochloric acid or sulfuric acid, serves as a catalyst for the reaction. The addition of water to ethene to form ethanol is a typical hydration reaction. The parts of ethanol that come from the addition of water are shown in blue in the equation below.

$$\underset{\text{Ethene}}{H_2C{=}CH_2} + \underset{\text{Water}}{H{-}OH} \xrightarrow[100^\circ C]{H^+} \underset{\text{Ethanol}}{H{-}CH_2{-}CH_2{-}OH}$$

Halogenation Reactions When the reagent X—Y is a halogen molecule such as chlorine or bromine, the product of the reaction is a disubstituted halocarbon. The addition of bromine to ethene to form the disubstituted halocarbon 1,2-dibromoethane is an example.

$$\underset{\text{Ethene (colorless)}}{H_2C{=}CH_2} + \underset{\text{Bromine (orange)}}{Br{-}Br} \longrightarrow \underset{\text{1,2-dibromoethane (colorless)}}{BrCH_2{-}CH_2Br}$$

The addition of bromine to carbon–carbon multiple bonds is often used as a chemical test for unsaturation in an organic molecule. Bromine has an orange color, but most bromine-substituted organic compounds are colorless. The test for unsaturation is performed by adding a few drops of a 1% solution of bromine in tetrachloromethane to the suspected alkene. As Figure 23.5 shows, the loss of the orange color is a positive test for unsaturation. If the orange color remains, the sample is completely saturated.

Hydrogen halides, such as HBr or HCl, also can add to a double bond. Because the product contains only one substituent, it is called a monosubstituted halocarbon. The addition of hydrogen chloride to ethene is an example.

$$\underset{\text{Ethene}}{H_2C{=}CH_2} + \underset{\text{Hydrogen chloride}}{H{-}Cl} \longrightarrow \underset{\text{Chloroethane}}{H{-}CH_2{-}CH_2{-}Cl}$$

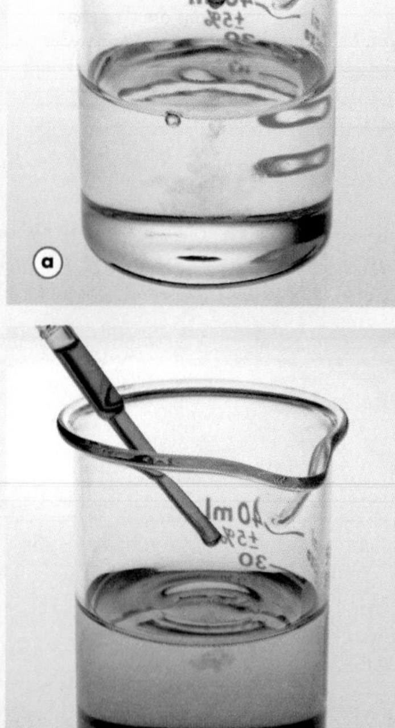

Figure 23.5 Test for Unsaturation
Bromine solution can be used to identify unsaturated organic compounds. **a.** After mixing, if the solution remains colorless, the compound is positive for unsaturation. **b.** If the solution remains orange, the compound is negative for unsaturation.

Professional Development Note: Trans Fats

Trans fats are produced through hydrogenation, the chemical process by which hydrogen is added to unsaturated fatty acids in vegetable oil. Hydrogenation converts the unsaturated bonds in the oil to saturated bonds, creating products such as margarine—a solid, spreadable fat with increased shelf life. Although hydrogenation eliminates some double bonds, it transforms others from the natural *cis* configuration to the *trans* configuration. Research indicates that eating trans fats is associated with an increased risk for heart disease.

Figure 23.6 Hydrogenation
The fats in cooking oils have double bonds. Oils are converted to solid spreads by adding hydrogen to the double bonds.
Interpret Photos ***What evidence of a chemical reaction do you see in the photos?***

Hydrogenation Reactions The addition of hydrogen to a carbon–carbon double bond to produce an alkane is called a **hydrogenation reaction.** Hydrogenation usually requires a catalyst. A platinum (Pt) or palladium (Pd) catalyst is often used. The manufacture of margarine or spreads from various oils is a common application of hydrogenation. As shown in Figure 23.6, adding hydrogen to unsaturated fats in oils results in the formation of saturated fats. Saturated fats have higher melting points than unsaturated fats. This is why margarine remains solid at room temperature.

The hydrogenation of a double bond is a reduction reaction. In the examples below, ethene is reduced to ethane, and cyclohexene is reduced to cyclohexane.

$$\mathrm{H_2C{=}CH_2 + H{-}H \xrightarrow{Pt} H_3C{-}CH_3}$$

Ethene Hydrogen Ethane

$$\mathrm{C_6H_{10} + H{-}H \xrightarrow{Pt} C_6H_{12}}$$

Cyclohexene Hydrogen Cyclohexane

See addition reactions animated online. KINETIC ART

Under normal conditions, benzene resists hydrogenation. It also resists the addition of a halogen or a hydrogen halide. However, under conditions of high temperatures and high pressures of hydrogen, and with certain catalysts, three molecules of hydrogen gas can reduce one molecule of benzene to form cyclohexane.

$$\mathrm{C_6H_6 + 3H{-}H \xrightarrow[Heat]{Pt} C_6H_{12}}$$

Benzene Hydrogen Cyclohexane

USE VISUALS Have students examine Figure 23.6. **Ask** What is the difference between a saturated fat and an unsaturated fat? *(Saturated fats consist of carbon atoms saturated with hydrogen atoms, and they are typically solid at room temperature; unsaturated fat is a form of fat that is liquid at room temperatures.)*

CRITICAL THINKING Discuss the health aspects of consuming the various types of fat. **Ask** What is better for you to consume, butter or margarine? Why? *(Sample answer: Margarine is made from vegetable oils, so it contains no cholesterol; butter is made from animal fat, so it contains cholesterol and high levels of saturated fat.)* **Ask** What are the health risks for consuming too many foods high in saturated fats? *(Eating foods that contain saturated fats raises the level of cholesterol in your blood. High levels of blood cholesterol increase your risk of heart disease and stroke.)* Point out that most foods contain several different kinds of fats, including saturated, polyunsaturated, monounsaturated, and trans fats, and that some kinds are better for your health than others. Have students research each type of fat and create a pros versus cons chart in regards to human consumption and health. Ask students to include an example of a chemical structure of each type of fat.

APPLY CONCEPTS Write several examples of addition reactions on the board and help students predict the structure of the expected product in each case.

Check for Understanding

The Essential Question How are chemical reactions used in organic chemistry?

Assess students' understanding of this Essential Question as it relates to addition reactions by asking students to write a one-minute summary explaining how addition reactions are used in organic chemistry.

ADJUST INSTRUCTION If students are having difficulty with their summaries, have them reread the Addition Reactions section and review the figures and associated reactions with them. Point out that in each example, a double bond is being reduced during the reaction. Then, have students revise their summaries.

Answers

FIGURE 23.6 a change in color

LESSON 23.2

Explain

Ethers

CHEMISTRY & YOU Menthol is an alcohol, as indicated by the *-ol* ending in cyclohexanol.

USE MODELS Describe the molecular structure of ethers. Prepare a ball-and-stick or space-filling model to demonstrate that ethers are slightly polar due to the bent geometry around the oxygen atom. Point out that ethers are characterized by a 120° bond angle. **Ask** Are ethers more or less soluble in water than alkanes of similar size? *(Ethers are generally more soluble due to their ability to participate in hydrogen bonding with water.)* **Ask** Would you expect an ether named butyl ether to be symmetric or nonsymmetric? How do you know? *(Symmetric; the prefix di- is often dropped when naming symmetric ethers.)*

APPLY CONCEPTS Discuss the naming of ethers using a number of specific examples. Explain that because ethers contain nonpolar and polar regions and because they are very inert to most chemical reagents, they make excellent solvents for reactions involving nonpolar and polar organic compounds.

CRITICAL THINKING Ask students whether it is possible to have a compound that contains both ether and alcohol functional groups. *(Yes, 4-methoxy-2-butanol is an example.)*

Extend

Explain that polyethers are compounds with more than one ether group. In small groups, have students research the structures of polyethylene glycol and polypropylene glycol, the characteristics of each compound, and some important uses of each. Ask students to display their findings in the class.

Q: *Menthol is the common name of the organic compound responsible for the cooling sensation in many consumer products. Menthol is a cyclohexanol with two alkyl substituents. Is menthol an alcohol or an ether? How can you tell?*

Ethers

What is the general formula of an ether?

Another class of organic compounds may sound familiar to you—ethers. An **ether** is an organic compound in which oxygen is bonded to two carbon groups. **The general formula of an ether is ROR.** Like an alcohol, an ether molecule is bent because of the unshared pairs of electrons on the oxygen atom.

R—Ö—R

Ether

To name an ether using the IUPAC system, first you need to identify the two R groups. The smaller R group is treated as part of the substituent and the *-ane* or *-ene* ending is replaced with *-oxy.* The larger R group is the parent hydrocarbon. In the common names of ethers, both R groups are treated as substituents. The *-ane* or *-ene* endings are replaced with *-yl* for both R groups. To form the common name, list the names of the two R groups in alphabetical order and follow with the word *ether.* Two simple ethers are shown below along with their IUPAC and common names.

$CH_3CH_2—O—CH_3$ — Methoxyethane (ethyl methyl ether)

$CH_3—O—C_6H_5$ — Methoxybenzene (methyl phenyl ether)

Ethyl methyl ether and methyl phenyl ether are nonsymmetric molecules. This is because the R groups attached to the oxygen atom are different. When both R groups are the same, the ether is symmetric. The same rules apply when naming symmetric ethers with the IUPAC system as shown in the examples below. The common names of symmetric ethers use the prefix *di-*. Sometimes, however, the prefix *di-* is dropped and a compound such as diethyl ether is simply called ethyl ether.

$CH_3CH_2—O—CH_2CH_3$ — Ethoxyethane (diethyl ether)

$C_6H_5—O—C_6H_5$ — Phenoxybenzene (diphenyl ether)

Figure 23.7 Isoflurane
Isoflurane ($C_3H_2ClF_5O$) is a halogenated ether that is used as an inhaled anesthetic. This veterinarian is administering an inhaled anesthetic to a koala.
Observe ***Is isoflurane a symmetric or a nonsymmetric ether?***

Diethyl ether was the first reliable general anesthetic. Doctors used it for more than a century. However, because diethyl ether is highly flammable and often causes nausea, it was eventually replaced by other anesthetics such as isoflurane shown in Figure 23.7. Diphenyl ether is used in the manufacture of perfumes and soaps for its characteristic geranium smell.

Ethers usually have lower boiling points than alcohols of comparable molar mass, but higher boiling points than comparable hydrocarbons and halocarbons. Unlike alcohols, ethers are not capable of forming hydrogen bonds with other ether molecules. The oxygen atom in an ether is a hydrogen acceptor. But recall that ethers have no hydroxy hydrogen atoms to donate in hydrogen bonding. Ethers can, however, form hydrogen bonds with water. So ethers are more soluble in water than hydrocarbons and halocarbons. Ethers are less soluble in water than alcohols because they form fewer hydrogen bonds than alcohols do when interacting with water.

Professional Development Note

Ethers as Solvents

Ethers are used as nonpolar solvents to extract natural products such as lipids. Low molar-mass ethers are partially soluble in water due to the ability of the oxygen atoms in the ether functional groups to form hydrogen bonds with water molecules. Solubility in water decreases as the mass of the carbon groups increases.

Amines

What is the general formula of an amine?

An **amine** is an organic compound in which nitrogen is bonded to a carbon group. Amines are similar to ammonia (NH_3). When one, two, or three of the hydrogens in ammonia are replaced by carbon groups, the compound is classified as an amine. **The general formula of an amine is RNH_2, R_2NH, or R_3N.** Amines can be classified according to the number of R groups attached to the nitrogen atom. An amine with the general formula RNH_2 is a primary amine because one R group is attached to the nitrogen atom. Amines with two and three R groups attached to the nitrogen atom are secondary and tertiary amines, respectively.

The IUPAC system for naming primary amines is similar to that of alcohols. The *-e* ending of the parent hydrocarbon is changed to *-amine*. For example, CH_3NH_2 is named methanamine. The common names of primary amines are similar to the IUPAC names. The alkyl or aryl group is named and followed by *-amine*. The common name for methanamine is methylamine. Figure 23.8 shows ball-and-stick models of the two simple primary amines ethanamine and benzenamine.

$CH_3CH_2NH_2$ — Ethanamine (ethylamine)

$C_6H_5NH_2$ — Benzenamine (phenylamine)

Like alcohols, primary amines form intermolecular hydrogen bonds. Because nitrogen is less electronegative than oxygen, the hydrogen bonds in amines are not as strong as those of alcohols. As a result, primary amines have lower boiling points than alcohols with a comparable number of carbons. Amines can also hydrogen bond with water. So smaller amines are soluble in water, but as the number of carbons increase, the solubility in water decreases.

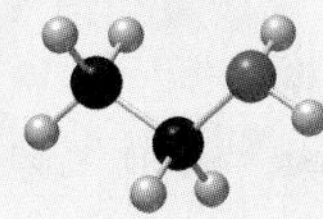

Ethanamine

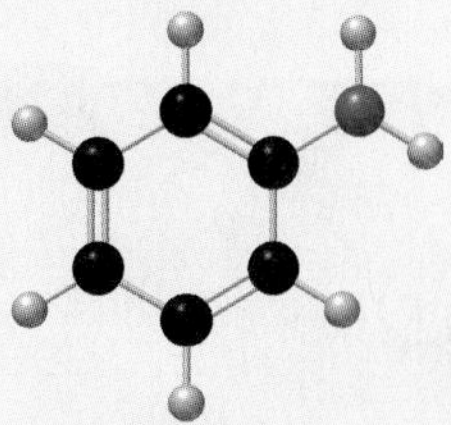

Benzenamine

Figure 23.8 Primary Amines Ethanamine is used to make plastics, pharmaceuticals, and pesticides. Benzenamine is used to make furniture foam and some of the dyes that give clothing their colors.
Explain ***Why are these compounds primary amines?***

23.2 LessonCheck

9. Review What is the general formula of an alcohol?

10. Review How are addition reactions commonly used in organic chemistry?

11. Describe Write the general formula of an ether.

12. List What are three possible general formulas of an amine?

13. Identify Write the IUPAC and common names of each compound below.

a. $CH_3CH_2CH(OH)CH_3$

b. $CH_3CH_2CH_2NH_2$

c. $CH_3CH_2CH_2OCH_2CH_2CH_2CH_3$

14. Predict Give the structure for the expected organic product from each of the following addition reactions.

a. $CH_3CH{=}CHCH_3 + H_2O \xrightarrow[100°C]{H^+}$

b. $CH_2{=}CHCH_3 + Cl_2 \longrightarrow$

c. $CH_3CH{=}CHCH_3 + HBr \longrightarrow$

d. $CH_3CH{=}CHCH_3 + H_2 \xrightarrow{catalyst}$

BIGIDEA REACTIONS

15. Write a short paragraph in which you compare and contrast substitution reactions and addition reactions. Include at least two examples of each reaction type, and describe your examples with balanced equations.

Lesson Check Answers

9. ROH
10. to introduce new functional groups to organic molecules and to convert alkenes to alkanes
11. ROR
12. RNH_2, R_2NH, R_3N
13. a. 2-butanol, *sec*-butyl alcohol
    b. propanamine, propylamine
    c. propoxybutane, butyl propyl ether
14. a. $CH_3CH_2CH(OH)CH_3$
    b. $CH_2ClCHClCH_3$
    c. $CH_3CH_2CHBrCH_3$
    d. $CH_3CH_2CH_2CH_3$
15. **BIGIDEA** Students' paragraphs should mention that substitution reactions involve atom substitution, and addition reactions involve atom addition. Check students' example reactions.

LESSON 23.2

Explain

Amines

MAKE A CONNECTION Explain that amines are known for their often unpleasant odor. Ask students if they are familiar with the odor of decomposing meat. Two of the compounds responsible for this odor are amines, and are appropriately named putrescine ($H_2N(CH_2)_4NH_2$) and cadaverine ($H_2N(CH_2)_5NH_2$). Point out the relationship between the name *putrescine* and the word *putrid*, and the name *cadaverine* and the word *cadaver*. Draw the structures of the two compounds on the board and challenge students to determine the IUPAC names for each. *(putrescine: 1,4-butanediamine; cadaverine: 1,5-pentanediamine)*

Evaluate

Informal Assessment

Have students draw the structural formulas for 3-pentanol, 2-methyl-2-butanol, 1-heptanol, ethyl propyl ether, and diphenyl ether. Have them classify each alcohol as primary, secondary, or tertiary. Have students name and describe a chemical test for unsaturation with an organic molecule, and write the chemical equation for the reaction. Write the structure of 2-butanol on the board. **Ask** What reactants could a chemist use to prepare this compound? *(2-butene and aqueous acid)* Then have students complete the 23.2 Lesson Check.

Reteach

Review the lesson and clarify that the two atoms being added to double-bonded carbons in an alkene can be identical or they can be different. The atoms added during hydrogenation are the same. Make sure students understand that the atoms added are different when HCl reacts with an alkene to produce an alkyl chloride or when water and an alkene react to produce an alcohol.

Answers

FIGURE 23.7 nonsymmetric

FIGURE 23.8 because only one R group is attached to the nitrogen atom

LESSON 23.3

Key Objectives

23.3.1 IDENTIFY the structural characteristic that an aldehyde and a ketone share.

23.3.2 IDENTIFY the general formula of a carboxylic acid.

23.3.3 EXPLAIN why dehydrogenation is classified as an oxidation reaction.

23.3.4 IDENTIFY the general formula of an ester.

Additional Resources

- Reading and Study Workbook, Lesson 23.3
- Probeware Laboratory Manual

PearsonChem.com

Engage

CHEMISTRY & YOU Have students study the opening text and photograph. **Ask** How do you think the natural flavors and odors of a real banana compare chemically to artificial banana flavoring? *(The natural odor and flavor of a real banana are likely due to a mix of organic chemicals, while artificial flavoring is generally made with a single organic compound.)*

Activate Prior Knowledge

Explain that many reactions involving carbonyl groups are reversible, and their results depend on equilibrium position. Have students review what they know about the factors that affect equilibrium.

Foundations for Reading

BUILD VOCABULARY Use the following terms and help students create a concept map: *carbonyl group, aldehydes, ketones, carboxylic acids, carboxyl group,* and *fatty acids.*

IN ACADEMIC STANDARDS for SCIENCE

- Develop explanations based on reproducible data and observations gathered during laboratory investigations.

C.4.5 Describe, classify and give examples of various kinds of reactions-synthesis (combination), decomposition, single displacement, double displacement and combustion.

23.3 Carbonyl Compounds

C.4.5 Classify types of reactions

CHEMISTRY & YOU

Q: *What gives a banana its characteristic smell?* Many organic molecules have pleasant aromas. These molecules provide the fragrances of certain foods and flowers. You will learn about a number of them in this lesson.

Key Questions

What structural characteristic do an aldehyde and a ketone share?

What is the general formula of a carboxylic acid?

Why is dehydrogenation classified as an oxidation reaction?

What is the general formula of an ester?

Vocabulary

- carbonyl group
- aldehyde
- ketone
- carboxyl group
- carboxylic acid
- fatty acid
- dehydrogenation reaction
- ester

Aldehydes and Ketones

What structural characteristic do an aldehyde and a ketone share?

Remember that in an alcohol, an oxygen atom is bonded to a carbon group and a hydrogen atom. In an ether, an oxygen atom is bonded to two carbon groups. An oxygen atom can also be bonded to a single carbon atom by a double covalent bond. Such an arrangement is called a carbonyl group. A **carbonyl group** is a functional group with the general structure C=O.

The C=O functional group is present in aldehydes and ketones. An **aldehyde** is an organic compound in which the carbon of the carbonyl group is joined to at least one hydrogen. The general formula of an aldehyde is RCHO. A **ketone** is an organic compound in which the carbon of the carbonyl group is joined to two other carbons. The general formula of a ketone is RCOR. The structures of an aldehyde and a ketone are summarized below.

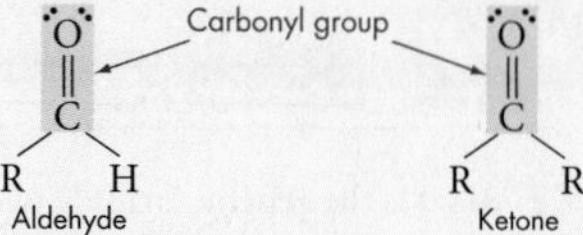

The IUPAC system may be used for naming aldehydes and ketones. For either class of compounds, first identify the longest carbon chain that contains the carbonyl group. Replace the *-e* ending of the parent structure with *-al* to designate an aldehyde. In the IUPAC system, the continuous-chain aldehydes are named methanal, ethanal, propanal, butanal, and so forth.

Ketones are named by changing the ending of the longest continuous carbon chain that contains the carbonyl group from *-e* to *-one.* If the carbonyl group of a ketone could occur at more than one place on the chain, then its position is designated by the lowest possible number. Table 23.5 illustrates the naming of some common aldehydes and ketones.

812 Chapter 23 • Lesson 3

Focus on ELL

1 CONTENT AND LANGUAGE Write the lesson objectives on the board, highlighting key terms in each objective. Provide students with a graphic organizer such as a Venn diagram so that they can compare and contrast aldehydes and ketones.

2 FRONTLOAD THE LESSON Divide the class into four groups, one for each key question. Have one group member read the question aloud slowly. The group should predict the answer and then try to find the answer by previewing the lesson. Ask all groups to present their findings to the class.

3 COMPREHENSIBLE INPUT Use ball-and-stick models to demonstrate key functional groups and reactions in the text. Have students view the Kinetic Art animation of oxidation reactions to supplement the discussion of the Oxidation-Reduction section.

Table 23.5

Some Common Aldehydes and Ketones

Compound type	IUPAC name (common name)	Structural formula	Ball-and-stick model
Aldehyde	Methanal (formaldehyde)	$H-\overset{\overset{O}{\|}}{C}-H$	
Aldehyde	Ethanal (acetaldehyde)	$CH_3-\overset{\overset{O}{\|}}{C}-H$	
Aldehyde	Benzaldehyde (benzaldehyde)	$C_6H_5-\overset{\overset{O}{\|}}{C}-H$	
Ketone	Propanone (acetone)	$CH_3-\overset{\overset{O}{\|}}{C}-CH_3$	
Ketone	Diphenylmethanone (benzophenone)	$C_6H_5-\overset{\overset{O}{\|}}{C}-C_6H_5$	

Uses of Aldehydes and Ketones The simplest aldehyde is methanal (HCHO), also called formaldehyde. Methanal is very important industrially. Its greatest use is in the manufacture of synthetic resins. Methanal is usually available as a 40% aqueous solution, known as formalin. Formalin can be used to preserve biological specimens. The methanal in solution combines with protein in tissues to make the tissues hard and insoluble in water. This prevents the specimen from decaying.

The most common industrial ketone is propanone, also called acetone. Propanone is a colorless, volatile liquid that boils at 56°C. Propanone is used in industry as a solvent for resins, plastics, and varnishes. Many nail-polish removers contain propanone, too.

LESSON 23.3

READING STRATEGY Have students write the general formula for an alcohol and an ether. As they read about aldehydes, ketones, and carboxylic acids, have them write those general formulas and compare them with the formulas of alcohols and ethers.

Explain

Aldehydes and Ketones

USE VISUALS Have students examine Table 23.5. Pronounce the IUPAC and the common name for each compound. **Ask** Are you familiar with any of these compounds? If so, in what respect? *(Answers will vary.)* Discuss how these compounds are named. Explain that the difference between an aldehyde and a ketone lies in whether a hydrogen atom is attached to the carbonyl group. Display the table using a transparency and overhead projector. Point out that in ketones, as in ethers, the functional group lies between two R groups. Except for methanal, the carbonyl group in aldehydes is attached to one R group and one hydrogen atom.

Explore

Class Activity

Have students gently sniff the contents of labeled test tubes containing small amounts of vanillin, isopentyl acetate, and benzaldehyde. **Ask** Which odor is that of vanilla ice cream? *(vanillin)* Bananas? *(isopentyl acetate)* Maraschino cherries? *(benzaldehyde)* Write structures for the compounds on the board. Refer to Table 23.5 for benzaldehyde and vanillin; isopentyl acetate is $CH_3COOCH_2CH_2CH(CH_3)_2$. Explain that compounds with carbonyl groups account for the odor and taste of many foods.

MAKING CONNECTIONS Explain that isopentyl acetate is referred to as banana oil, due to its banana-like flavor and smell. Mention to students that this compound was first discovered as one of the alarm pheromones of honeybees. Later it was discovered in yellow passion fruits, pears, and apples.

Vanilla Beans

Vanilla flavoring comes from the seed capsules, called "beans," of several species of orchid of the genus *Vanilla*. The capsules are collected, boiled, dried, and cured to produce the long, thin, brown bean-like pods that are familiar to cooks everywhere. The vanilla orchid was first used as a flavoring by the Aztecs and the Mayans in Mexico and Central America for their *xocolatl* (chocolate) drinks. Due to advances in pollination techniques, most of the natural vanilla produced today comes from Madagascar, Reunion Island, and Tahiti rather than from Central America.

Explore

Class Activity

Many students may find it helpful in understanding the differences between aldehydes and ketones if they build models of compounds containing these functional groups. Have students make models of propanal, propanone, and propanoic acid. Have them examine their models and make comparisons. If model-building kits are not available, toothpicks and gumdrops or marshmallows can be used. Direct students not to eat gumdrops or marshmallows used in model building.

Explain

START A CONVERSATION Explain that many familiar smells and flavors come from organic compounds. For example, many ketones and aldehydes, particularly those with high molar masses, have fragrant odors. **Ask** What are some functional groups other than aldehydes and ketones that can have an effect on the particular odor of an organic compound? *(alcohols, esters, amines)* Have students name some of their favorite scents. **Ask** What are the natural sources of these perfumes? *(animal oils, plant extracts)* Explain that perfume makers usually combine many different compounds to create a complex scent. Although these compounds may come from animal oils and plant extracts, they can be synthesized artificially. Citral, found in roses and citrus fruits, is an aldehyde. Civetone, the odor of musk, is a ketone.

USE VISUALS Direct students' attention to Table 23.6. Ask students to think of some examples of simple organic molecules that contain carbonyl groups. Have them write the structures and build models of these molecules. **Ask** What are some structural features of carbonyl-containing molecules? *(Carbonyl groups display trigonal planar geometry.)* Point out that a carbonyl carbon represents the positive center of aldehydes and ketones and that an oxygen atom represents the negative center. Discuss the types of intermolecular forces operating between the functional groups in aldehydes and ketones. **Ask** How does the shape of a molecule affect the strength of intermolecular attractions? *(If the shape causes bond polarities within the molecule to be unbalanced, the molecule is usually polar, and intermolecular forces will be stronger than if the molecule were nonpolar.)*

Interpret Data

Some Organic Compounds with Three Carbons

Compound	Formula	Boiling point (°C)	Primary intermolecular interactions
Propane	$CH_3CH_2CH_3$	−42	Dispersion forces
Propanal	CH_3CH_2CHO	49	Polar-polar interactions
Propanone	CH_3COCH_3	56	Polar-polar interactions
1-Propanol	$CH_3CH_2CH_2OH$	97	Hydrogen bonding

Table 23.6 These four organic compounds have the same number of carbon atoms. Yet their boiling points are different.
a. Classify Determine the type of each organic compound.
b. Compare Use your knowledge of intermolecular interactions to explain the similarities and differences in the boiling points of the four compounds.

A wide variety of aldehydes and ketones have been isolated from plants and animals. Many of them, particularly those with high molar masses, have fragrant odors. They are usually known by their common names, which can indicate their natural sources or perhaps a characteristic property. Benzaldehyde is the simplest aromatic aldehyde. It's also known as oil of bitter almond because it is largely responsible for the taste and aroma of almonds. Many flavoring agents, including vanilla bean extract and the cinnamon sticks shown in Figure 23.9, contain aromatic aldehydes.

Figure 23.9 Cinnamaldehyde
The cinnamon sticks in this drink contain an aldehyde with the common name cinnamaldehyde.

Properties of Aldehydes and Ketones Aldehydes and ketones cannot form intermolecular hydrogen bonds because they lack —OH and —NH groups. Consequently, they have boiling points that are lower than those of corresponding alcohols. Aldehydes and ketones can attract each other, however, through polar–polar interactions of their carbonyl groups. As a result, their boiling points are higher than those of the corresponding alkanes. These attractive forces account for the fact that nearly all aldehydes and ketones are either liquids or solids at room temperature. The exception is methanal, which is a gas. Table 23.6 compares the boiling points of an alkane, an aldehyde, a ketone, and an alcohol with the same number of carbon atoms.

Aldehydes and ketones can form weak hydrogen bonds between the carbonyl oxygen and the hydrogen atoms of water. The lower members of the series—up to three carbons—are soluble in water in all proportions. As the length of the hydrocarbon chain increases above four, however, water solubility decreases. When the carbon chain exceeds five or six carbons, the solubility of both aldehydes and ketones is very low. As might be expected, all aldehydes and ketones are soluble in nonpolar solvents.

Differentiated Instruction

LPR LESS PROFICIENT READERS Write the general structures of an aldehyde, a ketone, a carboxylic acid, and an ester on the board. Have students copy the structures in their notebooks, and label the carbonyl groups, hydroxy groups, aliphatic groups, and aryl groups. Have them indicate whether each structure is an aldehyde, ketone, ester, or carboxylic acid.

ELL ENGLISH LANGUAGE LEARNERS Arrange students in several small topic circles with proficient students in each group. Have them discuss aldehydes and ketones in a factual, detailed progression. Ask each group to summarize their discussion for you as you circulate between groups to assess understanding and correct misconceptions.

L3 ADVANCED STUDENTS Have students study the structure of cinnamaldehyde and other carbonyl-containing compounds that have carbon-carbon bonds conjugated with a carbon-oxygen double bond. Have students consider how the location of a carbonyl group next to an alkene group affects the stability of the molecule through resonance.

Carboxylic Acids

What is the general formula of a carboxylic acid?

A **carboxyl group** is a functional group that consists of a carbonyl group attached to a hydroxy group. It can be written as —COOH or —CO_2H. A **carboxylic acid** is an organic compound with a carboxyl group. **The general formula of a carboxylic acid is RCOOH.**

$$\underset{\text{Carboxylic acid}}{\mathrm{R{-}C({=}O){-}OH}}$$

(Carbonyl group: C=O; Hydroxy group: OH)

As you may have guessed from their name, carboxylic acids are acidic. Carboxylic acids are considered weak acids because they ionize weakly in solution. In water, they can lose a hydrogen ion and form a carboxylate ion, as shown in the following reaction.

$$\underset{\text{Carboxylic acid}}{\mathrm{R{-}\overset{O}{\overset{\|}{C}}{-}OH}} \rightleftharpoons \underset{\text{Carboxylate ion}}{\mathrm{R{-}\overset{O}{\overset{\|}{C}}{-}O^-}} + \underset{\text{Hydrogen ion}}{\mathrm{H^+}}$$

In the IUPAC system, carboxylic acids are named by replacing the *-e* ending of the parent structure with the ending *-oic acid*. Remember, the parent hydrocarbon of a carboxylic acid is the longest continuous carbon chain containing the carboxyl group. So, for example, the carboxylic acid CH_3COOH is named ethanoic acid. Figure 23.10 shows the ball-and-stick model of ethanoic acid. Table 23.7 lists the names and formulas of some common saturated aliphatic carboxylic acids.

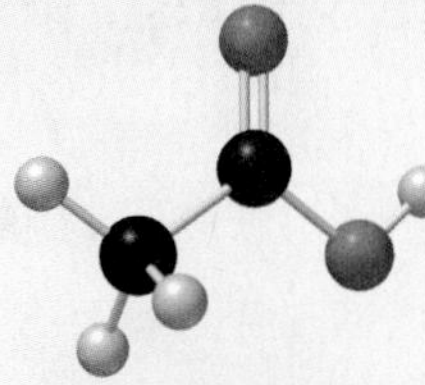

Figure 23.10 Ethanoic Acid Ethanoic acid is a simple carboxylic acid. It is a colorless, volatile liquid.
Identify ***What R group is attached to the carboxyl group in ethanoic acid?***

Table 23.7

Some Saturated Aliphatic Carboxylic Acids

Molecular formula	Carbon atoms	IUPAC name	Common name	Melting point (°C)
HCOOH	1	Methanoic acid	Formic acid	8
CH_3COOH	2	Ethanoic acid	Acetic acid	17
CH_3CH_2COOH	3	Propanoic acid	Propionic acid	−22
$CH_3(CH_2)_2COOH$	4	Butanoic acid	Butyric acid	−6
$CH_3(CH_2)_4COOH$	6	Hexanoic acid	Caproic acid	−3
$CH_3(CH_2)_6COOH$	8	Octanoic acid	Caprylic acid	16
$CH_3(CH_2)_8COOH$	10	Decanoic acid	Capric acid	31
$CH_3(CH_2)_{10}COOH$	12	Dodecanoic acid	Lauric acid	44
$CH_3(CH_2)_{12}COOH$	14	Tetradecanoic acid	Myristic acid	58
$CH_3(CH_2)_{14}COOH$	16	Hexadecanoic acid	Palmitic acid	63
$CH_3(CH_2)_{16}COOH$	18	Octadecanoic acid	Stearic acid	70

UbD Check for Understanding

What structural characteristics do an aldehyde and a ketone share?

Have students complete the following sentence:

The main structural difference between and aldehyde and a ketone is that in an aldehyde the carbon of the carbonyl group is joined to at least one ______, whereas in a ketone the carbon of the carbonyl group is joined to two ______.

(hydrogen; carbons)

ADJUST INSTRUCTION If students are having difficulty completing the sentence, briefly review the text leading up to Table 23.5. Then have students repeat the exercise.

Explain

Carboxylic Acids

START A CONVERSATION Write the structural formula for a carboxylic acid on the board or overhead projector. **Ask** What functional groups are present, and what characteristics would you expect the molecule to have? *(Students should recognize the hydroxy and carbonyl groups and may suggest that the molecule would have properties of both an alcohol and an aldehyde.)* **Ask** Why are carboxylic acids considered to be acids? *(They contain an ionizable proton.)* Explain that the presence of the double-bonded oxygen weakens the bond between the other oxygen atom and the hydrogen atom bonded to it, so that the hydrogen can ionize easily. Point out that the overall functional group is called a carboxyl group and the molecule containing it is called a carboxylic acid.

USE VISUALS Use an overhead projector to display a transparency of Table 23.7. Discuss the IUPAC rules for naming carboxylic acids. Remind students that the systematic names are derived from the parent alkanes. Discuss the basis for the observed trends in physical properties for the homologous series of aliphatic carboxylic acids. **Ask** What is the name of an aliphatic carboxylic acid that is a solid at room temperature? *(Answers will vary, but should be restricted to those acids with melting points higher than 25°C.)* **Ask** What is the name of an aliphatic carboxylic acid that is a liquid at room temperature? *(Answers will vary, but should be restricted to those acids with melting points lower than 25°C.)* **Ask** What is the common name of an 18-carbon aliphatic carboxylic acid? *(stearic acid)*

Extend

Have students consider why carboxylic acids are considered weak acids. Write the chemical equation for the ionization of dilute aqueous ethanoic acid on the board. Point out that the acid is only partly ionized at equilibrium. Challenge students to write the equilibrium constant expression for the reaction shown. Then have them write the chemical equation for the ionization of propanoic acid.

Answers

INTERPRET DATA

a. propane, hydrocarbon; propanal, aldehyde; propanone, ketone; propanol, alcohol

b. Hydrogen bonds are stronger than polar-polar interactions, which are stronger than dispersion forces. Thus, the alcohol has the highest boiling point, and the hydrocarbon has the lowest boiling point. The aldehyde and the ketone have similar boiling points because they experience the same types of interactions.

FIGURE 23.10 methyl group (CH_3)

Explain

USE VISUALS Have students study the photograph in Figure 23.11. Explain that carboxylic acids are found in many natural products. Ask students to name the acid in the lemons and limes shown. Write the structural formula for the acid on the board. Point out the carboxyl functional groups. Have students name other examples of food products that contain acids and describe the tastes and odors associated with carboxylic acids.

Oxidation-Reduction Reactions

APPLY CONCEPTS Go through the sequence of oxidations that occurs for alkanes. Explain that carbon dioxide is the most oxidized form of carbon, and an alkane is the most reduced form of carbon. Discuss each step as being an example of an oxidation or reduction. Remind students that the loss of hydrogen is oxidation, and the gain of hydrogen is reduction. In any series of oxidations or reductions involving hydrocarbons, an alkane is the least oxidized and an alkyne is most oxidized.

Figure 23.11 Citric Acid
A common carboxylic acid is citric acid, which is found in lemons and limes.
Use Models *How many carboxyl groups does a citric acid molecule have?*

Carboxylic acids are abundant and widely distributed in nature. The lemons and limes in Figure 23.11 contain citric acid, a carboxylic acid that gives citrus fruits their sour flavor. The IUPAC name for citric acid is 2-hydroxypropane-1,2,3-tricarboxylic acid. Many carboxylic acids have common names derived from a Greek or Latin word that describes their natural sources. For example, the common name for ethanoic acid is acetic acid, which comes from the Latin word *acetum*, meaning vinegar. Common household vinegar contains about 5% (v/v) acetic acid. Many carboxylic acids were first isolated from fats and are called **fatty acids.** Propionic acid, the three-carbon acid, literally means first fatty acid.

Like alcohols, carboxylic acids form intermolecular hydrogen bonds. Thus, carboxylic acids have higher boiling and melting points than other compounds of similar molar mass. The low-molar-mass members of the aliphatic carboxylic acid series are colorless, volatile liquids. The higher members of the series are nonvolatile, waxy solids with low melting points. All aromatic carboxylic acids are solids at room temperature.

The carboxyl group in carboxylic acids is polar and readily forms hydrogen bonds with water molecules. As a result, methanoic, ethanoic, propanoic, and butanoic acids are completely soluble in water. After four carbons, however, the solubility drops sharply. Most carboxylic acids are also soluble in organic solvents such as ethanol or propanone.

Oxidation-Reduction Reactions

Why is dehydrogenation classified as an oxidation reaction?

The classes of organic compounds you have just studied—aldehydes, ketones, and carboxylic acids—are related by oxidation and reduction reactions. Recall from Chapter 20 that oxidation is the gain of oxygen, loss of hydrogen, or loss of electrons and reduction is the loss of oxygen, gain of hydrogen, or gain of electrons. Also remember that one does not occur without the other.

In organic chemistry, the number of oxygen atoms and hydrogen atoms attached to carbon indicates the degree of oxidation of a compound. The fewer hydrogens on a carbon–carbon bond, the more oxidized the bond. Thus, a triple bond is more oxidized than a double bond, which is more oxidized than a single bond.

The loss of a molecule of hydrogen from an organic molecule is called a **dehydrogenation reaction.** It is the opposite of a hydrogenation reaction. Strong heating and a catalyst are usually needed to make dehydrogenation reactions occur. **Dehydrogenation is an oxidation reaction because the loss of each molecule of hydrogen involves the loss of two electrons from the organic molecule.** The remaining carbon electrons pair to make a second or third bond, as shown in the reactions below.

See oxidation reactions animated online.

$$H_3C{-}CH_3 \xrightarrow[\text{oxidation}]{\text{loss of hydrogen (dehydrogenation)}} H_2C{=}CH_2 \xrightarrow[\text{oxidation}]{\text{loss of hydrogen (dehydrogenation)}} H{-}C{\equiv}C{-}H$$

Least oxidized (most reduced) — Most oxidized (least reduced)

Differentiated Instruction

L1 STRUGGLING STUDENTS Students may find it difficult to understand that oxidation and reduction can be defined in terms of the loss or gain of hydrogen. Pair up students who are having trouble with this concept. Have students write out each series of oxidation-reduction reactions on this and the next few pages of the text, noting the number of hydrogen and oxygen atoms attached to carbon at each step.

L3 ADVANCED STUDENTS Have students assign oxidation numbers to carbon for each step in the sequence of oxidation reactions shown in the text in which methane is converted to carbon dioxide. **Ask** How many molecules of hydrogen are given up by methane? *(2)* How many electrons are lost? *(4)* How many molecules of H_2 would ethane yield? *(3)*

Oxidation in organic chemistry also involves the number and degree of oxidation of oxygen atoms attached to a carbon atom. For example, methane, a saturated hydrocarbon, can be oxidized in steps to carbon dioxide. This occurs if it alternately gains oxygen atoms and loses hydrogen atoms. Methane is oxidized to methanol, then to methanal, then to methanoic acid, and finally to carbon dioxide. The carbon dioxide is most oxidized or least reduced, and methane is least oxidized or most reduced.

$$\underset{\text{Methane (most energetic molecule)}}{CH_4} \xrightarrow[\text{oxidation}]{\text{gain of oxygen}} \underset{\text{Methanol}}{H{-}\overset{H}{\underset{H}{C}}{-}OH} \xrightarrow[\text{oxidation}]{\text{loss of hydrogen}} \underset{\text{Methanal}}{H{-}\overset{O}{\overset{\|}{C}}{-}H} \xrightarrow[\text{oxidation}]{\text{gain of oxygen}} \underset{\text{Methanoic acid}}{H{-}\overset{O}{\overset{\|}{C}}{-}OH} \xrightarrow[\text{oxidation}]{\text{loss of hydrogen}} \underset{\text{Carbon dioxide (least energetic molecule)}}{O{=}C{=}O}$$

The more reduced a carbon compound is, the more energy it can release upon its complete oxidation to carbon dioxide. The oxidation of organic compounds is exothermic. The energy-releasing properties of oxidation reactions are extremely important for the production of energy in living systems. To play a sport, such as the one shown in Figure 23.12, your body must produce energy by oxidation reactions.

Primary alcohols can be oxidized to aldehydes, and secondary alcohols can be oxidized to ketones by warming them with acidified potassium dichromate ($K_2Cr_2O_7$). Tertiary alcohols, however, cannot be oxidized because there is no hydrogen atom present on the carbon atom attached to the hydroxy group. Examples of oxidation of a primary and secondary alcohol are shown below.

$$\underset{\text{Ethanol (primary alcohol)}}{CH_3{-}\overset{O}{\overset{|}{\underset{H}{\underset{|}{C}}}}{-}H} \xrightarrow[H_2SO_4]{K_2Cr_2O_7} \underset{\text{Ethanal (aldehyde)}}{CH_3{-}\overset{O}{\overset{\|}{C}}{-}H}$$

$$\underset{\text{2-Propanol (secondary alcohol)}}{CH_3{-}\overset{OH}{\overset{|}{\underset{H}{\underset{|}{C}}}}{-}CH_3} \xrightarrow[H_2SO_4]{K_2Cr_2O_7} \underset{\text{Propanone (ketone)}}{CH_3{-}\overset{O}{\overset{\|}{C}}{-}CH_3}$$

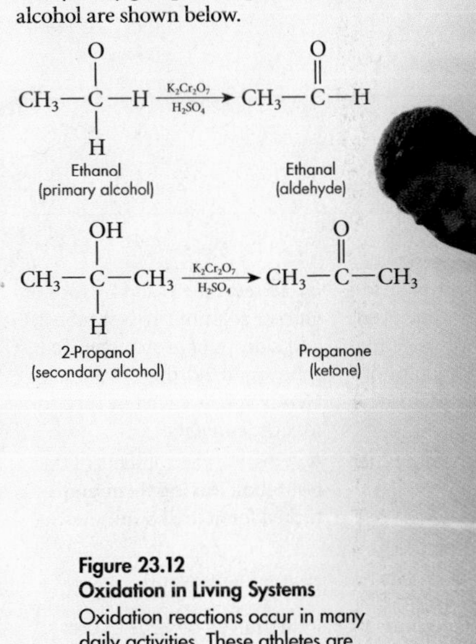

Figure 23.12
Oxidation in Living Systems
Oxidation reactions occur in many daily activities. These athletes are energized by oxidation reactions taking place within the cells of their bodies.

UbD Check for Understanding

The Essential Question What effect does a functional group have on an organic compound?

Ask When a carbonyl group is attached to a hydroxyl group, what is its effect on the organic compound? *(Sample answers: It is a weak acid and has higher boiling and melting points than other compounds of similar molar mass.)*

ADJUST INSTRUCTION If students are having difficulty answering, have them review the text before and after Table 23.7. Then allow them to revise their answers.

Explore

Teacher Demo

PURPOSE Students observe the oxidation of primary and secondary alcohols to carboxylic acids and ketones, respectively, by chromic acid.

MATERIALS CrO_3, concentrated sulfuric acid, water, 3 test tubes, acetone, ethanol, isopropanol, and *t*-butyl alcohol

PROCEDURE Explain how the loss of a molecule of hydrogen by an alcohol can lead to the formation of a carbonyl group. Prepare an acidic solution of CrO_3 by dissolving 0.25 g CrO_3 in 0.75 mL water and adding 0.25 mL concentrated sulfuric acid. A clear orange solution should be obtained. Set up three test tubes, each containing 1 mL acetone. Dissolve about 5 drops of ethanol, isopropanol, and *t*-butyl alcohol in separate tubes. Add 1–2 drops of chromic acid reagent to each tube and gently tap the tubes to mix. Note any changes that occur. A positive test for a primary or secondary alcohol is indicated by a change in color of the solution from orange to blue-green. No change should occur in the case of *t*-butyl alcohol. Have students record their observations and write the chemical equations for any reactions they observed. **Ask** Why was no reaction observed for the test tube containing t-butyl alcohol? *(There is no hydrogen on the carbon bearing the hydroxyl group.)*

Answers

FIGURE 23.11 three

Explore

Quick Lab

OBJECTIVE After completing this activity, students will be able to describe a positive test for an aldehyde.

TEACHING TIP It may be necessary to warm the test tube in a bath of warm water. *Note*: The silver nitrate solution in the *Materials* list is 5% (w/v).

PREP TIME 20 minutes

CLASS TIME 30 minutes

SAFETY Do not store Tollens' reagent for more than a few hours. It decomposes on standing and yields an explosive mixture. Neutralize it with a dilute acid and flush with plenty of water. Avoid skin contact with these chemicals.

EXPECTED OUTCOME The deposition of a silver mirror on the inner walls of the test tube is a positive indicator for the presence of an aldehyde functional group. (The silver mirror can be removed from the inside of the test tube with nitric acid.)

ANALYZE AND CONCLUDE

1. A layer of reflective silver metal is deposited on the inner walls of test tube 1. No reaction is observed in test tubes 2 or 3.
2. In an alkaline solution of silver nitrate, the aldehyde is oxidized to a carboxylic acid according to the following reaction:

$$RCHO(aq) + 2Ag(NH_3)_2^+(aq) + 3OH^-(aq) \rightarrow 2Ag(s) + RCOO^-(aq) + NH_4^+(aq) + 2H_2O(l) + 3NH_3(aq)$$

3. The test is useful for distinguishing aldehydes from alcohols and ketones.

Figure 23.13 Fehling's Test
When an aldehyde is mixed with Fehling's reagent (left test tube) and heated, the blue copper(II) ions in Fehling's reagent are reduced to form Cu_2O, a red precipitate (right test tube). Infer ***What is the oxidation state of copper in the product?***

Aldehydes are so easily oxidized that it is difficult to prevent further oxidation to carboxylic acids as shown in the reaction below.

$$\underset{\text{Aldehyde}}{R-\overset{\overset{\displaystyle O}{\|}}{C}-H} \xrightarrow[H_2SO_4]{K_2Cr_2O_7} \underset{\text{Carboxylic acid}}{R-\overset{\overset{\displaystyle O}{\|}}{C}-OH}$$

Further oxidation is avoided by removing them from the reaction mixture as they are formed. Unlike aldehydes, ketones are relatively resistant to further oxidation, so there is no need to remove them from the mixture during the reaction.

Tests for aldehydes make use of the ease with which these compounds are oxidized. Benedict's and Fehling's reagents are deep-blue alkaline solutions of copper(II) sulfate. Figure 23.13 illustrates Fehling's test for an aldehyde. When an aldehyde is oxidized with Benedict's or Fehling's reagent, a red precipitate of copper(I) oxide (Cu_2O) is formed. The aldehyde is oxidized to its acid, and copper(II) ions (Cu^{2+}) are reduced to copper(I) ions (Cu^+).

Quick Lab

Testing for an Aldehyde

Purpose To distinguish an aldehyde from an alcohol or a ketone using Tollens' reagent

Materials

- 1*M* sodium hydroxide
- 5% silver nitrate
- 6*M* aqueous ammonia
- 4 small test tubes
- test tube rack
- plastic droppers
- glucose solution
- propanone
- ethanol

Procedure

1. Add 1 drop of 1*M* sodium hydroxide to 2 mL of 5% silver nitrate in a test tube. Add 6*M* aqueous ammonia drop by drop, gently agitating the tube after each addition until the brownish precipitate dissolves. This will be your Tollens' reagent.
2. Place 10 drops of Tollens' reagent in each of three clean, labeled test tubes.
3. To test tube 1, add 2 drops of glucose solution. To test tube 2, add 2 drops of propanone. To test tube 3, add 2 drops of ethanol. Gently agitate each test tube to mix the contents.
4. Observe the contents of the test tubes, leaving them undisturbed for at least 5 minutes.

Analyze and Conclude

1. **Observe** What evidence of a chemical reaction did you observe in test tube 1? In test tube 2? In test tube 3?
2. **Describe** Write the equation for any chemical reaction you observed.
3. **Infer** If you observed a chemical reaction in one or more of the test tubes, what practical uses might the reaction have?

Focus on ELL

4 LANGUAGE PRODUCTION Have students work in groups of four to complete the lab. Make sure each group has ELLs of varied language proficiencies, so that more proficient students can help less proficient ones. Have students work according to their proficiency level.

BEGINNING: LOW/HIGH Restate the procedures in simpler terms. Allow students to answer in short written phrases or orally.

INTERMEDIATE: LOW/HIGH Draw a picture of the reaction observed in each of the test tubes.

ADVANCED: LOW/HIGH Discuss and help the group write the equations for each of the chemical reactions observed.

Esters

What is the general formula of an ester?

Esters are probably the most pleasant and delicious organic compounds one can study. Many esters have pleasant, fruity odors. Esters give blueberries, pineapples, apples, pears, bananas, and many other fruits their characteristic aromas. They also give many perfumes their fragrances. An **ester** is an organic compound in which the —OH of the carboxyl group has been replaced by an —OR from an alcohol. Esters contain a carbonyl group and an ether link to the carbonyl carbon. **The general formula of an ester is RCOOR.**

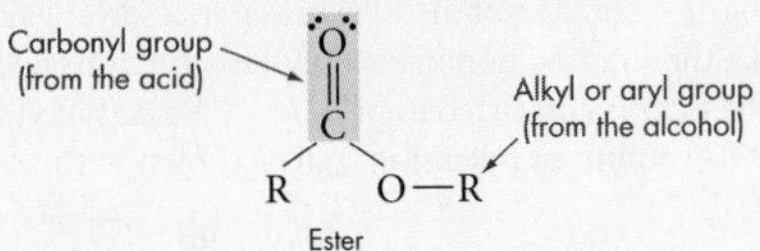

Figure 23.14 shows the esters that contribute to the characteristic odors of bananas and strawberries. Simple esters are neutral substances. Although the molecules are polar, they cannot form hydrogen bonds with one another because they do not contain hydrogen attached to oxygen or another electronegative atom. As a result, only weak attractions hold ester molecules to one another. As you might expect, esters have much lower boiling points than carboxylic acids. The low-molar-mass esters are somewhat soluble in water, but esters containing more than four or five carbons have very limited solubility.

Esters may be prepared from a carboxylic acid and an alcohol. The process is called esterification. The reactants, usually a carboxylic acid and a primary or secondary alcohol, are heated with an acid as a catalyst. The synthesis of ethyl ethanoate from ethanoic acid and ethanol is an example of esterification.

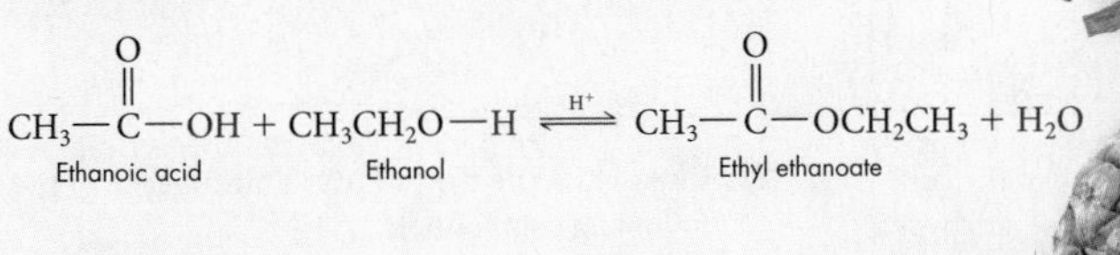

$$CH_3-C(=O)-OH + CH_3CH_2O-H \overset{H^+}{\rightleftharpoons} CH_3-C(=O)-OCH_2CH_3 + H_2O$$

CHEMISTRY & YOU

Q: *The smell of strawberries is largely due to benzyl acetate. The smell of almonds is largely due to benzaldehyde. How are the general formulas of these compounds similar? How are they different?*

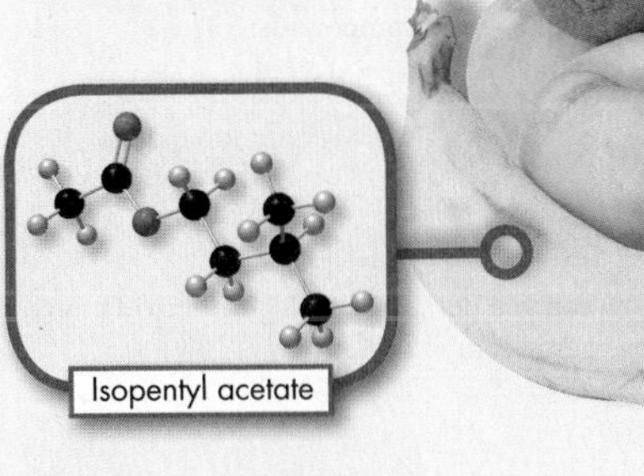

Figure 23.14 Esters
The characteristic aromas of many fruits are due to esters. Isopentyl acetate is an ester found in bananas. Strawberries contain the ester benzyl acetate.

Explore

CHEMISTRY & YOU Both compounds have a carbonyl group and a phenyl group. Benzyl acetate is a 9-carbon ester, and benzaldehyde is a 7-carbon aldehyde.

Teacher Demo

PURPOSE Students will observe the synthesis of an ester from an acid and an alcohol.

MATERIALS water bath, hot plate, test tubes, several carboxylic acids, several corresponding alcohols, concentrated sulfuric acid, filter paper, watch glass

SAFETY AND DISPOSAL Sulfuric acid is caustic and can burn skin and clothing. Neutralize the combined reaction mixtures and flush them down the drain.

PROCEDURE Explain that esters are derivatives of carboxylic acids in which the —OH group of the carboxyl group has been replaced by an —OR group from an alcohol. The synthesis of an ester from an acid and an alcohol is called esterification. Synthesize a variety of fragrant esters in class and give students a chance to smell the products. Set up a water bath on a hot plate to maintain the temperature at about 80°C. To synthesize the ester, mix 1–2 mL (or 0.2–0.6 g if the acid is a solid) of each carboxylic acid with 3–5 mL of the corresponding alcohol in a test tube. Add 1–2 drops of concentrated sulfuric acid and heat for 15 minutes in the water bath. Dip a piece of filter paper into the test tube, and place it on a clean watch glass. Have students record the esters you prepare and their associated odors. Ask them to write the chemical equation for each reaction. Point out that the only difference in these reactions is in the R groups attached to the carboxyl and alcohol functional groups.

EXPECTED OUTCOME

Acid	Alcohol	Odor of Ester
ethanoic	ethyl	apple
ethanoic	isopentyl	banana
ethanoic	pentyl	apricot
ethanoic	octyl	orange
butyric	ethyl	pineapple
salicylic	methyl	wintergreen

Professional Development Note

Scientists Growing Flavors

Scientists are learning how to produce flavor molecules and other naturally occurring compounds in plant cell cultures and genetically altered yeast cells. For example, Japanese scientists are using a tissue culture technique called root culture to produce ginseng for teas and medicines and for a red pigment used in cosmetics.

In the United States, scientists are obtaining vanilla flavor from cells surgically removed from a vanilla orchid and cultured in a glass bioreactor. Prior to any commercial production, a patent application is usually filed, and manufacturers must seek approval from the Food and Drug Administration to use the prepared flavor molecules in food products.

Answers

FIGURE 23.13 +1

LESSON 23.3

Explain

Esters

START A CONVERSATION Explain that the formation of an ester from a carboxylic acid and an alcohol is an excellent example of dynamic equilibrium. Under the proper conditions, the equilibrium can be shifted greatly in favor of the products. Other conditions favor the decomposition of the ester into the carboxylic acid and alcohol. Have students review Le Châtelier's principle and use it to predict conditions that would favor the reactants and conditions that would favor the products. **Ask** How can a chemist improve the yield of an ester from an esterification reaction? *(Use an excess of a reactant, or remove the water or ester by distillation as it is formed.)*

APPLY CONCEPTS Discuss the base-promoted hydrolysis of an ester. Write a number of examples of reactions on the chalkboard or overhead projector. Explain which part of the ester yields the alcohol and which part yields carboxylic acid products. Ask students about the difference in volatility of the products and how a chemist might separate them once they are formed. Write the molecular structure for propyl ethanoate on the board. **Ask** What products would be formed from the base-promoted hydrolysis of propyl ethanoate? *(propanol and a salt of ethanoic acid)*

Evaluate

Informal Assessment

Write the molecular structures of vanillin, 3-heptanone, ethyl butanoate, and pentanoic acid on the board. Have students classify each compound as an aldehyde, ketone, ester, or carboxylic acid. Then have students complete the 23.3 Lesson Check.

Reteach

Review the oxidation states of carbon in organic compounds. Remind students that the oxidation state of carbon can cycle from –4 to +4.

If an ester is heated with water for several hours, usually very little happens. In strong acid or base solutions, however, the ester breaks down. An ester is hydrolyzed by the addition of water to produce a carboxylic acid and an alcohol. The reaction is rapid in acidic solution.

$$\underset{\text{Ethyl ethanoate}}{CH_3-\overset{\overset{\displaystyle O}{\|}}{C}-OCH_2CH_3} + H_2O \xrightleftharpoons{H^+} \underset{\text{Ethanoic acid}}{CH_3-\overset{\overset{\displaystyle O}{\|}}{C}-OH} + \underset{\text{Ethanol}}{HOCH_2CH_3}$$

Hydroxide ions also promote this reaction. Usually aqueous solutions of sodium hydroxide or potassium hydroxide are the source of hydroxide ions. Because many esters do not dissolve in water, a solvent such as ethanol is added to make the solution homogeneous. The reaction mixture is usually heated. All of the ester is converted to products. The carboxylic acid product is in solution as its sodium or potassium salt as shown in the example below.

$$\underset{\text{Ethyl ethanoate}}{CH_3-\overset{\overset{\displaystyle O}{\|}}{C}-OCH_2CH_3} + NaOH \longrightarrow \underset{\text{Sodium ethanoate}}{CH_3-\overset{\overset{\displaystyle O}{\|}}{C}-O^-Na^+} + \underset{\text{Ethanol}}{HOCH_2CH_3}$$

If the reaction mixture is acidified, the carboxylic acid forms.

$$\underset{\text{Sodium ethanoate}}{CH_3-\overset{\overset{\displaystyle O}{\|}}{C}-O^-Na^+} + HCl \longrightarrow \underset{\text{Ethanoic acid}}{CH_3-\overset{\overset{\displaystyle O}{\|}}{C}-OH} + NaCl$$

23.3 LessonCheck

ONLINE PROBLEMS

16. Review Describe the structure of the carbonyl groups that are characteristic of aldehydes and ketones.

17. Identify What is the general formula of a carboxylic acid?

18. Explain Why is dehydrogenation an oxidation reaction?

19. Describe What is the general formula of an ester?

20. Predict What products are expected when the following compounds are oxidized?

a. $CH_3CH_2CH_2CH_2OH$

b. $CH_3CH_2\underset{}{\overset{\overset{\displaystyle OH}{|}}{C}}HCH_3$

c. $CH_3CH_2\underset{\underset{\displaystyle CH_3}{|}}{\overset{\overset{\displaystyle OH}{|}}{C}}CH_3$

21. Classify Write the IUPAC name for each of the following compounds:

a. CH_3CH_2CHO

b. $CH_3CH_2CH_2\overset{\overset{\displaystyle O}{\|}}{C}CH_2CH_3$

c. $CH_3CH_2CH_2CH_2COOH$

22. Describe Draw structural formulas for the following organic compounds:

a. hexanoic acid

b. butanal

c. 2-pentanone

BIGIDEA CARBON CHEMISTRY

23. How can you describe the degree of oxidation of an organic compound?

Lesson Check Answers

16. A carbonyl group is a carbon atom double bonded to an oxygen atom. aldehyde: carbon of the carbonyl group is always joined to at least one hydrogen (RCHO); ketone: carbon is joined to two other carbons (RCOR)

17. RCOOH or RCO_2H

18. The loss of hydrogen (H_2) also results in the loss of two electrons.

19. RCOOR or RCO_2R

20. **a.** butanal

b. 2-butanone

c. no reaction

21. **a.** propanal **b.** 3-hexanone **c.** pentanoic acid

22. **a.** $CH_3CH_2CH_2CH_2CH_2COOH$

b. $CH_3CH_2CH_2CHO$

c. $CH_3COCH_2CH_2CH_3$

23. **BIGIDEA** The fewer hydrogen atoms and the more oxygen atoms a compound has, the more oxidized it is.

Scratch and Sniff Stickers

When you were a kid, you may have been rewarded with a scratch and sniff sticker for an A+ assignment. Many scratch and sniff stickers smell like fruits such as cherries, strawberries, or grapes. Others smell like root beer, pickles, pizza, or even stinky sneakers. Have you ever wondered where the scents come from?

Esters are usually the smelly component used in scratch and sniff sticker technology. The esters used in the stickers are volatile liquids, which explains why you are able to smell them. A single ester or a mixture of esters with the desired odor is contained in numerous tiny brittle capsules that are glued onto the surface of the sticker paper. The capsules are so small that you can barely notice the rough texture of the sticker. When you scratch the sticker, you break some of the capsules and allow the ester molecules to make their way to your nose.

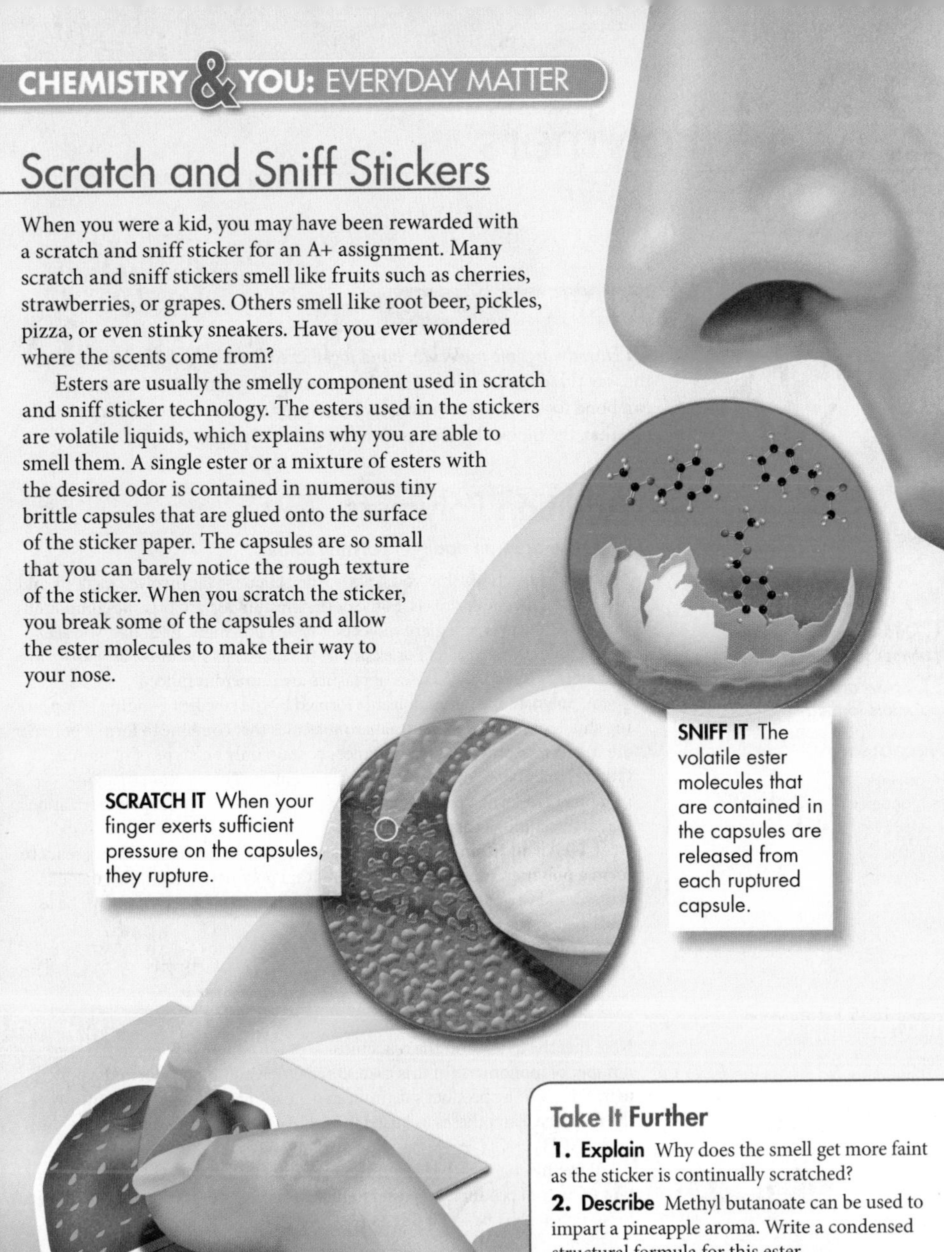

SCRATCH IT When your finger exerts sufficient pressure on the capsules, they rupture.

SNIFF IT The volatile ester molecules that are contained in the capsules are released from each ruptured capsule.

Take It Further

1. **Explain** Why does the smell get more faint as the sticker is continually scratched?
2. **Describe** Methyl butanoate can be used to impart a pineapple aroma. Write a condensed structural formula for this ester.
3. **Infer** Which carboxylic acid and alcohol would you use to prepare ethyl heptanoate, the ester that smells like grapes?

Differentiated Instruction

L1 STRUGGLING STUDENTS Have students write down their favorite scent and then research the IUPAC chemical name and the chemical structure. Then, assign them another scent based on their grasp of the concepts; an easier one for those struggling to gain confidence, and a more difficult one for others.

LPR LESS PROFICIENT READERS Have students handle oranges, lemons, or limes. Draw their attention to the oily residue that may linger on their skin and the aroma left from handling the fruit. For each type of fruit, have the IUPAC chemical name and chemical model available for them to visually associate with the scent.

L3 ADVANCED STUDENTS Have students research the chemical compounds in a favorite perfume or cologne. Challenge students to write the structural formulas for the compounds contained in the product.

CHEMISTRY & YOU

Volatility is a property shared by many organic substances. Volatile substances can vaporize at a rather low temperature, a characteristic that is desirable in the scent chemicals utilized in scratch and sniff applications. It is this characteristic that allows the human nose to detect the chemicals associated with various odors and fragrances as the airborne molecules interact with odorant receptors on olfactory cells in the nasal cavity. These cells send nerve impulses to the olfactory bulb of the brain, which interprets the impulses and identifies the scents. Although the exact mechanisms by which humans and other animals identify and create a memory bank of scents is not yet known, researchers do know that humans can recognize about 10,000 different scents. This is quite a feat, considering that most natural scents are the result of a mixture of chemicals, rather than a single chemical.

Some microencapsulation systems rely on protein polymers to form the capsules. *Challenge students to create their own complex scratch and sniff spots using unflavored gelatin and a variety of scents.* You may need to assist students in the following ways:

- Unflavored gelatin must be softened in cold water before being dissolved in hot water.
- The paper used must be soaked with the gelatin and allowed to dry before testing.

Have students exchange their spots and try to identify the scents used.

Extend

Connecting to PSYCHOLOGY

Psychologists use scents of pleasant things in the form of essential oils. The use of different scents has been used to aid in altering the physiological and mental state of humans and assist in the healing process of physical ailments. Pleasant aromas have been shown to alter moods. The use of this phenomenon is called "aromatherapy" and drives the industry of scented candles and other air fresheners or deodorizers. Some scents have been associated with a more energetic feeling, while others are promoted to provide a calm or relaxed feeling. Provide a selection of aromatherapy oils and ask students to explain the effect each has on their mood. Be aware, however, that some scents may trigger allergic reactions or severe headaches in some individuals. Ask students to identify any scents that might have this effect before beginning the activity.

Answers

TAKE IT FURTHER

1. More capsules are broken and depleted of the ester molecules within the capsules.
2. $CH_3CH_2CH_2COOCH_3$
3. heptanoic acid and ethanol

LESSON 23.4

Key Objectives

23.4.1 DESCRIBE how addition polymers are formed.

23.4.2 DESCRIBE how condensation polymers are formed.

Additional Resources

- Reading and Study Workbook, Lesson 23.4
- Probeware Laboratory Manual

PearsonChem.com

Engage

CHEMISTRY & YOU Relate the construction of a skydiver chain to the formation of a polymer. **Ask** What is the repeating unit, or monomer, in the chain? *(each person of the chain)* **Ask** What do you think are some of the characteristics of monomers? *(Answers will vary. Students may suggest that monomers must have the ability to be joined together, must be small molecules, or must be chemically reactive.)*

Build Background

Show students pictures of a disposable baby diaper, nylon stockings, and PVC pipe. Explain that each of the items are everyday items made of polymers.

Foundations for Reading

BUILD VOCABULARY The word *polymer* comes from the Greek word *poly*, meaning "many," and the Greek word *meros*, meaning "parts." A polymer is a compound that is formed by the covalent bonding of repeating small parts, called monomers.

23.4 Polymers

CHEMISTRY & YOU

Q: *How do organic molecules bond together to form long chains?* Similar to the way these skydivers are linked together in a chain, organic molecules can bond together to form long molecular chains. As more molecules bond together, the molecular chains grow longer and longer.

Key Questions

How does an addition polymer form?

How are condensation polymers formed?

Vocabulary

- polymer
- monomer

Addition Polymers

How does an addition polymer form?

Most of the reactions that you have learned about so far involve reactants and products of low molar mass. Some of the most important organic compounds that exist, however, are giant molecules called polymers. Each day, you see many different polymers. For example, the materials you know as plastics are polymers. The kinds and uses of plastics are numerous indeed!

A **polymer** is a large molecule formed by the covalent bonding of repeating smaller molecules. The smaller molecules that combine to form a polymer are called **monomers.** Some polymers contain only one type of monomer. Others contain two or more types of monomers. The reaction that joins monomers to form a polymer is called polymerization. Most polymerization reactions require a catalyst.

An addition polymer forms when unsaturated monomers react to form a polymer. Ethene undergoes addition polymerization. The ethene molecules bond to one another to form the long-chain polymer polyethylene as described in the equation below.

$$\underset{\text{Ethene (ethylene)}}{xCH_2{=}CH_2} \longrightarrow \underset{\text{Polyethylene}}{H{+}CH_2{-}CH_2{+}_xH}$$

Note that the letter *x* on the reactant side of the equation refers to the number of monomers (in this case, ethene) that combine to form the polymer. The *x* on the product side indicates the number of repeating units in the polymer. Parentheses are used to identify the repeating unit (in this case, $-CH_2-CH_2-$).

Polyethylene, which is chemically resistant and easy to clean, is an important industrial product. It is used to make plastic bottles, containers, and even toys, such as the one in Figure 23.15.

Figure 23.15 Polyethylene Many familiar household items, including this unsinkable duck, are made of polyethylene.

Focus on ELL

1 CONTENT AND LANGUAGE Have students identify the prefix for each vocabulary term. Ask students to list other words that use the same prefixes *(Sample answers: polyatomic, polygon, monoatomic, monorail)*. Ask students to predict the meaning of the key vocabulary using the list generated.

2 FRONTLOAD THE LESSON Create two chains of pop beads: one made from a solid color of beads and one of mixed colors. Explain that the two chains represent polymers. Ask students to study each chain and identify their similarities and differences. Ask students to predict what each chain has in common with polymers.

3 COMPREHENSIBLE INPUT Write the product of the first addition polymer on the board, and beside it, write the condensed structural formula for butane. Erase the parentheses and parenthetical subscripts for each, and write them in a contrasting color. Use the two formulas to demonstrate how the parentheses and their accompanying subscript indicate the presence of multiple identical units within each molecule.

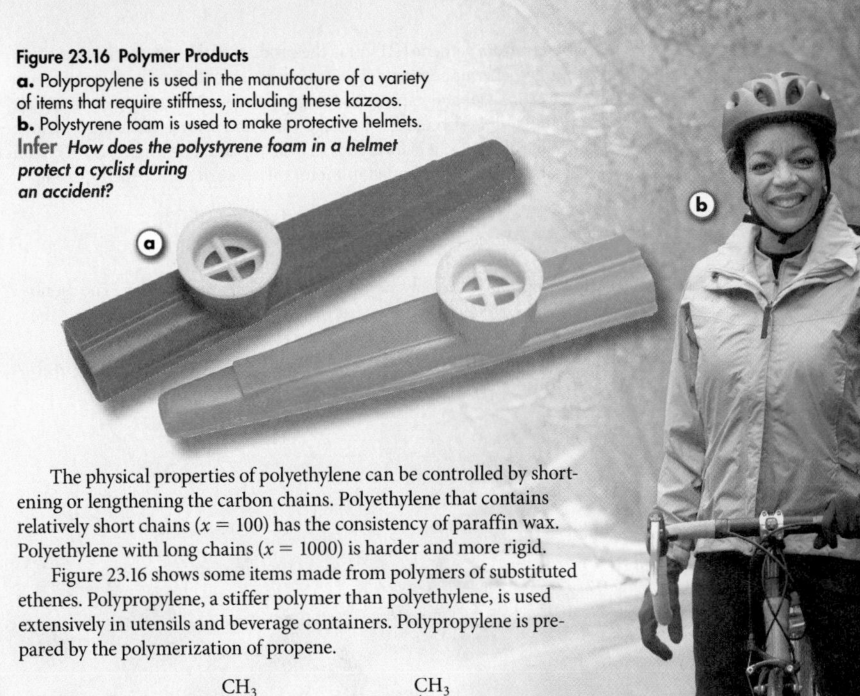

Figure 23.16 Polymer Products
a. Polypropylene is used in the manufacture of a variety of items that require stiffness, including these kazoos.
b. Polystyrene foam is used to make protective helmets.
Infer *How does the polystyrene foam in a helmet protect a cyclist during an accident?*

The physical properties of polyethylene can be controlled by shortening or lengthening the carbon chains. Polyethylene that contains relatively short chains ($x = 100$) has the consistency of paraffin wax. Polyethylene with long chains ($x = 1000$) is harder and more rigid.

Figure 23.16 shows some items made from polymers of substituted ethenes. Polypropylene, a stiffer polymer than polyethylene, is used extensively in utensils and beverage containers. Polypropylene is prepared by the polymerization of propene.

$$\underset{\text{Propene (propylene)}}{x\mathrm{CH_2{=}CH(CH_3)}} \longrightarrow \underset{\text{Polypropylene}}{\mathrm{\left(CH_2{-}CH(CH_3)\right)}_x}$$

Polystyrene, in the form of a rigid foam, is a poor heat conductor. This makes it useful for insulating homes and for manufacturing molded items such as coffee cups and picnic coolers. Polystyrene is prepared by the polymerization of styrene as shown in the reaction below.

$$\underset{\text{Styrene (vinyl benzene)}}{x\mathrm{CH_2{=}CH(C_6H_5)}} \longrightarrow \underset{\text{Polystyrene}}{\mathrm{\left(CH_2{-}CH(C_6H_5)\right)}_x}$$

Many halocarbon polymers, including polyvinyl chloride (PVC), have useful properties. For example, polyvinyl chloride is used to make plumbing pipes, rainwear, and garden hoses. It is also produced in sheets, sometimes with a fabric backing, for use as a tough plastic upholstery covering. Vinyl chloride is the monomer of polyvinyl chloride.

$$\underset{\text{Chloroethene (vinyl chloride)}}{x\mathrm{CH_2{=}CH(Cl)}} \longrightarrow \underset{\text{Polyvinyl chloride (PVC)}}{\mathrm{\left(CH_2{-}CH(Cl)\right)}_x}$$

The First Plastic—Celluloid

One of the first plastics, developed by an Englishman named Alexander Parkes, was a synthetic ivory named pyroxlin. Parkes marketed it under the trade name "Parkesine," and it won a bronze medal at the 1862 World's Fair in London. Parkesine was made from cellulose treated with nitric acid and a solvent. An American amateur inventor named John Wesley Hyatt later improved the product. Because cellulose was the main constituent used in the synthesis of his new material, Hyatt named it celluloid. A large variety of objects were made of celluloid, and it was a huge success. However, celluloid had a dangerous defect—it burned easily and tended to explode. Hyatt liked to tell stories about celluloid billiard balls exploding when struck very hard. Today, celluloid is used only to make table-tennis balls.

READING STRATEGY Students may have trouble understanding the concept of a polymer as an array of repeating units. Drawing a segment containing 3 or 4 monomers of each polymer will help students visualize the polymer.

Explain

Addition Polymers

START A CONVERSATION Discuss the similarities between the addition of hydrogen and halogens to alkenes and the formation of addition polymers. Explain that chemists have also found that when certain catalysts are used, alkenes will add to one another to form a long chain of hydrocarbon units. **Ask** What is a catalyst? *(A catalyst is a substance that increases the rate of a reaction without being used up in the reaction.)*

USE MODELS Write the formula for the simplest repeating unit of a polymer on the board. Ask students to identify the polymer and to write out the structure for the polymer showing how three successive monomers are linked together. **Ask** Why can't we write an exact formula for a polymer? *(The length of the carbon chain varies from polymer molecule to polymer molecule. The best that can be done is to write the formula for the simplest repeating unit, the monomer, in parentheses. The subscript x indicates a very large whole number.)*

Extend

Connect to MATHEMATICS

Have students write a paragraph comparing and contrasting the chemistry concept of polymers and monomers to the mathematical concept of polynomials and monomials. Encourage students to use diagrams as well as words to describe their analogy.

Answers

FIGURE 23.16 It absorbs a large portion of the force of impact (the energy is used to deform the foam), leaving only part of the force to be suffered by the head and brain.

Explore

Teacher Demo

PURPOSE Students will better understand how monomers link together to form polymers.

PROCEDURE Bring a length of metal-link chain to class. Use the metal chain as a model for a polymer. Explain that a polymerization reaction is a reaction that joins monomers to form a polymer. Point out the features in the chain that are similar to a polymer molecule. In addition, model a polymerization reaction using paperclips. Each paperclip represents a monomer. Explain that each end of the paperclip represents a reaction site. Build a chain by successively connecting one paperclip to another. The chain formed is analogous to a polymer. Point out that the main differences between polymers in organic chemistry are due to the type of monomers used to form the chains.

Explain

APPLY CONCEPTS Have students study the chemical equations for the formation of the addition polymers discussed in this section. Write the equations on the board and help students identify the repeating units. Discuss the features of an addition polymerization and how the name of a polymer is derived from the monomer units. Bring to class samples of items that are composed of addition polymers. Have students describe and compare their physical characteristics. **Ask** What are some of the advantages of synthetic polymers over metals and other natural materials? *(They are moldable, lightweight, chemically resistant, and easy to clean.)*

Polytetrafluoroethene (PTFE) is the product of the polymerization of tetrafluoroethene monomers. PTFE is very resistant to heat and chemical corrosion. You are probably familiar with this polymer as a coating on nonstick cookware. PTFE is also used to insulate wires and cables. Because PTFE is very durable and slick, it is formed into bearings and bushings used in chemical reactors. It also is suspended in motor oils as a friction-reducing agent.

$$x\text{CF}_2{=}\text{CF}_2 \longrightarrow \text{+CF}_2{-}\text{CF}_2\text{+}_x$$

Tetraflouroethene PTFE

Polyisoprene is the polymer that constitutes natural rubber. The monomer of polyisoprene, isoprene, is harvested from tropical plants such as the rubber tree. Polyisoprene is used to make rubber bands, soles of athletic shoes, and many other common items. Figure 23.17 shows how isoprene is used in the manufacture of tires.

Figure 23.17 Natural Rubber

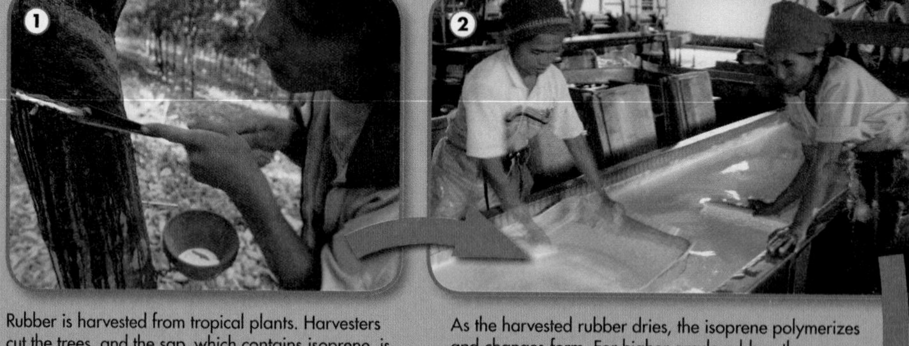

Rubber is harvested from tropical plants. Harvesters cut the trees, and the sap, which contains isoprene, is collected in a vessel such as a bowl.

As the harvested rubber dries, the isoprene polymerizes and changes form. For higher-grade rubber, the polymerization occurs under controlled conditions.

Finally, the manufacturer processes and molds the polymer to form the desired product.

Differentiated Instruction

LPR LESS PROFICIENT READERS Have students collect several household items composed of different polymer materials. Ask them to create a table listing the items and the plastic container codes found on the bottom of each item. Have students do research to find out how the codes are involved in recycling.

L1 STRUGGLING STUDENTS Have student pairs create ball-and-stick models for the reactants of each reaction described in the lesson. Have students manipulate the models to form the products of the reactions by removing the appropriate molecules and creating the new bonds.

L3 ADVANCED STUDENTS Have students do research on the pioneering work of Elias James Corey, a Harvard University professor and Nobel Prize-winning organic chemist. Tell students that his work led to the synthesis of more than 100 important drugs and other naturally occurring compounds. Ask students to write a short report on their findings.

Condensation Polymers

How are condensation polymers formed?

The formation of a polyester is a common example of condensation polymerization. **Condensation polymers are formed by the joining of monomers with the loss of a small molecule such as water.** Polyesters are polymers that consist of many repeating units of dicarboxylic acids and dihydroxy alcohols joined by ester bonds.

The formation of a polyester can be represented by a block diagram, which shows only the functional groups involved in the polymerization reaction. The squares and circles represent unreactive parts of the organic molecules. Condensation polymerization requires that there be two functional groups on each monomer molecule.

$$x\mathrm{HO{-}\overset{\displaystyle O}{\overset{\|}{C}}{-}\square{-}\overset{\displaystyle O}{\overset{\|}{C}}{-}OH} + x\mathrm{HO{-}\bigcirc{-}OH} \longrightarrow$$

Dicarboxylic acid — Dihydroxy alcohol

$$\left(\mathrm{\overset{\displaystyle O}{\overset{\|}{C}}{-}\square{-}\overset{\displaystyle O}{\overset{\|}{C}}{-}O{-}\bigcirc{-}O}\right)_x + 2x\mathrm{H_2O}$$

Representative polymer unit of a polyester

The polyester polyethylene terephthalate (PET) is formed from terephthalic acid and ethylene glycol as shown in the reaction below.

$$x\mathrm{HO{-}\overset{\displaystyle O}{\overset{\|}{C}}{-}C_6H_4{-}\overset{\displaystyle O}{\overset{\|}{C}}{-}OH} + x\mathrm{HO{-}CH_2CH_2{-}OH} \longrightarrow$$

Terephthalic acid — Ethylene glycol

$$\left(\mathrm{\overset{\displaystyle O}{\overset{\|}{C}}{-}C_6H_4{-}\overset{\displaystyle O}{\overset{\|}{C}}{-}O{-}CH_2CH_2{-}O}\right)_x + 2x\mathrm{H_2O}$$

Representative polymer unit of PET

Figure 23.18 shows two very different products made from PET: water bottles and a fleece jacket. But, in fact, the fabric of the jacket is made from recycled PET bottles. PET fibers form when the compound is melted and forced through tiny holes in devices called spinnerettes. The fibers are used for tire cord and permanent-press clothing. PET fibers are often blended with cotton to make clothing that is more comfortable on hot, humid days than those containing 100% polyester. These clothes retain the wrinkle resistance of 100% polyester. Woven PET fiber tubing can be used to replace major blood vessels. PET melts may also be forced through a narrow slit to produce sheets of film that are used extensively on credit cards and as coverings for frozen dinners.

Q: *Polymers make up most of the bottles, containers, and packaging that you see around you. Try to identify the polymers in some of these items. How would you classify them—as additional polymers or condensation polymers?*

Figure 23.18 Recycling PET
Plastic bottles from which you sip your drink may one day be part of someone else's wardrobe. It only takes about a dozen large bottles that are made from PET to make one fleece jacket.

LESSON 23.4

Explain

Condensation Polymers

APPLY CONCEPTS Explain that the key to polymerization is the ability of molecules to form repeating chains. For this to happen, each reacting molecule must have at least two reactive sites. In addition polymerization, the two active sites become available as double bonds are opened up. In condensation polymerization, the sites become available as two groups (such as —H and —OH groups) are split off the molecules.

Extend

Connect to MEDICINE

Have students research the use of synthetic polymers in medicine. Some applications include prosthetics, skin substitutes, artificial artery grafts, and heart valve replacements.

UbD Check for Understanding

The Essential Question How are chemical reactions used in organic chemistry?

Assess students' understanding of how addition and condensation reactions are used to create organic polymers by writing the following cloze item on the board and asking them to fill in the blanks:

________ reactions are used to form polymers from alkenes and alkynes.
________ reactions are used to form polymers when at least one reactant has a functional group containing –OH. *(addition; condensation)*

ADJUST INSTRUCTION If students are having difficulty distinguishing between the two types of reactions, remind students to think about what happens when condensation forms on a cold surface. *(Water collects on the surface.)* Then, have students review the text associated with the lesson's key questions and answers.

LESSON 23.4

Explore

Teacher Demo

PURPOSE Students observe the synthesis of nylon.

MATERIALS 5% aqueous solution of 1,6-hexanediamine, small beaker, 20% sodium hydroxide, dropper, 5% solution of adipoyl chloride in cyclohexane, forceps or tongs, water, paper towel

SAFETY Wear safety goggles, a chemical-resistant apron, and chemical-resistant gloves while preparing solutions and performing the demonstration. Use good ventilation. Rinse the nylon thoroughly before handling.

PROCEDURE Pour 5 mL of a 5% aqueous solution of 1,6-hexanediamine into a small beaker. Add 5 drops of 20% sodium hydroxide. Slowly pour 5 mL of a 5% solution of adipoyl chloride in cyclohexane down the wall of the beaker. A two-phase system will form, and a polymer film will appear at the interface. Use forceps or tongs to slowly draw out a long strand of nylon. Rinse thoroughly with water and set it on a paper towel to dry. Have students examine the color, hardness, and strength. Discuss the uses of nylon, and explain that it is a polyamide and a condensation polymer.

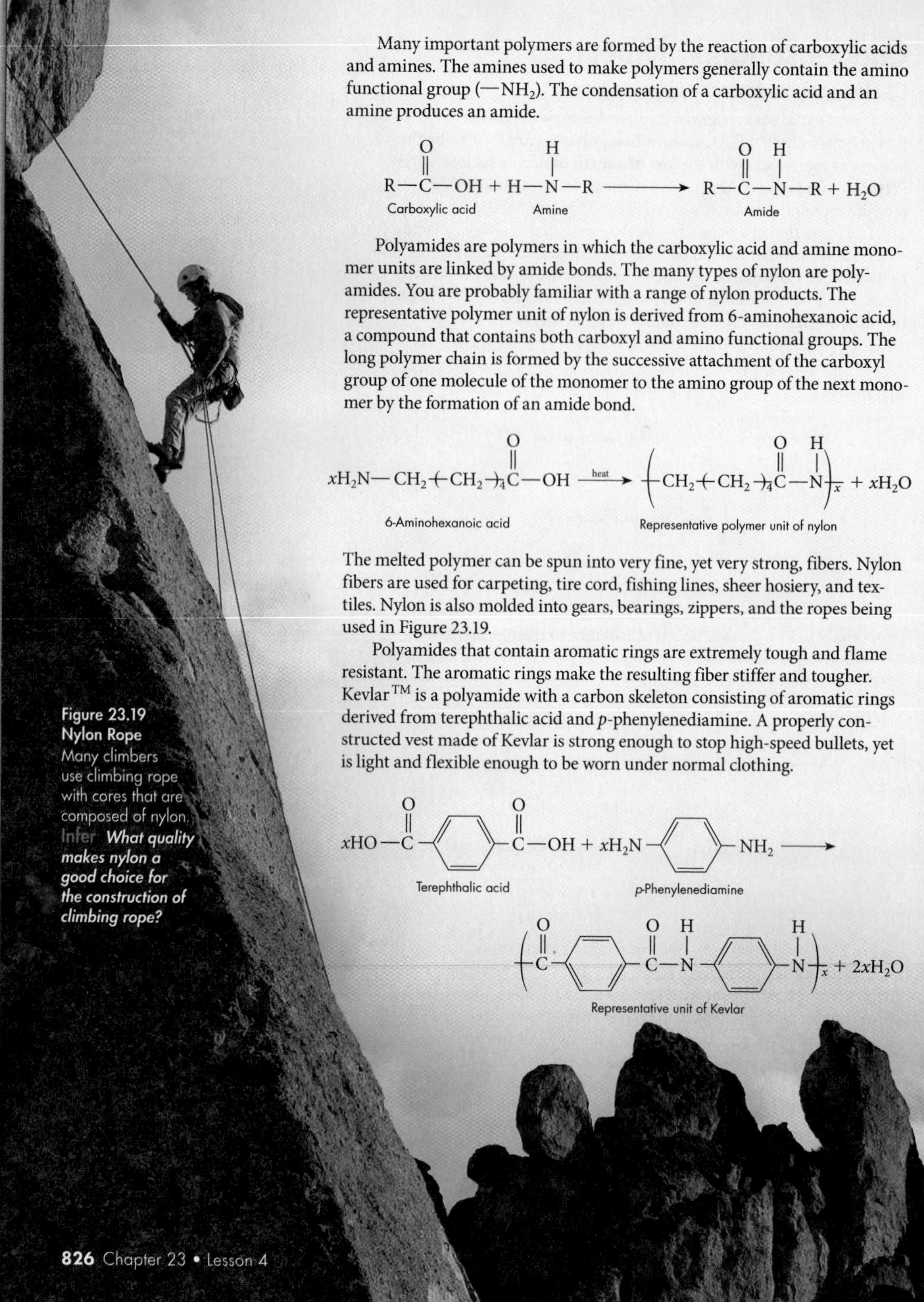

Many important polymers are formed by the reaction of carboxylic acids and amines. The amines used to make polymers generally contain the amino functional group ($—NH_2$). The condensation of a carboxylic acid and an amine produces an amide.

$$\underset{\text{Carboxylic acid}}{R—\overset{\overset{O}{\|}}{C}—OH} + \underset{\text{Amine}}{H—\overset{\overset{H}{|}}{N}—R} \longrightarrow \underset{\text{Amide}}{R—\overset{\overset{O}{\|}}{C}—\overset{\overset{H}{|}}{N}—R} + H_2O$$

Polyamides are polymers in which the carboxylic acid and amine monomer units are linked by amide bonds. The many types of nylon are polyamides. You are probably familiar with a range of nylon products. The representative polymer unit of nylon is derived from 6-aminohexanoic acid, a compound that contains both carboxyl and amino functional groups. The long polymer chain is formed by the successive attachment of the carboxyl group of one molecule of the monomer to the amino group of the next monomer by the formation of an amide bond.

$$\underset{\text{6-Aminohexanoic acid}}{xH_2N—CH_2(CH_2)_4\overset{\overset{O}{\|}}{C}—OH} \xrightarrow{\text{heat}} \underset{\text{Representative polymer unit of nylon}}{\left(CH_2(CH_2)_4\overset{\overset{O}{\|}}{C}—\overset{\overset{H}{|}}{N}\right)_x} + xH_2O$$

The melted polymer can be spun into very fine, yet very strong, fibers. Nylon fibers are used for carpeting, tire cord, fishing lines, sheer hosiery, and textiles. Nylon is also molded into gears, bearings, zippers, and the ropes being used in Figure 23.19.

Polyamides that contain aromatic rings are extremely tough and flame resistant. The aromatic rings make the resulting fiber stiffer and tougher. Kevlar™ is a polyamide with a carbon skeleton consisting of aromatic rings derived from terephthalic acid and *p*-phenylenediamine. A properly constructed vest made of Kevlar is strong enough to stop high-speed bullets, yet is light and flexible enough to be worn under normal clothing.

$$\underset{\text{Terephthalic acid}}{xHO—\overset{\overset{O}{\|}}{C}—C_6H_4—\overset{\overset{O}{\|}}{C}—OH} + \underset{p\text{-Phenylenediamine}}{xH_2N—C_6H_4—NH_2} \longrightarrow$$

$$\underset{\text{Representative unit of Kevlar}}{\left(\overset{\overset{O}{\|}}{C}—C_6H_4—\overset{\overset{O}{\|}}{C}—\overset{\overset{H}{|}}{N}—C_6H_4—\overset{\overset{H}{|}}{N}\right)_x} + 2xH_2O$$

Figure 23.19 Nylon Rope
Many climbers use climbing rope with cores that are composed of nylon.
Infer *What quality makes nylon a good choice for the construction of climbing rope?*

Focus on ELL

4 LANGUAGE PRODUCTION Have students work in groups of four to complete the Small-Scale Lab on page 828. Make sure each group has ELLs of varied language proficiencies, so that more proficient students can help less proficient ones. Have students work according to their proficiency level.

BEGINNING: LOW/HIGH Model the procedure for the beginning-level student. Give each student a part in the lab.

INTERMEDIATE: LOW/HIGH Rewrite the 3-step procedure into a series of simpler steps.

ADVANCED: LOW/HIGH Have students help students with lower proficiencies with the written answers and predictions.

Nomex™ is another polyamide that contains aromatic rings. It is used in the fabrication of flame-resistant building materials and in the manufacture of the flame-resistant clothing the firefighter wears in Figure 23.20. Nomex is a poor conductor of electricity. Because it is also rigid, it is used to make parts for electrical fixtures. Nomex is a polyamide made from isophthalic acid and *m*-phenylenediamine as shown in the reaction below.

Figure 23.20 Nomex Clothing Flame-resistant clothing is made of Nomex, a polyamide with aromatic rings.

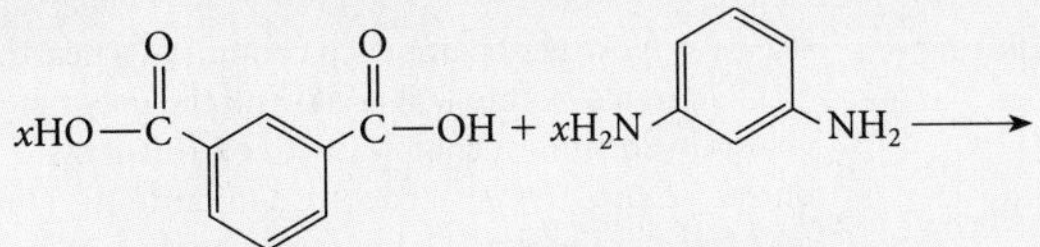

$$\left(C(=O)\text{—}C_6H_4\text{—}C(=O)\text{—}N(H)\text{—}C_6H_4\text{—}N(H) \right)_x + 2xH_2O$$

Representative unit of Nomex

Proteins, which are polyamides of naturally occurring molecules, rank among the most important of all biological molecules. You will learn about these polymers in more detail in Chapter 24.

ONLINE PROBLEMS 23.4 LessonCheck

24. Review Describe how addition polymers form.

25. Review How do condensation polymers form?

26. Compare What is a polymer? A monomer?

27. Describe What structure must a monomer have if it is to undergo addition polymerization?

28. Identify What is formed when a carboxylic acid and an amine combine? Give an example of the type of polymer that is formed by this reaction.

29. Explain How is water involved in the formation of a condensation polymer?

30. Use Analogies Write an analogy that you can use to describe to a middle-school student how a polymer is constructed from monomers. Use diagrams as well as words to describe your analogy.

31. Identify Give names and uses for three polymers you encounter in your home.

Lesson Check Answers

24. They form when unsaturated monomers react and link to one another.

25. They form by the joining of monomers, usually with loss of water.

26. A polymer is a large molecule formed by the covalent bonding of repeating small molecules. Monomers are the smaller molecules, or the repeating unit of a polymer.

27. an unsaturated carbon-carbon bond

28. an amide bond; Nylon, Kevlar, and Nomex are polyamides.

29. Water is a product of most condensation polymerizations.

30. Students' analogies might include putting together a toddler's plastic pop beads, stringing a bead necklace, or making a paperclip or popcorn chain.

31. Answers will vary.

Extend

Explain that synthetic polymers can be traced to the unique bonding ability of carbon. To show how small changes in the structures of monomers can lead to dramatic differences in physical properties of the polymers formed, ask students to compare polymers based on ethene and vinyl chloride.

Evaluate

Informal Assessment

Have students write sentences, in their own words, describing what takes place during the formation of a polymer. **Ask** What is the difference between an addition polymer and a condensation polymer? Give examples of each. *(An addition polymer is formed when unsaturated monomers react to form a polymer. One example is the reaction of ethene to form polyethylene. Condensation polymers are formed by the head-to-tail joining of monomer units. The other product of the condensation reaction is a small molecule such as water.)* Then, have students complete the 23.4 Lesson Check.

Reteach

Remind students that the area of chemistry devoted to the study of polymers involves more than just plastics. Synthetic fibers are used to create clothing, automobile tires, and materials used in the medical and dental fields. Guide students to examine the structures throughout the lesson and then have students describe the uses of some addition and condensation polymers.

Answers

FIGURE 23.19 It can be spun into fibers that are small enough to allow a firm grip with your hands and strong enough to support the weight of your body without breaking.

SMALL-SCALE LAB

Small-Scale Lab

OBJECTIVE After completing this activity, students will be able to use a borax solution to cross-link carbohydrate polymers, write structural formulas for cross-linked polymers, and relate the properties of some polymers to their structures.

Students tend to think of polymerization as a large-scale industrial process. This lab shows that polymers can be synthesized on a much smaller scale.

SKILLS FOCUS observing, inferring, drawing conclusions, interpreting diagrams, comparing and contrasting

PREP TIME 30 minutes

CLASS TIME 30 minutes

MATERIALS 3 1/2-oz plastic cups, soda straw, powdered guar gum, plastic spoons, 4% borax solution, pipet, corn starch or liquid laundry starch, paper, metric ruler, white school glue

ADVANCE PREPARATION Prepare the solution.

Solution	Preparation
4% borax	10 g $Na_2B_4O_7 \cdot 10H_2O$ in 250 mL

SAFETY AND DISPOSAL Students should not take home the polymers as the mixtures tend to develop mold in a few days. Dispose of these polymers in the waste basket. While the cross–linked polymers are non-toxic, borax is toxic if ingested. Remind students to wash their hands thoroughly when finished.

TEACHING TIPS School glue adheres to the stirring spoon as it cross-links. Decant any excess liquid that does not adhere. Wooden craft sticks can be substituted for plastic spoons if desired.

EXPECTED OUTCOME Students will make a polymer that has some liquid properties and some solid properties.

ANALYZE

1. The polymer is a viscous gel that oozes. It has properties of both a liquid and a solid.
2. 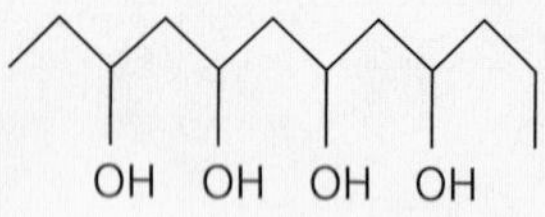 (zigzag chain with —OH groups) OH OH OH OH
3. 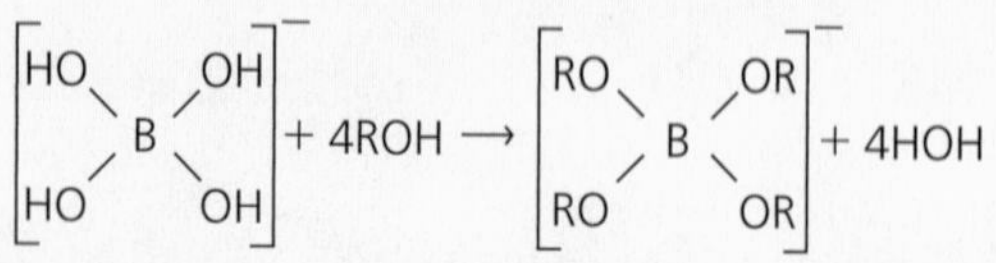 $[B(OH)_4]^- + 4ROH \longrightarrow [B(OR)_4]^- + 4HOH$
4. (two polymer chains cross-linked through B, with O O / B / O⁻ O groups)

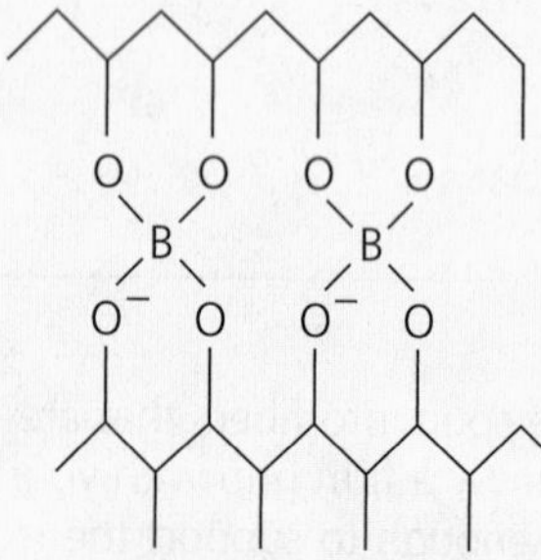

Small-Scale Lab

Polymers

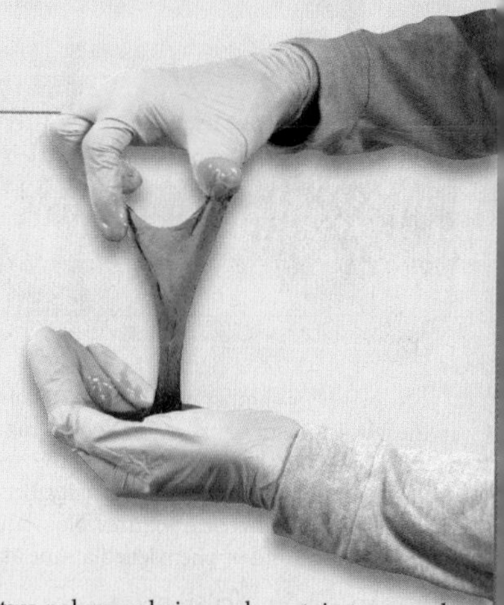

Purpose

To cross-link some polymers and examine their properties

Materials

- 3½-oz plastic cup
- soda straw
- guar gum powder
- plastic spoon
- 4% borax solution
- pipet
- food coloring (optional)

Procedure

1. Half fill a 3½-oz cup with water. Add food coloring to the water if you want the polymer to be colored.

2. Use a soda straw as a measuring scoop to obtain approximately 2 cm of powdered guar gum. **CAUTION** *Do not use your mouth to draw up the guar gum into the straw.* Gently sprinkle the guar gum powder into the water while stirring with a plastic spoon. Add the guar gum powder slowly to prevent it from clumping. Stir the mixture well.

3. While stirring, add one full pipet (about 4 mL) of borax solution. Continue to stir until a change occurs.

Analyze

1. Is the polymer you just made a liquid or a solid? What special characteristics does it have?

2. Guar gum is a carbohydrate, a polymer with many repeating alcohol functional groups (—OH). Draw a zigzag line to represent a crude polymer chain. Add —OH groups along the chain to represent the alcohol functional groups.

3. Borate ions combine with alcohol to form water and borate complexes of the alcohol as shown below. Write a similar equation that replaces all of the —OH groups on the borate with —OR groups.

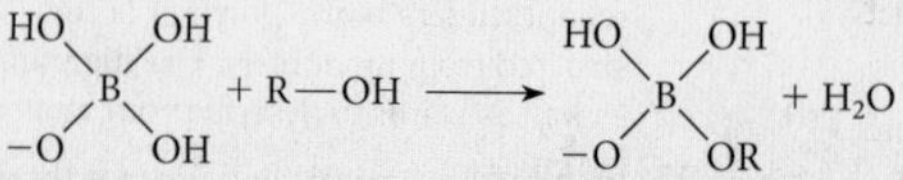

$$B(OH)_3(O^-) + R{-}OH \longrightarrow B(OH)_2(OR)(O^-) + H_2O$$

4. If two polymer chains each contain two nearby —OH groups, borate will cross-link the polymer chains by forming a complex with two alcohols on each chain. Draw a structure similar to the one you drew for Question 3, but replace your four R groups with two polymer chains.

You're the Chemist

1. Design an Experiment! Try using borax to cross-link other common carbohydrate polymers, such as cornstarch or liquid laundry starch. For each polymer, half fill a 3-oz cup with the chosen carbohydrate polymer and add enough water to bring the liquid to within about 1 cm of the rim. Stir carefully and thoroughly. Add one full pipet (about 4 mL) of borax solution while stirring. Describe the similarities and differences between this cross-linked polymer and the polymer you made previously. Compare the properties of these polymers.

2. Analyze Data! Cut a 1 cm × 15 cm strip of paper. Use a drop of glue or a stapler to fasten one end of the paper strip to the other end to form a ring. Now cut out some identical-sized strips of paper, and glue or staple them together into an interlocking chain of paper rings. Explain how this chain is like a polymer. Make another paper ring that cross-links your chain to one of your classmates' chains. Explain how these linked chains are like cross-linked polymers.

828 Chapter 23 • Small-Scale Lab

YOU'RE THE CHEMIST

1. Through experimentation, students are able to produce a variety of polymers with different properties. Students should compare and contrast the properties of the various polymers they make.
2. The chain is similar to a polymer because it contains many repeating units linked end to end.

FOR ENRICHMENT Have students quantify the synthesis of the polymer by varying the amount of water used in the reaction.

23 Study Guide

BIGIDEAS

CARBON CHEMISTRY AND REACTIONS

Functional groups affect the physical and chemical properties of organic compounds. Chemical reactions can be used to change the structure of organic molecules, often by introducing a new functional group. Substitution, addition, oxidation-reduction, and polymerization reactions are commonly used in organic chemistry.

23.1 Introduction to Functional Groups

Organic compounds can be classified according to their functional groups.

The general formula of a halocarbon is RX, where X is a halogen substituent.

Substitution reactions are an important method of introducing new functional groups to organic molecules.

- functional group (798)
- halocarbon (800)
- alkyl halide (800)
- aryl halide (800)
- substitution reaction (801)

23.2 Alcohols, Ethers, and Amines

The general formula of an alcohol is ROH.

Addition reactions are an important method of introducing new functional groups to organic molecules. They are also used to convert alkenes to alkanes.

The general formula of an ether is ROR.

The general formula of an amine is RNH_2, R_2NH, or R_3N.

- alcohol (804)
- hydroxy group (804)
- fermentation (807)
- addition reaction (807)
- hydration reaction (808)
- hydrogenation reaction (809)
- ether (810)
- amine (811)

23.3 Carbonyl Compounds

The C=O functional group is present in aldehydes and ketones.

The general formula of a carboxylic acid is RCOOH.

Dehydrogenation is an oxidation reaction because the loss of each molecule of hydrogen involves the loss of two electrons from the organic molecule.

The general formula of an ester is RCOOR.

- carbonyl group (812)
- aldehyde (812)
- ketone (812)
- carboxyl group (815)
- carboxylic acid (815)
- fatty acid (816)
- dehydrogenation reaction (816)
- ester (819)

23.4 Polymers

An addition polymer forms when unsaturated monomers react to form a polymer.

Condensation polymers are formed by the joining of monomers with the loss of a small molecule such as water.

- polymer (822)
- monomer (822)

What's Online

Review and Assessment Resources

CHAPTER PROBLEM SET Have students practice more functional group problems using the Chapter 23 online Problem Set.

DIRECTED VIRTUAL LAB Have students complete the Virtual Lab *Properties of Organic Molecules* to help reinforce the concepts of how functional groups change the properties of organic molecules. In the post-lab exercise, students will analyze the data and draw conclusions based on the results of the experiment.

KINETIC ART Direct students to the Kinetic Art animations *Hydration and Hydrogenation* and *Redox Reactions*, and have them answer the questions that accompany each animation.

Study Tip

IDENTIFY KEY CONCEPTS AND VOCABULARY For each lesson of the chapter, read the key concepts that are listed in the Study Guide. Then review the chapter text for any concepts that you do not fully understand. Next, define each vocabulary term in your own words and check definitions against the text's definitions. When you check the text, read all the related text, not just the sentence that defines the boldfaced vocabulary term.

Focus on ELL

5 ASSESS UNDERSTANDING Create a formative assessment for each lesson to determine students' level of understanding.

BEGINNING

LOW Create an assessment in which students match a picture or diagram to the corresponding word or concept.

HIGH Create an assessment in which students list characteristics of the different functional groups.

INTERMEDIATE: LOW/HIGH Create a multiple-choice assessment that measures students' understanding of the key concepts.

ADVANCED: LOW/HIGH Create an assessment that involves comparing and contrasting different functional groups.

UbD Performance Tasks

MAKING SUBSTITUTIONS Carboxylic acids react with thionyl chloride ($SOCl_2$) to form acid chlorides, releasing SO_2 and HCl as byproducts. Have students research the IUPAC naming conventions for acid chlorides. Then have them write an equation for the production of an acid chloride from the carboxylic acid of their choice and name the product.

POLYMERIZED Research one of the many items shown in Lesson 23.4. Explain the polymer used to create the item. Describe the structure of the polymer and how the structure helps to make the item usable. Determine whether the item is recyclable or not.

Evaluate

Answers

LESSON 23.1

32. any carbon-containing group

33. a. $ClCH_2CCl_2CH_3$ (Cl, Cl on C-2) **b.** benzene ring with Br at 1, 3, 5 **c.** cyclohexane ring with Cl on adjacent carbons

34 **a.** 3-chloropropene
b. 1,2-dichloro-4-methylpentane
c. 1,3-dibromobenzene

35. a.
$CHCl_2-CH_2-CH_3$ 1, 1-dichloropropane
$CH_2Cl-CHCl-CH_3$ 1, 2-dichloropropane
$CH_2Cl-CH_2-CH_2Cl$ 1, 3-dichloropropane
$CH_3-CCl_2-CH_3$ 2, 2-dichloropropane

b.
$CH_3-CH_2-CH_2-CH_2Br$ 1-bromobutane
$CH_3-CH_2-CHBr-CH_3$ 2-bromobutane
$CH_3-CH(CH_3)-CH_2Br$ 1-bromo-2-methylpropane
$CH_3-CBr(CH_3)-CH_3$ 2-bromo-2-methylpropane

36. a. phenol (benzene ring–OH) **b.** cyclohexane ring–OH
c. $CH_3-CH(CH_3)-OH$ **d.** benzene ring–Br

LESSON 23.2

37. a. 2-propanol
b. 1,2-propanediol
c. 2-methyl-2-propanol

38. a. CH_2H-CH_2Br bromoethane **b.** CH_2Cl-CH_2Cl 1, 2-dichloroethane
c. CH_2H-CH_2OH ethanol **d.** CH_2H-CH_2H ethane
e. CH_2H-CH_2Cl chloroethane

39. a. $CH_2BrCHBrCH_2CH_3$ **b.** $CH_3CHICHICH_3$ **c.** $CH_3CHHCHBrCH_3$
d. $CH_3CHHCHHCH_3$ **e.** cyclohexane ring with Cl on adjacent carbons

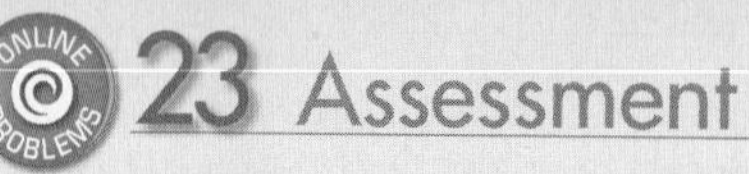

23 Assessment

★ Solutions appear in Appendix E

Lesson by Lesson

23.1 Introduction to Functional Groups

32. What does R in the formula RCH_2Cl represent?

★**33.** Write a structural formula for each compound.
a. 1,2,2-trichlorobutane
b. 1,3,5-tribromobenzene
c. 1,2-dichlorocyclohexane

34. Name the following halocarbons:
a. $CH_2{=}CHCH_2Cl$
b. $CH_3CH(CH_3)CH_2CH(Cl)CH_2Cl$
c. benzene ring with Br at positions 1 and 3

35. Write structural formulas and give IUPAC names for all the isomers of the following compounds:
a. $C_3H_6Cl_2$ **b.** C_4H_9Br

36. What organic products are formed in the following reactions?
a. benzene ring–Br + NaOH $\xrightarrow{heat}$ ___ + NaBr
b. cyclohexane ring–Cl + NaOH ⟶ ___ + NaCl
c. $CH_3CHCl(CH_3)$ + NaOH ⟶ ___ + NaCl
d. benzene + Br_2 $\xrightarrow{catalyst}$ ___ + HBr

23.2 Alcohols, Ethers, and Amines

★**37.** Give the IUPAC names for these alcohols.
a. $CH_3-CH(OH)-CH_3$
b. $CH_2(OH)-CH(OH)-CH_3$
c. $CH_3-C(CH_3)(OH)-CH_3$

★**38.** Write structures and names of the products obtained upon addition of each of the following reagents to ethene.
a. HBr **b.** Cl_2 **c.** H_2O **d.** H_2 **e.** HCl

39. Write the structure for the expected product from each reaction.
a. $CH_2{=}CHCH_2CH_3 + Br_2 \longrightarrow$
b. $CH_3CH{=}CHCH_3 + I_2 \longrightarrow$
c. $CH_3CH{=}CHCH_3 + HBr \longrightarrow$
d. $CH_3CH{=}CHCH_3 + H_2 \longrightarrow$
e. cyclohexene + $Cl_2 \longrightarrow$

40. Give the IUPAC and common names for the following ethers:
a. $CH_3OCH_2CH_3$
b. benzene ring$-O-CH_2CH_3$
c. $CH_2{=}CHOCH{=}CH_2$
d. $CH_3CH(CH_3)OCH(CH_3)CH_3$

41. Name the following amines:
a. $CH_3CH_2CH_2CH_2NH_2$
b. $CH_3CH_2NH_2$

23.3 Carbonyl Compounds

★**42.** Give the IUPAC names for the following carbonyl compounds:
a. $CH_3C(=O)CH_3$
b. CH_3CHO
c. $CH_3CH(CH_3)CH_2CHO$
d. $CH_3CH_2C(=O)CH_2CH_2CH_3$
e. benzene ring$-CH_2CHO$
f. $CH_3(CH_2)_3COOH$

43. Write the structure for the expected organic product from each reaction.
a. $CH_3CH_2OH \xrightarrow[H_2SO_4]{K_2Cr_2O_7}$
b. $CH_3CHO \xrightarrow[H_2SO_4]{K_2Cr_2O_7}$
c. $CH_3CH_2CH_2CHO \xrightarrow[H_2SO_4]{K_2Cr_2O_7}$
d. $CH_3-CH_2-CH(OH)-CH_3 \xrightarrow[H^+]{K_2Cr_2O_7}$

40. a. methoxyethane, ethyl methyl ether
b. ethoxybenzene, ethyl phenyl ether
c. ethenoxyethene, divinyl ether (or vinyl ether)
d. dimethylmethoxydimethyl-methane, diisopropyl ether (or isopropyl ether)

41. a. butanamine or butyl amine
b. ethanamine or ethylamine

23.4 Polymers

44. Different samples of a polymer such as polyethylene can have different properties. Explain.

***45.** Draw the structure of the repeating units in a polymer that has the following monomers:

a. 1-butene
b. 1,2-dichloroethene

Understand Concepts

46. Write a general formula for each type of compound.

a. halocarbon	d. ester
b. ketone	e. amide
c. aldehyde	f. ether

47. Place the following compounds in order from lowest boiling point to highest boiling point. Molar masses are given in parentheses.

a. CH_3CHO (44 g)
b. CH_3CH_2OH (46 g)
c. $CH_3CH_2CH_3$ (44 g)

48. Write the structure and name of the expected products for each reaction.

a. $CH_3COOH + CH_3OH \xrightarrow{H^+}$
b. $CH_3CH_2CH_2COOCH_2CH_3 + H_2O \xrightarrow{H^+}$
c. $CH_3CH_2OH \xrightarrow[H^+]{K_2Cr_2O_7}$

49. Explain why a carbon–carbon double bond is nonpolar, but a carbon–oxygen double bond is very polar.

***50.** Classify each compound as an alcohol, a phenol, or an ether.

a. (naphthalene ring)–OH
b. (benzene ring)–O–(benzene ring)
c. (benzene ring)–CH_2OH
d. (benzene ring)–OH
e. CH_3CH_2CHOH with CH_3 on the CHOH carbon
f. $CH_3CH_2OCH_3$

51. Explain why 2-methyl-1-propanol can be oxidized, but 2-methyl-2-propanol cannot be oxidized.

52. Write the structural formulas for the products of these reactions.

a. $CH_3CH_2CH{=}CH_2 + Cl_2 \longrightarrow$
b. $CH_3CH_2CH{=}CH_2 + Br_2 \longrightarrow$
c. (cyclohexene) + HBr $\longrightarrow$

***53.** For each compound pictured below, identify the functional group and name the compound. The red atoms represent oxygen.

a.
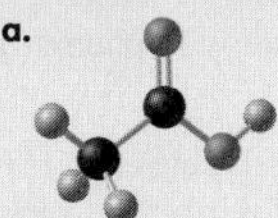

c.
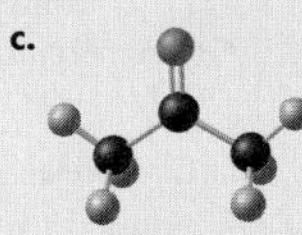

b.
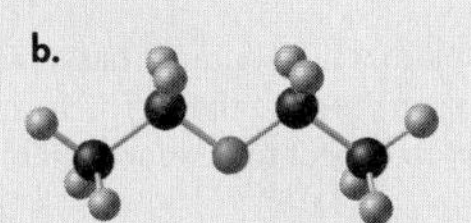

d.

54. Write the name and structure for the alcohol that must be oxidized to make each carbonyl compound.

a. HCHO
b. CH_3CCH_3 with =O on the central C
c. CH_3CHCHO with CH_3 on the CH carbon
d. (cyclohexanone)

55. Write the structures of the expected products for the following reactions:

a. $CH_3CH_2COOCH_2CH_3 + NaOH \longrightarrow$
b. CH_3COO–(benzene ring) + KOH $\longrightarrow$
c. $CH_3CH_2COOCH_2CH_2CH_3 + H_2O \xrightarrow{H^+}$
d. $CH_3CH_2COOCH_2CHCH_3 + H_2O \xrightarrow{HCl}$ with CH_3 on the CH carbon

LESSON 23.3

42. a. propanone
b. ethanal
c. 3-methylbutanal
d. 3-hexanone
e. 2-phenylethanal
f. pentanoic acid

43. a. CH_3CHO
b. CH_3COOH
c. $CH_3CH_2CH_2COOH$
d. $CH_3CH_2COCH_3$

LESSON 23.4

44. The properties of polyethylene vary with the length of the chains.

45. a. $-CH_2-CH-$ with CH_2CH_3 attached to the CH
b. $-CH-CH-$ with Cl on each C

UNDERSTAND CONCEPTS

46. a. RX b. RCOR
c. RCHO d. RCOOR
e. RCONHR f. ROR

47. c, a, b

48. a. CH_3COOCH_3 (methyl ethanoate or methyl acetate) + H_2O
b. $CH_3CH_2CH_2COOH$ (butanoic acid) + CH_3CH_2OH (ethanol)
c. CH_3CHO (ethanal)

49. Both atoms in a carbon-carbon double bond have the same electronegativity value, so the bond is nonpolar. Because oxygen is more electronegative than carbon, a carbon-oxygen bond is very polar.

50. a. phenol
b. ether
c. alcohol
d. phenol
e. alcohol
f. ether

51. 2-methyl-1-propanol is a primary alcohol, which is easily oxidized. The carbon atom attached to the hydroxyl group loses a hydrogen atom in the process. 2-methyl-2-propanol is a tertiary alcohol, which does not have a hydrogen atom present on the carbon atom attached to the hydroxyl group. Thus, 2-methyl-2-propanol cannot be oxidized.

52. a. $CH_3CH_2CH-CH_2$ with Cl on each of the last two C
b. $CH_3CH_2CH-CH_2$ with Br on each of the last two C
c. (cyclohexane ring) with H and Br on adjacent carbons

53 a. carboxyl group, ethanoic acid (acetic acid)
b. ether, diethyl ether (ethyl ether)
c. ketone (carbonyl group), propanone (acetone)
d. alcohol (hydroxyl group), ethanol (ethyl alcohol)

54. a. CH_3OH
methanol
b. $CH_3-CH-CH_3$ with OH on the central C
2-propanol
c. $CH_3-CH-C-OH$ with CH_3 on the CH and two H on the C
2-methyl-1-propanol
d. (cyclohexane ring) with OH and H on the same C
cyclohexanol

55. a. $CH_3CH_2COO^-Na^+$, CH_3CH_2OH
b. $CH_3COO^-K^+$, (benzene ring)–OH
c. CH_3CH_2COOH, $CH_3CH_2CH_2OH$
d. CH_3CH_2COOH, CH_3CHCH_2OH with CH_3 on the CH carbon

ASSESSMENT

ASSESSMENT

Evaluate

Answers

THINK CRITICALLY

56. a. $CH_3COOCH_3 + H_2O \xrightarrow{HCl}$
methyl ethanoate water

$CH_3-\overset{\overset{O}{\|}}{C}-OH + CH_3OH$
ethanoic acid methanol

b. $CH_3CH_2COOCH_2CH_2CH_3 + H_2O \xrightarrow{NaOH}$
propyl propanoate water

$CH_3CH_2-\overset{\overset{O}{\|}}{C}-O^-Na^+ + CH_3CH_2CH_2OH$
sodium propanoate 1-propanol

c. $HCOOCH_2CH_3 + H_2O \xrightarrow{KOH}$
ethyl methanoate water

$H-\overset{\overset{O}{\|}}{C}-O^-K^+ + CH_3CH_2OH$
potassium methanoate ethanol

57. The chemical properties (and toxicity) of organic compounds are determined by the compound as a whole. As a substituent in a molecule, a phenyl group does not have the same properties as benzene.

58. The oxygen atom in diethyl ether polarizes the small molecule. This enables diethyl ether to dissolve in water, which is also polar. The large dihexyl ether molecule has large nonpolar parts and does not dissolve. Because propane is nonpolar, it is less soluble in water than diethyl ether.

59. The alcohol molecules form hydrogen bonds with one another, resulting in a higher boiling point. They also form hydrogen bonds with water molecules, causing 1-butanol to be more soluble than diethyl ether. (Although diethyl ether is polar, 1-butanol has greater polarity.)

60. Acetaldehyde is polarized by its carbonyl oxygen, forming stronger intermolecular attractions. Propane, a nonpolar compound, has weak intermolecular attractions. Thus, propane molecules are more easily liberated from the liquid state.

61. Ethanoic acid (two carbons long) is more soluble in water than decanoic acid (ten carbons long).

62. Substitution of an alkane by a halogen, for example, usually gives a mixture of products. Addition to a double bond or a triple can give a single addition product.

63. cadaverine, $H_2N(CH_2)_5NH_2$;
putrescine, $H_2N(CH_2)_4NH_2$
Both compounds are amines.

Think Critically

56. Apply Concepts Draw the structures of the expected products and name the reactants and products for each reaction below.

a. $CH_3COOCH_3 + H_2O \xrightarrow{HCl}$

b. $CH_3CH_2COOCH_2CH_2CH_3 + H_2O \xrightarrow{H^+}$

c. $HCOOCH_2CH_3 + KOH \longrightarrow$

***57. Draw Conclusions** Benzene is poisonous and a proven carcinogen. Yet many compounds containing benzene rings, such as benzaldehyde, are common in the foods you eat. Why are some organic compounds with phenyl groups safe to eat?

58. Compare Explain why diethyl ether is more soluble in water than dihexyl ether. Would you expect propane or diethyl ether to be more soluble in water? Why?

59. Compare Explain why 1-butanol has a higher boiling point than diethyl ether. Which compound would you expect to be more soluble in water? Why?

60. Evaluate Propane ($CH_3CH_2CH_3$) and acetaldehyde (CH_3CHO) have the same molar mass, but propane boils at –42°C and acetaldehyde boils at 21°C. Account for this difference.

***61. Compare and Contrast** How would you expect the water solubility of ethanoic and decanoic acids to compare?

62. Infer The processes used to synthesize many organic compounds often use compounds that contain double or triple bonds as reactants. Explain why using unsaturated reactants might be advantageous over using saturated reactants.

***63. Classify** Cadaverine (1,5-pentanediamine) and putrescine (1,4-butanediamine) are unpleasant-smelling compounds that are formed by bacteria in rotting flesh. Draw their structures. What class of organic compounds are these two chemicals?

64. Graph Use the data in Table 23.3 to make a graph of boiling point versus number of chlorine atoms for the five compounds listed in the table. Is the graph a straight line? Use the graph to describe the relationship between boiling point and degree of halogen substitution.

65. Apply Concepts For the following structures, write a chemical equation showing how to produce the compound.

a. 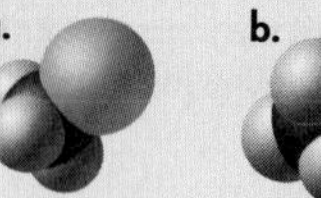

b. 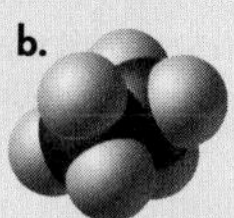

***66. Infer** Tetrahydrofuran (THF) is an important industrial organic solvent. THF is a cyclic ether containing four carbon atoms in the ring. Draw a structural formula of this cyclic ether.

67. Predict Write structures and names of the organic products obtained when the following compounds react with Fehling's reagent.

a. $C_6H_5-CH_2CHO$

b. $CH_3\underset{|}{\overset{CH_3}{C}}HCHO$

c. $CH_3\overset{CH_3}{C}HCH_2CHO$

68. Interpret Graphs The graph shows the boiling points of the C_2–C_{10} straight-chain 1-hydroxy alcohols and straight-chain 1-chloroalkanes plotted versus their molar masses.

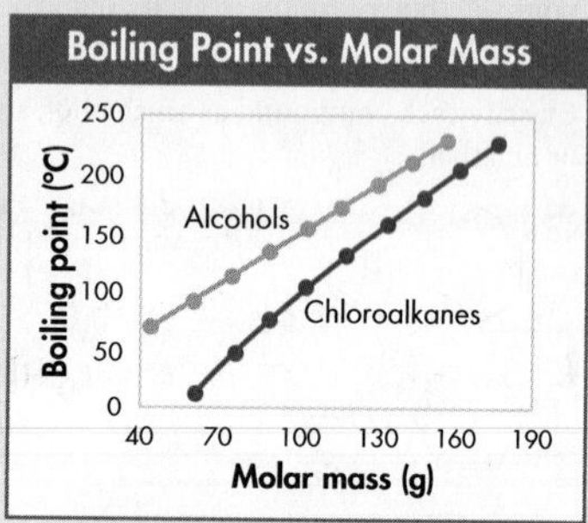

a. Explain why the boiling points of the alcohols are consistently higher than those of the chloroalkanes of similar molar mass.

b. Why does the gap in boiling point between the alcohols and chloroalkanes decrease with increasing chain length?

832 Chapter 23 • Assessment

64. No, the graph is not a straight line. Boiling point increases as the degree of halogen substitution increases, but less drastically for each subsequent halogen substitution.

65. a. $CH_4 + Cl_2 \xrightarrow{\text{UV light}} CH_3Cl + HCl$

b. $CH_2{=}CH_2 + H_2 \xrightarrow{Pt} CH_3CH_3$

66. (cyclic ether structure: five-membered ring containing O)

67. a. $C_6H_5-CH_2COOH$
2-phenylethanoic acid

b. $CH_3\overset{CH_3}{C}HCOOH$
2-methylpropanoic acid

c. 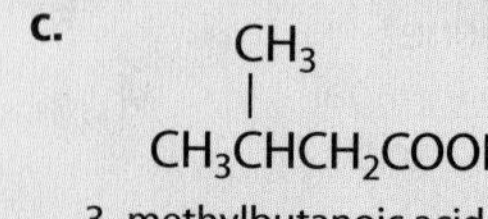
$CH_3\overset{CH_3}{C}HCH_2COOH$
3-methylbutanoic acid

68. a. Hydrogen bonding between alcohol molecules leads to higher boiling points relative to the chloroalkanes.

b. The boiling point gap decreases as the contribution of molar mass to the boiling points becomes more important.

Enrichment

*69. **Interpret Diagrams** Cholesterol is a compound in your diet and is also synthesized in the liver. Sometimes it is deposited on the inner walls of blood vessels, causing hardening of the arteries. Describe the structural features and functional groups of this important molecule.

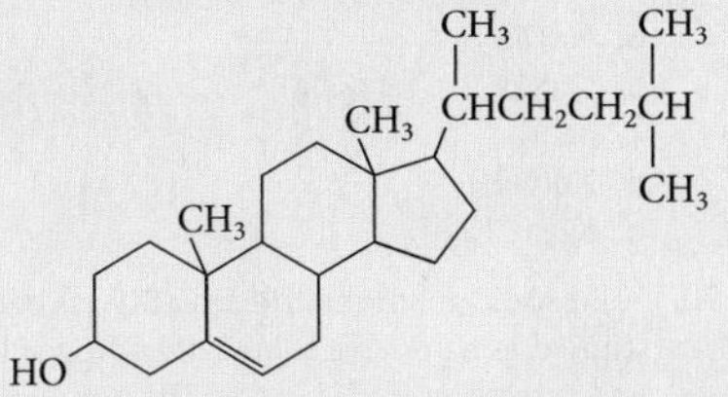

CH_3 CH_3
CH_3 $CHCH_2CH_2CH$
CH_3
CH_3
HO

70. **Propose a Solution** Hydrocarbons from petroleum are an important source of raw material for the chemical industry. Using reactions covered in this chapter, propose a scheme for the synthesis of ethylene glycol, a major component of antifreeze, from petrochemical ethene.

*71. **Draw Conclusions** Human hair is composed of long-chain polymers. Some of the monomers in these polymers contain sulfur atoms. When two sulfur atoms are adjacent to one another, they form strong, covalent disulfide (S—S) bonds that can link two polymer molecules together. The location of these bonds between the hair polymers affects how curly or straight the hair is and helps to hold the polymers in their shape. Waving lotion that is used in permanent waves is a reducing agent, and neutralizing agent is an oxidizing agent. Using this information, explain in terms of chemistry how a permanent wave can change the shape of hair. Describe what chemical steps should be taken to change the shape of someone's hair from straight to curly.

72. **Infer** Which of the following monomers can be used to produce the addition polymer $\text{+}CF_2\text{+}_x$: $CH_2{=}CF_2$, $CH_2{=}CHF$, CF_4, $CF_2{=}CF_2$, or $CHF{=}CHF$?

73. **Analyze Data** An unidentified compound of molecular formula $C_5H_{10}O$ produced a red precipitate when treated with Fehling's reagent. Further testing showed that the compound does not contain a hydrocarbon ring. Write structural formulas for all of the compounds that this unknown substance might be.

Write About Science

74. **Research a Problem** Methyl tert-butyl ether (MTBE) is a commonly used gasoline additive. Recently, regulatory agencies have begun to limit its use. Write a paragraph about MTBE, in which you explain the function of MTBE in gasoline and why its use is now limited.

75. **Connect to the BIGIDEA** Acetylsalicylic acid (aspirin) and ibuprofen are both active ingredients of over-the-counter pain relievers. They work by preventing certain proteins from sending pain signals to the brain. Look up the structures of these two active ingredients. Write a short report, in which you compare the structures and functional groups of the two compounds.

CHEMYSTERY

The Toothpaste Terror

Manufacturers include moistening agents in their toothpaste formulations. The most common moistening agent in toothpaste is glycerol. Usually glycerol is listed as glycerin on toothpaste ingredient labels. Glycerol is an alcohol with three hydroxy groups, which help keep the toothpaste moist by forming hydrogen bonds with water molecules in the air. Because glycerin has a sweet taste, it also sweetens the toothpaste.

*76. **Compare** Counterfeiters have been known to manufacture cheap toothpaste by substituting diethylene glycol for glycerol. Like glycerol, diethylene glycol is a moistening agent, but it is also toxic. Compare the structures of diethylene glycol and glycerol.

77. **Connect to the BIGIDEA** Often there is more than one moistening agent in toothpaste. Look at the ingredient list on your toothpaste container. Are there any other alcohols besides glycerol that may serve as moistening agents? Search the Internet to find the structures of each alcohol, and identify the number of hydroxy groups.

ASSESSMENT

75. Students' answers should compare the structures and functional groups of the following two compounds.

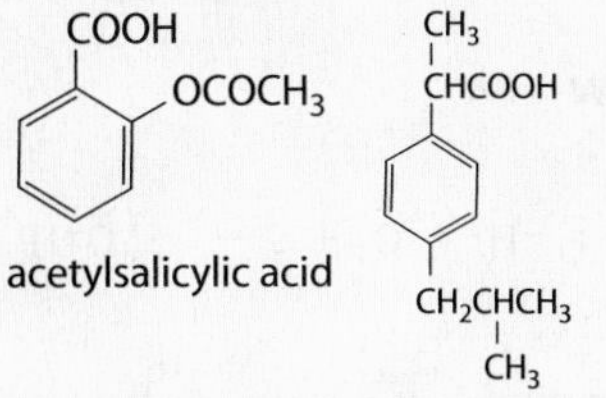

COOH, $OCOCH_3$ — acetylsalicylic acid

CH_3, CHCOOH, CH_2CHCH_3, CH_3 — ibuprofen

CHEMYSTERY **SUMMARIZE** After students have read through the CHEMystery, call on volunteers to summarize which properties of glycerol make it a practical application for keeping toothpaste moist. Have students classify glycerol according to its proper functional groups. **Ask** Given what you know about the functional groups of this product, how should Yani have cleaned the mirror in order to quickly dissolve the toothpaste? *(Sample answer: Clean it off with a wet cloth, since it is water soluble.)*

CHEMystery Answers

76. The two molecules are similar in size. Diethylene glycol has four carbons, and glycerol has three carbons. Diethylene glycol is an ether with two hydroxy groups. Glycerol has three hydroxy groups.

77. Two common alcohols found in toothpaste are sorbitol (6 hydroxyl groups) and propylene glycol (2 hydroxyl groups).

Answers

ENRICHMENT

69. Cholesterol is an alcohol with a hydroxy group on a cycloalkane. It has four nonaromatic rings. It has a double bond in one of its rings, as well as a large alkyl group, making the molecule nonpolar.

70. $CH_2CH_2(g) + Br_2(l) \rightarrow CH_2BrCH_2Br(l)$
$CH_2BrCH_2Br(l) + 2NaOH(aq) \rightarrow CH_2OHCH_2OH(l) + 2NaBr(aq)$

71. Waving lotion reduces —S—S— bonds to —SH bonds. Hair can be placed in curlers to form the hair in the desired shape. The neutralizing agent is an oxidizing agent that reforms —S—S— bonds, locking the hair into its curly shape. Similar steps could be used to straighten curly hair.

72. $CF_2{=}CF_2$

73. $CH_3CH_2CH_2CH_2CHO$; $CH_3CH_2CH(CH_3)CHO$; $CH_3CH(CH_3)CH_2CHO$; $CH_3C(CH_3)_2CHO$

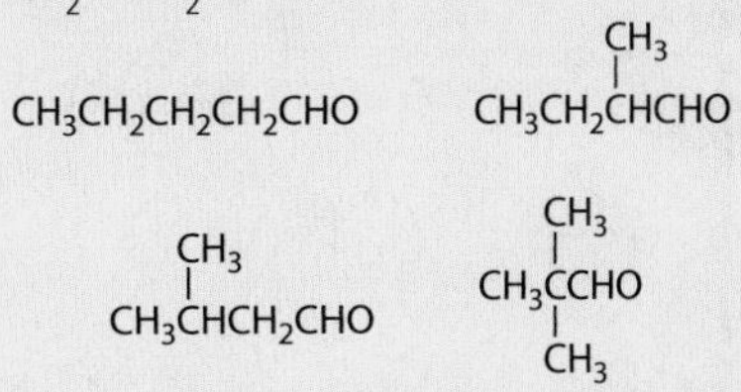

WRITE ABOUT SCIENCE

74. Engines knock when gasoline burns too soon or too quickly. MTBE replaced tetraethyl lead as an anti-knocking additive. MTBE from leaking storage tanks and fuel spills contaminates groundwater.

Evaluate

Answers

CUMULATIVE REVIEW

78. b, 3

79. 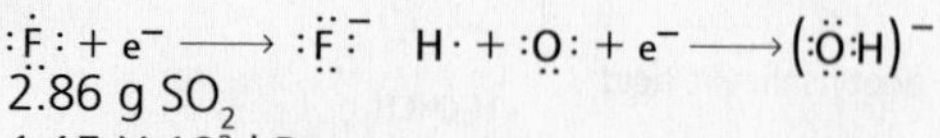 $:\ddot{F}: + e^- \longrightarrow :\ddot{F}:^-$ $H\cdot + :\ddot{O}: + e^- \longrightarrow (:\ddot{O}H)^-$

80. 2.86 g SO_2

81. 1.15×10^2 kPa

82. Anhydrous $Na_2CO_3(s)$ is the better value because the decahydrate is 63.0% water.

83. The amount of dissolved oxygen is greater entering the plant; solubilities of gases tend to decrease with increasing temperature.

84. 0.117*M* $Ca(NO_3)_2$

85. At any given moment, the rate of dissolving of solute is equal to the rate of precipitation of solute. As a result, the concentration of the solution remains constant.

86. 71 kJ

87. **a.** shift toward products
b. shift toward reactants
c. shift toward reactants
d. shift toward reactants

88. a, c, d, b

89. **a.** Na +1; N +3; O −2
b. Co +2; S +6; O −2
c. Se +4; O −2
d. Zn +2; O −2; H +1
e. K +1; Pt +2; Cl −1

90. oxidized: H of BH_4^-; reduced: H of H_2O; unaffected: Na, B, O

91. spontaneous redox reactions

92. Reduction always occurs at the cathode. In the electrolytic cell, the cathode is the negative electrode.

93. **a.** 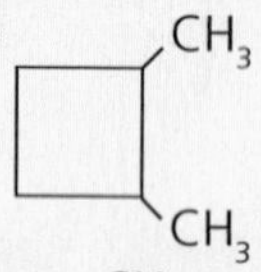 (cyclobutane ring with CH_3 on two adjacent carbons)

b. $CH_3C(CH_3){=}CHCH_2CH_3$

c. $CH_3CH{=}CHCH_3$

d. $CH_3C{\equiv}CCH_2CH_3$

e. $CH_3CH_2CH_2CH_2C(CH_3){=}CH_2$

94. coal

95. b and c

Cumulative Review

***78.** What is the maximum number of orbitals in the *p* sublevel of an atom?

a. 1 **b.** 3 **c.** 5 **d.** 9

79. Using electron dot structures, illustrate the formation of F^- from a fluorine atom and OH^- from atoms of hydrogen and oxygen.

80. Calculate the mass, in grams, of one liter of SO_2 at standard temperature and pressure.

81. A sample of 1.40 L of nitrogen gas in a sealed container at 25.0°C and 1.00×10^2 kPa is heated to 68.7°C. What is the new pressure?

***82.** Sodium carbonate is often sold as the anhydrous compound [$Na_2CO_3(s)$] or as the decahydrate [$Na_2CO_3{\cdot}10H_2O(s)$]. If the price per kilogram of the anhydrous and decahydrate forms are the same, which compound is the better value?

83. Assume that water enters a power plant at 20°C and leaves at 30°C. Will the amount of dissolved oxygen in the water be greater entering or leaving the plant?

84. A solution is made by diluting 250 mL of 0.210*M* $Ca(NO_3)_2$ solution with water to a final volume of 450 mL. Calculate the molarity of $Ca(NO_3)_2$ in the diluted solution.

85. In a saturated solution containing undissolved solute, the solute is continually dissolving, but the solution concentration remains constant. Explain this statement.

***86.** A 500-g aluminum tray at 22°C is heated to 180°C in an oven. How many kJ of heat does the aluminum tray absorb if the specific heat of aluminum is 0.90 J/(g·°C)?

87. Predict the direction of shift in the equilibrium position for each change in conditions.

$$2NO_2(g) \rightleftharpoons 2NO(g) + O_2(g)$$

a. O_2 partial pressure decrease
b. total pressure increase
c. O_2 partial pressure increase
d. NO partial pressure increase

***88.** List these K_a values for weak acids in order of increasing acid strength.

a. 3.5×10^{-6}
b. 2.7×10^{-3}
c. 1.5×10^{-5}
d. 6.6×10^{-5}

89. Assign an oxidation number to each atom in these compounds.

a. $NaNO_2$
b. $CoSO_4$
c. SeO_2
d. $Zn(OH)_2$
e. K_2PtCl_4

90. Solid sodium borohydride ($NaBH_4$) is being studied as a possible source of hydrogen fuel for hydrogen-powered vehicles. The borohydride reacts with water to produce hydrogen gas and sodium metaborate. Identify which atoms of the reactants are oxidized, which atoms are reduced, and which atoms are unaffected in this reaction.

$$NaBH_4(s) + 2H_2O(l) \longrightarrow 4H_2(g) + NaBO_2(aq)$$

91. What is the source of the electrical energy produced in a voltaic cell?

***92.** At which electrode in an electrolytic cell does reduction always occur? What is the charge on this electrode?

93. Draw a condensed structural formula for each compound.

a. 1,2-dimethylcyclobutane
b. 2-methyl-2-pentene
c. 2-butene
d. 2-pentyne
e. 2-methylhexene

***94.** Is petroleum or coal most likely to be a good source of aromatic compounds?

95. Which of these statements applies to ethene?

a. saturated hydrocarbon
b. H—C—H bond angle of 120°
c. alkene
d. aromatic compound

If You Have Trouble With . . .

Question	78	79	80	81	82	83	84	85	86	87	88	89	90	91	92	93	94	95
See Chapter	5	7	10	14	15	16	16	16	17	18	19	20	20	21	21	22	22	22

Standardized Test Prep for Indiana

Standard: C.4.5

Select the choice that best answers each question or completes each statement.

1. The acid-catalyzed hydrolysis of an ester gives a carboxylic acid and
   (A) an amine. (C) an alcohol.
   (B) an ether. (D) an alkene.
2. Ethane, methanal, and methanol have similar molar masses. Which series lists the compounds in order of increasing boiling point?
   (A) ethane, methanal, methanol
   (B) methanal, methanol, ethane
   (C) methanol, methanal, ethane
   (D) ethane, methanol, methanal
3. A carbonyl group is characterized by a
   (A) carbon–carbon double bond.
   (B) carbon–oxygen double bond.
   (C) carbon–nitrogen single bond.
   (D) carbon–oxygen single bond.

The lettered choices below refer to Questions 4–7. A lettered choice may be used once, more than once, or not at all.

(A) alcohol (C) carboxylic acid
(B) ketone (D) ether

To which class of organic compounds does each of the following compounds belong?

4. CH_3CH_2COOH
5. $CH_3CH_2CH_2OH$
6. $CH_3CH_2OCH_3$
7. CH_3COCH_3

Use the following models with Question 8.

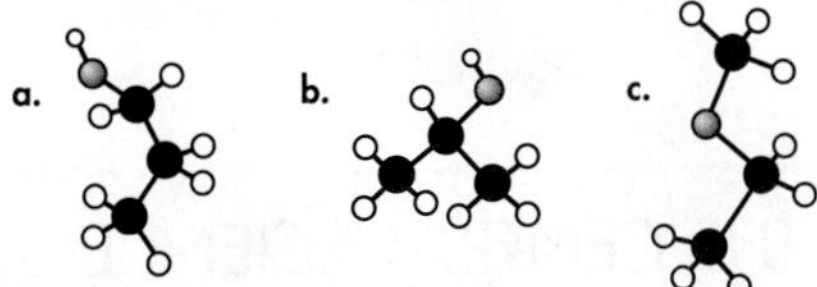

8. The molecular formula for each compound is C_3H_8O. Write the name for each compound.
9. There are two compounds with a carbonyl group that have the molecular formula C_3H_6O. Write a complete structural formula for each compound. Name each compound.

Tips for Success

Eliminate Wrong Answers If you don't know which choice is correct, eliminate those you know are wrong. If you can rule out some choices, you'll increase your chances of choosing the correct answer.

Characterize the reactions in Questions 10–14 as an addition, esterification, oxidation, polymerization, or substitution reaction.

10. $CH_3CHO \xrightarrow[H_2SO_4]{K_2Cr_2O_7} CH_3COOH$
11. $CH_2{=}CH_2 + HCl \longrightarrow CH_3CH_2Cl$
12. $CH_3CO_2H + CH_3CH_2OH \xrightarrow{H^+} CH_3COOCH_2CH_3 + H_2O$
13. $xCH_2{=}CH_2 \longrightarrow H\!\left(CH_2{-}CH_2\right)_x\!H$
14. $C_6H_6 + Br_2 \xrightarrow{catalyst} C_6H_5Br + HBr$

For each question there are two statements. Decide whether each statement is true or false. Then decide whether Statement II is a correct explanation for Statement I.

	Statement I		Statement II
15.	The addition of hydrogen to an alkene is a reduction reaction.	BECAUSE	The addition of hydrogen to any molecule is a reduction reaction.
16.	Aldehydes are easily oxidized.	BECAUSE	Oxidation of aldehydes produces alcohols.
17.	Ethanol (CH_3CH_2OH) is immiscible in water in all proportions.	BECAUSE	Ethanol molecules can form hydrogen bonds with other ethanol molecules.

If You Have Trouble With . . .

Question	1	2	3	4	5	6	7	8	9	10	11	12	13	14	15	16	17
See Lesson	23.3	23.3	23.3	23.3	23.2	23.2	23.3	23.2	23.3	23.3	23.2	23.3	23.4	23.1	23.2	23.3	23.2

ASSESSMENT

STP Answers

MULTIPLE CHOICE

1. C
2. A
3. B
4. C
5. A
6. D
7. B

OPEN RESPONSE

8. a. 1-propanol (or propyl alcohol)
   b. 2-propanol (or isopropyl alcohol)
   c. methoxyethane (or ethyl methyl ether)
9.

```
    H   H   H
    |   |   |
H — C — C — C = O
    |   |
    H   H
```

propanal

```
    H   O   H
    |   ||  |
H — C — C — C — H
    |       |
    H       H
```

propanone

10. oxidation
11. addition
12. esterification
13. polymerization
14. substitution
15. True, True, correct explanation
16. True, False
17. False, True

24 The Chemistry of Life Planning Guide

Introducing the BIGIDEA: CHEMISTRY AS THE CENTRAL SCIENCE

Chemistry overlaps with all of the other sciences, including biology.

IASS	Lessons and Objectives	Print Resources		
		For the Student	For the Teacher	
	24.1 A Basis for Life p 838–840 **24.1.1** Identify the two major cell types that occur in nature. **24.1.2** Identify the compound that is reduced during photosynthesis and the compounds formed.	Reading and Study Workbook Lesson 24.1 Lesson Assessment 24.1 p 840	Class Activity, p 839: Living and Nonliving Things	
	24.2 Carbohydrates p 841–843 **24.2.1** Construct the general formula for carbohydrates.	Reading and Study Workbook Lesson 24.2 Lesson Assessment 24.2 p 843	Teacher Demo, p 843: Benedict's Test	
	24.3 Amino Acids and Their Polymers p 844–848 **24.3.1** Diagram the general structure of an amino acid. **24.3.2** Identify what determines the differences in the chemical and physiological properties of peptides and proteins. **24.3.3** Describe how enzymes affect the rates of reactions in living things.	Reading and Study Workbook Lesson 24.3 Lesson Assessment 24.3 p 848 Small-Scale Lab: The Egg: A Biochemical Storehouse p 849	Teacher Demo, p 845: Peptide Models Teacher Demo, p 846: Protein in Hard Tissue	
	24.4 Lipids p 850–852 **24.4.1** Identify the physical property that distinguishes lipids from other biological molecules.	Reading and Study Workbook Lesson 24.4 Lesson Assessment 24.5 p 852	Student Activity, p 851: Better Butter	
	24.5 Nucleic Acids p 854–861 **24.5.1** Identify the functions of DNA and RNA. **24.5.2** Identify the number of bases of DNA required to specify one amino acid in a peptide chain. **24.5.3** Define gene mutations. **24.5.4** Identify two examples of DNA technology used today.	Reading and Study Workbook Lesson 24.5 Lesson Assessment 24.5 p 861 Quick Lab: A Model of DNA p 856	Class Activity, p 858: Genetic Code	
	24.6 Metabolism p 862–866 **24.6.1** Describe the function of ATP in the cell. **24.6.2** Describe what happens to biological molecules and energy during catabolism and anabolism. **24.6.3** Explain how nitrogen-fixing bacteria provide plants with a usable form of nitrogen.	Reading and Study Workbook Lesson 24.6 Lesson Assessment 24.6 p 866		

Assessing the BIGIDEA: CHEMISTRY AS THE CENTRAL SCIENCE

	Essential Questions 1. What are the characteristics of the four main types of biological molecules? 2. What is the function of anabolism and catabolism in a cell?	Study Guide p 868 STP p 873 Reading and Study Workbook Self-Check and Vocabulary Review Chapter 24		

	PearsonChem.com	
	Editable Worksheets	Online Activities
	Teaching Resources, Lesson 24.1 Review	24.1 Lesson Overview A Fish's Life
	Teaching Resources, Lesson 24.2 Review Small-Scale Lab Manual Lab 40: Reactions of Biomolecules	24.2 Lesson Overview
	Teaching Resources, Lesson 24.3 Review Small-Scale Lab Manual Lab 40: Reactions of Biomolecules	24.3 Lesson Overview Peptide Structures
	Teaching Resources, Lesson 24.4 Review Small-Scale Lab Manual Lab 40: Reactions of Biomolecules Lab 51: Preparation of Soap	24.4 Lesson Overview
	Teaching Resources, Lesson 24.5 Review	24.5 Lesson Overview
	Teaching Resources, Lesson 24.6 Review	24.6 Lesson Overview The Nitrogen Cycle Food Calorimetry
	Core Teaching Resources (includes editable worksheets) • Lesson Reviews • Practice Problems • Interpret Graphs • Chapter Quizzes and Tests	Chapter 24 Problem Set

Materials List

For the Student

Small-Scale Lab p 849
- chicken egg
- ruler
- balance

Quick Lab p 856
- cardboard tube from paper-towel roll
- felt-tip markers, two colors
- metric ruler
- thumbtack
- 10 toothpicks

For the Teacher

Class Activity p 839
- photographs of living and nonliving things

Teacher Demo p 842
- foam balls and toothpicks or molecule kit

Teacher Demo p 843
- hot-water bath
- 4–6 mL of 1% carbohydrate solution
- 2 test tubes
- distilled water
- 2–3 mL of Benedict's reagent
- dropper
- samples of mono- and polysaccharides

Teacher Demo p 845
- models of amino acids

Teacher Demo p 846
- thigh bone from cooked poultry
- 1*M* HCl
- beaker

Class Activity p 851
- butter
- regular margarine
- low-calorie margarine
- 3 test tubes
- balance
- warm-water bath

Additional Digital Resources

Online Student Edition
Online Teacher's Edition
Exam*View Assessment Suite CD-ROM*
Classroom Resources DVD

CHAPTER 24

What's Online

ONLINE PROBLEM SET Additional practice problems about biochemistry can be found in the Chapter 24 Online Problem Set.

DIRECTED VIRTUAL LAB Encourage students to review the energetic properties of different food and food additives using the animated virtual lab for Chapter 24. The animation is followed by analysis questions, where the student is guided through the lab concepts.

KINETIC ART Have students watch animations about protein folding and the nitrogen cycle. Each animation is followed by analysis questions.

CONCEPTS IN ACTION Students can watch an interactive animation about the similarities in the chemistry of breathing for fish and humans. The animation is followed by questions to explore other relationships in biochemistry.

24 The Chemistry of Life

INSIDE:

PearsonChem.com

All organisms, including you and your classmates, are made of the same types of molecules: carbohydrates, proteins, lipids, and nucleic acids.

IN ACADEMIC STANDARDS for SCIENCE

- Develop explanations based on reproducible data and observations gathered during laboratory investigations.
- Clearly communicate their ideas and results of investigations verbally and in written form using tables, graphs, diagrams, and photographs.

Focus on ELL

1 CONTENT AND LANGUAGE Write the chapter title on the board. Underline the word *Life*. Ask volunteers to explain their ideas of what life is. Have students skim through the lesson titles, essential questions, and CHEMystery and identify words that are unfamiliar to them. Group students who are unfamiliar with the same word. Have each group complete a Frayer model diagram for this word, with a definition, characteristics, an example, and a nonexample.

BEGINNING

LOW Create a symbol or drawing to represent the word.

HIGH Write the word and introduce it to the class by pointing to it and slowly saying it aloud.

INTERMEDIATE: LOW/HIGH Brainstorm examples and nonexamples with the class.

ADVANCED: LOW/HIGH Write the definition and present it to the class. Explain to the class how the examples represent the word.

BIGIDEA

CHEMISTRY AS THE CENTRAL SCIENCE

Essential Questions:

1. *What are the characteristics of the four main types of biological molecules?*
2. *What is the function of anabolism and catabolism in a cell?*

CHEMYSTERY

Phenyl-what?

Do you ever look at the labels on food and wonder what they mean? In the United States, the Food and Drug Administration (FDA) regulates the requirements for how foods are labeled. Information that you might see on foods includes a nutrition label, the quantity of the food, the ingredients the food contains, and an expiration date.

Some food labels also contain warnings about the ingredients in the food. Some labels warn you about something common, like nuts or milk. But other times the warning may leave you wondering what it means. The warning on the label shown here says the product "Contains phenylalanine." What is phenylalanine? Why would people need to know that phenylalanine is in their food?

► Connect to the BIGIDEA As you read about molecules involved in life's processes, think about how the chemistry of these molecules is central to living things.

Introduce the Chapter

CONNECTIONS TO THE REAL WORLD Students may not have thought about the chemical nature of the food they eat. Use the following activity to help students relate food and chemical concepts they will study in this chapter.

Activity *You will need to bring into class a variety of food nutrition panels and foods representing each of the four main types of biological compounds.* **Ask** Where does your body get the chemicals it needs to live and grow? *(from the food you eat)* Have students examine each food label and name different chemicals. Engage them in discussing possible benefits the body gets from the different chemicals. Discuss how the fresh fruits and vegetables or other foods you have brought in provide nutrients people need for biological processes.

UbD

Understanding by Design

Students are building toward understanding how chemistry is the central science by learning about the chemistry of life.

PERFORMANCE GOALS At the end of Chapter 24, students will be able to answer the essential questions by applying their knowledge of the chemistry of life.

ESSENTIAL QUESTIONS Read the essential questions aloud. **Ask** What do you think a biological molecule is? *(molecule produced by living things)* **Ask** What are some processes that occur in the human body that require energy? *(Sample answers: breathing, changing food into chemicals the body can use)*

BIGIDEA Use the photo of students to help students connect to this chapter's concepts. Activate students' prior knowledge by discussing ways the people and trees are alike. Explain that all organisms have properties that depend on chemical processes. **Ask** What are some ways that chemical processes help the people and trees survive? *(Sample answer: Trees and people grow and reproduce.)* Point out that these processes require chemicals obtained from the environment.

CHEMYSTERY Connect the CHEMystery to the Big Idea by engaging students in a discussion of food ingredient labels. Explain that understanding the chemistry of food is important. As a hint about why people might need to know that phenylalanine is in food, point out that the warning on the label is in bold type. Encourage students to think about why this information would be important to some people.

UntamedScience™

The Secret's in the Muscles

Suze learns how a chemical called myoglobin enables dolphins to dive deeper and longer than humans can dive.

Key Objectives

24.1.1 IDENTIFY the two major cell types that occur in nature.

24.1.2 IDENTIFY the compound that is reduced during photosynthesis and the compounds formed.

Additional Resources

- Reading and Study Workbook, Lesson 24.1

PearsonChem.com

Engage

CHEMISTRY & YOU **Ask** What gases may have dominated Earth's early atmosphere? *(Studies suggest the following: CO_2, CH_4, H_2, NH_3, N_2, and H_2O.)* **Ask** What life-supporting gas does photosynthesis produce? *(oxygen)* **Ask** What types of organisms photosynthesize? *(many types of bacteria and almost all plant species)*

Activate Prior Knowledge

Encourage students to share what they know about the flow of energy and matter through an ecosystem. Remind students that the movement of both through an ecosystem is the result of chemical processes that take place in the cells of living organisms.

Foundations for Reading

BUILD VOCABULARY The word *photosynthesis* can be broken down into its parts: *photo*, meaning "light," *syn*, meaning "together," and *thesis*, meaning "to put." Photosynthesis uses light to combine carbon dioxide and water into glucose, with oxygen as a by-product.

24.1 A Basis for Life

CHEMISTRY & YOU

Q: *Where do fish get their oxygen?* All animals need oxygen to survive. Most fish obtain oxygen as water flows across their gills. We breathe in oxygen from the air around us. The air you breathe is composed mainly of nitrogen (N_2) and oxygen (O_2). Earth's early atmosphere may have been very different, with very little oxygen, and most inhospitable to life. It is thought that the atmosphere changed over time. In this lesson, you will learn how oxygen in both the air and water is produced.

Key Questions

What are the two major types of cells that occur in nature?

What compound is reduced during photosynthesis? What compounds are formed?

Vocabulary

- photosynthesis

The Structure of Cells

What are the two major types of cells that occur in nature?

Life! You are certainly familiar with it, but what does it really mean? Until recently, life was defined as the ability of an organism to grow and to reproduce its own kind. However, recent discoveries made at the fringes of life seem to blur this simple definition. As difficult as it is to define life, you can generally regard tiny structures called cells as the fundamental units of life.

Organisms are composed of as few as one cell or as many as billions of cells. **Two major cell types occur in nature: prokaryotic cells and eukaryotic cells.** The prokaryotic cell is the more ancient of the two. Microscopic examination of fossilized remains shows that prokaryotic cells were present on Earth at least 3 billion years ago. Eukaryotic cells did not appear until about 1 billion years ago. Figure 24.1 shows both types of cells.

Figure 24.1 Cell Types
Typical prokaryotic and eukaryotic cells are shown here. Note that only the eukaryotic cell has a nucleus.
Compare and Contrast *How do prokaryotic and eukaryotic cells compare in size?*

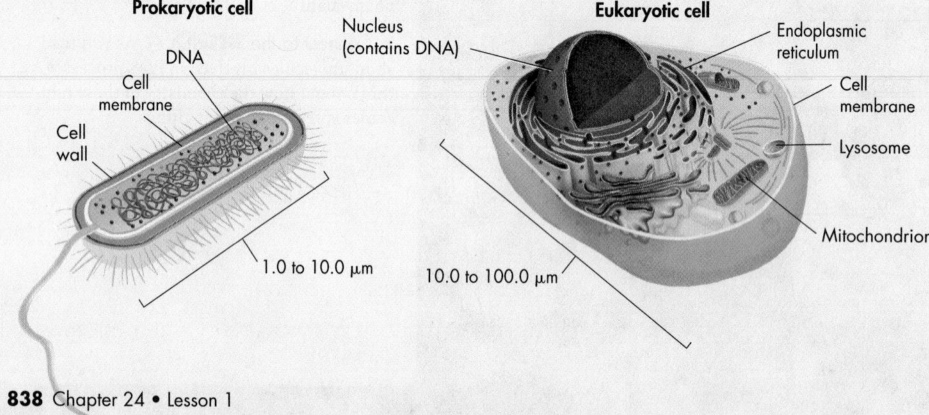

838 Chapter 24 • Lesson 1

Focus on ELL

1 CONTENT AND LANGUAGE Review vocabulary terms such as *organelles*, *prokaryotic cells*, *eukaryotic cells*, and *chloroplast* by assigning each term to a group. Have each group research the definition of the term, create a drawing of what the term refers to, and present the drawing to the class. Encourage students to use any one of the vocabulary methods they have learned in this course.

2 FRONTLOAD THE LESSON Access students' prior knowledge by asking them what they already know about prokaryotic and eukaryotic cells. Have several biology books available for students to reference. Discuss why understanding the structure of cells is important for a chemist that studies biological molecules.

3 COMPREHENSIBLE INPUT Project Figure 24.3 on a dry-erase board, and provide students with printed copies. Write the chemical formulas of reactants and products associated with each process beside its descriptive label.

Both eukaryotic and prokaryotic cells contain all the chemicals necessary for life, encased in a cell membrane. The cell membrane is a sack that holds the contents of a cell and acts as a selective barrier for the passage of substances into and out of the cell. Eukaryotic cells are considerably larger and more complex than prokaryotic cells, but the chemical processes carried out by both types of cells are very similar.

One major feature that distinguishes eukaryotic cells from prokaryotic cells is that eukaryotic cells contain membrane-enclosed organelles. Organelles, meaning little organs, are small structures suspended in the interior cellular fluid, or cytoplasm. The organelles are the sites of many specialized functions in eukaryotic cells. For example, the nucleus, a structure that is important in eukaryotic cell reproduction, is not present in prokaryotic cells. Mitochondria (singular: mitochondrion) are the source of cellular energy in eukaryotic cells that use oxygen. Mitochondria are often referred to as the powerhouses of the cell. Lysosomes are the sites for the digestion of substances taken into a cell. Yet another membrane-enclosed structure in eukaryotic cells is the highly folded, netlike endoplasmic reticulum (ER). Among its various functions, the ER serves as an attachment site for ribosomes. The ribosomes, small organelles that are not membrane-enclosed, are the sites where essential substances called proteins are made.

The Energy and Carbon Cycle

What compound is reduced during photosynthesis? What compounds are formed?

Organisms must have energy to survive. The ultimate source of this energy is the sun. Cells of green plants and certain algae contain organelles called chloroplasts that are able to capture solar energy and make food. Within a chloroplast is a light-capturing system of membranes, shown in Figure 24.2, that converts light energy into chemical energy by a process called **photosynthesis.** In addition to sunlight, photosynthetic organisms require carbon dioxide and water. **Photosynthesis uses the energy from sunlight to reduce carbon dioxide to compounds that contain C—H bonds, mainly in the form of glucose ($C_6H_{12}O_6$).** The following equation summarizes the process:

$$\underset{\text{Carbon dioxide (carbon in more oxidized state)}}{6CO_2} + \underset{\text{Water}}{6H_2O} + \underset{\text{from sunlight}}{\text{Energy}} \longrightarrow \underset{\text{Glucose (carbon in more reduced state)}}{C_6H_{12}O_6} + \underset{\text{Oxygen}}{6O_2}$$

Figure 24.3 on the next page illustrates the relationship between photosynthesis and the carbon compounds used by all organisms. In the energy and carbon cycle, photosynthetic organisms produce necessary carbon compounds. Animals, which do not carry out photosynthesis, get these carbon compounds by eating plants or by eating animals that feed on plants. Both plants and animals get energy by unleashing the energy stored in the chemical bonds of these carbon compounds. The nutrients are oxidized back to carbon dioxide and water in the process.

$$\underset{\text{Glucose (carbon in more reduced state)}}{C_6H_{12}O_6} + \underset{\text{Oxygen}}{6O_2} \longrightarrow \underset{\text{Carbon dioxide (carbon in more oxidized state)}}{6CO_2} + \underset{\text{Water}}{6H_2O} + \text{Energy}$$

Figure 24.2 Chloroplast
The reactions of photosynthesis take place on the inner membranes and in the spaces between membranes of a chloroplast. The pigments in chloroplasts are why leaves are green in color.

Professional Development Note

Detailing Photosynthesis

Melvin Calvin (1911–1997) and others in his laboratory were important participants in tracing the chemical reactions that take place in photosynthesis. Their work was made possible by two techniques: using radioactive carbon-14 to follow carbon's path through the complex process and using chromatography to identify the products at each step.

READING STRATEGY Instruct students to read the caption with each figure in this lesson. Have students read the caption and other text with Figure 24.3 and have them summarize how this diagram conveys the concept of photosynthesis and the carbon compounds used by all organisms.

LESSON 24.1

Explain

The Structure of Cells

USE VISUALS Direct students' attention to Figure 24.1. Note that eukaryotic cells can vary greatly in size and shape, but they are always larger than prokaryotic cells. Point out that many metabolic reactions in eukaryotic cells are compartmentalized. **Ask** What possible advantages would organelles provide for a cell? *(Many of the chemical activities that occur in cells involve similar reactants and products. By nature, they are incompatible. If these processes are physically separated, they can proceed concurrently without conflict.)*

Explore

Class Activity

Have teams of students create a two-column chart of characteristics that distinguish living from nonliving systems. *(Living things have one or more cells; they obtain and consume energy; they respond to changing conditions; they grow and develop, and they can reproduce.)* Then present photographs of living and nonliving things and help students use their lists to classify each subject as living or nonliving.

Answers

FIGURE 24.1 Eukaryotic cells are larger than prokaryotic cells.

LESSON 24.1

Extend

Have students explore the concept of why some large metropolitan cities are encouraging "green" roofs for office buildings. Students should be able to explain at least four reasons why a green roof is beneficial to the environment.

CHEMISTRY & YOU Photosynthesis

Evaluate

Informal Assessment

Write the equation for photosynthesis on the board. **Ask** Should this reaction be classified as acid–base, redox, or a combination? *(redox)* Why? *(Carbon dioxide is reduced and water is oxidized.)* **Ask** Is photosynthesis endothermic or exothermic? *(endothermic)* **Ask** Which part of the plant cell specializes in the light-harvesting reactions of photosynthesis? *(the chloroplast)* Then have students complete the 24.1 Lesson Check.

Reteach

Have students examine Figure 24.3. Explain that the diagram represents the flow of energy and nutrients through Earth's biosphere, the regions on Earth where living organisms exist. Point out that energy is required to maintain the highly organized chemical environments found in living cells. **Ask** Where does the energy come from that drives the cycle? *(the Sun)* **Ask** Why are plants and other photosynthetic organisms called producers and animals called consumers? *(Photosynthetic organisms produce high-energy forms of carbon compounds that are consumed by animals.)*

CHEMISTRY & YOU

Q: *Fish, like all animals, need oxygen to survive. What process produces oxygen?*

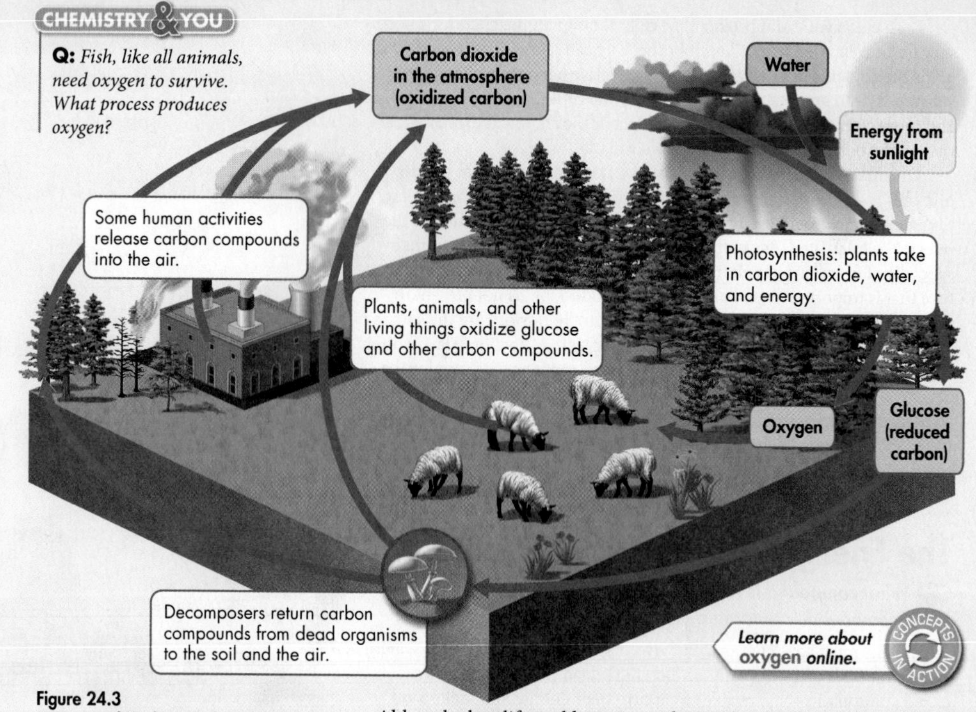

Learn more about oxygen online. CONCEPTS IN ACTION

Figure 24.3
Energy and Carbon Cycles
In the energy and carbon cycle, photosynthesis and the oxidation of glucose are responsible for the major transformations and movements of carbon. Plants release oxygen into the atmosphere through photosynthesis.
Interpret Diagrams ***In which parts of the cycle is oxygen consumed?***

Although plant life could survive without animals, animal life could never survive without plants. Without photosynthesis, the supply of carbon compounds that animals need to get energy would not exist.

Oxygen is another important product of photosynthesis. Photosynthetic land-dwelling and aquatic organisms produce the oxygen found in Earth's atmosphere, oceans, and lakes. Oxygen is needed for most organisms to live. The importance of photosynthetic organisms is a major reason for the concern about the loss of such organisms through the destruction of forests.

All biological processes, including photosynthesis, are based on certain essential kinds of chemical substances. Surprisingly, the great complexity of life arises from just a few types of biological molecules. In the remainder of this chapter, you will learn about the molecular structures of the great classes of biological molecules and the roles they play in living things.

24.1 LessonCheck

ONLINE PROBLEMS

1. **Identify** What two types of cells occur in nature?
2. **Review** What chemical changes occur during photosynthesis?
3. **Identify** What are the fundamental units of life?
4. **Describe** Describe the structure of a eukaryotic cell.
5. **Describe** What is the function of chloroplasts in green plants and algae?
6. **Identify** Write an equation that describes the oxidation of glucose.
7. **Explain** Explain how carbon moves through the environment.

Lesson Check Answers

1. eukaryotic cells and prokaryotic cells
2. CO_2 is reduced to glucose and oxygen is released.
3. cells
4. A eukaryotic cell has a nucleus and other membrane-enclosed organelles.
5. Chloroplasts capture solar energy and make food.
6. $C_6H_{12}O_6 + 6O_2 \rightarrow 6CO_2 + 6H_2O +$ energy
7. Carbon moves through the environment in the carbon and energy cycle: Through photosynthesis, plants convert carbon dioxide to carbon compounds used for energy and growth; animals ingest plant products for energy, producing carbon dioxide in the process, which returns to the atmosphere.

24.2 Carbohydrates

CHEMISTRY & YOU

Q: *Why does a cow chew all day?* Cows spend most of their day chewing their food. A cow's diet is rich in cellulose, which belongs to a class of organic molecules known as carbohydrates. In this lesson, you will learn about the structures and functions of carbohydrates.

Key Question

What is the general formula of carbohydrates?

Vocabulary

- carbohydrate
- monosaccharide
- disaccharide
- polysaccharide

Classifying Carbohydrates

What is the general formula of carbohydrates?

Long-distance runners often prepare for a big race by eating a great deal of bread and pasta, a process known as carbohydrate loading. Breads and pastas are excellent sources of the family of important molecules called carbohydrates. **Carbohydrates** are monomers and polymers of aldehydes and ketones that have numerous hydroxyl groups attached; they are made up of carbon, hydrogen, and oxygen. **Most carbohydrates have the general formula $C_n(H_2O)_n$.** The name *carbohydrate* comes from the early observation that because of the formula, $C_n(H_2O)_n$, the compounds appear to be hydrates of carbon. But, carbohydrates are not true hydrates.

Carbohydrates are the main source of energy for the body. They are found in most foods, including fruits, breads, and the pasta and legumes in Figure 24.4. Carbohydrates are also in many sweets, such as cookies and pies. In this lesson, you will learn about the similarities and differences among some well-known types of carbohydrates.

Monosaccharides The simplest carbohydrate molecules are called simple sugars, or **monosaccharides.** Glucose and fructose are examples of simple sugars. Glucose is abundant in plants and animals. Glucose is the primary energy source for our bodies. Depending on the source, glucose has also been called corn sugar, grape sugar, or blood sugar. Fructose occurs in a large number of fruits and in honey. Glucose and fructose both have the molecular formula $C_6H_{12}O_6$. However, glucose has an aldehyde functional group, whereas fructose has a ketone functional group. Therefore, glucose and fructose are constitutional isomers. Both undergo many of the same reactions as ordinary aldehydes and ketones.

Figure 24.4 Carbohydrates in Food Carbohydrates are the most abundant sources of energy in food. But some sources are healthier than others. Vegetables, fruits, beans, and whole-grain pastas are a healthier source of carbohydrates than white breads, cookies, and sodas.

841

Focus on ELL

1 CONTENT AND LANGUAGE Write the term *polysaccharide* on the board, and pronounce it for students. Explain that this term comes from the Greek word *poly*, meaning "full" or "many," and the Sanskrit word *sarkara* meaning "sugar"; it is a compound formed by the linking together of simple sugar molecules. Ask students to share the word for sugar in their language; most are derived from the Sanskrit word.

2 FRONTLOAD THE LESSON Ask students to write down everything they ate or drank the day before. As you discuss carbohydrates, have them circle the items consumed that contained carbohydrates.

3 COMPREHENSIBLE INPUT When discussing cellulose, create a short ball-and-stick model of three units of glucose. Manipulate the model to show how the glucose units in cellulose differ from those in starch.

Key Objective

24.2.1 CONSTRUCT the general formula for carbohydrates.

Additional Resources

- Reading and Study Workbook, Lesson 24.2

PearsonChem.com

Engage

CHEMISTRY & YOU Direct students to the photograph and read the opening text. Encourage them to think about the structure of plant cells as they predict an answer to the question. Have them write their predictions in their notebooks.

Activate Prior Knowledge

Remind students of what they know about polymers. As they read this section, they should use this prior knowledge to confirm differences among different types of polysaccharides.

Answers

FIGURE 24.3 Living things consume oxygen as they oxidize glucose and other carbon compounds.

Foundations for Reading

BUILD VOCABULARY Have students make a concept map of the relationships among carbohydrates, monosaccharides, disaccharides, and polysaccharides. Encourage students to personalize their concept maps with illustrations or photographs.

READING STRATEGY Preview the chemical structures shown throughout the lesson. Review key information about functional groups and the types of reactions they can undergo.

Explain

Classifying Carbohydrates

USE VISUALS Direct students to Figure 24.5. Point out that plants are mostly composed of a carbohydrate known as cellulose. **Ask** What are carbohydrates, and what is the general formula of these compounds? *(Carbohydrates are a group of organic compounds that includes simple and complex sugars. Most carbohydrates have the general formula $C_n(H_2O)_n$.)*

Explore

Teacher Demo

PURPOSE Students build models depicting the structures of glucose and fructose.

MATERIALS foam balls and toothpicks or molecule kit

PROCEDURE Construct ball-and-stick models of glucose and fructose. Point out the functional groups. Demonstrate the cyclization of the straight-chain forms. Connect the two cyclic forms to produce a disaccharide, and note the loss of a water molecule. Point out that if this condensation reaction is allowed to continue, many monosaccharide units could join to form a polysaccharide.

ANSWER TO READING SUPPORT Monosaccharides are the building blocks of more complex sugars. A disaccharide forms from two monosaccharides. A polysaccharide forms from many monosaccharide monomers.

In aqueous solution, simple sugars such as glucose and fructose exist in a dynamic equilibrium between straight-chain and cyclic forms. The cyclic form predominates. The structures for each sugar in both forms are below.

Straight-chain and cyclic forms of glucose

Straight-chain and cyclic forms of fructose

READING SUPPORT

Build Vocabulary: ***Prefixes*** The word *saccharide* is from the Sanskit *sarkara*, which means "sugar." The prefixes *mono-*, *di-*, and *poly-* mean one, two and many, respectively. ***How does knowing this help you remember the meanings of monosaccharide, disaccharaide, and polysaccharide?***

Note the aldehyde functional group (—CHO) on the straight-chain form of glucose and the ketone functional group (—C(=O)—) on the straight-chain form of fructose.

Disaccharides Simple sugars form the building blocks of more complex carbohydrates. The cyclic forms of two simple sugars can be linked by means of a condensation reaction. For example, the linking of glucose and fructose with the loss of a water molecule produces sucrose—common table sugar. Sugar cane plants, such as those in Figure 24.5, are a major source of sucrose. A sugar such as sucrose that forms from the condensation of two monosaccharides is known as a **disaccharide.** The reaction by which it forms is as follows:

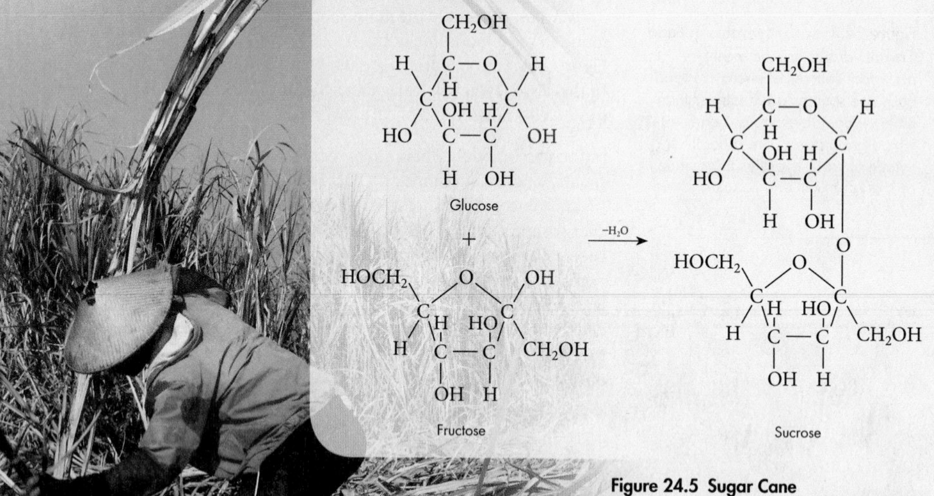

Figure 24.5 Sugar Cane
Sucrose is obtained commercially mainly from the juice of sugar cane and sugar beets. Sugar cane, shown here, grows as a tall grass and is often harvested by hand.

Glucose Testing

The ability to detect monosaccharides is an important function of clinical laboratories and at-home tests for those afflicted with diabetes mellitus. The Benedict's test is one way to test for monosaccharides. It is not specific to glucose, however, but tests for the total reducing substance, which may also include certain disaccharides and ketones. Available test strips allow more precise measurement of glucose levels. They combine enzymes specific to glucose with indicators whose colors depend on the concentration of glucose.

Polysaccharides The formation of a disaccharide is sometimes the first step in a condensation polymerization reaction that produces extremely large molecules. The polymers produced by the linkage of many monosaccharide monomers are called **polysaccharides.** Starches, the major storage form of glucose in plants, are polysaccharide polymers that consist of glucose monomers. Figure 24.6 shows a portion of a starch molecule.

A typical linear starch molecule contains hundreds of glucose monomers. Other starches are branched molecules, each branch containing about a dozen glucose units. Glycogen, the energy source stored in the liver and muscle cells of animals, is a more highly branched molecule than plant starches. Glycogen, too, consists of glucose monomers.

Cellulose is probably the most abundant biological molecule on Earth. As you can see in Figure 24.6, cellulose also is a polymer of glucose. The orientation of the bond that links the glucose monomers in cellulose is different from the bond orientation in starch and in glycogen. Starch can be digested by most organisms and is partially soluble in water. Cellulose, however, can be digested by only a few kinds of microorganisms, such as those that live in the digestive tracts of cattle and termites. Cellulose is insoluble in water and is an important structural polysaccharide that provides form, hardness, and rigidity in plants. Plant cell walls are made of cellulose. Cotton is about 80 percent cellulose.

Q: *A cow's diet consists mostly of cellulose. A cow's stomach contains several parts. Food enters the first part of the stomach and is then regurgitated so the cow can chew it again. The regurgitated food then goes to a later part of the stomach where special bacteria live. Why is chewing and regurgitating necessary for a cow to digest cellulose?*

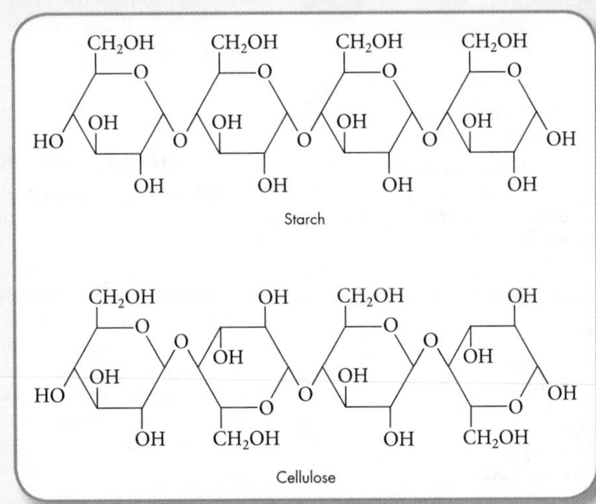

Figure 24.6 Polysaccharides
Starch and cellulose are similar polymers made up of hundreds of glucose monomers. They differ in the orientation of the bond between the glucose units. Because of this difference, starch is readily digestible, but cellulose is indigestible by most organisms.
Use Models *What are the differences between the complete structural formulas used in Figure 24.5 and the formulas used in Figure 24.6?*

24.2 LessonCheck

8. **Review** What is the general formula for carbohydrates?
9. **Identify** Where is glucose found in nature?
10. **Explain** How can the cyclic forms of two simple sugars be combined?
11. **Compare** Distinguish between the important structural features of sucrose, glucose, and fructose.
12. **Describe** Describe the main characteristics of monosaccharides, disaccharides, and polysaccharides.
13. **Compare** Starch and cellulose have different properties, but both are composed of glucose units. Explain what makes them different.
14. **Review** Name a source for each polysaccharide:
    a. starch
    b. cellulose
    c. glycogen
15. **Identify** What is the most abundant carbohydrate on Earth, and where is it found?

Lesson Check Answers

8. $C_n(H_2O)_n$
9. in plants and animals
10. via a condensation reaction
11. *sucrose*: disaccharide of glucose and fructose; *glucose*: monosaccharide with aldehyde functional group; *fructose*: monosaccharide with ketone functional group
12. Monosaccharides (simple sugars) can exist in straight-chain or cyclic form. Disaccharides form when two simple sugars condense. Polysaccharides are polymers of monosaccharides.
13. The orientation of the bonds that link the glucose monomers is different.
14. a. plants
    b. plant cell walls
    c. liver and muscle cells of animals
15. Cellulose; it is found in plant cell walls.

Explain

CHEMISTRY & YOU Chewing and regurgitating helps break down the cellulose, so that it can be digested by the microorganisms that live in a cow's digestive tract.

Explore

Class Activity

PURPOSE Students will classify sugars as reducing or nonreducing.

MATERIALS hot-water bath, 2–5 mL of 1% carbohydrate solution, 2 test tubes, distilled water, 4–6 mL of Benedict's reagent, dropper, samples of mono- and polysaccharides

PROCEDURE Place 2–5 mL of 1% carbohydrate solution (corn syrup) in one test tube; place distilled water in another test tube as a negative control. Add 2–3 mL of Benedict's reagent to each tube and heat in the hot-water bath for a few minutes. The Cu_{2+} ions in the reagent react with reducing sugars to form a brick-red cuprous oxide (Cu_2O) precipitate. The amount of precipitate formed depends on the amount of sugar present. Have students observe the formation of a red precipitate. Test a number of mono- and polysaccharides and have students classify them as reducing or nonreducing.

EXPECTED OUTCOME The precipitate indicates the presence of a simple sugar.

Evaluate

Informal Assessment

Write on the board the structure of mono-, di-, and polysaccharides and give a common example of each. Then have students identify the types of elements, functional groups, and linkages found in carbohydrates. Then have students complete the 24.2 Lesson Check.

Answers

FIGURE 24.6 The complete structural formulas in Figure 24.5 include every atom in the molecule. The structural formulas in Figure 24.6 include the functional groups, but the symbols for the carbons in the ring structures are omitted as well as the symbols for the hydrogen atoms that are bonded to these carbon atoms.

LESSON 24.3

Key Objectives

24.3.1 DIAGRAM the general structure of an amino acid.

24.3.2 IDENTIFY what determines the differences in the chemical and physiological properties of peptides and proteins.

24.3.3 DESCRIBE how enzymes affect the rates of reactions in living things.

Additional Resources

- Reading and Study Workbook, Lesson 24.3

PearsonChem.com

Engage

CHEMISTRY & YOU Direct students to the photograph and read aloud the text. Ask a student volunteer to explain his or her regimen for building strength. Encourage the volunteer to discuss his or her nutritional habits and strength training ritual.

Activate Prior Knowledge

Review topics that students need for success in this chapter, such as polarity, carboxyl groups, amino groups, aliphatic and aromatic hydrocarbons, and optical isomers. Then have class members discuss what they know about proteins.

24.3 Amino Acids and Their Polymers

CHEMISTRY & YOU

Q: *Why do your muscles need amino acids?* Your muscles are constantly in use. For example, muscles in your stomach aid in digestion of the food you eat, and muscles in your fingers allow you to turn pages in a book or manipulate a computer mouse. Strength-building exercises can cause your muscles to become larger and stronger. This could not happen without amino acids. In this lesson, you will learn about amino acids and some of their functions.

Key Questions

What is the general structure of an amino acid?

What determines the differences in the chemical and physiological properties of peptides and proteins?

How do enzymes affect the rates of reactions in living things?

Vocabulary

- amino acid
- peptide
- peptide bond
- protein
- enzyme
- substrate
- active site

Amino Acids

What is the general structure of an amino acid?

Many biological compounds contain nitrogen in addition to carbon, oxygen, and hydrogen. Some of the most important nitrogen-containing molecules in organisms are amino acids. In fact, the polymers of amino acids make up more than one half of the dry weight of your body.

An **amino acid** is any compound that contains an amino group ($—NH_2$) and a carboxyl group (—COOH) in the same molecule. For chemists and biochemists, however, the term is usually reserved for the 20 common amino acids that are formed and used by living organisms. **Amino acids consist of a carboxyl group, an amino group, a hydrogen, and an R group side chain that are all covalently bonded to a central carbon atom.**

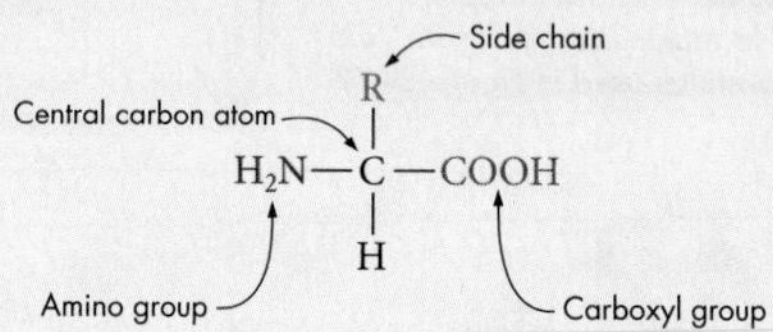

The chemical nature of the side-chain group accounts for the differences in properties of the 20 amino acids. In some amino acids, the side chains are nonpolar aliphatic or aromatic hydrocarbons. In other amino acids, the side chains are neutral but polar. In still others, the side chains are acidic or basic.

Because the central carbon of amino acids is asymmetric, these compounds can exist as enantiomers. As you may recall from Lesson 22.3, enantiomers may be right- or left-handed. Nearly all the amino acids found in nature are of the left-handed, or L, form.

Table 24.1 gives the names of amino acids with their three-letter abbreviations. Examine the abbreviations. You will use them as shortcuts when you read or write about protein structure.

Focus on ELL

1 CONTENT AND LANGUAGE Ask students to work in pairs and scan the text for chemistry vocabulary that they have heard before, such as *acid* and *protein*. Ask students to make a list of the words and quick descriptions of the contexts in which they have heard the words.

2 FRONTLOAD THE LESSON Have students preview the chemical structures shown in the lesson text and identify familiar functional groups and reactions. Tell students to ask themselves what they already know about these groups and reactions and to predict how they might be related to proteins.

3 COMPREHENSIBLE INPUT Use fan-folded computer paper and toy springs to model the three-dimensional structures of proteins illustrated in Figure 24.8.

Table 24.1

Common Amino Acids							
Name	**Symbol**	**Name**	**Symbol**	**Name**	**Symbol**	**Name**	**Symbol**
Alanine	Ala	Glutamine	Gln	Leucine	Leu	Serine	Ser
Arginine	Arg	Glutamic acid	Glu	Lysine	Lys	Threonine	Thr
Asparagine	Asn	Glycine	Gly	Methionine	Met	Tryptophan	Trp
Aspartic acid	Asp	Histidine	His	Phenylalanine	Phe	Tyrosine	Tyr
Cysteine	Cys	Isoleucine	Ile	Proline	Pro	Valine	Val

Peptides and Proteins

What determines the differences in the chemical and physiological properties of peptides and proteins?

A **peptide** is any combination of amino acids in which the amino group of one amino acid is united with the carboxyl group of another amino acid. The amide bond between the carboxyl group of one amino acid and the nitrogen in the amino group of the next amino acid in the peptide chain is called a **peptide bond.** Peptide bonds always involve the central amino and central carboxyl groups. The side chains are not involved in the bonding.

$$\underset{\text{Amino acid}}{H_2N-\overset{R}{\underset{H}{C}}-\overset{O}{\overset{\|}{C}}-OH} + \underset{\text{Amino acid}}{H-\underset{H}{N}-\overset{R}{\underset{H}{C}}-\overset{O}{\overset{\|}{C}}-OH} \longrightarrow \underset{\text{Peptide}}{H_2N-\overset{R}{\underset{H}{C}}-\overset{O}{\overset{\|}{C}}-\underbrace{\underset{H}{N}}_{\text{Peptide bond}}-\overset{R}{\underset{H}{C}}-\overset{O}{\overset{\|}{C}}-OH} + H_2O$$

A free amino group is at one end of the peptide. The convention is to write the peptide formula so that the free amino group is at the left end. There is also a free carboxyl group, which appears at the right end of the molecule.

More amino acids may be added to the peptide in the same fashion to form long chains by condensation polymerization. The order in which the amino acids of a peptide molecule are linked is called the amino acid sequence of that molecule. The amino acid sequence of a peptide is conveniently expressed using the three-letter abbreviations for the amino acids. For example, Asp—Glu—Gly represents a peptide containing three amino acids. This tri-peptide contains aspartic acid, glutamic acid, and glycine, in that order, with the free amino group assumed to be on the left end (on the Asp) and the free carboxyl group on the right end (on the Gly). Note that Asp—Glu—Gly is a different peptide from Gly—Glu—Asp because the order of amino acids is reversed, and thus the free amino group and free carboxyl group are on different amino acids.

In theory, the process of adding amino acids to a peptide chain can continue indefinitely. A peptide with more than ten amino acids is a polypeptide. A peptide with more than about 100 amino acids is a **protein.** On average, a molecule of 100 amino acids has a molecular mass of about 10,000 amu. Proteins are an important class of biomolecules. For example, your skin, hair, nails, and muscles are all made of proteins. Proteins are needed for almost all chemical reactions that occur in the body. We can make some of the amino acids that our cells use to make proteins. Other amino acids must be obtained by eating foods rich in proteins, such as the beans in Figure 24.7.

Figure 24.7
Amino Acids in Your Diet
Beans and brown rice are good sources of amino acids.

Differentiated Instruction

L1 STRUGGLING STUDENTS Many laundry detergents contain enzymes that are supposed to help remove stains. Ask students to read the labels of a number of laundry detergents and find out if enzymes are listed among the ingredients. Ask students to hypothesize how enzymes improve the cleaning performance of a detergent.

LPR LESS PROFICIENT READERS Help students create a flowchart to show the steps involved in the binding and subsequent activity of an enzyme.

L3 ADVANCED STUDENTS Have students determine how the amino acid sequence of a protein dictates its secondary and tertiary structure. Ask them to draw a simple protein (or part of one) to show how some of the bonds affect its shape.

Foundations for Reading

BUILD VOCABULARY Have students use a Venn diagram to compare peptides to proteins. They should include the vocabulary for the lesson and ask themselves a series of thought questions, such as, *What do peptides and proteins have in common? What is unique to each?*

READING STRATEGY Have students skim the lesson, paying particular attention to the tables and figures. Have students take notice of the arrangements of peptide molecules: straight chains, helical chains, and pleated sheets.

Explain

Amino Acids

START A CONVERSATION Discuss how to calculate the number of possible arrangements of amino acids in proteins. **Ask** How many different ways can 8 different amino acids be arranged to form a peptide? *(8! = 8 × 7 × 6 × 5 × 4 × 3 × 2 × 1 = 40,320)* **Ask** How many arrangements are possible for a peptide made of 10 different amino acids? *(10! = 3,628,800)* Point out to students that an average protein usually contains more than 100 amino acids.

Explore

Peptides and Proteins

Teacher Demo

PURPOSE Students will observe how amino acids link through amide bonds to form peptides.

PROCEDURE Create models of peptides by linking amino acids with amide bonds. Point out that peptide formation is another example of a condensation polymerization. Ask students to describe how continuing the polymerization process produces protein molecules.

EXPECTED OUTCOME Students should be able to describe peptide formation by manipulating the models.

Explain

Peptides and Proteins

USE VISUALS Point out that the types of amino acids in proteins have a large influence on the physical and chemical properties of a protein. The amino acids listed in Table 24.1 can be grouped into two major categories: polar and nonpolar. Explain that the ratio of polar to nonpolar amino acids profoundly affects the shape of a protein, as shown in Figure 24.8. Convey that when a protein folds, nonpolar amino acids are sequestered to the interior of the protein, away from the aqueous environment, and polar amino acid residues often predominate on the surface of the protein.

CHEMISTRY & YOU Our cells use amino acids to make proteins. Proteins are needed for almost all chemical reactions that occur in the body. One type of protein, known as myoglobin, stores oxygen in muscle cells.

Explore

Teacher Demo

PURPOSE Students will learn that some proteins play structural roles in hard tissue.

MATERIALS thighbone from cooked poultry, 1*M* HCl, beaker

PROCEDURE Students may be surprised to learn that protein is a major structural constituent of bone. Illustrate this fact by soaking a thighbone from cooked poultry in 1*M* HCl overnight. The next day, rinse the bone with water.

EXPECTED OUTCOME Upon visual inspection it may seem as if no reaction has occurred at all, but demonstrate that now the bone can be bent and twisted quite easily. Point out that the mineral salts of the bone were dissolved in the acid. The acid-insoluble portion remaining consists of a special protein called collagen. **Ask** What possible roles does collagen have in the formation and maintenance of healthy bone tissue? *(It provides a framework for the deposition of calcium phosphate, the mineral that makes up bone, and for the growth of bone marrow cells.)*

Figure 24.8 Peptide Structures
Peptides form three-dimensional shapes. **a.** This is a representation of amino acids in a peptide chain. **b.** The chain may coil into a helix. **c.** Two peptide chains may become arranged in a pleated, sheetlike structure.
Apply Concepts *What types of bonds determine the three-dimensional shape of a protein?*

See peptide structure animated online.

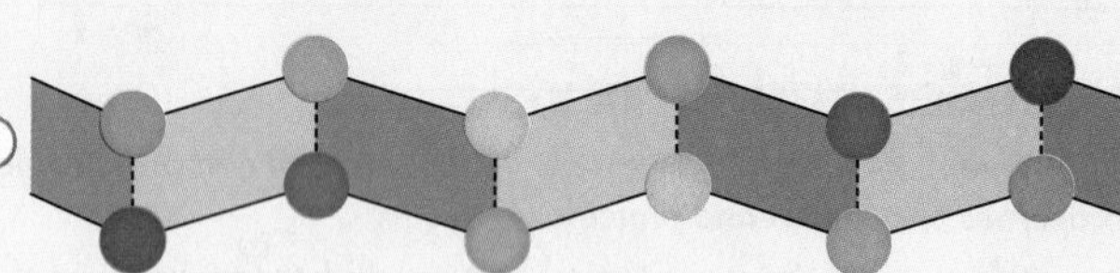

CHEMISTRY & YOU

Q: *Why do your muscles need amino acids?*

Differences in the chemical and physiological properties of peptides and proteins result from differences in the amino acid sequence. Twenty amino acids can be linked in an enormous number of ways in a protein molecule. As many as 20^{100} different amino acid sequences are possible for a protein of 100 amino acids containing a combination of the 20 different amino acids.

Protein molecules are folded into relatively stable three-dimensional shapes. Figure 24.8a represents a long peptide chain of a protein, and Figure 24.8b shows how sections of peptide chain may coil into a regular spiral, known as a helix. Peptide chains may also be arranged side by side to form a pleated sheet, as shown in Figure 24.8c. Irregular folding of the chains also can occur. The three-dimensional shape of a protein is determined by interactions among the amino acids in its peptide chains. Protein shape is partly maintained by hydrogen bonds between adjacent folded chains. Covalent bonds also form between sulfur atoms of cysteine side chains that are folded near each other. In that way, separate polypeptide chains may be joined into a single protein. Figure 24.9 traces the shape of myoglobin, a protein that stores oxygen in muscle cells. The peptide chains of most of the myoglobin molecule are twisted into helixes.

Figure 24.9 Myoglobin
The three-dimensional structure of myoglobin, the oxygen storage protein of muscle tissue, is shown here. Most of the peptide chain of myoglobin is wound into helixes. Myoglobin also contains a nonprotein structure called heme. Heme contains four linked rings with an iron(II) ion (Fe^{2+}) at the center. Molecular oxygen binds to the heme iron. Marine animals, such as this dolphin, have a large concentration of myoglobin in their muscles which allows them to store oxygen during long dives.

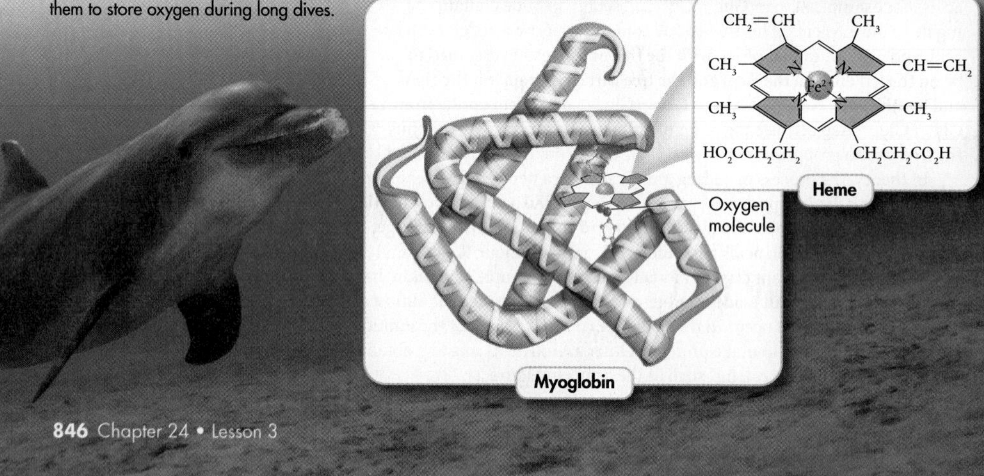

UbD Check for Understanding

The Essential Question What are the characteristics of the four main types of biological molecules?

Assess students' understanding of proteins by asking them to identify the main characteristics of proteins. *(Proteins are made of chains of over 100 amino acids joined by peptide bonds. Hydrogen bonding and covalent bonding between a protein's amino acids determine its three-dimensional structure.)*

ADJUST INSTRUCTION If students are having trouble answering, have them reread the text of the Peptides and Proteins section and re-examine Figure 24.8.

846 Chapter 24 • Lesson 3

Enzymes

How do enzymes affect the rates of reactions in living things?

Enzymes are proteins that act as biological catalysts. **Enzymes increase the rates of chemical reactions in living things.** In 1926, the American chemist James B. Sumner reported the first isolation and crystallization of an enzyme. The enzyme he isolated was urease. Urease hydrolyzes urea, a constituent of urine, into ammonia and carbon dioxide. The strong ammonia smell of wet diapers that sit for a long time is the result of the action of bacteria that contain this enzyme. The equation for the reaction is shown below.

$$\underset{\text{Urea}}{H_2N-\overset{\overset{\displaystyle O}{\|}}{C}-NH_2(aq)} + \underset{\text{Water}}{H_2O(l)} \xrightarrow{\text{urease}} \underset{\text{Ammonia}}{2NH_3(g)} + \underset{\text{Carbon dioxide}}{CO_2(g)}$$

Since the discovery of urease, thousands of enzymes have been isolated and structurally characterized as proteins.

In addition to being able to promote reactions, enzymes have two other properties of true catalysts. First, they are unchanged by the reaction they catalyze. Second, they do not change the normal equilibrium position of a chemical system. The same amount of product is eventually formed whether or not an enzyme is present. Few reactions in cells ever reach equilibrium, however. The products tend to convert rapidly to another substance in a subsequent enzyme-catalyzed reaction. According to Le Châtelier's principle, such removal of a product pulls the reaction toward completion.

How Enzymes Work Enzymes catalyze most of the chemical changes that occur in the cell. **Substrates** are the molecules on which an enzyme acts. In a typical enzymatic reaction, shown in Figure 24.10, the substrate interacts with side chains of the amino acids on the enzyme. These interactions cause the making and breaking of bonds. A substrate molecule must make contact with, and bind to, an enzyme molecule before the substrate can be transformed into product. The place on an enzyme where a substrate binds is called the **active site.** An active site is usually a pocket or crevice formed by folds in the peptide chains of the enzyme protein. The peptide chain of an enzyme is folded in a unique way to accommodate the substrate at the active site.

Figure 24.10 Enzyme and Substrate
A substrate fits into a distinctively shaped active site on an enzyme. Bond-breaking occurs at the active site to produce the products of the reaction.
Predict ***What would happen if access to the active site were blocked by another molecule?***

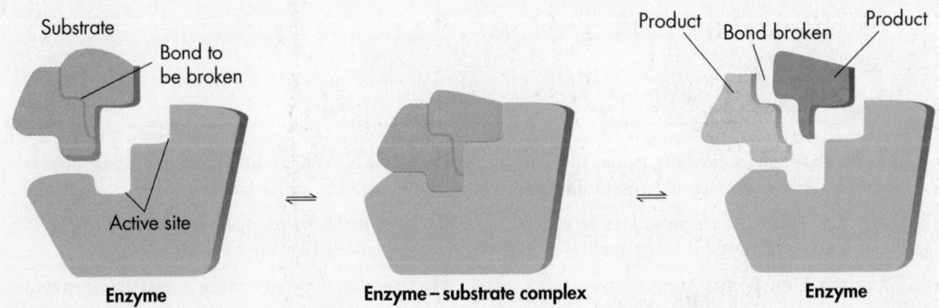

Explain

Enzymes

USE VISUALS Have students examine Figure 24.10. Compare the binding of a substrate and an enzyme to the way in which a key fits into a lock. **Ask** How does the enzyme exclude all but the correct substrate? *(The substrate has a matching, or complementary, shape.)* **Ask** What would happen to the system if you heated it until the enzyme shape changed? *(The substrate would no longer fit.)*

Extend

Ask for a show of hands of students who have seen the warning on gelatin packages regarding the use of fresh pineapple or papaya as add-ins. Explain that gelatin is derived from collagen, a connective protein in the skin and bones of animals. Tell students that pineapple and papaya contain enzymes that break down proteins, preventing the gelatin from setting. Challenge students to identify the protein-digesting enzyme in each fruit and to discover why it is safe to use the canned forms of the fruits in a gelatin mold, even though the fresh fruits cannot be used. *(The enzyme in papaya is papain, and the one in pineapple is bromelain. Even though these enzymes are active in fresh fruit, the high temperature used during the canning process inactivates them. Thus, a gelatin mold made with canned papaya or pineapple will set normally.)*

Focus on ELL

4 LANGUAGE PRODUCTION Have students work in groups of four to complete the Small-Scale Lab on page 849. Make sure each group has ELLs of varied language proficiencies so that more proficient students can help less proficient ones. Have students work according to their proficiency level.

BEGINNING: LOW/HIGH Provide students with a chart they can use to record the measurements and complete the calculations.

INTERMEDIATE: LOW/HIGH Model the calculations for students with a different-sized egg.

ADVANCED: LOW/HIGH Have students assist less proficient students with the calculations and answering the questions.

Answers

FIGURE 24.8 hydrogen bonds

FIGURE 24.10 Transformation of the substrate into product could not occur.

LESSON 24.3

Connect to NUTRITION

Explain that coenzymes are often vitamins; for example, B vitamins help release energy from foods. Lack of B vitamins usually produces malnourishment of some kind, since food can't be digested properly without them. Have students determine what occurs if a person is deficient in niacin and vitamin B_6. *(It can produce skin and nervous system disorders, such as headaches, depression, and memory loss.)*

Evaluate

Informal Assessment

Have students form into groups of four. Distribute a sketch of the general structure of a tripeptide, such as Gly—Val—Trp, using letter abbreviations in place of the R group. Ask students to label the peptide bonds and the carboxyl and amino ends. **Ask** Which element do proteins contain that carbohydrates do not? *(nitrogen)* **Ask** What is an enzyme? *(Enzymes are proteins that function as biological catalysts.)* Have students describe one enzyme and its function. *(Sample answer: urease, which catalyzes the hydrolysis of urea)* Then have students complete the 24.3 Lesson Check.

Reteach

Review secondary and tertiary protein structure. Point out that many of the amino acids in an enzyme are not directly involved in catalysis because they lie far from the active site. However, they do participate in intermolecular forces, which maintain the folded shape of the protein. This shape is a prerequisite for its biological activity.

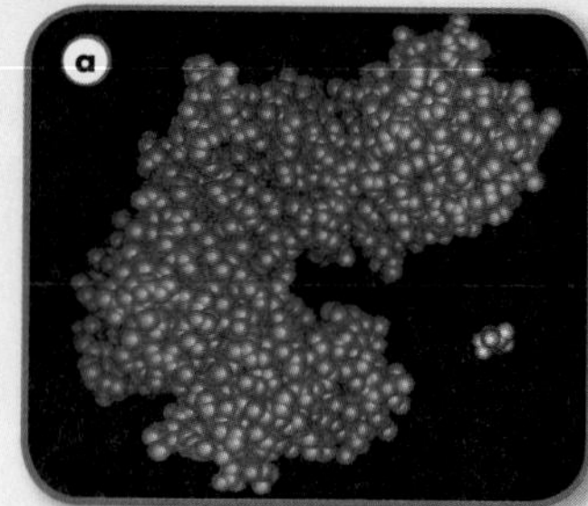

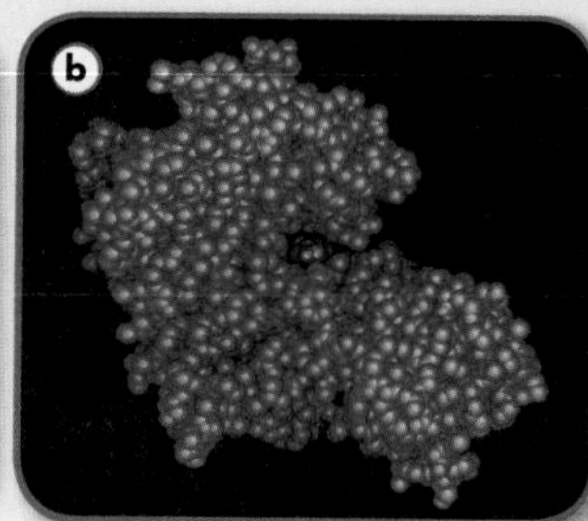

Figure 24.11 Carbonic Anhydrase
a. Carbonic anhydrase (blue) has only one substrate, carbonic acid (red). **b.** An enzyme-substrate complex of carbonic anhydrase and carbonic acid. Notice how the substrate fits snugly into the active site.

Since the active site of each enzyme has a distinctive shape, only a specific substrate molecule can fit into the enzyme, similar to how only one key shape will fit into a certain lock. Thus, each enzyme can catalyze only one chemical reaction at a time. An enzyme-substrate complex is formed when an enzyme molecule and a substrate molecule are joined. Figure 24.11 shows a model of the enzyme-substrate complex formed between the enzyme carbonic anhydrase and its substrate, carbonic acid.

To see the efficiency of enzymes, consider the effects of carbonic anhydrase on carbonic acid. Carbonic anhydrase catalyzes the reversible breakdown of carbonic acid to carbon dioxide and water. One molecule of carbonic anhydrase can catalyze the breakdown of 36 million molecules of carbonic acid in one minute!

$$\underset{\text{Carbonic acid}}{H_2CO_3(aq)} \xrightleftharpoons{\text{carbonic anhydrase}} \underset{\text{Carbon dioxide}}{CO_2(g)} + \underset{\text{Water}}{H_2O(l)}$$

Figure 24.12 Catalase
Liver contains high levels of the enzyme catalase. When a small amount of crushed liver cells is added to a solution of hydrogen peroxide, oxygen gas is rapidly evolved.

Coenzymes Some enzymes can directly catalyze the transformation of biological substrates without assistance from other substances. Other enzymes need nonprotein coenzymes, also called cofactors, to assist the transformation. Coenzymes are metal ions or small organic molecules that must be present for an enzyme-catalyzed reaction to occur. Many water-soluble vitamins, such as B vitamins, are coenzymes. Metal ions that act as coenzymes include the cations of magnesium, potassium, iron, and zinc. The enzyme catalase includes an iron(III) ion in its structure. Catalase catalyzes the breakdown of hydrogen peroxide to water and oxygen, as shown in Figure 24.12 and in the reaction below.

$$2H_2O_2(aq) \xrightarrow{\text{catalase}} 2H_2O(l) + O_2(g)$$

24.3 LessonCheck

ONLINE PROBLEMS

16. **Review** What are the four groups that surround the central carbon atom in an amino acid?
17. **Identify** What determines the differences in the properties of peptides and proteins?
18. **Explain** How do enzymes affect the reaction rates in living things?
19. **Identify** Which functional groups are always involved in amide bonds?
20. **Explain** What is meant by the amino acid sequence of a protein?
21. **Describe** Describe three properties of enzymes.

Lesson Check Answers

16. amino, carboxyl, R, and hydrogen
17. differences in the amino acid sequence.
18. Enzymes speed up the rate of reaction but do not affect the position of equilibrium.
19. The amino group of one amino acid combines with the carboxylic acid group of another to form an amide linkage.
20. the order of amino acids in the protein molecule
21. catalysts (they remain unchanged in a reaction); usually proteins; specific to the reactions they catalyze

Small-Scale Lab

The Egg: A Biochemical Storehouse

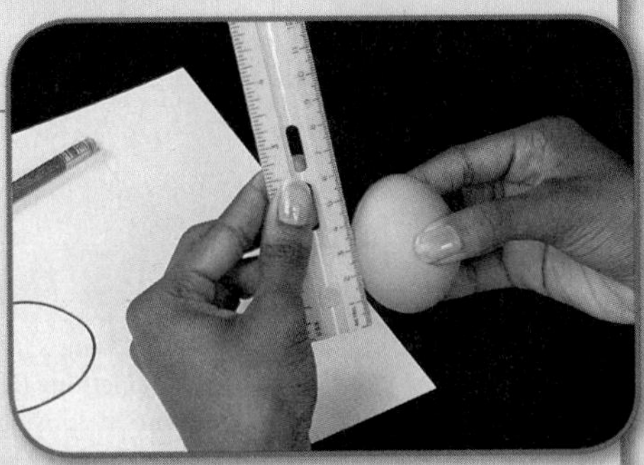

Purpose
To explore some physical and chemical properties of a chicken egg

Materials
- chicken egg
- ruler
- balance

Procedure
Obtain a chicken egg. Examine the egg's shape, and measure its length and width in centimeters. Measure the mass of the egg. Make an accurate, life-size sketch of your egg and record all of your data on the sketch.

Analyze and Conclude
Using your experimental data, record the answers to the following questions below your drawing.

1. Calculate One way to compare the shapes of eggs is by using a shape index. The shape index is the width (w) of an egg expressed as a percentage of its length (l). Calculate the shape index of your egg.

$$\text{Shape index} = \frac{w}{l} \times 100\%$$

2. Calculate The volume, original mass (when freshly laid), and surface area of an egg can easily be estimated by using the following equations.

$V = (0.5236)(lw^2)$ $m = (0.5632)(lw^2)$
$A = (3.138)(lw^2)^{2/3}$
V = volume m = original mass
A = surface area

Use your data to calculate the volume, original mass, and surface area of your egg. Show your work, and record your results.

3. Compare Which is greater, the measured mass or the calculated mass of your egg? Suggest why the mass of an egg might change over time.

4. Compare Using your measured mass and your calculated volume, calculate the density of your egg. Compare this value with the density of a freshly laid egg (density of freshly laid egg = 1.075 g/cm^3).

You're the Chemist
The following small-scale activities allow you to develop your own procedures and analyze the results.

1. Design an Experiment Design an experiment to answer the following question: Does the mass of an egg change over time?

2. Analyze Data Using your measured mass, your calculated original mass, and your experiments on the mass loss of an egg over time, estimate the age of your egg. What assumptions must you make?

3. Design an Experiment Design and carry out an experiment to measure the volume of your egg. Write down what you did and what you found.

4. Design an Experiment Carry out a series of experiments, or consult with your classmates and use their data, to determine if and how the shape index varies with the size of the egg (small, medium, large, extra large, jumbo).

5. Analyze Data Determine how the mass of an egg varies with its size (small, medium, large, extra large, jumbo).

6. Analyze Data An eggshell contains a calcium carbonate matrix with a protein cuticle. Place one drop of HCl on an eggshell and observe what happens. Write a chemical equation for this reaction. **CAUTION** *HCl is caustic and can burn skin.*

7. Analyze Data Proteins can be detected by adding aqueous solutions of copper(II) sulfate and sodium hydroxide to a sample. A violet color indicates the presence of protein. Test powdered milk and an eggshell for protein. What are your results?

8. Design an Experiment Design and carry out an experiment to answer the following question: Does temperature affect the mass of an egg over time?

Explore

Small-Scale Lab

OBJECTIVE Students examine the physical and chemical properties of a chicken egg.

SKILLS FOCUS Measuring, calculating, designing experiments

PREP TIME 20 minutes

CLASS TIME 20 minutes

MATERIALS Chicken egg, metric ruler, mass balance, NaOH, $CuSO_4$, HCl

ADVANCE PREPARATION Make up reagents as follows:

Solution	Preparation
0.5*M* NaOH	20.0 g in 1.0 L
0.2*M* $CuSO_4$	12.5 g $CuSO_4 \cdot 5H_2O$ in 250 mL
1.0*M* HCl	82 mL of 12*M* in 1.0 L

SAFETY Always add acid to water carefully and slowly!

TEACHING TIPS For each class, obtain at least a dozen eggs of two or three different sizes: medium, large, and extra large. Some markets also sell small or jumbo eggs. Students can then obtain data from classmates about eggs that are different in size from their own. Have students write their names on their eggs so they can identify them the next day. Store some eggs in a refrigerator and some at room temperature if students are going to do activities 1, 2, and 8 in the You're the Chemist section.

EXPECTED OUTCOME Sample data: length 5.90 cm; width 4.55 cm; mass 62.42 g

ANALYZE AND CONCLUDE

1. $\text{Shape index} = \frac{4.55\text{ cm}}{5.90\text{ cm}} \times 100 = 77.1\%$
2. $V = (0.5236)(5.90\text{ cm})(4.55\text{ cm})^2 = 64.0\text{ cm}^3$
   $m = (0.5632)(5.90\text{ cm})(4.55\text{ cm})^2 = 68.8\text{ g}$
   $A = 3.138[(5.90\text{ cm})(4.55\text{ cm})^2]^{2/3} = 77.3\text{ cm}^2$
3. The measured mass of 62.42 g is less than the calculated mass of 68.8 g by 6.4 g. The egg may have lost water.
4. $d = \frac{62.42\text{ g}}{64.0\text{ cm}^3} = 0.975\text{ g/cm}^3$
   This is less than the density of a freshly laid egg.

YOU'RE THE CHEMIST

1. Measure the mass each day for two or three days. Eggs lose 0.2 to 0.5 gram per day.
2. Assuming the egg loses 0.20 g per day, the age of the egg is:
   $(68.8\text{ g} - 62.42\text{ g}) \times \frac{1\text{ day}}{0.20\text{ g}} = 32\text{ days old}$
3. Measure the volume by water displacement.
4. Larger eggs have smaller shape indexes.
5. Extra-large eggs are usually more than 70 g, and medium eggs are less than 50 g.
6. $2HCl + CaCO_3 \rightarrow CO_2(g) + H_2O + CaCl_2$
7. Both powdered milk and the eggshell produce a violet color, indicating the presence of protein.
8. Measure the mass of a refrigerated egg for three days. Measure its mass for three more days while storing it at room temperature. The warmer the temperature, the greater the mass loss.

LESSON 24.4

Key Objective

24.4.1 IDENTIFY the physical property that distinguishes lipids from other biological molecules.

Additional Resources

- Reading and Study Workbook, Lesson 24.4

PearsonChem.com

Engage

CHEMISTRY & YOU Direct students to read the text that begins the lesson. As a class, brainstorm places in the body where fats are found.

Build Background

Display familiar food items such as a tub of margarine, a bottle of olive oil, a stick of butter. Ask students to predict which of these items are fats and which are oils. Students will have an opportunity later to discover the difference between the two.

Foundations for Reading

BUILD VOCABULARY Have students dissect the words *hydrophilic* and *hydrophobic*. Ask them for examples of words that use the same roots, to help distinguish between the meanings.

READING STRATEGY Have students look ahead through the lesson and predict what the properties of lipids are and what types exist. They should check their predictions after they read the lesson. Have students summarize how lipids are different from carbohydrates and proteins.

24.4 Lipids

CHEMISTRY & YOU

Q: *Why is fat an important part of our diet?* The media often portray fat as something that should be avoided because it is bad for you. It is probably not surprising that red meat can be high in fat. But you might not expect avocadoes like the one shown here to be high in fat. In this lesson, you will learn why fats are actually necessary in our diet.

Key Question

What physical property distinguishes lipids from other classes of biological molecules?

Vocabulary

- lipids
- triglyceride
- saponification
- phospholipid
- wax

Figure 24.13 Fats in Foods
Moderate levels of dietary fats and oils are essential to health.

Describing Lipids

What physical property distinguishes lipids from other classes of biological molecules?

Fats, oils, and other water-insoluble compounds are called **lipids.** **Carbohydrates and proteins tend to dissolve in water. Whereas, lipids tend to dissolve readily in organic solvents, such as ether and chloroform.** Most fats, such as the butter in Figure 24.13, are obtained from animals. The fats from palm kernels and coconuts, however, are exceptions. Most oils, such as olive oil, are plant products.

Although excessive dietary fat is harmful, you do need some lipids in your diet to stay healthy. Experts recommend that your lipid intake make up less than 30 percent of your daily caloric intake of food. Lipids provide an efficient way for your body to store energy. They are also needed to keep your cell membranes healthy.

Triglycerides Natural fats and oils exist as triesters of glycerol with fatty acids, which are long-chain carboxylic acids (C_{12} through C_{24}). This form of lipid is known as a **triglyceride.** Triglycerides are important as the long-term storage form of energy in the human body. The following equation shows the general reaction for the formation of triglycerides.

$$\begin{matrix} CH_2OH \\ | \\ CHOH \\ | \\ CH_2OH \end{matrix} + \begin{matrix} HO-\overset{\overset{O}{\|}}{C}-R \\ HO-\overset{\overset{O}{\|}}{C}-R \\ HO-\overset{\overset{O}{\|}}{C}-R \end{matrix} \longrightarrow \begin{matrix} CH_2-O-\overset{\overset{O}{\|}}{C}-R \\ | \\ CH-O-\overset{\overset{O}{\|}}{C}-R \\ | \\ CH_2-O-\overset{\overset{O}{\|}}{C}-R \end{matrix} + 3H_2O$$

Glycerol — 3 Fatty acid molecules — Triglyceride (triester of glycerol) — Water

850

Focus on ELL

1 CONTENT AND LANGUAGE Provide students with a compare/contrast table to organize the information they encounter about the three classes of lipids. If possible, pair beginning and advanced speakers of the same language and allow them to brainstorm in their native language as they work together on their tables.

2 FRONTLOAD THE LESSON Display a bottle of olive oil, a tube of wax lip balm, and a diagram of a cell membrane for students to examine. Explain to them that even though the olive oil, the lip balm, and the cell membrane seem completely unrelated, they all contain lipids. List the three types of lipids on the board, and have students predict which item contains which type of lipid. Have students revisit their predictions after completing the lesson.

3 COMPREHENSIBLE INPUT Explain Figure 24.16 and restate information in its caption as needed. Read the question out loud, and draw a picture on the board that assists students in answering the question.

A fat, such as beef tallow or coconut oil, is mixed with an excess of sodium hydroxide and heated.

Sodium chloride is added to the saponification mixture. This causes the sodium salts of the fatty acids to separate as a thick curd of crude soap.

The crude soap is purified and can then be processed as desired.

Figure 24.14 Soapmaking
These photographs illustrate soapmaking. Once the soap is formed, it is poured into molds. Later it may be milled, or shredded, with scent or color added, and then remolded to produce a finished product.

Like other esters, fats and oils are easily hydrolyzed in the presence of acids and bases. The hydrolysis of oils or fats by boiling with an aqueous solution of an alkali-metal hydroxide is called **saponification.** Saponification is used to make soap. Soaps are thus the alkali metal (Na, K, or Li) salts of fatty acids. A typical saponification reaction is shown below.

$$\begin{array}{l} CH_2{-}O{-}\overset{O}{\overset{\|}{C}}{-}(CH_2)_{16}CH_3 \\ | \\ CH{-}O{-}\overset{O}{\overset{\|}{C}}{-}(CH_2)_{16}CH_3 \\ | \\ CH_2{-}O{-}\overset{O}{\overset{\|}{C}}{-}(CH_2)_{16}CH_3 \end{array} + 3NaOH \longrightarrow \begin{array}{l} CH_2OH \\ | \\ CHOH \\ | \\ CH_2OH \end{array} + 3CH_3(CH_2)_{16}{-}\overset{O}{\overset{\|}{C}}{-}O^-Na^+$$

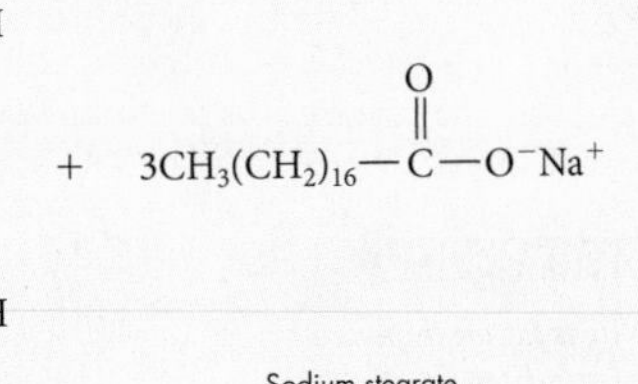

Tristearin (triester of glycerol and stearic acid) — Glycerol — Sodium stearate (a soap)

As shown in Figure 24.14, soap can be made from a fat, such as beef tallow or coconut oil. Glycerol is an important byproduct of saponification reactions. It is recovered by evaporating the water layer.

Phospholipids Lipids that contain phosphate groups are called **phospholipids.** Phospholipids are abundant in cells. Figure 24.15 shows a typical phospholipid molecule, lecithin. The lecithin molecule has a hydrophilic (water-loving) ionic head and oily or hydrophobic (water-hating) hydrocarbon tails. Recall the solubility rule, "Like dissolves like." Lecithin, which is partly hydrophobic and partly hydrophilic, behaves in water like both an insoluble hydrocarbon and a soluble ionic compound. In water, the hydrophobic carbon chains of lecithin aggregate to exclude water. The hydrophilic part is drawn to water, which can solvate it. In water, phospholipids spontaneously form a spherical double layer, called a lipid bilayer, in which the hydrophobic tails of phospholipid molecules are sandwiched between two layers of hydrophilic heads.

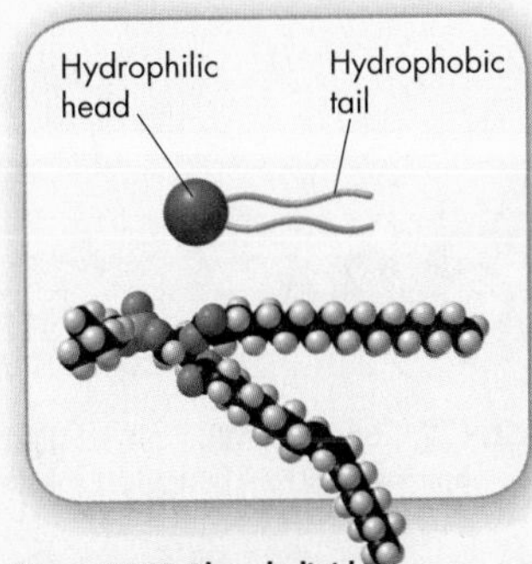

Figure 24.15 Phospholipid
In the simplified diagram, the hydrophilic head is shown as a sphere and the hydrophobic tails as wavy lines. The space-filling model is the phospholipid lecithin.

Explain

Describing Lipids

START A CONVERSATION Explain that lipids contain identifiable functional groups, but their most distinguishing property is their insolubility in water. **Ask** Why do you think lipids are insoluble? *(They are nonpolar.)* **Ask** What utility does this characteristic give lipids? *(They can be used to contain water-soluble products and processes.)*

Explore

Student Activity

Have students compare the composition of butter, regular margarine, and low-calorie margarine by melting the same mass of each kind of spread in separate test tubes. Put the test tubes in warm water to liquefy the samples. Then allow the samples to settle. The liquid will form two layers: the top layer is a lipid (butterfat or partially hydrogenated vegetable oil) and the bottom layer is water. Students can compare the samples quantitatively and qualitatively and calculate the percentage of water in each fat.

MAKE A CONNECTION Explain to the class that lipid bilayers are an important component of cells today and were probably essential to the origin of cellular life. Convey that before DNA and RNA existed to direct the formation of complex molecules, molecules had to be able to self-assemble. Point out that bipolar molecules like phospholipids can assemble themselves into membranes; a membrane-bound environment was an important step in providing a place for complex biological chemistry.

Differentiated Instruction

L1 SPECIAL NEEDS Create a model of a cell membrane by floating a layer of table-tennis balls in a tub of water. Add a few tennis balls. Explain that the water underneath the balls represents the cytoplasm of the cell, the table-tennis balls represent the phospholipids, and the tennis balls represent the proteins.

LPR LESS PROFICIENT READERS Write the words *triglyceride* and *saponification* on the board, along with pronunciation guides for them. Have students practice pronouncing the words, and have them write the pronunciation guides in their notebooks.

L3 ADVANCED STUDENTS Have students determine the structure of oleic acid, noting the hydrophobic region and the hydrophilic region.

LESSON 24.4

Explain

Describing Lipids

USE VISUALS Have students study Figure 24.16. **Ask** Why are the phospholipid tails facing each other? *(They are hydrophobic and the membrane is surrounded by water.)*

Waxes

APPLY CONCEPTS Point out that an important difference between waxes and oils is the length of the hydrocarbon backbone. Explain that waxes are derivatives of high molar mass carboxylic acids. Remind students that carboxylic acids react with alcohols to form esters. Thus, waxes are essentially high molar mass esterification products of carboxylic acids.

CHEMISTRY & YOU Fats provide an efficient way for your body to store energy. They are also needed to keep your cell membranes healthy.

Evaluate

Informal Assessment

Display samples of different types of lipids and ask students to decide what type each is, why they think so, and what that implies about the structure of the lipid. Then have students complete the 24.4 Lesson Check.

Reteach

Review the chemical structures on pages 850 and 851. Help students to understand that the fats and oils hydrolyze in the presence of acids or bases because lipids are esters.

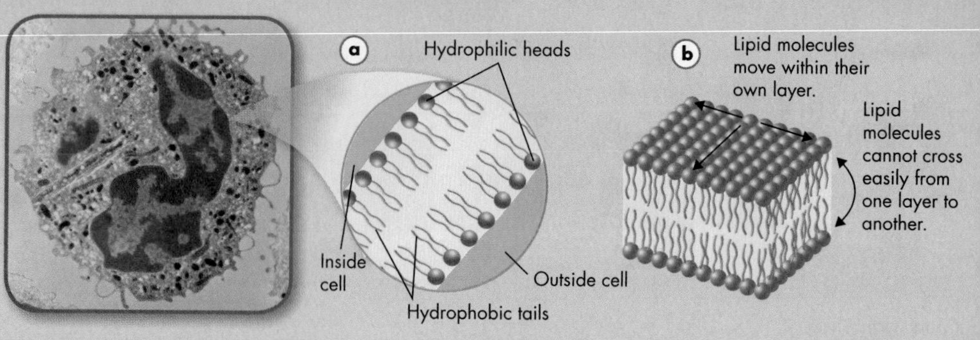

Figure 24.16 Lipid Bilayer
A cell membrane has a lipid bilayer structure. **a.** The hydrophilic heads are in contact with water, but the hydrophobic tails are not. **b.** The lipid molecules move easily within their own layer but do not readily move to the other layer.
Apply Concepts ***What prevents a lipid molecule from crossing to the opposite side of the bilayer?***

Cell membranes, such as the one shown in Figure 24.16, consist primarily of lipid bilayers. The lipid bilayer of a cell membrane acts as a barrier against the passage of molecules and ions into and out of the cell. However, cells do need to take in certain ions and molecules, such as nutrients, while excluding other materials. Selective absorption is accomplished by the protrusion of protein molecules through the lipid bilayer. These proteins form channels through which specific ions and molecules can selectively pass. Not all membrane proteins extend all the way through the membrane. Proteins, such as enzymes, may be bound to the interior surface of the membrane. Many membrane proteins have attached carbohydrate molecules. The carbohydrate portion is on the exterior of the lipid bilayer, where it can hydrogen-bond with water. The protein portion is on the interior of the lipid bilayer, so it does not contact the water.

Q: *Why is fat an important part of our diet?*

Waxes Another type of lipids are **waxes.** Waxes are esters of long-chain fatty acids and long-chain alcohols. The hydrocarbon chains for both the acid and the alcohol usually contain from 10 to 30 carbon atoms. Waxes are low-melting, stable solids. In many plants, a wax coat protects the surfaces of leaves from water loss and attack by microorganisms. For example, carnauba wax, a major ingredient in car wax and floor polish, is found on the leaves of a South American palm tree. In animals, waxes coat the skin, hair, and feathers and help keep these structures pliable and waterproof.

24.4 LessonCheck

22. Compare What physical property sets lipids apart from biological substances such as carbohydrates and proteins?

23. Compare Compare the molecular structures of the three main types of lipids.

24. Identify What are the products of a saponification reaction?

25. Explain What role do phospholipids and proteins play in cell membranes?

26. Explain How do phospholipids behave in water?

27. Identify What two classes of organic compounds combine to form a wax?

28. Identify What is the function of waxes in plants? In animals?

Lesson Check Answers

22. Lipids do not dissolve in highly polar solvents such as water.
23. Triglycerides are triesters of glycerol with long chains of fatty acids attached. Phospholipids are lipids that contain phosphate groups; they have a hydrophobic end and a hydrophilic end. Waxes are esters of long-chain fatty acids and long-chain alcohols.
24. glycerol and a soap
25. Phospholipids make up the lipid bilayer of the cell membrane. Protein molecules protrude from this layer, forming channels through which nutrients pass.
26. They spontaneously form a lipid bilayer, with hydrophobic tails of the phospholipids in the middle.
27. long-chain fatty acids and long-chain alcohols
28. Wax protects the surfaces of leaves from water loss and attack by microorganisms. Wax keeps skin, hair, and feathers pliable and waterproof.

Biochemists

Biochemistry is the chemistry of life, and biochemists combine the fields of chemistry, microbiology, cell biology, genetics, and physics to study the chemical and physical processes of cells and organisms.

Biochemists often work in medical, agricultural, and food-related fields. Those working in medical fields may investigate the causes of diseases and genetic disorders and develop new drugs and medications. They might also research the chemical changes that take place in cells to identify how substances such as drugs, hormones, and household chemicals affect cells, tissues, and body functions. Biochemists working in agriculture may develop new techniques for such tasks as crop cultivation and pest control. Biochemists working in food science may analyze the effects of cooking, canning, and processing on the nutritional value of foods, or they may study the effects of certain foods and nutrients on the human body.

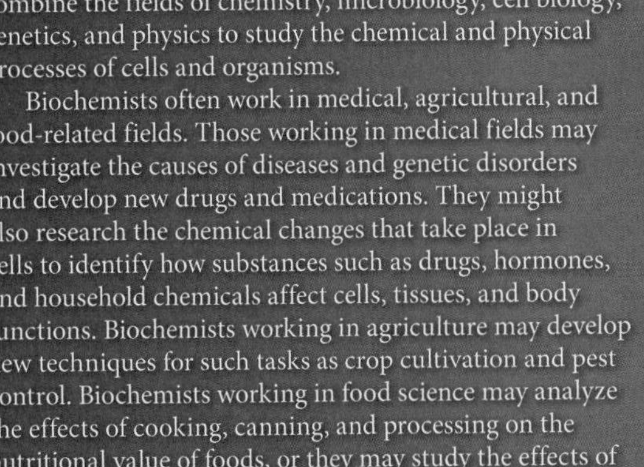

FLU FIGHTERS Biochemists investigate the chemical aspects of the immune system and viruses to create vaccines that help prevent viral diseases such as known strands of influenza viruses.

MONITORING THE ENVIRONMENT Biochemists analyze the effects of environmental conditions on plants and animals such as these corals in the Indian Ocean.

CREATING BIOFUELS Understanding both plant genetics and chemistry allows biochemists to research how plants such as corn and soy can be used as renewable fuel sources.

Take It Further

1. **Identify** What are three life processes that depend on chemistry? Explain your answer.
2. **Evaluate the Impact on Society** What is one way in which understanding the chemical processes involved in aging might help future generations?

Differentiated Instruction

L1 STRUGGLING STUDENTS Read aloud the description of Flu Fighters, and have students examine the photo. Lead students in discussing why this job is important, and then have them write a sentence describing the job. Repeat this process for Monitoring the Environment and Creating Biofuels. Discuss how learning about different careers can help students plan their futures.

ELL ENGLISH LANGUAGE LEARNERS Have students skim through the passages to find unfamiliar words before reading the text. For each of the three jobs represented in the photos, have students write a sentence about how working in one of these jobs in their native country would benefit that country. Allow time for students to share this information with the class.

L3 ADVANCED STUDENTS Have students conduct online research about the educational requirements for a career in biochemistry. Tell students to take on the role of a recruiter. Instruct them to create a technological presentation that persuades students to pursue the study of biochemistry after graduation from high school.

CHEMISTRY & YOU Draw students' attention to the photo labeled *Flu Fighters*. Remind students that a virus particle is basically a packet of DNA or RNA surrounded by a protein coat (a capsid) and possibly an additional outer envelope of lipids and proteins. The surface proteins help virologists classify viruses into one of three main types (A, B, or C). Type A influenzas are divided into subtypes, based on specific combinations of two proteins: hemagglutinin (HA) and neuraminidase (NA). Each protein has several variations. Pose the following question to students: *How would you characterize swine influenza?* You may need to assist students in the following ways:

- The natural hosts of type A influenzas are birds, but these influenzas can be transmitted to humans and other animals, such as horses and pigs.
- Humans are natural hosts of types B and C influenzas.
- Swine influenza is officially designated as H1N1 influenza.
- Type A influenzas are capable of causing deadly epidemics and pandemics, but type B is only associated with epidemics. Type C outbreaks are sporadic and mild.

Explain

APPLY CONCEPTS Point out that there are many fields in which biochemists are needed. Have students name several fields of scientific research or manufacturing. For each field they name, have students consider how a biochemist might be needed. Examples include studying ways to increase plant yields or insect resistance for crops and studying types of nutrients and medicines for farm animals.

Extend

Connect to SPACE SCIENCE

The first female commander of the International Space Station was Peggy Whitson, a biochemist. As part of her responsibilities on the space station, she conducted biochemical research into the effects of low-gravity conditions on plants. Have students find the results of her research and write a report detailing the results and her conclusions.

Answers

FIGURE 24.16 the hydrophobic center of a lipid bilayer

TAKE IT FURTHER

1. Possible answers include aging, reproduction, heredity, and the immune system.
2. Biochemists could develop drugs to treat diseases associated with aging such as Alzheimer's disease.

LESSON 24.5

Key Objectives

24.5.1 IDENTIFY the functions of DNA and RNA.

24.5.2 IDENTIFY the number of bases of DNA required to specify one amino acid in a peptide chain.

24.5.3 EXPLAIN gene mutations.

24.5.4 IDENTIFY two examples of DNA technology used today.

Additional Resources

- Reading and Study Workbook, Lesson 5

PearsonChem.com

Engage

CHEMISTRY & YOU Point out that some physical traits are inherited, while others are acquired during life. **Ask** What are some inherited traits? *(hair color, eye color, blood type)* **Ask** What are some acquired traits? *(muscle mass, hair length, skills)*

Build Background

Have students call out all possible three-letter arrangements of A, B, and C (allow double and triple letters as in AAA, AAB, and so on), and list the arrangements on the board. *(There should be 27 possible arrangements.)* Tell students that the body uses nucleic acids in a similar fashion to code for amino acid sequences in proteins.

IN ACADEMIC STANDARDS for SCIENCE

- Develop explanations based on reproducible data and observations gathered during laboratory investigations.

24.5 Nucleic Acids

CHEMISTRY & YOU

Q: *Why do children often look similar to their parents?* Maybe people have told you that you have your mother's eyes or your father's nose. Although this is not literally true (your eyes and nose are your own), you do inherit some traits from your parents. In this lesson, you will learn about molecules that are involved in the inheritance of traits from parents.

Key Questions

What are the functions of DNA and RNA?

How many bases of DNA are required to specify one amino acid in a peptide chain?

What are gene mutations?

What are two examples of DNA technologies used today?

Vocabulary

- nucleic acid
- nucleotide
- gene

DNA and RNA

What are the functions of DNA and RNA?

More than 100 years ago, a Swiss biochemist discovered a class of nitrogen-containing compounds in the nuclei of cells. The nuclei were first obtained from dead white blood cells in the pus of infected wounds. The eventual understanding of the biological role of the compounds has led to a revolution in biochemistry.

These nitrogen-containing compounds, called **nucleic acids,** are polymers that are found primarily in a cell's nucleus. They are indispensable components of every living thing. Two kinds of nucleic acids are in cells—*d*eoxyribo*n*ucleic *a*cid (DNA) and *r*ibo*n*ucleic *a*cid (RNA). **DNA stores the information needed to make proteins and governs the reproduction and growth of cells and new organisms. RNA has a key role in the transmission of the information stored in DNA and in the synthesis of proteins.**

The monomers that make up the DNA and RNA polymers are called **nucleotides.** Nucleic acids are therefore polynucleotides. As shown below, each nucleotide consists of a phosphate group, a five-carbon sugar, and a nitrogen-containing unit called a nitrogen base.

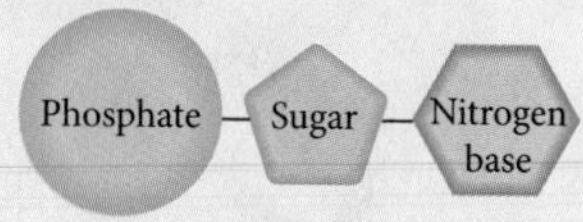

Nucleotide

The sugar unit in the nucleotides of DNA is the five-carbon monosaccharide known as deoxyribose. There are four different nitrogen bases in DNA—adenine, guanine, thymine, and cytosine. These four bases are abbreviated A, G, T, and C, respectively, and are shown in a short segment of a DNA molecule in Figure 24.17. Notice that adenine and guanine each contains a double ring and that thymine and cytosine each contains a single ring. Ribose, which has one more oxygen atom than deoxyribose, is the sugar found in the nucleotide monomers of RNA. The base thymine is never found in RNA. Instead, it is replaced by a fifth nitrogen base, called uracil, which is abbreviated U.

Focus on ELL

1 CONTENT AND LANGUAGE Explain what acronyms and abbreviations are and why people use them. Write *ASAP, SCUBA,* and *LASER* on the board and write below each word the meaning of each letter. Then write *DNA* and *RNA* on the board and discuss what the letters stand for.

2 FRONTLOAD THE LESSON As a class discuss the shortcuts students use when texting, instant messaging, or e-mailing, such as *LOL, BRB,* and *BBS*. Then tell students that, just as they use these three-letter codes to convey specific meanings in electronic conversation, DNA uses nucleic acids grouped in threes to code for specific amino acids.

3 COMPREHENSIBLE INPUT Manipulate a physical model of the DNA helix while discussing Figures 24.17 and 24.18. When discussing mutations, repeat the red cat exercise but add a letter to represent an insertion. Repeat again by changing a letter to represent a point mutation.

Figure 24.17 DNA
The nucleotide monomers of DNA are linked together through their sugar-phosphate groups. Two strands of DNA coil into a helix.

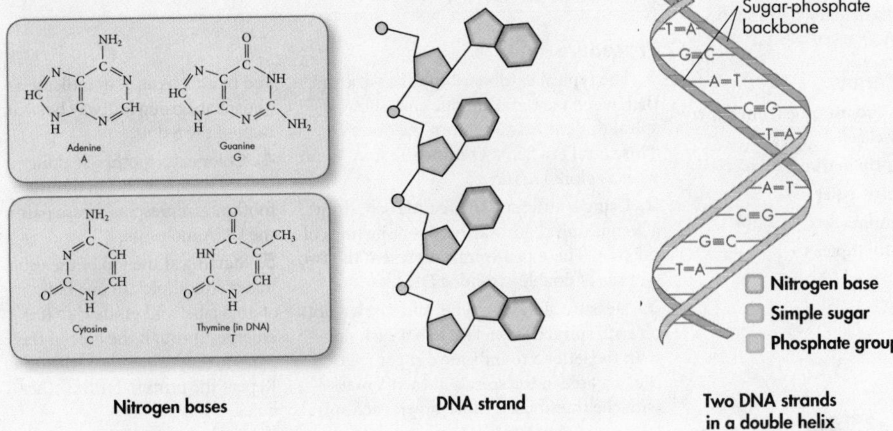

Chemists studying nucleic acids discovered that the amount of adenine in DNA always equals the amount of thymine (A = T). Similarly, the amount of guanine always equals the amount of cytosine (G = C). The significance of these facts was not apparent until 1953, when James Watson and Francis Crick proposed that the structure of DNA consists of two polynucleotide chains wrapped into a spiral shape, as in Figure 24.17. This spiral is the famous double helix of DNA. For the nitrogen bases to fit neatly into the double helix, every double-ringed base on one strand must be paired with a single-ringed base on the opposing strand. The pairing of A with T and G with C not only provides the best possible fit; it also allows the maximum number of hydrogen bonds to form between the opposing bases, as Figure 24.18 shows. Thus, the pairing of A and T (with two hydrogen bonds between the opposing bases) and of G and C (with three hydrogen bonds) makes for the most stable arrangement in the double helix.

Figure 24.18
Hydrogen Bonding in DNA
The two DNA strands in a double helix are held together by many hydrogen bonds; there are two hydrogen bonds between each thymine (T) and adenine (A) and three hydrogen bonds between each cytosine (C) and guanine (G).
Interpret Diagrams ***In the pairing of C with G, how many hydrogen bonds involve nitrogen? How many involve oxygen?***

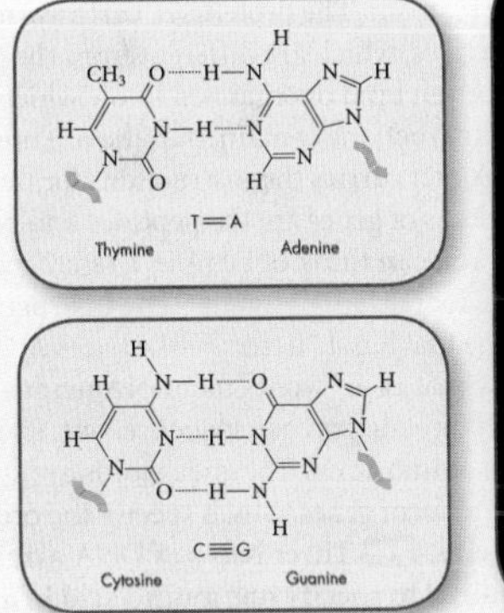

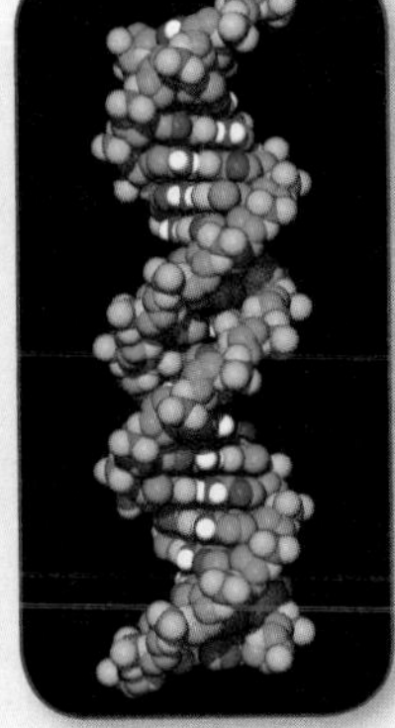

Foundations for Reading

BUILD VOCABULARY Explain to students that the root of many key words in this lesson relates to the nucleus of the cell. Point out that nucleic acid, for example, was first found in the nucleus of eukaryotic cells and that nucleotides are the compounds that make up these nucleus-inhabiting acids.

READING STRATEGY Have students identify the main idea under each red heading. Ask them to note details as they read that support the main idea for that portion of the text.

Explain

DNA and RNA

USE VISUALS Have students examine Figure 24.17. Point out the sequence complementarity between strands. **Ask** How does the experimental data that moles A = moles T, and moles G = moles C support the concept of base pairing? *(A and G are always hydrogen-bonded to T and C, respectively, in the opposite strand. Thus, there is a one-to-one mole correspondence.)*

Then direct students' attention to the molecular structures in Figure 24.18. **Ask** What features of the structure of the bases make this arrangement the best fit? *(two hydrogen bonds in the T-A connection, three for C-G, and length of bonds)*

UbD Check for Understanding

The Essential Question What are the characteristics of the four main types of biological molecules?

Have students write a brief summary of the characteristics of nucleic acids. *(There are five naturally occurring nucleic acids, each with a phosphate, a sugar, and a nitrogen base. The sugar in RNA is ribose; the sugar in DNA is deoxyribose. The five bases are adenine (A), guanine (G), cytosine (C), thymine (T), and uracil (U). Both RNA and DNA contain adenine, guanine, and cytosine. DNA also contains thymine; RNA contains uracil in place of thymine. The DNA helix is formed from two complementary strands of DNA, in which A bonds with T and C bonds with G.)*

ADJUST INSTRUCTION For the students struggling with this concept, have them reread the section on DNA and RNA, and then review Figures 24.17 and 24.18 with them. If possible, make DNA model kits available for students to manipulate.

Answers

FIGURE 24.18 nitrogen, 1; oxygen, 2

LESSON 24.5

Explore

Quick Lab

OBJECTIVE Students will construct a model of double-stranded DNA.

SKILLS FOCUS Using models, inferring

PREP TIME 10 minutes

CLASS TIME 20 minutes

MATERIALS Cardboard tube from paper-towel roll; felt-tip markers, two colors; metric ruler; thumbtack; 10 toothpicks

TEACHING TIPS To help students get parallel strands, have them cut a thin strip of paper that just wraps around the tube (following the spiral seam); then they can measure the length of the paper, marking points along it every 5 cm. Students can use the paper to mark points on the tube and connect points to draw the parallel strand.

EXPECTED OUTCOME Models should have an orderly column of bases on the inside, with phosphate and sugar groups on the outside. **Ask** What forces hold the two strands of DNA together? *(Primarily hydrogen bonding; dispersion forces also act between stacked bases.)* **Ask** What forces the two strands apart? *(Repulsion of negative charges on phosphate groups; two strands have greater entropy than a single helical structure.)*

ANALYZE AND CONCLUDE

1. Base pairs are neatly stacked inside the double helix.
2. the exterior
3. the exterior

FOR ENRICHMENT **Ask** What must occur to replicate base pairs of DNA? *(The two strands must separate.)* Instruct students to alter their model so that it can be used to demonstrate DNA replication.

Quick Lab

A Model of DNA

Purpose To construct a model of double-stranded DNA

Materials

- cardboard tube from paper-towel roll
- felt-tip markers (two colors)
- metric ruler
- thumbtack
- 10 toothpicks

Procedure

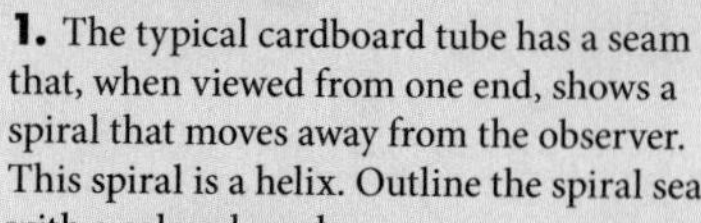

1. The typical cardboard tube has a seam that, when viewed from one end, shows a spiral that moves away from the observer. This spiral is a helix. Outline the spiral seam with a colored marker.
2. Using a different-colored marker, draw a second spiral midway between the lines of the first. These two spirals represent the two strands of double-stranded DNA.
3. Measure along the tube, and mark a dot on each spiral every 5 cm. Label each dot with the letter *S* to indicate a sugar unit. Make a hole in the spirals at each *S* mark with the thumbtack. Move down each spiral and mark a letter *P* to indicate a phosphate group halfway between each of the *S* dots.
4. Color each toothpick along half its length with a marker. A toothpick represents a base pair in the DNA molecule.
5. Starting at the top of the tube, insert a toothpick in one hole at an *S* label and guide it so it emerges through the hole in the *S* on the opposite side of the tube. Repeat the process for the other holes.

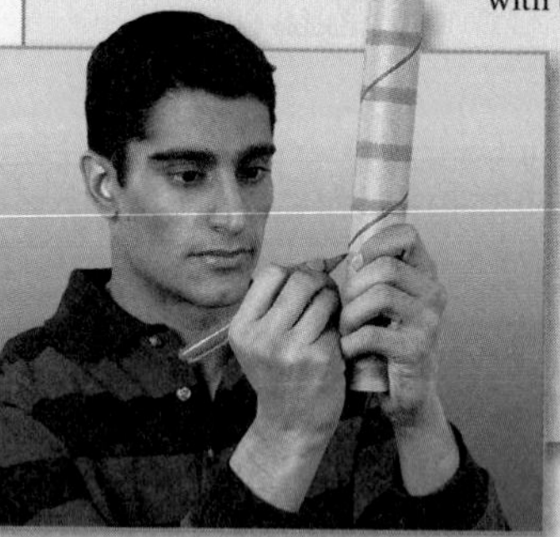

Analyze and Conclude

1. **Use Models** Are the bases on the interior or the exterior of the double helix? Are they randomly arranged or neatly stacked?
2. **Analyze Data** Are the phosphate groups on the exterior or the interior of the DNA structure?
3. **Analyze Data** Are the sugar groups on the interior or the exterior of the DNA molecule?

The Genetic Code

How many bases of DNA are required to specify one amino acid in a peptide chain?

An organism contains many proteins that are characteristic of that particular organism. The proteins of earthworms are different from the proteins of pine trees, which are different from the proteins of humans. How do cells in a given kind of organism know which proteins to make? The cells use the instructions contained in the organism's DNA. A **gene** is a segment of DNA that carries the instructions for making one peptide chain. Thus, the products of genes are the peptides and proteins found in an organism.

You can think of DNA as a reference manual that stores the instructions for building proteins. The instructions are written in a simple language that has 4 "letters"—the bases A, T, G, and C. Experimental data show that each "word" in a DNA manual is exactly three letters in length. Each three-letter base sequence, or triplet, codes for one of the 20 common amino acids. The code words are strung together in the DNA molecule to form genes, which specify the order of amino acids in peptides and proteins. **Three bases of DNA arranged in a specific sequence are required to specify one amino acid in a peptide or protein chain.**

Focus on ELL

4 LANGUAGE PRODUCTION Have students work in pairs to complete the lab. Make sure each pair has students of varied language proficiencies so that a proficient student can help a less proficient student. Have students work according to their proficiency level.

BEGINNING: LOW/HIGH Model the creation of the double helix by systematically showing students what to do at each step of the procedure. Have students write a description of each step of the procedure in their native language during your demonstration.

INTERMEDIATE: LOW/HIGH For their responses to the Analyze and Conclude questions, allow students to answer orally and to point to places on their models.

ADVANCED: LOW/HIGH Paraphrase the procedure and Analyze and Conclude questions for students with lower English proficiencies. Encourage students to use symbols and diagrams when applicable.

Table 24.2 provides the DNA code words for the 20 common amino acids. For example, you can see that the DNA code word AAA specifies the amino acid phenylalanine (Phe) and that the DNA code word CGA specifies the amino acid alanine (Ala). Note that most amino acids are specified by more than one code word, but a code word never specifies more than one amino acid. With DNA code words of three letters, 900 bases arranged in a specific sequence would be required to code for a peptide chain made up of 300 amino acids arranged in a specific sequence.

One of the the code words (TAC) signifies the initiation of a peptide. Three code words (ATT, ATC, and ACT) are reserved as end, or termination, code words. The translation of a base sequence of DNA in a gene into the amino acid sequence of a peptide begins with the initiation code word and runs continuously until a termination code word is reached. The termination code word signals a stop to the addition of amino acids in the production of the peptide. You can think of a termination code as being similar to the period at the end of a sentence.

The molecular masses of DNA molecules reach into the millions and possibly billions. Even with only four bases, the number of possible sequences of nucleotides in a DNA chain is enormous. The sequence of the nitrogen bases A, T, G, and C in the DNA of an organism constitutes the genetic plan, or blueprint, for that organism. This genetic plan is inherited from parents and passed to offspring. Differences in the number and sequence of the bases in DNA ultimately are responsible for the great diversity of living creatures found on Earth.

CHEMISTRY & YOU

Q: *Why do children often look similar to their parents?*

Table 24.2

Three-Letter DNA Code Words for the Amino Acids

First Letter in Code Word	Second Letter in Code Word: A	G	T	C	Third Letter in Code Word
A	AAA Phe	AGA Ser	ATA Tyr	ACA Cys	A
	AAG Phe	AGG Ser	ATG Tyr	ACG Cys	G
	AAT Leu	AGT Ser	ATT End	ACT End	T
	AAC Leu	AGC Ser	ATC End	ACC Trp	C
G	GAA Leu	GGA Pro	GTA His	GCA Arg	A
	GAG Leu	GGG Pro	GTG His	GCG Arg	G
	GAT Leu	GGT Pro	GTT Gln	GCT Arg	T
	GAC Leu	GGC Pro	GTC Gln	GCC Arg	C
T	TAA Ile	TGA Thr	TTA Asn	TCA Ser	A
	TAG Ile	TGG Thr	TTG Asn	TCG Ser	G
	TAT Ile	TGT Thr	TTT Lys	TCT Arg	T
	TAC Met	TGC Thr	TTC Lys	TCC Arg	C
C	CAA Val	CGA Ala	CTA Asp	CCA Gly	A
	CAG Val	CGG Ala	CTG Asp	CCG Gly	G
	CAT Val	CGT Ala	CTT Glu	CCT Gly	T
	CAC Val	CGC Ala	CTC Glu	CCC Gly	C

Explain

The Genetic Code

CHEMISTRY & YOU Proteins determine a variety of traits. DNA contains the instructions for which proteins cells should make. Offspring inherit DNA from their parents.

USE AN ANALOGY Display a copy of the Morse code. Tell students that the millions of possible sequences from the combination of only four nucleotides can be compared to the very large number of words that are formed from only two symbols—the dot and dash of Morse code. Moreover, both systems use patterns to communicate information. Point out, however, that Morse code is a relatively simple human invention that uses arbitrary combinations of dots and dashes to represent letters of the alphabet. The genetic code is a complex natural communication system that directs and controls all life processes. Make clear to students that while Morse code combinations vary from one to five symbols, all the base sequences in the genetic code are exactly three bases in length.

MAKE A CONNECTION Tell students that cracking the genetic code was one of the greatest achievements of modern science. Make sure students understand that the three-letter codes used to translate DNA to protein are the same in humans and in bacteria. Have students ponder this concept and then ask them to state what implications this has for the evolution of life on Earth. *(It supports the argument that life on Earth evolved from the same organisms.)*

USE VISUALS Direct students' attention to Table 24.2. **Ask** How many three-letter codes are there? *(64)* **Ask** Are all possible combinations of bases used? *(yes)* **Ask** Why are there more codes than amino acids? *(Some amino acids have more than one code.)* **Ask** Is there an advantage to redundant codes? *(Redundancy might protect against harmful mutations.)*

Differentiated Instruction

L3 **ADVANCED STUDENTS** Challenge students to write a computer program that converts any sequence of DNA bases into an amino-acid sequence by using the three-letter codes in Table 24.2.

LPR **LESS PROFICIENT READERS** Provide students with a copy of Table 24.1 from Lesson 24.3 so they can know what amino acids the abbreviations in Table 24.2 represent. Help students interpret the information present in both tables.

L1 **STRUGGLING STUDENTS** Review the difference between hydrogen bonding and chemical bonding. Reinforce the concept that DNA strands are held together by hydrogen bonds rather than as a result of chemical reactions.

The Chemistry of Life 857

Explore

Gene Mutations

Class Activity

Have students practice reading the genetic code. Write a number of short DNA sequences on the board and ask students to use Table 24.2 to translate the genetic information into an amino-acid sequence. *(Example: A-C-G-G-C-A-A-G-A-G-A-A-T-G-G-A-C-T Cys-Arg-Ser-Leu-Thr-End)*

Then, ask students to predict the effects of substitutions, additions, and deletions on the translation of the DNA sequence. **Ask** How might this sequence be affected if a thymine were substituted for the first guanine? *(This mutation would insert a stop code. The protein synthesis would prematurely terminate, and there would most likely be no biological activity.)*

Extend

Connect to BIOLOGY

A mutation is a change in the genetic material of a cell. If a mutation occurs in the sex cells of an organism, it can be passed on from one generation to the next. Mutations can have beneficial, negative, or neutral effects. Over time, accumulated mutations can lead to the formation of a new species. Scientists use molecular clocks to compare the DNA of species to estimate the last time the species shared a common ancestor—that is to say, to estimate when a speciation event occurred. Molecular clocks are based on an assumption that neutral mutations accumulate in the DNA of different species at about the same rate. So, the more differences there are between the DNA sequences of two species, the more time has elapsed since the speciation event.

Different molecular clocks "tick" at different rates—some genes accumulate mutations faster than others. Scientists therefore have to be careful when selecting which sections of DNA to use for comparison. Researchers can check the accuracy of molecular clocks by comparing the number of mutations in a particular gene in species whose age has been determined by other means, such as through fossil evidence.

Have students do research to find out more about DNA mutations and molecular clocks. Ask them to explain how changing the DNA sequence can lead to changes that can be passed from parents to offspring, and have them provide an example of how a molecular clock was used in scientific research.

Gene Mutations

What are gene mutations?

When a change occurs in a DNA code word, the result is a mutation in the DNA. **Substitutions, additions, or deletions of one or more nucleotides in the DNA molecule are called gene mutations.** The effect of the deletion of a single base from a gene can be illustrated by the following analogy. Suppose a string of letters of the alphabet goes as follows:

PATTHEREDCAT

The letters may not make sense at first glance. However, if you separate them into three-letter words, they form a perfectly sensible statement:

PAT THE RED CAT

Now delete the first letter, and again separate the string into three-letter segments:

ATT HER EDC AT

This last sequence is nonsensical. Similarly, the deletion of a base in the DNA base sequence can turn the information into nonsense. A sequence that once may have coded for the proper sequence of amino acids in a necessary protein may be replaced by a sequence that produces a useless or damaging amino acid sequence. The same sort of harmful effect may be produced by mutations involving substitutions or additions of nucleotides.

Such mutations might result in the production of a faulty protein or of no protein at all. Diseases that result from gene mutations are called genetic disorders. Thousands of genetic disorders have been identified. Galactosemia is an example of a genetic disorder that affects about 1 in 55,000 newborn babies. Galactosemia results from a mutation in an enzyme called GALT (galactose-1-phosphate uridyl transferase). GALT is needed to break down the sugar galactose into glucose, as shown in Figure 24.19. Without normal GALT, galactose can build up in the body and cause kidney failure, an enlarged liver, cataracts, and brain damage. Newborns are usually tested for galactosemia at birth. The only way to treat galactosemia is to avoid foods that contain galactose, such as milk, cheese, and dried beans.

Not all gene mutations are harmful. Occasionally, a mutation can result in the synthesis of a protein that is more efficient than the version that previously existed. Such a mutation could thus be beneficial to the survival of the affected organism.

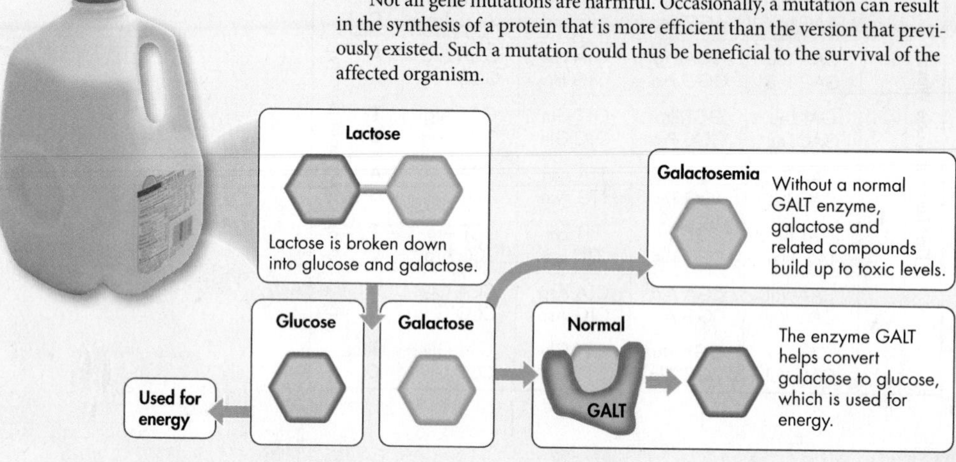

Figure 24.19 Galactosemia
Persons with galactosemia have mutations in the gene for the enzyme GALT. They cannot complete the breakdown of lactose. Any food containing galactose or lactose, including milk and products containing milk, should be avoided.

Caffeine Mutations?

Many people have looked for health risks associated with the consumption of caffeinated beverages. One concern raised is the potential mutagenic effects due to the misincorporation of caffeine into nucleic acids. Caffeine is a nitrogen compound that is structurally related to purine bases, especially guanine. Its resemblance may allow it to bind to DNA-synthesizing enzymes, which would incorporate it into a polynucleotide. However, caffeine lacks hydrogen for proper base pairing, and this would destabilize the DNA helix.

858 Chapter 24 • Lesson 5

DNA Technologies

What are two examples of DNA technologies used today?

Since DNA is such an important part of living things, it may not be surprising that DNA technology has had a significant impact on our society. In the following pages, you will learn about some of these technologies.

DNA Typing Only a small fraction of a human's DNA is used for coding the information needed for protein synthesis. The rest consists of repeating, noncoding base sequences that separate or sometimes interrupt gene coding sequences. The role of these stretches of noncoding DNA is unclear. The noncoding sequences are similar for members of the same family but are slightly different for almost every individual. Differences also exist in the coding portions of DNA. The base sequences of DNA are slightly different for different individuals, except for identical twins. Identical twins look similar because they have identical DNA.

DNA typing uses the variation in the DNA of individuals as a basis for creating DNA profiles to identify a person from samples of his or her hair, skin cells, or body fluid. Because DNA sequences, like fingerprints, are unique for each individual, DNA typing has also been called DNA fingerprinting. To construct a DNA profile, scientists first isolate the DNA in a sample. Only a tiny sample is needed. A sample can be anything that contains DNA, including teeth, fingernails, blood, hair, saliva, and skin cells. Figure 24.20 shows some items that investigators obtained from a crime scene.

Samples can be typed in several different ways, but the method used most commonly by the Federal Bureau of Investigation (FBI) is short tandem repeat (STR) analysis. A short tandem repeat is a short segment of DNA that is repeated several times. For example, one of the regions of DNA used by the FBI contains repeats of the sequence AGAT. To obtain a profile, the FBI looks at 13 different STR regions.

The DNA profile can then be compared with a sample of DNA from a known individual, as Figure 24.21 shows. The FBI has a technology system called the Combined DNA Indexing Systems (CODIS) that allows laboratories throughout the country to share and search DNA profiles. The chances of two people (except for identical twins) having the same DNA profile for these 13 regions is 1 in 1 billion.

Figure 24.20
Crime Scene Evidence
These scientists are analyzing items taken from a crime scene. DNA may be isolated from blood on the items, in this case clothing and an axe.

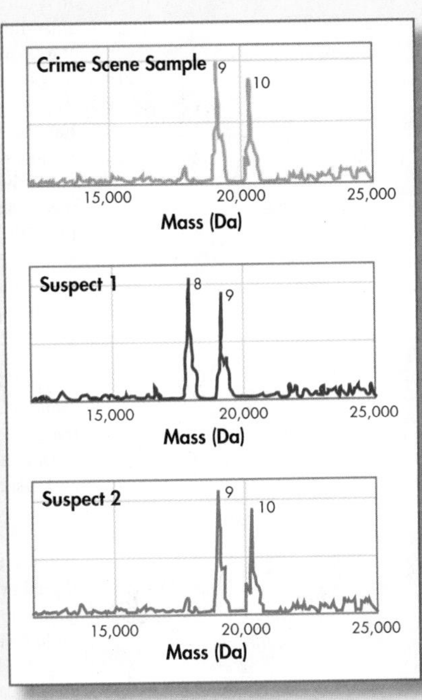

Figure 24.21 DNA Profile
This is an example of what a DNA profile for one STR region might look like. The DNA from the crime scene has a segement with 9 repeats and a segment with 10 repeats. Suspect 1 can be eliminated because the length of that person's segments does not match the crime scene sample.

Differentiated Instruction

L3 ADVANCED STUDENTS Have students research the legal programs started after 1992 when the National Research Council expressed support for DNA testing as a means for identifying criminals. Have students find the statistics showing the number of convictions overturned since the inception of these legal programs.

L1 SPECIAL NEEDS STUDENTS Using an ink pad and a sheet of plain paper, take one fingerprint each from two or three volunteers. Project the fingerprints on a screen. Ask students if they see any fingerprints that are exactly alike. Convey that no two people have the same fingerprint, and then relate this concept to DNA.

ELL ENGLISH LANGUAGE LEARNERS Write the word *profile* on the board. Explain that it has both common and scientific meanings. Assign students to small groups of varying proficiencies, and have each group look up the definitions of this term in a dictionary and identify the definition that most closely fits the context of DNA profiles.

Explain

DNA Technologies

APPLY CONCEPTS Forensic investigators try to understand a crime by using a variety of tactics. They often try to collect evidence that will allow them to create a DNA profile. Analysts then cross-check the profile with the ones in the FBI's Combined DNA Index System (CODIS). CODIS is a database of DNA profiles from crime scenes and known offenders. The profiles can also be compared to samples taken from victims and suspects.

START A CONVERSATION Tell students that DNA fingerprinting has been controversial in terms of its reliability. **Ask** What current events or past criminal trials have you read about that involve the reliability of DNA fingerprinting? *(Answers will vary.)* **Ask** Do you think that proven scientific discoveries should be allowed to exonerate people in jail for crimes that were judged using evidence examination techniques from the past? *(Generate a class discussion using a well-known criminal trial.)* Explain to the class that the closer the match between two samples of DNA, the more likely that the two samples are from one individual. Tell students that one of the reasons people are nervous about the reliability of DNA results is that the expense of scanning evidence increases with each region of DNA that is scanned. Explain that the reliability of DNA fingerprinting varies, depending on the number of regions scanned and the level of expertise of the interpreter. **Ask** If you were on a jury, which would you think is more reliable: an eyewitness or the results of DNA fingerprinting? *(Answers may vary, but inform students that the reliability of an eyewitness is 50:50.)*

LESSON 24.5

Explain

DNA Technologies

USE AN ANALOGY Relate splicing an electrical wire to the methodology of recombinant DNA. If possible, have a student familiar with splicing electrical wire demonstrate the process. Introduce the complex topic of recombinant DNA methodology as simply removing a damaged section and replacing it with a newer and more reliable section that will allow the intended process to continue.

MAKE A CONNECTION Point out that the treatment of human diseases with recombinant proteins, though effective, is a replacement therapy. Explain that patients must have a continuous supply of the protein to remain asymptomatic. Ultimately, genetic engineers would like to develop methods for treating the cause of a genetically linked disorder by enabling the patient's own cells to produce a missing or faulty protein, an area of research referred to as gene therapy.

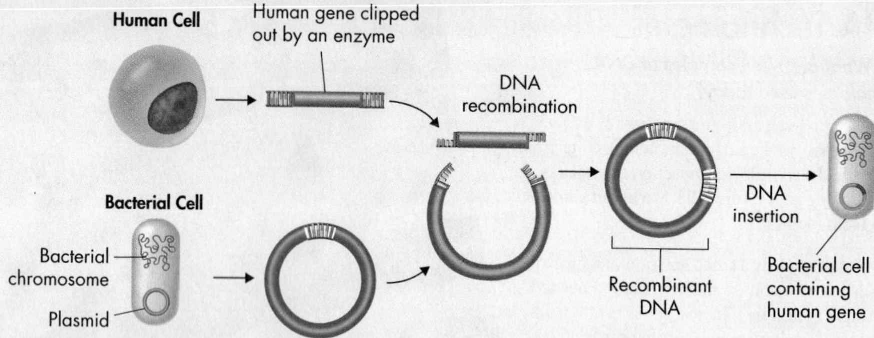

Figure 24.22
Recombinant DNA
Included here are the elements of an experiment involving recombinant DNA. In this experiment, DNA from one organism is inserted into the DNA of a different organism.

Recombinant DNA Technology Scientists have learned to manipulate genes by various methods. **Recombinant DNA technology consists of methods for cleaving a DNA chain, inserting a new piece of DNA into the gap created by the cleavage, and resealing the chain.** Figure 24.22 illustrates such a method. The altered DNA formed by this method is known as recombinant DNA.

Applications in Medicine The first practical application of recombinant DNA technology was to insert the gene for making human insulin into bacteria. Most people naturally make insulin, a polypeptide that controls levels of blood sugar. However, insufficient insulin production results in diabetes. The symptoms of diabetes can often be controlled by insulin injections. In the past, human insulin was not available for this purpose. Pig insulin, which is quite similar in structure to human insulin, was used as a substitute. Some patients, however, were allergic to pig insulin. Today, diabetic patients use the human form of insulin produced by bacteria that have been altered by recombinant DNA technology. Use of this insulin removes the need for the potentially dangerous use of pig insulin.

Other proteins produced by recombinant DNA technology are used as medicinal drugs. For example, an enzyme called tissue plasminogen activator (TPA) is used to dissolve blood clots in patients who have suffered heart attacks. Another protein, interferon, is thought to relieve or delay some of the debilitating effects of multiple sclerosis. Recombinant DNA technology is also being applied to the cure of genetic disorders in an experimental treatment known as gene therapy.

Applications in Agriculture In agriculture, new recombinant DNA techniques can make plants resistant to pests and weed killers and produce fruits and vegetables that are better suited for shipping and storage. The most common traits in genetically modified crops are herbicide resistance and insect resistance in corn, cotton, soybean, and canola.

Crops have also been genetically modified for pharmaceutical purposes. These so-called pharma crops are genetically modified to produce drugs to treat or prevent diseases such as cancer or AIDs. Genetically altered organisms have many potential benefits, but some people have concerns about their safety. There is also concern that genetically modified crops could contaminate other crops if they are grown and processed in close proximity to them.

Recombinant DNA

Recombinant DNA technology uses small circular DNA molecules called plasmids, which occur naturally in bacteria and in some yeasts. Once a human gene has been combined with the plasmid DNA, the recombinant molecule is reintroduced into bacteria. Because human DNA and bacterial DNA use the same bases and because the genetic code is universal, the bacteria translate the human gene in the recombinant DNA into protein exactly as they would their own genes. When grown in large incubators, the bacteria can produce massive quantities of the recombinant protein.

Cloning Ethical concerns were raised in 1997 when Scottish scientists announced the birth of a lamb named Dolly, shown in Figure 24.23. In normal animal reproduction, an offspring is a genetic mixture of the characteristics of both parents. Dolly, however, was a clone—an offspring of a single individual. A clone is an exact genetic copy of its parent because it is formed using the DNA of only that parent. Since the cloning of Dolly, other animals have also been successfully cloned, including cows, mice, and cats. The birth of cloned animals has raised the question of whether humans might eventually be cloned. Many people are concerned about some of the possible outcomes of cloning identical individuals. These situations are one aspect of more general concerns about the uniqueness of life.

Figure 24.23 Cloning
Polly and Dolly had no fathers. Dolly was cloned from a single cell taken from her mother. Polly was also cloned from a single cell that was genetically modified.
Infer ***How did Dolly's DNA compare with the DNA in her mother's cells?***

24.5 LessonCheck

ONLINE PROBLEMS

29. **Review** What functions are performed by DNA and RNA?

30. **Review** What does a three-letter base sequence of DNA specify?

31. **Identify** What are three types of gene mutations?

32. **Identify** What methods are used in recombinant DNA technology?

33. **Describe** Describe how scientific methods might be applied to the process of DNA fingerprinting.

34. **Evaluate** Why do you think cloning is controversial? What are your personal thoughts on this matter?

Lesson Check Answers

29. DNA stores genetic information; RNA transmits genetic information.

30. an amino acid

31. substitution, addition, and deletion of a nucleotide

32. Answers might include the ideas of cleaving a DNA chain, inserting a new piece of DNA into the gap created by the cleavage, resealing the chain, and expressing the new gene product.

33. Forensic investigators use scientific methods when trying to solve a crime. At the crime scene, they make observations and collect evidence, including DNA samples. The investigators make hypotheses about how the crime was committed and who may be responsible. They can test these hypotheses by questioning suspects and comparing DNA fingerprints of the suspects with those left at the crime scene.

34. People have different opinions about issues such as the dignity of human life and interfering with nature.

Extend

Connect to LITERATURE

Have students read the autobiography of the Nobel Prize–winning chemist Arthur Kornberg. The title of his autobiography is *For the Love of Enzymes: The Odyssey of a Biochemist*. Kornberg provides an exceptionally readable historical account of the discovery and characterization of DNA and nucleic acids. This book documents many of the important biological discoveries made in the twentieth century. Assign students to utilize technology to make a timeline of the history of DNA.

Interested students may also wish to read James Watson's *The Double Helix*, a first-person account of the discovery of the structure of DNA.

Informal Assessment

Have students write a short paragraph comparing and contrasting DNA and RNA and explaining their roles in the production of proteins. Then have students complete the 24.5 Lesson Check.

Reteach

Display Figure 24.17 on an overhead projector. Use it to review the chemical and functional differences between DNA and RNA. Have students identify a complementary RNA strand for one of the strands of DNA shown in the figure.

Answers

FIGURE 24.23 Their DNA was the same.

LESSON 24.6

Key Objectives

24.6.1 **DESCRIBE** the function of ATP in the cell.

24.6.2 **DESCRIBE** what happens to biological molecules and energy during catabolism and anabolism.

24.6.3 **EXPLAIN** how nitrogen-fixing bacteria provide plants with a usable form of nitrogen.

Additional Resources

- Reading and Study Workbook, Lesson 24.6

PearsonChem.com

Engage

CHEMISTRY & YOU Discuss with the class why a hummingbird's constant activity causes it to have a very high metabolic rate. **Ask** What is metabolic rate? *(how fast the body's chemical reactions proceed)*

Activate Prior Knowledge

Briefly review oxidation-reduction reactions. Then ask students to describe what they know about ATP from biology class. Provide biology textbooks for students to refer to for a more complete understanding of a living cell.

Foundations for Reading

BUILD VOCABULARY Explain that the suffix *-ism* means "process." Then have students write definitions of *metabolism, catabolism,* and *anabolism* in their own words.

24.6 Metabolism

CHEMISTRY & YOU

Q: *Why does a hummingbird eat so much?* A hummingbird eats more than its weight in food each day. Hummingbirds have a high body temperature, fast heart rate, and a fast breathing rate. All of these factors affect the hummingbird's metabolism. In this lesson, you will learn about the reactions that are part of an organism's metabolism.

Key Questions

What is the function of ATP in living cells?

What happens to biological molecules and energy during catabolism and anabolism?

How do nitrogen-fixing bacteria provide plants with a usable form of nitrogen?

Vocabulary

- adenosine triphosphate (ATP)
- metabolism
- catabolism
- anabolism

ATP

What is the function of ATP in living cells?

All living things need energy to function. **Adenosine triphosphate (ATP),** shown in Figure 24.24, is a molecule that transmits this energy in the cells of living organisms. The function of ATP can be compared to a belt connecting an electric motor to a pump. The motor generates energy capable of operating the pump. But if a belt does not connect the motor to the pump, the energy produced by the motor is wasted. You can think of ATP as the belt that connects the production and use of energy by cells. **In living cells, ATP is the energy carrier between the spontaneous reactions that release energy and nonspontaneous reactions that use energy.**

Recall that oxidation reactions, such as the combustion of methane in a furnace or the oxidation of glucose in a living cell, are spontaneous reactions that release energy. This energy can be captured when adenosine diphosphate (ADP) condenses with an inorganic phosphate group to become ATP. The addition of a phosphate group, called phosphorylation, occurs during certain biochemical oxidation reactions.

Figure 24.24 ATP
ATP is made of adenine, ribose, and three phosphate groups. ATP provides energy to muscles for moving the body.
Compare and Contrast *How is the structure of ATP similar to that of a DNA nucleotide?*

NH_2
N
N
N
N
Adenine
Adenosine
O O O
$^-O-P-O-P-O-P-O-CH_2$ O
O^- O^- O^-
Ribose
Triphosphate
OH OH

Focus on ELL

1 CONTENT AND LANGUAGE Ask students to state the key questions in their own words. Allow students with limited English to use their native languages for support.

2 FRONTLOAD THE LESSON Ask students what happens to their body temperature when they exercise or participate in a sport. *(Body temperature increases.)* Explain that this is a sign that their muscles and other tissues have increased their metabolic rate. Tell students that as they study the lesson, they should be alert for examples of chemical reactions that might be responsible for this increase.

3 COMPREHENSIBLE INPUT Explain Figure 24.27 and read aloud the caption.

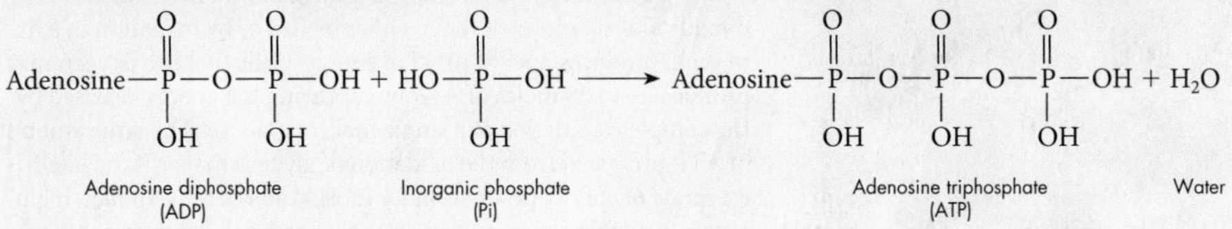

$$\text{Adenosine}-\underset{\text{OH}}{\overset{\text{O}}{\overset{\|}{\text{P}}}}-\text{O}-\underset{\text{OH}}{\overset{\text{O}}{\overset{\|}{\text{P}}}}-\text{OH} + \text{HO}-\underset{\text{OH}}{\overset{\text{O}}{\overset{\|}{\text{P}}}}-\text{OH} \longrightarrow \text{Adenosine}-\underset{\text{OH}}{\overset{\text{O}}{\overset{\|}{\text{P}}}}-\text{O}-\underset{\text{OH}}{\overset{\text{O}}{\overset{\|}{\text{P}}}}-\text{O}-\underset{\text{OH}}{\overset{\text{O}}{\overset{\|}{\text{P}}}}-\text{OH} + H_2O$$

Adenosine diphosphate (ADP) | Inorganic phosphate (Pi) | Adenosine triphosphate (ATP) | Water

The formation of ATP efficiently captures energy produced by the oxidation reactions in living cells. Every mole of ATP produced by the phosphorylation of ADP stores about 30.5 kJ of energy. The reverse happens when ATP is hydrolyzed back to ADP: Every mole of ATP that is hydrolyzed back to ADP releases about 30.5 kJ of energy. Cells use this released energy to drive processes that would ordinarily be nonspontaneous. Because of its ability to capture energy from one process and transmit it to another, ATP is sometimes referred to as a high-energy compound; however, the energy produced by the breakdown of ATP to ADP is not particularly high for the breaking of a covalent bond. ATP is important because it occupies an intermediate position in the energetics of the cell. It can be formed by using the energy obtained from a few higher-energy oxidation reactions. The energy that is contained in the bonds of ATP can then be used to drive other cellular processes.

CHEMISTRY & YOU

Q: *Why does a hummingbird eat so much?*

Metabolism Reactions

What happens to biological molecules and energy during catabolism and anabolism?

Thousands of chemical reactions take place in the cells of a living organism. The entire set of chemical reactions carried out by an organism is known as the organism's **metabolism.** The reactions that occur in metabolism can be divided into two main processes, catabolism and anabolism.

Catabolism In metabolism, unneeded cellular components and the nutrients in food are broken down into simpler compounds by chemical reactions collectively called **catabolism.** Catabolic reactions release energy as well as produce simple compounds. **The degradation of complex biological molecules such as carbohydrates, lipids, proteins, and nucleic acids during catabolism provides the energy and the building blocks for the construction of new biological compounds needed by the cell.** Through the formation of ATP, catabolic reactions provide the energy for such needs as body motion and the transport of nutrients to cells where they are required. The oxidation reactions of catabolism also provide energy in the form of heat. These reactions help keep your body temperature constant at 37°C.

The complete oxidation of glucose to carbon dioxide and water is one of the most important energy-yielding processes of catabolism. Study Figure 24.25, which summarizes the major steps in the degradation of one glucose molecule to six molecules of carbon dioxide. The complete oxidation actually involves many reactions that are not shown. As you can see in the figure, the major carbon-containing reactants and products are named, and they are also referred to according to the number of carbons they contain.

Figure 24.25 Glucose Catabolism
The breakdown of glucose to carbon dioxide and water is one of the most important energy-yielding processes of catabolism.
Make Generalizations ***What happens to the number of carbon-carbon bonds from one step to the next?***

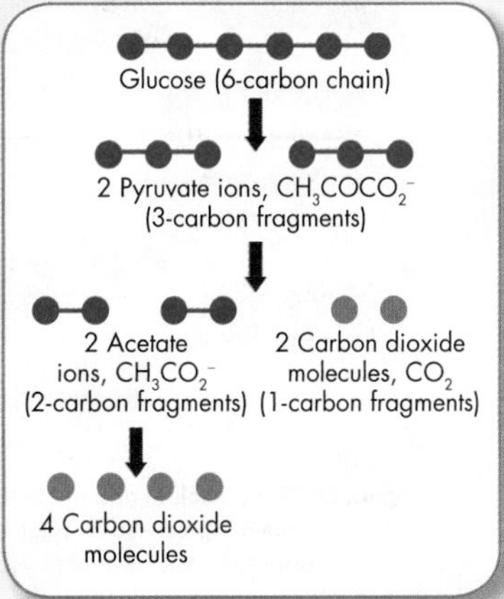

Check for Understanding

What is the function of ATP in living cells?

Assess students' understanding about ATP by asking them to write an analogy that can be used to convey the function of ATP. Tell students that they cannot use the same analogy given in the textbook.

ADJUST INSTRUCTION If students are having trouble answering, have them reread the section titled *ATP*. As a class, brainstorm various analogous situations. Then ask students to answer the question again.

READING STRATEGY After students read about catabolism and anabolism, ask them to create a diagram that shows how carbon moves from one type of metabolism to the other. Tell them to indicate where energy (ATP) comes into play.

Explain

ATP

APPLY CONCEPTS Have students compare the free energy changes associated with spontaneous and nonspontaneous processes. Explain that energy released by a reaction and available to do work is called free energy. Spontaneous processes, which favor formation of products, release free energy. Point out that it is the phosphate–oxygen anhydride bonds in ATP that store the chemical energy in ATP. When ATP is hydrolyzed, the new bonds that form release energy. This free energy is used to form new types of chemical bonds.

CHEMISTRY & YOU Hummingbirds need a lot of energy because they are in constant motion and have a high metabolism. They get this energy from food.

MAKE A CONNECTION Explain that all living things require a continuous input of free energy. Point out that more than 50 percent of the free energy released by cellular reactions is used to cause other chemical changes; the rest is lost as heat. Compare this efficiency to that of the internal combustion engine—at most, 30 percent of the free energy of combustion is used to propel a car.

Answers

FIGURE 24.24 Like the nucleotides of DNA and RNA, ATP has a basic structure consisting of a phosphate group, a five-carbon sugar, and a nitrogen base. The sugar unit in both ATP and RNA is ribose.

FIGURE 24.25 The number of C—C bonds decreases.

Explain

Metabolism Reactions

APPLY CONCEPTS Explain that in the human body, glycogen is stored in skeletal muscles and the liver. Tell students that liver glycogen is used to maintain normal blood sugar levels whereas muscle glycogen is used solely for muscle work. Explain that large amounts of carbohydrates, including glycogen, are used by the body during athletic events that require power or speed. **Ask** What compounds provide the necessary energy during these events? *(carbohydrates)* **Ask** Why do you suppose some athletes use interval training, a technique that alternates rest and work phases? *(to train the biochemical energy pathways and to train the muscles used in an athletic activity)*

Extend

Tell students that proper training and diet can increase by more than a thousandfold the amount of glycogen stored after a training session. Have students research the training activities of a successful or favorite athlete of their choice. Students should be prepared to explain their findings to the class.

USE VISUALS Have students examine Figure 24.27. **Ask** Are there any inputs created by catabolism that are not reflected in outputs? *(no)* Discuss why this is so for an animal, for example. *(We need all these substances to live, and there must to be a way to break them down and use them to build a body.)*

Figure 24.26 Animal Metabolism
Organisms such as this mouse use the energy stored in the chemical bonds of food molecules to power their body processes.

The combustion of one mole of glucose to six moles of carbon dioxide and six moles of water, either by fire or by oxidation in a living cell, produces 2.82×10^3 kJ of energy. Cells that use oxygen may produce up to 38 moles of ATP by capturing the energy released by the complete oxidation of a single mole of glucose! The large amount of ATP produced from the oxidation of glucose makes it the likeliest mode of energy production for most kinds of cells. In fact, if glucose is available, brain cells use no other source of carbon compounds for energy production. The need for energy and building blocks is the reason why all organisms, such as the field mouse shown in Figure 24.26, require food.

Anabolism Some of the simple compounds produced by catabolism are used to synthesize more-complex biological molecules—carbohydrates, lipids, proteins, and nucleic acids—necessary for the health and growth of an organism. The synthesis reactions of metabolism are called **anabolism.** Unlike catabolism, which releases energy, anabolism uses energy.

Figure 24.27 gives an overview of the relationship between catabolism and anabolism. Nutrients and unneeded cell components are degraded to simpler components by the reactions of catabolism. The oxidative reactions of catabolism yield energy captured in the formation of ATP. **In anabolism, the products and the energy of catabolism are used to make new compounds and cell parts needed for cellular life and growth.** You already know that energy produced by physical and chemical processes is of little value unless the energy can be captured to do work. If it is not captured, the energy is lost as heat. The chemical energy produced by catabolism must have some means of being used for the chemical work of anabolism. The ATP molecule is that means of transmitting energy.

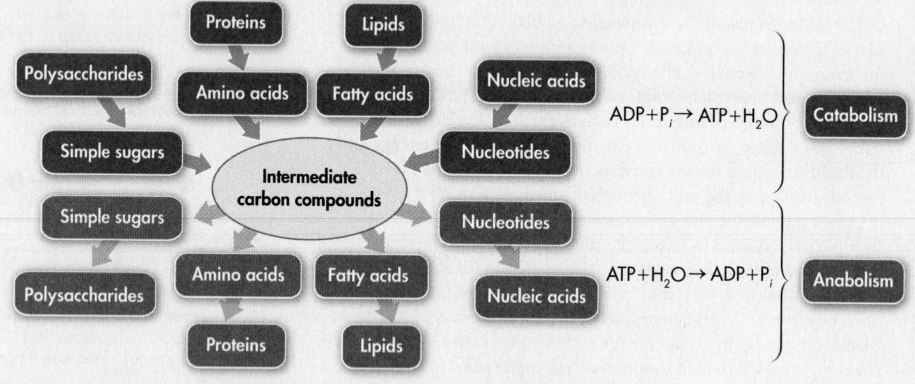

Figure 24.27 Catabolism and Anabolism
Simple compounds produced by catabolism are used in the synthesis reactions of anabolism.
Apply Concepts ***What part of metabolism releases energy? What part uses energy?***

Differentiated Instruction

LPR LESS PROFICIENT READERS Have students create separate cards for all the steps in glucose breakdown that are shown in Figure 24.25. Students should expand on the shorthand version in the visual, using words or chemical structures, so that each step makes sense to them. They can shuffle their cards and have another student put them into the correct sequence.

L1 STRUGGLING STUDENTS Have students create a T-chart to organize the similarities and differences between catabolism and anabolism in their own words.

ELL ENGLISH LANGUAGE LEARNERS Supply students with enlarged copies of Figure 24.27 and Figure 24.28. Pair beginning and intermediate speakers with advanced speakers of the same language. Encourage advanced speakers to explain points of confusion to their partners in their native language, and allow students to make native language notes on their handouts.

The Nitrogen Cycle

How do nitrogen-fixing bacteria provide plants with a usable form of nitrogen?

You have learned that the biological molecules taken into an organism's body as nutrients in food are broken down during catabolism. Food contains carbohydrates, proteins, lipids, nucleic acids, vitamins, and minerals. These nutrients are composed mainly of carbon, hydrogen, and oxygen atoms. Many biological compounds, such as proteins, contain nitrogen as well. Although Earth's atmosphere is 78 percent nitrogen gas, no animals and only a few plants can use this form of nitrogen to make nitrogen-containing compounds. However, certain bacteria can convert nitrogen gas into usable forms in a process called nitrogen fixation. **Nitrogen-fixing bacteria reduce atmospheric nitrogen ($N_2(g)$) to ammonia ($NH_3(g)$), a water-soluble form of nitrogen that can be used by plants.** In soil and biological fluids, most ammonia is present as ammonium ions.

Plants incorporate ammonia into biological nitrogen compounds such as proteins, nucleic acids, and ATP. Because animals cannot synthesize these compounds, they get them by eating plants or other animals that eat plants. When these plants and animals die, they decay with the aid of bacteria. Decaying matter returns nitrogen to the soil as ammonia, nitrite ions (NO_2^-), or nitrate ions (NO_3^-). Moreover, some nitrogen gas is returned to the atmosphere. This flow of nitrogen between the atmosphere and Earth and its living creatures is the nitrogen cycle, shown in Figure 24.28.

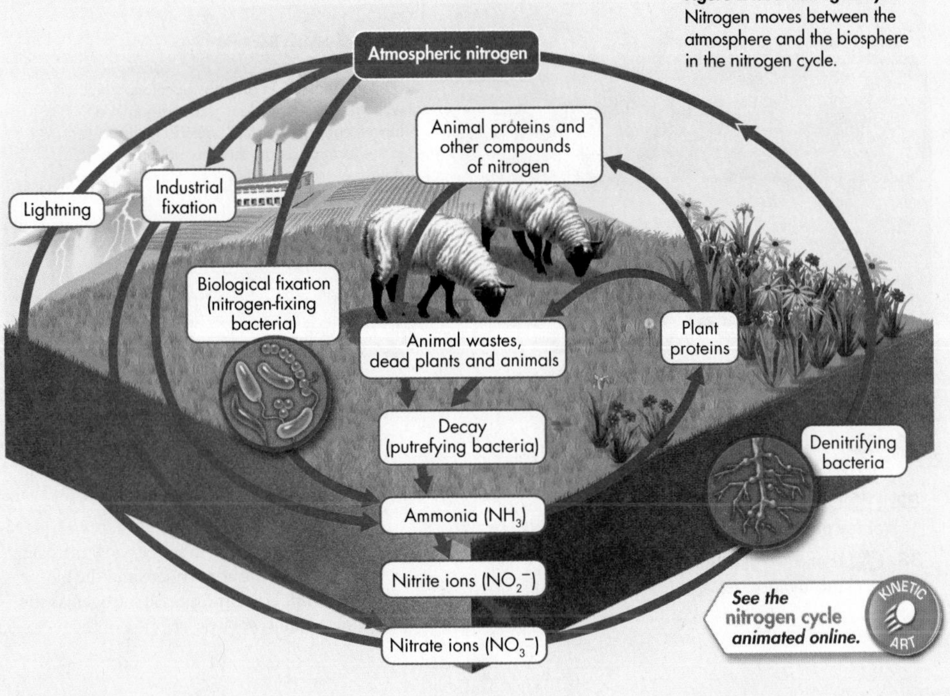

Figure 24.28 Nitrogen Cycle
Nitrogen moves between the atmosphere and the biosphere in the nitrogen cycle.

Explain

The Nitrogen Cycle

USE VISUALS Direct students' attention to Figure 24.28. Point out that nitrogen often is the limiting factor in how many plants can grow to maturity in a system. Explain that some plants are adapted to low-nitrogen environments, where faster-growing plants are unable to obtain adequate nitrogen to take over the habitat. Tell students that release of extra nitrogen from fertilizers would allow the faster-growing plants to outcompete the low-nitrogen specialists, eventually eliminating the low-nitrogen plants from the habitat.

UbD Check for Understanding

The Essential Question What are the functions of anabolism and catabolism in a cell?

Assess knowledge of metabolism by allowing students to write or verbalize a scenario that depicts the functions of anabolism and catabolism.

ADJUST INSTRUCTION If students are having difficulty distinguishing between catabolism and anabolism, have them read the sections titled *Catabolism* and *Anabolism* again. Then remind students that catabolic reactions make free energy available in the form of ATP. Anabolic reactions use energy from ATP to drive the formation of other products.

Answers

FIGURE 24.27 Catabolism releases energy; anabolism uses energy.

LESSON 24.6

Evaluate

Informal Assessment

Display the diagram of the carbon cycle, from page 840, and the diagram of the nitrogen cycle side-by-side. Ask students to compare the two cycles. **Ask** Where does energy, in either diagram, come into play? What is its source? *(Students should be able to indicate which pathways release energy and which require energy to proceed.)* Then have students complete the Lesson 24.6 Lesson Check.

Reteach

Draw a schematic diagram on the board that illustrates the exchange of energy between anabolic and catabolic reactions. Connect the terms *ATP* and *ADP* by up and down arrows. Label the up arrow *anabolism*. In anabolic reactions, small molecules are put together to form larger ones. Explain that these nonspontaneous processes must be coupled to the free energy of ATP hydrolysis; thus, ADP is produced during anabolic reactions. Label the down arrow *catabolism*. Explain that in catabolic reactions, large molecules are broken down; catabolic reactions most often release energy that can be used to drive the synthesis of ATP.

Biological Nitrogen Fixation There are two types of nitrogen-fixing bacteria: free-living and symbiotic. Free-living bacteria lead an independent existence in soil. Symbiotic bacteria, such as *Rhizobium*, live in a mutually beneficial arrangement with plants. Symbiotic bacteria live in nodules on the roots of legumes, such as alfalfa, clover, peas, and beans. These root nodules are shown in Figure 24.29. Soil fertility can be improved by plowing nitrogen-rich legumes back into the ground instead of harvesting them.

Industrial Nitrogen Fixation Modern agriculture uses an enormous quantity of nitrogen, which plays a role in the nitrogen cycle. For the past several years, the daily amount of atmospheric nitrogen fixed by industrial processes in the production of fertilizers has probably exceeded the amount fixed by living organisms in Earth's forests and oceans. Nitrogen fertilizers enter the biosphere when they are taken up by plants. In addition, a small amount of atmospheric nitrogen is fixed by lightning discharges, which produce the soluble nitrogen oxides (NO, NO_2, N_2O_4, and N_2O_5).

Figure 24.29
Nitrogen-Fixing Bacteria
The bumps on these plant roots contain bacteria that live in a symbiotic relationship with the plant. The plant gets nitrogen in a form it can use, and the bacteria get food in the form of sugars that the plant makes during photosynthesis.

24.6 LessonCheck

35. **Explain** What is the role of ATP in energy production and energy use in living cells?

36. **Explain** What is the function of catabolism in the cells of living organisms?

37. **Sequence** How does anabolism make use of the products of catabolism?

38. **Identify** What form of nitrogen is supplied to plants by nitrogen-fixing bacteria?

39. **Identify** How many moles of ATP are formed from the complete oxidation of one mole of glucose in a cell that uses oxygen?

BIGIDEA
CHEMISTRY AS THE CENTRAL SCIENCE

40. Write a paragraph that describes how nitrogen moves between the atmosphere and the biosphere. (*Hint:* Use atmospheric nitrogen as the starting point in your description of the process.)

Lesson Check Answers

35. ATP transmits energy in the body. Energy is captured when ATP forms and released when ATP is hydrolyzed.

36. Catabolism is the subset of metabolic reactions that involves degradation of complex molecules and production of energy.

37. Catabolic reactions provide the energy and the building blocks for the construction of new compounds needed by the cell in anabolism.

38. Atmospheric nitrogen is converted to ammonia and ammonium ions.

39. as many as 38 moles of ATP

40. The nitrogen cycle is the flow of nitrogen between the atmosphere and biosphere. Atmospheric nitrogen is converted to ammonia and ammonium ions by nitrogen-fixing bacteria. Plants and then plant-eating organisms synthesize nitrogen-containing compounds. The decay of these plants and animals returns nitrogen to the atmosphere.

CHEMISTRY & YOU: TECHNOLOGY

DNA Testing

Would you want to know if you are predisposed to a treatable disease, such as diabetes or breast cancer? What about an incurable disease, such as Huntington's disease? The answers to these questions, and many more, are only a drop of saliva away.

In the past few years, researchers have learned how to test for hundreds of genetic disorders using genetic testing. Genetic testing involves analyzing a person's DNA to determine changes in genes that may indicate a specific disorder. The DNA used for the tests can be obtained from a sample of blood, urine, or even saliva. In fact, today you can mail a saliva sample to a lab, and then view your test results online!

Genetic testing can be valuable in determining a person's likelihood of developing a treatable disease. However, learning of a genetic predisposition for a disease that currently has no treatments or cures can cause anxiety or despair. In addition, there are many ethical concerns involved in genetic testing, including who gets access to the test results and how many actual tests are performed on the DNA. The information in DNA is much greater than just the specific disease a lab could test for. Because of these concerns, it is important to consider all the consequences of knowing your genetic information before getting tested.

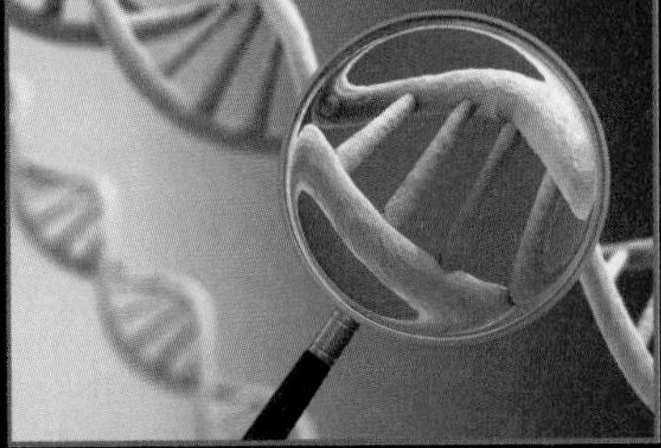

DNA MUTATIONS Alterations in the order of the bases in DNA cause gene mutations, which can lead to diseases.

TESTING FROM HOME At-home tests involve collecting saliva or blood samples. The sample is then sent to a lab, which analyzes the DNA for mutations known to correspond to different diseases and disorders.

Take It Further

1. **Identify** What are some of the possible benefits of genetic testing? What are some of the possible risks?
2. **Form an Opinion** Would you want to be tested for potential genetic disorders? Why or why not?

CHEMISTRY & YOU Have students read the passage and study the photos showing applications of DNA testing. Lead them in a discussion of current applications of DNA testing. Then, have students look back at DNA Typing on page 859 and have them compare and contrast DNA typing to testing individuals for genetic disorders.

Explain

USE AN ANALOGY Explain that DNA testing is similar to trying to solve a coded message. Like a coded message, DNA contains a lot of details that are not easily understood. Similar to the way that analyzing a code can help reveal those details, testing DNA and analyzing it can reveal a lot of information about an organism.

Extend

Connect to BIOLOGY

DNA testing is an important tool that biologists use to analyze the evolutionary path between ancient and modern species. By analyzing the DNA remains of ancient organisms, scientists can compare them with modern species to determine differences and similarities. Have students research the use of DNA testing in this manner, and have them identify a species that has been recently reclassified or an evolutionary relationship that has been discovered as a result of DNA testing. Have students share their findings with the class.

21st Century Learning To be successful in the 21st century, students need skills and learning experiences that extend beyond subject matter mastery. The project below helps students build the following 21st Century Skills: *Global Awareness; Financial, Economic, Business, and Entrepreneurial Literacy; Civic Literacy; Critical Thinking and Problem Solving; Communication and Collaboration; Flexibility and Adaptability; Social and Cross-Cultural Skills.*

DEBATING DNA Pose the following challenge to your students. *Suppose a bill has been introduced in Congress to require genetic testing of all newborn infants. Would you be in favor of its passage or against it? Debate the issue with others in your class.*

- Divide the class into teams of two students each.
- Each team should research both sides of the issue and should be prepared to argue in favor of either position and against the opposing viewpoint.
- Students should consider both the societal effects and the moral considerations of DNA testing, as well as financial effects.

Answers

TAKE IT FURTHER

1. Answers will vary. One benefit is that people can take preventative measures. The results must be kept private, though. People might not qualify for insurance coverage or employment opportunities could be adversely affected if the results are revealed.
2. Answers will vary.

Evaluate

Review and Assessment Materials

CHAPTER PROBLEM SET Have students practice more biochemistry problems using the Chapter 24 Online Problem Set.

DIRECTED VIRTUAL LAB Have students complete the Virtual Lab *Food Calorimetry* to help reinforce the understanding of the amounts of energy that humans can receive when eating different types of foods. In the post-lab exercise, students will analyze the data and draw conclusions based on the results of the experiment.

KINETIC ART Direct students to the Kinetic Art animations *Protein Folding* and *The Nitrogen Cycle*, and have them answer the questions that accompany each animation.

Study Tip

STUDY WITH A FRIEND Offer students the following advice: Quiz each other, compare notes, and discuss discrepancies. Organize the types of biomolecules you learned about in this chapter. You could make a chart that compares the different types of chemicals, or you could make a flowchart of processes. Work together to make sure you include all the important points.

24 Study Guide

BIGIDEA
CHEMISTRY AS THE CENTRAL SCIENCE

There are four main types of biological molecules. Most carbohydrates are polymers that release energy when broken down. Proteins are polymers of amino acids and are needed for most chemical reactions in cells. Lipids are water-insoluble and can be used for long-term energy storage. Nucleic acids are polymers of nucleotides. The nucleic acid DNA carries the instructions for a cell. Catabolic reactions break down biological molecules to provide energy and building blocks for the cell. Anabolic reactions build biological molecules to store energy and make new cell parts.

24.1 A Basis for Life

The two major cell types that occur in nature are prokaryotic cells and eukaryotic cells.

Photosynthesis uses sunlight to reduce CO_2 to compounds that contain C—H bonds, mainly in the form of glucose.

- photosynthesis (839)

24.2 Carbohydrates

Most carbohydrates have the general formula $C_n(H_2O)_n$.

- carbohydrate (841)
- monosaccharide (841)
- disaccharide (842)
- polysaccharide (843)

24.3 Amino Acids and Their Polymers

An amino acid has a carboxyl group, an amino group, a hydrogen atom, and an R group bonded to a central carbon atom.

Differences in the amino acid sequence result in differences in the properties of peptides.

Enzymes increase reaction rates.

- amino acid (844)
- peptide (845)
- peptide bond (845)
- protein (845)
- enzyme (847)
- substrate (847)
- active site (847)

24.4 Lipids

Lipids tend to dissolve readily in organic solvents, whereas carbohydrates and proteins tend to dissolve in water.

- lipid (850)
- triglyceride (850)
- saponification (851)
- phospholipid (851)
- wax (852)

24.5 Nucleic Acids

DNA stores information needed to make proteins and governs the reproduction of cells. RNA transmits information stored in DNA during protein synthesis.

A sequence of three bases of DNA is required to specify one amino acid in a peptide.

Gene mutations occur when one or more nucleotides in DNA are substituted, added, or deleted.

Examples of DNA technology include DNA typing, production of bacteria that make human proteins, genetically modifying foods and animals, and cloning.

- nucleic acid (854)
- nucleotide (854)
- gene (856)

24.6 Metabolism

In living cells, ATP is the energy carrier between the spontaneous reactions that release and the nonspontaneous reactions that use energy.

The degradation of biological molecules during catabolism provides the energy and the building blocks for making new compounds. In anabolism, new compounds needed for cellular life and growth are made from the products of catabolism.

Nitrogen-fixing bacteria reduce atmospheric nitrogen to ammonia, a water-soluble form of nitrogen that can be used by plants.

- adenosine triphosphate (ATP) (862)
- metabolism (863)
- catabolism (863)
- anabolism (864)

UbD Performance Tasks

DNA TECHNOLOGY Separate the class into two debate teams. Have them spend a class period debating DNA technology. One of the teams should create an argument for the technology, and the other should create an argument against the technology. Make sure each member of each team has a clearly defined role. Roles can be leader, researcher, presenter, artist, organizer, strategizer, and so on.

CATABOLISM VS. ANABOLISM Have students create a unique diagram that correctly depicts the functions of anabolism and catabolism. Diagrams should include how carbohydrates, amino acids, lipids, and nucleic acids relate to each function.

Focus on ELL

5 ASSESS UNDERSTANDING Have students create a portfolio for this chapter. Include various types of assignments, based on each student's language proficiency.

BEGINNING

LOW Include lists, models or pictures of models, or diagrams created by the student.

HIGH Include copies of the graphic organizers used to take notes or show how carbohydrates, amino acids, lipids, and nucleic acids relate to one another.

INTERMEDIATE

LOW Include any prereading predictions about lesson topics, as well as the results of those predictions.

HIGH Include any lab write-ups that were assigned during the chapter.

ADVANCED: LOW/HIGH Include an essay on the pros and cons of two uses of DNA technology.

24 Assessment

★ Solutions appear in Appendix E

Lesson by Lesson

24.1 A Basis for Life

41. What is the main difference between a prokaryotic and a eukaryotic cell?

★**42.** Explain what happens in photosynthesis.

43. Write a balanced equation for the complete oxidation of glucose.

44. Describe three organelles found in eukaryotic cells. Give a function of each organelle.

24.2 Carbohydrates

45. Name two important monosaccharides.

46. Where in nature are glucose and fructose found?

★**47.** How does the carbonyl functional group differ in glucose and fructose?

48. Which monosaccharides combine to form the disaccharide sucrose?

49. What is the product of the complete hydrolysis of starch? Of glycogen?

50. What product is formed when cellulose is broken down?

24.3 Amino Acids and Their Polymers

51. What is the name given to the bond connecting two amino acids in a peptide chain?

52. How many peptide bonds does the tripeptide Ser—Gly—Phe have? Explain.

★**53.** Describe two common patterns found in the folding of protein chains.

54. Are the structures of the following two tripeptides the same? Explain.

a. Ala—Ser—Gly **b.** Gly—Ser—Ala

55. Describe the function of an enzyme.

56. What is an enzyme-substrate complex? How does it form?

24.4 Lipids

57. Distinguish between a fat and an oil.

58. What is a triglyceride?

59. What is a soap?

60. Draw structural formulas for the products of the complete hydrolysis of tristearin.

61. Draw a simple representation of a lipid bilayer.

62. What two types of compounds combine to form a wax?

24.5 Nucleic Acids

63. What two types of nucleic acids do cells have?

★**64.** What are the components of a nucleotide?

65. What is the structural difference between the sugar unit in RNA and the sugar unit in DNA?

66. What type of bonding helps hold a DNA double helix together?

67. Which of the following base pairs are found in a DNA molecule: A—A, A—T, C—G, G—A, A—U, T—U?

68. How many bases specify an amino acid in the genetic code?

★**69.** What are the consequences of a substitution of one base for another in DNA? Give an example.

★**70.** What is the basis for identifying an individual by DNA profiling?

71. What is recombinant DNA?

24.6 Metabolism

72. Write an abbreviated, balanced equation for the hydrolysis of ATP to ADP.

73. Where do the complex biomolecules your body degrades during catabolism come from?

★**74.** How are catabolism and anabolism related?

75. Describe the nitrogen cycle in your own words.

76. What is the source of raw materials used in anabolic reactions?

77. What is meant by industrial nitrogen fixation? What is produced during this process?

Understand Concepts

★**78.** The formula for palmitic acid is $CH_3(CH_2)_{14}CO_2H$. A popular soap is mostly sodium palmitate. Draw a structural formula for sodium palmitate.

Evaluate

Answers

LESSON 24.1

41. A eukaryotic cell has a cell membrane and organelles. Prokaryotic cells have a cell membrane, but lack organelles.

42. Photosynthetic organisms use the sun's energy to synthesize carbon compounds from CO_2 and H_2O

43. $C_6H_{12}O_6 + 6O_2 \rightarrow 6CO_2 + 6H_2O$ + energy

44. *Sample answer:* nucleus (cellular reproduction); mitochondria (cellular energy production); lysosomes (digestion of substances)

LESSON 24.2

45. glucose and fructose

46. Glucose is found in blood, corn, and grapes. Fructose is found in honey and many fruits.

47. Glucose is an aldehyde; fructose is a ketone.

48. glucose and fructose

49. glucose; glucose

50. glucose

LESSON 24.3

51. peptide bond

52. two; one between the first and second amino acids, one between the second and third amino acids

53. Peptide chains fold into helixes or into pleated sheets in which peptide chains lie side by side.

54. No; Both peptides contain alanine, serine, and glycine, but the sequences are different.

55. Enzymes catalyze biological reactions.

56. It is a complex formed when an enzyme and substrate bind to each other.

LESSON 24.4

57. At room temperature, animal fats are solid; plant oils are liquid.

58. triester of glycerol and fatty acids

59. alkali metal salt of a fatty acid

60.

$$\begin{array}{l} CH_2OH \\ | \\ CHOH \\ | \\ CH_2OH \end{array} + 3CH_3(CH_2)_{16}-\overset{\overset{\displaystyle O}{\|}}{C}-O^-Na^+$$

Glycerol Sodium stearate

61.

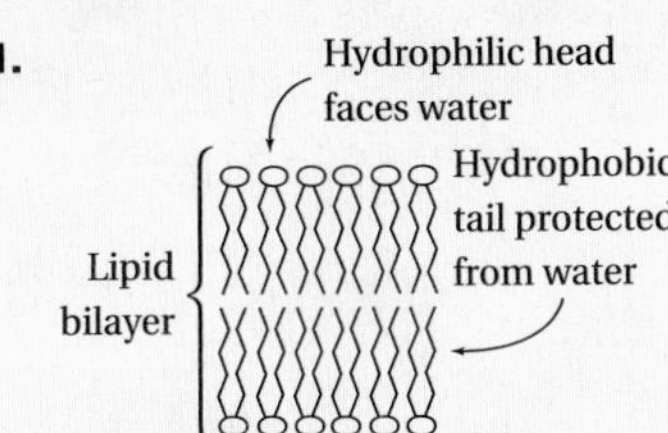

62. long-chain fatty acid and long-chain alcohol

LESSON 24.5

63. DNA and RNA

64. phosphate group, sugar unit, nitrogen base

65. Ribose (in RNA) has one more oxygen.

66. hydrogen bonding

67. A–T and C–G

68. Three

69. A substitution of one base for another may have no effect because the amino acid specified by the DNA code is not changed. Or a substitution of one base for another can result in a gene mutation in which the amino specified by the DNA code is changed.

70. Each individual's DNA is unique.

71. altered DNA that contains a foreign gene

LESSON 24.6

72. ATP + $H_2O \rightarrow$ ADP + P_i

73. foods or existing body tissues

74. In catabolism, molecules are broken down and energy is captured. These products and this free energy are used to make biomolecules in anabolism.

75. N_2 is converted to NH_3, which is taken up by plants, which are eaten by animals. Decaying organisms return N_2 to the air.

76. products of catabolism

77. Atmospheric nitrogen is reduced to ammonia by industrial processes.

UNDERSTAND CONCEPTS

78.

$$CH_3(CH_2)_{14}-\overset{\overset{\displaystyle O}{\|}}{C}-O^-Na^+$$

sodium palmitate

ASSESSMENT

Evaluate

Answers

79. The negatively charged hydrophilic phosphate heads interact favorably with water.

80. a. Arg-Gly-Cys-Asn **b.** Arg-Gly-Cys-Asn (same)

81. a. triglyceride **b.** glucose **c.** dipeptide

82. C-G-x-C-C-x-T-C-A(or G); There are multiple codes for the same amino acid.

83. G-C-T-A-G-G-T

84. a. amino acid
b. monosaccharide (simple sugar)
c. nucleotide

85. permit passage of nutrients into cells

86. No; three code words specify the termination of a peptide chain.

87. many applications in medicine and agriculture

88. 30.5 kJ/mol × 38 mol = 1159 kJ stored in ATP; 1159 kJ/2.82 × 10^3 kJ × 100% = 41.1%

89. about 275 mol ATP

90. The shape of the cellulose and starch molecules are different; humans lack enzymes necessary to cleave cellulose to glucose monomers.

91. A substrate molecule or molecules binds to the active site of the enzyme to form an enzyme-substrate complex.

92. Coenzymes must be present for an enzyme-catalyzed reaction to occur.

93. evidence from fossilized bacteria

94. Mutations could occur that would change how the organism constructs a protein.

95. Photosynthesis provides the carbon compounds that plants and animals need to exist.

96. monosaccharides are single simple sugars; examples are glucose, fructose. Disaccharides are two linked monosaccharide units; an example is sucrose. Polysaccharides are many linked monosaccharide units; examples are starch, glycogen, cellulose.

97. A carboxyl group of one amino acid and amino group of another amino acid undergo condensation to form a peptide bond.

98. hydrogen bonding between A-T (A-U in RNA) and G-C in DNA; helps hold DNA in double helix

79. Why are the hydrophilic heads located on the outsides of the cell membrane?

80. Consider the following sequence of DNA: GCC–CCA–ACG–TTA.

a. Using the code words for amino acids in Table 24.2 on page 857, write the amino acid sequence formed by translation of the DNA sequence into a peptide.

b. What amino acid sequence would result from the substitution of adenine (A) for the second cytosine (C)?

81. Identify or classify each of the following biological molecules.

a.

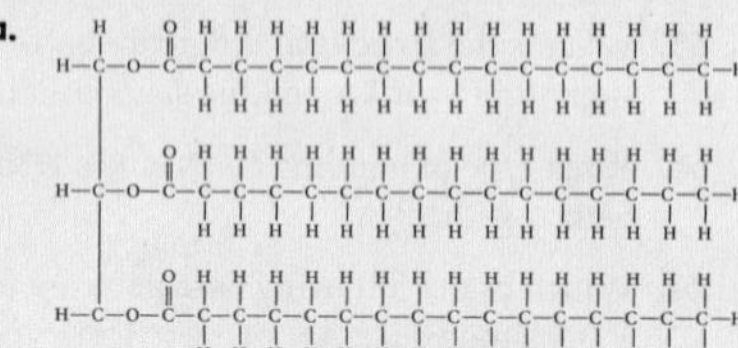

b.

c.

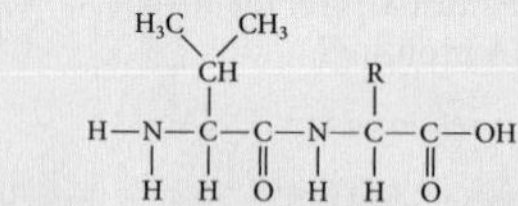

***82.** Use Table 24.2 on page 857 to write a base sequence for DNA that codes for the tripeptide Ala—Gly—Ser. Why might your answer be different from the answers of your classmates?

83. A segment of a DNA strand has the following base sequence: CGATCCA. Write the base sequence that would be found on the other strand in the double helix.

84. Which type of monomer produces each of the following polymers?

a. protein
b. polysaccharide
c. nucleic acid

85. What is one function of membrane proteins?

***86.** Does every code word in DNA specify an amino acid in protein synthesis? Explain.

87. What are some of the outcomes of recombinant DNA research?

***88.** The complete oxidation of glucose releases 2.82 × 10^3 kJ/mol of energy, and the formation of ATP from ADP requires 30.5 kJ/mol. What percent of the energy released in the complete oxidation of glucose is captured in the formation of ATP?

***89.** An average adult expends about 8400 kJ of energy every day. How many moles of ATP must be converted to ADP to provide this amount of energy?

90. Why can't humans digest cellulose, considering that it is made of the same monomers as starch?

91. How is an enzyme-substrate complex formed in the enzyme's active site?

92. What role do coenzymes play in metabolism?

93. Suggest a reason why prokaryotic cells are thought to be more ancient than eukaryotic cells.

94. What are the possible consequences of an error in DNA sequence?

95. Explain why photosynthesis might be considered the most important chemical process on Earth.

***96.** Describe the differences among monosaccharides, disaccharides, and polysaccharides. Give examples of each type of carbohydrate.

97. Describe how amino acids join to form a peptide bond.

***98.** What is base pairing? How does base pairing relate to the structure of DNA?

Think Critically

99. Explain Interpret this statement: "Carbon dioxide is an energy-poor molecule, but glucose is an energy-rich molecule."

100. Sequence In the DNA double helix, where are the base pairs located in relation to the backbone structure: inside the double helix or outside the double helix? What must happen before the protein-making machinery of the cell can "read" the code words formed by the DNA bases?

101. Explain Which type of gene mutation do you think will do more damage to an organism: a substitution mutation in which one base is substituted for another base or an addition mutation in which a base is added to a sequence of bases? Explain.

THINK CRITICALLY

99. CO_2 cannot produce energy through oxidation. Glucose can be oxidized to CO_2 and produce energy.

100. Inside the helix; strands must unwind

101. Substitution of one base could result in no change in the amino acid sequence of a protein (if there are multiple code words for the amino acid), or a different amino acid, which could be damaging or helpful to the organism. A base added to the start of the gene would result in a nonsense amino acid sequence for the protein produced by the gene. A base added to the end of the gene could make the termination codon untranslatable, and the protein product might not terminate. Therefore an addition mutation is more likely to cause damage.

102. **Predict** Suggest a reason why a bean plant might not grow well if planted in sterilized soil.

103. **Identify** What class of polymer is formed from each of the following monomers?

   a. amino acids
   b. monosaccharides
   c. nucleotides

*104. **Interpret Diagrams** Complete the following equation by drawing the structural formulas of the products:

(Structural formulas: glucose, ring with CH_2OH, H, O, OH, HO groups + fructose, ring with CH_2OH, O, H, HO, CH_2OH, OH groups)

Enrichment

105. **Compare and Contrast** Compare the structure of a DNA nucleotide with an RNA nucleotide.

106. **Compare** Describe the structural features all amino acids have in common. What structural features differ among the amino acids?

*107. **Identify** Using structural formulas, write a chemical equation for the formation of a dipeptide from two amino acids. What functional group does the reaction create?

108. **Cause and Effect** Explain why cell growth stops when the dietary intake of nutrients is insufficient.

109. **Interpret Diagrams** The following compound is hydrolyzed by boiling with sodium hydroxide. What are the saponification products?

$$CH_2-O-\overset{O}{\overset{\|}{C}}-(CH_2)_{14}CH_3$$
$$CH-O-\overset{O}{\overset{\|}{C}}-(CH_2)_{12}CH_3$$
$$CH_2-O-\overset{O}{\overset{\|}{C}}-(CH_2)_{16}CH_3$$

110. **Identify** What causes the spontaneous formation of a lipid bilayer?

111. **Explain** Peptide chains fold and bend into three-dimensional shapes. Suggest how a peptide chain is held in this 3-D shape.

*112. **Interpret Diagrams** A sequence of nine bases in a gene codes for the amino acid sequence Trp-Met-Met. What is the sequence of bases in this DNA fragment? Use Table 24.2 on page 857 to help you. Could you determine the base sequence for certain if the amino acid sequence were Trp-Met-Leu? Why or why not?

Write About Science

113. **Explain** Write a paragraph explaining how cells are able to selectively absorb certain ions and molecules while excluding other materials.

114. **Connect to the BIGIDEA** Choose one of the biological molecules discussed in this chapter. Write a paragraph describing how the elements and structure of the molecules relate to the function of the molecules.

CHEMYSTERY

Phenyl-what?

Phenylalanine is an amino acid, a monomer of proteins. So, how could phenylalanine be hazardous to your health? For most people, phenylalanine is not a health concern. But it is a concern for people who have the genetic disorder phenylketonuria, or PKU. Normally, phenylalanine is converted by the body to another amino acid, tyrosine. People with PKU lack the enzyme that is required to convert phenylalanine to tyrosine. When this happens, phenylalanine accumulates to toxic levels in the body. PKU is treated by eating a diet low in phenylalanine (the body still needs a small amount phenylalanine).

115. **Infer** Given that phenylalanine is an amino acid, which types of foods would you expect to be high in phenylalanine?

116. **Connect to the BIGIDEA** Why does a nutritionist need to understand chemistry?

ASSESSMENT

112. A-C-C-T-A-C-T-A-C; no, Leu has 6 possible code words.

WRITE ABOUT SCIENCE

113. The cell wall is selectively permeable meaning that only certain molecules and ions can enter or leave the cell. The three ways in which molecules and ions can enter or leave the cell are passive transport, active transport, and in a vesicle. Small molecules or ions cross by transport. Larger molecules cross within a vesicle. Answers should include a brief description of these three ways.

114. Students can write about carbohydrates, amino acids, lipids, nucleic acids, or ATP. Answers should relate structure and function.

CHEMYSTERY

USE AN ANALOGY Have students read over the CHEMystery. Engage them in a discussion about what they have learned about amino acids and proteins. Remind students that proteins are needed for almost all chemical reactions in the body. Help students understand phenylketonuria (PKU) by using an analogy of getting a sunburn. Explain that your body needs sunlight to produce vitamin D, but too much sunlight can damage your skin. Similarly, your body needs the amino acid phenylalanine, but for people with the PKU disorder, too much phenylalanine is harmful. It can lead to poor development and brain disorders.

CHEMYSTERY ANSWERS

115. foods that contain protein such as poultry, eggs, beans, and brown rice

116. Nutritionists treat and prevent illnesses by making modifications to a patient's diet. They need to know how the chemicals in food affect your health. They also need to know which biological molecules are contained in certain foods and how these biological molecules affect your health.

Answers

102. Bacteria necessary for nitrogen fixation might be killed by sterilization of the soil.

103. a. polypeptides and proteins
   b. polysaccharides
   c. polynucleotides (RNA and DNA)

104. (Structural formula: glucose ring with CH_2OH, H, O, OH, HO groups linked through O to fructose ring with $HOCH_2$, O, H, HO, CH_2OH, OH groups) $+ H_2O$

ENRICHMENT

105 A nucleotide consists of a phosphate group, a five-carbon sugar unit, and a nitrogen base. In DNA, the sugar is deoxyribose; in RNA, the sugar is ribose. In DNA, the bases may be A, G, T, or C; in RNA, the bases may be A, G, U, or C.

106 Every amino acid consists of a central carbon atom bonded to an amine group, a carboxyl group, a hydrogen atom, and a side chain. Different amino acids have different side chains.

107. An amide goup is formed by the reaction.

$$H_2N-\underset{H}{\overset{R}{C}}-\overset{O}{\overset{\|}{C}}-OH + H-\underset{H}{N}-\underset{H}{\overset{R}{C}}-\overset{O}{\overset{\|}{C}}-OH \longrightarrow$$

$$H_2N-\underset{H}{\overset{R}{C}}-\overset{O}{\overset{\|}{C}}-\underset{H}{N}-\underset{H}{\overset{R}{C}}-\overset{O}{\overset{\|}{C}}-OH + H_2O$$

108. Nutrients supply carbon compounds, ions, and energy needed for growth. The (insufficient) food primarily is used for energy production necessary to sustain the cell.

109. glycerol + a soap

110. Phospholipids have hydrophobic tails that exclude water and polar heads that attract water.

111. hydrogen bonds between adjacent parts of folded chains and covalent bonds between side-chain groups of cysteine

Evaluate

Answers

CUMULATIVE REVIEW

117. The particles in real gases have volumes and are attracted to one another.

118. a and b are electrolytes. c and d are nonelectrolytes.

119. a. 0.519°C **c.** 0.238°C
b. 0.209°C **d.** 0.227°C

120. 1.71×10^1 kJ heat absorbed

121. Besides being composed of highly combustible material, the needles have a large surface area that increases their rate of combustion.

122. The equivalents of acid and base must be equal.

123. a. 4.15 **c.** 12.79
b. 5.26 **d.** 10.36

124. a. fluorine **b.** oxygen **c.** chlorine

125. Oxidation always occurs at the anode. In the voltaic cell the anode is the negative electrode.

126. The nickel wire would be silver-plated.

127. a. Cu **c.** Ag **e.** Cu
b. Ni **d.** Fe **f.** Cu

128. a. $CH_3—CH_2—CH_2—CH_2—CH_2—CH_2—CH_3$

b.
$$\begin{array}{c}CH_3\\ |\\ CH_3—CH—CH{=}CH—CH_2—CH_3\end{array}$$

c.

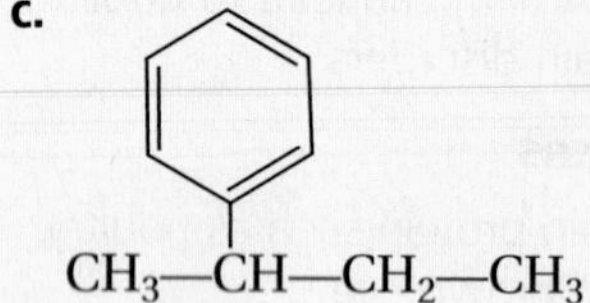

$CH_3—CH—CH_2—CH_3$

d. CH_2CH_3 (benzene ring) CH_2CH_3

129. a. 1-pentene
b. cyclononane
c. hexane
d. decane

130. a. cyclopentane
b. 2-methyl-2-propanol
c. 3-pentanone

131. a. $CH_3CO_2CH_3$ **b.** $CH_3CHOHCOOH$

132. a. polytetrafluoroethene (non-stick cookware)
b. polyethylene (plastic bags and containers)
c. polyvinyl chloride (PVC pipe)
d. polystyrene (electric insulation)

133.

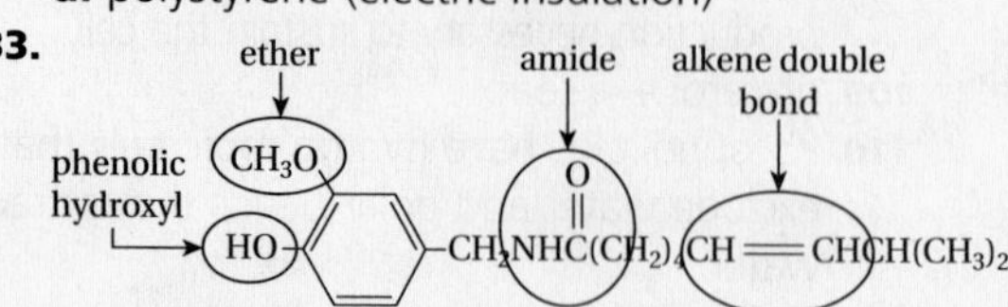

Cumulative Review

117. Describe two factors that cause real gases to depart from the ideal gas law.

118. Characterize these compounds as electrolytes or nonelectrolytes.
a. NaCl **c.** CCl_4
b. $CuSO_4$ **d.** H_2O

***119.** Calculate the boiling-point elevation for these aqueous solutions.
a. 0.507*m* NaCl **c.** 0.155*m* $CaCl_2$
b. 0.204*m* NH_4Cl **d.** 0.222*m* $NaHSO_4$

***120.** How much heat (in kJ) is released or absorbed when 0.265 mol of sodium bicarbonate is decomposed according to the reaction below?

$$2NaHCO_3(s) \rightarrow Na_2CO_3(s) + H_2O(g) + CO_2(g)$$
$$\Delta H = 129 \text{ kJ}$$

121. Explain why the needles on a dried-out fir tree can burn with almost explosive rapidity.

122. What must be true at the end point of an acid-base titration?

123. Calculate the pH of each of the following solutions:
a. $[H^+] = 7.0 \times 10^{-5}M$
b. $[OH^-] = 1.8 \times 10^{-9}M$
c. $[OH^-] = 6.1 \times 10^{-2}M$
d. $[H^+] = 4.4 \times 10^{-11}M$

124. Identify the oxidizing agent in each reaction.
a. xenon + fluorine ⟶ xenon tetrafluoride
b. sulfur + oxygen ⟶ sulfur trioxide
c. gaseous chlorine + aqueous sodium bromide ⟶ aqueous bromine + aqueous sodium chloride

125. At which electrode in a voltaic cell does oxidation always occur? What is the charge on this electrode?

126. What would you observe when a length of nickel wire is immersed in an aqueous solution of silver nitrate?

127. For each pair of metals shown below, decide which metal is more readily reduced.
a. Cu, Mg **c.** Ag, Sn **e.** Ni, Cd
b. Cd, Ni **d.** Zn, Fe **f.** Al, Cu

***128.** Write a molecular structure for each compound.
a. heptane **c.** 2-phenylbutane
b. 2-methyl-3-hexene **d.** 1,3-diethylbenzene

129. Name the next highest homolog of each of these compounds.
a. 1-butene **c.** pentane
b. cyclooctane **d.** nonane

***130.** Give the IUPAC name for these compounds.

a.
$$\begin{array}{c}CH_2\\ H_2C \quad\quad CH_2\\ H_2C—CH_2\end{array}$$

b.
$$\begin{array}{ccccc} & & H & & \\ & & | & & \\ H & H— & C & —H & H\\ | & & | & & |\\ H—C & — & C & — & C—H\\ | & & | & & |\\ H & & OH & & H\end{array}$$

c.

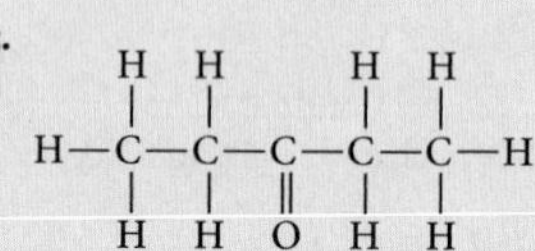

131. Write a molecular formula for each compound.
a. methyl acetate
b. 2-hydroxypropanoic acid

132. Name each polymer and state at least one of its uses.

a. $\text{+}CF_2—CF_2\text{+}_x$ **c.** $H\text{+}CH_2—CH_2\text{+}H$

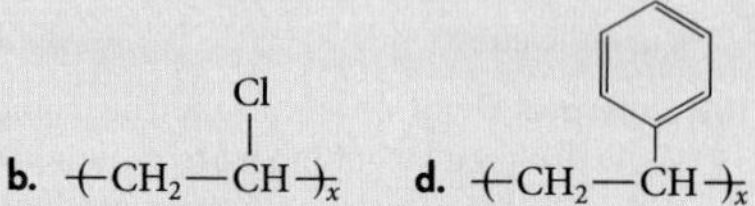

133. Capsaicin, shown below, is the major contributor to the heat of chili peppers. Circle and name the functional groups in capsaicin.

CH_3O, HO (benzene ring) $—CH_2NHC(CH_2)_4CH{=}CHCH(CH_3)_2$ (with C=O on the amide carbon)

If You Have Trouble With . . .

Question	117	118	119	120	121	122	123	124	125	126	127	128	129	130	131	132	133
See Chapter	14	15	16	17	18	19	19	20	21	21	21	22	22	23	23	23	23

Standardized Test Prep for Indiana

Tips for Success

Anticipating the Answer You can sometimes figure out the answer to a question before you look at the answer choices. Try to answer the question in your mind before you look at the answer choices. Then read the choices and choose the choice that agrees with the answer in your mind.

Select the choice that best answers each question or completes each statement.

1. What phrase best describes ATP?
   (A) energy producer
   (B) energy consumer
   (C) energy pump
   (D) energy transmitter

For Questions 2–5, match the category of organic compounds listed below that is most closely identified with each biological molecule.

I. monosaccharides
II. amino acids
III. fatty acids
IV. nucleotides

2. proteins
3. nucleic acids
4. lipids
5. carbohydrates

6. Which element is not found in amino acids?
   (A) phosphorus (C) oxygen
   (B) nitrogen (D) hydrogen

7. For any enzyme to function, the substrate must bind to the
   (A) product. (C) active site.
   (B) cofactor. (D) peptide.

Use the paragraph to answer Questions 8–10.
Because an amino acid contains both a carboxyl group and an amino group, it is amphoteric, that is, it can act as either an acid or a base. Crystalline amino acids have some properties—relatively high melting points and high water solubilities—that are more characteristic of ionic substances than of molecular substances.

8. Write an equation showing glycine acting as an acid in a reaction with water. (Glycine is the simplest amino acid. Its side chain is R = H.)
9. Write an equation showing glycine acting as a base in a reaction with water.
10. It is possible for glycine to undergo an internal Brønsted-Lowry acid-base neutralization reaction. Write the resulting structural formula. Explain how this reaction would account for the ionic properties of glycine.

For each question, there are two statements. Decide whether each statement is true or false. Then decide whether Statement II is a correct explanation for Statement I.

	Statement I		Statement II
11.	Lipids tend to be insoluble in water.	BECAUSE	Lipids have mainly nonpolar bonds.
12.	Starch and cellulose are both digestible by most organisms.	BECAUSE	Glucose is the monomer in both starch and cellulose.
13.	Many of the reactions in catabolism are oxidation reactions.	BECAUSE	Oxidation reactions tend to be energy-producing reactions.
14.	The sequence of bases in DNA contains the code for making proteins.	BECAUSE	Each pair of bases in DNA codes for a specific amino acid.

If You Have Trouble With . . .

Question	1	2	3	4	5	6	7	8	9	10	11	12	13	14
See Lesson	24.6	24.3	24.5	24.4	24.2	24.3	24.3	24.3	24.3	24.3	24.4	24.2	24.6	24.5

STP Answers

1. D
2. II
3. IV
4. III
5. I
6. A
7. C
8. $H_2NCH_2COOH(aq) + H_2O(l) \rightarrow H_2NCH_2COO^-(aq) + H_3O^+(aq)$
9. $H_2NCH_2COOH(aq) + H_2O(l) \rightarrow H_3N^+CH_2COOH(aq) + OH^-(aq)$
10. The carboxylic acid group can donate a proton to the amino group to form $H_3N^+CH_2COO^-$.
11. True, True, correct explanation
12. False, True
13. True, True, correct explanation
14. True, True, correct explanation

25 Nuclear Chemistry Planning Guide

Introducing the BIGIDEA: ELECTRONS AND THE STRUCTURE OF ATOMS

Elements are composed of particles called atoms, and every atom contains a nucleus and one or more electrons.

IASS	Lessons and Objectives	Print Resources		
		For the Student	For the Teacher	
C.2.7	**25.1** Nuclear Radiation p 876–879 **25.1.1** Explain the relationship between unstable isotopes and radioactivity. **25.1.2** Describe three main types of nuclear radiation.	Reading and Study Workbook Lesson 25.1 Lesson Assessment 25.1 p 879	Teacher Demo, p 878: An Effect of Radiation	
C.2.7, C.2.8	**25.2** Nuclear Transformations p 880–886 **25.2.1** Identify factors that determine the type of decay a radioisotope undergoes. **25.2.2** Solve problems that involve half-life. **25.2.3** Identify two ways transmutations can occur.	Reading and Study Workbook Lesson 25.2 Lesson Assessment 25.2 p 886 Small-Scale Lab: Radioactivity and Half-Lives p 887	Class Activity, p 885: Particle Acceleration	
C.2.7	**25.3** Fission and Fusion p 888–891 **25.3.1** Describe what happens in a nuclear chain reaction. **25.3.2** Describe what happens to spent fuel rods. **25.3.3** Distinguish fission reactions from fusion reactions.	Reading and Study Workbook Lesson 25.3 Lesson Assessment 25.3 p 891	Teacher Demo, p 889: Model a Chain Reaction	
C.2.7	**25.4** Radiation in Your Life p 894–897 **25.4.1** Compare three devices that are used to detect radiation. **25.4.2** Describe some practical uses of radioisotopes.	Reading and Study Workbook Lesson 25.4 Lesson Assessment 25.4 p 897 Quick Lab: Inverse-Square Relationships p 896	Teacher Demo, p 895: Background Radiation	

Assessing the BIGIDEA: ELECTRONS AND THE STRUCTURE OF ATOMS

	Essential Questions 1. What happens when an unstable nucleus decays? 2. How is the structure of atoms altered during fission and fusion? 3. How does nuclear chemistry affect your life?	Study Guide p 898 Math Tune-Up p 899 STP p 905 Reading and Study Workbook Self-Check and Vocabulary Review Chapter 25		

	PearsonChem.com	
	Editable Worksheets	Online Activities
	Teaching Resources, Lesson 25.1 Review	25.1 Lesson Overview Alpha and Beta Decay Radiation
	Teaching Resources, Lesson 25.2 Review Small-Scale Lab Manual Lab 41: Half-Lives and Reaction Rates	25.2 Lesson Overview Happy Birthday, Half-Life Using Half-Lives in Calculations
	Teaching Resources, Lesson 25.3 Review	25.3 Lesson Overview How a Nuclear Reactor Works
	Teaching Resources, Lesson 25.4 Review Lab 52: Radioactivity and Radiation Probeware Lab: Quick Lab: Studying Inverse-Square Relationships Lab Practical 25-1: Radioactivity and Radiation	25.4 Lesson Overview
	Core Teaching Resources (includes editable worksheets) • Lesson Reviews • Practice Problems • Interpret Graphs • Vocabulary Review • Chapter Quizzes and Tests • Lab Record Sheets	Chapter 25 Problem Set Exponential Decay

Materials List

For the Student

Small-Scale Lab p 887
- ruler
- penny
- graph paper

Quick Lab p 896
- flashlight
- strips of duct tape
- scissors
- poster board
- white (50 cm × 50 cm) meter ruler or tape measure flat surface long enough to hold the meter ruler
- graph paper
- light sensor (optional)

For the Teacher

Teacher Demo p 878
- photographic film
- radiation source
- key

Class Activity p 885
- reference materials and Internet access
- miscellaneous materials for models

Teacher Demo p 889
- wooden matchsticks

Teacher Demo p 895
- Geiger counter

Additional Digital Resources

Online Student Edition
Online Teacher's Edition
Exam*View Assessment Suite CD-ROM*
Classroom Resources DVD

CHAPTER 25

What's Online

CHEMISTRY TUTORIAL Students can work step-by-step with the online tutorial to use half-lives in calculations by accessing the Chapter 25 Chemistry Tutorial.

ONLINE PROBLEM SET Additional practice problems about nuclear chemistry can be found in the Chapter 25 Online Problem Set.

MATH TUTORIAL The Online Math Tutorial will help guide students through exponential decay.

DIRECTED VIRTUAL LAB Encourage students to review the properties of different types of radiation and study shielding types using the animated virtual lab for Chapter 25.

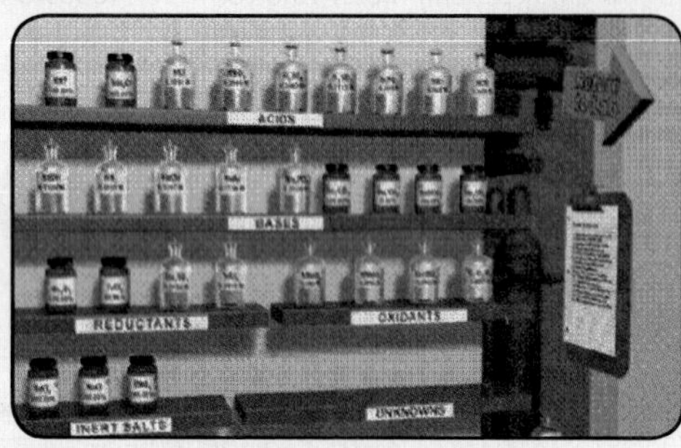

KINETIC ART Have students watch animations about alpha and beta decay and how a nuclear reactor works.

CONCEPTS IN ACTION Students can watch an interactive animation demonstrating the concept of half-life through an example of gradually eating a birthday cake.

IN ACADEMIC STANDARDS for SCIENCE

Nature of Science

- Develop explanations based on reproducible data and observations gathered during laboratory investigations.
- Clearly communicate their ideas and results of investigations verbally and in written form using tables, graphs, diagrams, and photographs.

Standard 2: Atomic Structure and the Periodic Table

C.2.7 Compare and contrast nuclear reactions. For nuclear reactions, describe how the fusion and fission processes transform elements present before the reaction into elements present after the reaction.

Also covered C.2.8

25 Nuclear Chemistry

INSIDE:

- **25.1 Nuclear Radiation**
- **25.2 Nuclear Transformations**
- **25.3 Fission and Fusion**
- **25.4 Radiation in Your Life**

PearsonChem.com

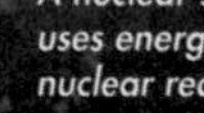

A nuclear submarine uses energy released by nuclear reactions.

874

Focus on ELL

1 CONTENT AND LANGUAGE Write the words *nucleus* and *nuclear* on the board. Ask students to recall the definition of *nucleus* based on what they learned in Chapter 4. Explain that this chapter is about the changes that can occur during nuclear reactions.

BEGINNING: LOW/HIGH Draw a diagram of a simple atom and identify the nucleus, and the subatomic particles. Use simple words and the diagram to illustrate that the composition of an atom's nucleus can change.

INTERMEDIATE: LOW/HIGH Create visual representations of the words *nuclear, radiation, transformations, fission,* and *fusion.* Then explain how the words are related.

ADVANCED: LOW/HIGH Read the lesson titles aloud. Have students predict how nuclear chemistry is related to each of these topics.

BIGIDEA

ELECTRONS AND THE STRUCTURE OF ATOMS

Essential Questions:

1. *What happens when an unstable nucleus decays?*
2. *How is the structure of atoms altered during fission and fusion?*
3. *How does nuclear chemistry affect your life?*

CHEMYSTERY

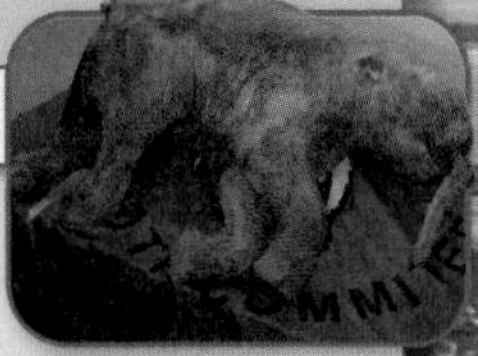

An Ice-Age Baby

In 2007, a reindeer herder in north-central Russia thought he spotted the frozen remains of a reindeer. What he found was a perfectly preserved baby wooly mammoth about the size of a large dog. The only thing missing was its shaggy coat. Scientists named the mammoth Lyuba, which means "love" in Russian.

Wooly mammoths are an extinct species related to modern elephants. These large mammals had curly tusks and long hair. They lived on Earth from about 1.8 million years ago until about 11,500 years ago. How do scientists know when Lyuba lived and died?

▶ Connect to the BIGIDEA As you read about nuclear reactions, think about which type of reaction could be used to find the age of a wooly mammoth fossil.

INDIANA'S ACADEMIC STANDARDS FOR SCIENCE

Atomic Structure and the Periodic Table C.2.7, C.2.8. See lessons for details.

Introduce the Chapter

IDENTIFYING MISCONCEPTIONS Students may have trouble understanding that radioactive decay takes place at a constant rate, or that the time period during which the decay occurs can be divided into intervals called half-lives. Use the activity to simulate the decay process.

Activity *You will need 40 jelly beans and 40 seeds or two other small, identical objects. Place each set of objects on a paper plate.* Have a volunteer replace 20 of the jelly beans with 20 seeds, one at a time. Time the process and say that this time is the half-life for jelly beans because half the original number of beans were replaced. Ask a second volunteer to show what will happen during the next half-life. Give this student the same amount of time as the first volunteer. *(Ten beans will be replaced, which is half of the remaining 20.)*

Understanding By Design

Students are building toward understanding nuclear chemistry using the structure of atomic nuclei.

PERFORMANCE GOALS At the end of Chapter 25, students will be able to answer the essential questions by applying their knowledge of nuclear chemistry.

ESSENTIAL QUESTIONS Read the essential questions aloud. **Ask** How may an atom make itself more chemically stable? *(It gains, loses, or shares valence electrons to form a stable octet.)* **Ask** How might an unstable atomic nucleus change to become stable? *(The numbers of protons and/or neutrons change.)* **Ask** How might these changes be related to fusion and fission? *(Fusion and fission are processes that involve unstable nuclei.)*

BIGIDEA Use the photo of the nuclear submarine to help students connect to this chapter's concepts. Activate prior knowledge by asking students if they know any ways that nuclear energy is used. Point out that nuclear energy, like other forms of energy, can be transformed into different forms of energy. The submarine in the photo transforms the energy generated by nuclear reactions into both electrical and mechanical energy. Tell students that in this chapter they will study how the structure of atoms is related to nuclear reactions.

CHEMYSTERY Have students read over the CHEMystery. Connect the CHEMystery to the Big Idea by explaining that nuclear reactions change an atom's structure. Have students predict how scientists know when the baby wooly mammoth lived and died.

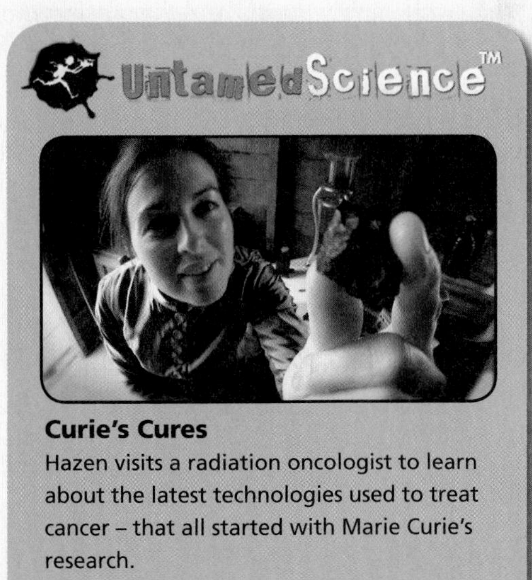

Curie's Cures

Hazen visits a radiation oncologist to learn about the latest technologies used to treat cancer – that all started with Marie Curie's research.

LESSON 25.1

Key Objectives

25.1.1 EXPLAIN how an unstable nucleus releases energy.

25.1.2 DESCRIBE the three main types of nuclear radiation.

Additional Resources

Reading and Study Workbook, Lesson 25.1

Available Online or on Digital Media:

- Teaching Resources, Lesson 25.1 Review

PearsonChem.com

Engage

CHEMISTRY & YOU Have students study the photograph and read the text that opens the lesson. **Ask** When you have an X-ray taken at the dental office, why does the technician drape a heavy shield over your body? *(the shield is a radiation barrier)* Explain that these X-ray shields are typically comprised of lead, which is why they are so heavy. Tell students that because of its high density and high atomic number, lead is effective at stopping radiation from X-rays.

Activate Prior Knowledge

Display the electromagnetic spectrum, given in Figure 5.8. Remind students that in Chapter 5 they learned that the electromagnetic spectrum consists of radiation over a broad range of wavelengths. Point out that electromagnetic radiation includes radio waves, microwaves, infrared waves, visible light, ultraviolet waves, X-rays, and gamma rays.

IN ACADEMIC STANDARDS for SCIENCE

C.2.7 Compare and contrast nuclear reactions with chemical reactions. For nuclear reactions, describe how the fusion and fission processes transform elements present before the reaction into elements present after the reaction.

25.1 Nuclear Radiation

C.2.7 Compare and contrast nuclear reactions with chemical reactions.

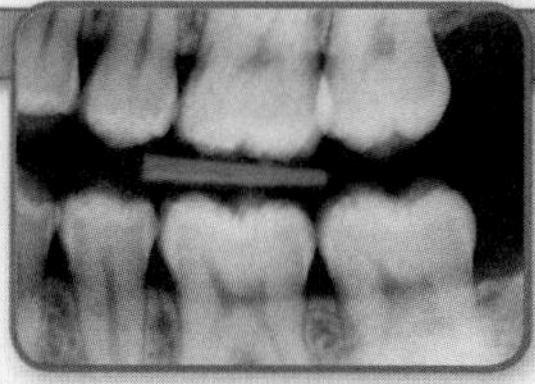

CHEMISTRY & YOU

Q: *What makes some types of radiation more dangerous than other types?* Atoms emit electromagnetic radiation when an electron moves from a higher energy level to a lower energy level. Most electromagnetic radiation, such as visible light, has low energy and is not dangerous. X-rays are an exception. Lengthy or frequent exposure to X-rays can damage cells in your body. This lesson will explain why exposure is also a concern with nuclear radiation.

Key Questions

How do nuclear reactions differ from chemical reactions?

What are three types of nuclear radiation?

Vocabulary

- radioactivity
- nuclear radiation
- radioisotope
- alpha particle
- beta particle
- gamma ray

Figure 25.1 Marie Curie
Marie Curie and her husband Pierre shared the 1903 Nobel Prize in physics with Becquerel for their pioneering work on radioactivity.

Radioactivity

How do nuclear reactions differ from chemical reactions?

In 1896, the French chemist Antoine Henri Becquerel made an accidental discovery. He was studying the ability of uranium salts that had been exposed to sunlight to fog photographic film plates. During bad weather, when Becquerel could not expose a sample to sunlight, he left the sample on top of the photographic plate. When he developed the plate, he discovered that the uranium salt still fogged the film. At the time, two of Becquerel's associates were Marie and Pierre Curie. The Curies were able to show that rays emitted by uranium atoms caused the film to fog. Marie Curie is shown in Figure 25.1. She used the term **radioactivity** to refer to the spontaneous emission of rays or particles from certain elements, such as uranium. The rays and particles emitted from a radioactive source are called **nuclear radiation.**

Radioactivity, which is also called radioactive decay, is an example of a nuclear reaction. In both chemical reactions and nuclear reactions, atoms become more stable. The word *stable* means "constant" or "not likely to change." In a chemical reaction, atoms tend to attain a more stable electron configuration by transferring or sharing electrons. Nuclear reactions begin with unstable isotopes, or **radioisotopes.** Atoms of these isotopes become more stable when changes occur in their nuclei. The changes are always accompanied by the emission of large amounts of energy. **Unlike chemical reactions, nuclear reactions are not affected by changes in temperature, pressure, or the presence of catalysts. Also, nuclear reactions of a given radioisotope cannot be slowed down, speeded up, or stopped.**

Radioactive decay is a spontaneous process that does not require an input of energy. If the product of a nuclear reaction is unstable, it will decay too. The process continues until unstable isotopes of one element are changed, or transformed, into stable isotopes of a different element. These stable isotopes are not radioactive.

Focus on ELL

1 CONTENT AND LANGUAGE Write *alpha*, *beta*, and *gamma* on the board and explain that these words are the names of Greek letters. Then, write the letters beside their names. Note that scientists often use Greek letters as symbols to represent sub-categories of substances.

2 FRONTLOAD THE LESSON Write the symbols for the isotopes of carbon-12 and carbon-14 on the board, and review what the numbers in each symbol represent. Then preview the equations in the lesson, and ask students how they differ from chemical equations.

3 COMPREHENSIBLE INPUT Explain the use of the terms *parent* and *daughter* with respect to nuclei. Then write the equation for the alpha decay of uranium-238 on the board, replacing the alpha particle with a question mark. Walk students through the steps of balancing a nuclear equation. Then have students examine Table 25.1 to locate a particle that will balance the equation.

Table 25.1

Characteristics of Some Types of Radiation

Type	Consists of	Symbol	Charge	Mass (amu)	Common source	Penetrating power
Alpha radiation	Alpha particles (helium nuclei)	α, ${}^{4}_{2}\mathrm{He}$	2+	4	Radium-226	Low (0.05 mm body tissue)
Beta radiation	Beta particles (electrons)	β, ${}^{0}_{-1}\mathrm{e}$	1–	1/1837	Carbon-14	Moderate (4 mm body tissue)
Gamma radiation	High-energy electromagnetic radiation	γ	0	0	Cobalt-60	Very high (penetrates body easily)

Types of Radiation

What are three types of nuclear radiation?

Radiation is emitted during radioactive decay. **Three types of nuclear radiation are alpha radiation, beta radiation, and gamma radiation.** Table 25.1 summarizes the characteristics of these three types of radiation.

Alpha Radiation Some radioactive sources emit helium nuclei, which are also called alpha particles. Each **alpha particle** contains two protons and two neutrons and has a double positive charge. In nuclear equations, an alpha particle is written ${}^{4}_{2}\mathrm{He}$ or α. The electric charge symbol is usually omitted.

The radioisotope uranium-238 emits alpha radiation and is transformed into another radioisotope, thorium-234. Figure 25.2 illustrates this process.

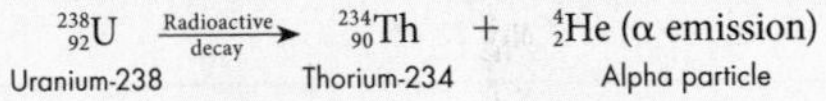

$$\underset{\text{Uranium-238}}{{}^{238}_{92}\mathrm{U}} \xrightarrow[\text{decay}]{\text{Radioactive}} \underset{\text{Thorium-234}}{{}^{234}_{90}\mathrm{Th}} + \underset{\text{Alpha particle}}{{}^{4}_{2}\mathrm{He}} \ (\alpha \text{ emission})$$

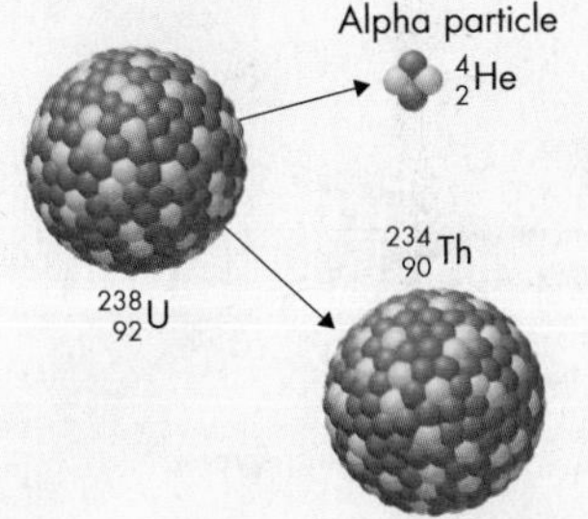

Figure 25.2 Alpha Decay
Uranium-238 decays and forms thorium-234. The radiation emitted is an alpha particle.
Interpret Diagrams
Describe the structure of an alpha particle.

See an online animation of alpha decay.

When an atom loses an alpha particle, the atomic number of the product is lower by two and its mass number is lower by four. In a balanced nuclear equation, the sum of the mass numbers (superscripts) on the right must equal the sum on the left. The same is true for the atomic numbers (subscripts).

Because of their large mass and charge, alpha particles do not travel very far and are not very penetrating. A sheet of paper or the surface of your skin can stop them. But radioisotopes that emit alpha particles can cause harm when ingested. Once inside the body, the particles don't have to travel far to penetrate soft tissue.

Nuclear Decay

Nuclear decay, unlike chemical reactions, is unaffected by temperature, pressure, or the presence of catalysts. Spontaneous nuclear decay cannot be slowed down, speeded up, or stopped. Nuclear decay illustrates the interconnected nature of matter and energy—matter is related to energy according to the equation $E = mc^2$, where c is a constant equal to the speed of light. The energy produced during nuclear decay can be released as electromagnetic radiation, such as gamma rays, or as kinetic energy, which is carried by particles, such as alpha particles.

Foundations for Reading

BUILD VOCABULARY Have students look up the meanings of the first three vocabulary terms. Then, have them infer the meaning of the prefix used in all three terms. Students should infer that the prefix *radio-* means "to send out in all directions."

READING STRATEGY Have students use context clues to infer the definition of *decay,* as it refers to radioactivity. Inferences should include that *decay* means "breaking down into smaller parts."

Explain

Radioactivity

START A CONVERSATION Ask What particles in an atom are involved in chemical reactions? *(valence electrons)* **Ask** How does a chemical reaction compare to a nuclear reaction?*(Chemical reactions involve the transfer or sharing of electrons; the nuclei remain unchanged. Nuclear reactions involve changes in the nuclei of atoms.)*

Misconception Alert

Students may think that the radioisotopes of an element are chemically different from other isotopes of that element. Point out that chemical reactivity depends on the electron structure of an atom, not its nuclear structure.

Types of Radiation

APPLY CONCEPTS Write the following statement on the board: *Nuclear chemistry is the study of changes in matter that originate in atomic nuclei.* **Ask** What types of nuclear radiation exist, and how harmful are they? *(The three most common types of radiation emitted by unstable nuclei are alpha (α), beta (β), and gamma (γ) radiation. Although all forms of nuclear radiation are somewhat harmful, gamma rays are particularly dangerous because they penetrate body tissues.)*

USE VISUALS Have students examine Table 25.1. Encourage them to memorize the names and symbols used to represent each type of radiation. Point out the differences in charge and mass. Students may wish to recreate Table 25.1 on an index card and use the information to help them write and balance nuclear equations. Point out the relative penetrating power of each type of radiation.

Answers

FIGURE 25.2 An alpha particle consists of two protons and two neutrons.

LESSON 25.1

Explore

Types of Radiation

Teacher Demo

PURPOSE Students will observe the effects of radiation on photographic film.

MATERIALS photographic film, radiation source, key

SAFETY Use only safely packaged radiation sources. Do not directly touch any radioactive material.

PROCEDURE Explain that some types of radiation can reduce silver in photographic film. For example, exposing photographic film to visible light or to radiation reduces the silver in the emulsion and forms an image. This property led to the discovery of radioactivity by Becquerel. Demonstrate this by exposing part of a sheet of photographic film to light and developing it. Next, place a metal key on top of unexposed photographic film that has been wrapped in black paper. Place a radiation source on top of the key. After one week, develop the film.

EXPECTED OUTCOME The area of the film that was beneath the key is unexposed, unlike the rest of the film. Remind students of this demonstration when they study film badges in Lesson 25.4.

CHEMISTRY & YOU The amount of energy determines the penetrating power of electromagnetic radiation.

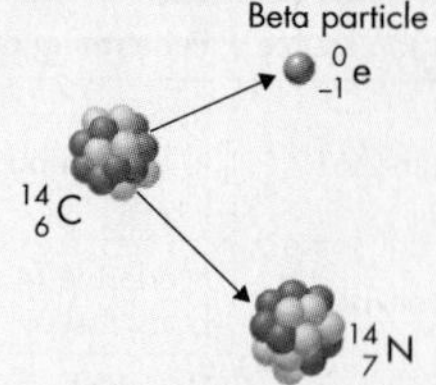

Figure 25.3 Beta Decay
When a carbon-14 atom decays, the products are nitrogen-14 and a beta particle.

Beta Radiation An electron resulting from the breaking apart of a neutron in an atom is called a **beta particle.** The neutron breaks apart into a proton, which remains in the nucleus, and a fast-moving electron, which is released.

$$^{1}_{0}n \longrightarrow ^{1}_{1}p + ^{0}_{-1}e$$

Neutron → Proton + Electron (beta particle)

The symbol for the electron has a subscript of –1 and a superscript of 0. The –1 represents the charge on the electron. The 0 represents the extremely small mass of the electron compared to the mass of a proton.

Carbon-14 is a radioisotope. It emits a beta particle as it decays and forms nitrogen-14. Figure 25.3 illustrates this reaction.

$$^{14}_{6}C \longrightarrow ^{14}_{7}N + ^{0}_{-1}e \ (\beta \text{ emission})$$

Carbon-14 (radioactive) → Nitrogen-14 (stable) + Beta particle

The nitrogen-14 atom has the same mass number as carbon-14, but its atomic number has increased by 1. It contains an additional proton and one fewer neutron. The nuclear equation is balanced.

A beta particle has less charge than an alpha particle and much less mass than an alpha particle. Thus, beta particles are more penetrating than alpha particles. Beta particles can pass through paper but are stopped by aluminum foil or thin pieces of wood. Because of their opposite charges, alpha and beta radiation can be separated by an electric field, as shown in Figure 25.4.

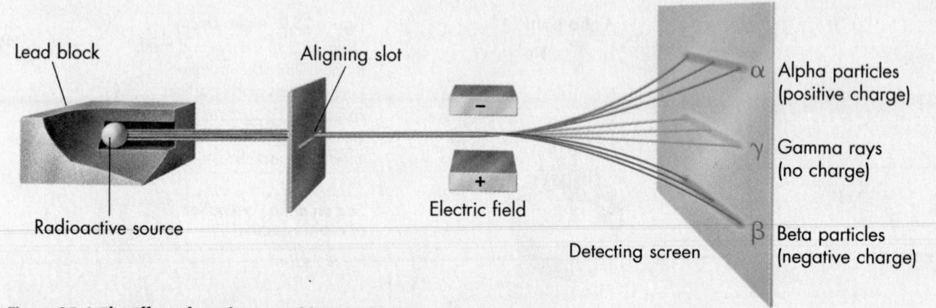

Figure 25.4 The Effect of an Electric Field on Radiation
An electric field has a different effect on each type of radiation. Alpha and beta particles move in opposite directions. Alpha particles move toward the negative plate and beta particles move toward the positive plate. Gamma rays are not deflected as they pass between the plates.
Applying Concepts ***Why are gamma rays not deflected?***

878 Chapter 25 • Lesson 1

UbD Check for Understanding

What are three types of nuclear radiation?

Have students list three types of nuclear radiation and explain how they differ from one another. *(Three types of nuclear radiation are alpha particles, beta particles, and gamma rays. Alpha particles are helium nuclei, which consist of two protons and two neutrons and have a charge of +2. Beta particles are electrons that are emitted from the nucleus when a neutron splits into a proton and an electron. Gamma rays are high-energy photons that are often emitted when alpha or beta particles are emitted, but have no effect on the mass number or atomic number of nuclei.)*

ADJUST INSTRUCTION Have students review Table 25.1 and examine the nuclear reaction examples shown in the Types of Radiation section. Then allow students an opportunity to revise their answers.

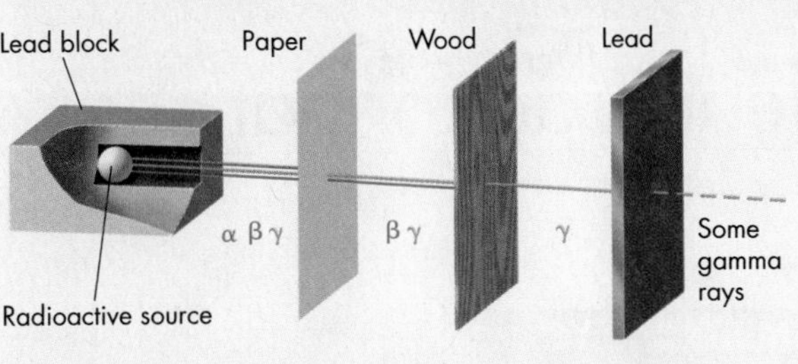

Figure 25.5 Relative Penetrating Power of Nuclear Radiation
Because of their large mass and charge, alpha particles (red) are the least penetrating of the three main types of nuclear radiation. Gamma rays (yellow) have no mass or charge and are the most penetrating.
Infer ***How penetrating are beta particles (green) compared to alpha particles and gamma rays?***

Gamma Radiation A high-energy photon emitted by a radioisotope is called a **gamma ray.** The high-energy photons are a form of electromagnetic radiation. Nuclei often emit gamma rays along with alpha or beta particles during radioactive decay. The following examples demonstrate this process.

$$\underset{\text{Thorium-230}}{{}^{230}_{90}\text{Th}} \longrightarrow \underset{\text{Radium-226}}{{}^{226}_{88}\text{Ra}} + \underset{\text{Alpha particle}}{{}^{4}_{2}\text{He}} + \underset{\text{Gamma ray}}{\gamma}$$

$$\underset{\text{Thorium-234}}{{}^{234}_{90}\text{Th}} \longrightarrow \underset{\text{Protactinium-234}}{{}^{234}_{91}\text{Pa}} + \underset{\text{Beta particle}}{{}^{0}_{-1}\text{e}} + \underset{\text{Gamma ray}}{\gamma}$$

Gamma rays have no mass and no electrical charge. So the emission of gamma radiation does not alter the atomic number or mass number of an atom. Because gamma rays are extremely penetrating, they can be very dangerous. For example, gamma rays pass easily through paper, wood, and the human body. They can be stopped, although not completely, by several meters of concrete or several centimeters of lead, as shown in Figure 25.5.

Q: *Gamma rays can be dangerous because of their penetrating power. What property determines the relative penetrating power of electromagnetic radiation?*

25.1 LessonCheck

1. **Compare** What factors do not affect nuclear reactions, but do affect chemical reactions?
2. **Describe** Briefly describe the three main types of nuclear radiation.
3. **Identify** What part of an atom undergoes change during radioactive decay?
4. **Compare and Contrast** How is the atomic number of a nucleus changed by alpha decay? By beta decay? By gamma decay?
5. **Describe** What two items must be equal for a nuclear equation to be balanced?
6. **Relate Cause and Effect** How does alpha decay affect the mass number of a nucleus? How does beta decay affect the mass number?
7. **Identify** Which of the three types of radiation described in this lesson is the most penetrating, and why?
8. **Predict** When polonium-210 decays by alpha radiation, what isotope is formed?

Lesson Check Answers

1. temperature, pressure, or the presence of catalysts
2. Alpha particles are helium nuclei with a positive charge. Beta particles are fast-moving electrons. Gamma rays are high-energy photons.
3. The nucleus undergoes change.
4. In alpha decay, the atomic number decreases by two. In beta decay, the atomic number increases by one. With gamma rays, the atomic number does not change.
5. The sum of the mass numbers and the sum of the atomic numbers on the right side of the equation must equal the corresponding sums on the left side of the equation.
6. In alpha decay, the mass number decreases by four. In beta decay, the mass number does not change.
7. Gamma radiation, because it has the highest energy
8. lead-206

Evaluate

Informal Assessment

Write several partial equations for nuclear decay involving alpha and beta particles using only the parent and daughter nuclei. Have students compare the two nuclei in each equation to identify the type of particle emitted and justify their conclusions. **Ask** What other type of emission is possible during nuclear decay? *(gamma rays)* Then have students complete the 25.1 Lesson Check.

Reteach

Remind students that when writing a nuclear equation, the sums of the mass numbers and atomic numbers of the reactants must equal the sums of the mass numbers and atomic numbers of the products. Students should check their equations by comparing the sums of the superscripts and subscripts on each side of the equation.

Answers

FIGURE 25.4 Gamma rays are not deflected by the electric field because they have no charge.

FIGURE 25.5 Beta particles are more penetrating than alpha particles but less penetrating than gamma rays.

Key Objectives

🔑 **25.2.1 DESCRIBE** the type of decay a radioisotope undergoes.

🔑 **25.2.2 SOLVE** problems that involve half-life.

🔑 **25.2.3 IDENTIFY** two ways transmutations can occur.

Additional Resources

Reading and Study Workbook, Lesson 25.2

Available Online or on Digital Media:

- Teaching Resources, Lesson 25.2 Review, Interpreting Graphics
- Small-Scale Chemistry Lab Manual, Lab 41

PearsonChem.com

Engage

CHEMISTRY & YOU Have students read the opening text. Tell students that radon-222 emits alpha radiation—the least harmful of the three main types of nuclear radiation. **Ask** So why does the level of radon need to be monitored? *(Because radon is a gas, it can be inhaled. Thus, the radiation that it emits can penetrate body tissues.)*

Build Background

Have students consider how stable a nucleus would be if it contained only protons. **Ask** In terms of stability, what purpose might the neutrons serve in an atom? *(The neutrons separate the positive charges in a nucleus and limit the repulsive forces between protons.)*

IN ACADEMIC STANDARDS for SCIENCE

C.2.7 Compare and contrast nuclear reactions with chemical reactions. For nuclear reactions, describe how the fusion and fission processes transform elements present before the reaction into elements present after the reaction.

C.2.8 Understand that the radioactive decay process is random for any given atom, but that this property leads to a predictable and measurable exponential decay of a sample of radioactive material. Calculate the initial amount, the fraction remaining, or the half-life of a radioactive sample given two of the three variables.

25.2 Nuclear Transformations

C.2.7 Compare and contrast nuclear reactions with chemical reactions; C.2.8 Radioactive decay

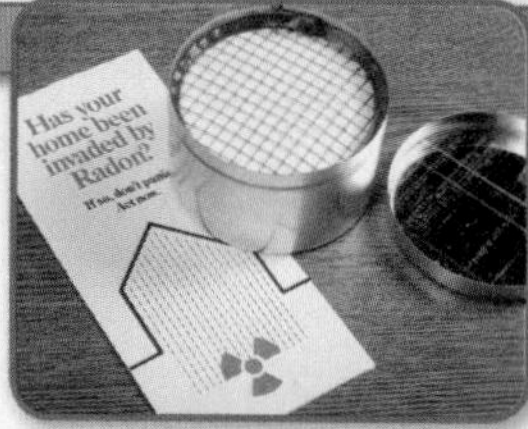

CHEMISTRY & YOU

Q: *What is the source of radon in homes?* All the isotopes of radon gas are unstable and emit radiation. Inhaled radon is the second leading cause of lung cancer in the United States. Radon may accumulate in a basement that is not well ventilated. Because radon is a colorless, odorless gas, people often do not know that they are being exposed to high levels of radon. Test kits are available to measure the levels of radon in a building. In this lesson, you will study the decay series that produces this hazardous gas.

Key Questions

🔑 ***What determines the type of decay a radioisotope undergoes?***

🔑 ***How much of a radioactive sample remains after each half-life?***

🔑 ***What are two ways in which transmutation can occur?***

Vocabulary

- nuclear force
- band of stability
- positron
- half-life
- transmutation
- transuranium elements

Nuclear Stability and Decay

🔑 ***What determines the type of decay a radioisotope undergoes?***

All atomic nuclei, except those of hydrogen atoms, consist of neutrons and two or more protons. If a force did not hold these subatomic particles together, the like-charged protons would repel one another and fly apart. The **nuclear force** is an attractive force that acts between *all* nuclear particles that are extremely close together, such as protons and neutrons in a nucleus. At these short distances, the nuclear force dominates over electromagnetic repulsions and holds the nucleus together.

More than 1,500 different nuclei are known. Only 264 of the known nuclei are stable and do not decay. The rest are unstable and will change over time. The stability of a nucleus depends on the ratio of neutrons to protons. Figure 25.6 shows a graph of the number of neutrons vs. the number of protons for all known stable nuclei. The region of the graph in which these points are located is called the **band of stability.** For elements of low atomic number (below about 20), this ratio is about 1. Above atomic number 20, stable nuclei have more neutrons than protons.

A nucleus may be unstable and undergo spontaneous decay for different reasons. 🔑 **The neutron-to-proton ratio in a radioisotope determines the type of decay that occurs.** Some nuclei are unstable because they have too many neutrons relative to the number of protons. When one of these nuclei decays, a neutron emits a beta particle (fast-moving electron) from the nucleus. A neutron that emits an electron becomes a proton.

$$^{1}_{0}n \longrightarrow {}^{1}_{1}p + {}^{0}_{-1}e$$

This process is known as beta emission. It increases the number of protons while decreasing the number of neutrons. Radioisotopes that undergo beta emission include the following.

$$^{66}_{29}Cu \longrightarrow {}^{66}_{30}Zn + {}^{0}_{-1}e$$

$$^{14}_{6}C \longrightarrow {}^{14}_{7}N + {}^{0}_{-1}e$$

Focus on ELL

1 CONTENT AND LANGUAGE Have students name things that may be stable or unstable in everyday life. Then have them describe how an atom can be unstable. Discuss the common use of the word *decay* to mean "break down" or "rot." Use fruit as an example to explain that decaying or rotting is a process that evolves over time.

2 FRONTLOAD THE LESSON Have students consider cell division as an example of exponential change. Point out that in cell division, the number of existing cells double each time the group of cells completes a cell cycle: two cells become four, four become eight, and so on. Tell students that radioactive decay is also exponential.

3 COMPREHENSIBLE INPUT When discussing Figure 25.8, project the figure on the board alongside an image of the periodic table. Help students trace the decay series on the periodic table to reinforce the concept that radioactive decay of one element's isotope results in the formation of another element's isotope.

InterpretGraphs

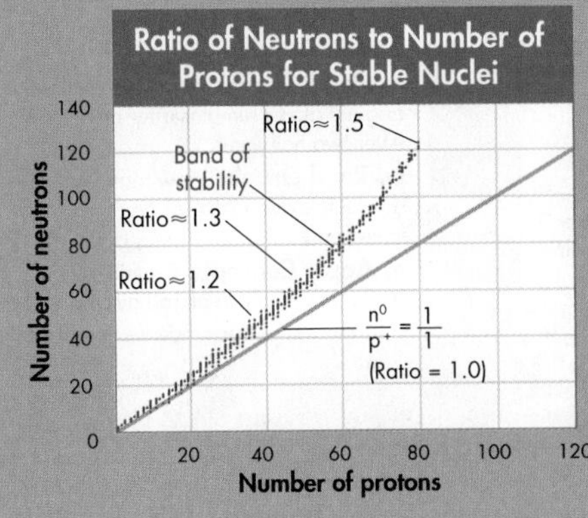

Figure 25.6 A plot of neutrons vs. protons for all stable nuclei forms a pattern called the band of stability, which is shown in purple. The green line shows what the pattern would be if the ratio were 1 for every nucleus.

a. Identify What does each dot represent?

b. Read Graphs What is the ratio of neutrons to protons for tin (Sn, atomic number = 50)?

c. Describe How does the neutron-to-proton ratio change as the number of protons increases in stable nuclei?

Other nuclei are unstable because they have too few neutrons relative to the number of protons. These nuclei increase their stability by converting a proton to a neutron. An electron is captured by a nucleus during this process, which is called electron capture. Here are two examples of electron capture.

$$^{59}_{28}Ni + ^{0}_{-1}e \longrightarrow ^{59}_{27}Co$$

$$^{37}_{18}Ar + ^{0}_{-1}e \longrightarrow ^{37}_{17}Cl$$

A **positron** is a particle with the mass of an electron but a positive charge. Its symbol is $^{0}_{+1}e$. During positron emission, a proton changes to a neutron, just as in electron capture. Here are two examples of positron emission.

$$^{8}_{5}B \longrightarrow ^{8}_{4}Be + ^{0}_{+1}e$$

$$^{15}_{8}O \longrightarrow ^{15}_{7}N + ^{0}_{+1}e$$

When a proton is converted to a neutron, the atomic number decreases by 1 and the number of neutrons increases by 1.

All nuclei that have an atomic number greater than 83 are radioactive. These nuclei have both too many neutrons and too many protons to be stable. Therefore, they undergo radioactive decay. Most of them emit alpha particles. Alpha emission increases the neutron-to-proton ratio, which tends to increase the stability of the nucleus. In alpha emission the mass number decreases by four and the atomic number decreases by two.

$$^{226}_{88}Ra \longrightarrow ^{222}_{86}Rn + ^{4}_{2}He$$

$$^{232}_{90}Th \longrightarrow ^{228}_{88}Ra + ^{4}_{2}He$$

Recall that conservation of mass is an important property of chemical reactions. In contrast, mass is not conserved during nuclear reactions. An extremely small quantity of mass is converted into the energy released during radioactive decay.

UbD Check for Understanding

BIGIDEA ELECTRONS AND THE STRUCTURE OF ATOMS Assess students' knowledge about nuclear stability and decay by asking them to complete the following analogy:

The structural stability that results from a proper ratio of mortar to bricks in a building is like the nuclear stability that results from a proper ratio of ________ to ________ in an atom. *(neutrons, protons)*

ADJUST INSTRUCTION If students are having trouble answering, have them re-read the first three paragraphs of Nuclear Stability and Decay section for review. Then ask them to complete the analogy again.

Foundations for Reading

BUILD VOCABULARY Have students look up the definitions of the vocabulary words. Then, have them paraphrase each definition.

READING STRATEGY As students read through the section, have them look for and list examples of cause and effect relationships. For example, *nuclei are unstable because their neutron-to-proton ratio is greater than one.*

Explain

Nuclear Stability and Decay

CRITICAL THINKING Explain that, for each element, there exists only a small range of neutron-to-proton ratios that produce stable nuclei. If a nucleus does not reflect a stable ratio, it spontaneously decays until a stable ratio of neutrons to protons results.

Extend

Explain that beta decay can result in either the emission of an electron (a β^- particle) or a positron (a β^+ particle). Write the partial equation for the decay of chlorine-32 into sulfur-32 on the board. **Ask** Which kind of particle would need to be emitted to balance the equation? *(a positron)* Explain that during positron emission, a proton is converted into a neutron. **Ask** What kind of unstable nuclei would be likely to emit positrons? *(one with too many protons for the number of neutrons)*

MAKE A CONNECTION Explain that emission of a positron also results in a neutrino, while electron emission is accompanied by an anti-neutrino. Neutrinos and anti-neutrinos have no charge and virtually no mass. Their emissions are not included in the student text because they are not chemically significant and do not react significantly with matter. However, their presence satisfies the laws of energy-mass conservation and momentum conservation.

Answers

FIGURE 25.6

a. Each dot represents a stable nucleus.

b. approximately 1.3

c. The ratio increases.

LESSON 25.2

Explain

Half-Life

INTERPRET GRAPHS Have students examine the graph in Figure 25.6. Discuss the role neutrons play in stabilizing the nuclei of atoms. Point out that nearly 85% of all known nuclei are unstable and the distribution of stable and unstable isotopes is not even. Because neutrons help bind protons together in the nucleus, the ratio of protons to neutrons is a major factor in determining stability. Encourage students to research the strong nuclear force, the force of attraction that holds nucleons together in a stable nucleus.

CRITICAL THINKING

Direct students' attention to Figure 25.7. **Ask** Will the curve on the graph ever reach zero? *(Sample answer: Theoretically, it could reach zero if the last radioactive atom in a sample decays. However, it would not be practical to graph the number of half-lives this event would take, and the amount of radioisotope present would be non-detectable.)* For advanced math students, **Ask** Which mathematical concepts can be used to explain the behavior of the graph? *(Sample answers: exponential function, limits, infinity, asymptote)*

USE VISUALS Refer students to Table 25.2 and have them note the range of half-life values. Remind students that these values do not indicate how long a given atom of an isotope will exist, but only how long it takes for half of the atoms in a sample to undergo radioactive decay.

CHEMISTRY & YOU One intermediate in the decay series for U-238 is Rn-222. This gas can seep into buildings and collect there if a building is not well ventilated.

Interpret Graphs

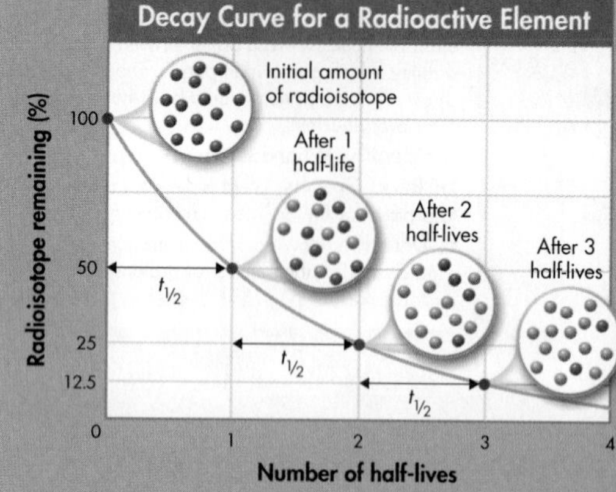

Figure 25.7 This decay curve shows that during each half-life, half of the radioactive atoms decay into atoms of another element.

a. Read Graphs What percent of the original atoms remains after one half-life? After two half-lives?

b. Read Graphs How many half-lives does it take for 12.5% of the radioisotope to remain?

c. Apply Concepts Explain why this graph can apply to all radioisotopes.

Half-Life

How much of a radioactive sample remains after each half-life?

Every radioisotope has a characteristic rate of decay, which is measured by its half-life. A **half-life** ($t_{\frac{1}{2}}$) is the time required for one-half of the nuclei in a radioisotope sample to decay to products, as shown in Figure 25.7. **During each half-life, half of the remaining radioactive atoms decay into atoms of a new element.**

CHEMISTRY & YOU

Q: *Uranium compounds are found in rocks and in soils that form from these rocks. How can these uranium compounds lead to a build up of radon in homes and other buildings?*

Comparing Half-Lives Half-lives can be as short as a second or as long as billions of years. Table 25.2 shows the half-lives of some radioisotopes that occur in nature. Scientists use the half-lives of some long-term radioisotopes to determine the age of ancient objects. Many artificially produced radioisotopes have short half-lives, which makes them useful in nuclear medicine. Short-lived isotopes are not a long-term radiation hazard for patients.

Table 25.2

Half-Lives of Some Naturally Occurring Radioisotopes

Isotope	Half-life	Radiation emitted
Carbon-14	5.730×10^3 years	β
Potassium-40	1.25×10^9 years	β, γ
Radon-222	3.8 days	α
Radium-226	1.6×10^3 years	α, γ
Thorium-234	24.1 days	β, γ
Uranium-235	7.0×10^8 years	α, γ
Uranium-238	4.5×10^9 years	α

Differentiated Instruction

LPR **LESS PROFICIENT READERS** Have students who struggle with reading comprehension draw inferences from tables and numbers. For example, have students compare and contrast the change in radioactivity levels over time for samples of the isotopes potassium-40 and thorium-234. Using Table 25.2, they should be able to infer that the radiation emitted by a sample of K-40 would diminish much less rapidly than the radiation from a sample of Th-234 because of the much longer half-life of K-40.

L1 **STRUGGLING STUDENTS** To reinforce the concept of half-life, have students consider a $1000 prize with the following condition: Half of the remainder of the prize must be spent each month. **Ask** How many months would pass before only $1 was left? *(10 months)*

L3 **ADVANCED STUDENTS** Have students plot a decay curve for one of the elements in Table 25.2. Students should use an initial amount of 100 g of radioisotope, and plot change in amount of radioisotope over time.

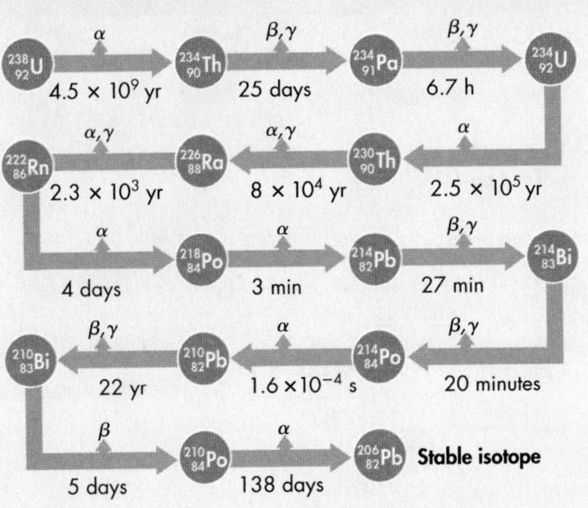

Figure 25.8 Decay Series of U-238
Uranium-238 decays through a series of radioactive intermediates, including radon (Rn) gas.
Interpret Diagrams ***What is the stable end product of this series?***

One isotope that has a long half-life is uranium-238. It decays through a complex series of unstable isotopes to the stable isotope lead-206. Figure 25.8 summarizes this process. The age of uranium-containing minerals can be estimated by measuring the ratio of uranium-238 to lead-206. Because the half-life of uranium-238 is 4.5×10^9 years, it is possible to use its half-life to date rocks as old as the solar system.

Radiocarbon Dating Scientists often find the age of an object that was once part of a living system by measuring the amount of carbon-14 ($^{14}_{6}C$) it contains. Carbon-14 has a half-life of 5730 years. Most of Earth's carbon, however, consists of the more stable isotopes $^{12}_{6}C$ and $^{13}_{6}C$. The ratio of $^{14}_{6}C$ to the other carbon isotopes in the environment is fairly constant because high-energy cosmic rays from space constantly produce $^{14}_{6}C$ in carbon dioxide in the upper atmosphere.

Plants use carbon dioxide to produce carbon compounds, such as glucose. In those compounds, the ratio of carbon isotopes is the same as in the air. The same ratio is maintained as animals consume the plants, and other animals. Thus, the ratio of carbon-14 to other carbon isotopes is constant during an organism's lifetime. When an organism dies, it stops exchanging carbon with the environment and its radioactive $^{14}_{6}C$ atoms decay without being replaced. Therefore, the ratio of $^{14}_{6}C$ to stable carbon in the remains of an organism changes in a predictable way. Archaeologists can use this data to estimate when an organism died.

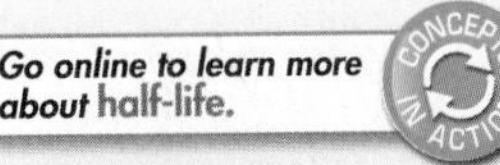

Exponential Decay Function You can use the following equation to calculate how much of an isotope will remain after a given number of half-lives.

$$A = A_0 \times \left(\tfrac{1}{2}\right)^n$$

In the formula, A stands for the amount remaining, A_0 for the initial amount, and n for the number of half-lives. The exponent n indicates how many times A_0 must be multiplied by $\frac{1}{2}$ to determine A. Table 25.3 shows examples in which $n = 1$ and $n = 2$.

Table 25.3

Decay of Initial Amount (A_0) of Radioisotope

Half-Life	Amount Remaining
0	$A_0 \times (\frac{1}{2})^0 = A_0$
1	$A_0 \times (\frac{1}{2})^1 = A_0 \times \frac{1}{2}$
2	$A_0 \times (\frac{1}{2})^2 = A_0 \times \frac{1}{2} \times \frac{1}{2}$

MAKING CONNECTIONS Emphasize that the rate of disintegration of the nuclei of an isotope is unaffected by factors such as heat, pressure, or chemical reactions. Point out that the half-life for a particular radioactive isotope is unique to that isotope. Guide students to understand that no two isotopes have exactly the same half-life.

USE VISUALS Have students study Figure 25.8, which charts the radioactive decay of U-238. Explain that many radioisotopes go through a complex series of nuclear reactions before a stable product is formed.

Extend

Explain to students that there are four fundamental, or universal forces—*strong nuclear forces, weak nuclear forces, electromagnetic forces,* and *gravitational forces.* Have students research these forces and then make a table to compare them in terms of relative strength and range (distance over which the force can act).

Force	Relative Strength	Range
strong nuclear	strongest	short distances
weak nuclear	intermediate	very short distances
electromagnetic	strong	long distances
gravitational	weakest	long distances

Professional Development Note

LIMITATIONS OF RADIOCARBON DATING There are two situations where radiocarbon dating is not an effective method of determining a fossil's age. First, not all fossils contain organic material, but are simply the imprint that remains after the organic material decomposed. Second, carbon-14 has a half-life of only 5730 years, meaning that very little of the isotope remains in items older than 60,000 years of age. In both cases, scientists can use radioisotopes of other elements to determine the fossil's age. This is usually accomplished by dating the rock in which the fossil is located.

Answers

INTERPRET GRAPHS

a. 50%; 25%
b. three half-lives
c. The percentage of a radioactive sample that remains after each half-life is the same for all isotopes.

FIGURE 25.8 lead-206

LESSON 25.2

Explain

Sample Practice Problems

A. How much of a 0.74-mg sample of U-235 will remain after 2.8×10^9 years? *(4.6×10^{-2} mg)*

B. A 0.456-mg sample of hydrogen-3 was collected. After 24.52 years, 0.114 mg of the sample remains. What is the half-life of hydrogen-3? *(12.26 years)*

C. Strontium-90 is a beta emitter with a half-life of 29 years. What is the mass of strontium-90 in a 5.0-g sample of the isotope at the end of 87 years? *(0.63 g)*

CRITICAL THINKING Have students write down procedure steps they can follow to work problems involving whole-number half-lives.

Extend

Connect to MATHEMATICS

Challenge advanced math students to expand their procedures for solving half-life problems to include partial half-lives. Point out that if k is the rate of decay, then $t_{\frac{1}{2}} = \frac{\ln 2}{k}$.

Challenge students to test their procedure by solving the following problem: A fossil was found to contain 60% of its original carbon-14. How old is the fossil?

($k = \ln 2/t_{1/2} = 0.693/5{,}730 \text{ yr} = 1.21 \times 10^{-4}\text{yr}^{-1}$;

$t = \frac{1}{k}\ln\frac{A_0}{A_t} = 1/(1.21 \times 10^{-4}\text{yr}^{-1}) \times \ln(1/0.60) =$ 4,200 yr)

SampleProblem 25.1

Using Half-lives in Calculations

Carbon-14 emits beta radiation and decays with a half-life ($t_{\frac{1}{2}}$) of 5730 years. Assume you start with a mass of 2.00×10^{-12} g of carbon-14.

a. How long is three half-lives?

b. How many grams of the isotope remain at the end of three half-lives?

1 Analyze List the knowns and the unknowns. To calculate the length of three half-lives, multiply the half-life by three. To find the mass of the radioisotope remaining, multiply the original mass by $\frac{1}{2}$ for each half-life that has elapsed.

KNOWNS	UNKNOWNS
$t_{\frac{1}{2}}$ = 5730 years	3 half-lives = ? years
initial mass(A_0) 2.00×10^{-12} g	mass remaining = ? g
number of half-lives (n) = 3	

2 Calculate Solve for the unknowns.

Multiply the half-life of carbon-14 by the total number of half-lives.

a. $t_{\frac{1}{2}} \times n = 5730 \text{ years} \times 3 = 17{,}190 \text{ years}$

The initial mass of carbon-14 is reduced by one half for each half-life. So multiply by $\frac{1}{2}$ three times.

b. Remaining mass $= 2.00 \times 10^{-12}\text{ g} \times \frac{1}{2} \times \frac{1}{2} \times \frac{1}{2}$

$= 0.250 \times 10^{-12}$ g

$= 2.50 \times 10^{-13}$ g

You can get the same answer by using the equation for an exponential decay function.

c. $A = A_0\left(\frac{1}{2}\right)^n = (2.00 \times 10^{-12}\text{ g})\left(\frac{1}{2}\right)^3$

$= (2.00 \times 10^{-12}\text{ g})\left(\frac{1}{8}\right)$

$= 0.250 \times 10^{-12}$ g

$= 2.50 \times 10^{-13}$ g

3 Evaluate Do the results make sense? The mass of carbon-14 after three half-lives should be one-eighth of the original mass. If you divide 2.50×10^{-13} g by 2.00×10^{-12} g, you will get 12.5%, or $\frac{1}{8}$.

For Problem 9, first figure out the number of half-lives.

9. Manganese-56 is a beta emitter with a half-life of 2.6 h. What is the mass of manganese-56 in a 1.0-mg sample of the isotope at the end of 10.4 h?

10. Thorium-234 has a half-life of 24.1 days. Will all the thorium atoms in a sample decay in 48.2 days? Explain.

884 Chapter 25 • Lesson 2

Foundations for Math

FRACTIONS AND EXPONENTS Students may be uncomfortable working with a fraction raised to an exponent in the exponential decay function. However, they may be comfortable working with whole numbers raised to an exponent.

Remind students that the rules of exponents state that $\left(\frac{1}{2}\right)^n$ can be rewritten as $\frac{1^n}{2^n}$.

Because 1 raised to any power is always 1, this equation can be further simplified to $\frac{1}{2^n}$. Demonstrate for students that if they know the value of 2^n, they can easily determine the value of $\left(\frac{1}{2}\right)^n$.

In **Sample Problem 25.1,** $n = 3$, so $\left(\frac{1}{2}\right)^n = \left(\frac{1}{2}\right)^3$. Because $2^3 = 8$, expanding this equation further gives $\frac{1^3}{2^3} = \frac{1}{2^3} = \frac{1}{8}$.

Transmutation Reactions

What are two ways in which transmutation can occur?

For thousands of years, alchemists tried to change lead into gold, an element which is more highly valued than lead. Despite much effort, they were not able to achieve their goal. What they wanted to achieve is **transmutation,** or the conversion of an atom of one element into an atom of another element. This change can occur in at least two ways. **Transmutation can occur by radioactive decay or when particles bombard the nucleus of an atom.** The particles may be protons, neutrons, alpha particles, or small atoms.

Transmutations are common in nature. The production of carbon-14 from nitrogen-14 that takes place in the upper atmosphere is one example. Recall the decay series of uranium-238, which was described in Figure 25.8. In this series, 14 transmutations occur before a stable isotope of lead is produced. Some transmutations that do not occur in nature can be forced to occur in a laboratory or in a nuclear reactor. Ernest Rutherford performed the earliest artificial transmutation in 1919. He bombarded nitrogen gas with alpha particles. The results of this action are shown in Figure 25.9. As the nitrogen atoms absorb the alpha particles, they form fluorine-18 atoms.

$$\underset{\text{Nitrogen-14}}{{}^{14}_{7}\text{N}} + \underset{\text{Alpha particle}}{{}^{4}_{2}\text{He}} \longrightarrow \underset{\text{Fluorine-18}}{{}^{18}_{9}\text{F}}$$

The unstable fluorine atoms quickly decay to form a stable isotope of oxygen and a proton.

$$\underset{\text{Fluorine-18}}{{}^{18}_{9}\text{F}} \longrightarrow \underset{\text{Oxygen-17}}{{}^{17}_{8}\text{O}} + \underset{\text{Proton}}{{}^{1}_{1}\text{p}}$$

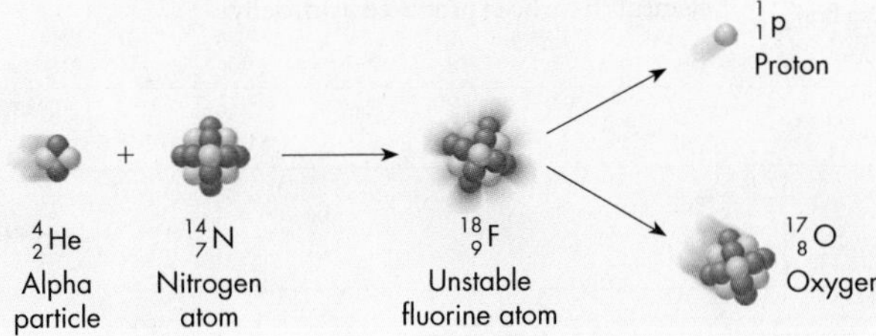

Figure 25.9 The Transmutation of Nitrogen-14
The first artificial transmutation reaction involved bombarding nitrogen gas with alpha particles.
Interpret Diagrams ***Which particle is the intermediate in this nuclear reaction?***

Rutherford's experiment eventually led to the discovery of the proton. He and other scientists noticed a pattern as they did different transmutation experiements. In every case, hydrogen nuclei were emitted. Scientists realized that these hydrogen nuclei (protons) must have a fundamental role in atomic structure. James Chadwick's discovery of the neutron in 1932 also involved a transmutation experiment. Neutrons were produced when beryllium-9 was bombarded with alpha particles.

$$\underset{\text{Beryllium-9}}{{}^{9}_{4}\text{Be}} + \underset{\text{Alpha particle}}{{}^{4}_{2}\text{He}} \longrightarrow \underset{\text{Carbon-12}}{{}^{12}_{6}\text{C}} + \underset{\text{Neutron}}{{}^{1}_{0}\text{n}}$$

Focus on ELL

4 LANGUAGE PRODUCTION Have students work in groups of four to complete the small-scale lab. Make sure each group has ELLs of varied language proficiencies, so that more proficient students can help less proficient ones. Have students work according to their proficiency level.

BEGINNING: LOW/HIGH Restate the procedures and analysis questions in simpler terms. Complete the You're the Chemist questions as a class and have students copy the answers into their notebooks.

INTERMEDIATE: LOW/HIGH Discuss the Analyze and You're the Chemist questions with each lab group before students attempt them. Provide simple examples to remind students of mathematical concepts involved with graphing.

ADVANCED: LOW/HIGH Have students help beginning and intermediate speakers complete their analysis questions.

Explain

Transmutation Reactions

MAKING CONNECTIONS Explain that one of the main goals of medieval alchemists was the transmutation of common metals to precious metals. No chemical reaction can achieve this goal. However, through transmutations, modern chemists can change one element into another. Point out that transmutation reactions also allow chemists to produce elements that do not occur naturally.

Explore

Class Activity

PURPOSE Students will learn about and model an accelerator.

MATERIALS reference materials and Internet access, materials for making models

PROCEDURE Particle accelerators, such as linear accelerators, cyclotrons, and synchrotrons, are used in transmutation experiments. Divide the class into groups of four to five. Have each group choose one type of accelerator and conduct library research on its design and function. The group should create a model of the accelerator, which students can display and explain.

EXPECTED OUTCOME Models and explanations should show understanding of how transmutation is accomplished.

Answers

Sample Problem 25.1

9. 0.063 mg Mn-56

10. No, $(1/2)^2$, or 1/4, of the sample will remain.

FIGURE 25.9 the fluorine-18 atom

Evaluate

Informal Assessment

Have students describe how the ratio of neutrons to protons changes in nuclei that undergo beta particle and alpha particle emission. Then, have students explain the concept of half-life using a numerical example. *(Sample answer: initial state = 256 atoms; one half-life = 128 atoms; two half-lives = 64 atoms, and so on.)* Write several balanced and unbalanced nuclear equations on the board and have students identify and correct the unbalanced equations. Then have students complete the 25.2 Lesson Check.

Reteach

Point out the three natural processes that can result in a nucleus attaining a stable ratio of neutrons to protons:

- A beta emission increases the number of protons and decreases the number of neutrons.
- A positron emission decreases the number of protons and increases the number of neutrons.
- In heavy elements, the nucleus may emit an alpha particle, which decreases the numbers of both neutrons and protons.

Have students organize their notes by these three processes, and encourage them to add equations and examples to help facilitate their understanding.

Extend

Connect to ASTRONOMY

Until recently, astronomers thought the universe consisted solely of the matter that they could observe directly. But this idea was disproved by American astronomer Vera Rubin. Rubin discovered that the matter the astronomers can see makes up as little as ten percent of the mass in the galaxies. The rest exists in the form of dark matter (matter that does not give off electromagnetic radiation). Dark matter cannot be seen directly, but its presence can be inferred by observing the effect of its gravity on visible objects.

In the late 1900s, astronomers observed that the expansion of the universe appeared to be accelerating. This observation was puzzling because no force could account for it. Astronomers infer that a mysterious new force, which they call dark energy, is causing the expansion of the universe to accelerate. Physicists are using the Large Hadron Collider to collect evidence of particles that are associated with dark matter and dark energy.

Figure 25.10 Particle Accelerator
The Large Hadron Collider is the most powerful accelerator in the world. It has a circumference of about 27 kilometers and is about 100 meters underground. The accelerator is designed to mimic the conditions that existed right after the Big Bang.

Elements with atomic numbers above 92, the atomic number of uranium, are called **transuranium elements.** All of these elements are radioactive. All transuranium elements undergo transmutation. These elements are synthesized in nuclear reactors and nuclear accelerators. Reactors produce beams of low-energy particles. Accelerators are used to increase the speed of bombarding particles to very high speeds. Sometimes particles must pass through a series of accelerators before they reach the desired speed. The European Oranization for Nuclear Research, known as CERN, has a number of accelerators at its site on the border between France and Switzerland. Figure 25.10 shows CERN's largest accelerator.

When uranium-238 is bombarded with the relatively slow neutrons from a nuclear reactor, some uranium nuclei capture these neutrons. The product is uranium-239.

$$^{238}_{92}U + ^{1}_{0}n \longrightarrow ^{239}_{92}U$$

Uranium-239 is radioactive and emits a beta particle. The other product is an isotope of the artificial radioactive element neptunium (atomic number 93).

$$^{239}_{92}U \longrightarrow ^{239}_{93}Np + ^{0}_{-1}e$$

Neptunium is unstable and decays, emitting a beta particle and a second artificial element, plutonium (atomic number 94).

$$^{239}_{93}Np \longrightarrow ^{239}_{94}Pu + ^{0}_{-1}e$$

Plutonium and neptunium are both transuranium elements. The majority of these elements do not occur in nature. Scientists in Berkeley, California, synthesized the first two artificial elements in 1940. Since that time, more than 20 additional transuranium elements have been produced artificially.

25.2 LessonCheck

11. Identify What factor determines the type of decay that occurs in a radioisotope?

12. Predict How much of a sample of radioisotope remains after one half-life? After two half-lives?

13. Explain How can transmutation occur in a stable isotope?

14. Apply Concepts Complete the following nuclear equations. Use what you know about balanced nuclear equations to identify the missing particles.

a. $^{27}_{13}Al + ^{4}_{2}He \longrightarrow ^{30}_{14}Si + ?$

b. $^{214}_{83}Bi \longrightarrow ^{4}_{2}He + ?$

c. $^{27}_{14}Si \longrightarrow ^{0}_{-1}e + ?$

d. $^{66}_{29}Cu \longrightarrow ^{66}_{30}Zn + ?$

15. Calculate A radioisotope has a half-life of 4 days. How much of a 20-gram sample of this radioisotope remains at the end of each time period?

a. 4 days

b. 8 days

16. Calculate The mass of cobalt-60 in a sample decreased from 0.800 g to 0.200 g over a period of 10.5 years. From this information, calculate the half-life of cobalt-60.

17. Interpret Graphs Suppose the variable on the *x*-axis in Figure 25.7 were time instead of number of half-lives. Could the same graph be used to represent all radioisotopes? Why or why not?

Lesson Check Answers

11. the neutron-to-proton ratio

12. 50% of the sample remains after one half-life; 25% remains after two half-lives.

13. Transmutation can occur in a stable isotope when particles bombard its nuclei.

14. a. $^{27}_{13}Al + ^{4}_{2}He \longrightarrow ^{30}_{14}Si + ^{1}_{1}H$

b. $^{214}_{83}Bi \longrightarrow ^{4}_{2}He + ^{210}_{81}Tl$

c. $^{27}_{14}Si \longrightarrow ^{0}_{-1}e + ^{27}_{15}P$

d. $^{66}_{29}Cu \longrightarrow ^{66}_{30}Zn + ^{0}_{-1}e$

15. a. 10 g

b. 5 g

16. 5.25 years

17. No, because radioisotopes have different half-lives.

Small-Scale Lab

Radioactivity and Half-Lives

Purpose

To simulate the transformation of a radioactive isotope over time, graph the data, and relate the data to radioactive decay and half-lives

Materials

- coin
- graph paper

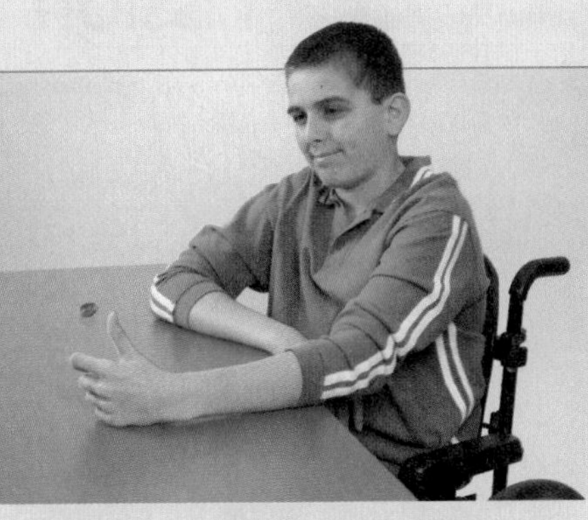

Procedure

1. On a sheet of paper, make a data table similar to the one below.

2. For trial 1, flip a coin 100 times. In your table, record the total number of heads that result.

3. For trial 2, flip the coin the same number of times as the number of heads in trial 1. Record the number of flips and the number of heads that result.

4. Continue the procedure until you obtain no more heads.

Trial	Number of flips	Number of heads
1	100	
2		
3		
4		
5		
6		
7		
8		

Analyze

1. Graph Use graph paper to plot the number of flips (*y*-axis) versus the trial number (*x*-axis). Draw a smooth line to connect the points.

2. Interpret Graphs Is the rate of the number of heads produced over time linear or nonlinear? Is the rate constant over time or does it change?

3. Relate Cause and Effect Why does each trial reduce the number of heads by about one-half?

4. Use Models A half-life is the time required for one-half of the atoms of a radioisotope to decay to products. What value represents one half-life for the process of flipping coins?

You're the Chemist

1. Design an Experiment Design and carry out an experiment using a single die to model radioactive decay. Plot your data.

2. Calculate Radon-222 undergoes alpha decay, emitting an alpha particle (helium nucleus),

$$^{222}_{86}\text{Rn} \longrightarrow ^{218}_{84}\text{Po} + ^{4}_{2}\text{He}$$

Find the half-life of radon-222 in Table 25.2 and determine how long it takes for only one eighth of a sample of radon-222 to remain.

3. Calculate Carbon-14 undergoes beta decay, emitting a beta particle (electron),

$$^{14}_{6}\text{C} \longrightarrow ^{14}_{7}\text{N} + ^{0}_{-1}\text{e}$$

Find the half-life of carbon-14 in Table 25.2 and determine what fraction of the carbon-14 in a sample will remain after 11,460 years.

Small-Scale Lab

OBJECTIVE After completing this activity, students will be able to simulate the transformation of a radioactive isotope and relate simulated data to radioactive decay and half-life.

PREP TIME 5 minutes

CLASS TIME 20 minutes

TEACHING TIPS Emphasize that the trials involve flipping a coin to simulate radioactivity. The appearance and removal of a "head" represents the decay of an unstable nucleus. The rate of removal is analogous to the half-life of a radioactive isotope—around 50%. Although the number of heads decreases over time, the percent of heads produced remains relatively steady until the sample becomes too small. Point out that in a sample as small as 100, the likelihood of producing a number of heads other than 50 is high. If the sample size were increased, the relative error would decrease. As the sample size approaches zero, the results are no longer statistically reliable. Predictions for a mole of atoms, which provides a large sample, can be quite accurate.

EXPECTED OUTCOME For each trial, the number of heads is approximately half the number of flips.

ANALYZE

1. Student graphs should resemble the graph in Figure 25.7.
2. Nonlinear; the rate decreases over time.
3. For each flip, the probability of a head is 0.50.
4. one trial

YOU'RE THE CHEMIST

1. *Sample answer:* Count the total number of even numbers that occur in 100 rolls of the die. Roll the die again a number of times equal to the number obtained on the first trial. Do trials until the number of events equals zero. Plot number of evens versus trials.
2. After 3.8 days, half the sample remains. After 7.6 days, one-fourth remains. After 11.4 days, one-eighth remains.
3. This time period is two half-lives (11,460 years/5730 years = 2) of carbon-14. After two half-lives, one-fourth of the sample remains.

FOR ENRICHMENT Repeat the procedure four more times, using a new table for each trial. Average the results in each cell of the table for the five trials. **Ask** Which is closer to the expected values, the results from a single trial or the averaged results? *(The average values should be closer.)* As a class, discuss the advantages of using multiple trials in an experiment.

LESSON 25.3

Key Objectives

25.3.1 **DESCRIBE** what happens in a nuclear chain reaction.

25.3.2 **DISTINGUISH** fission reactions from fusion reactions.

Additional Resources

Reading and Study Workbook, Lesson 25.3

Available Online or on Digital Media:

- Teaching Resources, Lesson 25.3 Review

PearsonChem.com

Engage

CHEMISTRY & YOU Explain that about 92% of all the sun's atoms are hydrogen. The second most abundant element is helium. **Ask** What can you infer about the reactant and product of the nuclear reaction that takes place in the sun? *(Hydrogen is the reactant and helium is the product.)*

Build Background

Have students brainstorm examples of chain reactions, such as a multi-car accident on an icy highway. Have students consider how these chain reactions often result in a significant number of collisions from the single initial collision, and the energy inputs and transfers required for the collisions to proceed.

25.3 Fission and Fusion

C.2.7 Compare and contrast nuclear reactions with chemical reactions

CHEMISTRY & YOU

Q: *Where does the sun's energy come from?* The sun appears as a bright, fiery ball in the sky. The sun is so bright that you should never look at it directly without eye protection. The sun is about halfway through its life cycle. It has been producing energy for about 5 billion years and is expected to continue to produce energy for about 5 billion more. In this lesson, you will study the nuclear reaction that takes place in the sun.

Key Questions

What happens in a nuclear chain reaction?

How do fission reactions and fusion reactions differ?

Vocabulary

- fission
- neutron moderation
- neutron absorption
- fusion

Nuclear Fission

What happens in a nuclear chain reaction?

When the nuclei of certain isotopes are bombarded with neutrons, the nuclei split into smaller fragments. This process is called **fission.** Uranium-235 and plutonium-239, for example, are fissionable isotopes. Figure 25.11 shows how uranium-235 breaks into two smaller fragments of roughly the same size when struck by a slow-moving neutron. At the same time, more neutrons are released by the fission. These neutrons strike the nuclei of other uranium-235 atoms, which causes a chain reaction. **In a chain reaction, some of the emitted neutrons react with other fissionable atoms, which emit neutrons that react with still more fissionable atoms.**

Nuclear fission can release enormous amounts of energy. The fission of 1 kg of uranium-235, for example, yields an amount of energy equal to that produced when 20,000 tons of dynamite explode. In an uncontrolled nuclear chain reaction, all the energy is released in fractions of a second. An atomic bomb is a device that can trigger an uncontrolled nuclear chain reaction.

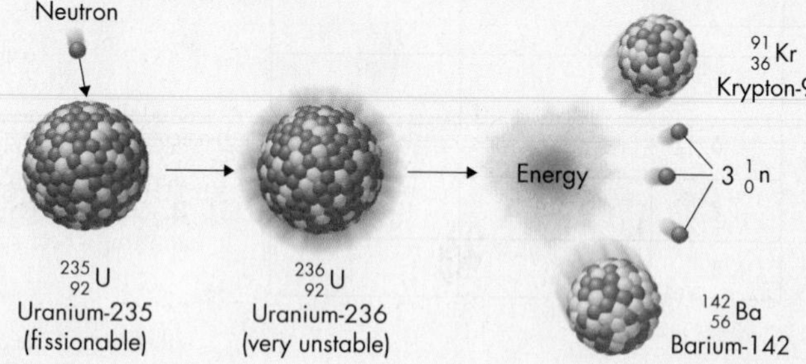

Figure 25.11 Fission of Uranium
When struck by a slow-moving neutron, a uranium-235 nucleus breaks into two smaller nuclei and releases three neutrons.
Predict *What happens when the released neutrons strike other uranium-235 nuclei?*

IN ACADEMIC STANDARDS for SCIENCE

C.2.7 Compare and contrast nuclear reactions with chemical reactions. For nuclear reactions, describe how the fusion and fission processes transform elements present before the reaction into elements present after the reaction.

Focus on ELL

1 CONTENT AND LANGUAGE Have students write the definitions of *fission* and *fusion* in their notebooks in their native language and in English. Have students prepare a two-column table to compare and contrast these two processes. Tell students to record sample reactions, and characteristics of each process as they progress through the lesson.

2 FRONTLOAD THE LESSON Have students preview Figures 25.11 and 25.14 and identify any similarities and differences that they find. Have students add these characteristics to their two-column table.

3 COMPREHENSIBLE INPUT Provide students with modeling clay. Have them work with a partner to model fission and fusion, using the diagrams in the lesson as a guide. Students should understand that fission is the breaking apart of a nucleus, and fusion is the combining of nuclei.

Fission can be controlled so energy is released more slowly. Nuclear reactors, such as the one shown in Figure 25.12, use controlled fission to produce useful energy. The reaction takes place within uranium-235 or plutonium-239 fuel rods. Much of the energy produced in this reaction is in the form of heat. A fluid, usually liquid sodium or water, removes heat from the core, or central part, of the reactor. Thus, the fluid is called a coolant. The heated fluid is used to change water to steam, which drives a turbine that generates electricity. The control of fission in a nuclear reactor involves two steps, neutron moderation and neutron absorption.

Neutron Moderation **Neutron moderation** is a process that slows down neutrons so the reactor fuel can capture them to continue the chain reaction. Moderation is necessary because most of the neutrons produced move so fast that they would pass right through a nucleus without being captured. Water and carbon in the form of graphite are good moderators.

Neutron Absorption To prevent the chain reaction from going too fast, some of the slowed neutrons must be trapped before they hit fissionable atoms. **Neutron absorption** is a process that decreases the number of slow-moving neutrons. Control rods, made of materials such as cadmium, are used to absorb neutrons. When the control rods extend almost all the way into the reactor core, they absorb many neutrons, and fission occurs slowly. As the rods are pulled out, they absorb fewer neutrons and the fission process speeds up. If the chain reaction were to go too fast, heat might be produced faster than the coolant could remove it. The reactor core would overheat, which could lead to mechanical failure and release of radioactive materials into the atmosphere. Ultimately, a meltdown of the reactor core might occur.

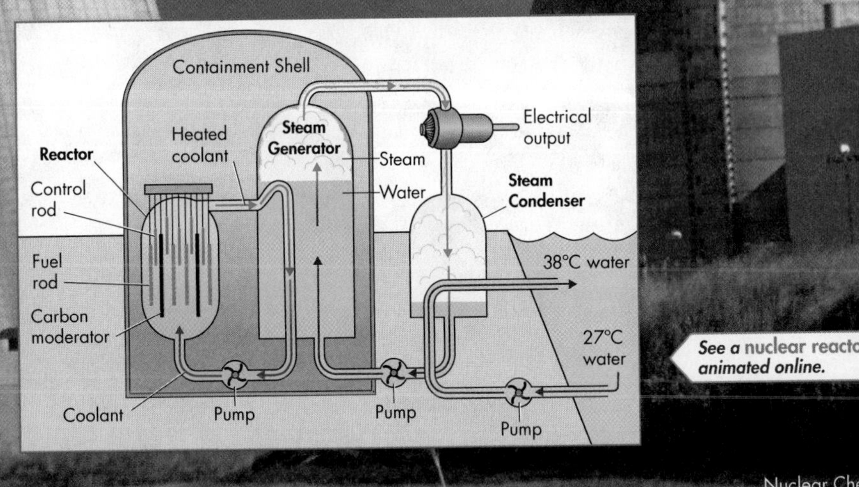

Figure 25.12 Nuclear Reactor
A nuclear reactor is used to produce electricity. A coolant absorbs heat produced by the controlled fission reaction and transfers the heat to water, which changes to steam. The steam drives a turbine, which drives a generator that produces electricity.
Interpret Diagrams ***What happens to the steam after it drives the turbine?***

See a nuclear reactor animated online.

Foundations for Reading

BUILD VOCABULARY Explain that *fusion* is the noun form of the verb *fuse*, from the Latin word *fusus* meaning "melt together." Explain that during fusion, two nuclei combine to form a nucleus of larger mass.

READING STRATEGY Pair students and have them read each section silently. Have them record vocabulary that they find confusing. After the pair has finished a section, have them work together to help clarify any confusing concepts.

Explain

Nuclear Fission

USE VISUALS Display Figure 25.11 on an overhead projector and explain that uranium-235 is called a fissionable material because its nucleus becomes unstable when it is struck by a neutron. Note that three neutrons are produced by each fission event. Explain that fission reactions also produce radioactive waste products such as krypton-91 and barium-142.

USE MODELS Have students draw a diagram showing how the neutrons from a single fission process can induce subsequent fissions in other uranium-235 nuclei. The diagrams should illustrate the exponential nature of fission.

Explore

Teacher Demo

PURPOSE Students will observe a chain reaction.

MATERIALS wooden matchsticks

SAFETY Keep all flammable materials away from the demo area. Have a fire extinguisher ready.

PROCEDURE Cut some matchsticks in half and arrange the heads in branching chains on a non-combustible surface. The end of each matchstick should touch the heads of two other matchsticks, forming a "Y." Ignite the first matchstick head.

EXPECTED OUTCOME The matches ignite in a chain reaction.

Differentiated Instruction

LPR LESS PROFICIENT READERS Explain that the term *fission* is generally used when an object splits into smaller parts. Demonstrate fission using rock candy crumbling into small pieces. Relate the concept of fission to fragmenting, breaking up, or crumbling. Explain that the term *fusion* is generally used when an object combine into a larger whole. Demonstrate fusion using popcorn pieces that are joined together to make a popcorn ball. Relate the concept of fusion to merging or joining together.

L3 ADVANCED STUDENTS Explain that the difference between the mass of a nucleus and the masses of the separated protons and neutrons is called the mass defect. Challenge students to use the mass defect and Albert Einstein's famous equation, $E = mc^2$, to calculate the energy released during a fission reaction.

L1 STRUGGLING STUDENTS After preparing the match demo, allow students to examine the matchstick chain. Project Figure 25.11 on the board, and identify parts of the chain that are equivalent to components of the U-235 reaction.

Answers

FIGURE 25.11 More neutrons are released which can then strike the nuclei of other fissionable atoms.

FIGURE 25.12 The steam is condensed into liquid water, which is pumped back into the steam generator.

LESSON 25.3

Explain

Nuclear Fission

MAKE A CONNECTION Explain that the wastes produced in fission reactors contain isotopes with half-lives measured in the thousands or hundreds of thousands of years. Many proposals for storing or disposing of these wastes involve methods and materials that may be highly unsuitable. For example, the placement of wastes in thick drums that would then be sunk in the oceans—one suggested method—may contain the waste for only decades or a few centuries. The containers would begin to leak long before the contents were safe. The U.S. Department of Energy is responsible for cleaning up 130 nuclear sites and safely managing their waste. The sites include locations where uranium was milled, research labs, and former nuclear weapons production facilities. Have students research how the cleanups are progressing and how and where the nuclear waste is being disposed.

Extend

Have students work in groups of four. Tell them that each group has to create and implement a simulation that is analogous to a uranium fission chain reaction. Groups should be prepared to show their simulation during class. *(Sample simulations: the "wave" at a sporting event, house of cards, and so on.)*

Explain

Nuclear Fusion

Generate a discussion about the limitations of using fusion as an alternative energy source. Point out that the only known way of generating the necessary heat to start a fusion reaction is to use a fission bomb—an uncontrolled reaction. Currently, no known materials or engineering designs exist that could withstand either the temperatures or the necessary fission-generated explosion. As a result, scientists have been pursuing the possibility of finding a situation in which fusion would occur at room temperature, a process called *cold fusion*. Have students research the history and current state of cold fusion research, and write a report explaining their findings.

CHEMISTRY & YOU Explain that the sun contains hydrogen nuclei. **Ask** How is the energy of the sun produced? *(Energy is released during the formation of helium nuclei.)*

READING SUPPORT

Build Reading Skills: ***Inference*** When you make an inference, you are really reading between the lines. An inference should be based on information in the text and any background knowledge you may have. After you read the Nuclear Waste section, use inference to explain how the storage of nuclear wastes could affect the environment.

Nuclear Waste Fuel rods from nuclear power plants are one major source of nuclear waste. The fuel rods are made from a fissionable isotope, either uranium-235 or plutonium-239. The rods are long and narrow—typically 3 meters long with a 0.5-cm diameter. In a typical reactor, three hundred fuel rods are bundled together to form an assembly, and one hundred assemblies are arranged to form the reactor core. During fission, the amount of fissionable isotope in each fuel rod decreases. Eventually the rods no longer have enough fuel to ensure that the output of the power station remains constant. The isotope-depleted, or spent, fuel rods must be removed and replaced with new fuel rods.

Spent fuel rods are classified as high-level nuclear waste. They contain a mixture of highly radioactive isotopes, including fission products and what remains of the nuclear fuel. Some of these fission products have very short half-lives, on the order of fractions of seconds. Others have half-lives of hundreds or thousands of years. All nuclear power plants have holding tanks, or "swimming pools," for spent fuel rods. Water cools the spent rods, and also acts as a radiation shield to reduce the radiation levels. The pools, like the one shown in Figure 25.13, are typically 12 meters deep. Storage racks at the bottom of these pools are designed to hold the spent fuel assemblies. The rods continue to produce heat for years after their removal from the core.

The spent fuel rods may spend a decade or more in a holding tank. In the past, plant operators expected spent fuel rods to be reprocessed. Any leftover fissionable isotope in the rods would be recycled in the manufacture of new fuel rods. However, with large deposits of uranium ore available—many in the United States—it is less expensive to mine new fuel than to reprocess depleted fuel. At some nuclear plants, the storage pool has no space left. In order to keep these plants open, their fuel rods must be moved to off-site storage facilities. Finding appropriate storage sites is difficult because high-level waste may need be stored for a long time. Plutonium-239, for example, will not decay to safe levels for 20,000 years. Often, people are concerned about having nuclear waste stored nearby or shipped through their communities.

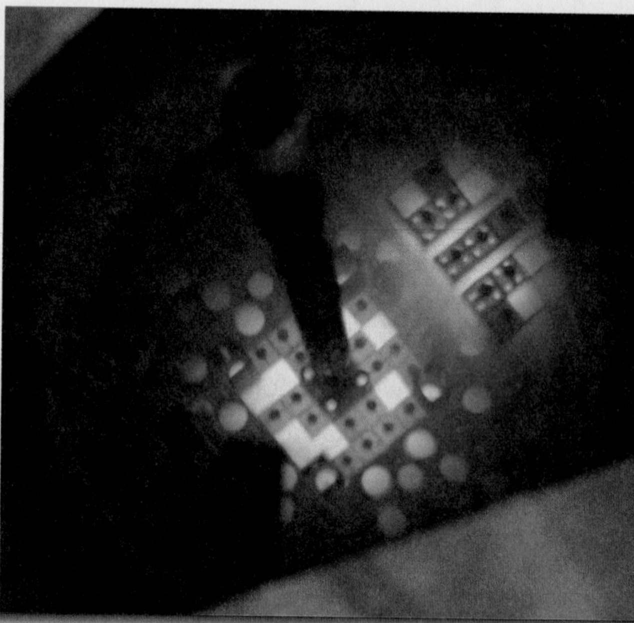

Figure 25.13 Disposal of Fuel Rods Racks at the bottom of this pool contain spent fuel rods. The blue glow is from beta particles that the rods emit into the water.

890 Chapter 25 • Lesson 3

Nuclear Waste

Nuclear waste began accumulating in the 1940s at nuclear weapon facilities, which no longer operate. Today, nuclear waste is created by power plants, and sectors of society, such as hospitals, that use radioisotopes. The nuclear waste accumulated over decades has been in temporary storage awaiting methods for permanent disposal. The waste is classified as low-, mid- or high-level. Transuranic waste is low- to -mid-level waste contaminated with alpha-emitting radioisotopes with half-lives greater than 20 years and atomic numbers greater than 92. This waste consists of disposable items such as tools or clothing. High-level nuclear waste consists of spent fuel rods from nuclear reactors and the waste from the production of nuclear weapons.

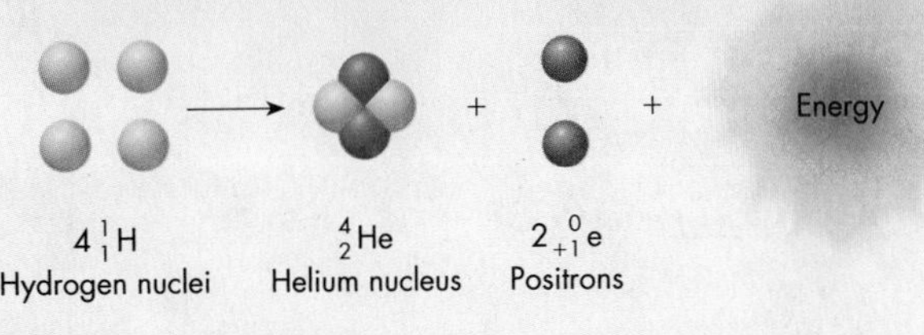

Figure 25.14 Fusion in the Sun
In the sun, hydrogen nuclei fuse to produce helium nuclei.
Interpret Diagrams ***What are the other products of this reaction?***

Nuclear Fusion

How do fission reactions and fusion reactions differ?

The sun, directly and indirectly, is the source of most energy used on Earth. The energy emitted by the sun results from nuclear fusion. **Fusion** occurs when nuclei combine to produce a nucleus of greater mass. In solar fusion, hydrogen nuclei (protons) fuse to make helium nuclei. Figure 25.14 shows that the reaction also produces two positrons. **Fusion reactions, in which small nuclei combine, release much more energy than fission reactions, in which large nuclei split apart and form smaller nuclei.** However, fusion reactions occur only at very high temperatures—in excess of 40,000,000°C.

The use of controlled nuclear fusion as an energy source on Earth is appealing. The potential fuels are inexpensive and readily available. Some scientists are studying a reaction in which a deuterium (hydrogen-2) nucleus and a tritium (hydrogen-3) nucleus combine to form a helium nucleus.

$$^2_1H + ^3_1H \longrightarrow ^4_2He + ^1_0n + \text{energy}$$

The problems with fusion lie in achieving the high temperatures needed to start the reaction and in containing the reaction once it has started. The high temperatures required to start fusion reactions have been achieved by using a fission bomb. Such a bomb is the triggering device used for setting off a hydrogen bomb, which is an uncontrolled-fusion device. This process is of no use, however, as a controlled generator of power.

Q: *The high temperatures needed to support fusion exist within the sun's core. Late in the sun's life cycle, other fusion reactions will occur. What element would form when two helium nuclei fuse?*

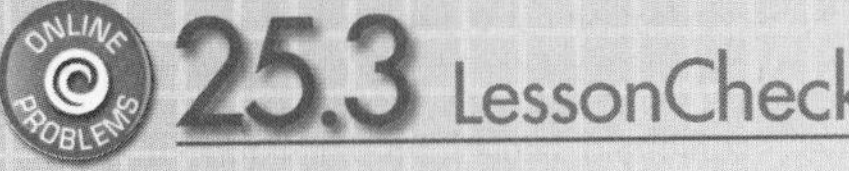

25.3 LessonCheck

18. **Relate Cause and Effect** Explain what happens in a nuclear chain reaction.
19. **Compare** How are fusion reactions different from fission reactions?
20. **Explain** What is neutron moderation, and why is it necessary in a nuclear reactor?
21. **Identify** What are two sources of the radioactive nuclei present in spent fuel rods?
22. **Evaluate** Suppose the technical problems with fusion reactors could be overcome. What are some advantages to using a fusion reactor to produce electricity?
23. **Interpret Diagrams** Review the diagram of a reactor in Figure 25.12. What role does water play in a typical nuclear reactor?
24. **Infer** Some nuclear waste is stored about 600 meters below the desert in New Mexico inside caverns dug out of an ancient bed of rock salt. The land above the storage site is owned by the federal government. Why do you think this location was chosen?
25. **Describe** Read about heavy water reactors on page R35 of the Elements Handbook. What is the advantage of using heavy water instead of ordinary water as a neutron moderator?

Lesson Check Answers

18. Fissionable atoms emit neutrons, which react with other fissionable atoms, which emit more neutrons.
19. In fission reactions, nuclei are split into smaller nuclei. In fusion reactions, small nuclei combine to form larger nuclei. Fusion reactions produce a greater amount of energy.
20. A neutron moderator slows down neutrons so that the nuclei in the reactor fuel can capture them.
21. unused nuclear fuel and fission products
22. The potential fuels are readily available and inexpensive.
23. Water absorbs heat from the reactor. The steam produced is used to generate electricity.
24. *Sample answer:* The site is below ground in an unpopulated area that is controlled by the government.
25. D_2O is a more efficient moderator than H_2O. So in heavy-water reactors, the fuel does not need to be enriched.

Evaluate

Informal Assessment

Ask students to write nuclear equations describing fission and fusion. **Ask** How might a meltdown of a nuclear reactor occur? *(Rapid removal of the control rods could allow too many neutrons to react with fissionable nuclei.)* **Ask** Describe the conditions under which nuclear fusion will occur. *(temperatures in excess of 4.00×10^7°C; methods to confine and control the reaction are needed)* Then have students complete the 25.3 Lesson Check.

Reteach

Explain that both nuclear fission and nuclear fusion produce energy by the conversion of matter to energy. In fission, heavy nuclei are split into lighter nuclei. In fusion, light nuclei combine to form heavier nuclei. Fission reactions are relatively easy to control but produce radioactive wastes. Fusion reactions are difficult to initiate and control but produce little radioactive waste. Guide students in creating a concept map to help them organize the content in this lesson.

Answers

READING SUPPORT *Sample answer:* Leakage from a storage site could contaminate groundwater and, subsequently, soil.

FIGURE 25.14 positrons

CHEMISTRY & YOU

CHEMISTRY & YOU

ACTIVATE PRIOR KNOWLEDGE Make sure students understand the role power plants play in an industrial society. **Ask** What is the purpose of a power plant? *(to generate electricity)* **Ask** What parts of a community need a reliable supply of electricity? *(homes, businesses, medical facilities, factories, and so on)* Ask students who have experienced a power outage to describe the effects of the outage.

Explain

USE VISUALS Have students use the illustrations and the table to make a list of similarities and differences between traditional nuclear power plants and small-scale nuclear power plants. *(Similarities: controlled fission reaction, steam used to generate electricity, nuclear waste. Differences: size of facility, construction time, location of reactors, method for circulating coolant, ability to continue operating if there is a problem with a reactor.)*

START A CONVERSATION Explain that problems with disposal of nuclear wastes have led some people to propose that nuclear power be abandoned as a source of energy. After students read the feature, ask students whether they think that the proposed small-scale reactors will change some people's minds.

CHEMISTRY & YOU: TECHNOLOGY

Small-Scale Nuclear Power

About 20 percent of the electricity generated each year in the United States comes from nuclear power. Operating costs for nuclear power plants are lower than for plants that burn fossil fuels. However, a nuclear power plant, like the one shown in the photo, costs more and takes longer to build than a coal-burning plant. Also, people have concerns about the safety of these plants.

Plants that burn fossil fuels also have issues. Concerns about global warming have made the reduction of carbon emissions a priority for many countries. To meet this goal, plants that burn fossil fuels must install technology to trap the carbon dioxide produced when coal or natural gas burn. A greater reliance on nuclear power can help reduce carbon emissions.

Researchers at Oregon State University have proposed a new kind of nuclear power plant to address some of the problems presented by traditional plants. The large above-ground reactor of a traditional plant would be replaced by self-contained modules, which would be located sixty-five feet below ground level. The plant could open with only a few modules. More modules can be added as needed.

Old vs. New

	Traditional Plant	Small-Scale Plant
Construction	Single-unit plants must be built onsite. Construction time is 7–10 years.	Modular units are factory-built and shipped to the site. Construction time is 3–3.5 years.
Coolant method	Valves and pumps circulate the coolant around the reactor core.	Gravity, condensation, evaporation, and convection circulate the coolant.
Safety	The reactor is above ground. A backup electricity source is needed for coolant systems. If the reactor must be shut down, no electricity is generated.	The modules are underground, in self-contained steel chambers, immersed in a pool of water. One module can be shut down without affecting the other modules.
Facility size	Many traditional single-unit plants require about 20 acres.	A plant that provides as much electricity as a small traditional plant would require about 4 acres.

Nuclear Power

As of 2010, 436 nuclear power plants were in operation worldwide and about 30 countries got some of their electricity from nuclear power. In France, over 75% of electricity is generated from nuclear power. By contrast, only about 20% of the electricity generated in the United states comes from nuclear power. The Nuclear Regulatory Commission (NRC), which was established in 1974, is the agency that ensures the safe use of nuclear materials for civilian purposes. It is the NRC that must approve any new design for a nuclear power plant. A design certification is valid for 15 years, but can be renewed for an additional 10 to 15 years. Approval of a design application is independent from approval of the construction or operation of a nuclear power plant.

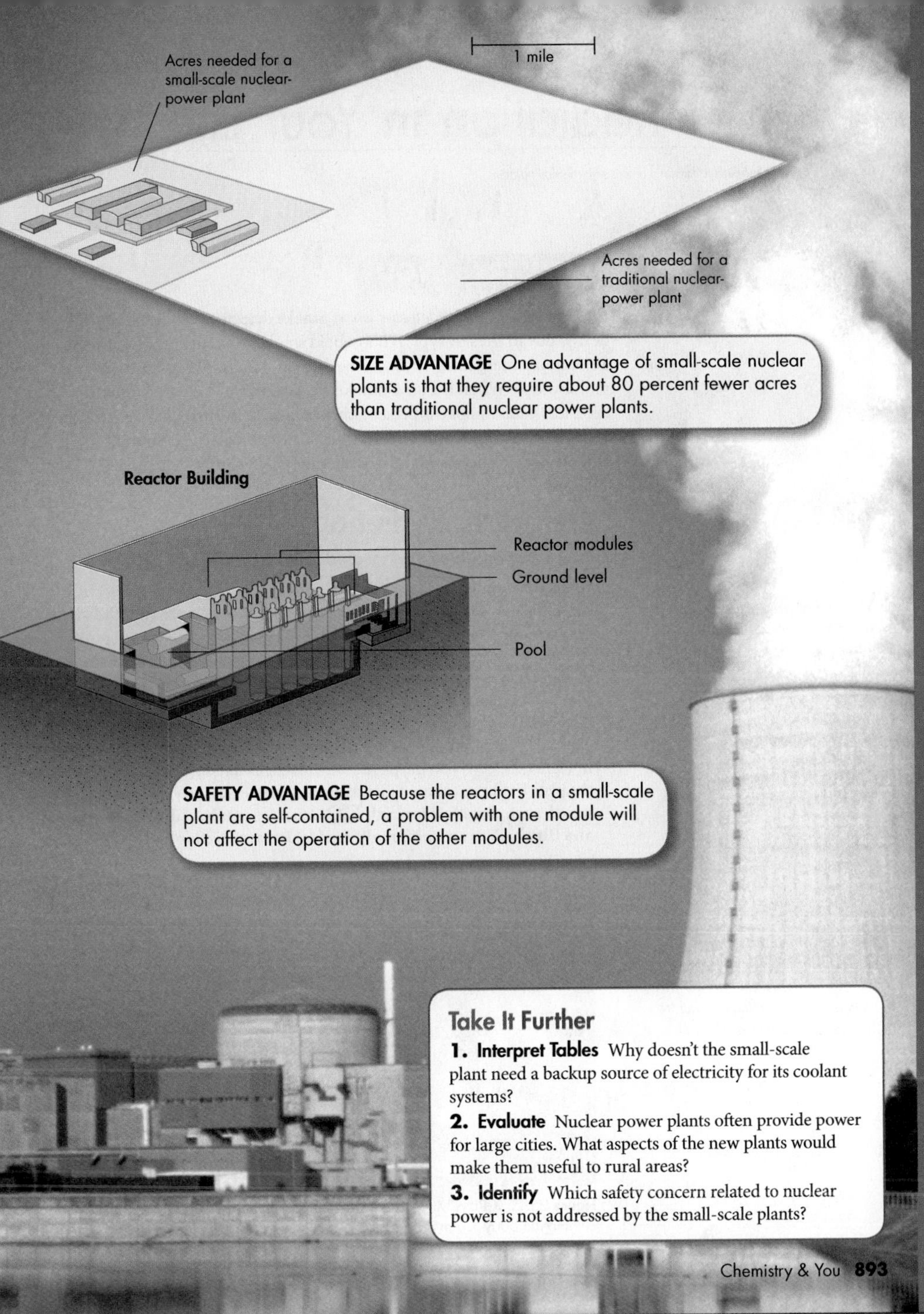

Extend

Connect to HISTORY

Tell students that the first large-scale nuclear power plant in the United States was built in Shippingport, Pennsylvania, a town about 35 miles northwest of Pittsburgh. The plant came on line December 18, 1957. One of the reasons that the plant was built in the middle of Pennsylvania coal country was that residents were not happy with the air pollution associated with coal-burning power plants. Engineers learned a lot from this first plant, including how to decommission a reactor when the plant was shut down in the 1980s.

ANALYZE DATA Display a copy of the map showing the locations of operating nuclear power reactors from the NRC web site. Have students compare the map with a population density map of the United States and identify the relationship between the data.

Answers

TAKE IT FURTHER

1. Pumps, which depend on electricity, are not used to circulate the coolant.
2. *Sample answer:* With a modular approach, a larger number of smaller plants could be constructed to serve a more scattered population.
3. The issue of nuclear waste is not addressed by the small-scale plants.

LESSON 25.4

Key Objectives

25.4.1 IDENTIFY three devices that are used to detect radiation.

25.4.2 DESCRIBE some practical uses of radioisotopes.

Additional Resources

Reading and Study Workbook, Lesson 25.4

Available Online or on Digital Media:

- Teaching Resources, Lesson 25.4
- Laboratory Manual, Lab 52

PearsonChem.com

Engage

CHEMISTRY & YOU Have students study the photograph and read the text that opens the section. Am-241 undergoes α decay. **Ask** What is the other product of the decay of an Am-241 nucleus? *(Np-237)* **Ask** Why do you have to exert special care in disposing of these smoke detectors? *(The detectors still contain radioactive materials.)*

Build Background

Lead a class discussion about the pros and cons of radiation exposure. Allow time for students to share personal knowledge involving radiation. Have students find recent articles in medical journals about the benefits of using radiation in treatments, as well as articles about the harmful consequences of radiation exposure.

IN ACADEMIC STANDARDS for SCIENCE

- Develop explanations based on reproducible data and observations gathered during laboratory investigations.
- Clearly communicate their ideas and results of investigations verbally and in written form using tables, graphs, diagrams, and photographs.

C.2.7 Compare and contrast nuclear reactions with chemical reactions. For nuclear reactions, describe how the fusion and fission processes transform elements present before the reaction into elements present after the reaction.

25.4 Radiation in Your Life

C.2.7 Compare and contrast nuclear reactions with chemical reactions

Q: *How does a smoke detector work?* Smoke detectors can limit injuries or deaths due to fires. A typical household smoke detector contains a small amount of americium, $^{241}_{95}Am$, in the form of AmO_2. Americium-241 is a radioisotope. When the air is smoke-free, a current flows through the smoke detector. When smoke is present, there is a drop in current. This drop is detected by an electronic circuit, which causes an alarm to sound. This lesson will help you understand the role of radiation in smoke detection.

Key Questions

What are three devices used to detect radiation?

What are some practical uses of radioisotopes?

Vocabulary

- ionizing radiation

Detecting Radiation

What are three devices used to detect radiation?

Radiation emitted by radioisotopes has enough energy to knock electrons off some atoms of a bombarded substance, producing ions. Thus, the radiation emitted by radioisotopes is called **ionizing radiation.** It is not possible for humans to see, hear, smell, or feel ionizing radiation. So people must rely on detection devices to alert them to the presence of radiation and to monitor its level. These devices work because of the effects of the radiation when it strikes atoms or molecules in the detector. For example, the radiation can expose a photographic plate, which produces an image such as the one shown in Figure 25.15. When the plate is developed, its darkened areas show where the plate has been exposed to radiation. Some devices rely on the current produced when atoms are ionized. **Geiger counters, scintillation counters, and film badges are commonly used to detect radiation.**

Q: *Radiation emitted in a smoke detector ionizes the nitrogen and oxygen in air, and a current flows. When smoke particles attach to the ions, the ions lose their charge. What happens next?*

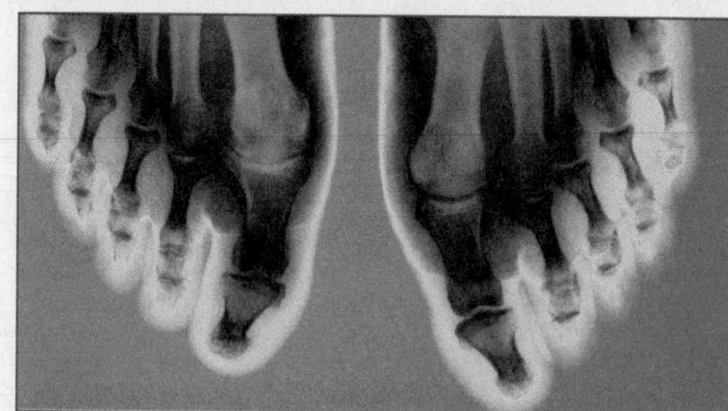

Figure 25.15 X-Rays X-rays allow doctors to see inside the body without having to cut into the body. Color was added to highlight parts of the image.

Focus on ELL

1 CONTENT AND LANGUAGE Discuss the term *ionizing radiation*. Explain that there are many types of radiation, such as visible light, microwaves, and gamma radiation. Explain that only high-energy radiation has enough energy to knock electrons off atoms.

2 FRONTLOAD THE LESSON Some students or their family members may have undergone medical tests or treatments that involved swallowing a radioactive substance. Some students may have relatives that work with radioactive materials. If they are willing, have students share their experiences. List the experiences on the board and refer back to them as you proceed with the lesson.

3 COMPREHENSIBLE INPUT Take students on a picture walk through the lesson. Have them look at each picture or diagram and read the caption. Afterwards, have them summarize what they will learn about in the lesson. If possible, provide a smoke detector, a Geiger counter and a film badge for students to examine.

Geiger Counter A Geiger counter uses a gas-filled metal tube to detect radiation. The tube has a central wire electrode that is connected to a power supply. When ionizing radiation penetrates a thin window at one end of the tube, the gas inside the tube becomes ionized. Because of the ions and free electrons produced, the gas is able to conduct electricity. Each time a Geiger tube is exposed to radiation, current flows. The bursts of current drive electronic counters or cause audible clicks from a built-in speaker. Geiger counters can detect alpha, beta, and gamma radiation. The first small, hand-held Geiger counters were developed in the 1930s.

Astronomers use Geiger counters to detect cosmic rays from outer space. Geologists use Geiger counters to search for radioactive minerals, such as uranium ores. These devices are also used to check for leaks in hospitals and other places that use radiation. Figure 25.16 shows one use for a Geiger counter.

Figure 25.16 Geiger Counters
This person is using a Geiger counter to check for pockets of radiation in contaminated dirt at a spill site.

Scintillation Counter A scintillation counter uses a phosphor-coated surface to detect radiation. When ionizing radiation strikes the surface, the phosphor produces bright flashes of light, or scintillations. The number of flashes and energies are detected electronically. The data is then converted into electronic pulses, which are measured and recorded. Scintillation counters are more sensitive than Geiger counters. This means that they can detect some radiation that would not be detected by a Geiger counter. Scintillation counters are used to track the path of radioisotopes through the body. They are also used to monitor the possible transport of radioactive materials across national borders and through airports.

Film Badge Figure 25.17 is a diagram of a typical film badge. The badge contains layers of photographic film covered with black light-proof paper. The film is sealed in a plastic or metal holder. To reach the film, radiation must pass through a filter, which absorbs some radiation, or a transparent area through which radiation can pass easily. People who work with or near ionizing radiation must wear a film badge to monitor their exposure while they are at work. At specific intervals, the film is removed and developed. The strength and type of radiation exposure are determined by comparing the darkness of the film in all the exposed areas. Records are kept of the results. Film badges do not protect a person from radiation, but they do monitor the degree of exposure. To protect themselves, workers must keep a safe distance from the source and use adequate shielding.

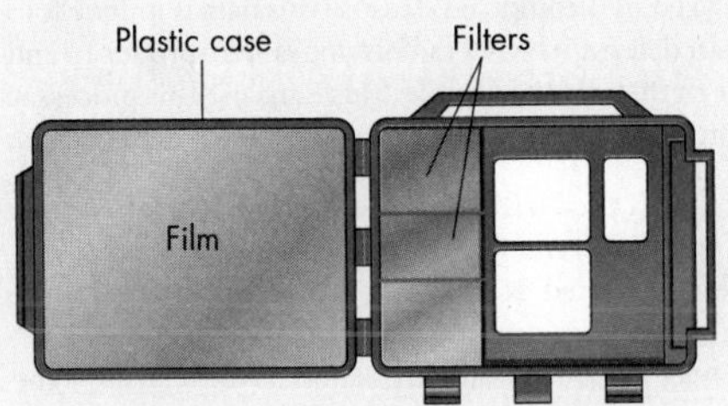

Figure 25.17 Film Badge
In a film badge, radiation passes through one of several filters or through a transparent area before it strikes the film. Different amounts of radiation pass through each area.

Gold-Foil Experiment

Hans Geiger (1882–1945) was a student of Ernest Rutherford. It was Geiger, along with another student named Marsden, who carried out the gold-foil, alpha-scattering experiment at Rutherford's direction. Geiger gained experience with the problem of detecting radiation, which ultimately led to his invention of the Geiger counter. Ask students to review the gold-foil experiment and describe the detection method used to measure the scattering of alpha particles. This method was a precursor of the scintillation counter.

Foundations for Reading

BUILD VOCABULARY Point out that the suffix *-ize* means "to make." Thus, *ionizing* radiation is able to make ions.

READING STRATEGY Encourage students to take a leadership role in reading this section. Have students take turns reading and conducting discussions about what was read.

Explain

Detecting Radiation

APPLY CONCEPTS Explain that a Geiger counter is mainly used to detect beta radiation because alpha particles cannot penetrate the tube and gamma rays pass through without causing many ionizations. By contrast, scintillation counters are designed to detect all types of radiation. However, they are not generally portable. The film badge is a updated version of Becquerel's photographic plate.

USE AN ANALOGY Given the changes in how computer and television screens display images, the only object that students may have seen that relies on a phosphor coating is a fluorescent light.

CHEMISTRY & YOU When ions in a smoke detector lose their charge, the current drops, which triggers an alarm.

Explore

Teacher Demo

PURPOSE Students will observe background radiation.

MATERIALS Geiger counter

PROCEDURE Using a Geiger counter equipped with a rate meter and a loudspeaker, have students listen to the random clicking noise. Be sure that there are no radioactive sources nearby. Point out that the counter is recording background radiation, which is always present. **Ask** What are possible sources of background radiation? *(cosmic, soil, and interior of Earth)* Explain that background radiation varies from location to location. To make an accurate measurement of radiation from a given source, local background radiation must be taken into account. Next, place the Geiger counter near an object containing a radioisotope, such as a smoke detector.

EXPECTED OUTCOME The Geiger counter indicates small amounts of background radiation and larger amounts near a source of radioactivity.

LESSON 25.4

Explore

Quick Lab

OBJECTIVE After completing this activity, students will be able to demonstrate the relationship between radiation intensity and the distance from the radiation source.

SKILLS FOCUS observing, measuring, using graphs, calculating

PREP TIME 20 minutes

CLASS TIME 30 minutes

EXPECTED OUTCOME As distance increases, intensity decreases and area increases.

ANALYZE AND CONCLUDE

1. Intensity is inversely related to the square of the distance.
2. When the distance is doubled, area increases by a factor of four, and intensity is one-fourth as much. When the distance is tripled, the area increases by a factor of nine, and the intensity is one-ninth as much.

ENRICHMENT Have students repeat the Quick Lab two times, changing the size of the hole from what it is in the original lab. On one trial, the hole should be slightly larger. On the other trial, the hole should be slightly smaller. **Ask** Does the size of the hole affect the results? *(No, the results are the same.)*

Quick Lab

Inverse-Square Relationships

Purpose To demonstrate the relationship between radiation intensity and the distance from the radiation source

Materials

- flashlight
- strips of duct tape
- scissors
- poster board, white (50 cm × 50 cm)
- meter ruler or tape measure
- flat surface, long enough to hold the meter ruler
- graph paper

Procedure

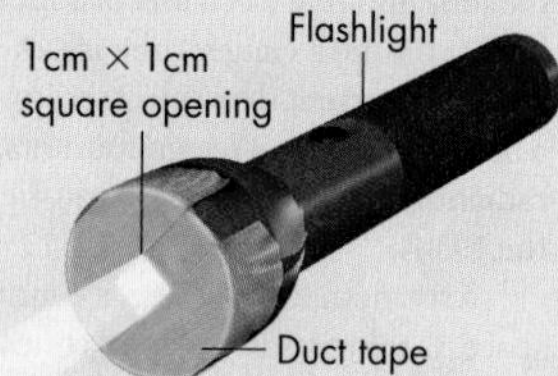

1. Estimate and record the distance (*A*) from the bulb filament to the front surface of the flashlight.
2. Cover the end of a flashlight with strips of duct tape. Leave a 1 cm × 1 cm square slit in the center of the tape.
3. Place the flashlight on its side on a flat surface. Turn on the flashlight. Darken the room.
4. Mount a large piece of white poster board in front of the flashlight, perpendicular to the horizontal surface.
5. Move the flashlight away from the board in short increments. At each position, record the distance (*B*) from the flashlight to the board and the length (*L*) of one side of the square image on the board.
6. On a sheet of graph paper, plot L on the y-axis versus $A + B$ on the x-axis. On another sheet, plot L^2 on the y-axis versus $A + B$ on the x-axis.

Analyze and Conclude

1. **Make Generalizations** As the flashlight is moved away from the board, what happens to the intensity of the light in the square image? Use your graphs to describe the relationship between intensity and distance.
2. **Explain** When the distance of the flashlight from the board (*B*) is doubled and tripled, what happens to the areas and intensities of the squares?

Using Radiation

What are some practical uses of radioisotopes?

Although radiation can be harmful, it can be used safely and has many important applications. **Radioisotopes are used to analyze matter, study plant growth, diagnose medical problems, and treat diseases.**

Analyzing Matter Scientists use radiation to detect trace amounts of elements in samples. The process is called neutron activation analysis. A sample is bombarded with neutrons from a radioactive source. Some atoms in the sample become radioactive. The half-life and type of radiation emitted can be detected and analyzed by a computer. Because this data is unique for each isotope, scientists can determine what radioisotopes were produced and infer what elements were in the original sample. Museums use this process to detect art forgeries. Crime laboratories use it to analyze gunpowder residues.

Using Tracers Radioisotopes called tracers are used in agriculture to test the effects of herbicides, pesticides, and fertilizers on plants. A tracer is introduced into the substance being tested. Next, plants are treated with the tagged substance. Devices that detect radioactivity are used to locate the substance in the plants. The tracer may also be monitored in animals that consume the plants, as well as in water and soil.

Focus on ELL

4 LANGUAGE PRODUCTION Have students work in groups of two to complete the lab. Make sure each group has ELLs of varied language proficiencies, so that more proficient students can help less proficient ones. Have students work according to their proficiency level.

BEGINNING: LOW/HIGH Restate each step of the Procedure in simpler words. Answer the Analyze and Conclude questions as a class.

INTERMEDIATE: LOW/HIGH Discuss the lab steps before students begin. Be sure they understand difficult words such as *filament, slit,* and *perpendicular.* Have students work with a partner to answer the Analyze and Conclude questions.

ADVANCED: LOW/HIGH Have students pre-read the lab and ask any questions before they begin. Have students help less proficient students plot the graph.

Diagnosing Medical Problems Radioisotopes can be used to detect disorders of the thyroid gland, which is located in the throat. The main function of this gland is to control the rate at which your cells release energy from food. The thyroid gland extracts iodide ions from blood and uses them to make the hormone thyroxine. To diagnose thyroid disease, the patient is given a drink containing a small amount of the radioisotope iodine-131. After about two hours, the amount of iodide uptake is measured by scanning the patient's throat with a radiation detector. Figure 25.18 shows the results of such a scan. In a similar way, the radioisotope technetium-99m is used to detect brain tumors and liver disorders. Phosphorus-32 is used to detect skin cancer.

Treating Diseases Radiation is one method used in the treatment of some cancers. Cancer is a disease in which abnormal cells in the body are produced at a rate far beyond the rate for normal cells. The mass of cancer cells that result from this runaway growth is called a tumor. Fast-growing cancer cells are more susceptible to damage by high-energy radiation such as gamma rays than are healthy cells. Thus, radiation can be used to kill the cancer cells in a tumor. Some normal cells are also killed, however, and cancer cells at the center of the tumor may be resistant to the radiation. Therefore, the benefits of the treatment and the risks to the patient must be carefully evaluated before radiation treatment begins. Cobalt-60 and cesium-137 are typical radiation sources for cancer therapy.

Salts of radioisotopes can also be sealed in gold tubes and directly inserted in tumors. This method of treatment is called seeding. The salts emit beta and gamma rays that kill the surrounding cancer cells. Because the radioisotope is in a sealed container, it is prevented from traveling to other parts of the body.

Prescribed drugs containing radioisotopes of gold, iodine, or phosphorus are sometimes used in radiation therapy. For example, a dose of iodine-131 larger than that used to detect thyroid diseases can be given to a patient to treat the disease. The radioactive iodine passes through the digestive system into the blood, which carries it to the thyroid. The iodine that collects in the gland emits beta particles and gamma rays , which provide therapy.

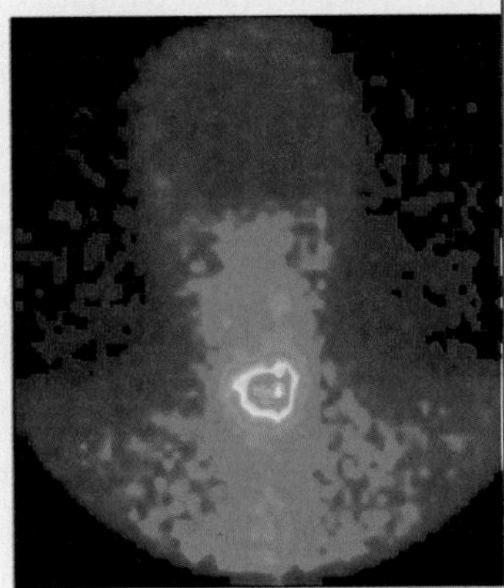

Figure 25.18 Radioactive Tracer
This scanned image of a thyroid gland shows where radioactive iodine-131 has been absorbed. Doctors use these images to identify thyroid disorders.

25.4 LessonCheck

ONLINE PROBLEMS

26. **Compare** In each of the three detection devices described in the lesson, what is used to detect the radiation?

27. **Review** What are two ways radioisotopes can be used in medicine?

28. **Define** Why is the radiation emitted by radioisotopes called ionizing radiation?

29. **Explain** Suppose you worked with or near a radiation source. Why might your employer use a film badge rather than a Geiger counter to monitor your exposure to radiation?

30. **Infer** Why do airports use scintillation counters and not Geiger counters to search for radioactive materials?

31. **Compare** Of Geiger counters, scintillation counters, and film badges, which device is most similar to a smoke detector? Explain your choice.

32. **Sequence** Briefly describe the three steps that occur when iodine-131 is used to diagnose thyroid disease?

33. **Explain** What is one advantage of using sealed tubes, or seeds, to treat a tumor?

Lesson Check Answers

26. Geiger counter: an inert gas; scintillation counter: phosphor-coated surface; film badge: photographic film
27. for diagnosis and treatment of diseases
28. It can knock electrons off atoms, causing them to become charged.
29. A Geiger counter measures the radiation emitted by a source at a given time. A film badge monitors exposure to radiation over time.
30. Scintillation counters can detect radiation that would not be detected by a Geiger counter.
31. A Geiger counter, because it also depends on a flow of electric current
32. Patient drinks mixture containing I-131; thyroid gland absorbs iodine; amount of uptake is measured.
33. The effect of the radioisotope is limited to a small area.

Extend

Connect to MEDICINE

The branch of medicine that deals with radioactive isotopes is called nuclear medicine. Tell students that many of the radioactive isotopes used in medicine are not naturally occurring; they are produced artificially by methods such as neutron bombardment. Explain that radioisotopes are used by physicians to diagnose and treat some forms of cancer. Point out that actively dividing cells are usually more sensitive to radiation than are normal cells. **Ask** What characteristic is important to consider when choosing a radioisotope as a treatment for cancer? *(The half-life should be long enough to kill the cancer cells but short enough to limit a patient's exposure.)* Explain that radioisotopes used in nuclear medicine typically have half-lives of a few days or less. Have interested students research the consequences of over-exposure to alpha, gamma, and beta radiation and present their findings in a poster.

Evaluate

Informal Assessment

Provide students with a two-column chart. Have students use the chart to compare and contrast detection of radiation by a film badge and detection of radiation by a scintillation counter. On the back of the chart, have students list or describe two practical uses of radioactive isotopes. *(Sample answer: as tracers to study plant growth and in the treatment of cancer)* Then have students complete the 25.4 Lesson Check.

Reteach

Review with students the three types of radiation detectors. Explain that, those devices do not protect the user from exposure to radiation. Note that workers who work with or near radioisotopes often wear special clothing, such as lead aprons or specialized self-contained suits, to shield themselves from certain types of radiation.

Evaluate

Review and Assessment Materials

CHEMISTRY TUTORIAL Encourage students to review solving half-life problems using the animated step-by-step tutorial for Chapter 25. Each Chemistry Tutorial is followed by a similar practice problem, where the student is guided through the problem-solving process.

CHAPTER PROBLEM SET Have students practice more nuclear chemistry problems using the Chapter 25 online Problem Set.

MATH TUTORIAL Struggling students can use a tutorial on exponential decay as a quick review. Ask students to answer the guided practice problem that follows the Math Tutorial.

DIRECTED VIRTUAL LAB Have students complete the Virtual Lab *Radiation* to help reinforce the concept of the penetration ability of alpha and beta radiation. In the post-lab exercise, students will analyze the data and draw conclusions based on the results of the experiment.

KINETIC ART Direct students to the Kinetic Art animations *Alpha and Beta Decay* and *How a Nuclear Reactor Works*, and have them answer the questions that accompany each animation.

CONCEPTS IN ACTION If students have trouble understanding the concept of half-life, ask them to view the Concepts in Action animation, which presents the material in a familiar context. Have students complete the questions after the animation.

Study Tip

Show all your calculations when working problems. Writing down the steps of a calculation helps you think through each step in the thought process of problem solving. Also, if you don't show your calculations, you will have more difficulty finding the source of your error if you made a mistake.

25 Study Guide

BIGIDEA ELECTRONS AND THE STRUCTURE OF ATOMS

Unstable atomic nuclei decay by emitting alpha or beta particles. Often gamma rays are emitted, too. During fission and fusion, atoms change their chemical identity as the number of protons in their nuclei change. In fission, large nuclei split into two or more smaller nuclei. In fusion, smaller nuclei combine to form larger nuclei at extremely high temperature and pressure. The ability to detect particles emitted when nuclei decay helps scientists study processes that take place in living organisms. This ability also allows scientists to determine the age of fossils and other objects.

25.1 Nuclear Radiation

Unlike chemical reactions, nuclear reactions are not affected by changes in temperature, pressure, or the presence of catalysts. Also, nuclear reactions of a given radioisotope cannot be slowed down, speeded up, or stopped.

Three types of nuclear radiation are alpha radiation, beta radiation, and gamma radiation.

- radioactivity (876)
- nuclear radiation (876)
- radioisotope (876)
- alpha particle (877)
- beta particle (878)
- gamma ray (879)

25.2 Nuclear Transformations

The neutron-to-proton ratio in a radioisotope determines the type of decay that occurs.

After each half-life, half of the original radioactive atoms have decayed into atoms of a new element.

Transmutation can occur by radioactive decay, or when particles bombard the nucleus of an atom.

- nuclear force (880)
- band of stability (880)
- positron (881)
- half-life (882)
- transmutation (885)
- transuranium elements (886)

Key Equation

$$A = A_0 \times \left(\frac{1}{2}\right)^n$$

25.3 Fission and Fusion

In a chain reaction, some of the emitted neutrons react with other fissionable atoms, which emit neutrons that react with still more fissionable atoms.

Fusion reactions, in which small nuclei combine, release much more energy than fission reactions, in which large nuclei split apart to form smaller nuclei.

- fission (888)
- neutron moderation (889)
- neutron absorption (889)
- fusion (891)

25.4 Radiation in Your Life

Geiger counters, scintillation counters, and film badges are commonly used to detect radiation.

Radioisotopes are used to analyze the composition of matter, study plant growth, diagnose medical problems, and treat diseases.

- ionizing radiation (894)

UbD

Understanding by Design

Performance Tasks

DECAYING GRACEFULLY Have students calculate the number of days needed for radon-222 to pass through 5 half-lives and the mass that would remain if a 4.00×10^{-10} g sample decayed for 5 half-lives. Then, have students plot a decay curve for radon-222 over the number of half-lives. *(19 days; 1.25×10^{-11} g)*

PICTURE DICTIONARY Have students create a picture dictionary of terms they have studied in this chapter. Instruct students to include a definition for each term listed in the Study Guide and a drawing that demonstrates the term.

Math Tune-Up: Nuclear Reactions

Problem	1 Analyze	2 Calculate	3 Evaluate
Plutonium-239 decays by emitting an alpha particle. What is the product of this reaction? $^{239}_{94}\text{Pu} \longrightarrow {}^{4}_{2}\text{He} + \text{X}$	Knowns: mass number of Pu = 239 mass number of α = 4 atomic number of Pu = 94 atomic number of α = 2 Unknowns: Mass number of X = ? Atomic number of X = ? Identity of X = ? In an equation for a nuclear reaction, mass numbers and atomic numbers must be balanced.	The mass number of X must equal the mass number of Pu minus the mass number of α. The atomic number of X must equal the atomic number of Pu minus the atomic number of α. $^{239}_{94}\text{Pu} \longrightarrow {}^{4}_{2}\text{He} + {}^{239-4}_{94-2}\text{X}$ $\longrightarrow {}^{4}_{2}\text{He} + {}^{235}_{92}\text{X}$ The element with atomic number 92 is uranium. $^{239}_{94}\text{Pu} \longrightarrow {}^{4}_{2}\text{He} + {}^{235}_{92}\text{U}$	The mass numbers total 239 on both sides of the equation. The atomic numbers total 94 on both sides of the equation. The isotope uranium-235 is a well-known radioisotope.
Thorium-234 has a half-life of 24.1 days. If a thorium-234 sample has a mass of 6.4×10^{-12} g, how much of the sample is left after 72.3 days?	Knowns: Original mass of Th = 6.4×10^{-12} g Decay time = 72.3 days $t_{\frac{1}{2}}$ = 24.1 days Unknowns: Mass of Th remaining = ? The mass of thorium-234 decreases by half with each half-life. Find the number of half-lives in 72.3 days and multiply the mass of thorium by $\frac{1}{2}$ for each half-life.	Divide the decay time by the half-life of thorium to find the number of half-lives. $\frac{72.3\ \cancel{\text{days}}}{24.1\ \cancel{\text{days}}/\text{half-life}} = 3$ half-lives Multiply the mass of thorium by $\frac{1}{2}$ three times. $6.4 \times 10^{-12}\ \text{g} \times \frac{1}{2} \times \frac{1}{2} \times \frac{1}{2} =$ 8.0×10^{-13} g	After three half-lives, the number of atoms of a radioisotope will decrease to $\frac{1}{8}$ of the original number.
Carbon-14 undergoes beta decay and produces nitrogen-14. The half-life for this process is 5730 years. In a sample from an ancient piece of pottery, the ratio of carbon-14 atoms to nitrogen-14 atoms is 25%. How old is the pottery?	Knowns: $t_{\frac{1}{2}}$ = 5730 yr Ratio of C-14 to N-14 = 25% Unknown: Age of sample = ? The number of C-14 atoms decreases by half every 5730 years. Find the number of half-lives that reduce the number of C-14 atoms to $\frac{1}{4}$ (25%) and multiply by 5730.	A ratio of C-14 atoms to N-14 atoms of $\frac{1}{4}$ means that C-14 has decayed for two half-lives: $\frac{1}{4} = \frac{1}{2} \times \frac{1}{2} = \left(\frac{1}{2}\right)^2$ Convert two half-lives to years. 5730 years $\times$ 2 = 11,460 years	The pottery sample is 11,460 years old, which is equal to two half-lives or the time required for 75% of the carbon-14 atoms to decay.

Hint: Review Sample Problem 25.1 for another calculation involving half-life.

Focus on ELL

5 ASSESS UNDERSTANDING Use a visually based assessment by choosing one or two diagrams from each lesson and having students explain the content.

BEGINNING:

LOW Ask students yes/no questions about the diagrams.

HIGH Ask students questions that they can answer with one or two words.

INTERMEDIATE: LOW/HIGH Ask students questions that they can answer with one or two simple sentences.

ADVANCED: LOW/HIGH Show each diagram to students and have them explain the content in several sentences.

Math Review

Example and sample practice problems related to nuclear reactions can be found on the following pages:

LESSON 25.2 Using Half-lives in Calculations—page 884

Evaluate

Answers

LESSON 25.1

34. A radioisotope is an isotope that is radioactive.

35. $^{226}_{88}\text{Ra} \longrightarrow ^{222}_{86}\text{Rn} + ^{4}_{2}\text{He}$

36. $^{210}_{82}\text{Pb} \longrightarrow ^{210}_{83}\text{Bi} + ^{0}_{-1}\text{e}$

37. **a.** α, +2 **b.** β, −1 **c.** γ, 0

38. **a.** $^{238}_{92}\text{U} \longrightarrow ^{234}_{90}\text{Th} + ^{4}_{2}\text{He}$;
thorium-234

b. $^{230}_{90}\text{Th} \longrightarrow ^{226}_{88}\text{Ra} + ^{4}_{2}\text{He}$;
radium-226

c. $^{235}_{92}\text{U} \longrightarrow ^{231}_{90}\text{Th} + ^{4}_{2}\text{He}$;
thorium-231

d. $^{222}_{86}\text{Rn} \longrightarrow ^{218}_{84}\text{Po} + ^{4}_{2}\text{He}$;
polonium-218

39. **a.** $^{14}_{6}\text{C} \longrightarrow ^{14}_{7}\text{N} + ^{0}_{-1}\text{e}$

b. $^{90}_{38}\text{Sr} \longrightarrow ^{90}_{39}\text{Y} + ^{0}_{-1}\text{e}$

c. $^{40}_{19}\text{K} \longrightarrow ^{40}_{20}\text{Ca} + ^{0}_{-1}\text{e}$

d. $^{13}_{7}\text{N} \longrightarrow ^{13}_{8}\text{O} + ^{0}_{-1}\text{e}$

40. **a.** mass number is unchanged; atomic number: increases by 1

b. mass number decreases by 4; atomic number decreases by 2

c. Mass number and atomic number are both unchanged.

41. **a.** $^{234}_{92}\text{U}$ **b.** $^{206}_{81}\text{Tl}$ **c.** $^{206}_{82}\text{Pb}$ **d.** $^{226}_{88}\text{Ra}$

LESSON 25.2

42. It undergoes radioactive decay.

43. $^{17}_{9}\text{F} \longrightarrow ^{17}_{8}\text{O} + ^{0}_{+1}\text{e}$

44. **a.** $^{13}_{6}\text{C}$ **b.** $^{1}_{1}\text{H}$ **c.** $^{16}_{8}\text{O}$ **d.** $^{14}_{7}\text{N}$

45. One half-life is the time required for one-half of the atoms of a radioisotope to emit radiation and decay.

46. so the person is exposed to radioactivity for a limited time

47. 0.63 mg

48. Transmutations that occur in nature are natural radioactivity. Transmutations that are forced to occur in reactors or accelerators are artificial radioactivity.

49. The elements with an atomic number greater than 92; none occur in nature and all are radioactive.

LESSON 25.3

50. Nuclear fission occurs when certain isotopes are bombarded with neutrons and split into smaller fragments. In a chain reaction, neutrons emitted during fission bombard other nuclei, which split and emit additional neutrons.

51. Fuel rods are removed when they can no longer sustain a constant output of electricity. The rods contain the remaining nuclear fuel and fission products. The rods are stored in pools of water.

52. Fusion requires extremely high temperatures, making it difficult to start or contain the reaction.

LESSON 25.4

53. Ionizing radiation, such as X-rays and gamma radiation, has sufficient energy to remove electrons from the atoms it hits.

54. Humans cannot use their senses to detect radiation.

55. People who work with or near ionizing radiation must wear a badge to monitor their exposure.

56. Cancer cells grow at a faster rate than do healthy cells.

ONLINE PROBLEMS

25 Assessment

* Solutions appear in Appendix E

Lesson by Lesson

25.1 Nuclear Radiation

34. Explain how radioisotopes are different from other isotopes.

35. The decay of radium-226 produces an isotope of the element radon and alpha radiation. The atomic number of radium (Ra) is 88; the atomic number of radon (Rn) is 86. Write a balanced equation for this transformation.

***36.** An isotope of the element lead (Pb) decays to an isotope of the element bismuth (Bi) by emission of a beta particle. Complete the equation for the reaction by supplying the missing atomic number and mass number.

$$^{210}_{?}\text{Pb} \longrightarrow ^{?}_{83}\text{Bi} + ^{0}_{-1}\text{e}$$

37. Write the symbol and charge for each item.

a. alpha particle
b. beta particle
c. gamma ray

***38.** Alpha radiation is emitted during the decay of the following isotopes. Write balanced nuclear equations to describe each decay process. Name the element produced in each case.

a. uranium-238 ($^{238}_{92}\text{U}$) **b.** thorium-230 ($^{230}_{90}\text{Th}$) **c.** uranium-235 ($^{235}_{92}\text{U}$) **d.** radon-222 ($^{222}_{86}\text{Rn}$)

39. The following radioisotopes are beta emitters. Write balanced nuclear equations to describe each decay process.

a. carbon-14 ($^{14}_{6}\text{C}$) **b.** strontium-90 ($^{90}_{38}\text{Sr}$) **c.** potassium-40 ($^{40}_{19}\text{K}$) **d.** nitrogen-13 ($^{13}_{7}\text{N}$)

***40.** How are the mass number and atomic number of a nucleus affected by the loss of the following?

a. beta particle
b. alpha particle
c. gamma ray

41. The following radioactive nuclei decay by emitting alpha particles. Write the product of the decay process for each isotope.

a. $^{238}_{94}\text{Pu}$ **b.** $^{210}_{83}\text{Bi}$ **c.** $^{210}_{84}\text{Po}$ **d.** $^{230}_{90}\text{Th}$

25.2 Nuclear Transformations

***42.** What happens to an atom with a nucleus that falls outside the band of stability?

43. Write an equation for the radioactive decay of fluorine-17 by positron emission.

44. Identify the more stable isotope in each pair.

a. $^{14}_{6}\text{C}$, $^{13}_{6}\text{C}$ **b.** $^{3}_{1}\text{H}$, $^{1}_{1}\text{H}$ **c.** $^{15}_{8}\text{O}$, $^{16}_{8}\text{O}$ **d.** $^{13}_{7}\text{N}$, $^{14}_{7}\text{N}$

45. Define *half-life*.

46. Why is it important that radioactive isotopes used for diagnosis or treatment of medical problems have relatively short half-lives?

***47.** A patient is given 20 mg of iodine-131. The half-life of iodine-131 is 8 days. How much of the isotope will remain in the patient's body after 40 days?

48. What is the difference between natural and artificial radioactivity?

49. What are the transuranium elements? Why are they unusual?

25.3 Fission and Fusion

50. Describe the process of nuclear fission, and define a nuclear chain reaction.

51. Why are spent fuel rods removed from a reactor core? What do they contain? What happens to them after they are removed?

52. Fusion reactions produce enormous amounts of energy. Why is fusion not used to generate electrical power?

25.4 Radiation in Your Life

53. Why are X-rays and the radiation emitted by radioisotopes called ionizing radiation?

54. Why must people rely on devices such as Geiger counters to detect radiation?

55. What type of people are required to wear a film badge and what is the purpose of this device?

56. Why are cancer cells more easily damaged by high-energy radiation than healthy cells are?

900 Chapter 25 • Assessment

Understand Concepts

* **57.** Write balanced nuclear equations for these transmutations.

a. $^{30}_{15}P$ to $^{30}_{14}Si$
b. $^{13}_{6}C$ to $^{14}_{6}C$
c. $^{131}_{53}I$ to $^{131}_{54}Xe$

58. How are the nuclear reactions that take place in the sun different from the nuclear reactions that take place in a nuclear reactor?

59. Complete these nuclear equations.

a. $^{32}_{15}P \longrightarrow \square + ^{0}_{-1}e$
b. $\square \longrightarrow ^{14}_{7}N + ^{0}_{-1}e$
c. $^{238}_{92}U \longrightarrow ^{234}_{90}Th + \square$
d. $^{141}_{56}Ba \longrightarrow \square + ^{0}_{-1}e$
e. $\square \longrightarrow ^{181}_{77}Ir + ^{4}_{2}He$

60. Write nuclear equations for the beta decay of the following isotopes.

a. $^{90}_{38}Sr$ **b.** $^{14}_{6}C$ **c.** $^{137}_{55}Cs$ **d.** $^{239}_{93}Np$ **e.** $^{50}_{22}Ti$

61. The graph shows the radioactive decay curve for thorium-234. Use the graph to answer the questions below.

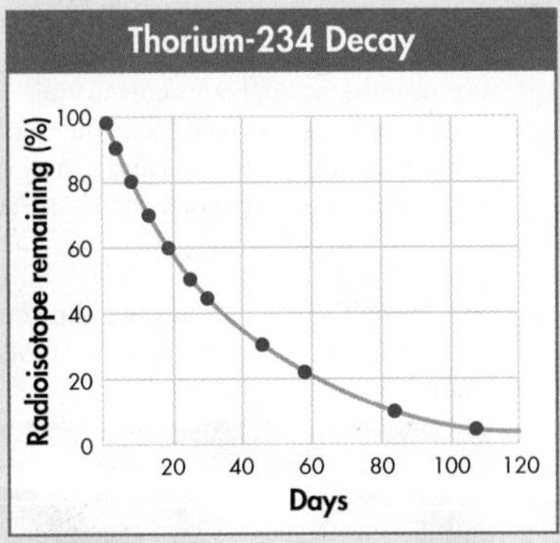

a. What percentage of the isotope remains after 60 days?
b. After 40 days have passed, how many grams of a 250-g sample of thorium-234 would remain?
c. How long would it take in days for 44 g of thorium-234 to decay to 4.4 g of thorium-234?
d. What is the half-life of thorium-234?

62. Write a balanced nuclear equation for each word equation.

a. Radon-222 emits an alpha particle to form polonium-218.
b. Radium-230 is produced when thorium-234 emits an alpha particle.
c. When polonium-210 emits an alpha particle, the product is lead-206.

63. Briefly describe the contributions the following people made to the study of radioactivity and nuclear chemistry.

a. Marie Curie
b. Antoine Henri Becquerel
c. James Chadwick
d. Ernest Rutherford

64. How many protons and how many neutrons are in each of the following nuclei?

a. $^{60}_{27}Co$ **b.** $^{206}_{82}Pb$ **c.** $^{233}_{90}Th$ **d.** $^{3}_{1}H$

* **65.** A sample of matter has 32 million radioactive atoms. How many of these atoms would be left after five half-lives?

66. Write balanced nuclear equations for alpha emission by each of these isotopes.

a. $^{231}_{91}Pa$ **b.** $^{241}_{95}Am$ **c.** $^{226}_{88}Ra$ **d.** $^{252}_{99}Es$

* **67.** Write balanced nuclear equations for beta emission by each of these isotopes.

a. $^{3}_{1}H$ **b.** $^{28}_{12}Mg$ **c.** $^{131}_{53}I$ **d.** $^{75}_{34}Se$

68. Use the concept of stability to compare chemical reactions with nuclear reactions.

69. The ratio of carbon-14 to carbon-12 in a chunk of charcoal from an archaeological dig is one-half the ratio of carbon-14 to carbon-12 in a piece of freshly cut wood. Hold old is the chunk of charcoal?

70. How are a positron and an electron similar? How are they different?

* **71.** Use what you know about balanced nuclear reaction equations to complete the following equations.

a. $^{38}_{19}K \longrightarrow ^{38}_{20}Ca + ?$
b. $^{242}_{94}Pu \longrightarrow ? + ^{4}_{2}He$
c. $^{68}_{31}Ga \longrightarrow ? + ^{0}_{1}e$
d. $^{68}_{32}Ge \longrightarrow ^{68}_{31}Ga + ?$

ASSESSMENT

Answers

UNDERSTAND CONCEPTS

57. a. $^{30}_{15}P + ^{0}_{-1}e \longrightarrow ^{30}_{14}Si$
b. $^{13}_{6}C + ^{1}_{0}n \longrightarrow ^{14}_{6}C$
c. $^{131}_{53}I \longrightarrow ^{131}_{54}Xe + ^{0}_{-1}e$

58. Nuclear fusion takes place in the sun. A nuclear reactor utilizes nuclear fission.

59. a. $^{32}_{16}S$ **b.** $^{14}_{6}C$ **c.** $^{4}_{2}He$ **d.** $^{141}_{57}La$ **e.** $^{185}_{79}Au$

60. a. $^{90}_{38}Sr \longrightarrow ^{90}_{39}Y + ^{0}_{-1}e$
b. $^{14}_{6}C \longrightarrow ^{14}_{7}N + ^{0}_{-1}e$
c. $^{137}_{55}Cs \longrightarrow ^{137}_{56}Ba + ^{0}_{-1}e$
d. $^{239}_{93}Np \longrightarrow ^{239}_{94}Pu + ^{0}_{-1}e$
e. $^{50}_{22}Ti \longrightarrow ^{50}_{23}V + ^{0}_{-1}e$

61. a. about 20% **c.** about 83 days
b. about 85 g **d.** about 25 days

62. a. $^{222}_{86}Rn \longrightarrow ^{218}_{84}Po + ^{4}_{2}He$
b. $^{234}_{90}Th \longrightarrow ^{230}_{88}Ra + ^{4}_{2}He$
c. $^{210}_{84}Po \longrightarrow ^{206}_{82}Pb + ^{4}_{2}He$

63. a. coined the term radioactivity and did pioneering work with Becquerel
b. discovered natural radioactivity from uranium ores
c. discovered the neutron
d. performed the first artificial transmutation

64. a. 27 protons and 33 neutrons
b. 82 protons and 124 neutrons
c. 90 protons and 143 neutrons
d. 1 proton and 2 neutrons

65. 1 million atoms

66. a. $^{231}_{91}Pa \longrightarrow ^{227}_{89}Ac + ^{4}_{2}He$
b. $^{241}_{95}Am \longrightarrow ^{237}_{93}Np + ^{4}_{2}He$
c. $^{226}_{88}Ra \longrightarrow ^{222}_{86}Rn + ^{4}_{2}He$
d. $^{252}_{99}Es \longrightarrow ^{248}_{97}Bk + ^{4}_{2}He$

67. a. $^{3}_{1}H \longrightarrow ^{3}_{2}He + ^{0}_{-1}e$
b. $^{28}_{12}Mg \longrightarrow ^{28}_{13}Al + ^{0}_{-1}e$
c. $^{131}_{53}I \longrightarrow ^{131}_{54}Xe + ^{0}_{-1}e$
d. $^{75}_{34}Se \longrightarrow ^{75}_{35}Br + ^{0}_{-1}e$

68. In chemical reactions, the transfer or sharing of electrons may make atoms more stable. In nuclear reactions, changes in their nuclei make atoms of unstable isotopes more stable.

69. 5730 years

70. A positron and an electron have the same mass but opposite charges.

71. a. $^{0}_{-1}e$ **b.** $^{238}_{92}U$ **c.** $^{68}_{32}Ge$ **d.** $^{0}_{+1}e$

ASSESSMENT

Evaluate

Answers

THINK CRITICALLY

72. **a.** platinum **e.** xenon
b. thorium **f.** californium
c. francium **g.** vanadium
d. titanium **h.** palladium
Thorium (b), francium (c), and californium (f) have no stable isotopes.

73. Zirconium-90 and neon-21 nuclei are stable; neodymium-130 nuclei are unstable.

74. In every round of the tournament, one-half the teams are eliminated; in every half-life, one-half the nuclei decay. In the tournament a single team eventually emerges as the winner and the tournament stops, but radioactive decay continues (almost) indefinitely.

75. $^{215}_{85}At$

76. 11,460 years

77. The large mass of the alpha particle and its charge make it more likely that it will collide with another particle and be stopped.

78. Radioactive isotopes of these elements can be incorporated into the body tissues of organisms. When the isotopes decay, they can damage cells within tissues.

79. The treatment dose is larger than the dose used for detection.

80. No, because the detector would not be effective after a relatively short period of time

81. one neutron

82. $^{25}_{12}Mg$

83. It would likely do the most harm if it were inhaled or ingested.

84. 24.6 years

85. The combination of half-life and type of radiation emitted is unique for each isotope.

86. The discovery of radioactivity contradicted Dalton's assumption that atoms of one element are never changed into atoms of another element.

87. $^{18}_{8}O$

88. Toppling one domino leads to the fall of other dominoes just as the emission of neutrons by one fissionable atom leads to the emission of more neutrons and more fission.

Think Critically

***72.** **Classify** Name the elements represented by the following symbols and indicate which of them would have no stable isotopes.

a. Pt **b.** Th **c.** Fr **d.** Ti
e. Xe **f.** Cf **g.** V **h.** Pd

73. **Interpret Graphs** Use the graph to determine which of these isotopes have stable nuclei: neon-21, zirconium-90, and neodymium-130.

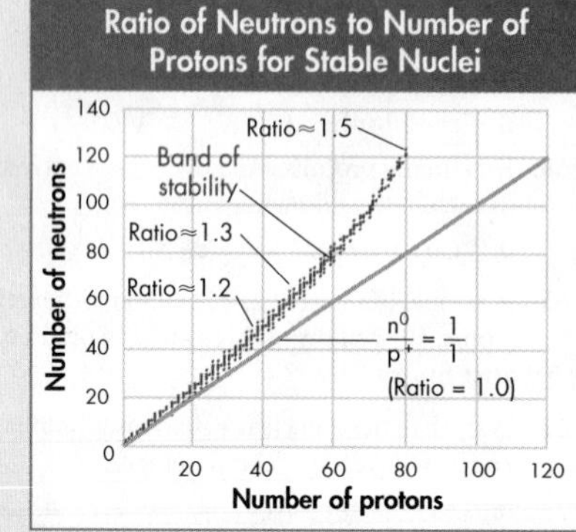

74. **Use Analogies** Compare the half-life of an element to a single-elimination sports tournament. *Hint:* What happens in each round of the tournament?

75. **Infer** A radioactive nucleus decays to give a bismuth-211 ($^{211}_{83}Bi$) nucleus and an alpha particle. What was the original nucleus?

***76.** **Calculate** The carbon-14 content of an object produces 4 counts per minute per gram of carbon. Living matter has a carbon-14 content that produces 16 counts per minute per gram of carbon. What is the age of the object?

77. **Relate Cause and Effect** Why is an alpha particle less penetrating than a beta particle?

78. **Infer** Why might radioisotopes of C, N, and O be especially harmful to living creatures?

79. **Compare and Contrast** Iodine-131 is used to diagnose and treat thyroid disorders. What is the main difference between the two processes?

80. **Apply Concepts** Should a radioisotope with a half-life measured in days be used in a smoke detector? Why or why not?

81. **Calculate** Bismuth-209 was bombarded with iron-58 for several days. Meitnerium ($^{266}_{109}Mt$) was produced. How many neutrons were released per atom?

$$^{209}_{83}Bi + ^{58}_{26}Fe \longrightarrow ^{266}_{109}Mt + ?^{1}_{0}n$$

***82.** **Predict** When neutrons strike magnesium-24 ($^{24}_{12}Mg$), a neutron is captured and photons are ejected. What new element is formed?

83. **Make Generalizations** Plutonium-239 emits alpha particles, which do not penetrate a thin sheet of paper or skin. Under what conditions would plutonium-239 be especially hazardous for organisms?

***84.** **Calculate** Tritium (hydrogen-3) has a half-life of 12.3 years. How old is a bottle of wine if the tritium content is 25% that of new wine?

85. **Apply Concepts** What properties of isotopes make it possible to use neutron activation analysis to identify the composition of matter?

86. **Relate Cause and Effect** What effect did the discovery of radioactivity have on Dalton's model of the atom?

87. **Analyze Data** A sample of californium-249 ($^{249}_{98}Cf$) was used as the target in the synthesis of seaborgium ($^{236}_{106}Sg$). Four neutrons were emitted for each transformed $^{249}_{98}Cf$. The result was a nucleus with 106 protons and a mass of 263 amu. What type of particle struck the target?

$$^{249}_{98}Cf + ? \longrightarrow 4^{1}_{0}n + ^{263}_{106}Sg$$

88. **Use Models** Explain how the falling dominoes in the photograph can be used as a model for a chain reaction.

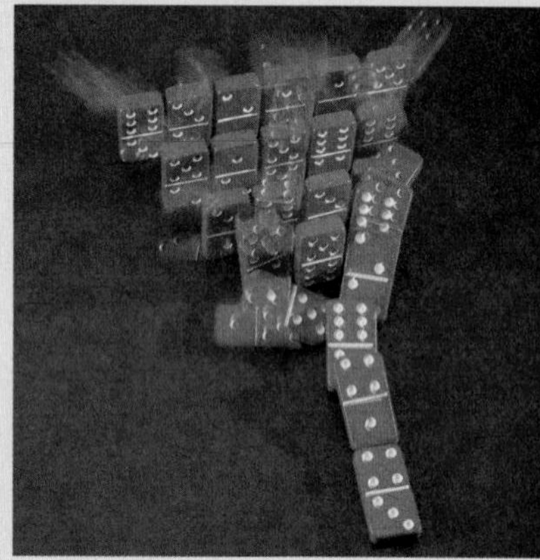

902 Chapter 25 • Assessment

Enrichment

89. Calculate The radioisotope cesium-137 has a half-life of 30 years. A sample decayed at the rate of 544 counts per minute in the year 1985. In what year will the decay rate be 17 counts per minute?

90. Interpret Graphs Use the graph below to answer the questions.

a. Describe the process that is being depicted in the graph.

b. Suggest an appropriate title for the graph.

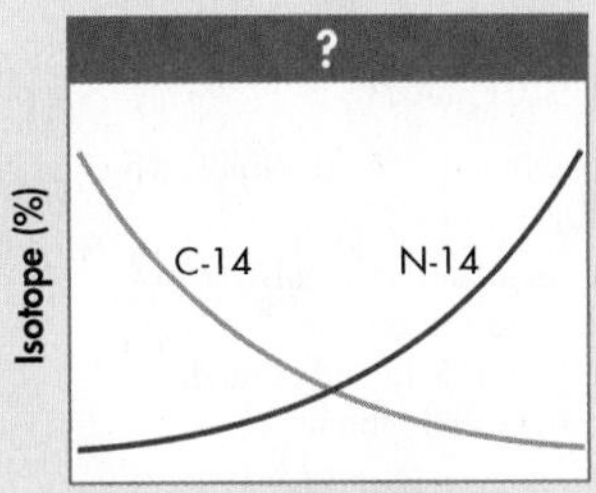

91. Apply Concepts Bismuth-211 decays by alpha emission to yield another radioisotope, which emits beta radiation as it decays to a stable isotope. Write equations for the nuclear reactions and name the decay products.

92. Analyze Data What isotope remains after three beta particles and five alpha particles are lost from a thorium-234 isotope?

93. Evaluate In the following argument, the third statement is based on the first two statements. Is the reasoning logical? Why or why not?

(1) Radiation kills fast-growing cells.
(2) Cancer cells are fast-growing.
(3) Therefore, radiation kills only cancer cells.

*** 94. Calculate** Uranium has a density of 19 g/cm^3. What volume does a mass of 8.0 kg of uranium occupy?

95. Make Generalizations Element 107 (Bh) is formed when nuclei of element 109 (Mt) each emit an alpha particle. Nuclei of element 107, in turn, each emit an alpha particle, forming an atom with a mass number of 262. Write balanced equations for these two nuclear reactions.

Write About Science

96. Summarize Research methods used to date materials such as pottery, coral, and stone. Prepare a written report that summarizes your findings on the radioisotopes used, their half-lives, and their limitations.

97. Explain Research how technetium-99m is produced. What does the letter *m* at the end of its name stand for? How is the isotope used in bone imaging?

CHEMYSTERY

An Ice-Age Baby

Scientists were able to use carbon-14 dating to determine the age of the baby mammoth discovered in 2007. During Lyuba's short time on Earth, the ratio of carbon-14 to other carbon isotopes in her body was constant. After she died, the unstable carbon-14 atoms began to decay. The ratio of carbon-14 atoms to other carbon atoms was no longer fixed. Based on the ratio of carbon isotopes in the preserved sample, Lyuba lived and died about 40,000 years ago.

Archaeologists use the same method to date artifacts left behind by ancient cultures. An artifact is an object made or shaped by humans. Examples of artifacts are tools, weapons, and ornaments.

98. Calculate Approximately how many half-lives does 40,000 years represent? The half-life for carbon-14 is 5730 years.

99. Apply Concepts Archaeologists find a wooden bowl, a stone arrowhead, and a bone bead at a site that they suspect is about 40,000 years old. Which objects can they use to test their hypothesis? Explain.

100. Connect to the BIGIDEA Describe what happens to the structure of carbon-14 atoms as this isotope decays.

CHEMYSTERY

APPLY CONCEPTS After students have read through the CHEMystery, discuss with students how scientists used a knowledge of nuclear reactions to determine when Lyuba lived. **Ask** What type of nuclear reaction is used in carbon-14 dating? *(radioactive decay)* **Ask** Did the fact that Lyuba was found frozen in ice affect the decay of the carbon-14? *(No. Radioactive decay does not depend on temperature.)*

Then, have students consider how the type of item to be dated determines what isotope a scientist uses to determine the item's age. **Ask** Suppose the reindeer herder who found Lyuba had instead found a rock with the imprint of a fossil that appeared extremely old. Could carbon-14 dating have been used to determine the age of the imprint? Explain. *(No. Carbon-14 dating is useful only for dating once-living things.)* Note that, because the half-life of carbon-14 is 5730 years, it can only be used for dating objects up to about 60,000 years old.

CHEMYSTERY ANSWERS

98. about 7 half-lives

99. They can use either the bowl or the bead because the material in both objects came from a once-living organism.

100. A neutron in the carbon-14 nucleus splits into a proton and an electron. The proton remains in the nucleus; the electron is emitted as a beta particle.

Answers

ENRICHMENT

89. 2135

90. a. the radioactive (beta) decay of carbon-14, along with the increase of the nitrogen product

b. Effect of Time on Carbon-14 and Nitrogen-14 Concentrations

91. $^{211}_{83}Bi \longrightarrow ^{207}_{81}Tl + ^{4}_{2}He$; thallium–207

$^{207}_{81}Tl \longrightarrow ^{207}_{82}Pb + ^{0}_{-1}e$; lead–207

92. Bismuth-214 remains.

93. The reasoning is not sound. Although it would affect cancer cells the most, it would also affect other cells.

94. 4.2×10^2 cm^3

95. $^{268}_{109}Mt \longrightarrow ^{264}_{107}Bh + ^{4}_{2}He$

$^{264}_{107}Bh \longrightarrow ^{262}_{105}Db + ^{4}_{2}He$

WRITE ABOUT SCIENCE

96. Students should note that the half-life of an isotope can limit its usefulness for dating certain objects.

97. Tc-99m is made from the decay of molybdenum-99, which also has a relatively short half-life. The *m* means "metastable," which indicates that the isotope might become more or less stable. Medical uses might include the following examples. To pinpoint a liver problem, doctors inject a mixture containing Tc-99m, which should be absorbed by the liver. If gamma radiation emitted by the Tc-99m atoms in the mixture is detected in the kidneys, there is likely a problem with the liver. If the mixture reaches the liver but does not reach the gallbladder, the duct connecting the organs is probably blocked. Tc-99m is also used to detect the presence of cancerous cells in bone marrow.

ASSESSMENT

Evaluate

Cumulative Review

101. **a.** 26 protons and electrons, and 33 neutrons
b. 92 protons and electrons, and 143 neutrons
c. 24 protons and electrons, and 28 neturons

102. The Pauli exclusion principle states that no two electrons in an atom can have the same quantum numbers. Hund's rule states that electrons that occupy orbitals of equal energy distribute with unpaired spins as much as possible.

103. **a.** covalent **b.** ionic **c.** covalent **d.** ionic

104. The oxygen atom carries a partial negative charge, and the hydrogen atoms carry partial positive charges. The water molecule's bent shape results in a net polarity, and the negative end of one molecule attracts the positive end of another molecule.

105. $9.22 \times 10^3\ cm^3\ H_2$; $0.412\ mol\ H_2$

106. **a.** $Ca(OH)_2 + 2HCl \rightarrow CaCl_2 + 2H_2O$
b. $Fe_2O_3 + 3H_2 \rightarrow 2Fe + 3H_2O$
c. $2NaHCO_3 + H_2SO_4 \rightarrow Na_2SO_4 + 2CO_2 + 2H_2O$
d. $2C_2H_6 + 7O_2 \rightarrow 4CO_2 + 6H_2O$

107. 6.7 mL

108. **a.**
```
      CH3
      |
CH3—C—CH2—CH2—CH2—CH3
      |
      CH3
```
b. (cyclopentane ring with CH3 on each of two adjacent carbons)
```
   CH3
   |
  /\
 /  \—CH3
 \__/
```
c.
```
      CH3
      |
CH3—C=CH—CH2—CH2—CH2—CH3
```
d. CH3—C≡C—CH3

e. CH3—(benzene ring)—CH3

f.
```
            CH2CH3
            |
CH3—CH2—CH—CH2—CH2—CH2—CH2—CH3
```

109. **a.** propanoic acid **b.** propanal **c.** 1-propanol **d.** 1-aminopropane **e.** 1-chloropropane **f.** ethylmethyl ether

110. **a.** ethanal **b.** ethene **c.** ethanoic acid **d.** ethyne

111. 1-propanol and ethanoic acid

Cumulative Review

101. How many protons, neutrons, and electrons are in an atom of each isotope?
a. iron-59
b. uranium-235
c. chromium-52

102. What is the Pauli exclusion principle? What is Hund's rule?

103. Identify the bonds between each pair of atoms as ionic or covalent.
a. carbon and silicon
b. calcium and fluorine
c. sulfur and nitrogen
d. bromine and cesium

104. The diagram below shows a water molecule. Identify the location of any partial positive and partial negative charges on the molecule. Then explain how the partial charges and their locations produce an attraction between different water molecules.

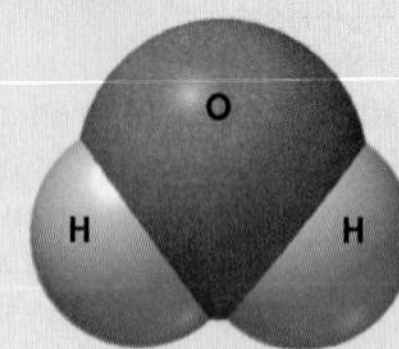

★105. A piece of magnesium with a mass of 10.00 g is added to sulfuric acid. How many cubic centimeters of hydrogen gas (at STP) will be produced if the magnesium reacts completely? How many moles of hydrogen gas are in this volume?

106. Balance the following equations.
a. $Ca(OH)_2 + HCl \longrightarrow CaCl_2 + H_2O$
b. $Fe_2O_3 + H_2 \longrightarrow Fe + H_2O$
c. $NaHCO_3 + H_2SO_4 \longrightarrow Na_2SO_4 + CO_2 + H_2O$
d. $C_2H_6 + O_2 \longrightarrow CO_2 + H_2O$

★107. You have a 0.30*M* solution of sodium sulfate. What volume (in mL) must be measured to give 0.0020 mol of sodium sulfate?

108. Draw the structural formula for each compound.
a. 2,2-dimethylhexane
b. 1,2-dimethylcyclopentane
c. 2-methyl-2-heptene
d. 2-butyne
e. 1,4-dimethylbenzene
f. 3-ethyloctane

109. Name each compound.
a. CH_3CH_2COOH
b. CH_3CH_2CHO
c. $CH_3CH_2CH_2OH$
d. $CH_3CH_2CH_2NH_2$
e. $CH_3CH_2CH_2Cl$
f. $CH_3CH_2OCH_3$

110. For each pair of compounds, which is the more highly oxidized?
a. ethanol and ethanal
b. ethane and ethene
c. ethanoic acid and ethanal
d. ethyne and ethane

111. What two compounds result from the acid-catalyzed hydrolysis of propyl ethanoate?

112. Which of these classes of compounds does not contain a carbon–oxygen double bond?
a. amide
b. ketone
c. aldehyde
d. carboxylic acid

113. Match a numbered item with each term.

a. amino acid	**(1)** carbohydrate
b. fat	**(2)** nucleic acid
c. monosaccharide	**(3)** lipid
d. peptide bond	**(4)** protein
e. sugar	
f. DNA	
g. saponification	
h. genetic code	
i. enzyme	
j. triglyceride	

If You Have Trouble With...

Question	101	102	103	104	105	106	107	108	109	110	111	112	113
See Chapter	4	5	8	8	12	11	16	22	23	23	23	23	24

112. They all contain a carbon-oxygen double bond.

113. **a.** (4) **b.** (3) **c.** (1) **d.** (4) **e.** (1) **f.** (2) **g.** (3) **h.** (2) **i.** (4) **j.** (3)

Standardized Test Prep for Indiana

Standard: C.2.8

Tips for Success

Anticipate the Answer Use what you know to decide what you think the answer should be. Then see if your answer, or one similar to it, is given as a choice.

Select the choice that best answers each question.

1. If a radioisotope undergoes beta emission,
   (A) the atomic number changes.
   (B) the number of neutrons remains constant.
   (C) the isotope loses a proton.
   (D) the mass number changes.

2. The radioisotope radon-222 has a half-life of 3.8 days. How much of an initial 20.0-g sample of radon-222 would remain after 15.2 days?
   (A) 5.00 g (C) 1.25 g
   (B) 12.5 g (D) 2.50 g

3. Spent fuel rods from nuclear reactors
   (A) are no longer radioactive.
   (B) are stored under water for at least a decade.
   (C) contain only one isotope of uranium, ^{238}U.
   (D) remain radioactive for less than 100 years.

4. What particle is needed to balance this equation?
   $^{27}_{13}Al + ^{4}_{2}He \longrightarrow ? + ^{30}_{15}P$

For each nuclear equation in Questions 5–8, name the particle that is being emitted or captured.

5. $^{59}_{26}Fe \longrightarrow ^{59}_{27}Co + ^{0}_{-1}e$
6. $^{185}_{79}Au \longrightarrow ^{181}_{77}Ir + ^{4}_{2}He$
7. $^{59}_{27}Co + ^{1}_{0}n \longrightarrow ^{60}_{27}Co$
8. $^{118}_{54}Xe \longrightarrow ^{118}_{53}I + ^{0}_{+1}e$

Use the drawings of atomic nuclei to answer Questions 9 and 10.

9. Write the name and symbol for each isotope.
10. Which isotope is radioactive?

(A) 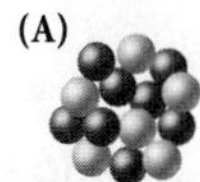 (B) 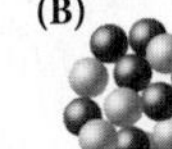 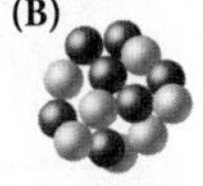 (C) 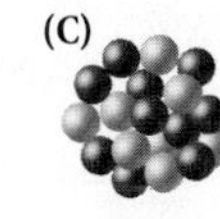

Proton Neutron

Use the graph to answer Questions 11–13.

Estimate the percent remaining of the radioisotope after the given number of half-lives.

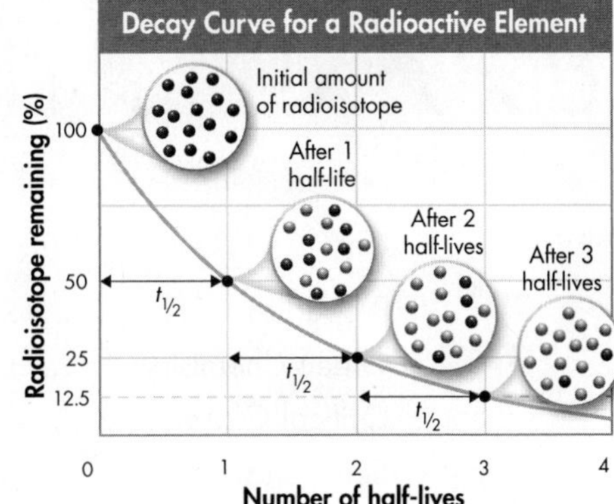

11. $0.5\ t_{1/2}$
12. $1.25\ t_{1/2}$
13. $3.75\ t_{1/2}$

The lettered choices below refer to Questions 14–17. A lettered choice may be used once, more than once, or not at all.

(A) film badge
(B) radioactive tracer
(C) radiation therapy
(D) neutron activation analysis
(E) Geiger counter

Which of the above items or processes is best described by each of the following applications?

14. treating some cancers
15. detecting ionizing radiation
16. monitoring exposure to radiation
17. diagnosing some diseases

If You Have Trouble With...

Question	1	2	3	4	5	6	7	8	9	10	11	12	13	14	15	16	17
See Lesson	25.1	25.2	25.3	25.3	25.1	25.1	25.3	25.2	25.1	25.2	25.2	25.2	25.2	25.4	25.4	25.4	25.4

STP Answers

1. A
2. C
3. B
4. neutron, $^{1}_{0}n$
5. beta emission
6. alpha emission
7. neutron capture
8. positron emission
9. a. carbon-14, $^{14}_{6}C$
   b. nitrogen-14, $^{14}_{7}N$
   c. oxygen-18, $^{18}_{8}O$
10. Carbon-14 is radioactive.
11. 71%
12. 42%
13. 7.5%
14. C
15. E
16. A
17. B

APPENDIX A

Contents

The Elements Handbook provides information that has traditionally been included within the text in a descriptive chemistry chapter. Beginning on page R2, the information in the handbook is organized by groups of elements. For most groups, there is a two-page spread that contains bulleted lists of properties. The data is divided into categories for easy access. A property spread also supplies information on sources of elements, a reminder of where a group fits into the periodic table, and data on the discovery of elements.

For most groups, there is a two-page spread that contains short articles about specific elements. The articles assume knowledge of the appropriate concepts from the text. How much background students will need depends on the topic. Students could read *Gold* on page R40 early in the year and understand most of the content, but *Swimming Pool Chemistry* on page R30 would be more challenging. The experience would be like reading an article in a newspaper that contains some unfamiliar terms and concepts. The articles are meant to introduce topics without attempting to cover all possible aspects of a topic. For many articles, you will find additional background in the TE wrap.

Appendix A Elements Handbook

ELEMENTS AROUND YOU AND WITHIN YOU

Elements in Earth's Crust	
Element	**Parts per million**
Oxygen	466,000
Silicon	277,000
Aluminum	82,000
Iron	41,000
Calcium	41,000
Sodium	23,000
Potassium	21,000
Magnesium	21,000
Titanium	4400
Hydrogen	1400

Elements in the Atmosphere	
Element	**Parts per million***
Nitrogen	780,900
Oxygen	209,500
Argon	9300
Neon	18
Helium	5.2
Krypton	1.14
Hydrogen	0.5
Xenon	0.086
Radon	Traces

*Data is for dry air.

Elements Dissolved in the Oceans	
Element	**Parts per million**
Chlorine	19,400
Sodium	10,800
Magnesium	1300
Sulfur	904
Calcium	411
Potassium	392
Bromine	67
Carbon	28
Strontium	8
Boron	5

Elements in the Human Body	
Element	**Percent of total body mass**
Oxygen	61
Carbon	23
Hydrogen	10
Nitrogen	2.6
Calcium	1.4
Phosphorus	1.1
Sulfur	0.2
Potassium	0.2
Sodium	0.14
Chlorine	0.12

Elements Around You

To Do

If students total the data in the tables, they will discover that the parts per million for elements in Earth's crust, elements dissolved in the oceans, and elements in the atmosphere add up to 977,800 ppm, 33,315 ppm, and 999,725 ppm, respectively. Students should be able to figure out that the discrepancy in the first and last categories are due to missing data on additional elements. The large discrepancy in the data for the oceans may seem puzzling at first, until students recall that a solution requires a solvent.

Elements Within You

After students study Chapter 24, they should be able to offer an informed explanation for the distribution of elements in the human body. For example, they can note the presence of nitrogen in proteins and nucleic acids, and the presence of phosphorus in nucleic acids, phospholipids, and the molecules that capture and store energy.

GROUP 1A

Group 1A

The meaning of the term *discovery* can vary because scientists sometimes knew that an element existed, and had named the element, long before the element could be isolated from its compounds. For example, Arfvedson discovered lithium during an analysis of ore containing $LiAl(Si_2O_5)_2$, but lithium was not isolated until W.T. Brande and Sir Humphrey Davy performed an electrolysis of lithium oxide.

Physical Properties

It isn't always easy to establish the melting point or boiling point for a substance. Energy added to a sample to cause a change of state can cause other changes. For example, a melting point cannot be determined for sucrose because it decomposes when heated. It is possible to melt iodine, but not at atmospheric pressure because at that pressure iodine will sublime. To determine the melting point of diamond, the sample must be heated in an inert atmosphere. Otherwise it reacts with oxygen in the air. Even in an inert atmosphere, it is not possible to determine a melting point for graphite because the graphite sublimes.

To Do

To see how physical properties change across the periods in the periodic table, students can compare the ranges on the *y*-axes of the melting and boiling point graphs and the density graphs for Groups 1A–8A.

GROUP 1A

Lithium discovered in 1817 by Johan August Arfvedson

Sodium discovered in 1807 by Sir Humphry Davy

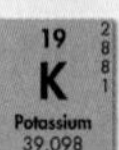

Potassium discovered in 1807 by Sir Humphry Davy

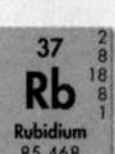

Rubidium discovered in 1861 by Robert Bunsen & Gustav Kirchhoff

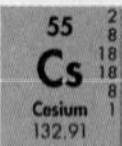

Cesium discovered in 1860 by Robert Bunsen & Gustav Kirchhoff

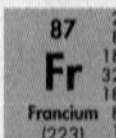

Francium discovered in 1939 by Marguerite Perey

Physical Properties

- Alkali metals are silver-gray solids that are soft enough to cut with a knife. They are soft because they have only one valence electron.
- The presence of a single valence electron also explains the low melting points and boiling points of alkali metals.

Pure alkali metals are stored under oil because they will react with oxygen and moisture in air.

POTASSIUM

Melting and Boiling Points

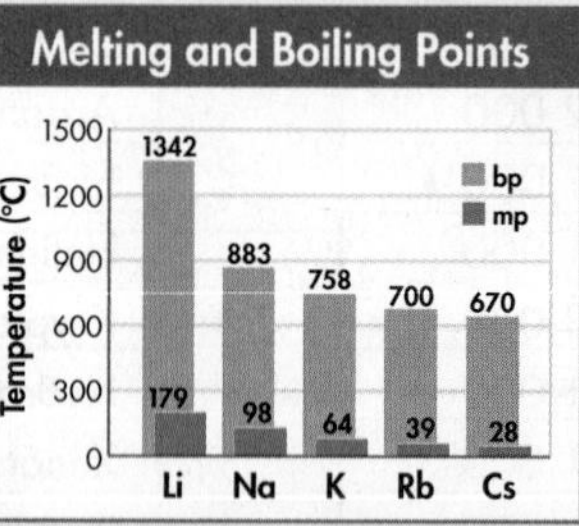

Density

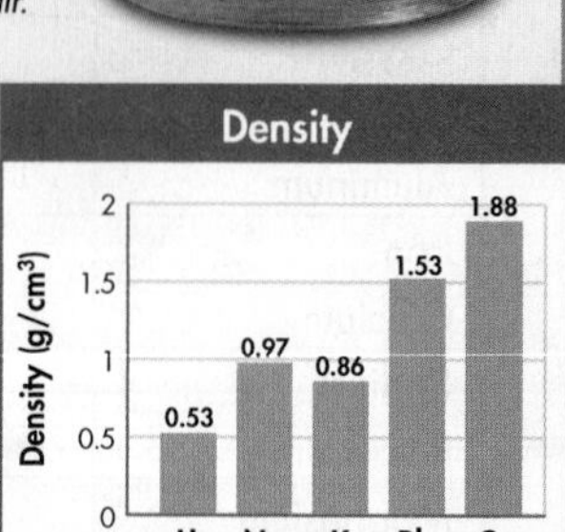

Sources

- Alkali metals do not occur uncombined in nature because they are highly reactive.
- Sodium occurs widely as sodium chloride in underground salt and brine deposits. It is a major component of seawater.
- Sodium is the only alkali metal manufactured on a large scale. It is generally produced by the electrolysis of molten sodium chloride.

  $2NaCl(l) \longrightarrow 2Na(l) + Cl_2(g)$

  Lithium and potassium are produced by a similar process.
- The mineral sylvite, KCl, is a source of potassium.

Manufacture of Sodium

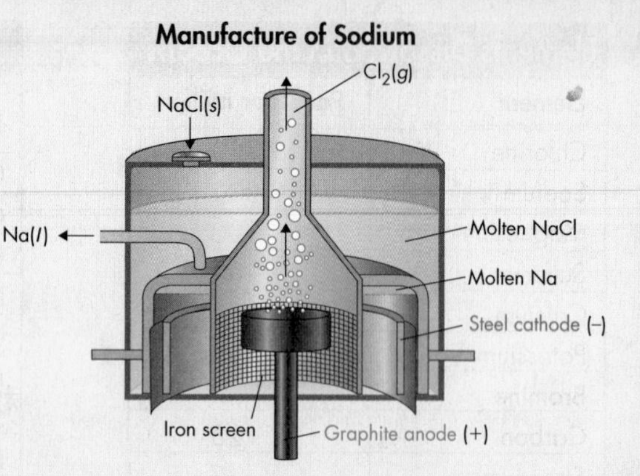

Sources

As new sources are discovered, the relative abundance of an element may need to be revised. For example, rubidium is now considered more abundant in Earth's crust than zinc or copper. However, there is little market demand for rubidium. Uses for rubidium include fiber optics, night-vision devices, PET scans, and photoelectric cells. Some rubidium is produced as a by-product of processing lithium and cesium ores.

To Do

Some students may want to research how rubidium atoms have been used to produce a Bose-Einstein condensate at a temperature only a few billionths of a degree above absolute zero.

Atomic Properties

- Alkali metals have an electron configuration that ends in ns^1.
- The alkali metals are the most reactive metals.
- Alkali metals form ions with a 1+ charge.
- The atoms of alkali metals are the largest in their periods.
- Cesium is a good reducing agent because its first ionization energy is very low.

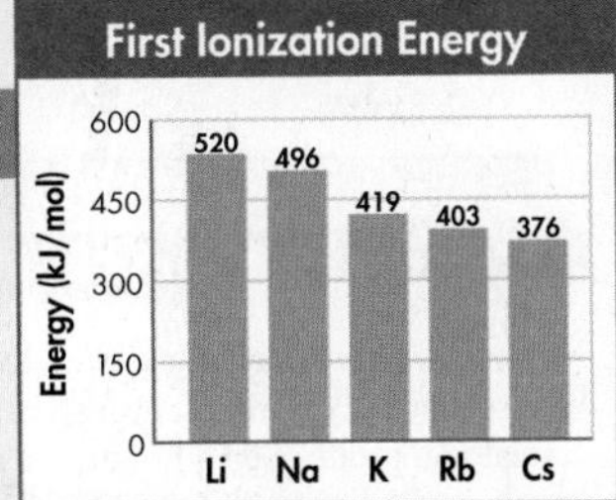

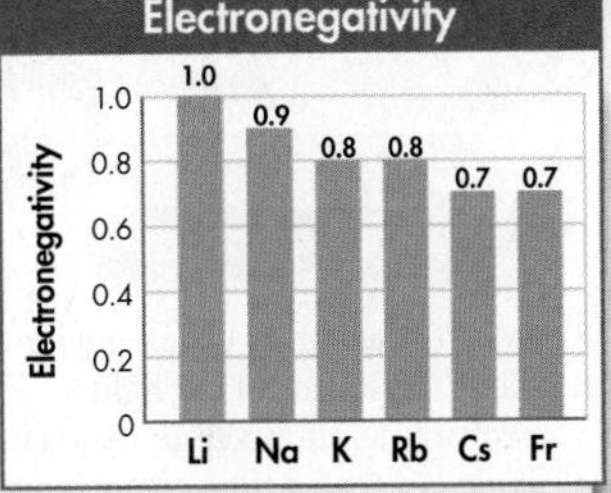

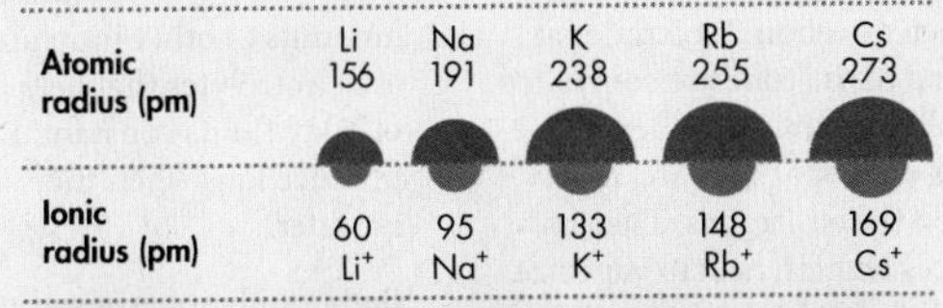

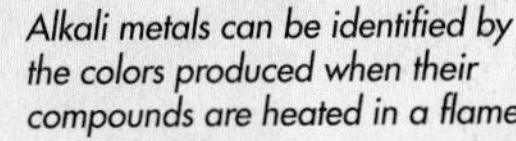

Alkali metals can be identified by the colors produced when their compounds are heated in a flame.

Lithium Sodium Potassium Rubidium Cesium

Important Compounds and Reactions

- All alkali metals react with water to form an alkaline solution. Example:

$$2K(s) + 2H_2O(l) \longrightarrow 2KOH(aq) + H_2(g)$$

- All alkali metals react with halogens to form an ionic halide. Example:

$$2Cs(s) + Cl_2(g) \longrightarrow 2CsCl(s) \qquad \Delta H = -442.8 \text{ kJ/mol}$$

- When heated, sodium hydrogen carbonate (baking soda) produces $CO_2(g)$ and steam, which causes baked goods to rise.

$$2NaHCO_3(s) \xrightarrow{\Delta} Na_2CO_3(s) + CO_2(g) + H_2O(g) \qquad \Delta H = 129 \text{ kJ/mol}$$

- Lithium hydride is a convenient source for preparing hydrogen.

$$LiH(s) + H_2O(l) \longrightarrow LiOH(aq) + H_2(g)$$

- Potassium superoxide, KO_2, is a source of oxygen in submarines. It removes CO_2 from the atmosphere as it produces oxygen.

$$4KO_2(s) + 2CO_2(g) \longrightarrow 2K_2CO_3(s) + 3O_2(g).$$

- Sodium hypochlorite, NaClO, is used as a bleach and to disinfect swimming pools.
- Sodium hydrogen carbonate, $NaHCO_3$, is used in fire extinguishers.
- Potassium nitrate, KNO_3, is used in matches and chemical fertilizers.
- Sodium hydroxide, NaOH, is used as a drain cleaner. It is also used to produce other chemicals and in many industrial processes.

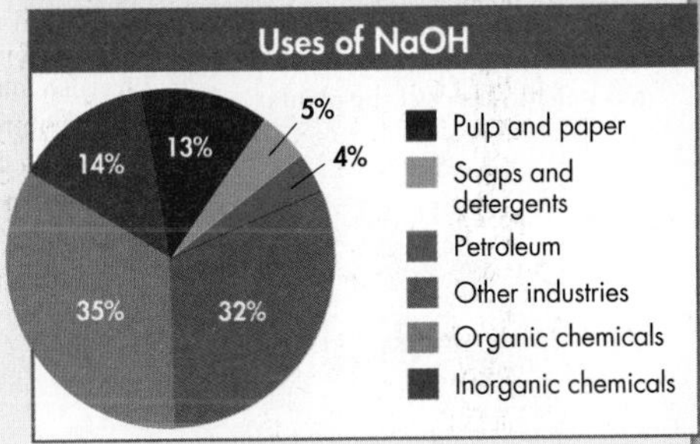

Atomic Properties

The source for most of the atomic radii data is *Bonding and Structure* by N. W. Alcock (Ellis Horwood, 1999).

Of all elements with atomic numbers between 1 and 100, atoms of francium are the least stable. The longest-lived isotope of francium, francium-233, has a half-life of only 22 minutes.

To Do

Comparing the ranges on the *y*-axes of the first ionization energy graphs for Groups 1A–8A provides another way to emphasize the variation of properties across periods in the periodic table.

Important Compounds and Reactions

To Do

The U.S. chemical industry produces large amounts annually (by mass) of sulfuric acid, phosphoric acid, ammonia, sodium hydroxide, nitric acid, ammonium nitrate, hydrogen chloride, and ammonium sulfate. As a long-term class project, students could collect data from the Elements Handbook and other sources to explain this production.

When students study organic chemistry, they could do similar research on ethylene, propylene, ethylene dichloride, urea, ethylbenzene, styrene, ethylene oxide, cumene, and 1,3-butadiene.

Group 1A

Na Vapor Lamps

mercury to a sodium vapor lamp helps with the problem of color recognition because of the strong blue and green lines in mercury's emission spectrum. However, because the tubes contain mercury, used lamps must be handled according to the rules of disposal for hazardous wastes or the tubes must be recycled.

The sodium vapor in a lamp can be at low pressure or at high pressure. Low-pressure sodium vapor lamps that do not contain mercury cause the least light pollution and are less likely to interfere with astronomical observations.

Because sodium vapor can etch glass, the tube in a sodium vapor lamp is made from a very pure form of aluminum oxide (Al_2O_3) called polycrystalline alumina (PCA). PCA doesn't react with the sodium, and it allows light to pass through.

To Do

Students could collect data on the background colors and the color of lettering used for traffic signs (e.g., stop, yield, wrong way, and do not enter signs; speed limits and mileposts; parking and construction signs; and information on local tourist attractions). Then students could consider how a sodium vapor lamp might affect the visibility of the signs.

GROUP 1A

Na Vapor Lamps

When streetlights have a golden glow, the source of the light is probably sodium vapor. Inside the lamp is a sealed tube containing a sodium-mercury alloy and a starter gas, such as xenon. Electrodes at each end of the tube are connected to an electric circuit. When the lamp is on, a spark, or arc, forms between the electrodes. The arc produces enough heat to vaporize the sodium and mercury atoms. Within the arc, atoms ionize. Outside the arc, ions recombine with electrons and light is emitted—yellow for sodium and blue-green for mercury.

A sodium vapor lamp uses less energy than most other light sources and costs less to operate. But the lamp isn't a perfect light source. The color of an object is visible only when light of that color is reflected off the object. So yellow lines on the pavement appear yellow under a sodium vapor lamp, but a red stop sign appears gray.

People who grow plants under artificial light may use sodium vapor lamps to stimulate the production of flowers and fruits. The vapor lamps must be combined with other light sources. Otherwise, the plant stems will be too weak to support the plants.

Na K Restoring Electrolytes

"Don't sweat it" may be good advice for handling stress, but not for maintaining a healthy body. The sweat you produce on a hot day or during exercise cools your body as it evaporates. Sweat consists mainly of water, sodium chloride, and small amounts of other inorganic salts. The salts are electrolytes that help keep the volume of body fluids constant. Electrolytes produce ions when they dissolve in water.

Body fluids contain sodium ions and potassium ions. Potassium ions are the principal cations inside cells. Sodium ions are the principal cations in the fluids outside of cells. The transmission of nerve impulses depends on the movement of sodium and potassium ions across the membranes of nerve cells. Potassium ions cause the heart muscle to relax between heartbeats.

Replacing the water lost during exercise is important, but not sufficient. Electrolytes must be replaced too. Some signs of electrolyte depletion are muscle cramps, nausea, and an inability to think clearly. Many athletes use sports drinks to replace electrolytes. Some experts recommend these drinks for people who lose more than 8 liters of sweat daily or exercise continuously for more than 60 minutes.

Bananas account for about one-third of the fresh fruit consumed in the United States.

Table salt is the chief source of sodium in the diet. But large amounts are also found in unexpected places, such as eggs. For a healthy adult, the recommended daily intake of sodium chloride is about 5 grams per day—about half the amount many people consume. The daily recommended amount of potassium is about 1 gram. The word *daily* is important because your kidneys excrete potassium even when the supply is low. Eating foods high in potassium and low in sodium is ideal. Such foods include bananas, chicken, and orange juice.

Na K Restoring Electrolytes

To Do

Relying more on fresh foods than processed foods can reduce the total amount of sodium chloride in a person's diet. Students could look at the labels for some processed foods and compare the sodium chloride content.

Students might also want to estimate how much sodium chloride they ingest daily from processed foods. Remind them that this quantity is in addition to the sodium chloride that is present in unprocessed foods.

R4 Elements Handbook

Cs Cesium Atomic Clock

Some watches contain a quartz crystal that vibrates at a constant rate. The vibration provides the "beat" that is translated into the time you see displayed. A clock with a quartz crystal is more accurate than a mechanical clock, which has moving parts that can be worn down by friction. But a quartz crystal isn't accurate enough for modern communication and navigation systems.

For greater accuracy, you need an atomic clock, which may gain or lose only one second in a million years! In most atomic clocks, cesium-133 atoms provide the "beat." Unlike quartz crystals, all cesium-133 atoms are identical and they don't wear out with use. The clock is designed so that the atoms repeatedly absorb and emit radiation. The emitted radiation has a frequency of exactly 9,192,631,770 cycles per second. The cycles are counted and translated into seconds, minutes, and hours.

There are four cesium atomic clocks at the National Institute of Standards and Technology in Fort Collins, Colorado. Short-wave radio waves are useds to transmit accurate time signals from these clocks to most of North America. Clocks and watches advertised as "radio controlled" contain a tiny antenna and receiver that pick up the signal and decode the information. The owner sets the time zone. The radio signal sets the time.

Global Positioning System (GPS) receivers can pinpoint any location on Earth to within a few meters. The system depends on accurate time signals from atomic clocks in 24 Earth-orbiting satellites.

The GPS receiver in this car compares signals sent at exactly the same time from three different satellites to determine a location.

Na Salt of the Earth

In the ancient world, table salt (NaCl) was an extremely valuable commodity. Before refrigeration, salt was used to preserve foods such as meats and fish. As salt draws the water out of bacteria, they shrivel up and die. Sailors ("salty dogs") especially depended on salt pork and fish to survive long voyages.

Trade routes were established and roads were built to transport salt. In ancient China, coins were made of salt and taxes were levied on salt. Roman soldiers were paid an allotment called a *salarium argentum*, or "salt silver," from which the word *salary* is derived.

Li Bipolar Disorder

Millions of people in the United States experience the extreme mood swings of bipolar disorder. During a manic phase, they think they can conquer the world. During a depression, they may feel hopeless. The ionic compound lithium carbonate can be used to control these symptoms.

Exactly how the treatment works is not known. Scientists do know that lithium ions can increase the level of serotonin. Serotonin is one of the chemical messengers (neurotransmitters) that transmit messages between brain cells.

Did You Know?

Most alkali metals are stored under oil. But **lithium** has such a low density that it floats on oil. So a coat of petroleum jelly is applied to lithium before it is stored.

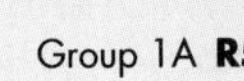

Cs Cesium Atomic Clock

A quartz crystal can keep accurate time to about one second every ten years. Over time, the vibrations of the crystal are affected by changes in temperature, impurities in the crystal, and the accumulated effects of the vibrations.

In a cesium clock, the valence electron in a cesium-133 atom oscillates between two hyperfine states. In one state, the magnetic field of the valence electron is in the same direction as the magnetic field of the nucleus. In the other state, the magnetic field is in the opposite direction.

GPS satellites transmit data that indicates their location and current time. 24 satellites are operating at all times with 3 satellites in reserve. Signals that are released simultaneously from different satellites will arrive at a receiver at slightly different times. A minimum of three satellites is needed. Using additional satellites increases the precision of the data.

To Do

Ask students to interpret the following statement. A person with one watch knows the time very well, but a person with two watches is not so certain.

Na Salt of the Earth

A good person is sometimes described as the "salt of the earth" and an unworthy person as "not worth his salt," or not worth the wages he receives.

In non-industrial countries, the principal use of NaCl is still food preparation and preservation. In the United States, 40% of NaCl is used for the production of chlorine and sodium hydroxide, and another 40% in de-icing roads. Table salt accounts for only 1% of usage.

Li Bipolar Disorder

Dr. John Cade discovered the mood-altering effect of lithium ions in 1949. His discovery had a major impact on the field of psychiatry. Up to that point, the accepted view was that mental health problems were always psychological. Talk therapy was the approved method of treatment. But many patients did not respond to this approach. Dr. Cade's work demonstrated that a mental illness could be caused by the wrong amount of a chemical in the body.

In the case of depression, the chemical in question is serotonin, which is manufactured in the brain. Scientists agree that a serotonin deficiency and depression are related. The level of serotonin in the blood is lower in people who are diagnosed with depression. What scientists have not yet determined is whether the decrease in serotonin is a cause or a symptom of depression.

Group 2A

Vauquelin discovered an element he called gluinium as an oxide in beryl. The name reflected the sweet taste of the oxide. When Friedrich Wöhler and A. Bussy isolated the element in 1828, Wohler renamed it beryllium. Adair Crawford noted the existence of strontium in minerals in 1790.

Physical Properties

Beryllium's strength and rigidity make it an ideal material for mid-range and tweeter diaphragms. In most loud speakers, a current alternating at the frequency of the sound that generated it flows through a coil. As the coil is attracted to and repelled by a permanent magnet, the coil and the cone-shaped diaphragm attached to the coil vibrate. The vibration of the diaphragm reproduces the original sound. A beryllium diaphragm can reproduce frequencies that cannot be heard when other materials are used.

Sources

To Do

Help students write balanced equations and classify the reactions involved in the production of magnesium.

(1) Decomposition

$$CaCO_3(s) \xrightarrow{\Delta} CaO(s) + CO_2(g)$$

(2) Combination

$$CaO(s) + H_2O(l) \rightarrow Ca(OH)_2(aq)$$

(3) Double-replacement

$$Ca(OH)_2(aq) + MgCl_2(aq) \rightarrow Mg(OH)_2(s) + CaCl_2(aq)$$

(4) Double-replacement/neutralization

$$Mg(OH)_2(s) + 2HCl(aq) \rightarrow MgCl_2(aq) + 2H_2O(l)$$

(5) Decomposition

$$MgCl_2(aq) \xrightarrow{elec.} Mg(s) + Cl_2(g)$$

(6) Combination

$$Cl_2(g) + H_2(g) \rightarrow 2HCl(l)$$

Students may wonder what the point is of reactions (3) and (4). The precipitation of $Mg(OH)_2$ is used to concentrate the magnesium. If you add 250 mL of 0.1M $Ca(OH)_2$ to 250 mL of seawater, students can observe the precipitate, which can be separated from the remaining solution by decanting or filtration.

GROUP 2A

Beryllium discovered in 1798 by Nicholas Vauquelin

Magnesium isolated in 1808 by Sir Humphry Davy

Calcium discovered in 1808 by Sir Humphry Davy

Strontium discovered in 1808 by Sir Humphry Davy

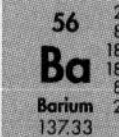

Barium discovered in 1808 by Sir Humphry Davy

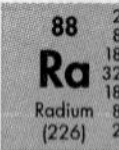

Radium discovered in 1898 by Marie Curie & Pierre Curie

Physical Properties

- Alkaline earth metals are relatively soft, but harder than alkali metals.
- Alkaline earth metals have a gray white luster when freshly cut. When exposed to air, they quickly form a tough, thin oxide coating.
- Densities, melting points, and boiling points tend to be higher than for the alkali metal in the same period.
- Magnesium alloys are strong and lightweight. They are used in cameras, lawnmowers, aircraft, and automobiles.

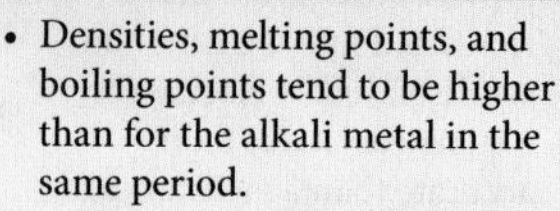

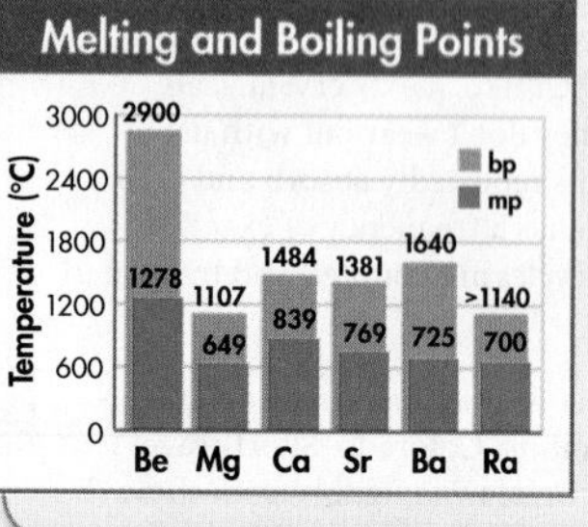

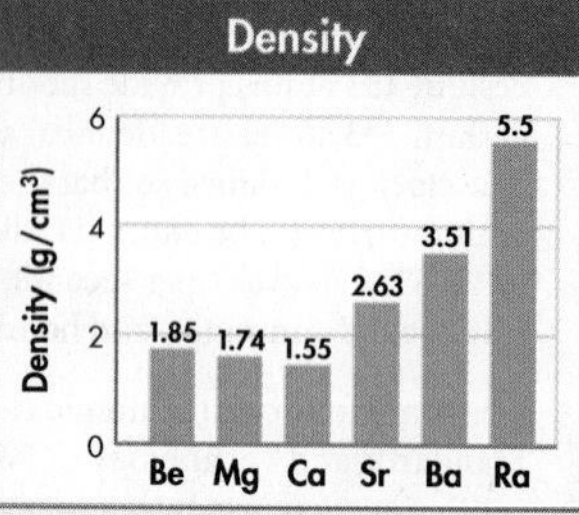

Sources

- Alkaline earth metals are not found in nature in the elemental state.
- Many mountain ranges contain alkaline earth carbonates—limestone ($CaCO_3$) and dolomite ($CaCO_3 \cdot MgCO_3$).
- Barium is produced by reducing its oxide with aluminum at high temperature.

$$3BaO(s) + 2Al(s) \longrightarrow 3Ba(l) + Al_2O_3(s)$$

- Salts of highly radioactive radium are by-products of uranium refining.

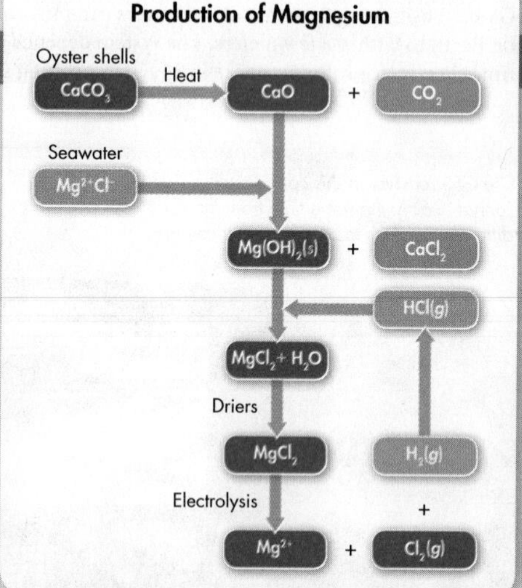

Oyster shells containing $CaCO_3$ are used to extract magnesium from seawater. The chlorine gas produced during the electrolysis of magnesium chloride is fed back into the process.

Atomic Properties

- Alkaline earth metals have an electron configuration that ends in ns^2.
- The alkaline earth metals are strong reducing agents, losing 2 electrons and forming ions with a 2+ charge.
- Because radium is luminous, it was once used to make the hands and numbers on watches glow in the dark.
- The ratio of ^{87}Sr to ^{86}Sr varies with location. This data is used to solve puzzles, such as the source of timber used in prehistoric buildings.

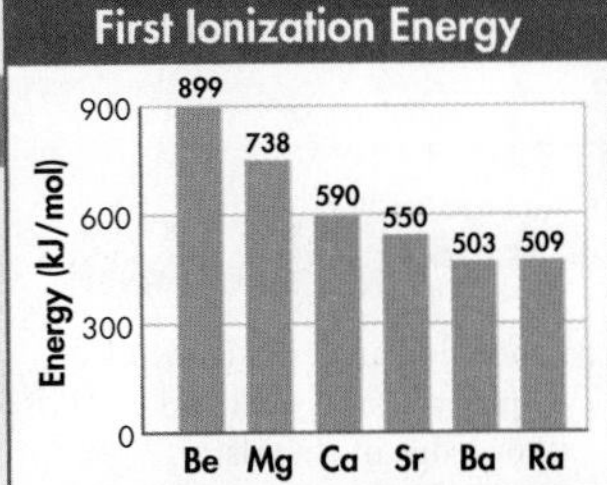

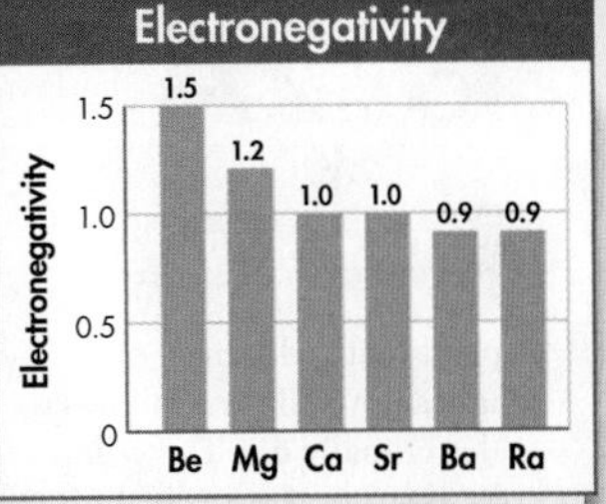

	Be	Mg	Ca	Sr	Ba	Ra
Atomic radius (pm)	113	160	197	215	224	223
Ionic radius (pm)	44 Be^{2+}	66 Mg^{2+}	99 Ca^{2+}	112 Sr^{2+}	134 Ba^{2+}	143 Ra^{2+}

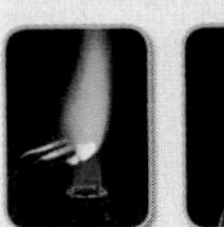
Calcium

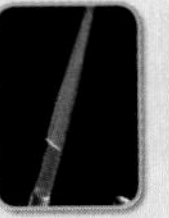
Strontium

Barium

Calcium, strontium, and barium can be identified by the colors produced when their compounds are heated in a flame.

Important Compounds and Reactions

- Alkaline earth metals are less reactive than alkali metals.
- Alkaline earth metals react with halogens to form ionic halides. Example:

  $Mg(s) + Br_2(l) \longrightarrow MgBr_2(s)$

- All Group 2A metals (except Be) react with water to form an alkaline solution. Example:

  $Sr(s) + 2H_2O(l) \longrightarrow Sr(OH)_2(aq) + H_2(g)$

- Alkaline earth metals react with oxygen to form binary oxides. Example:

  $2Ca(s) + O_2(g) \longrightarrow 2CaO(s)$ $\quad \Delta H = -635.1$ kJ/mol

- Heating limestone produces lime, CaO.

  $CaCO_3(s) \xrightarrow{\Delta} CaO(s) + CO_2(g)$ $\quad \Delta H = 176$ kJ/mol

- Slaked lime, $Ca(OH)_2$, reacts with carbon dioxide to form limestone.

  $Ca(OH)_2(s) + CO_2(g) \longrightarrow CaCO_3(s) + H_2O(g)$

- Barium peroxide is used as a dry powdered bleach. It reacts with water to form the bleaching agent, hydrogen peroxide.

  $BaO_2(s) + 2H_2O(l) \longrightarrow H_2O_2(aq) + Ba(OH)_2(aq)$

- Gypsum, calcium sulfate dihydrate, $CaSO_4 \cdot 2H_2O$, is used to make plasterboard.
- Calcium phosphate, $Ca_3(PO_4)_2$, is the major component of bone and tooth enamel.

Slaked lime is an ingredient in plaster, cement, and the mortar used in this stone wall.

Atomic Properties

One of the puzzles that the ratio of ^{87}Sr to ^{86}Sr was used to solve was the source of timber for the prehistoric great houses in Chaco Canyon, New Mexico. Of the more than 200,000 trees that were used to build the houses, at least one-fifth were spruce or fir trees that were carried from mountainous sites 75–100 km away. Scientists can identify specific sites by comparing $^{87}Sr/^{86}Sr$ ratios in beams to ratios in trees from the sites. The ratios show that the timber for construction beams came from the Chuska and San Mateo mountains, but not from the San Pedro Mountains.

Important Compounds and Reactions

To Do

Students could look at the relationship between solubility and the properties and uses of calcium oxide, calcium carbonate, calcium sulfate, and calcium phosphate. Refer to Table B.7 on page R48.

Group 2A

Mg Chlorophyll

Chlorophyll is an example of a chelate, a substance in which a central metal ion is bonded to a large organic molecule. There are two types of chlorophyll. In chlorophyll *b,* CHO replaces the methyl group at the upper right in the formula. Chlorophyll absorbs light in the red and blue-violet regions of the visible spectrum. In chlorophyll *a* and chlorophyll *b,* the peaks are at different wavelengths. Thus, chlorophyll absorbs a broader range of wavelengths than it would if there were only one type.

Sr Ba Fireworks

The first fireworks were probably invented in China more than a thousand years ago. They were a mixture of sulfur, charcoal, and potassium nitrate, which burned quickly with a flash. This black powder is still a major component of modern fireworks. When iron filings or copper or zinc were added to this gunpowder, early versions of fireworks with limited colors (orange, yellow, white) were produced. The rest of the colors were not added to the palette until the 1800s.

The pattern produced by a shell depends on how the stars are arranged within the shell. Shells are named for their patterns—*chrysanthemum, weeping willow, or strobe.* Strobes are clusters of flickering silvery lights that drift slowly to the ground. The gold sparks in a display come from burning iron filings or charcoal. Brilliant white flames are produced when magnesium, titanium, and aluminum burn at very high temperatures.

To Do

Because fireworks displays are often a central part of public celebrations, students may not realize that fireworks are classified as hazardous materials. Suggest that students investigate the hazards associated with the misuse of fireworks, and the regulations that are in place in your state to prevent injuries.

GROUP 2A

Mg Chlorophyll

A plant that is deprived of magnesium will turn yellow and eventually die. The yellow color is a sign that the plant is not producing enough of the green pigment chlorophyll, which is found in structures called chloroplasts. One square millimeter of a leaf's surface contains about half a million chloroplasts. In chloroplasts, light energy is changed to chemical energy.

Chlorophyll molecules absorb wavelengths of blue light and red light, and reflect green light. There is a magnesium ion at the center of each chlorophyll molecule. It is embedded like a jewel in a crown-like ring. The magnesium ion forms a bond with each nitrogen atom.

Chlorophyll a

Chlorophyll is typical of a family of compounds called porphyrins, which contain a central metal ion. In heme, the ion is Fe^{2+}. There are four heme molecules in hemoglobin, which transports oxygen in the blood.

Sr Ba Fireworks

Producing an aerial fireworks display requires skill and knowledge of chemistry. Shells are stored in steel pipes anchored in sand. A typical shell contains two charges, one to launch the shell and one to cause the shell to burst. When the lifting charge explodes, the shell is blown out of the pipe and the fuse attached to the bursting charge ignites. The shell travels for a few seconds before it bursts.

The name of this fireworks effect is chrysanthemum.

The explosions that launch the shells and cause them to burst are exothermic redox reactions. The reducing agent, or fuel, may be aluminum, magnesium, or sulfur. The oxidizing agent may be a nitrate, a chlorate, or a perchlorate, such as potassium perchlorate ($KClO_4$).

When a shell bursts, it releases tiny pellets called "stars," which produce the colors. Only a few elements are needed to produce these colors. Two are alkaline earth metals, strontium for bright red and barium for light green. These same colors are produced when strontium and barium compounds are heated in a flame. Flame tests can identify elements because an element emits characteristic wavelengths of light when its electrons absorb energy and then return to the ground state.

Labels for shells used in a fireworks display describe the color and effect produced by the shell.

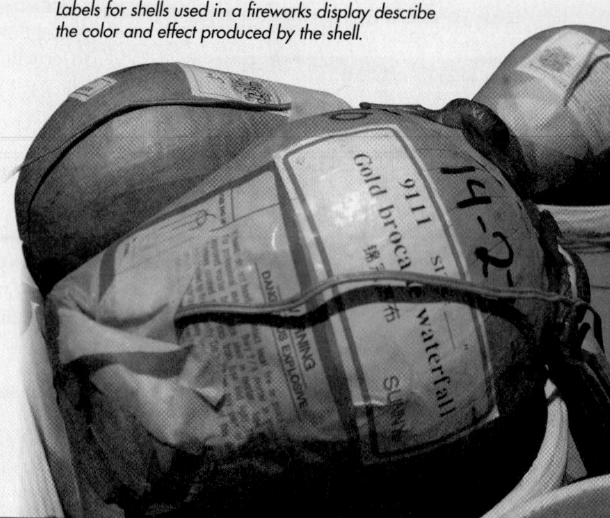

Ca Limestone Caves

The chemical reactions that form a limestone cave are simple. Carbon dioxide in the air dissolves in rain to form weak carbonic acid, H_2CO_3. As rain passes through soil, it dissolves carbon dioxide produced by decaying plants and becomes even more acidic. The rainwater seeps into limestone, $CaCO_3$, beneath the soil. The $CaCO_3$ dissolves in the carbonic acid, forming a solution of calcium hydrogen carbonate, $Ca(HCO_3)_2$.

$$CaCO_3(s) + H_2CO_3(aq) \longrightarrow Ca^{2+}(aq) + 2HCO_3^-(aq)$$

Over millions of years, as more and more limestone dissolves, a cave forms and slowly grows in size. Once the cave stops growing, another process may occur. Calcium hydrogen carbonate solution drips through the cave's roof. Carbon dioxide is released from solution, leaving behind a tiny deposit of solid calcium carbonate.

$$Ca^{2+}(aq) + 2HCO_3^-(aq) \longrightarrow CaCO_3(s) + CO_2(g) + H_2O(l)$$

The deposit slowly grows into a stalactite, which hangs from the ceiling like an icicle. Drops of solution that fall from the ceiling form stalagmites on the cave floor. Calcium carbonate is white. So any color in the deposits is due to traces of metal ions, such as copper (blue-green) and iron (red-brown). The growth rate of the formations depends mainly on the volume of water that drips through the roof and the concentration of calcium hydrogen carbonate in the water. In many caves, the rate of growth is measured in centimeters per hundreds or thousands of years.

Stalactites and stalagmites may join together to form columns.

Ca Building Strong Bones

As a young adult, you can have a major effect on the health of your bones later in life. You acquire 90% of the bone tissue in your skeleton before age 18 as females and age 20 as males. Physical activity and the amount of calcium in your diet are factors that affect the buildup of bone mass.

To maintain a constant level of calcium in your blood, your body may release calcium from bone tissue. If enough calcium is lost from bones, they become brittle and tend to break easily. This condition, known as osteoporosis, is most common in older people. However, it can occur in young and middle-aged adults.

The recommended daily intake of calcium from age 9 to age 18 is 1.3 g per day. This is the amount of calcium in a liter of milk. Calcium-fortified foods, such as orange juice, are a good choice for those who cannot digest the lactose in milk.

Vitamin D helps your body absorb calcium. The vitamin D your skin produces when exposed to sunlight is usually not enough to meet your daily requirement. So orange juice is also fortified with vitamin D.

Did You Know?

Calcium oxide (lime) emits a bright white light when it is heated. Before electric lights, theaters could spotlight a single actor by focusing the light from heated lime. Thus, the expression *in the limelight* describes a person in a prominent position.

Ca Limestone Caves

Limestone caves tend to be longer and more complex than other types of caves. Limestone is a sedimentary rock, which generally forms in warm, shallow seas. When the water table drops in a limestone cave, dry spaces that can be explored are left. It is in these dry spaces that stalactites and stalagmites form.

Ca Building Strong Bones

In recent years there has been a debate about how the increased consumption of soft drinks by teens has affected the calcium content of their bones. One hypothesis is that acids in soft drinks interfere with absorption of calcium. Another is that the consumption of soft drinks reduces the consumption of calcium sources such as milk.

Group 3A

Physical Properties

To Do

To celebrate its 80th anniversary, *Chemical & Engineering News* published an issue on the elements (September 8, 2003). Go to their Web site for instructions on accessing the issue online, purchasing multiple copies, and ordering a free issue for yourself via e-mail. Students are likely to enjoy the article on gallium by Oliver Sacks.

Sources

In the Elements Handbook, students will often see references to minerals and ores. A mineral is a naturally occurring, inorganic, crystalline solid with a characteristic chemical composition. Some properties used to identify minerals are crystal structure, color, luster, density, hardness, and how its crystals fracture. An ore is a material from which a valuable mineral can be profitably extracted.

GROUP 3A

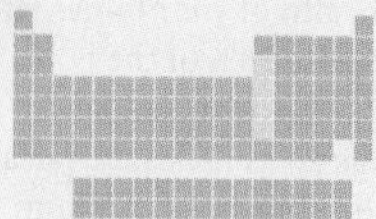

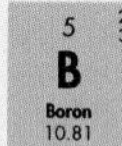

Boron discovered in 1808 by Sir Humphry Davy and by Joseph-Louis Gay-Lussac & Louis-Jacques Thénard

Aluminum discovered in 1825 by Hans Christian Oersted

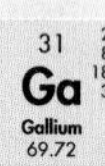

Gallium discovered in 1875 by Paul-Emile Lecoq de Boisbaudran

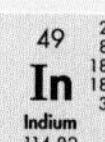

Indium discovered in 1863 by Ferdinand Reich & Hieronymus T. Richter

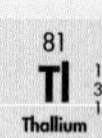

Thallium discovered in 1861 by Sir William Crookes

Physical Properties

- Boron is a metalloid. The rest of the Group 3A elements are metals.
- Aluminum is a valuable structural material because of its strength, especially in alloys with silicon or iron. These alloys have a low density and resist corrosion.
- Gallium has an extremely wide liquid temperature range (30°C to 2204°C). Solid gallium floats in liquid gallium, which is unusual for a metal.

Boron is black, lustrous, and extremely hard, but brittle.

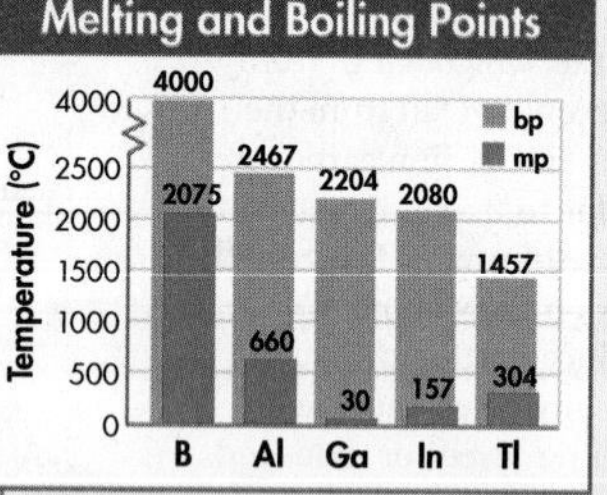

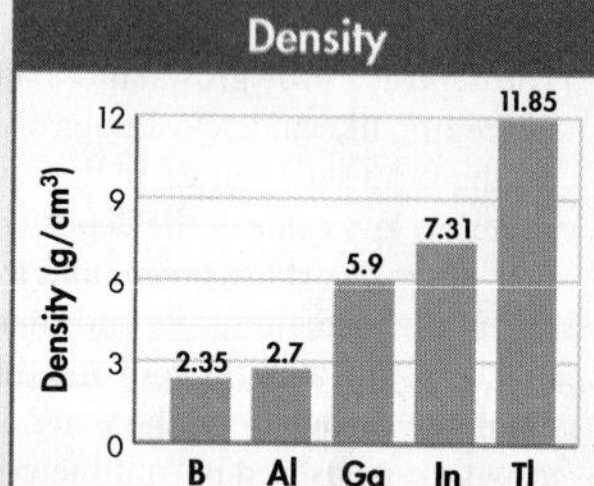

Sources

- Boron is always combined with oxygen in nature. Boron can be prepared by the reaction of its oxide with magnesium metal.

$$B_2O_3(s) + 3Mg(s) \longrightarrow 2B(s) + 3MgO(s)$$

- Bauxite is a common ore of aluminum. The primary mineral in bauxite is alumina, Al_2O_3, which is reduced to aluminum by electrolysis.
- Gallium, indium, and thallium are quite rare. They are typically extracted from ores being processed to extract other metals.

The conveyor belt is being used to unload bauxite from a train. Bauxite is a major source of aluminum. It is also a source of gallium.

Atomic Properties

- Group 3A elements have an electron configuration that ends in ns^2np^1.
- The most common oxidation number for boron, aluminum, gallium, and indium is +3. For thallium, it is +1.
- Group 3A elements become more metallic from top to bottom within the group.
- Radioactive thallium-201 is injected into patients taking a stress test used in the diagnosis of heart disease.

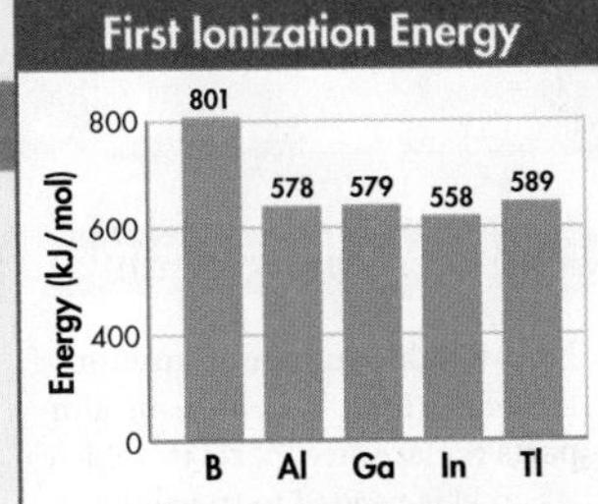

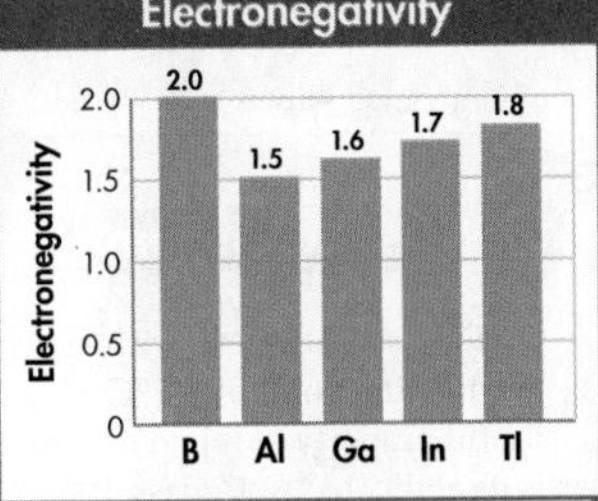

	B	Al	Ga	In	Tl
Atomic radius (pm)	83	143	141	166	172
Ionic radius (pm)	23 B^{3+}	51 Al^{3+}	62 Ga^{3+}	81 In^{3+}	95 Tl^{3+}

Important Compounds and Reactions

- Group 3A elements react with halogens to form halides. Example:

  $2Al(s) + 3Cl_2(g) \longrightarrow 2AlCl_3(s)$

  Aluminum chloride is used as a catalyst in organic reactions.
- Group 3A elements react with oxygen to form oxides. Example:

  $4Al(s) + 3O_2(g) \longrightarrow 2Al_2O_3(s)$

  $\Delta H = -1676$ kJ/mol

  Because this reaction is so exothermic, powdered aluminum is a component of some explosives, fireworks, and rocket fuels.
- Aluminum sulfate (alum), $Al_2(SO_4)_3 \cdot 18H_2O$, is used as a coagulant in water treatment plants.
- Gallium arsenide, GaAs, converts electric current to light in light-emitting diodes (LEDs). It is produced as follows.

  $(CH_3)_3Ga(g) + AsH_3(g) \longrightarrow GaAs(s) + 3CH_4(g)$

This LED is shown at about five times actual size.

Corundum is a mineral form of aluminum oxide. Rubies are corundum in which a few aluminum ions have been replaced by chromium ions.

- Borax, $Na_2B_4O_7 \cdot 10H_2O$, is used to soften water and is found in glasses and glazes. When a mixture of borax and hydrochloric acid is heated, boric acid is produced.

  $Na_2B_4O_7(aq) + 2HCl(aq) + 5H_2O(l) \longrightarrow 4H_3BO_3(aq) + 2NaCl(aq)$
- Boric acid is poisonous if ingested. The solid is used as an insecticide against cockroaches. A dilute solution of boric acid is an eyewash.
- Boron carbide, B_4C, is almost as hard as diamond. It is used for items that must resist wear, such as cutting tools.
- Laboratory glassware is made of heat-resistant glass containing 12% to 15% boric oxide.

Atomic Properties

To Do

Challenge students to figure out why thallium commonly has an oxidation number of +1 in addition to +3 when other elements in the group do not. Thallium can lose an electron from a 6*p* orbital more easily than the other elements can lose an electron from their highest occupied *p* orbital because the energy difference between the 6*s* and 6*p* orbitals is greater than the difference between the *s* and *p* orbitals in the lower principal energy levels. The occupied 4*f* sublevel in thallium is responsible for the difference.

Important Compounds and Reactions

The compositions of sapphires and rubies are almost identical. They are mainly crystals of colorless aluminum oxide (Al_2O_3). Traces of cobalt make sapphires blue and traces of chromium make rubies red.

Crystalline boron nitride is an industrial abrasive, which is used to cut and polish metals. But when finely ground boron nitride is added to face powder, it provides a soft, silky texture.

Group 3A

In Uses of Indium

Traditionally, solder was an alloy of tin and lead, which melts at about 180°C. Nowadays many solders are lead-free and some low-temperature solders do not even contain any tin. One mixture used to solder glass and ceramics is 97% indium and 3% silver. Often a flux is used when metals are being soldered. A flux is used to remove oxides from the surface of metals before they are joined.

To Do

You might want to arrange for a demonstration of soldering by a teacher in your school or another expert. Ask the person doing the demonstration to discuss how the composition of solder affects its properties and its uses.

Al Recycling Aluminum

When old bridges are repaired, the steel or concrete decks may be replaced with aluminum decks. The aluminum decks require much less maintenance because a thin oxide layer protects the underlying layers of aluminum. The aluminum decks can carry equal or greater loads than steel decks because aluminum has a high strength to weight ratio.

Strength is defined as the ability of a material to withstand forces of compression, tension, and shear. At the limit of its compressive strength, a material will tend to compact, i.e., its volume will be reduced. At the limit of its tensile strength, a material will be stretched to the point where it breaks. At the limit of its shear strength, a material will slide along a plane parallel to the force. With a deck on a bridge, the major concern is compressive strength.

GROUP 3A

In Uses of Indium

Indium is a soft metal with a low melting point of 157°C. One useful property of liquid indium is its ability to "wet" glass. It spreads out and forms a thin layer on the glass instead of beading up. This property allows indium to be deposited on glass to make mirrors that are as reflective, but more corrosion-resistant, than silvered mirrors.

Low-melting alloys that contain indium are used as solders to join glass to glass. They also can join a metal to a metal at low temperatures. There are products, such as electronics components, that could be damaged if joined at a high temperature. The seals of some fire sprinkler heads are held in place by indium alloys. When heat from a fire melts the alloy, the seal is released, which allows water to pour from the sprinkler head.

Al Recycling Aluminum

Lightweight, durable aluminum has many uses. When aluminum parts replace steel parts in vehicles, less fuel is needed to travel the same distance. About 20% of the aluminum produced is used for packaging, including foil and cans.

Aluminum is used in durable consumer goods such as baseball bats.

In the United States, over 80 billion aluminum soft drink cans are sold each year. More than 50% of these cans are recycled. The energy saved by recycling just one of these cans could be used to operate a television for three hours. The energy needed to recycle aluminum is 5% of the energy needed to obtain new aluminum from ore. An added benefit of recycling is a reduction in solid waste.

Recycled aluminum cans are shredded, crushed, and heated to remove materials other than aluminum. Aluminum pieces about the size of a potato chip are placed in furnaces where they are mixed with new aluminum and melted. Then, the molten aluminum is poured into 7.6-m ingots that have a mass of about 13,600 kg. The ingots pass through rolling mills that reduce their thickness from about 0.5 m to about 0.25 mm. The thin sheets are coiled and shipped to a manufacturer who produces the bodies and tops of the cans. Aluminum from a recycled can is part of a new can within 60 days.

Aluminum cans are compressed before being shipped to aluminum companies for recycling.

Al Manufacturing Aluminum

Aluminum is the most abundant metal in Earth's crust (8.3% by mass). It is found in minerals such as bauxite (impure aluminum oxide, Al_2O_3). But for years after its discovery, there wasn't a practical way to extract aluminum from its ores. This rare and expensive metal was used, like gold, mainly for decoration.

A professor at Oberlin College in Ohio challenged his class to find an inexpensive way to produce aluminum. In 1885, Charles Hall responded to the challenge. He set up a laboratory in a woodshed. He knew that other chemists had tried to decompose aluminum oxide by electrolysis. This method was not practical because the melting point of aluminum oxide is quite high (2045°C). Hall found that mixing aluminum oxide with cryolite (Na_3AlF_6) produced a mixture that melted at a much lower temperature of 1012°C.

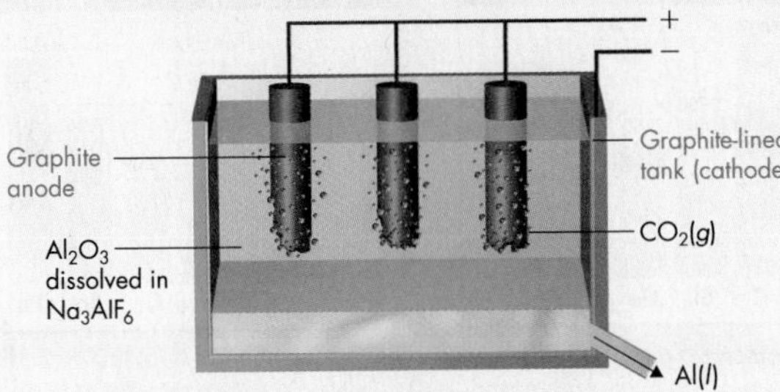

The process Hall invented is often called the Hall-Heroult process. Paul Heroult, a 23-year-old Frenchman, developed the same process almost simultaneously. It is still used today. Pure aluminum oxide is extracted from bauxite through heating to a temperature above 1000°C. The aluminum oxide is dissolved in molten cryolite and contained in a graphite-lined iron tank. The graphite rods used as an anode are consumed during the process.

$$\text{Anode: } C(s) + 2O^{2-}(l) \longrightarrow CO_2(g) + 4e^-$$

$$\text{Cathode: } 3e^- + Al^{3+}(l) \longrightarrow Al(l)$$

The products are carbon dioxide and molten aluminum metal. Because the aluminum is more dense than the aluminum oxide-cryolite mixture, it collects at the bottom of the tank and is drawn off periodically.

Al Anodized Aluminum

If an aluminum object, such as a flashlight, has a color other than silver, it was anodized before it was dyed. The main goal of anodizing is to protect the aluminum from corrosion by coating it with aluminum oxide. The ability to be dyed is an added benefit.

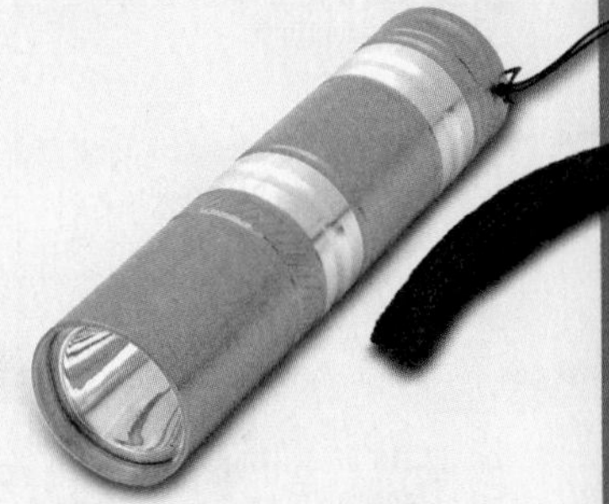

The aluminum object becomes the anode in an electrolytic cell. The electrolyte is a dilute acid. When an electric current flows through the cell, aluminum oxide forms on the surface of the aluminum.

The oxide layer is thin, hard, and dense. It contains tiny pores, which can absorb an organic dye. If a dyed piece is placed in boiling water, the oxide layer absorbs water and swells. The swelling closes the pores and seals in the dye. Because the oxide layer is so thin, the silver of the aluminum base is visible through the dye and gives the object a metallic sheen.

Did You Know?

Indium's name comes from indigo, a bright line in its emission spectrum. Indium will emit a high-pitched sound when it is bent. Tin and gallium also "cry" when bent.

Al Manufacturing Aluminum

Hall was able to convince a group in Pittsburgh to invest in his process. The investors formed the Pittsburgh Reduction Company, which became the largest producer of aluminum in the United States. The company was later renamed the Aluminum Company of American (Alcoa). Hall bequeathed much of the fortune he made to various universities. Oberlin College received three million dollars in 1914, which is equivalent to more than 50 million dollars today.

Al Anodized Aluminum

Untreated elemental aluminum cannot absorb enough dye, even if the metal is immersed in a coloring bath, to color the aluminum.

To Do

If you or one of your students has access to an aluminum object that has been anodized, bring the object to class for other students to observe.

Group 4A

See Roald Hoffmann's article on silicon from the 80th anniversary issue of *Chemical & Engineering News* for an explanation for why silicon plays so minor a role in biological processes in comparison to carbon.

Physical Properties

Based on their macroscopic properties, it isn't obvious that diamond and graphite have the same chemical composition. In 1772, Lavoisier demonstrated that diamond was carbon by analyzing the products produced when equal masses of diamond and charcoal were burned.

Sources

The orthosilicate ion, SiO_4^{4-}, is found in minerals such as zircon, $ZrSiO_4$. Several minerals contain the disilicate ion, $Si_2O_7^{6-}$, in which an oxygen atom is shared by two silicon atoms. In most silicates, the tetrahedra are linked together in chains or sheets, as in talc, $Mg_3(Si_2O_5)_2(OH)_2$. The repeating unit in a chain is $Si_2O_6^{4-}$, or SiO_3^{2-}. The repeating unit in a sheet is $Si_2O_5^{2-}$. Silicates that contain more than one metal can have a variable composition as long as the total charge on the cations is constant.

To Do

Andalusite belongs to a subclass of silicates called nesosilicates. Pyrophyllite belongs to the subclass phyllosilicates. Students could research the six subclasses of silicates and explain the basis of the classification system. The subclasses are nesosilicates (single unit), sorosilicates (double unit), inosilicates (single and double chains), cyclosilicates (rings), phyllosilicates (sheets), and tectosilicates (frameworks).

GROUP 4A

Carbon known since ancient times

Silicon discovered in 1824 by Jöns Jacob Berzelius

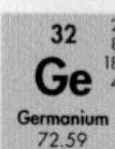

Germanium discovered in 1886 by Clemens Winkler

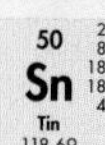

Tin known since ancient times

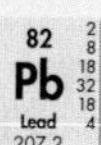

Lead known since ancient times

Physical Properties

- Group 4A elements are all solids at room temperature.
- The metallic properties of Group 4A elements increase from carbon to lead.
- Diamond, graphite, and buckminsterfullerene are three allotropes of carbon.

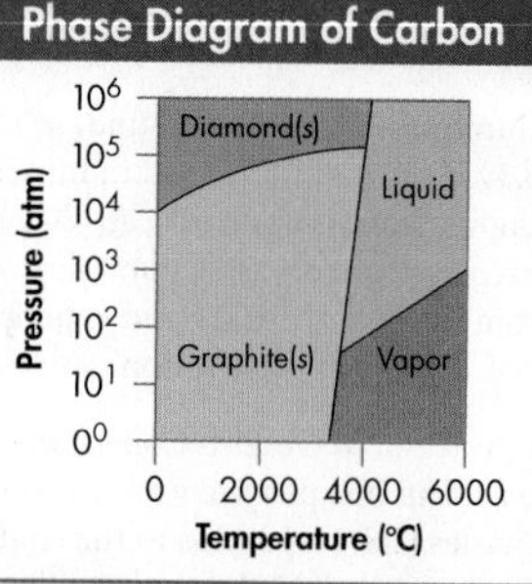

Graphite is more stable than diamond at STP, but the activation energy is too high for diamond to change to graphite at these conditions.

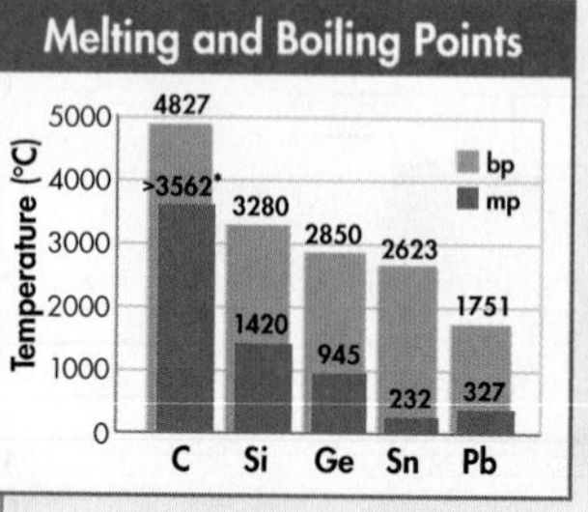

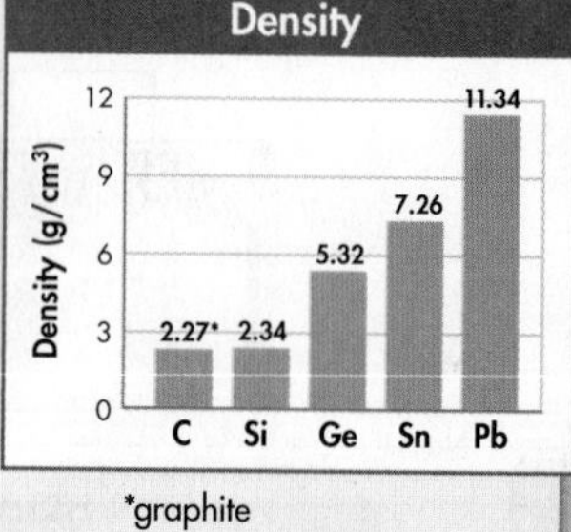

Sources

- Carbon is found in nature as an element, in Earth's atmosphere as carbon dioxide, in Earth's crust as carbonate minerals, and in organic compounds produced in cells.
- Silicon can be produced by the reduction of silicon dioxide (silica) with magnesium, carbon, or aluminum. Example:

$$SiO_2(s) + 2Mg(s) \longrightarrow Si(s) + 2MgO(s)$$

- Tin is prepared by reduction of the mineral cassiterite, SnO_2.

$$SnO_2(s) + 2C(s) \longrightarrow 2CO(g) + Sn(s)$$

- Lead is refined from the mineral galena, PbS. Galena is heated in air to form a mixture of PbO and $PbSO_4$. Lead is produced through further reaction of these compounds with PbS.

About 90% of the minerals in Earth's crust are silica and silicates. In silicates, each silicon atom is surrounded by three or four oxygen atoms. These units can be linked together in chains, sheets, rings, or crystals.

Atomic Properties

- Group 4A elements have an electron configuration that ends in ns^2np^2.
- For Group 4A elements, the most common oxidation numbers are +4 and +2. For carbon, −4 is also common.
- Silicon and germanium are semiconductors.

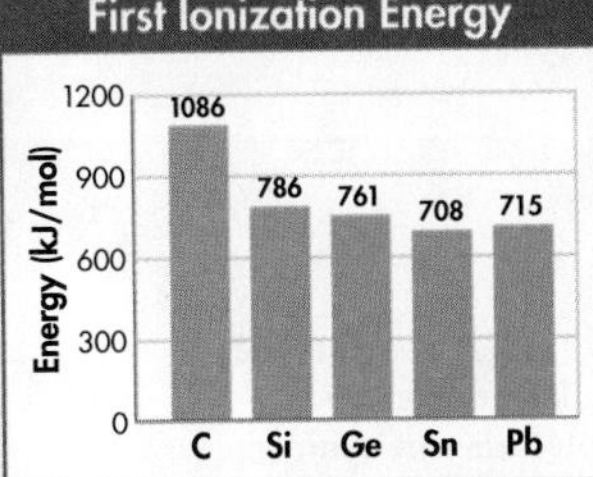

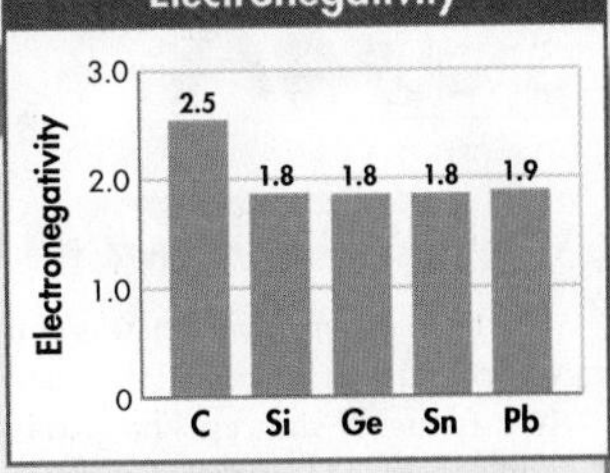

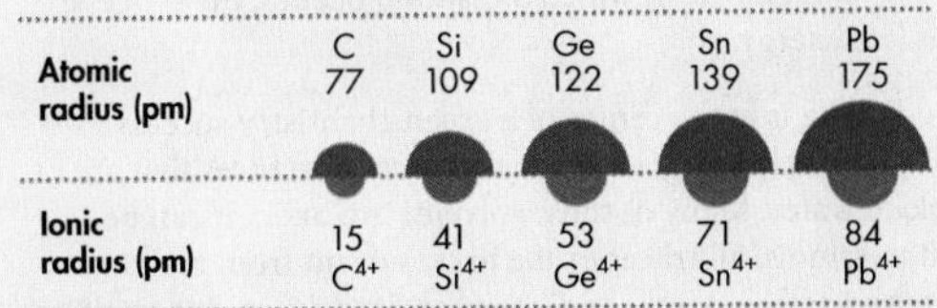

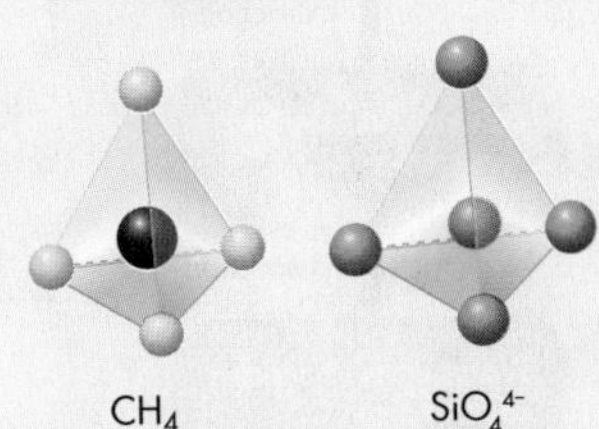

When carbon and silicon form four covalent bonds, there is often sp^3 hybridization. The result is compounds and ions with tetrahedral structures.

Important Compounds and Reactions

- Group 4A elements are oxidized by halogens.

  $Ge(s) + 2Cl_2(g) \longrightarrow GeCl_4(l)$
- Group 4A elements combine with oxygen to form oxides. Example:

  $Sn(s) + O_2(g) \longrightarrow SnO_2(s)$
- Complete combustion of hydrocarbons yields carbon dioxide and water. Example:

  $CH_4(g) + 2O_2(g) \longrightarrow CO_2(g) + 2H_2O(l)$

  $\Delta H = -890$ kJ/mol
- Plants use carbon dioxide to produce carbohydrates and oxygen.
- Aqueous sodium silicate, Na_2SiO_3, is used as an adhesive for paper, as a binder in cement, and to stabilize shale during oil drilling.

  $SiO_2(s) + 2NaOH(aq) \longrightarrow Na_2SiO_3(aq) + H_2O(l)$
- Acetylene is a fuel used for welding. It forms when calcium carbide and water react.

  $CaC_2(s) + 2H_2O(l) \longrightarrow C_2H_2(g) + Ca(OH)_2(aq)$
- Tungsten carbide, WC, is used on the cutting surfaces of drill bits and saw blades.
- Tin(II) fluoride, SnF_2, is used in some toothpastes to prevent tooth decay.

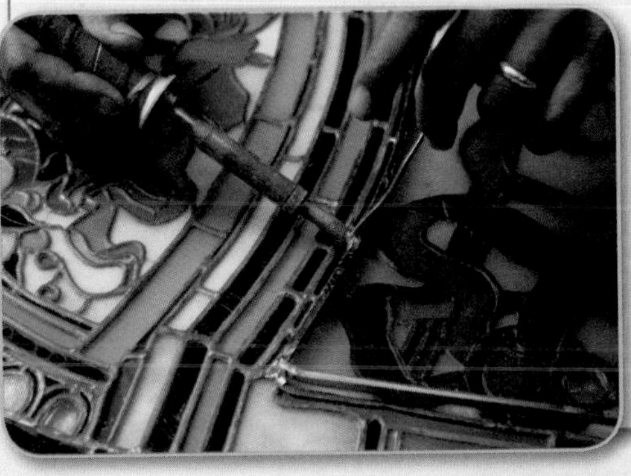

Silicon dioxide, SiO_2, is the sand on many beaches. It is used to make glass, including glass that is "stained" with metallic salts. The pieces of glass in this window are held together with strips of lead.

Atomic Properties

The graphs for first ionization energy and electronegativity can help you discuss the fact that the first element in a group of representative elements is often the least representative of the group as a whole. The elements in period 2 have only the electrons in the 1*s* orbital to shield the valence electrons from the attraction of the nucleus.

Important Compounds and Reactions

A mixture of sand (SiO_2), lime (CaO), and soda ash (Na_2CO_3) is used to make glass. The addition of small amounts of transition metal oxides can change the color of glass. Adding iron(III) oxide (Fe_2O_3) gives glass the green tint used in some beverage bottles and tinted windows. Blue glass contains CuO or CoO, orange glass contains Cr_2O_3, and red glass contains colloidal gold.

GROUP 4A

Group 4A

C Green Chemistry

The phrase "green chemistry" was coined in 1992 to define an approach to developing chemical technology. The green chemistry mission statement of the EPA is "to promote innovative chemical technologies that reduce or eliminate the use or generation of hazardous substances in the design, manufacture, and use of chemical products." The EPA program supports research and education through partnerships with industries, academia, other government agencies, and non-government agencies. Los Alamos National Laboratory is one important partner. Los Alamos was instrumental in the development of supercritical carbon dioxide as a recyclable cleaning solvent. Los Alamos has collaborated with industries on projects ranging from fuel cells to a process for producing ethylene that will reduce carbon dioxide emissions and conserve natural gas.

Caffeine is an alkaloid. The amount of caffeine in a coffee bean is typically 1% to 2% by mass. Around 1900, a German chemist, Ludwig Roselius, became interested in extracting caffeine from coffee because he blamed his father's death on the drinking of excessive amounts of coffee. There were a number of solvents that could remove the caffeine from coffee, but most were known to be toxic. By the 1980s, the solvent used for 70 years, methylene chloride, was suspected to be a carcinogen. Ethyl acetate and water are also used to decaffeinate coffee.

GROUP 4A

C Green Chemistry

The term *green chemistry* was coined in 1992. It describes the effort to design chemical processes that don't use or produce hazardous substances. The goal is to protect the environment and conserve resources. For example, if a catalyst is used to reduce the temperature at which a reaction occurs, the process requires less energy.

Carbon dioxide is at the center of a green chemistry success story. Organic solvents are used to dissolve substances that are insoluble in water. Many of these solvents are toxic. It can be difficult to remove all traces of the toxic solvent from reaction products and safely recycle or dispose of the solvent. Supercritical carbon dioxide can replace some organic solvents.

A gas becomes a supercritical fluid at a temperature and pressure called its critical point. For carbon dioxide, this occurs at 31.1°C and about 100 atmospheres. At its critical point, carbon dioxide is in a hybrid state. It has a high density (like a liquid) but it is easily compressed (like a gas). Many organic compounds dissolve in supercritical carbon dioxide. The solvent is easily separated from a reaction mixture because it evaporates at room temperature and atmospheric pressure. It is also used to separate substances from mixtures. It can extract caffeine from coffee beans, dry-clean clothes, or clean circuit boards.

Decaffeinating Coffee

Caffeine dissolves inside coffee beans soaked in water.

↓ Extraction

Caffeine diffuses into supercritical CO_2.

↓ Absorption

Water droplets leach caffeine from CO_2.

After the caffeine is extracted, the coffee beans are dried and roasted. The aqueous solution of caffeine is sold to soft drink manufacturers.

Si Optical Glass

Glass is a material with the structure of a liquid, but the hardness of a solid. In most solids, the particles are arranged in an orderly lattice. In solid glass, the molecules remain disordered, as in a liquid. The main ingredient in most glass is silica (SiO_2), which is one of the few substances that can cool without crystallizing.

The glass used in eyeglasses, microscopes, and telescopes is called optical glass. This type of glass is purer than window glass and transmits more light. Optical glass can be drawn into long fibers that are used like tiny periscopes to view tissues deep within the human body.

In an optical fiber, light travels through a thin glass center called the core. A second glass layer reflects light back into the core. An outer plastic layer protects the fiber from damage. The transmission of light signals through glass is called fiber optics.

When the fibers are bundled into cables, they often replace electrical cables in computer networks. They are also used to transmit television signals and phone calls over long distances.

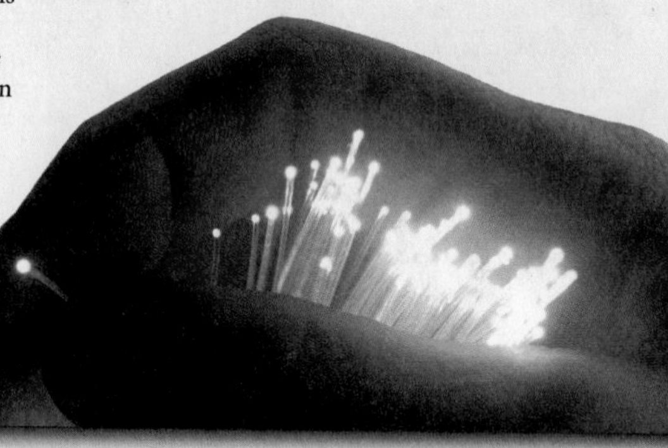

Si Optical Glass

A light signal can travel a great distance through the core of an optical fiber because the cladding doesn't absorb any light. Impurities in the cladding cause the signal to degrade, which is why the glass must be optical glass. Optical fibers are cheaper than copper wires of equivalent length. Because an optical fiber can have a smaller diameter than a copper wire, more fibers can be bundled into a cable with the same diameter.

To Do

Students may want to research what is happening in eyeglass lenses that darken when exposed to light. When lenses containing silver chloride are exposed to bright light, the silver chloride breaks down, releasing opaque silver atoms. These atoms reduce the amount of light that can penetrate the lens. When the amount of available light is reduced, the silver and chlorine atoms re-associate and the glass becomes clear again.

Si Semiconductors

With a cellular phone, you can call or text your friends from almost any location. Most phones also let you read and send e-mails, get directions, take a photograph, listen to music, or access the latest news. How can such a complex device be small enough to fit in your pocket? Semiconductor technology is responsible.

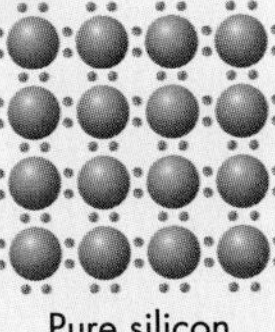

Pure silicon

Silicon is a semiconductor. In its pure form, it conducts an electric current better than most nonmetals but not as well as metals. But its ability to conduct can be changed dramatically by doping, or adding traces of other elements, to the silicon crystal.

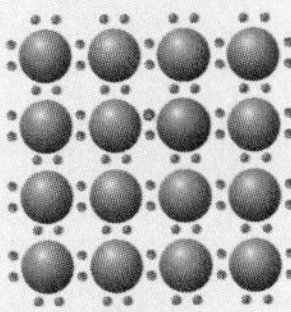

n-type (with arsenic)

Doping with arsenic produces a donor, or n-type, semiconductor. Each arsenic atom has five valence electrons, compared with four for silicon. So there are extra electrons in the crystal. Doping with boron produces an acceptor, or p-type, semiconductor. Because boron has only three valence electrons, there is a positive "hole" in the crystal for every boron atom. The extra electrons or holes are free to move and conduct an electric current.

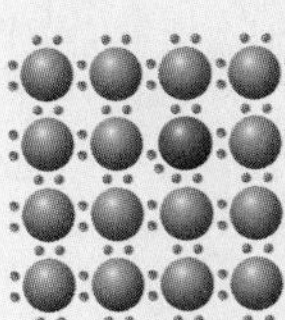

p-type (with boron)

Combinations of n-type and p-type semiconductors are used to build tiny electronic components. An integrated circuit containing millions of components can fit on a small semiconductor wafer. The resulting "chip" can be used to control devices such as computers, calculators, DVD players, televisions, and phones.

A cell phone's integrated circuit must process more and more data as features are added to the phone.

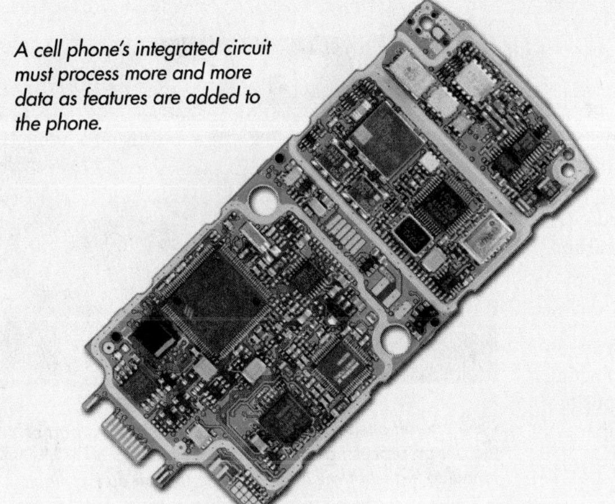

C Beyond Buckyballs

Buckminsterfullerene (C_{60}) is one member of a family of fullerenes. These structures are closed-cage spherical or nearly spherical forms of elemental carbon. The cages are networks of 20 to 600 carbon atoms.

Scientists have verified the existence of nesting spheres of fullerenes. C_{60} can be nested inside C_{240}, and this pair can be nested inside C_{540}. These nesting structures are sometimes called bucky-onions because they resemble the layers of an onion.

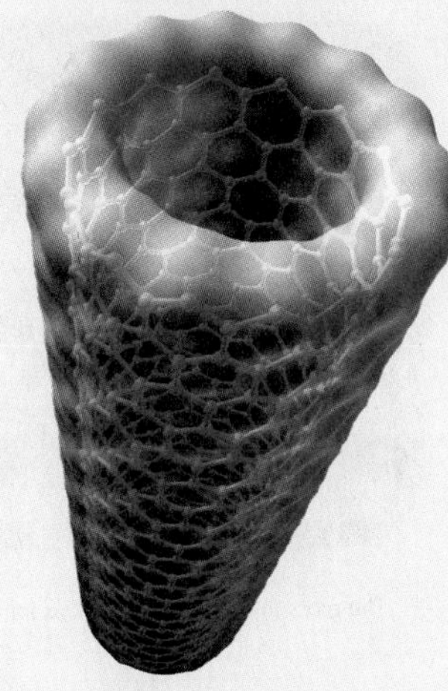

Dr. Sumio Iijima discovered a tubular fullerene, or carbon nanotube, in Japan in 1991.

Did You Know?

One name for diamonds is "ice." A diamond can quickly draw heat from your hand when you touch it. Such a high thermal conductivity is unusual for a substance containing covalent bonds.

Si Semiconductors

The essential components of a cell phone are a circuit board, antenna, liquid crystal display, keyboard, microphone, speaker, and battery. On the circuit board are analog-to-digital and digital-to-analog chips to translate the outgoing and incoming signals. There is a system processor that performs millions of calculations per second for the manipulation of the signals and a microprocessor that handles the keyboard, display, and all other functions.

C Beyond Buckyballs

The intensive properties of a material are degraded by structural defects. For example, if there is a tiny crack in a steel wire, the strength of the wire will be significantly reduced. Because tubular fullerenes are molecules, they don't have the structural defects associated with multi-wall carbon fibers. With molecules, it is also possible to reliably predict chemical changes.

Group 4A

C Greenhouse Gases

The Intergovernmental Panel on Climate Change (IPCC) was founded in 1988 to review the evidence and assess the impact of climate change. The IPCC published major reports in 1990, 1995, 2001, and 2007. The fifth report is planned for 2013 or 2014. (See the IPCC Web site for a statement on IPCC principles and procedures.) The IPCC has concluded that Earth's temperature increased by 0.6°C since the late 1880s and is likely to increase by another 1.9–2.9°C by 2100.

The source for the data in the graph is the Goddard Institute for Space Studies. The 10 warmest years in the 1900s occurred in the last 15 years of the century. 1998 was the warmest year since scientists began recording global mean surface temperatures. During the 1900s, sea level rose 4–8 inches, and there was a decrease in both snow cover in the Northern Hemisphere and floating ice in the Arctic Ocean. Precipitation over land increased worldwide by about 1%.

Carbon dioxide is not the only greenhouse gas. Methane (CH_4), nitrous oxide (N_2O), sulfur hexafluoride (SF_6), and chlorofluorocarbons (CFCs) exist in much smaller quantities but produce a more powerful greenhouse effect than CO_2. For example, PFCs have a global warming potential of up to 7000 times that of CO_2. The atmospheric content of these gases is also rising as a result of human activities.

GROUP 4A

C Greenhouse Gases

There are gases in Earth's atmosphere that are called greenhouse gases because they act like the glass in a greenhouse. Sunlight easily passes through these gases to Earth's surface. Some of the solar energy is reflected off the surface as infrared radiation. This radiation is absorbed by greenhouse gases and radiated back to Earth. By trapping infrared radiation, the greenhouse gases keep Earth's surface about 33°C warmer than it would be otherwise.

Average Temperature at Earth's Surface

Temperature (°C): 13.5, 13.7, 13.9, 14.1, 14.3, 14.5

Year: 1910, 1930, 1950, 1970, 1990, 2010

With an increase in greenhouse gases, more infrared radiation is trapped, causing global warming.

The glass in a greenhouse traps infrared radiation.

Carbon dioxide (CO_2) is the most abundant greenhouse gas. It is released into the air as a product of cellular respiration and removed from the air during photosynthesis. Such interactions normally keep the amount of atmospheric CO_2 in check. But the burning of fossil fuels releases more than 20 billion metric tons of CO_2 every year. Also, as forests are cleared for agriculture, the ability of plants to remove CO_2 from the atmosphere is reduced.

Scientists agree that a rise of only a few degrees in Earth's temperature could cause problems. They disagree on how severe the problems could be. Could climates change so that farmlands become deserts? Could the melting of ice caps cause sea levels to rise until coastal cities are under water?

Si Composite Materials

Most composites contain two distinctly different materials. The materials can be arranged in layers as when a sheet of plastic is sealed between panes of glass. Or a composite may consist of a matrix in which fibers of a second material are embedded. Often, the matrix is plastic. The fibers can be carbon.

Composites reinforced with carbon fibers are stronger than steel, yet light in weight. These composites are used in sports equipment, such as hockey sticks and golf clubs. It is less tiring to swing a tennis racket made from a carbon-fiber composite than one made from wood or metal. Carbon-fiber baseball bats act more like wood bats than do aluminum bats.

New aircraft often rely on composite materials. As part of the design process, a model of the plane is placed in a wind tunnel to test what will happen to the aircraft during flight.

Si Composite Materials

Composites are heterogeneous mixtures. Most composites consist of a binder, or matrix, and a reinforcement, which is stronger and stiffer than the matrix. The matrix holds the reinforcement, which can be in the form of particles or fibers, together. In concrete, cement is the matrix and gravel the reinforcement.

To Do

Students who play a string instrument may want to research and report on the properties of bows made from carbon-fiber composites. (If students research composites they are likely to see the terms *carbon fiber* and *graphite fiber* used interchangeably. They are also likely to see *bonded* used to describe how the reinforcement is held in a matrix.)

C Recycling Plastics

At about 2 kg of waste per person per day, the United States leads the world in the production of solid waste. Luckily, the United States is also a leader in recycling. It is important to recycle plastics because they are made from crude oil, a non-renewable resource. Also, some plastics release toxic gases, such as hydrogen cyanide (HCN) and hydrogen chloride (HCl), when they burn in an incinerator. Finally, plastics are used for packaging material because they do not decay when exposed to sunlight, water, or microorganisms. The downside of this resistance to decay is that plastics can remain unchanged in dumps and landfills for decades.

Plastics are usually sorted by type before they are melted and reprocessed. The plastics industry has a code to identify types of plastics. The numeral 1 is assigned to polyethylene terephthalate (PET), which is used in soft-drink bottles. The numeral 2 refers to high-density polyethylene (HDPE), which is used in milk jugs and shampoo bottles. These are the two types in greatest demand. Carpets and clothing are made from recycled PET fibers. Recycled HDPE is used as a wood substitute for decks and benches.

Fleece clothing often contains post-consumer recycled (PCR) plastic.

Si Silicone Polymers

If you have worn hard contact lenses or used shaving cream, you have used a silicone. Silicone polymers have chains in which silicon and oxygen alternate. The properties of silicones depend on the groups that are bonded to the silicon atoms and the length of the chains.

In silicone rubber and resins, there are cross-links between the chains. These silicones repel water and remain elastic, even at low temperatures. They are used in space suits, as gaskets in airplane windows, and as sealants that are squeezed into place and left to harden.

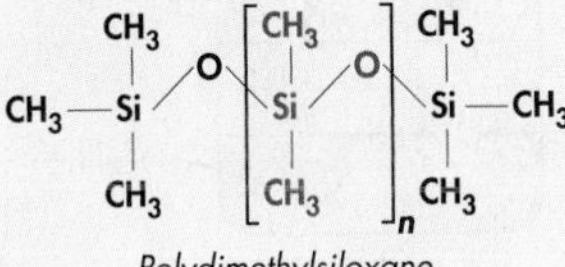

Polydimethylsiloxane

In polydimethylsiloxane, two methyl groups are bonded to each silicon atom in the chain. Polydimethylsiloxane is used as a lubricant in skin and suntan lotions.

C Carbon Monoxide

It is hard to detect colorless, odorless carbon monoxide gas. When it is inhaled, its molecules bind to the hemoglobin in red blood cells. They bind about 200 times more effectively than oxygen molecules do. So less oxygen reaches body tissues. Headaches, dizziness, nausea, and drowsiness are symptoms of low-level carbon monoxide poisoning. Higher levels of carbon monoxide are fatal.

The incomplete combustion of fuel in gas furnaces and space heaters produces carbon monoxide. It also forms, to some degree, in all internal combustion engines. In the United States, cars have catalytic converters, which convert carbon monoxide to carbon dioxide. In most communities, home owners are required to install devices that monitor the carbon monoxide concentration.

Cigarette smoke contains carbon monoxide. The carbon monoxide from one cigarette can remain in a smoker's blood for several hours. Smoking increases the risk of heart attacks because the heart must pump harder to deliver oxygen to cells when the level of oxygen in the blood is reduced.

Did You Know?

Members of the Scott expedition to the South Pole in 1912 may have died because of **tin.** Their supply of paraffin fuel leaked out through tiny holes in the tin-soldered joints of the storage cans because tin slowly changes to a powder below 13°C.

GROUP 4A

C Recycling Plastics

The numeral 3 is used for polyvinyl chloride, the numeral 4 for low-density polyethylene (LDPE), the numeral 5 for polypropylene, the numeral 6 for polystyrene, and the numeral 7 for all other plastics.

According to the EPA, Americans generated 250 million tons of trash in 2008. About 61 million tons were recovered through recycling and another 22.1 million tons through composting. About 32 million tons were burned with energy recovery. The remaining 54.2% was discarded. Visit the EPA Web site for more data on trends in municipal solid waste in 2008 and in subsequent years.

To Do

Students could research recycling programs in your community. They could find out which items are recycled and where these items are collected (curbside or at designated drop-off sites).

C Carbon Monoxide

The Consumer Product Safety Commission recommends that households install a CO_2 detector with an audible alarm near bedrooms and a second monitor near the heating source.

To Do

Students might want to research the relationship between the level of carbon monoxide in the air near highways and emissions from vehicles. In urban areas, the concentration tends to peak during morning and afternoon rush hours. Carbon monoxide tends to accumulate more in cold weather when engines run less efficiently. Other factors to consider are the presence or absence of wind to disperse pollutants or the presence of inversion layers in the atmosphere.

Si Silicone Polymers

Students will also be familiar with some properties of silicone polymers if they have bounced a superball, stretched Silly Putty, removed a peel off label, placed weather-stripping caulk around a door frame, polished a car, or ingested a remedy for an upset gaseous stomach.

Group 5A

Physical Properties

Cryogen comes from the Greek *kryos,* meaning "icy cold." Foods can be packed, sealed, and sprayed with liquid nitrogen, which absorbs heat as it evaporates. Rapid freezing is also used to preserve tissue samples, such as blood and bone marrow. Surgeons use liquid nitrogen to remove warts and skin cancers.

Sources

Roasting involves heating an ore to a high temperature in a furnace, which causes a reaction of the ore with a gaseous component of the atmosphere in the furnace. The oxidation of sulfide ores is an example of roasting. The products of roasting will be a metal oxide and sulfur dioxide or the elemental metal and sulfur dioxide (depending on the activity of the metal). Roasting of antimony produces the oxide, which can then be reduced with carbon. A flux of sodium sulfate is used to prevent evaporation of molten antimony.

$2Sb_2S_3 + 9O_2 \rightarrow 2Sb_2O_3 + 6SO_2$

$2Sb_2O_3 + 3C \rightarrow 4Sb + 3CO_2$

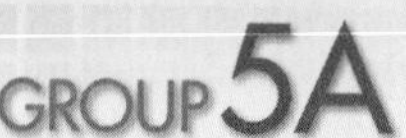

GROUP 5A

Nitrogen discovered in 1772 by Daniel Rutherford

Phosphorus discovered in 1669 by Hennig Brand

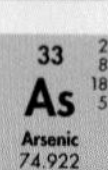

Arsenic discovered in 1250 by Albertus Magnus

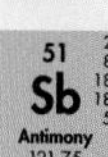

Antimony discovered in 1600s or earlier

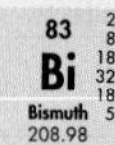

Bismuth described in 1450 by Basil Valentine; shown to be a distinct element in 1753 by Claude-François Geoffroy

Physical Properties

- Except for nitrogen gas, Group 5A elements are solid at room temperature.
- The metallic properties of Group 5A elements increase from top to bottom within the group. N and P are nonmetals. As and Sb are metalloids. Bi is a metal.
- Liquid nitrogen is a cryogen, a liquid refrigerant that boils below −190°C.
- Phosphorus has 10 allotropes including white and red.

White phosphorus is very reactive and must be stored under water. Red phosphorus is much more stable.

Melting and Boiling Points

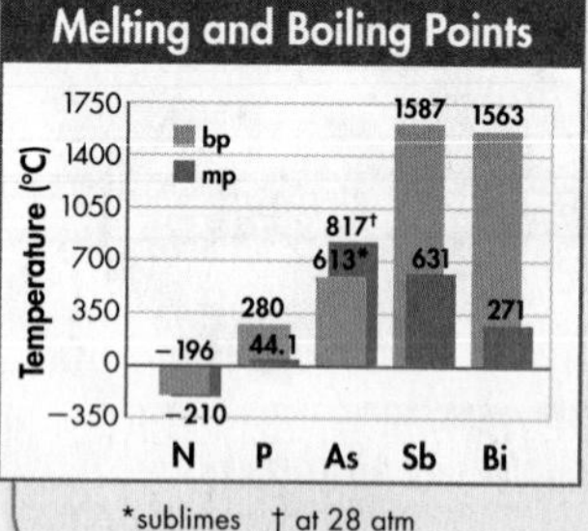

*sublimes † at 28 atm

Density

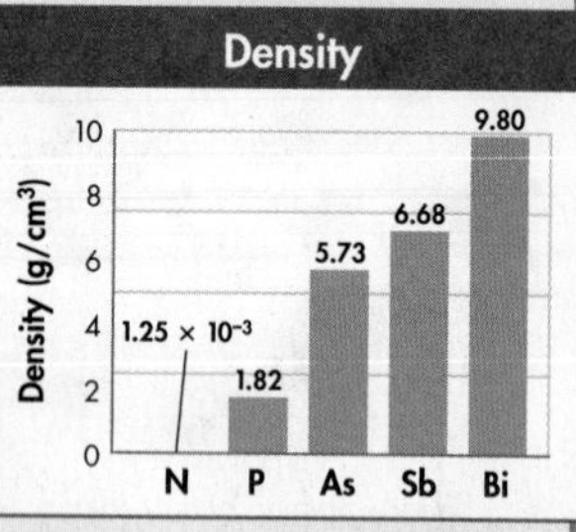

Sources

- Nitrogen is obtained from the fractional distillation of liquefied air.
- Phosphorus is derived from phosphate minerals. Example:

$$Ca_3(PO_4)_2(s) + 3SiO_2(s) + 5C(s) \longrightarrow 2P(l) + 3CaSiO_3(s) + 5CO(g)$$

- Arsenic is prepared by heating a mixture of $FeAs_2$ and FeS_2 in the absence of air.
- Antimony is prepared by roasting the ore stibnite, Sb_2S_3.

Fractional Distillation of Air

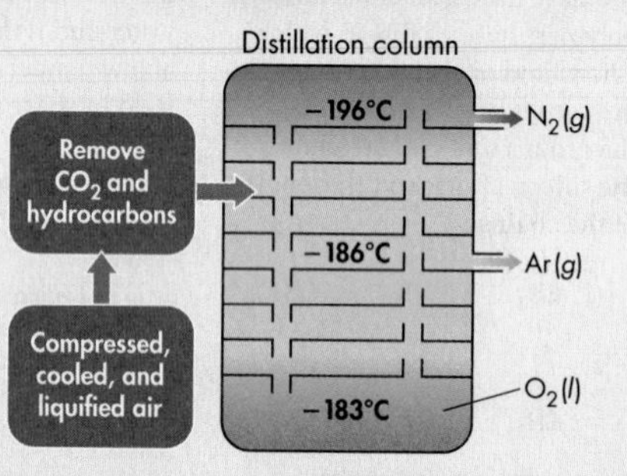

R20 Elements Handbook

Atomic Properties

- Group 5A elements have an electron configuration that ends in ns^2np^3.
- The most common oxidation numbers for Group 5A elements are +3, +5, and −3.
- Nitrogen has oxidation numbers from −3 to +5 (and all numbers in between) in a variety of stable compounds.
- Elemental nitrogen, N_2, is highly unreactive due to its strong N-to-N triple bond.

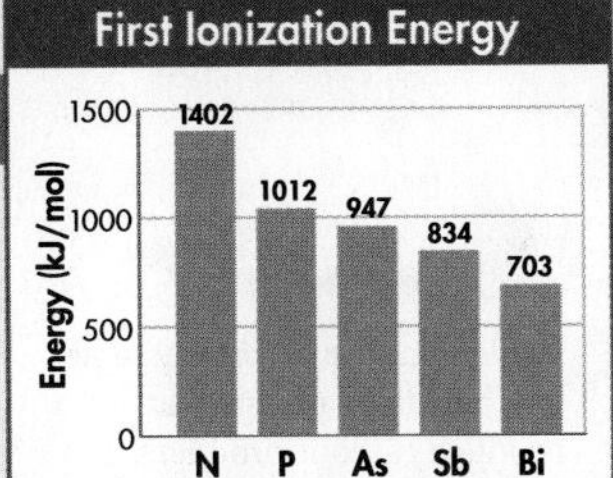

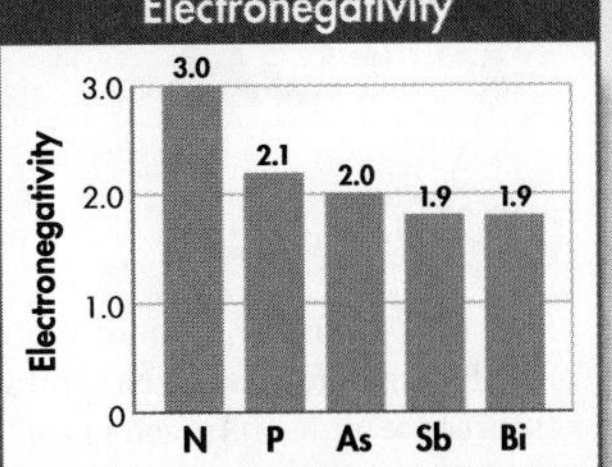

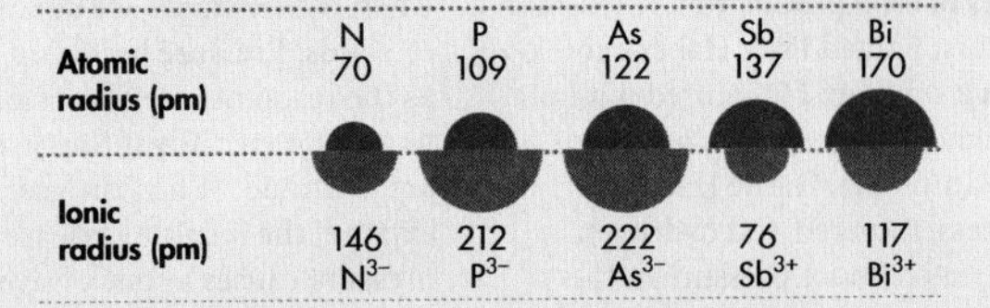

Important Compounds and Reactions

- Nitrous oxide, N_2O, (laughing gas) is an anesthetic. It is made from ammonium nitrate.

$$NH_4NO_3(s) \xrightarrow{\Delta} N_2O(g) + 2H_2O(g)$$

- Nitrogen dioxide, NO_2, is an air pollutant produced when fossil fuels burn at high temperatures.
- Lightning causes the nitrogen and oxygen in air to react and form nitric oxide, NO.
- Poisonous hydrazine, N_2H_4, is used as a rocket fuel. It is prepared by the following reaction.

$$2NH_3(aq) + OCl^-(aq) \longrightarrow N_2H_4(aq) + Cl^-(aq) + H_2O(l)$$

The reaction is complex. One intermediate is chloramine, NH_2Cl, which is also poisonous. The labels on bottles of household ammonia and chlorine bleach warn you not to mix these solutions. If you ignore the warning, chloramine will be produced.

- Nitric acid, HNO_3, is used to make fertilizers and explosives. It is produced by the Ostwald process.
- Phosphoric acid is used in soft drinks and fertilizers. It is made by a double-replacement reaction.

$$Ca_3(PO_4)_2(s) + 3H_2SO_4(aq) \longrightarrow 3CaSO_4(aq) + 2H_3PO_4(aq)$$

Dentists use phosphoric acid to etch the exposed dentine layer of a drilled tooth to help a filling adhere to the tooth.

- Ammonia is synthesized directly from its elements by the Haber-Bosch process.

$$3H_2(g) + N_2(g) \longrightarrow 2NH_3(g)$$

- Amino acids such as glycine, H_2NCH_2COOH, are the building blocks of proteins.
- Bismuth subsalicylate, $BiO(C_7H_5O_3)$, is the active ingredient of a pink liquid antacid.
- Arsenic trioxide, As_2O_3, is powerful poison. Its use as a weed killer and insecticide has been restricted.
- Antimony is added to alloys to increase their hardness.

Ostwald Process

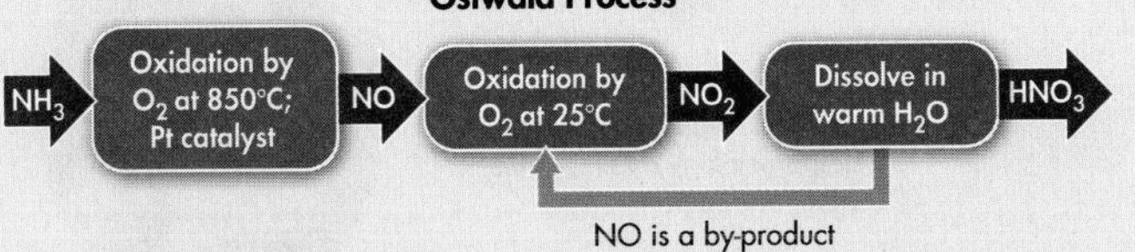

Ammonia is oxidized to nitrogen dioxide in two stages. The nitrogen dioxide forms nitric acid as it passes through warm water.

Atomic Properties

The table compares the bond lengths and bond dissociation energies for single, double, and triple nitrogen-nitrogen bonds.

Bond	Length	Dissociation Energy
single	1.47 Å	145 kJ/mol
double	1.24 Å	456 kJ/mol
triple	1.10 Å	945 kJ/mol

To Do

Students can use the bond lengths and bond dissociation energies to show the relationship between the number of shared electron pairs, bond length, and bond strength.

Important Compounds and Reactions

The approach to healing by traditional physicians during the renaissance was based on the existence of four humors (blood, phlegm, yellow bile, and black bile). Most traditional remedies were derived from plants. Apothecaries and other followers of Paracelsus used inorganic compounds as treatments. For example, a mixture used to treat fever contained powdered nitre (potassium nitrate), carbonate of potash (potassium carbonate), antimonial wine (hydrated potassium antimonyl tartrate), and sweet spirits of nitre (ethyl nitrate).

To Do

Help students write balanced equations for the reactions in the Ostwald Process.

$$4NH_3(g) + 5O_2(g) \rightarrow 4NO(g) + 6H_2O(g)$$

$$2NO(g) + O_2(g) \rightarrow 2NO_2(g)$$

$$3NO_2(g) + H_2O(l) \rightarrow 2H^+(aq) + 2NO_3^-(aq) + NO(g)$$

Group 5A

N Ammonia

Haber reacted three-to-one molar mixtures of hydrogen and nitrogen at different conditions of temperature and pressure. The yield of ammonia at equilibrium decreased with an increase in temperature and increased with an increase in pressure. Because the boiling point of ammonia (−33°C) is much higher than the boiling points of nitrogen (−196°C) or hydrogen (−253°C), it is possible to remove the ammonia from the equilibrium mixture and drive the reaction to completion.

The Haber process is sometimes called the Haber-Bosch process. Karl Bosch implemented Haber's process on an industrial scale. It was Bosch who found a catalyst that was cheaper and more available than the osmium and uranium that Haber had used. His team had to develop a furnace that could operate safely at the required temperature and pressure. In 1931, Bosch and Friedrich Bergius were awarded a Nobel Prize for the "invention and development of chemical high pressure methods."

Before the Haber Process, deposits of dried guano in Chile were the main source of fixed nitrogen. (The guano was deposited by fish-eating sea birds.) During World War I, a naval blockade of South America prevented Germany from importing the nitrates it needed to manufacture explosives. It has been argued that World War I would have ended sooner if Haber had not developed his process. Because Haber (as chief of Germany's Chemical Warfare Service) permitted the use of chlorine gas as a weapon, the awarding of his Nobel Prize in 1918 was controversial.

GROUP 5A

N Ammonia

The nitrogen in air is not in a form that is useful for plants. Natural sources of usable nitrogen in soil aren't sufficient to support current levels of crop production. A German chemist named Fritz Haber came up with a solution. He figured out how to convert atmospheric nitrogen into ammonia. In the Haber process, nitrogen and hydrogen are heated under pressure in the presence of iron.

$$N_2(g) + 3H_2(g) \longrightarrow 2NH_3(g) \quad \Delta H = 46.19 \text{ kJ/mol}$$

The ammonia is liquefied. Liquid ammonia, aqueous solutions of ammonia, and ammonium salts are used as fertilizers. Liquid ammonia is also used as a refrigerant. Many cleaning products contain aqueous ammonia, which is a weak base. Ammonia is also used to manufacture explosives.

Haber Process

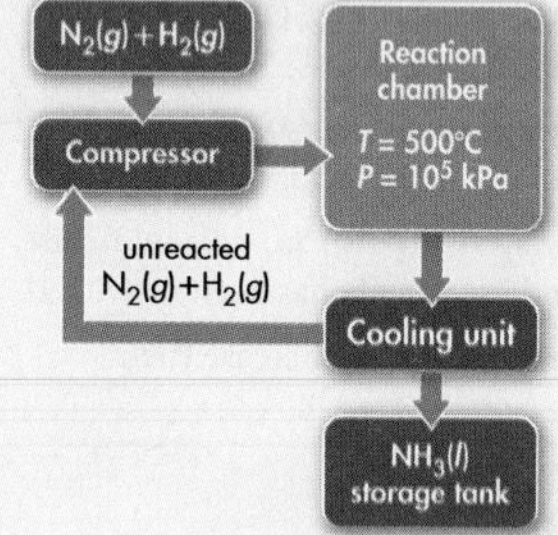

When the reaction mixture is cooled, the liquified ammonia can be separated from the nitrogen and hydrogen.

N Explosives

Explosions are extremely rapid exothermic reactions that produce gaseous products. The most forceful explosions occur when the reactants are liquids or solids. Pressure builds up as the reactants are converted to gases, especially if the gases are confined. When the gases expand, the resulting release of pressure causes a shock wave. The kinetic energy of this wave, the wind that follows it, and the heat is the destructive force of the explosion.

Engineers use a series of small controlled explosions to force a building to collapse inward, or implode.

An explosion requires fuel and an oxidizer. If the oxidizer and fuel are separate, the reaction is a combustion reaction. This type of explosion could occur if natural gas leaking in a building was ignited by a spark or flame.

$$CH_4 + 2O_2(g) \longrightarrow CO_2(g) + 2H_2O(g) \quad \Delta H = 890 \text{ kJ/mol}$$

If the fuel is its own oxidizer, a decomposition reaction takes place. Nitroglycerine, $C_3H_5(NO_3)_3$, is a thick, pale, oily liquid. It decomposes to form a mixture of gaseous products.

$$4C_3H_5(NO_3)_3(l) \longrightarrow 6N_2(g) + O_2(g) + 12CO_2(g) + 10H_2O(g) \quad \Delta H = -1427 \text{ kJ/mol}$$

Self-oxidizing explosives are often unstable. Jarring them may be enough to cause detonation. Alfred Nobel, a Swedish chemist and inventor, found a way to use nitroglycerine with less risk. After his family's nitroglycerine factory exploded in 1864, he moved his experiments to a barge in the middle of a lake. One day he found a cask of nitroglycerine that had leaked. Luckily, the diatomite in which the cask was packed had absorbed the liquid. Nobel found that the mixture was stable until detonated by a blasting cap. Plus, the mixture was as explosive as the pure liquid. Nobel named his invention dynamite.

In his will, Nobel established a fund to provide annual prizes in chemistry, physics, physiology and medicine, literature, and peace.

N Explosives

An explosion is highly exothermic when the bonds in the explosive are relatively weak and the bonds in the products are relatively strong. Because the bonds in N_2, CO, and CO_2 are very strong, many explosives are organic molecules with nitro or nitrate groups, which can react to produce N_2, CO, or CO_2.

Nitroglycerin is used to treat angina. Its effect is to cause blood vessels to dilate. The nitroglycerin used in pills, sprays, and patches is very dilute. Nitroglycerin has a short shelf life. It is unstable and must be stored in containers that prevent exposure to light.

N Acid Rain

Normal rainfall has a pH of about 5.6. It is mildly acidic because carbon dioxide in the air dissolves in tiny water droplets and forms carbonic acid (H_2CO_3). In acid rain, the pH is lower due to the emission of nitrogen oxides and sulfur oxides into the atmosphere. Some natural sources, such as volcanoes, emit these oxides. But most come from the burning of fossil fuels. In the atmosphere, the oxides form nitric acid (HNO_3) and sulfuric acid (H_2SO_4), which fall to Earth in rain or snow.

Because acid rain dissolves and washes away nutrients from soil, trees are dying in the Appalachian Mountains. Because aquatic species are highly sensitive to changes in pH, some lakes once full of fish and frogs are nearly lifeless.

Power plants are one source of sulfur oxides and nitrogen oxides. Devices called scrubbers are used to remove these pollutants from smoke. In one type of scrubber, smoke passes through an aqueous suspension of lime (CaO). The lime reacts with SO_2 to form solid calcium sulfite ($CaSO_3$).

Another type of scrubber can be used to remove nitrogen oxides. Ammonia is sprayed onto a surface covered with a mixture of catalysts. As smoke passes over the surface, nitrogen oxides react with ammonia and oxygen, forming nitrogen and water.

Another way to deal with nitrogen oxides is to keep them from forming in the first place. The method involves burning a fuel, such as coal, at a high temperature. When the burner contains more fuel than air, most of the oxygen in the air reacts with the fuel rather than the nitrogen.

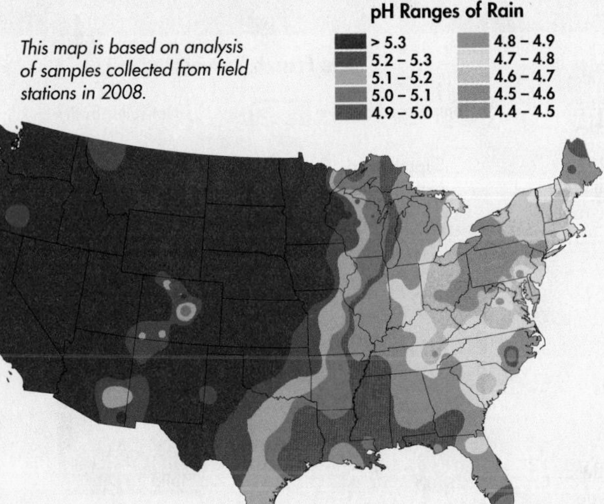

This map is based on analysis of samples collected from field stations in 2008.

N P Fertilizers

Most fertilizers contain salts of nitrogen, phosphorus, and potassium. These elements are essential to plant growth. On the label, the percents by mass of these nutrients are always listed in the order N-P-K.

Because the salts in a fertilizer can vary, there are rules for reporting the content. Nitrogen is always reported as the percent by mass of elemental nitrogen. Phosphorus and potassium are reported as the percent by mass of phosphorus pentoxide, P_2O_5, and potassium oxide, K_2O.

This system makes it easy to compare fertilizers. For example, a fertilizer labeled 20-10-10 has twice the mass of nitrogen as one labeled 10-10-10. But they both have the same mass of phosphorus and potassium. The numbers may not add up to 100% because fertilizers always contain ingredients not included in the N-P-K analysis.

Did You Know?

If nail polish or lipstick contains **bismuth** oxychloride (BiOCl), it appears lustrous and pearly.

N Acid Rain

Acid rain has had a major impact on the ecology of lakes. Most lakes in which organisms live have a pH between 6.5 and 8.5. All of the vertebrates, most of the invertebrates, and many microorganisms cannot survive at a pH below 4.0. Lakes that are most likely to have a serious drop in pH due to acid rain are those with low concentrations of ions such as HCO_3^-, which can act as a buffer.

To Do

Have students use the isopleth map to determine the pH of rainfall in your region as of 2008. (An isopleth is a line connecting points on a graph or map that have equal values for a given variable.) Students can research trends by viewing the isopleth maps produced annually by the National Atmospheric Deposition Program (NADP).

N P Fertilizers

Mineral plant nutrients can be classified as primary nutrients (N, P, and K), secondary nutrients (Ca, Mg, and S), and micronutrients (B, Cu, Fe, Cl, Mn, Mo, and Zn). Primary and secondary nutrients tend to be less available to plants in soils with a low pH, but micronutrients tend to be more available to plants in soils with a high pH. Most plants do well in soil with a pH between 6.0 and 7.5.

To Do

Invite an expert to talk to the class about why fertilizers with different percent compositions are used with different types of plants or at different stages in plant growth.

Group 5A **R23**

GROUP 6A

Group 6A

Physical Properties

When molten sulfur cools below 119°C, it forms crystals of monoclinic sulfur. This allotrope contains eight-member rings of covalently bonded sulfur atoms (S_8). Below 95.5°C, the sulfur changes to rhombic sulfur, which is also composed of S_8 units. If molten sulfur is poured into cold water, an amorphous form of sulfur results. Within hours, the amorphous sulfur loses its elasticity as it is converted to rhombic crystals, which are more stable. Sulfur can be purified through boiling and condensation. Tiny rhombic crystals in flowerlike patterns form on the wall of the condensation chamber. The powder is known as flowers of sulfur.

Sources

Some of the world's richest deposits of elemental sulfur are on the Gulf Coast in Louisiana and Texas. These deposits are not easily mined because they are buried under several hundred feet of quicksand. Herman Frasch (1851-1914) devised an ingenious method for getting sulfur out of the ground. The method takes advantage of sulfur's low melting point.

Wells are drilled into the sulfur bed. Three concentric tubes (with 2.5-cm, 7.5-cm, and 15-cm diameters) are installed. Compressed air and superheated water are pumped through the inner and outer tubes. The molten sulfur that emerges from the center tube is pumped into large storage vats where it cools and solidifies into huge blocks, which can be 130 m long, 70 m wide, and 30 m high. Dynamite is used to break up the blocks of sulfur for shipping.

GROUP 6A

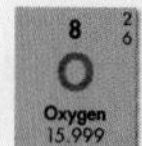

Oxygen discovered in 1772 by Carl Scheele and in 1774 by Joseph Priestly

Sulfur known since ancient times

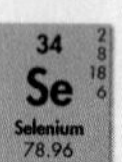

Selenium discovered in 1817 by Jöns Jacob Berzelius

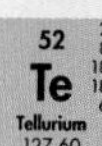

Tellurium discovered in 1782 by Franz Joseph Müller von Reichenstein

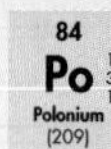

Polonium discovered in 1898 by Marie Curie

Physical Properties

- Except for oxygen gas, O_2, Group 6A elements are solid at room temperature.
- The metallic properties of Group 6A elements increase from top to bottom within the group.
- Polonium is a radioactive metal.

monoclinic

orthorhombic

The unit cell in crystalline sulfur is an S_8 molecule.

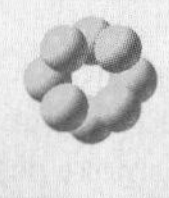

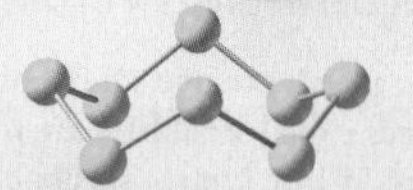

Melting and Boiling Points

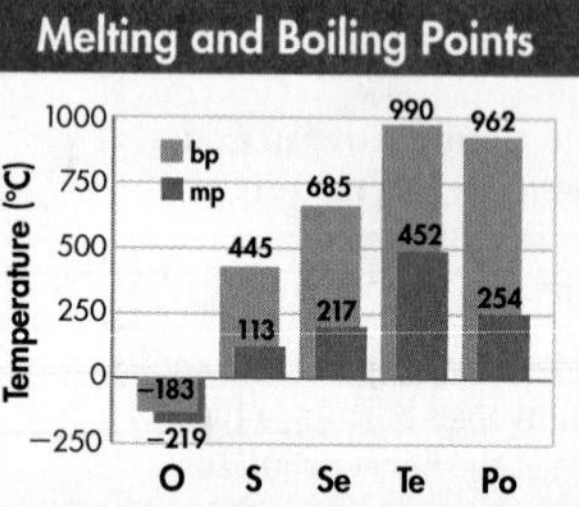

Density

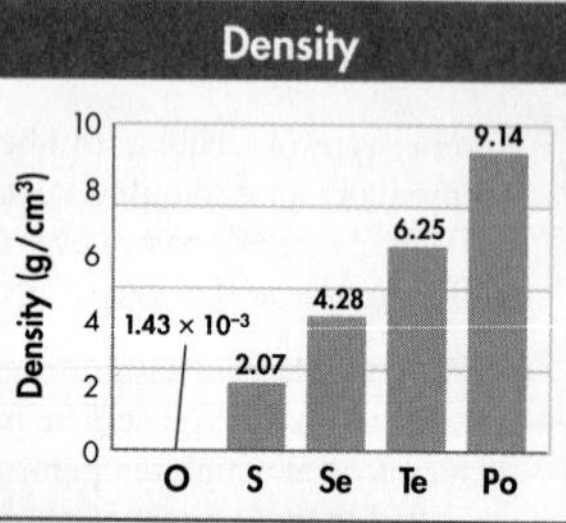

Sources

- Large-scale production of oxygen is by fractional distillation of liquid air. Liquid oxygen is stored and shipped at its boiling point of −183°C in vacuum-walled bottles.
- The Frasch process is used to mine sulfur from underground deposits. A well is drilled into a sulfur bed and a set of concentric tubes installed. Superheated water melts the sulfur. Compressed air forces it to the surface.
- Sulfur is also produced from hydrogen sulfide, H_2S, and sulfur dioxide, SO_2.

$$2H_2S(g) + SO_2(g) \longrightarrow 2H_2O(l) + 3S(s)$$

- Selenium and tellurium are by-products of the processing of sulfide ores for other metals.
- Polonium is formed by the radioactive decay of radium in minerals such as pitchblende.

The Frasch Process

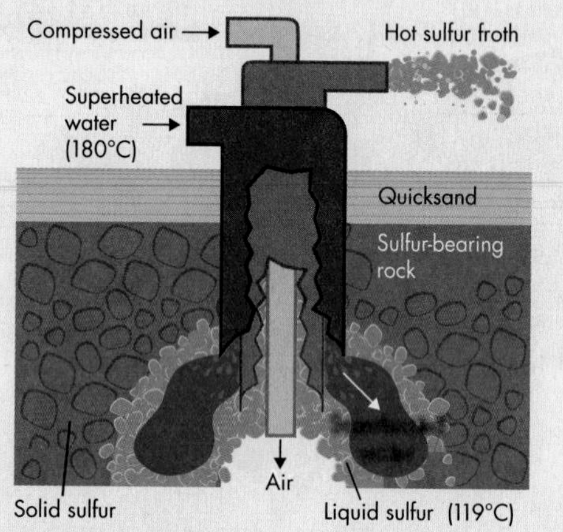

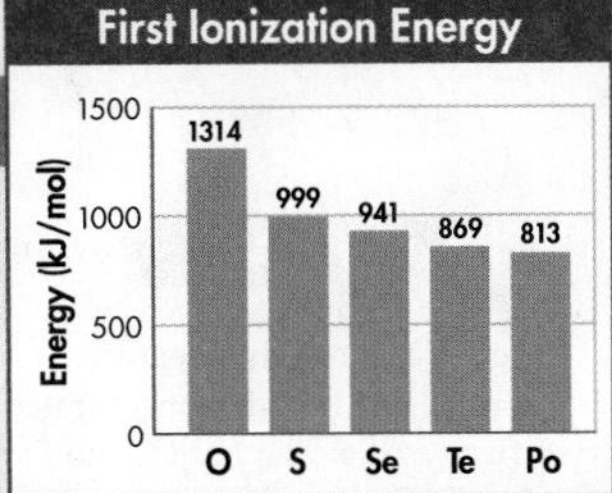

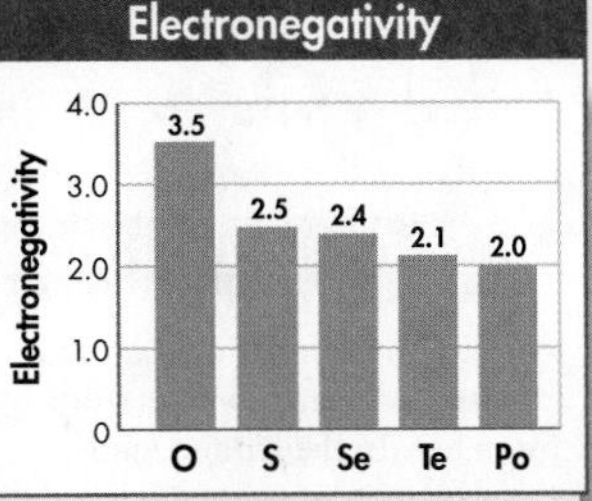

Atomic Properties

- Group 6A elements have an electron configuration that ends in ns^2np^4.
- For Group 6A elements, the most common oxidation numbers are +4, +6, and −2.
- Oxygen is paramagnetic because there are unpaired electrons in O_2 molecules.

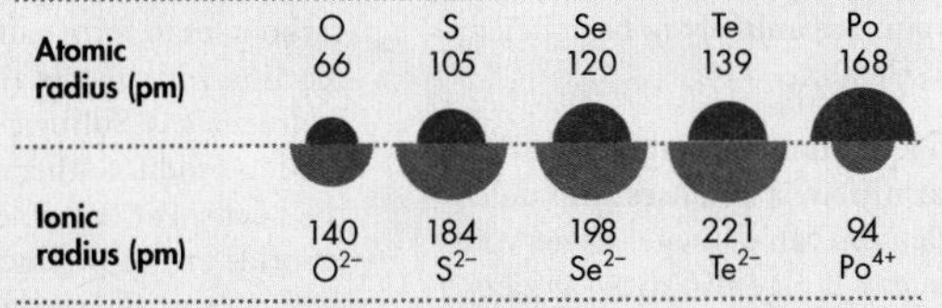

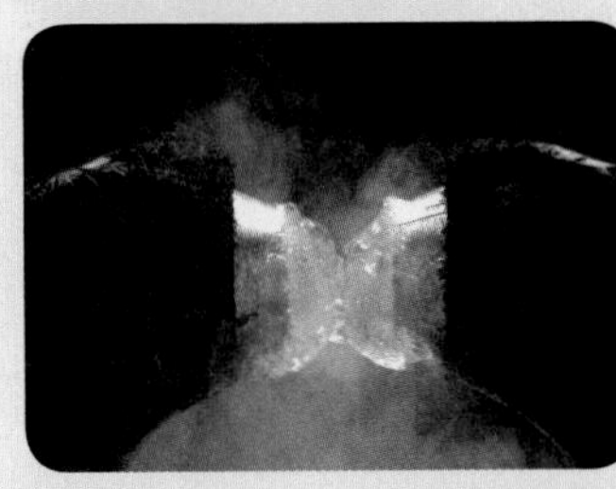

Liquid oxygen is held between the poles of a magnet because of its attraction to the magnet.

Important Compounds and Reactions

- Oxygen reacts with almost all other elements to form oxides. Example:
  $4K(s) + O_2(g) \longrightarrow 2K_2O(s) \quad \Delta H = -363.2 \text{ kJ/mol}$
- Ozone, O_3, is produced directly from oxygen, O_2, during lightning strikes.
  $3O_2(g) \longrightarrow 2O_3(g) \quad \Delta H = +285 \text{ kJ/mol}$
- Oxygen is necessary for releasing energy from fuels, such as glucose, in organisms.
  $C_6H_{12}O_6(s) + 6O_2(g) \longrightarrow 6CO_2(g) + 6H_2O(g) \quad \Delta H = -2808 \text{ kJ/mol}$
- Oxygen is used to produce steel and to oxidize hydrogen in fuel cells.
- Sulfur compounds often have unpleasant odors. Hydrogen sulfide, H_2S, smells like a rotten egg. It forms when metallic sulfides and hydrochloric acid react. Example:
  $FeS(s) + 2HCl(aq) \longrightarrow H_2S(g) + FeCl_2(aq)$
- Concentrated sulfuric acid, H_2SO_4, is a strong dehydrating agent. Example:
  $C_{12}H_{22}O_{11}(s) \xrightarrow{H_2SO_4} 12C(s) + 11H_2O(g)$

When concentrated sulfuric acid is added to sucrose, water vapor and carbon are produced. The release of the vapor causes the carbon to expand.

- Ethyl mercaptan, CH_3CH_2SH, is commonly called "stench." It is added to supplies of odorless natural gas so people will know when the gas is leaking.
- Sodium thiosulfate, $Na_2S_2O_3$, also known as *hypo*, is used in the development of dental X-rays.
- The addition of cadmium selenide, CdSe, gives glass a beautiful ruby color.

Atomic Properties

Students may think that only metals can be affected by a magnetic field. Explain that substances in which all electrons are paired are diamagnetic, i.e., weakly repelled by an external magnetic field. Substances in which there are unpaired electrons are paramagnetic, i.e., attracted to an external magnetic field. The more unpaired electrons, the stronger the force of attraction.

Important Compounds and Reactions

Concentrated sulfuric acid dissolves readily in water. The process is highly exothermic. The ΔH_{soln} for sulfuric acid is −95.28 kJ/mol. The strong affinity of sulfuric acid for water makes sulfuric acid a powerful dehydrating agent, as demonstrated by its effect on sucrose.

Each odorant has a detection threshold, or concentration at which 50% of people can identify the presence of the odor. Ethyl mercaptan's detection threshold is 0.0026 ppm, which is relatively low. Because olfactory acuity, or the ability to detect odors, tends to decrease with age, many elderly people are unable to detect ethyl mercaptan at its concentration in natural gas. Sulfur compounds with unpleasant smells are generally classified as putrid.

Group 6A

Se Selenium in Food

Keratin is the principal protein in the hoofs and hair of horses, as it is in human hair. Sulfur-containing thiol groups in the amino acid cysteine form the disulfide bonds between adjacent protein chains. Selenium can be substituted for sulfur when present in excess. Chronic selenium poisoning in horses is often called "bob-tail disease" because hair in the mane and tail breaks at the location in the hair shaft where selenium is substituted for sulfur.

S Sulfuric Acid

For a long time, researchers were unable to determine how the conversion of sulfur dioxide to sulfur trioxide, which is normally a slow process, occurred so rapidly in air. Research has since shown that dust and other solid particles in the air act as a catalyst for the process.

GROUP 6A

Se Selenium In Food

Selenium is an antioxidant that protects cell membranes from damage. In the United States, grain is grown on selenium-rich soils. Livestock and people who eat those grains are unlikely to be deficient in selenium.

An essential nutrient may be harmful in large doses. Too much selenium can damage the nervous system. It may also cause anxiety and fatigue.

This milkvetch (Astragalus bisulcatus) accumulates high levels of selenium.

Soils in the Great Plains and Rocky Mountain regions often contain high levels of selenium. If cattle graze on plants that grow in those soils, they may develop chronic selenium poisoning. The symptoms include loss of hair, sore hoofs, lameness, and a lack of energy.

Acute selenium poisoning causes cattle to lose their vision and stumble aimlessly before dying from respiratory failure. Cowboys called this condition the "blind staggers."

S Sulfuric Acid

When sulfur burns in air, the product is the irritating gas sulfur dioxide, SO_2.

Pure sulfuric acid is a dense, colorless, oily liquid. Concentrated sulfuric acid is 98% H_2SO_4 and 2% H_2O. Dilute sulfuric acid reacts with metals, oxides, hydroxides, or carbonates to form sulfates. The reaction with metals also releases hydrogen gas. Sulfuric acid can be used to produce other acids from their salts. For example, hydrogen chloride can be produced from sulfuric acid and sodium chloride.

$$H_2SO_4(l) + 2NaCl(s) \longrightarrow Na_2SO_4(s) + 2HCl(g)$$

Sulfuric acid is produced mainly from sulfur dioxide. The process is called the contact process because the key reaction takes place when the reactants are in contact with the surface of the solid catalyst.

(1) Melted sulfur is burned in air.

$$S(l) + O_2(g) \longrightarrow SO_2(g)$$

(2) Sulfur dioxide is oxidized in the presence of a vanadium oxide catalyst, V_2O_5.

$$2SO_2(g) + O_2(g) \xrightarrow{V_2O_5} 2SO_3(g)$$

(3) Sulfur trioxide dissolves in water and forms sulfuric acid.

Much of the sulfuric acid produced in North America is used to make fertilizers. Sulfuric acid is also used in petroleum refining, the production of other chemicals, and for pickling iron and steel. During pickling, oxides are removed from the surface of a metal.

This marine slug of the species Berthella martensi produces sulfuric acid, which discourages predators.

O Ozone

Near Earth's surface, ozone (O_3) is a pollutant. In the stratosphere, ozone is literally a lifesaver. The ozone layer in the stratosphere absorbs 99% of the sun's harmful ultraviolet (UV) radiation. In the 1970s, scientists began to suspect that the ozone layer might be threatened. They based their concerns on laboratory models. In 1985, their suspicions were confirmed when British researchers discovered a "hole" in the ozone layer over Antarctica. In the winter, the amount of ozone was reduced by almost one half.

The scientists traced the thinning of the ozone layer to chemicals called chlorofluorocarbons (CFCs). These chemicals were used mainly as propellants in aerosol spray cans and as coolants in refrigerators and air conditioners. CFCs are highly stable and inert in the lower atmosphere. Over time, they drift up into the stratosphere, where they are broken down by solar radiation. Through a repeated cycle of reactions (2) and (3), a single chlorine atom can destroy as many as 100,000 molecules of ozone.

(1) $CCl_3F \longrightarrow Cl\cdot + \cdot CCl_2F$

(2) $Cl\cdot + O_3 \longrightarrow ClO\cdot + O_2$

(3) $ClO\cdot + O \longrightarrow Cl\cdot + O_2$

In 1978, the United States banned the use of CFCs in aerosols. In 2007, most countries agreed to stop producing CFCs by 2020. The ban has had an effect. Concentrations of CFCs have begun to level off in the stratosphere and even decline in the lower atmosphere.

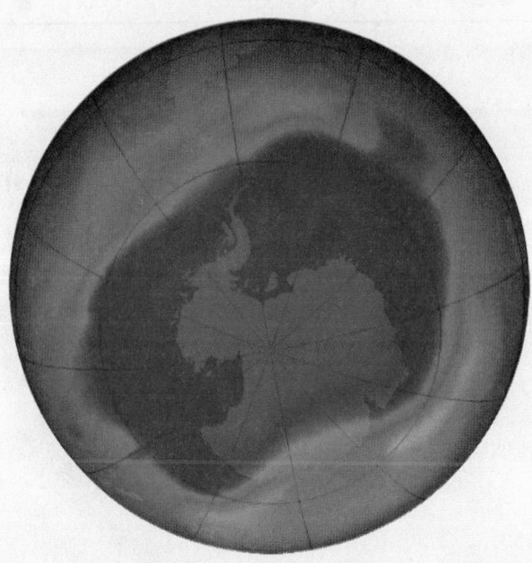

In September 2006, the "hole" over Antarctica was record size. The amount of ozone is low in the blue areas. It may take decades for the ozone layer to reach pre-1980 levels.

S Thiols

Skunks use a foul-smelling liquid to repel predators. The compounds in the mixture are mainly sulfur-containing molecules. One type of sulfur compound called a thiol is responsible for the foul odor. Thiols are organic compounds in which the oxygen of an alcohol has been replaced by sulfur. The general formula for a thiol is RSH.

The foul liquid is secreted and stored in glands until the skunk feels threatened. When it is threatened, the skunk contracts the muscles around the glands and sprays the liquid in the direction of the threat. A skunk can propel the liquid for about three meters! Other than smelling bad for a while, the "attacker" is not harmed.

Did You Know?

Organic compounds containing **sulfur** give onions their taste and smell. When an onion is cut, reactions occur that produce propanethial-S-oxide. When receptors in your eyes are exposed to this irritating gas, they trigger the production of tears.

O Ozone

As sunlight travels through the ozone layer, ozone absorbs light of known wavelengths. The amount of light of these wavelengths that passes through the atmosphere is an indication of the amount of ozone present. In 2003, researchers at the University of Alabama in Huntsville used such data collected by NASA Earth-observing satellites to show that the rate of ozone depletion in the upper stratosphere had slowed since 1997.

When CFCs are broken down by solar radiation, they release chlorine atoms. A chlorine atom can destroy as many as 100,000 molecules of ozone because it is a free radical. A free radical is very reactive because it contains one or more unpaired electrons.

To Do

Encourage interested students To Do additional reading on the research that led to the Montreal Protocol, which established an international ban on CFC-based products.

S Thiols

Ask students if they have had to deal with a pet that was sprayed by a skunk, and, if so, what approach they used. Students may say that they washed the pet with tomato juice, followed by a vinegar rinse. This traditional remedy doesn't actually remove the source of the odor. Rather, it masks the odor.

A better approach is based on an industrial method for scrubbing sulfur-containing compounds from waste gas streams in coal-burning power plants. The pet is washed with a mixture of hydrogen peroxide, baking soda, and liquid soap. The soap breaks up the oils in the spray. The hydrogen peroxide and baking soda react and produce oxygen, which reacts with the thiols.

Group 7A

Physical Properties

Astatine comes from the Greek *astatos,* meaning "unstable." It was synthesized by bombarding bismuth with alpha particles. Astatine-210 is the longest-lived isotope with a half-life of 8.3 hours. Despite the small quantities available for study, some properties of astatine have been demonstrated, including its ability to dissolve more easily in organic solvents than in water.

Sources

It was difficult for chemists to isolate fluorine because fluorine is such as strong oxidizing agent. The task was further complicated by the toxicity of hydrogen fluoride. Many chemists, including Davy and Gay-Lussac, suffered health problems from their attempts to prepare fluorine. The process Henri Moissan used to prepare fluorine (electrolysis of a solution of potassium fluoride in hydrogen fluoride) is the process used today.

To Do

Invite a mineral collector to class to show off samples of fluorite, which is second only to quartz in its popularity with collectors. You could use the visit to discuss crystal systems.

GROUP 7A

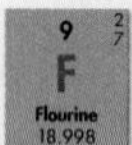

Fluorine discovered in 1886 by Henri Moissan

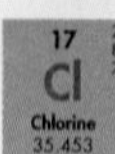

Chlorine discovered in 1774 by Carl Wilhelm Scheele

Bromine discovered in 1826 by Antoine-Jérôme Balard

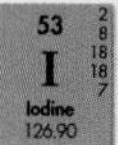

Iodine discovered in 1811 by Bernard Courtois

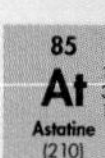

Astatine discovered in 1940 by Dale R. Corson, K. R. Mackenzie, & Emilio Segrè

Physical Properties

- Halogens are nonmetals. At room temperature, fluorine and chlorine are gases and bromine is a liquid. Iodine and astatine are solids.
- Halogens are very reactive. The reactivity decreases from fluorine to astatine. Halogens do not exist in the elemental form in nature.
- Astatine isotopes are radioactive with short half-lives.

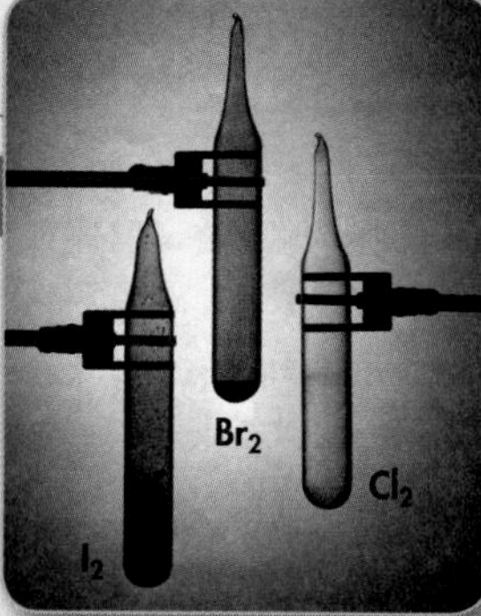

The colorful vapors of bromine and iodine are visible because bromine is volatile and iodine sublimes easily at room temperature.

Melting and Boiling Points

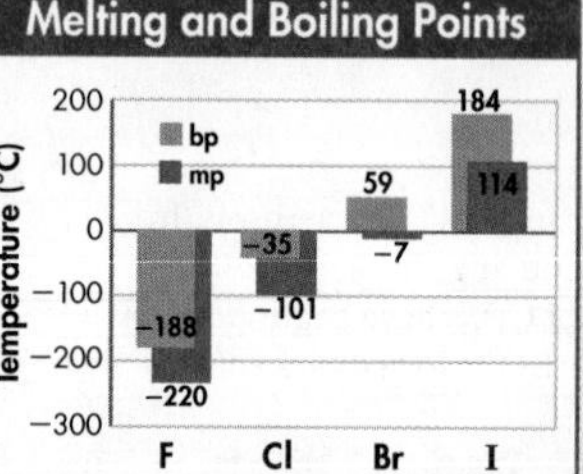

Density

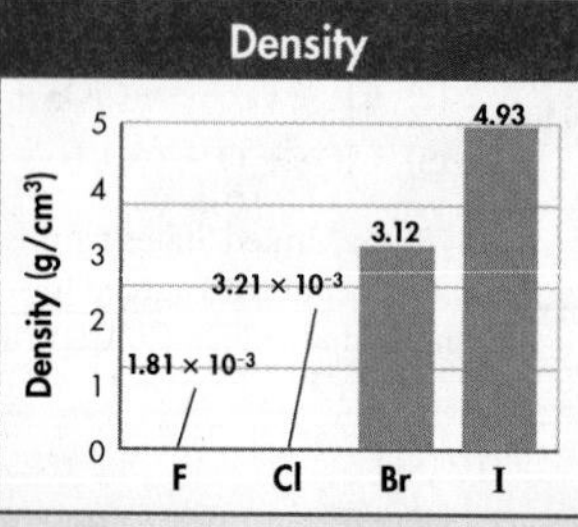

Sources

- Chlorine gas is made commercially by the electrolysis of brine.

$$2NaCl(aq) + 2H_2O(l) \longrightarrow Cl_2(g) + H_2(g) + 2NaOH(aq)$$

- Bromine is obtained from seawater by a displacement reaction with chlorine.

$$2NaBr(aq) + Cl_2(g) \longrightarrow 2NaCl(aq) + Br_2(l)$$

- Iodine is found in brine and in sodium iodate, $NaIO_3$, in deposits of sodium nitrate. Iodine is produced from $NaIO_3$ by this redox reaction.

$$2NaIO_3(aq) + 5NaHSO_3(aq) \longrightarrow I_2(g) + 2Na_2SO_4(aq) + 3NaHSO_4(aq) + H_2O(l)$$

- Fluorine is manufactured by the electrolysis of potassium fluoride, KF, dissolved in liquid hydrogen fluoride, HF.

Fluorite, CaF_2, is the principal mineral of fluorine. The term fluorescent comes from this mineral, which glows in the presence of UV radiation.

Atomic Properties

- Group 7A elements have an electron configuration that ends in ns^2np^5.
- Halogens exist as diatomic molecules.
- Each halogen has the highest electronegativity in its period.
- The most common ionic charge for halogens is 1−. Except for fluorine, halogens also have positive oxidation numbers of +1, +3, +5, and +7.

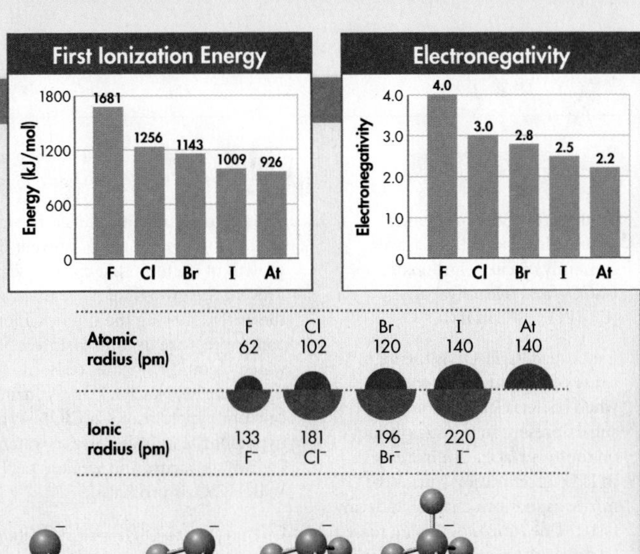

Chlorine forms four anions with oxygen. The oxidation number of chlorine is different in each anion.

Hypochlorite ion Chlorite ion Chlorate ion Perchlorate ion

Important Compounds and Reactions

- Halogens form metal halides. Example:

  $2Na(s) + Cl_2(g) \longrightarrow 2NaCl(s) \quad \Delta H = -411 \text{ kJ/mol}$

- Halogens form hydrogen halides. Example:

  $H_2(g) + Cl_2(g) \longrightarrow 2HCl(g) \quad \Delta H = -92.3 \text{ kJ/mol}$

- Dry bleach is a mixture of compounds represented by the formula CaCl(ClO). Dry bleach is used to bleach paper and textiles. It also removes stains and disinfects laundry.

  $Ca(OH)_2(aq) + Cl_2(g) \longrightarrow CaCl(ClO)(aq) + H_2O(l)$

- Small amounts of oxygen are produced in the laboratory by heating potassium chlorate, $KClO_3$. Potassium chlorate is an oxidizing agent in fireworks, matches, and explosives.

  $2KClO_3(s) \longrightarrow 2KCl(s) + 3O_2(g)$

- Canisters of sodium chlorate are used on submarines to produce oxygen. They are carried on airplanes in case of an emergency.
- Chlorine is used to make the monomer vinyl chloride, $CH_2{=}CHCl$, which reacts to form the polymer polyvinyl chloride (PVC).

Chlorine reacts vigorously with sodium to form solid sodium chloride.

- Hydrofluoric acid (HF) is made from the mineral fluorite and sulfuric acid. Although HF is extremely dangerous, it has many uses, including cleaning metals and frosting glass.

  $CaF_2(s) + H_2SO_4(aq) \longrightarrow 2HF(g) + CaSO_4(s)$

- Nonstick pans are coated with a polymer of tetrafluoroethene, $F_2C{=}CF_2$.
- Tincture of iodine is a solution of iodine, I_2, and potassium iodide, KI, in alcohol. It is an example of an iodine-based skin disinfectant.

Atomic Properties

The ball-and-stick models of the anions containing chlorine and oxygen could prove useful when students study VSEPR theory in Chapter 8.

To Do

Ask students to calculate the oxidation number of chlorine in the hypochlorite, chlorite, chlorate, and perchlorate ions.

Important Compounds and Reactions

Finding a container to store elemental fluorine was a challenge because fluorine attacks glass and causes most metals to burst into flame. Alloys of copper and nickel are used because when fluorine reacts with the alloy, it forms a protective (passive) fluoride layer on the surface.

Fluorine is used to enrich uranium. Fluorine reacts with uranium to produce the gas uranium hexafluoride (UF_6). Because the rate of diffusion of the gas varies with the isotope, uranium-235 (the fissionable isotope) can be separated from uranium-238.

To Do

Challenge students to explain why HCl, HBr, and HI form strong acids when dissolved in water, but HF forms a weak acid. (The strong hydrogen bonds between the polar HF molecules limit the ionization of hydrogen fluoride in water.) Then challenge students to find out why concentrated hydrofluoric acid, is extremely dangerous. (Hydrofluoric acid causes severe painful burns on contact with body tissues. Because of the small size of its molecules, hydrofluoric acid can penetrate skin and fat quickly. It decalcifies bone and acts as a poison once it is in the blood.)

GROUP 7A

Group 7A

F Tooth Decay

Low concentrations of fluoride ions are present in all natural water sources, but the level may not be sufficient for the prevention of tooth decay. In the United States, the optimum concentration for fluoride in water supplies is in the range of 0.7 to 1.2 ppm.

Because fluoride ions are toxic in high concentrations, FDA regulations limit the amount of fluoride in toothpaste. Even so, people should rinse away excess toothpaste. There is a warning on toothpaste labels about the use of toothpaste by children under 6 years of age. Young children need to be supervised until they have learned how to minimize the ingestion of toothpaste.

Cl Swimming Pool Chemistry

Students may wonder why the chlorine that kills bacteria doesn't harm humans who use a swimming pool. The answer lies with the concentration.

To Do

In the early 1900s, health officials in developed countries began to use chlorine to treat drinking water. Students could research the effect that adding chlorine to drinking water had on outbreaks of diseases such as cholera, typhoid fever, and dysentery, which are caused by waterborne pathogens.

GROUP 7A

F Tooth Decay

Your teeth have a hard outer layer called enamel. This layer is mainly calcium carbonate, $CaCO_3$, and hydroxyapatite, $[Ca_3(PO_4)_2]_3 \cdot Ca(OH)_2$.

Lactic acid, $C_3H_6O_3$, is the main cause of tooth decay. It forms when bacteria in saliva feed on sugars present in the sticky plaque on tooth surfaces. An increase in H^+ concentration causes the minerals in tooth enamel to decay faster. One source of this increased acidity is the phosphoric acid in soft drinks.

Fluoride ions are added to the water supply in many cities. Most toothpastes contain fluoride ions. The ions replace hydroxide ions in hydroxyapatite to form fluoroapatite, $[Ca_3(PO_4)_2]_3 \cdot CaF_2$. This replacement makes the enamel more resistant.

Fluoride ions alone will not prevent tooth decay. You need to brush your teeth and floss to keep plaque from building up on your tooth enamel.

Cl Swimming Pool Chemistry

The person who maintains a swimming pool has two main goals. Prevent the growth of bacteria that cause diseases. Prevent the growth of algae that can foul the water and clog the filters. Chlorine compounds are used to disinfect pool water. "Liquid chlorine" contains sodium hypochlorite, NaClO. "Dry chlorine" is calcium hypchlorite, $Ca(ClO)_2$. When hypochlorite ions dissolve in water, hydrolysis occurs and weak hypochlorous acid, HClO, is produced.

$$ClO^-(aq) + H_2O(l) \rightleftharpoons HOCl(aq) + OH^-(aq)$$

The amount of undissociated hypochlorous acid in the pool water depends on the pH. If the pH is too high, the hydrolysis reaction will shift toward the reactants and reduce the concentration of HClO. If the pH is too low, too much acid will form. A high concentration of acid can cause eye irritation, damage plaster, and corrode the metal piping and filters in the pool.

If the pH of the pool water is too high, solid sodium hydrogen sulfate can be used to react with the OH^- ions.

$$NaHSO_4(s) + OH^-(aq) \longrightarrow Na^+(aq) + SO_4^{2-}(aq) + H_2O(l)$$

If the pH is too low, sodium carbonate can be used to neutralize some of the acid.

$$Na_2CO_3(s) + 2H^+(aq) \longrightarrow 2Na^+(aq) + H_2O(l) + CO_2(aq)$$

The use of fluorides in drinking water and toothpastes has caused such a drastic reduction in tooth decay that many young adults have never had a cavity.

R30 Elements Handbook

F Blood Substitutes

The most important function of blood is its ability to deliver oxygen from the lungs to cells throughout the body and carry carbon dioxide from the cells to the lungs. Blood transfusions have been the traditional response to blood loss due to severe injuries or surgery. Blood transfusions have saved millions of lives. But it takes time to determine the recipient's blood type and less common blood types may not be available. Also, blood can carry disease-causing bacteria and viruses.

Scientists have tried for years to develop an effective and profitable blood substitute. One approach focused on perfluorocarbons (PFCs). PFCs are organic compounds in which all the hydrogen has been replaced by fluorine. PFCs can dissolve and transport oxygen. They are fairly inexpensive to manufacture, their purity can be controlled, and they don't react with other substances in the body.

Some problems with PFCs were revealed during clininal trials. Because PFCs are insoluble in water, they must be mixed with lipids to form an emulsion. The emulsion can be unstable in blood, which means that it is not effective for long. Because PFCs carry less oxygen than hemoglobin does, the patient must breathe oxygen-rich air.

Because blood is so complex, scientists may never find a substitute to perform all its functions. Therefore, there is still a great need for donations of human blood.

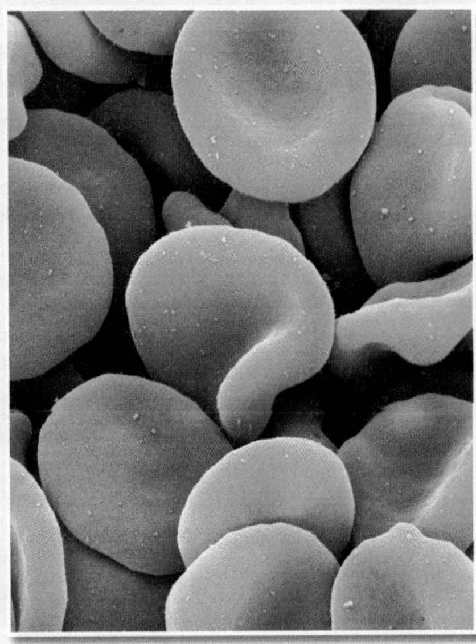

Hemoglobin is the molecule in red blood cells that transports oxygen. For a blood substitute to be effective, it has to perform this same function.

I Iodized Salt

The thyroid gland produces hormones that help to control the body's growth and the energy produced by cells. Trace amounts of iodine are needed to produce thyroid hormones. An adult needs about 150 mg of iodine daily.

Ocean fish are a good source of iodine. When people used to get most of their food from local sources, a person who lived far from the ocean often had an iodine deficiency. To compensate for the lack of iodine, the thyroid gland might enlarge. A severe deficiency causes some forms of mental disability. Adding potassium iodide to table salt proved to be a simple solution to this public health problem.

In the United States, iodized salt was first sold in 1924. There is about 400 mg of iodine in a teaspoon of iodized salt. The use of iodized salt has virtually eliminated the problem of iodine deficiency in the United States.

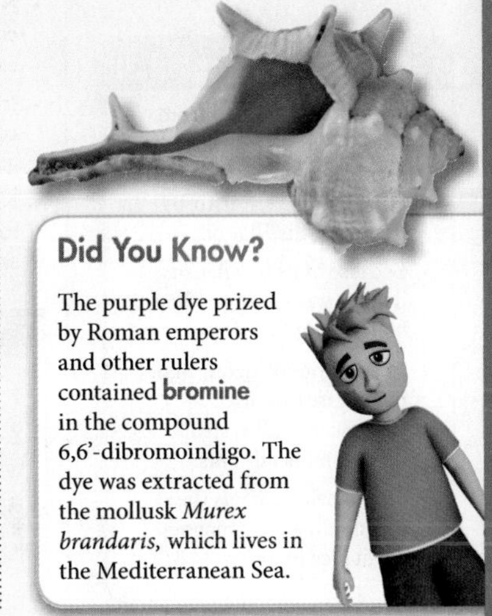

Did You Know?

The purple dye prized by Roman emperors and other rulers contained **bromine** in the compound 6,6'-dibromoindigo. The dye was extracted from the mollusk *Murex brandaris,* which lives in the Mediterranean Sea.

F Artificial Blood

The term *blood substitute* is a bit misleading because the substitutes generally perform only one function of blood. A more accurate description would be "Cell-free oxygen carriers." The ideal oxygen carrier (1) would not require refrigeration, (2) could be used with all blood types, (3) would be free of infectious agents, and (4) would be easy to use.

To Do

Use the discussion of blood substitutes as an antidote to the idea that scientific research is a series of success stories. So far, blood substitutes, whether PFC-based or hemoglobin-based, have had limited success despite the time and money invested.

I Iodized Salt

Recipes that call for large quantities of table salt often specify non-iodized salt. Some cooks claim large amounts of potassium iodate affect the flavor of food. In other cases, the issue is not the iodine, but other additives, e.g., non-caking agents, that the user is trying to avoid.

To Do

Table salt manufacturers in the United States use potassium iodide. But in countries where the climate is hot and humid, potassium iodate may be used instead. Challenge students to find data to demonstrate why potassium iodate might be preferable to potassium iodide in hot, humid climates. (The solubility of potassium iodide is 148 g/100 g H_2O at 25°C; the solubility of potassium iodate is 9.22 g/100g H_2O at 25°C.)

Group 8A

The names of all the noble gases except radon come from Greek words. Helium comes from *helios,* meaning "sun." Xenon comes from *xenos,* meaning "strange." Krypton comes from *kryptos,* meaning "hidden." Argon comes from *argos,* meaning "inert." Neon comes from *neos,* or "new."

Physical and Chemical Properties

More than 80 compounds of xenon have been produced. Krypton forms the colorless solid krypton(II) fluoride, KrF_2. The compound of argon, argon fluorohydride (HArF), produced in 2000 exists only at temperatures below −246°C.

To Do

Show students how to draw an electron dot structure for xenon(II) fluoride (XeF_2). Then have students draw electron dot structures for xenon(IV) fluoride (XeF_4), xenon(IV) oxide (XeO_3), and xenon(VIII) oxide (XeO_4). As a further challenge, students could try drawing the electron dot structures for XeO_2F_2 or $XeOF_4$.

Sources

Although there are measurable amounts of radon in most regions of the United States, the highest levels are recorded in regions where the bedrock is granite. Radon is a product of the natural decay of uranium-238. Radon can seep through cracks in foundations and basements floors. Because it is denser than air, radon tends to collect in the lower levels of a house. Inhaling radon may increase the risk of lung cancer. High levels of radon can be reduced by improved ventilation in basements and other closed areas where radon collects.

GROUP 8A

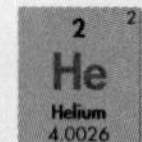

Helium discovered in 1868 by Pierre Janssen

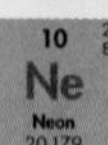

Neon discovered in 1898 by Sir William Ramsay & Morris Travers

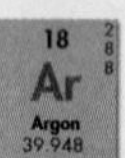

Argon discovered in 1894 by Lord Rayleigh & Sir William Ramsay

Krypton discovered in 1898 by Sir William Ramsay & Morris Travers

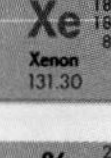

Xenon discovered in 1898 by Sir William Ramsay & Morris Travers

86 Rn Radon (222)

Radon discovered in 1900 by Friedrich E. Dorn

Physical and Chemical Properties

- All Group 8A elements are monatomic gases at STP.
- Noble gases are colorless, odorless, and tasteless.
- The first compound of a noble gas, $XePtF_6$, was made in 1962. More than 100 compounds of fluorine and xenon are now known.

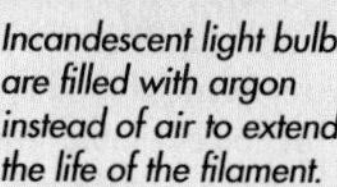

Incandescent light bulbs are filled with argon instead of air to extend the life of the filament.

- A compound of argon, HArF, exists only at temperatures below −246°C.

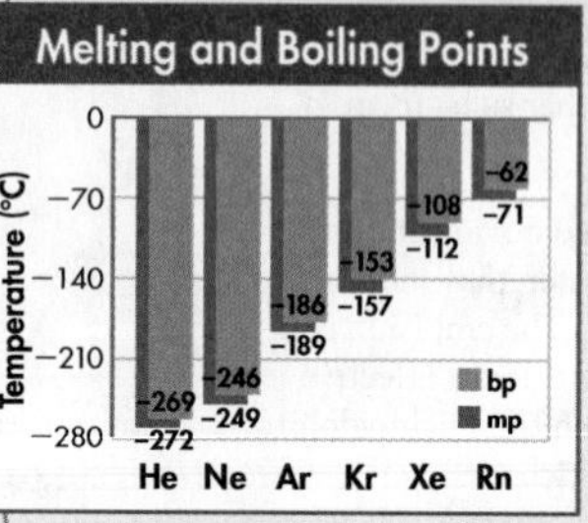

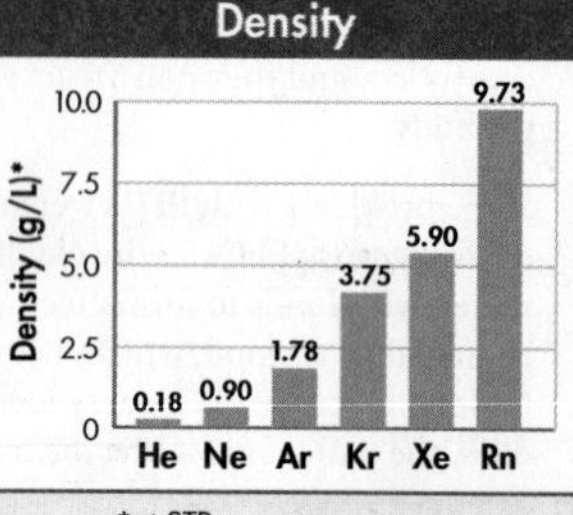

*at STP

Sources

- Helium is separated from natural gas deposits. Neon, argon, krypton, and xenon are separated from air by fractional distillation.
- Because of its low density, helium is used in weather balloons and airships.
- In addition to "neon" lights, noble gases are used in fluorescent bulbs, strobe lights, and headlights.
- Liquid helium cools the magnets used for magnetic resonance imaging (MRI).

Atomic Properties

- Noble gases have an electron configuration that ends in ns^2np^6, except for helium ($1s^2$).
- In noble gas compounds, the most common oxidation number for the gas is +2.
- Noble gases have the highest ionization energies because their energy levels are filled.

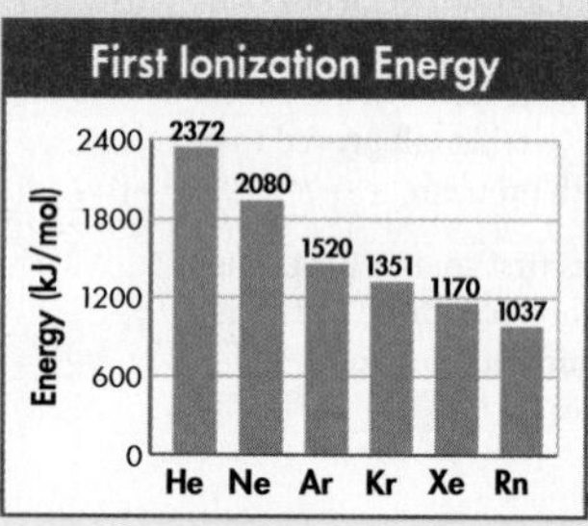

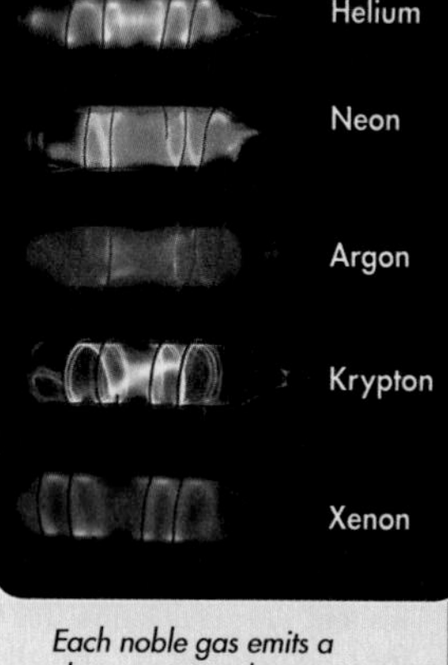

Each noble gas emits a characteristic color in a gas discharge tube.

Atomic Properties

The noble gases that form compounds most readily have larger atoms in which there are more electrons to shield the electrons in the highest occupied energy level from the nuclear charge.

To Do

Encourage students to research the role of krypton-86 in the development of a definition for a meter. Ask them to determine why a definition based on krypton-86 was adopted in 1960 and why it was replaced in 1983.

Ne Neon Lights

By 1855, scientists could produce light by passing an electric current through a gas under low pressure in a sealed glass tube. With the discovery of the noble gases, a new technology emerged. In 1910, George Claude displayed the first neon lamp in Paris, France.

In 1923, a car dealer from Los Angeles bought two signs that spelled out "Packard" for $24,000 (about $250,000 in today's dollars). When he displayed the signs in Los Angeles, people described the light as "liquid fire." By the 1930s, businesses were using neon lights to draw the attention of customers.

Neon and argon are the gases most often used in neon lights. Orange-red lights contain only neon. Other colors are produced by adding a bit of mercury to the noble gas. The tube is coated with a material that glows when exposed to UV light emitted by mercury vapor.

Ar Taken For a Ride By Argon

Whether riding on a paved city street or on an unpaved mountain trail, a bicyclist is likely to find rough patches. When faced with rough terrain, the cyclist may worry about the tires, but probably not about the bicycle frame. The metal frames are made of steel, aluminum alloy, or titanium tubes that are joined together.

The tubes are joined together by Tungsten Inert Gas (TIG) welding. An electric arc is struck between a tungsten electrode and the parts to be welded. Heat from the arc melts the ends of the tubes and fuses them together. Filler may be placed between the ends of the tubes to increase the strength of the joint or to produce a smoother joint.

During welding in air, there is a danger that the metal tubes or electrode will oxidize. To prevent oxidation, the area around the arc is filled with an inert gas, most often argon. Welding with argon has an added benefit. Because argon is a poor conductor of heat, the arc that forms is narrow. This narrow arc produces a weld that is both neat in appearance and mechanically strong.

Xe Xenon-Ion Engine

The signals for a television program may bounce off a communication satellite. The satellite is in orbit above the equator. The position of the satellite may be maintained by a xenon-ion propulsion system.

When electrons strike xenon atoms in a xenon-ion engine, the atoms lose electrons and form positive ions. The ions are accelerated by a charged grid and shot from the engine at about 105 km/h. This action pushes, or thrusts, the satellite in the opposite direction. With multiple engines facing in different directions, a satellite can be moved in any direction.

Although a xenon-ion engine produces a relatively small amount of thrust, it can provide thrust for months or years. This makes xenon-ion engines a good choice for lengthy space missions. In addition, an inert gas poses no hazard for the satellite or the people who handle the propellant tanks.

Did You Know?

When liquid **helium** is cooled to below 2 K, its viscosity drops to zero. It will escape from an unsealed container by flowing up the sides of the container.

Ne Neon Lights

The development of neon lights can be traced back to an observation made by the French astronomer Jean Picard in 1675. He observed a faint glow in the tube of a mercury barometer when the tube was shaken.

Los Angeles in the 1930s was a low-rise city. Businesses used elaborate, art deco neon signs to attract motorists, who could see the signs from miles away. Many of the vintage signs have been restored. In 1949, Raymond Chandler wrote, "There ought to be a monument to the man who invented neon lights." (A monument to Claude might be a problem given his political choices. He was imprisoned from 1945–1949 for his support of the Vichy government in occupied France.)

Xe Xenon-Ion Engines

Because it completes one orbit in 24 hours, a communication satellite appears to be stationary above Earth. The altitude at which a satellite orbits Earth depends on its task. Stationary weather and communication satellites orbit at an altitude of about 22,300 miles. The region in which they orbit is called the Clarke Belt in honor of the science-fiction writer Arthur C. Clarke. Countries are assigned positions in the Clarke Belt for their communications and weather satellites.

Ar Taken For a Ride By Argon

Welders who use TIG must be aware of a safety issue. Breathing argon-rich air can deprive the body of oxygen. A person could pass out or even die from asphyxiation. So argon must be used in a well-ventilated room.

Hydrogen

Hydrogen was prepared long before it was recognized as a distinct substance. The Swiss alchemist Theophrastus Paracelsus probably handled hydrogen in the 1500s. He produced a flammable gas when he dissolved a metal in acid. In 1671, Boyle published a paper in which he described a reaction between iron filings and acids that produced "an inflammable gas." Cavendish described hydrogen as "inflammable air from metals." Lavoisier named hydrogen from the Greek *hydro,* meaning "water" and *genes,* meaning "generate."

Atomic and Physical Properties

Giving unique names to isotopes is limited to the isotopes of hydrogen. Hydrogen is an exception to the rule that there are only minor differences in the physical properties of an element's isotopes. When a neutron is added to an atom with a single proton, the mass is doubled (or tripled when a second neutron is added). These relatively large differences in atomic mass contribute to greater differences in the physical properties of hydrogen isotopes and their compounds. For example, H_2O melts at 0°C and boils at 100°C; D_2O melts at 3.81°C and boils at 101.42°C.

Sources

Much hydrogen is also obtained from petroleum refining. During refining, hydrogen is a by-product of the catalytic break down of large hydrocarbon molecules into smaller molecules.

Small quantities of hydrogen are prepared in the laboratory by the reaction of dilute acid with an active metal, such as zinc.

$$Zn(s) + 2HCl(aq) \rightarrow H_2(g) + ZnCl_2(aq)$$

Hydrogen

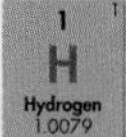

Hydrogen discovered in 1766 by Henry Cavendish

Sources

- Hydrogen is rarely found on Earth in an uncombined state.
- Electrolysis of water produces the purest hydrogen, but the process requires too much energy to be economical.

$$2H_2O(l) \longrightarrow 2H_2(g) + O_2(g) \quad \Delta H = +572 \text{ kJ}$$

- Hydrogen is produced when methane and steam react at 1100°C over a nickel catalyst.

$$H_2O(g) + CH_4(g) \xrightarrow{Ni} CO(g) + 3H_2(g) \quad \Delta H = +206 \text{ kJ}$$

The products pass over a metal oxide catalyst at 400°C. As carbon monoxide reacts with added steam, more hydrogen is produced.

$$CO(g) + H_2O(g) \longrightarrow CO_2(g) + H_2(g) \quad \Delta H = -41 \text{ kJ}$$

Carbon dioxide is removed as the gases flow through a basic solution.

$$CO_2(g) + 2OH^-(aq) \longrightarrow CO_3^{2-}(aq) + H_2O(l)$$

Atomic and Physical Properties

- Hydrogen has an electron configuration of $1s^1$.
- The most common oxidation numbers for hydrogen are +1 and −1.
- Most hydrogen (99.985%) is protium, or hydrogen-1.
- The other stable isotope is deuterium (hydrogen-2), which has the symbol D. Harold Urey discovered heavy hydrogen, D_2, in 1931.
- Tritium (hydrogen-3) was discovered in 1934. Its half-life is 12.3 years.

Properties of Hydrogen

Property	Value
Density at STP	0.09 g/L
Melting point	−259°C
Boiling point	−253°C
Ionization energy	1.312×10^3 kJ/mol
Electronegativity	2.1

Atomic radius (pm) H 30

Ionic radius (pm) 1.2 H^+

Atomic and Physical Properties

- Hydrogen forms molecular binary hydrides with nonmetals. Example:

$$H_2(g) + Cl_2(g) \longrightarrow 2HCl(g)$$

- Two thirds of the hydrogen produced in the U.S. is used to synthesize the molecular binary compound ammonia.
- Hydrogen forms ionic hydrides with alkali metals and alkaline earth metals. These hydrides are powerful reducing agents. Example:

$$Ca(s) + H_2(g) \longrightarrow CaH_2(s)$$

- Hydrogen is used to make methanol, CH_3OH. The reaction takes place at 200–300 atm and 400°C in the presence of a metal oxide catalyst.

$$CO(g) + 2H_2(g) \longrightarrow CH_3OH(g)$$

Methanol is an industrial solvent. It is used to make formaldehyde, CH_2O, which is used to make plastics.

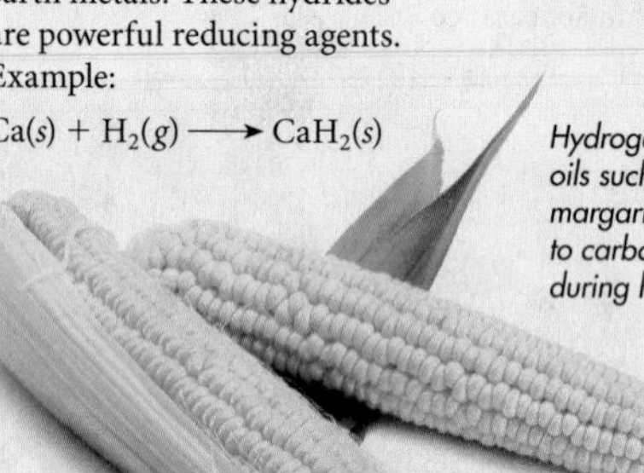

Hydrogen is used to turn liquid oils such as corn oil into solid margarine. Hydrogen is added to carbon–carbon double bonds during hydrogenation.

Important Compounds and Reactions

The oxidation number of hydrogen in covalent binary hydrides is +1. With the exception of water, most low molar mass covalent binary hydrides are gases or liquids with low boiling points. The oxidation number of hydrogen in ionic binary hydrides is −1. These hydrides are white crystalline solids with high melting points. They are formed by direct reaction of alkali metals and alkaline earth metals with hydrogen at about 400°C.

To Do

Ask students to write equations for the dissociation of a covalent binary hydride and an ionic binary hydride in water. Then have them identify the reaction in which hydrogen is acting as a base (in the ionic hydride) and the reaction when hydrogen is acting as an acid. Note that the hydride ion is one of the strongest bases known.

H Heavy Water Reactors

Because 1 in 6400 hydrogen atoms is deuterium, 1 in 41 million water molecules is D_2O. D_2O is called heavy water because it is about 10% heavier than ordinary water. In heavy water nuclear reactors, D_2O is used in place of H_2O as a neutron moderator. Both types of water are good moderators, but D_2O is more efficient than H_2O because D_2O absorbs fewer neutrons. So in a heavy water reactor, the uranium used to make the fuel doesn't have to be enriched. The initial cost of separating the heavy water from light water is offset by the lower cost of the uranium fuel.

Water passes through a series of extraction towers in which D_2O is separated from H_2O.

H Hydrogen Economy

Hydrogen is a pollution-free fuel. When it burns in air, the only product is water with trace amounts of nitrogen oxides. No carbon dioxide, carbon monoxide, oxides of sulfur, or unburned hydrocarbons are emitted. In a *hydrogen economy*, hydrogen would replace fossil fuels as the energy source for heating, transportation, and industrial processes. Achieving this goal will require new technology for the production, distribution, and storage of hydrogen.

Cars that run on hydrogen are being built. In one model, hydrogen is stored under pressure in a tank. The hydrogen is fed into a fuel cell, where it combines with oxygen to produce electricity. The electricity runs an electric motor that in turn drives the wheels. The valves to the hydrogen tank are designed to shut down if sensors detect a leak.

Large-scale production of hydrogen currently begins with fossil fuels. But in a true hydrogen economy, the source of the hydrogen would not be fossil fuels. Some scientists are working on a process that uses enzymes to extract hydrogen from wood chips or grass. Others are focused on algae that produce hydrogen in the presence of sunlight.

At the Los Angeles auto show in 2008, one auto manufacturer presented this design concept for a hydrogen-powered sports car.

H Hydrogen Peroxide

The strips used to whiten teeth usually contain hydrogen peroxide, H_2O_2. It is a powerful oxidizer that doesn't produce toxic gases or unwanted residues. A 3% aqueous solution is safe for use at home. Stronger concentrations are used to treat wastewater, bleach paper, and make germ-fighting detergents.

Hydrogen peroxide can inhibit the growth of bacteria in water pipes and increase the growth of bacteria that clean up polluted soils. How is this possible? The answer is *selectivity*. Scientists can adjust variables such as pH, temperature, and concentration so that hydrogen peroxide oxidizes one pollutant and not another.

Did You Know?

Hydrogen in the center of the sun has a density of about 200 g/mL. The temperature is about 13 million degrees Celsius. Radiation released when hydrogen nuclei fuse takes about a million years to reach the sun's surface.

H A Hydrogen Economy

Metal hydrides provide a compact way to store hydrogen. Metal hydrides are formed by the reaction of hydrogen with inner transition metals and some transition metals. The smaller hydrogen atoms occupy holes in the crystal lattice of the much larger metal atoms. The reaction takes place under pressure. The stored hydrogen is released through a drop in pressure or an increase in temperature.

The nature of the bonding in metal hydrides isn't well understood. They have the general formula MH_x, where x represents the number of hydrogen atoms in the simplest formula. Because of the variable number of holes in the parent alloy, many metallic hydrides are non-stoichiometric. For example, a titanium hydride has the formula $TiH_{1.7}$.

H Heavy Water Reactors

Due to the high neutron absorption factor of D_2O over other moderators, heavy water reactors produce larger numbers of radioisotopes than other types of nuclear reactors. Many of these isotopes are used in medicine and agriculture. Heavy water reactors also permit a higher number of uranium-238 atoms to absorb neutrons, which leads to the production of plutonium-239. If a heavy water reactor ran at full power, it could produce 8–10 kg of plutonium per year, which is enough to build a nuclear weapon. But, a large plutonium separation facility would be needed to produce this much weapons-grade plutonium.

H Hydrogen Peroxide

Hydrogen peroxide is a natural metabolite of many organisms, which decompose the H_2O_2 they produce into water and oxygen. Hydrogen peroxide is often used to control odors. It may be applied directly to aqueous wastes containing odorants (such as hydrogen sulfide) or to wet scrubbers that are used to remove pollutants from airstreams. In the case where the odors are the result of biological activity, hydrogen peroxide may be added as a preventive measure to eliminate anoxic (oxygen deficient) conditions, which favor the generation of bad odors.

In 2000, IUPAC published a paper in which *selectivity* was distinguished from *specificity* in relation to analytical tests. A specific test is one that occurs only with the substance of interest. A selective test can occur with multiple substances, but exhibits a degree of preference for the substance of interest. Few tests are specific; many are selective.

Hydrogen **R35**

Transition Metals

Physical Properties

The "sea of electrons" model can explain such properties of transition metals as conductivity and malleability. But it is not adequate to explain the periodic trend in melting points, which reflects an increase in the strength of metallic bonding until Group 6B and a subsequent decrease. To explain this trend, chemists apply molecular orbital bonding theory to metallic bonding; i.e., the existence of bonding and antibonding interactions within a metal.

Sources

An ore must contain enough of the desired metal so that the market value of the metal offsets the cost of finding, removing, and refining the ore. For many metals, the ore must contain at least 30% of the metal. However, with high value metals, such as gold, it may be profitable to mine ore containing a much smaller percentage of the metal.

To Do

Students can use sources such as the business sections of newspapers to research which metals are traded on stock exchanges, and to rank the metals from least expensive to most expensive. Ask students to use what they have learned about different metals to propose reasons for the rankings. (Remind students of how the invention of the Hall-Heroult process affected the value of aluminum.)

Transition Metals

Before 1700

Au gold
Ag silver
Cu copper
Fe iron
Hg mercury
Zn zinc

1700–1799

Co cobalt (1735)
Pt platinum (1735)
Ni nickel (1751t)
Mn manganese (1774)
Mo molybdenum (1778)
W tungsten (1783)
Ti titanium (1791)
Y yttrium (1794)
Cr chromium (1797)

1800–1899

V vanadium (1801)
Nb niobium (1801)
Ta tantalum (1802)
Pd palladium (1803)
Rh rhodium (1803)
Os osmium (1803)
Ir iridum (1803)
Cd cadmium (1817)
Zr zirconium (1824)
Ru ruthenium (1844)
Sc scandium (1878)

After 1900

Lu lutetium (1907)
Hf hafnium (1923)
Re rhenium (1925)

Physical Properties

Except for copper and gold, transition metals, including platinum, have a silvery luster.

- Most transition metals are ductile, malleable, and good conductors of heat and electric current.
- For transition metals, density tends to increase across a period, while melting point increases to a peak in Group 6B and then decreases.
- Compounds of transition metals tend to have color.

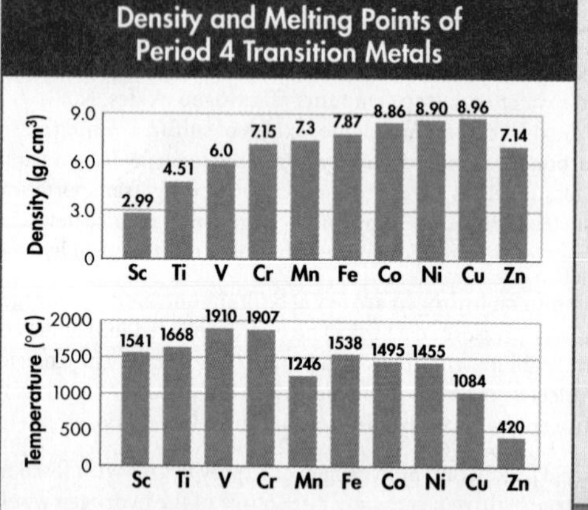

Sources

- Transition metals come from mineral deposits in Earth's crust. Minerals that are used for the commercial production of metals are called ores.
- For centuries, people have developed techniques for separating metals from ores. The ore is concentrated and the metal removed by reduction. Then the metal is refined and purified.

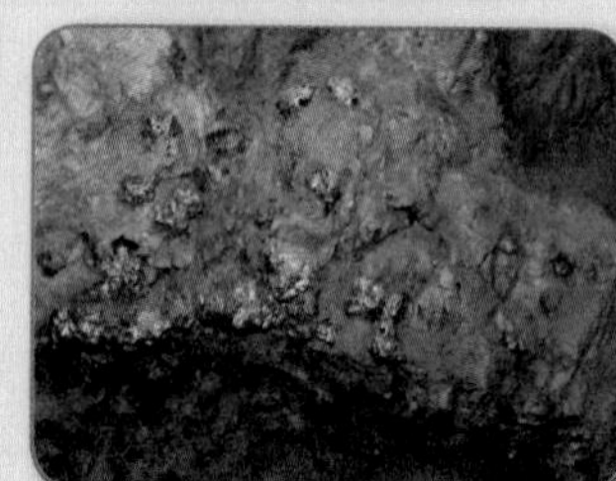

Gold exists as an element in nature. But its ore needs to be concentrated before the gold can be extracted and purified.

Atomic Properties

- Among the transition metals, as atomic number increases, there is an increase in the number of electrons in the second-to-highest occupied energy level.
- In periods 5 and 6, transition metals in the same group have identical or almost identical atomic radii. Thus, these pairs of elements have very similar chemical properties. They tend to occur together in nature and are difficult to separate.

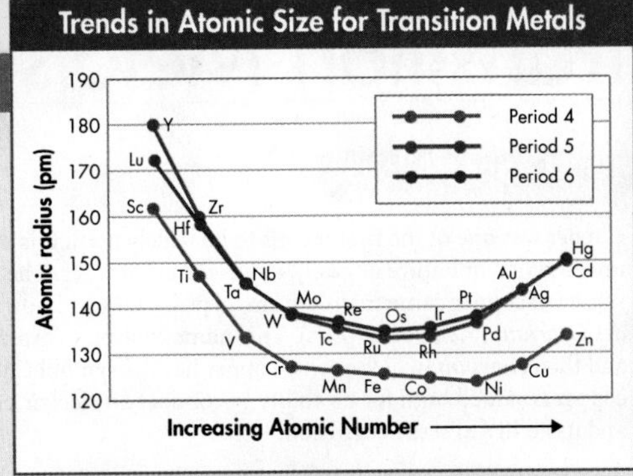

Chemical Properties

- There is great variation in reactivity among transition metals. Scandium and yttrium are similar to Group 1A and 2A metals. They are easily oxidized on exposure to air and react with water to release hydrogen. Platinum and gold are extremely unreactive and resist oxidation.
- In general, transition metals have multiple oxidation states. Compounds in which these elements are in their highest oxidation states are powerful oxidizing agents.
- Most transition metals form compounds with distinctive colors. The color of a transition metal compound or solution can indicate the oxidation state of the metal.

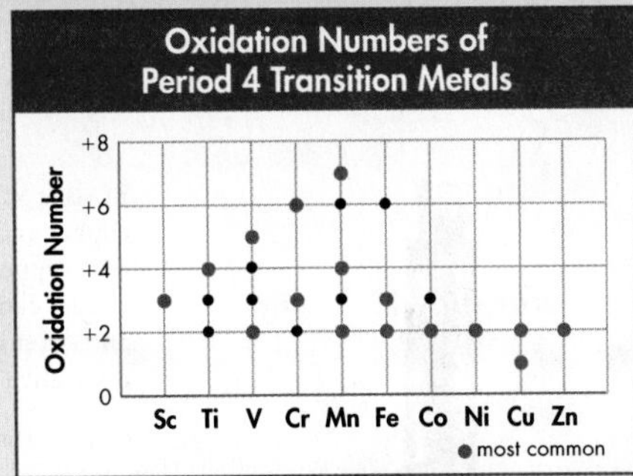

In these aqueous solutions, vanadium, chromium, and manganese are in their highest oxidation states.

The oxidation number of vanadium is +5 in the yellow solution, +4 in the blue solution, +3 in the green solution, and +2 in the purple solution.

Atomic Properties

Metal ions that have a partially filled *d* sublevel are usually colored; metal ions with a completely filled or completely empty *d* sublevel are colorless.

Chemical Properties

The transition metals vary greatly in their reactivity. For example, scandium and yttrium are similar to Group 1A metals in that they are easily oxidized on exposure to air and react with water to produce hydrogen. Yet platinum and gold are extremely unreactive and resist oxidation.

Transition Metals

Cu Copper Alloys

The Bronze Age reigned supreme for many centuries. Because bronze could be cast, it was used for making tools, plowshares, weapons, armor, and decorative objects. About 1200 B.C., there was a disruption of international trade routes and the supply of tin dried up. Metalworkers began to use iron instead of bronze because iron ore was widely available for smelting with charcoal. The substitution of iron for bronze in tools and weapons was the result of iron's abundance rather than any inherent advantages of iron.

Because only 0.13 mm of the Statue of Liberty's copper skin was weathered and oxidized during the statue's first century, the copper skin was one of the few parts of the statue that didn't need to be replaced or rebuilt when the statue was renovated for its centennial. The torch section was the only copper part that required work. It was rebuilt with new copper and patinated before installation to match the rich, green color of the rest of the Statue.

To Do

Encourage students to research the chemistry of patina formation. Have them look at the different compounds that form at various stages and how environmental conditions affect the rate at which a patina forms.

To Do

Beginning in the thirteenth century, some cultures used monumental brasses to commemorate the dead. These large brass plates were engraved with a likeness of the deceased, decorated with heraldic devices and inscriptions, and set on the surface of the tomb. Show students reproductions of rubbings made from plates in the United Kingdom. Compare the rubbings to those made of granite tombstones in New England burial grounds.

Transition Metals

Cu Copper Alloys

Copper was one of the first metals to be widely used. It is found uncombined in nature or easily reduced from its ores. The Roman supply of copper came mainly from Cyprus and was known as *aes Cyprium* (metal of Cyprus). This name evolved to *cyprium* and then *cuprium*, which is why copper has the symbol Cu. Pure copper is valued both for its ability to conduct an electric current and its ability to resist corrosion.

Around 3500 BC, people began to add tin to copper to form bronze. This alloy is harder than pure copper and easier to melt for casting. Metalworkers could produce bronze with different properties by varying the amount of tin. Bronze used to make statues might contain as little as 10% tin by mass. Bronze used to make bells would contain 13% to 25% tin. Most copper coins are bronze with 4% tin and 1% zinc.

When a bronze bell is struck, the clear, loud tone lasts for several seconds. This ancient Chinese bell was probably part of a set of graduated chimes.

Brass is an alloy of copper and zinc. Brass is harder than pure copper and more malleable than bronze. Brass containing at least 65% copper can be worked when it is cold. Brass with 55% to 65% copper can be worked when it is hot. Before large amounts of gold and silver reached Europe in the 1500s, brass was the metal used for decorative items.

Copper that is exposed to oxygen and water forms a patina of basic copper salts. This thin film protects the underlying metal from further oxidation. The composition of the patina, its color, and the rate at which it forms vary with the climate. The rate is faster near the ocean.

Historically, brass was used to make high-quality scientific instruments like this microscope.

The exterior of this science museum near the harbor in Amsterdam is copper, which has formed a patina.

Fe Iron and Steel

Carbon is used to extract iron from its ores. At first, people used charcoal from burnt wood. In 1709, Abraham Darby invented a process that used coke instead of charcoal. Coke is almost pure carbon. It is produced when coal is heated in the absence of air and the impurities removed as gases. With coke, producing iron became less costly and more efficient.

Iron ore is reduced to metallic iron in a blast furnace. Ore, coke, and limestone are added at the top of the furnace. Molten iron and slag collect at the bottom. The "pig" iron produced contains 3% to 5% carbon and smaller amounts of other impurities, which make the iron brittle. Pig iron can't be rolled or welded, but it can be cast. Cast iron is used to make stoves and engine blocks for cars.

Most pig iron is used to make steel. The methods for making steel differ, but they all lower the carbon content to less than 2% and remove other impurities. About 90% of the steel produced is carbon steel, which contains no other metals. Mild steel, which is malleable and ductile, contains less than 0.2% carbon. Medium steel (0.2% to 0.6% carbon) is used for structural components, such as beams and girders. Because high-carbon steel (0.8% to 1.5% carbon) is harder than other carbon steels, it is used to make items such as drill bits and knives.

Transition metals are used to produce alloy steels with a specific set of properties. The most common stainless steel contains about 18% chromium and 8% nickel.

Blast Furnace

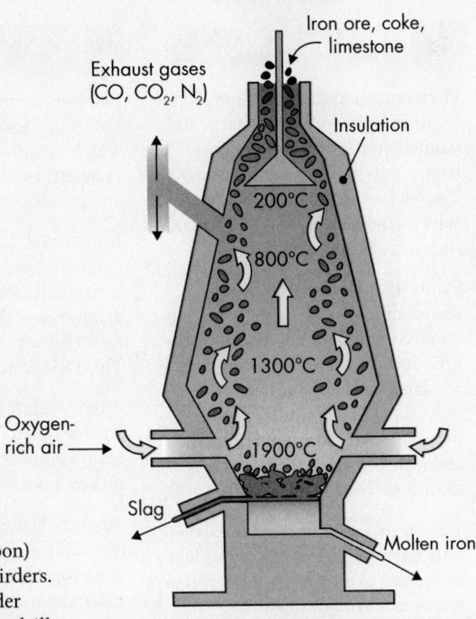

Iron ore typically contains Fe_2O_3, and SiO_2. When heated, limestone produces CaO, which reacts with SiO_2 to form slag. Slag is used to manufacture Portland cement.

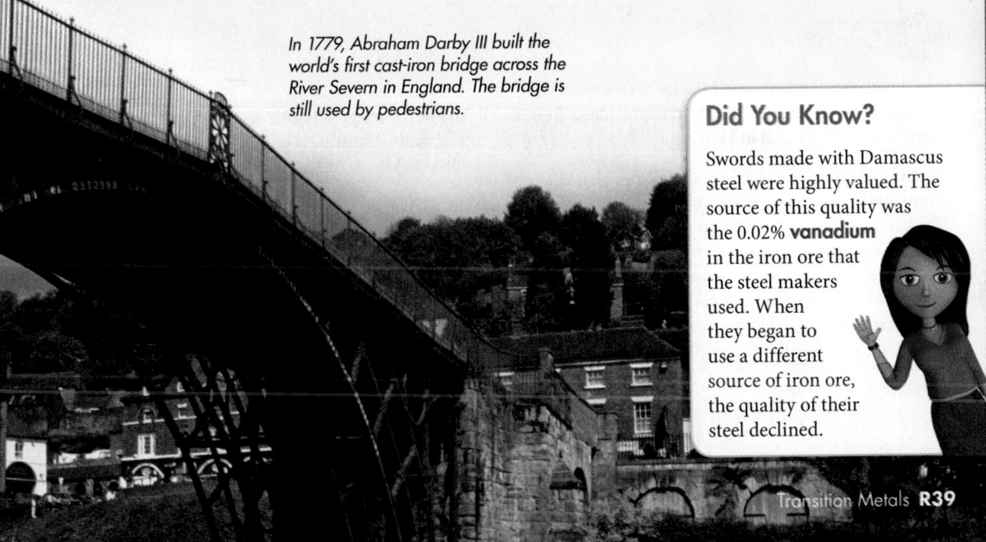

In 1779, Abraham Darby III built the world's first cast-iron bridge across the River Severn in England. The bridge is still used by pedestrians.

Did You Know?

Swords made with Damascus steel were highly valued. The source of this quality was the 0.02% **vanadium** in the iron ore that the steel makers used. When they began to use a different source of iron ore, the quality of their steel declined.

Fe Iron and Steel

Various ancient cultures developed techniques for making steel. By 500 B.C., carbon steel was being manufactured in China. There is evidence of ancient forced-draft furnaces on the shores of Lake Victoria, and oral histories indicate that steel was being produced in Africa about 2000 years ago.

In the Bessemer process for making steel, a blast of air is bubbled through molten pig iron in a huge egg-shaped container. The carbon reacts vigorously with the oxygen in the air. With this process, it is not possible to control the final carbon content or the amount of other metals in the alloy.

The open-hearth method uses a shallow lined vessel to produce high quality steel in five to eight hours. The chief advantage of the open-hearth method is that each batch can be analyzed and alloying metals added as needed. When the steel is ready, the furnace is tapped.

In the basic oxygen process, about 300 metric tons of steel can be produced in one hour. The furnace vessel is lined with a basic oxide, such as calcium oxide, and pure oxygen is blown into the molten mass through a narrow tube called a lance at about 10 atmospheres pressure.

TRANSITION METALS

Transition Metals

Cd Phytoremediation

Plants used for phytoremediation are called hyperaccumulators because the concentration of a target substance in the plant is greater than its concentration in the soil. One disadvantage of using plants to clean up soil is that the process usually takes longer than the alternatives. One advantage is that the contaminated soil is cleaned up on site. The soil does not have to be dug up and transported to another location for treatment.

Fe Ni Co Permanent Magnets

Currently, neodymium–iron–boron magnets are the most compact magnetic materials available; that is, they exhibit considerable magnetic strength for their density. These magnets are susceptible to corrosion with the main product of surface corrosion being Nd_2O_3. Because the rate of corrosion increases as the temperature increases, the maximum temperature at which the magnets are useful is 120°C. The magnets are often plated with nickel to prevent corrosion.

Au Gold

A gold content of 18k and above is rarely used for jewelry because the high gold content makes the gold alloy soft enough to be easily dented. Various metals are added to gold to form alloys. The most common is copper, which preserves the gold color. Metals such as nickel and zinc are used to produce white gold.

Transition Metals

Cd Phytoremediation

Phytoremediation uses plants such as sunflowers, Indian mustard, and dandelions to remove pollutants from contaminated soil and water. The contaminants include organic solvents, pesticides, and toxic metals such as cadmium and chromium.

Plants have a natural ability to absorb nutrients through their roots. Often a plant does not distinguish a toxic metal such as cadmium from a nutrient such as zinc because these metals have similar chemical properties. So cadmium is absorbed and transported to the leaves and stems, where it accumulates.

The plants are composted or burned after harvesting. The metal residues are buried in an approved landfill or recovered through smelting.

Fe Ni Co Permanent Magnets

Refrigerator magnets contain a barium ferrite, $BaO{\cdot}6Fe_2O_3$, or strontium ferrite, $SrO{\cdot}6Fe_2O_3$, powder, which is embedded in plastic or rubber. Horseshoe magnets usually contain an alloy of aluminum, nickel, and cobalt.

Iron, nickel, and cobalt are strongly attracted to magnetic fields. When these metals are exposed to a magnetic field, their cations line up in an orderly arrangement. When the field is removed, the ions remain lined up, and the material can act as a magnet. This type of magnetism is called ferromagnetism.

A magnet retains its strength unless it is heated past a point called the Curie temperature. For iron, this temperature is 1043 K. For cobalt, it is 1388 K. For nickel, it is 627 K.

Magnets made from a neodymium, boron, and iron alloy are very powerful. If they are allowed to fly together, they will shatter. They are used to check for counterfeit bills because they can detect tiny magnetic particles placed in the ink of genuine bills.

The wrench is strongly attracted to the tiny neodymium-iron-boron cube.

Au Gold

Gold occurs chiefly as small flecks of free metal in veins of quartz. About 5 g of gold is produced from a metric ton (10^6 g) of gold-bearing rock.

Gold can be pounded into sheets so thin that they will transmit light. These sheets, called gold leaf, are used for lettering and decoration in general. Gold is used on the outside surfaces of satellites because it resists corrosion. Its high electrical and thermal conductivity make gold a good choice to plate contacts in microcircuits.

Pure gold is alloyed to make it harder and more durable. Gold alloys are safe to use as fillings for teeth because gold is highly unreactive.

Units called karats (k) are used to describe the purity of gold. Pure gold is 24k or 100% gold. Gold in coins is usually 22k or 92% gold. Gold in rings is often 14k (58% gold).

Connectors in electronics cables may be gold-plated to enhance signal transfer and minimize corrosion.

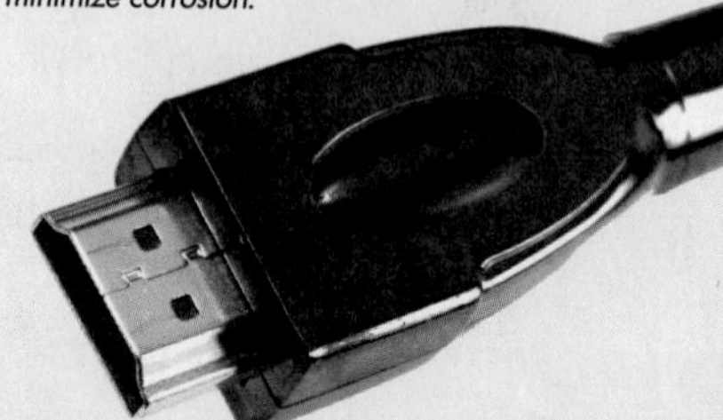

Micronutrients

Trace amounts of some transition metals are essential for human health.

Iron is found mainly in hemoglobin and myoglobin. Hemoglobin is the protein that transports oxygen in blood. Myoglobin is the protein that stores oxygen in muscle tissue. Vitamin C helps the absorption of iron by promoting the reduction of Fe^{3+} ions to Fe^{2+} ions.

Zinc is a cofactor in many enzymes. It helps protect the immune system. Hormones that control growth and reproduction do not function properly without zinc. A lack of zinc impairs the sense of taste and reduces the appetite.

Copper is a component of enzymes that control the synthesis of melanin, hemoglobin, and phospholipids in the sheath that protects nerves.

Molybdenum affects the absorption of copper. It is also needed for the oxidation of lipids and the metabolism of sulfur and nitrogen.

Chromium assists in the metabolism of glucose and may help to control adult-onset diabetes. A lack of chromium may affect growth.

Manganese is required for the proper function of the nervous system and the thyroid gland. It is needed for glucose metabolism. It helps maintain healthy bones and cartilage.

Cobalt is a component of vitamin B_{12}, which is required for the synthesis of red blood cells.

Transition Metal Micronutrients

Element	RDA or AI*	Dietary Sources
Iron	10 mg RDA (M) 20 mg RDA (F)	liver, green vegetables, egg yolk, fish, whole wheat, nuts, oatmeal, molasses, and beans
Zinc	11 mg RDA (M) 8 mg RDA (F)	liver, eggs, meat, milk, whole grains, and shellfish
Copper	900 μg RDA	beans, peas, and shellfish
Molybdenum	45 μg RDA	beans, peas, and whole grains
Chromium	35 μg AI (M) 25 μg AI (F)	meat and whole grains
Manganese	2.3 mg AI (M) 1.8 mg AI (F)	nuts, whole grains, dried fruits, and green leafy vegetables

*Recommended Dietary Allowance or Adequate Intake

Ti Sunscreens

You need some exposure to sunlight so your skin cells can make vitamin D, which is needed for healthy bones and teeth. Yet the UV radiation in sunlight can damage skin cells and even lead to skin cancer. The best way to protect your skin is to limit your time in the sun. The next best way is to use a sunscreen.

All the active ingredients in sunscreens protect against UVB light (280–320 nm), which is the primary cause of sunburn. Some ingredients protect against UVA light (320–400 nm), which penetrates deeper and causes long-term damage. A sun protective factor rating (SPF) measures only how effective a sunscreen is against UVB, not UVA.

Titanium dioxide, TiO_2, can reflect and scatter UV light. This stable and nonirritating oxide has one drawback. It looks like white paint on the skin. One manufacturer has addressed this problem by decreasing the size of the TiO_2 particles to a diameter of about 21 nm. At this size, TiO_2 appears transparent because its particles are smaller than wavelengths of visible light and light isn't reflected by the particles.

Did You Know?

An octopus has blood that is blue, not red, because the compound that transports oxygen in an octopus contains **copper**, not iron. Snails, oysters, and spiders are also bluebloods.

TRANSITION METALS

Micronutrients

There are probably more than seven transition metal micronutrients. For example, nickel may affect cell membranes in the liver and vanadium may affect growth. Fe^{2+} ions are absorbed in the stomach and small intestine. Vitamin C helps the absorption by promoting the reduction of Fe^{3+} ions to Fe^{2+} ions.

To Do

Students could research the difference between an RDA and an AI; how these quantities are determined; and why the recommended amounts vary by gender.

Ti Sunscreens

Skin responds to UV exposure by producing more melanin, which absorbs some of the damaging radiation. But damage can still occur even if a person has naturally dark skin. An active ingredient in sunscreen absorbs, reflects, or scatters radiation in the UV range of 280–400 nm. Examples of organic active ingredients are avobenzene and oxybenzene, which absorb UV light and dissipate the energy as heat. Inorganic products act by reflecting and scattering UV light. Sunscreens containing both organic and inorganic ingredients can have a very high SPF rating.

To Do

Suggest that students compare the active ingredients and ratings for different sunscreens. Have them check to see if the label claims that the product protects against both UVA and UVB radiation.

Appendix B Reference Tables

Table B.1

Some Properties of the Elements							
Element	Symbol	Atomic number	Atomic mass	Melting point (°C)	Boiling point (°C)	Density (g/cm³) (gases at STP)	Oxidation numbers
Actinium	Ac	89	(227)	1050	3200	10.07	+3
Aluminum	Al	13	26.98154	660.37	2467	2.6989	+3
Americium	Am	95	243	994	2607	13.67	+3, +4, +5, +6
Antimony	Sb	51	121.75	630.74	1587	6.691	−3, +3, +5
Argon	Ar	18	39.948	189.2	185.7	0.0017837	
Arsenic	As	33	74.9216	817	613	5.73	−3, +3, +5
Astatine	At	85	(210)	302	337	—	
Barium	Ba	56	137.33	725	1640	3.5	+2
Berkelium	Bk	97	(247)	986	—	14.78	
Beryllium	Be	4	9.01218	1278	2970	1.848	+2
Bismuth	Bi	83	208.9804	271.3	1560	9.747	+3, +5
Bohrium	Bh	107	(264)	—	—	—	
Boron	B	5	10.81	2075	3675	2.34	+3
Bromine	Br	35	79.904	7.2	58.78	3.12	−1, +1, +5
Cadmium	Cd	48	112.41	320.9	765	8.65	+2
Calcium	Ca	20	40.08	839	1484	1.55	+2
Californium	Cf	98	(251)	900	—	14	
Carbon	C	6	12.011	3550	4827	2.267	−4, +2, +4
Cerium	Ce	58	140.12	799	3426	6.657	+3, +4
Cesium	Cs	55	132.9054	28.40	669.3	1.873	+1
Chlorine	Cl	17	35.453	100.98	34.6	0.003214	−1, +1, +5, +7
Chromium	Cr	24	51.996	1907	2672	7.18	+2, +3, +6
Cobalt	Co	27	58.9332	1495	2870	8.9	+2, +3
Copernicium	Cn	112	(277)	—	—	—	
Copper	Cu	29	63.546	1083.4	2567	8.96	+1, +2
Curium	Cm	96	(247)	1340	—	13.51	+3
Darmstadtium	Ds	110	(269)	—	—	—	
Dubnium	Db	105	(262)	—	—	—	
Dysprosium	Dy	66	162.50	1412	2562	8.550	+3
Einsteinium	Es	99	(252)	—	—	—	
Erbium	Er	68	167.26	159	2863	9.066	+3
Europium	Eu	63	151.96	822	1597	5.243	+2, +3
Fermium	Fm	100	(257)	—	—	—	
Fluorine	F	9	18.998403	219.62	188.54	0.00181	−1
Francium	Fr	87	(223)	27	677	—	+1
Gadolinium	Gd	64	157.25	1313	3266	7.9004	+3
Gallium	Ga	31	69.72	29.78	2204	5.904	+3
Germanium	Ge	32	72.59	937.4	2830	5.323	+2, +4
Gold	Au	79	196.9665	1064.43	2856	19.3	+1, +3
Hafnium	Hf	72	178.49	2227	4602	13.31	+4
Hassium	Hs	108	(265)	—	—	—	
Helium	He	2	4.00260	272.2	268.934	0.001785	
Holmium	Ho	67	164.9304	1474	2695	8.795	+3
Hydrogen	H	1	1.00794	259.14	252.87	0.00008988	−1, +1
Indium	In	49	114.82	156.61	2080	7.31	+1, +3
Iodine	I	53	126.9045	113.5	184.35	4.93	−1, +1, +5, +7
Iridium	Ir	77	192.22	2410	4130	22.42	+3, +4
Iron	Fe	26	55.847	1535	2750	7.874	+2, +3
Krypton	Kr	36	83.80	156.6	152.30	0.003733	
Lanthanum	La	57	138.9055	921	3457	6.145	+3
Lawrencium	Lr	103	(262)	—	—	—	+3
Lead	Pb	82	207.2	327.502	1740	11.35	+2, +4

R42 Appendix B

REFERENCE TABLES

Table B.1

Some Properties of the Elements (cont.)							
Element	**Symbol**	**Atomic number**	**Atomic mass**	**Melting point (°C)**	**Boiling point (°C)**	**Density (g/cm^3) (gases at STP)**	**Oxidation numbers**
Lithium	Li	3	6.941	180.54	1342	0.534	+1
Lutetium	Lu	71	174.967	1663	3395	9.840	+3
Magnesium	Mg	12	24.305	648.8	1107	1.738	+2
Manganese	Mn	25	54.9380	1244	1962	7.32	+2, +3, +4, +7
Meitnerium	Mt	109	(268)	—	—	—	
Mendelevium	Md	101	257	—	—	—	+2, +3
Mercury	Hg	80	200.59	38.842	356.58	13.55	+1, +2
Molybdenum	Mo	42	95.94	2617	4612	10.22	+6
Neodymium	Nd	60	144.24	1021	3068	6.90	+3
Neon	Ne	10	20.179	248.67	246.048	0.0008999	
Neptunium	Np	93	(237)	640	3902	20.25	+3, +4, +5, +6
Nickel	Ni	28	58.69	1453	2732	8.902	+2, +3
Niobium	Nb	41	92.9064	2468	4742	8.57	+3, +5
Nitrogen	N	7	14.0067	209.86	195.8	0.0012506	−3, +3, +5
Nobelium	No	102	(259)	—	—	—	+2, +3
Osmium	Os	76	190.2	3045	5027	22.57	+3, +4
Oxygen	O	8	15.9994	218.4	182.962	0.001429	−2
Palladium	Pd	46	106.42	1554	2970	12.02	+2, +4
Phosphorus	P	15	30.97376	44.1	280	1.82	−3, +3, +5
Platinum	Pt	78	195.08	1772	3627	21.45	+2, +4
Plutonium	Pu	94	(244)	641	3232	19.84	+3, +4, +5, +6
Polonium	Po	84	(209)	254	962	9.32	+2, +4
Potassium	K	19	39.0982	63.25	760	0.862	+1
Praseodymium	Pr	59	140.9077	931	3512	6.64	+3
Promethium	Pm	61	(145)	1168	2460	7.22	+3
Protactinium	Pa	91	231.0359	1560	4027	15.37	+4, +5
Radium	Ra	88	(226)	700	1140	5.5	+2
Radon	Rn	86	(222)	71	61.8	0.00973	
Rhenium	Re	75	186.207	3180	5627	21.02	+4, +6, +7
Rhodium	Rh	45	102.9055	1966	3727	12.41	+3
Roentgenium	Rg	111	(272)	—	—	—	
Rubidium	Rb	37	85.4678	38.89	686	1.532	+1
Ruthenium	Ru	44	101.07	2310	3900	12.41	+3
Rutherfordium	Rf	104	(261)	—	—	—	
Samarium	Sm	62	150.36	1077	1791	7.520	+2, +3
Scandium	Sc	21	44.9559	1541	2831	2.989	+3
Seaborgium	Sg	106	(263)	—	—	—	
Selenium	Se	34	78.96	217	684.9	4.79	−2, +4, +6
Silicon	Si	14	28.0855	1410	2355	2.33	−4, +2, +4
Silver	Ag	47	107.8682	961.93	2212	10.50	+1
Sodium	Na	11	22.98977	97.81	882.9	0.971	+1
Strontium	Sr	38	87.62	769	1381	2.63	+2
Sulfur	S	16	32.06	112.8	444.7	2.07	−2, +4, +6
Tantalum	Ta	73	180.9479	2996	5425	16.654	+5
Technetium	Tc	43	(98)	2172	4877	11.50	+4, +6, +7
Tellurium	Te	52	127.60	449.5	989.8	6.24	−2, +4, +6
Terbium	Tb	65	158.9254	1356	3123	8.229	+3
Thallium	Tl	81	204.383	303.5	1457	11.85	+1, +3
Thorium	Th	90	232.0381	1750	4790	11.72	+4
Thulium	Tm	69	168.9342	1545	1947	9.321	+3
Tin	Sn	50	118.69	231.968	2270	7.31	+2, +4
Titanium	Ti	22	47.88	1660	3287	4.54	+2, +3, +4
Tungsten	W	74	183.85	3410	5660	19.3	+6
Uranium	U	92	238.0289	1132.3	3818	18.95	+3, +4, +5, +6
Vanadium	V	23	50.9415	1890	3380	6.11	+2, +3, +4, +5
Xenon	Xe	54	131.29	111.9	107.1	0.005887	
Ytterbium	Yb	70	173.04	819	1194	6.965	+2, +3
Yttrium	Y	39	88.9059	1522	3338	4.469	+3
Zinc	Zn	30	65.38	419.58	907	7.133	+2
Zirconium	Zr	40	91.22	1852	4377	6.506	+4

Reference Tables **R43**

Table B.2

Electron Configuration of the Elements																				
Element		Sublevels																		
		1s	*2s*	*2p*	*3s*	*3p*	*3d*	*4s*	*4p*	*4d*	*4f*	*5s*	*5p*	*5d*	*5f*	*6s*	*6p*	*6d*	*7s*	*7p*
1	Hydrogen	1																		
2	Helium	2																		
3	Lithium	2	1																	
4	Beryllium	2	2																	
5	Boron	2	2	1																
6	Carbon	2	2	2																
7	Nitrogen	2	2	3																
8	Oxygen	2	2	4																
9	Fluorine	2	2	5																
10	Neon	2	2	6																
11	Sodium	2	2	6	1															
12	Magnesium	2	2	6	2															
13	Aluminum	2	2	6	2	1														
14	Silicon	2	2	6	2	2														
15	Phosphorus	2	2	6	2	3														
16	Sulfur	2	2	6	2	4														
17	Chlorine	2	2	6	2	5														
18	Argon	2	2	6	2	6														
19	Potassium	2	2	6	2	6		1												
20	Calcium	2	2	6	2	6		2												
21	Scandium	2	2	6	2	6	1	2												
22	Titanium	2	2	6	2	6	2	2												
23	Vanadium	2	2	6	2	6	3	2												
24	Chromium	2	2	6	2	6	5	1												
25	Manganese	2	2	6	2	6	5	2												
26	Iron	2	2	6	2	6	6	2												
27	Cobalt	2	2	6	2	6	7	2												
28	Nickel	2	2	6	2	6	8	2												
29	Copper	2	2	6	2	6	10	1												
30	Zinc	2	2	6	2	6	10	2												
31	Gallium	2	2	6	2	6	10	2	1											
32	Germanium	2	2	6	2	6	10	2	2											
33	Arsenic	2	2	6	2	6	10	2	3											
34	Selenium	2	2	6	2	6	10	2	4											
35	Bromine	2	2	6	2	6	10	2	5											
36	Krypton	2	2	6	2	6	10	2	6											
37	Rubidium	2	2	6	2	6	10	2	6			1								
38	Strontium	2	2	6	2	6	10	2	6			2								
39	Yttrium	2	2	6	2	6	10	2	6	1		2								
40	Zirconium	2	2	6	2	6	10	2	6	2		2								
41	Niobium	2	2	6	2	6	10	2	6	4		1								
42	Molybdenum	2	2	6	2	6	10	2	6	5		1								
43	Technetium	2	2	6	2	6	10	2	6	5		2								
44	Ruthenium	2	2	6	2	6	10	2	6	7		1								
45	Rhodium	2	2	6	2	6	10	2	6	8		1								
46	Palladium	2	2	6	2	6	10	2	6	10										
47	Silver	2	2	6	2	6	10	2	6	10		1								
48	Cadmium	2	2	6	2	6	10	2	6	10		2								
49	Indium	2	2	6	2	6	10	2	6	10		2	1							
50	Tin	2	2	6	2	6	10	2	6	10		2	2							
51	Antimony	2	2	6	2	6	10	2	6	10		2	3							
52	Tellurium	2	2	6	2	6	10	2	6	10		2	4							
53	Iodine	2	2	6	2	6	10	2	6	10		2	5							
54	Xenon	2	2	6	2	6	10	2	6	10		2	6							
55	Cesium	2	2	6	2	6	10	2	6	10		2	6			1				
56	Barium	2	2	6	2	6	10	2	6	10		2	6			2				

R44 Appendix B

Table B.2

Electron Configuration of the Elements (cont.)																				
Element		**Sublevels**																		
		1s	**2s**	**2p**	**3s**	**3p**	**3d**	**4s**	**4p**	**4d**	**4f**	**5s**	**5p**	**5d**	**5f**	**6s**	**6p**	**6d**	**7s**	**7p**
57	Lanthanum	2	2	6	2	6	10	2	6	10		2	6	1		2				
58	Cerium	2	2	6	2	6	10	2	6	10	1	2	6	1		2				
59	Praseodymium	2	2	6	2	6	10	2	6	10	3	2	6			2				
60	Neodymium	2	2	6	2	6	10	2	6	10	4	2	6			2				
61	Promethium	2	2	6	2	6	10	2	6	10	5	2	6			2				
62	Samarium	2	2	6	2	6	10	2	6	10	6	2	6			2				
63	Europium	2	2	6	2	6	10	2	6	10	7	2	6			2				
64	Gadolinium	2	2	6	2	6	10	2	6	10	7	2	6	1		2				
65	Terbium	2	2	6	2	6	10	2	6	10	9	2	6			2				
66	Dysprosium	2	2	6	2	6	10	2	6	10	10	2	6			2				
67	Holmium	2	2	6	2	6	10	2	6	10	11	2	6			2				
68	Erbium	2	2	6	2	6	10	2	6	10	12	2	6			2				
69	Thulium	2	2	6	2	6	10	2	6	10	13	2	6			2				
70	Ytterbium	2	2	6	2	6	10	2	6	10	14	2	6			2				
71	Lutetium	2	2	6	2	6	10	2	6	10	14	2	6	1		2				
72	Hafnium	2	2	6	2	6	10	2	6	10	14	2	6	2		2				
73	Tantalum	2	2	6	2	6	10	2	6	10	14	2	6	3		2				
74	Tungsten	2	2	6	2	6	10	2	6	10	14	2	6	4		2				
75	Rhenium	2	2	6	2	6	10	2	6	10	14	2	6	5		2				
76	Osmium	2	2	6	2	6	10	2	6	10	14	2	6	6		2				
77	Iridium	2	2	6	2	6	10	2	6	10	14	2	6	7		2				
78	Platinum	2	2	6	2	6	10	2	6	10	14	2	6	9		1				
79	Gold	2	2	6	2	6	10	2	6	10	14	2	6	10		1				
80	Mercury	2	2	6	2	6	10	2	6	10	14	2	6	10		2				
81	Thallium	2	2	6	2	6	10	2	6	10	14	2	6	10		2	1			
82	Lead	2	2	6	2	6	10	2	6	10	14	2	6	10		2	2			
83	Bismuth	2	2	6	2	6	10	2	6	10	14	2	6	10		2	3			
84	Polonium	2	2	6	2	6	10	2	6	10	14	2	6	10		2	4			
85	Astatine	2	2	6	2	6	10	2	6	10	14	2	6	10		2	5			
86	Radon	2	2	6	2	6	10	2	6	10	14	2	6	10		2	6			
87	Francium	2	2	6	2	6	10	2	6	10	14	2	6	10		2	6		1	
88	Radium	2	2	6	2	6	10	2	6	10	14	2	6	10		2	6		2	
89	Actinium	2	2	6	2	6	10	2	6	10	14	2	6	10		2	6	1	2	
90	Thorium	2	2	6	2	6	10	2	6	10	14	2	6	10		2	6	2	2	
91	Protactinium	2	2	6	2	6	10	2	6	10	14	2	6	10	2	2	6	1	2	
92	Uranium	2	2	6	2	6	10	2	6	10	14	2	6	10	3	2	6	1	2	
93	Neptunium	2	2	6	2	6	10	2	6	10	14	2	6	10	4	2	6	1	2	
94	Plutonium	2	2	6	2	6	10	2	6	10	14	2	6	10	6	2	6		2	
95	Americium	2	2	6	2	6	10	2	6	10	14	2	6	10	7	2	6		2	
96	Curium	2	2	6	2	6	10	2	6	10	14	2	6	10	7	2	6	1	2	
97	Berkelium	2	2	6	2	6	10	2	6	10	14	2	6	10	9	2	6		2	
98	Californium	2	2	6	2	6	10	2	6	10	14	2	6	10	10	2	6		2	
99	Einsteinium	2	2	6	2	6	10	2	6	10	14	2	6	10	11	2	6		2	
100	Fermium	2	2	6	2	6	10	2	6	10	14	2	6	10	12	2	6		2	
101	Mendelevium	2	2	6	2	6	10	2	6	10	14	2	6	10	13	2	6		2	
102	Nobelium	2	2	6	2	6	10	2	6	10	14	2	6	10	14	2	6		2	
103	Lawrencium	2	2	6	2	6	10	2	6	10	14	2	6	10	14	2	6	1	2	
104	Rutherfordium	2	2	6	2	6	10	2	6	10	14	2	6	10	14	2	6	2	2	
105	Dubnium	2	2	6	2	6	10	2	6	10	14	2	6	10	14	2	6	3	2	
106	Seaborgium	2	2	6	2	6	10	2	6	10	14	2	6	10	14	2	6	4	2	
107	Bohrium	2	2	6	2	6	10	2	6	10	14	2	6	10	14	2	6	5	2	
108	Hassium	2	2	6	2	6	10	2	6	10	14	2	6	10	14	2	6	6	2	
109	Meitnerium	2	2	6	2	6	10	2	6	10	14	2	6	10	14	2	6	7	2	
110	Darmstadium	2	2	6	2	6	10	2	6	10	14	2	6	10	14	2	6	9	1	
111	Roentgenium	2	2	6	2	6	10	2	6	10	14	2	6	10	14	2	6	10	1	
112	Copernicium	2	2	6	2	6	10	2	6	10	14	2	6	10	14	2	6	10	2	

Reference Tables **R45**

Table B.3

Physical Constants	
Atomic mass unit	1 amu = 1.6605×10^{-24} g
Avogadro's number	$N = 6.0221 \times 10^{23}$ particles/mol
Gas constant	$R = 8.31$ L·kPa/K·mol
Ideal gas molar volume	$V_m = 22.414$ L/mol
Masses of subatomic particles	
Electron (e^-)	$m_e = 0.0005486$ amu $= 9.1096 \times 10^{-28}$ g
Proton (p^+)	$m_p = 1.007277$ amu $= 1.67261 \times 10^{-24}$ g
Neutron (n^0)	$m_n = 1.008665$ amu $= 1.67492 \times 10^{-24}$ g
Speed of light (in vacuum)	$c = 2.997925 \times 10^8$ m/s

Table B.4

SI Units and Equivalents

Quantity	SI unit	Common equivalents
Length	meter (m)	1 meter = 1.0936 yards 1 centimeter = 0.39370 inch 1 inch = 2.54 centimeters 1 mile = 5280 feet = 1.6093 kilometers
Volume	cubic meter (m^3)	1 liter = 10^{-3} m^3 = 1.0567 quarts 1 gallon = 4 quarts = 8 pints = 3.7854 liters 1 quart = 32 fluid ounces = 0.94635 liter
Temperature	kelvin (K)	1 kelvin = 1 degree Celsius °C = $\frac{5}{9}$(F − 32) K = °C + 273.15
Mass	kilogram (kg)	1 kilogram = 1000 grams = mass weighing 2.2046 pounds 1 amu = 1.66057×10^{-27} kilograms
Time	second (s)	1 hour = 60 minutes 1 hour = 3600 seconds
Energy	joule (J)	1 joule = 1 kg·m^2/s^2 (exact) 1 joule = 0.23901 calorie 1 calorie = 4.184 joules
Pressure	pascal (Pa)	1 atmosphere = 101.3 kilopascals = 760 mm Hg (torr) = 14.70 pounds per square inch

R46 Appendix B

Table B.5

Other Symbols and Abbreviations

Symbol	Meaning
α	alpha rays
β	beta rays
γ	gamma rays
Δ	change in
$\delta+$, $\delta-$	partial ionic charge
λ	wavelength
π	pi bond
σ	sigma bond
ν	frequency
amu	atomic mass unit
(*aq*)	aqueous solution
atm	atmosphere (*pressure*)
bp	boiling point
°C	degree Celsius (*temperature*)
c	speed of light in a vacuum
cm	centimeter (*length*)
E	energy
e^-	electron
fp	freezing point
G	Gibbs free energy
g	gram (*mass*)
(*g*)	gas
gfm	gram formula mass
H	enthalpy

Symbol	Meaning
ΔH_f	heat of formation
h	hour
h	Planck's constant
Hz	hertz (*frequency*)
J	joule (*energy*)
K	kelvin (*temperature*)
K_a	acid dissociation constant
K_b	base dissociation constant
K_b	molal boiling point elevation constant
K_{eq}	equilibrium constant
K_f	molal freezing point depression constant
K_w	ion product constant for water
K_{sp}	solubility product constant
kcal	kilocalorie (*energy*)
kg	kilogram (*mass*)
kPa	kilopascal (*pressure*)
L	liter (*volume*)
(*l*)	liquid
M	molarity
m	meter (*length*)
m	mass

Symbol	Meaning
m	molality
mL	milliliter (*volume*)
mm	millimeter (*length*)
mol	mole (*amount*)
mp	melting point
N	normality
n^0	neutron
n	number of moles
n	principal quantum number
P	pressure
p^+	proton
Pa	pascal (*pressure*)
R	ideal gas constant
S	entropy
s	second
(*s*)	solid
SI	International System of Units
STP	standard temperature and pressure
T	temperature
$t_{\frac{1}{2}}$	half-life
V	volume
v	velocity

Table B.6

Common Polyatomic Ions

Charge	Name	Formula	Charge	Name	Formula
1–	Chlorate	ClO_3^-	2–	Carbonate	CO_3^{2-}
	Chlorite	ClO_2^-		Chromate	CrO_4^{2-}
	Cyanide	CN^-		Dichromate	$Cr_2O_7^{2-}$
	Dihydrogen phosphate	$H_2PO_4^-$		Oxalate	$C_2O_4^{2-}$
	Ethanoate	CH_3COO^-		Peroxide	O_2^{2-}
	Hydroxide	OH^-		Silicate	SiO_3^{2-}
	Hydrogen carbonate	HCO_3^-		Sulfate	SO_4^{2-}
	Hydrogen sulfate	HSO_4^-		Sulfite	SO_3^{2-}
	Hydrogen sulfite	HSO_3^-		Thiosulfate	$S_2O_3^{2-}$
	Hypochlorite	ClO^-			
	Nitrate	NO_3^-	3–	Phosphate	PO_4^{3-}
	Nitrite	NO_2^-		Phosphite	PO_3^{3-}
	Perchlorate	ClO_4^-			
	Permanganate	MnO_4^-	1+	Ammonium	NH_4^+
	Thiocyanate	SCN^-			

Reference Tables **R47**

Table B.7

Solubilities of Compounds at 25°C and 101.3 kPa

	ethanoate	bromide	carbonate	chlorate	chloride	hydroxide	iodide	nitrate	oxide	perchlorate	phosphate	sulfate	sulfide
aluminum	S	S	X	S	S	I	S	S	I	S	I	S	d
ammonium	S	S	S	S	S	X	S	S	X	S	S	S	S
barium	S	S	I	S	S	S	S	S	sS	S	I	I	d
calcium	S	S	I	S	S	S	S	S	sS	S	I	sS	I
copper(II)	S	S	X	S	S	I	S	S	I	S	I	S	I
iron(II)	S	S	I	S	S	I	S	S	I	S	I	S	I
iron(III)	S	S	X	S	S	I	S	S	I	S	I	sS	d
lithium	S	S	sS	S	S	S	S	S	S	S	sS	S	S
magnesium	S	S	I	S	S	I	S	S	I	S	I	S	d
potassium	S	S	S	S	S	S	S	S	S	S	S	S	S
silver	sS	I	I	S	I	X	I	S	I	S	I	sS	I
sodium	S	S	S	S	S	S	S	S	S	S	S	S	S
strontium	S	S	I	S	S	S	S	S	S	S	I	I	I
zinc	S	S	I	S	S	I	S	S	I	S	I	S	I

Key: S = soluble
sS = slightly soluble
I = insoluble
d = decomposes in water
X = no such compound

R48 Appendix B

Appendix C Safety in the Chemistry Lab

The experiments in this book have been carefully designed to minimize the risk of injury. However, safety is also your responsibility. The following rules are essential for keeping you safe in the laboratory. The rules address pre-lab preparation, proper laboratory practices, and post-lab procedures.

Pre-Lab Preparation

1. Read the entire procedure before you begin. Listen to all of your teacher's instructions. When in doubt about a procedure, ask your teacher.

2. Do only the assigned experiments. Do any experiment only when your teacher is present and has given you permission to work.

3. Know the location and operation of the following safety equipment: fire extinguisher, fire blanket, emergency shower, and eye wash station.

4. Know the location of emergency exits and escape routes. To make it easy to exit quickly, do not block walkways with furniture. Keep your work area orderly and free of personal belongings such as coats and backpacks.

5. Protect your clothing and hair from chemicals and sources of heat. Tie back long hair and roll up loose sleeves when working in the laboratory. Avoid wearing bulky or loose-fitting clothing. Remove dangling jewelry. Wear closed-toe shoes at all times in the laboratory.

Proper Laboratory Practices

6. Even with well-designed and tested laboratory procedures, an accident may occur while you are working in the lab. Report any accident, no matter how minor, to your teacher.

7. Wear chemical splash goggles at all times when working in the laboratory. These goggles are designed to protect your eyes from injury. While working in the lab, do not rub your eyes, because chemicals are easily transferred from your hands to your eyes.

⚠ If, despite these precautions, a chemical gets in your eye, remove any contact lenses and immediately wash your eye with a continuous stream of lukewarm water for at least 15 minutes.

8. To reduce danger, waste, and cleanup, always use the minimal amounts of chemicals specified for an experiment.

9. Never taste any chemical used in the laboratory, including food products that are the subject of an investigation. Treat all items as though they are contaminated with unknown chemicals that may be toxic. Keep all food and drink that is not part of an experiment out of the laboratory. Do not eat, drink, or chew gum in the laboratory.

⚠ If you accidentally ingest a substance, notify your teacher immediately.

10. Don't use chipped or cracked glassware. Don't handle broken glass. If glassware breaks, tell your teacher and nearby classmates. Discard broken glass as instructed by your teacher.

⚠ If, despite these precautions, you receive a minor cut, allow it to bleed for a short time. Wash the injured area under cold, running water and notify your teacher. More serious cuts or puncture wounds require immediate medical attention.

11. Do not handle hot glassware or equipment. You can prevent burns by being aware that hot and cold equipment can look exactly the same.

⚠ If you are burned, immediately run cold water over the burned area for several minutes until the pain is reduced. Cooling helps the burn heal. Ask a classmate to notify your teacher.

12. Recognize that the danger of an electrical shock is greater in the presence of water. Keep electrical appliances away from sinks and faucets to minimize the risk of electrical shock. Be careful not to spill water or other liquids in the vicinity of an electrical appliance.

⚠ If, despite these precautions, you spill water near an electrical appliance, stand back, notify your teacher, and warn other students in the area.

13. Report any chemical spills immediately to your teacher. Follow your teacher's instructions for cleaning up spills. Warn other students about the identity and location of spilled chemicals.

Safety in the Chemistry Lab R49

⚠ If, despite these precautions, a corrosive chemical gets on your skin or clothing, notify your teacher. Then wash the affected area with cold, running water for several minutes.

Post-Lab Procedures

14. Dispose of chemicals in a way that protects you, your classmates, and the environment. Always follow your teacher's directions for cleanup and disposal. Clean your small-scale reaction surface by draining the contents onto a paper towel. Then wipe the surface with a damp paper towel and dry the surface completely. Dispose of the paper towels in the waste bin.

15. Wash your hands thoroughly with soap and water before leaving the laboratory.

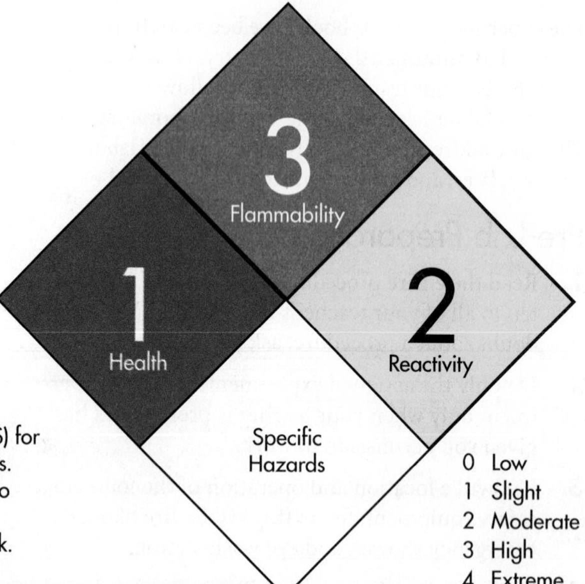

A Materials Safety Data Sheet (MSDS) for a chemical describes any safety issues. A diagram summarizes risks related to flammability, health, and reactivity. A number scale indicates the level of risk.

Take appropriate precautions when any of the following safety symbols appears in an experiment.

Safety Symbols

Eye Safety Wear safety goggles.

Clothing Protection Wear a lab coat or apron when using corrosive chemicals or chemicals that can stain clothing.

Skin Protection Wear plastic gloves when using chemicals that can irritate or stain your skin.

Broken Glass Do not use chipped or cracked glassware. Do not heat the bottom of a test tube.

Open Flame Tie back hair and loose clothing. Never reach across a lit burner.

Flammable Substance Do not have a flame near flammable materials.

Corrosive Substance Wear safety goggles, an apron, and gloves when working with corrosive chemicals.

Poison Don't chew gum, drink, or eat in the laboratory. Never taste a chemical in the laboratory.

Fume Avoid inhaling substances that can irritate your respiratory system.

Thermal Burn Do not touch hot glassware or equipment.

Electrical Equipment Keep electrical equipment away from water or other liquids.

Sharp Object To avoid a puncture wound, use scissors or other sharp objects only as intended.

Disposal Dispose of chemicals only as directed.

Hand Washing Wash your hands thoroughly with soap and water.

Appendix D Problem Bank

Chapter 2

1. Are intensive or extensive properties most helpful in identifying a substance? Explain your choice.

2. Classify the following mixtures as homogeneous or heterogeneous.

a. a toaster

b. a calculator

c. the air on a clear day

d. a copper coin

3. Identify the following as a mixture or a substance.

a. tomato juice

b. a rusty, iron fence post

c. a drop of mercury

d. a milkshake

4. Classify the following as a physical change or a chemical change.

a. salad dressing separates into layers after standing

b. spilled acid burns a hole in cotton jeans

c. alcohol freezes

d. ice sublimes

5. Classify the following properties of an aluminum bar as either physical or chemical.

a. burns in pure oxygen

b. melts at 660°C

c. bends easily

d. is nonmagnetic

Chapter 3

6. How many significant figures are in each measurement?

a. 786.32 mg	e. 0.0500 s
b. 0.0004 s	f. 66.066 mg
c. 5.060 kg	g. 4000 m
d. 7006.0 g	h. 40.0×10^3 m

7. Write each of the measurements in Problem 6 in scientific notation.

8. What is the total mass of three gold bars that weigh 5543 mg, 23.45 mg, and 697.4 mg?

9. The normal daily high temperature on the planet Zork is −85°C. Express this temperature in kelvins.

10. Methane, a gas that contributes to global warming, has a density of 0.714 g/L. What is the mass, in grams, of 25.0 L of methane?

11. The density of zinc is 9.394 g/cm³ at 20°C. What is the volume (in cubic centimeters) of a sphere of zinc metal that has a mass of 15.6 g?

12. Calculate the mass in kilograms of 964 mL of the element mercury. The density of mercury is 13.6 g/mL.

13. Make the following conversions:

a. 55 mg to grams

b. 5.76 dL to liters

c. 0.96 m to micrometers

d. 5.26 ns to seconds

e. 87 kg to milligrams

f. 846 mmol to centimoles

g. 3.4 nm to picometers

h. 6.66×10^3 kg to megagrams

i. 2.34×10^{-5} mL to microliters

14. When you donate a unit of blood to the Red Cross, you "give" about 0.55 L of blood. How many cubic centimeters (cm³) of blood is this?

15. The recommended daily amount (RDA) of vitamin B_6 for adults is 0.2 cg. How many micrograms of vitamin B_6 should be consumed each day?

16. A person develops jaundice (characterized by yellowing of the skin) when the concentration of bilirubin in his or her blood is 18 mg bilirubin per liter of blood. Assuming a total blood volume of 5.2 L, what is the number of grams of bilirubin in the person's blood?

Appendix D Problem Bank

Answers

Chapter 2

1. Intensive properties: because they depend on the composition of the sample, not the amount of matter in the sample.

2. a. heterogeneous b. heterogeneous c. homogeneous d. homogeneous

3. a. mixture b. mixture c. substance d. mixture

4. a. physical change b. chemical change c. physical change d. physical change

5. a. chemical b. physical c. physical d. physical

Chapter 3

6. a. 5 b. 1 c. 4 d. 5 e. 3 f. 5 g. 1 h. 3

7. a. 7.8632×10^2 b. 4×10^{-4} c. 5.060×10^0 d. 7.0060×10^3 e. 5.00×10^{-2} f. 6.6066×10^1 g. 4×10^3 h. 4.00×10^4

8. 6264 mg

9. 188 K

10. 17.9 g

11. 1.66 cm³

12. 13.1 kg

13. a. 5.5×10^{-2} g b. 5.76×10^{-1} L c. 9.6×10^5 µm d. 5.26×10^{-9} s e. 8.7×10^7 mg f. 8.46×10^1 cmol g. 3.4×10^3 pm h. 6.66 Mg i. 2.34×10^{-2} µL

14. 5.5×10^2 cm³

15. 2×10^3 µg

16. 9.4×10^{-2} g

APPENDIX D

17. 4.3×10^2 µg

18. 4.26 bottles

19. 2.4×10^5 mg

20. 3.8×10^7 beats/year

21. 2.42 kg

22. 11.6 days

23. **a.** 8.0×10^{-5} cm **b.** 8.0×10^{-1} µm **c.** 8.0×10^{-6} dm **d.** 8.0×10^{-4} mm

24. 0.43 s

25. 2.50×10^3 kg

26. 26 m

27. 4 days

28. 3 capsules/dose

29. **a.** 1.03×10^3 packs **b.** 1.41 years

30. **a.** 18.0 g **b.** $31.50

31. 52.8 cm³ sodium

32. 4.95×10^3 kg

33. 3.93 kg

34. 71.5 kg of air

35. **a.** 1×10^6 cm³ **b.** 1×10^3 cm³ **c.** 5×10^{-21} cm³ **d.** 2×10^{12} cm³

Chapter 4

36. **a.** not equal **b.** equal **c.** equal **d.** not equal

37. **a.** 23 protons, 23, electrons, 28 neutrons

b. 13 protons, 13 electrons, 14 neutrons

c. 50 protons, 50 electrons, 70 neutrons

d. 72 protons, 72 electrons, 106 neutrons

17. Mites are small eight-legged "bugs" in the same family as spiders. A particularly small mite has a mass of 0.0043 dg. What is the mass of this mite expressed in micrograms?

18. In the United States, a typical can of soda holds 355 mL. How many 2.00-L bottles could be filled from a 24-can case of soda?

19. You are going to carry out a chemical reaction in which you need 16 g of oxygen for every 7.0 g of nitrogen that will be used. If you have 0.554 kg of oxygen, how many milligrams of nitrogen do you need?

20. If your heart beats at an average rate of 72 times per minute, how many times will your heart beat each year?

21. Four empty beakers weigh a total of 1.84 kg. Each beaker when full holds 0.75 kg of water. How much do two full beakers of water weigh?

22. How many days would it take you to count a million pennies if you could count one penny each second? Express the answer to 3 significant figures.

23. A soap bubble film is 8.0×10^2 nm thick. Express this thickness in the following units:

a. centimeters **c.** decimeters
b. micrometers **d.** millimeters

24. The pitcher's mound on a regulation baseball field is 60 feet 6 inches from home plate. How many seconds does it take a 96 mph fastball to reach home plate? (5280 ft = 1 mile)

25. Gemstones such as diamonds are measured in carats, where 1 carat = 2.00 dg. How many kilograms of diamonds were produced if 12.5 million carats of diamonds were mined in a recent year?

26. How many meters does a car moving at 95 km/hour travel in 1.0 second?

27. A milliliter of water is equal to 20 drops of water. If water is dripping from a faucet at the rate of 7 drops per minute, how many days will it take to completely fill a 2.00-L soda bottle?

28. A prescription for a certain drug calls for a dose of 0.200 mg/kg of body weight, four times a day. The drug is packaged in capsules of 5 mg. How many capsules per dose should be given to a patient who weighs 75 kg?

29. A certain low-tar cigarette contains 11.0 mg of tar per cigarette.

a. If all the tar gets into the lungs, how many packs of cigarettes (20 cigarettes per pack) would have to be smoked to produce 0.500 lb of tar? (454 g = 1 lb)

b. If a person smoked two packs per day, how many years would it take to accumulate 0.500 lb of tar?

30. A chemist needs 25.0 mL of a liquid compound.

a. What mass of the compound is necessary if the density is 0.718 g/cm³?

b. If the compound costs $1.75/gram, what is the cost of this amount of the compound?

31. What volume of sodium has the same mass as 22.0 cm³ of silicon? The density of sodium is 0.97 g/cm³; the density of silicon is 2.33 g/cm³.

32. What is the mass, in kilograms, of a block of platinum that measures 23.0 cm by 78.4 cm by 122 cm? The density of platinum is 22.5 g/cm³.

33. Sulfuric acid sold for laboratory use consists of 96.7% sulfuric acid, H_2SO_4, by mass. The density of the solution is 1.845 g/cm³. Compute the number of kilograms of H_2SO_4 in a 2.20-L bottle of laboratory sulfuric acid.

34. How many kilograms of dry air are in a room that measures 15.0 ft by 18.0 ft by 8.00 ft? Use an average density of air of 1.168 g/L. There are 30.48 cm in one foot.

35. Calculate the number of cubic centimeters in each of the following:

a. 1 m³ **c.** 5 nm³
b. 1 dm³ **d.** 2×10^{-3} km³

Chapter 4

36. In which of these atom(s) is the number of protons equal to the number of neutrons?

a. germanium-72 **c.** silicon-28
b. calcium-40 **d.** hydrogen-1

37. Give the total number of subatomic particles (protons, electrons, and neutrons) in each atom.

a. vanadium-51 **c.** tin-120
b. aluminum-27 **d.** hafnium-178

38. Identify the element name and mass number of an atom with the given composition.

a. 42 protons, 56 neutrons, 42 electrons

b. 2 protons, 1 neutron, 2 electrons

c. 76 protons, 113 neutrons, 76 electrons

d. 31 protons, 40 neutrons, 31 electrons

39. Use the mass and percent abundance of the four isotopes of strontium to calculate the atomic mass of strontium.

Isotope	Mass (amu)	Abundance (%)
Strontium-84	83.193	0.560
Strontium-86	85.909	9.86
Strontium-87	86.908	7.00
Strontium-88	87.906	82.58

40. An atom of carbon and an atom of element Z together weigh 6 amu less than double the weight of an atom of oxygen. If an atom of oxygen weighs 16 amu and an atom of carbon weighs 12 amu, what does an atom of element Z weigh?

Chapter 5

41. What is the maximum number of electrons in each of the following?

a. the fourth energy level

b. the 5*p* energy sublevel

c. a single 4*f* orbital

d. the first three energy levels

42. Write the electron configuration of each atom.

a. nickel

b. sulfur

c. arsenic

d. rubidium

43. Identify the symbols of the elements with the following electron configurations:

a. $1s^2 2s^2 2p^6 3s^2 3p^6 3d^{10} 4s^1$

b. $1s^2 2s^2 2p^6 3s^2 3p^5$

c. $1s^2 2s^2 2p^6 3s^2 3p^6 3d^{10} 4s^2 4p^6 4d^2 5s^2$

d. $1s^2 2s^2 2p^6 3s^2 3p^6 3d^{10} 4s^2 4p^6$

44. How many electrons are in the

a. third energy level of an indium atom?

b. second energy level of an oxygen atom?

c. third energy level of a vanadium atom?

d. first energy level of a barium atom?

45. Calculate the wavelength (in meters) of each of these frequencies of electromagnetic radiation.

a. 9.82×10^{19}/s

b. 2.24×10^{14}/s

c. 5.31×10^{7}/s

d. 7.78×10^{10}/s

46. Order the wavelengths in Problem 45 from highest to lowest energy.

Chapter 6

47. Based on their relative positions on the periodic table, which atom in each pair has the smaller atomic radius?

a. Na, K

b. Cl, Br

c. K, Br

d. Ne, Na

48. Based on their relative positions on the periodic table, which atom in each pair has the greater electronegativity?

a. B, C

b. Na, Al

c. Li, Cs

d. As, F

49. Based on their relative positions on the periodic table, which atom in each pair has the highest first ionization energy?

a. F, Br

b. Li, F

c. Ca, Be

d. K, Ar

38. a. molybdenum-98 **b.** helium-3
c. osmium-189 **d.** gallium-71

39. 87.62 amu

40. 14 amu

Chapter 5

41. a. 32 electrons **b.** 6 electrons
c. 2 electrons **d.** 28 electrons

42 a. $1s^2 2s^2 2p^6 3s^2 3p^6 3d^8 4s^2$

b. $1s^2 2s^2 2p^6 3s^2 3p^4$

c. $1s^2 2s^2 2p^6 3s^2 3p^6 3d^{10} 4s^2 4p^3$

d. $1s^2 2s^2 2p^6 3s^2 3p^6 3d^{10} 4s^2 4p^6 5s^1$

43. a. Cu **b.** Cl
c. Zr **d.** Kr

44. a. 18 electrons **b.** 6 electrons
c. 11 electrons **d.** 2 electrons

45. a. 3.05×10^{-12} m **b.** 1.34×10^{-6} m
c. 5.65 m **d.** 3.85×10^{-3} m

46. highest to lowest: a, b, d, c

Chapter 6

47. a. Na **b.** Cl
c. Br **d.** Ne

48. a. C **b.** Al
c. Li **d.** F

49. a. F **b.** F
c. Be **d.** Ar

APPENDIX D

50. Element X, group 2A; element Y, group 1A

51. a. smaller **b.** smaller
c. larger **d.** larger

Chapter 7

52. The number of valence electrons remains the same.

53. a. 2 **b.** 3
c. 3 **d.** 2

54. K^+ and Ca^{2+}

55. MX_2 (M with 2 dots) and MX_3 (M with 3 dots) are valid formulas

56. a. 78 **b.** 21
c. 54 **d.** 10

Chapter 8

57. a. H:T̈e:H (with a pair of dots above and below Te)

b. H:Äs:H with H bonded below As

c. Si at the centre with Br above and below it, written Br:S̈i:Br

d. :Ï:Ï: (two I atoms, each with dot pairs)

58. a. 10 **b.** 26
c. 26 **d.** 18

59. a. bent **b.** pyramidal
c. tetrahedral **d.** linear

60. a. polar **b.** polar
c. nonpolar **d.** nonpolar

Chapter 9

61. a. molecular **b.** ionic
c. ionic **d.** ionic
e. ionic **f.** molecular
g. molecular **h.** ionic

62. a. silicon tetrachloride **b.** PI_3
c. dibromine heptoxide **d.** IF
e. bromine pentafluoride **f.** As_2O_3
g. nitrogen trichloride **h.** P_2O_5

50. Here are the first, second, and third ionization energies (kJ/mol) respectively for the representative elements "X" and "Y."

Element	First ionization energy (kJ/mol)	Second ionization energy (kJ/mol)	Third ionization energy (kJ/mol)
X	738	1450	7732
Y	496	4565	6912

In what group on the periodic table would these elements most likely be found?

51. Would the ion formed from each element be larger or smaller than the atom from which it was formed?

a. calcium

b. aluminum

c. bromine

d. nitrogen

Chapter 7

52. In any group of representative elements on the periodic table, how does the number of valence electrons vary as the elements within the group increase in mass?

53. How many valence electrons are lost by the metallic element when forming each of these ionic compounds?

a. BaS

b. In_2Se_3

c. GaP

d. SrI_2

54. Write the formulas for two cations of representative elements that have the electron configuration $1s^22s^22p^63s^23p^6$.

55. If "X" is the formula for any halogen and "M" is the formula for any metal, which of these is a valid formula of an ionic compound formed between "M" and "X"? What is the electron dot structure for "M" in each of the compounds that can be formed?

a. MX_2

b. M_2X_2

c. MX_3

d. M_2X_3

56. How many electrons are in each ion?

a. Pb^{4+} **c.** Te^{2-}

b. Cr^{3+} **d.** C^{4-}

Chapter 8

57. Draw an electron dot structure for each substance.

a. H_2Te

b. AsH_3

c. $SiBr_4$

d. I_2

58. How many electrons are in the electron dot structure for each of these polyatomic ions?

a. cyanide, CN^-

b. bromate ion, BrO_3^-

c. phosphite ion, PO_3^{3-}

d. nitrite, NO_2^-

59. Use VSEPR theory to predict the shape of each molecule in Problem 57.

60. Classify each of the molecules in Problem 57 as polar or nonpolar.

Chapter 9

61. Classify each of these compounds as molecular or ionic.

a. CF_4

b. PtO_2

c. SrI_2

d. NH_4Br

e. K_2CO_3

f. NI_3

g. $C_5H_{10}O_5$

h. $Ba(OH)_2$

62. Name or write the formulas for these molecular compounds.

a. $SiCl_4$

b. phosphorus triiodide

c. Br_2O_7

d. iodine monofluoride

e. BrF_5

f. diarsenic trioxide

g. NCl_3

h. diphosphorus pentoxide

63. Write the formulas for these ionic compounds.

a. barium iodide

b. iron(III) acetate

c. potassium dichromate

d. ammonium bromide

e. cesium nitride

f. cobalt(III) nitrate

g. aluminum oxalate

h. mercurous chloride

64. Name these ionic compounds.

a. Rb_2S e. $HgCl_2$

b. LiI f. $CuClO_3$

c. $Pb(C_2H_3O_2)_2$ g. NaCN

d. Mg_3N_2 h. $Cr(ClO_4)_3$

65. Name these compounds.

a. Cs_2O e. H_2CrO_4

b. SnS_2 f. CaC_2O_4

c. N_4S_4 g. $(NH_4)_3PO_4$

d. B_2O_3 h. As_4O_{10}

66. Write the formulas for these compounds.

a. calcium oxide

b. sulfurous acid

c. diboron tetrachloride

d. calcium hydrogen phosphate

e. tin(II) chromate

f. ferric hydroxide

g. manganese(II) chlorite

h. iodine monochloride

67. Explain why it is not fair to be asked to write a formula for each named "compound."

a. iron bromide

b. sulfur oxide

c. lead hypochlorite

d. phosphorus chloride

68. Iron forms two compounds with oxygen. One compound consists of 1.396 g of iron and 0.400 grams of oxygen. The other has 0.582 g iron and 0.250 g oxygen. Show by calculation whether this pair of compounds obeys the law of multiple proportions.

Chapter 10

69. How many of each kind of atom are in a formula unit of each compound?

a. $(NH_4)_2SO_3$

b. $AlPO_4$

c. $Ca(C_2H_3O_2)_2$

d. $Fe_2(SO_4)_3$

70. How many of each kind of atom are in a molecule of each compound?

a. $C_3H_7O_2$

b. $C_3H_5(OH)_3$

c. $C_2H_4(COOH)_2$

d. $C_7H_5(NO_3)_3$

71. Calculate the molar mass of each of these binary ionic compounds.

a. MgO c. Hg_2I_2

b. $AlCl_3$ d. Sr_3N_2

72. Calculate the molar mass of each of these ionic compounds.

a. $(NH_4)_2C_2O_4$

b. $Ca(OH)_2$

c. Na_2HPO_4

d. $Mg(HSO_4)_2$

73. Calculate the molar mass of each of these molecular compounds.

a. N_2O_5

b. C_3H_7OH

c. SO_3

d. XeF_6

74. Calculate the molar mass of each of these compounds.

a. DEET, $C_{12}H_{17}ON$, an insect repellent

b. aspartame, $C_{14}H_{18}N_2O_5$, a sugar substitute

c. codeine, $C_{18}H_{21}NO_3$, an analgesic (painkiller)

d. sodium benzoate, $NaC_7H_5O_2$, a food preservative

75. What is the mass, in grams, of each of the following?

a. 5.000 mol Ar

b. 1.64 mol $NaNO_2$

c. 0.886 mol $(NH_4)_2SO_4$

d. 18.3 mol SiF_4

63. a. BaI_2 **b.** $Fe(C_2H_3O_2)_3$
c. $K_2Cr_2O_7$ **d.** NH_4Br
e. Cs_3N **f.** $Co(NO_3)_3$
g. $Al_2(C_2O_4)_3$ **h.** Hg_2Cl_2

64. a. rubidium sulfide **b.** lithium iodide
c. lead(II) acetate **d.** magnesium nitride
e. mercury(II) chloride **f.** copper(I) chlorate
g. sodium cyanide **h.** chromium(III) perchlorate

65. a. cesium oxide **b.** tin(II) sulfide
c. tetranitrogen tetrasulfide **d.** diboron trioxide
e. chromic acid **f.** calcium oxalate
g. ammonium phosphate **h.** tetraarsenic decoxide

66. a. CaO **b.** H_2SO_3
c. B_2Cl_4 **d.** $CaHPO_4$
e. $SnCrO_4$ **f.** $Fe(OH)_3$
g. $Mn(ClO_2)_2$ **h.** ICl

67. a. The ionic charge of iron must be given by a Roman numeral.
b. Prefixes must be used in the name of a molecular compound.
c. The ionic charge of lead must be given by a Roman numeral.
d. Prefixes must be used in the name of a molecular compound.

68. The ratio of masses of iron per gram of oxygen is 3:2, so these compounds obey the law of multiple proportions.

Chapter 10

69. a. N, 2; H, 8; S, 1; O, 3 **b.** Al, 1; P, 1; O, 4
c. Ca, 1; C, 4; H, 6; O, 4 **d.** Fe, 2; S, 3; O, 12

70. a. C, 3; H, 7; O, 2 **b.** C, 3; H, 8; O, 3
c. C, 4; H, 6; O, 4 **d.** C, 7; H, 5; N, 3; O, 9

71. a. 40.3 g **b.** 133.5 g
c. 655.0 g **d.** 217.2 g

72. a. 124.0 g **b.** 74.1 g
c. 142.0 g **d.** 218.5

73. a. 108.0 g **b.** 60.0 g
c. 80.1 g **d.** 245.3 g

74. a. 191.0 g **b.** 294.0 g
c. 299.0 g **d.** 144.0 g

75. a. 199.5 g **b.** 113 g
c. 117 g **d.** 1.91×10^3 g

APPENDIX D

76. a. 2.97 mol b. 9.26×10^{-4} mol
c. 0.692 mol d. 29.5 mol

77. a. 2.65×10^{24} molecules
b. 1.71×10^{23} formula units
c. 9.75×10^{23} molecules
d. 7.71×10^{24} formula units

78. a. 0.121 mol b. 2.46 mol
c. 0.664 mol d. 15.0 mol

79. a. 81.5 L b. 1.45 L
c. 189 L d. 28.2 L

80. a. 2.52 mol b. 0.341 mol
c. 3.96×10^{-2} mol d. 5.63 mol

81. a. 2.00×10^{23} molecules
b. 2.16×10^{23} atoms
c. 6.84×10^{21} formula units
d. 2.37×10^{23} formula units

82. a. 1.86×10^{-12} g b. 6.43 g
c. 55 g d. 1.33×10^{-2} g

83. a. 3.49 g b. 555 g
c. 279 g d. 5.7×10^{-3} g

84. a. 0.148 L b. 33.5 L
c. 0.813 L d. 54.9 L

85. a. 1.72×10^{24} molecules
b. 9.03×10^{22} molecules
c. 1.28×10^{27} molecules
d. 1.85×10^{21} atoms

86. a. 0.136 L b. 2.27 L
c. 432 L d. 167 L

87. a. 63 oxygen atoms
b. 3.61×10^{24} oxygen atoms
c. 1.08×10^{23} oxygen atoms
d. 1.82×10^{23} oxygen atoms

88. a. 23 g b. 2.50 g
c. 222 g d. 7.72 g

89. a. 86.6% Pb, 13.4% O
b. 62.1% C, 10.3% H, 27.6% O
c. 18.3% K, 59.3% I, 22.4% O
d. 29.1% Na, 40.6% S, 30.3% O
e. 57.2% I, 42.8% F
f. 0.69% H, 55.1% Br, 44.2% O
g. 56.4% P, 43.6% O
h. 54.5%C, 9.1% H, 36.4 % C

76. How many moles is each of the following?
a. 579 g Pt
b. 0.0426 g NO_2
c. 56.8 g H_2SO_3
d. 6.78×10^3 g CsH_2PO_4

77. Find the number of representative particles in each of the following:
a. 4.40 mol Pd
b. 0.284 mol NaI
c. 1.62 mol NH_3
d. 12.8 mol $Fe(C_2H_3O_2)_2$

78. How many moles is each of the following?
a. 7.26×10^{22} atoms Zr
b. 1.48×10^{24} molecules C_2H_6O
c. 4.00×10^{23} formula units $KClO_3$
d. 9.02×10^{24} molecules OF_2

79. Calculate the volume, in liters, of each of these gases at STP.
a. 3.64 mol H_2
b. 0.0648 mol C_2H_6
c. 8.44 mol SO_3
d. 1.26 mol Xe

80. How many moles is each of the following at STP?
a. 56.4 L He
b. 7.64 L N_2
c. 0.888 L CO
d. 126 L SO_2

81. Calculate the number of representative particles in each mass.
a. 14.6 g CO_2
b. 68.3 g Os
c. 0.847 g KCl
d. 174 g Au_2O_3

82. Calculate the mass of each of the following samples.
a. 7.00×10^9 molecules Br_2
b. 9.22×10^{22} formula units NaF
c. 4.8×10^{24} atoms Li
d. 2.66×10^{20} molecules H_2CO

83. Find the mass of each of the gases at STP.
a. 2.44 L O_2 c. 78.0 L SO_3
b. 777 L CH_4 d. 0.0642 L H_2

84. Calculate the volume of each of these gases at STP.
a. 0.469 g Cl_2
b. 44.8 g NO
c. 2.76 g N_2O_3
d. 93.2 g F_2

85. Calculate the number of representative particles in each volume.
a. 64.0 L H_2S
b. 3.36 L C_3H_8
c. 4.78×10^4 L HF
d. 6.88×10^{-2} L Kr

86. Find the volume at STP of the following:
a. 3.66×10^{21} molecules F_2
b. 6.11×10^{22} molecules PH_3
c. 1.16×10^{25} atoms Ne
d. 4.48×10^{24} molecules C_2H_2

87. Calculate the number of oxygen atoms in each of the following:
a. 7 molecules of the explosive nitroglycerine, $C_3H_5(NO_3)_3$
b. 3.00 mol of the antiseptic hydrogen peroxide, H_2O_2
c. a balloon filled with 2.00 L O_2
d. 8.04 g of the fertilizer, NH_4NO_3

88. Calculate the number of grams of hydrogen in each of the following.
a. a balloon filled with 7.06×10^{24} hydrogen molecules
b. a balloon filled with 14.0 L of methane, CH_4, at STP
c. a 2.00-L bottle of water (density of H_2O = 1.00 g/mL)
d. a 69.5-g ice cube (density of ice = 0.917 g/cm^3)

89. Calculate the percent composition of each compound.
a. PbO_2
b. $(CH_3)_2CO$
c. KIO_3
d. $Na_2S_2O_3$
e. IF_5
f. $HBrO_4$
g. P_4O_6
h. C_3H_7COOH

R56 Appendix D

90. Use the answers from Problem 89 to calculate the number of grams of the indicated element in the compound.

a. lead in 63.8 g PbO_2
b. carbon in 1.664 g $(CH_3)_2CO$
c. oxygen in 36.8 g KIO_3
d. sulfur in 6.26 g $Na_2S_2O_3$
e. fluorine in 594 g IF_5
f. bromine in 82.7 g $HBrO_4$
g. phosphorus in 2.66 g P_4O_6
h. carbon in 55.0 g C_3H_7COOH

91. Which of these are empirical formulas?

a. $Al_2(SO_4)_3$
b. $C_6H_4Cl_2$
c. $C_2H_4(OH)_2$
d. $K_2Cr_2O_7$

92. What is the empirical formula of

a. $C_6H_{16}N_2$, a compound used to make nylon?
b. $C_6H_8N_2$, a component of chocolate?
c. C_8H_8, used to make polystyrene foam plastics?
d. C_3H_7OH, rubbing alcohol?

93. Determine the empirical formula for each compound from the percent composition data.

a. 85.71% C, 14.29% H
b. 60.94% Ba, 10.65% C, 28.41% O
c. 37.50% C, 12.50% H, 50.00% O
d. 27.87% P, 72.13% S
e. 67.61% U, 32.39% F
f. 74.19% Na, 25.79% O
g. 32.43% C, 5.41% H, 43.24% O, 18.92% N
h. 18.70% Li, 16.26% C, 65.04% O

94. Find the molecular formula from the given empirical formula and molar mass.

Empirical formula	Molar mass (g/mol)	Molecular formula
C_2H_3	54.0	a. ______
C_2H_2Cl	123.0	b. ______
$C_3H_4O_3$	176.0	c. ______
C_5H_7N	162.0	d. ______

95. A compound with a molar mass of 312.2 g/mol contains 69.23% C, 3.85% H, and 26.92% N. What is the molecular formula of this compound?

96. The molar mass of caffeine, the stimulant found in coffee, is 194.0 g/mol. The percent composition of caffeine is 49.48% C, 5.19% H, 28.85% N, and 16.48% O. What is the molecular formula of caffeine?

97. Linoleic acid, which has a molar mass of 280.0 g/mol, is found in many vegetable oils. The percent composition of this compound is 77.1% carbon, 11.4% hydrogen, and 11.4% oxygen. Find the empirical formula and molecular formula of this compound.

98. A 2.716-g sample of a compound of C, H, N, and O was found to contain 0.7580 g C, 0.0633 g H, and 0.8843 g N. The molar mass of the compound is 129 g/mol. Calculate the compound's empirical and molecular formula.

Chapter 11

99. What is the function of the element platinum in this reaction?

$$2H_2 + O_2 \xrightarrow{Pt} 2H_2O$$

100. Balance the following equations:

a. $Hg(NO_3)_2 + NH_4SCN \longrightarrow Hg(SCN)_2 + NH_4NO_3$
b. $CH_4O + O_2 \longrightarrow CO_2 + H_2O$
c. $Ca + Cl_2 \longrightarrow CaCl_2$
d. $Na_3PO_4 + CoCl_2 \longrightarrow Co_3(PO_4)_2 + NaCl$
e. $Fe + AgNO_3 \longrightarrow Fe(NO_3)_2 + Ag$
f. $N_2H_4 \longrightarrow NH_3 + N_2$
g. $C_{12}H_{26} + O_2 \longrightarrow CO_2 + H_2O$
h. $CuCl + Mg \longrightarrow Cu + MgCl_2$

101. Classify each of the equations in Problem 100 by type.

102. Write balanced equations for each of these reactions. Indicate states of matter in your equations.

a. Potassium metal reacts with water to form hydrogen gas and aqueous potassium hydroxide.
b. Nitrogen monoxide gas reacts with gaseous carbon monoxide to form carbon dioxide gas and nitrogen gas.

90. a. 55.3 g Pb b. 1.03 g C
c. 8.24 g O d. 2.54 g S
e. 254 g F f. 45.6 g Br
g. 1.50 g P h. 30.0 g C

91. a. yes b. no
c. no d. yes

92. a. C_3H_8N b. C_3H_4N
c. CH d. C_3H_7OH

93. a. CH_2 b. BaC_2O_4
c. CH_4O d. P_2S_5
e. UF_6 f. Na_2O
g. $C_2H_4O_2N$ h. Li_2CO_3

94. a. C_4H_6 b. $C_4H_4Cl_2$
c. $C_6H_8O_6$ d. $C_{10}H_{14}N_2$

95. $C_{18}H_{12}N_6$

96. $C_8H_{10}N_4O_2$

97. empirical formula, $C_9H_{16}O$; molecular formula, $C_{18}H_{32}O_2$

98. empirical formula, CHNO; molecular formula, $C_3H_3N_3O_3$

Chapter 11

99. acting as a catalyst

100. a. $Hg(NO_3)_2 + 2NH_4SCN \rightarrow Hg(SCN)_2 + 2NH_4NO_3$
b. $2CH_4O + 3O_2 \rightarrow 2CO_2 + 4H_2O$
c. $Ca + Cl_2 \rightarrow CaCl_2$
d. $2Na_3PO_4 + 3CoCl_2 \rightarrow Co_3(PO_4)_2 + 6NaCl$
e. $Fe + 2AgNO_3 \rightarrow Fe(NO_3)_2 + 2Ag$
f. $3N_2H_4 \rightarrow 4NH_3 + N_2$
g. $2C_{12}H_{26} + 37O_2 \rightarrow 24CO_2 + 26H_2O$
h. $2CuCl + Mg \rightarrow 2Cu + MgCl_2$

101. a. double-replacement b. combustion
c. combination d. double-replacement
e. single-replacement f. decomposition
g. combustion h. single replacement

102. a. $2K(s) + 2H_2O(l) \rightarrow H_2(g) + 2KOH(aq)$
b. $2NO(g) + 2CO(g) \rightarrow 2CO_2(g) + N_2(g)$

c. $4HCl(aq) + O_2(g) \rightarrow 2H_2O(l) + 2Cl_2(g)$

d. $Ca(OH)_2(aq) + 2HC_2H_3O_2(aq) \rightarrow 2H_2O(l) + Ca(C_2H_3O_2)_2(aq)$

e. $3O_2(g) + 2PbS(s) \rightarrow 2SO_2(g) + 2PbO(s)$

f. $Li_2O(s) + H_2O(l) \rightarrow 2LiOH(aq)$

g. $MnO_2(s) + H_2C_2O_4(aq) \rightarrow MnO(s) + H_2O(l) + 2CO_2(g)$

h. $B_2H_6(g) + 3O_2(g) \rightarrow 3H_2O(l) + B_2O_3(s)$

103. a. $2HCl(aq) \xrightarrow{electricity} H_2(g) + Cl_2(g)$

b. $3Br_2(l) + 2AlI_3(aq) \rightarrow 2AlBr_3(aq) + 3I_2(aq)$

c. $2Na(s) + S(s) \rightarrow Na_2S(s)$

d. $Ba(OH)_2(aq) + 2HNO_3(aq) \rightarrow 2H_2O(l) + Ba(NO_3)_2(aq)$

e. $2C_7H_{14}O_2(l) + 19O_2(g) \rightarrow 14H_2O(l) + 14CO_2(g)$

f. $Ni(NO_3)_2(aq) + Na_2CO_3(aq) \rightarrow NiCO_3(s) + 2NaNO_3(aq)$

104. a. $MnO_2 + 4HCl \rightarrow MnCl_2 + Cl_2 + 2H_2O$

b. $PCl_5 + 4H_2O \rightarrow H_3PO_4 + 5HCl$

c. $Ca_3P_2 + 6H_2O \rightarrow 2PH_3 + 3Ca(OH)_2$

d. $Li_3N + 3H_2O \rightarrow 3LiOH + NH_3$

e. $2H_2O_2 + N_2H_4 \rightarrow N_2 + 4H_2O$

f. $SiCl_4 + 2Mg \rightarrow 2MgCl_2 + Si$

g. $V_2O_5 + 2H_2 \rightarrow V_2O_3 + 2H_2O$

h. $HBr + KHSO_3 \rightarrow KBr + H_2O + SO_2$

105. a. $BaSO_4$ b. $CrCO_3$
c. no precipitate formed d. HgS

106. a. $3Ni^{2+}(aq) + 6Cl^-(aq) + 6K^+(aq) + 2PO_4^{3-}(aq) \rightarrow Ni_3(PO_4)_2(s) + 6Cl^-(aq) + 6K^+(aq)$

b. $2H^+(aq) + 2C_2H_3O_2^-(aq) + Ca^{2+}(aq) + 2OH^-(aq) \rightarrow 2H_2O(l) + Ca^{2+}(aq) + 2C_2H_3O_2^-(aq)$

c. $Ca^{2+}(aq) + 2I^-(aq) + 2Na^+(aq) + SO_4^{2-}(aq) \rightarrow CaSO_4(s) + 2I^-(aq) + 2Na^+(aq)$

d. $2Na^+(aq) + 2OH^-(aq) + Pb^{2+}(aq) + 2NO_3^-(aq) \rightarrow Pb(OH)_2(s) + 2Na^+(aq) + 2NO_3^-(aq)$

107. a. $Cl^-(aq) + K^+(aq)$ b. $Ca^{2+}(aq) + C_2H_3O_2^-(aq)$
c. $I^-(aq) + Na^+(aq)$ d. $Na^+(aq) + NO_3^-(aq)$

108. a. $3Ni^{2+}(aq) + 2PO_4^{3-}(aq) \rightarrow Ni_3(PO_4)_2(s)$

b. $2H^+(aq) + 2OH^-(aq) \rightarrow 2H_2O(l)$

c. $Ca^{2+}(aq) + SO_4^{2-}(aq) \rightarrow CaSO_4(s)$

d. $Pb^{2+}(aq) + 2OH^-(aq) \rightarrow Pb(OH)_2(s)$

c. Hydrochloric acid reacts with oxygen gas to form liquid water and chlorine gas.

d. Aqueous calcium hydroxide reacts with acetic acid to form water and aqueous calcium acetate.

e. Oxygen gas reacts with solid lead(II) sulfide to form sulfur dioxide gas and lead(II) oxide.

f. Solid lithium oxide reacts with water to form aqueous lithium hydroxide.

g. Solid manganese dioxide reacts with oxalic acid to form solid manganese(II) oxide, water, and gaseous carbon dioxide.

h. Gaseous diboron hexahydride reacts with oxygen gas to form liquid water and solid diboron trioxide.

103. Complete and then balance each of these equations.

a. $HCl(aq) \xrightarrow{electricity}$

b. $Br_2(l) + AlI_3(aq) \longrightarrow$

c. $Na(s) + S(s) \longrightarrow$

d. $Ba(OH)_2(aq) + HNO_3(aq) \longrightarrow$

e. $C_7H_{14}O_2(l) + O_2(g) \longrightarrow$

f. $Ni(NO_3)_2(aq) + Na_2CO_3(aq) \longrightarrow$

104. Balance each of these equations.

a. $MnO_2 + HCl \longrightarrow MnCl_2 + Cl_2 + H_2O$

b. $PCl_5 + H_2O \longrightarrow H_3PO_4 + HCl$

c. $Ca_3P_2 + H_2O \longrightarrow PH_3 + Ca(OH)_2$

d. $Li_3N + H_2O \longrightarrow LiOH + NH_3$

e. $H_2O_2 + N_2H_4 \longrightarrow N_2 + H_2O$

f. $SiCl_4 + Mg \longrightarrow MgCl_2 + Si$

g. $V_2O_5 + H_2 \longrightarrow V_2O_3 + H_2O$

h. $HBr + KHSO_3 \longrightarrow KBr + H_2O + SO_2$

105. Use Table 11.3 to predict whether a precipitate will form when aqueous solutions of these pairs of salts are mixed. If a precipitate forms, write its formula.

a. ammonium sulfate and barium bromide

b. chromium(II) chloride and lithium carbonate

c. potassium nitrate and sodium chloride

d. sodium sulfide and mercury(II) nitrate

106. Write a balanced complete ionic equation for each of these double-replacement reactions. All the reactants are in aqueous solution.

a. nickel(II) chloride + potassium phosphate

b. acetic acid + calcium hydroxide

c. calcium iodide + sodium sulfate

d. sodium hydroxide + lead(II) nitrate

107. Identify the spectator ions in each of the reactions in Problem 106.

108. Write net ionic equations for each of the reactions in Problem 106.

Chapter 12

109. Interpret each equation in terms of interacting particles.

a. $H_2 + F_2 \longrightarrow 2HF$

b. $2K_3PO_4 + 3CoCl_2 \longrightarrow Co_3(PO_4)_2 + 6KCl$

c. $2PbS + 3O_2 \longrightarrow 2PbO + 2SO_2$

d. $Fe + S \longrightarrow FeS$

110. Write all possible mole ratios for these equations.

a. $2NO + Cl_2 \longrightarrow 2NOCl$

b. $2KClO_3 \longrightarrow 2KCl + 3O_2$

c. $3N_2H_4 \longrightarrow 4NH_3 + N_2$

d. $2Na + O_2 \longrightarrow Na_2O_2$

111. Show by calculation that the following equations obey the law of conservation of mass:

a. $3NO_2 + H_2O \longrightarrow 2HNO_3 + NO$

b. $4HCl + O_2 \longrightarrow 2H_2O + 2Cl_2$

c. $2Li + S \longrightarrow Li_2S$

d. $2CH_4O + 3O_2 \longrightarrow 2CO_2 + 4H_2O$

112. Nitric acid, HNO_3, is produced by a process that allows nitrogen dioxide to react with water.

$$3NO_2(g) + H_2O(l) \longrightarrow 2HNO_3(aq) + NO(g)$$

a. How many moles of nitrogen dioxide, NO_2, are required to produce 3.56 mol of nitric acid?

b. How many moles of water react with 0.946 mol of nitrogen dioxide?

113. Calcium hydroxide reacts with nitric acid to produce an aqueous solution of calcium nitrate.

$$Ca(OH)_2(aq) + 2HNO_3(aq) \longrightarrow 2H_2O(l) + Ca(NO_3)_2(aq)$$

a. How many moles of calcium hydroxide are needed to react with 5.88 mol of nitric acid?

b. How many moles of calcium nitrate are produced when 2.30 mol of water are made in this reaction?

114. Chromium combines with oxygen to form chromium(III) oxide.

$$4Cr(s) + 3O_2(g) \longrightarrow 2Cr_2O_3(s)$$

a. How many moles of chromium are needed to react with 45.6 g of oxygen?

b. How many moles of chromium(III) oxide are produced when 2.86 g of chromium react?

115. Sodium hydroxide is formed when sodium oxide reacts with water.

$$Na_2O(s) + H_2O(l) \longrightarrow 2NaOH(aq)$$

a. Calculate the grams of sodium hydroxide formed when 2.24 moles of sodium oxide react with water.

b. What mass of water (in centigrams) is needed to react with 0.126 mol of sodium oxide?

116. The reaction of nitrogen monoxide with carbon monoxide produces carbon dioxide and nitrogen.

$$2NO(g) + 2CO(g) \longrightarrow 2CO_2(g) + N_2(g)$$

a. How many liters of nitrogen monoxide at STP are needed to produce 3.40 mol of nitrogen gas?

b. When 2.18 moles of nitrogen are made in this reaction, how many liters of carbon dioxide at STP are produced?

117. Hydrogen fluoride gas is produced directly from its component elements.

$$H_2(g) + F_2(g) \longrightarrow 2HF(g)$$

a. When 40.0 L of fluorine at STP reacts with an excess of hydrogen, how many moles of hydrogen fluoride are made?

b. How many moles of fluorine are needed to make 8.04 L of hydrogen fluoride at STP?

118. Rust (iron(III) oxide) is formed by the reaction of oxygen with iron.

$$4Fe(s) + 3O_2(g) \longrightarrow 2Fe_2O_3(s)$$

a. Calculate the mass of oxygen required to react with 10.0 g of iron.

b. How many grams of rust form when 2.48 g of iron reacts with an excess of oxygen?

119. Silver chloride precipitates when aqueous solutions of calcium chloride and silver nitrate are mixed.

$$CaCl_2(aq) + 2AgNO_3(aq) \longrightarrow 2AgCl(s) + Ca(NO_3)_2(aq)$$

a. How many grams of calcium nitrate are formed when 0.500 g of calcium chloride reacts with an excess of silver nitrate?

b. How many grams of calcium chloride are required to react completely with 34.8 g of silver nitrate?

120. The complete combustion of octane, a component of gasoline, forms carbon dioxide and water.

$$2C_8H_{18}(l) + 25O_2(g) \longrightarrow 16CO_2(g) + 18H_2O(l)$$

a. How many grams of C_8H_{18} must be reacted to give 5.00 g of CO_2?

b. How many liters of oxygen gas at STP are required to burn 2.20 g C_8H_{18}?

121. A precipitate of nickel(II) carbonate forms when aqueous solutions of sodium carbonate and nickel(II) nitrate are mixed.

$$Ni(NO_3)_2(aq) + Na_2CO_3(aq) \longrightarrow NiCO_3(s) + 2NaNO_3(aq)$$

a. How many grams of each reactant must be used to form 67.2 g of the precipitate?

b. When 1.88 g of nickel(II) carbonate are formed, how many grams of sodium nitrate are produced?

122. One way to make ethanol is to react ethene with water at high pressure.

$$C_2H_4(g) + H_2O(g) \longrightarrow C_2H_6O(l)$$

a. How many grams of each reactant are needed to produce 8.84 g of ethanol?

b. How many liters of ethanol are produced when 1.00 kg of ethene is reacted with an excess of water? The density of ethanol is 0.789 g/mL.

Chapter 12

109. a. One molecule of hydrogen reacts with one molecule of fluorine to form two molecules of hydrogen fluoride.

b. Two formula units of potassium phosphate react with three formula units of cobalt(II) chloride to form one formula unit of cobalt(II) phosphate and six formula units of potassium chloride.

c. Two formula units of lead(II) sulfide react with three molecules of oxygen to form two formula units of lead(II) oxide and two molecules of sulfur dioxide.

d. One atom of iron reacts with one atom of sulfur to form one formula unit of iron(II) sulfide.

110. a. 2 mol NO/1 mole Cl_2, 2 mol NO/2 mol NOCl. 1 mol Cl_2/2 mol NO, 1 mol Cl_2/2 mol NOCl, 2 mol NOCl/2 mol NO, 2 mol NOCl/1 mol Cl_2

b. 2 mol $KClO_3$/2 mol KCl, 2 mol $KClO_3$/3 mol O_2, 2 mol KCl/2 mol $KClO_3$, 2 mol KCl/3 mol O_2, 3 mol O_2/2 mol KCl, 3 mol O_2/2 mol $KClO_3$

c. 3 mol N_2H_4/4 mol NH_3, 3 mol N_2H_4/1 mol N_2, 4 mol NH_3/3 mol N_2H_4, 4 mol NH_3/1 mol N_2, 1 mol N_2/3 mol N_2H_4, 1 mol N_2/4 mol NH_3

d. 2 mol Na/1 mol O_2, 2 mol Na/1 mol Na_2O_2, 1 mol O_2/2 mol Na, 1 mol O_2/1 mol Na_2O_2, 1 mol Na_2O_2/1 mol O_2, 1 mol Na_2O_2/2 mol Na

111. a. 138.0 g + 18.0 g = 126.0 g + 30.0 g
156.0 g = 156.0 g

b. 146.0 g + 32.0 g = 36.0 g + 142.0 g
178.0 g = 178.0 g

c. 13.8 g + 32.1 g = 45.9 g
45.9 g = 45.9 g

d. 64.0 g + 96.0 g = 88.0 g + 72.0 g
160.0 g = 160.0 g

112. a. 5.34 mol NO_2 **b.** 0.315 mol H_2O

113. a. 2.94 mol $Ca(OH)_2$ **b.** 1.15 mol $Ca(NO_3)_2$

114. a. 1.90 mol Cr **b.** 0.0275 mol Cr_2O_3

115. a. 179 g NaOH **b.** 227 cg H_2O

116. a. 152 L NO **b.** 97.7 L CO_2

117. a. 3.57 mol HF **b.** 0.179 mol F_2

118. a. 4.30 g O_2 **b.** 3.55 g Fe_2O_3

119. a. 0.739 g $Ca(NO_3)_2$ **b.** 11.4 g $CaCl_2$

120. a. 1.62 g C_8H_{18} **b.** 5.40 L O_2

121. a. 103 g $Ni(NO_3)_2$ and 60.0 g Na_2CO_3

b. 2.69 g $NaNO_3$

122. a. 5.38 g C_2H_4, 3.46 g H_2O

b. 2.08 L C_2H_6O

Problem Bank **R59**

123. $Al_2(SO_4)_3(aq) + 6NaOH(aq) \rightarrow 2Al(OH)_3(s) + 3Na_2SO_4(aq)$

a. 4.36 g NaOH **b.** 20.8 g $Al(OH)_3$

124. a. 10.6 g H_2O_2 **b.** 5.53 L O_2

125. a. 5.11 kg HNO_3 **b.** 154 g NO

126. a. 169 g Al **b.** 7.11 mL $Fe(l)$

127. a. 45.2 g HCl, 9.92 g O_2

b. 19.2 L O_2

128. a. 1.46 g HF, 4.99 g $CaSO_4$

b. 1.95 kg CaF_2

129. a. $2N_2O_5 \rightarrow O_2 + 4NO_2$

b. 0.593 g O_2, 3.41 g NO_2

130. a. 2.38×10^3 g Cl_2 **b.** 201 dL Cl_2

131. a. $2C_2H_4O + O_2 \rightarrow 2C_2H_4O_2$

b. 197 g O_2

Chapter 12 (continued)

123. Balance the equation for the formation of aluminum hydroxide, a common ingredient in some antacid tablets.

$$Al_2(SO_4)_3(aq) + NaOH(aq) \longrightarrow Al(OH)_3(s) + Na_2SO_4(aq)$$

a. How many grams of sodium hydroxide are required to react completely with 6.22 g of aluminum sulfate?

b. When 32.0 grams of sodium hydroxide reacts with an excess of aluminum sulfate, how many grams of aluminum hydroxide are formed?

124. One source of elemental oxygen in the laboratory is the decomposition of hydrogen peroxide.

$$2H_2O_2(l) \longrightarrow 2H_2O(l) + O_2(g)$$

a. How many grams of hydrogen peroxide are needed to produce 5.00 g of oxygen?

b. When 16.8 g of hydrogen peroxide are decomposed, how many liters (at STP) of oxygen are produced?

125. One source of acid rain is the production of nitric acid from nitrogen dioxide and water in the atmosphere.

$$3NO_2(g) + H_2O(l) \longrightarrow NO(g) + 2HNO_3(aq)$$

a. How many kilograms of nitric acid are produced when 5.60 kg of nitrogen dioxide reacts with an excess of water?

b. Calculate the mass in grams of nitrogen monoxide produced when 0.648 kg of nitric acid is formed by this reaction.

126. Steel rails for trains are welded together with the liquid (molten) iron formed by the immense heat generated by this reaction.

$$2Al(s) + Fe_2O_3(s) \longrightarrow Al_2O_3(s) + 2Fe(l)$$

a. How many grams of aluminum are needed to react completely with 0.500 kg of iron(III) oxide?

b. How many milliliters of molten iron are produced when 80.0 g of iron(III) oxide are reacted with an excess of aluminum? The density of iron is 7.87 g/cm^3. Assume the densities of molten iron and solid iron are the same.

127. Chlorine gas is made by reacting oxygen with hydrochloric acid.

$$4HCl(aq) + O_2(g) \longrightarrow 2Cl_2(g) + 2H_2O(l)$$

a. How many grams of each of the reactants are required to produce 44.0 g of chlorine?

b. At STP, how many liters of oxygen are needed to react completely with 125 g of hydrochloric acid?

128. Hydrogen fluoride is made by reacting sulfuric acid with calcium fluoride.

$$H_2SO_4(l) + CaF_2(s) \longrightarrow 2HF(g) + CaSO_4(s)$$

a. How many grams of hydrogen fluoride and calcium sulfate are produced when 2.86 g of calcium fluoride reacts with an excess of sulfuric acid?

b. Calculate the number of kilograms of calcium fluoride that must react with an excess of sulfuric acid to produce 1.00 kg of hydrogen fluoride.

129. When solid dinitrogen pentoxide is heated, it produces oxygen and nitrogen dioxide.

a. Write the balanced equation for this reaction.

b. How many grams of each product are formed when 4.00 g of dinitrogen pentoxide is completely decomposed?

130. Laundry bleach (NaClO) is made by reacting chlorine with sodium hydroxide.

$$Cl_2(g) + 2NaOH(aq) \longrightarrow NaClO(aq) + NaCl(aq) + H_2O(l)$$

a. How many grams of chlorine must react with an excess of sodium hydroxide to produce 2.50 kg of sodium hypochlorite?

b. At room temperature, chlorine gas has a density of 2.95 g/L. How many dL of chlorine gas are needed to react completely with 66.8 g of sodium hydroxide?

131. Bubbling oxygen gas through liquid acetaldehyde (C_2H_4O) forms a single product, acetic acid.

a. Write a balanced equation for this reaction.

b. How many grams of oxygen are needed to react completely with 542 g of acetaldehyde?

132. Water is decomposed into its elements by an electric current.

a. Write a balanced equation for the reaction.

b. What is the total volume (in liters at STP) of gases produced when 222 g of water are decomposed?

133. Oxygen is generated in a rebreathing gas mask by a reaction of water vapor with potassium superoxide, KO_2.

$$4KO_2(s) + 2H_2O(l) \longrightarrow 3O_2(g) + 4KOH(s)$$

a. How many liters of oxygen gas at STP are produced when 56.0 g of potassium superoxide react completely with water vapor?

b. How many grams of potassium hydroxide are produced when the 56.0 g of KO_2 react with an excess of water?

134. Find the limiting reagent for each set of reactants. Then calculate the number of moles of each reactant remaining and the amount of each product formed after the reaction.

$$4NH_3(g) + 3O_2(g) \longrightarrow 2N_2(g) + 6H_2O(l)$$

a. 4.00 mol NH_3 + 4.00 mol O_2

b. 2.00 mol NH_3 + 1.00 mol O_2

c. 7.00 mol NH_3 + 5.00 mol O_2

d. 3.25 mol NH_3 + 2.75 mol O_2

135. Diboron trioxide is formed by reacting 14.0 g of diboron hexahydride with 68.0 g of oxygen.

$$B_2H_6(g) + 3O_2(g) \longrightarrow 3H_2O(l) + B_2O_3(s)$$

a. Identify the limiting reagent.

b. Calculate the mass of diboron trioxide produced.

136. When hydrochloric acid is added to calcium carbonate, bubbles of carbon dioxide gas are produced.

$$CaCO_3(s) + 2HCl(aq) \longrightarrow CaCl_2(aq) + H_2O(l) + CO_2(g)$$

a. What is the limiting reagent when 1.68 g of HCl is added to 4.82 g $CaCO_3$?

b. In the reaction, how many milliliters of water are produced? (density = 1.00 g/cm^3)

c. What is the volume, in liters, of carbon dioxide produced? Assume STP.

137. The elements phosphorus and chlorine react to form phosphorus trichloride.

$$P_4(s) + 6Cl_2(g) \longrightarrow 4PCl_3(l)$$

a. What is the limiting reagent when 100.0 g of phosphorus reacts with 200.0 g of chlorine?

b. How many grams of phosphorus trichloride are formed?

138. Hydrogen gas is one of the products of the reaction of aluminum with hydrochloric acid.

$$2Al(s) + 6HCl(aq) \longrightarrow AlCl_3(aq) + 3H_2(g)$$

How many grams of H_2 are produced when 20.0 g of Al reacts with 60.0 g of HCl?

139. When copper(I) oxide, Cu_2O, is heated in oxygen, copper(II) oxide is formed.

$$2Cu_2O(s) + O_2(g) \longrightarrow 4CuO(s)$$

When 4.00 mol of copper(I) oxide reacts with 2.00 mol of oxygen, 7.44 mol of CuO is obtained. What is the percent yield of this reaction?

140. When 7.00 mol C reacts with 5.00 mol SO_2, 1.80 mol CS_2 is formed.

$$3C(s) + 2SO_2(g) \longrightarrow CS_2(l) + 2CO_2(g)$$

What is the percent yield of this reaction?

141. An excess of water is reacted with 25.0 g of calcium carbide. A mass of 7.20 g of C_2H_2 is obtained.

$$CaC_2(s) + 2H_2O(l) \longrightarrow C_2H_2(g) + Ca(OH)_2(aq)$$

What is the percent yield of C_2H_2?

142. An excess of sulfur dioxide was reacted with 0.150 g of oxygen gas. A mass of 0.725 g of sulfur trioxide was recovered.

$$2SO_2(g) + O_2(g) \longrightarrow 2SO_3(g)$$

What is the percent yield of sulfur trioxide?

143. When 30.0 g CH_4 reacts with 90.0 g O_2 and 30.0 g NH_3, 94.4 g H_2O is formed.

$$2CH_4(g) + 3O_2(g) + 2NH_3(g) \longrightarrow 2HCN(g) + 6H_2O(l)$$

What is the percent yield of this reaction?

132. **a.** $2H_2O \rightarrow 2H_2 + O_2$
b. 829 L

133. **a.** 13.2 L **b.** 44.2 g KOH

134. **a.** limiting reagent is NH_3; after reaction: 0.0 mol NH_3, 1.00 mol O_2, 2.00 mol N_2, and 6 mol H_2O

b. limiting reagent is O_2; after reaction: 0.67 mol NH_3, 0.0 mol O_2, 0.67 mol N_2, 2.00 mol H_2O

c. limiting reagent is O_2; after reaction: 0.33 mol NH_3, 0.0 mol O_2, 3.33 mol N_2, 10.00 mol H_2O

d. limiting reagent is NH_3; after reaction: 0.0 mol NH_3, 0.31 mol O_2, 1.63 mol N_2, and 4.88 mol H_2O

135. **a.** B_2H_6 is limiting **b.** 35.3 g B_2O_3

136. **a.** HCl is limiting **b.** 0.414 mL H_2O
c. 0.516 L CO_2

137. **a.** chlorine is limiting **b.** 258 g PCl_3

138. 1.64 g H_2

139. 93.0%

140. 77.3%

141. 71.3%

142. 96.5%

143. 99.1%

Problem Bank **R61**

APPENDIX D

Chapter 13

144. a. 3.59 atm **b.** 109 kPa

c. 198 mm Hg **d.** 2.05 atm

e. 644 mm Hg **f.** 44.6 kPa

145. The intermolecular forces are greater in water.

146. The water will evaporate faster in the pie pan because of its greater surface area.

147. The relative amount of liquid and solid would remain unchanged.

Chapter 14

148. 295 k pa

149. 521 L

150. 3.03 mL

151. 427 mL

152. 398°C

153. 922 kPa

154. 36.8 kPa

155. 7.4 kL

156. 1.13 mol N_2

157. a. 44.6 L Ne **b.** no

158. 116 kPa

159. 24.6 L F_2

160. 24.1 kPa

161. P_{Ne} = 101 kPa; P_{Ar} = 505 kPa

162. 3.29 L

163. a. 743 kPa

b. The pressure would not change.

c. The pressure would double to 1.49×10^3 kPa

Chapter 13

144. Make the following pressure conversions:

a. 364 kPa to atm

b. 815 mm Hg to kPa

c. 0.260 atm to mm Hg

d. 1555 mm Hg to atm

e. 85.8 kPa to mm Hg

f. 0.440 atm to kPa

145. Water evaporates much more slowly at room temperature than acetone. How does the relative strength of intermolecular forces compare in these two compounds?

146. In the same location, equal quantities of water are poured into a drinking glass and a glass pie pan. In which container will the water evaporate first? Explain your choice.

147. Equal masses of liquid and solid wax are placed in an oven maintained at a temperature exactly at the melting point of the wax. How would the relative amounts of liquid and solid wax change over time?

Chapter 14

148. A sample of gas at a pressure of 124 kPa has a volume of 3.00 L. If the gas is compressed to a volume of 1.26 L, what is its new pressure? (Assume constant temperature.)

149. A scuba tank has a volume of 11.0 L. What volume of gas in liters at 0.950 atm is required to completely fill the tank to a pressure of 45.0 atm, assuming no change in the temperature of the gas?

150. A syringe contains 2.60 mL of gas at 20.0°C. What is the volume of gas after the temperature is increased to 68.0°C?

151. A contained gas has a volume of 120.0 mL at −183°C. What volume does this gas occupy at 47.0°C?

152. To what temperature must a contained gas at a pressure of 464 mm Hg and a temperature of 40.0°C be raised to increase the pressure to 994 mm Hg?

153. The pressure of a gas in a cylinder at 27.0°C is 846 kPa. What is the pressure in the cylinder when the temperature is increased to 54.0°C?

154. Calculate the final pressure of a gas initially at 122 kPa pressure that is expanded from 4.50 L at 56°C to 18.0 L at 124°C.

155. A weather balloon has a volume of 3.5 kL at 1.01 atm and 18°C. What is the balloon's volume at a pressure of 0.420 atm and −18°C?

156. A cylinder contains 4.50 L of nitrogen at 35°C and a pressure of 644 kPa. How many moles of N_2 are in the cylinder?

157. A balloon containing 1.46 mol of neon gas has a volume of 36.2 L.

a. Under the same conditions, what is the volume of the balloon if an additional 0.34 mol of Ne is added to the balloon?

b. Would the answer change if 0.34 mol of He were added instead of neon?

158. What is the pressure (in kPa) in a 5.00-L tank containing 0.240 mol of oxygen gas at a temperature of 17°C?

159. Calculate the volume of 0.880 mol of fluorine gas at 26°C and 88.8 kPa.

160. A metal cylinder contains 0.440 mol of nitrogen gas at a pressure of 34.0 kPa. What is the pressure in the container after 0.128 mol of nitrogen are removed?

161. All the neon gas from a 10.0-L container at a pressure of 202 kPa is added to a 20.0-L container of argon at a pressure of 505 kPa. After the transfer, what are the partial pressures of neon and argon?

162. A child buys a balloon filled with 3.50 L of helium on a very hot day when it's 39.0°C outside. Assuming a constant pressure, what is the volume of the balloon when the child brings the balloon home to an air-conditioned house at 20.0°C?

163. Suppose you have a 0.500-L cylinder that contains 0.150 mol of oxygen gas, O_2, at 25°C.

a. What is the pressure inside the cylinder?

b. How would the pressure inside the cylinder change if you substituted 0.150 mol of sulfur dioxide gas, SO_2, for the 0.150 mol of oxygen gas?

c. How would the pressure inside the cylinder change if you added 0.150 mol of sulfur dioxide gas, SO_2, to the oxygen already in the cylinder?

R62 Appendix D

164. In a typical automobile engine, the gas mixture in a cylinder is compressed and the pressure increases from 1.00 atm to 9.50 atm. If the uncompressed volume of the cylinder is 755 mL, what is the volume when fully compressed? (Assume constant temperature.)

165. What is the new pressure when an aerosol can with an initial pressure of 4.50 atm at 25°C is heated in a fire to 650°C?

166. How many moles of air are in the lungs of an average person with a total lung capacity of 3.8 L? Assume that the person is at sea level (1.00 atm) and has a normal body temperature of 37°C.

167. Two containers of equal size are filled with 4.0 g of He and 32.0 g of O_2, respectively. Assuming a constant temperature, would you expect the pressures of these two gases to be identical? Explain your answer.

168. Lithium nitride is formed from its elements.

$$6Li(s) + N_2(g) \longrightarrow 2Li_3N(s)$$

How many milliliters of nitrogen gas at STP are needed to react with 0.246 g of lithium?

169. Nitrogen and hydrogen react to form ammonia.

$$3H_2(g) + N_2(g) \longrightarrow 2NH_3(g)$$

How many liters of hydrogen gas measured at 86.4 kPa pressure and 245°C are needed to react completely with 6.44 g N_2?

170. Auto air bags are inflated by nitrogen gas formed through this decomposition reaction:

$$2NaN_3(s) \longrightarrow 2Na(s) + 3N_2(g)$$

How many grams of NaN_3 are needed to inflate an air bag to a volume of 10.6 L, assuming STP?

Chapter 15

171. Which of these molecules can form hydrogen bonds with water?

a. H_2 c. HCl

b. CH_3OH d. C_2H_6

172. Classify each substance as an electrolyte or a nonelectrolyte.

a. NH_4NO_3 c. $NaBr_2$

b. C_2H_6O d. Cl_2

173. Calculate the percent by mass of water in lithium perchlorate trihydrate.

174. An experiment requires making a solution that contains 34.6 g of $CaCl_2$. Your only source is the hydrate, $CaCl_2 \cdot 2H_2O$. How many grams of the hydrate do you need to use to obtain the required mass of $CaCl_2$?

175. A 19.97-g sample of a hydrate contains 5.08 g Cu, 2.57 g S, 5.12 g O, and 7.20 g H_2O. What is the empirical formula of this hydrate?

Chapter 16

176. The solubility of carbon dioxide gas at 50°C and 1.00 atm pressure is 7.6×10^{-2} g/100 g H_2O. Assuming constant temperature, calculate the solubility of CO_2 when the pressure is increased to 2.50 atm.

177. Calculate the molarity of each of these solutions.

a. 4.24 mol NaCl in 2.00 L of solution

b. 0.164 mol of $C_5H_{10}O_5$ in 125 mL of solution

c. 0.0056 mol CsBr in 50.0 mL of solution

d. 2.84 mol C_2H_6O in 0.650 L of solution

178. What is the molarity of each of these solutions?

a. 3.34 g $CuNO_2$ in 0.150 L of solution

b. 0.0688 g CoF_2 in 20.0 mL of solution

c. 88.8 g KOH in 0.755 L of solution

d. 1.66 g $LiNO_3$ in 455 mL of solution

179. Find the moles of solute in the following solutions:

a. 650 mL of 0.28*M* $NaNO_3$

b. 1.4 L of 0.35*M* KI

c. 0.340 L of 2.22*M* $CaCl_2$

d. 148 mL of 0.0068*M* LiF

180. Calculate the mass of solute in each of these solutions.

a. 2.00 L of 0.440*M* MgF_2

b. 6.80 dL of 1.88*M* CH_4O

c. 65.0 mL of 0.0360*M* $NaNO_3$

d. 5.00 mL of 1.48*M* HCl

164. 79.5 mL

165. 13.9 atm

166. 0.15 mol

167. Yes, because the masses are 1.00 mol of each gas, and equal moles of gases at the same temperature have the same volume.

168. 133 mL N_2

169. 34.4 L H_2

170. 20.5 g NaN_3

Chapter 15

171. a. cannot **b.** can **c.** can **d.** cannot

172. a. electrolyte **b.** nonelectrolyte **c.** electrolyte **d.** nonelectrolyte

173. 33.7%

174. 45.8 g $CaCl_2 \cdot 2H_2O$

175. $CuSO_4 \cdot 5H_2O$

Chapter 16

176. 0.19 g/100 g H_2O

177. a. 2.12*M* **b.** 1.31*M* **c.** 0.11*M* **d.** 4.37*M*

178. a. 0.203*M* **b.** 3.55×10^{-2}*M* **c.** 2.10*M* **d.** 5.29×10^{-2}*M*

179. a. 0.18 mol $NaNO_3$ **b.** 0.49 mol KI **c.** 0.755 mol $Ca(OH)_2$ **d.** 1.01×10^{-3} mol LiF

180. a. 54.8 g MgF_2 **b.** 40.9 g CH_3OH **c.** 0.199 g $NaNO_3$ **d.** 0.270 g HCl

181. **a.** 20.0 mL **b.** 160 mL
c. 75 mL **d.** 38.0 mL

182. 63 mL of alcohol

183. **a.** 2.39% CsI **b.** 17.6% NaOH
c. 4.75% K_2CO_3 **d.** 37.5% $NaNO_3$

184. **a.** 0.309 CH_4O and 0.691 C_2H_6O
b. 0.342 H_2O and 0.658 HCl
c. 0.484 C_2H_6O and 0.516 H_2O
d. 0.120 C_2H_6O and 0.880 CH_4O

185. **a.** true **b.** false
c. false

186. **a.** 0.148*m* KCl **b.** 0.137*m* $LiNO_3$
c. 5.34×10^{-2}*m* CsI **d.** 2.66*m* $MgBr_2$

187. **a.** f.p. = −8.70°C; b.p. = 102.40°C
b. f.p. = −6.53°C; b.p. = 101.80°C
c. f.p. = −5.36°C; b.p. = 101.47°C
d. f.p. = −2.19°C; b.p. = 100.60°C

188. **a.** 76.2 mL of 0.366*M* $Ni(NO_3)_2$
b. 3.31 g $NiCO_3$

189. **a.** 38.2 mL $Al_2(SO_4)_3(aq)$
b. 0.811 g $Al(OH)_3$

190. **a.** KCl and NaOH, because they both dissociate to give two particles/formula unit in solution.

b. Since $AlCl_3$ dissociates to give 4 particles/formula unit and KCl dissociates to give 2 particles/formula unit, the conductivity of the aluminum chloride solution should be about twice that of the potassium chloride solution.

181. How many milliliters of a stock solution of 2.50*M* $SrCl_2$ solution are required to make each diluted solution?

a. 50.0 mL of 1.00*M* $SrCl_2$
b. 1.0 L of 0.40*M* $SrCl_2$
c. 750 mL of 0.25*M* $SrCl_2$
d. 65.0 dL of 0.146*M* $SrCl_2$

182. An aqueous solution is 65% (v/v) rubbing alcohol. How many milliliters of alcohol are in a 97-mL sample of this solution?

183. Calculate the percent by mass of each of these solutions.

a. 6.50 g CsI in 266 g H_2O
b. 246 g NaOH in 1.40 kg H_2O
c. 0.428 g K_2CO_3 in 8.58 g H_2O
d. 1.20 kg $NaNO_3$ in 2.00 kg H_2O

184. Calculate the mole fraction of each component of the following solutions:

a. 2.40 mol CH_4O and 5.36 mol C_2H_6O
b. 1.25 mol H_2O and 87.6 g HCl
c. 24.0 g C_2H_6O and 10.0 g H_2O
d. 0.464 g C_2H_6O and 2.36 g CH_4O

185. Potassium bromide is dissolved in water. Which statements are true when comparing the solution to pure water?

a. The boiling point of the solution is higher.
b. The vapor pressure of the solution is higher.
c. The freezing point of the solution is higher.

186. Calculate the molality of these solutions.

a. 0.246 mol KCl in 1.66 kg solvent
b. 0.116 mol $LiNO_3$ in 844 g solvent
c. 56.6 mmol CsI in 1.06 kg solvent
d. 6.66 mol $MgBr_2$ in 2.50 kg solvent

187. Calculate the freezing and boiling points of each of these aqueous solutions.

a. 2.34*m* NH_4Br
b. 1.17*m* $CaCl_2$
c. 24.4 g LiCl in 0.400 kg H_2O
d. 44.8 g $MgCl_2$ in 1.20 kg H_2O

188. When aqueous solutions of sodium carbonate and nickel(II) nitrate are mixed, nickel(II) carbonate precipitates.

$$Ni(NO_3)_2(aq) + Na_2CO_3(aq) \longrightarrow NiCO_3(s) + 2NaNO_3(aq)$$

a. What volume of 0.366*M* $Ni(NO_3)_2$ is required to completely react with 55.8 mL of 0.500*M* Na_2CO_3?
b. How many grams of nickel(II) carbonate precipitate in this reaction?

189. Aluminum hydroxide precipitates when aqueous solutions of aluminum sulfate and sodium hydroxide are mixed.

$$Al_2(SO_4)_3(aq) + 6NaOH(aq) \longrightarrow 2Al(OH)_3(s) + 3Na_2SO_4(aq)$$

a. What volume of 0.136*M* $Al_2(SO_4)_3$ is needed to completely react with 26.0 mL of 1.20*M* NaOH?
b. What mass of aluminum hydroxide precipitates in this reaction?

190. A conductivity meter quantitatively measures the ability of an aqueous solution to conduct electrical current. The magnitude of the conductivity value is proportional to the number of ions in the solution. Data from an experiment are given in the table.

Solutions (0.2*M*)	Conductivity (μS/cm)
KCl	2050
$AlCl_3$	4500
$CaCl_2$	3540
NaOH	2080
C_2H_6O	0

a. Which two of the solutions have similar conductivities? Why would you expect this?
b. The ratio of the conductivity of the aluminum chloride solution to the conductivity of the potassium chloride solution is approximately two to one. Why would you expect this result?

Chapter 17

191. As the mass of a substance increases,

a. does its heat capacity increase, decrease, or remain constant?

b. does its specific heat increase, decrease, or remain constant?

192. The temperature of a 6.42-gram piece of glass is 15°C. How many calories would it take to increase the temperature of the glass to 96°C? The specific heat of glass is 0.12 cal/(g·°C).

193. Ethanol has a specific heat of 2.43 J/(g·°C). If 468 J of heat is added to 29.0 g of ethanol initially at 25.0°C, what is the final temperature of the liquid?

194. When 1564 J of energy is added to a sample of gold at 25.0°C, the temperature of the gold increases by 424.0°C. What is the mass of the gold? The specific heat of gold is 0.129 J/(g·°C).

195. Identical masses of aluminum and lead at the same temperature absorb identical amounts of heat energy. The specific heat of aluminum is 0.901 J/(g·°C); the specific heat of lead is 0.129 J/(g·°C). Which gets hotter, the aluminum or the lead?

196. Suppose your diet provides 2100 Cal (kcal) in a day, and your body weight is 68 kg. Start with an initial body temperature of a normal 37°C. Calculate the maximum temperature that your body would reach by absorbing all 2100 kcal at once. For purposes of this problem, assume that your body is 100% water. The specific heat of water is 1.00 cal/(g·°C).

197. Nitrogen monoxide is formed from its elements.

$$N_2(g) + O_2(g) \longrightarrow 2NO(g) \quad \Delta H = 181 \text{ kJ/mol}$$

a. Is this reaction exothermic or endothermic?

b. How many kilojoules of energy are needed to form 8.70 mol NO?

198. Carbon dioxide and water are produced by the complete combustion of propane, C_3H_8.

$$C_3H_8(g) + 5O_2(g) \longrightarrow 3CO_2(g) + 4H_2O(g) + 526 \text{ kcal}$$

a. Is this reaction exothermic or endothermic?

b. How many kcal of energy are produced when 14.4 g C_3H_8 burns in an excess of oxygen?

199. The following reaction was used to fuel the rockets in the Apollo mission landing module.

$$2N_2H_4(l) + N_2O_4(l) \longrightarrow 3N_2(g) + 4H_2O(g) \quad \Delta H = -1049 \text{ kJ}$$

a. Is this reaction endothermic or exothermic?

b. How many grams of N_2H_4 must be reacted with an excess of N_2O_4 to produce 645 kJ of energy?

c. How many kilojoules of energy are produced when 5.40 g N_2O_4 reacts with an excess of N_2H_4?

200. The heat of fusion of mercury is 2.30 kJ/mol. How much heat (in J) is released when 24.0 g Hg changes from the liquid state to the solid state at its freezing point?

201. How much heat energy is required to change 50.0 g of liquid water at 100°C to water vapor at 100°C? The molar heat of fusion is 6.01 kJ/mol, and the molar heat of vaporization is 40.7 kJ/mol.

202. There is a dramatic temperature change when solid ammonium nitrate is dissolved in water.

$$NH_4NO_3(s) \longrightarrow NH_4^+(aq) + NO_3^-(aq) \quad \Delta H_{soln} = 25.7 \text{kJ/mol}$$

a. Is there a temperature increase or decrease when ammonium nitrate dissolves in water?

b. Calculate the heat change when 55.0 g $NH_4NO_3(s)$ dissolves in water.

203. Use the data in Table 17.4 and the additional values for $\Delta H°$ given below to calculate the standard heat of reaction ($\Delta H°$) for each of these reactions.

Substance	$\Delta H_f°$ (kJ/mol)
$N_2H_4(l)$	50.63
$HNO_3(aq)$	−207.4

a. $2SO_2(g) + O_2(g) \longrightarrow 2SO_3(g)$

b. $3N_2H_4(l) \longrightarrow 4NH_3(g) + N_2(g)$

c. $3NO_2(g) + H_2O(g) \longrightarrow 2HNO_3(aq) + NO(g)$

d. $2NO(g) + 2CO(g) \longrightarrow 2CO_2(g) + N_2(g)$

Chapter 17

191. a. increase **b.** remain constant

192. 62 cal

193. 31.6°C

194. 28.6 g

195. lead

196. 68°C

197. a. endothermic **b.** 787 kJ

198. a. exothermic **b.** 172 kcal

199. a. exothermic **b.** 39.4 g N_2H_4 **c.** 61.6 kJ

200. 275 J

201. 113 kJ

202. a. decrease **b.** 17.7 kJ

203. a. −197.8 kJ **b.** −336.7 kJ **c.** −184.2 kJ **d.** −746.7 kJ

Problem Bank **R65**

APPENDIX D

204. –44 kJ

205. –188 kJ/mol

206. 82.3 g H_2O

Chapter 18

207. rate = $[NO]^3[H_2]$

208. a. reactants **b.** products
c. reactants **d.** products

209. a. $K_{eq} = \dfrac{[POCl_3]^2}{[PCl_3]^2 \times [O_2]}$

b. $K_{eq} = \dfrac{[Cl_2O] \times [H_2O]}{[HOCl]_2}$

c. $K_{eq} = \dfrac{[BrF_5]^2}{[Br_2] \times [F_2]^5}$

d. $K_{eq} = \dfrac{[NO_2]^2 \times [H_2O]^8}{[N_2H_4] \times [H_2O_2]}$

210. a. $K_{eq} = 2.26$ **b.** $K_{eq} = 659$
c. $K_{eq} = 4.32 \times 10^{-2}$ **d.** $K_{eq} = 3.60 \times 10^6$

211. add more CO; add more H_2, remove CH_4O as it forms, cool the reaction.

212. a. the color turns more yellowish
b. the color turns more yellowish

213. a. the color turns more bluish
b. the color turns more pinkish

214. a. yes **b.** yes
c. yes **d.** no

204. Ethanol is manufactured by reacting water with ethane, C_2H_4.

$$C_2H_4(g) + H_2O(l) \longrightarrow C_2H_6O(l)$$

Use the following equations to calculate the $\Delta H°$ for this reaction.

$$C_2H_6O(l) + 3O_2(g) \longrightarrow 2CO_2(g) + 3H_2O(l) \quad \Delta H = -1367 \text{ kJ}$$

$$C_2H_4(g) + 3O_2(g) \longrightarrow 2CO_2(g) + 2H_2O(l) \quad \Delta H = -1411 \text{ kJ}$$

205. Use the following equations to calculate the standard heat of formation, in kJ/mol, of magnesium nitrate:

$$2MgO(s) \longrightarrow 2Mg(s) + O_2(g) \quad \Delta H = 1203 \text{ kJ}$$

$$Mg_3N_2(s) \longrightarrow 3Mg(s) + N_2(g) \quad \Delta H = 463 \text{ kJ}$$

$$Mg(NO_3)_2(s) + 8Mg(s) \longrightarrow 6MgO(s) + Mg_3N_2(s) \quad \Delta H = -3884 \text{ kJ}$$

206. The heat of sublimation of dry ice (solid CO_2) is 25.2 kJ/mol. How many grams of water at 0.0°C would be frozen by the complete sublimation of 48.0 g of dry ice that is dropped into the water? The heat of fusion of water is 6.01 kJ/mol.

Chapter 18

207. Nitric oxide reacts with hydrogen to form nitrogen gas and water.

$$2NO(g) + 2H_2(g) \longrightarrow N_2(g) + 2H_2O(g)$$

In an experiment, doubling the H_2 concentration causes the rate of reaction to double. When the NO concentration is doubled, the rate of the reaction increases by a factor of eight. Write the rate law for the reaction.

208. At equilibrium, are the reactants or the products favored for reactions that have the following equilibrium constants?

a. $K_{eq} = 5.6 \times 10^{-7}$ **c.** $K_{eq} = 5.6 \times 10^{-14}$
b. $K_{eq} = 5.6 \times 10^{21}$ **d.** $K_{eq} = 5.6 \times 10^{5}$

209. Write the expression for the equilibrium constant for each reaction.

a. $2PCl_3(g) + O_2(g) \rightleftharpoons 2POCl_3(g)$
b. $2HOCl(g) \rightleftharpoons Cl_2O(g) + H_2O(g)$
c. $Br_2(g) + 5F_2(g) \rightleftharpoons 2BrF_5(g)$
d. $N_2H_4(g) + 6H_2O_2(g) \rightleftharpoons 2NO_2(g) + 8H_2O(g)$

210. Using the equations in Problem 209, calculate the value of K_{eq} when the following amounts of reactants and products are present in a 1-L container at equilibrium.

a. 1.44 mol PCl_3, 1.44 mol O_2, and 2.60 mol $POCl_3$
b. 0.220 mol HOCl, 4.68 mol Cl_2O, and 6.82 mol H_2O
c. 0.0500 mol Br_2, 1.00 mol F_2, and 0.0465 mol BrF_5
d. 0.400 mol N_2H_4, 0.100 mol H_2O_2, 1.20 mol NO_2, and 1.00 mol H_2O

211. List three ways to cause a shift in this equilibrium to the right, forming more $CH_4O(g)$.

$$CO(g) + 2H_2(g) \rightleftharpoons CH_4O(g) + \text{heat}$$

212. A yellow gas (Y) reacts with a colorless gas (C) to produce a blue gas (B), according to this equation:

$$C(g) + 3Y(g) \rightleftharpoons 2B(g) + \text{heat}$$

The system is initially at equilibrium and has a green color. What happens to the color of the system if the following stresses are placed on the system? Note: Blue and yellow mix to form green.

a. A large amount of colorless C is removed from the reaction container.
b. The reaction container is heated.

213. Assume that the following chemical system is originally at equilibrium and the color of the liquid is purplish (a mixture of the colors pink and blue):

$$\text{heat} + \underset{\text{pink}}{[Co(H_2O)_6]^{2+}(aq)} + 4\,Cl^-(aq) \rightleftharpoons \underset{\text{blue}}{[CoCl_4]^{2-}(aq)} + 6\,H_2O(l)$$

a. How does the color change if chloride ion is added to the system?
b. How does the color change if the reaction mixture is cooled?

214. Use Table 18.1 to predict whether a precipitate will form when the following pairs of substances are mixed.

a. $K_2S(aq) + Cu(NO_3)_2(aq)$
b. $NH_4Cl(aq) + Pb(NO_3)_2(aq)$
c. $Na_2CO_3(aq) + ZnCl_2(aq)$
d. $KNO_3(aq) + BaCl_2(aq)$

R66 Appendix D

215. Find the equilibrium concentrations of zinc and sulfide ions in a saturated solution of zinc sulfide with $K_{sp} = 3.0 \times 10^{-23}$.

216. For each change, does the entropy increase or decrease?

a. A gold nugget melts.

b. Liquid wax solidifies.

c. Liquid water forms from water vapor.

d. Liquid water forms from hydrogen and oxygen gas.

Chapter 19

217. Identify each of the species in each equation as a Brønsted-Lowry acid or a Brønsted-Lowry base.

a. $H_2O(aq) + CN^-(aq) \rightleftharpoons OH^-(aq) + HCN(aq)$

b. $HClO_3(aq) + H_2O(aq) \rightleftharpoons ClO_3^-(aq) + H_3O^+(aq)$

c. $C_5H_5NH^+(aq) + OH^-(aq) \rightleftharpoons C_5H_5N(aq) + H_2O(aq)$

d. $HSO_4^-(aq) + H_3O^+(aq) \rightleftharpoons H_2SO_4(aq) + H_2O(aq)$

218. Which of the following are conjugate acid-base pairs?

a. NH_4^+, NH_3

b. H_3PO_4, $H_2PO_4^-$

c. HSO_4^-, SO_4^{2-}

d. H_3O^+, OH^-

219. Ammonia can act as a Lewis base toward which of these compounds?

a. CH_4

b. BCl_3

c. NF_3

d. OF_2

220. Calculate the pH of each solution and classify it as acidic or basic.

a. $[H^+] = 4.6 \times 10^{-4}M$

b. $[H^+] = 1.2 \times 10^{-8}M$

c. $[OH^-] = 8.3 \times 10^{-4}M$

d. $[OH^-] = 2.8 \times 10^{-11}M$

e. $[H^+] = 3.9 \times 10^{-2}M$

f. $[OH^-] = 1.5 \times 10^{-9}M$

221. Calculate the $[H^+]$ and the $[OH^-]$ from the pH of each solution.

a. pH = 6.03

b. pH = 1.18

c. pH = 12.68

d. pH = 4.33

e. pH = 9.16

f. pH = 3.46

222. Rank these acids from strongest to weakest.

a. HX, $K_a = 1 \times 10^{-4}$

b. HY, $K_a = 1 \times 10^{-11}$

c. HP, $K_a = 1 \times 10^{-2}$

d. HQ, $K_a = 1 \times 10^{-9}$

223. Acetylsalicylic acid (aspirin) has a K_a of 3×10^{-4}. A 0.00056*M* solution of aspirin would be best described by which terms: weak, strong, dilute, concentrated?

224. A 0.10000*M* solution of an unknown acid, HX, has a hydrogen ion concentration of $3.65 \times 10^{-4}M$. Calculate the value of K_a for this acid.

225. Name and write the formula for the salt formed in each of the following acid-base neutralizations.

a. aluminum hydroxide with phosphoric acid

b. oxalic acid with magnesium hydroxide

c. sulfurous acid with lithium hydroxide

d. sodium hydroxide with carbonic acid

226. How many moles of sulfuric acid are required to neutralize 1.40 mol of potassium hydroxide?

$$2KOH(aq) + H_2SO_4(aq) \longrightarrow K_2SO_4(aq) + 2H_2O(l)$$

227. What is the molarity of a hydrochloric acid solution if 25.0 mL of the solution reacts completely with 1.66 g $NaHCO_3$?

$$HCl(aq) + NaHCO_3(s) \longrightarrow NaCl(aq) + H_2O(l) + CO_2(g)$$

228. In an acid base titration, how many mL of 0.180*M* HCl are required to neutralize 20.0 mL of 0.220*M* NaOH?

229. How many milliliters of 0.456*M* $Ca(OH)_2$ are needed to neutralize 25.0 mL of 0.300*M* HCl?

215. $[Zn^{2+}] = [S^{2-}] = 5.5 \times 10^{-12}\ M$

216. **a.** increase **b.** decrease **c.** decrease **d.** decrease

Chapter 19

217. **a.** $H_2O(aq) + CN^-(aq) \rightleftharpoons OH^-(aq) + HCN(aq)$
acid base base acid

b. $HClO_3(aq) + H_2O(aq) \rightleftharpoons ClO_3^-(aq) + H_3O^+(aq)$
acid base base acid

c. $C_5H_5NH^+(aq) + OH^-(aq) \rightleftharpoons C_5H_5N(aq) + H_2O(aq)$
acid base base acid

d. $HSO_4^-(aq) + H_3O^+(aq) \rightleftharpoons H_2SO_4(aq) + H_2O(aq)$
acid base base acid

218. **a.** yes **b.** yes **c.** yes **d.** no

219. **a.** no **b.** yes **c.** no **d.** no

220. **a.** 3.34, acidic **b.** 7.92, basic **c.** 10.92, basic **d.** 3.45, acidic **e.** 1.41, acidic **f.** 5.18, acidic

221 **a.** $[H^+] = 9.3 \times 10^{-7}M$, $[OH^-] = 1.1 \times 10^{-8}M$

b. $[H^+] = 6.6 \times 10^{-2}M$, $[OH^-] = 1.5 \times 10^{-13}M$

c. $[H^+] = 2.1 \times 10^{-13}M$, $[OH^-] = 4.8 \times 10^{-2}M$

d. $[H^+] = 4.7 \times 10^{-5}M$, $[OH^-] = 2.1 \times 10^{-10}M$

e. $[H^+] = 6.9 \times 10^{-10}M$, $[OH^-] = 1.4 \times 10^{-5}M$

f. $[H^+] = 3.5 \times 10^{-4}M$, $[OH^-] = 2.9 \times 10^{-11}M$

222. **c.** > **a.** > **d.** > **b.**

223. a dilute solution of a weak acid

224. 1.34×10^{-6}

225. **a.** aluminum phosphate, $AlPO_4$

b. magnesium oxalate, MgC_2O_4

c. lithium sulfite, Li_2SO_3

d. sodium carbonate, Na_2CO_3

226. 0.700 mol H_2SO_4

227. 0.790*M* HCl

228. 24.4 ml HCl

229. 8.22 mL $Ca(OH)_2$

230. **a.** $BrO^- + H^+ \rightarrow HBrO$

b. $HBrO + OH^- \rightarrow BrO^- + H_2O$

Chapter 20

231. **a.** +6 **b.** +2
c. +4 **d.** +6
e. 0 **f.** +6

232. **a.** –1 **b.** +5
c. +3 **d.** 0
e. +1 **f.** +1

233. **a.** Fe **b.** $CuCl_2$
c. Fe **d.** Fe

234. **a.** oxidized **b.** oxidized
c. reduced **d.** reduced

235. **a.** $HgS(s) \rightarrow Hg(l) + S(s)$

b. $2S(s) + 3O_2(g) \rightarrow 2SO_3(g)$

c. $H_2SO_4(aq) + 4Ca(s) + 8H^+(aq) \rightarrow H_2S(g) + 4Ca^{2+}(aq) + 4H_2O(l)$

d. $2Al(s) + 3S(s) \rightarrow Al_2S_3(s)$

236. **a.** $Mg(s) + 2H_2O(l) \rightarrow Mg(OH)_2 + H_2(g)$
single replacement

b. $PF_3(g) + F_2(g) \rightarrow PF_5(g)$
combination

c. $C_2H_2(g) + 2H_2(g) \rightarrow C_2H_6(g)$
combination

d. $2NaNO_3(s) \rightarrow 2NaNO_2(s) + O_2(g)$
decomposition

237. **a.** Fe, +2; Mn, +7 **b.** S, +6, O, –2
c. MnO_4^- **d.** Mn ion in MnO_4^-

238. **a.** $2NO + 2H_2 \rightarrow N_2 + 2H_2O$
nitrogen is reduced, hydrogen is oxidized

b. $2KMnO_4 + 10FeSO_4 + 8H_2SO_4 \rightarrow K_2SO_4 + 2MnSO_4 + 5Fe_2(SO_4)_3 + 8H_2O$
manganese is reduced, iron is oxidized

c. $P_4 + 6NO \rightarrow P_4O_6 + 3N_2$
nitrogen is reduced, phosphorus is oxidized

d. $3SO_2 + 2HNO_3 + 2H_2O \rightarrow 3H_2SO_4 + 2NO$
nitrogen is reduced, sulfur is oxidized

239. **a.** $2KMnO_4(aq) + 3NaNO_2(aq) + H_2O(l) \rightarrow 2MnO_2(s) + 3NaNO_3(aq) + 2KOH(aq)$
oxidizing agent is $KMnO_4$;
reducing agent is $NaNO_2$

b. $I_2(s) + 2Na_2S_2O_3(aq) \rightarrow Na_2S_4O_6(aq) + 2NaI(aq)$
oxidizing agent is I_2;
reducing agent is $Na_2S_2O_3$

c. $6HCl(aq) + 2NH_4Cl(aq) + K_2Cr_2O_7(aq) \rightarrow 2CrCl_3(aq) + 2KCl(aq) + N_2(g) + 7H_2O(l)$
oxidizing agent is $K_2Cr_2O_7$;
reducing agent is NH_4Cl

d. $2FeCl_2(aq) + H_2O_2(aq) + 2HCl(aq) \rightarrow 2FeCl_3(aq) + 2H_2O(l)$
oxidizing agent is H_2O_2;
reducing agent is $FeCl_2$

240. **a.** $Cr_2O_7^{2-}(aq) + 6I^-(aq) + 14H^+(aq) \rightarrow 2Cr^{3+}(aq) + 3I_2(s) + 7H_2O(l)$
$Cr_2O_7^{2-}$ is reduced; I^- is oxidized

b. $2MnO_4^-(aq) + 5SO_3^{2-}(aq) + 6H^+ \rightarrow 2Mn^{2+}(aq) + 5SO_4^{2-}(aq) + 3H_2O(l)$
MnO_4^- is reduced; SO_3^{2-} is oxidized

c. $3C_2O_4^{2-}(aq) + 2MnO_4^-(aq) + 4OH^-(aq) \rightarrow 6CO_3^{2-}(aq) + 2MnO_2(s) + 2H_2O(l)$
$C_2O_4^{2-}$ is oxidized; MnO_4^- is reduced

d. $3CN^-(aq) + 2MnO_4^-(aq) + H_2O(l) \rightarrow 2MnO_2(s) + 3CNO^-(aq) + 2OH^-(aq)$
CN^- is oxidized; MnO_4^- is reduced

230. A buffer ($HBrO/BrO^-$) solution is made by mixing together equal quantities of hypobromous acid (HBrO) and sodium hypobromite (NaBrO).

a. Write an equation for the reaction that occurs when an acid is added to this buffer.

b. Write an equation for the reaction that occurs when a base is added to this buffer.

Chapter 20

231. What is the oxidation number of sulfur in each of these species?

a. SF_6 **d.** SO_3
b. CaS_2O_3 **e.** S
c. K_2SO_3 **f.** H_2SO_4

232. What is the oxidation number of bromine in each of the following?

a. CsBr **d.** Br_2
b. $NaBrO_3$ **e.** BrCl
c. BrO_2^- **f.** NaBrO

233. Consider the following reaction:

$CuCl_2(aq) + Fe(s) \longrightarrow Cu(s) + FeCl_2(aq)$

a. Which reactant lost electrons?

b. What is the oxidizing agent?

c. Which reactant was oxidized in this reaction?

d. What is the reducing agent?

234. In each of these reactions, is sulfur oxidized or reduced?

a. $HgS(s) \longrightarrow Hg(l) + S(s)$

b. $S(s) + O_2(g) \longrightarrow SO_3(g)$

c. $H_2SO_4(aq) + Ca(s) \longrightarrow H_2S(g) + Ca^{2+}(aq)$

d. $Al(s) + S(s) \longrightarrow Al_2S_3(s)$

235. Balance each of the equations in Problem 234.

236. Balance each of the following redox reactions and classify each as a combination, decomposition, or single-replacement reaction.

a. $Mg(s) + H_2O(l) \longrightarrow Mg(OH)_2 + H_2(g)$

b. $PF_3(g) + F_2(g) \longrightarrow PF_5(g)$

c. $C_2H_2(g) + H_2(g) \longrightarrow C_2H_6(g)$

d. $NaNO_3(s) \longrightarrow NaNO_2(s) + O_2(g)$

237. Iron(II) sulfite reacts with an acid solution of permanganate ion.

$5FeSO_3 + 14H^+ + 3MnO_4^- \longrightarrow 3Mn^{2+} + 5Fe^{3+} + 5SO_4^{2-} + 7H_2O$

a. What is the oxidation number of iron and of manganese in the reactants?

b. What is the oxidation number of sulfur and of oxygen in SO_4^{2-}?

c. Identify the oxidizing agent in this reaction.

d. What is reduced in this reaction?

238. Write a balanced chemical equation for each of these reactions, and identify the element oxidized and the element reduced.

a. Nitrogen monoxide reacts with hydrogen to form nitrogen and water.

b. Potassium permanganate, iron(II) sulfate, and sulfuric acid react, producing manganese(II) sulfate, iron(III) sulfate, potassium sulfate, and water.

c. Elemental phosphorus (P_4) and nitrogen monoxide react to form tetraphosphorus hexoxide and nitrogen.

d. Sulfur dioxide, nitric acid, and water react to produce sulfuric acid and nitrogen monoxide.

239. Balance each redox equation by the oxidation-number-change method. Identify the oxidizing agent and the reducing agent.

a. $KMnO_4(aq) + NaNO_2(aq) + H_2O(l) \longrightarrow MnO_2(s) + NaNO_3(aq) + KOH(aq)$

b. $I_2(s) + Na_2S_2O_3(aq) \longrightarrow Na_2S_4O_6(aq) + NaI(aq)$

c. $HCl(aq) + NH_4Cl(aq) + K_2Cr_2O_7(aq) \longrightarrow CrCl_3(aq) + KCl(aq) + N_2(g) + H_2O(l)$

d. $FeCl_2(aq) + H_2O_2(aq) + HCl(aq) \longrightarrow FeCl_3(aq) + H_2O(l)$

240. Use the half-reaction method to write a balanced ionic equation for each reaction. Identify the species oxidized and the species reduced.

a. $Cr_2O_7^{2-}(aq) + I^-(aq) \longrightarrow Cr^{3+}(aq) + I_2(s)$ (in acidic solution)

b. $MnO_4^-(aq) + SO_3^{2-}(aq) \longrightarrow Mn^{2+}(aq) + SO_4^{2-}(aq)$ (in acidic solution)

c. $C_2O_4^{2-}(aq) + MnO_4^-(aq) \longrightarrow CO_3^{2-}(aq) + MnO_2(s)$ (in basic solution)

d. $CN^-(aq) + 2MnO_4^-(aq) \longrightarrow MnO_2(s) + CNO^-(aq)$ (in basic solution)

Chapter 21

241. What happens when a strip of zinc is dipped into a solution of aluminum chloride?

242. A strip of aluminum is dipped into a solution of nickel(II) sulfate. Explain the result.

243. A voltaic cell is constructed using the following half-reactions:

$Ag^{+}(aq) + e^{-} \longrightarrow Ag(s) \qquad E^{\circ}_{Ag^{+}} = +0.80\ V$

$Al^{3+}(aq) + 3e^{-} \longrightarrow Al(s) \qquad E^{\circ}_{Al^{3+}} = -1.66\ V$

Determine the cell reaction.

244. Calculate the standard cell potential for the voltaic cell described in Problem 243.

245. What process, reduction or oxidation, always occurs at the anode of an electrolytic cell?

Chapter 22

246. What bonding properties make carbon uniquely suited to make a very large number of organic compounds?

247. Draw structural formulas for these compounds.

a. 3-phenylpentane

b. 2-phenyl-1-butene

248. Describe the structural features of an alkene molecule that permits the existence of *cis-trans* isomers.

249. How is an asymmetric carbon produced in an organic molecule?

250. Which of these structures are *cis* isomers?

a. H, CH_2CH_3 / C=C / CH_3CH_2, H

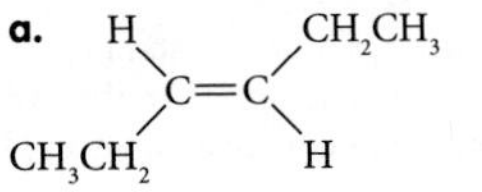

b. CH_3, CH_3 / C=C / H, H

c. H, CH_3 / C=C / H, CH_3

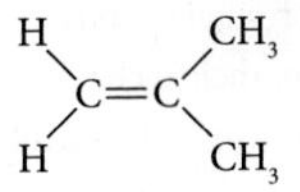

d. CH_3, (phenyl) / C=C / H, H

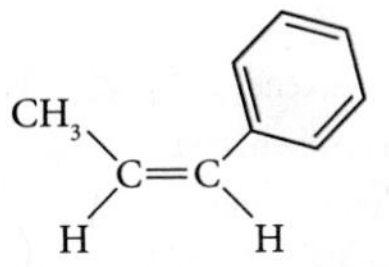

e. H, $CH_2CH_2CH_3$ / H—C—C—H / CH_3CH_2, H

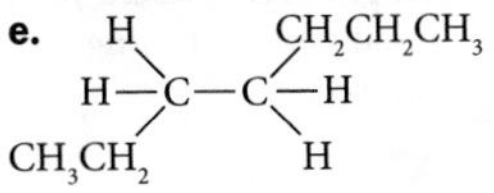

251. Write the IUPAC names for these structures. Use *cis* and *trans* prefixes when appropriate.

a. H, H / C=C / CH_3CH_2, CH_3

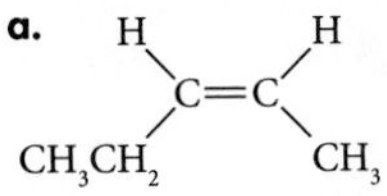

b. (cyclooctane)—CH_3

c. CH_3, CH_3 / C=$\underset{|}{C}$HCH$_2$CH$_2$CH$_3$ with CH_2CH_3 on the CH

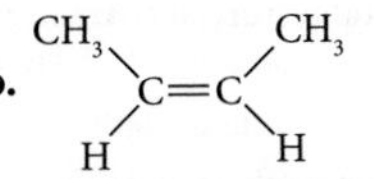

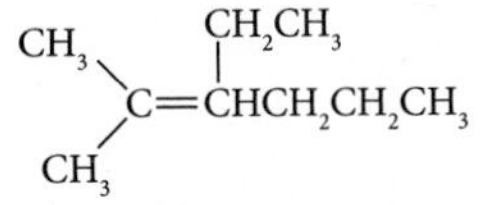

d. $CH_3CH(CH_2)_7CH_3$ with CH_3 on the CH

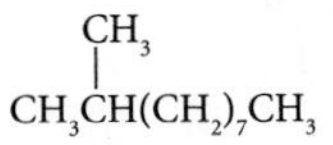

e. CH_3, CH_2CH_3 / C=C / CH_3, H

PROBLEM BANK

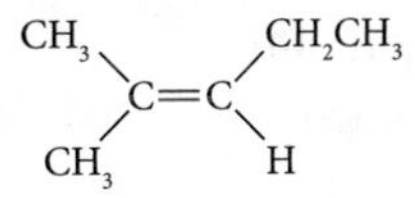

Chapter 21

241. nothing

242. The aluminum strip will become nickel plated.

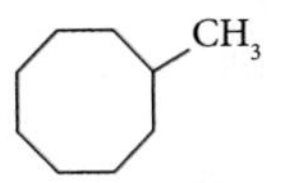

243. $Al(s) + 3Ag^{+}(aq) \rightarrow Al^{3+}(aq) + 3Ag(s)$

244. Standard cell potential is: +9.66 V

245. oxidation

Chapter 22

246. the ability of carbon atoms to form four stable covalent bonds with other carbon atoms

247. a. benzene ring with CH_2CH_3 and CH_2CH_3 **b.** $CH_3CH_2CHCH_2CH_3$ with benzene ring on the CH

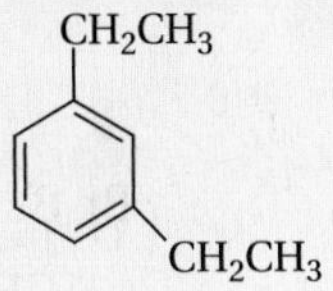

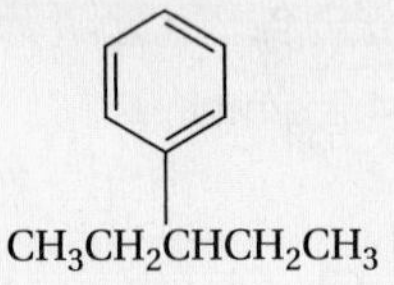

c. benzene ring with CH_2CH_3 and $CH_2CH_2CH_3$ **d.** $CH_2{=}CCH_2CH_3H$ with benzene ring on the C

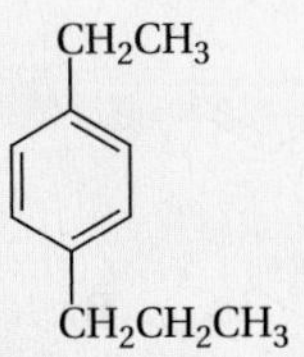

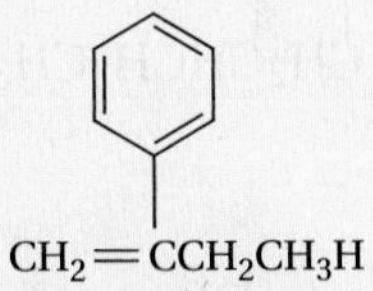

e. benzene ring with CH_2CH_3, CH_3CH_2, CH_2CH_3

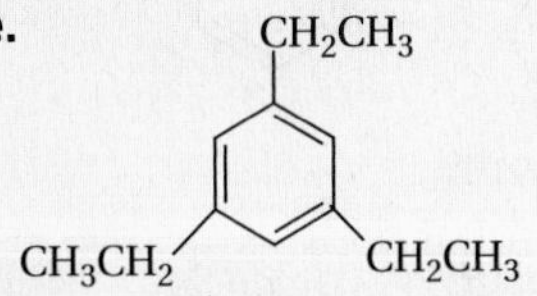

248. carbon–carbon double bond

249. four different groups attached to a carbon atom

250. b., d.

251. a. cis-2-pentene **b.** methylcyclooctane
c. 2-methyldecane **d.** 2-methyl-2-pentene
e. 2-methyl-3-ethyl-2-hexene

APPENDIX D

252. phenyl

253. more resistant because the benzene ring is stabilized by resonance

254. a., c., d., e.

256. coal, because it contains aromatic compounds from which benzene can be readily made.

Chapter 23

257. **a.** halogen **b.** carboxyl **c.** amino **d.** carbonyl **e.** ether

258. b., d, e.

259. 84.2 g/mol

260.

a. 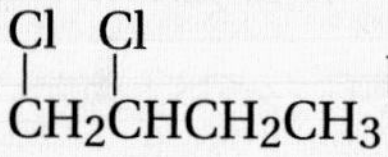 $CH_2ClCHClCH_2CH_3$ (Cl Cl on adjacent carbons)

b. 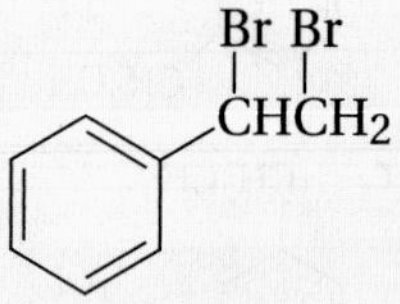 (benzene ring)–CHBrCH$_2$Br (Br Br)

c. 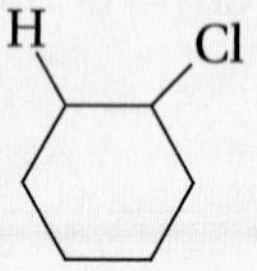 (cyclohexane ring with H and Cl)

d. 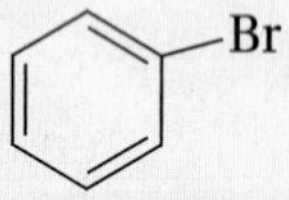 (benzene ring)–Br

252. What name is given to a benzene ring when it is a substituent in an organic molecule?

253. Are benzene rings more resistant or less resistant to chemical reactions than the carbon-carbon double bond in an alkane? Explain your answer.

254. Which structural feature(s) characterize alkanes?

a. carbon-hydrogen single bonds
b. carbon-carbon double bonds
c. carbon-carbon single bonds
d. may contain branched chains
e. may contain phenyl substituents

255. Draw structural formulas for these compounds.

a. 1,3-diethylbenzene
b. 1,4-ethylpropylbenzene
c. 1,3,5-triethylbenzene

256. If you want to devise an industrial process to prepare methylbenzene, would you prefer petroleum or coal as your starting material? Explain your choice.

Chapter 23

257. Name the functional group in these molecular structures.

a. CH_3CH_2Br

b. 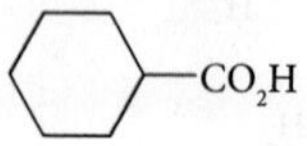 (cyclohexane ring)–CO_2H

c. 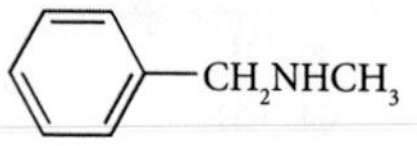 (benzene ring)–CH_2NHCH_3

d. 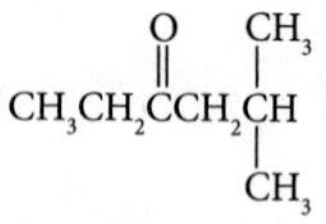 $CH_3CH_2C(=O)CH_2CH(CH_3)_2$

e. $CH_3CH_2CH_2OCH_3$

258. Which of these structures represent aryl halides?

a. 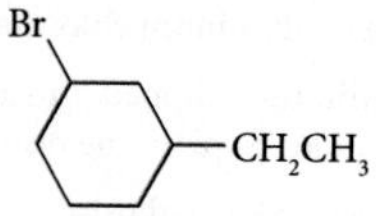 (cyclohexane ring with Br)–CH_2CH_3

b. 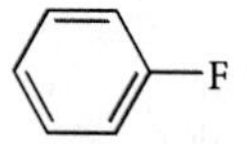 (benzene ring)–F

c. 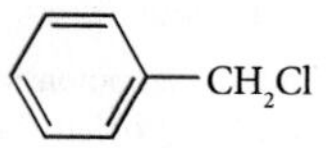 (benzene ring)–CH_2Cl

d. 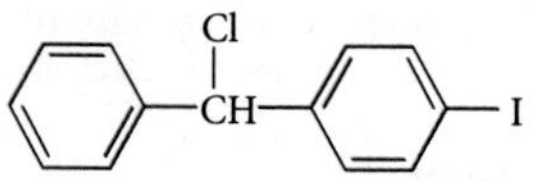 (benzene ring)–CH(Cl)–(benzene ring)–I

e. 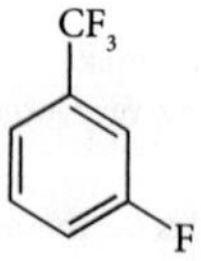 (benzene ring with CF_3 and F)

259. If 5.23 g of an alkene is required to completely decolorize 9.92 g of bromine, what is the molar mass of the alkene?

260. Write structural formulas for the organic product of these reactions.

a. $CH_2{=}CHCH_2CH_3 + Cl_2 \longrightarrow$

b. 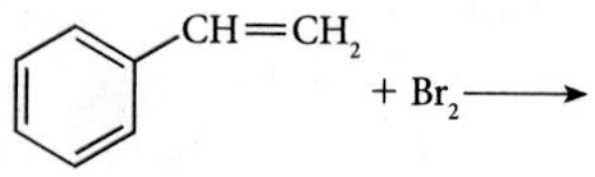 (benzene ring)–$CH{=}CH_2$ + $Br_2 \longrightarrow$

c. 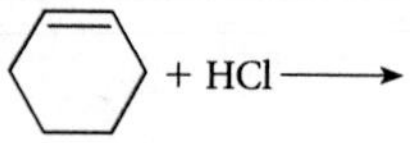 (cyclohexene ring) + HCl $\longrightarrow$

d. 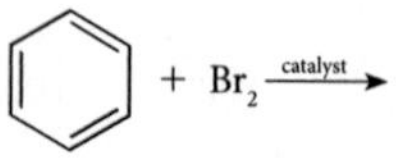 (benzene ring) + $Br_2 \xrightarrow{\text{catalyst}}$

261. Treatment of 7.57 g of pentanoic acid with a large excess of ethanol in the presence of a catalytic amount of hydrochloric acid produced 8.21 g of ethyl pentanoate after purification. Write a balanced equation for the reaction and calculate the percent yield of the ester.

262. Classify these reactions as oxidation or reduction.

a. $C_6H_5CH(OH)CH_3 \xrightarrow[H_2SO_4]{K_2Cr_2O_7} C_6H_5COCH_3$

b. $CH_3CH_2CH_2CH_2CHO \xrightarrow{CuSO_4} CH_3CH_2CH_2CH_2CO_2H$

c. $C_6H_5CH_2CH_3 + 3H_2 \xrightarrow{Pt} C_6H_{11}CH_2CH_3$ (cyclohexyl)

d. $CH_4 + 2O_2 \longrightarrow CO_2 + 2H_2O$

263. How many liters of hydrogen gas at STP are required to saturate 0.150 mole of benzene?

264. Write complete equations for the following organic reactions. Be sure to include all reactants and catalysts required for the transformations.

a. 1-pentene to pentane

b. 2,3-dimethyl-2-butene to 2-chloro-2,3-dimethylbutane

c. ethane to chloroethane

d. cyclohexene to 1,2-dibromocyclohexane

265. A sample of benzaldehyde is oxidized to produce the crystalline white solid, benzoic acid. Write the structural formula of benzoic acid.

266. Complete these reactions.

a. $CH_3CH_2CH_2I + KOH \xrightarrow[100°C]{H_2O}$

b. $C_6H_5{-}CH{=}CH{-}C_6H_5 + H_2O \xrightarrow[100°C]{H^+}$

c. $CH_3CH_2CHO \xrightarrow{K_2Cr_2O_7}$

d. $C_6H_5{-}CH(OH){-}CH_2{-}C_6H_5 \xrightarrow{K_2Cr_2O_7}$

267. Classify each of these reactions as a hydration, hydrogenation, hydrolysis, or substitution.

a. $C_5H_9{-}CH_2Cl + KOH \xrightarrow[100°C]{H_2O} C_5H_9{-}CH_2OH + KCl$ (cyclopentyl)

b. $C_6H_{11}{-}COOCH_3 + H_2O \xrightarrow{H^+} C_6H_{11}{-}COOH + CH_3OH$ (cyclohexyl)

c. cyclohexene $+ H_2O \xrightarrow[100°C]{H^+}$ $C_6H_{11}{-}OH$

d. $C_6H_{11}{-}C{\equiv}CH + 2H_2 \xrightarrow{Pt} C_6H_{11}{-}CH_2CH_3$

Chapter 24

268. Saliva contains an enzyme called amylase, which catalyzes the breakdown of starch to its monomers. Suggest a reason why a piece of bread begins to taste sweet when it has been chewed for a short time.

261. $CH_3CH_2CH_2CH_2COOH + CH_3CH_2OH \xrightarrow{HCl} CH_3CH_2CH_2CH_2COOCH_2CH_3$

85.1% yield of ethyl pentanoate

262. **a.** oxidation **b.** oxidation
c. reduction **d.** oxidation

263. 10.1 L

264. **a.** $CH_2{=}CHCH_2CH_2CH_3 + H_2 \xrightarrow{Pt} CH_3CH_2CH_2CH_2CH_3$

b. $(CH_3)_2C{=}C(CH_3)_2 + HCl \longrightarrow CH_3{-}CH(CH_3){-}CCl(CH_3){-}CH_3$

c. $CH_3CH_3 + Cl_2 \xrightarrow{UV\ light} CH_3CH_2Cl + HCl$

d. cyclohexene $+ Br_2 \longrightarrow$ 1,2-dibromocyclohexane

265. $C_6H_5{-}CO_2H$

266. **a.** $CH_3CH_2CH_2OH + KI$

b. $C_6H_5{-}CH(OH){-}CH_2{-}C_6H_5$

c. CH_3CH_2COOH

d. $C_6H_5{-}CO{-}CH_2{-}C_6H_5$

267. **a.** substitution **b.** hydrolysis
c. hydration **d.** hydrogenation

Chapter 24

268. Action of amylase breaks down starch to its glucose monomers; glucose tastes sweet.

269. a., b., c.

270. 1.67 x 10^{17} minutes; 2.78 x 10^{15} hours

271. 4.1 g

272. **a.** 342.0 g/mol **b.** 2.08 x $10^{-2}M$
c. $C_{12}H_{22}O_{11} + 12O_2 \rightarrow 12CO_2 + 11H_2O$

273. **a.** 23.0 g **b.** 8.67 x 10^2 kJ
c. 2.07 x 10^2 Cal **d.** 10.4%

274. 1.87 km

275. 1.5 x 10^8 turns

276. Add the equations and the free energies to obtain

phosphoenolpyruvic acid + ADP ⟶ ATP + pyruvic acid

$\Delta G = -30.9$ kJ/mol

Transfer of a phosphoryl group from phosphoenolpyruvate to ADP to form ATP is spontaneous.

277. 2.33 mol ATP

Chapter 25

278. The mass number of atom C is 4 less than that of atom A; the atomic number of C is 1 less than that of atom A.

279. ${}^{133}_{54}Xe \rightarrow {}^{133}_{55}Cs + {}^{0}_{-1}\beta$

280. **b.** and **d.** stable; neutron-to-proton ratio in Band of Stability.

281. **a.** ${}^{26}_{11}Na \rightarrow {}^{0}_{-1}\beta + {}^{26}_{12}Mg$
b. ${}^{234}_{92}U \rightarrow {}^{4}_{2}He + {}^{230}_{90}Th$

282. 1.69 x 10^5 years

269. Which of these statements apply to enzymes?
a. do not change position of equilibrium
b. catalyze biological reactions
c. bind substrates and sometimes cofactors at active sites
d. change position of equilibrium to favor products
e. are almost always nucleic acids

270. If one molecule of the enzyme catalase can break down 3.60 × 10^6 peroxide molecules in one minute, how many minutes would it take this enzyme molecule to break down 1 mole of peroxide? How many hours?

271. The level of glucose in the blood is normally 70–120 mg/dL. If a student's volume of blood is 4.5 L and her blood glucose level is 90 mg/dL, what is the total number of grams of glucose in her blood?

272. Sucrose, ordinary table sugar, has the molecular formula $C_{12}H_{22}O_{11}$.
a. What is the molar mass of sucrose?
b. What is the molarity of an aqueous solution containing 7.12 g/L of sucrose?
c. Write a balanced equation for the complete combustion of sucrose in air.

273. A student eats a 115-g hamburger that contains 20.0% by mass of fat. The chemical potential energy of fat is 37.7 kJ/gram.
a. How many grams of fat does the hamburger contain?
b. How many kilojoules of energy are stored in the fat?
c. What is this amount of energy in dietary Calories?
d. What percentage of total Calories does this amount of energy represent in a 2.00 × 10^3 Calorie/day diet?

274. Walnuts, rich in fats and oils and with a chemical potential energy of 25.8 kJ/g, are a nutritious treat for many people. Suppose you eat 27.3 g of walnuts and then take a brisk walk. If you burn 3.76 × 10^2 kJ of kinetic energy for each kilometer you walk, how far will you need to walk to use up the energy provided by the walnuts?

275. There are 3.4 base pairs in every complete turn of the double helix in a DNA molecule. How many turns are in a DNA molecule that contains 5.0 × 10^8 base pairs?

276. ATP is not the only biological molecule capable of transmitting energy. For example, hydrolysis of the molecule phosphoenolpyruvic acid releases even more free energy than the hydrolysis of ATP. Use the equations below to write a net equation showing that the energy released by the hydrolysis of phosphoenolpyruvic acid can be coupled to the phosphorylation of ADP to form ATP in a spontaneous reaction.

Phosphoenolpyruvic acid + H_2O ⟶ pyruvic acid + P_i
$\Delta G = -61.4$ kJ/mol

ADP + P_i + ⟶ ATP + H_2O
$\Delta G = 30.5$ kJ/mol

277. The hydrolysis of ATP to ADP and inorganic phosphate releases 30.5 kJ/mol of free energy. If all of this energy is wasted as heat, how many moles of ATP must be hydrolyzed to raise the temperature of 1 L of water at 20°C to 37°C?

Chapter 25

278. How do the mass number and the atomic number of radioactive atom A change if it first emits an alpha particle and the resulting atom then emits a beta particle to finally give atom C?

279. A radioactive active atom undergoes beta decay to yield cesium-133. Write a balanced equation for this nuclear reaction.

280. Which nuclei would you predict to be stable? Explain your answer.
a. ${}^{9}_{3}Li$ **c.** ${}^{20}_{8}O$
b. ${}^{59}_{27}Co$ **d.** ${}^{146}_{60}Nd$

281. Write a balanced nuclear equation for the following:
a. beta decay of sodium-26
b. alpha decay of ${}^{234}_{92}U$

282. Plutonium-239 presents a serious nuclear waste disposal problem. If seven half-lives are required for the radioactivity from the waste to reach a tolerable level and if Pu-239 has a $t_{1/2} = 2.41 \times 10^4$ years, how long must the Pu-239 be stored?

Appendix E Selected Solutions

Chapter 1

25. $12\ \cancel{\text{blocks}} \times \dfrac{1\ \cancel{\text{mi}}}{10\ \cancel{\text{blocks}}} \times \dfrac{20\ \text{min}}{1\ \cancel{\text{mi}}} = 24\ \text{min}$

26. $48\ \cancel{\text{min}} \times \dfrac{1\ \cancel{\text{mi}}}{20\ \cancel{\text{min}}} \times \dfrac{10\ \text{blocks}}{1\ \cancel{\text{mi}}} = 24\ \text{blocks}$

27. Algebra II would be scheduled during the 2nd period instead of 1st period.

28. No, Manny's schedule would not change. Algebra II would not be scheduled during 2nd period because Art must be scheduled during this time. Algebra II would not be scheduled during 3rd period because Chemistry must be scheduled during this time. Therefore, Algebra II would still be scheduled during 1st period.

35. Chemistry is concerned with the changes that matter undergoes.

37. A scientist who is studying cancer with the goal of finding an effective treatment is doing applied research because the research is directed toward a practical goal or application.

38. A firefighter needs to know which chemicals to use to fight different types of fires; a knowledge of chemistry will help a reporter gather information during an interview with a chemist.

42. Scientists can study the composition of distant stars by analyzing the light they transmit to Earth.

44. The most powerful tool that any scientist can have is the scientific method.

46. c

48. Repeat the experiment. If you get the same result, you must propose a new hypothesis.

49. A scientific law summarizes the results of many experiments; a theory explains the results of the experiments.

51. Good problem solvers do b, c, and d.

54. Two games won out of three games played means that one game is lost out of three games played:

$162\ \cancel{\text{games played}} \times \dfrac{1\ \text{loss}}{3\ \cancel{\text{games played}}} = 54\ \text{losses}$

56. $1{,}000{,}000\ \cancel{\text{pennies}} \times \dfrac{1\ \cancel{\text{sec}}}{1\ \cancel{\text{penny}}} \times \dfrac{1\ \cancel{\text{min}}}{60\ \cancel{\text{sec}}} \times \dfrac{1\ \cancel{\text{h}}}{60\ \cancel{\text{min}}} \times$

$\dfrac{1\ \text{day}}{24\ \cancel{\text{h}}} = 11.6\ \text{days (about 12 days)}$

57. Answers vary; possible answers: 1c, 2d, 3e, 4b, 5a

58. One possible answer is the development of materials to produce artificial limbs.

61. The doctor's hypothesis is that the sore throat is the result of bacteria that cause strep throat. The experiment she performs to test the hypothesis is to test the sample for the presence of the bacteria that causes strep.

62. Your experiment may be correct, but your hypothesis may be wrong. You should reexamine your hypothesis and repeat the experiment.

65. **a.** Independent variable: amount of salt added

b. Dependent variable: freezing point of the salt water

c. Changing the volume of the water would also change the relative amount of salt per volume of water and would affect the results.

d. Yes, to a point. The freezing point appears to be leveling off at about 15°C.

66. $40\ \cancel{\text{mi by bicycle}} \times \dfrac{30\ \text{mi by car}}{4\ \cancel{\text{mi by bicycle}}} = 300\ \text{mi by car}$

70. A person who is educated in the theories and practice of chemistry is more likely to recognize the significance of an accidental discovery and have the means and motivation to develop that accidental discovery into an important scientific contribution.

71. Method 1: Divide the weight of four beakers by 2; 2.0 lb/2 = 1.0 lb. Method 2: Multiply the weight of one beaker by 2; 0.5 lb × 2 = 1.0 lb. The answers are the same.

72. A theory can never be proven. It is a well-tested explanation of a broad set of observations. A theory may need to be changed in the future to explain new observations.

73. c

75. Number of crates = 5 × 6 × 5 = 150 crates

$150\ \cancel{\text{crates}} \times \dfrac{4\ \cancel{\text{boxes}}}{1\ \cancel{\text{crate}}} \times \dfrac{20\ \cancel{\text{cartons}}}{1\ \cancel{\text{box}}} \times \dfrac{12\ \text{eggs}}{1\ \cancel{\text{carton}}}$

$= 144{,}000\ \text{eggs}$

77. **a.** $1\ \cancel{\text{package}} \times \dfrac{1\ \cancel{\text{carton}}}{6\ \cancel{\text{packages}}} \times \dfrac{1\ \cancel{\text{box}}}{12\ \cancel{\text{cartons}}} \times \dfrac{1\ \cancel{\text{crate}}}{8\ \cancel{\text{boxes}}} \times$

$\dfrac{\$576.00}{1\ \cancel{\text{crate}}} = \$1.00$

b. number of envelopes in a package

Selected Solutions **R73**

80. Answers will vary, but possible answers are as follows: Factors: (1) PLA is made from natural resources; (2) producing PLA requires less energy than producing petroleum-based plastics; (3) compost facility availability. Factors (1) and (2) make PLA products a good choice, but factor (3) may make it difficult to choose PLA products.

Chapter 2

10. Iron is magnetic; table salt is not. Table salt will dissolve in water; iron will not.

11. By lowering the temperature to below the boiling point of each gas, you could condense each substance and separate the gases.

20. Liquid A is probably a substance. Liquid B is a mixture because a solid remained after evaporation.

21. The liquid was not an element because a solid was left when the liquid evaporated. A physical process, such as evaporation, cannot be used to break down a compound. Therefore, the liquid was a mixture.

42. State; both are gases.

44. A vapor; the term *vapor* is used to refer to the gaseous state of a substance which normally exists as a liquid or solid at room temperature.

47. Sharpening a pencil is an irreversible change. Making ice cubes is a reversible change.

50. **a.** heterogeneous
b. homogeneous
c. depends on how well the batter is mixed
d. homogeneous
e. heterogeneous
f. homogeneous
g. heterogeneous
h. homogeneous

53. **a.** Sodium chloride (table salt) is a compound because it is made of two elements, sodium and chloride.

b. Salt water is a mixture because it is made of two compounds, water (H_2O) and NaCl.

c. Sodium is an element because it is not combined with any other element.

57. The compound water contains two parts hydrogen to one part oxygen.

60. **a.** physical
b. chemical (color change)
c. chemical (production of a gas)
d. physical

62. $40\text{ g } NH_4NO_3 - 14\text{ g N} - 8\text{ g O} = 18\text{ g } H_2O$

66. Sulfur, which is the only substance in the table that is a solid at room temperature

67. Mass is an extensive property, which depends only on the amount of matter in the sample, not on the composition of the sample.

70. The particles in solids are packed tightly together so it is difficult to compress them. The particles in a gas are spaced relatively far apart.

76. **a.** gas produced
b. formation of a precipitate
c. color and texture change
d. energy change, odor change
e. gas produced
f. color change, odor change

80. The wax appears to disappear because the products of the reaction—carbon dioxide and water vapor—are colorless gases.

81. Add sufficient water to dissolve all of the sugar. Separate the charcoal and sand from the sugar water by filtration. Large pieces of charcoal could be separated on the basis of color or size. Small pieces of charcoal could be burned.

85. **a.** Yes; because the graph is a straight line, the proportion of iron to oxygen is a constant, which is true for a compound.

b. No; a point for the values given wouldn't fall on the line. The mass ratio of iron to oxygen is different.

88. **a.** oxygen and calcium
b. silicon, aluminum, and iron
c. Different; the second most abundant element in Earth's crust, silicon, is not abundant in the human body, and the second most abundant element in the human body, carbon, is not among the most abundant elements in Earth's crust. If the elements are different, then the compounds must also be different.

Chapter 3

1. **a.** $(6.6 \times 10^{-8}) + (5.0 \times 10^{-9}) =$
$(6.6 \times 10^{-8}) + (0.5 \times 10^{-8}) =$
$(6.6 + 0.5) \times 10^{-8} = 7.1 \times 10^{-8}$

b. $(9.4 \times 10^{-2}) - (2.1 \times 10^{-2}) =$
$(9.4 - 2.1) \times 10^{-2} = 7.3 \times 10^{-2}$

2. $\dfrac{6.6 \times 10^{6}}{(8.8 \times 10^{-2}) \times (2.5 \times 10^{3})} = 0.30 \times 10^{5} = 3.0 \times 10^{4}$

R74 Appendix E

3. Error = experimental value − accepted value = 2.04 m − 2.00 m = 0.04 m

$\% \text{ error} = \dfrac{|\text{error}|}{\text{accepted value}} \times 100\% = \dfrac{|0.04 \text{ m}|}{2.00 \text{ m}} \times 100\%$

= 2%

4. **a.** 4 **b.** 4 **c.** 2 **d.** 5

5. **a.** 3 **b.** 2 **c.** 4 **d.** 4

6. **a.** 8.71×10^1 m **b.** 4.36×10^8 m **c.** 1.55×10^{-2} m **d.** 9.01×10^3 m **e.** 1.78×10^{-3} m **f.** 6.30×10^2 m

7. **a.** 9×10^1 m **b.** 4×10^8 m **c.** 2×10^{-2} m **d.** 9×10^3 m **e.** 2×10^{-3} m **f.** 6×10^2 m

8. **a.** 61.2 m + 9.35 m + 8.6 m = 79.15 m = 79.2 m

b. 9.44 m − 2.11 m = 7.33 m

c. 1.36 m + 10.17 m = 11.53 m

d. 34.61m − 17.3 m = 17.3 m

9. 14.2 g + 8.73 g + 0.912 g = 23.842 g = 23.8 g

10. **a.** $8.3 \text{ m} \times 2.22 \text{ m} = 18.4 \text{ m}^2 = 18 \text{ m}^2 = 1.8 \times 10^1 \text{ m}^2$

b. $8432 \text{ m}^2 \div 12.5 \text{ m} = 675 \text{ m} = 6.75 \times 10^2 \text{ m}$

c. $35.2 \cancel{s} \times \dfrac{1 \text{ min}}{60 \cancel{s}} = 0.587 \text{ min} = 5.87 \times 10^{-1} \text{ min}$

11. $V = l \times w \times h = 22.4 \text{ m} \times 11.3 \text{ m} \times 5.2 \text{ m} = 1316.2 \text{ m}^3 = 1.3 \times 10^3 \text{ m}^3$

19. mp: K = °C + 273 = 960.8 + 273 = 1234 K
bp: K = °C + 273 = 2212 + 273 = 2485 K

20. °C = K − 273 = 77.2 − 273 = −196°C

21. $\text{Density} = \dfrac{\text{mass}}{\text{volume}} = \dfrac{612 \text{ g}}{245 \text{ cm}^3} = 2.50 \text{ g/cm}^3$

No, because the density of aluminum is 2.7 g/cm^3

22. $\text{Density} = \dfrac{\text{mass}}{\text{volume}} = \dfrac{68.0 \text{ g}}{6.48 \text{ cm}^3} = 10.5 \text{ g/cm}^3$

36. $1 \cancel{\text{wk}} \times \dfrac{7 \cancel{d}}{1 \cancel{\text{wk}}} \times \dfrac{24 \cancel{h}}{1 \cancel{d}} \times \dfrac{60 \text{ min}}{1 \cancel{h}} = 10{,}080 \text{ min} = 1.0080 \times 10^4 \text{ min}$

37. $40 \cancel{h} \times \dfrac{60 \cancel{\text{min}}}{1 \cancel{h}} \times \dfrac{60 \text{ s}}{1 \cancel{\text{min}}} = 144{,}000 \text{ s} = 1.44000 \times 10^5 \text{ s}$

38. $570 \cancel{\text{cm ribbon}} \times \dfrac{1 \text{ student}}{8.5 \cancel{\text{cm ribbon}}} = 67 \text{ students}$

39. $48.0 \cancel{\text{°C change}} \times \dfrac{1.80\text{°F change}}{1.00 \cancel{\text{°C change}}} = 86.4\text{°F change}$

40. $5.00 \cancel{\text{g Au}} \times \dfrac{1 \text{ atom Au}}{3.271 \times 10^{-22} \cancel{\text{g Au}}} = 1.53 \times 10^{22} \text{ atoms Au}$

41. **a.** $0.044 \cancel{\text{km}} \times \dfrac{10^3 \text{ m}}{1 \cancel{\text{km}}} = 44 \text{ m}$

b. $4.6 \cancel{\text{mg}} \times \dfrac{1 \text{ g}}{10^3 \cancel{\text{mg}}} = 4.6 \times 10^{-3} \text{ g}$

c. $0.107 \cancel{g} \times \dfrac{10^2 \text{ cg}}{1 \cancel{g}} = 10.7 \text{ cg}$

42. **a.** $15 \cancel{\text{cm}^3} \times \dfrac{1 \text{ L}}{10^3 \cancel{\text{cm}^3}} = 0.015 \text{ L} = 1.5 \times 10^{-2} \text{ L}$

b. $7.38 \cancel{g} \times \dfrac{1 \text{ kg}}{10^3 \cancel{g}} = 7.38 \times 10^{-3} \text{ kg}$

c. $6.7 \cancel{s} \times \dfrac{10^3 \text{ ms}}{1 \cancel{s}} = 6.7 \times 10^3 \text{ ms}$

d. $94.5 \cancel{g} \times \dfrac{10^6 \text{ µg}}{1 \cancel{g}} = 9.45 \times 10^7 \text{ µg}$

43. **a.** $14.8 \cancel{\text{g B}} \times \dfrac{1 \text{ cm}^3}{2.34 \cancel{\text{g B}}} = 6.32 \text{ cm}^3$

b. $4.62 \cancel{\text{g Hg}} \times \dfrac{1 \text{ cm}^3}{13.5 \cancel{\text{g Hg}}} = 0.342 \text{ cm}^3 = 3.42 \times 10^{-1} \text{ cm}^3$

44. **a.** $\text{Volume} = \dfrac{\text{mass}}{\text{density}} = \dfrac{14.8 \cancel{\text{g B}}}{2.34 \cancel{\text{g B}}/\text{cm}^3} = 6.32 \text{ cm}^3$

b. $\text{Volume} = \dfrac{\text{mass}}{\text{density}} = \dfrac{4.62 \cancel{\text{g Hg}}}{13.5 \cancel{\text{g Hg}}/\text{cm}^3} = 0.342 \text{ cm}^3$

$= 3.42 \times 10^{-1} \text{ cm}^3$

45. $50.0 \cancel{\text{cm}^3} \times \dfrac{0.950 \text{ g}}{1 \cancel{\text{cm}^3}} = 47.5 \text{ g}$

46. $0.227 \cancel{\text{nm}} \times \dfrac{1 \text{ m}}{10^9 \cancel{\text{nm}}} = 0.227 \times 10^{-9} \text{ m} = 2.27 \times 10^{-8} \text{ m}$

47. $1.3 \times 10^4 \cancel{\text{km}} \times \dfrac{10^3 \cancel{m}}{1 \cancel{\text{km}}} \times \dfrac{10 \text{ dm}}{1 \cancel{m}} = 1.3 \times 10^{4+3+1} \text{ dm}$

$= 1.3 \times 10^8 \text{ dm}$

48. $\dfrac{19.3 \cancel{g}}{1 \cancel{\text{cm}^3}} \times \dfrac{10^6 \cancel{\text{cm}^3}}{1 \text{ m}^3} \times \dfrac{1 \text{ kg}}{10^3 \cancel{g}} = 19.3 \times 10^{6-3} \text{ kg/m}^3$

$= 19.3 \times 10^3 \text{ kg/m}^3 = 1.93 \times 10^4 \text{ kg/m}^3$

49. $\dfrac{7.0 \times 10^6 \text{ RBCs}}{1 \cancel{\text{mm}^3}} \times \dfrac{10^9 \cancel{\text{mm}^3}}{1 \cancel{\text{m}^3}} \times \dfrac{1 \cancel{\text{m}^3}}{10^3 \cancel{\text{dm}^3}} \times \dfrac{1 \cancel{\text{dm}^3}}{1 \text{ L}}$

$= 7.0 \times 10^{6+9-3} \text{ RBCs/L} = 7.0 \times 10^{12} \text{ RBCs/L}$

59. **a.** 43 g **b.** 7.3 cm^2 **c.** 225.8 L **d.** 92.0 m **e.** 32.4 m^3 **f.** 104 m^3

60. (58) **a.** 9.85×10^1 L **b.** 7.63×10^{-4} cg **c.** 5.70×10^1 m **d.** 1.22×10^1°C **e.** 7.50×10^{-3} mm **f.** 1.76×10^3 mL

(59) **a.** 4.3×10^1 g **b.** 7.3×10^0 cm^2 **c.** 2.258×10^2 L **d.** 9.20×10^1 m **e.** 3.24×10^1 m^3 **f.** 1.04×10^2 m^3

Selected Solutions R75

63. pm, nm, μm, mm, cm, dm, m, km; $1\text{ pm} = 10^{-12}\text{ m}$, $1\text{ nm} = 10^{-9}\text{ m}$, $1\ \mu\text{m} = 10^{-6}\text{ m}$, $1\text{ mm} = 10^{-3}\text{ m}$, $1\text{ cm} = 10^{-2}\text{ m}$, $1\text{ dm} = 10^{-1}\text{ m}$, $1\text{ km} = 10^{3}\text{ m}$

66. $K = °C + 273 = 962 + 273 = 1235\text{ K}$

69. $\text{Density} = \frac{\text{mass}}{\text{volume}} = \frac{57.3\text{ g}}{4.7\text{ cm}^3} = 12\text{ g/cm}^3$

No; the density of the metal bar is 12 g/cm^3, but the density of gold is 19 g/cm^3.

74. **a.** $157\text{ cs} \times \frac{1\text{ s}}{100\text{ cs}} = 1.57\text{ s}$

b. $42.7\text{ L} \times \frac{10^3\text{ mL}}{1\text{ L}} = 42{,}700\text{ mL} = 4.27 \times 10^4\text{ mL}$

c. $261\text{ nm} \times \frac{1\text{ m}}{10^9\text{ nm}} \times \frac{10^3\text{ mm}}{1\text{ m}} = 261 \times 10^{3-9}\text{ mm}$

$= 261 \times 10^{-6}\text{ mm} = 2.61 \times 10^{-4}\text{ mm}$

d. $0.065\text{ km} \times \frac{10^3\text{ m}}{1\text{ km}} \times \frac{10\text{ dm}}{1\text{ m}} = 650\text{ dm} =$

$6.5 \times 10^2\text{ dm}$

e. $642\text{ cg} \times \frac{1\text{ g}}{10^2\text{ cg}} \times \frac{1\text{ kg}}{10^3\text{ g}} = \frac{642\text{ kg}}{10^{2+3}} = \frac{642\text{ kg}}{10^5}$

$= 642 \times 10^{-5}\text{ kg} = 6.42 \times 10^{-3}\text{ kg}$

f. $8.25 \times 10^2\text{ cg} \times \frac{1\text{ g}}{10^2\text{ cg}} \times \frac{10^9\text{ ng}}{1\text{ g}} = 8.25 \times 10^9\text{ ng}$

75. **a.** $\frac{0.44\text{ mL}}{1\text{ min}} \times \frac{1\text{ L}}{10^3\text{ mL}} \times \frac{10^6\ \mu\text{L}}{1\text{ L}} \times \frac{1\text{ min}}{60\text{s}}$

$= (7.3 \times 10^{-3}) \times 10^3\ \mu\text{L/s} = 7.3\ \mu\text{L/s}$

b. $\frac{7.86\text{ g}}{1\text{ cm}^2} \times \frac{10^4\text{ cm}^2}{1\text{ m}^2} \times \frac{1\text{ m}^2}{10^6\text{ mm}^2} \times \frac{10^3\text{ mg}}{1\text{ g}}$

$= 7.86 \times 10^1\text{ mg/mm}^2 = 78.6\text{ mg/mm}^2$

c. $\frac{1.54\text{ kg}}{\text{L}} \times \frac{10^3\text{g}}{1\text{ kg}} \times \frac{1\text{ L}}{10^3\text{ mL}} \times \frac{1\text{ mL}}{1\text{ cm}^3} = 1.54\text{ g/cm}^3$

77. **a.** $28.3\text{ cg} \times \frac{1\text{ g}}{10^2\text{ cg}} \times \frac{10^3\text{ mg}}{\text{g}} = 28.3 \times 10^1\text{ mg}$

$= 283\text{ mg}$

b. $283\text{ mg} \times \frac{1\text{ g}}{10^3\text{ mg}} = 0.283\text{ g}$

c. $0.283\text{ g} \times \frac{1\text{ kg}}{10^3\text{ g}} = 0.000283\text{ kg}$

d. $6.6 \times 10^3\text{ mg} \times \frac{1\text{ g}}{10^3\text{ mg}} = 6.6\text{ g}$

e. $6.6\text{ g} \times \frac{10^2\text{ cg}}{1\text{ g}} = 660\text{ cg} = 6.6 \times 10^2\text{ cg}$

f. $6.6\text{ g} \times \frac{1\text{ kg}}{10^3\text{ g}} = 6.6 \times 10^{-3}\text{ kg}$

g. $2.8 \times 10^{-4}\text{ g} \times \frac{10^3\text{ mg}}{\text{g}} = 2.8 \times 10^{-1}\text{ mg}$

h. $2.8 \times 10^{-4}\text{ g} \times \frac{10^2\text{ cg}}{\text{g}} = 2.8 \times 10^{-2}\text{ cg}$

i. $2.8 \times 10^{-4}\text{ g} \times \frac{1\text{ kg}}{10^3\text{ g}} = 2.8 \times 10^{-7}\text{ kg}$

86. $125\text{ kg coal} \times \frac{1.30\text{ kg C}}{2.00\text{ kg coal}} = 81.25\text{ kg C} = 81.3\text{ kg C}$

87. $50\text{ g air} \times \frac{1\text{ cm}^3\text{ air}}{1.19 \times 10^{-3}\text{ g air}} = 42.0 \times 10^3\text{ cm}^3\text{ air}$

$= 4.20 \times 10^4\text{ cm}^3\text{ air}$

89. $\text{Volume} = 158\text{ g } H_2O \times \frac{1\text{ cm}^3}{1.000\text{ g } H_2O} = 158\text{ cm}^3$

$\text{Density} = \frac{\text{mass}}{\text{volume}} = \frac{127\text{ g}}{158\text{cm}^3} = 0.804\text{ g/cm}^3$

90. $\frac{0.15\text{ s}}{1\text{ min}} \times \frac{60\text{ min}}{1\text{ h}} \times \frac{24\text{ h}}{1\text{ day}} \times \frac{1\text{ min}}{60\text{ s}}$

$= 0.15 \times 24\text{ min/day} = 3.6\text{ min/day lost}$

91. $V = 28.6\text{ cm} \times (73.0 \times 10^{-1}\text{ cm}) \times (0.72 \times 10^2\text{ cm})$
$= 1.50 \times 10^4\text{ cm}^3$

$\text{Density} = \frac{\text{mass}}{\text{volume}} =$

$\frac{1.38 \times 10^4\text{ g}}{1.50 \times 10^4\text{ cm}^3} \times \frac{1\text{ kg}}{10^3\text{ g}} \times \frac{10^3\text{ cm}^3}{1\text{ L}} = 0.92\text{ kg/L}$

93. $1.5 \times 10^8\text{ km} \times \frac{10^3\text{ m}}{1\text{ km}} \times \frac{1\text{ s}}{3.0 \times 10^8\text{ m}} \times \frac{1\text{ min}}{60\text{ s}}$

$= 8.3\text{ min}$

94. $\frac{5.52\text{ g}}{1\text{ cm}^3} \times \frac{1\text{ kg}}{10^3\text{ g}} \times \frac{10^6\text{ cm}^3}{1\text{ m}^3} \times \frac{1\text{ m}^3}{10^3\text{ dm}^3} = \frac{5.52 \times 10^6\text{ kg}}{10^6\text{ dm}^3}$

$= 5.52\text{ kg/dm}^3$

97. $\text{Mass}_{\text{amalgam}} = 26.0\text{ g} + 10.8\text{ g} + 2.4\text{ g} + 0.8\text{ g} = 40.0\text{ g}$

$25.0\text{ g amalgam} \times \frac{26.0\text{ g Ag}}{40.0\text{ g amalgam}} = 16.3\text{ g Ag}$

98. $\frac{112\text{ km}}{1\text{ hr}} \times \frac{10^3\text{ m}}{1\text{ km}} \times \frac{1\text{ hr}}{60\text{ min}} \times \frac{1\text{ min}}{60\text{ s}} =$

$0.0311 \times 10^3\text{ m/s} = 31.1\text{ m/s}$

103. $\text{Volume of Fe} = 355\text{ g Fe} \times \frac{1\text{ cm}^3\text{ Fe}}{7.87\text{ g Fe}} = 45.1\text{ cm}^3\text{ Fe}$

$\text{Mass} = \text{volume} \times \text{density} = 45.1\text{ cm}^3 \times \frac{11.3\text{ g Pb}}{1\text{ cm}^3}$

$= 510\text{ g Pb}$

R76 Appendix E

104. $\frac{8.0 \times 10^{-1}\text{ cg Sr}}{1\ \cancel{\text{kg seawater}}} \times \frac{1\ \cancel{\text{kg seawater}}}{10^3\ \cancel{\text{g seawater}}} \times \frac{1.0\ \cancel{\text{g seawater}}}{1\ \cancel{\text{cm}^3\text{ seawater}}} \times \frac{10^6\ \cancel{\text{cm}^3\text{ seawater}}}{1\text{ m}^3\text{ seawater}} \times \frac{1\text{ g Sr}}{10^2\ \cancel{\text{cg Sr}}} = 8.0\text{ g Sr/m}^3\text{ seawater}$

107. $34.5\ \cancel{\text{g Au}} \times \frac{1\ \cancel{\text{cm}^3\text{ Au}}}{19.3\ \cancel{\text{g Au}}} \times \frac{1\text{ mL Au}}{1\ \cancel{\text{cm}^3\text{ Au}}} = 1.79\text{ mL Au}$

Chapter 4

16. **a.** 9 protons and 9 electrons
b. 20 protons and 20 electrons
c. 13 protons and 13 electrons
d. 19 protons and 19 electrons

17. **a.** 16 **b.** 16 **c.** 23 **d.** 23 **e.** B **f.** 5 **g.** 5

18. **a.** neutrons = mass number − atomic number = 80 − 35 = 45
b. neutrons = mass number − atomic number = 32 − 16 = 16
c. neutrons = mass number − atomic number = 108 − 47 = 61
d. neutrons = mass number − atomic number = 207 − 82 = 125

19. **a.** $^{12}_{6}\text{C}$ **b.** $^{11}_{5}\text{B}$ **c.** $^{9}_{4}\text{Be}$ **d.** $^{16}_{8}\text{O}$

20. $^{16}_{8}\text{O}$, $^{17}_{8}\text{O}$, $^{18}_{8}\text{O}$

21. neutrons = mass number − atomic number = 50 − 24 = 26; chromium-50 has 26 neutrons.

neutrons = mass number − atomic number = 52 − 24 = 28; chromium-52 has 28 neutrons.

neutrons = mass number − atomic number = 53 − 24 = 29; chromium-53 has 29 neutrons.

22. Boron's atomic mass (10.81 amu) is closer to 11 than it is to 10, so boron-11 is more abundant than boron-10.

23. Silicon's atomic mass (28.086 amu) is closest to 28 and closer to 29 than it is to 30. Therefore, silicon-28 is more abundant than silicon-29, and silicon-29 is more abundant than silicon-30.

24. for ^{63}Cu: 62.93 amu × 0.692 = 43.5 amu
for ^{65}Cu: 64.93 amu × 0.308 = 20.0 amu
atomic mass = 43.5 amu + 20.0 amu = 63.5 amu

25. for ^{79}Br: 78.92 amu × 0.5069 = 40.00 amu
for ^{81}Br: 80.92 amu × 0.4931 = 39.90 amu
atomic mass = 40.00 amu + 39.90 amu = 79.90 amu

40. repel

44. He did not expect alpha particles to be deflected at a large angle.

51. **a.** 19 **b.** 9 **c.** 14 **d.** 29
e. 22 **f.** 22 **g.** 25 **h.** 30

53. for ^{6}Li: 6.015 amu × 0.075 = 0.45 amu
for ^{7}Li: 7.016 amu × 0.925 = 6.49 amu
atomic mass = 0.45 amu + 6.49 amu = 6.94 amu

54. because of the existence of isotopes

60. 5 protons and 6 neutrons in the nucleus; 5 electrons outside the nucleus

64. $[(82 + 122) \times 0.014] + [(82 + 124) \times 0.241] + [(82 + 125) \times 0.221] + [(82 + 126) \times 0.524] = 207$ amu

66. **a.** 8,289,000 + 502,570 + 120,570 + 7800 + 1435 + 477 + 211 + 198 = 8,922,261
8,289,000 ÷ 8,922,261 = 0.929 = 92.9%
b. (8,289,000 + 502,570 + 120,570) ÷ 8,922,261 = 0.9989 = 99.89%
c. (477 + 211 + 198) ÷ 8,922,261 = 0.0000993 = 0.00993%

67. $^{14}_{7}\text{N}$: 14.003 amu; 99.63%
$^{15}_{7}\text{N}$: 15.000 amu; 0.37%
average atomic mass = 14.01 amu

74. The following are reasonable hypotheses: (*i*) The space in an individual atom is large relative to the volume of the atom but very small relative to an object the size of a hand. (*ii*) There are many layers of atoms in a wall or a desk. The space that exists is distributed evenly throughout the solid, similar to the distribution of air pockets in foam insulation.

78. $6.941 = [6.015 \times (1 - A)] + (7.016 \times A)$
$6.941 = 6.015 - 6.015A + 7.016A$
$6.941 - 6.015 = 7.016A - 6.015A$

$$A = \frac{0.926}{1.001} = 0.925 = 92.5\%$$

79. $17(1.67 \times 10^{-24}\text{ g}) + 18(1.67 \times 10^{-24}\text{ g}) + 17(9.11 \times 10^{-28}\text{ g}) = 5.857 \times 10^{-23}\text{ g}$
$5.857 \times 10^{-23}\text{ g} - (5.81 \times 10^{-23}\text{ g}) = 4.70 \times 10^{-25}\text{ g}$

86. $54\text{ g H}_2\text{O} - 6\text{ g H} = 48\text{ g O}$

88. $4.42\ \cancel{\text{cm}^3} \times \frac{22.5\text{ g Pt}}{1\ \cancel{\text{cm}^3}} = 99.5\text{ g Pt}$

Chapter 5

8. **a.** $1s^2 2s^2 2p^2$
b. $1s^2 2s^2 2p^6 3s^2 3p^6$
c. $1s^2 2s^2 2p^6 3s^2 3p^6 3d^8 4s^2$

9. **a.** $1s^2 2s^2 2p^1$; 1 unpaired electron
b. $1s^2 2s^2 2p^6 3s^2 3p^2$; 2 unpaired electrons
c. $1s^2 2s^2 2p^6 3s^2 3p^4$; 2 unpaired electrons

15. $\lambda = \frac{c}{\nu} = \frac{2.998 \times 10^8\text{ m}/\cancel{\text{s}}}{1.50 \times 10^{13}/\cancel{\text{s}}} = 2.00 \times 10^{-5}\text{m}$; longer wavelength than red light

Selected Solutions **R77**

16. $\nu = \frac{c}{\lambda} = \frac{2.998 \times 10^8 \text{ m/s}}{5.00 \times 10^{-8} \text{ m}} = 6.00 \times 10^{15}\text{/s}$;

ultraviolet

17. $E = h\nu = (6.626 \times 10^{-34} \text{ J}\cdot\text{s}) \times (5.00 \times 10^{11}\text{/s}) = 3.31 \times 10^{-22} \text{ J}$

18. $\nu = \frac{c}{\lambda} = \frac{2.998 \times 10^8 \text{m/s}}{260 \text{ nm}} \times \frac{10^9 \text{ nm}}{1 \text{ m}} = 1.2 \times 10^{15}\text{/s}$

$E = h\nu = (6.626 \times 10^{-34} \text{ J}\cdot\text{s}) \times (1.2 \times 10^{15}\text{/s})$
$= 8.0 \times 10^{-19} \text{ J}$

28. Bohr proposed that electrons traveled in circular paths around the nucleus.

30. An electron is found 90% of the time inside this boundary.

33. The $2p$ sublevel contains three orbitals: $2p_x$, $2p_y$, and $2p_z$.

34. a. 1 ($1s$) b. 2 ($2s$, $2p$) c. 3 ($3s$, $3p$, $3d$) d. 4 ($4s$, $4p$, $4d$, $4f$)

35. Aufbau principle: Electrons occupy the lowest possible energy levels. Pauli Exclusion Principle: An atomic orbital can hold at most two electrons. Hund's Rule: One electron occupies each of a set of orbitals with equal energies before any pairing of electrons occurs.

36. $2s$, $3p$, $4s$, $3d$

37. a. valid b. invalid c. invalid d. valid

39. The p orbitals in the third quantum level have three electrons.

42. a. 2 b. 3 c. 1 d. 6

44. a. $1s^2 2s^2 2p^6 3s^2 3p^6 3d^{10} 4s^2 4p^4$
b. $1s^2 2s^2 2p^6 3s^2 3p^6 3d^2 4s^2$
c. $1s^2 2s^2 2p^6 3s^2 3p^6 3d^3 4s^2$
d. $1s^2 2s^2 2p^6 3s^2 3p^6 4s^2$

47. a. v, vi, iv, iii, i, ii
b. It is the reverse.

50. A quantum is a discrete amount of energy. Photons are light quanta.

51. A photon of ultraviolet light has a higher frequency (smaller wavelength) than a photon of infrared light. Therefore, a photon of ultraviolet light has more energy than a photon of infrared light.

52. $E = h\nu = (6.626 \times 10^{-34} \text{ J}\cdot\text{s}) \times (5.80 \times 10^{14}\text{/s}) = 3.84 \times 10^{-19} \text{ J}$

53. Classical physics views energy changes as continuous. In the quantum concept, energy changes occur in tiny discrete units called quanta.

54. The electron of the hydrogen atom is raised (excited) to a higher energy level.

57. $1s^2 2s^2 2p^6 3s^2 3p^6 3d^{10} 4s^2 4p^3$; $n = 1$, 2 electrons; $n = 2$, 8 electrons; $n = 3$, 18 electrons; $n = 4$, 5 electrons; the fourth energy level ($n = 4$) is not filled.

59. $1s^2 2s^2 2p^3$; nitrogen (7 electrons); 3 unpaired electrons (in the $2p$ sublevel)

60. a. Na, sodium (11 electrons)
b. N, nitrogen (7 electrons)
c. Si, silicon (14 electrons)
d. O, oxygen (8 electrons)
e. K, potassium (19 electrons)
f. Ti, titanium (22 electrons)

63. a. $\lambda = 4.36 \times 10^{-7} \text{ m} \times \frac{10^2 \text{ cm}}{1 \text{ m}} = 4.36 \times 10^{-5} \text{ cm}$
b. ultraviolet
c. $\nu = \frac{c}{\lambda} = \frac{2.998 \times 10^8 \text{ m/s}}{4.36 \times 10^{-7} \text{ m}} = 6.88 \times 10^{14}\text{/s}$

65. a. Electrons with a low velocity will be emitted.
b. More electrons will be emitted but with a low velocity.
c. Electrons will be emitted with a higher velocity.

66. $\nu = \frac{c}{\lambda} = \frac{2.998 \times 10^8 \text{ m/s}}{6.45 \times 10^{-7} \text{ cm}} \times \frac{10^2 \text{ cm}}{1 \text{ m}}$
$= 4.65 \times 10^{14}\text{/s}$

$E = h\nu = (6.626 \times 10^{-34} \text{ J}\cdot\text{s}) \times (4.65 \times 10^{14}\text{/s})$
$= 3.08 \times 10^{-19} \text{ J}$

Red light is lower energy than green light.

69. a. emits energy
b. requires the absorption of energy
c. requires the absorption of energy
d. requires the absorption of energy

70. The outermost electron of sodium absorbs photons of wavelength 589 nm as it jumps to a higher energy level, but the electron is not energetic enough to emit a photon at this wavelength. Therefore, the white light spectrum exhibits a dark line at 589 nm.

72. a. $E = h\nu = h \times \frac{c}{\lambda} = 6.626 \times 10^{-34} \text{ J}\cdot\text{s} \times \frac{2.998 \times 10^8 \text{ m/s}}{1.2 \times 10^{-4} \text{ m}} = 1.7 \times 10^{-21} \text{ J}$

b. $E = h\nu = h \times \frac{c}{\lambda} = 6.626 \times 10^{-34} \text{ J}\cdot\text{s} \times \frac{2.998 \times 10^8 \text{ m/s}}{5.1 \times 10^{-7} \text{ m}} = 3.9 \times 10^{-19} \text{ J}$

R78 Appendix E

c. $E = h\nu = h \times \dfrac{c}{\lambda} = 6.626 \times 10^{-34}\ \text{J}\cdot\text{s} \times \dfrac{2.998 \times 10^{8}\ \cancel{\text{m}}/\cancel{\text{s}}}{1.4 \times 10^{-8}\ \cancel{\text{m}}} = 1.7 \times 10^{-17}\ \text{J}$

The energy of the photon of light increases as its wavelength decreases.

73. An orbit confines the electron to a fixed circular path around the nucleus; an orbital is a region around the nucleus in which electrons are likely to be found.

75. Answers will vary. Students may note that radio waves have the lowest energy in the electromagnetic spectrum and, thus, would not be energetic enough to cook food. Others may reason that if microwaves cook food faster than infrared radiation, then radio waves would cook food even faster.

78. **a.** atom with electron in $n = 1$ level

b. atom with electron in $n = 4$ level

c. atom with electron in $n = 4$ level

d. atom with electron in $n = 1$ level

79. The element is potassium (19 electrons).

a. excited state, valence electron has been promoted from 4*s* to 5*p*

b. ground state, lowest energy electron configuration

c. impossible configuration, 3*p* orbitals can hold a maximum of 6 electrons, not 7

81. **a.** $\nu_1 = \dfrac{c}{\lambda} = \dfrac{2.998 \times 10^{8}\ \cancel{\text{m}}/\text{s}}{5.77 \times 10^{-3}\ \cancel{\text{cm}}} \times \dfrac{10^{2}\ \cancel{\text{cm}}}{1\ \cancel{\text{m}}}$
$= 5.20 \times 10^{12}/\text{s}$

$\nu_2 = \dfrac{c}{\lambda} = \dfrac{2.998 \times 10^{8}\ \cancel{\text{m}}/\text{s}}{6.82 \times 10^{-4}\ \cancel{\text{cm}}} \times \dfrac{10^{2}\ \cancel{\text{cm}}}{1\ \cancel{\text{m}}}$
$= 4.40 \times 10^{13}/\text{s}$

$\nu_3 = \dfrac{c}{\lambda} = \dfrac{2.998 \times 10^{8}\ \cancel{\text{m}}/\text{s}}{3.16 \times 10^{-4}\ \cancel{\text{cm}}} \times \dfrac{10^{2}\ \cancel{\text{cm}}}{1\ \cancel{\text{m}}}$
$= 9.49 \times 10^{13}/\text{s}$

$\nu_4 = \dfrac{c}{\lambda} = \dfrac{2.998 \times 10^{8}\ \cancel{\text{m}}/\text{s}}{1.76 \times 10^{-4}\ \cancel{\text{cm}}} \times \dfrac{10^{2}\ \cancel{\text{cm}}}{1\ \cancel{\text{m}}}$
$= 1.70 \times 10^{14}/\text{s}$

$\nu_5 = \dfrac{c}{\lambda} = \dfrac{2.998 \times 10^{8}\ \cancel{\text{m}}/\text{s}}{1.36 \times 10^{-4}\ \cancel{\text{cm}}} \times \dfrac{10^{2}\ \cancel{\text{cm}}}{1\ \cancel{\text{m}}}$
$= 2.20 \times 10^{14}/\text{s}$

$\nu_6 = \dfrac{c}{\lambda} = \dfrac{2.998 \times 10^{8}\ \cancel{\text{m}}/\text{s}}{6.38 \times 10^{-5}\ \cancel{\text{cm}}} \times \dfrac{10^{2}\ \cancel{\text{cm}}}{1\ \cancel{\text{m}}}$
$= 4.70 \times 10^{14}/\text{s}$

b.

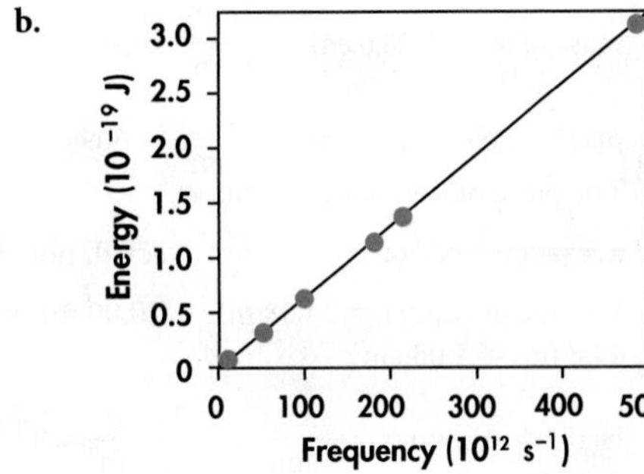

c. $6.63 \times 10^{-34}\ \text{J}\cdot\text{s}$

d. The slope is Planck's constant.

83. Hydrogen atom ($Z = 1$), $n = 1$:

$E = 1^2 \times \dfrac{2.18 \times 10^{-18}\ \text{J}}{1^2} = 2.18 \times 10^{-18}\ \text{J}$

Hydrogen atom ($Z = 1$), $n = 2$:

$E = 1^2 \times \dfrac{2.18 \times 10^{-18}\ \text{J}}{2^2} = 5.45 \times 10^{-19}\ \text{J}$

Li^{2+} ion ($Z = 3$), $n = 1$:

$E = 3^2 \times \dfrac{2.18 \times 10^{-18}\ \text{J}}{1^2} = 1.96 \times 10^{-17}\ \text{J}$

85. Two magnets would push each other apart. In the same way, electrons with the same spin would push apart and be unable to occupy the same orbital.

88. The light emitted from an incandescent bulb has wavelengths from 300 nm to 700 nm, which corresponds to a frequency range of about $4 \times 10^{14}\ \text{s}^{-1}$ to $1 \times 10^{15}\ \text{s}^{-1}$:

$\nu = \dfrac{c}{\lambda} = \dfrac{2.998 \times 10^{8}\ \cancel{\text{m}}/\text{s}}{700\ \cancel{\text{nm}}} \times \dfrac{10^{9}\ \cancel{\text{nm}}}{1\ \cancel{\text{m}}} = 4 \times 10^{14}\ \text{s}^{-1}$

$\nu = \dfrac{c}{\lambda} = \dfrac{2.998 \times 10^{8}\ \cancel{\text{m}}/\text{s}}{300\ \cancel{\text{nm}}} \times \dfrac{10^{9}\ \cancel{\text{nm}}}{1\ \cancel{\text{m}}} = 1 \times 10^{15}\ \text{s}^{-1}$

This means that the energy absorbed by the photons is in the range of about 3×10^{-19} J to 7×10^{-19} J:

$E = h\nu = (6.626 \times 10^{-34}\ \text{J}\cdot\cancel{\text{s}}) \times (4 \times 10^{14}/\cancel{\text{s}})$
$= 3 \times 10^{-19}\ \text{J}$

$E = h\nu = (6.626 \times 10^{-34}\ \text{J}\cdot\cancel{\text{s}}) \times (1 \times 10^{15}/\cancel{\text{s}})$
$= 7 \times 10^{-19}\ \text{J}$

89. **a.** heterogeneous

b. heterogeneous

c. homogeneous

91. A compound has constant composition; the composition of a mixture can vary.

93. $77\ \cancel{\text{pm}} \times \dfrac{1\ \cancel{\text{m}}}{10^{12}\ \cancel{\text{pm}}} \times \dfrac{10^{6}\ \mu\text{m}}{1\ \cancel{\text{m}}} = 7.7 \times 10^{-5}\ \mu\text{m}$

Selected Solutions **R79**

95. mass of lead $= 28.0 \cancel{cm^3} \times \frac{11.3\ g}{1 \cancel{cm^3}} = 316\ g$

mass of gold $= 16.0 \cancel{cm^3} \times \frac{19.3\ g}{1 \cancel{cm^3}} = 309\ g$

The piece of lead has more mass.

97. a. exact b. exact c. not exact d. not exact

99. Volume of copper = 24.08 mL − 20.00 mL = 4.08 mL = 4.08 cm^3

Density of copper $= \frac{mass}{volume} = \frac{36.4\ g}{4.08\ cm^3} = 8.92\ g/cm^3$

101. Helium gas is less dense than the nitrogen gas and oxygen gas in the air.

102. Accuracy is a measure of how close the value is to the true value; precision is a measure of how close a series of measurements are to one another.

105. Neon-20 has 10 neutrons in the nucleus; neon-21 has 11 neutrons in the nucleus.

Chapter 6

9. a. $1s^22s^22p^2$
b. $1s^22s^22p^63s^23p^63d^{10}4s^24p^65s^2$
c. $1s^22s^22p^63s^23p^63d^34s^2$

10. a. B, Al, Ga, In, Tl
b. F, Cl, Br, I, At
c. Ti, Zr, Hf, Rf

27. The close match between the predicted properties and the actual properties of gallium helped gain wider acceptance for Mendeleev's periodic table.

34. helium

38. a. sodium c. germanium
b. strontium d. selenium

43. The ionic radius of a metal cation is smaller than the atomic radius of the metal atom.

47. a. O b. F c. O d. S

51. b; nitrogen and phosphorus are in the same group (Group 5A).

56. a. H, Li, Na, K, Rb, Cs, Fr
b. O, S, Se, Te, Po
c. Zn, Cd, Hg, Uub

60. It is relatively easy to remove two electrons from magnesium; it is much more difficult to remove a third electron. It is relatively easy to remove three electrons from aluminum; it is much more difficult to remove a fourth electron.

65. a. The electrons in calcium are removed from the same energy level. In potassium, the second electron is removed from a lower energy level.

b. Because cesium has a larger atomic radius than lithium, the nuclear charge in a cesium atom has a smaller effect on the electrons in the highest occupied energy level.

c. The third electron removed from a magnesium atom is in a lower energy level.

66. Zinc has a greater nuclear charge (more protons) than calcium.

68. a. $1s^22s^22p^63s^23p^63d^{10}4s^24p^6$
b. $1s^22s^22p^6$
c. $1s^22s^22p^6$
d. $1s^22s^22p^6$

75. Electron affinity increases (becomes more negative) from left to right across a period because the nuclear charge increases and the shielding effect is constant.

78. a. Possible cations are Rb^+ and Sr^{2+}; possible anions are Br^-, Se^{2-}, and As^{3-}.

b. No, a cation is isoelectronic with the noble gas in the preceding period, and an anion is isoelectronic with the noble gas in the same period.

85. a. physical change c. physical change
b. chemical change d. chemical change

90. a. $2.24 \cancel{nm} \times \frac{1\ m}{10^9 \cancel{nm}} = 2.24 \times 10^{-9}\ m$

b. $8.13 \cancel{cm} \times \frac{1\ m}{10^2 \cancel{cm}} = 8.13 \times 10^{-2}\ m$

c. $7.4 \cancel{pm} \times \frac{1\ m}{10^{12} \cancel{pm}} = 7.4 \times 10^{-12}\ m$

d. $9.37 \cancel{mm} \times \frac{1\ m}{10^3 \cancel{mm}} = 9.37 \times 10^{-3}\ m$

93. The density of the olive is 1.05 g/cm^3. The olive will sink because its density is greater than that of water.

98. a. 48 b. 44 c. 114 d. 110

100. a. none b. one, $2p$ c. none d. none

Chapter 7

10. a. In order to have a completely filled valence shell, the iodine atom must gain 1 electron. This electron comes from 1 potassium atom, which loses 1 electron. The formula is KI.

b. Each oxygen atom needs 2 electrons to have an octet, but each aluminum atom loses 3 electrons; so, 2 aluminum atoms are needed for every 3 oxygen atoms. The formula is Al_2O_3.

11. A calcium atom loses 2 valence electrons to form a Ca^{2+} cation. A chlorine atom gains 1 electron to form a Cl^- anion. The formula of the compound that forms is $CaCl_2$.

R80 Appendix E

28. a. Group 5A; 5 valence electrons
b. Group 1A; 1 valence electron
c. Group 5A; 5 valence electrons
d. Group 2A; 2 valence electrons
e. Group 7A; 7 valence electrons
f. Group 4A; 4 valence electrons

31. a. 2; a calcium atom has 2 valence electrons, which it can lose.
b. 3; an aluminum atom has 3 valence electrons, which it can lose.
c. 1; a lithium atom has 1 valence electron, which it can lose.
d. 2; a barium atom has 2 valence electrons, which it can lose.

34. a. 3; a nitrogen atom has 5 valence electrons. It achieves an octet by gaining 3 electrons.
b. 2; a sulfur atom has 6 valence electrons. It achieves an octet by gaining 2 electrons.
c. 1; a chlorine atom has 7 valence electrons. It achieves an octet by gaining 1 electron.
d. 3; a phosphorus atom has 5 valence electrons. It achieves an octet by gaining 3 electrons.

35. a. A sulfur atom has 6 valence electrons and gains 2 electrons to attain a noble-gas configuration. The formula of the ion formed is S^{2-}.
b. A sodium atom has 1 valence electron and loses 1 electron to attain a noble-gas configuration. The formula of the ion formed is Na^{+}.
c. A fluorine atom has 7 valence electrons and gains 1 electron to attain a noble-gas configuration. The formula of the ion formed is F^{-}.
d. A phosphorus atom has 5 valence electrons and gains 3 electrons to attain a noble-gas configuration. The formula of the ion formed is P^{3-}.

40. a, c, e (ionic compounds form between atoms of metals and nonmetals)

43. a. Ca^{2+}, F^{-}
b. Al^{3+}, Br^{-}
c. Li^{+}, O^{2-}
d. Al^{3+}, S^{2-}
e. K^{+}, N^{3-}

47. Atoms in metals are arranged in a compact and orderly manner.

49. body-centered cubic: Na, K, Fe, Cr, or W; face-centered cubic: Cu, Ag, Au, Al, or Pb; hexagonal close-packed: Mg, Zn, or Cd

55. It has lost valence electrons.

58. a. sulfide ion (anion), S^{2-}
b. aluminum ion (cation), Al^{3+}
c. nitride ion (anion), N^{3-}
d. calcium ion (cation), Ca^{2+}

59. a. $1s^2 2s^2 2p^6 3s^2 3p^6 3d^6$
b. $1s^2 2s^2 2p^6 3s^2 3p^6 3d^7$
c. $1s^2 2s^2 2p^6 3s^2 3p^6 3d^8$

62. All have the noble-gas configuration of $1s^2 2s^2 2p^6 3s^2 3p^6$.

67. a, c, e, f (these substances are not composed of metals and nonmetals)

68. a. $BaBr_2$ b. Al_2S_3 c. K_3N

72. 12

75. a. Cu, Zn
b. Ag, Cu
c. Cu, Sn
d. Fe, Cr, Ni, C

76. Each dot in the electron dot structure represents a valence electron in the electron configuration diagram.

79. a. 6A (the Ca^{2+} cation must be balanced by an anion with a charge of 2−)
b. 7A (the Mg^{2+} cation must be balanced by 2 anions with a charge of 1−)
c. 1A (the N^{3-} anion must be balanced by 3 cations with a charge of 1+)
d. 6A (the 2 Al^{3+} cations must be balanced by 3 anions with a charge of 2−)
e. 1A (the F^{-} anion must be balanced by a cation with a charge of 1+)
f. 2A (the S^{2-} anion must be balanced by a cation with a charge of 2+)

83. Both metals and ionic compounds are composed of ions. Both are held together by electrostatic attractions. Metals always conduct an electric current, and ionic compounds conduct an electric current only when melted or dissolved in water. Ionic compounds are composed of cations and anions, but metals are composed of cations and free-floating valence electrons. Metals are ductile, but ionic compounds are brittle.

86. a. Lithium is a metal; a lithium atom will form a cation by losing 1 electron.
b. Sodium is a metal; a sodium atom will form a cation by losing 1 electron.
c. Neon is a noble gas; it is chemically nonreactive.
d. Chlorine is a nonmetal; a chlorine atom will form an anion by gaining 1 electron.
e. Magnesium is a metal; a magnesium atom will form a cation by losing 2 electrons.

89. a. CaO; the electrostatic forces of attraction between the Ca^{2+} cations and O^{2-} anions in CaO are much stronger than the forces between the Na^{+} cations and Cl^{-} anions in NaCl.
b. Yes; CaO has stronger electrostatic attractions than NaCl, which must be overcome to melt the compound.

Selected Solutions **R81**

92. The charge on the copper cation is 2+. Its electron configuration is $1s^2 2s^2 2p^6 3s^2 3p^6 3d^9$.

98. a. chemical c. physical
b. chemical d. chemical

100. a. liquid, vapor c. liquid, vapor
b. vapor d. liquid, vapor

102. a. $6 \times 10^4 \cancel{cm} \times \dfrac{1 \text{ m}}{10^2 \cancel{cm}} = 600 \text{ m}$

b. $6 \times 10^6 \cancel{mm} \times \dfrac{1 \text{ m}}{10^3 \cancel{mm}} = 6000 \text{ m}$

c. $0.06 \cancel{km} \times \dfrac{10^3 \text{ m}}{1 \cancel{km}} = 60 \text{ m}$

d. $6 \times 10^9 \cancel{nm} \times \dfrac{1 \text{ m}}{10^9 \cancel{nm}} = 6 \text{ m}$

Distance (b) is the longest.

104. $\text{volume} = \dfrac{\text{mass}}{\text{density}} = \dfrac{62.9 \cancel{g}}{2.33 \cancel{g}/\text{cm}^3} = 27.0 \text{ cm}^3$

108. a. 1 b. 3 c. 1 d. 5

111. a. $500 \cancel{nm} \times \dfrac{1 \text{ m}}{10^9 \cancel{nm}} = 5 \times 10^{-7} \text{ m}$

b. ultraviolet

112. a. K, $1s^2 2s^2 2p^6 3s^2 3p^6 4s^1$
b. Al, $1s^2 2s^2 2p^6 3s^2 3p^1$
c. S, $1s^2 2s^2 2p^6 3s^2 3p^4$
d. Ba, $1s^2 2s^2 2p^6 3s^2 3p^6 3d^{10} 4s^2 4p^6 4d^{10} 5s^2 5p^6 6s^2$

Chapter 8

7. a. :C̤̈l:C̤̈l: b. :B̤̈r:B̤̈r: c. :Ï:Ï:

8. a. H:Ö:Ö:H b. :C̤̈l:P̤̈:C̤̈l: with :C̤̈l: below P

9. $[\text{H:Ö:}]^-$

10. $[\text{:Ö:S:Ö:}]^{2-}$ (with :Ö: above and below S) $[\text{:Ö:C::Ö}]^{2-}$ (with :Ö: above C)

29. a. 0.7; moderately polar covalent
b. 2.2; ionic
c. 1.0; moderately to very polar covalent
d. 1.0; moderately to very polar covalent
e. 2.5; ionic
f. 0; nonpolar covalent

30. c and d (tie at 0.4), b (0.7), a (0.9)

48. One atom contributes both electrons to a coordinate covalent bond, as in CO.

50. An unshared pair of electrons is needed for a coordinate covalent bond. There are no unshared pairs in compounds with only C—H and C—C bonds.

54. b and c; assuming only single bonds, the P and S atoms each have 10 valence electrons.

58. a. H:N̤̈:H with H below N b. :B̤̈r:C̤̈l: c. H:Ö:Ö:H d. H:S̤̈i:H with H above and below Si

60. a. linear d. bent
b. tetrahedral e. linear
c. trigonal planar f. bent

63. a. sp^3 b. sp^2 c. sp d. sp

65. c (1.9), d (1.4), a (.09), f (0.5), b (0.4), e (0.0)

69. $1 \cancel{\text{mol } CH_4} \times \dfrac{393 \text{ kJ}}{1 \cancel{\text{mol } CH_4}} \times 4 = 1572 \text{ kJ}$

71. The 3*s* and three 3*p* orbitals of phosphorus hybridize to form four sp^3 atomic orbitals. The resulting shape is pyramidal with a bond angle of 107° between the sigma bonds.

78. a (the phosphorus atom in PBr_5 has 10 valence electrons)

82. Ethyl alcohol can form intermolecular hydrogen bonds between its polar —OH groups, but dimethyl ether cannot form hydrogen bonds.

84. a. bent b. tetrahedral c. pyramidal

91. a. :F̤̈—Be—F̤̈: (Be has only 4 valence electrons.)

b. S bonded to six :F̤̈: atoms (S has 12 valence electrons.)

c. :Ö: bonded to Cl, Cl bonded to :Ö: (Cl has only 7 valence electrons.)

d. B bonded to three :F̤̈: atoms (B has only 6 valence electrons.)

e. :F̤̈—Ẍ̤e—F̤̈: (Xe has 10 valence electrons.)

97. a. $66.5 \cancel{mm} \times \dfrac{1 \cancel{m}}{10^3 \cancel{mm}} \times \dfrac{10^6 \text{ µm}}{1 \cancel{m}} = 6.65 \times 10^4 \text{ µm}$

b. $4 \times 10^{-2} \cancel{g} \times \dfrac{100 \text{ cg}}{1 \cancel{g}} = 4 \text{ cg}$

c. $\dfrac{5.62 \cancel{mg}}{1 \cancel{mL}} \times \dfrac{1 \cancel{g}}{10^3 \cancel{mg}} \times \dfrac{10 \text{ dg}}{1 \cancel{g}} \times \dfrac{10^3 \cancel{mL}}{1 \text{ L}} = 56.2 \text{ dg/L}$

d. $\dfrac{85 \cancel{km}}{1 \cancel{h}} \times \dfrac{1 \cancel{h}}{60 \cancel{min}} \times \dfrac{1 \cancel{min}}{60 \text{ s}} \times \dfrac{10^3 \text{ m}}{1 \cancel{km}} = 2.4 \times 10^1 \text{ m/s}$

101. Protons (positive charge) and electrons (negative charge) must be equal.

109. a. barium b. silicon c. sodium

R82 Appendix E

111. All have the same number of electrons as a noble gas.

Chapter 9

1. a. selenide ion c. phosphide ion
b. barium ion d. iodide ion

2. a. three electrons lost c. one electron lost
b. two electrons gained d. two electrons lost

10. a. BaS b. Li_2O c. Ca_3N_2 d. CuI_2

11. a. NaI b. $SnCl_2$ c. K_2S d. CaI_2

12. a. zinc sulfide c. barium oxide
b. potassium chloride d. copper(II) bromide

13. a. calcium oxide c. iron(II) sulfide
b. copper(I) selenide d. aluminum fluoride

14. a. $(NH_4)_2SO_3$ b. $Ca_3(PO_4)_2$

15. a. $LiHSO_4$ b. $Cr(NO_2)_3$

16. a. calcium oxalate
b. potassium hypochlorite
c. potassium permanganate
d. lithium sulfite

17. a. aluminum hydroxide c. tin(II) phosphate
b. sodium chlorate d. sodium chromate

27. a. oxygen difluoride c. sulfur trioxide
b. disulfur decafluoride d. sulfur hexafluoride

28. a. N_2O_4 b. XeF_4 c. S_2F_{10} d. IF_7

37. a. hydrofluoric acid
b. nitric acid
c. sulfurous acid

38. a. $HClO_4$ b. HI c. $HClO_2$

39. a. barium hydroxide
b. calcium hydroxide
c. rubidium hydroxide

40. a. CsOH b. $Be(OH)_2$ c. $Mn(OH)_3$

48. $\frac{2.98 \text{ g Pb}}{0.461 \text{ g O}} = \frac{x}{1.00 \text{ g O}}$

$$x = \frac{2.98 \text{ g Pb} \times 1.00 \cancel{\text{g O}}}{0.461 \cancel{\text{g O}}} = 6.46 \text{ g Pb}$$

$$\frac{9.89 \text{ g Pb}}{0.763 \text{ g O}} = \frac{y}{1.00 \text{ g O}}$$

$$y = \frac{9.89 \text{ g Pb} \times 1.00 \cancel{\text{g O}}}{0.763 \cancel{\text{g O}}} = 13.0 \text{ g Pb}$$

$$\text{mass ratio} = \frac{6.46 \cancel{\text{g Pb}}}{13.0 \cancel{\text{g Pb}}} = 1{:}2$$

49. Since $\frac{7 \text{ Fe}}{3 \text{ O}} = 2.3$ and $\frac{23 \text{ g Fe}}{10 \text{ g O}} = 2.3$, the compound must be iron(III) oxide.

58. a. 2+ b. 2+ c. 3+ d. 1+

64. Determine the charge of the anion, and then work the formula backward to find the charge of the transition metal cation needed to give a net charge of zero for the formula unit.

70. a. tri- c. di- e. penta-
b. mono- d. hexa- f. tetra-

76. a. lead(II) hydroxide c. $Cu(OH)_2$
b. cobalt(II) hydroxide d. $Fe(OH)_2$

79. No, the ratio of nitrogen to oxygen is 42:26, which is not a 7:4 ratio.

82. a. sodium chlorate
b. mercury(I) bromide
c. potassium chromate
d. perchloric acid
e. tin(IV) oxide
f. iron(III) acetate
g. potassium hydrogen sulfate
h. calcium hydroxide
i. barium sulfide

86. binary molecular compound

88. $SnCl_4$

96. a. N_2O, dinitrogen monoxide
b. NO_2, nitrogen dioxide
c. NO, nitrogen monoxide
d. N_2O_4, dinitrogen tetroxide

98. a. The charges do not balance; CsCl.
b. Neon does not form compounds.
c. The charges do not balance; ZnO.
d. The subscripts are not the lowest whole-number ratio; BaS.

100. binary ionic compounds: d and g; binary molecular compounds: a and f; compounds with polyatomic ions: b, c, e, h, and i; acids: b and e; base: c

101. a. 3.60 g solid/1.40 g gas = 2.57 and
6.48 g solid/2.52 g gas = 2.57
The compound obeys the law of definite proportions.
b. % gas = 1.40/5.00 and 2.52/9.00 = 28.0% gas
0.28 × 14.0 g = 3.92 g colorless gas

102. law of conservation of mass

Selected Solutions **R83**

106. **a.** Potassium carbonate has a greater water solubility than $CaCO_3$, so you could see which one dissolves in water more readily.

b. The copper compound is blue; the iron compound is white.

c. Add water to dissolve the NH_4Cl, and then filter out the insoluble $BaSO_4$.

d. chlorine (nonmetal), sulfur (nonmetal), bromine (nonmetal), barium (metal), iodine (nonmetal), mercury (metal)

e. barium sulfate, calcium carbonate, potassium carbonate, copper(II) sulfate pentahydrate, iron(II) sulfate pentahydrate, ammonium chloride

f. mass = density × volume

$\text{mass} = 47.0\ \cancel{\text{cm}^3} \times 13.59\ \text{g}/\cancel{\text{cm}^3} = 639\ \text{g Hg}$

g. $\text{volume} = \dfrac{\text{mass}}{\text{density}} = \dfrac{16.6\ \cancel{\text{g}}}{2.07\ \cancel{\text{g}}/\text{cm}^3} = 8.02\ \text{cm}^3$

h. color, density, melting point, and boiling point

111. ionic; $Ca(OH)_2$

113. **a.** 4 **b.** 2 **c.** 2 **d.** 4 **e.** 2 **f.** 1

118. **a.** 2 **b.** protons **c.** electrons **d.** neutrons

119. **a.** neon **b.** carbon **c.** boron **d.** helium

124. **a.** 1 **b.** 6 **c.** 8 **d.** 2 **e.** 7 **f.** 5

127. b and c; molecular compounds formed by two nonmetals have covalent bonds.

Chapter 10

1. $0.50\ \cancel{\text{bushel apples}} \times \dfrac{1\ \cancel{\text{dozen apples}}}{0.20\ \cancel{\text{bushel apples}}} \times \dfrac{2.0\ \text{kg apples}}{1\ \cancel{\text{dozen apples}}} = 5.0\ \text{kg apples}$

2. $14\ \cancel{\text{kg apples}} \times \dfrac{1\ \cancel{\text{dozen apples}}}{2.0\ \cancel{\text{kg apples}}} \times \dfrac{12\ \cancel{\text{apples}}}{1\ \cancel{\text{dozen apples}}} \times \dfrac{8\ \text{seeds}}{1\ \cancel{\text{apple}}} = 670\ \text{seeds}$

3. $2.80 \times 10^{24}\ \cancel{\text{atoms Si}} \times \dfrac{1\ \text{mol Si}}{6.02 \times 10^{23}\ \cancel{\text{atoms Si}}} = 4.65\ \text{mol Si}$

4. $2.17 \times 10^{23}\ \cancel{\text{molecules Br}_2} \times \dfrac{1\ \text{mol Br}_2}{6.02 \times 10^{23}\ \cancel{\text{molecules Br}_2}} = 0.360\ \text{mol Br}_2$

5. $1.14\ \cancel{\text{mol SO}_3} \times \dfrac{6.02 \times 10^{23}\ \cancel{\text{molecules SO}_3}}{1\ \cancel{\text{mol SO}_3}} \times \dfrac{4\ \text{atoms}}{1\ \cancel{\text{molecule SO}_3}} = 2.75 \times 10^{24}\ \text{atoms}$

6. $2.12\ \cancel{\text{mol C}_3\text{H}_8} \times \dfrac{6.02 \times 10^{23}\ \cancel{\text{molecules C}_3\text{H}_8}}{1\ \cancel{\text{mol C}_3\text{H}_8}} \times \dfrac{3\ \text{C atoms}}{1\ \cancel{\text{molecule C}_3\text{H}_8}} = 3.83 \times 10^{24}\ \text{C atoms}$

$2.12\ \cancel{\text{mol C}_3\text{H}_8} \times \dfrac{6.02 \times 10^{23}\ \cancel{\text{molecules C}_3\text{H}_8}}{1\ \cancel{\text{mol C}_3\text{H}_8}} \times \dfrac{8\ \text{H atoms}}{1\ \cancel{\text{molecule C}_3\text{H}_8}} = 1.02 \times 10^{25}\ \text{H atoms}$

7. $1\ \cancel{\text{mol P}} \times \dfrac{31.0\ \text{g P}}{1\ \cancel{\text{mol P}}} = 31.0\ \text{g P}$

$3\ \cancel{\text{mol Cl}} \times \dfrac{35.5\ \text{g Cl}}{1\ \cancel{\text{mol Cl}}} = 106.5\ \text{g Cl}$

mass of 1 mol PCl_3 = 31.0 g P + 106.5 g Cl = 138.0 g

molar mass of PCl_3 = 138 g/mol

8. $1\ \cancel{\text{mol Na}} \times \dfrac{23.0\ \text{g Na}}{1\ \cancel{\text{mol Na}}} = 23.0\ \text{g Na}$

$1\ \cancel{\text{mol H}} \times \dfrac{1.0\ \text{g H}}{1\ \cancel{\text{mol H}}} = 1.0\ \text{g H}$

$1\ \cancel{\text{mol C}} \times \dfrac{12.0\ \text{g C}}{1\ \cancel{\text{mol C}}} = 12.0\ \text{g C}$

$3\ \cancel{\text{mol O}} \times \dfrac{16.0\ \text{g O}}{1\ \cancel{\text{mol O}}} = 48.0\ \text{g O}$

mass of 1 mol $NaHCO_3$ = 23.0 g + 1.0 g + 12.0 g + 48.0 g = 84.0 g

molar mass of $NaHCO_3$ = 84.0 g/mol

16. $20\ \cancel{\text{mol C}} \times \dfrac{12.0\ \text{g C}}{1\ \cancel{\text{mol C}}} = 240.0\ \text{g C}$

$42\ \cancel{\text{mol H}} \times \dfrac{1.0\ \text{g H}}{1\ \cancel{\text{mol H}}} = 42.0\ \text{g H}$

$1\ \text{mol C}_{20}\text{H}_{42} = 240.0\ \text{g C} + 42.0\ \text{g H} = 282.0\ \text{g C}_{20}\text{H}_{42}$

$4.52 \times 10^{-3}\ \cancel{\text{mol C}_{20}\text{H}_{42}} \times \dfrac{282.0\ \text{g C}_{20}\text{H}_{42}}{1\ \cancel{\text{mol C}_{20}\text{H}_{42}}} = 1.27\ \text{g}$

17. $1\ \cancel{\text{mol Fe}} \times \dfrac{55.8\ \text{g Fe}}{1\ \cancel{\text{mol Fe}}} = 55.8\ \text{g Fe}$

$2\ \cancel{\text{mol O}} \times \dfrac{16.0\ \text{g O}}{1\ \cancel{\text{mol O}}} = 32.0\ \text{g O}$

$2\ \cancel{\text{mol H}} \times \dfrac{1.0\ \text{g H}}{1\ \cancel{\text{mol H}}} = 2.0\ \text{g H}$

1 mol $Fe(OH)_2$ = 55.8 g Fe + 32.0 g O + 2.0 g H = 89.8 g $Fe(OH)_2$

$2.50\ \cancel{\text{mol Fe(OH)}_2} \times \dfrac{89.8\ \text{g}}{1\ \cancel{\text{mol Fe(OH)}_2}} = 225\ \text{g Fe(OH)}_2$

18. $3.70 \times 10^{-1}\ \cancel{\text{g B}} \times \dfrac{1\ \text{mol B}}{10.8\ \cancel{\text{g B}}} = 3.43 \times 10^{-2}\ \text{mol B}$

R84 Appendix E

19. $2\ \cancel{\text{mol N}} \times \frac{14.0\text{ g N}}{1\ \cancel{\text{mol N}}} = 28\text{ g N}$

$3\ \cancel{\text{mol O}} \times \frac{16.0\text{ g O}}{1\ \cancel{\text{mol O}}} = 48.0\text{ g O}$

$1\text{ mol } N_2O_3 = 28\text{ g N} + 48.0\text{ g O} = 76.0\text{ g } N_2O_3$

$75.0\ \cancel{\text{g } N_2O_3} \times \frac{1\text{ mol } N_2O_3}{76.0\ \cancel{\text{g } N_2O_3}} = 0.987\text{ mol } N_2O_3$

20. a. $3.20 \times 10^{-3}\ \cancel{\text{mol } CO_2} \times \frac{22.4\text{ L } CO_2}{1\ \cancel{\text{mol } CO_2}} = 7.17 \times 10^{-2}\text{ L } CO_2$

b. $3.70\ \cancel{\text{mol } N_2} \times \frac{22.4\text{ L } N_2}{1\ \cancel{\text{mol } N_2}} = 82.9\text{ L } N_2$

c. $0.960\ \cancel{\text{mol } CH_4} \times \frac{22.4\text{ L } CH_4}{1\ \cancel{\text{mol } CH_4}} = 21.5\text{ L } CH_4$

21. a. $67.2\ \cancel{\text{L } SO_2} \times \frac{1\text{ mol } SO_2}{22.4\ \cancel{\text{L } SO_2}} = 3.00\text{ mol } SO_2$

b. $0.880\ \cancel{\text{L He}} \times \frac{1\text{ mol He}}{22.4\ \cancel{\text{L He}}} = 0.039\text{ mol He}$

c. $1.00 \times 10^3\ \cancel{\text{L } C_2H_6} \times \frac{1\text{ mol } C_2H_6}{22.4\ \cancel{\text{L } C_2H_6}} = 44.6\text{ mol } C_2H_6$

22. $\frac{3.58\text{ g}}{\cancel{\text{L}}} \times \frac{22.4\ \cancel{\text{L}}}{1\text{ mol}} = 80.2\text{ g/mol}$

23. molar mass of Kr = 83.8 g/mol

$\frac{83.8\text{ g}}{1\ \cancel{\text{mol Kr}}} \times \frac{1\ \cancel{\text{mol Kr}}}{22.4\text{ L}} = 3.74\text{ g/L}$

33. mass of compound = 9.03 g + 3.48 g = 12.51 g

$\%\text{Mg} = \frac{\text{mass of Mg}}{\text{mass of compound}} \times 100\% = \frac{9.03\ \cancel{\text{g}}}{12.51\ \cancel{\text{g}}} \times 100\% = 72.2\%\text{ Mg}$

$\%\text{N} = \frac{\text{mass of N}}{\text{mass of compound}} \times 100\% = \frac{3.48\ \cancel{\text{g}}}{12.51\ \cancel{\text{g}}} \times 100\% = 27.8\%\text{ N}$

34. mass of oxygen = 14.2 g − 13.2 g = 1.0 g

$\%\text{Hg} = \frac{\text{mass of Hg}}{\text{mass of compound}} \times 100\% = \frac{13.2\ \cancel{\text{g}}}{14.2\ \cancel{\text{g}}} \times 100\% = 93.0\%\text{ Hg}$

$\%\text{O} = \frac{\text{mass of O}}{\text{mass of compound}} \times 100\% = \frac{1.0\ \cancel{\text{g}}}{14.2\ \cancel{\text{g}}} \times 100\% = 7.0\%\text{ O}$

35. a. mass of N in 1 mol NH_3 = $1\ \cancel{\text{mol}} \times 14.0\text{ g/}\cancel{\text{mol}} = 14.0\text{ g}$

molar mass of NH_3 = 14.0 g + 3 × 1.0 g = 17.0 g

$\%\text{N} = \frac{\text{mass of N in 1 mol } NH_3}{\text{molar mass of } NH_3} \times 100\% = \frac{14.0\ \cancel{\text{g}}}{17.0\ \cancel{\text{g}}} \times 100\% = 82.4\%\text{ N}$

b. mass of N in 1 mol NH_4NO_3 = $1\ \cancel{\text{mol}} \times 14.0\text{ g/}\cancel{\text{mol}} = 14.0\text{ g}$

molar mass of NH_4NO_3 = 14.0 g + 4 × 1.0 g + 14.0 g + 3 × 16.0 g = 80.0 g

$\%\text{N} = \frac{\text{mass of N in 1 mol } NH_4NO_3}{\text{molar mass of } NH_4NO_3} \times 100\% = \frac{28.0\ \cancel{\text{g}}}{80.0\ \cancel{\text{g}}} \times 100\% = 35.0\%\text{ N}$

36. a. mass of C in 1 mol C_2H_6 = $2\ \cancel{\text{mol}} \times 12.0\text{ g/}\cancel{\text{mol}} = 24.0\text{ g}$

mass of H in 1 mol C_2H_6 = $6\ \cancel{\text{mol}} \times 1.0\text{ g/}\cancel{\text{mol}} = 6.0\text{ g}$

molar mass of C_2H_6 = 24.0 g + 6.0 g = 30.0 g

$\%\text{C} = \frac{\text{mass of C in 1 mol } C_2H_6}{\text{molar mass of } C_2H_6} \times 100\% = \frac{24.0\ \cancel{\text{g}}}{30.0\ \cancel{\text{g}}} \times 100\% = 80.0\%\text{ C}$

$\%\text{H} = \frac{\text{mass of H in 1 mol } C_2H_6}{\text{molar mass of } C_2H_6} \times 100\% = \frac{6.0\ \cancel{\text{g}}}{30.0\ \cancel{\text{g}}} = 20.0\%\text{ H}$

b. mass of Na in 1 mol $NaHSO_4$ = $1\ \cancel{\text{mol}} \times 23.0\text{ g/}\cancel{\text{mol}} = 23.0\text{ g}$

mass of H in 1 mol $NaHSO_4$ = $1\ \cancel{\text{mol}} \times 1.0\text{ g/}\cancel{\text{mol}} = 1.0\text{ g}$

mass of S in 1 mol $NaHSO_4$ = $1\ \cancel{\text{mol}} \times 32.1\text{ g/}\cancel{\text{mol}} = 32.1\text{ g}$

mass of O in 1 mol $NaHSO_4$ = $4\ \cancel{\text{mol}} \times 16.0\text{ g/}\cancel{\text{mol}} = 64.0\text{ g}$

molar mass of $NaHSO_4$ = 23.0 g + 1.0 g + 32.1 g + 64.0 g = 120.1 g

$\%\text{Na} = \frac{\text{mass of Na in 1 mol } NaHSO_4}{\text{molar mass of } NaHSO_4} \times 100\% = \frac{23.0\text{ g}}{120.1\text{ g}} = 19.2\%\text{ Na}$

$\%\text{H} = \frac{\text{mass of H in 1 mol } NaHSO_4}{\text{molar mass of } NaHSO_4} \times 100\% = \frac{1.0\text{ g}}{120.1\text{ g}} = 0.80\%\text{ H}$

$\%\text{S} = \frac{\text{mass of S in 1 mol } NaHSO_4}{\text{molar mass of } NaHSO_4} \times 100\% = \frac{32.1\text{ g}}{120.1\text{ g}} = 26.7\%\text{ S}$

$\%\text{O} = \frac{\text{mass of O in 1 mol } NaHSO_4}{\text{molar mass of } NaHSO_4} \times 100\% = \frac{64.0\text{ g}}{120.1\text{ g}} = 53.3\%\text{ O}$

37. a. $125\ \cancel{\text{g } NH_3} \times \frac{82.4\text{ g N}}{100\ \cancel{\text{g } NH_3}} = 103\text{ g N}$

b. $125\ \cancel{\text{g } NH_4NO_3} \times \frac{35.0\text{ g N}}{100\ \cancel{\text{g } NH_4NO_3}} = 43.8\text{ g N}$

Selected Solutions **R85**

38. a. $350\ \text{g C}_2\text{H}_6 \times \dfrac{2.0\ \text{g} \times 10^1\ \text{g H}}{100\ \text{g C}_2\text{H}_6} = 7.0 \times 10^1\ \text{g H}$

b. $20.2\ \text{g NaHSO}_4 \times \dfrac{0.83\ \text{g H}}{100\ \text{g NaHSO}_4} = 0.17\ \text{g H}$

39. a. $94.1\ \text{g O} \times \dfrac{1\ \text{mol O}}{16.0\ \text{g O}} = 5.88\ \text{mol O}$

$5.9\ \text{g H} \times \dfrac{1\ \text{mol H}}{1.0\ \text{g H}} = 5.9\ \text{mol H}$

The mole ratio of O to H is $\text{O}_{5.88}\text{H}_{5.9}$

$\dfrac{5.88\ \text{mol O}}{5.88} = 1\ \text{mol O}$

$\dfrac{5.9\ \text{mol H}}{5.88} = 1\ \text{mol H}$

The mole ratio of O to H is OH.
The empirical formula is OH.

b. $67.6\ \text{g Hg} \times \dfrac{1\ \text{mol Hg}}{200.6\ \text{g Hg}} = 0.337\ \text{mol Hg}$

$10.8\ \text{g S} \times \dfrac{1\ \text{mol S}}{32.1\ \text{g S}} = 0.336\ \text{mol S}$

$21.6\ \text{g O} \times \dfrac{1\ \text{mol O}}{16.0\ \text{g O}} = 1.35\ \text{mol O}$

The mole ratio of Hg to S to O is $\text{Hg}_{0.337}\text{S}_{0.336}\text{O}_{1.35}$.

$\dfrac{0.337\ \text{mol Hg}}{0.336} = 1\ \text{mol Hg}$

$\dfrac{0.336\ \text{mol S}}{0.336} = 1\ \text{mol S}$

$\dfrac{1.35\ \text{mol O}}{0.336} = 4\ \text{mol O}$

The mole ratio of Hg to S to O is HgSO_4.
The empirical formula is HgSO_4.

40. $62.1\ \text{g C} \times \dfrac{1\ \text{mol C}}{12.0\ \text{g C}} = 5.18\ \text{mol C}$

$13.8\ \text{g H} \times \dfrac{1\ \text{mol H}}{1.0\ \text{g H}} = 13.8\ \text{mol H}$

$24.1\ \text{g N} \times \dfrac{1\ \text{mol N}}{14.0\ \text{g N}} = 1.72\ \text{mol N}$

$\dfrac{5.18\ \text{mol C}}{1.72} = 3\ \text{mol C}$

$\dfrac{13.8\ \text{mol H}}{1.72} = 8\ \text{mol H}$

$\dfrac{1.72\ \text{mol N}}{1.72} = 1\ \text{mol N}$

The empirical formula is $\text{C}_3\text{H}_8\text{N}$.

41. efm of CClN = 12.0 g/mol + 35.5 g/mol + 14.0 g/mol = 61.5 g/mol

$\dfrac{\text{molar mass}}{\text{efm}} = \dfrac{184.5\ \text{g/mol}}{61.5\ \text{g/mol}} = 3$

$(\text{CClN}) \times 3 = \text{C}_3\text{Cl}_3\text{N}_3$.

42. efm = 12.0 g/mol + 3(1.0 g/mol) + 16.0 g/mol = 31.0 g/mol

$\dfrac{\text{molar mass}}{\text{efm of CH}_3\text{O}} = \dfrac{62.0\ \text{g/mol}}{31.0\ \text{g/mol}} = 2$

$(\text{CH}_3\text{O}) \times 2 = \text{C}_2\text{H}_6\text{O}_2$

51. a. molecule
b. formula unit
c. molecule
d. atom

52. a. 3
b. 2
c. 9
d. 10

54. a. $2.41 \times 10^{24}\ \text{formula units of NaCl} \times \dfrac{1\ \text{mol NaCl}}{6.02 \times 10^{23}\ \text{formula units NaCl}} = 0.400 \times 10^1 =$ 4.00 mol NaCl

b. $9.03 \times 10^{24}\ \text{atoms of Hg} \times \dfrac{1\ \text{mol Hg}}{6.02 \times 10^{23}\ \text{atoms Hg}}$ = 15.0 mol Hg

c. $4.65 \times 10^{24}\ \text{molecules of NO}_2 \times \dfrac{1\ \text{mol NO}_2}{6.02 \times 10^{23}\ \text{molecules NO}_2} = 0.772 \times 10^1 =$ 7.72 molecules NO_2

57. a. $3.00\ \text{mol Sn} \times \dfrac{6.02 \times 10^{23}\ \text{atoms Sn}}{1\ \text{mol Sn}} =$ 1.81×10^{24} atoms Sn

b. $0.400\ \text{mol KCl} \times \dfrac{6.02 \times 10^{23}\ \text{formula units KCl}}{1\ \text{mol KCl}}$ $= 2.41 \times 10^{23}$ formula units KCl

c. $7.50\ \text{mol SO}_2 \times \dfrac{6.02 \times 10^{23}\ \text{molecules SO}_2}{1\ \text{mol SO}_2} =$ 4.52×10^{24} molecules SO_2

d. $4.80 \times 10^{-3}\ \text{mol NaI} \times \dfrac{6.02 \times 10^{23}\ \text{formula units NaI}}{1\ \text{mol NaI}} =$ 2.89×10^{21} formula units NaI .

60. a. (3 × 1.0 g/mol) + (1 × 31.0 g/mol) + (4 × 16.0 g/mol) = 98.0 g/mol
b. (2 × 14.0 g/mol) + (3 × 16.0 g/mol) = 76.0 g/mol

R86 Appendix E

c. $(1 \times 40.1\text{ g/mol}) + (1 \times 12.0\text{ g/mol}) + (3 \times 16.0\text{ g/mol}) = 100.1\text{ g/mol}$

d. $2 \times [(1 \times 14.0\text{ g/mol}) + (4 \times 1.0\text{ g/mol})] + (1 \times 32.1\text{ g/mol}) + (4 \times 16.0\text{ g/mol}) = 132.1\text{ g/mol}$

e. $(4 \times 12.0\text{ g/mol}) + (9 \times 1.0\text{ g/mol}) + (2 \times 16.0\text{ g/mol}) = 89.0\text{ g/mol}$

f. $2 \times 79.9\text{ g/mol} = 159.8\text{ g/mol}$

63. **a.** mass C in 1 mol $C_{12}H_{22}O_{11} = 12\ \cancel{\text{mol}} \times 12.0\text{ g}/\cancel{\text{mol}} = 144.0\text{ g}$

mass H in 1 mol $C_{12}H_{22}O_{11} = 22\ \cancel{\text{mol}} \times 1.0\text{ g}/\cancel{\text{mol}} = 22.0\text{ g}$

mass O in 1 mol $C_{12}H_{22}O_{11} = 11\ \cancel{\text{mol}} \times 16.0\text{ g}/\cancel{\text{mol}} = 176.0\text{ g}$

molar mass $C_{12}H_{22}O_{11} = 144.0\text{ g} + 22.0\text{ g} + 176.0\text{ g} = 342.0\text{ g}$

$$0.250\ \cancel{\text{mol } C_{12}H_{22}O_{11}} \times \frac{342.0\text{ g } C_{12}H_{22}O_{11}}{1\ \cancel{\text{mol } C_{12}H_{22}O_{11}}} = 85.5\text{ g } C_{12}H_{22}O_{11}$$

b. mass Na in 1 mol NaCl $= 1\ \cancel{\text{mol}} \times 23.0\text{ g}/\cancel{\text{mol}} = 23.0\text{ g}$

mass Cl in 1 mol NaCl $= 1\ \cancel{\text{mol}} \times 35.5\text{ g}/\cancel{\text{mol}} = 35.5\text{ g}$

molar mass of NaCl $= 23.0\text{ g} + 35.5\text{ g} = 58.5\text{ g}$

$$0.250\ \cancel{\text{mol NaCl}} \times \frac{58.5\text{ g NaCl}}{1\ \cancel{\text{mol NaCl}}} = 14.6\text{ g NaCl}$$

c. mass K in 1 mol $KMnO_4 = 1\ \cancel{\text{mol}} \times 39.1\text{ g}/\cancel{\text{mol}} = 39.1\text{ g}$

mass Mn in 1 mol $KMnO_4 = 1\ \cancel{\text{mol}} \times 54.9\text{ g}/\cancel{\text{mol}} = 54.9\text{ g}$

mass O in 1 mol $KMnO_4 = 4\ \cancel{\text{mol}} \times 16.0\text{ g}/\cancel{\text{mol}} = 64.0\text{ g}$

molar mass $KMnO_4 = 39.1\text{ g} + 54.9\text{ g} + 64.0\text{ g} = 158.0\text{ g}$

$$0.250\ \cancel{\text{mol } KMnO_4} \times \frac{158.0\text{ g } KMnO_4}{1\ \cancel{\text{mol } KMnO_4}} = 39.5\text{ g } KMnO_4$$

64. **a.** $$1.00 \times 10^2\ \cancel{\text{g } C_{12}H_{22}O_{11}} \times \frac{1\text{ mol } C_{12}H_{22}O_{11}}{342.0\ \cancel{\text{g } C_{12}H_{22}O_{11}}} = 0.292\text{ mol } C_{12}H_{22}O_{11}$$

b. $$1.00 \times 10^2\ \cancel{\text{g NaCl}} \times \frac{1\text{ mol NaCl}}{58.5\ \cancel{\text{g NaCl}}} = 1.71\text{ mol NaCl}$$

c. $$1.00 \times 10^2\ \cancel{\text{g } KMnO_4} \times \frac{1\text{ mol } KMnO_4}{158.0\ \cancel{\text{g } KMnO_4}} = 0.633\text{ mol } KMnO_4$$

67. **a.** $$7.64\ \cancel{\text{mol Ar}} \times \frac{22.4\text{ L Ar}}{1\ \cancel{\text{mol Ar}}} = 171\text{ L Ar}$$

b. $$1.34\ \cancel{\text{mol } SO_2} \times \frac{22.4\text{ L } SO_2}{1\ \cancel{\text{mol } SO_2}} = 30.0\text{ L } SO_2$$

c. $$0.442\ \cancel{\text{mol } C_2H_6} \times \frac{22.4\text{ L } C_2H_6}{1\ \cancel{\text{mol } C_2H_6}} = 9.90\text{ L } C_2H_6$$

d. $$2.45 \times 10^{-3}\ \cancel{\text{mol } H_2S} \times \frac{22.4\text{ L } H_2S}{1\ \cancel{\text{mol } H_2S}} = 5.49 \times 10^{-2}\text{ L } H_2S$$

68. $$\frac{0.902\text{ g}}{1\ \cancel{\text{L}}} \times \frac{22.4\ \cancel{\text{L}}}{1\text{ mol}} = 20.2\text{ g/mol}$$

72. Total mass $= 13.3\text{ g} + 5.7\text{ g} = 19.0\text{ g}$

$$\%\text{ Fe} = \frac{13.3\ \cancel{\text{g}}}{19.0\ \cancel{\text{g}}} \times 100\% = 70.0\%\text{ Fe}$$

$$\%\text{ O} = \frac{5.7\ \cancel{\text{g}}}{19.0\ \cancel{\text{g}}} \times 100\% = 3.0 \times 10^1\%\text{ O}$$

73. **a.** $$\%\text{ H} = \frac{2.0\ \cancel{\text{g}}}{34.1\ \cancel{\text{g}}} \times 100\% = 5.9\%\text{ H}$$

$$\%\text{ S} = \frac{32.1\ \cancel{\text{g}}}{34.1\ \cancel{\text{g}}} \times 100\% = 94.1\%\text{ S}$$

b. $$\%\text{ N} = \frac{28.0\ \cancel{\text{g}}}{124.0\ \cancel{\text{g}}} \times 100\% = 22.6\%\text{ N}$$

$$\%\text{ H} = \frac{8.0\ \cancel{\text{g}}}{124.0\ \cancel{\text{g}}} \times 100\% = 6.5\%\text{ H}$$

$$\%\text{ C} = \frac{24.0\ \cancel{\text{g}}}{124.0\ \cancel{\text{g}}} \times 100\% = 19.4\%\text{ C}$$

$$\%\text{ O} = \frac{64.0\ \cancel{\text{g}}}{124.0\ \cancel{\text{g}}} \times 100\% = 51.6\%\text{ O}$$

c. $$\%\text{ Mg} = \frac{24.3\ \cancel{\text{g}}}{58.3\ \cancel{\text{g}}} \times 100\% = 41.7\%\text{ Mg}$$

$$\%\text{ O} = \frac{32.0\ \cancel{\text{g}}}{58.3\ \cancel{\text{g}}} \times 100\% = 54.9\%\text{ O}$$

$$\%\text{ H} = \frac{2.0\ \cancel{\text{g}}}{58.3\ \cancel{\text{g}}} \times 100\% = 3.4\%\text{ H}$$

d. $$\%\text{ Na} = \frac{69.0\ \cancel{\text{g}}}{164.0\ \cancel{\text{g}}} \times 100\% = 42.1\%\text{ Na}$$

$$\%\text{ P} = \frac{31.0\ \cancel{\text{g}}}{164.0\ \cancel{\text{g}}} \times 100\% = 18.9\%\text{ P}$$

$$\%\text{ O} = \frac{64.0\ \cancel{\text{g}}}{164.0\ \cancel{\text{g}}} \times 100\% = 39.0\%\text{ O}$$

74. **a.** $$3.54\text{ g } \cancel{H_2S} \times \frac{94.1\text{ g S}}{100\text{ g } \cancel{H_2S}} = 3.33\text{ g S}$$

b. $$25.0\text{ g } \cancel{(NH_4)_2C_2O_4} \times \frac{22.6\text{ g N}}{100\text{ g } \cancel{(NH_4)_2C_2O_4}} = 5.65\text{ g N}$$

c. $$97.4\text{ g } \cancel{Mg(OH)_2} \times \frac{41.7\text{ g Mg}}{100\text{ g } \cancel{Mg(OH)_2}} = 40.6\text{ g Mg}$$

d. $$804\text{ g } \cancel{Na_3PO_4} \times \frac{18.9\text{ g P}}{100\text{ g } \cancel{Na_3PO_4}} = 152\text{ g P}$$

76. An empirical formula has the lowest whole-number ratio of elements.

a. molecular **c.** molecular and empirical

b. molecular **d.** molecular and empirical

Selected Solutions **R87**

79. **a.** efm of CH_2O = 12.0g/mol + 2 × (1.0 g/mol) + 16.0 g/mol = 30.0 g/mol

$$\frac{\text{molar mass}}{\text{efm}} = \frac{90 \text{ g/mol}}{30.0 \text{ g/mol}} = 3$$

$(CH_2O) \times 3 = C_3H_6O_3$

b. efm of HgCl = 200.6 g/mol + 35.5 g/mol = 236.1 g/mol

$$\frac{\text{molar mass}}{\text{efm}} = \frac{472.2 \text{ g/mol}}{236.1 \text{ g/mol}} = 2$$

$(HgCl) \times 2 = Hg_2Cl_2$

80. **a.** $1.00 \text{ mol } C_{12}H_{22}O_{11} \times \frac{6.02 \times 10^{23} \text{ molecules } C_{12}H_{22}O_{11}}{1 \text{ mol } C_{12}H_{22}O_{11}} \times \frac{45 \text{ atoms}}{1 \text{ molecule } C_{12}H_{22}O_{11}} = 2.71 \times 10^{25} \text{ atoms}$

b. $2.00 \text{ mol } C_{12}H_{22}O_{11} \times \frac{6.02 \times 10^{23} \text{ molecules } C_{12}H_{22}O_{11}}{1 \text{ mol } C_{12}H_{22}O_{11}} \times \frac{12 \text{ C atoms}}{1 \text{ molecule } C_{12}H_{22}O_{11}} = 1.44 \times 10^{25} \text{ C atoms}$

c. $2.00 \text{ mol } C_{12}H_{22}O_{11} \times \frac{6.02 \times 10^{23} \text{ molecules } C_{12}H_{22}O_{11}}{1 \text{ mol } C_{12}H_{22}O_{11}} \times \frac{22 \text{ H atoms}}{1 \text{ molecule } C_{12}H_{22}O_{11}} = 2.65 \times 10^{25} \text{ H atoms}$

d. $3.65 \text{ mol } C_{12}H_{22}O_{11} \times \frac{6.02 \times 10^{23} \text{ molecules } C_{12}H_{22}O_{11}}{1 \text{ mol } C_{12}H_{22}O_{11}} \times \frac{11 \text{ O atoms}}{1 \text{ molecule } C_{12}H_{22}O_{11}} = 2.42 \times 10^{25} \text{ O atoms}$

82. **a.** efm of CH_2O = 12.0 g/mol + 2 × (1.0 g/mol) + 16.0 g/mol = 30.0 g/mol

Compound A: $\frac{\text{molar mass of A}}{\text{efm}} = \frac{60.0 \text{ g/mol}}{30.0 \text{ g/mol}} = 2$

$(CH_2O) \times 2 = C_2H_4O_2$

Compound D: $\frac{\text{molar mass of D}}{\text{efm}} = \frac{150.0 \text{ g/mol}}{30.0 \text{ g/mol}} = 5$

$(CH_2O) \times 5 = C_5H_{10}O_5$

Compound E: $\frac{\text{molar mass of E}}{\text{efm}} = \frac{180.0 \text{ g/mol}}{30.0 \text{ g/mol}} = 6$

$(CH_2O) \times 6 = C_6H_{12}O_6$

b. $\text{slope} = \frac{\Delta \text{ molar mass}}{\Delta \text{ mass of C}} = \frac{150.0 - 60.0}{60.0 - 24.0} = \frac{2.5}{1}$

The slope is the ratio of the empirical formula mass to the mass of carbon in the empirical formula: 30.0 g/12.0 g = 2.5/1.

c. The two other data points occur when molar mass/efm = 3 and 4. These data points correspond to compounds with molecular formulas $C_3H_6O_3$ and $C_4H_8O_4$, respectively. Thus, the *x, y* values are (36, 90) and (48, 120).

85. molar mass of Cl_2O = 2 × (35.5 g/mol) + 16.0 g/mol = 87.0 g/mol

$$90.0 \text{ g } Cl_2O \times \frac{1 \text{ mol } Cl_2O}{87.0 \text{ g } Cl_2O} \times \frac{1 \text{ mol O}}{1 \text{ mol } Cl_2O} \times \frac{16.0 \text{ g O}}{1 \text{ mol O}} = 16.6 \text{ g O}$$

87. $150.0 \text{ mg } C_{17}H_{18}FN_3O_3 \times \frac{1 \text{ g } C_{17}H_{18}FN_3O_3}{1000 \text{ mg } C_{17}H_{18}FN_3O_3} \times \frac{1 \text{ mol } C_{17}H_{18}FN_3O_3}{331.0 \text{ g } C_{17}H_{18}FN_3O_3} \times \frac{6.02 \times 10^{23} \text{ molecules } C_{17}H_{18}FN_3O_3}{1 \text{ mol } C_{17}H_{18}FN_3O_3} \times \frac{1 \text{ F atom}}{1 \text{ molecule } C_{17}H_{18}FN_3O_3} = 2.73 \times 10^{20} \text{ F atoms}$

89. **a.** $94.1 \text{ g O} \times \frac{1 \text{ mol O}}{16.0 \text{ g O}} = 5.88 \text{ mol O}$

$5.9 \text{ g H} \times \frac{1 \text{ mol H}}{1.0 \text{ g H}} = 5.9 \text{ mol H}$

$\frac{5.88 \text{ mol O}}{5.88} = 1.00 \text{ mol O}; \frac{5.9 \text{ mol H}}{5.88} = 1.00 \text{ mol H};$

The empirical formula is HO.

$$\frac{\text{molar mass}}{\text{efm}} = \frac{34 \text{ g/mol}}{17.0 \text{ g/mol}} = 2$$

$(HO) \times 2 = H_2O_2$

b. $50.7 \text{ g C} \times \frac{1 \text{ mol C}}{12.0 \text{ g C}} = 4.23 \text{ mol C}$

$4.2 \text{ g H} \times \frac{1 \text{ mol H}}{1.0 \text{ g H}} = 4.2 \text{ mol H}$

$45.1 \text{ g O} \times \frac{1 \text{ mol O}}{16.0 \text{ g O}} = 2.83 \text{ mol O}$

$\frac{4.23 \text{ mol C}}{2.83} = 1.49 \text{ mol C}; \frac{4.2 \text{ mol H}}{2.83} = 1.5 \text{ mol H};$

$\frac{2.83 \text{ mol O}}{2.83} = 1.00 \text{ mol O}$

1.49 mol C × 2 = 3 mol C; 1.5 mol H × 2 = 3 mol H; 1.00 mol O × 2 = 2 mol O. The empirical formula is $C_3H_3O_2$.

$$\frac{\text{molar mass}}{\text{efm}} = \frac{142 \text{ g/mol}}{71.0 \text{ g/mol}} = 2$$

$(C_3H_3O_2) \times 2 = C_6H_6O_4$

R88 Appendix E

c. $56.6\ g\ K \times \frac{1\ mol\ K}{39.1\ g\ K} = 1.45\ mol\ K$

$8.7\ g\ C \times \frac{1\ mol\ C}{12.0\ g\ C} = 0.73\ mol\ C$

$34.7\ g\ O \times \frac{1\ mol\ O}{16.0\ g\ O} = 2.17\ mol\ O$

$\frac{1.45\ mol\ K}{0.73} = 2.00\ mol\ K; \frac{0.73\ mol\ C}{0.73} = 1.00\ mol\ C;$

$\frac{2.17\ mol\ O}{0.73} = 3.00\ mol\ O$

The empirical formula is K_2CO_3.

$\frac{\text{molar mass}}{\text{efm}} = \frac{138.2\ g/mol}{138.2\ g/mol} = 1$

$(K_2CO_3) \times 1 = K_2CO_3$

90. **a.** Atomic mass of boron = 10.8 amu

$15\ atoms\ B \times \frac{10.8\ amu}{1\ atom\ B} = 162\ amu$

$\frac{162\ amu}{6} = 27.0\ amu$

b. aluminum

91. $6.02 \times 10^{23}\ particles \times \frac{5 \times 10^{-6}\ cm}{1\ particle} \times \frac{1\ m}{100\ cm} \times \frac{1\ km}{1000\ m} = 3 \times 10^{13}\ km$

94. $5.50\ L \times \frac{1\ mol\ He}{22.4\ L} \times \frac{4.0\ g\ He}{1\ mol\ He} = 0.98\ g\ He$

96. Let the number of carbon atoms = x. Then, the empirical formula is $C_xH_{3x}O_{x/2}$. The smallest value of x that will give whole number subscripts is 2. Therefore, the empirical formula is C_2H_6O.

100. **a.** $68\ g\ C \times \frac{1\ mol\ C}{12.0\ g\ C} = 5.7\ mol\ C$

$7\ g\ H \times \frac{1\ mol\ H}{1.0\ g\ H} = 7.0\ mol\ H$

$20\ g\ O \times \frac{1\ mol\ O}{16.0\ g\ O} = 1.3\ mol\ O$

$9\ g\ N \times \frac{1\ mol\ N}{14.0\ g\ N} = 0.64\ mol\ N$

$\frac{5.7\ mol\ C}{0.64} = 9\ mol\ C; \frac{7\ mol\ H}{0.64} = 11\ mol\ H;$

$\frac{1.3\ mol\ O}{0.64} = 2\ mol\ O; \frac{0.6\ mol\ N}{0.64} = 1\ mol\ N$

The empirical formula is $C_9H_{11}O_2N$.

b. $\frac{\text{molar mass}}{\text{efm}} = \frac{165.2\ g/mol}{165.0\ g/mol} = 1$

The molecular formula is $C_9H_{11}O_2N$.

101. From the given information, the ratio of atoms is $C_{0.6}H_1O_{1.8}N_{0.6}$. The empirical formula is $(C_{0.6}H_1O_{1.8}N_{0.6}) \times 5 = C_3H_5O_9N_3$.

efm = $3 \times (12.0\ g/mol) + 5 \times (1.0\ g/mol) + 9 \times (16.0\ g/mol) + 3 \times (14.0\ g/mol) = 227.0\ g/mol$

$\text{molar mass} = \frac{1.00\ g}{0.00441\ mol} = 227\ g/mol$

$\frac{\text{molar mass}}{\text{efm}} = \frac{227}{227.0} = 1$

The molecular formula is also $C_3H_5O_9N_3$.

103. $75.0\ g\ air \times \frac{1\ L\ air}{1.19\ g\ air} \times \frac{20.95\ L\ O_2}{100\ L\ air} \times \frac{1\ mol\ O_2}{22.4\ L\ O_2} \times \frac{6.02 \times 10^{23}\ molecules\ O_2}{1\ mol\ O_2} = 3.54 \times 10^{23}\ molecules\ O_2$

105. Consider a cube containing 1 mol of NaCl. The mass of the cube (to 4 significant figures) would then be 22.99 g + 35.45 g = 58.44 g. The volume of the cube can be calculated using the density of solid NaCl:

$58.44\ g\ NaCl \times \frac{1\ cm^3}{2.165\ g\ NaCl} = 26.99\ cm^3$

The length of one side of the cube is then

$\sqrt[3]{26.99\ cm^3} = 3.000\ cm$

The number of ions per side is as follows:

$3.000\ cm \times \frac{1\ ion}{2.819 \times 10^{-8}\ cm} = 1.064 \times 10^8\ ions$

The total number of ions in the cube is then $(1.064 \times 10^8\ ions)^3 = 1.205 \times 10^{24}\ ions$

To calculate Avogadro's number, determine the number of formula units in the cube:

$1.205 \times 10^{24}\ ions \times \frac{1\ formula\ unit}{2\ ions} = 6.025 \times 10^{23}\ formula\ units$

108. $30.0\ mg \times \frac{1\ g}{1000\ mg} \times \frac{1\ mol\ of\ compound}{312\ g} \times \frac{6.02 \times 10^{23}\ molecules}{1\ mol\ of\ compound} = 5.79 \times 10^{19}\ molecules$

110. chemical change: wax burning
physical changes: wax melting, wax vaporizing

111. **a.** physical change
b. chemical change
c. chemical change
d. physical change
e. chemical change
f. physical change

Selected Solutions **R89**

114. $\frac{84.0\text{ g}}{2.75\text{ cm} \times 4.80\text{ cm} \times 7.5\text{ cm}} = 0.85\text{ g/cm}^3$

The block will float because its density (0.85 g/cm^3) is less than the density of water (1.00 g/cm^3).

115. a. $4.72\cancel{\text{g}} \times \frac{1000\text{ mg}}{1\cancel{\text{g}}} = 4.72 \times 10^3\text{ mg}$

b. $\frac{2.7 \times 10^3\cancel{\text{cm}}}{\cancel{\text{s}}} \times \frac{1\cancel{\text{m}}}{100\cancel{\text{cm}}} \times \frac{1\text{ km}}{1000\cancel{\text{m}}} \times \frac{60\cancel{\text{s}}}{1\cancel{\text{min}}} \times$

$\frac{60\cancel{\text{min}}}{1\text{ h}} = 97\text{ km/h}$

c. $4.4\cancel{\text{mm}} \times \frac{1\cancel{\text{m}}}{1000\cancel{\text{mm}}} \times \frac{10\text{ dm}}{1\cancel{\text{m}}} = 4.4 \times 10^{-2}\text{ dm}$

116. a. 40 protons, 40 electrons, 50 neutrons
b. 46 protons, 46 electrons, 62 neutrons
c. 35 protons, 35 electrons, 46 neutrons
d. 51 protons, 51 electrons, 72 neutrons

117. a. $1s^22s^22p^5$
b. $1s^22s^1$
c. $1s^22s^22p^63s^23p^63d^{10}4s^24p^65s^1$

119. Cr, Cd, Cu, Co

126. d. CaS_2 f. Ba(OH)

127. a. iron(III) hydroxide c. sodium carbonate
b. ammonium iodide d. carbon tetrachloride

128. a. KNO_3 b. CuO c. Mg_3N_2 d. AgF

Chapter 11

1. When solid sodium is dropped in water, hydrogen gas and aqueous sodium hydroxide are produced.

2. $S(s) + O_2(g) \longrightarrow SO_2(g)$

3. $3CO + Fe_2O_3 \longrightarrow 2Fe + 3CO_2$

4. $2C + O_2 \longrightarrow 2CO$

5. a. $FeCl_3 + 3NaOH \longrightarrow Fe(OH)_3 + 3NaCl$
b. $CS_2 + 3Cl_2 \longrightarrow CCl_4 + S_2Cl_2$
c. $2KI + Pb(NO_3)_2 \longrightarrow PbI_2 + 2KNO_3$
d. $2C_2H_2 + 5O_2 \longrightarrow 4CO_2 + 2H_2O$

6. a. $Ca(OH)_2 + H_2SO_4 \longrightarrow CaSO_4 + 2H_2O$
b. $Na + H_2O \longrightarrow NaOH + H$

12. HBr

13. $2HI \longrightarrow H_2 + I_2$

14. $3Mg + N_2 \longrightarrow Mg_3N_2$

15. a. $Fe(s) + Pb(NO_3)_2(aq) \longrightarrow Fe(NO_3)_2(aq) + Pb(s)$
b. $Cl_2(aq) + 2NaI(aq) \longrightarrow 2NaCl(aq) + I_2(aq)$
c. $Ca(s) + 2H_2O(l) \longrightarrow Ca(OH)_2(aq) + H_2(g)$
d. $Zn(s) + H_2SO_4(aq) \longrightarrow ZnSO_4(aq) + H_2(g)$

16. a. $3NaOH(aq) + Fe(NO_3)_3(aq) \longrightarrow Fe(OH)_3(s) + 3NaNO_3(aq)$
b. $3Ba(NO_3)_2(aq) + 2H_3PO_4(aq) \longrightarrow Ba_3(PO_4)_2(s) + 6HNO_3(aq)$
c. $FeS(s) + 2HCl(aq) \longrightarrow H_2S(g) + FeCl_2(aq)$

17. a. $3KOH(aq) + H_3PO_4(aq) \longrightarrow K_3PO_4(aq) + 3H_2O(l)$
b. $AgNO_3(aq) + NaCl(s) \longrightarrow AgCl(s) + NaNO_3(aq)$
c. $3Ca(OH)_2(aq) + 2H_3PO_4(aq) \longrightarrow Ca_3(PO_4)_2(aq) + 6H_2O(l)$
d. $2KI(aq) + Pb(NO_3)_2(aq) \longrightarrow 2KNO_3(aq) + PbI_2(s)$
e. $3H_2SO_4(aq) + 2Al(OH)_3(aq) \longrightarrow Al_2(SO_4)_3(aq) + 6H_2O(l)$

18. a. $CH_2O(g) + O_2(g) \longrightarrow CO_2(g) + H_2O(g)$
b. $C_7H_{16}(l) + 11O_2(g) \longrightarrow 7CO_2(g) + 8H_2O(g)$
c. $2C_6H_6(l) + 15O_2(g) \longrightarrow 12CO_2(g) + 6H_2O(g)$

19. a. $C_6H_{12}O_6(s) + 6O_2(g) \longrightarrow 6CO_2(g) + 6H_2O(g)$
b. $C_3H_6O(l) + 4O_2(g) \longrightarrow 3CO_2(g) + 3H_2O(g)$
c. $2C_5H_{12}O(l) + 15O_2(g) \longrightarrow 10CO_2(g) + 12H_2O(g)$

25. $OH^-(aq) + H^+(aq) \longrightarrow H_2O(l)$

26. Complete ionic equation:
$3Ca^{2+}(aq) + 6OH^-(aq) + 6H^+(aq) + 2PO_4^{3-}(aq) \longrightarrow Ca_3(PO_4)_2(s) + 6H_2O(l)$
Net ionic equation: same as complete ionic equation

27. The precipitate formed is lead chloride.
$Pb^{2+}(aq) + Cl^-(aq) \longrightarrow PbCl_2(s)$

28. Complete ionic equation:
$Fe^{3+}(aq) + NO^{3-}(aq) + 3Na^+(aq) + 3OH^-(aq) \longrightarrow 3Na^+(aq) + NO^{3-}(aq) + Fe(OH)_3(s)$
Net ionic equation:
$Fe^{3+}(aq) + OH^-(aq) \longrightarrow Fe(OH)_3(s)$

37. a. $2PbO_2(s) \longrightarrow 2PbO(s) + O_2(g)$
b. $2Fe(OH)_3(s) \longrightarrow Fe_2O_3(s) + 3H_2O(l)$
c. $(NH_4)_2CO_3(s) \longrightarrow 2NH_3(g) + H_2O(g) + CO_2(g)$
d. $CaCl_2(aq) + H_2SO_4(aq) \longrightarrow CaSO_4(s) + 2HCl(aq)$

38. a. $2Mg(s) + O_2(g) \longrightarrow 2MgO(s)$
b. $4P(s) + 5O_2(g) \longrightarrow 2P_2O_5(s)$
c. $Ca(s) + S(s) \longrightarrow CaS(s)$

46. a. $H^+(aq) + OH^-(aq) \longrightarrow H_2O(l)$
b. $Ag^+(aq) + Cl^-(aq) \longrightarrow AgCl(s)$

51. a. $Na_2O(s) + H_2O(l) \longrightarrow 2NaOH(aq)$
b. $H_2(g) + Br_2(g) \longrightarrow 2HBr(g)$
c. $Cl_2O_7(l) + H_2O(l) \longrightarrow 2HClO_4(aq)$

R90 Appendix E

53. **a.** tube A

b. $2Na(s) + 2H_2O(l) \longrightarrow 2NaOH(aq) + H_2(g)$
single-replacement

57. **a.** $CdS(s)$

b. $Na^+(aq)$ and $NO_3^-(aq)$

c. $Cd^{2+}(aq) + S^{2-}(aq) \longrightarrow CdS(s)$

60. **a.** single-replacement

b. $Cl_2(g) + 2I^-(aq) \longrightarrow I_2(aq) + 2Cl^-(aq)$

62. **a.** $2K(s) + 2H_2O(l) \longrightarrow 2KOH(aq) + H_2(g)$

b. $C_2H_5OH(l) + 3O_2(g) \longrightarrow 2CO_2(g) + 3H_2O(g)$

c. $2Bi(NO_3)_3(aq) + 3H_2S(g) \longrightarrow Bi_2S_3(s) + 6HNO_3(aq)$

d. $2Al(s) + 3Br_2(l) \longrightarrow 2AlBr_3(s)$

65. Smoking is not permitted near an oxygen source because a fire will burn faster in an area of high oxygen concentration. However, if a match were struck in a room full of oxygen and isolated from combustible material, it would only burn more vigorously.

70. 22 protons, 28 neutrons, and 22 electrons

74. **a.** K_2CrO_4 **b.** $NaHSO_3$ **c.** permanganic acid **d.** potassium oxalate

77. First, determine the empirical formula.

$$49.5\ \cancel{g\ C} \times \frac{1\ \text{mol C}}{12.0\ \cancel{g\ C}} = 4.13\ \text{mol C}$$

$$5.20\ \cancel{g\ H} \times \frac{1\ \text{mol H}}{1.01\ \cancel{g\ H}} = 5.15\ \text{mol H}$$

$$16.5\ \cancel{g\ O} \times \frac{1\ \text{mol O}}{16.0\ \cancel{g\ O}} = 1.03\ \text{mol O}$$

$$28.9\ \cancel{g\ N} \times \frac{1\ \text{mol N}}{14.0\ \cancel{g\ N}} = 2.06\ \text{mol N}$$

Then, divide each value by 1.03. The empirical formula is $C_4H_5ON_2$. The mass of the empirical formula is 97 g.

$$\frac{\text{gfm}}{\text{efm}} = \frac{194.1\ g}{97\ g} = 2$$

$$(C_4H_5ON_2) \times 2 = C_8H_{10}O_2N_4$$

Chapter 12

1. $$288\ \cancel{ESW_3HP_2} \times \frac{1\ \text{seat}}{1\ \cancel{ESW_3HP_2}} = 288\ \text{seats}$$

$$288\ \cancel{ESW_3HP_2} \times \frac{3\ \text{wheels}}{1\ \cancel{ESW_3HP_2}} = 864\ \text{wheels}$$

$$288\ \cancel{ESW_3HP_2} \times \frac{2\ \text{pedals}}{1\ \cancel{ESW_3HP_2}} = 576\ \text{pedals}$$

2. Answers will vary but should include the correct number of "parts" to make the product. For example, 1 deck + 2 trucks + 4 wheels $\longrightarrow DT_2W_4$. Note: The truck (T) of a skateboard holds the wheels (W). The deck (D) of a skateboard is the part that a skateboarder stands on when riding.

3. 2 molecules H_2 + 1 molecule $O_2 \longrightarrow$ 2 molecules H_2O

2 moles H_2 + 1 mole $O_2 \longrightarrow$ 2 moles H_2O

44.8 L H_2 + 22.4 L $O_2 \longrightarrow$ 44.8 L H_2O

4. $C_2H_4(g) + 3O_2 \longrightarrow 2CO_2(g) + 2H_2O(g)$;

1 mol C_2H_4 + 3 mol $O_2 \longrightarrow$ 2 mol CO_2 + 2 mol H_2O;

22.4 L C_2H_4 + 67.2 L $O_2 \longrightarrow$ 44.8 L CO_2 + 44.8 L H_2O;

$$\left(1\ \cancel{mol} \times \frac{28.0\ g}{\cancel{mol}}\right) + \left(3\ \cancel{mol} \times \frac{32.0\ g}{\cancel{mol}}\right) \longrightarrow \left(2\ \cancel{mol} \times \frac{44.0\ g}{\cancel{mol}}\right) + \left(2\ \cancel{mol} \times \frac{18.0\ g}{\cancel{mol}}\right)$$

28.0 g C_2H_4 + 96.0 g $O_2 \longrightarrow$ 88.0 g CO_2 + 36.0 g H_2O

124 g = 124 g

11. **a.** $$\frac{4\ \text{mol Al}}{3\ \text{mol } O_2} \quad \frac{3\ \text{mol } O_2}{4\ \text{mol Al}} \quad \frac{4\ \text{mol Al}}{2\ \text{mol } Al_2O_3}$$

$$\frac{2\ \text{mol } Al_2O_3}{4\ \text{mol Al}} \quad \frac{3\ \text{mol } O_2}{2\ \text{mol } Al_2O_3} \quad \frac{2\ \text{mol } Al_2O_3}{3\ \text{mol } O_2}$$

b. $$3.7\ \cancel{\text{mol } Al_2O_3} \times \frac{4\ \text{mol Al}}{2\ \cancel{\text{mol } Al_2O_3}} = 7.4\ \text{mol Al}$$

12. **a.** $$14.8\ \cancel{\text{mol Al}} \times \frac{3\ \text{mol } O_2}{4\ \cancel{\text{mol Al}}} = 11.1\ \text{mol } O_2$$

b. $$0.78\ \cancel{\text{mol } O_2} \times \frac{2\ \text{mol } Al_2O_3}{3\ \cancel{\text{mol } O_2}} = 0.52\ \text{mol } Al_2O_3$$

13. $$5.00\ \cancel{g\ CaC_2} \times \frac{1\ \cancel{\text{mol } CaC_2}}{64.1\ \cancel{g\ CaC_2}} \times \frac{1\ \cancel{\text{mol } C_2H_2}}{1\ \cancel{\text{mol } CaC_2}} \times \frac{26.0\ g\ C_2H_2}{1\ \cancel{\text{mol } C_2H_2}} = 2.03\ g\ C_2H_2$$

14. $$49.0\ \cancel{g\ H_2O} \times \frac{1\ \cancel{\text{mol } H_2O}}{18.0\ \cancel{g\ H_2O}} \times \frac{1\ \text{mol } CaC_2}{2\ \cancel{\text{mol } H_2O}} = 1.36\ \text{mol } CaC_2$$

15. $$6.54\ \cancel{g\ KClO_3} \times \frac{1\ \cancel{\text{mol } KClO_3}}{122.6\ \cancel{g\ KClO_3}} \times \frac{3\ \cancel{\text{mol } O_2}}{2\ \cancel{\text{mol } KClO_3}} \times \frac{6.02 \times 10^{23}\ \text{molecules } O_2}{1\ \cancel{\text{mol } O_2}} = 4.82 \times 10^{22}\ \text{molecules } O_2$$

16. $$5.00 \times 10^{22}\ \cancel{\text{molecules NO}} \times \frac{1\ \cancel{\text{mol NO}}}{6.02 \times 10^{23}\ \cancel{\text{molecules NO}}} \times \frac{3\ \cancel{\text{mol } NO_2}}{1\ \cancel{\text{mol NO}}} \times \frac{46.0\ g\ NO_2}{1\ \cancel{\text{mol } NO_2}} = 11.5\ g\ NO_2$$

Selected Solutions R91

17. $3.86 \text{ L CO} \times \frac{1 \text{ mol CO}}{22.4 \text{ L CO}} \times \frac{1 \text{ mol } O_2}{2 \text{ mol CO}} \times \frac{22.4 \text{ L } O_2}{1 \text{ mol } O_2}$
$= 1.93 \text{ L } O_2$

18. $0.42 \text{ L } H_2 \times \frac{1 \text{ mol } H_2}{22.4 \text{ L } H_2} \times \frac{4 \text{ mol } PH_3}{6 \text{ mol } H_2} \times \frac{22.4 \text{ L } PH_3}{1 \text{ mol } PH_3}$
$= 0.28 \text{ L } PH_3$

19. $27.9 \text{ mL } O_2 \times \frac{2 \text{ mL } SO_2}{3 \text{ mL } O_2} = 18.6 \text{ mL } SO_2$

20. $0.38 \text{ L } SO_2 \times \frac{1 \text{ L } CO_2}{2 \text{ L } SO_2} \times \frac{10 \text{ dL } CO_2}{1 \text{ L } CO_2} = 1.9 \text{ dL } CO_2$

26. $2.70 \text{ mol } C_2H_4 \times \frac{3 \text{ mol } O_2}{1 \text{ mol } C_2H_4} = 8.10 \text{ mol } O_2$

8.10 mol O_2 are needed to react with 2.70 mol C_2H_4, but there are only 6.30 mol O_2 given for the reaction. Therefore, O_2 is the limiting reagent.

27. $6.00 \text{ g HCl} \times \frac{1 \text{ mol HCl}}{36.5 \text{ g HCl}} = 0.160 \text{ mol HCl}$

$5.00 \text{ g Mg} \times \frac{1 \text{ mol Mg}}{24.3 \text{ g Mg}} = 0.210 \text{ mol Mg}$

$0.16 \text{ mol HCl} \times \frac{1 \text{ mol Mg}}{2 \text{ mol HCl}} = 0.080 \text{ mol Mg}$

HCl is the limiting reagent.

28. **a.** $2.70 \text{ mol } C_2H_4 \times \frac{2 \text{ mol } O_2}{1 \text{ mol } C_2H_4} = 5.40 \text{ mol } O_2$

C_2H_4 is the limiting reagent.

b. $2.70 \text{ mol } C_2H_4 \times \frac{2 \text{ mol } H_2O}{1 \text{ mol } C_2H_4} = 5.40 \text{ mol } H_2O$

29. $2.40 \text{ mol } C_2H_2 \times \frac{5 \text{ mol } O_2}{2 \text{ mol } C_2H_2} = 6.00 \text{ mol } O_2$

C_2H_2 is the limiting reagent.

$2.40 \text{ mol } C_2H_2 \times \frac{2 \text{ mol } H_2O}{2 \text{ mol } C_2H_2} \times \frac{18.0 \text{ g } H_2O}{1 \text{ mol } H_2O}$
$= 43.2 \text{ g } H_2O$

30. $84.8 \text{ g } Fe_2O_3 \times \frac{1 \text{ mol } Fe_2O_3}{159.9 \text{ g } Fe_2O_3} \times \frac{2 \text{ mol Fe}}{1 \text{ mol } Fe_2O_3}$
$\times \frac{55.9 \text{ g Fe}}{1 \text{ mol Fe}} = 59.3 \text{ g Fe}$

31. $Cu(s) + 2AgNO_3(aq) \longrightarrow 2Ag(s) + Cu(NO_3)_2(aq)$

$5.00 \text{ g Cu} \times \frac{1 \text{ mol Cu}}{63.6 \text{ g Cu}} \times \frac{2 \text{ mol Ag}}{1 \text{ mol Cu}} \times \frac{107.9 \text{ g Ag}}{1 \text{ mol Ag}}$
$= 17.0 \text{ g Ag}$

32. $50.0 \text{ g } SiO_2 \times \frac{1 \text{ mol } SiO_2}{60.1 \text{ g } SiO_2} \times \frac{1 \text{ mol SiC}}{1 \text{ mol } SiO_2} \times \frac{40.1 \text{ g SiC}}{1 \text{ mol SiC}}$
$= 33.36 \text{ g SiC}$

$\% \text{ yield} = \frac{27.9 \text{ g SiC}}{33.4 \text{ g SiC}} \times 100\% = 83.5\%$

33. $N_2 + 3H_2 \longrightarrow 2NH_3$

$15.0 \text{ g } N_2 \times \frac{1 \text{ mol } N_2}{28.0 \text{ g } N_2} \times \frac{3 \text{ mol } H_2}{1 \text{ mol } N_2} \times \frac{2.0 \text{ g } H_2}{1 \text{ mol } H_2} = 3.2 \text{ g } H_2$

N_2 is the limiting reagent, so use the given mass of nitrogen to find the theoretical yield of NH_3.

$15.0 \text{ g } N_2 \times \frac{1 \text{ mol } N_2}{28.0 \text{ g } N_2} \times \frac{2 \text{ mol } NH_3}{1 \text{ mol } N_2} \times \frac{17.0 \text{ g } NH_3}{1 \text{ mol } NH_3}$
$= 18.2 \text{ g } NH_3$

$\% \text{ yield} = \frac{10.5 \text{ g } NH_3}{18.2 \text{ g } NH_3} \times 100\% = 57.7\%$

45. **a.** $2.7 \text{ mol C} \times \frac{1 \text{ mol } CS_2}{5 \text{ mol C}} = 0.54 \text{ mol } CS_2$

b. $5.44 \text{ mol } SO_2 \times \frac{5 \text{ mol C}}{2 \text{ mol } SO_2} = 13.6 \text{ mol C}$

c. $0.246 \text{ mol } CS_2 \times \frac{4 \text{ mol CO}}{1 \text{ mol } CS_2} = 0.984 \text{ mol CO}$

d. $118 \text{ mol } CS_2 \times \frac{2 \text{ mol } SO_2}{1 \text{ mol } CS_2} = 236 \text{ mol } SO_2$

46. **a.** $3.60 \times 10^2 \text{ g } CH_3OH \times \frac{1 \text{ mol } CH_3OH}{32.0 \text{ g } CH_3OH}$
$\times \frac{1 \text{ mol CO}}{1 \text{ mol } CH_3OH} = 11.3 \text{ mol CO}$

$3.60 \times 10^2 \text{ g } CH_3OH \times \frac{1 \text{ mol } CH_3OH}{32.0 \text{ g } CH_3OH}$
$\times \frac{2 \text{ mol } H_2}{1 \text{ mol } CH_3OH} = 22.5 \text{ mol } H_2$

b. $4.00 \text{ mol } CH_3OH \times \frac{1 \text{ mol CO}}{1 \text{ mol } CH_3OH} \times \frac{28.0 \text{ g CO}}{1 \text{ mol CO}}$
$= 112 \text{ g CO}$

$4.00 \text{ mol } CH_3OH \times \frac{2 \text{ mol } H_2}{1 \text{ mol } CH_3OH} \times \frac{2.0 \text{ g } H_2}{1 \text{ mol } H_2}$
$= 16 \text{ g } H_2$

c. $2.85 \text{ mol CO} \times \frac{2 \text{ mol } H_2}{1 \text{ mol CO}} \times \frac{2.0 \text{ g } H_2}{1 \text{ mol } H_2} = 11 \text{ g } H_2$

50. **a.** $32.9 \text{ g } Li_3N \times \frac{1 \text{ mol } Li_3N}{34.7 \text{ g } Li_3N} \times \frac{3 \text{ mol } H_2O}{1 \text{ mol } Li_3N}$
$\times \frac{18.0 \text{ g } H_2O}{1 \text{ mol } H_2O} = 51.2 \text{ g } H_2O$

R92 Appendix E

b. $32.9\ \text{g Li}_3\text{N} \times \dfrac{1\ \text{mol Li}_3\text{N}}{34.7\ \text{g Li}_3\text{N}} \times \dfrac{1\ \text{mol NH}_3}{1\ \text{mol Li}_3\text{N}} \times \dfrac{6.02 \times 10^{23}\ \text{molecules NH}_3}{1\ \text{mol NH}_3}$

$= 5.71 \times 10^{23}\ \text{molecules NH}_3$

c. $15.0\ \text{L NH}_3 \times \dfrac{1\ \text{mol NH}_3}{22.4\ \text{L NH}_3} \times \dfrac{1\ \text{mol Li}_3\text{N}}{1\ \text{mol NH}_3} \times \dfrac{34.7\ \text{g Li}_3\text{N}}{1\ \text{mol Li}_3\text{N}} = 23.2\ \text{g Li}_3\text{N}$

53. **a.** $3.0\ \text{mol Al} \times \dfrac{3\ \text{mol Cl}_2}{2\ \text{mol Al}} = 4.5\ \text{mol Cl}_2$

Al is the limiting reagent.

b. $3.0\ \text{mol Al} \times \dfrac{2\ \text{mol AlCl}_3}{2\ \text{mol Al}} = 3.0\ \text{mol AlCl}_3$

c. $5.3\ \text{mol Cl}_2 - 4.5\ \text{mol Cl}_2 = 0.80\ \text{mol Cl}_2$

54. $15.0\ \text{g Sb}_2\text{S}_3 \times \dfrac{1\ \text{mol Sb}_2\text{S}_3}{339.9\ \text{g Sb}_2\text{S}_3} \times \dfrac{2\ \text{mol Sb}}{1\ \text{mol Sb}_2\text{S}_3} \times \dfrac{121.8\ \text{g Sb}}{1\ \text{mol Sb}} = 10.8\ \text{g Sb}$

$\%\ \text{yield} = \dfrac{9.84\ \text{g Sb}}{10.8\ \text{g Sb}} \times 100\% = 91.1\%$

57. **a.** $1.49\ \text{g HNO}_3 \times \dfrac{1\ \text{mol HNO}_3}{63.0\ \text{g HNO}_3} \times \dfrac{4\ \text{mol Zn}}{10\ \text{mol HNO}_3} \times \dfrac{6.02 \times 10^{23}\ \text{atoms Zn}}{1\ \text{mol Zn}} = 5.70 \times 10^{21}\ \text{atoms Zn}$

b. $29.1\ \text{g NH}_4\text{NO}_3 \times \dfrac{1\ \text{mol NH}_4\text{NO}_3}{80.0\ \text{g NH}_4\text{NO}_3} \times \dfrac{4\ \text{mol Zn}}{1\ \text{mol NH}_4\text{NO}_3} \times \dfrac{65.4\ \text{g Zn}}{1\ \text{mol Zn}} = 95.2\ \text{g Zn}$

60. **a.** $1.0\ \text{kg N}_2\text{H}_4 \times \dfrac{10^3\ \text{g N}_2\text{H}_4}{1.0\ \text{kg N}_2\text{H}_4} \times \dfrac{1\ \text{mol N}_2\text{H}_4}{32.0\ \text{g N}_2\text{H}_4} \times \dfrac{1\ \text{mol N}_2}{1\ \text{mol N}_2\text{H}_4} \times \dfrac{22.4\ \text{L N}_2}{1\ \text{mol N}_2} = 7.0 \times 10^2\ \text{L N}_2$

$1.2\ \text{kg O}_2 \times \dfrac{10^3\ \text{g O}_2}{1\ \text{kg O}_2} \times \dfrac{1\ \text{mol O}_2}{32.0\ \text{g O}_2} \times \dfrac{1\ \text{mol N}_2}{1\ \text{mol O}_2} \times \dfrac{22.4\ \text{L N}_2}{1\ \text{mol N}_2} = 8.4 \times 10^2\ \text{L N}_2$

Since $7.0 \times 10^2\ \text{L N}_2 < 8.4 \times 10^2\ \text{L N}_2$, the limiting reagent is N_2H_4 and O_2 is in excess.

b. $1.0\ \text{kg N}_2\text{H}_4 \times \dfrac{10^3\ \text{g N}_2\text{H}_4}{1\ \text{kg N}_2\text{H}_4} \times \dfrac{1\ \text{mol N}_2\text{H}_4}{32.0\ \text{g N}_2\text{H}_4} \times \dfrac{1\ \text{mol O}_2}{1\ \text{mol N}_2\text{H}_4} \times \dfrac{32.0\ \text{g O}_2}{1\ \text{mol O}_2} = 1.0 \times 10^3\ \text{g O}_2\ \text{used}$

$1.0 \times 10^3\ \text{g O}_2 \times \dfrac{1\ \text{kg O}_2}{10^3\ \text{g O}_2} = 1.0\ \text{kg O}_2$

$1.2\ \text{kg O}_2 - 1.0\ \text{kg O}_2 = 0.2\ \text{kg O}_2$

The excess remaining reagent is $0.2\ \text{kg O}_2$.

64. $158\ \text{g CH}_4 \times \dfrac{1\ \text{mol CH}_4}{16.0\ \text{g CH}_4} \times \dfrac{3\ \text{mol H}_2}{1\ \text{mol CH}_4} \times \dfrac{6.02 \times 10^{23}\ \text{molecules H}_2}{1\ \text{mol H}_2} = 1.78 \times 10^{25}\ \text{molecules H}_2$

67. **a.** $SF_2(l) + 2F_2(g) \longrightarrow SF_6(g)$

b. $5.00\ \text{mg SF}_2 \times \dfrac{1\ \text{g SF}_2}{10^3\ \text{mg SF}_2} \times \dfrac{1\ \text{mol SF}_2}{70.1\ \text{g SF}_2} \times \dfrac{2\ \text{mol F}_2}{1\ \text{mol SF}_2} \times \dfrac{6.02 \times 10^{23}\ \text{molecules F}_2}{1\ \text{mol F}_2}$

$= 8.59 \times 10^{19}\ \text{molecules}$

c. $6.66\ \text{g SF}_2 \times \dfrac{1\ \text{mol SF}_2}{70.1\ \text{g SF}_2} \times \dfrac{2\ \text{mol F}_2}{1\ \text{mol SF}_2} \times \dfrac{22.4\ \text{L F}_2}{1\ \text{mol F}_2}$

$= 4.26\ \text{L F}_2$

69. B ⟶ C:

$\%\ \text{yield} = \dfrac{\text{actual yield}}{\text{theoretical yield}} \times 100\% = \dfrac{2.00\ \text{mol C}}{4.00\ \text{mol C}} \times 100\% = 50.0\%$

C ⟶ D:

Let x represent the actual yield of compound D.

$\%\ \text{yield} = 25.0\% = \dfrac{x}{2\ \text{mol D}} \times 100\%$

$x = 0.250 \times 2\ \text{mol D}$
$x = 0.500\ \text{mol D}$

D ⟶ E:

Let y represent the actual yield of compound E.

$\%\ \text{yield} = 10.0\% = \dfrac{y}{0.500\ \text{mol E}} \times 100\%$

$y = 0.100 \times 0.500\ \text{mol E}$
$y = 0.0500\ \text{mol E}$

E ⟶ F:

$\%\ \text{yield} = \dfrac{\text{actual yield}}{\text{theoretical yield}} \times 100\% = \dfrac{0.0100\ \text{mol F}}{0.0500\ \text{mol F}} \times 100\% = 20.0\%$

Selected Solutions **R93**

72. **a.** $2Ca_3(PO_4)_2 + 6SiO_2 \longrightarrow P_4O_{10} + 6CaSiO_3$
$P_4O_{10} + 10C \longrightarrow P_4 + 10CO$

b. $5.5 \times 10^5\ \text{g}\ Ca_3(PO_4)_2 \times \frac{1\ \text{mol}\ Ca_3(PO_4)_2}{310.3\ \text{g}\ Ca_3(PO_4)_2} \times \frac{1\ \text{mol}\ P_4O_{10}}{2\ \text{mol}\ Ca_3(PO_4)_2} = 8.9 \times 10^2\ \text{mol}\ P_4O_{10}$

$2.3 \times 10^5\ \text{g}\ SiO_2 \times \frac{1\ \text{mol}\ SiO_2}{60.1\ \text{g}\ SiO_2} \times \frac{1\ \text{mol}\ P_4O_{10}}{6\ \text{mol}\ SiO_2} = 6.4 \times 10^2\ \text{mol}\ P_4O_{10}$

Since $6.4 \times 10^2\ \text{mol}\ P_4O_{10} < 8.9 \times 10^2\ \text{mol}\ P_4O_{10}$, SiO_2 is the limiting reagent.

c. $6.4 \times 10^2\ \text{mol}\ P_4O_{10} \times \frac{1\ \text{mol}\ P_4}{1\ \text{mol}\ P_4O_{10}} \times \frac{124.0\ \text{g}\ P_4}{1\ \text{mol}\ P_4} = 7.9 \times 10^4\ \text{g}\ P_4$

d. $6.4 \times 10^2\ \text{mol}\ P_4O_{10} \times \frac{10\ \text{mol C}}{1\ \text{mol}\ P_4O_{10}} \times \frac{12.0\ \text{g C}}{1\ \text{mol C}} = 7.7 \times 10^4\ \text{g C}$

75. The balanced equation is:
$C_6H_{12}O_6 \longrightarrow 2C_2H_5OH + 2CO_2$

$1.0 \times 10^3\ \text{kg}\ C_6H_{12}O_6 \times \frac{10^3\ \text{g}\ C_6H_{12}O_6}{1\ \text{kg}\ C_6H_{12}O_6} \times \frac{1\ \text{mol}\ C_6H_{12}O_6}{180.0\ \text{g}\ C_6H_{12}O_6} \times \frac{2\ \text{mol}\ C_2H_5OH}{1\ \text{mol}\ C_6H_{12}O_6} \times \frac{46.0\ \text{g}\ C_2H_5OH}{1\ \text{mol}\ C_2H_5OH} \times \frac{1\ \text{kg}\ C_2H_5OH}{10^3\ \text{g}\ C_2H_5OH} \times \frac{5.0\ \text{h}}{8\ \text{kg}\ C_2H_5OH} \times \frac{1\ \text{day}}{24\ \text{h}} = 13\ \text{days}$

78. First calculate the amount of $CaCO_3$ needed to produce 81.8 g of $CaCl_2$:

$81.8\ \text{g}\ CaCl_2 \times \frac{1\ \text{mol}\ CaCl_2}{111.0\ \text{g}\ CaCl_2} \times \frac{1\ \text{mol}\ CaCO_3}{1\ \text{mol}\ CaCl_2} \times \frac{100.1\ \text{g}\ CaCO_3}{1\ \text{mol}\ CaCO_3} = 73.8\ \text{g}\ CaCO_3$

Then calculate % $CaCO_3$ in the limestone:

$\frac{73.8\ \text{g}\ CaCO_3}{84.4\ \text{g limestone}} \times 100\% = 87.4\%\ CaCO_3$

82. **a.** $4NH_3 + 5O_2 \longrightarrow 4NO + 6H_2O$
$2NO + O_2 \longrightarrow 2NO_2$
$3NO_2 + H_2O \longrightarrow 2HNO_3 + NO$

b. $88.0\ \text{g}\ NH_3 \times \frac{14.0\ \text{g N}}{17.0\ \text{g}\ NH_3} \times \frac{63.0\ \text{g}\ HNO_3}{14.0\ \text{g N}} = 326\ \text{g}\ HNO_3$

c. 70.0% of 1 kg HNO_3 = 700 g HNO_3

$700\ \text{g}\ HNO_3 \times \frac{88.0\ \text{g}\ NH_3}{326\ \text{g}\ HNO_3} = 189\ \text{g}\ NH_3$

88. **a.** 22, 22, 25 **b.** 50, 50, 70 **c.** 8, 8, 10 **d.** 12, 12, 14

99. $(3 \times 9.0) + (2 \times 27.0) + (6 \times 28.1) + (18 \times 16.0) = 537.6$

$147\ \text{g}\ Be_3Al_2Si_6O_{18} \times \frac{27.0\ \text{g Be}}{537.6\ \text{g}\ Be_3Al_2Si_6O_{18}} = 7.38\ \text{g Be}$

101. $90\ \text{g} \times 0.267 = 24\ \text{g C}$

$24\ \text{g C} \times \frac{1\ \text{mol C}}{12.0\ \text{g C}} = 2.0\ \text{mol C}$

$90\ \text{g} \times 0.022 = 2.0\ \text{g H}$

$2.0\ \text{g H} \times \frac{1\ \text{mol H}}{1.0\ \text{g H}} = 2.0\ \text{mol H}$

$90\ \text{g} \times 0.711 = 64\ \text{g O}$

$64\ \text{g O} \times \frac{1\ \text{mol O}}{16.0\ \text{g O}} = 4.0\ \text{mol O}$

The empirical formula is $C_2H_2O_4$.

Chapter 13

1. $385\ \text{mm Hg} \times \frac{101.3\ \text{kPa}}{760\ \text{mm Hg}} = 51.3\ \text{kPa}$

$51.3\ \text{kPa} \times \frac{1\ \text{atm}}{101.3\ \text{kPa}} = 0.507\ \text{atm}$

2. $33.7\ \text{kPa} \times \frac{1\ \text{atm}}{101.3\ \text{kPa}} = 0.33\ \text{atm} > 0.25\ \text{atm}$

31. In an elastic collision, energy is transferred between particles.

35. **a.** $190\ \text{mm Hg} \times \frac{101.3\ \text{kPa}}{760\ \text{mm Hg}} = 25\ \text{kPa}$

b. $190\ \text{mm Hg} \times \frac{1\ \text{atm}}{760\ \text{mm Hg}} = 0.25\ \text{atm}$

40. Since the Kelvin temperature is directly proportional to the average kinetic energy and the temperature increased from 300 K to 900 K, then the average kinetic energy triples as well.

43. Two opposing processes are occurring at identical rates.

49. Escaping molecules have more kinetic energy than the average. Thus, the average kinetic energy and temperature of the remaining molecules are lower.

52. The intermolecular attractions between molecules are weaker than the attractions between ions.

53. Water from the food sublimed and then condensed on the lid.

55. The average kinetic energy of the molecules is greater because, by definition, a fever is a state of increased body temperature.

R94 Appendix E

SELECTED SOLUTIONS

60. Decrease; as the attractions become stronger, it becomes more difficult for molecules to overcome the attractions and vaporize.

63. The Kelvin temperature is directly proportional to the average kinetic energy. As the temperature rises, the air particles speed up and increase in kinetic energy which causes the raft to expand. As the temperature drops, the air particles slow down resulting in a decrease in kinetic energy, which causes the raft to not be fully inflated.

65. Possible answer: Since the beaker is an open container, the water should boil at 100°C at or close to sea level. Your partner probably misread the thermometer and should recheck the value.

72. No; at 15 kPa, water would boil at a temperature of about 50°C, which is much higher than room temperature.

76. No; if (a) > (b) then water vapor will condense at a greater rate than the liquid evaporates.

79. a. body-centered cubic

b. 8

c. CsCl (one Cl^- ion and $8 \times \frac{1}{8}$ equals one Cs^+ ion)

85. a. $1s^2 2s^2 2p^6 3s^2 3p^6$

b. $1s^2 2s^2 2p^6 3s^2 3p^6$

c. $1s^2$

93. a. $56.2\ g\ HClO_4 \times \frac{1\ mol\ HClO_4}{100.5\ g\ HClO_4} \times \frac{1\ mol\ Cl_2O_7}{2\ mol\ HClO_4} \times \frac{183.0\ g\ Cl_2O_7}{1\ mol\ Cl_2O_7} = 51.2\ g\ Cl_2O_7$

b. $3.40\ mol\ HClO_4 \times \frac{1\ mol\ H_2O}{2\ mol\ HClO_4} \times \frac{22.4\ L\ H_2O}{1\ mol\ H_2O} \times \frac{1000\ mL\ H_2O}{1\ L\ H_2O} = 3.81 \times 10^4\ mL\ H_2O$

95. $H_2S(aq) + Cd(NO_3)_2(aq) \longrightarrow 2HNO_3(aq) + CdS(s)$

99. a. $1\ mol\ C_{12}H_{22}O_{11} \times \frac{11\ mol\ H_2O}{1\ mol\ C_{12}H_{22}O_{11}} \times \frac{18.0\ g\ H_2O}{1\ mol\ H_2O} = 198\ g\ H_2O$

b. $11\ mol\ H_2O + 12\ mol\ C = 23\ mol$

c. $1\ mol\ C_{12}H_{22}O_{11} \times \frac{12\ mol\ C}{1\ mol\ C_{12}H_{22}O_{11}} \times \frac{12.0\ g\ C}{1\ mol\ C} = 144\ g\ C$

100. $40.0\ g\ C_2H_4 \times \frac{1\ mol\ C_2H_4}{28.0\ g\ C_2H_4} \times \frac{1\ mol\ H_2}{1\ mol\ C_2H_4} \times \frac{2.0\ g\ H_2}{1\ mol\ H_2} = 2.86\ g\ H_2$

C_2H_4 is the limiting reagent.

Chapter 14

9. $V_2 = \frac{P_1 \times V_1}{P_2} = \frac{105\ kPa \times 2.50\ L}{40.5\ kPa} = 6.48\ L$

10. $P_2 = \frac{P_1 \times V_1}{V_2} = \frac{205\ kPa \times 4.00\ L}{12.0\ L} = 68.3\ kPa$

11. $T_1 = 325°C + 273 = 598\ K$

$T_2 = 25°C + 273 = 298\ K$

$V_2 = \frac{V_1 \times T_2}{T_1} = \frac{6.80\ L \times 298\ K}{598\ K} = 3.39\ L$

12. $T_1 = -50.0°C + 273 = 223K$

$T_2 = 100.0°C + 273 = 373\ K$

$V_2 = \frac{V_1 \times T_2}{T_1} = \frac{5.00\ L \times 373\ K}{223\ K} = 8.36\ L$

13. $T_1 = 41°C + 273 = 314\ K$

$T_2 = 22°C + 273 = 295\ K$

$P_2 = \frac{P_1 \times T_2}{T_1} = \frac{108\ kPa \times 295\ K}{314\ K} = 101\ kPa$

14. $T_1 = 27°C + 273 = 300\ K$

$T_2 = \frac{P_2 \times T_1}{P_1} = \frac{225\ kPa \times 300\ K}{198\ kPa} = 341\ K\ (68°C)$

15. $T_1 = 25°C + 273 = 298\ K$

$T_2 = 125°C + 273 = 398\ K$

$V_2 = \frac{P_1 \times V_1 \times T_2}{T_1 \times P_2} = \frac{155\ kPa \times 1.00\ L \times 398\ K}{298\ kPa \times 605\ K} = 0.342\ L$

16. $T_1 = -50°C + 273 = 223\ K$

$T_2 = 102°C + 273 = 375\ K$

$P_2 = \frac{P_1 \times V_1 \times T_2}{T_1 \times V_2} = \frac{107\ kPa \times 5.00\ L \times 375\ K}{223\ K \times 7.00\ L} = 129\ kPa$

26. $n = \frac{P \times V}{R \times T}$

$n = \frac{1.89 \times 10^3\ kPa \times 685\ L}{8.31 \frac{L \cdot kPa}{K \cdot mol} \times 621\ K} = 251\ mol\ He$

27. $T = 25°C + 273 = 298\ K$

$P = \frac{n \times R \times T}{V}$

$P = \frac{0.450\ mol \times 8.31 \frac{L \cdot kPa}{K \cdot mol} \times 298\ K}{0.650\ L} = 1.71 \times 10^3\ kPa$

Selected Solutions **R95**

28. $T = 37°C + 273 = 310 K$

$$n = \frac{P \times V}{R \times T} = \frac{102 \text{ kPa} \times 2.20 \text{ L}}{8.31 \frac{\text{L}\cdot\text{kPa}}{\text{K}\cdot\text{mol}} \times 310 \text{ K}}$$

$$= 0.0871 \text{ mol air}$$

$$0.0871 \text{ mol air} \times \frac{29 \text{ g air}}{1 \text{ mol air}} = 2.5 \text{ g air}$$

29. $T = 25°C + 273 = 298 K$

$$n = 12.0 \text{ g } O_2 \times \frac{1 \text{ mol } O_2}{32.0 \text{ g } O_2} = 0.375 \text{ mol } O_2$$

$$V = \frac{n \times R \times T}{P}$$

$$V = \frac{0.375 \text{ mol} \times 8.31 \frac{\text{L}\cdot\text{kPa}}{\text{K}\cdot\text{mol}} \times 298 \text{ K}}{52.7 \text{ kPa}} = 17.6 \text{ L}$$

37. $P_{total} = P_{O_2} + P_{N_2} + P_{CO_2}$

$$P_{CO_2} = P_{total} - (P_{O_2} + P_{N_2})$$

$$P_{CO_2} = 32.9 \text{ kPa} - 6.6 \text{ kPa} - 23.0 \text{ kPa}$$

$$P_{CO_2} = 3.3 \text{ kPa}$$

38. $P_{total} = P_{O_2} + P_{N_2} + P_{He}$

$$P_{total} = 20.0 \text{ kPa} + 46.7 \text{ kPa} + 26.7 \text{ kPa}$$

$$P_{total} = 93.4 \text{ kPa}$$

39. $$\frac{\text{Rate}_{H_2}}{\text{Rate}_{CO_2}} = \sqrt{\frac{\text{molar mass}_{CO_2}}{\text{molar mass}_{H_2}}} = \sqrt{\frac{44.0 \text{ g}}{2.0 \text{ g}}} = \sqrt{22} = 4.7$$

The ratio is 4.7:1.

54. $$P_2 = \frac{P_1 \times T_2}{T_1} = \frac{300 \text{ kPa} \times 101 \text{ K}}{303 \text{ K}} = 100 \text{ kPa}$$

57. $T_1 = 150.0°C + 273 = 423 K$

$$T_2 = \frac{T_1 \times V_2}{T_1} = \frac{423 \text{ K} \times 600 \text{ mL}}{300 \text{ mL}} = 846 \text{ K } (573°C)$$

58. $T_1 = 327.0°C + 273 = 600 K$

$$T_2 = \frac{T_1 \times V_2}{T_1} = \frac{600 \text{ K} \times 5 \text{ L}}{15 \text{ L}} = 200 \text{ K } (-73°C)$$

61. $$P_2 = \frac{P_1 \times T_2}{T_1} = \frac{6.58 \text{ kPa} \times 211 \text{ K}}{539 \text{ K}} = 2.58 \text{ kPa}$$

65. $T = 35°C + 273 = 308 K$

$$V = \frac{n \times R \times T}{P}$$

$$V = \frac{1.24 \text{ mol} \times 8.31 \frac{\text{L}\cdot\text{kPa}}{\text{K}\cdot\text{mol}} \times 308 \text{ K}}{96.2 \text{ kPa}} = 33.0 \text{ L}$$

67. $T = 35°C + 273 = 308 K$

$$n = 4.50 \text{ g } CH_4 \times \frac{1 \text{ mol } CH_4}{16.0 \text{ g } CH_4} = 0.281 \text{ mol } CH_4$$

$$P = \frac{n \times R \times T}{V}$$

$$P = \frac{0.281 \text{ mol} \times 8.31 \frac{\text{L}\cdot\text{kPa}}{\text{K}\cdot\text{mol}} \times 308 \text{ K}}{2.00 \text{ L}}$$

$$= 360 \text{ kPa} = 3.60 \times 10^2 \text{ kPa}$$

69. $T = 0°C + 273 = 273 K$

$$n = \frac{P \times V}{R \times T} = \frac{99 \text{ kPa} \times 240 \text{ L}}{8.31 \frac{\text{L}\cdot\text{kPa}}{\text{K}\cdot\text{mol}} \times 273 \text{ K}} = 10.5 \text{ mol He}$$

$$10.5 \text{ mol He} \times \frac{4.0 \text{ g He}}{1 \text{ mol He}} = 42 \text{ g He}$$

73. $$\frac{\text{Rate}_{He}}{\text{Rate}_{Ne}} = \frac{\sqrt{\text{molar mass}_{Ne}}}{\sqrt{\text{molar mass}_{He}}} = \frac{\sqrt{20.2 \text{ g}}}{\sqrt{4.0 \text{ g}}} = \sqrt{5.05} = 2.25$$

The ratio is 2.25:1.

83. $T_1 = 20°C + 273 = 293 K$

$$T_2 = \frac{T_1 \times P_2 \times V_2}{P_1 \times V_1}$$

$$T_2 = \frac{293 \text{ K} \times 56.7 \text{ kPa} \times 8.00 \text{ L}}{86.7 \text{ kPa} \times 3.50 \text{ L}} = 438 \text{ K } (165°C)$$

87. Let g represent a certain gas.

$$\text{Rate}_g = 4 \times \text{Rate}_{O_2}$$

$$\frac{\text{Rate}_g}{\text{Rate}_{O_2}} = \sqrt{\frac{\text{molar mass}_{O_2}}{\text{molar mass}_g}}$$

$$\frac{4 \times \text{Rate}_{O_2}}{\text{Rate}_{O_2}} = \sqrt{\frac{32.0 \text{ g}}{\text{molar mass}_g}}$$

$$4^2 = \left(\sqrt{\frac{32.0 \text{ g}}{\text{molar mass}_g}}\right)^2$$

$$16 = \frac{32.0 \text{ g}}{\text{molar mass}_g}$$

$$\text{molar mass}_g = \frac{32.0 \text{ g}}{16} = 2.0 \text{ g}$$

88. Let g represent an unknown gas.
Let n = the number of moles of gas.

$$\frac{\text{Rate}_g}{\text{Rate}_{O_2}} = \sqrt{\frac{\text{molar mass}_{O_2}}{\text{molar mass}_g}}$$

$$\frac{\frac{n}{75 \text{ s}}}{\frac{n}{30 \text{ s}}} = \sqrt{\frac{32.0 \text{ g}}{\text{molar mass}_g}}$$

$$\left(\frac{30}{75}\right)^2 = \frac{32.0 \text{ g}}{\text{molar mass}_g}$$

$$900 \times \text{molar mass}_g = 5625 \times 32.0 \text{ g}$$

$$\text{molar mass}_g = \frac{180{,}000 \text{ g}}{900} = 200 \text{ g}$$

R96 Appendix E

94. **a.** $T = 120°C + 273 = 393\ K$

$$34.0\ \cancel{g\ NH_3} \times \frac{1\ mol\ NH_3}{17.0\ \cancel{g\ NH_3}} = 2\ mol\ NH_3$$

$$96.0\ \cancel{g\ O_2} \times \frac{1\ mol\ O_2}{32.0\ \cancel{g\ O_2}} = 3\ mol\ O_2$$

$$2\ \cancel{mol\ NH_3} \times \frac{5\ mol\ O_2}{4\ \cancel{mol\ NH_3}} = 2.5\ mol\ O_2$$

$2.5\ mol\ O_2 < 3\ mol\ O_2$, so NH_3 is limiting reagent.

$$n_{NO} = 2\ \cancel{mol\ NH_3} \times \frac{4\ mol\ NO}{4\ \cancel{mol\ NH_3}} = 2\ mol\ NO$$

$$P = \frac{n \times R \times T}{V}$$

$$P_{NO} = \frac{2\ \cancel{mol} \times 8.31\ \frac{\cancel{L}\cdot kPa}{\cancel{K}\cdot \cancel{mol}} \times 393\ \cancel{K}}{40.0\ \cancel{L}} = 163\ kPa$$

b. $n_{H_2O} = 2\ \cancel{mol\ NH_3} \times \dfrac{6\ mol\ H_2O}{4\ \cancel{mol\ NH_3}} = 3\ mol\ H_2O$

$$P_{H_2O} = \frac{3\ \cancel{mol} \times 8.31\ \frac{\cancel{L}\cdot kPa}{\cancel{K}\cdot \cancel{mol}} \times 393\ \cancel{K}}{40.0\ \cancel{L}} = 245\ kPa$$

O_2 is in excess by 0.5 mol.

$$P_{O_2} = \frac{0.5\ \cancel{mol} \times 8.31\ \frac{\cancel{L}\cdot kPa}{\cancel{K}\cdot \cancel{mol}} \times 393\ \cancel{K}}{40.0\ \cancel{L}} = 41\ kPa$$

$$P_{total} = P_{O_2} + P_{NO} + P_{H_2O}$$
$$= 41\ kPa + 163\ kPa + 245\ kPa = 449\ kPa$$

96. **b.** 700 mm Hg

c. directly proportional

d. The pressure rises 2.4 mm of Hg for every 1°C.

$$\text{slope of line } (m) = \frac{750\ mm\ Hg - 726\ mm\ Hg}{20°C - 10°C}$$
$$= \frac{24\ mm\ Hg}{10°C} = 2.4\ mm\ Hg/°C$$

e. $m = 2.4$; y-intercept $= 700$ mm Hg;
equation of line:
$P = (2.4\ mm\ Hg/°C)T + 700\ mm\ Hg$

f. Gay-Lussac's law; sample data points:
$(T_1, P_1) = (10°C, 726\ mm\ Hg)$
$(T_2, P_2) = (20°C, 750\ mm\ Hg)$

$T_1 = 10°C + 273 = 283\ K$
$T_1 = 20°C + 273 = 293\ K$

$$\frac{P_1}{T_1} = \frac{P_2}{T_2}$$

$$\frac{726\ mm\ Hg}{283\ K} = \frac{750\ mm\ Hg}{293\ K} = 2.6\ mm\ Hg/K$$

99. Let $n_{initial}$ represent moles of methane/ethyne.

$n_{initial} = n_{CH_4} + n_{C_2H_2}$ [equation 1]

Let n_{CO_2} represent final moles of CO_2. $PV = nRT$, so $V = nRT/P$. At constant volume and temperature:

$$V = \frac{n_{CO_2} \times \cancel{R} \times \cancel{T}}{25.2\ kPa} = \frac{n_{initial} \times \cancel{R} \times \cancel{T}}{16.8\ kPa}$$

$$n_{CO_2} \times 16.8\ kPa = 25.2\ kPa \times n_{initial}$$

$$n_{CO_2} = \frac{25.2\ \cancel{kPa}}{16.8\ \cancel{kPa}} \times n_{initial}$$

$n_{CO_2} = 1.5n_{initial}$ [equation 2]

Chemical equations for combustion:

$CH_4 + 2O_2 \longrightarrow CO_2 + 2H_2O$
(Each mol CH_4 burned yields 1 mol CO_2.)

$C_2H_2 + \frac{5}{2}O_2 \longrightarrow 2CO_2 + H_2O$
(Each mol C_2H_2 burned yields 2 mol CO_2.)

Therefore, $n_{CO_2} = n_{CH_4} + 2n_{C_2H_2}$ [equation 3]

Substitute equation 2 into equation 3:
$1.5n_{initial} = n_{CH_4} + 2n_{C_2H_2}$

Substitute for $n_{initial}$ using equation 1:

$1.5(n_{CH_4} + n_{C_2H_2}) = n_{CH_4} + 2n_{C_2H_2}$

$n_{CH_4} = n_{C_2H_2}$

So, in terms of moles, the initial mixture is equal parts methane and ethyne, or 50% methane (CH_4).

100. Let x be the percentage of total gas volume occupied by its molecules.

$$V_{H_2\ molecules} = 3.0 \times 10^{20}\ \cancel{molecules\ H_2} \times \frac{6.7 \times 10^{-24}\ \cancel{mL}}{1\ \cancel{molecule\ H_2}} \times \frac{1\ L}{10^3\ \cancel{mL}} = 2.0 \times 10^{-6}\ L$$

a. $\dfrac{V_{molecules}}{V_{gas}} \times 100\% = \dfrac{2.0 \times 10^{-6}\ \cancel{L}}{0.10\ \cancel{L}} \times 100\%$
$= 2.0 \times 10^{-3}\%$

b. $\dfrac{V_{molecules}}{V_{gas}} \times 100\% = \dfrac{2.0 \times 10^{-6}\ \cancel{L}}{1 \times 10^{-4}\ \cancel{L}} \times 100\% = 2.0\%$

107. $h = 1.60\ mm = 0.160\ cm$

$V = l \times w \times h = 4.50\ cm \times 1.30\ cm \times 0.160\ cm = 0.936\ cm^3$

$$\text{Density} = \frac{\text{mass}}{\text{volume}} = \frac{9.92\ g}{0.936\ cm^3} = 10.6\ g/cm^3$$

113. **a.** tin(II) bromide
b. barium sulfate
c. magnesium hydroxide
d. iodine pentafluoride

Selected Solutions **R97**

115. **a.** molar mass of $Ca(CH_3CO_2)_2$ = (1 × 40.1 g/mol) + (4 × 12.0 g/mol) + (6 × 1.0 g/mol) + (4 × 16.0 g/mol) = 158.1 g/mol

b. molar mass of H_3PO_4= (3 × 1.0 g/mol) + (1 × 31.0 g/mol) + (4 × 16.0 g/mol) = 98.0 g/mol

c. molar mass of $C_{12}H_{22}O_{11}$ = (12 × 12.0 g/mol) + (22 × 1.0 g/mol) + (11 × 16.0 g/mol) = 342.0 g/mol

d. molar mass of $Pb(NO_3)_2$ = (1 × 207.2 g/mol) + (2 × 14.0 g/mol) + (6 × 16.0 g/mol) = 331.2 g/mol

117. **a.** efm of C_2H_4O = (2 × 12.0g/mol) + (4 × 1.0 g/mol) + (1 × 16.0 g/mol) = 44.0 g/mol

$$\frac{\text{molar mass}}{\text{efm}} = \frac{88\ \cancel{\text{g/mol}}}{44.0\ \cancel{\text{g/mol}}} = 2$$

$(C_2H_4O) \times 2 = C_4H_8O_2$

b. efm of CH = 12.0 g/mol + 1.0 g/mol = 13.0 g/mol

$$\frac{\text{molar mass}}{\text{efm}} = \frac{104\ \cancel{\text{g/mol}}}{13.0\ \cancel{\text{g/mol}}} = 8$$

$(CH) \times 8 = C_8H_8$

c. $\%C = 26.7\% = \frac{\text{mass of C}}{90\ \text{g}} \times 100\%$

mass of C = 0.267 × 90 g = 24 g

$$24\ \cancel{\text{g C}} \times \frac{1\ \text{mol C}}{12.0\ \cancel{\text{g C}}} = 2.0\ \text{mol C}$$

$\%O = 71.1\% = \frac{\text{mass of O}}{90\ \text{g}} \times 100\%$

mass of O = 0.711 × 90 g = 64 g

$$64\ \cancel{\text{g O}} \times \frac{1\ \text{mol O}}{16.0\ \cancel{\text{g O}}} = 4.0\ \text{mol O}$$

$\%H = 2.2\% = \frac{\text{mass of H}}{90\ \text{g}} \times 100\%$

mass of H = 0.022 × 90 g = 2.0 g

$$2.0\ \cancel{\text{g H}} \times \frac{1\ \text{mol H}}{1.0\ \cancel{\text{g H}}} = 2.0\ \text{mol H}$$

The empirical formula is $C_2H_2O_4$. Since molar mass/efm = 1, $C_2H_2O_4$ is both the empirical formula and molecular formula

121. **a.** $4Al(s) + 3O_2(g) \longrightarrow 2Al_2O_3(s)$

b. $$583\ \cancel{\text{g }Al_2O_3} \times \frac{1\ \text{mol }Al_2O_3}{102\ \cancel{\text{g }Al_2O_3}} = 5.72\ \text{mol }Al_2O_3$$

$$5.72\ \cancel{\text{g }Al_2O_3} \times \frac{4\ \cancel{\text{mol Al}}}{2\ \cancel{\text{mol }Al_2O_3}} \times \frac{27.0\ \text{g Al}}{1\ \cancel{\text{mol Al}}} = 309\ \text{g Al}$$

$$5.72\ \cancel{\text{g }Al_2O_3} \times \frac{3\ \cancel{\text{mol }O_2}}{2\ \cancel{\text{mol }Al_2O_3}} \times \frac{32.0\ \text{g }O_2}{1\ \cancel{\text{mol }O_2}} = 275\ \text{g }O_2$$

Chapter 15

8. mass of 5 mol H_2O = 5[(2 × 1.0 g) + 16.0 g] = 90.0 g

molar mass of $CuSO_4 \cdot H_2O$ = 63.5 g + 32.1 g + (4 × 16.0 g) + 90.0 g = 249.6 g

$$\%\ \text{by mass } H_2O = \frac{\text{mass of water}}{\text{mass of hydrate}} \times 100\% =$$

$$\frac{90.0\ \cancel{\text{g}}}{249.6\ \cancel{\text{g}}} \times 100\% = 36.1\%$$

9. $$5.00\ \cancel{\text{g }Na_2CO_3} \times \frac{100.0\ \text{g }Na_2CO_3 \cdot 10\ H_2O}{37.06\ \cancel{\text{g }Na_2CO_3}} =$$

$$13.5\ \text{g }Na_2CO_3 \cdot 10H_2O$$

25. Surface molecules are attracted to the liquid molecules below but not to the air. Molecules inside the liquid are attracted in all directions.

28. A surfactant is a wetting agent such as soap or detergent. A surfactant interferes with hydrogen bonding between water molecules and reduces surface tension.

31. Water has low vapor pressure.

35. Bodies of water would freeze from the bottom up. This would kill many forms of aquatic life.

38. Solvent: water; solute: sugar

39. No; the molecules and ions are smaller than the pores of the filter and would therefore pass through the filter.

40. Solvent molecules surround positively charged and negatively charged ions.

43. **a.** HCl (polar) dissolves.

b. K_2SO_4 (ionic) dissolves.

c. NaI (ionic) dissolves.

d. C_2H_6 (nonpolar) will not dissolve.

e. NH_3 (polar) dissolves.

f. $CaCO_3$ (strong ionic forces) will not dissolve.

45. Its ions are free to move toward positively and negatively charged electrodes.

48. a. $Na_2SO_4 \cdot 10H_2O$

b. $CaCl_2 \cdot 2H_2O$

c. $Ba(OH)_2 \cdot 8H_2O$

49. a. tin(IV) chloride pentahydrate

b. iron(II) sulfate heptahydrate

c. barium bromide tetrahydrate

d. iron(III) phosphate tetrahydrate

50. $MgSO_4 \cdot 7H_2O(s) \longrightarrow MgSO_4 \cdot H_2O(s) + 6H_2O(g)$

52. Hygroscopic substances absorb water vapor from the air and create a dry environment in a sealed container.

54. solutions, colloids, suspensions

55. Colloids and suspensions exhibit the Tyndall effect, but solutions do not. The particles in a suspension will settle out over time.

59. Brownian motion and repulsion between like-charged ions adsorbed on the surfaces of colloidal particles.

62. hexane, ethanol, water

63. a. 1.0000 g/mL b. 4°C

c. No; there would be a break in the curve at 0°C as liquid water at 0°C changes to ice at 0°C.

65. Hydrobromic acid disassociates into hydrogen and bromide ions when dissolved in water, but methanol does not.

66. a. Water expands when it freezes to ice.

b. Water is polar, and wax is nonpolar; water has a higher surface tension.

c. Water has a lower vapor pressure than alcohol.

68. a. gasoline c. water

b. gasoline d. water

69. a. No, both form clear, colorless solutions.

b. Evaporate the water to examine the crystals; test for electrical conductivity; do a flame test.

72. a. $NH_4Cl(s) \longrightarrow NH_4^+(aq) + Cl^-(aq)$

b. $C_2H_4O_2(s) \longrightarrow H^+(aq) + C_2H_3O_2^-(aq)$

c. $Cu(NO_3)_2(s) \longrightarrow Cu^{2+}(aq) + 2NO_3^-(aq)$

d. $HgCl_2(s) \longrightarrow Hg^{2+}(aq) + 2Cl^-(aq)$

73. a. sodium carbonate monohydrate;

mass of 1 mol $H_2O = (2 \times 1.0\ g) + 16.0\ g = 18.0\ g$

molar mass of $Na_2CO_3 \cdot H_2O = (2 \times 23.0\ g) + 12.0\ g + (3 \times 16.0\ g) + 18.0\ g = 124.0\ g$

$$\%\ \text{by mass}\ H_2O = \frac{18.0\ \cancel{g}}{124.0\ \cancel{g}} \times 100\% = 14.5\%$$

b. magnesium sulfate heptahydrate;

mass of 7 mol $H_2O = 7[(2 \times 1.0\ g) + 16.0\ g] = 126.0\ g$

molar mass of $MgSO_4 \cdot 7H_2O = 24.3\ g + 32.1\ g + (4 \times 16.0\ g) + 126.0\ g = 246.4\ g$

$$\%\ \text{by mass}\ H_2O = \frac{126.0\ \cancel{g}}{246.4\ \cancel{g}} \times 100\% = 51.14\%$$

75. a. $Ba(OH)_2(s) + 8H_2O(l) \longrightarrow Ba(OH)_2 \cdot 8H_2O(s)$

b. mass of 8 mol $H_2O = 8[(2 \times 1.0\ g) + 16.0\ g] = 144.0\ g$

molar mass of $Ba(OH)_2 \cdot 8H_2O = 137.3\ g + 2(16.0\ g + 1.0\ g) + 144.0\ g = 315.3\ g$

$$\%\ \text{by mass}\ H_2O = \frac{144.0\ \cancel{g}}{315.3\ \cancel{g}} \times 100\% = 45.67\%$$

76. From Practice Problem 8, the percent H_2O in $CuSO_4 \cdot 5H_2O$ is 36.1%. This means that for every 100.0 g of hydrate, there are 100.0 g − 36.1 g = 63.9 g of anhydrous $CuSO_4$.

$$10.0\ g\ \cancel{CuSO_4} \times \frac{100.0\ g\ CuSO_4 \cdot 5H_2O}{63.9\ \cancel{g\ CuSO_4}} = 15.6\ g\ CuSO_4 \cdot 5H_2O$$

78. a. (1), (3), (6), (8)

b. (1), (2), (5), (6), (7), (9)

c. (2), (4), (5)

80. a. nonelectrolyte

b. weak electrolyte

c. strong electrolyte

81. The container would break because water expands as it freezes.

83. The surface tension of water keeps the strider from sinking. The surfactant would reduce the surface tension and the strider would shrink.

85. The hydrogen-bonded structures in liquid water are disrupted when ethyl alcohol is added because the alcohol competes for hydrogen bonds with water molecules and the water structure collapses. Thus, mixtures of water and ethyl alcohol have less volume than the sum of the volumes of the components. Mixing two liquids could result in a volume greater than the sum of the volumes of the components if the structural ordering in the mixture is greater than in the separated components.

86. Most of the important chemical reactions of life take place in aqueous solutions inside cells.

Selected Solutions **R99**

89. a. pink b. pink c. blue

d. mass of 6 mol H_2O = $6[(2 \times 1.0\text{ g}) + 16.0\text{ g}] =$ 108.0 g

molar mass of $CoCl_2 \cdot 6H_2O = 58.9\text{ g} + (2 \times 35.5\text{ g}) + 108.0\text{ g} = 237.9\text{ g}$

$$\%\text{ by mass } H_2O = \frac{108.0\text{ g}}{237.9\text{ g}} \times 100\% = 45.40\%$$

e. water or water vapor

91. In spring, when the ice melts and the temperature of the surface water increases to 4°C, it becomes heavier than the water below it and sinks. The downward movement of surface water forces water in the deeper parts of the lake upward, where it is warmed.

92. A surfactant helps to wet the burning material, so less water is needed to put out the fire. Thus, less water carries pollutants into the environment.

93. Water enters cracks in pavement and expands when it freezes, creating larger cracks. Continuous freeze-thaw cycles cause pavement to break up and form potholes.

97. The molecules in the dirt and grease dissolve in the water.

100. a. 5 b. 2 c. 2 d. 4

102. $H^+ + H{:}\ddot{O}{:}H \longrightarrow H{:}\ddot{O}{:}H^+$ (with H bonded below O)

104. a. $6CO_2(g) + 6H_2O(l) \longrightarrow C_6H_{12}O_6(s) + 6O_2(g)$

b. $2Na(s) + 2H_2O(l) \longrightarrow Na^+(aq) + 2OH^-(aq) + H_2(g)$

106.
$$2.00 \times 10^{-3}\text{ mol } H_2O_2 \times \frac{2\text{ mol } H_2O}{2\text{ mol } H_2O_2} \times \frac{18.0\text{ g } H_2O}{1\text{ mol } H_2O} = 0.0360\text{ g } H_2O$$

$$2.00 \times 10^{-3}\text{ mol } H_2O_2 \times \frac{1\text{ mol } O_2}{2\text{ mol } H_2O_2} \times \frac{22.4\text{ L } O_2}{1\text{ mol } O_2} = 0.0224\text{ L } O_2$$

107.
$$2.60 \times 10^2\text{ g } H_2O \times \frac{1\text{ mol } H_2O}{18.0\text{ g } H_2O} \times \frac{1\text{ mol } C_2H_4O}{1\text{ mol } H_2O} \times \frac{44.0\text{ g } C_2H_4O}{1\text{ mol } C_2H_4O} = 636\text{ g } C_2H_4O$$

109. a.
$$40\text{ cm}^3 \times \frac{1\text{ mL}}{1\text{ cm}^3} \times \frac{1\text{ L}}{1000\text{ mL}} \times \frac{1\text{ mol } O_2}{22.4\text{ L}} = 1.8 \times 10^{-3}\text{ mol } O_2$$

$$60\text{ cm}^3 \times \frac{1\text{ mL}}{1\text{ cm}^3} \times \frac{1\text{ L}}{1000\text{ mL}} \times \frac{1\text{ mol } H_2}{22.4\text{ L}} = 2.7 \times 10^{-3}\text{ mol } H_2$$

$$1.8 \times 10^{-3}\text{ mol } O_2 \times \frac{2\text{ mol } H_2}{1\text{ mol } O_2} = 3.6 \times 10^{-3}\text{ mol } H_2$$

Since 2.7×10^{-3} mol H_2 < 3.6×10^{-3} mol H_2, hydrogen is the limiting reagent.

b.
$$2.7 \times 10^{-3}\text{ mol } H_2 \times \frac{2\text{ mol } H_2O}{2\text{ mol } H_2} \times \frac{18.0\text{ g } H_2O}{1\text{ mol } H_2O} = 0.049\text{ g } H_2O$$

c. oxygen

d.
$$2.7 \times 10^{-3}\text{ mol } H_2 \times \frac{1\text{ mol } O_2}{2\text{ mol } H_2} = 1.4 \times 10^{-3}\text{ mol } O_2$$

Excess O_2 = $(1.8 \times 10^{-3} - 1.4 \times 10^{-3})$ mol O_2 = 0.4×10^{-3} mol O_2

$$0.4 \times 10^{-3}\text{ mol } O_2 \times \frac{22.4\text{ L}}{1\text{ mol } O_2} = 9 \times 10^{-3}\text{ L}$$

111. $T_1 = 100°C + 273\text{ K} = 373\text{ K}$

$T_2 = 200°C + 273\text{ K} = 473\text{ K}$

$$P_2 = \frac{P_1 \times T_2}{T_1} = \frac{1.00\text{ atm} \times 473\text{ K}}{373\text{ K}} = 1.27\text{ atm}$$

Chapter 16

1. $S_2 = \dfrac{S_1 \times P_2}{P_1} = \dfrac{0.16\text{ g/L} \times 288\text{ kPa}}{104\text{ kPa}} = 0.44\text{ g/L}$

2. $P_2 = \dfrac{P_1 \times S_2}{S_1} = \dfrac{1.0\text{ atm} \times 9.5\text{ g/L}}{3.6\text{ g/L}} = 2.6\text{ atm}$

10. $\dfrac{36.0\text{ g } C_6H_{12}O_6}{2.0\text{ L}} \times \dfrac{1\text{ mol } C_6H_{12}O_6}{180\text{ g } C_6H_{12}O_6} = 0.10\text{ mol/L} = 0.10M$

11. $\dfrac{0.70\text{ mol NaCl}}{250\text{ mL}} \times \dfrac{10^3\text{ mL}}{1\text{ L}} = 2.8\text{ mol/L} = 2.8M$

12. $335\text{ mL} \times \dfrac{1\text{ L}}{10^3\text{ mL}} \times \dfrac{0.425\text{ mol } NH_4NO_3}{1\text{ L}} = 0.142\text{ mol } NH_4NO_3$

13. $250\text{ mL} \times \dfrac{1\text{ L}}{10^3\text{ mL}} \times \dfrac{2.0\text{ mol } CaCl_2}{1\text{ L}} = 0.50\text{ mol } CaCl_2$

$$0.50\text{ mol } CaCl_2 \times \frac{111.1\text{ g } CaCl_2}{1\text{ mol } CaCl_2} = 56\text{ g } CaCl_2$$

14. $V_1 = \dfrac{M_2 \times V_2}{M_1} = \dfrac{0.760M \times 250\text{ mL}}{4.00M} = 47.5\text{ mL}$

15. $V_2 = \dfrac{M_1 \times V_1}{M_2} = \dfrac{0.20\,M \times 250\text{ mL}}{1.0\,M} = 50\text{ mL}$

Use a pipet to transfer 50 mL (calculation above) of the 1.0*M* solution to a 250-ml volumetric flask. Then add distilled water up to the mark.

16. $\dfrac{10\text{ mL propanone}}{200\text{ mL}} \times 100\% = 5.0\text{ \% propanone (v/v)}$

17. $3.0\%(\text{v/v}) = \frac{V_{H_2O_2}}{V_{soln}} \times 100\%$

$V_{H_2O_2} = \frac{3.0\%}{100\%} \times 400.0\text{ mL} = 12\text{ mL}$

18. $250\ \cancel{\text{g solution}} \times \frac{10\text{ g MgSO}_4}{100\ \cancel{\text{g solution}}} = 25\text{ g MgSO}_4$

34. $750\ \cancel{\text{g H}_2\text{O}} \times \frac{0.400\ \cancel{\text{mol NaF}}}{10^3\ \cancel{\text{g H}_2\text{O}}} \times \frac{42.0\text{ g NaF}}{1\ \cancel{\text{mol NaF}}} = 12.6\text{ g NaF}$

35. $\frac{10.0\ \cancel{\text{g NaCl}}}{600\ \cancel{\text{g H}_2\text{O}}} \times \frac{1\text{ mol NaCl}}{58.5\ \cancel{\text{g NaCl}}} \times \frac{10^3\ \cancel{\text{g H}_2\text{O}}}{1\text{ kg H}_2\text{O}} = 0.285m\text{ NaCl}$

36. $n_{C_2H_6O} = 300\ \cancel{\text{g C}_2\text{H}_6\text{O}} \times \frac{1\text{ mol C}_2\text{H}_6\text{O}}{46.0\ \cancel{\text{g C}_2\text{H}_6\text{O}}} = 6.52\text{ mol C}_2\text{H}_6\text{O}$

$n_{H_2O} = 500\ \cancel{\text{g H}_2\text{O}} \times \frac{1\text{ mol H}_2\text{O}}{18.0\ \cancel{\text{g H}_2\text{O}}} = 27.8\text{ mol H}_2\text{O}$

$X_{C_2H_6O} = \frac{n_{C_2H_6O}}{n_{C_2H_6O} + n_{H_2O}} = \frac{6.52\ \cancel{\text{mol}}}{27.8\ \cancel{\text{mol}} + 6.52\ \cancel{\text{mol}}} = 0.190$

$X_{H_2O} = \frac{n_{H_2O}}{n_{C_2H_6O} + n_{H_2O}} = \frac{27.8\ \cancel{\text{mol}}}{27.8\ \cancel{\text{mol}} + 6.52\ \cancel{\text{mol}}} = 0.810$

37. $n_{CCl_4} = 50.0\ \cancel{\text{g CCl}_4} \times \frac{1\text{ mol CCl}_4}{153.8\ \cancel{\text{g CCl}_4}} = 0.325\text{ mol CCl}_4$

$n_{CHCl_3} = 50.0\ \cancel{\text{g CHCl}_3} \times \frac{1\text{ mol CHCl}_3}{119.4\ \cancel{\text{g CHCl}_3}} = 0.419\text{ mol CHCl}_3$

$X_{CCl_4} = \frac{n_{CCl_4}}{n_{CCl_4} + n_{CHCl_3}} = \frac{0.325\ \cancel{\text{mol}}}{0.325\ \cancel{\text{mol}} + 0.419\ \cancel{\text{mol}}} = 0.437$

$X_{CHCl_3} = \frac{n_{CHCl_3}}{n_{CCl_4} + n_{CHCl_3}} = \frac{0.419\ \cancel{\text{mol}}}{0.325\ \cancel{\text{mol}} + 0.419\ \cancel{\text{mol}}} = 0.563$

38. $10.0\ \cancel{\text{g C}_6\text{H}_{12}\text{O}_6} \times \frac{1\text{ mol C}_6\text{H}_{12}\text{O}_6}{180.0\ \cancel{\text{g C}_6\text{H}_{12}\text{O}_6}} = 0.0556\text{ mol C}_6\text{H}_{12}\text{O}_6$

$m = \frac{0.0556\text{ mol C}_6\text{H}_{12}\text{O}_6}{50.0\ \cancel{\text{g H}_2\text{O}} \times \frac{1\text{ kg H}_2\text{O}}{10^3\ \cancel{\text{g H}_2\text{O}}}} = 1.11m\text{ C}_6\text{H}_{12}\text{O}_6$

$\Delta T_f = K_f \times m = 1.86°\text{C}/\cancel{m} \times 1.11\cancel{m} = 2.06°\text{C}$

39. $200\ \cancel{\text{g C}_3\text{H}_6\text{O}} \times \frac{1\text{ mol C}_3\text{H}_6\text{O}}{58.0\ \cancel{\text{g C}_3\text{H}_6\text{O}}} = 3.45\text{ mol C}_3\text{H}_6\text{O}$

$m = \frac{3.45\text{ mol C}_3\text{H}_6\text{O}}{400\ \cancel{\text{g benzene}} \times \frac{1\text{ kg benzene}}{10^3\ \cancel{\text{g benzene}}}} = 8.63m\text{ C}_3\text{H}_6\text{O}$

$\Delta T_f = K_f \times m = 5.12°\text{C}/\cancel{m} \times 8.62\cancel{m} = 44.2°\text{C}$

40. $m = \frac{1.25\text{ mol CaCl}_2}{1400\ \cancel{\text{g H}_2\text{O}} \times \frac{1\text{ kg H}_2\text{O}}{10^3\ \cancel{\text{g H}_2\text{O}}}} = 0.893m\text{ CaCl}_2$

Each formula unit of $CaCl_2$ disassociates into 3 particles, so molality of total particles is:

$3 \times 0.893m = 2.68m$

$\Delta T_b = K_b \times m = 0.512°\text{C}/\cancel{m} \times 2.68\cancel{m} = 1.37°\text{C}$

$T_b = 100°\text{C} + 1.37°\text{C} = 101.37°\text{C}$

41. $m = \frac{\Delta T_b}{K_b} = \frac{2.00\cancel{°\text{C}}}{0.512\cancel{°\text{C}}/m} = 3.91m$ (molality of total particles)

Each formula unit of NaCl disassociates into 2 particles, so the solution concentration is:

$3.91m/2 = 1.96m$

Mass of NaCl needed per 1 kg of H_2O is:

$1.96\ \cancel{\text{mol NaCl}} \times \frac{58.5\text{ g NaCl}}{1\ \cancel{\text{mol NaCl}}} = 115\text{ g NaCl}$

55. a. $S_2 = \frac{S_1 \times P_2}{P_1} = \frac{0.026\text{ g/L} \times 0.60\ \cancel{\text{atm}}}{1.00\ \cancel{\text{atm}}} = 0.016\text{ g/L}$

b. $S_2 = \frac{S_1 \times P_2}{P_1} = \frac{0.026\text{ g/L} \times 1.80\ \cancel{\text{atm}}}{1.00\ \cancel{\text{atm}}} = 0.047\text{ g/L}$

58. $V_1 = \frac{M_2 \times V_2}{M_1} = \frac{0.100\cancel{M} \times 100.0\text{ mL}}{0.500\cancel{M}} = 20.0\text{ mL}$

59. $\frac{0.50\ \cancel{\text{g NaCl}}}{100\ \cancel{\text{mL}}} \times \frac{1\text{ mol NaCl}}{58.5\ \cancel{\text{g NaCl}}} \times \frac{10^3\ \cancel{\text{mL}}}{1\text{ L}} = 0.085M$

61. a. $2500\ \cancel{\text{g solution}} \times \frac{0.90\text{ g NaCl}}{100\ \cancel{\text{g solution}}} = 23\text{ g NaCl}$

b. $0.050\ \cancel{\text{kg solution}} \times \frac{1000\ \cancel{\text{g solution}}}{1\ \cancel{\text{kg solution}}} \times \frac{4.0\text{ g MgCl}_2}{100\ \cancel{\text{g solution}}} = 2.0\text{ g MgCl}_2$

63. a. $\frac{25\ \cancel{\text{mL}}\text{ ethanol}}{150\ \cancel{\text{mL}}} \times 100\% = 17\%\ (\text{v/v})\text{ ethanol}$

b. $\frac{175\ \cancel{\text{mL}}\text{ isopropyl alcohol}}{275\ \cancel{\text{mL}}} \times 100\% = 63.6\%\ (\text{v/v})\text{ isopropyl alcohol}$

Selected Solutions **R101**

72. **a.** Molality of $Na_2SO_4 = \dfrac{1.40 \text{ mol } Na_2SO_4}{1750 \text{ g } H_2O \times \dfrac{1 \text{ kg } H_2O}{10^3 \text{ g } H_2O}}$

$= 0.800m \; Na_2SO_4$

Each mol of Na_2SO_4 that dissolves yields 3 mol of particles.

Molality of total particles $= 3 \times 0.800m = 2.40m$

$\Delta T_f = K_f \times m = 1.86°C/m \times 2.40\,m = 4.46°C$

$T_f = 0°C - 4.46°C = -4.46°C$

b. Molality of $MgSO_4 = \dfrac{0.060 \text{ mol } MgSO_4}{100 \text{ g } H_2O \times \dfrac{1 \text{ kg } H_2O}{10^3 \text{ g } H_2O}}$

$= 0.60m \; MgSO_4$

Each mol of $MgSO_4$ that dissolves yields 2 mol of particles.

Molality of total particles $= 2 \times 0.60m = 1.2m$

$\Delta T_f = K_f \times m = 1.86°C/m \times 1.2m = 2.2°C$

$T_f = 0°C - 22°C = -2.2°C$

76. $12.0 \text{ g } C_{10}H_8 \times \dfrac{1 \text{ mol } C_{10}H_8}{128.0 \text{ g } C_{10}H_8} = 0.0938 \text{ mol } C_{10}H_8$

$m = \dfrac{0.0938 \text{ mol } C_{10}H_8}{50.0 \text{ g } C_6H_6 \times \dfrac{1 \text{ kg } C_6H_6}{1000 \text{ g } C_6H_6}} = 1.88m \; C_{10}H_8$

$\Delta T_f = K_f \times m = 5.12°C/m \times 1.88m = 9.63°C$

$\Delta T_b = K_b \times m = 2.53°C/m \times 1.88m = 4.76°C$

85. **a.** At 20°C, the solubility of KCl in water is 34.0 g/100 g H_2O, or 44.2 g/130 g H_2O. Therefore, 44.2 g KCl remain dissolved.

b. 50.0 g KCl − 44.2 g KCl = 5.8 g KCl

87. $S = \dfrac{36.0 \text{ g NaCl}}{100 \text{ g } H_2O} \times \dfrac{0.750}{0.750} = \dfrac{27.0 \text{ g NaCl}}{75.0 \text{ g } H_2O}$

The solution contains 26.5 g NaCl/75.0 g H_2O and is therefore unsaturated.

91. $m = \dfrac{\Delta T_f}{K_f} = \dfrac{0.460°C}{5.12°C/m} = 0.898m$

Let x = molar mass of the nondissociating solute.

$\dfrac{5.76 \text{ g}/x}{750 \text{ g benzene}} \times \dfrac{1000 \text{ g benzene}}{1 \text{ kg benzene}} = 0.898m$

$7.68 \text{ g}/x = 0.0898 \text{ mol}$

$x = 7.68 \text{ g}/0.0898 \text{ mol} = 85.5 \text{ g/mol}$

92. $1000 \text{ g } H_2O \times \dfrac{1 \text{ mol}}{18.0 \text{ g } H_2O} = 55.6 \text{ mol } H_2O$

$m_{C_{12}H_{22}O_{11}} = 1.62m = 1.62 \text{ mol } C_{12}H_{22}O_{11}/\text{kg } H_2O$

$X_{H_2O} = \dfrac{55.6 \text{ mol } H_2O}{55.6 \text{ mol } H_2O + 1.62 \text{ mol } C_{12}H_{22}O_{11}} = 0.972$

$X_{C_{12}H_{22}O_{11}} = \dfrac{1.62 \text{ mol } C_{12}H_{22}O_{11}}{55.6 \text{ mol } H_2O + 1.62 \text{ mol } C_{12}H_{22}O_{11}}$

$= 0.0283$

100. $Na_2SO_4(aq) + BaCl(aq) \longrightarrow BaSO_4(s) + 2NaCl(aq)$

$5.28 \text{ g } BaSO_4 \times \dfrac{1 \text{ mol } BaSO_4}{233.4 \text{ g } BaSO_4} \times \dfrac{1 \text{ mol } Na_2SO_4}{1 \text{ mol } BaSO_4}$

$= 0.0226 \text{ mol } Na_2SO_4$

$\text{Molarity} = \dfrac{0.0226 \text{ mol } Na_2SO_4}{250 \times 10^{-3} \text{ L}} = 0.090M \; Na_2SO_4$

103. $V_1 = \dfrac{M_2 \times V_2}{M_1} = \dfrac{0.50M \times 100 \text{ mL}}{2.0M} = 25 \text{ mL}$

Step 1: Start with a stock solution that is 2.0*M* KCl.

Step 2: Transfer 25 mL of 2.0*M* KCl to a 100-mL volumetric flask.

Step 3: Add distilled water up to the 100-mL mark.

108. atomic mass of Rb = (0.72165)(84.912 amu) + (0.27835)(86.909 amu) = 61.277 amu + 24.191 amu = 85.468 amu

114. **a.** 1 mol Fe = 55.8 g Fe

1 mol Cu = 63.5 g Cu

1 mol Hg = 200.6 g Hg

1 mol S = 32.1 g S

b. Each sample contains 6.02×10^{23} atoms.

c. $25.0 \text{ g Fe} \times \dfrac{1 \text{ mol Fe}}{55.8 \text{ g Fe}} = 0.448 \text{ mol Fe}$

$25.0 \text{ g Fe} \times \dfrac{1 \text{ mol Cu}}{63.5 \text{ g Cu}} = 0.394 \text{ mol Cu}$

$25.0 \text{ g Fe} \times \dfrac{1 \text{ mol Hg}}{200.6 \text{ g Hg}} = 0.125 \text{ mol Hg}$

$25.0 \text{ g Fe} \times \dfrac{1 \text{ mol S}}{32.1 \text{ g S}} = 0.779 \text{ mol S}$

115. $1500 \text{ g } H_2 \times \dfrac{1 \text{ mol } H_2}{2.0 \text{ g } H_2} \times \dfrac{22.4 \text{ L } H_2}{1 \text{ mol } H_2} = 1.7 \times 10^4 \text{ L}$

120. $T_1 = 25°C + 273 = 298 \text{ K}$

$T_1 = 45°C + 273 = 318 \text{ K}$

$P_2 = \dfrac{P_1 \times T_2}{T_1} = \dfrac{101.3 \text{ kPa} \times 318 \text{ K}}{298 \text{ K}} = 108 \text{ kPa}$

Chapter 17

1. Heat flows from the system (wax) to the surroundings (air). The process is exothermic.

2. Since the beaker becomes cold, heat is absorbed by the system (chemicals within the beaker) from the surroundings (beaker and surrounding air). The process is endothermic.

3. $\Delta T = 85°C - 21°C = 64°C$

$C_{\text{olive oil}} = \dfrac{q}{m \times \Delta T} = \dfrac{435 \text{ J}}{3.4 \text{ g} \times 64°C} = 2.0 \text{ J/(g} \cdot °C)$

R102 Appendix E

4. $q = C_{Hg} \times m \times \Delta T = (0.14\ J/(g \cdot °C))(250.0\ g)(52°C)$
$= 1800\ J = 1.8\ kJ$

12. $V_{water} = 50.0\ mL + 50.0\ mL = 100.0\ mL$

$m_{water} = 100.0\ mL \times \frac{1.00\ g}{1\ mL} = 100.0\ g$

$\Delta T = 26.0°C - 22.5°C = 3.5°C$

$\Delta H = -q_{surr} = -m_{water} \times C_{water} \times \Delta T = -(100.0\ g)$
$\times (4.18\ J/(g \cdot °C))(3.5°C) = -1460\ J = -1.46\ kJ$
(1.46 kJ of heat was released)

13. $\Delta T = 26.4°C - 25.0°C = 1.4°C$

$\Delta H = -q_{surr} = -m_{water} \times C_{water} \times \Delta T = -(25.0\ g)$
$\times (4.18\ J/(g \cdot °C))(1.4°C) = -150\ J$
(150 J of heat is released by the pebble)

14. $\Delta H = 3.40\ mol\ Fe_2O_3 \times \frac{26.3\ kJ}{1\ mol\ Fe_2O_3} = 89.4\ kJ$

15. $\Delta H = 5.66\ g\ CS_2 \times \frac{1\ mol\ CS_2}{76.2\ g\ CS_2} \times \frac{89.3\ kJ}{1\ mol\ CS_2} = 6.63\ kJ$

22. $m_{H_2O(s)} = 0.400\ kJ \times \frac{1\ mol\ H_2O(s)}{6.01\ kJ} \times \frac{18.0\ g\ H_2O(s)}{1\ mol\ H_2O(s)}$
$= 1.20\ g\ H_2O(s)$

23. $\Delta H = 50.0\ g\ H_2O(l) \times \frac{1\ mol\ H_2O(l)}{18.0\ g\ H_2O(l)} \times \frac{-6.01\ kJ}{1\ mol\ H_2O(l)}$
$= 16.7\ kJ$

24. $\Delta H = 63.7\ g\ H_2O(l) \times \frac{1\ mol\ H_2O(l)}{18.0\ g\ H_2O(l)} \times \frac{40.7\ kJ}{1\ mol\ H_2O(l)}$
$= 144\ kJ$

25. $\Delta H = 0.46\ g\ C_2H_5Cl \times \frac{1\ mol\ C_2H_5Cl}{64.5\ g\ C_2H_5Cl} \times \frac{24.7\ kJ}{1\ mol\ C_2H_5Cl}$
$= 0.18\ kJ$

26. $\Delta H = 0.677\ mol\ NaOH(s) \times \frac{-44.5\ kJ}{1\ mol\ NaOH(s)} = -30.1\ kJ$

27. moles of $NH_4NO_3(s) = 88.0\ kJ \times \frac{1\ mol\ NH_4NO_3(s)}{25.7\ kJ}$
$= 3.42\ mol\ NH_4NO_3(s)$

35. $\Delta H_f°(\text{reactants}) = 1\ mol\ Br_2(g) \times \frac{30.91\ kJ}{1\ mol\ Br_2(g)}$
$= 30.91\ kJ$

$\Delta H_f°(\text{products}) = 1\ mol\ Br_2(l) \times \frac{0\ kJ}{1\ mol\ Br_2(l)} = 0\ kJ$

$\Delta H° = \Delta H_f°(\text{products}) - \Delta H_f°(\text{reactants})$
$= 0\ kJ - 30.91\ kJ = -30.91\ kJ$

36. $2NO(g) + O_2(g) \longrightarrow 2NO_2(g)$

$\Delta H_f°(\text{reactants}) = 2\ mol\ NO(g) \times \frac{90.37\ kJ}{1\ mol\ NO(g)}$
$+ 1\ mol\ O_2(g) \times \frac{0\ kJ}{1\ mol\ O_2(g)} = 180.7\ kJ$

$\Delta H_f°(\text{products}) = 2\ mol\ NO_2(g) \times \frac{33.85\ kJ}{1\ mol\ NO_2(g)}$
$= 67.70\ kJ$

$\Delta H° = \Delta H_f°(\text{products}) - \Delta H_f°(\text{reactants})$
$= 67.70\ kJ - 180.7\ kJ = -113.0\ kJ$

47. **a.** exothermic

b. The immediate surroundings are the glass beaker and the air. If one or more of the substances is in water, the water is also considered part of the surroundings.

48. **a.** exothermic **c.** exothermic
b. endothermic **d.** endothermic

51. **a.** $8.50 \times 10^2\ cal \times \frac{1\ Cal}{1000\ cal} = 0.85\ Cal$

b. $444\ cal \times \frac{4.18\ J}{1\ cal} = 1.86 \times 10^3\ J$

c. $1.8\ kJ \times \frac{1000\ J}{1\ kJ} = 1.8 \times 10^3\ J$

d. $45 \times 10^{-1}\ kJ \times \frac{1000\ J}{1\ kJ} \times \frac{1\ cal}{4.18\ J} = 1.1 \times 10^2\ cal$

53. $q = C \times m \times \Delta T = (0.24\ J/g \cdot °C)(400.0\ g)(45°C)$
$= 4.3 \times 10^3\ J$

59. $\Delta H = 0.75\ mol\ Mg \times \frac{-1204\ kJ}{2\ mol\ Mg} = -4.5 \times 10^2\ kJ$

62. **a.** $\Delta H = 3.50\ mol\ H_2O(l) \times \frac{-6.01\ kJ}{1\ mol\ H_2O(l)} = -21.0\ kJ$

b. $\Delta H = 0.44\ mol\ H_2O(g) \times \frac{-40.7\ kJ}{1\ mol\ H_2O(g)} = -18\ kJ$

c. $\Delta H = 1.25\ mol\ NaOH \times \frac{-44.5\ kJ}{1\ mol\ NaOH} = -55.6\ kJ$

d. $\Delta H = 0.15\ mol\ C_2H_6O \times \frac{5.8\ kJ}{1\ mol\ C_2H_6O} = 6.5\ kJ$

65. **a.** $\Delta H = 1\ mol\ Al_2O_3(s) \times \frac{3352\ kJ}{2\ mol\ Al_2O_3(s)} = 1676\ kJ$

b. ΔH is positive so the reaction is endothermic.

66. Reverse the second equation and change the sign of ΔH. Then add the equations and the values of ΔH.

$Pb(s) + 2Cl_2 \longrightarrow PbCl_4(l)$ $\quad \Delta H = -329.2\ kJ$
$PbCl_2(s) \longrightarrow Pb(s) + Cl_2(g)$ $\quad \Delta H = 359.4\ kJ$

$PbCl_2(s) + Cl_2(g) \longrightarrow PbCl_4(l)$ $\quad \Delta H = 30.2\ kJ$

Selected Solutions **R103**

70. Substance B; for equal masses, the substance with the greater heat capacity undergoes the smaller temperature change.

74. $C_{\text{stainless steel}} = \dfrac{q}{m \times \Delta T} = \dfrac{141\ \text{J}}{1.55\ \text{g} \times 178°\text{C}} = 0.511\ \text{J/(g·°C)}$

76. **a.** $\Delta H_f°(\text{reactants}) = 1\ \cancel{\text{mol } CH_4(g)} \times \dfrac{-74.86\ \text{kJ}}{1\ \cancel{\text{mol } CH_4(g)}} + \dfrac{3}{2}\ \cancel{\text{mol } O_2(g)} \times \dfrac{0\ \text{kJ}}{1\ \cancel{\text{mol } O_2(g)}} = -74.86\ \text{kJ}$

$\Delta H_f°(\text{products}) = 1\ \cancel{\text{mol } CO(g)} \times \dfrac{-110.5\ \text{kJ}}{1\ \cancel{\text{mol } CO(g)}} + 2\ \cancel{\text{mol } H_2O(l)} \times \dfrac{-285.8\ \text{kJ}}{1\ \cancel{\text{mol } H_2O(l)}} = -682.1\ \text{kJ}$

$\Delta H° = \Delta H_f°(\text{products}) - \Delta H_f°(\text{reactants})$
$= (-682.1\ \text{kJ}) - (-74.86\ \text{kJ}) = -607.2\ \text{kJ}$

b. $\Delta H_f°(\text{reactants}) = 2\ \cancel{\text{mol } CO(g)} \times \dfrac{-110.5\ \text{kJ}}{1\ \cancel{\text{mol } CO(g)}} + 1\ \cancel{\text{mol } O_2(g)} \times \dfrac{0\ \text{kJ}}{1\ \cancel{\text{mol } O_2(g)}} = -221.0\ \text{kJ}$

$\Delta H_f°(\text{products}) = 2\ \cancel{\text{mol } CO_2(g)} \times \dfrac{-393.5\ \text{kJ}}{1\ \cancel{\text{mol } CO(g)}} = -787.0\ \text{kJ}$

$\Delta H° = \Delta H_f°(\text{products}) - \Delta H_f°(\text{reactants})$
$= (-787.0\ \text{kJ}) - (-221.0\ \text{kJ}) = -566.0\ \text{kJ}$

78. Multiply the first equation by 2, reverse it, and change the sign of ΔH.

$2PCl_3(g) + 2Cl_2(g) \longrightarrow 2PCl_5(s)$	$\Delta H = (2)(-87.9\ \text{kJ})$
$2P(s) + 3Cl_2(g) \longrightarrow 2PCl_3(s)$	$\Delta H = -574\ \text{kJ}$
$2P(s) + 5Cl_2(g) \longrightarrow 2PCl_5(s)$	$\Delta H = -750\ \text{kJ}$

81. $q = 106\ \cancel{\text{Cal}} \times \dfrac{1000\ \text{cal}}{1\ \cancel{\text{Cal}}} = 1.06 \times 10^5\ \text{cal}$

$\Delta T = 100.0°\text{C} - 25.0°\text{C} = 75.0°\text{C}$

$m_{\text{water}} = \dfrac{q}{C_{\text{water}} \times \Delta T} = \dfrac{1.06 \times 10^5\ \cancel{\text{cal}}}{(1.00\ \cancel{\text{cal}}/(\text{g}·\cancel{°\text{C}}))(75.0\cancel{°\text{C}})} = 1.41 \times 10^3\ \text{g}$

85. Calculate the grams of ice that melt when 3.20 kcal is absorbed.

$3.20\ \cancel{\text{kcal}} \times \dfrac{4.18\ \cancel{\text{kJ}}}{1\ \cancel{\text{kcal}}} \times \dfrac{1\ \cancel{\text{mol } H_2O}}{6.01\ \cancel{\text{kJ}}} \times \dfrac{18.0\ \text{g } H_2O}{1\ \cancel{\text{mol } H_2O}} = 40.1\ \text{g ice melts}$

The amount of ice remaining is:

$1.0\ \text{kg} - 0.401\ \text{kg} = 0.96\ \text{kg ice} = 9.6 \times 10^2\ \text{g ice}$

89. **a.** This equation is the original equation multiplied by 2, so multiply ΔH by 2:

$\Delta H = 2 \times (-92.38\ \text{kJ}) = -184.76\ \text{kJ}$

b. This equation is the original equation multiplied by $\frac{3}{2}$, so multiply ΔH by $\frac{3}{2}$:

$\Delta H = \frac{3}{2} \times (-92.38\ \text{kJ}) = -138.6\ \text{kJ}$

c. This equation is the original equation multiplied by $\frac{1}{2}$, so multiply ΔH by $\frac{1}{2}$:

$\Delta H = \frac{1}{2} \times (-92.38\ \text{kJ}) = -46.19\ \text{kJ}$

91. **a.** Calculate the heat absorbed by the melting ice:

$40.0\ \cancel{\text{g } H_2O(l)} \times \dfrac{1\ \cancel{\text{mol } H_2O(s)}}{18.0\ \cancel{\text{g } H_2O(s)}} \times \dfrac{6.01\ \cancel{\text{kJ}}}{1\ \cancel{\text{mol } H_2O(l)}} \times \dfrac{1000\ \text{J}}{1\ \cancel{\text{kJ}}} = 1.34 \times 10^4\ \text{J}$

Convert to calories:

$1.34 \times 10^4\ \text{J} \times \dfrac{1\ \text{cal}}{4.18\ \text{J}} = 3.21 \times 10^4\ \text{cal}$

Convert to kilocalories:

$3.21 \times 10^3\ \cancel{\text{cal}} \times \dfrac{1\ \text{kcal}}{1000\ \cancel{\text{cal}}} = 3.21\ \text{kcal}$

b. $m_{\text{water}} = \dfrac{q}{C_{\text{water}} \times \Delta T} = \dfrac{1.34 \times 10^4\ \text{J}}{4.18\ \text{J/(g·°C)} \times (25.0°\text{C})} = 128\ \text{g}$

94. $\Delta H_f°(\text{reactants}) = 1\ \cancel{\text{mol } C_6H_{12}O_6(s)} \times \dfrac{-1260\ \text{kJ}}{1\ \cancel{\text{mol } C_6H_{12}O_6(s)}} + 6\ \cancel{\text{mol } O_2(g)} \times \dfrac{0\ \text{kJ}}{1\ \cancel{\text{mol } O_2(g)}} = -1260\ \text{kJ}$

$\Delta H_f°(\text{products}) = 6\ \cancel{\text{mol } CO_2(g)} \times \dfrac{-393.5\ \text{kJ}}{1\ \cancel{\text{mol } CO_2(g)}} + 6\ \cancel{\text{mol } H_2O(l)} \times \dfrac{-285.8\ \text{kJ}}{1\ \cancel{\text{mol } H_2O(l)}} = -4075.8\ \text{kJ}$

$\Delta H° = \Delta H_f°(\text{products}) - \Delta H_f°\ (\text{reactants}) =$
$(-4075.8\ \text{kJ}) - (-1260\ \text{kJ}) = -2820\ \text{kJ}$

97. The heat released as the glass cools equals the heat absorbed by the water.

$q_{\text{glass}} = -q_{\text{water}}$ and $T_{f,\text{glass}} = T_{f,\text{water}} = T_f$, so:

$C_{\text{glass}} m_{\text{glass}}(T_f - T_{i,\text{glass}}) = -C_{\text{water}} m_{\text{water}}(T_f - T_{i,\text{water}})$

$T_f(C_{\text{glass}} m_{\text{glass}} + C_{\text{water}} m_{\text{water}}) = C_{\text{water}} m_{\text{water}} T_{i,\text{water}} + C_{\text{glass}} m_{\text{glass}} T_{i,\text{glass}}$

$T_f = \dfrac{C_{\text{water}} m_{\text{water}} T_{i,\text{water}} + C_{\text{glass}} m_{\text{glass}} T_{i,\text{glass}}}{C_{\text{glass}} m_{\text{glass}} + C_{\text{water}} m_{\text{water}}} =$

$\dfrac{(1.00\ \text{cal}/(\cancel{\text{g}}·°\text{C}))(175\ \cancel{\text{g}})(21°\text{C}) + (2.1\ \text{cal}/(\cancel{\text{g}}·°\text{C}))(41.0\ \cancel{\text{g}})(95°\text{C})}{(2.1\ \text{cal}/(\cancel{\text{g}}·°\text{C}))(41.0\ \cancel{\text{g}}) + (1.00\ \text{cal}/(\cancel{\text{g}}·°\text{C}))(175\ \cancel{\text{g}})}$

$= \dfrac{3700\ \text{cal} + 8200\ \text{cal}}{86\ \text{cal/°C} + 175\ \text{cal/°C}} = \dfrac{11900\ \cancel{\text{cal}}}{261\ \cancel{\text{cal}}/°\text{C}} = 45.6°\text{C}$

R104 Appendix E

102. Evaporation is an endothermic process. As the water evaporates, it absorbs heat from the trees and fruit, causing their temperature to decrease. This can lead to the branches, leaves and fruit freezing.

106. a. 6.99 m   c. 3.6×10^2 m/s
b. 10.68 g   d. 4.44°C

108. $\lambda = \frac{c}{\nu} = \frac{2.998 \times 10^8\ \text{m/}\cancel{\text{s}}}{93.1 \times 10^6\ \cancel{\text{s}^{-1}}} = 3.22\ \text{m}$

112. a. K_3N c. $Ca(NO_3)_2$
b. Al_2S_3 d. $CaSO_4$

113. $44.8\ \cancel{\text{L}} \times \frac{1\ \cancel{\text{mol H}_2}}{22.4\ \cancel{\text{L}}} \times \frac{6.02 \times 10^{23}\ \text{H}_2\ \text{molecules}}{1\ \cancel{\text{mol H}_2}}$
$= 1.20 \times 10^{24}\ \text{H}_2$ molecules

115. $N_2(g) + O_2(g) \longrightarrow 2NO(g)$
$2NO(g) + O_2(g) \longrightarrow 2NO_2(g)$

118. $V_2 = \frac{V_1 \times T_2}{T_1} = \frac{8.57\ \text{L} \times 355\ \cancel{\text{K}}}{273\ \cancel{\text{K}}} = 11.1\ \text{L}$

Chapter 18

7. $\text{Rate}_1 = 0.5\ \text{mol/(L·s)}$
$\text{Rate}_2 = x$
$\frac{\text{Rate}_2}{\text{Rate}_1} = \frac{x^1}{0.5^1} = \left(\frac{1}{2}\right)^1$
$2x = 0.5\ \text{mol/(L·s)}$
$x = \frac{0.5\ \text{mol/(L·s)}}{2} = 0.25\ \text{mol/(L·s)}$;
$\frac{\text{Rate}_2}{\text{Rate}_1} = \frac{x^1}{0.5^1} = \left(\frac{1}{4}\right)^1$
$4x = 0.5\ \text{mol/(L·s)}$
$x = \frac{0.5\ \text{mol/(L·s)}}{4} = 0.125\ \text{mol/(L·s)}$

8. $\text{Rate} = k[\text{A}]$
$k = \frac{\text{Rate}}{[\text{A}]} = \frac{\frac{\cancel{\text{mol}}}{(\cancel{\text{L}}\cdot\text{s})}}{\frac{\cancel{\text{mol}}}{\cancel{\text{L}}}} = \frac{1}{\text{s}} = \text{s}^{-1}$

17. a. favors products   c. favors reactants
b. favors reactants   d. favors reactants

18. a. favors reactants   c. favors products
b. favors reactants   d. favors products

19. $K_{eq} = \frac{[NH_3]^2}{[N_2][H_2]^3} = \frac{(0.10M)^2}{(0.25M) \times (0.15M)^3} = 12$

20. $K_{eq} = \frac{[N_2][H_2]^3}{[NH_3]^2} = \frac{(0.25M) \times (0.15M)^3}{(0.10M)^2} = 8.4 \times 10^{-2}$;
One is the inverse of the other.

21. $K_{eq} = \frac{[NO]^2}{[N_2][O_2]} = \frac{(0.02M)^2}{0.50M \times 0.50M} = 1.6 \times 10^{-3}$

22. $0.047\ \cancel{\text{mol H}_2\text{O}} \times \frac{1\ \text{mol H}_2}{1\ \cancel{\text{mol H}_2\text{O}}} = 0.047\ \text{mol H}_2$
$0.10 - 0.047 = 0.053$ mol H_2 at equilibrium (mol H_2 = mol CO_2)
$K_{eq} = \frac{[H_2O][CO]}{[H_2][CO_2]} = \frac{(0.047\cancel{M}) \times (0.047\cancel{M})}{(0.053\cancel{M}) \times (0.053\cancel{M})} = 0.79$

23. $K_{eq} = \frac{[N_2O_4]}{[NO_2]^2}$
$5.6 = \frac{0.66M}{[NO_2]^2}$
$[NO_2]^2 = \frac{0.66M}{5.6}$
$[NO_2] = \sqrt{\frac{0.66M}{5.6}} = 0.34M$

24. $[H_2] = [I_2] = 0.50M$;
$K_{eq} = \frac{[H_2][I_2]}{[HI]^2}$
$0.020 = \frac{0.50M \times 0.50M}{[HI]^2}$
$[HI]^2 = \frac{0.50M \times 0.50M}{0.020} = \frac{0.25M^2}{0.020} = 12.5M^2$
$[HI] = \sqrt{12.5M^2} = 3.5M$

33. $K_{sp} = [Pb^{2+}] \times [S^{2-}]$
$3.0 \times 10^{-28} = [Pb^{2+}]^2$
$[Pb^{2+}] = \sqrt{3.0 \times 10^{-28}M^2} = 2 \times 10^{-14}M$

34. $K_{sp} = [Ca^{2+}] \times [CO_3^{2-}]$
$4.5 \times 10^{-9} = [Ca^{2+}]^2$
$[Ca^{2+}] = \sqrt{4.5 \times 10^{-9}M^2} = 6.7 \times 10^{-5}M$

35. $K_{sp} = [Pb^{2+}] \times [S^{2-}]$
$[S^{2-}] = \frac{K_{sp}}{[Pb^{2+}]} = \frac{(8 \times 10^{-19})}{0.04} = 2 \times 10^{-17}M$

36. $K_{sp} = [Sr^{2+}] \times [SO_4^{2-}]$
$[SO_4^{2-}] = \frac{K_{sp}}{[Sr^{2+}]} = \frac{(3.2 \times 10^{-7})}{0.10} = 3.2 \times 10^{-6}M$

54. Atoms, ions, or molecules can react to form products, or they can bounce apart unchanged.

58. A catalyst increases the rate of reactions by providing an alternative reaction mechanism with a lower activation energy.

Selected Solutions **R105**

61.

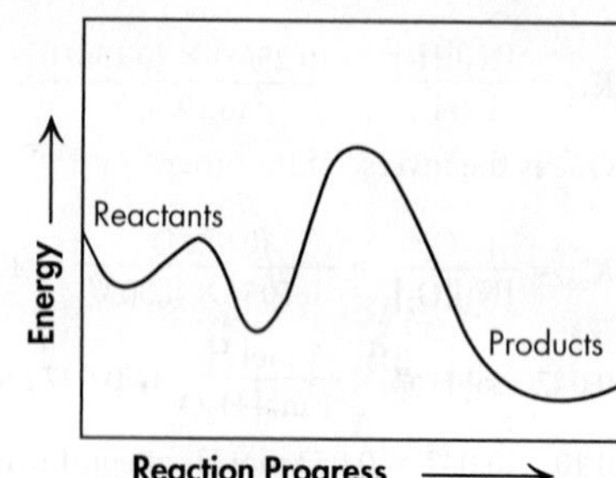

The intermediate is $N_2O_2(g)$.

66. a. $K_{eq} = \frac{[CH_4] \times [H_2S]^2}{[H_2]^4 \times [CS_2]}$

b. $K_{eq} = \frac{[PCl_3] \times [Cl_2]}{[PCl_5]}$

71. c, b, d, a

77. a. Entropy increases. b. Entropy decreases.

81. c

85. Since the total energy change for the forward reaction (−20 kJ) is negative, energy is released, and the reaction is exothermic. The reverse reaction is endothermic (energy is absorbed). The reaction progress curve for the reverse reaction is as follows:

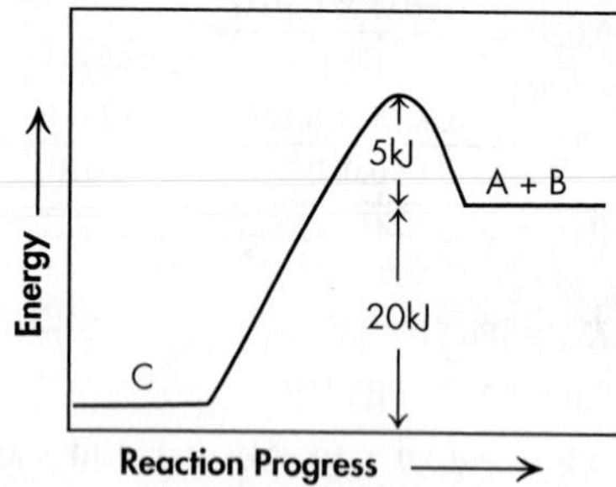

The activation energy of the reverse reaction is the total height of the peak (20 kJ + 5 kJ = 25 kJ).

89. increase in products

91. $K_{eq} = \frac{[CO] \times [H_2O]}{[CO_2] \times [H_2]} = \frac{0.448 \times 0.448}{0.552 \times 0.552} = 6.59 \times 10^{-1}$

92. The product of the concentrations of the ions must be greater than the ion-product constant (K_{sp}).

97. a. favors products c. favors reactants
b. no effect d. no effect

100. a. IO^-
b. two
c. the slow reaction
d. $2H_2O_2 \longrightarrow 2H_2O + O_2$
e. No, the I^- is changed in the reaction. A catalyst would not appear in the reaction as a reactant, an intermediate, or a product.

103. first order in NO_2^-, first order in NH_4^+, second order overall

105. Increase pressure, cool the reaction mixture, and remove NH_3

109. a. about 3 g
b. about 1.3 g
c. The rate of decomposition slows down over time.

115. Potassium chloride is an ionic compound, not a molecular compound.

117. a. sodium perchlorate, 1−
b. potassium permanganate, 1−
c. calcium phosphate, 3−
d. magnesium carbonate, 2−
e. sodium sulfate, 2−
f. potassium dichromate, 2−

121. a. $2KClO_3(s) + \text{heat} \longrightarrow 2KCl(s) + 3O_2(g)$

b. molar mass = 39.1 + 35.5 + (3 × 16.0) = 122.6 g

$$4.88\ \cancel{g\ KClO_3} \times \frac{1\ \cancel{mol\ KClO_3}}{122.6\ \cancel{g\ KClO_3}} \times \frac{3\ \cancel{mol\ O_2}}{2\ \cancel{mol\ KClO_3}} \times \frac{32\ g\ O_2}{1\ \cancel{mol\ O_2}} = 1.91\ g\ O_2$$

129. $2.40\ \cancel{L\ KCl} \times \frac{0.66\ mol\ KCl}{1\ \cancel{L\ KCl}} = 1.58\ mol\ KCl$

133. The process is exothermic because heat is released.

Chapter 19

1. a. H^+ is the Lewis acid; H_2O is the Lewis base.
b. $AlCl_3$ is the Lewis acid; Cl^- is the Lewis base.

2. Lewis base; it has a nonbonding pair of electrons that it can donate.

10. a. basic c. acidic
b. basic d. neutral

11. $[H^+] = \frac{K_w}{[OH^-]} = \frac{1.0 \times 10^{-14}}{1 \times 10^{-3}M} = 1 \times 10^{-11}M$; basic

12. a. $pH = -\log[H^+] = -\log(0.045M) = 1.35$
b. $pH = -\log[H^+] = -\log(8.7 \times 10^{-6}M) = 5.06$
c. $pH = -\log[H^+] = -\log(0.0015M) = 2.82$
d. $pH = -\log[H^+] = -\log(1.2 \times 10^{-3}M) = 2.92$

13. a. $pH = -\log[H^+] = -\log(1.0 \times 10^{-12}M) = 12$
b. $pH = -\log[H^+] = -\log(1 \times 10^{-4}M) = 4$

14. a. $pH = -\log[H^+]$
$5.00 = -\log[H^+]$
$-5.00 = \log[H^+]$
$[H^+] = 10^{-5.00} = 1.00 \times 10^{-5}M$

R106 Appendix E

b. $pH = -\log[H^+]$
$12.83 = -\log[H^+]$
$-12.83 = \log[H^+]$
$[H^+] = 10^{-12.83} = 1.48 \times 10^{-13}M$

15. a. $pH = -\log[H^+]$
$4.00 = -\log[H^+]$
$-4.00 = \log[H^+]$
$[H^+] = 10^{-4.00} = 1.00 \times 10^{-4}M$

b. $pH = -\log[H^+]$
$11.55 = -\log[H^+]$
$-11.55 = \log[H^+]$
$[H^+] = 10^{-11.55} = 2.82 \times 10^{-12}M$

16. a. $[H^+] = \dfrac{K_w}{[OH^-]} = \dfrac{1.0 \times 10^{-14}}{4.3 \times 10^{-5}M} = 0.23 \times 10^{-9}M$
$pH = -\log[H^+] = -\log(0.23 \times 10^{-9}M) = 9.64$

b. $[H^+] = \dfrac{K_w}{[OH^-]} = \dfrac{1.0 \times 10^{-14}}{4.5 \times 10^{-11}M} = 0.22 \times 10^{-3}M$
$pH = -\log[H^+] = -\log(0.22 \times 10^{-3}M) = 3.66$

17. a. $[H^+] = \dfrac{K_w}{[OH^-]} = \dfrac{1.0 \times 10^{-14}}{5.0 \times 10^{-9}M} = 0.20 \times 10^{-5}M$
$pH = -\log[H^+] = -\log(0.20 \times 10^{-5}M) = 5.70$

b. $[H^+] = \dfrac{K_w}{[OH^-]} = \dfrac{1.0 \times 10^{-14}}{8.3 \times 10^{-4}M} = 0.12 \times 10^{-10}M$
$pH = -\log[H^+] = -\log(0.12 \times 10^{-10}M) = 10.91$

25. $0.1000M - 4.2 \times 10^{-3}M = 0.0958M$
$K_a = \dfrac{(4.2 \times 10^{-3}) \times (4.2 \times 10^{-3})}{(0.0958)} = 1.8 \times 10^{-4}$

26. $0.2000M - 9.86 \times 10^{-4}M = 0.199M$
$K_a = \dfrac{(9.86 \times 10^{-4}) \times (9.86 \times 10^{-4})}{(0.199)} = 4.89 \times 10^{-6}$

35. $H_3PO_4(aq) + 3KOH(aq) \longrightarrow K_3PO_4(aq) + 3H_2O(l)$
$1.56 \cancel{\text{mol } H_3PO_4} \times \dfrac{3 \text{ mol KOH}}{1 \cancel{\text{mol } H_3PO_4}} = 4.68 \text{ mol KOH}$

36. $HNO_3(aq) + NaOH(aq) \longrightarrow NaNO_3(aq) + H_2O(l)$
$0.20 \cancel{\text{mol } HNO_3} \times \dfrac{1 \text{ mol NaOH}}{1 \cancel{\text{mol } HNO_3}} = 0.20 \text{ mol NaOH}$

37. $25.0 \cancel{\text{mL KOH}} \times \dfrac{1.00 \cancel{\text{mol KOH}}}{1000 \cancel{\text{mL KOH}}} \times \dfrac{1 \cancel{\text{mol HCl}}}{1 \cancel{\text{mol KOH}}}$
$\times \dfrac{1000 \text{ mL HCl}}{0.45 \cancel{\text{mol HCl}}} = 56 \text{ mL HCl}$

38. $38.5 \cancel{\text{mL NaOH}} \times \dfrac{0.150 \cancel{\text{mol NaOH}}}{1000 \cancel{\text{mL NaOH}}} \times \dfrac{1 \text{ mol } H_3PO_4}{3 \cancel{\text{mol NaOH}}}$
$= 0.00193 \text{ mol } H_3PO_4$
$\dfrac{0.00193 \text{ mol } H_3PO_4}{0.0150 \text{ L } H_3PO_4} = 0.129M\ H_3PO_4$

44. a. $HPO_4^{2-}(aq) + H^+(aq) \longrightarrow H_2PO_4^-(aq)$
b. $H_2PO_4^-(aq) + OH^-(aq) \longrightarrow HPO_4^{2-}(aq) + H_2O(l)$

45. $HCOO^-(aq) + H^+(aq) \longrightarrow HCOOH(aq)$

52. Acids ionize to give hydrogen ions in aqueous solution. Bases ionize to give hydroxide ions in aqueous solution.

55. a. $2Li(s) + 2H_2O(l) \longrightarrow 2LiOH(aq) + H_2(g)$
b. $Ba(s) + 2H_2O(l) \longrightarrow Ba(OH)_2(aq) + H_2(g)$

63. a. $pH = -\log[H^+] = -\log\left(\dfrac{K_w}{[OH^-]}\right)$
$= -\log\left(\dfrac{1 \times 10^{-14}}{1 \times 10^{-2}M}\right) = 12$; basic
b. $pH = -\log[H^+] = -\log(1 \times 10^{-2}M) = 2$; acidic

66. a. strong base
b. weak base
c. strong acid
d. strong acid

73. a. $NaOH(aq) + HCl(aq) \longrightarrow NaCl(aq) + H_2O(l)$
$28.0 \cancel{\text{mL HCl}} \times \dfrac{1 \cancel{\text{L HCl}}}{10^3 \cancel{\text{mL HCl}}} \times \dfrac{1 \cancel{\text{mol HCl}}}{1 \cancel{\text{L HCl}}}$
$\times \dfrac{1 \text{ mol NaOH}}{1 \cancel{\text{mol HCl}}} = 0.028 \text{ mol NaOH}$
$\dfrac{0.028 \text{ mol NaOH}}{20.0 \cancel{\text{mL}}} \times \dfrac{10^3 \cancel{\text{mL}}}{1 \text{ L}} = 1.40M$

b. $3NaOH(aq) + H_3PO_4(aq) \longrightarrow Na_3PO_4(aq) + 3H_2O(l)$
$17.4 \cancel{\text{mL } H_3PO_4} \times \dfrac{1 \cancel{\text{L } H_3PO_4}}{10^3 \cancel{\text{mL } H_3PO_4}} \times \dfrac{1 \cancel{\text{mol } H_3PO_4}}{1 \cancel{\text{L } H_3PO_4}}$
$\times \dfrac{3 \text{ mol NaOH}}{1 \cancel{\text{mol } H_3PO_4}} = 0.0522 \text{ mol NaOH}$
$\dfrac{0.0522 \text{ mol NaOH}}{20.0 \cancel{\text{mL}}} \times \dfrac{10^3 \cancel{\text{mL}}}{1 \text{ L}} = 2.61M$

75. $HCO_3^-(aq) + H_2O(l) \longrightarrow H_2CO_3(aq) + OH^-(aq)$

81. $HPO_4^{2-}(aq) \longrightarrow H^+(aq) + PO_4^{3-}(aq)$
(acting as an acid)
$HPO_4^{2-}(aq) + H^+(aq) \longrightarrow H_2PO_4^-(aq)$
(acting as a base)

85. a. $HClO_2$, chlorous acid
b. H_3O^+, hydronium ion
c. H_3PO_4, phosphoric acid
d. NH_4^+, ammonium ion

91. a. $2HCl(aq) + Mg(OH)_2(aq) \longrightarrow MgCl_2(aq) + 2H_2O(l)$
b. $2HCl(aq) + CaCO_3(aq) \longrightarrow H_2O(l) + CO_2(g) + CaCl_2(aq)$
c. $3HCl(aq) + Al(OH)_3(aq) \longrightarrow AlCl_3(aq) + 3H_2O(l)$

95. b, c, d, a

100. $HOCN(aq) + OH^-(aq) \longrightarrow H_2O(l) + OCN^-(aq)$
$OCN^-(aq) + H^+(aq) \longrightarrow HOCN(aq)$

Selected Solutions **R107**

101. The *y*-axis might correspond to $[H^+]$ because HCl is a strong acid.

106. $pOH = -\log[OH^-]$

110. $K_w = K_aK_b = \frac{[H^+]\cancel{[A^-]}}{\cancel{[HA]}} \times \frac{\cancel{[HA]}[OH^-]}{\cancel{[A^-]}} = [H^+][OH^-]$

113. Rapid breathing releases more CO_2. The shift in equilibrium causes H_2CO_3 and H^+ concentrations to decrease. The loss of H^+ increases the basicity of the blood and alkalosis results. Slow breathing does not release enough CO_2. The CO_2 buildup increases the H_2CO_3 and H^+ concentrations. The H^+ concentration increases the acidity of the blood and acidosis results.

118. neutralization; raised

121. liquid

126. hydrogen bond

129. $V_1 = \frac{M_2 \times V_2}{M_1} = \frac{2.5\cancel{M} \times 1.50\ L}{8.0\cancel{M}} = 0.47\ L$

132. **a.** $34.5\ \cancel{cal} \times \frac{1\ J}{0.239\ \cancel{cal}} = 144\ J$

b. $250\ \cancel{Cal} \times \frac{10^3\ \cancel{cal}}{1\ \cancel{Cal}} \times \frac{1\ \cancel{J}}{0.239\ \cancel{cal}} \times \frac{1\ kJ}{10^3\ \cancel{J}} = 1.05 \times 10^3\ kJ$

c. $0.347\ \cancel{kJ} \times \frac{10^3\ \cancel{J}}{1\ \cancel{kJ}} \times \frac{0.239\ cal}{1\ \cancel{J}} = 82.9\ cal$

137. **a.** $NaCl(aq)$ **b.** $CO_2(g)$ **c.** hot water

Chapter 20

1. **a.** Na: oxidized, loses electrons (reducing agent); S: reduced, gains electrons (oxidizing agent)

b. Al: oxidized, loses electrons (reducing agent); O_2: reduced, gains electrons (oxidizing agent)

2. **a.** oxidation (loss of 1 electron)

b. oxidation (loss of 2 electrons)

c. reduction (gain of 2 electrons)

d. reduction (gain of 2 electrons)

10. **a.** S, +3; O, −2 **c.** P, +5; O, −2

b. Na, +1; O, −1 **d.** N, +5; O, −2

11. **a.** $\overset{+1+5-2}{KClO_3}$; +5 **c.** $\overset{+2\ +7\ -2}{Ca(ClO_4)_2}$; +7

b. 0 **d.** $\overset{+1\ -2}{Cl_2O}$; +1

12. **a.** H_2 is oxidized (0 to +1); O_2 is reduced (0 to −2).

b. N is reduced (+5 to +3); O is oxidized (−2 to 0).

13. **a.** H_2 is the reducing agent; O_2 is the oxidizing agent.

b. N is the oxidizing agent; O is the reducing agent.

14. **a.** N in NH_4^+ is oxidized (−3 to 0); H is unchanged; N in NO_2^- is reduced (+3 to 0); O is unchanged.

b. Pb is reduced (+4 to +2); O is unchanged; H is unchanged; I is oxidized (−1 in HI to 0 in I_2).

15. **a.** N in NH_4^+ is the reducing agent; N in NO_2^- is the oxidizing agent.

b. Pb is the oxidizing agent; I is the reducing agent.

21. **a.** redox reaction: Mg is oxidized (0 to +2), Br_2 is reduced (0 to −1)

b. not a redox reaction

22. **a.** not a redox reaction

b. redox reaction: H_2 is oxidized (0 to +1), Cu is reduced (+2 to 0)

23. **a.**

(1)(−6) = −6

$\overset{+1+5-2}{KClO_3(s)} \longrightarrow \overset{+1\ -1}{KCl(s)} + \overset{0}{O_2(g)}$

(3)(+2) = +6

One K atom must be reduced for every 3 oxygen atoms oxidized.

$KClO_3(s) \longrightarrow KCl(s) + O_2(g)$

Balance by inspection; put the coefficient 3 in front of O_2, and the coefficient 2 in front of $KClO_3$ and KCl:

$2KClO_3(s) \longrightarrow 2KCl(s) + 3O_2(g)$

b. $2HNO_2(aq) + 2HI(aq) \longrightarrow 2NO(g) + I_2(s) + 2H_2O(l)$

24. **a.**

(2)(−3) = −6

$\overset{+3\ -2}{Bi_2S_3(s)} + \overset{+1+5-2}{HNO_3(aq)} \longrightarrow \overset{+3\ +5\ -2}{Bi(NO_3)_3(aq)} + \overset{+2\ -2}{NO(g)} + \overset{0}{S(s)} + \overset{+1\ -2}{H_2O(l)}$

(3)(+2) = +6

2 N atoms must be reduced for every 3 sulfur atoms oxidized.

$Bi_2S_3(s) + 2HNO_3(aq) \longrightarrow Bi(NO_3)_3(aq) + 2NO(g) + 3S(s) + H_2O(l)$

Balance by inspection; put the coefficient 2 in front of $Bi(NO_3)_3$, the coefficient 8 in front of HNO_3, and the coefficient 4 in front of H_2O:

$Bi_2S_3(s) + 8HNO_3(aq) \longrightarrow 2Bi(NO_3)_3(aq) + 2NO(g) + 3S(s) + 4H_2O(l)$

b. $SbCl_5(aq) + 2KI(aq) \longrightarrow SbCl_3(aq) + 2KCl(aq) + I_2(s)$

R108 Appendix E

25. Oxidation half-reaction: $\overset{+2}{Sn^{2+}} \longrightarrow \overset{+4}{Sn^{4+}}$

Reduction half-reaction: $\overset{+6}{Cr_2O_7^{2-}} \longrightarrow \overset{+3}{Cr^{3+}}$

Oxidation: $Sn^{2+}(aq) \longrightarrow Sn^{4+}(aq)$ (atoms balanced)

Reduction: $Cr_2O_7^{2-}(aq) + 14H^+ \longrightarrow 2Cr^{3+}(aq) + 7H_2O(l)$ (atoms balanced)

Oxidation: $Sn^{2+}(aq) \longrightarrow Sn^{4+}(aq) + 2e^-$ (charges balanced)

Reduction: $Cr_2O_7^{2-}(aq) + 14H^+ + 6e^- \longrightarrow 2Cr^{3+}(aq) + 7H_2O(l)$ (charges balanced)

Oxidation: $3Sn^{2+}(aq) \longrightarrow 3Sn^{4+}(aq) + 6e^-$

Reduction: $Cr_2O_7^{2-}(aq) + 14H^+ + 6e^- \longrightarrow 2Cr^{3+}(aq) + 7H_2O(l)$

$3Sn^{2+}(aq) + Cr_2O_7^{2-}(aq) + 14H^+ + \cancel{6e^-} \longrightarrow 3Sn^{4+}(aq) + \cancel{6e^-} + 2Cr^{3+}(aq) + 7H_2O(l)$

$3Sn^{2+}(aq) + Cr_2O_7^{2-}(aq) + 14H^+ \longrightarrow 3Sn^{4+}(aq) + 2Cr^{3+}(aq) + 7H_2O(l)$

26. $Zn(s) + NO_3^-(aq) + H_2O(l) + OH^-(aq) \longrightarrow Zn^{2+} + OH^-(aq) + NH_3(aq)$

Oxidation half-reaction: $\overset{0}{Zn} \longrightarrow \overset{+2}{Zn^{2+}}$

Reduction half-reaction: $\overset{+5}{NO_3^-} \longrightarrow \overset{-3}{NH_3}$

Oxidation: $Zn(s) \longrightarrow Zn^{2+}(aq)$ (atoms balanced)

Reduction: $NO_3^-(aq) + 6H_2O(l) \longrightarrow NH_3(aq) + 9OH^-(aq)$ (atoms balanced)

Oxidation: $Zn(s) \longrightarrow Zn^{2+}(aq) + 2e^-$ (charges balanced)

Reduction: $NO_3^-(aq) + 6H_2O(l) + 8e^- \longrightarrow NH_3(aq) + 9OH^-(aq)$ (charges balanced)

Oxidation: $Zn(s) \longrightarrow Zn^{2+}(aq) + 2e^-$ (charges balanced)

Reduction: $NO_3^-(aq) + 6H_2O(l) + 8e^- \longrightarrow NH_3(aq) + 9OH^-(aq)$ (charges balanced)

Oxidation: $4Zn(s) \longrightarrow 4Zn^{2+}(aq) + 8e^-$ (charges balanced)

Reduction: $NO_3^-(aq) + 6H_2O(l) + 8e^- \longrightarrow NH_3(aq) + 9OH^-(aq)$ (charges balanced)

$4Zn(s) + NO_3^-(aq) + 6H_2O(l) + \cancel{8e^-} \longrightarrow 4Zn^{2+}(aq) + \cancel{8e^-} + NH_3(aq) + 9OH^-(aq)$

$4Zn(s) + NO_3^-(aq) + 6H_2O(l) \longrightarrow 4Zn^{2+}(aq) + NH_3(aq) + 9OH^-(aq)$

$4Zn(s) + NO_3^-(aq) + 6H_2O(l) + 16OH^-(aq) \longrightarrow 4Zn^{2+}(aq) + NH_3(aq) + 9OH^-(aq) + 16OH^-(aq)$ (spectator ions added)

$4Zn(s) + NO_3^-(aq) + 6H_2O(l) + 16OH^-(aq) \longrightarrow 4Zn(OH)_4^{2-}(aq) + NH_3(aq) + 9OH^-(aq)$

$4Zn(s) + NO_3^-(aq) + 6H_2O(l) + 7OH^-(aq) \longrightarrow 4Zn(OH)_4^{2-}(aq) + NH_3(aq)$

34. **a.** $2Ba(s) + O_2(g) \longrightarrow 2BaO(s)$; barium is oxidized
b. $CuO(s) + H_2(g) \longrightarrow Cu(s) + H_2O(l)$; copper is reduced
c. $C_2H_4(g) + 3O_2(g) \longrightarrow 2CO_2(g) + 2H_2O(l)$; carbon is oxidized
d. $3CaO(s) + 2Al(s) \longrightarrow Al_2O_3(s) + 3Ca(s)$; calcium is reduced

36. **a.** oxidizing agent **b.** reducing agent **c.** oxidizing agent **d.** oxidizing agent

37. **a.** H_2 is oxidized; S is reduced.
b. N_2 is reduced; H_2 is oxidized.
c. S is oxidized; O_2 is reduced.
d. H_2 is oxidized; O_2 is reduced.

42. **a.** O, −2; H, +1 **b.** P, +5; O, −2 **c.** I, +5; O, −2 **d.** H, +1; P, +5; O, −2 **e.** H, +1; S, +6; O, −2

44. **a.** $2Al(s) + 3Cl_2(g) \longrightarrow 2AlCl_3(s)$
b. $2Al(s) + Fe_2O_3(s) \longrightarrow Al_2O_3(s) + 2Fe(s)$
c. $3Cl_2(g) + 6KOH(aq) \longrightarrow KClO_3(aq) + 5KCl(aq) + 3H_2O(l)$
d. $2HNO_3(aq) + 3H_2S(aq) \longrightarrow 3S(s) + 2NO(g) + 4H_2O(l)$
e. $KIO_4(aq) + 7KI(aq) + 8HCl(aq) \longrightarrow 8KCl(aq) + 4I_2(s) + 4H_2O(l)$

45. redox: a, b, c, d, e

46. **a.** $CuS(s) + 8NO_3^-(aq) + 8H^+(aq) \longrightarrow Cu(NO_3)_2(aq) + SO_2(g) + 6NO_2(g) + 4H_2O(l)$
b. $6I^-(aq) + 2NO_3^-(aq) + 8H^+(aq) \longrightarrow 3I_2(s) + 2NO(g) + 4H_2O(l)$

47. **a.** $4MnO_4^-(aq) + 3ClO_2^-(aq) + 2H_2O(l) \longrightarrow 4MnO_2(s) + 3ClO_4^-(aq) + 4OH^-(aq)$
b. $2Cr^{3+}(aq) + 3ClO^-(aq) + 10OH^-(aq) \longrightarrow 2CrO_4^{2-}(aq) + 3Cl^-(aq) + 5H_2O(l)$
c. $6Mn^{3+}(aq) + I^-(aq) + 6OH^-(aq) \longrightarrow 6Mn^{2+}(aq) + IO_3^-(aq) + 3H_2O(l)$

50. **a.** +4 **b.** +5 **c.** +5 **d.** +3 **e.** +5 **f.** +3

54. **a.** $16H^+(aq) + 2Cr_2O_7^{2-}(aq) + C_2H_5OH(aq) \longrightarrow 4Cr^{3+}(aq) + 2CO_2(g) + 11H_2O(l)$
b. oxidizing agent

59. **a.** $Ba(ClO_3)_2(s) \longrightarrow BaCl_2(s) + 3O_2(g)$
b. $2PbS(s) + 3O_2(g) \longrightarrow 2PbO(s) + 2SO_2(g)$

63. $1s^22s^22p^63s^23p^5$; a chlorine atom can "lose" its 7 valence electrons or it can gain 1 electron to fill the last $3p$ orbital.

Selected Solutions **R109**

66. In every redox reaction one species loses one or more electrons and is a reducing agent. Another substance gains one or more electrons and is an oxidizing agent.

70. **a.** $Rb(s) + I_2(s) \longrightarrow RbI_2(s)$; oxidizing agent is I

b. $Ba(s) + 2H_2O(l) \longrightarrow Ba(OH)_2(aq) + H_2(g)$; oxidizing agent is H

c. $2Al(s) + 3FeSO_4(aq) \longrightarrow Al_2(SO_4)_3(aq) + 3Fe(s)$; oxidizing agent is Fe

d. $C_4H_8(g) + 6O_2(g) \longrightarrow 4CO_2(g) + 4H_2O(l)$; oxidizing agent is O

e. $Zn(s) + 2HBr(aq) \longrightarrow ZnBr_2(aq) + H_2(g)$; oxidizing agent is H

f. $Mg(s) + Br_2(l) \longrightarrow MgBr_2(s)$, oxidizing agent is Br

73. $2AgNO_3(aq) + Cu(s) \longrightarrow Cu(NO_3)_2(aq) + 2Ag(s)$

$$85.0\ \cancel{mL} \times \frac{1\ \cancel{L}}{1000\ \cancel{mL}} \times \frac{0.150\ \cancel{mol\ AgNO_3}}{1\ \cancel{L\ AgNO_3}} \times \frac{1\ \cancel{mol\ Cu}}{2\ \cancel{mol\ AgNO_3}} \times \frac{63.5\ g\ Cu}{1\ \cancel{mol\ Cu}} = 0.405\ g\ Cu$$

80. **a.** $Cr_2O_7^{2-}(aq) + 14H^+ + 6e^- \longrightarrow 2Cr^{3+}(aq) + 7H_2O(l)$

$Cr_2O_7^{2-}$ is an oxidizing agent.

b. $S_2O_3^{2-}(aq) + 5H_2O(l) \longrightarrow 2SO_4^{2-}(aq) + 10H^+ + 8e^-$

$S_2O_3^{2-}$ is a reducing agent.

c. $CrO_4^{2-}(aq) + 4H_2O(l) + 3e^- \longrightarrow Cr(OH)_3(aq) + 5OH^-(aq)$

CrO_4^{2-} is an oxidizing agent.

d. $MnO_4^-(aq) + 8H^+ + 5e^- \longrightarrow Mn^{2+}(aq) + 4H_2O(l)$

MnO_4^- is an oxidizing agent.

e. $C_2O_4^{2-}(aq) \longrightarrow 2CO_2(g) + 2e^-$

$C_2O_4^{2-}$ is a reducing agent.

f. $2H_2O(l) + MnO_4^-(aq) + 3e^- \longrightarrow MnO_2(s) + 4OH^-(aq)$

MnO_4^- is an oxidizing agent.

86. **a.**

Alkane burned	O_2 used (mol)	CO_2 produced (mol)	H_2O produced (mol)
CH_4	2	1	2
C_2H_6	3.5	2	3
C_3H_8	5	3	4
C_4H_{10}	6.5	4	5
C_5H_{12}	8	5	6
C_6H_{14}	9.5	6	7

b. $C_xH_y + [x + (y/4)]O_2 \longrightarrow xCO_2 + (y/2)H_2O$

88. $T = 25°C + 273 = 298\ K$

$$P = \frac{n \times R \times T}{V}$$

$$= \frac{\left(13.8\ \cancel{g\ N_2} \times \frac{1\ \cancel{mol\ N_2}}{28.0\ \cancel{g\ N_2}}\right) \times 8.31 \left(\frac{\cancel{L} \cdot kPa}{\cancel{K} \cdot \cancel{mol}}\right) \times 298\ \cancel{K}}{6.8\ \cancel{L}}$$

$= 1.8 \times 10^2\ kPa$

91. $V_2 = \frac{M_1 \times V_1}{M_2} = \frac{1.5\cancel{M\ HCl} \times 440\ mL\ HCl}{6.0\cancel{M\ HCl}}$

$= 110\ mL\ 6.0M\ HCl$

Dilute 110 mL of 6.0*M* HCl to 440 mL total volume.

94. $PbBr_2(s) \longrightarrow Pb^{2+} + 2Br^-$

$K_{sp} = [Pb^{2+}][Br^-] = 2.1 \times 10^{-6}$

Let $x = [Pb^{2+}]$.

Then $2x = [Br^-]$.

$(x)(2x)^2 = 2.1 \times 10^{-6}$

$4x^3 = 2.1 \times 10^{-6}$

$x = [Pb^{2+}] = 8.1 \times 10^{-3}$

The solubility of $PbBr_2$ is $8.1 \times 10^{-3}M$.

100. **a.** $pH = -\log[H^+] = -\log(0.000010) = -\log(10^{-5})$
$= 5.00$ (acidic)

b. $[H^+] = \frac{K_w}{[OH^-]} = \frac{1.0 \times 10^{-14}M^2}{1 \times 10^{-4}M} = 1.0 \times 10^{-10}M$

$pH = -\log[H^+] = -\log(1.0 \times 10^{-10}M)$
$= 10.00$ (basic)

c. $[H^+] = \frac{K_w}{[OH^-]} = \frac{1.0 \times 10^{-14}M^2}{1 \times 10^{-1}M} = 1.0 \times 10^{-13}M$

$pH = -\log[H^+] = -\log(1.0 \times 10^{-13}M)$
$= 13.00$ (basic)

d. $pH = -\log[H^+] = -\log(3.0 \times 10^{-7})$
$= 6.50$ (acidic)

Chapter 21

8. The half-reactions are

Oxidation: $Cr(s) \longrightarrow Cr^{3+}(aq) + 3e^-$

Reduction: $Zn^{2+}(aq) + 2e^- \longrightarrow Zn(s)$

Writing both half-cells as reductions:

$Cr^{3+}(aq) + 3e^- \longrightarrow Cr(s)$ $\qquad E°_{Cr^{3+}} = -0.74\ V$

$Zn^{2+}(aq) + 2e^- \longrightarrow Zn(s)$ $\qquad E°_{Zn^{2+}} = -0.76\ V$

$E°_{cell} = E°_{red} - E°_{oxid} = E°_{Zn^{2+}} - E°_{Cr^{3+}}$

$= -0.76\ V - (-0.74\ V) = -0.02\ V$

$E°_{cell} < 0$, so the reaction is not spontaneous.

R110 Appendix E

9. The half-reactions are

Oxidation: $Fe(s) \longrightarrow Fe^{2+}(aq) + 2e^-$

Reduction: $Co^{2+}(aq) + 2e^- \longrightarrow Co(s)$

Writing both half-cells as reductions:

$Fe^{2+}(aq) + 2e^- \longrightarrow Fe(s)$ $\quad E°_{Fe^{2+}} = -0.44\ V$

$Co^{2+}(aq) + 2e^- \longrightarrow Co(s)$ $\quad E°_{Co^{2+}} = -0.28\ V$

$E°_{cell} = E°_{red} - E°_{oxid} = E°_{Co^{2+}} - E°_{Fe^{2+}}$

$= -0.28\ V - (-0.44\ V) = +0.16\ V$

$E°_{cell} > 0$, so the reaction is spontaneous.

10. Cu^{2+} is reduced, and Al is oxidized.

Oxidation: $Al(s) \longrightarrow Al^{3+}(aq) + 3e^-$

Reduction: $Cu^{2+}(aq) + 2e^- \longrightarrow Cu(s)$

Writing both half-cells as reductions:

$2[Al(s) \longrightarrow Al^{3+}(aq) + 3e^-]$

$3[Cu^{2+}(aq) + 2e^- \longrightarrow Cu(s)]$

$2Al(s) \longrightarrow 2Al^{3+}(aq) + \cancel{6e^-}$

$3Cu^{2+}(aq) + \cancel{6e^-} \longrightarrow 3Cu(s)$

$2Al(s) + 3Cu^{2+}(aq) \longrightarrow 2Al^{3+}(aq) + 3Cu(s)$

11. Ag^+ is reduced, and Cu is oxidized.

The half-reactions are

Oxidation: $Cu(s) \longrightarrow Cu^{2+}(aq) + 2e^-$

Reduction: $Ag^+(aq) + e^- \longrightarrow Ag(s)$

Writing both half-cells as reductions:

$Cu(s) \longrightarrow Cu^{2+}(aq) + 2e^-$

$2[Ag^+(aq) + e^- \longrightarrow Ag(s)]$

$Cu(s) \longrightarrow Cu^{2+}(aq) + \cancel{2e^-}$

$2Ag^{2+}(aq) + \cancel{2e^-} \longrightarrow 2Ag(s)$

$Cu(s) + 2Ag^{2+}(aq) \longrightarrow Cu^{2+}(aq) + 2Ag(s)$

12. $E°_{cell} = E°_{red} - E°_{oxid} = E°_{Cu^{2+}} - E°_{Al^{3+}}$

$= 0.34\ V - (-1.66\ V) = +2.00\ V$

13. $E°_{cell} = E°_{red} - E°_{oxid} = E°_{Ag^+} - E°_{Cu^{2+}}$

$= 0.80\ V - (+0.34\ V) = +0.46\ V$

28. Using Table 21.1, the more active element is the most readily oxidized.

a. Cu **b.** Ca **c.** Mg **d.** Sn **e.** Zn **f.** Al

33. A zinc container (anode) filled with electrolyte paste; the cathode is a graphite rod embedded in the paste; Zn is oxidized; MnO_2 is reduced.

36. $Pb(s)\ |\ PbSO_4(aq)\ ||\ PbO_2(s)\ |\ PbSO_4(s)$

40. The standard reduction potential of a half-cell is a measure of the tendency of a given half-reaction to occur as a reduction under standard conditions. The difference between the standard reduction potentials of the two half-cells is called the standard cell potential.

44. a. Cu is oxidized, and H^+ is reduced.

Writing both half-reactions as reductions:

$Cu(s) \longrightarrow Cu^{2+}(aq) + 2e^-$ $\quad E°_{Cu^{2+}} = +0.34\ V$

$2H^+(aq) + 2e^- \longrightarrow H_2(g)$ $\quad E°_{H^+} = 0.000\ V$

$E°_{cell} = E°_{red} - E°_{oxid} = E°_{H^+} - E°_{Cu^{2+}}$

$= 0.000\ V - (+0.34\ V) = -0.34\ V$

$E°_{cell} < 0$, so the reaction will not occur spontaneously.

b. Ag is oxidized, and Fe^{2+} is reduced.

Writing both half-reactions as reductions:

$Ag(s) \longrightarrow Ag^+(aq) + e^-$ $\quad E°_{Ag^+} = +0.80\ V$

$Fe^{2+}(aq) + 2e^- \longrightarrow Fe(s)$ $\quad E°_{Fe^{2+}} = -0.44\ V$

$E°_{cell} = E°_{red} - E°_{oxid} = E°_{Fe^{2+}} - E°_{Ag^+}$

$= -0.44\ V - (+0.80\ V) = -1.24\ V$

$E°_{cell} < 0$, so the reaction will not occur spontaneously.

46. Voltaic cells convert chemical energy into electrical energy. Electrolytic cells use electrical energy to cause a chemical reaction.

50. Chloride ions are oxidized to produce chlorine gas, and water is reduced to produce hydrogen gas.

55. The iron nail will become copper-plated as Fe is oxidized to Fe^{2+} and Cu^{2+} is reduced to Cu.

Oxidation: $Fe(s) \longrightarrow Fe^{2+}(aq) + \cancel{2e^-}$

Reduction: $Cu^{2+}(aq) + \cancel{2e^-} \longrightarrow Cu(s)$

Overall cell reaction:

$Fe(s) + Cu^{2+}(aq) \longrightarrow Fe^{2+}(aq) + Cu(s)$

57. a. $Sn(s) + Pb^{2+}(aq) \longrightarrow Sn^{2+}(aq) + Pb(s)$

Oxidation: $Sn(s) \longrightarrow Sn^{2+}(aq) + \cancel{2e^-}$

Reduction: $Pb^{2+}(aq) + \cancel{2e^-} \longrightarrow Pb(s)$

Overall cell reaction:

$Sn(s) + Pb^{2+}(aq) \longrightarrow Sn^{2+}(aq) + Pb(s)$

$E°_{cell} = E°_{red} - E°_{oxid} = E°_{Pb^{2+}} - E°_{Sn^{2+}}$

$= -0.13\ V - (-0.14\ V) = +0.01\ V$

b. $H_2(g) + Br_2(l) \longrightarrow 2H^+(aq) + 2Br^-(aq)$

Oxidation: $H_2(g) \longrightarrow 2H^+(aq) + \cancel{2e^-}$

Reduction: $Br_2(l) + \cancel{2e^-} \longrightarrow 2Br^-(aq)$

Overall cell reaction:

$H_2(g) + Br_2(l) \longrightarrow 2H^+(aq) + 2Br^-(aq)$

$E°_{cell} = E°_{red} - E°_{oxid} = E°_{Br^-} - E°_{H^+}$

$= +1.07\ V - 0.00\ V = +1.07\ V$

60. If the cell potential for a redox reaction is positive, the reaction is spontaneous as written.

Selected Solutions **R111**

62. a. oxidation: $2Cl^-(l) \longrightarrow Cl_2(g) + 2e^-$ (at anode)

reduction: $Al^{3+}(l) + 3e^- \longrightarrow Al(l)$ (at cathode)

b. Multiply the oxidation half-reaction by 3 and the reduction half-reaction by 2 to balance the electrons:

$6Cl^-(l) \longrightarrow 3Cl_2(g) + \cancel{6e^-}$

$2Al^{3+}(l) + \cancel{6e^-} \longrightarrow 2Al(l)$

Overall cell reaction:

$2AlCl_3(l) \longrightarrow 2Al(l) + 3Cl_2(g)$

c. Chlorine gas is produced at the anode. Liquid aluminum is produced at the cathode.

68. a. Possible oxidation reactions at anode:

(*i*) $2Cl^-(aq) \longrightarrow Cl_2(g) + 2e^-$

(*ii*) $2H_2O(l) \longrightarrow O_2(g) + 4H^+(aq) + 4e^-$

b. Possible reduction reactions at cathode:

(*i*) $Na^+(aq) + e^- \longrightarrow Na(s)$

(*ii*) $2H_2O(l) + 2e^- \longrightarrow H_2(g) + 2OH^-(aq)$

c. (*i*) Chloride ions are more readily oxidized to chlorine gas than water molecules are oxidized to oxygen.

d. (*ii*) Water molecules are more easily reduced than sodium ions.

69. The stronger oxidizing agent will have the more positive standard reduction potential ($E°$).

a. Ca^{2+} b. Hg_2^{2+} c. Cu^+ d. Hg^{2+}

71. A direct current flows in one direction only. If an alternating current is used, the reactions at the anode and cathode would constantly be reversed.

73. The cathode reaction in a dry cell is not reversible, so dry cells are not rechargeable. The anode and cathode reactions in a lead storage battery are reversible. Therefore, lead storage batteries are rechargeable when electrical energy is applied.

75. The spoon is being electroplated with silver in an electrolytic cell. The spoon serves as the cathode, and the anode is metallic silver. The electrolyte is a solution of silver cyanide. When a direct current is applied, silver ions move from the anode to the spoon, where they are reduced to silver metal.

79. d; the voltage of a dry cell decreases steadily.

80. a. The standard reduction potential for AgCl (+0.22 V) is more positive than the standard reduction potential for Ni^{2+} (−0.25 V), so AgCl is reduced and Ni is oxidized.

Oxidation: $Ni(s) \longrightarrow Ni^{2+}(aq) + 2e^-$

Reduction: $2[AgCl(s) + e^- \longrightarrow Ag(s) + Cl^-(aq)]$

Overall cell reaction:

$Ni(s) + 2AgCl(s) \longrightarrow 2Ag(s) + NiCl_2(aq)$

$E°_{cell} = E°_{red} - E°_{oxid} = E°_{AgCl} - E°_{Ni^{2+}}$

$= +0.22\text{ V} - (-0.25\text{ V}) = +0.47\text{ V}$

b. The standard reduction potential for Cl^- (+1.36 V) is more positive than the standard reduction potential for Al^{3+} (−1.66 V), so Cl^- is reduced and Al^{3+} is oxidized.

Oxidation: $2[Al(s) \longrightarrow Al^{3+}(aq) + 3e^-]$

Reduction: $3[Cl_2(g) + 2e^- \longrightarrow 2Cl^-(aq)]$

Overall cell reaction:

$2Al(s) + 3Cl_2(s) \longrightarrow 2AlCl_3(s)$

$E°_{cell} = E°_{red} - E°_{oxid} = E°_{Cl^-} - E°_{Al^{3+}}$

$= +1.36\text{ V} - (-1.66\text{ V}) = +3.02\text{ V}$

84. The battery output would not be 12 V.

88. Connect the positive electrode of the power source to the gold object and the negative electrode to another metal. The gold object will then serve as the anode and the gold will be oxidized.

91. $T_1 = 30°C + 273 = 303\text{ K}$

$T_2 = 60°C + 273 = 333\text{ K}$

$$V_2 = \frac{V_1 \times T_2}{T_1} = \frac{425\text{ mL} \times 333\text{ }\cancel{K}}{303\text{ }\cancel{K}} = 467\text{ mL}$$

93. a. $250\text{ }\cancel{g} \times \dfrac{0.90\text{ g NaCl}}{100\text{ }\cancel{g}} = 2.3\text{ g NaCl}$

b. $500\text{ }\cancel{mL} \times \dfrac{1\text{ }\cancel{L}}{1000\text{ }\cancel{mL}} \times \dfrac{2\text{ }\cancel{mol\ KNO_3}}{1\text{ }\cancel{L}} \times \dfrac{101.1\text{ g KNO}_3}{1\text{ }\cancel{mol\ KNO_3}} = 101\text{ g KNO}_3$

97. $4.80\text{ }\cancel{g\ CH_4} \times \dfrac{1\text{ }\cancel{mol\ CH_4}}{16.0\text{ }\cancel{g\ CH_4}} \times \dfrac{-890\text{ kJ}}{1\text{ }\cancel{mol\ CH_4}} = -267\text{ kJ}$

98. a and c; $K_{eq} < 1$

100. a. $pH = -\log[H^+] = -\log(1.0 \times 10^{-8}M) = 8.00$

b. $pH = -\log[H^+] = -\log(0.000010M) = 5.00$

c. $[H^+] = \dfrac{K_w}{[OH^-]} = \dfrac{1.0 \times 10^{-14}}{1.0 \times 10^{-4}M} = 1.0 \times 10^{-10}M$

$pH = -\log[H^+] = -\log(1.0 \times 10^{-10}M) = 10.00$

d. $[H^+] = \dfrac{K_w}{[OH^-]} = \dfrac{1.0 \times 10^{-14}}{1.0 \times 10^{-9}M} = 1.0 \times 10^{-5}M$

$pH = -\log[H^+] = -\log(1.0 \times 10^{-5}M) = 5.00$

103. a. +6 b. −2 c. +4 d. +2 e. 0 f. +4

105. b; Ca is oxidized to Ca^{2+}, and Cl_2 is reduced to Cl^-.
c; Ca is oxidized to Ca^{2+}, and H^+ is reduced to H_2.

R112 Appendix E

Chapter 22

1.

```
    H  H  H  H  H
    |  |  |  |  |
 H—C—C—C—C—C—H
    |  |  |  |  |
    H  H  H  H  H

    H  H  H  H  H  H
    |  |  |  |  |  |
 H—C—C—C—C—C—C—H
    |  |  |  |  |  |
    H  H  H  H  H  H
```

2. 10 single bonds

3. **a.** 3-ethylhexane **b.** 2-methylbutane

4. 4-ethyl-2-methylheptane

5.

```
  CH3
  |
CH3CHCHCH2CH2CH3
     |
     CH3
```

6.

```
    CH3  CH3  CH3
    |    |    |
CH3—CH—CH—C—CH2—CH2—CH2—CH3
              |
              CH2
              |
              CH3
```

19. **a.**

```
       H
       |
CH3—(C)—CHO
       |
       Cl
```

b. no asymmetric carbon

20. **a.**

```
          CH3
          |
CH3CH2—(C)—Br
          |
          F
```

b. no asymmetric carbon

42. pentane: $CH_3CH_2CH_2CH_2CH_3$
hexane: $CH_3CH_2CH_2CH_2CH_2CH_3$

45. **a.** 2-methylbutane

b. 2,3-dimethylbutane

c. 3-methylhexane

47. **a.** unsaturated, because there is a double carbon-carbon bond

b. saturated, because the compound contains the maximum number of hydrogen atoms per carbon atom

52. No, only molecules with at least one asymmetric carbon can have enantiomers.

55. **a.**

```
  CH2CH3
  |
(benzene ring)
  |
  CH2CH3
```

c.

```
                      CH3
                     /
CH3—(benzene ring)
```

b.

```
     CH3
     |
CH3—CH—CH—CH2—CH3
          |
   (benzene ring)
```

58. The combustion of sulfur in coal produces the air pollutants SO_2 and SO_3.

60. **a.** Ethyne (acetylene) has one triple carbon-carbon bond and two single carbon-hydrogen bonds.

b. In methylbenzene, there are hybrid bonds within the ring. All other bonds are single bonds.

c. All the bonds in propane are single bonds.

62. propane, butane, pentane

66. The middle structure is most stable due to resonance within the ring.

68. No, the structures are identical; one has been flipped over.

71. **a.** 4 **b.** 2 **c.** 3 **d.** 1

72. The amount of heat per carbon is higher for methane (−890 kJ/mol of carbon burned) than for benzene (−3268 ÷ 6 carbons = −545 kJ/mol of carbon burned). Burning aromatic compounds produces more soot.

76. $CH_2{=}C{=}CH_2$

Selected Solutions **R113**

78. **a.** 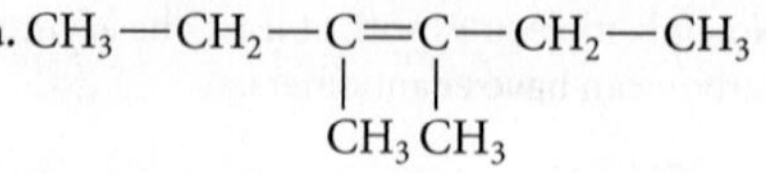 $CH_3—CH_2—C{=}C—CH_2—CH_3$ (with CH_3 on each of the two double-bonded carbons)

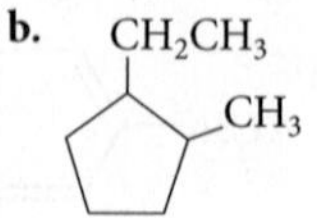

$CH_3\ CH_3$

b. CH_2CH_3, CH_3 (on a cyclopentane ring)

c.

CH_3—CH_2—CH_2—CH_2—C(—CH_2—CH_2—CH_3)(—CH_2—CH_2—CH_3)—CH_2—CH_2—CH_2—CH_3

(as drawn: CH_3, CH_2, CH_2 above the central C; CH_3, CH_2 above the left CH_2; CH_2—CH_2—C—CH_2—CH_2; CH_2, CH_2, CH_3 below the central C; CH_2, CH_2, CH_3 below the right CH_2)

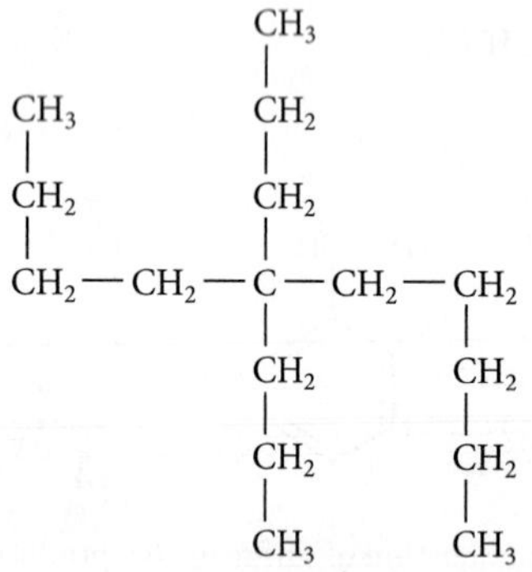

82. **a.**

$$CH_3CH_2CH_2CH_2CH_3 \rightleftharpoons CH_3\overset{\displaystyle CH_3}{\overset{|}{C}}HCH_2CH_3$$

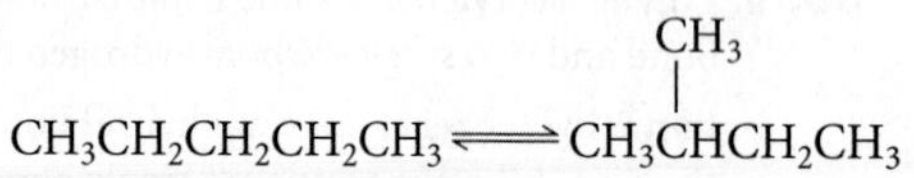

b. constitutional

c. 2-methylbutane

86. **a.** $C_6 = 5, C_7 = 9, C_8 = 18, C_9 = 35, C_{10} = 75$

b. As the size of the alkane molecule gets larger, the number of different ways that the carbon atoms can be bonded together (to form constitutional isomers) increases dramatically.

90. No, enantiomers have identical physical properties.

92. **a.** $6.20 \times 10^{-1}\ \cancel{\text{mol Cl}_2} \times \dfrac{22.4\ \text{L}}{1\ \cancel{\text{mol Cl}_2}} = 13.9\ \text{L}$

b. $V_2 = \dfrac{P_1 \times V_1}{P_2} = \dfrac{0.5\ \cancel{\text{kPa}} \times 6\ \text{L}}{3\ \cancel{\text{kPa}}} = 1\ \text{L}$

c. $P_X + P_Y + P_Z = P_{total}$

$P_X + 30\ \text{kPa} = 50\ \text{kPa}$

$P_X + 30\ \text{kPa} - 30\ \text{kPa} = 50\text{kPa} - 30\text{kPa}$

$P_X = 20\ \text{kPa}$

93. $750\ \cancel{\text{mL KNO}_3} \times \dfrac{1\ \cancel{\text{L KNO}_3}}{1000\ \cancel{\text{mL KNO}_3}} \times \dfrac{1.50\ \text{mol KNO}_3}{1\ \cancel{\text{L KNO}_3}}$

$= 1.13\ \text{mol KNO}_3$;

$1.13\ \cancel{\text{mol KNO}_3} \times \dfrac{101.1\ \text{g KNO}_3}{1.000\ \cancel{\text{mol KNO}_3}} = 114\ \text{g KNO}_3$

98. **a.** favors reactants **b.** favors products

100. **a.** $[H^+] = \dfrac{1.0 \times 10^{-14}}{[OH^-]} = \dfrac{1.0 \times 10^{-14}}{1.0 \times 10^{-4}} = 1.0 \times 10^{-10}M$

$pH = -\log[H^+] = -\log(1.0 \times 10^{-10}) = 10.00$

b. $[H^+] = \dfrac{1.0 \times 10^{-14}}{[OH^-]} = \dfrac{1.0 \times 10^{-14}}{3.9 \times 10^{-7}} = 2.6 \times 10^{-8}M$

$pH = -\log[H^+] = -\log(2.6 \times 10^{-8}) = 7.59$

c. $[H^+] = \dfrac{1.0 \times 10^{-14}}{[OH^-]} = \dfrac{1.0 \times 10^{-14}}{0.010} = 1.0 \times 10^{-12}M$

$pH = -\log[H^+] = -\log(1.0 \times 10^{-12}) = 12.00$

d. $[H^+] = \dfrac{1.0 \times 10^{-14}}{[OH^-]} = \dfrac{1.0 \times 10^{-14}}{5.0 \times 10^{-3}} = 2.0 \times 10^{-12}M$

$pH = -\log[H^+] = -\log(2.0 \times 10^{-12}) = 11.70$

102. **a.** H_3PO_4 **c.** H_2CO_3

b. CsOH **d.** $Be(OH)_2$

104. **a.** $2 + C + 3(-2) = 0$

$-4 + C = 0$

$C = +4$

Ca, +2; C, +4 ; O, −2

b. Cl, 0

c. $1 + I + 3(-2) = 0$

$I - 5 = 0$

$I = +5$

Li, +1; I, +5; O, −2

d. $2(+1) + S + 3(-2) = 0$

$S - 4 = 0$

$S = +4$

Na, +1; S, +4; O, −2

110. The reaction is nonspontaneous.

Chapter 23

33. **a.** 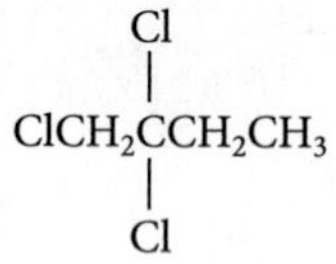 $ClCH_2CCl_2CH_2CH_3$ (drawn as $ClCH_2CCH_2CH_3$ with Cl above and below the second carbon)

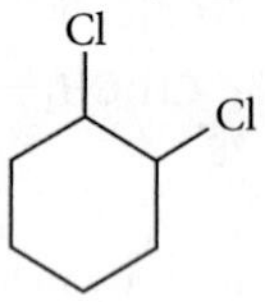

b. 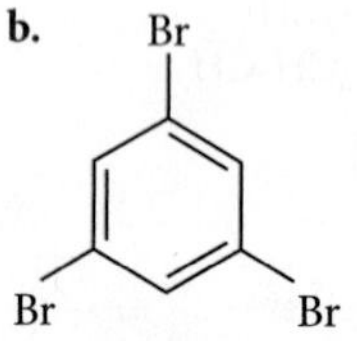 Br, Br, Br (1,3,5-tribromobenzene ring)

c. Cl, Cl (on adjacent carbons of a cyclohexane ring)

37. **a.** 2-propanol

b. 1,2-propanediol

c. 2-methyl-2-propanol

R114 Appendix E

38. a. 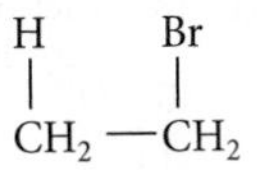
```
H       Br
|       |
CH2 — CH2
```
bromoethane

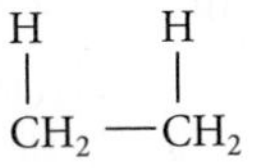

b. 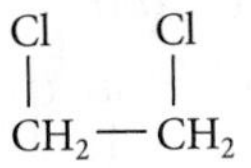
```
Cl      Cl
|       |
CH2 — CH2
```
1,2-dichloroethane

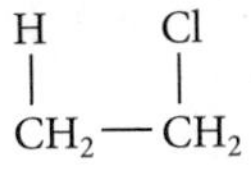

c. 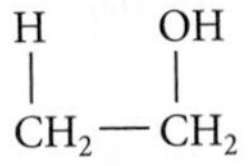
```
H       OH
|       |
CH2 — CH2
```
ethanol

d.
```
H       H
|       |
CH2 — CH2
```
ethane

e.
```
H       Cl
|       |
CH2 — CH2
```
chloroethane

42. a. propanone
b. ethanal
c. 3-methylbutanal
d. 3-hexanone
e. 2-phenylethanal
f. pentanoic acid

45. a. 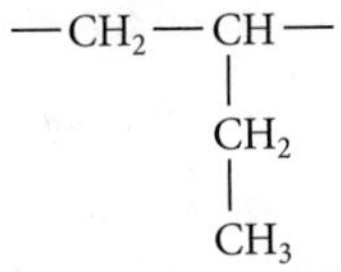
```
—CH2—CH—
      |
      CH2
      |
      CH3
```
b. 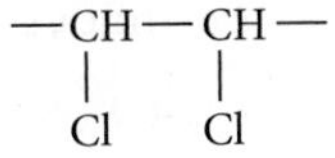
```
—CH—CH—
  |   |
  Cl  Cl
```

50. a. phenol
b. ether
c. alcohol
d. phenol
e. alcohol
f. ether

53. a. carboxyl group, ethanoic acid (acetic acid)
b. ether, diethyl ether (ethyl ether)
c. ketone (carbonyl group), propanone (acetone)
d. alcohol (hydroxy group), ethanol (ethyl alcohol)

57. The chemical properties (and toxicity) of organic compounds are determined by the compound as a whole. As a substituent in a molecule, a phenyl group ring does not have the same properties as benzene.

61. Ethanoic acid (two carbons long) is more soluble in water than decanoic acid (ten carbons long).

63. cadaverine, $H_2N(CH_2)_5NH_2$;
putrescine, $H_2N(CH_2)_4NH_2$
Both compounds are amines.

66. 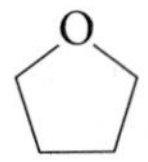 (five-membered ring containing O)

69. Cholesterol is an alcohol with a hydroxy group on a cycloalkane. It has four nonaromatic rings. It has a double bond in one of its rings, as well as a large alkyl group, making the molecule nonpolar.

71. Waving lotion reduces —S—S— bonds to —SH bonds. Hair can be placed in curlers to form the hair in the desired shape. The neutralizing agent is an oxidizing agent that reforms —S—S— bonds, locking the hair into its curly shape. Similar steps could be used to straighten curly hair.

76. The two molecules are similar in size. Diethylene glycol has four carbons, and glycerol has three carbons. Diethylene glycol is an ether with two hydroxy groups. Glycerol has three hydroxy groups.

diethylene glycol
```
CH2—CH2—O—CH2—CH2
|                |
OH               OH
```

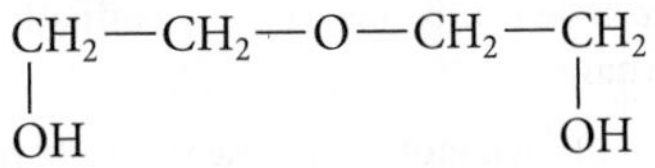

glycerol
```
CH2—CH—CH2
|    |    |
OH   OH   OH
```

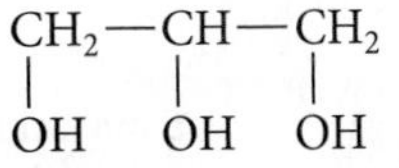

78. b, 3

82. Anhydrous $Na_2CO_3(s)$ is the better value because the decahydrate is 63.0% water.

86. $\Delta T = 180°C - 22°C = 158°C$

$q = C_{Al} \times m \times \Delta T$

$$q = \frac{0.90\,\text{J} \times \frac{1\text{ kJ}}{10^3\text{ J}} \times 500\,\text{g} \times 158°\text{C}}{(\text{g}\cdot°\text{C})} = 71\text{ kJ}$$

88. a, c, d, b

92. Reduction always occurs at the cathode. In an electrolytic cell, the cathode is the negative electrode.

94. coal

Chapter 24

42. Photosynthetic organisms use the sun's energy to synthesize carbon compounds, such as glucose, from CO_2 and H_2O.

47. Glucose is an aldehyde; fructose is a ketone.

53. Peptide chains fold into helixes or into pleated sheets in which peptide chains lie side by side.

64. phosphate group, sugar unit, nitrogen base

69. A substitution of one base for another may have no effect because the amino acid specified by the DNA code is not changed. Or a substitution of one base for another can result in a gene mutation in which the amino specified by the DNA code is changed.

70. Each individual's DNA is unique.

Selected Solutions **R115**

74. In catabolism, biological molecules are broken down and energy is released. In anabolism, energy and the products of catabolism are used to make biological molecules.

78.
```
                  O
                  ‖
CH3(CH2)14—C—O⁻Na⁺
```
(sodium palmitate)

82. C-G-x-C-C-x-T-C-A(or G)
Answers vary because there are multiple codes for the same amino acid.

86. No; three code words specify the termination of a peptide chain.

88. Oxidation of each mol of glucose yields 38 mol ATP, so:

$$\frac{38\ \cancel{\text{mol ATP}} \times 30.5\ \cancel{\text{kJ}}/\cancel{\text{mol ATP}}}{2.82 \times 10^3\ \cancel{\text{kJ}}} \times 100\% = 41.1\%$$

89. $8400\ \cancel{\text{kJ}} \times \frac{1\ \text{mol ATP}}{30.5\ \cancel{\text{kJ}}} = 275\ \text{mol ATP}$

96. A monosaccharide (such as glucose and fructose) consists of a single simple sugar unit. A disaccharide (such as sucrose) consists of two linked monosaccharide units. A polysaccharide (such as starch or glycogen) consists of many linked monosaccharide units.

98. Base-pairing describes the hydrogen bonding that occurs between thymine (T) and adenine (A) and between cytosine (C) and guanine (G) in DNA. Base-pairing helps hold together the two DNA strands in a double helix.

104.
```
           CH2OH
            |
     H      C—O        H
      \   / |    \    /
       C    H      C
      / \ OH H    /  |
    HO    C—C        |
          |  |       |
          H  OH      |     + H2O
                     O
                     |
  HOCH2       O      |
        \   /   \    |
         C        C
        / \H  HO / \
       H   |  |     CH2OH
           C—C
           |  |
           OH H
```

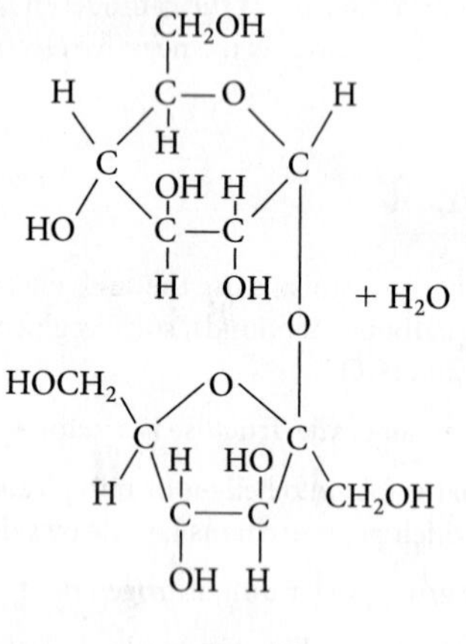

107. An amide group is formed by the reaction.
```
     R  O              R  O
     |  ‖              |  ‖
H2N—C—C—OH + H—N—C—C—OH ⟶
     |              |  |
     H              H  H

       R  O     R  O
       |  ‖     |  ‖
  H2N—C—C—N—C—C—OH + H2O
       |     |  |
       H     H  H
```

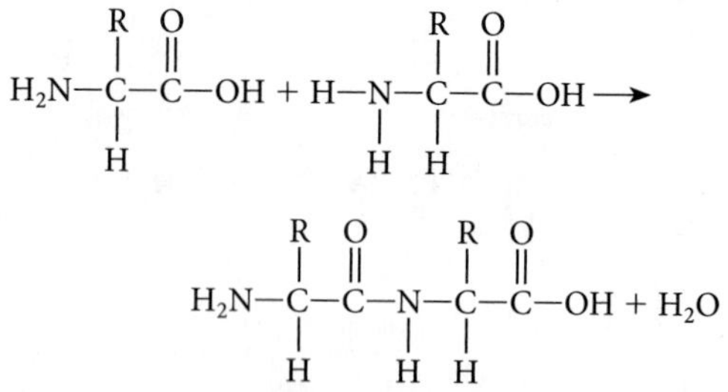

112. A-C-C-T-A-C-T-A-C
No; Leu has 6 possible code words.

119. **a.** Each formula unit of NaCl disassociates into 2 particles, so $0.507m \times 2 = 1.014m$.

$$\Delta T_b = K_b \times m = \frac{0.512°\text{C}}{\cancel{m}} \times 1.014\cancel{m} = 0.519°\text{C}$$

b. Each formula unit of NH_4Cl disassociates into 2 particles, so $0.204m \times 2 = 0.408m$.

$$\Delta T_b = K_b \times m = \frac{0.512°\text{C}}{\cancel{m}} \times 0.408\cancel{m} = 0.209°\text{C}$$

c. Each formula unit of $CaCl_2$ disassociates into 3 particles, so $0.155m \times 3 = 0.465m$.

$$\Delta T_b = K_b \times m = \frac{0.512°\text{C}}{\cancel{m}} \times 0.465\cancel{m} = 0.238°\text{C}$$

d. Each formula unit of $NaHSO_4$ disassociates into 2 particles, so $0.222m \times 2 = 0.444m$.

$$\Delta T_b = K_b \times m = \frac{0.512°\text{C}}{\cancel{m}} \times 0.444\cancel{m} = 0.227°\text{C}$$

120. $\Delta H = 0.265\ \cancel{\text{mol NaHCO}_3} \times \frac{129\ \text{kJ}}{2\ \cancel{\text{mol NaHCO}_3}}$

$= 17.1$ kJ heat absorbed

128. **a.** $CH_3—CH_2—CH_2—CH_2—CH_2—CH_2—CH_3$

b.
```
      CH3
      |
CH3—CH—CH=CH—CH2—CH3
```

c.
```
   (benzene ring)
        |
CH3—CH—CH2—CH3
```

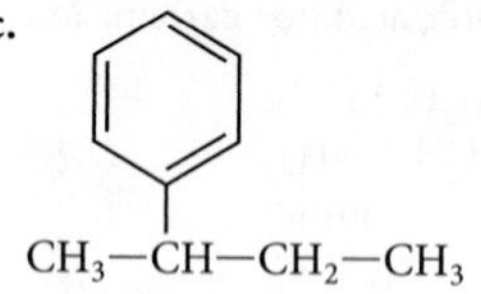

d.
```
  CH2CH3
  |
(benzene ring)
        \
         CH2CH3
```

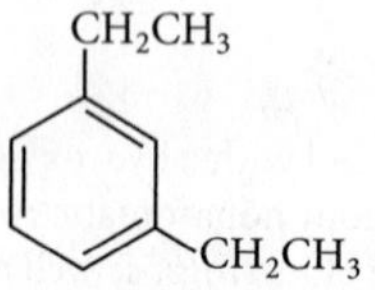

130. **a.** cyclopentane

b. 2-methyl-2-propanol

c. 3-pentanone

R116 Appendix E

Chapter 25

9. $10.4\ \cancel{h} \div 2.6\ \cancel{h}/\text{half-life} = 4$ half-lives

$A = A_0\left(\frac{1}{2}\right)^4 = 1.0\text{ mg} \times \frac{1}{16} = 0.063\text{ mg Mn-56}$

10. $48.2\ \cancel{\text{days}} \div 24.1\ \cancel{\text{days}}/\text{half-life} = 2$ half-lives

$(\text{Th-234 atoms}) \times \left(\frac{1}{2}\right)^2 = (\text{Th-234 atoms}) \times \frac{1}{4}$

No, $\frac{1}{4}$ of the Th-234 sample will remain.

36. ${}^{210}_{82}\text{Pb} \longrightarrow {}^{210}_{83}\text{Bi} + {}^{0}_{-1}\text{e}$

38. **a.** ${}^{238}_{92}\text{U} \longrightarrow {}^{234}_{90}\text{Th} + {}^{4}_{2}\text{He}$; thorium-234

b. ${}^{230}_{90}\text{Th} \longrightarrow {}^{226}_{88}\text{Ra} + {}^{4}_{2}\text{He}$; radium-226

c. ${}^{235}_{92}\text{U} \longrightarrow {}^{231}_{90}\text{Th} + {}^{4}_{2}\text{He}$; thorium-231

d. ${}^{222}_{86}\text{Rn} \longrightarrow {}^{218}_{84}\text{Po} + {}^{4}_{2}\text{He}$; polonium-218

40. **a.** mass number is unchanged; atomic number increases by 1

b. mass number decreases by 4; atomic number decreases by 2

c. Mass number and atomic number are both unchanged.

42. The atom undergoes radioactive decay.

47. $n = 40\ \cancel{\text{days}} \div 8\ \cancel{\text{days}}/\text{half-life} = 5$ half-lives

$A = A_0\left(\frac{1}{2}\right)^n = (20\text{ mg})\left(\frac{1}{2}\right)^5 = \frac{20}{32} = 0.625\text{ mg}$

0.625 mg of I-131 remains.

57. **a.** ${}^{30}_{15}\text{P} + {}^{0}_{-1}\text{e} \longrightarrow {}^{30}_{14}\text{Si}$

b. ${}^{13}_{6}\text{C} + {}^{1}_{0}\text{n} \longrightarrow {}^{14}_{6}\text{C}$

c. ${}^{131}_{53}\text{I} \longrightarrow {}^{131}_{54}\text{Xe} + {}^{0}_{-1}\text{e}$

65. $A = A_0\left(\frac{1}{2}\right)^n = (3.2 \times 10^7\text{ atoms})\left(\frac{1}{2}\right)^5$

$= \frac{3.2 \times 10^7\text{ atoms}}{32} = 1 \times 10^6\text{ atoms}$

67. **a.** ${}^{3}_{1}\text{H} \longrightarrow {}^{3}_{2}\text{H} + {}^{0}_{-1}\text{e}$ **c.** ${}^{131}_{53}\text{I} \longrightarrow {}^{131}_{54}\text{Xe} + {}^{0}_{-1}\text{e}$

b. ${}^{28}_{12}\text{Mg} \longrightarrow {}^{28}_{13}\text{Al} + {}^{0}_{-1}\text{e}$ **d.** ${}^{75}_{34}\text{Se} \longrightarrow {}^{75}_{35}\text{Br} + {}^{0}_{-1}\text{e}$

71. **a.** ${}^{0}_{-1}\text{e}$ **b.** ${}^{238}_{92}\text{U}$ **c.** ${}^{68}_{32}\text{Ge}$ **d.** ${}^{0}_{+1}\text{e}$

72. **a.** platinum **d.** titanium **g.** vanadium

b. thorium **e.** xenon **h.** palladium

c. francium **f.** californium

Thorium (b), francium (c), and californium (f) have no stable isotopes.

76. Carbon-14 has decayed $\frac{4\ \cancel{\text{counts/min}}}{16\ \cancel{\text{counts/min}}} = \frac{1}{4}$.

Since $\frac{1}{4} = \left(\frac{1}{2}\right)^2$, the artifact is 2 half-lives old.

One half-life is 5.73×10^3 years.

2 half-lives $= (5.73 \times 10^3) \times 2 = 11{,}460$ years.

82. ${}^{25}_{12}\text{Mg}$

84. $25\% = \frac{25}{100} = \frac{1}{4}$.

$\left(\frac{1}{2}\right)^n = \frac{1}{4}$, so $n = 2$ half-lives.

12.3 years × 2 half-lives = 24.6 years.

94. $\text{volume} = \frac{\text{mass}}{\text{density}} = \frac{8.0\ \cancel{\text{kg}} \times \frac{1000\ \cancel{\text{g}}}{1\ \cancel{\text{kg}}}}{\frac{19\ \cancel{\text{g}}}{1\text{ cm}^3}} = 4.2 \times 10^2\text{ cm}^3$

105. $10.00\ \cancel{\text{g Mg}} \times \frac{1\ \cancel{\text{mol Mg}}}{24.3\ \cancel{\text{g Mg}}} \times \frac{1\text{ mol H}_2}{1\ \cancel{\text{mol Mg}}} = 0.412\text{ mol H}_2$

$0.412\ \cancel{\text{mol H}_2} \times \frac{22.4\ \cancel{\text{L H}_2}}{1\ \cancel{\text{mol H}_2}} \times \frac{1000\ \cancel{\text{mL H}_2}}{1\ \cancel{\text{L H}_2}} \times \frac{1\text{ cm}^3\text{ H}_2}{1\ \cancel{\text{mL H}_2}}$

$= 9.22 \times 10^3\text{ cm}^3\text{ H}_2$

107. $\text{volume} = \frac{\text{moles}}{\text{molarity}} = \frac{0.0020\ \cancel{\text{mol Na}_2\text{SO}_4}}{\frac{0.30\ \cancel{\text{mol Na}_2\text{SO}_4}}{1\ \cancel{\text{L}}} \times \frac{1\ \cancel{\text{L}}}{1000\text{ mL}}}$

$= 6.7\text{ mL}$

Selected Solutions **R117**

Glossary

A

absolute zero: the zero point on the Kelvin temperature scale, equivalent to −273.15°C *(79)*
cero absoluto: punto cero en la escala de temperatura Kelvin; equivale a −273.15°C

accepted value: a quantity used by general agreement of the scientific community *(65)*
valor aceptado: cantidad que se usa por acuerdo general de la comunidad científica

accuracy: the closeness of a measurement to the true value of what is being measured *(64)*
exactitud: qué tan cerca está una medición del valor real de lo que se mide

acid: a compound that produces hydrogen ions in solution; see also hydrogen-ion donor, Lewis acid *(285)*
ácido: compuesto que, en solución, produce iones hidrógeno; ver también donante iones hidrógeno, ácido de Lewis

acid dissociation constant (K_a): the ratio of the concentration of the dissociated form of an acid to the undissociated form; stronger acids have larger K_a values than weaker acids *(665)*
constante de disociación ácida (K_a): razón de la concentración de la forma disociada de un ácido a la concentración de la forma no disociada; los ácidos fuertes tienen valores K_a más altos que los ácidos débiles

acidic solution: any solution in which the hydrogen-ion concentration is greater than the hydroxide-ion concentration *(654)*
solución ácida: cualquier solución en la que la concentración de iones hidrógeno es mayor que la de iones hidróxido

activated complex: an unstable arrangement of atoms that exists momentarily at the peak of the activation-energy barrier; an intermediate or transitional structure formed during the course of a reaction *(596)*
complejo activado: acomodo inestable de átomos que existe momentáneamente en el punto más alto de la barrera de energía de activación; estructura intermedia o de transición que se forma en el curso de una reacción

activation energy: the minimum energy colliding particles must have in order to react *(596)*
energía de activación: energía mínima que deben tener las partículas para que, al chocar, reaccionen

active site: a groove or pocket in an enzyme molecule into which the substrate (reactant molecule) fits; where the substrate is converted to products *(847)*
sitio activo: hendidura o bolsa en una molécula de enzima, en la que embona el sustrato (molécula que reacciona); donde el sustrato se convierte en productos

activity series: a list of elements in order of decreasing activity; the activity series of halogens is Fl, Cl, Br, I *(360)*
serie de actividad: lista de elementos en orden de actividad decreciente; la serie de actividad de los halógenos es F, Cl, Br, I

actual yield: the amount of product that forms when a reaction is carried out in the laboratory *(405)*
rendimiento real: cantidad de producto que se forma cuando se lleva a cabo una reacción en el laboratorio

addition reaction: a reaction in which a substance is added at the double bond of an alkene or at the triple bond of an alkyne *(807)*
reacción de adición: reacción en la que una sustancia se añade al doble enlace de un alqueno o al triple enlace de un alquino

adenosine triphosphate (ATP): a molecule that transmits the energy needed by cells of all living things *(862)*
trifosfato de adenosina (ATP): molécula que transmite la energía que necesitan las células de todos los seres vivos

alcohol: an organic compound having an —OH (hydroxy) group; the general structure is R—OH *(804)*
alcohol: compuesto orgánico que posee un grupo —OH (hidroxilo); su estructura general es R—OH

aldehyde: an organic compound in which the carbon of the carbonyl group is joined to at least one hydrogen; the general formula is RCHO *(812)*
aldehído: compuesto orgánico en el que el carbono del grupo carbonilo está unido a por lo menos un hidrógeno; su fórmula general es RCHO

aliphatic hydrocarbon: any straight-chain or branched-chain alkane, alkene, or alkyne *(780)*
hidrocarburo alifático: cualquier alcano, alqueno o alquino de cadena lineal o cadena ramificada

alkali metal: any metal in Group 1A of the periodic table *(167)*
metal alcalino: cualquier metal del grupo 1A de la tabla periódica

alkaline earth metal: any metal in Group 2A of the periodic table *(167)*
metal alcalinotérreo: cualquier metal del grupo 2A de la tabla periódica

alkane: a hydrocarbon containing only single covalent bonds; alkanes are saturated hydrocarbons *(764)*
alcano: hidrocarburo que sólo contiene enlaces covalentes sencillos; los alcanos son hidrocarburos saturados

R118 Glossary

alkene: a hydrocarbon containing one or more carbon–carbon double bonds; alkenes are unsaturated hydrocarbons *(772)*

alqueno: hidrocarburo que contiene uno o más enlaces dobles carbono–carbono; los alquenos son hidrocarburos insaturados

alkyl group: a hydrocarbon substituent; the methyl group ($—CH_3$) is an alkyl group (768)

grupo alquilo: un hidrocarburo sustituto; el grupo metilo ($—CH_3$) es un grupo alquilo

alkyl halide: a halocarbon in which one or more halogen atoms are attached to the carbon atoms of an aliphatic chain *(800)*

haluro de alquilo: compuesto halocarbonado en el que uno o más átomos de halógeno están unidos a los átomos de carbono de una cadena alifática

alkyne: a hydrocarbon containing a carbon–carbon triple bond; alkynes are unsaturated hydrocarbons *(773)*

alquino: hidrocarburo que contiene un triple enlace carbono–carbono; los alquinos son hidrocarburos insaturados

allotrope: one of two or more different molecular forms of an element in the same physical state; oxygen (O_2) and ozone (O_3) are allotropes of the element oxygen *(434)*

alótropo: una de dos o más formas moleculares distintas de un elemento en el mismo estado físico; el oxígeno (O_2) y el ozono (O_3) son alótropos del elemento oxígeno

alloy: a mixture composed of two or more elements, at least one of which is a metal *(211)*

aleación: mezcla formada por dos o más elementos, donde al menos uno de ellos es un metal

alpha particle: a positively charged particle emitted from certain radioactive nuclei; it consists of two protons and two neutrons and is identical to the nucleus of a helium atom *(877)*

partícula alfa: partícula con carga positiva emitida por ciertos núcleos radiactivos; consta de dos protones y dos neutrones, y es idéntica al núcleo de un átomo de helio

amine: an organic compound in which nitrogen is bonded to a carbon group *(811)*

amina: compuesto orgánico en el cual el nitrógeno se enlaza a un grupo de carbonos

amino acid: an organic compound having amino ($—NH_2$) and carboxyl ($—COOH$) groups in the same molecule; proteins are made from the 20 naturally occurring amino acids *(844)*

aminoácido: compuesto orgánico que posee grupos amino ($—NH_2$) y carboxilo ($—COOH$) en la misma molécula; las proteínas se forman a partir de los 20 aminoácidos naturales

amorphous solid: describes a solid that lacks an ordered internal structure; denotes a random arrangement of atoms *(434)*

sólido amorfo: describe un sólido que carece de una estructura interna ordenada; denota un acomodo aleatorio de átomos

amphoteric: a substance that can act as both an acid and a base *(651)*

anfótero: una sustancia que puede actuar como ácido y también como base

amplitude: the height of a wave's crest *(138)*

amplitud: altura de la cresta de una onda

anabolism: synthesis processes in the metabolism of cells; these processes usually require the expenditure of energy *(864)*

anabolismo: procesos de síntesis dentro del metabolismo de las células; por lo regular, esos procesos requieren gasto de energía

analytical chemistry: the area of chemistry that focuses on the composition of matter *(3)*

química analítica: rama de la química que estudia la composición de la materia

anhydrous: describes a substance that does not contain water *(498)*

anhidro: se refiere a una sustancia que no contiene agua

anion: any atom or group of atoms with a negative charge *(176)*

anión: cualquier átomo o grupo de átomos que posee carga negativa

anode: the electrode at which oxidation occurs *(730)*

ánodo: electrodo en el que hay oxidación

applied chemistry: research that is directed toward a practical goal or application *(3)*

química aplicada: investigaciones que tienen una meta o aplicación práctica

aqueous solution: water that contains dissolved substances *(494)*

solución acuosa: agua que contiene sustancias disueltas

aromatic compound: an organic compound that contains a benzene ring or other ring in which the bonding is like that of benzene; aromatic compounds are also known as arenes *(780)*

compuesto aromático: compuesto orgánico que contiene un anillo bencénico u otro anillo con enlaces similares a los del benceno; los compuestos aromáticos también se conocen como arenos

aryl halide: a halocarbon in which one or more halogens are attached to the carbon atoms of an arene ring *(800)*

haluro de arilo: compuesto en el que uno o más átomos de halógeno están unidos a átomos de carbono de un anillo de areno

Glossary **R119**

asymmetric carbon: a carbon atom that has four different atoms or groups attached *(776)*
carbono asimétrico: átomo de carbono unido a cuatro átomos o grupos distintos

atmospheric pressure: the pressure exerted by atoms and molecules in the atmosphere surrounding Earth, resulting from collisions of these particles with objects *(421)*
presión atmosférica: presión ejercida por átomos y moléculas de la atmósfera que rodea a la Tierra y que resulta de los choques de dichas partículas con los objetos

atom: the smallest particle of an element that retains its identity in a chemical reaction *(102)*
átomo: partícula más pequeña de un elemento que conserva su identidad en una reacción química

atomic emission spectrum: the pattern formed when light passes through a prism or diffraction grating to separate it into the different frequencies of light it contains *(140)*
espectro de emisión atómica: patrón que se forma cuando la luz atraviesa un prisma o una rejilla de difracción que la separa en las dife-rentes frecuencias de luz que contiene

atomic mass: the weighted average of the masses of the isotopes of an element *(117)*
masa atómica: promedio ponderado de las masas de los isótopos de un elemento

atomic mass unit (amu): a unit of mass equal to one-twelfth the mass of a carbon-12 atom *(116)*
unidad de masa atómica (uma): unidad de masa igual a un doceavo de la masa de un átomo de carbono 12

atomic number: the number of protons in the nucleus of an atom of an element *(112)*
número atómico: número de protones que hay en el núcleo del átomo de un elemento

atomic orbital: a mathematical expression describing the probability of finding an electron at various locations; usually represented by the region of space around the nucleus where there is a high probability of finding an electron *(131)*
orbital atómico: expresión matemática que describe la probabilidad de hallar un electrón en diversos lugares; se suele representar como la región del espacio en torno al núcleo donde hay una probabilidad elevada de hallar un electrón

atomic radius: one-half the distance between the nuclei of two atoms of the same element when the atoms are joined *(174)*
radio atómico: mitad de la distancia entre los núcleos de dos átomos del mismo elemento cuando dichos átomos están unidos

aufbau principle: the rule that electrons occupy the orbitals of lowest energy first *(134)*
principio de aufbau: regla según la cual los electrones primero ocupan los orbitales de energía más baja

Avogadro's hypothesis: equal volumes of gases at the same temperature and pressure contain equal numbers of particles *(320)*
hipótesis de Avogadro: volúmenes iguales de gases a la misma temperatura y presión contienen el mismo número de partículas

Avogadro's number: the number of representative particles contained in one mole of a substance; equal to 6.02×10^{23} particles *(308)*
número de Avogadro: número de partículas representativas contenidas en un mol de una sustancia; es igual a 6.02×10^{23} partículas

B

balanced equation: a chemical equation in which mass is conserved; each side of the equation has the same number of atoms of each element *(350)*
ecuación balanceada: ecuación química en la que se conserva la masa; cada lado de la ecuación tiene el mismo número de átomos de cada elemento

band of stability: the location of stable nuclei on a neutron-vs.-proton plot *(880)*
banda de estabilidad: región ocupada por los núcleos estables en un diagrama neutrones-protones

barometer: an instrument used to measure atmospheric pressure *(421)*
barómetro: instrumento que sirve para medir la presión atmosférica

base: a compound that produces hydroxide ions in solution; see also hydrogen-ion acceptor, Lewis base *(287)*
base: compuesto que, en solución, produce iones hidróxido, ver también receptor de iones hidrógeno, base de Lewis

base dissociation constant (K_b): the ratio of the concentration of the conjugate acid times the concentration of the hydroxide ion to the concentration of the base *(668)*
constante de disociación básica (K_b): razón de la concentración del ácido combinado multiplicada por la concentración del ion hidróxido, a la concentración de la base

basic solution: any solution in which the hydroxide-ion concentration is greater than the hydrogen-ion concentration *(654)*
solución básica: cualquier solución en la que la concentración de ion hidróxido es mayor que la concentración de ion hidrógeno

battery: a group of voltaic cells that are connected to one another *(733)*
batería: grupo de celdas voltaicas conectadas entre sí

beta particle: an electron resulting from the breaking apart of neutrons in an atom *(878)*
partícula beta: electrón que se produce al descomponerse los neutrones de un átomo

R120 Glossary

binary compound: a compound composed of two elements; NaCl and Al_2O_3 are binary compounds *(272)*
compuesto binario: compuesto integrado por dos elementos; NaCl y Al_2O_3 son compuestos binarios

biochemistry: the area of chemistry that focuses on processes that take place in organisms *(3)*
bioquímica: rama de la química que se concentra en los procesos que se dan en los organismos

boiling point (bp): the temperature at which the vapor pressure of a liquid is just equal to the external pressure on the liquid *(428)*
punto de ebullición (p. eb.): temperatura en la que la presión de vapor de un líquido es apenas igual a la presión externa sobre el líquido

boiling-point elevation: the difference in temperature between the boiling point of a solution and the boiling point of the pure solvent *(537)*
incremento del punto de ebullición: diferencia de temperatura entre el punto de ebullición de una solución y el punto de ebullición del disolvente puro

bond dissociation energy: the energy required to break the bond between two covalently bonded atoms; this value is usually expressed in kJ per mol of substance *(236)*
energía de disociación de enlaces: energía requerida para romper el enlace entre dos átomos unidos de forma covalente; este valor suele expresarse en kJ por mol de sustancia

bonding orbital: a molecular orbital that can be occupied by two electrons of a covalent bond *(240)*
orbital de enlace: orbital molecular que puede ser ocupado por los dos electrones de un enlace covalente

Boyle's law: for a given mass of gas at constant temperature, the volume of the gas varies inversely with pressure *(456)*
ley de Boyle: para una masa dada de gas a temperatura constante, el volumen del gas varía en proporción inversa con la presión

branched-chain alkane: an alkane with one or more alkyl groups attached to the parent structure *(768)*
alcano de cadena ramificada: alcano con uno o más grupos alquilo unidos a la estructura madre

Brownian motion: the chaotic movement of colloidal particles, caused by collision with particles of the solvent in which they are dispersed *(506)*
movimiento browniano: movimiento caótico de partículas coloidales, debido a los choques con las partículas del disolvente en el que están dispersas

buffer: a solution in which the pH remains relatively constant when small amounts of acid or base are added; a buffer can be either a solution of a weak acid and the salt of a weak acid or a solution of a weak base with the salt of a weak base *(678)*
solución amortiguadora: solución cuyo pH permanece relativamente constante si se le añaden pequeñas cantidades de ácido o base; una solución amortiguadora puede ser una solución de un ácido débil y la sal de un ácido débil o una solución de una base débil y la sal de una base débil

buffer capacity: a measure of the amount of acid or base that may be added to a buffer solution before a significant change in pH occurs *(679)*
capacidad amortiguadora: medida de la cantidad de ácido o base que se puede añadir a una solución amortiguadora sin que haya un cambio importante del pH

C

calorie (cal): the quantity of heat needed to raise the temperature of 1 g of pure water 1°C *(77)*
caloría (cal): cantidad de calor necesaria para elevar 1°C la temperatura de 1 g de agua pura

calorimeter: an insulated device used to measure the absorption or release of heat in chemical or physical processes *(562)*
calorímetro: aparato con material aislante que sirve para medir la absorción o desprendimiento de calor durante procesos químicos o físicos

calorimetry: the precise measurement of heat flow out of a system for chemical and physical processes *(562)*
calorimetría: medición precisa del cambio del calor durante procesos químicos y físicos

carbohydrate: the name given to monomers and polymers of aldehydes and ketones that have numerous hydroxyl groups; sugars and starches are carbohydrates *(841)*
carbohidrato: nombre dado a monómeros y polímeros de aldehídos y cetonas que tienen muchos grupos hidroxilo; los azúcares y almidones son carbohidratos

carbonyl group: a functional group having a carbon atom and an oxygen atom joined by a double bond; it is found in aldehydes, ketones, esters, and amides *(812)*
grupo carbonilo: grupo funcional que consiste en un átomo de carbono y uno de oxígeno unidos por un doble enlace; se le encuentra en aldehídos, cetonas, ésteres y amidas

carboxyl group: a functional group consisting of a carbonyl group attached to a hydroxyl group; it is found in carboxylic acids *(815)*
grupo carboxilo: grupo funcional que consiste en un grupo carbonilo unido a un grupo hidroxilo; se le encuentra en los ácidos carboxílicos

Glossary **R121**

carboxylic acid: an organic acid containing a carboxyl group; the general formula is RCOOH *(815)*
ácido carboxílico: ácido orgánico que contiene un grupo carboxilo; su fórmula general es RCOOH

catabolism: the reactions in living cells in which substances are broken down and energy is produced *(863)*
catabolismo: reacción, dentro de las células vivas, por la que diversas sustancias se descomponen y producen energía

catalyst: a substance that increases the rate of reaction by lowering the activation-energy barrier; the catalyst is not used up in the reaction *(348)*
catalizador: sustancia que aumenta la velocidad de reacción disminuyendo la barrera de energía de activación; el catalizador no se consume en la reacción

cathode: the electrode at which reduction occurs *(730)*
cátodo: electrodo en el que hay reducción

cathode ray: a stream of electrons produced at the negative electrode (cathode) of a tube containing a gas at low pressure *(105)*
rayo catódico: haz de electrones producido en el electrodo negativo (cátodo) de un tubo que contiene un gas a baja presión

cation: any atom or group of atoms with a positive charge *(176)*
catión: cualquier átomo o grupo de átomos que posee carga positiva

cell potential: the difference between the reduction potentials of two half-cells *(737)*
potencial de celda: diferencia entre los potenciales de reducción de dos medias celdas

Celsius scale: the temperature scale on which the freezing point of water is 0°C and the boiling point is 100°C *(78)*
escala Celsius: escala de temperatura en la que el punto de congelación del agua es 0°C y el punto de ebullición del agua es 100°C

Charles's law: the volume of a fixed mass of gas is directly proportional to its Kelvin temperature if the pressure is kept constant *(458)*
ley de Charles: el volumen de una masa fija de gas es directamente proporcional a su temperatura Kelvin si la presión se mantiene constante

chemical change: a change that produces matter with a different composition than the original matter *(43)*
cambio químico: cambio que produce materia con una composición diferente que la de la materia original

chemical equation: an expression representing a chemical reaction; the formulas of the reactants (on the left) are connected by an arrow with the formulas for the products (on the right) *(348)*
ecuación química: expresión que representa una reacción química; las fórmulas de los reactantes (a la izquierda) se unen mediante una flecha a las fórmulas de los productos (a la derecha)

chemical equilibrium: a state of balance in which the rates of the forward and reverse reactions are equal; no net change in the amount of reactants and products occurs in the chemical system *(610)*
equilibrio químico: estado de equilibrio en el que las velocidades de la reacción de evolución y la reacción inversa son iguales; no hay un cambio total en la cantidad de reactantes y productos en el sistema químico

chemical formula: an expression that indicates the number and type of atoms present in the smallest representative unit of a substance *(202)*
fórmula química: expresión que indica el número y tipo de átomos que están presentes en la unidad más pequeña representativa de una sustancia

chemical potential energy: energy stored in chemical bonds *(556)*
energía potencial química: energía almacenada en los enlaces químicos

chemical property: the ability of a substance to undergo a specific chemical change *(48)*
propiedad química: capacidad de una sustancia para sufrir un cambio químico específico

chemical reaction: a change in which one or more reactants change into one or more products; characterized by the breaking of bonds in reactants and the formation of bonds in products *(48)*
reacción química: cambio en el que uno o más reactantes se convierten en uno o más productos; se caracteriza por la ruptura de enlaces en los reactantes y la formación de enlaces en los productos

chemical symbol: a one- or two-letter representation of an element *(45)*
símbolo químico: representación de un elemento que emplea una o dos letras

chemistry: the study of the composition of matter and the changes that matter undergoes *(2)*
química: estudio de la composición de la materia y los cambios que ésta sufre

***cis* configuration:** the configuration in which substituent groups are on the same side of a double bond *(776)*
configuración *cis*: configuración en la cual los grupos sustitutos están del mismo lado de un doble enlace

***cis-trans* isomers:** compounds that have atoms in the same order, but differ in the orientation of groups around a double bond *(776)*
isómeros *cis-trans*: compuestos cuyos átomos tienen el mismo orden, pero difieren con respecto a la orientación de los grupos alrededor de un enlace doble

coefficient: a small whole number that appears in front of a formula in a balanced chemical equation *(350)*
coeficiente: número entero pequeño que aparece antepuesto a una fórmula en una ecuación química balanceada

R122 Glossary

coenzyme: a small organic molecule or metal ion necessary for an enzyme's biological activity *(848)*
coenzima: pequeña molécula orgánica o ion metálico que se requiere para que una enzima tenga actividad biológica

colligative property: a property of a solution that depends only upon the number of solute particles, and not upon their identities; boiling-point elevation, freezing-point depression, and vapor-pressure lowering are colligative properties *(534)*
propiedad coligativa: propiedad de una solución que depende únicamente del número de partículas de soluto, y no del tipo de soluto; el incremento del punto de ebullición, la disminución del punto de congelación y el descenso de la presión de vapor son propiedades coligativas

collision theory: atoms, ions, and molecules can react to form products when they collide, provided that the particles have enough kinetic energy *(596)*
teoría de choques: los átomos, iones y moléculas pueden reaccionar para formar productos cuando chocan, siempre que las partículas tengan suficiente energía cinética

colloid: a mixture whose particles are intermediate in size between those of a suspension and a solution *(505)*
coloide: mezcla cuyas partículas tienen un tamaño intermedio entre las de una suspensión y una solución

combination reaction: a chemical change in which two or more substances react to form a single new substance; also called a synthesis reaction *(356)*
reacción de combinación: cambio químico en el que dos o más sustancias reaccionan para formar una sola sustancia nueva; también llamado reacción de síntesis

combined gas law: the law that describes the relationship among the pressure, temperature, and volume of an enclosed gas *(462)*
ley combinada de los gases: ley que describe las relaciones entre la presión, la temperatura y el volumen de un gas encerrado

combustion reaction: a chemical change in which an element or a compound reacts with oxygen, often producing energy in the form of heat and light *(363)*
reacción de combustión: cambio químico en el que un elemento o un compuesto reacciona con oxígeno y por lo regular produce energía en forma de luz y calor

common ion: an ion that is common to both salts in a solution; in a solution of silver nitrate and silver chloride, Ag^+ would be a common ion *(624)*
ion común: ion que es común a dos sales disueltas en una solución; en una solución de nitrato de plata y cloruro de plata, Ag^+ sería un ion común

common ion effect: a decrease in the solubility of an ionic compound caused by the addition of a common ion *(624)*
efecto de ion común: disminución en la solubilidad de un compuesto iónico debida a la adición de un ion común

complete ionic equation: an equation that shows dissolved ionic compounds as dissociated free ions *(370)*
ecuación iónica completa: ecuación que muestra los compuestos iónicos disueltos en forma de iones disociados libres

compound: a substance that contains two or more elements chemically combined in a fixed proportion *(42)*
compuesto: sustancia que contiene dos o más elementos combinados químicamente en una proporción fija

compressibility: a measure of how much the volume of matter decreases under pressure *(450)*
compresibilidad: medida de cuánto disminuye el volumen de la materia cuando se le aplica presión

concentrated solution: a solution containing a large amount of solute *(525)*
solución concentrada: solución que contiene una gran cantidad de soluto

concentration: a measurement of the amount of solute that is dissolved in a given quantity of solvent; usually expressed as mol/L *(525)*
concentración: medida de la cantidad de soluto que está disuelto en una cantidad específica de disolvente; suele expresarse en mol/L

condensed structural formula: a structural formula that leaves out some bonds and/or atoms; the presence of these atoms or bonds is understood *(766)*
fórmula estructural condensada: fórmula estructural que no muestra algunos enlaces o átomos; se sobreentiende la presencia de estos enlaces o átomos

conjugate acid: the particle formed when a base gains a hydrogen ion; NH_4^+ is the conjugate acid of the base NH_3 *(650)*
ácido conjugado: partícula que se forma cuando una base gana un ion hidrógeno; NH_4^+ es el ácido conjugado de la base NH_3

conjugate acid-base pair: two substances that are related by the loss or gain of a single hydrogen ion; ammonia (NH_3) and the ammonium ion (NH_4^+) are a conjugate acid-base pair *(650)*
par conjugado ácido-base: dos sustancias relacionadas entre sí por la pérdida o ganancia de un solo ion hidrógeno; el amoníaco (NH_3) y el ion amonio (NH_4^+) son un par conjugado ácido-base

conjugate base: the particle that remains when an acid has donated a hydrogen ion; OH^- is the conjugate base of the acid water *(650)*
base conjugada: partícula que queda cuando un ácido transfiere un ion hidrógeno; OH^- es la base conjugada del ácido agua

constitutional isomers: compounds that have the same molecular formula, but whose atoms are bonded in a different order *(775)*
isómeros constitucionales: compuestos que tienen la misma fórmula molecular, pero cuyos átomos están enlazados en distinto orden

Glossary **R123**

conversion factor: a ratio of equivalent measurements used to convert a quantity from one unit to another *(84)*

factor de conversión: razón de medidas equivalentes usadas para convertir una cantidad de una unidad a otra

coordinate covalent bond: a covalent bond in which one atom contributes both bonding electrons *(232)*

enlace covalente coordinado: enlace covalente en el que un átomo aporta dos electrones de enlace

coordination number: the number of ions of opposite charge that surround each ion in a crystal *(205)*

número de coordinación: número de iones de carga opuesta que rodean a cada ion en un cristal

covalent bond: a bond formed by the sharing of electrons between atoms *(215)*

enlace covalente: enlace que se forma cuando dos átomos comparten electrones

cracking: the controlled process by which hydrocarbons are broken down or rearranged into smaller, more useful molecules *(783)*

pirólisis: proceso controlado por el cual los hidrocarburos se descomponen o reacomodan para obtener moléculas más pequeñas y útiles

crystal: a solid in which the atoms, ions, or molecules are arranged in an orderly, repeating, three-dimensional pattern called a crystal lattice *(432)*

cristal: sólido en el que los átomos, iones o moléculas están dispuestos en un patrón tridimensional ordenado y repetitivo llamado red cristalina

cycloalkane: cyclic hydrocarbon that contains only single bonds *(779)*

cicloalcano: hidrocarburo cíclico que solamente contiene enlaces simples

cyclic hydrocarbon: an organic compound that contains a hydrocarbon ring *(779)*

hidrocarburo cíclico: compuesto orgánico que contiene un anillo de hidrocarburo

D

Dalton's atomic theory the first theory to relate chemical changes to events at the atomic level *(103)*

teoría atómica de Dalton: primera teoría en relacionar los cambios químicos con sucesos a nivel atómico

Dalton's law of partial pressures: at constant volume and temperature, the total pressure exerted by a mixture of gases is equal to the sum of the partial pressures of the component gases *(470)*

teoría de Dalton de las presiones parciales: a volumen y temperatura constantes, la presión total ejercida por una mezcla de gases es igual a la suma de las presiones parciales de los gases componentes

decomposition reaction: a chemical change in which a single compound is broken down into two or more simpler products *(358)*

reacción de descomposición: cambio químico en el que un solo compuesto se descompone en dos o más productos más simples

dehydrogenation reaction: a reaction in which hydrogen is lost *(816)*

reacción de deshidrogenación: reacción en la que se pierde hidrógeno

deliquescent: describes a substance that removes sufficient water from the air to form a solution; the solution formed has a lower vapor pressure than that of the water in the air *(501)*

delicuescente: término que describe una sustancia que absorbe suficiente humedad del aire como para formar una solución; la solución formada tiene una presión de vapor más baja que la de la humedad del aire

density: the ratio of the mass of an object to its volume *(80)*

densidad: razón de la masa de un objeto a su volumen

dependent variable: the variable that is observed during an experiment; also called responding variable *(16)*

variable dependiente: variable que se observa durante un experimento; también llamada variable de repuesta

desiccant: a hygroscopic substance used as a drying agent *(499)*

desecante: sustancia higroscópica empleada como agente secante

diatomic molecule: a molecule consisting of two atoms *(223)*

molécula diatómica: molécula que cons-ta de dos átomos

diffusion: the tendency of molecules to move toward areas of lower concentration until the concentration is uniform throughout *(472)*

difusión: tendencia de las moléculas a moverse hacia áreas de baja concentración hasta que la concentración es uniforme en todo el medio

dilute solution: a solution that contains a small amount of solute *(525)*

solución diluida: solución que contiene muy poco soluto

dimensional analysis: a technique of problem-solving that uses the units that are part of a measurement to help solve the problem *(86)*

análisis dimensional: técnica para resolver problemas que se apoya en las unidades de las mediciones para resolver el problema

dipole: a molecule that has two poles, or regions, with opposite charges *(249)*

dipolo: molécula que tiene dos polos o regiones de carga opuesta

R124 Glossary

dipole interactions: intermolecular forces resulting from the attraction of oppositely charged regions of polar molecules *(250)*
 interacción dipolar: fuerzas intermoleculares que resultan de la atracción de regiones de moléculas polares que tienen cargas opuestas

diprotic acid: any acid that contains two ionizable protons (hydrogen ions); sulfuric acid (H_2SO_4) is a diprotic acid *(647)*
 ácido diprótico: cualquier ácido que contenga dos protones (iones hidrógeno) ionizables; el ácido sulfúrico (H_2SO_4) es un ácido diprótico

disaccharide: a carbohydrate formed from two monosaccharide units; common table sugar (sucrose) is a disaccharide *(842)*
 disacárido: carbohidrato formado por dos unidades de monosacárido; el azúcar de mesa común (sacarosa) es un disacárido

dispersion forces: attractions between molecules caused by the electron motion on one molecule affecting the electron motion on the other through electrical forces; these are the weakest interactions between molecules *(251)*
 fuerzas de dispersión: atracciones entre moléculas que se dan cuando el movimiento de los electrones de una molécula afecta el movimiento de los electrones de la otra mediante fuerzas eléctricas; se trata de las interacciones más débiles entre moléculas

displacement reaction: see single-replacement reaction
 reacción de desplazamiento: véase reacción de sustitución sencilla

distillation: a process used to separate components of a mixture using differences in boiling points *(40)*
 destilación: proceso que se emplea para separar las sustancias de una mezcla por medio de diferentes puntos de ebullición

double covalent bond: a bond in which two atoms share two pairs of electrons *(230)*
 enlace covalente doble: enlace en el que dos átomos comparten dos pares de electrones

double-replacement reaction: a chemical change that involves an exchange of positive ions between two compounds *(362)*
 reacción de sustitución doble: cambio químico que implica un intercambio de iones positivos entre dos compuestos

dry cell: a commercial voltaic cell in which the electrolyte is a moist paste; despite their name, the compact, portable batteries used in flashlights are dry cells *(732)*
 pila seca: celda voltaica comercial en la que, a pesar del nombre, el electrolito es una pasta húmeda; las baterías compactas y portátiles que se usan en las linternas son pilas secas

E

effloresce: to lose water of hydration; the process occurs when the hydrate has a vapor pressure higher than that of water vapor in the air *(499)*
 eflorecerse: perder agua de hidratación; el proceso se presenta cuando la presión de vapor del hidrato es más alta que la del vapor de agua en el aire

effusion: the process that occurs when a gas escapes through a tiny hole in its container *(472)*
 efusión: proceso en el cual un gas escapa por un agujero diminuto en su recipiente

electrical potential: the ability of a voltaic cell to produce an electric current *(737)*
 potencial eléctrico: capacidad de una celda voltaica para producir corriente eléctrica

electrochemical cell: any device that converts chemical energy into electrical energy or electrical energy into chemical energy *(730)*
 celda electroquímica: cualquier dispositivo que convierte energía química en energía eléctrica o energía eléctrica en energía química

electrochemical process: the conversion of chemical energy into electrical energy or electrical energy into chemical energy; all electrochemical processes involve redox reactions *(728)*
 proceso electroquímico: conversión de energía química en energía eléctrica o energía eléctrica en energía química; en todos los procesos electroquímicos intervienen reacciones redox

electrode: a conductor in a circuit that carries electrons to or from a substance other than a metal *(730)*
 electrodo: en un circuito, un conductor que transporta electrones hacia o desde una sustancia que no es un metal

electrolysis: a process in which electrical energy is used to bring about a chemical change; the electrolysis of water produces hydrogen and oxygen *(745)*
 electrolisis: proceso en el que se usa energía eléctrica para realizar un cambio químico; la electrolisis del agua produce hidrógeno y oxígeno

electrolyte: a compound that conducts an electric current when it is in an aqueous solution or in the molten state; all ionic compounds are electrolytes, but most covalent compounds are not *(496)*
 electrolito: compuesto que conduce una corriente eléctrica cuando está en solución acuosa o está derretido; todos los compuestos iónicos son electrolitos, pero muy pocos compuestos covalentes lo son

electrolytic cell: an electrochemical cell used to cause a chemical change through the application of electrical energy *(745)*
 celda electrolítica: celda electroquímica que se usa para efectuar un cambio químico mediante la aplicación de energía eléctrica

Glossary **R125**

electromagnetic radiation: energy waves that travel in a vacuum at a speed of 2.998×10^8 m/s; includes radio waves, microwaves, infrared waves, visible light, ultraviolet waves, X-rays, and gamma rays *(139)*
radiación electromagnética: ondas de energía que viajan en el vacío a una velocidad de 2.998×10^8 m/s; incluye las ondas de radio, microondas, ondas infrarrojas, luz visible, ondas ultravioleta, rayos X y rayos gamma

electron: a negatively charged subatomic particle *(105)*
electrón: partícula subátomica con carga negativa

electron configuration: the arrangement of electrons of an atom in its ground state into various orbitals around the nuclei of atoms *(134)*
configuración electrónica: distribución de los electrones de un átomo en su estado basal, en diversos orbitales alrededor del núcleo del átomo

electron dot structure: a notation that depicts valence electrons as dots around the atomic symbol of the element; the symbol represents the inner electrons and atomic nucleus; also called Lewis dot structure *(195)*
estructura de punto electrón: notación que muestra los electrones de valencia como puntos alrededor del símbolo atómico del elemento; el símbolo representa los electrones internos y el núcleo atómico; también se conoce como estructura de puntos de Lewis

electronegativity: the ability of an atom to attract electrons when the atom is in a compound *(181)*
electronegatividad: capacidad de un átomo para atraer electrones cuando el átomo está en un compuesto

element: the simplest form of matter that has a unique set of properties; an element cannot be broken down into simpler substances by chemical means *(42)*
elemento: forma más simple de materia que posee un conjunto único de propiedades; un elemento no puede descomponerse en sustancias más simples usando métodos químicos

elementary reaction: a reaction in which reactants are converted to products in a single step *(607)*
reacción básica: reacción en la que los reactantes se convierten en productos en un solo paso

empirical formula: a formula with the lowest whole-number ratio of elements in a compound; the empirical formula of hydrogen peroxide (H_2O_2) is HO *(330)*
fórmula empírica: fórmula que muestra las proporciones de los elementos en un compuesto con los números enteros más pequeños posibles; la fórmula empírica del peróxido de hidrógeno (H_2O_2) es HO

emulsion: the colloidal dispersion of one liquid in another *(507)*
emulsión: dispersión coloidal de un líquido en otro

enantiomers: molecules that differ from one another in the way that four different groups are arranged around a carbon atom *(777)*
enantiómero: moléculas que se diferencian entre sí por la forma en que cuatro grupos diferentes están dispuestos alrededor de un átomo de carbono

endothermic process: a process that absorbs heat from the surroundings *(557)*
proceso endotérmico: proceso en el que se absorbe calor del entorno

end point: the point in a titration at which the indicator changes color *(674)*
punto final: punto de una valoración química en cual el indicador cambia de color

energy: the capacity for doing work or producing heat *(77)*
energía: capacidad para efectuar trabajo o producir calor

energy level: the specific energies an electron in an atom or other system can have *(129)*
nivel energético: las energías específicas que puede tener un electrón en un átomo u otro sistema

enthalpy (*H*): the heat content of a system at constant pressure *(562)*
entalpía (*H*): cantidad de calor en un sistema a presión constante

entropy (*S*): a measure of the disorder of a system; systems tend to go from a state of order (low entropy) to a state of maximum disorder (high entropy) *(630)*
entropía (*S*): medida del desorden de un sistema; los sistemas tienden a pasar de un estado ordenado (baja entropía) a un estado de máximo desorden (alta entropía)

enzyme: a protein that acts as a biological catalyst *(847)*
enzima: proteína que actúa como catalizador biológico

enzyme–substrate complex: the structure formed when a substrate molecule joins an enzyme at its active site *(847)*
complejo enzima-sustrato: estructura que se forma cuando una molécula de sustrato se une a una enzima en su sitio activo

equilibrium constant (K_{eq}): the ratio of product concentrations to reactant concentrations at equilibrium, with each concentration raised to a power equal to the number of moles of that substance in the balanced chemical equation *(616)*
constante de equilibrio (K_{eq}): razón de las concentraciones de los productos a las concentraciones de los reactantes en el equilibrio, con cada concentración elevada a una potencia igual al número de moles de esa sustancia en la ecuación química balanceada

equilibrium position: the relative concentrations of reactants and products of a reaction that has reached equilibrium; indicates whether the reactants or products are favored in the reversible reaction *(611)*
posición de equilibrio: las concentraciones relativas de reactantes y productos de una reacción que ha alcanzado el equilibrio; indica si se favorecen los reactantes o productos en la reacción reversible

R126 Glossary

equivalence point: the point in a titration where the number of moles of hydrogen ions equals the number of moles of hydroxide ions *(674)*

punto de equivalencia: punto de una valoración química en la que el número de moles de iones hidrógeno es igual al número de moles de iones hidróxido

error: the difference between the accepted value and the experimental value *(65)*

error: diferencia entre el valor aceptado y el valor experimental

ester: a derivative of a carboxylic acid in which the —OH of the carboxyl group has been replaced by the —OR from an alcohol; the general formula is RCOOR *(819)*

éster: derivado de un ácido carboxílico en el que el —OH del grupo carboxilo ha sido sustituido por el —OR de un alcohol; la fórmula general es RCOOR

ether: an organic compound in which oxygen is bonded to two carbon groups; the general formula is R—O—R *(810)*

éter: compuesto orgánico en el que el oxígeno está unido a dos grupos carbono; la fórmula general es R—O—R

evaporation: vaporization that occurs at the surface of a liquid that is not boiling *(426)*

evaporación: vaporización que se da en la superficie de un líquido que no está en ebullición

excess reagent: a reagent present in a quantity that is more than sufficient to react with a limiting reagent; any reactant that remains after the limiting reagent is used up in a chemical reaction *(401)*

reactivo excesivo: reactivo que está presente en una cantidad más que suficiente para reaccionar con un reactivo limitante; cualquier reactante que queda después de que se ha usado todo el reactivo limitante en una reacción química

exothermic process: a process that releases heat to its surroundings *(557)*

proceso exotérmico: proceso en el que se desprende calor hacia el entorno

experiment: a repeatable procedure that is used to test a hypothesis *(16)*

experimento: procedimiento repetido que sirve para probar una hipótesis

experimental value: a quantitative value measured during an experiment *(65)*

valor experimental: valor cuantitativo que se mide durante un experimento

extensive property: a property that depends on the amount of matter in a sample *(34)*

propiedad extensiva: propiedad que depende de la cantidad de materia en una muestra

F

fatty acid: the name given to continuous-chain carboxylic acids that were first isolated from fats *(816)*

ácido graso: nombre que se da a los ácidos carboxílicos de cadena continua que se aislaron originalmente de las grasas

fermentation: the production of ethanol from sugars by the action of yeast or bacteria *(807)*

fermentación: producción de etanol a partir de azúcares por la acción de levaduras o bacterias

filtration: a process that separates a solid from the liquid in a heterogeneous mixture *(40)*

filtración: proceso para separar un sólido de un líquido en una mezcla heterogénea

first-order reaction: a reaction in which the reaction rate is proportional to the concentration of only one reactant *(605)*

reacción de primer orden: reacción cuya velocidad de reacción es proporcional a la concentración de un solo reactante

fission: the splitting of a nucleus into smaller fragments, accompanied by the release of neutrons and a large amount of energy *(888)*

fisión: división de un núcleo en fragmentos más pequeños, acompañada por desprendimiento de neutrones y una gran cantidad de energía

formula unit: the lowest whole-number ratio of ions in an ionic compound; in magnesium chloride, the ratio of magnesium ions to chloride ions is 1:2 and the formula unit is $MgCl_2$ *(202)*

unidad de fórmula: razón más baja, expresada en números enteros, de los iones de un compuesto iónico; en el cloruro de magnesio, la razón de iones magnesio a iones cloruro es de 1:2, así que la unidad de fórmula es $MgCl_2$

free energy: the energy available to do work *(627)*

energía libre: energía que está disponible para realizar trabajo

freezing point: the temperature at which a liquid changes to a solid *(431)*

punto de congelación: temperatura a la cual un líquido se convierte en un sólido

freezing-point depression: the difference in temperature between the freezing point of a solution and the freezing point of the pure solvent *(536)*

disminución del punto de congelación: diferencia de temperatura entre el punto de congelación de una solución y el del disolvente puro

frequency (ν): the number of wave cycles that pass a given point per unit of time; frequency and wavelength are inversely proportional to each other *(138)*

frecuencia (ν): número de ciclos de onda que pasan por un punto específico en la unidad de tiempo; la frecuencia y la longitud de onda son inversamente proporcionales

Glossary **R127**

fuel cell: a voltaic cell that does not need to be recharged; the fuel is oxidized to produce a continuous supply of electrical energy *(734)*
celda de combustible: celda voltaica que no necesita recargarse; el combustible se oxida para producir un suministro continuo de energía eléctrica

functional group: a specific arrangement of atoms in an organic compound that is capable of characteristic chemical reactions; the chemistry of an organic compound is determined by its functional groups *(798)*
grupo funcional: distribución específica de átomos en un compuesto orgánico que puede participar en reacciones químicas características; la química de un compuesto orgánico está determinada por sus grupos funcionales

fusion: the process of combining nuclei to produce a nucleus of greater mass *(891)*
fusión: proceso en el que se combinan núcleos para producir un núcleo con mayor masa

G

gamma ray: a high-energy photon emitted by a radioisotope *(879)*
rayo gamma: fotón de alta energía emitido por un radioisótopo

gas: a form of matter that takes the shape and volume of its container; a gas has no definite shape or volume *(37)*
gas: estado de la materia que adopta la forma y el volumen del recipiente que la contiene; los gases no tienen forma ni volumen definidos

gas pressure: results from the force exerted by a gas per unit surface area of an object; due to collisions of gas particles with the object *(421)*
presión de gas: resultado de la fuerza que ejerce un gas por unidad de área total de un objeto; se debe a los choques de las partículas de gas contra el objeto

Gay-Lussac's law: the pressure of a gas is directly proportional to the Kelvin temperature if the volume is constant *(460)*
ley de Gay-Lussac: la presión de un gas es directamente proporcional a su temperatura Kelvin si se mantiene constante el volumen

gene: a segment of DNA that codes for a single peptide chain *(856)*
gen: segmento de ADN que contiene el código para una sola cadena péptida

glass: transparent fusion product of inorganic materials that have cooled to a rigid state without crystallizing *(434)*
vidrio: producto transparente que resulta de la fusión de materiales inorgánicos que se han enfriado hasta solidificarse sin cristalizarse

Graham's law of effusion: the rate of effusion of a gas is inversely proportional to the square root of its molar mass; this relationship is also true for the diffusion of gases *(472)*
ley de efusión de Graham: la velocidad de efusión de un gas es inversamente proporcional a la raíz cuadrada de su masa molar; esta relación también se cumple en la difusión de gases

gram (g): a metric mass unit equal to the mass of 1 cm^3 of water at 4°C *(77)*
gramo (g): unidad métrica de masa equivalente a la masa de 1 cm^3 de agua a 4°C

ground state: the lowest possible energy of an atom described by quantum mechanics *(145)*
estado fundamental: energía más baja que puede tener un átomo descrito por la mecánica cuántica

group: a vertical column of elements in the periodic table; the constituent elements of a group have similar chemical and physical properties *(47)*
grupo: columna vertical de elementos en la tabla periódica; los elementos de un grupo tienen propiedades físicas y químicas similares

H

half-cell: the part of a voltaic cell in which either oxidation or reduction occurs; it consists of a single electrode immersed in a solution of its ions *(730)*
semicelda: parte de una celda voltaica en la que se lleva a cabo la oxidación o reducción; consta de un solo electrodo sumergido en una solución de sus iones

half-life ($t_{1/2}$): the time required for one-half of the nuclei of a radioisotope sample to decay to products *(882)*
semivida ($t_{1/2}$): tiempo que tarda en desintegrarse la mitad de los núcleos de una muestra de un radioisótopo

half-reaction: an equation showing either the oxidation or the reduction that takes place in a redox reaction *(712)*
semirreacción: ecuación que muestra la oxidación o bien la reducción que se da en una reacción redox

half-reaction method: a method of balancing a redox equation by balancing the oxidation and reduction half-reactions separately before combining them into a balanced redox equation *(712)*
método de semirreacción: método para balancear una ecuación redox equilibrando por separado las semirreacciones de oxidación y reducción antes de combinarlas para obtener una ecuación redox balanceada

halide ion: a negative ion formed when a halogen atom gains an electron *(199)*
ion haluro: ion negativo que se forma cuando un átomo de halógeno gana un electrón

R128 Glossary

halocarbon: any member of a class of organic compounds containing covalently bonded fluorine, chlorine, bromine, or iodine *(800)*

compuesto halocarbonado: cualquier miembro de una clase de compuestos orgánicos que contienen flúor, cloro, bromo o yodo unidos mediante enlaces covalentes

halogen: a nonmetal in Group 7A of the periodic table *(167)*

halógeno: no metal del grupo 7A de la tabla periódica

heat (*q*): energy that transfers from one object to another because of a temperature difference between the objects *(556)*

calor (*q*): energía que fluye de un objeto a otro debido a la diferencia de temperatura entre los objetos

heat capacity: the amount of heat needed to increase the temperature of an object exactly 1°C *(559)*

capacidad calorífica: cantidad de calor necesaria para elevar exactamente 1°C la temperatura de un objeto

heat of combustion: the heat of reaction for the complete burning of one mole of a substance *(568)*

calor de combustión: calor de reacción al quemarse totalmente un mol de una sustancia

heat of reaction: the enthalpy change for a chemical equation exactly as it is written *(565)*

calor de reacción: cambio de entalpía correspondiente a una ecuación química en la forma exacta en que está escrita

Heisenberg uncertainty principle: it is impossible to know exactly both the velocity and the position of a particle at the same time *(148)*

principio de incertidumbre de Heisenberg: es imposible conocer con exactitud la velocidad y la posición de una partícula al mismo tiempo

Henry's law: at a given temperature the solubility of a gas in a liquid is directly proportional to the pressure of the gas above the liquid *(523)*

ley de Henry: a una temperatura determinada, la solubilidad de un gas en un líquido es directamente proporcional a la presión del gas sobre el líquido

hertz (Hz): the unit of frequency, equal to one cycle per second *(138)*

hertz (Hz): unidad de frecuencia, equivalente a un ciclo por segundo

Hess's law of heat summation: if you add two or more thermochemical equations to give a final equation, then you also add the heats of reaction to give the final heat of reaction *(578)*

ley de Hess de la suma de los calores: ley según la cual, si se suman dos o más ecuaciones termoquímicas para obtener una ecuación final, también se suman los calores de reacción para obtener el calor de reacción final

heterogeneous mixture: a mixture that is not uniform in composition; components are not evenly distributed throughout the mixture *(39)*

mezcla heterogénea: mezcla cuya composición no es uniforme; sus componentes no están distribuidos de forma equitativa en toda la mezcla

homogeneous mixture: a mixture that is uniform in composition; components are evenly distributed and not easily distinguished *(39)*

mezcla homogénea: mezcla cuya composición es uniforme; sus componentes están distribuidos de forma equitativa y no es fácil distinguirlos

homologous series: a group of compounds in which there is a constant increment of change in molecular structure from one compound in the series to the next *(765)*

serie homóloga: grupo de compuestos en el que se observa un incremento constante de cambio en la estructura molecular de un compuesto al siguiente

Hund's rule: electrons occupy orbitals of the same energy in a way that makes the number or electrons with the same spin direction as large as possible *(134)*

regla de Hund: los electrones ocupan orbitales de la misma energía haciendo que el número de electrones cuyo espín tiene la misma dirección sea lo más grande posible

hybridization: the mixing of several atomic orbitals to form the same total number of equivalent hybrid orbitals *(244)*

hibridización: combinación de varios orbitales atómicos para formar el mismo número total de orbitales híbridos equivalentes

hydrate: a compound that has a specific number of water molecules bound to each formula unit *(498)*

hidrato: compuesto que tiene un número específico de moléculas de agua unidas a cada unidad de fórmula

hydration reaction: a reaction in which water is added to an alkene *(808)*

reacción de hidratación: reacción en la que se añade agua a un alqueno

hydrocarbon: an organic compound that contains only carbon and hydrogen *(762)*

hidrocarburo: compuesto orgánico que contiene sólo carbono e hidrógeno

hydrogenation reaction: a reaction in which hydrogen is added to a carbon–carbon double bond to give an alkane *(809)*

reacción de hidrogenación: reacción en la que se añade hidrógeno a un doble enlace carbono–carbono para obtener un alcano

GLOSSARY

Glossary **R129**

hydrogen bonds: attractive forces in which a hydrogen covalently bonded to a very electronegative atom is also weakly bonded to an unshared electron pair of another electronegative atom *(251)*
enlaces de hidrógeno: fuerzas de atracción en las que un átomo de hidrógeno, unido por un enlace covalente a un átomo muy electronegativo, también se une débilmente a un par no compartido de electrones de otro átomo electronegativo

hydrogen-ion acceptor: a base, according to the Brønsted-Lowry theory; ammonia acts as a base when it accepts hydrogen ions from water *(649)*
receptor de iones hidrógeno: una base, según la teoría de Brønsted-Lowry; el amoniaco actúa como base cuando acepta iones hidrógeno del agua

hydrogen-ion donor: an acid, according to the Brønsted-Lowry theory *(649)*
donador de iones hidrógeno: un ácido, según la teoría de Brønsted-Lowry

hydronium ion (H_3O^+): the positive ion formed when a water molecule gains a hydrogen ion *(647)*
ion hidronio (H_3O^+): ion positivo que se forma cuando una molécula de agua gana un ion hidrógeno

hydroxy group: the —OH functional group in alcohols *(804)*
grupo hidroxilo: el grupo funcional —OH de los alcoholes

hygroscopic: a term describing salts and other compounds that remove moisture from the air *(499)*
higroscópico: se llama así a las sales y otros compuestos que absorben humedad del aire

hypothesis: a proposed explanation for an observation *(16)*
hipótesis: explicación propuesta para una observación

I

ideal gas constant: the constant in the ideal gas law with the symbol *R* and the value 8.31 (L · kPa)/(K · mol) *(465)*
constante del gas ideal: constante de la ley del gas ideal; se representa con el símbolo *R* y tiene un valor de 8.31 (L · kPa)/(K · mol)

ideal gas law: the relationship $PV = nRT$, which describes the behavior of an ideal gas *(465)*
ley del gas ideal: relación $PV = nRT$, que describe el comportamiento del gas ideal

immiscible: describes liquids that are insoluble in one another; oil and water are immiscible *(521)*
inmiscible: se dice de los líquidos que son insolubles uno en el otro; el aceite y el agua son inmiscibles

independent variable: the variable that is changed during an experiment; also called manipulated variable *(16)*
variable independiente: variable que cambia durante un experimento; también se llama variable manipulada

inhibitor: a substance that interferes with the action of a catalyst *(601)*
inhibidor: sustancia que interfiere la acción de un catalizador

inner transition metal: an element in the lanthanide or actinide series; the highest occupied *s* sublevel and nearby *f* sublevel of its atoms generally contain electrons; also called inner transition element *(172)*
metal de transición interna: elemento de las series de los lantánidos o los actínidos; el subnivel *s* más alto ocupado y el subnivel *f* cercano de sus átomos contienen electrones; también se llama elemento de transición interna

inorganic chemistry: the study of substances that, in general, do not contain carbon *(3)*
química inorgánica: estudio de sustancias que, en general, no contienen carbono

intensive property: a property that depends on the type of matter in a sample, not the amount of matter *(34)*
propiedad intensiva: propiedad que depende del tipo de materia de una muestra, no de la cantidad de materia

intermediate: a product of one of the steps in a reaction mechanism; it becomes a reactant in the next step *(607)*
intermediario: producto de uno de los pasos de un mecanismo de reacción; se convierte en reactante en el siguiente paso

International System of Units (SI): the revised version of the metric system, adopted by international agreement in 1960 *(74)*
Sistema Internacional de Unidades (SI): versión modificada del sistema métrico, adoptada por acuerdo internacional en 1960

ion: an atom or group of atoms that has a positive or negative charge *(176)*
ion: átomo o grupo de átomos que tiene carga positiva o negativa

ionic bond: the electrostatic attraction that binds oppositely charged ions together *(201)*
enlace iónico: atracción electrostática que une a iones con carga opuesta

ionic compound: a compound composed of positive and negative ions *(201)*
compuesto iónico: compuesto formado por iones positivos y negativos

ionization energy: the energy required to remove an electron from an atom in its gaseous state *(177)*
energía de ionización: energía necesaria para sacar un electrón de un átomo en su estado gaseoso

ionizing radiation: radiation with enough energy to knock electrons off some atoms of a bombarded substance to produce ions *(894)*
radiación ionizante: radiación que tiene la energía suficiente para desprender electrones de algunos átomos de una sustancia bombardeada, produciendo así iones

R130 Glossary

ion-product constant for water (K_w): the product of the concentrations of hydrogen ions and hydroxide ions in water; it is 1×10^{-14} at 25°C *(654)*
constante de producto iónico del agua (K_w): producto de las concentraciones de iones hidrógeno y de iones hidróxido del agua; es 1×10^{-14} a 25°C

isomers: compounds that have the same molecular formula but different molecular structures *(775)*
isómeros: compuestos que tienen la misma fórmula molecular, pero diferentes estructuras moleculares

isotopes: atoms of the same element that have the same atomic number but different atomic masses due to a different number of neutrons *(114)*
isótopos: átomos del mismo elemento que tienen el mismo número atómico pero diferentes masas atómicas porque tienen un distinto número de neutrones

J

joule (J): the SI unit of energy; 4.184 J equal one calorie *(77)*
julio (J): unidad de energía en el SI; 4.184 J equivalen a una caloría

K

Kelvin scale: the temperature scale in which the freezing point of water is 273 K and the boiling point is 373 K; 0 K is absolute zero *(78)*
escala Kelvin: escala de temperatura en la que el punto de congelación del agua es 273 K, y el de ebullición, 373 K; 0 K es el cero absoluto

ketone: an organic compound in which the carbon of the carbonyl group is joined to two other carbons; the general formula is RCOR *(812)*
cetona: compuesto orgánico en el que el carbono del grupo carbonilo está unido a otros dos carbonos: la fórmula general es RCOR

kilogram (kg): the mass of 1 L of water at 4°C; it is the base unit of mass in SI *(77)*
kilogramo (kg): masa de 1 L de agua a 4°C; es la unidad base de masa en el SI

kinetic energy: the energy an object has because of its motion *(420)*
energía cinética: energía que tienen los objetos de acuerdo con su movimiento

kinetic theory: a theory explaining the states of matter, based on the concept that all matter consists of tiny particles that are in constant motion *(420)*
teoría cinética: teoría que explica los estados de la materia basándose en el concepto de que toda la materia está formada por pequeñas partículas que están en constante movimiento

L

law of conservation of energy: in any chemical or physical process, energy is neither created nor destroyed *(557)*
ley de conservación de la energía: ley según la cual en ningún proceso químico o físico se crea ni se destruye energía

law of conservation of mass: in any physical change or chemical reaction, mass is conserved; mass can be neither created nor destroyed *(50)*
ley de conservación de la masa: en cualquier cambio físico o reacción química, la masa se conserva; la masa no puede crearse ni destruirse

law of definite proportions: in samples of any chemical compound, the masses of the elements are always in the same proportion *(289)*
ley de las proporciones definidas: en muestras de cualquier compuesto químico, las masas de los elementos siempre están en la misma proporción

law of disorder: it is a natural tendency of systems to move in the direction of maximum chaos or disorder *(630)*
ley del desorden: tendencia natural de los sistemas a desplazarse en la dirección hacia el máximo caos o desorden

law of multiple proportions: whenever two elements form more than one compound, the different masses of one element that combine with the same mass of the other element are in the ratio of small whole numbers *(290)*
ley de las proporciones múltiples: siempre que dos elementos forman más de un compuesto, las diferentes masas de un elemento que se combinan con la misma masa del otro elemento están en razón de números enteros pequeños

Le Châtelier's principle: when a stress is applied to a system in dynamic equilibrium, the system changes in a way that relieves the stress *(612)*
principio de Le Châtelier: cuando se aplica una tensión a un sistema que está en equilibrio dinámico, el sistema cambia a modo de aliviar dicha tensión

Lewis acid: any substance that can accept a pair of electrons to form a covalent bond *(651)*
ácido de Lewis: cualquier sustancia capaz de aceptar un par de electrones para formar un enlace covalente

Lewis base: any substance that can donate a pair of electrons to form a covalent bond *(651)*
base de Lewis: cualquier sustancia capaz de ceder un par de electrones para formar un enlace covalente

limiting reagent: any reactant that is used up first in a chemical reaction; it determines the amount of product that can be formed in the reaction *(401)*
reactivo limitante: cualquier reactante que se haya consumido primero en una reacción química; determina la cantidad de producto que se puede formar en la reacción

Glossary **R131**

lipid: a member of a large class of relatively water-insoluble organic compounds; fats, oils, and waxes are lipids *(850)*
lípido: miembro de una clase amplia de compuestos orgánicos relativamente insolubles en agua; las grasas, aceites y ceras son lípidos

liquid: a form of matter that flows, has a fixed volume, and an indefinite shape *(36)*
líquido: forma de materia que fluye; tiene volumen fijo pero forma indefinida

liter (L): the volume of a cube measuring 10 centimeters on each edge (1000 cm^3); it is the common unprefixed unit of volume in the metric system *(76)*
litro (L): volumen de un cubo cuyas aristas miden 10 centímetros cada una (1000 cm^3); es la unidad común de volumen en el sistema métrico

M

manipulated variable: *see* independent variable *(16)*
variable manipulada: *véase* variable independiente

mass: a measure of the amount of matter that an object contains; the SI base unit of mass is the kilogram *(34)*
masa: medida de la cantidad de materia contenida en un objeto; la unidad base de masa en el SI es el kilogramo

mass number: the total number of protons and neutrons in the nucleus of an atom *(113)*
número de masa: número total de protones y neutrones que contiene el núcleo de un átomo

matter: anything that has mass and occupies space *(2)*
materia: todo lo que tiene masa y ocupa espacio

measurement: a quantitative description that includes both a number and a unit *(62)*
medición: descripción cuantitativa que incluye tanto números como unidades

melting point (mp): the temperature at which a substance changes from a solid to a liquid; the melting point of water is 0°C *(431)*
punto de fusión (p.f.): temperatura a la que una sustancia cambia del estado sólido al líquido; el punto de fusión del agua es 0°C

metabolism: all the chemical reactions carried out by an organism; includes energy-producing (catabolism) reactions and energy-absorbing (anabolism) reactions *(863)*
metabolismo: todas las reacciones químicas llevadas a cabo por los organismos; incluyen reacciones que producen energía (catabolismo) y reacciones que consumen energía (anabolismo)

metal: one of a class of elements that are good conductors of heat and electric current; metals tend to be ductile, malleable, and shiny *(165)*
metal: miembro de una clase de elementos que son buenos conductores del calor y la electricidad; los metales suelen ser dúctiles, maleables y brillantes

metallic bond: the force of attraction that holds metals together; it consists of the attraction of free-floating valence electrons for positively charged metal ions *(209)*
enlace metálico: fuerza de atracción que mantiene unidos los átomos de un metal; se debe a la atracción entre los electrones de valencia, que flotan libremente, y los iones metálicos de carga positiva

metalloid: an element that tends to have properties that are similar to those of metals and nonmetals *(166)*
metaloide: elemento cuyas propiedades son similares a las de los metales y de los no metales

meter (m): the base unit of length in SI *(75)*
metro (m): unidad base de longitud en el SI

miscible: describes liquids that dissolve in one another in all proportions *(521)*
miscible: se les llama así a los líquidos que se disuelven uno en el otro en todas las proporciones

mixture: a physical blend of two or more substances that are not chemically combined *(38)*
mezcla: incorporación física de dos o más sustancias que no se combinan químicamente

model: a representation of an object or event *(16)*
modelo: representación de un objeto o evento

molal boiling-point elevation constant (K_b): the change in boiling point for a 1-molal solution of a nonvolatile molecular solute *(543)*
constante molal de la elevación del punto de ebullición (K_b): cambio en el punto de ebullición de una solución 1-molal de un soluto molecular no volátil

molal freezing-point depression constant (K_f): the change in freezing point for a 1-molal solution of a nonvolatile molecular solute *(542)*
constante molal de la disminución del punto de congelación (K_f): cambio en el punto de congelación de una solución 1-molal de un soluto molecular no volátil

molality (m): the concentration of solute in a solution expressed as the number of moles of solute dissolved in 1 kilogram (1000 g) of solvent *(538)*
molalidad (m): concentración de soluto en una solución expresada como el número de moles de soluto disueltos en 1 kilogramo (1000 g) de disolvente

molar heat of condensation (ΔH_{cond}): the amount of heat released by one mole of a vapor as it condenses to a liquid at a constant temperature *(572)*
calor molar de condensación (ΔH_{cond}): cantidad de calor que un mol de vapor desprende al condensarse, convirtiéndose en líquido, a temperatura constante

molar heat of fusion (ΔH_{fus}): the amount of heat absorbed by one mole of a solid substance as it melts to a liquid at a constant temperature *(569)*
calor molar de fusión (ΔH_{fus}): cantidad de calor que un mol de una sustancia sólida absorbe al fundirse, convirtiéndose en líquido, a temperatura constante

R132 Glossary

molar heat of solidification (ΔH_{solid}): the amount of heat lost by one mole of a liquid as it solidifies at a constant temperature *(569)*
calor molar de solidificación (ΔH_{solid}): cantidad de calor que un mol de un líquido pierde al solidificarse a temperatura constante

molar heat of solution (ΔH_{soln}): the enthalpy change caused by the dissolution of one mole of a substance *(574)*
calor molar de disolución (ΔH_{soln}): cambio de calor debido a la disolución de un mol de una sustancia

molar heat of vaporization (ΔH_{vap}): the amount of heat absorbed by one mole of a liquid as it vaporizes at a constant temperature *(571)*
calor molar de vaporización (ΔH_{vap}): cantidad de calor absorbida por un mol de un líquido al evaporarse a temperatura constante

molarity (M): the concentration of solute in a solution expressed as the number of moles of solute dissolved in 1 liter of solution *(525)*
molaridad (M): concentración de soluto en una solución expresada como el número de moles de soluto disueltos en 1 litro de solución

molar mass: a term used to refer to the mass of a mole of any substance *(313)*
masa molar: término empleado para referirse a la masa de un mol de cualquier sustancia

molar volume: the volume occupied by 1 mole of a gas at standard temperature and pressure (STP); 22.4 L (320)
volumen molar: volumen ocupado por 1 mol de un gas a temperatura y presión estándar (TPE); 22.4 L

mole (mol): the amount of a substance that contains 6.02×10^{23} representative particles of that substance *(308)*
mol: cantidad de una sustancia que contiene 6.02×10^{23} partículas representativas de esa sustancia

molecular compound: a compound that is composed of molecules *(223)*
compuesto molecular: compuesto formado por moléculas

molecular formula: a chemical formula of a molecular compound that shows the kinds and numbers of atoms present in a molecule of a compound *(223)*
fórmula molecular: fórmula química de un compuesto molecular que indica los tipos y números de átomos presentes en una molécula de un compuesto

molecular orbital: an orbital that applies to the entire molecule *(240)*
orbital molecular: orbital que abarca toda la molécula

molecule: a neutral group of atoms joined together by covalent bonds *(215)*
molécula: grupo neutro de átomos unidos por enlaces covalentes

mole fraction: the ratio of the moles of solute in solution to the total number of moles of both solvent and solute *(540)*
fracción molar: razón de los moles de soluto en solución al número total de moles de disolvente y soluto

mole ratio: a conversion factor derived from the coefficients of a balanced chemical equation interpreted in terms of moles *(390)*
razón molar: factor de conversión derivado de los coeficientes de una ecuación química equilibrada interpretada en términos de moles

monatomic ion: a single atom with a positive or negative charge resulting from the loss or gain of one or more valence electrons *(264)*
ion monoatómico: un solo átomo con carga positiva o negativa debido a la pérdida o ganancia de uno o más electrones de valencia

monomer: a simple molecule that repeatedly combines to form a polymer *(822)*
monómero: molécula sencilla que se combina repetidamente para formar un polímero

monoprotic acid: any acid that contains one ionizable proton (hydrogen ion); nitric acid (HNO_3) is a monoprotic acid *(647)*
ácido monoprótico: ácido que sólo contiene un protón (ion hidrógeno) ionizable; el ácido nítrico (HNO_3) es un ácido monoprótico

monosaccharide: a carbohydrate consisting of one sugar unit; also called a simple sugar *(841)*
monosacárido: carbohidrato que consta de una sola unidad de azúcar; también llamado azúcar simple

N

net ionic equation: an equation for a reaction in solution showing only those particles that are directly involved in the chemical change *(370)*
ecuación iónica neta: ecuación de una reacción en solución que sólo muestra las partículas que intervienen directamente en el cambio químico

network solid: a solid in which all of the atoms are covalently bonded to each other *(252)*
sólido en cadena: sólido en el que todos los átomos están unidos entre sí por enlaces covalentes

neutralization reaction: a reaction in which an acid and a base react in an aqueous solution to produce a salt and water *(672)*
reacción de neutralización: reacción en la que un ácido y una base reaccionan en una solución acuosa para producir una sal y agua

Glossary **R133**

neutral solution: an aqueous solution in which the concentrations of hydrogen and hydroxide ions are equal; it has a pH of 7.0 *(653)*
solución neutral: solución acuosa en la que las concentraciones de iones hidrógeno e iones hidróxido son iguales; tiene un pH de 7.0

neutron: a subatomic particle with no charge and a mass of 1 amu; found in the nucleus of an atom *(107)*
neutrón: partícula subatómica sin carga que tiene una masa de 1 uma; se le encuentra en el núcleo de los átomos

neutron absorption: a process that decreases the number of slow-moving neutrons in a nuclear reactor; this is accomplished by using control rods made of a material such as cadmium, which absorbs neutrons *(889)*
absorción de neutrones: proceso que reduce el número de neutrones lentos en un reactor nuclear; esto se logra mediante el uso de varillas de control hechas con un material como el cadmio, que absorbe neutrones

neutron moderation: a process used in nuclear reactors to slow down neutrons so the reactor fuel captures them to continue the chain reaction *(889)*
moderación de neutrones: proceso empleado en reactores nucleares para frenar los neutrones de modo que el combustible del reactor los capture para continuar la reacción en cadena

noble gas: an element in Group 8A of the periodic table; the *s* and *p* sublevels of the highest occupied energy level are filled *(170)*
gas noble: elemento del grupo 8A de la tabla periódica; los subniveles *s* y *p* del nivel energético ocupado más alto están totalmente llenos

nonelectrolyte: a compound that does not conduct an electric current in aqueous solution or in the molten state *(496)*
no electrolito: compuesto que no conduce una corriente eléctrica en solución acuosa ni en estado fundido

nonmetal: an element that tends to be a poor conductor of heat and electric current; nonmetals generally have properties opposite to those of metals *(165)*
no metal: elemento que suele ser mal conductor del calor y la electricidad; las propiedades de los no metales generalmente son opuestas a las de los metales

nonpolar covalent bond: a covalent bond in which the electrons are shared equally by the two atoms *(247)*
enlace covalente no polar: enlace covalente en el que los dos átomos comparten equitativamente los electrones

nonspontaneous reaction: a reaction that does not favor the formation of products at the specified conditions *(628)*
reacción no espontánea: reacción que no favorece la formación de productos en las condiciones especificadas

normal boiling point: the boiling point of a liquid at a pressure of 101.3 kPa or 1 atm *(430)*
punto normal de ebullición: el punto de ebullición de un líquido a una presión de 101.3 kPa o 1 atm

nuclear force: an attractive force that acts between all nuclear particles that are extremely close together, such as protons and neutrons in a nucleus *(880)*
fuerza nuclear: fuerza de atracción que actúa entre todas las partículas nucleares que están extremadamente cerca unas de otras, como los protones y los neutrones en un núcleo

nuclear radiation: the penetrating rays and particles emitted by a radioactive source *(876)*
radiación nuclear: rayos y partículas penetrantes que una fuente radiactiva emite

nucleic acid: a polymer of ribonucleotides (RNA) or deoxyribonucleotides (DNA) found primarily in cell nuclei; nucleic acids play an important role in the transmission of hereditary characteristics, protein synthesis, and the control of cell activities *(854)*
ácido nucleico: polímero de ribonucleótidos (ARN) o desoxirribonucleótidos (ADN) que se encuentra primordialmente en el núcleo de las células; los ácidos nucleicos desempeñan un papel importante en la transmisión de las características hereditarias, en la síntesis de proteínas y en el control de las actividades celulares

nucleotide: one of the monomers that make up DNA and RNA; it consists of a nitrogen-containing base (a purine or pyrimidine), a sugar (ribose or deoxyribose), and a phosphate group *(854)*
nucleótido: uno de los monómeros que constituyen el ADN y el ARN; consiste en una base nitrogenada (una purina o una pirimidina), un azúcar (ribosa o desoxirribosa) y un grupo fosfato

nucleus: the tiny, dense central portion of an atom, composed of protons and neutrons *(108)*
núcleo: la diminuta porción central densa de un átomo; se compone de protones y neutrones

observation: information obtained through the senses; observation in science often involves a measurement *(15)*
observación: información obtenida a través de los sentidos; en la ciencia, la observación suele implicar la medición

octet rule: atoms react by gaining or losing electrons so as to acquire the stable electron structure of a noble gas, usually eight valence electrons *(195)*
regla del octeto: los átomos reaccionan ganando o perdiendo electrones a modo de adquirir la estructura electrónica estable de un gas noble, que por lo regular consta de ocho electrones de valencia

R134 Glossary

organic chemistry: the study of compounds containing carbon *(3)*
química orgánica: estudio de los compuestos que contienen carbono

oxidation: a process that involves complete or partial loss of electrons or a gain of oxygen; it results in an increase in the oxidation number of an atom *(694)*
oxidación: proceso que implica la pérdida total o parcial de electrones o la ganancia de oxígeno; conduce a un aumento en el número de oxidación de un átomo

oxidation number: a positive or negative number assigned to an atom to indicate its degree of oxidation or reduction; the oxidation number of an uncombined element is zero *(701)*
número de oxidación: número positivo o negativo que se asigna a un átomo para indicar su grado de oxidación o reducción; el número de oxidación de un elemento no combinado es cero

oxidation-number-change method: a method of balancing a redox equation by comparing the increases and decreases in oxidation numbers *(710)*
método de cambio del número de oxidación: método para balancear una ecuación redox comparando los incrementos y reducciones de los números de oxidación

oxidation-reduction reaction: a reaction that involves the transfer of electrons between reactants *(693)*
reacción de oxidación–reducción: reacción en la que hay transferencia de electrones entre los reactantes

oxidizing agent: the substance in a redox reaction that accepts electrons; in the reaction, the oxidizing agent is reduced *(695)*
agente oxidante: en una reacción redox, la sustancia que acepta electrones; en la reacción, el agente oxidante se reduce

P

partial pressure: the contribution each gas in a mixture of gases makes to the total pressure *(469)*
presión parcial: contribución de cada gas de una mezcla de gases a la presión total

pascal (Pa): the SI unit of pressure *(422)*
pascal (Pa): unidad de presión en el SI

Pauli exclusion principle: an atomic orbital may describe at most two electrons, each with opposite spin direction *(134)*
principio de exclusión de Pauli: orbital atómico puede describir como máximo a dos electrones, los cuales deben tener espín opuesto

peptide: an organic compound formed by a combination of amino acids in which the amino group of one acid is united with the carboxyl group of another through an amide bond *(845)*
péptido: compuesto orgánico formado por la combinación de aminoácidos de modo que el grupo amino de un ácido se une al grupo carboxilo de otro creando un enlace amida

peptide bond: the bond between the carbonyl group of one amino acid and the nitrogen of the next amino acid in the peptide chain; the structure is

$$\overset{\displaystyle O}{\overset{\|}{-C}}-\overset{\displaystyle H}{\overset{|}{N}}-$$

(845)
enlace péptido: enlace que hay entre el grupo carbonilo de un aminoácido y el nitrógeno del siguiente aminoácido de la cadena péptida; la estructura es

$$\overset{\displaystyle O}{\overset{\|}{-C}}-\overset{\displaystyle H}{\overset{|}{N}}-$$

percent composition: the percent by mass of each element in a compound *(325)*
composición porcentual: porcentaje en masa de cada elemento de un compuesto

percent error: the percent that a measured value differs from the accepted value *(65)*
error porcentual: porcentaje en que un valor medido difiere del valor aceptado

percent yield: the ratio of the actual yield to the theoretical yield for a chemical reaction expressed as a percentage; a measure of the efficiency of a reaction *(405)*
rendimiento porcentual: razón del rendimiento real al rendimiento teórico de una reacción química, expresado como porcentaje; es una medida de la eficiencia de la reacción

period: a horizontal row of elements in the periodic table *(46)*
periodo: fila horizontal de elementos en la tabla periódica

periodic law: when the elements are arranged in order of increasing atomic number, there is a periodic repetition of their physical and chemical properties *(162)*
ley periódica: si los elementos se acomodan en orden de menor a mayor número atómico, se observa una repetición periódica de sus propiedades físicas y químicas

periodic table: an arrangement of elements in which the elements are separated into groups based on a set of repeating properties *(46)*
tabla periódica: distribución de los elementos dividiéndolos en grupos según un conjunto de propiedades repetidas

Glossary **R135**

pH: a number used to denote the hydrogen-ion concentration, or acidity, of a solution; it is the negative logarithm of the hydrogen-ion concentration of a solution *(656)*

pH: número empleado para denotar la concentración de ion hidrógeno (acidez) de una solución; es el logaritmo negativo de la concentración de ion hidrógeno en una solución

phase: any part of a sample with uniform composition and properties *(39)*

fase: cualquier parte de una muestra que tiene composición y propiedades uniformes

phase diagram: a graph showing the conditions at which a substance exists as a solid, liquid, or vapor *(438)*

diagrama de fases: gráfica que muestra las condiciones en las que una sustancia existe como sólido, líquido o vapor

phospholipid: a lipid that contains a phosphate group; because phospholipids have hydrophilic heads and hydrophobic tails, they form the lipid bilayers found in cell membranes *(851)*

fosfolípido: lípido que contiene un grupo fosfato; como los fosfolípidos tienen una cabeza hidrofílica y una cola hidrofóbica, pueden formar las bicapas lípidas de las membranas celulares

photoelectric effect: the ejection of electrons by certain metals when they absorb light with a frequency above a threshold frequency *(143)*

efecto fotoeléctrico: liberación de electrones por algunos metales cuando absorben la luz con una frecuencia superior a un umbral (o frecuencia mínima)

photon: a quantum of light; a discrete bundle of electromagnetic energy that interacts with matter similarly to particles *(144)*

fotón: cuanto de luz; paquete discreto de energía electromagnética que interactúa con la materia de forma similar a como lo hacen las partículas

photosynthesis: the process by which green plants and algae use radiant energy from the sun to synthesize glucose from carbon dioxide and water *(839)*

fotosíntesis: proceso por el cual las plantas y algas verdes aprovechan la energía radiante del Sol para sintetizar glucosa a partir de dióxido de carbono y agua

physical change: a change during which some properties of a material change, but the composition of the material does not change *(37)*

cambio físico: cambio durante el cual se alteran algunas propiedades de un material, pero sin que se altere la composición del material

physical chemistry: the area of chemistry that deals with the mechanism, the rate, and the energy transfer that occurs when matter undergoes a change *(3)*

fisicoquímica: área de la química que se relaciona con el mecanismo, la velocidad y la transferencia de energía que ocurre cuando la materia sufre un cambio

physical property: a quality or condition of a substance that can be observed or measured without changing the substance's composition *(35)*

propiedad física: cualidad o condición de una sustancia que se puede observar o medir sin alterar la composición de la sustancia

pi bond (π bond): a covalent bond in which the bonding electrons are most likely to be found in sausage-shaped regions above and below the bond axis of the bonded atoms *(241)*

enlace pi (enlace π): enlace covalente en el que hay una alta probabilidad de encontrar los electrones de enlace en regiones alargadas que están arriba y abajo del eje de enlace de los átomos enlazados

Planck's constant (*h*): a number used to calculate the radiant energy *(E)* absorbed or emitted by a body based on the frequency of radiation (143)

constante de Planck (*h*): número que se usa para calcular la energía radiante (*E*) que un cuerpo absorbe o emite basándose en la frecuencia de radiación

polar covalent bond (polar bond): a covalent bond between atoms in which the electrons are shared unequally *(248)*

enlace covalente polar (enlace polar): enlace covalente entre átomos que no comparten equitativamente sus electrones

polar molecule: a molecule in which one side of the molecule is slightly negative and the opposite side is slightly positive *(249)*

molécula polar: molécula que tiene un lado ligeramente negativo y el otro ligeramente positivo

polyatomic ion: a tightly bound group of atoms that behaves as a unit and has a positive or negative charge *(232)*

ion poliatómico: grupo fuertemente enlazado de átomos, que se comporta como una unidad y tiene carga positiva o negativa

polymer: a very large molecule formed by the covalent bonding of repeating small molecules, known as monomers *(822)*

polímero: molécula muy grande formada por la unión, mediante enlaces covalentes, de moléculas pequeñas repetidas, llamadas monómeros

polysaccharide: a complex carbohydrate polymer formed by the linkage of many monosaccharide monomers; starch, glycogen, and cellulose are polysaccharides *(843)*

polisacárido: carbohidrato complejo formado por el encadenamiento de muchos monómeros monosacáridos; el almidón, el glucógeno y la celulosa son polisacáridos

positron: a particle with the mass of an electron but a positive charge *(881)*

positrón: partícula con la misma masa que un electrón pero con carga positiva

R136 Glossary

precipitate: a solid that forms and settles out of a liquid mixture *(49)*
precipitado: sólido que se forma a partir de una mezcla líquida y se asienta

precision: describes the closeness, or reproducibility, of a set of measurements taken under the same conditions *(64)*
precisión: cifra que describe la variabilidad de una serie de mediciones efectuadas en las mismas condiciones

product: a substance produced in a chemical reaction *(48)*
producto: sustancia que se obtiene en una reacción química

protein: any peptide with more than 100 amino acids *(845)*
proteína: cualquier péptido que tiene más de 100 aminoácidos

proton: a positively charged subatomic particle found in the nucleus of an atom *(107)*
protón: partícula subatómica con carga positiva que se encuentra en el núcleo de los átomos

pure chemistry: the pursuit of chemical knowledge for its own sake *(3)*
química pura: búsqueda de conocimientos químicos por sí mismos

pure substance: *see* substance
sustancia pura: *véase* sustancia

Q

quantum: the amount of energy needed to move an electron from one energy level to another *(129)*
cuanto: cantidad de energía necesaria para desplazar un electrón de un nivel energético a otro

quantum mechanical model: the modern description, primarily mathematical, of the behavior of electrons in atoms *(130)*
modelo según la mecánica cuántica: descripción moderna, primordialmente matemática, del comportamiento de los electrones en los átomos

R

radioactivity: the process by which nuclei emit particles and rays *(876)*
radiactividad: proceso por el cual los núcleos emiten partículas y rayos

radioisotope: an isotope that has an unstable nucleus and undergoes radioactive decay *(876)*
radioisótopo: isótopo cuyo núcleo es inestable y sufre desintegración radiactiva

rate: describes the speed of change over an interval of time *(595)*
velocidad (de reacción): cifra que describe la velocidad de cambio a lo largo de un intervalo de tiempo

rate law: an expression relating the rate of a reaction to the concentration of the reactants *(604)*
ley de velocidad de reacción: expresión que relaciona la velocidad de una reacción con la concentración de los reactantes

reactant: a substance present at the start of a reaction *(48)*
reactante: sustancia presente al principio de una reacción

reaction mechanism: a series of elementary reactions that take place during the course of a complex reaction *(607)*
mecanismo de reacción: serie de reacciones básicas que se dan durante el curso de una reacción compleja

reducing agent: the substance in a redox reaction that donates electrons; in the reaction, the reducing agent is oxidized *(695)*
agente reductor: en una reacción redox, la sustancia que cede electrones; en la reacción, el agente reductor se oxida

reduction: a process that involves a complete or partial gain of electrons or the loss of oxygen; it results in a decrease in the oxidation number of an atom *(694)*
reducción: proceso que implica una ganancia total o parcial de electrones o pérdida de oxígeno; provoca una disminución en el número de oxidación de un átomo

reduction potential: a measure of the tendency of a given half-reaction to occur as a reduction (gain of electrons) in an electrochemical cell *(737)*
potencial de reducción: medida de la tendencia que tiene una semirreacción específica de efectuarse como reducción (con ganancia de electrones) en una celda electroquímica

representative element: an element in an "A" group in the periodic table; as a group these elements display a wide range of physical and chemical properties. In their atoms, the *s* and *p* sublevels in the highest occupied energy level are partially filled *(171)*
elemento representativo: elemento de un grupo "A" de la tabla periódica; en conjunto, estos elementos exhiben una amplia gama de propiedades físicas y químicas. En sus átomos, los subniveles *s* y *p* del nivel energético ocupado más alto están parcialmente llenos

representative particle: the smallest unit into which a substance can be broken down without a change in composition, usually atoms, molecules, or ions *(308)*
partícula representativa: unidad más pequeña en que puede dividirse una sustancia sin que cambie su composición; por lo regular es un átomo, molécula o ion

resonance structure: one of the two or more equally valid electron dot structures of a molecule or polyatomic ion *(237)*
estructura de resonancia: una de las dos o más estructuras electrón-punto igualmente válidas de una molécula o ion poliatómico

responding variable: *see* dependent variable *(16)*
variable de respuesta: *véase* variable dependiente

Glossary **R137**

reversible reaction: a reaction in which the conversion of reactants into products and the conversion of products into reactants occur simultaneously *(609)*

reacción reversible: reacción en la que se da en forma simultánea la conversión de reactantes en productos y la conversión de productos en reactantes

S

salt bridge: a tube containing a strong electrolyte used to separate the half-cells in a voltaic cell; it allows the passage of ions from one half-cell to the other but prevents the solutions from mixing completely *(730)*

puente salino: tubo que contiene un electrolito fuerte y se usa para separar las semiceldas de una celda voltaica; permite el paso de iones de una semicelda a la otra, pero impide que las soluciones se mezclen totalmente

salt hydrolysis: a process in which the cations or anions of a dissociated salt accept hydrogen ions from water or donate hydrogen ions to water *(677)*

hidrólisis de sales: proceso por el cual los cationes o aniones de una sal disociada aceptan iones hidrógeno del agua o ceden iones hidrógeno al agua

saponification: the hydrolysis of fats or oils by a hot aqueous alkali-metal hydroxide; soaps are made by saponification *(851)*

saponificación: hidrólisis de grasas o aceites con una solución acuosa caliente de un hidróxido de metal alcalino; los jabones se hacen mediante la saponificación

saturated compound: an organic compound in which all carbon atoms are joined by single covalent bonds; it contains the maximum number of hydrogen atoms per carbon atom *(772)*

compuesto saturado: compuesto orgánico en el que todos los átomos de carbono están unidos unos a otros por enlaces covalentes sencillos; contiene el número máximo de átomos de hidrógeno por átomo de carbono

saturated solution: a solution containing the maximum amount of solute for a given amount of solvent at a constant temperature and pressure; an equilibrium exists between undissolved solute and ions in solution *(520)*

solución saturada: solución que contiene la cantidad máxima de soluto para una cantidad dada de disolvente a temperatura y presión constantes; existe equilibrio entre el soluto no disuelto y los iones en solución

scientific law: a concise statement that summarizes the results of many observations and experiments *(17)*

ley científica: expresión concisa que resume los resultados de muchas observaciones y experimentos

scientific method: a logical, systematic approach to the solution of a scientific problem; steps in the scientific method include making observations, testing hypotheses, and developing theories *(15)*

método científico: enfoque lógico y sistemático para resolver un problema científico; los pasos del método científico incluyen hacer observaciones, probar hipótesis y desarrollar teorías

scientific notation: an expression of numbers in the form $m \times 10^n$, where m is equal to or greater than 1 and less than 10, and n is an integer *(62)*

notación científica: convención por la cual los números se expresan en la forma $m \times 10^n$, donde m es un número mayor o igual que 1 y menor que 10, y n es un entero

self-ionization: a term describing the reaction in which two water molecules react to produce ions *(653)*

autoionización: reacción en la que dos moléculas de agua reaccionan para producir iones

sigma bond (σ bond): a bond formed when two atomic orbitals combine for form a molecular orbital that is symmetrical around the axis connecting the two atomic nuclei *(240)*

enlace sigma (enlace σ): enlace que se forma cuando dos orbitales atómicos se combinan para formar un orbital molecular que es simétrico respecto al eje que conecta a los dos núcleos atómicos

significant figures: all the digits that can be known precisely in a measurement, plus a last estimated digit *(66)*

dígitos significativos: todos los dígitos de una medición que se pueden conocer con precisión, más un último dígito estimado

single covalent bond: a bond formed when two atoms share a pair of electrons *(226)*

enlace covalente sencillo: enlace que se forma cuando dos átomos comparten un par de electrones

single-replacement reaction: a chemical change in which one element replaces a second element in a compound; also called a displacement reaction *(360)*

reacción de sustitución sencilla: cambio químico en el que un elemento reemplaza a un segundo elemento en un compuesto; también llamado reacción de desplazamiento

skeleton equation: a chemical equation that does not indicate the relative amounts of reactants and products *(348)*

ecuación esqueleto: ecuación química que no indica las cantidades relativas de los reactantes y productos

solid: a form of matter that has a definite shape and volume *(36)*

sólido: estado de la materia que tiene forma y volumen definidos

R138 Glossary

solubility: the amount of a substance that dissolves in a given quantity of solvent at specified conditions of temperature and pressure to produce a saturated solution *(520)*
solubilidad: cantidad de una sustancia que se disuelve en una cantidad dada de disolvente, bajo condiciones específicas de temperatura y presión, para producir una solución saturada

solubility product constant (K_{sp}): an equilibrium constant applied to the solubility of electrolytes; it is equal to the product of the concentrations of the ions each raised to a power equal to the coefficient of the ion in the dissociation equation *(622)*
constante del producto de solubilidad (K_{sp}): constante de equilibrio aplicada a la capacidad disoluble de electrolitos; es igual al producto de las concentraciones de los iones, cada una elevada a una potencia igual al coeficiente que tiene ese ion en la ecuación de disociación

solute: dissolved particles in a solution *(494)*
soluto: partículas disueltas en una solución

solution: a homogeneous mixture; consists of solutes dissolved in a solvent *(39)*
solución: mezcla homogénea que consiste en solutos disueltos en un disolvente

solvation: a process that occurs when an ionic solute dissolves; in solution, solvent molecules surround the positive and negative ions *(495)*
solvatación: proceso que tiene lugar cuando se disuelve un soluto iónico; en solución, las moléculas de disolvente rodean a los iones positivos y negativos

solvent: the dissolving medium in a solution *(494)*
disolvente: medio dispersor en una solución

specific heat: the amount of heat needed to increase the temperature of 1 g of a substance 1°C; also called specific heat capacity *(559)*
calor específico: cantidad de calor requerida para elevar 1°C la temperatura de 1 g de una sustancia

specific rate constant: a proportionality constant relating the concentrations of reactants to the rate of the reaction *(604)*
constante específica de velocidad de reacción: constante de proporcionalidad que relaciona las concentraciones de los reactantes con la velocidad de la reacción

spectator ion: an ion that is not directly involved in a chemical reaction; an ion that does not change oxidation number or composition during a reaction *(370)*
ion espectador: ion que no interviene directamente en una reacción química; ion que no cambia de número de oxidación ni de composición durante una reacción

spectrum: wavelengths of visible light that are separated when a beam of light passes through a prism; range of wavelengths of electromagnetic radiation *(139)*
espectro: longitudes de onda de la luz visible que se separan cuando un haz de luz atraviesa un prisma; gama de longitudes de onda de radiación electromagnética

spin: a quantum mechanical property of electrons that may be thought of as clockwise or counterclockwise *(134)*
espín: propiedad de los electrones según la mecánica cuántica en la cual la rotación se considera en sentido de las agujas del reloj o en sentido contrario a las agujas del reloj

spontaneous reaction: a reaction that favors the formation of products at the specified conditions; spontaneity depends on enthalpy and entropy changes *(628)*
reacción espontánea: reacción que favorece la formación de productos, bajo las condiciones especificadas; la espontaneidad depende de los cambios de entalpía y de entropía

standard atmosphere (atm): a unit of pressure; it is the pressure required to support 760 mm of mercury in a mercury barometer at 25°C *(422)*
atmósfera estándar (atm): unidad de presión; es la presión necesaria para sostener 760 mm de mercurio en un barómetro de mercurio a 25°C

standard cell potential (E°_{cell}): the measured cell potential when the ion concentration in the half-cells are 1.00*M* at 1 atm of pressure and 25°C *(737)*
potencial estándar de celda (E°_{cell}): potencial de celda que se mide cuando las concentraciones de los iones en las semiceldas son 1.00*M* a 1 atm de presión y 25°C

standard heat of formation (ΔH_f°): the change in enthalpy that accompanies the formation of one mole of a compound from its elements with all substances in their standard states at 25°C *(580)*
calor estándar de formación (ΔH_f°): cambio de entalpía que acompaña a la formación de un mol de un compuesto a partir de sus elementos, estando todas las sustancias en su estado estándar a 25°C

standard hydrogen electrode: an arbitrary reference electrode (half-cell) used with another electrode (half-cell) to measure the standard reduction potential of that cell; the standard reduction potential of the hydrogen electrode is assigned a value of 0.00 V *(738)*
electrodo estándar de hidrógeno: electrodo (semicelda) arbitrario de referencia que se usa junto con otro electrodo (semicelda) para medir el potencial estándar de reducción de esa celda; se asigna al potencial estándar de reducción del electrodo de hidrógeno el valor de 0.00 V

standard solution: a solution of known concentration used in carrying out a titration *(674)*
solución estándar: solución cuya concentración se conoce; se usa para efectuar valoraciones químicas

standard temperature and pressure (STP): the conditions under which the volume of a gas is usually measured; standard temperature is 0°C, and standard pressure is 101.3 kPa, or 1 atmosphere (atm) *(320)*
temperatura y presión estándar (TPE): las condiciones en las que normalmente se mide el volumen de un gas; la temperatura estándar es 0°C y la presión estándar es 101.3 kPa, o sea, 1 atmósfera (atm)

Glossary **R139**

stereoisomers: molecules that have atoms in the same order, but which differ in the arrangement of the atoms in space *(776)*

estereoisómeros: moléculas cuyos átomos están en el mismo orden, pero que difieren en la distribución de los átomos en el espacio

stoichiometry: that portion of chemistry dealing with numerical relationships in chemical reactions; the calculation of quantities of substances involved in chemical equations *(386)*

estequiometría: rama de la química que se ocupa de las relaciones numéricas en las ecuaciones químicas; el cálculo de las cantidades de sustancias presentes en las ecuaciones químicas

straight-chain alkane: a saturated hydrocarbon that contains any number of carbon atoms arranged one after the other in a chain *(764)*

alcano de cadena lineal: hidrocarburo saturado que contiene cualquier número de átomos de carbono acomodados uno tras otro en una cadena

strong acid: an acid that is completely (or almost completely) ionized in aqueous solution *(664)*

ácido fuerte: ácido que se ioniza casi totalmente en solución acuosa

strong base: a base that completely dissociates into metal ions and hydroxide ions in aqueous solution *(668)*

base fuerte: base que se disocia totalmente en una solución acuosa para dar iones metálicos y iones hidróxido

strong electrolyte: a solution in which a large portion of the solute exists as ions *(497)*

electrolito fuerte: solución en la que una porción considerable del soluto existe en forma de iones

structural formula: a chemical formula that shows the arrangement of atoms in a molecule or a polyatomic ion; each dash between a pair of atoms indicates a pair of shared electrons *(227)*

fórmula estructural: fórmula química que indica la distribución de los átomos en una molécula o ion poliatómico; cada raya entre un par de átomos indica un par de electrones compartidos

sublimation: the process in which a solid changes to a gas or vapor without passing through the liquid state *(436)*

sublimación: proceso por el cual un sólido cambia a gas o vapor sin pasar por el estado líquido

substance: matter that has a uniform and definite composition; either an element or a compound; also called pure substance *(35)*

sustancia: materia que tiene una composición uniforme y definida; puede ser un elemento o un compuesto; también llamada sustancia pura

substituent: an atom or group of atoms that can take the place of a hydrogen atom on a parent hydrocarbon molecule *(767)*

sustituto: átomo o grupo de átomos que puede ocupar el lugar de un átomo de hidrógeno en una molécula precursora de hidrocarburo

substitution reaction: a common type of organic reaction; involves the replacement of an atom or group of atoms by another atom or group of atoms *(801)*

reacción de sustitución: tipo común de reacción orgánica; implica el reemplazo de un átomo o grupo de átomos por otro átomo o grupo de átomos

substrate: a molecule on which an enzyme acts *(847)*

sustrato: molécula sobre la que actúa una enzima

supersaturated solution: a solution that contains more solute than it can theoretically hold at a given temperature; excess solute precipitates if a seed crystal is added *(522)*

solución sobresaturada: solución que contiene más soluto del que en teoría puede contener a una temperatura específica; el soluto en exceso se precipita si se añade un cristal que actúa como semilla

surface tension: an inward force that tends to minimize the surface area of a liquid; it causes the surface to behave as if it were a thin skin *(490)*

tensión superficial: fuerza que tiende a reducir al mínimo la superficie total de un líquido y actúa hacia el seno de éste; hace que la superficie se comporte como si fuera una membrana elástica

surfactant: any substance that interferes with the hydrogen bonding between water molecules and thereby reduces surface tension; soaps are surfactants *(490)*

tensoactivo: cualquier sustancia que perturba la formación de enlaces de hidrógeno entre las moléculas de agua y así reduce la tensión superficial; los jabones y detergentes son tensoactivos

surroundings: everything in the universe outside of the system *(557)*

entorno: todo lo que no forma parte del sistema, es decir, el resto del universo

suspension: a mixture from which some of the particles settle out slowly upon standing *(504)*

suspensión: mezcla de la que se separan lentamente algunas partículas por asentamiento cuando no se agita

synthesis reaction: *see* combination reaction

reacción de síntesis: *véase* reacción de combinación

system: a part of the universe on which you focus your attention *(557)*

sistema: parte del universo en la que centramos nuestra atención

R140 Glossary

T

technology: the means by which a society provides its members with those things needed and desired *(8)*
tecnología: los medios por los cuales una sociedad proporciona a sus miembros las cosas que necesitan y desean

temperature: a measure of the average kinetic energy of particles in matter; temperature determines the direction of heat transfer *(78)*
temperatura: medida de la energía cinética promedio de las partículas de la materia; la temperatura determina la dirección de la transferencia de calor

tetrahedral angle: a bond angle of 109.5° that results when a central atom forms four bonds directed toward the center of a regular tetrahedron *(242)*
ángulo tetraédrico: ángulo de enlace de 109.5° que se forma cuando un átomo central forma cuatro enlaces dirigidos hacia el centro de un tetraedro regular

theoretical yield: the amount of product that could form during a reaction calculated from a balanced chemical equation; it represents the maximum amount of product that could be formed from a given amount of reactant *(405)*
rendimiento teórico: cantidad de producto que podría formarse durante una reacción, calculada a partir de una ecuación química balanceada; representa la cantidad máxima de producto que podría formarse a partir de una cantidad determinada de reactantes

theory: a well-tested explanation for a broad set of observations *(17)*
teoría: explicación, probada exhaustivamente, de un conjunto amplio de observaciones

thermochemical equation: a chemical equation that includes the enthalpy change *(565)*
ecuación termoquímica: ecuación química que incluye el cambio de calor

thermochemistry: the study of energy changes that occur during chemical reactions and changes in state *(556)*
termoquímica: estudio de los cambios de calor que acompañan a las reacciones químicas y a los cambios de estado físico

titration: process used to determine the concentration of a solution (often an acid or base) in which a solution of known concentration (the standard) is added to a measured amount of the solution of unknown concentration until an indicator signals the end point *(673)*
valoración química: proceso empleado para determinar la concentración de una solución (a menudo un ácido o base) por el cual una solución de concentración conocida (solución estándar) se añade a una cantidad medida de una solución cuya concentración se desconoce, hasta que un indicador marca el punto final

***trans* configuration:** the configuration in which substituent groups are on the opposite sides of a double bond *(776)*
configuración *trans*: configuración en la que los grupos sustitutos están en lados opuestos de un doble enlace

transition metal: one of the Group B elements in which the highest occupied *s* sublevel and a nearby *d* sublevel generally contain electrons *(172)*
metal de transición: uno de los elementos del grupo B en el que el subnivel *s* ocupado más alto y un subnivel *d* cercano generalmente contienen electrones

transmutation: the conversion of an atom of one element to an atom of another element *(885)*
transmutación: conversión de un átomo de un elemento en un átomo de otro elemento

transuranium element: any elements in the periodic table with atomic number above 92, the atomic number of uranium *(886)*
elemento transuránico: cualquier elemento de la tabla periódica cuyo número atómico es mayor que 92, el número atómico del uranio

triglyceride: an ester in which all three hydroxyl groups on a glycerol molecule have been replaced by long-chain fatty acids; fats are triglycerides *(850)*
triglicérido: éster en el que los tres grupos hidroxilo de una molécula de glicerol han sido sustituidos por ácidos grasos de cadena larga; las grasas son triglicéridos

triple covalent bond: a covalent bond in which three pairs of electrons are shared by two atoms *(230)*
enlace covalente triple: enlace covalente en el que dos átomos comparten tres pares de electrones

triple point: the point on a phase diagram that represents the only set of conditions at which all three phases exist in equilibrium with one another *(402)*
punto triple: punto de un diagrama de fases que representa el único conjunto de condiciones en el que las tres fases existen en equilibrio

triprotic acid: any acid that contains three ionizable protons (hydrogen ions); phosphoric acid (H_3PO_4) is a triprotic acid (647)
ácido triprótico: ácido que contiene tres protones (iones hidrógeno) ionizables; el ácido fosfórico (H_3PO_4) es un ácido triprótico

Tyndall effect: scattering of light by particles in a colloid or suspension, which causes a beam of light to become visible *(506)*
efecto Tyndall: dispersión de la luz por las partículas de un coloide o una suspensión, que hace que un haz de luz se vuelva visible

Glossary **R141**

U

unit cell: the smallest group of particles within a crystal that retains the geometric shape of the crystal *(433)*
celda unitaria: grupo más pequeño de partículas dentro de un cristal que conserva la forma geométrica del cristal

unsaturated solution: a solution that contains less solute than a saturated solution at a given temperature and pressure *(520)*
solución insaturada: se dice de una solución que contiene menos soluto que una solución saturada a una temperatura y presión específicas

unsaturated compound: an organic compound with one or more double or triple carbon–carbon bonds *(772)*
compuesto insaturado: compuesto orgánico que tiene uno o más dobles o triples enlaces carbono–carbono

unshared pair: a pair of valence electrons that is not shared between atoms *(227)*
par no compartido: par de electrones de valencia que no es compartido por dos átomos

V

vacuum: a space where no particles of matter exist *(421)*
vacío: espacio en el que no existen partículas de materia

valence electron: an electron in the highest occupied energy level of an atom *(194)*
electrón de valencia: electrón que está en el nivel energético ocupado más alto de un átomo

van der Waals forces: the two weakest intermolecular attractions—dispersion interactions and dipole forces *(250)*
fuerzas de van der Waals: las dos atracciones intermoleculares más débiles—interacciones de dispersión y fuerzas dipolares

vapor: describes the gaseous state of a substance that is generally a liquid or solid at room temperature *(37)*
vapor: estado gaseoso de una sustancia que suele ser líquida o sólida a temperatura ambiente

vaporization: the conversion of a liquid to a gas or a vapor *(426)*
vaporización: conversión de un líquido en gas o vapor

vapor pressure: a measure of the force exerted by a gas above a liquid in a sealed container; a dynamic equilibrium exists between the vapor and the liquid *(427)*
presión de vapor: medida de la fuerza que ejerce un gas sobre un líquido en un contenedor sellado; equilibrio dinámico que existe entre el vapor y el líquido

voltaic cell: an electrochemical cell used to convert chemical energy into electrical energy; the energy is produced by a spontaneous redox reaction *(730)*
celda voltaica: celda electroquímica empleada para convertir energía química en energía eléctrica; la energía se produce por una reacción redox espontánea

volume: a measure of the space occupied by a sample of matter *(34)*
volumen: medida del espacio ocupado por una muestra de materia

VSEPR theory: valence-shell electron-pair repulsion theory; because electron pairs repel, molecules adjust their shapes so that valence electron pairs are as far apart as possible *(242)*
teoría RPENV: teoría de repulsión de pares de electrones del nivel de valencia; como los pares de electrones se repelen, las moléculas ajustan su forma de modo que los pares de electrones de valencia estén lo más alejados posible entre sí

W

water of hydration: water molecules that are an integral part of a crystal structure *(498)*
agua de hidratación: moléculas de agua que forman parte integral de una estructura cristalina

wavelength (λ): the distance between adjacent crests of a wave *(138)*
longitud de onda (λ): distancia entre crestas adyacentes de una onda

wax: an ester of a long-chain fatty acid and a long-chain alcohol *(852)*
cera: éster de un ácido graso de cadena larga y un alcohol de cadena larga

weak acid: an acid that is only slightly ionized in aqueous solution *(664)*
ácido débil: ácido que se ioniza poco en solución acuosa

weak base: a base that reacts with water to form the hydroxide ion and the conjugate acid of the base *(668)*
base débil: base que reacciona con agua para formar el ion hidróxido y el ácido combinado de la base

weak electrolyte: a solution that conducts electricity poorly because only a fraction of the solute exists as ions *(497)*
electrolito débil: solución que apenas conduce la electricidad porque sólo una fracción del soluto existe en forma de iones

weight: a force that measures the pull of gravity on a given mass *(77)*
peso: fuerza que mide la atracción de la gravedad sobre una masa específica

R142 Glossary

Index

The page on which a term is defined is indicated in **boldface** type. Page numbers for appendices begin with *R.*

A

Index **R143**

R144 Index

Index **R145**

D

R146 Index

Index **R147**

G

H

R148 Index

Index **R149**

R150 Index

Index **R151**

P

R152 Index

Q

R

S

Index **R153**

T

R154 Index

Index **R155**

Credits

Staff Credits
Art/Design—Courtenay Kelley, Beth Kun, Russ Lappa, Rich McMahon, Deborah Nicholls, Michelle Reyes, Stephen Rider, Rashid Ross, Laurel Smith, Lisa Smith-Ruvalcaba, Ted Smykal, Emily Soltanoff, Heidi Wilson, Heather Wright; *Composition/Content Management*—Lisa Bailey, Penny Baker, Larry Berkowitz, Lisa Carrillo, Phil Gagler, Mary Jean Jones, Kathy Martin, Kathy Mercandetti, Ken Myett, Karyn Mueller, Kim Ortell; *Digital/Multimedia*—Joan Archer-Scott Neil Benjamin, Meredith Glassman, Meytal Kotik, Debbie Munson, Hugh Rutledge; *Editorial*—Diane Braff, Ken Chang, Allison Cook Bellistri, Jody Dogra, Kimberly Fekany Lee, Kathryn Fobert, Jay Grasso, Kathy Immel, Anne McLaughlin; *Production*—Lori-Anne Cohen

Additional Credits
James Ackroyd, Jessica Beauchemin, Jim Glynn, Mark Hughes, Beth Kelly, Margy Kuntz, Steve Miller, Heather Myler, Kathleen Naughton, Laura Prescott, Karen Rowan, Cecelia Schneider, Cheryl Steinecker

Illustrations
Courtenay Kelley, Rich McMahon, Precision Graphics, Laurel Smith, Ted Smykal

Photographs
Every effort has been made to secure permission and provide appropriate credit for photographic material. The publisher deeply regrets any omission and pledges to correct errors called to its attention in subsequent editions.

Unless otherwise acknowledged, all photographs are the property of Pearson Education, Inc.

Photo locators denoted as follows: Top (T), Center (C), Bottom (B), Left (L), Right (R), Background (Bkgd)

Cover and Title Page: hundreddays/iStockphoto **x** (T) Claudio Bertoloni/Shutterstock, (B) Murat Boylu/Shutterstock; **xi** (BL) Harry Taylor/Courtesy of the Natural History Museum, London/©DK Images, (BR) Steve Mason/Getty Images, (T) Traveler1116/iStockphoto; **xii** (T) Don Johnston/Stone/Getty Images, (B) ©DK Images; **xiv** Fundamental Photographs; **xv** linhof/Shutterstock; **xxiv–1** (C) Alexander Raths/Shutterstock; **2** (T) Eric Isselée/Shutterstock; **3** (Inset) Andraz Cerar/Shutterstock, (Bkgd) Mitchell Funk/Getty Images; **4** Mitchell Funk/Getty Images; **5** Steve Mason/Getty Images; **6** (T) M. Kornmesser (ESA/Hubble)/NASA Image Exchange, (B) Traveler1116/iStockphoto; 7 (TR) Goodman Photography/Shutterstock, (TL) Reuters Media, (B) Spencer Platt/Staff/Getty Images; **9** (Inset, L) Cathleen Clapper/Shutterstock, (Bkgd) Paul Clarke/Shutterstock, (Inset, R) Sheila Terry/Photo Researchers, Inc.; **10** (R) Hank Morgan/AGE Fotostock, (L) Kenneth Eward/Photo Researchers, Inc.; **11** NASA Image Exchange; **12** (T) blackpixel/Shutterstock, (Bkgd) cla78/Shutterstock, (Inset, B) Karin Lau/Shutterstock, (Inset, T) WitR/Shutterstock; **14** (T) Photo Researchers, Inc., (B) Valentin Agapov/Shutterstock; **15** (T) The Metropolitan Museum of Art/Art Resource, NY/Art Resource, NY, (B) zumawireworldphotostwo471970/NewsCom; **16** MARK DUNCAN/©AP Images; **17** Ken O'Donoghue/Pearson Education; **18** (T) Bruce Ely/©AP Images, (B) Image Source/Christopher Robbins/Zuma Press, Inc.; **19** Bob Daemmrich/PhotoEdit, Inc.; **22** (T) Ashley Whitworth/Shutterstock, (B) Jupiter Images/Jupiter Images; **27** zumawireworldphotostwo471970/NewsCom; **29** Phil Cole/Getty Images; **30** John Moore/Pearson Education; **32** (Bkgd) Foodcollection/Getty Images; **33** (Inset) thumb/Shutterstock; **34** (BR) Aptyp_koK/Shutterstock, (BL) FloridaStock/Shutterstock, (T) tororo reaction/Shutterstock; **35** (TL) bunhill/iStockphoto, (TR) Leifr/Shutterstock, (TL) Ronen/Shutterstock; **36** (R, L, C) Geoffrey Nilsen Photography; **37** Lawrence Migdale/Photo Researchers, Inc.; **38** (T) Getty/Jupiter Images; **39** (R, L, C) Ken O'Donoghue; **42** (T) Dana Bartekoske/Shutterstock, (BR) David Lawrence, (BL) David lawrence, (BC) David Lawrence; **43** (L) Alamy, (C Inset) Charles D. Winters/Photo Researchers, Inc., (L inset) David Lawrence, (C) Getty Images, (R) Pearson Photo Studio, (R Inset) Viktor/Shutterstock; **44** (L) ©Andraz Cerar/Shutterstock, (CR) 300dpi/Shutterstock, (R) DK Images; **46** Theodore Clutter/Photo Researchers, Inc.; **48** (T) devi/Shutterstock, (B Inset) elena moiseeva/Shutterstock; **49** (C) Andrew Lambert Photography/Photo Researchers, Inc., (R) Stefano Scata/Foodcollection/Getty Images, (L) Gusto Productions/Photo Researchers, Inc.; **50** (R, L) Richard Megna/Fundamental Photographs; **51** (T) Ken O'Donoghue; **52** (L) JGI/Jamie Grill/Blend Images/Getty Images; **53** (TL) bunhill/iStockphoto, (Bkgd) Huguette Roe/Shutterstock, (BL) mrfotos/Shutterstock, (CC) prism68/Shutterstock, (BR) Sergio Schnitzler/Shutterstock, (TR) TheClarkester/Shutterstock; **54** (T) tororo reaction/Shutterstock; **57** (R, L) Fundamental Photographs; **58** (R) thumb/Shutterstock; **60** Gordon Wiltsie/Getty Images; **61** (Inset) David Toase/Getty Images; **62** (B) Steve Gschmeissner/Photo Researchers, Inc., (T) Technotr/iStockphoto; **66** (B) Alexander Kalina/Shutterstock, (T) Gjermund Alsos/Shutterstock; **72** Ken O'Donoghue/Pearson Education; **73** (BR) KPA/Hecht/Alamy Images, (TR) moodboard/Alamy Images, (Bkgd) photodisc/AGE Fotostock, (CR) Travel Ink/Getty Images; **77** NASA/NASA; **78** (B) Anthony Blake/PhotoLibrary Group, Inc., Ilya Andriyanov/Shutterstock; **80** Lijuan Guo/Shutterstock; **81** Pearson Education Curriculum Group; **83** (Inset, R) Richard B. Levine/PhotoLibrary Group, Inc., (Inset, L) Shutterstock, (Bkgd) White/Photolibrary Group, Inc.; **84** cheyennezj/Shutterstock; **87** Tetra Images/SuperStock; **92** (R) PhotoSpin/AGE Fotostock, (L) Richard Cano/iStockphoto; **97** NASA; **98** David Toase/Getty Images; **100** (Bkgd) IBM Research/Almaden Center/IBM Almaden Research Center; **101** (Inset) ©The Trustees of The British Museum; **102** motorolka/Shutterstock; **103** Harry Taylor/Courtesy of the Natural History Museum, London/©DK Images; **104** Yasuhiro Shirai/News & Media Relations, Rice University; **105** Claudio Bertoloni/Shutterstock; **106** (B) Richard Megna/Fundamental Photographs; **107** Wendy Kaveney Photography/Shutterstock; **110** (Bkgd) Andrew Syred/Photo Researchers, Inc., (BR) Science Photo Library/AGE Fotostock, (BL) Science Source/Photo Researchers, Inc.; **111** (CR, CL) Micrograph courtesy Obducat CamScan Ltd. Superior Electron Optics By Design./Obducat CamScam Ltd., (TR) Photolibrary/Getty Images, (BR) Volker Steger/Photo Researchers, Inc.; **112** PK-Photos/iStockphoto; **113** Murat Boylu/Shutterstock; **117** Murat Boylu/Shutterstock; **120** Ken O'Donoghue; **121** rubberball/Getty Images; **123** Richard Megna/Fundamental Photographs; **124** (Inset) ©The Trustees of The British Museum; **126** Alexey Stiop/Shutterstock; **128** (B) Adrian Sherratt/Alamy Images, (T) Philippe Plailly/Photo Researchers, Inc.; **130** Chris Curtis/Shutterstock; **133** (T) Boyer/Roger Violle/Getty Images, (C) Photo courtesy Riken/Riken; **134** Ayakovlev/Shutterstock; **137** Leifr/Shutterstock; **138** Karin Lau/Shutterstock; **140** (L) H. Mark Weidman Photography/Alamy Images, (R) scoutingstock/Shutterstock; **142** Richard Megna/Fundamental Photographs; **146** (BL) ©AP Images, (T) Design Pics Inc./Alamy, (BR) Dgrilla/Shutterstock; **147** Susumu Nishinaga/Photo Researchers, Inc.; **150** H. Mark Weidman Photography/Alamy Images; **153** scoutingstock/Shutterstock; **156** Colin Keates/©DK Images; **158** (Bkgd) artpartner-images/Alamy Images; **160** (T) Tatiana Popova/Shutterstock, (B) ©sciencephotos/Alamy; **161** (L) ©The Granger Collection, NY, (R) IgorGolovniov/Shutterstock; **163** (Bkgd) ©DK Images, (CR) ©John Durham/Photo Researchers, Inc., (CL) Abhishek Dogra, (BR) Omikron/Photo Researchers, Inc., (L) PhotoSpin/Alamy, (TR) Sandra Caldwell/Shutterstock; **164** (BC) Ken Davies/Masterfile Corporation, (BL) Vaughan Ryall/Alamy; **165** (T) ©DK Images, (C) slavchovr/Shutterstock, (B) Thomas Barrat/Shutterstock; **166** (L) Ilan Rosen/Alamy Images, (R) Volker Steger/Siemens/Photo Researchers, Inc.; **170** (BR) Michael Ledray/Shutterstock, (Bkgd) Vibrant Image Studio/Shutterstock; **171** (Inset) ©AP Images, (B) Igumnova Irina/Shutterstock, (CR) movit/Shutterstock; **172** (T) Paolo Koch/Photo Researchers, Inc.; **183** (R) Philip Dowell/©DK Images, (CL) Photo Researchers, Inc., (T) Sebastian Kaulilzki/Shutterstock; **184** Ken O'Donoghue/Prentice Hall; **185** Vaughan Ryall/Alamy; **192** ©Carsten Peter/Speleoresearch & Films/National Geographic Image Collection; **193** (Inset) Hakimata Photography/Shutterstock; **194** (BR) Phillip Hayson/Photo Researchers, Inc., (BL) Yurchyks/Shutterstock; **196** (Bkgd) A.T. Willett/Alamy Images; **199** (T) Carlos Caetano/Shutterstock; **201** (T) Carlos Caetano/Shutterstock; **202** (T) Charles D. Winters/Photo Researchers, Inc., (B) Charles Falco/Photo Researchers, Inc.; **206** (Bkgd) Kevin Foy/Alamy Images; **207** (T) Ken O'Donoghue/Prentice Hall; **209** (TL) Christina Richards/Shutterstock; **210** (B) Flickr/Getty Images; **212** (R) Aaron Amat/Shutterstock, (L) Emily Lai/Alamy, (C) Tyler Olson/Shutterstock; **213** (BL) Yurchyks/Shutterstock; **217** (Inset) Hakimata Photography/Shutterstock; **220** (Bkgd) Vera Volkova/Shutterstock; **221** (Inset) Danny E Hooks/Shutterstock; **222** (TL) Ivonne Wierink/Shutterstock, (BC) Jupiter Images/Jupiter Images, (BL) Jupiter Images/Jupiter Images, (BR) Michael Newman/PhotoEdit, Inc.; **223** (TR) Susa Tusa/NewsCom; **225** (R) Christopher Elwell/Shutterstock, (L) Nubephoto/Shutterstock; **226** (T) NASA; **228** (BL) Jupiterimages/Jupiter Images; **230** (Bkgd) Jupiterimages/Jupiter Images; **232** (BL) Editorial Image, LLC/Alamy Images; **234** (Bkgd) Jhaz Photography/Shutterstock; **237** Deborah Davis/Getty Images; **238** (Inset) Ken O'Donoghue; **239** (CR) magicinfoto/Shutterstock, (TR) Mark Sykes/Alamy Images; **240** (T) U.S. Geological Survey; **247** (B) David Young-Wolff/PhotoEdit, Inc., (T) Yaroslav/Shutterstock; **251** (BR) Runk/Schoenberger/Alamy Images; **252** (B) U. Bellhaeuser/Getty Images; **254** (T) Ken O'Donoghue; **255** U.Bellhaeuser/Getty Images; **259** (Inset) Danny E Hooks/Shutterstock; **262** (Bkgd) Mitshu/Shutterstock; **263** (Inset) Thinstock Images/Getty Images; **264** Santje/Shutterstock; **266** Fundamental Photographs; **269** (B) Vladimir Mucibabic/Shutterstock, (T) volschenkh/iStockphoto; **270** (Inset, B) blickwinkel/Alamy Images, (Bkgd) NewsCom, (Inset, T) Suzanne Long/The Image Works, Inc.; **271** (T) Elena Schweitzer/Shutterstock, (B) Gary Ombler/©DK Images; **272** Geoffrey Kuchera/Shutterstock; **274** (R) Blend Images/Getty Images, (L) Lincoln Museum and Gallery/Eileen Tweedy/The Art Archive; **276** (T) Albert Lozano/Shutterstock, Erhan Dayi/Shutterstock; **278** Ken O'Donoghue; **280** Brand X/Getty Images; **281** AFP/Getty Images; **284** (B) Chassenet/AGE Fotostock, (Bkgd) Christian Kover/PhotoLibrary Group, Inc.; **285** (B) Alamy Images, (T) Denis Tabler/Shutterstock; **287** Workbook Stock/Getty Images; **289** Westend61/SuperStock; **290** Fundamental Photographs; **292** Steve Bower/Shutterstock; **301** Thinkstock/Getty Images; **304** Robert Stone/Dreamstime LLC; **305** (R) Nicholas Monu/iStockphoto; **306** (B) Arkady/Shutterstock, (T) Eldad Yitzhak/Shutterstock; **308** (T) Nataliya Hora/Shutterstock; **310** (B) Getty Images/PhotoLibrary Group, Inc., (TL) Ken Karp/Addison Wesley; **316** (BL) arlindo71/iStockphoto; **317** Kinetic Imagery/Shutterstock; **324** Ken O'Donoghue; **325** (T) GIPhotoStock/Photo Researchers, Inc.; **328** Ken O'Donoghue; **330** (TL) Lisa F. Young/Shutterstock, (BL) Sebastian Crocker/Shutterstock; **334** (Inset) Steve Parsons/PA Photos/Landov LLC, (Bkgd) The Star-Ledger/Polaris Images; **336** (B) Arkady/Shutterstock; **341** Nicholas Monu/iStockphoto; **342** Ronald van der Beek/Shutterstock; **344** (Bkgd) Michel Gunther/Peter Arnold, Inc.; **345** (R) Image Source/Getty Images; **346** (T) Alamy Images, (B) Stefano Tiraboschi/Shutterstock; **347** (C) David Young-Wolff/PhotoEdit, Inc., (L) LAimages/Shutterstock, (R) Olga Utlykova/Shutterstock; **348** (T, B) Ken Karp; **354** Ken O'Donoghue; **356** Jupiter Images; **357** (R) Fundamental Photographs; **358** (L, CR, C) Ken Karp/Addison Wesley; **360** (R, L, C) David Lawrence/Addison Wesley; **362** (R, L) Richard Megna/Fundamental Photographs; **364** Richard Megna/Fundamental Photographs; **366** (BR, BL, BC) David Lawrence/Addison Wesley, (TR) Fundamental Photographs, (CR, CM, CL) Ken Karp/Addison Wesley; **367** (TR, TL, B) Richard Megna/Fundamental Photographs; **368** (T) Bettmann/Corbis, (B) EFE/Corbis; **369** (T) M. Wisshak/AGE Fotostock, Richard Megna/Fundamental Photographs; **370** (T) Patrick Hermans/Shutterstock; **372** David Lawrence/Addison Wesley; **375** Patrick Hermans/Shutterstock; **378** (TR) David Lawrence/Addison Wesley, (TL, BR) Richard Megna/Fundamental Photographs; **379** Image Source/Getty Images; **382** (Bkgd) Shutterstock; **383** (Inset) Alistair Scott/Shutterstock; **384** (T) Maxim Petrichuk/Shutterstock; **386** (Bkgd) B.S.Karan/Shutterstock; **390** Andy Sotiriou/Getty Images; **392** Erich Karkoschka/NASA; **397** (Bkgd) Don Johnston/Stone/Getty Images; **400** (TL) Aleksi Markku/Shutterstock; **404** (BL) PeteMuller/iStockphoto; **407** (C) Alan C. Heison/Shutterstock, (B) Joel Shawn/Shutterstock, (T) Maksud/Shutterstock; **409** B.S.Karan/Shutterstock; **411** Odin M. Eidskrem/Shutterstock; **414** Micargi Bicycles, Inc.; **415** Alistair Scott/Shutterstock; **418** (Bkgd) TechWizard/Shutterstock; **419** (Inset) SilviaJansen/iStockphoto; **421** Richard Megna/Fundamental Photographs; **424** J. Biretta/Hubble Heritage Team/NASA; **425** (BR, BL) Richard Megna/Fundamental Photographs, Volodymyr Krasyuk/Shutterstock; **426** (L) Ivan Demyanov/Shutterstock, (R) James King-Holmes/Photo Researchers, Inc.; **430** Francis Hammond/Getty Images; **432** (L) ©Colin Keates/Courtesy of the Natural History Museum, London/DK Images, (R) Colin Keates/©DK Images, (C) Harry Taylor/©DK Images; **433** Mark Schneider/Getty Images (CL) ©dimitriyd/Shutterstock, (CR) Joel Arem/Photo Researchers, Inc., (R) Jiri Slama/Shutterstock; **435** Ken O'Donoghue/Prentice

R156 Credits

Hall; **436** (T) bierchen/Shutterstock, (B) Richard Megna/Fundamental Photographs; **437** Jens Meyer/©AP Images; **438** Richard Megna/Fundamental Photographs; **439** John Kelly/Getty Images; **440** (Bkgd) Dgrilla/Shutterstock; **441** (TR) Alex Segre/Alamy Images, (TL) Epic Stock/Shutterstock, (BL) GA Tech/©Georgia Institute of Technology; **443** (Inset) ©Wolfgang Amri/iStockphoto; **445** SilviaJansen/iStockphoto; **450** Bob Daemmrich/PhotoEdit, Inc.; **451** Robert Llewellyn/PhotoLibrary Group, Inc.; **452** Magna Carta/Shutterstock; **453** bramalia/Shutterstock; **455** (T) NOAA Earth System Research Laboratory, Chemical Sciences Division, (B) steinphoto/iStockphoto, (Bkgd) Thomas Prior/Getty Images; **456** Carlos Caetano/Shutterstock; **458** (TL, TC, R) Matt Meadows/Peter Arnold, Inc.; **460** (Bkgd) lakov Kalinin/Shutterstock; **462** (TL) MikeE/Shutterstock; **463** justin maresch/Shutterstock, NASA; **464** ©Kelly Redinger/AGE Fotostock; **467** Simon Fraser/Newcastle General Hospital/Photo Researchers, Inc., themoog/iStockphoto; **469** maxstockphoto/Shutterstock; **470** ©Jim Lopes/Shutterstock; **472** (L, C) ©DK Images; **473** ©Larry Lee/Larry Lee Photography/Corbis; **475** Ken O'Donoghue/Pearson Education; **476** eVox/Alamy Images; **477** (BR) Antonio Scorza/AFP/Getty Images, (TR, CR) imagebroker/Alamy; **478** Simon Fraser/Newcastle General Hospital/Photo Researchers, Inc.; **481** (R) Andrew Lambert Photography/Photo Researchers, Inc., (L) Charles D. Winters/Photo Researchers, Inc.; **483** ©Wolfgang Amri/iStockphoto; **486** Serg64/Shutterstock; **487** (Inset) Pascal Rondeau/Getty Images; **488** (B) ©DK Images, (T) NASA Image Exchange; **490** (T) Denis Vrublevski/Shutterstock; **491** Ken O'Donoghue/Pearson Education; **492** Ted Kinsman/Photo Researchers, Inc.; **493** Joyce Photographics/Photo Researchers, Inc.; **495** Eunice Harris/Photo Researchers, Inc.; **496** (R, C) Richard Megna/Fundamental Photographs, (L) Richard Megna/Fundamental Photographs; **497** (BR) ERproduction Ltd/Getty Images; **498** (B) Kristen Brochmann/Fundamental Photographs, (TR) Richard Megna/Fundamental Photographs, (TL) Richard Megna/Fundamental Photographs; **499** Photo courtesy Cheryl Steinecker; **502** (Inset) Anthony Jay D.Villalon/Shutterstock; **503** (CR) Eightfish/Getty Images; **504** (BL) Russ Lappa/Pearson Education; **506** Ackab Photography/Shutterstock; **508** Ken O'Donoghue/Pearson Education; **509** Ted Kinsman/Photo Researchers, Inc.; **513** Pascal Rondeau/Getty Images; **516** (Bkgd) Robert Glusic/Corbis/Photolibrary Group, Inc.; **517** (Inset) Craig Lovell/AGE Fotostock; **518** (BR, BL, BC) Ken Karp/Addison-Wesley, (T) Photo Researchers, Inc.; **519** Charles D. Winters/Photo Researchers, Inc.; **520** (BL) Lilly Dong/Getty Images; **521** (T) Fillpe B. Varela/Shutterstock; **523** (TR, TL, TCL, CR) Charles D. Winters/Photo Researchers, Inc.; **525** (T) J. Herlgason/Shutterstock, (BL) Ken Karp/Addison Wesley, (BR, BC) Ken Karp/Addison-Wesley; **528** Ken Karp/Addison-Wesley, (BR, BL, BC) Richard Hutchings/Addison-Wesley; **532** (TL) Cathleen Clapper/Shutterstock, (Bkgd) Christian Ostrosky/Alamy Images, (CL, BL) DK; **534** (T) Suzanne Kischnick/Alamy Images; **535** (TR, TL, TC) Ken Karp/Addison-Wesley; **536** (Inset) Florida Images/Alamy Images, (Bkgd) Thomas Cooper/Getty Images; **537** jane/iStockphoto; **538** (T,) MIXA/Getty Images; **545** (T) Photo Studio; **551** Craig Lovell/AGE Fotostock; **552** (BL) Ken O'Donoghue/Prentice Hall; **554** (Bkgd) Alamy Images; **555** (Inset) Tony Ranze/Getty Images; **556** (T) Corbis/SuperStock, (B) Getty Images; **557** (R) AGE Fotostock, (L) BananaStock/AGE Fotostock, (Background, R) ilker canikligil/Shutterstock; **559** Unarmed/Shutterstock; **560** (B) JG Photography/Alamy; **562** Roman Sigaev/Shutterstock; **565** tinabelle/iStockphoto; **566** Alamy; **568** D. Parker/Photo Researchers, Inc.; **569** Blend Images/Patrik Giardino/Getty Images, Image Source/SuperStock; **571** JanPietruszka/iStockphoto; **572** (B) Ken Karp/Addison Wesley; **574** Ken O'Donoghue, (TL) Pyastolova Nadya/Shutterstock; **576** (Bkgd) robas/iStockphoto; **577** (Inset) AlenaderLaws/iStockphoto; **578** James Steidl/Shutterstock; **583** (TR) Ken O'Donoghue/Prentice Hall , Thinkstock; **589** Tony Ranze/Getty Images; **590** pakowacz/Shutterstock; **592** (Bkgd) Lex Van Lieshout/AGE Fotostock; **593** (Inset) Stepen Morton/©AP Images; **594** (B) Juice Images169/Alamy, (TL) zumawirewestphotos/NewsCom; **598** (BL) Ken Kalp/Addison Wesley, (BR) Ken Kalp/Addison Wesley; **599** (B) Stephen Morton/©AP Images; **600** (T) Ken O'Donoghue; **602** (Bkgd) Car Culture/Getty Images; **603** (TR) Pearson ; **604** (TL) CHRISTOPHE ENA/©AP Images; **609** (TL) David Frazier/Corbis; **613** (B) Steve Baccon/Getty Images; **616** (B) Geoffrey Nilson/Addison/Wesley; **620** (Bkgd) Epic Stock/Shutterstock, (Inset) Science Source/Photo Researchers, Inc.; **621** (TL) CNRI/Photo Researchers, Inc.; **622** (TL) Ken Karp/Addison Wesley; **627** (TL) Pyastolova Nadya/Shutterstock; **628** (TR, B) Samot/Shutterstock; **629** (T) Pakhnyushcha/Shutterstock; **640** (CR, CL) ©DK Images, (T) Joel Arem/Photo Researchers, Inc., (B) Lihee Avidan/The Image Bank/Getty Images; **641** (CR) Stepen Morton/©AP Images; **644** (Bkgd) Susan Van Etten/PhotoEdit, Inc.; **645** (Inset) Chet Gordon/The Image Works, Inc.; **646** (Inset, B) Iconica/Getty Images, (BL) Igor Dutina/Shutterstock, (T) Rolf Nussbaumer/Nature Picture Library; **648** Ken Karp/Addison-Wesley; **649** Fundamental Photographs; **653** Sarah Holmlund/Shutterstock; **661** (TL) Kojiro/Shutterstock, (B) Michael Howes/AGE Fotostock, (TR) UVimages/Photolibrary Group, Inc.; **663** (Inset) PhotoLibrary Group, Inc., (Bkgd) Sean Sprague/The Image Works, Inc.; **664** Foodcollection/AGE Fotostock; **666** (Bkgd) Dhoxax/Shutterstock, (Inset) Dimaparunin/Dreamstime LLC; **670** Jon Chonity; **671** (Inset) Serg64/Shutterstock, (Bkgd) Sylvain Sonnet/Getty Images; **672** Geoff Kidd/AGE Fotostock; **674** (R, L, C) Richard Megna/Fundamental Photographs; **676** (BR, BL, BC) Ken Karp/Addison-Wesley, (T) London Scientific Films/PhotoLibrary Group, Inc.; **678** (R, L) Ken Karp/Addison-Wesley; **681** Specta/Shutterstock; **685** Richard Megna/Fundamental Photographs; **687** Chet Gordon/The Image Works, Inc.; **690** (Bkgd) Henryk T. Kaiser/AGE Fotostock; **691** (Inset) ©DK Images; **692** (TL) Julija Sapic/Shutterstock, (BR) Richard Megna/Fundamental Photographs; **693** (L) Bluefly/Alamy, (R) Gregory Davies/Alamy; **694** (R, C) Ken Karp/Addison-Wesley, (L) Paul Silverman/Fundamental Photographs; **696** (TL) Blair Seitz/Shutterstock; **697** (T) Jorge Salcedo/Shutterstock; **698** (Bkgd) Kevin Fleming/Corbis, (TL) Jose Gil/Shutterstock, (BL) Shenval/Alamy; **700** (Bkgd) Alexey Stiop/Shutterstock, (Inset) MaxFX/Shutterstock; **701** (T) LWA/AGE Fotostock; **704** (TR, TL, BL) Ken Karp/Addison-Wesley; **707** (TL) Kokhanchikov/Shutterstock, (BL) Richard Megna/Fundamental Photographs, (BR) Richard Megnal/Fundamental Photographs; **708** (T) 2009fotofriends/Shutterstock; **710** (T) James Hardy/AGE Fotostock; **712** (T) Lipsky/Shutterstock; **717** Richard Megna/Fundamental Photographs; **718** (T) Kevin Fleming/Corbis; **721** (R) Kai Schwabe/AGE Fotostock; **723** ©DK Images; **724** (R, L, C) Ken Karp/Addison-Wesley; **726** (Bkgd) Andreas Gradin/Shutterstock; **727** (Inset) linhof/Shutterstock; **728** (T) oktobernight/iStockphoto, (BR, BL) Richard Megna/Fundamental Photographs; **730** John Moore/Pearson Education; **733** futureimage/iStockphoto; **735** Jim West/AGE Fotostock; **736** (L) ifong/Shutterstock, (R) Mary Evans Picture Library/The Image Works, Inc.; **737** ifong/Shutterstock; **745** (B) Drozdowski/Shutterstock, (T) Peter Gudella/Shutterstock; **747** Charles D. Winters/Photo Researchers, Inc.; **751** John Gress/Reuters Media; **753** ifong/Shutterstock; **757** linhof/Shutterstock; **760** Christopher Herwig/Getty Images; **761** Imago/Zuma Press, Inc.; **762** Imago/Zuma Press, Inc.; **764** (T) Charles D.Winters/Photo Researchers, Inc., (C) Matt Cardy/Getty Images, (B) Mike Liu/Shutterstock; **768** Tim Boyle/Getty Images; **772** Juanmonino/iStockphoto; **774** (Inset) Billy Hustace/Getty Images, (Bkgd) Don Seabrook/©AP Images; **775** Norman Pogson/Shutterstock; **778** Ken O'Donoghue; **779** Tomtsya/Shutterstock; **781** Bedolaga/Shutterstock; **782** Stephen Bay/Alamy; **784** (Inset, L) HENRY LAPO/Xinhua/Landov LLC, (Inset, R) James Cavallini/Photo Researchers, Inc., (Bkgd) Jon Hrusa/©AP Images; **785** (L, CR) Andrew J. Martinez/Photo Researchers, Inc., (R) Martin Shields/Photo Researchers, Inc., (CL) Ted Kinsman/Photo Researchers, Inc.; **786** Oded Balilty/©AP Images; **787** Ken O'Donoghue; **793** Imago/Zuma Press, Inc.; **796** (Bkgd) foodfolio/Alamy Images; **797** (CR) Aleksandr Stennikov/Shutterstock; **798** (B) Luxe/Shutterstock, (TL) Thinkstock; **799** (CR, Background, R) Thinkstock, (BL, BC) Yasonya/Shutterstock; **803** USTIN/Shutterstock; **804** (TR) Yuri Arcurs/Shutterstock; **806** (Bkgd) Ralf-Finn Hestoft/PhotoLibrary Group, Inc., (CL, BL) Thinkstock; **807** (L, C) Ian O'Leary/©DK Images, (R) Lauree Feldman/AGE Fotostock; **808** (T, B) Richard Megna/Fundamental Photographs; **809** (R) Diane Diederich/iStockphoto, (L) Maram/Shutterstock; **810** Lisa Maree Williams/Getty Images; **812** Picsfive/Shutterstock; **814** Olga Lyubkina/Shutterstock; **816** David Toase/Getty Images; **817** (BR) afaizal/Shutterstock; **818** (TL) Fundamental Photographs; **819** (BR) Pixelbliss/Shutterstock; **822** (T) Darryl Leniuk/Getty Images, (B) KariDesign/Shutterstock; **823** (TR) Garry Wade/Getty Images, (L) Photolibrary/AGE Fotostock; **824** (TL) Getty Images, (B) Getty Images\Staff/Getty Images, (TR) Jordi Cami/AGE Fotostock; **825** (BR) Angela Hampton/AGE Fotostock; **826** Danny Warren/Shutterstock; **827** Clinton Lewis/©AP Images; **828** Ken O'Donoghue ; **829** Garry Wade/Getty Images; **833** Aleksandr Stennikov/Shutterstock; **836** (Bkgd) Yellow Dog Productions/Shutterstock; **838** (T) Levent Konuk/Shutterstock; **839** (BR,) Serp/Shutterstock; **841** (B) Geoffrey Lawrence/Shutterstock, (T) Kojot/Shutterstock; **842** (B) Hywit Dimyadi/Shutterstock; **844** (T) Jules Studio/Shutterstock; **845** Dorling Kindersley; **846** (B) tubuceo/Shutterstock; **849** Ken O'Donoghue/Prentice Hall; **850** (B) David P. Smith/Shutterstock, (T) Shutterstock; **851** (L, C) Richard Huchings/Addison-Wesley, (R) Richard Hutchings/Addison-Wesley; **852** (TL) Steve Gschmeissner/Photo Researchers, Inc.; **853** (Bkgd) Alexis Rosenfield/Photo Researchers, Inc., (B) iStockphoto, (T) Marc Phares/Photo Researchers, Inc.; **854** (T) Dorgie Productions/Getty Images; **855** (BR, B) Dr. Tim Evans/Photo Researchers, Inc.; **856** Ken O'Donoghue/Prentice Hall; **858** J. Gatherum/Shutterstock; **859** (T) Peter Menzel/Photo Researchers, Inc.; **861** John Chadwick/©AP Images; **862** (T) Kenmcgrath/Dreamstime LLC; **864** (T) John Shaw/Photo Researchers, Inc.; **866** U.S. Department of Agriculture; **867** (C) Peter Dazeley/Getty Images; **874** (Bkgd) U.S. Navy/AGE Fotostock; **875** (Inset) RIA Novosti/Topham/The Image Works, Inc.; **876** (T) d_michel/iStockphoto, (B) Rue des Archives/©The Granger Collection, NY; **880** Blair Seitz/Photo Researchers, Inc.; **886** Thomas Pflaum/The Image Works, Inc.; **888** SOHO Extreme Ultraviolet Imaging Telescope/NASA; **889** (Bkgd) Joe Sohm/Photo Researchers, Inc.; **890** Gavrilov/PhotoLibrary Group, Inc.; **892** (Bkgd) Olivier Voisin/Photo Researchers, Inc.; **894** (B) Corbis, (T) S. Wanke/PhotoLink/Getty Images; **895** SuperStock; **897** Oullette/Theroux/Publiphoto/Photo Researchers, Inc.; **902** Ken Karp/Addison-Wesley; **903** RIA Novosti/Topham/The Image Works, Inc.; **906** ©Stephen Wilkes/Getty Images; **R02** Andrew Lambert Photography/Photo Researchers, Inc.; **R03** (L) David Taylor/Photo Researchers, Inc., (CL, C) Fundamental Photographs, (R, CR) Lester V. Bergman/Corbis; **R04** Chris Bence/Shutterstock; **R05** (B) Michael Ventura/Alamy Images, (T) Mike Danton/Alamy Images; **R07** (TR, TL, TC) Fundamental Photographs, (B) James Dawson/Alamy; **R08** (T) batoots/Shutterstock, (B) David Greedy/Stringer/Getty Images; **R09** andrej pol/Shutterstock; **R10** (B) Doug Steley A/Alamy Images, (T) Lester V. Bergman/Corbis; **R11** (B) Andrew Syred/Photo Researchers, Inc., (TR) ProArtWork/iStockphoto, (TL) Gary Ombler/©DK Images; **R12** (B) Patti McConville/Alamy Images, (T) Peter Kim/Shutterstock; **R13** Artur Synenko/Shutterstock; **R14** (R) Harry Taylor/©DK Images, (L) Jacana/Photo Researchers, Inc.; **R15** James L. Amos/Getty Images; **R16** S. Wanke/PhotoLink/Getty Images; **R17** (R) Alfred Pasieka/Photo Researchers, Inc., (L) David J. Green - technology/Alamy Images; **R18** (B) NASA Image Exchange, (L) Nigel Cattlin/Photo Researchers, Inc.; **R19** (R) Ingram Publishing/AGE Fotostock, (L) Richard Megna/Fundamental Photographs; **R20** (T) Richard Megna/Fundamental Photographs; **R22** (T) AfriPics/Alamy Images, (B) Bettmann/Corbis; **R24** (B) Natural History Museum, London/Alamy Images, (TL) Tom Bochsler/Prentice Hall; **R25** (BR, BL, BC) Ken Karp/Addison-Wesley, (T) Charles D. Winters/Photo Researchers, Inc.; **R26** (B) Andre Seale/AGE Fotostock, (T) Fundamental Photographs, (C) Photo Researchers, Inc.; **R27** (R) Big Pants Production/Shutterstock, (L) NASA Image Exchange; **R28** (T, B) Fundamental Photographs; **R29** Fundamental Photographs; **R30** (T) Corbis/AGE Fotostock, (B) Gallo Images/AGE Fotostock; **R31** (B) Alamy Images, (T) Juegen Berger/Photo Researchers, Inc.; **R32** (T) Anton Prado Photo/Shutterstock, (B) Juergen Bauer/Smart Elements; **R33** (T) Shutterstock, (B) The Image Works, Inc.; **R34** Maksymilian Skolik/Shutterstock; **R35** (B) Gabriel Bouys/NewsCom, (T) Keith Binns/iStockphoto; **R36** (B) Egmont Strigl/AGE Fotostock, (T) Harry Taylor/©DK Images; **R37** (R) Fundamental Photographs; **R38** (T) Burstein Collection/Corbis, (BL) Synthetic Alan King/Alamy Images, (BR) Terry Smith/Alamy Images; **R39** Getty Images; **R40** (T) Alamy Images, (B) Ingvar Bjork/Shutterstock.

Credits **R157**

Periodic Table of the Elements

Representative Elements
- Alkali Metals
- Alkaline Earth Metals
- Other Metals
- Metalloids
- Nonmetals
- Noble Gases

Transition Elements
- Transition metals
- Inner transition metals

- C Solid
- Br Liquid
- He Gas
- Tc Not found in nature

Key: 13 (Atomic number); 2 8 3 (Electrons in each energy level); Al (Element symbol); Aluminum (Element name); 26.982 (Atomic mass†)

†The atomic masses in parentheses are the mass numbers of the longest-lived isotope of elements for which a standard atomic mass cannot be defined.

	1 1A	2 2A	3 3B	4 4B	5 5B	6 6B	7 7B	8 (8B)	9 (8B)	10 (8B)	11 1B	12 2B	13 3A	14 4A	15 5A	16 6A	17 7A	18 8A
1	1 **H** Hydrogen 1.0079 — 1																	2 **He** Helium 4.0026 — 2
2	3 **Li** Lithium 6.941 — 2 1	4 **Be** Beryllium 9.0122 — 2 2											5 **B** Boron 10.81 — 2 3	6 **C** Carbon 12.011 — 2 4	7 **N** Nitrogen 14.007 — 2 5	8 **O** Oxygen 15.999 — 2 6	9 **F** Fluorine 18.998 — 2 7	10 **Ne** Neon 20.179 — 2 8
3	11 **Na** Sodium 22.990 — 2 8 1	12 **Mg** Magnesium 24.305 — 2 8 2											13 **Al** Aluminum 26.982 — 2 8 3	14 **Si** Silicon 28.086 — 2 8 4	15 **P** Phosphorus 30.974 — 2 8 5	16 **S** Sulfur 32.06 — 2 8 6	17 **Cl** Chlorine 35.453 — 2 8 7	18 **Ar** Argon 39.948 — 2 8 8
4	19 **K** Potassium 39.098 — 2 8 8 1	20 **Ca** Calcium 40.08 — 2 8 8 2	21 **Sc** Scandium 44.956 — 2 8 9 2	22 **Ti** Titanium 47.90 — 2 8 10 2	23 **V** Vanadium 50.941 — 2 8 11 2	24 **Cr** Chromium 51.996 — 2 8 13 1	25 **Mn** Manganese 54.938 — 2 8 13 2	26 **Fe** Iron 55.847 — 2 8 14 2	27 **Co** Cobalt 58.933 — 2 8 15 2	28 **Ni** Nickel 58.71 — 2 8 16 2	29 **Cu** Copper 63.546 — 2 8 18 1	30 **Zn** Zinc 65.38 — 2 8 18 2	31 **Ga** Gallium 69.72 — 2 8 18 3	32 **Ge** Germanium 72.59 — 2 8 18 4	33 **As** Arsenic 74.922 — 2 8 18 5	34 **Se** Selenium 78.96 — 2 8 18 6	35 **Br** Bromine 79.904 — 2 8 18 7	36 **Kr** Krypton 83.80 — 2 8 18 8
5	37 **Rb** Rubidium 85.468 — 2 8 18 8 1	38 **Sr** Strontium 87.62 — 2 8 18 8 2	39 **Y** Yttrium 88.906 — 2 8 18 9 2	40 **Zr** Zirconium 91.22 — 2 8 18 10 2	41 **Nb** Niobium 92.906 — 2 8 18 12 1	42 **Mo** Molybdenum 95.94 — 2 8 18 13 1	43 **Tc** Technetium (98) — 2 8 18 14 1	44 **Ru** Ruthenium 101.07 — 2 8 18 15 1	45 **Rh** Rhodium 102.91 — 2 8 18 16 1	46 **Pd** Palladium 106.4 — 2 8 18 18	47 **Ag** Silver 107.87 — 2 8 18 18 1	48 **Cd** Cadmium 112.41 — 2 8 18 18 2	49 **In** Indium 114.82 — 2 8 18 18 3	50 **Sn** Tin 118.69 — 2 8 18 18 4	51 **Sb** Antimony 121.75 — 2 8 18 18 5	52 **Te** Tellurium 127.60 — 2 8 18 18 6	53 **I** Iodine 126.90 — 2 8 18 18 7	54 **Xe** Xenon 131.30 — 2 8 18 18 8
6	55 **Cs** Cesium 132.91 — 2 8 18 18 8 1	56 **Ba** Barium 137.33 — 2 8 18 18 8 2	71 **Lu** Lutetium 174.97 — 2 8 18 32 9 2	72 **Hf** Hafnium 178.49 — 2 8 18 32 10 2	73 **Ta** Tantalum 180.95 — 2 8 18 32 11 2	74 **W** Tungsten 183.85 — 2 8 18 32 12 2	75 **Re** Rhenium 186.21 — 2 8 18 32 13 2	76 **Os** Osmium 190.2 — 2 8 18 32 14 2	77 **Ir** Iridium 192.22 — 2 8 18 32 15 2	78 **Pt** Platinum 195.09 — 2 8 18 32 17 1	79 **Au** Gold 196.97 — 2 8 18 32 18 1	80 **Hg** Mercury 200.59 — 2 8 18 32 18 2	81 **Tl** Thallium 204.37 — 2 8 18 32 18 3	82 **Pb** Lead 207.2 — 2 8 18 32 18 4	83 **Bi** Bismuth 208.98 — 2 8 18 32 18 5	84 **Po** Polonium (209) — 2 8 18 32 18 6	85 **At** Astatine (210) — 2 8 18 32 18 7	86 **Rn** Radon (222) — 2 8 18 32 18 8
7	87 **Fr** Francium (223) — 2 8 18 32 18 8 1	88 **Ra** Radium (226) — 2 8 18 32 18 8 2	103 **Lr** Lawrencium (262) — 2 8 18 32 32 9 2	104 **Rf** Rutherfordium (261) — 2 8 18 32 32 10 2	105 **Db** Dubnium (262) — 2 8 18 32 32 11 2	106 **Sg** Seaborgium (263) — 2 8 18 32 32 12 2	107 **Bh** Bohrium (264) — 2 8 18 32 32 13 2	108 **Hs** Hassium (265) — 2 8 18 32 32 14 2	109 **Mt** Meitnerium (268) — 2 8 18 32 32 15 2	110 **Ds** Darmstadtium (269) — 2 8 18 32 32 17 1	111 **Rg** Roentgenium (272) — 2 8 18 32 32 18 1	112 **Cn** Copernicium (277) — 2 8 18 32 32 18 2	113 ***Uut** Ununtrium (284) — 2 8 18 32 32 18 3	114 ***Uuq** Ununquadium (289) — 2 8 18 32 32 18 4	115 ***Uup** Ununpentium (288) — 2 8 18 32 32 18 5	116 ***Uuh** Ununhexium (293) — 2 8 18 32 32 18 6	117 ***Uus** Ununseptium Classification pending	118 ***Uuo** Ununoctium (299) — 2 8 18 32 32 18 8

Elements 104–118 are the transactinide elements.

*Discovery reported but not verified

Lanthanide Series

57	58	59	60	61	62	63	64	65	66	67	68	69	70
La Lanthanum 138.91 — 2 8 18 18 9 2	**Ce** Cerium 140.12 — 2 8 18 20 8 2	**Pr** Praseodymium 140.91 — 2 8 18 21 8 2	**Nd** Neodymium 144.24 — 2 8 18 22 8 2	**Pm** Promethium (145) — 2 8 18 23 8 2	**Sm** Samarium 150.4 — 2 8 18 24 8 2	**Eu** Europium 151.96 — 2 8 18 25 8 2	**Gd** Gadolinium 157.25 — 2 8 18 25 9 2	**Tb** Terbium 158.93 — 2 8 18 27 8 2	**Dy** Dysprosium 162.50 — 2 8 18 28 8 2	**Ho** Holmium 164.93 — 2 8 18 29 8 2	**Er** Erbium 167.26 — 2 8 18 30 8 2	**Tm** Thulium 168.93 — 2 8 18 31 8 2	**Yb** Ytterbium 173.04 — 2 8 18 32 8 2

Actinide Series

89	90	91	92	93	94	95	96	97	98	99	100	101	102
Ac Actinium (227) — 2 8 18 32 18 9 2	**Th** Thorium 232.04 — 2 8 18 32 18 10 2	**Pa** Protactinium 231.04 — 2 8 18 32 20 9 2	**U** Uranium 238.03 — 2 8 18 32 21 9 2	**Np** Neptunium (237) — 2 8 18 32 22 9 2	**Pu** Plutonium (244) — 2 8 18 32 24 8 2	**Am** Americium (243) — 2 8 18 32 25 8 2	**Cm** Curium (247) — 2 8 18 32 25 9 2	**Bk** Berkelium (247) — 2 8 18 32 27 8 2	**Cf** Californium (251) — 2 8 18 32 28 8 2	**Es** Einsteinium (252) — 2 8 18 32 29 8 2	**Fm** Fermium (257) — 2 8 18 32 30 8 2	**Md** Mendelevium (258) — 2 8 18 32 31 8 2	**No** Nobelium (259) — 2 8 18 32 32 8 2